Chordate Zoology

For B.Sc. (Pass & Hons.) Classes of Indian Universities

E.L. Jordan
M.Sc., M.A. (Toronto)
Formerly Head
Department of Zoology
Lucknow Christian College, Lucknow

Dr. P.S. Verma
M.Sc., Ph.D., FESI, FAZ
Formerly Reader
Department of Zoology
Meerut College, Meerut

S. CHAND
PUBLISHING

S Chand And Company Limited
(ISO 9001 Certified Company)

S Chand And Company Limited

(ISO 9001 Certified Company)

Head Office: Block B-1, House No. D-1, Ground Floor, Mohan Co-operative Industrial Estate, New Delhi – 110 044 | Phone: 011-66672000

Registered Office: A-27, 2nd Floor, Mohan Co-operative Industrial Estate, New Delhi – 110 044 Phone: 011-49731800

www.**schandpublishing.com**; e-mail: **info@schandpublishing.com**

Branches

Ahmedabad	:	Ph: 27542369, 27541965; ahmedabad@schandpublishing.com
Bengaluru	:	Ph: 22354008, 22268048; bangalore@schandpublishing.com
Bhopal	:	Ph: 4274723, 4209587; bhopal@schandpublishing.com
Bhubaneshwar	:	Ph: 2951580; bhubaneshwar@schandpublishing.com
Chennai	:	Ph: 23632120; chennai@schandpublishing.com
Guwahati	:	Ph: 2738811, 2735640; guwahati@schandpublishing.com
Hyderabad	:	Ph: 40186018; hyderabad@schandpublishing.com
Jaipur	:	Ph: 2291317, 2291318; jaipur@schandpublishing.com
Jalandhar	:	Ph: 4645630; jalandhar@schandpublishing.com
Kochi	:	Ph: 2576207, 2576208; cochin@schandpublishing.com
Kolkata	:	Ph: 23357458, 23353914; kolkata@schandpublishing.com
Lucknow	:	Ph: 4003633; lucknow@schandpublishing.com
Mumbai	:	Ph: 25000297; mumbai@schandpublishing.com
Nagpur	:	Ph: 2250230; nagpur@schandpublishing.com
Patna	:	Ph: 2260011; patna@schandpublishing.com
Ranchi	:	Ph: 2361178; ranchi@schandpublishing.com
Sahibabad	:	Ph: 2771238; info@schandpublishing.com

First Edition 1965
Subsequent Editions and Reprints 1968, 70, 73, 74, 76, 77, 78, 80, 81, 83, 86, 87, 88, 89, 90, 91, 93, 95, 96, 97, 98, 2000, 2001, 2002, 2003, 2004, 2005, 2006, 2008, 2009, 2010, 2011, 2012
Fourteenth Revised and Enlarged Edition 2013
Reprints 2014, 2015, 2016 (Twice), 2018, 2019 (Twice)
Reprint 2020

ISBN: 978-81-219-1639-4 **Product Code:** H6ZOC68ZOOL10ENAN13O

PRINTED IN INDIA

By Vikas Publishing House Private Limited, Plot 20/4, Site-IV, Industrial Area Sahibabad, Ghaziabad – 201 010 and Published by S Chand And Company Limited, A-27, 2nd Floor, Mohan Co-operative Industrial Estate, New Delhi – 110 044

DEDICATED

to

My Respected Mother

Late Smt. Gomti Devi

who could not survive

to see the fruits of

this labour

Dr. Prem Singh Verma

PREFACE TO THE FOURTEENTH EDITION

The vistas of biology are fast increasing with more and more emphasis on cell biology, molecular biology, genetics, biochemistry, physiology, electron microscopy, biotechnology and development biology (embryology), etc. Several topics have found their way into syllabi of universities. The importance of basic zoology, however, cannot be ignored; most teachers and researchers rightly feel the need for a strong base in general zoology as a prerequisite for all fields of specialisations. Jordan's *Chordate Zoology* has served a great deal towards the fulfilment of this aim. It has been a national bestseller for more than four decades. It was, therefore, found necessary to bring out the same book in a more useful and enlarged form. The book has been thoroughly revised and enhanced in this edition. I have retained all that was judged good and useful in the previous edition. However, a number of changes have been made in this edition. Subject matter has been revised and updated in the light of recent researches and new topics have been added at appropriate places keeping in view the changes prescribed in the syllabi of Indian Universities.

New chapters and topics which have been added in the present edition include : (i) Pisces (Fishes) : *Labeo rohita* (Type study); Comparison of Chondrichthyes and Osteichthyes; Dipnoi; Fin System; Scales; Air bladder or Swim bladder; Accessary respiratory organs; Parental care; Migration; and Economic importance. (ii) Amphibia : Parental care; Neoteny and Paedogenesis; Origin and Ancestry. (iii) Reptilia : *Calotes versicolar* (Type study); Snakes : General account; Reptilia: General account. (iv) Aves : Common Flightless birds (Ratitae); Types of beaks and feet or claws; Types of skull palate; *Archaeopteryx* and its significance; Birds are glorified reptiles; Birds as a flying machine; Aerial or flight adaptations; Migration; and Economic importance. (v) Mammalia : Prototheria (Order Monotremata); Metatheria; Dentition; Aquatic mammals and adaptations; Flying (Volant) mammals and adaptations; Adaptive radiations; Adaptive convergence; Origin and Ancestry; and Economic importance. (vi) Vertebrate Embryology and Comparative Anatomy: Several new topics have been added. (vii) Wildlife in India. (viii) Glossary.

In addition to this, general topics which are usually asked in the examinations have been thoroughly revised and some new topics have also been added in all the classes of chordates. In the last few years, most of the universities have introduced long answer type, short answer type, very short answer type and multiple choice type (objective type) questions in the examination papers. To cater to this requirement, various types of questions have been added at the end of each chapter.

The examination-oriented text is intelligible to for all students, even to those who do not have a good background in the subject. The language is simple, explanations are clear, and presentation is very systematic. To improve scientific accuracy, about 706 well-labelled diagrams have been included at appropriate places. Important words are given in bold face and definitions in italics. Subject matter is error-free. Notwithstanding several Indian books on Chordate Zoology or Vertebrate Zoology, I hope the present edition of Chordate Zoology shall serve the purpose of academicians and students.

I am thankful to the Management Team and the Editorial Department of S. Chand & Company Pvt. Ltd. for all the help and support in the publication of this book.

I would like to express my thanks to esteemed teachers for sending their valuable suggestions and opinions and shall hope to receive the same in future.

"PREETI" **Dr. P.S. VERMA**

121, New Mohanpuri Colony

Meerut- 250 001

Preface To The First Edition

There are some excellent textbooks of comparative anatomy and anatomy of vertebrates. It would be presumptive to present yet another book on chordate zoology without any justification. Yet the author feels justified in presenting this book to the students. Firstly it lays down the basic principles of the B.Sc. Zoology course of Indian Universities, secondly it is written in a manner which takes into account the morphology, physiology, histology, embryology, ecology, palaeontology and evolution of chordates. The book is a comprehensive guide though it does not burden the student with unnecessary details, yet it is a clear and complete study of Chordates and it includes not only the study of comparative anatomical structure, but also the origin of structure in terms of its embryology and evolution. The book contains numerous clear and well-labelled figures which make an understanding of the text an easier task.

The first chapter takes into account the geological time chart and the origin and characters of chordates. Then it deals with Protochordata, but it does not take into account the Hemichordata. The Hemichordata in modern zoological literature are now regarded as a separate phylum of invertebrates. Then the Agnatha are described in outline. The next part is a comprehensive account of the comparative anatomy of vertebrates dealing with the development, morphology, and evolution of various systems. This major portion is followed by a classification of vertebrates along with a description of many representative types from each class. Hardly any two books have an identical scheme of classification of animals; in this book the best points of various classifications are taken on the basis of morphology and evolution, and are coordinated into a wholesome system. The last part of the book is devoted to general biology dealing with Heredity, Mendelism, and Evolution.

E.L. JORDAN

Lucknow,

Ist January, 1965

Contents

1 Phylum Chordata : An Introduction

Chapter

Contents of this Chapter

WHAT IS CHORDATA?

The name of phylum Chordata is derived from two Greek words, the *chorde* (= a string or cord) and *ata* (= bearing). The common characteristic feature of phylum Chordata is the presence of a stiff, supporting rod-like structure along the back, the **notochord** (Gr., *noton* = back; L. *chorda* = cord) which is found in all the members of the phylum at some stage of their lives. Thus, chordates are those animals which have a cord, *i.e.,* notochord. The animals belonging to all other phyla of the Animal Kingdom are often called "the nonchordates" or "the invertebrates" since they do not have notochord or backbone in their body structure. The phylum Chordata was created by **Balfour** in 1880. The phylum Chordata includes primarily the **Protochordata (Acraniata)** and **Vertebrata (Craniata)**.

CHORDATE ANIMALS

Phylum Chordata is the largest of the deuterostome phyla. It is the highest and most important phylum comprising a vast variety of living and extinct animals. Most of the living chordate animals are well-known familiar vertebrate animals such as the fishes, amphibians, reptiles, birds and mammals. The less well-known animals include a number of marine forms such as the tunicates and lencelets.

Diversity of Chordates

The chordates exhibit great diversity of form (anatomy), physiology and habit. The phylum Chordata includes a vast group of diverse animals ranging from Ascidian to Man.

Number (Numerical strength). The number of chordate species is not large. About 49,000 species are on record which are only half of the living species of molluscs, and less than one-tenth those of arthropods. The two subphyla Urochordata and Cephalochordata include nearly 2,500 species. The subphylum Vertebrata includes 46,500 species. Among the vertebrates, fishes are the most numerous with approximately 25,000 species. The approximate number of species of amphibians is about 2,500, reptiles 6,000, birds 9,000 and mammals 4,500.

Size. The urochordates and cephalochordates are mostly of small size. The vertebrates include animals of medium to large size, notably some huge sharks and whales. The gigantic blue whale (*Balaenoptera musculus*), which is about 35 metres in length and 120 tons in weight, is the biggest known animal. The whale shark (*Rhineodon typus*) attains a length of 15 metres and is the second biggest vertebrate after whales. The smallest fish is a Philippine goby (*Pandaka*) measuring only 10 mm long.

Ecology. The chordates are not only the largest animals existing today, but ecologically they are among the most successful in the Animal Kingdom. They occupy various kinds of habitats and have adapted themselves to various modes of living than any other group, including the arthropods. Chordates are found in the sea, in freshwater, in the air and all parts of land from the equator to the poles. Birds and mammals are found in cold climates because they have a constant body temperature, which other animals do not have.

All lower chordates (protochordates) are marine, fishes are aquatic (freshwater and marine both) and higher chordates are predominantly terrestrial. Amphibians are not found in salt water. Birds do not live permanently in water, while some amphibians, reptiles and mammals permanently occupy the water. Most tunicates are sessile, all chordates are free-living and none is strictly parasitic.

THREE FUNDAMENTAL CHORDATE CHARACTERS

All the chordates possess three common fundamental characteristics at some stage in their life history. These three common fundamental morphological features are as follows :

1. Notochord (a longitudinal supporting rod-like structure);
2. Dorsal tubular (hollow) nerve cord; and
3. Pharyngeal gill-slits.

These three distinctive characteristics are unique for the phylum Chordata (Fig. 1.1). The existence of such common characteristics is considered as a result of inheritance from a common ancestry. Besides these three common characteristics, there are few other characteristics which have less diagnostic value.

1. Notochord or chorda dorsalis. The notochord (Gr., *noton* = back; *chorda* = cord) is a rod-like, elongated, elastic structure situated just above the alimentary canal and immediately beneath the dorsal tubular nerve cord. It is composed of large vacuolated **notochordal cells** enclosed in an outer fibrous and an inner elastic sheath of connective tissue (Fig. 1.2). It serves as a primitive internal skeleton and acts as a rigid axis, but permits movements of the body. It may persist throughout life, as in lancelet, lamprey and some fishes, or it may be replaced partially or completely by a backbone or vertebral column. The notochord originates from the endodermal roof of the embryonic archenteron.

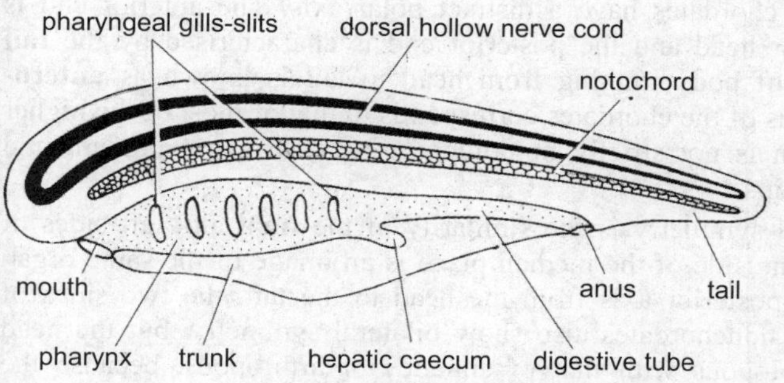

Fig. 1.1. Diagrammatic side view of a chordate showing three fundamental chordate characters.

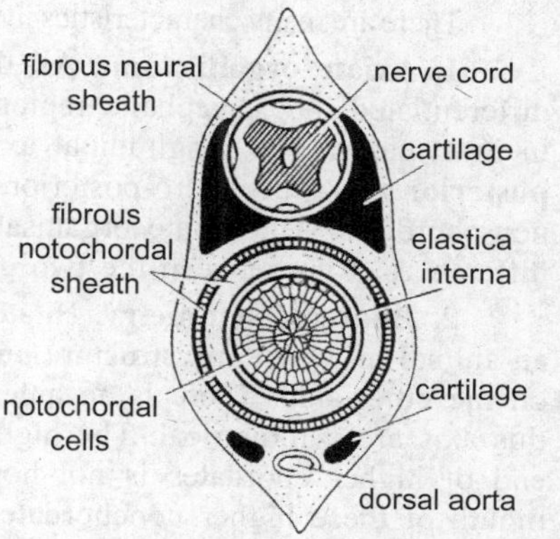

Fig. 1.2. Notochord of young dogfish.

2. Dorsal tubular nerve cord. In chordates, there is a dorsal, hollow or tubular, fluid-filled nerve cord lying just above the antero-posterior axis of the body (Fig. 1.2). The nerve cord is formed by an infolding of mid-dorsal strip of **neural ectoderm** and encloses a cavity or canal, the **neurocoel**. The nerve cord lies above the notochord and outside the coelom, extending lengthwise in the body. This tubular nerve cord persists throughout the life in chordates.

In vertebrates, the anterior region of nerve cord becomes specialised to form a cerebral vesicle or brain which is enclosed in a protective cartilaginous or bony cranium. The posterior part of nerve cord transforms into **spinal cord** which is protected within the neural canal of the vertebral column.

3. Pharyngeal gill-slits. The gill-slits have many alternative names, such as gill-clefts, pharyngeal or branchial-clefts, visceral-clefts, visceral or branchial-pouches. In all the chordates, at some stage of their life history, a series of paired internal gill-slits or gill-clefts perforate through the pharyngeal wall of the gut behind the mouth. Gill-slits are paired openings leading from the pharynx to the exterior. Lateral outpushing of the endodermal lining of pharynx meet and fuse with ectodermal inpushings from the exterior, the intervening walls being broken down, thus, gill-clefts are formed. Such gill-clefts appear during the development of every chordate, but in many aquatic forms they are lined with vascular lamellae which form gills for respiration. In terrestrial chordates which never breathe by gills, traces of gill-clefts are present during early development but disappear later in adults. Within these clefts are the gills in many aquatic animals. In many animals equipped with lungs, branchial-clefts, or branchial-grooves are always found in the embryo. In lower chordates, the visceral-clefts are used as a feeding mechanism throughout life, but in higher chordates, the gill-clefts often form some endocrine glands.

The gill-slits serve primarily for the passage of the water from the pharynx to outside, thus, bathing the gills for respiration. The water current secondarily helps in filter feeding by retaining food particles in the pharynx.

The above three common characters appear during early embryonic life of all the chordates. But all the above three characters rarely persist in the adult (*e.g., Branchiostoma*). Only one of these three primary characters of chordates is found in the adults of most of them, this is the dorsal tubular nerve cord, even this has its hollow lumen greatly reduced in some, and in Urochordata the central nervous system degenerates in the adult. The other two characters completely disappear in the adults of most chordates. However, the three primary characters are possessed only by chordates.

The three common chordate characters were probably characteristics of the ancestral chordates. These characters distinguish chordates from all other animals and appear to reveal their common ancestry.

CHARACTERS COMMON TO CHORDATES AND HIGHER NONCHORDATES

There are many characteristics in which chordates resemble the higher nonchordates or invertebrates.

1. Axiate organisation. All the chordates have a distinct polar axis. The anterior end is differentiated into a cephalic region or head and the posterior end is characterised by the **tail** in most cases. The longitudinal axis of body running from head to tail is known as **antero-posterior axis**. The antero-posterior axis of the chordates corresponds to that of most of the higher non-chordates. The axiate organisation is not strictly homologous, because many fundamental differences exist between the two groups.

2. Bilateral symmetry. Bilateral symmetry is the similarity of the right and left sides of an animal in which any structure on one side of the median place is an image of the same organ on the other side. There is an antero-posterior axis from the head to the tail, the two sides of this axis are symmetrical. The higher nonchordates also show bilateral symmetry, but the head end of higher chordates is not homologous with that of annelids or arthropods, because the mouth of these higher nonchordates is derived from the blastopore of the embryo, whereas in vertebrates the anus is formed near the blastopore, and the mouth is formed by an inpushing into archenteron at the opposite end.

3. Triploblastic condition. All Metazoa above the Coelenterata are triploblastic having three germinal layers called ectoderm, endoderm, and mesoderm, and it is assumed that the mesoderm of these groups is homologous. But in higher invertebrates the mesoderm arises as solid cord-like outgrowth from the junction of ectoderm and endoderm in gastrula stage, whereas in brachiopods, echinoderms and *Branchiostoma*, etc., the mesoderm is formed as lateral out-pushings of the archenteron of gastrula.

4. Coelom. Chordata are coelomate animals; they have a true coelom lined entirely by mesoderm. There are two distinct ways in which the coelom can be formed. In lower chordata (Cephalochordata) the coelom arises as direct paired pouches or outgrowths of the archenteron which fuse in a linear series (**enterocoelic coelom**), in Urochordata mesoderm arises as a solid mass in the same position as outgrowths of archenteron of Cephalochordata, but they have no coelom; but in higher forms such as higher chordates and higher nonchordates. In annelids, molluscs, arthropods and higher chordates, the coelom is formed from the lateral plate mesoderm splitting into an outer somatic and inner splanchnic layers which enclose the coelom (**schizocoelic coelom**).

5. Metamerism. When certain structures of an animal which resemble each other are repeated regularly one after the other, then each such division of the body is called a **metamere** (segment or somite), and the animal having a series of repeated metameres is said to be metameric or to be metamerically segmented. In higher chordates metamerism is not present externally but is seen only internally in nerves, blood vessels, muscles, vertebrae, ribs and excretory system, at least in their origin. Animals in which various metameres are almost alike are said to be **homonomous** in their segmentation, but this is a primitive condition and the various segments are more or less independent of each other. In the majority of animals the metameres differ from each other in several structures, such animals are said to have **heteronomous** segmentation in which the segments are dissimilar and there is a division of labour among the segments, this condition is an advanced one and has arisen from the homonomous condition. Vertebrates show only internal heteronomous segmentation, and that too during development. Metamerism in chordates is not homologous with that in metameric invertebrates (annelids, arthropods) but is only analogous. Metamerism is clearly seen in the embryo, but it becomes obscured by condensation in the adult in some organs, such as the kidneys and limbs, and it is completely lost in the head. Metameric condition is well marked in embryonic chordates in the arrangement of muscles and in the adult vertebrates in the arrangement of vertebrae.

6. Organ-systems. In an organ-system, several organs work together for the same function, such as digestion, circulation and respiration, etc. Organ-system grade of organisation is exhibited by all the chordates and all the nonchordates from nemertean worms onwards. However, the vertebrates exhibit a greater stage of development and fundamental unity of organ-system grade of organisation than even the highest invertebrates.

These structural similarities shared by the chordates and the nonchordates probably reflect upon their remote common ancestry. Although it is not possible to trace exact lineage, but all available evidences indicate that chordates have evolved from the invertebrates.

COMPARISON (DIFFERENCES) BETWEEN CHORDATES AND NONCHORDATES

So far we have seen that chordates share many structural similarities with the nonchordates (invertebrates) such as axiate organisation, bilateral symmetry, triploblastic condition, coelomate condition, metameric segmentation and organ-systems, etc. However, the chordates have several fundamental differences, the most striking have been shown in the Table 1.1.

TABLE 1.1. COMPARISON (DIFFERENCES) OF CHORDATA AND NONCHORDATA.

Features	Chordata	Nonchordata
1. Symmetry	Bilateral	Radial, biradial, bilateral or absent
2. Metamerism	True metamerism	True or pseudometamerism or absent
3. Postanal tail	Usually present	Absent
4. Grade of organisation	Organ-system	Protoplasmic to organ-system
5. Germ layers	Triploblastic (Three germ layers)	Diploblastic (2 germ layers), triploblastic (3 germ layers) or absent
6. Coelom	Truly coelomate	Acoelomate, pseudocoelomate or truly coelomate
7. Notochord	Present at some stage of life or replaced by a backbone made of ring-like vertebrae	Notochord or backbone absent
8. Pharyngeal gill-slits	Present at some stage of life	Absent
9. Blood vascular system	Closed	Open, closed or absent
10. Heart	Ventrally placed	Dorsal lateral or absent
11. Dorsal blood vessel	Blood flows posteriorly	Blood flows anteriorly
12. Hepatic portal system	Present	Absent
13. Haemoglobin	Contained in red blood corpuscles (RBCs)	Contained in plasma or absent
14. Respiration	Through gills or lungs	Through body surface, gills or tracheae
15. Nervous system	Hollow	Solid
16. Nerve cord	Single, hollow, tubular and dorsally placed	Double, solid and ventrally placed
17. Roots of segmental nerves	Dorsal and ventral separate	Dorsal and ventral roots not separate
18. Reproduction	Sexual reproduction predominant	Asexual reproduction predominant
19. Regeneration power	Usually poor	Usually good
20. Body temperature	Cold or warm blooded	Cold blooded

ORIGIN AND ANCESTRY OF CHORDATA

The chordates include organisms having a notochord, a dorsal tubular (hollow) nerve cord, pharyngeal gill-slits and a few features such as axiate organisation, bilateral symmetry, triploblastic condition, coelom, metamerism (segmentation), organ-systems, etc., that are common with the nonchordates. The question of the origin of chordates still remains unanswered and considerable controversy exists on this issue. Scientists have not succeeded in determining which lower forms have given rise to chordates. Their early ancestors most likely were soft-bodied and left no definite fossil remains. The geological records established beyond doubt that the chordates originated prior to Cambrian period as the oldest fossils of known vertebrates have been discovered in Cambrian strata. This fact has been emphasised by many workers on this line. Most scientists consider that the chordates have originated from invertebrates. Several theories have been proposed to explain the origin of chordates from nonchordate groups, but these theories have serious drawbacks and far from satisfactory. One theory advocates the origin of chordates from Echinodermata as such. Echinoderms are radially symmetrical animals and show no typical chordate characters, yet they seem to be most closely related to chordates. The auricularia larva of echinoderms is a free-swimming bilaterally symmetrical larva, and it very closely resembles the tornaria larva of hemichordates. But in recent years the ancestry of chordates from echinoderms source is not accepted.

Most zoologists, such as **Romer, Berril** and **Barrington,** etc., now favour the deuterostome line of chordate evolution, according to which the phyla Echinodermata, Hemichordata and Chordata show common ancestry on embryological and biochemical evidences. The protochordates provide the connecting link between early chordate ancestors and vertebrates. The differentiation probably occurred much earlier than Cambrian period. The earliest traces of vertebrates have been found in the rocks of the late Cambrian period and Ordovician period. A number of fishes followed in Silurian period and became abundant in the Devonian period. The subsequent periods show the evolution of amphibians, reptiles, birds and mammals.

MAJOR SUBDIVISIONS OF PHYLUM CHORDATA

(Outline Classification of Phylum Chordata)

Phylum Chordata is a heterogeneous assemblage of groups which differ widely from one another and exhibit various degrees of relationships to each other. In a classification of the phylum, these groups are customarily arranged in larger functional divisions or subdivisions based on specific structures or features. These subdivisions or taxa have been given different ranks under different systems of classification. The following terminology includes major subdivisions of phylum Chordata.

1. Subphyla and classes. Phylum Chordata is first conveniently divided into 3 or 4 primary subdivisions called **subphyla** based on the character of notochord. These are :

Subphylum 1. Hemichordata (or Adelochordata)

Subphylum 2. Urochordata (or Tunicata)

Subphylum 3. Cephalochordata

Subphylum 4. Vertebrata

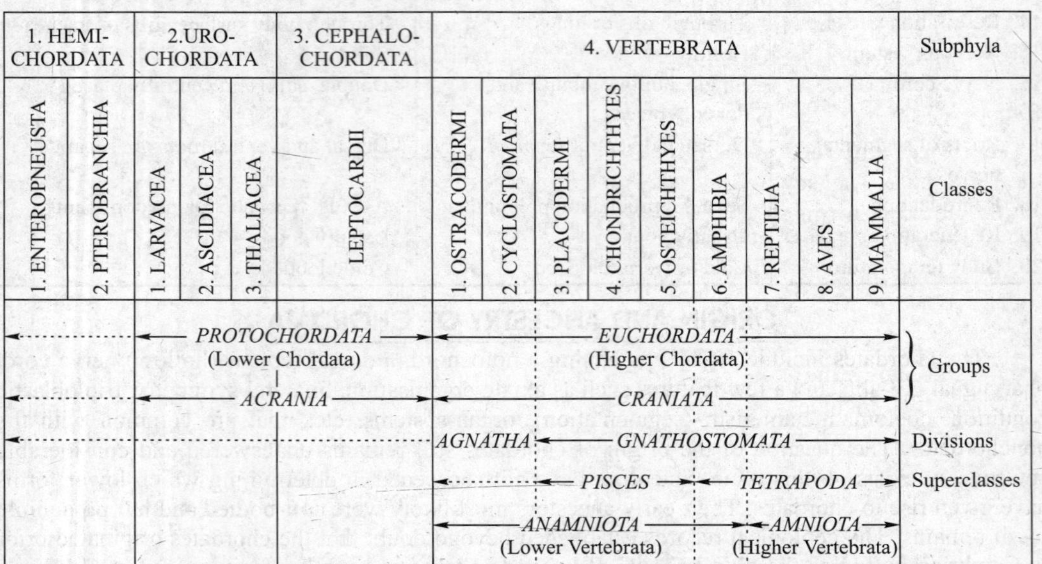

Fig. 1.3. Subdivisions of the phylum Chordata.

Subphylum Hemichordata (Gr., *hemi* = half; *chorde* = cord) have long been considered to be the lowest chordates. But recent workers consider the so called notochord of hemichordates not a true notochord but **stomochord**. In recent years the hemichordates have been separated from the phylum Chordata and given the status of a separate independent invertebrate phylum.

Subphylum **Urochordata** (Gr., *oura* = a tail; L., *chorda* = cord) includes 3 classes (Larvacea, Ascidiacea and Thaliacea), subphylum **Cephalochordata** (Gr., *kephale* = head; L., *chorde* = cord) includes a single class (Leptocardii), while the subphylum **Vertebrata** (L., *vertebratus* = backbone) is divided into 9 classes (Ostracodermi, Cyclostomata, Placodermi, Chondrichthyes, Osteichthyes, Amphibia, Reptilia, Aves and Mammalia).

2. Protochordata and Euchordata. The first two subphyla under phylum Chordata are Urochordata and Cephalochordata. They are all marine, relatively small and without a vertebral column or backbone. They are often collectively referred to as the **invertebrate chordates** or **nonvertebrates** or **protochordates** (Gr., *protos* = first; *chorde* = cord). They are regarded to be early, primitive or first chordates closely allied with the ancestral chordate stock. Earlier, the **Protochordata** were treated as a distinct phylum with the Hemichordata included as a third sub-phylum. At present, the Hemichordata is considered as an independent invertebrate phylum, while the Urochordata and Cephalochordata are considered to be true chordate subphyla. Therefore, the term "Protochordata" no longer includes any zoological taxonomic position, but is merely descriptive.

The third subphylum Vertebrata, possessing a vertebral column, is considered to be more advanced and belongs to the subdivision **Euchordata** of the phylum Chordata. Sometimes, the protochordates are known as the **lower chordates** or the euchordates as the **higher chordates**.

3. Acrania and Craniata. The protochordate subphyla lack a head and cranium, therefore, they are known as **Acrania** (Gr., *a* = absent, *kranion* = head). On the other hand, the subphylum Vertebrata (subdivision Euchordata) possess a distinct head and a cranium, therefore, they are known as **Craniata**. Typically the body of Craniata comprises a head, neck, trunk and tail.

4. Agnatha and Gnathostomata. The Vertebrata (or Craniata) are further subdivided into two unequal sections : Agnatha and Gnathostomata. **Agnatha** (Gr., *a* = not; *gnathos* = jaw) lack true jaws and paired appendages. Agnatha include a small number of primitive but highly specialised fish-like forms, the extinct ostracoderms and modern cyclostomes. All other vertebrates have true jaws and paired appendages and are called **Gnathostomata** (Gr., *gnathos* = jaw; stoma = mouth). In the past Agnatha and Gnathostomata were considered as subphyla of phylum Chordata. In the recent years, they are considered as branches or groups or superclasses of subphylum Vertebrata.

5. Pisces and Tetrapoda. Basically Gnathostomata is divided into two superclasses : Pisces and Tetrapoda. The superclass **Pisces** (L., *piscis* = fish) includes all the fishes which are strictly aquatic forms with paired fins. The superclass **Tetrapoda** (Gr., *tetra* = four; *podos* = foot) includes four-legged land vertebrates including amphibians, reptiles, birds and mammals.

6. Anamnia and Amniota. This method of grouping the vertebrates is based on the presence of a membrane, the **amnion**. The amnion holds the developing embryo in a reservoir of fluid and allows the laying of eggs on land. The animals that possess the amnion belong to **Amniota** which includes the classes Reptilia, Aves and Mammalia. The animals without amnion are **Anamniota** which includes cyclostomes, fishes and amphibians. Sometimes the anamniotes are referred to as the **lower vertebrates** and amniotes as **higher vertebrates**.

GENERAL CHARACTERS OF PHYLUM CHORDATA

1. Chordates are aquatic, aerial or terrestrial. All are free-living with no fully parasitic forms.
2. Body small to large, bilaterally symmetrical and metamerically segmented.

3. A **postanal tail** usually projects beyond the anus at some stage of life and may or may not persist in the adult.

4. Exoskeleton often present; well-developed in most vertebrates.

5. Body wall **triploblastic**, *i.e.,* presence of three germ layers: ectoderm, mesoderm and endoderm.

6. **Coelomate** animals, *i.e.,* a well developed true coelom is always present which is enterocoelic or schizocoelic in origin.

7. **Notochord** is always essentially present at some stage of life cycle. It is completely or partly replaced by vertebral column in majority of animals.

8. A cartilaginous or bony, living and jointed **endoskeleton** present in the majority of members (vertebrates).

9. Paired **pharyngeal gill-slits** are present on either side of the pharynx at some stage of life, may or may not be functional.

10. Digestive system is complete with digestive glands.

11. **Blood vascular system** is closed. Heart ventral with dorsal and ventral blood vessels. Hepatic portal system is present which is well developed.

12. **Excretory system** comprising proto-or meso-or metanephric kidneys.

13. **Nerve cord** is dorsal and tubular. Anterior end usually enlarged to form brain.

14. Sexes separate with rare exceptions.

BRIEF CLASSIFICATION OF PHYLUM CHORDATA WITH CHARACTERS

The chordates form a large heterogeneous group of members differing widely from one another in many respects. Due to great diversity in chordate forms, different schemes of classification have been proposed by a number of taxonomists from time to time. In the present text, the classification followed is simplified and is a synthesis of most recent classifications. Table 1.2 gives an outline classification of the phylum Chordata.

TABLE 1.2. OUTLINE CLASSIFICATION OF PHYLUM CHORDATA.

Phylum	Group	Subphylum	Division	Superclass	*Classes*	
Phylum Chordata	*Group* Acrania or Protochordata (Lower Chordata)	Subphylum *Hemichordata			1. Enteropneusta 2. Pterobranchia	
		Subphylum Urochordata			1. Ascidiacea 2. Thaliacea 3. Larvacea	
		Subphylum Cephalochordata			1. Leptocardii	
	Group Craniata or Euchordata (Higher Chordata)	*Subphylum* Vertebrata	*Division* Agnatha		1. Ostracodermi 2. Cyclostomata	**Anamniota** (Lower Vertebrata)
			Division Gnathostomata	*Superclass* Pisces	3. Placodermi 4. Chondrichthyes 5. Osteichthyes	
				Superclass Tetrapoda	6. Amphibia	
					7. Reptilia 8. Aves 9. Mammalia	**Amniota** (Higher Vertebrata)

**Subphylum Hemichordata is now considered to be an invertebrate group.*

PHYLUM CHORDATA

Widely diversified (differing) in size, habits and habitat. Bilaterally symmetrical, metamerically segmented, triploblastic, coelomate deuterostomes. All the chordates possess a supporting skeletal rod or notochord, a hollow dorsal nerve cord and paired gill-slits at some stage of their life history which may persist, change or disappear in adults. Cambrian to Recent. About 50,000 species.

Phylum Chordata can be divided into two groups : **Acrania** (Protochordata) and **Craniata** (Euchordata) which show contrasting characters.

Group A. Acrania (Protochordata)

(Gr., *a* = absent; *kranion* = head or Gr., *protos* = first; *chorde* = cord). All are marine, small, primitive or lower chordates. No cranium, jaws, vertebral column, paired appendages. About 2,000 species. The Acrania is divided into three subphyla : Hemichordata, Urochordata and Cephalochordata.

Subphylum I. Hemichordata

(Gr., *hemi* = half; *chorde* = cord). Body divided into 3 regions : proboscis, collar and trunk. Notochord doubtful, short confined to proboscis and non-homologous with that of chordates.

Class 1. Enteropneusta. (Gr., *enteron* = gut; *pneustos* = breathed). Body large and worm-like. Gill-slits numerous and paired. Alimentary canal straight. Acorn or tongue worms. Enteropneusts include 3 families, 15 genera and 70 species. **Examples:** *Balanoglossus, Saccoglossus, Ptychodera.*

Class 2. Pterobranchia. (Gr., *pteron* = feather; *branchion* = gill). Body small and compact. Gill-slits one pair or none, never U-shaped. Alimentary canal U-shaped. Pterobranchs include 2 orders, 3 genera and 20 species. **Examples:** *Cephalodiscus, Rhabdopleura.*

Class 3. Planctosphaeroidea. Transparent, round and specialised tornaria larva possessing extensively branched ciliary bands. Alimentary canal L-shaped. *Planctosphaera pelagica* is a representative of this class. This form is supposed to be specialised tornaria larva of some unknown hemichordate.

Class 4. Graptolita. The fossil graptolites (*e.g., Dendrograptus*) were abundant in Ordovician and Silurian periods. These are often considered as an extinct colonial class of Hemichordata. Their tubular chitinous skeleton and colonial habits show an affinity with *Rhabdopleura.*

Subphylum II. Urochordata or Tunicata

(Gr., *oura* = a tail; L., *chorda* = cord). Notochord and nerve cord in tail only in minute free-swimming tadpole larva. Adult sac-like, covered with tunic (test), often transparent; usually no notochord, nerve cord reduced to ganglion; no coelom, segmentation or nephridia.

Class 1. Ascidiacea. Sessile tunicates with scattered muscles in tunic. Solitary, colonial or compound. Gill-slits many; tunic well developed, permanent. Ascidians or sea squirts. Include 3 orders, 12 families, 37 genera and 1,200 species. **Examples:** *Herdmania, Ciona* and *Molgula.*

Class 2. Thaliacea. Free-swimming or pelagic tunicates with circular muscle bands in tunic. Salpians and chain tunicates. Include 3 orders, 5 families, 9 genera and 30 species. **Examples:** *Salpa, Doliolum, Pyrosoma.*

Class 3. Larvacea or Appendicularia. Tiny transparent, free-floating. Adults retain many larval features including tail. Only two gill-slits. Tunic not persistent. Include 2 orders, 2 families, 5 genera and 3 species. **Examples:** *Oikopleura, Appendicularia.*

Subphylum III. Cephalochordata

(Gr., *kephale* = head; L., *chorda* = cord). Notochord and nerve cord present throughout life along entire length of body. Include 1 class, 1 family, 2 genera and 30 species.

Class Leptocardii. Body fish-like, segmented with numerous gill-slits. Free-swimming and burrowing. Lancets. **Examples:** *Branchiostoma* (= *Amphioxus*), *Asymmetron.*

TABLE 1.3. CONTRASTING CHARACTERS OF PROTOCHORDATA AND EUCHORDATA.

Group A. Acrania (Protochordata)	Group B. Craniata (Euchordata) or Higher Chordata
1. Exclusively marine, small-sized chordates.	1. Aquatic or terrestrial, mostly large-sized vertebrates.
2. Appendages, head and exoskeleton absent.	2. Usually 2 pairs of appendages, well-developed head and exoskeleton present.
3. Coelom enterocoelic, arising from embryonic archenteron.	3. Coelom schizocoelic, arising by splitting mesoderm.
4. Notochord persistent. Skull, cranium and vertebral column absent.	4. Notochord covered or replaced by vertebral column. Skull and cranium well developed.
5. Pharynx having permanent gill-slits or gill-clefts. Endostyle present.	5. Pharyngeal gill-slits or gill-clefts persist or disappear. Endostyle absent.
6. Heart without chambers when present. Red blood corpuscles (RBCs) absent in blood.	6. Heart made of 2, 3 or 4 chambers. Blood contains RBCs.
7. Kidneys protonephridia.	7. Kidneys meso- or metanephridia.
8. Sexes separate or united. Reproduction asexual as well as sexual. Gonoducts usually absent.	8. Sexes separate. Only sexual reproduction Gonoducts always present.
9. Development indirect with a free-swimming larval stage.	9. Development indirect or direct, with or without a larval stage.

Group B. Craniata

Aquatic or terrestrial, usually large-sized, higher chordates or vertebrates with distinct head, a vertebral column, jaws and brain protected by a cranium or skull. The craniata includes a single subphylum **Vertebrata**.

Subphylum IV. Vertebrata

(L., *vertebratus* = backbone). Notochord supplemented or replaced by a vertebral column or backbone comprising of overlapping vertebrae. Body divisible into head, neck, trunk and tail. Usually dioecious. Vertebrata is the largest chordate subphylum including about 45,000 species.

The Subphylum Vertebrata is divided into two divisions : **Agnatha** and **Gnathostomata** having contrasting characters given in Table 1.4.

TABLE 1.4. CONTRASTING CHARACTERS OF AGNATHA AND GNATHOSTOMATA.

Division I. Agnatha	Division II. Gnathostomata
1. True jaws absent.	1. True jaws present.
2. Paired appendages absent.	2. Paired appendages (pectoral and pelvic) present.
3. Internal ear with 2 semicircular canals.	3. Internal ear with 3 semicircular canals.
4. Notochord persistent in adults.	4. Notochord persists or replaced by vertebrae.

Division I. Agnatha

(Gr., *a* = not; *gnathos* = jaw). Jawless fish-like vertebrates without true jaws and paired limbs.

Class 1. Ostracodermi. (Gr., *ostrakon* = shell; *derma* = skin). Several extinct orders of ancient primitive heavily armoured, Palaeozoic, world's first vertebrates, collectively called the ostracoderms. **Examples :** *Cephalaspis, Pteraspis, Birkenia, Thelodus.*

Class 2. Cyclostomata. (Gr., *cyclos* = circular; *stoma* = mouth). Body eel-like, long, cylindrical, skin smooth; without scales, jaws and lateral fins. Mouth rounded and suctorial. Gills 5 to 16 pairs. Heart 2-chambered. Parasites and scavangers. About 50 species. **Examples :** Lampreys (*Petromyzon*) and hagfishes (*Myxine*).

Division II. Gnathostomata

(Gr., *gnathos* = jaw; *stoma* = mouth). Jawed vertebrates having true jaws and paired limbs.

Gnathostomata has been further divided into two superclasses: Pisces and Tetrapoda. All the fishes and fish-like aquatic gnathostomes in the superclass **Pisces**, whereas all the four-footed terrestrial gnathostomes in the superclass **Tetrapoda**.

TABLE 1.5. CONTRASTING CHARACTERS OF SUPERCLASS PISCES AND TETRAPODA.

Superclass 1. Pisces	Superclass 2. Tetrapoda
1. Exclusively aquatic gnathostome vertebrates.	1. Aquatic or terrestrial. Some arboreal and aerial.
2. Paired limbs, if present, as fins.	2. Paired pentadactyle limbs present.
3. Median fins present.	3. Median fins absent.
4. Skin usually moist and scaly.	4. Skin usually dry not cornified (without scales).
5. Respiration aquatic, by gills.	5. Respiration aerial, by lungs.
6. Sense organs functional in water.	6. Sense organs function in air.

Superclass 1. Pisces

(L., *piscis* = fish). Fishes or fish-like aquatic forms with paired as well as median fins, gills and scaly skin.

Class 1. Placodermi. Several extinct orders of primitive earliest jawed fishes of Palaeozoic with bony head shield. Placoderms. A full-sized functional gill-arch precedes hyoid arch. **Examples :** *Climatius, Dinichthys.*

Class 2. Chondrichthyes. (Gr., *chondros* = cartilage; *ichthys* = fish). Mostly marine. Cartilaginous endoskeleton. Skin with minute placoid scales. Gill-slits not covered with operculum. Mouth and 2 nostrils ventral. Males with claspers. Cartilaginous fishes. Approximately 600 species. **Examples :** *Scoliodon* (dogfish), *Chimaera* (ratfish).

Class 3. Osteichthyes. (Gr., *osteon* = bone; *ichthys* = fish). Freshwater and marine. Bony endoskeleton. Skin having various types of scales (cycloid, ctenoid) other than placoid. Gill-slits covered with operculum. Males without claspers. Bony fishes. Approximately 20,000 species. **Examples :** *Labeo* (rohu), *Protopterus* (lungfish), *Hippocampus* (sea horse).

Superclass 2. Tetrapoda

(Gr., *tetra* = four; *podos* = foot). Land vertebrates with two pairs of pentadactyle limbs, cornified skin with lungs.

Class 1. Amphibia. (Gr., *amphi* = both; *bios* = life). Larval stage usually aquatic and breathes by gills. Adults typically terrestrial and respires by lungs. Living forms with moist glandular skin, no external scales. Typically 2 pairs of limbs, nostrils connecting mouth cavity. Heart 3-chambered. Amphibians. Approximately 2,500 species. **Examples :** *Rana* (frog), *Bufo* (toad), *Ambystoma* (salamander).

Class 2. Reptilia. (L., *reptilis* = covering). Terrestrial and aquatic tetrapods. Skin dry covered by ectodermal horny scales or scutes. Limbs typically 4, each with 5 clawed toes. Skeleton bony. Heart incompletely 4-chambered. Cold blooded. Respiration by lungs. Approximately 7,000 species.

Examples : *Hemidactylus* (wall lizard), *Uromastix* (spiny-tailed lizard), *Naja* (cobra), *Sphenodon* (tuatara), *Crocodilus* (crocodile).

Class 3. Aves. (L., *avis* = bird). Body covered with feathers. Forelimbs modified as wings usually adapted for flight. Hindlimbs for walking, perching or swimming. No teeth in beak. Heart 4-chambered. Lungs compact, with air-sacs. Warm blooded. Birds. About 9,000 species. **Examples :** *Struthio* (African ostrich), *Columba* (pigeon), *Gallus* (fowl).

Class 4. Mammalia. (L., *mamma* = breast). Body usually covered with hair. Skin with various glands. Female with mammary glands which secrete milk for suckling the young. Heart 4-chambered. Warm blooded, air breathing vertebrates. Mammals. Approximately 4,500 species. **Examples :** *Echidna* (spiny anteater), *Macropus* (kangaroo), *Rattus* (rat), *Homo* (man).

IMPORTANT QUESTIONS

Long Answer Questions

1. Classify phylum Chordata up to classes giving the important characters and examples of each group.
2. Discuss the features in which chordates show an advancement over nonchordates.
3. Classify phylum Chordata up to classes, giving the important characters and examples of each class.

Short Answer Questions

1. Write the main characteristics of phylum Chordata.
2. Compare the chordates with nonchordates.
3. Differentiate between :
   (*i*) Protochordata and Euchordata,
   (*ii*) Acrania and Craniata,
   (*iii*) Agnatha and Gnathostomata,
   (*iv*) Pisces and Tetrapoda.
4. Write short notes on :
   (*i*) Chorda dorsalis,
   (*ii*) Hemichordata,
   (*iii*) Pharyngeal gill-slits,
   (*iv*) Urochordata,
   (*v*) Tetrapoda.

Very Short Answer Questions

1. Name the chordate group in which solenocyte is found.
2. Name the chordate group in which branchial basket is found.
3. Name the chordate group in which Bursa fabricii is found.
4. What is chorda dorsalis?

Multiple Choice Questions

1. Which is not a chordate character?
   (*a*) Dorsal hollow tubular nerve cord
   (*b*) Longitudinal supporting rod-like notochord
   (*c*) A series of pharyngeal gill-slits
   (*d*) Diploblastic.
2. First jawed vertebrates belong to class
   (*a*) Cyclostomata
   (*b*) Placodermi
   (*c*) Ostracodermi
   (*d*) Chondrichthyes.
3. Endoskeleton of chordate animal is
   (*a*) Living structure
   (*b*) Living structure without growth
   (*c*) Living structure with growth
   (*d*) Non-living structure.
4. Notochord is confined to proboscis in
   (*a*) Urochordata
   (*b*) Hemichordata
   (*c*) Chordata
   (*d*) Cephalochordata.
5. Which of the following is present in all chordates?
   (*a*) Umbilical cord
   (*b*) Ventral tubular nerver cord
   (*c*) Dorsal tubular nerve cord
   (*d*) Pseudocoelom.

6. Placoid scales are present in
   (*a*) *Labeo* (*b*) *Scoliodon*
   (*c*) *Hippocampus* (*d*) *Protopterus*.
7. Amniota includes
   (*a*) Pisces, *Amphioxus* and reptiles
   (*b*) Amphibians, reptiles and mammals
   (*c*) Reptiles, Aves and mammals
   (*d*) Pisces, reptiles and mammals.
8. Which of the following is a chordate but not a vertebrate?
   (*a*) *Catla* (*b*) *Hippocampus*
   (*c*) *Amphioxus* (*d*) *Hemidactylus*.
9. Name the animal group whose members are considered doubtful chordates
   (*a*) Urochordata
   (*b*) Hemichordata
   (*c*) Cephalochordata
   (*d*) All of the above.
10. One of the following primitive class of chordates containing *Amphioxus* and a few related forms which are all marine
   (*a*) Urochordata
   (*b*) Cephalochordata
   (*c*) Hemichordata
   (*d*) Vertebrata.
11. Which of the following is not a characteristic of members of Urochordata or Tunicata?
   (*a*) Notochord and hollow central nervous system is present only in the larval stage
   (*b*) The body is externally covered by a thick cellulose cuticle or tunic
   (*c*) The coelom is absent
   (*d*) None of the above.
12. Name the primitive worm-like animals having certain chordate features such as gill-clefts and a small notochord in the anterior region of the head only
   (*a*) Urochordata
   (*b*) Cephalochordata
   (*c*) Hemichordata
   (*d*) None of the above.
13. Which of the following is a characteristic of members of the subphylum Urochordata?
   (*a*) Worm-like animals that live in burrows in the muddy bottom of the shore
   (*b*) Both the notochord and the nerve cord are well developed
   (*c*) Notochord found in the larva and is confined to the tail only
   (*d*) Cranium and brain box distinct.
14. Chordates with a backbone are called
   (*a*) Protochordates (*b*) Invertebrates
   (*c*) Hemichordates (*d*) Vertebrates.
15. Which of the following animal is not a vertebrate?
   (*a*) Frog (*b*) Fish
   (*c*) Sea horse (*d*) Sea squirt.
16. Which of the following is anamniote animal?
   (*a*) *Labeo* (*b*) *Crocodilus*
   (*c*) *Columba* (*d*) *Rattus*.

ANSWERS

1. (*d*)	2. (*b*)	3. (*c*)	4. (*b*)	5. (*c*)	6. (*b*)	7. (*c*)	8. (*c*)
9. (*b*)	10. (*b*)	11. (*d*)	12. (*c*)	13. (*c*)	14. (*d*)	15. (*d*)	16. (*a*).

2

Chapter

Type 1. *Balanoglossus* (A Tongue Worm)

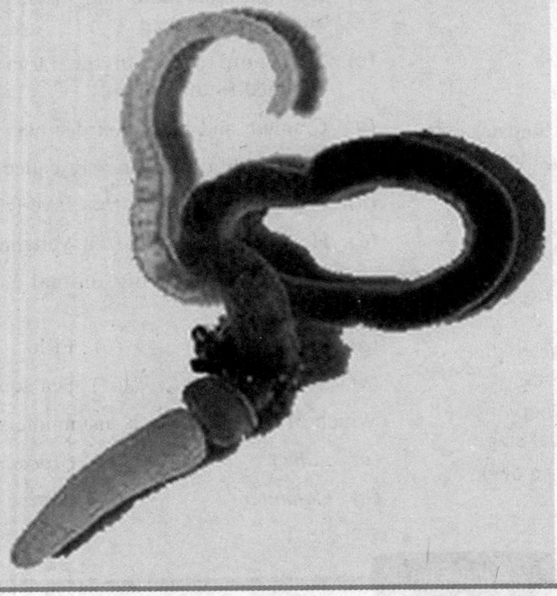

Contents of this Chapter

Hemichordata (Gr., *hemi* = half; *chorde* = cord) or **Adelochordata** are tongue worms which have been included for long as a subphylum or class of Phylum Chordata in the lower chordates or Protochordata, but now they are placed as an independent phylum of invertebrates close to Echinodermata, since their supposed affinities with chordates are doubtful. However, for convenience of the comparative study of the protochordates, the Hemichordata is being retained here as a chordate subphylum. Hemichordata are a small group of marine, solitary or colonial, worm-like, enterocoelous animals, most of which live in tubes. The soft, fragile body breaks readily in collecting. The body and enterocoelous coelom are divisible into three unequal regions, *viz,* **proboscis**, **collar** and **trunk**. In most forms one to several pairs of gill-clefts lead from the pharynx to the exterior. Nervous system is a network of nerve cells and fibres on the surface of the body, embedded in the epidermis. They have no bony tissue, but a buccal diverticulum is present in the preoral region, which is often regarded as a notochord. There is no tail. Development is direct in some, while in others there is a **tornaria larva**. The phylum has 100 species.

The most familiar hemichordate genus is *Balanoglossus* which belongs to the class Enteropneusta. Other closely related genera are *Saccoglossus* (= *Dolichoglossus*), *Glossobalanus*, *Ptychodera, Spengelia*, etc.

SYSTEMATIC POSITION

Phylum	...	Chordata
Subphylum	...	Hemichordata
Class	...	Enteropneusta
Family	...	Ptychoderidae
Genus (Type)	...	*Balanoglossus* (Tongue worm)

Balanoglossus (Gr., *balanos* = acorn; *glossa* = tongue) belongs to class **Enteropneusta** which have a straight digestive tube with mouth and anus at opposite ends. They have numerous gill-clefts but have no anus. They are burrowing animals ranging in size from about 2 cm to 50 cm. *Balanoglossus* has about 20 species.

HABITS AND HABITAT

Balanoglossus is a tubicolous or burrowing and exclusively marine animal. It is found in shallow waters between tide marks along the coast of warm and temperate oceans. *Balanoglossus* is world-wide in distribution.

Burrow. *Balanoglossus clavigerus* is **tubicolous** living in U-shaped burrows excavated in the sandy bottom. The burrows has two vertical limbs 50 – 70 cm deep and the two openings 10 – 30 cm apart. The burrows are open at both ends, and spiral coils of faeces like the castings of earthworms may be seen at the posterior opening which is rounded and hidden below the spirally coiled faeces. The anterior opening of the burrow is funnel-shaped and exposed. In its burrow *Balanoglossus* lies in a twisted condition but its anterior and posterior extremities are straight. **Knight Jones** (1952) reported that the animal moves in its burrow with the help of cilia present all over the body.

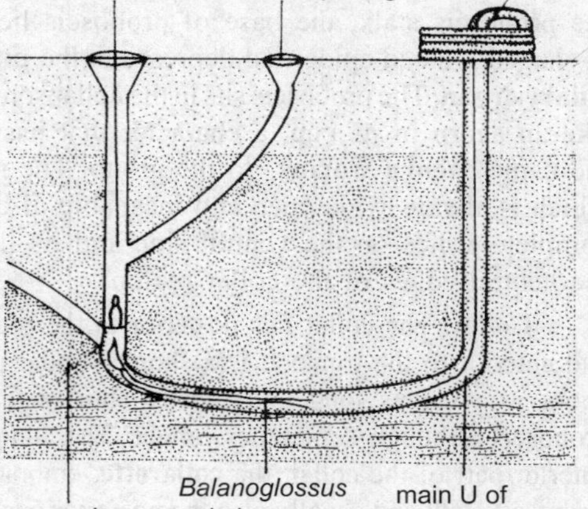

Fig. 2.1. *Balanoglossus* in U-shaped burrow. (After Young, 1962).

Protective device. The sandy walls of the tube are lined with mucus secreted by the skin mucous glands of the animal. This prevents the collapse of the tube protecting the animal from burial in sand. The animal also secretes a foul-smelling iodoform-like secretion which probably serves as a protective device. In *B. biminiensis*, 2.6–dibromophenol is secreted which may serve as a disinfectant of the burrow. Iodine is present in the hepatic regions, probably organically bound (**Barrington**, 1974). One species of *Balanoglossus* is phosphorescent.

Movements. Knight Jones (1952) reported that the animal moves in its burrow with the help of cilia present all over the body. Muscular contraction and expansion of the proboscis helps in burrowing. After the tide recedes, the animal protrudes its proboscis out of the burrow to explore the surroundings for food.

EXTERNAL MORPHOLOGY

Shape, size and colouration. The body of *Balanoglossus* is soft, elongated, cylindrical, bilaterally symmetrical, being richly ciliated all over and covered with mucus. The length of animal varies from 2 cm to 2.5 meters according to species. Most forms are drab coloured, though reddish tints are present, several species are luminiscent due to mucus. They have an offensive odour.

Division of body. The body is unsegmented and divided into three regions, *viz,* **proboscis** or **protosome**, **collar** or **mesosome**, and **trunk** or **metasome** (2.2).

Proboscis. The proboscis forms the anterior part of the body and is either rounded or conical in shape. It is continued posteriorly into a short, narrow **neck** or **proboscis stalk** which is attached with the collar. The proboscis is hollow and has thick muscular walls. Its cavity the proboscis coelom opens to the outside by means of a small opening called the **proboscis-pore** situated mid-dorsally near its base. In certain cases there are two proboscis-pores. In some species the proboscis-pore does not communicate with the outside environment, but terminate blindly. Beneath the proboscis stalk, the base of proboscis lies in a U-shaped ciliated epidermal depression, the **pre-oral ciliary organ**. The proboscis sits in the collar somewhat like an acorn in its cup, a character that has given the name **"acorn worms"** to the group. The mouth, which is always wide open and incapable of closing completely, lies on the ventral side and its lips are the ventral edges of the collar region.

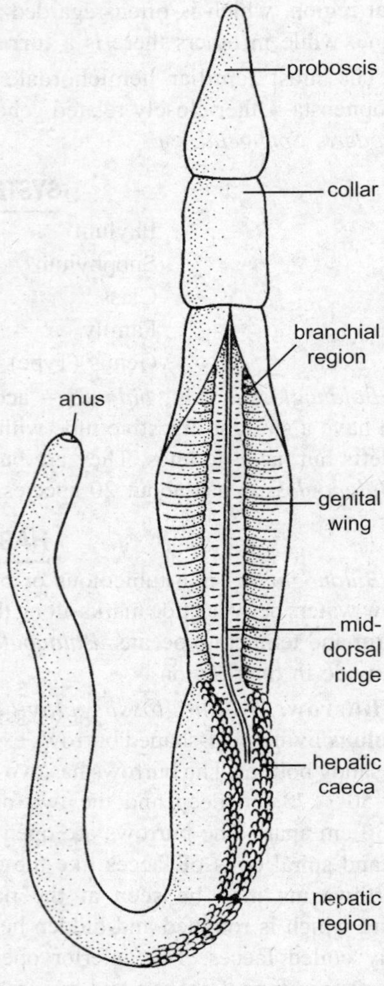

Fig. 2.2. *Balanoglossus*. External features in dorsal view.

Collar. The collar lies posterior to the proboscis and anterior to the trunk. It is a short cylinder usually about as wide as long and mostly shorter than the proboscis although sometimes longer. The funnel-like anterior part of the collar, the **collarette**, embraces the proboscis stalk and usually also the posterior part of the proboscis. Posteriorly the collar is sharply demarcated from the trunk by a circular indentation. The surface of the collar is often marked with elevations, depressions, and specially circular grooves. The collar is also muscular and possesses two coelomic cavities, called the **collar coelom**. The right and left coelomic cavities are separated from one another by dorsal and ventral mesenteries. The coelomic cavities of collar are completely cut off from the proboscis cavity. The collar cavity as well as the proboscis cavity are crossed by numerous strands of connective tissue which give the region a spongy appearance. The collar cavity communicateds with the exterior by a pair of short ciliated collar tubes (canals) leading into the first pair of gill-pouches.

Locomotion. The functional significance of the cavities and water pores in the proboscis and collar may best be explained through a description of the burrowing habits. When on the surface of the sandy bottom *Balanoglossus* pushes the tip of the proboscis into the sand, moving it around by muscular contractions until a shallow, cylindrical hole is made. Then the proboscis empties

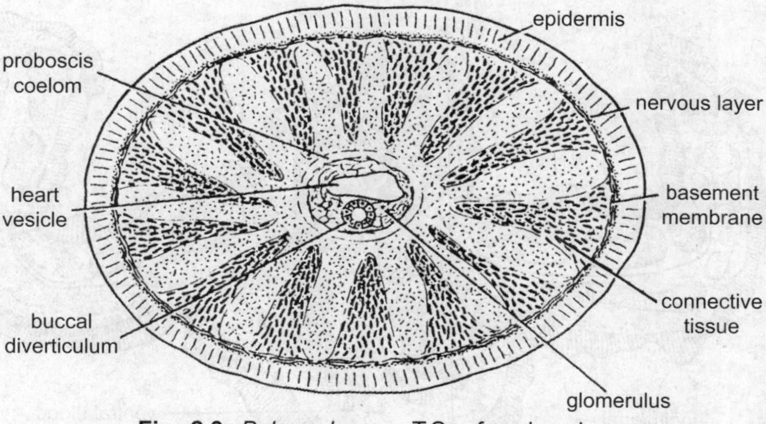

Fig. 2.3. *Balanoglossus*. T.S. of proboscis.

its water content through its pore and collapses. This allows the collar to enter the hole. By taking in water through the pores the collar expands so as to fit tightly into the hole like a cork in a bottle. The well-filled collar then gives a point of resistance for further rooting movements of the refilled proboscis, which loosens sand and stores it into the scoop-shovel mouth. Then both proboscis and collar relax and the latter squirms deeper into the hole before tightening its hold again. Once the collar gets a firm grip, the animal makes rapid progress and soon buries itself. The tail end is left near the surface, and at intervals comes out and deposits a pile of castings somewhat after the fashion of earthworms.

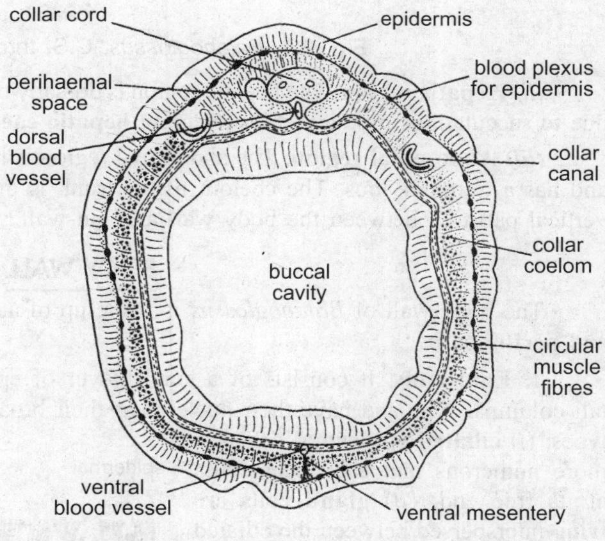

Fig. 2.4. *Balanoglossus*. T.S. of collar.

Trunk. The trunk is the elongated posterior part of the body. It is somewhat flat and annulated on the surface. It has a mid-dorsal and a mid-ventral longitudinal ridge along its entire length for the accommodation of corresponding nerve and blood vessel. The trunk is divisible into three parts : an anterior **branchiogenital region**, a middle **hepatic region**, and a posterior **abdominal** or **post-hepatic region**.

(i) **Branchiogenital region.** On the dorsal surface of the branchiogenital region of the trunk is a double row of small pore—the **branchial apertures,** one on either side of mid-dorsal ridge. Each row is situated in a long furrow. These pores increase in number during growth. In some species the most anterior are overlapped by a posterior prolongation of the collar called the **operculum**. The branchiogenital region has a pair of lateral, thin, flat and longitudinal flaps. In these the genital ridges, gonads are situated. In some genera, the genital ridges are so prominent that they form a pair of wing-like lateral folds, the **genital wings**, but in other genera folds are absent.

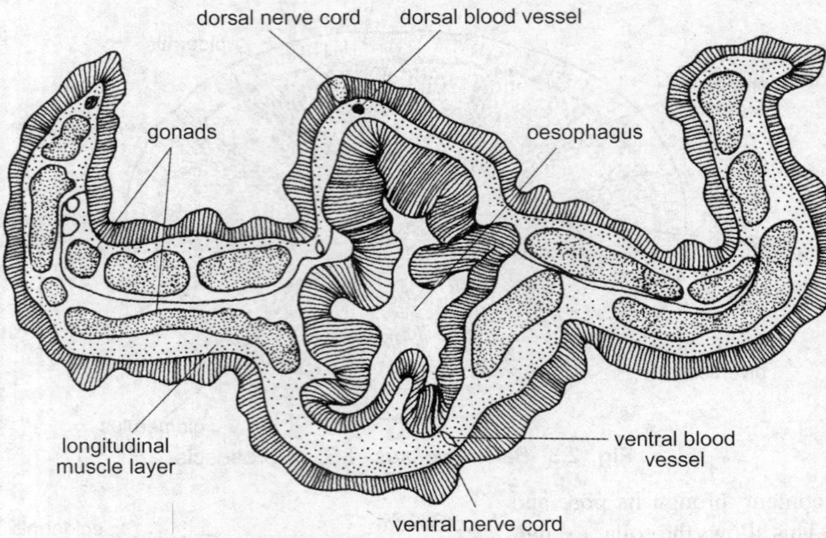

Fig. 2.5. *Balanoglossus*. C.S. through genital wings.

***(ii)* Hepatic region.** The hepatic region is dorsally marked externally with irregular elevations due to sacculations produced by projecting **hepatic caeca** of the intestine along with body wall.

***(iii)* Abdominal region.** The abdominal region is longest and cylindrical. It tapers gradually and has a terminal anus. The coelom of the trunk is divided into two lateral closed cavities by vertical partition between the body wall and gut-wall.

BODY WALL

The body wall of *Balanoglossus* is made up of an outer **epidermis**, an inner **musculature** and **peritoneum**.

1. Epidermis. It consists of a single layer of epithelial cells. The epithelial cells are of tall columnar type and have their nuclei near their broader bases. These cells are mainly of two types: (*i*) **ciliated epidermal** cells are more numerous and each bears cilia at its free end, (*ii*) **gland cells** are lying interspersed between the ciliated epidermal cells and are further of three kinds: (*a*) **goblet cells** are flask-shaped and secrete mucus; (*b*) **reticulate cells** are long cells with vacuolated cytoplasm which also secrete mucus; (*c*) **mulberry cells** are long cells containing coarse cytoplasmic granules and, hence, are also called **granular gland cells**. They secrete amylase. The mucus, secreted by gland cells, covers the animal and lines its burrow. The mucus has an obnoxious smell. In addition to these cells, the body wall of proboscis and anterior part of the collar also contain **neurosensory cells** which take darker stain than the rest. There is no dermis.

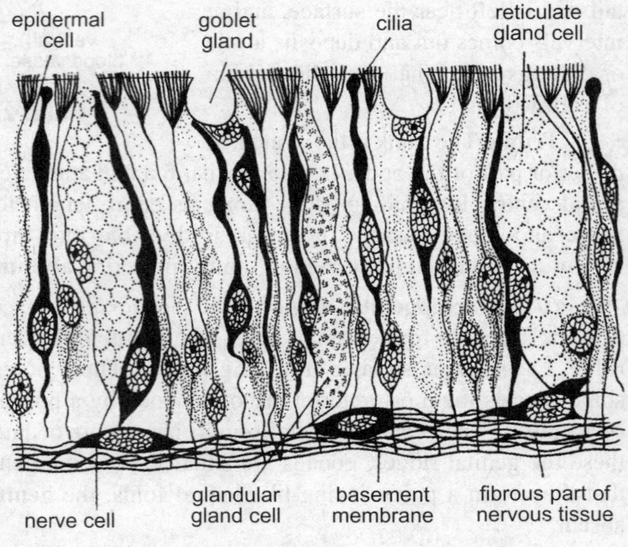

Fig. 2.6. *Balanoglossus*. V.S. of body wall.

Immediately below the epidermis is a thick **nervous layer** consisting of bipolar and quadripolar nerve cells and fibres which form a network lying in close contact with the epidermal cells. This layer is traversed by the filamentous bases of the epidermal cells that are connected with the basement membrane. The fibres of sensory epidermal cells synapse with the fibres of nerve cells.

Below the nervous layer is a thick **basement membrane** made up of two lamellae pressed together. The basement membrane supports the epidermis and serves for attachment of underlying muscles.

2. Musculature. The musculature of typical body wall and gut wall is greatly reduced and more or less replaced by muscles arising from the coelomic epithelium. The muscle fibres are smooth and of circular, longitudinal and diagonal types. The muscle layer lies below the basement membrane.

The **proboscis musculature** comprises a thin layer of circular muscle fibres and a thick layer of longitudinal muscle fibres. The longitudinal muscles fibres obliterate the proboscis coelom and some of the fibres cross one another diagonally. The **collar musculature** is confined to the collarette and consists of an inconspicuous layer of circular fibres and prominent bands of longitudinal and diagonal fibres. The longitudinal and diagonal fibres, along with connective tissue, also traverse the collar coelom in a criss-cross pattern. The **trunk musculature** consists mainly of moderately developed longitudinal muscle fibres which are better developed on the central side. The muscle layer is interrupted by the dorsal and ventral mesenteries and the lateral septa. Several radial muscle fibres are also found in the trunk region. The radial muscle fibres extend between the digestive tract and the body wall and traverse the trunk coelom.

3. Peritoneum. It is found just beneath the musculature and outside the coelom. It is also called the coelomic epithelium.

Functions of body wall. (*i*) The body wall protects the tender internal organs from mechanical injuries. (*ii*) Mucus secreted by gland cells adhere sand particles of burrow formed by the animal. (*iii*) Foul smell of mucus is protective. (*iv*) Neurosensory cells receive and transmit the stumuli. (*v*) Musculature helps in locomotion.

COELOM

The coelom is enterocoelous having been formed as outgrowths of the enteron. Corresponding with the three body regions the coelom is divided into three portions which are completely separated from each other by septa. The coelom is lined with coelomic epithelium or peritoneum. But enteropneusts are peculiar in that their coelomic epithelium has connective tissue and muscle fibres which fill much of the original coelomic cavities, and a distinct peritoneal lining has disappeared, moreover the coelomic musculatutre largely replaces the body wall muscles. The three parts of the coelom are an unpaired proboscis coelom, a pair of collar coeloms, and a pair of trunk coeloms.

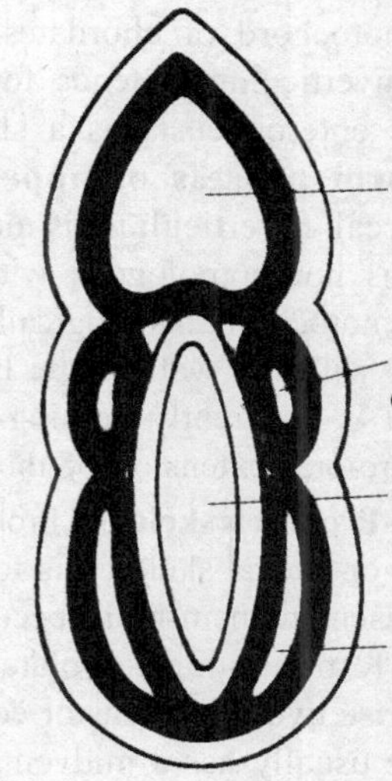

Fig. 2.7. *Balanoglossus.* Diagram of tripartite embryo showing coelomic cavities.

1. Proboscis coelom. The proboscis coelom or **protocoel** is a single space in the proboscis which is largely occupied by muscles and connective tissue and a few structures like buccal diverticulum, glomerulus and central sinus or heart. Dorsally, towards the posterior side, the proboscis coelom is divided by a dorsal mesentery into right and left dorso-lateral compartments which extend into the proboscis stalk; the left compartment is larger than the right and communicates with the exterior through the **proboscis pore** situated mid-dorsally at the base of the posterior stalk. Ventrally the proboscis coelom is divided by a ventral mesentery into right and left ventro-lateral compartments which are continuous behind the mesentery.

2. Collar coelom. The collar coelom or **mesocoel** has two cavities lying side by side in the collar, one on each side between the collar wall and buccal cavity. The two cavities are partitioned by incomplete mid-dorsal and mid-ventral mesenteries. The collar coelom does not communicate with the proboscis coelom, but posteriorly, its each cavity opens into the first gill sac of its side by a canal called **collar canal**. Each collar coelom opens to the exterior by a **collar pore**. The collar coelom is greatly obliterated by the collar musculature and connective tissue.

3. Trunk coelom. The trunk coelom or **metacoel** has two closed cavities lying between the body wall and alimentary canal. The two cavities are separated by an incomplete dorsal and a complete ventral mesentery. In the branchio-genital region each cavity is further divided by a **lateral septum** into a dorso-lateral and ventro-lateral compartment. The trunk coelom is separated from the collar coelom by a **collar-trunk septum**. The trunk coelom is obliterated by the trunk musculature.

Coelomic fluid. The proboscis and collar coeloms communicate with the exterior and get filled with sea water through their pores, which keeps them turgid. The trunk coelom is filled with a watery coelomic fluid having amoeboid coelomocytes. The coelomocytes originate from the coelomic epithelium. Each coelomocyte possesses a single large vacuole. According to **Spengel,** they behave like leucocytes by secreting a membrane around any foreign body that may invade the animal.

ENDOSKELETON

In *Balanoglossus,* there is no definite endoskeleton but there are four structures of a supporting nature, they are a **buccal diverticulum**, **proboscis skeleton**, **branchial skeleton**, and a **pygochord**.

1. Buccal diverticulum. Buccal diverticulum is a hollow pre-oral outgrowth, extending from the roof of the buccal cavity into the proboscis. It was for a long time also called a **notochord** or a **stomochord** on the assumption that it represents the anterior portion of the notochord of chordates. The buccal diverticulum extends forward in some enteropneusts as a slender **vermiform process** or **appendix**. The buccal diverticulum is neither analogous nor homologous with the chordate notochord; histologically it is identical with the wall of the buccal cavity, it is apparently nothing more than a pre-oral extension of the gut.

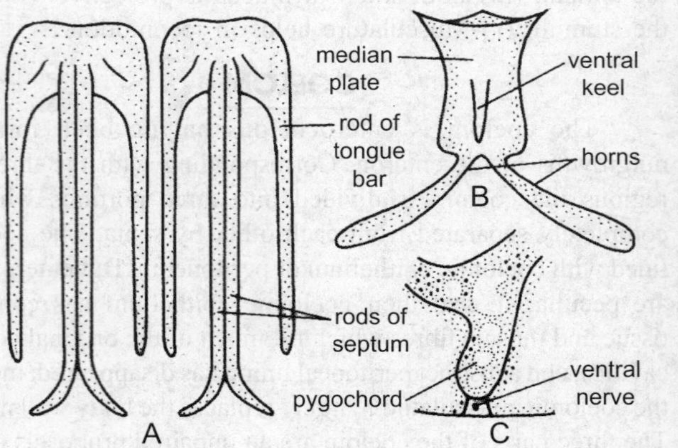

Fig. 2.8. *Balanoglossus*. Skeletal elements. A–Branchial skeleton; B–Proboscis skeleton; C–Pygochord.

2. Proboscis skeleton. Proboscis skeleton or nuchal skeleton is formed by the basement mem-brane becoming thick to form a laminated plate from which arise two thin horns or cornua, the plate usually has a midventral keel. The proboscis skeleton lies in the proboscis stalk, while its horns extend into the roof of the buccal cavity.

3. Branchial skeleton. The walls of the U-shaped gill-clefts are supported by skeletal rods called primary and secondary gill-rods formed by thickening of the basement membrane.

4. Pygochord. Pygochord is longitudinal rod-like structure extending from the ventral side of the intestine to the body wall, its function is not known but it may support the soft abdominal region.

DIGESTIVE SYSTEM

Alimentary Canal

In *Balanoglossus*, the alimentary canal is a straight tube. Its anterior opening, the **mouth**, is wide and circular and situated on the ventral side in a groove between the proboscis stalk and collarette. The mouth remains open constantly. The posterior opening or the anus is a circular aperture at the extreme posterior end of the trunk. Between the mouth and anus, the alimentary canal can be distinguished into four regions : **buccal tube**, **pharynx**, **oesophagus**, and **intestine**. Their walls are composed of ciliated epithelium lined externally by basement membrane and devoid of muscle fibres.

1. Mouth. It is a wide gap at the base of proboscis, lying ventrally between collarette and the proboscis stalk. According to **Knight Jones** (1952), it can be closed or opened and does not remain open permanently. The mouth is provided with two sets of muscle fibres, the radial fibres to open it and the concentric fibres to closed it. The mouth leads into buccal cavity.

2. Buccal cavity. The short buccal cavity lies in the collar region. It is lined with ciliated and glandular cells. From its roof arises a stiff, hollow buccal diverticulum extending into the proboscis coelom. Posteriorly it extends up to the collar-trunk septum and then extends into the trunk as pharynx.

3. Pharynx. Pharynx lies in the branchial region of the trunk. Its wall bears a longitudinal constriction along each lateral side. These lateral constrictions project into its lumen as ridges, called **parabranchial ridges** consisting of tall columnar cells. These ridges and constrictions incompletely divide the pharynx into a dorsal **branchial portion** (pore pharynx) and a ventral **digestive portion** (digestive pharynx).

(*i*) Branchial portion of pharynx. The dorsal branchial portion of pharynx is perforated dorso-laterally by two rows of U-shaped gill-slits. It is concerned with respiration.

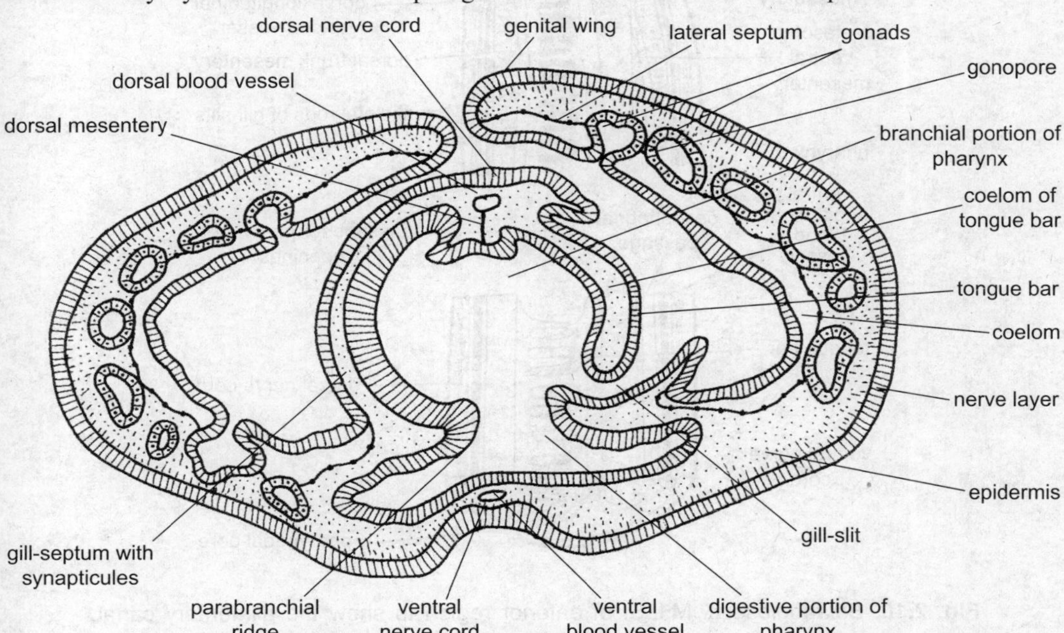

Fig. 2.9. *Balanoglossus.* T.S. through pharyngeal or branchiogenital region.

(*ii*) Digestive portion of pharynx. The digestive portion of pharynx is concerned with the food concentration, digestion and absorption of food. Its ciliated epithelial wall contains gland cells.

4. Oesophagus. Behind the last pair of gill-slits the pharynx continues into the oesophagus. The dorsal and ventral division of the pharynx continue for some distance into the oesophagus;

in this region, the dorsal part of the oesophagus is called **post-branchial canal** which possesses thick, folded and glandular epithelium. The posterior part of the oesophagus reduce in diameter and has deeply furrowed epithelium.

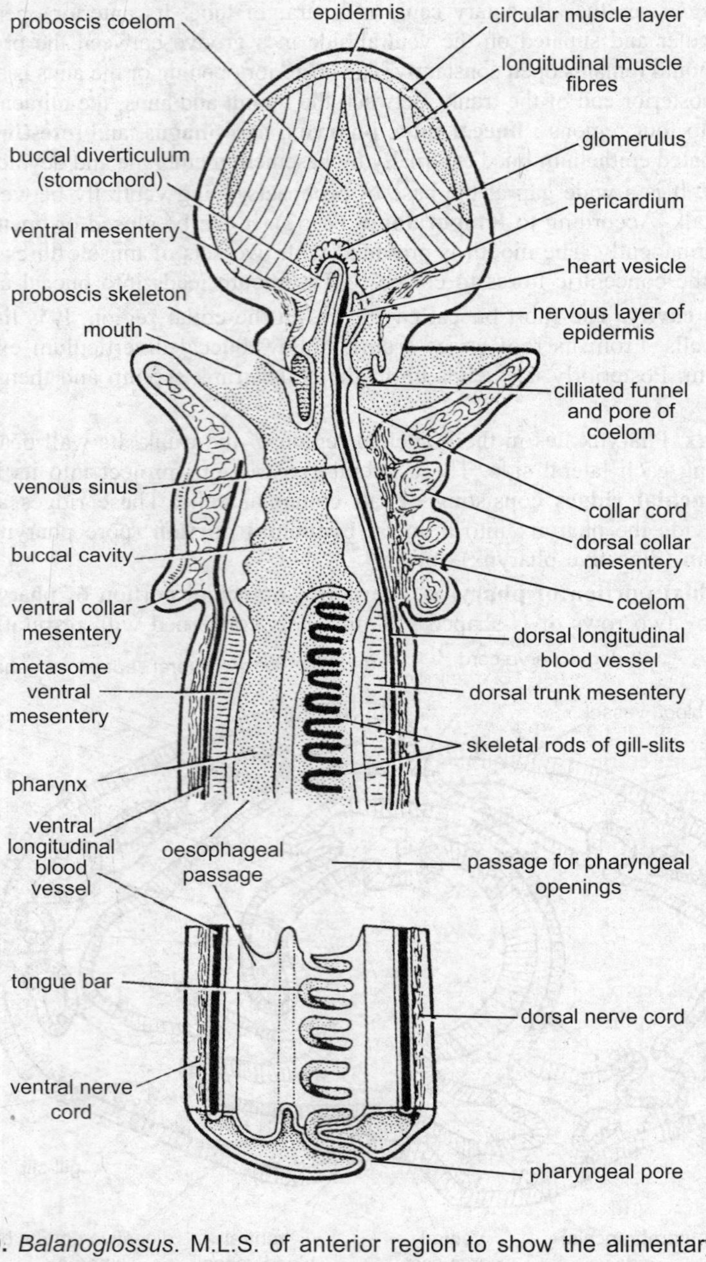

Fig. 2.10. *Balanoglossus*. M.L.S. of anterior region to show the alimentary canal.

5. Intestine. Behind the oesophagus is an intestine. It occupies the hepatic and post-hepatic regions of the trunk. The **hepatic region** of the intestine is highly vascular. Its epithelial cells are dark green or dark brown and its dorsal wall forms numerous prominent sacculations called **hepatic caeca** which push the body wall outwards and are, thus, visible externally. The **post-hepatic region** of the intestine is connected with the ventral body wall by the pygochord. The intestine has the form of a simple tube and bears a pair of dorso-lateral grooves lined by tall epithelial cells bearing long cilia.

6. Anus. Posteriorly the intestine opens out through the **anus** situated at the extreme hind end of the body. The anus often possesses sphincter muscles.

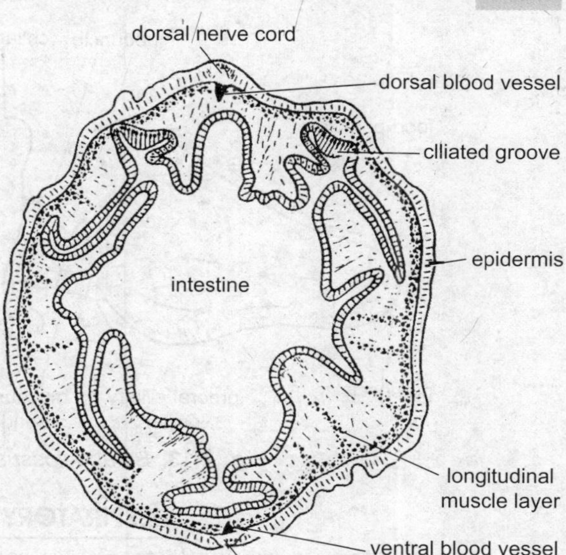

Fig. 2.11. *Balanoglossus*. T.S. through post-hepatic intestine.

Food, Feeding and Digestion

Balanoglossus is a "**ciliary feeder**". Its food comprises microscopic organisms and minute organic particles present in water and mud of bottom sand in which it makes its burrows. The lateral cilia lining the gill-slits set up a current of water directed backward which enters through the mouth, takes its course through the buccal tube, pharynx, gill-slits and branchial sacs, and leaves through gill-pores. This is called **respiratory-cum-food current**. Some food particles directly enter the mouth with this current, while some come in contact with the proboscis and are entangled in the mucus that covers it. The mucus is secreted by the epithelial gland cells of the proboscis epithelium. Cilia covering the probosocis direct the mucous string, containing food particles towards the pre-oral ciliary organ at the base of the proboscis. From here the mucous string is passed back into the mouth by the action of the proboscis cilia, assisted by the main water current entering the mouth. Organic particles present in the mud or sand are ingested directly along with mud or sand at the time of burrowing.

At the base of the proboscis, on the ventral side, there is a U-shaped depression bordered by long epidermal cells bearing long cilia. This structure is called the **pre-oral ciliary organ**. It tests the quality of food and the water entering the mouth. Undesirable substances are prevented from entering the mouth by the ventral part of the collarette which does so by covering the mouth. Thus, the rejected particles, instead of entering the mouth, pass back over the collar.

Backward movement of food through the alimentary canal is maintained by the cilia lining its walls. In the pharynx, the food passes through the ventral digestive portion. The exact process of digestion in *Balanoglossus* is not known with certainty, however, the digestion of food is brought about by the enzymes secreted by proboscis, gland cells of the pharynx, oesophagus and hepatic region of the intestine. Proboscis secretes mucus, which contains amylase is ingested with the food. The gland cells of pharynx and oesophagus also secrete enzymes. It is also claimed that hepatic caeca secrete amylase, maltase, lipase and weak protease. The enzymes digest the organic particles in the mud or sand. Undigested substances, along with mud or sand, pass out through the anus in large quantities which forms piles of "**castings**" at the posterior opening of the burrow.

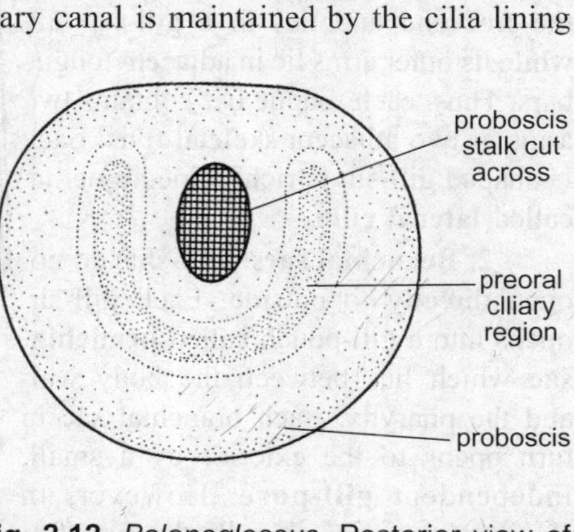

Fig. 2.12. *Balanoglossus.* Posterior view of proboscis showing the preoral ciliary organ.

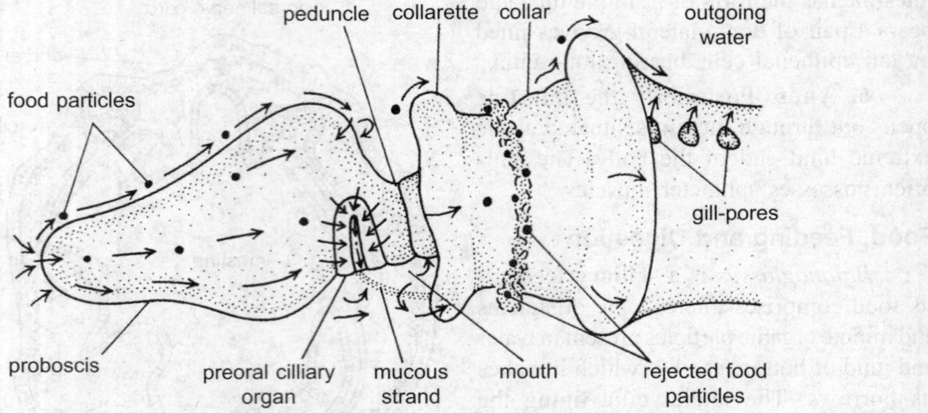

Fig. 2.13. *Balanoglossus*. Feeding current.

RESPIRATORY SYSTEM

The respiratory organs of *Balanoglossus* comprise: (*i*) the branchial portion of pharynx bearing gill-slits, and (*ii*) the branchial sacs that open out through gill-pores.

1. Branchial pharynx. As already described, two lateral longitudinal **para-branchial ridges** divide the pharyngeal cavity into a dorsal respiratory or **branchial** portion and a ventral digestive portion. Dorso-laterally, on each side, the branchial portion of pharynx is perforated by a longitudinal series of numerous U-shaped openings, the gill-slits. The number of gill-slits varies and increases as the animal grows older. Each gill-slit is a broad oval slit in the beginning, but later, a hollow projection of dorsal pharyngeal wall, called **tongue bar**, grows into the slit making it U-shaped. The hollow tongue bars enclose coelomic cavity and do not touch the ventral side of the gill-slits. The portions of the pharyngeal wall between two adjacent U-shaped gill-slits are called **gill-septa**. The gill-septa are solid and do not enclose coelom. A tongue bar is connected with its adjacent gill-septa by longitudinal connections are called **synapticula**. The tongue-bars and gill-septa are supported by M-shaped skeletal rods. The middle arm of an M-shaped rod is bifurcated at the free end and lies in a gill-septum while its outer arms lie in adjacent tongue bars. Thus, each tongue bar contains two arms of two adjacent skeletal rods. Each U-shaped gill-slit is richly lined by cilia called **lateral cilia**.

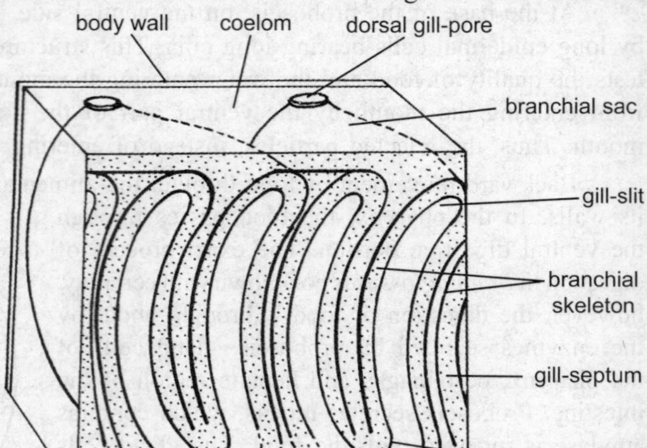

Fig. 2.14. *Balanoglossus*. Diagrammatic three-dimensional view of two gill-slits and two branchial sacs.

2. Branchial sacs. Gill-slits do not open directly to outside. Each gill-slit opens into a gill-pouch called **branchial sac** which lies between the body wall and the pharynx. Each branchial sac in turn opens to the exterior by a small, independent **gill-pore**. However, in *Balanoglossus misakiensis* the first four branchial sacs become united to open by a common gill-pore to outside. The collar coelom also communicates with the common branchial sac of its side through a **collar canal**. The gill-pores are visible externally in two longitudinal rows, one on each side of the mid-dorsal ridge in the branchiogenital region of the trunk.

Mechanism of respiration. The lateral cilia lining the gill-slits create a current of water (food-cum-respiratory current) that enters the pharynx through mouth, then passes through gill-slits into the branchial sacs and finally leaves through the gill-pores. The tongue bars are richly supplied with blood capillaries and take part in respiration. The blood of their capillary networks takes up oxygen dissolved in water and diffuses carbon dioxide to it.

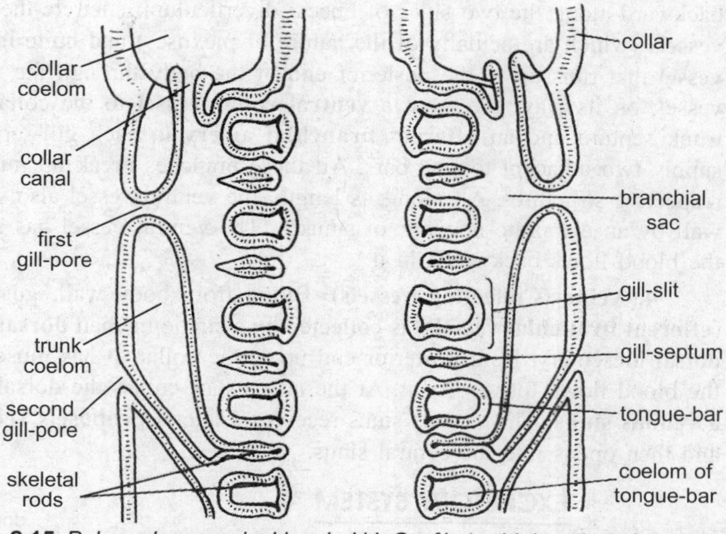

Fig. 2.15. *Balanoglossus misakiensis.* H.L.S. of branchial portion of pharynx showing first four gill-silts operating by a common gill-pore.

BLOOD VASCULAR SYSTEM

The blood vascular system of *Balanoglossus* is the **open** or **lacunar** type. It consists of : (*i*) colourless **blood**, (*ii*) **central sinus** and a **heart vesicle**, (*iii*) distributing vessels or **arteries** and **sinuses**, and collecting vessels or **veins**.

1. Blood. The blood is colourless fluid containing few white corpuscles which are possibly detached endothelial cells. The respiratory pigment is probably absent.

2. Central sinus and heart vesicle. The central sinus is a small elongated non-contractile sinus located in the proboscis just above the buccal diverticulum. Just above it is a closed triangular **cardiac sac** or **heart vesicle**. Its ventral wall is muscular and contracts rhythmically. The central sinus receives blood from collecting vessels that open into its posterior end. Anteriorly, it pumps blood into several **afferent vessels** which form a plexus in the **glomerulus** lying in front of it. In glomerulus the blood gets rid of excretory wastes.

3. Arteries (Distributing vessels). Blood from the glomerulus is carried away by four arteries. Of these two arteries, a **mid-dorsal probosci**ɜ **artery** and a **mid- ventral proboscis artery**, supply the proboscis. The other two, **efferent glomerular arteries**, run

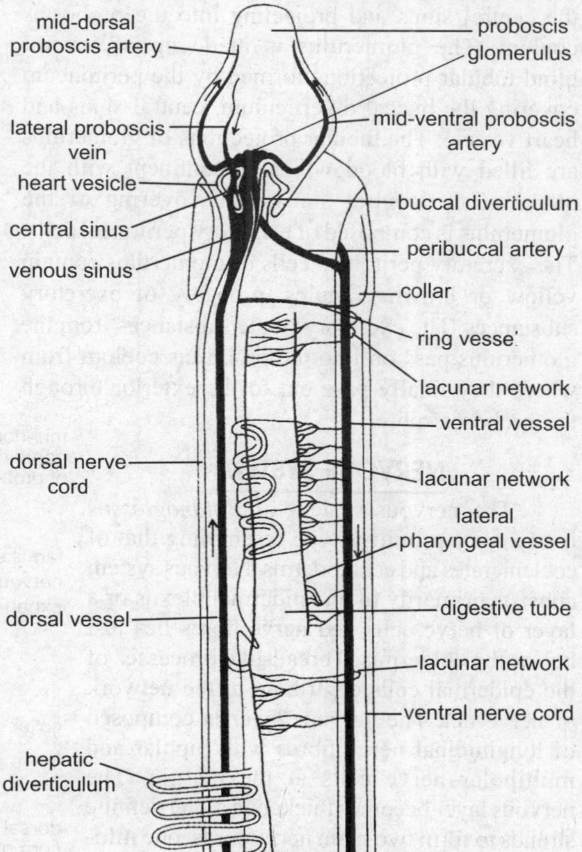

Fig. 2.16. *Balanoglossus.* Blood vascular system in anterior end in lateral view.

backward along the two sides of buccal diverticulum, encircle the buccal tube as **peribranchial vessels** (which are actually of the nature of plexuses) and unite in a single longitudinal **ventral vessel** that runs up to the posterior end of the body through the ventral mesentery. The ventral vessel, on its way, gives out a **ventral collar vessel** to the collar, a **ring vessel** to the collar-trunk septum and an **afferent branchial artery** to each gill-septum in which it bifurcates to supply two adjacent tongue bars. All these branches break up into a system of sinuses in their respective structures. All along its length, the ventral vessel also supplies the body wall and gut wall by an elaborate network of sinuses. The ventral vessel has muscular contractile walls and the blood flows backwards in it.

4. Veins (Collecting vessels). Blood from body wall, gut wall and branchial apparatus (**efferent branchial vessels**) is collected by a single median **dorsal vessel** which runs through the dorsal mesentery, from posterior end up to the collar. It has muscular and contractile walls and the blood flows forward in it. At the anterior of collar, the dorsal vessel dilates a little to form a **venous sinus**. The venous sinus receives a lateral **proboscis vein** from each side of proboscis and then opens into the central sinus.

EXCRETORY SYSTEM

In *Balanoglossus*, the excretory organ is **glomerulus** or **proboscis gland** lying in front of the central sinus and projecting into the proboscis coelom. The glomerulus is made up of several blind tubular projections formed by the peritoneum covering the buccal diverticulum, central sinus and heart vesicle. The tubular projections of glomerulus are filled with blood which is confluent with the blood of the central sinus. The covering of the glomerulus is composed of excretory peritoneal cells. The excretory peritoneal cells of glomerulus contain yellow or brown granules, probably of excretory substances. The excretory waste substances from the glomerulus pass on into the proboscis coelom from where they finally pass out to the exterior through the proboscis pore.

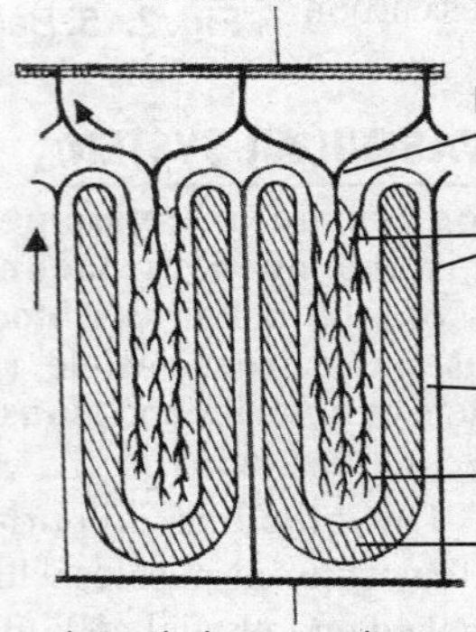

Fig. 2.17. *Balanoglossus*. Branchial vessels in portion of the branchial wall.

NERVOUS SYSTEM

The nervous system of *Balanoglossus* is of a very primitive type resembling that of coelenterates and echinoderms. Nervous system consists primarily of an epidermal plexus or a layer of nerve cells and nerve fibres lies just below the epidermis. Thread-like processes of the epidermal cells contribute to the network or nerve net. The nervous layer is composed of longitudinal nerve fibres with bipolar and multipolar nerve cells at the margin. The nervous layer becomes thickened along definite strands to form two main nerve cords, one **mid-dorsal** and the other **mid-ventral**, which run along the entire length of the trunk. **Ventral cord** extends up to collar-trunk septum where

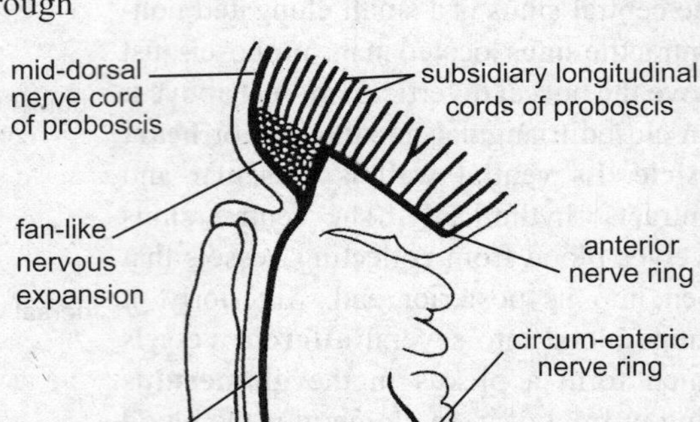

Fig. 2.18. *Balanoglossus*. Nerve cords in the anterior region of the body.

it is connected with the dorsal cord by a circular strand, called **circumenteric nerve ring**. **Dorsal cord** extends anteriorly up to the base of proboscis where it is connected with another circular stand called **anterior nerve ring**. From the anterior nerve ring longitudinal nerve fibres are given out, these nerve fibres are called subsidiary longitudinal cords of proboscis. In the collar region, dorsal cord leaves the epidermis and projects into the collar coelom as the **collar cord**. The collar cord contains a cavity called **neurocoel**. The collar cord arises from the epidermis, but has sunk in to take a deeper position. Its similarity to chordate nerve cord, formed by the invagination of the nervous ectoderm at the dorsal mid-line, is evident. The collar cord and sometimes the anterior part of the dorsal cord in the trunk or nerve ring, contains giant neurons. Each gives off a single large nerve fibre which crosses to the other side of the body and runs through the circumenteric ring to the ventral nerve cord. The number of giant neurons varies from about 10 to 160. They are responsible for rapid conduction of stimuli leading to quick retraction of body parts. Although the collar cord is the most complex part of the nervous system, it is no more than a conduction path and the site of giant neuron formation.

SENSE ORGANS

In *Balanoglossus*, the sense organs are poorly developed. In the epidermis are numerous **neurosensory cells** which are connected to the nerve net, they are more numerous on the proboscis. It is claimed that in some species a few neurosensory cells form **photoreceptors** sensory to light. On the ventral side of the base of the proboscis is a U-shaped depression called a **preoral ciliary organ**. It has ciliated cells joined to the nerve net, it is a chemoreceptor.

REPRODUCTIVE SYSTEM

Sexual Reproduction

In *Balanoglossus*, the sexes are separate and are indistinguishable externally except in case of the colour of the ripe gonads shown through the body wall in the living animal. The gonads occur in one or more longitudinal rows to the sides of the alimentary canal lying within the genital pleurae in the anterior part of the trunk. Gonads develop from the coelomic wall, though they have no connection with coelom in the adult. The gonads are generally sacciform bodies but may be elongated or lobulated and secondary gonads may arise by subdivision of the primary ones through lobulation. Each gonad is a sac, it continues into a short ductule which opens to the exterior by a **gonopore**. The gonopores are generally located to the lateral (external) side of the gill-pores in the same branchiogenital groove. The saccular gonads are lined with germinal epithelium which is continuous with the ectoderm. By the proliferation of cells from the germinal epithelium **sperms** or **ova** are produced. The mature sperms and ova are discharged outside through the **genital pores**. The sperm has a rounded head and a flagella-like tail but the ova are of two types. The small ovum measures about 0.06 mm in diameter and undergoes indirect development with a pelagic larva known as tornaria larva, while the larger ovum measures about 0.4 mm in diameter and undergoes direct development without larval stage. Fertilisation is external.

DEVELOPMENT

In *Balanoglossus*, during sexual reproduction zygote develops either **directly** or **indirectly** depending on the size. The small fertilised eggs develop indirectly but large fertilised eggs develop directly. The indirect development is through a larval form called **tornaria larva**. In both the cases early phases of development are same.

Fertilisation. During breeding season (May to June) mature ova and sperms are discharged in the surrounding water where fertilisation takes place. First the ova, egg-mass, are discharged by the female from its burrow and then the sperms are discharged by the male from its burrow. The number of eggs discharged at a time varies from few dozens to more than a thousand. Normally one to three hundred eggs are shed at a time. According to available evidences maturation starts some four hours before ovulation and that the egg is generally in the metaphase of the first meiotic division when shed. It is at this condition the egg is fertilisable. Fertilisation of eggs within 6 to 7 hours after shedding yields a high percentage of normal development. The spermatozoon is able to enter the eggs at any point over the entire surface. Thus, fertilisation is external. After fertilisation, the cleavage starts.

Indirect Development

Indirect type of development is studied in *Balanoglossus clavigerus* by workers like **Heider** (1909), **Stiansy** (1914) and **Payne** (1937).

Embryonic or Prelarval Development

Cleavage. The zygote, produced as a result of fertilization, undergoes cleavage. The cleavage is **holoblastic**, almost **equal** and mostly of the **radial** type. The first cleavage starts about two hours after fertilisation and produces two generally, but not invariably, equal cells. The second cleagave is like the first and produces usually (but not invariably) four approximately equal cells. As a result of third and subsequent cleagaves a sphere of equal blastomeres is produced, it is called **morula**. The morula undergoes the reorganisation of its blastomeres and takes the form of a single-layered hollow and spherical **blastula** or **coeloblastula**. Its central fluid-filled cavity is called the blastocoel. As the cells multiply the volume of blastula increases. Blastula results in about 6–15 hours after fertilization. Within 12–24 hours, an invagination starts in the blastula which deepens to form the **archenteron**. The archenteron opens to the outside through a **blastopore**. The blastopore marks the posterior end of the embryo. The blastopore soon closes and the embryo now called **gastrula**. The gastrula elongates along the antero-posterior axis. Now the anterior tip of the archenteron is differentiated as a coelomic vesicle called the **protocoel**. Thus, origin of coelom is **enterocoelic**. The remaining posterior part of the archenteron marks the future gut or alimentary canal. The protocoel becomes triangular in shape. One end of the protocoel gets attached to the underside of the apical thickening and another end opens to outside through an aperture, the **hydropore**, towards the dorsal side of the embryo. The protocoel and hydropore represent the future **proboscis coelom** and **proboscis pore** respectively. The collar and trunk coelom develop as solid invaginations of the hindgut, independent of the formation of protocoel.

Larval development. After the formation of the protocoel, the inner end of the early gut moves towards the ventral surface and opens to the outside through a mouth. The gut is now differentiated into the oseophagus, stomach and intestine. The intestine opens to outside through an **anus**, formed at the place of closed blastopore. By this time (after a day or so) the embryo becomes uniformly **ciliated** and escapes from fertilisation membrane to lead a free swimming larval life. It is called **tornaria larva**.

Tornaria larva. Tornaria larva was first of all discovered by **J. Muller** in 1850 and was considered by him as the larva of echinoderms. Later on in 1869 it was **Metschnikoff** who established that it is a larva of *Balanoglossus clavigerus*. The name tornaria is given to it because of its habit of rotating in circles. Tornaria larva is usually oval in shape and is excessively transparent. The size of tornaria larva varies from 1 mm to 3 mm. It has a ventral mouth near the equatorial plane of the body, a posterior terminal anus and gut differentiated into an oesophagus, stomach and intestine. The cilia form two bands on the body surface. The **anterior ciliary band** or **circumoral band** takes up a winding course over the preoral surface and forms a postoral loop. Its cilia are short and serve to collect the food. The **posterior ciliated band** or **circumanal ring** or **telotroch** lies as a ring in front of the anus. The cilia in this band are long, powerful and act as chief locomotor organ of tornaria. A ciliary wave passing along the telotroch causes the larva to rotate constantly in swimming. At the anterior end is an **apical plate** of thickened epidermal cells. The apical plate bears a pair of **eye spots** or **ocelli** and a tuft of sensory cilia called **apical tuft** or **ciliary organ**. The **protocoel** (proboscis coelom) is present in the form of a thin-walled sac and opens to the exterior through a **hydropore** (proboscis pore). To the right of the hydropore lies a pulsating heart vesicle which develops in the later stages of tornaria larva. The collar and trunk coeloms appear in the older larva.

Metamorphosis. The tornaria larva swims freely, leads a planktonic life feeding on minute organisms. After swimming for some time the tornaria larva sinks to the bottom and metamorphoses into an adult. During metamorphosis, the size of larva is reduced probably due to loss of water. Transparency, ciliary bands, sensory cilia and eye spots are lost. The body elongates and is distinguished into proboscis, collar and trunk by the appearance of two constrictions, and the trunk region is elongated. The hydropore persists as proboscis pore. Simultaneously the buccal diverticulum and gill-slits appear as outgrowth of the alimentary canal. The reproductive organs make their appearance, probably from the mesoderm. Thus, the larva gradually changes into the adult. The adult leads a benthonic life.

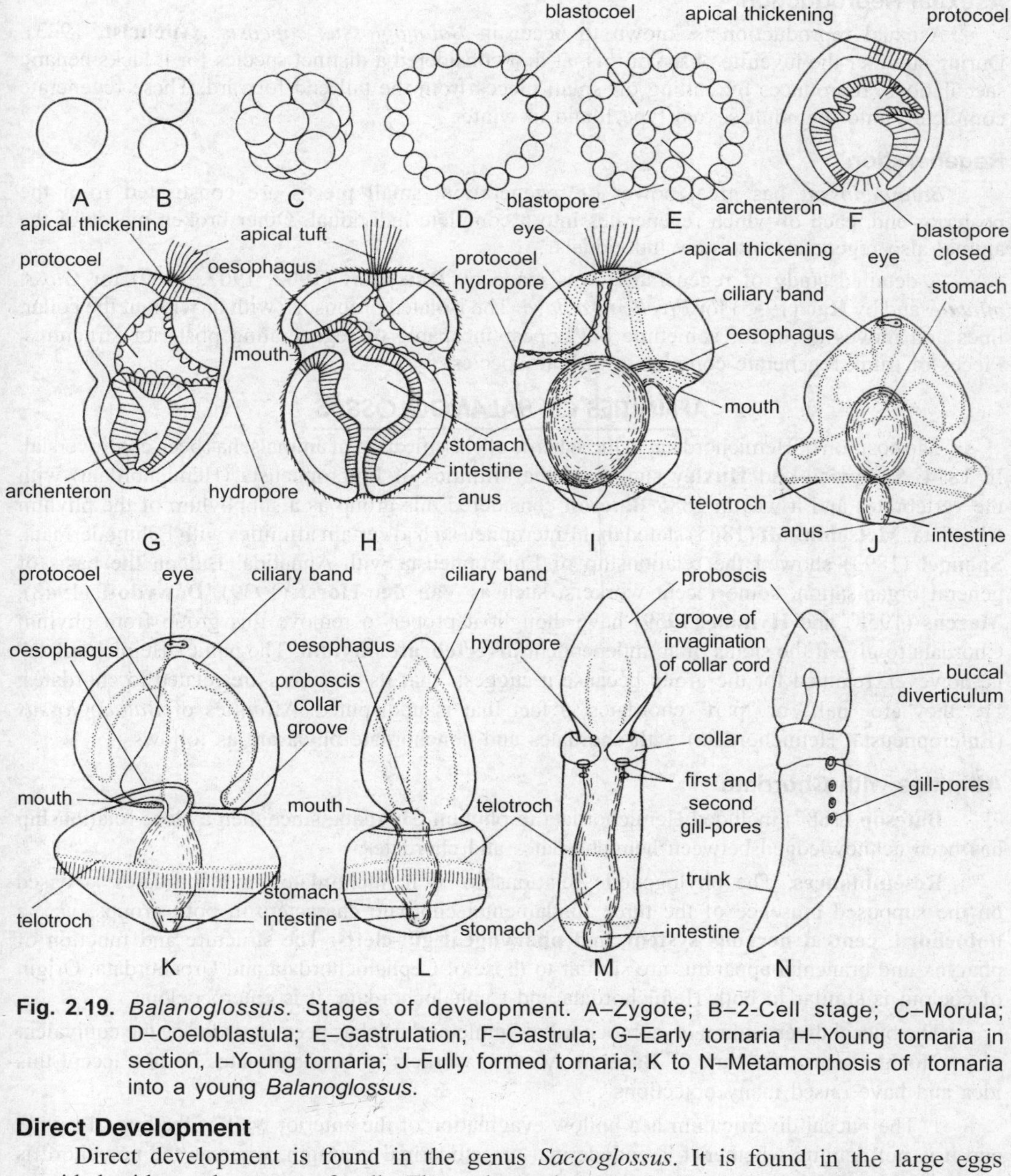

Fig. 2.19. *Balanoglossus*. Stages of development. A–Zygote; B–2-Cell stage; C–Morula; D–Coeloblastula; E–Gastrulation; F–Gastrula; G–Early tornaria H–Young tornaria in section, I–Young tornaria; J–Fully formed tornaria; K to N–Metamorphosis of tornaria into a young *Balanoglossus*.

Direct Development

Direct development is found in the genus *Saccoglossus*. It is found in the large eggs provided with good amount of yolk. The embryo elongates in the future antero-posterior axis, develops cilia, and after rotating for a time inside the egg membrane, escapes at about 24 to 30 hours and leads a planktonic existence for a time. It is provided with an apical tuft of cilia and the telotroch. *Saccoglossus kowalevskii*, however, does not hatch until the seventh day and at once takes up a benthonic life. After hatching the larva swims for a short time (24–26 hours) and soon elongates and develop in equatorial constriction marking the proboscis-collar boundary. The second constriction marks the collar-trunk boundary. The larva soon elongates and leads a benthonic life. It looses its apical tuft of cilia and the telotroch (posterior band of cilia). The larva now becomes worm-like juvenile *Balanoglossus*.

Asexual Reproduction

Asexual reproduction is known to occur in *Balanoglossus capensis* (**Gilchrist**, 1923). During summer the juvenile phase of this, at first considered a distinct species for it lacks hepatic sacculations, reproduces by cutting off small pieces from the tail end forward. These regenerate completely into the adult sexual type found in winter.

Regeneration

Balanoglossus has great power of regeneration, small pieces are constricted from the posterior end, each of which regenerates into a complete individual. Other broken pieces of the animal also regenerate into new individuals.

A detailed study of regeneration was made by **Dawydoff** (1902, 1907, 1909) for *Gloss, minutus* and by **Rao** (1955) for *Ptychodera flava*. The isolated proboscis, with or without the collar, lines and moves about for sometime but appear incapable of regenerating posterior structures. Pieces of trunk regenerate completely in both species.

AFFINITIES OF *BALANOGLOSSUS*

The position of Hemichordata, in the scheme of classification of animals, has been controversial. In 1814, **Sedgwick** and **Huxley** suggested the affinities of Enteropneusta (Hemichordata) with the vertebrates and it was in 1885 **Bateson** considered this group as a subphylum of the phylum Chordata. **Metschinkoff** (1865) stated that Enteropneusta had certain affinities with Echinodermata. **Spengel** (1893) showed the relationship of Enteropneusta with Annelida. But on the basis of general organisation, some recent workers, such as **Van der Horst** (1939), **Dawydoff** (1948), **Marcus** (1958) and **Hyman** (1959) have thought it proper to remove this group from phylum Chordata to give it the status of an independent **invertebrate phylum**. The name "Hemichordata" is, however, retained for the group because it suggests that its members are related to chordates, *i.e.,* they are "half" or "part" chordates, a fact that is undisputed. Affinities of *Balanoglossus* (Enteropneusta, Hemichordata) with chordates and nonchordate phyla are as follows :

Affinities with Chordata

Bateson (1887) included Hemichordata in phylum Chordata, since then a close relationship has been acknowledged between hemichordates and chordates.

Resemblances. The phylogenetic relationship of hemichordates and chordates is based on the supposed presence of the three fundamental chordate characters in both groups, *viz.,* a **notochord**, **central nervous system**, and **pharyngeal gill-clefts**. The structure and function of pharynx and branchial apparatus are similar to those of Cephalochordata and Urochordata. Origin of coelom is similar in both Hemichordata and Cephalochordata, it is enterocoelous.

The buccal diverticulum or stomochord of hemichordates has been regarded as the equivalent of a notochord since the time of **Bateson**. Modern workers of hemichordates do not accept this idea and have raised many objections :

1. The buccal diverticulum is a hollow evagination of the anterior wall of the buccal cavity, and it is not definite whether it is endodermal or ectodermal in origin, whereas the notochord is a long solid rod formed from the roof of the archenteron.

2. The buccal diverticulum is generally made of ordinary epithelial cells, while the notochord of vertebrates consists of large vacuolated cells.

3. The buccal diverticulum has no enclosing sheath as found around the notochord.

4. The buccal diverticulum lies ventral to the dorsal blood vessel, whereas the vertebrate notochord is always dorsal to the main dorsal blood vessel.

5. The buccal diverticulum is small and confined to the proboscis, while the notochord extends far backwards. It can be safely concluded that there is no representative of the notochord in hemichordates.

There are certain **resemblances** between the nervous system of hemichordates and chordates, such as (*i*) its position, and (*ii*) formation of the dorsal nerve cord from the dorsal epidermis, and the collar cord which often has a neuropore and is comparable with the brain of vertebrates. But there are major **differences**, such as (*i*) its superficial position in contact with the epidermis, (*ii*) possession of a main ventral nerve cord, and (*iii*) a circumenteric nerve ring, and collar cord is solid not hollow as found in true chordates, in these features the nervous system is distinctly invertebrate. Thus, the invertebrate features of the nervous system of hemichordates outweigh its chordate characters.

The chief link between the hemichordates and chordates lies in the pharynx and its gill-clefts. The details of the branchial apparatus having tongue bars, M-shaped skeletal rods and synapticula are exactly like those of *Amphioxus*. But the endostyle and epibranchial groove are absent from the pharynx of hemichordates. Such similarity can be only due to common ancestry, and phylogenetic relationship of hemichordates and chordates cannot be denied.

Differences. But the inclusion of hemichordates in phylum Chordata cannot be justified on the basis of a few similarities which are more than outweighed by important differences. The main differences are :

1. Chordates do not have the body and coelomic regions corresponding to those of hemichordates.

2. The circulatory and nervous systems of hemichordates are like those of invertebrates.

3. There is no post-anal tail in hemichordates.

4. Chordates are metamerically segmented animals, this segmentation is clearly shown by the muscular, nervous, circulatory and excretory systems, whereas there is a total absence of segmentation in hemichordates.

Affinities with Annelida

Spengel (1893) first suggested the relationships between hemichordata and Annelida.

Resemblances. The main resemblances of Hemichordata with Annelida are as follows :

1. The general body form and burrowing habit of tubicolous forms such as *Balanoglossus* are alike and mud is ingested in burrowing. It is passed out from the anus as castings.

2. The collar of *Balanoglossus* resembles with the clitellum of earthworm.

3. The proboscis of *Balanoglossus* and prostomium of earthworm are similar and both are pre-oral.

4. The vascular system of most hemichordates is like that of annelids with blood flowing anteriorly in the dorsal vessel and posteriorly in the ventral vessels.

5. The hemichordate **tornaria larva** appears like a modified **trochosphere larva** of polychaete worms.

Differences. The differences between two groups are so great that there can be no phylogenetic relationship between them. The differences are as follows :

1. Pharyngeal gill-slits, buccal diverticulum and dorsal tubular nerve cord found in *Balanoglossus*.

2. Paired ventral nerve cord present in annelids is absent in *Balanoglossus*. Nephridia found in annelids are absent in *Balanoglossus*.

3. The larva of Hemichordata and Annelida also differ in the following ways :

(*i*) Nephridia present in trochophore larva are absent in tornaria larva.

(*ii*) Pre-oral coelom found in tornaria larva is absent in trochophore larva.

(*iii*) In trochosphere larva blastopore becomes the mouth, while in tornaria larva it becomes the anus.

Affinities with Echinodermata

The adult hemichordate and adult echinoderm are so different that one cannot suspect any relationship between them, the only anatomical similarity between them is their nervous system which in both cases consists of nerve net lying near the surface embedded in the epidermis. But there is a strong affinity between the two phyla on embryological evidence, the method of formation of the gastrula and the coelom is very similar in the two phyla and for years the tornaria larva was considered to be the larva of an echinoderm. The **tornaria larva** shows a very striking resemblance with the **auricularia larva** and specially with **bipinnaria** of Asteroidea.

Resemblances. The resemblance extends into the following details :

1. The ciliated band is identical and follows the same course in the tornaria and the auricularia and bipinnaria, though the telotroch and eye spots of the tornaria are absent in echinoderm larvae.

2. The alimentary canal has the same shape and the same divisions into foregut, stomach and intestine in hemichordate and echinoderm larvae.

3. In both groups the blastopore becomes larval anus.

4. The cleavage and gastrulation follow the same pattern in both.

5. The greatest and the most convincing resemblance lies in the method of formation and arrangement of coelomic cavities. In both the coelom is of enterocoelous origin and it divides into three antero-posterior parts, which in hemichordates are called proboscis coelom (protocoel), collar coelom (mesocoel), and trunk coelom (metacoel), while in echinoderms the three parts of the coelom are axocoel, hydrocoel, and somatocoel. Moreover, the proboscis coelom and collar coelom of hemichordates open to the exterior by pores through short hydroporic canals, as does the hydrocoel in echinoderms.

6. The heart vesicle of hemichordates is related to the proboscis coelom and is homologous with the madreporic vesicle of echinoderm larvae and both these structures are closely connected to the glomerulus of hemichordates and the axial gland of echinoderms which have a combine vascular and excretory functions.

7. Proteins and phosphagens present in adult hemichordates and echinoderms are very similar.

Differences. There are following differences between two groups :

1. Eye spot present in tornaria is absent in bipinnaria.

2. The apical plate and telotroch present in tornaria are absent in bipinnaria.

3. The protocoel is paired in echinoderms, while unpaired in tornaria larva.

The many embryonic resemblances between hemichordates and echinoderms are merely due to convergent evolution due to similar mode of habits and habitat.

The many embryonic resemblances between hemichordates and echinoderms cannot possibly be accidental or due to convergent evolution. The only infallible conclusion is that the two groups are closely related and that they arose from a common ancestor. Echinoderms have deviated greatly from the ancestral type, while hemichordates are closer to it. The common ancestor gave rise to echinoderms as a blind side branch, while the main line of evolution produced the hemichordates and chordates.

Conclusion. The above affinities have led to the conclusion that echinoderms, hemichordates and chordates have arisen from a common ancestral stock, the dipleurula larva. Further, the echinoderms deviated greatly from the ancestral stock and formed blind branch in the main line of evolution. The main line of evolution continued to give rise to hemichordates and chordates.

It appears most reasonable to place them in the invertebrates as an independent phylum which has arisen from an ancestral stock that has given rise, on the one hand, to echinoderms and, on the other hand, to hemichordates and chordates.

SYSTEMATIC POSITION AND PHYLOGENY

Bateson (1885) placed the hemichordates in a separate subphylum under the phylum Chordata only due to their peculiar anatomical organisation. It represents the lowest group of Chordata. But they have only one character of Chordata, *i.e.,* presence of pharyngeal gill-slits. But the recent workers such as **Van der Host** (1939), **Dawydoff** (1948), **Marcus** (1958) and **Hyman** (1959) have thought it proper to remove the group from Phylum Chordata to give it the status of an independent invertebrate phylum. Since the group comprises only about 80 species, it is included in the category of a **minor phylum**.

The echinoderms and hemichordates, both have arisen from a common ancestor probably the dipleurula larva (**Bather** 1900). The dipleurula larva is very similar to the adult stage of tornaria and echinoderm larvae. But this idea was not accepted.

Groben (1923) established that the echinoderms are derived from pterobranchs of Hemichordata. Pterobranchs are more primitive than those of enteropneusts. They possess tentaculated arms, absent in enteropneusts, probably they are lost. The post-anal tail in juvenile enteropneusts, which is a relic of the stalk of pterobranch ancestor.

The close affinities of Hemichordata, Echinodermata and Chordata, etc., have led to the conclusion that they have common ancestor from which they have arisen independently. Probably the common ancestor was dipleurula larva of Bather. Recently **Barrington** (1965) expressed that Echinodermata deviated greatly from the ancestral stock and formed a blind branch. Hemichordata also did not stand on the direct line of ancestory but formed a divergent offshoot from the main line of chordate evolution. Thus, it may be suggested that *Balanoglossus* and other chordates are running on the same field but not on the same track. Since the hemichordates arose from the ancestral line after the divergence of the ancient Echinodermata but before the rise of true chordates, they are often called **prechordates**.

IMPORTANT QUESTIONS

Long Answer Questions

1. Describe the general organisation of hemichordates.
2. Discuss the affinities and phylogeny of Hemichordata.

Short Answer Questions

1. State the position and functions of the gonads in *Balanoglossus*.
2. Draw a labelled diagram of the vertical longitudinal section through the anterior half of *Balanoglossus*. (No description is required).
3. Discuss the systematic position of *Balanoglossus*.
4. Give an account of digestive system of *Balanoglossus*.
5. Describe the blood vascular system of *Balanoglossus*.
6. Write short notes on :
   (*i*) Coelom of *Balanoglossus*,
   (*ii*) Body wall of *Balanoglossus*,
   (*iii*) Morphology of *Balanoglossus*, and
   (*iv*) Endoskeleton of *Balanoglossus*.

Very Short Answer Questions

1. What is pygochord and where is it located?
2. What are the three divisions of coelom in *Balanoglossus*?

3. Epidermis in *Balanoglossus* is ______ layered.
4. In *Balanoglossus* body wall ______ is absent.
5. In *Balanoglossus* nervous layer is ______.
6. The common name of *Balanoglossus* is ______ or ______.
7. Protosome, mesosomes and metasomes are parts of ______ in *Balanoglossus*.
8. *Balanoglossus* is a ______ feeder.
9. In *Balanoglossus* the excretory organ is ______.

Multiple Choice Questions

1. *Balanoglossus* is commonly known as
   (*a*) Tonque worm (*b*) Acorn worm
   (*c*) Both (*a*) and (*b*) (*d*) Corn worm.
2. Larval form of hemichordates is
   (*a*) Tornaria (*b*) Nauplius
   (*c*) Amphiblastula (*d*) Bipinnaria.
3. Which of the following is a connecting link between chordates and nonchordates?
   (*a*) *Herdmania* (*b*) *Balanoglossus*
   (*c*) *Amphioxus* (*d*) *Asterias*.
4. Notochord is restricted only in the proboscis in
   (*a*) Hemichordata
   (*b*) Urochordata
   (*c*) Cephalochordata
   (*d*) All of the above.
5. Hemichordates have one chordate feature
   (*a*) Perforated pharynx
   (*b*) Buccal diverticulum
   (*c*) Three types of coelom
   (*d*) Ciliated integument.
6. Hemichordates do not resemble chordates with respect to
   (*a*) Notochord
   (*b*) Absence of dorsal tubular nerve cord
   (*c*) Presence of pharyngeal gill-cleft
   (*d*) None of the above.
7. Coelom in *Balanoglossus* is
   (*a*) Schizocoel
   (*b*) Enterocoel
   (*c*) Holocentric
   (*d*) None of the above.
8. Blood vascular system in *Balanoglossus* is
   (*a*) Colourless
   (*b*) Open type
   (*c*) Closed type
   (*d*) (*a*) and (*b*).

Answers

1. (*c*) 2. (*a*) 3. (*b*) 4. (*a*) 5. (*a*) 6. (*a*) 7. (*a*) 8. (*d*).

3 Chapter

Subphylum 1. Hemichordata : General Characters, Classification and Representative Types

A Generalised Enteropneust Hemichordate

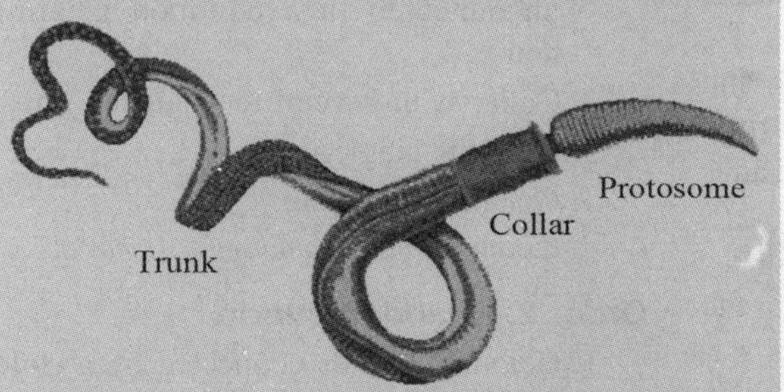

Contents of this Chapter

DEFINITION

Usually vermiform, solitary or colonial enterocoelous coelomate animals with intra-epidermal nervous system and a pre-oral gut with or without gill-slits and without typical nephridia.

GENERAL CHARACTERS

1. Solitary and colonial, mostly tubicolous, exclusively marine.

2. Body soft, fragile, vermiform and divisible into **proboscis, collar** and **trunk**.

3. Body wall with a single-layered epidermis.

4. Coelom enterocoelous, divisible into protocoel, mesocoel and metacoel.

5. Buccal diverticulum, earlier considered as notochord, present in the proboscis.
6. Digestive tract complete; in the form of straight or U-shaped tube.
7. Gill-slits, when present, are paired and one to numerous.
8. Circulatory system simple and well developed; closed type; usually with a contractile heart vesicle and two longitudinal vessels, one dorsal and one ventral, interconnected by lateral vessels and sinuses.
9. Excretion by a single glomerulus situated in the proboscis.
10. Nervous system primitive comprising mainly of an intra-epidermal nerve plexus.
11. Reproduction mostly sexual. Sexes separate or united. Gonads one to several pairs.
12. Fertilisation external. Development mostly indirect through a free swimming tornaria larva. Direct development is also found in some forms.

CLASSIFICATION

Phylum Hemichordata has been divided into following four classes :

Class 1. Enteropneusta

1. Commonly known as "acorn" or "tongue worms".
2. Solitary and burrowing worm-like marine animals.
3. Body consists of proboscis, collar and trunk; collar without tentaculated arms.
4. Alimentary canal straight; mouth and anus at opposite ends.
5. Numerous pairs of U-shaped gill-slits.
6. Two pairs of hepatic caeca present in the middle of the trunk.
7. Sexes separate; gonads numerous, sac-like.
8. Development with or without tornaria larva.

Examples : *Balanoglossus, Saccoglossus (= Dolichoglossus), Ptychodera.*

Class 2. Pterobranchia

1. Sedentary, solitary or colonial, tubicolous marine animals.
2. Proboscis with ciliated tentacles to produce ciliary feeding currents of water.
3. Collar with two or more tentaculated arms bearing tentacles.
4. One pair of gill-slits or none, never U-shaped.
5. Alimentary canal U-shaped with dorsal anus situated near the mouth at the same end.
6. Sexes separate or united; single or one pair of gonads.
7. Development direct, may or may not include a free swimming larval stage.
8. Asexual reproduction by budding in some.

Order 1. Cephalodiscida

1. Solitary or several individuals living unconnected in a common gelatinous house.
2. Collar with several tentaculated arms.
3. Gill-slits single pair.
4. Gonads single pair.

Examples : *Cephalodiscus, Atubaria.*

Order 2. Rhabdopleurida

1. Colonial, zooids connected by a stolon.
2. Collar with two tentaculated arms.
3. Gill-slits absent.
4. Gonad single.

Examples : *Rhabdopleura* (Single genus).

Class 3. Planctosphaeroidea

This class is represented by a few small rounded, transparent and pelagic larvae, supposed to be specialised tornaria of some unknown hemichordate *Planctosphaera pelagica*. The larva has branching arborescent ciliated bands on the surface. The alimentary canal of larva is U-shaped. The adult form is yet unknown.

Class 4. Graptolita

These are extinct colonial hemichordates, mainly known from the fossil structures of their tubes. Each animal is housed in a zooid. These were abundant in the Ordovician and Silurian periods.

Example : *Dendrograptus.*

REPRESENTATIVE TYPES

1. *Saccoglossus (=Dolichoglossus).* *Saccoglossus* (Fig. 3.1) is a typical enteropneust very much similar to *Balanoglossus* in habits, habitat and structure. It is a marine, slender soft-bodied tongue worm living in spirally twisted furrows. Body is divisible into proboscis, collar and a trunk. Proboscis or protosome is exceptionally elongated and pointed than in other tongue worms. Collar slightly overhangs the beginning of the trunk covering first three or four pairs of gill-pores. Trunk is differentiated into anterior branchial region and posterior branchial region. Genital wings and hepatic caeca are absent. Alimentary canal is straight and anus lies at the posterior end of the body. Mature gonads are yellow in male and gray in female and their position is marked externally by dorso-lateral genital folds in the middle part of the trunk. Synapticula are absent so that tongue bars hang freely in their gill-slits. Sexes are separate and fertilisation is external. Development is direct without a free-swimming tornaria larva. *Saccoglossus* is found in the Indo-Pacific Atlantic coasts, New Zealand and Australia.

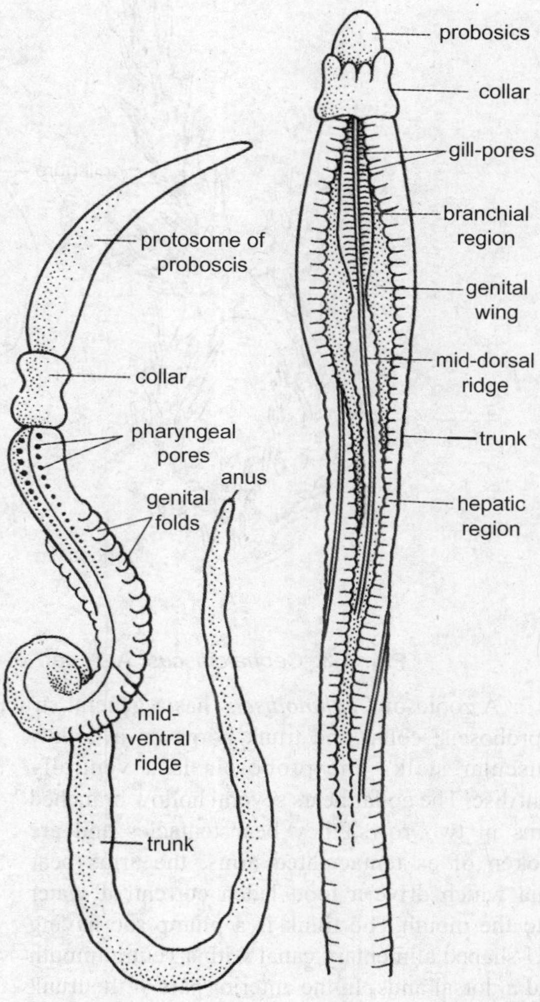

Fig. 3.1. *Saccoglossus.* **Fig. 3.2.** *Ptychodera.*

2. *Ptychodera.* *Ptychodera* (Fig. 3.2) bears a close resemblance to *Balanoglossus* in habits, habitat and structure. It is soft-bodied, slender marine burrowing animal. The general colour of the animal is yellowish. Its body is elongated and divisible into proboscis, collar and trunk. Proboscis and collar are short, while trunk is quite elongated. Gonads are situated in the genital wings. Hepatic caeca are distinct. Sexes are separate. Development is indirect involving a free-swimming tornaria larva. It exhibits bioluminescence. *Ptychodera* is found in the sea of tropical and subtropical regions.

3. *Cephalodiscus.* *Cephalodiscus* (Fig. 3.3) individuals or zooids occur together in large aggregations attached to objects on the sea bottom in tropical and temperate oceans. The zooids live in separate tubes embedded in a common matrix called a **coenoecium** but the various zooids are free from each other.

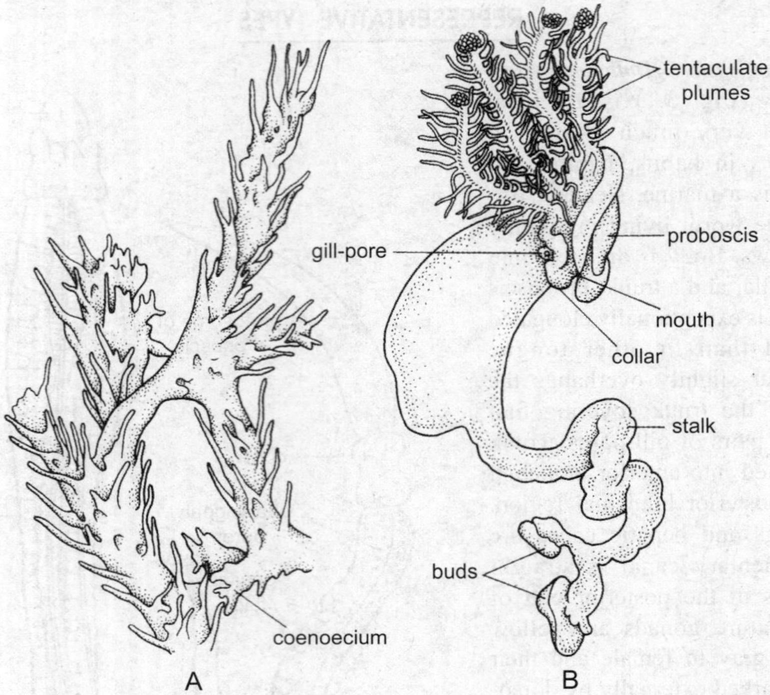

Fig. 3.3. *Cephalodiscus*. A–A part of colony; B–An individual zooid.

A zooid of *Cephalodiscus* has a general structure like that of *Balanoglossus*. The body has a proboscis, collar and trunk borne on a hollow muscular stalk. The proboscis is a ventrally bent disc. The collar bears several hollow branched arms in two rows, they bear tentacles and are spoken of as tentaculated arms, the arms bear cilia which drive a food laden current of water into the mouth. The trunk is a plump sac having a U-shaped alimentary canal with a ventral mouth and a dorsal anus. In the anterior part of the trunk is a single pair of gill-clefts. Sexes are separate, each has a pair of gonads opening by separate gonopores situated dorsally on the front part of the trunk. Fertilisation and direct development with no larval stage occur in the coenoecium. The original individual is produced sexually, on its trunk is a peduncle bearing buds, the buds become free and each forms a zooid which secretes a tube.

4. *Atubaria*. *Atubaria* (Fig. 3.4) was first described by **Sato** in 1936. It is sedentary and solitary genus found attached with hydroid colonies by its long stalk and closely resembling with *Cephalodiscus*. A coenoecium is absent. The zooid is about 1.5 mm in length. It has three

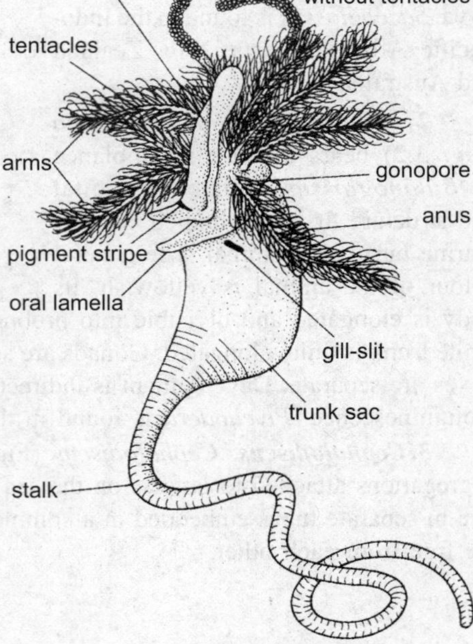

Fig. 3.4. *Atubaria*.

divisions of body. The collar bears four pairs of tentaculated arms of which second pair distally has rod-like terminations devoid of tentacles. Only one pair of pharyngeal gill-slits are present.

5. *Rhabdopleura.* *Rhabdopleura* (Fig. 3.5) is a true colonial form. The colonies are attached to stones, corals and sessile marine animals mostly in the North Atlantic. The colony has a thin horizontal branching tube called a coenoecium from which short erect tubes arise, each containing a zooid. A zooid is minute, its general structure is like an enteropneust with a proboscis, collar and trunk, attached to the trunk is a stalk. On the collar is a pair of hollow branching arms having cilia which collect the food of the animal. Alimentary canal is U-shaped so that the anus lies close to the mouth. There are no gill-clefts and a glomerulus is lacking.

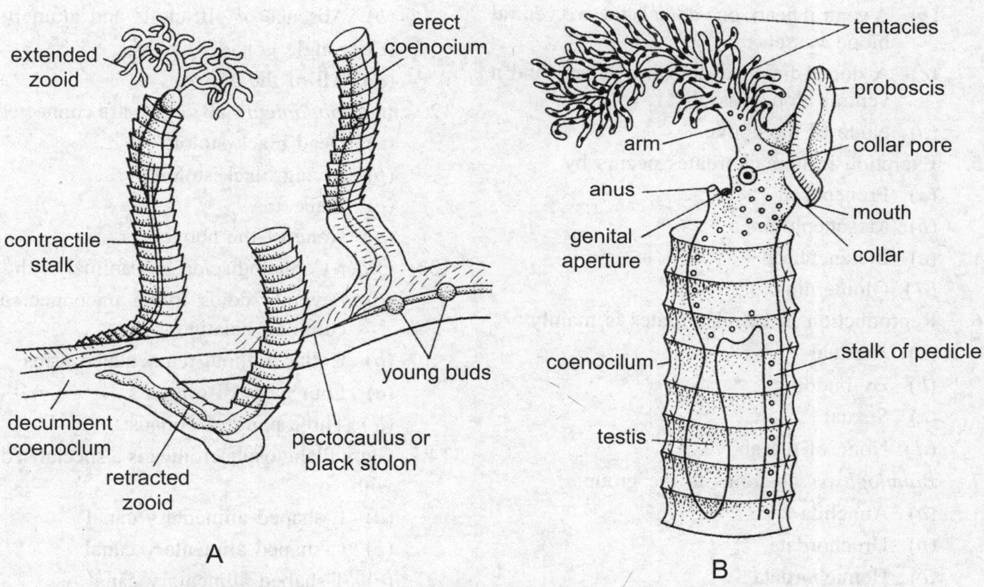

Fig. 3.5. *Rhabdopleura*. A–A portion of a colony; B–An individual zooid.

Sexes are separate, each sex has a single gonad with a gonopore on the right side. Large yolky eggs are laid. An original individual is produced sexually, on its trunk is a peduncle which bears buds, the buds do not become free and each forms a zooid. Each zooid secretes its own ringed vertical tube or case which is formed ring by ring. The horizontal part of each tube has a black stolon which connects the various zooids to the parent so that a colony of connected zooids is formed.

IMPORTANT QUESTIONS

Long Answer Questions

1. Give characters and classification of phylum Hemichordata up to orders.

Short Answer Questions

1. Classify *Balanoglossus* with full justification.
2. Write note on (*i*) Enteropneusta and (*ii*) Pterobranchia.

Multiple Choice Questions

1. In hemichordates body is divided into
   (*a*) Head, neck and trunk
   (*b*) Head, trunk and tail
   (*c*) Proboscis, collar and trunk
   (*d*) Proboscis, collar and tail.
2. Notochord in hemichordates is actually
   (*a*) Buccal diverticulum
   (*b*) True notochord

(*c*) Derivative of nerve cord
(*d*) Foregut.

3. Hemichordates are
(*a*) Filter feeder
(*b*) Ciliary filter feeder
(*c*) Ciliary feeder
(*d*) Bottom feeders.

4. Circulatory system in hemichordates includes
(*a*) A dorsal heart, a dorsal and a ventral blood vessels
(*b*) A ventral heart, one dorsal and two ventral blood vessels
(*c*) A dorsal heart, a dorsal, a lateral and a ventral blood vessels
(*d*) None of the above.

5. Excretion in hemichordates occurs by
(*a*) Pronephron
(*b*) Mesonephron
(*c*) Metanephron
(*d*) Glomerulus.

6. Reproduction in hemichordates is mainly
(*a*) Asexual
(*b*) By budding
(*c*) Sexual
(*d*) None of the above.

7. *Balanoglossus* belongs to the group
(*a*) Annelida
(*b*) Urochordata
(*c*) Hemichordata
(*d*) Cephalochordata.

8. Coelom in hemichordates is divided into ________, ________, and ________.

9. Body wall of hemichordates lacks ________.

10. Hemichordates are
(*a*) Marine
(*b*) Freshwater
(*c*) Both (*a*) and (*b*)
(*d*) Brackish water forms.

11. Genus *Rhabdopleura* is characterised by
(*a*) A pair of hollow branching arms
(*b*) Absence of gift-clefts and glomerulus
(*c*) Single gonad
(*d*) All of the above.

12. In *Rhabdopleura* all zooids are connected with
(*a*) Dead black stolon
(*b*) Living black stolon
(*c*) Stalk
(*d*) None of the above.

13. Order Cephalodiscida has animal with
(*a*) Several zooids living unconnected in a common gelatinous case
(*b*) Collar without tentaculated arms
(*c*) Four pairs of gill-slits
(*d*) Three pairs of gonads.

14. Class Planctosphaeroidea is a specialised larva with
(*a*) L-shaped alimentary canal
(*b*) C-shaped alimentary canal
(*c*) T-shaped alimentary canal
(*d*) None of the above.

Answers

1. (*c*) 2. (*a*) 3. (*b*) 4. (*a*) 5. (*d*) 6. (*c*) 7. (*c*)
8. proboscis coelom, collar coelom, trunk coelom 9. dermis 10. (*a*)
11. (*a*) 12. (*b*) 13. (*a*) 14. (*a*).

4

Chapter

Type 2. *Herdmania* (A Sea Squirt)

Contents of this Chapter

- *Geographical Distribution*
- *Systematic Position*
- *Habits and Habitat*
- *External Morphology*
- *General Anatomy*
- *Digestive System*
- *Respiratory System*
- *Circulatory System*
- *Circulation*
- *Nervous System*
- *Receptors*
- *Excretory Organ*
- *Reproductive Organs*
- *Development*
- *Affinities and Systematic Position of Herdmania (Urochordata)*

Subphylum **Urochordata** (Gr., *uros* = tail; *chorde* = cord) includes a peculiar group of widely distributed marine animals called **sea squirts** or ascidians and their allies. In most of them several chordate characters are lost in the adults, but their chordate affinities are clearly seen in their free-swimming larvae which have pharyngeal gill-clefts, a dorsal tubular central nervous system, and a notochord which is confined only to the tail (hence, the name **Urochordata**). During metamorphosis the chordate characters are lost except gill-clefts. Gill-clefts do not open to the exterior but into an ectoderm-lined atrium. The body becomes surrounded by a coat called test or tunic (hence, the name **Tunicata**). The changes in metamorphosis are a specialisation due to a sedentary life in which ciliary feeding is perfected, due to which locomotory organs and neuro-sensory system are simplified and reduced. They have no metameric segmentation and a coelom is absent. They are also hermaphrodite. About 2,200 Urochordata are known, they show a great diversity of structure, habits and habitats. They may be sessile or free-swimming. They are found in all seas, being found along the shores and up to a depth of more than 3 kms.

The genus *Rhabdocynthia* was first established by **Herdman** in 1891. But, in 1910, **Hartmeyer** changed it to *Herdmania* according to the law of priority as it was originally proposed by **Lahille** in 1888. It is now prescribed for study in almost all the universities of India and neighbouring countries. The description that follows is based mainly on the work of **S.M. Das** of *Herdmania pallida* first published in 1936.

GEOGRAPHICAL DISTRIBUTION

The Urochordata or Tunicata (L., *tunica* = outer covering) are marine animals found in all seas and at all depths, extending from the arctic and the antarctic regions to the tropics and from the littoral zone down to the abyssal depths of over 5 km. The Urochordata are derived from a common sessile ancestor and the most generalised condition of the class is found in sea-squirts to which *Herdmania* belongs. *Herdmania* is the commonest monascidian found in the Indian seas. Of the twelve known species of *Herdmania,* however, only four have been recorded from the Indian Ocean, *i.e., H. pallida, H. ceylonica, H. mauritiana* and *H. ennurensis*. Out of these again only the first two have been recorded from the coastal waters, while the remaining two are usually found in deeper waters. *Herdmania pallida* is described here in detail.

SYSTEMATIC POSITION

Phylum	...	Chordata
Subphylum	...	Urochordata
Class	...	Ascidiacea
Subclass	...	Pleurogona
Order	...	Stolidobranchia
Family	...	Pyuridae
Genus	...	*Herdmania* (Sea squirt)

HABITS AND HABITAT

Herdmania pallida is a solitary marine form found in shallow waters along the Indian sea coast. Each animal is found attached to the substratum usually separately, at its postero-ventral end by means of a **foot**. When the substratum is sandy, an expanded or elongated foot is formed which remains imbedded in the sand and keeps the animal fixed to the sea bottom. A large number of organisms inhabit the test of *Herdmania*; some of these are merely attached to its surface, while others are more or less imbedded within its substance. Sometime an individual is found attached to a living gastropod shell (*e.g., Turbinella pyrum, T. rapa, Xancus,* etc.) showing commensalism. *Herdmania* protects the gastropod from enemies since it is unpalatable due to its spicules, and gastropod takes it from place to place so that it may get good food and oxygen, etc. The food of *Herdmania* consists of microscopic plants and animals, *e.g.,* diatoms, algae and infusorians. *Herdmania* is more or less widely distributed, though it cannot be called a cosmopolitan species. *H. pallida* having been recorded from the Indian, the Pacific and the Atlantic Oceans, as also from Malaya and the West Indies. *Herdmania* is hermaphrodite and oviparous. Fertilisation takes place in the sea and the tadpole larvae emerging from the eggs lead a short free-swimming life before they settle down to the sea-bottom to lead a fixed existence after going retrogressive metamorphosis.

EXTERNAL MORPHOLOGY

Shape, Size and Colouration

The body of *Herdmania pallida* is roughly oblong in outline, narrower at its attached than at its free end. At its free end, it is provided with two external openings—the **branchial** and **atrial apertures**. The average size of the adult is about 9.5 cm long, 7 cm broad and 4 cm thick. Older animals may even attain to a size of 12 × 8 × 4 cm, while an exceptionally large measures 13 × 8 × 4.5 cm. The foot, when present, may attain a length of as much as 3 to 4 cms. The general colour of the body in a fresh specimen is pink. The presence of bright red patches, formed of terminal knobs (ampullae) in the blood vessels of the test, is a characteristic feature of *Herdmania*. The test is soft and leathery. It is more or less transparent in a young animal, but in an adult becomes usually opaque. The general surface of the test is much corrugated all over with lines, some shallow and others fairly deep, running in a criss-cross manner.

Body Divisions

The body is divided into two parts, **body proper** and **foot**, and the entire animal is covered by the test.

1. Body proper. It is the distal free portion of the body. The branchial aperture marks the **anterior end** of the animal; consequently the opposite end attached to the substratum is the **posterior end**. The side on which the atrial aperture is placed marks the dorsal side of the body, which is very limited. The side opposite to that on which the atrial opening is placed and which is partly attached to the substratum is ventral side which is extensive. The branchial and atrial apertures are situated on short protuberances of the body called **branchial** and **atrial siphons** respectively. When fully extended, the atrial siphon is longer than the branchial. In a large-sized animal the atrial siphon measures about 1.5 cm, while the branchial siphon is only about 1 cm in length. The atrial siphon is almost always directed upwards, the atrial aperture being more or less upright, while the branchial siphon is always bent a little outwards and the branchial aperture opens more or less laterally, always directed away from the atrial aperture. The average diameter of the branchial aperture, when fully expanded, is about 2 cm and that of the atrial aperture about 1.2 cm. At the base of branchial siphon, there is a ring of long **branchial tentacles**, while at the base of the atrial siphon there is a ring of slightly serrated folds which constitute the **atrial tentacles**. The test of the siphons is very elastic and can contract to close the apertures at the slightest disturbance of sea water.

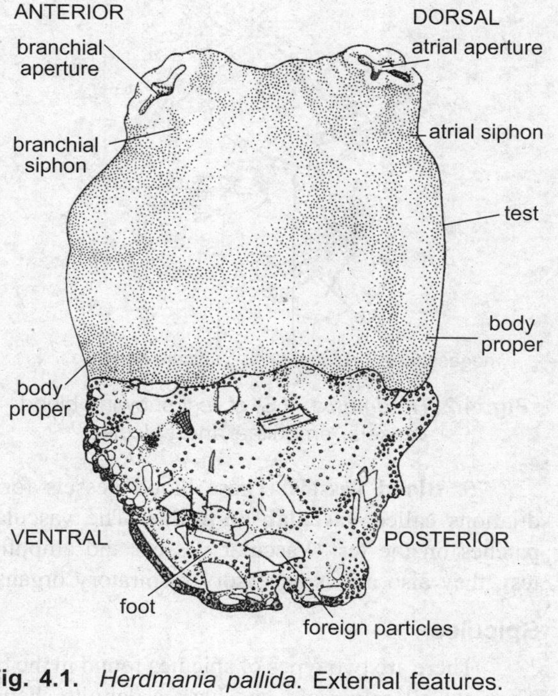

Fig. 4.1. *Herdmania pallida*. External features.

Foot. The foot, when present varies in character according to the nature of the substratum which the animal inhabits. If the substratum is of fine sand, the foot has an oval shape and a smooth surface and the test is quite hard in consistency. But if the substratum consists of coarse and the broken shell-pieces, the foot is irregular in outline and more or less soft in consistency.

Test or Tunic

The external covering surrounding the animal is a leathery, translucent test or tunic composed of **tunicin**, a substance akin to cellulose of plants. The foot is made entirely of test. It acts as a receptor as well as a respiratory organ. It is about 4–8 mm thick. It is ectodermal in origin. The test has (*i*) a clear matrix in which are embedded, (*ii*) cells of various shapes, (*iii*) interlacing fibrils, (*iv*) calcareous spicules, and (*v*) branching vascular vessels.

1. Matrix. It is gelatinous and made of a polysaccharide called **tunicine**.

2. Cells. These are mesodermal in origin and have migrated into the test. These are of 6 or 7 different types : (*i*) large eosinophilous cells, (*ii*) small eosinophilous cells, (*iii*) small amoeboid cells, (*iv*) granular cells, (*v*) round vacuolated cells, (*vi*) nerve cells with several processes, and (*vii*) squamous epithelial cells.

3. Interlacing fibrils. These form a fine network in the test and resemble with smooth muscle fibres.

4. Calcareous spicules. *Herdmania* has large number of calcareous spicules of two types. All the spicules bear several equidistant rings of minute spines all pointing in the same direction all along their length (Fig. 4.2).

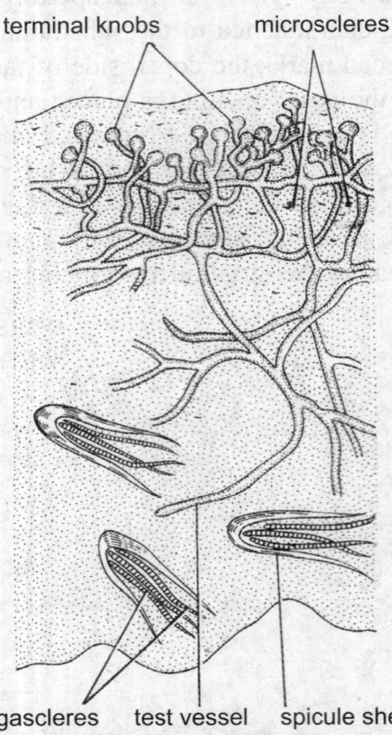

Fig. 4.2. *Herdmania*. V.S. of test showing blood vessels, ampullae and spicules.

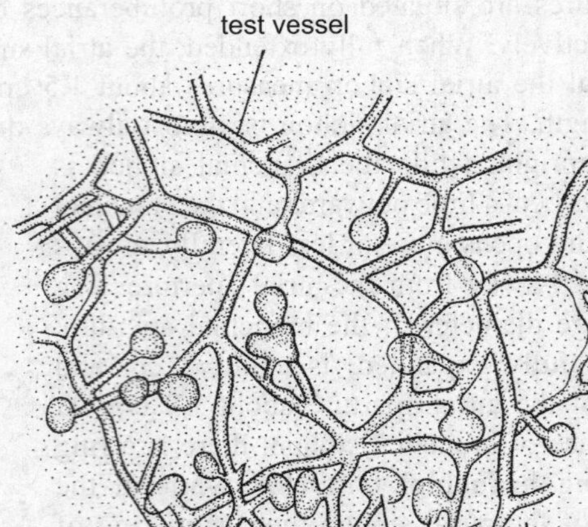

Fig. 4.3. *Herdmania*. Surface section of test showing blood vessels and ampullae.

5. Blood vessels. These blood vessels form a network in the test and end in bulb-like dilations called **vascular ampullae**. The vascular ampullae form vascular areas of bright red patches on the test. Vascular vessels and ampullae transport blood and, thus bring food to the test, they also act as accessory respiratory organs. The ampullae are also receptor organs.

Spicules

There are two types of spicules found in the body : small **microscleres** and large **megascleres**. They are all calcareous and have a definite shape.

1. Microscleres. The microscleres are found only in the test and lie scattered throughout its substance. Each spicule consists of a rounded knob-like head and an elongated tapering body. The head is generally smooth, rarely with a few spines on its sides but the body bears a large number of rings of spines ranging from 5–20. These rings of spines run round the body of the spicule, nearly equidistant from one another. The spines always pointing towards the head of the spicule. The size of the spicule varies according to the stage of its growth, the average size being 50 microns, while very large ones measure about 80 microns.

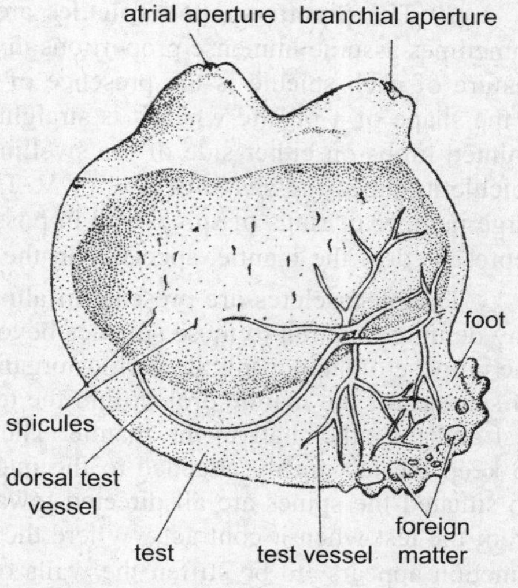

Fig. 4.4. *Herdmania*. Side view of the test with the right half out to expose the test vessels.

2. Megascleres. The megascleres are of two kinds: (*i*) spindle-shaped, and (*ii*) pipette-shaped spicules. Both these types of spicules are much larger than the microscleres.

(*i*) The **spindle-shaped spicules** are always found enclosed in a connective tissue sheath and lie scattered throughout the body of the animal in most of its tissues; but they are regularly arranged in the posterior half of the test. They are also very abundant in the mantle but are not uniformly distributed. They are usually very dense in the region of the stomach and the gonads and also at the bases of siphons, and are quite well represented in the region of the intestine but are sparse at the base of the longitudinal muscles. They are usually arranged in linear rows in sheaths, each row having the appearance of a string. Like the microscleres, each spicule has a large number of rings of spines ranging from 20 to 60, all pointing in the same direction. Their size varies according to their stage of growth; an average size is 1.5 mm, but some attain a size of 2.5 mm.

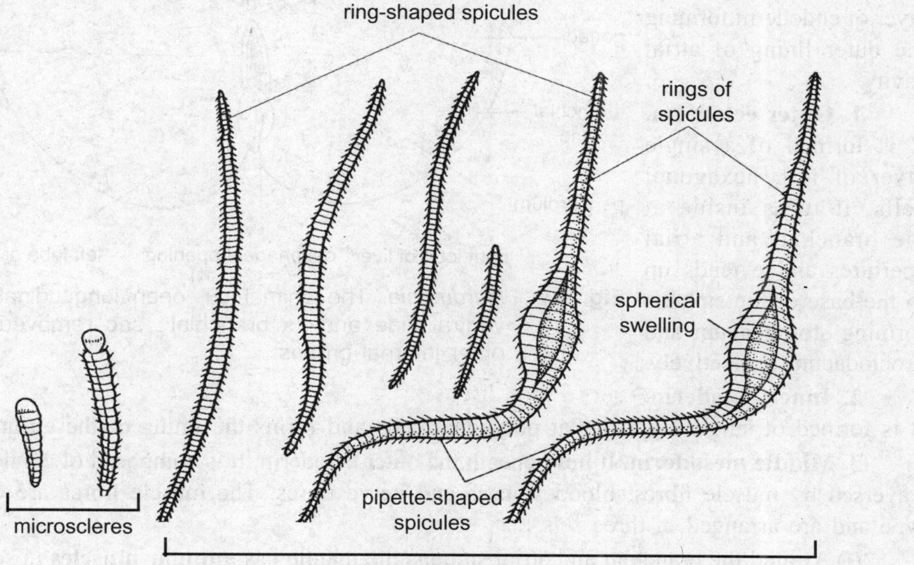

Fig. 4.5. *Herdmania*. Different types of spicules.

(ii) The **pipette-shaped spicules** are larger than those of the spindle-shaped variety and sometimes assume immense proportions in size, being as much as 3.5 mm long. The characteristic feature of each spicule is the presence of a large spherical swelling in the middle which gives it the shape of a pipette when it is straight. But very often the spicules are not straight, the two pointed limbs on either side of the swelling being bent one way or the other with the result the spicule assumes the shape of a U or V. They resemble the spindle-shaped spicules in having a large number of rings of spines and in possessing connective tissue sheaths. They are abundantly represented in the mantle, specially in the region of the gonads and lobes of the liver.

The megascleres are presents in almost every organ of the body. It is difficult to attribute any definite function to these spicules beyond saying that they probably form a skeletal tissue for the purpose of supporting the various organs of the body. It has been observed that a large number of spicules of the mantle project into the test, so that one-half of the spicule lies imbedded in the test and the other half in the mantle. The function of some of these spicules, therefore, seems to keep the test firmly attached to the mantle. This view is confirmed by the fact that they are so situated the spines are all directed towards the mantle, so that mantle cannot be drawn away from the test when it contracts. Where the spicules form a sheath around the blood-vessels, their function appears to be stiffen the walls of the vessels and prevent their collapse.

Mantle or Body Wall

Inside the test is the body wall or **mantle**. The mantle secretes the test. The mantle is present beneath the test and is attached only at the branchial and atrial apertures where it forms branchial and atrial siphons. The mantle encloses a large **atrial cavity** or **atrium**, containing water. Histologically the mantle has an outer layer of ectoderm, middle layer of mesoderm, and an inner layer of endoderm forming the outer lining of atrial cavity.

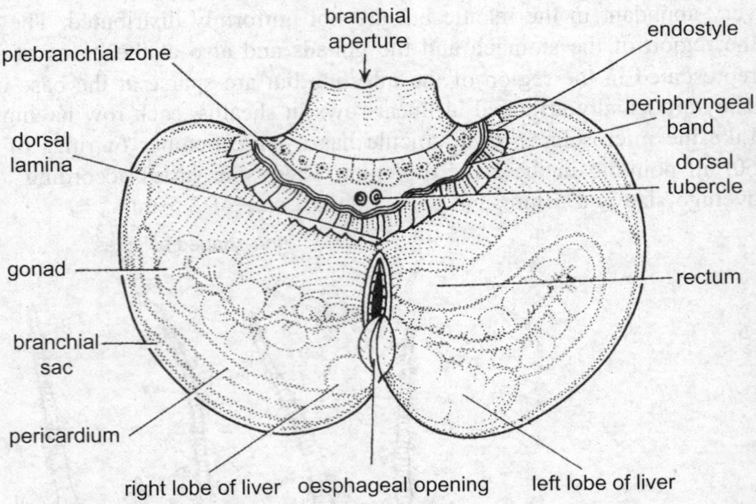

Fig. 4.6. *Herdmania*. The animal cut open longitudinally from the ventral side and its branchial sac removed to expose other internal organs.

1. Outer ectoderm. It is formed of a single layer of flat, hexagonal cells. It turns inside at the branchial and atrial apertures and extends up to the base of the siphons forming stomodaeum and proctodaeum respectively.

2. Inner ectoderm. It is formed of single layer of flat polygonal cells and forms the lining of the atrium.

3. Middle mesoderm. It lies beneath the outer ectoderm. It is composed of connective tissue traversed by muscle fibres, blood sinuses and nerve fibres. The muscle fibres are of unstriped type and are arranged in three sets :

(i) Around the branchial and atrial siphons the mantle has **annular muscles** in several rings.

(ii) Below the annular muscles are **longitudinal muscles** which start from branchial and atrial apertures, then fan out up to the middle of the body on each side.

(iii) **Branchioatrial muscles** found in between two siphons.

GENERAL ANATOMY

The body of *Herdmania* can be roughly divided into two parts : the **body proper** which projects into the sea water and the **foot** which lies imbedded in the sea-bottom. All the soft parts of the animal lie in the body proper. The foot is entirely composed of the test. If the test is removed from the left side of the body proper, many of the internal organs of the animal can be seen through the more or less transparent mantle; or the test can be removed altogether and the body examined.

When the body is examined from the left side one can easily identify the following structures :

1. The two projecting **siphons** with their annular and longitudinal muscles, and the respective apertures each guarded by four lips.

2. The stomach, intestine and rectum, the **intestine** forming a wide loop.

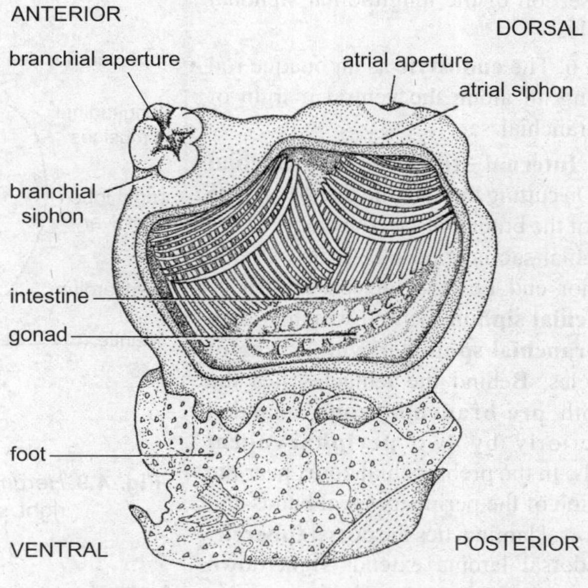

Fig. 4.7. *Herdmania.* Side view of the animal with the test cut open to the left to expose internal structures.

3. A part of the left lobe of the liver appearing as a chocolate-coloured body in front of stomach.

4. The **left gonad** forming a lobulated pinkish mass lying in the loop of the intestine.

5. Endostyle along the ventral margin of pharynx.

If the body is examined from the right, one can see the following structures :

1. The two siphons along with their musculature and apertures.

2. The **neural (hypophysial) gland** forming a dark elevated area in the intersiphonal region.

3. The **pericardium** appearing as a long transparent tube of uniform thickness, within which lies the **heart**.

4. The **right lobe of the liver** forming a more or less oval chocolate-coloured mass near the anterior end of the pericardium.

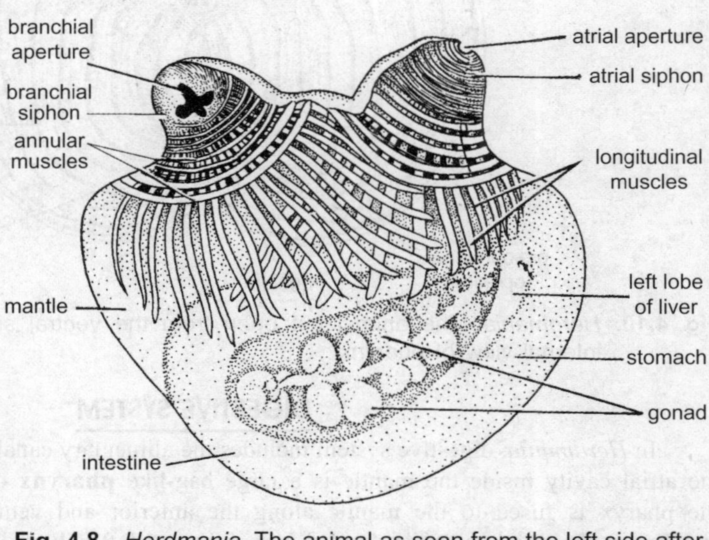

Fig. 4.8. *Herdmania.* The animal as seen from the left side after removal of the test.

5. The **right gonad** forming a lobulated pinkish mass close to the line of insertion of the longitudinal siphonal muscles.

6. The **endostyle** as an opaque rod running all along the ventral margin of the branchial sac.

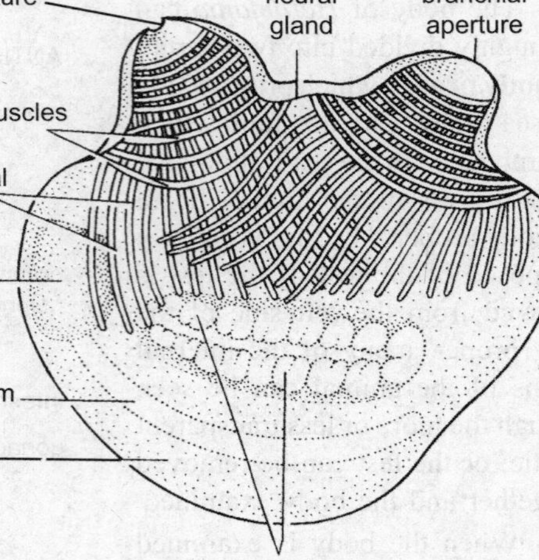

Fig. 4.9. *Herdmania*. The animal as seen from the right side after removal of the test.

Internal structure of branchial sac. On cutting through the mantle and the wall of the branchial sac, the cavity of the branchial sac is exposed. At its extreme anterior end lies the incised **mouth** or **branchial siphon**, behind which is seen the **branchial sphincter** with its ring of tentacles. Behind the tentacles lies the smooth **pre-branchial zone** bounded posteriorly by two **peripharyngeal bands**. In the prebranchial zone, near the junction of the peripharyngeal bands and the dorsal lamina, lies the dorsal **tubercle**. The dorsal lamina extends right down to the anterior margin of the **oesophageal opening** surrounded by the so-called **oesophageal area**. The exposed internal surface of the **branchial sac** shows a number of folds on both sides.

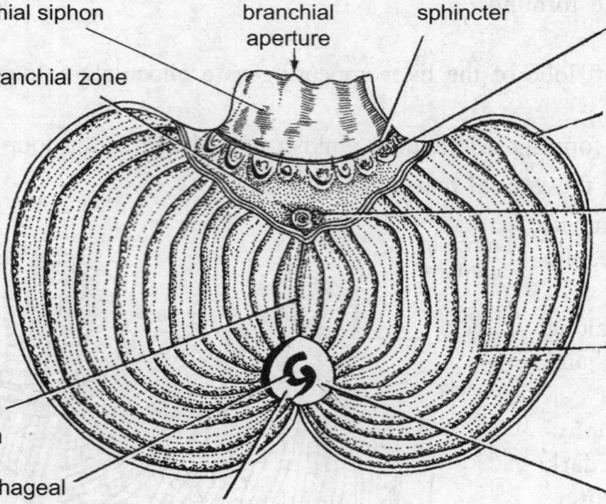

Fig. 4.10. *Herdmania*. The animal cut open from the ventral side longitudinally to show the internal view of pharynx.

DIGESTIVE SYSTEM

In *Herdmania*, digestive system includes the alimentary canal and digestive glands. Lying in the atrial cavity inside the mantle is a large bag-like **pharynx** or **branchial sac**. The wall of the pharyx is fused to the mantle along the anterior and ventral sides, on account of this attachment the atrial cavity is divided into right and left halves which are, however, continuous dorsally.

I. Alimentary Canal

The alimentary canal is coiled, beginning from the mouth and ending in anus. It has the following parts :

1. Mouth. The mouth or branchial aperture lies at the top of branchial siphon, which is designated as the anterior end of the body. It is guarded by four lips (lobes) of the test. The mouth leads into a branchial siphon.

2. Branchial siphon. It is a narrow and tubular cavity lined by ectoderm and is called the **stomodaeum** or **buccal cavity**. At the base of the stomodaeum is a ring of branching **tentacles** which act as a sieve allowing only minute food particles to go in. These are richly innervated and their number is about 64. Below the tentacles is a smooth **prebranchial (prepharyngeal) zone** having a swollen **dorsal tubercle** made of two spiral coils. Then there are two thin ciliated ridges, the **peripharyngeal bands** lying parallel and encircling the upper end of the pharynx. The posterior one of the two peripharyngeal bands is joined mid-dorsally to a **dorsal lamina**, and mid-ventrally with an **endostyle**.

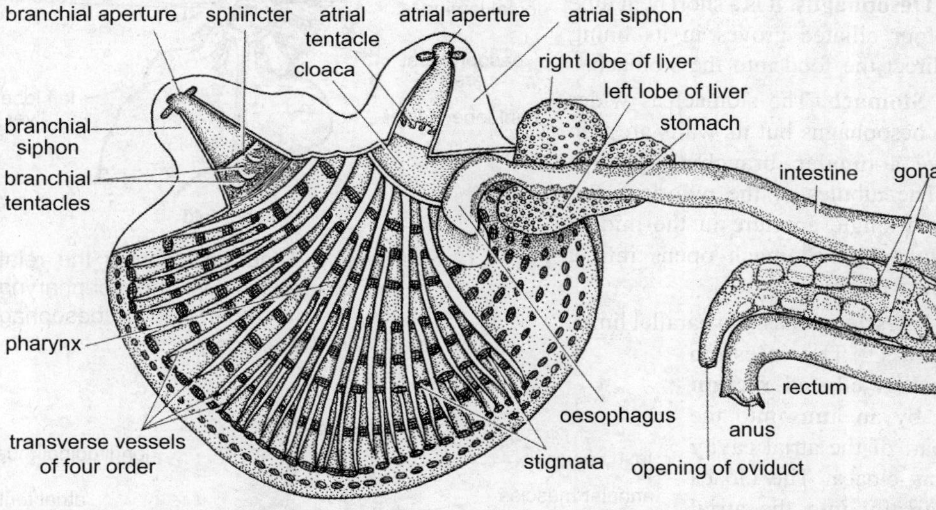

Fig. 4.11. *Herdmania*. The alimentary canal as seen from the left side after removal of the mantle.

3. Pharynx. It occupies the major part of the body cavity and is differentiated into two: prebranchial zone and branchial sac.

***(i)* Prebranchial zone.** It is the smaller anterior region having smooth walls without folds, cilia and stigmata or gill-slits.

***(ii)* Branchial sac or pharynx.** It is the posterior larger part of the pharynx. The pharynx almost fills, the atrial cavity. The wall of the pharynx is perforated by several rows of **stigmata** arranged in transverse rows, through these the cavity of the pharynx communicates with the atrial cavity. The edges of stigmata are beset with cilia which drive currents of water from the pharynx into the atrial cavity. The stigmata are not gill-clefts, but they are formed by sub-divisions of a few original **gill-clefts**. Between the stigmata the wall of the pharynx has transverse and longitudinal bars containing blood vessels, thus, the pharynx appears like a basket. The lining of the pharynx is raised into ten longitudinal folds on each side.

***(a)* Dorsal lamina** is a thin flap lying inside the mid-dorsal roof of the pharynx. It bears a number of conical, ciliated projections called **languets**. The dorsal lamina runs from the posterior peripharyngal band to the opening of the oesophagus. The dorsal lamina and its languets can bend to one side to form a sort of tube through which food-laden mucus passes.

(*b*) **Endostyle.** Lying in the mid-ventral floor of the pharynx is an **endostyle**. It is a longitudinal groove with four longitudinal rows of gland cells with ciliated cells between them. The middle row of cells has long cilia. The gland cells of endostyle secrete mucus which is driven laterally by ciliated cells. Endostyle is homologus with thyroid gland of vertebrates.

The posterior region of the pharynx is known as **oesophageal area**, around which all the folds of the pharynx converge. It has two semi-circular lips enclosing an aperture which leads into an **oesophagus**.

4. Oesophagus. It is a short bent tube having four ciliated groves in its lining which direct the food into the stomach.

5. Stomach. The stomach is wider than the oesophagus but its walls are thin and have a tubular, branching **pyloric gland**. The tubules of the pyloric gland open by a single aperture in the middle of the intestine. Stomach opens into a thin-walled intestine.

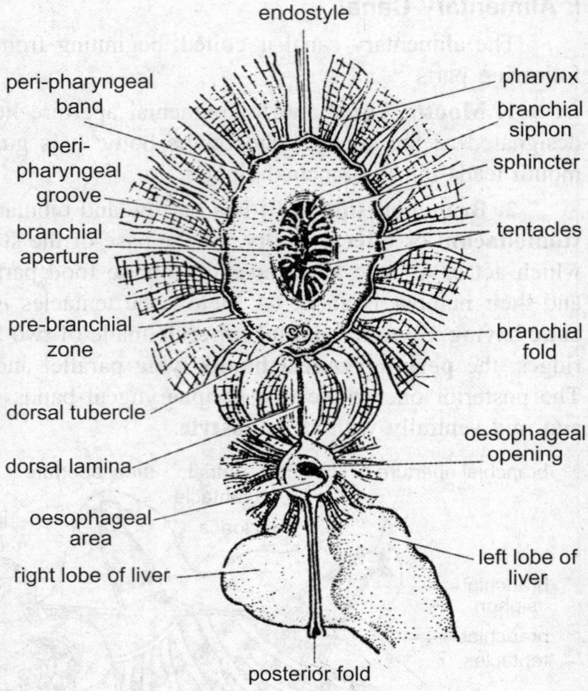

Fig. 4.12. *Herdmania*. Diagram showing the relation of the branchial folds, the peripharyngeal bands, the dorsal lamina, the oesophageal area and the endostyle.

6. Intestine. It has two parallel limbs forming a U. The intestine joins a short ciliated **rectum** opening by an **anus** into the dorsal part of the atrial cavity known as cloaca. The cloaca leads dorsally into the atrial siphon, opening outside through **atriopore** or **atrial aperture**. Like branchial siphon, the atrial siphon is also lined by ectoderm and, thus, represents the **proctodaeum**.

II. Digestive Glands

1. Liver. Lying against the stomach is a dark brown bilobed **liver**, the left lobe being larger. The liver is formed of a large number of fine tubules embedded in the matrix of connective tissue having blood sinuses. These tubules unite to form 11 or 12 hapatic duct which open separately into the

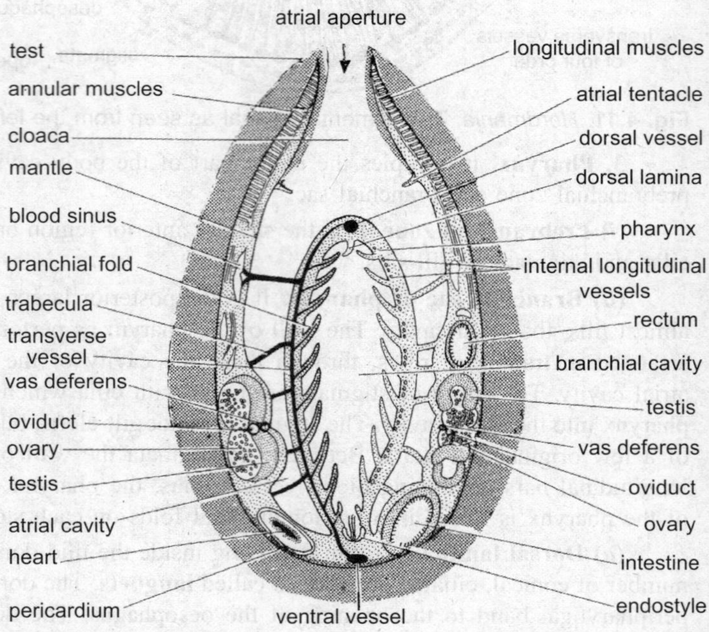

Fig. 4.13. *Herdmania*. Diagrammatic transverse section of the animal through the atrial siphon.

stomach. The hepatic secretion has a powerful amylase, protease and weak lipase. Bile pigments are also present. In the liver and stomach are starch-like granules which are probably stored food.

2. Pyloric gland. It is located in the walls of the posterior part of stomach and intestine. Pyloric gland is branched and its ductules unite to form a common duct which opens in the intestine. Its secretion is pancreatic in nature. It also probably acts as an excretory organ.

III. Food, Feeding and Digestion

(*i*) Food. Microorganisms such as protozoans, pieces of decaying animals, zooplanktons and algae are its food.

(*ii*) Feeding. Ciliary feeding takes place because of sedentary habit. Cilia bordering the stigmata keep up a constant current of water entering the mouth, and passing into the pharynx from where it goes out through the stigmata into the atrial cavity, and then to the exterior through the atrial aperture. The current of water carries in minute organic food particles which pass into the pharynx, where they settle down on the pharyngeal walls. Gland cells of the endostyle secrete mucus (which is not carried forward by cilia as was originally believed). The mucus is lashed out transversely from the endostyle by its cilia, at right angles to the endostyle. Food particles lying on the lining of the pharynx are caught in the mucus and generally transported upwards along the pharyngeal wall to the dorsal lamina, in passing the food particles and mucus are rolled into a cylindrical mass. The dorsal lamina cilia convert the cylindrical mass into a string in the tube formed by curving upwards of languets. This food-laden mucus string is carried into the oesophagus and then into the stomach.

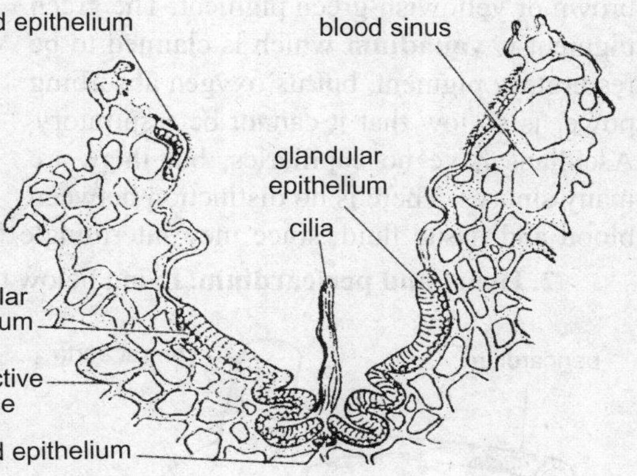

Fig. 4.14. *Herdmania.* T.S. of endostyle.

(*iii*) Digestion. The liver secretes a yellowish-brown digestive fluid into the stomach, it has many enzymes, an **amylase** which splits carbohydrates into maltose, a **protease** which breaks down proteins, and a weak **lipase** which probably acts on fats. The liver also stores starch as reserve food material. The pyloric gland is an accessory digestive organ, it is probably pancreatic in nature. Food is digested mainly in the stomach and absorption occurs in the intestine. The ciliated rectum expels the much-coiled excreta with such force that it is shot out about 10 cm through the atrial siphon.

RESPIRATORY SYSTEM

The current of sea water coming into the pharynx brings dissolved oxygen. The pharyngeal wall is very thin and has a rich blood supply in blood vessels and blood sinuses which provide large areas in which probably some carbon dioxide is exchanged for oxygen. The respiratory pigment of blood is practically incapable of absorbing any oxygen. Interchange of gases occurs in the atrium where **vascular trabeculae** are present. The test has **vascular ampullae** which act as accessory respiratory organs and bring about an exchange of gases through the surface of the test.

CIRCULATORY SYSTEM

The circulatory system of *Herdmania* includes heart and pericardium, blood vessels and blood.

1. Blood. The blood is slightly reddish and transparent. It has colourless **plasma**, many pigmented and non-pigmented corpuscles, and a few colourless amoeboid leucocytes. **Leucocytes** are nucleated, amoeboid and colourless. Other corpuscles are round, of various sizes, some have vacuoles, and they may or may not be nucleated. They may have no pigment, while others may contain orange, brown or yellowish-green pigment. The green pigment is **vanadium** which is claimed to be respiratory pigment, but its oxygen absorbing power is so low that it cannot be respiratory. Ascidians have no capillaries, but there are many sinuses. There is no distinction between blood and tissue fluid, since they intermingle freely.

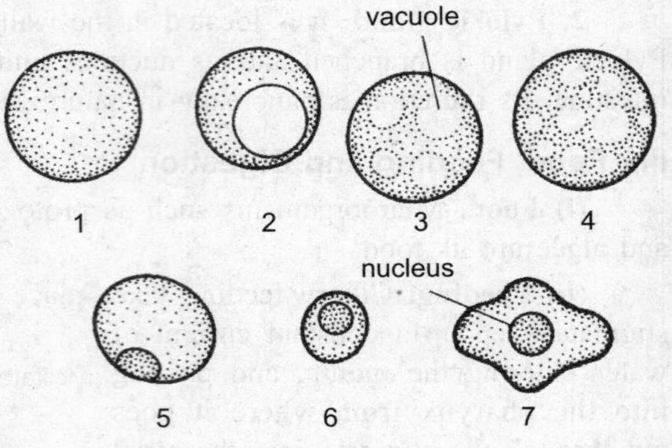

Fig. 4.15. *Herdmania*. The blood corpuscles.

2. Heart and pericardium. Lying below the right gonad is a tubular, elongated **pericardium** with thick non-contractile wall. It is about 7 cm long and 3 mm wide. It is closed at both the ends and contains a colourless pericardial fluid having corpuscles similar to those of blood. Its wall is composed of connective tissue having blood sinuses. It is internally lined by squamous epithelium. Inside the pericardium is a thin-walled, tubular, contractile **heart** with striated muscles. The heart is attached along its whole length to the pericardium by a thin connective tissue flap. The heart is opened at both the ends. A pear-shaped body is present in between the heart and pericardium which probably regulates the blood flow in the heart.

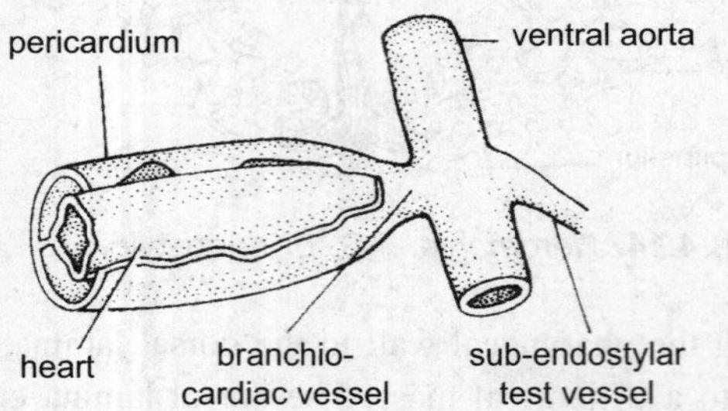

Fig. 4.16. *Herdmania*. Wall of the pericardium cut open to show the attachment of the heart.

3. Blood vessesls. There are no true blood vessels in most ascidians, but they are well developed in *Herdmania*. Ventrally the heart opens into a **ventral aorta** which lies in the pharyngeal wall along the whole length below the endostyle. The ventral aorta near the heart gives out a **ventral test vessel** supplying blood to the test. Throughout its length the ventral aorta gives out paired **transverse vessels** which run in the pharynx wall on each side between rows of stigmata. From the transverse vessels arise smaller branches which pass between the stigmata of each row. The transverse vessels traverse the wall of the pharynx and then join a **dorsal aorta** lying above the dorsal lamina in the dorsal wall of the pharynx. It is not connected with the heart. It sends branches to the neural complex, anteriorly to the dorsal tubercle and

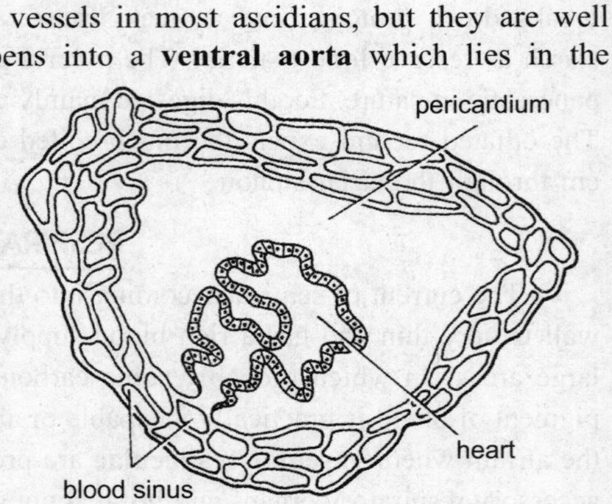

Fig. 4.17. *Herdmania*. T.S. of the pericardium and the heart.

branchial siphon. Then it leads posteriorly into a **branchio-visceral vessel**. The branchio-visceral vessel immediately divides into a **right oesophageal vessel** for the right liver lobe and right portion of oesophagus, and a **ventro-intestinal vessel** supplying blood to the left portion of oesophagus, stomach, intestine, rectum, left liver lobe and left gonad.

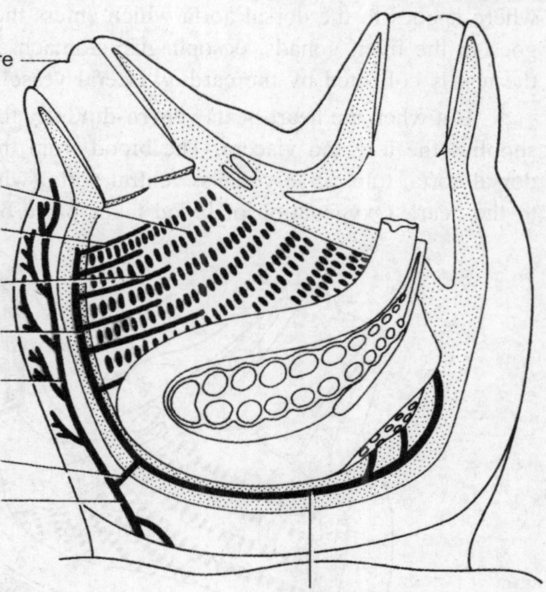

Fig. 4.18. *Herdmania*. Ventral aorta and its branches.

Cardio-visceral vessel. From the dorsal end of the heart arises a **cardio-visceral vessel** which near its origin gives a **right hepatic vessel** to right liver lobe and an **oesophageo-test vessel** to the oesophagus and test. The main dorsal branch gives a **test vessel**, a **left oesophageal vessel** and **right gonadial vessel** to corresponding organs.

(i) The main dorsal vessel further extends and gives off circular **subtentacular vessel** at the base of atrial tentacles and siphonal vessels to atrial siphon.

(ii) The middle **left gonadial vessel** to the left gonad, and

(iii) **Ventral vessel** soon divides into a **dorso-intestinal**, a **gastric** and a **dorsal test** vessels takes blood through its branches to the test, liver, gonads, oesophagus, stomach and intestine.

Circulation

Peristaltic contractions of the heart and small pear-shaped body within pericardium pass in waves from one end of the heart to the other for several beats, then there is a short pause after which contractions begin again and pass in waves in the opposite direction, thus, the flow of blood is alternately in opposite directions. Nowhere else in the animal kingdom this peculiar reversal in the flow of blood is seen.

When the heart beats **dorso-ventrally** the blood is driven into the ventral aorta from where it goes to the test and to the pharynx through the transverse vessels. The blood is oxygenated in the pharynx from

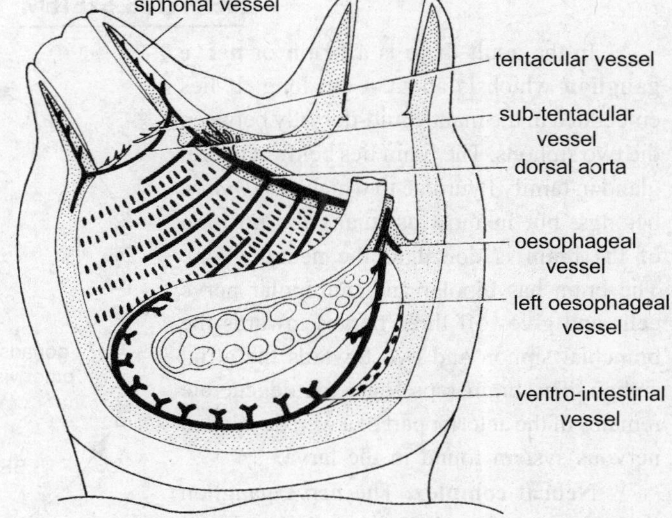

Fig. 4.19. *Herdmania*. Dorsal aorta and its branches.

where it goes to the dorsal aorta which enters the branchio-visceral vessel. From here the blood goes to the liver, gonads, oesophagus, stomach and intestine. The blood from these organs and the test is collected by the cardio-visceral vessel and returned to the heart.

But when the heart-beats **ventro-dorsally** the blood goes to the cardio-visceral vessel which supplies the test and viscera. The blood from the viscera goes to the branchio-visceral vessel, dorsal aorta, transverse vessels, ventral aorta (which also receives blood from the test) and then to the heart. Oxygenation of blood takes place both in the pharynx and the test.

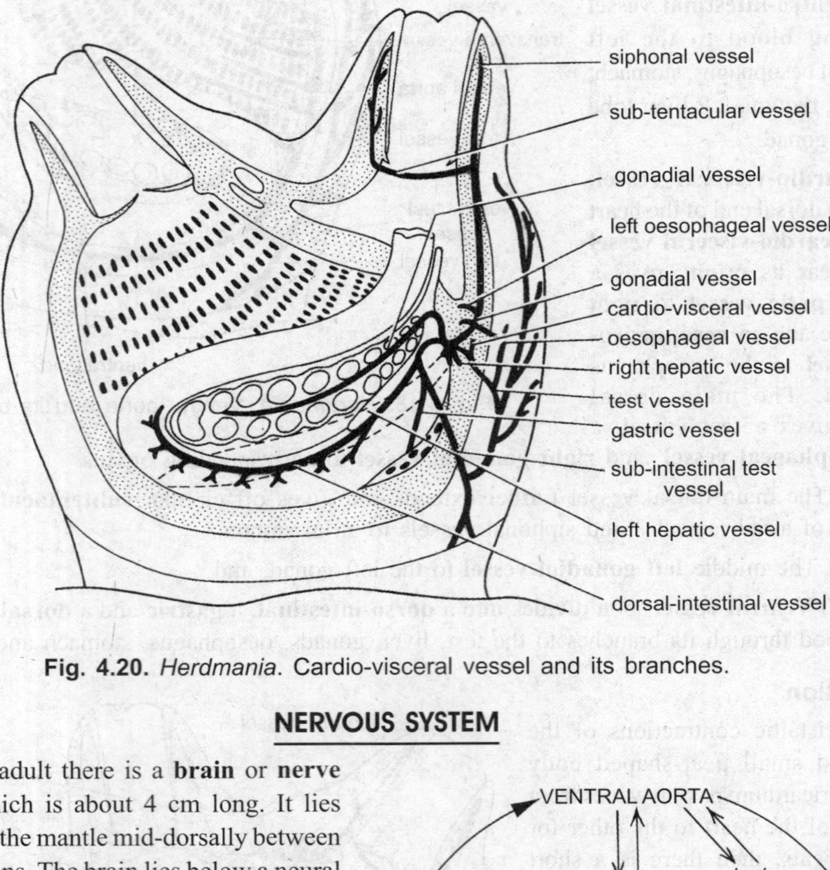

Fig. 4.20. *Herdmania*. Cardio-visceral vessel and its branches.

NERVOUS SYSTEM

In the adult there is a **brain** or **nerve ganglion** which is about 4 cm long. It lies embedded in the mantle mid-dorsally between the two siphons. The brain lies below a neural gland in family Pyuridae to which *Herdmania* belongs, but in most ascidians the position of the brain is dorsal to the neural gland. The brain has bipolar and multipolar nerve cells and gives off three nerves towards the branchial siphon and two towards the atrial siphon. The brain represents the degenerate remains of the anterior part of a well developed nervous system found in the larva.

Neural complex. The nerve ganglion or brain, the neural gland and the dorsal tubercle together form the **neural complex**.

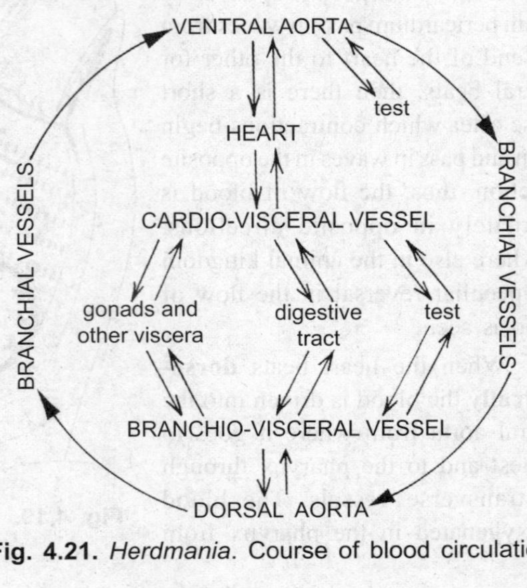

Fig. 4.21. *Herdmania*. Course of blood circulation.

RECEPTORS

There are no special sense organs. Many cells receive a nerve, especially in the test and the siphons. 1. Red pigmented cells situated on the margins of siphons and vascular ampullae are **photoreceptors** being sensitive to light. 2. Sensory cells of the test, margins of siphons and tentacles are **tangoreceptors** being sensory to contact. 3. The cells on the margins of siphons are **rheoreceptors**. They are sensitive to water currents. 4. Cells lining the siphons are **thermoreceptors**. They are sensitive to changes of temperature. 5. The tentacles and dorsal tubercle are the **chaemoreceptors**.

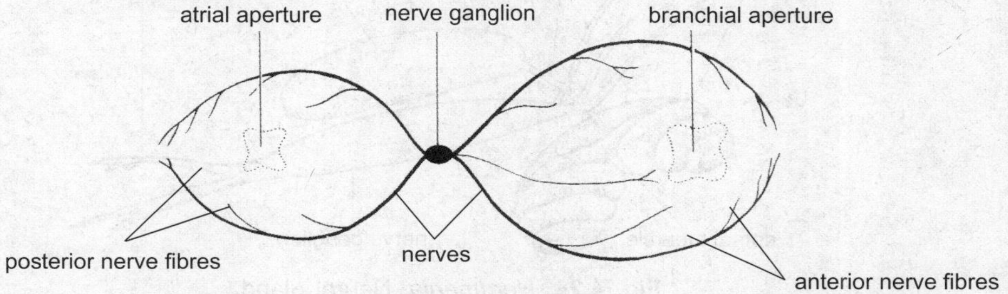

Fig. 4.22. *Herdmania*. The nerve ganglion with its nerves.

Dorsal tubercle. Lying in the pre-branchial zone, near the junction of dorsal lamina with the peripharyngeal bands is a **dorsal tubercle**. It has a broad base with two spirally coiled lobes. Each lobe has three coils and its surface has ciliated cells. Each lobe has a open coiled, ciliated narrow channel running from its base to the tip of the lobe. The channels of both the lobes unite at the dome-like base of the dorsal tubercle. The dorsal tubercle is richly supplied with the blood in blood sinuses present in the loose connective tissue and nerves. It is an **olfactory** and **gustatory** organ.

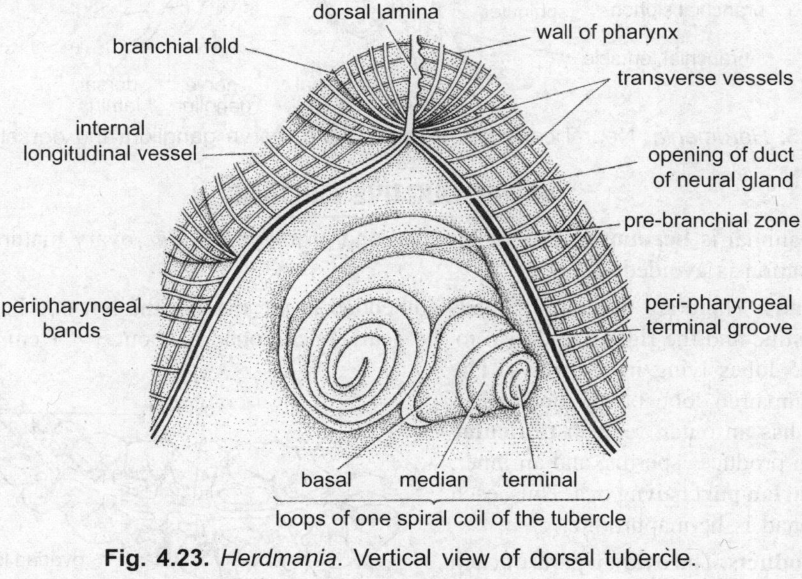

Fig. 4.23. *Herdmania*. Vertical view of dorsal tubercle.

EXCRETORY ORGAN

There is an oval **neural gland** of brown colour lying immediately above the brain. It is embedded in the mantle in the intersiphonal region and is 4 × 1 mm in size. It has several branching tubules, which open in a non-ciliated canal running along the entire length of the gland.

It leads into a short duct which opens by a ciliated funnel at the base of the dorsal tubercle in between two lobes. **Nephrocyte** cells of the blood collect waste particles, mainly xanthine and urates. They enter the lumen of the neural gland from where they pass out into the pre-branchial zone of pharynx. The neural gland also secretes a hormone which aids in oviposition and metamorphosis. Thus, the neural gland is probably homologous with the pituitary gland of vertebrates (**Jullin** 1881, **Bacq, Florkin** and **Das** 1956).

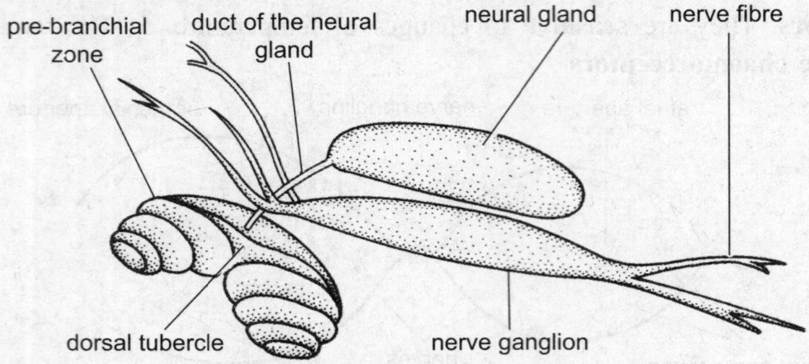

Fig. 4.24. *Herdmania*. Neural gland.

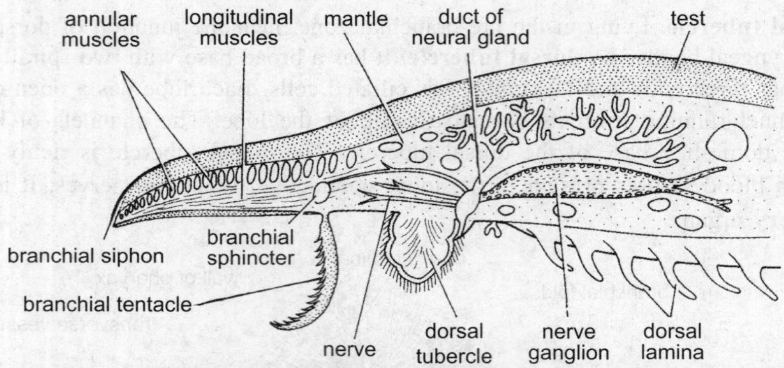

Fig. 4.25. *Herdmania*. Neural complex. (Neural gland, nerve ganglion and dorsal tubercle).

REPRODUCTIVE ORGANS

The animal is **hermaphrodite** or bisexual and protogynous, *i.e.,* ovary matures first, thus, self-fertilisation is avoided.

Gonads. There are two large gonads embedded in the mantle, the left one lies in the loop of the intestine and the right one dorsal to the heart. Each gonad is about 3 × 1 cm, and consists of 10 to 25 lobes lying in two rows. The posterior unpaired lobe being the largest. Each lobe has an outer, reddish **testicular part** which produces sperms, and an inner, pinkish **ovarian part** haivng ova. Thus, each lobe of gonad is hermaphrodite.

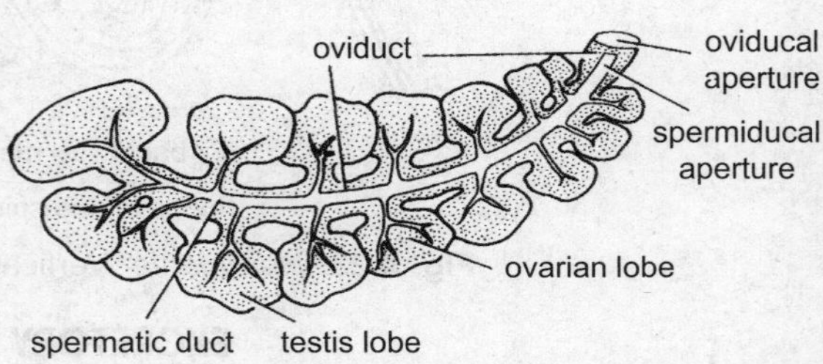

Fig. 4.26. *Herdmania*. Left gonad viewed from its inner face.

Gonoducts. The ovarian parts of each gonad are joined by narrow **ovarian ductules** to a thick oviduct, while the testicular parts are connected to a thin **spermatic duct** or **vas deferens** by **spermatic ductules**. The oviduct and vas deferens run parallel and open by their own apertures (oviducal and spermiducal apertures respectively) into the atrial cavity where ova and sperms are discharged. Oviducal aperture lies a little behind the anus and spermiducal aperture is little behind the oviducal aperture. Both the gonoducts are ciliated internally. The sex cells pass out of the atrial siphon into the sea. Generally cross fertilisation occurs in sea water.

It is suggested that the neural gland secretes a hormone similar to the one produced by the anterior lobe of the pituitary of vertebrates. The hormone influences the brain which stimulates the gonads to release their gametes. In this way the animals fixed close to each other are stimulated to spawn at the same time, thus, ensuring cross fertilisation.

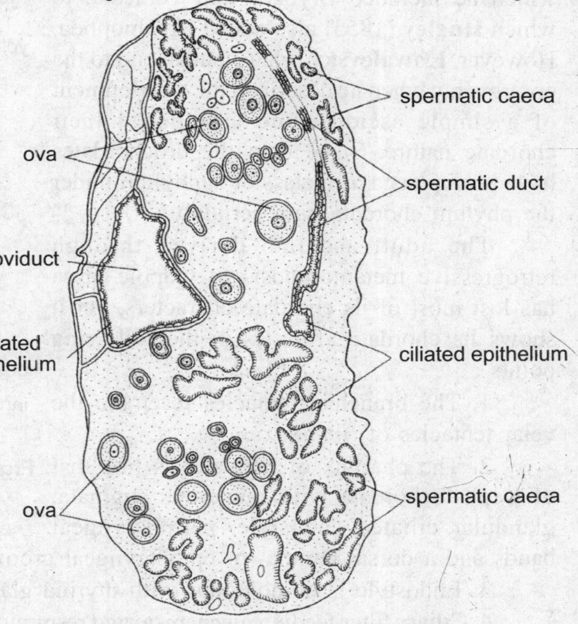

Fig. 4.27. *Herdmania*. T.S. gonad.

Gametes. A mature **spermatozoan** is about 4 mm in length. Each sperm has ovoid head bearing a long beak-shaped acrosome, a middle piece and a long tail. Sperms are of three types: (*i*) beak and head are of equal length, (*ii*) beak three times longer than the neck, and (*iii*) beak is three times shorter than the neck.

A mature **ovum** (Fig. 4.29) is about 0.3 mm in diameter. Its cytoplasm contains well distributed yellowish granulated yolk. Nucleus is eccentric with a conspicuous eccentric nucleolus. The ovum is surrounded by a thin **vitelline membrane**, which is further enveloped by two **chorion membranes** secreted by the follicular cells of the ovary. In between the vitelline membrane and inner chorion is a perivitelline space filled with perivitelline fluid. A number of follicle cells are found attached with the vitelline membrane and a few are found floating in the fluid. In between the outer and inner chorion membranes is found in the inter-chorional fluid. Follicle cells are also remain attached with the outer surface of the outer chorion. These cells keep the ovum floating in water.

Fertilisation. Fertilisation is external. Since the animal is protogynous, *i.e.,* ova matures earlier than the sperms, only cross fertilisation occurs.

DEVELOPMENT

Development of *Herdmania* has been described in a chapter in Vertebrate Embryology section.

AFFINITIES AND SYSTEMATIC POSITION OF *HERDMANIA* (UROCHORDATA)

The systematic position of Urochordata had been remained controversial. **Lamarck** (1816) first called them **Tunicata**. **Cuvier** (1817) classified them along with **Mollusca** but **Lamarck** placed them between **Radiata** and **Vermes**. **Milne Edwards**

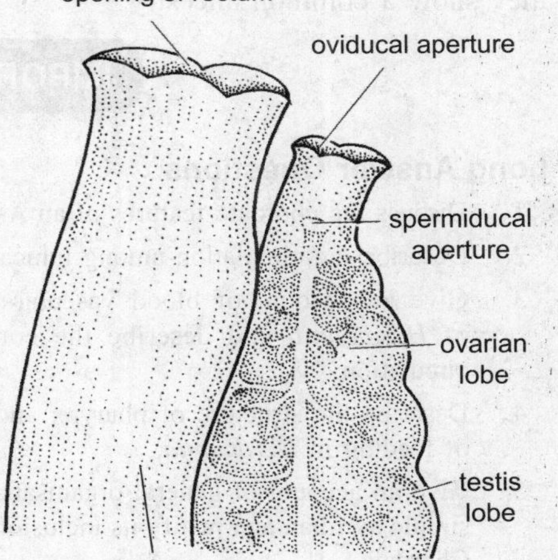

Fig. 4.28. *Herdmania*. Diagram showing openings of rectum, oviduct and spermatic duct.

(1843) established the class **Molluscoidea** in which he included Bryozoa and Tunicata, to which **Huxley** (1853) also added Brachiopoda. However, **Kowalevsky** (1886) put an end to the controversy when he described the development of a simple ascidian and established their chordate nature. Since then the urochordates have been placed as a class or suphylum under the phylum chordata with certainty.

The adult ascidian derived through retrogressive metamorphosis of tadpole larva has lost most of its chordate characters, yet it shows its chordate affinities in the following points :

1. The branchial tentacles resemble the velar tentacles of *Amphioxus*.

2. The pharynx of tunicates is like that of Cephalochordata with numerous stigmata, glandular ciliated endostyle, peripharyngeal bands and a dorsal lamina (or epipharyngeal groove).

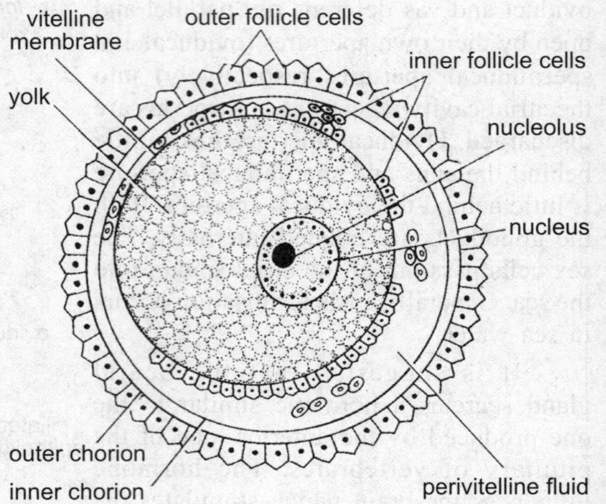

Fig. 4.29. *Herdmania*. Structure of a mature ovum.

3. Endostyle is homologous with thyroid gland of vertebrates.

4. Ciliary filter feeding mechanism and respiratory system in urochordates and cephalochordates are similar.

5. Origin and position of neural gland suggest its homology with the pituitary gland of vertebrates. Even more than the adult, the tadpole larva shows its chordate characteristics in its nervous system, stigmata and notochord. Besides these it has: (*a*) a muscular postanal tail with tail-fin, (*b*) cleavage and gastrulation in ascidian development are similar to those of chordates.

The urochordate tadpole larva is the relic of free-swimming chordate ancestor. Its degeneration is a secondary modification due to sedentary mode of life of adult.

Thus, Urochordata have been derived from a free-swimming chordate ancester and their closest affinities are with the Cephalochordata. The two classes have such close homologies that they show a common ancestry.

IMPORTANT QUESTIONS

Long Answer Questions

1. Discuss the chordate features of an Ascidian.
2. Describe the circulation among tunicates.
3. Give an account of blood vascular system of *Herdmania* and describe the course of circulation.
4. Describe the structure of pharynx and mode of feeding in *Herdmania*.
5. Give an account of general organisation of a simple ascidian and justify its inclusion in the Chordata.
6. Describe the external morphology of *Herdmania*.
7. Discuss the affinities of Urochordata in detail.
8. Describe the structure and function of the tunic of *Herdmania* in detial.
9. Describe the digestive system of *Herdmania*.

Short Answer Questions

1. With reference to *Herdmania* write note on the following :

   (*a*) Neural gland complex,

   (*b*) Spicules,

   (*c*) Blood corpuscles,

   (*d*) Gonads, and

   (*e*) Dorsal tubercle.

2. Give an account of the structure of the pharynx and the process of respiration in *Herdmania*.
3. Discuss the affinities of Urochordata.
4. Write note on the following of *Herdmania* :
   (*i*) Respiratory system,
   (*ii*) Excretory system,
   (*iii*) Nervous system, and
   (*iv*) Reproductive system.

Very short Answer Questions

1. Write the common name of *Herdmania*.
2. Write the names of various blood corpuscles in *Herdmania*.
3. What are the two types of spicules in *Herdmania*?
4. Give the name of cavity that encloses viscera.
5. What is the function of endostyle in *Herdmania*?
6. What is the type of feeding of *Herdmania*?
7. Write the name of the structure which regulates the flow of blood in heart of *Herdmania*.
8. What is the function of vanadocytes blood corpuscles in *Herdmania*?
9. Write the name of excretory organ in *Herdmania*.

Multiple Choice Questions

1. Urochordates inhabit
   (*a*) Marine habitat
   (*b*) Streams
   (*c*) Estuarine habitat
   (*d*) Freshwater lakes.
2. Which of the following terms is not used for *Herdmania*?
   (*a*) Sea skirt (*b*) Sea squirt
   (*c*) Undapasi (*d*) Mulaikanna.
3. Fertilisation in *Herdmania* is
   (*a*) External and cross (*b*) Internal
   (*c*) Internal and self (*d*) External.
4. In *Herdmania* matrix is formed of
   (*a*) Tunicine (*b*) Starch
   (*c*) Metacine (*d*) None.
5. Microscleres are restricted only in
   (*a*) Test (*b*) Matrix
   (*c*) Test and matrix (*d*) None.
6. Megascleres occur throughout the body in *Herdmania* except in
   (*a*) Heart (*b*) Liver
   (*c*) Pharynx (*d*) Gonads.
7. True coelom in *Herdmania* is greatly reduced due to the development of
   (*a*) Atrium
   (*b*) Body wall
   (*c*) Mesenteries
   (*d*) V-shaped intestine.
8. Most tunicates, except *Herdmania* are able to extract rare metal from sea water
   (*a*) Mercury (*b*) Silver
   (*c*) Platinum (*d*) Vanadium.
9. Pyloric gland in *Herdmania* are present in
   (*a*) Stomach and pharynx
   (*b*) Stomach and intestine
   (*c*) Pharynx and intestine
   (*d*) Oesophagus and stomach.
10. Intestine in *Herdmania* is :
    (*a*) V-shaped (*b*) U-shaped
    (*c*) L-shaped (*d*) C-shaped.
11. Branchial sac in *Herdmania* is called
    (*a*) Atrial basket
    (*b*) Branchial feather
    (*c*) Atrial flower
    (*d*) Branchial basket.
12. Accessory respiratory organ in urochordates is
    (*a*) Pharynx (*b*) Test
    (*c*) Stomach (*d*) Liver.
13. Body of *Herdmania* is divided into
    (*a*) Head, body proper and foot
    (*b*) Body proper and foot
    (*c*) Head, abdomen and foot
    (*d*) Body, tail and foot.
14. Annular muscles in *Herdmania* are found near
    (*a*) Pharynx (*b*) Heart
    (*c*) Siphon (*d*) Liver.
15. The aperture of siphons in *Herdmania* is guarded by
    (*a*) 4 lips (*b*) 3 lips
    (*c*) 2 lips (*d*) 1 lip.

Answers

1. (*a*)	2. (*a*)	3. (*a*)	4. (*a*)	5. (*a*)	6. (*a*)	7. (*a*)	8. (*d*)
9. (*b*)	10. (*b*)	11. (*d*)	12. (*b*)	13. (*b*)	14. (*b*)	15. (*a*).	

5 Chapter Subphylum 2. Urochordata : General Characters, Classification and Representative Types

Contents of this Chapter

GENERAL CHARACTERS

1. Exclusively marine and cosmopolitan, commonly known as **sea squirts**.

2. Solitary or colonial.

3. Fixed or free-swimming and pelagic.

4. Body is covered by a cuticular tunic or test in adult stage composed of tunicin, similar to cellulose.

5. An anterior branchial aperture and a dorsal atrial aperture are commonly present.

6. Coelom is completely absent, even the pericardial cavity is not coelomic. Atrium is lined by ectoderm and opens outside by atriopore.

7. Notochord present in the larval stages and absent in the adults.

8. Pharynx large with two or several pairs of gill-slits. Endostyle present. Feeding ciliary.
9. Respiration by gill-slits and test.
10. Dorsal tubular nerve cord is present in the larval forms, while degenerates in the form of small ganglion in adults.
11. Blood vascular system open. Heart is ventral, simple and tubular. Blood without respiratory pigment.
12. Excretion by neural gland, nephrocytes, and pyloric gland.
13. Sexes are united, *i.e.,* hermaphrodite.
14. Fertilisation external and cross.
15. Development indirect with tadpole larva, having chordate characters.

CLASSIFICATION

Urochordata includes about 2000 fixed and 100 pelagic species. It has been divided into the following three classes :

Class 1. Larvacea (Appendicularia)

1. Small (5 mm long solitary) free-swimming pelagic forms.
2. Neotenic forms which retain the larval form throughout adult life with persistent tail, nerve cord, notochord.
3. Test is temporary and renewed periodically.
4. Single pair of gill-slits is present opening directly to outside.
5. Anus opens ventrally on the surface of the body.
6. Sexes are united, *i.e.,* hermaphrodite.
7. No metamorphosis.

Order 1. Endostylophora

1. House bilaterally symmetrical, with separate inhalent and exhalent apertures.
2. Pharynx with endostyle.

Examples: *Oikopleura, Appendicularia.*

Order 2. Polystylophora

1. House bilaterally symmetrical with single aperture.
2. Pharynx with endostyle and four longitudinal ciliated rows.

Example: *Kowalevskia.*

Class 2. Ascidiacea

1. Fixed or free-swimming marine forms.
2. Simple or compound, solitary or colonial.
3. Adults are never provided with a locomotory appendage or tail and have no trace of notochord and nerve cord.
4. Test is permanent, well developed and thick.
5. Branchial sac is large and well developed with its walls perforated by numerous gill-slits. Atrium opens dorsally by atriopore.
6. Sexes united. Larva free-swimming and highly developed and undergoes retrogressive metamorphosis.
7. Colonial forms reproduces by gemmation

Order 1. Enterogona

1. Body sometimes divided into thorax and abdomen.
2. Neural gland usually ventral to nerve ganglion.
3. Single gonad lying in or behind intestinal loop.
4. Larva with ocelli and otolith.

Suborder 1. Phlebobranchia

1. Pharynx with acessory longitudinal vessels.
2. Budding or gemmation is rare.

Examples: *Ascidia, Ciona, Phallusia.*

Suborder 2. Aplousobranchia

1. Body elongated with distinct abdominal region.
2. Pharynx with bars, longitudinal vessels and transverse ridges.
3. Budding is common.

Example: *Clavelina.*

Order 2. Pleurogona

1. Body compact and not divided.
2. Neural gland dorsal or lateral to nerve ganglion.
3. Gonads two or more embedded in mantle wall.
4. Larva with otolith. Separate eye absent.

Examples: *Herdmania, Botryllus, Molgula, Styela.*

Class 3. Thaliacea

1. Free-swimming pelagic forms.
2. Solitary or colonial.
3. Test is permanent thin and transparent, with circular muscle bands.
4. Atriopore posterior.
5. Branchial sac has either two large or many small gill-slits.
6. Sexes united.
7. Tail, nerve cord and notochord absent in adults.
8. Life history exhibits an alternation of generations.

Order 1. Pyrosomida

1. Free-swimming, pelagic colonial form. Colony closed at one end.
2. Zooids of the colony are embedded in a common test. Branchial apertures of the zooids on the outer side and atrial apertures on the inner side.
3. Muscle bands confined to body ends.
4. Phosphorescent.
5. Free-swimming larval stage absent.

Example: *Pyrosoma.*

Order 2. Cyclomyaria (= Doliolida)

1. Free-swimming pelagic forms.
2. Body is cask-shaped with branchial and atrial apertures at opposite ends.
3. Test is moderately well developed, never much thickened.
4. Muscle-bands are always complete surrounding the body.
5. Stigmata small, few to many.
6. Hermaphrodite. Life history exhibits an alternation of generation.
7. Sexual generation is always polymorphic.
8. Larva possesses notochord and tail.

Examples : *Doliolum, Dolchinia, Doliopsis.*

Order 3. Hemimyaria or Desmomyaria (= Salpida)

1. Free-swimming pelagic forms.
2. Body is more or less fusiform with the branchial and atrial apertures nearly terminal and opposite.
3. Test is well developed and transparent.
4. Muscle-bands are always incomplete ventrally.
5. Pharynx communicate freely with the atrial cavity through a large gill-slit. stolon is long.
6. Life history exhibits an alternation of generations. Development occurs inside the pouch of the body and the embryo is nourished by placenta.
7. Tailed larva absent.

Examples : *Salpa, Scyclosalpa.*

REPRESENTATIVE TYPES

1. *Molgula.* *Molgula* (Fig. 5.1) is a round, solitary and sessile ascidian, about 2.5 cm long. It is found abundantly in the shallow water of Atlantic and less widely in the Indian Ocean, being fixed to wharfs, piers and rocks. There is no foot. Test is thin and transparent, it is often encrusted with shell pieces and sand. The siphons are well formed, the branchial aperture has 6 lips and the atrial aperture has 4 lips. In the branchial siphon are 16 branching tentacles. Pharynx is large with narrow stigmata arranged spirally. Dorsal lamina is C-shaped and has no languets. Alimentary canal forms a curved loop. There is a curved **renal vesicle** or kidney above the heart. There is a pair of protandrous gonads. *Molgula* is usually oviparous. The tadpole larva has no ocelli, and in some species there is no tail. Some species of *Molgula* are viviparous giving birth to a larva which may or may not have a tail in different species.

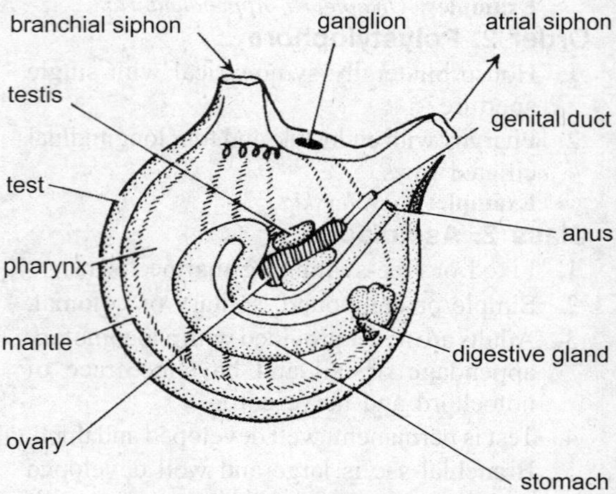

Fig. 5.1. *Molgula.*

2. *Botryllus.* *Botryllus* (Fig. 5.2) is widely distributed in the Atlantic and the Mediterranean, being found fixed to wharfs, piers and ships in shallow water. It is a colonial ascidian. The test forming a flat encrusting mass with individuals or **zooids** grouped together in radiating star-shaped fashion. The branchial apertures of zooids are at their outer ends, while their atrial apertures open into a common cloacal chamber opening in the centre of the group of zooids. The test of all zooids is common forming a gelatinous mass. The structure of each zooids is like that of a simple ascidian. Dorsal lamina has no languets, neural gland is dorsal to the brain. Pharynx has rectangular stigmata. There is a pair of hermaphrodite gonads, reproduction is sexual and also by budding, the buds remain permanently in the colony.

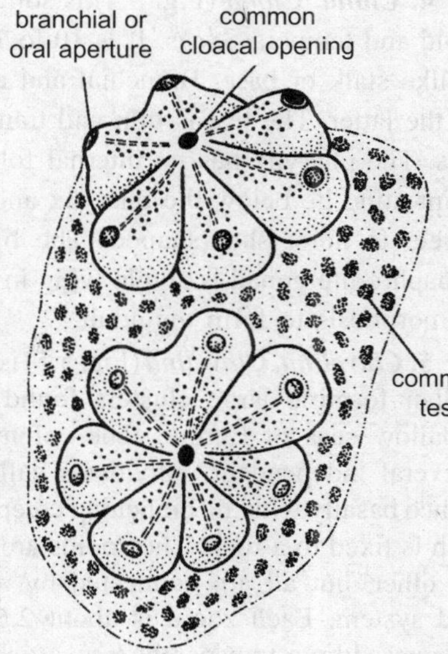

Fig. 5.2. Part of *Botryllus* colony.

3. *Ascidia.* *Ascidia* (Fig. 5.3) is common in the Atlantic Ocean and Mediterranean sea being fixed to rocks. It is 10 to 12 cm long. The test is thick and cartilaginous. The body is cylindrical with a broad base by which it is fixed. At the free end is a branchial aperture, and some distance from it on one side is an atrial aperture. The siphons are short. The pharynx is large with numerous stigmata but there are no internal longitudinal folds. The intestine contains a typhlosole. There is no liver. Dorsal tubercle is horse-shoe-shaped. Dorsal lamina is a longitudinal ridge with no languets. Brain is dorsal to the neural gland. Excretory renal vesicle lie on the stomach and intestine. There is a single hermaphrodite gonad in the intestinal loop.

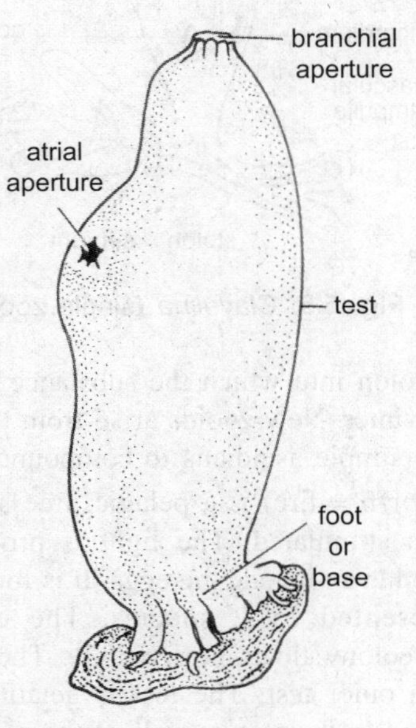

Fig. 5.3. *Ascidia.*

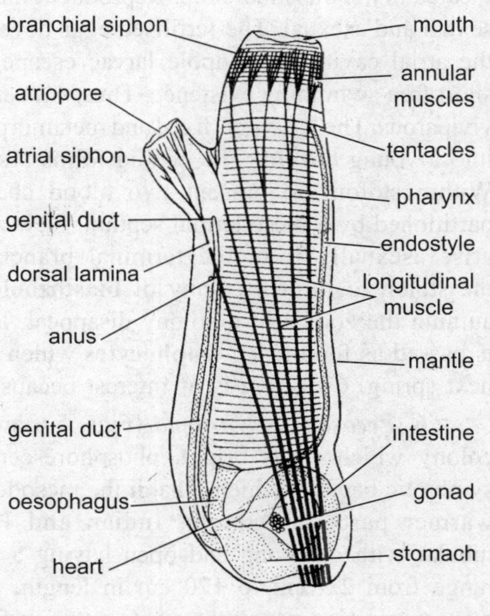

Fig. 5.4. *Ciona.*

4. *Ciona.* *Ciona* (Fig. 5.4) is solitary and is found in great abundance on rooks, ships, etc., in cold and temperate seas. It is 10 to 12 cm long with a cylindrical body. It is fixed by a broad foot-like stalk or base. Branchial and atrial siphons are cylindrical, the former being higher up than the latter. The test is thin and transparent. The mantle has longitudinal and circular muscle bands. The pharynx has no internal folds, the dorsal lamina has languets. Stomach, oesophagus and intestine lie below the pharynx and not on one side. In the intestine is a typhlosole. Dorsal tubercle is horse-shoe-shaped. The brain is dorsal to the neural gland. Heart is V-shaped. Hermaphrodite gonad is single lying in the intestinal loop. Tadpole larva undergoes retrogressive metamorphosis to form the adult.

5. *Clavelina.* *Clavelina* (Fig. 5.5) is a colonial ascidian forming large clusters found on rocks in shallow seas of Europe. The colony consists of several independent individuals called **zooids** attached basally with the branching creeping stolon which is fixed to a rock. The zooids are free from each other, but all are connected by a common blood system. Each zooid is about 2.5 cm long and covered by a transparent test. A zooid has an upper **thorax** and a lower **abdomen** connected by a middle narrow waiste. The thorax has two short siphons, a pharynx with endostyle, dorsal lamina with languets and a circular dorsal tubercle. The abdomen is narrow and long and contains the rest of the alimentary canal, heart, gonad and stolon.

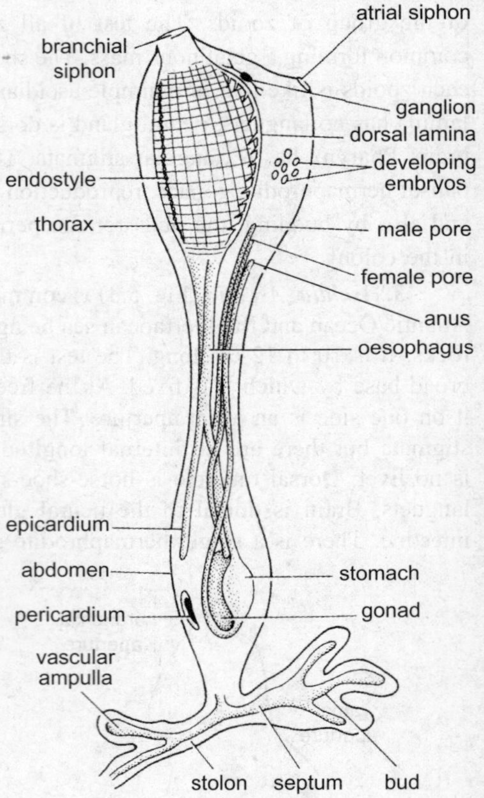

Fig. 5.5. *Clavelina* (single zooid).

A narrow tubular outgrowth extends from the pharynx through the abdomen into the stolon, it is called **epicardium**. Gonad is hermaphrodite located in the intestinal loop. Reproduction is both sexual and asexual. The fertilised eggs develop in the atrial cavity, the tadpole larvae escape for a short free-swimming existence. Thus, *Clavelina* is viviparous. The larva gets fixed and metamorphoses into a young **oozoid**. The oozoid forms a stolon. Within stolon are present two blood channels partitioned by a mesodermal septum. New zooids arise asexually from the terminal branches of the stolon to form a colony of **blastozooids**. In autumn the zooids of a colony disappear, leaving the stolon into which the substance of zooids is passed as food-laden **trophocytes** which survive the winter. New zooids arise from this stolon next spring. *Clavelina* is of interest because it links the simple ascidians to compound ones.

6. *Pyrosoma.* *Pyrosoma* (Figs. 5.6 and 5.7) (Gr., *pyro* = fire) is a pelagic, free swimming colony which emits bright, phosphorescent light when stimulated. The light is produced by symbiotic bacteria which inhabit the mesodermal cells and even invade the eggs. It is found in the warmer parts of Atlanic, Indian and Pacific represented by 2 species. The colony is tubular with only one end open having a **velum**. The colony floats horizontally. The colonies range from 2.5 cm to 120 cm in length, they have an outer test. The test is gelatinous with many projecting processes on its outer surface. Embedded in the test or wall of the cylinder are numerous zooids called **blastozooids**, they are arranged in such a way that the branchial aperture of each opens outwards guarded by a tongue-like appendage and the atrial aperture into a common cloacal cavity of the cylinder which opens outside by a cloacal aperture regulated by a diaphragm or **velum**. Colony locomotes by jet propulsion since a stream of water is shot out from the cloacal aperture. A blastozooid has a structure like an ascidian. It has a pharynx with many stigmata, an endostyle and a dorsal lamina with languets. At the inner end of the pharynx is an atrium. Heart lies behind the pharynx. Subneural gland and ocellus are also present. There are circum-oraland atrial muscles forming complete but feeble rings at the two ends. There is a hermaphrodite gonad with a single yolky ovum and a lobed testis. Near the endostyle is a **stolon** which forms additional blastozooids by budding. Each blastozooid sieves water from the sea into the common cloacal cavity of the colony, then a stream of water is shot out from the open posterior end moving the colony forward by propulsion. Speed is increased by the velum narrowing the posterior opening in ejecting water.

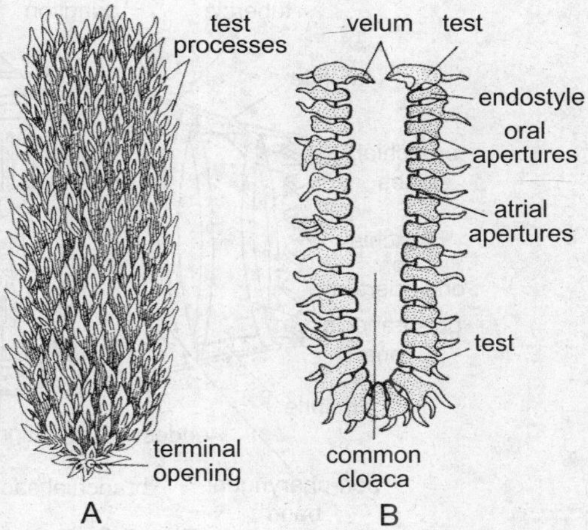

Fig. 5.6. *Pyrosoma*. A–Colony; B–Sagittal section of colony.

In *Pyrosoma* there is no tailed larva, development is direct and takes place with in the body of the parent. The fertilised egg undergoes meroblastic cleavage and forms an **oozooids** or **cyathozooid** having a stolon. It has a primitive gut, a tubular nervous system and a pair of laterally placed atrial tubes. Within maternal body it feeds upon kalymnocytes (follicle cells). (A zooid arising from an ovum is an oozooid, while the one formed by budding is a blastozooid). The stolon (anterior part of body) forms four buds called **blastozooids**, each of which has a stolon. The posterior part of the cyathozooids eventually atrophies. This colony of four individuals (**tetrazooid colony**) gets enclosed in a common test, it passes out through the cloaca of the parent colony. The cyathozooid begins to atrophy, but more blastozooids are produced by budding from the stolons of tetrazooid colony to form an adult *Pyrosoma* colony.

7. *Doliolum*. *Doliolum* (Fig. 5.8) is pelagic, free-swimming and transparent creature. It is found in cold, tropical and subtropical seas.

Structure. It is one to two cm long with a transparent barrel-shaped body covered over by a thin transparent test. The mouth and the atriopore are at the opposite ends. Both the apertures are surrounded by numerous lobes. Its life cycles includes two phases : a solitary sexual gonozooid and a colonial asexual oozoid.

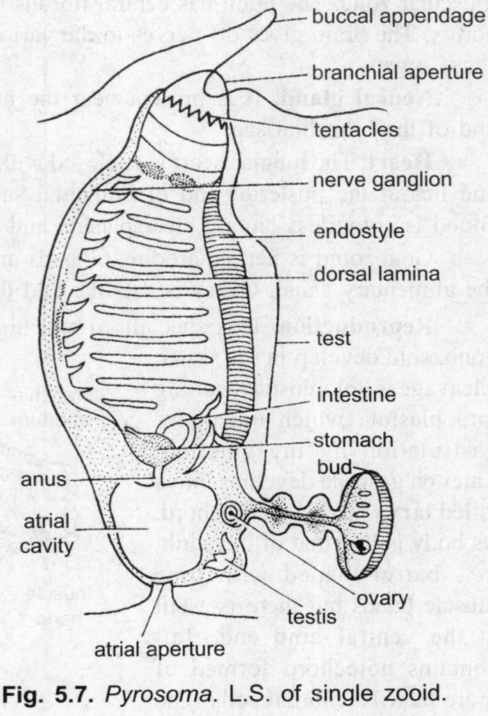

Fig. 5.7. *Pyrosoma*. L.S. of single zoid.

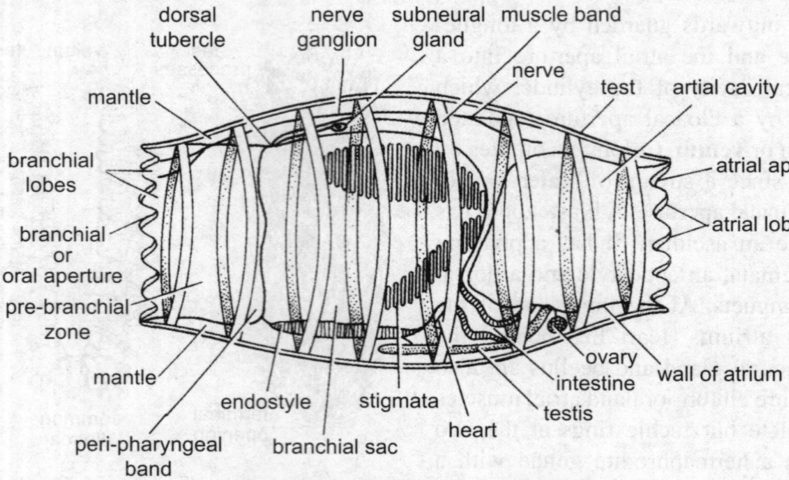

Fig. 5.8. *Doliolum*. (Sexual form).

Gonozoid. The **sexual gonozooid** or **solitaria phase** has a thin transparent test. The mantle or body wall has a single layer of epidermis and beneath it lies the mesenchyme which forms eight complete **muscle bands**, the first and last of which act as sphincters. Contracticns of muscles drive water through the posterior end, thus, the animal moves by propulsion. The mouth and atrial apertures are at opposite ends of the barrel-shaped body, each has 10 or 12 sensory lobes.

Alimentary canal. Mouth leads into a pharynx having several elongated stigmata in its posterior wall, a ventral endostyle, but there is no dorsal lamina. Pharynx leads into oesophagus, ventrally it leads into a dilated stomach, tubular intestine which finally opens into the atrial cavity.

Nervous system. There is a dorsal brain (cerebral ganglion) in the third or fourth inter-muscular zone. The brain has central fibrous medulla surrounded by the ganglionic cells forming cortex. The brain gives off nerves to the various organs. The branchial and atrial lobes has tactile sense organs.

Neural gland. It is present near the nerve ganglion and its duct opens into the anterior end of the branchial sac.

Heart. The tubular heart is enclosed within a transparent, non-contractile tubular pericardium and lies at the posterior end of branchial sac. The heart gives off vessels from its both ends. Blood is colourless having clear plasma and a less number of amoebocytes.

Gonozooid is hermaphrodite. Gonads are found on the left side of the middle line close to the alimentary canal. Ovary is sac-like and the testis tubular. Gonoducts open into the atrium.

Reproduction. It is specialised and highly complicated. The fertilised eggs of the sexual gonozooid develop in sea water. Cleavage is holoblostic resulting into blastula which undergoes gastrulation by invagination. Later on gastrula develops into a tailed **larva** having a notochord. Its body is like that of the adult, *i.e.,* barrel-shaped with nine muscle bands but there is a tail at the ventral hind end. Tail contains notochord formed of more than 10 chordal cells. The body has two small processes, a

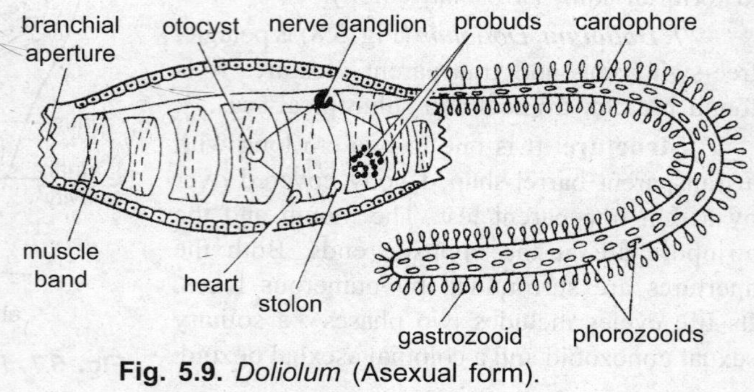

Fig. 5.9. *Doliolum* (Asexual form).

postero-dorsal **cadophore** and a ventral **stolon**. In its pharynx only four gill-slits are present on each side. It possesses the statocyst on the lateral side. Oozooid (asexual phase) also possesses a **dorsal appendix** (cadophore) in between seventh and eighth muscle bands. It is immovable elongated structure and encloses a part of coelom. Stolon is present on the ventral side. The larva after a little metamorphosis loses its tail but does not get fixed, it changes into an **asexual oozooid** stage or **gregaria phase** which acts as a float and **reproduces asexually**.

Asexual reproduction by budding. In the **asexual oozooid** the cadophore and stolon increase. The stigmata, endostyle and alimentary canal are lost. There are nine complete muscle bands which increase in thickness, and nervous system attains a higher development. Stolon's free end undergoes segmentation to form small constricted bodies which get detached from the stolon as **probuds**. The probuds migrate along the body and get fixed to the cadophore. The probuds divide several times and get fixed as lateral and median rows of buds on the cadophore. These buds are of three kinds. The lateral buds develop into small *Doliolum*-like zooids with rudiments of gonads that soon atrophy. Alimentary canal develops. These are called **trophozooids** or **gastrozooids** and these are helpful in the nutrition and respiration of the colony. Some of the median zooids called **phorozooids** act as nurses and become detached. These are barrel-shaped in structure with 8 muscle bands and vascular structures. They have no gonads. The other median zooids called **gonozooids** are fixed on the stalks of phorozooids and carried for a time. The gonozooids acquire gonads, separate from phorozooids and form the adult **sexual stage** or **solitaria phase**.

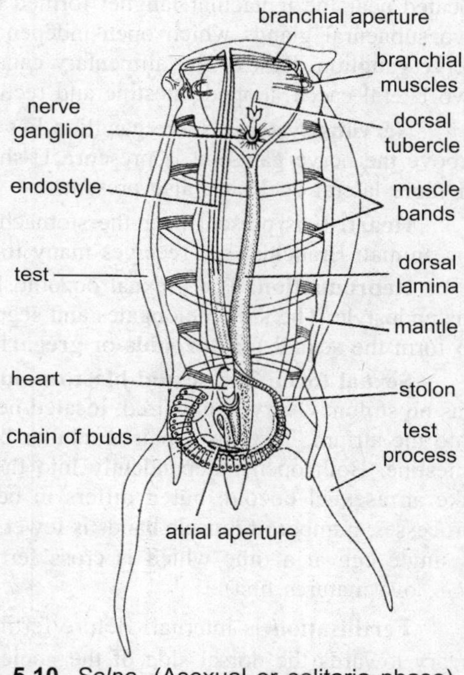

Fig. 5.10. *Salpa*. (Asexual or solitaria phase).

Doliolum exhibits polymorphism and alternation of generations in its complicated life history. The sexual *Doliolum* phase reproduces sexually and develops asexual phase which is polymorphic: nutritive gastrozooids, foster phorozooids and sexual gonozooids.

8. *Salpa*. *Salpa* (Fig. 5.10 and 5.11) is found in almost all seas as a free-swimming pelagic animal. It occurs in two phases : an **asexual oozooid** or **solitaria** form and a **sexual blastozooid** or **gregaria** form.

Structure of asexual *Salpa*. *Salpa* is a transparent, small pelagic marine creature. Its body is covered by a soft gelatinous test closely attached with the mantle. The body is ovoidal having a branchial and atrial apertures

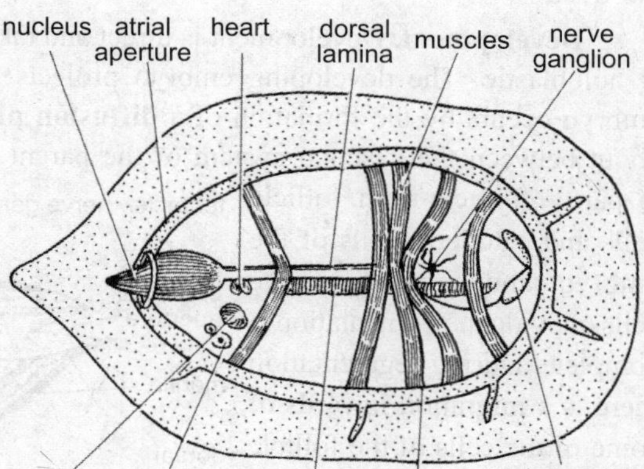

Fig. 5.11. *Salpa*. (Sexual or gregaria phase).

at the opposite ends. The nucleus is opaque which contains the heart and the alimentary canal. Muscles bands are six to nine C-shaped, incomplete ventrally. Atrial band is complete, while the oral band is prolonged into lips and acts as sphincter. Their contractions eject water through the atrial aperture and the animal is propelled forwards.

Digestive system. Branchial aperture is wide and guarded by movable upper and lower lips. It leads into a tubular buccal cavity which is followed by prebranchial zone of pharynx and then large elongated pharynx. It has a dorsal lamina and a central endostyle. The side walls of pharynx have no stigmata and freely communicate with the atrial cavity on each side. Dorsal tubercle is located near the tentacular languet formed by the anterior end of dorsal lamina. Beneath it lies two subneural glands which open independently into the pharynx. Above these is present the nerve ganglion. Rest of the alimentary canal includes the oesophagus, saccular stomach having two lateral caeca, looped intestine and rectum opening into atrium through anus.

Nervous system. Nerve ganglion lies near the anterior end gill and gives off many nerves. Above the nerve ganglion is present a U-shaped pigmented ridge-like eye. In some species one or more lateral ocelli are also present.

Heart. It is present near the stomach. It gives off many vessels to the various organs of the animal. Branchial sac receives many transverse vessels from the dorsal and central aorta.

Reproduction. The asexual oozooid has no gonads, but it has a **stolon** arising from near the endostyle. The stolon elongates and segments into a chain of buds, which break off in groups to form the sexual **blastozooids** or **gregaria stage** which later separate.

Sexual form. The **sexual blastozooid** has a testis and an ovary with a single ovum, but it has no stolon. Ovary is unpaired, located near the right side of nucleus. It opens through oviduct into the atrium. Testis is a paired tubular branched structure near the nucleus, on either side of intestine. Both open independently into the atrium. Sexual blastozooid has a general structure like an asexual oozoid, but it differs in being smaller. The body is asymmetrical with no test processes, number of muscle bands is fewer, gonads are present and there is no stolon. It produces a single egg at a time which is cross fertilised internally because the animal is **protogynous** (*i.e.,* ova matures first).

Fertilisation is internal. Before fertilisation the stalk-like oviduct shortens and draws the ovary towards the dorsal side of the nucleus. The ovary is now enclosed in a lumen having a relation with the atrium. Fertilisation of a single ovum occurs within this lumen. After fertilisation the oviducal opening becomes closed and ovarian epithelium forms a sac-like structure around the ovum.

Development. Development is direct and takes place within the body of the parent. Cleavage is holoblastic. The developing embryo projects into the branchial cavity. The nourishment of embryo occurs by the formation of a **diffusion placenta** through which a close union is brought about between the vascular system of the parent and that of the embryo. The placenta of *Salpa* is partly formed from follicle cells and ectoderm cells of the embryo, and partly from the cells of oviduct. Segmentation is complete. During segmentation there is a migration inwards of some of the cells of the follicle and of the wall of the oviduct. These cells enter the segmenting ovum and pass among the blastomeres. These cells are

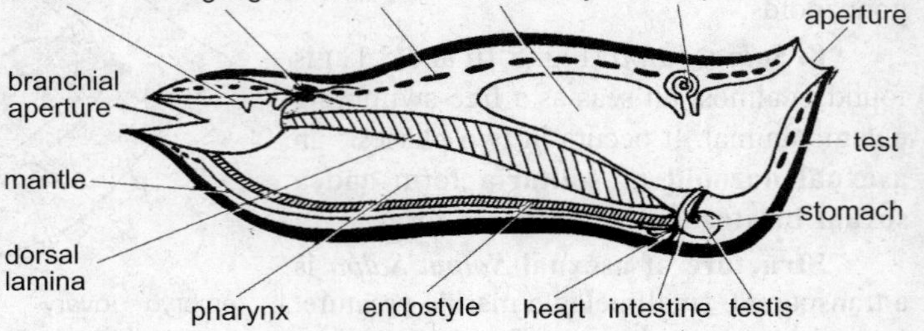

Fig. 5.12. *Salpa*. Diagrammatic sagittal section.

called kalymnocytes and possibly they carry nourishment to the developing embryo. These are broken up and eventually completely absorbed.

There is no tailed larval stage. The embryo develops the muscle bands and all the characteristic parts of the adult while present within the parents body. The sexually produced embryo grows into a solitary asexual oozoid which escapes from the parent and becomes free-swimming. After a time there grows a stolon, on the surface of which are formed a number of bud-like projections. These increase in size as the stolon elongates and each takes the form of a sexual *Salpa*. The chain of zooids formed on stolon breaks off in groups and swim as such while reproductive organs develop in each.

Salpa is, thus, dimorphic and it exhibits an alternation of generations in its life history.

9. *Oikopleura*. *Oikopleura* (Fig. 5.13) is pelagic free-swimming and neotenous found in almost all seas except polar regions.

House. The house is formed of a gelatinous transparent test which is cuticular and not made of tunicin.

Body. The animal is about 2 mm long and surrounded by a large house (loose test) which does not adhere to the animal. The large "house" has a pair of funnel-shaped **incurrent apertures** having a sieve of fine threads which allows only minute particles to enter along with water. Movements of the tail cause food and water to enter the house through the incurrent apertures. Inside the house is an elaborate **filtering apparatus** formed of sieve tubes which filters minute organisms which are sucked into the mouth due to ciliary action of stigmata. Food particles are entangled in mucus produced by the endostyle, the food-laden mucus cords are drawn into the stomach. When pressure of water increases in the house, a door opens at the tail end and water passes out, this jet of water also propels the house forwards.

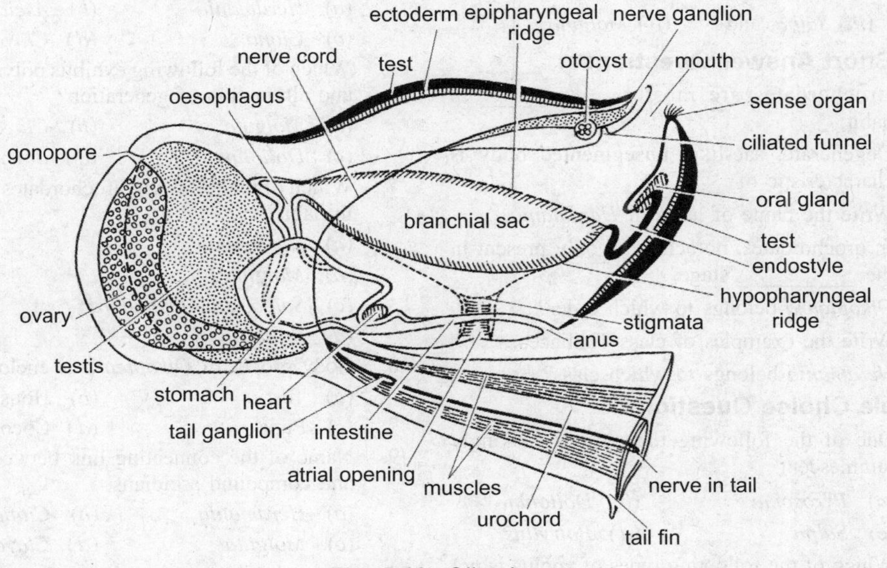

Fig. 5.13. *Oikopleura*.

The house also has an emergency **exit door** through which the animal can escape, it never re-enters the house but its ectoderm secretes a new house with great rapidity. The house is protective, hydrostatic and for filtering food. The animal has an oval trunk and a laterally compressed tail attached to the ventral side, near the hind end. An anterior mouth leads into a large pharynx having a short endostyle, two peripharyngeal bands, and two ventral ciliated stigmata which open directly to the exterior because there is no atrium. There is no dorsal lamina. Rest of the gut has an oesophagus, stomach and intestine opening to the exterior by an anus in front of the stigmata. A cerebral ganglion lies near the mouth, close to it is a statocyst. Neural gland is absent. Heart lies beneath the stomach. At the posterior side of the body is a large hermaphrodite gonad, and the animal is protandrous. (*Oikopleura dioica* is not hermaphrodite, it has separate males and females). The vibratile tail is laterally compressed and twisted. It is surrounded by a wide tail fin. The tail is supported by notochord and also contains a ganglionated nerve cord and unsegmented muscles. The larva does not undergo metamorphosis since it is neotenous. It becomes sexually mature due to the formation of gonad.

IMPORTANT QUESTIONS

Long Answer Questions

1. Give an account of the general organisation of *Appendicularia.*
2. Classify the subphylum Urochordata up to orders, giving characters of each group.

Short Answer Questions

1. Write comprehensive note on the *Pyrosoma.*
2. Write brief note on *Oikopleura* larva.
3. What is the significance of Bidder's organ?
4. Explain the difference between Larvacea and Thaliacea.
5. Write short notes on:
   (*i*) *Botryllus*, (*ii*) *Doliolum*,
   (*iii*) *Salpa*, and (*iv*) *Molgula.*

Very Short Answer Questions

1. Urochordates are mostly __________ in habit.
2. Degenerate, sac-like, unsegmented body is characteristic of __________.
3. Write the name of larva of *Herdmania.*
4. In urochordates, notochord is only present in the __________ stage.
5. *Oikopleura* belongs to which order?
6. Write the examples of class Thaliacea.
7. *Herdmania* belongs to which class?

Multiple Choice Questions

1. One of the following tunicates is strikingly luminescent
   (*a*) *Pyrosoma* (*b*) *Doliolum*
   (*c*) *Salpa* (*d*) *Botryllus.*
2. Which of the following types of zooids is not produced by the asexual form of *Salpa*?
   (*a*) Gastrozooid (*b*) Phorozoid
   (*c*) Gonozooid (*d*) Dactylozoid.
3. Excretion in urochordates occurs through
   (*a*) Neural glands
   (*b*) Glomerulus
   (*c*) Nephrocytes
   (*d*) Neural glands, pyloric glands and nephrocytes.
4. Dorsal tubular nerve cord of the larval stage is changed to
   (*a*) A single dorsal nerve ganglion in adult
   (*b*) A pair of nerve ganglion in adult
   (*c*) A single ventral nerve ganglion in adult
   (*d*) None of the above.
5. Which of the following urochordates is viviparous?
   (*a*) *Herdmania* (*b*) *Ascidia*
   (*c*) *Ciona* (*d*) *Clavellina.*
6. Which of the following exhibits polymorphism and alternation of generation?
   (*a*) *Molgula* (*b*) *Clavellina*
   (*c*) *Doliolum* (*d*) *Botryllus.*
7. Which of the following urochordates is pelagic in habit?
   (*a*) *Herdmania*
   (*b*) *Molgula*
   (*c*) *Salpa* and *Oikopleura*
   (*d*) *Botryllus.*
8. Body proper of *Oikopleura* is enclosed in
   (*a*) Test (*b*) House
   (*c*) Pupa (*d*) Cocoon.
9. Name of the connecting link between simple and compound ascidians
   (*a*) *Herdmania* (*b*) *Ciona*
   (*c*) *Molgula* (*d*) *Clavellina.*
10. In urochordates respiration occurs through
    (*a*) Test (*b*) Test and gill-slits
    (*c*) Gill-slits (*d*) Branchial basket.

ANSWERS

1. (*a*) 2. (*d*) 3. (*d*) 4. (*a*) 5. (*d*) 6. (*c*) 7. (*c*) 8. (*b*)
9. (*d*) 10. (*b*).

6

Chapter

Type 3. *Branchiostoma* (= *Amphioxus*) The Lancelet

Contents of this Chapter

Cephalochordata includes only two genera, *Asymmetron* and *Branchiostoma (Amphioxus)*. The generic name *Branchiostoma* has priority over *Amphioxus*, hence, *Amphioxus* is now used as a common name for any Cephalochordata.

Cephalochordata are small, marine, fish-like animals showing the main chordate characters. The notochord extends the entire length of the body projecting beyond the nervous system to the tip of the snout, (hence, Cephalochordata). They have a dorsal, tubular neural tube without a defined brain. Pharynx is large with numerous gill-clefts opening not to the exterior, but into an ectoderm-lined atrium. The gill-clefts are equivalent to stigmata or urochordates. The pharynx has an endostyle and ciliated tracts which are concerned with ciliary feeding. Metamerism is well marked, even the gonads are segmental. There is a definite coelom. Excretory organs are segmental nephridia akin to those of Annelida. Cephalochordata are closely related to Urochordata, so they cannot be assigned to two separate subphyla, as is often done, they should be placed in the same subphylum Acrania. Both have been derived from a common ancestral stock.

Branchiostoma = (*Amphioxus*) is a most extensively studied cephalochordate type. *Branchiostoma* = (*Amphioxus*) is commonly called **lancet** or **lancelet**. It was first described by the German scientist **Pallas** in 1778. Pallas considered it to be a slug (Phylum Mollusca) and named it *Limax lanceolatus*. An Italian scientist **Costa** (1834) first recognised its chordate nature and described it *Branchiostoma lanceolatum*. Two years later in 1836 **Yarrel** named it *Amphioxus lanceolatus*.

SYSTEMATIC POSITION

Phylum	...	Chordata
Group	...	Acrania
Subphylum	...	Cephalochordata
Class	...	Leptocardii
Family	...	Branchiostomidae
Genus (Type)	...	*Branchiostoma* = (*Amphioxus*) (The Lancelet)

GEOGRAPHICAL DISTRIBUTION

Branchiostoma is found in British seas and even as far north as the coast of Norway. It is much more common in warmer waters, such as the Mediterranean, and is also found in the Indian ocean. It is abundant in the Bay of Naples, and lives and breeds in great numbers in the salt lagoon the "Pantano" near Messina. On the Chinese coast it is found in such abundance that it is sold as food.

HABITS AND HABITAT

Branchiostoma lives in shallow waters, burrows in sand near the shore with greater part of the body buried and only a fringed anterior oral hood projecting into water. It can swim actively and it can burrow with great rapidity with either head or tail foremost, often it lies on one side on the sand. These animals are most active at night.

EXTERNAL FEATURES

Shape, size and colour. *Branchiostoma* is about 3.5 to 6.0 cm in length. The body is slender, somewhat translucent, superficially fish-like, laterally compressed and pointed at both ends, hence, the name lancelet. The anterior two-thirds of the body is roughly triangular in transverse section. The posterior third is nearly oval in section. The pointed anterior end of trunk projects in front as a snout or **rostrum**. Below the rostrum is an **oral hood** formed by dorsal and lateral projections of the body. It bears twenty or more stiff **oral cirri** or tentacles.

Body divisions. The body is distinguished in two regions : the trunk and tail. The true head is lacking.

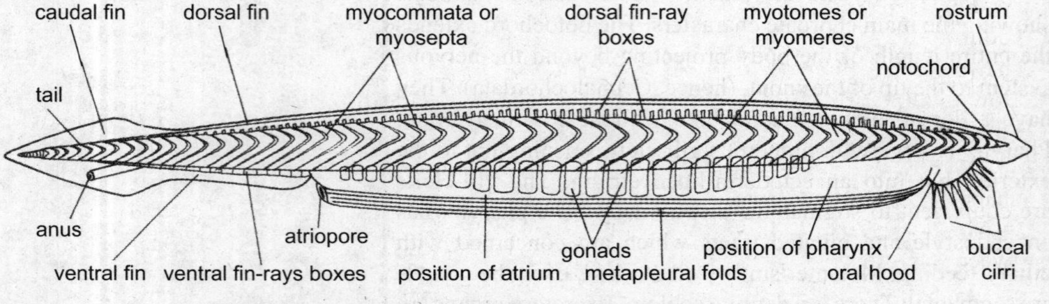

Fig. 6.1. *Branchiostoma*. Entire animal in right side view.

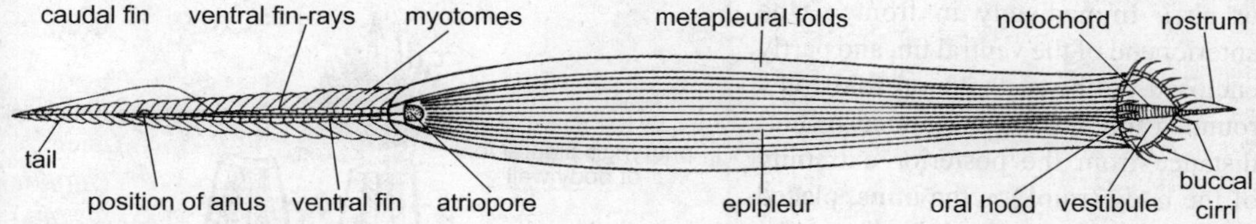

Fig. 6.2. *Branchiostoma*. Entire animal in ventral view.

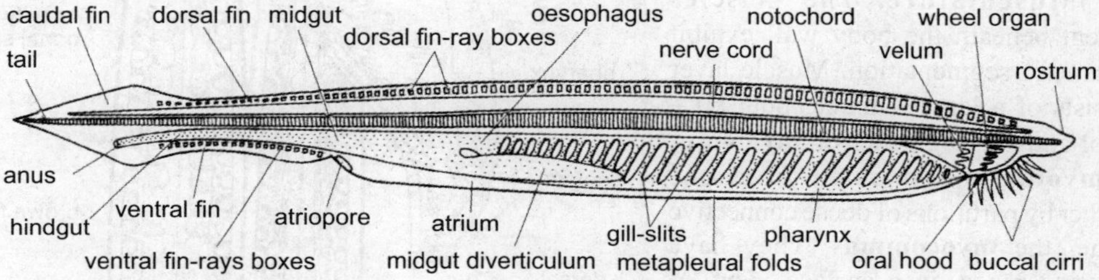

Fig. 6.3. *Branchiostoma*. General anatomy dissected in left lateral view.

Fins. *Branchiostoma* has three unpaired fins : dorsal, caudal and ventral. Paired fins are absent. Along the mid-dorsal side of the body is a **dorsal fin** running along the whole length. It is joined to a somewhat broader **caudal fin** around the tail. Mid-ventrally is a **ventral fin** running from the caudal fin to an atriopore. The dorsal fin is supported by a series of **fin rays**, and the ventral fin by a series of paired fin rays. The fin rays are made of a gelatinous substance and connective tissue. Fin rays are lacking in caudal fin. It is important to note that the fins do not have the same structure as the fins of fishes.

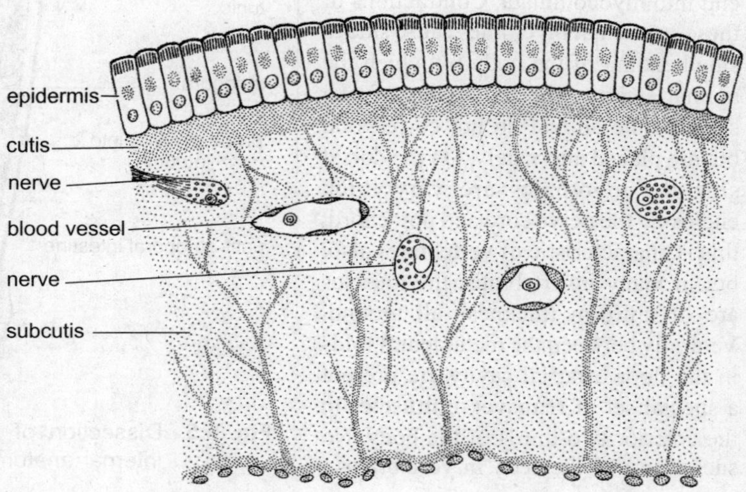

Fig. 6.4. *Branchiostoma*. V.S. of body wall.

Folds. The oral hood is continuous posteriorly with two lateral membranous hollow **metapleural folds** running ventrally up to the atriopore. Both the folds are joined by a horizontal fold of body wall, the **epipleur**. It forms the floor of the atrium internally.

Myotomes and gonads. The body wall shows metameric segmentation, being made of <-shaped blocks of muscles called **myotomes** or **myomeres**. These are found just beneath the transparent body wall. Beneath the myotomes on both sides are present a series of gonads between mouth and atriopore.

Apertures. Below the pointed anterior extremity is a large median aperture, the **mouth** surrounded by a frill-like membrane, the **oral hood**. Its edge is beset with numerous tentacles or cirri. Immediately in front of the anterior end of the ventral fin, and partly enclosed by the metapleural folds, is a rounded aperture, the **atriopore**. A short distance from the posterior extremity of the body (trunk) is the **anus**, placed asymmetrically on the left side at the base of the caudal fin. The post-anal portion of the body is called the tail.

Musculature. The muscles present beneath the body wall exhibit metameric segmentation. Muscle layer consists of a large number about 60 – 62 of muscle segments or **myomeres** or **myotomes**, separated from one another by partitions of dense connective tissue, the **myocommas**. They have the appearance, in a surface view, of a series of very open "V"s with their apices directed forwards. The myotomes have striated muscle fibres lying longitudinally, they are inserted at each end into myocommata. Contractions of these segmental myotomes cause side to side bending of the body by which the animal swims. The myotomes of two sides of the animal do not correspond but are placed alternately. On the dorsal side the myotomes are very thick enclosing the nerve cord and notochord, this is characteristic of vertebrates also, but in many invertebrates the muscles are of uniform thickness all around. Ventrally between the metapleural folds in the floor of atrial cavity are present a special set of **transverse muscles** in the anterior two-thirds of the body. No such muscles are seen in vertebrates.

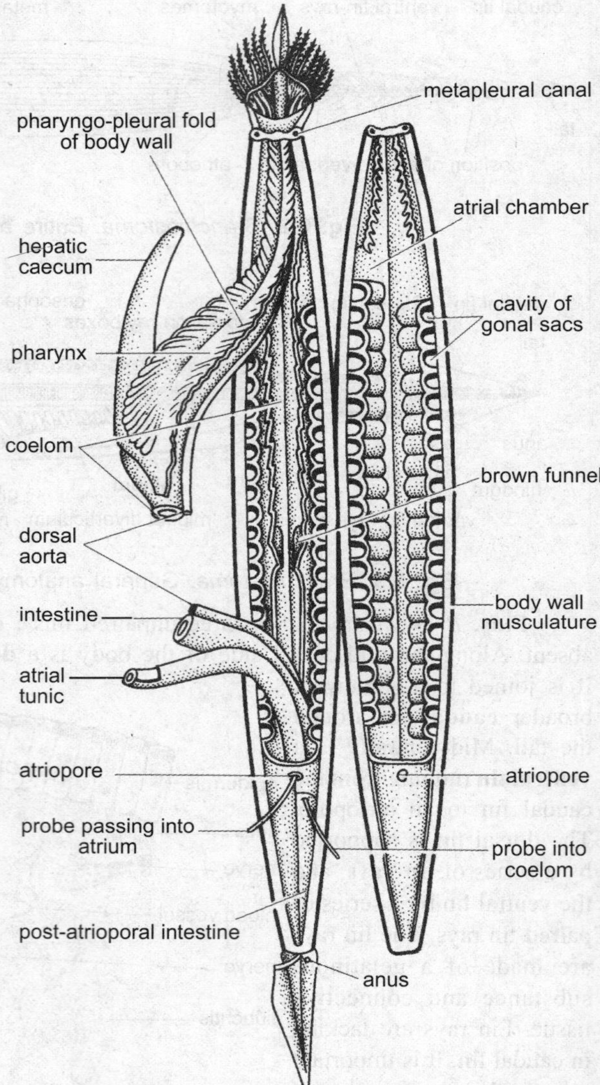

Fig. 6.5. Dissection of *Branchiostoma* to show the internal anatomy.

BODY WALL

The body wall (Fig. 6.4) is formed of outermost thin, transparent skin, well-developed musculature (already described) and a parietal peritoneum.

Skin. There is a thin, outermost, perforated **cuticle** covering an **epidermis** which, unlike vertebrates, is made of a single layer of columnar epithelial cells. These are placed on a basement membrane. Cuticle is secreted by the epidermal cells. In between the epidermal cells are some sensory cells, but there are no glands or chromatophores. A thin layer of tough fibrous connective tissue forms a **cutis** below the epidermis. Below the cutis is a **subcutis** made of a gelatinous material containing nerve fibres cells and blood vessels. The integument is like that of invertebrates, being made only of epidermis. A dermis is lacking, which is present in vertebrates. The cutis and subcutis are traversed by a system of canals. Below the subcutis are myotomes lined internally by a thin layer of **parietal peritoneum**, except in the region of the pharynx where this peritoneum is restricted to small tracts only.

SKELETON

In *Branchiostoma*, exoskeleton is not found. But a few endoskeletal structures are present which are not cartilaginous or bony.

Notochord. There is a cylindrical and tapering **notochord** at both the ends, remarkable for its extension from the tip of the rostrum to the end of the tail. It lies immediately above the alimentary canal and between the left and right myotomes. It is formed of large alternate fibrous and gelatinous, vacuolated cells and having the nuclei confined to its dorsal and ventral regions. Around these cells is a structureless layer, secreted by the cells, enclosed in a **notochordal sheath** of connective tissue. Notochordal sheath is produced dorsally into an investment for the canal enclosing the central nervous system. The turgid cells and notochordal sheath give the notochord its stiff elasticity enabling it to prevent any shortening of the body when myotomes contract. In *Branchiostoma* alone among chordates, the notochord extends anteriorly in front of cerebral vesicle. This prolongation is probably an adaptation to its habit of burrowing rapidly into the sand.

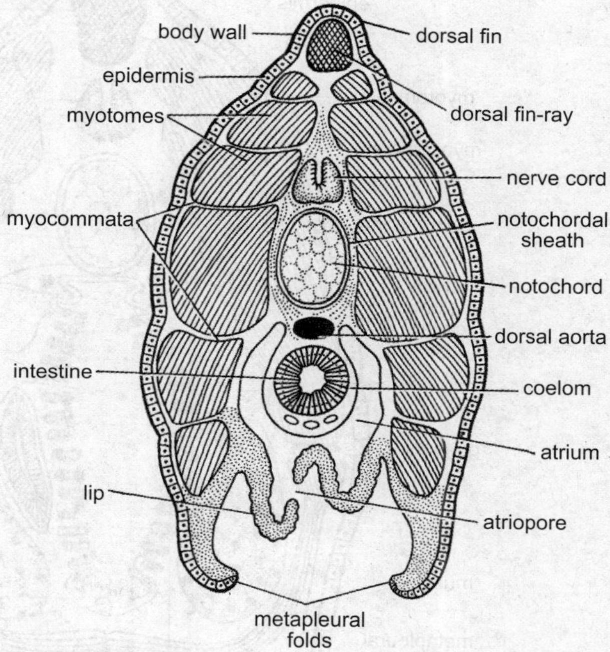

Fig. 6.6. T.S. of *Amphioxus* through atriopore.

Oral ring. The oral hood is supported by a gelatinous ring of cartilaginous consistency made of separate rod-like pieces lying end to end, and corresponding in number with the cirri. Each piece gives out a rod forming the axis of an oral cirrus.

Gill-rods. Gill-bars of the pharynx and endostyle are supported by skeletal rods made of elastic, gelatinous substance. Gill-rods are composed of agglutinated elastic fibres. These will be discussed along with the pharynx.

Fin-rays. Dorsal fin is supported by a single row of fin-rays, while the ventral fin has a double row of supporting fin-rays. These fin-rays are made of connective tissue containing a gelatinous substance inside. Fin-rays are separated from one another by small cavities.

ATRIUM

The atrial cavity or atrium is an extensive space lined with ectoderm. It is formed by a pair of metapleural folds one on either side above gill-slits of embryo, growing down and getting united ventrally by a transverse shelf so that a portion of the out side space becomes enclosed within the body (Fig. 6.6.). The atrium forms a large cavity which surrounds the pharynx and anterior part of the intestine both laterally and ventrally (Figs. 6.7. and 6.19). Thus, the gill-slits open into the atrium and not directly outside as in other chordates. The atrium opens outside through a small rounded, mid-ventral **atriopore** placed just in front of the ventral fin. Posteriorly it also extends back on the right side of the intestine as a blind pouch which runs almost up to the anus. From near the hind end of the pharynx the atrium gives out two conical pouches, each projecting in front into the coelom on either side of the pharynx, these pouches are **brown**

funnels or **atrio-coelomic canals** of unknown function (Fig 6.10). The atrium lined with ectoderm, protects the pharyngeal region of the animal.

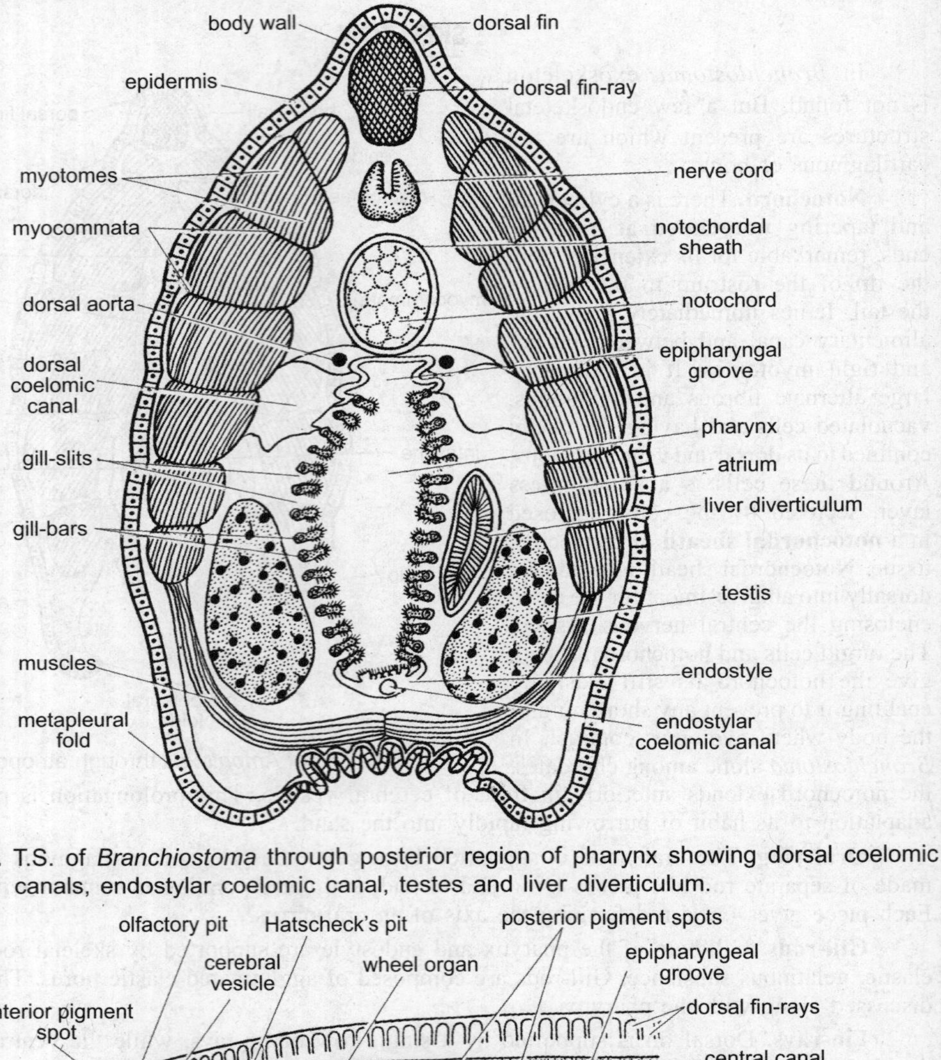

Fig. 6.7. T.S. of *Branchiostoma* through posterior region of pharynx showing dorsal coelomic canals, endostylar coelomic canal, testes and liver diverticulum.

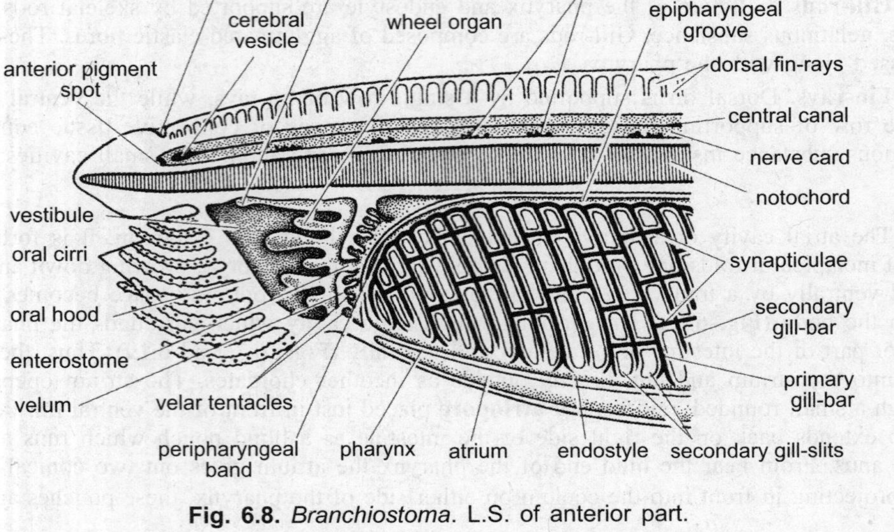

Fig. 6.8. *Branchiostoma*. L.S. of anterior part.

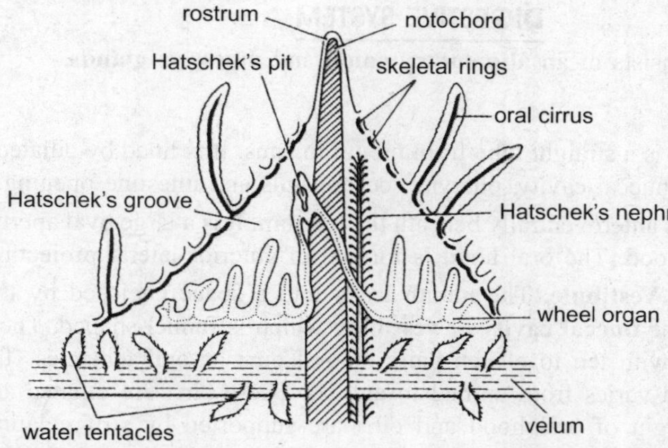

Fig. 6.9. *Branchiostoma*. Ventral view of anterior end showing Hatscheck's pit and groove.

COELOM

The body cavity is a well-developed coelom lined with somatic and splanchnic mesodermal epithelium. It contains a lymph-like coelomic fluid. It is enterocoelic in origin. Behind the pharynx the coelom is a spacious cavity around the intestine being suspended in the coelom by a dorsal mesentery. But it is reduced on the right side of the hindgut by a posterior extension of the atrium. Coelom also surrounds the midgut diverticulum lying within the atrium on the right side of the pharynx. But in the region of the pharynx the coelom is much reduced due to formation of gill-clefts, though in the larva it surrounds the pharynx all around except mid-dorsally. Thus, in the pharyngeal region of the adult it is reduced to three types of spaces, a mid-ventral **sub-endostylar coelom** running longitudinally below the endostyle, two **dorsal longitudinal canals** lying above the pharynx, one on either side and enclosing the brown funnels, and vertical **coelomic canals** in the primary gill-bars which connect the subendostylar coelom with a dorsal longitudinal coelomic canal on each side of the pharynx. Small coelomic spaces are also present within the gonads, these are the **gonocoels**. In higher chordates the coelom is completely lost in the region of the pharynx.

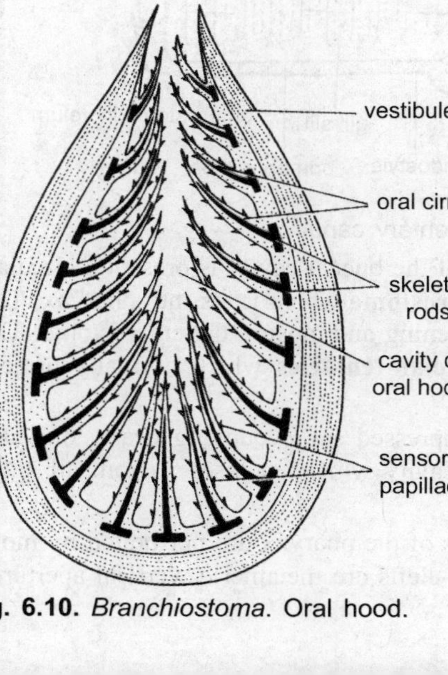

Fig. 6.10. *Branchiostoma*. Oral hood.

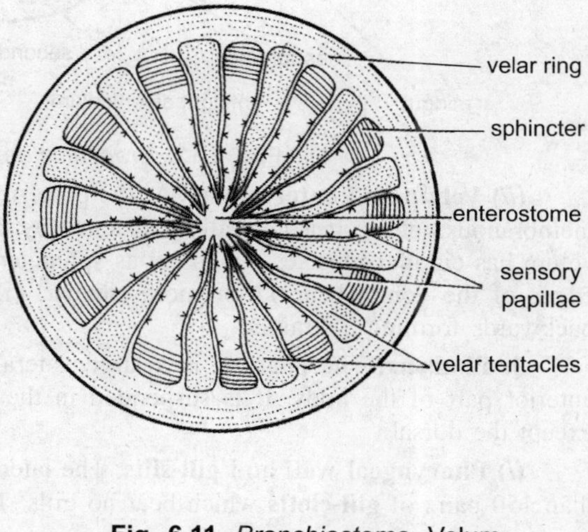

Fig. 6.11. *Branchiostoma*. Velum.

DIGESTIVE SYSTEM

Digestive system consists of an **alimentary canal** and **digestive glands**.

Alimentary Canal

The alimentary canal is a straight tube from mouth to anus. It is lined by ciliated epithelium. It is divisible into mouth, buccal cavity, pharynx, oesophagus and intestine opening by an anus.

1. Mouth. Mouth lies anteroventrally beneath the rostrum. It is a large oval aperture bordered by the membranous oral hood. The oral hood is formed of anterior lateral projection of trunk.

2. Buccal cavity or Vestibule. The mouth leads into a space enclosed by the oral hood lined by ectoderm called the **buccal cavity** or **vestibule** which is funnel-shaped. The free margin of the oral hood is beset with ten to eleven pairs of oral cirri or oral tentacles. Their number increases with the age and varies from animal to animal. These cirri are ciliated and also bear sensory papillae. The margin of oral hood and cirri are supported by stiff gelatinous skeletal rods. The oral cirri remain converged inward to form a sieve over the mouth at the time of feeding. Thus, they check the entry of larger particles of food into the vestibule. At the back of vestibule is a vertical membranous **velum** having a hole in the middle known as **enterostome**. The enterostome is also called a mouth. Since the enterostome leads into an endoderm-lined pharynx and not into an ordinary ectoderm-lined vestibule or stomodaeum, it cannot correspond to the mouth of chordates. Hence, the front opening into the vestibule is a true mouth.

***(i)* Muller organ.** The epithelium (ectoderm) of the oral hood is projected to form six to eight pairs of finger-like folds, each having a ciliated groove borded by ciliated ridge. These folds are collectively known as **wheel organ** or **Muller's organ** or **rotatory organ**. Outer of these, the mid-dorsal groove is the largest and ends in a pit or depression in the roof of buccal cavity. This is called **Hatscheks groove** and **Hatscheks pit** respectively. Both are ciliated and glandular, and secretes mucus.

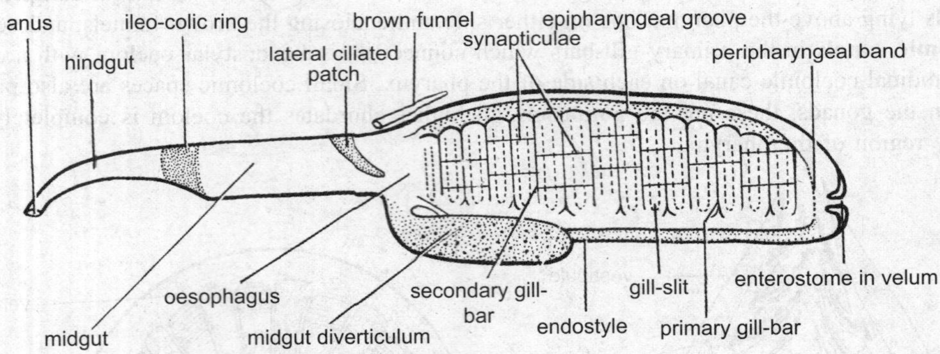

Fig. 6.12. *Branchiostoma.* Alimentary canal.

***(ii)* Velum and enterostome.** At the posterior end of the buccal cavity is present a circular, membranous **velum**. It is centrally perforated by the **enterostome** which leads into pharynx. The velum has circular muscles which act as sphincter for opening and closing the enterostome. The edges of the velum bear 12 or more slender, ciliated **velar tentacles** which normally project backwards forming a strainer.

3. Pharynx. The pharynx is a large, laterally compressed sac occupying nearly one-half anterior part of the body. It is suspended in the atrial cavity which encloses it from all sides except the dorsal.

***(i)* Pharyngeal wall and gill-slits.** The lateral walls of the pharynx are perforated by more than 150 pairs of **gill-clefts** which bear no gills. The gill-clefts are metameric, vertical apertures when first formed in the larva, but each gets divided into two. New gill-clefts are added with age at the posterior end of the pharynx, hence, their number varies in different specimens. Between the gill-clefts the wall of the pharynx is known as **gill-bars** or **gill-lamellae**. The pharyngeal gill-bars are of two types : **primary gill-bars** and **secondary gill-bars** or **tongue-bars**. They alternate regularly and differ in their structure and mode of development. A primary gill-bar is formed of the tissue between two successive gill-clefts after they have perforated to the exterior. It is composed of the wall of the pharynx and the body wall of the larva. The secondary gill-bars arise as downgrowths of the dorsal wall of larval gill-clefts, the wall grows downwords dividing the original gill-cleft into two halves vertically. Both primary and secondary gill-bars are covered on their external or outer surface by sparsely ciliated ectodermal or **atrial epithelium**, but on their inner, anterior and posterior surfaces by endodermal **pharyngeal epithelium** which is heavily ciliated. The cilia on the anterior and posterior endodermal surfaces of gill-bars are long and called the **lateral cilia** which propel water, and the cilia on the inner endodermal surface of each gill-bar form a long but narrow tract of **frontal cilia** which propel mucus. In the middle each gill-bar has a mesodermic core of connective tissue, blood vessels and gill-rods.

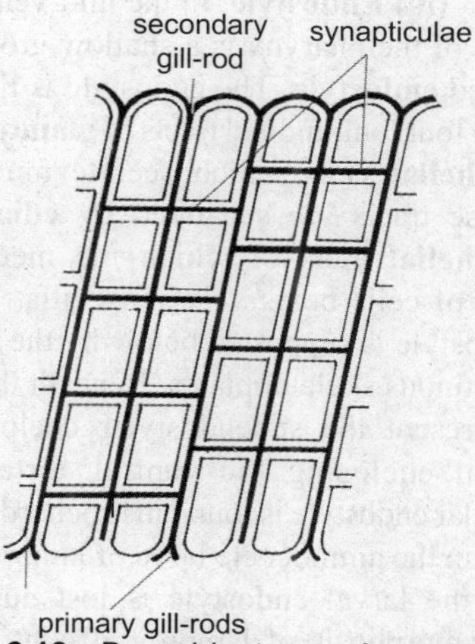

Fig. 6.13. *Branchiostoma*. A part of pharyngeal-wall.

The gill-bars are supported internally by gelatinous skeletal **gill-rods**. All gill-rods are united dorsally, but ventrally their free ends are forked in the primary gill-bars, and unforked or simple in the secondary gill-bars. The primary gill-bars are connected to each other by transverse junctions called **synapticula** which also contain gelatinous rods and blood vessels. The synapticula develop only after the gill-clefts are completed, they make the pharynx look like a basket, somewhat like the branchial sac of urochordates. The primary gill-bars contain a narrow **coelomic canal** running throughout the length of each and communicating dorsally and ventrally with other corresponding coelomic spaces, they also have three blood vessels in each, running lengthwise. The secondary gill-bars with simple gill-rods have no coelom and only two blood vessels run through each of them.

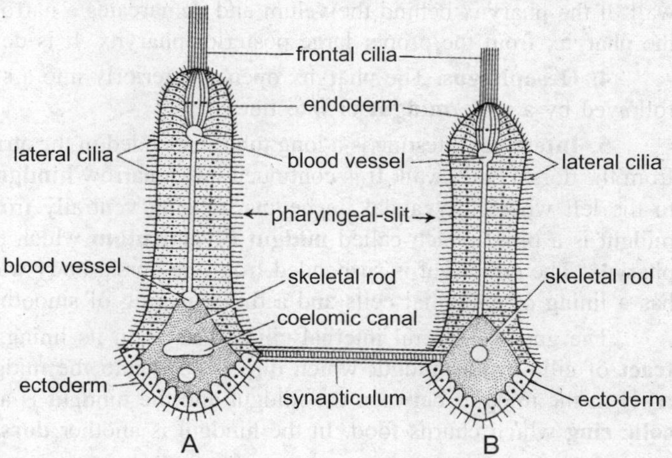

Fig. 6.14. *Branchiostoma*. T.S. of gill-bars. A–Primary gill-bar, B–Secondary gill-bar.

***(ii)* Epipharyngeal groove.** Besides the ciliated gill-bars, there are other ciliated tracts in the pharynx. Running mid-dorsally is a ciliated **epipharyngeal groove** which leads into the opening of oesophagus at the posterior end of the pharynx

(iii) Endostyle. In the mid-ventral wall of the pharynx is a shallow groove called **endostyle**. The endostyle is lined with four longitudinal tracts of **glandular epithelial cells** which secrete mucus. These tracts are separated by **ciliated epithelial cells** of which the median row of cells bears very long cilia. The endostyle is supported below by the two gelatinous skeletal plates. Beneath these is present the subendostylar coelomic canal enclosing the ventral aorta. A similar endostyle is found in urochordates and in the ammocoete larva of lampreys, but the larval endostyle is lost during metamorphosis of lamprey, it contributes to the formation of a thyroid gland in the adult. Like thyroid gland of vertebrates, it also concentrates radioactive iodine.

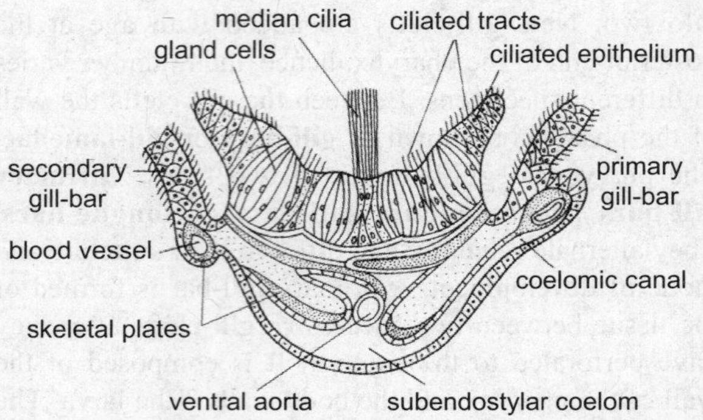

Fig. 6.15. *Branchiostoma*. T.S. of endostyle.

(*iv*) Peripharyngeal bands. The epipharyngeal groove and endostyle are joined to each other anteriorly by two narrow ciliated **peripharyngeal bands**. Each running on the lateral side wall of the pharynx behind the velum and demarcates a narrow anterior **prebranchial region** of the pharynx from the proper large posterior pharynx. It is devoid of cilia and gill-slits.

4. Oesophagus. The pharynx opens posteriorly into a straight narrow ciliated **oesophagus** followed by a wide **midgut** or intestine.

5. Intestine. Intestine is a long tube suspended in the atrium by the dorsal mesentery arising from the dorsal body wall. It is continued into a narrow **hindgut** opening by an anus lying slightly to the left where the caudal fin begins. Arising ventrally from the junction of oesophagus and midgut is a blind pouch called **midgut diverticulum** which extends in front on the right of the pharynx. The entire gut is suspended from the dorsal body wall by a **dorsal mesentery**. The gut has a lining of epithelial cells and a thin covering of smooth muscles.

The gut has several internal ciliated tracts in its lining, there is a crescent-shaped **lateral tract** of cilia in the midgut which directs food into the midgut diverticulum; there is a **dorsal tract** in the midgut; between the midgut and the hindgut is a large ciliated **ileo-colonic** or **ileo-colic ring** which churns food. In the hindgut is another **dorsal tract** of cilia.

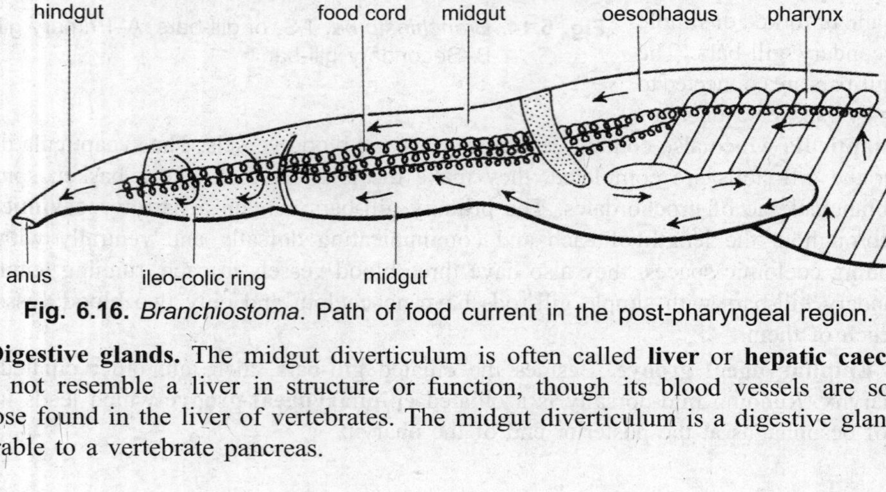

Fig. 6.16. *Branchiostoma*. Path of food current in the post-pharyngeal region.

Digestive glands. The midgut diverticulum is often called **liver** or **hepatic caecum**, but it does not resemble a liver in structure or function, though its blood vessels are somewhat like those found in the liver of vertebrates. The midgut diverticulum is a digestive gland and is comparable to a vertebrate pancreas.

Physiology of Digestion

Food. *Branchiostoma* feeds on protozoans, diatoms, algae, desmids and other organic particles suspended in sea water.

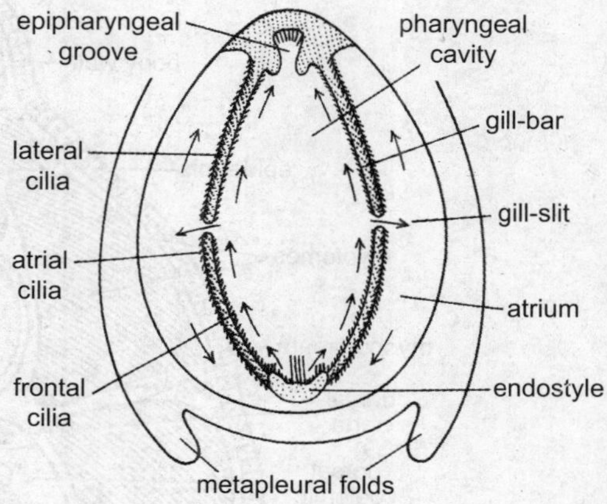

Fig. 6.17. *Branchiostoma*. Course of feeding current in diagrammatic T.S. of pharynx and atrium.

Feeding. *Branchiostoma* is a **ciliary feeder**. Action of cilia of the pharynx causes a current of water containing food. The current of water enters the mouth and goes to the pharynx from where it passes through gill-clefts into an atrial cavity, and then it goes out through an aperture of the atrial cavity called **atriopore**. In feeding the oral hood is extended and oral cirri are turned inwards, they prevent sand from entering the mouth. Rotary movements of cilia of wheel organ direct water towards the pharynx, but some food particles fall out of the water current, they are mixed with mucus secreted by groove of Hatschek and passed back into the pharynx. When the main water current passes through the enterostome into the pharynx, mainly due to lateral cilia of gill-bars, the suspended food particles fall on the gill-bars where they get entangled in mucus secreted mainly by the endostyle and to some extent by the pharyngeal epithelium. Mucus secreted by endostyle is transferred to the lateral wall of the pharynx by its lateral rows of cilia, the median row of long endostylar cilia keeps supplying mucus to the lateral rows of cilia of the endostyle. These cilia lash outwards throwing the muccus on the lateral wall of pharynx. The frontal cilia of gill-bars beat along the length of the bars in such a way that they propel mucus from the lateral to the mid-dorsal side of the pharynx. In this way a stream of mucus with food particles passes from the lower-side into epipharyngeal groove. The cilia of epipharyngeal groove beat backwards moving the cord of mucus and food into the oesophagus.

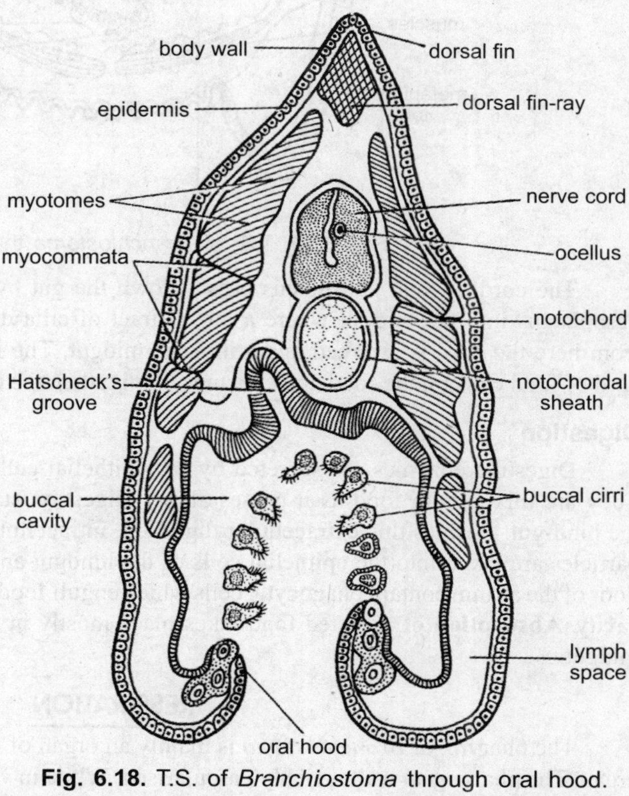

Fig. 6.18. T.S. of *Branchiostoma* through oral hood.

The peripharyngeal bands also collect and pass to the epipharyngeal groove any food particles which fall out of the water current at the extreme anterior end of the pharynx (where no gill-clefts are present). Food and mucus are not transferred from the endostyle into the peri-pharyngeal bands, as is often stated.

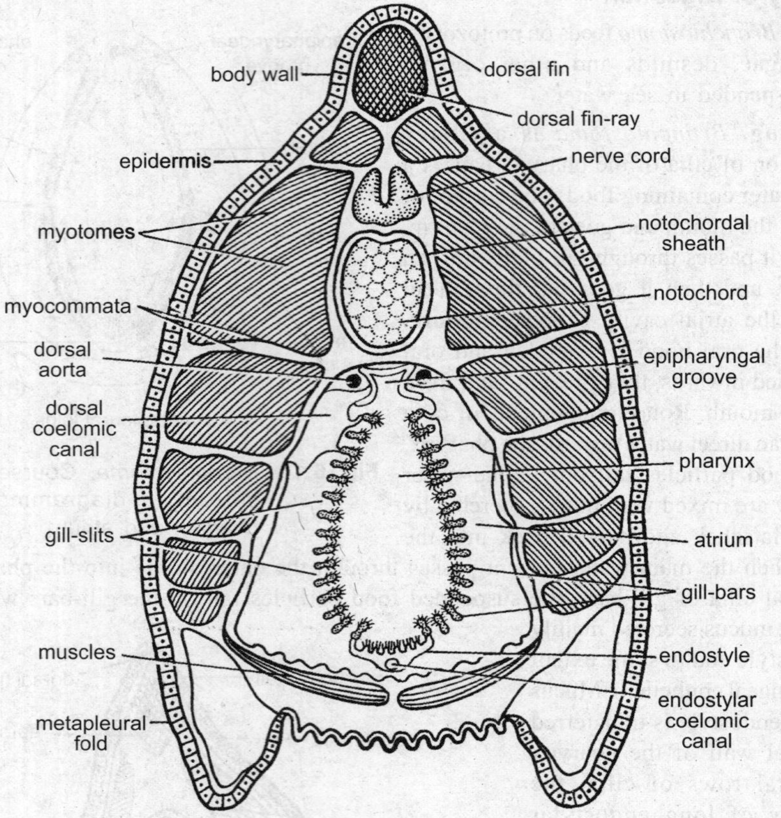

Fig. 6.19. T.S. of *Branchiostoma* through pharynx.

The cord of food and mucus passes down the gut by action of cilia. It is moved from the oesophagus into the midgut where a lateral tract of cilia directs it into the midgut diverticulum, from here the cord is returned again into the midgut. The iliocolic or iliocolonic ring rotates the cord of food causing the food and enzymes to mix, then the cord of food is moved into the hindgut.

Digestion

Digestive enzymes are secreted by the epithelial cells of the gut and midgut diverticulum. They are mixed with food as it passes along. Digestion starts in the midgut and is continued in the hind-gut. Besides this extracellular digestion, intracellular digestion also occurs in which food particles are taken into the epithelial cells of the hindgut and digested there. Some papillae on the floor of the atrium contain phagocytic cells which engulf food particles which may pass into the atrial cavity. **Absorption** of digested food takes place mostly in the hindgut and to lesser extent in the midgut.

RESPIRATION

The pharynx of *Branchiostoma* is mainly an organ of feeding, but it also acts as a respiratory organ. The pharyngeal wall is richly vascularised. While in vertebrates the pharynx is not concerned with food collection, and in some it is specialised for respiration by gills. In *Branchiostoma* some exchange of O_2 and CO_2 occurs between the water current and blood flowing through the blood vessels of gill-bars. But this appears doubtful since the blood contains no respiratory pigment. It appears more probable that an exchange of gases occurs over the whole surface of the body wall of the atrium and the metapleural folds.

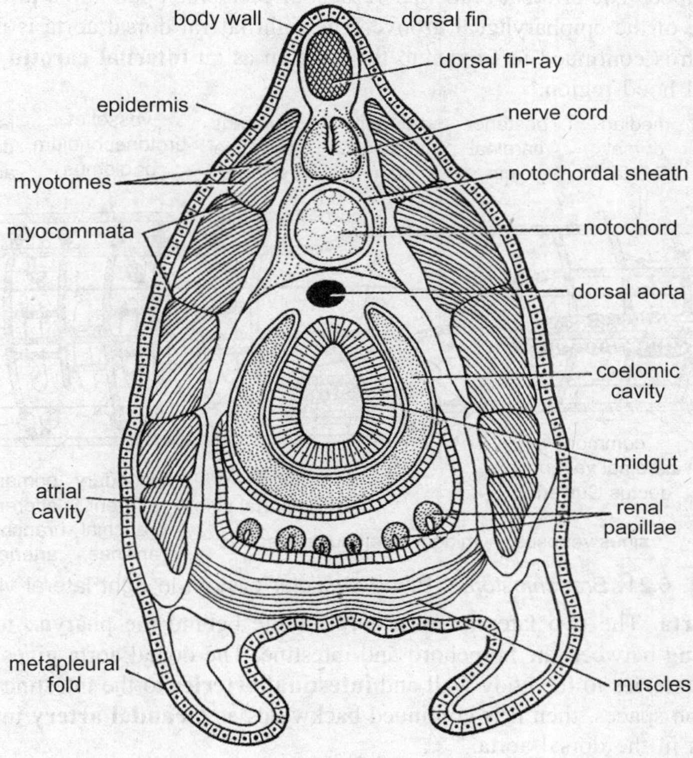

Fig. 6.20. T.S. of *Branchiostoma* through midgut.

CIRCULATORY SYSTEM

The blood vascular system of *Branchiostoma* is of closed type. The blood is colourless and devoid of corpuscles and respiratory pigment. Blood is found not only in blood vessels but also in lymph spaces, around fin-rays, and in metapleural folds. Its main function is to transport the food and excretory wastes.

Heart is not found in *Branchiostoma*. Most blood vessels are similar, but due to their homologies with blood vessels of vertebrates they are called **arteries** or **veins**. The principal arteries have muscular walls and only dorsal aorta has an endothelial lining.

Sinus venosus. It is a thin-walled sac below the hind end of pharynx. It is formed by the union of numerous vessels (veins) which collect blood from the entire body. It anteriorly leads into the ventral aorta.

Ventral aorta. Lying in the subendostylar coelom, below the pharynx, is a contractile **ventral aorta** which pumps blood forwards by peristaltic contractions. At the anterior end the ventral aorta passes into **two external carotid arteries**. From the ventral aorta paired **afferent branchial arteries** arise and pass upwards in the pharyngeal wall on both sides. The afferent branchial arteries at their commencement have contractile dilations called **bulbilli** which also pump blood, then they pass through the primary gill bars as three branches and through the synapticula they give two branches to each secondary gill-bar. They form vascular plexuses which supply nephridia. Blood is collected from the gill-bars and nephridia by aortic arches or paired **efferent branchial vessels**. In the afferent and efferent vessels the blood is exposed to a respiratory water current, but there is no proof of its oxygenation since there is no respiratory pigment in the blood. The efferent branchial vessels of each side open into a **lateral dorsal aorta** lying on the side of the epipharyngeal groove. The right lateral dorsal aorta is more dilated than the left one, each is continued in front into the rostrum as an **internal carotid artery** to supply the blood to oral hood region.

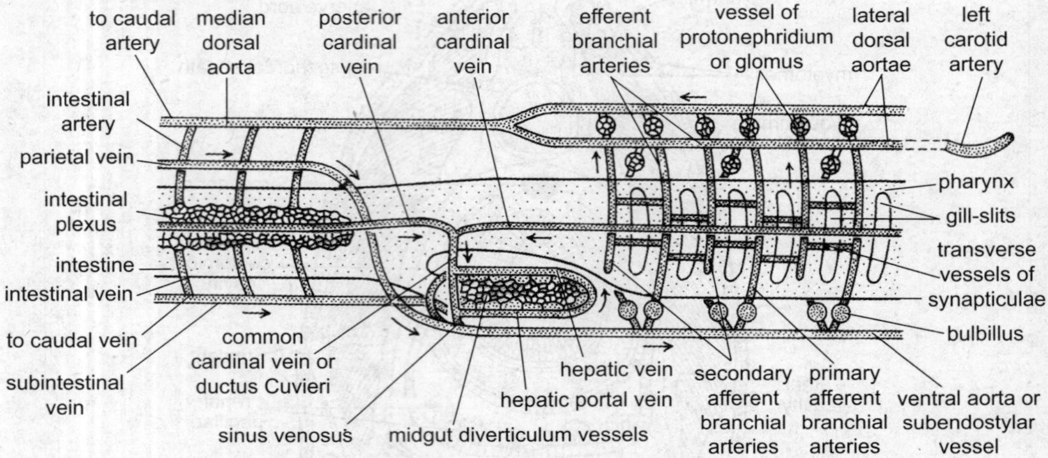

Fig. 6.21. *Branchiostoma*. Blood vascular system in right lateral view.

Dorsal aorta. The two lateral dorsal aortae unite behind the pharynx to form a median **dorsal aorta** lying between the notochord and intestine. The dorsal aorta gives out many small, paired **parietal arteries** to the body wall and **intestinal arteries** to the intestine where they form plexuses in lymph spaces, then it is continued backwards as a **caudal artery** into the tail. Blood flows backwards in the dorsal aorta.

Subintestinal vein. From the lymph spaces of the intestine blood is collected into a median longitudinal a subintestinal vein. For a greater part of its length the subintestinal is not a single vein but a plexus of small vessels, it runs below the intestine, and blood flows forwards in it. The blood from the tail is collected by a **caudal vein** which in the intestinal region joins the plexuses of the subintestinal vein.

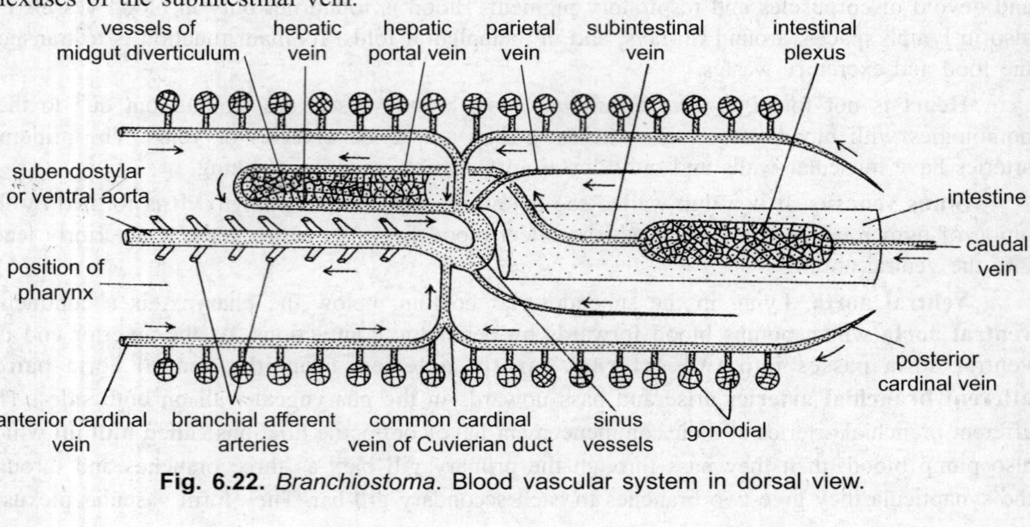

Fig. 6.22. *Branchiostoma*. Blood vascular system in dorsal view.

Hepatic portal system. The subintestinal vein in the anterior part of the midgut form a short but wide **hepatic portal vein** which runs along the ventral border of the midgut diverticulum and ramifies in its wall, thus, forming a system which is not strictly comparable with, but which foreshadows the **hepatic portal system** of vertebrates. Small blood vessels from the midgut diverticulum unite to form a short **hepatic vein** running along the dorsal border of the midgut diverticulum. The hepatic vein enters a sac-like **sinus venosus** from which arises the ventral aorta.

Parietal veins. A pair of **parietal veins** lying on either side of the gut receive blood from the dorsal body wall. They run dorsal to the intestine for a short distance and then run ventrally to open into the sinus venosus.

Cardinal veins. Running in the body wall at the level of gonads on each side are an **anterior cardinal** and a **posterior cardinal veins** which receive blood through small segmental veins coming from the body wall, myotomes and gonads. The anterior and posterior cardinal viens of each side unite just behind pharynx to form a **ductus Cuvieri** or **common cardinal vein**. The two ductus Cuvieri pass inwards through the atrium to join the sinus venosus.

A renal portal and a true hepatic portal system are absent because of the absence of kidneys and a true liver, yet the circulatory system is of the pattern found in vertebrates.

Course of blood circulation. Blood flows anteriorly in the caudal, subintestinal, parietals, posterior cardinals and ventral aorta. Whereas in the laterals and median dorsal aorta and anterior cardinals, it flows backwards.

Lymphatic system. Lymphatic spaces or sinuses are present in the cavities containing fin rays of the dorsal and ventral fins, and paired canals in the metapleural folds. These spaces are filled with colourless blood having no leucocytes. The meagure amount of O_2 required by *Branchiostoma* may be taken into the blood through these superficial areas.

EXCRETORY SYSTEM

The excretory organs of *Branchiostoma* are ectodermal in origin and comparable with the nephridia of annelids and platyhelminthes. The excretory organs of *Branchiostoma* are different from the kidney of vertebrates.

1. Protonephridia. The protonephridia of *Branchiostoma* are simple, closed ciliated sac-like and thin-walled ectodermal tubules. These are about 90 to 100 pairs arranged segmentally on dorso-lateral pharyngeal wall, one above each gill-slit on either side. Each protonephridium is a bent nephridial tube having a small upper posterior horizontal limb and a long lower anterior vertical limb. The vertical limb lies in the coelomic canal of the primary gill-bar and ends in a large group of solenocytes or terminate blindly. The horizontal limb has numerous smaller groups of the solenocytes. On the ventral surface of the horizontal limb, opposite a secondary gill-bar, is an aperture called **nephridiopore**. It bears long cilia and opens into the atrium. A nephridium contains about 500 solenocytes. Each solenocyte has a round nucleated head leading into a thin tubule opening into the lumen of nephridium through independent aperture. A long vibratile flagellum arise from the basal body of the cell (head) and extends into the tubule and project into the nephridium lumen. Tufts of solenocytes project into the dorsal longitudinal coelomic canals and are bathed in coelomic fluid. The nephridia are also supplied with small blood vessels. The nephridial tube opens by **nephridopore** into the atrium at the upper end of a gill-cleft lying against a secondary gill-bar. Thus, one end of the protonephridium lies in the coelom and the other end opens to the exterior (atrium). Protonephridia absorb nitrogenous waste from the blood sinuses and coelomic fluid by diffusion, and discharge it into the atrium.

2. Hatschek's nephridium. Besides the paired series of protonephridia there is a single large **nephridium of Hatschek** above the velum, slightly to the left of the notochord. It is like a protonephridium with a bent tube having numerous tubules ending in solenocytes. Its blind anterior end is just in front of Hatschek's pit, while the posterior end opens into the prebranchial region of pharynx just behind the velum. The Hatschek's nephridium is associated with a network of fine blood vessels and each solenocyte is surrounded by a small coelomic sac, it absorbs nitrogenous waste.

3. Brown funnels. There are two sac-like brown funnels which are situated dorsally upon the posterior end of pharynx, one on either side. The narrow anterior end of each funnel opens into the epibranchial or dorsal coelomic canal of its side, while the broad posterior end opens into the atrium. Most zoologists consider the funnels to be excretory, while some regard them to be receptor organs.

4. Renal papillae. In the floor of the atrium are numerous small **renal papillae**, they are probably excretory.

The nephridia of *Branchiostoma* develop from ectoderm, so that in their development and structure they resemble the nephridia of polychaete Annelida, but more recent views discount these similarities and the resemblance is only a case of parallelism since the relationship of the two groups is very remote. The excretory organs of chordates are mesodermal kidneys which require a high arterial blood pressure for filtration, in *Branchiostoma* kidneys and heart are lacking, the low blood pressure is sufficient for excretion by protonephridia.

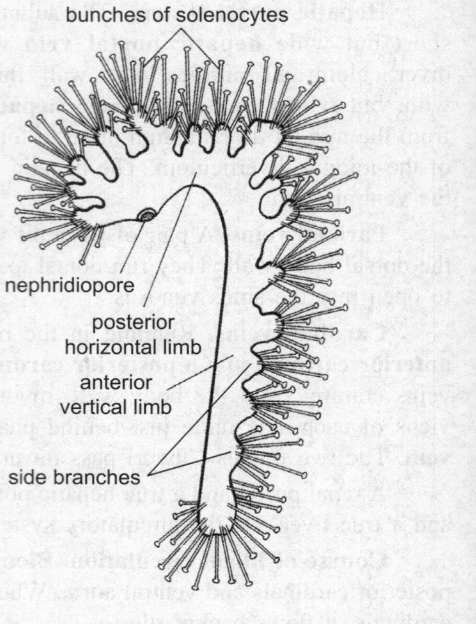

Fig. 6.23. *Branchiostoma*. Entire protonephridium.

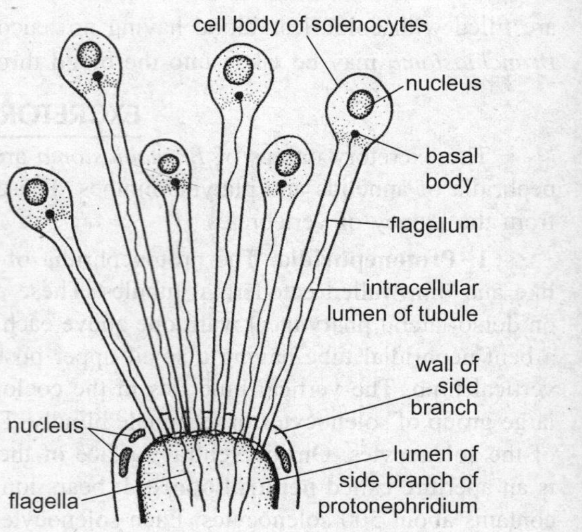

Fig. 6.24. *Branchiostoma*. An enlarged view of solenocytes.

NERVOUS SYSTEM

In *Branchiostoma,* nervous system is simple, having no brain as found in higher chordates.

1. Nerve cord. There is a mid-dorsal, hollow, tubular **neural tube** or nerve cord lying above the notochord. It starts behind the anterior end of the notochord and tapers to a point posteriorly and ends a short way in front of the hinder end of the notochord. The neural tube contains a narrow **central canal** or **neurocoele** throughout its length and filled with cerebrospinal fluid. **Nerve cells** are grouped around the central canal, and fibres arising from nerve cells are placed superficially, while in invertebrates the nerve cells are superficial and nerve fibres are placed centrally. Like annelids, the neural tube on the dorsal side has some giant neurons with giant nerves fibres running lengthwise.

Anteriorly the central canal dilates within the **cerebral vesicle** forming its ventricle. It gives out a dorsal **diverticulum** which extends behind over the central canal for some distance. The cerebral vesicle has two receptor organs, a **pigment eye spot** in its anterior wall and an **infundibular organ** on its floor.

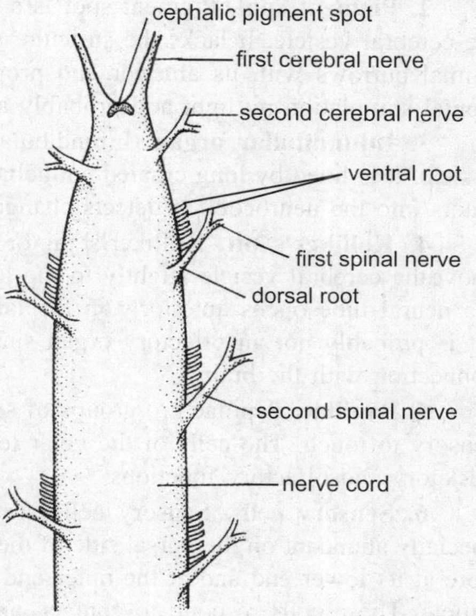

Fig. 6.25. *Branchiostoma.* Anterior part of nerve cord in dorsal view.

2. Nerves. From the anterior part of the neural tube (cerebral vesicle) arise two pairs of sensory nerves going to the oral hood cirri and sense organs. These nerves are called **cerebral nerves** and lack ventral roots. The body behind the cerebral vesicle gives off **spinal nerves** in segmental pairs. Each spinal nerve has a **dorsal root** with **afferent** sensory fibres entering, and a **ventral root** made of several separate **efferent** motor fibres leaving the neural tube. The dorsal roots come from the skin and is of mixed type (*i.e.,* sensory and motor) whereas, the ventral roots are motor and go to myotomes. The dorsal and ventral roots do not unite to form a mixed spinal nerve, as in vertebrates. Moreover, a dorsal root of a spinal nerve is slightly posterior to its ventral root. The dorsal roots also contain fibres running to the wall of the gut. The spinal nerves of the two sides do not correspond, the dorsal roots of one side are opposite the ventral roots of the other side. These spinal nerves have no myelin sheath, they are primitive.

3. Autonomic nervous system. There is an **autonomic nervous system** having two nerve plexuses in the smooth muscles of the wall of the intestine. These are connected to the neural tube by visceral nerves (preganglionic) through the dorsal roots. The autonomic nervous system is apparently represented only by the parasympathetic system, it controls the unstriped muscles of the gut.

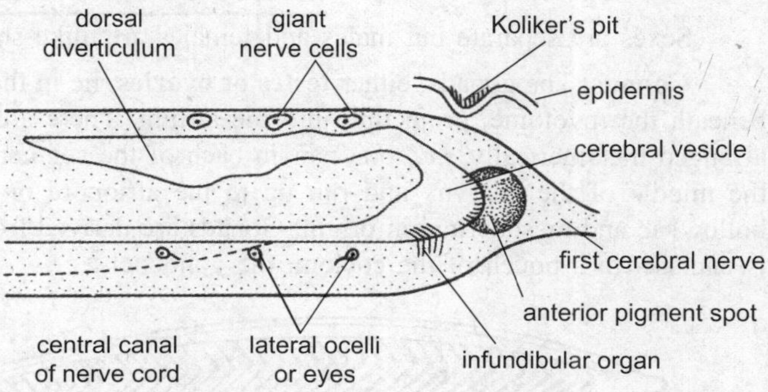

Fig. 6.26. *Branchiostoma.* V.L.S. of anterior part of nerve cord.

SENSE ORGANS (RECEPTORS)

Amphioxus is richly supplied with simple receptors.

1. Eye spots or ocelli. Simple eyes or ocelli are rows of black pigmented spots along the entire length of the ventro-lateral sides of the neural tube. They are arranged in definite tracts and are sensitive to light. Each eye has an outer pigment cell or melanocyte and a lens-like photosensitive cell with a striated apical border or vitreous body which secretes a pigmented cap. It acts as a lens. The photosensitive cell receives one nerve fibre from the nerve cord. The light-sensitive eyes are responsible for orientation of the animal as it burrows in sand.

2. Pigment spot. Pigment spot is a large pigmented spot in the extreme anterior wall of the cerebral vesicle. It lacks the structure of an eye and has no sensory function. But when the animal burrows with its anterior end projecting, then the pigment spot shields the eyes from frontal stimulation by light and probably acts as a thermoreceptor.

3. Infundibular organ. Infundibular organ is a depression in the floor of the cerebral vesicle. It is lined by long ciliated epithelial cells. Internally it gives out a Reissner's fibre which enters into the neurocoel. It detects changes in the pressure of the fluid in the neural tube.

4. Kolliker's pit. Kolliker's pit or olfactory pit is a pocket of ciliated ectoderm cells above the cerebral vesicle slightly to the left side. It marks the position of neuropore by which the neural tube opens anteriorly in the larva, but the neuropore closes in the adult. Kolliker's pit is probably not an olfactory organ since it has no specialised sensory cells, nor it has any connection with the brain.

5. Papillae. Papillae are groups of sensory cells on oral cirri and velar tentacles. They are sensory to touch. The cells of the velar tentacles and oral cirri are also chemoreceptors having gustatory and olfactory functions.

6. Sensory cells. Sensory cells are found scattered all over in the epidermis, they are specially abundant on the dorsal side of the body and on oral cirri. Each sensory cells has a nerve fibre at its lower end and at the outer end it has a hair-like sensory process projecting from the cuticle. These cells respond to contact, and some of them are concerned with determining the nature of the sand in which the animal will burrow, for the animal avoids sand which is too fine.

The above mentioned receptors in or near the integument are collectively known as **exteroceptors** or external receptors.

7. Free nerve endings. Besides the above external receptors, there are sensory nerve endings in myotomes which respond to internal stimuli, such as muscular contractions. These internal receptors in muscles are called **proprioceptors**.

REPRODUCTIVE SYSTEM

Sexes are separate but males and females are alike showing no sexual dimorphism.

Gonads. The gonads, either **testes** or **ovaries**, lie in the ventro-lateral part of the body wall beneath the myotome, projecting into the atrium. There are about 26 pairs of testes or ovaries arranged metamerically, *i.e.,* one pair in each of the segments 25 to 51. They begin from about the middle of the pharynx and run up to the atriopore on each side. Each gonad is a simple hollow sac and mesodermal in origin. Gonads are derived from the walls of the coelom and each is enclosed in a pouch of the coelom, the gonocoel.

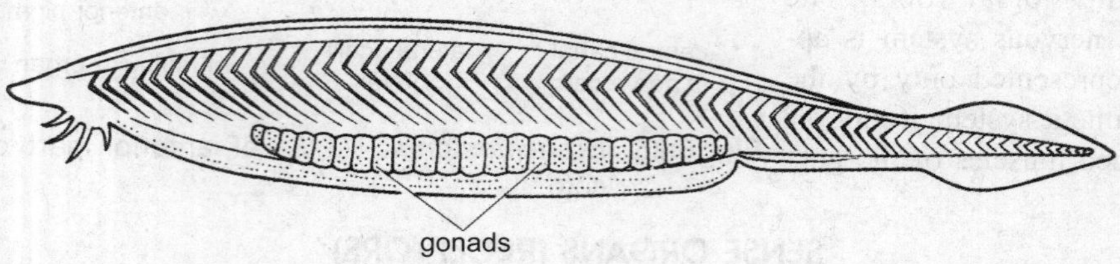

Fig. 6.27. *Branchiostoma*. Diagram showing the location of gonads.

Gametes. In sections, ovaries have cells with large nuclei but testes present a streaky appearance. The mature sperm of *Branchiostoma* is about 18 μ in length having a nucleated head with acrosome, a middle piece and a long tail. The ovum is spherical about 0.1 mm in diameter with a large nucleus. It is microlecithal type.

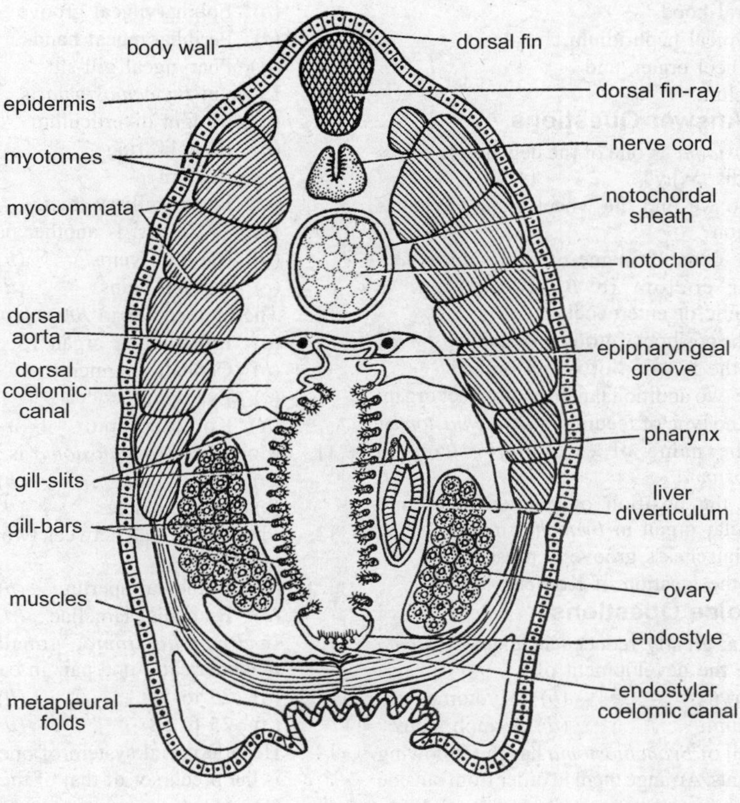

Fig. 6.28. T.S. of *Branchiostoma* passing through posterior region of pharynx showing gonads.

Gonoducts. There are no genital ducts. When gametes mature the walls of the gonads burst and ova or sperms escape into the atrium from where they escape through the atriopore.

DEVELOPMENT

Development in *Branchiostoma* is indirect, *i.e.,* having a larval stage. Fertilisation is **external** taking place in sea water. Development have been discussed in a Chapter in a section of Embryology.

IMPORTANT QUESTIONS

Long Answer Questions

1. Why would *Branchiostoma* starve to death without its endostyle? Explain.
2. What is the habitat of *Branchiostoma*? How does it move about in this habitat?
3. Describe the structure of notochord in *Branchiostoma*. Discuss its homology in the Hemichordata and Tunicata.
4. Describe in brief the internal anatomy of *Branchiostoma* and show whether it is primitive or secondary of degenerated.
5. Describe the blood vascular system of *Branchiostoma* and explain its primitive features.

Short Answer Questions

1. Draw a neat labelled sketch of the T.S. through tail region of *Branchiostoma*.
2. Write notes on the feeding mechanism in *Branchiostoma*.
3. Give an account of the mode of feeding in *Branchiostoma*.
4. Describe pharynx in *Branchiostoma*.
5. Describe the brain and sense organs in *Branchiostoma*.
6. Write notes on
   (*i*) Atrium,
   (*ii*) Hatschek's nephridium,

(*iii*) Oral hood,
(*iv*) Typical nephridium,
(*v*) Wheel organ, and
(*vi*) Velum.

Very Short Answer Questions

1. *Branchiostoma* is one of the degenerate forms of Agnatha. Why?
2. Where do we find the wheel organ? What is its function?
3. Write the common name of *Branchiostoma.*
4. Whether coelom in *Branchiostoma* is schizocoelic or enterocoelic?
5. What is the position of atrium in *Branchiostoma*?
6. What is the function of wheel organ?
7. Write the two additional names of wheel organ.
8. What is the type of feeding in *Branchiostoma*?
9. Write the name of excretory organs in *Branchiostoma.*
10. What is the position of pigment spot and infundibular organ in *Branchiostoma*?
11. Where Hatschek's groove is present?
12. What is the location of Kollicker's pit?

Multiple Choice Questions

1. Coelom is greatly reduced in *Branchiostoma* owing to the development of
   (*a*) Pharynx (*b*) Myotomes
   (*c*) Atrium (*d*) Lymph sinus.
2. Body wall of *Branchiostoma* has the following components. Arrange them in order from outside to inside.
   (*a*) Epidermis (*b*) Cuticle
   (*c*) Myomeres (*d*) Dermis.
3. Coelom in *Branchiostoma* is
   (*a*) Enterocoelic in origin
   (*b*) Schizocoelic in origin
   (*c*) Pseudocoelom
   (*d*) None of the above.
4. Atrium in *Branchiostoma* is formed by
   (*a*) A pair of metapleural folds
   (*b*) Folding of endoderm
   (*c*) Folding of mesoderm
   (*d*) None.
5. In *Branchiostoma* gill-slits open into
   (*a*) Coelom (*b*) Atrium
   (*c*) Exterior (*d*) Pharynx.
6. Wheel organ is found in
   (*a*) Oral hood (*b*) Mouth
   (*c*) Pharynx (*d*) Velum.
7. One of the following is converted into thyroid gland of vertebrates
   (*a*) Endostyle
   (*b*) Epipharyngeal groove
   (*c*) Peripharyngeal bands
   (*d*) Pharyngeal gill-slits.
8. Liver in *Branchiostoma* is
   (*a*) Midgut diverticulum
   (*b*) Iliocolic ring
   (*c*) Rectum
   (*d*) Lateral ciliary tract.
9. Ductus Cuvieri is another name for
   (*a*) Cardinal veins (*b*) Sinus venosus
   (*c*) Parietal veins (*d*) Aorta.
10. Thermoreceptor in *Branchiostoma* is
    (*a*) Infundibular organ
    (*b*) Cephalic pigments
    (*c*) Eye spots
    (*d*) Kollicker's pit.
11. Mouth in *Branchiostoma* is bordered by
    (*a*) Oral hood (*b*) Oral cirri
    (*c*) Oral tentacles (*d*) Oral frill.
12. Pharyngeal wall between two adjacent gill-slits is called
    (*a*) Branchial aperture (*b*) Gill-clefts
    (*c*) Branchial lamellae (*d*) Gill-lamellae.
13. In *Branchiostoma*, gonads are arranged metamerically one pair in each segment from
    (*a*) 25 to 51 (*b*) 29 to 51
    (*c*) 25 to 58 (*d*) 29 to 58.
14. Hepatic portal system of one of the following is the precursor of that of the higher chordates
    (*a*) *Herdmania*
    (*b*) *Branchiostoma*
    (*c*) *Labeo*
    (*d*) *Petromyzon.*
15. The intestine of *Branchiostoma* is divided into
    (*a*) Midgut, lateral ciliary tract and diverticulum
    (*b*) Lateral ciliary tract, diverticulum and rectum
    (*c*) Midgut, hindgut and rectum
    (*d*) Midgut, iliocolic ring and hindgut.
16. Primary gill-rod encloses
    (*a*) A blood vessel only
    (*b*) A coelomic canal
    (*c*) A blood vessel and coelomic canal
    (*d*) A coelomic canal and a central cavity.
17. Dorsal nerve root arising from the spinal cord in *Branchiostoma*
    (*a*) Is only sensory in function
    (*b*) Supplies visceral organs
    (*c*) Is only motor in function
    (*d*) Both sensory and motor in function.

Answers

1. (*c*) 2. (*b, a, d, c*) 3. (*a*) 4. (*a*) 5. (*b*) 6. (*a*) 7. (*a*)
8. (*a*) 9. (*a*) 10. (*b*) 11. (*a*) 12. (*c*) 13. (*a*) 14. (*b*) 15. (*d*)
16. (*c*) 17. (*d*).

7 Chapter

Subphylum 3. Cephalochordata : General Characters, Primitive, Degenerate and Specialised Characters and Affinities

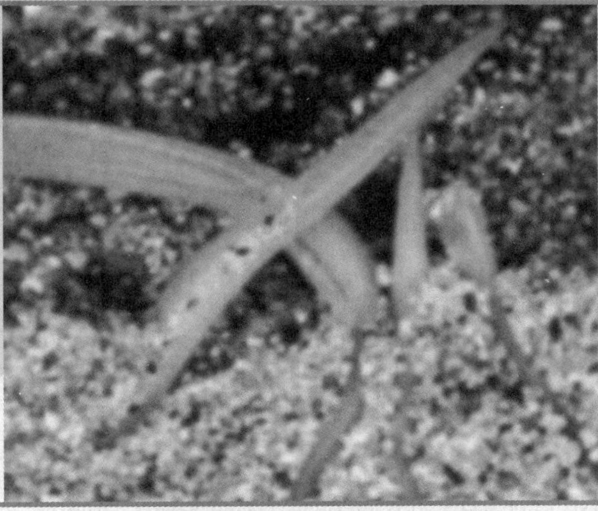

Contents of this Chapter

GENERAL CHARACTERS

1. Marine, widely distributed in shallow waters.
2. Mostly sedentary and buried with only anterior body end, projecting above bottom sand.
3. Body small, 5 to 8 cm long, slender, fish-like, segmented and transparent.
4. Body has trunk and tail. Head lacking.
5. Paired appendages lacking. Median fins present.
6. Exoskeleton absent. Epidermis single-layered.
7. Muscles dorso-lateral, segmented into myotomes.

8. Coelom enterocoelous, reduced in the pharyngeal region by development of atrial cavity.

9. Notochord rod-like, persistent, extending from rostrum to tail, hence, the name Cephalochordata.

10. Digestive tract complete. Pharynx large, perforated by numerous persistent gill-slits opening into atrium.

11. Respiration through general body surface. Special organs for respiration absent.

12. Circulatory system well developed, closed without heart and respiratory pigment. Hepatic portal system developed.

13. Excretion by protonephridia with solenocytes.

14. Nerve cord dorsal, tubular without ganglia and brain. Dorsal and ventral nerve roots separate.

15. Sexes separate. About 25 pairs of gonads. Gonoducts lacking. No asexual reproduction.

16. Fertilisation external in sea water.

17. Development indirect, including a free-swimming larva.

18. Cephalochordata has about 30 species of the genus *Branchiostoma* and all put in the class Leptocardii.

PRIMITIVE, DEGENERATE AND SPECIALISED CHARACTERS OF *BRANCHIOSTOMA* (CEPHALOCHORDATA)

Primitive Characters

Primitive characters are relics from its original ancestors. They are those characters which are developed and perfected in the course of evolution and they generally show progress. Primitive characters of *Branchiostoma* are as follows :

1. There is no specialised head.

2. Absence of paired limbs or fins.

3. The epidermis is made of single row of cells and a true dermis is absent.

4. Coelom enterocoelous.

5. A complete notochord is persistent throughout life with no vertebral column.

6. There is a complete metameric segmentation of myotomes from one end to the other, as in the embryo of vertebrates, and this obvious metamerism is retained throughout life.

7. Alimentary canal is a straight tube without loops, and the midgut diverticulum is a hollow tube. Jaws absent.

8. A peculiar pharynx specialised for ciliary feeding in which sorting of food takes place after a food-laden current of water enters the pharynx, this involves taking a very large volume of water and its continuous removal through numerous gill-clefts.

9. Blood vascular system is diagrammatically vertebrate-like but without a heart and practically no distinction between arteries and veins.

10. Special respiratory organs and respiratory pigment absent.

11. Excretory organs are segmental protonephridia having solenocytes. Protonephridia are not coelomoducts.

12. The neural tube does not form an anterior brain. The dorsal and ventral nerve roots do not unite to form spinal nerves. Dorsal roots have no ganglia, consequently impulses pass directly from the skin to the neural tube.

13. Paired sense organs such as eyes, ears and nose are absent. The receptors present are of a primitive type.

14. Gonads segmentally arranged and without gonoducts.

15. Eggs are small with almost no yolk. Blastula is hollow, spherical and single-layered.

Degenerate Characters

Branchiostoma may be a specialised and degenerate descendant of Agnatha. The degenerate characters include :

1. Poorly developed brain (cerebral vesicles) and simple sensory organs.
2. Lack of any cartilaginous or bony endoskeleton.
3. Lack of gonoducts.

Specialised or Secondary Characters

The following specialised or secondary characters present in *Branchiostoma* are peculiar adaptations to its mode of life. They also probably prevented them to undergo further evolution. Their degenerate and specialised characters do not consider *Branchiostoma* in direct line of chordate evolution. Cephalochordates arose as a side offshoot.

1. There is an oral hood enclosing the mouth which is provided with oral cirri bearing sensory organs.
2. Elaborate velum with sensory velar tentacles.
3. An extreme anterior extension of notochord into rostrum seen in no other chordate making it stronger for burrowing.
4. A very large and elaborate pharynx with an accompanying multiplication of ciliated gill-clefts. The extra number of gill-clefts is unrelated to body segmentation.
5. Wheel organ and Hatscheck's groove and pit develop which help in ciliary feeding.
6. Formation of a large atrium with an atriopore.
7. Peculiar reduction and distortion of coelom by the atrium.
8. Curious asymmetries seen in the adult and in early development.

Branchiostoma has a simple organisation compared to vertebrates because many important craniate structures are lacking in it. But it is definitely a simple chordate having a large number of primitive characters, such as notochord, dorsal hollow nerve cord, and gill-clefts. Moreover, it has a post-anal tail, hepatic diverticulum, hepatic portal system, and the beginning of a ventral heart. These characters show the probable structure of the original chordate ancestor to which *Branchiostoma* is related. However, its secondary characters show that it is not on the direct line of chordate descent in evolution.

It is possible that *Branchiostoma* and its allies resemble in many fundamental respects the fish-like creature that existed in Silurian seas, and it is possible that the modern fishes of today evolved from animals of much the same kind.

AFFINITIES OF CEPHALOCHORDATA

Cephalochordata show affinities with non-chordate as well as chordates. Cephalochordates have been regarded to be phylogenetically related to several non-chordate groups at one time or other. But these can be overlooked in favour of the more chordate-like characteristics of cephalochordates.

Chordate Affinities

Cephalochordata (*Branchiostoma*) exhibits the three basic chordate features such as notochord, dorsal tubular nerve cord and pharyngeal gill-slits in the most typical manner and there is no doubt about its chordate nature. However, it exhibits relationships with all the major groups of phylum chordata, and its real status in the phylum remains uncertain.

Affinities with Hemichordata. Hemichordata and Cephalochordata resemble in having similar (*i*) pharyngeal apparatus with numerous gill-slits and gill-bars, (*ii*) filter feeding mechanism, (*iii*) respiratory mechanism, (*iv*) enterocoelous coelom, and (*v*) numerous gonads without gonoducts.

Objections. But, muscles in Hemichordata are unsegmented, nervous system distinctly of non-chordate type, gill-slits dorsal in position instead of lateral, and a post-anal tail is lacking. Moreover, inclusion of Hemichordata under chordata is also uncertain because of doubtful nature of notochord. This shows Hemichordata is more primitive than Cephalochordata.

Affinities with Urochordata. *Branchiostoma* (Cephalochordata) and *Herdmania* (Urochordata) are regarded to be very closely related because of (*i*) the primitive ciliary feeding and respiratory mechanisms, (*ii*) a large pharynx having many gill-clefts (stigmata), an endostyle which produces mucus, peripharyngeal bands and an epipharyngeal groove (dorsal lamina), (*iii*) an ectoderm-lined atrial cavity opening to the exterior by an atriopore or its homologue, the atrial siphon, (*iv*) the mantle of *Herdmania* is equivalent to the atrial folds of *Branchiostoma*. Besides the homologies of adult anatomy, there are embryonic similarities, (*v*) there is a very striking similarity of the eggs of some tunicates to those of *Branchiostoma*, in both there is an identical arrangement of future parts in the cytoplasm, and (*vi*) the early stages of development, especially in the formation of nervous system and notochord are practically identical. Such vast similarities afford convincing evidence of a close relationship between Cephalochordata and Urochordata and they show that both have been derived from a common ancestor.

Objections. The adult Urochordata are extremely degenerate, sedentary animals having several characters not found in *Branchiostoma*. For example: (*i*) body unsegmented and covered by a cellulose test, (*ii*) without hollow nerve cord and notochord, (*iii*) with liver, (*iv*) with a heart covered by peritoneum, (*v*) absence of nephridia, (*vi*) sexes united and gonads hermaphrodite, and (*vii*) larva undergoes retrogressive metamorphosis.

These differences show that in spite of close similarities reflecting probable common ancestry, the cephalochordates are better evolved than the urochordates.

Affinities with Cyclostomata. *Branchiostoma* and **ammocoete larva** of lamprey (Cyclostomata) are similar in many characters such as : (*i*) elongated, slender fish-like body, (*ii*) continuous dorsal median fin, (*iii*) mouth surrounded by an oral hood, (*iv*) presence of a velum, (*v*) pharynx with an endostyle and gill-slits, and (*vi*) presence of dorsal hollow nerve cord and notochord. Besides these fundamental chordate characters, their adults have persistent velum, gill-slits, metameric myotomes and a post-anal tail.

Affinities with other vrtebrates. *Branchiostoma* also show similarities with other vertebrates in several ways such as : (*i*) post-anal tail, (*ii*) metameric myotomes, (*iii*) coelom lined by mesodermal epithelium, (*iv*) mid-gut diverticulum comparable with liver, (*v*) well formed hepatic portal system, and (*vi*) with forward flow of blood in ventral and backward flow of blood in dorsal longitudinal blood vessels.

Objections. Cephalochordates differ from cyclostomes and other vertebrates in most of their primitive characters, such as : (*i*) absence of head, paired limbs, skull, vertebral column, muscular true heart, blood without RBCs, brain, sense organs and gonoducts, etc., and (*ii*) in possessing nephridia, atrium, numerous gonads, asymmetry, etc.

IMPORTANT QUESTIONS

Long Answer Questions

1. Discuss the affinities of Cephalochordata.
2. Enlist the primitive, degenerate and specialised characters of cephalochordates.

Short Answer Questions

1. Write the general characters of Cephalochordata.
2. Write the specialised characters of Cephalochordata.

Multiple Choice Questions

1. Cephalochordates live in
   (*a*) Marine shallow waters
   (*b*) Marine deep waters
   (*c*) Brackish waters
   (*d*) Freshwater of rivers.
2. Body of *Branchiostoma* is divided into
   (*a*) Head and tail
   (*b*) Head, trunk and tail
   (*c*) Head and trunk
   (*d*) Trunk and tail.
3. Fins in *Branchiostoma* are
   (*a*) Median fins
   (*b*) Paired fins
   (*c*) Median and paired fins both
   (*d*) Median fins absent.
4. In cephalochordates notochord extends from
   (*a*) Pharynx to tail
   (*b*) Rostrum to pharynx
   (*c*) Rostrum to tail
   (*d*) Entirely absent.
5. In cephalochordates, gill-slits open into
   (*a*) Atrium
   (*b*) Pharynx
   (*c*) Coelom
   (*d*) Exterior.
6. Cephalochordates are mostly sedentary and are buried
   (*a*) Completely under the bottom sand
   (*b*) With only anterior end projecting above bottom sand
   (*c*) Under bottom and laterally
   (*d*) With only posterior end projecting above bottom sand.
7. Exoskeleton in cephalochordates is
   (*a*) Well-developed
   (*b*) Absent
   (*c*) Cartilaginous
   (*d*) Bony.
8. Muscles in cephalochordates are
   (*a*) Dorso-ventral
   (*b*) Dorso-lateral
   (*c*) Ventro-lateral
   (*d*) Antero-posterior.
9. In cephalochordates respiration takes place
   (*a*) Through nostrils
   (*b*) Through gills
   (*c*) Through general body surface
   (*d*) Through lungs.
10. Circulatory system in cephalochordates is
    (*a*) Well-developed, closed with a heart and lacks a respiratory pigment
    (*b*) Well-developed, open with a heart and lacks a respiratory pigment
    (*c*) Poorly-developed, open with a heart and lacks a respiratory pigment
    (*d*) Well-developed, closed without a heart and lacks a respiratory pigment.
11. In cephalochordates, excretion takes place by
    (*a*) Protonephridia with solenocytes
    (*b*) Mesonephridia with solenocytes
    (*c*) Protonephridia without solenocytes
    (*d*) Metanephridia without solenocytes.
12. In cephalochordates nerve cord is
    (*a*) Dorsal tubular with ganglia and brain
    (*b*) Dorsal tubular without ganglia and brain
    (*c*) Ventral tubular with ganglia and brain
    (*d*) Ventral tubular without ganglia and brain.
13. In cephalochordates
    (*a*) Sexes are separate
    (*b*) Organisms are hermaphrodite
    (*c*) One pair of gonads are present
    (*d*) Organisms are hermaphrodite with metamerically arranged gonads.
14. Development in Cephalochordata is
    (*a*) Indirect including a parasitic larva
    (*b*) Direct including a parasitic larva
    (*c*) Indirect including a free-swimming larva
    (*d*) Direct including a free-swimming larva.

ANSWERS

1. (*a*) 2. (*b*) 3. (*a*) 4. (*c*) 5. (*a*) 6. (*b*) 7. (*b*) 8. (*b*)
9. (*c*) 10. (*d*) 11. (*a*) 12. (*b*) 13. (*a*) 14. (*c*).

8

Chapter

Protochordata : General and Comparative

Contents of this Chapter

WHAT ARE PROTOCHORDATES ?

Protochordata (Gr., *protos* = first, *chorde* = cord) or protochordates are lower primitive chordates. Protochordata comprise three major divisions or subphyla : Hemichordata, Urochordata and Cephalochordata. This division is mainly based on the notochord found in three subphyla. In **Hemichordata**, the notochord is represented by a short stomodaeal diverticulum or stomochord to the proboscis. This type of notochord in Hemichordata is of doubtful existence. Due to this reason, the Hemichordata is now-a-days treated as an independent invertebrate phylum. The **Urochordata** have notochord only in the tail of the tadpole larva. **Cephalochordata** have a persistent notochord extending along the whole length of their body throughout life.

At one time, Protochordata was considered to be a distinct phylum, but now it has no official recognition. Even then, the term Protochordata has been retained, for the convenience of study.

SIGNIFICANCE OF PROTOCHORDATES

The significance of protochordates is of little economic importance. But protochordates have great phylogenetic significance to the zoologists. They exhibit great affinities and perhaps, common origin with the living vertebrates. They retain the three basic chordate characteristics (notochord, hollow dorsal nerve cord and pharyngeal gill-slits) throughout the life. As such, protochordates are considered most primitive and borderline chordates closer to the ancestor which probably gave rise to the final and largest group of chordates, the vertebrates.

COMPARISON OF THREE PROTOCHORDATE SUBPHYLA

Although Protochordata show three chief characteristic features (notochord, hollow dorsal nerve cord and pharyngeal gill-slits) but they differ from one another in several ways. Their chief contrasting characters have been summarised in Tables 8.1 to 8.10.

TABLE 8.1. GENERAL COMPARISON OF THREE PROTOCHORDATE SUBPHYLA AT A GLANCE.

Characters	Hemichordata (*Balanoglossus*)	Urochordata (*Herdmania*)	Cephalochordata (*Branchiostoma*)
1. **Distribution**	Worldwide, marine.	Worldwide, marine.	Worldwide, marine.
2. **Habits and Habitat**	Solitary or colonial, mostly tubicolous.	Solitary or colonial, fixed or free swimming.	Solitary, burrowing or free swimming.
3. **Shape**	Worm-like, elongated, cylindrical body.	Sac-like, degenerate body.	Fish-like laterally compressed body.
4. **Body divisions**	Proboscis, collar and trunk.	No such division.	Trunk and tail, no distinct head.
5. **Post-anal tail**	Absent.	Absent.	Present.
6. **Fins**	Absent.	Absent.	Median fin present.
7. **Test**	Absent, or test composed of tunicine.	Body covered with a tunic.	Absent.
8. **Coelom**	Enterocoelous, divisible into proboscis, collar and trunk coeloms.	Absent, replaced by atrial cavity.	Reduced by the development of atrial cavity.
9. **Atrial cavity**	Absent.	Well developed.	Well developed.
10. **Notochord**	Absent.	Absent in adult but present only in larval tail.	Present, it is rod-like, persists throughout life.
11. **Muscles**	Unsegmented.	Unsegmented.	Metamerically segmented as myotomes.
12. **Alimentary canal**	Complete, straight or coiled.	Complete and coiled.	Complete and straight.
13. **Oral hood**	Absent.	Absent.	Present, bordering the mouth aperture.
14. **Pharyngeal apparatus**	Large, simple with 2 dorsal linear rows of gill-slits.	Large, complex, with 2 to numerous pairs of lateral gill-slits.	Large, complex with several pairs of lateral gill-slits.
15. **Mode of feeding**	Ciliary or filter feeder.	Ciliary feeder.	Ciliary feeder.
16. **Respiratory organs**	Absent, respiration occurs within gill-slits.	Absent, respiration occurs through the pharyngeal wall.	Absent, respiration occurs through pharyngeal wall.

(*Contd.*)

Characters	Hemichordata (*Balanoglossus*)	Urochordata (*Herdmania*)	Cephalochordata (*Branchiostoma*)
17. **Heart**	A dorsal heart present.	Well developed ventral muscular heart present.	Absent.
18. **Excretory organs**	Glomerulus in proboscis.	Neural gland, pyloric gland and nephrocytes.	Protonephridia.
19. **Nervous system**	Intraepidermal.	Only dorsal nerve ganglion in adult.	Nerve cord dorsal, tubular without ganglia, simple brain present.
20. **Sexes**	Separate.	Hermaphrodite	Separate.
21. **Gonads**	One to several pairs.	One pair only.	Many pairs, metamerically arranged.
22. **Gonoducts**	Absent.	Present.	Absent.
23. **Development**	Direct or indirect, with or without metamorphosis.	Indirect with retrogressive metamorphosis.	Indirect with metamorphosis.

TABLE 8.2. COMPARISON : HABITS AND HABITAT OF THREE PROTOCHORDATE TYPES.

Characters	*Balanoglossus* (Hemichordata)	*Herdmania* (Urochordata)	*Branchiostoma (Amphioxus)* (Cephalochordata)
1. **Occurrence**	Marine, shallow coastal water of intertidal zones.	Marine, shallow waters with rocky bottoms.	Marine, shallow waters with sandy bottoms.
2. **Distribution**	Worldwide in all seas.	Worldwide, particularly in Indian, Pacific and Carribean seas.	Worldwide in all seas.
3. **Mode of life**	Solitary, burrowing, nocturnal and sluggish.	Solitary, attached to some substratum, diurnal and sluggish.	Solitary, free-swimming or burrows in sand, nocturnal and active.
4. **Burrow**	U-shaped at least with two openings lined by sand grains cemented togather by mucus.	Burrows are not formed.	Usually remains burried in sand with anterior end projecting out.
5. **Food and feeding habit**	Microorganisms and organic particles in sand and water. Microphagus.	Planktonic microorganisms. Microphagus.	Planktonic microorganisms. Microphagus.
6. **Mode of feeding**	Ciliary filter-feeder as well as swallowing with mud.	Ciliary filter-feeder, food particles are filtered in pharyngeal sac from food-respiratory water current.	Ciliary filter-feeder, food particles are filtered from food-laden water current in the pharyngeal sac.
7. **Water current**	Mouth → branchial sac → gill slits → out through gill pores.	Mouth → branchial sac → gill slits → atrial cavity → atriopore → out.	Mouth → branchial sac → gill slits → atrial cavity → atriopore → out.
8. **Reproduction**	Only sexual, sexes separate, exhibits great power of regeneration.	Only sexual, sexes united. Some reproduce asexually by budding.	Only sexual, sexes separate. No asexual reproduction.

TABLE 8.3. COMPARISON : EXTERNAL FEATURES OF THREE PROTOCHORDATE TYPES.

Characters	*Balanoglossus* (Hemichordata)	*Herdmania* (Urochordata)	*Branchiostoma (Amphioxus)* (Cephalochordata)
1. **Shape**	Worm-like, soft, elongated, cylindrical, bilaterally symmetrical body	Bag-like, laterally compreseed, oblong or rectangular body.	Fish-like, small, narrow, laterally compressed body with both the ends pointed.
2. **Size**	10 to 50 cm in length.	Usually about 9.5 cm long, 7 cm broad and 4 cm thick.	5 to 8 cm in length.
3. **Colour**	Bright with reddish or orange tints.	Fresh animal is pinkish.	Whitish, somewhat translucent.
4. **External ciliation**	Body uniformly ciliated.	—	Youngs ciliated. Adults non-ciliated.
5. **External covering**	—	A test or tunic which is largely made of tunicine forms a protective jacket or covering around the body.	—
6. **Body divisions**	Proboscis, collar and trunk, trunk is further differentiated into branchiogenital region, hepatic region and post-hepatic region.	Body proper and foot, the foot is entirely made of the test.	Trunk and tail, a true head is absent, trunk projects anteriorly into rostrum or snout.
7. **External apertures**	Proboscis pore, mouth, collar pores, numerous dorsal gill pores and terminal anus. representing cloacal aperture.	Branchial aperture at the tip of branchial siphon representing mouth and atrial aperture at the tip of atrial siphon	Opening of oral hood representing mouth, atriopore and sub-terminal anus and numerous lateral gill-slits.

TABLE 8.4. COMPARISON : DIGESTIVE SYSTEM (ALIMENTARY CANAL AND ASSOCIATED GLANDS) OF THREE PROTOCHORDATE TYPES.

Characters	*Balanoglossus* (Hemichordata)	*Herdmania* (Urochordata)	*Branchiostoma (Amphioxus)* (Urochordata)
1. **Alimentary canal**	Complete, straight narrow tube-like extending from mouth to anus.	Complete, curved tube-like having intestinal loop.	Complete, straight tube-like extending from mouth to anus.
2. **Mouth opening**	Wide, circular permanently open lying mid-ventrally between proboscis and collar.	Four-sided branchial opening at the tip of branchial siphon bordered by four. lips	Wide opening situated antero-ventrally bordered by oral hood having oral or buccal cirri.
3. **Buccal cavity**	Short passage without tentacles.	The cavity of branchial siphon is the buccal cavity, at its base a circle of branchial tentacles working as a sieve.	It is spacious funnel-shaped cavity of oral hood called vestibule having wheel organ.

(Contd.)

Characters	*Balanoglossus* (Hemichordata)	*Herdmania* (Urochordata)	*Branchiostoma (Amphioxus)* (Cephalochordata)
4. **Pharynx**	Large tubular, lateral parabranchial ridge divides it into a dorsal branchial portion with gill-slits and a ventral digestive portion.	Large, bag-like divided by two ciliated peripharyngeal ridges or bands into an anterior small prebranchial region and a large posterior branchial sac having numerous gill-slits.	Large tubular, lateral oblique ciliated peripharyngeal bands divide it into a small anterior prebranchial region and a large posterior branchial sac.
5. **Velum**	Absent.	Absent.	Present. It is a circular ring-like vertical membrane with velar tentacles between vestibule and pharynx.
6. **Endostyle**	Absent.	Situated mid-ventrally along the floor of pharynx.	Same like *Herdmania*.
7. **Hyperpharyngeal groove**	Absent.	It is the dorsal lamina with tongue-like processes called languets situated at the mid-dorsal border of the roof of branchial sac.	It is the epipharyngeal ciliated groove present mid-dorsally along the roof of pharynx, it is without languets.
8. **Gill-slits**	Two rows of dorso-lateral U-shaped gill-slits open into branchial sacs and finally to exterior by gill-pores.	Numerous about, 2,00,000 gill-slits open into atrial cavity.	150–200 pairs of gill-slits open into atrial cavity.
9. **Stomach**	Not differentiated from intestine.	Differentiated from intestine and sphinctered at both the ends.	Not differentiated from intestine.
10. **Intestine**	Straight tube, its anterior hepatic region possesses numerous dorsal sacculations called hepatic caeca and posterior post-hepatic region bears a pair of dorsolateral grooves having epithelial cells with long cilia.	U-shaped tube formed by descending and ascending limbs, left gonad is enclosed in the intestinal loop, thus, formed.	Straight tube, ciliated internally possesses lateral ciliary tract and iliocolic ring.
11. **Rectum**	Not demarcated from intestine.	Hind part of intestine demarcated as rectum which is ciliated internally.	It is clearly demarcated from intestine and heavily ciliated internally.
12. **Anus**	It is terminal circular sphinctered aperture at the tip of the trunk.	Rectum curves dorsally to open into atrium or cloaca through anus which is bounded by four lips.	It is a circular, sphinctered aperture at the base of caudal fin.
13. **Atrial cavity**	Not found.	Spacious, gill-slits and anus open into it which leads to the exterior through atrial aperture situated at the tip of atrial siphon.	Spacious around the pharynx, gill-slits open into it which leads to the exterior through atriopore, anus does not open into it.
14. **Digestive glands**	No definite digestive glands.	Two (right and left) lobes of liver, pyloric glands in the wall of stomach and intestine are supposed to be digestive.	Midgut diverticulum works like liver, no pyloric glands.

TABLE 8.5. COMPARISON : PHARYNX OF THREE PROTOCHORDATE TYPES.

Characters	*Balanoglossus* (Hemichordata)	*Herdmania* (Urochordata)	*Branchiostoma (Amphioxus)* (Cephalochordata)
1. **Situation**	It lies in the branchio-genital region of the trunk.	It occupies the major part of body cavity.	It occupies nearly one-half anterior part of the body.
2. **Shape**	Elongated, tubular.	Large, bag-like.	Large, spacious, laterally compressed chamber.
3. **Attachment**	By dorsal and ventral mesenteries to the body wall.	To the mantle mid-ventrally.	To the body wall dorsally.
4. **Atrial cavity**	Absent.	It surrounds the pharynx dorsolaterally and opens to the exterior through atrial aperture.	It surrounds the pharynx from all sides except dorsal and opens to the exterior through atriopore lying posteriorly.
5. **Relation with mouth**	Wide mid-ventral mouth leads to pharynx through short buccal cavity.	Four-sided mouth aperture at the tip of branchial siphon opens into pharynx through tubular passage of branchial siphon (buccal cavity) which at the base bears branched branchial tentacles.	Wide antero-ventral mouth aperture bordered by oral hood with oral cirri leads to the pharynx through vestibule (buccal cavity) and enterostome of the velum bordered by velar tentacles.
6. **Pharyngeal cavity**	Simple and uniformly lined by cilia and mucous cells.	Internal pharyngeal wall complicated, longitudinally folded with tracts of cilia and gland cells.	It is very much complicated with definite ciliary and glandular tracts.
7. **Division**	Pharynx is incompletely divided into a dorsal branchial or respiratory chamber and a ventral digestive chamber by two lateral parabranchial ridges.	Divided into an anterior smaller prebranchial zone and a larger posterior branchial sac or basket by ciliated peripharyngeal bands.	A pair of peripharyngeal bands divide it into a small anterior prebranchial zone and a large posterior pharynx proper also referred to as branchial basket.
8. **Gill-slits**	Branchial part of pharynx bears many U-shaped gill-slits leading into branchial sacs which open out by gill-pores.	Branchial sac is perforated by numerous, small, elongated gill-slits or stigmata which open into atrial cavity.	Lateral walls of pharynx proper are perforated by nearly 150 – 200 pairs of vertically placed oblique gill-slits or stigmata which open into atrial cavity.
9. **Endostyle**	Absent.	Present mid-ventrally as a shallow longitudinal groove at the floor of branchial sac. It is made of five ciliary tracts alternating with four glandular tracts.	Present, similar to *Herdmania*, made of alternating ciliary and mucus secreting cells.
10. **Dorsal lamina**	Absent.	Present, it hangs mid-dorsally having tongue-like languets in the branchial sac.	A distinct ciliated epipharyngeal groove present mid-dorsally in the pharynx proper, languets, absent.

(Contd.)

Characters	*Balanoglossus* (Hemichordata)	*Herdmania* (Urochordata)	*Branchiostoma (Amphioxus)* (Cephalochordata)
11. **Periphary-ngeal bands**	Absent.	Present; dorsal lamina and endostyle are connected on either lateral side by a pair of ciliated peripharyngeal bands enclosing a narrow peripharyngeal groove.	Present; endostyle and epipharyngeal groove are connected anteriorly by a pair of lateral peripharyngeal ciliated bands.
12. **Relation with intestine**	It opens directly into intestine.	It leads into intestine through oesophagus and stomach.	It leads into intestine through a short oesophagus and midgut.
13. **Functions**	The lateral cilia lining the gill-slits set up a respiratory-cum-food current, hence, it helps in gaseous exchange and also brings food particles.	Feeding and respiration.	Feeding and respiration.

TABLE 8.6. COMPARISON : FOOD AND FEEDING MECHANISM OF THREE PROTOCHORDATE TYPES.

Characters	*Balanoglossus* (Hemichordata)	*Herdmania* (Urochordata)	*Branchiostoma (Amphioxus)* (Cephalochordata)
1. **Food**	Microorganisms and organic food particles in sand and water.	Planktonic microorganisms.	Planktonic microorganisms.
2. **Feeding habit**	Microphagus.	Microphagus.	Microphagus.
3. **Mode of feeding**	Ciliary filter-feeding and swallowing mud.	Ciliary filter-feeding.	Like *Herdmania*.
4. **Collection of food particles**	External minute food particles are trapped outside in mucus secreted by proboscis.	Internal food particles are gathered inside pharynx in mucus secreted from glandular tracts of endostyle.	Internal food particles are gathered inside pharynx in mucus secreted by the glandular tracts of endostyle and by pharyngeal epithelium.
5. **Cause of water current production**	Lashing of lateral cilia on gill-slits of pharynx cause a constant water current which enters through mouth.	Beating of lateral cilia lining the gill-slits set up a constant current of water which enters through the mouth.	Rotatory movements of the cilia of wheel organ cause a constant flow of water current through mouth.
6. **Course of water current**	Mouth → buccal cavity → pharynx → gill slits → branchial sacs → gill pores → exterior.	Mouth → buccal cavity → pharynx → gill slits of stigmata → atrial cavity → atriopore → exterior.	Mouth → vestibule → enterostome → pharynx → atrial cavity → atriopore → exterior.
7. **Sieves to strain larger food particles**		Branched branchial tentacles of the branchial sphincter work as sieve to strain larger food particles.	Buccal cirri and velar tentacles work as sieves preventing larger food and sand particles to enter into the pharynx.

(*Contd.*)

Characters	*Balanoglossus* (Hemichordata)	*Herdmania* (Urochordata)	*Branchiostoma (Amphioxus)* (Cephalochordata)
8. **Fate of food particles**	Ciliary actions of pharynx cause food particles to move backwards through ventral digestive part of pharynx and then into oesophagus for further course.	Ciliary actions of pharynx cause mucus-entangled food particles from ventral endostyle to pass dorsal along pharyngeal wall and then to the dorsal lamina, and finally into the oesophagus for further course.	Ciliary actions cause food particles entangled in mucus to pass dorsally in the epipharyngeal groove and finally into the oesophagus for further course as food mucus cord.
9. **Reverse current**	Not noticed.	Usually noticed; larger food particles and unwanted water current are regulated out through mouth (branchial aperture) by producing strong reverse current by muscular contraction.	Most often, the atriopore closes and atrial floor suddenly contracts, forcing the larger food particles and water to pass out through mouth.
10. **Egestion**	Sand, mud and undigested food pass out through anus as castings like those of earthworms	Undigested matter is expelled out through atrial aperture by sudden contraction of body muscles.	Undigested matter is expelled out directly through anus.

TABLE 8.7 COMPARISON : EXCRETORY SYSTEM OF THREE PROTOCHORDATE TYPES.

Characters	*Balanoglossus* (Hemichordata)	*Herdmania* (Urochordata)	*Branchiostoma (Amphioxus)* (Cephalochordata)
1. **Excretory organs**	Glomerulus or proboscis gland is the only excretory organ.	Neural gland is the main excretory organ.	90 – 100 pairs of protonephridia are the main excretory organs.
2. **Situation**	It is situated at the base of proboscis in front of central sinus and projects into proboscis coelom.	It lies mid-dorsally embedded in mantle above the nerve ganglion.	These are present segmentally on dorso-lateral pharyngeal wall one above each gill-slit on either side.
3. **Structure**	Glomerulus is made of several blind tubular projections lined by peritoneum and being filled with blood.	Neural gland is an elliptical structure and consists of branched peripheral tubules and central tubule leading into a small duct which opens anteriorly by a large ciliated funnel at the base of dorsal tubercle.	Each nephridium is a bent tube-like structure being made of a vertical limb terminating blindly and a horizontal limb opening into atrium by a small nephridiopore.
4. **Solenocytes**	Absent.	Absent.	The outer surface of nephridium is thrown out into several small projections, each receiving a tuft of solenocytes or flame cells.

(*Contd.*)

Characters	*Balanoglossus* (Hemichordata)	*Herdmania* (Urochordata)	*Branchiostoma (Amphioxus)* (Cephalochordata)
5. **Physiology**	The excretory wastes from the blood are extracted by the excretory peritoneal cells of the glomerulus and pass on into the proboscis coelom and finally to the exterior by proboscis pore.	The nephrocytes, main excretory cells collect excretory wastes from the blood and finally discharge into the prebranchial zone of pharynx through the duct of nerural gland.	Nitrogenous wastes are extracted by solenocytes from blood and coelomic fluid discharged by nephridiopores into the atrium and passed out of body through atriopore and outgoing water current.
6. **Secretion of hormones**	—	Neural gland secretes a hormone that controls oviposition, development and metamorphosis; considered homologous to vertebrate pituitary gland.	—
7. **Accessory excretory organs**	Absent.	Absent.	Hatschek's nephridium lies in the roof of oral hood. A pair of brown funnels located dorso-laterally upon pharynx and renal papillae present over atrial floor.

Table 8.8. Comparison : Nervous System of Three Protochordate Types.

Characters	*Balanoglossus* (Hemichordata)	*Herdmania* (Urochordata)	*Branchiostoma (Amphioxus)* (Cephalochordata)
1. **Nervous system in general**	It is primitive type, similar to those of coelenterates and echinoderms.	It is simple and degenerate in adult; comparatively well built in larval stage.	It is simple but better developed than the first two types.
2. **Composition**	It consists of sub-epidermal plexus or layers of nerve cells and nerve fibres, forming two main nerve cords, one mid-dorsal and other mid-ventral.	It consists of a solid nerve ganglion lying below the neural gland mid-dorsally in the mantle between the two siphons. Nerve cord is not found in adult instead it is present in the larva.	It consists of a hollow nerve cord lying mid-dorsally above the notochord from anterior to the posterior end of the body.
3. **Brain**	Absent.	Nerve ganglion represents the so-called brain or cerebral ganglion.	Nerve cord is slightly dilated anteriorly to form the so-called brain or cerebral vesicle.
4. **Nerves**	No definite nerves.	The nerve ganglion gives out 3 nerves to branchial siphon region and 2 nerves to atrial siphon region. Spinal nerves not found.	2 pairs of sensory cerebral nerves and several pairs of spinal nerves having separate dorsal and ventral roots.
5. **Autonomic nervous system**	Absent.	Absent.	Present. It consists of two nerve plexus in the gut wall and controls the involuntary muscles of the gut wall. No sympathetic ganglia.

TABLE 8.9. COMPARISON : CIRCULATORY SYSTEM OF THREE PROTOCHORDATE TYPES.

Characters	*Balanoglossus* (Hemichordata)	*Herdmania* (Urochordata)	*Branchiostoma (Amphioxus)* (Cephalochordata)
1. **Status**	Open or lacunar type.	Closed type and highly specialised.	Closed type.
2. **Composition**	It consists of blood, a central sinus and a heart vesicle, arteries and sinuses and veins.	It consists of blood, heart and pericardium and blood vessels.	It consists of blood, sinus venosus, ventral aorta, dorsal aorta and arteries and veins.
3. **Blood**	Colourless fluid, respiratory pigment probably absent.	It is slightly reddish, transparent having a few colourless amoeboid leucocytes and many types of coloured corpuscles.	It is colourless.
4. **Central sinus**	Small elongated non-contractile structure situated in proboscis.	—	—
5. **Heart and pericardium**	There is a triangular cardiac sac or heart vesicle just above buccal diverticulum in the proboscis.	Heart is a thick-walled, tubular muscular structure opening at both the ends. It is enclosed in a non-contractile, elongated transparent pericardium.	Heart absent, blood vessels are muscular and contractile.
6. **Sinus venosus**	—	—	Small, thin-walled, sac-like, situated just below the posterior end of midgut diverticulum.
7. **Blood vessels**	Distributing vessels are mid-dorsal and mid-ventral proboscis arteries, and a ventral blood vessel. Collecting vessel is a single median dorsal vessel which forms ventral sinus at the anterior end of collar.	Ventral aorta, dorsal aorta, branchio-visceral vessel and cardiovisceral vessels are the four main blood vessels. All these work as distributing as well as collecting vessels.	Ventral aorta and dorsal aorta are distributing vessels, while anterior and posterior cardinals, parietal and hepatic portal veins are collecting vessels.
8. **Ductus Cuvieri**	Absent.	Absent.	Present.
9. **Portal systems**	Absent.	Absent.	A primitive type of only hepatic portal system present.
10. **Flow of blood**	Unidirectional.	Periodically reversed.	Unidirectional.

Table 8.10. Comparison : Reproduction of Three Protochordate Types.

Characters	*Balanoglossus* (Hemichordata)	*Herdmania* (Urochordata)	*Branchiostoma (Amphioxus)* (Cephalochordata)
1. **Modes of reproduction**	Asexual reproduction by regeneration and sexual reproduction.	Only sexual reproduction.	Only sexual reproduction.
2. **Sexes**	Separate, *i.e.*, dioecious.	Hermaphrodite.	Dioecious but no sexual dimorphism.
3. **Gonads**	Gonads, testes of male and ovaries of female are similar. They are sac-like bodies occurring in several longitudinal rows in the genital wings, on either side of the branchial region.	There are two large gonads; right gonad lies parallel and dorsal to pericardium, while left gonad within the intestinal loop. Each gonad is an elongated, lobulated and hermaphrodite. There are 10–25 distinct oval lobes arranged in two rows one on either side of a central axis. Each lobe is made of a larger, bright-red, cortical, peripheral testicular region and a smaller, pinkish, medullary inner ovarian region. Each lobe of gonad is, thus, hermaphrodite.	There are 26–27 pairs of similar gonads arranged-metamerically in two rows, one pair in each segment from 25 to 51. These are situated ventro-laterally from middle of pharyngeal region to the atriopore. The gonads are simple hollow sac-like.
4. **Gonoduct**	Each gonad has a narrow neck or ductule that opens out through a gonopore.	Each gonad has two gonoducts —oviduct and spermatic duct running along the central axis.	Absent, mature gametes are discharged into the atrium by dehiscence of gonadial wall at cicatrices (certain weak spots).
5. **Fertilisation**	External.	External.	External.
6. **Development**	Indirect, free-swimming tornaria larva.	Indirect, free-swimming tadpole larva.	Indirect, free-swimming larva.
7. **Metamorphosis**	Progressive.	Retrogressive.	Progressive.

IMPORTANT QUESTIONS

Long Answer Questions

1. Describe a colonial tunicate.
2. Describe asexual reproduction in ascidians.
3. Give a comparative account of the excretory system of protochordates studied by you.
4. Describe the structure and function of pharynx in *Herdmania* and compare it with that of *Balanoglossus* and *Branchiostoma*.
5. Compare the food and feeding habits of protochordates.

Short Answer Questions

1. Write the three common chordate characters.
2. What are protochordates? Discuss their affinities.
3. Give a general comparison of three protochordata subphyla in a tabular form.

Multiple Choice Questions

1. Post-anal tail is present in
   (*a*) Urochordates
   (*b*) Cephalochordates
   (*c*) Hemichordates
   (*d*) All protochordates.

2. Test is present in
   (*a*) Urochordates
   (*b*) Cephalochordates
   (*c*) Hemichordates
   (*d*) All protochordates.
3. Metamerically segmented muscles are present in
   (*a*) Cephalochordates
   (*b*) Urochordates
   (*c*) Hemichordates
   (*d*) All chordates.
4. Autonomic nervous system is present in
   (*a*) Cephalochordates
   (*b*) Hemichordates
   (*c*) Urochordates
   (*d*) All protochordates.
5. In *Balanoglossus* each branchial sac opens to the exterior
   (*a*) Directly
   (*b*) By a single gill-pore
   (*c*) Through the pharynx
   (*d*) By a single gill-slit.
6. In *Herdmania* the opening of branchial aperture into the pharynx is guarded by
   (*a*) Branched branchial tentacles
   (*b*) Unbranched tentacles
   (*c*) Branched and ciliated tentacles
   (*d*) Unbranched pharyngeal tentacles.
7. Development in Urochordata is
   (*a*) Direct
   (*b*) Indirect with retrogressive metamorphosis
   (*c*) Indirect with metamorphosis
   (*d*) Indirect without metamorphosis.
8. In Urochordata, the pyloric gland is
   (*a*) Digestive as well as excretory
   (*b*) Only respiratory
   (*c*) Only excretory
   (*d*) Only digestive.
9. In *Branchiostoma* rectum is
   (*a*) Not differentiated
   (*b*) Differentiated but not ciliated
   (*c*) Differentiated and sparsely ciliated
   (*d*) Differentiated and heavily ciliated.
10. In Hemichordata, the function of excretion is chiefly performed by
    (*a*) Pyloric gland
    (*b*) Proboscis gland
    (*c*) Liver
    (*d*) Neural gland.
11. In cephalochordates
    (*a*) Only cerebral nerves are present
    (*b*) Only spinal nerves are present
    (*c*) Both cerebral and spinal nerves are present
    (*d*) Nerves are absent.
12. In cephalochordates, the brown funnels situated near pharynx are
    (*a*) Respiratory (*b*) Excretory
    (*c*) Sensory (*d*) Circulatory.

ANSWERS

1. (*b*) 2. (*a*) 3. (*a*) 4. (*a*) 5. (*b*) 6. (*a*) 7. (*b*) 8. (*a*)
9. (*d*) 10. (*b*) 11. (*c*) 12. (*b*).

Chapter 9

Subphylum 4. Vertebrata : General Characters, Origin and Ancestry of Vertebrates

Contents of this Chapter

WHAT ARE VERTEBRATES ?

All the vertebrates are placed under the subphylum Vertebrata of the phylum Chordata. The name of Vertebrata is derived from the presence of serially arranged **vertebrae** (L., *vertebratus = jointed*) which constitute a major part of their axial exoskeleton, vertebral column or the backbone. Another important feature, that all the vertebrates share as a common diagnostic character is the presence of anterior skeletal **cranium** or **skull** which houses brain and various sense organs. The presence of cranium gives another name, the **Craniata** which is also sometimes used for the group. There is reason to believe that the distinctive vertebrate cranium and brain evolved even before the vertebral column and are, therefore, more fundamentally characteristic of vertebrates than the backbone.

A vertebrate may be defined as a special kind of chordate animal that has a cartilaginous or bony endoskeleton consisting of a cranium, housing a brain and a vertebral column through which the nerve cord passes. These two fundamental and related characters which have existed in vertebrates are not found in any other group of animals except the vertebrates since the late Cambrian and Ordovician.

CHORDATE *VERSUS* VERTEBRATES

Many people usually think chordates as synonymous with vertebrates. Now it is clear that the terms chordate and vertebrate are not synonymous. All the vertebrates are included in the subphylum Vertebrata. The vertebrates make up the vast majority of animals in the phylum Chordata. The vertebrates possess three important characteristics : dorsal hollow nervous system, notochord and pharyngeal gill-slits, with some inconspicuous and primitive chordates belonging to the subphyla Urochordata and Cephalochordata and doubtfully the Hemichordata. Therefore, the vertebrates and the protochordates together comprise the chordates. The terms 'Euchordata' and 'Craniata' can be regarded equivalent to Vertebrata.

GENERAL CHARACTERS OF SUBPHYLUM VERTEBRATA

All vertebrates have at least traces of **vertebrae**. This is one of the most important characteristic feature that gives the subphylum its name. Another important diagnostic feature is the presence of a skull or **cranium** from which its another name **Craniata** is derived for the subphylum. Besides, the vertebrates have also possess the three fundamental or diagnostic chordate characteristics : the **notochord**, **pharyngeal gill-slits** and **dorsal hollow nerve cord**. These are the **five** important diagnostic characteristics of vertebrates. The others are only satellite characteristics not necessarily unique among vertebrates.

1. Lower vertebrates aquatic, higher vertebrates predominantly terrestrial.

2. Body size medium to large, bilaterally symmetrical and metametrically segmented.

3. Body typically consists of head, trunk and a postanal tail. Especially in terrestrial forms a neck may also be present.

4. Trunk bears typically two pairs of jointed lateral appendages, which may be reduced or absent in some forms, serve for support, locomotion and other special functions.

5. Integument or body covering is a stratified epithelium consisting of an outer epidermis and an inner dermis; with many mucous glands in aquatic species.

6. Skin covered by a protective exoskeleton comprising scales, feathers, hairs, claws, nails, horns, etc.

7. Coelom large, nearly always developed as a shizocoel, and largely filled with visceral organs.

8. Notochord invested by cartilage or bone or replaced by a vertebral column.

9. Endoskeleton living, jointed; formed of bone or cartilage or both; consists of skull, vertebral column, girdles and limb bones.

10. Several muscles are attached to the endoskeleton for motion and locomotion.

11. Digestive canal more or less convoluted. Liver massive and tubular. Pharyngeal gill-slits not more than 7 pairs except in some cyclostomes.

12. Respiration by gills in lower aquatic forms and by lungs in terrestrial forms.

13. Blood vascular system closed. Heart ventral, muscular, contractile and consists of two, three or four chambers. Blood plasma contains both white and red corpuscles. The red blood corpuscles contain the respiratory pigment, haemoglobin.

14. Excretion by paired kidneys, mesonephric or metanephric, segmental or non-segmental, and discharging through ducts into cloacal or anal region.

15. Anterior end of dorsal nerve cord enlarges to form a complex brain which is protected by skull. Remaining nerve cord forming the spinal cord which is surrounded and protected by vertebrae, 10 to 12 pairs of cranial nerves in the head. Except in cyclostomes, the spinal nerves formed by the union of dorsal and ventral nerve roots, which are given out in each segment. In lower vertebrates, the dorsal and ventral nerve roots are separate like cyclostomes. In higher vertebrates, the dorsal and ventral nerve roots are united to form a common spinal nerve. The ventral nerve is efferent or motor in nature, it carries nerve impulse from the central nervous system to the efferent organs. The dorsal nerve is mixed and it bears a swelling called spinal ganglion.

16. Special sensory organ include a pair of eyes and a pair of auditory organs, derived in part from brain.

17. An endocrine system of ductless glands scattered through the body, regulate body processes such as growth and reproduction.

18. Sexes separate. Gonads paired discharging sex cells through genital ducts opening into or near the anus.

19. Development direct or indirect. There is never a typical invaginate gastrula. Mesoderm arises as paired longitudinal bands, which subsequently become segmented.

DIVERSITY OF VERTEBRATES

In many respects, vertebrates are the most important and most progressive of all groups of organisms. They are highly diverse group of organisms showing many interesting features. They also play important ecological roles.

Numerical strength. Subphylum Vertebrata is the largest of all chordate groups. About 49,000 species of animals which exhibit vertebral columns. Of these 40,000 species are living, the remaining extinct.

Kinds of vertebrates. The vertebrates are of different kinds. These are familiar backboned animals such as fishes, amphibians, reptiles, birds and mammals. They also include man, who is the supreme animal and the most potent force on the earth today.

Categories of vertebrates. The vertebrates may be grouped in various ways. The general categories of vertebrates are given in Table 9.1. Commonly nine classes of vertebrates are recognised, including two (Ostracodermi and Placodermi) that are extinct. The first two classes of vertebrates (Ostracodermi and Cyclostomata) lack jaws and are placed in the group Agnatha. The

TABLE 9.1. GENERAL CATEGORIES OF VERTEBRATES.

Groups	Superclasses	Classes with examples
I. Agnatha		1. Ostracodermi–extinct armoured fishes
		2. Cyclostomata–lampreys and hagfishes
II. Gnathostomata	(*a*) Pisces	3. Placodermi–extinct fishes with primitive jaws
		4. Chondrichthyes–cartilaginous sharks, rays and skates
		5. Osteichthyes–bony fishes
	(*b*) Tetrapoda	6. Amphibia–frogs, toads, salamanders
		7. Reptilia–turtles, lizards and snakes
		8. Aves–birds
		9. Mammalia–mice, whales, men

remaining classes have jaws and they form the group **Gnathostomata**. The gnathostomes are further grouped into two superclasses: **Pisces** (fishes) and **Tetrapoda** (four legged land vertebrates).

TABLE 9.2. SELECTED CHARACTERISTICS OF LIVING VERTEBRATE CLASSES.

<table>
<tr><th rowspan="2">Classes</th><th rowspan="2">Jaws</th><th rowspan="2">Endoskeleton</th><th rowspan="2">Locomotory appendages</th><th rowspan="2">Respiratory surface</th><th rowspan="2">Extra-embryonic membranes</th><th rowspan="2">Body temperature energetics</th><th colspan="2">Integument</th></tr>
<tr><th>Glands</th><th>Exoskeleton</th></tr>
<tr><td>Cyclostomata</td><td>AGNATHA (jawless)</td><td rowspan="2">CARTILAGE</td><td rowspan="3">FINS (Fishes)</td><td rowspan="2">GILLS</td><td rowspan="4">ANAMNIOTA (yolk sac and chorion)</td><td rowspan="5">ECTOTHERMAL</td><td rowspan="4">GLANDULAR mucous secretions</td><td>naked</td></tr>
<tr><td>Chondrichthyes</td><td rowspan="6">GNATHOSTOMATA (jawed)</td><td>placoid scales</td></tr>
<tr><td>Osteichthyes</td><td rowspan="5">BONE</td><td>Primarily gills</td><td>dermal scales</td></tr>
<tr><td>Amphibia</td><td rowspan="4">LIMBS (Tetrapodas)</td><td>Primarily lungs</td><td>naked</td></tr>
<tr><td>Reptilia</td><td rowspan="3">LUNGS</td><td rowspan="3">AMNIOTA (yolk sac, chorion, allantois and amnion)</td><td rowspan="2">AGLANDULAR dry</td><td>epidermal scales</td></tr>
<tr><td>Aves</td><td rowspan="2">ENDOTHERMAL</td><td>feathers and epidermal scales</td></tr>
<tr><td>Mammalia</td><td>SECONDARY GLANDULAR oily and watery secretions</td><td>hair</td></tr>
</table>

SELECTED CHARACTERS OF CLASSES

All the nine classes of vertebrates share the common chordate-vertebrate characters. Nevertheless, each of these classes differs from the others in some fundamental ways, exhibiting a progressive series of changes from the primitive jawless fishes to the highly complex mammals. Some of the selected characteristics of the seven living classes of vertebrates have been given in Table 9.2.

PHYLOGENY OR EVOLUTIONARY HISTORY OF VERTEBRATES

Phylogeny is derived from Greek roots which mean "to beget a race": (Gr., *phylon* = race or tribe + *gennao* = to bring forth). The evolutionary history of a group of organisms is termed **phylogeny**. The concept of phylogeny is used to place animal groups in proper evolutionary sequence. Thus, in phylogenetic order, fishes came before reptiles and reptiles before birds and mammals.

The evolutionary history of vertebrates is shown by a figure 9.1 which appears to be a good model to record the evolutionary history of vertebrates. The vertical dimension or height of the tree represent geological time. The diversification into groups is represented by branches. The width of each branch indicates the relative abundance of each group in time. Figure 9.2 exhibits various diagrams representing a phylogenetic tree, reflecting different ideas of interrelationship of animal groups.

Cambrian and Ordovician periods. The first fossils of vertebrates were found in the rocks of the Ordovician period in the form of **Ostracoderms**. These were small jawless, bony fish-like forms related to cyclostomes, which lived some 480 million years ago. This shows that their chordate ancestors must have existed much before, in the late Cambrian period. The scarcity of early vertebrate fossils is probably due to the fact that they evolved mainly in freshwater and did not have as much chance to become fossilised as marine forms did. The Ostracodermi became extinct but some Cyclostomata (modern lampreys and hagfishes) are still existing.

Silurian and Devonian periods. In Silurian period some fossil fish are found but far more are found in the succeeding Devonian period. Devonian period is known as the **Age of Fishes**. Ostracoderms were jawless fish, but during Devonian period the first jawed fish, the **placoderms**, arose. The **Placodermi** became extinct without leaving living representatives. It is likely that early placoderms were ancestors of cartilaginous and bony fishes, both of which have true jaws, but we do not have any direct evidence. Contrary to the former belief, cartilaginous fishes (**Chondrichthyes**) did not give rise to bony fishes (Osteichthyes).

Carboniferous period. In the late Devonian or early Carboniferous the lobed finned bony fishes (Cropssopterygii) gave rise to **labyrinthodonts** or primitive **stem Amphibia**. They were the first vertebrates to walk on land. The Amphibia became abundant and mutated in many directions during Carboniferous period which is usually known as the **Age of Amphibians**.

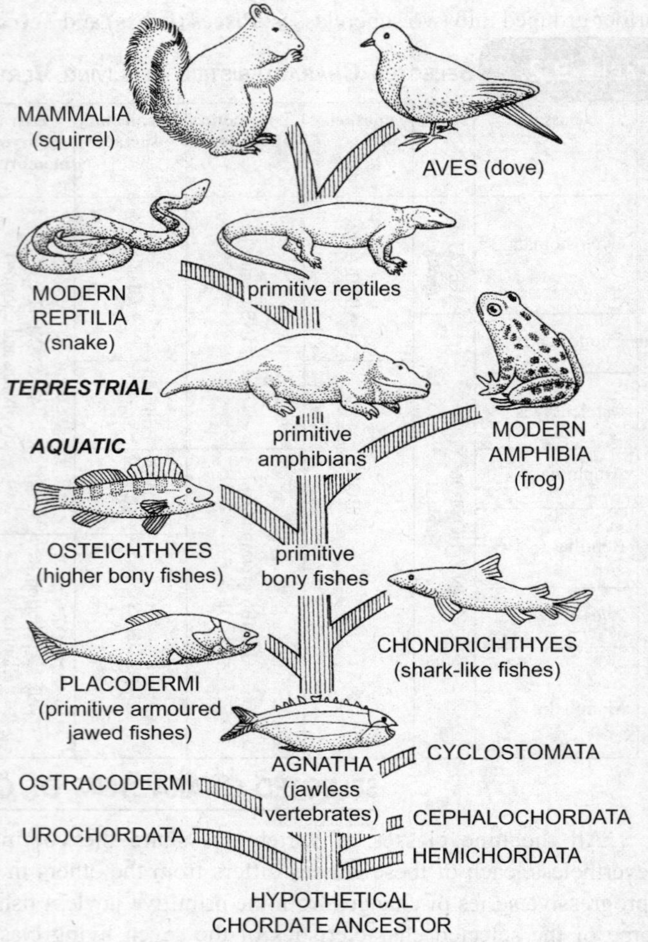

Fig. 9.1. A simple phylogenetic tree of vertebrates showing classes.

Mesozoic Era. During the early Carboniferous period, the very primitive amphibians also gave rise to the **primitive reptiles**. They reached their peak and great abundance during Mesozoic era. Mesozoic era which is appropriately known as the **Age of Reptiles**. They included among others the famed dinosaurs, the ichthyosaurs and the pterodactyles or flying reptiles. They dominated the world for nearly 130 million years until the end of Mesozoic era when most of them suddenly became extinct. Today only 4 orders of reptiles are living, while as many as 10 orders are represented by their fossil remains.

The **ancestral mammals** were derived from the primitive reptiles during Triassic period. The **first birds** also appeared in the late Jurassic period and one of their fossils, the *Archaeopteryx*, had both reptilian as well as avian characteristics.

Cenozoic era. After the decline of the reptiles during the late Mesozoic era, both birds and mammals started flourishing. The mammals became the most diversified of all animals during Cenozoic era. Cenozoic era is also called the **Age of Mammals**. This era started nearly 70 million years ago and is divided into two periods : Tertiary and Quaternary. After reaching their peaks in Tertiary, the mammals are steadily declining in numbers of species even though man, a member of this group is considered to be the dominant form of life on earth today.

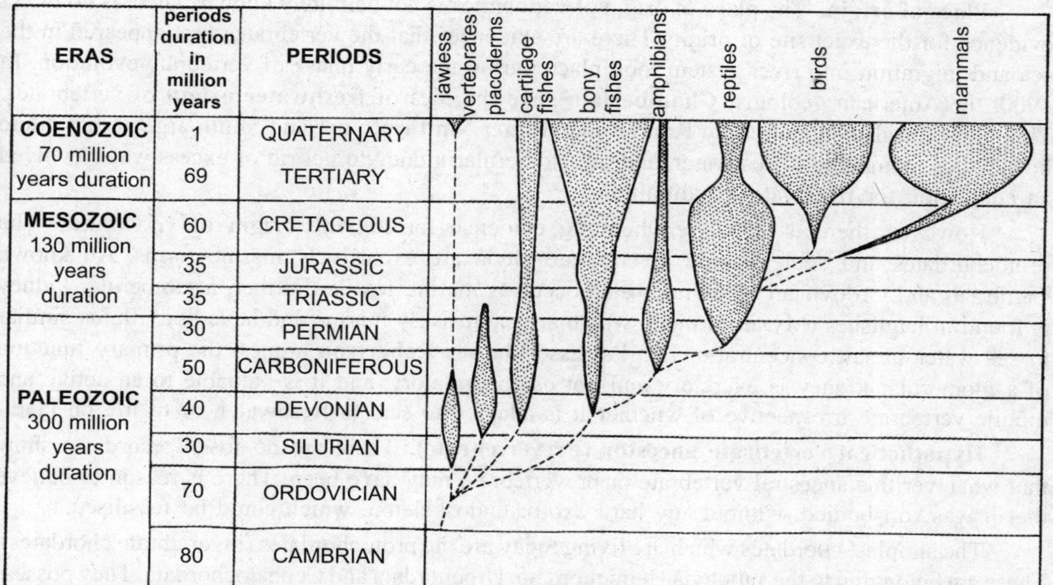

Fig. 9.2. Geological chart with a simple family tree of vertebrate classes showing their relative abundance through time.

ORIGIN AND ANCESTRY OF VERTEBRATES (CHORDATES)

The problem of searching into origin and ancestry of vertebrates has long been of particular interest to zoologists. But, like that of most of the animal phyla, the origin of vertebrates also remains obscure. Over the years, several hypothesis have been proposed to explain the origin of vertebrates, but none could stand the test of close scrutiny.

Time of origin. The chordates constitute very ancient forms. The earliest known truly vertebrate animals were freshwater forms which were abundant during the late Silurian and middle Devonian periods. Their fossils are collectively known as **ostracoderms.** Ostracoderms are placed with living cyclostomes (lampreys and hagfishes) in the jawless group called **Agnatha**. Ostracoderms, as their name refers to, have their body covered by a dermal bony armour which forms an elaborate rounded solid shield on the head. Like cyclostomes, they had presumably a persistent notochord and no vertebrae. The mouth was anterior, ventral and devoid of jaws and teeth. They had no paired appendages homologous with those of vertebrates. Paired lateral eyes, median nostril and a median pineal eye were present on the top of the head. A variable number of pharyngeal gill-pouches opened by lateral common or separate gill-openings. They were adapted for filter feeding.

The fragmentary remains of the invertebrate chordates have been recorded in the Cambrian strata and that of ostracoderms in the middle Ordovician time. The absence of any vertebrate fossils in rocks older than the Cambrian period, permits only speculation about the earlier history of the vertebrates. Which group was ancestral to the first true vertebrates (ostracoderms)? In fact, there have been no fossils intermediate between ostracoderms, which are already vertebrates, and any other earlier group of animals. As a result, there has been a great deal of speculation about the time of origin and the early progenitors of the vertebrates (chordates). Probably the vertebrate organisation had been evolving for several millions of years before the appearance of the first, late Cambrian fossils. It can be said that the time of origin was not later than the beginning of the Ordovician, if not long before.

Place of origin. The place of origin of vertebrates is not definitely known. There is no direct evidence for the exact site of origin. There are evidences that the vertebrates first appeared in the sea and migration into river system took place during the early phase of vertebrate evolution. In 1900, the American geologist **Chamberlain** gave the idea of **freshwater origin** of vertebrates. This idea was also supported by **Romer** and **Homer Smith**. Romer and Smith argued that dilute body fluids, compared to sea water, and the glomerular kidney to get rid of excess water evolved an adaptation for freshwater conditions.

However, there is also overwhelming evidence for **marine origin** of vertebrates. The protochordates and deuterostome invertebrate phyla are exclusively marine forms. All known Cambrian and Ordovician vertebrates also occur as marine fossils. Further, a glomerular kidney is found in hagfishes (Myxiniformes), which are exclusively marine and have body fluids similar to sea water in salt concentration. As Professor **James Robertson** argues, the primary function of a glomerular kidney is excretory and not osmoregulatory, and it is valuable to an active and mobile vertebrate irrespective of whether it is adapted to sea, to freshwater, or to life on land.

Hypothetical vertebrate ancestor (Prevertebrate). There are no fossil records to show that whatever this ancestral vertebrate or prevertebrate may have been. There is reason to believe that it was soft-bodied, without any hard exo-or endoskeleton, which could be fossilised.

The simplest chordates which are living today are the protochordates (invertebrate chordates). These are belonging to the subphyla Hemichordata, Urochordata and Cephalochordata. They possess the notochord, dorsal nerve cord, pharyngeal gill-slits and post-anal tail. They lack the vertebrae and some other features of the earliest as well as the living vertebrates, but they exhibit closest affinities and certainly a common origin with the vertebrates. Thus, it seems most reasonable and logical to draw inferences about the imaginary, generalised or ancestral vertebrate among them.

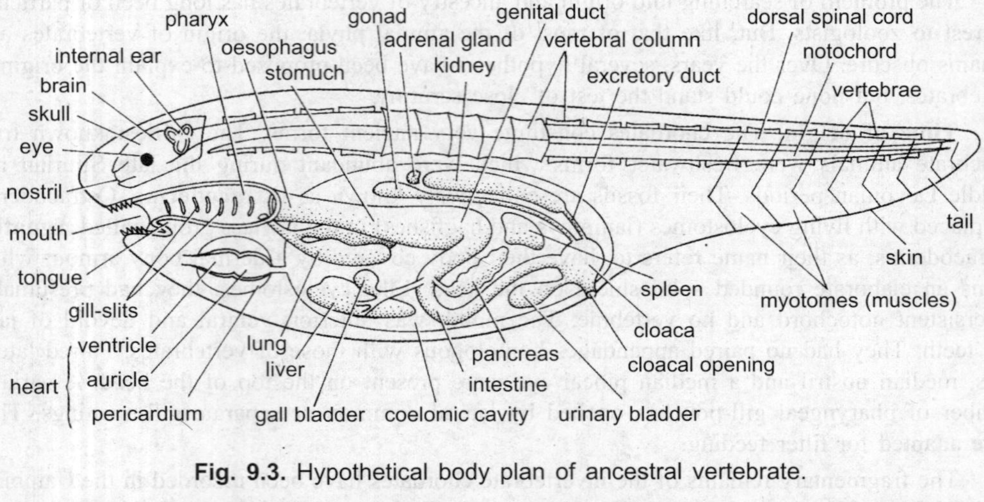

Fig. 9.3. Hypothetical body plan of ancestral vertebrate.

The American geologist **Chamberlain,** who proposed the theory of freshwater origin of vertebrates in 1900, also proposed the plan of a hypothetical protovertebrate (Fig. 9.3). According to Chamberlain, it was an aquatic, motile, actively swimming fish-like animal having a bilaterally symmetrical body with definite head and tail ends. As in higher invertebrates and chordates, the basic internal organisation would be some sort of modified tube-within-a-tube arrangement with the major internal organs present inside a large body cavity or coelom. It would also possess all the diagnostic chordate-vertebrate features. It possessed an internal skeleton in the form of mid-dorsal, longitudinal flexible rod, the notochord, surrounded by the vertebrae. There were internal supports in fins, bony plates in the skin and a rigid cranium or skull encasing the brain and sense organs. V-shaped muscles were arranged segmentally along the sides of the body, particularly the tail. These muscles were forming the myotomes which were used in locomotion. There was a simple anteroventral opening without jaws called the mouth. Probably, it fed on microorganisms, filtered from the water or from bottom detritus. The water taken in through the mouth passed outwards through paired lateral pharyngeal gill-slits. The passing out water bathing the external gills for aquatic respiration. A liver and primitive kidneys were present. The presence of a circulatory system with a single differentiated heart in prevertebrate is uncertain. The sense organs including the lateral line organs were well developed. The endocrine system was probably already well developed in the prevertebrate. The gonads were paired. The fertilisation of eggs was external.

It is thought that from such beginnings the vertebrates were evolved.

Origin of chordates. Here we shall discuss the origin of earlier chordate ancestors of the vertebrates. That chordates have originated from the invertebrates is not questioned by most zoologist now-a-days. Since the earlier chordate ancestors were soft-bodied forms, they left no fossil remains to us clues as to their origin. Therefore, only basis for finding out the origin of earlier chordates is available from the resemblance between the lower chordates (protochordates) and the invertebrates. There are some structural features shared by them, such as bilateral symmetry, antero-posterior body axis, triploblastic coelomate condition, metameric segmentation, etc., may be considered as basis of their common ancestry.

Theories of invertebrate ancestry of chordates. Similarities existing between some invertebrates and the chordates have led to the enunciation of several theories on the origin of chordates. All these theories postulate that the chordates originated either directly from some invertebrates or through the intervention of some invertebrate chordates (protochordates). Almost every invertebrate phylum—Coelenterata, Nemertean, Phoronida, Annelida, Arthropoda and Echinodermata—has been suggested. But these theories are far from being satisfactory and convincing and have only a historical value. Only the **echinoderm theory** has received some attention and acceptance and shall be considered and evaluated under deuterostome line of chordate ancestry.

Division of Bilateria. The Bilateria is divided into two major subdivisions: **Protostomia** and **Deuterostomia**. The basis of division is the basic difference in embryonic and larval developments. Probably this division represent two main lines of evolution within the Animal Kingdom. The main differences between Protostomia and Deuterostomia are given in Table 9.3.

TABLE 9.3. BASIC DIFFERENCES BETWEEN PROTOSTOMIA AND DEUTEROSTOMIA.

Character	Protostomia	Deuterostomia
1. Cleavage	Spiral and determinate	Mostly radial, and indeterminate
2. Blastopore	Forms mouth	Forms anus
3. Mesoderm formation	By cell cleavage between ectoderm and endoderm	By outpocketing from dorso-lateral endodermal wall of archenteron
4. Coelom formation	Schizocoelous, by split of mesoderm	Enterocoelous, by fusion of gut pouches (except vertebrates)
5. Type of larva	Trochophore	Tornaria or Bipinnaria (except vertebrates)
6. Phosphagen	Arginine	Creatine
7. Major phyla included	Annelida, Mollusca and Arthropoda	Echinodermata, Pogonophora, Hemichordata and Chordata.

Deuterostome line of Chordate Evolution. Table 9.3 exhibits common features of all deuterostomia. These common features suggest strong evidence of embryological and biochemical nature of a closer evolutionary relationship between three main deuterostome phyla: Echinodermata, Hemichordata and Chordata. These common features are as follows :

1. Early cleavages of zygote are indeterminate, *i.e.*, each early blastomere is capable of developing into a whole adult if separated.

2. Blastopore of gastrula forms the anus, while mouth is formed as a secondary opening.

3. Folds or pockets arise from the endoderm of developing archenteron of the embryo. The fusion of spaces in the pockets forms the **coelom** (enterocoelous, except in vertebrates) and their walls become the mesoderm.

4. The pelagic larva of echinoderms and hemichordates exhibit a close structural resemblance. The vertebrates, however, do not have floating larvae, having been lost in the course of evolution.

5. Biochemically, all deuterostome use an identical phosphagen, the **creatine**, in the energy cycle of their muscular contraction. The invertebrates use the phosphagen, the **arginine**. However, certain hemichordates as well as echinoids use both arginine phosphate as well as creatine phosphate. From these facts, it is interpreted that hemichordates are connecting link between chordates and nonchordates.

6. Serological tests demonstrate that the proteins of the three deuterostome phyla (Echinodermata, Hemichordata and Chordata) are more closely related to one another than those of any other phyla.

The exact relatonship of the three deuterostome phyla (Echinodermata, Hemichordata and Chordata) is still unknown, but there is little doubt that they have a common evolutionary history. Several workers have tried to explain the deuterostome line of chordate evolution. Some of the proposals are as follows :

1. Echinoderm ancestry (Echinoderm theory). On the basis of anatomical, embryological, palaeotological, biochemical and serological evidences, various workers had tried to establish that the chordates probably had originated directly from some primitive echinoderm or some echinoderm larva. The tornaria larva of *Balanoglossus* (Hemichordata) and the larvae of echinoderms (bipinnaria or dipleurula) exhibit close similarity (Fig. 9.4). **Johannes Muller** and **Bateson** suggested that the tornaria larva (Hemichordata) and dipleurula larva (Echinodermata) have evolved from a common ancestral source. The common features are : (*i*) The body is minute and transparent. (*ii*) Both have a dorsal pore. (*iii*) Almost identical twisted external ciliated bands are present. (*iv*) Formation and disposition of coelom are similar. (*v*) Both are free-swimming forms. (*vi*) Body is bilaterally symmetrical. But the presence of apical plate with eye spots in tornaria larva raises doubts about the common ancestry for the echinoderms and the hemichordates. **Garstang** and **de Beer** proposed the **neotenous larva theory**. They suggested that probably the auricularis larva of echinoderms became sexually mature and later this neotenic larva gave rise to the chordates.

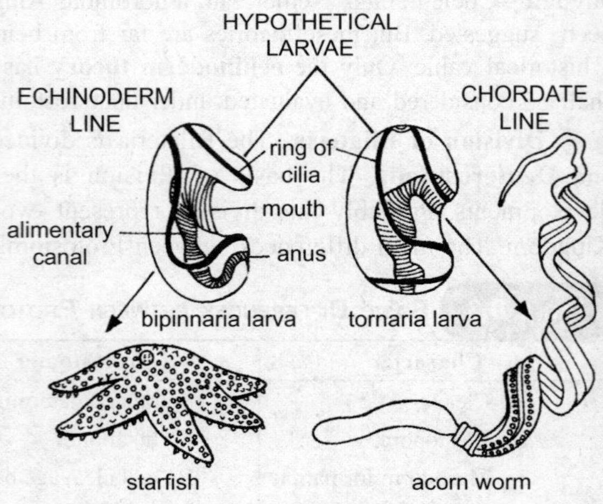

Fig. 9.4. Similarity of larval forms of echinoderms and hemichordates has supported the idea that both came from a common ancestor.

Garstang (1894) also imagined that if the ciliated bands together with the underlying nervous tissue of auricularia larva of echinoderms, concentrates to form ridges leaving a groove between them and if the lips of the groove fuses subsequently, it will give rise tube. It will resemble with the nervous system of chordates.

The most primitive echinoderms are that of Cambrian and Ordovician Carpoid echinoderms. **Torsten** and **Gislen** assumed that the Carpoid echinoderms might have evolved from tornaria-like creatures which have begun to settle down to lead a sedentary life. The water vascular system might have developed out of the ciliated grooves of the tornaria-like creatures. It has also been claimed that in the lower Silurian, one of the Carpoid echinoderms had the calyx perforated by a series of 16 small apertures. These apertures are compared with the gill-slits of *Branchiostoma*.

The phylogenetic relationship between the echinoderms and the chordates has been shown by biochemical studies. Most of the nonchordates conduct energy transfer with arginine phosphate, but ophiuroids, cephalochordates, ascidians and vertebrates use creatine phosphate. The hemichordates and echinoids use both arginine phosphate as well as creatine phosphate. **J. Needham** (1932) has shown the presence of phosphagen as the phosphorus-carrier in the echinoderms and chordates. **Wilhelmi** (1942) has shown the affinity between two groups by serological tests. All these isolated biochemical studies put weight on the concept of derivation of chordates from echinoderms.

The descent of chordates from the echinoderms by the direct transformation of echinoderm or its neotenous larva into a chordate is no longer accepted now-a-days. The view most widely accepted at present is that the living echinoderms and the living chordates had in common an immediate ancestor.

2. Hemichordate ancestry (Balanoglossus ancestry). There is a strong suggestive evidence that the early evolutionary stage of Deuterostomia group was sessile or sedentary. The pharynx perforated by gill-slits, a characteristic feature of chordates, is also a likely adaptation to sedentary habit. The hemichordates are sedentary and possess pharyngeal gill-slits and a hollow dorsal nerve cord. But the presence of a true notochord is doubtful and their adult body plan is quite different from vertebrates. Therefore, the hermichorates as a likely ancestor of vertebrates seems to be impossible. Due to this fact, the hemichordates are put under separate phylum of their own.

3. Urochordate ancestry. The urochordate or ascidian theory of vertebrate origin was first proposed by **W. Garstang** in 1928. It was later elaborated by **N.J. Berrill** (1955) in his book "*Origin of Vertebrates,* **Romer** (1959) and others. The adult tunicates or ascidians exhibit the primitive sessile, marine and filter feeding condition of the ancestral chordates. But their body plans are so divergent that it is impossible to imagine a direct evolutionary transformation of an adult ascidian into a vertebrate. On the other hand, the ascidian larva are tadpole-like, elongated, bilaterally symmetrical and free-swimming forms with pharyngeal gill-slits, notochord, dorsal tubular nerve cord, and a muscular post-anal tail. They represent slightly modified living structure of the ancestral chordate that gave rise to the vertebrate line of evolution. According to this theory, some of these larvae of ascidians failed to metamorphose into adults, but became **neotenous**, *i.e.*, sexually mature by developing gonads precociously, and later evolved into cephalochordates and vertebrates.

However, the ascidian theory of chordate origin does not seem to be perfect. The main draw back is that this theory considers sessile nonchordates to be ancestral to chordates. Whereas, they are highly specialised because sessile condition is a specialised condition wherever it occurs in the Animal Kingdom.

4. Cephalochordate ancestry (Branchiostoma ancestry). The theory of origin of vertebrates from cephalochordates appears to be convincing, because they fulfil the theoretical definition of typical chordate organisation. The cephalochordates, particularly the lancelets (*Branchiostoma lanceolatum*), are an interesting group of animals. They possess the three basic chordate features: notchord, dorsal tubular nerve cord and the pharyngeal gill-slits. According to **Colbert**, the living *Amphioxus* (= *Branchiostoma*) possesses the logical structure of a model prevertebrate. **Homer Smith's** reconstruction of the hypothetical protovertebrate in his book *"From Fish to Philosopher"* (1953) also resembles *Amphioxus*. But the excretory system of cephalochordates consisting of flame cells called **solenocytes** is altogether different from that of vertebrates. The solenocytes are ectodermal in origin and, therefore, not homologous with the mesodermal origin of vertebrate kidneys. Further, cephalochordates lack strong cephalisation and sense organs and the unique extension of notochord indicate that the cephalochordate may provide hint about the likely ancestral body plan of vertebrates, but they are not themselves ancestral. Both may represent divergent path of evolution from a common ancestor.

5. Barrington's hypothesis. It is a most plausible hypothesis proposed by **E.J.W. Barrington** in 1965. This hypothesis is based on the deuterostome line of chordate evolution. The common echinoderm-chordate ancestor was in all probability a small, sessile or semisessile **arm feeding creature**. It fed by ciliary method by trapping food particles in a set of waving tentacles. From this ancestral stalk were derived early stalked echinoderms and pogonophores. The next was the derivation of a sessile **filter feeder** or **stem chordate**. The external tentacles were replaced by an internal filtering apparatus in which food is entrapped inside pharynx which develops external gill-slits and a mucus-secreting endostyle. *Cephalodiscus*, a living pterobranch hemichordate, shows the transitional stage between the two modes of feeding because it has a single pair of gill-slits besides the crown of tentacles. In course of time and space a group evolved which had specialised pharyngeal apparatus to assist internal food collection. This group holds the key of the origin of urochordates, cephalochordates and vertebrates. The hemichordates developed **pharyngotremy** that is perforated pharynx with internal food trapping mechanism. This resulted in the evolution of free-living hemichordates on one hand and the sessile ancestral urochordates (tunicates) on the other. Some ancestral tunicates, instead of producing ciliated larvae common to the earliest groups, formed tadpole larvae with all the typical somatic features of chordates. According to **Garstang**, the larva became elongated and increased in size, the longitudinal ciliary bands shifted mid-dorsally and changed to the hollow nerve cord, the adoral cilia developed into the endostyle, and the muscle fibres evolved in the tail. This typical chordate larva by paedogenesis suppressed the sessile adult stage, developed gonads precociously (premature) and became the ancestor of cephalochordates *(Branchiostoma)*, vertebrates and the larvaceans probably representing three cases of parallel evolution.

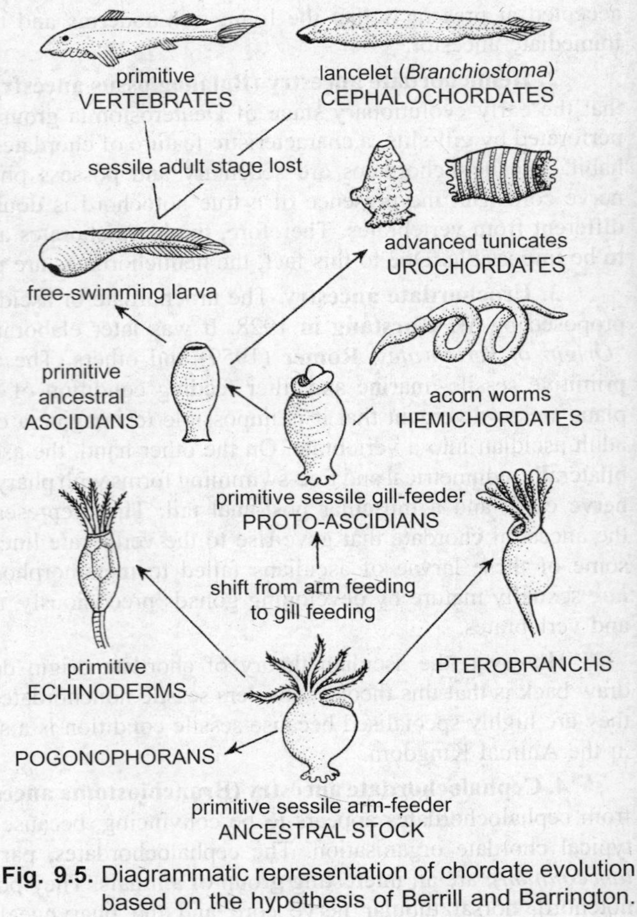

Fig. 9.5. Diagrammatic representation of chordate evolution based on the hypothesis of Berrill and Barrington.

IMPORTANT QUESTIONS

Long Answer Questions

1. Write an essay on origin and ancestry of vertebrates.
2. Give the evolutionary history of vertebrates.
3. What are vertebrates? Describe their general characters.

Short Answer Questions

1. Differentiate the lower vertebrates from higher vertebrates.
2. Write five diagnostic vertebrate characteristics.
3. Differentiate between protostomia and deuterostomia.

Very Short Answer Questions

1. In which class of vertebrate Ductus Botalli is found?
2. Name the chordate group in which epipubic bone is found.
3. Name the chordate group in which floccular is found.
4. Name the chordate group in which zygosphene is found.
5. What is the name of blood vessel which connects carotid and systemic arteries?

Multiple Choice Questions

1. Which of the following artery is absent in vertebrates other than pigeon?
   (*a*) Carotid (*b*) Branchial
   (*c*) Pulmonary (*d*) Pectoral.
2. Metanephros is the adult kidney of
   (*a*) Amniotes (*b*) Anamniotes
   (*c*) Agnatha (*d*) Gnathostomata.
3. The term vertebrate is synonymous to
   (*a*) Craniata (*b*) Chordata
   (*c*) Urochordata (*d*) Protochordata.
4. In vertebrates pharyngeal gill-slits are not more than
   (*a*) 6 pairs (*b*) 7 pairs
   (*c*) 8 pairs (*d*) 9 pairs.
5. Heart in vertebrates is
   (*a*) Dorsal (*b*) Ventral
   (*c*) Dorsolateral (*d*) Ventrolateral.
6. Respiratory pigment in vertebrates is
   (*a*) Haemocyanin (*b*) Haemoglobin
   (*c*) Chloroquin (*d*) Heparin.
7. Brain in vertebrates is
   (*a*) Dorsal to alimentary canal
   (*b*) Ventral to alimentary canal
   (*c*) Ventral to notochord
   (*d*) Dorsal to alimentary canal and notochord.
8. Period of Age of fishes was
   (*a*) Devonian (*b*) Silurian
   (*c*) Cambrian (*d*) Ordovician.
9. First jawed-vertebrates were
   (*a*) Placoderms (*b*) Ostracoderms
   (*c*) Amphibians (*d*) All of the above.
10. Era of Age of reptiles was
    (*a*) Mesozoic (*b*) Archaeozoic
    (*c*) Proterozoic (*d*) Paleozoic.
11. Connecting link between reptiles and birds was
    (*a*) *Apteryx* (*b*) *Archaeopteryx*
    (*c*) *Crocodilus* (*d*) None.
12. Era of age of mammals was
    (*a*) Cenozoic (*b*) Mesozoic
    (*c*) Paleozoic (*d*) Archaeozoic.
13. The first fossil records of vertebrates were found in the rocks of
    (*a*) Ordovician (*b*) Cambrian
    (*c*) Jurassic (*d*) Devonian.
14. The first tetrapods arose in
    (*a*) Ordovician period (*b*) Devonian period
    (*c*) Silurian (*d*) Carboniferous.
15. The "age of amphibians" is used for
    (*a*) Ordovician period
    (*b*) Silurian period
    (*c*) Carboniferous period
    (*d*) Devonian period.
16. Mammals arose during the
    (*a*) Cretaceous period (*b*) Jurassic period
    (*c*) Permian period (*d*) Triassic period.
17. Which phosphagen is utilised by vertebrates in energy cycles of muscle contraction?
    (*a*) Arginine (*b*) Creatine
    (*c*) Guanine (*d*) Cysteine.
18. In deuterostomes, blastopore forms
    (*a*) The mouth (*b*) Brain
    (*c*) The anus (*d*) Kidney.
19. Spinal nerves in vertebrates are formed by the union of
    (*a*) A pair of ventral roots
    (*b*) A pair of dorsal roots
    (*c*) Two dorsal and one ventral nerve root
    (*d*) A dorsal and a ventral nerve root.
20. Excretion of vertebrates takes place by
    (*a*) Paired mesonephros kidneys
    (*b*) Unpaired kidneys
    (*c*) Unpaired pronephron
    (*d*) Paired mesonephric and metanephric.

ANSWERS

1. (*a*)	2. (*a*)	3. (*a*)	4. (*b*)	5. (*b*)	6. (*b*)	7. (*d*)	8. (*a*)
9. (*a*)	10. (*a*)	11. (*b*)	12. (*a*)	13. (*a*)	14. (*d*)	15. (*c*)	16. (*d*)
17. (*b*)	18. (*c*)	19. (*d*)	20. (*d*).				

10 Chapter — Class 1. Ostracodermi (Extinct) : General Characters, Classification and Representative Types

Contents of this Chapter

The earliest known vertebrates to appear in fossil record were jawless primitive fish-like animals collectively known as the **ostracoderms**. These were placed under the class **Ostracodermi**. The class Ostracodermi is represented by the fossil vertebrates of late Cambrian and middle Ordovician periods. Although the fossils are frequent in the sea, there are ample reasons to believe that their progenitors were freshwater forms living near estuaries. The ostracoderms resembled the present day cyclostomes (lampreys and hagfishes) in many respects and together with them constitute a special group of jawless vertebrates, the **Agnatha**.

OCCURRENCE

Ostracoderms were first observed as fragmented fossils occurring in rocks of late Cambrian and middle Ordovician periods approximately 500 million years ago. They were quite abundant during the upper Silurian and Devonian periods. Most of fossils of Ostracodermi were preserved in the bottom sedıments of freshwater streams. However, the opinion is sharply divided as to whether their habitat was freshwater or marine.

IMPORTANT FEATURES

The ostracoderms were primitive vertebrates, small to medium in size. Their body form was fish-like, usually flattened dorsoventrally, with a huge head and gill region, a tapering but muscular trunk and some sort of tail fin. They had no jaws and no pectoral or pelvic fins but had only median fins. These earliest vertebrates were bony and heavily armoured. The bony plates and the scales of the ostracoderms were well preserved. The head was covered in a solid shield made of broad bony dermal plates, while the rest of the body surrounded by a series of smaller plates often called dermal scales. This has led to their names "ostracoderms" "armoured fishes" or bony skinned animals (Gr., *ostrakon* = shell; *derma* = skin). It has been suggested that the heavy endoskeleton served as a protection against the giant scorpion-like arthropods, the eurypterids which were the dominant predators of Cambrian, Ordovician and Silurian periods. Later, when these enemies disappeared, the jawed descendants of ostracoderms also lost their heavy armour which only hindered rapid progress.

The ostracoderm head had a pair of large lateral eyes and a median penial eye on the top. A single median nostril was located anterior to penial eye. The mouth was anteroventral, small and without jaws or teeth. The gill-slits were round and all had similar gill-pouches. Sensory fields on head were probably a part of the lateral line system. The flattened body and feeble fins suggest that they were probably bottom dwellers and filter feeders like most of the present day lower chordates. Very little is known about ostracoderm internal anatomy. The endoskeleton was moderately ossified. They had no axial endoskeleton or vertebrae. An internal ear with two semicircular canals was present.

CLASSIFICATION

All the fossils of ostracoderms have been put under a single class **Ostracodermi**. The Ostracodermi is divided into 2 subclasses and 5 suborders which are as follows :

Subclass I. Monorhina

A single, large, slit-like median nostril on the top of the head between the eyes. The Monorhina has been divided into three orders.

Order 1. Euphanerida. Silurian. This order is represented by a single genus and species *Jamoytius kerwoodi*, which was previously placed under order Anaspida by **Stensio** and **Ritche** (1960). **Newth** regarded it as ammocoete larva of an ostracoderm, discovered in the Silurian strata of England. *Jamoytius kerwoodi* measures about 18 cm in length. It had a very primitive body construction. The body was unarmoured, fusiform with a blunt head. The two lateral folds and a median dorsal fin, notochord and muscle myotomes were exactly like those of *Branchiostoma*. No gill-slits. Regarded as an ancestor of ostracoderms.

Order 2. Anaspida. This order includes the small fish-like ostracoderms. The fossils were recorded from Silurian to Devonian period. Body small about 15 cm long. Head naked or covered by a complex of small plates. Arrangement of scales was complicated. In the trunk region, the scales were oriented in longitudinal rows. Mouth was terminal. Eyes laterally placed. Nostril single median. Paired appendages were represented by a pair of pectoral spines. A series of spines was present in the mid-dorsal line of the trunk. An anal spine might have also been present. About 8 gill-slits were present in front of the pectoral spines in a slanting fashion. The tail hypocercal (lower lobe larger). They were active swimmers. Examples : *Birkenia*, *Lasanius*, *Pterolepis*, *Rhyncholepis* and *Pharyngolepis*.

Order 3. Osteotraci (Cephalaspida). These fishes were recorded from Silurian to Devonian period. The head was expanded, dorsoventrally flattened and was covered by a thick continuous cephalic shield (carapace) which was prolonged laterally into two lateral horns. The surface of the carapace was ornamented with spines and tubercles. The trunk was covered by elongated vertical scales. Well-formed dorsal and pectoral fins were present. Tail heterocercal (upper lobe larger). Several pairs of gill-slits. Sensory fields on head. Lateral eyes close together. Lampreys (Petromyzontiformes) Probably were derived from Osteotraci. Examples : *Cephalaspis*, (Fig. 10.1), *Hemicyclaspis*, *Tremataspis Kiaeraspis*, *Didymaspis*, *Benneviaspis* and many others.

Subclass II. Diplorhina

Two separate nasal openings, where known.

Order 1. Heterostraci (Pteraspida). The members of this order were heavily armoured ostracoderms of middle Ordovician to Devonian periods. Body large. Head was flattened and was protected by a bony cephalic shield. The fins were absent. The tail was laterally compressed, covered with small scales, hypocercal. Mouth slit-like aperture. Anterior portion of the snout is prolonged into rostrum. Lateral eyes widely separated. A single large gill-slit on each side. No sign of sensory organ. Lateral line was distinct. A backwardly directed dorsal spine projected upwards from the posterior part of the cephalic shield. The trunk was covered over by rectangular scales. Hagfishes (Myxiniformes) were probably derived from Heterostraci. Examples : *Pteraspis* (Fig. 10.2), *Drepanaspis*.

Order 2. Coelolepida. The members of this order were found from Ordovician to Devonian periods. Body was small (10 to 12 cm long) and fusiform or torpedo-like. Surface of the body was covered by numerous minute placoid scales like denticles. Tail hypocercal or heterocercal. Lateral eyes widely separated. Broad flangs project anterolaterally from body. Examples : *Phlebolepis*, *Lanarkia*, *Coelolepis*.

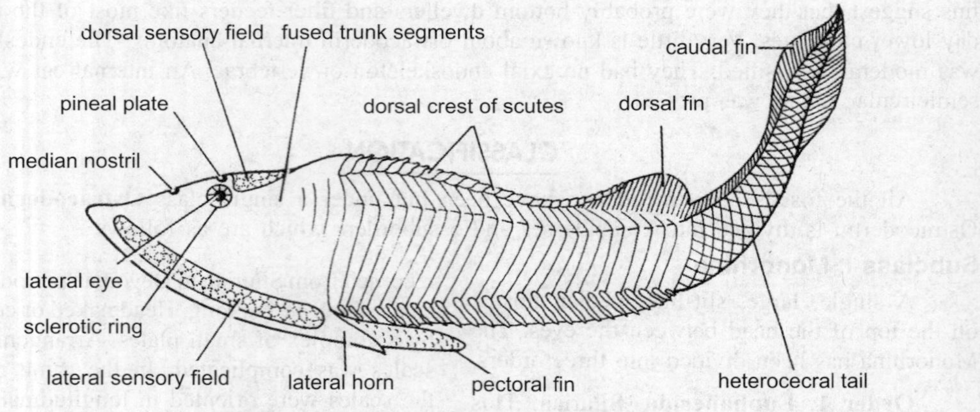

Fig. 10.1. Restoration of *Cephalaspis*.

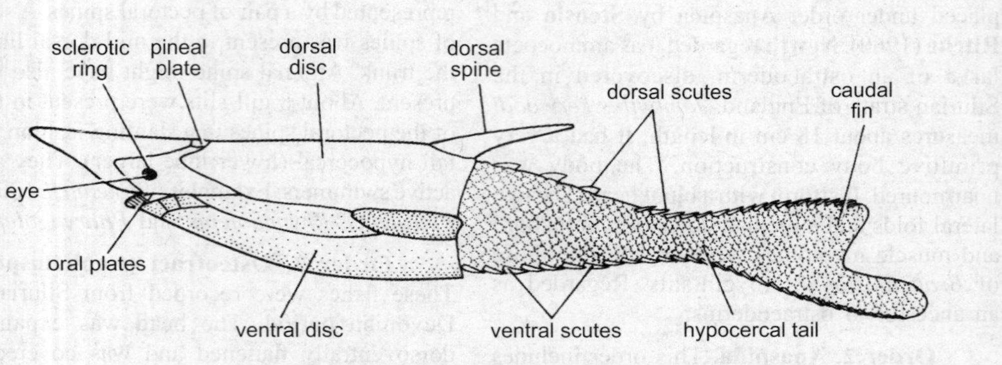

Fig. 10.2. Restoration of *Pteraspis rostrata*.

BIOLOGICAL SIGNIFICANCE OF OSTRACODERMS

The ostracoderms were the primitive vertebrates recorded in the geological history. The biological significance of ostracoderms have been discussed as follows :

1. The ostracoderms are specially interesting because they represent the oldest known vertebrate fossils in the late Cambrian and Ordovician rocks dating back to nearly 500 million years. They are the remote ancestors of all the vertebrates including man.

2. The microscopic studies of their fossilised bony tissues reveals a great complexity of structure. This implies that these vertebrates were far advanced and had undergone a considerable period of evolution before becoming fossilised.

3. The lack of earlier vertebrate fossils shows that they had perhaps evolved in a habitat (freshwater?) which was inimical to fossilisation. It is also likely that the earliest ancestors lacked hard skeletal materials as bones.

4. The ostracoderms had developed heavy bony armour perhaps for survival against the attacks of contemporary giant arachnid predators, the eurypterids. After the disappearance of giant arachnids (eurypterids) the descendants of ostracoderms, the cyclostomes, also lost the univanted heavy armour which was a hinderance in rapid progression.

5. The cartilage of cyclostomes and sharks and skates (Chondrichthyes) was previously considered a precursor to bone and more primitive. Since the ostracoderms had bony skeletons, the bone is now considered more primitive and the cartilage is interpreted as a degenerate condition.

INTER-RELATIONSHIP AND AFFINITIES

The fossil ostracoderms, as we find them, were specialised and products of a long evolutionary past. The ostracoderms had probably evolved from unarmoured ancestral forms such as *Jamoytius* and became diverged to various modes of life. Now the question arises whether the ostracoderms should be placed in the direct line of descent of vertebrates. It is certain that the group shows many structural similarities with the cyclostomes (the most primitive surviving vertebrates) are the degenerated descendants of some forms of ostracoderms. There are strong evidences in favour of this view. But the relationship of ostracoderms with gnathostomes is still obscure. The gnathostomes (jawed vertebrates) evolved simultaneously with the ostracoderms during the Devonian period and no fossil link is so far discovered depicting the line of transformation.

The ostracoderms could not compete with the jawed fish that evolved in such diversity during Devonian and became extinct. Before extinction the ostracoderms gave rise to the first bony fishes, the placoderms, and the cartilaginous chondrichthyes.

IMPORTANT QUESTIONS

Long Answer Questions

1. Write the characters, classification and affinities of ostracoderms.

Short Answer Questions

1. Write the biological importance of ostracoderms.
2. Write notes on:
   (*i*) *Ostracoderms*, (*ii*) *Cephalaspis* and (*iii*) *Pteraspis*.

Multiple Choice Questions

1. The earliest known vertebrates appeared in fossil records were
   (*a*) Placoderms
   (*b*) Ostracoderms
   (*c*) Tetrapods
   (*d*) Reptites.
2. Ostracoderms are placed in
   (*a*) Pisces
   (*b*) Tetrapoda
   (*c*) Aquatha
   (*d*) Gnathostomata.
3. Fossils of ostracoderms were preserved in

(*a*) Bottom sediments of oceans
(*b*) Bottom sediments of freshwater streams
(*c*) Bottom sediments of large lakes
(*d*) Gondvana land.

4. Ostracoderms date back almost
(a) 500 million years in rocks of late Cambrian and middle ordovician
(*b*) 300 million years
(*c*) 400 million years
(*d*) 600 million years.

5. Ostracoderm had no
(*a*) jaws and median fins
(*b*) Median fins
(*c*) jaws and paired fins
(*d*) jaws.

6. Exoskeleton of ostracoderms was
(*a*) Ctenoid scales
(*b*) Derwal scales
(*c*) Placoid scales
(*d*) cycloid scales.

7. Sense organs in ostracoderms were
(*a*) A pair of large lateral eyes, a median pineal eye and a median nostril
(*b*) A pair of nostrils
(*c*) A single lateral and a median eye
(*d*) Three nostrils.

Answers

1. (*b*) 2. (*c*) 3. (*b*) 4. (*a*) 5. (*c*) 6. (*b*) 7. (*a*).

11 Chapter

Type 4. *Petromyzon* (The Lamprey)

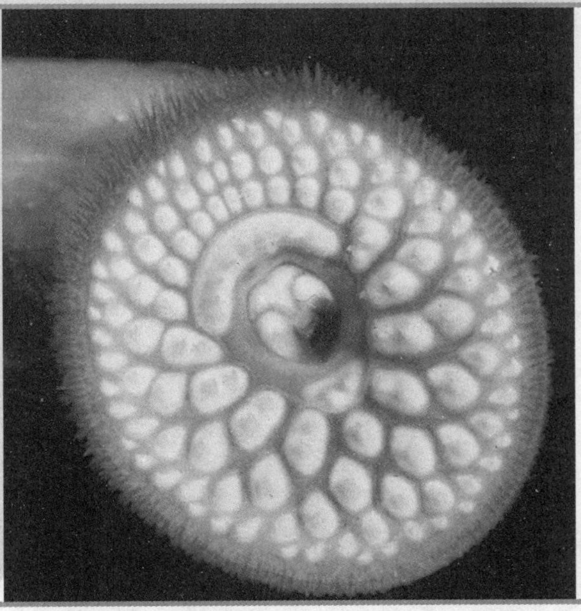

Contents of this Chapter

The Cyclostomata (Gr., *cyklos* = circular + *stoma* = mouth) are the living agnathans, they are primitive in many respects, but specialised in others. They are a modified and degenerate offshoot of the primitive vertebrate stalk, arose in the Devonian. In the adult state they are parasitic or scavengers on fishes. They have round bodies with laterally compressed or diphycercal tail. They resemble eels superficially. The suctorial mouth is ventral and round (hence, Cyclostomata). Buccal cavity has a muscular tongue bearing epidermal teeth by which they rasp the flesh of fishes. They are the only living vertebrates without jaws. They have 6–14 pairs of internal gills in different species. Gill-chambers are round pouches (hence, Marsipobranchii). The skin is soft and devoid of scales, paired appendages are absent, though median fins are present and supported by cartilaginous fin rays. Exoskeleton is lacking. Endoskeleton is cartilaginous with no bones and the vertebral column is primitive. There is a single median nostril, and only one or two semicircular canals are present in each auditory organ. Heart is enclosed in cartilage derived from the hinder visceral arch. They have no spleen. Cyclostomata include lampreys and hagfishes. Its two order are not closely related, because they have evolved from different orders of ostracoderms.

SYSTEMATIC POSITION

Phylum	...	Chordata
Subphylum	...	Vertebrata
Group	...	Agnatha
Class	...	Cyclostomata
Order	...	Petromyzontiformes
Family	...	Petromyzonidae
Type	...	*Petromyzon* (Lamprey)

DISTRIBUTION

Lampreys have almost world-wide distribution being present in both salt and freshwaters of North America, Europe, West Africa, Japan, Chile, Australia, New Zealand and Tasmania.

HABITS AND HABITAT

Lampreys (*Petromyzon*) are both freshwater and marine forms. Its larval stage called **ammocoete** is a freshwater, sedentary, and filter feeding creature like *Amphioxus*. While some adult freshwater forms are free-living but the others are ectoparasitic on fishes and turtles. The lampreys are all carnivores, and although they are sometimes spoken as quasi-parasitic, they should properly be regarded as true predators, for they actually catch and eat their prey. The method of feeding is peculiar and quite specialised. Lampreys are swift swimmers, capable of overtaking slower prey. The animal attacks fishes, turtles, etc., of many sorts, usually larger than itself. It attaches itself by means of its circular mouth.

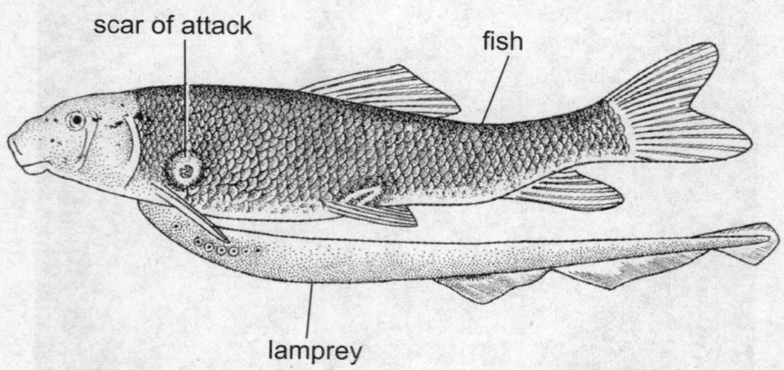

Fig. 11.1. Lake lamprey attached to a bony fish which also shows the scars of the attacks of other lampreys.

The mouth operates as a sucker or vacuum cup and has very tenacious holding power. While attached to the prey, it rasps or shreds the flesh of the prey with its tongue apparatus and horny teeth on the buccal funnel.

Lampreys usually live in sea and migrate to the rivers for the purpose of spawning. After spawning they usually die. In the river the lampreys do not eat anything, but live on the accumulated subcutaneous fat that provides the nourishment. They migrate to the river usually in autumn, while the sexual maturity attained in winter. But the breeding occurs in spring. A freshwater lamprey, *Petromyzon marinus unicolor* migrates a few miles upstreams for the purpose of spawning. The brook lampreys (represented by the genus *Lampetra*) have the peculiar habit of nest-building at the bottom of the river where the eggs are laid. They select the bottom where both sand and stones are present. By the action of the suctorial mouth, they arrange the stones in such a fashion that a small depression is produced at the centre. Fertilisation is external.

EXTERNAL FEATURES

Shape and Size

The body of adult *Petromyzon* is elongated, cylindrical and eel-like. *Petromyzon marinus* (sea-lamprey) attains a length of a metre, *Lampetra fluviatilis* (Lampern or common freshwater lamprey) reaches a length of about 90 cm, and *Lampetra planeri* (Sandpride or lesser freshwater lamprey) does not exceed 45 cm in length.

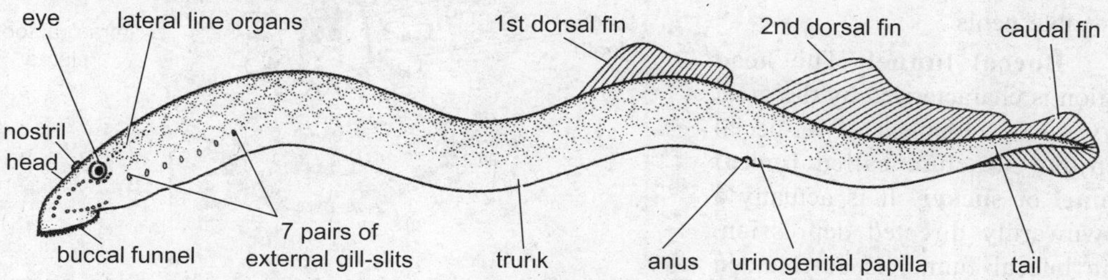

Fig. 11.2. *Petromyzon marinus.* External features.

Colour

The upper surface of the body is usually dark mottled greenish-brown. The surface of the body is smooth and slimy due to secretory epiderwal glands. The scales are absent.

Division of Body

The body of *Petromyzon* is divisible in three regions, *viz.*, **head**, **trunk**, and **tail**, which are not clearly demarcated. The head and trunk are nearly cylindrical and the tail is more or less laterally flattened.

Fins. Paired fins are absent. There is a median **dorsal fin** which is divided into two unequal parts by a notch are placed near the posterior end. A **caudal fin** is present around the tail which is continuous with the posterior dorsal fin. The fins are supported by thin cartilaginous fin-rays. (In *Lampetra planeri* the female possesses an anal fin. This fin is reduced to a copulatory papilla in male).

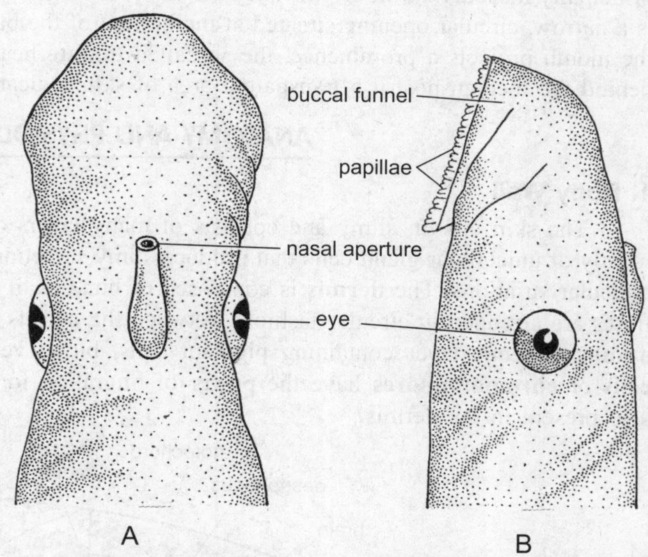

Fig. 11.3. *Petromyzon*. A–Dorsal view of head showing the position of nostril; B–Lateral view of head.

Eyes. On either lateral side of head is present a well developed eye having no eyelids, but are covered by a transparent area of skin.

Apertures. (*i*) **Mouth** lies at the apex of the buccal funnel bordered by a ring of cartilage. (*ii*) **Nostril** or nasohypophyseal aperture is single located mid-dorsally over the head in between the two eyes. (*iii*) Just behind each eye are present a row of seven rounded **external gill-slits**. (*iv*) At the junction of trunk and tail, on the ventral side is present a slit-like depression, the **cloaca**. Protruding through the cloaca is a urinogenital papilla bearing at its tip a minute urinogenital aperture. **Anus** lies just in front of papilla within the cloacal depression. (*v*) Along each lateral side of the body and below the head are present numerous small **sensory pores** of the lateral line septum. In male *Petromyzon* is present an eversible penis.

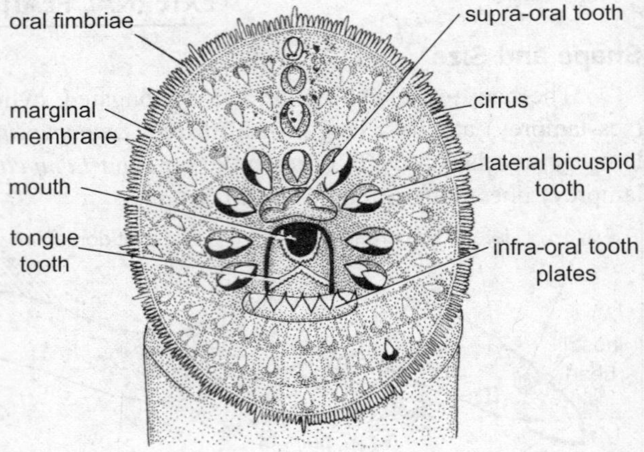

Fig. 11.4. *Petromyzon.* Ventral view of the head showing details of buccal funnel and position of mouth and tongue.

Buccal funnel. The head region is characterised by the great forward development of a special cup-like structure called **buccal funnel** or **sucker**. It is actually a downwardly directed depression. The buccal funnel is surrounded by a marginal membrane which is provided with soft small projection called the **oral fimbriae** or **papillae** and the longer **sensory cirri** in between the papilla. The inside of the buccal funnel is beset with radiating rows of yellow, epidermal, **horny teeth**. The teeth rest on cartilaginous pads and are principally mesodermal in origin. They are not homologous with true vertebrate teeth. The mouth is a narrow, circular opening situated at the bottom of the buccal funnel. From below and behind the mouth projects a prominence, the so-called **tongue** bearing larger horny teeth. Immediately behind the median nostril a transparent area of skin indicates the position of the pineal organ.

ANATOMY AND PHYSIOLOGY

1. Body Wall

The skin is soft, slimy and consists of many layers of cells. The **epidermis** is composed mostly of unicellular gland cells that produce slimy secretion and its outermost cells border show cuticular striations. The **dermis** is composed of bundles of collagenous and elastic fibres. These fibres are arranged in circular fashion. Between the dermis and bodywall musculature, there lies a **subcutaneous** layer containing pigment cells, blood vessels and fatty tissue. The **pigment cells** or **chromatophores** have the power of migration for changing the skin colour. They are also present in the dermis.

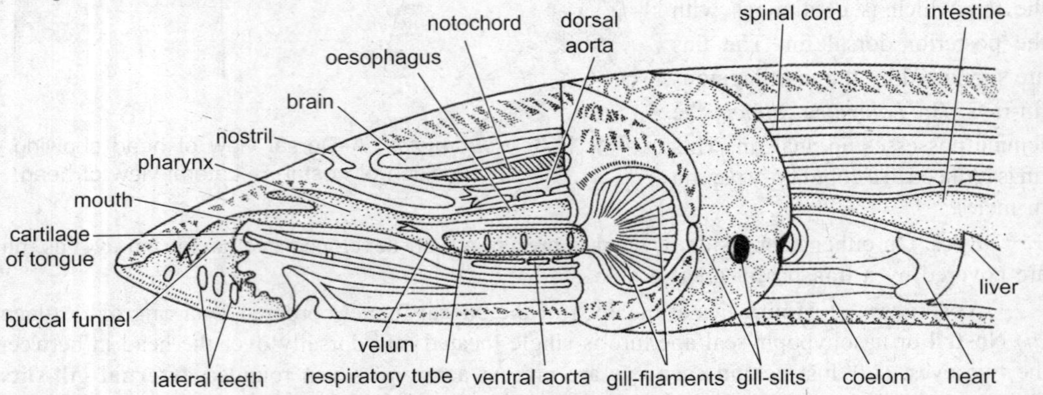

Fig. 11.5. *Petromyzon.* Longitudinal sectional view of the anterior portion showing the internal structure. A portion of the body is kept intact to show the position of a gill-slit. (After Storer and Usinger).

2. Muscular System

The muscles of trunk and tail are arranged in E-shaped myotomes, separated by **myocommas**. The myotomes are W or E-shaped muscle blocks and the muscle fibres composing the myotomes run longitudinally. The myotomes of one side alternate with those of the other. The muscle fibres are striated. Contraction of muscles of one side bends the body to that side. These muscles are helpful in swimming of the fish. Radial muscles operate the buccal funnel and the tongue is moved by stout retractor and smaller protractor muscles.

3. Skeleton

It includes the notochord and cartilaginous endoskeleton. The **axial skeleton** comprises a lowly developed **skull**, persistent **notochord** and collection of **cartilages**. The **notochord** is composed of large vacuolated **notochordal cells** enveloped by a thick fibrous **notochordal sheath.** This is further surrounded by another connective tissue layer which covers the spinal cord and remains connected with the myocommas. Connective tissue layer contains scattered thickening of cartilaginous elements. Besides notochord, body skeleton is formed of cartilage. The **cartilage** is composed of large cells embedded in a matrix of **chondrin**.

(*i*) Numerous slender cartilaginous rods extend dorsally and ventrally to support the median fins.

(*ii*) The **skull** is very peculiar and primitive in structure. Its floor is formed by a basal plate made by the union of parachordals and trabeculae surrounding posteriorly the anterior end of the notochord. Attached with the basal plate there are pieces which form an incomplete cartilaginous box enclosing the brain and the special sense organs. Immediately anterior to the notochord is a large aperture, the **basi-cranial fontanelle** due to the non-union of the posterior ends of the trabeculae. Through this aperture passes the naso-hypophysial sac. The side walls of the cranium are strong but the roof is exclusively composed of tough membranous fibro-cartilage excepting a narrow transverse bar. Two compact auditory capsules are united with the posterior end of the basal plate. The olfactory capsule is an unpaired concave-convex plate which supports the posterior wall of the olfactory sac and is pierced by the paired apertures for the olfactory nerves. This capsule is united with the cranium by fibrous tissue. On each side of the basal plate, there lies the **subocular arch** to support the eye. A slender **styloid process** hangs from the subocular plate. Each styloid process is connected with a small **cornual cartilage**. The cranium is attached with the skeleton that supports the gills and buccal funnel.

(*iii*) The **branchial basket** is composed of nine irregularly curved vertical bars of cartilage on each side. The first vertical bar is placed almost immediately posterior to the styloid cartilage and is homologous with the extra-hyal cartilage of the elasmobranchs. The second lies immediately in

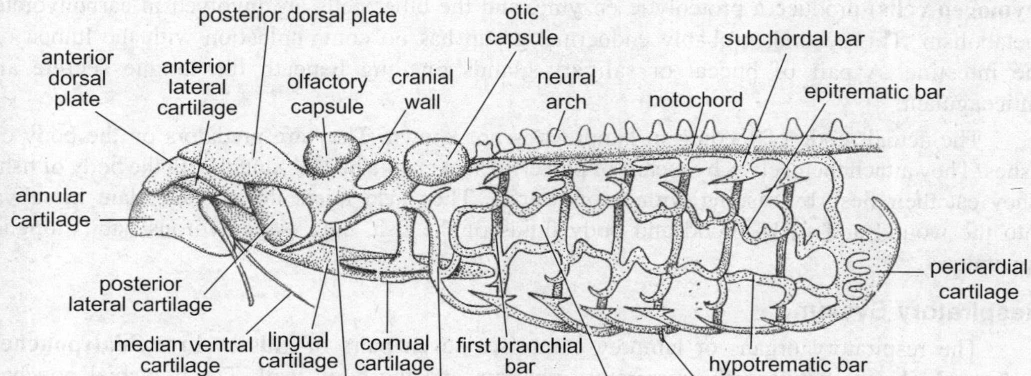

Fig. 11.6. *Petromyzon.* Skull and branchial basket of an adult animal.

front of the first gill-cleft, and the remaining seven just behind the seven gill-clefts. These bars are united together by four longitudinal rods.

(*iv*) The posterior part of the branchial basket is extended to form a cup-like **pericardial cartilage** to accommodate the heart. The whole branchial basket lies external to the gill pouches and branchial arteries, not like typical visceral arches, in the walls of the pharynx.

(*v*) The buccal funnel is supported by a ring-like **annular cartilage**.

(*vi*) An elongated **lingual cartilage** supports the tongue.

Digestive System

The **mouth** is situated just above the tongue at the base of the buccal funnel. The opening and closing of the mouth is done by the fore-and- aft piston-like movement of the tongue. The **buccal funnel** is a specialised cup-like depression which acts as a **sucker**. Numerous teeth are present in the buccal funnel which are laminated horny cones. Beneath them lie mesodermal papillae (covered with ectoderm). The teeth that are borne on the tongue are elongated and sharply pointed. The teeth are capable of being replaced, if lost. Special set of muscles operate the buccal funnel and tongue apparatus. Lampreys can attach with the body of other fishes by producing a vacuum cup inside the buccal funnel. The vacuum cup is produced inside the oral cavity by the pulling on the tongue backwards by the contraction of the **cardio-apicalis muscle**. The release of sucker is effected by the contraction of an **annular muscle** running along the margin of the buccal funnel.

The mouth leads into a **buccal cavity** formed from the stomodaeum of the embryo. It communicates behind with two tubes a dorsal **oesophagus** and the ventral **respiratory tube** or **pharynx** having linearly arranged 7 gill-slits on either side. Guarding the entrance of the respiratory tube is a curtain-like fold, the **velum** with velar tentacles. The velar apparatus keeps the digestive part completely separated from the respiratory tube when the animal feeds. The **oesophagus** leads directly into a straight **intestine** by a vulvular aperture. True stomach is absent in lampreys. The posterior end of the intestine is widened to form the **rectum** which opens through anus in the **cloaca**. The lumen of the intestine is crescentic due to the presence of a spiral longitudinal fold, the **typhlosole**. The typhlosole takes a slightly spiral course along the intestine and is, hence, known as **spiral valve.**

The digestive glands are well-developed in lampreys. The **liver** is a bilobed organ and surrounds the anterior part of intestine. Gall-bladder and bile-duct occur in the larval stage, but are absent in the adult. There is no pancreas as such. In both larval and adult lampreys, however, there occur in the gut epithelium aggregations of seemingly secretory cells of two different kinds : one kind (**zymogen cells**) produce a proteolytic enzyme, and the other cells are involved in carbohydrate metabolism. This second, probably endocrine, group has no communication with the lumen of the intestine. A pair of buccal or salivary glands opening beneath the tongue secrete an anticoagulant.

The actual mode of feeding in lampreys is not known. They are predators on the body of fishes. They attach themselves by means of powerful sucker-like buccal funnel with the body of fish. They eat their flesh by rasping with the powerful. They also inject the anticoagulant in saliva into the wound and sucks blood and body fluids of the fish, etc., (sanguivorous habit) tongue apparatus.

Respiratory System

The respiratory organs of lamprey comprises seven pairs of **gills** or **branchial pouches** or **branchial sacs** between the respiratory pharynx and the body wall. The branchial pouches open directly into the respiratory tube and have no direct connection with alimentary canal. The gill-pouches have numerous vascular gill-lamellae developed on their inner surfaces and are separated from one another by wide membranous **interbranchial septa**. Each gill-pouch opens independently to the exterior by an external gill-slit and in the respiratory pharynx by an internal gill-slit.

The water enters into the respiratory pharynx through the 7 pairs of gill-pouches and passes out through the same route. But in larval stage, water current enters through mouth and goes out through the gill-slits. The elastic branchial basket is compressed by the contraction of the muscles to throw the water like jets through the gill-openings, while during inspiration the branchial basket expands. The gaseous exchange takes place inside the gill-pouches bearing vascular gill-filaments or lamellae.

Circulatory System

In lamprey the circulatory system is well developed and similar to *Amphioxus*. The **heart** is an S-shaped structure situated behind the last pair of gill-pouches. It is enclosed in a thick-walled pericardium supported by a cartilaginous plate. The heart consists of three chambers– a **sinus venosus**, an **auricle** and a **ventricle**. The sinus venosus is a thin-walled chamber and leads into a thin-walled auricle lying above the sinus venosus. The auricle opens into the thick-walled ventricle. There is no conus arteriosus. The ventricle leads into the ventral aorta whose proximal end presents a slight dilation or **bulbous anteriosus**. The ventral aorta gives off 7 pairs of efferent branchial arteries to 7 pairs of gills. Oxygenation of blood occurs here. From gills blood enters into 7 pairs of efferent branchial arteries, which unite to form a median **dorsal aorta** distributing the blood to the various parts of the body. There is no renal portal system but a hepatic portal vein is present. The blood of lamprey is red in colour due to the presence of haemoglobin. It contains both red blood corpuscles and white blood cells. The blood forming tissue is present in the kidneys and the spinal cord. Lymphatic system is also present.

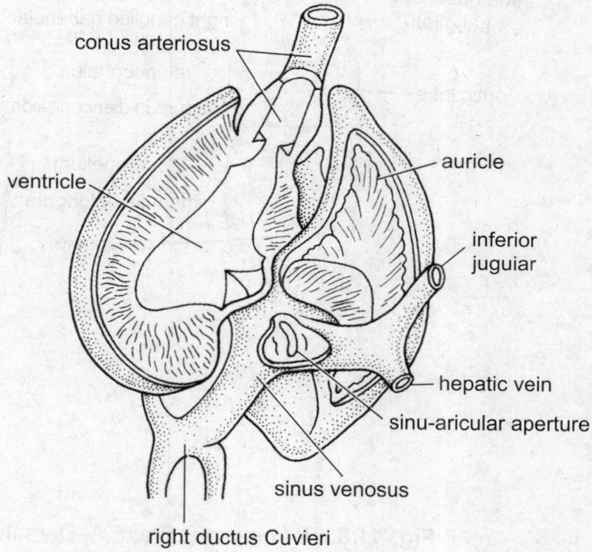

Fig. 11.7. *Petromyzon*. Diagrammatic longitudinal section of the heart. The chambers are exposed showing the internal details (after Jollie).

Nervous System

The nervous system of lamprey is quite well-developed. The brain exhibits the typical pattern that is found throughout the vertebrate series. Some of the structures that are well-developed in the higher forms occur in a relatively primitive condition. The brain is divisible into three primary parts, *viz*., **forebrain** or **prosencephalon, midbrain** or **mesencephalon** and **hindbrain** or **rhombencephalon**. The forebrain includes large paired olfactory lobes followed by a pair of small cerebral hemispheres attached to the diencephalon. A broad **infundibulum** lies attached beneath the diencephalon and it dorsally gives off a **pineal** or **median eye**. The midbrain possesses a pair of large **optic lobes** rather dorsally placed. The hindbrain is differentiated into a dorsal small transverse band, the **cerebellum** and a much larger dorsal **medulla oblongata**. Between the optic lobes and the medulla oblongata extends the choroid plexus. The ventricles within the brain are four as in other vertebrates. The cranial nerves are ten pairs.

The spinal cord is band-like structure which is dorso-ventrally flattened. It is of transparent grey colour. The spinal nerves are like those of *Amphioxus* in that the dorsal and ventral roots do not join. The sympathetic nervous system is poorly developed.

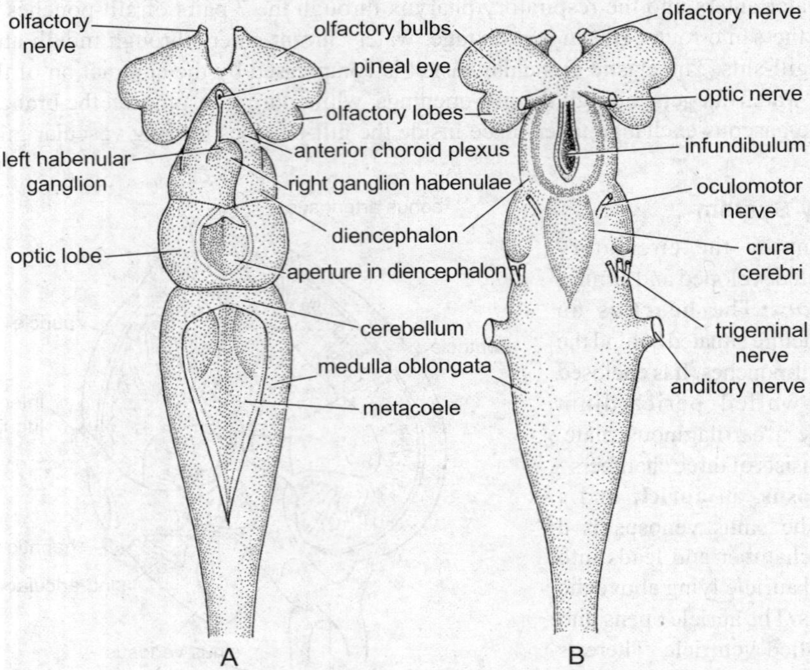

Fig. 11.8. *Petromyzon.* Brain. A–Dorsal view; B–Ventral view.

Sense Organs

The organs of special sense are well-developed. The paired eyes are primitive and having no suspensory ligament and ciliary apparatus. They are formed, like the pineal eyes, by evaginations of the wall of the diencephalon. Accomodation is affected by a process found in no other vertebrates. Cornea consists of two distinct layers, separated by a gelatinous substance. Iris is present around pupil which does not change in diameter under different illuminations. Median eye or pineal organ is better developed. It contains a pigmented retina with sensory cells and a flattened imperfect lens. It is photosensitive. The membranous labyrinth has only two semicircular canals. The single external nostril leads by a short passage into a rounded olfactory sac lined by olfactory mucous membrance. It gives off a large naso-hypophyseal sac which ends blindly below the anterior end of notochord. The lateral line sense organs are relatively simple and are exposed to the surrounding water and innervated by cranial nerves.

Excretory System

In lamprey, the adult functional kidney is of **mesonephric** type. The kidney develops from the nephrogenous tissue (nephrotome). Each kidney is long, strap-shaped body developed from the mesonephros of the embryo. The mesonephric tubules become greatly elongated winding tubule and the anterior end of each tubule bears a double-walled **Bowmann's capsule** containing glomerulus. Each kidney is attached along one edge to the dorsal wall of the body cavity by part of the peritoneum. Along the other (free) edge runs the mesonephric duct which opens posteriorly into a small **urinogenital sinus** placed just behind the rectum. This opens by a urinogenital papilla into a pit, the cloaca in which the alimentary canal also ends. The side walls of the sinus are pierced by a pair of **genital pores** which place its cavity in communication with the coelom. Kidneys contain lymphoid tissue where blood cells may be formed. This tissue is also phagocytic–destroys old blood cells.

Reproductive System

The sexes are separate in the adult. The gonad is a large unpaired occupying almost the whole of the body cavity and lack reproductive duct. The testis consists of many follicles containing male gametes which are liberated into the body cavity. The ovary is composed of eggs which are covered by a layer of follicular epithelium. The eggs also rupture in the body cavity. From the body cavity both the eggs and sperms escape by way of abdominal pores present in the lateral walls of urinogenital sinus and escape outside the body through urinogenital opening. Fertilisation is, thus, external, *i.e.*, occurs in water.

Life History

Lampreys become sexually nature in spring or early summer and migrate into freshwater streams or rivers for breading. On return to the rivers they stop feeding and start building nests of small stones in river beds. The males remove pebbles from river beds by their buccal funnel and make a small pit. In mating, the male holds the female near her head by its buccal funnel, then the two intertwine their bodies. Eggs are squeezed out through the cloaca into the nest and sperms are deposited on them. Fertilisation is, thus, external. The eggs and sperms are not laid at once; mating is repeated several times, after which the animals are exhausted and soon die. Fertilised eggs are covered up with mud or sand so that they are not carried away by the current. The eggs are small, being 1 mm in diameter, they are telolecithal but cleavage is holoblastic and unequal. Gastrulation is by invagination. In 3 to 4 weeks the young hatches from the egg, it is called an **ammocoete larva** which is at first transparent and about 1 cm long. The larva lives in U-shaped tunnels of mud where it feeds on organic ooze at night. It emerged from its tunnel at night only to change its feeding ground. It swims with its head downwards and then buries itself again. The larval period is of long duration, it is 3 to 4 years, but in some lampreys it may be 7 years. During this period the ammocoete larva becomes opaque and grows to 170 mm in length.

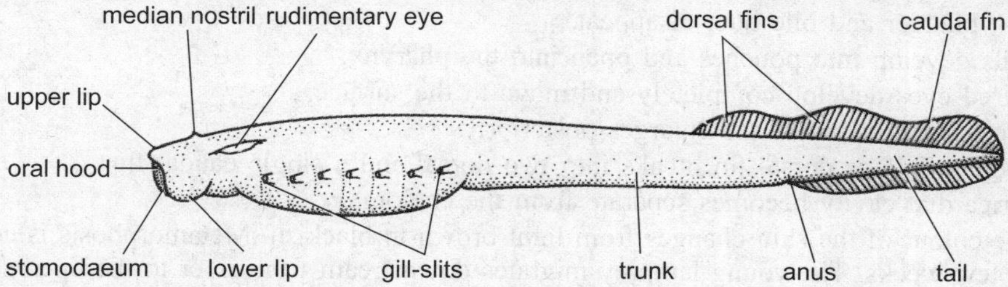

Fig. 11.9. Ammocoete larva. External features in lateral view.

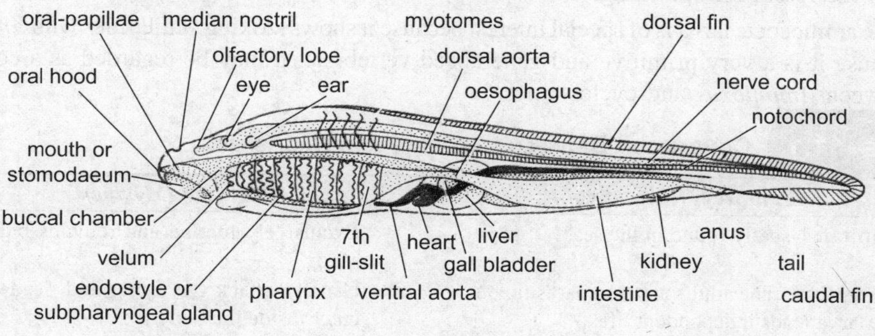

Fig. 11.10. Ammocoete larva showing internal structure.

Ammocoete Larva

Ammocoete larva has an eel-like body, but it differs from the adult in several respects. It has a single continuous median **dorsal fin**. There is no buccal funnel, the mouth is surrounded by a semicircular upper lip of **oral hood** like that of *Amphioxus*. Below the mouth is a short transverse lower lip, there are no teeth. The paired eyes are covered by muscles and skin. Branched **buccal tentacles** lie at the entrance to the alimentary canal. Behind these tentacles is a **velum** made of a pair of cup-shaped structures attached to anterior wall of pharynx. The pharynx has seven pairs of gill-pouches. The pharynx is continued posteriorly into the oesophagus. On the tail are **photoreceptors** which ensure that the larva lies completely buried. Below the pharynx is a short sac-like **endostyle** having glandular cells. It is different from that of an *Amphioxus* in not being an open groove. In the floor of the pharynx, *i.e.*, on the roof of endostyle, however, is a ciliated groove. Mucus from the endostyle goes to the groove in the pharynx, and then to a pair of **peripharyngeal grooves** behind the velum. The velum maintains a stream of water into the pharynx, in this it is helped by muscular movements of the pharynx. Food particles consisting of unicellular algae and bacteria come through the mouth with the current of water and are entangled in mucus. In the pharynx the food-laden mucus forms a cord which is carried into the oesophagus. Cilia in the oesophagus now propel the food into the gut. Thus, feeding in the ammocoete larva is by muscular movements of velum and pharynx, whereas in *Amphioxus* it is brought about by ciliary action, this is a forward step in evolution. It enables a quicker and larger intake of food.

Metamorphosis

After a prolonged larval period (*i.e.*, 3 to 7 years) metamorphosis occurs.

(*i*) The buccal region elongates so that the oral hood becomes the buccal funnel with horny teeth, tongue, and a round mouth.

(*ii*) The endostyle becomes modified to form a **thyroid gland** which contains iodine.

(*iii*) The velum disappears leaving a rudiment from which skeletal rods develop to form a filter guarding the pharynx.

(*iv*) The oesophagus separates from the pharynx and comes to lie dorsal to it.

(*v*) Gall-bladder and bile duct disappear.

(*vi*) Gills develop into pouches and open into the pharynx.

(*vii*) Paired eyes develop completely and move to the surface.

(*viii*) The nasohypophysial sac enlarges posteriorly.

(*ix*) The continuous dorsal fin breaks into two dorsal and a single caudal fin.

(*x*) Pericardial cavity becomes separate from the coelom.

The colour of the skin changes from light brown to blackish. Metamorphosis is completed within a few weeks. The young lamprey migrates downstream from river to the open sea where it remains for 3 or 4 years and becomes carnivorous adult. But the gonads ripen only when it returns to the rivers for spawning.

The ammocoete larva is of special interest because it shows striking similarities with *Amphioxus*, and because it is a very primitive and generalised vertebrate, it may be regarded as a connecting link between *Amphioxus* and cyclostomes.

TABLE 11.1. COMPARISON BETWEEN LAMPREYS AND HAGFISHES.

Lampreys *(Petromyzon)*	Hagfishes *(Myxine)*
1. **Habitat:** Usually found in the sea.	Exclusively marine and remains buried in the sand.
2. **Parasitism:** The adults are semiparasitic but the larva leads independent life.	Usually attack the dying and dead fishes and enter inside the body.
3. **Breeding:** Migrate to the rivers for spawning (anadromous).	Spawn on ocean floor.

(*Contd.*)

Lampreys *(Petromyzon)*	Hagfishes *(Myxine)*
4. **Body:** The body is cylindrical and stout. About 1 metre long.	The body is slender "eel-like" and feeble. Less than 1 metre long.
5. **Fins:** The dorsal fin is divided into two by a notch. The caudal fin is well-represented.	The dorsal fin is ill-developed. The caudal fin is slightly developed.
6. **Nostril:** Single nostril lies on the dorsal side of the head and occupies a median position.	Single nostril lying very close to the mouth and opens terminally.
7. **Eyes:** Paired eyes are functional and well-developed.	Paired eyes are rudimentary and sunk beneath the skin. The eyes are devoid of lenses and eye-muscles.
8. **Pineal:** Pineal eye present.	Pineal eye absent.
9. **Mouth:** The mouth is subterminal in position due to the over development of the upper lip region to form a funnel-like structure.	Mouth is terminal is position and is surrounded by soft lips.
10. **Oral tentacles:** Tentacles around mouth are absent.	Three or four pairs of tentacles are present around the edge of the mouth.
11. **Buccal funnel:** Buccal funnel present.	Buccal finnel absent.
12. **Tongue:** Tongue feebly developed.	Tongue strongly developed.
13. **Salivary glands:** Paired salivary glands are present.	Paired salivary glands are absent.
14. **Teeth:** Horny teeth are present in the buccal funnel and tongue. The teeth present on the tongue are larger in size.	Small teeth are present on the tongue along two rows.
15. **Gill-pouches:** Gill-pouches seven pairs which do not open into the gut, but into the respiratory tube.	The number of gill-pouches varies from six to fifteen pairs. *Myxine* possesses six pairs of gill-pouches, while *Eptatretus* bears thirteen to fifteen pairs. Gill-pouches open directly into the gut.
16. **External gill-slits:** Seven pairs of external gill-slits in the adult. An additional pair may be present in the larva.	One pair of external gill-slits is present in *Myxine* but 13–15 pair in *Eptatretus*.
17. **Aortic arch:** Each aortic arch divides and sends branches to the posterior hemibranch and anterior hemibranch of the adjacent gill-pouches.	Each aortic arch supplies the hemibranch of a single gill-pouch.
18. **Brain:** Brain is comparatively well-developed.	Brain is simpler and degenerated.
19. **Cranial nerves:** Ten pairs of cranial nerves are present.	Only oculomotor, trochlear and abducens nerves are present.
20. **Nasohypophysial duct:** Nasohypophysial duct ends blindly.	Nasohypophysial duct open posteriorly into the mouth cavity.
21. **Kidney:** The pronephros does not persist in the adult. The mesonephros is the functional adult kidney. The mesonephric tubules are complicated and lack segmental disposition.	The pronephros is retained in the adult. The functional kidney is of mesonephric type and simpler. The mesonephric tubules have segmental arrangement.
22. **Urinogenital sinus:** Urinogenital sinus is present.	Urinogenital sinus is absent.
23. **Sexes:** The sexes are separate in the **adult,** but the ova may be present in the testis.	The sexes are united. The anterior position of the gonad acts as the ovary and posterior part is testis-like.
24. **Eggs:** The eggs are small and are devoid of **horny** shell. The eggs are ovoid in shape and are essentially isolecithal.	The eggs are large and enveloped by cylindrical horny shell. The shell bears terminal hook-like processes. The eggs are elongated and typically telolecithal.
25. **Cleavage:** Segmentation holoblastic.	Segmentation meroblastic in *Eptatretus*.
26. **Development:** The development is accompanied by metamorphosis of the Ammocoetes larva.	The development is direct, *i.e.*, the metamor-metamorphosis is known.

SUMMARY OF THE CHARACTERS OF CYCLOSTOMES

AFFINITIES OF CYCLOSTOMES

Cyclostomes are evidently chordate. They are primitive vertebrates. Their ammocoete larva resembles in most characters with that of *Branchiostoma*, which shows primitive relationship. Whereas, adult cyclostomes possess specialised as well as degenerate characteristics.

I. Primitive Characters of Cyclostomes

1. Characters resembling those of *Amphioxus*

1. Absence of jaws, exoskeleton and paired fins.
2. Continuous notochord (but with an added sheath).
3. Segmental musculature (myotomes) but little modified from head to tail.
4. Ciliated alimentary tract straight and without much regional specialisation.
5. Relatively large numbers of gill-slits.
6. Endostyle in lamprey larva.
7. Gonads without gonoducts.

Besides these, the ammocoete larva of lampreys resembles *Amphioxus* as follows :

1. Fish-like body. 2. Oral hood anterior to mouth. 3. Continuous dorsal and caudal fins. 4. Ciliated digestive tract. 5. Filter feeding habit and 6. Endostyle functions in feeding.

2. Characters more primitive than in fishes (Differences from fishes)

1. No biting jaws, Scales, true teeth, true fin rays, girdles, ribs, stomach, spleen and gonoducts.
2. Continuous median dorsal fin.
3. Diphycercal caudal fin.
4. Single median nostril, instead of paired.
5. Cranium incomplete or poorly developed.
6. No vertebrae or poorly developed vertebrae.
7. Rundimentary pancreas.
8. No spinal valve, or only slightly developed spiral valve, in intestine.
9. Brain relatively small or generalised.
10. Ninth and tenth cranial nerves not enclosed in the cranium. Absence of medullated nerves.
11. Sympathetic nervous system very primitive and poorly developed.
12. Heart a rather loosely twisted S-shaped tube without conus arteriosus.
13. Lateral line organs poorly developed and in isolated pits.
14. Hypophysial duct rather large, open to the exterior and not connected with the pituitary body.

3. Affinities with Ostracoderms

Cephalaspids and anaspids are fossil agnathans that show similarity to the modern cyclostomes than pteraspids (Neterostraci or Pteraspida). Ostracoderms belonging to Ordovician are the Oldest fossil vertebrates. They were abundant in Silurian period and become extinct in Devonian. Probably they were the forerunners of higher fishes. The fossil ostracoderms and present cyclostomes are kept in Agnatha due to the following similarities :

1. Presence of a median pineal eye.
2. Presence of velar pump-like lamprey.
3. Endostyle sac-like.
4. Single nasal opening though nasal sacs are paired.
5. Brain is like that of lamprey.
6. Two semicircular canals in the ear.
7. Dorsal and ventral nerve roots separate up to 15.
8. Pairs of branchial pouches surrounded by a branchial basket.
9. Continuous uncostricted notochord.
10. Absence of jaws.

Stensio holds that the pteraspids have given rise to the myxinoids, and the cephalaspids to the lampreys. The Agnatha were the first animals of the chordate type to become large. They feed on detritus on the bottom. The lampreys and hagfishes have been derived from early Agnatha by the evolution of a sucking mouth, perhaps with loss of the bony skeleton and paired limbs.

II. Specialised Characters

1. Sucking mouth buccal funnel and horny teeth (in lampreys) for attachment.
2. Powerful tongue armed with sharp horny teeth works as a rasping organ.
3. Secretion of anticoagulants in saliva to feed on blood of prey.
4. Sac-like gill-pouches. Located far behind head. It is probably an adaptation to burrowing.
5. Complete separation of lower sac-like respiratory pharynx from upper digestive pharynx.
6. Water entering gill-pouches and also leaving them through external gill openings and not through mouth.
7. In hagfishes, presence of large mucus secreting mucous glands.
8. Dorsal position of nostril on head in lampreys.
9. Large, heavy-yolked egg, with meroblastic cleavage and no larval stage in hags.

III. Degenerate Characters

1. Simple elongated eel-like body, more marked in hags. Whereas ostracoderms body is broad fish-like.
2. Rudimentary paired eyes covered by thick skin in hags.
3. Lack of exoskeleton or bony armour.
4. Absence of paired fins and girdles.
5. Lack of ossification of endoskeleton, it is cartilaginous.
6. Reduced liver and lack of gall-bladder and bile-duct in adult lamprey.

IMPORTANT QUESTIONS

Long Answer Questions

1. Give a detailed account of the external features of *Petromyzon.*
2. Describe the digestive system and feeding in *Petromyzon.*
3. Mention the differences between *Petromyzon* and *Myxine.*

Short Answer Questions

1. Write note on buccal funnel of *Petromyzon.*
2. Give an account of skeleton in *Petromyzon.*
3. Describe branchial basket of *Petromyzon.*
4. Describe the arrangement of teeth in the buccal funnel of *Petromyzon.*

5. Describe the structure of ammocoete larva.
6. Write about the metamorphosis of ammocoete larva.
7. Describe the circulatory system in *Petromyzon*.

Very Short Answer Questions

1. *Petromyzon* is included under Agnatha because it is a jawless vertebrate. (True/False)
2. *Petromyzon* has a pair of nostrils. (True/False)
3. *Petromyzon* has both median and paired fins (True/False)
4. *Petromyzon* has a .... buccal funnel.
5. Inside the buccal funnel are present radiating rows of .... which are .... and ....
6. What is the position of mouth in *Petromyzon*?
7. What are the supra-oral and infra-oral tooth plates in *Petromyzon*?
8. What is the type of kidney in *Petromyzon*?
9. Exoskeleton in *Petromyzon* is ....
10. Name the cartilage which surrounds the buccal funnel.

Multiple Choice Questions

1. *Petromyzon* belongs to the group
   (*a*) Agnatha (*b*) Pisces
   (*c*) Gnathostomata (*d*) Tetrapoda.
2. Larva of *Petromyzon* is called
   (*a*) Ammocoete (*b*) Tornaria
   (*c*) Nauplius (*d*) Crinoid.
3. In lampreys paired appendages are
   (*a*) Absent (*b*) 2 pairs
   (*c*) 3 pairs (*d*) 1 pair.
4. Marginal membrane around buccal funnel is be set with
   (*a*) Oral fimbrae (*b*) Oral cirri
   (*c*) Buccal cirri (*d*) None.
5. The pineal organ is
   (*a*) Present behind the eye
   (*b*) Present in between the eye and nostril
   (*c*) Absent.
6. A urinogenital papilla is present
   (*a*) Near anus
   (*b*) Near caudal fin
   (*c*) Protrudes through cloaca
   (*d*) Absent.
7. Sucking action of buccal funnel is operated by
   (*a*) Radial muscles
   (*b*) Circular muscles
   (*c*) Longitudinal muscles
   (*d*) All.
8. Buccal funnel is surrounded by a cartilage in *Petromyzon* called
   (*a*) Annular cartilage
   (*b*) Linqual cartilage
   (*c*) Branchial cartilage
   (*d*) None.
9. Kidneys in lampreys are
   (*a*) Mesonephric (*b*) Metanephric
   (*c*) Pronephric (*d*) Glomerular.
10. Period of metamorphosis of ammocoetes larva is
    (*a*) 3 to 7 years (*b*) 3 to 10 years
    (*c*) 1 to 5 years (*d*) 6 years.
11. Eggs of lampreys are
    (*a*) Isolecithal (*b*) Alecithal
    (*c*) Polylecithal (*d*) Telolecithal.
12. Heart of *Petromyzon* contains
    (*a*) A sinus venosus, an auricle, a ventricle and conus arteriosus
    (*b*) A pair of auricle, a ventricle and sinus venosus
    (*c*) Sinus venosus, auricle and 2 ventricles
    (*d*) None of the above.
13. Blood from various parts of the body in
    (*a*) Aorta
    (*b*) Sinus venosus
    (*c*) Auricle
    (*d*) Ventricle.
14. In *Petromyzon* ventral aorta gives off seven pairs of afferent branchial arteries to
    (*a*) Kidney (*b*) Brain
    (*c*) Liver (*d*) Gills.
15. Each gill-pouch in *Petromyzon* communicates with the respiratory pharynx by
    (*a*) Gill-arches
    (*b*) Gill-lamella
    (*c*) External gill-slits
    (*d*) Internal gill-slits.
16. Exoskeleton in lampreys is
    (*a*) Bony
    (*b*) Cartilaginous
    (*c*) Scaly
    (*d*) Absent.
17. The teeth in upper and lower side of mouth of lampreys are grouped
    (*a*) Preoral and supraoral tooth plates
    (*b*) Preoral and postoral tooth plates
    (*c*) Supraoral and infraoral tooth plates
    (*d*) Suboral and supraoral tooth plates.

18. Two tubes are given from the buccal early in *Petromyzon*
    (*a*) Oesophagus and respiratory pharynx
    (*b*) Eustachian tube and intestine
    (*c*) Oesophagus and Eustachian tube
    (*d*) Nasopharynx and intestine.
19. In *Petromyzon* water cavity by
    (*a*) Endostyle
    (*b*) Velum
    (*c*) Typhlosole
    (*d*) Spiral valve.
20. The outermost layer of cutaneous epidermis
    (*a*) Have a ciliated cuticular border
    (*b*) Have a squamous cuticular border
    (*c*) Have a glandular cuticular border
    (*d*) Have a striated cuticular border.
21. In *Petromyzon* dermis is composed of
    (*a*) Collagen and elastin fibres
    (*b*) Only elastin fibres
    (*c*) Only collagen fibres
    (*d*) Longitudinal muscles.
22. In *Petromyzon* muscles of trunk and tail are arranged in
    (*a*) C-shaped myotomes
    (*b*) D-shaped myotomes
    (*c*) E-shaped myotomes
    (*d*) F-shaped myotomes.
23. In *Petromyzon* the cartilage surrounding the buccal funnel is
    (*a*) Branchial basket
    (*b*) Lingual cartilage
    (*c*) Annular cartilage
    (*d*) Pericardium.
24. In *Petromyzon* passage of tube in the respiratory tube is prevented by
    (*a*) Oral hood
    (*b*) Velum
    (*c*) Sphincter muscles
    (*d*) Scroll valve.
25. In *Petromyzon* anticogulant is secreted by
    (*a*) Goblet cells
    (*b*) Mucous glands
    (*c*) Zymogen cells
    (*d*) Salivary glands.
26. In *Petromyzon* optic lobes are present in
    (*a*) Forebrain
    (*b*) Midbrain
    (*c*) Hindbrain
    (*d*) Medulla oblongata.
27. Intermediates between cephalochordates and vertebrates are
    (*a*) Ostracoderms
    (*b*) Placoderms
    (*c*) *Branchiostoma*
    (*d*) Ammocoete larva.

ANSWERS

1. (*a*)	2. (*a*)	3. (*a*)	4. (*a*)	5. (*a*)	6. (*c*)	7. (*a*)	8. (*a*)
9. (*a*)	10. (*a*)	11. (*d*)	12. (*a*)	13. (*b*)	14. (*d*)	15. (*d*)	16. (*d*)
17. (*c*)	18. (*a*)	19. (*b*)	20. (*d*)	21. (*a*)	22. (*c*)	23. (*c*)	24. (*b*)
25. (*d*)	26. (*b*)	27. (*d*)					

12

Chapter

Class 2. Cyclostomata : General Characters, Classification and Representative Types

Contents of this Chapter

GENERAL CHARACTERS

1. Body long, rounded and eel-like.

2. Median fins with cartilaginous fin rays, but no paired appendages. Tail diphycercal.

3. Skin soft, smooth, containing unicellular mucous glands but without scales.

4. Trunk and tail muscles segmented into myotomes separated by mycommata.

5. Endoskeleton fibrous and cartilaginous. Notochord persist throughout life. Imperfect neural arches (arcualia) over notochord represent rudimentary vertebrae.

6. Jaws absent (Group Agnatha).

7. Mouth ventral, suctorial and circular. Due to circular mouth, the class name Cyclostomata (Gr., cyclos = circular, stoma = mouth).

8. Nostril is single and median.

9. Digestive system without stomach. Intestine with a fold, typhlosole.

10. Gills 5 to 16 pairs in lateral sac-like pouches of pharynx, hence, another name of class **Marsipobranchii**. Gill-slits 1 to 16 pairs.

11. Heart 2 chambered with 1 auricle and 1 ventricle, with a conus arteriosus anteriorly. Many aortic arches in gill region. Hepatic portal system present. Blood with leucocytes and nucleated circular erythrocytes. Body temperature variable (poikilothermal).

12. Two metanephric kidneys with ducts to urinogenital papilla.

13. Dorsal nerve cord with differentiated brain. 8 to 10 pairs of cranial nerves.

14. Single median olfactory sac. Auditory organ with 1 or 2 semicircular canals.

15. Sexes separate or united. Gonad single, large, without gonoduct.

16. Fertilisation external. Development direct or with a prolonged larval stage.

CLASSIFICATION

About 50 species of cyclostomes are recognised. They belong to two major divisions (**Petromyzontiformes** and **Myxiniformes**). They are termed variously as subclasses, orders or families. Because they possess a round jawless mouth, they are combined in the class Cyclostomata. The similarity of these two groups is probably the result of convergent evolution. However, they show important and basic morphological differences which can be attributed to their long phylogenetic separation and different habits and habitats.

Order 1. Petromyzontiformes (Gr., *petros* = stone; *myzon* = suck): Members of this order are called lampreys or lamper eels or lamperns or sand pride, etc.

1. Mouth ventral, suctorial with rasping tongue beset with many horny teeth.
2. Nostril dorsal. Nasohypophyseal sac terminates posteriorly in a blind sac, *i.e.*, it does not communicate with the pharynx.
3. 7 pairs of gill-pouches and gill-slits which open into a separate respiratory pharynx below the oesophagus.
4. Dorsal fin well developed.
5. Branchial basket complete.
6. Dorsal and ventral roots of spinal nerves remain separate.
7. Ear with 2 semicircular canals.
8. Eggs numerous, small. Development indirect with a long larval stage and metamorphosis.
9. Both marine and freshwater forms.

Examples : Lampreys. Over 30 species. *Petromyzon*, *Lampetra*, *Ichthyomyzon*.

Order 2. Myxiniformes (Gr., *myxa* = slime; *oidea* = type of): Representatives of order are called hagfishes. They are exclusively marine.

1. Mouth terminal and surrounded by 8 small tentades. Teeth few. No buccal funnel.
2. Nostril terminal. Nasohypophyseal sac opens posteriorly in the pharynx.
3. Gill-pouches and gill-slits 6 to 14 pairs.
4. Dorsal fin feeble or absent.
5. Branchial basket poorly developed.
6. Dorsal and ventral roots of spinal newes united.
7. Ear with only 1 semicircular canal.
8. Eggs few, large. Development dark.
9. Hagfishes are all marine species.

Examples : Hagfishes. About 15 species *Myxine*, *Eptatretus* (= Bdellostoma), *Paramyxine*.

REPRESENTATIVE TYPES

1. *Myxine.* Members belonging to the order Myxiniformes are commonly known as **hagfishes**. They are exclusively marine. *Myxine* (Fig. 12.1) is found buried in the sea bottom. *Myxine* has a wide distribution along sea coasts of both Atlantic and pacific Oceans, occurring in the waters of Northern Europe, North Atlantic, America, Chili, Africa and Japan, etc. Body is eel-like, measuring about 2 feet (50–60 cm) in length and differentiated into head, trunk and tail. The surface of the body is soft and smooth without scales. The mouth is terminal and surrounded by soft lips. Buccal funnel and jaws are absent. Branchial basket is also reduced. Lateral to the mouth are four pairs of short tentacles supported by skeletal rods. Nostril is single, lies very close to the mouth and opens terminally. Single pineal eye is visible on the top of the head. Paired eyes are vestigial or degenerated due to bottom dwelling habit. Six pairs of gills which do not open separately to the outside but open by a single pair of **external gill openings**. Single median ventral fin runs from about the middle of the ventral surface extending around the tail region. Large mucus glands are present opening by **mucous pores** along both the sides of the body and secrete mucus. Hermaphroditic with single **ovotestis,** anterior part being ovary and posterior testis. These animals are parasitic or quasi-parasitic because they are sometimes found within the bodies of their prey, which are fishes of various types. Nocturnal feeders. During the day time they live buried in the sea bottom mud at depths of over 2,000 feet. Hagfishes do not migrate to freshwater to spawn. Development is direct.

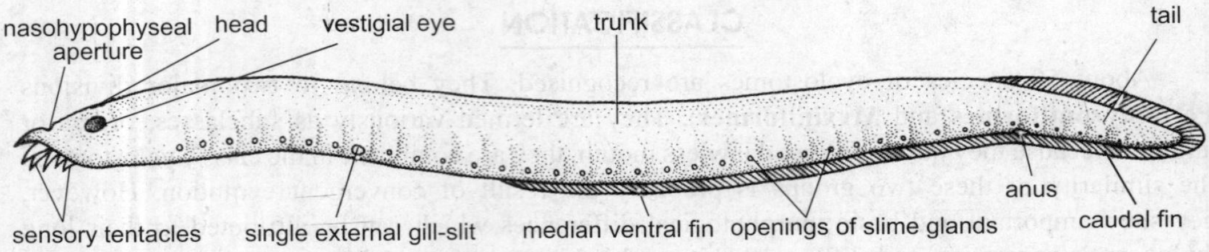

Fig. 12.1. *Myxine.*

2. ***Eptatretus* (= *Bdellostoma*).** *Bdellostoma* (Fig. 12.2) is also commonly known as **hagfish**. It is found buried in the bottom mud of sea. It occurs off the Pacific coasts of both North and South America, South Africa and New Zealand. The long eel-like body has a soft smooth integument without scales. It is about one metre in length. The mouth is terminal surrounded by soft lips. Buccal funnel and jaws are absent. Four pairs of short tentacles supported by skeletal rods are present on the lateral sides of the mouth. The single nostril lies very close to the mouth and opens terminally. Single pineal eye is present on the top of the head. Paired eyes are vestigial or degenerated due to the bottom dwelling habit. The **gill-openings** are 6–14 in number which all open independently to the exterior by round pores. The opening of pharyngocutaneous duct lies behind the last gill-slit of left side. It opens into the pharynx. The median fin is confined to the caudal region. Large slime or mucous glands are present opening by mucous pores on both sides of the body. Hermaphroditic. Single **ovotestis,** the anterior part being ovary and the posterior testis. It is parasitic or quasi-parasitic. Nocturnal feeders. During the day time they live buried in the sea bottom mud.

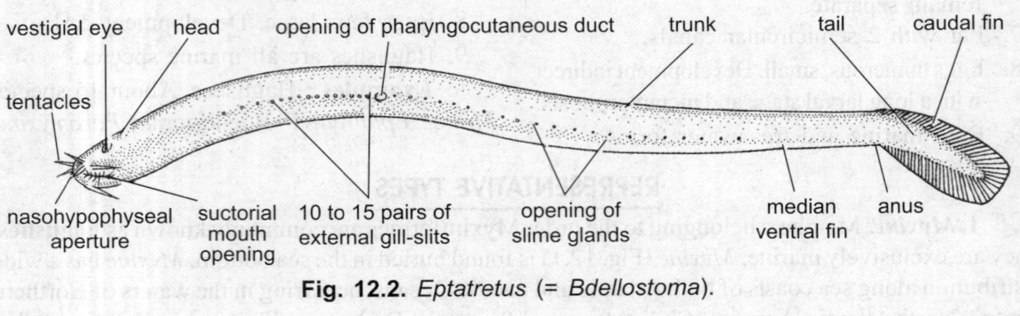

Fig. 12.2. *Eptatretus* (= *Bdellostoma*).

IMPORTANT QUESTIONS

Long Answer Questions

1. Give the important characters of Cyclostomata and classify the group.
2. Give characters of Cyclostomata.

Short Answer Questions

1. Certain fishes do not die when taken out of water. Explain.
2. Write brief note on external features of *Myxine*.
3. What is the food of *Myxine*? What is its importance to fisheries?
4. Give the external features of *Myxine*.
5. Write notes on : (*i*) *Eptatretus*, (*ii*) *Myxine*.

Very Short Answer Questions

1. Which protochordate do cyclostomes resemble most?
2. The presence of single common branchial aperture on each side of *Myxine*. Explain in two lines.

Multiple Choice Questions

1. The larva of *Petromyzon* lasts for about
   (*a*) 5 to 6 days (*b*) 3 to 7 years
   (*c*) 4 to 5 weeks (*d*) 4 to 5 months.
2. Tail in cyclostomes is
   (*a*) Diphycercal (*b*) Homocercal
   (*c*) Heterocercal (*d*) Hypocercal.
3. In cyclostomes scales are
   (a) Placoid (*b*) Absent
   (*c*) Cycloid (*d*) Ctenoid.
4. Intestinal fold in cyclostomes is
   (*a*) Typhlosole (*b*) Scroll valve
   (*c*) Endostyle (*d*) Velum.
5. Cranial nerves in cyclostomes are
   (*a*) 8 to 10 pairs (*b*) 4 to 6 pairs
   (*c*) 6 to 8 pairs (*d*) 10 pairs.
6. Nasohypophyseal sac in Petromyzontiformes is
   (*a*) Closed and not connected to pharynx
   (*b*) Open not connected to pharynx
   (*c*) Close connected to pharynx
   (*d*) Open connected to pharynx.
7. Mouth in Myxiniformes is
   (*a*) Terminal with four pairs of tentacles
   (*b*) Subterminal with 4 pairs of tentacles
   (*c*) Lateral with 4 pairs of tentacles
   (*d*) Ventral with 4 pairs of tentacles.
8. Myxiniformes animals are
   (*a*) Marine
   (*b*) Freshwater
   (*c*) Estuarine
   (*d*) Some freshwater and some marine.
9. Ear in Petromyzontiformes possesses
   (*a*) No Semicircular canals
   (*b*) 1 semicircular canal
   (*c*) 2 semicircular ducts
   (*d*) 3 semicircular ducts.
10. Ear in Myxiniformes are
    (*a*) No semicircular duct
    (*b*) 1 semicircular duct
    (*c*) 2 semicircular ducts
    (*d*) 3 semicircular ducts.
11. Members of the order Myxiniformes are commonly known as
    (*a*) Sharks (*b*) Lung fishes
    (*c*) Hagfishes (*d*) Lampreys.

ANSWERS

1. (*b*) 2. (*a*) 3. (*b*) 4. (*a*) 5. (*a*) 6. (*a*) 7. (*a*) 8. (*a*)
9. (*c*) 10. (*b*) 11. (*c*)

13 Chapter

Class 3. Placodermi (Extinct) : General Characters and Classification

Contents of this Chapter

THE EARLIEST JAWED VERTEBRATES

In the Silurian, some 450 million years ago, as the ostracoderms were disappearing, a host of more efficient and jawed fishes appeared. Earlier grouped in a single category, they are now placed into two separate classes : **Placodermi** and **Acanthodii**. According to taxonomists, these classes are placed lower to Chondrichthyes in the evolutionary scale.

Acanthodians were freshwater, superficially shark-like fishes. Usually they had elongated fusiform bodies almost eel-like. Their bodies were covered by diamond-shaped scales. A common fossil genus *Climatius* was about 8 cm long. Between larger pectoral and pelvic fins, they had extra smaller paired fins some of which in the form of stout spines, hence, the name 'spiny shark'. Some of them had **true teeth**, especially on the lower jaw. The acanthodians survived into earlier Permian before they became extinct.

OCCURRENCE

Placodermi were the earliest jawed vertebrates of fossil record. They appeared in Silurian, flourished in Devonian and Carboniferous and became extinct in Permian. They probably lived both in freshwater as well as sea water. Some primitive agnath ostracoderms were probably the ancestors of placoderms. But their fossil record does not exhibit any connecting link between the jawless and jawed fishes. Like the ostracoderms, the jawed fishes also appear fully formed without intermediates. Fossil evidence for ancestry of placodermi simply does not exist.

IMPORTANT FEATURES

The placoderms possess the heavy defensive armour of bony plates with powerful jaws and efficient fins.

1. Bony armour. The name placodermi means 'armoured fish' or 'plate skinned' (Gr., *plakos* = plates; *derma* = skin). The dermal armour of placodermi links them genetically with their predecessors, the osteoderms. The placoderms were highly diversified. They were ranging in length from a few centimeters to three meters or more. The giant predator, *Dunkleosteus*, grew to ten meters. Despite the differences among them, all were characterised by the presence of a bony skeleton. Some in particular *Dunkleosteus* exhibited a heavy armour of bony plates over the head and anterior part of trunk, while the rest of the body was nearly naked.

2. Jaws. All placoderms possess jaws. The jaws are supposed to have originated from the first pair of gill-bars (mandibular arch) in front of the first gill-slit. Hyoid gill-arch persisted and as a result of which spiracles had not arisen. Hyoid arch remained unmodified and did not support the jaws. This type of primitive jaw suspension called **aphetohyoidean** type. The placoderm jaw is **autostylic**, *i.e.,* it is articulated by its own processes to the cranium. No placoderms ever developed teeth of a modern type as did the acanthodians. The jaws of placoderms were immovably bound to cranium or rest of the head. These handicaps may have been involved in their eventual extinction.

Fig. 13.1. *Dinichthyes.*

3. Paired fins. Most placoderms possess paired fins. These are essentially adaptive organs. In an aquatic environment, development of strong mobile fins was coincident with the evolution of jaws, for swimming faster. The lateral fins served to produce turning movements in any direction (right, left, up or down) and to prevent roll, pitch and yaw when swimming in a straight path.

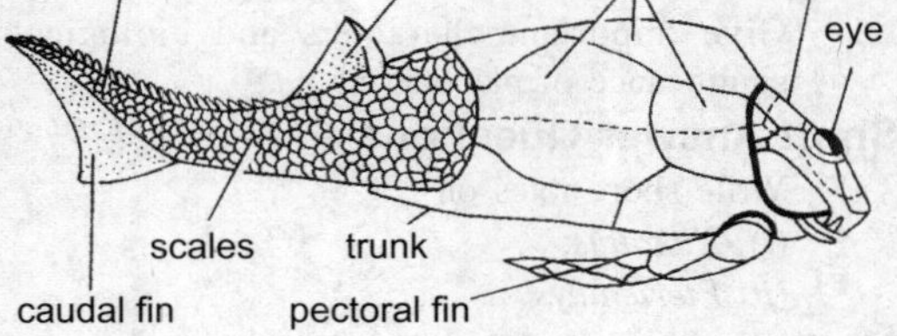

Fig. 13.2. *Pterichthyodes.*

Placoderms survived for short period only and are often considered as 'unsuccessful ancient experiment' in the evolution of gnathostomes.

CLASSIFICATION

Various schemes of classification have been forwarded by different workers. **Berg** (1940) and **Romer** (1959) forwarded detailed classification of fishes, which have been generally accepted. A new scheme of classification, often called modern classification of fishes, have been forwarded by **Bertin** and **Armbourg** (1958) and **Greenwood** *et.al.*, (1989).

All the extinct earliest jawed fishes were once believed to be generally related and, therefore, placed in a single group or class, the **Placodermi**. But the spiny sharks are now placed in a separate class **Acanthotii**. Two groups, subclasses or orders of class **Placodermi** are generally recognised. These are : Arthrodiriformes and Antiarchiformes.

Order 1. Arthrodiriformes. This group flourished in the Devonian. Earliest placoderms. Resembled with ostracoderms in appearance and habitat. Heavy bony armour, shields of head and trunk meeting in a movable joint (Gr., *arthros* = joint). Powerful gaping jaws with sharp shearing blades. Violently predaceous. Example : *Dunkleosteus*, *Coccosteus*, *Dinichthyes* (Fig. 13.1).

Order 2. Antiarchiformes. This group was confined to the Devonian period. Bottom-dwellers and mud-feeders in freshwater. Head and fore part of trunk was covered by a depressed bony shield. These cephalic plates had a different arrangement and the mode of attachment of head with body shield was also different. Pectoral fins long. Example : *Bothriolepes*, *Pterichthyodes* (=*Pterichthys*) (Fig. 13.2).

BIOLOGICAL SIGNIFICANCE

Placoderms were the earliest of the known fossil jawed vertebrates. They appeared in the Silurian, flourished in the Devonian and Carboniferous and became extinct in the Permian. Their bony armour links them genetically with their predecessors, the ostracoderms. Due to the presence of bony jaws and paired fins, they were more specialised than ostracoderms. A new era in the history of fishes started with the advent of jaws. Prof. A.S. Romer in his book '**The Vertebrate Body**' recalls that "Perhaps the greatest of all advances in vertebrate history was the development of jaws and the consequently revolution in the mode of life of early fishes. There exists no fossil evidence for the ancestry of placoderms. Some primitive ostracoderms were probably their ancestors. According to one view, the placoderms all became extinct by the end of Palaeozoic era, without giving rise to any surviving forms. According to another view, the cartilaginous fishes or **Chondrichthyes** and the bony fishes or **Osteichthyes**, both arose in the early Devonian from some primitive group of placoderms. Contrary to the former belief, the cartilaginous fishes did not give rise to the bony fishes. Thus bone, rather than cartilage, is to be considered a primitive characteristic.

IMPORTANT QUESTIONS

Long Answer Questions

1. Give important characters and biological significance of placoderms.

Short Answer Questions

1. Write short notes on :
   (*i*) *Climatius*;
   (*ii*) *Pterichthys*

Multiple Choice Questions

1. Earliest jawed vertebrates of fossil record are
   (*a*) Placoderms (*b*) Ostracoderms
   (*c*) Acanthodii (*d*) Cyclostomes.
2. Placoderms had
   (*a*) Movable jaws
   (*b*) Upper jaw movable
   (*c*) Immovable jaws
   (*d*) Only lower jaw movable.
3. What is true for placoderms?
   (*a*) Plate skinned
   (*b*) Spiny sharks
   (*c*) Lampreys
   (*d*) Hagfishes.
4. Jaw suspension in placoderms is
   (*a*) Holostylic
   (*b*) Autostylic
   (*c*) Hyoidean type
   (*d*) Aphetohyoidean type.
5. Placoderms flourished in
   (*a*) Devonian and Carboniferous
   (*b*) Silurian
   (*c*) Permian
   (*d*) None.

ANSWERS

1. (*a*) 2. (*c*) 3. (*a*) 4. (*d*) 5. (*a*).

14 Chapter

Type 5. *Scoliodon*: A Cartilaginous Fish (The Dogfish)

Contents of this Chapter

Pisces (Fishes) are aquatic, cold blooded vertebrates that breathe by means of gills and live in water. There are about 30,000 to 40,000 species of fishes differing widely from each other in shape, size, habits and habitat. They live in all the seas, rivers, canals, lakes, dams, ponds and in almost every place where there is water. Fishes usually have a streamlined body but some are elongated snake-like and few are dorso-ventrally flattened. They have paired and unpaired fins supported by soft or spiny fin rays. Dorsal, anal and caudal fins are unpaired, while the pectorals and pelvics are paired. There is always an exoskeleton of dermal scales. Respiratory organs are chiefly gills. Besides the olfactory, photic and auditory organs they have a lateral-line system or receptors. Sexes are always separate. Economically the fishes constitute a very important group of animals. Besides being used as food, fish liver is an important source of oil. Fishes also provide fish-meal, fish-manure, isinglass and several other products of commerce. It is, therefore, natural that man has paid a considerable attention to the study of habits and habitat, anatomy and physiology of fishes.

Living fishes with jaws fall into two well marked classes : **Chondrichthyes**, (Gr., *chondros* = cartilage; *ichthys* = fish), the cartilaginous fishes such as sharks and rays and **Osteichthyes** (Gr., *osteon* = bone; *ichthys* = fish), the bony fishes such as ray-finned fishes and lung fishes.

The Chondrichthyes, also called elasmobranchs, comprise sharks, rays, skates, chimaeras, etc. Dogfish sharks are extensively studied nearly all over the world. The description included here belongs to the common Indian dogfish shark, *Scoliodon sorrakowah*, worked out by **E.M. Thillayampallam** and first published in 1928 in the series of Indian Zoological Memoirs.

SYSTEMATIC POSITION

Phylum	...	Chordata
Subphylum	...	Vertebrata
Division	...	Gnathostomata
Superclass	...	Pisces
Class	...	Chondrichthyes
Subclass	...	Selachii
Order	...	Squaliformes (Pleurotremata)
Family	...	Carcharinidae
Genus	...	*Scoliodon*
Species	...	*sorrakowah*

DISTRIBUTION

Scoliodon has wide geographical distribution. It has been recorded from Zanzibar to Sri Lanka, and Sri Lanka to Malay Archipelago in the Indian Ocean, from the Bay of Bengal, the East Indies and the Philippine Islands, from Mexico to Panama in the Eastern Pacific, from Labrador to Brazil in the Atlantic, from Cuba in West Indies, and also off the Eastern coasts of South America. Four species of *Scoliodon* are found in Indian waters and these are *S. sorrakowah*, *S. dumerilii*, *S. palasorrah* and *S. walbeehmi*. The common Indian dogfish is *Scoliodon sorrakowah* which means 'black shark' in Tamil (*sorra* = a shark + *kowah* = black). Fossil records of *Scoliodon* have been in the geological strata from the lower Eocene to later periods.

HABITS AND HABITAT

Scoliodon is marine like most sharks. It is found in all parts of the open sea, *i.e.*, Indian, Pacific and Atlantic Oceans. It is carnivorous feeding on crabs, lobsters, worms and fishes, etc. It is a fast swimmer as the body is perfectly stream-lined without having rigid projections. Sexes are separate. Fertilisation is internal and development is direct. It is viviparous, *i.e.,* producing living young ones that develop inside the uteri.

EXTERNAL FEATURES

Shape, Size and Colour

Scoliodon has a long, laterally compressed spindle-shaped body tapering at both ends. The full grown specimen measures from 30 to 60 cm in length. The colour of the body is dark grey above and pale white beneath, while the portions of the caudal fin are more or less dark. Body surface is rough due to backwardly directed spines of placoid scales embedded in the dermis.

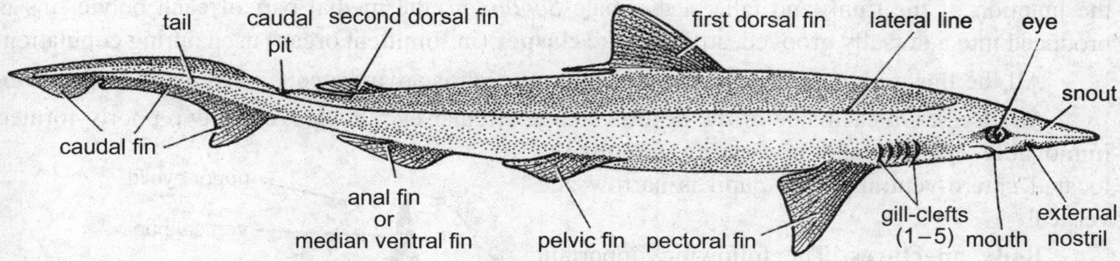

Fig. 14.1. *Scoliodon.* External features.

Division of Body

The body is divisible into **head**, **trunk** and **tail**, though there are no distinct boundaries between these regions.

***(i)* Head.** The head is strongly compressed dorso-ventrally and is produced in front into a wedge-shaped snout or rostrum.

***(ii)* Trunk.** The trunk is almost elliptical in transverse section. Its thickest part lying in front of the middle of the body. The trunk gradually tapers behind into the tail.

***(iii)* Tail.** The tail is laterally compressed and is bent upwards at a small angle and fringed with a caudal fin. Such a tail is known as **heterocercal tail**.

Fins. *Scoliodon* is provided with two sets of fins which are flattened expansions of the skin supported by cartilaginous rods and horny fin-rays : these are unpaired or **median fins** and **paired lateral fins**.

***(i)* Median fins.** The median fins are two **dorsal fins,** a **ventral** or **anal fin** and a **caudal fin**. The **first dorsal fin** is large and triangular in shape and is situated a little in front of the middle of the body. It has a basal lobe. The **second dorsal fin** is also triangular in outline but is very small and is situated midway between the first dorsal and the tip of the tail. The **caudal fin** extends along the dorsal and ventral surfaces of the tail in the median line and forms a dorsal and ventral lobe. The dorsal lobe is very much reduced and forms only a low ridge along the greater part of the upper surface but the ventral lobe is well developed and is divided into two parts, the anterior part being much larger and more extensive than the posterior. The **ventral** or **anal fin** is situated in the mid-ventral line about 5.0 cm in front of the caudal fin, more or less opposite the second dorsal. Each of the two dorsals and the ventral fin are produced behind into long and narrow fleshy processes called **basal lobes**.

***(ii)* Lateral fins.** The lateral fins are paired **pectoral** and **pelvic fins**. The large **pectoral fins** originate from the ventro-lateral margins of the body immediately behind the gill-clefts. The **pelvic fins** are much smaller than the pectoral fins and arise close together from the ventral surface at

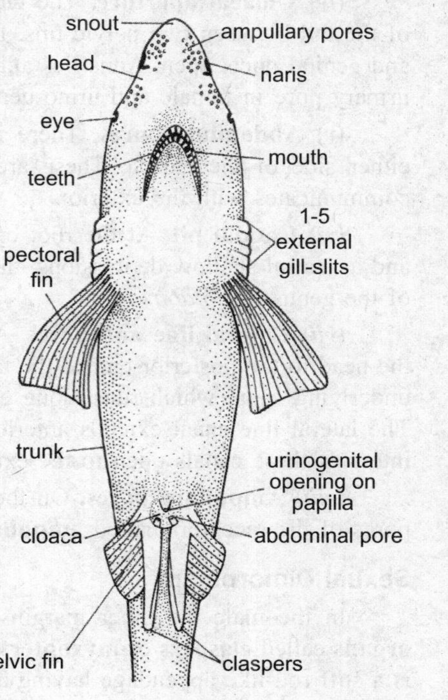

Fig. 14.2. Male Indian dogfish in ventral view.

the junction of the trunk and tail. In the male *Scoliodon*, the medial part of each pelvic fins is produced into a dorsally grooved, stiff rod-like **clasper** (intromittent organ) used during copulation.

All the fins are directed backwards a feature which is advantageous in forward progression.

(*iii*) Eyes. On each side of the head is a large circular eye. Each eye has two poorly-formed immovable **eye-lids** and **nictitating membrane** located antero-ventrally. The pupil is narrow and vertical.

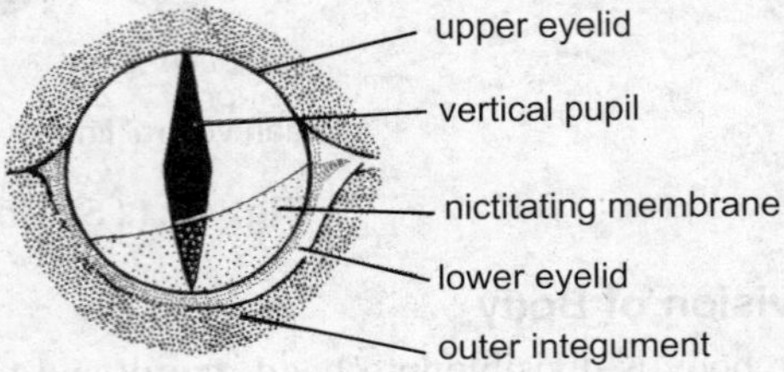

Fig. 14.3. *Scoliodon.* Right Eye.

Body apertures. The following important apertures are present on the body surface.

(*i*) Mouth. On the ventral side of the head is a wide crescentic mouth. It is bounded by upper and lower jaws. Each jaw is armed with one or two rows of sharp backwardly directed teeth, having smooth not-serrated edges for holding and tearing the prey. In catching its prey the shark brings its jaws into play by raising the snout and thrusting the mouth forward.

(*ii*) Nostrils. On the anterior side of the mouth are present two crescentic apertures the nostrils leading into olfactory sacs which do not open into the mouth cavity. A small fold of skin from the anterior edge partially covers each nostril. The nostrils are only olfactory and have no respiratory function.

(*iii*) Gill-clefts. Behind the eyes are situated a series of vertical slits, five on each side, called **branchial** or **gill-clefts**. These apertures lead into the gill-pouches and thence into pharynx. These are respiratory in function.

(*iv*) Cloacal aperture. The cloacal aperture is an elongated median opening at the root of the tail between two pelvic fins. It leads into the cloaca into which the intestine and urinary and genital ducts open. Anus lies anteriorly and in the cloaca and a cone-shaped papilla bearing urinary pore in female and urino-genital pore in male located behind the anus.

(*v*) Abdominal pores. There is a pair of openings situated on the elevated papillae on either side of the cloaca. These are the **abdominal pores**. Through these pores the coelom communicates with the exterior.

(*vi*) Caudal pit. At the root of the tail, just in front of the caudal fins, is a small dorsal and a ventral shallow depressions, the **caudal pits**. The caudal pits are the characteristic feature of the genus *Scoliodon*.

(*vii*) Lateral line and pores. A faint line runs on either side of the body extending from the head to the posterior end of the tail, this is called the lateral line. It marks the position of an underlying canal which runs along each side of the body and contains special receptor organs. The lateral line canal extends anteriorly into the head where it branches into several canals; at intervals these canals open to the exterior through minute pores.

(*viii*) Ampullary pores. On the head and snout open several groups of minute ampullary pores of the receptors called **ampullae of Lorenzini**. When pressed they exude mucus.

Sexual Dimorphism

In the male the inner margins of the pelvic fins bear a pair of rod-shaped **copulatory organs** called **claspers** or **myxopterygia**, they are modified parts of the pelvic fins. Each clasper is a stiff rod-like appendage having a groove on its dorsal surface which leads into a cavity, the **siphon,** beginning at the base of the clasper.

SKIN

The body is invested by an outer leathery covering called **skin** or integument. In *Scoliodon*, the skin or integument consists of two layers : an outer ectodermal **epidermis** and an inner mesodermal **dermis**.

1. Epidermis. The epidermis is composed of many layers of epithelial cells amongst which are interspersed numerous unicellular **mucous glands** secreting mucus for lubricating the body surface. Innermost layer of cells rest on a basement membrane and called stratum germinativum.

2. Dermis. The dermis comprises dense areolar connective tissue mixed with smooth muscle fibres, blood capillaries, pigment cells and nerves. Dermis or corium is divided into an outer layer having few loose fibres called the **stratum laxum**, and an inner layer having compact fibres called **stratum compactum**. Just below the epidermis are found pigment cells, the **melanophores** or **chromatophores**. The dermis is firmly attached to the underlying muscles. In a fresh specimen the skin is slimy, but in preserved specimens the slimy mucus is generally removed and the skin becomes rough. This roughness of the skin is due to the presence of closely lying minute dermal denticles called **placoid scales** which are arranged in regular oblique rows and form the exoskeleton of the shark covering the entire surface of the body and even parts of the fins.

Functions of skin. In *Scoliodon*, the skin serves the following functions :

(*i*) Like a wrapper, it protects internal organs against mechanical injuries. (*ii*) Secretion of slimy mucus makes body surface slippery and difficult to catch by predators, minimises friction in locomotion and resists entry of microorganisms into body. (*iii*) Its colouration imparts camouflage, as the fish blends with silvery surface of water when seen from below and with the dark bottom when seen from above, by predators. (*iv*) Skin receptors help in reacting to changes in the surrounding.

EXOSKELETON

In *Scoliodon*, the exoskeleton comprises the placoid scales, embedded in the skin.

1. Structure of a placoid scale. A typical placoid scale consists of mainly two parts : the **basal plate** and the flat trident **spine**. The **basal plate** is diamond-shaped and lies embedded in the dermis and firmly placed by Sharpey's and other connective tissue fibres. The basal plate is formed of the trabecular calcified tissue closely allied to the cement. The inner surface of the basal plate bears an opening which leads into the **pulp cavity**. During life the pulp cavity is filled with vascular connective tissue called the **pulp** containing numerous **odontoblasts (dentine-forming cells)**, **blood vessels**, **nerves**, and **lymph channels**. The **spine** is flat and trident and projecting out of the skin and backwardly directed. The surface of the spines is not smooth but presents a stratified appearance under the microscope. The spine is composed of a hard calcareous substance, the **dentine**. The dentine is traversed by minute nearly parallel **canaliculi** with delicate branches containing long protoplasmic processes of odontoblasts. It is coated externally with a layer of hard shiny enamel-like **vitrodentine** secreted by ameloblasts of the overlying ectoderm. The placoid scales are derived partly from the dermis and partly from the epidermis. The basal plate and the dentine of the spine are derived from mesoderm, while the enamel is secreted by the ectoderm.

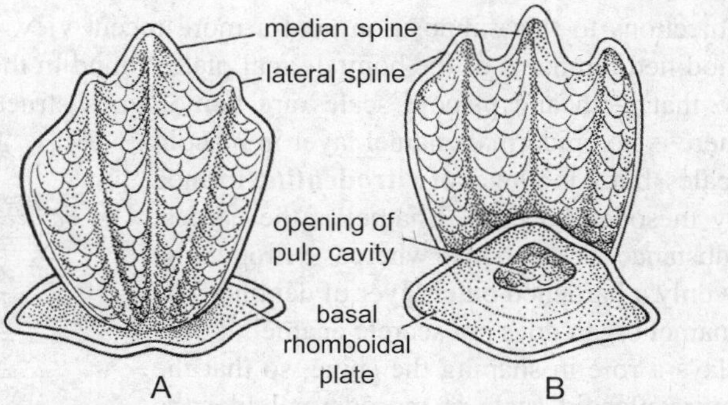

Fig. 14.4. *Scoliodon.* Placoid scales. A–Dorsal view; B–Ventral view.

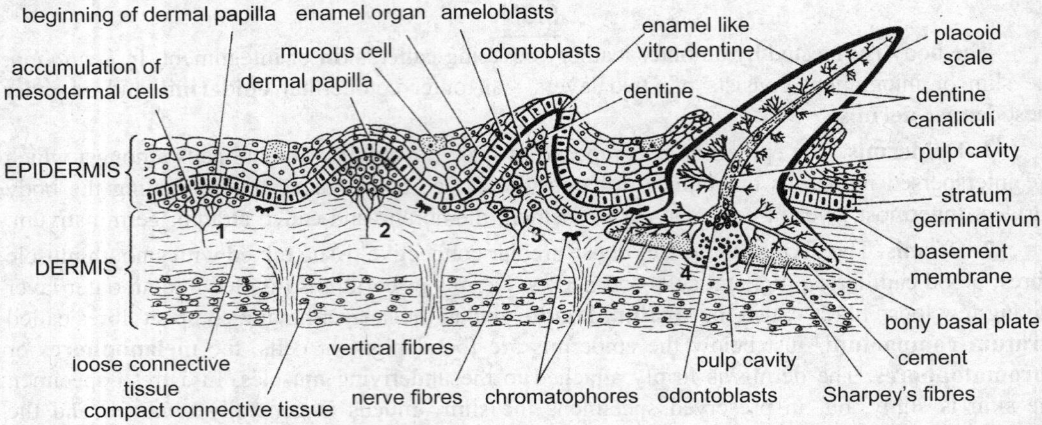

Fig. 14.5. *Scoliodon*. V. S. of skin showing development of placoid scale. 1. Accumulation of dermal cells; 2. Formation of dermal papilla; 3. Growing spine; 4. V. S. of a fully formed scale.

2. Development of placoid scale. A group of mesodermal cells of dermis collect in the dermis just beneath the epidermis to form a **dermal papilla**. The outermost cells of dermal papilla become the **odontoblasts**. The dermal papilla grows upwards pushing the epidermis, then it takes the shape of a basal plate and spine. The odontoblasts secrete dentine all around forming the basal plate and spine of the placoid scale. The dentine of the basal plate becomes calcified. The Malpighian layer of the epidermis overlying the dermal papilla is now called the **ameloblasts** form the **enamel organ**, which form the vitrodentine over the dentine of the spine. The dermal papilla in a growing scale forms pulp in the pulp cavity of the scale. The overlying epidermis moves away and also wears off so that the spine projects above, while the basal plate remains embedded in the dermis.

3. Homology of placoid scale. Placoid scales are the forerunners of vertebrate teeth because the two have essentially the same form, structure and embryonic development. The gradation from placoid scales to teeth is seen in the mouth of a shark. Shark teeth are enlarged placoid scales formed in the skin of jaws and are, thus, homologous with placoid scales. But there are objections to this supposition, and a more recent view is that both placoid scales and teeth are modified remnants of the bony dermal plates found in the ancestral ostracoderms and placoderms, so that teeth and placoid scales are homologous structures. Moreover, it has been shown that there is no epidermal enamel layer in placoid scales, but it is a layer of **vitrodentine** formed by mesodermal cells. Enamel is the hardest substance in the body, whereas vitrodentine is only a hardened outer layer of dentine. The enamel organ does not secrete enamel, it only plays a role in shaping the spine, so that the entire placoid scale is mesodermal like the scales of bony fishes, whereas a tooth has an enamel covering derived from ectoderm. The claim that a gradation from placoid scales to teeth is observed in the mouth of a shark is interpreted as a divergence shown between vitrodentine-covered placoid scales, on the one hand, and enamel-covered teeth on the other.

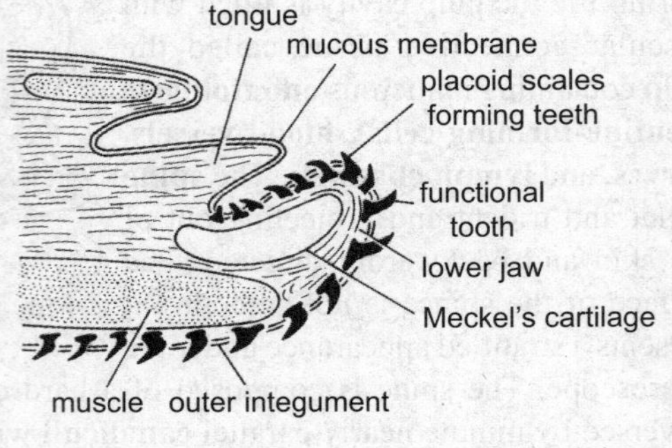

Fig. 14.6. *Scoliodon*. Section of lower jaw showing formation and replacement of teeth from placoid scale.

ENDOSKELETON

The endoskeleton is composed entirely of cartilage hardened at places by calcifications. The true bones are entirely absent. The endoskeleton can be divided into : **(*i*) axial skeleton** comprising the vertebral column, the skull and **(*ii*) appendicular skeleton** including the pectoral and pelvic girdles and the skeletal elements of fins.

Axial Skeleton

Vertebral column. The vertebral column comprises a chain of about 130 **vertebrae**, which develop around an elastic axis of vacuolated cells, the **notochord**. The vertebral column is clearly divisible into two regions : trunk and tail.

1. Trunk vertebra. A trunk vertebra consists of ventral, cylindrical **acentrum** surrounding the slender notochord, a dorsal neural arch enclosing the neural canal for the spinal cord, and a pair of **transverse processes** projecting ventro-laterally from the centrum. The vertebrae of trunk region are **amphicoelous**, *i.e.,* the centrum bears concavities on anterior as well as posterior surfaces. In transverse section the centrum shows four wedge-shaped calcified fibro-cartilages, arranged in the form of a cross which is called **Maltese cross**. These cartilages also leave some uncalcified area between them, hence, the vertebra or centrum is known as **asterospondylus**. Through the series of centra are found remnants of the notochord. The concave anterior and posterior surfaces of the centra are covered by a dense calcified layer. The centrum bears on its dorsal side two neural arches which meet mid-dorsally to form the **neural spine**. Each neural arch consists on each side a small cartilage, the neural plate (basidorsal) which becomes completely fused with the neural process in the adult. Between successive neural plates is present a series of **interdorsal** or **interneural plates**. Small median cartilages, the **neural spines** fit in between both neural and inter-neural plates of opposite sides and form the complete arch. From the ventrolateral side of the centrum there arise a pair of slender **transverse processes** which are directed outwards. The transverse processes of the anterior vertebrae are short and extend outward, but in the posterior trunk vertebrae they bend downward and enclose an open groove.

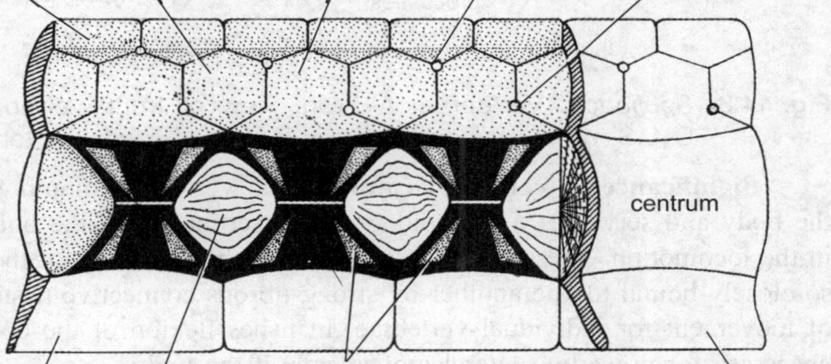

Fig. 14.7. *Scoliodon.* A portion of the vertebral column of the trunk.

2. Caudal vertebra. The **caudal vertebra** is slightly modified in comparison to the trunk vertebra. The neural arch is short, while the neural spine is very much large and pointed. The transverse processes converge to meet and fuse together forming a **haemal arch** which encloses a haemal canal through which caudal artery and vein pass. The haemal arch is produced into a long **haemal spine** pointing backwards. The haemal spines of the posterior caudal vertebrae support the large ventral lobes of the caudal fin, while the dorsal lobe is supported by neural spines. The posterior part of the vertebral column is bent dorsally to form an asymmetrical or **heterocercal tail**.

The **trunk vertebrae** are **monospondylous** but caudal vertebrae are **diplospondylous**, there being two complete vertebrae to each mytome.

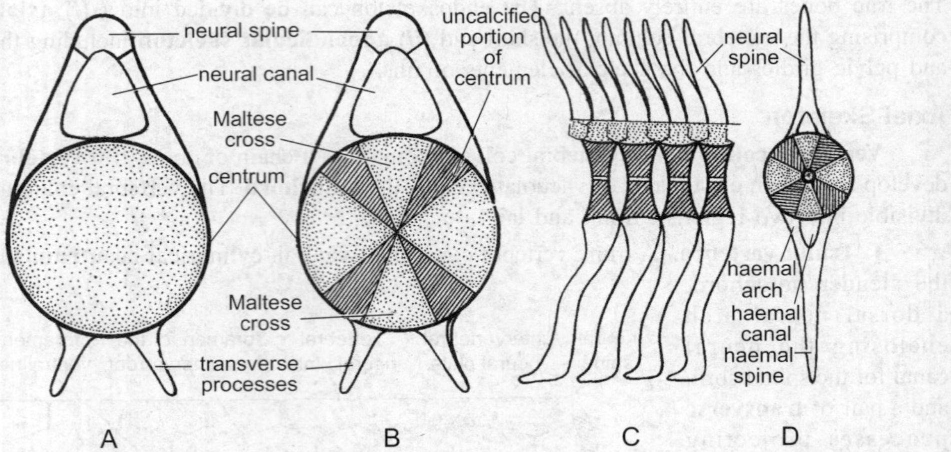

Fig. 14.8. *Scoliodon.* Vertebrae. A–Front view of trunk vertebra; B–T.S. of trunk vertebra; C–L.S. of four caudal vertebra; D–T.S. of caudal vertebra.

Significance of vertebral column. The vertebral column forms the axial scaffolding of the body and serves as a foothold for the insertion of all the body and tail muscles concerned in the locomotion. The segmentation increases the flexibility of the column but the vertebrae are so closely bound to one another by strong fibrous connective tissue that there is a little freedom of movement for individual vertebrae. In fishes flexion of the column as a whole is all that is necessary for prevailing lateral movements of the body.

Skull. The skull of *Scoliodon* consists of : (*i*) the **cranium** surrounding and protecting the brain, (*ii*) the **sense capsules** closing the olfactory and auditory organs, and (*iii*) the **visceral** skeleton forming the jaws and the skeletal supports of the pharynx and gills.

(*i*) Cranium and sense capsules. The cranium is simple and cartilaginous. Its general shape is very much like that of a violin box, open in front and behind, with an arched roof and a flattened floor. Two pairs of sense capsules, the olfactory and auditory, are intimately fused to it. In the hinder region the massive auditory capsules project on either side into a deep concavity for lodging the membranous labyrinth. The anterior region, besides presenting swelling on each side is formed by the olfactory capsule. Anteriorly the cranium gives off three cartilaginous bars which are prolonged forward and unite to form **rostrum**. The cranium may be roughly divided into four regions : (*a*) the **occipital region**, (*b*) the **auditory region**, (*c*) the **orbital region** and (*d*) the **ethmoidal region.**

(*a*) Occipital region. The occipital region forms the posterior part of the cranium. It is perforated behind by a large median opening, the **foramen magnum** through which the brain is continued into the spinal cord. Just below the foramen magnum there is a cup-shaped concavity which encloses the notochord and lies opposite to the concave centrum of the first vertebra. On either side of the foramen magnum, there is a prominent rounded **occipital condyle** for articulation with the first vertebra. Above the foramen magnum, the roof of the occipital region carries a prominent median crest called the **occipital crest**. External to the occipital condyle, on either side lies a large foramen for the exit of the tenth or vagus nerve.

(*b*) Auditory region. The auditory region lies in front of the occipital region and comprises two auditory capsules enclosing internal ears and the part of the cranium with which they are intimately fused. The auditory capsules remain separate from the cranium in the embryo but become firmly united with it in the adult. On the roof of the cranium, in between the two auditory capsules there is a marked oval depression, the **parietal fossa**. The parietal fossa possesses two pairs of apertures. The anterior pair of apertures are smaller through which the endolymphatic duct of each internal ear pierces the cranium. The posterior pair is comparatively larger and forms the openings of the perilymphatic spaces of the two capsules. Each auditory capsule possesses three ridges which mark the position of three **semicircular canals** of the internal ear. Hyomandibula of hyoid arch (2nd) articulates with in the post-orbital groove of both sides.

***(c)* Orbital region.** The orbital region lies anterior to the auditory region almost in the middle of the cranium. It is hollow, called the orbit and lodges the eyes, their muscles, the large orbital blood sinuses and a large number of nerves. Each orbit is bordered dorsally by a **supra-orbital ridge** or **crest**, ventrally by **infra-orbital** or **sub-orbital ridge**, anteriorly and posteriorly by **pre-orbital** and **post-orbital processes** respectively. These processes are the extensions of supra-orbital ridge. Within the orbit in the cranial wall are present a number of foramina for cranial nerves. The anterior boundary of the orbit is formed by a posteriorly directed slender process which arises from the posterior part of the olfactory capsule and is called the **pre-orbital cartilage**. A similar process arising from the anterior part of the auditory capsule directed forwards marks the posterior boundary. This process is known as **post-orbital cartilage**. A lateral outgrowth of the floor of the orbit forms the **suborbital ridge** which is produced anteriorly into the ant-orbital process forming a place for the insertion of ligaments of the upper jaw.

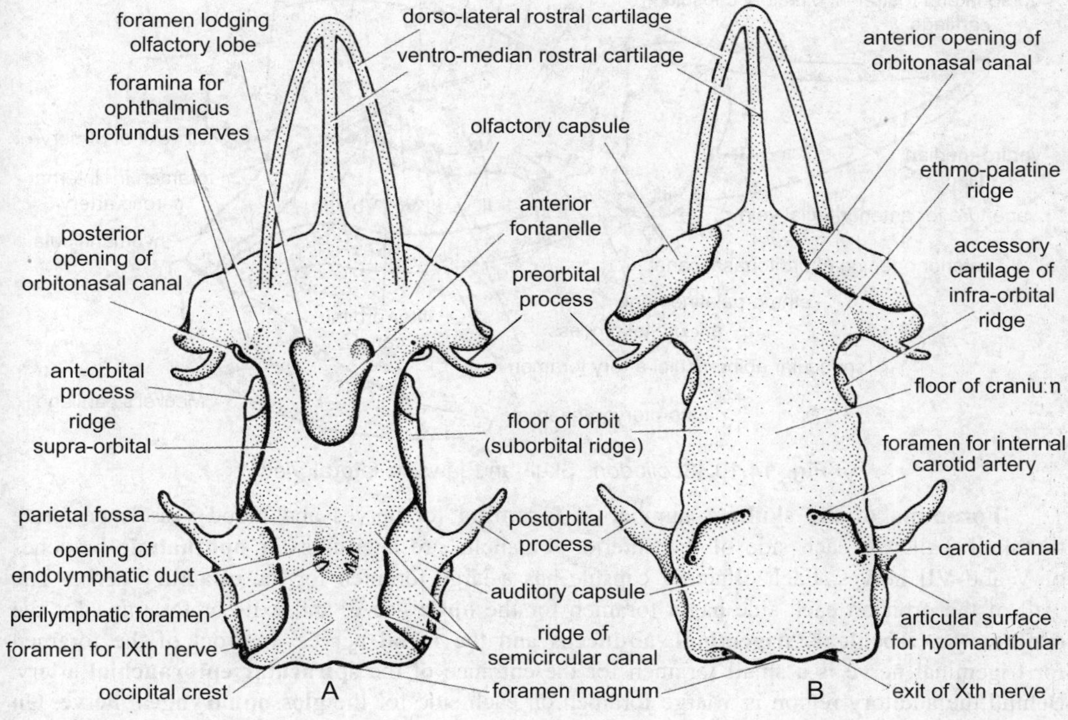

Fig. 14.9. *Scoliodon*. Skull. A–Dorsal view; B–Ventral view.

***(d)* Ethmoidal region.** The ethmoidal region consists of two olfactory capsules and the rostrum. The lateral olfactory capsules are large oval cartilaginous cups at the anterior end of the skull which are firmly united to the cranium in the adult. Each capsule lodges an olfactory sac. Postero-ventral surface of each capsule has a short, stout **ethmopalatine ridge**. A thin median cartilage, the **internasal septum** or **mesethmoid** separates the two olfactory capsules from each other. The cranial roof in this region shows a depression, the **anterior fontanelle** covered over by a sheet of connective tissue. With in each olfactory capsule there is a large opening leading into the cranial cavity, through which the olfactory nerve enters the olfactory sac of its own side. The rostrum or **snout** is formed by the fusion of three cartilaginous bars. In front of the anterior fontanelle arise a pair of dorso-lateral cartilages, one from the roof of each olfactory capsule. These cartilages run forward to converge and meet in front with a **median ventral cartilage** which projects from the base of the cranium.

The **floor of the cranium** is broad and flat and bears towards its hind end two obliquely transverse grooves called the **carotid canals**. The anterior half of each carotid canal bears two apertures, one leading into the cranium transmitting the **internal carotid artery** and the other leading into the orbit transmits the **stapedial artery**. Immediately behind the olfactory capsules lie two large prominent articular surfaces for the attachment of the **ethmopalaline ligaments** of the upper jaw, while in front of these articular surfaces lie the **anterior openings of the orbito-nasal canals**.

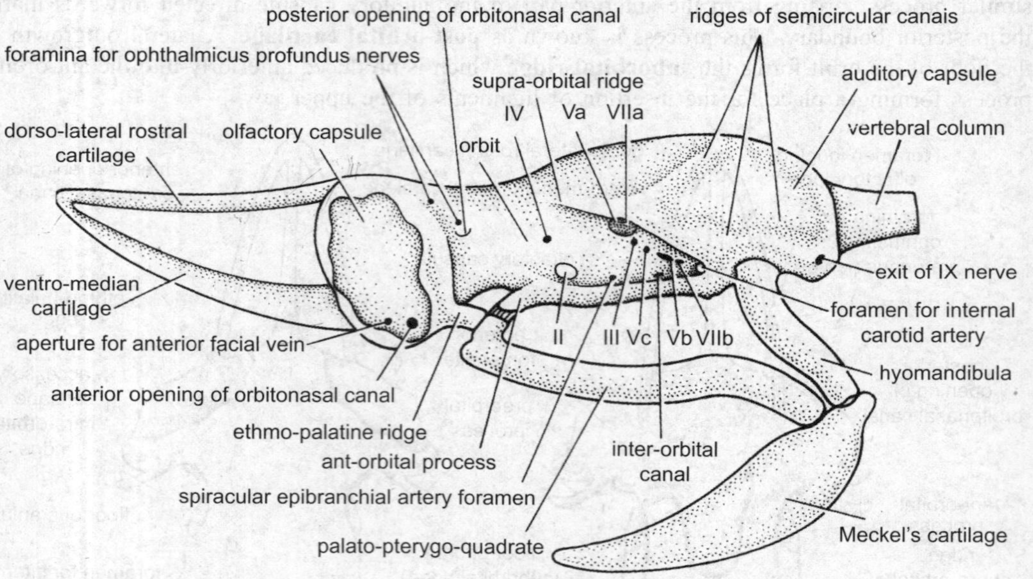

Fig. 14.10. *Scoliodon*. Skull and jaws in lateral view.

Foramina of the skull. A number of **foramina** for nerves and blood vessels perforate the skull wall. On each side of the anterior fontanelle are apertures for **ophthalmic** branches of V and VII nerves. Each olfactory capsule has a large foramen for the **olfactory** nerve. The wall of the orbit on each side has a foramen for the **optic** nerve, and exits or foramina for the **oculomotor**, **trochlear**, **trigeminal**, **abducens** and the **facial** nerves. In front of the foramen for trigeminal nerve is a small foramen for the entrance of the **spiracular epibranchial** artery. Behind the auditory region is a large foramen on each side for the **glossopharyngeal** nerve. On the outer side of each occipital condyle is a large foramen for the **vague** nerve. The floor of the skull in the posterior part of the cranium has two large **carotid canals**, each has two foramina, one of the **internal carotid** artery, and the other for the **stapedial** artery.

Visceral skeleton. The visceral skeleton or the splanchnocranium is closely connected with the cranium and forms the jaws and pharyngeal skeleton. It consists of a series of seven pairs of U-shaped **visceral arches** which lie in the sides and floor of the pharynx.

(*i*) The first **visceral arch** or **mandibular arch**, forms the upper and lower jaws bearing teeth. The upper jaw or **palatoquadrate** has two cartilages joined in front. Near its anterior end it has an orbital process from which ligaments arise to join the cranium. The lower jaw has two **Muckel's cartilages** united anteriorly. At its posterior end each Meckel's cartilage has an articular surface for a movable joint with the upper jaw.

(*ii*) The **second visceral arch** or **hyoid arch** is made of five cartilages, a lower median **basihyal** in the floor of the pharynx supporting the buccal floor and tongue, and a pair of **ceratohyal** in the buccal floor and a pair of dorsal **hyomandibular** which bear slender cartilaginous branched **gill-rays**. Each hyomandibular is connected at its upper end articulates with the cranium, and at its lower end with the palatoquadrate (upper jaw) and Meckel's cartilages (lower jaw). The jaws are not directly joined to the cranium but are only suspended by the hyomandibular, this type of suspensorium is **hyostylic** and is found in most Chondrichthyes. The remaining hyoid arch supports a gill.

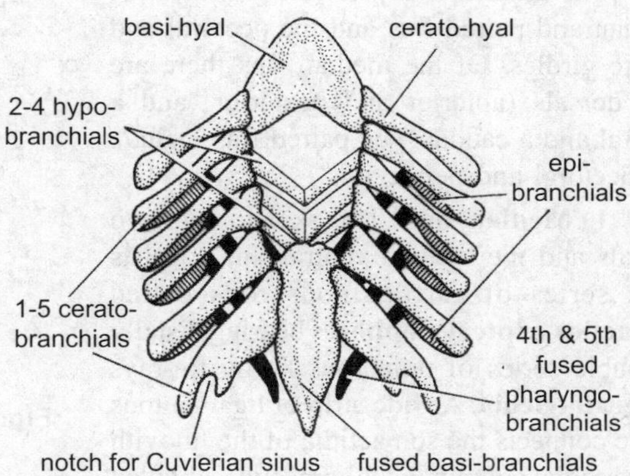

Fig. 14.11. *Scoliodon*. Visceral skeleton in ventral view.

(*iii*) **Branchial arches.** The remaining posterior five visceral arches are called **branchial arches** because they support the gills and the wall of pharynx allowing muscular movements of the pharyngeal wall for a respiratory current. Typically each branchial arch is made of nine pieces of cartilages, a lower median rod-like **basibranchial** to which are attached on each side a ventral **hypobranchial**, a lateral **ceratobranchial**, and **epibranchial**, and a dorsal **pharyngobranchial** cartilages. The ceratobranchials and epibranchials bear slender cartilaginous unbranched **gill-rays** which support the gills. Besides these there are four pairs of **extra-branchial cartilages** which lie at right angles to the first four pairs of branchial arches.

Variations in the branchial arches. In *Scoliodon* some of the cartilages are lost or fused with others. It has all five basibranchials fused to form a single cartilage. The hypobranchials are very small in the first branchial arch and absent in the fifth. All five ceratobranchials are present, the fifth one being very large. The first four epibranchials are present but the fifth one is lost. There are four pharyngobranchials, those of the fourth and fifth branchial arches having fused into a single piece.

The skull is, thus, formed from the **chondrocranium**, **first visceral arch**, and the **dorsal part** of the second visceral arch. This condition is found in all fishes and tetrapoda, except in ungulates and primates in which the middle part of the hyoid arch also fuses with the skull in the auditory region as the **styloid process**. The skull forms the support of the head with protective coverings for the brain and sense organs, and it forms the skeleton of the jaws and internal support of the pharynx and gills. The presence of jaws

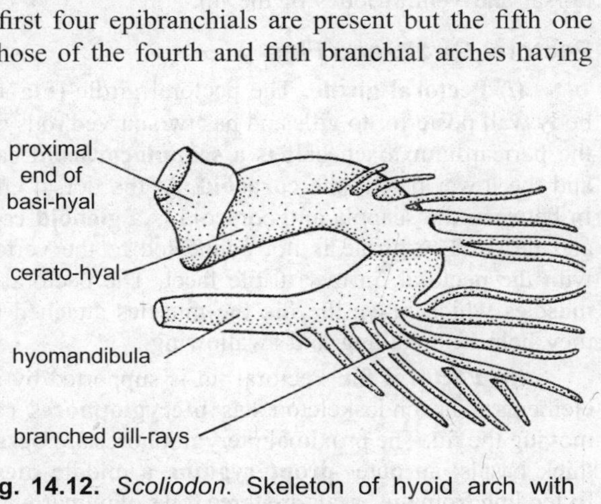

Fig. 14.12. *Scoliodon*. Skeleton of hyoid arch with gill-rays in ventral view.

is a great advance in evolution, they support the mouth which is used not only in feeding and biting but it is also used for handling the environment, as in nest building in many fishes, it is useful for holding objects.

Appendicular Skeleton

The appendicular skeleton includes the median and paired fins and the pectoral and pelvic girdles. Of the median fins there are two dorsals (anterior and posterior) and a ventral and a caudal. The paired fins include the pectoral and pelvic fins.

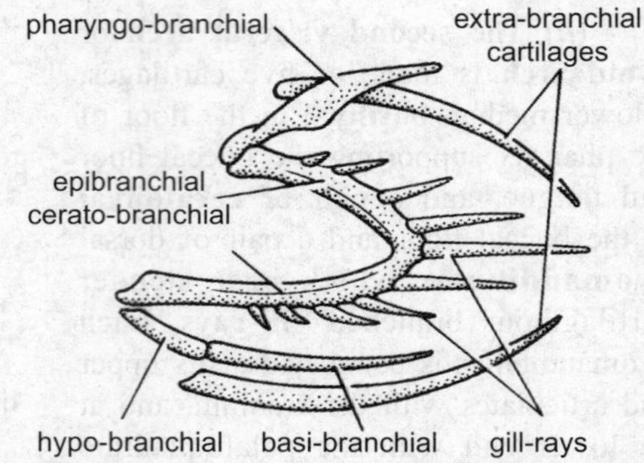

Fig.14.13. *Scoliodon*. Skeleton of the first branchial arch of left side.

1. Median fins. The skeleton of two dorsals and the median ventral fins consists of a series of cartilaginous rods called **somactids** or **pterygiophores** bearing distally a double series of numerous horny fin-rays or **ceratotrichia**. A wide strip of ligamentous tissue connects the somactidia of the fin with the vertebral column. Each somactid is divided typically into three segments, the proximal, mesial and distal segments. The segments of the adjoining somactidia tending to fuse at their bases. The **first dorsal fin** is supported by twenty-one somactidia, the proximal segments of which are fused along the posterior border of the fin to form an obliquely horizontal axis in this region. The **second dorsal fin** and median **ventral** or **anal fin** are built on the same plan as the first dorsal. In the **caudal fin** there are no somactidia but the neural and haemal spines of the vertebral column are elongated and flattened to support the dorsal and ventral lobes of the fin.

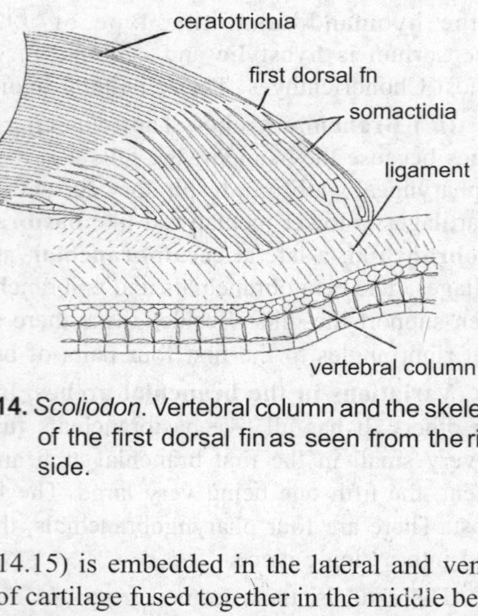

Fig. 14.14. *Scoliodon*. Vertebral column and the skeleton of the first dorsal fin as seen from the right side.

Pectoral Girdle and Fins

(*i*) Pectoral girdle. The pectoral girdle (Fig 14.15) is embedded in the lateral and ventral body wall posterior to gills and has two curved rods of cartilage fused together in the middle below the pericardium. Each half is a **scapulocoracoid** cartilage, its upper curved part is the **scapula** and the lower part is the **coracoid**. At the dorsal end of scapula is a distinct suprascapular part. In between the scapula and coracoid is a **glenoid region** with three foramina for branchial artery and nerve. The girdle is not connected to the vertebral column. The glenoid region articulates with the pectoral fin by a triple facet. The pectoral girdle provides a fulcrum for attachment of muscles which move the fin, the muscles attached to coracoid control the visceral arches, thus, they help in breathing and swallowing.

(*ii*) Pectoral fin. Pectoral fin is supported by a series of both endoskeletal and exoskeletal elements. The endoskeleton has **pterygiophores** cartilages to which muscles are attached for moving the fin. The proximal pterygiophores are **basals** and distal ones are **radials**. There are three thick basals, an outer **propterygium**, a middle **mesopterygium**, and an inner **metapterygium**. Extending from the basals are three rows of numerous radials, *i.e.*, proximal, mesial, and distal radial cartilages collectively called **somactidia**. Arising from distal somactidia are horny, unjointed exoskeletal **ceratotrichia** or fin rays. They support the fin blade.

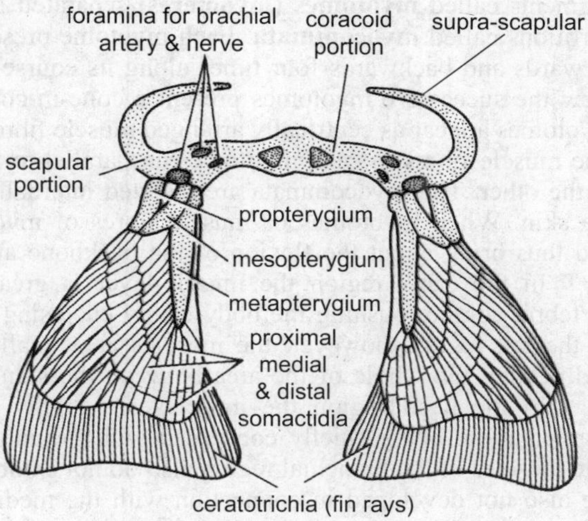

Fig. 14.15. *Scoliodon*: Pectoral girdle and fins in dorsal view.

Pelvic Girdle and Fins

(*i*) Pelvic girdle. Pelvic girdle (Fig. 14.16) is a flat transverse rod of cartilage called **ischiopubic bar**. It lies transversely with its ends bent, near each end is a nerve foramen. On each side of the girdle is an **acetabular region** for attachment of basal cartilage of pelvic fins. The pelvic girdle is more primitive than the pectoral girdle. In fishes it is never connected to vertebral column and lies embedded in the abdominal wall in front of cloaca.

(*ii*) Pelvic fin. Like pectoral fin, the pelvic fin also has the same endoskeletal and exoskeletal structures. There is a single curved basal cartilage called the **basipterygium** to which several **somactidia** (radials) are attached. The fin blade is supported by exoskeletal **ceratotrichia** or fin rays attached to somactidia.

The pelvic fin shows sexual dimorphism, in the male a part of each fin is modified to form a **clasper** or **myxopterygium**. It arises from the distal end of the basipterygium. It is an elongated tubular axial cartilage which is dorsally grooved. At its distal end is a sharp pointed **style** enclosed within a dorsal and a ventral plate and at the upper end of the style is a small **accessory cartilage**. The clasper has an anterior aperture, the **apopyle** which communicates with cloaca and the **siphon** and a posterior **hypopyle** opening on a sharp, pointed style. The siphon lies beneath the skin of the ventral surface of the body and communicated with the claspers by means of paired ducts. The claspers are co-vered with skin. They are erectile and used for transporting sperms during copulation.

The skeleton of **median fins** is similar to that of paired fins. The endoskeletal pterygiophores form a series of radials **(somactidia)** which bear a double series of distal exoskeletal **ceratotrichia**. In the caudal fin there are no somactidia, their place being taken by neural and haemal spines of caudal vertebrae.

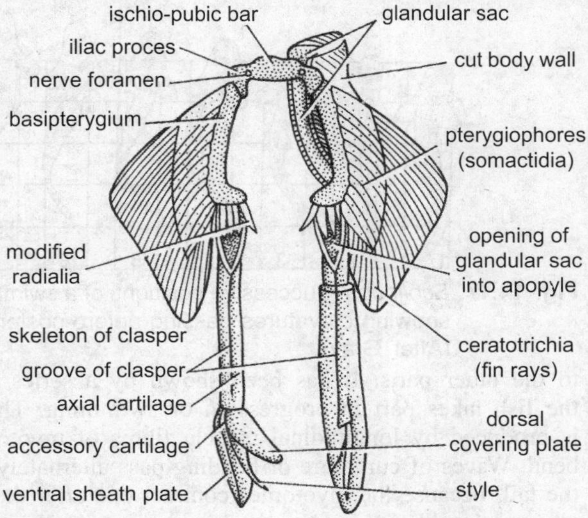

Fig. 14.16. *Scoliodon*. Pelvic girdle and fins of a male fish in dorsal view.

MUSCULATURE

The greater part of the body-bulk of *Scoliodon* consists of muscles closely packed together between the skin and the skeleton. In the region of the trunk and tail the muscles are highly developed and present a perfectly segmental arrangement comprising a paired series of muscle segments called **myotomes (myomeres)** separated from one another by tough connective tissue partitions called **myocommata**. Each myotome presents a zig-zag appearance being sharply bent forwards and backwards four times along its course from its dorsal to its ventral end. In surface view the successive myotomes present a cone-in-cone arrangement but in transverse section the myotomes appear as centrically arranged muscle fibres placed one below the other dorso-ventrally. The muscle fibres in each myotome run parallel to the length of the body from one myocomma to the other. The myocommata are inserted internally on the vertebral column and externally on the skin. When myotomes or muscle fibres of myotomes contract, they pull the myocommata and thus bring about the flexion of the backbone and the body.

In the trunk region the muscle-layer is greatly thickened dorsally on each side of the vertebral column pushing the body-cavity enclosing the internal organs towards the ventral side. In the tail region, however, the muscles are equally developed all round the vertebral column and occupy the whole of the area between the skin and the vertebral column.

In the head region the muscles are highly specialised and do not show any trace of segmentation, being chiefly concerned with the movements of the eyes, jaws and the pharynx. The muscles of the branchial region also do not exhibit any marked segmentation. Special muscles are also not developed in connection with the median and lateral fins. Thus, it is clear that the primitive segmental arrangement of the myotomes is lost in the head and is greatly modified in the region of the gills and the fins.

LOCOMOTION

The locomotion of fishes has been explained by **Breder** (1926) and **Gray** (1933). By taking a series of photographs at definite intervals of time of a swimming fish against a background marked with a graduated scale, **Gray** has been able to record : (*a*) the motion of the body of the fish relative to the surrounding water, and (*b*) the movements of each part of the body relative

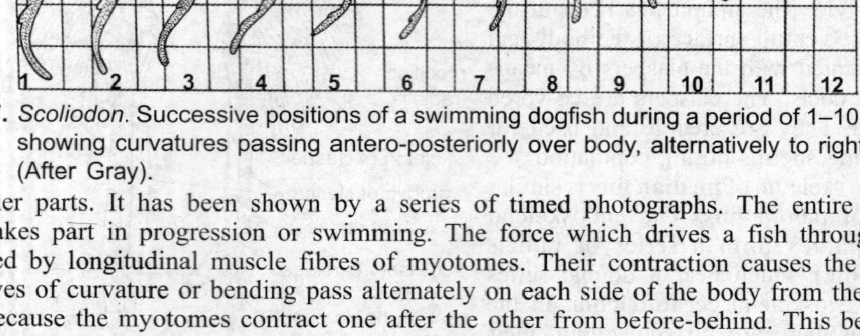

Fig. 14.17. *Scoliodon*. Successive positions of a swimming dogfish during a period of 1–10 seconds showing curvatures passing antero-posteriorly over body, alternatively to right and left (After Gray).

to the other parts. It has been shown by a series of **timed photographs**. The entire body of the fish takes part in progression or swimming. The force which drives a fish through water is produced by longitudinal muscle fibres of myotomes. Their contraction causes the body to bend. Waves of curvature or bending pass alternately on each side of the body from the head to the tail, because the myotomes contract one after the other from before-behind. This bending is more noticeable behind the head region, especially in the tail, because of a progressively greater transverse amplitude of the waves as they pass backwards. As one wave of contraction passes backwards on one side, it is followed by a similar second wave on the other side of the body. This contraction of successive myotomes causing waves of bending is called **metachronal contraction**. In metachronal contractions each segment or part of the body moves in a transverse direction to its path of swimming in such a way that it is bent at an angle or obliquely backwards, so that this part transmits a backward momentum to push the water backwards. Thus, the whole body operates as a single self-propelling system, and the fish is thrust forward.

The role of paired and median fins in swimming is to give stability to the fish and to maintain equilibrium, so that they enable the fish to keep a constant course or to change its course. The body is subjected to three forces while swimming in water, (*a*) **rolling** or turning side-ways or deviating from the longitudinal path of motion, this is called **drag**, (*b*) **pitching** or deviating from a transverse plane, this is called **lateral force**, (*c*) **yawing** or deviating from a vertical axis by raising or lowering the head, this is called **lift**.

The dorsal fins which are fixed in elasmobranchs prevent **rolling**, thus, keeping the fish steady in a horizontal plane. **Equilibrium** in a vertical plane is controlled by the moveable pectoral fins, the smallest movement of these fins prevents **yawing** and restores the original direction of locomotion. The paired fins and caudal fin prevent **pitching**. The powerful movements of the heterocercal tail not only give a forward thrust to the fish, but they also make the fish dive head downwards, this is counterbalanced by the paired fins, chiefly the pectoral fins (the pelvic fins have no influence or stability and they are insignificant). The pectoral fins lie in front of the centre of gravity and act as **elevators**, forcing the head upwards. Hence, the heterocercal tail and pectoral fins neutralise each others effects and maintain the fish in a constant plane. **Turning** is produced either by a wave of myotomal contraction only on one side of the body, or by asymmetrical braking with the pectoral fins.

COELOM AND VISCERA

In *Scoliodon*, the **coelom** is specious and is divided into two unequal cavities, the small anterior **pericardial** and the large posterior **abdominal**, separated from each other by a membranous partition, the **septum transversum**.

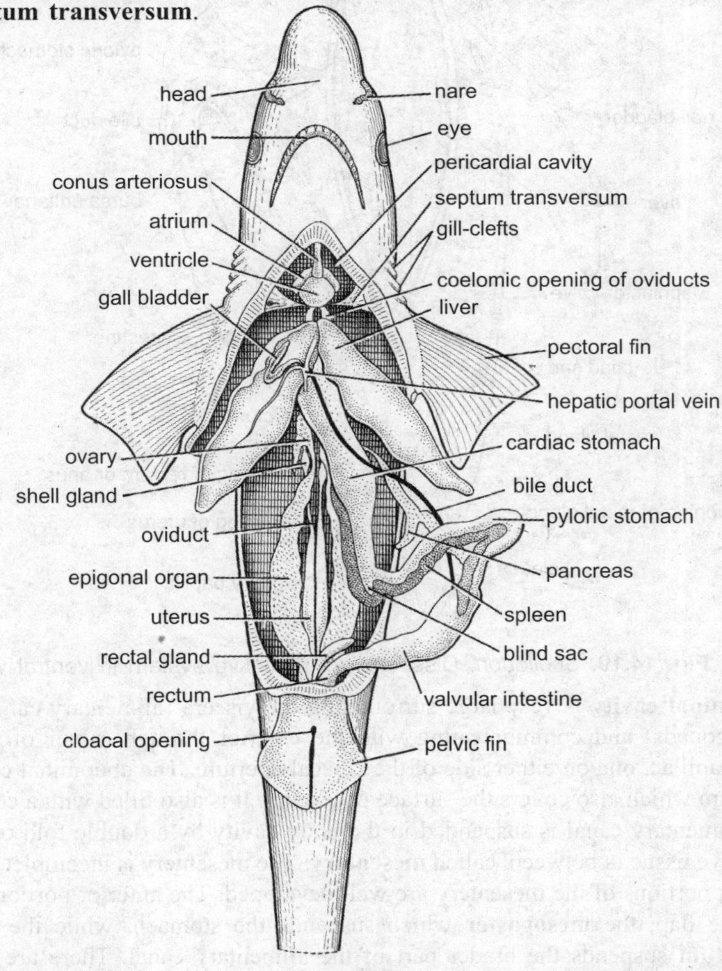

Fig. 14.18. *Scoliodon*. Ventral view of the internal viscera in a female animal.

Pericardial cavity is comparatively smaller in size and triangular in shape and lies beneath the pharynx enclosing the heart. It is enclosed between a tightly fitting smooth layer of peritoneum lining the outer wall of the cavity and the inner pericardial layer which adheres closely to the heart itself. The pericardial cavity contains a clear colourless fluid, the **pericardial fluid**. It communicates with the abdominal cavity through aperture in the septum transversum called the **pericardio-peritoneal canal**.

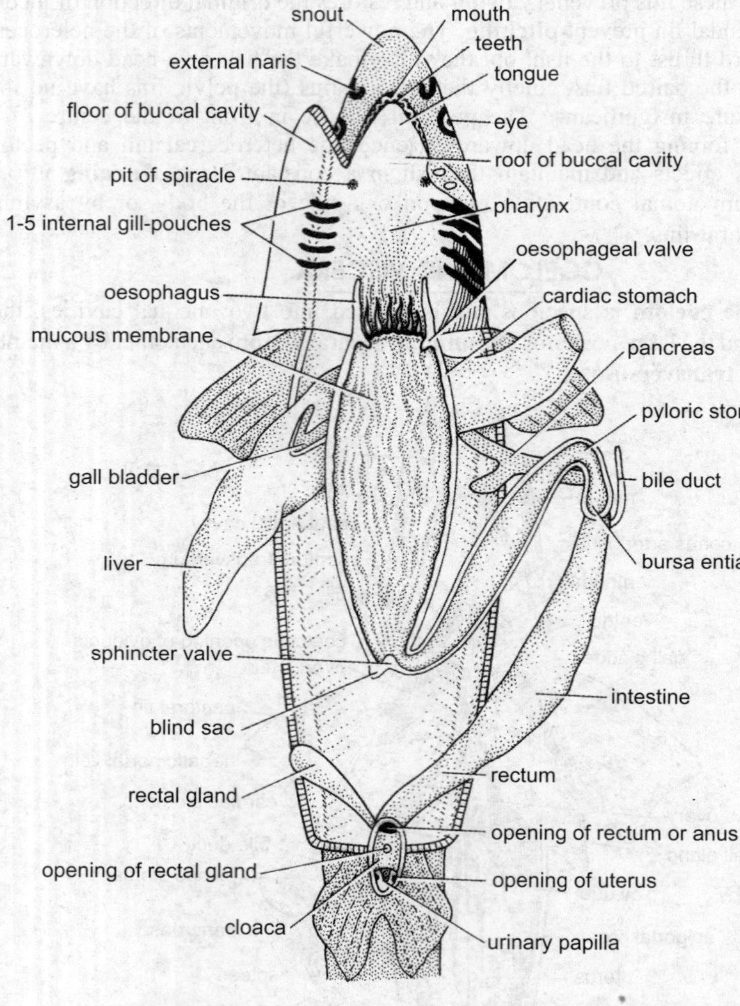

Fig. 14.19. *Scoliodon*. Dissection of digestive system in ventral view.

Abdominal cavity is very large surrounding the viscera (alimentary canal, liver, pancreas, spleen and gonads) and communicating with the exterior through a pair of **abdominal pores** situated on papillae, one on either side of the cloacal aperture. The abdominal cavity is lined with the peritoneum which also covers the surface of viscera. It is also filled with a colourless coelomic fluid. The alimentary canal is suspended in the body cavity by a double fold of peritoneum with thin connective tissue in between, called **mesentery**. The mesentery is incomplete, only the anterior and posterior portions of the mesentery are well developed. The anterior portion of the mesentery forms a large flap, the **mesogaster** which suspends the stomach, while the posterior portion, the **mesorectum** suspends the hinder part of the alimentary canal. There are other membranes which attach different organs to one another and are called **omenta** (singular omentum). The membrane which connects the stomach with the liver is knows as **gastro-hepatic omentum**, while the membrane connecting the spleen with the stomach is the **gastro-splenic omentum**.

DIGESTIVE SYSTEM

The digestive system of *Scoliodon* can be described under the following heads :

1. Alimentary canal,
2. Glands of alimentary canal and
3. Physiology of digestion.

(*i*) Alimentary Canal

The alimentary canal of *Scoliodon* comprises the mouth, buccal cavity, pharynx, oesophagus, stomach, intestine and rectum opening in the cloaca through anus.

1. Mouth. It is a ventral crescentic opening guarded by upper and lower lips which are folds of integument.

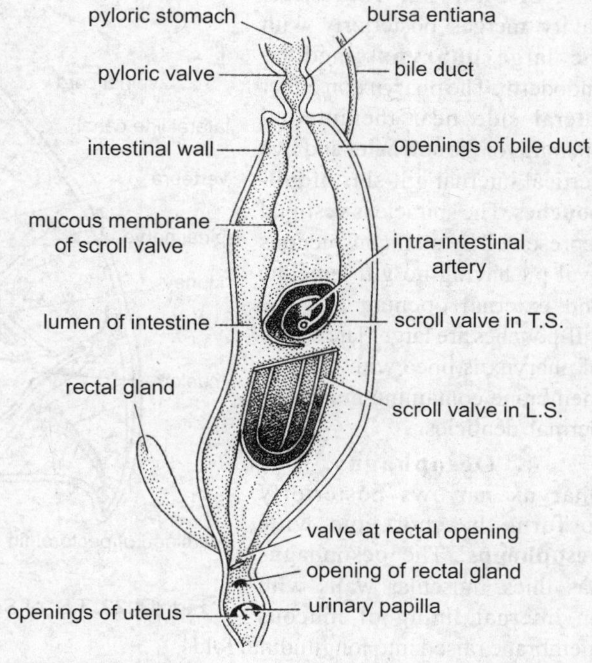

Fig. 14.20. *Scoliodon*. Intestine cut open to show scroll valve.

2. Buccal cavity. The mouth leads into a spacious dorso-ventrally compressed buccal cavity. It is bordered with jaws. The buccal cavity is lined with a thick mucous membrane. It is raised ventrally into a thick fold to form the so-called **tongue** which is non-muscular, non-glandular and non-protrusible. It is internally supported by basihyal cartilage. The mucous membrane is rough due to the presence of dermal denticles or teeth. The teeth are oblique and have sharp more or less compressed cusps, the edges of which are smooth and non-serrated. They are all alike in shape, **homodont**, and are borne in several parallel rows on the inner margin of the upper and lower jaws. They are not attached to the jaw cartilages and are simply embedded in the skin like scales. The teeth are used to catch the prey and prevent its escape but not to crush or masticate it. There are several rows of teeth over the jaws embedded in the mucous membrane covering the jaws. These are replaced as they become severely wornout or broken during life time **(polyphyodont)**. There are no glands in the buccal cavity comparable to the salivary glands of higher vertebrates.

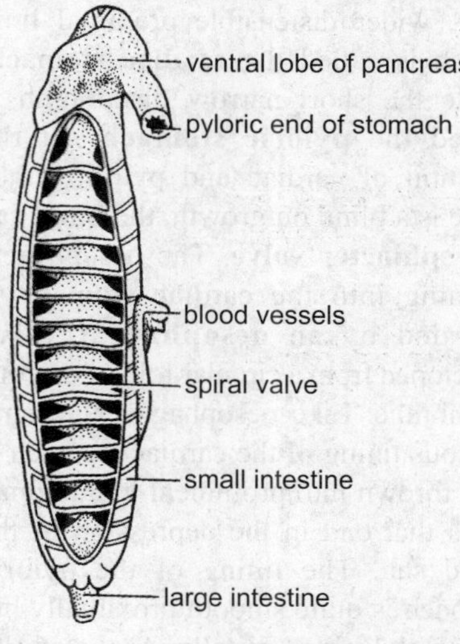

Fig. 14.21. Small intestine of *Squalus acanthias* (spiny dogfish) cut open to show spiral valve.

3. Pharynx. The buccal cavity merges posteriorly with the large **pharynx** lined by endoderm. The pharynx on either lateral side bear the internal openings of the **spiracles** and five vertical internal gill-slits of **gill-pouches**. The spiracle is vestigial represented by an inconspicuous oval pit having no gill lamellae and external opening and the gill-pouches are large. The cavity of pharynx is lined with mucous membrane containing numerous dermal denticles.

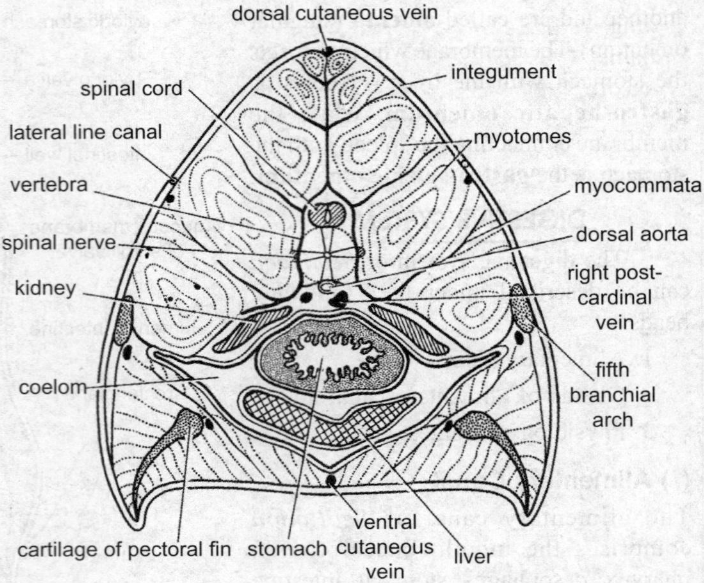

Fig. 14.22. T.S. of *Scoliodon* passing through stomach and liver.

4. Oesophagus. The pharynx narrows posteriorly to form the short and wide **oesophagus**. The oesophagus has thick muscular walls with an internal lining of mucous membrane raised into longitudinal folds.

5. Stomach. The oesophagus widens posteriorly to form the large muscular **stomach**. The stomach is bent on itself and forms a J-shaped organ, the long, wider distensible proximal limb which is called the **cardiac stomach**, while the short narrow distal limb is called the **pyloric stomach**. At the junction of cardiac and pyloric limbs there is a blind outgrowth, the **blind sac** and **sphincter valve**. The oesophageal opening into the cardiac stomach is guarded by an **oesophageal valve** developed from a circular fold of mucous membrane. Like oesophagus, the inner mucous lining of the cardiac stomach is also thrown into prominent longitudinal folds that end in the depression of the blind sac. The lining of the pyloric stomach is quite smooth proximally but is slightly folded distally. At the end of pyloric stomach there is a muscular **bursa entiana**. The opening of pyloric stomach into the bursa entiana is guarded by a circular band of muscle-fibres called the **pyloric valve**.

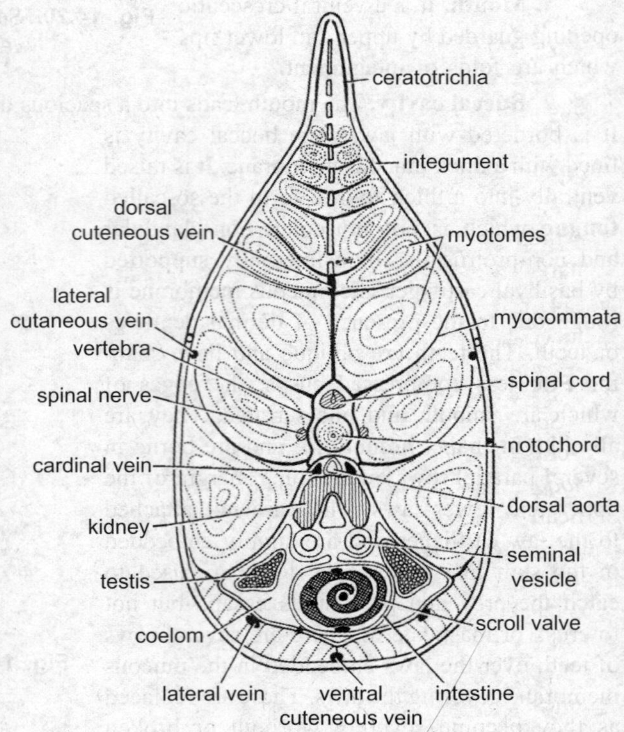

Fig. 14.23. T.S. of the body of male *Scoliodon* passing through the intestine, "scroll-valve" and testes.

6. Intestine. The bursa entiana continues into the **intestine**. The intestine is a wide tube running straight backwards into abdominal cavity and its middle region is like cardiac stomach in diameter. Its narrow anterior part receives the bile and pancreatic ducts. The internal mucous living of the intestine is folded anti-clockwise into a characteristic fold the **scroll valve**. Its one edge attached to the inner wall of the intestine and the other rolled up longitudinally on itself into a scroll, making an anti-clockwise spiral of about two and a half turns (Fig. 14.20). In a transverse section (Fig. 14.23) the scroll (spiral) valve looks like a watch-spring. The scroll valve serves not only to increase the extent of the absorptive surface of the intestine but also prevents the rapid flow of food through the intestine. In *Scylliorhinus* and *Squalus*, etc., the scroll valve takes about 14 to 15 turns arranged like a spiral-staircase of a tower (Fig. 14.21).

7. Rectum. The short and narrow **rectum** is the last part of the alimentary canal. The tubular **rectal (caecal) gland** opens dorsally into the rectum. Its function is not known. The rectum opens into the **cloaca** through **anus**. Into the cloaca also open the urinogenital ducts.

(*ii*) Glands of the Alimentary Canal

The glands of alimentary canal (digestive glands) are **liver** and **pancreas**.

1. Liver, rectal gland, and spleen. The **liver** is an elongated yellowish gland, consisting of two lobes which extend back along the greater part of the abdominal cavity. The two lobes are united anteriorly and attached to septum transversum by a ligament, but remain apart posteriorly. The right lobe of the liver carries a V-shaped thin-walled sac, the **gall bladder**, which stores the bile secreted by the liver. The gall bladder communicates with the anterior end of the intestine through **bile-duct**. In a full grown specimen the bile duct is about 3 cm long. The liver produces bile, stores glycogen and fat. It also destroys wornout red blood corpuscles since Kupffer cells are present.

2. Pancreas. The **pancreas** is a compact bilobed gland situated in the angle between the cardiac and pyloric stomach. It comprises a longer **dorsal lobe** running parallel to the posterior part of the cardiac stomach and a smaller **ventral lobe** lying closely applied to the pyloric stomach. The **pancreatic duct** opens into the intestine just opposite to the opening of the bile-duct.

3. Rectal gland. The **rectal** or **caecal gland** is a short thick hollow diverticulum arising from the dorsal wall of the rectum. It is richly vascularised and formed of lymphoid tissue. It discharges a fluid into the intestinal lumen but the physiological effect of the fluid is unknown.

4. Spleen. It is a large lymphoid organ attached with the cardiac and pyloric stomach. It produced lymphocytes and, thus, has no physiological relation with the alimentary canal.

(*iii*) Food and Physiology of Digestion

Scoliodon is **carnivorous** and feeds mainly on other fishes, but the diet may also include rock-crabs, lobsters and spider-crabs.. Food is swallowed without mastication. The teeth of *Scoliodon* only prevent the escape of prey from the mouth and do not perform the function of mastication. Probably the teeth may be used for tearing the food. Since there are no salivary glands in the mouth there is no digestion in the buccal cavity. The wall of the pharynx possesses numerous mucous glands. The secretion of these glands has no digestive function but simply helps in the passage of food. The oesophagus also has no digestive function. The digestion starts only in the cardiac stomach. The mucous membrane of cardiac stomach secretes the gastric juice which contains **pepsin** and **hydrochloric acid** that converts proteins into syntonin, proteoses and peptones. The gastric juice is not able to digest chitin. The pyloric stomach and scroll-valve have no digestive activity of their own but they activate the pancreas. The **pancreas** secretes **trypsin**, **amylopsin**, and **lipase**. As the semi-digested food enters the intestine it is acted upon by the bile and the pancreatic juice. The **bile** makes the food alkaline and, thus, helps the action of pancreatic juice. The trypsin acts on the remaining proteins, the amylopsin converts starch into sugar and lipase turns fats into fatty acids. The digested food is absorbed into the blood over the extensive surfaces of the intestine and scroll-valve.

RESPIRATORY SYSTEM

In *Scoliodon*, the respiration is **aquatic,** since the animal resides in water. It breathes by means of gills borne in a series of gill-pouches on either lateral side of the pharynx. Water enters the mouth and after passing through the buccal cavity, pharynx, gill-pouches bearing gill-lamallae, goes out through the external gill-slits after bathing the gills.

(*i*) Respiratory Organs

There are five pairs of **gill-pouches** bearing gills, arranged in a series behind the hyoid arch in the lateral walls of the pharynx. Each gill-pouch is compressed antero-posteriorly and communicates with the cavity of the pharynx through a large **internal branchial aperture** and with the exterior through a narrow **external branchial aperture** (commonly called gill-slit). The endodermal mucous membrane of gill-pouches or interbranchial septa is raised into a series of horizontal folds to form lamelliforms **branchial lamellae** or gill-filaments. The branchial lamellae have a rich blood supply, and they have a very thin covering membrane through which blood is exposed to sea water for an exchange of gases. Each gill-pouch has two sets of gill-lamellae, one on its anterior wall and the other on the posterior. Each set of lamellae is a half gill or **hemibranch,** so that gill-pouch has two hemibranches. The **gill-pouches** are separated from each other by fibro-muscular partition called the **inter-branchial septa** (they should more correctly be called intra-branchial septa because each lies between two successive gill-pouches). The interbranchial septa extend well beyond the branchial lamellae, then each bends posteriorly to form a flap which protect the lamellae as well as external gill-slit. The inner part of each interbranchial septum has a supporting cartilaginous **visceral arch** with slender **gill-rays**. Visceral arches also give out rigid comb-like **gill-rakers** which project inwards to protect the internal branchial apertures from entering the food.

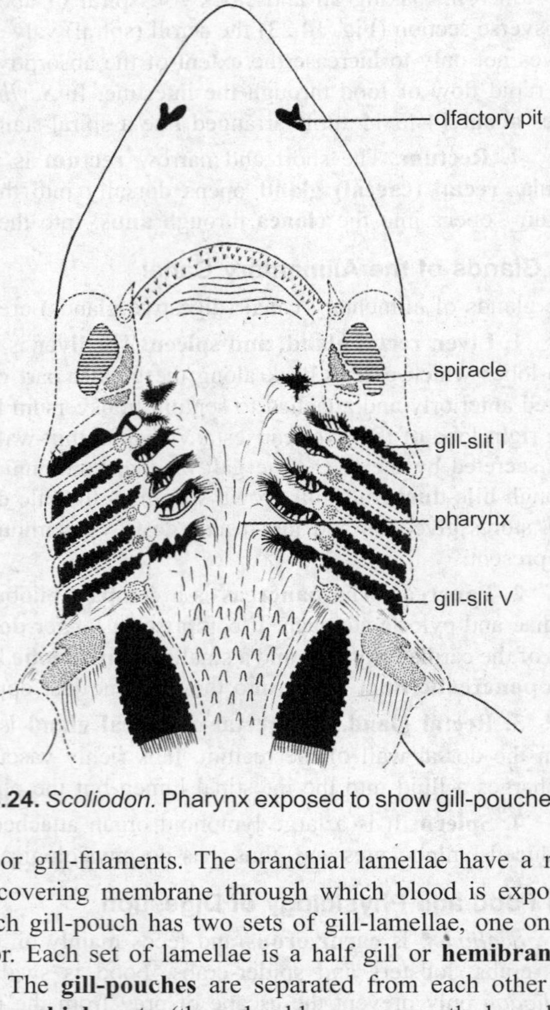

Fig. 14.24. *Scoliodon.* Pharynx exposed to show gill-pouches.

Each visceral arch supports the posterior branchial lamellae or hemibranch of posterior gill-pouch and the anterior branchial lamellae of the anterior gill-cleft. These two hemibranchs or demibranchs with their interbranchial septum and the visceral arch constitute a complete gill or **holobranch**. The posterior hemibranch of a septum has longer lamellae than the anterior one. Thus, a gill-pouch has two demibranchs belonging to two adjacent gills. In *Scoliodon*, the hyoid arch bears only a hemibranch on its posterior surface, the first four branchial arches has holobranchs and the fifth branchial arch is without any gill **(abranch)**. Thus, it has nine hemibranchs.

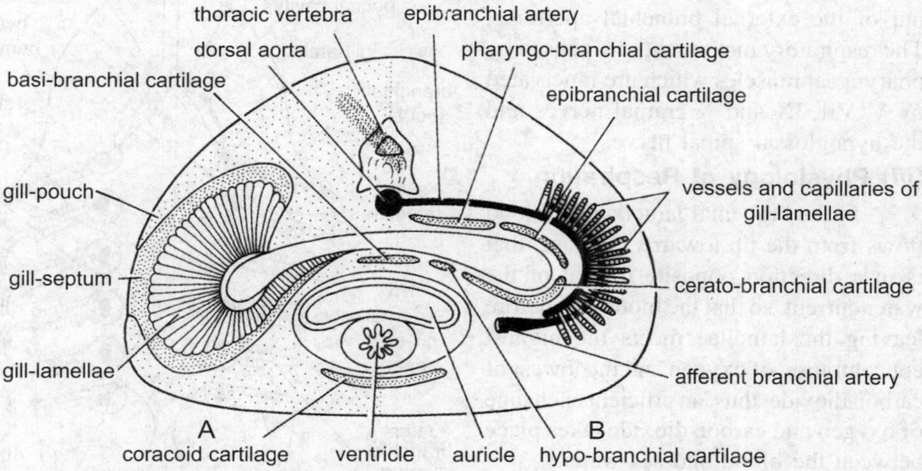

Fig. 14.25. *Scoliodon*. Diagrammatic respresentation of respiratory system in T.S. of body. A–A gill-pouch opened to show a gill in surface view; B–Parts of branchial arch and blood supply in gill-lamellae.

Between the mandibular and hyoid arches or the first internal gill-slit on either lateral side of pharynx is a **spiracle**. In most elasmobranchs the spiracle bears minute branchial lamellae forming a false gill or psendobranch and opens to the exterior by an external branchial aperture. It is supplied with arterial blood and plays no part in respiration. But in *Scoliodon* the spiracles are vestigial pits in the pharynx with no lamellae and external branchial apertures.

(*ii*) Mechanism of Respiration

1. Inspiration. The floor of the buccopharyngeal cavity is depressed by the contraction of hypobranchial (hypoglossal) muscles due to which the visceral arches expand the wall of the pharynx, so that sea-water containing dissolved oxygen rushes in through the opened mouth. Entry of water into external branchial apertures is prevented by an anterior fold of skin on each gill-pouch.

2. Expiration. During expiration the mouth becomes closed by the action of adductor muscle. Now the buccopharyngeal floor is raised, and contractions of the wall of the pharynx force the water into internal

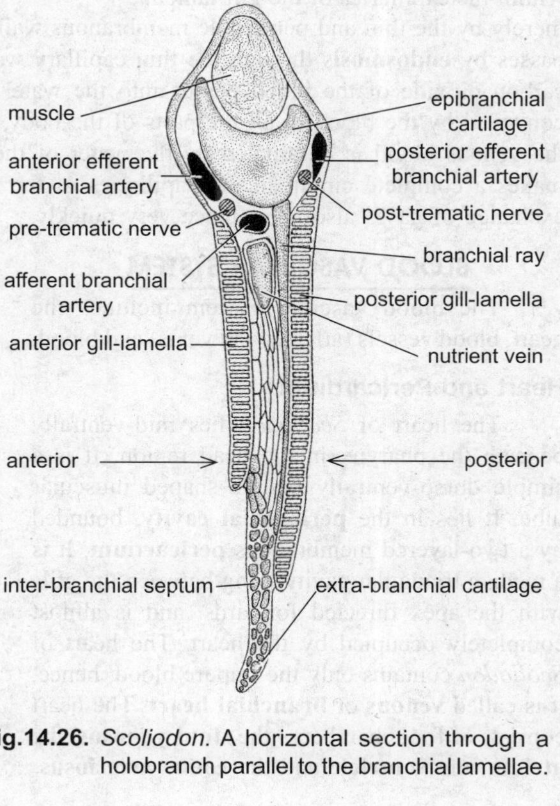

Fig.14.26. *Scoliodon.* A horizontal section through a holobranch parallel to the branchial lamellae.

branchial apertures, the oesophagus being closed, and then into gill-clefts, where it washes the branchial lamellae and goes out of the external branchial apertures. The respiratory movements are caused by pharyngeal muscles which are innervated by V, VII, IX and X cranial nerves and the hypoglossal spinal nerve.

(*iii*) Physiology of Respiration

In the branchial lamellae the blood flows from the tip towards the base, that is in a direction opposite to that of the water current, so that the blood just before leaving the lamellae meets the highest concentration of oxygen and the lowest of carbon dioxide, thus, an efficient exchange of oxygen and carbon dioxide takes place between the blood and sea-water.

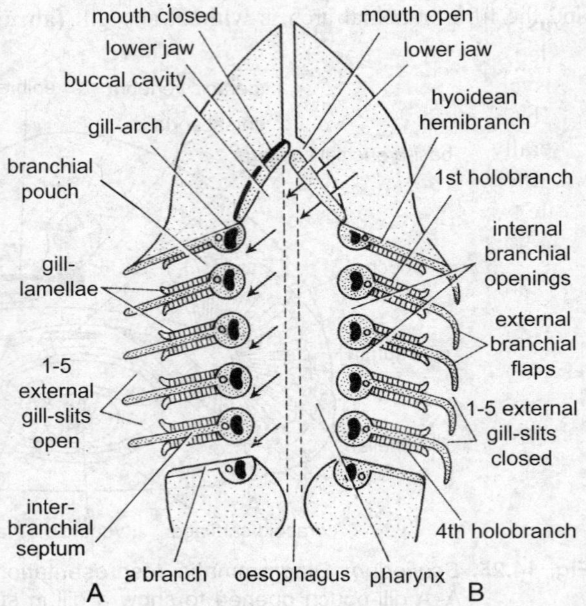

Fig. 14.27. *Scoliodon.* Respiratory mechanism. A–Expiration; B–Inspiration.

Fresh sea-water entering the gill-pouches with the respiratory current contains oxygen dissolved in it. This water is separated from the blood contained within the capillaries of the gill-lamellae merely by the thin and permeable membranous walls of the capillaries. The oxygen of the water passes by **endosmosis** through the thin capillary walls into the blood, and at the same time the carbon dioxide of the blood passes into the water by a process of **exosmosis**. The oxygen is conveyed by the blood to all the parts of the body, while carbon dioxide brought to the gills in the venous blood is eliminated by the water of the outgoing respiratory current. As the blood makes a complete circuit in the capillaries of the gills in a very short time, it is evident that exchange of gases also takes place very quickly.

BLOOD VASCULAR SYSTEM

The blood vascular system includes the heart, blood vessels (arteries and veins) and blood.

Heart and Pericardium

The heart of *Scoliodon* lies mid-ventrally beneath the pharynx in the head region. It is a simple dorso-ventrally bent S-shaped muscular tube. It lies in the **pericardial cavity**, bounded by a two-layered membranous **pericartium**. It is a median triangular cavity lying between the gills with the apex directed forwards, and is almost completely occupied by the heart. The heart of *Scoliodon* contains only the impure blood, hence, it is called **venous** or **branchial heart**. The heart consists of four chambers : the **sinus venosus**, the **atrium**, the **ventricle** and the **conus arteriosus**.

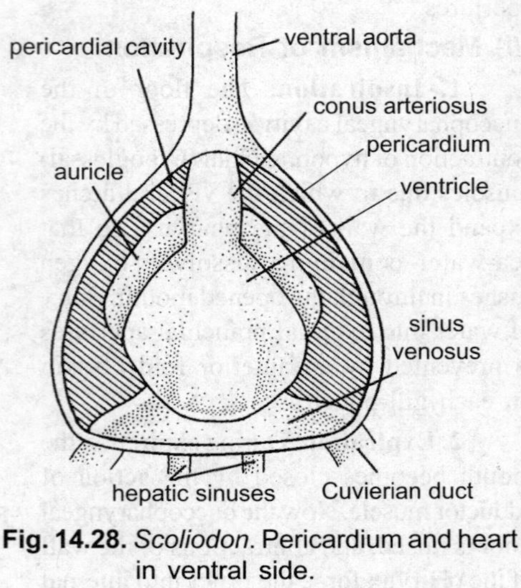

Fig. 14.28. *Scoliodon.* Pericardium and heart in ventral side.

(*a*) **Sinus venosus.** The **sinus venosus** is a triangular, thin-walled posterior chamber extended transversely and lying fused along the base of the pericardial cavity. Laterally it receives two large veins, the **ducti Cuvieri**, one on each side, while two **hepatic sinuses** open into it in the postero-median line. The sinus venosus opens anteriorly into the atrium through the **sinu-atrial sinu-auricular aperture,** guarded by a pair of membranous valves, the **sinu-atrial valves**. These valves prevent the beckward flow of the blood.

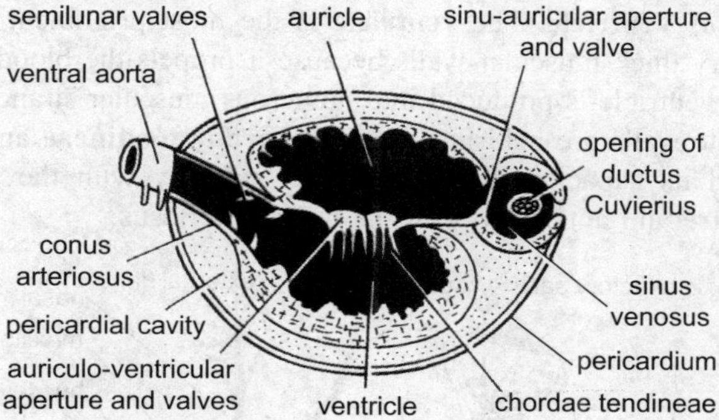

Fig. 14.29. *Scoliodon.* Pericardium and heart in sagittal section.

(*b*) **Auricle.** The **atrium (auricle)** is a large triangular sac, lying in front of the sinus venosus and dorsally to the ventricle. It occupies the dorsal half of the pericardial cavity. Its walls are somewhat thicker than those of the sinus venosus. Its lateral posterior angles produced into processes which project laterally at the sides of the ventricle like ears. It opens into the ventricle through the **atrio-ventricular aperture**, guarded by a **bilabiate valve** which prevents the backward flow of the blood.

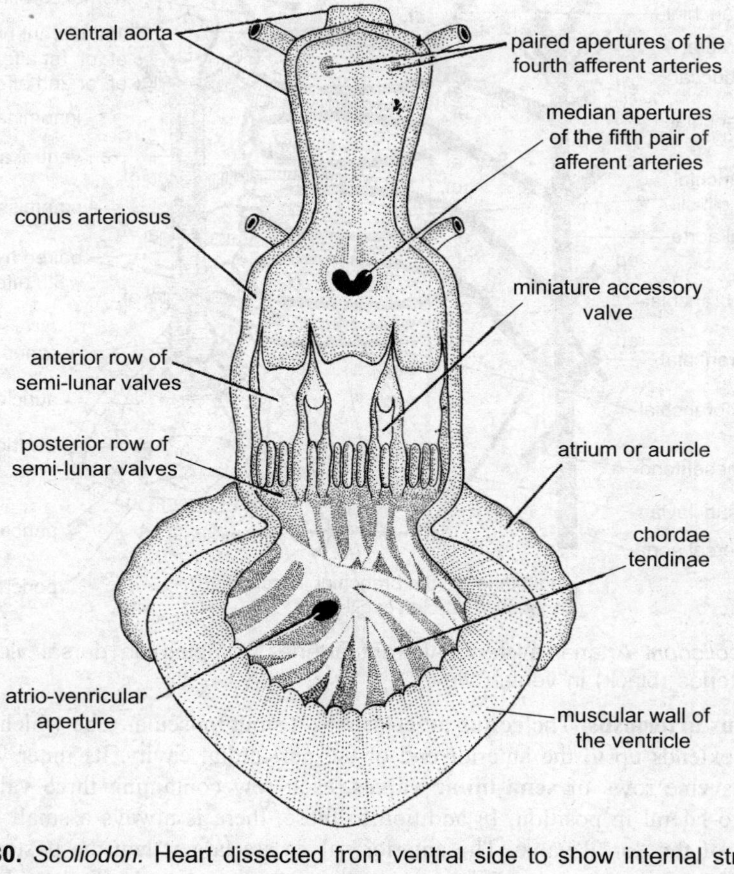

Fig. 14.30. *Scoliodon.* Heart dissected from ventral side to show internal structure.

(*c*) **Ventricle.** The **ventricle** is the most prominent pear-shaped chamber of the heart. It has very thick muscular walls because it propels the blood to the entire body. The inner surface of the ventricle is produced into numerous muscular strands which give it a spongy texture. The opposite walls are held in place by **chordae tendineae** and also protect the ventricle to expand beyond its capacity. It communicates dorsally with the atrium through the **atrio-ventricular aperture** and anteriorly with the conus arteriosus.

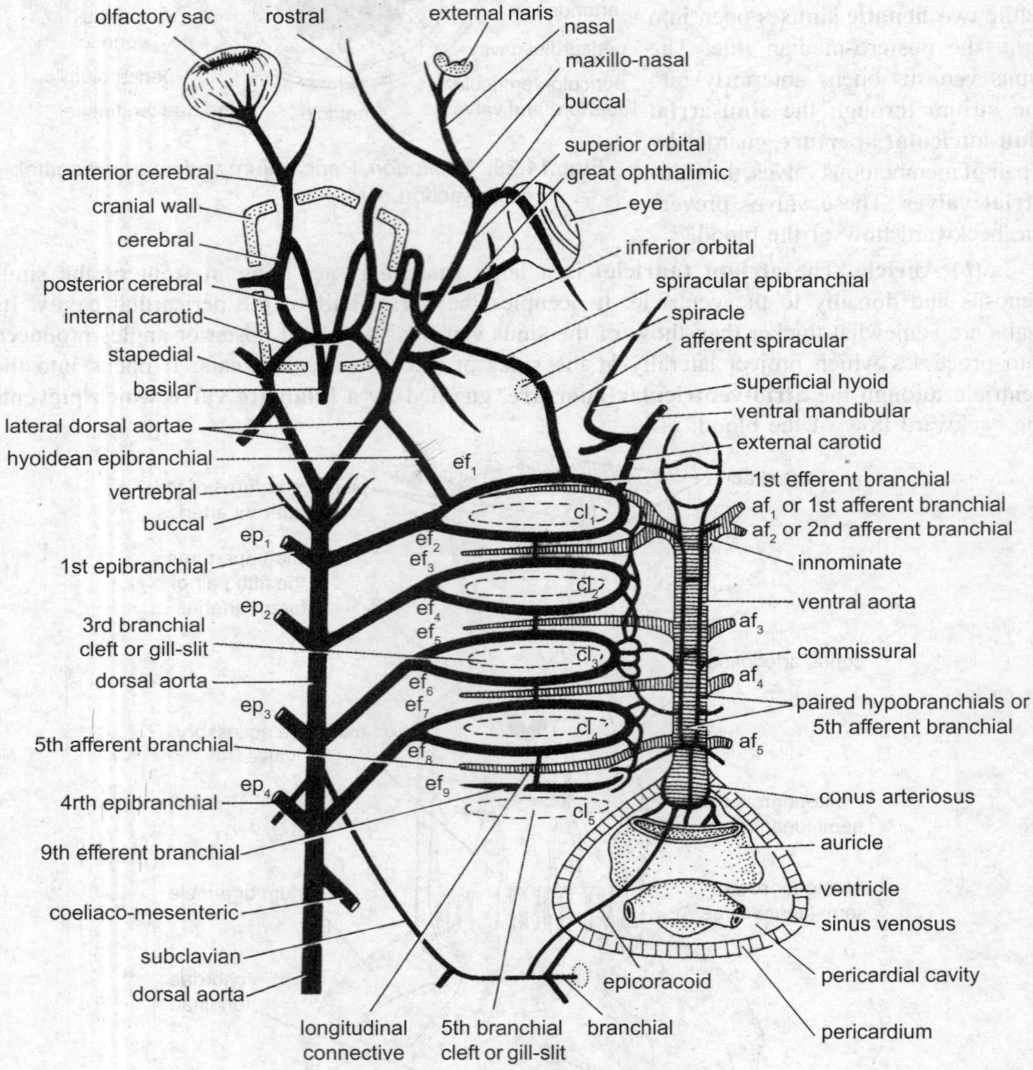

Fig. 14.31. *Scoliodon.* Arterial system. Afferent arteries (stippled) in dorsal view and efferent arteries (black) in ventral view.

(*d*) **Conus arteriosus.** The **conus arteriosus** is a stout muscular tube which arises from the ventricle and extends up to the anterior end of the pericardial cavity. Its inner wall is provided with two transverse rows of **semi-lunar valves,** each row containing three valves, one dorsal and two ventro-lateral in position. In addition to these, there is always a small accessory valve on either side of the dorsal valve. The anterior valves are larger than the posterior valves. The free-ends of the valves are connected to the walls of the ventricle by fine tendinous threads to keep the valves in position. These valves prevent the backward flow of blood into the ventricle. The conus arteriosus is continued forward through the wall of the pericardium as the **ventral aorta**. The ventricle and conus constitute the forwarding pump for the blood.

Working. The heart of dogfish is venous type since it receives only venous or deoxygenated blood. The blood also flows through the heart only once (single circulation). Heart pumps the venous blood into the gills for aeration. For this purpose, the different chambers of the heart rhythmically contract at regular intervals in a definite sequence—first sinus venosus→ auricle → ventricle → conus arteriosus. Contraction of the heart is called **systole** and relaxation is called **diastole**. The valves at different places prevent the backward flow of blood into the preceding chambers. The walls of the heart are also supplied oxygenated blood through coronary vessels.

Arterial System

Arteries have more muscular walls than the veins. The heart pumps the blood into the arteries which are of following types :

1. Afferent branchial arteries. The **ventral aorta** arises from the conus arteriosus and runs forwards along the ventral surface of pharynx right up to the posterior border of the hyoid arch, where it bifurcates into two short branches, the **innominate arteries**. Each innominate artery divides into **first** and **second afferent branchial arteries**. The first afferent branchial artery runs along the posterior border of hyoid arch and supplies branches to all the gill-lamellae of the hyoidean demibranch. The second afferent branchial artery runs along the posterior border of the first branchial arch and supplies arterial branches to the anterior and posterior gill-lamellae of the first branchial arch. The **third, fourth and fifth afferent branchial arteries** arise independently from the ventral aorta, almost equidistant from one another but the fifth branchials of both sides arise from a common median aperture from ventral aorta. These run along the outer borders of the second, third and fourth branchial arches. They supply the blood to the third, fourth and fifth gill-arches respectively. These all break up into capillaries in the gill-lamellae.

2. Efferent branchial arteries. The blood from the capillaries of gill-lamellae is collected by a series of blood vessels called the **efferent branchial arteries**. There are nine efferent branchial vessels on each side–pretrematic is along the anterior border of gill-pouch and the post-trematic along the posterior border of gill-pouch. Of these the first eight join in pairs to form four complete loops around the first four gill-pouches. The ninth runs along the anterior border of the fifth gill-pouch. The four loops and ninth efferent branchial artery are joined by short **longitudinal connectives** running across the interbranchial septa. The loops are further connected with one another by a network of longitudinal commissural vessels between their ventral extremities. From the upper end of each of the four efferent branchial loop arises an **epibranchial artery**. These run backwards and inwards to the mid-dorsal line, and all the four pairs of **epibranchial arteries** unite in the middle in the roof of the pharynx to form the **median dorsal aorta**. The half-loop of ninth efferent branchial has no epibranchial of its own.

3. Arteries of the head. The **first efferent branchial** or **hyoidean efferent** and a small part from the dorsal aorta supply blood to the head. The first efferent branchial vessel or hyoidean gives rise to three branches–the **external carotid**, **afferent spiracular** and **hyoidean epibranchial**.

(a) The **external carotid artery** arises from the antero-ventral corner of the first efferent branchial vessel and runs along the outer surface of the hyoid arch. It divides into two branches: (*i*) the **ventral mandibular** supplying the coraco-mandibular muscles and muscles of lower jaw, and (*ii*) **superficial hyoid** supplying the skin and sub-cutaneous tissues over the ventral part of the hyoid arch.

(b) The **afferent spiracular artery** arises at about the middle of the hyoidean efferent. It runs forward on the outer side of the hyomandibular and surrounds the spiracle by the spiracular epibranchial artery across the floor of the orbit and then enters the cranium through a small foramen. Just before entering the cranium it gives off the **great ophthalmic artery** to the eye-ball. Within the cranium it unites immediately with a branch of internal carotid to form the **cerebral artery** which is a short vessel dividing immediately into **anterior** and **posterior cerebral arteries**.

(*c*) The **hyoidean epibranchial artery** arises from the dorsal part of the first efferent and runs forward and inward. At the level of the orbit it receives a branch from the dorsal aorta and then divides into a **stapedial** and an **internal carotid**. (*i*) The **stapedial artery** runs forward and enters the orbit where it gives two branches, one supplying the eye-muscles and the superficial tissues in the region above the auditory capsules and other supplying the anterior boundary of the orbit. The former is called the **inferior orbital**, while the latter is called the **superior orbital**. (*ii*) The **internal carotid** passes inward along a groove in the roof of the buccal cavity and enters the cranium through a foramen. Within the cranial cavity it divides into two branches one of which unites with its fellow of the opposite side, while the other unites with the stapedial to form the anterior and posterior **cerebral arteries**.

4. Dorsal aorta and its branches. The **dorsal aorta** is formed by the union of four pairs of epibranchial arteries. It runs backwards along the whole length of the body, lying beneath the vertebral column in the trunk. In the tail region it continues within the haemal canal of the tail vertebrae as the **caudal artery**. The dorsal aorta supplies oxygenated blood through its paired and unpaired branches to all the structures behind the gills. It gives off the following branches from anterior to posterior sides :

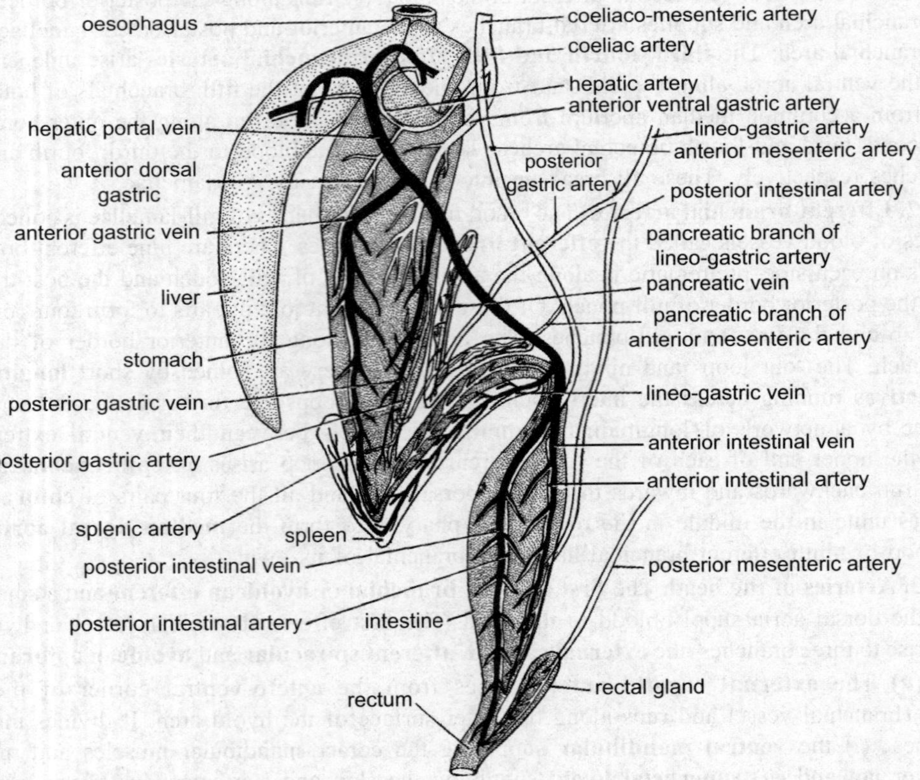

Fig. 14.32. *Scoliodon*. Vascular supply of the alimentary canal.

(*i*) **Lateral aortae.** These arise from the anterior end of dorsal aorta and divides into two. Each branch unites with the hyoidean epibranchial of its side.

(*ii*) Buccal. A pair to the roof of buccal cavity.

(*iii*) Vertebral. Paired vessels to the vertebral column.

(*iv*) Subclavian. Close to the union of the fourth epibranchial arteries, the dorsal aorta gives off a pair of small vessels, the **subclavian arteries**. Each subclavian passes outwards and backwards to the pectoral girdle and the pectoral fin of one side, which it supplies.

(*v*) Coeliaco-mesenteric. It is a large median artery arising from the dorsal aorta slightly behind the junction of the fourth pair of epibranchial arteries. It divides into two unequal branches: the smaller **coeliac** and the larger **anterior mesenteric**. The coeliac artery supplies the stomach and liver, etc., and the anterior mesenteric artery supplies the pancreas, intestine and the rectum.

(*vi*) Lieno-gastric artery. It arises from the dorsal aorta a short distance behind the origin of coeliaco-mesenteric. It is also a median blood vessel giving off branches supplying the genital organs, stomach, spleen and posterior part of the intestine.

(*vii*) Posterior mesenteric artery. It is a small median vessel which supplies the posterior end of the gonads and finally ends in the rectal gland.

(*viii*) Parietal arteries. These are a series of paired vessels arising at intervals along the whole length of the dorsal aorta and supplying the body wall. The **renal arteries** are small paired vessels arising from the parietal arteries and supplying the kidneys.

(*ix*) Iliac arteries. These are a pair of arteries similar to the parietals, each of them extending into the pelvic as a **femoral artery** and there breaks up into capillaries.

5. Hypobranchial blood plexus. It includes a pair of longitudinal median **hypobranchial arteries** present in the ventral wall of ventral aorta. These are connected with each other by transverse vessels. Four **commissural vessels** connect hypobranchial with the **lateral hypobranchial** chain of its side. The lateral hypobranchial chain of either side is connected with the ventral ends of the efferent branchial loops. Both the hypobranchial arteries unite posteriorly to form a **median coracoid artery** which divides into a **coronary artery** for heart and a **pericardial artery** for pericardium.

Venous System

In *Scoliodon*, the venous blood from the entire body is returned to the heart by the veins. The veins differ in structure from the arteries in possessing thin walls and frequently contain valves which prevent backward flow of blood. Many veins form wide irregular spaces without definite walls called **sinuses** during their course. It is characteristic of venous system of Chondrichthyes (elasmobranchs). The venous system of *Scoliodon* can be divided into the following systems, *viz*., (*i*) anterior cardinal system, (*ii*) posterior cardinal system, (*iii*) hepatic portal system, and (*iv*) cutaneous system.

1. Anterior cardinal system. The anterior cardinal system collects blood from the part of the body which lies anterior to the heart, *i.e.*, head region. It consists of a (*i*) pair of large vessels or sinuses the **internal jugular veins**. Each internal jugular comprises the **olfactory sinus**, **orbital sinus**, **post-orbital sinus** and the **anterior cardinal sinus**. The **olfactory sinus** receives blood from the rostral region by an **anterior facial vein**. The **orbital sinus** lies in the orbital region and is connected with its fellow of the opposite side by an **inter-orbital vein**. The orbital sinus opens into the **post-orbital sinus** and finally into the anterior cardinal sinus. The **anterior cardinal** is a large sinus and lies in between the dorsal ends of the gill-pouches and the muscles of the body wall. It opens into the **ductus Cuvieri** behind and receives a **hyoidean sinus** from inferior jugular sinus. It also receives blood from the five **dorsal nutrient branchial sinuses** of the gills. (*ii*) A pair of **inferior jugular sinus.** These are small median ventral sinuses, one on either side. These collect blood from the bucco-pharyngeal region, gill-pouches and heart region and then open laterally into the ductus Cuvieri.

2. Posterior cardinal system or renal portal system. The posterior cardinal system collects blood from the part of the body which lies posterior to the heart. It consists of a **median caudal vein**, two **renal veins** and two large **posterior cardinal sinuses**. (*i*) The **caudal vein** runs below the caudal artery and lies in the haemal arch of the caudal vertebrae. It receives numerous branches on either side from the myotomes of the tail. On entering the abdominal cavity dorsally to the cloaca, the caudal artery bifurcates into two branches, the right and left **renal portal veins**, (*ii*) The **renal portal veins** continue forward along the dorso-lateral margins of the kidneys and give out many **afferent renal veins** which break up into capillaries in the kidney. The renal portal

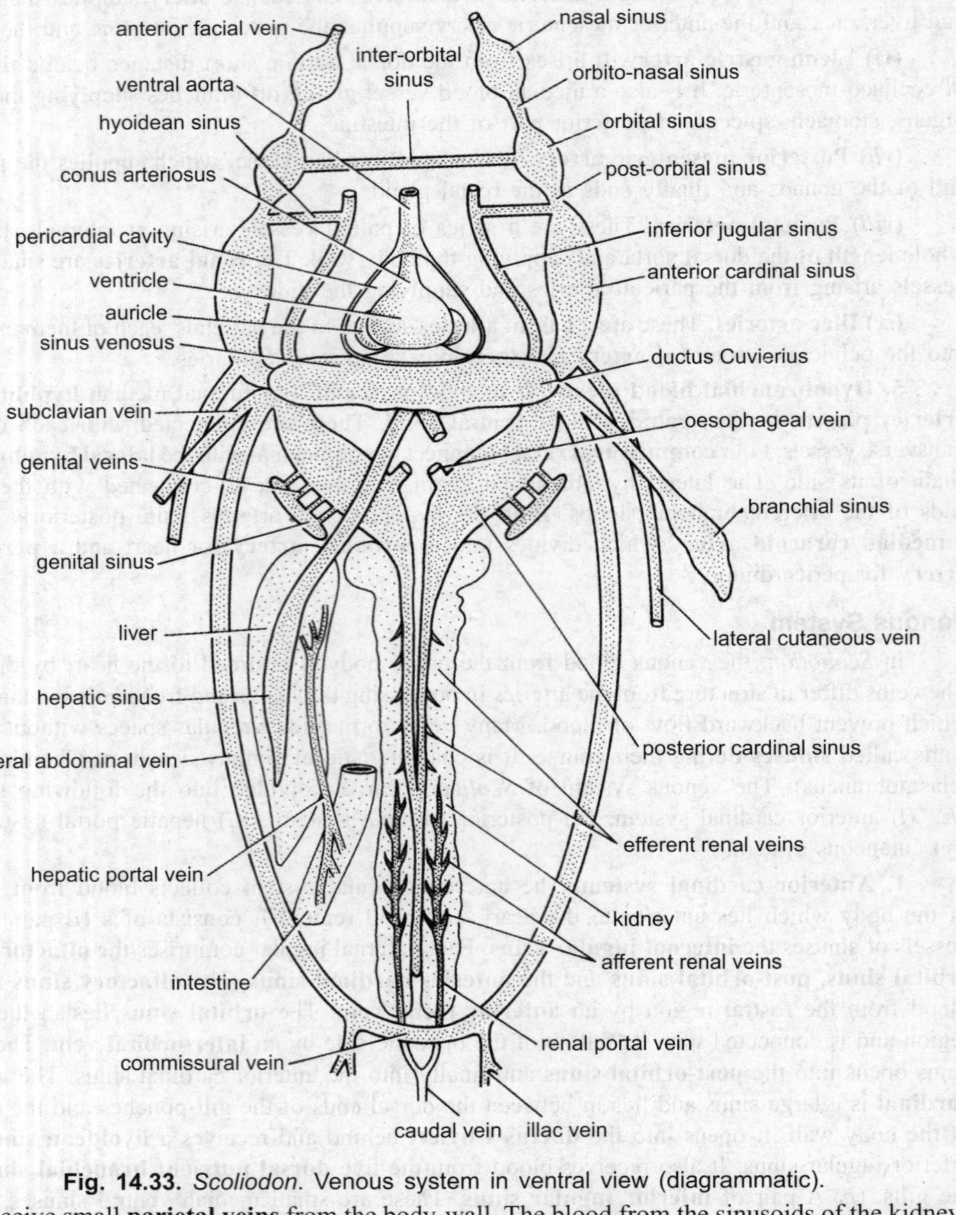

Fig. 14.33. *Scoliodon*. Venous system in ventral view (diagrammatic).

veins receive small **parietal veins** from the body-wall. The blood from the sinusoids of the kidneys is collected by the **efferent renal veins** which join together to form a single vessel. Anteriorly it divides into two **posterior cardinal sinuses**. The two posterior cardinal sinuses lie close together along the roof of the abdominal cavity. In the region of the oesophagus, the posterior cardinals expand into wide thin-walled sacs and each receives an **oesophageal vein** from **oesophagus** and numerous **genital veins** from the **genital sinus**. Finally, each posterior cardinal opens into the **Cuvierian duct** by a relatively small aperture opposite the opening of the anterior cardinal sinus.

3. Hepatic portal system. The blood from the alimentary canal and its associated glands is collected by a number of veins which unite to form the **hepatic portal vein**. The hepatic portal vein is formed by the union of the **anterior** and **posterior intestinal veins**. The large vein so formed receives a **lineo-gastric vein**, and **anterior** and **posterior gastrics** and then runs forward to divide into two branches which enter the right and left lobes of the liver. The blood brought to the liver is distributed in the lobes of the liver by fine capillaries. It is collected by a second set of capillaries which together form two large thin-walled sinuses, the **hepatic sinuses** which open into the **sinus venosus** by two small apertures.

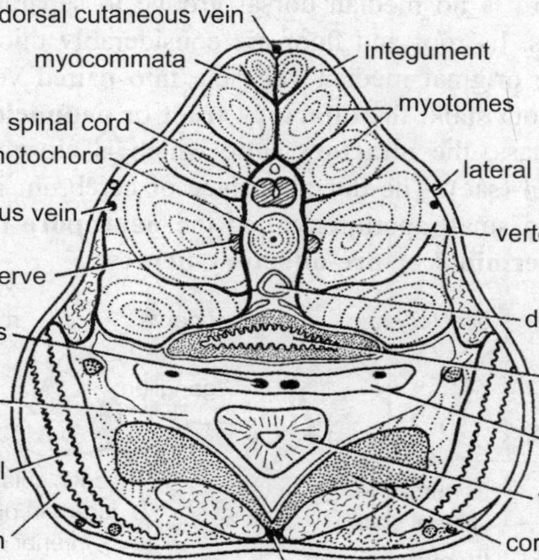

Fig. 14.34. T.S. of *Scoliodon* passing through the oesophagus and pericardial cavity.

4. Lateral abdominal veins. There are two **lateral abdominal veins**. Each collect blood from body wall, cloacal region and paired fins of its own side. In the posterior region both are connected by a **commissural vein** and each receives an **iliac vein** from the pelvic fin formed by the union of an outer femoral and inner cloacal vein. Each lateral abdominal vein anteriorly receives a **subclavian vein** from pectoral fin and then empties into the ductus Cuvierius laterally.

Lateral vein instead of joining with the sobclavian, opens separately into the precaval. The **genital sinus** of either side opens into the posterior cardinal sinus of their own side.

5. Cutaneous system. The cutaneous system includes a mid-dorsal, a mid-ventral, and two lateral cutaneous veins. The **dorsal cutaneous vein** runs beneath the skin along the mid-dorsal line and collects blood from the skin of the dorsal side. The **ventral cutaneous vein** runs along the mid-ventral line beneath the skin and joins the lateral abdominal in front of the cloacal vein behind. The **lateral cutaneous vein** runs along each side just beneath the lateral line canal. Each lateral cutaneous empties its blood into the **branchial vein** in front.

Blood

The blood of *Scoliodon* consists of a colourless plasma and corpuscles suspended in it. Corpuscles are of two types : RBCs (erythrocytes) are oval and nucleated bodies and contain respiratory pigment, haemoglobin and WBCs (leucocytes) and amoeboid cells resembling with lymphocytes of other vertebrates.

NERVOUS SYSTEM

The nervous system of *Scoliodon* comprises (*i*) the **central nervous system** including the brain and spinal cord, and (*ii*) the **peripheral nervous system** comprising the cranial and spinal nerves and (*iii*) the **autonomic nervous system**.

I. Central Nervous System

The central nervous system is essentially tubular and encloses a narrow central canal within the spinal cord which widens out anteriorly to form the ventricles of the brain.

1. Brain. The brain of *Scoliodon* is large and considerably advance to that of the Cyclostomata. It is divided into a **forebrain**, a **midbrain** and a **hindbrain**.

(*i*) Forebrain. The **forebrain** comprises the **cerebrum** and the **diencephalon**.

(*a*) Cerebrum. The **cerebrum** forms the anterior part of the brain and is an undivided mass of nervous tissue. There is no median dorsal groove to separate the cerebrum into right and left cerebral hemispheres. Its roof and floor are considerably thickened and meet internally in the mid-line, dividing the original median ventricle into paired ventricles. From either side of its anterior end arises a stout stalk, the **olfactory tract** or **peduncle**. Each olfactory peduncle ends into a large bilobed mass, the **olfactory lobe** which is closely applied to the posterior surface of the olfactory organ (sac). The dorsal surface of cerebrum is quite smooth, but on its mid-ventral surface there is a small aperture, called the **neuropore** through which come out a pair of delicate nerves, the **terminal** or **pre-olfactory nerves**.

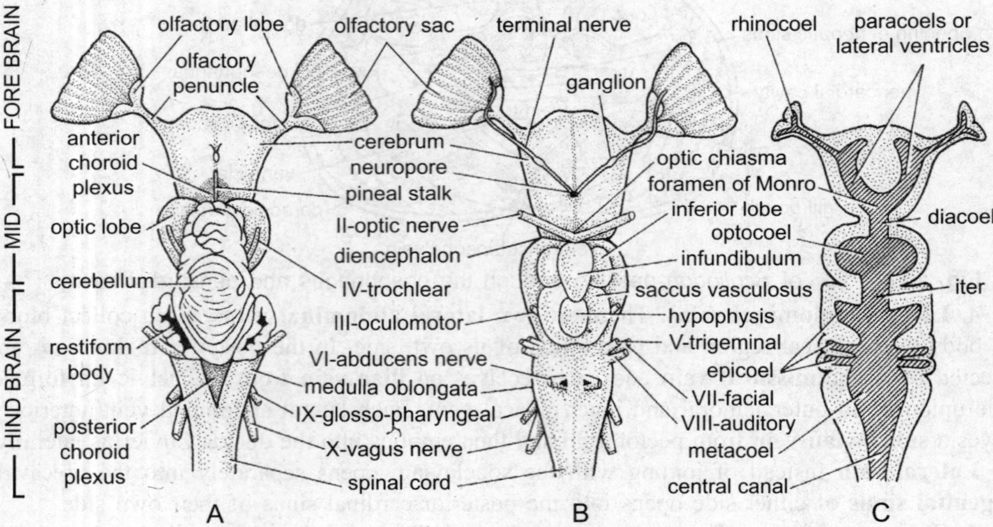

Fig. 14.35. *Scoliodon*. Structure of brain. A–Dorsal view; B–Ventral view; C–Horizontal longitudinal section.

(*b*) Diencephalon. The cerebrum is continued posteriorly into the narrow **diencephalon** which is extremely short and is completely covered by the forward prolongation of the cerebellum. Its lateral walls are composed of two thick masses, the **thalami**. The roof of the diencephalon is extremely thin and membranous which is highly vascular and is known as **anterior choroid plexus**. The postero-dorsal end of the diencephalon gives rise to a pineal stalk or **epiphysis cerebri** which ends in a rounded knob called **pineal body** which is fixed to the membranous part of the roof of skull. It has no visual or secretory function. The floor of the diencephalon gives off a hollow outgrowth, the **infundibulum** or **hypothalamus**, which projects downward and backward.

The infundibulum possesses at its distal end a rounded body called the **hypophysis**. On either side of the infundibulum lie the two **lobi inferiores** which are the dilations of the infundibulum. The distal end of each is continued into a thin-walled, pigmented vascular and glandular sac, the **saccus vasculosus**. It may be pressure receptor centre. In front of the infundibulum lies the **optic chiasma** formed by the decussation of the fibres of the two optic nerves.

(*ii*) Midbrain. The **midbrain** consists of two **optic lobes** only. The optic lobes are rounded swellings and are almost completely covered over dorsally by the cerebellum and ventrally by the infundibular outgrowths. The optic lobes receive the optic tracts and in addition fibres which transmit a variety of afferent impulses unconnected with optical stimulation. The third cranial nerve arises from the floor of the midbrain, while the fourth cranial nerve arises from the roof between the optic lobes and the cerebellum.

(*iii*) Hindbrain. The **hindbrain** comprises the **cerebellum** and the **medulla oblongata**.

(*a*) The cerebellum. The cerebellum is large and elongated. Its anterior portion overlaps the optic lobes in front and part of medulla behind. The cerebellum possesses numerous irregular convolutions on its dorsal surface. Two deep transverse furrows divide the cerebellum into three lobes, while a median longitudinal furrow dividing it into right and left halves. The sensory fibres from the ear and others from the facial and glossopharyngeal and vagus nerves enter into it. These carry afferent impulses from the lateral line system.

(*b*) Medulla oblongata. The **medulla oblongata** is almost triangular in shape and forms the posterior part of the brain. Its posterior narrow part pass into the spinal cord. From its anterior end arise a pair of hollow outgrowths, the **corpora restiformia** which lie on the dorsolateral aspects of the medulla and are slightly overlapped by the cerebellum. The corpora restiformia of two sides are connected with the help of a band of nervous tissue. The floor and sides of the medulla oblongata are thick, while the roof is extremely thin, non-nervous and vascular called the **posterior choroid plexus**. The fifth, sixth, seventh, eighth, ninth and tenth cranial nerves arise from the medulla oblongata. The branchial respiration and cardio-vascular activities are controlled by the medulla. Its cavity is called the **fourth ventricle** and is continuous behind with the central canal of the spinal cord.

Cavities of the brain. Vertical and horizontal longitudinal sections of the brain reveal various ventricles or cavities of the brain. The cavity of each olfactory lobe is known as **rhinocoel** which communicate with the lateral ventricles of the cerebrum behind. The lateral ventricles open into the large **third ventricle (diacoel)** of the diencephalon behind, each by the **foramen of Monro**. The cavity of the third ventricle extends into the infundibulum of the pituitary body and also into the base of pineal stalk. The optic lobes possess **optocoele** within them. The **fourth ventricle** is the cavity of the medulla into which also opens the cavity of the cerebellum called the **epicoels**. A common space connecting the third and fourth ventricles, into which also open the optocoeles, is called the **iter**. Behind, the fourth ventricle is continued into the central canal of the spinal cord and in front into the unusually wide **iter** or **aqueductus Sylvii** of midbrain.

Spinal Cord

The brain is continued behind into the spinal cord which extends to the end of the tail. It lies protected within the neural canal of vertebrae. It is similar to the spinal cord of higher vertebrates in having dorsal and ventral fissures, central canal and outer white and inner grey matter surrounding the central canal.

II. Peripheral Nervous System

The peripheral nervous system comprises the **cranial** and **spinal nerves** and the **autonomic nervous system**.

A. Cranial Nerves

In *Scoliodon*, ten pairs of cranial nerves arise from the brain, besides a pair of anterior terminal nerves.

O Nerve. The **terminal** or **pre-olfactory** nerve arises from the ventral surface of the cerebrum through the neuropore, runs along the olfactory peduncle and innervate the **olfactory septum** or **nasal septum** and external nostrils. The nerve is sensory in function.

1. First or olfactory nerve. It arises as a number of fibres from each olfactory lobe and innervate the olfactory sac of its side. It is sensory in function.

2. Second or optic nerve. The optic nerve arises from the optic thalamus, *i.e.,* on the ventral side of deencephalon. The nerve of each side crosses the other to form the **optic chiasma**. It innervates the retina of the eye. It is sensory in function.

3. Third or oculomotor nerve. It is a slender nerve arising from the ventral surface of the midbrain. It divides into branches which supply the anterior, superior and inferior recti muscles, and the inferior oblique muscles of the eye-ball. It is motor in function and controls the movements of eye-ball, iris and lens.

4. Fourth or trochlear nerve. It arises from the dorso-lateral surface of the midbrain, between the optic lobes and the cerebellum. It innervates the superior oblique muscle of the eye-ball. It is motor in function and controls the rotation of eye-ball.

5. Fifth or trigeminal nerve. It arises from the antero-lateral side of the medulla oblongata. It bears a Gasserian ganglion within the cranium. It has originally three main branches : (*i*) the **ophthalmicus superficialis**, (*ii*) the **maxillaris**, and (*iii*) the **mandibularis**. In addition to these there is one secondarily associated nerve, the **ophthalmicus profundus.**

(*i*) Ophthalmicus superficialis. It is a small branch entering the orbit along with the ophthalmicus superficialis of the seventh nerve. It innervates the skin of the snout and is sensory in nature.

(*ii*) Maxillaris. It forms two branches, the **maxillaris superior** and **maxillaris inferior**. The maxillaris superior is ribbon-like and runs along the floor of the orbit along with buccal branch of VII. It innervates the skin of the upper jaw, while the **maxillaris inferior** innervates the posterior part of the upper jaw. It is sensory in function.

(*iii*) Mandibularis. It runs along the posterior wall of orbit and innervates the muscles of the lower jaw, tongue and gill-region. It is a mixed nerve.

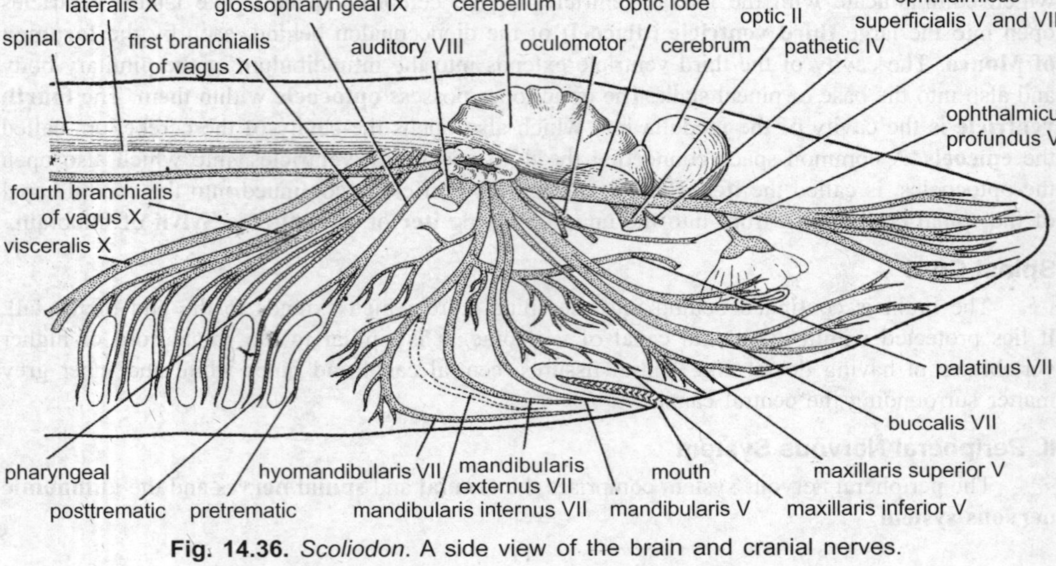

Fig. 14.36. *Scoliodon*. A side view of the brain and cranial nerves.

***(iv)* Ophthalmicus profundus.** It passes through the inner side of orbit, sends a branch to the eye-ball and goes to the skin of the dorsal surface of the snout. It is sensory in nature.

6. Sixth or abducens nerve. It is a slender nerve arising mid-ventrally from the medulla oblongata. It innervates the posterior rectus muscle of the eye-ball. It is sensory in function.

7. Seventh or facial nerve. It is very large nerve and emerges from the side of medulla oblongata in two bundles: (*i*) The first bundle is a thick broad **ophthalmicus superficialis** running forward along the upper border of eye orbit. It along with ophthalmic branch of the fifth nerve goes to the lateral line receptors and ampullae of Lorenzini on the dorsal side of the snout. The second bundle divides into three branches called **ramus buccalis**, **ramus hyomandibularis** and **ramus palatinus**. The facial nerve is a mixed type of nerve having sensory as well as motor fibres.

***(i)* Ophthalmicus superficialis.** It innervates the lateral line receptors and ampullae of Lorenzini of the dorsal side of snout.

***(ii)* Ramus buccalis.** It runs along the floor of eye orbit and innervates the infra-orbital canal of the snout and its associated groups of ampullae of Lorenzini.

***(iii)* Ramus hyomandibularis.** It runs along the posterior wall of orbit and gives out a small branch to the lateral line receptors and then divides into following three branches :

***(a)* Mandibularis externus.** It innervates the mandibular canal.

***(b)* Mandibularis internus.** It innervates the mucous membrane of the buccal floor.

***(c)* Hyoidean or hyomandibular.** It innervates the muscles of the throat (hyoid arch).

***(iv)* Ramus palatinus.** It runs along the floor of orbit and innervates the roof of the buccal and pharyngeal cavity.

8. Eighth or auditory nerve. It arises close to the origin of fifth and seventh cranial nerves from the side of the medulla and divides immediately into two branches : the **vestibular** and **saccular**. It is sensory in function.

***(i)* Vestibular.** It innervates the membranous labyrinth.

***(ii)* Saccular.** It innervates the cochlea.

9. Ninth or glossopharyngeal nerve. It arises from the ventro-lateral surface of the medulla behind the origin of sixth nerve. It runs obliquely backwards and divides into three branches : **pre-trematic** or **hyoid branch**, **post-trematic** and **pharyngeal**. This nerve is mixed in function.

***(i)* Pre-trematic.** It runs along the anterior border of the first gill-pouches and supplies the mucous membrane around the first gill-cleft.

***(ii)* Post-trematic.** It runs along the posterior border of the first gill-pouches.

***(iii)* Pharyngeal.** It innervates the mucous membrane of pharynx.

10. Tenth or vagus nerve. Vagus nerve is very large and arises from many roots from the postero-lateral side of medulla oblongata behind the ninth. It divides into three branches : **branchialis**, **visceralis** and **lateralis**.

***(i)* Branchialis.** Branchialis consists of four main nerves going around the second, third, fourth and fifth gill-pouches. Each nerve consists of a **pre-trematic**, **post-trematic** and **pharyngeal** branches. The branchial nerve-branches supply the mucous membrane around the gill-pouches and pharynx.

***(ii)* Visceralis.** It is a large nerve which runs backward into the body cavity and innervates the viscera including the alimentary canal, liver, lungs and the heart.

***(iii)* Lateralis.** Lateralis runs parallel but below the lateral line canal giving branches to the receptors of the lateral line along the length of the body.

An **occipital nerve** arises behind the vagus and joins the first two spinal nerves to form a **hypobranchial nerve** which goes to muscles in the floor of the buccal cavity.

The cranial nerves of *Scoliodon*, as in all fishes, are suited to an aquatic mode of life and are related to the gills, lateral line receptors, and ampullae of Lorenzini, for their innervation the trigeminal, facial, glossopharyngeal, and vagus are specialised. In land vertebrates the branches of trigeminal and facial which go to the lateral line receptors and various groups of ampullae of Lorenzini have disappeared, as also the branches of the glossopharyngeal and vagus which go to the gills because gills are lost, the lateralis branch of the vagus is also lost with the disappearance of lateral line receptors in terrestrial vertebrates.

B. Spinal Nerves

The spinal nerves arise in pairs from the sides of the spinal cord at regular intervals along its length. The number of paired spinal nerves is very large, approximately corresponding to the number of vertebrae. Each spinal nerve of each side arises by two roots, a **dorsal** or **sensory root** and a **ventral** or **motor root**. The ventral root always arises in front of the dorsal root. The dorsal root has a ganglion. Both the roots run backward for a short distance within the neural canal and perforate the cartilages of the neural arch separately, but unite outside to form a common mixed spinal nerve. Each spinal nerve divides into three branches : (*i*) **ramus dorsalis** supplying the skin and muscles of the dorsal body-wall, (*ii*) **ramus ventralis** supplying the skin and muscles of the ventral body-wall, and (*iii*) **ramus communicans** containing visceral sensory and motor fibres and joining the autonomic nervous system. The dorsal root carries **somatic sensory**, **visceral sensory** and **visceral motor** fibres, while the ventral root has **somatic motor** and **visceral motor** fibres.

In the region of the pectoral fin spinal nerves three to six join each other near the pectoral girdle to form a **cervicobrachial plexus** going to the pectoral fin. Spinal nerves seven to eleven pass downwards and send branches to the muscles of ventral body wall and to the pectoral fin. All spinal nerves from the twelfth to the anus go to body wall muscles. Then nine or ten spinal nerves go to the pelvic fin, some of them joining each other to form a **lumbosacral plexus**. The remaining spinal nerves go to the caudal region.

III. Autonomic Nervous System

The autonomic nervous system comprises a paired series of irregularly arranged ganglia situated anteriorly in the dorsal wall of the posterior cardinal sinuses and posteriorly in the dorsal part of the kidney on each side of the mid-dorsal line. The first ganglion is small. The second or **gastric ganglion** is fairly large, being formed by the fusion of a number of ganglia. It lies immediately behind the post-branchial plexus. It is connected with numerous fibres from spinal nerves and gives off branches to the viscera. The succeeding ganglia are small. There is always at least one ganglion in each segment, but often there are two or three in a segment. Sometimes the ganglia of successive segments are joined by longitudinal connectives, but there is no definite continuous chain. The posterior ganglia innervate the genital ducts, kidneys, urinary sinus, intestine and rectum.

RECEPTOR ORGANS (SENSE ORGANS)

The receptor organs or sense organs of *Scoliodon* include the paired **olfactory**, **optic** or **photoreceptor**, **statoacoustic organs**, the **lateral line organs** or **neuromasts** and the **ampullae of Lorenzini**.

1. Olfactory Organs (Olfactoreceptors)

The olfactory organs of *Scoliodon* are a pair of blind olfactory sacs lying on the ventral side of the head in front of the mouth. They are invaginations of the ectoderm and do not communicate with the buccal cavity by an internal opening. Each olfactory sac is ellipsoid in shape and proportionately very large. It is covered externally by a thin membrane and is lodged on each side in the cartilaginous olfactory capsule of the skull. The mucous membrane lining the inner wall of the olfactory sac is thrown into a double series of folds, the **Schneiderian folds**. The Schneiderian folds are held in place by a median band of connective tissue, the **median raphe**. Internally olfactory epithelium consists of two types of cells, the **olfactory receptor cells** and **supporting cells**. The olfactory receptor cells bear internally sensory hairs and their basal ends are connected with the fibres of the olfactory nerve which passes from each olfactory sac with olfactory lobe of its side of the brain. The olfactory sacs open outside ventro-laterally by external nares.

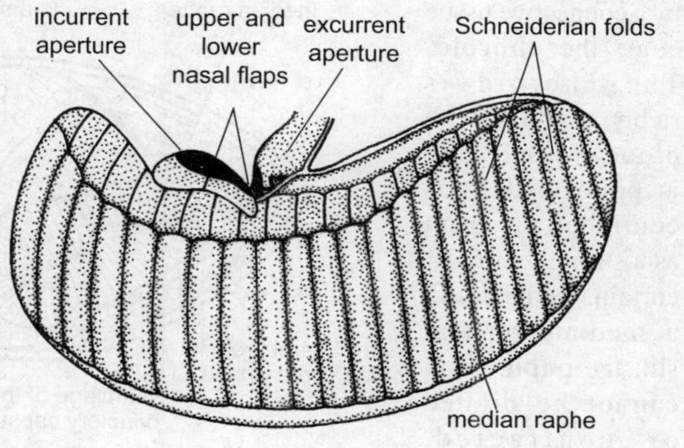

Fig. 14.37. *Scoliodon.* Olfactory sac dissected out to show its relation to the inner nasal flaps.

Each external nostril is more or less completely divided into two, a lateral **incurrent siphon** and a median **excurrent siphon** by three nasal flaps. Of these flaps two project inwards fitting closely, one above the other, while the third hangs down from the dorsal side slightly overlapping the other two. The median raphe of the Schneiderian folds lies parallel to the long axis of the nasal aperture, and, therefore, the current of water that enters through the incurrent siphon is at first directed against the oval wall of the olfactory sac, then turns internally to the nasal valve in the direction of the median raphe and finally passes out through the median excurrent siphon. The current of water enters at the lateral end of the nasal aperture and, thus, takes a zig-zag course through the olfactory sac finally leaving by the mesial end of the nasal aperture.

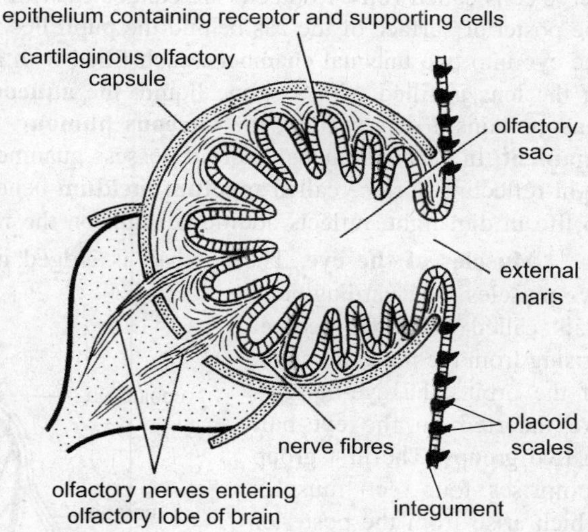

Fig. 14.38. *Scoliodon.* T.S. of olfactory capsule with olfactory sac.

The great size of the olfactory sacs is characteristic of elasmobranchs and is correlated with a highly developed sense of smell. The highly developed sense of smell enables them to detect food. If the external narial openings are plugged the dogfish will swim over food material without detecting it, on the other hand with one nostril open it will detect even concealed food.

2. Eyes (Photoreceptors)

In *Scoliodon*, the eyes are very large and housed in orbits on the lateral sides of cranium. Each eye is elliptical and supported not only by the recti and oblique muscles but also by a cartilaginous **optic pedicel** which connects the eye-ball to the orbit. The eye-ball consists of three concentric layers, *viz.*, **sclerotic, choroid** and **retina**. The **sclerotic** is cartilaginous and opaque and forms the outer protective layer of the eye. The sclerotic of the exposed part of the eye is transparent and is known as **cornea** through which the light enters. Cornea is flat and fused with an outer **conjunctiva** that is continuous with the living of immovable eyelids. The sclerotic is lined internally by a connective tissue coat, the **choroid**. The choroid is richly supplied with blood vessels and is pigmented. It is continued in front as a strongly circular curtain, the **iris**, with a median vertical slit, the **pupil**. Pupil cannot be dilated or contracted. The **retina** is the innermost layer of the posterior region of the eye-ball and forms its sensitive portion. It is made up of special elongated nerve cells called **rods**. Rod cells are connected with the fibres of optic nerve. Closely attached to the posterior surface of the iris behind the pupil lies the spherical **crystalline lens** which divides the eye into two unequal chambers, each filled with a semifluid substance. The chamber in front of the lens is filled with a saline liquid, the **aqueous humour**, while the chamber behind the lens contains a denser fluid, the **vitreous humour**. The lens is kept in place by a **suspensory ligament**. In elasmobranchs, choroid possess guanine crystals bearing cells which form a silvery light-reflecting surface called **tapetum lucidum** beneath the rods of retina. This is an adaptation to life in dim light, reflects additional light on the retinal cells.

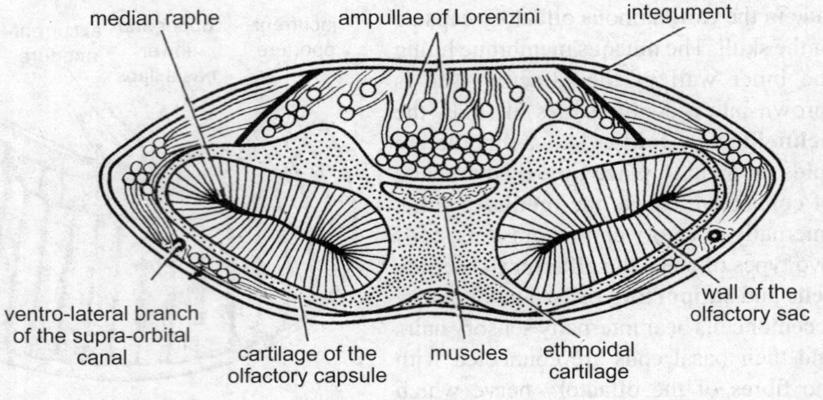

Fig. 14.39. T.S. of *Scoliodon* passing through the region of the olfactory sacs.

Muscles of the eye. The eye-ball is lodged in the orbit and is held in position by six eye-muscles and a cartilaginous stalk called the **optic pedicel** arising from the posterior angle of the orbit. The eye-muscles are inserted on the eye-ball in two groups. The first group comprises four recti muscles which arise from the posterior median angle of the orbit and are called the **superior rectus**, **inferior rectus, anterior rectus** and **posterior rectus** muscles. The **superior rectus** is inserted on the postero-dorsal surface of the eye-ball and the **inferior rectus** is inserted on the postero-ventral surface of the eye-ball. These rotate the eye-ball in vertical plane. The **anterior rectus** is inserted on the anterior surface and the **posterior rectus** is inserted on the posterior surface of the eye-ball and rotate it in horizontal plane. The second group comprises two **oblique muscles** which arise close together from the anterior angle of the orbit and run backward and

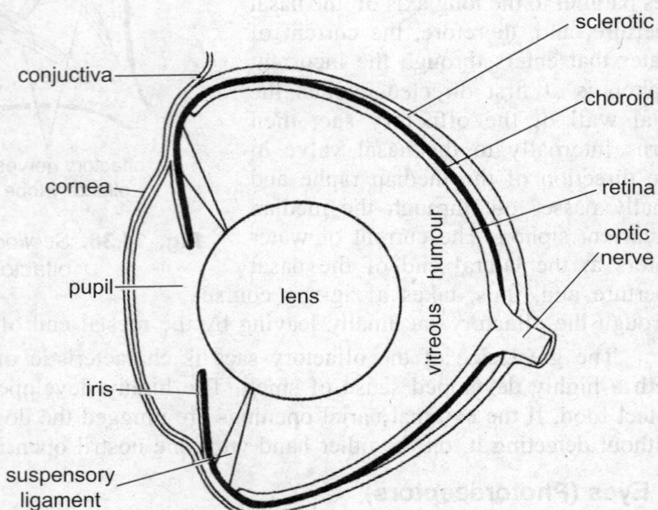

Fig. 14.40. Sagittal section through the eye of an elasmobranch.

outward to the eye-ball. The **superior oblique** is inserted on the antero-dorsal surface of the eye-ball and the **inferior oblique** is inserted on the antero-ventral surface of the eye-ball in front of the inferior rectus. These muscles twist the eye-ball.

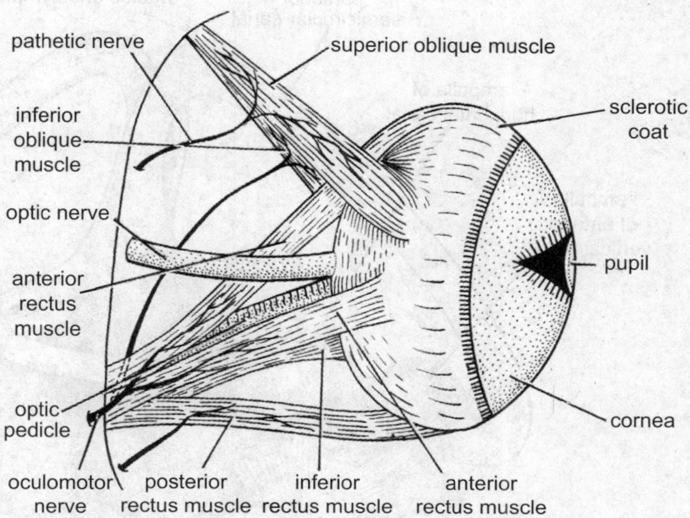

Fig. 14.41. *Scoliodon*. Eye muscles and their nerve supply.

Working. The eyes of *Scoliodon* though large yet are placed apart as to render it impossible for both of them to focus together on a given point. There is, therefore, no binocular vision and each eye has its own range of vision. Further, since little adjustment of the shape of the lens is possible and the pupil is non-contractile, the power of estimation is also poor. But the lens can project through the pupil against the rounded cornea, so that even rays of light entering at a very wide angle can be caught and focussed on the retina. Since the retina is without cones, the fish is colour blind and the eye cannot discriminate minute details, but the eye is adapted for near vision in dim light and fish can see its prey.

3. Stato-acoustic Organs (Internal Ears)

Structure. The stato-acoustic organs of *Scoliodon* are the **internal ears** or **membranous labyrinths**. The internal ear of each side is a closed ectodermal sac, the **membranous labyrinth** which is enclosed within the cartilaginous auditory capsule one on either postero-lateral side of cranium. It consists of mainly two parts—the central **otosaccus** and three peripheral **semicircular canals**. The otosaccus is laterally compressed and is differentiated into a dorsal and anterior part, the **utriculus**, and a ventral and posterior part, the **sacculus**. The posterior outgrowth of sacculus is called **lagena cochleae**. From the utriculus arises an anterior outgrowth called **recessus utriculi**. The three slender semicircular canals arise from the otosaccus and also open into it. From the top of the utriculus arise the **anterior vertical** and **horizontal semicircular canals** which form their independent ampullae before opening into the middle of the utriculus above the recessus utriculus. The **posterior vertical canal** arises from the dorsal surface of the sacculus, forms the complete circle, and opens by its ampulla into the lagena cochleae. The three semicircular canals are arranged at right angles to one another in the three planes of the space. The cavity of the sacculus communicates with the exterior through a long slender tube, the **ductus endolymphaticus**, which runs upward and pierces the roof of the cranium within the parietal fossa to open to the exterior by a minute pore in front of the commissural occipital canal.

The interior of membranous labyrinth, including the semicircular canals, is filled with a fluid, the **endolymph** which is sea water. The endolymph contains the calcareous particles the **otoliths**. The space between the membranous labyrinth and the wall of the auditory capsule is the perilymphatic space, filled with a **perilymph** which is actually the cerebro-spinal fluid. The perilymphatic space opens to the exterior by a large aperture, the **fenestra** lying immediately behind the opening of the ductus endolymphaticus.

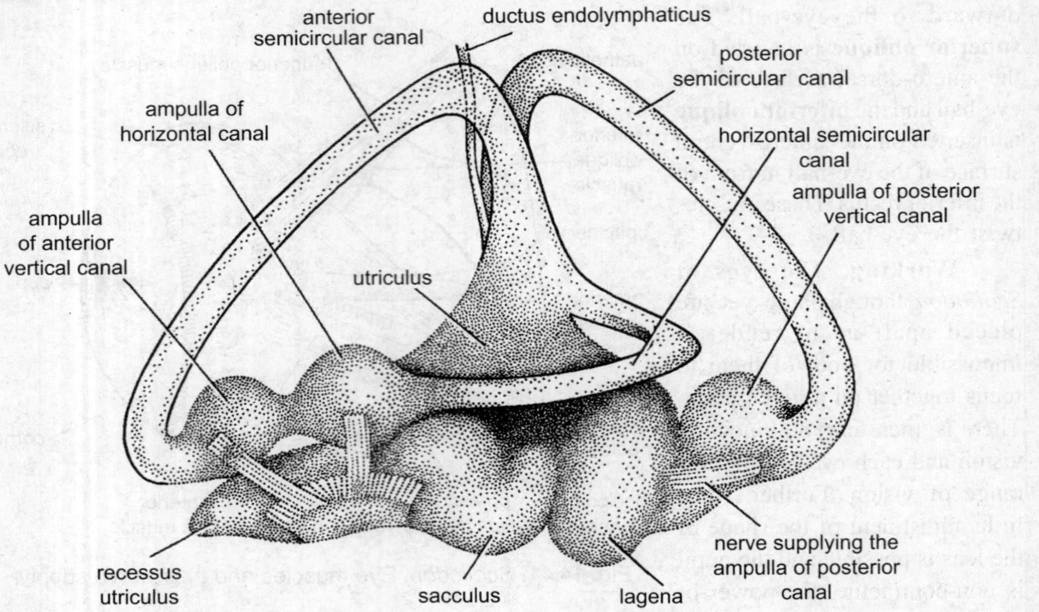

Fig. 14.42. *Scoliodon*. An outer view of the left membranous labyrinth.

The membranous labyrinth is innervated by the auditory nerve. In utriculus, sacculus, recessus utriculi, lagena cochleae as well as in the ampullae of the semicircular canals, there are groups of delicate receptor cells bearing stiff hairs. Groups of receptor cells located in the utriculus and sacculus are called **maculae**, and those found in the ampullae are called **cristae**.

Functions. In fishes the internal ear is not concerned with hearing but it performs the following functions : (*i*) it maintains muscle-tone, (*ii*) it detects acceleration of speed, change of direction and orientation, (*iii*) it is concerned with balance or equilibrium with regard to gravity. Movements of endolymph and otoliths stimulate sensory nerve endings in vestibule and ampullae, thus, informing the fish position in water and (*iv*) it may detect low frequency vibrations of water but there is no proof that fishes can hear. Sacculus and lagena probably receive auditory stimuli.

4. Neuromast Organs (Rheoreceptors)

The neuromast organs consist of (*i*) **lateral line receptors** and (*ii*) **pit-organs**.

(*i*) Lateral line receptors. The lateral line receptors lie in the **lateral line canal** which runs on each side of the trunk and tail. The lateral line canal is a slender mucus filled canal embedded in the dermis. It opens on the surface by minute pores found at intervals through a series of vertical tubes. Each lateral line canal is continued into the **cephalic canal** on the head. These canals have groups of epidermal receptors or **neuromasts** embedded in their walls. The neuromasts are innervated by the branches of seventh nerve and the lateralis branch of the vagus.

In the head region two lateral canals are joined by a transverse **commissural occipital canal** above the head, then each lateral line canal runs forward as a **post-orbital canal** which divides into two branches, a **supra-orbital canal** above the orbit and an **infra-orbital canal** below the orbit, both run up to the snout. The supra-orbital canal further divides into **anterior** and **dorsal branches**. The dorsal branch continues as **ventro-lateral branch** and joins with the infra-orbital canal. Arising from the infra-orbital canal is a **jugal canal** below the eye which runs back up to the first gill-cleft. The jugal canal gives off a **mandibular canal** to the lower jaw. The canals are lined with epithelium having many mucous gland cells which secrete mucus. The canals open at intervals on the surface by vertical tubes. The canals are filled with fluid and mucus.

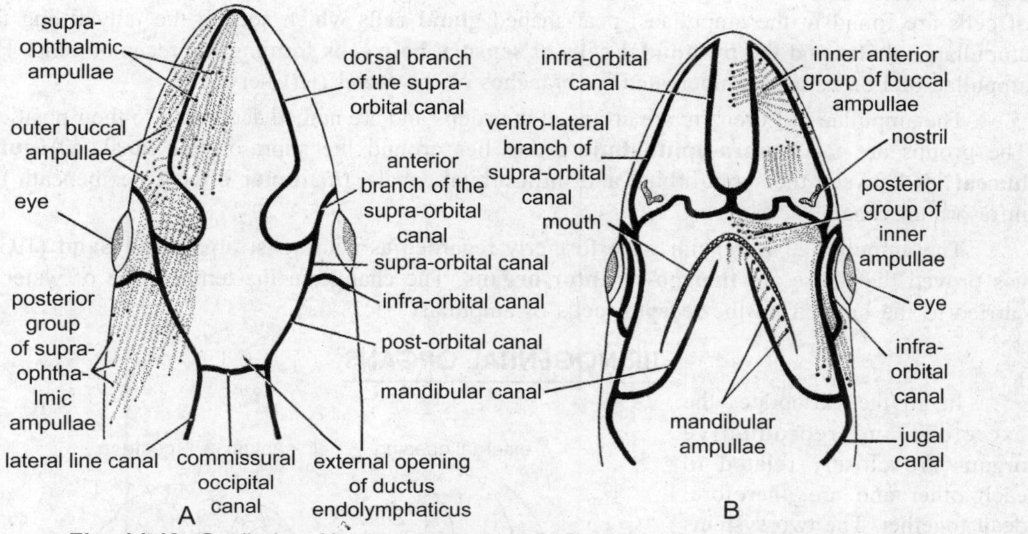

Fig. 14.43. *Scoliodon*. Neuromast organs. A–Dorsal surface; B–Ventral surface.

Neuromast organs. In the canals are found scattered sensory receptor organs, the **neuromasts**. Each made of a group of sensory receptor cell and supporting ectodermal cells. Each receptor cell has a stiff sensory hair projecting into the canal and a nerve fibre at the other end. The hairs of receptor cells are tipped with a heavy gelatinous substance. The lateral line neuromasts are current receptors **(rheoreceptors)** detecting any type of vibrations of water.

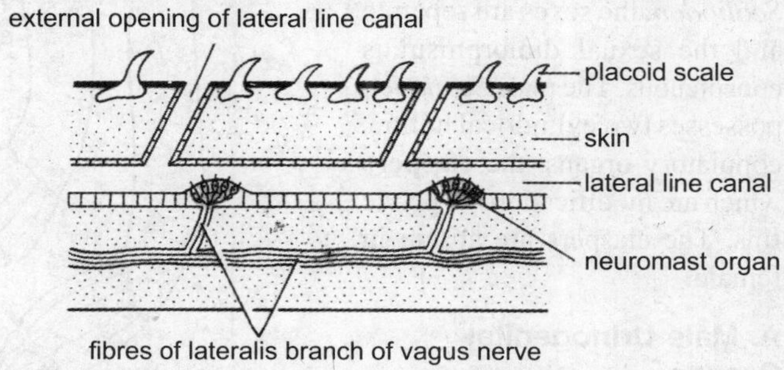

Fig. 14.44. *Scoliodon*. L.S. of lateral line canal showing neuromast organs.

***(ii)* Pit organs.** Pit organs are found scattered on the dorsal and lateral surfaces of the head. Pit organs consist of sensory hair cells and supporting ectodermal cells innervated by nerve-fibres of seventh cranial nerve. Pit organs are scattered individual neuromasts found in all fishes. They are **rheoreceptors**.

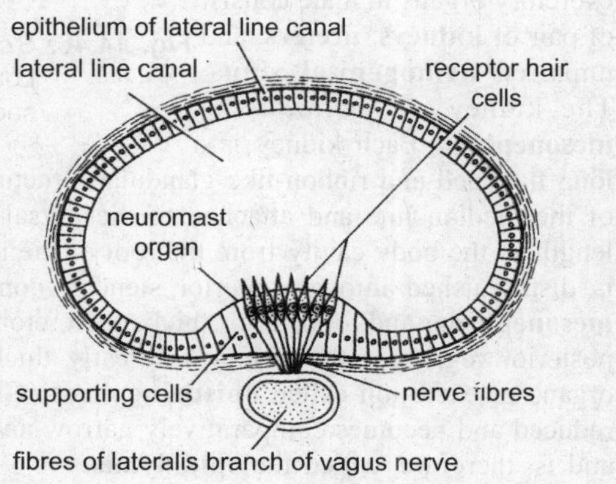

Fig. 14.45. *Scoliodon*. T.S. of lateral line canal through a neuromast organ.

5. Ampullae of Lorenzini

The **ampullae of Lorenzini** are found in clusters on the dorsal and ventral surfaces of the head embedded below the skin but opening externally on the surface of the skin. Each ampulla has a pore opening on the surface, the pore leads into an elongated mucous-filled **tubule** which ends in a radially septate **ampullary sac** lying deeply beneath the skin. Each ampullary sac consists of eight to nine radially dilated chambers arranged around a central core, the **centrum**. Two kinds of cells are found in the ampullae : pear-shaped **gland cells** which secrete the jelly filling the ampullary tubules and the **pyramidal cells** or **sensory hair cells** forming the receptor cells. The ampullae of Lorenzini are innervated by branches of the facial (*vii*) nerve.

The ampullae of Lorenzini are arranged in groups and are named according to their position. The groups are : (*i*) **supra-ophthalmic** group lies around the supra-orbital canal, (*ii*) **outer buccal** lies between the supra-orbital and infra-orbital canals, (*iii*) **inner buccal** lies beneath the infra-orbital canal.

The ampullae of Lorenzini were formerly regarded as neuromast organs but **Sand** (1938) has proved that these are **thermo-receptor organs**. The change in the temperature of water is carried to the brain from the receptor cells of ampullary sac.

URINOGENITAL ORGANS

In all the vertebrates the excretory and reproductive organs are closely related to each other and are, therefore, dealt together. The two systems together are known as **urino-genital system** and the organs as **urinogenital organs**. In *Scoliodon*, the sexes are separate and the sexual dimorphism is conspicuous. The male *Scoliodon* possesses two cylindrical hollow copulatory organs, the **claspers** which are modifications of pelvic fins. The claspers are absent in female.

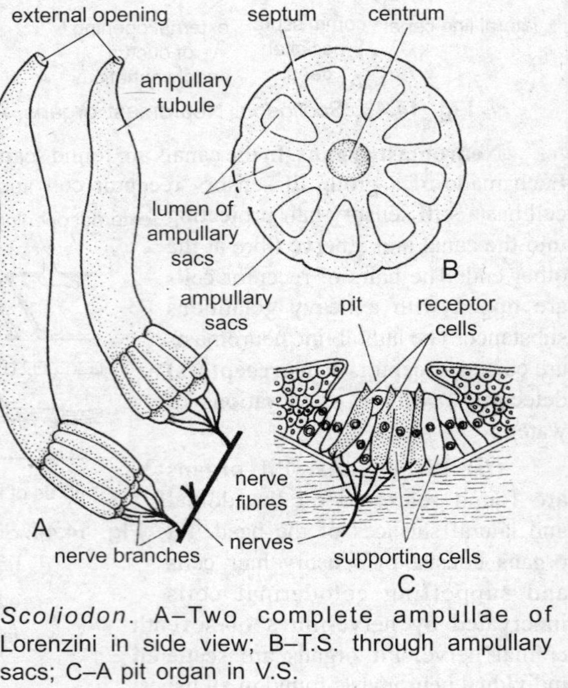

Fig. 14.46. *Scoliodon*. A–Two complete ampullae of Lorenzini in side view; B–T.S. through ampullary sacs; C–A pit organ in V.S.

A. Male Urinogenital Organs

1. Excretory organs. The excretory organs in male consist of pair of **kidneys**, **ureters**, and unpaired **urinogenital sinus**. The kidney of *Scoliodon* is **mesonephric**. Each kidney is a long flattened and ribbon-like glandular structure lying dorsally to the peritoneum on either side of the median line and attached to the dorsal abdominal wall. They extend through the entire length of the body cavity from the root of the liver in front up to the cloaca behind. Each kidney is distinguished into an anterior slender, non-excretory portion, the **genital kidney (cranial mesonephros)** and a posterior thicker, excretory part, **renal kidney (caudal mesonephros)**. The posterior region of the kidney is greatly thickened, laterally compressed and forms the chief organs of excretion called **opisthonephros** (**Graham Kerr** 1919). While, the anterior region is reduced and becomes comparatively narrow and takes the function of conveying genital products and is, therefore, called the **epididymis**.

The substance of the opisthonephric kidney is made up of coiled glandular uriniferous tubules. Each **uriniferous tubule** comprises a **Bowman's capsule** enclosing the **glomerulus** and a coiled **renal tubule**, several of which open into a common **collecting tubule**. The renal tubules have a special urea-absorbing segment. The collecting tubules of each kidney open independently into a common duct, the **ureter** which finally opens posteriorly into a wide median chamber, the **urinogenital sinus**. The urinogenital sinus opens into the **cloaca** through an aperture located at the tip of a small **urinogenital papilla**.

2. Reproductive organs. The reproductive organs include a pair of **testes**, a pair of **vas deferens**, and a pair of **seminal vesicles**.

Testes. There are two elongated **testes** lying one on either side of the middle line. Each testis extends from the base of the liver in front to the caecal gland behind. The testes are attached to the dorsal body wall by a fold of peritoneum called the **mesorchium** and to the caecal gland behind by a non-glandular tissue. Each testis is made up of numerous narrow seminiferous tubules lined with germinal cells which produce sperms by spermatogenesis. From each testis arise several thin tubular **vasa efferentia** which open into the anterior end of a large, narrow closely coiled tube occupying the entire part of the anterior part of the kidney or epididymis but becomes very much enlarged posteriorly to form the wide straight **vesicula seminalis** or **seminal vesicle** in which sperms are stored. The seminal vesicles of both the sides open behind independently into a large triangular chamber, the **urinogenital sinus** which in its turn opens into the cloaca on an elevated **urinogenital papilla**. On either side of the urinogenital sinus arises a club-shaped **sperm-sac**. The function of the sperm-sac is unknown.

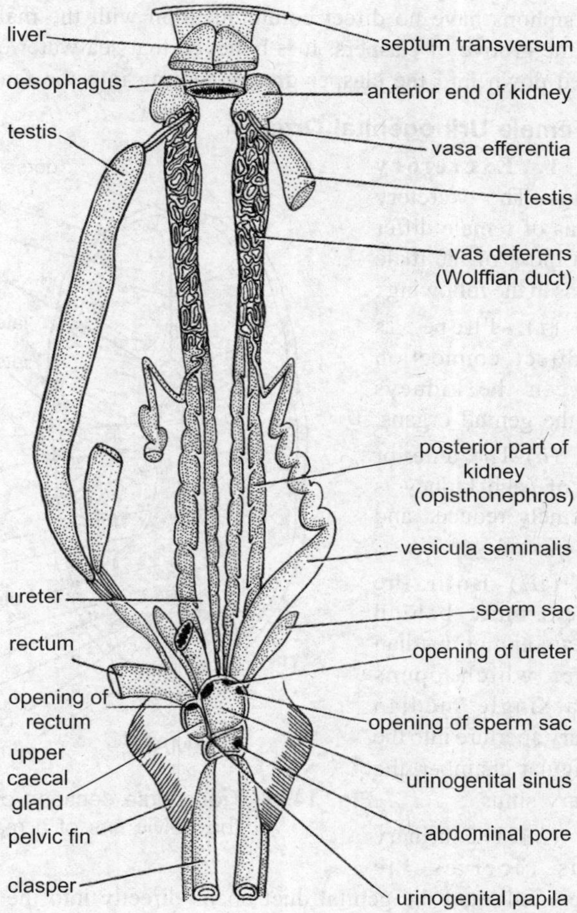

Fig. 14.47. *Scoliodon*. Male urinogenital organs.

Accessory reproductive organs. These are the **claspers** and **siphons**. The **claspers** are grooved elongations of the pelvic fins. The sperms from the cloaca pass into the claspers and are then transferred to the female, thus, acting as intromittent organs in coition. Each clasper has a closed groove with an anterior opening, **apopyle** lying near the cloaca and a posterior opening called the **hypopyle**. Sperms enter apopyle from the urinogenital papilla and then transferred to the cloaca of female during coitus. On the ventral side of the body just beneath the skin there is a pair of elongated glandular and muscular sacs, the **siphons**. These sacs extend anteriorly almost to the level of the posterior margin of the pectoral fins where they end blindly. Posteriorly these sacs extend as **siphon tubes**, each of which opens into the anterior opening of clasper of its side. The siphons have no direct communication with the male urinogenital organs except they open into the groove of claspers. It is believed that sea water enters the siphons during copulation and forced down into the clasper groove forcing into the female cloaca.

B. Female Urinogenital Organs

1. Excretory organs. The excretory organs of female differ from those of the male organs in the following :

(*i*) There is no direct connection between the kidneys and the genital organs.

(*ii*) The anterior part of each kidney is extremely reduced and non-excretory.

(*iii*) Both the ureters unite behind into a common median ureter which opens by a single median urinary aperture into the triangular chamber, the urinary sinus.

(*iv*) The urinary sinus receives the ureters only and the genital duct opens directly into the cloaca.

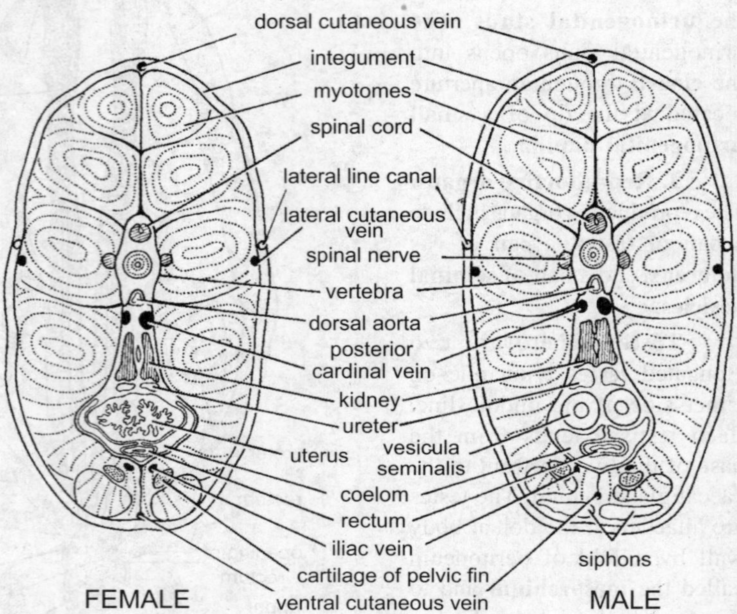

Fig. 14.48. Transverse sections of the body passing through the base of the pelvic fins of a male and a female *Scoliodon*.

Each **kidney** of the female also has two parts, an anterior part which is degenerate and reduced to a narrow tapering strand which is non-renal and non-functional, and a broad posterior part which contains masses of coiled uriniferous tubules. From the inner margin of each kidney arises a **ureter** or **Wolffian duct** into which open the uriniferous tubules. The two ureters unite to form a common ureter which opens into a large median **urinary sinus**. The urinary sinus is formed by the enlarged ends of the Wolffian duct. The urinary sinus opens into the cloaca by an aperture on a urinary papilla.

2. Reproductive organs. The female reproductive organs comprise a pair of **overies**, a pair of **oviducts**, **shell gland** and a pair of **uteri**.

Ovaries. The **ovaries** are paired organs varying in form and size according to the age of *Scoliodon*. The lobulated overies are situated, one on either side of the vertebral column, just behind the base of liver. They are suspended in the body cavity, from the dorsal abdominal wall by a fold of peritoneum, the **mesovarium**. Round follicles project from each ovary, each follicle contains an ovum. Between the ovary in front and caecal gland behind extends a long tubular strand of tissue, the **epigonal organ**.

Oviducts. the two **oviducts** are the well developed **Mullerian ducts** extending the entire length of the body cavity. Each oviduct is a long thick-walled muscular tube and their anterior narrow ends unite mid-ventrally beneath the oesophagus near septum transversum and opens into the coelom by a single longitudinal slit, called the **oviducal funnel** or **ostium**. Each oviduct runs backward for a short distance and then enlarges to form the **shell gland** or **oviducal gland**. It secretes a thin membrane over the descending fertilised eggs. The oviduct then again narrows and then it again dilates to form a sac-like **uterus**. The uteri of both the sides unite to form a

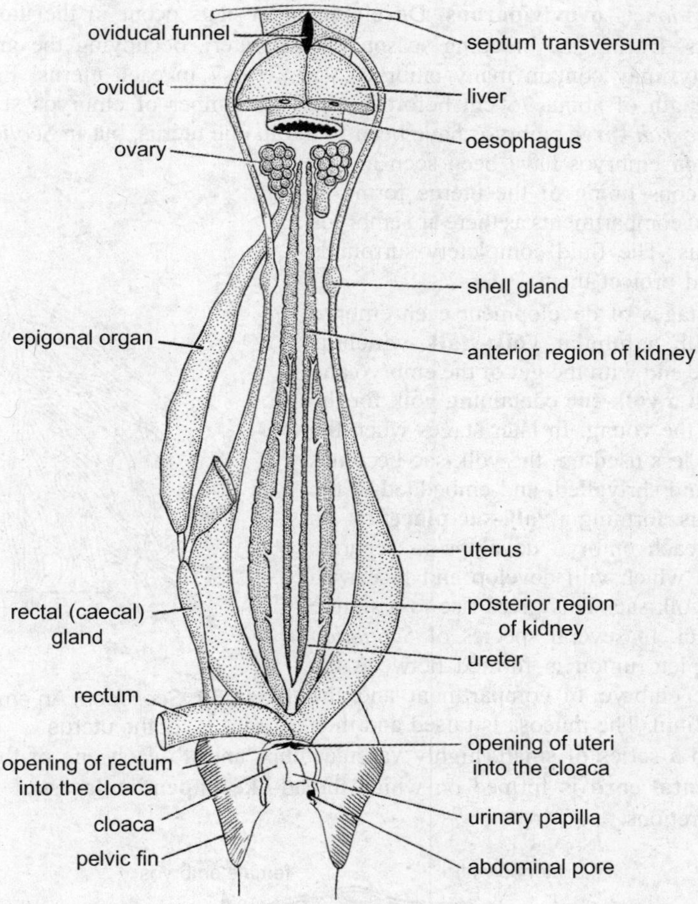

Fig. 14.49. *Scoliodon*. Female urinogenital organs.

short **vagina** which opens into the cloaca by a large aperture. A fold of mucous membrane of this region separates the vagina from the cloaca, and acts as a valve which closes the vagino-cloacal aperture during the development of the embryo within the uterus. The oviducts have no direct communication with the ovaries so that mature ova are at first shed into the abdominal cavity and later forced by the action of the body-muscles into the oviducts through the single oviducal funnel. Fertilisation occurs in the section of oviduct between the oviducal funnel and the shell-gland. The fertilised ova come into the uterus.

REPRODUCTION

Copulation

During copulation one or both the claspers of the male are inserted into the cloaca of the female and are fixed in position by the erection of terminal pieces. The spermatic fluid is forced through the grooves of the claspers into the oviducts through the vagina.

Fertilisation

In *Scoliodon*, the fertilisation is **internal**. Fertilisation of mature eggs takes place in the section of the oviduct between the oviducal funnel and the shell-gland.

Development

The *Scoliodon* is **ovoviviparous**. Development of eggs occur in the uteri and give birth to living youngs. During the breeding season the two uteri, occupying the greater part of the abdominal cavity, may contain many embryos, *i.e.*, 3 to 7 in each uterus. Each embryo may attain a total length of about 75 cm before birth. The number of embryos seems to vary : in *Scoliodon palasorrah* **three** embryos have been found in one uterus, but in *Scoliodon sorrakowah* as many as **seven** embryos have been seen in one uterus. The mucous lining of the uterus forms as many fluid-filled compartments as there are embryos within the uterus. The fluid completely surrounds the embryos and protect them.

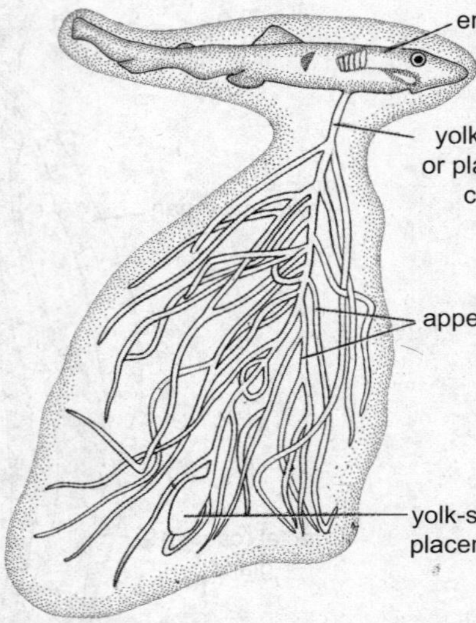

Fig. 14.50. *Scoliodon*. An embryo taken out of of the uterus.

In early stages of development each embryo is provided with a tubular **yolk-stalk** which is connected at one end with the gut of the embryo and at the other with a **yolk-sac** containing yolk for the nourishment of the young. In later stages when the yolk is more or less used up, the yolk-sac becomes greatly folded and shrivelled, and embedded in the uterine wall, thus, forming a **yolk-sac placenta**. As we know that each embryo develops in separate compartment in which villi develop and fuse with grooves in the yolk-sac. This gives rise to a simple yolk-sac placenta. In several species of *Scoliodon* an almost complete union is formed between the yolk-sac of the embryo in compartment and the maternal epithelium. The mucosa is raised and then invaginated into a series of small, highly vascular cups and, to each one of these a yolk-sac is joined. A **placental cord** is formed on which thread-like **appendicula** assist in the absorption of maternal secretions.

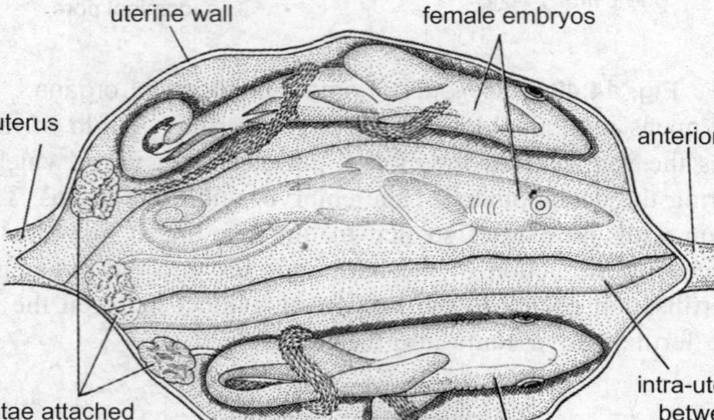

Fig. 14.51. *Scoliodon*. Uterus cut open to show the embryo within.

Some sharks and rays are oviparous. In these cases the impregnated ovum becomes surrounded by a layer of semi-fluid albumen and enclosed in a shell of keratin secreted by the shell gland. Shell in many dogfishes is four cornered, with twisted filamentous appendages at the angles, by means of which it becomes attached to sea-weeds and the like. In the rays the filaments are absent. In *Heterodontus,* it is an ovoid body whose wall presents a broad, spiral flange. In *Rhinobatus* and *Trygonorhina*, which are viviparous, each shell encloses three or four eggs.

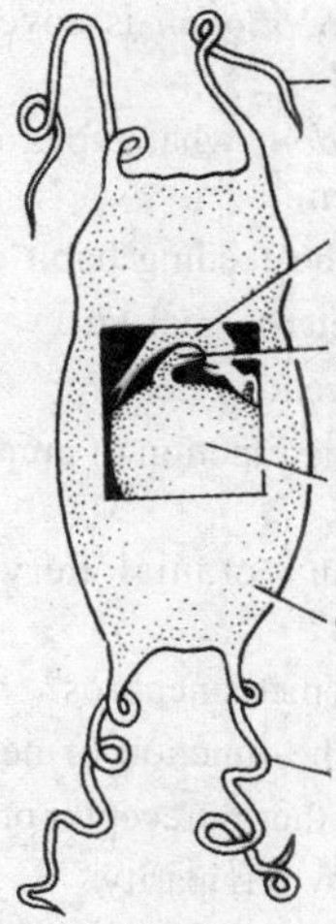

Fig.14.52. The egg case or mermaid's purse of dogfish *Scyllium*.

ECONOMIC IMPORTANCE

Sharks are predaceous carnivores feed on lobsters, crabs and fishes. Poor people living along the sea coasts eat dogfishes. Its flesh is not delicious but it is highly nutritious. Its dried skin called **shagreen** is used as an abrasive for polishing furniture, etc. It is used for dissection in laboratories of the colleges and universities.

IMPORTANT QUESTIONS

Long Answer Questions

1. Give an account of habits, habitat, geographical distribution and external features of *Scoliodon*.
2. Describe the structure of pectoral and pelvic girdles of *Scoliodon*.
3. Give an account of the digestive system and mechanism of feeding in *Scoliodon*.
4. Describe the structure of gill and mode of respiration in *Scoliodon*.
5. Describe the blood vascular system of *Scoliodon* and also describe the course of circulation of blood.
6. Describe the structure of heart and course of blood circulation through it in *Scoliodon*.
7. What is blood circulation? Describe the afferent branchial arterial system in *Scoliodon*.
8. Describe the urinogenital system in male and female *Scoliodon*.
9. Give an account of the central nervous system of *Scoliodon*.
10. Describe the cranial nerves of *Scoliodon*.

Short Answer Questions

1. Draw a placoid scale, label it and explain its structure in ten sentences.
2. Describe the skin of *Scoliodon* and compare it with the skin of a mammal.
3. Enumerate the muscles which bring about forward and backward rotation of eyeball in *Scoliodon*.
4. Give an account of internal ear of *Scoliodon*.
5. Give the origin, distribution and nature of cranial nerves in *Scoliodon*.
6. Describe the lateral line system in *Scoliodon*.
7. Describe the structure of brain of *Scoliodon*, mention the function of the various parts.
8. Describe the process of respiration in *Scoliodon*.
9. Write short notes on :
   (*i*) Ampullae of Lorenzini;
   (*ii*) Neuromast organs;
   (*iii*) Placoid scales;
   (*iv*) Spiral valve;
   (*v*) Mermaid's purse
10. Draw well labelled diagrams of the following of *Scoliodon* : (*i*) Arterial or venous system; (*ii*) Brain and cranial nerves; (*iii*) Male or female urinogenital system; (*iv*) Internal ear; (*v*) T.S. through pharyngeal region; (*vi*) T.S. through intestinal region; (*vii*) V.L.S. of a holobranch; (*viii*) Heart dissected from ventral side.

Very Short Answer Questions

1. Body of *Scoliodon* is covered by what type of scales?
2. In *Scoliodon*, what types of fins are there? Name them.
3. What is the feeding habit of *Scoliodon*?
4. Define heterocercal tail.
5. What is venous heart?
6. What is the function of ampullae of Lorenzini in shark?
7. How many cranial nerves are found in *Scoliodon*?
8. What is opisthonephros?
9. What is the function of neuromast organs?
10. Name the thermoreceptor organs of *Scoliodon*.
11. Explain ovoviviparity.

Multiple Choice Questions

1. *Scoliodon's* body is covered by
   (*a*) Placoid scales
   (*b*) Ctenoid scales
   (*c*) Dermal scales
   (*d*) Cycloid scales.
2. Fins in *Scoliodon* are
   (*a*) Median fins and paired fins
   (*b*) Only median fins
   (*c*) Only paired fins
   (*d*) Fins are lacking.
3. Median fins in *Scoliodon* are
   (*a*) Two dorsal, one caudal and one ventral fin
   (*b*) One dorsal, one caudal and one ventral fin
   (*c*) One dorsal, two caudals and one ventral
   (*c*) One dorsal, one caudal and two ventral fins.
4. Tail in *Scoliodon* is
   (*a*) Homocercal
   (*b*) Heterocercal
   (*c*) Hypocercal
   (*d*) Hypercereal.
5. Intromittent organ in *Scoliodon* is
   (*a*) A part of pectoral fin
   (*b*) A part of first dorsal fin
   (*c*) Medial part of pelvic fin
   (*d*) Medial part of ventral fin.
6. The eyes in *Scoliodon* are protected by
   (*a*) Eyelids
   (*b*) Immovable nictitating membrane
   (*c*) Cuticle
   (*d*) Movable nictitating membrane.
7. Nostrils in *Scoliodon* are
   (*a*) Dorsal and olfactory
   (*b*) Ventral and olfactory
   (*c*) Dorsal and respiratory
   (*d*) Ventral and respiratory.
8. Spine of placoid scale is made up of
   (*a*) Calcite
   (*b*) Dentine
   (*c*) Calcium carbonate
   (*d*) Tunicine.
9. The pulp in placoid scale is formed by
   (*a*) Odontoblasts
   (*b*) Ameloblasts
   (*c*) Scleroblasts
   (*d*) Dermal papilla.
10. The teeth in dogfish are modified
    (*a*) Placoid scales
    (*b*) Bony scales
    (*c*) Cycloid scales
    (*d*) Ctenoid scales.
11. The teeth of *Scoliodon* are meant for
    (*a*) Biting and tearing
    (*b*) Only biting
    (*c*) Only chewing
    (*d*) Biting, tearing and chewing.
12. In *Scoliodon* mucus is secreted by
    (*a*) Cells in stratum laxum
    (*b*) Cells in stratum germinativum
    (*c*) Cell in stratum compactum
    (*c*) Melanophores.
13. Myotomes in dogfish are separated by
    (*a*) Myocommata
    (*b*) Adipose tissue
    (*c*) Malpighian layer
    (*d*) Odontoblasts.
14. The rhythm of contraction of myomeres is governed by
    (*a*) Myocommata
    (*b*) Spinal cord
    (*c*) Brain
    (*d*) Medulla oblongata.

15. The pericardial and abdominal cavities in dogfish is separated by
    (*a*) Septum transversum
    (*b*) Septum longitudinum
    (*c*) Muscle sheath
    (*d*) Diagonal septum.
16. The pericardial and abdominal cavities in dogfish communicate with the another through
    (*a*) Pericardio-periabdominal canal
    (*b*) Pericardio-peritoneal canal
    (*c*) Peritoneal canal
    (*d*) Abdominal canal.
17. Dentition in dogfish is
    (*a*) Homodont, polyphyodont and lyodont
    (*b*) Heterodont, polyphyodont and lyodont
    (*c*) Heterodont, diphyodont and lyodont
    (*d*) Homodont, diphyodont and lyodont.
18. In dogfish, the different parts of alimentary canal are bound to one another by
    (*a*) Circular muscle fibres
    (*b*) Incomplete mesenteries or omenta
    (*c*) Complete mesenteries
    (*d*) Substances produced by cement glands.
19. In *Scoliodon*, the oesophageal opening into cardiac stomach is guarded by
    (*a*) Blind sac
    (*b*) Oesophageal valve
    (*c*) Sphincter valve
    (*d*) Scroll valve.
20. Sphincter valve is present
    (*a*) At the junction of pharynx and cardiac stomach
    (*b*) At the junction of cardiac stomach and pyloric stomach.
    (*c*) At the junction of pyloric stomach and oesophagus
    (*d*) At the junction of oesophagus and intestine.
21. The scroll valve serves to
    (*a*) Delay the passage of food
    (*b*) Speed up the passage of food
    (*c*) Mix the food with bile juice
    (*d*) Digest the food quickly.
22. Liver of dogfish is
    (*a*) Single lobed
    (*b*) Bilobed
    (*c*) Trilobed
    (*d*) Four lobed.
23. In *Scoliodon*, sinusoids in the gill-lamella receive venous blood from
    (*a*) Afferent branchial artery
    (*b*) Efferent branchial artery
    (*c*) Epibranchial artery
    (*d*) Ventral aorta.
24. Heart in fishes is
    (*a*) Single chambered
    (*b*) Two chambered
    (*c*) Three chambered
    (*d*) Four chambered.
25. In *Scoliodon*, the heart receives
    (*a*) Only arterial blood
    (*b*) Only venous blood
    (*c*) Both arterial and venous blood
    (*d*) Blood only from gills.
26. The S-shaped heart of *Scoliodon* is differentiated into
    (*a*) One auricle and one ventricle
    (*b*) Auricle, ventricle and sinus venosus
    (*c*) Auricle, ventricle and conus arteriosus
    (*d*) Sinus venosus, auricle, ventricle and conus arteriosus.
27. Sinus venosus receives blood from
    (*a*) Ventral aorta
    (*b*) Ductus Cuvieri
    (*c*) Ductus Cuvieri and hepatic sinuses
    (*d*) Hepatic sinuses.
28. Brain in *Scoliodon* is enclosed in
    (*a*) Cranium
    (*b*) Muscular structure
    (*c*) Chondrocranium
    (*d*) Bony skull.
29. In *Scoliodon*, the restiform body is present in
    (*a*) Medulla oblongata
    (*b*) Cerebrum
    (*c*) Spinal cord
    (*d*) Cerebellum.
30. In *Scoliodon*, the foramen of Monro is the opening between
    (*a*) First and second ventricle of cerebrum and third ventricle
    (*b*) Olfactory sacs
    (*c*) Cavities of cerebrum
    (*d*) Third and fourth ventricle.

31. In *Scoliodon*, Gasserian ganglion is present in
(*a*) Trigeminal
(*b*) Oculomotor
(*c*) Olfactory
(*d*) Optic.

32. The swimming movement of *Scoliodon* are controlled by
(*a*) Cerebrum
(*b*) Midbrain
(*c*) Hindbrain
(*d*) Spinal cord.

33. The cranial nerves arising from the midbrain of *Scoliodon* are
(*a*) III and IV
(*b*) IV and V
(*c*) V and VI
(*d*) VI and VII.

34. In *Scoliodon*, the cranial nerve numbered as 'O' is
(*a*) Olfactory
(*b*) Optic
(*c*) Oculomotor
(*d*) Terminal or pre-olfactory.

35. In *Scoliodon*, ampullae of Lorenzini are supplied by
(*a*) Buccalis
(*b*) Palatinus
(*c*) Hyomandibular
(*c*) Ophthalmicus superficialis.

36. In *Scoliodon*, otoliths are found in
(*a*) Aqueous humor
(*b*) Vitreous humor
(*c*) Recessus utriculi
(*d*) Endolymph of membranous labyrinth.

ANSWERS

1. (*a*)	2. (*a*)	3. (*a*)	4. (*b*)	5. (*c*)	6. (*d*)	7. (*b*)	8. (*b*)
9. (*d*)	10. (*a*)	11. (*a*)	12. (*b*)	13. (*a*)	14. (*b*)	15. (*a*)	16. (*b*)
17. (*a*)	18. (*b*)	19. (*b*)	20. (*b*)	21. (*a*)	22. (*b*)	23. (*a*)	24. (*d*)
25. (*b*)	26. (*d*)	27. (*c*)	28. (*c*)	29. (*a*)	30. (*a*)	31. (*a*)	32. (*c*)
33. (*a*)	34. (*d*)	35. (*d*)	36. (*d*).				

15

Chapter

Type 6. *Labeo rohita* : Rohu (A Bony Fish)

Contents of this Chapter

Class Osteichthyes (Gr., *osteon* = bone + *ichthyes* = fish) includes the ray-finned "bony fishes", both marine (*e.g.*, cod, sole, herring, eels) and freshwater. Subclass Actinopterygii of class Osteichthyes are the commonest vertebrates of today. Its another subclass Crossopterygii includes today only four genera having a few species. The Actinopterygii get their name from the structure of paired fins. Today's actinopterygians are more advanced having streamlined body. The structure of scales is ganoid. The tail is homocercal or more rarely diphycercal. More advanced species have lost spiracles, abdominal pores and a spiral valve in the gut. The actinopterygians in general show no trace of internal nostrils. Gills are filamentous. In heart, elastic **bulbus arteriosus** has been evolved in exchange of conus arteriosus. The conus becoming reduced to one or, at most, two rows of valves.

The actinopterygians seem to have arisen in freshwater, as did other bony fishes, and later some invaded the sea in Palaeozoic times. *Labeo rohita* is a common Indian freshwater bony fish. The following description relates to *Labeo rohita*. This fish is studied as the type specimen of bony fishes in many Indian Universities.

SYSTEMATIC POSITION

Phylum	...	Chordata
Subphylum	...	Vertebrata
Division	...	Gnathostomata
Superclass	...	Pisces
Class	...	Osteichthyes
Subclass	...	Actinoptergii
Superorder	...	Teleostei
Order	...	Cypriniformes
Type	...	*Labeo rohita* (Rohu)

DISTRIBUTION

Labeo is a tropical genus of carps distributed in tropical Africa and East Indies. Roughly two dozen species of *Labeo* are found in India. The most common is *Labeo rohita* (rohu) and *Labeo calbasu* (kala bans) which inhabit throughout India, Bangladesh and Pakistan.

HABITS AND HABITAT

Labeo rohita (Rohu) is commonly found in freshwater ponds, lakes, rivers and streams. It is principally herbivorous and bottom feeder, feeding on algae and aquatic plants. It breathes by means of gills and also frequently comes to water surface for gulping air into the air bladder. It is oviparous, breeding occurs in running water in July and August. Fertilisation is external, *i.e.*, in water.

EXTERNAL FEATURES

Shape, size and colour. The rohu has a spindle-shaped body. Its body colour is blackish on the dorsal side and silvery on the ventro-lateral sides. It measures up to 1 metre in length and weighs about 20 to 25 kg. The body is divisible into head, trunk and tail.

Head. Head extends from tip of snout to the posterior margin of operculum. Snout is depressed, obtuse and project beyond the jaws. A pair of **nostrils** are present on the dorsal surface of snout. **Mouth** is a crescentic transverse subterminal slit bounded by thick, fleshy upper and lower lips. **Eyes** are large, laterally situated and have no eyelids. One pair of small delicate **maxillary barbels** are usually present on the dorso-lateral sides of the mouth. **Rostral barbels** are usually absent. **Operculum** is thin, large, and plate-like gill cover on either lateral, posterior side of the head.

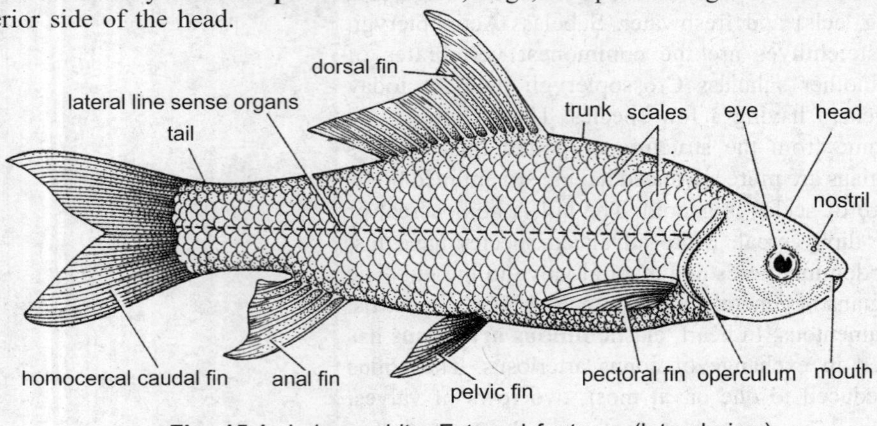

Fig. 15.1. *Labeo rohita*. External features (lateral view).

Trunk. It is elongated and oval in cross section, *i.e.,* broad in the middle and narrow towards the dorsal and ventral sides. The trunk is covered with thin, dermal plate-like overlapping **cycloid scales**. On either lateral side of the trunk is present the **lateral line**, which also extends over the tail. The scales along the lateral line contain pores which communicate with the tubular canal. The canal possesses the sensory organs, the **rheoreceptors** which are sensitive to current of water. At the posterior end of trunk, mid-ventrally there are three small apertures in a line: anterior **anus**, middle **genital** and posterior **urinary**.

Trunk also bears paired and upaired fins. **Paired fins** are the pectorals and the pelvics. **Pectoral fins** are placed at the antero-lateral sides of the trunk just behind the operculum. Each pectoral fin is supported by 19 fin-rays. **Pelvic fins** are placed on the ventral side behind the pectorals, and each pelvic fin is supported by 9 fin-rays.

Unpaired fins are (*i*) a **dorsal fin** located on the mid-dorsal line of trunk, roughly over the middle of the body, (*ii*) **anal fin** lies posterior to anus in the mid-ventral line, and (*iii*) **tail fin** or **caudal fin** over the tail. It is homocercal with two equal lobes. Dorsal fin is supported by 13 fin-rays, anal fin has 4 to 6 fin-rays and tail fin is supported by several fin-rays.

Tail. The part of the body behind the urinary aperture is the tail. It is laterally compressed and narrow behind. The tip of tail bears **homocercal caudal fin** which is deeply clefted into two similar lobes. Tail with caudal fin acts as a locomotor organ.

BODY WALL OR INTEGUMENT

Body wall is composed of outer **epidermis** and the inner **dermis**. The **epidermis** is a squamous stratified epithelium and contains mucous glands and pigment cells. Mucous cells secrete mucus making the body of fish slippery. **Dermis** is like that of the dogfish. Beneath the epidermis the scales are lodged in pouches of the dermis. Each scale is a flat, nearly circular plate of bone. Over the scales (circular plate of bone) are present concentric rings which represent the annual growth stages. These rings reveal the age of a fish. Scales are arranged overlapping one another from before backwards. A small three-sided portion of each scale comes to lie immediately beneath the epidermis, while the rest is hidden beneath the scales immediately anterior to it.

Below the integument are present the muscles. The muscles of the trunk and tail are arranged in a zig-zag fashion as in dogfish. The muscles are lined internally by peritoneum, which also lines the coelom or body cavity. The visceral organs are also covered over by peritoneum.

EXOSKELETON

Exoskeleton of *Labeo rohita* comprises scales and fin-rays. Scales are cycloid type which are thin overlapping bony plates, partly covered by skin. Centrally each scale has an area called focus. Around focus there are concentric rows of ridges called circulii which are used to count the age of fishes. Moreover, there are several radial lines, starting from focus and extending up to the periphery, called radii. Fin-rays are fin-supporting structures. They are of two types spines (single ray) and soft rays (segmented rays). Spines in the fins of *Labeo* are unsegmented, uniserial structures. Soft rays are segmented, often branched and biserial.

ENDOSKELETON

The endoskeleton of *Labeo* is completely ossified and forms the framework of the body. It is divided into an **axial skeleton** which includes the skull, vertebral column and ribs, and **appendicular skeleton** that includes the girdles and supporting elements of fins.

Skull

Skull of *Labeo* (Fig. 15.2) is an extremely complex structure, composed of many investing membrane or dermal bones and cartilage bones. The skull is divided into **cranium** enclosing the brain with paired olfactory, optic and auditory capsules for the respective sense organs and the **visceral** or **branchial arches**. The visceral arches are loosely attached with the skull.

1. Cranium and sense capsules. The cranium and sense capsules are firmly united together. The cranium of *Labeo rohita* is divided into four regions : (*i*) Ethmoidal region, (*ii*) Orbito-temporal region, (*iii*) Otic region, (*iv*) Occipital region.

(*i*) Ethmoidal region. It forms the anterior part of the cranium and is formed of paired **nasals**, **ectoethmoids**, **lacrymals** and unpaired median **mesethmoid**, **vomer** and **rostral**. The **mesethmoid** is a transversely placed bone, in front of the frontal bones. The **rostral** fits into a notch in the mesethmoid. The **vomer** is a quadrangular bone placed beneath the mesethmoid in front of the parasphenoid on the ventral side of the skull. The **nasals** articulate with the postero-lateral sides of the mesethmoid and forms the roof of the olfactory capsules. The **ectoethmoids** also take part in the formation of olfactory capsules. Each has a deep nasal pit which lodges the nasal sac. The **lacrymals** are attached with the antero-lateral borders of mesethmoid on the antero-dorsal surface of the skull.

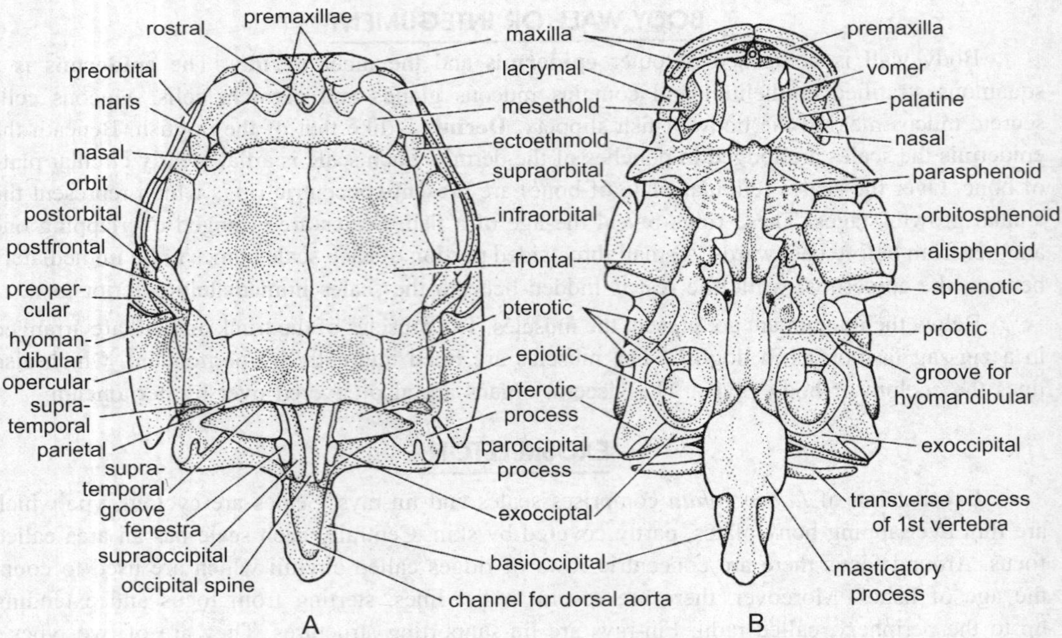

Fig. 15.2. *Labeo*. Skull. A–Dorsal view; B–Ventral view.

(*ii*) Orbito-temporal region. In front of auditory region the cranium is excavated on each side by a large **orbit**. A vertical plate or **inter-orbital septum** separates the two cavities from one another. Anterior boundary of each orbit is formed by **ectoethmoid,** posterior boundary is formed by the **sphenotic**, dorsal boundary is made by **frontal** and ventral boundary is formed by **alisphenoid** and **orbitosphenoid**. Five small orbital bones form the orbital ring. These are anterior **preorbital**, posterior **postorbital**, **supraorbital** and **postfrontal** are of the dorsal side bones and **infraorbital** forms the anteroventral part of the ring.

Temporal region or **sphenoidal region** is further divided into (*a*) anterior **frontal region** and (*b*) posterior **parietal region.**

(*a*) Frontal region is formed of **frontal bones** on the dorsal side (also forms the roof of cranium). **Orbitosphenoids** and a median **parasphenoid** form the floor of this region. Frontals articulate anteriorly with the median **mesethmoid** and laterally and mesially with the **nasals** and **ectoethmoids**. Frontals articulate antero-laterally with the **supraorbitals** and postero-laterally with the **postfrontals**.

(*b*) Parietal region is formed of **parietals** on the dorsal side, **alisphenoids** form its lateral sides, and **parasphenoid** on the postero-ventral side. **Basisphenoids** are absent.

(*iii*) Otic region. Otic or auditory region is found behind the orbital region. It is formed of (*a*) anterior **prootic** which unites with its fellow of the opposite side in the floor of the brain case, just in front of the basioccipital; (*b*) **opisthotic** in the posterior part of the capsule, external to the exoccipital; (*c*) **sphenotic** above the pro-otic and forming part of the boundary of the orbit; (*d*) **pterotic** above the exoccipital and opisthotic, forming a distinct lateral ridge and produced behind into a prominent **pterotic process**; and (*e*) **epitoic** is a small bone wedged in between the **supra-** and **exoccipitals** and **pterotic**, and produced into a short **epiotic process**. Opisthotic is absent in adult *Labeo rohita*. All these bones form an inverted cup-like structure that lodges the internal ear.

(*iv*) Occipital region. It forms the posteriormost part of the skull. In the occipital region are four bones. These are the **basioccipital**, forming the greater part of the occipital condyle and the hinder region of the skull-floor, the paired **exoccipitals** placed one on each side of the **foramen magnum** and meeting both above and below it, and the **supraoccipital**, forming the occipital crest on the dorsal side. Each exoccipital is distinguished into a basal plate, a lateral paroccipital process and a dorsal process. Basioccipital is a long, tubular bone. It is thick in mid region.

2. Visceral skeleton. In fishes, it forms the part of the skull. It includes seven paired arches uniting in the mid-ventral line.

(*i*) The first or **mandibular arch** forms the jaws. It is formed of a **palato-pterygoquadrate** which forms the upper jaw and a **Meckel's cartilage** that forms the lower jaw in the embryonic stage. **Palatoquadrate** is homologous with the upper jaw of dogfish. However, instead of remaining cartilaginous, it is ossified by three replacing bones. The toothed **palatine** lies in front, articulating with the olfactory capsule. **Metapterygoid** projecting upwards from the **quadrate**, but between quadrate and the palatine are two investing bones, the **ectopterygoid** on the ventral, and the **entopterygoid** on the dorsal edge of the original cartilaginous bar. The **quadrate** at the posterior end of the entopterygoid furnishes a condyle for the articulation of the lower jaw. These bones do not, however, enter into the gape, and do not form the actual upper jaw of the adult fish. External to them are two large investing bones, the **premaxilla** and the **maxilla**, which together form the actual or secondary upper jaw. Both these bones bear many teeth.

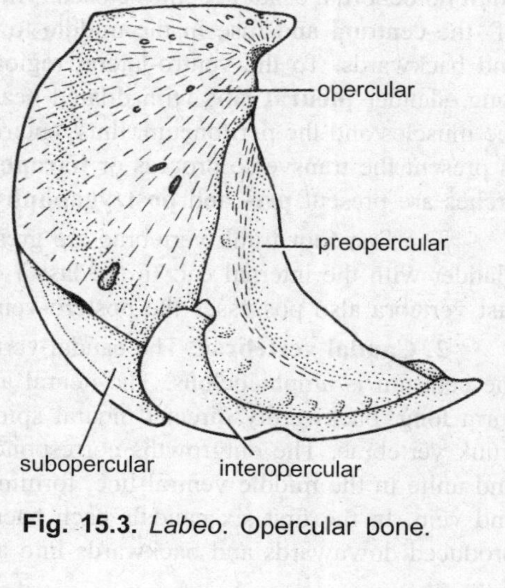

Fig. 15.3. *Labeo*. Opercular bone.

The lower jaw is formed of **dentaries**, **angulars** and **articulars**. Articular is formed by the ossification of Meckel's cartilage and articulates with the quadrate of the upper jaw. Dentary meets with its fellow of the other side in the middle. All these bones bear no teeth.

(*ii*) The second or **hyoid arch** forms the suspensorium and **hyoid cornu**. **Suspensorium** is formed of two bones, the **hyomandibular** and **symplectic**. **Hyomandibular** articulates with the auditory capsule and on the other side with the symplectic. **Hyoid cornu** is composed of paired **interhyal**, **epihyal**, **ceratohyal**, **hypohyal** and a median unpaired **basihyal**. The epihyal and ceratohyal are provided with three sabre-shaped **branchiostegal rays**. Beneath the basihyal is present a median **urohyal**. Basihyal supports the tongue.

Operculum. The operculum is formed of four investing bones; opercular, preopercular, subopercular and interopercular. The **opercular** is a large, flat bone with a condyle which articulates with the hyomandibular. The **subopercular** lies below and internal to the opercular. The **interopercular** pits between the lower portions of the three preceding bones, and is attached by ligament to the angle of the mandible. The sabre-shaped **branchiostegal rays** are attached along the posterior border of the epi- and ceratohyal.

The remaining five arches are called **branchial** or **gill-arches**, diminishing in size from before backwards. First three branchial arches are composed of pharyngobranchial above, then **epibranchial**, then a large **ceratobranchial**, and a small **hypobranchial** below. The right and left hypobranchials of each arch are connected by an unpaired median **basibranchial**. All these segments are ossified by replacing bones. The basibranchials are connected with one another and with the basihyal by cartilage so as to form a median ventral bar in the floor of the pharynx. In the fourth arch hypobranchial is absent. The fifth arch is reduced to a single bone on each side. Small spine-like ossifications are attached in a single or double row along the inner aspect of each of the first four arches. These are the **gill-rakers**. They serve as a sieve to prevent the escape of food through the gill-slits and to protect the gill filaments.

Vertebral Column

The vertebral column is differentiated into distinct bony **vertebrae**. It is divisible into an anterior or **trunk region** and a posterior **caudal region**.

1. Trunk vertebra. Trunk vertebrae are 21 in number. A typical trunk vertebra consists of a **centrum** with deeply concave anterior and posterior faces. In the embryonic stage the concavities are communicated by a narrow **notochordal canal** perforating the centrum. In the adult notochordal canal becomes closed. The low neural arches arise from the latero-dorsal surface of the centrum and fuse in the middle to form a long, slender **neural spine** directed upwards and backwards. To the ventro-lateral region of the vertebra are attached by ligament a pair of long, slender **pleural ribs** with dilated heads. These curve downwards and backwards between the muscles and the peritoneum, thus, encircling the abdominal cavity. On either side of centrum is present the transverse process or **parapophysis**. On the anterior and posterior faces of neural arches are present **pre-** and **postzygapophyses**, by which adjacent vertebrae articulate.

The first four trunk vertebrae are greatly changed since these vertebrae connect the swim-bladder with the internal ear. In the last 3 or 4 trunk vertebrae the parapophyses are fused. The last vertebra also possesses the postero-ventral process.

2. Caudal vertebrae. The caudal vertebrae are 16 – 17 in number. In the caudal vertebrae the centrum is amphicoelous. The neural arches arising from the centrum are fused dorsally to form long, backwardly directed neural spine. Pre- and postzygapophyses are also present as in trunk vertebrae. The outgrowths corresponding to the parapophyses are fused with the centrum and unite in the middle ventral line, forming a **haemal arch**. Through this run the caudal artery and vein. In the first six caudals each haemal arch bears a pair of ribs. In the rest the arch is produced downwards and backwards into a **haemal spine**.

The posterior end of the caudal region is enormously modified for the support of the tail fin. The hindmost centra have their axes deflected upwards and the last centrum is rod-like structure, the **urostyle**. The neural and haemal spines of the last five vertebrae are very broad and closely connected with one another. They are more numerous than the centra, and three or four haemal arches are attached to the urostyle. In this way a firm vertical plate of bone is formed, to the edge of which the caudal fin rays are attached fanwise in a symmetrical manner.

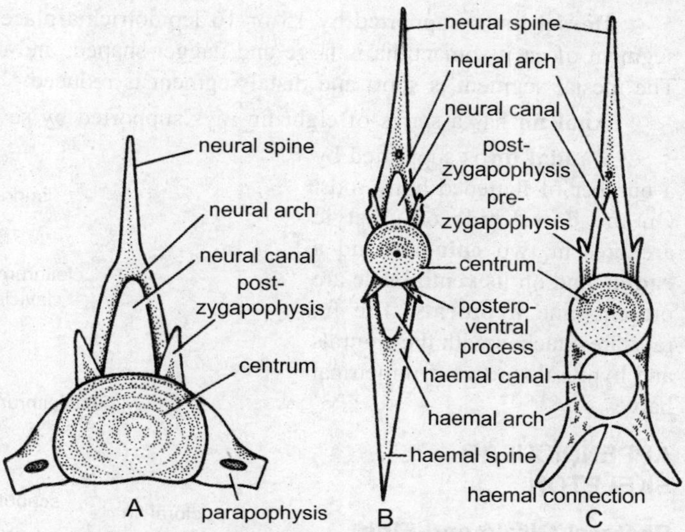

Fig. 15.4. *Labeo*. Vertebrae. A–Trunk vertebra in posterior view; B–Typical caudal vertebra in posterior view; C–First caudal vertebra in anterior view.

3. Weberian ossicle. A more sensitive apparatus exists in the carps (*e.g.*, *Labeo*) and siluroids (order Ostariophysi), in which a chain of bones connects the swim-bladder with the internal ear on either side, forming the **Weberian apparatus**. It is supposed that this apparatus is a modification of processes of certain vertebrae, which carries pressure stimuli anteriorly to the perilymph. It was discovered by **Weber** in 1820 who regarded them to be ear ossicles.

The first four trunk vertebrae have no parapophyses. The neural spine of the first vertebra is formed of two small pieces called **claustrum** and **scaphium**. These pieces are the two anteriormost elements of Weberian apparatus. The centra of second and third vertebrae are completely fused to form a large centrum that carries a flattened triangular plate, the **tripus**. Tripus is the modification of transverse process of third vertebra, and it is the last element of Weberian apparatus. In between scaphium and tripus is present a stout ligament. Embedded in this ligament is a small bone with a backwardly directed spinous process, which is called **interclarium**. It lies in between scaphium and tripus. Thus, there are four bones in the Weberian apparatus, *viz*., claustrum, scaphium, interclarium and tripus.

Skeleton of median fins. The skeleton of median fins (dorsal and anal) consists of a series of (*i*) **somactidia** or radials and (*ii*) the **dermotrichia** or dermal fin rays.

Somactidia are parallel bony rods found embedded within the body muscles. Each somactid has a proximal, mesial and distal segments. **Dermotrichia** support the fold of the fin. In *Labeo rohita*, the dermotrichia or **lepidotrichia** which lie in the substance of the fin are slender bones, jointed together and mostly branched. At the free edges of fins are also present unbranched horny **actinotrichia**.

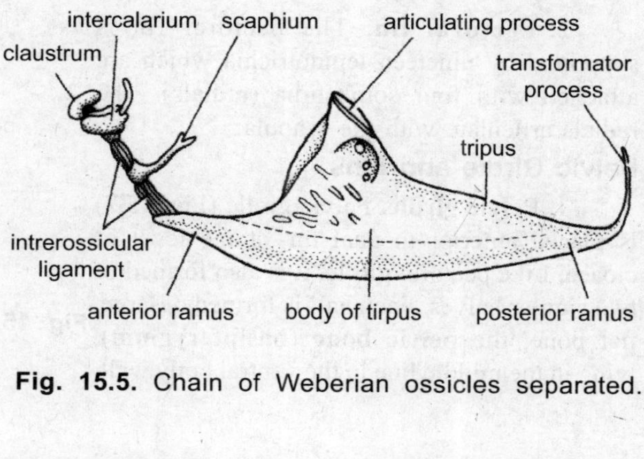

Fig. 15.5. Chain of Weberian ossicles separated.

Dorsal fin is supported by 15 or 16 lepidotrichia placed on 14 somactidia. The proximal segment of each somactidia is large and dagger-shaped, and called **axonast** (interspinous bone). The mesial segment is short and distal segment is reduced.

Anal fin has a series of eight fin rays supported by seven somactidia.

Caudal fin is supported by a number of flattened bony rods. On the dorsal side of urostyle are present two **epiurals** and a **radial**, and on its ventral side are present nine **hypiurals**. The fin rays are attached with the epiurals and hypiurals in two symmetrical halves.

APPENDICULAR SKELETON

Pectoral Girdle and Fins

1. Pectoral girdle. Each half of pectoral girdle (Fig. 15.6) is formed of (*i*) a **scapula**, situated dorsally to the glenoid facets and developed partly as a replacing, partly as an investing bone; (*ii*) a **coracoid** situated ventrally to the glenoid facet and (*iii*) a **mesocoracoid** situated above the coracoid and anterior to the scapula. Externally to these is a very large investing bone, the **cleithrum**, extending downwards under the throat. Its distal end is connected by means of a **supraclavicle** to a forked bone, the **posttemporal**, one branch of which articulates with the epiotic and the other with the pterotic process. To the inner surface of the cleithrum are attached two flat scale-like bones with a slender rod-like **postclavicle** passing backwards and downwards among the muscles.

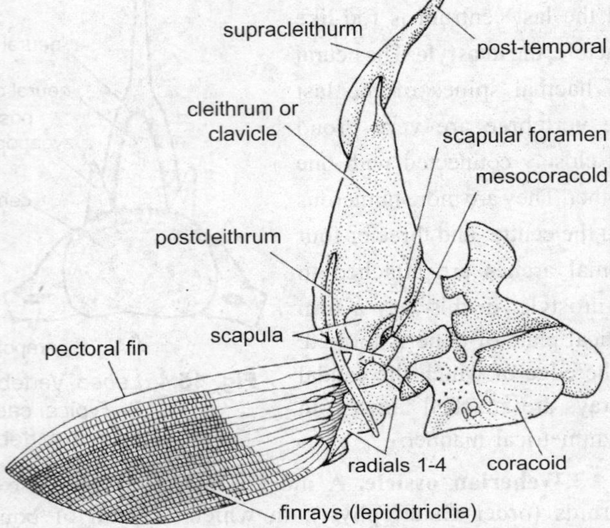

Fig. 15.6. *Labeo*. Left half of pectoral girdle with attached pectoral fin.

2. Pectoral fin. The pectoral fin is supported by nineteen lepidotrichia which are attached with four somactidia (radials). The radials articulate with the scapula.

Pelvic Girdle and Fins

1. Pelvic girdle. Pelvic girdle (Fig. 15.7) is placed anterior to anal fin, at the level of cloaca. Like pectoral girdle, it is also formed of two similar halves. Each half is formed of large, flat bone, the **pelvic bone** (**basipterygium**), lying in the middle line in the ventral body wall.

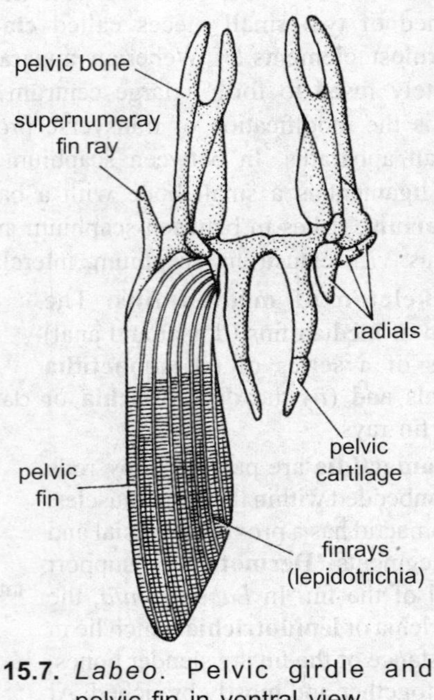

Fig. 15.7. *Labeo*. Pelvic girdle and right pelvic fin in ventral view.

The pelvic bone has an anterior elongated part which is deeply forked at the tip, and a posterior rod-like cartilaginous part. The anterior forked end is connected with the rib of the twelfh trunk vertebra by ligaments. The posterior rod-like parts of both the halves are united in the mid-line.

2. Pelvic fin. Each pelvic fin is supported by nine lepidotrichia or fin rays and three somactidia or radials. The radials articulate with the posterior border of the pelvic bone. The first and second radials, each bear two fin rays, and the third largest radial bears the remaining five fin-rays.

DIGESTIVE SYSTEM

Digestive system includes the alimentary canal and the associated digestive glands.

Alimentary Canal

The alimentary canal is a long tube comprising mouth, buccal cavity, pharynx, oesophagus, stomach, intestine and rectum.

1. Mouth. The mouth is a subterminal crescentic slit bounded by fleshy upper and lower lips. These lips bear four or five rows of conical papillae.

2. Buccal cavity. The mouth leads into a broad, dorso-ventrally compressed buccal cavity. The mucous membrane lining the buccal cavity bears minute papillae. In *Labeo* teeth and tongue are not present. The floor of the buccal cavity is provided with thick muscles.

3. Pharynx. The buccal cavity leads into a dorsoventrally compressed pharynx. The pharynx is laterally bounded by gill-arches and is well differentiated into a broad anterior respiratory part and a narrow posterior masticatory part. The anterior respiratory part is perforated laterally by 4 pairs of gill-slits. From gill-arches project into pharyngeal cavity small spiny **gill-rakers** which prevent the passage of food through gill-slits. The posterior masticatory part of pharynx bears closely set 3 rows of homodont pharyngeal teeth on its ventrolateral walls for masticating the food. Each tooth has a narrow basal root and a conical crown. The floor of this part is highly folded transversely.

4. Oesophagus. The pharynx leads into a short, narrow, tubular oesophagus. Its mucosal lining is longitudinally folded. The pneumatic duct **(ductus pneumaticus)** from the air bladder opens dorsally into the oesophagus.

5. Stomach/Intestine. In *Labeo*, true stomach is absent. The oesophagus opens into a swollen sac-like anterior part of intestine. This sac-like part is called **intestinal bulb** in which food is stored. The oesophageal opening into the intestinal bulb is guarded by an **oesophageal valve** to prevent regurgitation of digested food. Gastric glands are lacking in intestinal bulb. The mucosa of the intestinal bulb contains absorptive and mucous cells. The intestinal bulb is also differentiated into an anterior broader **cardiac part** into which open the pancreatic and bile ducts, and a posterior narrow **pyloric part**. Pyloric caeca are absent. The mucosa of cardiac part form honey-comb-like folds and that of pyloric part forms longitudinal folds.

Intestinal bulb is followed by a thin-walled, narrow, elongated and convoluted **intestine**. Its mucosa forms oblique transverse folds in the anterior region and longitudinal folds in the posterior region. Since *Labeo* is a herbivorous fish, its intestine is longer in comparison of carnivorous fishes. Intestinal villi and scroll valve are absent in *Labeo.*

6. Rectum. The intestine is followed terminally into a thin-walled, slightly dilated **rectum**. Its mucosal lining shows distinct oblique transverse folds. The rectum opens outside through **anus** located just in front of the urinogenital opening.

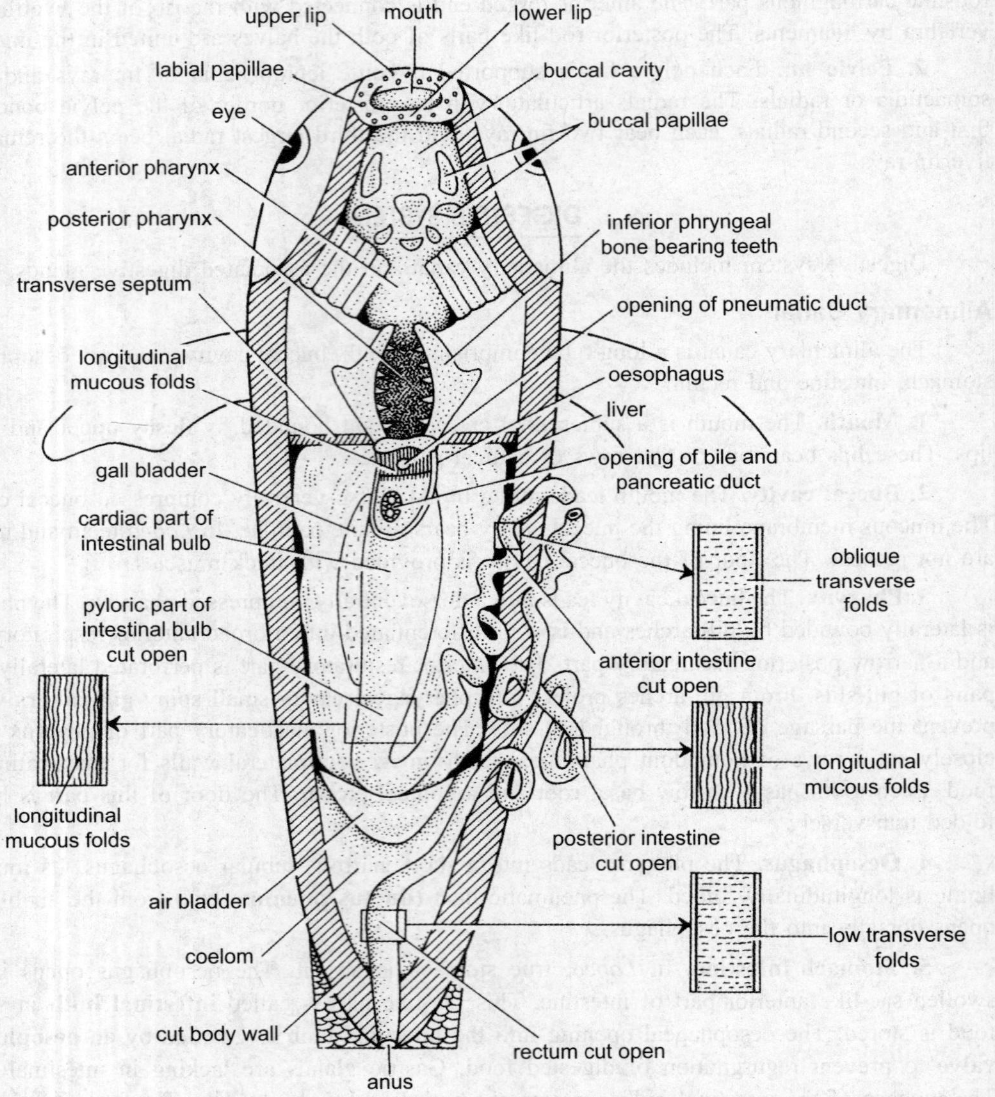

Fig. 15.8. *Labeo*. Dissection of digestive system. Some regions of alimentary canal cut open to show folds of mucous lining.

Digestive Glands

Digestive glands of *Labeo* are liver and pancreas.

1. Liver. It is a large, dark brown bilobed gland. Right liver lobe is narrower than that of left liver lobe. Both the lobes are interconnected at three regions, *viz.*, anterior, median and posterior, by median connecting lobes. **Gall bladder** is an elongated sac of about 8 cm long and 2.5 cm in diameter, and is placed between the two liver lobes. A **cystic duct** arises from the anterior end of the gall bladder and receives three hepatic ducts from the liver lobes and then open into the cardiac part of the intestinal bulb.

2. Pancreas. It is a diffused structure found among the coils of intestine and also extends into the liver. Its exocrine part formed of acini are also found in spleen tissue in rohu, mrigal and catla fishes. Pancreatic duct opens separately into the cardiac part of intestinal bulb.

Digestion. The mechanism of digestion is not clear. Intestinal mucosa secretes trypsin, erepsin and enterokinase. Pancreas secretes amylase. Pepsin and HCl are absent since stomach is lacking in *Labeo*.

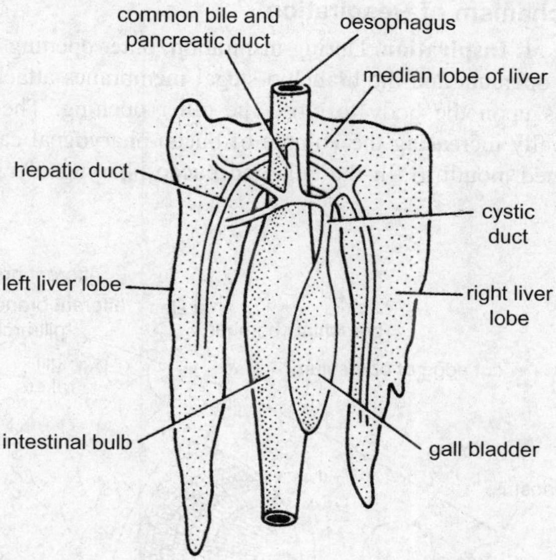

Fig. 15.9. *Labeo*. Liver and gall bladder in dorsal view.

AIR-BLADDER OR SWIM-BLADDER

Air-bladder or swimbladder is a hydrostatic organ. It lies within coelom, dorsal to alimentary canal. It is a shiny thin-walled sac filled with air of high oxygen content. The air-bladder is divided into an anterior and a posterior chamber. The anterior chamber communicates with the oesophagus dorsally through a short, thin **pneumatic duct**. Air-bladder functions like a **hydrostatic organ**. It regulates specific gravity of body by secretion or absorption of gas in the air-bladder. The inner lining of air-bladder is highly vascularised, forming red bodies or rete mirabile. The rete mirabile located in the anterior chamber possesses extraordinary power of isolating oxygen, carbon dioxide and nitrogen from the blood and to fill the chamber with these secreted gases. But the rete mirabile of posterior chamber serves reverse role, *i.e.*, it absorbs these gases. Besides acting as hydrostatic organ, the air-bladder also performs a variety of other functions such as gas secretion, gas resorption, respiratory function, sensory function, auditory function and sound production, etc.

RESPIRATORY SYSTEM

In *Labeo*, respiration is aquatic, occurring with the help of gills.

Respiratory Organs

In *Labeo*, there are four pairs of **gills** present in the branchial chambers covered by operculum. Each gill has a double row of **gill-filaments** or **branchial-filaments** supported by gill-arch. Each gill filament bears numerous minute transverse **lamellae** covered with thin epithelium. These lamellae contain capillaries of afferent and efferent branchial arteries. The four pairs of gills are borne on the first four branchial arches. In *Labeo* and other teleosts the gill-filaments are attached to a very much reduced interbranchial septum. Thus, this type of gill is called **pectinate** or **filiform**. The fifth gill arch bears no gill-filaments. On the inner surface of operculum is a comb-like body, the **pseudobranchia**, formed of a single row of branchial-filaments, and representing either the vestigial gill (hemibranch) of the mandibular arch or the spiracular hemibranch. Thus, there are 4 holobranchs attached to first four branchial arches and 5 gill-slits on either side covered by the operculum. The pharyngeal (inner) border of each gill-arch has spinous processes, called the **gill-rakers** which do not allow the food particles to enter into the branchial chamber.

Mechanism of Respiration

1. Inspiration. During inspiration, outer opening of the gill-chamber remains tightly closed, *i.e.*, opercula and the branchio-stegal membranes attached to the posterior margin of operculum, press upon the body to close the other opening. The branchial-arches and the opercula bulge laterally increasing the volume of bucco-pharyngeal cavity. Thus, the water flows in through the opened mouth to fill the enlarged bucco-pharyngeal cavity.

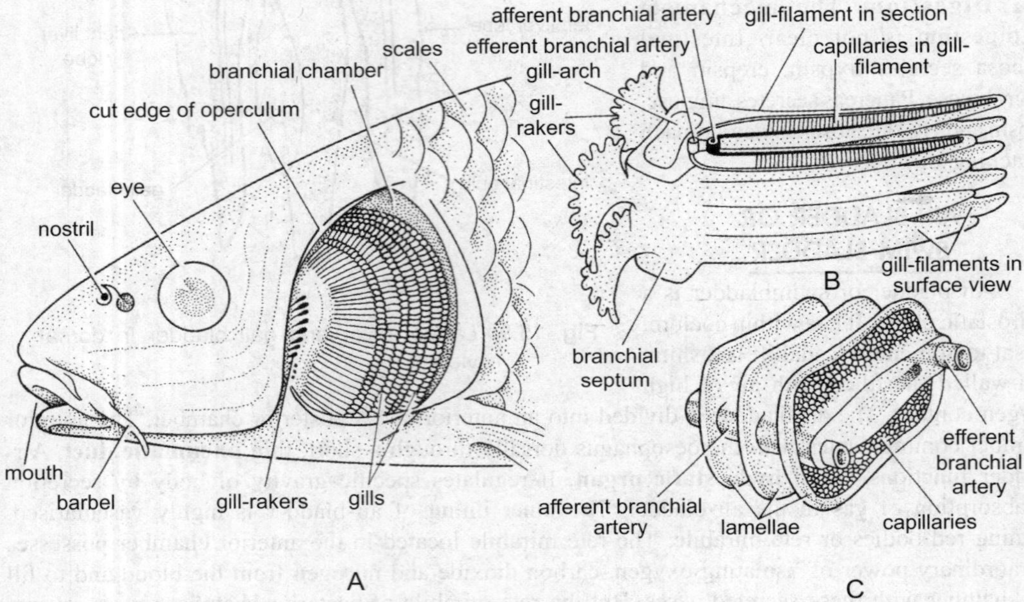

Fig. 15.10. Gills of a bony fish. A – Operculum cut away showing gills in the left gill-chamber; B – Part of gill showing gill-rakers and filaments; C–Three lamellae of gill-filament showing blood supply.

2. Expiration. During expiration, mouth becomes closed by oral valves, and the bucco-pharyngeal cavity contracts exerting pressure on the contained water. In the meantime the opercula and branchio-stegal membranes are lifted off, thus, the water from the bucco-pharyngeal cavity passes out through the gill-chamber, bathing the gill-filaments. Contraction and expansion of pharyngeal cavity occurs by alternate retraction and protraction of the hyoid arch which supports the bucco-pharyngeal cavity.

Physiology of Respiration

The lamellae of gill-filaments are richly vascularised by the capillaries of **afferent** and **efferent branchial arteries**. Afferent branchials bring deoxygenated blood into the lamellae. Oxygen of water diffuses through the thin walls of blood capillaries into their blood and CO_2 of the blood of these capillaries diffuses into the water. Thus, gaseous exchange occurs when blood passes out through the gill-chambers. The oxygenated blood is carried by the **efferent branchial arteries** and is supplied to the entire body.

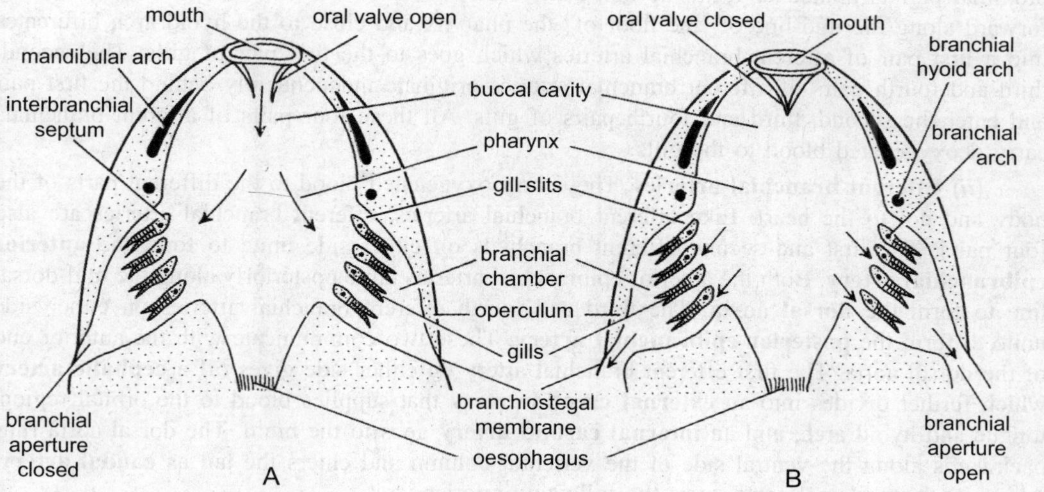

Fig. 15.11. Respiratory mechanism of a bony fish. Arrows show direction of respiratory water currents. A–Inspiration; B–Expiration.

Asphyxiation occurs when water is deficient in oxygen content or the fish is kept out of water for a long time. Due to asphyxiation the fish dies.

BLOOD VASCULAR SYSTEM

Like that of the cartilaginous fishes, blood vascular system consists of a "single circulation." Blood vascular system of *Labeo* consists of (1) blood, (2) heart, (3) arteries and (4) veins.

1. Blood. It consists of the fluid part, plasma and blood corpuscles. Erythrocytes are elliptical and nucleated with haemoglobin. Besides erythrocytes, nucleated reticulocytes and thrombocytes, as well as coarse and fine granulocytes and various lymphocytes occur. Haematopoiesis occurs chiefly in teleosts in the kidney intertubular tissue and in the spleen.

2. Heart. Heart is placed in the pericardial coelom ventral to the oesophagus and enclosed within thin-walled pericardium. Heart is 2– chambered, consisting of **sinus venosus, atrium** and **ventricle**. **Sinus venosus** is a thin-walled sac placed dorsally over the atrium and ventricle. It receives deoxygenated blood from the entire body by right and left **ductus Cuvieri**. **Auricle** lies ventral to the sinus venosus and is a thin-walled single chamber. It receives blood from the sinus through a valved **sinu-auricular aperture**. The auricle opens into a thick-walled ventricle by atrio-ventricular aperture guarded by valves. These valves prevent the flow of blood in reverse direction. There is no conus arteriosus. The proximal end of the ventral aorta is dilated to form a thin-walled elastic **bulbus arteriosus**. It differs from a conus in being part of the aorta, and not the heart. The bulbus arteriosus in teleosts dilates and contracts as a response to the beat of the ventricle, and by this means pressure through the short aorta to the gills is maintained. Like elasmobranchs the heart action is regulated by a depressor nerve of vagal origin. Sympathetic innervation is lacking. The heart is called the venous heart since only deoxygenated blood circulates through the various parts of the heart.

3. Arterial System. There are two types of arteries : afferent branchial arteries carrying the deoxygenated blood from the heart to the gills and the efferent branchial arteries carrying the oxygenated blood from the gills to the entire body.

(*i*) Afferent branchial arteries. From the ventricle arises a stout **ventral aorta** whose proximal part is dilated to form the non-contractile **bulbus arteriosus**. The ventral aorta runs forward along the mid-line of the floor of the pharynx and close to the hyoid arch bifurcates into a first pair of afferent branchial arteries which goes to the first pair of gills. The second, third and fourth pairs of afferent branchial arteries originate independently behind the first pair and enter the second, third and fourth pairs of gills. All these four pairs of afferent branchials carry deoxygenated blood to the gills.

***(ii)* Efferent branchial arteries.** They carry oxygenated blood to the different parts of the body and not to the heart. Like afferent branchial arteries, efferent branchial arteries are also four pairs. The first and second efferent branchials on either side unite to form the **anterior epibranchial artery**. Both the anterior epibranchial arteries unite posteriorly along the mid-dorsal line to form the **dorsal aorta**. The third and fourth efferent branchial arteries on either side unite to form the **posterior epibranchial artery**. These two communicate with the anterior end of the dorsal aorta. The first efferent branchial artery on either side gives off a **cephalic artery** which further divides into an **external carotid artery** that supplies blood to the orbital region, tongue and hyoid arch, and an **internal carotid artery** goes to the brain. The dorsal aorta runs backwards along the ventral side of the vertebral column and enters the tail as **caudal artery**. Along its route, dorsal aorta gives the following arteries :

(*a*) A **coeliaco-mesenteric artery** for swim-bladder, stomach, intestine, liver, spleen, etc.

(*b*) A pair of **subclavian arteries**, each dividing into a **pectoral** and a **pelvic arteries** for pectoral and pelvic fins.

(*c*) A number of paired **segmental arteries** for segmental muscles.

(*d*) A few pairs of **renal arteries** to the kidneys.

(*e*) A pair of **genital arteries** to the gonads.

4. Venous System. Veins of the venous system carry deoxygenated blood to the sinus venosus. The blood is carried to the sinus venosus by right and left **ductus Cuvieri**. Each Cuverian duct is formed by three veins, **anterior cardinal sinus**, a **jugular sinus** and a **posterior cardinal sinus**. The posterior cardinal sinus receive segmental veins, renal veins and genital veins, etc. Ductus Cuvieri also receive pectoral and pelvic veins, and a hepatic vein.

The **caudal vein** from the tail after entering into the trunk region bifurcates into two branches : right posterior cardinal sinus passes through the substance of right kidney and opens into the right ductus Cuvieri. The left posterior cardinal sinus arises from the capillaries of renal portal veins.

Portal System. A vein that originates from capillaries and ends up again into capillaries in other organ before going to the heart is called portal vein and the system is called portal system. There are two portal systems : (*i*) renal portal system and (*ii*) hepatic portal system.

(*i*) Renal portal system. The left branch of the caudal vein enters into the left kidney and breaks up into capillaries and forms the renal portal system. These capillaries rejoin forming the left posterior cardinal sinus.

(*ii*) Hepatic portal system. The veins from the various parts of the alimentary canal unite to form the **hepatic portal vein**. It enters into the liver and breaks up into capillaries and these capillaries reunite forming the hepatic vein that communicates with the ductus Cuvieri.

NERVOUS SYSTEM

Brain. The brain of *Labeo*, though in general not unlike that of elasmobranchs, is very different in detail and specialised. The **cerebellum** is very large. Its anterior part does not bulge outward. Instead it pushes forwards under the roof of mesencephalon to form **valvula cerebelli**. It is characteristic of teleost fishes. The **optic lobes** are also large with two large **lobi inferiores** on the ventral side. The **optic nerves** do not form a chiasma, but simply cross one another or decussate. The **diencephalon** is much reduced and dorsally indicated only as the place of origin of the pineal body. The infundibulum between the lobi inferiores gives attachment to the pituitary body (hypophysis). The small undivided **prosencephalon** comes immediately in front of the midbrain. **Olfactory bulbs** are situated in close apposition with the forebrain, without intervening olfactory tracts. Each contains a cavity (rhinocoele) communicating with the undivided ventricle of the forebrain. Three transverse bands of fibres connect the right and left halves of the forebrain, an **anterior commissure** joining the corpora striata, a **posterior commissure** situated just behind the origin of the pineal body, and an **inferior commissure** in front of the infundibulum. The **pineal body** is rounded and placed at the end of a hollow stalk. No trace of any optic structure remains within. A shorter offshoot of the root of the diencephalon may perhaps represent a rudimentary pineal eye. Behind the pituitary body is a **saccus vasculosus**. It may be a pressure receptor.

Cranial nerves. There are ten pairs of cranial nerves and a pair of **terminal nerves** (cranial nerve, "O"). Their origin, distribution and nature are like that of *Scoliodon*.

Spinal nerves. These are paired and each like that of *Scoliodon* arises as a dorsal sensory root and a ventral motor root.

Autonomic Nervous System

In teleosts paired symmetrical chains of **sympathetic ganglia** begin at the terminal level and run backwards to the end of the tail. A ganglion is connected with each cranial dorsal root. These receive pre-ganglionic fibres from the ventral roots in the trunk region and pass forwards in the sympathetic chain. The sympathetic trunk is connected with the mixed spinal nerves by **rami communicantes**. Each ganglion receives a white ramus containing pre-ganglionic fibres from its spinal nerve and gives a grey ramus to the same nerve. The grey ramus carries post-ganglionic fibres, some of which are concerned with melanophore contraction. Thus, the autonomic nervous system is far more advanced than that of cartilaginous fishes. It is comparable to the basic plan in tetrapods.

Sense Organs

Olfactory receptors are paired **olfactory pits**, each of which opens to the external by means of two apertures but they do not communicate to the buccal cavity. Olfactory pits are lined by the olfactory cells. Their function is olfactory and has no role in respiration.

Taste buds (chemoreceptors) are found in many parts of the body. Taste buds are present over the lips, in the epithelial lining of the first three gill-slits and on the barbels.

Tactile receptors are abundant over the lips and barbels and are also found over the entire body.

Lateral line system is well developed in actinopterygians. Each sensory organ lies in a pit and communicates with surrounding water through a pore. These pits are linked by canals and innervated by the **lateral line** branch of the vagus. Lateral line system helps the fish to perceive low frequency vibrations in water and also apprises the fish of the approach of predator or prey.

Eye. A pair of eyes over the head are the photoreceptors. Each eye has three layers : (*i*) outer cartilaginous **sclerotic layer**, (*ii*) a median vascular **choroid layer** and an innermost photosensitive **retinal layer**. In between the sclerotic and choroid is a silvery layer or **argentea** which gives its colour. The **cornea** is flat with which the globular **lens** is almost in contact, so that the anterior chamber of the eye is extremely small. There are no choroid processes. In the posterior part of the eye, between the choroid and the argentea, is a thickened ring-shaped structure, the **choroid gland** which surrounds the optic nerve. It is not glandular, but it is a complex network of blood vessels or **rete mirabile**. Close to the entrance of the optic nerve a vascular fold of the choroid, the **falciform**

process, pierces the retina, and is continued to the back of the lens. Here it ends in a muscular knob, the **campanula Halleri** or **retractor lentis**. The falciform process and campanula Halleri takes an important part in the process of **accommodation** by which the eye becomes adapted to forming and receiving images of objects at various distances. Accommodation is also effected by shifting the position of lens and not by changing the shape of lens as occurs in higher vertebrates. The pupil size appears to alter very little or not at all. Vision is **monocular**, each eye has different visual field.

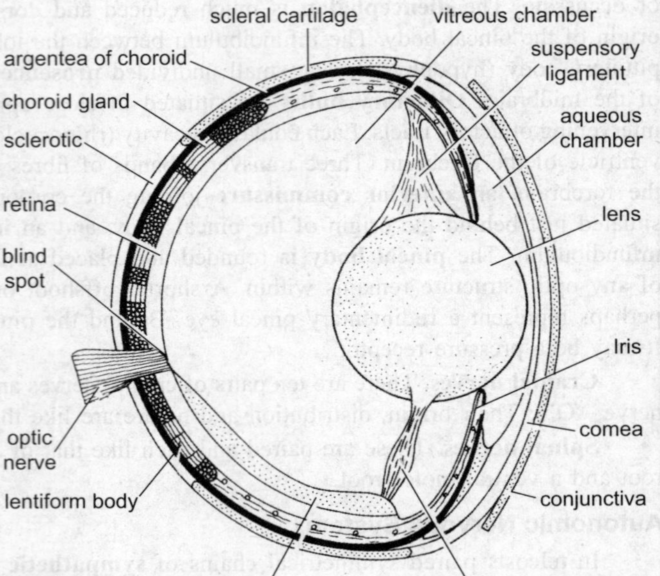

Fig. 15.12. V.S. eye of a bony fish.

Ear or organ of audio-equilibrium. In fishes, the external and middle ears are absent, only internal ear or **membranous labyrinth** is found. The membranous labyrinth is formed of an upper **utriculus** and a lower **sacculus**. Three semicircular canals open into the utriculus. The sacculus is sac-like and its floor gives rise to a **lagena**. The endolymph present in the membranous labyrinth contains **otoliths** or ear-stones, which are of three types: **sagitta** is relatively large and it almost fills the sacculus, **asteriscus** is a small granule lying in the lagena, and the **lapillus** is found in utriculus close to the ampullae of the anterior and horizontal canals. Bony fishes hear with great discrimination, despite the absence of organ of Corti. The utriculus and the semicircular canals help the fish to maintain equilibrium. The sacculus and lagena perceive the sound waves. The body surface transmits the vibrations to the membranous labyrinth.

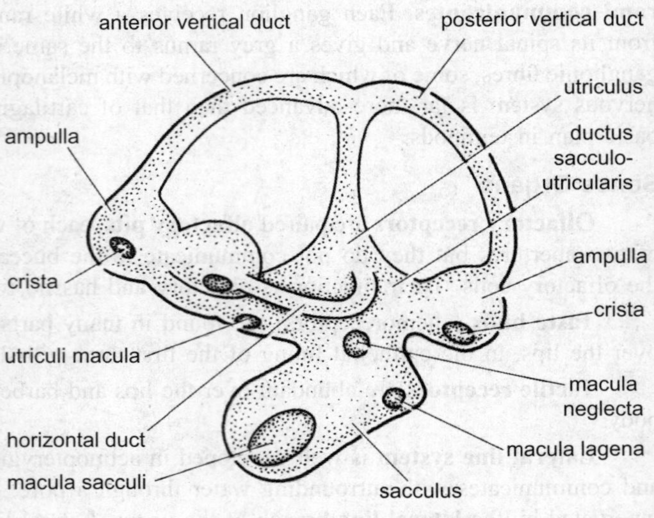

Fig. 15.13. *Labeo*. Internal ear.

URINOGENITAL SYSTEM

Sexes are separate. The excretory and reproductive organs in male and female are quite separate in *Labeo*.

Excretory Organs

The **kidneys** are of large size, extending the whole length of the abdominal cavity. They are placed dorsally above the air bladder, and partly fused together in the middle region. Anteriorly both the kidneys are free. Their anterior ends are much dilated and consist of lymphatic tissue in the adult, thus, ceasing to discharge a renal function. They are mesonephric type (derived from the mesonephros of the embryo). The mesonephric ducts or ureters arising from the kidneys unite into a single tube, which is dilated to form a **urinary bladder**. It opens into the urinogenital sinus. Urinary bladder is mesodermal in origin and is not homologous with the endodermal bladder of higher vertebrates. Fishes are **ammonotelic**, *i.e.*, they excrete mitrogenous metabolites largely in the form of ammonia, though some urea is formed and discharged.

Reproductive Organs

The **gonads** are of great size in the sexually mature fish. In the male, the **testes** are paired, extending the whole length of the abdominal cavity beneath the air bladder. Each testis is continued posteriorly into a **genital duct** which opens into the urinogenital sinus. The spermatozoa are shed into the external medium.

In the female, **ovaries** are also paired, running the whole length of the abdominal cavity beneath the air bladder. They are much wider than the testes and are covered with peritoneum only on their inner faces. The oviducts are not found. The ova are discharged when ripe from their outer faces into the coelom. The anterior wall of the urinogenital sinus is pieread by a pair of **genital pores** formed temporarily through which ova escape out into the water. Genital pores are not homologous with the abdominal pores of the cartilaginous fishes.

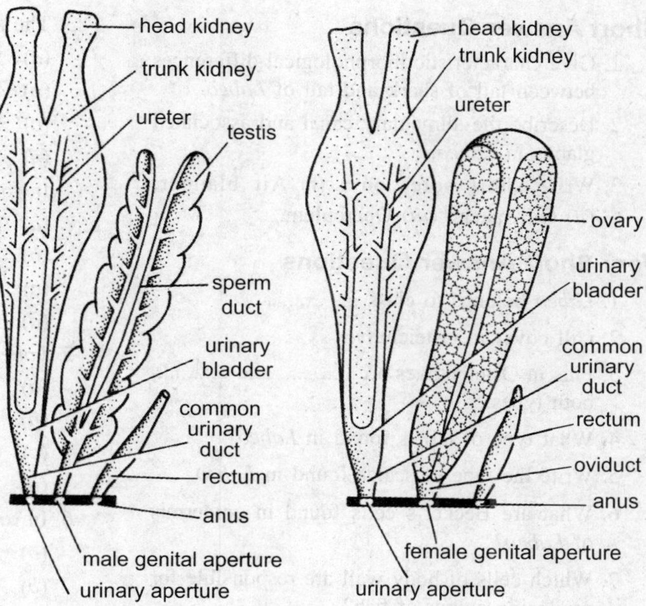

Fig. 15.14. *Labeo*. Urinogenital organs. A–Male; B–Female.

Fertilisation. *Labeo* is oviparous. Fertilisation is external. The male sheds its seminal fluid on the new-laid eggs. The ovum is covered by a thick membrane, the **zona radiata**. This is perforated by an aperture, the **micropyle**, through which a sperm enters. Hatching occurs within 2 to 15 hours. Hatchlings have a prominent yolk sac attached to the ventral side of the body. Yolk sac is absorbed in 5–7 days and the new young ones called **fry** begin to feed. They start feeding with fringed lips and grow into **fingerlings** measuring about 5–15 mm in length. They attain sexual maturity in about two years.

ECONOMIC IMPORTANCE

Labeo is a large genus having several species which are of considerable importance as an article of food. *Labeo rohita* (rohu) is perhaps the commonest of Indian fish and most generally esteemed for eating purposes. It is also a much domesticated freshwater fish because of its excellence as food.

IMPORTANT QUESTIONS

Long Answer Questions

1. Describe the organs and mechanism of respiration in bony fish.
2. Describe the arterial system of *Labeo*.
3. Give an account of the urinogenital system of a teleostean fish.

Short Answer Questions

1. Give characteristic morphological differences between tail of shark and tail of *Labeo*.
2. Describe the alimentary canal and associated glands in *Labeo*.
3. Write short notes on : (*i*) Air bladder; (*ii*) Gill rakers, (*iii*) Operculum.

Very Short Answer Questions

1. *Labeo* belongs to class ....
2. Gill cover in Osteichthyes is ....
3. Fins in Osteichthyes are .... and ...., both types.
4. What type of tail is found in *Labeo*?
5. Write the type of scales found in *Labeo*.
6. What are Becker's cells found in epidermis of *Labeo*?
7. Which cells of body wall are responsible for producing colour of fish?
8. .... and tongue are absent in rohu.
9. What is the function of gill-rakers present in pharyngeal cavity?
10. What is intestinal bulb in rohu?
11. What is the function of air or swim-bladder?
12. Which fish is called physostomous?
13. How many pairs of gills are found in *Labeo*?
14. Anterior knob-like, non-renal part of kidneys is ....

Multiple Choice Questions

1. The principal locomotory organ in rohu is
   (*a*) Pectoral fins (*b*) Pelvic fins
   (*c*) Dorsal fins (*d*) Tail.
2. Circulii of cycloid scales are used to determine the
   (*a*) Age of fish
   (*b*) Nutritional value of fish
   (*c*) Viability of fish
   (*d*) Sex of fish.
3. The skin epidermis of rohu lacks
   (*a*) Stratum germinativum
   (*b*) Stratum corneum
   (*c*) Stratum laxum
   (*d*) Stratum compactum.
4. In rohu, homodont teeth are present in
   (*a*) Posterior pharyngeal chamber
   (*b*) Anterior pharyngeal chamber
   (*c*) Buccal cavity
   (*d*) Oesophagus.
5. *Labeo rohita* is chiefly
   (*a*) Herbivorous
   (*b*) Carnivorous
   (*c*) Parasitic
   (*d*) Predator.
6. In rohu, stomach is
   (*a*) J-shaped
   (*b*) U-shaped
   (*c*) Cylindrical
   (*d*) Absent.
7. In rohu gill-rakers
   (*a*) Promote the entry of food into gill-chambers
   (*b*) Prevent the entry of food into gill-chambers
   (*c*) Prevent regurgitation of food into buccal chamber
   (*d*) Promote regurgitation of food into buccal chamber.
8. Air-bladder in rohu is connected to the roof of oesophagus by
   (*a*) Pneumatic duct
   (*b*) Oesophageal duct
   (*c*) Glandular duct
   (*d*) Pharyngeal duct.

ANSWERS

1. (*d*) 2. (*a*) 3. (*b*) 4. (*a*) 5. (*a*) 6. (*d*) 7. (*b*) 8. (*a*)

16 Chapter

Superclass Pisces : General Characters, Classification and Representative Types

Contents of this Chapter

The Superclass **Pisces** (L. *Piscis* = fish) are the truly jawed vertebrates. They have organs of respiration and locomotion related to a permanently aquatic life. The respiratory organs are the gills and the organs of locomotion are paired and unpaired fins. All are poikilothermous.

GENERAL CHARACTERS

1. Aquatic, either freshwater or marine, herbivorous or carnivorous, cold blooded, oviparous or ovoviviparous vertebrates.

2. Body usually streamlined, spindle-shaped, some are elongated snake-like and a few are dorsoventrally compressed, and differentiated into head, trunk and tail.

3. Locomotion by paired pectoral and pelvic fins along with median dorsal and caudal fins, supported by true dermal fin-rays. Muscular tail used in propulsion.

4. Exoskeleton of dermal scales, denticles or bony plates (in Placodermi) covering body surface. Placoid in Chondrichthyes and ganoid, cycloid or ctenoid in Osteichthyes.

5. Endoskeleton is cartilaginous or bony. The notochord in usually replaced by vertebrae, either bone or cartilage. Presence of well develoed skull and a system of visceral arches, of which the first pair forms the upper and lower jaws, the latter movably articulated with the skull.

6. Muscles arranged into segments called myotomes, with separate dorsal and ventral parts.

7. Alimentary canal with definite stomach and pancreas and terminates into cloaca or anus.

8. Organs of respiration are gills. Gill-slits 5 to 7 pairs, naked or covered by an operculum.

9. Heart is venous and two chambered, *i.e.*, one auricle and one ventricle. Sinus venosus and renal and portal systems present. Erythrocytes nucleated. Poikilothermous.

10. Kidneys mesonephros. Excretions ureotelic.

11. Brain with usual five parts. Cranial nerves ten pairs.

12. Nostrils are paired but do not open into pharynx except Dipnoi. Nasal capsules are partly separate in Chondrichthyes and completely separate in Osteichthyes.

13. Tympanic cavity and ear ossicles are absent.

14. Internal ear with three semicircular canals.

15. Lateral line system is well developed.

16. Sexes separate. Gonads typically paired. Gonoducts open into cloaca or independently.

17. Fertilisation internal or external. Females of Chondrichthyes are oviparous or ovoviviparous and of Osteichthyes are mostly oviparous and rarely ovoviviparous or viviparous. Eggs with large amount of yolk. Cleavage meroblastic.

18. Extra-embryonic membranes are absent.

19. Development usually direct without or with little metamorphosis.

CLASSIFICATION

About 40,000 species of fishes are known. Various workers have proposed different schemes of classification of fishes. However, no classification has been universally accepted because of confusion due to large number of fishes and great diversity in their shape, size, habits and habitat.

J. Muller (1844) gave first classification of the lower vertebrates and divided **Pisces** into six subclasses :

1. Dipnoi
2. Teleostei,
3. Ganoidei,
4. Elasmobranchii,
5. Marsipobranchii (Cyclostomi) and
6. Leptocardii (Amphioxini).

He included cyclostomes in Marsipobranchii and cephalochordates in Leptocardii. Subsequently. **Agassiz** (1857) separated lampreys and hagfishes into a separate class Myzontes and divided fishes into four classes. **Boulenger** (1924) classified the teleostean fishes into thirteen orders. **Regan** (1906–1929) gave an elaborate classification of fishes in his several papers. **Jordan** (1923) divided fish-like vertebrates into six classes. Among the more recent authors **Goodrich** (1930), **Berg** (1940), **Grasse** (1958) and **Romer** (1959) have given a detailed classification of fishes.

Romer (1959) classified the fishes into two classes : **Chondrichthyes** (includes all cartilaginous fishes) and **Osteichthyes** (includes all bony fishes Dipnoi and Teleostomi). Osteichthyes is divided into two subclasses : Sarcopterygii and Actinopterygii. **Parker** and **Haswell** (1962) have further combined all the extinct jawed fishes under a single class : **Placodermi** or **Aphetohyoidea**. This simple division of superclass Pisces into three classes : Placodermi, Chondrichthyes and

Osteichthyes, has been followed more or less by all the eminent authors. In the present text, the authors have also adopted this scheme of classification with slight modifications because of its simple and direct approach to our readers, while ignoring in other merits and demerits.

Class 4. Chondrichthyes (Gr., *chondros* = cartilage; *ichthys* = fish)

1. Mostly marine and predaceous.
2. Body fusiform and spindle-shaped.
3. Fins both median and paired, all supported by fin-rays. Pelvic fins bear claspers in male. Tail heterocercal.
4. Skin covered with placoid scales or naked and bears mucous glands.
5. Endoskeleton entirely cartilaginous, often calcified, without true bones. Notochord persistent. Vertebrae complete and separate. Pectoral and pelvic girdles present.
6. Mouth ventral. Jaws present. Teeth are modified placoid seales. Stomach J-shaped. Intestine with spiral valve.
7. Respiration by gills. Gill-slits 5 to 7 on each side. Operculum or gill-cover absent. No air-bladder and lungs.
8. Heart two chambered (one auricle and one ventricle) Sinus venosus and conus arteriosus present. Both renal and hepatic portal system present.
9. Kidneys opisthonephros (only posterior part of kidney is excretory in function). Excretion ureotelic.
10. Cloaca present. Its aperture serves as a common outlet for rectum, renal and reproductive ducts.
11. Brain with large olfactory lobes and cerebellum. Cranial nerves ten pairs.
12. Olfactory sacs do not open into pharynx. Membranous labyrinth (internal ear) with three semicircular canals. Lateral line system present.
13. Sexes separate. Male usually possesses claspers. Gonads paired. Gonoducts open into cloaca.
14. Fertilisation internal. Oviparous or ovoviviparous. Eggs large, yolky. Cleavage meroblastic. Development direct, without metamorphosis.

CLASSIFICATION OF CHONDRICHTHYES

The class **Chondrichthyes** (Gr., *chondros* = cartilage; *ichthys* = fish) also called **Elasmobranchii** (Gr., *elasmo* = plate; *branchia* = gills), including sharks, rays, skates and chimaeras. According to **Schultz** there are about 600 living species of cartilaginous fishes. The classification followed here is based after that of **Romer** (1959).

Subclass 1. Selachii
(Gr., *selachos* = a shark)

1. Paired fins without a median axis.
2. Base of pectoral fin constricted and pectoral radials jointed.
3. Gill-slits numerous on either side and each is protected by independent skin flaps.
4. Presence of a spiracle behind each eye.
5. Vertebral centra usually differentiated.
6. Cloaca present. Males with claspers.
7. Tail heterocercal with large lobe.

Order 1. Squaliformes or Pleurotremata
(Gr., *pleuro* = side; *trema* = opening)

1. Body typically spindle-shaped.
2. Gill-slits lateral, 5–7 pairs. Spiracles small.
3. Pectoral fins moderate, constricted at base. Anterior margin of pectoral fin free.
4. Hyomandibular bearing branchial rays.
5. Right and left halves of pectoral girdle dorsally separated.
6. Tail heterocercal.

Examples : True sharks. About 250 living species. Dogfishes (*Scoliodon, Chiloscyllium, Mustelus, Carcharinus*), Spiny dogfish (*Squalus*), Seven-gilled shark (*Heptanchus*), Tiger shark or Zebra shark (*Stegostoma*), Hammer-headed shark (*Zygaena* or *Sphyrna*), Whale shark (*Rhineodon*).

Order 2. Rajiformes or Hypotremata
(Gr., *hypo* = below; *trema* = opening)

1. Body depressed, flattened dorsoventrally.
2. Gill-slits ventral, 5 pairs.
3. Pectoral fins enlarged, fused with sides of the head and body.
4. Spiracles large and functional.
5. Two halves of pectoral girdle united either to each other or to the vertebral column.
6. No branchial rays on hyomandibular.

Examples : Skates and rays. About 300 species. Skate (*Raja*), Stingray (*Trygon*), Electric ray (*Torpedo*), Eagle ray (*Myliobatus*), Guitar fish (*Rhinobatus*), Sawfish (*Pristis*).

Subclass 2. Holocephali

(Gr., *holos* = entire; *kephale* = head)

1. Single gill opening on either side covered by a fleshy operculum.
2. No scales, spiracles and cloaca.
3. Jaws with tooth plates.
4. Nasal opening single
5. Lateral line system with open groove.
6. Males with claspers. Cephalic clasper on the head of male.
7. Tail long and thin.

Examples : Rat fishes or chimaeras. About 25 species. *Hydrolagus* (= *Chimaera*).

SOME REPRESENTATIVE TYPES OF CHONDRICHTHYES

Sharks and Rays

1. *Heptanchus*. *Heptanchus* is regarded as the most primitive shark. It is long bodied and slow moving. There are seven pairs of gill-clefts. *Hexanchus,* another primitive shark, has six pairs of gill-slits. Mouth is sub-terminal and there is a single soft dorsal fin. There is a long heterocercal tail. Vertebral column is entirely cartilaginous and does not constrict the notochord. Jaw suspension is amphistylic, and the upper jaw articulates with a postorbital process of the cranium. The pectoral girdle has a median ventral element as an indication of a sternum. It is viviparous but there is no placenta. It is found in the warm parts of the Atlantic and the Mediterranean.

2. *Scyliorhinus* (= *Scyllium*). *Scyllium* is a marine true dogfish. Its body is elongated, laterally compressed and of moderate size. The body is distinguished into head, trunk and tail. Eyes without nictitating membrane. Mouth and nasal openings are ventral spiracles and present behind the eyes. Teeth are small provided with median cusps. Two dorsal fins, small and without spines. The first lies above or behind the pelvic fins and the second behind the anal fin. Tail bent slightly upward. Fertilisation is internal. Oviparous; egg cases are large and quadrate with long tendrils for attachment. Sexes are separate, male possesses claspers. *Scyllium* is widely distributed in temperate and tropical seas. *S. canicula* and *S. catulus* are common on British coasts.

3. *Sphyrna* (= *Zygaena*) or *Raniceps*. *Sphyrna* (Fig. 16.1) is a marine shark, commonly called **hammer-headed shark**. Body is elongated, measuring about 4.0–5.0 metres in length. The head is flattened in front and expanded sideways into two conspicuous lateral lobes, hence, resembling a hammer. The eyes lie at the tips of lateral lobes. Eyes with 3 eyelids and nictitating membrane. Mouth is crescentic and ventral in position. Nostrils lie ventrally at the base of lateral lobes. Five pairs of lateral gill-slits are present which are uncovered. Two dorsal fins are present. The first dorsal fin is situated in front the pelvic fin and the second opposite the anal fin. Spiracles are absent. Vertebrae are asterospondylus. Viviparous. It is ferocious and attacks on its prey with its head. These sharks are usually adapted for living in deep waters. *Sphyrna* is distributed in tropical and subtropical seas. It is abundant in Indian Ocean.

4. *Pristis*. *Pristis* (Fig. 16.1) is commonly known as **saw fish**. Body is elongated, depressed and shark-like. It may attain the considerable length of 3–6 metres or even longer. Head and skull prolonged into a long flattened saw-like **rostrum**, the lateral margins of which are provided with a series of strong tooth-like denticles giving it the appearance of a saw. No rostral tentacles. Teeth in jaws are minute and obtuse. Spiracles are present behind the eyes. Dorsal fins are large, the first dorsal is opposite to the pelvic fin. Tail is well developed and terminating in heterocercal caudal fin. Predaceous. Food chiefly comprises small fishes and fleash of whales and other marine animals. They are viviparous. Their rostrum is formed while the young is in the mother's uterus. The fish is invaluable for liver oil of high vitamin value and skin for scale boards. *Pristis* is found in Mediterranean and Atlantic Oceans particularly in America, West Indies, China and Gulf of Mexico. There are two species in India, *Pristis cuspidatus* and *P. microdon.*

Skates and Rays

The **skates** (rays) are more diverse than the sharks and comprise about 400 living species. The guitar-fishes have the body dorso-ventrally flattened and it is an adaptation to a bottom-dwelling existence. They are supposed to be confined to shallow waters, some inhabit great depths and are blind. Skates are also of grotesque shapes. Some rays attain great bulk. Stingray (*Bathytoshia*) of Pacific reaches a length of 6.0 metres, a width of about 2.0 to 3.0 metres and a weight of about 385 kg. Devil rays may grow even bigger and are wider than the body length. *Manta* reaches a width of about 6.0 metres across its diamond-shaped body and wings (pectoral fins). In the rays the pectoral fins are extremely large and much bigger than the pelvic fins. The pectoral fins fringe the greater part of the length of the flattened body, and are prolonged on either side and even in front of head. Spiracles are present on the dorsal surface behind eyes. In bottom-living rays, the oxygenated water is taken in through the spiracles. The water is driven across the gills surface by the closure of a special valve. Along each side on the ventral surface in rays, there is a row of 5 gills or branchial-slits or clefts. The cloacal opening lies just in front of root of tail. A pair of small **abdominal pores** are placed close to which lead to abdominal cavity.

1. *Raja* or *Raia*. *Raja* (Fig. 16.1) or *Raia* is commonly known as **skate**. Body is dorso-ventrally depressed. Large pectoral fins extend along the lateral margins of the trunk and head from the pelvic fins to the snout, and are confluent there, forming the lateral portions of a large rhombic disc. The mouth is on the ventral side and provided with numerous sharp, rasping teeth with which they tear up their prey. Spiracles are present behind the eyes. Tail is slender and sharply marked off from the trunk. Two small dorsal fins are present on the tail. Caudal fin is very small. Caudal electric organs (modified caudal muscles) are present on each side of the terminal portion of the tail. Larger or smaller spines are generally present on the skin. Sexual dimorphism is well marked. Male is provided with a pair of **claspers** present near the pelvic fin. Carnivorous, food chiefly comprises crustaceans and fishes. They catch their prey by dropping down over them and covering them with the broad body and fins. Sluggish bottom dwellers. Oviparous, egg cases four horned without tendrils. *Raja* or *Raia* is found in temperate seas but is more abundant in northern hemisphere.

2. *Torpedo*. *Trpedo* or *Astrape* (Fig. 16.1) is commonly called **electric ray**. Body is flattened dorsoventrally and the anterior part is expanded laterally. There are no scales. Pectoral fins are large and extend in front so as to be continuous with the head, and the front end is rounded. It swims by flapping its broad pectoral fins up and down. Eyes and spiracles are dorsal, mouth, gill-clefts, and nostrils are ventral. In respiration water is drawn in through spiracles. Tail is fairly thick and bears two dorsal fins and a caudal fin. *Torpedo* has vesicles of Savi in the ventral epidermis, and it has a pair of **electric organs** formed from metamorphosed muscles, they lie between the head and pectoral fins and run through the entire thickness of the body. They receive nerves from an electric lobe of the medulla. They give a powerful electric shock of about 100 volts. *Torpedo* is found in Mediterranean, Red Sea, Atlantic and Pacific Oceans. *T. marmorata* is found in Indian Ocean.

3. *Chimaera*. *Chimaera* (Fig. 16.1) is commonly known as **rat fish** or **king of herrings**. They belong to the genus *Hydrolagus* (= *Chimaera*) having 25 very poorly known species. It is about 60 cm long with a skeleton of cartilage and a large compressed head and a blunt snout with a small mouth having lips. Mouth has crushing plates, two pairs in the upper jaw and one pair in the lower, they have no enamel but vitrodentine and reduced pulp cavities. There is a single nasal aperture, only four pairs of gills, they are covered by cartilaginous operculum on each side opening by a single branchial aperture. Eyes are very large and diencephalon is long and thin. There is no spiracle, the intestine has a spiral valve. The skin is naked with placoid scales restricted to certain parts only. Pectoral fins are very large, the first dorsal fin has an anterior cartilaginous spine, the second dorsal fin is elongated, the ventral fin is very small, the tail is produced into a long whip-like filament with a narrow caudal fin. The lateral line is an open groove. Endoskeleton is cartilaginous. Vertebrae are reduced to separate nodules and the notochord is unconstricted. Intestine has a spiral valve. In the male there are five claspers covered with placoid scales, one normal, pelvic pair, a second pair, in front of the pelvic fins in pockets of the skin, they also project backwards, and an unpaired frontal or cephalic clasper or spine on the head between the eyes. It is not known how these claspers are used. *Chimaera* occurs off the coasts of Europe from Norway to Portugal including the Mediterranean and also in the neighbourhood of Azores, Cape of Good Hope and coasts of Japan.

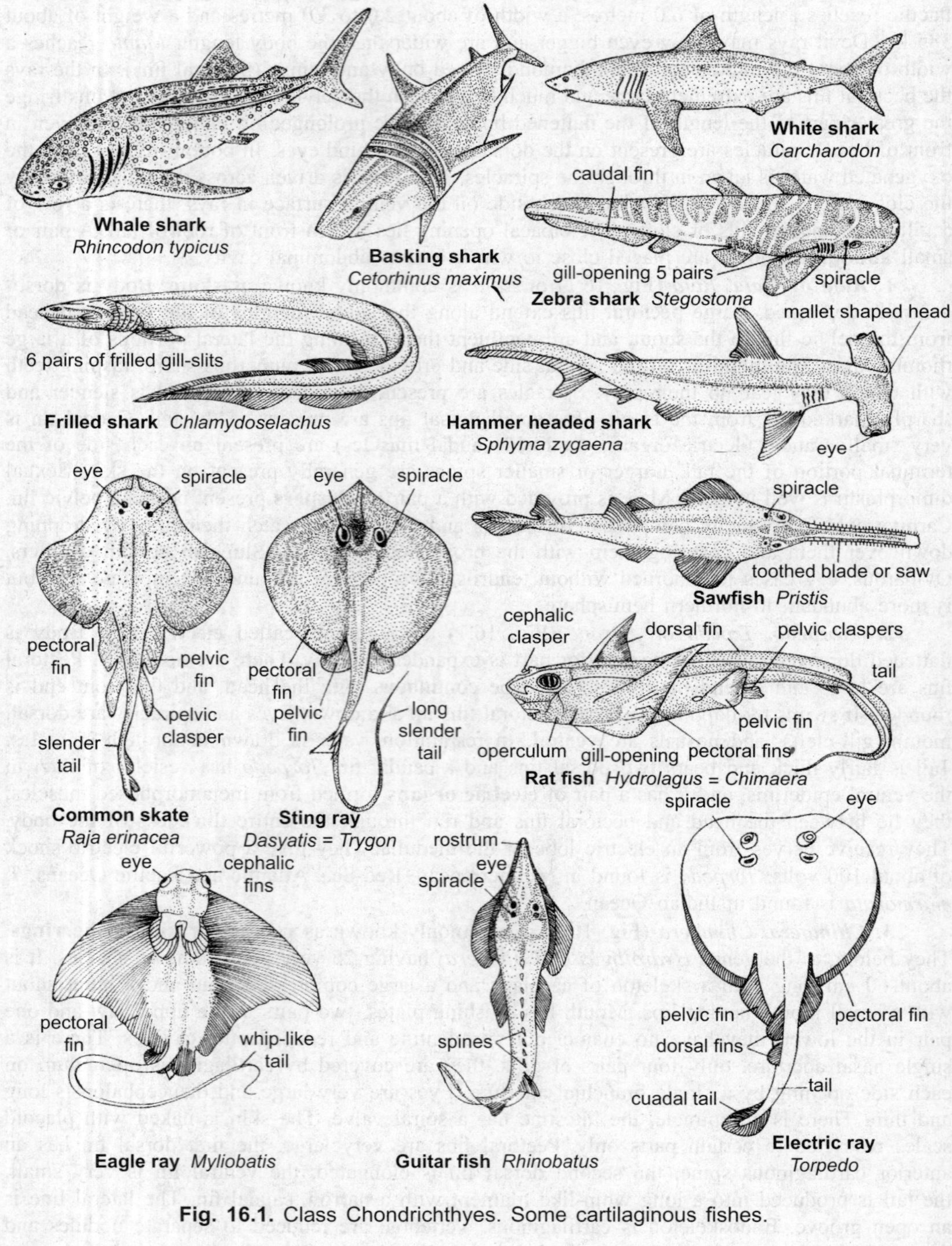

Fig. 16.1. Class Chondrichthyes. Some cartilaginous fishes.

Class 5. Osteichthyes

(Gr., *osteon* = bone; *ichthyes* = fish)

1. Inhabit all types of water–fresh, brackish, or salt; worm or cold.
2. Body spindle-shaped and streamlined.
3. Fins both median and paired, supported by fin rays of cartilage or bone. Tail usually homocercal.
4. Skin with many mucous glands, usually embedded dermal scales of three types : ganoid, cycloid and ctenoid. Some fishes without scales. No placoid scales.
5. Endoskeleton chiefly of bone. Cartilage in sturgeons and some others. Notochord replaced by distinct vertebrae. Pelvic girdle usually small and simple or absent. Claspers absent.
6. Mouth terminal or subterminal. Jaws usually with teeth. Cloaca absent, anus present.
7. Respiration by gills. Four pairs of gills on bony gill-arches, covered by a common operculum on either side.
8. An air-bladder or swim-bladder often present with or without duct connected to pharynx. Lung-like in some (Dipnoi).
9. Ventral heart 2-chambered (1 auricle + 1 ventricle). Sinus venosus and conus arteriosus present. Aortic arches four pairs. Erythrocytes oval, nucleated. Temperature variable (poikilothermal).
10. Adult kidneys mesonephric. Excretion ureotelic.
11. Brain with very small olfactory lobes, small cerebrum and well-developed optic lobes and cerebellum. Cranial nerves ten pairs.
12. Lateral line system well developed. Internal ear with three semicircular canals.
13. Sexes separate. Gonads paired.
14. Fertilisation usually external. Mostly oviparous rarely ovoviviparous. Egg minute to 12 mm in size, cleavage meroblastic. Development direct, rarely with metamorphosis.

CLASSIFICATION OF OSTEICHTHYES

The class Osteichthyes includes a large assemblage of true bony fishes. There are well over 30,000 to 40,000 living species, both freshwater and marine. Some of the freshwater forms are the carp, perch, bass, trout, catfish, sucker, etc., while the marine fishes are the tarpon, meckerel, tuna, sailfish, barracula, flying fish, etc. The classification of class Osteichthyes described here has been largely followed after **A.S. Romer** (1966) which has also been followed by most authors including **Storer** and **Usinger**. However, several new schemes of classification, such as **Greenwood**, *et.al.*, (1966), **Lander** and **Liem** (1983), **Nelson** (1984) and **Pough**, *et.al.*, (1989) are also known. Two subclasses are recognised: Sarcopteygii and Actionopterygii. Only important orders of each subclass have been described here.

Subclass I. Sarcopterygii

(Gr., *sarcos* = fleshy; *pterygium* = fin)

1. Paired fins leg-like or lobed, with a fleshy, bony central axis covered by scales.
2. Dorsal fins two, caudal fin heterocercal with an epichordal lobe.
3. Olfactory sacs usually connected to mouth cavity by internal nostrils or **choanae**.
4. Popularly called **fleshy** or **lobe finned**, or **air breathing fish**.

Sarcopterygii has been divided into two superorders or orders: Crossopterygii and Dipnoi.

Order 1. Crossopterygii

(Gr., *crossoi* = a fringe; *pteryx* = fin)

1. Paired fins lobed covered with scales. Pectoral fins are supported by a jointed median axis bearing radials. Caudal fin three-lobed.
2. Scales covered by a layer of cosmine.
3. Premaxillae, maxillae and squamosal bones present.
4. Internal nares present or absent.
5. Spiracles present.
6. Air-bladder vestigial.

Examples : *Latimeria, Coelacanthus* (Extinct).

Order 2. Dipnoi

(Gr., *di* = double; *pnoe* = breathing)

1. Cycloid scales covering the skin.
2. Single gill-slit on either side present and covered by operculum.
3. Paired fins lobed.
4. Tail fin symmetrical (diphycercal), with no trace of separate dorsal fins.
5. Internal nares present opening into mouth; spiracles absent.
6. Air-bladder single or paired, lung-like.
7. Vertebrae cartilaginous arches, notochord remaining as an unconstricted rod.
8. Premaxillae and maxillae absent.
9. Lower jaw (palato-quadrate) firmly fused to neurocranium (skull autostylic).
10. Stomach absent, intestine ciliated. No hepatic caeca. Spiral valve present.
11. Auricle partly divided into two, ventricle completely divided by a ridge.
12. Ventral aorta shortened into a spirally twisted muscular bulbus cordis provided with valves.
13. Pulmonary and systemic circulation separate.
14. Cleavage holoblastic.

Examples : Lung fishes. Only three living genera : *Neoceratodus*, *Protopterus* and *Lepidosiren*.

Subclass II. Actinopterygii

(Gr., *actis* = ray; *pteryx* = fin)

1. Includes all ray-finned fishes.
2. Paired fins thin, broad, without fleshy basal lobes and supported by dermal fin-rays.
3. Radials of paired fins not arranged biradially.
4. One dorsal fin, may be divided.
5. Caudal fin without epichordal lobe.
6. Tail generally homocercal, in a few heterocereal or semi-heterocercal.
7. Scales either ganoid or reduced to thin horny structures or completely absent in some.
8. Gill-slits covered by operculum. Spiracles absent.
9. Olfactory sacs not connected to mouth cavity. Internal nares absent.
10. Squamosal bone absent.

Actinopterygii has been divided into three infraclasses or superorders : Chondrostei, Holostei and Teleostei.

Superorder A. Chondrostei

1. Primitive ray-finned fish or cartilaginous ganoids.
2. Tail fin heterocercal.
3. Scales usually ganoid
4. Mouth opening large.

Order 1. Polypteriformes

1. Typical rhomboid ganoid scales.
2. Dorsal fin of eight or more finlets.
3. Pectoral fins with a small prominent scale covered fleshy lobe.
4. Ossified skeleton.
5. Spiracles present.
6. Air-bladder bilobed opening into the intestine ventrally.

Examples : *Polypterus* (Bichir).

Order 2. Acipenseriformes

1. Body covered with five rows of bony scutes.
2. Snout elongated, having barbles on the ventral surface.
3. Caudal fin heterocercal.
4. Skeleton largely cartilaginous. Endocranium cartilaginous.
5. Jaws without teeth.

Example : *Acipenser* (Sturgeon), *Polydon* (Paddle-fish).

Superorder B. Holostei

(Gr., *holos* = entire; *osteon* = bone)

1. Intermediate ray-finned fish, transitional between Chondrostei and Teleostei.
2. Ganoid or cycloid scales.
3. Tail fin heterocercal.
4. Mouth opening small.

Order 1. Amiiformes

1. Thin, overlapping cycloid scales.
2. Caual fin abbreviate heterocercal.
3. Long dorsal fin.
4. Pectoral radials are attached to the scapulocoracoid cartilage.
5. Vertebral centra non-opisthocoelous.
6. Premaxillary not protractile, firmly articulated with the cranium.
7. Snout normal rounded.
8. Spircles and clavicles absent.
9. Presence of a single swim-bladder.

Examples : *Amia* (Bowfin).

Order 2. Semionotiformes or Lepidosteiformes

1. Scales rhomboidal ganoid and in oblique rows.
2. Body elongated.
3. Nasal opening at the end of much elongated snout.
4. Caudal fin abbreviate heterocercal.
5. Vertebrae completely ossified, opisthocoelous.
6. Air-bladder cellular.

Example : *Lepidosteus* or *Lepisosteus* (Garpike).

Superorder C. Teleostei

(Gr., *teleos* = complete; *ostgon* = bone)

1. Scales cycloid ctenoid or absent.
2. Mouth terminal, small.
3. Tail fin mostly heterocercal.
4. Endoskeleton more or less bony.
5. Single external gill-slit on each side of the head covered over by operculum.
6. Cloaca and claspers absent.
7. Air-bladder usually present.
8. Spiracle is lost.
9. Spiral valve in the intestine absent.
10. Conus arteriosus greatly reduced. There is an enlarged bulbus arteriosus.
11. Advanced or modern ray-finned fishes.

Order 1. Clupeiformes

1. Scales cycloid and well developed.
2. Head and operculum not scaled.
3. Single dorsal and a small ventral fin without spines. Ventral fin may be absent. Pelvic fins abdominal.
4. Caudal fin homocercal.
5. Air-bladder communicate with the pharynx.
6. Vertebral centra completely ossified.
7. No auditory vesicles.
8. Weberian apparatus absent.

Examples : *Clupea* (Herringer), *Salmo* (Atlantic salmon), *Sardinops* (Pacific sardine), *Esox* (Pike), *Notopterus* (Chital fish), *Elops*, *Gadusia*, *Ilisha*, etc.

Order 2. Scopeliformes

1. Deep sea fishes having phosphorescent organs.
2. Dorsal and anal fins without spines. An adipose fin present.
3. Mouth wide and provided with numerous small teeth.
4. Air-bladder absent.

Example : *Harpodon* (Mumbai duck).

Order 3. Cypriniformes or Ostariophysi

1. Fins either without spines or dorsal, anal and pectoral have a spine each.
2. Ventral (pelvic) fins abdominal.
3. Air-bladder connected with the pharynx by a duct.
4. A peculiar Webarian apparatus, connecting the internal ear with the air-bladder, present.

Representatives of this order are grouped in two divisions : Cyprini, Siluri.

Division I. Cyprini

1. Body covered with scales or naked. Never covered with bony plates.
2. Third and fourth vertebra not fused with each other.

Examples : *Labeo* (Rohu), *Cirrhina, Botia, Nemachilus, Cyprinus* (Carp) *Puntius, Tor, Esomus, Oxygaster,* etc.

Division II. Siluri

1. Body naked, not covered by scales.
2. Maxillary bone reduced supporting the barbules.
3. Second, third, fourth and sometimes the fifth vertebrae are generally fused.

Examples : *Arius, Heteropneustes* or *Sacobranchus* (Singhi) *Clarius* (Magur), *Wallago* (Lachi), *Mystus* (Tengra), *Ompok*, *Ailia*, *Silonia*, *Bagarius*, etc.

Order 4. Anguilliformes

1. Body elongated eel-like or snake-like.
2. Scales vestigial or absent.
3. Dorsal and anal fins long and confluent behind.
4. Pelvic fins, if present, abdominal.
5. Fins devoid of spines.
6. Air-bladder with duct.

Examples : *Anguilla* (Freshwater eel), *Muraena* (Moray).

Order 5. Beloniformes (Synentognathi)

1. Body elongated covered with cycloid scales.
2. Fins without spines.
3. Pectoral fins large and high on body.
4. Ventral (pelvic) fins abdominal.
5. Some of them are capable of jumping into the air and glide with the help of enlarged pectoral fins.

Examples : *Belone* or *Xenentodon* (Garfish), *Hemirhamphus* (Half beak), *Exocoetus* and *Cypselurus* (Flying fishes),

Order 6. Syngnathiformes (Solenichthyes)

1. Body, covered with protective layer of scales or bony rings.
2. Snout tubular with suctorial mouth.
3. Pectoral fins small, pelvics absent and a single dorsal fin present.
4. Fin-rays of dorsal, pectoral and pelvic fins not branched.
5. Tail prehensile in sea horse, not in pipe fish.
6. Air-bladder closed.
7. Males possess brood pouch for the development of the young.
8. Physoclistic.

Examples : *Hippocampus* (Sea horse), *Fistularia* (Flute fish), *Syngnathus* (Pipe fish).

Order 7. Ophiocephaliformes or Channiformes

1. Body covered with cycloid scales.
2. Head depressed, covered with plate-like-scales.
3. Fins without spines.
4. Air-bladder very long and without duct.
5. Accessory respiratory organs present.
6. Physoclistic.

Example : *Ophiocephalus* or *Channa* (Snake head fish).

Order 8. Symbranchiformes

1. Body elongated eel-like or snake-like devoid of scales.
2. Dorsal, caudal and anal fins continuous. Pectoral fins absent.
3. Fins without spines
4. Gill-slits single or join to form a transverse ventral slit.
5. Air-bladder absent.

Examples : *Amphipnous*, *Symbranchus* (Eels).

Order 9. Mastacembeliformes

1. Freshwater eel-like fishes.
2. Dorsal, caudal and anal fins confluent. Sometimes a small fin separate.
3. Some free spines present in front of dorsal fin. Anal fin with three spines.
4. Ventral (pelvic) fins absent, but pectoral fins present.
5. Nostrils on tubular tentacles at the end of snout.
6. Buccal cavity enlarged for air breathing.
7. Physoclistic.

Examples : *Mastacembelus*, *Macrognathus*.

Order 10. Perciformes or (Percomorphi)

1. Two dorsal fins, ventral (pelvic) fins thoracic with not more than 6 rays.
2. Fins usually with spines.
3. Weberian apparatus absent.
4. Air bladder without duct.
5. Plysoclistic.

Examples : *Anabas* (Climbing perch), *Perca* (Yellow perch), *Lates* (Bhetki).

Order 11. Scorpaeniformes

1. Enlarged heads and pectoral fins.
2. Projecting spines from gill-covering.

Example : *Pterois* (Scorpion fish).

Order 12. Pleuronectiformes

1. Bottom dwellers.
2. Body flat, lying on one side, adapted for bottom living.
3. Head asymmetrical.
4. Both eyes present on the upper or dorsal side and close to each other.
5. Dorsal and anal fins fringing the body and along the caudal encircle the body. Fins usually without spines.
6. Air-bladder absent in adults.

Examples : *Synaptura*, *Pleuronectes*, *Solea* (Flat fishes).

Order 13. Echeneiformes (Discocephali)

1. Body covered with cycloid scales.
2. First dorsal fin modified into a flat oval adhesive disc or sucker on head for attachment. It possesses 12–28 transverse ridges which are modified spines.

3. No spines in second dorsal and anal fins.
4. Air-bladder absent.

Example : *Echeneis* or *Remora* (Sucker fish).

Order 14. Tetradontiform (Plectognathi)

1. Body usually globular.
2. Body scales modified into spines. Scutes or bony plates cover the body.
3. Strong jaws with a sharp beak.
4. Gill-slits small like a hole on either side of fish in front of pectorals.
5. Ventral fines thoracic or subthoracic.
6. Air-bladder present or absent.
7. Some inflate by swallowing water.

Examples : *Diodon* (Porcupine fish), *Terodon* (Globe fish), *Ostracion* (Coffer fish or Trunk fish).

Order 15. Lophiiformes (Pediculati)

1. First ray of spinous dorsal fin placed on the head is transformed into a fishing organ consisting of a rod (**illicium**) and a lure called **esca**.
2. Mouth large with long pointed teeth.
3. Body with minute scales or scaleless.
4. Pelvic fins present or absent. Pectorals well developed.
5. Air-bladder absent.
6. Luminescent organs present.

Examples : *Lophius* and *Antennarius* (Angular fishes).

Other Osteichthyes

The class Osteichthyes of bony fishes is truly the most successful group of aquatic vertebrates. This group includes the dominant fishes of seas today. No other group surpasses these fishes in evolutionary diversity and nowhere else do we see better example of adaptive radiation. There are probably 30,000–40,000 living species of bony fishes. Of these, the superorder Teleostei, comprising about 20,000 described species. Thus, Teleostei constitutes perhaps the single largest group of living vertebrates. For convenience, these fishes may be considered into 4 groups as under :

I. Lobe-finned Fish

Latimeria. Crossopterygians or lobe-finned fish are now represented only by fossils. They were very dominant in Devonian times about 280 million years or more ago and became extinct by the end of Cretaceous. However, several living specimens of a specialised side branch, the Coelacanths, have been found in recent years since 1938 near Comoro Islands between Africa and Madagascar. **Dr. J.L.B. Smith** of Rhodes University named this **"living fossil"** *Latimeria chalumnae.* The generic name was given in honour of **Miss Courteny Latimer**, curator of East London Museum (South Africa), who recognised it as unusual. The specific name refers to the first site of capture; offshore the mouth of Chalumna river of South Africa.

Latimeria (Fig. 16.2) is steely blue-gray with irregular white spots and highly reflective golden eyes. Size ranges from 0.75 to slightly over 2.0 metres and weighing about 90 kg. The entire body is covered with cosmoid scales. The scales have hollow tubercles, overlap each other to form a strong, thick protective covering. The paired fins are

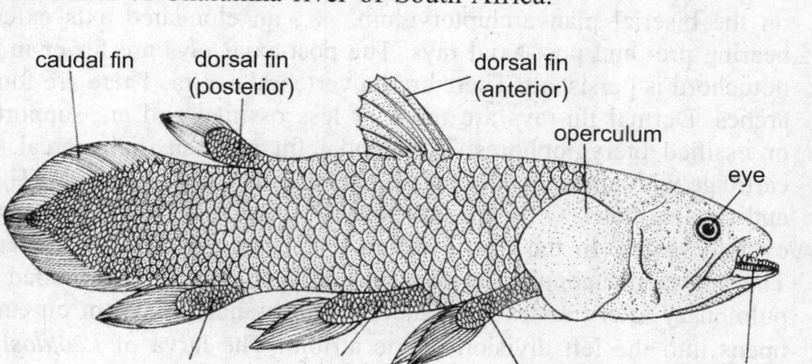

Fig. 16.2 *Latimeria chalumnae*. The living coelocanth.

lobed and have a series of jointed bones. It is believed that the limbs of land animals arose from such lobed fins. There are two dorsal fins and an anal fin. The anteriorly situated first dorsal fin is unlobed, the second dorsal fin and anal fin are lobed. The caudal fin is diphycercal with distinct 3 lobes. The front part of the head moves on the hind part at a joint. There are numerous small conical teeth and it feeds on fishes. Notochord is unconstricted, vertebral column is cartilaginous, but both neural and haemal spines are ossified. Spines present on the back of fish have coelom, thus, the name Coelocanth is given. Below the lower jaw a pair of bony plates called gular plates are found. Gill-arches are 5. Internal nostrils are absent. Alimentary canal consists of muscular oesophagus, followed by well developed stomach having gastric glands. Intestine is a long tube having spiral valve. It opens into rectum, opening into cloaca. Rectal gland is present. Liver is bilobed, with large gall-bladder. Pancreas is present. Air-bladder is covered with fat and is a hydrostatic organ. The heart is primitive. It is S-shaped with the atrium and sinus venosus behind the ventricle, the conus arteriosus has four rows of valves. Near its well-developed olfactory capsules are present a pair of sac-like rostral organs opening to the surface through a series of six pores. They are perhaps electro-receptors. They have a well ossified internal skeleton and small conical teeth suited for seizing prey. Kidneys are fused and are attached to the ventral wall of abdominal cavity which is peculiar. Each kidney has a ureter which enlarges to form a bladder. In this way two urinary bladders are present. A pair of testes lies posterior to anus. Urinogenital opening is situated between pelvic fins. Copulatory organs are absent. Fertilisation is internal. In the absence of claspers how male *Latimeria* achieve copulation is not known.

Evidences show that the crossopterygians were not only ancestral to all bony fish, but they also gave rise to the earliest amphibians and other vertebrates that were to master the land.

II. Lung Fish

The Dipnoi, commonly known as **lung-fish**, are considered to be specialised or degenerate descendants of the primitive lobe fins which they closely resemble. Dipnoi are Devonian offshoot from the crossopterygian stem. They still survive as three degenerate and specialised forms in widely separated tropical or subtropical, freshwater habitats of a special kind. Lung-fishes appear first in mid-Devonian, flourished moderately in the Permian and Triassic, and then became rare. The living Dipnoi are *Neoceratodus* of South-Eastern Queensland, *Protopterus* of Africa, and *Lepidosiren* of South America. They possess internal nostrils. They breathe by means of gills and a highly developed lung. *Neoceratodus* is able to live in stagnant water and never dies if removed from water. *Protopterus* and *Lepidosiren* can survive the drying of the environment by aestivating in mud with small aperture to allow respiration. *Protopterus* secretes slimy protective cocoon. In both *Protopterus* and *Lepidosiren* an epiglottis occurs. Lung fishes have bony scales and dermal fin-rays. Paired fins are unlike those of other fishes and are constructed on the biserial plan–archipterygium, *i.e.*, an elongated axis extending out from the body and bearing pre- and post-axial rays. The post-axial rays are fewer in number than the preaxial. The notochord is persistent. There are no vertebral centra. There are four to six cartilaginous branchial arches. Dermal fin-rays are more or less ossified and are supported by numerous cartilaginous or ossified pterygiophores. The caudal fin is often diphycercal. The pectoral arch is a single cartilage with a pair of superficial investing bones on each side. The pelvic arch is well developed and cartilaginous. Gills are covered by operculum. Dermal skeleton is formed by over-lapping cycloid scales. In the heart, atrium and sinus venosus are imperfectly divided into two parts. The contractile conus arteriosus is partly or completely divided by a longitudinal partition. A pulmonary artery arises from the efferent branchial system on each side, and a pulmonary vein opens into the left division of the atrium. The larva of *Lepidosiren* and *Protopterus* possesses both suckers and external gills. Adults of some species of *Protopterus* possess remains of these gills. Paired oviducts open anteriorly into coelom. Male vasa efferentia carry sperms through the excretory part of the mesonephros.

1. *Neoceratodus.* *Neoceratodus* (Fig. 16.3) is commonly known as **Australian lung fish** or **Burnett salmon**. Body is elongated and compressed measuring about 1.5 metres or more. Scales are large, thin and covered with spines which are not regarded as denticles. Paired fins are rather leaf-like or paddle-like. Caudal fin is symmetrical and pointed. Dental plates are oval crescentic or triangular terminating in a smooth or feebly denticulated biting margins. Lower jaw with a small toothless dentary on each side. Branchial arches are five in number and bisegmented. Hyobranchial-cleft is open and associated with pseudobranch. The first four branchial-arches carry holobranchs. Air-bladder is single. It cannot survive out of water. The lateral line system is somewhat degenerated. Inactive and sluggish in habit, usually lying motionless on the bottom. Carnivorous. Feeds on freshwater crustaceans, worms and molluscs. *Neoceratodus* is found in Burnett and Mary rivers in Queensland, Australia.

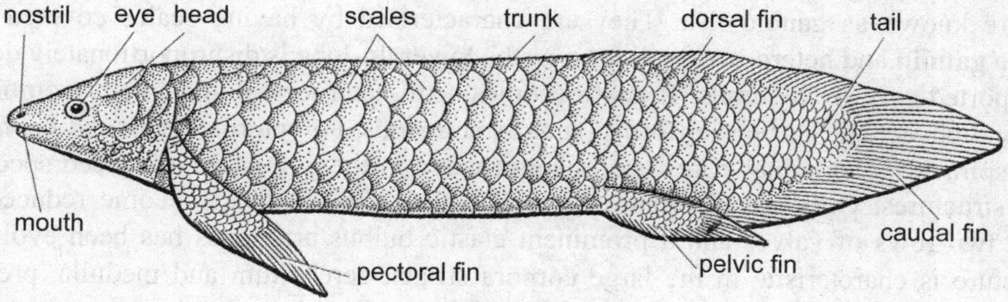

Fig. 16.3. *Neoceratodus.*

2. *Protopterus.* *Protopterus* (Fig. 16.4) is commonly known as **Nile lung fish** or **African Lung fish.** Body is elongated, cylindrical and more or less eel-like and grows to a length of 2 metres. Small cycloid scales are completely enclosed in the skin. The paired fins are long, thin and filamentous. Dorsal and caudal fins are united. Tail is principal organ of locomtion. Six branchial arches and five pairs of gill-slits are covered by operculum. Larval gills are usually retained as vestiges throughout life. Air-bladder is double lung. It is said to breathe by its lungs as well as by its gills. Carnivorous and voracious feeder. During unfavourable seasons it undergoes **summer sleep** and burrow into the soil to a depth of about 60.0 cm. *Protopterus* has a wide distribution over the middle portion of great African continent ranging from the river Senegal and the White Nile on the north to the Congo basin, Lake Tanganiyika and Zambesi on the south.

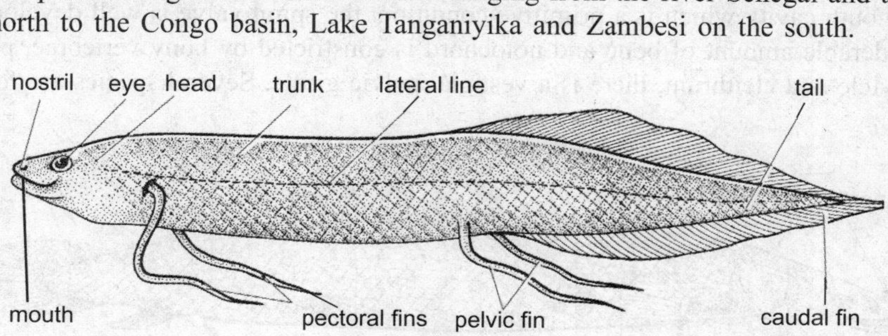

Fig. 16.4. *Protopterus.*

3. *Lepidosiren.* *Lepidosiren* (Fig. 16.5) is commonly known as **South American lung fish**. Body is elongated, cylindrical and more or less eel-like. The skin covering the body contains the small cycloid scales. Paired fins are thin and filamentous. Dorsal and caudal fins are united. Tail is principal organs of locomotion. Five branchial arches and four pairs of gill-slits are covered by operculum. Cutaneous gills disappear soon after the larval metamorphosis. Air-bladder is double lung. It is not exclusively carnivorous. Food chiefly comprises fresh snails and mass of algae. *Lepidosiren* occurs along the course of main Amazon river.

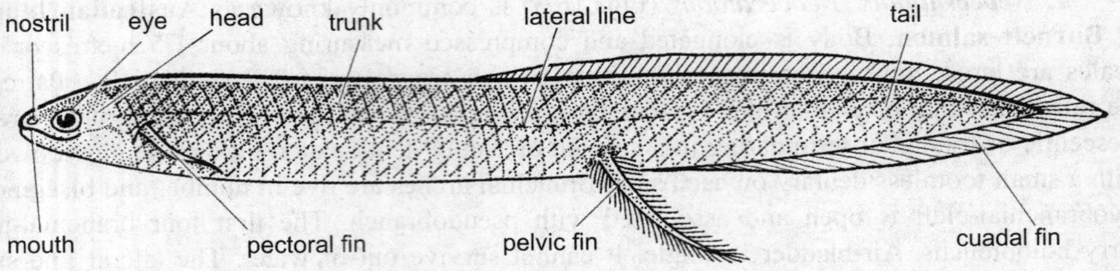

Fig. 16.5. *Lepidosiren.*

III. Ganoid Fish

The actinopterygians (modern living fish) belonging to the superorders Chondrostei and Holostei are known as ganoid fish. They are characterised by having scales covered with an enamel-like **ganoin** and heterocercal tail fin, *i.e.*, the lower fin-lobe is disproportionately developed and is supported only by hypurals. It next becomes semi-heterocercal and finally homocercal or more rarely diphycercal. The earlier forms (such as sturgeons, Polypterini, *Amia* and *Lepidosteus*), possessed spiracles, abdominal pores, and a spiral valve in the gut. More advanced species have lost these structures. Gills are filamentar type. The conus in the heart become reduced to one or at most two rows of valves and a prominent elastic bulbus arteriosus has been evolved. The brain structure is charcteristic in the large corpora striata, cerebellum and medulla, presence of valvula cerebelli and in the absence of cerebral lobes. The actinopterygians in general show no trace of internal nostrils, (except the stargazer, *Astrocopus*).

1. *Polypterus.* *Polypterus* (Fig. 16.6) along with other ganoid fishes has survived from ancient times. The dorsal fin is divided into a number of finlets, hence, the name. The pectoral fin has acquired a thick scale-covered basal lobe, hence, it was once placed in Crossopterygii, but its endoskeleton is different. Body is covered with thick ganoid scales, dermal denticles like those of elasmobranchs are found on the scales and lepidotrichia. The tail is no longer heterocercal but appears somewhat diphycercal. Nostrils are dorsolateral, and spiracles persist. It expels water through the spiracles. There is a paired ventral air-bladder which is like a lung, its pneumatic duct arises from the ventral side of the pharynx, the extent to which the air-bladder functions as a lung is uncertain, because *Polypterus* cannot survive out of water. The pituitary gland opens into the mouth cavity which is a primitive condition, the spiral valve is well developed. Skeleton has considerable amount of bone and notochord is constricted by bony vertebrae, pectoral girdle has a clavicle and cleithrum, there is a vestigial pelvic girdle. Several species are found in rivers of Africa.

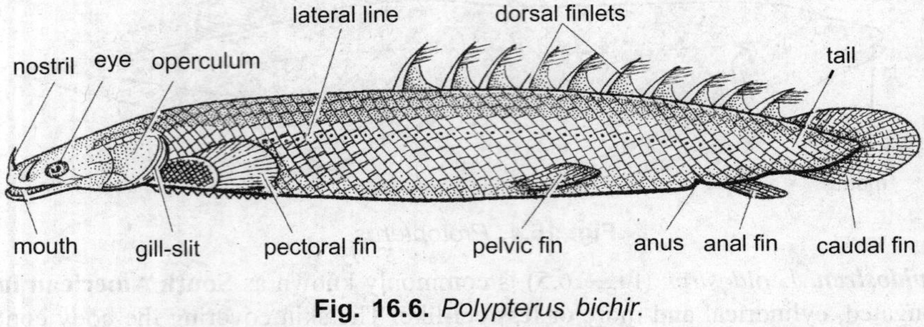

Fig. 16.6. *Polypterus bichir.*

2. *Acipenser.* *Acipenser* (Fig. 16.7) is commonly known as **sturgeon**. It has an elongated snout bearing barbels, mouth is ventral and tubular, jaws are reduced and toothless, sturgeons stir up the mud for suctorial feeding on small invertebrates. The jaws hang free from the hyomandibular. The body has no scales but is covered with five rows of bony dermal scutes which are grooved and bear spines, the heterocercal tail has some ganoid scales. The spiraches persist, and there are 5 pairs of gill-clefts. There is a reduction of dermal bones and skeleton is almost entirely cartilaginous, but dermal bones form a shield on the cranium, vertebrae are incomplete and notochord is unconstricted. Sturgeons show some primitive characters of elasmobranchs, such as spiracles, unconstricted notochord, heterocercal tail, conus arteriosus, and a pericardio-peritoneal canal. They also show some degenerate characters in the jaws, scales, and loss of ossification.

Some sturgeons are found in freshwater but most live in the sea but come to rivers for breeding, they are the largest freshwater fishes, *Acipenser huso* being 24 ft. and weighing 25 maunds. Sturgeons are used as food and their preserved roe is eaten as caviar.

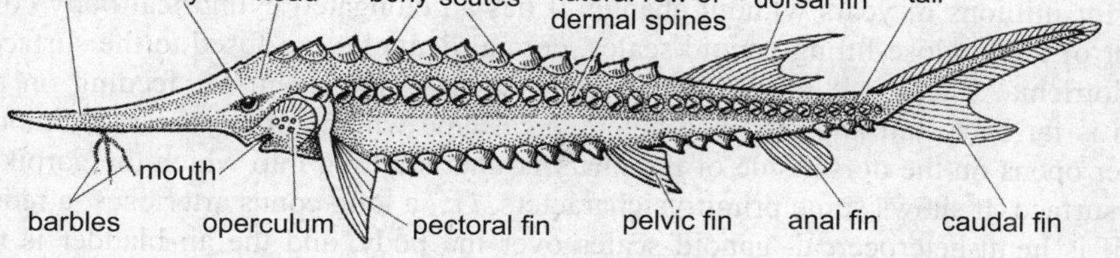

Fig. 16.7. *Acipenser sturio* (sturgeon).

3. *Polyodon.* *Polyodon* (Fig. 16.8) is commonly known as paddle-fish or spoon bill. Body is elongated and fusiform measuring up to one and a half to two metres in length. Skin is naked. Bony scutes are absent and ganoid scales are present only on the upper lobe of the tail. The most outstanding feature is the great elongation of snout into a flat oar-like rostrum. Contrary to the common belief that the paddle or shovel-like rostrum is used to stir food from muddy river bottom. Paddle-fishes are planktivorous. Mouth is wide and ventral. Barbels are absent. Eyes small and open spiracles are present. A small operculum is present. Tail is heterocercal. *Polyodon* exhibits intriguing geographical distribution. Of the two species, one occurs in the Yantze River Valley of China, the other in Mississippi River Valley of North America.

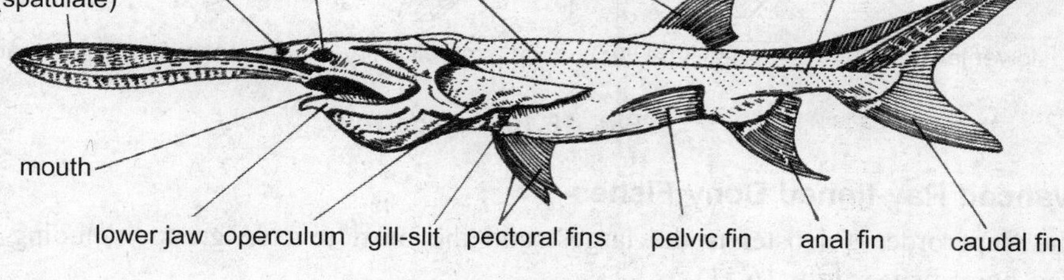

Fig. 16.8. *Polyodon.*

4. *Amia.* *Amia calva* (Fig. 16.9) is commonly known as **bow-fin.** It looks like and has nearly achieved the teleostean condition. It has a long, continous dorsal fin, hence, the name bow-fin. Mouth is wide with sharp teeth. The scales covering the body have been reduced to cycloid type which overlap slightly but there are still some ganoid scales on the head.

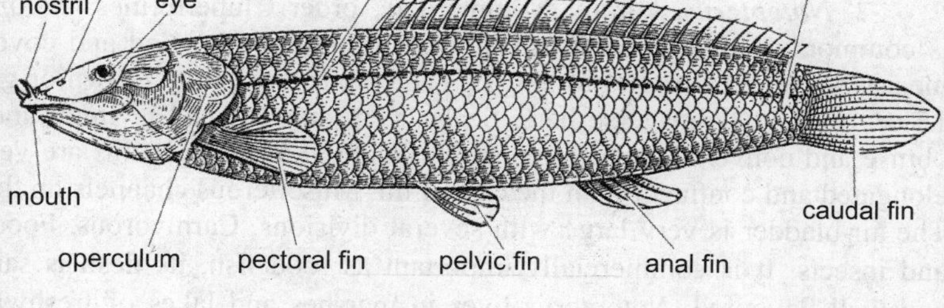

Fig. 16.9. *Amia calva.*

The tail appears homocercal in which the posterior end of the vertebral column is turned up, but externally it has become symmetrical. The air-bladder opens dorsally into the pharynx and it functions as a lung. *Amia* shows some primitive characters, *i.e.*, presence of a conus arteriosus and a bulbus arteriosus, a vestigial spiral valve and the air-bladder is respiratory. It shows some advanced characters, spiracles have disappeared, branchiostegal rays are present, and vertebrae are amphicoelous. It is 60.0 cm long and is carnivorous in the rivers and lakes of North America. Eggs are laid in nests and guarded by the male.

5. *Lepidosteus*. *Lepidosteus* (Fig. 16.10) is commonly known as **garpike.** It arose in the Permian, flourished in the seas in the Jurassic, and died out in the Cretaceous. *Lepidosteus* has survived for millions of years without change. It has an elongated cylindrical body covered with an armour of thick, close-fitting ganoid scales, dermal denticles are fused to the surface of scales and lepidotrichia. There is a long snout or beak having sharp teeth for feeding on fishes. The dorsal fin is far back, tail is a modified heterocercal type which appears nearly homocercal. The air-bladder opens on the dorsal side of the pharynx and is a lung into which the garpike gulps air from the surface. It shows some primitive characters, *i.e.*, a long conus arteriosus, a reduced spiral valve, tail is hemi-heterocercal, ganoid scales over the body, and the air-bladder is respiratory. On the other hand it shows some advanced characters, *i.e.*, spiracle is absent, pyloric caeca are present, clavicle is lost and maxillae are reduced. The skeleton is ossified, there are dermal bony plates, vertebrae are bony but they are opisthocoelous which is rare in fishes. *Lepidosteus* is a predator on other fishes, one species in the Mississipi is about 3.0 metres long. They are found in rivers of America and Cuba.

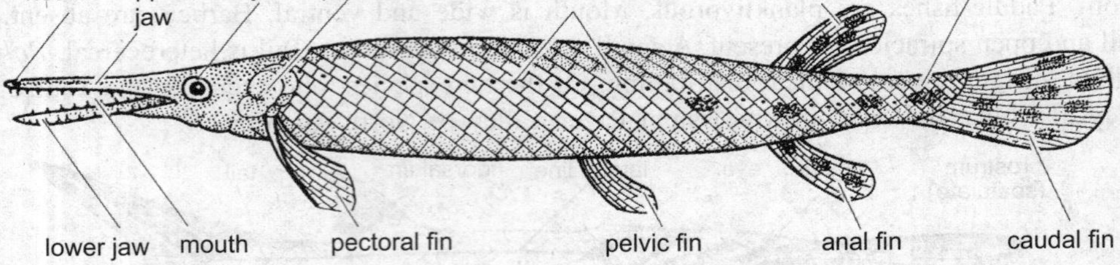

Fig. 16.10. *Lepidosteus.*

IV. Advanced Ray-finned Bony Fishes

The Superorder Teleostei is the largest and the most diverse group including advanced ray-finned bony fishes (Fig. 16.11).

1. *Salmo*. Commonly called **salmon** or **trout**, belongs to the family Salmonidae of the order Clupeiformes. A freshwater and marine fish found in temperate and arctic zones of Northern Hemisphere. It is introduced in Kashmir and Nilgiris. It has a small adipose fin on back. It is much prized for food and game.

2. *Notopterus*. Family Notopteridae, order Clupeiformes. *Notopterus chitala* (Fig. 16.11) is commonly known as **chital**. Body is strongly compressed and covered with minute scales. It measures up to 1.5 metres in length. The colour is coppery brown or greyish along the back with 15 or 16 silvery transverse bars. Scales very small. Head is small and mouth is large. Snout is obtuse and convex. Dorsal fin is small. Pectoral and pelvic fins are very small. Anal fin is much elongated and confluent with the caudal fin. Musciferous channels on the head are well developed. The air-bladder is very large with several divisions. Carnivorous. Food chiefly comprises worms and insects. It is commercially important as food fish, its flesh is said to be uncommonly rich and well flavoured. *Notopterus* lives in marshes and lakes of freshwater and brackish water of West Africa, India, Myanmar and Malaya. *Notopterus chitala* is found exclusively in freshwaters of India.

3. *Clarias.* Family Claridae, order Cypriniformes. *Clarias batrachus* (Fig. 16.11) called **mangur** in Hindi. Body is elongated, scaleless and measuring up to 45.0 cm in length. The general colour of the body is uniform brown or greyish black. Head depressed with top and sides covered with osseus plates. Sensory barbels are four pairs. Dorsal fin is long and without spines, extending from the neck to the caudal fin. Anal fin also long. No adipose fin. Caudal fin is more or less rounded. Pectoral fins are provided with spines. Accessory respiratory organs are branched tree-like, specially designed to take in oxygen from the air. The air-bladder is connected with internal ear by Weberian ossicles. It is highly nourishing and esteemed as food. It is also used in laboratories for experimental purposes. *Clarias* is found in Africa and South and West Asia. *Clarias batrachus* occurs in fresh and brackish waters throughout India.

4. *Heteropneustes* (= *Saccobranchus*). Family Heteropneustidae, order Cypriniformes. *Heteropneustes fossilis* (Fig. 16.11) is commonly called **singhi** in Hindi. Body is elongated and laterally compressed, measuring about 30 cm in length. Skin without scales. Head is flattened. Eyes with free circular margins. Barbels are long and four pairs. Dorsal fin short without spines, ventral fin situated at the level of the dorsal fin. Pectoral fins are strong with poison spine. Anal fin is elongated, reaches up to the caudal fin separated from it by a notch. Gill-opening is wide, the membranes not being confluent with the skin of isthumus and separated by a deep notch. Accessory breathing organs are present. Air-bladder is also present. *Heteropneustes* (=*Saccobranchus*) is found in freshwaters of India, Pakistan and Sri Lanka.

5. *Wallago* (= *Wallagonia*). Family Siluridae, order Cypriniformes. Common names are **lachi**, **mullee** or **boalli**. Body is elongated and laterally compressed with a straight back. They may reach in length up to 2 metres but the specimens generally available are 60–90 cm in length. Head is very large, trunk small and tail long and tapering. Mouth is wide extending below and even behind eyes. Jaws are provided with teeth. Barbels 4, two maxillary and two mandibular. Eyes above the level of the mouth and not covered with skin. Dorsal fin is short and spineless. Anal fin is very long. Caudal fin bilobed. Skin scaless. Predatory in habit. It predates on fishes particularly young carps. *Wallago attu* (Fig. 16.11) is found in freshwater rivers and lakes throughout India.

6. *Anguilla.* Family Muraenidae, order Anguilliformes. *Anguilla* (Fig. 16.11) is commonly called **eel**. It has a long snake-like body, more than a metre. long with an excessive number of vertebrae. There are no visible scales but minute ones are hidden in the dermis. Pectoral fins are reduced and pelvic fins are absent, the dorsal fin is soft, long, and continuous with the caudal fin. Air-bladder has a pneumatic duct. The skin is respiratory both in water and in air, and eels are noted for their overland journeys by wriggling on damp grass. There are no oviducts, the eggs are discharged through genital pores. The electric eel *Gymnotus* and *Eelctrophorus* have electric organs formed from metamorphosed muscles, they can give a very powerful shock which in *Electrophorus* is of 370 volts. Moray are marine eels of bright colour patterns and savage, predaceous habits. *Ophichthys* is an eel of the Indian coast found in estuaries and it comes up into rivers.

Eels are noted for their migration which is opposite of other fishes, they migrate from rivers into the sea. Species of *Anguilla* found in rivers of Europe and America can live both in marine and freshwaters, the adults migrate from the rivers far into the Atlantic Ocean where they change their yellow-green colour to a silvery one. They spawn in the region of the West Indies, one eel laying up to twenty million eggs, after spawning only once in a life time both males and females die. The young which hatch from eggs are called **leptocephalus** which are transparent, very thin laterally and leaf-like in shape, they are so transparent that they are known as glass fish. They are pelagic and float on the surface of water from one to three years reaching a length of 7.5 cm. They undergo metamorphosis becoming shorter but cylindrical, now they are called **elvers**.

The elvers march upstream into rivers, sometimes crawling overland through moist grass, they reach rivers, ponds and lakes where they become adults in a few years.

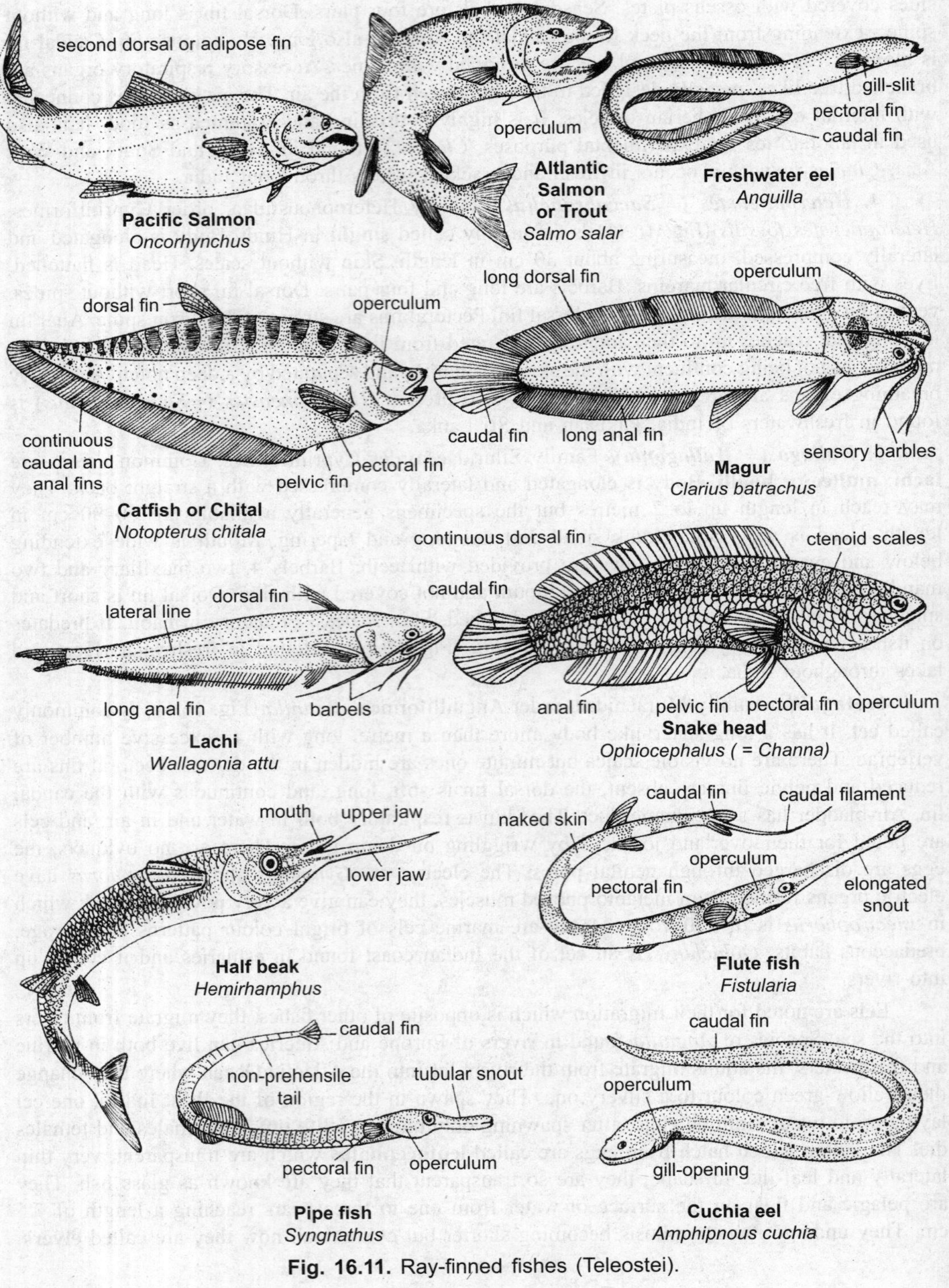

Fig. 16.11. Ray-finned fishes (Teleostei).

7. *Amphipnous.* Family Amphipnidae, order Symbranchiformes. *Amphipnous cuchia* (Fig. 16.11) is commonly known as **blind serpent**. Body is elongated, eel-like or snake-like and grows up to 60.0 cm in length. Skin appears almost naked because the minute scales are buried deep in it. Head bears the nostrils and eyes. Mouth is small. The gills are rudimentary and the fish has evidently lost the power of aquatic respiration. The air-breathing organs consists of a pair of sacs growing out of the pharynx above the gills. They spend most of their time in grass on the banks of ponds and rivers. When in water it constantly rises to the surface for respiration. The body is dark green above and dirty pale red below. No swim bladder. *Amphipnous cuchia* is found in the rivers and ponds of Bihar and Bengal and also found in Myanmar.

8. *Ophiocephalus (= Channa).* Family Ophiocephalidae, order Ophiocephaliformes. *Ophiocephalus* (= *Channa*) *punctatus* (Fig. 16.11) and *O. striatus* are commonly called live-fish and popularly known as **snake-heads.** Body is elongated, anteriorly cylindrical and posteriorly compressed and covered with cycloid scales. They may grow to the length of 45.0 cm. Head is depressed and covered with large cycloid scales. Dorsal and anal fins are single, long and without spines. Gill-opening is wide and lateral in position. Accessory breathing organs are present. On the roof of its pharynx, the fish has a pair of cavities which have folded linings richly supplied with blood vessels for taking the air. The accessory respiratory organe enable these fishes to survive out of water for a few hours or migrate from one pool to another. Air-bladder is present and elongated. Carnivorous. Eaten mostly by poor classes. *Ophiocephalus (= Chana)* is found in the freshwaters of India, Myanmar, Sri Lanka, Malaya, Thailand, China, Formosa and Islands of East Archipelago.

9. *Hemirhamphus.* Family Hemirhamphidae, order Beloniformes. Body is elongated and sub-cylindrical, measuring up to 30 cm in length. The skin is covered with cycloid scales. The colour of the body is greenish blue above with two bands along each side, one silvery and other black called half-beaked fish. Upper jaw is short, more or less triangular in shape. Lower jaw is elongated and prominent projected beyond the upper jaw, both jaws having pointed teeth. Eyes are lateral in position. Gill-opening is wide. Tail is homocercal. They are mainly herbivorous, feeding on green algae. It is used as food fish. *Hemirhamphus* (Fig. 16.11) is found in seas of India and China. It is quite abundant in south-east coast of India.

10. *Exocoetus.* Family Scombresocidae, order Beloniformes. *Exocoetus* (Fig. 16.12) is commonly known as **flying-fish**. Body is moderately elongated, measuring 30 to 45 cm, compressed and covered with cycloid scales. The colour is bluish and silvery below. Pectoral fins are large wing-like. They serve as parachute to sustain the fish in its gliding leaps. Dorsal fin opposite the anal find. Mouth is wide, both the jaws bear teeth. Tail is hypobatic, *i.e.*, ventral lobe of tail fin is large. Air-bladder is present. Oviparous. It is a pelagic shoaling fish, carnivorous feeding on prawns and young teleosts and their eggs. Excellent food fish. It is considered as good edible fish and provide fishery in certain seasons. *Exocoetus* and a related genera *Cypselurus* inhabit tropical and subtropical regions of Atlantic, Pacific and Indian Oceans. *E. poecilopterus* is found from Indian Ocean to China seas.

11. *Hippocampus.* Family Syngnathidae, order Syngnathiformes. *Hippocampus* (Fig. 16.12) is commonly called **sea horse**. Size variable 5 to 20 cm. The head is prolonged into a tubular rostrum or snout at the end of which is a small mouth, so the head looks like that of a horse, hence, the common name **sea horse**. There are no teeth in the adult. Body is enclosed in closely fitting bony scales which form a continuous armour. The tail is long and has no caudal fin. It is prehensible and wrapped around stems of sea weeds for attachment. On the back is a dorsal fin with soft fin-rays. Gills lophobranchs made of several small rounded lobes. Operculum is large, plate-like structure. Muscles are feebly developed. It swims in an erect position by waves passing along the dorsal fin. Pelvic fins are absent but small pectoral fins lie behind the operculum. From the head and body project filamentous or leaf-like processes. Colour of the body and processes resembles that of sea weeds in which it lives, and on which it also feeds. On the abdomen of the male is a **brood pouch** formed from combined pelvic fins, fertilised eggs are placed in the brood pouch by the male, the young ones develop in the brood pouch. *Hippocampus* occurs in tropical and temperate seas including Indian Ocean.

12. *Syngnathus.* Family Syngnathidae, order Syngnathiformes. *Syngnathus* (Fig. 16.11) is commonly known as **pipe-fish**. Body is more or less elongated, 45 cm long, having an exoskeleton of dermal rings. Snout is smaller and tubular. The mouth is small, terminal, toothless and suctorial, lies at the extremity of the snout. Eyes are lateral in position. Gill-slit is reduced to a very small opening near the upper posterior angle of the operculum. Pectoral fins are very small. Single dorsal fin. Ventral fins are absent. Tail is long and not prehensile, with a poorly developed caudal fin. Male possesses a **brood pouch** on the ventral side of the abdomen. In the brood pouch eggs are retained until they hatch as young ones. *Syngnathus* is found mainly in the Oceans of India and China.

13. *Fistublaria.* Family Syngnathidac, order Syngnathiformes. *Fistularia* (Fig. 16.11) is commonly known as **flute-fish** or **flute-mouth**. Body is greatly elongated and naked. It measures up to 2 metres in length. Snout forming a long tube which terminates in a narrow mouth with minute teeth. Eyes are large and lateral in position. Dorsal and anal fins are opposite to each other. Anal fin with six soft rays. Pelvic fins are very small. Caudal fin is forked. The middle rays of caudal fin are produced into a long caudal filament. Air-bladder is physocleistous. *Fistularia* is found in tropical, subtropical and temperate waters of Atlantic and Indo-Pacific Oceans from Africa to China. It is very common in coastal waters of Chennai.

14. *Anabas.* Family Anabantidae, order Perciformes. *Anabas* (Fig. 16.12) is commonly known as climbing perch. It has an oblong, laterally compressed body about 20.0 cm long. The dorsal fin is elongated and has many spines. The anal fin is also long and bears spines. The pelvic fins lie close to the pectoral fins. On the operculum are backwardly directed spines. Eyes are large and body is covered with large, overlapping cycloid scales. The lateral line is discontinuous. In front of the gills are accessory respiratory organs made of folded plates covered with vascular mucous membrane, they are called labyrinthiform organs which lie in a suprabranchial cavity, they are used for breathing air, thus, *Anabas* can remain outside water for long periods, it is so dependent on atmospheric oxygen that is asphyxiated if immersed in water for long. It can also retain water in its pharyngeal bones when on excursions on land. It moves on land by its fins and opercular spines. The belief that *Anabas* can climb trees is not true. It is found in Indian rivers and South-East Asia.

15. *Pleuronectes* or *Synaptura.* Family Pleuronectidae, order Pleuronectiformes. *Pleuronectes* or *Synaptura* (Fig. 16.12) is commonly known as **flat fish**. Body is thin, laterally compressed and flat. Scales covering the body are usually imbricate, ctenoid or cycloid scales. The side of the body bearing the eyes turned upwards being coloured, while the lower side is white. Head asymmetrical. Both the eyes are situated on the upper side. Mouth is more or less protractile. Dorsal and anal fins are long without spines and confluent with the caudal fin. Caudal fin is well developed. Air-bladder is absent in the adult. Gills four, a slit behind the fourth, pseudobranchiae present. Adapted for bottom living. It is economically important as food fish. *Synaptura* is found in Atlantic, Pacific and Indian Oceans. Common flat fishes of Indian coasts are *Pleuronectes* or *Synaptura*, *Psettodes*, *Zebra* and *Solea.*

16. *Echeneis* or *Remora.* Family Echeneidae, order Echeneiformes. *Echeneis* (Fig. 16.12) is commonly called **remora** or **sucker fish**. Body is elongated, fusiform and covered with small scales. Size up to 50 cm. Head is depressed and furnished above with an adhesive organ. Eyes are lateral in position. Mouth cleft is wide and deep. First dorsal fin is modified into an **adhesive disc.** Adhesive disc is flat, oval and transversely furrowed and is an effective organ for attachment. Sucker represents modified anterior dorsal fin. Second dorsal fin and anal fins are elongated without spines and opposed to each other. Tail is homocercal. Air-bladder is absent. It feeds on other fishes and attaches itself by means of its adhesive organ to boats, sharks, turtles, cetaceans and other large swift swimming animals. On the east coast of Africa, it is employed for catching the turtles and fishes with a remora tied with a cord. Remora eats smaller fishes. *Echeneis* is found in the tropical and subtropical seas including seas of India and China.

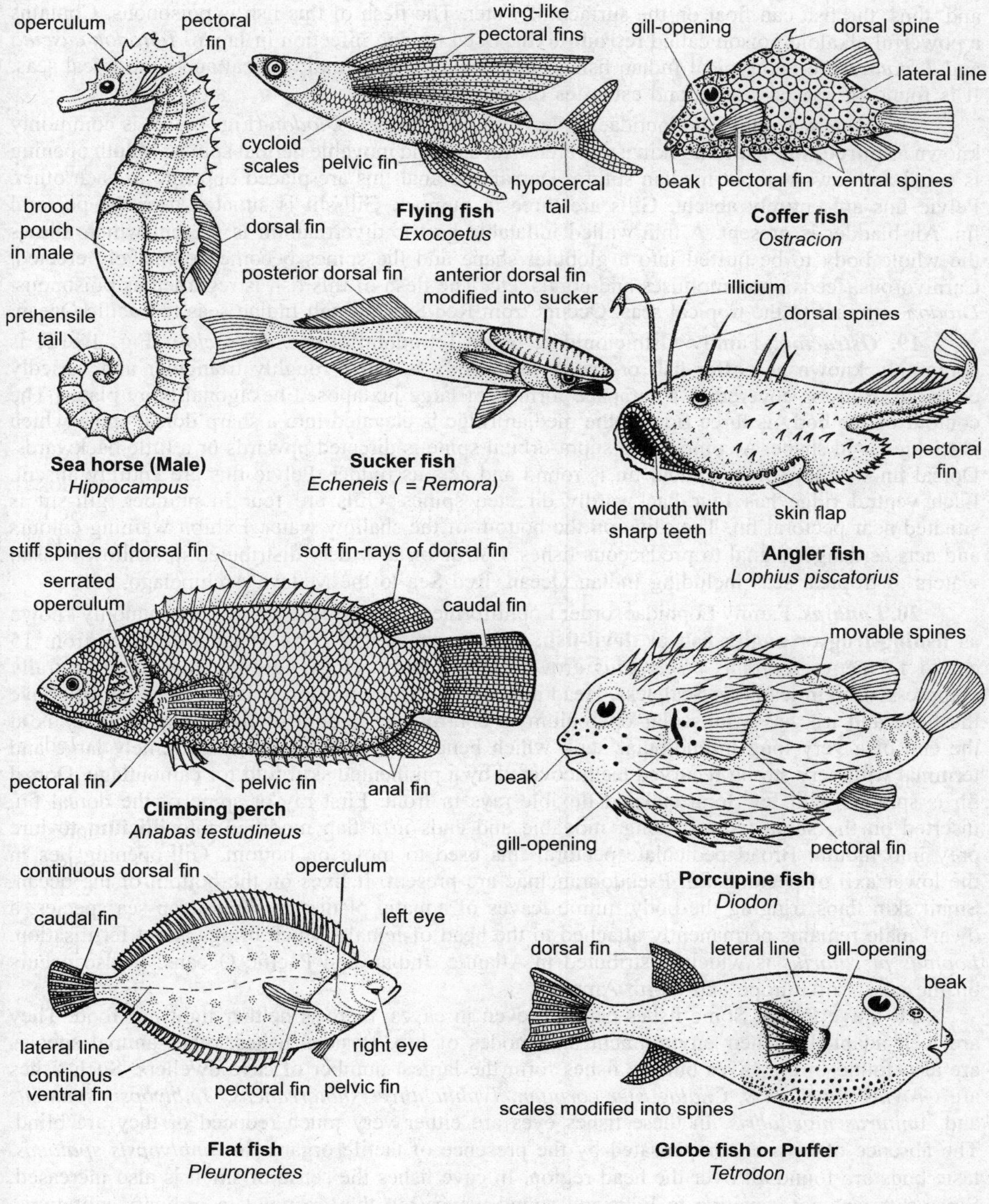

Fig. 16.12. (Continued) Ray-finned fishes (Teleostei).

17. *Tetrodon.* Family Tetrodontidae, order Tetrodontiformes. *Tetrodon* (Fig. 16.12) is commonly known as **globe-fish** or **puffer**. Body is spherical and covered with rough scales. The colour is light brown with dark brown spots throughout the body. Dorsal and anal fins have fin-rays. Pelvic fins are entirely absent. Teeth are fused to form plates and beak with a median suture. Eyes large and protruding. Gills are three. Gill-slit or gill-opening is situated near the pectoral fin. Air-bladder is horse-shoe shaped. The oesophagus is dilatable. The air is engulfed and, thus, the fish can float on the surface of water. The flesh of this fish is poisonous. Contains a powerful alkaloid poison called **tetrodotoxin**. Used in lung infection in Japan. *Tetrodon cuscuta* and *T. patoca* are two typical Indian fishes. *Tetrodon* is found in all tropical and subtropical seas. It is found in coastal waters and estuaries of Indian rivers.

18. *Diodon.* Family Diodontidae, order Tetrodontiformes. *Diodon* (Fig. 16.12) is commonly known as **porcupine-fish**. The skin is covered with stiff and movable dermal spines. Mouth opening is small and jaws without median suture. Dorsal and anal fins are placed opposite to each other. Pelvic fins are entirely absent. Gills are three in number. Gill-slit is situated near the pectoral fin. Air-bladder is present. A thin-walled inflatable gastric diverticulum is present which allows the whole body to be puffed into a globular shape and the spines become defensively erected. Carnivorous, feeds upon molluscs and corals, etc. The flesh of this fish is regarded as poisonous. *Diodon* is found in the tropical seas. Occurs from Red Sea through Indian seas to Pacific Ocean.

19. *Ostracion.* Family Ostracionidae, order Tetrodontiformes. *Ostracion* (Fig. 16.12) is commonly known as **coffer-fish** or **trunk-fish.** Body is short, roughly triangular and variedly coloured. Body is covered in a carapace formed of large juxtaposed hexagonal bony plates. The contour of the body is three ridged, the median ridge is elevated into a sharp dorsal spine which is triangular in shape. A compressed supra-orbital spine is directed upwards or a little backwards. Dorsal fin without spine. Caudal fin is round and acts as rudder. Pelvic fins are entirely absent. Each ventral ridge has four backwardly directed spines. Gills are four in number, gill-slit is situated near pectoral fin. They live on the bottom of the shallow water. Exhibit warning colours and acts as danger signal to predaceous fishes. *Ostracion* is widely distributed in shallow coastal waters of tropical seas including Indian Ocean, Red Sea to the Malaya Archipelago, etc.

20. *Lophius.* Family Lophidae, order Lophiiformes. *Lophius* (Fig. 16.12) is commonly known as **fishing-frog** or **angler-fish** or **devil-fish.** Various species of *Lophius* range in sizes from 15 cm to 1.5 metres. *Lophius piscatorius* grows to 1.5 metres. Body ugly-looking dorso-ventrally compressed. Skin leathery. Scaleless. Head and anterior trunk portion is large. Eyes dorsal above the head and iris has a muscular operculum to control the amount of light. The pituitary is at the end of a very long infundibular stalk which bends forwards. Mouth is extremely large and terminal with very strong recurved teeth covered by a pigmented skin fold for camouflage. Dorsal fin is spinous with few long isolated flexible rays in front. First ray or spine of the dorsal fin, inserted on the snout, is very long, movable and ends in a flap modified into **illicium** to lure prey into mouth. Broad pediculate pectoral fins used to move on bottom. Gill-opening lies in the lower axil of pectoral fin. Pseudobranchiae are present. It lives on the bottom of the ocean. Small skin flaps fringing the body mimic leaves of aquatic plants. In some deep sea species, a dwarf male remains permanently attached to the head of female which brings about fertilisation. *Lophius piscatorius* is widely distributed in Atlantic, Indian and Pacific Oceans. It also occurs on the coasts of Europe and North America.

21. Cave fishes. Some fishes can live even in caves, there is neither light nor food. They are well adapted to their environment and modes of life. Besides fishes, other animal species are also found in the caves but the fishes form the largest number of cave dwellers. Such fishes are *Grovias migrilabris*, *Cholagaster cornium*, *Typhlichthyes subterraneus, Amblyopsis spalaeus* and *Amiturus nigrilabris*. In these fishes eyes are either very much reduced or they are blind. The absence of eyes is compensated by the presence of tactile organs. In *Amblyopsis spalaeus*, taste buds are found all over the head region. In cave fishes the sense of smell is also increased. Since they are not sensitive to light and sound waves but they respond to ordinary vibrations. Due to scarcity of food the organs of digestion are modified to utilise all the nutrients of food.

IMPORTANT QUESTIONS

Long Answer Questions

1. Give the distinctive features of class Chondrichthyes (= Elasmobranchii) and classify it up to orders, giving examples of each group.
2. Give the distinctive characters of class Osteichthyes with important features of each group, giving examples of each group.

Short Answer Questions

1. Mention eight characteristic features of rays and skates.
2. Write the names of three ray-fishes. Give three interesting features about each.
3. Write short notes on : (*i*) *Chimaera*, (*ii*) *Pristis*, (*iii*) *Torpedo*, (*iv*) *Rhinobatus*, and (*v*) *Sphyrna* or *Zygaena*.
4. Compare pleurotremata and hypotremata.
5. Write the zoological importance of Crossopterygii.
6. Differentiate between subclass Sarcoptergii and Actinoptergii.
7. Write about *Latimeria.*
8. Write short notes on : (*i*) *Anabas*, (*ii*) *Anguilla*, (*iii*) *Echeneis*, (*iv*) *Exocoetus*, (*iv*) *Hippocampus*, (*v*) *Amia*, (*vi*) *Polypterus*, (*vii*) *Polyodon.*

Very Short Answer Questions

1. Write common name of *Chimaera.*
2. Give the scientific name of hammer-headed shark.
3. Operculum in Holocephali is cartilaginous or bony or fleshy.
4. What is spiracle?
5. Define cloaca.
6. In which subclass gill-slits are multiple and each is covered by individual skin flap?
7. In which order gill-slits are ventral?
8. In which subclass of Chondrichthyes cloaca is absent.
9. Dorso-ventrally flattened body is found in .... and rays.
10. In skates and rays .... are fused laterally with head and trunk.
11. In *Chimaera* teeth are .... and jaws are covered with .... for crushing and eating molluscs.
12. Write the names of lung fish and their distribution.
13. What are ganoid fish?
14. Ganoid fish belong to which superorders?
15. Write the names of ray-finned bony fishes.
16. Write the names of pipe-fish, flying-fish, eel, sea-horse, flute-fish, flat-fish and sucker-fish.
17. *Tetrodon* is commonly called ....
18. Porcupine fish is known as ....
19. Write the common name of *Ostracion.*

Multiple Choice Questions

1. Chondrichthyes are mostly marine and
   (*a*) Herbivorous (*b*) Predaceous
   (*c*) Sanguivorous (*d*) Parasitic.
2. Body of cartilaginous fishes is streamlined and
   (*a*) Circular (*b*) Cylindrical
   (*c*) Spindle-shaped (*d*) Oval.
3. Tail in chondrichthyes is
   (*a*) Homocercal (*b*) Heterocercal
   (*c*) Hypocercal (*d*) Hypercercal.
4. The number of gills in Chondrichthyes are
   (*a*) 5 to 7 pairs (*b*) 3 to 5 pairs
   (*c*) 2 to 4 pairs (*d*) 4 to 6 pairs.
5. Respiration in Chondrichthyes is by
   (*a*) 2 to 4 pairs of gills
   (*b*) 4 to 6 pairs of gills
   (*c*) 3 to 5 pairs of gills
   (*d*) 5 to 7 pairs of gills.
6. Kidney in Chondrichthyes is
   (*a*) Opisthonephric (*b*) Mesonephric
   (*c*) Pronephric (*d*) Metanephric.
7. In Chondrichthyes
   (*a*) Both renal and renal portal system are present
   (*b*) Only hepatic portal system is present
   (*c*) Only renal portal system is present
   (*d*) Portal system absent altogether.
8. In subclass Holocephali gill-opening on either side is
   (*a*) Single covered by operculum
   (*b*) Single uncovered
   (*c*) Two covered by operculum
   (*d*) Two uncovered.

9. Tail in bony fishes is usually
   (*a*) Homocercal (*b*) Heterocercal
   (*c*) Hypocercal (*d*) Hypercercal.
10. In bony fishes, cloaca is
   (*a*) Single (*b*) Absent
   (*c*) One pair (*d*) Two pairs.
11. Aortic arches in class Osteichthyes are
   (*a*) 2 pairs (*b*) 3 pairs
   (*c*) 4 pairs (*d*) 5 pairs.
12. Excretion in class Osteichthyes is
   (*a*) Ammonotelic (*b*) Aminotelic
   (*c*) Ureotelic (*d*) Uricotelic.
13. Internal ears of bony fishes possess
   (*a*) 1 semicircular canal
   (*b*) 2 semicircular canals
   (*c*) 3 semicircular canals
   (*d*) 0 semicircular canal.
14. Subclass Sarcopterygii includes
   (*a*) Lobe-finned fishes
   (*b*) Ray-finned fishes
   (*c*) Cat-fishes
   (*d*) Snake-like fishes.
15. Lung fishes included in the order
   (*a*) Crossopterygii
   (*b*) Dipnoi
   (*c*) Polypteriformes
   (*d*) Acipenseriformes.
16. Members of superorder Chondrostei possess
   (*a*) Placoid scales (*b*) Cycloid scales
   (*c*) Ctenoid scales (*d*) Ganoid scales.
17. *Fistularia* is commonly known as
   (*a*) Flute fish (*b*) Dog fish
   (*c*) Sword fish (*d*) Saw fish.
18. Cow fish is the common name of
   (*a*) *Ostracion* (*b*) *Lophius*
   (*c*) *Diodon* (*d*) *Tetrodon.*
19. *Tetrodon* is commonly known as
   (*a*) Cow fish (*b*) Devil fish
   (*c*) Cave fish (*d*) Puffer.
20. Pipe fish is the common name of
   (*a*) *Syngnathus* (*b*) *Hippocampus*
   (*c*) *Remora* (*d*) *Anabas.*

ANSWERS

1. (*b*)	2. (*c*)	3. (*b*)	4. (*a*)	5. (*d*)	6. (*a*)	7. (*a*)	8. (*a*)
9. (*a*)	10. (*b*)	11. (*c*)	12. (*c*)	13. (*c*)	14. (*a*)	15. (*b*)	16. (*d*)
17. (*a*)	18. (*a*)	19. (*d*)	20. (*a*).				

17

Chapter

Superclass Pisces and Fish : General Account

Contents of this Chapter

The superclass *Pisces* (L., *piscis* = fish) of the truly jawed vertebrates (Gnathostomata) includes all the fishes. The fishes are essentially aquatic forms with paired fins for swimming and gills for respiration. Fishes have undergone extensive adaptive modifications and exhibit an endless variety of forms in different environments. The organisation of the fishes is so diverse that it is difficult to give a complete and comprehensive account of the group.

COMPARISON OF CHONDRICHTHYES AND OSTEICHTHYES

Fishes are usually described under two broad categories depending on the nature of endoskeleton. The fishes having cartilaginous endoskeleton are usually regarded as the cartilaginous fishes or Chondrichthyes (Gr., *chondros* = catilage, *ichthys* = fish), while the bony fishes are called

Osteichthyes (Gr., *osteo* = bony, *ichthys* = fish). The cartilaginous and bony fishes have a common fish-like body organisation, but they show many differences which are summarised in Table 17.1.

TABLE 17.1. COMPARISON OF CHONDRICHTHYES AND OSTEICHTHYES.

Features (Characteristics)	Chondrichthyes (Cartilaginous fishes)	Osteichthyes (Bony fishes)
1. Habitat	Mostly marine	Both marine and freshwater
2. Shape	Usually dorso-ventrally flattened	Usually bilaterally flattened
3. Caudal fin	Heterocercal	Homocercal or diphycercal
4. Pelvic fins	Usually posterior. In male form claspers for transferring sperms into genital tract of female	Usually anterior, sometimes posterior. Claspers absent whenever present not formed by pelvic fins
5. Mouth opening	Ventral on head. Large and crescentic	Terminal on head. Variable in size
6. Gill-openings	Usually five pairs of naked gill-slits. No operculum	5 pairs of gill-slits covered by a lateral flap of skin called operculum, so that a single gill-opening on either side
7. Spiracles	Usually 1st gill-slit become spiracles which open just behind eyes	Spiracles absent
8. Cloaca	Between two pelvic fins lies mid-ventrally common cloacal opening for alimentary canal, urinary bladder and genital products	Cloaca absent. Anus and urinary and genital opertures open separately
9. Exoskeleton	Separate dermal placoid scales	Overlapping dermal cosmoid, ganoid, cycloid or ctenoid scales
10. Endoskeleton	Wholly cartilaginous	Mostly bony
11. Jaw suspension	Hyostylic	Hyostylic and autostylic
12. Stomach	Typically J-shaped	Shape variable. Absent in some
13. Intestine	Short with an internal fold or scroll valve in lumen	Long without scroll valve
14. Rectal gland	Present	Absent
15. Liver	Generally has 2 lobes	Generally has 3 lobes
16. Type of gills	Lemellibranch with long interbranchial septum	Filiform with reduced interbranchial septum
17. Air (Swim) bladder	Absent	Usually present
18. Conus arteriosus	Present in heart	Absent
19. Afferent branchial vessels	5 pairs from ventral aorta to gills	Only 4 pairs
20. Efferent branchial vessels	9 pairs	4 pairs
21. Brain	Primitive with larger olfactory lobes and cerebrum and smaller optic lobes and cerebellum. Cerebral hemispheres undivided	Advanced with smaller olfactory lobes and cerebrum and larger optic lobes and cerebellum. Cerebral hemi-spheres divided–two in number
22. Restiform bodies	Present in brain	Absent
23. Ductus endolymphaticus	Open on top of head	Do not open to exterior
24. Retina	Lacks cones	Cones present
25. Accomodation of eye	Lens moved forward by protractor lentis muscle	Lens moved by retractor lentis muscle

(Contd.)

Features (Characteristics)	Chondrichthyes (Cartilaginous fishes)	Osteichthyes (Bony fishes)
26. Ampullae of Lorenzini	Present	Absent
27. Male genital tract	Connected to anterior genital part of kidney	Not connected to kidney
28. Oviducts	Not connected to ovaries	Connected to ovaries
29. Urinary and genital apertures	United and urinogenital apertures lead into common cloaca	Separate and open indendently to exterior
30. Fertilisation	Internal	External in water
31. Eggs	Few, large with much yolk	Numerous, small with less yolk
32. Development	Internal in ovoviviparous types. Externally inside egg cases in oviparous types	Usually external without an egg case

DIPNOI

The dipnoans, belonging to the order Dipnoi of the subclass Sarcopterygii of class Osteichthyes, are generally called **"lung fishes"**. The name Dipnoi (Gr., *di* = two, *pnoe* = breathing) means "double breathers" as they respire through gills as well as lungs. The dipnoans evolved during the middle Devonian and flourished well in the Permian and Triassic periods. They became rare after Triassic and are represented now by three specialised genera.

Geographical Distribution

Fossil dipnoans appeared in mid-Devonian probably as an offshot of the crossopterygian stem. They flourished moderately in Permian and Triassic and later became rare. At present, the Dipnoi are represented by three genera occurring in widely separated tropical and subtropical freshwater habitats.

There are only 3 living genera of lung fishes, one in each of the 3 continents of the Southern Hemisphere.

1. *Neoceratodus (= Epiceratodus).* Single species *Neoceratodus forsteri* found in Australia in Burnett and Mary rivers of Queensland. A direct descendant of ancient lung fish. Attain a length of 1.5 metres and a weight of 45 kg.

2. *Protopterus.* 4 or so species found in tropical Africa ranging from the rivers Senegal, White Nile and Zambesi and lake Tanganayika. Grow up to 1 or 2 metres and weight about 40 kg.

3. *Lepidosiren.* Single species *Lepidosiren paradoxa* found in tropical South America in river Amazon and its tributaries.

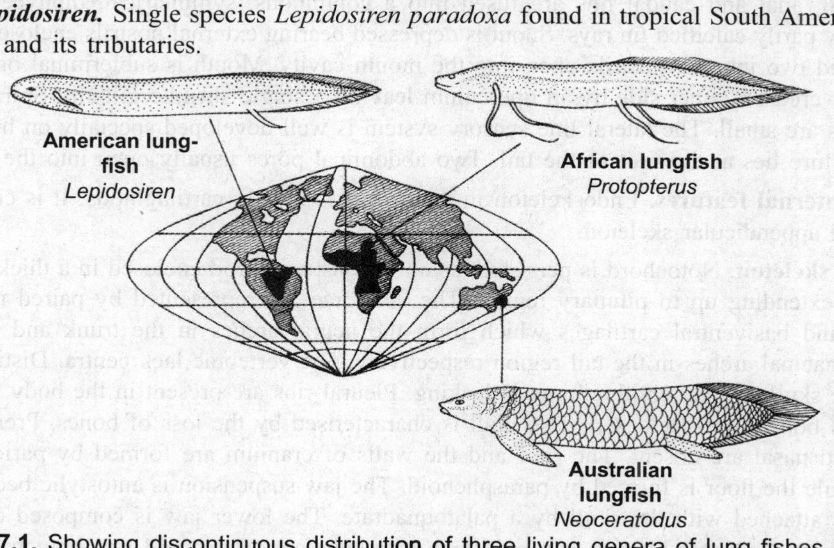

Fig. 17.1. Showing discontinuous distribution of three living genera of lung fishes (Dipnoi).

Habits and Habitat

Lung fishes are large, bizarre fishes, inhabiting semipermanent freshwaters and swamps in Africa, South America and Australia. *Protopterus* lives in large lakes and rivers of tropical Africa. It can survive even when the rivers become completely dried up in summer. It can dig into the mud and aestivate (summer sleep) there for at least six months in a cocoon made out by clay and mucus. During the period of aestivation, nutrition is derived from the stored fat. *Lepidosiren* has also the property of aestivation to escape death during summer. It lives in rivers which become shallow and stagnant in summer, but never dry up completely. It lives mostly at the bottom of the river. *Neoceratodus* thrives well by switching over to pulmonary respiration. When the water is plenty and oxygen is easily available, *Neoceratodus* adopts gill-respiration. During summer when the water-level becomes low and toxic due to the decomposition of organic materials, *Neoceratodus* adopts well to pulmonary respiration.

All the surviving genera of the lung fishes are sluggish in nature and are bottom dwellers. They are carnivorous, although *Lepidosiren* sometimes feeds on plant materials. It lives mainly on snails. *Protopterus* is predaceous in nature. It feeds on worms, crustaceans, insects, frogs and many other small animals. *Protopterus* is often found to attack its own species and bite off portions of the limbs, tail and other parts of the body. The lost parts of the body are regenerated by the these fishes. *Neoceratodus* is also a carnivorous fish which feeds on molluscs, crustaceans and other worms.

In *Lepidosiren* and *Protopterus*, parental care is present. The eggs are laid in muddy nests. The nest is simply a tunnel of about 30 cm deep. The male usually guard the nest till the development is complete. The parental care is highly specialised in *Lepidosiren* because its pelvic fins become specially modified during breeding season. The pelvic fins become greatly enlarged and highly vascularised. The significance of such modification is either to release excess of oxygen to the surrounding water for respiration of the developing youngs or possibly to act as accessory respiratory structure to compensate aerial respiration while remaining around the nest.

Peculiarties

1. External features. Lung fishes are slender fish-like or eel-like creatures, 1 to 2 metres in length. Body is covered by overlapping cycloid scales. Paired pectoral and pelvic fins, often designated as the limbs, are narrow lobe-like or filamentous, with a central axis of flesh and bone. Dorsal, anal and caudal fins are fused into a continuous, symmetrical, diphycercal tail, supported by partly calcified fin rays. Snout is depressed bearing external nostrils enclosed within upper lip and two internal nostrils open into the mouth cavity. Mouth is subterminal or ventral. Gills are covered on either side by an operculum leaving a single slit-like gill-slit. Spiracles are absent. Eyes are small. The lateral line sensory system is well developed specially on head. The cloacal aperture lies at the root of the tail. Two abdominal pores usually open into the cloaca.

2. Internal features. Endoskeleton in dipnonans is mostly cartilaginous. It is composed of axial and appendicular skeleton.

Axial skeleton. Notochord is persistent as an unconstricted rod enclosed in a thick, fibrous sheath and extending up to pituitary region. The vertebrae are represented by paired neural or basidorsal and basiventral cartilages which form the neural arches in the trunk and both the neural and haemal arches in the tail region respectively. The vertebrae lack centra. Distinct joint between the skull and vertebral column is lacking. Pleural ribs are present in the body wall like those of the bony fishes (teleosts). The skull is characterised by the loss of bones. Premaxillae, maxillae and nasal are absent. The roof and the walls of cranium are formed by parietals and frontals, while the floor is formed by parasphenoid. The jaw suspension is autostylic because the mandible is attached with the skull by a palatoquadrate. The lower jaw is composed of paired Meckel's cartilages, tooth-bearing coronoid and supraangulars. The hyoid and branchial arches are all cartilaginous. There are 4 to 6 cartilaginous branchial arches. The branchial arches bear gill-rakers.

Appendicular skeleton. The pectoral girdle consists of a stout cartilage with a pair of investing bones the cleithra and infraclavicle. The pelvic girdle is composed of a cartilaginous plate with a long epipubic and prepubic processes. The skeleton of the fin consists of a central axis of small cartilaginous rods with radials on both preaxial and postaxial sides. This is called an archipterygium.

Digestive system. The teeth form characteristic tooth plates for crushing the molluscan shells. The alimentary canal is a simple tube. The pharynx leads into an oesophagus. The lung-fishes lack distinct stomach. The posterior region of the oesophagus shows slight dilations which is regarded as a part of the stomach by many zoologists. The cavities between the stomach and intestine are separated by a flap-like **pyloric valve**. The intestine is ciliated and contains a **spiral valve** running along the entire length of the intestine and makes about six and a half turns. The liver is a single massive gland. The gall-bladder is large and situated on the left margin of the liver.

Respiratory system. Both gill and pulmonary respiration take place in the lung-fishes. Although the dipnoans possess the gills as well as lungs, they are mostly the lungs. The nostrils help in aerial respiration. The external nostrils lie at the margin of the mouth and the internal nostrils open into the buccal cavity. A slit-like **glottis** is present in the floor of the oesophagus which opens into a short trachea. The trachea passes into the lungs. The glottis is provided with a fibrocartilaginous plate which resembles the epiglottis.

Air-bladder (swim-bladder) is modified into the **"lung"**, which is similar to that of higher vertebrates in structure and function. The internal cavity of lung is produced into numerous **alveoli** which lead into minute alveolar sacs.

Circulatory system. The circulatory system is well developed. The heart is enclosed in a stiff pericardium and is situated somewhat posterior to the gills. The heart of lung fishes consists of four chambers the **sinus venosus**, **auricle**, **ventricle** and **conus arteriosus**. The sinus venosus is incompletely divided into two parts. It opens into the auricle by a broad **sinuaricular opening**. The atrium is divided by a complete septum. Thus, the heart is incompletely divided into arterial and venous channels. Venous blood runs thorugh the right side and oxygenated blood through the left side of the heart. The **atrioventricular opening** is closed or opened by means of a fibrous plug. Pulmonary vein bringing oxygenated blood from the lungs, passes through the sinus venosus and opens into the left auricle. The deoxygenated blood from the rest of the body is poured into the right side of the sinus venosus passes to the right auricle. Ventricle is undivided and leads into the conus which is divided into two channels by means of valves. The arrangement of the valves is such that the oxygenated blood flows through the first two arches into the carotid and cerebral circulation. Venous blood tends to go into the posterior branchial arches and is finally carried to the lungs.

The ventral aorta is small and gives off four pairs of afferent branchial arteries close to the conus. In the arches bearing gills, paired efferent branchial vessels are present and join to form the dorsal aorta. Pulmonary arteries arise from the paired dorsal aortae and carry blood to the lungs.

The conus arteriosus become spirally twisted and the cavity becomes complicated by the presence of valves. A **spiral valve** is present. A pulmonary artery is given off on either side from efferent branchial system. A pulmonary vein opens into left side of auricle. There are two precavals (ductus Cuvieri) and a large posterior vena cava (posterior cardinal vein) as in tetrapods.

Urinogenital organs. The excretory system is like that of *Scoliodon* (dog-fish). It consists of a pair of mesonephric kidneys which discharge into the cloaca through the ureters. In the female, a pair of ovaries are present close to the kidneys. A pair of oviducts open by funnel-like apertures into the abdominal cavity and join at their hinder ends before opening into the cloaca. In the male, a pair of elongated testes are present close to the kidneys. The anterior part of the testis serves for spermatogenesis and the posterior part acts as vesicula seminalis. The two vasa deferentia join to open into the cloaca. Vestigial Mullerian ducts are also present. Vasa efferentia carry sperms through excretory part of kidney in male. Fertilisation is external. Development includes metamorphosis. Larvae may develop suckers and external gills.

Nervous system. The brain of lung-fishes shows many characteristics resembling that of amphibians. A pair of well developed cerebral hemispheres are present in the brain. Cerebellum is small and the optic lobes are fused to form a single oval body. A pineal body is present. The brain, thus, lacks the special features of elasmobranchs or teleosts.

Classification

Order **Dipnoi** is divided into 2 suborders :

Suborder 1. Monopneumona

Lung single. Lateral jointed rays of archipterygium (paired fins) are well developed.

Examples : Living Australian *Neoceratodus forsteri* and extinct *Ceratodus*.

Suborder 2. Dipneumona

Lung double. Lateral rays of archipterygium (paired fins) are vestigial or absent.

Examples : African *Protopterus* and Australian *Lepidosiren paradoxa*.

Affinities of Dipnoi

With special features of their own, the Dipnoi combine characteristics in which they resemble different groups of fishes as well as Amphibia.

1. Affinities with fishes (in general). Lung-fishes are true fishes beyond doubt as they resemble them in general in the following features :

(*i*) Body spindle-shaped and streamlined.
(*ii*) Locomotory appendages fins.
(*iii*) Diphycercal tail fin.
(*iv*) Largely ossified, slender dermal fin rays.
(*v*) Body covered by overlapping cycloid scales.
(*vi*) Notochord persistent.
(*vii*) Vertebrae without centra.
(*viii*) Skull with little ossification and with many investing bones.
(*ix*) 4 to 6 pairs of branchial arches present.
(*x*) Aquatic respiration by gills.
(*xi*) Lateral line sense organs present.

2. Affinities with Elasmobranchii. Lung-fishes resemble the cartilaginous fishes (elasmobranchs) in the following primitive characters :

(*i*) Spiral valve in the intestine.
(*ii*) Similar conus arteriosus.
(*iii*) No nephrostome in the kidney tubules.
(*iv*) Similar diencephalon.
(*v*) Similar female reproductive system.
(*vi*) Each gill-arch with 2 efferent arteries.

The above similarities are not sufficient enough to establish any relationship. The main differences from elasmobranchs are presence of diphycercal tail, opercula and lungs, absence of claspers and external fertilisation.

3. Affinities with Holocephali. According to **Jarvik** (1964, 1967), dipnoans and holocephalians resemble remarkably with each other as follows :

(*i*) Teeth fused into dental plates on jaws.

(*ii*) Gills covered by operculum.

(*iii*) Operculo-gular membranes of both sides fused.

(*iv*) Jaw suspension autostylic.

(*v*) Lack of stomach. Intestine with a spiral valve.

(*vi*) Excurrent nostrils opening into the mouth cavity.

(*vii*) Similar kidneys, gonads and ducts.

(*viii*) Each gill-arch with 2 efferent arteries.

(*ix*) Similar cranial nerves.

Despite such similarities it is extremely difficult to establish any relationship between dipnoans and holocephalians. The main differences from holocephalians are presence of lungs, absence of claspers and external fertilisation.

4. Affinities with Actinopterygii. Lung-fishes resemble subclass Actinopterygii in the following characters :

(*i*) Blunt snout.

(*ii*) Body covered with cycloid scales.

(*iii*) Powerful palatine and splenial teeth.

(*iv*) Presence of inferior jugular veins.

(*v*) Presence of swim-bladder (air-bladder).

(*vi*) Presence of operculum over gills.

However, the Actinopterygii belong to a separate evolutionary line. They have thin, broad fins modified for swimming, and the external nostrils never penetrate into the mouth. Most of them are small in size, have reduced snout, large eyes, single separate dorsal fin and a homocercal caudal fin.

5. Affinities with Crossopterygii. The two orders Dipnoi and Crossopterygii were included under the subclass Sarcopterygii by **Romer** (1959). Instead of specialising for aquatic habitats (as did the Actinopterygii) they have adapted for semiaquatic existence. They show many close similarities as follows :

(*i*) Powerful leg-like lobate fins.

(*ii*) Caudal fin diphycercal.

(*iii*) Presence of cosmine covering the cycloid scales.

(*iv*) Presence of blunt snout with ventral nostrils.

(*v*) Presence of powerful palatine and splenial bones.

(*vi*) Vertebral column reaching up to the end of caudal fin.

(*vii*) Paired inferior jugular veins are present.

(*viii*) Presence of air-bladder (swim-bladder) although modified into lungs with good pulmonary circulation.

(*ix*) Presence of operculum.

(*x*) Contractile conus arteriosus.

(*xi*) Larval forms in some cases with external gills as accessory respiratory organs.

Besides, the fossil crossopterygians (rhipidistians such as Devonian *Osteolepis*) and fossil dipnoans (such as Devonian *Dipterus*) show even closer affinity in :

(*i*) Similar body shapes and sizes (20 to 70 cm).
(*ii*) Separate 2 dorsal, 1 anal and 1 heterocercal caudal fin supported by dermal bony rays.
(*iii*) Paired fins somewhat lobate with a fleshy scaly central axis. Pectoral fins placed high.
(*iv*) Presence of internal nostrils.
(*v*) Cycloid scales modified from cosmoid type.
(*vi*) Similar skull bones.
(*vii*) No vertebral centra.
(*viii*) Similar opercular and gular bones.
(*ix*) Comparable lower jaw.
(*x*) Similar lateral line sensory system.
(*xi*) Presence of swim-bladder (air-bladder) functioning as the lung.
(*xii*) Intestine contains spiral valves.
(*xiii*) Conus arteriosus is contractile.
(*xiv*) External gills are present in the larvae.

This led to the belief that the dipnoans are degenerate descendants of the crossopterygians, which the early dipnoans closely resembled. But **Jarvik** (1968) and others throw doubt upon this belief. According to them, certain structures are more specialised in dipnoans than the crossopterygians. Their basic **differences** are in structural organisation of food crushing apparatus, fin skeleton, vertebral column, visceral skeleton, neural endocranium, snout, division of heart, atrium, blood supply of swim-bladder, etc.

Affinities with Amphibia. Lung-fishes (dipnoans) resemble the amphibians in several features, such as :

(*i*) Semiaquatic and marshy hebitat.
(*ii*) Skin glands multicellular.
(*iii*) Presence of dermal scales in Gymnophiona.
(*iv*) Internal nostrils piercing roof of mouth cavity.
(*v*) Presence of vomerine teeth.
(*vi*) Lungs capable of pulmonary respiration. A pulmonary artery and a pulmonary vein are present.
(*vii*) Spiracles lacking.
(*viii*) Auricle and sinus venosus partially divided into right and left halves.
(*ix*) Conus arteriosus spirally twisted and divided into two by a longitudinal partition.
(*x*) Pericardium is thin-walled.
(*xi*) Ventral aorta short.
(*xii*) Presence of anterior abdominal vein, posterior vena cava.
(*xiii*) Jaw suspension autostylic.
(*xiv*) Brain similar in structure of cerebrum and cerebellum.
(*xv*) Sperms carried through excretory part of mesonephric kidney.
(*xvi*) Similar structure and development.
(*xvii*) Larval stages possess external gills and sucker.

The close similarities led early workers to conclude that lung-fishes (dipnoans) gave rise to Amphibians, a view no longer held now-a-days. According to **Dolo**, these similarities probably were due to convergent evolution and have possibly from adaptive modifications on account of similar habits and habitat. On the other hand, lung-fishes have many specialised features by which they **differ** from amphibians. These are as follows :

(*i*) Lobate fins instead of limbs for locomotion.

(*ii*) Fin skeleton not like that of primitive tetrapods.

(*iii*) Skull mainly cartilaginous with little ossification.

(*iv*) Presence of characteristic tooth-plates instead of teeth.

(*v*) Maxillae and premaxillae absent.

(*vi*) Some anterior vertebrae fused with skull.

(*vii*) Lungs located dorsal to gut.

(*viii*) Urinary bladder develops from dorsal wall of cloaca in Dipnoi but from ventral wall in Amphibia.

Phylogeny and Significance

The origin, evolution and affinities of Dipnoi are still an unsolved problem due to diverse opinions. They combine characteristics in which they resemble almost all the other groups of fishes as well as Amphibia. Fossil primitive Dipnoi (*e.g.*, *Dipterus*), shows greater similarity with fossil crossopterygians (*e.g.*, *Osteolepis*), than do their living members. During the course of their evolution, the modern Dipnoi have undergone several changes or specialisations which are as follows :

(*i*) Anterior dorsal fin was reduced and eventually lost.

(*ii*) Remaining median fins elongated and fused so that originally heterocercal tail became symmetrically diphycercal.

(*iii*) Reduction in number of dermal plates or bones of skull and operculum.

(*iv*) Thick bony cosmoid scales modified into thin cycloid scales.

(*v*) Extensive sheet of cosmine covering head and body was lost.

(*vi*) Fusion of conical teeth to form characteristic crushing tooth-plates with ridges to feed effectively upon shell fish.

(*vii*) Air-bladder became functional lungs not unlike those of higher vertebrates.

Evidence indicates that Dipnoi are degenerate descendants of Crossopterygii which early dipnoans resembled. **Romer** (1945) thought that Dipnoi and rhipidistian Crossopterygii had a common ancestor. On the other hand close similarity between Dipnoi and Amphibia led early workers to conclude that dipnoans gave rise to amphibians, a view no longer held now. However, it is universally accepted that Amphibia have originated either directly from rhipidistian Crossopterygii or from some common ancestor.

FIN SYSTEM

Fins are the chief organs of locomotion in fishes. These are either folds of skin or projections from the body surface. The fins are supported by fin-rays. These supporting rays may be bony, cartilaginous, fibrous or horny. There are mainly two types of fins in fishes : (1) **Unpaired** or **median fins** and (2) **Paired lateral fins**.

1. Unpaired median fins. These include 1 or 2 **dorsal fins** along mid-dorsal line, a ventral **anal fin** behind anus or vent (cloaca) and a **tail** or **caudal fin** around the tip of tail. Dorsal fins may be in a series or reduced or absent. Anal fins may be absent especially in bottom dwellers.

2. Paired lateral fins. The paired fins are the pectorals and pelvics corresponding to the fore- and hindlimbs of the terrestrial vertebrates. These include **pectoral fins** anteriorly and

pelvic fins posteriorly. Pelvic fins are called thoracic when placed below the pectoral fins and abdominal when situated just in front of anus. In some cases they are absent.

Both the unpaired median and paired fins are supported by skeletal rods called radials and dermal fin rays. A great variety of fins is observed in fishes. The diversity in fin system in fishes is due to their adaptive responsiveness.

Use of fins. Fish swim mainly by lateral movement of tail and tail fin. Other fins are mainly used as steering devices and rudders. When the body is at rest, the paired lateral fins serve to maintain equilibrium. Fins are also modified to serve other purposes. Lung-fishes use them as legs in walking. The flying fish use their large and extended pectoral fins for gliding. Pelvic fins in some male Chondrichthyes become modified as claspers. In remoras, anterior dorsal fin forms an adhesive disc or sucker on head. In some teleosts the anal fin forms an intromittent organ or ovipositor.

Origin of Fins

The first chordates lacked paired appendages but it is universally accepted that tetrapod limbs arose during evolution from the fins of fishes. However, clues to the origin of fins themselves have remained obscure and unresolved. The median or unpaired fins in fishes are believed to have originated from a continuous fold of tissue. This fold extends from the posterior region of the head and continues posteriorly around the tail and extend up to anus. This fold of skin is supported by a series of parallel cartilaginous rods. During the course of development each supporting rod divides into a lower **basal** piece, embedded in the body wall and an upper **radial** piece, lying in the fin fold. From such a continuous fin-fold, the dorsal, caudal and anal fins have been evolved by restriction of radials at certain areas and the progressive degeneration of fold between them. Several theories have been advanced to explain the source and manner of origin of fins (Figs. 17.2 and 17.3).

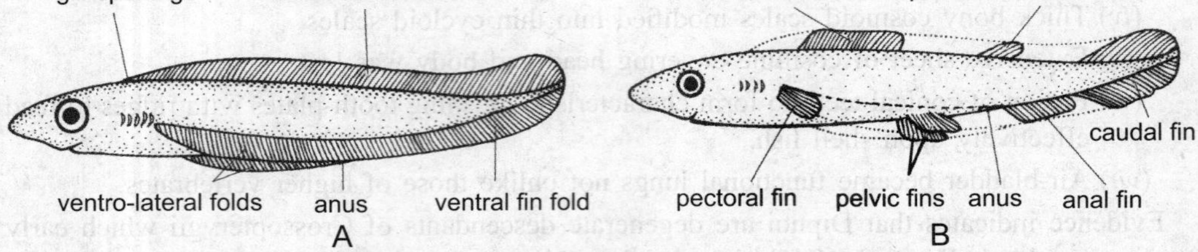

Fig. 17.2. Diagram illustrating the origin of fins according to Balfour's fin fold theory.

1. Gill-arch theory. According to the gill-arch theory of **Gegenbaur** proposed in the ninteenth century, the paired fins are modified gill structures, the girdles represent the gill-arches and the fin folds with their skeleton represent gill-flaps or septa. The position of the pelvic fins can be explained on the assumption that some of the posterior gill-arches have been shifted posteriosly. This idea finds no support from morphological, embryological and palaeontological studies.

2. External gill theory. According to **Graham Kerr**, paired fins and their skeleton are derived from external gills which occur temporarily in some larval forms. However, this theory has no supporting evidence.

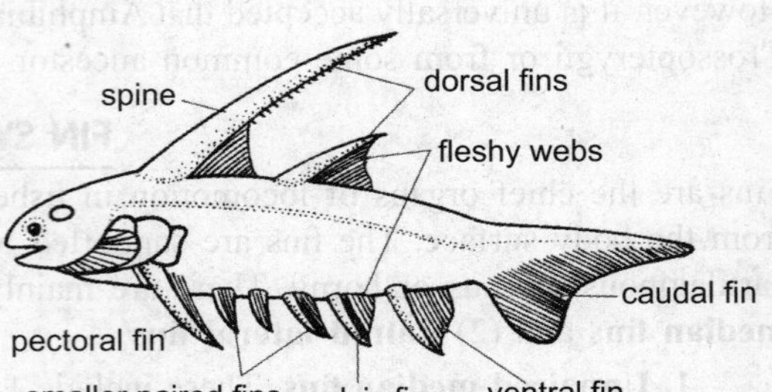

Fig. 17.3. Schematic derivation of fins according to the fin-spines theory.

3. Fin-fold theory. This theory was first proposed by **Balfour** and **Thacher** and later supported by **Mivart**. A number of eminent vertebrate morphologists like **Wiedersheim**, **Haswell**, **Regan**, **Sewertzoff Osburn** and **Goodrich** have presented several evidences in support of this theory. According to this theory, the paired fins have originated from a paired **lateral** continuous **fin-folds** running down each side of the body behind the gill-openings up to the end of the tail. These lateral fin-folds were separate in the anterior part but fused in the caudal region behind the anus, forming a ventral fin-fold. Besides these, there was a dorsal median fold. The condition of the lateral fin-folds in the ancestral forms was presumably like the metapleural folds of *Branchiostoma*. There are many palaeontological and embryological evidences in favour of origin of the paired fins from the lateral fin-folds. Some of the evidences are : (*i*) Basic skeletal structures of paired and unpaired fins are similar indicating a common mode of origin. (*ii*) In the embryos of certain elasmobranchs, early stages show continuous series of muscle bands which disappear, except those forming the isolated paired and unpaired fins. (*iii*) In some extinct Devonian acanthodian sharks, a row of numerous small accessory spiny fins on either side between pectoral and pelvic fins, have been interpreted as remnants of fin-folds. (*iv*) In the extinct shark *Cladoselache*, the pectoral and pelvic fins are broad, without basal notches and supported by parallel cartilaginous rods (pterygiophores) suggesting a primitive fin-folds-like origin. (*v*) In *Climatius*, there was a series of smaller spines and finlets between the pectoral and pelvic fins. This suggests that they were originally joined by a continuous fold.

The idea of the fin-fold origin of the paired and unpaired fins in fishes is supported by anatomical, embryological and palaeontological evidences. This theory was accepted by most of the workers in this field of research.

4. Fin spines theory. According to this theory, recent workers claim that the so-called "several pairs of fins" in acanthodians indicate nothing but a special kind of multiplication of the defensive spines, which may often support membranous structures. Subsequently all spines and webs may have been lost except one anterior and one posterior pair which later evolved into pectoral and pelvic fins. Furthermore, no true primitive fossil fish with completely continuous fins is still known. This observation compels the recent workers to discard the fin-fold theory. They suggest that the fin appeared probably in connection with some paired and unpaired median spines, as found in many ostracoderms. In course of evolution membranous structures appeared between these spines and the adjacent body wall.

5. Ostracoderm theory. Some ostracoderms possessed lateral fleshy lobes, projecting from either lateral side. They may have been the precursors of pectoral fins. Other ostracoderms had ventro-lateral rows of dermal spines similar to extra fins of acanthodians. Most of these spines were lost while some were retained in the pectoral and pelvic regions, thus, accounting for the origin of paired fins (and limbs) from ostracoderm ancestors.

Caudal Fins or Tails

Of the unpaired fins, the caudal fin is well developed in most fishes because it plays the most important role in forward propulsion during swimming. The caudal fin in most of the fishes is highly developed with few exceptions like *Hippocampus* and some of the eels. In *Hippocampus* (sea horse) the tail is prehensile. In bottom dwelling rays, the caudal fin tends to become reduced and in the sting rays the caudal fin is lacking. The caudal fin exhibits different shapes in different groups of fishes correlated with their habits, so that it is of great importance in classification of fishes. Three main types of caudal fins are found in fishes. These are : diphycercal (protocercal), heterocercal and homocercal (Fig. 17.4).

1. Diphycercal (Protocercal). This type of caudal fin is regarded to be the most primitive type. It is not exhibited by many living fish. The vertebral column extends up to the tip of the tail and divides the caudal fin into two equal halves. The dorsal half is called **epichordal lobe** and the ventral half is known as **hypochordal lobe**. The epichordal and hypochordal parts of the caudal fin are equal in size and symmetrical. Diphycercal caudal fin occurs in modern cyclostomes, primitive sharks, Holocephali (*Chimaera*), living Dipnoi (lung-fishes), living crossopterygii (*Latimeria*), many larval teleosts and deep sea fishes. In *Chimaera* and some deep sea fishes, the caudal fin is called **isocercal** (Gr., *iso* = equal) which is very much elongated and symmetrical. *Latimeria* and extinct coelacanths have unique symmetrical three-lobed tail with a marked median lobe. In the living dipnoans, the caudal fin is of **protocercal** type. Presence of protocercal tail in some highly developed fishes is possibly due to its secondary modification.

2. Heterocercal. It is the intermediate type in which vertebral column bends upwards and reaches up to the tip of the more prominent dorsal lobe, thus, making the caudal fin strongly asymmetrical (Gr., *heteros* = different, other). This type of caudal fin is found in modern elasmobranchs, extinct osteolepid crossopterygian (*Osteolepis*), extinct dipnoans (*Dipterus*) and living holosteans (*Acipenser*, *Polyodon*). Heterocercal caudal fin is characteristic of bottom feeders with ventral mouth and without swim-bladder (air-bladder). The strokes of larger dorsal lobe in swimming serve to direct fish towards bottom.

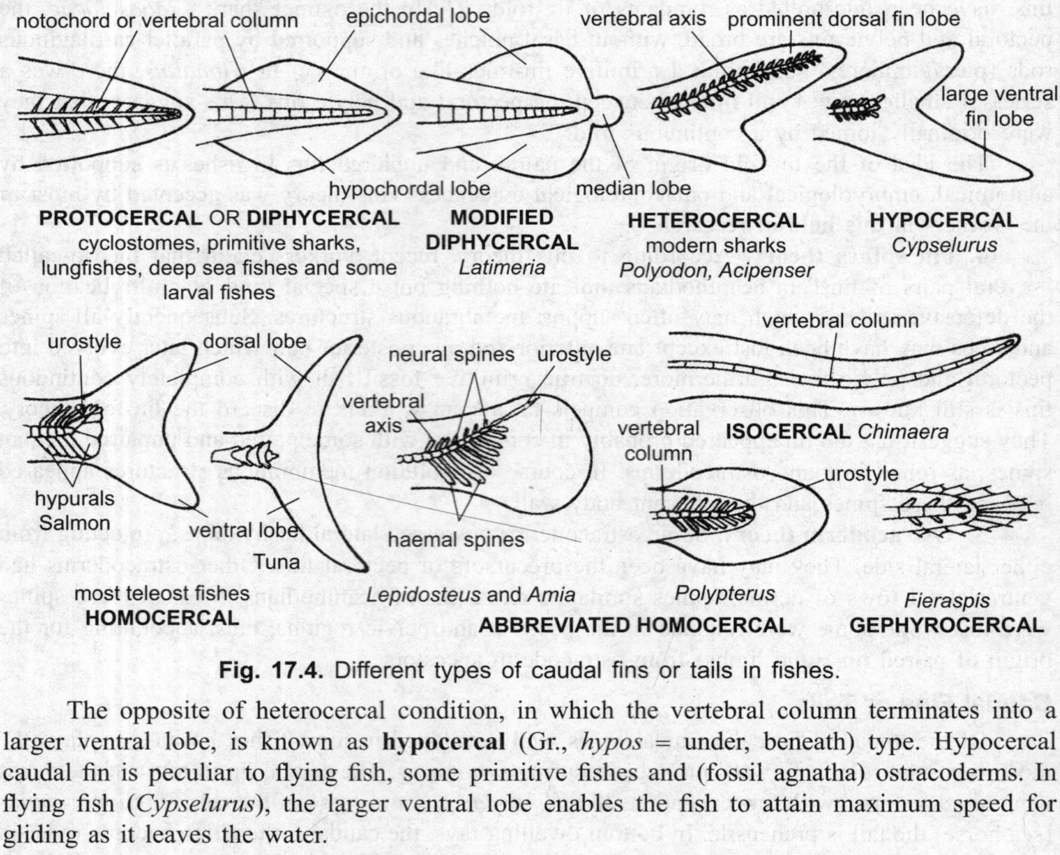

Fig. 17.4. Different types of caudal fins or tails in fishes.

The opposite of heterocercal condition, in which the vertebral column terminates into a larger ventral lobe, is known as **hypocercal** (Gr., *hypos* = under, beneath) type. Hypocercal caudal fin is peculiar to flying fish, some primitive fishes and (fossil agnatha) ostracoderms. In flying fish (*Cypselurus*), the larger ventral lobe enables the fish to attain maximum speed for gliding as it leaves the water.

3. Homocercal. This is the advanced and most common type of caudal fin. The homocercal (Gr., *homos* = common, alike) caudal fin is the characteristic of the higher bony fishes (teleosts). It is symmetrical externally but internally it is asymmetrical. The posterior end of vertebral column is turned upwards and becomes greatly reduced. The tip of vertebral column do not reach the posterior limit of fin. There is no apparent dorsal lobe but the ventral lobe is greatly enlarged and divided into two equal superficial lobes. Most of the teleosts retain the typical **homocercal** condition, but in some forms this condition is slightly modified.

Homocercal type has several variations. In cod and tuna, the homocercal condition is modified. The upward terminal portion of vertebral column is withdrawn. In some deep sea fishes, the terminal part of vertebral column is straight and greatly elongated to form **isocercal** caudal fin or tail. In *Fieraspis* and *Orthagoriscus*, the caudal fin completely disappears forming a **gephyrocercal** caudal fin. In chondrosteans (*Polypterus*, *Amia*, *Lepidosteus*) the vertebral column is a bit elongated and upturned, also with or without an upturned fleshy lobe. This is called **abbreviated homocercal** caudal fin. Most of these variations represent an intermediate stage between heterocercal and homocercal types.

The protocercal or diphycercal type of caudal fin is considered as the most primitive type, the heterocercal as the intermediate stage and the homocercal condition represents the advanced stage. In many teleosts, the caudal fin starts as diphycercal then becomes heterocercal and finally assumes the homocercal condition. The transition between three types of caudal fins in the developmental history of fishes is significant from the phylogenetic point of view.

SCALES OF FISHES

In many vertebrates, the exoskeletal covering of body is made of two types of scales : epidermal and dermal. **Epidermal scales** are cornified derivatives of the Malpighian layer of epidermis. They are well developed in terrestrial vertebrates such as reptiles, birds and mammals. **Dermal scales** are mesenchymal in origin and especially developed in the fishes. They are small, thin, cornified, calcareous or bony plates which fit closely together or overlap. As regards the arrangement of scales on fish body is concerned, they are most often imbricated and thus, overlap like shingles on the roof, with their free margins directed towards the tail, so as to minimise the friction of water. But some times total reversal of the pattern of arrangement is seen in some fishes. Among barbot (*Lota*) and freshwater eel (*Anguilla*) the pattern is mosaic rather than overlapping one another, they are separated minutely or meet their neighbours only at their margins. Scales vary in size and shape in different species. The body of all fishes except members of family Siluridae and a few bottom dwellers is covered by scales. Scales are usually found covering entire body surface, but in some fishes such as *Chimaeras*, *Polyodon* and *Acipenser* the scales are present in localised regions or areas of the body.

Primitive fossil fishes of Silurian and Devonian periods possess exoskeleton in the form of plates and scales which consist of three distinct layers. The innermost layer consisted of a compact bone, **isopedine**, the intermediate or middle layer of **spongy vascular bone** and the outer layer is of **dentine**.

According to the mode of the their origin, there are two types of scales : (*i*) those which are formed due to the secretory activity of both epidermis and dermis, as the placoid scales of elasmobranchs and (*ii*) non-placoid scales that are derived from the dermis only as the scales of teleosts. Structurally, the scales are classified as cosmoid, ganoid or rhomboid, placoid, cycloid and ctenoid, the last two are also called the bony ridge scales.

1. Cosmoid scales. The cosmoid scales are found in the extinct Crossopterygii and Dipnoi. The external layer of scale is thin and enamel-like and is called the **vitrodentine**. The middle layer is made up of hard, non-cellular, dentine-like material called the **cosmine** and contains a number of branching tubules and chambers. The inner layer is made up of vascularised bony substance, **isopedine**. These scales grew by the addition of new isopedine material from below, along the edges. Cosmoid scales are not found in the living fishes. The living Dipnoi possess thin cycloid scales and have lost the cosmine layer.

2. Ganoid scales. These scales are characteristic of the primitive actinopterygians called the ganoid fishes, and are of various forms and structure. These scales are heavy and have an outer layer of hard inorganic, enamel-like material called **ganoine**. The middle layer is **cosmine**

containing numerous branching tubules. The innermost layer is thickest and is made up of lamellar bone, **isopedine**. These scales grow by the addition of new layers to lower as well as upper surface. These scales are usually rhomboid in shape and articulate by peg and socket joints.

The ganoid scales are best developed in the chondrosteans (*Polypterus*, *Acipenser*) and holosteans (*Lepidosteus*) so that these are often called **ganoid fishes**. In *Acipenser*, they are in the form of large, isolated, bony scutes and are present in five longitudinal rows at places where there is greater wear and tear. In *Lepidosteus*, the scales are hard, polished, rhombic plates, fitting edge to edge, thus, forming a complete armour. Scales of this fish have lost the middle cosmine layer, thus, reducing their thickness and rigidity. The scales of *Amia* are even thinner, do not possess the ganoine layer and resemble typical cyloid scales.

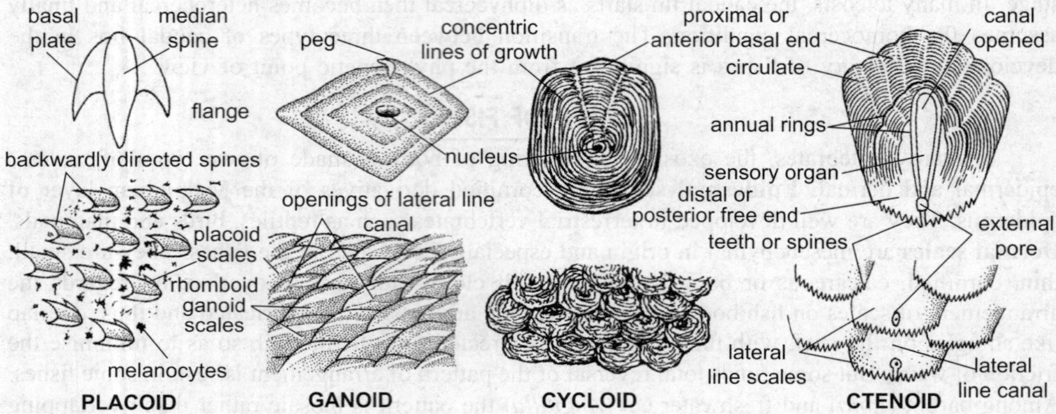

Fig. 17.5. Different types of dermal scales found in fishes. Lower row shows parts of skin with numerous scales. Upper row shows single scales.

3. Placoid scales. These scales are characteristic of elasmobranch fishes (sharks) only. Each scale has a disc-like basal plate embedded in the dermis and a spine projecting out through the epidermis. In structure, a placoid scale resembles a tooth. The spine has an external covering of enamel-like, hard, transparent material called vitrodentine. This is followed by a layer of dentine enclosing a pulp cavity from which several branching dentine tubules radiate in different directions. The centre of the basal plate is perforated by an aperture to provide entrance to the blood vessels and nerves from the dermis. The placoid scales are closely set in skin but do not overlap each other and giving a sand paper-like quality to the skin.

4. Cycloid scales. The cycloid scales are thin, flexible, transluscent plates, rather circular in outline, thicker in the centre and marked with several concentric lines of growth which can be used for determining the age of the fish. They are found in a large number of teleostean fishes having soft rayed fins, such as *Labeo*, *Catla*, *Barbus*, *Cirrhina*, etc. The central part of the scale is called the focus and is the first part to develop. In many species, oblique grooves or radii run from the focus towards the margin of the scale. They form a protective covering over the skin and project diagonally in an imbricating pattern. The part of the scale which is exposed to view *in situ* condition (posterior area) generally shows less distinct ridges or circuli and chromatophores are also sometimes attached to it. The anterior area lies embedded in the skin.

5. Ctenoid scales. The ctenoid scales are also circular and can be distinguished from the cycloid scales by having a more or less serrated free edge. Moreover, several spines are present on the surface of the posterior area of the scale. These scales are characteristic of modern higher teleosteans such as perch, sunfish, etc. They are more firmly attached and their exposed free hind parts which are not overlapped, bear numerous small comb-like teeth or spines. Intermediate types between cycloid and ctenoid scales are also found on different parts of the body. Certain fishes, such as flounders, may bear both types, ctenoid scales dorsally and cycloid scales ventrally.

Modifications and variations of scales. Some fishes are totally scaleless or naked such as *Torpedo* (electric-ray) and catfishes. In eels, they are minute and deeply embedded in dermis. In some fishes (chimaeras) they become localised. Jaw teeth in the sharks are regarded as modified scales. The teeth forming the saw of *Pristis* and the sting-ray are modifications of the scales. The scales at the base of tail in the sturgeon fish, *Acanthurus*, are modified to forming sharp cutting blades. In coffer fish, *Ostracion*, the scales form polygonal bony plates articulating with one another to form a rigid protective box. In the globe-fish, *Tetrodon* and the porcupine fish, *Diodon*, the scales are elongated to form spines for protection. The jointed fin-rays, leptotrichia of the bony fishes are also considered to be modified scales. In the basking shark, *Cetorhinus*, the scales become modified to form gill-rakers. The scales along the lateral line of the fish become perforated to provide an exit for the canal. In the sea-horse, *Hippocampus*, and the pipe-fish, *Syngnathus*, scales are fused to form protective bony rings around the body.

Uses of scales. Scales form a protective covering of exoskeleton on the body. The structure of scales is useful in broad classification of the group. Thus, among the fishes, placoid scales are characteristic of Chondrichthyes, ganoid scales are found in the primitive bony fishes, while the cycloid and ctenoid scales are found in the higher teleosts. Sometimes, the scales are useful in separating orders and families. Thus, siluroid fishes are scaleless and can be distinguished from cyprinoids. Scales form an important source of information about fossil fishes and in classifying kitchen middens of extinct men. Scales are useful in studying the food habits of piscivorous animals. In taxonomic work, several species are identified by counting the scales along the lateral line and round the body.

Cycloid and ctenoid scales are of considerable help in calculating the age of and growth rate of fishes. Many species undergo seasonal growth which is apparent from the lines of growth on the scale. In some species like *Salmo*, spawning marks can be seen on the scales, so that it is possible to find how many times a fish has spawned.

Scales grow throughout life in size with the fish. Growth results in concentric lines which make age determination possible in salmon, trout, bass and several other species. For every species, its scale pattern is rather constant. Thus, arrangement, number, form and structure of scales play an important role in identification and classification of fish species.

AIR-BLADDER OR SWIM-BLADDER

The air-bladder or swim-bladder is more or less a sac-like structure lying between the alimentary canal and the kidneys. It is a characteristic organ of Osteichthyes (bony fishes). It is a gas-filled pneumatic sac, called air-bladder or swim-bladder. Air-bladder does not occur in elasmobranchs. However, it is found in all Osteichthyes (bony fishes) except a few bottom dwellers (*Lophius*, *Pleuronectes*, etc.). It is vestigial in *Latimeria*, the only living crossopterygian. Air-bladder shows a number of structural modifications in various groups of bony fishes.

Origin of air-bladder. The air-bladder is generally considered to be homologous with the lungs of Dipnoi, Osteichthyes and Tetrapods. Opinions differ as regard the development of air-bladder in fishes. In teleosts, it originates as an unpaired dorsal and dorso-lateral diverticulum of the oesophagus. The diverticulum with an opening in the oesophagus becomes subsequently divided into two halves. The left half often atrophies and the right half becomes well developed and take a median position. In dipnoans and Polypteridae (*Polypterus*), the air-bladder is modified into lungs and originates as the downgrowths from the floor of the pharynx. These outgrowths have been rotated around the right side of the alimentary canal to occupy the dorsal position. As a consequence of shifting of the position, the original right lung becomes the left one.

But the air-bladder arises from the dorsal wall of the foregut, whereas the lungs develop from the ventral wall of the pharynx. They differ in blood supply and functions also. **Goodrich** (1930), concludes that both these structures have developed from the posterior pair of gill-pouches. According to **Jones** (1957), air-bladder of teleosts may have evolved independently of the lungs. **Spengel** advocates the view that the air-bladder in fishes originates from the posterior pair of gill-pouches, but definite embryo-logical evidence in support of this idea is lacking.

Generally speaking, the air-bladder arises as an outgrowth from the oesophageal region of the alimentary canal. It shows a great diversity in mode of development, structure and function in different fishes. It lies ventral to alimentary canal in *Polypterus*, laterally in Dipnoi and dorsally or dorsoventrally in teleosts (Fig. 17.6).

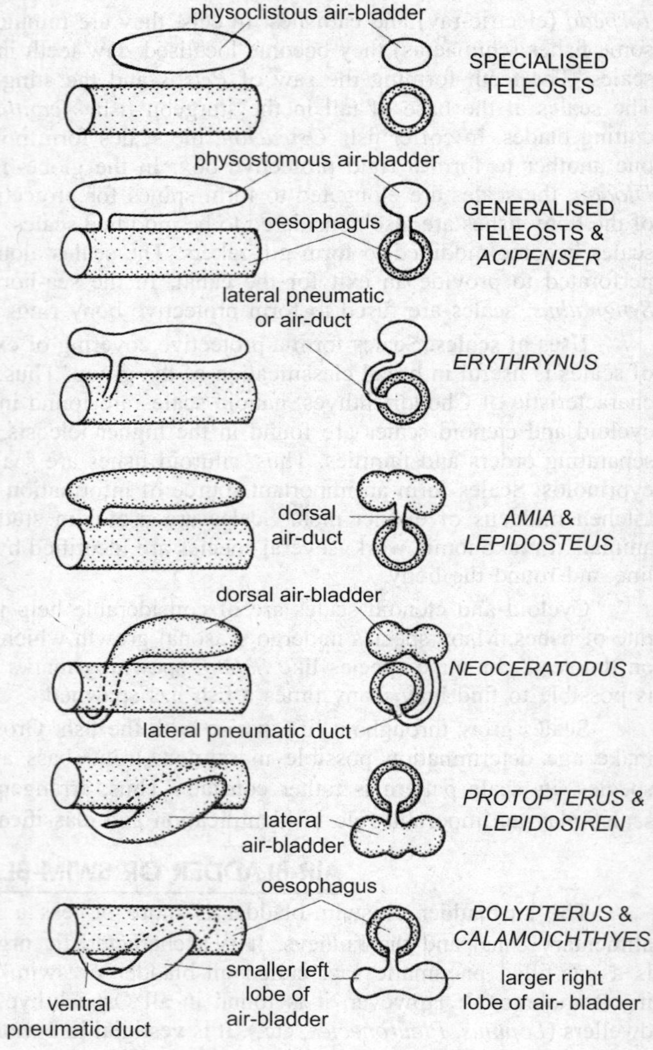

Fig. 17.6. Various types of air-bladders and lungs in fishes shown in L.S. (left side) and T.S. (right side).

Basic structure of air-bladder. The air-bladder in fishes varies greatly in structure, shape and size. It is essentially a tough sac-like structure with an over-lying capillary network. Beneath the capillary system there is a connective tissue layer called **tunica externa**. Below this layer lies the **tunica interna** consisting primarily of smooth muscle fibres and epithelial gas-gland. The air-bladder may be composed of one or two or many chambers. In teleosts, the air-bladder lies below the kidneys between the gonads and above the gut. The connection with the oesophagus may be retained throughout life or may be lost in the adult. **Blot** (1807) and **Morean** (1876) have shown that the gas secreted by the air-bladder is mostly oxygen. Nitrogen and little quantity of carbon dioxide are also present.

The air-bladder receives its blood from branches of the coeliaco-mesenteric artery or directly from the posterior branches of dorsal aorta. The venous blood is then drained into a vessel that joins the hepatic portal system, while in some species the air-bladder vein joins the posterior cardinal vein. The air-bladder also shows differences in its degree of vascularity in various teleosts and in the formation of **'red bodies'** or **'red glands'**. In some species (Clupeidae and Salmonidae), the capillaries are uniformly distributed all over the surface of the air-bladder and do not form a **'retia mirabilia'**, while in other Physostomes as the carps (*Cyprinus*, *Labeo*, *Tortor*) the blood vessels are arranged in a fan-like manner and are concentrated at one or more points on the inner surface of the air-bladder, forming red masses of various shapes, called the **'red bodies'**. These are essentially **'retia mirabilia'** consisting of numerous arterial and venous capillaries, running parallel to one another and carrying blood to and from the gas gland. These constitute a **'wonder net'** of capillaries which do not communicate until they reach the epithelium of the air-bladder. In the physostomous fishes, this structure is more primitive, being covered with a simple flat epithelium and is called the **'red body'**. In the physoclistous fishes, the capillaries are covered with a thick glandular folded epithelium and is called the **'red gland'**. Among the Physostomi, the eels (*Anguilla anguilla*) shows a close resemblance to the condition found in physoclisti.

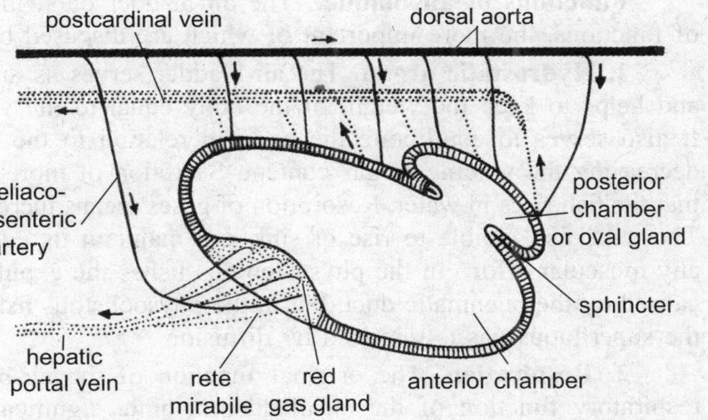

Fig. 17.7. Structure and blood supply of a ductless teleostean air-bladder in section.

Gases in the air-bladder come from blood secreted by the red gland. The posterior chamber is thin-walled and forms the oval gland which permits reabsorption of gases by the blood. Secretion and resorption are under the control of autonomic nervous system. In physostome fishes with air-bladder connected to pharynx by a duct, air can also be gulped or bubbled through the mouth (Fig. 17.7).

Types of air-bladder. Depending on the presence of the duct (**ductus pneumaticus**) between the air-bladder and the oesophagus, the air-bladder in fishes can be divided into two broad categories : **physostomus** and **physoclistous** types. Depending on the conditions of air-bladder, the teleosts are classified by older taxonomists into two groups : **Physostomi** and **Physoclisti**. A transitional condition is observed in eels.

Condition in ganoids and dipnoi. In most primitive fishes, the air-bladder serves as an accessory respiratory organ or lung which seems to have been its original function. In *Polypterus* (Chondrostei), one of the most primitive bony fishes living today, it is a smooth walled bilobed sac with a short left and a long right lobe, opening on the floor of the pharynx below the gill-slits. The two sacs are fused into one at the proximal end and the opening (glottis) is provided with a muscular sphincter. In *Acipenser*, the air-bladder is short and oval in shape, its walls are smooth and the glottis is a wide opening into the oesophagus. In the holosteans *Amia* and *Lepidosteus*, the air-bladder is a single large and highly vascular lobe, lying dorsal to oesophagus and its pneumatic duct also opening dorsally into oesophagus. Thus, in *Amia* and *Lepidosteus*, the air-bladder serves as both a respiratory organ (lung) and a hydrostatic organ. In the Dipnoi (lung-fishes), the air-bladder resembles closely with the lung of amphibians in structure. In *Protopterus* and *Lepidosiren*, the condition is similar but the two lobes are equal in size and have a thick, vascular walls with alveoli and septa. In the Australian lung fish *Neoceratodus*, air-bladder consists of a single lobe lying dorsal to the gut, but still connected ventrally to the oesophagus by a narrow pneumatic duct, passing down along the right side of the gut.

Condition in teleosts. The air-bladder, although present in many teleosts, is completely absent in several groups of fishes as the flat fishes (Pleuronectiformes), Giganturiformes, Saccophyringiformes, Pegasiformes, Echeineiformes, Gobeisociformes and Symbranchiformes. The air-bladder assumes various shapes in different teleosts and may be tubular, fusiform, oval, heart-shaped, dumb-bell shaped or horse-shoe shaped. The air-bladder in higher teleosts or bony fishes is the most specialised, playing little or no part in respiration, and primarily serving as a hydrostatic organ. Two types of air-bladders are known. In the more generalised groups of teleosts, (salmon, eel) the air bladder retains connection with the gut via a pneumatic duct, just as in ganoids and dipnoi. Such an open air-bladder is called **physostomous**. In the teleost *Erythrinus* the air-bladder has a lateral attachment to gut. In the more specialised teleosts (perch, cod) the duct becomes atrophied or lost. Such a closed or ductless air-bladder is called **physoclistous**.

Functions of air-bladder. The air-bladder or swim-bladder in fishes performs a number of functions, the more important of which are discused below :

1. Hydrostatic organ. The air-bladder serves as an important hydrostatic organ in fishes and helps to keep the weight of the body equal to the volume of the water the fish displaces. It also serves to equilibrate the body in relation to the surrounding medium by increasing or decreasing the volume of gas content. Secretion of more gases means lower specific gravity so that the fish rises in water. Resorption of gases means increased specific gravity and the fish sinks. Thus, the fish is able to rise or sink and maintain its equilibrium or position in water without any muscular effort. In the physostomous fishes the expulsion of the gas from the air-bladder is caused by the pneumatic duct, but in the physoclistous fishes where the pneumatic duct is absent the superfluous gas is removed by diffusion.

2. Respiration. The original function of the air-bladder was probably respiratory. The respiratory function of the air-bladder is quite significant. A cellular lung-like air-bladder is present in the teleostomes such as *Polypterus*, *Amia* and *Lepidosteus* in which it is primarily respiratory in function. In Dipnoi, the air-bladder resembles closely with an amphibian lung. In certain teleosts as *Megalops*, *Chirocentrus*, *Gymnarchus*, *Erythrinus*, *Umbra* and some cyprinoids, the air-bladder works like a lung. These fishes live in swamps or pools where the carbon dioxide tension is high and that of oxygen is low. These teleosts come to the surface, swallow air and pass it back to the air-bladder which is highly vascular. It appears that the lung-like function has been reacquired in these teleosts as a special adaptation for living in foul water.

Air-bladder can also be used for storing oxygen to be used in emergency. This has been demonstrated both in physoclistous and physostomous fishes as *Perca fluviatilis*, *Tinca tinca*, *Opsanus tau*, *Cyprinous carpio*, *Carassius auratus*, etc. Although it has been shown that in many species the amount of oxygen stored in the bladder would enable the fish to survive for a few minutes only, the air-bladder may be more important in deep water fishes where the amount of oxygen stored is much higher. Studies have shown that in fishes living nearer the surface the gas in air-bladder is much like the air, while in those living at greater depths, oxygen forms a major component, extending up to 75% in some species.

3. Sound production. The air-bladder plays an important role in sound production. Some fishes are able to produce sounds with the gases inside air-bladder by the use of special muscles attached to the air-bladder. But the actual mechanism is not understood. Many fishes as *Doras*, *Platystoma*, *Malapterurus*, *Trigla* can produce grunting or hissing or drumming sound. The circulation of air inside the air-bladder causes the vibrations of the incomplete septa, which in turn, produce sound. The sound may also be produced by compression of extrinsic and intrinsic musculature of air-bladder. *Polypterus*, *Protopterus* and *Lepidosiren* can produce sound by compression and forceful expulsion of gases in the air-bladder. In *Cynoscion* male, the compression of the air-bladder is achieved by contraction of specialised muscles, **musculus sonorificus**. The sound production serves to startle the enemies or to attract mates.

4. Auditory function. Air-bladder serves to transmit sound waves to the ear especially in the Ostariophysi, more efficiently than in the species in which a connection with the ear is missing. In Cypriniformes, a series of small bones, the Weberian ossicles, connect the air-bladder and perilymph cavity containing internal ear. Low frequency vibrations of the contained gas, induced by noises in water, are transmitted by the ossicles to the membranous labyrinth. Thus, these fishes can hear.

5. Sensory function. When the fish is subjected to pressure changes by moving into different depths of water, compression of the wall of the air-bladder which functions as a pressure receptor like a manometer, barometer or a hydrophone. It has been suggested that the air-bladder acting as a sense organ enables a fish to maintain a steady depth. If the fish moves above or below a certain depth at which it is in equilibrium, changes in the tension in the bladder wall lead to compensatory swimming movements and the fish returns to the original level. However, there is not much evidence in support of this hypothesis.

In Cypriniformes (Ostariophysi), the connection of the air-bladder (gas bladder) with the membranous labyrinth through the Weberian ossicles, increases the sensitivity of the fish to volume changes in the air-bladder and the fish is able to control the escape of gas via the pneumatic duct.

ACCESSORY RESPIRATORY ORGANS

Generally adult fishes depended chiefly on pharyngeal gills for aquatic respiration. However, other devices also occur to supplement or replace gill respiration. All such additional respiratory organs, other than gills, are known as **accessory respiratory organs**. The development of accessory respiratory organs is found mostly in freshwater fishes of tropical region and very rarely in marine fishes. Sometimes the fishes of the tropical freshwaters and hill-streams, develop accessory respiratory organs to meet extra demand for oxygen, because depletion of oxygen occurs during summers as the water level falls to a considerable degree. Accessory respiratory organs enable the fishes to live in oxygen-deficient water, to aestivate over prolonged droughts in dry summer, to take excursions on land or simply to meet extra demand for oxygen. To overcome these adverse situations, accessory respiratory organs functionable in aquatic and/or aerial environment have been developed in fishes. So the development of such structures is essentially adaptive in nature. Some accessory organs subserve aquatic respiration, while others aerial respiration (Fig. 17.8).

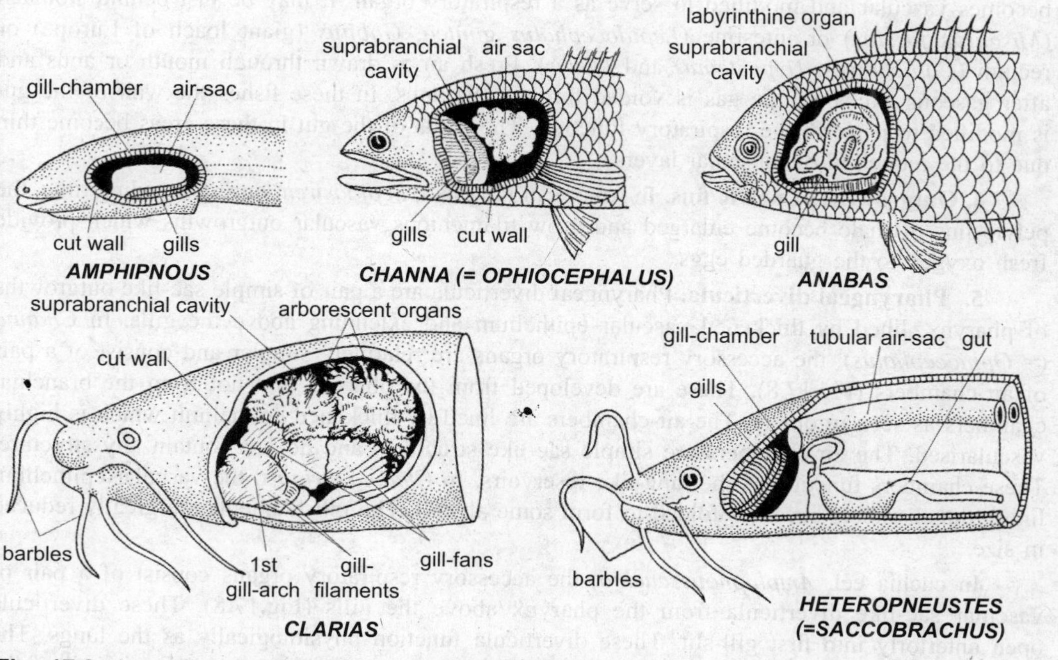

Fig. 17.8. Accessory respiratory organs dissected on the left side in some air-breathing teleost fishes.

Several types of accessory respiratory organs have been evolved in different species of fishes. These accessory respiratory organs of fishes are as follows :

1. Skin or integument. In the eel, *Anguilla anguilla, Amphipnous cuchia* and in *Periophthalmus* and *Boleophthalmus*, the skin is highly vascular and serves for exchange of gases as in frog, when the fish is out of water. These fishes habitually leave the water and migrate from one place to another through damp vegetation. During this period, the moist skin serves as an important organ in respiration. They can respire cutaneously both in air and in water.

Since *Amphipnous* and *Mastacembelus* live in oxygen deficient stagnant water, the skin is of little use for respiration but it plays an important role in extracting oxygen from air, when the fishes are exposed in drying up muddy ponds, or when fish is moving out of water. The glandular secretions of the skin protect it from desiccation in the air.

Median fin folds of many fishes are supplied with numerous blood vessels that help in cutaneous respiration. Besides, the highly vascular opercular folds of *Sturgeous* and many cat fishes serve as accessory respiratory structures.

2. Bucco-pharyngeal epithelium. In most of the fishes, the epithelial lining of buccal cavity and pharynx is usually highly vascular and permeable to gases in water. It may remain simple or may develop folds, pleats or tongues projecting into the buccal cavity and pharynx to make it an efficient respiratory organ. But in mudskippers (*Periophthalmus* and *Boleophthalmus*) the highly vascularised buccopharyngeal epithelium helps in absorbing oxygen directly from the atmosphere. These tropical fishes leave water and spend most of the time skipping or walking about through dampy areas particularly round the roots of the mangroove trees. The old idea that the mudskippers use the vascular tail as the respiratory organ is not supported by recent ichthyologists.

3. Gut epithelium. In several fishes epithelial lining of certain parts of alimentary canal becomes vascular and modified to serve as a respiratory organ. It may be just behind stomach (*Misgurus fossilis*) or intestine (*Lepidocephalus guntea*, *Gobitus* (giant loach of Europe) or rectum (*Callichthyes*, *Hypostomus* and *Doras*). Fresh air is drawn through mouth or anus and after gaseous exchange the gas is voided through the anus. In these fishes the wall of the gut is modified to perform the respiratory function. The walls of the gut in these areas become thin due to the reduction of muscular layers.

4. Outgrowths of pelvic fins. In American lung fish, *Lepidosiren*, during breeding time, the pelvic fins of male become enlarged and grow filamentous vascular outgrowths which provide fresh oxygen to the guarded eggs.

5. Pharyngeal diverticula. Pharyngeal diverticula are a pair of simple sac-like outgrowths of pharynx, lined by thickened vascular epithelium and extending above the gills. In *Channa* (= *Ophiocephalus*), the accessory respiratory organs are relatively simpler and consist of a pair of air-chambers (Fig.17.8). These are developed from the pharynx and not from the branchial chambers as seen in others. The air-chambers are lined by thickened epithelium which is highly vascularised. The air-chambers are simple sac-like structures and do not contain any structure. These chambers function as the lung-like reservoirs. In *Channa striatus*, the vascular epithelium lining the chambers becomes folded to form some alveoli. The gill-filaments are greatly reduced in size.

In cuchia eel, *Amphipnous cuchia*, the accessory respiratory organs consist of a pair of vascular sac-like diverticula from the pharynx above the gills (Fig.17.8). These diverticula open anteriorly into first gill-slit. These diverticula function physiologically as the lungs. The gills are greatly reduced and a few rudimentary gill-filaments are present on the second of the three remaining gill-arches. The third gill-arch is found to bear fleshy vascular (respiratory) epithelium.

In *Periophthalmus* also, a small, shallow pharyngeal diverticulum lined with respiratory epithelium (vascular epithelium) is present on each side of the roof of the pharynx.

6. Opercular chamber modified for aerial respiration. In some species, the inhaled air is passed through the gill-slits into the opercular chamber where it is stored for sometime. The opercular chamber becomes bulged out in the form of two little balloons in the hinder region of the head and after sometimes its walls collapse and the air is passed out through the small external branchial opening. The membrane lining the opercular chamber becomes thin and highly vascular to allow exchange of gases. This is seen in *Periophthalmus* and *Boleophthalmus*.

7. Branchial diverticula. The outgrowths from gill-chambers form more complicated aerial accessory respiratory organs than the simpler pharyngeal outgrowths in other fishes. Such air breathing organs are present in *Heteropneustes*, *Clarias*, *Anabas*, *Trichogaster*, *Macropodus*, *Betta*, etc. Important modifications in some of these species are described below :

(a) Heteropneustes fossilis (= Saccobranchus). This Indian catfish has a pair of long, tubular and dorsally situated **air-sacs**, arising posteriorly from gill-chambers and extending almost up to the tail. They are highly vascular. The air is drawn in and expelled out through pharynx.

(b) Anabas testudineus. The Indian climbing perch has two, spacious, suprabranchial cavities as dorsal outgrowths of the two gill-chambers. Each cavity contains a special **labyrinthine organ** formed of much folded, concentric bony plates developed from the first epibranchial bone and covered with thin vascular mucous membrane. Margins of these plates are wavy and the plates are covered with vascular gill-like epithelium. Each branchial outgrowth communicates freely not only with the opercular cavity but also with buccopharyngeal cavity. Air is drawn through mouth into suprabranchial cavities and expelled through opercular opening. The fish is so dependent on atmospheric oxygen that it will drown if denied access to surface to gulp air.

(c) Trichogaster fasciatus. The accessory respiratory organs in this species consist of a suprabranchial chamber, a labyrinthine organ and the respiratory membrane. The suprabranchial chamber is situated above the gills on either side as in *Anabas*, communicates with the pharynx by means of inhalent aperture and with the exterior through the opercular chamber by means of an exhalent aperture. The labyrinthine organ develops from the epibranchial of the first gill-arch and is simpler in structure than that of *Anabas*. It is in the form of a spiral organ possessing two leaf-like expansions and is composed of loose connective tissue covered by a vascular epithelium. The respiratory membrane lining the air-chamber and covering the labyrinthine organs consist of vascular and non-vascular areas, of which the former possesses a large number of 'islets' containing parallel blood capillaries. The islets are believed to be derived from the secondary lamellae of a typical gill-filament.

(d) Clarias batrachus. The Indian cat fish, *Clarias batrachus* has the most complicated accessory respiratory organs. The accessory air-breathing organs of this fish consists of (*i*) the suprabranchial cavity or chamber, (*ii*) the two beautiful 'rosettes' or air-trees or arborescent organs or dendritic organs, (*iii*) the 'fans' and (*iv*) the respiratory membrane. The suprabranchial chamber lies above the gills and is divided into two cup-like compartments and is lined by a highly vascular respiratory membrane. Two beautiful 'rosettes' or dendritic organs are present on each side and are supported by epibranchials of the second and the fourth branchial arches. The first of these is smaller in size and lies in the anterior compartment. Each is a highly branched tree-like structure supported by cartilaginous internal skeleton. The terminal knobs or bulbs of each dendritic organ consists of a core of cartilage covered by vascular epithelium showing eight folds in it. According to **Datta Munshi**, each knob represents eight abbreviated and fused gill-filaments. Some of the primary gill-lamellae of each gill-arch are fused so as to form a 'fan' or gill-plate. Hence, there are four 'fans' on either side and each consists of vascular and non-vascular areas. The respiratory membrane lining the suprabranchial chamber is also composed of vascular and non-vascular areas, of which the former show a large number of 'islets'. According to **Datta Munshi** (1961), the respiratory membrane has developed by the abbreviation and fusion of the primary gill-lamellae and shortening of secondary gill-lamellae. Well developed exhalent and inhalent apertures are present for the suprabranchial chamber. The fish rises to the surface of water and gulps in air, which from the pharynx enters into the suprabranchial chamber through the inhalent aperture. When the air enters the opercular cavity, it is directed into the suprabranchial chamber by the action of the two fans on the second and third gill-arches. The exhalation of the air is effected by the contraction of suprabranchial chamber and the movement of the fans. This creates a partial vacuum in the suprabranchial chamber. The mouth is opened and the buccopharyngeal cavity is enlarged to inhale air.

8. Air-bladders. Swim-bladder of higher bony fishes (teleosts) is essentially a hydrostatic organ. But in lower bony fishes (dipnoans and ganoids), the air-bladder acts like a lung to breathe air and is truly an accessory respiratory organ. The wall of bladder is vascular and sacculated with alveoli. In *Amia* and *Lepidosteus*, the wall of the swim-bladder is sacculated and resembles lung. In *Polypterus*, the swim-bladder is more lung-like and gets a pair of pulmonary arteries arising from the last pair of epibranchial arteries. The swim-bladder in dipnoans resembles strikingly the tetrapod lung in structure as well as in function. In *Neoceratodus*, it is single but in *Protopterus* and *Lepidosiren* it is bilobed. The inner surface of the 'lung' is increased by spongy alveolar structures. In these fishes, the lung is mainly respiratory in function during aestivation because the gills become useless during this period. Like that of *Polypterus*, the 'lung' in dipnoans gets the pulmonary arteries from the last epibranchial arteries. The swim-bladder of feather tail, *Notopterus notopterus* has a wide pneumatic duct and a network of blood capillaries covered by a thin epithelium in its wall. This helps in exchange of gases.

Functions of accessory respiratory organs. The accessory respiratory organs contain a higher percentage of oxygen. The fishes possessing such respiratory organs are capable of living in water where oxygen concentration is very low. Under this condition these fishes come to the surface of water to gulp in air for transmission to the accessory respiratory organs. If these fishes are prevented from coming to the surface, they will die due to asphyxiation for want of oxygen. So the acquisition of accessory respiratory organs in fishes is an adaptive feature.

Further it has been observed that the rate of absorption of oxygen in such organs is much higher than the rate of elimination of carbon dioxide. Hence, it is natural that the gills excrete most of the carbon dioxide. Absorption of oxygen appears to be the primary function of the accessory respiratory organs.

Origin and significance of acessory respiratory organs. During development, the fifth gill-arch does not develop gill-lamellae, and its embryonic gill material forms rudiments of the gill-arch, and aggregates to form a structure called the 'gill-mass'. The air-breathing organs or accessory respiratory organs develop from gill-mass (**Singh**, 1993). In some species, the gill-arches other than the fifth arch, also take part in the formation of accessory respiratory organs. The gill-lamellae which normally develop on gill-arches for aquatic respiration, become modified to form the respiratory epithelium of the suprabranchial chamber, dendritic organs and the air-sacs for aerial respiration. According to **Singh** (1993), the air-sacs have evolved from the same basic material which has given rise to the gill in teleosteam fishes. The accessory respiratory organs of *Heteropneustes* and *Clarias*, are modifications of the gills (**Munshi**, 1985). In these species, swim-bladder is either absent or greatly reduced. During the Tertiary and Quaternary period of the Cenozoic era, the oxygen level of the atmosphere and of water was reduced. Due to depletion of oxygen in rivers and swamps, the gills were unable to cope with the requirements of the body. Hence, several teleostean species developed accessory respiratory organs to absorb oxygen from air.

Most of the fishes possessing air-breathing organs or accessory respiratory organs are capable of living in highly deoxygenated water of the swamps and muddy ponds infested with weeds. They have been observed to gulp in air from the surface and to pass it to the accessory respiratory organs. If prevented from reaching the surface, these fishes die due to asphyxiation. This shows that the accessory respiratory organs are capable of maintaining life of the fish in oxygen deficient water. It has been shown that the absorption of oxygen in the accessory respiratory organs is much greater than the excretion of carbon dioxide. Hence, most of the carbon dioxide is excreted in the gills and the chief function of the accessory respiratory organs is the absorption of oxygen required for sustenance of life.

PARENTAL CARE IN FISHES

Parental care can be defined as an association between the parents and the offsprings, so as to increase the chances of the survival of the young ones, and in fishes it includes all the post-spawning care of the offsprings by the parents. Most fishes do not care for their eggs or youngs and leave the spawning grounds soon after fertilisation. The lack of parental behaviour is correlated with production of great numbers of eggs and sperms. But there are many fishes where definite parental care has been evolved. Various devices have been adopted to ensure proper development of the eggs into adults. One or both the sexes may participate in the process. These include selection of a suitable site, nest building and various other methods of protection of the larvae.

Species which do not exhibit any special device for safety of the ova, generally produce a very large number of eggs to increase the chances of survival of atleast a few of them. Eggs of many species possess various mechanisms for attachment to stones, pebbles or aquatic vegetation, so that they are prevented from being washed away with the current of water.

1. Nest building. Some fishes prepare crude nests for egg laying. At first a suitable place for preparing the nest is selected and some species may defend the place till death. Males of many species like the Darters (*Etheostoma*), sunfishes and cichlids, prepare a shallow basin-like nest for laying eggs by females. The stones and pebbles are removed from such nest and male keeps close watch over the eggs till hatching. A few species, however, leave the nest unprotected. Many freshwater fishes prepare crude nest with aquatic vegetation where eggs are laid. *Protopterus* and *Lepidosiren* prepare deep hole into which the females lay eggs. Males protect the nest till development is complete. *Amia calva* (bowfin) prepare a crude circular nest among aquatic vegetation. The fertilised ova are protected by male who keeps guard over the nest till the young ones are hatched. The young ones are allowed to leave the nest in a body under the protection of father. Both the male and female of some cat fishes of North America prepare a crude nest in the mud for egg-laying. The nest is sometimes provided with protective cover of logs, stones, etc.

Most interesting example is provided by the male stickleback *Gasterosteus aculeatus*, a small freshwater fish of North American lakes and ponds. The male fish actually builds a nest of dead aquatic weeds which are joined together by a sticky secretion produced from the kidneys. When the nest assumes a considerable size, the male makes a small tunnel. After the formation of tunnel and an elaborate courtship ritual, the male drags a mature female into the tunnel for laying eggs. After laying eggs, the female swims away and the male keeps watch over the fertilised eggs till development is over. In addition, foamy nest prepared by blowing of bubbles of air and sticky mucus are also encountered in many fishes. The bubbles of air and mucus adhere to form a floating mass of foam. The eggs are collected by the male in his mouth cavity and he throws them in such a way that the eggs can adhere to the lower surface of foamy nest. This type of caring for eggs is found in *Betta*, *Macropodus* and many other fishes.

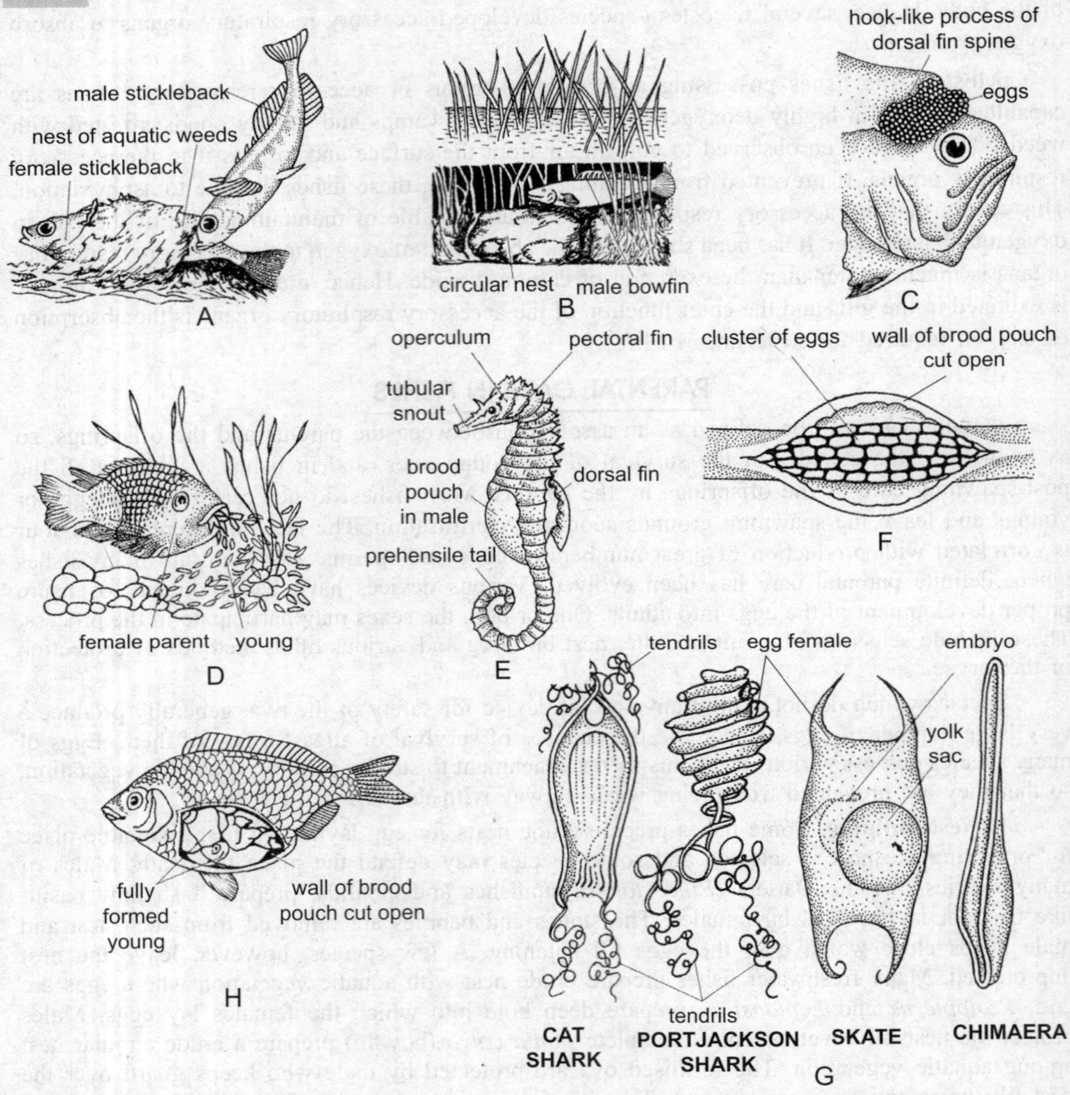

Fig. 17.9. Some interesting examples of parental care in fishes. A–Male stickleback (*Gasterosteus aculeatus*) muzzles female at the base of tail to stimulate her to lay eggs in a nest of dead aquatic plants; B–Male bowfin (*Amia calva*) guarding over circular nest; C–The male Australian *Kurtus* incubates eggs on its forehead; D–Very young of *Tilapia massambica* take refuge in female parents buccal cavity in times of danger; E–A male *Hippocampus* carrying brood pouch; F–Brood pouch of male *Syngnathus* opened to show eggs; G–Mermaids purses for eggs; H–Body cavity of *Cymatogaster aggregatus* cut open to show a fully formed young ready for hatching.

2. Mouth cavity as shelter. In some species, eggs develop within the mouth of the parent. In many cichlids, the female carries the eggs in her oral cavity. After hatching, the young fry do not leave the shelter for some time, and swim about in water very near the mouth, so that they can return to it in case of danger (found in *Tilapia*). In the cat fish, *Arius* the male carries the eggs and young ones in his mouth, and does not take food during this period.

3. Coiling round eggs. In butter fish, *Pholis* rolls the eggs into a rounded ball and then one of the parents remains on guard, possibly male, guards the egg-ball by coiling round it.

4. Attachment to body. In *Kurtus indicus* (Perciformes) the male develops a bony hook projecting from the forehead and is supported by a special process of skull bone. The eggs are grouped in two bunches with the help of filamentous processes of the egg membrane. The eggs are attached to the hook of the forehead, in such a way that one bunch of eggs lies on either side of the head of the male, as he swims in water.

5. Formation of integumentary cups. In a cat fish, *Platystacus* of Brazil, shows an interesting method of parental case. During breeding season, the skin of lower surface of the body of the female fish becomes soft and spongy. Immediately after fertilisation of the eggs, the female presses her body against the eggs in such a way that each egg becomes lodged in a small integumentary depression. Each egg is attached inside the cup by an inconspicuous stalk. The eggs remain in this position till hatching.

6. Development of brood pouches. In the pipe fish, *Syngnathus* and the sea horse, *Hippocampus* the eggs develop within the broad pouch of the male. The eggs are transferred into the brood pouch by the female and development takes place within the brood pouch.

7. Mermaid's purses. Oviparous sharks (*e.g., Scyllium*) lay fertilised eggs inside the protective horny egg capsules or mermaid's purses, which remain anchored to the sea weeds by their long tendrils. The young hatch out after rupturing the egg case.

8. Viviparity. The highest degree of parental care is found in viviparous fishes where young develop within the oviduct of the female. A few species are viviparous, such as the dog-fish, *Scoliodon* and the surf fish *Cymatogaster aggregatus*. Both fertilisation and development are internal. Developing embryos are nourished mostly by a yolk sac placenta and the young are born with the characteristic of the adult. Viviparity provides maximum protection and represents the highest degree of parental case.

MIGRATION IN FISHES

Generally fishes restrict their movements within small territorial limits and do not go out of their home ranges. However, a few species travel long distances moving from place to place in search of food or for breeding. This movement of a large number of fishes for the purpose of feeding or spawning, is known as migration. It may take place in vertical direction, as from the deeper to the surface water, or it may be in horizontal direction, either upstream or downstream. **Thompson** (1952) defines a true migration as seasonal movement that implies return to the starting point. **Harden Jones** (1968) has defined migration as a class of movement which impels migrants to return to the region from which they have migrated. This definition envisages a return to the original place. **Baker** (1978) has given a more general definition, and defines migration 'as the act of moving from one spatial unit to another'. There is no restriction of return journey. More recently, **Dingle** (1980) has defined the term migration as a 'specialised behaviour especially evolved for displacement of the individual in space'. This definition implies evolutionary changes leading to specific migratory patterns. Dingle has opined that accidental or unintentional movements cannot be included in the definition of migration.

There are two opposing trends of migration in fishes : **1. Catadromous migration**, *i.e.*, movement from freshwater to salt water (sea) for spawning. The most famous example of catadromous fish is the freshwater eel, *Anguilla*. **2. Anadromous migration** meaning the reverse movement of the fishes from the salt water (sea) to the freshwater (river system) for spawning. Examples of anadromous fishes are salmon, shad, striped bass, sturgeon, *Alosa, Hilsa*, and some trout. Completely free movement between fresh and marine water without the purpose of breeding is called **amphidromous migration** exhibited by fishes like *Megalopa*, *Chanos*, etc. (Fig. 17.10).

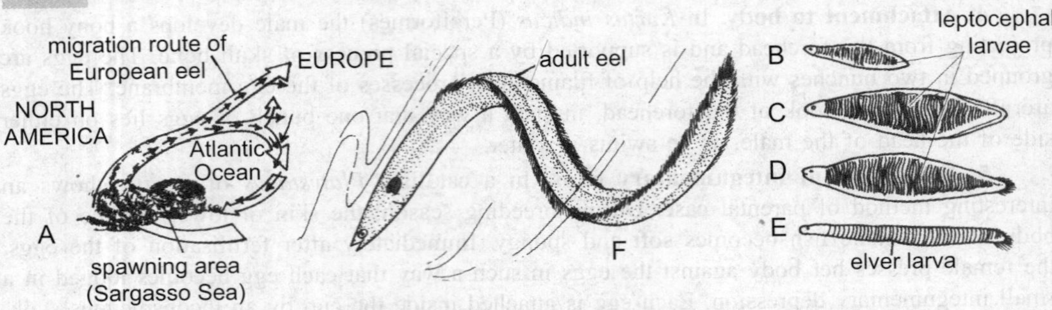

Fig. 17.10. Migration of freshwater eel, *Anguilla*. A–Migration route and breeding place; B, C, D–Leptocephali larvae; E–Elver or glass eel; F–Adult eel.

1. Eels. The best example of **catadromous migration** is furnished by two common species of eels, *Anguilla rostrata* of European freshwater rivers and *Anguilla vulgaris* of America. With the advent of autumn, their colour changes from yellow to metalic silver. Feeding stops with the shrinking of their digestive tract. Eyes become larger, snout becomes sharper with thinner lips and the gonads are fully mature. The silvery eels then enter the sea and migrate about 4,500 kilometres westward from Europe or eastward from America. Reaching their breeding place in the Surgasso sea of Bermuda, the adults die immediately after spawning in deep waters. The eggs hatch into little transparent, leaf-like flattened pelagic larvae, called **leptocephalia**, less than 6 mm long. They have sharp needle-like teeth for feeding. During their long homeward journey, they grow into **elvers** or **glass eels** about 8 cm long, with cylindrical bodies. On reaching land, the males remain behind in brackish waters near coasts, while the females ascend freshwater streams and rivers. The elvers feed and grow to become yellow eels in some years. How the elvers without parents are able to find their way across the sea towards homes of their parents remains an unsolved mystery. Until their real identity was discovered, the leptocephali larvae were called glass fishes and placed in the genus *Leptocephalus*. The biggest larva of some unknown species was 184 cm long which was captured in 1930 by Dana Expedition west of the Cape of Good Hope (Africa).

2. Salmon. There is a single species of Atlantic Salmon (*Salmo salar*) and five species of Pacific salmon (*Oncorhynchus*). They furnish the best example of **anadromous migration**. In winter, both the sexes have their feeding grounds at sea to ascend the freshwater mountain streams, reaching the identical spot where they originally grew some years before. They stop feeding, change to a dull reddish brown from silver, and excavate shallow saucer-like pits in bottom gravel. After spawning the adults die, but some of the Atlantic species (*Salmo*) may survive, return to sea and spawn for a second or third time in life. After hatching, the larvae fish feed and grow out for some time in the streams before going out to sea. The young salmons grow faster, in the ocean because of abundant food there. Experimental evidence shows that some strong olfactory sense of salmon determines its homing into original birth place, for different streams have different odours. Many marine fishes migrate to the river from sea for spawning. This anadromous migration is observed in *Acipenser*, *Alosa*, *Salmo*, *Hilsa*. These fishes usually produce huge numbers of eggs to compensate the risk to which eggs are subjected.

Causes of migration. Several authors have given various reasons as to why fish migrate. According to **Northcote** (1978) this is : (*i*) to optimise feeding, (*ii*) to avoid unfavourable conditions, (*iii*) to enhance reproductive success, and (*iv*) possibly to promote colonisation. Specific fish migrations may not fulfil all these requirements, and might be for any one of the above reasons. Migrations are brought about because reproductive habitats are not likely to be the best areas for other activities such as feeding and over wintering. Unfavourable chemical, physiological and biological conditions may occur in any environment at different times due to seasonal changes, and migration is a means to escape such conditions. The timing of migration pattern appears to have evolved in such a way that the young fish arrive at feeding areas when the environmental conditions are optical for food production. The strategy of the fish is to exploit rich food source, to enhance food intake which is necessary for increased growth rate, fecundity and survival.

Advantages of migrations. According to **Nikolsky**, migration is an adaptation towards abundance. The spawning or nursery grounds may not have enough food to maintain both the mature and immature members of a large population. Hence, it would be an advantage to have separate spawning, nursery and feeding grounds. The fact that many commercial species are migratory, supports the view that migration is an adaptation towards abundance. Further, there appears to be some advantage to a species whose adults return to spawn in an area where the environmental conditions were similar to those under which they themselves survived when young. A return to the parent spawning ground provides a means by which these favourable conditions may be exploited. Thus, a better egg and larvae survival would lead to a greater number of spawners on a particular ground.

Fish are often dispersed widely over large areas while feeding, but congregate at specific spawning grounds to enhance reproductive success. Precise timing of spawning migration ensures simultaneous arrival of sufficient members of both the sexes at one place to ensure reproductive success of the group.

ECONOMIC IMPORTANCE OF FISH

Fishes form one of the most important group of vertebrates for man, influencing his life in various ways. Millions of human beings suffer due to hunger and malnutrition, and fishes form a rich source of food and provide a means to tide over the nutritional difficulties of man. In addition to serving as an important item of food, fishes provide several byproducts to us. Fish have considerable economic importance are useful as well as harmful to man.

Useful Fishes

1. Fish as human food. Fish have formed an important item of human diet. Nearly all fish, freshwater and marine are edible and have been an important source of protein, fat and vitamins A and D since time immemorial. In most fishes, the flesh is white, contains about 13 to 20% of protein and has a food value of 300 to 1600 calories per pound. According to **Pottinger** and **Baldwin**, fish meat contains atleast 5 of the esxsential amino acids. Besides this, it contains vitamin A and D together with large amount of phosphorus present in it. Important marine fishes include salmon, cod, halibut, herring, eels, tuna, mackerel, and sardines. Important freshwater food fishes are cat fish, trout, bass, perch and mullet. Even eggs of certain fishes, such as Russian sturgeon, are eaten as cavier. The major food fishes of India include *Labeo, Catla, Cirrhina, Mystus, Wallago, Notopterus, Ophiocepholus*, etc. The cartilaginous sharks, skates and rays are also used as human food in several countries. They are eaten by poorer classes of people living along the sea coasts of India and Sri Lanka. The canned meat of sharks is sold commercially under the name of gray fish. In south-east Asian countries, shark fins are dried and boiled to get a gelatinous material used as soups.

Fresh fish meat is usually cooked for human consumption. However, large quantities are refrigerated, salted, smoked, canned or pickled. Today, fisheries of the world carry on business worth several hundred thousand rupees annually and also provide employment to thousands of people.

2. Fish as food of cattle. The scrap from canneries, as well as entire fishes that are not relished by man, are dried and ground in mill. This is called **fish meal** and is used as artificial food for poultry, pig and cattle. Fish meal is produced in several states like Maharashtra, Andhra Pradesh, Tamil Nadu, Bengal and Kerala chiefly from sardines, meckerels, ribbon fish, etc. The fish is first cooked in large pots containing sufficient quantity of water, on fire or on steam. The cooked material is then pressed to remove moisture and dried in the sun on suitable platforms. The resultant product is then stored, and if preserved in airtight containers after sterilisation retains its nutritive value for a long time. The fish meal contains about 60% protein and high percentage of calcium phosphate so that it is very valuable for cattle and poultry. The manufacture of fish meal can be undertaken as a cottage industry requiring little expenditure.

3. Fish manure. Fish that are unfit for human consumption are used to prepare fish manure for the fields. During peak season, when there is a large supply of fish, or they are landed in spoiled condition, they are sun dried on the beach. The dried fish is ground and converted to manure, which contains a high percentage of nitrogen and phosphate. Mostly sardines are used for preparing fish meal and the waste material forms the fish guano. It contains 8.9% nitrogen and phosphate, and when mixed with soil, forms a rich fertiliser for plants. It is several times richer than ordinary cattle manure.

4. Fish oils. The most important fishery by-product is that of the fish oil, which is of two kinds : body oil and liver oil. The oil extracted from the whole body of the fish is called fish body oil, while that obtained from liver of certain fish is called the fish liver oil. Liver oil contains vitamin A and D, while the body oil contains them in traces. The refined oil from the liver of fishes has a medicinal use, being the source of vitamin A and D. The body oil from fish has many uses, such as in painting, varnishing, soap, candle, leather and steel industries. For preparing body oil, fish are boiled in large quantity of water. Oil is removed quickly and washed in boiling salt water. Liver oil is prepared from the liver of several species, including sharks and rays. Oil is extracted from the liver soon after the fish are caught to avoid action of enzymes. Liver is cut into small pieces and boiled in sufficient quantity of water. Oil is skimmed from the surface of water and sent for purification. Liver oil contains 55–75% fat, 5–10% protein and the rest water. It is of considerable medicinal value.

5. Fish skin and leather. The skin of several fishes like the sharks and rays are used for making polishing and smoothing material. Shark skin leather is of some commercial importance in the manufacture of many useful articles such as shoes and hand bags, etc. In Japan, lantern are prepared from the skin of puffer fishes (*e.g., Tetrodon*). Some tribal people used skins of puffer and porcupine fishes (*e.g., Diodon*) for war helmets. Crude skins of sharks and rays are used by carpenters and metal workers. Shark skin tanned with placoid scales on it, is called **shagreen**. It has been used as an abrasive for polishing wood and ivory and also for covering jewel boxes, fine books and sword handles.

6. Fish glue. Liquid glues are prepared from skin, head and other trimmings of certain fishes. This glue has an adhesiveness of great power for paper, wood, leather and glass.

7. Isinglass. It is a gelatinous product obtained from the air-bladders of certain fishes such as sturgeons, carps, perches, salmons, cat fishes, cods, etc. The isinglass is a shining powder and is used for clearing wine, beer, making edible jelly and in the preparation of adhesive material. The air-bladder is removed from fish, washed in cold water and flattened by beating it on a piece of wood. The bladder is then dried in the sun, and is exported for the preparation of isinglass. The finest quality of isinglass is obtained from Russia.

8. Fish fin. The fins of sharks are exported to China where they are used for preparing soup.

9. Medicines and disease control. The refined oils extracted from livers of cods and sharks are rich in vitamins A and D. Pituitary glands yield important extracts for medicines. Fishes like top minnows (*Gambusia affinis*), *Trichogaster*, *Chela, Puntius, Barilius, Danio, Colisa, Rasbora, Esomus, Ambassis, Aplocheilus, Barbus*, *Panchax*, etc., feed voraciously on mosquito larvae. These larvicidal fishes are propagated and distributed widely into ponds, lakes and tanks to destroy mosquito that transmit malaria, yellow fever and other dreadful diseases of tropical countries. Certain fishes and their bye-products contribute to useful Ayurvedic and Unani medicines for treatment of duodenal ulcers, skin diseases, night blindness, general weakness, loss of appetite, colds, coughs, bronchitis, ashthma, tuberculosis, etc.

10. Sports and recreation. Sport fishing by individuals and fishing parties is a popular recreation of million of people, as well as a source of food, all over the world. The most commonly hunted fishes are the freshwater perch and trout and the marine tarpon. However, some of the best game fish is most famous for sport that it provides, but its flesh is not palatable. Many people's hobby is to cultivate certain tropical fishes as pets. Both native as well as foreign fishes are displayed in home aquaria for their beauty and graceful movements. Common aquarium fishes are gold fish (*Carassius auratus*), angel fish (*Pterophyllum*), sword tail guppy (*Xiphophorus*), minnow (*Gambusia affinis*), siamese fighter (*Betta splendens*), paradise fish (*Macropodus*), *Hemigrammus*, *Aphyocharax*, loach (*Botia*), *Trichogaster, Tilapia*, etc. Goldfish cultured and not found in nature and the Japanese have produced their several curious artificial varieties. Pet shops now-a-days stock many kinds of fishes for hobbyists and scientists.

11. Fancy articles. Scales of garpike (*Lepidosteus*) are used for jewelry and novelties. From scales of some fish is secured a pigment whose water suspension is known as **pearl essence**. It is used in the manufacture of artificial pearl in Europe, especially in France.

12. Scientific study. Fishes have considerable use as experimental animals, especially in the fields of Genetics, Embryology, Animal Behaviour and Pharmacology. Certain fishes such as *Latimeria* and dipnoans have anatomical features of great zoological interest. Fishes like dogfish (*Scoliodon*), perch (*Perca*) and carp (*Labeo*), etc., are dissected for anatomical study in zoological laboratories. Researches in ichthyology are conducted for the benefit of fisheries and mankind.

13. Industries. As the fish forms a rich source of food, millions of people are engaged in fishing industry and depends on fisheries for their livelihood in various ways. Besides those who directly catch the fish for marketing, there are equally large number of people engaged in subsidiary industries like refrigeration, preservation, canning, and in the manufacture of fish products and by-products.

Harmful Fishes

1. Destructive. All the cartilaginous fishes are predaceous and feed chiefly on large quantities of crabs, lobsters, squids and other valuable marine animals. They cause great harm by destroying eggs, young and even adults of our food fishes. Sharks are extremely dangerous in the sea and injure fisherman and damage their nets, steal baits, devour captured fish and drive away squids and fishes to be netted. The damage done by these sharks amounts to several hundred thousand rupees.

2. Injurious. Some larger sharks and sword fish may capsize small boats and injure or even kill fisherman. All sharks, small or big, are a menace to bathers and skin divers in shallow waters. The pirhanas of Central and South America have very strong teeth and are dangerous even to man.

Some fishes like cartilaginous electric ray (*Torpedo*) have electric organs which give powerful shocks to swimmers and fishermen. An extreme case of specialisation is the bony electric eel (*Electrophorus electricus*) of the Amazon. It is nearly 1.5 metres long and the postero-ventral four-fifth part of the body is occupied by the electric organ. It can generate electricity up to 600 volts in potential with a maximum output of 1000 watts. Carnivorous fishes eat away the larvae of useful insects.

3. Poisonous. Poisonous glands are found in many cartilaginous fishes such as sting ray (*Trygon*) and eagle ray (*Myliobatis*). They can inflict painful wounds, sometimes fatal by the poisonous stings or spines. Poisonous glands are also found in *Squalus* (dogfish), *Heterodontos*, *Chimaera*, *Diodon* (porcupine fish), *Tetrodon* (globe fish), scorpion fish, etc. These are also capable of causing wounds by their spines or spiny opercula. Flesh of some tropical fishes (*e.g.*, *Tetrodon*) is also poisonous and may prove fatal to man.

IMPORTANT QUESTIONS

Long Answer Questions

1. Differentiate between Chondrichthyes and Osteichthyes.
2. Give a general account of Dipnoi and discuss their affinities.
3. Discuss the various theories regarding the origin of paired fins in fishes.
4. Discuss the different types of scales found in fishes.
5. Write an essay on air-bladder or swim-bladder in fishes and discuss the modifications of air-bladder in different groups of fishes.
6. Write an essay on accessory respiratory organs in fishes.
7. Discuss the parental case in fishes.
8. Write an essay on migration in fishes.
9. Write an essay on economic importance of fishes.

Short Answer Questions

1. Make a tabular comparison between Chondrichthyes and Osteichthyes.
2. Briefly describe the following :

   (*i*) Dipnoi; (*ii*) Fin system in fishes; (*iii*) Scales in fishes; (*iv*) Air-bladder or swim-bladder in fishes; (*v*) Accessory respiratory organs in teleosts; (*vi*) Parental care in fishes; (*vii*) Migration in fishes.
3. Briefly describe the economic importance of fishes.

Very Short Answer Questions

1. Write short notes on the following :

   (*i*) Dipnoi; (*ii*) Heterocercal and homocercal tail; (*iii*) Cycloid and ctenoid scales; (*iv*) *Neoceratodus*; (*v*) *Protopterus*; (*vi*) *Lepidosiren*; (*vii*) Types of swim-bladder; (*viii*) *Salmon*.

Multiple Choice Questions

1. The fish body is divided into
   (*a*) Head and trunk
   (*b*) Head trunk and tail
   (*c*) Head, neck and tail
   (*d*) Head, neck, trunk and tail.
2. Superclass Pisces includes organisms which are essentially aquatic and use
   (*a*) Paired fins for swimming and gills for respiration.
   (*b*) Paired fins for locomotion and gills for sound production
   (*c*) Paired fins for respiration and gills for swimming
   (*d*) Paired fins for sound prodúction and gills for respiration.
3. The superclass Pisces is divided into three subclasses
   (*a*) Placodermi, Chondrichthyes and Osteichthyes
   (*b*) Elasmobranchi, Chondrichthyes and Osteichthyes
   (*c*) Actinopterygii, Chondrichthyes and Osteichthyes
   (*d*) Pterichthyes, Acanthodi and Coccostei.
4. In fishes, the paired nasal sacs
   (*a*) Open into mouth
   (*b*) Open into pharynx
   (*c*) Open into internal ear
   (*d*) Do not open into mouth.
5. Rectal glands are present in
   (*a*) Both Chondrichthyes and Osteichthyes
   (*b*) Only Chondrichthyes
   (*c*) Only Osteichthyes
   (*d*) Neither Chondrichthyes nor Osteichthyes.
6. Liver in Osteichthyes is
   (*a*) 1-lobed (*b*) 2-lobed
   (*c*) 3-lobed (*d*) 4-lobed.
7. The eggs of fishes are
   (*a*) Microlecithal, extra-embryonic membranes absent
   (*b*) Mesolecithal, extra-embryonic membranes absent
   (*c*) Megalecithal, extra-embryonic membranes present
   (*d*) Megalecitha!, extra-embryonic membranes absent.
8. Members of order Dipnoi are commonly known as
   (*a*) Cowfishes (*b*) Pipefishes
   (*c*) Lungfishes (*d*) Catfishes.
9. *Protopterus* is commonly known as
   (*a*) African lungfish
   (*b*) Australian lungfish
   (*c*) American lungfish
   (*d*) Asian lungfish.

10. *Neoceratodus* is found in
(*a*) Asia (*b*) Australia
(*c*) South Africa (*d*) South America.

11. The body of lung-fishes is covered with
(*a*) Placoid scales (*b*) Cycloid scales
(*c*) Ctenoid scales (*d*) Ganoid scales.

12. The internal nostrils in lung fishes
(*a*) Open into mouth cavity
(*b*) Open into lungs
(*c*) Do not open into mouth
(*d*) Open into pharynx.

13. The centrum in vertebrae in lung fishes is
(*a*) Procoelous (*b*) Amphicoelous
(*c*) Acoelous (*d*) Absent.

14. The accessory respiratory organs in lung fishes is
(*a*) Swim-bladder
(*b*) Lungs
(*c*) Gills
(*d*) Urinary bladder.

15. Pelvic fins in fishes are called thoracic when
(*a*) They are situated anterior to pectoral fins
(*b*) They are situated just above the pectoral fins
(*c*) They are situated just below the pectoral fins
(*d*) They are situated just in front of anus.

16. Lateral and median fins are used as
(*a*) Swimming organs
(*b*) Defensive organs
(*c*) Steering devices
(*d*) Sound producing organs.

17. The caudal fin in *Latimeria* is
(*a*) Single lobed (*b*) Bilobed
(*c*) 3-lobed (*d*) 4-lobed.

18. Swim-bladder is absent in
(*a*) Elasmobranchs (*b*) Teleosts
(*c*) Acanthodians (*d*) Placoderms.

19. Which of the following is not a function of air-bladder?
(*a*) Respiration
(*b*) Reproduction
(*c*) Hydrostasis
(*d*) Sound production.

20. Which of the following is not an accessory respiratory organs in fishes?
(*a*) Skin or integument
(*b*) Pectoral fins
(*c*) Pelvic fins
(*d*) Gut epithelium.

21. In *Hippocampus* the brood pouches are developed in
(*a*) On the back of female
(*b*) On the belly of male
(*c*) On the belly of female
(*d*) On the back of male.

22. Catadromous migration is concerned with movement of fishes
(*a*) From freshwater to freshwater
(*b*) From salt water to salt water
(*c*) From freshwater to salt water
(*d*) From saltwater to freshwater.

23. Anadromous migration refers to movement of fishes
(*a*) From saltwater to freshwater
(*b*) From freshwater to saltwater
(*c*) From saltwater to saltwater
(*d*) From freshwater to freshwater.

Answers

1. (*a*)	2. (*a*)	3. (*a*)	4. (*d*)	5. (*b*)	6. (*c*)	7. (*d*)	8. (*c*)
9. (*a*)	10. (*b*)	11. (*b*)	12. (*a*)	13. (*d*)	14. (*b*)	15. (*c*)	16. (*c*)
17. (*c*)	18. (*a*)	19. (*b*)	20. (*c*)	21. (*b*)	22. (*c*)	23. (*a*).	

18 Chapter

Type 7. *Rana tigrina* (Common Indian Bull Frog)

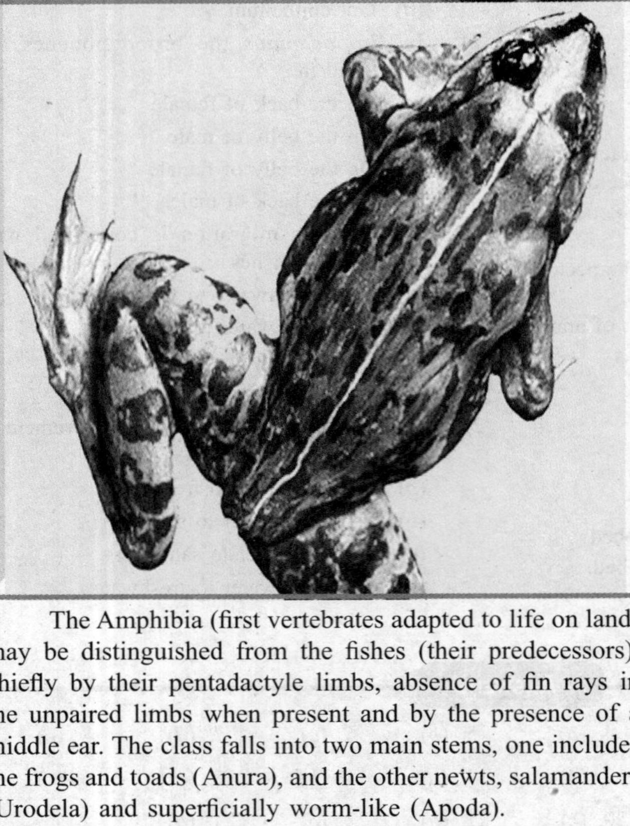

Contents of this Chapter

The Amphibia (first vertebrates adapted to life on land) may be distinguished from the fishes (their predecessors), chiefly by their pentadactyle limbs, absence of fin rays in the unpaired limbs when present and by the presence of a middle ear. The class falls into two main stems, one includes the frogs and toads (Anura), and the other newts, salamanders (Urodela) and superficially worm-like (Apoda).

The earliest known anuran ancestors, *Protobatrachus*, did not appear until the early Mesozoic. This was frog-like, possessed a tail in the adult, but the skull was similar to those of modern frogs, except that an opisthotic was retained. The ribs were short. The presacral vertebrae were reduced in number (more than eight). Free caudal vertebrae were present. The elongated ilium and fairly long femur were frog-like, but the radius and ulna, and tibia and fibula were separate. The earliest true frog remains occur in the upper Jurassic. The urodels also have not been found before upper Jurassic and no fossil Apoda has been discovered. Today the most terrestrial Amphibia are the Anura. The Urodela have retreated once more to the water, and the degenerate Apoda into moist holes in the ground.

Amphibia breathe by means of gills in the various larval stages, and by lungs when adult. The gills are retained by the adults of some urodels, and the skin, which is usually naked, often plays an important part in the respiration. The skull is **autostylic** (*i.e.*, upper jaw is fused with the skull and the free hyomandibula has been converted into a **columella auris**).

The class **Amphibia** (Gr., *amphi* = both + bios = life) includes all those animals which live partly in freshwater as well as partly on land. They are the lowest and earliest tetrapods which first of all successfully descended on land from some fish-like ancestors possibly in Devonian time. The living salamanders, the toads, the frogs and limbless tropical caecilians are the most important representatives of this class. They are characterised by their outstanding features and thus stand in between the fishes and reptiles. They are cold-blooded or **poikilothermal** or **ectothermal** animals as the body temperature, like fishes, directly fluctuates according to the temperature of the environment. They possess four pentadactyle limbs which are supported by girdles instead of fins of fishes and are adapted for locomotion both on land and water as well. Their skin is usually soft, glandular and moist lacking in the protective external scales. They have two nostrils connected to mouth cavity. Their skull is provided with two occipital condyles and the heart is three-chambered. They respire by gills, lungs, skin or the mouth lining. Fertilisation is either external or internal. Mostly are oviparous. Eggs are with some yolk and enclosed in gelatinous coverings. Development is always indirect as larva is formed during life cycle which is generally known as tadpole larva.

SYSTEMATIC POSITION

Phylum	...	Chordata
Group	...	Vertebrata
Subphylum	...	Gnathostomata
Class	...	Amphibia
Order	...	Anura
Suborder	...	Diplasiocoela
Genus	...	*Rana*
Species	...	*tigrina*

HABIT AND HABITAT

Rana tigrina, the common **Indian bull frog**, generally lives in or near water but in rains it may be found far away from its natural habitat, the aquatic home. It lives near water probably (*i*) to carry out the **cutaneous respiration** by keeping the skin moist as soon as it gets dry and (*ii*) to escape from the man and a number of other enemies by slipping or jumping into the water. During the breeding season frogs are more closely confined to the water than other times of the year.

Habits

Frogs are harmless, non-poisonous and quick-moving active animals. It rests on land by keeping its short angular forelimbs upright and the long hindlimbs folded by the side of the body; when disturbed, it jumps into water by suddenly extending its hindlimbs in order to escape.

1. Locomotion. Frogs move by leaping and swimming, and in both of these operations the long hindlegs play the chief part.

2. Feeding. It is **carnivorous** in its feeding habit and feeds on small insects, spiders and earthworms which are caught by throwing out the extensible sticky mucous-coated tongue from the mouth and swallowed entire organisms. The tongue is attached anteriorly, while the forked posterior end lies free. It always rejects motionless animals and undesirable food.

Fig. 18.1. Frog. Stages of leaping on land.

(*i*) Leaping. In the ordinary resting position the body is inclined upward in front, being supported on the forelegs and its thumb points nearly backward. The posterior part of the body rests upon the ground and the hindlimbs are folded up ready for a spring. By sudden extension of the hindlegs the body is propelled through the air. In leaping, the forelimbs are used more to hold up the anterior part of the body and to point the animal in the desired direction of movement than as actual organs for propulsion. It may leap from 2 to 3 feet.

(*ii*) Swimming. In both operations the hindlegs are alternately drawn up in the form of a "Z" and quickly extended. As the hindlegs are pushed back during swimming, the toes are spread apart, the webs between them affords resistance to passing thorugh the water. This motion gives the body a forward impulse. The forelimbs are also used in swimming, taking strokes sometimes together and sometimes alternately. When the animal swims downward, the forelegs beat backward and upward. When it swims upward, forelegs beat downward elevating the anterior part of the body, which is then pushed upward by the strokes of the hindlegs. When frog comes to surface to breathe or simply to float, only the tip of the nose remains exposed. The hindlegs hang obliquely in a state of moderate extension and the forelegs are held out from the body.

3. Croaking sound. The sound produced by the frog is called the **croaking**. It is more often heard during the breeding season (rainy season), when they are supposed to serve the purpose of a sex call. The sound of male is louder and deeper than that of the female. In large frogs the voice are deeper than in small ones. Frogs croak under water as well as on land. It is produced by forcing the air from the lungs over vocal cords into the mouth, the external nares remain closed to prevent its escape. Then the buccal cavity contracts, forcing the air back into the lungs again. The same process is repeated again and again.

4. Seasonal changes. It is cold-blooded or **poikilothermal** or **ectothermal** animal as its body temperature does not remain constant but fluctuates with that of environment. In winter the temperature of the body becomes so much low that its all body activities get ceased down and it becomes sluggish. In this condition it cannot live more on the land, so it digs down into damp earth at the bottom of ponds to pass the adverse conditions of winter and rest. This going of underground is known as **winter sleep** or **hibernation**. During winter sleep or hibernation lung breathing is stopped, while skin breathing continues which suffice the need of the oxygen. Moreover it does not feed but to keep up its vital activities it consumes the reserve food stored in the form of glycogen and fat. It remains underground till the rain falls. Similarly during the summer it once again goes underground to sleep as its all body activities are slowed down due to high temperature. This going of underground is known as **summer sleep** or **aestivation**. The spring and rainy seasons are the periods of great activity for the frogs.

5. Colour changes. Frogs have the power of changing their colour in harmony with their surroundings to protect from enemies. The power of the skin to change its colour in relation to surrounding conditions depends upon changes in chromatophores (pigment cells). The colour change of the skin depends upon the difference in the distribution of pigments (black and yellow). When the pigment of black chromatophores is expanded, the skin becomes dark in colour and when it becomes contracted, the skin becomes light coloured.

6. Breeding. In mating or copulation the male clasps the female by its forelegs from its back just behind her forelegs. The nuptial pads on the bases of inner fingers of male also help in firmly clasping the female. This sexual embrace is called **amplexus** which continues for several days until the female deposits the eggs through cloaca in water. The male also discharges his spermatic fluid over them to fertilise. The male then loses his clasping instinct and is totally indifferent to the other sex. Egg's mass called frogs **spawn** is formed by adhering the jelly surrounding the eggs. The jelly of the eggs absorbs water and swells and adheres with each other. The fertilised eggs (zygotes) develop into free-swimming **tadpole** within a fortnight and then undergoes metamorphosis to become miniature adult.

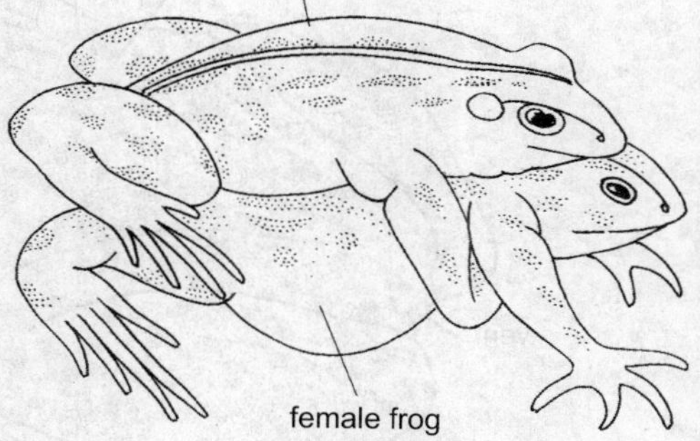

Fig. 18.2. Frogs in amplexus.

7. Enemies. Frogs are the most defenceless animals and fall easy prey to a variety of carnivorous animals, who devour them in great numbers. These animals are snakes, turtles, mongooses, racoons, cranes, herons, hawks and crows. Man is also their enemy who is very fond of their legs. The frog is also afflicted by a large number of parasitic forms, such as leeches remain attached to frog for a long period, larvae of blowflies infest the intestine of frogs, but usually in great numbers in toads. Nematodes such as round worms are found in the intestine of frogs. Trematodes or flukes are also found in the intestine and bladder, and sometimes in lungs are also found. Among protozoans are *Opalina*, *Balantidium*, *Copromonas* found in rectum; *Entamoeba* found in the intestine; *Eimeria* infest intestinal epithelum; *Isospora* spp. in the cells of kidneys.

8. Economic importance. Frog has remained a favourite object for the study of animal structure and function since a long time. It is very commonly used in the physiological researches. It is due to its easy availability. It is also widely used in instruction, since its structure is more or less similar to that of the human body. It is a good friend of farmers because it feeds on insects harmful to crops. Its muscular hindlegs are eaten in China and so they are reared in ponds, etc.

The frogs, generally, breed in the rainy season, appearing with the first shower. As soon as the rains fall the frogs come out from their hidden places (hibernation and aestivation) and start croaking. Male frogs, by croaking, attract the female frogs for breeding or copulation. Frogs, generally, seek shallow quiet water as a breeding place where they lay down their eggs for fertilisation. The parents never pay any attention to them but lead a carefree life of excessive feeding. The eggs develop into embryos that hatch out as tadpoles or larvae in about two weeks.

There are so many natural enemies of frog such as snakes, turtles, herons, raccoons and man. To escape from these enemies frog tries to jump into water or conceal itself in the bottom of debris or green vegetation as its body colour coincides with that of environment.

EXTERNAL FEATURES

Shape and size. Besides aerial mode of life, frog also leads aquatic mode of life. Therefore, it has streamlined body which is the characteristic of the aquatic animals and assits in swimming in water. The two ends, the anterior and the posterior, of the body are pointed and the triangular flattened head, with its blunt apex directed forward, is broadly united to the trunk. Thus, there is no neck to connect the head and trunk together and no tail.

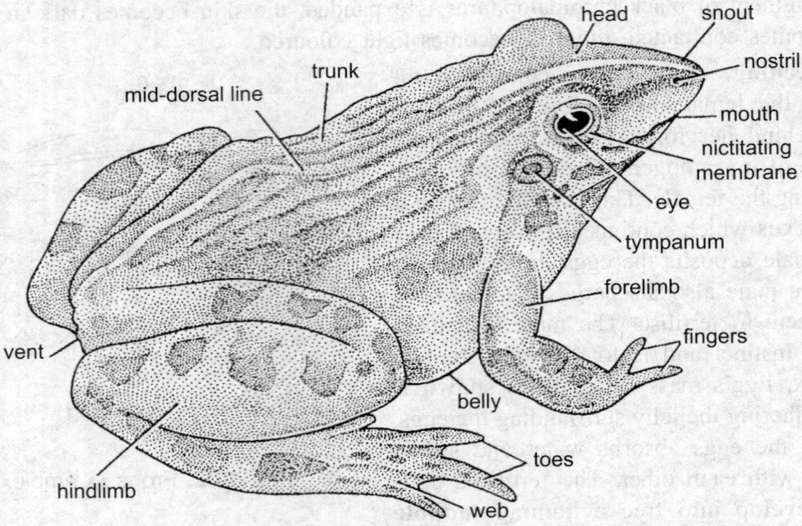

Fig. 18.3. An Indian frog, *Rana tigrina*, in sitting posture.

The size of the frog varies from species to species or even in the same species depending upon the age of the individual. The size may range from few centimetres to many centimetres.

Colour. The colour of the body at the dorsal side is green with black spots and streaks but ventrally it is paler. This type of colouration harmonises with that of surrounding environment.

Skin is smooth, thin, moist and slimy, and fits loosely on the body. Skin of back is folded or thickened longitudinally called **dermal plicae**.

Division of body. The body of the frog is divided into two parts, the **head** and **trunk**, the true neck and tail of tadpole being absent. The head and trunk are broadly joined.

Head. The head is almost triangular and somewhat flattened. Its anteriorty directed blunt apex is known as **snout** which terminates into a large, transverse mouth. At the tip of the snout are two laterally placed **nostrils** or **external nares** communicating with the buccal cavity through internal nares, serving in respiration. The head dorsolaterally bears two large prominent bulging eyes. This position enables the frog to see in all the directions and, thus, compensate the disadvantage on land due to the absence of the neck. The eyes are protected by two eyelids, the upper eyelid is thick, fleshy, opaque and almost immovable but the lower one is thin, transparent and movable, capable to cover the eye. Beneath it there is a transparent third eyelid or **nictitating membrane** which is merely an outgrowth of the lower eyelid that can cover the eyeball in water and also keep it moist in the air. The frog can see through it. In the middle of the head, just in front of the eyes, there is a light coloured patch–the **brow spot** which represents the vestigial pineal eye. There are no external ears but behind and below each eye there is a nearly circular obliquely placed a tough transparent membrane–the **tympanic membrane** or **ear drum**. In the male frog under the head on either side are placed two bluish wrinkled patches of skin–the **vocal sacs** which are used to produce croaking sound to attract the females for copulation.

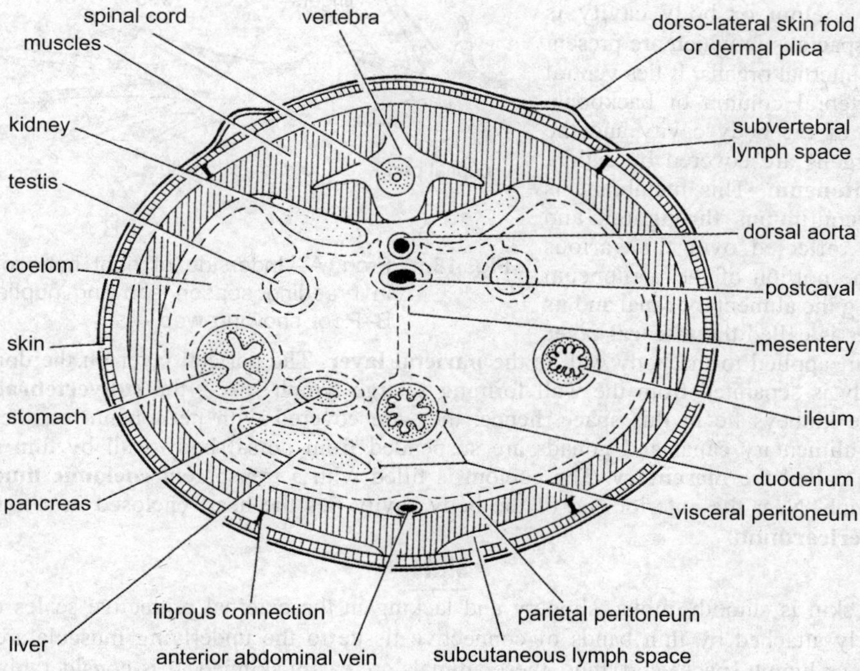

Fig. 18.4. Frog. Diagrammatic T.S. of trunk showing coelom, peritoneum (broken lines) and viscera.

Trunk. The head is broadly joined with short somewhat flattened ovoid trunk. At its dorsal side in the middle region in the resting stage there is a characteristic **sacral hump** which is due to the linking of the hip girdle to the vertebral column. At the posterior end of trunk, in between the hindlimbs is present the **cloacal opening** or **vent** through which foecal matter, urine and reproductive bodies (sperms and ova) are discharged. Attached to the trunk are two pairs of limbs. The **forelimbs** are shorter, while the **hindlimbs** are larger. The forelimbs are meant to hold and support the front part of the body at the time of jumping but the hindlimbs assist in jumping and swimming as the webs are present in between the toes. Each forelimb comprises an **upper arm** (brachium), **forearm** (antebrachium), **wrist** and **hand** (manus) with four fingers (digits) and a vestigial **"thumb"** or pollex. In male the base of the first (inner) finger is thickened especially in the breeding season, forming the **nuptial pad** for clasping the female at the time of amplexus. Each hindlimb comprises an upper **thigh**, **shank** or **lower leg**, **ankle** (tarsus) and long **foot**. The latter has a narrow **sole** and five slender toes connected by broad thin webs of skin which help in swimming.

TABLE 18.1. SEXUAL DIMORPHISM.

Male	Female
1. Smaller in size and darker in colour.	1. Larger in size and less dark in colour.
2. Slim.	2. Stout in breeding period.
3. Possess vocal sacs.	3. Vocal sacs absent.
4. Possess nuptial or copulatory pads on inner fingers.	4. Nuptial or copulatory pads absent.

Coelom and Viscera

The coelom or body cavity is large and spacious in which are present viscera or internal organs. It lies ventral to the vertebral column or backbone. The walls of the body cavity and the visceral organs are covered by a thin, moist **peritoneum**. This membrane is perfectly continuous throughout and is simply reflected over the various organs. The portion of the peritoneum surrounding the alimentary canal and its appendages is called the **visceral layer** and the part applied to the body wall is the **parietal layer**. The parietal layer on the dorsal side of the body is separated from the wall forming a large lymph space, the **subvertebral lymph space**. The kidneys lie in this space, hence, they are covered with peritoneum on the neutral side. The alimentary canal and gonads are suspended from dorsal body wall by thin sheet of membrane called the **mesentery**. The coelom is filled with a transparent **coelomic fluid** which is like lymph. Near the anterior end of the body cavity lies the heart enclosed in a transparent sac, the **pericardium**.

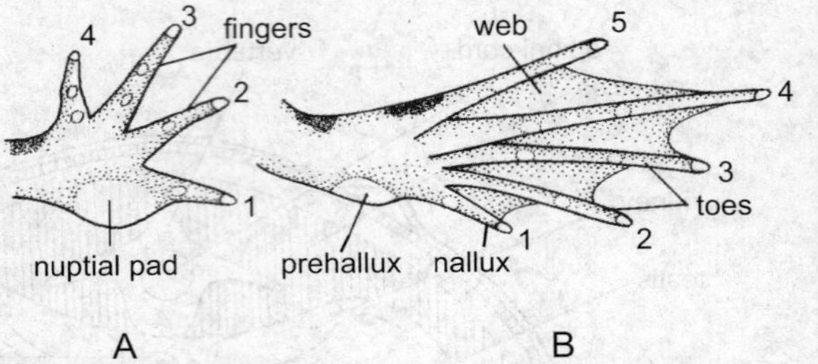

Fig. 18.5. Frog. A–Underside of right hand of male in breeding season showing nuptial pad; B–Root showing web.

SKIN

The skin is smooth, moist, slippery and lacking in the external protective scales or hairs. It is loosely attached by thin bands of connective tissue to the underlying musculature due to subcutaneous lymph spaces and, thus, these animals are easily skinned. It is considerably thicker on the dorsal side of the body than it is below. At the dorsal side of the body it is thrown into a number of folds which extend from behind the eyes. The ridges, thus, formed by the thickening of the skin are known as **dorsolateral dermal plicae.** Its colour on the back and the limbs is dark green with dark coloured streaks and patches, while on the ventral side it is pale yellow.

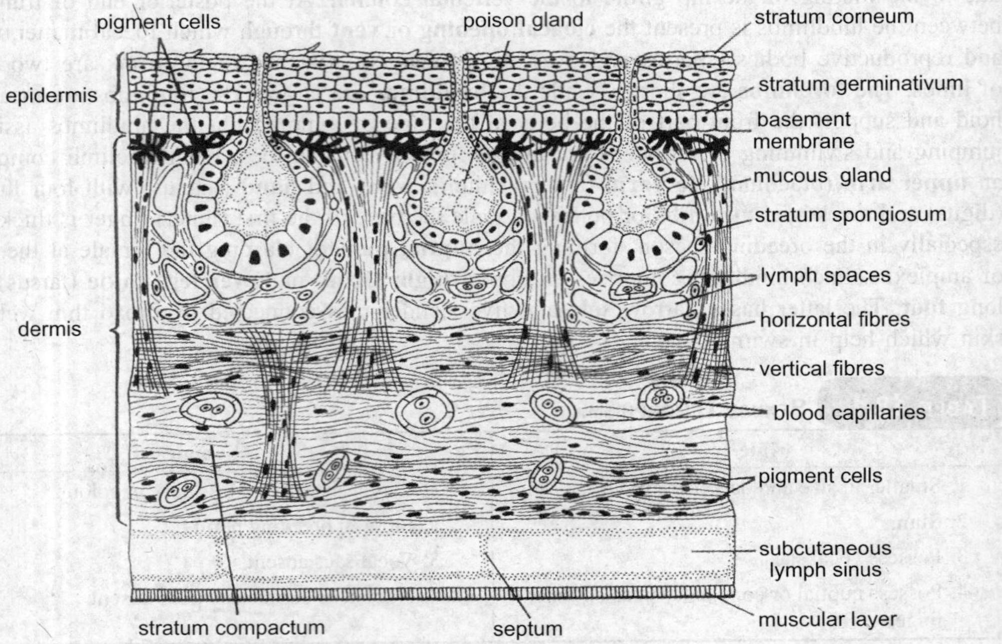

Fig. 18.6. Frog. V.S. of skin.

Structure. Structurally, like other vertebrates, the skin is composed of two layers, the **epidermis** and the **dermis**.

Epidermis. The epidermis is an outer layer which is non-vascular, stratified and further composed of several layers of epithelial cells. It is usually shed and renewed at regular intervals by a process of moulting. The outermost layer is keratinized and made up of flattened, squamous epithelial cells. It is known as **stratum corneum**. The innermost layer called **stratum germinativum** or **stratum Malpighii** is made up of active columnar epithelial cells which are capable in producing the new cells that pass towards the outer surface and become more and more flattened and ultimately lose their columnar shape as they reach the surface. It is due to the gradual change of protoplasm of these cells into a horny substance called **keratin**. As the old cells are worn out due to friction, they are replaced by new ones formed by the cells of the layer stratum germinativum or stratum Malpighii. The **stratum corneum** is shed off from time to time and eaten by frog. The shedding of stratum corneum is due to the secretions of thyroid and pituitary glands. This activity is known as **moulting**.

Dermis. The **dermis** forms a tough, flexible and somewhat elastic layer just underneath the epidermis. The dermis is separable into two layers, an outer comparatively loose layer (**stratum spongiosum**), which contains most of the glands, and an inner layer (**stratum compactum**) formed of dense connective tissue. The stratum spongiosum consists of a loose network of fibrous connective tissue, richly supplied with lymph spaces and blood vessels. Just beneath the epidermis it forms a thin layer which contains numerous pigment cells. Each cell is irregular in shape with branched processes. They have black and yellow pigments and impart colouration to the skin. In the deeper portion are embedded the glands.

The stratum compactum is composed of a dense layer of connective tissue whose fibres run in a wavy course parallel to the surface of the skin. At intervals this layer is crossed by vertical strands, which often extend through the stratum spongiosum into the epidermis. In addition to fibrous connective tissue, these strands contain smooth muscle fibres, elastic fibres, nerves and blood vessels.

The subcutaneous connective tissue forms a loose layer beneath the stratum compactum and a second very thin layer next to the muscles. The two layers are separated by large lymph spaces except in the septa, where they become continuous.

Glands. In the skin of frog two types of glands are found—the **mucous glands** and the **poison glands**. Actually these glands are the derivatives of the epidermis but they lie in the stratum spongiosum of the dermis.

The **mucous glands** are somewhat smaller, flask-shaped found in abundance practically over the entire surface of the body. Their ducts are narrow and lined with a layer of small flattened epithelial cells. The body of the gland is also lined by a single layer of epithelial cells except near the opening of the neck, where there are two layers. It is this epithelium which forms the mucus which is discharged into the lumen of the gland, and poured out through the neck over the surface of skin. The mucus is a colourless watery fluid which keeps the skin moist, glistening and sticky.

Outside of the epithelium of glands is a muscular coat of smooth muscle cells. The function of the muscle cells is the expulsion of the mucus of the glands.

They are more numerous on the dorsal side of the body and hindlegs, and they are especially abundant, and large in the dermal plicae.

These glands are lined with a single layer of epithelial cells and communicate with the exterior through their respective fine ducts which are narrow and lined with a layer of small flattened epithelial cells. Their epithelial cells are cylindrical nearly filled with granules. Outside the epithelium like mucous glands, is a muscular coat and a connective tissue coat. The secretion of the poison glands is a whitsh fluid with a burning taste. It protects the animal in some degree from the enemies.

Functions of the skin. The skin of frog performs the following functions :

1. It gives definite shape and texture to the body and also acts as a protective covering over the body.

2. It protects the body against the invasion of foreign bodies and fungal spores.

3. The mucous glands keep the skin moist, glistening and sticky. The mucus also prevents the invasion of the water and other harmful materials dissolved in water.

4. It forms a chief respiratory organ as its moist surface brings about an exchange of respiratory gases (O_2 and CO_2) in between the body of the animal and the environment.

5. Being devoid of sweat glands it acts as an excretory organ as the shedding of stratum corneum from time to time helps in removing the excretory wastes which are no longer needed for the body.

6. Due to presence of nerve endings it acts as an important sensory organ.

7. The frog never drinks the water through buccal cavity but absorbs through skin and, thus, compensates the loss of water from body. The skin is loosely attached to the body, and a considerable quantity of water may collect in the large subcutaneous lymph spaces.

8. The skin of frog larva produces **hatching enzymes** which dissolve the egg membrane so that hatching may occur.

ENDOSKELETON

The skeleton that supports the soft parts lies internally and is designated as the endoskeleton. It is chiefly made of bones and cartilages. These two structures are joined with one another to form the internal framework. The endoskeleton is described under two broad heads : (*i*) the **axial skeleton** and (*ii*) the **appendicular skeleton**. The axial skeleton comprises the **skull, vertebral column** and **sternum** and the appendicular skeleton consists of **pectoral** and **pelvic girdles**, and the **skeleton** of **paired limbs**.

Axial Skeleton

1. Skull. The skull of frog is broad and flat and consists of a narrow **cranium** or brain box, paired **sense capsules**, large **orbits**, the **jaws, hyoid** and **cartilages of larynx**.

2. Cranium. The cranium is the narrow cavity lodging the brain, hence, it is also called brain box. In the front part of the cranium is a **sphenethmoid** enclosing the forebrain and olfactory sacs.

Skull is divided by a transverse partition into an anterior ethmoidal region and a posterior sphenoidal region which encloses the forebrain. The ethmoidal region encloses the olfactory sacs. Sphenethmoid is only exposed laterally. It is dorsally covered by a pair

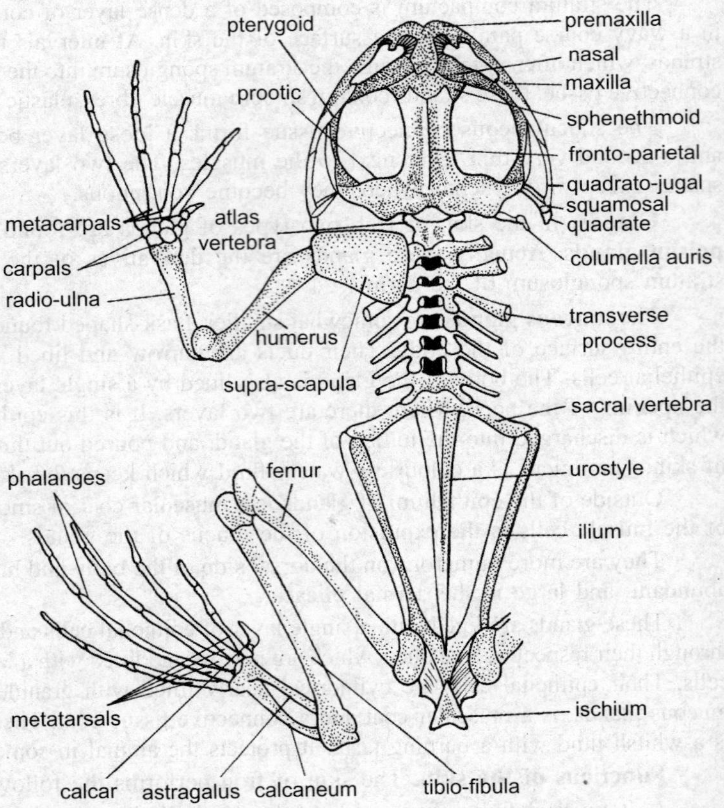

Fig. 18.7. Frog. Skeleton (Left side only).

of nasals and a fronto-parietal except a small diamond-shaped area. It is ventrally covered by a dagger-shaped parasphenoid.

Occipital segment. At the posterior end of the cranium is a **foramen magnum** surrounded by two **exoccipitals**. Each exoccipital bears at its posterior end a convexity, the **occipital condyle** which articulates with the concavity of the atlas vertebra. The auditory capsule is fused on the outer side of each exoccipital. Each auditory capsule has a **prootic** in front, the capsule has an aperture, the **fenestra ovalis** into which a cartilaginous **stapedial plate** fits. The stapedial plate is a part of the **columella** derived from the hyomandibular.

Olfactory capsules. The olfactory capsules have two **nasals** dorsally and two **vomers** ventrally, the vomers bear vomerine teeth. A pair of special bones called **septomaxillary** (ethmoids) form the boundary of nostrils. They are associated with and surround the Jacobson's organ.

Optic capsules. They enclose the eyes and are not fused with the skull.

Visceral Skeleton

Visceral skeleton includes the upper and lower jaws, the hyoid apparatus, the columella auris and cartilages of larynx.

Upper jaw. The upper jaw has two halves, each half has an anterior **premaxilla** followed by a long **maxilla**, both bear teeth. The posterior part of the upper jaw has a small **quadratojugal**. Its broad posterior end unites with quadrate cartilage, which is a small thin rod forming the suspensorium. The mandible articulates with the quadrate cartilage. Ventral anterior to the orbit is a slender, rod-like **palatine**. At the posterior lateral end of cranium is present a large 3-rayed or Y-shaped **pterygoid**. It articulates anteriorly with the maxilla and palatine and on the inner side with the parasphenoid and auditory capsule, and posteriorly with the quadrato-jugal and quadrate cartilage. At the posterior dorsolateral end of cranium is the hammer-shaped bone, the **squamosal**. It lies above the pteryoid. Its anterior limb or head is free and the short posterior limb articulates with the auditory capsule and prootic. Its handle joins with the quadrate cartilage.

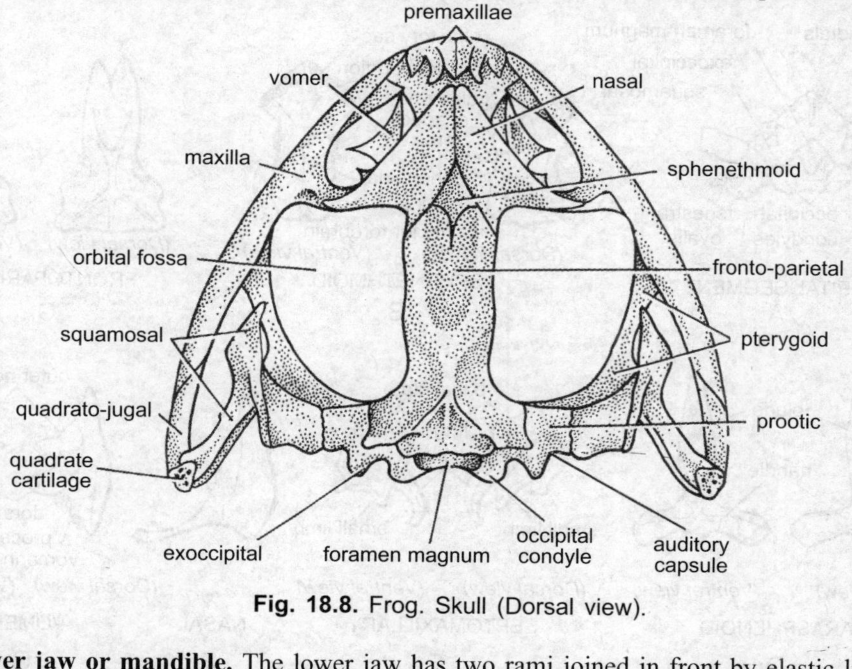

Fig. 18.8. Frog. Skull (Dorsal view).

Lower jaw or mandible. The lower jaw has two rami joined in front by elastic ligament. Each half has a core of **Meckel's cartilage** covered over by an **angulosplenial** forming the inner and posterior portion of each ramus. Its anterior end is tapering and posterior end possesses dorsally

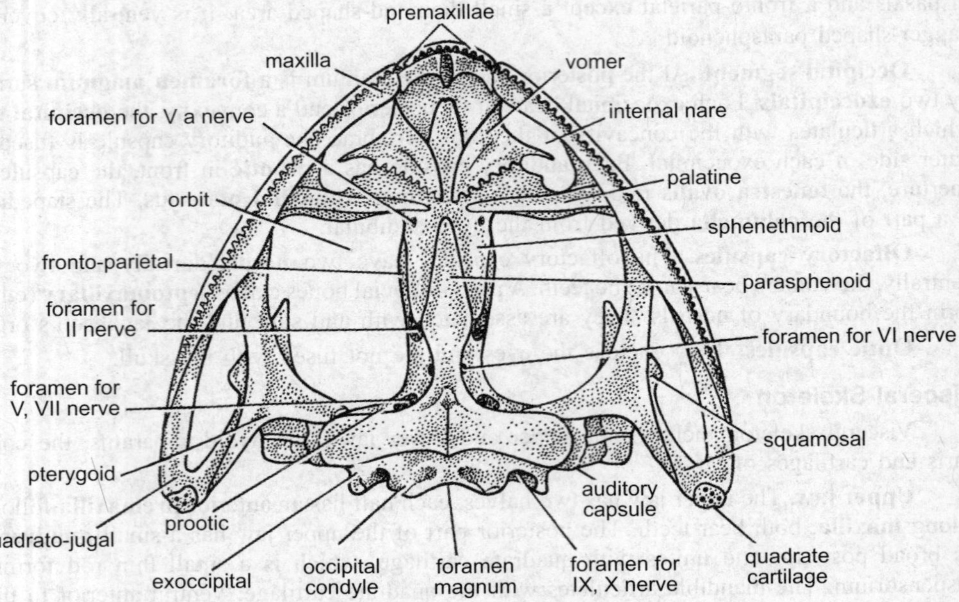

Fig. 18.9. Frog. Skull (ventral view).

a condyle for the articulation with the quadrate cartilage. Just anterior to the condyle is present a small dorsal projection, the **coronary process**. Anterior outer surface of Meckel's cartilage is covered by a small, flat, dogger-like **dentary**. It extends anteriorly up to Mentomeckelian bone and its posterior part is attached with the outer side of angulosplenial. At the extereme anterior end of Meckel's cartilage is a small cartilage bone, the **Mentomeckelian**.

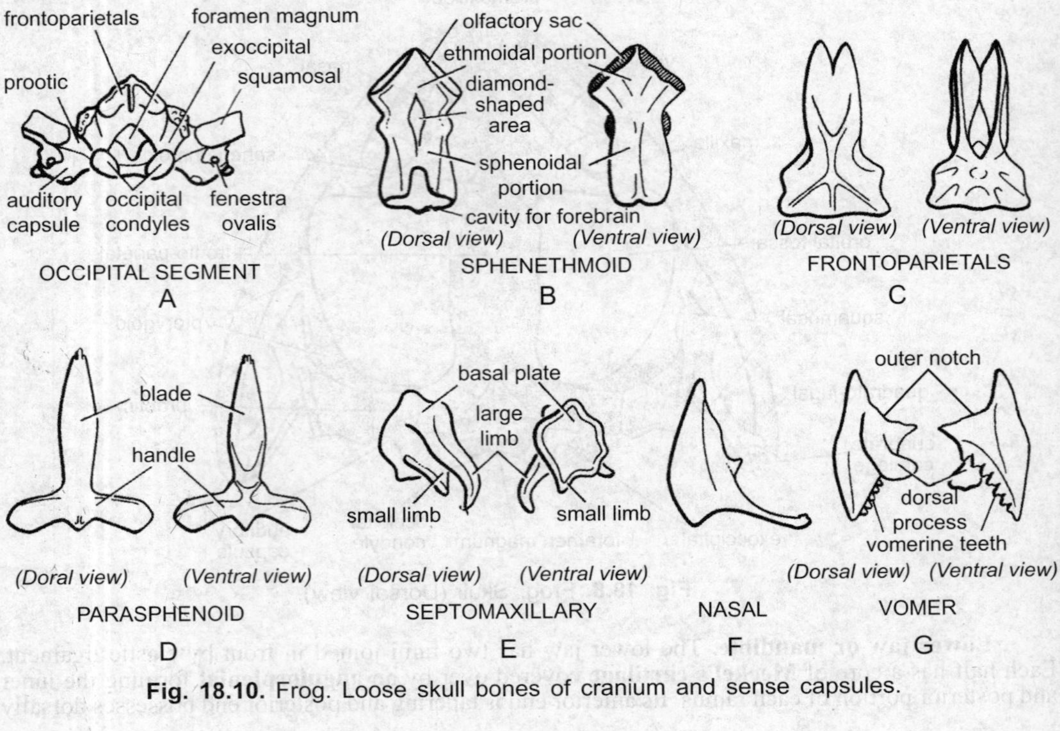

Fig. 18.10. Frog. Loose skull bones of cranium and sense capsules.

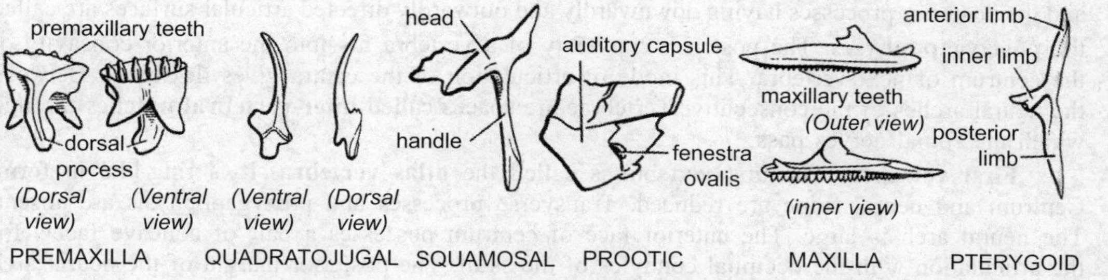

Fig. 18.11. Frog. Loose skull bones of upper jaw.

Hyoid. Hyoid has a thin body made of cartilage. It is produced in front into two **alar processes** and behind into two posterior lateral processes. In front are also two long slender anterior cornua of cartilage, they curve backwards and join the auditory capsules. Posteriorly is a pair of strong bony posterior cornua or **thyrohyals** between which lies the glottis. The hyoid lies in the floor of the pharynx, beneath the tongue.

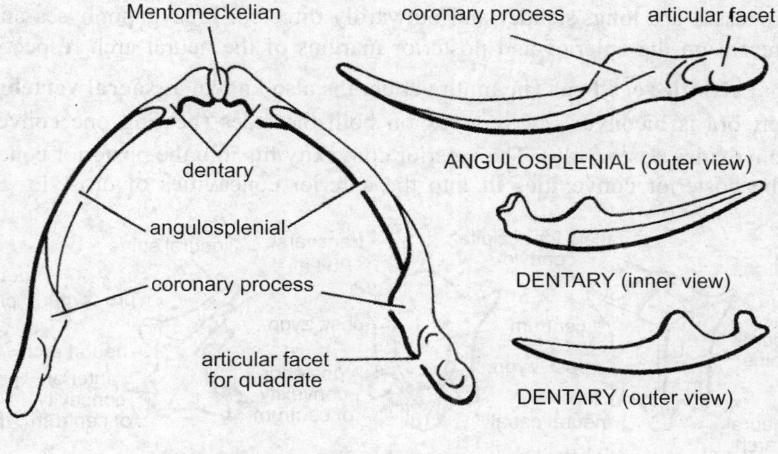

Fig. 18.12. Frog. Bones of mandible (lower jaw).

Vertebral Column

The vertebral column or backbone of frog encloses and protects the spinal cord. It is very much short due to the absence of tail. It is composed of nine vertebrae and a terminal rod-like structure called the **urostyle**. The first, eighth and ninth vertebrae are peculiar, while vertebrae from second to seventh are almost similar in structure.

Typical vertebra. A typical vertebra has a solid cylindrical part known as the **centrum**. The centrum is **procoelous**, *i.e.*, it is concave in front and convex behind. On the dorsal side, the centrum bears a ring-like **neural arch** which encloses the **neural canal**. The neural arch possesses a backwardly directed spinous process or **neural spine**. The lateral sides of the neural arch carries transverse processes. The neural arch possesses two articulating processes. The anterior processes having upwardly and inwardly directed articular surfaces are known as the **prezygapophyses**

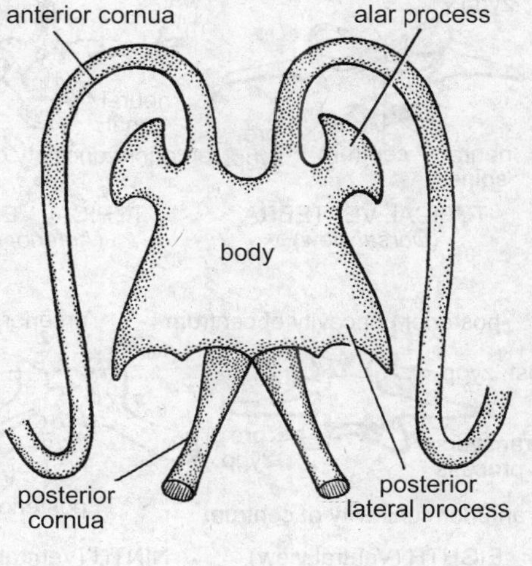

Fig. 18.13. Frog. Hyoid apparatus.

and the posterior processes having downwardly and outwardly directed articular surfaces are called the **postzygapophyses**. The posterior convexity of a vertebra fits into the anterior concavity of the centrum of next vertebra. This mode of articulation of the centra gives flexibility. Between the neural arches of the consecutive vertebrae are spaces called **inter-vertebral notches** through which the spinal nerves pass.

First vertebra. The first vertebra is called the **atlas vertebra**. It is ring-like in form. Centrum and neural spine are reduced. Transverse processes and prezygapophysis are absent. The neural arch is large. The anterior face of centrum possesses a pair of concave facets for the articulation with the occipital condyles of the skull. The posterior margin of the neural arch bears a pair of **postzygapophyses**.

Eighth vertebra. The centrum of eighth vertebra is **amphicoelous**, *i.e.*, concave on both the sides. The anterior concavity receives the posterior convexity of the VIIth vertebra. Transverse processes are long, slender and outwardly directed. Prezygapophyses and postzygapophyses are present on the anterior and posterior margins of the neural arch respectively.

Ninth vertebra. The ninth vertebra is also known as **sacral vertebra**. The centrum of ninth vertebra is biconvex, *i.e.*, convex on both the sides (bearing one convexity anteriorly and two convexities posteriorly). The anterior convexity fits into the posterior concavity of eighth vertebra. The posterior convexities fit into the anterior concavities of urostyle. Transverse processes are

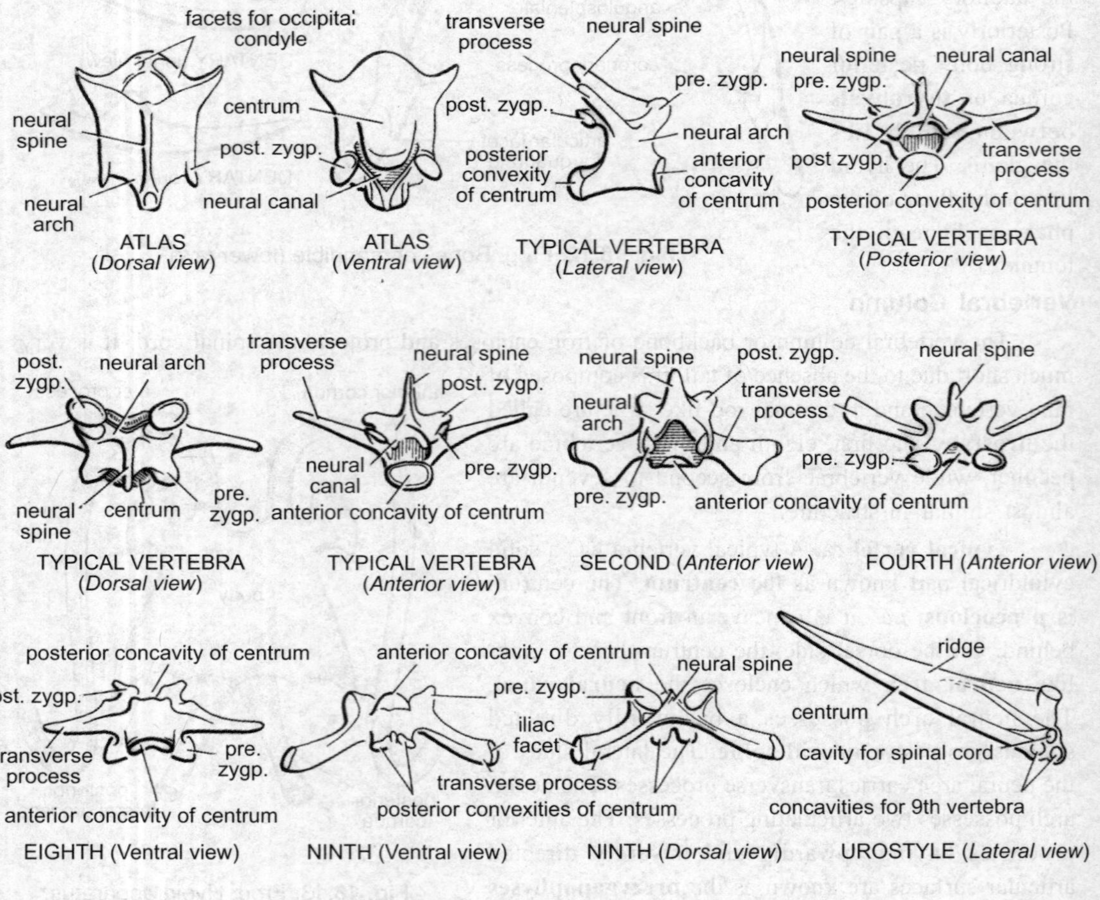

Fig. 18.14. Frog. Vertebrae.

cylindrical, stout and backwardly directed. **Iliac facet** is present at the tip of each transverse process for the articulation of ilium bone of pelvic girdle. Neural spine is inconspicuous, *i.e.*, greatly reduced. Prezygapophyses are well developed along the anterior end of neural arch, while the postzygapophyses are entirely absent.

Functions of vertebral column. The vertebral column serves the following functions :

1. It supports the trunk region.
2. It encloses and protects the spinal cord.
3. In front it supports the head which is held slightly above the ground.
4. It acts as a body-axis from which viscera are suspended in the body cavity by mesenteries.
5. It helps in locomotion by providing attachment in the pelvic girdle and hindlimbs.

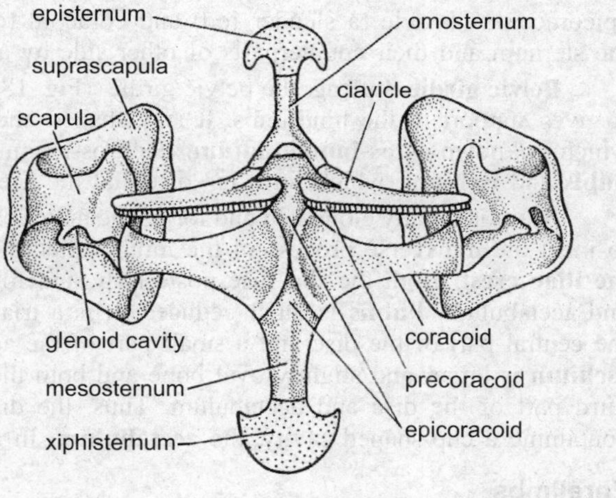

Fig. 18.15. Frog. Pectoral girdle and sternum (Ventral view).

Sternum. It lies midventrally connected between the two halves of pectoral girdle. It is composed of four parts. Anterior to the clavicle lies an inverted Y-shaped bony **omosternum** which is anteriorly attached with the rounded, flat cartilaginous **episternum**. Posterior to the epicoracoid and coracoids is a bony rod-like **sternum proper** or **mesosternum** to which is attached a broad cartilaginous **xiphisternum** posteriorly. Ribs are absent.

Appendicular Skeleton

Pectoral girdle. The pectoral girdle or shoulder girdle (Fig. 18.15) is present in the thoracic region and provides attachment to the forelimbs and their muscles. It also protects the inner softer parts of the thorax. It consists of two similar halves united mid-ventrally and separated dorsally. Each half is divided into a dorsal **scapular portion** and a ventral **coracoid portion**.

Scapular region. The scapular portion comprises the **suprascapula** and **scapula**. **Suprascapula** is a thin, flat, somewhat rectangular cartilaginous plate on the dorsal side. It covers dorsally the first four vertebrae. Its free margin is calcified and the lower part articulating with the scapula is bony. At its posterior end is present a glenoid cavity into which articulates the head of humerus. **Scapula** is a strong, flat bony plate broad towards the end and narrower in the middle.

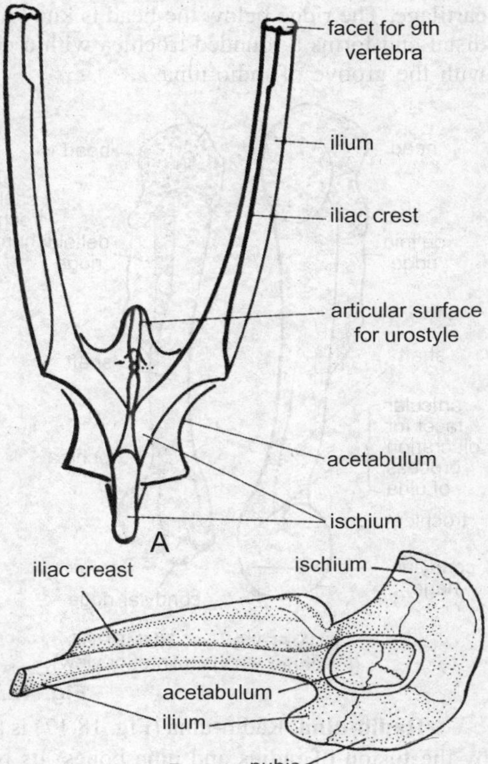

Fig. 18.16. Frog. Pelvic girdle, A–Dorsal view; B–One half (os-innominatum) in lateral view.

Coracoid portion. The coracoid portion comprises the clavicle, coracoid, precoracoid and epicoracoid. Clavicle (a slender rod) and coracoid (dumb-bell-shaped) meet mid-ventrally with the sternum and their counterparts of other side by a strip of cartilage, the epicoracoid.

Pelvic girdle. In frog, the pelvic girdle (Fig. 18.16) lies in the posterior region of the trunk. It gives support to the hindlimbs. It is V-shaped and composed of two similar halves, each of which is known as **os-innominatum**. Each os-innominatum is composed of three bones, **ilium, pubis** and **ischium**, which form the disc and the acetabulum.

Ilium is greatly elongated and forms the major part of each os-innominatum. It runs forwards to meet the transverse process of the ninth vertebra. It bears a prominent dorsal vertical ridge, the **iliac crest**. Both the ilia fuse posteriorly and form the anterior and upper half of the disc and acetabulum. **Publis** is much reduced. It is a triangular piece of calcified cartilage, forming the central part of the disc and a small part of the acetabulum. Both the pubes are also fused. **Ischium** is larger and slightly oval bone and both the ischia fused in the middle and form one-third part of the disc and acetabulum. Thus, the disc is formed by the union of three bones containing a cup-shaped cavity, the **acetabulum**. In acetabulum the head of femur articulates.

Forelimbs

The bones of the forelimbs include **humerus**, **radio-ulna** and the **bones of hand**.

Humerus. Humerus (Fig. 18.17) is the bone of the upper arm of forelimb. It is a short, stout and cylindrical bone with a slightly curved shaft. Its proximal rounded end is known as the **head** which fits into the glenoid cavity of pectoral girdle. The head is covered with calcified cartilage. The ridge below the head is known as **deltoid ridge** to which muscles are attached. The distal end forms a rounded trochlea with a condylar ridge on either side. The trochlea articulates with the groove of radio-ulna.

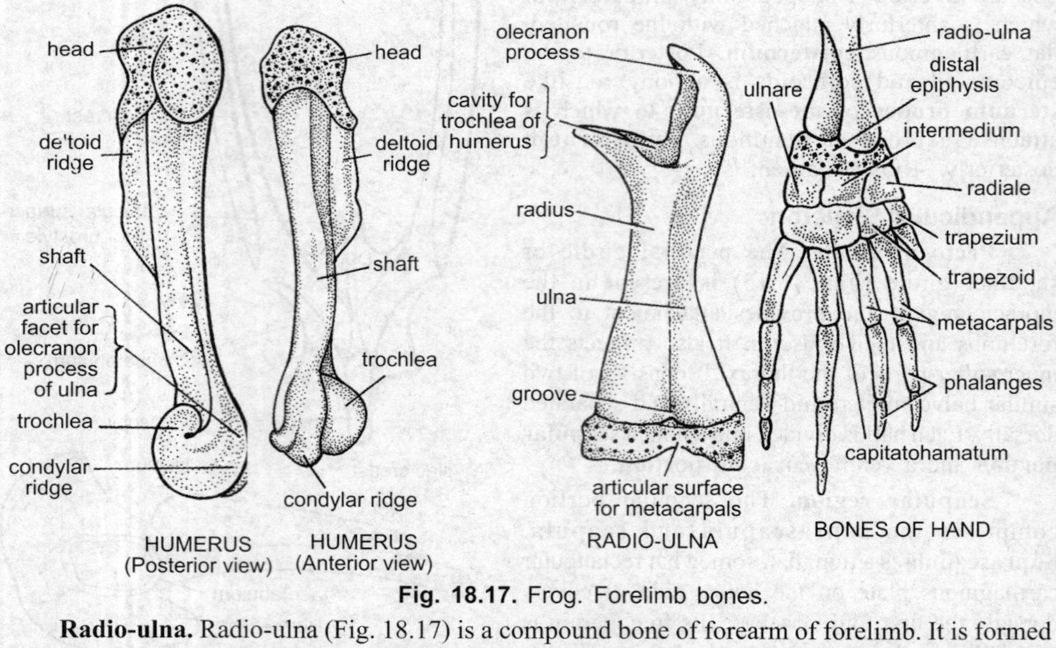

Fig. 18.17. Frog. Forelimb bones.

Radio-ulna. Radio-ulna (Fig. 18.17) is a compound bone of forearm of forelimb. It is formed by the fusion of radius and ulna bones. Its proxinial end has a concavity to receive the trochlea of humerus. The proximal portion of ulna projects into an **olecranon process** forming the elbow joint. The distal portion of radio-ulna is somewhat flat having a median groove dividing it into an anterior (radial) part and a posterior (ulnar) part terminating into a facet providing articular surface for the proximal row of carpals.

Bones of hand. The bones of wrist are called **carpals**. The carpal bones are six in number and are arranged in two rows of three each. The bones of the proximal row are called **ulnare**, **intermedium** (centrale) and **radiale**. These bones articulate with the radio-ulna. The bones of the distal row are called **capitohamatum, trapezoid** and **trapezium**. These bones articulate with the metacarpals.

The hand or **manus** is provided with five slender metacarpals. The first metacarpal is rudimentary. The digit corresponding to thumb is absent, the remaining four metacarpals are supported byphalanges. The second digit bears two phalanges. The third and fourth digits bear three phalanges each.

Hindlimbs

The bones of handlimbs include **femur**, **tibio-fibula**, **astragalus-calcaneum** and **bones of the foot**.

Femur. Femur (Fig. 18.18) is the bone of thigh region of hindlimb. It is long and slender having a slightly curved shaft. The proximal swollen end is called the **head.** Head fits into the acetabulum of pelvic girdle forming a ball and socket joint. The distal end forms a condyle which articulates with the tibio-fibula. The head and condyle are covered by calcified cartilage.

Tibio-fibula. Tibio-fibula (Fig. 18.18) is a compound bone of shank region of hindlimb. It is formed by the fusion of inner tibia and outer fibula bones forming a single bone called the **tibio-fibula**. In between the two is a median longitudinal groove. The proximal and distal ends are covered by calcified cartilage. Near the proximal end tibia bears an **cnemial** or **tibial crest**. The proximal end articulates with the femur, while the distal end articulates with the **astragalus-calcaneum**.

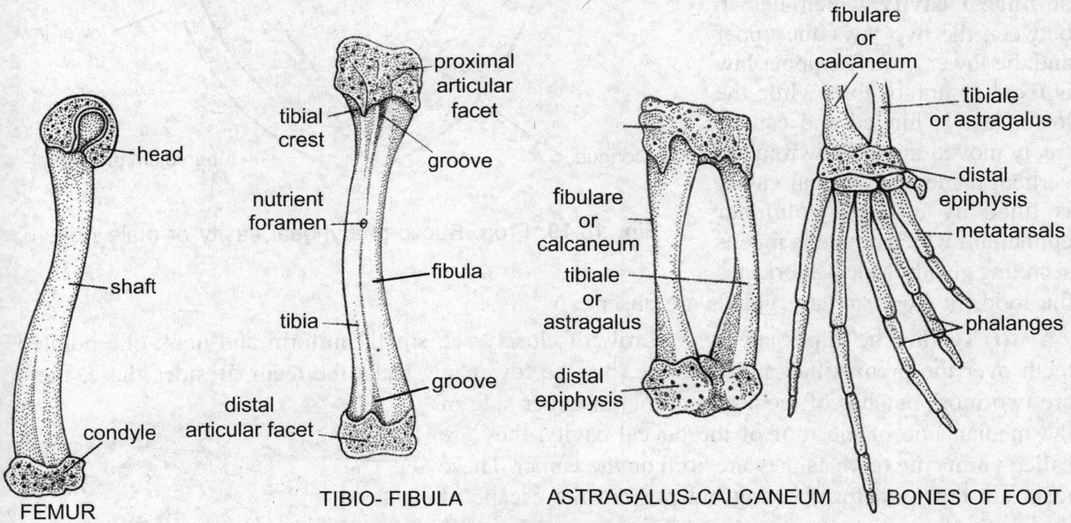

Fig. 18.18. Frog. Hindlimb bones.

Astragalus-calcaneum. Astragalus-calcaneum (Fig. 18.18) is a compound bone of ankle of hindlimb. The ankle consists of two rows of four tarsal bones. The proximal row consists of two long bones fused together at their proximal and distal ends with a wide gap in the middle. The inner bone is thinner and slightly curved, called the **astragalus** or **tibiale**, while the outer bone is thicker and straight, called the **calcaneum** or **fibulare**. The proximal and distal ends are covered by epiphyses of calcified cartilage. Distal row of tarsals has two very small bones.

Bones of foot. The foot or **pes** (Fig. 18.18) is supported by five long and slender metatarsals bearing five true toes, having 2, 2, 3, 4 and 3 phalanges respectively. A small preaxial sixth toe composed of 2 or 3 bones is present on the inner side of the first toe or **hallux**. Preaxial sixth toe is called the **prehallux** or **calcar** and does not project from the toe.

DIGESTIVE SYSTEM

The digestive system of frog mainly includes the **alimentary canal** and the **digestive glands**. In the alimentary canal the processes of mastication, digestion and absorption take place, while the digestive glands secrete certain enzymes which bring about the digestion of the ingested food.

Alimentary Canal

The alimentary canal of the frog is essentially a coiled tube of varying diameter that extends from mouth to cloacal aperture. It consists of the **mouth, buccal cavity**, **pharynx**, **oesophagus**, **stomach, intestine** and **cloaca**.

1. Mouth. The alimentary canal starts with an aperture which is known as **mouth opening**. Mouth is a very wide gape bounded by upper and lower jaws covered by immovable lips.

2. Buccal cavity. Mouth leads into a wide and broad **oral** or **buccal cavity** which lies in between the two jaws–the upper and the lower jaw. The upper jaw is fixed or immovable, while the lower jaw is hinged and can be freely moved up and down in the vertical plane. The buccal cavity is lined by ciliated, columnar epithelium which contains mucus secreting glands. Mucus lubricates the food. In frog, salivary glands are absent.

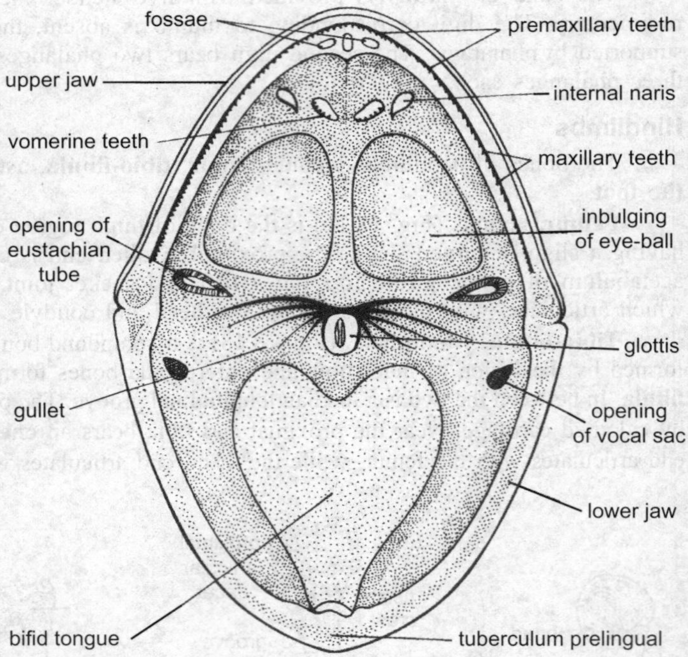

Fig. 18.19. Frog. Bucco-pharyngeal cavity of male frog.

(*i*) Teeth. The upper jaw bears a row of closely set, small, uniform and hook-like pointed teeth over the premaxillae and maxillae, but the lower jaw lacks the teeth. Besides these, there are two more patches of teeth found one on either side of the median line of the roof of the buccal cavity, they are called **vomerine teeth** as they are born on the vomer. These teeth are not used for chewing but prevent the escape of the captured prey.

The teeth are **homodont**, *i.e.*, similar in shape. They are curved backwards and attached to the bones **(acrodont)** instead of being set in sockets. Each tooth is conical in its shape and consists of two parts–the **base** and the **crown**. The **base** is attached to the jaw bone and is made of a bone-like substance. The **crown** is the free end of tooth and is formed of **dentine** traversed by numerous fine branching canals or **canaliculi** leading from the interior of the tooth. The tip of the crown is coated with a very hard, resistant, shining white layer of **enamel** substance. In the interior of each tooth there is a cavity which is called

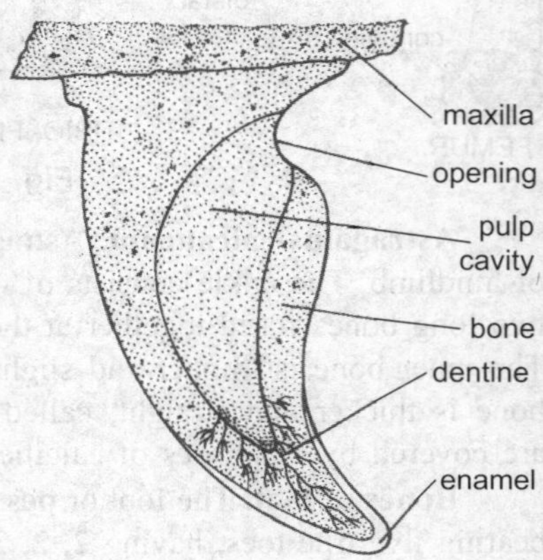

Fig. 18.20. Frog. L.S. of a tooth.

pulp cavity. It is filled with a highly vascular and nutritive tissue or **pulp** which also contains **odontoblast cells** which produce new material for the growth of teeth. During lifetime the old and worn out teeth are being continuously replaced by new ones which are produced below the old ones which are cast out or lost. As the teeth are replaced several times during the lifetime of the frog, so they are called **polyphyodont**.

(*ii*) Internal nostrils. The buccal cavity in its roof near the vomerine teeth has two openings, the **internal** or **posterior nares** connecting with the nasal cavities through which the respiratory gases pass to and from the buccal cavity during respiration.

(*iii*) Tongue. The large, thick, fleshy and protrusible tongue attached in front to the inner border of lower jaw, and free and notched behind lies on the floor of the buccal cavity. Its upper surface bears **taste buds** in the form of small papillae and **mucous glands** of which the secretions make the tongue sticky. Neither taste buds nor mucous glands produce any digestive enzymes. The tongue can be thrown out and retracted suddenly to capture and engulf insects. According to **Hartog**, the throwing out of the tongue is brought about by the sudden flowing of squeezed lymph from the lymph sac to another due to muscular contraction but the swallowing is accomplished simply by raising the floor of the buccal cavity in which a flat **hyoid cartilage** is embedded.

(*iv*) Bulgings of orbits. The roof of buccal cavity behind the vomerine teeth has two large, oval pale areas, the bulgings of eyeballs. During swallowing of food, eyes are depressed into the buccal cavity which pushes the food into the pharynx.

3. Pharynx. The buccal cavity narrows behind as the pharynx which in turn opens into the oesophagus through **gullet**. The buccal cavity and pharynx are sometimes called the **bucco-pharyngeal cavity**. In the roof of the pharynx on either lateral side is present a wide eustachean opening which communicates with the middle ear. The **glottis** which is a median slit in the pharynx behind the tongue guards the entrance to the lungs. It is always opened during breathing but closes while food is being swallowed. In male frog two openings of **vocal sacs** are also formed in the angle of the lower jaw on the floor of the pharynx. These act as resonators at the time of croaking.

4. Oesophagus. The gullet leads into a short, broad and muscular part of the alimentary canal called **oesophagus**. This part of the alimentary canal is very short due to the absence of neck but highly distensible as its inner lining is thrown into a large number of longitudinal folds which allow the sufficient expansion of the oesophagus during the passage of the ingested food through it to the stomach. It opens into the stomach in such a way that no demarcation line is formed between the oesophagus and stomach.

5. Stomach. It is the most important part of the alimentary canal where the digestion of ingested food takes place with the help of certain digestive enzymes secreted by the digestive glands situated in its wall. It is in the form of wide and curved tube and lies between the oesophagus and intestine. It is suspended by a mesentery, the **mesogaster**, into the left side of the body cavity or the coelom. It is distinguished into two parts – the large wider anterior **cardiac stomach** and the posterior, short narrow **pyloric stomach**.

The internal lining of the stomach has numerous longitudinal folds which may allow the expansion of the stomach whenever needed. Its mucous epithelium possesses multicellular **gastric glands** which secrete **pepsinogen** enzyme, and unicellular **oxyntic glands** which secrete hydrochloric acid. The pyloric end of stomach is constricted and its opening into small intestine is guarded by **sphincter** (a circular ring-like muscle). It regulates the passage of food from the stomach into the intestine.

6. Intestine. The stomach leads into the long, tubular and coiled intestine. It is also attached to the dorsal body wall by mesentery. It consists of two parts : the **small** and **large intestine**.

(*i*) **Small intestine.** The small intestine lies in several loops supported by a fan-like membrane, the **mesentery**. The anterior region of the small intestine which curves upwards to form a U with the stomach is the **duodenum**, the rest part of it continues as the coiled **ileum**. In the duodenum opens a common **hepatopancreatic duct** from liver and pancreas bringing the bile and pancreatic juice. The internal mucous lining is thrown into low transverse folds.

The **ileum** is a long, narrow, very much coild tube and its lower end leads into **rectum** (large intestine). The internal mucous lining of ileum forms a number of longitudinal folds. True villi, glands and crypts of higher vertebrates are absent. Digestion and absorption of food takes place in this part. Its lower end opens into cloaca by spinctered anus. Its mucosal lining forms low longitudinal folds.

7. Cloaca. It is a small sac-like structure and receives the openings of anus and urinogenital apertures. Cloaca opens towards the exterior through **vent** or **cloacal spening** situated at the posterior end of the body.

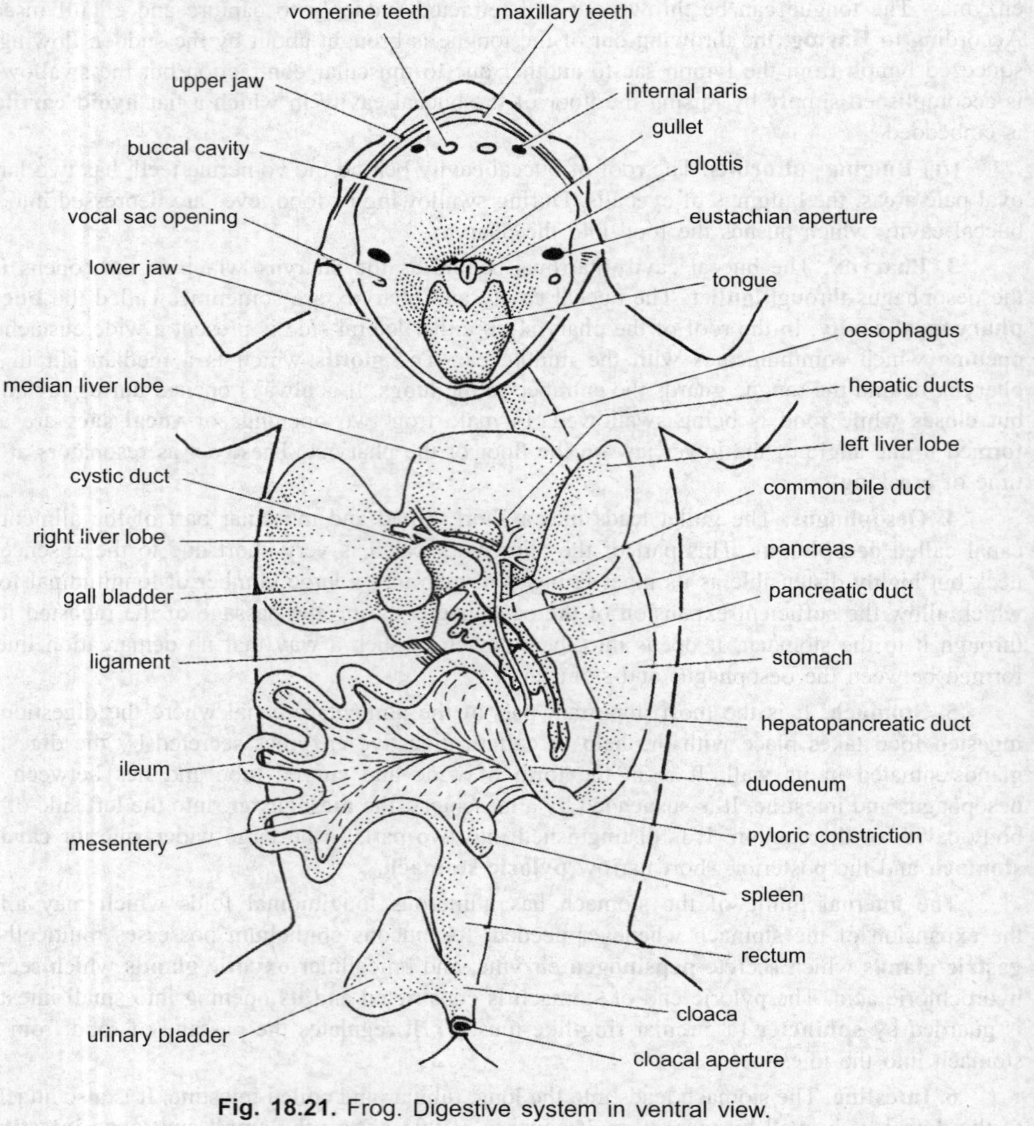

Fig. 18.21. Frog. Digestive system in ventral view.

Digestive Glands

Two large digestive glands, in addition to the glands located in the stomach and small intestine, are also found in the frog. These glands are the **liver** and the **pancreas**.

1. Liver. It is the largest reddish brown gland found in the anterior region of the body cavity close to the heart and lungs. It consists of two lobes–the right and the left. These two lobes remain connected with each other by a narrow bridge of liver tissue. The left lobe is again subdivided into two lobes. Between the right and left lobes, a thin-walled, round, large, greenish sac, the **gall bladder**, is found. It serves as a reservoir for the bile constantly secreted by the liver cells. The bile passes into the gall bladder through **cystic ducts** as well as directly into the **bile duct** by way of minute **hepatic ducts.** The hepatic and crystic ducts join to form a **common bile duct** which runs through the pancreas and opens into the duodenum. The common bile duct is also known as **hepatopancreatic duct** as it receives the fine ducts of the pancreas on its way to the duodenum. Bile lacks digestive juices. Bile only emulsifies fats, thus, liver is not a true digestive gland.

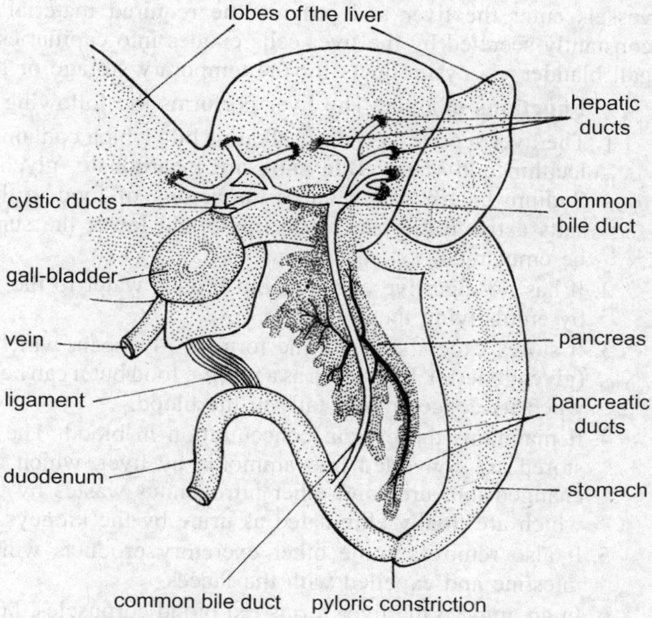

Fig. 18.22. Frog. Digestive organs and glands.

Histology of liver. The liver is composed of numerous **lobules** or **tubules** which not only branch but also anastomose with another to form a complex network due to which it is sometimes called as a **reticular gland**. These lobules are separated from each other by the presence of connective tissue containing bile capillaries, hepatic ducts, blood sinuses and blood capillaries. Each lobule is made up of numerous polyhedral, glandular hepatic cells which contain nuclei and cytoplasm along with protein granules, droplets of fats, glycogen and often black or dark brown pigment granules. The hepatic cells are arranged in columns between the bile capillaries which unite to form larger hepatic ducts. Hepatic ducts ultimately open into cystic duct directly attached with the gall bladder. The hepatic ducts of different lobes and cystic duct unite to form the **bile duct**.

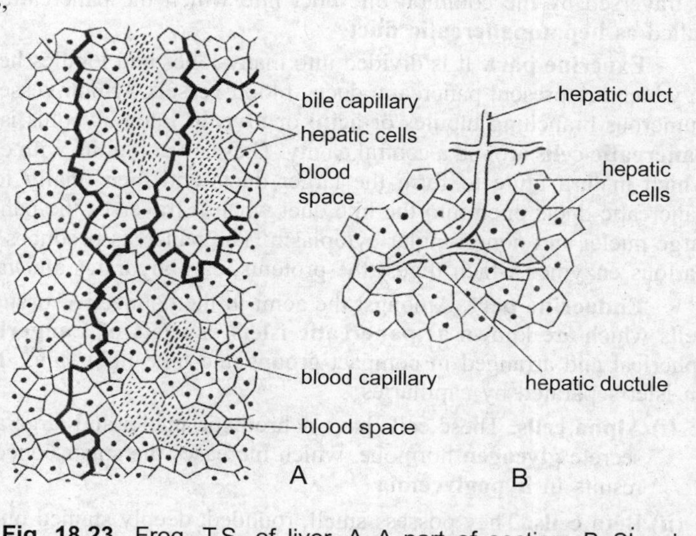

Fig. 18.23. Frog. T.S. of liver. A–A part of section; B–Showing hepatic ductule.

The liver receives blood from the hepatic artery and the hepatic portal vein. These blood vessels enter the liver and provide the required material for the formation of bile. The bile constantly secreted by the liver cells exudes into capillaries from which it passes either into the gall bladder via cystic ducts for the temporary storage or the intestine through the bile duct.

Functions of liver. The liver performs the following important functions :

1. The liver secretes a watery, alkaline bile which contains bile salts, bile pigments, cholesterol, lecithin and water. **Bile salts** are bicarbonate, glycocholate and taurocholate of sodium. Sodium bicarbonate reduces the acidity of food in the intestine, while the other two bile salts activate the pancreatic lipase and lower the surface tension of fats so that they can be emulsified.
2. It has no digestive enzymes but it adds water to the food and aids in the digestion of fats by emulsifying them.
3. It stores excess sugar in the form of glycogen, which is formed by the change of glucose **(glycogenesis)**. It is stored as a reserve food but it can be changed into glucose **(glycogenolysis)** when its concentration falls in the blood.
4. It maintains the protein concentration in blood. The excess amino acids instead of being stored are converted into ammonia by liver, which combines with carbon dioxide and is changed into urea and other nitrogenous wastes by the action of enzymes **(deamination)** which are finally eliminated as urine by the kidneys.
5. It also removes some other excretory products which are discharged with bile into the intestine and expelled with the faeces.
6. In an embryo the liver forms red blood corpuscles, but in an adult it destroys old and worn out erythrocytes by its Kupffer cells.
7. It stores copper and iron and forms vitamin A.
8. It produces **fibrinogen** and **prothrombin** which are essential for the clotting of blood. It also produces **heparin** which prevents clotting of blood in blood vessels.
9. It kills bacteria and also eliminates foreign substances from the blood.
10. **Prussic acid** is formed in the body as a by-product and it is harmful. Liver converts it into harmless potassium sulphocyanide.

2. Pancreas. It is an irregular, branched, flattened and pale coloured gland of exocrine and endocrine function, which is held by mesentery between the stomach and duodenum. It is traversed by the common bile duct into which the pancreatic ducts also open which is now called as **hepatopancreatic duct**.

Exocrine part. It is divided into many lobes and lobules held together by connective tissue in which are present pancreatic ducts, blood vessels, lymph vessels and nerves. The lobules have numerous branching tubules or **acini** or **alveoli**. Each alveolus is formed of pyramidal glandular **pancreatic cells** around a central cavity. These alveoli unite with each other through their ductules, which in turn unite to form the larger ducts and then finally form the **pancreatic ducts**. The pancreatic ducts open into the bile duct when it traverses the pancreas. The pancreatic cells have large nuclei and non-granular cytoplasm. The pancreas produces pancreatic juice which contains various enzymes which digest the proteins, carbohydrates and fats of the ingested food.

Endocrine part. Amongst the acini in the connective tissue are present compact groups of cells which are known as **pancreatic islets** or **islets of Langerhans**. These cells are somewhat spherical and arranged in compact groups and take light stain. There are three kinds of cells in an islet separated by capillaries.

(*i*) **Alpha cells.** These cells have achromatic nuclei and large acidophilic granules. These cells secrete **glycogen** hormone, which increases the sugar concentration in blood. Its deficiency results in **hypoglycemia**.

(*ii*) **Beta cells.** They possess small, rounded, deeply stained nuclei and orange-brown granules. They secrete insulin hormone.

***(iii)* D cells.** These cells possess vesicular nuclei and basophilic granules.

Insulin is essential for metabolism of carbohydrates, regulating the storage of glycogen in liver and muscles, concentration of sugar in the blood and for increasing the ability of tissues to oxidise glucose as a source of energy. Deficiency of insulin causes a disease called diabetes (hyperglycemia).

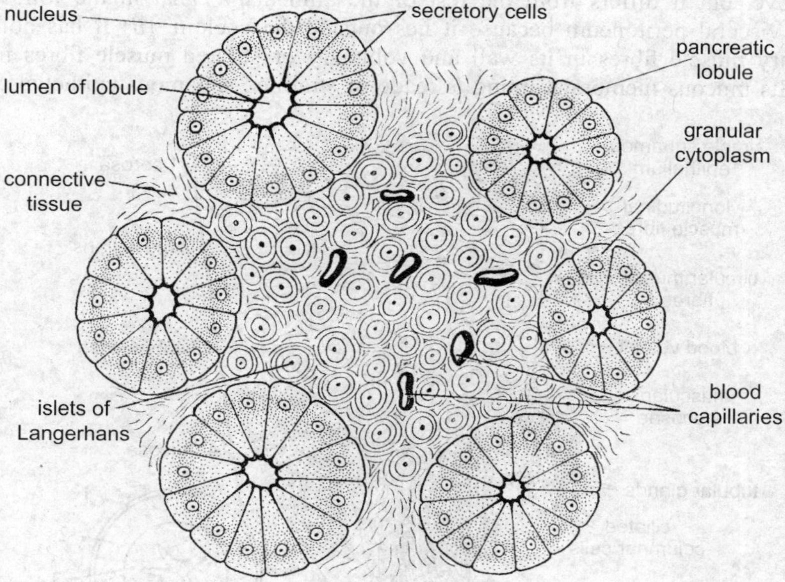

Fig. 18.24. Frog. T.S. of pancreas.

Histology of the Alimentary Canal

Histologically, the wall of alimentary canal of frog and other vertebrates is made up of four distinct concentric layers. These are present in the following order from within-outwards : (1) The mucosa; (2) The submucosa; (3) The muscularis; and (4) the visceral peritoneum or serosa.

1. Mucosa. It is the innermost layer or mucous membrane. It remains folded forming various pits and different types of glands. It is concerned with secretion and absorption of digested food, etc. It is further composed of the following layers :

***(i)* Epithelium.** It is the innermost layer formed of simple columnar epithelium (glandular and ciliated) based on thin basement membrane.

***(ii)* Lamina propria.** It is a thin connective tissue layer having blood capillaries, lymph and nerves.

***(iii)* Muscularis mucosae.** It is a narrow smooth muscle layer containing inner circular and outer longitudinal muscles.

2. Submucosa. It is a thin protective layer composed of coarse connective tissue, elastic fibres, fat, blood and lymph vessels and nerve cells. It contains glands in mammal. It also possesses Meissner plexus formed of nerve cells and fibres.

3. Muscularis. It is composed of outer longitudinal and inner circular smooth muscle fibres which are in spirals. Between these two layers of muscles is a thin layer of connective tissue having a network of nerve cells and nerve fibres of the autonomic ganglionated myocentric Auerbach plexus.

4. Visceral peritoneum or serosa. It is the outermost thin layer which lacks in the oesophagus. It is composed of very much thin connective tissue layer and an outer layer of flattened cells. It is continuous with the peritoneum lining of the body cavity called mesentery.

Certain variations are found in different parts of the alimentary canal because the above said layers, particularly the mucosa, become greatly modified in different parts of the alimentary canal to enable that part to carry on certain specific functions.

1. Oesophagus. Histologically, the oesophagus more or less has the same structure as described above, but it differs from the rest of the alimentary canal in the following facts : (*i*) It has no visceral peritoneum because it lies outside the coelom. (*ii*) It has both voluntary and involuntary muscle fibres in its wall and voluntary or striated muscle fibres in the upper portion. (*iii*) Its mucous membrane lining is made of stratified squamous epithelial cells instead

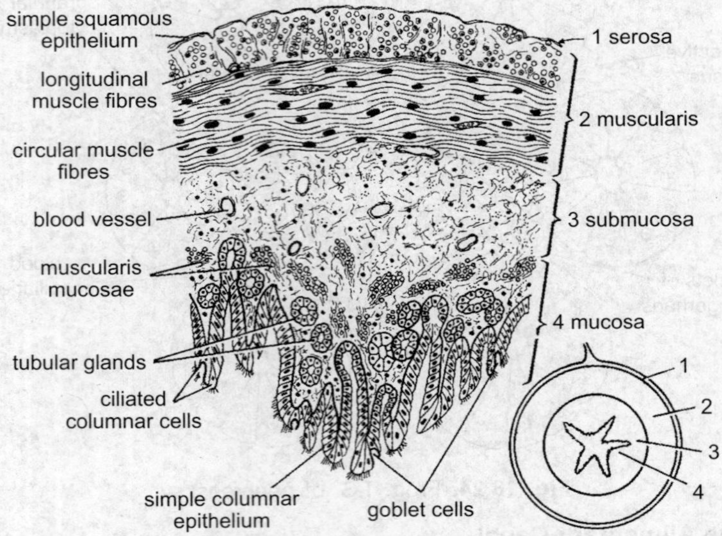

Fig. 18.25. Frog. T.S. of oesophagus.

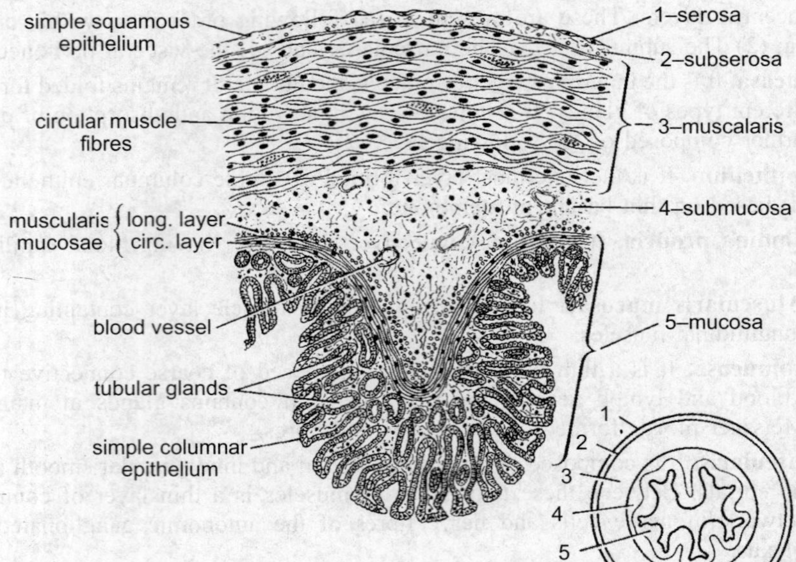

Fig. 18.26. Frog. T.S. of stomach.

of columnar cells. It has some goblet cells which produce **mucus** making the food material slippery. Mucous epithelium is invaginated into submucosa forming tubular branched glands secreting an enzyme **pepsin** for the digestion of proteins.

2. Stomach. The wall of the stomach (Fig. 18.26) is thick and made of the same typical parts of the alimentary canal, *i.e.*, mucosa, submucosa, muscularis and serosa, but it has two peculiarities :

(*i*) Stomach wall is thick and longitudinally folded internally. Folds disappear when stomach is distended. The mucous epithelium is formed of simple columnar mucous secreting gland cells. These glands are embedded in the connective tissue of lamina propria. Glands represent invaginations of the mucous epithelium. They are elongated tubular structures set very closely together, and frequently more or less branched, called **gastric glands**.

(*ii*) These glands differ in their structure at different regions of the stomach. The gastric glands of the cardiac region of the stomach are called **cardiac glands**, while of the fundus and pyloric region are called as **fundic** and **pyloric** respectively.

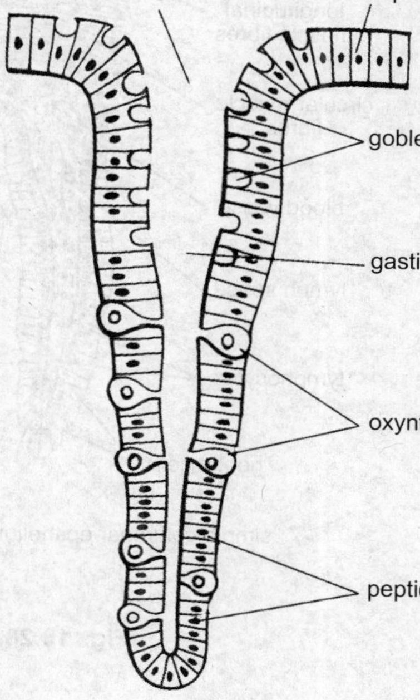

Fig. 18.27. Frog. V.S. of epithelium of a fundic gastric gland.

The **cardiac glands** are very long with deep set mouth but the **fundic** and **pyloric** glands comparatively are less deep and smaller. The cardiac and pyloric glands secrete only mucus from the surface cells. **Fundic glands** (or cardiac glands in some) have three kinds of cells, **mucus neck cells** produce mucus, **oxyntic cells** produce hydrochloric acid, they may be present in the cardiac region also, **zymogen cells** or **peptic cells** produce pepsin. Pepsin is produced in the form of inactive propepsin or pepsinogen which is soon converted into active pepsin by the hydrochloric acid.

(*iii*) Villi are absent.

3. Small intestine or duodenum. It has all the four usual coats of the alimentary canal, but its mucosa is quite thick which forms irregular, branched transverse folds which increase the absorptive surface of the alimentary canal. It is formed of columnar epithelial cells having mucous secreting gland cells. Mucosa of small intestine is thrown into numerous folds, but there are no true villi nor definite glands nor crypts of higher vertebrates. Muscularis mucosae is thin. Remaining layers are of typical type.

Ileum. It also contains usual four coats. Mucosa is thrown into several folds of various sizes and these are produced into the intestinal lumen decreasing its diameter. Mucosa has tall columnar epithelial cells in a single row within which **goblet** and **absorptive cells** are scattered. The muscle layers are less developed. Muscularis mucosa is thin and comprises only one layer of muscle fibres. True villi, crypts are lacking.

4. Large intestine or rectum. It is also composed of the same four coats of the alimentary canal. The muscularis mucosae is less developed, muscular coat is thick and both contain voluntary muscle fibres. The mucosa is thick and forms longitudinal folds and comprises numerous tubular goblet cells, secreting mucus. Mucosa epithelium is stratified near anus.

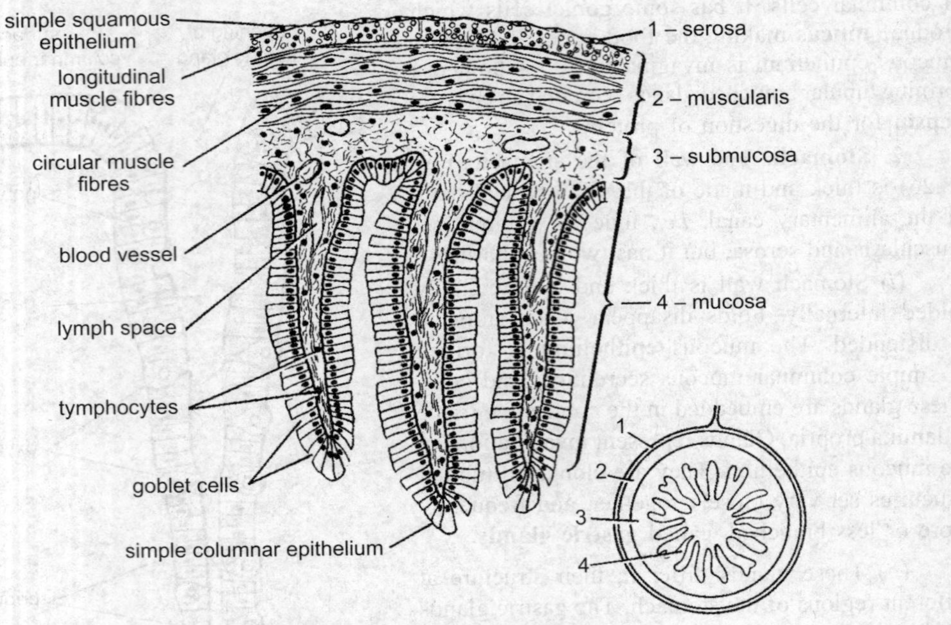

Fig. 18.28. Frog. T.S. of intestine.

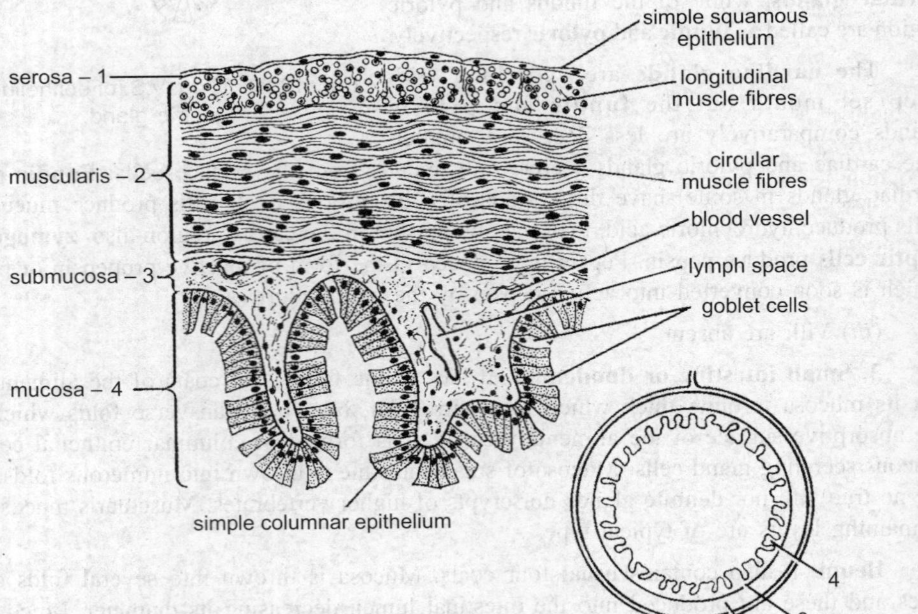

Fig. 18.29. Frog. T.S. of rectum.

Physiology of Digestion

1. Food. The frog is carnivorous, feeds chiefly on earthworms, spiders, snails, fishes, smaller frogs and other small insects which it captures and swallows whole directly into the stomach with the help of protrusible tongue.

2. Ingestion of food. Its mode of catching of its prey is remarkable. When feeding, the frog sits at a suitable place frequented by insects. When any prey comes near to it, it opens its mouth and suddenly flicks out its sticky tongue and strikes the prey. As soon as the prey comes in contact with the tongue it immediately adheres to it. Now the tongue is withdrawn into the buccal cavity. Once the prey is caught into the buccal cavity it would not be allowed to escape due to the presence of the hook-like inwardly directed maxilliary and vomerine teeth. From the buccal cavity it is pushed into oesophagus by the contraction of the pharyngeal wall and thence by peristalsis caused by the contraction and dilation of muscular wall of oesophagus, into the stomach.

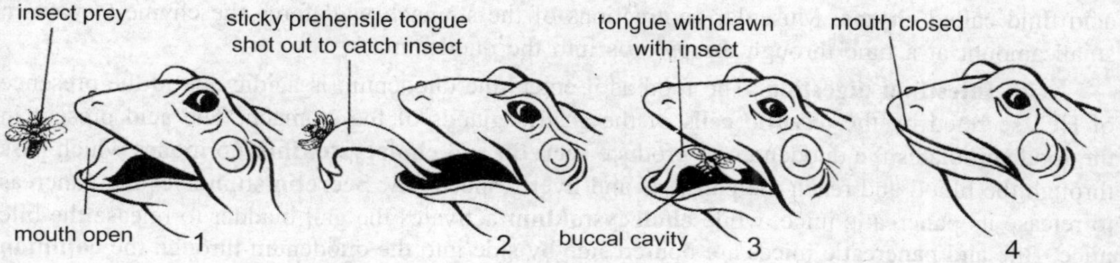

Fig. 18.30. Frog. Mechanism of capturing prey with sticky prehensile tongue.

3. Digestion of food. During feeding whatever the food is ingested by the animal that contains complex organic substances which are not of immediate use as they are insoluble and cannot diffuse through the mucous membrane lining of the alimentary canal, therefore, they must be subjected to the physical and chemical changes of the digestion so that they may be transformed into soluble forms for the immediate use of the body. The physical changes are brought about by the peristaltic movements of the alimentary canal, while the chemical changes are brought about by the organic catalysts called **enzymes** which only hasten the chemical reactions without being changed themselves. They are always specific in their property and are complex proteins which are produced by the exocrine glands and they always act at the optimum body temperature. They can also reverse the reaction, *i.e.*, the substances they had changed can be reformed. They are of different types subject to the type of food on which they act. Thus, proteins are digested by proteolytic enzymes, carbohydrates by diastatic (amylolytic) enzymes and fats by lipolytic enzymes.

***(i)* Buccal digestion.** In frog the captured prey is neither subjected to any physical change (mastication) nor any chemical action in the buccal cavity as the buccal epithelium does not have any digestive gland.

From the buccal cavity the prey is directly pushed into oesophagus where it undergoes physical changes due to constant peristaltic movement of its wall. Besides mucus, the glands of oesophagus also secrete an enzyme called **pepsin** but no digestion occurs as it does not become active till it reaches the stomach, while the mucus simply makes the active food inactive and soft and, thus, makes the passage easier.

***(ii)* Gastric digestion.** Due to oesophageal contractions and relaxations the food comes down to stomach. The stomach performs the following three main functions : (*a*) Storage; (*b*) Mechanical mixing; and (*c*) Chemical modifications.

As soon as food reaches the stomach, its wall shows a series of muscular contractions, the so-called peristaltic movements which not only allow the food to go down but also break down into small pieces and mix it thoroughly with gastric juice secreted by the gastric glands situated in the internal lining of stomach. The gastric glands secrete their secretions when they are activated by **gastrin** hormone which is produced by the stomach wall as soon as the food comes down to stomach. The gastric juice secreted by the gastric glands contains a large amount of **water**, inactive **pepsinogen** enzyme and free **hydrochloric acid**. Inactive pepsinogen changes into active **pepsin** on being mixed with hydrochloric acid. The acid also prevents the bacterial decomposition and dissolves the inorganic salts as well as makes the food soft. The pepsin of stomach along with pepsin of oesophagus acts on proteins of food and changes them into peptones and proteoses.

Pepsinogen + HCl → Active pepsin

Pepsin + Proteins → Peptones + Proteoses

After being remained for about 2–3 hours in the stomach the food is thoroughly churned and mixed by the muscular contraction of the stomach wall and takes the form of a thick creamy acid fluid called **chyme**. Muscular contractions of the stomach wall force the chyme to pass in small amount at a time through the pylorus into the duodenum.

***(iii)* Intestinal digestion.** The food as it enters the duodenum is acidic due to the presence of HCl secreted by the **oxyntic cells** of the gastric glands of the stomach. The acid present in the food stimulates the duodenum to produce **secretin** and **cholecystokinin** hormones which pass through the blood and reach the pancreas and liver respectively. **Secretin** stimulates the pancreas to release its pancreatic juice, while **cholecystokinin** activates the gall bladder to release the bile juice. Bile and pancreatic juices are poured side by side into the duodenum through the **common hepatopancreatic duct**. At the same time the intestinal mucosa also secretes the intestinal juice called **succus entericus** with the help of **enterocrinin** hormone. Thus, three substances mix up with the food in intestine, *i.e.*, bile, pancreatic juice and succus entericus. All these act on food to digest it thoroughly.

(*a*) Bile. It is a greenish alkaline fluid which contains no digestive enzymes so that it does not take any part in the digestion of food. It is alkaline because it contains certain inorganic salts like sodium bicarbonate, sodium glycocholate and sodium torocholate which not only neutralise the acidity of the semidigested food, chyme but also prepares it for the activity of enzymes of pancreatic juice which can act only in the alkaline medium. They also break down fats to form small globules which can be emulsified. Bile juice also activates the fat digesting enzyme of the pancreas, the lipase.

(*b*) Pancreatic juice. It is also a watery alkaline fluid containing three powerful enzymes called **trypsinogen**, **amylopsin** (amylase) and **steapsin** (lipase), all of which act in an alkaline medium and hydrolyse all the three types of food substances.

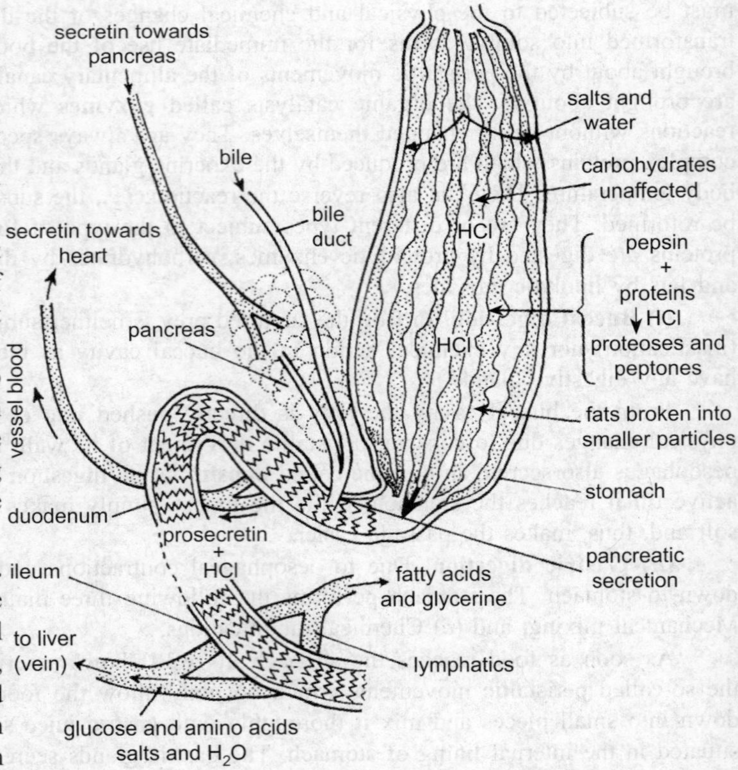

Fig. 18.31. Frog. Diagrammatic representation of the internal lining and physiology of the gut.

(*i*) **Trypsinogen** is an inactive enzyme unless it is mixed with enterokinase enzyme of succus entericus. Enterokinase, thus, activates the trypsinogen to form active **trypsin**. **Trypsin** acts on proteins, peptones and proteoses changing them into simple amino acids. (*ii*) **Amylopsin** (amylase) acts on starches reducing them into maltose. (*iii*) **Steapsin** (lipase) acts on emulsified fat to form fatty acids and glycerol.

1. Trypsinogen + Enterokinase → Trypsin (active)
   Trypsin + Protein → Peptones + Proteoses → Amino acids
2. Amylopsin + Starch → Maltose (Disaccharides)
3. Steapsin + Emulsified fats → Fatty acids + Glycerol.

(*c*) Succus entericus. It is also an alkaline fluid contains many enzymes known as enterokinase, peptidases (erepsin), lipase, maltase, invertase, lactase, ribonuclease and deoxyribonuclease. (*i*) **Enterokinase** activates the inactive trypsinogen of pancreatic juice to form active trypsin. (*ii*) Peptidases includes proteolytic enzymes like **erepsin** which act on polypeptides breaking them into amino acids. (*iii*) **Lipase** and **maltase** along with the same enzyme of the pancreatic juice act on emulsified fats and maltose and convert them into fatty acids and sugars respectively. (*iv*) Invertase and lactase act on sucrose and lactose reducing them to glucose. (*v*) Ribo- and deoxyribonuclease change the nucleic acids into nucleotides which are used in the synthesis of DNA, RNA and ATP.

Thus, the food is thoroughly digested with the help of these enzymes. As a result, all the proteins are reduced to amino acids, the carbohydrates to glucose and similar sugars and fats to glycerol and fatty acids.

4. Absorption and Assimilation of Digested Food

(*i*) Absorption. Absorption is that process by which the digested food is taken into blood. It takes place mainly in the duodenum and ileum as they are very much suited for this due to the development of various folds with villi-like processes, which increase the absorptive surface of these two regions of the alimentary canal. Each villus is richly supplied with blood capillaries and lymph vessel or lacteal. Water, mineral salts and other nutrients are directly absorbed through the mucosa.

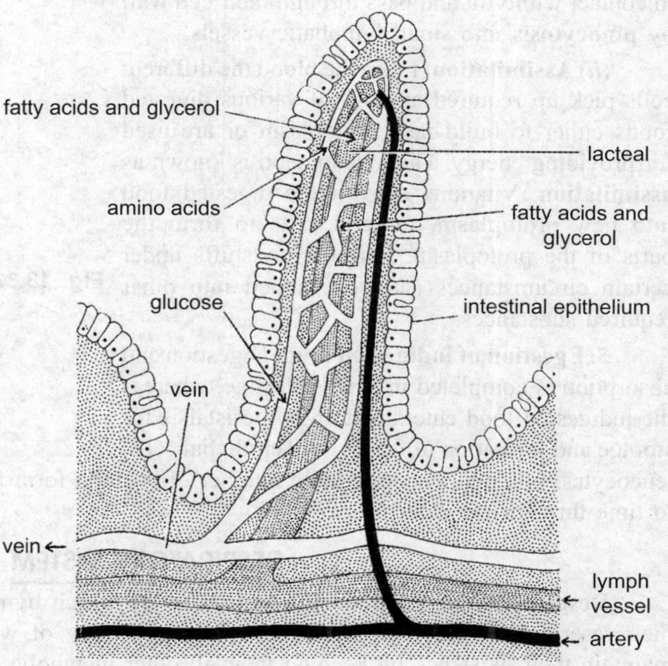

Fig. 18.32. Frog. Absorption of digested food in the intestine.

The amino acids and glucose and fructose pass by diffusion from the mucosa into the blood capillaries of the intestine and finally reach the liver through the **hepatic portal vein**. The liver maintains the steady supply of the required amount of sugars and amino acids into blood. The excess sugar is stored as glycogen, whereas the excess amino acids cannot be stored in reserve but are converted into urea by liver cells which is eliminated as urine from the body by kidneys. If sugar concentration falls from the normal in the blood, the reserve glycogen is changed into glucose by the liver cells and sent to the blood. From the blood the cells pick up the required amino acids to synthesise the proteins to form the protoplasm.

The glycerol and fatty acids pass into lymph vessels called lacteals. The glycerol can be absorbed easily as it is soluble in water, whereas the fatty acids as such cannot be absorbed as they are insoluble in water. So before absorption the fatty acids mix with bile salts to make themselves soluble so that they can be absorbed. After absorption in the lacteals, the glycerol and fatty acids are again converted into fat globules of much smaller molecules. Thus, fats enter as glycerol and fatty acids into the blood through lymph vessels.

Recent researches show that it is not necessary for all fats to be changed into glycerol and fatty acids for absorption, it is seen that fat droplets come in contact with villi and pass through their cell wall by **pinocytosis** into small lymphatic vessels.

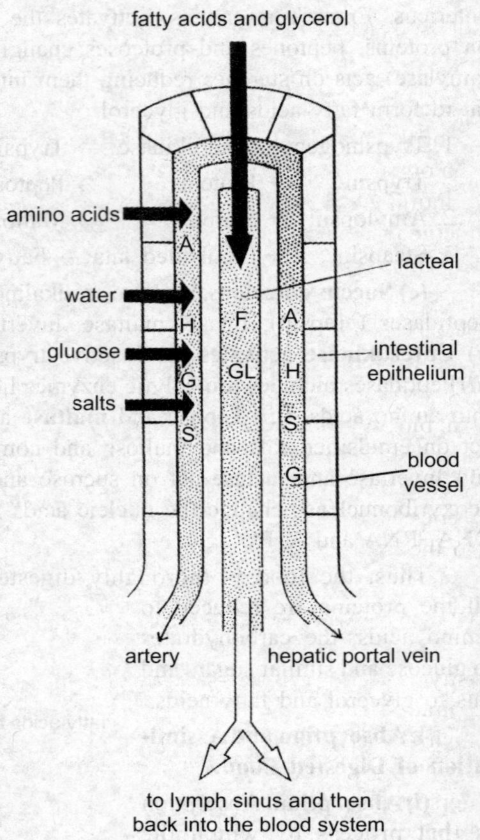

Fig. 18.33. Frog. Diagrammatic representation of absorption of food in the intestinal wall. A–Amino acids; H–Water; S–Salts; G–Glucose; F–Fatty acids; GL–Glycerol.

***(ii)* Assimilation.** From the blood the different cells pick up required amount of various digested foods either to build new protoplasm or are used for providing energy. This phenomeon is known as **assimilation**. Vitamins convert the digested food into new protoplasm, mineral salts to form the parts of the protoplasm. Various foodstuffs under certain circumstances can be changed into other required substances.

5. Egestion of indigested food. Digestion and absorption is completed in small intestine, whereas the indigested food enters rectum by peristalsis for storage and formation of faeces. Old epithelial cells, leucocytes, bile pigments and large number of bacteria form faeces which is removed from time to time through the cloacal opening.

RESPIRATORY SYSTEM

Respiration can be defined as a process by which living beings obtain oxygen to oxidise their reserve digested food in order to release energy of which they are in constant need to maintain their existence by keeping their all other metabolic activities in order. In the course of this reaction carbon dioxide is also formed which is removed regularly from the body due to its injurious property. Organs which help in the intake of oxygen and supplying it to the body tissues and removal of carbon dioxide constitute the respiratory system. Respiration in tadpole of frog occurs by means of gills as in fishes. In adult frog, due to its amphibian life, respiration occurs through skin (cutaneous respiration), lining of the bucco-pharyngeal cavity (buccal respiration) and the lungs (pulmonary respiration). Ordinary respiratory requirements are met by the skin and bucco-pharyngeal cavity, lungs are used only when the need of oxygen is great.

1. Cutaneous respiration. The skin of frog is an important organ of respiration. The frog, due to amphibious mode of life, passes most of the time of its life in water. During this period the skin only serves as an organ of respiration for gaseous exchange. Similarly, when frog undergoes summer sleep **(aestivation)** and winter sleep **(hibernation)**, the skin is the only organ of respiration.

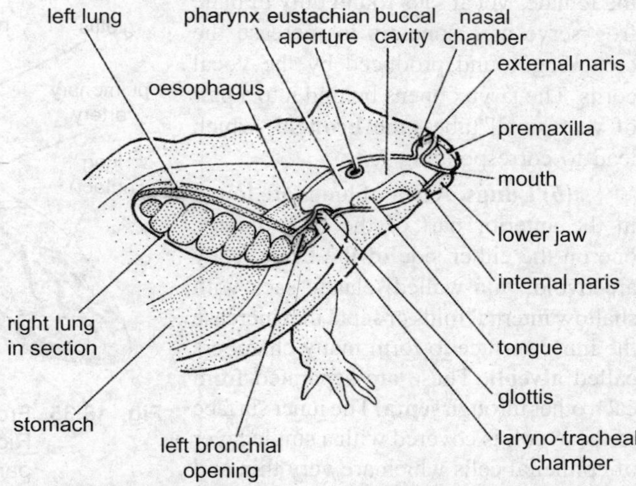

Fig. 18.34. Frog. Diagrammatic L.S. of respiratory system.

The skin of frog is very much suited for the respiratory function as it is very thin and richly supplied with blood capillaries and remains moist with the water and also mucus, secreted by mucous glands. During gaseous exchange the oxygen first dissolves in the moisture present over the body and then diffuses into the blood circulating in the blood capillaries, while the resultant carbon dioxide passes out from the blood into the surrounding medium (water) by diffusion. In cutaneous respiration, no movements are needed because skin always remains exposed to air or water.

2. Bucco-pharyngeal respiration or Buccal respiration. The mucous lining of the buccal cavity is richly supplied with blood capillaries and remains moist by the mucus. The buccal respiration occurs by lowering and raising of the floor of the buccal cavity, during the course of which the air is constantly sucked into the buccal cavity and is drawn out through the external and internal nares. In this type of respiration, the mouth and glottis remain closed. Thus, no air enters or goes out from the lungs. When the floor of the buccal cavity is lowered, the air enters the buccal cavity through the nostrils or the nares. The oxygen of air dissolves in the layer of mucus and then goes into blood. At the same time carbon dioxide is given out into the buccal cavity from the blood which is expelled along with residual air through the nostrils when the floor of the buccal cavity is raised.

3. Pulmonary respiration. Respiration on land in air with the help of lungs is the **pulmonary respiration**. In frog, lungs are poorly developed. The intake of oxygen by lungs is not sufficient to the body. Therefore, oxygen intake through moist skin and buccal cavity is needed.

1. Organs of respiration. The organs of aerial respiration are a pair of **lungs**. The lungs are not only the organs of respiration but are also hydrostatic organs as they enable frog to float in water when they are inflated. The air enters and leaves the lungs through the respiratory tract.

(*a*) Respiratory tract. It includes the external nostrils, nasal chambers, internal nostrils, bucco-pharyngeal cavity, glottis, laryngo-tracheal chamber and a pair of bronchi. The median slit-like glottis on the floor of pharynx opens into larynx (laryngo-tracheal chamber). Larynx is a small sac whose walls are supported by two arytenoid and one cricoid cartilages. Cricoid cartilage is a slender ring surrounding the larynx. The arytenoid cartilages are a pair of semilunar valves, which rest upon the cricoid cartilage. Their upper edges form the lateral margins of the glottis. They afford attachment to muscles by which glottis may be opened or closed. The true sound producing organs are a pair of elastic bands, the **vocal cords**, extending longitudinally across the larynx. Their median edges are thickened and lie near each other in the middle line. Sound is produced by the expulsion of air from the lungs which set the free edges of the vocal cords in vibration. Vibrations in the sound are caused by altering the tension on the cords through the action of laryngeal muscles. The vocal apparatus of the male frog is much larger than that of the female. Vocal sacs found only in male frog serve as resonators to increase the croacking sound produced by the vocal cords. The larynx opens behind into a pair of very small tubes, the **bronchi**, which lead to corresponding lung.

(*b*) **Lungs.** A pair of lungs are found in the anterior part of the body cavity, one on the either side of the heart. They are ovoid, thin-walled, elastic sacs with shallow internal folds or septa that increase the inner surface to form many chambers called **alveoli**. These are separated from each other through **septa**. The inner surface of the aveoli is covered with a single layer of epithelial cells which are very thin and flattened except on the edges of the septa where they are ciliated and cylindrical. On the inner side of the epithelium there is aerolar type of connective tissue comprising blood and lymph vessels and unstriped muscle fibres which give remarkable power of contraction and expansion to the lungs. The outer surface of the lung is coated with coelomic epithelium called peritoneum.

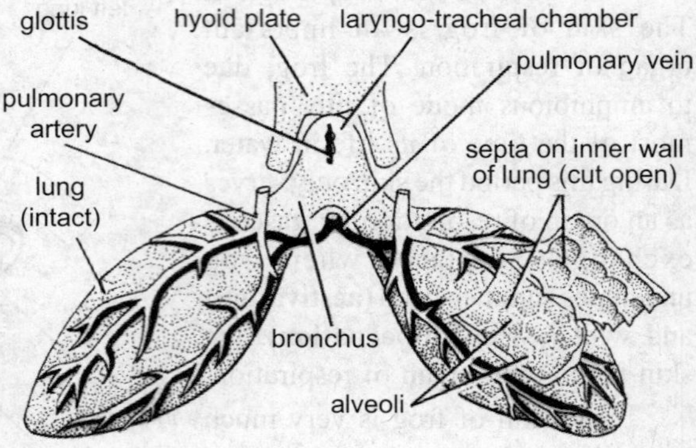

Fig. 18.35. Frog. Respiratory organs in dorsal view. Right lung partly cut open to show inner partitions and alveoli.

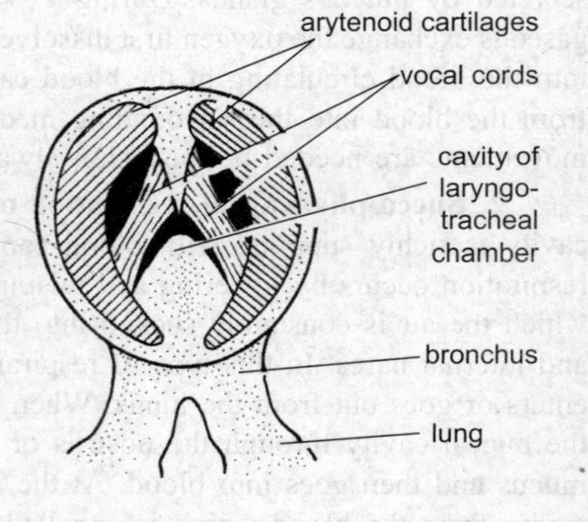

Fig. 18.36. Frog. Laryngo-tracheal chamber opened to show vocal cords.

4. Mechanism of pulmonary respiration. The incoming and outgoing of the air from the lung is brought about by the action of the floor of the buccal cavity which actually acts as a **force pump**. The actions of the floor of the buccal cavity are brought by two sets of muscles, the sternohyal and the petrohyal muscles. (*i*) **Sternohyal muscles** arise from the coracoid and clavicle or sternum and attached to the lower surface of the hyoid apparatus located in the floor of the buccal cavity. (*ii*) **Petrohyal muscles** are attached on one end with the squamosal bone above and on the other side with the upper surface of hyoid apparatus. The respiratory mechanism involves two phases: inspiration and expiration. During pulmonary respiration the air is forced into the lungs. This is aided by the movable premaxillae bones of the upper jaw situated just below the external nares and the foresaid muscles.

(*a*) **Inspiration.** During inspiration the frog closes the glottis and mouth, and the nostrils remain open. On the contraction of the sternohyal muscles the floor of the buccal cavity along with hyoid is lowered increasing the volume of the buccal cavity. Thus, the air enters the cavity through the external nares. Now the glottis opens and the Mentomeckelian

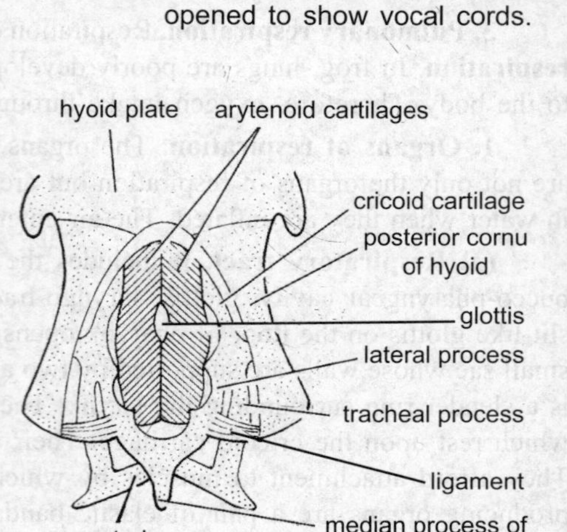

Fig. 18.37. Frog. Skeletal cartilages of laryngo-tracheal chamber.

bones of lower jaw push the premaxillae bones of the upper jaw upwards closing the nostrils. Now the petrohyal muscles contract, this results in lifting the hyoid apparatus along with the floor of the buccal cavity forcing the air to enter into the lung through the glottis. This drawing in of air into the lungs is called **inspiration**.

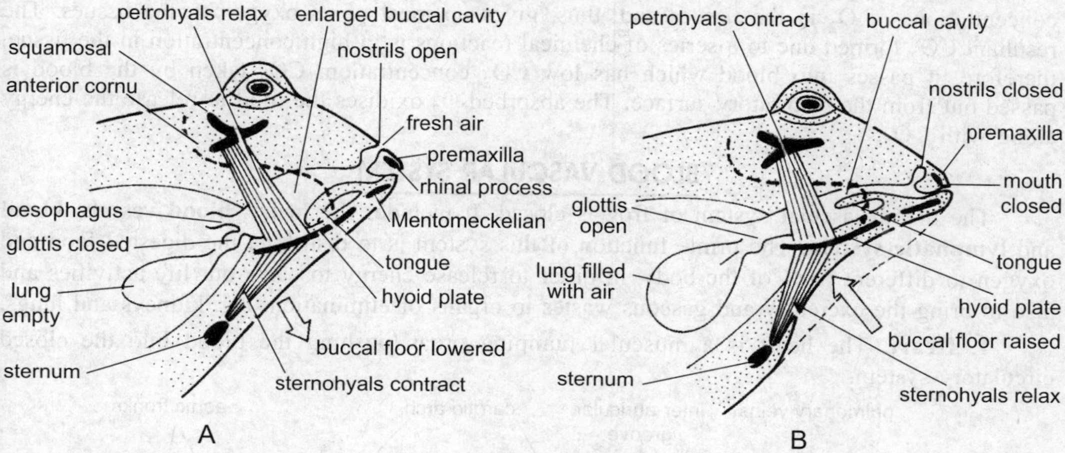

Fig. 18.38. Frog. Diagrammatic representation of respiratory mechanism involving two sets of muscles. The two stages bring respiration. Reverse sequence will result in expiration.

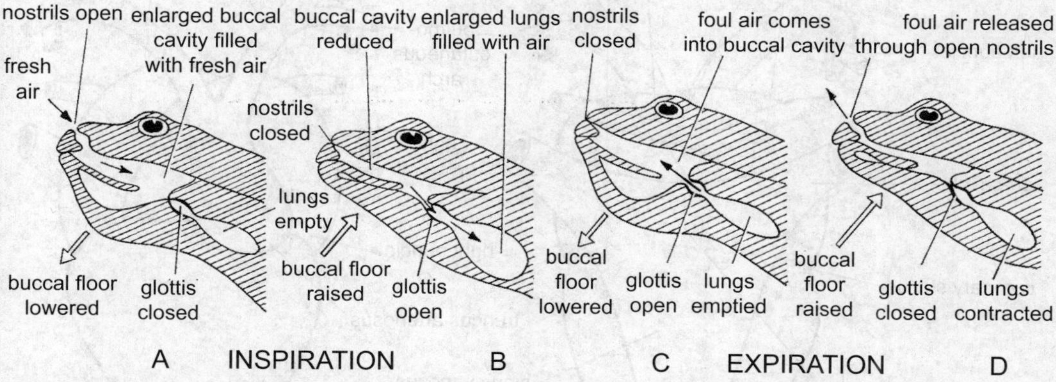

Fig. 18.39. Frog. Breathing movements during ventilation of lungs. Black arrows indicate pathway of air in and out of lungs. White arrows show up and down movements of buccal floor.

***(b)* Expiration.** Before expiration, when lungs are filled with air, the glottis closes and the air is kept in the lungs for a short time. During this period buccal respiration occurs. Soon the glottis becomes opened and the air from the lungs is expelled into the buccal cavity by the contraction of the lungs and the abdominal muscles and by lowering the floor of the buccal cavity. Now the buccal floor is raised again, the glottis closes and external nares are opened, forcing the air out through the external nares.

Recently a large number of experiments have been conducted on the mechanism of the pulmonary respiration in frog. These experiments show that when the floor of the buccal cavity is lowered the fresh air is not only rushed into the buccal cavity but the air from the lungs is also drawn into the buccal cavity where the two airs mix. On raising of the floor of the buccal cavity the mixed air goes outside the body through the external nares as well as goes into the lungs through the glottis.

5. Physiology of respiration. The erythrocytes (R.B.Cs.) of frog's blood contain a respiratory pigment, the **haemoglobin** to which the respiratory gases (O_2 and CO_2) can combine temporarily only in high partial pressure. In the respiratory surfaces the O_2 in the air is found in high concentration due to which it combines with haemoglobin of the blood to form **oxyhaemoglobin**. It is transported by blood to the different tissues of the body where it dissociates due to less concentrations of O_2 in the tissues and, thus, gives up most of its oxygen to the tissues. The resultant CO_2 formed due to a series of chemical reactions is in high concentration in the tissue, therefore, it passes into blood which has low CO_2 concentration. CO_2 taken by the blood is passed out from the respiratory surface. The absorbed O_2 oxidises the food to release the energy along with CO_2.

BLOOD VASCULAR SYSTEM

The blood vascular system of frog is closed. It includes the **heart**, **blood vessels**, **blood** and **lymphatic system**. The prime function of this system is to distribute the digested food and oxygen to different parts of the body, in order to release energy to carry out life activities and also to bring the excretory and gaseous wastes to organs of elimination, *i.e.*, kidneys and lungs.

1. Heart. The heart is a muscular pumping organ pushing the blood into the closed circulatory system.

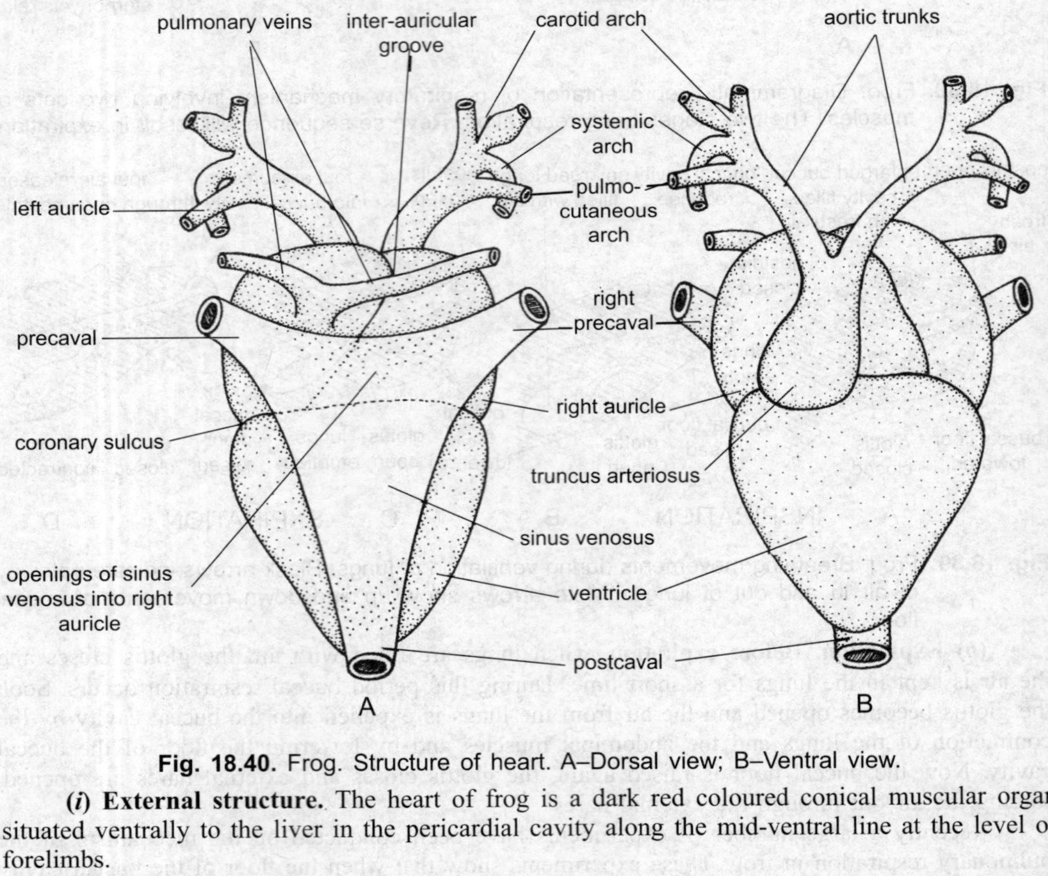

Fig. 18.40. Frog. Structure of heart. A–Dorsal view; B–Ventral view.

***(i)* External structure.** The heart of frog is a dark red coloured conical muscular organ situated ventrally to the liver in the pericardial cavity along the mid-ventral line at the level of forelimbs.

***(a)* Pericardium.** The heart is enclosed within a sac formed of two membranes, an outer **pericardium** and an inner **epicardium** which closely invests the heart. Between these two membranes a **serous** or **pericardial fluid** is found which prevents friction and also keeps the heart moist. It also protects the heart from outer shocks.

(*b*) Chambers of heart. The heart looks like an angular structure with broader anterior part and narrow posterior part. The heart is 3-chambered. The broader part of the heart contains two **atrium** or **auricles**, whereas the posterior part has a single **ventricle**. The atrium or auricle lies anterior to the ventricle. Both the auricles are externally demarcated by a faint longitudinal **inter-auricular groove**. The two auricles, however, are clearly separated from the ventricle by a transverse **coronary sulcus** or auriculo-ventricular groove. The **ventricle** is the most conspicuous and prominent part of the heart. It is conical in shape with thick muscular walls. Besides auricles and ventricle, the heart has two more chambers—truncus arteriosus and sinus venosus. **Truncus arteriosus** arises from the ventral upper right side of the ventricle as a cylindrical structure, which extends forward obliquely across the right auricle and finally divides into two branches or trunks, each further divides into three arches—carotid, systemic and pulmocutaneous. **Sinus venosus** lies on the dorsal surface of the heart. It is a thin-walled dark-coloured somewhat triangular chamber which is formed by the union of three caval veins two anterior, precavals and one posterior, postcaval.

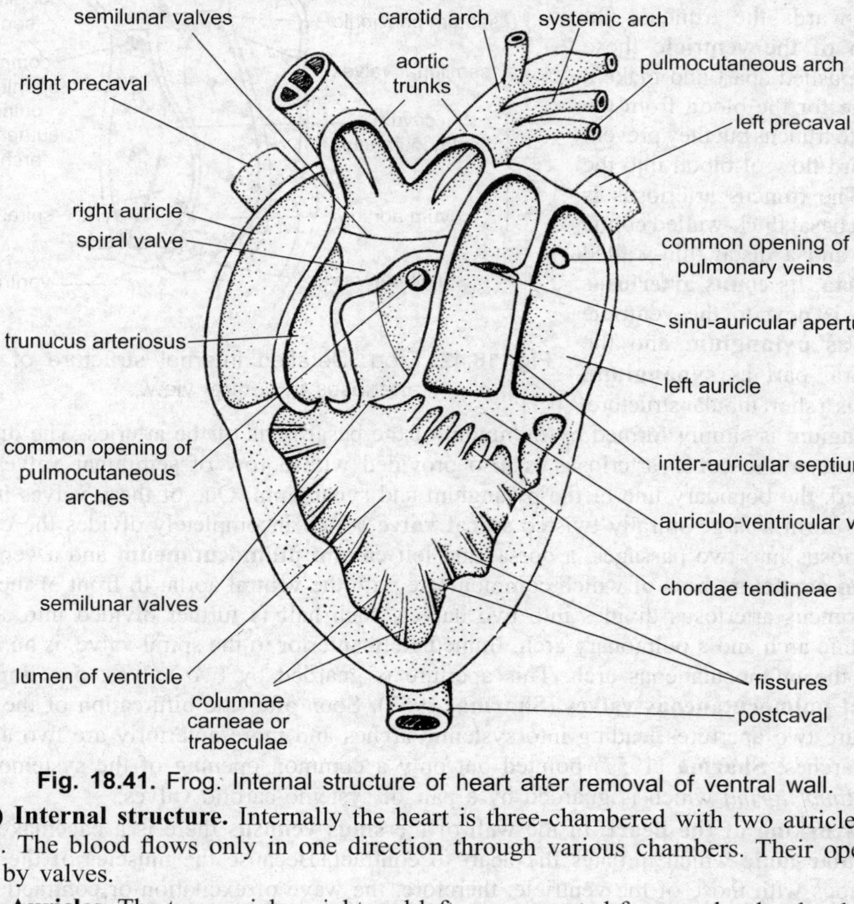

Fig. 18.41. Frog. Internal structure of heart after removal of ventral wall.

(*ii*) Internal structure. Internally the heart is three-chambered with two auricles and one ventricle. The blood flows only in one direction through various chambers. Their openings are guarded by valves.

(*a*) Auricles. The two auricles, right and left, are separated from each other by thin vertical **inter-auricular septum**. Left auricle is smaller than the right. In the right auricle close to the septum there is a transverse oval opening called **sinu-auricular aperture** through which blood enters into the auricle from the sinus-venous. It is guarded by two lip-like **sinu-auricular valves**, one arising from the dorsal edge and the other from the ventral. These valves allow the free flow of blood only into the right auricle but prevent the backward flow of the blood. In the left auricle slightly anterior to the sinu-auricular aperture but close to the septum there is a small opening of pulmonary vein which has no valve. The two auricles open into a single ventricle by an **auriculo-ventricular aperture** which is bounded by two pairs of **auriculo-ventricular valves**, one arising from the dorsal edge and the other from the ventral edge of this aperture.

***(b)* Ventricle.** The ventricle is a most conspicuous triangular chamber of the heart with muscular walls. Its inner surface has irregular ridges, the **columnae carnae** with deep pockets between them, which to some extent prevent the mixing of the blood from the two auricles. These ridges reduce the lumen of the ventricle. The flaps of auriculo-ventricular valves are attached to the wall of ventricle by thread-like **chordae tendinae**.

***(c)* Truncus arteriosus.** From the upper right side of the ventricle arises a tubular **truncus arteriosus**. Its opening is guarded by three **semilunar valves**, whose edges are directed towards the truncus. On contraction of the ventricle these valves are pushed apart and make a free passage for the blood from the ventricle into truncus but they prevent the backward flow of blood into the ventricle. The truncus arteriosus is formed of a basal thick-walled **conus arteriosus** and a distal thin-walled **ventral aorta**. Its **conus arteriosus** part which is next to the ventricle is known as **pylangium** and the distal ventral part as **synangium**. Pylangium is a short tubular structure, while synangium is simply formed by the union of the basal parts of the arteries. The distal end of pylangium or the conus arteriosus is also provided with a row of semilunar valves which actually mark the boundary line of the pylangium and synangium. One of these valves has been modified to form a large spirally twisted **spiral valve** which incompletely divides the cavity of conus arteriosus into two passages, a dorsal and left **cavum pulmocutaneum** and a ventral and right **cavum aorticum**, both of which communicate with the ventral aorta. In front of the ventral aorta the truncus arteriosus divides into two halves, each half is further divided into a carotid arch, systemic arch and a pulmonary arch. Immediately anterior to the spiral valve, is an aperture leading to the pulmocutaneous arch. This aperture is guarded by two valves, one **dorsal** and one **ventral pulmocutaneous valves** (**Sharma**, 1957). Soon after the bifurcation of the truncus arteriosus are two apertures leading into systemic arches and more anteriorly are two apertures of carotid arches. **Sharma** (1957) pointed out only a common opening of the systemo-carotid arches in *Rana tigrina* which is guarded by a pair of systemo-carotid valves.

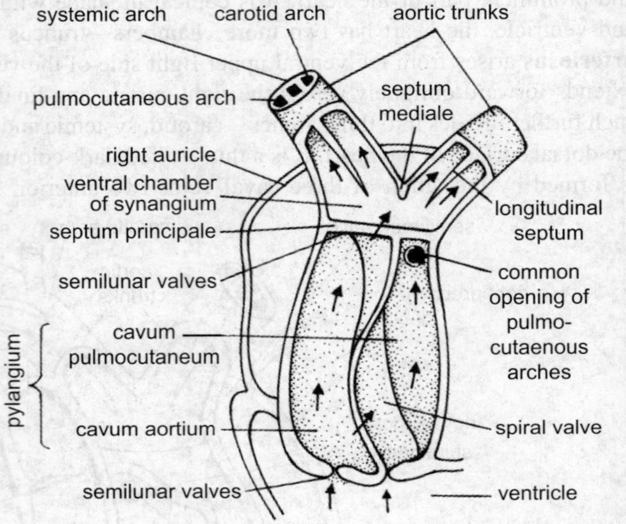

Fig. 18.42. Frog. Detailed internal structure of truncus arteriosus in ventral view.

***(iii)* Working of the heart.** In the wall of the **sinus venosus** there is a pacemaker called **sinu-auricular node** which initiates the heart to contract. Because the muscles of the auricles are continuous with those of the ventricle, therefore, the wave of excitation or contraction starts from the sinus venosus and ends at the truncus, thus, these chambers contract in sequence. As soon as sinus venosus contracts, its impure blood is pumped into the right auricle through sinu-auricular aperture. At the same time the left auricle receives blood from the lungs via pulmonary veins. Both auricles contract almost simultaneously driving blood into the ventricle through the auriculo-ventricular aperture. The flow of blood in the heart is explained by two theories.

***(a)* Older view.** According to old theory of **Brucke** (1851), and confirmed by **Sabatier** (1873), the left side of ventricle receives oxygenated blood from the left auricle, while the right side receives deoxygenated blood from right auricle. The mixing of the blood is avoided in the ventricle because of the viscous nature of blood and the spongy texture of the ventricle due to the presence of columnae carneae. Thus, there is an impure blood in the right side, pure in the left side and in the middle, however, mixing of two bloods takes place. On contraction of the ventricle, first the impure blood from the right side goes into the truncus arteriosus. The spiral valve in the conus arteriosus is so placed that it allows the deoxygenated blood to go into the pulmo-cutaneous arches through their common opening and carried to lungs and skin for oxygenation. Spiral valve then closes the common opening of pulmo-cutaneous arches. Now the mixed blood from the middle of the ventricle is pushed through cavum aorticum into the systemic arches and oxygenated blood from the left side of the ventricle into the carotid arches. Thus, spiral valve in truncus arteriosus directs the flow of blood into various arches. This explanation of the flow of the blood from the ventricle into the arches is now discarded.

(*b*) Modern view. According to new theory propounded by **Vandervael** (1933) and **Foxon** (1953), the blood coming into auricles whether from lungs or sinus venosus is all oxygenated because oxygenation of blood takes place not only in the lung but also in the skin and the buccal cavity. It means the blood which comes into sinus venosus from skin and buccal cavity is equally oxygenated, if not more. On contraction of the two auricles the blood comes into ventricle where the two get completely mixed up. This mixed blood is distributed to all parts of the body through the common **carotid**, **systemic**, and **pulmocutaneous** arches. The spiral valve takes no part in turning the blood, it only supports the conus arteriosus and prevents it from collapsing. The carotid labyrinth does not raise the pressure of blood but it is sensory and detects pressure changes of oxygen in the blood. The above explanation of flow of the blood in the heart is widely accepted now-a-days.

DeLong (1962) indicated that the carotid arches get oxygenated blood, pulmocutaneous arches lesser oxygenated blood and the systemic arches mixed blood. The precise mechanism is not understood.

The flow and pressure is regulated by the extrinsic nerves of the heart. Nerve fibres from the vagus slow down the speed of beat, whereas fibres from sympathetic nerves accelerate the speed of beat.

Histology of Blood Vessels

1. Arteries. The blood vessels which carry oxygenated blood (in most of the cases) away from the heart to different parts of the body are called **arteries**. They have thick and more elastic walls. Each artery is made of three concentric layers : (*i*) **Tunica intima.** It is the innermost layer which is made of flat endothelial cells. Internally it is lined by an elastic membrane. (*ii*) **Tunica media.** It is the middle layer which is simply made of circularly disposed off smooth or unstriated muscle fibres and elastic connective fibres. It is greatly developed, therefore, the arteries are non-collapsible. (*iii*) **Tunica externa** or **adventitia**. It is the outermost layer which is made of loosely arranged elastic and collagen fibres.

As these blood vessels do not have any valves, therefore, blood flows in them at a relatively greater speed. Moreover, they are relatively deep-seated in the body, therefore, in their course in the body tissues, they gradually decrease in size and pass over into **arterioles**, **arterial capillaries** and finally into **true capillaries** which form a network between the **arterial capillaries** and **venous capillaries.**

2. Capillaries. They are found abundantly in those tissues or organs where extensive metabolism occurs, such as muscles and glands but few or even altogether absent in tissues where metabolism does not occur or is very slow such as cartilage and epithelia. The structure of the capillaries depends on their function. Because these bring blood into the closest possible relation with the surrounding tissues, their walls are extremely thin. They lack both tunica media and tunica externa or adventitia, having only the innermost layer of flat endothelial cells called **tunica intima**. Their walls are so thin that an exchange of food material, gases and waste takes place between the blood and the tissue under pressure.

3. Veins. From the fine capillaries the blood passes into thin **venules**, thence into veins by which it is finally returned to the pumping centre, the heart. The veins, thus, start with capillaries. They are always thin-walled and collapsible. They are (in most of the cases) provided with semi-lunar valves which prevent the backward flow of the blood. Generally the flow of blood in the veins is always much slower than in the arteries because the blood, as it comes into the veins, the pressure is very much reduced. So in veins the onward flow of the blood is maintained by the semi-lunar valves, contraction of the muscles of the body and by the sucking activity of the heart.

The veins have almost the same structure as that of arteries. It means they are also made of all the three layers but these layers are very thin, consequently the walls of the veins are comparatively thin than the arteries of the same size, but they are more elastic. In veins the muscles and elastic connective tissues are less developed, while the collagen fibres of the outer layer are greatly developed in arteries. In most veins the tunica media is found without muscles, while tunica externa makes up the greater part of the wall.

TABLE 18.2. DIFFERENCES BETWEEN ARTERIES AND VEINS.

Arteries	Veins
1. The lumen of an artery is small.	1. The vein of the same size has the lumen of great size.
2. They have thick muscular and highly elastic walls.	2. They have thin, fibrous and scarcely elastic walls.
3. When blood goes out they retain their shape.	3. They collapse.
4. They are devoid of valves.	4. They are provided with valves to maintain onward flow.
5. In them blood flows fast and with jerks.	5. In them blood flows evenly and less fast.
6. They carry oxygenated blood away from the heart to different parts of the body.	6. They collect the deoxygenated blood from different parts of the body and finally return to the heart.

ARTERIAL SYSTEM

The blood vessels which carry the blood away from the heart to different parts of the body constitute the **arterial system**. In frog arterial system begins with the truncus arteriosus which divides into two large right and left branches or trunks. Each branch divides into three aortic arches : an anterior **carotid arch**, a middle **systemic arch** and a posterior **pulmocutaneous arch**.

1. Carotid arch. The carotid arch of each side also divides into two, an **external carotid** or **lingual** going to the lower jaw and tongue and an **internal carotid** to the orbit and brain. Each internal carotid at its commencement bears a swollen **carotid body** or **labyrinth** formed from the second pair of gill-slit. Carotid bodies or labyrinths have a network of capillaries and chromoffin cells. They detect the pressure of oxygen and CO_2 in the blood.

2. Systemic arch. The two systemic arches as pass outward and curve around the oesophagus to join to form a median **dorsal aorta** going backward beneath the vertebral column. Each systemic arch gives out three arteries : (*i*) **oesophageal** artery to the oesophagus, (*ii*) **occipitovertebral** artery to the head, vertebral column and spinal cord and (*iii*) a large **subclavian** to the forelimbs. The oesophageal artery very often arises from the occipitovertebral artery.

3. Pulmocutaneous arch. Each pulmocutaneous arch divides into two arteries : a **pulmonary artery** going to a lung and a **cutaneous artery** to the skin and buccal cavity.

Dorsal aorta. It is an unpaired artery formed by the union of two systemic arches. It runs posteriorly mid-dorsally just beneath the vertebral column. It gives off a number of arteries.

(*i*) Coeliaco-mesenteric artery. A large but unpaired coeliaco-mesenteric artery arises at the point where the dorsal aorta is formed by the union of the right and left systemic arteries. Coeliaco-mesenteric passing through the mesentery divides into two main branches : a small **coeliac** artery going to the stomach (gastric) and liver (hepatic) and a large **anterior mesenteric** sending branches to the duodenum (duodenal), spleen (splenic) and ileum (intestinal). From the anterior mesenteric artery arises an unpaired posterior **mesenteric** artery going to the large intestine. It sometimes arises from the posterior end of dorsal aorta.

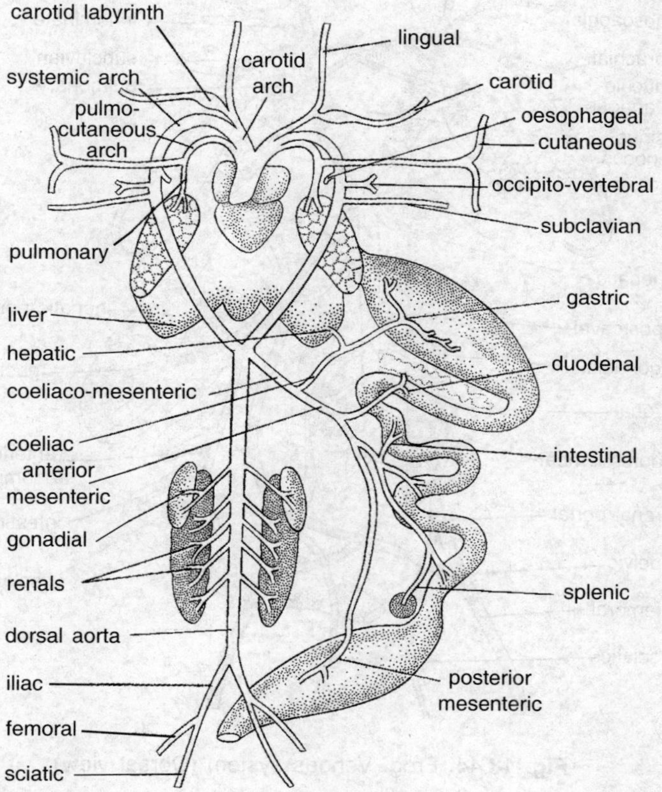

Fig. 18.43. Frog. Arterial system (Ventral view).

(*ii*) Gonadial. A pair of small gonadial arteries to the gonads, called **spermatic** in male and **ovarian** in female.

(*iii*) Renal. Dorsal aorta runs backward between the kidneys and gives off 5 to 6 pairs of **renal arteries** to both the kidneys.

(*iv*) Common iliacs. Posteriorly dorsal aorta divides into two large **iliac arteries** supplying the hindlimbs. Each iliac as it enters the leg gives off a branch which divides into an **epigastric** supplying the central body wall, and a **recto-visceral** artery to the rectum and bladder. A **femoral** artery to the skin and the muscles of the anterior part of the thigh and an **iliac** artery continues further as the **sciatic** artery going to the shank.

VENOUS SYSTEM

The blood vessels which carry the blood towards the heart constitute the venous system. The venous system in frog constitutes the following main veins :

1. Pulmonary veins. The oxygenated blood from two lungs is returned by two **pulmonary** veins, which open dorsally into the left auricle after uniting with each other.

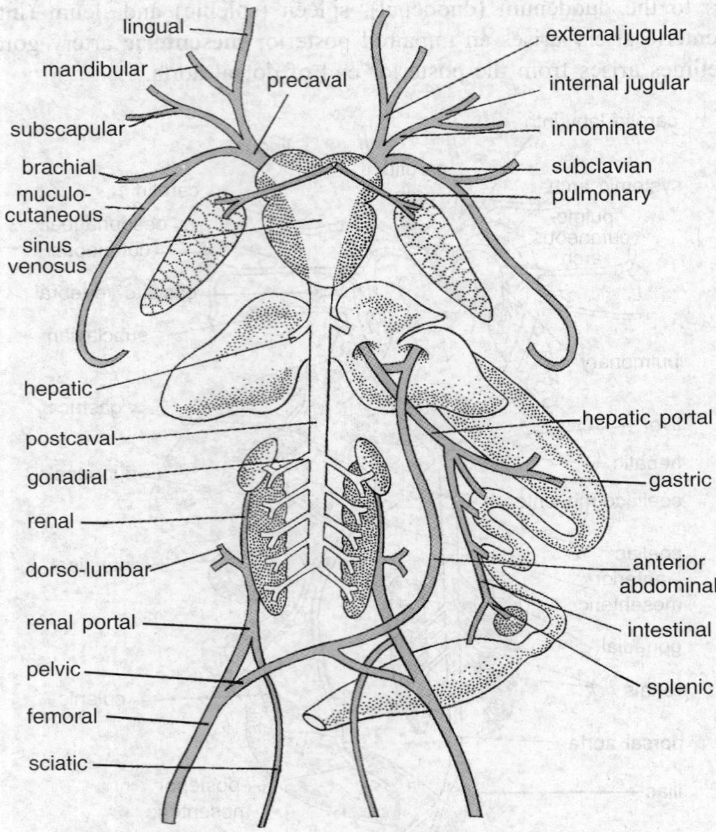

Fig. 18.44. Frog. Venous system (Dorsal view).

2. Caval veins. The deoxygenated blood from the rest of the body comes to the sinus venosus through two **precavals (anterior venae cavae)** and one **postcaval vein (posterior vena cavae)**.

(*i*) Precavals or anterior vena cavae. Each precaval or anterior vena cava is formed by the union of three veins : **(*a*) External jugular** receiving branches from the tongue **(lingual)** and floor of the mouth (**mandibular**). **(*b*) Innominate** receiving branches from the brain and orbit **(internal jugular)**, and from shoulder and back of arm **(subscapular)**. **(*c*) Subclavian** vein receiving branches from arm **(brachial)** and from skin and muscles of the abdomen and also mucous membrane of mouth and head muscles **(musculo-cutaneous)**.

(*ii*) Postcaval or posterior vena cava. The postcaval vein receives blood from the kidneys by 5 to 6 pairs of **renal veins** and a pair of **gonadial veins** (**spermatic** in male and **ovarian** in female) from gonads. It arises between the kidneys and runs forward ventral to the dorsal aorta in the mid-ventral line. Before it is communicated with the sinus venosus it also receives a pair of **hepatic** veins from the liver.

3. Renal portal vein. The veins which collect the blood from the posterior side of the body constitutes **renal portal system.** Two large veins, the **femoral** and **sciatic**, return the blood from each leg. On entering the body cavity each femoral vein divides into a ventral **pelvic** and a dorsal **renal portal vein**. The pelvic meets its fellow from the opposite side forming the **anterior abdominal vein**. It receives blood from the bladder and central abdominal wall and runs forward to enter into liver. Before entering the liver, it is connected with the hepatic portal vein. The renal portal unites with the sciatic and runs along the outer margin of the kidneys sending branches to it. The renal portal vein receives the blood from the body wall of lumbar region through **dorso-lumbar** vein.

4. Hepatic portal system. Hepatic portal system collects the blood from the alimentary canal through many branches and carries it to the liver where the veins break up into capillaries and the blood is collected by hepatic veins to pour into the postcaval.

LYMPHATIC SYSTEM

The blood is distributed to different parts of the body by the arteries. During its course of distribution certain portions of fluid plasma along with leucocytes escape into spaces (intercellular spaces) in the tissues through the capillaries as they are having very thin walls. Most of the fluid is absorbed by blood capillaries but some of it passes by diffusion into special vessels called the lymph vessels. This squeezed fluid from the blood is known as **lymph**, while the special vessels are the **lymphatic vessels**, both constitute the **lymphatic system**.

The lymphatic system is an open system because it has lymph spaces between tissue cells and it does not form a complete circuit. Since lymph flows only in one direction, that is, from the tissues towards the heart, it has only capillaries and lymph vessels which are equivalent to veins.

1. Lymph. It is a colourless fluid which is a part of blood. Actually it is blood minus its red blood corpuscles and plasma proteins. It has less concentration of calcium and phosphorus than in blood. It has the power to coagulate slowly. It fills body spaces and bathes the tissue cells where it is known as **tissue fluid**. It is a medium of exchange between the blood and tissue cells. It carries food and oxygen to the cells and takes away water and other waste substances from the cells.

2. Lymph vessels. The lymph capillaries found between the tissue cells are interwoven with but independent of blood capillaries. Lymph capillaries form an anastomosing network in all the body organs except in the nervous system. The diameter of lymph capillaries is larger and less uniform than that of blood capillaries. At their commencement the closed ends of lymph capillaries have small swollen knobs which help in collecting lymph from tissues. The lymph capillaries unite to form larger lymph vessels or lymphatics which finally discharge their contents into veins. Thus, lymphatics are efferent vessels draining lymph into veins.

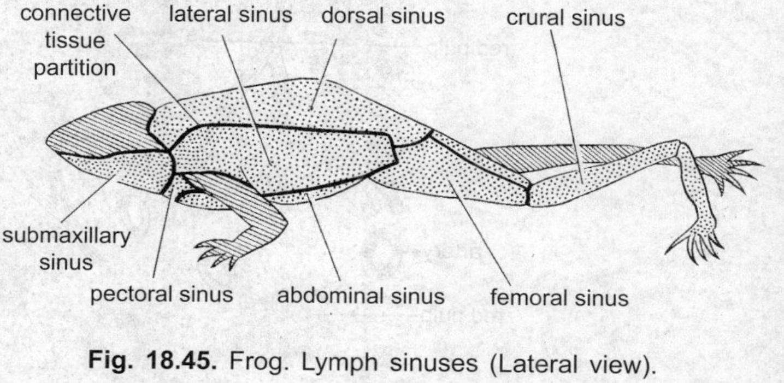

Fig. 18.45. Frog. Lymph sinuses (Lateral view).

Lymphatics are made of connective tissue fibres with some unstriped muscles and are lined with endothelial cells. They are provided with valves present in pairs ; they prevent backward flow of lymph.

3. Lymph spaces. In frog in the tissues there are numerous lymph spaces from which arise lymph capillaries, the lymph capillaries unite to form lymph vessels, some of which dilate to form large lymph channels or **sinuses** which are as follows :

(*i*) **Subcutaneous lymph sinuses.** They are large spaces found on the dorsal, ventral and lateral sides between the skin and muscles. They are separated from each other by fibrous membranes (septa).

(*ii*) **Subvertebral sinuses.** They are found around the dorsal aorta and also enclose the kidney.

(*iii*) **Pericardial sinuses.** They enclose the heart, their lymph is called **pericardial fluid**.

(*iv*) **Coelom.** It is a body cavity containing lymph which is commonly known as **coelomic fluid**.

4. Lymph hearts. From the diffuse lymphatic system lymph is pumped back into veins by two pairs of **lymph hearts**. One pair is situated just behind the transverse processes of the third vertebrae or below the scapulae, opening into the subscapular veins. The second pair is found on either side at the end of the urostyle, they open into the femoral veins.

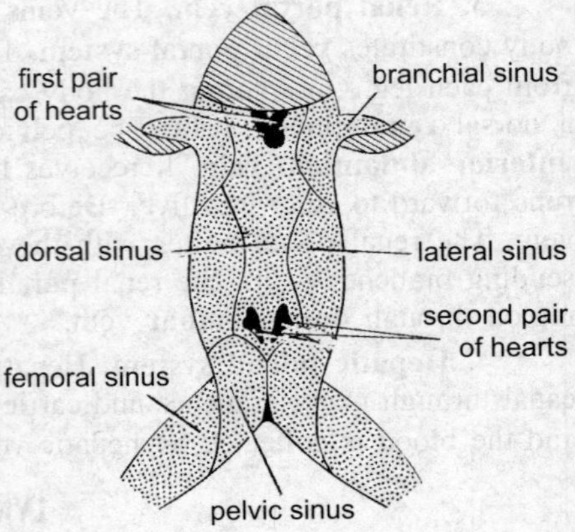

Fig. 18.46. Frog. Lymph sinuses and hearts (Dorsal view).

5. Spleen. It lies near the rectum in the mesentery and is a dark red coloured body. It acts as a reservoir of blood, it destroys the worn out R.B.Cs., and produces antibodies, new R.B.Cs., and W.B.Cs., It is made of connective tissue and muscle fibres covered with fibrous capsule. It has many compartments containing splenic pulp (lymphatic tissue) which includes mainly the R.B.Cs., and W.B.Cs.

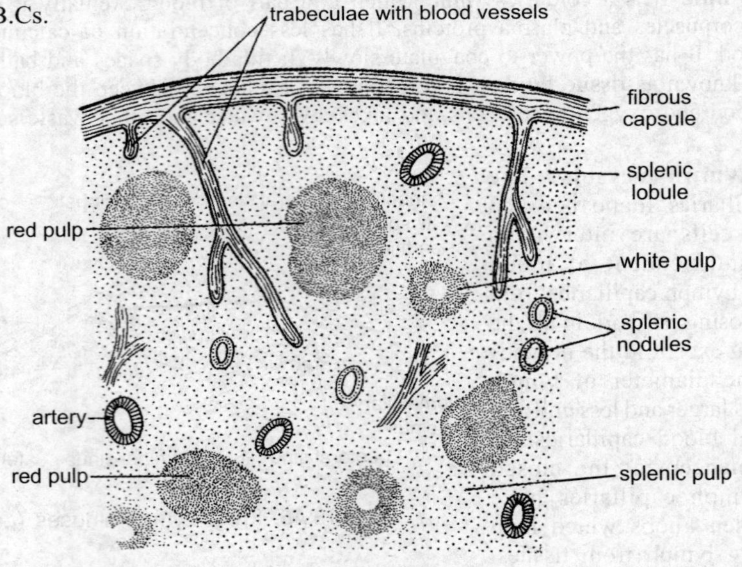

Fig. 18.47. Frog. T.S. of spleen.

BLOOD

The blood is really a fluid connective tissue containing fluid plasma in which red and white blood cells (corpuscles) are suspended. Plasma, of course, is non-living and constitutes about 60% of the blood, while blood corpuscles form the remaining 40%. Blood corpuscles are living cells.

1. Plasma. Plasma is a liquid composed of 90% water and less than 6% of various inorganic and organic substances in solution. Inorganic substances are chloride, bicarbonates, sulphates, and phosphates of Na, K, Ca, Fe and Mg. Organic substances include plasma proteins like fibrinogen (coagulated substance), glucose, amino acids, hormones, enzymes and waste products. The amount of various constituents of plasma is fairly constant, though substances leave the plasma all the time, and some others enter it.

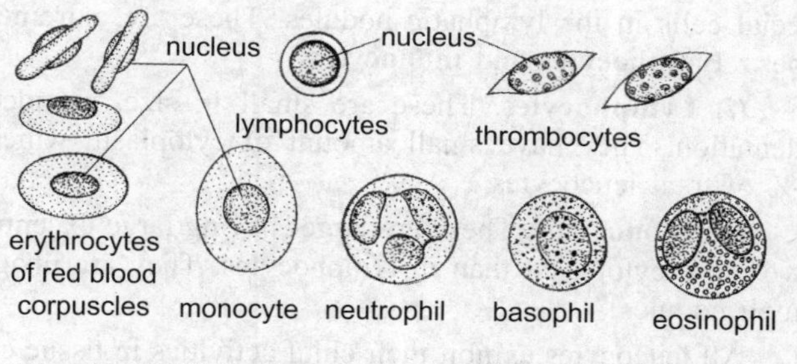

Fig. 18.48. Frog. Various types of blood corpuscles.

2. Blood cells or corpuscles. The blood cells or corpuscles include the **red blood corpuscles (erythrocyte), white blood corpuscles (leucocytes)** and **thrombocytes**, all are nucleated. None of these divides or multiplies in the blood. Certain of them are formed in the red bone marrow and others in the nodules of lymphatics.

***(i)* Red blood corpuscles or erythrocytes.** In frog these are typically flattened cells of more or less elliptical shape but appear biconvex when seen along their edges. Each erythrocyte is yellowish in colour but when numerous are present they impart a red colour to the blood, therefore, they are popularly called red blood corpuscles. Each is filled with faint stain colour cytoplasm and contains a bulging biconvex nucleus. They contain iron, water, various inorganic salts and respiratory pigment **haemoglobin**. Haemoglobin is a protein pigment which gives red colour to the corpuscles. It is a complex protein made of 90% globin (a colourless protein) and 5% hematin (red compound of iron).

The haemaglobin has a great affinity to oxygen, when it comes in contact, it becomes oxyhaemaglobin which breaks up into oxygen and haemaglobin in the tissue.

An erythrocyte is $23 \times 14\,\mu$ in size. Their number ranges from 250,000 to 450,000 per cubic millimeter of blood. In mammal they are circular, biconcave and non-nucleated and measure 8μ in their diameter. The life of erythrocytes is limited, in frog they live for about 100 days but in man for about 120 days.

***(ii)* White blood corpuscles or leucocytes.** These are small nucleated and mostly amoeboid cells. Their number ranges from 5,000 to 7,000 per cubic millimeter. These are sometimes called **phagocytes** as they engulf bacteria and other foreign substances, and old tissue cells through the process of **phagocytosis**. They have short life and can disintegrate in a day liberating their substances into blood. They are produced constantly in the liver, yellow bone marrow and spleen of frog but in mammals in red bone marrow, spleen, thymus and lymph nodes.

These are of two types depending upon the presence or absence of granules in their cytoplasm, (*i*) granulocytes and (*ii*) non-granulocytes.

1. Granulocytes or polymorpholeucocytes. These are of irregular shape having granular cytoplasm and lobed nucleus. These form 70% of total leucocytes and are produced in red bone marrow.

According to their staining property these are of three types :

***(i)* Basophils.** These cells are stained readily with basic dyes such as methylene blue or haematoxylin. They have a few large granules covering the two or three lobed nucleus. These are non-phagocytic and constitute 8% of the total leucocytes.

***(ii)* Eosinophils.** These are stained readily with basic dyes like eosin. They have many smaller granules and a nucleus of two lobes connected by a thread. These constitute about 3% of total leucocytes. These are also non-phagocytes.

***(iii)* Neutrophils (Heterophils).** These cells are stained with neutral dyes. They have numerous fine granules and a nucleus having many lobes. These form 67% of total leucocytes. These are phagocytic and engulf foreign bacteria and destroy them.

2. Non-granulocytes. These have a non-granular cytoplasm and are produced from the special cells in the lymphatic nodules. These are extremely variable in size. These are of two types : **lymphocytes** and **monocytes**.

(*i*) Lymphocytes. These are small in size, rounded and have a large nucleus with an indentation. They have small amount of cytoplasm which surrounds the nucleus. These form 28% of total leucocytes.

(*ii*) Monocytes. These are larger having large excentric nucleus indented on one side. They have more cytoplasm than in lymphocytes. They are phagocytes and are produced in spleen or lymph nodules.

All leucocytes exhibit their chief activities in tissue cells, they transport substances, remove dead cells and engulf bacteria. Whenever bacteria or foreign particles enter the body or any part is injested, the phagocytic neutrophils and monocytes squeeze out from the endothelium to engulf the bacteria and foreign particles. These cells produce antibodies to neutralise the infection caused by bacteria. The foreign particles or substances which cause the production of **antibodies** are called **antigens**.

3. Thrombocytes (Spindle cells). These are small spindle-shaped nucleated cells which are known as thrombocytes.

In mammals instead of thrombocytes blood platelets are found which are round, colourless, biconvex and non-nucleated.

Clotting of Blood

Clotting of blood is its most important property due to which blood is never escaped from the body as it clots and seals the site of injury. Blood moving normally through the circulating system does not clot because of the presence of antithrombin, an organic compound called **heparin** produced in liver. During clotting the **fibrinogen** protein present in the plasma of blood is precipitated by the action of enzyme **thrombin** to form a dense meshwork of fine threads of **fibrin** in which various types of corpuscles become entangled. This results in the formation of **clot**. **Thrombin** does not occur in an active form in the blood but it occurs as a precursor called **prothrombin**. Prothrombin in the presence of calcium ions is converted into active thrombin. When a blood vessel is injured a substance called **thromboplastin** or **thrombokinase** is released from the disintegrating thrombocytes (blood platelets) and from the injured tissues. This substance neutralises the effect of hepagin, and thrombin causes the precipitation of the fibrinogen proteins into fibrin. It is, thus, apparent that clotting of blood depends on the presence of prothrombin, thromboplastin and calcium ions.

Functions of Blood

The general functions of the blood can be summarised as follows :

1. Transport of oxygen. Blood contains haemoglobin in its erythrocytes which with oxygen forms oxyhaemoglobin and transports it to different parts of the body for the oxidation to release energy.

2. Removal of carbon dioxide. After the oxidation of food material in the tissue cells, carbon dioxide is formed which is picked up by plasma and erythrocytes which take it to the lungs or skin from where it is released.

3. Transport of food materials. All the digested food materials are absorbed through the intestinal wall and find their way into the blood and are carried in the plasma until they are taken by the tissue cells.

4. Removal of unwanted substances (metabolic wastes). Plasma of the blood picks up all the metabolic wastes like urea and uric acid from various parts of the body and takes them first to the liver and then to the kidneys from where these are eliminated.

5. Transport of other substances. There are many other substances such as hormones and **antitoxins** which are also transported by the blood to different parts of the body in order to bring harmonious cooperation and integration of various parts of the body.

6. Clotting of blood. It prevents loss of much blood from an injury by forming a clot from its fibrinogen over the site of injury.

7. Equilibration of water content. It maintains the equilibration of water contents in the tissue cells.

8. Regulation of body temperature. It regulates and equalises body temperature by carrying heat away from the region where it is produced.

9. Protection from diseases. Phagocytic leucocytes combat bacteria and foreign bodies, thus, prevent diseases. The plasma contain antitoxins and antibodies which neutralise the harmful effects of bacteria and toxins.

10. Healing of wounds. Blood brings all the essential substances to the site of injury for healing wounds.

NERVOUS SYSTEM

The nervous system includes (*i*) a **central nervous system** comprising the brain and spinal cord, (*ii*) a **peripheral nervous system** consisting of cranial and spinal nerves arising from the brain and spinal cord respectively and (*iii*) an **autonomic nervous system** made of two ganglionated sympathetic nerves. The autonomic nervous system is often regarded as a part of the peripheral nervous system because the two are connected.

1. Central Nervous System

A. Brain. The brain of frog is elongated, bilaterally symmetrical, white coloured structure safely situated in the cranial cavity of the skull. It is surrounded by a thin, pigmented and vascular connective tissue membrane, the **piamater**, which is closely applied with the brain. Outside this membrane is a tough, fibrous membrane lining the interior of the cranial cavity called **duramater**. These two membranes are called **meninges** (singular, menix). The space between the piameter and duramater is known as **subdural** which is filled with a kind of shock absorbing watery, clear lymphatic **cerebro-spinal fluid**. It is also found in the cavities of brain and central canal of spinal cord. The brain of frog is divisible into three main parts :

(*i*) Forebrain or Prosencephalon;

(*ii*) Midbrain or Mesencephalon;

(*iii*) Hindbrain or Rhombencephalon.

(*i*) Forebrain. It is the largest part of the brain consisting of a pair of anteriorly directed **olfactory lobes**, a pair of **cerebral hemispheres**, and a **diencephalon**. **(*a*) Olfactory lobes.** The olfactory lobes are anterior small, spherical structures which are fused together in the median plane. Each lobe gives off an olfactory nerve and possesses a small cavity **rhinocoel** or olfactory autricle. **(*b*) Cerebral hemispheres.** The two cerebral hemispheres are long, oval structures separated from olfactory lobes by a slight constriction. They are wider behind and narrower in front. They are separated from each other by a deep median longitudinal fissure. Each cerebral hemisphere has a large **lateral ventricle** or **paracoel** which is continuous anteriorly with the **rhinocoel**. Posteriorly the lateral ventricles communicate with each other and with the ventricle of diencephalon called **diocoel** by an opening, the **interventricular foramen** or **foramen of Monro**. The nerve cell bodies form masses around the lateral ventricles and lie in layers. Fibres of the olfactory, tactile and optic impulse reach the cerebral hemisphere which may act as correlating centres but the hemispheres are largely olfactory in function.

On ventro-lateral side of each of the cerebral hemispheres there is a thick fibrous tract called the **corpus striatum** containing a network of white medullated nerve fibres and nerve cells. The corpora striata of two hemispheres are joined by a transversely running tract of fibres called **anterior commissure** and above which is another commissure (upper line) partly representing the hippocampal commissure of the brain of reptiles and mammals. They have a thick roof called the **pallium** in which more nerve cells have moved to the periphery.

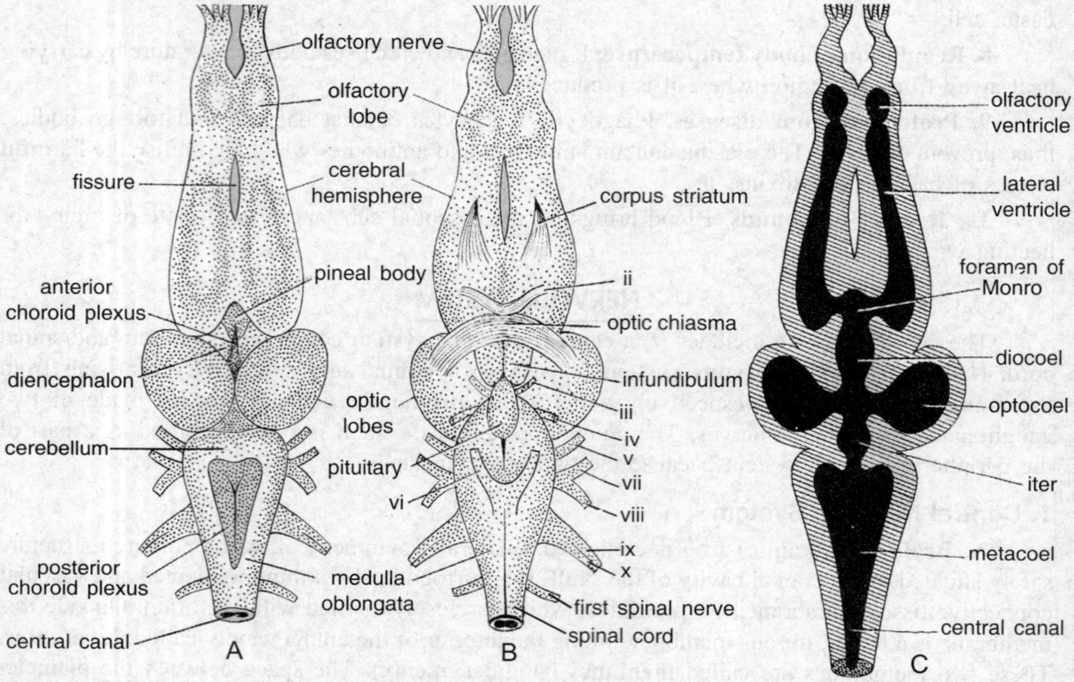

Fig. 18.49. Frog. Brain. A–Dorsal view; B–Ventral view; C–Longitudinal horizontal section of brain showing ventricles.

(*c*) Diencephalon or thalamencephalon is a short unpaired structure of the forebrain situated behind the cerebral hemispheres. Its lateral walls are thick called **optic thalami** (singular **thalamus**) and its thick floor is called the **hypothalamus**. Its roof is thin and lined with a vascular membrane, the **anterior choroid plexus**. Behind it arises a hollow, thin-walled stalk, called the **pineal stalk** which terminates dorsally at the brow spot. The pineal stalk or epiphysis, which originally was continuous with the brow spot, becomes constricted off from it in early larval life. On the ventral side of diencephalon is the **optic chiasma** or crossing of the optic nerves which go the eyes. Just behind the optic chiasma is a flattened bilobed **infundibular lobe** or **infundibulum** extending posteriorly and divided by a median longitudinal groove. It is formed of nervous tissue and contains a cavity which is continuous with the III **ventricle** or **diocoel** of diencephalon. The **hypothalamus cerebri** or pituitary body lies behind and partly covered by infundibulum. The hypothalamus is an important centre regulating the whole endocrine system as well as other parts of the brain. It is composed of anterior and posterior parts. The anterior part of the hypophysis has no connection with the brain.

(*ii*) Midbrain. It is well developed consisting of two dorso-lateral large rounded **optic lobes**. The optic lobes are centres for impulses coming from the eyes. Their cavities are called **optocoel** or **optic ventricles** communicating with each other and the fourth ventricle behind through a narrow cavity, the **iter** or **aqueduct of Sylvius**. Below the optic lobes are present two thick longitudinal bands of nerve fibres, called the **crura cerebri**. These connect diencephalon and medulla. These form the floor of midbrain. Lying transversely between the diencephalon and optic lobes is a band of nerve fibres called **posterior commissure**.

(*iii*) Hindbrain. It consists of the **cerebellum** and the **medulla oblongata**. (*a*) The **cerebellum** is a rudimentary narrow transverse solid band lying dorsally immediately behind the optic lobes. Its function is probably to regulate the vestibulo-oculomotor system controlling movements of the eyes. (*b*) **Medulla oblongata** is short and somewhat triangular structure which is simply a widening of the spinal cord. Its cavity is also triangular called **fourth ventricle** or **metacoel** which is joined in front to the **iter** but posteriorly it is continuous with the central cavity of the spinal cord. Its roof is thin, vascular and thrown into folds called the **posterior choroid plexus**. From the sides of medulla arises several pairs of cranial nerves. Its lower surface is divided by a median fissure which is continuous with the ventral fissure of the spinal cord. On the dorsal surface of the medulla oblongata there is a triangular area of reddish brown colour which is called **posterior choroid plexus**.

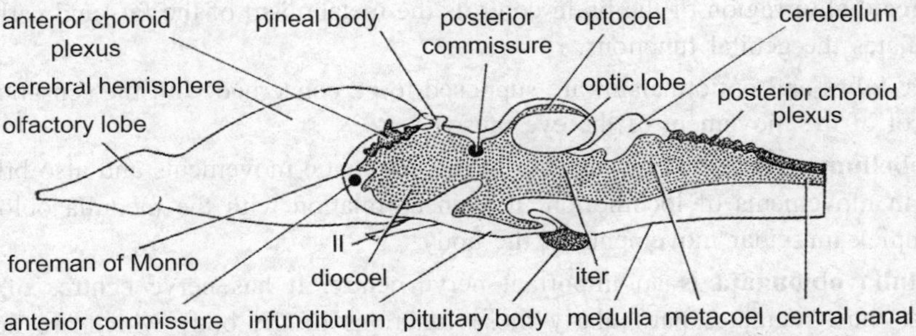

Fig. 18.50. Frog. Brain. Longitudinal vertical section of brain.

B. Spinal cord. The medulla oblongata continues behind as spinal cord, lying in the neural canal of the vertebral column. It is short, somewhat flattened dorsoventrally and terminates behind the lumbar swelling in a tapering narrow thread called **filum terminale** lying in the urostyle. The **filum terminale** with the nerve roots on either side is sometimes called **cauda equina** as it looks like a horse tail. It presents two enlargements during its course, one at the level of the forelimbs where nerves for arms arise and one far behind at the level of hindlimbs where nerves for hindlimbs arise. The anterior enlargement of the spinal cord is known as **brachial** or **cervical** swelling, while the posterior as **lumbar** or **sciatic** swelling. It is traversed throughout its length by two longitudinal grooves, one on the dorsal side and the other on the ventral side, known as **dorsal** and **ventral fissuree** respectively. These two fissures almost completely divide the spinal cord into two symmetrical halves. Like the brain the spinal cord is also formed of **gray matter** with ganglion cells and non-medullated nerve cells surrounding the central canal, and **white matter** with ganglion cells

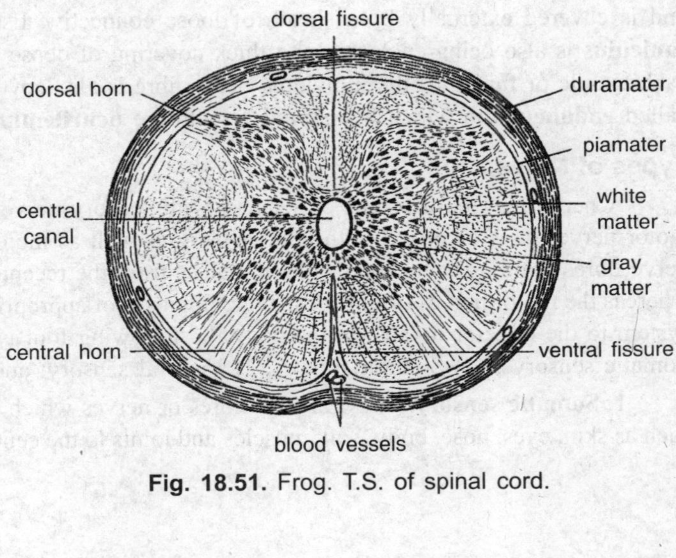

Fig. 18.51. Frog. T.S. of spinal cord.

and medullated nerve fibres surrounding the grey matter. Grey matter is produced at the corners into dorsal and ventral columns or horns, while the white matter is divided by grey matter into four zones or **funicules**, the dorsal, the ventral and two lateral funicules. It is covered by two protective membranes ; the **duramater** lines the neural canal and the vascular, thin and pigmented **piamater** closely covers the cord.

Functions of Different Parts of the Brain

Brain is the only centre for the immediate control of all vital activities as it receives impulses from different parts of the body through sensory nerves and issues orders through motor fibres to different parts of the body for appropriate action.

1. Cerebrum. The **pallium** of cerebral hemispheres controls the activities of the olfactory, tactile and optic organs, whereas the cerebral hemispheres coordinate the activities of the **neuro-muscular** mechanism of the body, but these are supposed to be the seat of intelligence and voluntary control in higher animals.

2. Diencephalon region of the brain controls the metabolism of the fats and carbohydrates and also regulates the genital functions.

3. Optic lobes and optic thalami are supposed to be concerned with the sensation of sight and the control of the movement of the eye muscles.

4. Cerebellum controls the mechanism of the automatic movements and also brings about coordination in movements of locomotion. It is in correlation with the medulla oblongata and regulates complex muscular movements of the body.

5. Medulla oblongata is an important nerve centre. It has nerve centres of all reflex functions and, thus, regulates particularly those functions of the body which are not directly under the control of the will like heart beating, respiration, swallowing, taste, hearing, sound production and secretions of various digestive juices.

2. Peripheral Nervous System

The peripheral nervous system includes the nerves arising from the brain and spinal cord. Those nerves which arise directly from the brain are called **cranial nerves**, while those arising from the spinal cord are called **spinal nerves**.

Structure of nerve. The nerves are solid structures looking like white threads. Each nerve is composed of several bundles of **nerve fibres** called the **funiculi** (singular **funiculus**) and is covered externally by a sheath of loose connective tissue called the **epineurium**. Each **funiculus** is also being enclosed by a thick covering of dense tissue called the **perineurium**. In each bundle or funiculus the individual nerve fibre is also covered by a connective tissue sheath called **endoneurium** which is continuous with the **neurilemma** of nerve fibres.

Types of Nerves

A nerve may be **afferent** or sensory with sensory nerve fibres or **efferent** or motor with motor nerve fibres or mixed with both sensory as well as motor nerve fibres. Sensory nerves or nerve fibres are those which carry the impulses from the receptors to the central nervous system, whereas the motor nerves or fibres carry the impulses of appropriate order from the central nervous system to the effector organs. In the body the following four types of fibres are recognised : (1) Somatic sensory ; (2) Somatic motor; (3) Visceral sensory; and (4) Visceral motor.

1. Somatic sensory. These are the fibres or nerves which carry the impulses from receptors such as skin, eyes, nose, body wall, muscles and joints to the central nervous system. The dendrons of these nerve fibres are very long, starting from the receptors pass to dorsal root ganglion in which their cell body lies.

2. Somatic motor. Such fibres or nerves carry impulses from the central nervous system to the effector organs which include mainly the muscles. Their dendrites and cell bodies are always found in grey matter but their long axons pass through the ventral roots to the muscles (effector organ).

3. Visceral sensory. Such fibres or nerves carry sensations from the receptors situated in the wall of alimentary canal to the central nervous system. Their dendrites start from the receptors located in the wall of alimentary canal and pass to the cell bodies situated in the dorsal root ganglia from where the axons pass into the grey matter.

4. Visceral motor. These fibres are able to convey impulses from the central nervous system to the involuntary muscles of the alimentary canal, glands and other visceral organs. Their dendrites and cell bodies are found in the grey matter, while the axons pass out through the ventral root and end in the autonomic ganglia (preganglionic) from where the second neuron starts whose axon extends to the involuntary muscles or glands (post-ganglionic).

Cranial Nerves

In an adult frog ten brain pairs of **cranial nerves** are found which emerge from the brain through various foramina of the cranium to supply the different organs of the body. Each cranial nerve originates from the brain by two roots, a dorsal and a ventral root, but these two roots do not unite with each other and, thus, look like separate nerves. These are numbered I to X, some of which are purely motor, some are sensory, while others are mixed. The cranial nerves with their names are as follows :

I. Olfactory nerve. It arises from the anterior end of olfactory lobe and innervates the cells of olfactory sac. It is **sensory** in nature.

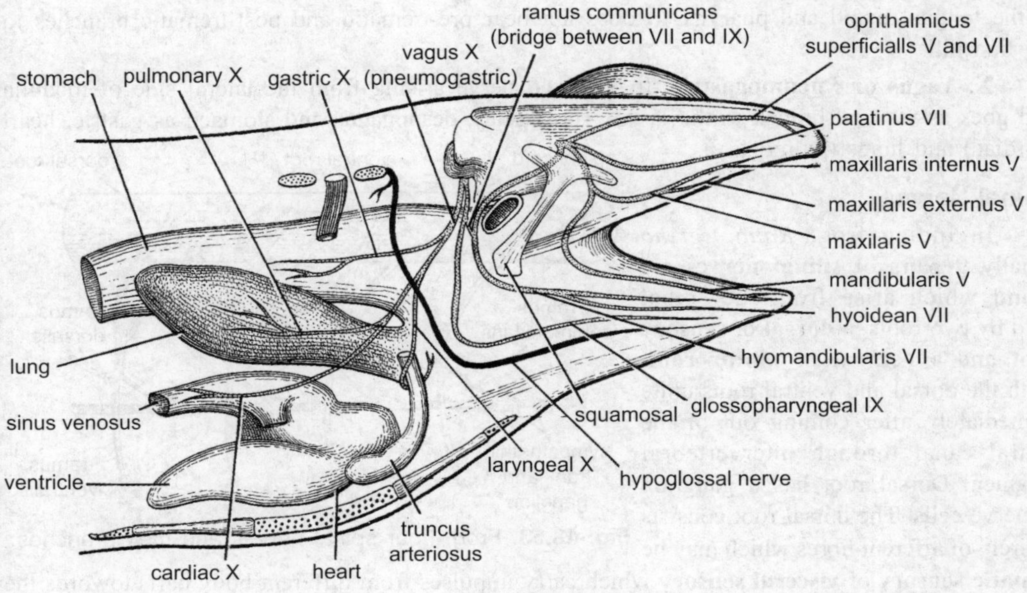

Fig. 18.52. Frog. Cranial nerves.

II. Optic nerve. Nerve fibres arise from the retina of the eye. The fibres of the two sides generally cross or decussate out the optic chiasma and then enter the optic thalamus of the opposite side, finally terminating in the thalamencephalon. It is also purely **sensory**.

III. Oculomotor nerve. It is small nerve arising from the ventral side of the midbrain **(crura-cerebri)**. It divides into branches which supply the anterior, superior and inferior recti muscles and the inferior oblique muscle of the eye ball. It is exclusively **motor**.

IV. Trochlear or pathetic nerve. It is also a small nerve arising from the dorsal side of the brain between the optic lobes and cerebellum and going to the superior oblique muscle of the eye ball. it is exclusively **motor**.

V. Trigeminal nerve. It is the largest of the cranial nerves arising from the sides of the anterior end of the medulla oblongata. Before it emerges from the skull it bears **Gasserian ganglion**. It divides into three branches : (*a*) **Ophthalmic superficialis** passes along the dorsal border of the orbit and goes to the skin of the snout. It is **somatic sensory**. (*b*) **Mandibular** and (*c*) **Maxillary** arise from a common stem and then separate. The mandibular goes to the muscles of lower jaw and the maxillary forms the two branches going to the skin of the upper jaw and upper lip. Maxillary is a **somatic sensory**, whereas the mandibular is **visceral motor nerve**. Thus, trigeminal is a **mixed** nerve.

VI. Abducens nerve. It arises from the ventral side of the medulla oblongata and enters the orbit and goes to the posterior rectus muscle of the eye ball. It is a **motor** nerve.

VII. Facial nerve. It arises from the antero-lateral side of medulla oblongata close behind the fifth. It is **mixed nerve** as having both visceral sensory and visceral motor fibres. It is divided into two branches : (*a*) **Palatine** going to the roof of the buccal cavity. (*b*) **Hyomandibular** going to the tongue and muscles of the lower jaw.

VIII. Auditory nerve. It is **somatic sensory** arising from the medulla oblongata behind the seventh and goes to internal ear.

IX. Glossopharyngeal. It is **mixed** nerve arising from the lateral side of medulla and goes to the tongue, hyoid and pharynx. It does not bear pre-trematic and post-trematic branches to the first gill.

X. Vagus or Pneumogastric. It is **mixed** nerve arising from the lateral side of medulla and goes as visceral branch to the larynx (laryngeal), oesophagus and stomach as gastric, heart (cardiac) and lungs (pulmonary).

Spinal Nerves

In Indian frog, *Rana tigrina* usually 9 pairs of **spinal nerves** are found which arise from the spinal cord by two roots, a **dorsal** or **sensory root** and a **ventral** or **motor root**. Both the dorsal and ventral roots unite immediately after coming out of the neural canal through intervertebral foramen. Dorsal root has a ganglion of nerve cells. The dorsal root consists entirely of afferent fibres which may be **somatic sensory** or **visceral sensory** which carry impulses from different body parts towords the spinal cord. The **ventral root** consists of **somatic motor** or **visceral motor** fibres of which the cell bodies together with dendrons are situated in the ventral part of the grey matter of spinal cord.

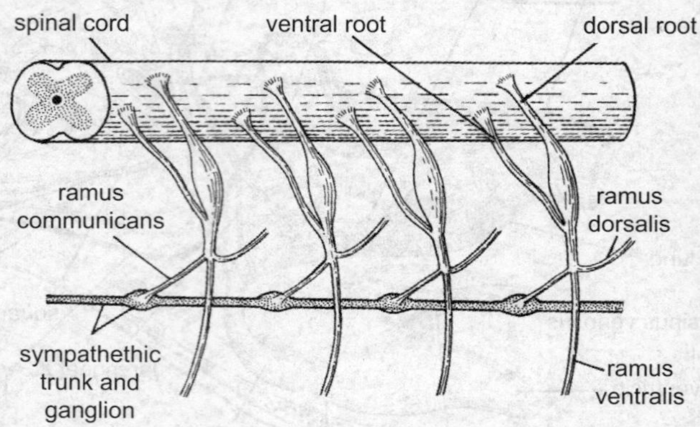

Fig. 18.53. Frog. Four spinal nerves and their branches.

These carry responses from the spinal cord to the body tissues. Thus, in the spinal nerves these different kinds of fibres are mixed but as the roots are approached, they are segregated.

The dorsal root ganglia are covered by white soft chalky masses, calcareous bodies or glands of **Swammerdam** or **periganglionic glands**. It is probably the reserve supply of calcium to the body. Each nerve immediately after its origin gives off a short dorsal branch, **ramus dorsalis** to the dorsal skin and muscles of the back, a large ventral branch, **ramus ventralis** to the ventral skin and muscles of the body and a very short **ramus communicans** which unite to the nearest sympathetic ganglion. Ramus dorsalis contains only somatic sensory fibres, whereas the ramus ventralis contains somatic motor fibres, while ramus communicans contains both visceral sensory and visceral motor fibres.

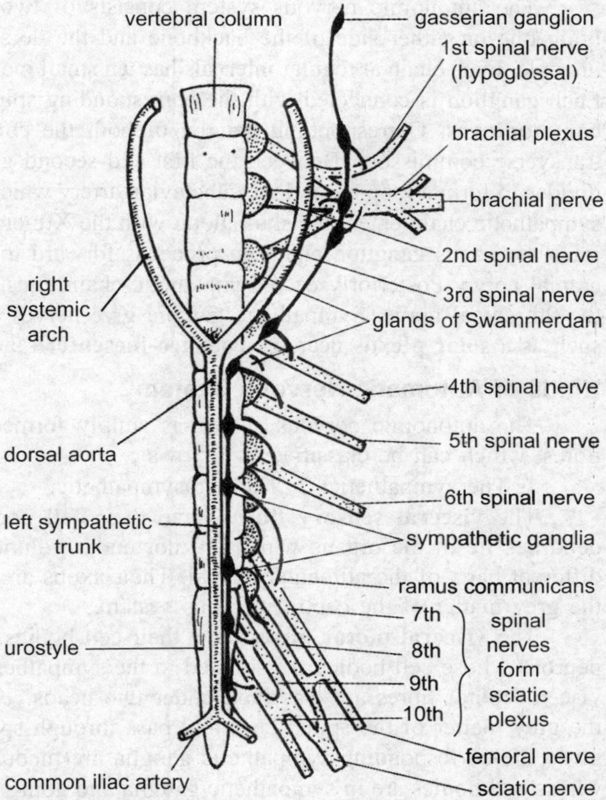

Fig. 18.54. Frog. Dissection showing spinal nerves 1–10 and sympathetic trunk on the **left side** only.

The first spinal nerve (hypoglossal) comes out between the first and second vertebrae and innervates the tongue and several muscles attached to the hyoid. The second spinal nerve is quite large and emerges between the second and third vertebrae. It receives the branches of the first and third spinal nerves to form a **brachial plexus**, then it proceeds as the **brachial nerve** to the skin and muscles of the forelimbs. The fourth, fifth and sixth nerves are small going obliquely to the skin and muscles of the abdomen. The seventh, eighth and ninth spinal nerves pass almost directly backward and anastomose to form the **sciatic** or **lumbo-sacral plexus** from which a large **sciatic nerve** and several small nerves enter the hindlimb. The most important one is **sciatic** nerve which divides into **tibialis** and **peroneus**. The tibialis gives branches to the gastrocnemius, tibialis posticus and numerous muscles of the plantar surface of the foot. The **peroneus** supplies the peroneus muscle tibialis. The tenth spinal nerve is not found in *Rana tigrina* but in other frogs it is present. When it is present it emerges from a foramen in the anterior part of the urostyle and goes to the cloaca and urinary bladder.

The roots of the last four pairs of spinal nerves are elongated forming a bundle of nerves called **cauda equina** which lies inside the vertebral column along the filum terminale.

3. Autonomic Nervous System

The autonomic nervous system is partly independent and not under voluntary control. Though it is involuntarily controlled by the nerve centres located in the central nervous system, it is also connected to spinal nerves and some cranial nerves. It is simply concerned with the intestinal regulation of the body with the central nervous system together with its spinal and cranial nerves and is concerned with the external regulations.

The autonomic nervous system consists of two delicate longitudinal chains of ganglia, lying one on either side of the backbone and the dorsal aorta from the brain to the end of the urostyle. Each chain at regular intervals has ten small metamerically arranged **sympathetic ganglia**. Each ganglion is connected with the corresponding spinal nerve by a small branch called **ramus communicans**. Corresponding ganglia of both the cords are also connected together by small transverse commissure. Between the first and second ganglia the sympathetic chain of each side divides to form a loop around the subclavian artery which is known as **annulus of Vieussens.** Each sympathetic chain enters the skull along with the Xth cranial nerve through the vagus foramen and joins the vagus ganglion and then proceeds forward to join the **Gasserian ganglion** of the fifth cranial nerve. Posteriorly each sympathetic chain ends by joining with one or more branches of the 9th spinal nerve. Sympathetic ganglia give nerves to the viscera, where they form plexuses, such as a **solar plexus** near the **coeliaco-mesenteric artery** and a **cardiac plexus** on the heart.

Kinds of Automatic Nervous System

The autonomic nervous system is simply formed of visceral motor and visceral sensory fibres, which can be classified as follows :

1. The sympathetic; 2. The parasympathetic.

The **visceral sensory fibres** have their cell bodies in the dorsal root ganglia and their dendrites lie in the organs which are not under voluntary control like the heart, blood vessels, different parts of the alimentary canal. Their axons are extended from the dorsal root ganglia to the grey matter of the central nervous system.

The **visceral motor fibres** have their cell bodies in the spinal cord forming synapses with neurons whose cell bodies are situated in the sympathetic ganglia. Because of these synapses the visceral motor fibres are classified under two heads : **1. Preganglionic fibres** which arise from the grey matter of the spinal cord and pass through spinal nerves and rami communicantes and go to the corresponding sympathetic ganglia are medullated. **2. Postganglionic fibres** are those whose cell bodies are in sympathetic ganglia and going in organs they supply are non-medullated. Thus, sympathetic nerves are made of both visceral sensory and visceral motor fibres which on stimulation secrete a chemical substance called **sympathin** which generally stimulate the organs.

Parasympathetic system includes parasympathetic nerves and ganglia. The fibres of parasympathetic nerves come from cranial and spinal nerves, while parasympathetic ganglia are situated in the organs innervated by parasympathetic nerves. Their **preganglionic** fibres are very long and extend from the central nervous system to a ganglion in or near the organ which they innerverate. Here they are connected to the short **postganglionic** fibres. These fibres on stimulation secrete a chemical substance called **acetylcholine** which has an inhibitory effect on the organs. The parasympathetic nerve fibres are included in **oculomotor**, **trigeminal**, **facial** and **vagus** cranial nerves.

The visceral and some other organs receive fibres both from sympathetic and parasympathetic nerves. These are antagonistic in effect as sympathin of sympathetic fibres brings an action, whereas acetylcholine of parasympathetic fibres retards it.

The autonomic nervous system simply regulates the functions of certain organs which are involuntary. It is not independent because it is intimately bound structurally and functionally with central and peripheral nervous system. Its nerve centres are located in the central nervous system, while its most fibres are parts of the peripheral nervous system.

RECEPTORS

In frog various types of **receptors** or **organs of special sense** are found which are supplied with nerve fibre and, thus, convey the stimulus to the central nervous system. These receptors can be grouped under two heads :

1. External or Exteroceptors. External receptors are those which receive impulses from the external environment. External receptors can be grouped under following heads :

(*i*) **Tangoreceptors** or organs of touch;
(*ii*) **Olfactoreceptors** or organs of smell;
(*iii*) **Gustatoreceptors** or organs of taste;
(*iv*) **Photoreceptors** or organs of sight;
(*v*) **Statoacoustic receptors** or organs of hearing and balance.

(*i*) Tangoreceptors or organs of touch. The entire skin serves as organs of touch as it is abundantly supplied with sensory nerve endings situated in the spaces between the cells. At places compact groups of cells form **corpuscles** which project into papillae of epidermis. These are supplied with sensory nerve endings. Such groups of cells are called **tactile organs** or **patches**. These are very much sensitive to touch and also to temperature. The nerve endings never reach the cells of the outermost layer of epidermis. Consequently the stimuli which produce the sensation of touch are not directly received by the tactile organs. The tactile organs make the skin of the frog sensitive to touch, heat, cold and the effects of the chemicals. In tadpole larva a lateral line system is found which is absent in adult frog.

(*ii*) Olfactoreceptors or organs of smell. These are the organs which are simply concerned with distinguishing the various kinds of smell given off from different substances or things. They include a pair of **olfactory** or **nasal sacs** located in the olfactory capsules of the skull. Each nasal sac communicates with the outside by the external nares and with the buccal cavity through the internal nares. These are internally lined with columnar epithelial cells out of which certain are special modified cells called **neurosensory** or **olfactory cells** which are bipolar in shape. Their deeper ends are connected to the fibres of the **olfactory nerves** which are directly connected with the brain (olfactory lobes), while their other ends are produced into **sensory hairs** projecting into the lumen of the nasal sac. The mucous lining of the sac also has supporting and mucus secreting cells whose secretion keeps the nasal epithelium moist in order to make it more effective. The odours of a substance are brought to these organs either through the medium of air or water. From these organs the stimuli travel along the olfactory nerves to the brain.

(*iii*) Gustatoreceptors or organs of taste. Gustatoreceptors or organs of taste are found in the form of **taste buds** which are confined chiefly over the tongue and the floor and roof of the buccal cavity. Each **taste bud** is more or less spherical body consisting of a group of barrel-shaped columnar cells, some of which are **neurosensory** and others are supporting cells. The neurosecretory cells are slender and elongated and found in the middle and are covered all around by supporting cells. The free ends of neurosensory cells are produced into delicate fine **taste hairs** projecting above the surface, while other ends are innervated with sensory nerve fibres. The supporting cells are larger in size but lack sensory hairs. Taste buds of frog are supplied by the branches of the VII and IX cranial nerves.

The mucous membrane of the tongue is produced into two kinds of papillae : the conical **filiform papillae** and rounded knob-like **fungiform papillae**. The taste buds are confined only in fungiform papillae and stimulates whenever their taste hairs come in contact with a substance in solution.

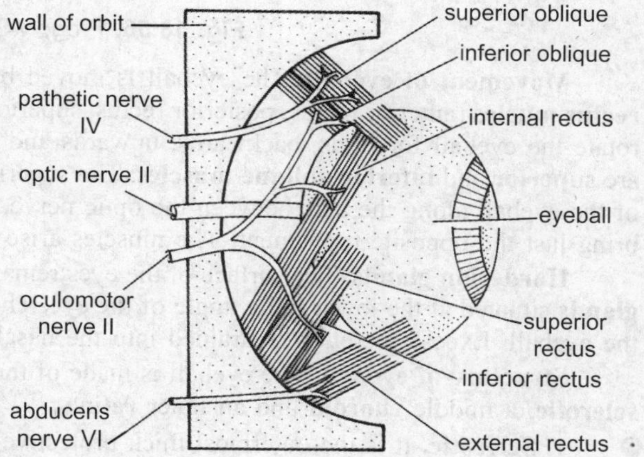

Fig. 18.55. Frog. Dissection of right eyeball in dorsal view showing eye muscles and their nerve supply.

(iv) **Photoreceptors or organs of sight.** The organs of sight are two spherical eyes which are situated in the orbital fossae on either side of the head. Their structure and function is like that of other vertebrates. Their one-third part is visible externally but remaining part lies hidden in the orbit. The orbit has no bone at its bottom, therefore, the eyeballs can be seen projecting on the roof of the buccal cavity as spherical prominences. The eyes are protected with two eyelids–the upper is immovable and lower one is movable. There is a transparent **nictitating membrane** which covers the eyes when the frog goes inside the water or underground. The nictitating membrane is simply a transparent part of the upper border of the lower eyelid which lies folded beneath the opaque part of the lower eyelid. Its movement is controlled by means of two types of muscles, the **retrator bulbi** and the **levator bulbi**. On the contraction of the **retrator bulbi** muscles the eyeball is drawn deep into the orbit due to which the eyelids come nearer and the nictitating membrane is unfolded over the eye. On the contraction of the **levator bulbi** muscles the eyelids are separated that results in the folding back of the nictitating membrane to its original position.

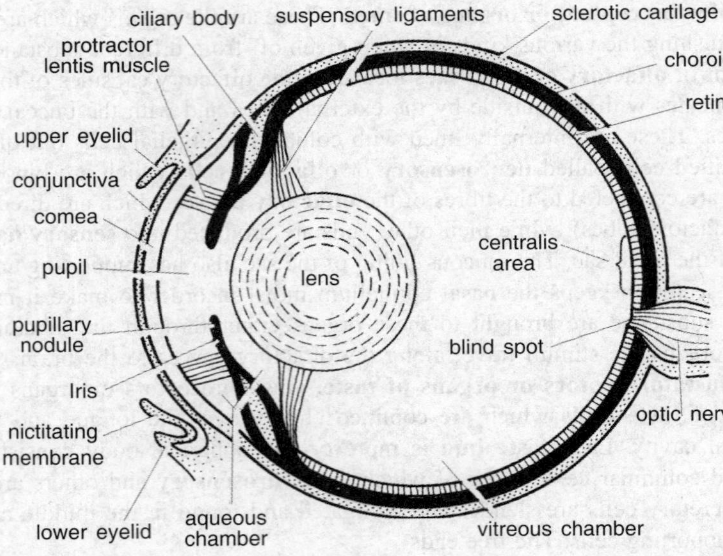

Fig. 18.56. Frog. V.S. of eye.

Movement of eyeball. The eyeball is moved by a set of six extrinsic muscles : *(i)* Four **rectus mucles** (anterior rectus, posterior rectus, superior rectus and inferior rectus). These muscles rotate the eyeball forwards, backwards, upwards and downwards respectively. *(ii)* Two muscles are **superior** and **inferior oblique muscles**. The superior oblique muscles bring about the rotation of the eyeball along the axis between the optic nerve and cornea, while inferior oblique muscles bring just the opposite movement. The muscles arise from the front part of the orbit.

Harderian glands. The surface of the eyes remains moist due to the presence of **Harderian glands** situated at the lower inner angle of the eye. These glands secrete a liquid which lubricates the eyeball. Excess secretion is drained into the nasal sac through a fine **naso-lachrymal duct**.

Structure of eyeball. The eyeball is made of three concentric layers or coats, an outermost **sclerotic**, a middle **choroid** and an inner **retina**.

1. Sclerotic. It is the outermost thick protective layer which is made of fibrous connective tissue and cartilage. It is actually the optic capsule which has not fused with the skull but fits closely to the eye. It occupies about two-thirds of the entire circumference of the eyeball and is mostly out of sight being within the orbits. The remaining one-third of it is continued in front as an arched transparent **cornea**. It simply maintains the shape of the eyeball and also protects the eyeball and also provides surface for attachment of extrinsic eye muscles. Cornea is covered externally by a thin transparent membrane called **conjunctiva** which is the continuation of skin. It is also continued with the inner lining of the eyelids. Nictitating membrane which is the continuation of the inner membrane of lower eyelid protects the outer exposed surface of the eye. It has thin blood capillaries.

2. Choroid. It is the middle vascular and pigmented layer made of loose connective tissue fibres. Towards the anterior side it thickens as a ring-like **ciliary body** having ciliary muscles and then separates from the sclerotic forming a circular pigmented **iris** with a central aperture, the **pupil**, which looks like a black spot. Within the iris are two sets of involuntary muscles, circular and radials, which regulate the opening of pupil.

Immediately behind the pupile iris is a large, circular and transparent crystalline lens which is enclosed in a delicate **lens capsule**. It is suspended behind the pupil by a membranous **suspensory ligament** from ciliary body. The iris divides the hollow of the eyeball into a small anterior **aqueous chamber** and a large posterior **vitreous chamber**. The aqueous chamber is filled with a watery **aqueous humour**, whereas the vitreous chamber is filled with **gelatinous vitreous humour**.

3. Retina. It is the innermost sensitive layer present only in the posterior part of eyeball behind ciliary body. It is made of several layers. The outermost layer is formed of pigment cells lining the choroid and beneath it is the sensory layer of rods and cones, which is followed by several layers of sensory cells. The innermost layer is of nerve fibres which are connected to the **optic nerve**. Towards the posterior side the optic nerve leaves the eyeball through the three coats. This site is called the **blind spot** which has no rods and cones. Close to it is the **yellow spot** where a distinct image is formed but at blind spot no image is formed.

Microscopic structure of retina. Microscopically the retina is a complicated coat consisting of several layers of different kinds of sensory cells. Next to the outer pigmented layer of the retina, there is a layer of visual cells consisting of **rods** and **cones** followed by a layer of **bipolar nerve cells** and an outer layer of **ganglionic cells**. The rods and cones are modified sensory cells which have been so named due to their shape. These cells are placed at right angle to the surface of the retina and their thin ends are embedded in a layer of pigment cells. The rods are cylindrical cells containing a purple pigment and the cones are conical tapering cells. The rods are meant simply to perceive the amount of light, while the cones for distinguishing colours.

The bipolar nerve cells on one side form synapses with the rods and cones, while on the other side with the ganglionic cells. The axons of ganglionic cells spread over the inner surface of the retina, and converge at the back of the eyeball where they pierce the retina and come together to form the **optic nerve** going to the brain.

The posterior part of the retina just opposite to the lens is called **area centralis** or **yellow spot** which contains only cones and has yellow pigment, the images are normally focussed on this area.

Working of the eye. Frog has a **monocular vision** as the two eyes are situated far away from each other over the head and their images also do not coincide. The eyes function like photographic camera. The eyelids function like the shutter of camera, the iris like diaphragm which regulates the amount of light entering the eye through the pupil, the **lens** like camera's lens and the sensitive **retina** like the film or the plates of the camera.

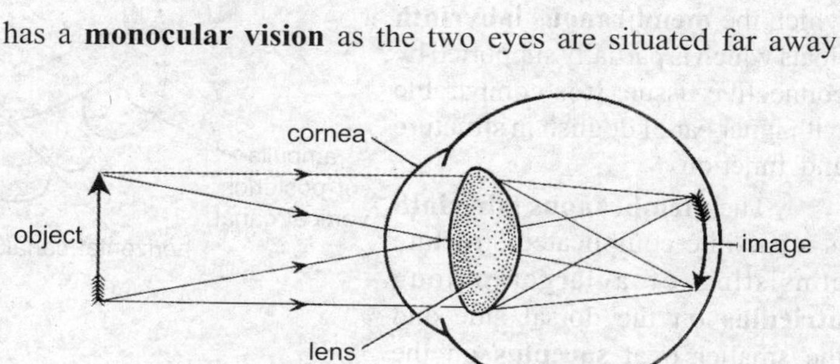

Fig. 18.57. Frog. Image formation in the eye.

When light from any object in front of the eye falls upon it, the rays pass through the cornea and aqueous humour and reach the spherical lens through the pupil. The cornea and lens both in combination focus the rays on the retina to stimulate the rods and cones. The impulses then pass along the bipolar and ganglionic cells to reach the brain through the optic nerve. Light rays falling on the retina produce an **inverted** and **reduced image** of the object which is rightly corrected by the brain but never reinverted in the brain as it is often supposed. The rays passing through the pupil converge due to their passage through the cornea and aqueous humour which together act as a convex lens but the rays as pass through the **crystalline lens** are further refracted in order to form an inverted and diminished image of the object on the retina which is rightly corrected by the brain.

During the image formation the iris acts as a diaphragm. If the object is well illuminated the pupil closes down to a pin point in order to check the excess of light intensity inside the eyeball and also to prevent the disintegration of the retinal cells, but when the object is poorly illuminated the pupil is widely open in order to allow as much light as possible to enter the eyeball to form the clear image of the object. In this way the iris helps in the formation of a well defined image after regulating the amount of light entering the eye.

(*v*) Statoacoustic receptors or Organs of hearing (Ear). The statoacoustic organ is for hearing and equilibrium. It includes a pair of ears present on the postero-lateral sides of skull enclosed in auditory capsules. The ear of frog includes a **middle ear** and an **internal ear**. The external ears are absent.

1. Middle ear. The middle ear consists of all those structures which are related simply in transmitting the sound waves to the internal ear which acts as suspensory apparatus. The cavity of middle ear is called the tympanic cavity lined by a membrane. It is an air-filled chamber communicating with the pharynx by a slender **eustachian tube**, the opening is normally kept closed by a valve. Externally the cavity of middle ear is bounded by a circular patch of dark skin, the **tympanic membrane** or **eardrum** which is tightly stretched over a ring of cartilage, the **annulus tympanicus**, it is simply a modified skin. It is vibratile in nature and from the centre of which a club-shaped rod, the **columella auris**, is extended across the tympanic cavity and attached internally to a membrane and a cartilaginous small nodule, the **stapedial plate** fused over a small window-like oval aperture, **fenestra ovalis** (a hole in the auditory capsule). The columella auris is partly made of bone and partly of cartilage. The columella auris is equivalent to hymandibula of dogfish.

2. Internal ear. The internal ear consists of a bony **auditory capsule** which is cartilaginous in the beginning but later in the adult is bony. It is formed by the **pro-otic** and **exoccipital** bones. The auditory capsule is filled with a watery fluid called **perilymph** in which the **membranous labyrinth** floats which is partially supported by connective tissue. It is comparable with inner ear of dogfish in structure and function.

The **membranous labyrinth** is a sac-like complicated structure consisting of a larger oblong **utriculus** on the dorsal side and the smaller oval **sacculus** on the ventral side.

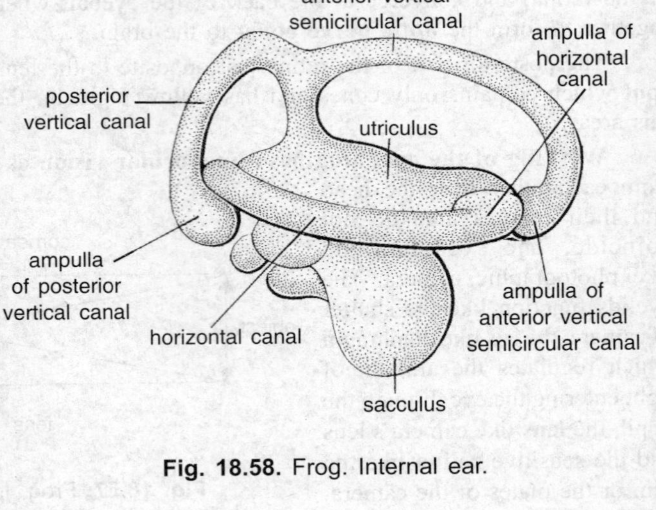

Fig. 18.58. Frog. Internal ear.

From the posterior side of the sacculus arise two small dilations, the **lagena** or **cochlea** and the **pars basilaris** which is now regarded as a part of lagena. Lagena is the forerunner of cochlea of higher vertebrates.

Dorsal to these is the protuberance of utriculus, called the **pars neglecta**. From the inner dorsal side of the sacculus arises a narrow tube, the **ductus endolymphaticus** and terminates in a dilated thin-walled endolymphatic sac present over the hindbrain within skull.

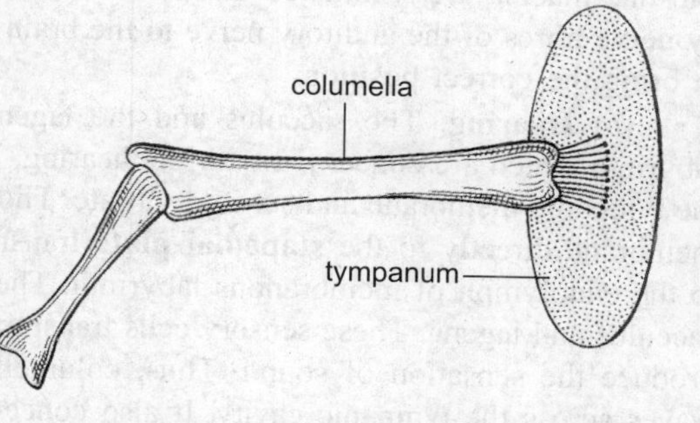

Fig. 18.59. Frog. Columella auris.

The three semicircular canals arise from the utriculus, are placed at right angles to one another, these are anterior and posterior vertical semicircular canals, while the third is a horizontal semicircular canal. The anterior and posterior canals have their adjacent limbs dorsally united and open into the utriculus through a common opening, whereas the horizontal semicircular canal opens into the utriculus at either end. Each semicircular canal at its distal end is dilated to form a small round **ampulla**. The ampullae of the anterior and horizontal semicircular canals are found at their anterior ends and that of the posterior canal is at its posterior end.

The entire **membranous labyrinth** is hollow and is filled with a fluid called **endolymph** which contains pieces of calcium carbonate forming **otoliths** or **ear stones**. The various parts of the membranous labyrinth are innervated by the fibres of the auditory nerve.

Histology of membranous labyrinth. The wall of entire membranous labyrinth is made of dense fibrous connective tissue and is lined with cubical epithelial cells. The epithelial lining is modified at certain places to form receptors of the labyrinth which are called **sensory patches** or **acoustic spots**. There is one acoustic spot in each ampulla, one spot in the utriculus, one in the sacculus and one in the lagena. The acoustic spots of the ampullae are called **cristae**, while those of the utriculus, lagena and sacculus, **maculae**. The macula of utriculus is a large **pars neglecta** and that of the lagena is called **basilar papilla**. The **cristae** and **maculae** have the same structure as they are made of **sensory hair cells** and the **supporting cells**. The sensory cells have stiff and tapering hair-like processes at their inner free ends, while their lower ends are connected with nerve fibres of the auditory nerve.

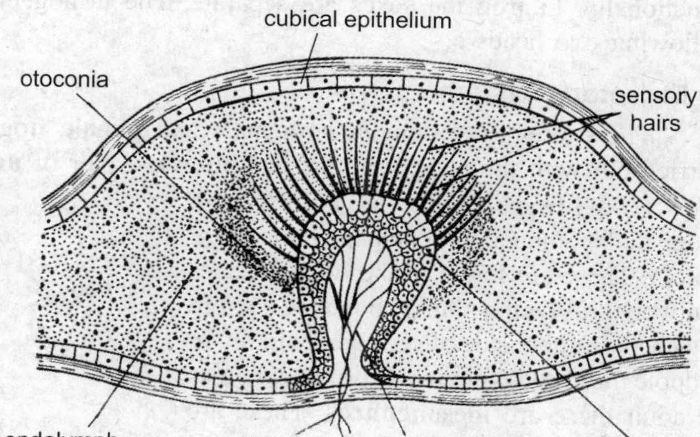

Fig. 18.60. Frog. V.S. of ampulla through an acoustic ridge or crista.

Working of the ear. Ears have two functions, hearing and equilibrium. The membranous labyrinth of the frog functions primarily as the organ of balancing and secondarily as the organ of hearing.

(*a*) Equilibrium. The semicircular canals along with their ampullae and the utriculus are concerned with the balancing of the body. On any change in the position of the animal the endolymph and the otoliths start movement and also exert pressure due to which the three cristae and one macula are stimulated. From which the impulses of change of position are transmitted by nerve fibres of the auditory nerve to the brain which sends the impulses to muscles concerned to bring the correct position.

***(b)* Hearing.** The sacculus and the lagena are the main structures of the membranous labyrinth which are concerned with the hearing. In hearing the sound waves strike the surface of the tympanic membrane and set it to vibrate. The columella which is extended from the tympanic membrane directly to the **stapedial plate** transmits these vibrations to the perilymph and then to the endolymph of membranous labyrinth. These vibrations stimulate the sensory hair cells of sacculus and lagena. These sensory cells transmit the impulses to the brain by the auditory nerve, produce the sensation of sound. Thus, columella is meant only for the transmission of sound waves across the tympanic cavity. It also concentrates the sound waves to a point.

2. Interoceptors. These are certain muscles and tendons of the body and also the alimentary canal which respond to stimuli as they are very much supplied with nerve fibres called **interoceptors**. The interoceptors found in the alimentary canal and viscera are called **interoceptors**, whereas those found in the muscles and joints are called **proprioceptors**.

Interoceptors provide information about the hunger, thirst, pain or comfort in the alimentary canal. **Proprioceptors** provide information about pains in viscera and tension in the muscles of the body.

URINOGENITAL SYSTEM

Since the excretory and reproductive systems are closely associated, hence, it is customary to call the two systems together as a **urogenital** or **urinogenital system**, though both are unrelated functionally. In frog the sexes are separate. The urinogenital organs can be studied under the following two heads :

1. Excretory System

The excretory system in both male and female frog is similar. The excretion is mainly carried out with the help of a pair of **kidneys**, a pair of **ureters**, a urinary bladder and cloaca.

***(i)* Kidneys.** Both the kidneys are elongated, compact, flattened and dark red in colour. These are found in the lymph spaces (subvertebral lymph sinus) above the coelom attached on either side of vertebral column. In tadpole the kidneys are **pronephros**, whereas in adult these are **mesonephros**. These are covered ventrally by peritoneum.

***(ii)* Ureters.** From the outer smooth convex posterior side of each kidney arises a **mesonephric** or **Wolffian duct** or **ureter** which passes backwards to open into dorsal side of the cloaca. The openings of the ureters are placed over a separate papilla on the dorsal side of cloaca. In male frog the ureters dilate just posterior to the kidney to form a **vesicula seminalis** in which sperms are stored. In male frog the ureters convey the sperms and urine, and, hence, are called **urinogenital ducts.** The ventral surface of each kidney has a yellow coloured **adrenal** or

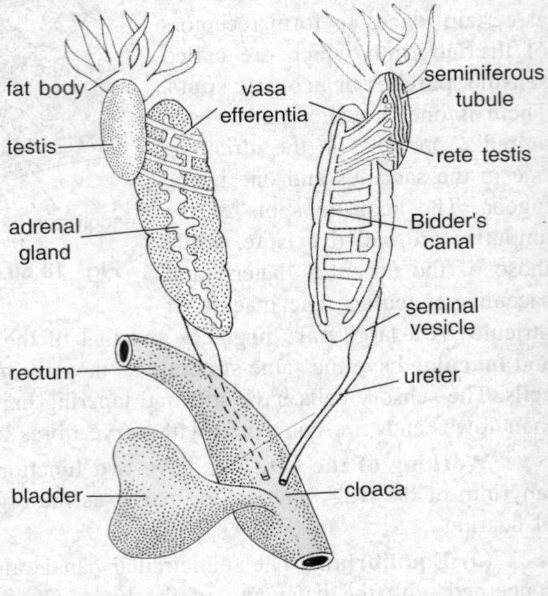

Fig. 18.61. Frog. Male urinogenital organs.

supra renal gland of endocrine function. To the anterior of each kidney are attached numerous finger-like fat bodies, a testis in male and ovary in female. Fat bodies are reserves for nourishment.

***(iii)* Urinary bladder.** It is large, thin-walled bilobed distensible structure. It also opens into the ventral wall of cloaca by a sphinctered aperture. Its aperture lies below and opposite to the openings of ureters. The inner surface of bladder is lined with a layer of epithelium about three cells thick. The middle layer of the bladder consists of a network of smooth muscle fibres and outside this layer is a thin sheet of connective tissue covered externally by the peritoneum.

***(iv)* Cloaca.** It is a small, medium sac receiving the anus, urinogenital apertures and the opening of urinary bladder. Cloaca opens outside by a cloacal aperture placed at the posterior end of the body between the two hindlimbs.

Histology of kidney. The kidneys are haemopoietic organs forming most of the red blood corpuscles. Each kidney is composed of a number of (about 2000) twisted **renal** or **uriniferous tubules** held together by connective tissue and richly supplied with blood vessels and their capillaries. Each **renal** or **uriniferous tubule** starts as a thin, double-walled ciliated cup, the **Bowman's capsule** enclosing a bunch of blood capillaries, the **glomerulus** which receives blood from an **afferent arteriole** of the renal artery. Uriniferous tubules are the functional units of kidney. These are lined with a glandular and at places with ciliated epithelium. From the glomerulus blood is collected by **efferent arteriole** which joins with a renal vein. The tubule gets the blood supply from the capillaries of efferent arteriole and renal portal vein. The Bowman's capsule along with its glomerulus is called **Malpighian body** or **corpuscle tubules**. Each tubule opens into the collecting tubule which runs transversely across the kidney towards the dorsal surface. In turn all the transverse collecting tubules open into a longitudinal **Bidder's canal** lying towards the inner margin of kidney and towards outer margin into the ureter. Ventral ciliated funnel-shaped

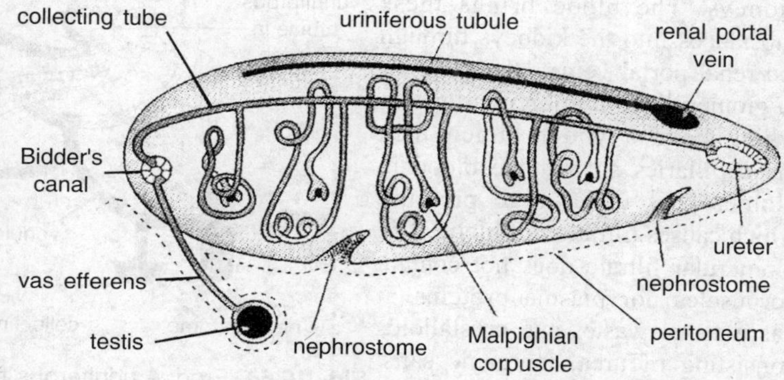

Fig. 18.62. Frog. Diagrammatic representation of kidney.

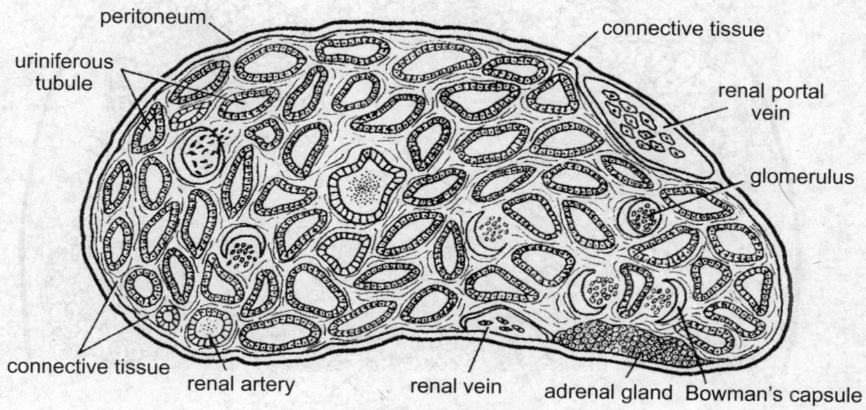

Fig. 18.63. Frog. T.S. of kidney.

nephrostomes. They carry wastes from the coelom into renal veins in frog or in uriniferous tubules in tadpoles.

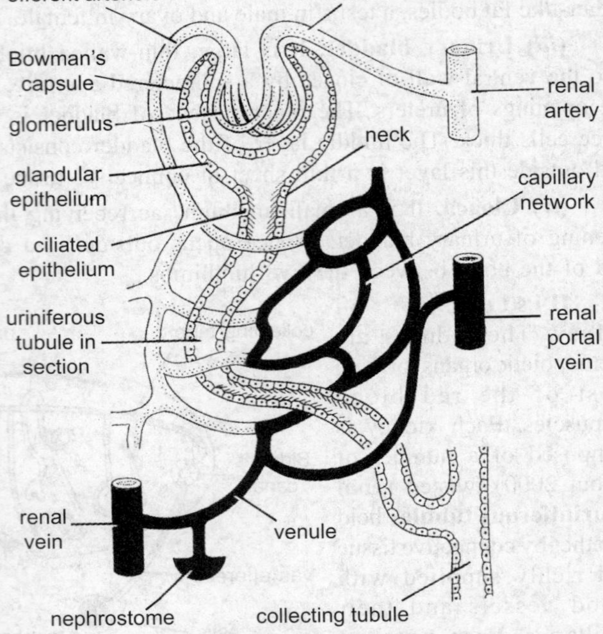

Fig. 18.64. Frog. A uriniferous tubule in section with its blood capillaries.

Physiology of Excretion

Due to continuous catabolic activities certain substances are formed which are harmful to body, hence, their elimination is very essential which is done by the kidneys. The blood brings these substances into the kidneys through the renal portal veins. Blood comes to glomeruli from afferent arterioles which have a wider lumen then the capillaries so that the blood in glomeruli is under high pressure which causes filtration of blood. The glomerular filtrate does not contain corpuscles nor plasma proteins, it has plasma, wastes and crystalloids consisting of urea, inorganic salts, glucose and large quantities of water. The glomerular filtrate passes through the thin wall of the Bowman's capsules into the lumen of uriniferous tubules and the filtered blood goes into renal vein through the efferent arteriole.

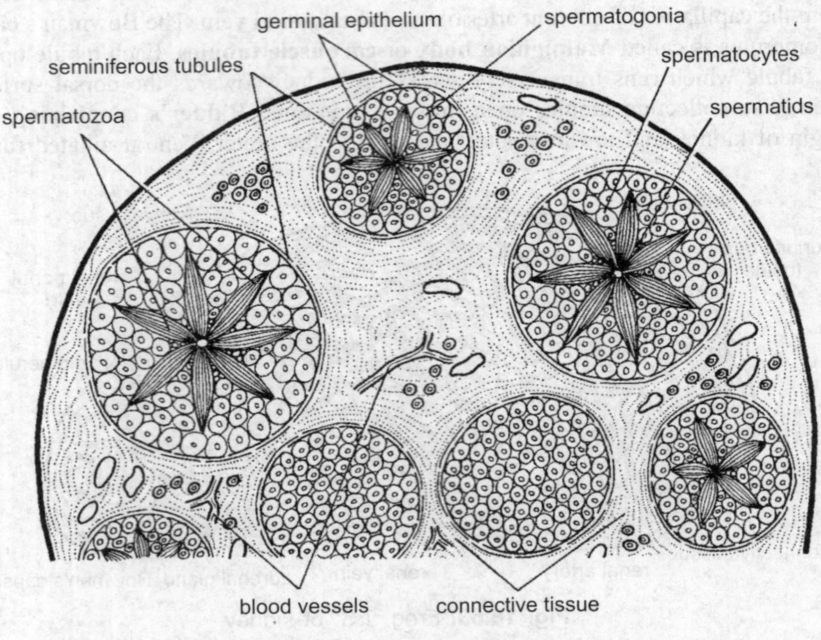

Fig. 18.65. Frog. T.S. of a part of testis.

In the convoluted tubules selective reabsorbtion of useful substances takes place. The substances that are useful, such as glucose, amino acids, inorganic chlorides and some water are reabsorbed from the filtrate and put back into the blood of the capillaries of renal veins, while harmful substances such as urea and other salts along with water form urine which passes through the collecting tubes into ureters and goes to the cloaca. It is stored in the urinary bladder for a time. When bladder is full, its involuntary muscles contract, the urine is forced into the cloaca which is voided at intervals through the cloacal aperture.

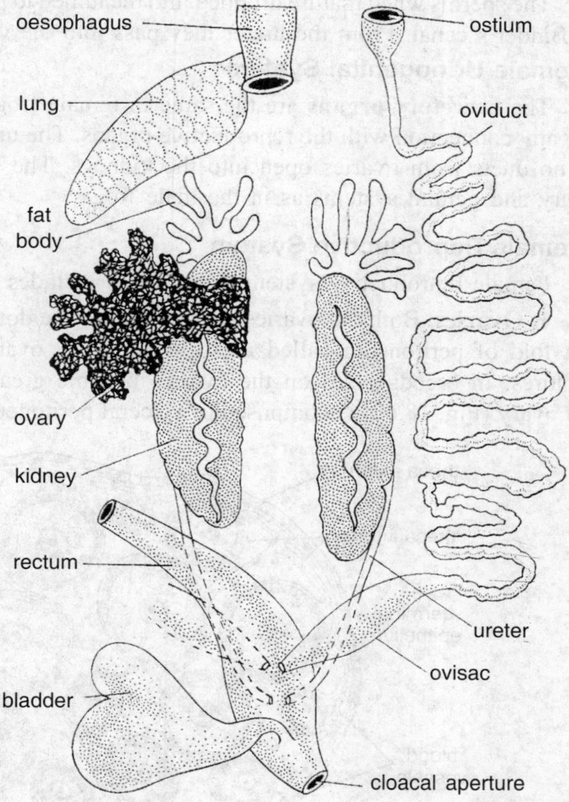

Fig. 18.66. Frog. Female urinogenital organs.

2. Male Reproductive System

Male reproductive system (Fig. 18.61) includes a pair of testes attached to kidneys, vasa efferentia and a pair of urinogenital ducts. Copulatory organs are absent.

1. Testis. The testis are rounded or ovoid, **light yellow** bodies attached to the antero-ventral surfaces of the kidneys by a double fold of peritoneum, the **mesorchium**. Actually each testis is surrounded by peritoneum, which is extended dorsally as a double membrane, the **mesorchium**, to the dorsal side of the body cavity, where its becomes continuous with the general coelomic lining. Just in front of each testis are found attached yellowish finger-like processes. They serve as a sort of storehouse of nutrients which are provided to the developing spermatozoa and during hibernation.

2. Vasa efferentia. The vasa efferentia consist of a variable number of slender tubes arising from the inner margin of testis and extend within the mesorchium and then enter the inner margin of the kidney to open into the Bidders, canal. The Bidders' canal communicates with the ureter through collecting tubules of kidney. In this way sperms enter the ureter of kidney through vasa efferentia, Bidders' canal and collecting tubules. The vasa efferentia are originally outgrowths of the walls of the Malpighian corpuscles which become connected with the testis.

3. Urinogenital duct. Ureter in male frog is a urinary duct as well as a vas deferens to convey the urine and spermetozoa. Hence, it is called a urinogenital duct. Both the ureters open into the dorsal wall of cloaca separately on urinogenital papillae.

Histology of testis. Histologically, each testis (Fig. 18.65) is composed of a large number of **seminiferous tubules** held together by connective tissue. In the connective tissue are found blood capillaries and lymph vessels, nerves, muscle fibres and groups of **interstitial cells** in between the lubules. These cells secrete a hormone testosterone which bring the secondary sexual characters of the individual. The wall of tubule is lined by germinal cells which produce sperms by spermatogenesis. The mature sperms are found in bundles in the lumen of tubules floating in the spermatic fluid.

The sperms when mature dropped into the lumen to pass into the ureter through vasa efferentia and Bidder's canal. From the ureter they pass into the vesicula seminalis where they are stored.

3. Female Urinogenital System

The **excretory organs** are the same in female frog as found in male frog, but they do not have any connection with the reproductive organs. The ureter does not dilate as vesicula seminalis and no ducts from ovaries open into the kidneys. The cloaca serves as a common passage for urinary and genital systems as in the male frog.

4. Female Reproductive System

Female reproductive system (Fig. 18.66) includes a pair of ovaries and a pair of oviducts,

1. Ovaries. Both the ovaries are attached to the dorsal abdominal wall, close to the kidneys, by a fold of peritoneum called **mesovarium**. The ovaries are large, lobulated hollow sac-like structures. In breeding season the ovaries become greatly enlarged. Histologically, the wall of each ovary (Fig. 18.67) is composed of visceral peritoneum which forms germinal epithelium and

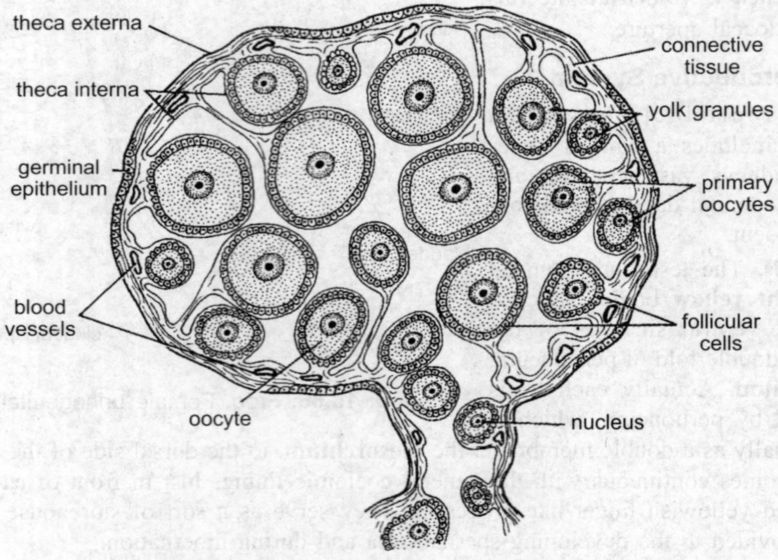

Fig. 18.67. Frog. T.S. of a part of ovary.

internal to it is the fibrous connective tissue having blood vessels, muscle fibres and nerves. The germinal epithelium gives off several small groups of cells or **oogonia** which form ovarian follicles. Within each follicle a central cell enlarges forming **ovum**. It contain a nucleus and granular yolky cytoplasm. The remaining follicle cells form the follicular epithelium around the ovum. Later with the maturation of ovum, follicular epithelium is replaced by **vitelline membrane** secreted by the ovum. In front of each ovary are attached **fat bodies**.

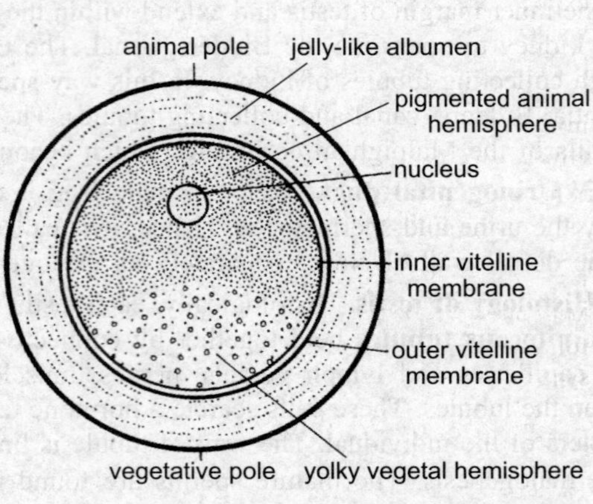

Fig. 18.68. An unfertilised ripe egg.

2. Oviducts. On each side is a long and much coiled glandular and ciliated **oviduct** or **Mullerian duct**. It starts near the base of the lung by a thin-walled ciliated **coelomic** or **oviducal funnel**. At the posterior end near cloaca, each oviduct dilates to form a thin-walled **ovisac** called **uterus** which opens by a narrow aperture on a papilla in the cloaca. The cilia of the oviduct direct the eggs posteriorly and the glands secrete albuminous coat around each egg during their descent. Oviducts become much enlarged and coiled just before the breeding season. The eggs escape from the surface of the ovary into the coelom and are directed by cilia into the oviduct and are temporarily stored in the ovisacs.

Each egg is spherical and about 1.75 mm in diameter. Its upper black pigmented half is the animal hemisphere containing nucleus and the lower half containing white yolk is the vegetal hemisphere. Egg is enveloped into a thin vitelline membrane and outer to it is the thick albuminous coat.

IMPORTANT QUESTIONS

Long Answer Questions

1. Discuss the food, feeding habits and the digestive organs of *Rana tigrina.*
2. Explain the structure and functions of the liver of frog.
3. Describe the various modes of respiration in frog.
4. Describe the structure and working of heart of frog.
5. Give an account of arterial and venous system of frog.
6. What is excretion? Describe the excretory organs and mechanism of excretion in frog.
7. Give a detailed account of the structure of brain of frog.
8. Describe the ventricles of an amphibian brain.
9. Give an account of the urinogenital system of frog.

Short Answer Questions

1. Describe the following organs in frog and explain how frog is adapted to conditions in water and on land : (*a*) organs of hearing and (*b*) organs of vision.
2. Describe the function of liver and pancreas in frog.
3. Give an account of lymphatic system of frog.
4. What is valve? Show the utility of various valves present in the heart of frog.
5. Describe the structure of truncus arteriosus in frog's heart.
6. Make well labelled diagrams to illustrate the following in frog : (*a*) Arterial or venous system; (*b*) T.S. of kidney; (*c*) V.S. of skin; (*d*) T.S. of testis or ovary; (*e*) V.L.S. of brain.
7. Describe the structure of kidney of frog.
8. Write about the ventricles found in the brain of frog.
9. Write short notes on :
   (*i*) Amplexus;
   (*ii*) Hibernation;
   (*iii*) Laryngotracheal chamber;
   (*iv*) Renal portal system;
   (*v*) Truncus arteriosus;
   (*vi*) Sympathetic nervous system;
   (*vii*) Reflex arc.
10. Describe in tabular form, the origin, the distribution and the nature of cranial nerves of frog.

Very Short Answer Questions

1. What is hibernation and aestivation?
2. What is the function of nuptial pads which develop during breeding season on the bases of inner fingers of male frog?
3. Frog is a .... animal. (cold blooded).
4. What is amplexus?
5. On what food substances frog depends during dormancy? (glycogen and fat).
6. Frogs change their colour of skin according to the environment to protect themselves from enemies. This protective colouration is called .... (camouflage).
7. Sexual embrace of female frog by male frog is called .... (amplexus).

Multiple Choice Questions

1. *Rana tigrina* is a member of the order
   (*a*) Lepospondyli (*b*) Urodela
   (*c*) Apoda (*d*) Anura.
2. The tongue of frog is
   (*a*) Short, protrusible and attached at the front end
   (*b*) Short, protrusible and attached at the distal end
   (*c*) Long, non-protrusible and attached at the front end
   (*d*) Long, protrusible and attached at the front end.

3. The frog swims in water by
   (*a*) Powerful forward thrusts of its forelimbs
   (*b*) Powerful backward thrusts of its forelimbs
   (*c*) Powerful backward thrusts of its hindlimbs
   (*d*) Powerful forward thrusts of its hindlimbs.
4. During hibernation or asestivation frog lives up on
   (*a*) Stored glucose and fat
   (*b*) Stored glycogen and fat
   (*c*) Stored starch and fat
   (*d*) Stored galactose and fat.
5. The third eyelid in frog is called
   (*a*) Pineal eye
   (*b*) Upper eyelid
   (*c*) Lower eyelid
   (*d*) Nictitating membrane.
6. Salivary glands in frog are
   (*a*) 1 pair (*b*) 2 pair
   (*c*) 3 pairs (*d*) Absent.
7. Teeth in frog are meant for
   (*a*) Preventing prey from slipping out
   (*b*) Chewing
   (*c*) Cutting
   (*d*) Tearing.
8. The number of voluntary muscles in frog is about
   (*a*) 200 (*b*) 300
   (*c*) 400 (*d*) 500.
9. During which period does the frog show aestivation
   (*a*) Autumn (*b*) Spring
   (*c*) Winter (*d*) Summer.
10. Stomach in frog is attached to the dorsal body wall by means of mesentery called
    (*a*) Endogaster (*b*) Mesogaster
    (*c*) Megagaster (*d*) Oxygaster.
11. The hepatopancreatic duct opens into
    (*a*) Buccal cavity (*b*) Oesophagus
    (*c*) Stomach (*d*) Duodenum.
12. Common bile duct in frog is
    (*a*) Combined cystic ducts from gall bladder
    (*b*) Combined cystic ducts from gall bladder and hepatic ducts from liver
    (*c*) Combined hepatic ducts from liver
    (*d*) Combined hepatic and pancreatic ducts.
13. The gastric glands are activated for the secretion of HCl by
    (*a*) Gastrin (*b*) Insulin
    (*c*) Renin (*d*) Enterogastrone.
14. The liquefied semidigested acidic food in stomach is known as
    (*a*) Bolus (*b*) Bile
    (*c*) Chyme (*d*) Rhyme.
15. During buccal respiration in frog
    (*a*) Nostrils remain open and glottis remains closed
    (*b*) Nostrils remain closed and glottis remains open
    (*c*) Both nostrils and glottis remain open
    (*d*) None of these.
16. Which of the following is not an intestinal hormone in frog
    (*a*) Enterogasterone
    (*b*) Cholecystokinin
    (*c*) Gastrin
    (*d*) Secretin.
17. Which of the following is not a function of bile
    (*a*) Hydrolysis of carbohydrates
    (*b*) Neutralysing acidity of chyme
    (*c*) Activation of pancreatic lipase
    (*d*) Emulsification of fat.
18. Auricles in frog are marked off externally from the ventricle by
    (*a*) Interauricular groove
    (*b*) Interventricular groove
    (*c*) Auriculoventricular septa
    (*d*) Coronary sulcus.
19. Sinus venosus in frog is formed by the union of
    (*a*) 2 precavals and 2 postcavals
    (*b*) 1 precaval and 2 postcavals
    (*c*) 1 precaval and 1 postcaval
    (*d*) 2 precavals and 1 postcaval.
20. Truncus arteriosus of amphibians is divided into
    (*a*) 3 main branches
    (*b*) 2 main branches
    (*c*) 4 main branches
    (*d*) None of the above.
21. In frog the sexual embrance between male and female is called
    (*a*) Amplexus (*b*) Aplexus
    (*c*) Spawning (*d*) Swarming.
22. The three rectal ciliates in frog are
    (*a*) *Nyctotherus, Balantidium* and *Trichomonas*
    (*b*) *Balantidium*, *Opalina* and *Entamoeba*
    (*c*) *Nyctotherus*, *Balantidium* and *Opalina*
    (*d*) *Balantidium*, *Opalina and Amoeba.*

Answers

1. (*d*)	2. (*d*)	3. (*c*)	4. (*b*)	5. (*d*)	6. (*d*)	7. (*a*)	8. (*a*)
9. (*d*)	10. (*b*)	11. (*d*)	12. (*b*)	13. (*a*)	14. (*c*)	15. (*a*)	16. (*c*)
17. (*a*)	18. (*d*)	19. (*d*)	20. (*b*)	21. (*a*)	22. (*c*).		

19

Chapter

Class 6. Amphibia : General Characters, Classification and Representative Types

Contents of this Chapter

GENERAL CHARACTERS

1. Aquatic (freshwater) or semiaquatic, air and water breathing, carnivorous, cold-blooded, oviparous vertebrates.

2. Body is divisible into head and long trunk. Neck and tail may be present or absent.

3. Limbs tetrapodus pentadactyle type. Some limbless. Digits clawless.

4. Skin smooth or rough, rich in glands which kept it moist.

5. Scales, if present, are hidden in the skin.

6. Endoskeleton mostly bony. Skull with two occipital condyles.

7. Vertebrae are acentrous, pseudocentrous and notocentrous.

8. Heart is divided into two atria and one ventricle (3-chambered) and it has a conus provided with valves. Red blood corpuscles (RBCs) are biconvex, oval and nucleated.

9. Respiration by lungs, skin and buccal cavity lining. Gills are present at least, during some early stages of development. Gills present in some aquatic adults.

10. Kidneys are mesonephric and provided with persistent nephrostome. Excretion ureotelic. Ureters open into cloaca.

11. Brain not well developed. Cranial nerves 10 pairs.

12. Lateral sense organs are present during some stages of development.

13. Sexes separate. No copulatory organ in male. Genital ducts open into the cloaca. Fertilisation external. Eggs with gelatinous covering, usually laid in water.

14. Development indirect. The larvae pass through an aquatic stage before metamorphosing into adult.

CLASSIFICATION

Different authors have classified Class Amphibia with their own conception. **Parker** and **Haswell** divided the class into eight orders, five extinct and three living. It has been customary to place the extinct orders together in a subclass **Stegocephalia (Adam Sedgwick)** and all living groups in **Lissamphibia**. **G. Kingsley Noble** (1924) classified class Amphibia into six orders, 3 orders of extinct and 3 orders of living. The classification followed under these lines is of **G. Kingsley Noble** (1924).

Extinct Amphibia

Subclass 1. Stegocephalia (Extinct) Stegocephalia are extinct lizard-like or salamander-like Amphibia. Dorsal surface of skull is completely covered by dermal bones. A parietal foramen is present between the parietal bones. Limbs pentadactylous. Skin with scales and bony dermal plates. Appeared in Lower Carboniferous to the Upper Triassic in which they disappeared.

Fig. 19.1. *Eryops*, a small fossil labyrinthodont.

Order 1. Labyrinthodontia

1. These were crocodile or salamander-like fish-eating Amphibia.
2. They lived from the Lower Carboniferous to the Triassic period.
3. They had a skull completely roofed over with bones.
4. Many more bony elements were present in their skulls than occur today in the modern Amphibia.
5. Enlarged teeth with greatly folded dentine found in both the jaws, similar to their crossopterygian ancestors.
6. Many species were aquatic or semi-aquatic but others may have been terrestrial.

Example : *Eryops.*

Order 2. Phyllospondyli

1. They were small, salamander-like Amphibia. Found in the Carboniferous and Permian deposits.
2. Four fingers and five toes.
3. The skull roof had a separate quadrato-jugal and a lacrimal. Presence of a preopercular bone showing traces of fish ancestry.
4. Vertebrae tubular with spinal cord and notochord lying in one cavity.
5. Well marked transverse proceses with stout ribs.
6. Pubis cartilaginous.

7. A clavicle, cleithrum and scapula were present but the coracoid remained cartilaginous.

Example : *Branchiosaurus (Ichthyostega).*

1. Salamander-like small amphibians. Carboniferous and Permian Amphibia.
2. Vertebrae were single rings around the continuous notochord.
3. Ribs articulate with the column intervertebrally.

Example : *Diplocaulus* had horned skulls. Limbs small.

Subclass 2. Lissamphibia

It includes modern Amphibia having small, simple teeth and no dermal bony skeleton.

Living Amphibia

Order 1. Gymnophiona or Apoda (Gr., *gymnos* = naked + *ophioneos* = serpent-like)

1. Primitive blind burrowing tropical forms.
2. Body is elongated eel-type.
3. Limbs, girdles and tail are absent.
4. The skin is provided with numerous transverse grooves or wrinkles.
5. Minute scales are found embedded in the grooves of the skin in some forms.
6. Copulatory organs are present in males.

Examples : About 55 species. *Ichthyophis, Uraeotyphlus, Siphonops.*

Order 2. Caudata or Urodela (Gr., *ura* = tail + *delos* = visible) or (L., *cauda* = tail)

1. Body with distinct head, trunk and tail.
2. Limbs are of equal size and usually weak.
3. Gills permanent or lost in adult and gill-slits in the adult.
4. Skin devoid of scales and tympanum.
5. Teeth are present in both the jaws.
6. Males lacking copulatory organs. Females usually oviparous.
7. Larvae are aquatic like adults or resemble adults in form and with teeth.

Suborder 1. Cryptobranchoidea

1. Permanently aquatic and primitive.
2. Body is depressed.
3. Skin is soft and flabby.
4. Eyelids are absent.
5. Gills are absent in adults.
6. Fertilisation is external.

Examples : *Cryptobranchus, Hynobius.*

Suborder 2. Ambystomoidea

1. Adults terrestrial with eyelids.
2. Palatal teeth present in transverse rows or converge across the posterior margins of vomer. Parasphenoid toothless.
3. Vertebrae amphicoelous.
4. Fertilisation is internal.
5. Exhibits the phenomena of **neoteny** or **paedogenesis**, *i.e.*, larva becomes sexually mature and reproduces.

Examples : *Ambystoma,* Axolotl larva.

Suborder 3. Salamandroidea

1. Eyelids are present or absent.
2. Usually either gills or lungs are present but in some forms respiration by skin.
3. Palatal teeth in two longitudinal rows. Parasphenoid toothless.
4. Vertebrae are opisthococlous.
5. Usually viviparous.
6. Fertilisation is internal.

Examples : *Salamandra, Amphiuma, Triton.*

Suborder 4. Proteida

1. Body is depressed.
2. Permanent larval forms.
3. Tail provided with fin.
4. Three pairs of external gills and two gill-openings persist throughout life.
5. Eyelids are absent.
6. Vertebrae amphicoelous.
7. Premaxillaries, vomer and mandible toothed.
8. Permanently aquatic.

Examples : *Proteus, Necturus.*

Suborder 5. Meantes

1. Body is slender eel-type.
2. Permanent larval forms.
3. Hindlimbs are absent.
4. External gills (3 pairs) are well developed and persist throughout life.
5. Eyelids are absent.
6. Jaws with horny covering.
7. Fertilisation is external.
8. Completely aquatic.

Example : *Siren.*

Order 6. Anura or Salientia

1. Body is short and broad.
2. Head and trunk are fused. Neck and tail are absent.
3. Forelimbs are short, while hindlimbs are long.
4. External gills or gill-slits are absent. Respiration is by lungs.
5. Eyelids well developed.
6. Tympanum present.
7. Mandible toothless.
8. Fertilisation is usually external.

Suborder 1. Amphicoela

1. Vertebrae are amphicoelous.
2. Two relict tail muscles, absent in other anurans.
3. Free ribs and abdominal ribs.
4. Fertilisation is internal.

Examples : *Ascaphus* with reduced lungs; *Liopelma.*

Suborder 2. Opisthocoela

1. Tongueless.
2. Vertebrae are opisthocoelous.
3. Free ribs either in larva or adult.

Examples : *Alytes, Pipa, Bombinator, Xenopus, Discoglossus.*

Suborder 3. Anomocoela

1. Vertebrae are procoelous or amphicoelous.
2. Sacral vertebrae are fused with urostyle.
3. No free ribs.
4. Upper jaw toothed.
5. Skin is rough.

Examples : *Pelobates, Scaphiopus.*

Suborder 4. Procoela

1. Vertebrae are procoelous, without ribs.
2. Urostyle articulates with the last vertebra by two condyles.
3. Teeth absent in *Bufo*.

Examples : *Bufo, Hyla* (upper jaw toothed).

Suborder 5. Diplasiocoela

1. First seven vertebrae are procoelous.
2. Eighth vertebra is amphicoelous.
3. Sacral vertebra acoelous and bears two condyles posteriorly for the attachment with the urostyle.
4. Pectoral girdle is usually fused with sternum.
5. Ribs are absent.

Examples : *Rana, Rhacophorus.*

REPRESENTATIVE TYPES

Ichthyophis. *Ichthyophis* (blindworm) (Fig. 19.2) is a worm-like caecilian with ring-like grooves on the body. It is about 30 cm long. It burrows in moist earth. In the skin are numerous minute concealed scales (absent in *Siphonops*). Head is small and depressed, with solidly built bony skull and microscopic eyes are covered with opaque skin so that it is blind. On the head, between the eyes and nostrils, is a pair of special protrusible sensory tentacles lodged in pits, they are kept moist by the secretion of Harderian glands. The tentacles are used for tasting food. Middle ear and tympanum are absent, so it cannot hear. There is a long trachea, the right lung is long but the left lung is reduced. There are no limbs or girdles and a sternum is lacking. Vertebrae are amphicoelous. Anus is sub-terminal so that there is very short tail. The male has an eversible copulatory organ bearing hooks. *Ichthyophis* is primitive but its tentacles, absence of tail, and copulatory organ are specialised features. It is found underground in India, Sri Lanka, Malaya and Siam.

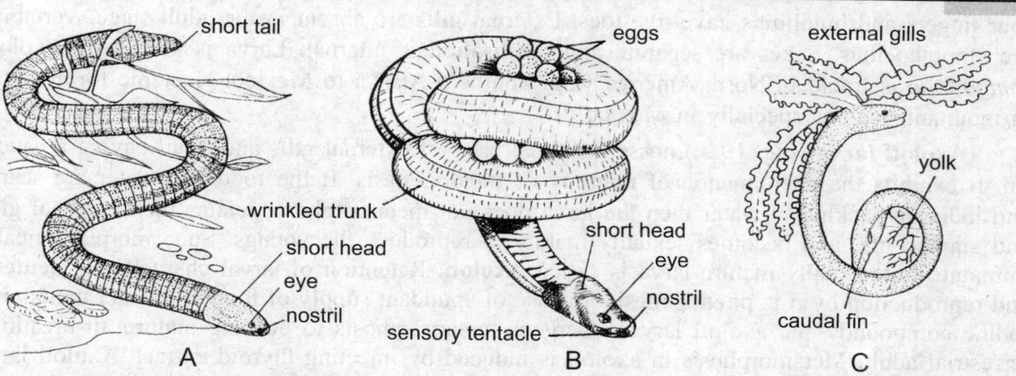

Fig. 19.2. *Ichthyophis glutinosa*, A–Male; B–Female guarding her eggs; C–Gilled-larva.

The female lays 20 eggs in a burrow near water. (Some genera of Apoda are viviparous, such as *Geotrypetes* and *Typhlonectes*). The female coils around the fertilised eggs. They are large and yolky and covered with albumen having twisted chalazae, the chalazae join the eggs in a bunch. Cleavage is meroblastic. Embryos develop around yolk sacs. The gilled-stage is passed within the egg. Each embryo having three pairs of long, plumed gills. The gills are lost before the embryo hatches, obviously these external gills are not respiratory but they probably absorb nourishment. The female then takes the eggs to shallow water where larvae hatch. The larva has one pair of open gill-clefts and a caudal fin, it swims about coming to the surface to take air into its lungs. After a long period the gill-clefts close, tail fin is lost and the adult come to land for a burrowing life. *Typhlonects* is viviparous.

Salamanders and Newts

They belong to the order Urodela. The adult and larval urodeles differ little from each other. Larva has gills and gill-slits which may or may not persist in the adult. They are mostly aquatic animals, but a few are purely terrestrial in the adult state, merely return to the water to lay their eggs or deposit their larvae. In adults eyes are small, sometimes functionless, and without lids except in **Salamandride**, Tympanic membrane and tympanic cavity are not developed. They are usually oviparous, rarely viviparous. In a few cases the larvae have the power of devloping sexual organs and reproduce like adults. There are about 100 species.

Urodeles generally belong to two groups :

1. Perennibranchs. These retain gills and are permanently aquatic.

Examples : *Proteus, Siren, Necturus.*

2. Caducibranchs. Adults have no gills and so they become terrestrial.

Examples : *Amphiuma Cryptobranchus, Megalobatrachus* possess gill-slits in adult state, whereas *Ambystoma, Triturus, Salamandra* possess neither external gills nor gilt-slits in adult stage.

1. *Cryptobranchus.* *Cryptobranchus* (Fig. 19.3.) is commonly known as American **hell-bender**. It is a giant, fully aquatic salamander. It is semi-larval (incompletely metamorphosed) in adult stage. Body is large reaching a length of 60 cm or more. Head is broad, bears a pair of small lidless eyes. Gills are absent in the adult but the first gill-cleft or spiracle remains open and acts as an outlet for water taken into the mouth during buccal respiration. Lateral line organs are well developed. Predaceous. Chiefly feeds on fishes. Sexes are separate. Eggs are laid in gelatinous sacs pulled out into two long strings and are fertilised when the males deposit sperm masses near the egg laying females. *Cryptobranchus* is widely distributed over Eastern United States from the Great lakes to Georgia and Louisiana.

2. *Ambystoma.* *Ambystoma* or *Amblystoma tigrinum* (Fig. 19.3) is commonly known as **tiger salamander**. Body is elongated lizard-like, measures about 12 cm in length. The colour of the body is black with yellow patches. Skin is poisonous. Limbs are well developed, forelimbs bear four fingers and hindlimbs have five toes. External gills are absent in the adult stage. Vertabrae are amphicoelous. Sexes are separate and fertilisation is internal. Larva is known as Axolotl. *Ambystoma* is found in North America from Southern Alaska to Mexico. Neotenic forms occur in mountain regions specially in Mexico.

Axolotl larva (Fig. 19.3.) possesses three pairs of external gills and a tail having a caudal fin. It exhibits the phenomenon of **neoteny** or **paedogenesis**. If the food and water are scarce and iodine is lacking in water then the larva does not metamorphose, retains their external gills and aquatic life, and becomes sexually mature to reproduce the youngs. Such morphologically immature but sexually mature larva is called **Axolotl**. Retention of larval characters is **neoteny** and reproduction by it is **paedogenesis**. In case of abundant supply of food and water containing iodine compounds, the axolotl larva undergoes metamorphosis to become mature air-breathing terrestrial adult. Metamorphosis in axolotl is induced by injecting thyroid extract. Axolotl larva is found in the mountain regions of Mexico.

3. *Salamandra.* *Salamandra* (Fig. 19.3) is a terrestrial salamander with a lizard-like body having strong limbs and a nearly cylindrical tail. Salamanders have no tympanic membrane or middle ear and tympanum, hence, they cannot hear. The adult has no gills, eyelids are movable. They have teeth on both jaws and also vomerine teeth, Vertebrae are opisthocoelous. The interauricular septum is perforated. *Salamadra maculosa*, the European fire salamander, is black and yellow in colour, about 15 embryos develop in the oviducts. It is viviparous and when the young are born, they still have external gills, metamorphosis occurs later. *Salamander atra* of Switzerland lives in rapid Alpine streams. It is black in colour and is ovoviviparous, two eggs develop in the oviducts, when larvae hatch they have no external gills and they take to water.

4. *Amphiuma.* *Amphiuma means* (Congo eel) (Fig. 19.3) has a long cylindrical body about 90 cm in length. Limbs are very small and rudimentary with two or three digits, they are of no value in locomotion. It is a semi-larval form of salamandrid derivation. The eyes are small and lidless. There are no gills in the adult but one pair of gill-clefts is retained. It has lungs with a long trachea supported by cartilages. There is no sternum. *Amphiuma* has the largest erythroeytes of any animal, they are 80μ in diameter. Eggs are laid in water and the female coils itself around them for protection. It is a salamander found in American swamps and muddy waters.

5. *Triton.* *Triton* (Fig. 19.3) is commonly known as **European salamander**. Body is elongated and cylindrical. Skin is soft and covered with slime. Scales are entirely absent. Fore- and hindlimbs are well developed. Forelimbs bear four fingers and hindlimbs bear five toes. Tail is provided with a median fin above and below. Fin is devoid of fin-rays. Mouth is wide having the paired external nostrils just above it. Eyes are small having the upper and lower lids. Tympanum or eardrum is absent. In adults the gills are absent and respiration is by skin. Lateral line organs are found in the larva. Alimentary canal, excretory ducts and the gonadial ducts all open into the cloaca. It exhibits the marked **sexual dimorphism**. Male possesses a dorsal median **crest**. It is terrestrial animal. *Triton* is distributed in Europe and U.S.A.

6. *Proteus.* *Proteus* (Fig. 19.3) is commonly known as **olm** or **European blind cave salamander**. It has an elongated eel-like body reaching up to the length of 5 cm. The skin has no pigment so the colour is whitish. Head is broad, eye rudiments are sunk deep in the head so that it is blind. It is neotenic permanent larva with three pairs of red-coloured branching external gills and two pairs of open gill-clefts, lungs are also present. The tail is compressed and has a caudal fin. Limbs are weakly developed, the hindlimbs have two digits and are smaller than the forelimbs which have three digits. The skull is embryonic with persistent cartilaginous trabeculae. Pelvic girdle has long anteriorly directed pubo-ischium. It has no organ of Jacobson. It lives in the dark caves of central European mountains. If *Proteus* is kept in light for long, it first becomes grayish and finally black in colour. If its young ones are reared in red light, well developed eyes appear and the skin becomes pigmented showing that the inherent basis for these characters has not been lost in this species.

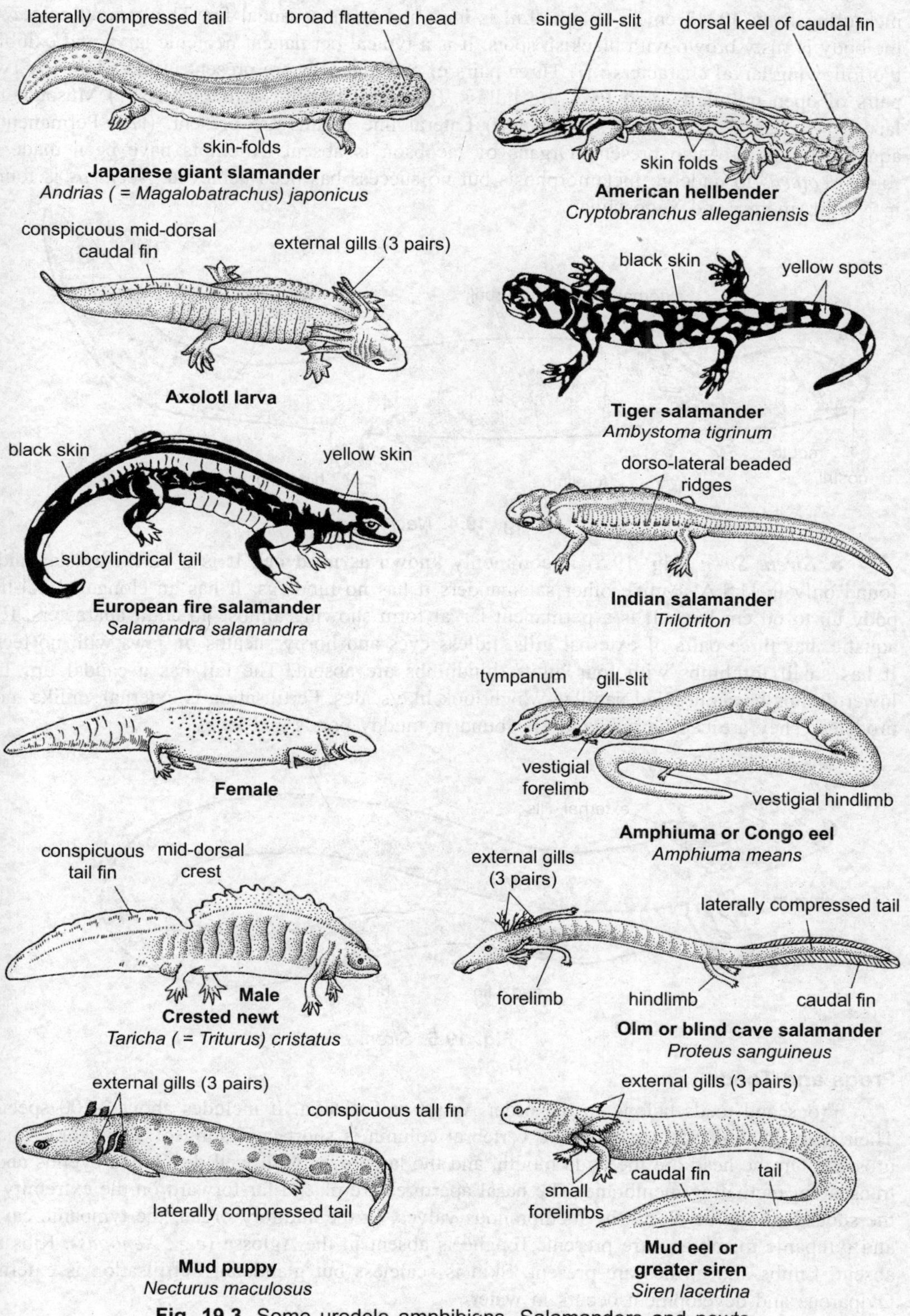

Fig. 19.3. Some urodele amphibians. Salamanders and newts.

7. *Necturus.* *Necturus* (Fig. 19.4) is commonly known as **mud puppy**. Body is elongated measuring from 30–42 cm in length. Tail is long and bears a caudal fin. The general colour of the body is rusty brown with blackish spots. It is a typical permanent neotenic larva and exhibits the following larval characters : (*i*) Three pairs of external gills are present in the adult. (*ii*) Two pairs of open gill-clefts. (*iii*) Eyes are lidless. (*iv*) Skull largely cartilaginous. (*v*) Musculature larval. (*vi*) Larval circulatory system. (*vii*) Lateral line organs are present. (*viii*) Permanently aquatic. Lungs are also present. Organs of Jacobson is absent. Attempts have been made to force *Necturus* to undergo metamorphosis but no success has been achieved. *Necturus* is found in North America and Yugoslavia.

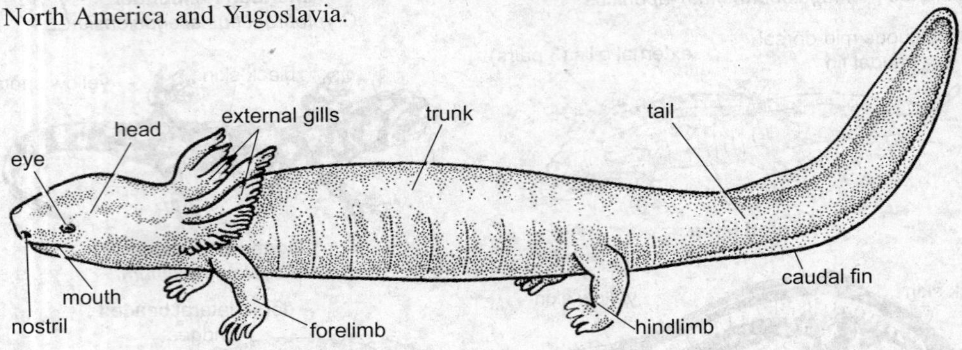

Fig. 19.4. *Necturus.*

8. *Siren*. *Siren* (Fig. 19.5) is commonly known as **mud eel**. It is a blackish salamander found only in U.S.A. Unlike other salamanders it has no hindlegs. It has an elongated eel-like body up to 60 cm long. It is a permanent larval form showing almost no adult characters. It is aquatic, has three pairs of external gills, lidless eyes and horny sheaths on jaws with no teeth. It has small forelimbs with four digits, hindlimbs are absent. The tail has a caudal fin. The lower dermis has concealed papillae which look like scales. Fertilisation is external, unlike most urodeles. They are carnivorous and are found in muddy ponds and lakes.

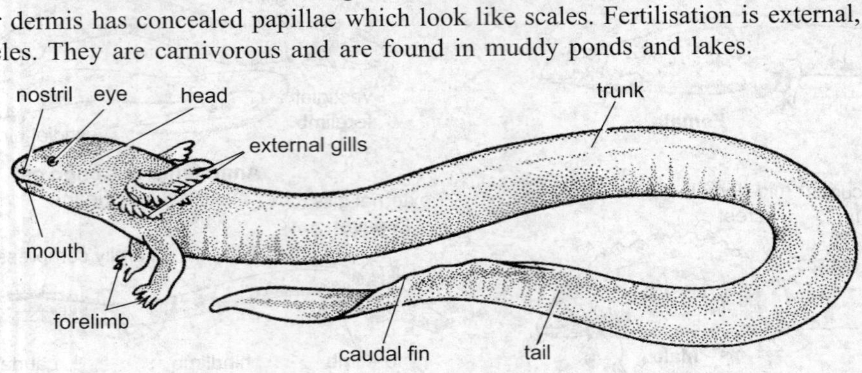

Fig. 19.5. *Siren.*

Frogs and Toads

Frogs and toads belong to the order Anura or Salientia. It includes about 2,200 species. Their body is short without tail. The vertebral column is short consisting of 9 vertebrae and urostyle. On the head are the wide mouth, and the large eyes with well developed eyelids and transparent nictitating membrane. The nasal apertures are placed far forward on the extremity of the snout and can be closed by membranous valves. In the auditory organ, the tympanic cavity and tympanic membrane are present. Tongue is absent in the Aglossa (*e.g., Xenopus*). Ribs are absent. Limbs and girdles are present. Skin is scaleless but glandular. Fertilisation is external. Oviparous and development occurs in water.

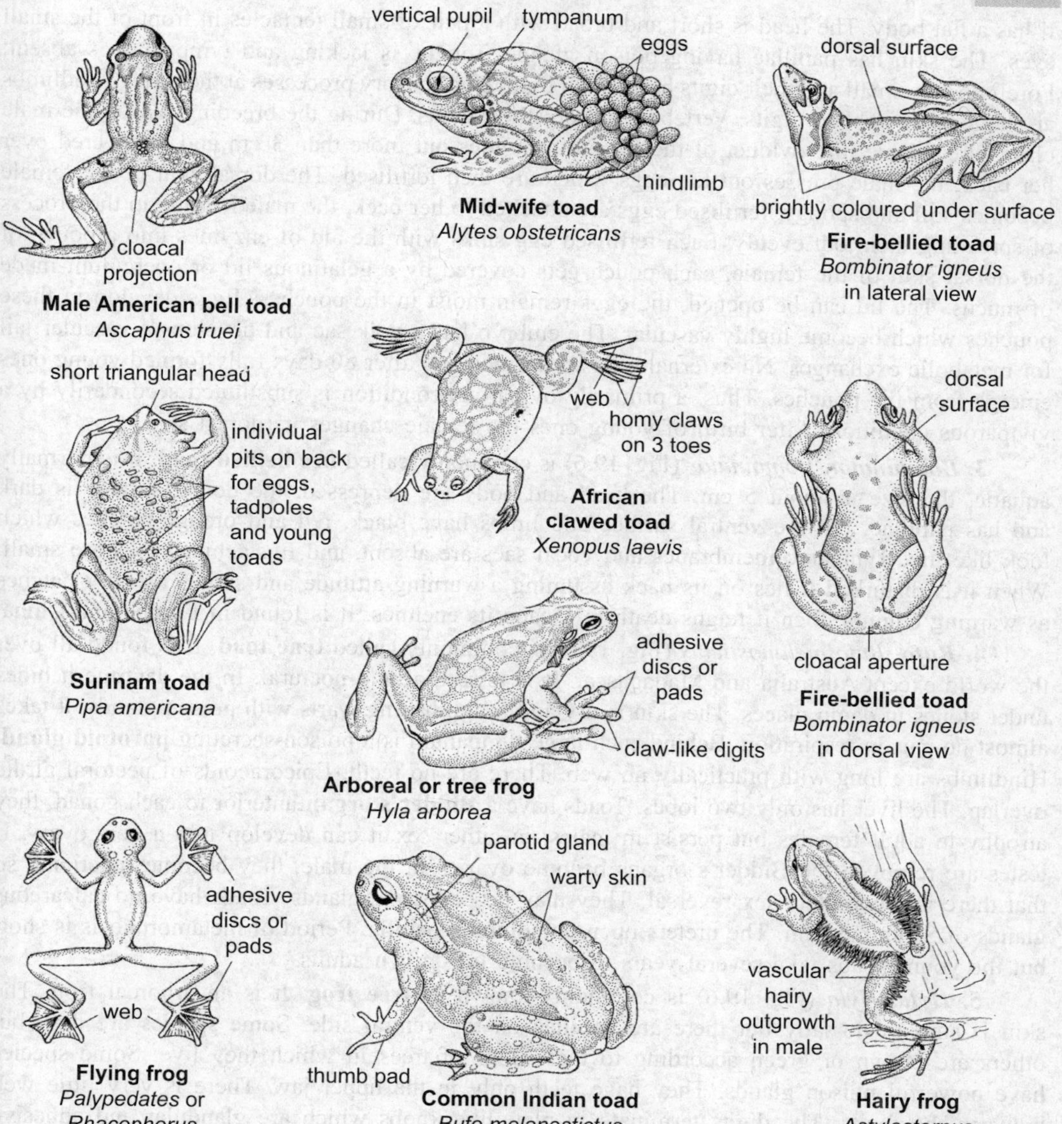

Fig. 19.6. Some anuran amphibians. Frogs and toads.

1. *Alytes.* *Alytes obstetrican* (Fig. 19.6) is commonly known as **mid-wife toad.** It has papillae on the skin both on the dorsal and ventral side. Tympanum is large, vocal sacs are absent. Vertebrae are opisthocoelous. It has a particular breeding habit. The male massages the cloaca of the female and eggs are laid in one string and fertilised by the male. The male *Alytes* winds the string of eggs around its posterior body and thighs, then it buries itself in moist earth. It makes occasional trips to water to keep the eggs wet. When tadpoles are in an advanced stage and ready to hatch, the male plunges itself into water where tadpole larvae hatch and undergo metamorphosis. It is found in France and Italy.

2. *Pipa.* *Pipa pipa* (Fig. 19.6) is commonly known as **Surinam toad**. It has several species found in Brazil, such as *Pipa dorsigera* and *Pipa americana*. It is an ugly toad which is aquatic.

It has a flat body. The head is short and broad with a pair of small tentacles in front of the small eyes. The skin has papillae having poison glands. Tongue is lacking and tympanum is absent. Forelimbs are small and their digits have small branching sensory processes at their tips. Hindlimbs are large with webbed digits. Vertebrae are opisthocoelous. During the breeding season the male clasps are female, the oviduct of the female protrudes out more than 3 cm and is covered over her back, the male presses out the eggs which are then fertilised. The dorsal skin of the female becomes soft and spongy, fertilised eggs are attached to her back, the male helping in the process of spreading them out evenly. Each fertilised egg sinks with the aid of enzymes into a pouch of the dorsal skin of the female, each pouch gets covered by a gelatinous lid or operculum made of mucus. The lid can be opened, the eggs remain moist in the pouches. Eggs develop in these pouches which become highly vascular. The embryo has a yolk sac and develops a vascular tail for metabolic exchanges. No external gills are formed and after 80 days fully formed young ones emerge from the pouches. Thus, a primarily oviparous condition is substituted secondarily by a viviparous condition. After birth of young ones the female changes its dorsal skin.

3. *Bombinator.* *Bombinator* (Fig. 19.6) is commonly called **fire bellied toad**. It is normally aquatic, the size is about 5 cm. The head and body are depressed, the dorsal surface is dark and has papillae, but the ventral surface and limbs have black, red and orange patches which look like fire. Tympanic membranes and vocal sacs are absent, and Eustachian tubes are small. When it is disturbed it lies on its back assuming a warning attitude and shows its fiery patches as warning colours, then it feigns death to escape its enemies. It is found in Europe and China.

4. *Bufo.* *Bufo melanostictus* (Fig. 19.6) is commonly called **true toad**. It is found all over the world except Australia and Madagascar. It is terrestrial and noctural. In the daytime it hides under stones in damp places. The skin is dry and rough having warts with poison glands, it takes almost no part in respiration. Behind each large tympanum is a poison-secreting **parotoid gland**. Hindlimbs are long with practically no web. There are no teeth. Epicoracoids of pectoral girdle overlap. The liver has only two lobes. Toads have a **Bidder's organ** anterior to each gonad, they atrophy in adult females but persist in males, in either sex it can develop into a new ovary. If testes are removed the Bidder's organs become ovaries in the male, they become functional so that there is a complete sex reversal. They may be endocrine glands. Toads have no calcareous glands of Swammerdam. The ureters open on common papilla. Period of metamorphosis is short but the young toads take several years to become full-grown adults.

5. *Hyla.* *Hyla* (Fig. 19.6) is commonly known as **tree frog**. It is an arboreal frog. The skin is smooth dorsally but there are papillae on the ventral side. Some species are dark but others are brown or green according to the colour of trees in which they live. Some species have powerful poison glands. They have teeth only in the upper jaw. There is very little web between the digits. The digits terminate in plate-like knobs which are glandular and adhesive discs. They are not suctorial but have a moist, corrugated lower surface with mucus. An extra joint or **intercalary cartilage** between the last two phalanges of each digit facilitates adjustments of the adhesive discs for climbing trees. *Hyla* is cosmopolitan found in damp forests, but it is not found in India and Africa. Several species are confined only to the New World. *Hyla goeldii* female carries eggs on her back. The Jamaican *Hyla* places its eggs in small pools of rain which have collected in large leaves.

6. *Rhacophorus.* *Rhacophorus* (Fig. 19.6) is commonly known as **flying frog**. The body is slender. The limbs are thin and long with large feet having well developed web between the digits of all four limbs, and the digits bear rounded adhesive pads as in *Hyla*. It lives in trees and can glide with outstretched limbs from tree to tree or from a tree to the ground. It also exhibits parental care by depositing eggs in the nests near water. It has the capacity to change the colour rapidly. *Rhacophorus maculatus* and *Rhacophorus malabaricus* are found in India. *Rhacophorus pardalis* is a Javanese flying frog.

IMPORTANT QUESTIONS

Long Answer Questions

1. Give the diagnostic characters of class Amphibia. Classify modern Amphibia giving salient features and examples of each order.
2. Explain in detail how Apoda can be distinguished from Urodela and why they are put together under the same class.
3. Write the salient features of Gymnophiona.
4. Discuss the effects of terrestrialisation on the anatomy of frog.

Short Answer Questions

1. Name any two amphibians where gills are present in adults. Classify one of them.
2. Differentiate between frog and toad.
3. Write short notes on : (*i*) *Ambystoma*; (*ii*) Axolotl; (*iii*) *Amphiuma*; (*iv*) *Ichthyophis*; (*v*) Stegocephalia; (*vi*) *Bufo*; (*vii*) Urodela.
4. Differentiate between Stegocephalia and Lissamphibia.

Very Short Answer Questions

1. Write about Gymnophiona.
2. Write an example of Apoda.
3. Write three examples of Amphibia in which gills present throughout life.
4. In which order of Amphibia tympanum is not found?
5. Gills are not found in .... (Anura).
6. Anurans have no .... (tails).
7. To which amphibian Axolotl larva belongs?
8. Write the scientific names of two salamanders.
9. What is the common name of *Amphiuma*?
10. Write the scientific names of blind cave salamander and mud eel.
11. Lungs and gills are not found in .... (*Plethodon*)

Multiple Choice Questions

1. In amphibians respiration takes place by
   (*a*) Gills, lungs, skin and buccal lining
   (*b*) Skin and gills
   (*c*) Gills and lungs
   (*d*) Gills, lungs and buccal lining.
2. Erythrocytes in amphibians are
   (*a*) Oval and enucleated
   (*b*) Oval and nucleated
   (*c*) Convex and nucleated
   (*d*) Concave and nucleated.
3. Excretion in amphibians is
   (*a*) Ammonotelic (*b*) Aminotelic
   (*c*) Uricoletic (*d*) Ureotelic.
4. Ear ossicles present in the middle ear of frog is
   (*a*) Iris (*b*) Columella auris
   (*c*) Auris (*d*) Columella.
5. Caecilians belong to the order
   (*a*) Anura (*b*) Opisthocoela
   (*c*) Gymnophiona (*d*) Urodela.
6. Extinct amphibians are placed in the subclass
   (*a*) Lissamphibia (*b*) Stegocephalia
   (*c*) Lepospondyli (*d*) Cryptobranchoidea.
7. In *Plethodon* respiration occurs through
   (*a*) Lungs
   (*b*) Gills
   (*c*) Skin
   (*d*) Skin and lining of buccopharyngeal cavity.
8. Neoteny refers to
   (*a*) Retention of rudimentary organs
   (*b*) Retention of larval characters in adults
   (*c*) Reproduction in larval forms
   (*d*) Metamorphosis.
9. *Pipa pipa* is also called
   (*a*) Horned toad (*b*) Midwife toad
   (*c*) Desert toad (*d*) Surinam toad.
10. Axolotl is the larva of
    (*a*) *Ambystoma* (*b*) *Salamandra*
    (*c*) *Siren* (*d*) *Proteus*.
11. To which of the following order does the toad belong
    (*a*) Urodela (*b*) Apoda
    (*c*) Anura (*d*) Labyrinthodontia.
12. Tadpole larva is a connecting link between
    (*a*) Fishes and amphibians
    (*b*) Amphibia and reptilians
    (*c*) Amphibia and Aves
    (*d*) Amphibians and mammals.

ANSWERS

1. (*a*) 2. (*b*) 3. (*d*) 4. (*b*) 5. (*c*) 6. (*b*) 7. (*d*) 8. (*b*)
9. (*d*) 10. (*a*) 11. (*c*) 12. (*a*).

20 Chapter

Amphibia : General Account

Contents of this Chapter

PARENTAL CARE IN AMPHIBIA

In amphibians there are many devices for the protection of the eggs during the early stages of development and the youngs. In this way nature has practised economy in the number of eggs, which varies in direct proportion to the chances of destruction. Parental care is the care of the eggs or the youngs until they become able to protect themselves from the predators.

These devices fall under two heads : (1) Protection by the parents by means of nests, nurseries, or shelters and (2) Direct caring or nursing by parents. The different modes of protection are given below in the three important orders of class Amphibia.

1. Protection by Means of Nests, Nurseries and Shelters

A number of different species of frogs and toads construct nests or shelters of leaves or other materials in which the eggs are deposited and the youngs are developed.

A. In enclosures in the water (mud nests). A large tree frog (*Hyla faber*) known in Brazil as the **"Ferreiro"**. It protects its progeny by building a basin-shaped nest or nursery in shallow water on the border of the pond. The female scoops mud to a depth of 7.5 or 10 cm and with the mud, thus, removed a circular wall is built around the nest, which emerges above the surface of the water. The inside wall is smoothened by the flattened webbed hands and the bottom is also levelled by belly and hands. The eggs and early larvae are, thus, protected from predators (insects and fishes, etc.) until they are able to defend themselves. Heavy rains later on destroy the wall and larvae go to the water.

B. In holes near water (foam nests). A still better mode of protecting the offspring during the early stages of development has been adopted by a Japanese tree frog *Rhacophorus schlegelii*. The male and female in embrace bury themselves in the damp earth on the edge of ditch or flooded rice field, and make a hole or chamber, a few centimetres above water level. The walls of this chamber are polished and during this process the gallery by which they enter into that chamber gets obliterated and then oviposition begins. The female first produces a secretion from cloaca which is beaten into a froth. The eggs are deposited into the froth. Now the inactive male impregnates them, and then both of them separate and make an exit gallery towards the ditch. It is obliquely downwards towards the water, later on this is used by the larvae who come to the water to complete the development. The bubbles collapse, the froth liquefies and this liquid acts as an efficient vehicle for transporting the larvae down the tunnel into the water. Similarly female of South American tree frog, *Leptodactylus mystacinus* stirs up a frothy mass of mucus which is filled up in holes near water and then eggs are laid in it. The tadpoles from these nests easily enter water. Some anuran females discharge huge mucus and beat it into a foam with their hindlegs and then eggs are laid. Later on hatching tadpoles drop into water from the foam.

C. In nests on trees (tree nests). Some tree frogs like *Phyllomedusa* in South America, *Rhacophorus malabaricus* in India, and *Chiromantis* in tropical Africa glues the eggs to foliage hanging over water, and after hatching, the tadpoles drop straight into the water. *Hyla resinfictrix* (tree frog) lines a shallow cavity of the tree by bees wax brought from the hives of stingless bees. Eggs are laid there when it is filled with rain water. Tadpoles develop here safely.

Autodax (Urodela) lay 10–20 eggs in a dry hole in ground or in a hole on a tree, up to 10 metres above the ground. Both parents remain in the hole to protect the eggs and larvae and also provide them moisture. Youngs remain within the hole for a considerable period with their parents.

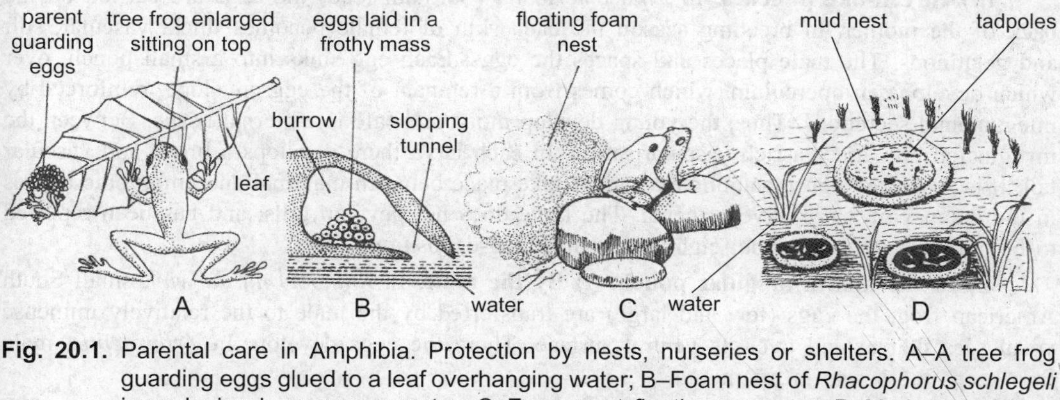

Fig. 20.1. Parental care in Amphibia. Protection by nests, nurseries or shelters. A–A tree frog guarding eggs glued to a leaf overhanging water; B–Foam nest of *Rhacophorus schlegeli* in a sloping burrow near water; C–Foam nest floating on water; D–Mud nest of *Hyla faber*.

D. In transparent gelatinuous bags. The eggs of *Phrynixalus biroi* are large which are enclosed in sausage-shaped transparent common membranous bag secreted by the female and is left in the mountain streams. The whole development takes place within the eggs and little frogs go out in perfect condition. No gills have been observed and the large tail serves as a breathing organ of young ones. *Salamandrella keyserlingi* (urodele) deposits its small eggs in a gelatinous bag which is attached to an aquatic plant below the water level.

E. On trees or in moss away from water. Several species of tropical American genus *Hylodes* lay their large eggs in damp places under stones or moss or plant leaves. The metamorphosis is hurried up within the egg. Due to plenty of yolk in the egg the entire development takes place within the egg and young frogs hop out as an air breather with a vestige of tail.

2. Direct Nursing by the Parent

A. Tadpoles transported from one place to another. Small South American frogs *Phyllobates* and *Dendrobates* and tropical African frogs *Arthroleptis* and *Pelobates* lay their eggs on ground. The hatched tadpoles adhere by their sucker-like lips and flattened abdomen to the back of one of their parents and, thus, they are carried from one place to the other and in this way they can even go from one pool to the other and this is particularly when one pond is to dry up.

B. Eggs protected by male. The eggs (17 in number) of *Mantophryne robusta* are strung together by an elastic gelatinous envelope forming a clump over which the male sits for development. It may be outside water. The larvae have no gills, well developed legs, large tail which is vascular and respiratory.

C. Eggs carried by the parents. In Obstetric toad (*Alytes obstetricans*) of Europe, the male winds the strings of eggs–formed by adhesion of their gelatinuous investment–round his body and hindlegs. Here they are retained until the tadpoles are ready to be hatched.

Female *Rhacophorus reticulatus* (Sri Lankan tree frog) carries the eggs glued to her belly.

In *Desmognathus fusca* (urodele) the eggs are laid in the form of rosary-like strings. The string is bound round the body and the female nourishes them at a comparatively dry spot.

D. Eggs in Back Pouches

(*i*) Exposed. In a Brazilian tree-frog, *Hyla goeldii*, the female carries the eggs on the back within an incipient brood pouch in which the eggs remain exposed. How they reached there is not known but probably male does it. In *Nototrema* also the eggs are placed over the back in a single large brood pouch covered by the skin and opened posteriorly in front of cloacal aperture.

(*ii*) In cell-like pouches. In *Pipa americana* (Surinam toad) the eggs are carried on the back of the mother. In breeding season the back skin of female becomes thick, vascular, soft and gelatinous. The male places and spaces the eggs. Each egg sinks into a small pouch, over which develops an operculum, which comes from a remnant of the egg envelope, reinforced by integumental secretions. Thus, the young develop moist and safe in maternal tissue. Between the invaginated pits arises a rich vascularisation. In each larva there develops a broad and vascular tail. It is suspected that metabolic exchanges take place between maternal and embryonic tissues in the manner of a primitive placenta. The larva does not develop gills, and has been reported to be born as a tadpole about eighty days after egg-deposition.

E. In the mouth or gular pouch. (*i*) By the male. In *Rhinoderma darwini*, small South American frog, the eggs (few and large) are transferred by the male to the relatively immense vocal sacs that extend over its ventral surface. There the eggs develop. In *Arthroleptis*, male frog keeps the larvae in his mouth.

(*ii*) By the female. The female of a West African tree-frog, *Hylambates breviceps*, carries the eggs in her mouth. Female *Rheobatrachus silus* (Australian frog) keeps her eggs in her stomach. The tadpoles are expelled through mouth after metamorphosis.

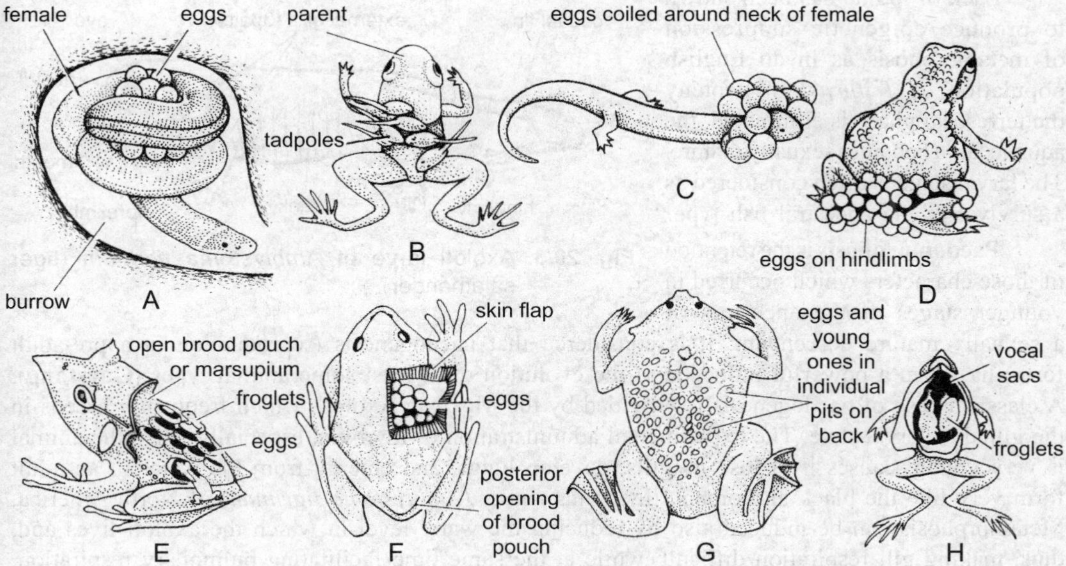

Fig. 20.2. Direct parental care in Amphibia. A–Female *Ichthyophis* coiling round eggs; B–Transportation of tadpoles attached to back of a parent; C–*Desmognathus fuscus* with eggs; D–*Alytes obstetricans* carrying eggs around his thighs; E–A marsupial frog with eggs exposed in open brood pouch on back; F–*Nototrema* or *Gastrotheca*, with flap of dorsal brood sac cut open to show eggs; G–In *Pipa*, eggs develop completely into individual capsules on back of female; H–Froglets inside vocal sacs cut open of female *Rhinoderma darwinii*.

F. Coiling around eggs. In *Plethodon* (urodele) the eggs are laid in small packages of about five beneath the stones or in the hollow of rotten log, and the mother coils round them. In *Megalobatrachus maximus* (urodele) the male coils round the eggs.

Female *Amphiuma* (urodele) also coils round the eggs laid in burrows in damp soil.

Coecilians *Ichthyophis* and *Hypogeophis* are oviparous, lay eggs in burrows in damp soil and coil round them until they hatch.

G. Viviparous or viviparity. Two small East African toads, *Pseudophryne vivipara* and *Nectophryne tornieri*, are known to be viviparous, but no observations have yet been made on them beyond the fact that larvae are found in the uteri. Caecilians like *Typhlonectes, Geotrypetes, Schistometopum, Chthonerpeton, Gymnopis* are ovoviviparous.

NEOTENY AND PAEDOGENESIS

In the genus *Amblystoma* (urodele) the animal sometimes becomes sexually mature in the larval condition and does not metamorphose (*Siredon*, the axolotl). This phenomenon of the retention of larval characters in the sexually mature state has been called **neoteny** or **paedogenesis**. It is not confined to *Amblystoma* but is found occasionally in the genus *Triton*. It is a physiological adaptation. It also happens in *Necturus* and *Amphiuma* (obligate neoteny) and in these metamorphosis cannot be induced by thyroxine T_4. While in axolotl, *Amblystoma*, some or all of the individuals of a given population metamorphose. In these **facultative neotenes** thyroxine T_4 generally induces metamorphosis. Neoteny is due to a low level of thyrotropin releasing hormone (TRH) probably due to a combination of genetic predisposition with environmental factors.

Lack of iodine has been shown to produce epigenetic suppression of metamorphosis as in an English population of *Triturus*. In neoteny the terrestrial phase is eliminated, the aquatic larva becomes sexually mature. The larval stage is often considered as a survival of the ancestral fish type.

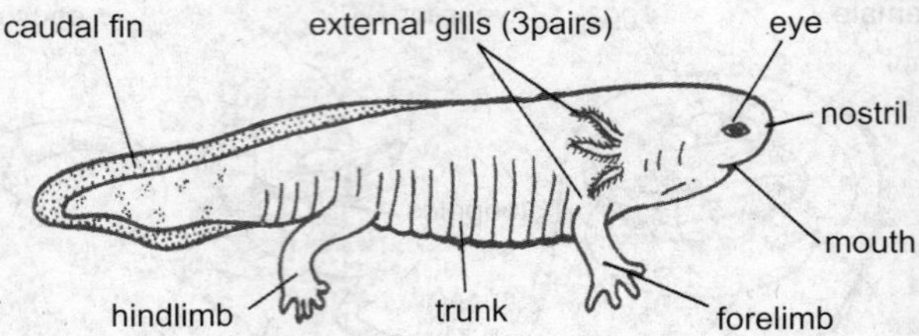

Fig. 20.3. Axolotl larva of *Amblystoma tigrinum* (tiger salamander).

Paedomorphosis is the retention of those characters which occurred in younger stages of the ancestors of a sexually mature descendant. It is considered that paedogenesis (reproduction in a pre-adult form) has been a powerful influence in the evolution of at least some animal types (**Garstang**). A classical case of paedogenesis is provided by the Mexican Axolotl, which frequently breeds in the gilled or larval state. The experimental administration of thyroxin, especially when the animal is very young, causes it to lose its gills, develop lungs, and emerge from the water in an adult form very like the black and orange tiger salamander (*Amblystoma tigrinum*) of North America. Metamorphosis can be induced also by reducing the water level in which the axolotl lives and, thus, making gill respiration difficult, while at the same time facilitating pulmonary respiration. It is to the branchiate stage that the name Axolotl properly applies.

ORIGIN AND ANCESTRY OF AMPHIBIA (TETRAPODA)

During the late Devonian period osteolepid fishes such as *Eusthenopteron* lived in the freshwater systems. It is supposed that some of these animals, first crawling from pool to pool and then spending more time on the land, gave rise to the terrestrial Amphibia. They already possessed lungs and stoutly-constructed fins. The pace of evolution was very slow as it is known that the Devonian Amphibia, and also many of the Carboniferous Amphibia, still looked and behaved very like fishes. Nearly all of these early amphibians were wholly or largely aquatic, feeding on fish or an aquatic invertebrates. The crocodile-like forms such as *Eryops* and small terrestrial amphibians such as *Cacops* were evolved not before Permian.

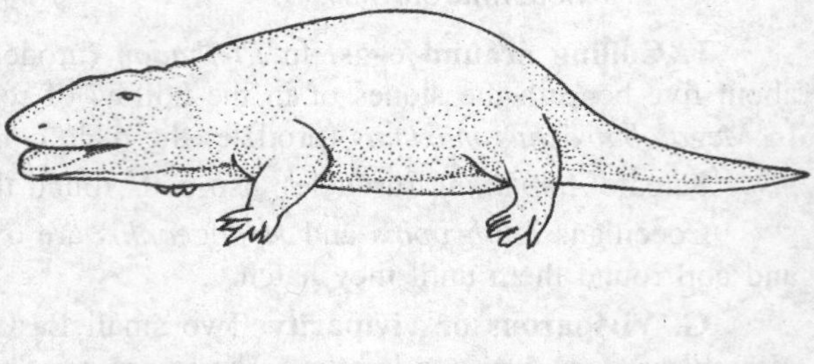

Fig. 20.4. *Eryops.* An early Permian form having well developed limbs, probably living in swampy areas.

Earliest Amphibia. The fossils that appear to be closest to the possible tetrapod ancestor are the osteolepids of the Lower and Middle Devonian periods about 375 million years ago. They were definitely fishes. They may have used air as well as water for obtaining oxygen, as do modern lungfishes. *Elpistostege* is a single Upper Devonian skull which is intermediate in proportions between such fishes and the earliest undoubted tetrapods, *Ichthyostega* and similar forms found in freshwater beds of Greenland. They belonged to late Devonian or early Carboniferous, about 350 million years ago. They are the oldest members of Labyrinthodontia having teeth with folded dentine. The enamel and dentine surrounding the pulp cavity at the base of tooth was folded into a labyrinthine pattern. They were all fish eating. This condition was also present in their crossopterygian fish ancestors. Throughout the succeeding 100 million years of the Carboniferous and Permian periods they flourished and developed many different lines, one giving rise to the reptiles. They were mainly aquatic or semi-aquatic forms, and only a few seem to have been completely terrestrial.

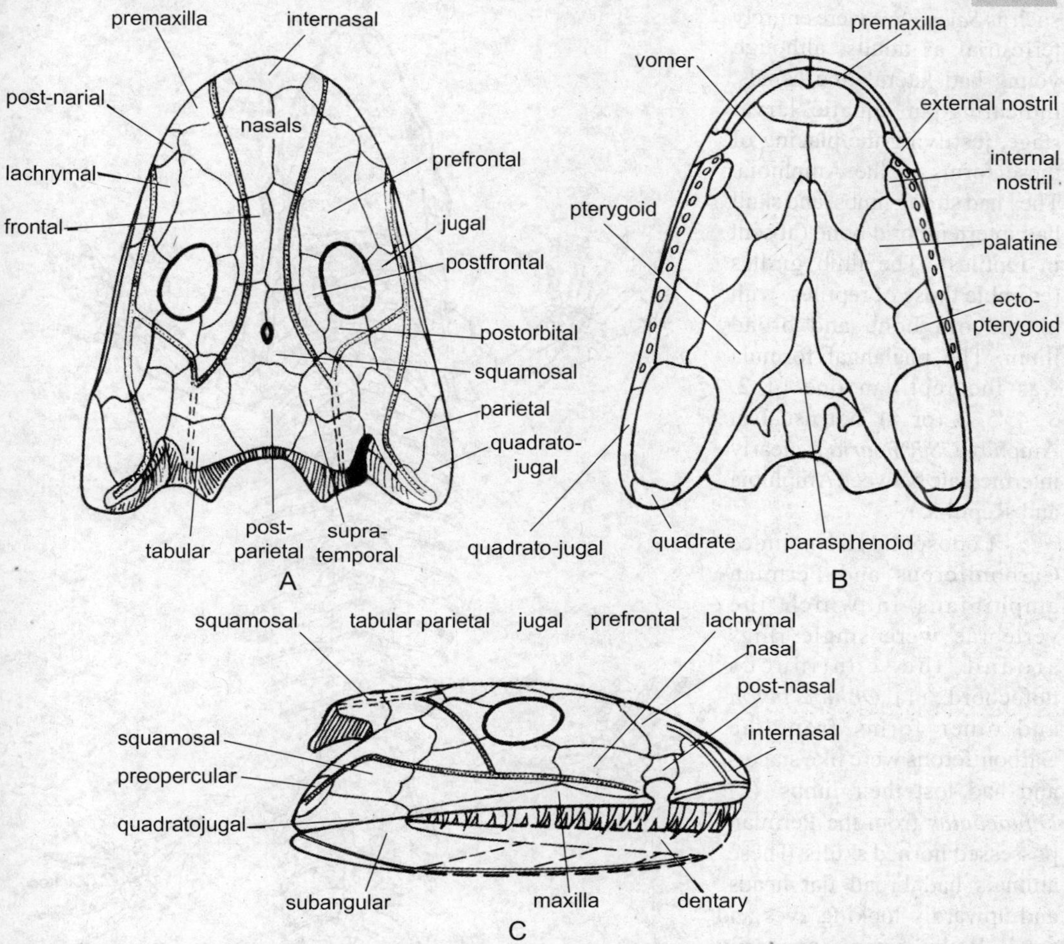

Fig. 20.5. Skull of *Ichthyostege* (labyrinthodont). A–Dorsal view; B–Ventral view; C–Lateral view

The earliest labyrinthodonts were definitely tetrapods and showed sharp changes from the fish type. Some had scales, while others had a dry leathery skin. The ventral side of body was sometimes protected by ventral plates. In *Ichthyostega* the skull shows amphibian features but retains traces of fish ancestry in its shape, with a short, wide snout and long posterior region, and presence of a preopercular bone. The nostril lies on the edge of upper lip. Gills and operculum were absent. These creatures were about 100 cm long and possessed a long tail with a dorsal fin, but had strong legs.

In the earliest Palaeozoic Amphibia (*Ichthyostegalia*) the vertebrae were like those of crossopterygian, composed of three parts, a dorsal neural arch and a centrum of two parts—anterior intercentrum and posterior pleurocentrum. The intercentrum carries a ventral arch and rib. In many other aquatic amphibians the vertebrae are reduced to simple cylinders of bone surrounding the persistent notochord, a condition called **lepospondylous**, which is found also in modern forms.

Loxomma, a Carboniferous Amphibia, were probably largely aquatic. Later members such as *Eryops* and *Cacops* of the early Permian had stout legs and were probably able to live in swamps or on firm ground. *Palaeogyrinus* of the Carboniferous (earliest members) were close to the ancestry of all Amphibia. These were largely with small legs and long tail, and vertebrae retained both pleurocentrum and intercentrum, the two-ringed condition. The later Permian Amphibia such as *Seymouria* were entirely terrestrial as adults, although young had lateral line canals, indicating an aquatic larval stage, justifying the placing of these forms in the Amphibia. They had strong limbs, and skull had intertemporal bone (absent in reptiles). The limb girdles resemble those of reptiles, with a coracoid bone and broad ilium. The phalangal formula was the reptilian one of 2, 3, 4, 5, 3 (or 4), not usual in Amphibia. *Seymouria* is clearly intermediate between Amphibia and Reptilia.

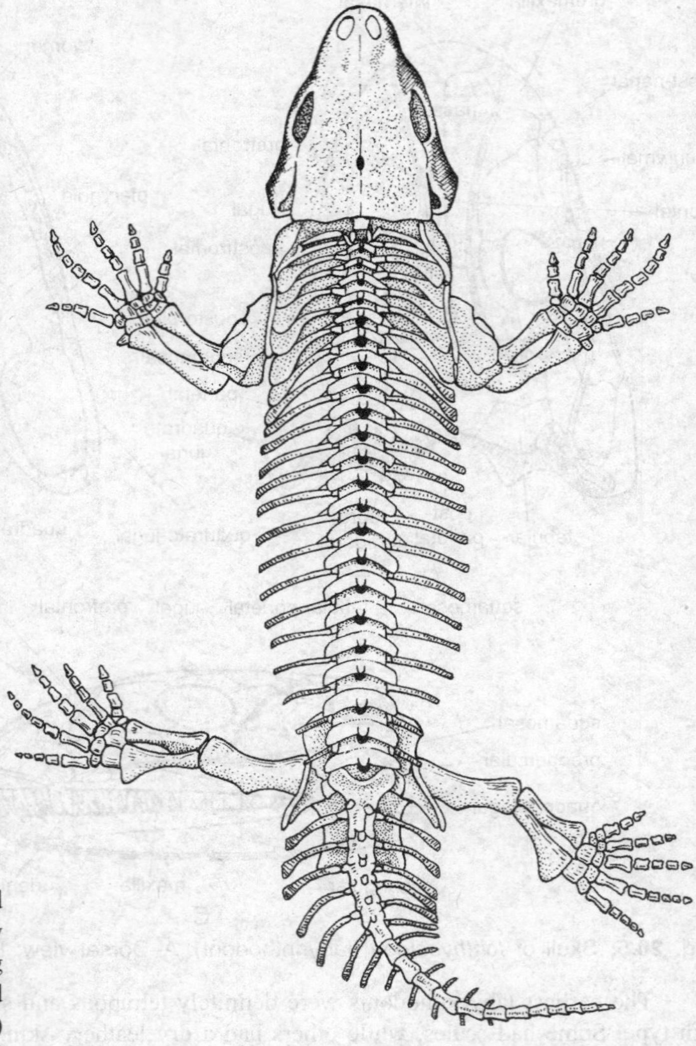

Fig. 20.6. Skeleton of *Seymouria*.

Lepospondyli includes Carboniferous and Permian amphibians in which the vertebrae were single rings around the continuous notochord. (1) *Ophiderpeton* and other forms from the Carboniferous were like snakes and had lost their limbs. (2) *Diplocaulus* from the Permian possessed horned skulls. These animals had broad flat heads and upwardly-looking eyes and small limbs were secondarily aquatic carnivores, with long bodies and vertebrae adapted for bending movements. (3) *Microbrachis* had more normal limbs and bodies. Some were aquatic, others fully terrestrial, some used burrows and might have been the ancestors of the Apoda.

Origin of Modern Amphibia

The modern forms are generally considered to be a single monophylatic group, but this is not certain and their relationship to the Palaeozoic Amphibia is unclear. They show flattening of the skull and loss of ossification in the roof and the palate. The vertebrae are simple ring-like those of temnospondyls (*e.g., Loxomma, Eryops, Cacops*). The earliest known anuran *Protobatrachus* (= *Triadobatrachus*) from lower Triassic has a frog-like skull but many vertebrae with ribs, and a segmented tail. The ilia and torsal bones are long. The urodeles are superficially more like early amphibians. Fossil salamanders of Jurassic resemble modern forms. **Jarvik** (1980) has suggested their origin from the fish *Porolepis* which is supposed to have an internal nostril. Fossil apodes are not known, but their retention of scales suggests an early separation from the rest and they may be derived from the lepospondyls rather than from labyrinthodonts.

IMPORTANT QUESTIONS

Long Answer Questions

1. Write an essay on parental care in Amphibia.
2. What is neoteny? Describe the external features of Axolotl and *Ambystoma*.
3. Give an account of the general organisation of Amphibia with reference to their emergence to the terrestrial life.
4. Discuss the origin and ancestry of Amphibia.

Short Answer Questions

1. Write what you know about neoteny?
2. Give a brief account of parental care in Amphibia.
3. Differentiate between neoteny and paedogenesis.
4. Write about adaptations of amphibians to amphibious life.
5. Write short notes on : (*i*) Environmental factors affecting neoteny; (*ii*) Origin of modern Amphibia.

Very Short Answer Questions

1. What is neoteny?
2. What is paedogenesis?
3. In which amphibian neoteny and paedogenesis is found?
4. Where Axolotl larva is found?
5. Why larva of *Ambystoma* does not metamorphose?
6. What is parential care in Amphibia?

Multiple Choice Questions

1. Partial neoteny refers to
   (*a*) Sexual reproduction in larvae, capable of metamorphosis under suitable conditions
   (*b*) Delayed metamorphosis due to physiological or ecological changes in environment
   (*c*) Animal fails to hibernate
   (*d*) Animal remains larval throughout.
2. Paedogenesis refers to
   (*a*) Precocious development of gonads
   (*b*) Retention of larval characters by adults
   (*c*) Retrogressive metamorphosis
   (*d*) Retention of rudimentary characters in adults.
3. Direct development occurs in
   (*a*) *Rana clamitans*
   (*b*) *Rhacophorus*
   (*c*) *Hylodes*
   (*d*) *Leptodactylous*.
4. Which of the following does carry eggs in its back pouches?
   (*a*) *Hylambates breviceps*
   (*b*) *Nototrema*
   (*c*) *Pipa*
   (*d*) *Cryptobatrachus evansi*.
5. Which of the following exhibits viviparity?
   (*a*) *Pseudophryne*
   (*b*) *Hyla goeldii*
   (*c*) *Nototrema*
   (*d*) All.
6. In which amphibian eggs develop within vocal sacs?
   (*a*) *Rhinoderma darwinii*
   (*b*) *Salamandra salamandra*
   (*c*) *Pelobates*
   (*d*) *Phrynixalus*.

Answers

1. (*b*) 2. (*a*) 3. (*c*) 4. (*a*) 5. (*a*) 6. (*a*).

21

Chapter

Type 8. *Uromastix hardwickii* (The Spiny-Tailed Lizard)

Contents of this Chapter

The reptiles arose in the Carboniferous from labyrinthodont amphibian stock. They are fully adapted to maintain life on land. Many features show a considerable advance over the Amphibia, for example, dryness of the skin, method of respiration and the devices for economising in the use of water. The living reptiles today have 5,000 known species. Reptiles are ectothermal, terrestrial or aquatic, carnivorous, air-breathing, cold-blooded, oviparous tetrapodes. Generally they have an exoskeleton of dry horny epidermal scales, shields, plates and scutes below which in some there are bony dermal plates also. Endoskeleton is more completely bony than in amphibians. The skull has one condyle, vertebral column is differentiated into regions, vertebrae are gastrocentrous, *i.e.,* anterior surface is concave, the posterior convex, and in living forms two vertebrae form a sacrum, ilio-sacral articulation is postacetabular. Limbs are pentadactyl ending in claws. Lungs are the sole respiratory organs (except in some aquatic Chelonia in which cloaca is respiratory). Ventricle begins to divide into two parts, conus arteriosus has disappeared but the sinus venosus persists, the ventral aorta is divided into three trunks. Kidneys are metanephric with no nephrostomes. They are associated with water conservation. Excretory waste is mostly uric acid. There are 12 pairs of cranial nerves. There is a cloaca which is often complicated. Fertilisation is internal because male has a single or double penis. Reptiles are the first true land vertebrates because, along with birds and mammals, they have evolved an egg which is laid and develops on land, the embryo developing within embryonic membranes consisting of amnion and allantois which bring about embryonic respiration and excretion. A yolk sac provides nourishment, consequently the three classes together are known as Amniota. Even aquatic reptiles come on land to lay eggs; gill-pouches are present in the embryo but gills are never formed.

The reptiles flourished and radiate in the Mesozoic era with giant dinosaurs and plesiosaurs dominating the earth and seas, they were terrestrial, aquatic, and aerial forms in abundance, hence, the Mesozoic is called the age of reptiles. Some of the earliest reptiles, the **cotylosaurs** (*e.g., Rameriscus* and *Solenodonsaurus*) called the "stem reptiles", so closely resembled the labyrinthodont amphibians that it is not easy to distinguish between the two groups. By the Permian the first great radiation of reptiles was under way with numerous Anapsid and Synapsid forms predominating. By the late Triassic cotylosaurs had disappeared, but from them further reptilian radiation occurred and in the Mesozoic, one of the most spectacular and fascinating faunal group developed which ever lived. In late Triassic appeared the dinosaurs and pterosaurs belonging to Archosauria, some of which were very large and spectacular and dominated the earth, their only survivors are the crocodilians. Some of these archosaurs were the ancestors of birds and mammals, and some had given rise to the present-day reptiles. Most of these giant reptiles died out in late Cretaceous or at the latest the early Coenozoic. It is not known what caused their disappearance. Certainly it is believed that there had been a widespread elevation of the earth's surface together with changes in temperature and vegetation. This would, no doubt, affect a cold blooded population. But most of the marine reptiles vanished as well. There is evidence of an increased aridity, which would certainly embarrass the gigantic swamp-dwelling dinosaurs. Doubtless climatic changes affected many successful reptiles by extinguishing the food upon which they were irrevocably adapted to feed. It has been suggested too, that many of the small-brained (*e.g., Stegosaurus*) in Mesozoic reptiles perhaps disappeared because they evolved no means of defending their eggs from predaceous ancestral mammals. This may have been a factor. Vague notions concerning genetic instability leading to over-specialisation and phylogenetic sequence have also been put forward. The best that can be said at present is that a combination of climatic and other factors, possibly including competition by the arising Mammalia in the beginning of Jurassic or even earlier, and perhaps by the birds as well, caused the relatively sudden extinction of numerous highly successful types. Thus, most of the reptiles disappeared, a few have persisted to the present day with little change, others evolved further and gave rise to modern living reptiles which live mostly in the tropical and sub-tropical regions, and yet others gave rise to birds on the one hand and to mammals on the other. The reptiles living today are lizards, snakes and lizard-like Tuatara, turtles and tortoises, and crocodiles, gharials, alligators and caimans.

SYSTEMATIC POSITION

Phylum	....	Chordata
Subphylum	...	Vertebrata
Division	...	Gnathostomata
Superclass	....	Tetrapoda

Class	....	Reptilia
Subclass	....	Diapsida
Superorder	....	Lepidosauria
Order	.....	Squamata
Suborder	...	Lacertilia
Genus	...	*Uromastix*
Species	....	*hardwickii*

HABIT AND HABITAT

Distribution

There are 11 species of the genus *Uromastix* inhabiting the desert regions of North Africa and North-Western Asia. Only one species, *Uromastix hardwickii*, is represented in India. It is found almost throughout North-Western India (Punjab, Haryana, Rajasthan, Uttar Pradesh) and Gujarat. It has not been reported from South India. It is also found in Sindh and Baluchistan in Pakistan. The lizard is commonly known as "Sanda" in Hindi and "Sanha" in Punjabi.

Habitat

Uromastix inhabits dry sandy tracts such as deserts with scanty vegetation. It lives in individual burrows and prefers dry places. It is usually found in soft yellowish soil. It is also called "Sand lizard" because of its sandy habitat.

Habits

Uromastix is gentle and quiet in habit and it can be easily tamed in captivity. The mode of progression is of "walking type". It is a burrowing lizard. It digs its own burrow in the ground with its powerful claws. It is diurnal. It retreats into burrow at night or in the day during very hot or cold weather. The burrow is also used for shelter and protection against enemies and for laying eggs. At night, the lizard closes the entrance of the burrow for safety. It is chiefly herbivorous and feeds on grasses, flowers, fruits and succulent leaves of wild shrubs and trees. It sometimes preys on insects. In captivity, it can be fed on pieces of bread and moistened or roasted gram. During winter, usually from December to March, *Uromastix* undergoes hibernation in order to avoid the cold season. It suspends feeding, slows down all activities and passes through a period of dormancy inside its burrow. During hibernation, the lizard lives on the accumulated fat of the body.

The breeding season of *Uromastix* begins in March and April in Northern India. During breeding season, the males become agile and aggressive and resort to fighting for the females. Courtship precedes copulation which takes place generally between 10 a.m. and 3 p.m. on a clear warm and sunny day. During mating, the male everts both its hemipenes, one of which is inserted into the cloaca of the female to discharge the spermatic fluid. Fertilisation is internal. *Uromastix* is oviparous. The female lizard lays about 10 to 15 large eggs which possess thin membranous shell. The eggs are laid in the burrow of the female in late April or early May. The eggs measure 20 to 30 mm in size, are dusty white in colour, rich in yolk and develop with the heat of the sun. Cleavage is meroblastic. The newly hatched young resemble the parents.

Economic Importance

The flesh of *Uromastix*, particularly the tail muscles, is eaten by some tribal people and considered a delicacy. The oil extracted from its fat bodies is sold by roadside quacks, is claimed as cure for impotence and used for massage. The powder of its faeces mixed with "kajal" is supposed to cure corneal opacity. The skin of this lizard is used for making fancy leather articles and even exported. Nowadays these articles are not much in demand. *Uromastix* is also used as a suitable material for dissection in the laboratories and also studied in the classroom.

EXTERNAL FEATURES

Shape, Size and Colour

The body of *Uromastix* is long, massive and depressed. The length of the body may vary from 20 to 30 cm. The general colour of the body on the dorsal surface is yellow-brown with dark spots, while the ventral surface is light pale.

The whole body surface is covered with granular epidermal scales. The crest of spines present on the neck and back found in some members of family Agamidae, *e.g., Calotes* to which *Uromastix* also belongs, is in this species are entirely absent.

Division of Body

The body of *Uromastix* is distinctly divisible into four regions, *viz.,* **head, neck, trunk** and **tail**.

Head. The head is small, depressed and roughly triangular and produced in front into a short and blunt snout. The mouth is anterior and like a wide slit bounded by immovable lips covered with scales. Lower lip is rounded but the margin of upper lip is sharp covering the lower lip. A pair of oval openings, the **external nares**, are located at the anterior end of snout above the mouth. During digging, the nostrils become closed by a prominent swelling of their posterior mucous linings which serve as valves. A pair of small bead-like eyes is situated dorso-laterally almost in the middle of the head. The pupil is round and the iris is light orange in colour. Each eye has two thick movable eye-lids and a nictitating membrane. Eye-lids are well developed. The lower eye-lid is much more mobile than the upper one. The two eye-lids, completely close the eye. In addition to the two eye-lids, the nictitating membrane is a transparent thin membrane and is placed in the anterior corner of the eye. It also acts like a third eye-lid. The nictitating membrane cleans the cornea at the time of need. Behind the eyes are two shallow pits called the **external ear openings** or **auditory meatus**. At the base of each auditory meatus lies an oval tympanic membrane. The meatus has a denticulated anterior border but the posterior border is not denticulated.

Neck. Head is followed by the neck. The neck is short and devoid of gular pouches. There are prominent folds of skin on the ventral and lateral sides of the neck when the animal is at rest but these folds disappear when the animal stretches out its neck.

Trunk. The neck is followed by the long, broad and depressed trunk to which are attached two pairs of limbs. The trunk is smooth and flattened ventrally but rough and convex dorsally. Its lateral sides skin is loosely folded. On the ventral surface a thick-lipped transverse **cloacal aperture** is placed at the junction of the trunk and the tail.

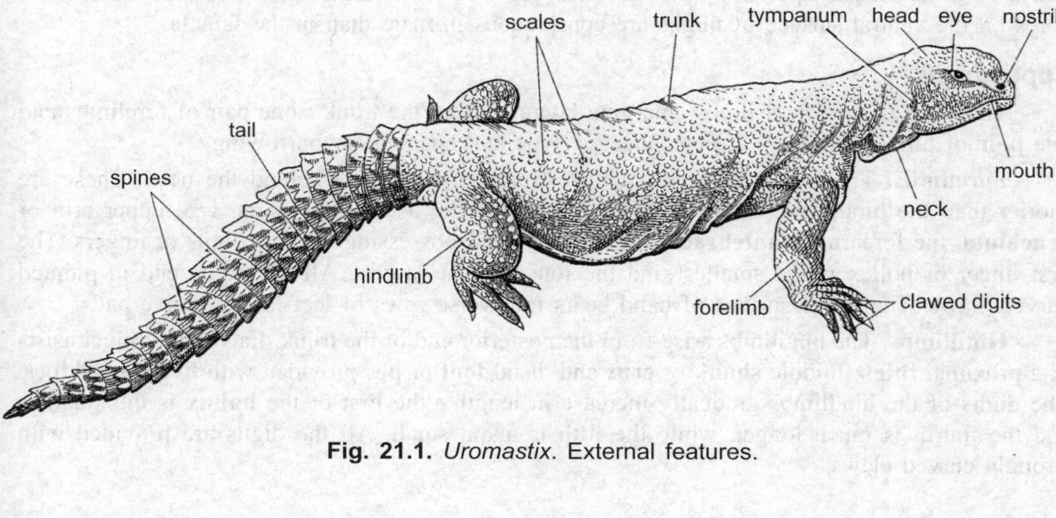

Fig. 21.1. *Uromastix*. External features.

Tail. The tail is long, thick and strongly depressed, widest at the base and tapers posteriorly. There are about twenty transverse rows of hard and strongly keeled scales covering the dorsal surface of the tail. These scales form a sort of armour for self-defence and, thus, check the loss of water. Due to the presence of spiny scales on the tail, the *Uromastix* is also called the "spiny-tailed lizard". Caudal autotomy is lacking. Just behind the cloacal aperture, on the ventral side at the broad base of the tail is present a pair of nodule-like swellings, which are due to the presence of hemipenes in male.

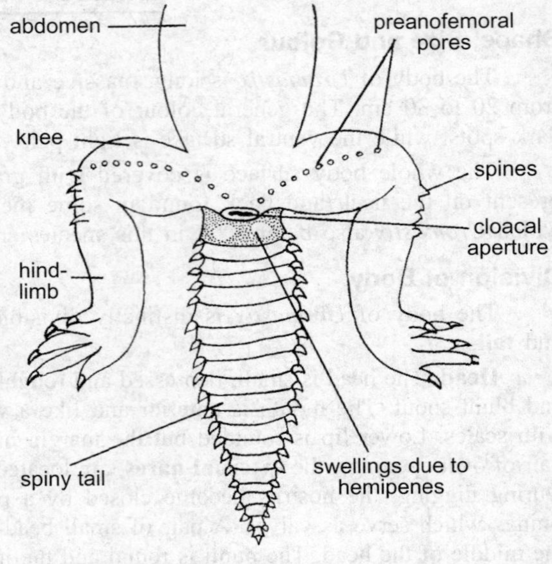

Fig. 21.2. *Uromastix*. Posterior part of the body to show the cloacal aperture and preanofemoral pores.

Keith (1955) has given an interesting account of scales found on the dorsal, ventral and lateral surfaces of the body. According to her, the whole body is covered with scales. The scales present on the dorsal surface of the head are unequal, smooth or keeled. These are larger on the snout but become smaller towards the eye. The scales on the cheeks are oval and 12–14 in number. These are called **upper labials**.The scales present on the dorsal surface are small, numerous with or without larger scales lying between them. The ventral scales are subquadrangular in shape and they are bigger even than the bigger dorsal scales. The gular scales are rounded and smaller than ventrals. On the tail are present **spinose scales**. Due to the presence of spinose scales *Uromastix* is called **spiny-tailed lizard**.

Sexual Dimorphism

In the male, the cloaca possesses a pair of protrusible copulatory organs, called **hemipenes** (**Majupuria,** 1962). These open separately into the proctodaeum. These are absent in females. In the male, in breeding season, a pair of nodule-like swellings appear ventrally at the base of the tail just behind cloacal aperture. These are also absent in females. Besides these, preanofemoral pores on the ventral surface of thighs are conspicuous in male than in the female.

Appendages

There are two pairs of limbs emerging laterally from the trunk : one pair of forelimbs and one **pair** of hindlimbs. They are pentadactyle type and adapted for burrowing.

Forelimbs. The forelimbs arise laterally from the trunk just behind the neck. These are shorter than the hindlimbs. Each forelimb consists of the usual three parts, *viz.,* upper arm or **brachium**, the forearm or **antebrachium** and the **hand** possessing the five digits or fingers. The first finger or pollex is the smallest and the fourth is the largest. All the digits end in pointed curved claws. The lower surface of hand bears transverse rows of leaf-like adhesive pads.

Hindlimbs. The hindlimbs arise from the posterior end of the trunk. Each hindlimb consists of a proximal **thigh**, middle **shank** or **crus** and distal **foot** or **pes** provided with five clawed toes. The digits of the hindlimbs gradually increase in length : the first or the **hallux** is the smallest and the fourth is much longer, while the fifth is again small. All the digits are provided with strongly curved claws.

Preanofemoral pores. There are about 9 to 18 prominent pores arranged in a curved row on the ventral surface of both thighs from cloaca to the knee-joint. These are called **preanofemoral pores** and more prominent in male. Each pore is surrounded by a ring of scales, amongst which one is larger than the others. Each pore leads into a tubular gland producing a horny secretion which foams temporary spines which help in copulation.

SKIN OR INTEGUMENT

The skin of *Uromastix* is rough, dry and scaly and has almost no glands. It prevents the loss of water. This is an adaptation to prevent evaporation of water. The histological structure of skin shows two layers : an outer **epidermis** and an inner **dermis**.

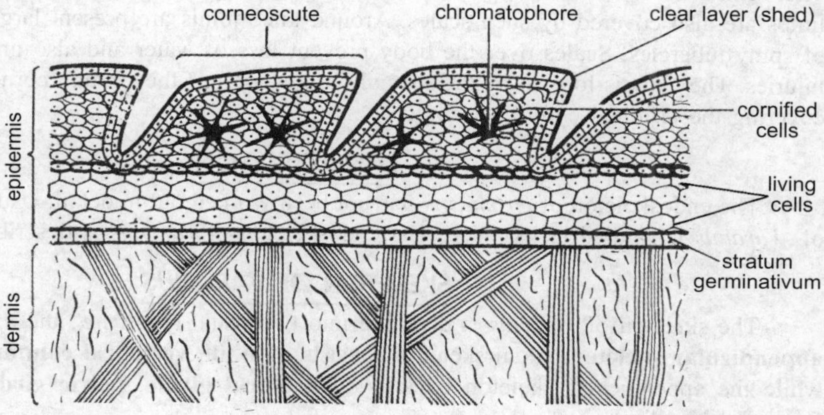

Fig. 21.3. *Uromastix*. V.S. of integument. (Diagrammatic).

1. Epidermis. The epidermis is stratified and derived from ectoderm. The epidermis comprises a well developed **stratum corneum** or horny layer which serves admirably to protect the lizard from desiccation. The stratum corneum is composed of thin, flattened and dead cells due to the deposition of horny scleroprotein or keratin. Due to accumulation of keratin in the outermost stratum corneum cells, they become keratinized or cornified and then they die. The epidermis produces **horny scales** covering the body. This layer is sloughed off periodically in flakes (ecdysis). Below this is a layer of **chromatophores**, which are not found in the skin of those areas which are covered with scales. At the base of epidermis lies a single-layered **germinal epithelium** with prominent nuclei. The germinal epithelium proliferates cells which move outward and form the stratum corneum. The horny layers covering the terminal ends of digits form the claws.

2. Dermis. The **dermis** lies beneath the epidermis. It is a well developed thick layer of dense fibrous connective tissue. It contains muscle-fibres, blood vessels, nerves and chromatophores. The scales, covering the entire surface of the animal, are formed from special folds of dermis. There are no glands in the skin except a few **femoral glands** which are found on the ventral surface of the thigh. These glands secrete a substance in males that hardens into temporary spines which help in clasping the female at the time of copulation.

3. Muscles. Below the body wall or skin is present a layer of muscles which help in locomotion, etc. Beneath the muscle layer is the **peritoneum** which is formed of closely fitting mesodermal cells. It encloses the body cavity.

EXOSKELETON

The exoskeleton of *Uromastix* are overlapping epidermal horny scales covering the body and claws. The entire body is covered with small horny overlapping epidermal scales. These are shed off at regular intervals. Periodically a new set of scales forms beneath the old and when this happens the outer and older series is shed. Each scale projects backward to ovelap the one behind. The scales are formed from thickened areas of the integument which grow up and backward and become cornified. They are continuous with each other at their bases. The scales present on the dorsal surface of the head are unequal, obtusely keeled or smooth. The labials are small, supra-occulars are smallest and scales on the snout are the largest. There is a simple arrangement of scales on the trunk : dorsals, ventrals and preanals. The dorsals (dorsal scales) are small, numerous and arranged in cross series. Ventrals (ventral scales) are sub-quadrangular and are larger than dorsals. The dorsal or upper surface of the tail is covered by large keeded spinose scales arranged in twenty transverse rows. Thus, tail is used as a weapon of defence. The ventral or lower surface of the tail is covered by sub-caudals or sub-caudal scales. The limbs are also covered by such scales. Around knee joints are present larger scales in the form of spiny tubercles. Scales over the body prevent loss of water and also protects the body from injuries. The claw is formed by the special modification of the stratum corneum of the epidermis covering the terminal end of the digit.

ENDOSKELETON

The endoskeleton of *Uromastix* has not been worked out in detail. Usually the endoskeleton of *Varanus* is studied in its place. The endoskeleton of *Varanus* is described here in detail.

SKELETON OF *VARANUS*

The skeleton of *Varanus* is divisible into two usual types, *i.e.,* the **axial skeleton** and the **appendicular skeleton**. Axial skeleton consists of **skull**, **vertebral column**, **sternum** and **ribs**, while the appendicular skeleton consists of **pectoral girdle**, **pelvic girdle** and the **bones** of **fore-** and **hindlimbs**.

Characteristics of Skull

1. It is **monocondylic**, *i.e.,* single occipital condyle formed by the basioccipital.
2. It is **tropibasic**, *i.e.,* presence of interorbital septum separating the two orbits.
3. Alisphenoids, orbitosphenoids and presphenoid are absent.
4. Prefrontals, supraorbitals and postorbitals are present.
5. A pair of parietals are fused and perforated by a median parietal foramen.
6. Both the premaxillae are also fused to form a single bone.
7. Presence of three vacuities (temporal fossae).
8. Presence of a single ear ossicle or **columella** in the tympanic cavity.
9. Teeth polyphyodont, homodont and pleurodont.
10. Autostylic suspensorium, *i.e.,* lower jaw articulates with the quadrate.

1. Skull. The skull of *Varanus* may be divided into four main regions, *viz.,* the **cranium**, **sense capsules** (auditory, optic and olfactory) and the **visceral skeleton** comprising the jaws and hyoid apparatus.

Cranium. The cranium proper forms the posterior part of the skull and lodges the brain. The cranium comprises three segments, *i.e.,* the posterior occipital segment, the middle parietal segment and the anterior frontal segment.

(*a*) Occipital segment. The occipital segment is the posteriormost part of the cranium. It encloses a large aperture **foramen magnum** in its centre. It consists of four bones, *i.e.,* the dorsal **supraoccipital**, the lower **basioccipital** and paired lateral **exoccipitals**. Supraoccipital is more or less rectangular forming the roof of the hindermost part of the cranium. It articulates anteriorly with the parietals and laterally with the prootics forming the post-temporal fossa. Basioccipital forms the floor of the hindermost portion of the cranium and posteriorly bears a single occipital condyle. It articulates anteriorly with the basisphenoid, laterally with the exoccipitals and prootics.

Exoccipitals surround the foramen magnum from its lateral sides. Each exoccipital is produced on its outer side into a paroccipital process articulating with the supratemporal, squamosal, parietal and quadrate. Each exoccipital has a combined juglar and condylar foramen for the exit of X, XI and XII cranial nerves and artery and vein. Each exoccipital articulates with the supraoccipital, prootic and the basioccipital. Occipital segment articulates posteriorly with the atlas vertebra by the occipital condyle of the basioccipital. All the bones of this region are fused forming a ring-like structure.

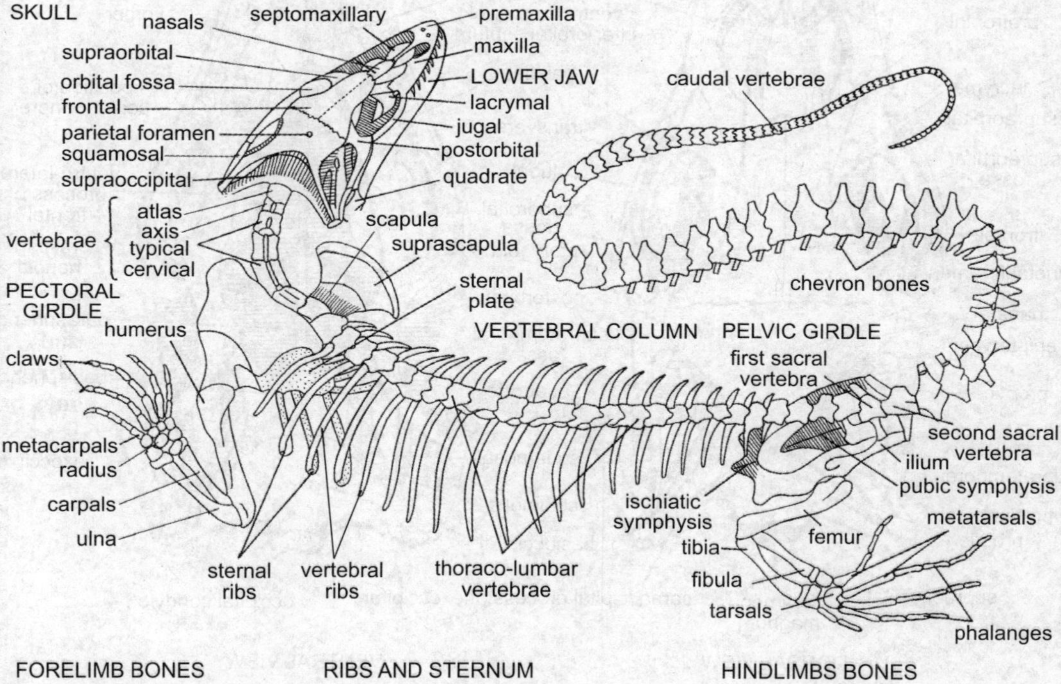

Fig. 21.4. *Varanus*. Complete skeleton in dorso-lateral view.

(*b*) Parietal region. The parietal segment comprises only two bones, parietals and a basisphenoid. **Parietal**. The parietal is a large flat bone on the postero-dorsal side of the cranium. It forms the roof of the cranium in the parietal region or segment. The two parietals are fused to form a single bone. It is broad in front and narrow behind. It has a **parietal foramen** in its centre. Its postero-lateral angles are produced into **supratemporal processes**, each of which articulates with the quadrate, squamosal, supratemporal and paroccipital process of the exoccipital of its side. The supratemporal processes form the outer margins of the post-temporal fossae. The parietals articulate anteriorly with the frontals, posteriorly with the supraoccipital, on the sides with the post-orbitals and below with the epipterygoid. **Basisphenoid**. The basisphenoid forms the floor of the cranium in the parietal region. It is a broad, flat and somewhat rectangular bone. On its antero-lateral sides it gives off a pair of basipterygoid processes which unite with the pterygoids at their free ends. At the anterior margin it bears a median knob-like process which receives the hinder end of parasphenoid. Basisphenoid articulates posteriorly with the anterior margin of the basioccipital.

(*c*) Frontal region. The frontal segment is composed of only three bones, a pair of frontals and parasphenoid. **Frontals.** The frontals are large triangular bones lying on the dorsal side with their apex forwards. They meet one another along the median line by a **frontal suture** and with the parietal behind by a **coronal suture**. The frontals form the roof of the cranium in the frontal

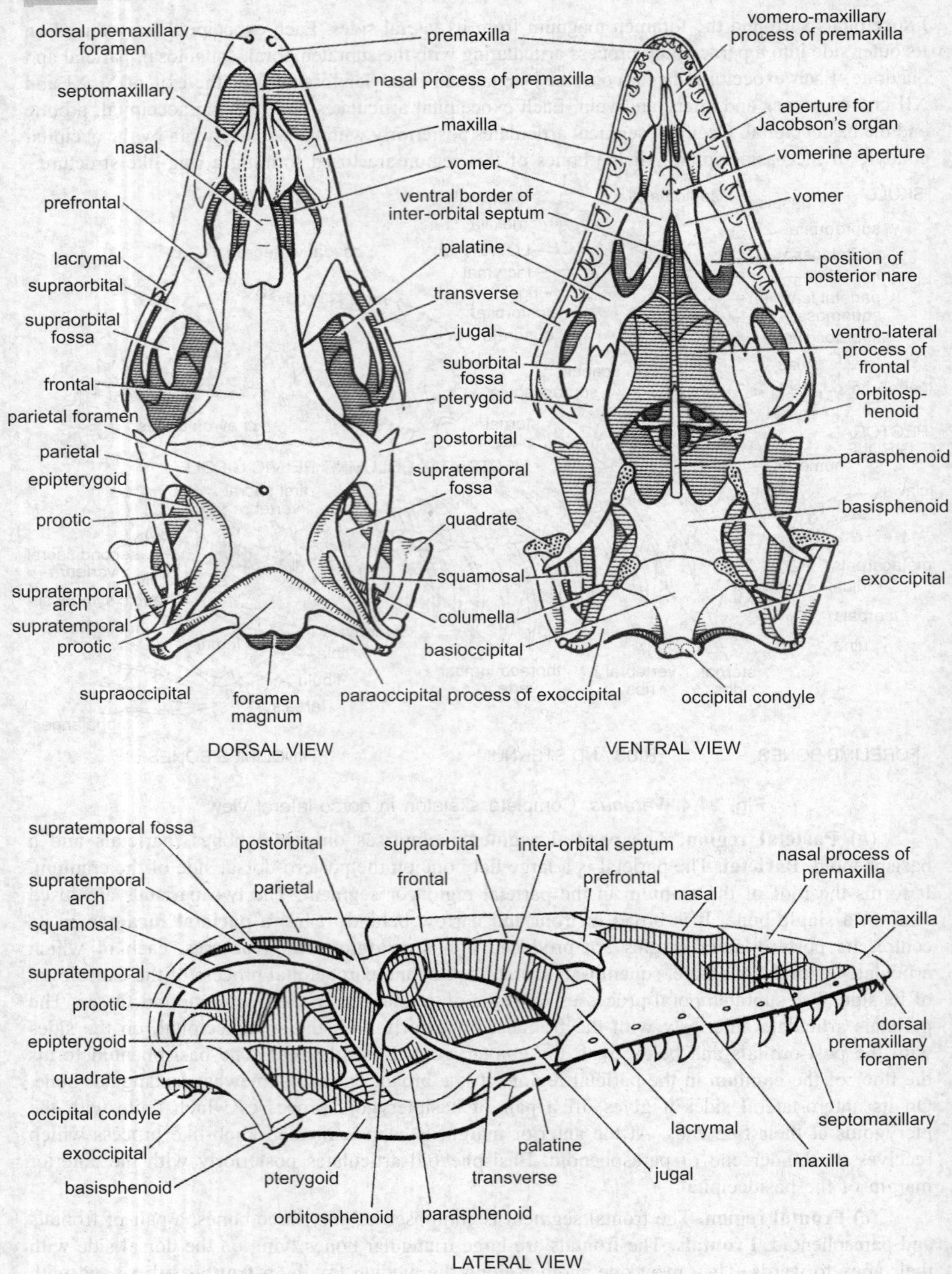

Fig. 21.5. *Varanus*. Skull in dorsal, ventral and lateral view.

segment. Frontals articulate in front with nasals, with prefrontals, palatines and postorbitals on the outer sides, with the parietals behind and below with the parasphenoid. The **parasphenoid** is a long, narrow, rod-like bone in the mid-ventral line.

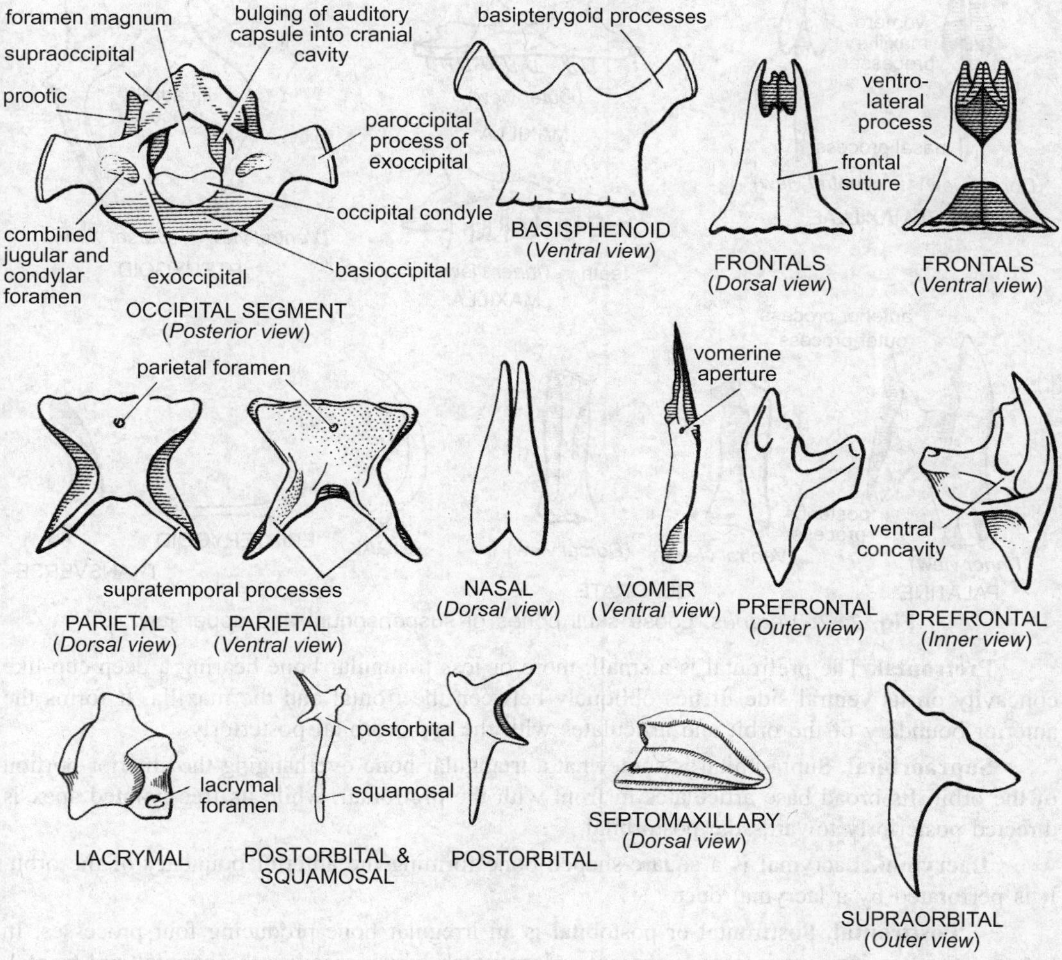

Fig. 21.6. *Varanus*. Loose skull bones of cranium and sense capsules.

2. Sense capsules. The sense capsules protect and lodge the organs of special sense. There are three pairs of sense capsules : auditory, optic and olfactory.

(*a*) Auditory capsules. The auditory capsules enclose the internal ears, the organs of hearing. They are situated on the lateral sides of the posterior end of cranium. Each capsule comprises a single small irregular vertical bone, the **prootic** lying just outside the supraoccipital and **epiotic** and **opisthotic** are indistinguishable. The epiotic is fused with the supraoccipital, while the opisthotic is fused with the exoccipital.

(*b*) Optic capsules. The optic capsules or orbits are a pair of large excavations on the lateral sides of the cranium. They enclose the eyes. The two orbits are separated from one another by a thin, longitudinal vertical partition, the **interorbital septum**. Each orbit is bounded by five bones: prefrontal, supra-orbital, lacrymal, post-frontal or post-orbital and jugal.

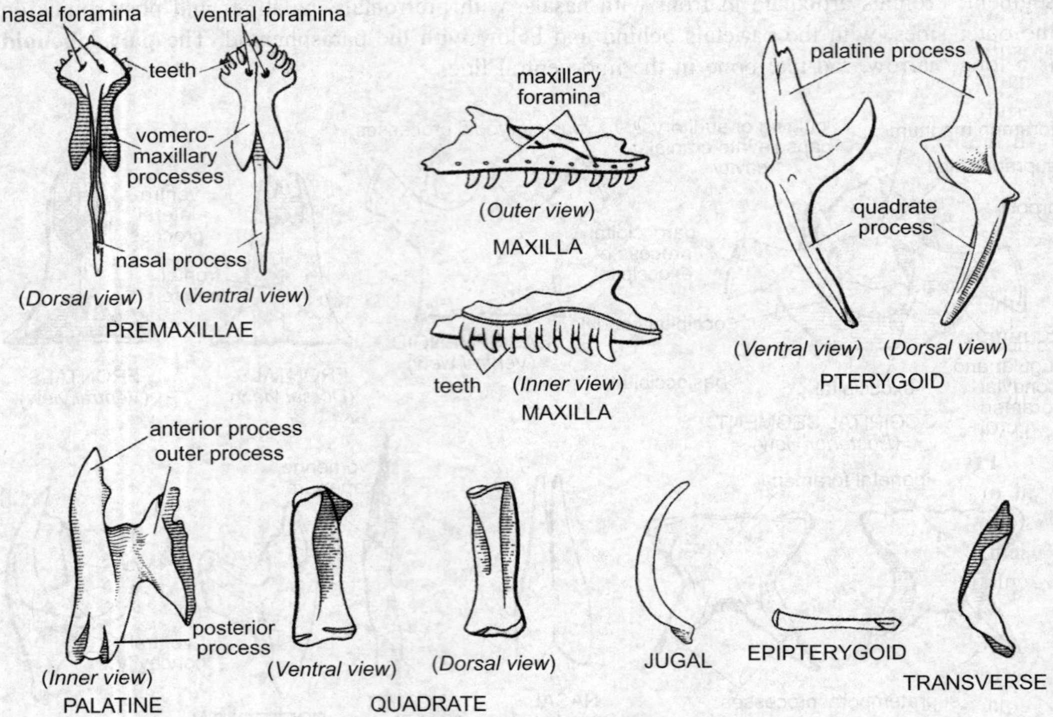

Fig. 21.7. *Varanus.* Loose skull bones of suspensorium and upper jaw.

Prefrontal. The prefrontal is a small, more or less triangular bone bearing a deep cup-like concavity on its ventral side. It lies obliquely between the frontal and the maxilla. It forms the anterior boundary of the orbit and articulates with the suprarorbital posteriorly.

Supraorbital. Supraorbital is somewhat a triangular bone overhanging the anterior portion of the orbit. Its broad base articulates in front with the prefrontal, while its free pointed apex is directed posteriorly towards the postorbital.

Lacrymal. Lacrymal is a square-shaped bone forming the anterior boundary of the orbit. It is perforated by a lacrymal duct.

Postfrontal. Postfrontal or postobital is an irregular bone producing four processes. In between the two inner processes is present a deep notch which receives the parietal and frontal. The posterior longest process articulates with the squamosal, whereas the antero-lateral process remains free. Postfrontal name is given because it is present behind the frontal, and postorbital name is given because it forms postero-dorsal boundary of orbit. It forms the posterior boundary of the orbital capsule.

Jugal. Jugal is a slender, curved, rod-like bone forming the ventral border of the orbit.

***(c)* Olfactory capsules.** The olfactory capsules lie side by side in front of the cranium and contain the organs of smell. Each capsule is composed of dorsal nasal, septomaxillary and ventral vomer.

Nasals. Nasals form the roof of the olfactory capsules. Two nasals fused in the mid-line into a single compound bone. Each nasal is a flat triangular bone, narrow anteriorly and broad behind. Nasals articulate in front with the nasal process of premaxilla and two septomaxillaries, and behind with the frontals.

Vomer. The vomer is an elongated rod-shaped bone lying on the floor of the olfactory capsule. It meets with its counterpart in the mid-line in the anterior half and their posterior half remains free. It forms the inner margin of the posterior nare and is perforated by a vomerine aperture in its centre. It articulates with the vomero-maxillary process of premaxilla in front, with the palatine behind and Jacobson's organ outside in the anterior region.

3. Visceral skeleton. It comprises the jaws, suspensorium and hyoid apparatus.

Jaws. The jaws support the borders of the mouth. There are two jaws, *i.e.,* upper jaw and lower jaw.

(*a*) Upper jaw. The upper jaw consists of two similar halves or rami which are fused together anteriorly in the middle line but diverge posteriorly. It is intimately fused with the cranium. Each half or ramus consists of nine bones divisible into two sets : the outer and inner. The outer set includes four bones : premaxilla, maxilla, jugal, septomaxillary, and quadrate, while the inner set includes five bones : pterygoid, palatine, transverse or transpalatine or ectopterygoid, epipterygoid and squamosal.

Premaxillae. Two premaxillae are fused in the mid-line into a single bone and form the anterior limit of the snout. On its ventral surface it possesses 6–8 small conical teeth along its margins. Posteriorly it is produced into a long, narrow dorsal nasal process which is lodged into the groove present on the anterior part of the nasals. A pair of wing-like ventral vomero-maxillary processes articulating with the vomers behind. Premaxillae articulate on either lateral sides with the maxillae.

Maxilla. The maxilla forms the major portion of the upper jaw on either side. It is a long irregular bone lying on either side behind the premaxillae. Outer surface of each maxilla is perforated by a number of **maxillary foramina**. The main body of each maxilla is known as alveolar part which bears a row of 8–10 teeth along its outer margin. The teeth are small, conical and pointed backward, and pleurodont type. The palatine process present on the inner side is poorly developed due to which a gap is present between maxilla and palatine. Maxilla articulate in front with the premaxillae, septomaxillary and vomer. Its upper process articulates with the prefrontal, supraobital, nasal and lacrymal, and behind with palatine, jugal and transverse bones.

Jugal. Jugal is a slender curved, rod-like bone behind the maxilla. Besides forming a part of the upper jaw, it also bounds the orbit ventrally. It articulates anteriorly with the maxilla and lacrymal and on the inner side with the transverse, but free posteriorly.

Palatine. Palatine is the bone of the roof of the buccal cavity. It is small, flat and more or less irregular bone. It is produced into three processes. The anterior long process joins with the vomer, the posterior small process with the pterygoid and the outer broad process with the maxilla and transverse. Palatines form the posterior boundary of inner nares.

Pterygoid. Pterygoid is a large elongated and irregular bone. It has an anterior palatine process and a posterior quadrate process. It articulates anteriorly with the palatine on the inner side and transverse on the outer side, posteriorly with the basipterygoid process of basisphenoid on the inner side and with the quadrate on the outer side. Pterygoids are present on the roof of mouth.

Transverse. The transverse is a small curved bone connecting the pterygoid with the jugal and the maxilla. It forms the floor of the orbit. It is also called the transpalatine or ectopterygoid.

Epipterygoid. Epipterygoid is very slender rod-like bone extending almost vertically from the upper surface of the pterygoid to the prootic. It is also called columella crani.

Squamosal. The squamosal is a curved rod-like bone attached to the posterior end of the arcade. Its anterior end meets a backward process of the post-frontal to form the supra-temporal arch, and its posterior end bends downwards to articulate with quadrate, supra-temporal, parietal and exoccipital.

Septomaxillary. A pair of small, irregular flat bones are present on the dorsal side of vomers and one on either side of nasal process of premaxillae in the nasal region. These articulate with maxillae and nasals.

Suspensorium. It includes quadrate and squamosal bones which suspend the lower jaw with the cranium and upper jaw.

Quadrate. The quadrate is a small, thick, rod-like bone of the suspensorium. It lies obliquely in the postero-lateral side of the hind region of the cranium. Posteriorly it articulates with the squamosal, supra-temporal, supra-temporal process of the parietal and paroccipital process of exoccipital. Anteriorly it articulates with the quadrate process of pterygoid and lower jaw.

Lower jaw. The lower jaw or mandible consists of two rami meeting anteriorly. Each ramus is composed of six bones, *i.e.,* articular, angular, supra-angular, coronoid, splenial and dentary around an axial Meckel's cartilage.

Articular. The articular is the posteriormost bone of the ramus. On its dorsal surface it bears an articular surface for the quadrate and extends posteriorly terminating into an articular cartilage.

Angular. Angular is a small splint-like bone fitted in between the dentary and articular and perforated by an angular foramen.

Supra-angular. Supra-angular is a flat, elongated and more or less rectangular bone lying in the middle of the ramus. It carries a pair of mandibular foramina for the mandibular nerves. It is present over the upper half of the outer surface of articular.

Coronoid. Coronoid is a small more or less conical bone present above the supra-angular. It forms the dorsal side of the middle portion of ramus. It is produced into an upwardly directed coronoid process just behind the last tooth.

Splenial. Splenial is a flat, irregular membranous bone lying on the inner side of the dentary.

Dentary. Dentary is the largest bone forming the distal half portion of the ramus. It bears a row of 8 to 10 small, conical mandibular teeth which are of **pleurodont type**. Its outer surface also carries numerous foramina for the mandibular nerves. Both the dentaries unite suturally at their anterior ends.

Hyoid apparatus. The hyoid apparatus lies embedded in the floor of the buccal cavity. It is cartilaginous structure which supports the tongue. It consists of a basihyal (main body) and two pairs of cornua arising from basihyal.

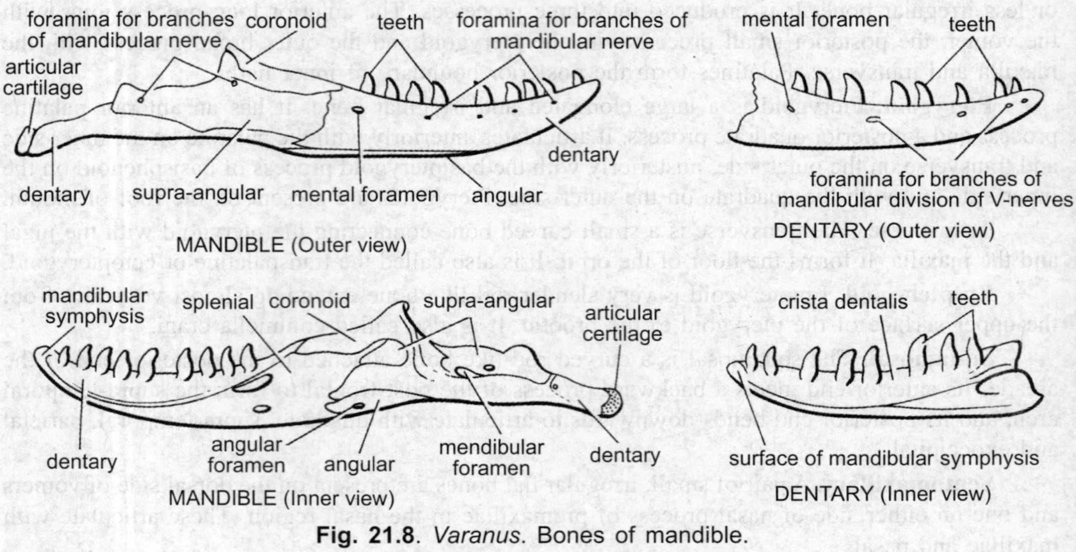

Fig. 21.8. *Varanus.* Bones of mandible.

Basihyal. Basihyal is in the form of an elongated, median, rod-like structure forming the body proper of hyoid apparatus. Its tapers anteriorly into os-entoglossus or lingual process.

Anterior cornua. The anterior cornua are elogated cartilaginous rods connected ventrally with the basihyal and running upwards curving round the gullet and ending on the ventral surface of the auditory capsule of their side. Each anterior cornua consists of a proximal and a distal piece, both joined together by ligaments.

Posterior cornua. The posterior cornua are also cartilaginous two segmental rods arising from the posterior end of the basihyal and running towards the hind end. Posterior cornua arise from branchial arches, while anterior cornua arise from hyoid arch.

Vertebral Column

The vertebral column of *Varanus* consists of a large number (83–85) of vertebrae. Epiphyses are absent and centra of vertebrae are strongly procoelous. The vertebral column is divisible into four regions : the **cervical** containing 8 vertebrae, the **thoraco-lumbar** including 22 vertebrae, the **sacral** having only 2 vertebrae, and the caudal with about 51–53 vertebrae.

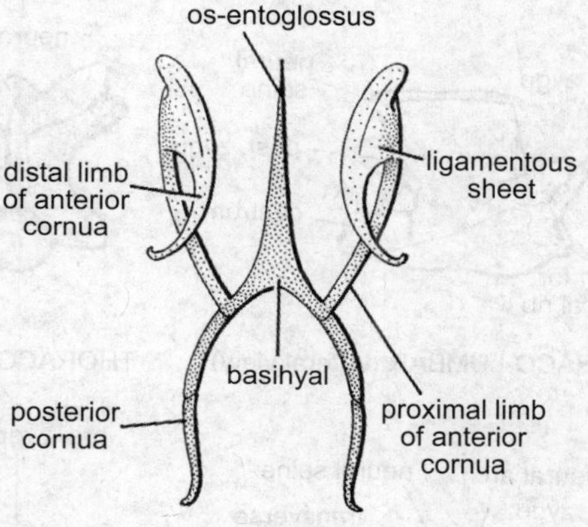

Fig. 21.9. *Varanus.* Hyoid apparatus.

1. Cervical vertebrae. A typical cervical vertebra consists of a stout and elongated centrum which is concave in front and convex behind, *i.e.,* **procoelous.** Neural spine is crest-like. Prezygapophyses present at the anterior surface of neural arch are directed upwards and outwards, while postzygapophyses present on the posterior surface of neural arch are directed downwards and inwards. On the ventral surface, the posterior region of the centrum carries a backwardly directed hypapophysis. On the outer anterior surface the centrum carries articular facets for cervical ribs. Articular facets are present behind the third cervical vertebra.

2. Atlas vertebra. Atlas vertebra is the first vertebra of cervical region. It is small and ring-like in shape. It has no distinct centrum. It is composed of three distinct bony pieces of which two are dorso-lateral and one ventral around the neural canal. The neural canal in living animals is divided by a ligament into the dorsal and ventral portion which is downwardly projected into a small spine. Ventral piece occupies the position of centrum. The dorso-lateral pieces have a space filled by a membrane at the point of their union in the mid-dorsal line. Along the anterior face of the ventral piece is present a cavity for the articulation of occipital condyle of the basioccipital of skull. On the posterior side of the ventral piece is also present an articular facet for the odontoid process of the axis vertebra. Pre- and postzygapophyses are absent. Transverse processes are also entirely absent.

3. Axis vertebra. Axis vertebra is the second vertebra of the cervical region. It resembles the typical cervical vertebra in structure but it is slightly larger and bears a posteriorly directed hook-like spine below the odontoid process and arising from the centrum. The neural spine is vertical crest-like. It is characterised by the presence of an odontoid process at the anterior face of centrum. Below the odontoid process, from the anterior end of centrum arises a backwardly directed hook-like **spine**. It represents the intercentrum or hypapophysis of atlas vertebra. A hypapophysis is also present on the ventral surface of the centrum in the posterior region. Prezygapophyses are rudimentary represented by simple notches but postzygapophyses are well developed.

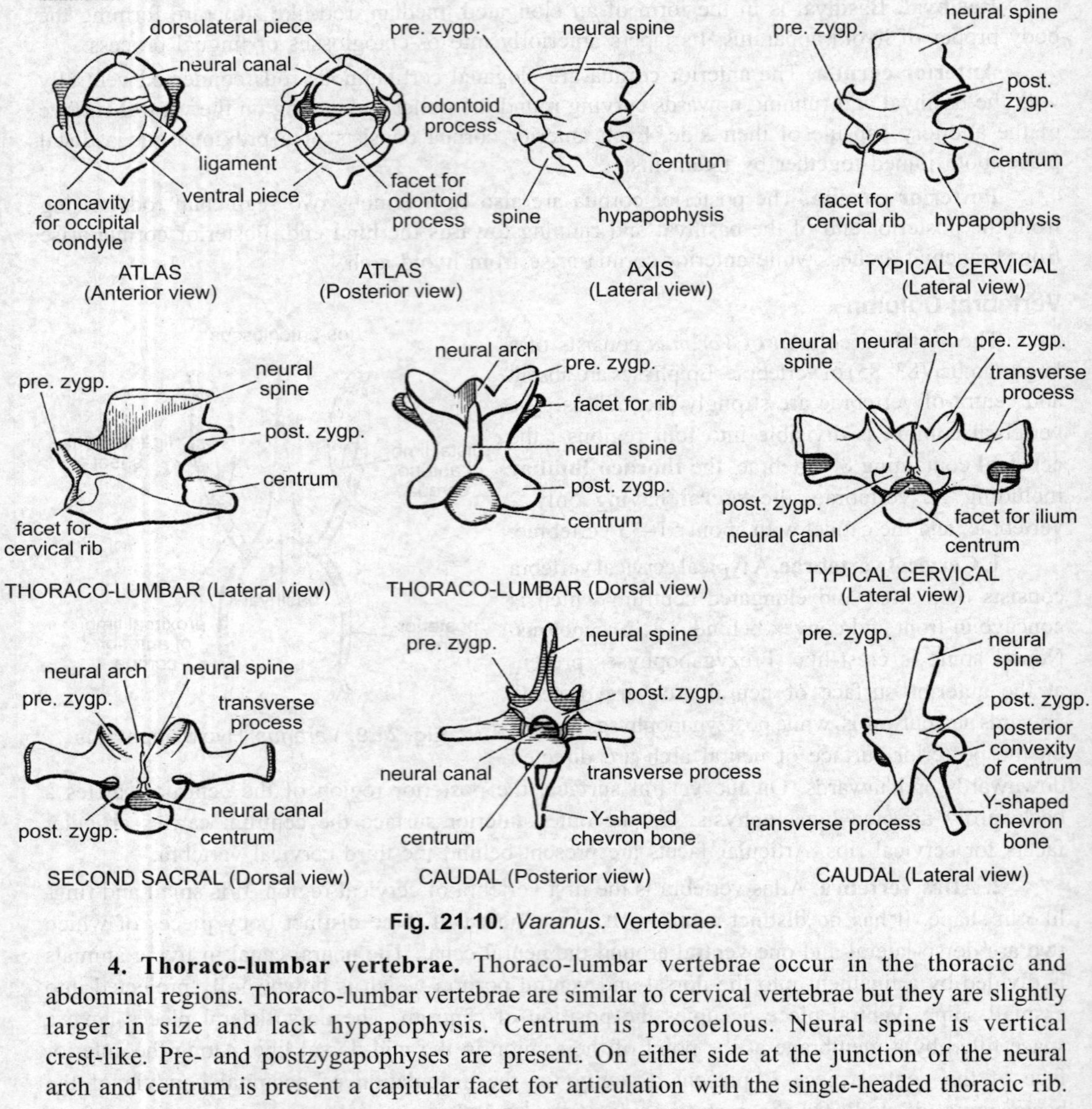

Fig. 21.10. *Varanus*. Vertebrae.

4. Thoraco-lumbar vertebrae. Thoraco-lumbar vertebrae occur in the thoracic and abdominal regions. Thoraco-lumbar vertebrae are similar to cervical vertebrae but they are slightly larger in size and lack hypapophysis. Centrum is procoelous. Neural spine is vertical crest-like. Pre- and postzygapophyses are present. On either side at the junction of the neural arch and centrum is present a capitular facet for articulation with the single-headed thoracic rib.

5. Sacral vertebrae. Both the sacral vertebrae are firmly united. These are stout and strong giving support to the pelvic girdle. **Anterior** or **first sacral vertebra** is more stout than the second. Its centrum is short and strongly procoelous. The neural spine is small crest-like. Pre- and postzygapophyses are well developed. Hypapophysis is absent. Transverse processes are strong and expanded at the tips which are also notched to receive a part of the ilia bones of the pelvic girdle. **Posterior** or **second sacral** has also short, stout and strongly procoelous centrum, neural spine is low and crest-like, and zygapophyses well developed. Unlike first sacral, transverse processes though expanded but not notched at the tips.

6. Caudal vertebrae. The caudal vertebrae progessively become smaller towards the posterior end, finally getting reduced to mere bony rod-like centrum. **Anterior caudal vertebrae.** In the anterior caudal vertebrae the centrum is longer and strongly procoelous. The neural spine is long, pointed and backwardly directed. Pre- and postzygapophyses are well developed. Transverse processes are long, slender, outwardly and downwardly directed. The most characteristic feature of the vertebra is the presence of Y-shaped **chevron bone** attached to the ventral surface of the centrum which bears two nodules for this purpose. Y-shaped chevron bone has two haemal arches attached to the ventral surface of centrum and a long downwardly directed haemal spine. It encloses the haemal canal through which passes the artery and vein. **Posterior caudal vertebrae.** Posterior caudal vertebrae resemble the anterior caudal vertebrae in structure except the absence of chevron bone. The various processes also become gradually reduced.

Sternum. The sternum or breast bone is in the form of a rhomboidal plate of calcified cartilage lying enbedded in the ventral thoracic wall. Its antero-lateral borders articulate with the coracoids and epicoracoids of the pectoral girdle. Along its each postero-lateral border it bears two small facets for the articulation with the 2 sternal ribs. At its posterior end it also bears two sternal ribs. The posterior elongated stem of T-shaped interclavicle is lodged in a groove on the mid-ventral surface of sternal plate. A pair of slender curved rods, the clavicles, are applied to the anterior cross-piece of the interclavicle.

Ribs. Ribs are **unicephalous**, *i.e.,* single-headed. Their upper ends articulate with the capitular facets of corresponding vertebrae. Tubercular facets are lacking. Ribs are slender and curved formed of bone or bone and cartilage. Except first three cervicals, all the cervicals, thoracic and lumbars bear ribs.

Thoracic ribs. These are larger and differentiated into a dorsal bony **vertebral portion** articulating with the vertebra and a ventral cartilaginous **sternal portion**. First three thoracic ribs articulate with the sternum by their cartilaginous sternal parts. Rest thoracic ribs do not articulate with the sternum and, thus, remain free.

Cervical ribs. Leaving first three cervicals, all the remaining ones have cervical ribs. These are shorter and remain free ventrally and have no cartilaginous sternal portion

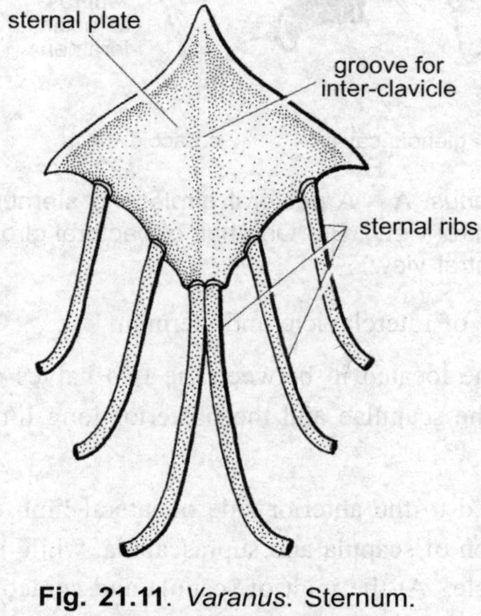

Fig. 21.11. *Varanus*. Sternum.

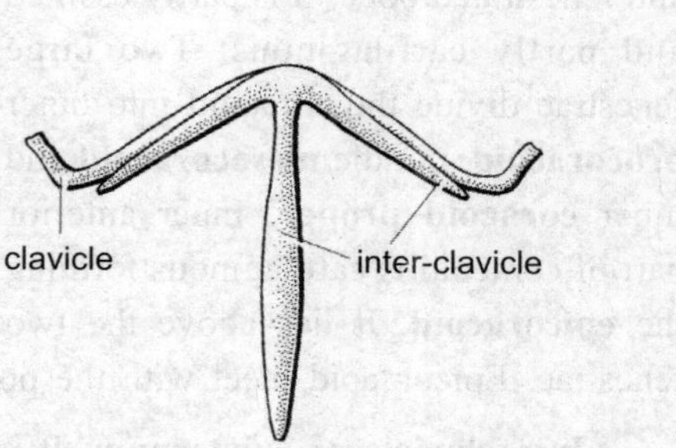

Fig. 21.12. *Varanus*. Inter-clavicle and clavicle.

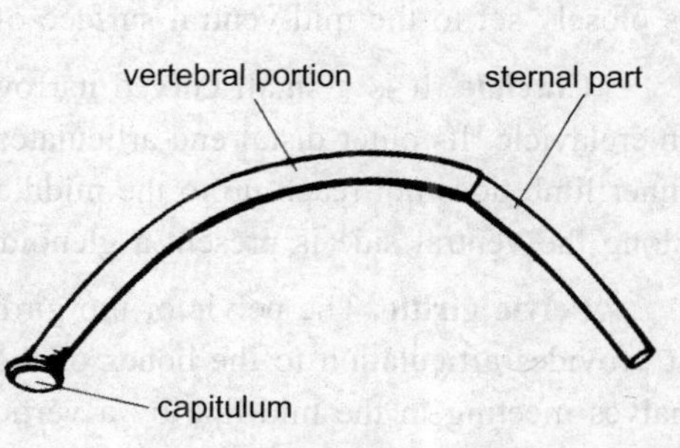

Fig. 21.13. *Varanus*. A thoracic rib.

Appendicular Skeleton

Pectoral girdle. The pectoral or shoulder girdle (Fig. 21.14) is situated at the anterior end of the trunk. It protects the heart and lungs along with the sternum and ribs and also provides articulation to the bones of forelimbs. The pectoral girdle is composed of two similar halves one lying on either side of the T-shaped interclavicle and sternum. Each half of the pectoral girdle consists of scapula, suprascapula, coracoid and epicoracoid.

Scapula. Scapula is bony, oblong and flat plate which is narrow in the middle. Its outer broader end articulates with the suprascapula, while its inner narrow end unites with the coracoid. Its lower posterior end forms a part of the **glenoid cavity** and its dorsal anterior end gives out an ossified process, the **mesoscapula**.

Suprascapula. Suprascapula is more or less rectangular thin plate of calcified cartilage with a free distal border. It articulates proximally with the scapula and its distal border is free.

Coracoid. Coracoid is large, flat and fenestrated bone. It is partly ossified and partly cartilaginous. Two large fenestrae divide the coracoid into outer **procoracoid**, middle **mesocoracoid** and inner **coracoid proper**. Inner anterior part of coracoid is cartilaginous forming the **epicoracoid**. It lies above the two fenestrae. Epicoracoid meet with the posterior long arm of interclavicle and sternum.

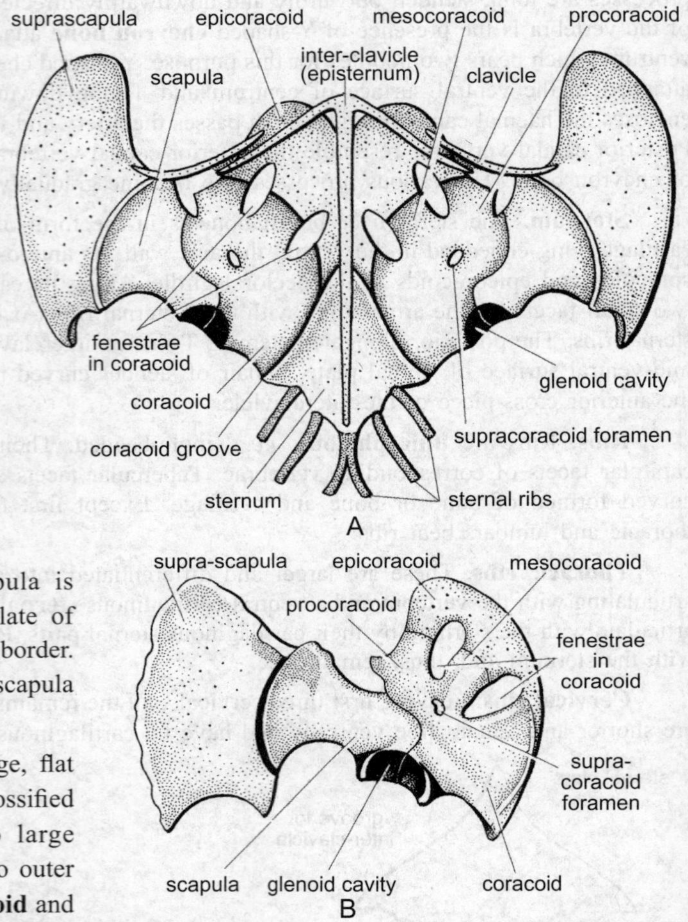

Fig. 21.14. *Varanus*. A – A pectoral girdle and sternum (Ventral view); B – One half of pectoral girdle (Ventral view).

Interclavicle or episternum. It is a T-shaped bone located in between the two halves of pectoral girdle. Its lateral curved limbs lie anterior to the scapulae and the posterior long limb is closely set to the mid-ventral surface of sternum.

Clavicle. It is a small curved narrow bone attached to the anterior side of lateral limb of interclavicle. Its outer distal end articulates with the union of scapula and suprascapula, while its inner limb does not reach up to the middle of interclavicle. At the joint of scapula and coracoid along the ventral side is present a glenoid cavity for the articulation of head of humerus.

Pelvic girdle. The pelvic or hip girdle (Fig. 21.15) is situated at the hind end of the trunk. It provides articulation to the bones of the hindlimbs. Pelvic girdle also consists of two similar halves meeting in the mid-line by a vertical ligament. Each half is known as **os-innominatum** and composed of ilium, pubis and ischium. All these three bones are not fused with each other. At their meeting points on the outer surface is a concave acetabulum for the head of humerus.

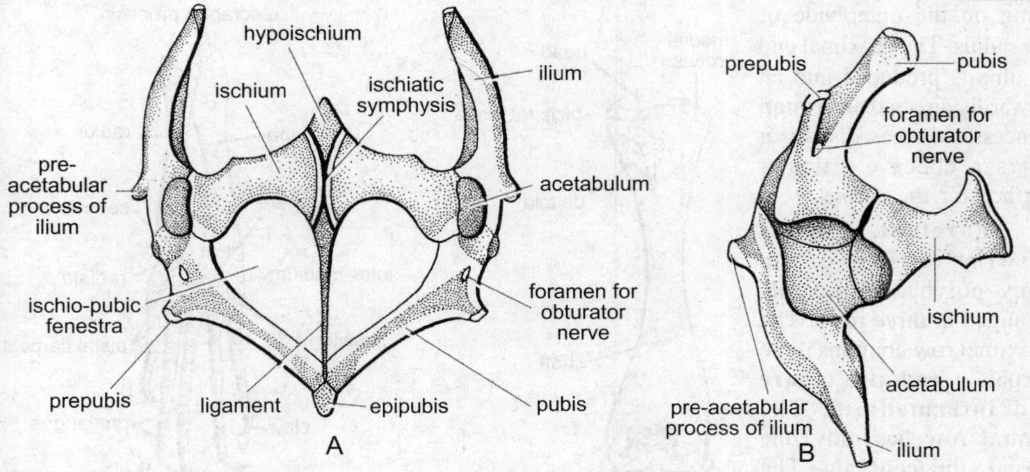

Fig. 21.15. *Varanus.* A – A pelvic girdle (Ventral view); B – One half of pelvic girdle.

Ilium. Ilium is strong, compressed, rod-like bone directed upwards and backwards to articulate with the sacral vertebrae (within the groove of transverse process of I sacral). On the outer side it is produced into a **pre-acetabular process** in front of the acetabulum and also contributes in the formation of about one-third part of acetabulum.

Pubis. Pubis is flat and slightly curved bone, passes downwards to meet in front with its fellow of the other side in the middle line at the pubic symphysis. Between the anterior ends of two pubes is present a backwardly directed nodule of calcified cartilage, the **epipubis**. On the anterior end near its fusion with the ilium and ischium it has an oval foramen for the obturator nerve. Just external and slightly posterior to the foramen is a small rod-like process, the **prepubis**, which is directed outwards. Pubis also contributes to about one-third of the acetabulum.

Ischium. Ischium is a flat and slightly curved bone, runs inward to meet its fellow of the other side at the ischiatic symphysis. It articulates on the outer side with the pubis and ilium of its side. A small rhomboidal piece of calcified cartilage, **hypoischium**, is present in between the two ischia at the anterior face of ischiatic symphysis and gives support to the ventral wall of cloaca. Ischium also forms about one-third of the acetabulum. The acetabulum is a cup-shaped cavity situated on the outer side at the point where all the three bones of each half of pelvic girdle join with each other. It provides articular surface for the head of femur.

A wide space is present between the pubes and ischia of both sides which is divided into two lateral **ischio-pubic fenestrae** by a median ligament. The ligament is often lost in the dried girdle.

Bones of forelimb. The forelimb contains humerus, radius and ulna, and the bones of forefoot.

Humerus. Humerus (Fig. 21.16) is the bone of the upper arm of the forelimb. It has an elongated shaft, both the ends of which are expanded and covered by epiphyses of calcified cartilage. The proximal end bears the rounded head which fits into the glenoid cavity of the pectoral girdle. A prominent crest-like **deltoid ridge** is present below the head. A **bicipital fossa** is enclosed between the medial process of proximal end and head. The distal end has a pulley-like articular surface, the **trochlea**, divisible into radial condyle and ulnare condyle by which it articulates with the radius and ulna.

Radius and ulna. Radius and u!na (Fig. 21.16) are the bones of forearm of the forelimb. **Radius** is slender and consists of a shaft and two epiphyses. At the distal end of radius is present a concave articular facet for the carpus and a preaxial **styloid proces**. **Ulna** is somewhat stouter lying on the outer side of the radius. The proximal end of ulna is produced into an upwardly directed **olecranon process**, whereas distally it bears a concave articular surface for the carpus.

Forefoot. Carpus or wrist consists of ten bony polyhedral **carpals** arranged in three rows. The proximal row contains three carpals : **radiale**, **ulnare** and **intermedium**. The central row has only one carpal, the **centrale**. The distal row has five small carpals. Besides the above, a pisciform is attached to the distal epiphysis of the ulna on its post-axial side.

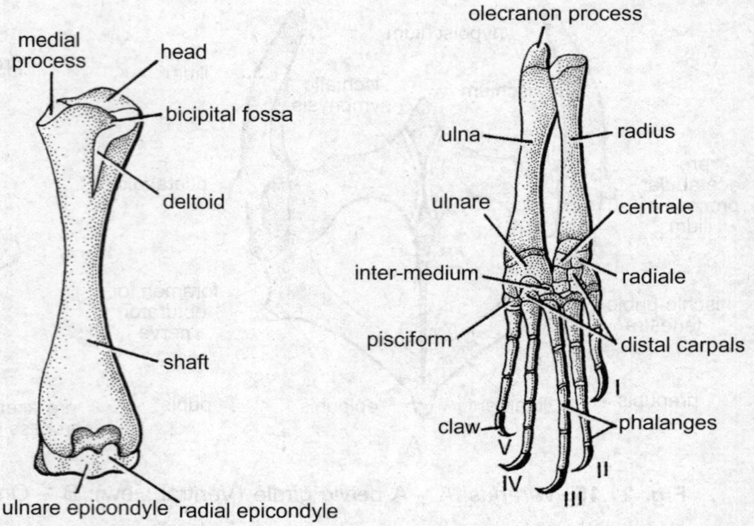

HUMERUS

RADIUS, ULNA AND BONES OF HAND

Fig. 21.16. *Varanus*. Forelimb bones.

Manus. The digits are five each consisting of a **metacarpal** and varying number of phalanges, *i.e.,* 2, 3, 4, 5 and 3. The terminal phalanx has a horny claw.

Bones of hindlimb. The hindlimb (Fig. 21.17) contains femur, tibia and fibula, and the bones of hindfoot.

Femur. Femur is the bone of thigh of hindlimb. It is a stout bone having an elongated shaft and two epiphyses. The proximal end has a rounded head which fits into the acetabulum of pelvic girdle. A bit posterior to the head, on the pre-axial side is the prominent **lesser trochanter** and the **greater trochanter** lying on the post-axial side is greatly reduced. The distal epiphysis is pulley-shaped having two condyles for the articulation of tibia and fibula and a tuberosity in between the two.

Tibia and fibula. Tibia and fibula are the bones of the shank of hindlimb. **Tibia** is a stout bone and bears a longitudinal ridge, the **enemial crest**, along the anterior or dorsal surface. The proximal end of tibia bears two concave facets for the articulation with the condyles of femur. **Fibula** is relatively slender bone proximally articulating with the distal tuberosity of the femur. The distal end of both tibia and fibula provide surface for the articulation with the tarsals.

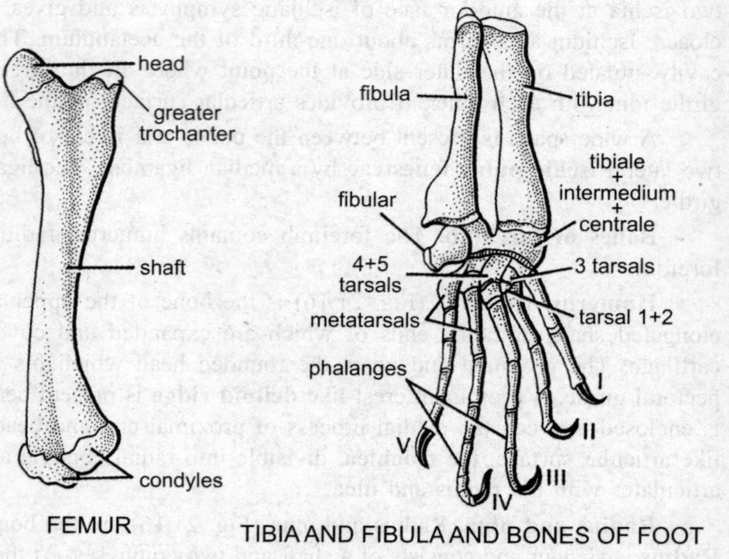

FEMUR

TIBIA AND FIBULA AND BONES OF FOOT

Fig. 21.17. *Varanus*. Hindlimb bones.

Foot bones. Tarsus contains only five tarsal bones, two in the proximal row and three in the distal row. Two tarsals of the proximal row are suturally fused and articulate with tibia and fibula. The distal row consists of three cuboid tarsals articulating with the five metatarsals of the foot. The number of phalanges in each toe is variable : the first toe has two, second toe has three, third toe has four, fourth toe has five and fifth toe has three phalanges. The terminal phalanx of each toe has a horny claw.

DIGESTIVE SYSTEM

The digestive system of *Uromastix* exhibits a number of unique features like all other lizards. The digestive system includes the **alimentary canal** and **digestive glands**.

Alimentary Canal

The alimentary canal of *Uromastix* is a long and convoluted tube. It can be divided into mouth, buccal cavity, pharynx, oesophagus, stomach, duodenum, ileum or small intestine,\colon, rectum or large intestine, cloaca.

1. Mouth. The mouth opening is fairly a wide gape bounded by the upper and lower immovable and muscular lips. Both the lips are covered with horny scales. The sharp edge of upper lip covers the rounded edge of lower lip. The mouth opens into the buccal cavity.

2. Buccal cavity. It is narrow in front and broad towards the posterior side. It is lined with mucous membrane and contains tongue, internal nares and teeth over the jaws. The mucous membrane of the buccal cavity possesses the mucous glands in four groups.

Tongue. The buccal cavity has a well developed muscular tongue on its floor. It is attached to the floor of the buccal cavity along the median ventral line. The tongue is long, anteriorly bifid and protrusible. Its broad posterior end is fixed and also bifid. The tongue bears the voluntary muscles, taste buds and mucous glands. Taste buds are present over the glandular papillae found in the mid-dorsal groove.

Internal nares. These are present in the roof of the buccal cavity near the anterior end. These lead into nasal chambers and open outside by external nares located at the anterior end of snout.

Teeth. Teeth are arranged over the premaxillae and maxillae of the upper jaw and dentaries of the lower jaw. The teeth are small, conical and **acrodont**, *i.e.,* they are attached to the outer borders of the bones of jaws. The palate bears no teeth.

3. Pharynx. The buccal cavity is continuous with the pharynx. The pharynx lies posterior to the tongue. The mucous lining of the pharynx is thrown into longitudinal folds. The floor of pharynx contains a median longitudinal slit called the **glottis** leading into trachea. It is present in between the posterior bifurcation of the tongue. Near the jaw angles are present rounded apertures of eustachian tubes, one on either side. Pharynx leads posteriorly into the oesophagus through a wide **gullet**. It only opens when food is swallowed.

4. Oesophagus. The oesophagus is a long, narrow, straight cylindrical muscular tube. It is capable of distension at the time of food swallowing. There

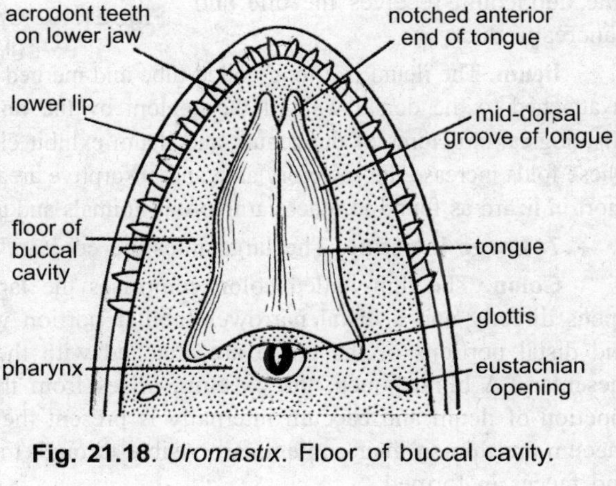

Fig. 21.18. *Uromastix*. Floor of buccal cavity.

are numerous longitudinal folds present on the entire inner surface of the oesophagus. It leads into the wide cylindrical stomach through long neck.

5. Stomach. The stomach is a long cylindrical sac-like structure, wider than the oesophagus. It has thick muscular walls and lies in the left half of the body and is roughly U-shaped. The stomach is suspended in the body cavity by a fold of peritoneum called the **mesogaster**. The stomach is differentiated into two parts : anterior part is known as **cardiac stomach** which lies dorsal to the left lobe of the liver, and the posterior part is known as **pyloric stomach** which lies slightly to the right side. The liver is attached to the stomach by a thin fold of peritoneum called the **hepatic omentum**.

Histological studies of the stomach reveal that the walls of the stomach are much thicker than that of the oesophagus and the intestine. Numerous well developed longitudinal folds of mucosa are seen in the lumen of both cardiac and pyloric stomach but these are well developed in the cardiac stomach. The distal end of pyloric stomach is constricted and internally bears a **pyloric sphincter** or **valve** (a ring of circular muscles). It allows only the digested food to pass into the duodenum. The stomach is the seat of digestion.

6. Small intestine. It is a long, narrow, coiled tube and comprises an anterior **duodenum** and a posterior **ileum**.

Duodenum. The duodenum with the stomach forms a U-shaped structure. In between the duodenum and stomach is present the pancreas. The duodenum receives the bile and pancreatic duct.

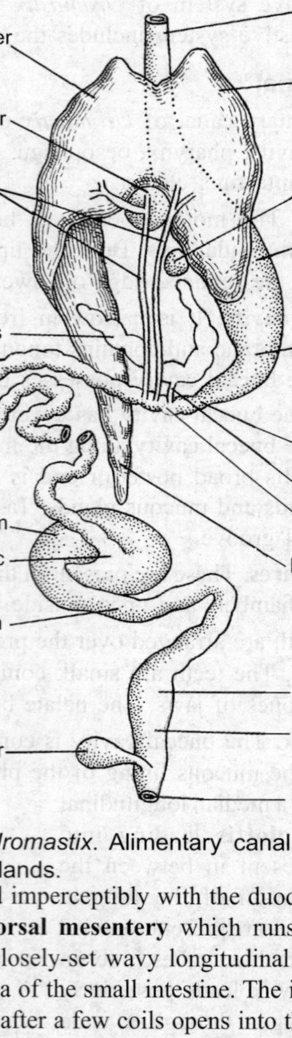

Fig. 21.19. *Uromastix*. Alimentary canal and digestive glands.

Ileum. The ileum is a long coiled tube and merged imperceptibly with the duodenum. The ileum is attached to the dorsal wall of the coelom by the **dorsal mesentery** which runs along the coils. Histological structures of duodenum and ileum exhibit closely-set wavy longitudinal folds of mucosa. These folds increase the secretory area and absorptive area of the small intestine. The ileum is relatively short in lizard as found in other carnivorous animals and after a few coils opens into the large intestine.

7. Large intestine. The large intestine consists of a proximal **colon** and a distal **rectum**.

Colon. The thin-walled colon comprises the large proximal portion into which the ileum opens directly and a distal narrower tubular portion which leads into the rectum. The proximal and distal portions of the colon are attached with the dorsal wall of the coelom by the dorsal mesentery. A blind pouch, the **caecum**, arises from the junction of the ileum and colon. At the junction of ileum and caecum internally is present the **ileo-coecal valve**. The walls of colon and caecum are thin so their contents are visible through them. Within the colon the water is absorbed and faeces in formed.

Rectum. The colon is continued into the tubular and narrower rectum. The rectum is very much thick-walled. It takes a sharp curve upwards towards the left and again turns sharply in the posterior direction. The faeces is stored in the rectum. The rectum opens into the **cloaca** and both are separated by a sphincter. The cloaca opens to the exterior by the **cloacal opening**.

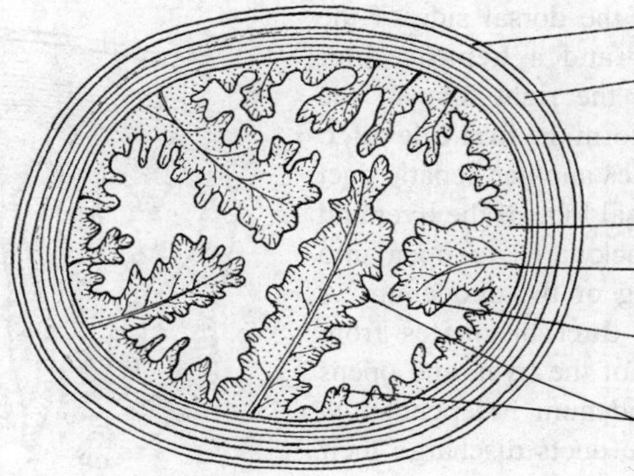

Fig. 21.20. *Uromastix.* T. S. of cardiac stomach.

8. Cloaca. The cloaca is divisible internally into three chambers, the **coprodaeum**, the **urodaeum** and the **proctodaeum**. The anterior chamber coprodaeum receives the rectum. The middle chamber urodaeum dorsally receives a pair of ureters and gonoducts and ventrally the urinary bladder. These two chambers open into the proctodaeum, closed by a cloacal sphincter. This division of the cloaca is associated with the necessity for the retention of water. The posterior chamber proctodaeum opens externally by the **cloacal aperture**. It is a transverse slit located ventrally at the junction of trunk and tail.

The cloaca serves for reabsorption of water from the faeces and urine. Such conservation of water is a terrestrial adaptation as the lizard lives in an arid land with scarcity of water. The faeces and urine are converted into dehydrated cylindrical castings or **pellets** (Fig. 21.23.) which are egested through the cloacal aperture.

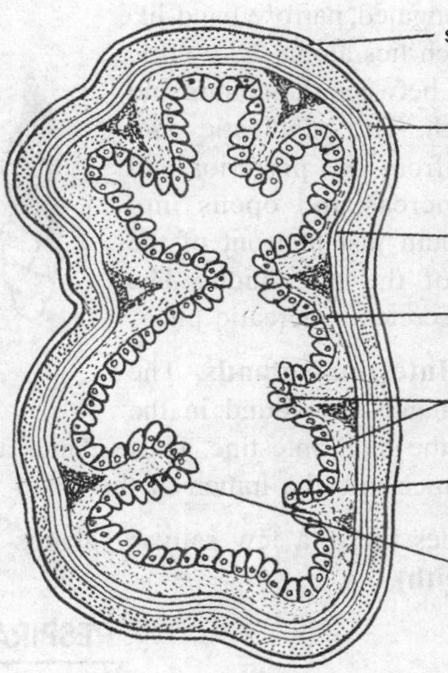

Fig. 21.21. *Uromastix.* T. S. of stomach.

Digestive Glands

In *Uromastix*, the digestive glands associated with the alimentary canal are : ***(i)* gastric glands**, **(*ii*) liver**, **(*iii*) pancreas**, and **(*iv*) intestinal glands**.

(*i*) Gastric glands. The gastric glands are microscopic, simple or branched pits in the mucous membrane of the stomach. Since they are microscopic, they are, therefore, not visible. These glands discharge **gastric juice** into the cavity of stomach.

(*ii*) Liver. The liver is a large dark brownish-red gland situated a little posterior to the heart between the lungs. It is connected with the stomach by a thin membranous fold, the **gastrohepatic omentum**. The gastrohepatic omentum arises from the dorsal surface of the liver and is continued backward as the **duodeno-hepatic omentum** connecting the liver with the duodenum. The liver consists of three lobes : right, left and dorsal. The **right lobe** is long and narrow extending up to the right gonad. The **left lobe** is short and broad lying ventral to the cardiac stomach. The **dorsal lobe** is small, found attached to the dorsal side of the left lobe. The **gall bladder** is a small rounded sac situated ventrally on the posterior end of the junction between the left and right lobes of the liver. The liver cells secrete bile which is stored in gall bladder. A **cystic duct**

arising from the dorsal side of the gall bladder and a **hepatic duct** arising from the right lobe of the liver join to form the **first bile duct**. It also receives another hepatic duct from the dorsal lobe of the liver and then runs backwards to open into the beginning of the duodenum. A **second bile duct** originates from the left lobe of the liver, and opens into the duodenum independently. Both the bile ducts discharge their contents into the duodenum behind the opening of the pancreatic duct.

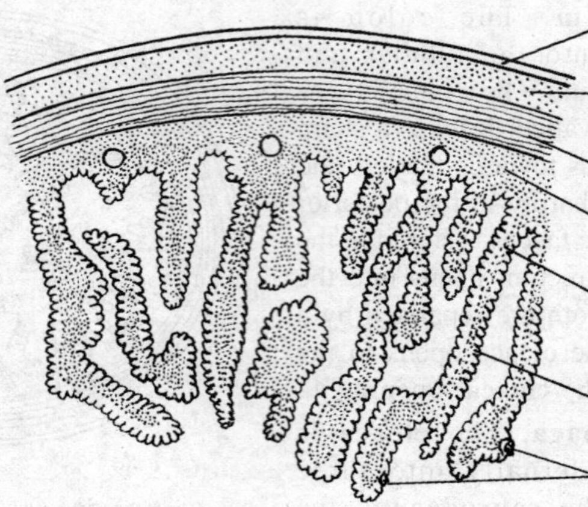

Fig. 21.22. *Uromastix*. T. S. of duodenum.

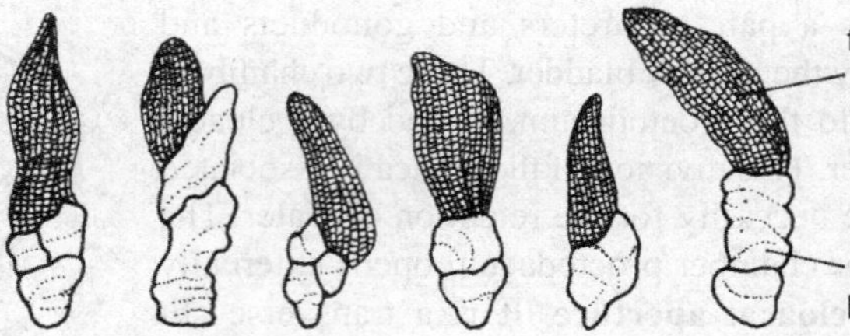

Fig. 21.23. *Uromastix*. Faecal pellets.

***(iii)* Pancreas.** The pancreas is a white, elongated, narrow, band-like gland, which lies along the pyloric stomach in between duodenum and the stomach. The pancreatic duct originates from the posterior end of the pancreas and opens into the duodenum just in front of the openings of the bile ducts. The pancreas secretes pancreatic juice.

***(iv)* Intestinal glands.** The intestinal glands are found in the mucosa of the small intestine. These glands are numerous, microscopic and invisible. They secrete intestinal juice into the lumen of the small intestine.

Besides these, a few salivary glands are also present in the buccal cavity which secrete mucus **(Keith)**.

RESPIRATORY SYSTEM

In amphibians, the respiratory interchange of gases takes place through the delicate moist skin, but this is not possible in reptiles whose dry skin possesses tougher horny layer. The modern reptiles do not possess gills at any stage of life, therefore, respiration by gills is not possible. However, on the walls of the temporary embryonic pharyngeal clefts have been observed minute papillae, which have been interpreted as vestiges of gill-filaments. The functional gills do not occur in any reptilian embryo. Therefore, the process of respiration in adult reptiles is entirely **pulmonary** except in some cases (probably sea turtles) where the **buccopharyngeal** respiration also takes place.

The respiratory system of *Uromastix* consists of two main portions, *viz.,* **respiratory tract** and **respiratory organs**.

Respiratory Tract

The respiratory tract starts from the **external nares**. These are a pair of small oval apertures on the snout slightly above the mouth. On its inner side each naris possesses a prominent swelling of mucous membrane which serves as a valve to close the aperture during burrowing. The external nares lead into small sacs, the **nasal chambers** or **olfactory sacs**, lying in the skull. The nasal chambers open posteriorly by a pair of small apertures, the **internal nares**, into the anterior part of a spacious **buccopharyngeal cavity**. On the floor of buccopharyngeal cavity between the basal bifurcations of the tongue lies a small median longitudinal slit called the **glottis** which leads into a short median chamber called **larynx**. The wall of larynx is supported by a pair of **arytenoid** and a median **cricoid cartilages**. The arytenoid cartilages are situated in the antero-lateral wall of the larynx and give support to the lips of glottis. Posteriorly these two cartilages articulate with the antero-lateral sides of the cricoid cartilage. In *Uromastix*, the vocal cords are lacking. The larynx is provided with two pairs of muscles : the inner **musculus compressor laryngis** which serves as a sphincter of the glottis and the outer **musculus dilator laryngis** which serves as a dilator of glottis. The larynx leads posteriorly into the trachea; which is an elongated cylindrical tube running backwards through the neck ventral to oesophagus. The wall of the trachea is supported by a series of complete cartilaginous rings throughout its length which prevent the trachea to collapse. Some of the tracheal rings are bifurcated. After entering the thorax the trachea almost reaches the base of the heart, where it bifurcates into two very short, narrow tubes, the **bronchii**, which are also supported by complete cartilaginous rings. Each bronchus enters the lung of its side and does not branch into **intrapulmonary extensions** or **bronchioles**.

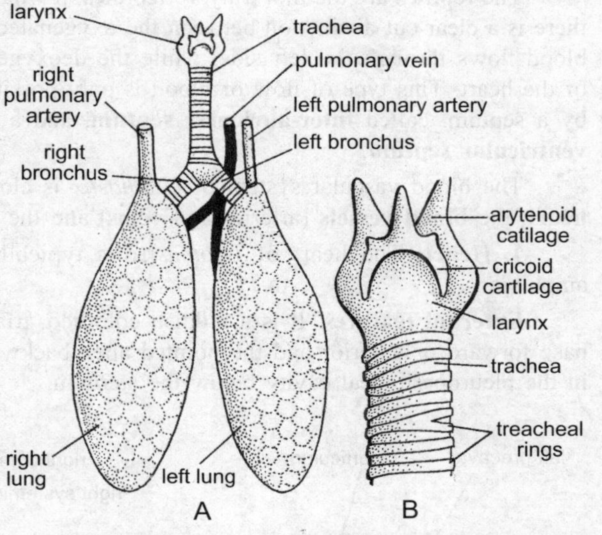

Fig. 21.24. *Uromastix.* A – Respiratory system in ventral view; B – Larynx and trachea in dorsal view.

Respiratory Organs

The respiratory organs of *Uromastix* are a pair of **lungs**. They are situated one on either side of the heart in the thoracic cavity. Both the lungs are symmetrical in shape and size. They are elongated, fusiform, thin-walled elastic hollow sacs of orange colour. The posterior ends of lungs appear pointed when they are empty but rounded when they are inflated. The inner lining of the lungs is raised into a network of low ridges which enclose shallow depressions or alveoli. The cavity of the lung is continuous. These ridges are more numerous and closer in the anterior part of the lung than in the posterior part. The ridges increase the respiratory surface of lungs. The walls of the lungs and alveoli have a rich supply of blood-capillaries. The blood supply is more abundant in the anterior part of the lungs than in the posterior part. Therefore, the anterior parts of the lungs are more respiratory than the posterior parts. As earlier described, the bronchi do not branch into bronchioles in the lungs. The blood entering the lungs through pulmonary artery is deoxygenated and the oxygenated blood is carried to the heart by pulmonary vein.

Mechanism of Respiration

In *Uromastix*, the mechanism of respiration is different from that of the frog. The process of respiration is effected only by ribs and their inter-costal muscles. The inter-costal muscles, which extend between the ribs, pull the ribs backward and forward and, thus, the pleuro-peritoneal cavity is decreased or increased in size. When the cavity is increased, the lungs expand and fresh air rushes into the lungs from outside through the nostrils, buccal cavity, glottis, trachea and bonchii. When the cavity is decreased, the lungs are pressed and the air is forced out through the same above passage. Exchange of gases takes place in the alveoli through their thin moist walls. In *Uromastix*, the respiration is only pulmonary.

BLOOD VASCULAR SYSTEM

The reptiles are the first truly vertebrates in which pulmonary respiration occurs. Therefore, there is a clear-cut distinction between the oxygenated and deoxygenated blood. The oxygenated blood flows through the left side, while the deoxygenated blood passes through the right side of the heart. This type of flow of blood is only possible by the complete division of the auricle by a septum, called **inter-auricular septum**, and a partial division of ventricle by the **inter-ventricular septum**.

The blood vascular system of *Uromastix* is closed and consists of three main parts : the **heart**, the **blood vessels** (arteries and veins) and the **blood**.

1. Heart. The heart of *Uromastix* is typically reptilian and more advanced than the amphibians.

External features. It is a reddish coloured, triangular and muscular organ with its broad base forward or anterior and the pointed apex backward. It lies forward in the mid-ventral line in the pleuroperitoneal cavity below the sternum.

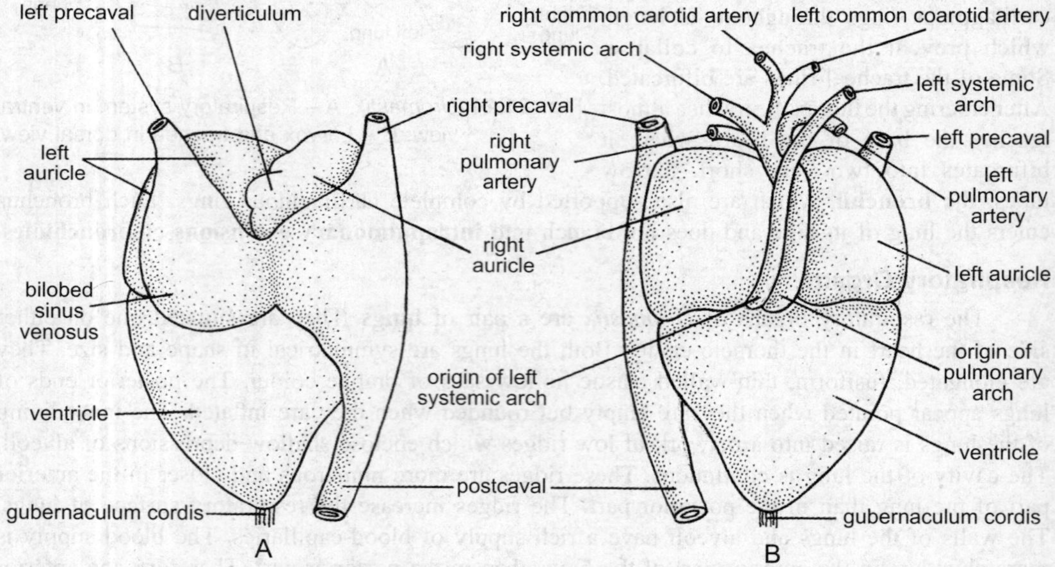

Fig. 21.25. *Uromastix*. External features of heart. A – Dorsal view, B – Ventral view.

(*i*) Pericardium. The heart is enclosed in a thin, firm, two-layered transparent sac, the **pericardium**, to which it is attached at its apex by a small cord, the **gubernaculum cordis**. It keeps the heart in position within the pericardium. The outer layer of pericardium is called the **parietal pericardium**, while the inner layer is termed **visceral pericardium**. The parietal and visceral pericardium are continuous at the anterior end of the heart. Between the two layers of pericardium is a narrow space, the **pericardial cavity**, which contains a watery **pericardial fluid**. The pericardial fluid protects the heart from mechanical injuries and shocks and also allows it free movement.

(*ii*) Heart chambers. The heart consists of four chambers : a large dorsal **sinus venosus**, two **auricles** (right and left) and a **ventricle**.

(*a*) **Sinus venosus.** The sinus venosus is a large thin-walled and bilobed chamber situated on the dorsal side of the heart covering both the auricles from above. It is formed by the union of a pair of precavals and a single postcaval veins. By a deep dorsal groove the sinus venosus is divided into two lobes, a **larger right** and a **smaller left**. The sinus venosus receives the deoxygenated blood from the body through the two precavals and a postcaval which open into the sinus by separate apertures.

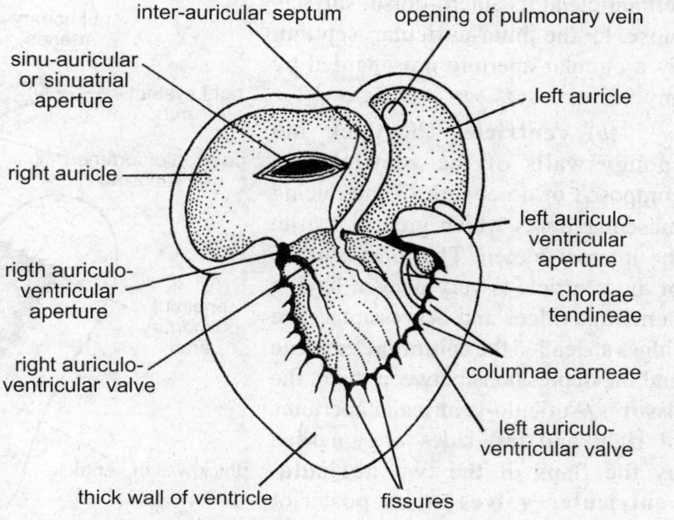

Fig. 21.26. *Uromastix*. Internal structure of heart in ventral view.

(*b*) **Auricles.** There are two auricles : the **right** and the **left**. The two auricles are externally separated by a longitudial **interauricular groove**. The right auricle is larger than the left. From its antero-dorsal edge, a small sac-like diverticulum is given off. This is said to be a normal feature of the heart in lizards. Its function is not known. The internal lining of the right auricle is raised into numerous interlacing strands of muscular fibres called **musculi pectinati**. The right auricle opens into the lumen of the right side of the ventricle through the right side of the **auriculo-ventricular opening** which is guarded by the **auriculo-ventricular valves**. A small posterior inner margin of the right auricle extends into the lumen of the ventricle. This portion of the right auricle which projects into the right side of the ventricle is called the **atrium dextrum intraventriculare**. The sinus venosus opens into the right auricle through a transversely elongated aperture called **sinu-auricular aperture** which is situated in the median dorsal wall of the right auricle. The sinu-auricular aperture is guarded by two lip-like **sinu-auricular valves** which project into the lumen of the right auricle.

According to **Bhatia** (1929), sinu-auricular valves are absent in *Uromastix hardwikii*. Only thick muscular lips keep the aperture ordinarily closed. It only opens when blood is forced through it into the right auricle.

(*c*) **Ventricle.** The ventricle is a conical chamber whose walls are thick and spongy. It is separated from the auricles by a transverse **auriculo-ventricular groove** or **coronary sulcus**. Its apex is bluntly rounded and points a little to the right of the median axis of the heart and is attached to the pericardium by the **gubernaculum cordis**. The base of the ventricle is more or less straight. Its dorsal surface is slightly concave but the ventral is greatly convex.

Internal structure. It can be studied by cutting the section of heart.

(*a*) **Auricles.** The auricles are thin-walled. Both the auricles are separated from each other by a thin, muscular, unperforated and slightly curved partition, the **inter-auricular septum**. It is attached at its anterior end to the wall of auricles a little left of the diverticulum. At its posterior end the septum is free and reaches the lumen of the ventricle dividing the auriculo-ventricular aperture into right and left portions.

The **left auricle** is smaller than the righ. It is also drawn out posteriorly into the lumen of the ventricle to form the **atrium sinistrum intraventriculare**. The pulmonary vein opens into the left auricle at its antero-dorsal surface, close to the inter-auricular septum, by a circular aperture not guarded by any valve.

***(b)* Ventricle.** The thick and spongy walls of the ventricle are composed of a network of interlacing muscular fibres which greatly restrict the internal lumen. The inner surface of the ventricle is very uneven having numerous ridges and depressions. The ridges are called the **columnae carneae** and the depressions between them, the **fissures**. Auriculo-ventricular apertures of right and left sides are guarded by the flaps of the two **auriculo-ventricular valves** at its posterior end. These two valves guard their respective portions of the **auriculo-ventricular apertures**. The free edges of the auriculo-ventricular valves are connected to the wall of the ventricle by firm cords called the **chordae tendineae**. A conspicuous flap, called the **muscular ridge** or **interventricular septum**, arises from the ventral wall of the ventricle on the left near the apex and extends horizontally forwards towards the right. It divides the cavity of the ventricle into a large dorsal part, the **cavum dorsale**, and a small ventral part, the **cavum pulmonale**. The two parts keep communicating with one another over the margin of the ridge. Three aortic arches originate from the ventricle : a **pulmonary arch** and two **systemic arches** (right and left). The pulmonary arch takes its origin from the cavum pulmonale, while the systemic arches arise from the cavum dorsale. Each of the three arches has a pair of **semilunar valves** at its base. The cavities of these valves face away from the ventricle to prevent the back-flow of blood.

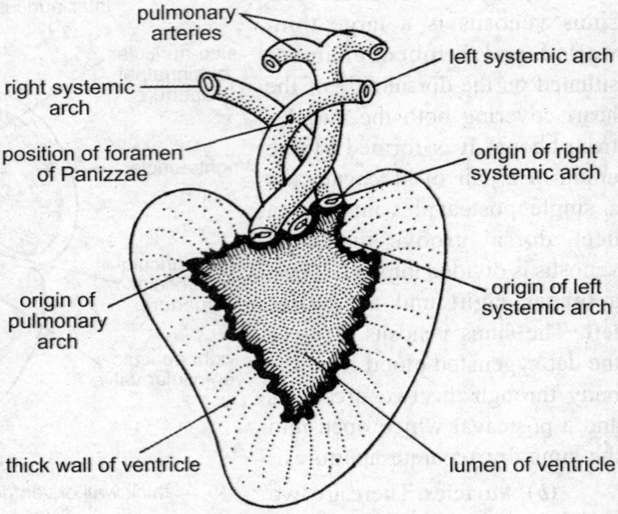

Fig. 21.27. *Uromastix*. Origin of aortic arches from ventricle.

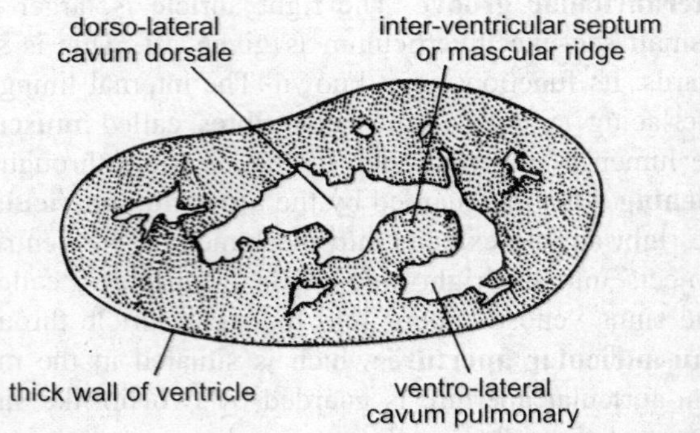

Fig. 21.28. *Uromastix*. T.S. of ventricle to show muscular ridge.

Working of Heart

The sinus venosus receives deoxygenated blood from the whole body by way of three venae cavae. When the sinus venosus contracts, driving its blood into the right auricle through the sinu-auricular aperture. Meanwhile, the left auricle receives oxygenated blood from the lungs through the pulmonary vein. Now the auricles contract almost simultaneously forcing the deoxygenated blood from the right auricle into the cavum pulmonale (right side of the ventricle). During this process, return of blood to the sinus venosus is prevented by the sinu-auricular valves. From the left auricle the oxygenated blood enters the cavum dorsale (left side of the ventricle). In the middle portion of the ventricle, some mixing of pure and deoxygenated blood takes place. Now the ventricle contracts in such a way that the muscular ridge keeps the deoxygenated and oxygenated blood separate. The contraction of ventricle sends the deoxygenated blood from the cavum pulmonale into the pulmonary arch and the oxygenated and partly mixed blood from the cavum-dorsale into the systemic arches which is distributed to the entire body. The head receives the purest blood. The return of blood from the ventricle into the auricles, during the contraction of ventricle, is prevented by the auriculo-ventricular valves. The return of blood from the aortic arches in the ventricle is checked by the semilunar valves at the base of arches. Therefore, it can be said that the heart of *Uromastix* is functionally four-chambered.

The blood circulation in the body is as follows :

Systemic arches → Body → Sinus venosus
↓
Sinu-auricular aperture
↓
Right auricle
↓
Auriculo-ventricular aperture
↓
Cavum pulmonale (right-ventricle) → Pulmonary arch → Lungs → Pulmonary vein
↑ (up from Pulmonary vein)
Sinu-pulmonary aperture
↑
Left auricle
↑
Auriculo-ventricular aperture
↑
Cavum dorsale
↑
Systemic arches

HEART

2. Blood Vessels. There are two types of blood vessels, *viz.,* **arteries** and **veins**.The arteries carry blood from the heart to the various organs of the body and collectively form the **arterial system**. The veins collect blood from various organs of the body and return it to the heart and collectively form the **venous system**. The arteries and veins are connected by way of capillaries.

I. Arterial System

There are three aortic arches which arise independently from the ventricle, a little to the right of the middle line, *viz.,* the **pulmonary arch** and a pair of **systemic arches** (right and left systemic arches).

1. Pulmonary arch. The pulmonary arch lies ventrally and arises from the right ventral side of the ventricle (cavum pulmonale). As it travels forwards, it curves and becomes dorsal to the left systemic arch, where it soon divides into **right pulmonary** and **left pulmonary arteries** going to the right and left lungs respectively.

2. Systemic arches. Both the systemic arches arise directly from the cavum dorsale of the ventricle carrying oxygenated blood. The **left systemic arch** is fully visible in the ventral view of the heart. It arises from the right side of the ventricle but curves later to the left and carries mixed blood. The point of origin of the **right systemic arch** is not visible from the ventral side but can be seen in the dorsal view. The **right systemic arch** arises from the left side of the ventricle and lies dorsal to both the pulmonary and the left systemic arches. Later it curves to the right side. The two systemic arches communicate with one another by a **foramen of Panizzae**, situated where the arches cross each other anterior to the heart. Both the systemic arches then turn upwards, backwards and inwards and join posterior to the heart in the mid-dorsal line to form **dorsal aorta**.

The right systemic arch gives off a very short artery before it starts curving to the right in the notch between the two auricles which is called the **carotis primaria** or **innominate artery**.The right systemic arch also gives off a **common subclavian artery** before joining the left systemic arch.

(*a*) **Common carotid arteries.** The innominate arises from the right systemic arch before curving to the right side. It bifurcates to form the **right** and **left common carotid arteries**. Each common carotid runs forward for some distance parallel to the systemic arch of its side and then divides into the **external** and the **internal carotid arteries**. The external and the internal carotid arteries extend forwards through the neck, both of which branch into numerous arteries supplying oxygenated blood to the head region. Each internal carotid artery is connected with the systemic arch of its side by a vessel called the **ductus caroticus**. It is a remnant of embryonic lateral aorta. Since carotid arteries arise from the right systemic arch, so they are also known as **carotico-systemic arches**. The **ductus arteriosus** or the **ductus Botalli** (which usually connects the pulmonary artery and the systemic arch) is absent in *Uromastix hardwickii*.

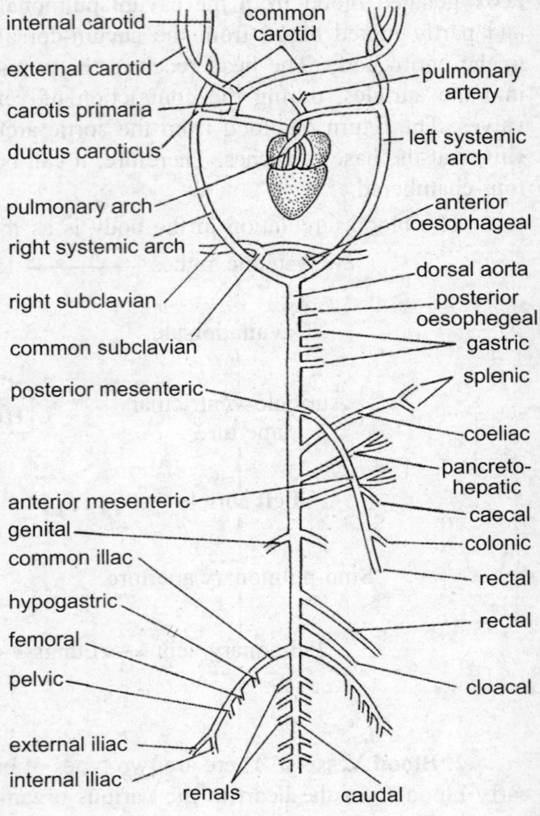

Fig. 21.29. *Uromastix*. Arteries.

(*b*) **Common subclavian artery.** It arises from the right systemic arch before it joins the left systemic arch. It soon divides into the **right** and the **left subclavian arteries** which run outwards to supply the respective forelimb. Each subclavian artery gives off three vessels, *i.e.,* **scapular** to the scapular region, **coracoid** to the coracoid muscles and **brachial** to the arm.

(*c*) **Anterior oesophageal artery.** The left systematic arch also gives off a small **anterior oesophageal artery** which supplies the anterior part of the oesophagus. **Shah** also described posterior oesophageal artery which arises from the dorsal aorta. However, **Bhatia** has figured only one oesophageal artery.

3. Dorsal aorta. The dorsal aorta is formed by the union of the right and left systemic arches. It extends straight backwards in the mid-dorsal line beneath the vertebral column, giving off the following arteries :

(*i*) **Posterior oesophageal artery.** Immediately after the formation of the dorsal aorta a small unpaired artery arises from it which is called the posterior oesophageal artery. It supplies blood to the posterior part of the oesophagus.

(*ii*) **Gastric arteries.** Three small unpaired gastric arteries arise behind the posterior oesophageal artery. These arteries supply blood to the anterior, middle and posterior part of stomach and are called **anterior**, **middle** and **posterior gastric arteries**.

(*iii*) **Posterior mesenteric artery.** The posterior mesenteric artery is large and unpaired, arises far forward at the level of the fifth parietal, just in front of the origin of coeliac. It runs backwards and distributes blood to the large intestine by three branches : the **caecal**, **colic** (colon) and **rectal**.

***(iv)* Coeliac artery.** It arises behind the posterior mesenteric and runs forwards, crosses the posterior mesenteric and supplies blood to the spleen and to stomach by two branches, *viz.*, the **spleenic** and **gastric**.

***(v)* Pancreo-hepatic artery.** It is a fairly large unpaired vessel supplying blood to the duodenum, liver and pancreas.

***(vi)* Anterior mesenteric or intestinal artery.** It arises behind the pancreo-hepatic artery and supplies blood to the entire small intestine by a number of branches.

***(vii)* Genital or gonadial arteries.** These are paired, arise near the tenth parietal and supply blood to the gonads and suprarenal glands. They are called **spermatic arteries** in the male and **ovarian arteries** in the female.

***(viii)* Rectal artery.** It is an unpaired artery, arises in between the thirteenth and fourteenth parietals. It passes down along the dorsal wall of the rectum and supplies blood to it.

***(ix)* Cloacal artery.** It is an unpaired median artery which arisés close to the common iliac arteries. It supplies blood to the lateral wall of the cloaca and the terminal portion of the rectum.

***(x)* Common iliac arteries.** A pair of large common iliac arteries arise near the middle region of the kidneys and enter the hindlimbs. Before entering the limbs, each common iliac artery gives off a **hypogastric artery** to the urinary bladder, a **femoral artery** to the thigh muscles and a **pelvic artery** to the **pelvic girdle** and the fat bodies. It supplies the hindlimbs by two branches, *viz.*, **external** and **internal iliac arteries**.

***(xi)* Renal arteries.** Three pairs of small renal arteries supply blood to the kidneys.

***(xii)* Caudal artery.** The posterior end of the dorsal aorta continues behind as the caudal artery. The caudal artery runs through the haemal canal of the caudal vertebrae.

***(xiii)* Parietal arteries.** During its course, the dorsal aorta also gives off about 15 pairs of small parietal arteries, supplying blood to the trunk muscles and caudal muscles.

II. Venous System

The venous system of *Uromastix* is most primitive. The blood from the lungs is returned to the left auricle by two pulmonary veins. The blood from the other parts of the body is returned by three large veins. The two **precavals** (right and left) return the blood from the anterior side, while the **postcaval veins** return the blood from the posterior side and enter into the bilobed sinus venosus which joins the right auricle. Thus, the venous system of *Uromastix* has the following main veins :

1. Precaval veins;
2. Postcaval vein;
3. Hepatic portal vein;
4. Renal portal vein.

1. Precaval veins. There are two precaval veins, the right and the left, which drain the blood from the anterior body region—head, neck, shoulders, forearms and thoracic wall. Each precaval is formed by the union of four veins : ***(i)* common jugular, *(ii)* subclavian, *(iii)* intercostal** and ***(iv)* laryngo-tracheal**.

***(i)* Common jugular vein.** The common jugular vein is formed by the union of **external** and **inernal jugular veins** at the level of the tympanum. The **external jugular** vein receives blood from the upper jaw by a **maxillary vein** and from the lower jaw by a **mandibular vein**. The **internal jugular** receives blood from the brain by a **cerebral vein** and from the orbital region by an **orbital vein**. The common jugular runs back into the neck where it receives blood from the auditory region of the head by a **temporal vein**. A little behind the junction of common jugular vein and temporal vein, there is a large sinus (expansion) called **jugular sinus**. **Das** (1960) has reported displacement of some elements of the external jugular. This view is also confirmed by the embryological studies of **Grossar** and **Brezina**. However, some authors including **Bhatia** (1929) described that there is no external jugular on the left side and according to them the left precaval is formed by the union of an internal jugular and a subclavian only.

***(ii)* Subclavian vein.** The large **subclavian vein** runs inwards from the forelimb and joins the posterior end of jugular sinus. The subclavian vein receives blood from the arm through three small veins: **(*a*) scapular** from the shoulder region, **(*b*) brachial** from the arm, and **(*c*) axillary** from the arm-pit and thorax.

***(iii)* Intercostal vein.** The intercostal vein collects blood from the skin and muscles of the chest and runs forwards to meet the precaval vein internal to subclavian. By the union of common jugular, temporal, subclavian and intercostal, a large precaval vein is formed that opens into the sinus venosus.

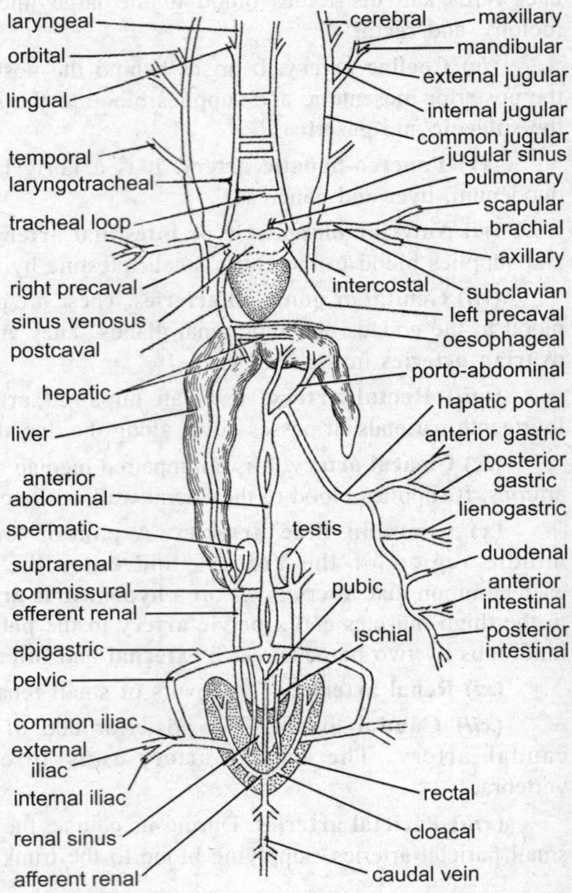

Fig. 21.30. *Uromastix*. Veins.

***(iv)* Laryngo-tracheal vein.** The laryngo-tracheal vein is a very slender vessel. It receives a very fine **lingual vein** from the tongue and a **laryngeal vein** from the larynx and trachea. It then passes backwards alongside of the trachea and joins the precaval close to its entry into the sinus venosus. The two laryngo-tracheal veins are symmetrical running parallel to the trachea and are inteconnected by four or five transverse tracheal loops. **Bhatia** (1929) described that the left laryngo-tracheal vein does not unite with the left precaval.

2. Postcaval vein. The postcaval vein opens into the right posterior angle of the sinus venosus. It collects blood from the posterior region of the body, *i.e.,* kidneys, gonads and liver. It arises from the region of kidneys. Blood from the kidneys is collected through capillaries in a thin-walled **renal sinus** lying in the middle region of the kidneys. A pair of **efferent renal veins** commence from the renal sinus and pass forwards along the sides of the vertebral column up to the gonads, being separated from one another by a pad of connective tissue. The right efferent renal vein is thicker than the left. Both the efferent renal veins receive blood from dorsal bodywall through a few small veins in the way. Just behind the gonad, the left efferent renal vein sharply bends medially and joins the right efferent renal vein to form the **commissural vein**. A small genital vein from the gonads and a fine supra-renal vein from the supra-renal gland on each side pours into the commissural vein or the efferent renal vein of that side. Only the right efferent renal vein proceeds as the **postcaval vein**, which passes through the right lobe of the liver and opens into the sinus venosus. The postcaval vein during its course also receives two or more hepatic veins from the liver and a small **vertebral vein** from the backbone. A small **oesophageal vein** from the oesophagus joins the postcaval after its emergence from the liver.

3. **Pulmonary vein.** There is a single pulmonary vein formed by the union of several veins from both the lungs. It runs along the ventro-lateral border of the trachea and passes dorsally to the sinus venosus before opening into the left auricle close to the inter-auricular septum. It is the only vein which carries oxygenated blood.

4. **Hepatic portal vein.** Hepatic portal vein is a large vein which collects the blood from the alimentary canal by a number of fine branches and carries it to the liver. It is formed by the union of numerous veins from the different regions of the alimentary tract such as the **anterior** and **posterior gastric veins** from the stomach, **lienogastric vein** from the spleen and a part of the stomach, **duodenal vein** from the duodenum, **anterior intestinal** from the ileum and **posterior intestinal** from the caecum, colon, and rectum. The hepatic portal vein and all its tributaries run in the mesentery. Anteriorly, the hepatic portal vein receives a branch of the anterior abdominal vein called the **porto-abdominal vein** and enters the middle of the left lobe of the liver where it capillarises. Blood from the liver as described earlier is collected by the hepatic veins and poured into the postcaval. The hepatic portal system is formed by the hepatic portal vein and porto-abdominal vein.

5. **Renal portal and anterior abdominal veins.** The **caudal vein** starts from the tip of the tail and runs in the haemal canal of the caudal vertebrae receiving on its way several paired branches from the tail muscles. The caudal vein after reaching the kidneys divides into a pair of diverging **renal portal** or **afferent renal veins**. The afferent renal veins pass over the ventral surface of the kidneys partly buried in their substance. They receive **cloacal** and **rectal veins** from the cloaca and rectum respectively. The renal portal vein of each side continues forward into the kidney, its side and ramifies into capillaries.

Blood from each hindlimb is collected by the **internal** (sciatic) and **external iliac** (femoral) **veins**. These two unite after entering the abdominal cavity to form a **common iliac vein** which is connected to the renal portal vein by a transverse connection (iliac-afferent union) near the middle of the kidney. Each common iliac vein extends forwards as the **pelvic** or **lateral vein** along the dorsal surface of the large yellow fat body. The pelvic vein receives in the way fine **adipose veins** from the fat body, **ischial vein** and **epigastric vein** from the dorsal bodywall, **vesicular vein** from the urinary bladder and **parietal veins** from the posterior muscles. A fine median pubic vein joins both the pelvic veins. The two pelvic veins converge in front of the kidneys and unite in the mid-ventral line to form the **anterior abdominal vein**. The anterior abdominal vein is formed by the imperfect union of two pelvic veins. In *Uromastix*, the anterior abdominal vein is double at places indicating a primitive feature of fishes where two lateral abdominal veins have not fused along their entire course. Anteriorly the two components of the anterior abdominal vein remain separate, one of the anterior abdominal vein joins with the hepatic portal vein and enters the left lobe of the liver as **porto-abdominal vein**, while another enters the left lobe of liver directly.

Blood

Uromastix (reptiles) are cold-blooded or **poikilothermous**, like fishes and amphibians. Red cells are nucleated and survive for much longer than those of mammals. The total haemoglobin concentrations and oxygen carrying power of the blood are only half those of mammals. The white cells of reptiles include the same types as in other vertebrates.

NERVOUS SYSTEM

The nervous system in *Uromastix* is generally on the amphibian plan. Many differences in the plan of nervous system between the two groups are not apparent anatomically but reptiles, however, indicate a higher degree of nervous efficiency. One of the major differences is the absence of lateral line system in reptiles. The absence of this system has been fully compensated by improvements in the sensory organs. Like other vertebrates, the nervous system of *Uromastix* consists of the following divisions :

1. Central nervous system includes the brain and spinal cord.
2. Peripheral nervous system includes the cranial and spinal nerves.
3. Autonomic nervous system includes a pair of ganglionated lateral trunks.

Central Nervous System

The central nervous system comprises the **brain** and **spinal cord**.

1. Brain

In *Uromastix,* the brain is very simple and similar to that of the frog, however, it shows some advancement over that of the frog. The brain lies quite loosely in the cranial cavity lined with a tough fibrous membrane, **duramater**. Another membrane which surrounds the brain is called **piamater**. In between the two layers **arachnoid fluid** is present which protects the brain. The brain of *Uromastix* is divisible into the three usual parts : the **forebrain**, **midbrain** and **hindbrain**.

(*i*) Forebrain. The forebrain forms the anterior part of the brain. It consists of large swollen and elongated **cerebral hemispheres**, **olfactory bulbs** and **diencephalon**.

(*a*) Cerebral hemispheres. The cerebral hemispheres are two oval bodies, somewhat narrow in front but broad behind closely applied together in the median line but a dorsal **median longitudinal fissure** forms the demarcation between the two.

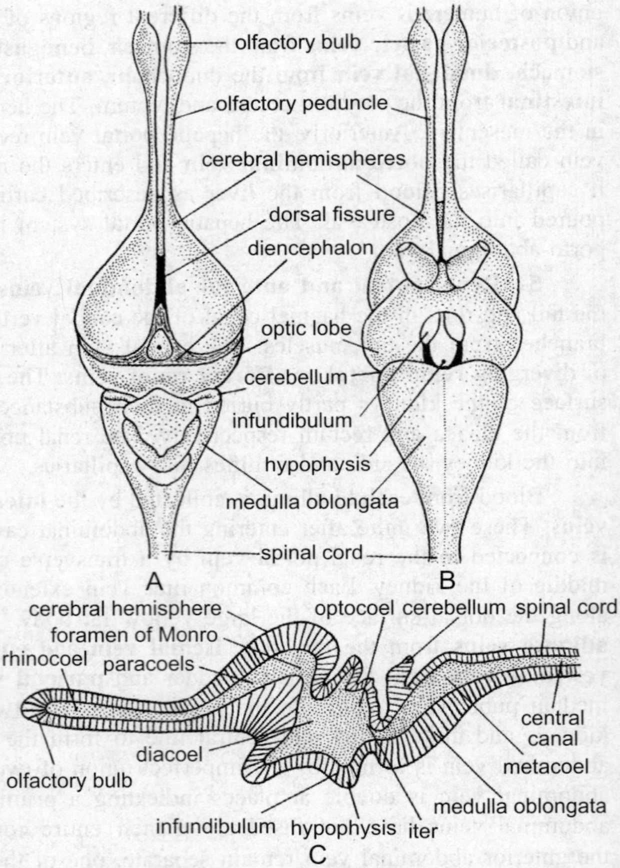

Fig. 21.31. *Uromastix.* Structure of brain. A – Dorsal view; B – Ventral view; C – Sagittal section showing ventricles.

(*b*) Olfactory bulbs. A pair of smaller olfactory bulbs and their posterior narrow stalks, the **olfactory peduncles** by which they are attached with the corresponding cerebral hemispheres, are the anteriormost parts of the forebrain. From olfactory bulbs originate the olfactory nerves. The cavity of each olfactory bulb is called **rhinocoel**. The roof of each cerebral hemisphere is thin and called the **pallium** and the thick ventral and lateral walls together constitute the **corpus striatum**. The pallium in *Uromastix* is, however, thicker than that of the amphibians. In *Uromastix* the cerebral hemispheres are large due to large fibrous tracts called **corpora striata**. The two corpora striata are connected by a tranverse **anterior commissure** which receives many efferent fibres from the side-walls of the diencephalon. Above the anterior commissure lies the **hippocampal commissure** which connects the hippocampal regions of the cerebral hemispheres. Each cerebral hemisphere encloses a cavity, the **lateral ventricle** or **paracoel** which is continuous in front with the **rhinocoel** of the olfactory lobe of its side.

(*c*) Diencephalon. The diencephalon is a small rounded area situated between the cerebral hemispheres and the midbrain. It is hardly visible from the dorsal side.The lateral walls of the diencephalon are thick forming the **optic thalami**. The thick floor forms the **hypothalamus**. The roof of the diencephalon is thin-walled and highly vascular forming the **anterior choroid plexus**. Behind the anterior choroid plexus lie the **habenular commissure** and **posterior commissure** which is a transverse band of nerve fibres. From the thin dorsal surface of the diencephalon arises a median outgrowth, the **pineal apparatus**, which consists of two parts : the anterior **parietal organ** or **parietal eye** lying in the parietal foramen, and the posterior **pineal body** or **epiphysis**.The parietal organ or parietal eye is considered to be functional eye. From the thick ventral surface of the diencephalon arises an **infundibulum** behind which is attached the **pituitary body** or **hypophysis**. **Optic chiasma** is situated anterior to the infundibulum. Optic chiasma is formed by the crossing of two optic nerves. Diencephalon encloses a laterally compressed cavity, the **diacoel** or **third ventricle**, which communicates with the lateral ventricles of the cerebral hemispheres by a small common aperture, the **foramen of Monro**.

(*ii*) Midbrain. The midbrain consists of two dorso-laterally placed rounded **optic_lobes**. On the ventral surface of the midbrain lie two bands of longitudinal, nerve fibres, stretching between the forebrain in front and the medulla oblongata behind, these are the **crura cerebri**. Each optic lobe contains a ventricle, the **optocoel**.The two optocoels open into a narrow median canal, the **iter**, which connects the diacoel (cavity of diencephalon) in front and the metacoel (cavity of cerebellum) behind.

(*iii*) Hindbrain. The hindbrain comprises the **cerebellum** and the **medulla oblongata**.

(*a*) Cerebellum. The cerebellum is poorly developed as in the frog. It is merely a flattened semicircular ridge at the anterior dorsal surface of the medulla oblongata.

(*b*) Medulla oblongata. The medulla oblongata is the hindermost part of the brain. It is broad in front and narrow behind. It has a prominent **ventral flexure** where it passes into the spinal cord. The roof of the medulla is thin, vascular and folded forming the **posterior choroid plexus.** It encloses a cavity, the **fourth ventricle** or **metacoel** or **myelocoel.** Its cavity is connected with the diocoel through a narrow median passage, the **iter**.

All the ventricles of the brain are continuous and posteriorly connected with the cavity of the spinal cord and full of a watery fluid, the **cerebro-spinal fluid** secreted by the choroid plexuses.

The brain of *Uromastix* shows relative increase in size of the cerebral hemispheres. But the olfactory bulbs are reduced to a smaller size. The increase in the size of the cerebral hemispheres indicates an enlargement of instinctive activities. The behaviour of reptiles is certainly on a higher level than that of the amphibians in general. The reptiles respond more quicker than the amphibians.

Histology of Brain

The brain is composed of two types of nervous tissue, *viz.,* the **grey matter** and **white matter.** The grey matter is composed of nerve-cells and the white matter consists of nerve fibres. In the forebrain the grey matter is external and white matter is internal, while in the midbrain and medulla oblongata the arrangement is reverse of this.

Functions of Brain

The **olfactory lobes** control the sense of smell.

The **cerebral hemispheres** get the impulses from the sense organs and initiate motor impulses for voluntary movements. The cerebral hemispheres are the seat of intelligence, memory, emotions and will.

The **diencephalon** acts as a relay station in conveying the impulses to the cerebral hemispheres and as integrating centre for the autonomic nervous system.

The **optic lobes** are concerned with the sense of sight. The crura cerebri carry the sensory impulses from the medulla oblongata and the spinal cord to the optic thalami and thence to the cerebral hemispheres.

The **cerebellum** coordinates the voluntary movements and also maintains the equilibrium.

The **medulla oblongata** controls the involuntary movements. Therefore, it is regarded as the most important part of the brain.

2. Spinal Cord

The spinal cord is a long, whitish, somewhat dorso-ventrally flattened tube. It is situated in tbe neural canal of the vertebral column. It is also covered by the same meninges which cover the brain. A very shallow groove, the **dorsal fissure,** occurs along the mid-dorsal line of the cord. A prominent mid-ventral groove called the **ventral fissure** extends along the entire length of the spinal cord. A thin longitudinal vertical septum of connective tissue lies in the thickness of the spinal cord beneath the dorsal fissure. The spinal cord encloses a narrow central cavity known as **central canal**. The central canal opens anteriorly into the fourth ventricle of the medulla oblongata but is closed posteriorly. It is filled with the cerebro-spinal fluid.

The spinal cord is composed of two types of nervous tissue : the inner **grey matter** and the outer **white matter**. In section the grey matter is H-shaped with prominent **dorsal** and **ventral** columns or cornua from which arise the dorsal and ventral roots of the spinal nerves. The grey matter is mainly composed of nerve cells and non-medullated nerve-fibres. The white matter is divisible into the **dorsal, ventral** and **lateral funiculi**. It comprises medullated nerve fibres.

The spinal cord conducts sensory and motor impulses to and from the brain. It is also a centre of reflex action.

Peripheral Nervous System

The peripheral nervous system consists of twelve pairs of **cranial nerves** from the brain and sixteen pairs of **spinal nerves** from the spinal cord. The first ten pairs of the cranial nerves correspond with the ten pairs of cranial nerves of fishes and amphibians. The eleventh or **accessory** and the twelfth or **hypoglossal** originate from the medulla behind the tenth pair and innervate the muscles of larynx, neck and tongue. The sixteen pairs of **spinal nerves** originate by two roots, dorsal and ventral, as in other vertebrates.

Autonomic Nervous System

The autonomic nervous system consists of two conspicuous white ganglionated sympathetic trunks which are continued along each side of the vertebral column. Each of these autonomic trunk ganglion, is connected with the corresponding spinal nerve by **ramus communicans**. The autonomic nervous system performs both sensory and motor functions and helps to regulate involuntary reaction.

SENSE ORGANS

The sense organs of *Uromastix* have not been studied. The following description is of *Lacerta*. The external organs of taste are completely lacking in Reptilia. The sense of taste is confined to the mouth only.

1. Olfactory organs. The olfactory organs are more complex than those of Anura. The nasal cavities open at the end of the snout by the **external nares** and into the cavity of the mouth by a pair of slit-like **internal nares** situated near the middle line of the palate. The external nare opens itno the vestibule through which air passes to the sensory epithelium of the nasal or olfactory cavity proper. The nasal cavity contains a convoluted **turbinal bone** over which the mucous membrane extends. Its cell are sensitive to chemicals in solution. The organs of chemical sense are better developed in reptiles. In reptiles, there are a pair of **vomero-nasal organs** or **organs of Jacobson**. The organs of Jacobson are well developed sac-like chambers lying below the nasal cavity but above the buccal cavity. Each organ consists of a blind sac lined with olfactory epithelium. It develops as a ventral hollow outgrowth of the nasal cavity. They are lined with olfactory epithelium. Each sac opens into the buccal cavity by a narrow duct a little in front of the internal nare.

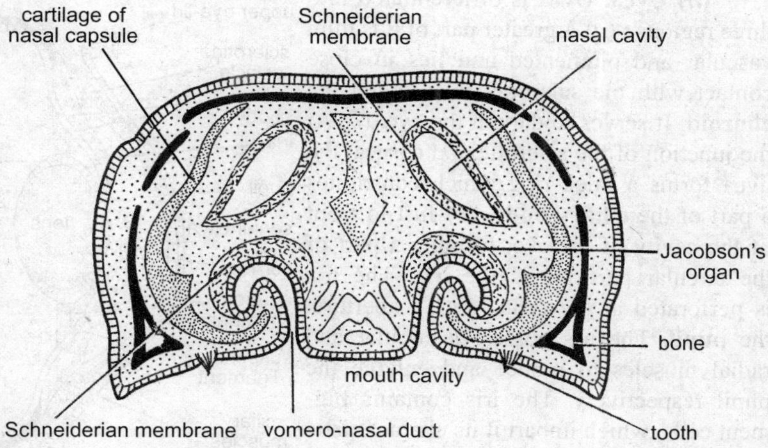

Fig. 21.32. *Uromastix*. T. S. of head showing Jacobson's organ and nasal sacs.

2. Jacobson's organs. The organ of Jacobson is innervated by branches of the olfactory and the trigeminal nerves. This sense organ is important in most lizards and snakes and appreciates scent particles introduced into it by the tongue tips. These organs serve to smell the food when it has been taken into the buccal cavity. In some reptiles, they play a part in activities such as trailing prey and locating members of the opposite sex. Paired ducts communicate with the buccal cavity, probably enabling the olfactory appreciation of substances held in the mouth.

3. Eyes. The eyes of *Uromastix* exhibit some advancement over those of the amphibians due to the transition from water to land. Although both the eyelids are movable but the lower eyelid is more movable in comparison to the upper eyelid. The eyelids close the eyes to protect them from the dust, heat and rain. There is a third eyelid, the **nictitating membrane**, in the anterior corner of the eye which develops and lies folded beneath the lower eyelid. It is really a double fold of conjunctiva and is not homologous with that of the frog. The nictitating membrane can be drawn rapidly over the moist cornea and, thus, it protects and cleans the eye. There are glands which keep the eyelids and cornea moist and clean. The **Harderian gland** lying on the anterior side of the eyeball lubricates the nictitating membrane, and the **lacrymal gland** lying on the_posterior side of the eyelid keeps the eye moist and clean. The tears (water fluid) are passed from the eye into the nose through a lacrymal duct.

The eye of *Uromastix* consists of the usual three layers, *viz.,* (*i*) the outer **fibrous tunic** or **sclerotic**, (*ii*) the middle **uvea** or **choroid** and (*iii*) the inner **retina**.

***(i)* Fibrous tunic.** The thick and tough fibrous tunic protects the eyeball and maintains its form. The fibrous tunic is distinguished into two distinct regions : an anterior, small, transparent and exposed portion called the **cornea** and a posterior, large, opaque cartilaginous portion called the **sclerotic** lying hidden in the orbit. A small anterior portion of the sclerotic, the cornea is, however, visible and is commonly called the **white** of the eye. The cornea is supported by a ring of eleven small **sclerotic bones**, or **ossicles**, present around the iris. The cornea is curved and provides the main refracting surface. A thin transparent epithelial layer is fused to the outer surface of the cornea. It is known as **conjunctiva**. The conjunctiva is continued over the inner surface of the eyelids.

(*ii*) **Uvea.** Uvea is differentiated into three regions : (*a*) A greater part of it is thin, vascular and pigmented and lies in close contact with the sclerotic. It is called the **choroid**. It serves to darken the eye. (*b*) At the junction of the sclerotic and cornea, the uvea forms a thick ring which constitutes a part of the **ciliary body**. (*c*) Just in front of the cavity of the eye, forming a part of the circular partition is the **iris**. The iris is perforated at the centre by an aperture, the **pupil**. The iris possesses circular and radial muscles to reduce and enlarge the pupil respectively. The iris contains pigment cells, which impart it its characteristic light-orange colour.

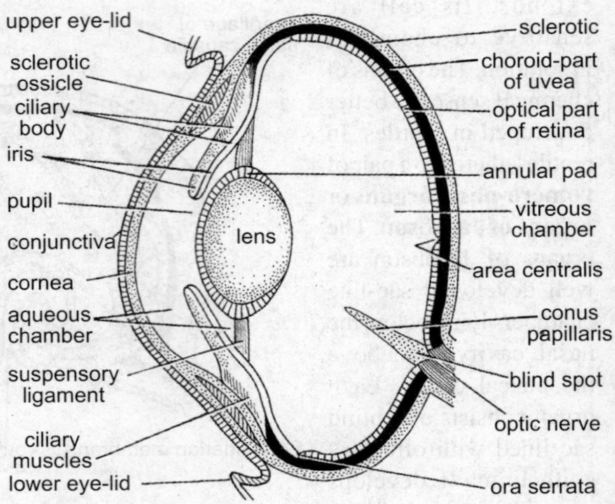

Fig. 21.33. *Uromastix*. V.S. of eye.

(*iii*) **Retina.** Retina lies inside the entire uvea and is differentiated into three distinct regions: (*a*) Its portion lying in contact with the choroid is thick and sensitive and is called the **optical part of retina**. It contains more cones than rods, thus, there is a good daylight vision and probably good colour perception. This condition is found in most lizards (diurnal types) and many turtles. The double cones of turtles and lizards may serve to detect polarized light (**Underwood**, 1970). This portion is protected from excessive exposure to light by yellow droplets found over this portion. (*b*) Remaining retina is thin and non-sensory. It lines the iris and ciliary body. The part forming the ciliary body is known as the **ciliary part** and that to the iris as the **iridial part**. The junction of the optical and ciliary parts of the retina is irregular and is called **ora serrata**.

The **lens** is a biconvex, transparent body lying immediately behind the iris. It is more convex behind than in front. A ring of soft tissue, the **annular pad**, surrounds the circumference of the lens. The lens is suspended and held in position by radially arranged fibres which arise from the ciliary body and are attached to the annular pad and form the **suspensory ligament**. The suspensory ligament is formed of ciliary muscles and ciliary processes. The lens and its suspensory ligament divide the cavity of the eye into two unequal compartments, *viz.,* the anterior small **aqueous chamber** and the posterior large **vitreous chamber**. The aqueous chamber is further divided by iris into anterior and posterior parts containing a clear watery fluid, the **aqueous humour**, and are continuous through the pupil. The aqueous humour keeps the eyeball taut or stretehed for distinct vision.

The optic nerve arises from the inner surface of the retina and joins the brain after piercing through all the three eye-coats at the back of the eyeball. The point of origin of optic nerve is called the **blind spot**. No image is formed at this point. A small part of the retina opposite the centre of the lens is the place of the most distinct vision. It is called the **area centralis** or **yellow spot**. A peculiar pigmented, blackish brown, highly vascular, cushion-like rod, the **conus papillaris**, projects into the vitreous humour from the blind spot. Probably it provides nutrition, like the pecten of birds.

Accommodation for near vision is usually produced by the striated ciliary muscles, so arranged that they cause the ciliary process to squeeze the lens, making its anterior surface more rounded.

Ear. The ear of *Uromastix* consists of two principal parts : the **middle ear** and the **internal ear**. External ear is absent.

Middle ear. The middle ear comprises an air-filled cavity called the **tympanic cavity**. It is bounded externally by the tympanum and internally by the auditory capsule. The tympanus lies behind the jaws, sunk a little below the surface. From the lower part of the tympanic cavity a narrow passage, the **Eustachian tube**, extends downwards and inwards to open into the posterior part of the pharynx. A slender rod-like ear ossicle, the **columella auris**, stretches across the tympanic cavity. The columella auris is divided into two parts : the inner bony **stapes** fitting into a small membrane covered aperture, the **fenestra ovalis**, and the outer cartilaginous **extrastapes** is attached to the inner surface of the tympanum.

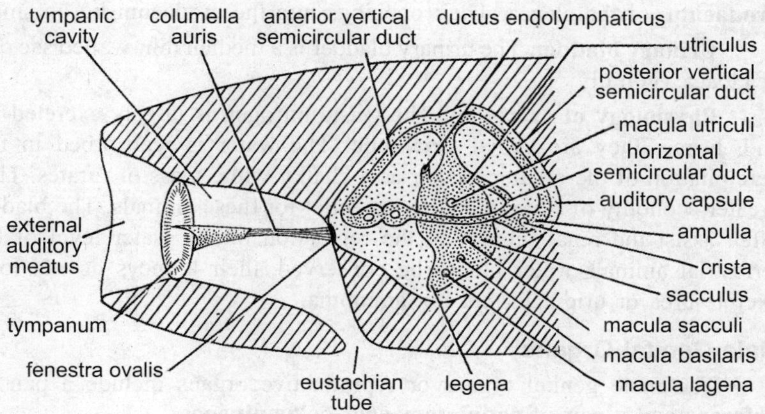

Fig. 21.34. *Uromastix*. Ear.

Internal ear. The internal ear is situated in the auditory capsule. It is a soft, delicate and complicated structure known as **membranous labyrinth** or **vestibule**. Its upper cylindrical part is known as the **utriculus** and the lower large and rounded **sacculus**.Three semi-circular ducts (2 vertical and 1 horizontal) open into the utriculus at both of their ends. The two vertical ducts join together by their adjacent ends before opening into the utriculus. One end of each semicircular duct is enlarged to form an ampulla. The sacculus gives rise to a small lobe, the **pars lagena**, which represents the cochlea of mammalian ear. It is an uncoiled tube and hearing is performed by it. Its wall, known as the limbus, is strengthened by a peculiar form of connective tissue. The receptor hairs (organ of Corti) are carried on a basilar membrane and are in contact with a tectorial membrane attached to the limbus. Hearing is good in some lizards which also produce sounds for communication. Snakes respond mainly to earth-borne vibrations through the quadrate attached to the stapes (tympanum is absent in snakes and some lizards also). A narrow **ductus endolymphaticus** extends dorsally from the saculus and ends in a small blind sac over the hindbrain.

URINOGENITAL SYSTEM

In *Uromastix*, the excretory and reproductive systems are closely associated as in other vertebrates. Therefore, the two systems are described together under the heading **urinogenital system**. The sexes are separate but the sexual dimorphism is not well developed.

Excretory System

The excretory system is similar in both the sexes and consists of a pair of **kidneys**, a pair of **ureters** and an unpaired **urinary bladder**.

Kidneys. The kidneys are elongated, compact, dark-red bodies situated in the posterior portion of the body cavity, one on either side of and close to the median line. The kidneys are attached with the dorsal body wall outside the body cavity and are covered with peritoneum on their ventral surface only. Kidneys are **metanephric** in nature. Each kidney is roughly divided into two lobes: the anterior broad and posterior narrow. The broad anterior lobes of both the kidneys are free and slightly apart, while the narrow posterior lobes are united so that they appear V-shaped

Histology. The kidneys are composed of narrow, much coiled uriniferous tubules, which are embedded in connective tissue having blood vessels, nerves and muscle fibres. These tubules communicate with their corresponding ureters.

Ureters. From the ventral side of each kidney arises a narrow, short, delicate **metanephric duct** or **ureter**. Both the ureters open behind on the dorsal side into the middle chamber, the **urodaeum**, of the cloaca. Their openings into the urodaeum lie on small papillae.

Urinary bladder. The urinary bladder is a median thin-walled sac opening into the urodaeum of cloaca ventrally.

Physiology of excretion. The waste nitrogen is largely excreted as uric acid in Squamata and, hence, they are called **uricotelic**. The water is reabsorbed in the urodaeum, with the percipitation of the organic matter as a chalky white mass of urates. Thus, this method allows a greater economy of water which is essential for these animals. The bladder and cloaca of reptiles often assist the renal tubules in the regulation of the water level of the blood. In completely terrestrial animals water has to be conserved, their kidneys do not lose much water and they excrete urea or uric acid, but not ammonia.

Male Genital Organs

The male genital organs or reproductive organs include a pair of **testes**, a pair of **vas deferens** and a pair of **copulatory sacs** or **hemipenes**.

Testes. The testes are two white oval bodies situated in the body cavity quite ahead of kidneys. Both the testes are not on the same level. Generally, the right testis is slightly ahead of the left. Both the testes are suspended from the dorsal bodywall by the peritoneal fold, the **mesorchium**. The adrenal gland is attached to the posterior side of each testis.

Vasa deferentia. The vasa deferentia (Wolffian ducts) are narrow convoluted tubes running backwards from the epididymis of testes along the dorsal bodywall. Posteriorly the vasa deferentia pass over the ventral surface of the kidneys and open into the urodaeum of the cloaca on the dorsal side in front of the opening of the ureters. From each testis several delicate vasa efferentia formed by the union of seminiferous tubules of the testis pass through the mesorchium and form epididymis present on the outer posterior surface of the testis. Vasa efferentia are outgrowths of mesonephric uriniferous tubules, the epididymis is the remains of the mesonephros.

Das (1960) described that the vestigial **Mullerian ducts** persist as a pair of thickened peritoneal strands, running parallel to the vasa deferentia from testes to the urodaeum.

Hemipenes. The hemipenes are the copulatory organs. These are two eversible hollow sacs lying under the skin behind the cloacal aperture at the base of tail. They appear as pocket-like swellings when withdrawn and almost cylindrical when everted. Each hemipenis consists of a proximal hard **stalk** or **pedicel** and a distal soft **head** or **glans**. A longitudinal **sulcus** or **groove** extends along the outer side of the stalk and ends in a depression on the head of the hemipenis. During copulation, the seminal fluid flows through the sulcus. Proximally the

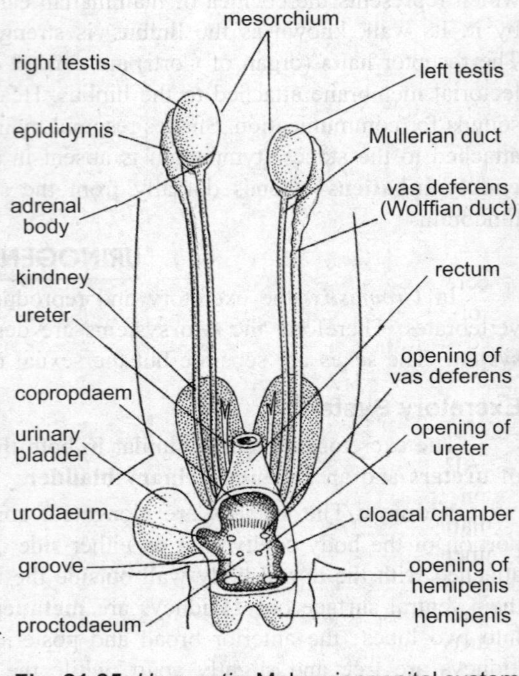

Fig. 21.35. *Uromastix*. Male urinogenital system.

hemipenes communicate with the opening of vas deferens in the urodaeum of the cloaca. During copulation, only one hemipenis is inserted into the cloaca of the female for the discharge of spermatozoa. Erection of the hemipenes is due to muscular action and filling with blood, then they are everted and become cylindrical and project beyond the cloaca.

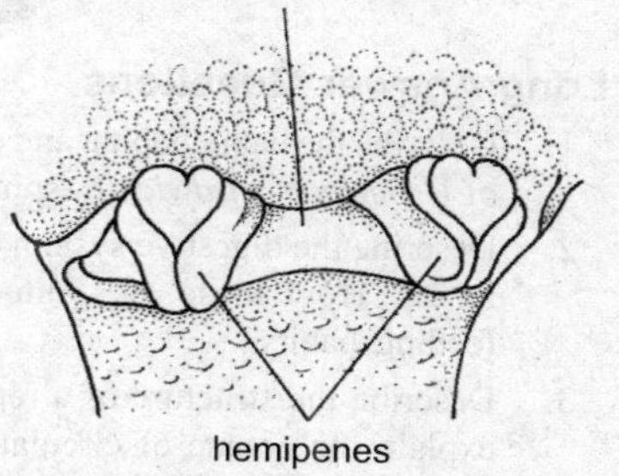

Fig. 21.36. *Uromastix.* Hemipenes of a lizard projecting through cloacal aperture.

Female Genital Organs

The female genital organs or reproductive organs include a pair of **ovaries** and a pair of **oviducts.**

Ovaries. The ovaries are yellowish-white and irregular in shape. The ovaries are situated in the posterior part of the body cavity anterior to the kidneys. The right ovary is somewhat anterior to the left. Each ovary is held to the dorsal bodywall by a double fold of peritoneum called the **mesovarium**. The outer surface of each ovary is produced into several rounded raised up areas or follicles. The follicles contain oocytes which are in different stages of development. The ovary is formed of several lobules lined by germinal epithelium. Its cells undergo oogenesis to form the ova. The mature ova are released into the body cavity by the rupture of follicles. They float in the coelomic fluid of the body cavity.

Oviducts. The oviducts (**Mullerian ducts**) are long, thin-walled coiled tubes lying outside the ovaries. Each oviduct is attached to the dorsal bodywall by a broad ligament or **mesotubarium**. Each oviduct opens into the anterior part of the body cavity by a wide externally directed ciliated aperture, the **ostium**, situated at the end of a large **oviducal funnel**. Posteriorly the oviducts open independently in the cloaca (urodaeum) in its dorsal wall, their openings lying in front of those of the ureters. While passing along the ventral surface of the kidney, the oviduct is enlarged so as to form a **shell-gland** or **ovisac**, whose function is to secrete **egg-shell** around the fertilised eggs and to store them. The terminal part of the oviduct is called **vagina**. It opens dorsally into the urodaeum of cloaca.

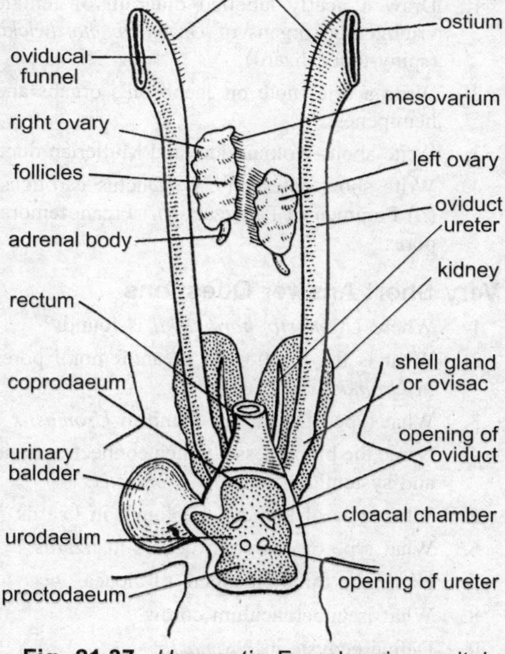

Fig. 21.37. *Uromastix.* Female urinogenital system.

COPULATION AND FERTILISATION

The sexes are separate but the sexual dimorphism is not well marked. During breeding season, *i.e.,* March and April, the male and female undergo courtship which is followed by copulation. During copulation the two individuals lie side by side opposing their cloacae. The hemipenes of male is everted and inserted into the cloaca of the female and, thus, transference of sperms takes place. The fertilisation is internal and takes place in the anterior part of oviduct, and the eggs are laid in burrows. The eggs are large, measuring 42 by 22 mm in size. The eggs are dirty-white in colour and develop with the heat of sun. A single female lays about 10 to 15 eggs. Since the eggs of *Uromastix* contains a large amount of yolk, the cleavage is **meroblastic**. The development takes place on land.

IMPORTANT QUESTIONS

Long Answer Questions

1. Describe the habit, habitat and external features of *Uromastix hardwickii*. (spiny-tailed lizard).
2. Describe the digestive system of a spiny-tailed lizard. How these are suited to take their feeding habits?
3. Describe the structure of a reptilian heart and explain the course of circulation through it.
4. Describe the arterial and venous system of *Uromastix hardwickii* (spiny-tailed lizard).
5. Give an account of brain of *Uromastix hardwickii* (spiny-tailed lizard).

Short Answer Questions

1. Draw a neatly labelled diagram of female urinogenital organs of *Uromastix hardwickii* (spiny-tailed lizard).
2. Write a brief note on Jacobson's organs and hemipenes.
3. Write about Wolfian duct and Mullerian duct.
4. Write short notes on : (*i*) Ductus caroticus; (*ii*) Foramen of Panizzae; (*iii*) Preanofemoral pores.

Very Short Answer Questions

1. Where *Uromastix hardwickii* is found ?
2. What is the location of preanofemoral pores in *Uromastix hardwickii* ?
3. What type of teeth are found in *Uromastix* ?
4. Name the blood vessel which connects carotid and systemic arteries in *Uromastix*.
5. What type of kidneys are found in lizards ?
6. What type of excretion occurs in lizards ?
7. Name the three chambers of cloaca.
8. What is gubernaculum cordis ?
9. Define ecdysis in *Uromastix*.

Multiple Choice Questions

1. *Uromastix hardwickii* is commonly known as
   (*a*) Spiny-tailed lizard (*b*) Girgit
   (*c*) Wall-lizard (*d*) Flying lizard.
2. The body of *Uromastix* is divisible into
   (*a*) Head, neck, trunk and tail
   (*b*) Head, neck and tail
   (*c*) Head, neck and trunk
   (*d*) Head, trunk and tail.
3. Shedding of scaly covering periodically in flakes is
   (*a*) Ecdysis (*b*) Moulting
   (*c*) Pupation (*d*) Metamorphosis.
4. Lips guarding mouth in *Uromastix* are
   (*a*) Thin and movable
   (*b*) Fleshy and movable
   (*c*) Fleshy and immovable
   (*d*) Thin and immovable.
5. In *Uromastix* pharynx leads to oesophagus through an aperture, the
   (*a*) Gullet (*b*) Glottis
   (*c*) Cloaca (*d*) Nares.
6. In *Uromastix* liver is attached to the body wall by a peritoneal fold called
   (*a*) Mesogaster (*b*) Hepatic omentum
   (*c*) Pyloric sphincter (*d*) Temporal fossae.
7. In *Uromastix* cloaca is divided into
   (*a*) Coprodaeum and urodaeum
   (*b*) Coprodaeum and proctodaeum
   (*c*) Coprodaeum, urodaeum and proctodaeum
   (*d*) Urodaeum and proctodaeum.
8. Walls of larynx are supported by
   (*a*) Arytenoid cartilage
   (*b*) Cricoid cartilage
   (*c*) Coracoid cartilage
   (*d*) Arytenoid and cricoid cartilage.
9. Right and left systemic arches communicate through
   (*a*) Foramen of Panizzae
   (*b*) Foramen of Monro
   (*c*) Foramen Magnum
   (*d*) Fenestra ovalis.
10. Fertilised egg is provided with shell by
   (*a*) Oviducal funnel (*b*) Ovisac
   (*c*) Vagina (*d*) Ostium.

Answers

1. (*a*) 2. (*a*) 3. (*a*) 4. (*b*) 5. (*a*) 6. (*b*) 7. (*c*) 8. (*d*)
9. (*a*) 10. (*b*).

22

Chapter

Type 9. *Calotes versicolor* (The Garden Lizard)

Contents of this Chapter

SYSTEMATIC POSITION

Phylum	...	Chordata
Subphylum	...	Vertebrata
Division	...	Gnathostomata
Superclass	...	Tetrapoda
Class	...	Reptilia
Subclass	...	Diapsida
Order	...	Squamata
Suborder	...	Lacertilia
Family	...	Agamidae
Type	...	*Calotes versicolor*

NATURAL HISTORY

Distribution. *Calotes* is a common garden lizard and is commonly known as "girgit" in North India. It is also known as oriental "blood sucker" because of red colour of its throat. *Calotes versicolor* is the most commonly distributed Indian species.

Habits and habitat. *Calotes* is a common arboreal lizard of gardens and fields. It is diurnal, often found sitting on tree branches, shrubs and hedges and mimics with the surroundings. It is very swift in running over the ground and also swims in water. It is insectivorous but also feeds on earthworms, small crabs, lizards and frogs. Sexes are separate showing no sexual dimorphism externally. Its breeding period is from May to November. It lays eggs in crevices or in debris. Development is direct without any larval form.

EXTERNAL FEATURES

Shape, size and colour. Body is elongated, slender, lizard-like and divisible into **head**, **neck**, **trunk** and **tail**. Male is brightly coloured, golden yellow with a greenish tinge. Throat is scarlet red, often with a black transverse bar. Female is not so brightly coloured. Male is larger and stronger than female, measuring about 35 cm in length including tail (Fig. 22.1).

Head. The head is large and somewhat pyramidal in shape with a short conical snout which bears a wide, slit-like aperture, the mouth running round the anterior border of the head and a pair of small external nares situated above the anterior extremity. At the sides of the head are the eyes, each provided with upper and lower opaque, movable eyelids and with a transparent third eyelid or **nictitating membrane**, which, when withdrawn, lies in the anterior angle of the orbit. Behind the eye is a circular patch of skin-the **tympanic membrane** somewhat sunk below the general level of the skin.

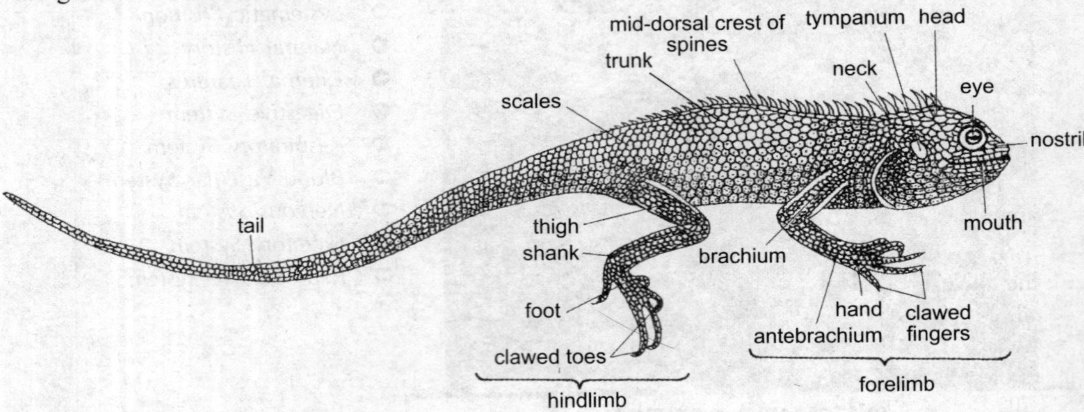

Fig. 22.1 *Calotes versicolor*. External features.

Trunk. Head is connected with the trunk with a short neck. The trunk is elongated, strongly convex dorsally, laterally compressed and flattened ventrally. At the junction of trunk and tail, on the ventral surface is a slit-like transverse aperture, the cloacal aperture. The trunk bears two pairs of limbs. Each forelimb has an upperarm or brachium, the forearm or antebrachium, the hand or manus and five horny clawed digits; while each hindlimb comprises the thigh or femur, shank or crus, foot or pes and five clawed digits, of which the hallux or first is the smallest.

Tail. Tail is cylindrical, tapering showing alternate dark and light annuli and is nearly twice as long as the head and trunk together. In adult male the base of tail just behind the cloacal aperture becomes quite swollen due to hemipenes.

Skin and exoskeleton. Skin is dry and devoid of glands. The body is entirely covered by epidermal horny, imbricate and backwardly directed scales of unequal size which are periodically shed in flakes. On the head, a prominent shield covers the parietal foramen and two spines are present above each ear opening. Over the head and trunk, along the mid-dorsal line, is a characteristic crest of large spine-like movable scales directed backward and gradually diminishing in size posteriorly. There are melanophores lying deep in the dermis and iridocytes and lipophores more superficially. It is able to change colour. The colour may change with the environment, serving the purpose of concealment. Melanophores are controlled partly or entirely by the sympathetic nervous system. Colour is largely controlled through the eyes. Histology of skin is similar to that of *Uromastix*.

Scales and claws form the exoskeleton of most of the reptiles including *Calotes*.

Scales. Scales form the continuous covering over the body of *Calotes*, but becomes thinner in the grooves between the scales. Scales on the body of *Calotes* are of two types : large scales and small scales. Sensory structures in the form of bristles are present on both the scales called **prototrichus**. Scales are epidermal in origin, develop from Malpighian layer. Each scale is a plate-like structure supported by a **bony plate** or the **ossicle**.

Claws. Tips of digits of *Calotes* are provided with sharp claws. Each claw is made up of a dorsal and ventral scale-like horny plate. Dorsal plate is called **unguis** and ventral plate is called **subunguis**. The latter is flattened. Both these plates are derived from the Malpighian layer and placed in such a way that their coverage should form a sharp claw. Cellular structure of skin is similar to that of *Uromastix*.

Endoskeleton

Endoskeleton of *Calotes* is very much similar to that of *Varanus* or monitor lizard which has been already described in the Chapter of *Uromastix*. The endoskeleton of *Varanus* is more convenient to study because of its larger size.

DIGESTIVE SYSTEM

The digestive system of *Calotes* is formed of the alimentary canal and digestive glands.

I. Alimentary Canal

The alimentary canal is a convoluted tube beginning from mouth and ends in a cloaca.

1. Mouth. It is a wide transverse slit bounded by the upper and lower jaws, bearing teeth. Teeth are small, conical and similar **(homodont)**, and directed backward. They are attached with the inner lateral sides of jaw bones. Such type of attachment of teeth is called **pleurodont**. They are adapted to catching and holding and not for chewing.

2. Buccal cavity. Mouth leads into a large and dorsoventrally compressed **buccal cavity**. In its roof open two small openings of **internal nares** and on its floor is present a narrow, elongated and fleshy protrusible tongue which is bifid in front.

3. Pharynx. It is the posterior part of the buccal cavity. At the back of the tongue is present the **glottis**. In the pharynx are also present a pair of lateral openings of **eustachian tubes** and a posterior wide opening the **gullet**, leading to oesophagus.

4. Oesophagus. It is a straight, narrow tube passing through the neck dorsal to trachea and opens into stomach.

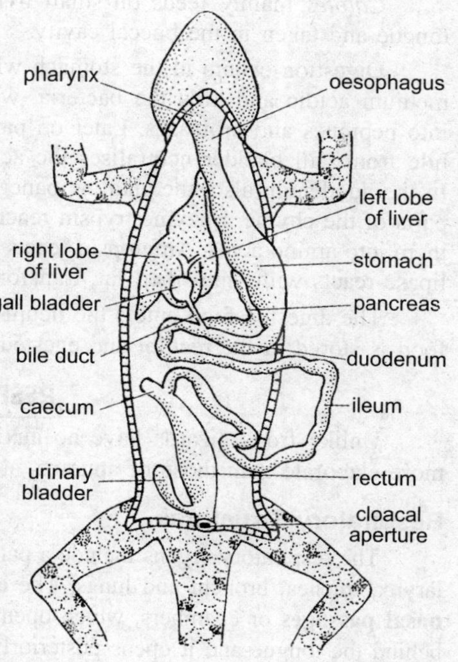

Fig. 22.2. *Calotes*. Digestive system.

5. Stomach. It is a muscular, cylindrical tube, which, though expansible, is little wider than the oesophagus. The stomach is not quite distinct into cardiac and pyloric portion. The posterior pyloric stomach leads into small intestine through pyloric sphincter.

6. Small intestine. The small intestine is a long, narrow and coiled tube. Its small anterior, U-shaped part is the **duodenum** and the posterior, much coiled part is the **ileum**. The duodenum receives the bile and pancreatic ducts. The ileum leads into the large intestine or rectum. At the point where the small intestine joins the rectum, the latter is produced into a short **caecum**.

7. Large intestine. It is small, sac-like, thick-walled tube, opening into **cloaca** by a sphinctered anus. The colon is absent.

8. Cloaca. The cloaca is linearly subdivided into **coprodaeum**, **urodaeum** and **proctodaeum**. The last part opens externally through a sphinctered cloacal aperture.

II. Digestive Glands

Digestive glands include mucous and salivary glands, gastric glands, liver, pancreas and intestinal glands. In the buccal cavity are present few unicellular mucous and salivary glands. The **liver** is a large, bilobed, brownish gland and placed on the dorsal side of stomach, behind the heart. The **gall bladder** is a small, rounded sac lying in a cleft at the lower margin of the right liver lobe. It acts as a reservoir for bile secreted by the liver cells. The bile duct from the gall bladder opens into the upper part of duodenum. **Pancreas** is a whitish elongated structure situated in between the stomach and duodenum. It opens into the duodenum by a separate pancreatic duct. It secretes the pancreatic juice which is alkaline in nature and contains three enzymes, trypsin, amylopsin and lipase. The unicellular gastric glands are present in the inner wall of stomach which secretes gastric juice containing digestive enzyme **pepsin**. Besides this, unicellular parietal glands are also present in its wall that secrete hydrochloric acid.

Food, Feeding Mechanism and Digestion

Calotes mainly feeds on small living insects. It catches its prey with the help of sticky tongue and taken in the buccal cavity.

Digestion occurs in the stomach where HCl and pepsin react on the food. HCl makes the medium acidic and kills the bacteria, while pepsin reacts with the food protein and breaks it into peptones and proteases. Later on partly digested food (chyme) enters the duodenum where bile from gall bladder neutralises the acidic chyme and emulsifies the fat portion of the food. In the duodenum also, the alkaline pancreatic juice reacts with the protein, carbohydrate and fat parts of the chyme. Enzyme trypsin reacts with the protein, peptones and proteases and converts them into amino acids. Amylopsin reacts with the carbohydrates and transforms into glucose, and lipase reacts with emulsified fat, transforming into fatty acid and glycerol.

The digested food enters the ileum where the digested food is absorbed and the undigested food is stored in the rectum and egested periodically through cloaca.

RESPIRATORY SYSTEM

Unlike frogs, lizards have no auxiliary means of respiration. Lungs of *Calotes* are little more elaborate than those of anurans.

Respiratory Structures

The respiratory organs include a pair of external nares, nasal chambers, internal nares, glottis, larynx, trachea, bronchi and lungs. The external nares lie a little in front of eyes. They lead into **nasal passages** or chambers, which open into the roof of the buccal cavity. The **glottis** is located behind the tongue and it opens posteriorly into a short chamber, the **larynx**. It is less prominently developed than in many of the Amphibia. Its walls are supported by a **cricoid** and a pair of **arytenoid** cartilages. The larynx opens into a narrow, elongated cylindrical tube, the **trachea**. Its wall is supported by a large number of small cartilaginous rings, the **tracheal rings**. The trachea is bifurcated into two narrow tubes in the thorax, the **bronchi**. Each bronchus enters into a lung.

Lungs. The lungs are elastic, elongated sacs, the right lung is slightly larger than the left one. The walls of the lungs are folded into ridges giving the appearance of a honeycomb. These ridges are much closer and more numerous towards the anterior end than towards the posterior end of the lung. These chambers are called **alveoli** where gaseous exchange occurs. In the distal part of the lungs, such chambers are absent. This posterior part of the lung is considered as **reservoir** for the residual air.

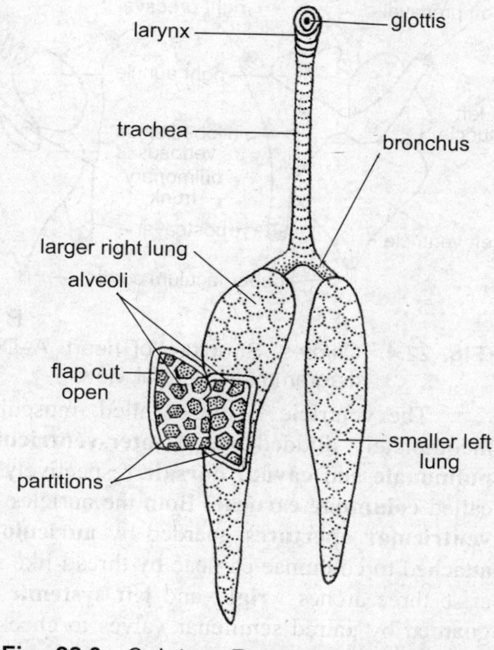

Fig. 22.3. *Calotes*. Respiratory system. A portion of the right lung is cut open to show the internal partitions and alveoli.

Respiratory Mechanism

In most reptiles including *Calotes* the pleural and peritoneal cavities communicate. The respiratory movements are performed by the intercostal muscles attached to the ribs. **Inspiration** is caused by the movement of intercostal muscles, raising the ribs that increases the volume of the thorax and reduces the lung pressure causing the inflow of air into the lung. Oxygen of the air enters the blood of the blood capillaries and CO_2 of the blood enters the alveoli. **Expiration** is done by lowering the ribs due to contraction of the intercostal muscles that decreases the volume of the thoracic cavity, pressing also the lungs. Thus, air having CO_2 flows back to the exterior.

BLOOD VASCULAR SYSTEM

Heart

1. External features. Heart of *Calotes* is placed in the anterior part of the pleuroperitonial cavity. The heart is enclosed in a thin transparent membrane, the **pericardium**, which is lined with **endocardium**, thus, it is two-layered. The space between the heart and pericardium is filled with **pericardial fluid**. The heart is triangular in shape and three-chambered–two **auricles** and a **ventricle**. The right auricle is larger than the left auricle. Both the auricles are clearly marked from the posterior ventricle by a transverse **auriculo-ventricular groove**. Truncus arteriosus is absent. The thin-walled and bilobed **sinus venosus** is reduced and placed transversely and dorsal to the right auricle. It is formed by the fusion of the three **venae cavae**. These three venae cavae empty independently into the sinus venosus. There are no valves. The smaller left lobe of sinus venosus is formed by the left precaval, while the larger right lobe is formed by the union of right precaval and the postcaval. The apex of the ventricle is attached with the liver by a thin, white cord of tissue, the **gubernaculum cordis**.

2. Internal structure. Both the auricles are internally separated by a thin, complete **inter–auricular septum**. The left auricle is smaller and receives the single **pulmonary vein** on its dorsal wall. Its aperture has no valve. The right auricle is larger and its inner wall is raised internally into smaller ridges, the **musculi pectinati**. Sinus venosus opens into the dorsal wall of right auricle by a semicircular **sinu-auricular aperture** guarded by sinu-auricular valves. These valves develop from the upper and lower margins of the aperture and their frilled free ends project into the lumen of right auricle.

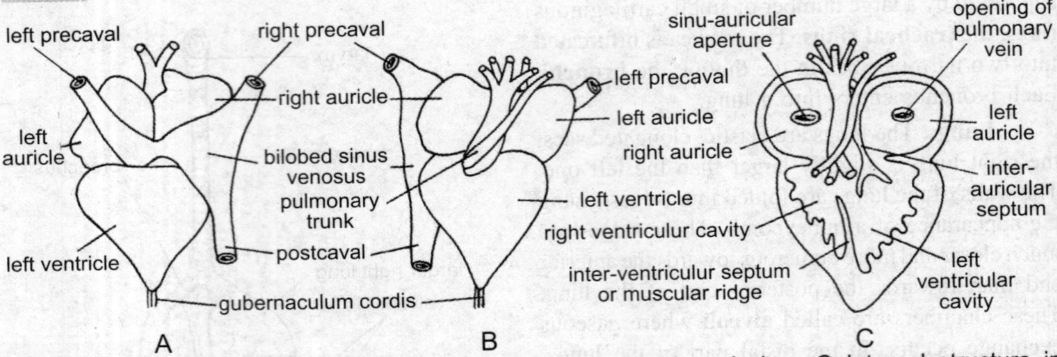

Fig. 22.4. *Calotes*. Structure of heart. A–Dorsal view; B–Ventral view; C–Internal structure in diagrammatic ventral view.

The ventricle is thick-walled, muscular, spongy and triangular in shape. Its cavity is incompletely divided by an **inter-ventricular septum** into right and left halves or **cavum pulmonale** and **cavum dorsale** respectively. Its inner layer is projected into interlacing ridges called **columnae carneae**. Both the auricles open into ventricle through right and left **auriculo-ventricular apertures** guarded by **auriculo-ventricular valves**. The flaps of these valves are attached to columnae carneae by thread-like muscles, the **chordae tendineae**. From the ventricle arise three arches : **right** and **left systemic** and **pulmonary**. The openings of these arches are guarded by paired semilunar valves to check the return of the blood.

Working of heart. In *Calotes*, the circulation of blood is double. The deoxygenated blood of the sinus venosus enters the right auricle through sinu-auricular aperture. The oxygenated blood from the lungs enters the left auricle by the pulmonary vein. The left auricle pours its blood into the left part of cavum dorsale of the ventricle through the auriculo-ventricular aperture. The right auricle also pours its blood into the right portion of cavum pulmonale or ventrale of the ventricle through the auriculo-ventricular aperture.

The oxygenated blood from the cavum dorsale goes principally through both carotid arches and the right systemic arch. Some of the right systemic blood may be added via the **ductus caroticus** to that of the internal carotid. While the deoxygenated blood from the cavum pulmonale passes into the pulmonary arteries for reoxygenation in the lungs. The blood in the right part of the cavum dorsale goes into the left systemic arch and probably passes through the ductus caroticus on that side into the carotid.

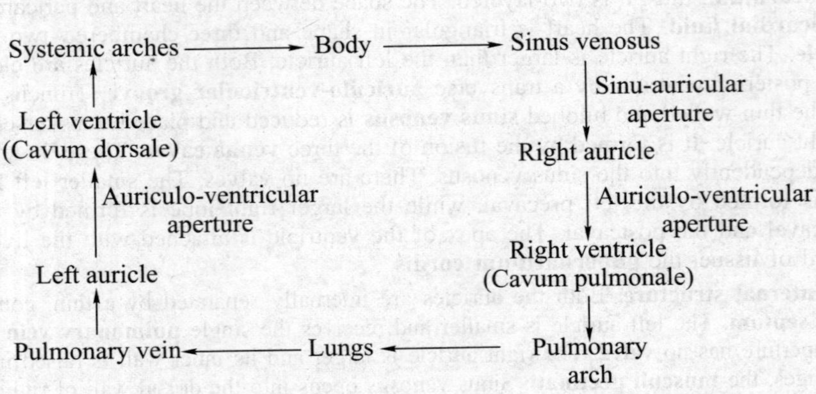

Arterial System

From the ventricle of *Calotes* arise three aortic arches :

1. A pulmonary arch,
2. A pair of systemic arches.

These three arches are twisted around each other at the base and covered by a fibrous sheath.

1. Pulmonary arch. It arises from the right portion of the ventricle (*i.e.*, cavum pulmonale) and soon splits into two pulmonary arteries, each entering into a lung. It carries deoxygenated blood.

2. A pair of systemic arches. Right systemic arch arises from the left side of the ventricle (*i.e.*, left part of cavum dorsale) and carries pure blood. Left systemic arch originates from the right side of the ventricle and carries mixed blood. Both the systemic arches cross each other, curve backwards round the oesophagus and unite behind the heart to form the **dorsal aorta**.

Right systemic arch. (*i*) Innominate. From the apex curvature of right systemic arch arises a single common **carotid artery** or **innominate**. It runs anteriorly and soon divides into right and left carotid arteries, that run parallel to the systemic arch of that side. Each carotid artery again divides into **internal** and **external carotid** arteries supplying blood to the head region and before division each carotid gives off a **thyroid artery**. Each external carotid artery also gives off one **laryngo-tracheal artery** and three **buccal arteries**. Each internal carotid artery at its tip bifurcates into **palatine artery** and **stapedial artery**. The internal carotids supply blood to the brain. The internal carotids are also connected with the systemic arches of their side by a small **ductus caroticus** which are the remnants of embryonic lateral aorta.

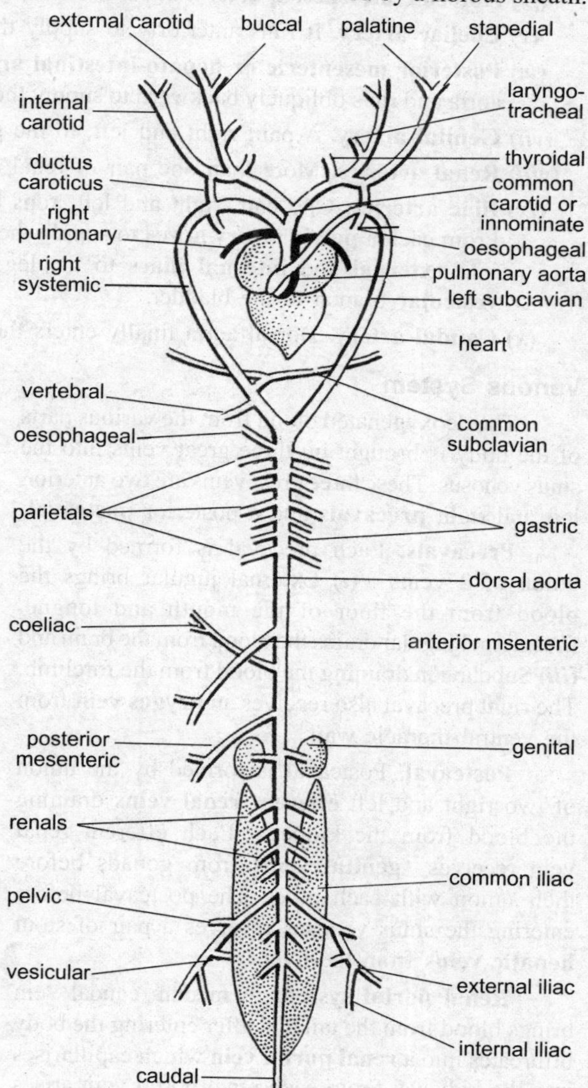

Fig. 22.5. *Calotes*. Arterial system.

(*ii*) Subclavian. The common subclavian arises from the right systemic arch and it divides into **right and left subclavian arteries** which supply blood to the forelimbs.

(*iii*) Vertebral. It also arises from the right systemic arch supplying blood to the vertebral column.

(*iv*) Oesophageal. These arise from both right and left systemic arches. Right systemic arch gives off 3 and left systemic arch gives off 4 oesophageals.

3. Dorsal aorta. It runs backwards mid-dorsally and gives off the following branches :

(*i*) Anterior oesophageal artery. It arises from thr central surface of the dorsal aorta.

(*ii*) **Parietal arteries.** Several pairs to dorsal muscles and vertebral column.

(*iii*) **Gastric arteries.** These vary from 4 to 8 pairs to stomach.

(*iv*) **Anterior mesenteric artery.** It runs obliquely backward to supply the intestine.

(*v*) **Coeliac artery.** It runs anteriorly to supply the pyloric stomach.

(*vi*) **Posterior mesenteric or hepato-intestinal artery.** It arises from the right side of dorsal aorta and runs obliquely backward to supply the gall-bladder and intestine after bifurcation.

(*vii*) **Genital artery.** A pair, right and left, to the gonads (testes or ovaries).

(*viii*) **Renal arteries.** More than one pair of renal arteries to the kidneys.

(*ix*) **Iliac arteries.** One pair, right and left, runs backward to the hindlimb of its own side. From each a **pelvic artery** arises to supply the pelvic girdle. Later, each branch bifurcates into **external** and **internal iliacs** to the leg. Before bifurcation it gives off a slender **vesicular** branch to the bladder.

(*x*) **Caudal artery.** Dorsal aorta finally enters the tail as caudal artery.

Venous System

The deoxygenated blood from the various parts of the body is brought by three great veins into the sinus venosus. These three great veins are two anterior, left and right **precavals** and a posterior **postcaval**.

Precavals. Each precaval is formed by the union of 2 veins : (*i*) External jugular brings the blood from the floor of the mouth and tongue, (*ii*) Internal jugular drains the blood from the brain and (*iii*) Subclavian draining the blood from the forelimb. The right precaval also receives an **azygos vein** from the ventral thoracic wall.

Postcaval. Postcaval is formed by the union of two right and left **efferent renal veins** draining the blood from the kidneys. Each efferent renal vein receives **genital veins** from gonads before their union with each other. The postcaval before entering the sinus venosus receives a pair of stout **hepatic veins** from the liver.

Renal portal system. A median caudal vein brings blood from the tail and after entering the body bifurcates into a **renal portal vein** which capillarises into the kidneys. From each renal portal vein arises a **pelvic vein**, which receives **femoral** and **sciatic** veins from the hindlimb. Both the pelvic veins unite in front of kidneys to form a median **epigastric** or **anterior abdominal vein** that enters into liver and breaks up into capillaries.

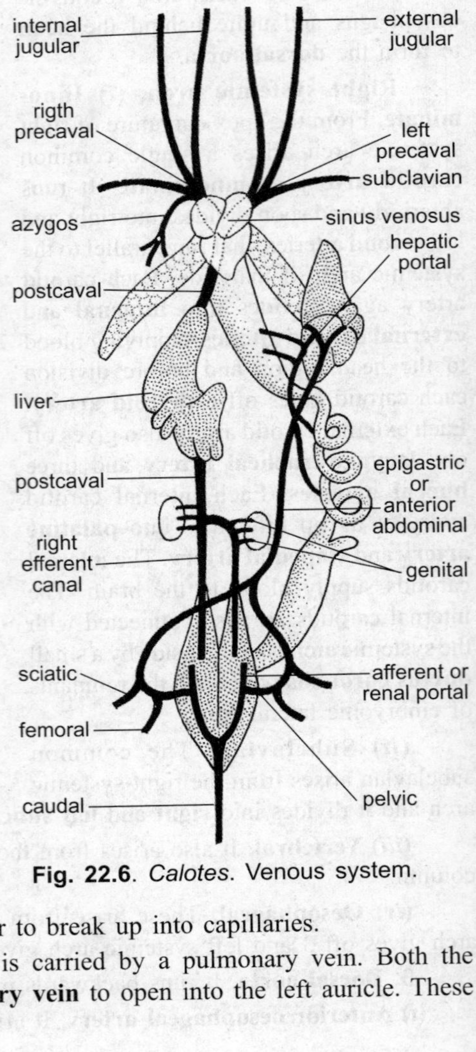

Fig. 22.6. *Calotes*. Venous system.

Hepatic portal system. The **hepatic portal vein** is formed by the branches from stomach, intestine, pancreas, etc., and it enters into the left lobe of liver to break up into capillaries.

Pulmonary vein. From each lung the blood is carried by a pulmonary vein. Both the pulmonary veins unite to form a **common pulmonary vein** to open into the left auricle. These veins carry oxygenated blood to the heart.

Blood

The blood of *Calotes* is red in colour and contains **plasma** and **blood corpuscles**. The red blood corpuscles are elliptical, biconvex and nucleated. The white blood corpuscles are like other vertebrates. The red cells of reptiles survive for much longer than those of mammals. The total haemoglobin concentrations and oxygen carrying power of the blood are only half those of mammals.

The *Calotes,* like other reptiles, is cold blooded or poikilothermous.

Lymphatic System

The lymphatic system is exceptionally well developed and with large vessels. There are no lymph nodes but large cisterns are found at the sites where nodes are found in birds and mammals. Lymph is pumped by paired lymphatic hearts to a large cisterna chyli in the abdomen and from there passes through thoracic ducts, forming sheaths around the aortae, to enter the base of the subclavian and jugular veins.

NERVOUS SYSTEM

The nervous system of *Calotes* consists of the following parts :

1. Central nervous system includes brain and spinal cord.

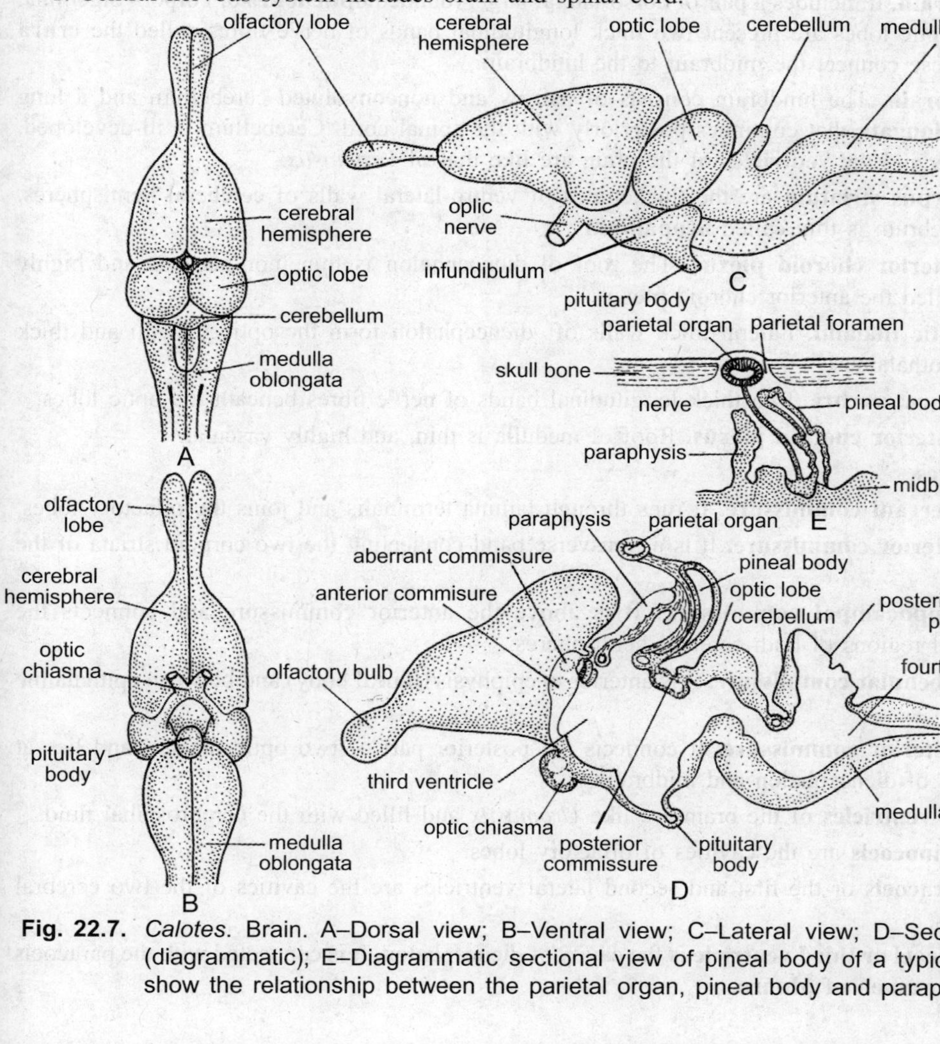

Fig. 22.7. *Calotes*. Brain. A–Dorsal view; B–Ventral view; C–Lateral view; D–Sectional view (diagrammatic); E–Diagrammatic sectional view of pineal body of a typical lizard to show the relationship between the parietal organ, pineal body and paraphysis.

2. Peripheral nervous system includes the cranial and spinal nerves.

3. Autonomic or sympathetic nervous system.

1. Central Nervous System

Brain. The brain is lodged in the cranium. The brain as well as the spinal cord are enveloped by two meninges, the **piamater** and **duramater**. The piamater is vascular and closely applied with the brain, while duramater is fibrous and lies outside the piamater. In between these two membranes is a space, **subdural space**.

In *Calotes*, brain is differentiated into (*i*) forebrain, (*ii*) midbrain and (*iii*) hindbrain.

Forebrain. The forebrain consists of olfactory bulbs, cerebral hemispheres and diencephalon.

(*i*) Olfactory bulbs. The anterior part of telencephalon is in the form of a pair of sac-like **olfactory bulbs** or lobes which are connected posteriorly with the posterior part of telencephalon, the **cerebral hemispheres** through narrow, elongated **olfactory peduncle**. The diencephalon bears on the dorsal surface two outgrowths, anterior **parietal organ** and posterior **pineal body**. **Paraphysis** is reduced and anterior to parietal organ. From the floor of diencephalon arises a funnel-like structure, the **infundibulum**, at the apex of which is placed the **hypophysis** or **pituitary body**.

Midbrain. It includes a pair of dorso-lateral large, rounded **optic lobes** or corpora bigemina. Ventral to optic lobes are present two thick longitudinal bands of nerve fibres called the **crura carebri**. These connect the midbrain to the hindbrain.

Hindbrain. The hindbrain consists of narrow and nonconvoluted cerebellum and a long **medulla oblongata** that continues posteriorly with the spinal cord. Cerebellum is ill-developed. **Thickenings** and the **ventricles** of the brain are like that of *Uromastix*.

1. Corpus striatum are the thickenings of ventro-lateral walls of cereberal hemispheres. Roof of cerebrum is thin called **neopallium**.

2. Anterior choroid plexus. The roof of diencephalon is thin, non-nervous and highly vascular, called the anterior choroid plexus.

3. Optic thalami. Lateral thick walls of diencephalon form the optic thalami and thick floor is hypothalamus.

4. Crura cerebri. Two thick longitudinal bands of nerve fibres beneath the optic lobes.

5. Posterior choroid plexus. Roof of medulla is thin, and highly vascular.

Commissures

1. Aberrant commissure. It runs through lamina terminalis and joins the olfactory lobes.

2. Anterior commissure. It is a transverse band connecting the two corpora striata of the cerebrum.

3. Hippocampal commissure. It is above the anterior commissure and connects the hippocampal regions of both cerebral hemispheres.

4. Habenular commissure. It is anterior to epiphysis (pineal body) and between epithalamic ganglia.

5. Posterior commissure. It connects the posterior part of two optic thalami and lies at the junction of diencephalon and midbrain.

Cavities or ventricles of the brain are like *Uromastix* and filled with the cerebrospinal fluid.

1. Rhinocoels are the cavities of olfactory lobes.

2. Paracoels or the first and second lateral ventricles are the cavities of the two cerebral hemispheres.

3. Diacoel or third ventricle is the cavity of diencephalon. It is connected with the paracoels through a **foramen of Monro.**

4. Metacoel or fourth ventricle is the cavity of medulla oblongata which is connected with the diacoel through a narrow **iter**.

5. Optocoels are the cavities of optic lobes and are also connected with the **iter**.

All these cavities of the brain are connected with each other and also connected with the cavity of spinal cord.

Spinal Cord

It lies in the neural canal of the vertebral column. It is anteriorly connected with the brain. It is thick-walled and bears a furrow in its roof and floor. Histologically, it is made up of white matter externally and grey matter internally. White matter is formed of medullated nerve fibres and grey matter of nerve cells.

Pineal Eye in Reptiles

In most reptiles (*Lacerta, Calotes, Varanus, Anguis, etc.*) and particularly in *Sphenedon* the roof of diencephalon is produced into a hollow median outgrowth, in front of epiphysis or pineal organ. The **parietal organ** or **pineal eye** is situated in the parietal foramen of the cranial roof just under the integument and covered in the young by a transparent scale. The distal end of this becomes constricted off as a hollow sphere, while the remainder is converted into a nerve. The wall of the hollow sphere becomes divergently modified on opposite sides. The distal side gives rise to a lens-like thickening. The proximal side forms a membrane several layers in thickness–the **retina**. The whole is enclosed in a capsule of connective tissue. The nerve usually degenerates before the animal reaches maturity.

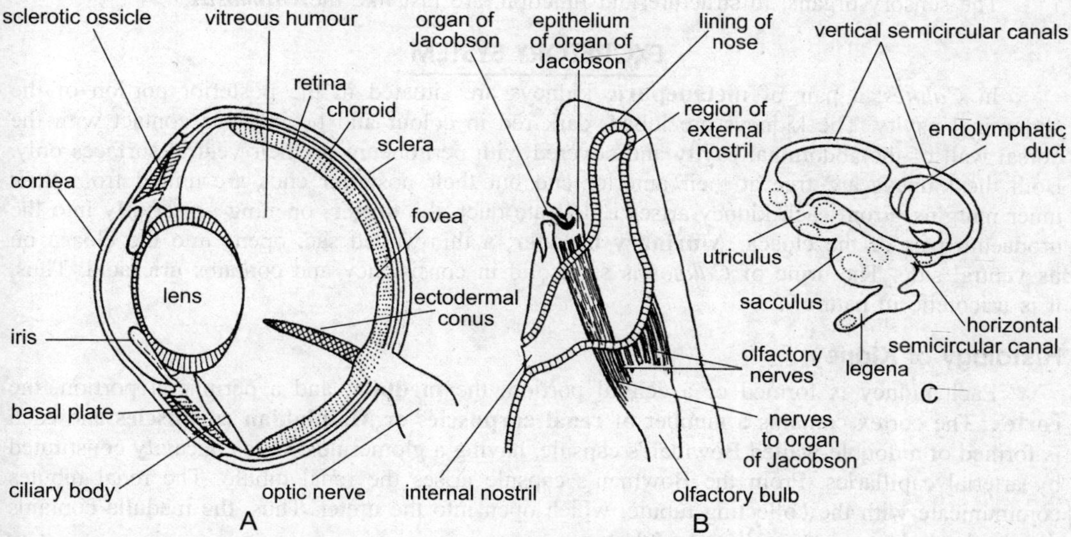

Fig. 22.8. *Calotes.* Sense organs. A–Section of eye; B–Showing the relationship of the organ of Jacobson with the nasal cavity; C–Internal ear.

The function of the reptilian pineal eye is still obscure. The skin over the top of the pineal eye in some reptiles becomes depigmented and takes the shape of cornea. The parietal organ is though eye-like in appearance, but it has no trace of any visual function. Histological evidences have not proved that this organ is a vestige of once functional median eye.

2. Peripheral Nervous System

The cranial and spinal nerves form the peripheral nervous system.

Cranial nerves. Like *Uromastix*, in *Calotes* also cranial nerves are 12 pairs besides the terminal nerve. The origin, distribution and biological nature of the first to tenth pairs of cranial nerves is exactly similar to frog and fishes. The **spinal accessory** (11th) and **hypoglossal** (12th) cranial nerves arise from the medulla behind 10th pair and innervate the muscles of larynx, neck and tongue.

Spinal nerves. Sixteen pairs of spinal nerves arise from the spinal cord and come out from it between the vertebrae. These are mixed types of nerves. Each nerve originates by two roots, a dorsal and a ventral, and both unite before emerging from the vertebral column. After coming out from the vertebral column, each divides into **four branches** or **rami** : **(*i*) Dorsal branch** innervates sense organs, glands and muscles; **(*ii*) Ventral branch** is long and innervates hypoaxial organs; **(*iii*) Meningeal branch** innervates the neural canal; and **(*iv*) Ramus communicans** communicates with the autonomic nervous system. Several spinal branches often unite and form the **plexus**.

3. Autonomic Nervous System

A chain of ganglia on either side of dorsal aorta, beneath the vertebral column, form the autonomic nervous system. Each ganglion is connected with adjacent spinal nerves by a small nerve, the **ramus communicans**. Corresponding ganglia of both the chains are also connected by small transverse commissures. On either side the chain of ganglia or sympathetic cord enters the skull to unite with the vagus ganglion of Xth nerve and then again runs forward to join the gasserian ganglion of Vth nerve. The branches from these ganglia go to muscles of heart, lungs, alimentary canal and glands.

Sense Organs

The sensory organs, in structure and function, are just like the *Uromastix*.

EXCRETORY SYSTEM

In *Calotes*, a pair of **metanephric** kidneys are situated in the posterior portion of the abdominal cavity. The kidneys are lobed, dark red in colour and lie in close contact with the dorsal wall of the abdominal cavity and covered with peritoneum on their ventral surfaces only. Both the kidneys are free at their anterior end but their posterior ends are united from their inner margins. From each kidney arises a delicate duct, the **ureter**, opening posteriorly into the urodaeum part of the cloaca. A **urinary bladder**, a thin-walled sac, opens into the cloaca on its ventral side. The urine of *Calotes* is semisolid in consistency and contains uric acid. Thus, it is uricotelic in nature.

Histology of Kidney

Each kidney is formed of a central portion, the **medulla**, and a peripheral portion, the **cortex**. The cortex contains a number of **renal corpuscles** or **Malpighian corpuscles** and each is formed of a double-walled Bowman's capsule, having a glomerulus in its concavity constituted by arterial capillaries. From the Bowman's capsule arises the renal tubule. The renal tubules communicate with the collecting tubule, which open into the ureter. Thus, the medulla contains the renal tubules and the collecting tubules.

REPRODUCTIVE SYSTEM

In *Calotes*, sexes are separate but not easily distinguishable externally.

Male Reproductive System

In male, testes are two oval white bodies. The right testis is situated just posterior to the right lobe of the liver, and the left testis is located somewhat farther posterior. Each testis is attached to the dorsal bodywall by a fold of the peritoneum, the **mesorchium**.The testes undergo drastic change in size. In the breeding period they increase in size and in non-breeding period they reduce in size. From the inner side of each testis arises the **epididymis**, which extends backwards as a narrow convoluted **vas deferens** or spermiduct. Both the vasa deferentia open independently into the cloaca over a small papilla, and before opening into the cloaca, each receives the ureter of its own side. Thus, this common duct is called the urino-genital duct through which passes the urine as well as sperms.

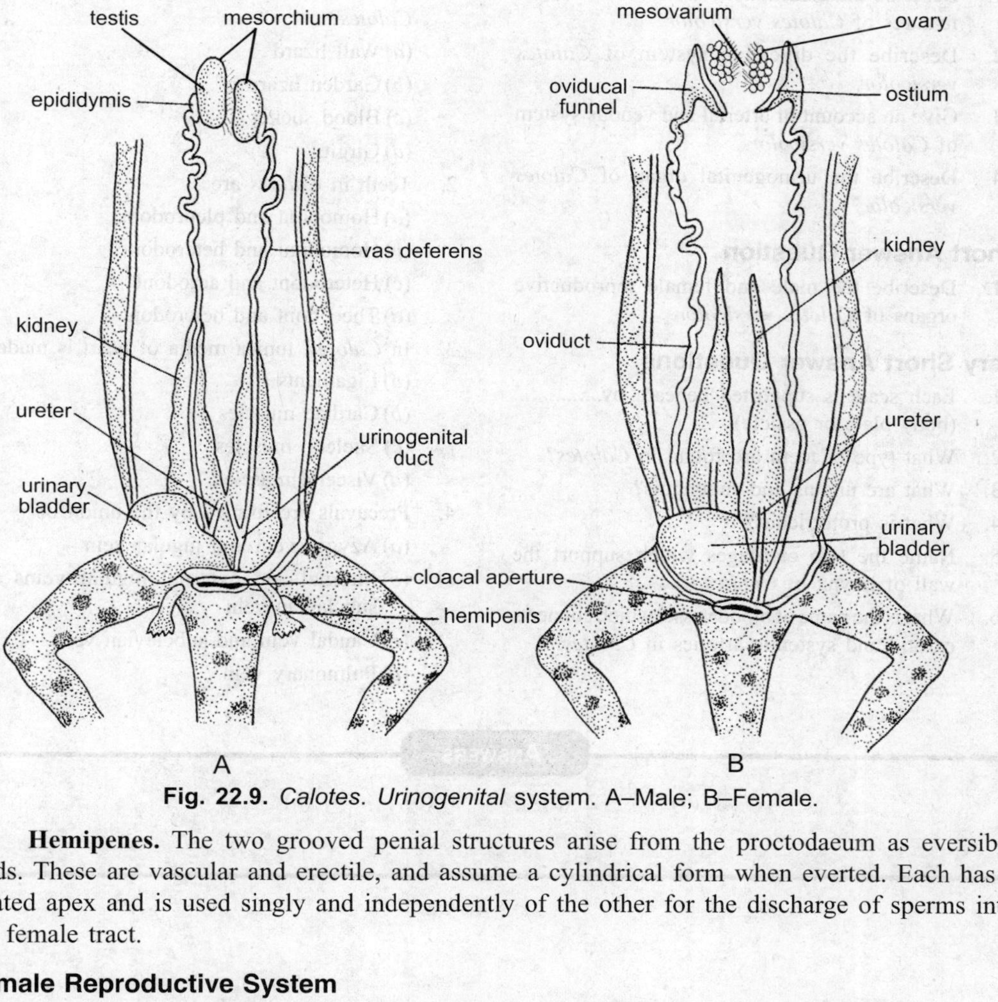

Fig. 22.9. *Calotes*. *Urinogenital* system. A–Male; B–Female.

Hemipenes. The two grooved penial structures arise from the proctodaeum as eversible folds. These are vascular and erectile, and assume a cylindrical form when everted. Each has a dilated apex and is used singly and independently of the other for the discharge of sperms into the female tract.

Female Reproductive System

This includes a pair of irregularly oval-shaped **ovaries**. Their surfaces are raised up into rounded elevations marking the position of the **oocytes**. They are situated in the abdominal cavity and each is attached to the dorsal bodywall by a fold of peritoneum, called the **mesovarium**. A pair of thin-walled, wide **oviducts** are also present and these are attached to the dorsal body-wall by a fold of peritoneum, the **broad ligament**. The anterior end of each oviduct is wide, funnel-shaped, ciliated and lies close to the corresponding ovary. Both the oviducts open into the urodaeum, a little in front of the ureters. The eggs are extruded into the body cavity and pass into the anterior ends of the oviducts. The lower part of each oviduct is slightly dilated, called the **uterus**. Here the eggs are stored for a short time before laying. The walls of the oviducts are glandular, their anterior end bears **albumen glands** and the posterior part bears **shell glands**.

Fertilisation. Fertilisation of eggs takes place in the anterior part of oviduct. The eggs are heavily yolked and are enveloped by calcareous shell.

IMPORTANT QUESTIONS

Long Answer Questions

1. Describe the distribution, habitat and external features of *Calotes versicolor*.
2. Describe the digestive system of *Calotes versicolor*.
3. Give an account of arterial and venous system of *Calotes versicolor*.
4. Describe the urinogenital organ of *Calotes versicolor*.

Short Answer Question

1. Describe the male and female reproductive organs of *Calotes versicolor*.

Very Short Answer Questions

1. Each scale is supported beneath by.............. (bony plate or ossicle).
2. What type of teeth are found in *Calotes*?
3. What are unguis and subunguis?
4. What is prototrichus?
5. Name the two cartilages which support the wall of larynx of *Calotes*.
6. What is the name of blood vessel which connects carotid and systemic arteries in *Calotes*?

Multiple Choice Questions

1. Which of the following is not a name of *Calotes*?
   (*a*) Wall lizard
   (*b*) Garden lizard
   (*c*) Blood sucker
   (*d*) Girgit.
2. Teeth in *Calotes* are
   (*a*) Homodont and pleurodont
   (*b*) Homodont and heterodont
   (*c*) Heterodont and acrodont
   (*d*) Thecodont and heterodont.
3. In *Calotes*, tunica media of heart is made of
   (*a*) Ligaments
   (*b*) Cardiac muscles
   (*c*) Skeletal muscles
   (*d*) Visceral muscles.
4. Precavals are formed by the union of
   (*a*) Azygos vein and jugular vein
   (*b*) Internal and external jugular veins and subclavian vein
   (*c*) Caudal vein and subclavian vein
   (*d*) Pulmonary vein.

Answers

1. (*a*) 2. (*a*) 3. (*b*) 4. (*b*)

23 Chapter

Snakes : General Account

Contents of this Chapter

Snakes are obviously descended from lizards of some kind in Mesozoic times, but their precise mode of origin is obscure. Possibly snake-like reptiles occurred in the Lower Cretaceous, but they were not abundant until the beginning of Tertiary. Some apparently marine forms (*e.g., Palaeophis*) occurred in Eocene, and large constricting snakes like the modern boas and pythons are also found in deposits of Eocene. The Colubridae, containing the majority of smaller, harmless snakes (*e.g.,* grass snakes) did not expand until the Oligocene. The fanged, venomous snakes are certainly identified from Miocene onwards. The venomous Elapidae and Viperidae are first seen as recently as the Miocene-Pleistocene.

EXTERNAL FEATURES

The body of snake is elongated, narrow and cylindrical, usually tapering towards the posterior end, and sometimes with or without a constriction behind the head. The short caudal region is indicated by the position of the cloacal opening. Forelimbs are never represented even by vestiges. But there are inconspicuous vestiges of hindlimbs in the form of small claw-like processes in pythons and other snakes.

The mouth of a snake is capable of being very widely opened by the free articulation of the lower jaw. It mainly distinguishes it from the snake-like lizard. In snakes, the eyelids are immovable and the tympanum is absent.

Size. Reticulated python (*Python reticulatus*) of India and Malaya (25 feet) and Anaconda (*Eunectes murinus*) of South America (25 feet and 300 lb) are the largest living snakes, Indo-Malayan King Cobra (*Naja hannah*) may grow up to 18 feet. Longest sea snakes grow to a length of 8 feet.

Locomotion

Snakes move on land or in water by the movement of spinal column and muscles. This movement pushes the sinuous body horizontally against the resistant surrounding medium. Thick bodied snakes (*e.g.,* boas, pythons, vipers) sometimes progress slowly in a straight line by means of contraction waves which pass along the costocutaneous musculature from head to tail. These specialised muscles are attached to the ventral scales which engage projections in the ground and so draw the animal along.

The North American side-winders (*Crotalus*) and Egyptian sand-vipers (*Cerastes*) move over sand by means of a series of horizontal double loops which engage the earth and propel the animal along sideways.

Various tropical and subtropical tree snakes glide across leaves and twigs from one tree to another.

The Asiatic *Chrysopelea* can ascend vertical walls and tree-trunks aided by laterally keeled ventral scales which engage very tiny projections.

Hearing

In snakes, tympanum, tympanic cavity and eustachean tube are absent. In spite of this, snakes appear to have a good sense of hearing. The columella auris occurs embedded in muscular and fibrous tissue and extends from the stapedial plate to the quadrate, against which it abuts by a cartilaginous epiphysis. In some snakes stapes is a bony plate enclosing the fenestra ovalis and without a shaft-like columella. The snake is probably insensitive to air-borne sound. The rattle snake cannot hear its own rattling. The attachment of columella to quadrate undoubtedly makes the snake highly sensitive to earth-borne vibrations through quadrate, such as approaching feet. They are also quite sensitive to a narrow waveband of low frequency air-borne vibrations.

In snakes the tongue is long, forked at the tip and highly mobile and lodged in a sac in the floor of mouth. It is well provided with sense organs and used as a tactile organ. It can be protruded through an indentation at the extremity of the snout even when mouth is closed. Besides a tactile organ, the tongue of snakes presumably also acts as an auditory organ for receiving earth-borne vibrations.

Feeding Mechanism

Snakes do not "slime" their victims before swallowing. Teeth are sharp-pointed, curved backwards and ankylosed to the jaws. They are present on the maxillae, palatines, pterygoids and on the dentaries. Teeth occasionally absent from pterygoid. They chiefly serve to hold the prey

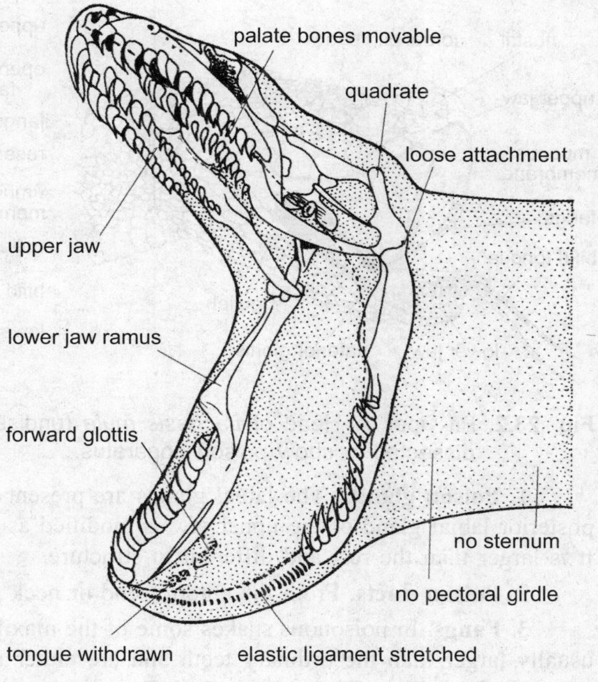

Fig. 23.1. Snake. Mechanism of swallowing large objects.

while it is being swallowed. Snakes feed exclusively on living animals, both warm- and cold-blooded. They swallow the prey without mastication. Swallowing is effected thus : teeth on the lower jaw are alternately hooked further and further forwards into the body of prey (the two halves of the mandible moving forwards alternately), as a result of which mouth and pharynx of the snake are gradually drawn over the animal, the surface of which is at the same time made slippery by the secretion of the buccal glands. During this process the larynx is projected forwards between the rami of the jaws, so that respiration can be maintained. After the completion of the laborious process of swallowing, the animal appears to be entirely prostrated and passes a long period in inactivity, during which slow digestion takes place.

Some snakes kill their prey by crushing, *e.g.,* Python; some by poison and others, the majority, swallow their prey directly. Digestion begins in the stomach and is rapid. When a large animal is being slowly engulfed, the first part of the prey is partly digested before the hind-parts have been swallowed. After taking in a large meal, pythons are able to do without another for more than a year. Besides pepsin secreted by fundic region of stomach, snake venom also assist in digestion. Swallowed rat by a boa constrictor, bones of its head have disappeared in two days and the whole skeleton in five days. *Dasypeltis*, the egg-eating snake, swallows the eggs whole and crushes them with special tooth-like processes of the neck vertebrae.

Poison Apparatus

The poisonous snakes possess poison apparatus in their heads. It is absent in non-poisonous snakes.

The poison apparatus consists of a pair of **1. poison glands**, **2.** their **ducts**, **3.** a pair of **fangs** and muscles. The poison glands are situated one on either side of the upper jaw. Each poison gland is sac-like (oval) in sea snakes or large and tubular in vipers. In *Naja naja*, the poison gland is about the shape and size of an almond kernel. The gland is thickly encapsulated with fibrous connective tissue. It is composed of a body and a neck. The roughly fan-shaped capito-mandibularis superficial muscle, originating on the post-frontal (post-orbital) bone and parietal ridges, embraces much of the body of the poison gland. Its contraction during biting, swiftly and synchronously squeezes poison into the poison duct. The duct passes forward along the side of the upper jaw and loops over itself just in front of the fang and opens either at the base of the fang or at the base of the tunnel on the fang.

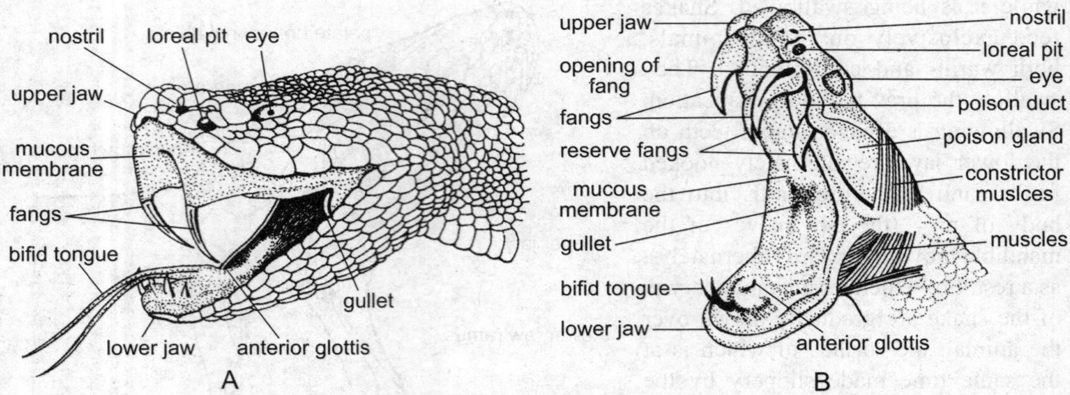

Fig. 23.2. Pit viper. A–Head of *Lachesis muta* (undissected); B–Head of rattle snake, *Crotalus*, dissected to show poison apparatus.

1. Poison glands. The labial glands are present in a row in the upper and lower jaw. The posterior labial gland of the upper jaw is modified as the poison gland in the poisonous snakes. It is larger than the rest and different in structure.

2. Poison ducts. From the anterior end or neck of each poison gland arises a poison duct.

3. Fangs. In poisonous snakes some of the maxillary teeth become poison fangs. These are usually larger than the ordinary teeth and are either grooved or perforated by a canal for the passage of the duct of the poison gland. Fangs are long, curved and pointed. The fangs regenerate by one of the small reserve fangs at its base (vipers). On the basis of structure and position, three types of fangs occur in poisonous snakes. These are as follows :

(a) Solenoglyphous (*solen* = pipe + *glyph* = hollowed). In Viperidae (vipers and rattle snakes), there is a single, large, curved, poison fang on the maxilla with small reserve-fangs at its base. These are the only teeth borne by the maxilla, which is very short. The large poison fang is capable of being rotated through a considerable angle, and moved nearly horizontal position (in which it lies along the roof of the mouth embedded in folds of the mucous membrane) to a nearly vertical position, when snake opens its mouth to strike its prey. Each poison fang contains a canal open at each end.

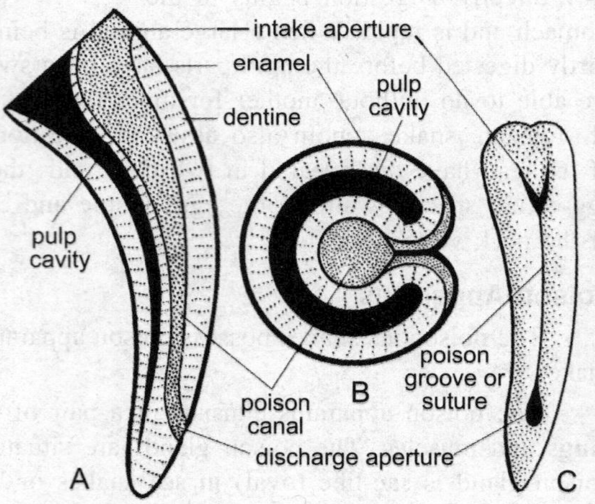

Fig. 23.3. Snake fangs. A–Solenoglyphous fang in L.S; B–Solenoglyphous fang in T.S.; C–Entire grooved fang.

(b) Opisthoglyphous (*opistho* = behind). In some poisonous snakes of the family Colubridae (subfamily Homalopsinae, Dipsadomorphinae and Elachiostodontinae) such as *Dryophis*, *Elapops*, *Lycognathus*, *Elachistodon westermanni* found in Bengal, etc., one or more of the posterior maxillary teeth grooved along its posterior border. There are more or less poisonous and poison is weak.

(c) Proteroglyphous (*protero* = first). In sea snakes, *Bungarus* (krait), cobras (*Naja*), fangs (anterior maxillary teeth) are small, permanently erect and grooved along its anterior face.

(d) Aglyphous. Fangs are lacking, so such snakes are non-poisonous.

4. Muscles. The poison apparatus is associated with specialised bands of three types of muscles, *viz.,* ***(i)*** **digastric** ***(ii)*** **sphenopterygoid** or **protractor-pterygoid** and ***(iii)*** **anterior** and **posterior temporalis. Digastric muscle** is attached to the squamosal of the skull at one end and the articular of the lower jaw. **Sphenopterygoid** is attached anteriorly to the sphenoidal region and posteriorly to the dorsal surface of the pterygoid. It helps in pulling the pterygoid forward. **Anterior** and **posterior temporalis muscles** are attached to the sidewalls of the cranium and the lower jaw. They help in closing the lower jaw.

Biting Mechanism

The skull and jaw bones in poisonous snakes are loosely and movably articulated, thus, allowing an enormous gape and swallowing whole of large prey. In cobras fangs are small and remain permanently erect, but in vipers the fangs are large and curved and lie against the roof of mouth cavity when closed. Premaxilla, usually toothless and the bones of the upper jaw are loosely attached to rest of the skull. Quadrate jointed to the squamosal. There are movable joints between the frontals behind and prefrontals and nasals in front and also between several other bones of brain case, palate and jaws. These joints have loose ligaments and allow movement in several directions and so permit a huge gape. The two halves of the lower jaw are connected together by elastic ligamentous tissue. So they are capable of being widely separated from one another.

The mechanism of biting is a complicated process and it can be described in the following three steps :

(i) **Opening of the mouth.** By the contraction of digastric muscles the mouth is opened (lower jaw moves down).

(ii) **Rotation of maxilla.** As the mouth opens the lower jaw moves down and the lower end of quadrate moves forward. Quadrate and squamosal are very movable. The pterygoid is movably attached to the palatine. Quadrate pushes the pterygoid forward and the pterygo-palatine joint bent. This forward movement of the pterygoid is conveyed by the transpalatine bone to the maxilla and causes it to rotate through about 90° upon its prefrontal articulation in such a way that the surface to which the fang is attached is carried forwards and ventralwards, and the fang is erected, *i.e.,* is made to project downwards at the front end of the mouth. The contraction of sphenopterygoid muscles also helps in the movement of pterygoid forward.

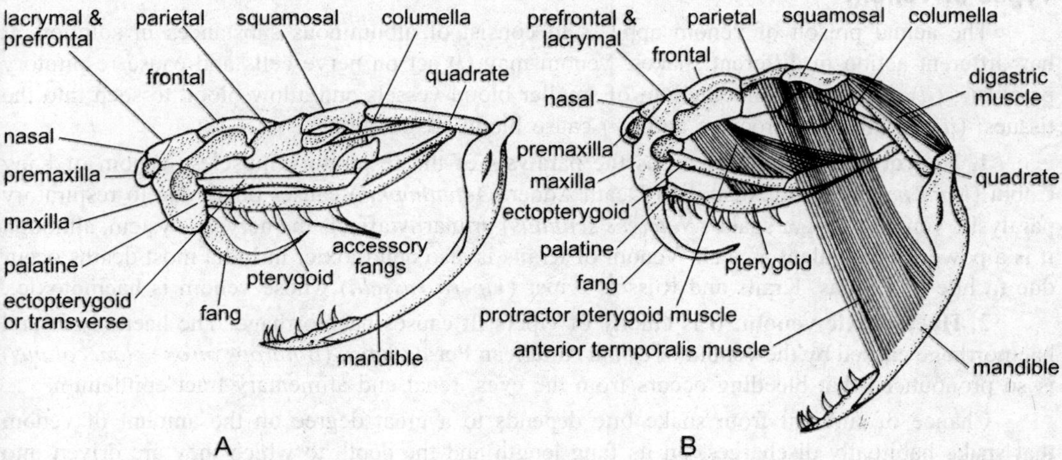

Fig. 23.4. Skull of a viper showing biting mechanism. A–Mouth closed at rest; B–Mouth opened when striking the prey.

***(iii)* Closing of mouth.** The closing of the mouth is brought about by the contraction of the temporalis and sphenopterygoid muscles. The point of fang is directed backward while the mouth is closed. It takes longer time to open the mouth than to close it.

***(iv)* Transference of venom.** During the contraction of the digastric muscle the posterior ligament is relaxed and during the rotation of the squamosal bone the fan-shaped ligaments are stretched to squeeze the wall of the poison gland. This makes the poison to come out of the poison gland through the poison duct and the fang.

Swallowing Mechanism

The rami of the lower jaw are separate in front but are connected by an elastic ligament which allows them to come apart. The upper jaw is loosely attached to the skull. A long quadrate articulates movably with the squamosal forming a lever which, when straightened, separates the lower jaw from the base of the skull, thus, the mouth and throat are opened so wide that the snake can swallow very large animals. Swallowing is done very slowly, and to enable breathing, the glottis is pushed forwards between the two rami of the lower jaw. Absence of pectoral girdle and sternum is an aid to swallowing large animals.

Non-poisonous snakes. Fangs are not found. Teeth rarely developed on the premaxillae in snakes. But teeth are present on the maxillae, palatines and pterygoids, as well as the dentary of the mandible. Teeth are solid, elongated, sharp-pointed and usually strongly recurved. The function of such teeth is to hold prey and prevent it from struggling free while it is being swallowed.

Snake Venom

Snake venom is secreted by poison glands which are the modified supralabial or parotid glands present in the head of snakes. Venom is injected in the body of prey at the time of bite through the grooved maxillary fangs. Snake venoms vary in composition, most are mixtures of enzymes and proteins. They often include polypeptides that poison the respiratory muscles of the prey. Many include proteases, which help in digestion. Venom is a clear sticky fluid of very light yellow or greenish colour. It is acidic in reaction and tasteless and odourless. It causes no harm if swallowed but there should be no scratch in the interior of alimentary canal. It is only fatal when mixed with blood. It is soluble in water, glycerine or salt solution and in this condition also it is poisonous. It can be precipitated in silver nitrate and potassium permanganate.

Types of Venom

The actual poison of venom appears to consist of albuminous substances in solution. It has different action in different snakes. Venom may (*i*) act on nerve cells and cause respiratory paralysis; (*ii*) destroy the endothelium of smaller blood vessels and allow blood to seep into the tissues; (*iii*) destroy erythrocytes and (*iv*) cause blood coagulation.

1. Neurotoxic venom. It causes the paralysis of the respiratory muscles. Venom of King Cobra (*Naja hannah*) and Australian Death Adder (*Acanthophis*) causes death due to respiratory paralysis. Venom of Tiger snake (*Notechis scutatus*) primarily affects the nervous system, although it is a powerful coagulant as well. Venom of Kraits is also neurotoxic. In India most deaths occur due to bite of Cobras, Kraits and Russels' viper (*Vipera russelli*) whose venom is haemotoxic.

2. Haemotoxic venom. It is chiefly of vipers. It causes haemorrhage. The haemolysis and haemorrhage caused by the venom of South American Fer-de-lance (*Bothrops atrox = lanceolatus*) is so pronounced that bleeding occurs from the eyes, renal and alimentary tract epithelium.

Chance of survival from snake bite depends to a great degree on the amount of venom that snake habitually discharges, on its fang length and the depth to which they are driven into the tissues. It has been suggested that about 2 milligrams of neurotoxic venom of Tiger snake and Blue krait of India (*Bungarus candidus*) are toxic to the man.

Treatment

The best treatment is to inject the antivenom serum or antivenin (serum of animals which have been rendered partially immune by repeated doses of the venom). Failing this, the best plan is to apply a ligature above the wound, twisting the string with a stick, and then to make an incision of the wound. Then bandage the limb downwards towards the wound repeating this several times. Direct application into the widened wound of bleaching powder (calcium hypochlorite) or of a 1 per cent solution of Condy's fluid is good. Amputation is of course the best remedy, if a very deadly snake has bitten the part. Alcohol and injection of ammonia are useless, and sucking the wound is dangerous. Many mammals are said to be immune, *e.g.*, mongoose, hedgehog and pig.

DISTINCTION BETWEEN POISONOUS AND NON-POISONOUS SNAKES OF INDIA

The majority of snakes are non-poisonous and harmless animals. Some of the common poisonous snakes of India are : cobra, king cobra, krait, pit-viper, Indian viper, rattle snake, russel's snake, coral snake, and sea-snakes. The identification of poisonous and non-poisonous snakes is very simple. The following key will serve to distinguish **poisonous snakes** from **non-poisonous snakes** :

1. If it is a **marine snake** with the tail laterally compressed, it is **poisonous**.

2. In a terrestrial snake the tail is rounded or cylindrical and not laterally compressed. It may be both poisonous and non-poisonous. Examine its ventral scales.

(*i*) If all the ventral scales are small or the ventral scales are somewhat broad and continuous with dorsals, then it is a non-poisonous snake.

(*ii*) If the ventral scales are large transverse plates extending fully across the ventral side or belly, the snake may be poisonous or non-poisonous. To find out examine the dorsal surface of the head.

(*a*) If all the dorsal scales of the head are small, then it may be a pitless viper (poisonous).

(*i*) If there is a **loreal pit** between the nostril and the eye, and scales over the head are small, then it is a pit viper (poisonous).

(*ii*) If the **subcaudals** are double and there is a loreal pit, then it is a russel's viper (*Vipera russelli*) and when subcaudals are single, then it is *Echis carinata*.

(*b*) If dorsal side of the head has both small scales and large shields, the snake may or may not be poisonous; to ascertain look shields present at the side of the head.

3. If the **third supra-labial** shield touches the nostril and eye, then it is a **poisonous snake**, may be cobra, king cobra or coral snake.

4. If the upper side of the head has both small scales and large shields but there is no loreal pit, and the third supra-labial shield does not touch the eye, then examine the back shields of the snake and lateral side of the iower jaw.

(*i*) The middle row of scales on the back called **vertebrals** may be larger than other laterals.

(*ii*) Ventral side of lower jaw has fourth infra-labial shield larger than the others. If both (*i*) and (*ii*) characters are present, then it is a **krait (poisonous)**.

5. If the snake has small scales and large shields on the head but does not have the characters of cobra, krait or coral snake, then it is **non-poisonous**.

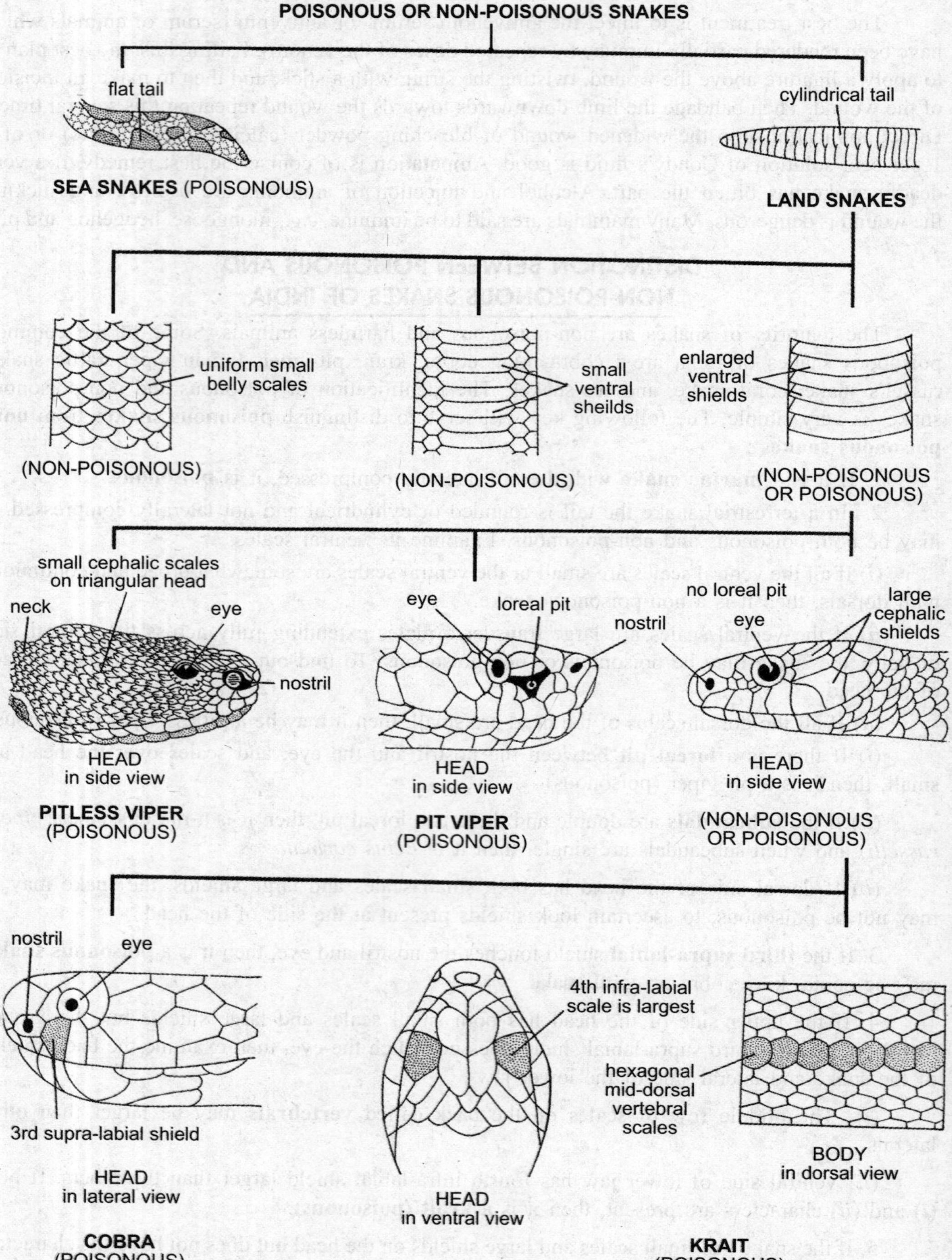

Fig. 23.5. Diagram for identification of poisonous and non-poisonous snakes.

KEY FOR IDENTIFICATION OF POISONOUS AND NON-POISONOUS SNAKES OF INDIA

Snakes

- I—Tail flat and laterally compressed
  - Sea snakes (*Hydrophis*)
    - → **Poisonous**
- II—Tail cylindrical, rounded and not laterally compressed
  - Terrestrial or land snakes
    - → **Poisonous or Non-poisonous**
      - A. Small scales on the belly continuous with dorsals and on the back
        - → **Non-poisonous**
      - B. Belly scales not broad enough to extend right across it
        - → **Non-poisonous** (Python)
      - C. Ventrals broad covering the entire width of the belly
        - Small scales on the head. Loreal pit absent.
          - Pitless viper (Subcaudal double) Subcaudal single *Echis carinata*
            - → **Poisonous**
        - Small scales or shields on the head and a loreal pit between the nostril and eye
          - Pit viper
            - → **Poisonous**
        - Shields on the head. No loreal pit.
          - Ventrals not enlarged. Third supra-labial shield (upper lip shield) touches the eye and nose shield.
            - **Poisonous**
              - Neck with hood and spectacle marking
                - Cobra (*Naja*)
                  - → **Poisonous**
              - Neck without hood, coral spots on the belly
                - Coral snakes (*Callophis*)
                  - → **Poisonous**
          - Vertebrals (*i.e.,* scales on the middle of the back) are enlarged, subcaudals undivided and bands and half rings across the back. Fourth infra-labial is largest.
            - Krait
              - → **Poisonous**
          - None of the characters of cobra, coral snake or krait
            - → **Non-poisonous**

IMPORTANT QUESTIONS

Long Answer Questions

1. How will you distinguish poisonous snakes from non-poisonous snakes?
2. Give distinguishing characters of the following: (*a*) *Echis carinata*; (*b*) *Bungarus caerulens*; (*c*) *Naja naja.*
3. How are the snakes able to swallow animals much larger than the normal size of their mouth?
4. How are the bones of skull fitted in snakes? With a labelled sketch, state how they operate in the biting action of a poisonous snake.
5. How will you distinguish between a poisonous and a non-poisonous snake? Describe the biting mechanism of a poisonous snake.

Short Answer Questions

1. What is the advantage of the lower jaw of snake joined only by an elastic ligament? Explain.
2. Write a short note on the external characters of poisonous snakes.
3. How will you differentiate a krait from a viper? Give sketches to support your arguments.
4. Describe the poison apparatus of any poisonous snake.
5. Compare the fangs of the cobra with those of viper.
6. How would you distinguish *Typhlops*?
7. What are the scales on the head of a pit viper?
8. Describe some important snakes of South India.
9. What are the three types of fangs? Describe each with example.

Very Short Answer Questions

1. Poison glands of snakes are modified.................
2. Write about masseter (temporal) muscle in poisonous snakes.
3. What is solenoglyphous type of fang?
4. What is the effect of neurotoxin on prey?
5. Write about the effect of haemotoxin on prey.
6. Which snakes possess haemotoxic venom?
7. Which snake's venom affect blood vascular system and nervous system?

Multiple Choice Questions

1. A viper is easily identified by the
   (*a*) Flattened tail
   (*b*) Small scales on head
   (*c*) Enlarged vertebrals
   (*d*) Broad ventrals.
2. A villager has been bitten by a snake and you rush to help. If at first sight you cannot identify the snake, you try to find out if it is poisonous at all by quickly looking at its
   (*a*) Tail (*b*) Ventral side
   (*c*) Markings on body (*d*) Head shields.
3. Having ascertained whether it is poisonous or non-poisonous and if it happens to be poisonous you would proceed with other identifying marks. If the head is covered with scales and not shields, then it is
   (*a*) Krait (*b*) Viper
   (*c*) Cobra (*d*) Rat snake.
4. If the vertebrals are enlarged on its back, it is
   (*a*) Cobra (*b*) Viper
   (*c*) Krait (*d*) Natrix.
5. If the tail is laterally compressed and not cylindrical, it is
   (*a*) Cobra (*b*) Krait
   (*c*) Viper (*d*) Sea snake.
6. In *Bungarus* there are
   (*a*) 4 infra-orbitals and all are of the same size
   (*b*) 4 infra-labials and 4th is the largest
   (*c*) 5 infra-labials and 4th is the largest
   (*d*) 5 infra-labials and all are of the same size.
7. Snakes are sensitive to
   (*a*) Earth-borne vibrations
   (*b*) Noises made by birds
   (*c*) Thunder
   (*d*) Air-borne vibrations.
8. Solenoglyphous is found in
   (*a*) Family Colubridae
   (*b*) Viper and rattle snake
   (*c*) Coral snakes and Sea snakes
   (*d*) Cobra and Kraits.

Answers

1. (*b*)	2. (*a*)	3. (*b*)	4. (*c*)	5. (*d*)	6. (*b*)
7. (*a*)	8. (*b*)				

Class 7. Reptilia : General Characters, Classification and Representative Types

24

Chapter

Contents of this Chapter

Reptiles are cold-blooded, usually scaly vertebrates, with a right and a left aortic arch, a single occipital condyle and pulmonary respiration. The ovum is large and meroblastic, and the embryo has an amnion and allantois. The name Reptilia refers to the mode of locomotion (L., *repere* or *reptum* = to creep or crawl). The study of reptiles is called **Herpetology** (Gr., *herpeton* = reptiles).

GENERAL CHARACTERS

1. Reptiles are cold-blooded, terrestrial or aquatic air-breathing animals.

2. Body is divisible into head, neck, trunk and tail, and covered with horny scales or scutes.

3. Limbs tetrapodus (2 pairs), pentadactyle type. Digits with horny claws. However, limbs are absent in some lizards (*Ophiosaurus*) and snakes.

4. Mouth terminal, jaws provided with simple conical teeth. Teeth are replaced by horny beaks in turtles and tortoises. A typical cloaca is present.

5. Skin is dry cornified and skin glands are absent.

6. Endoskeleton bony, skull with single median occipital condyle. Interclavicle is T-shaped.

7. Mandible consists of many pieces and articulates with the cranium through the quadrate bone.

8. Vertebrae are gastrocentrous.

9. Ribs single-headed.

10. Respiration by lungs throughout life.

11. Heart is three-chambered, divided into two auricles and incompletely divided ventricle. In crocodiles heart is four-chambered.

12. Right and left aortic arches are complete and functional.

13. Red blood corpuscles (R.B.Cs) are oval and nucleated.

14. Kidneys without nephrostomes and metanephric. Each kidney is provided with separate ureter. Excretion uricotelic.

15. Cerebrum of brain is well developed than that of amphibians. Twelve pairs of cranial nerves.

16. Sexes separate. Male usually with a pair of copulatory organs (snakes and lizards) and unpaired in Chelonia and Crocodilia.

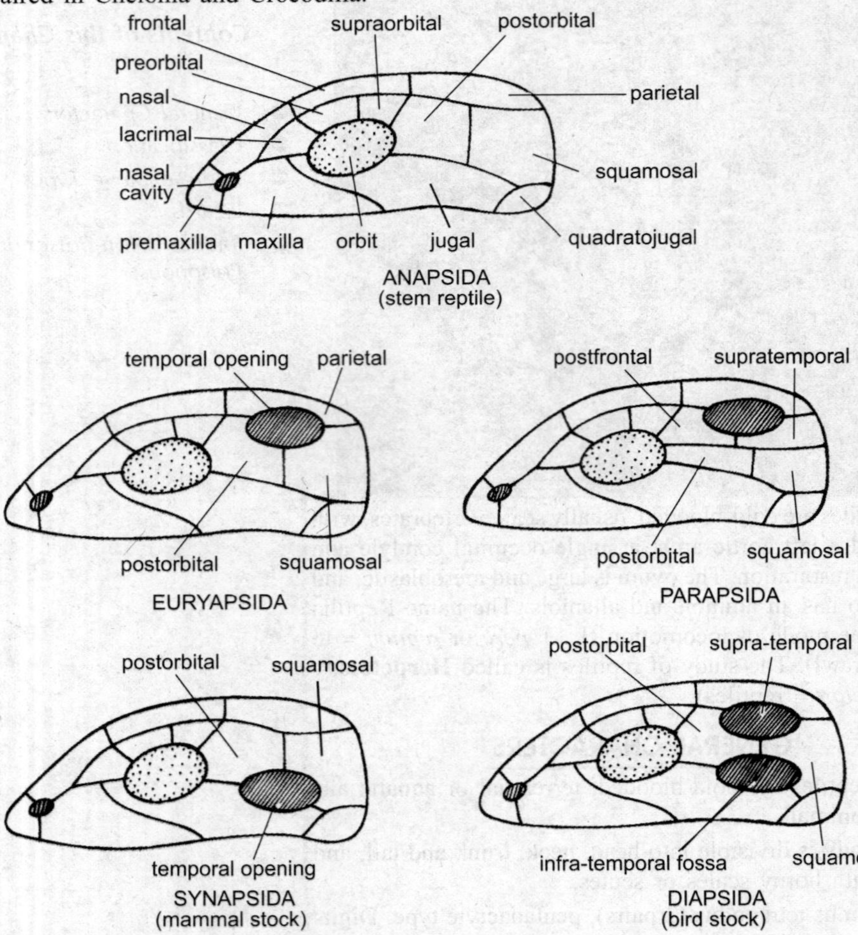

Fig. 24.1. Five types of skulls in lateral view in five subclasses of reptiles.

17. Lateral line sense organs are absent. Jacobson's organs are present in the roof of the buccal cavity.

18. Fertilisation is internal. Eggs are laid on land which are large and covered with leathery shells.

19. Embryonic membranes (amnion and allantois) are present.

20. Parental care is absent.

CLASSIFICATION

Class Reptilia has about 7,000 living and several extinct species (**Bogert**). These are grouped into about 16 orders, of which only four are living. The class Reptilia has been divided into the following five subclasses on the basis of presence or absence of temporal vacuities in the skull.

1. Anapsida
2. Euryapsida
3. Parapsida
4. Diapsida
5. Synapsida.

SUBCLASS 1. ANAPSIDA

1. Solid roof of the skull.
2. No temporal vacuity or fossa in the skull.
3. Quadrate is connected to the otic bones.
4. Limbs are generally strong.
5. Body is enclosed in the dorsal carapace and ventral plastron.

Order 1. Cotylosauria (stem reptiles)

1. Extinct reptiles having complete cranial roof in the temporal region.
2. Exhibit resemblances with Labyrinthodont amphibians.
3. Jaws and teeth are poorly developed.
4. Limbs are feeble and project laterally.
5. Pelvic girdle is flat and plate-like.
6. Range in size from one to ten feet.
7. Become extinct in the Triassic.

Example : *Seymouria.*

Order 2. Chelonia or Testudinata

(Gr., *chelone* = turtle; L., *testudo* = turtle).

1. Terrestrial or aquatic.
2. Body is enclosed in a bony shell composed of dorsal carapace and ventral plastron, which may or may not be covered with horny shields. Nasal opening single at the front end of snout.
3. Tetrapoda pentadactyle with clawed walking limbs and/or webbed paddles.
4. Tail is always present.
5. Skull anapsid without a parietal foramen. Quadrate bone is fixed. Sternum absent.
6. Jaws without teeth but with horny beaks.
7. Thoracic vertebrae and ribs usually fused with the carapace. Two sacral vertebrae.
8. Humerus with entepicondylar foramen.
9. Pubes and ischia form symphyses.
10. Cloacal opening is longitudinal.
11. Copulatory organ is unpaired and simple.
12. Oviparous.

It includes about 400 species of marine turtles, freshwater terrapins and terrestrial tortoises.

Examples : *Chelone, Chrysemis, Testudo, Kachuga, Trionyx, Dermochelys.*

SUBCLASS 2. EURYAPSIDA (EXTINCT)

Skull with a single pair of temporal openings (one on either dorso-lateral side) bounded below by postorbital and squamosal bones.

SUBCLASS 3. PARAPSIDA (EXTINCT)

1. Skull with one pair of temporal vacuities, one on either side. Each temporal vacuity is bounded by post-frontal and supra-temporal below and parietal above.
2. Includes all extinct forms, from Mesozoic to Cretaceous.

Examples : *Ichthyosaurus, Plegiosaurus Protosaurus, Ophthalmosaurus.*

SUBCLASS 4. DIAPSIDA

Two temporal vacuities on each side of the skull which are separated by a bar of post-orbital and squamosal bones.

SUPERORDER I–LEPIDOSAURIA (SCALY LIZARDS)

1. Two temporal vacuities are present (diapsid skull).
2. Anterior orbital vacuities are absent.
3. Post-temporal fenestrae are usually present.
4. Humerus with two foramina.

Order 1. Rhyncocephalia (L., *rhynchos* = snout + Gr., *kephale* = head)

1. Two vacuities are present in the skull.
2. Teeth acrodont.
3. An epipterygoid is present.
4. Quadrate bone is fixed or immovable.
5. Large parietal foremen is present. Non-functional median eye is present in the living form. No antorbital foramen.
6. Vertebrae amphicoelus and intercenta present. Limbs pentadactylous adapted for walking.
7. External copulatory organs are entirely absent.
8. Ribs single-headed and have uncinate processes. Abdominal ribs and sternum present.

Example : *Sphenodon* (lizard-like Tuatara).

Order 2. Squamata (L., *squama* = scale or *squamatus* = scaly)

1. Body is covered by horny epidermal scales.
2. Single (supra or upper) temporal vacuity in lizards and no temporal vacuity in snakes.
3. Teeth pleurodont or acrodont.
4. Quadrate is movably articulated with skull.
5. Quadratojugal is absent.
6. Vertebrae are usually procoelous. Sacrum with two vertebrae.
7. Limbs clawed, present or absent in snakes and some lizards.
8. Cloacal opening transverse. Jacobson's organs paired.
9. Male possesses a pair of eversible copulatory organs (penes).

It includes about 6800 species of lizards and snakes which are divided into two following suborders :

Suborder I. Lacertilia (= Sauria)

1. Terrestrial, arboreal, burrowing and aquatic forms.
2. Carnivorous and some are herbivorous.
3. Single (supra or upper) temporal vacuity is present.
4. Limbs pentadactyle and are usually present. Girdles also present.
5. The rami of mandibles are fused in front. Mouth non-expandable. Premaxillae bear teeth.
6. Sternum, episternum and urinary bladder present.
7. Eyelids movable. Nictitating membrane present.
8. Ear openings and tympanum present.
9. Both lungs equally developed.

Examples: *Hemidactylus, Calotes, Chaemeleon, Draco, Phrynosoma, Varanus, Crotaphytus, Mabuiya, Gecko, Ophisaurus, Rhineura, Iguana, Heloderma, Molloch.*

Suborder 2. Ophidia (= Serpentes)

1. Terrestrial or aquatic, arboreal or burrowing.
2. Body slender and narrow.
3. Temporal vacuities are entirely absent.
4. Limbs and girdles absent. Vestigial hindlimbs and pelvic girdle in boa and python.
5. Eyelids immovable. Nictitating membrane absent.
6. Ear openings and tympanum absent.
7. Tongue is bifid and protrusible.
8. The rami of mandibles are united by a ligament and can be separated at the time of swallowing.
9. Sternum, episternum and urinary bladder absent.
10. Maxillae, palatines and pterygoids freely movable.
11. Premaxillae toothless.
12. Zygosphene and zygantra are present in the vertebra.
13. Left lung small.

Examples : *Typhlops, Python, Eryx johni, Lycodon, Bungarus Naja naja, Enhydrina, Hydrophis, Viper, Ancistrodon, Natrix, Zaemenis, Dendrophis, Crotalus, Elaps.*

SUPERORDER II—ARCHOSAURIA (RULING REPTILES)

Archosaurs were the dominant land animals of late Mesozoic and include dinosaurs and pterosaurs. Crocodiles are the only living descendants of the group.

1. Skull diapsid with closed upper temporal vacuity but possesses both the temporal arches.

   Interparietal, tabular bones or a parietal foramen absent.
2. Anterior orbital vacuities are usually present behind the nostrils.
3. Teeth are thecodont.
4. Vertebrae amphicoelous or procoelous.
5. Humerus with foramina. Hindlegs longer and powerful.

Order. Crocodila

(Gr., *krokodeilos* = crocodile)

1. Freshwater and predatory carnivorous forms.
2. Body is covered with an exoskeleton of thick horny epidermal scales, bony plates and scutes.
3. Tail is long, strong and laterally compressed.
4. Limbs short, powerful, clawed and webbed.
5. Maxillae, palatines and pterygoids are united along the middle to form a secondary palate.
6. Quadrate immovable. Parietal foramen absent.
7. Both supra and infra temporal arches present.
8. Teeth are thecodont lodged in sockets.
9. Vertebrae procoelous. Cervical vertebrae with two-headed ribs.
10. Sternum is present.
11. Thoracic ribs double-headed and possess uncinate processes. Abdominal ribs are present.
12. Prepubis present. Pubis does not share in the formation of acetabulum.
13. Heart is four-chambered and ventricle is completely divided into two by a septum.
14. Lungs are spongy.
15. Cloacal opening longitudinal.
16. Males possess a median unpaired erectile and grooved penis.

Examples: *Crocodilus, Alligator, Gavialis.*

SUBCLASS 5. SYNAPSIDA (EXTINCT)

1. Skull with a pair of temporal vacuities, one on either side, each bounded by the post-orbital and squamosal above and by the jugal and squamosal below.
2. Teeth usually heterodont.
3. Dentary large except in the most primitive members.
4. Pectoral girdle has a coracoid and precoracoid.
5. Includes all extinct forms from Carboniferous to Permian.

Examples : *Dimetrodon, Cynognathus.*

REPRESENTATIVE TYPES

Turtles and Tortoises

The term **turtle** usually includes the aquatic and semiaquatic forms, **tortoise** includes the land forms and terrapin is used for some edible freshwater species. The characteristic of chelonian organisation is the shortening and broadening of the body, together with the development of bony plates, forming a box into which the head and limbs can be withdrawn. The total number of segments is only about 8 in the neck, 10 in the trunk and a series of reduced caudals. The shell usually includes a dorsal carapace and ventral plastron. Each of these is made up of inner plates of bone, covered by separate outer plates of horny material. The carapace includes five rows of bony plates, namely, median **neurals**, and paired **costals** and **marginals**. These plates are ossifications in the dermis, attached to the vertebrae and ribs. The plastron is developed from the expanded dermal bones of the pectoral girdle together with dermal ossifications. The whole is covered in most chelonians by rows of special smooth epidermal plates forming the "tortoise" shell. A new and larger layer is added to each of these plates each year, thus, making a number of "growth rings" from which age of the tortoise can be calculated, though the outer layers often become rubbed off.

The limb girdles lie inside the encircling ribs. The pectoral girdle's scapula meets the carapace dorsally and carries a long acromial process and a backwardly directed coracoid, both being attached by ligaments to the plastron. The ilia are attached to two sacral vertebrae. Limbs are stout and with five digits. In marine turtles they are transformed into paddles.

In recent Chelonia teeth are totally absent, both in the herbivorous and carnivorous forms. The edges of the jaws form sharp ridges, covered with horny beak. Skull is completely roofed in **Chelonia**, dermal bones are widely separated from the brain case. In other groups of Chelonia, the dermal roofing has been reduced or emarginated.

The heart has a muscular sinus, separated auricles, a partly divided ventricle and two equal aortic arches. Lungs are spongy attached to the dorsal surface of the shell, sometimes enclosed in a separate pleural cavity (*Testudo*). Some aquatic forms (*Emys*) also respire by taking water into special vascularised diverticuli of urodaeum. The metabolism of tortoises is low and they can remain for long periods without breathing. For this purpose they obtain energy by anaerobic glycolysis. Further, the tissues including the brain are unusually tolerant of anoxia. In temperate climates all species hibernate regularly.

The kidney is metanephric and the nitrogenous waste vary between urea and uric acid. Cloaca is typically subdivided and reabsorption of water occurs in cloaca to form a solid whitish excretory product. In male the copulatory organ is single. A female may lay 10 clutches of 100 eggs in a year. Turtles all come ashore to breed. The marine *Chelonia mydas* (green turtle) make migrations of more than 1500 km, perhaps by sun compass. They place their eggs carefully in holes made by scooping away the sand with the hind feet and then the traces are covered. *Geochelone gigas*, the giant tortoise of the Aldabra Islands, also digs a nest for its eggs.

Chelonia mydas is also found in Bay of Bengal and is about 1 metre long. *Dermochelys*, the leathery turtle having no horny shell, is over 2 metre long and weighs over 500 kg.

Chelonian history starts back from the Triassic. *Proganochelys* (*Triassochelys*) was an early turtle, with a shell like that of modern forms, but still possessing teeth on the palate. The head, tail and limbs could not be withdrawn into the shell and were protected by spines. In later evolution *Pleurodira*, retraction of head became possible. In side-neck turtle, the neck is folded side ways. This group was world-wide in Cretaceous and survived today in Tropical Africa (*Chelus*), South America and Australia. In **Cryptodira**, the neck is curved in a dorso-ventral plane. This type is also known from Jurassic and Cretaceous and includes most of the modern types, *Dermochelys,* the leathery turtle, has a carapace consisting only of a mosaic of small bony plates beneath its leathery skin. *Trionyx* is a freshwater turtle with a soft shell and no horny plates. Chelonia includes more than 200 species which are quite varied and widespread. These include terrestrial animals, such as *Testudo graeca* and *T. imarginata* tortoises of Southern Europe, the freshwater tortoises such as *Chrysemys* and other American terrapins; and *Emys*, the European water tortoise. *Trionyx gangeticus* is found in rivers of North India and *T. laithi* in South India. *Chelydra serpentina* (snapping turtles) and *Macrolemys* (alligator turtles) are carnivorous feeding on fish, frogs, water-fowls, etc. *Chrysemys picta* (painted turtle or terrapin) feeds on insects. *Malaclemys* (diamond-back terrapin) is used as food by man. *Kachuga* found in Indian rivers have carapace with spinous plates. *Chelodina longicollis* (side-necked turtle of Australia) has a S-shaped folded neck. *Chelus* (mata-mata) of America and Africa is a curious side-necked freshwater turtle. *Testudo* is a common land tortoise. *Testudo abingdoni* (giant land tortoise) of Galapagos islands and Indian Ocean islands weighs over 100 kg and its age varies from 200 to 400 years.

1. *Chelone.* *Chelone mydas* (Fig. 24.2) is a marine form commonly called **green turtle**. It is edible and is found in Indian, Pacific and Atlantic Oceans. Forelimbs form wing-like paddles with only a single claw found on the first digit of the paddle, digits are not visible externally, hindlimbs are clawed. Head and neck are not completely retractile, jaws have denticulate edges.

Skull and shell are light and porous. Tail is very short. Carapace is flat and heart-shaped. Plastron is joined to carapace only by ligaments, there being no bony ridges, osteoderms of plastron leave a large unossified gap in the middle. Shell has osteoderms and horny scales. Marine turtles do not drink water but extract salt from the sea by means of a pair of glands situated above the eyes. It lives mostly on seaweeds but also eats fishes. It comes ashore for breeding and lays up to 200 eggs. *Chelone mydas* and *Chelone virgata* are found in the Bay of Bengal, the upper surface is marbled olive and the lower is pale yellow. In the plastron there is an intergular scute between two gulars.

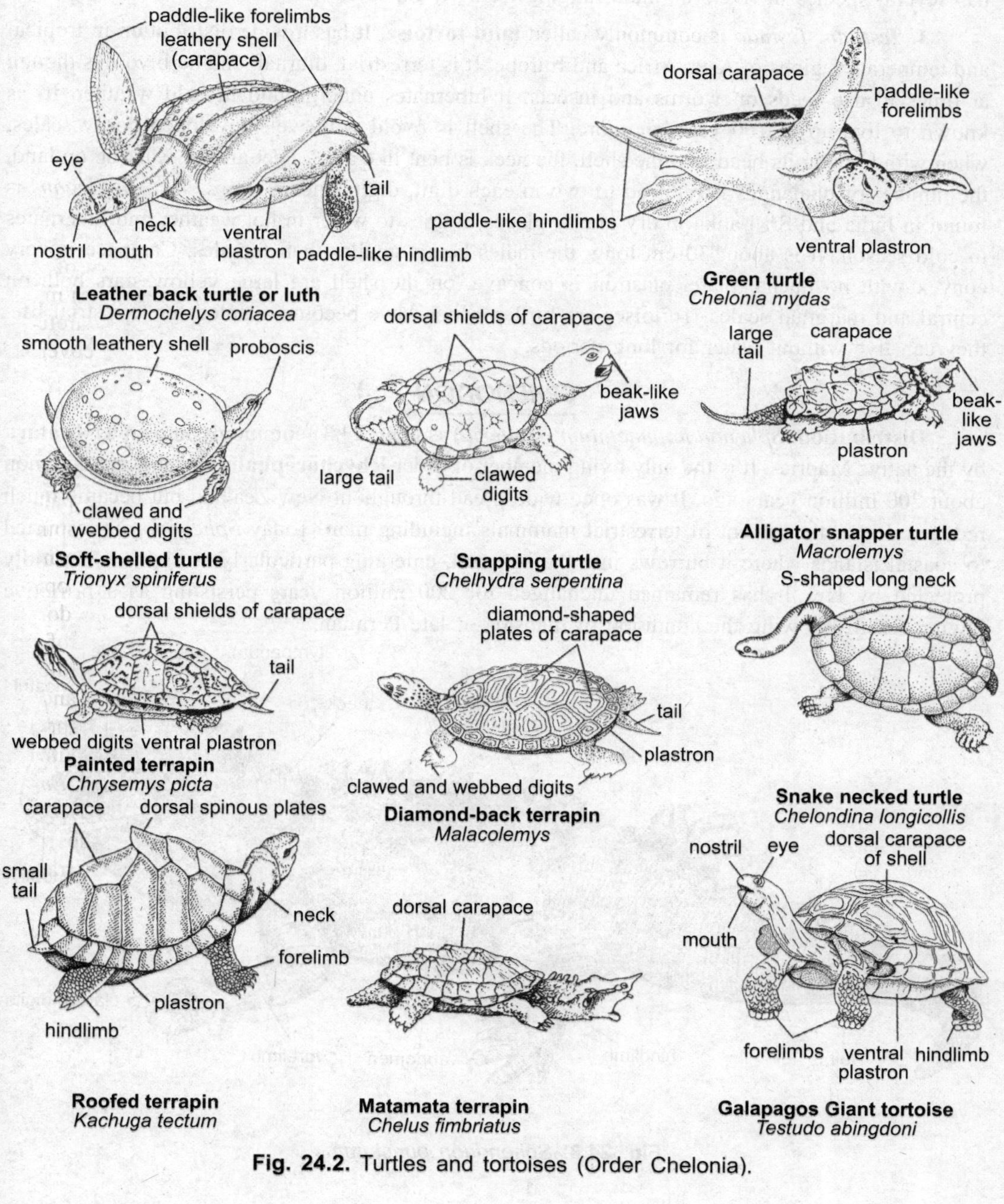

Fig. 24.2. Turtles and tortoises (Order Chelonia).

2. *Trionyx.* *Trionyx* (Fig. 24.2) is commonly known as **tortoise** or **soft river terrapin**. It has many species distributed in rivers of Asia, Africa and N. America. The shell has no horny scales but a covering of soft skin. Carapace and plastron are not joined. *Trionyx gangeticus* is found in the rivers of N. India, its shell is 60 cm long and is olive above and yellow below. Head is pointed having a black longitudinal stripe, there is no horny beak but lips are fleshy. Carapace has two nuchal bony plates. Each limb bears three claws. *Emyda* is a soft-shelled terrapin of Indian rivers in which the ribs extend beyond the costal plates of carapace. *Kachuga* has several species in rivers of India ranging from 15 cm to 55 cm.

3. *Testudo.* *Testudo* is commonly called **land tortoise**. It has a wide distribution in tropical and temperate regions of Asia, Africa and Europe. It is terrestrial, diurnal, and herbivorous though at times it also feeds on worms and insects. It hibernates underground in cold weather. It is known to live up to 100 years or more. The shell is ovoid with well developed horny scales, when withdrawing its head into the shell, the neck is bent like an S. Feet are for walking on land, the number of phalanges is reduced to two in each digit, digits end in claws. *Testudo elegans* is found in India and Sri Lanka in dry grassy places, it goes to water in hot weather and hibernates in cold season. It is about 30 cm long, the males being smaller than females. Carapace is very convex with no nuchal plate, plastron is concave, on the shell are large yellow stars both on central and marginal scales. Tortoises are turtles which have become adapted for terrestrial life, they can live without water for long periods.

SPHENODON

Distribution. *Sphenodon punctatum* (*Hatteria*) is (Fig. 24.3) commonly known as **Tuatara** by the native Maories. It is the only living member of order **Rhychocephalia** which were common about 200 million years ago. It was once widespread throughout New Zealand but became much reduced due to the advent of terrestrial mammals including man. Today *Sphenodon* is restricted to coastal islands where it burrows into the soft soil, emerging particularly at night. It is rigidly protected by law. It has remained unchanged for 200 million years persisting as a primitive "living fossil" showing the condition of diapsids of late Permian.

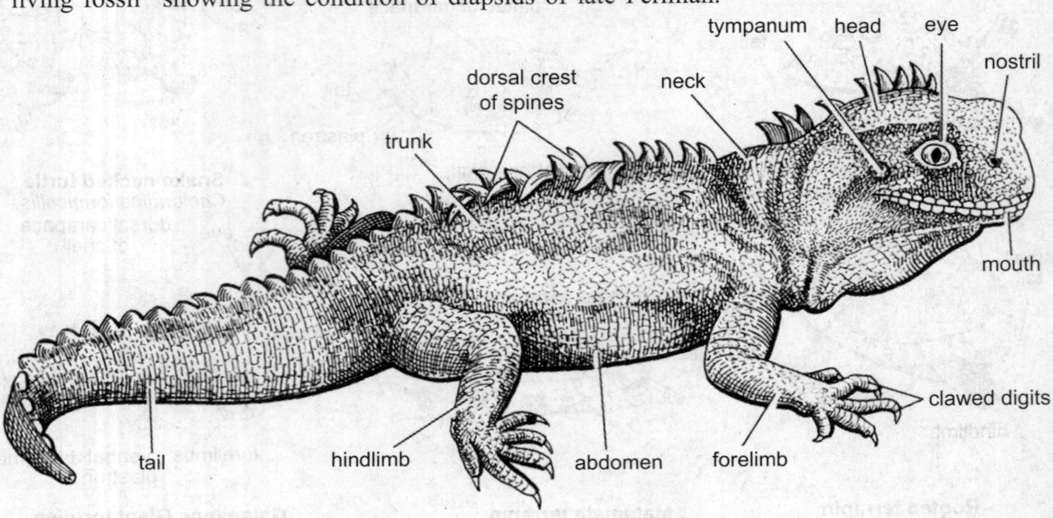

Fig. 24.3. *Sphenodon punctatum.*

Habits and habitat. *Sphenodon* inhabits burrows made in the soft soil and in rocks, often in association with petrels, and is nocturnal in habit. They are carnivorous, feeding on invertebrates and fish. They are slow in their movements. However, they can run fast and can defend themselves with some vigour. They lay eggs from November to February. The eggs take over a year (13 months) to hatch. The survival of *Sphenodon* in captivity is twenty-eight years.

External features. *Sphenodon* is a lizard-like reptile measuring about 60 cm in length. It possesses a scaly skin, a long laterally compressed tail and four-clawed pentadactyle limbs adapted for walking. Its upper surface has granular scales, there is a median row of scales forming spines from the head to the end of the tail, except on the neck, lower surface has transverse rows of large squarish plates. Its colour is dull olive-green with white and yellow spots. Its eyes are large and dark brown, and the pupil is vertical like other lizards. Vision is monocular. Anal opening of *Sphenodon* is transverse. There are no hemipenes (copulatory organs) in male.

Anatomical Features

Endoskeleton. Vertebrae are **amphicoelous**, with remnants of the notochord in the intercentral spaces. Intercentra may be represented by bony elements formed by ossification of ventral portions of the intervertebral discs. Chevron bones also occur in the caudal vertebrae. There is a median bone, proatlas, intercalated between the atlas and the occipital region of the skull. Ribs (single-headed) are present with all the vertebrae of the pre-sacral region. In *Sphenodon*, caudal ribs are also present which become fused with the vertebrae. Uncinate process is also present in the ribs.

Skull is lizard-like, but there is a complete lower temporal arch. The quadrate is fixed, wedged in by the quadrato-jugal, squamosal and pterygoid. Premaxillae are not fused together, but separated by a suture. There is a broad palate formed by the plate-like vomers, palatines and pterygoids. **Humerus** of *Sphenodon* has a foramen above the outer and one above the inner condyle. Carpals are eleven – 4 including pisciform in the proximal row, 2 centrales and 5 in the distal row. **Pubes** of pelvic girdle are united in a symphysis, in front of which is a cartilaginous epipubis. A large oval foramen is present between the ischium and pubis. A cartilaginous **hypoischium** is attached to the ischia behind. In the **tarsus**, tibial and fibular elements are distinct but firmly united. Intermedium and centrale are firmly fixed to the tibiale. There are 3 distal tarsal bones.

Alimentary canal. *Sphenodon* possesses pointed, triangular, laterally-compressed teeth in two parallel rows, one along the maxilla, the other along the palatine. Lower jaw also possesses similar teeth, bite in between these two upper rows. All the rows becoming worn down in the adult to form continuous ridges. In young *Sphenodon* a tooth has been found on each vomer, a condition unusual among reptiles.

Respiratory system includes an elongated trachea whose anterior part is dilated to form the **larynx**. Its wall is supported by cricoid and arytenoid cartilages. *Sphenodon* emit a soft frog-like croak. Lungs are lizard-like.

Heart. In *Sphenodon,* sinus venosus is not easily distinguished externally as visible in other reptiles. Other structure is similar to that of other lizards. Jacobson's organs in *Sphenodon* is in a rather primitive form.

Parietal or Pineal eye. In *Sphenodon* is found the median **pineal eye** developed in close relation to the epiphysis. It is situated in the parietal foramen of the cranial roof just beneath the integument, and covered in the young by a transparent scale. It has a lens and retina supplied by a nerve from the brain. In other reptiles nerve usually degenerates before the animal reaches maturity.

Tympanic membrane. In *Sphenodon*, the tymparic membrane is not visible externally but it is visible after removing the skin in the aural region. Tympanic cavity is represented by a large pharyngeal recess.

Systematic position. *Sphenodon* (Rhynchocephalia) make their appearance in the Permian and, therefore, they are the oldest known reptiles. For this reason and also due to resemblances to the lizards, Crocodilia, Dinosauria and Chelonia they have been regarded as an ancestral group or closely allied to the ancestors of reptiles. No fossil specimen of *Sphenodon* has been found. It is now referred as "living fossil" because it has many primitive characteristics of stem reptiles of the Permian period and has undergone a relatively less change than most others. It may be significant that it is found in an isolated island region. *Sphenodon* is a highly modified animal which persisted from Permian to recent times. It shows a number of similarities and dissimilarities to the other groups of reptiles, *i.e.,* Squamata and Crocodilia, and so it is placed in a separate order Rhynchocephalia.

Lizards

1. *Hemidactylus*. *Hemidactylus* (Fig. 24.4) is commonly known as wall lizard or house lizard. It is found in homes in India, Africa, Sri Lanka and China. Other genera are common all over the warmer countries. It is not a lover of sunshine. It is about 25 cm long, of pale-green colour. Head is broad and flat, ear opening is vertical, tail is moderately long, scales are minute and smooth. It has a short, sticky, slightly notched tongue with which it feeds on insects at night. The digits are dilated with a double series of ridged lamellae on the ventral side. They have sharp claws, which enable the lizard to walk on ceilings and smooth surfaces. The tail can break off at an unossified part of caudal vertebrea. The broken part of the tail wriggles actively to distract a predator, while the lizard escapes. The tail regenerates, but the new tail is not exactly the same as the old one because the lost vertebrae are not regenerated. *Hemidactylus* is one of geckos which are the only Squamata which lay round hard-shelled eggs, in all others the eggs are leathery. Male has femoral pores on the thighs.

2. *Calotes*. *Calotes versicolor* (Fig. 22.1) is commonly called **garden lizard** or **girgit**. It is also known as **blood sucker** because of the red colour of the throat. It is found in India, China and Malaysia and is an arboreal insectivorous agamid, as also *Uromastix*. It is about 30 cm long and has coarse rough scales. The tail is very long and slender. There is a crest of sharp spines on the neck and back. The Indian *Calotes versicolor* is known for its colour changes, the normal colour is olive-green but in courtship or threat the body becomes yellow, and the neck and sides of the head become red. Colour changes are due to temperature and environment. The colour is present in melanophores of the skin which are controlled by hormones of the pituitary. It has no gular sac found on the neck in some species. It nods its head up and down. Body is covered with strongly keeled scales. The tail does not break. It lives on trees and is insectivorous and oviparous.

3. *Draco*. *Draco* (Fig. 24.4) is commonly called **flying dragon** or **flying lizard**. The body is compressed dorso-ventrally, on the neck are three soft hooks. Teeth are acrodont. Five or six ribs on each side of the body are elongated and support a wide fan-shaped flap of skin between the fore- and hind limbs. When ribs are raised, the skin is stretched to form wing-like parachutes which enable the lizard to glide downwards from branch to branch but does not fly. The hooks on the neck help in alighting on branches. *Draco* is arboreal and beautifully coloured like the flowers of trees in which it lives. It is insectivorous and is found in the Indo-Malayan peninsula. *Draco volans* is found in Sumatra, Java and Borneo. *Draco dussumieri* occurs in Chennai.

4. *Iguana*. *Iguana* (Fig. 24.4) is a lizard of South and Central America and West Indies. It has a laterally compressed body and long tail. It is about 2 metre long. Along the dorsal side of the neck and body is a crest of long spines. On the throat is present pendulous and expansible folds of skin which play a part in courtship and rivalry. Teeth are acrodont. It is both herbivorous and carnivorous. Another iguanid, *Amblyrhynchus*, of Galapagos Islands is the only living marine lizard, though it spends much time on the shore basking in the sun and feeding on weeds.

5. *Phrynosoma*. *Phrynosoma* (Fig. 24.4) is commonly known as **horned toad**. It grows to a length of 7.5 cm. Body is much flattened and broadened and covered with larger and smaller

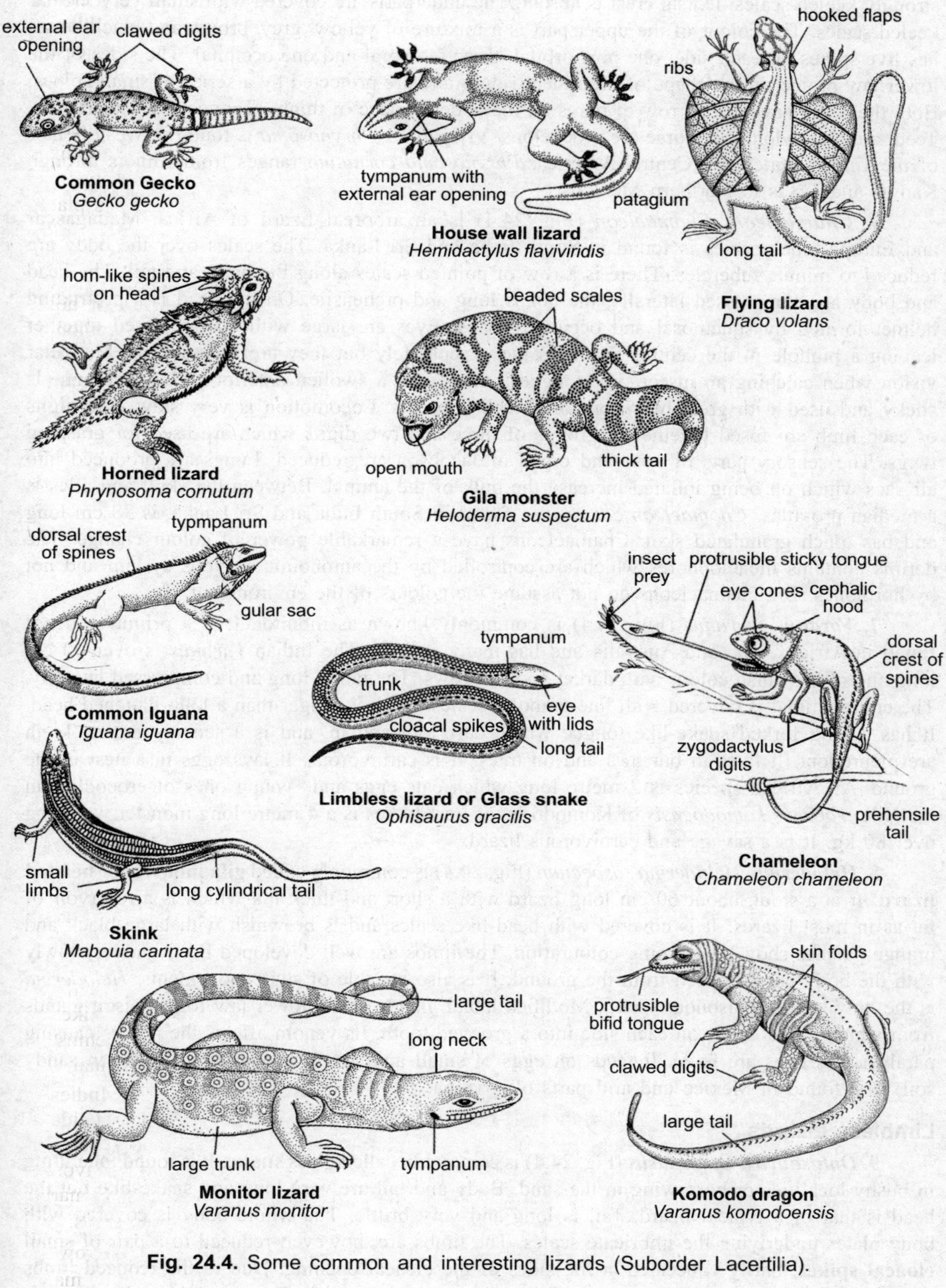

Fig. 24.4. Some common and interesting lizards (Suborder Lacertilia).

strongly keeled scales. Dorsal crest is absent. The underparts are covered with small very regular keeled scales. The colour of the upper part is a mixture of yellow, grey, brown and black. Head has five spikes on each side, one post-orbital, three temporal and one occipital. The sides of the lower jaw project in the shape of prominent ridges and are protected by a series of small spines. Both the sexes have a long row of pores on the under surface of thighs. Tongue is short and flat. Teeth are pleurodont. Vertebrae are procoelous. Viviparous. *Phrynosoma* is found in Western half of the United States and Central America. *Phrynosoma cornutum* ranges from Illinois through Kansas and Texas to Northern Mexico.

6. *Chamaeleon.* *Chamaeleon* (Fig. 24.4) is an arboreal lizard of Africa, Madagascar and Europe, one species is found in South India and Sri Lanka. The scales over the body are reduced to minute tubercles. There is a row of pointed scales along the body and tail. The head and body are compressed laterally, the tail is long and prehensile. On the head is a protruding helmet formed by squamosal and occipital bones. Eyes are large with eyelids fused together leaving a pinhole in the centre. The eyes move separately but they are focussed for binocular vision when catching an insect. The tongue is long with a swollen anterior end. The tongue is sticky and used with great precision for catching insects. Locomotion is very slow. The digits of each limb are fused together in groups of three and two digits which are used for grasping twigs. The sensory parts of nose and organ of Jacobson are reduced. Lungs are produced into air sacs which on being inflated increase the bulk of the animal. Between the skull and atlas is a median pro-atlas. *Chamaeleon calcaratus*, found in South India and Sri Lanka, is 38 cm long and has much granulated skin. Chamaeleons have a remarkable power of colour change. The dermis contains melanophores which are controlled by the autonomic nervous system and not by hormones. But chamaeleons do not assume the colours of the environment.

7. *Varanus.* *Varanus* (Fig. 24.4) is commonly known as monitor. It is a primitive lizard found in Africa, Asia, and Australia and has many species. The Indian *Varanus* is over 60 cm long and of brownish colour with darker spots in rows. The tail is long and compressed laterally. The entire animal is covered with fine granular scales. Neck is longer than a long flattened head. It has a long forked snake-like tongue which darts out and in, and is a sensory organ. Teeth are pleurodont. It lives in burrows and on trees. It is carnivorous. It lays eggs in a nest in the ground. An African species is 2 metre long which eats eggs and young ones of crocodiles in the Nile. *Varanus komodoensis* of Komodo Is. of East Indies is a 4 metre long monster weighing over 80 kg. It is a savage and carnivorous lizard.

8. *Heloderma.* *Heloderma suspectum* (Fig. 24.4) is commonly called **gila monster** or **beaded lizard**. It is a stout, about 60 cm long lizard with a short and thick tail which is a reservoir of fat as in most lizards. It is covered with bead-like scales and is brownish with large black and orange patches showing warning colouration. The limbs are well developed but it moves slowly with the body hardly raised from the ground. It is also capable of swift movements. *Heloderma* is the only known poisonous lizard. Modified labial glands of the lower jaw form poison glands from which a duct leads on each side into a grooved tooth. Its venom affects the nerves causing paralysis. Its eyes are fixed. It feeds on eggs of small animals. It lays eggs in a nest in sandy soil. It is found in Mexico and arid parts of Arizona.

Limbless Lizards

9. *Ophisaurus.* *Ophisaurus* (Fig. 24.4) is commonly called **glass snake**. It is found inhabiting in bushy localities or burrowing in the sand. Body and tail are very long and snake-like but the head is that of a typical lizard. Tail is long and very brittle. The whole body is covered with bony plates underlying the imbricate scales. The limbs are, however, reduced to a pair of small **cloacal spikes**, partly concealed at the sides of the cloacal opening. Due to the reduced limbs it is also known as **"limbless lizard"**. Eyes with movable eyelids. Ear opening is also present.

Tongue composed of two distinct portions, of which the anterior is thin emarginate, extensile and retractile into the posterior thicker portion. Teeth are pleurodont and solid. Pectoral and pelvic girdles are present. Vertebrae are procoelous. Presence of longitudinal fold of skin on the lateral sides of the body. Oviparous. Carnivorous, feed chiefly on snails, insects, worms, mice and small lizards. *Ophisaurus* is found in North America, South Russia, Asia Minor and Morocco. *Ophisaurus gracilis* is found in the Eastern Himalayas and Myanmar.

SNAKES

Snakes probably derived from a lizard ancestry in Mesozoic times. They have long, narrow bodies, devoid of limbs. Moreover, some snakes have traces of hindlimbs placed at the root of the tail and carry a conical claw projecting at the sides of anus. The mouth and pharynx is extraordinarily distensible for swallowing their prey larger than the mouth. The rami of the mandible are connected together only by elastic fibres at the symphysis, so that they can be widely separated. The maxillae, palatines and pterygoids are freely movable. Sternum and episternum are absent. As in lizards, a lachrymal duct carries away the tears and opens into the mouth in close relationship with the duct of Jacobson's organ. Tympanum and tympanic cavity are absent. Eyelids are replaced by transparent spectacles developed from the eyelid rudiments. Skin is scaly which vary much in form, number and arrangement. When they are small and overlap, they are called **scales**, but when they are large and only touch by their edges, they are termed as **shields**. Snakes moult (ecdysis) several times in the course of the year. They strip off the whole of the scaly epidermis. Vertebrae are very numerous and procoelous divided only into caudal and precaudal. Precaudal vertebrae except atlas carry ribs. Ribs articulate to the transverse processes of the trunk vertebrae and have only capitula. They are very movable in an antero-posterior direction. They are usually hollow and end ventrally in a cartilage attached to the connective tissue underlying the ventral shields. Snakes run on the extreme points of their ribs which are moved forwards, carrying with them the ventral shields to which they are attached. Their tongue is long, narrow and forked, and retractile into a basal sheath. It is well provided with sense organs and is exceedingly protractile. It is used as a tactile organ. Majority of living snakes are non-poisonous.

A. Non-Poisonous Snakes

1. *Typhlops.* *Typhlops* (Fig. 24.5) is commonly known as **blind snake**. It has a large number of species in various tropical and marshy regions of the earth. It lives underground and looks like an earthworm. It has a brown cylindrical body generally 25 cm long, but various species range from 17.5 cm to 60 cm in length. The body is covered uniformly with smooth, rounded, overlapping scales. Head is small and non-expansible. There is no distinct neck. Eyes are covered by scales so that it is blind. Tail is short and stumpy. It differs from other snakes in having a few teeth on the upper jaw only. Squamosal and transverse bones of the skull are absent. Quadrate is fixed, rami of the lower jaw are joined, and mouth is non-expansible. It has vestiges of the pelvic girdle. It is insectivorous and oviparous.

2. *Python.* *Python* (Fig. 24.5) is a large, massive, non-poisonous snake covered with small scales and shields on the head and large plates below. Tail is prehensile. It has a vertical pupil. Premaxilla also bears teeth. It constricts warm-blooded animals in its coils and kills them due to suffocation, then it slowly swallows the prey. Pythons and Boas have vestigial pelvic girdle which articulates with a small femur bearing a claw. The claws are seen on either side of the cloaca, they play a part in copulation. They are oviparous. *Python molurus*, the Indian python, is over 8 metres long. It is grayish-brown with red and black spots and a mark like a spearhead between the eyes, ventral side is yellowish. Scales of upper lip have pits. It is nocturnal and very often waits on trees for its prey. It loves to lie in water. *Python reticulatus* of Myanmar, Indo-China and Malaya is 10 metres long and is the longest known snake. Pythons are found in Asia and Africa.

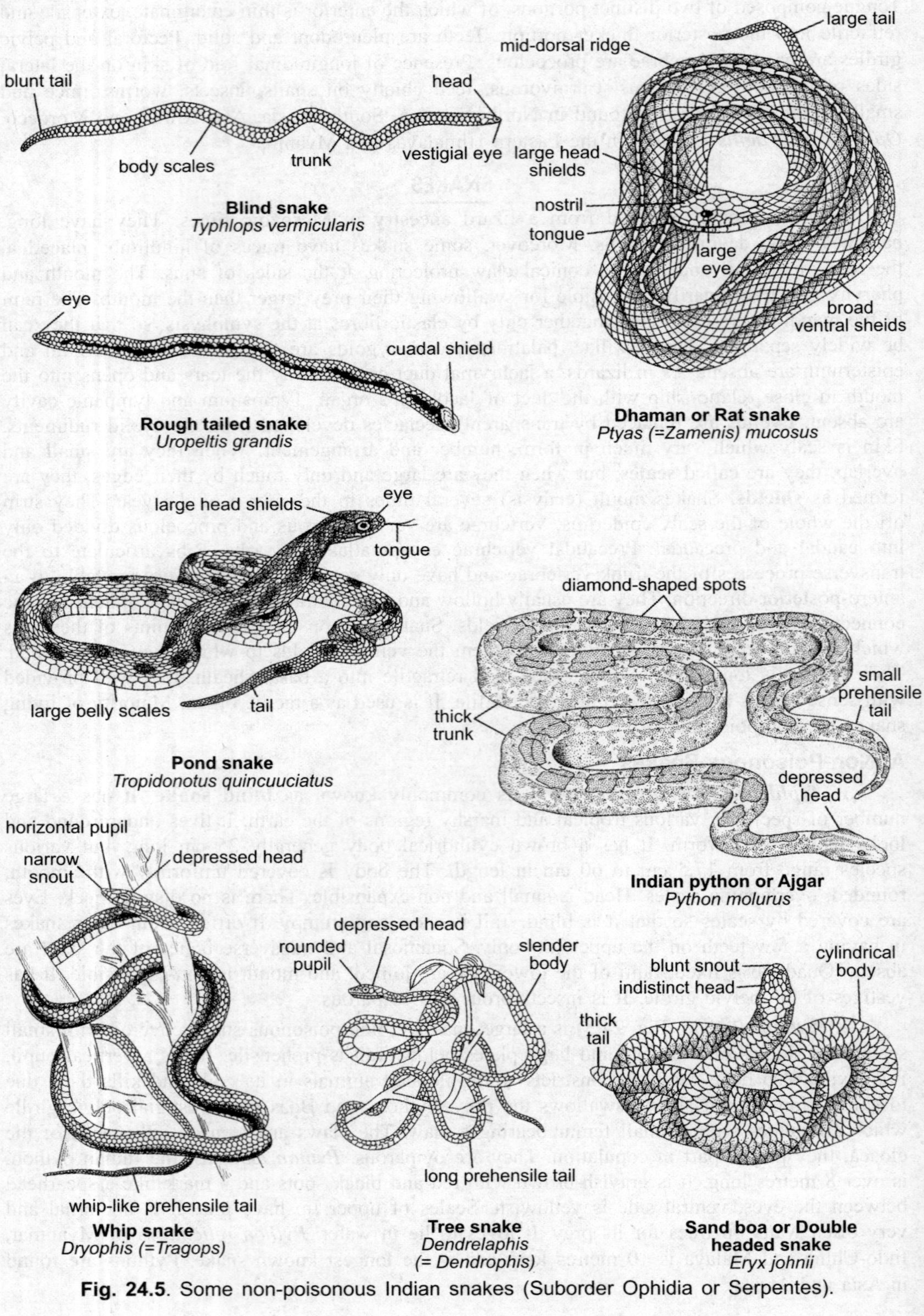

Fig. 24.5. Some non-poisonous Indian snakes (Suborder Ophidia or Serpentes).

3. *Eryx.* *Eryx johnii* (Fig. 24.5) is commonly called **Indian** or **John Sand Boa** or **Dumuhi**. It is a uropeltid, found in dry, hot, sandy parts of India, Africa and Sri Lanka. The body is short and flat with a stumpy tail, eyes are minute. The head is pointed and resembles the tail, hence, it is often called a two-head boa. It has a uniformly sandy brown colour, it may have cross bars and dark spots, with small smooth scales above and slightly enlarged plates below. *Eryx* is 1 metre long. It is a harmless, sub-terranean snake. It lives in burrows of small rodents on which it feeds, it also feeds on lizards and worms.

4. *Zamenis* (*Ptyas*). *Zamenis mucosus* (Fig. 24.5) is commonly called **dhaman** or **rat-snake of India**. It is a highly active non-poisonous snake which feeds on rats, small rodents, and frogs. It grows to a length of 2.5 metres. and has an elongated tail. Scales above are smooth with pits, below they are large transverse bands. Colour is brownish above with black bands on the tail in some, it is pale yellow ventrally. The eyes are large with round pupil. It has a sharp ridge along the backbone. *Ptyas* is very swift, ill-tempered and bites viciously. It has a nervous temperament and is untamable. It is oviparous. It is regarded as a friend of the farmers because it destroys rats. It is found in India, Myanmar and Java.

B. Poisonous Snakes

The poisonous snakes are kraits, cobras, pitless vipers and pit vipers, sea snakes, coral snakes and rattle snakes. These possess poison glands and fangs–modified maxillary teeth. Their tails except sea snakes are cylindrical. Heads are covered either with small scales (vipers and sea snakes) or with large shields. Ventrals cover the entire width of the belly so that laterals are not visible from the ventral side.

1. *Bungarus.* *Bungarus* (Fig. 24.6) is commonly called **krait**. It has several species in India and Malaya. It is slender and about 1 metre long, of pale brown colour with darker bluish-black markings having white cross bars, ventral surface is white. Fangs are small. Scales are smooth. The backbone is ridged having a central row of enlarged scales. It is oviparous and the female incubates the eggs. It feeds on toads, mice and smaller snakes. The common krait, *Bungarus caeruleus*, comes into homes, is slow moving, but is highly poisonous. Its venom is more poisonous than that of a cobra. *Bungarus fasciatus* is the banded krait about 1.20 metres long. It has black and yellow subequal cross bars and the back is keeled. It is found in Indonesia and India.

2. *Naja.* *Naja* (Fig. 24.6) is commonly called **Indian Cobra** or **nag**. *Naja naja* is also called *Naja tripudians*. It is a highly poisonous snake growing up to 2 metres in length, of brown or blackish colour. It lives in holes and under stones, but also comes into human dwellings. It feeds on frogs, lizards, rats and other snakes. Scales are smooth, dorsal surface of the head has small scales and large shields. The head is small. The cervical ribs when raised form a hood on the sides of the neck, its hood spread is more than in other cobras. Pupil is round. Fangs are small and relatively non-movable. On each side is a large poison gland enclosed in a fibrous capsule, its duct runs into a canal in the fang. When annoyed, it raises the front part of the body about 30 cm, spreads its hood and hisses through the nose. It bites with a chewing motion making several wounds. The venom is neurotoxic and fatal. It is more deadly than that of vipers. There are three common varieties of cobras in India according to the makings on the hood. On the ventral side all have two black and white marks. 1. But on the dorsal side of the hood the **diocellate** cobra has a black and white spectacle mark. 2. The **monocellate** cobra has a single black ring enclosing a pale area having a black spot in the centre. 3. The **black** cobra has no mark but the scales are barred. The cobra is oviparous and the female incubates the eggs. Cobra is found also in Malaya, Iran, Myanmar, and southern China.

Naja hannah (**king cobra** also called **Hamadryad**), is found in India, Malaya, and southern China. It grows up to 6 metres, but has a diameter of only 7.5 cm, the average length is 3.5 metres. It is the world's largest poisonous snake and the most deadly and venomous of all snakes. The

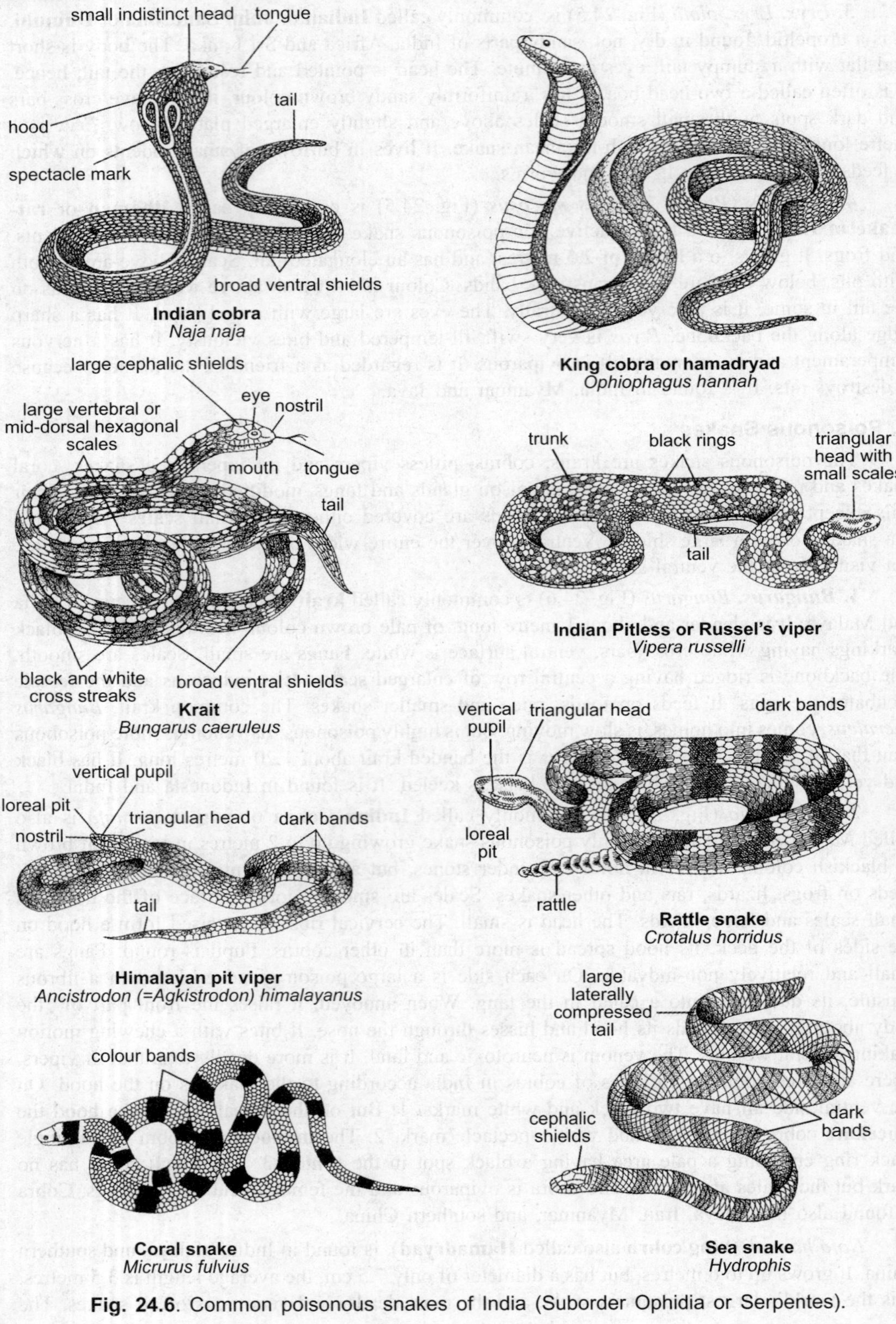

Fig. 24.6. Common poisonous snakes of India (Suborder Ophidia or Serpentes).

venom injected is large in quantity and causes quick death due to respiratory paralysis. The king cobra has a small head, the hood spread is only about 10 cm, on the hood are transverse bands and white marks. The colour is variable, it has a pale gray, fawn, olive, or black colour with dark cross bands which appear like rings, the end of the tail may be yellowish. It is oviparous and feeds on non-poisonous snakes, carefully avoiding poisonous ones. The king cobra is very intelligent, dangerous in habits, and bad-tempered, it is bold and often attacks without provocation.

3. *Vipera.* *Vipera* (viper) (Fig. 24.6) has thick body up to 4.3 cm long. Head is large, flat, and covered with small scales. Nostrils are lateral, eyes are far forwards and there is a sensory pit between the eye and the nostril in some. Pits are sensory organs with which the snake can detect its prey. The scales on the body are keeled. Tail is short and tapers abruptly. The Indian Russel's viper, *Vipera russelli*, has enlarged nostrils unlike other vipers. It is pale brown above with three longitudinal series of black rings, each ring is bordered with white, ventrally it is yellowish. Scales on the head are small and keeled. In vipers, the maxilla is small and bears long, movable fangs with canals. The fangs are erected for biting and folded back against the roof of the mouth when not in use. Behind the fangs there are no teeth in the upper jaw, but there are two rows of teeth on the palatines. Vipers are nocturnal. In biting it opens its mouth very wide and strikes like lightning, thrusting its long fangs, the poison is deadly. Vipers are viviparous and are found in India, Myanmar and Sri Lanka, other vipers have species in Europe, Africa and Asia.

4. *Ancistrodon* or *Agkistrodon.* *Ancistrodon himalayanus* (Fig. 24.6) is commonly called **Himalayan pit viper**. The general colour is bluish-brown above with dark brown or black spots appearing like cross bars and the ventral side is brown with black and white spots. Body is not much elongated, measuring about 60 cm in length. The scales covering the body have apical sensory pits. Mouth is grey with black spots. The upper lip is slightly raised in the rostral region. Supra-labials are 5–7, 3rd and 4th are enlarged but not touching the eye. Post-oculars are two, the last greatly enlarged and pre-oculars are three. A **loreal pit** is present in between the eye and the nostril on each side. **Loreal pits** function as thermoreceptors. Each pit has two cavities separated by a membrane. These organs (and the pits in the marginal scales along the mouth of pythons and some boas) are supplied by branches of the trigeminal nerve. This apparatus perhaps helps the animals to locate precisely and attack warm-blooded prey. Eyes are big with golden iris and vertical pupils. Tail is cylindrical without rattles. Viviparous. **Poisonous** and its poison is **haemotoxic**. *Ancistrodon* is quite a timid snake, remains hidden under stones, etc. *Ancistrodon himalayanus* is distributed in hilly regions of Eastern parts of India, Sri Lanka and Java. It is very common in Kashmir and is found even at an altitude of about 3658 metres.

5. *Hydrophis.* *Hydrophis* (Fig. 24.6) is commonly called **Sea snake**. The body is long and laterally compressed posteriorly. Head and neck are very slender. The general colour is dark olive green above with yellowish cross bars and white below. The ventral scales are small. Tail is **laterally compressed** and acts like paddle in swimming. Eyes are small with rounded pupil. Loreal shield is absent. One pre-ocular, two post-oculars and 7–8 supra-labials of which 3rd and 4th touching the eye. Presence of 14–18 maxillary teeth behind the poison fangs. Viviparous. Carnivorous, feeding on fishes, etc. *Hydrophis* is **deadly poisonous** and its venom is **neurotoxic**. *Hydrophis* occurs in the Bay of Bengal and Malay Archipelago.

6. *Crotalus.* *Crotalus* (Fig. 24.6) is commonly called **rattle snake**. It has several species in America and Mexico. It grows up to 2.5 metres and has a prominent sensory pit between the eye and nostril. Upper side of head has small scales. Fangs are long and movable, they are folded back when not in use. At the end of the tail is a rattle having a small hard button at the tip and usually 10 to 12 dry horny rings, the rings are formed as hollow segments by the end of the moulted skin being left attached to the tail. The rattle is always held up vertically. It makes a grating sound in locomotion and is probably for warning. The rattle snake is highly poisonous. It is carnivorous and viviparous.

Crocodilia

The amphibious crocodiles, gharials, alligators and caimans are the largest living reptiles. The dorsal surface or both dorsal and ventral surfaces are covered with rows of sculptured bony scutes which are covered by corresponding horny epidermal scales. Bony plates of the back are pitted and sculptured and frequently keeled. Ventral plates are less sculptured and not keeled. On the tail these plates form a dentated dorsal crest, paired in front, but single behind. The nostrils are placed at the end of long snout and can be opened and closed by longitudinal dilator muscle and constrictor muscle encircling the dilator muscle respectively. Tympanic membranes are exposed, but can be covered over by a cutaneous fold. The eyes have both the eyelids and a nictitating membrane. There is a bony plate in the upper eyelid. Pupil is vertical. There is a small pecten-like *Uromastix*. They have two pairs of limbs, of which the forelimbs are shorter and have five digits and the hindlimbs have four digits. The three preaxial digits in each limb possess nails. The hindlimbs are more or less webbed. Tail is laterally compressed and elongated.

Endoskeleton. In **skull** there is a low cartilaginous interorbital septum with a membranous interspace. The presphenoid and orbitosphenoid region remains cartilaginous. Alisphenoid and a separate prootic is present. The epiotic fuses with the supraoccipital and opisthotic with the exoccipital. Occipital condyle is formed by the basioccipital only. Parietals and frontals are unpaired in the adult. Premaxillaries surround the nostrils. Nasals are long and usually reach the premaxillaries, but in *Gavialis* nasals are separated from the premaxillaries by the maxillaries. Paired vomers and their opposed vertical plates separate the narial passages. Premaxillae, maxillae and palatines develop palatal plates which unite in the middle line and form a hard palate. Pterygoids are modified due to which internal narial opening is placed far back on the base of the skull. An ectopterygoid or transpalatine connects the pterygoid with the maxilla and jugal. Prefrontals, unperforated lacrymals and postfrontls are present. The orbit is separated from the temporal fossa by the postorbital bar formed by processes of the postfrontal and jugal. Both supratemporal and infratemporal arcade are present. Supratemporal arcade is formed by postfrontals and squamosals and the second one is formed by the jugal and quadratojugal. Quadrate is large and immovably fixed to the skull.

Vertebrae. There are 24 or 25 presacral vertebrae, 35 or more caudal and two sacral. Presacral vertebrae are 9, cervical 11 or 12 dorsals and 3 or 4 lumbars. All the vertebrae except atlas and axis, second sacral and the first caudal are **procoelous**. Second sacral is flat in front and concave behind and first caudal is biconvex. Cervical vertebrae possess ribs. These ribs are two-headed. Dorsal vertebrae also possess two-headed ribs. Lumbar vertebrae are without ribs. Sacral vartebrae have ribs attached to the centrum and neural arch. Chevron bones are present in the caudal vertebrae except on the first and posterior vertebrae.

Sternum consists anteriorly of a rhomboidal plate of cartilage to the sides of which are articulated two thoracic ribs. Sternal plate posteriorly bifurcated to which are attached 5–7 ribs. Interclavicle is slender placed on the ventral surface of sternal plate. Thoracic ribs possess uncinate processes in the form of small cartilaginous plate. Abdominal ribs are membrane bones placed on the ventral surface of rectus abdominis muscles. Each rib has two curved rods not united in the middle. Ribs are bicephalous. **Columella** is bony, extends from fenestra ovalis to articulate with a trifid cartilage (extra-columellar cartilage) which is attached to the tympanum. Tympanum is protected by scaly movable flaps. Eustachean tubes are very complicated and lagena is large.

Hyoid has a cartilaginous body, a pair of partly ossified cornua.

Lower jaw has the usual lacertian six bones–dentary, splenial, angular, supra-angular, articular and coronoid.

Pectoral girdle (one half) has a scapula with a dorsal cartilaginous suprascapula, marginal and unfenestrated coracoid. **Clavicle** is absent. Hand has five clawed digits. In the **pelvis** are present the ilia attached to two sacral ribs, large both ischia united in a ventral symphysis and pubes which do not take part in the formation of acetabulum. Ventral end of pubis extends forward as an epipubis (cartilaginous). Both the epipubes are united behind by a broad ligament. Pes has four clawed digits. Number 5 being absent.

Skin. It is horny on the scales but soft between. Each scale has a small pit in which epidermis is not cornified. At the bottom of each pit has a number of tactile bodies in the dermis. There are two pairs of musk secreting glands, one pair on the throat and another pair within lips of cloaca.

Alimentary canal. The teeth are thecodont, hollow, conical and without roots, attached over the premaxillae, maxillae and dentaries. They are shed at intervals and replaced by new teeth formed behind them. IV mandibular tooth is the largest. Tongue is flat and non-protrusible. Salivary glands are absent. There is a transverse fold of mucous membrane of the palate just in front of choanae, which with the help of flap behind tongue can shut off the mouth from pharynx. Oesophagus can be distended to store food. Stomach is a gizzard-like sac with muscular walls and tendinous discs. It contains stones (gastroliths). Pyloric region is glandular. Gall bladder is present but caecum is absent. Cloaca is divided into two parts by a muscular fold. Anterior chamber receives the paired openings of urinary and generative ducts. Posterior chamber, on the ventral side, has a grooved penis developed from the ventral wall of cloaca which ends in a glans penis. It consists of a fibrous body and some erectile tissue around the groove and in the glans. Urinary bladder is absent.

Respiratory organs. Trachea is long and provided with complete rings. Larynx is provided with vocal chords. Epiglottis is absent. Lungs are spongy and well developed. Tracheal division bronchi enter into lungs of its side from the middle and continues to its hind end. It looses its tracheal rings and beset with openings which lead into pouches whose walls are like honeycomb. Lungs lie in the anterior part of the body cavity. It is partitioned off from the rest by a partly muscular and partly membranous septum. It assists in the inspiratory movements.

Vascular system. Sinus venosus is closely applied to the heart and receives three great systemic veins. Ventricle is double. Right aortic arch proceeds from the left ventricle and gives off arteries to head and forelimbs. Left aortic arch arises from the right ventricle and gives off a large visceral artery and is connected by a narrow vessel with the right aortic arch. Pulmonary artery also arises from right ventricle. Opening of each of these three great arteries are guarded by two semilunar valves. **Foramen of Panizzae** is also present at the crossing point of two aortic arches. Thus, the arterial and venous blood are not completely separated.

Nervous system. Cerebellum is more developed than in other reptiles and possesses a median vermis and two small lateral lobes.

Urinogenital system. Kidneys are lobed. Testes are oval and ovaries are elongated. In male a median erectile penis is present. In female clitoris occurs. Male crocodiles defend territories by fighting and roaring. Courtship is complex and fertilisation is internal. The eggs are hard-shelled and are laid in the sand or in holes on the banks. Nile crocodile buries her eggs in a nest, then stays near them throughout incubation, helps the young to hatch, and escorts them to the water. Crocodiles first appeared in the late Triassic. There were numerous types of crocodile in the Jurassic and Cretaceous, living both in freshwater and in the sea. The present living forms dates from the Cretaceous period. Today many species of crocodiles are seriously threatened by the demands of the skin trade. So conservation measures, in association with farming, are being taken by different countries.

1. *Crocodylus.* *Crocodylus porosus* (Fig. 24.7) is commonly called **Mugger of India** or **Marsh Crocodile**. Body is covered with an exoskeleton of horny thick epidermal scales provided with dermal bony plates. The general colour of the upper parts is dark olive brown. The average length of this species is 4 metres. The head is triangular and rough at the top. The snout is stout rather broad. Limbs are pentadactyle with five fingers and four toes. The fingers are webbed at the base and the outer toes are broadly webbed. The tail is long and laterally compressed. Mandibular symphysis is short, extending only to the level of fourth or fifth tooth. There are upper and lower jaws, each contains 19 teeth on either side. The fourth mandibular tooth fits into a notch in the upper jaw. The fifth upper tooth is the largest. Vertebrae are procoelous. The nuchal scutes are six in number which are packed closely together forming a square. The dorsal shield is composed of four, sometimes six, rows of larger scutes of which the central pair is the broadest. The ears are covered by movable flaps of skin Oviparous. Predators. *Crocodylus porosus* has a wide distribution. It inhabits the rivers, ponds, tanks and marshes of India, Sri Lanka, Myanmar, Malacca, Malay Islands and Baluchistan.

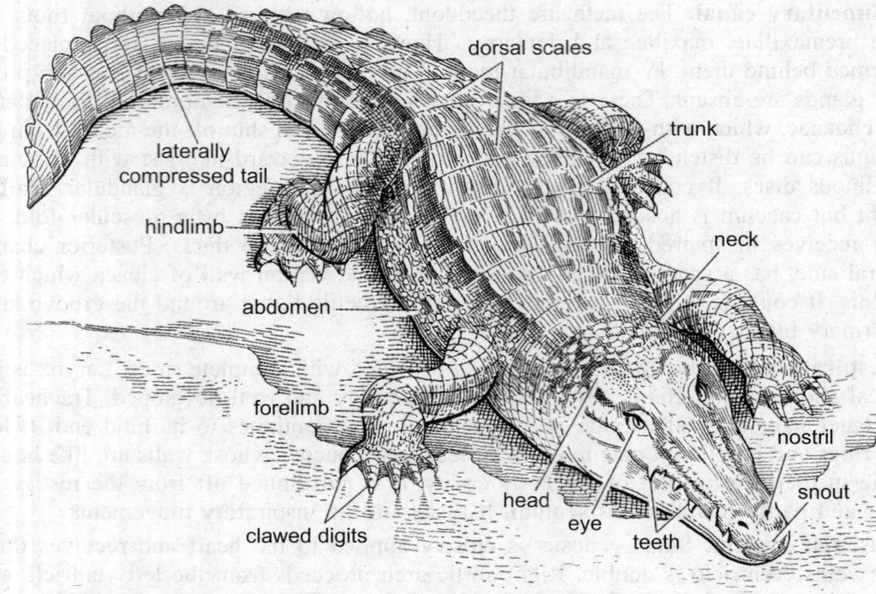

Fig. 24.7. *Crocodylus porosus* (Indian freshwater crocodile).

2. Alligator. Body of **Alligator** (Fig. 24.8) is covered with an exoskeleton of horny epidermal scales. The general colour is greenish black or dark brown above and yellowish below. It measures

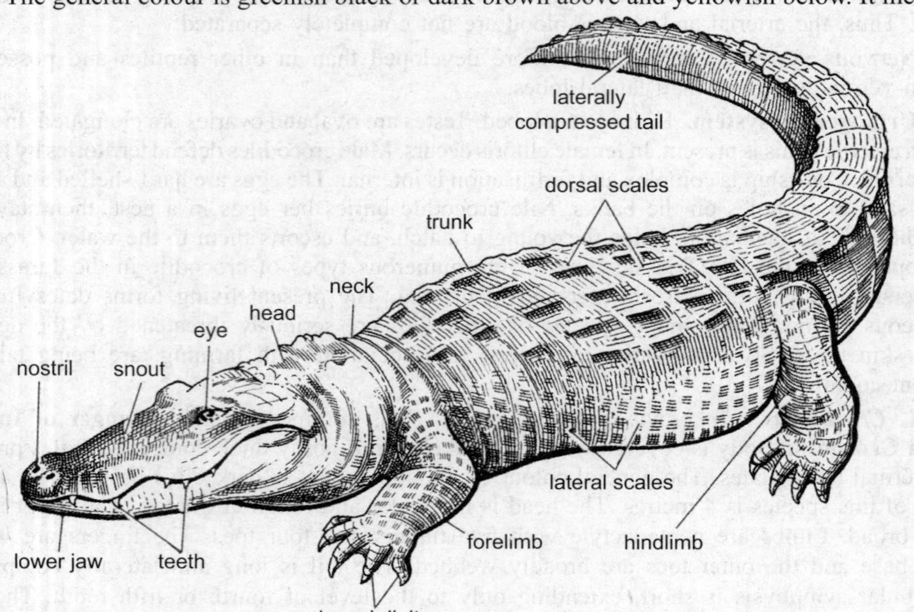

Fig. 24.8. *Alligator mississipiensis* (American alligator).

about 3 metres in length. Head is broad with short, broad and rounded snout. Limbs pentadactyle with five fingers and four toes. The fingers are half webbed and the toes are two-third webbed. The tail is laterally compressed. Mandibular symphysis is short extending only to the level of fourth or fifth tooth. Fourth mandibular tooth fits into a pit in the upper jaw. On either side 17–20 teeth in the upper jaw and 18–20 teeth in the lower jaw. The nasal bones divide the nasal groove by a median bony septum. The dorsal shield is formed by 6 or 8 longitudinal series of keeled bony scutes. The neck is protected by two pairs of large scutes. Predators. *Alligator mississipiensis* is found in the rivers of Southern States of North America and *A. sinensis* occurs in the rivers of China.

3. *Gavialis.* *Gavialis gangeticus* (Fig. 24.9) is commonly called **gharial**. It may reach the length of about 6 metres. The general colour is dark olive brown above with dark markings. Body is covered with an exoskeleton of epidermal horny scales. The snout is extremely long and slender. The limbs are pentadactyle with five clawed fingers and four clawed toes. Tail is laterally compressed. The mandibular symphysis is so long that it comprises a great portion of the splenial bones and extends backwards almost to the level of the last teeth. Nasal bones are very short and are separated from the premaxillaries. On either side 28 teeth in the upper jaw and 25 teeth in the lower jaw. First and fourth lower teeth fit into grooves in the upper jaw. Vertebrae are procoelous. In male, the nose is very much swollen and can be inflated like a bag when the nostrils are closed. *Gavialis* is a harmless creature and feeds entirely on fish. *Gavialis gangeticus* is found in India inhabiting chiefly the basin of Ganges, Brahmaputra, Indus, Mahanadi and Arakan.

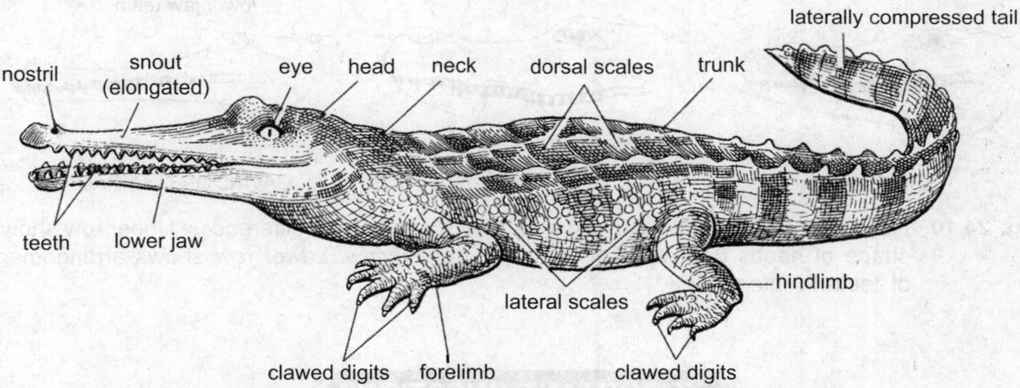

Fig. 24.9. *Gavialis gangeticus* (Gavial or gharial).

TABLE 24.1. DISTINCTION BETWEEN *CROCODYLUS*, *ALLIGATOR* AND *GAVIALIS*.

	Characters	*Crocodylus*	*Alligator*	*Gavialis*
1.	Habits	More aggressive and dangerous to man.	Less aggressive and attack when provoked.	Can feed only on fish due to narrow throat.
2.	Length of the body	8 metres.	3 metres.	6 metres.
3.	Colour	Olive green with black spots or bands.	Steel grey.	Dark olive green.
4.	Head and snout	Long and pointed.	Short and broad.	Very long and slender.

(*Contd.*)

	Characters	*Crocodylus*	*Alligator*	*Gavialis*
5.	Halves of the lower jaw	Fused not farther back than the 8th tooth.	Fused up to 5th tooth.	Fused in front as far back as the 15th tooth.
6.	Teeth	$\frac{16-19}{14-15}$ teeth on each side. Teeth unequal.	$\frac{17-20}{17-22}$ on each side, very unequal.	27–29 similar teeth on each side of upper jaw.
7.	Fourth mandibular tooth	Fits into a notch in the side of upper jaw. Externally visible.	Fits into the pit of the upper jaw. Not visible externally.	—
8.	Distribution	Africa, South Asia, North Australia, Tropical America.	North America and China.	India.

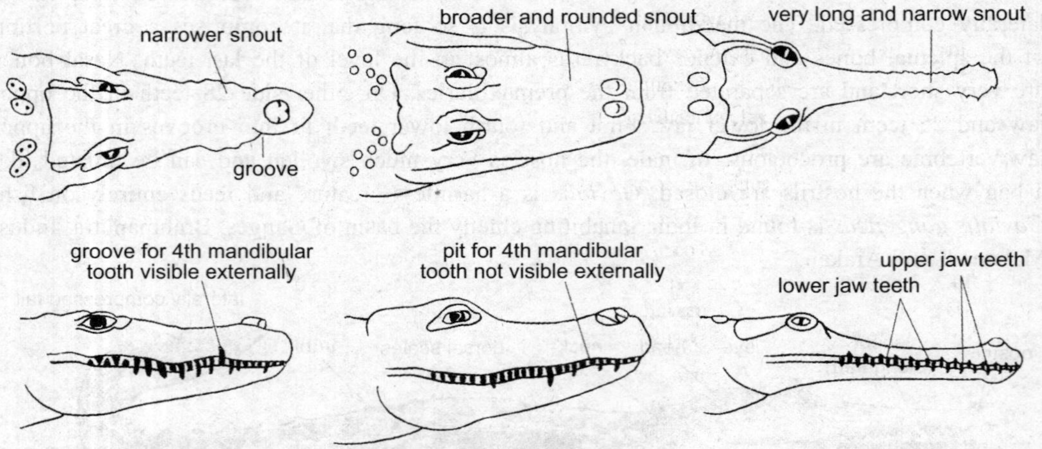

Fig. 24.10. Heads of crocodile, alligator and gharial to illustrate their differences. Upper row shows shape of heads and width of snouts in dorsal view. Lower row shows arrangement of teeth in lateral view.

IMPORTANT QUESTIONS

Long Answer Questions

1. Classify Reptilia up to orders giving important characters and suitable examples.
2. What are temporal vacuities? Explain their taxonomic importance in reptilian groups.
3. Describe principal anatomical peculiarities of *Sphenodon* and discuss its affinities.
4. Discuss that *Sphenodon* is the most primitive and the crocodilians the most advanced of all living reptiles.
5. Describe the general characters, affinities and distribution of Crocodilia.
6. Give a general account of Crocodilia.
7. Draw a neat and well labelled diagram of dorsal and ventral views of the skull of Crocodilia.
8. How would you distinguish a lizard from a snake? Give reasons for placing them in the same order. Could *Sphenodon* be included in the same order?
9. What adaptive function is shown by temporal vacuities in the skull of reptiles? Classify reptiles on the basis of temporal vacuities and stress on the evolutionary significance.

Short Answer Questions

1. Describe distinctive characters of class Reptilia.
2. Give distinguishing characters of Chelonia and Ophidia.
3. Differentiate between : (*i*) Lizards and snakes; (*ii*) Limbless lizard and snake; (*iii*) Limbless amphibian and snake; (*iv*) Lizard and Salamander; (*v*) Turtle and tortoise; (*vi*) Crocodile, alligator and gharial.
4. Name four crocodilians and their distribution.
5. *Chamaeleon* is well adapted for arboreal mode of life. Give special features for this adaptation.
6. Discuss biology of *Phrynosoma*.
7. Write a short note on the uropeltids of South India.
8. Mention the peculiarities of skull of crocodiles.
9. What features of "Horned toad" show that this is a reptile and not a toad?
10. Write short notes on : (*i*) Anapsida; (*ii*) *Chamaeleon*; (*iii*) *Draco*; (*iv*) *Heloderma*; (*v*) *Sphenodon*; (*vi*) Coral snake: (*vii*) Cobra; (*viii*) Krait; (*ix*) *Python*.

Very Short Answer Questions

1. Which two subclasses of Reptilia have living reptiles?
2. What is the meaning of the term Rhyncocephalia?
3. Name a living fossil which exists in New Zealand.
4. To which order turtles and tortoises belong?
5. Write the name of that order in which only one living species exists. Name the animal and where it is found.
6. Name living reptiles, one each for (*a*) diapsid skull and (*b*) anapsid skull.
7. In which order only bicephalous ribs are found?
8. Write the name of a poisonous lizard.
9. Which reptiles possess 4-chambered heart?
10. Draw a neat and labelled diagram of plastron.

Multiple Choice Questions

1. The study of reptiles is known as
   (*a*) Herpetology (*b*) Ornithology
   (*c*) Osteology (*d*) Ichthyology.
2. Which of the following groups shows close affinity with dinosaurs?
   (*a*) Turtles (*b*) Crocodiles
   (*c*) Lizards (*d*) Snakes.
3. Tuatara belongs to the order
   (*a*) Rhyncocephalia (*b*) Chelonia
   (*c*) Crocodilia (*d*) Squamata.
4. Turtles have
   (*a*) Horny beak (*b*) Homodont teeth
   (*c*) Heterodont teeth (*d*) Pleurodont teeth.
5. Vestigial hindlimbs and pelvic girdle are present in
   (*a*) Krait (*b*) Boa
   (*c*) Cobra (*d*) Turtle.
6. Interclavicle in reptiles is
   (*a*) W-shaped (*b*) T-shaped
   (*c*) H-shaped (*d*) A-shaped.
7. Turtles are included in the order
   (*a*) Rhyncocephalia (*b*) Chelonia
   (*c*) Crocodilia (*d*) Squamata.
8. Which of the following gland is responsible for moulting of snake?
   (*a*) Posterior lobe of pituitary
   (*b*) Thyroid
   (*c*) Adrenals
   (*d*) Thymus.
9. Embryonic membranes present in reptiles are
   (*a*) Amnion and chorion
   (*b*) Chorion and yolk
   (*c*) Yolk sac and allantois
   (*d*) Amnion, chorion, yolk sac and allantois.
10. *Ophisaurus* belongs to
    (*a*) Chelonia (*b*) Lacertilia
    (*c*) Rhynchocephalia (*d*) Ophidia.
11. Heart in crocodiles is
    (*a*) Incompletely 4-chambered
    (*b*) Completely 4-chambered
    (*c*) 3-chambered
    (*d*) 2-chambered.

Answers

1. (*a*) 2. (*b*) 3. (*a*) 4. (*a*) 5. (*b*) 6. (*b*) 7. (*b*) 8. (*b*)
9. (*d*) 10. (*b*) 11. (*b*).

25

Chapter

Reptilia : General Account

Contents of this Chapter

ORIGIN OF REPTILES

The origin of reptiles is well suited to maintain life on land. Many features show a considerable advance in this respect over the Amphibia. These are the (*i*) dryness of the skin, (*ii*) reproduction, and (*iii*) devices for economising the use of water. Reptiles have a dramatic evolution. They had an obscure beginning in the Palaeozoic, they enjoyed a dominant status in the Mesozoic and then in the Coenozoic they faded out of scene.

Amphibian origin. It is generally accepted that primitive reptiles originated from some primitive labyrinthodont Amphibia in the beginning of Carboniferous period. The emergence of such reptiles from labyrinthodont Amphibia was so gradual that, on the basis of skeletal remains alone, it is difficult to ascertain whether some of them are reptiles or Amphibia on way to transform to reptiles. The labyrinthodonts possessed characteristically folded or labyrinthine teeth, similar to their crossopterygian ancestors. They flourished through Carboniferous and Permian periods before extinction in Triassic. We cannot, however, point out to a single ancestor of reptiles. Probably they arose polyphyletically along a dozen or more independent lines.

Stem reptiles (Cotylosauria). During Carboniferous period of late Palaeozoic Era, about 250 million years ago, some labyrinthodont amphibians gradually took on reptilian characters. These earliest reptiles are called the **stem reptiles**. They belong to the order **Cotylosauria** of the subclass **Anapsida**. The transition was so gradual that often it is difficult to decide whether some fossil skeletons are those of advanced amphibians or primitive reptiles. They were contemporary of the primitive amphibians of late Pennsylvanian and early Permian times. In the later part of the Permian, however, the amphibians were outnumbered by these reptiles.

Ancestry through *Seymouria*. One of the members of the Cotylosauria was *Seymouria* (Fig. 25.1) found in the Lower Permian sediments in Texas (U.S.A.), perhaps 250 million years old. It was a lizard-like animal about 60 cm long, with a comparatively thick body, relatively small pointed head with dorsally placed nostrils, and a short tail. It had a few reptilian features like anapsid skull, single occipital condyle, large parietal eye and five digits. Structure of *Seymouria* was intermediate between the amphibians of that time and the early reptiles.

Fig. 25.1. *Seymouria* (A cotylosaur). Anapsid.

Seymouria leads us to certain logical conclusions. It is not directly ancestral to all reptiles. At the time it was living, the reptiles had already been present for some 50 million years. It is so perfectly intermediate between an amphibian and a reptile that its true position remains uncertain. **Romer** treats it as a reptile under the order Cotylosauria, whereas others classify it with primitive Amphibia under the order Seymouria-morpha. Pehaps *Seymouria* is a connecting link between Labyrinthodontia and Cotylosauria.

Ancestry through *Limnoscelis*. *Limnoscelis* was a genuine reptile from the late Carboniferous or early Permian of New Mexico. **Romer** (1946) critically studied the *Limnoscelis* and suggested that *Limnoscelis*, a Captorhinomorph Cotylosaur, is the primitive reptile. *Limnoscelis* was about 5 feet in length, some half of which was made up by tail, with an elongated body, low slung on short stubby legs which sprawled outward from the sides. Like *Seymouria*, it was also aquatic in habitat. Its skull was of anapsid type and was compressed from side to side and dorsoventrally flattened. The otic notch present in Labyrinthodonts had disappeared leaving an indication at the back of the skull in the region of its closure. The premaxillary teeth were enlarged and overhung the front teeth in the lower jaw, which is one of the most common specialisation of early reptiles but seldom seen in amphibians. Other members of the primitive reptiles are *Captorhinus* and *Hylonomus*.

Ancestry through *Diadectes*. *Diadectes* was contemporary to *Limnoscelis*, but the didectomorphs represents different evolutionary line from *Limnoscelis*. The didectomorphs retained the otic notch at the back of skull. *Diadectes* had developed a specialised dentition. The front teeth were chisel-shaped and back teeth had broad ridged crowns.

The first really large reptiles to make their appearance in the later Permian were the *Pareiasaurus*, related to *Diadectes*. These lumbering animals ran to spiny excrescences on the head and spiny armour of bony plates along back, which was presumably a defence against carnivorous mammals like reptiles.

From the foregoing discussion, it is clear that the Cotylosaurs were, broadly speaking, ancestral to all reptiles. Some workers including **Watson** believe that these Cotylosaurs separated into two divergent evolutionary lines. One of these lines is represented by Captorhinomorph (suborder) of which *Limnoscelis* was primitive member, led to mammal-like reptiles and ultimately to mammals and has been termed **theropsida**. The other line, the **Sauropsida**, however, is represented by Didectomorph Cotylosaurs, the remaining reptiles and birds. In this group the otic notch tends to persist.

Although, the authorities have not yet agreed, as at what precise stage of evolution these two Theropsida and Sauropsida lines separated from one another. It seems clear, therefore, that the modern reptiles are not intermediate in the evolutionary sense between amphibians and mammals. And to find out a common ancestor for a creature like a lizard and a mammal, one has to go back to an ancient reptilian or possibly to an amphibian, according to **Watson**.

ADAPTIVE RADIATION IN REPTILES

One of the many features that make the study of reptiles interesting is what **Osborn** has called adaptive radiation. By the term adaptive radiation is meant the differentiation of animal form and construction which, form a common starting point, follows lines of adaptations in diverse directions in response to the different needs demanded by different kinds of environment. The ancient reptiles gave trials to all sorts of haunts and ways of living. Some of them like Pterosaurs became aerial, Ichthyosaurs and Plesiosaurs were aquatic. Amphibians were the Sauropod Dinosaurs. Cursorial adaptation was shown by many Dinosaurs. It is assumed that some of the Mesozoic reptiles became adapted to fossorial and arboreal life. Adaptive radiation is a dynamic aspect of animal existence and as animals branch out into each available environment, they assume characteristic adaptations to meet their specific needs—a fact noticeable in the modifications associated principally with locomotion. The noteworthy feature in the radiation of reptiles is the fact that most of their adventure, specially on land, was successful as there was no vertebrate competitor.

DINOSAURS

The stars amongst the extinct reptiles were undoubtedly the Dinosaurs and the age of reptiles often is called as the age of Dinosaurs. At the end of Triassic, **Thecodontia** or **Pseudosuchia**, the early descendants of Archosauria, gave rise to the most fantastic Mesozoic reptiles, the **Dinosaurs** which means "terrible lizards" (Gr., *dinos* = terrible + *saurus* = lizard). They subdivided early into two orders: Saurischia and Ornithischia, depending on the structure of their pelvis (Figs. 25.2 and 25.3).

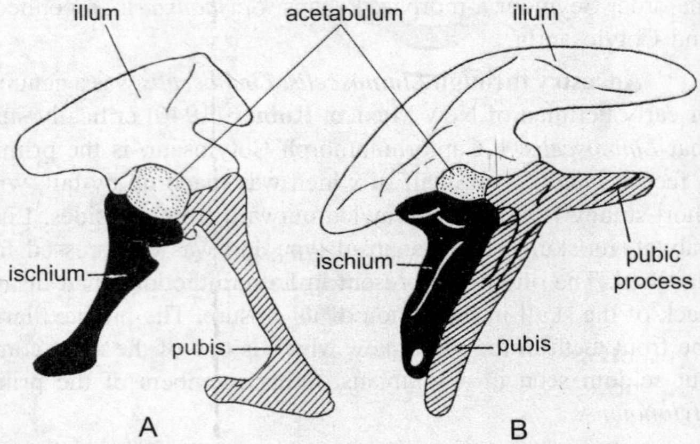

Fig. 25.2. Structure of pelvis in two orders of dinosaurs. A–Saurischia; B–Ornithischia.

Saurischia had a reptilian triradiate pelvis with pubis entirely separate and anterior to ischium. The early Saurischians were bipedal with short forelimbs, and four or five clawed digits, they progressed on land with hindlimbs. Some of the later forms became quadrupedal and herbivorous. The bipedal carnivorous forms comprise the suborder **Theropoda**. Suborder Theropoda included all flesh-eating and bipedal carnivores. Smaller Cretaceous ostrich-like forms, such as *Struthiomimus* and *Ornithomimus*, walked on 3 toes of large hindfeet. Their forelimbs also had 3 fingers, of which one was opposable like a thumb and used for grasping. Jurassic *Allosaurus*, a monster carnivore, was 10 metres long. But the largest and the most fearful predator that ever walked the face of earth was *Tyrannosaurus rex* from Cretaceous of North America. It was 15 metres long and stood 6 metres high. Its head was disproportionately great with large jaws armed with dagger-like teeth 15 cm long. The 3-toed massive hindlegs were adapted for running, but extremely short forelimbs were almost useless. Suborder Sauropoda included huge herbivorous quadrupedal dinosaurs. Some of them were the largest and heaviest of all terrestrial and amphibious vertebrates that ever lived. *Apatosaurus* (= *Brontosaurus*), *Diplodocus*, and *Branchiosaurus* were enormous Jurassic reptiles, each more than 25 metres long and weighing over 50 tons. They probably lived in swamps where their body would be supported partly by the buoyancy of water. They had long necks and tails, small head with exceptionally small brains and weak jaws.

Ornithischia are bird-like dinosaurs. They had a typical tetraradiate bird-like pelvic girdle with pubis directed backwards parallel to ischium. They were all herbivorous; some of them became quadrupedal. A good many of them developed horns and protective plates on their body. Bipedal *Iguanodon* (Lower Cretaceous) grew to 10 metres. Its sharp dagger-like thumb was probably used for defence. Quadrupedal *Stegosaurus* (Jurassic) measured 8 metres and weighed 10 tons. Forelimbs were much shorter than hindlimbs. Skull was much small and brain much smaller than the lumbar swelling of spinal cord. It had a parapet of heavy triangular plates on its neck, back and tail. The tail was also furnished with formidable sharp spikes or spines. *Triceratops* (Late Cretaceous) was 6 metres long and stood 3 metres high and had quadrupedal gait. Its enormous head carried 3 huge horns projecting from a large bony frill or collar projecting the vulnerable skull (Fig. 25.3).

Causes of Extinction

Dinosaurs had a majestic rise and a dramatic fall. After thriving and dominating the earth for 130 million years, the great dinosaurs and their contemporaries became suddenly extinct by the end of Cretaceous period. The extinction of this mighty race which underwent an explosive evolution is still a matter difficult to explain. Various factors have been suggested for their total extinction such as catastrophism, epidemic, food poisoning, racial senescence, climatic changes, over-specialisation, interspecific warfare or competition with ancestral mammals. None of these has been accepted as being completely satisfactory. Probably a combination of several factors are responsible for their extinction.

Now **Dr. T. Swain** of the New Royal Botanical Gardens in Britain says that dinosaurs were poisoned and murdered in cold blood. According to Dr. Swain it was nature which killed the reptiles by introducing into their diet a new class of plants containing "aromatic type of alkaloids". In support of his theory Dr. Swain says there is evidence of increasing physiological trouble in dinosaurs towards the end of their reign. He further said that there was an increase in the size of their hypothalamus and certain dinosaur fossils have been found in contorted positions suggestive of alkaloid poisoning. Dinosaurs were victims of changes in the ecological relationships between them and plants. Such changes hold the real key to the nature of extinction and to the whole evolutionary process. Dr. Swain goes on to say that demise of dinosaurs is a lesson which all animals including ourselves should heed.

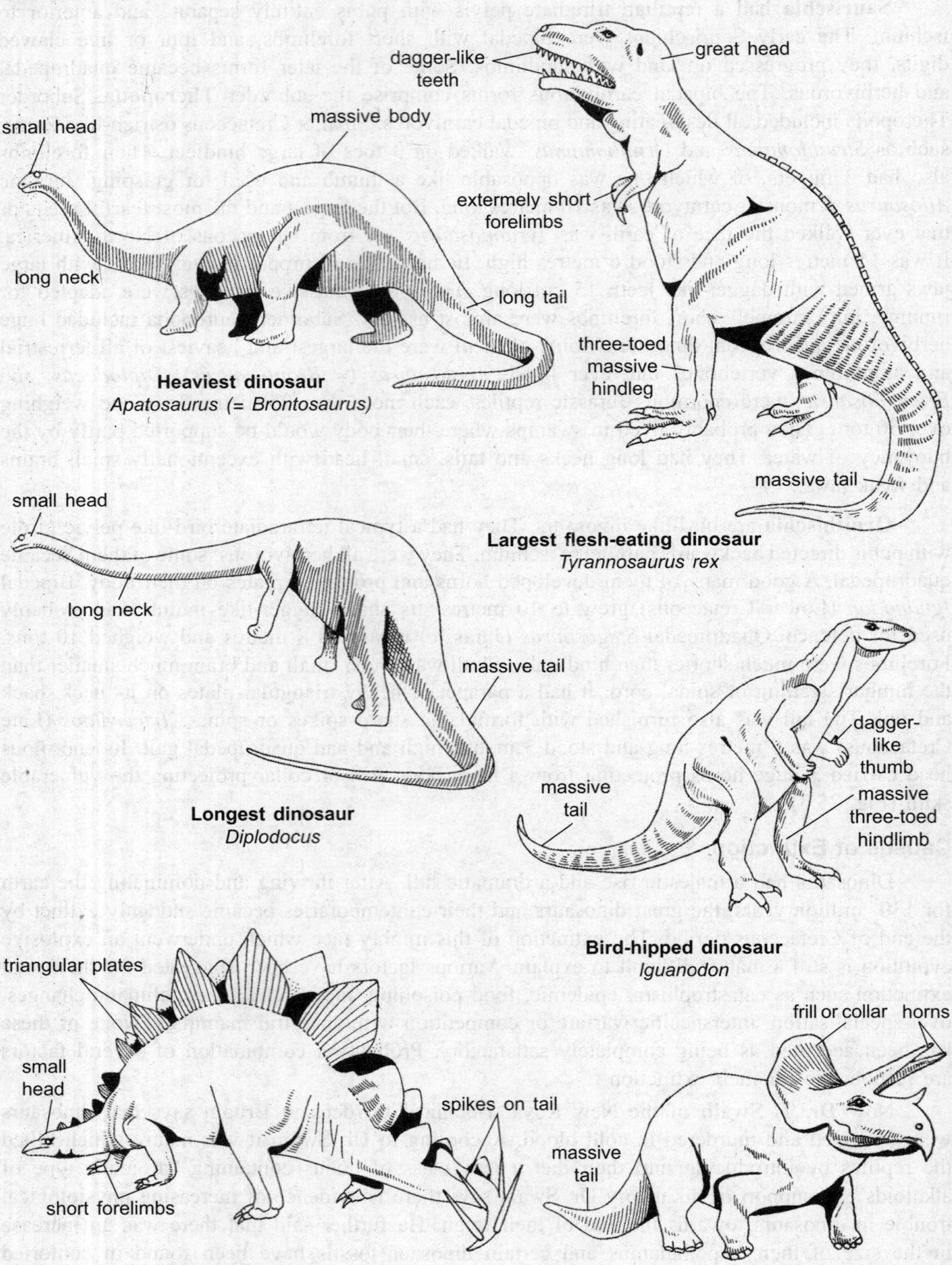

Fig. 25.3. Varied types of Mesozoic dinosaurs.

IMPORTANT QUESTIONS

Long Answer Questions

1. Discuss the origin and evolution of reptiles.
2. Write about the adaptive radiation in reptiles.
3. Reptiles are the first true land vertebrates. Discuss.
4. What are the physiological changes associated with the emergence of land vertebrates? Discuss.
5. Discuss the evolutionary trends among extinct reptiles.

Short Answer Questions

1. Write a few lines on the biological significance of the Mesozoic reptiles.
2. Write short notes on : (*i*) *Seymouria*; (*ii*) Dinosaurs; (*iii*) Age of reptiles.

Very Short Answer Questions

1. The number of occipital condyle in reptiles is only one. True/False.
2. Reptilian vertebrae are acoelous. True/False.
3. Archosauria were the reptiles dominating the Mesozoic era. True/False.
4. Archosauria are represented by extinct Pterosauria and Dinosaurs and modern crocodiles. True/False.
5. *Tyrannosaurus rex* was the longest dinosaur. True/False.

Multiple Choice Questions

1. Which of the following is not a reptilian character of *Seymouria*?

   (*a*) Skull flat with reduced ossification

   (*b*) Muscular limbs arise midventrally

   (*c*) Pelvic girdle attached to vertebral column by sacral vertebrae

   (*d*) Anapsid and monocondylic skull.

2. Which of the following is not true for *Seymouria*?

   (*a*) Skull is flat with high degree of ossification

   (*b*) An intertemporal bone is present

   (*c*) Palate is primitive

   (*d*) Traces of lateral line canals in the head region.

3. Age of reptiles is

   (*a*) Mesozoic era

   (*b*) Palaeozoic era

   (*c*) Carboniferous era

   (*d*) Permian era.

4. Dinosaurs dominated the earth for about

   (*a*) 130 million years

   (*b*) 150 million years

   (*c*) 160 million years

   (*d*) 155 million years.

5. Mammal-like reptiles belong to

   (*a*) Synapsid line

   (*b*) Anapsid line

   (*c*) Eurypsid line

   (*d*) Parapsid line.

ANSWERS

1. (*a*) 2. (*a*) 3. (*a*) 4. (*a*) 5. (*a*).

26 Type 10. *Columba livia* (The Common Rock Pigeon)

Chapter

Contents of this Chapter

EARLY HISTORY (ORIGIN)

Aves or birds arose in the Jurassic from ornithischian dinosaurs and have many characters resembling those of reptiles, consequently the two classes together are called **Sauropsida**. Ornithischia was the second main group of dinosaurs and appeared later than sauropods. They possessed 4-radiate pelvis, with the pubis directed backwards and an extra prepubic bone pointing forwards. Teeth were restricted to the hind part of the jaws, the front bearing a beak. At the front end of the lower jaw there was an extra bone (predentary). These were herbivorous forms and appeared in the Jurassic and achieved their maximum in the Cretaceous, by which time the sauropods had become less common.

The Triassic archosaurian reptiles (Subclass Archosauria–ruling reptiles) gave rise to two independent stocks that took to the air, called the pterodactyls and the birds. Both of these appear first in the Jurassic. The early archosaurs were bipedal animals, and the forelimbs were, therefore, free and available for use as wings. Other reptiles, such as *Draco*, develop a membrane between the limbs and the body to assist them in making soaring jumps. The flight of pterodactyl and birds may have originated like *Draco* or by flapping of the forelimbs during rapid running on the ground, then they became airborne for longer and longer periods (**Nopcsa**, 1907, 1923). Evolution of flight may have been different in both cases–the birds are fully bipedals, while the pterodactyls probably could not walk on their hindlegs and may have used the wing more for soaring than for flapping flight. Inspite of great differences there are interesting parallelism in the structure of the two groups, *viz.*, limb-bones became hollowed and light, skull bones fused and the jaws toothless and beaked. This parallelism in lives is known to be distinct showing remote common origin.

The pterodactyls are most commonly found in the Jurassic strata, less often in Cretaceous. The characteristic features that have produced the pterodactyl structure from a thecodont ancestry may be described as a lengthening of the head and neck, shortening of the body and the tail, lengthening of arms and especially of the fourth digit, shortening of the hindlimbs, and development of the ventral parts of the limb girdles. There are many others in the soft parts which also occur during the evolution of birds, for instance, the pterodactyls were warm-blooded, fur on the legs and the wings. However, there is no evidence that they possessed feathers. The wing was a membrane (patagium). Olfactory lobes were reduced but large cerebral hemispheres, optic lobes and cerebellum as in birds.

Rhamphorhynchus of Jurassic is still of archosaurian structure, *i.e.,* skull has two fossae and large teeth; humerus was short and strong, radius and ulna long, carpus with an extra "pteroid bone" in front, first three digits were short and hooked, fourth long supporting the wing and fifth absent. Hindlimb slender with five hooked digits. Tail was long ending in an expanded fin. Both girdles had well developed ventral regions and large sternum keeled in front. Scapula articulated with the vertebral column.

Pteranodon of Cretaceous showed further modifications, *e.g.,* trunk became shortened, forelimbs further lengthened, carpus being long, fourth digit much longer than the other three. Hindlimbs remained shortened and tail became very short. Teeth gradually lost and beak developed. In latest forms skull was drawn out backwards into a process. *Pteranodon* of late Cretaceous had a wing span of 8 metres *Pterodactyl* had a huge continuous membrane supported by a single finger. Pterodactyls were mostly marine. The largest flying birds alive today are the albatrosses, which use their great weight to gain height with the increasing velocity of wind a few metres above the sea. It is possible that the pterodactyls used a similar method of soaring. They were presumably unable to compete with the birds, and died out at the end of Cretaceous, along with many other reptiles.

Nearly all the orders of present-day living birds had appeared by the end of Eocene. It is estimated that the total number of bird species that have existed is 154,000, of which 9,600 are alive today (**Brodkorb**, 1971). The Pleistocene seems to have been a time of rapid formation of new species. It was not until the Cretaceous that birds became modernised, this involved both physiological and morphological changes, such as constant temperature, fusion of skull bones, formation of pneumatic bones, fusion of vertebrae to form a synsacrum and a suppression of the tail. But birds have become highly modified due to an aerial life and show many differences from reptiles. Present-day living birds number 8,590 species, they are aerial, terrestrial and aquatic.

CHARACTERISTICS OF AVES

1. Aves are bipedal, endothermal craniates with an exoskeleton of epidermal feathers on the greater part of the body.

2. Body temperature is high and constant in correlation with their energy.

3. Beak and claws have horny sheath, and feet are covered with scales.

4. Bones of the skull undergo early ankylosis.

5. There is a single occipital condyle.

6. The upper jaw is formed largely by premaxillae, and the lower jaw has 5 or 6 bones.

7. Forelimbs are modified to form wings with a carpo-metacarpus and only the first three digits of the hand.

8. Hindlimbs articulate far forwards to support the body weight in standing.

9. Head of the femur is at right angles to the shaft, there is a tibiotarsus and a tarso-metatarsus, there are never more than four digits, which are clawed.

10. Long bones are light with air spaces and no marrow.

11. The cervical and free thoracic vertebrae are usually heterocoelous. Several vertebrae fuse to form a synsacrum which fuses with the pelvic girdle.

12. Ribs are double-headed and have uncinate processes, both vertebral and sternal parts of ribs are bony.

13. The sternum has a ventral keel in flying birds.

14. Clavicle and interclavicle unite to form a furcula.

15. The ilium is very large, pubis and ischium are directed backwards, except in one case each they do not form pubic and ischiac symphyses.

16. The gullet is dilated into a crop, and the stomach is divided into a gizzard and a proventriculus.

17. Lungs are spongy and non-distensible, they are continued into air sacs.

18. Heart is four chambered and only the right aortic arch is present.

19. The skin is devoid of glands except an oil gland on the tail, and the tail is much reduced.

20. Cerebrum has large corpora striata, and optic lobes are lateral. The lumbar region of the spinal cord has a sinus rhomboidalis.

21. Eyes are well developed with a large pecten, they have sclerotic bones.

22. There are three-lobed metanephric kidneys, a urinary bladder is absent.

23. Right ovary and oviduct are usually atrophied.

24. They are oviparous with large yolky eggs enclosed in shell membranes and a hard shell, the embryo has an amnion, allantois, and yolk sac.

25. The young when hatched may be **precocial** or **nidifugous** being able to fend for themselves, or they may be **altricial** or **nidicolous** being naked and dependent on the parents for food for some time, in which case there is much parental care. Birds are noted for courtship, attachment to mates, nest building, and habit of migration in many.

SYSTEMATIC POSITION

Phylum	....	Chordata
Group	....	Vertebrata
Subphylum	....	Gnathostomata
Superclass	....	Tetrapoda
Class	....	Aves
Subclass	....	Neornithes
Superorder	....	Neognathae
Order	....	Columbiformes
Genus	....	*Columba*
Species	....	*livia*

HABIT AND HABITAT

The Blue Rock Pigeon lives in perfect freedom in ledges, fissures and holes of rocks, forts, old buildings and side walls of wells. It prefers to live in those places of towns and cities which have plenty of coarse grains. Thus, their favourite resorts include big buildings, godowns, grain markets, temples, mosques, churches, tombs, railway stations and office buildings. They never nest on trees.

Nature

The pigeons are inoffensive, harmless, timid and gregarious birds. During breeding season solitary couples make a simple, flat and artless nest of small sticks and thin roots, etc., at all sorts of places where there is some shelter from rain and sun. The eggs are laid in these nests and further development also occurs there. During winter, the pigeons collect into flocks which may be composed of several hundred individuals.

Food and Feeding

The pigeons are vegetarians, feeding on grains, pulses, seeds of fruits and grasses. Sometimes they feed on insects, snails and slugs probably mistaken for seeds. They regularly leave their places of retreats and settlings during mornings and evenings, and collect into flocks to plunder the nearby fields.

Locomotion

The pigeons are provided with long powerful wings which are well adapted for swift and strong flight. They walk on their two legs and such kind of walking is called **bipedal gait**. They walk on ground in search of food with great rapidity. When startled, they rise suddenly by striking the ground with their wings producing a crackling sound.

Sound

The pigeons do not sing, chirp or screech but produce a characteristic sound which resembles the syllables gootur-goon, gootur-goon.

Family Life

They lead a **monogamous** life, *i.e.,* one male lives and copulates with only one female throughout the life.

Reproduction and Incubation

The **internal fertilisation** is performed by copulation in which temporary union of male and female takes place at their cloacal ends, *i.e.,* cloacae of the two oppose each other, and transfer of sperms occurs directly into the urodaeum of female. The pigeons are oviparous, the eggs are laid in the nest and are incubated by the warmth of the parent's body and hatching occurs after a fortnight.

Parental Care

The newly hatched youngs are immature, helpless and featherless and are nourished by both parents by a fatty, curdy secretion, the **pigeon's milk**, which is secreted in their crop. The parental care and homing instinct are well developed in pigeons.

DISTRIBUTION

The Blue Rock Pigeons are widely distributed in Palaearctic (Europe) and Oriental regions (Asia) and North Africa. They are especially plentiful in Palestine city of Israel. In India, two subspecies of *Columba livia* are namely, *Columba livia neglecta* is found up to 13,000 feet on the Himalaya. Another *Columba livia intermedia* is smaller and darker race, which occurs throughout India. The Indian wild pigeon differs from that of European in having the rump or lower part of the back ash-coloured, while Indian pigeon is white.

EXTERNAL FEATURES

Shape and Size

The spindle-shaped or fusiform body of Blue Rock Pigeon is about 33 cm in length and is well adapted for rapid movement through the air.

Colouration

Except the eyes and the feet which are pink, rest of the body of pigeon is a slaty gray with glistening metallic green and purple sheen on the upper breast and around the neck. The wing has two black bars.

Body Divisions

The bilaterally symmetrical and compactly set body of pigeon is divisible into four regions: head, neck, trunk and tail. All of these body divisions are invested in a close covering of **feathers** which are directed backward and overlapping one another.

1. Head. The head is small, rounded and mobile. ***(i)* Beak.** Head is prolonged in front into a short, pointed, **bill** or **beak**, which is formed by an elongated upper jaw and comparatively smaller lower jaw. Both jaws lack teeth and remain ensheathed by a horny sheath or **rhamphotheca**. The terminal mouth is thus guarded by horny beaks and forms a wide gape. ***(ii)* Nostrils.** At the base of upper beak occurs a patch of naked, whitish, swollen area of soft skin, called **operculum** or **cere**. The cere overhangs on two oblique, slit-like **external nares** or **nostrils** which can be closed by the cere. ***(iii)* Eyes.** On either lateral side of the head is located a large eye. The eyes are round and are guarded by upper and lower eyelids and a transparent third eyelid or nictitating membrane. The nictitating membrane can be drawn across the eyeball from the anterior upper corner of the eye. The eyelids lack eyelashes and eyebrows. ***(iv)* Ear apertures.** Below and behind each eye is an external **ear opening** which leads to a short tube, the external auditory meatus, closed by the tympanic membrane. These apertures are hidden under special auricular feathers.

2. Neck. The neck is long, flexible and well demarcated from head and trunk. It helps in handling of food and compensates the forelimbs which have modified into wings.

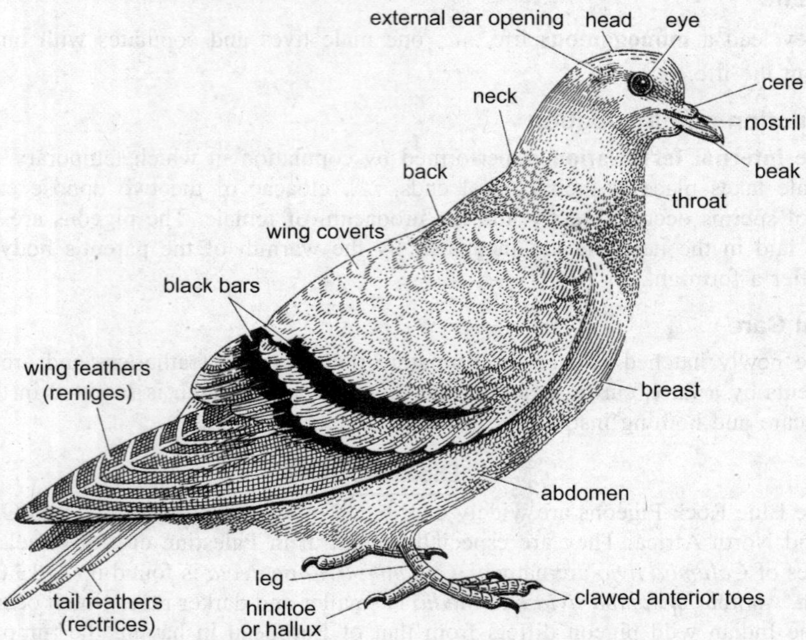

Fig. 26.1. Pigeon. (*Columba livia*). External features.

3. Trunk. The trunk is compact and spindle-shaped, bearing a prominent bony ridge, the **keel** or **carina**, at its mid-ventral line. There occurs a large, transversely elongated **cloacal aperture** on the ventral surface at the junction of the uropygium (tail) with the trunk. The trunk bears two pairs of pentadactyl type limbs : one pair of wings and one pair of legs.

(i) **Wings.** In pigeons, two forelimbs are modified into two powerful organs of flight, the **wings**. Each wing remains attached high on the antero-dorsal side of the trunk. Each wing is elongated, flattened and distally pointed with its longitudinal axis at right angles to that of the trunk. It can be folded close to the body in the form of Z during rest and extended during flight.

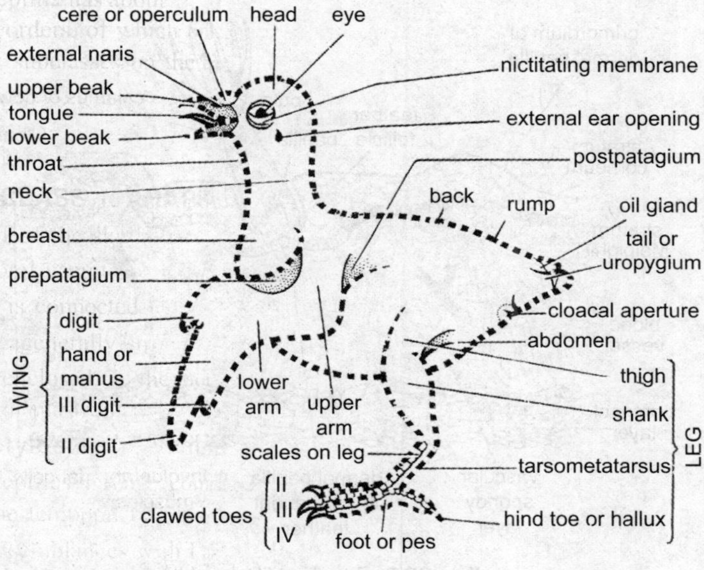

Fig. 26.2. Pigeon. Defeathered bird in left lateral view.

The wings show the three typical divisions of the forelimb : upper-arm or **brachium**, forearm or **ante-brachium** and the hand or **manus**, all closely bound together by skin. The manus or hand bears three, imperfectly-marked and clawless **digits**, of which the second is the largest. On the anterior or pre-axial border of the limb a fold of skin stretches between the upper-arm and the forearm called the **alar membrane** or **pre-patagium**. A similar but much smaller fold extends, post-axially, between the proximal portion of the upper-arm and the trunk, called the **post-patagium**.

(ii) **Legs.** The hindlimbs or legs arise somewhat anteriorly from the trunk to balance and support the entire weight of the body at rest. Each hindleg comprises three parts : the thigh, shank and foot. The **thigh** is short, closely bound to the trunk, and directed downwards and forwards. The **shank** is long and extends from the knee downwards and backwards. The **foot** is clearly divisible into a proximal portion, the **traso-metatarsus**, and **four digits**. The digits are long and clawed. The first digit is called **hallux** and is directed backwards. The remaining three digits, the 2nd, 3rd and 4th, extend forward. The legs are efficient organs of perching and locomotion with long toes and curved claws.

4. Tail. After removal of the feathers it can be seen that the true tail is short, conical projection of the trunk, known as **uropygium**. It bears a group of elongated tail-feathers or **rectrices**. On the dorsal surface of uropygium is a papilla bearing on its summit the opening of a **preen**, **coccygeal**, or **oil-gland**. It is the only integumentary or cutaneous gland found in birds. Its oily secretion is used for lubricating or dressing the feathers and beak (makes them water-proof and shining). It produces a specific recognition scent and it elaborates ergosterol.

which, when exposed on the feathers, is transformed by sunlight to vitamin D. It is assumed that the vitamin would be ingested during preening. It is also probably used for keeping the horny beaks from becoming brittle. The tail with its feathers is used as a rudder during flight and a balancer in perching.

SKIN

The skin of pigeon is dry, loose, hard and thin. Histologically, it consists of an outer **epidermis** of ectodermal origin and an inner **dermis** of mesodermal origin.

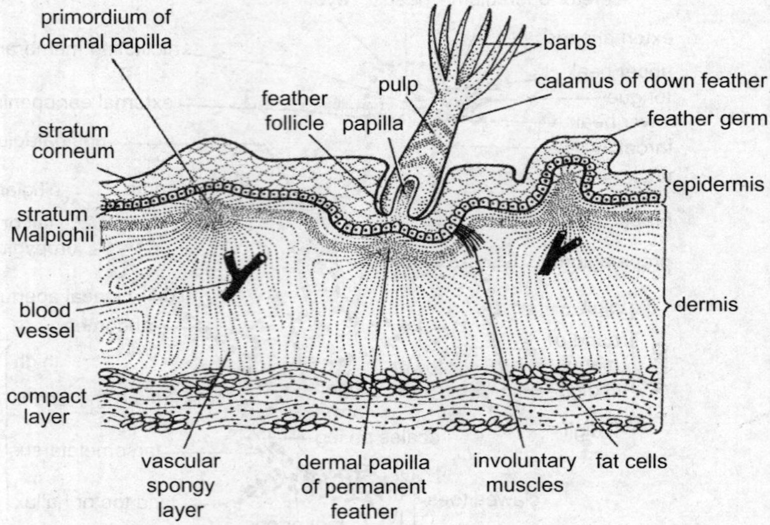

Fig. 26.3. Pigeon. V.S. of skin.

(*i*) Epidermis. The epidermis is several layers thick and composed of an outer **epitrichium**, a middle **stratum corneum** and an inner **stratum germinativum** or **stratum Malpighii.** The epitrichium is composed of a single row of flattened delicate cells. The stratum corneum is horny and protective and made of flat cornified cells. It is modified to form several exoskeletal structures, such as the **corneoscutes** or **scales** on the feet, **rhamphotheca** or horny covering on the beaks, claws and toes and feathers which covers the most of the body. The stratum germinativum is made up of large, cylindrical and continuously dividing cells. It gives origin to the cells of epitrichium and stratum corneum.

(*ii*) Dermis. The stratum germinativum of epidermis is followed by dermis. The dermis consists of muscle fibres, nerve fibres, blood capillaries, and connective tissue.

The epidermis and dermis have no pigmented cells, however, pigments occur in scales and feathers only. The skin lacks in sweat and any other type of cutaneous gland, except the **uropygial** or **preen gland** present dorsally at the base of tail.

EXOSKELETON

The pigeon has well developed exoskeletal system which serves many functions such as protection of body from temperature fluctuations of environment, radiations and mechanical injuries of various sorts. Further it makes the body of pigeon well adapted for aerial mode of life.

The exoskeleton of birds is derived from epidermis and occurs in the form of horny claws, beaks, spurs, webs, scales, feathers. **Spurs** are lacking in pigeons. Spur is a bony outgrowth of tarsometatarsus in the male fowl and some other birds. It is covered by a horny, scale-like epidermal sheath and pointed in shape. Male fowl uses it for fighting. In certain birds (plover,etc.), spurs may be found on carpometacarpus. **Webs** are modified fold of skin found on the feet of aquatic birds, such as ducks, geese and swans. It is found stretched between the toes. In cormorants, it includes all the four toes. Sometimes it leaves the hallux free. The horny claws and beaks have already been referred earlier in this chapter. Here, only scales and feathers will be discussed.

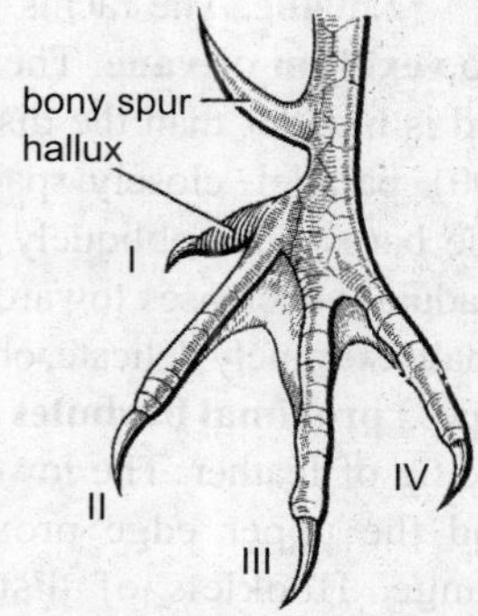

Fig. 26.4. Leg of fowl showing spur.

Scales

In pigeon, the scales are confined to the shanks and feet and some at the base of the beak. The scales are cornified derivatives of the stratum germinativum of epidermis which form a protective covering of body parts and are shed and replaced by moulting. The scales of pigeons and other birds resemble with reptilian scales in every respect, which indicates the origin of birds from reptiles.

Feathers

The entire body of pigeon is covered with a close and continuous covering of **feathers**, which constitutes the plumage. The feathers are found only in birds and are modified reptilian scales. They are light, strong, elastic, waterproof and show many colours due to melanin pigment of various shades as well as due to irridescence. Melanins range from black through brown to yellow, and laid down in the feathers by special cells in the papilla. Carotenoid pigments are also found, such as the yellow zooxanthin, the red astaxanthin.

Structure of a Typical Feather in Pigeon

To understand the structure of a feather, a contour feather which occurs on general body, wing and tail, can be considered as typical. A typical contour feather consists of a **central axis, main stem** or **scapus** and an expanded distal portion, the **vexillum** or **vane**.

1. Scapus (Axis). The scapus is divided into a basal portion, the **calamus** and an upper **shaft** or **rachis.**

***(a)* Calamus.** The calamus is hollow, tubular and semitransparent. The base of calamus remains inserted into a pit or **follicle** of the skin, from which non-striated muscle fibres pass to the feather and provide movement to each contour feather. The calamus opens below by a small opening called **inferior umbilicus**, which receives a small, conical, nutritive **dermal papilla** from the dermis. The nutrients and pigments are passed through the dermal papilla into the feather from the dermis, during the development of feather. Another pore, called **superior umbilicus**, occurs on the ventral side of junction of calamus and rachis. In flightless birds like cassowary, emu and extinct *Dinornis*, the aftershaft is as long as the main feather, from which it arises. In some pigeons and many other birds, a small tuft of soft down feather, called **after shaft** or **hyporachis**, occurs near the superior umbilicus and covers it.

***(b)* Rachis.** The part of the scapus above the calamus is a **rachis**. It forms the longitudinal axis of the vane. It is solid, opaque, roughly quadrangular in transverse section and filled with a closely packed mass of **pith cells**. A longitudinal furrow, the **umbilical groove**, runs along the ventral or inner surface of the rachis throughout its length.

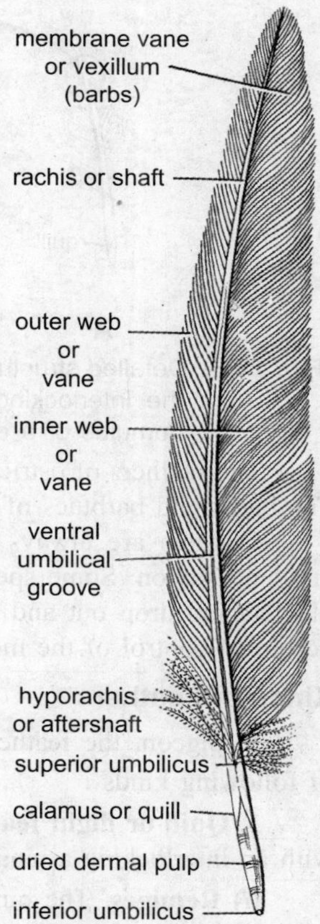

Fig. 26.5. Pigeon. A typical feather in ventral view.

2. Vane. The rachis bears a fan-like, webbed or expanded membranous part of the feather, the **vexillum** or **vane**. The vane is divided by rachis into two unequal lateral halves. Its proximal end is broader than the distal end. Each half of vane is composed of a series of numerous (about 600), parallel, closely spaced, delicate, lateral, thread-like structures called the **barbs** or **rami**. The barbs arise obliquely outwards from the two lateral sides of the rachis. The size of barbs gradually decreases towards both the ends of the rachis. Each barb on either side bears a fringe of small, extremely delicate, oblique filamentous processes, the **radii** or **barbules**. The barbules are of two types : **proximal barbules** directed towards the base of feather and **distal barbules** directed towards the tip of feather. The lower edge of distal barbules bears minute, **hamuli**, **barbicels** or **hooklets** and the upper edge proximal barbules are deeply curled or rolled to form a groove and flange. Hooklets of distal barbules hook over the grooved edges of proximal barbules binding the barbs together. With this limited sliding interlocking arrangement, all the barbs and barbules. are loosely held together, so that the vane forms a flexible, firm, wide, flat and continuous surface for striking the air during flight. This interlocking mechanism can be broken down so that the barbs become separate, but can be joined again by "preening" the whole feather.

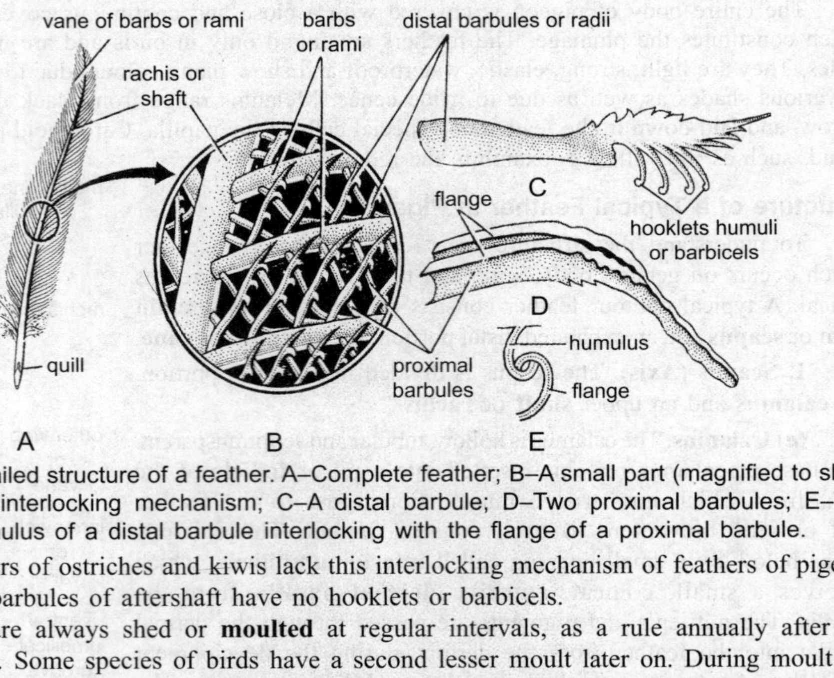

Fig. 26.6. Detailed structure of a feather. A–Complete feather; B–A small part (magnified to show the interlocking mechanism; C–A distal barbule; D–Two proximal barbules; E–The humulus of a distal barbule interlocking with the flange of a proximal barbule.

The feathers of ostriches and kiwis lack this interlocking mechanism of feathers of pigeon. The barbs and barbules of aftershaft have no hooklets or barbicels.

Feathers are always shed or **moulted** at regular intervals, as a rule annually after the breeding season. Some species of birds have a second lesser moult later on. During moult the old feathers drop out and new ones are formed from the same papillae. There is an elaborate hormonal control of the moult, basically by the thyroid.

Kinds of Feathers

In pigeon, the feathers are variously modified to serve different functions. They may be of following kinds :

1. Quill or flight feathers. The quill feathers have a strong rachis or shaft having barbules with an interlocking arrangement. They are classified into following types :

(*i*) Remiges. The quill feathers occurring on the wings and serving the purpose of flight are called **pinions, wing quills** or **remiges** (singular, **remex**). In remiges the posterior half of the vane is slightly broader than the outer or anterior half. Each wing of pigeon has 23 remiges which remain attached to its hinder border. Out of 23 remiges, eleven remiges are attached to the hand and are called **primaries** or **manuals**. The seven of these are attached to the metacarpal region and are called **metacarpals. The** remaining 4 are attached to the phalanxes of second and third digits and are called **digitals** which are further distinguished into : one **ad-digital** connected with the single phalanx of 3rd digit, 2 **mid-digitals** attached with the proximal phalanx of 2nd digit, and 2 **pre-digitals** with the distal phalanx of 2nd digit. The remaining 12 remiges are attached with the ulna of forearm and are called **secondaries** or **cubitals**. At the anterior border of the first digit (pollex) are attached a tuft of feathers called **ala spuria** or **bastard wing**.

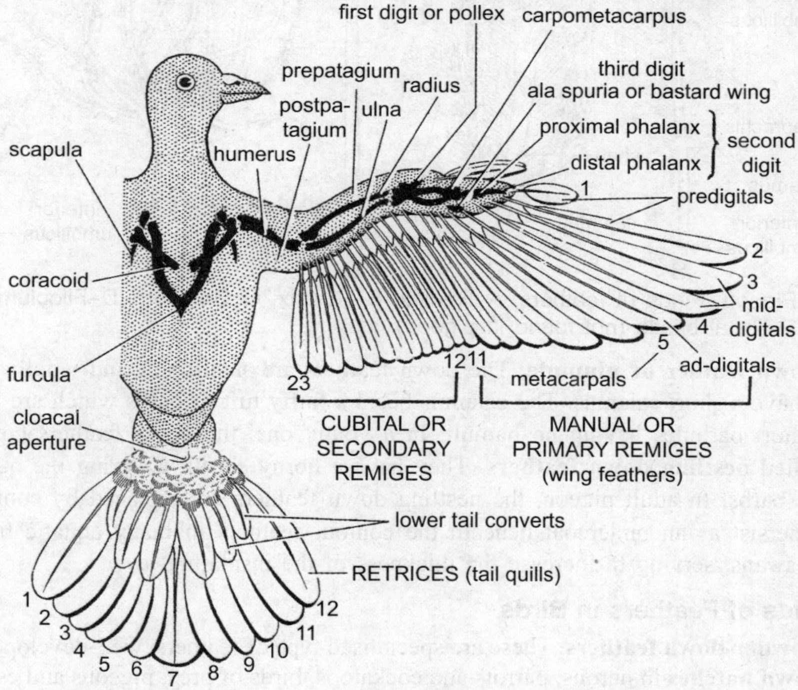

Fig. 26.7. Pigeon. Arrangement of flight feathers in ventral view.

***(ii)* Rectrices.** The quill feathers occurring around the uropygium to form the tail of pigeon are called **tail-quills** or **rectrices**. In pigeon, twelve long rectrices are arranged in semicircle or fan-like manner on the tail or uropygium. In rectrices two halves of the vane are almost equal in size. The rectrices act as a brake in alighting and as a rudder in vertical or lateral steering.

***(iii)* Coverts.** The quill feathers covering the bases of wing quills and tail quills are called **coverts**. The bases of the wing-quills are covered by several rows of **upper** and **under wing-coverts**, and the bases of the tail quills by **upper** and **under tail-coverts**. They are of smaller size than the quill feathers but both are structurally similar. They close the interstices between the calamuses (quills) of remiges and rectrices and, thus, presenting a continuous area to oppose the buoyancy of the air.

***(iv)* Contours.** The quill feathers forming the general covering of the body are called **contours** or **pennae**. They are smaller and woolly feathers having poorly developed barbules due to which barbs can be easily separated. These provide warmth and the smooth air flow, without turbulence.

2. Filoplumes (hair feathers or pin feathers). The filoplumes are small, delicate, hair-like feathers which remain sparsely distributed over the body among the contour feathers. A filoplume consists of a short calamus and a long thread-like rachis with a few weak terminal barbs, and barbules without any hamuli.

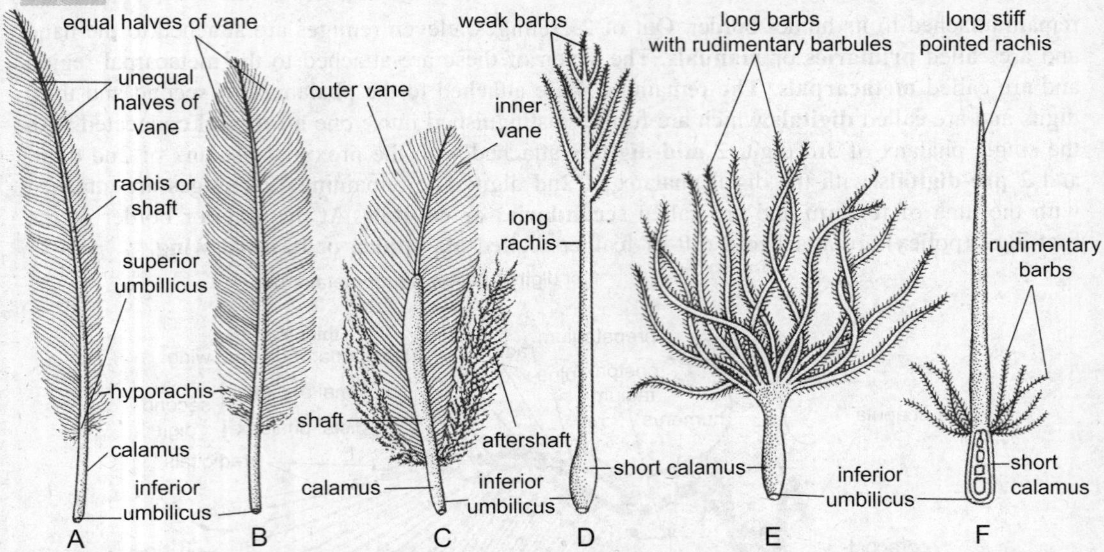

Fig. 26.8. Pigeon. Kinds of feathers. A–Remex; B–Rectrix; C–Contour; D–Filoplume; E–Down; F–Rictal bristle (not found in pigeon).

3. Down feather or plumule. The down feathers are small, soft and woolly and lack the rachis but have a short calamus. The calamus bears a fluffy tuft of barbs which are long, flexible and with short barbules having no hamuli. In a young one, the down feathers cover the body and are called **nestling down feathers**. They have a horny sheath covering the quill and basal portions of barbs. In adult pigeon, the nestling down feathers are replaced by contour feathers. But, they persist as an undercoat beneath the contour feathers in many aquatic birds, such as ducks and swans, serving to increase the thickness of the insulating layer.

Other Kinds of Feathers in Birds

4. Powder-down feathers. These are specialised type of feathers, well-developed in tracts or **powder-down patches** in herons, parrots and cockatoos, birds of prey, pigeons and especially **frog mouths** (*Podargus*). These feathers produce some powdery fragments for cleaning the plumage. The powder is a derivative of the innermost walls of Malpighian layer of the feather papilla.

5. Rictal bristles. Some birds, such as fly catchers, goat-suckers and whippoorwills, have stiff hair-like feathers called **rictal bristles** at the base of bill (rictus) and eyes. Each rictal bristle has a short calamus, and a slender rachis with a few rudimentary barbs at its base. They do not occur in pigeons.

6. Other types. There are certain other kinds of feathers which do not occur in pigeons, but occur in different birds. For example, **tactile feathers** or **vibrissae** occur at the root of the beaks or round the eyes of nocturnal birds, such as owl. Other peculiar feathers, such as **bristles** at the gape of Night jars, **eyelashes** of Hornbills, **wires** of birds of paradise and **ornamental plumes** of many birds, are modified contour feathers.

Pterylosis

The arrangement or distribution of feathers on the body is called **pterylosis**. In pigeons and majority of birds, except penguins, the feathers are not uniformly distributed over the whole body but are arranged in distinct patches or tracts called **pterylae**. The pterylae are followed by featherless areas or **apteria** which are covered with filoplumes in pigeons and by down feathers in ducks and many other birds. In flightless birds apteria are usually found only in the young, the adult having a uniform covering of feathers. The principal feather tracts or pterylae of pigeon are following :

(*i*) **Head or cephalic.** Covering the head.

(*ii*) **Neck or cervical.** On the neck.

(*iii*) **Shoulder or humeral.** Across the upper arm or humerus.

(*iv*) **Spinal.** Extending from neck to tail along the vertebral column.

(*v*) **Ventral.** It is a double tract with one branch running along each side of the breast.

(*vi*) **Wing or alar.** Composed of remiges with their coverts.

(*vii*) **Tail or caudal.** Composed of rectrices with their coverts.

(*viii*) **Femoral or lumbar.** Spreading obliquely on the outer side of the thigh.

(*ix*) **Crural.** Covering the shank of leg.

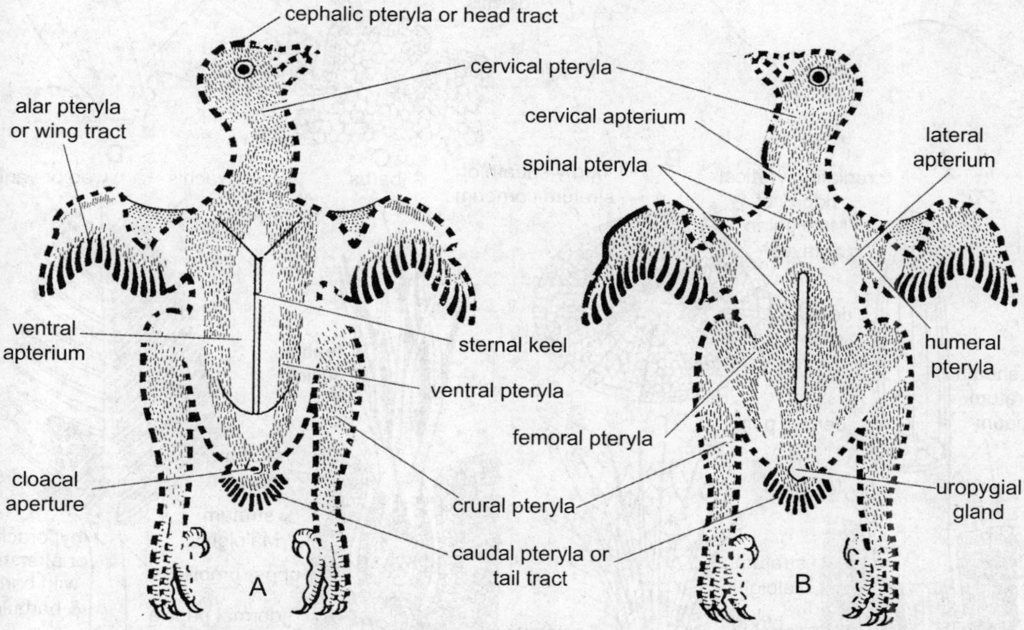

Fig. 26.9. Pigeon. Pterylosis (feather tracts). A–Ventral view; B–Dorsal view.

Development of Feathers

Feathers, like scales, arise in the embryo from papillae of the skin formed of dermis with an epidermal covering. The down feathers and contour feathers have different types of development, therefore, should be discussed separately as under.

(a) Development of a Down Feather

The mesodermal tissue of the dermis of delicate and semi-transparent skin of five-or six-day old embryo gather at a place to form a **dermal papilla** externally covered by epidermis. As the dermal papilla grows it takes the shape of a cone whose apex is directed backward. The papilla then becomes sunk in a sac called the **feather follicle**, from which it protrudes as an elongated **feather germ**. Its vascular dermal interior having the dermal blood capillaries, nerves and connective tissues is called **feather pulp**. The feather pulp feeds the growing feather.

The Malpighian layer of distal part of feather germ proliferates in such a way as to form a series of vertical or longitudinal radiating thickenings or ridges, which later separate from one another converting into the **rami** or **barbs**. The Malpighian layer of the epidermis at the base of the feather germ becomes horny or cornified to form calamus. The stratum corneum of the epidermis forms a horny, transparent and coherent cylindrical sheath, called **periderm**, around the growing feather. As the feather grows and expands, the pulp of the distal part of the feather germ is resorbed, the periderm splits and peels off liberating the barbs which are almost parallel to one another and slightly divergent. These are attached by their bases to the calamus. On the barbs, the barbules develop by secondary splitting. The feather pulp dries inside the calamus to form pith. Thus, a rudiment of **down feather** or **plumule** is formed. A dermal papilla persists at the base of the feather, to form a new down feather after the shedding of old down feather during ecdysis.

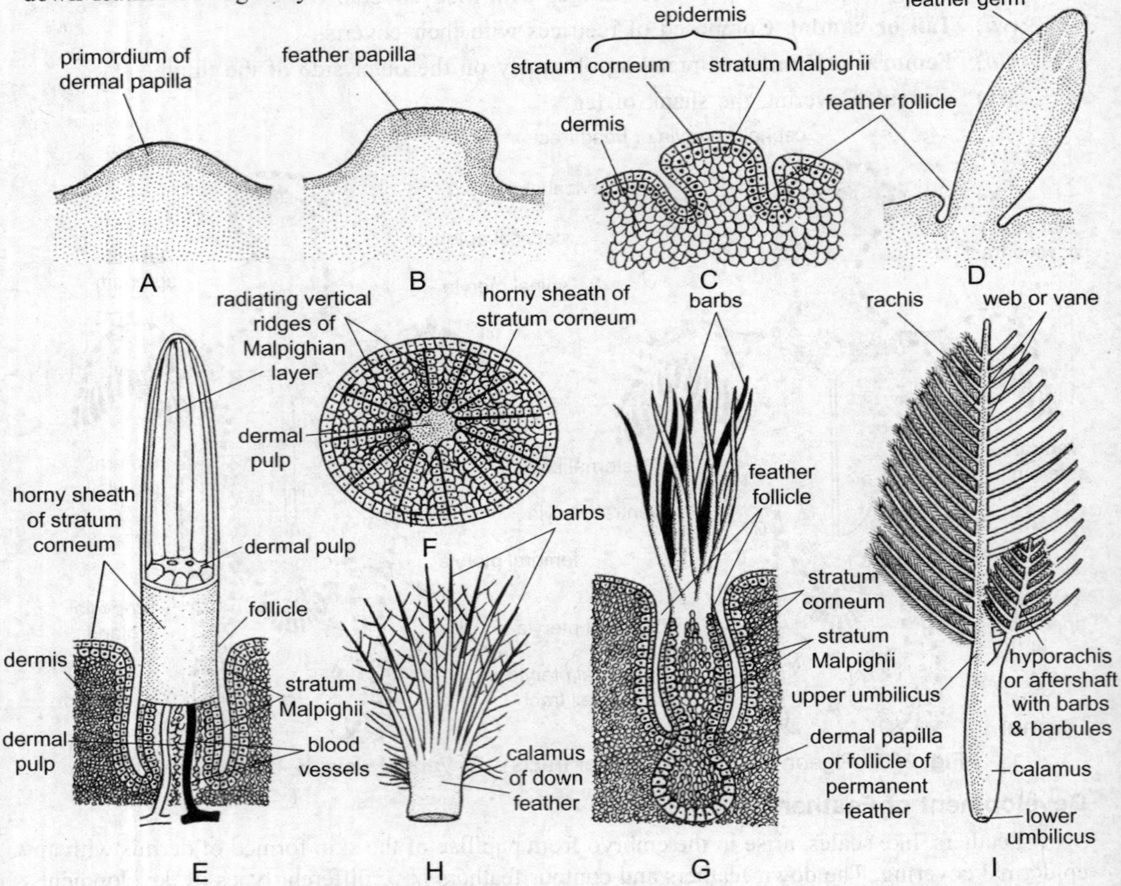

Fig. 26.10. Stages in the development of a contour feather. A–Beginning of feather papilla; B–Feather papilla arising above skin surface; C–Same in V.S; D–Feather germ; E–Feather germ in V.S; F–Feather germ in T.S; G–Down feather in its follicle; H–Mature down feather of a newly hatched bird; I–Young contour feather.

(b) Development of a Contour Feather

In the development of contour feather, the feather papilla is formed from the lower or deep end of that of the down feather, so that the earlier down feather drops off from its apex. Its early development is the same as in down feather. After the feather germ with its feather pulp, longitudinal ridges are not of the same size. Instead, two mid-dorsal Malpighian ridges thicken and fuse together to form a solid **rachis** which elongates and grows outwards more rapidly than the rest due to differential growth. As the rachis elongates, it carries up with it the remaining ridges which become the lateral barbs. New barbs appear below. A similar fusion of two ventral barbs with differential growth results in the formation of **hyporachis** or **aftershaft**.

Later barbules arise from both sides of the barb. The periderm splits and dries, the developing feather splits along the mid-ventral line and the barbs flatten out to form the **vane**. The portion of shaft which remains in the skin becomes **calamus**. The original basal aperture of feather germ becomes inferior umbilicus. The superior umbilicus develops later. The feather pulp inside the calamus dries to form the partition like pith.

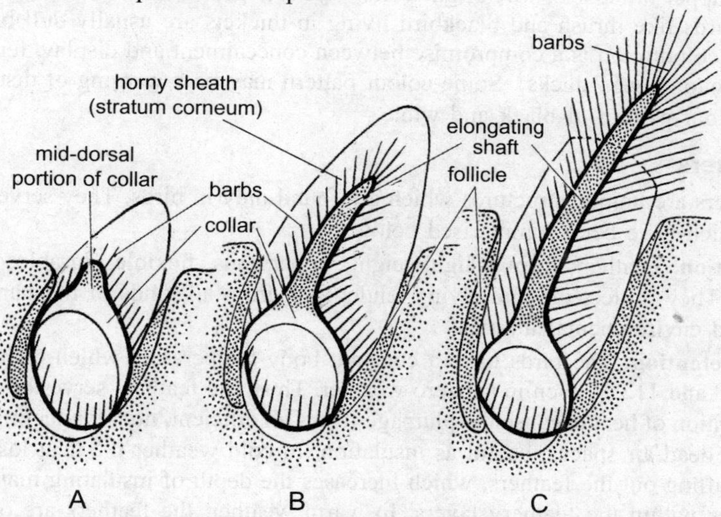

Fig. 26.11. Diagrammatic representation of a developing contour feather.

Moulting or Ecdysis

Shedding and replacement of feathers is **moulting** or **ecdysis** which takes place gradually, one moulting usually takes an average time of six weeks. At the base of each feather follicle, a dermal papilla persists from which new feathers will form. Thus, there is a continuous replacement of feathers throughout life. The replacement of feathers is seasonal in some birds such as peacock, while in other birds such as pigeon it is gradual throughout the year.

Pigmentation of Feathers

The characteristic pigmentation or colouration of pigeon's body is due to pigments in feathers and also due to characteristic arrangement of feathers. The feathers of pigeon and other birds show a varied and often brilliant colourations. The **colours** are produced partly by pigments and partly by reflection and diffraction effects (structural colouration). The most common pigments are melanins, ranging from black through brown to yellow. The **colours** are due to three factors: (*a*) Pigment is deposited in the feathers during development by special cells in the papilla. The processes of these amoeboid chromatophores convey granules of pigment into the barbs and barbules, where they are laid down in layers between those of keratin. Carotenoid pigments are also found such as the yellow zooxanthin and red astaxanthin. The colour that is seen, is due to absorption of some wavelengths of light by the pigment, thus, black, red, brown, yellow and orange colours are seen. White colour is not due to white pigment but is caused due to reflection of light from the feather without absorption of any wavelengths of light rays. In blue colours incident light is reflected from a layer of spongy keratin, pierced by holes 1–2 μm in diameter which absorbs the red and reflects blue light. (*b*) Structural arrangement or striations of feather surface are prismatic, these cause iridescence due to reflection of light, thus, producing iridescent hues, metallic colours, gray and some shades of blue. In iridescent feathers interference of light in thin surface films gives colours like those of soap bubbles. (*c*) A combination of pigments and prismatic striations of the feather produces green in which the yellow pigment combines with the structural blue. The colouration of birds are for concealment, recognition, and sexual stimulation and, hence, as the basis of their social life.

The actual colour patterns vary with the habits of the bird. Concealing (cryptic) colouration is very common. Even the bright colours may serve this purpose, by breaking up the outline of the bird when at rest or in motion. Most birds are dark above and pale below. The feathers often show mottled or speckled patterns rather than homogeneous colour. Finches and other birds living in sunlit upper branches show bright yellow, yellow-green, and blue colours, either singly or combined. Birds like thrush and blackbird living in thickets are usually dull brown or black. In most species colouration is a compromise between concealment and display, female is cryptic, the male conspicuous (*e.g.,* ducks). Some colour pattern may be a warning of distasteful quality, *e.g.,* magpie colour pattern of black and white.

Uses of Feathers

The feathers are unique structures which are found only in birds. They serve a varied array of functions which have been summarised below.

1. Protection. Feathers form a lightweight, impervious, flexible, durable and waterproof body covering. They protect the underlying tender skin from all kinds of mechanical, chemical, pathological and environmental injuries.

2. Heat retention. The birds have a constant body temperature which commonly remains in between 104° and 112°F, even in subzero weather. Thus, the feathers serve the most important function of retention of heat because the plumage forms an efficient, non-conducting covering with its innumerable dead air spaces, useful as insulation. In cold weather the heat loss is reduced to minimum by fluffing out the feathers, which increases the depth of insulating material by adding to the air spaces within the feathery layers. In warm weather, the feathers are often held close to the body to allow some escape of body heat.

3. Flight adaptation. Feathers make the bird's body well adapted for aerial mode of life. They are light, elastic, horny structures which make the body of birds of quite light weights. Further, the thin, flat and overlapping wing and tail feathers, with their close almost airtight linkage due to the interlocking arrangement of barbules, form surfaces to support the bird in flight.

4. Camouflaging (Protective colouration). The feathers of different birds have quite characteristic protective colouration like the colouration of their surroundings, which make them indistinguishable from their habitual surroundings and, thus, serve to protect them from their enemies.

5. Sexual dimorphism. Feathers provide protective colouration and also sexual display. Sexual dimorphism is common in monogamous as well as polygamous species. Colours and erectile plumes become **sign stimuli** which evoke or release specific reactions and whole pattern of behaviour in rivals and mates. Ornaments like bars and spots on the wings and tail, fully exposed only during flight in many gregarious birds, and often widely different in closely allied species, are probably **recognition marks** which enable stragglers to distinguish between a flock of their own and other species. For example, on many birds the crown feathers are modified into crests (peacock), topknots (tufted titmouse), ornamental plumes (California quail), etc. Some birds have special ruffs (ruffed grouse) or pinnae (prairie chicken) about the neck. Breeding plumes (cigrettes) in certain herons and egrets, upper tail, coverts in peacocks and elongated decoration in the streamers of some birds of paradise, serve the similar secondary sexual function.

6. Formation of nest. Some birds such as eider-duck and long-tailed-tit, use the feathers of other birds in the formation of warm and comfortable nests for their nestlings.

7. Other uses. The powdery secretions of powdery down feathers of many birds (*e.g.,* herons, bitterus etc.) help in keeping the plumage clean and free from ectoparasites. The feathers also provide a characteristic individuality to each species of bird, thus, have significant role in speciation. The spine-tipped rectrices of chimney-swifts and woodpeckers help them in clinging to vertical surfaces. Markings on feathers are recognition marks to their fellow birds.

Feathers are used for stuffing cushions and mattresses. In some birds patches of special feathers without a rachis break up to make a greasy "powder down".

MUSCULAR SYSTEM

The muscular system of birds is efficient and well developed. It is well specialised to meet the different needs of the bird. The diaphragm of pigeon is rudimentary. The muscles of back are atrophied due to inactivity or immobility of the trunk vertebrae since few thoracic vertebrae are fused and also the lumbars, sacrals and few anterior caudals are fused. However, the muscular system of neck, wings, tail, legs and ventral side of the body is well developed. Feathers are provided with non-striated muscles at the base, with an autonomic innervation serve to move the feathers. Striated subcutaneous muscles move groups of feathers. Control of feather position is important for regulation of heat loss, for flight and in sexual display, etc. Feathers can be studied under following heads :

Appendicular Muscles

The muscles which provide movement to forelimbs, wings and hindlimbs are called **appendicular muscles**. Both types of appendicular muscles (flight muscles and perching muscles) are well developed and can be studied separately as follows :

(A) Flight Muscles

The muscles which operate the forelimbs (wings) during flight are called **flight muscles**. These are pectoral, accessory and tensor.

***(i)* Pectoral muscles.** The most significant flight muscles of birds (pigeons) are pectoral muscles. These muscles remain attached to the keel of the sternum and to the wings, and provide up and down movements to the wings. They are of following two types :

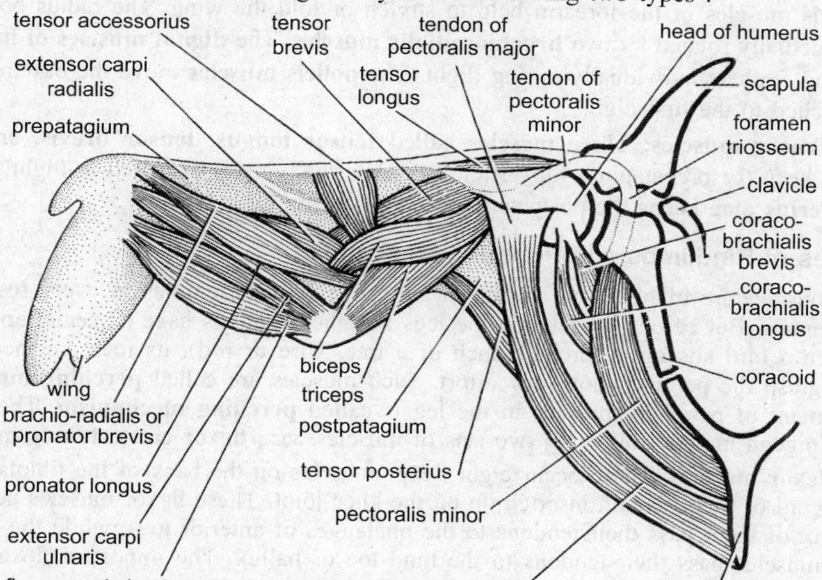

Fig. 26.12. Pigeon. Diagrammatic arrangement of flight muscles.

***(a)* Pectoralis major.** The pectoralis major is a very large, triangular and most powerful flight muscle which arises from the ventral side of the sternal keel and clavicle, one on each side, and forms the so-called 'breast'. This muscle weighs about one-fifth as much as the entire body and has dark red colour due to rich blood supply. In front each muscle is attached to the keel of sternum and the clavicle and anteriorly its fibres join the ventral side of the head of humerus by a broad and flat tendon. When the pectoralis major muscle contracts, the wing is pulled downwards and forwards, so that the body of pigeon is lifted up and propels itself through the air, because it causes the down stroke of wing, so also called **depressor muscle**.

***(b)* Pectoralis minor.** Pectoralis minor, also called **deep pectoralis**, **supracoracoideus** or **subclavius**, is a small and elongated muscle which elevates the wing during flight. It lies deep to the pectoralis major. It arises from the anterior part of the sternum, dorsal to the pectoralis major. Its tendon passes through the **foramen triosseum**, an aperture occurring in between the junction of the clavicle, scapula and coracoid; and is inserted on the dorsal side of the head of humerus. Pectoralis minor is an **elevator muscle** and it causes the upstroke of the wing. When it contracts, the foramen triosseum acts like a pulley for its tendon, pulling the humerus backwards and upwards, and, thus, raising the wing during flight. In pigeon, the pectoralis minor is especially developed and causes quick takeoff of the bird during flight.

***(ii)* Accessory muscles.** Besides pectoral muscles, the accessory muscles also elevate or depress the wing during flight. A **coraco-brachialis longus** or **coraco-humeral** lies beneath the pectoral muscles. It arises from the coracoid and the costal process of sternum, and its tendon is attached to the posterior side of the head of humerus. It lowers the hinder aspect of the wing. Another smaller and narrow muscle is **coraco-brachialis brevis** or **scapulo-humeral** lies in front of the longus. It also extends from girdle to humerus and raises the hinder edge of the wing. Both muscles rotate the wing in the glenoid cavity.

The **biceps** and **triceps** are the intrinsic muscles of the upper arm which operate the elbow and perform adjustments during flight. The **extensor carpi radialis** and the **extensor carpi ulnaris** muscles of the forearm help to stretch or fold the wing. The radius bone of the forearm is medially rotated by two **brachioradialis muscles**. The **digital muscles** of hand move the digits and feathers individually during flight. The **policis muscles** move the bastard wing or alula is attached to the first digit.

***(iii)* Tensor muscles.** Three muscles called **tensor longus**, **tensor brevis,** and **tensor accessorius** keep the prepatagium fully stretched when the wing is extended in flight. A similar **tensor posterius alae** keeps the postpatagium tensed during flight.

(B) Muscles of Hindlimbs and Perching Mechanism

The musculature of hindlimb of pigeon and other birds resembles in every respect with that of mammals. But certain muscles in the legs and their tendons have a special arrangement so that when a bird sits on a perch (branch of a tree, wire or rod), its toes are mechanically flexed and grasp the perch without any effort. Such muscles are called **perching muscles** and the arrangement of perching muscles in the leg is called **perching mechanism**. The perching muscles of pigeon include following two sets of muscles, *viz.*, **flexor** and **extensor** muscles.

1. Flexor muscles. There occur eight flexor muscles on the back of the tibiotarsus bone of hindlimb, all of them remain inserted up on the knee-joint. These flexor muscles perform the perching. Six of them pass their tendons to the phalanges of anterior toes, while the remaining two flexor muscles pass their tendons to the hind toe or hallux. The important flexor muscles are following :

***(i)* Ambiens.** In some birds, an ambiens muscle arises from the ilium and passes along the inner surface of thigh. Its long tendon runs beneath the patella bone round to the outer side of the knee enclosed in a special sheath and passes to the outer side of the tibiotarsus to join the upper end of the flexor muscle of the second and third toes. This muscle has little role in perching.

***(ii)* Peroneus medius.** This muscle occurs singularly on the anterior aspect of the shank, attached to the upper part of the tibiotarsus bone. Its tendon divides into three tendons, going to the three front digits.

***(iii)* Gastrocnemius.** It is a calf muscle mainly concerned with producing flexion of the toes in the act of perching and occurs on the back of tibiotarsus. Its tendon passes behind the ankle and trifurcate to supply the three anterior toes along with peroneus medius muscles. These tendons often act as a single unit.

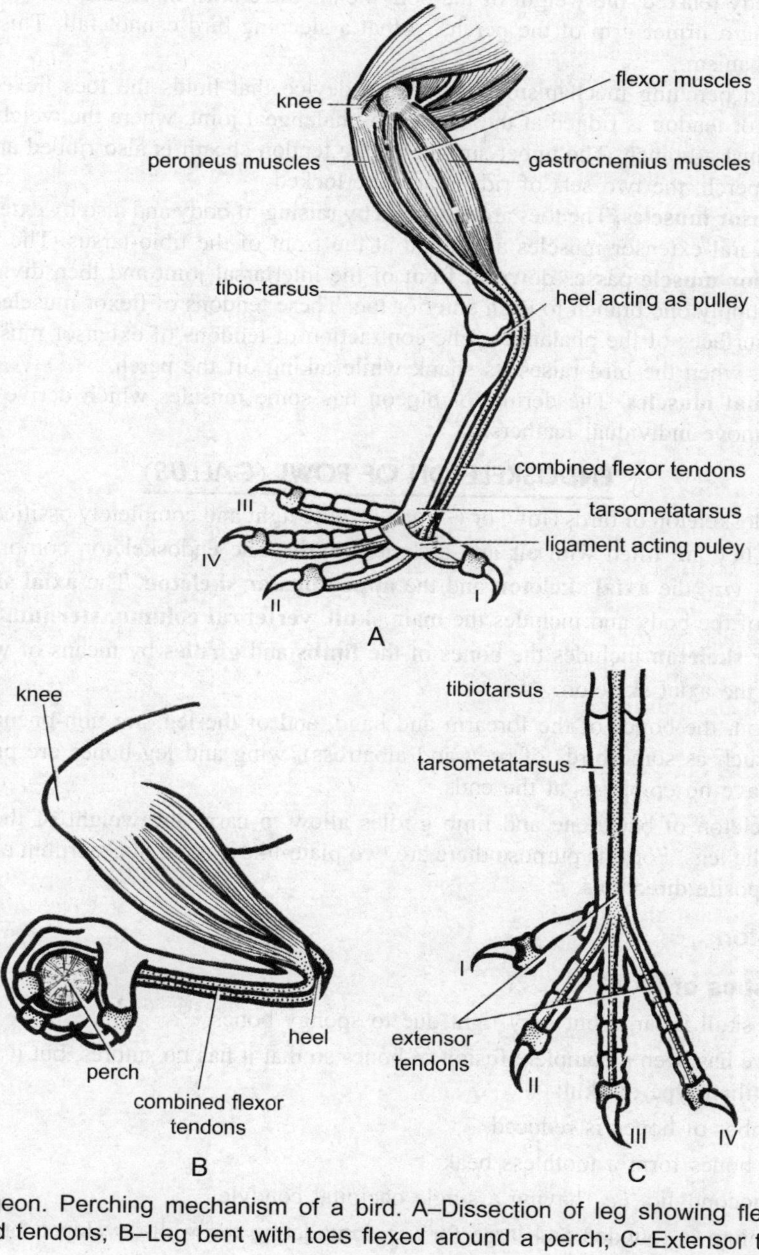

Fig. 26.13. Pigeon. Perching mechanism of a bird. A–Dissection of leg showing flexor muscles and tendons; B–Leg bent with toes flexed around a perch; C–Extensor tendons.

(*iv*) Flexor perforans. This muscle is attached to the upper part of the tibiotarsus (above the knee). Its tendon going to the hallux and is joined by a slip with the peroneus medius. Thus, a pull up on one tendon flexes all the toes.

When a bird sits on its perch the legs are bent at the knee and the ankle, thus, the tendons are flexed and the digits are automatically closed around the perch. A pull on any tendon will close all four digits. Flexing of digits is due to a **grip reflex** which is inflated by a **sensory receptor** on the lower surface of the foot, the moment the foot touches the perch. When a bird sleeps with body relaxed, the weight of the body bends the ankles more and the tendons become tighter ensuring a firmer grip of the perch, so that a sleeping bird cannot fall. This is automatic perching mechanism.

A second perching mechanism is a locking device that holds the toes flexed. The lower surface of flexor tendon is ridged at the metatarso-phalangeal joint, where the weight of the body presses it against a branch. The upper surface of the tendon sheath is also ribbed and as the bird settles on its perch, the two sets of ridges are interlocked.

2. Extensor muscles. The toes are unlocked by raising of body and also by extensor muscles. In pigeon, several extensor muscles are found at the front of the tibio-tarsus. The tendon of the **tibialis anterior muscle** passes down in front of the intertarsal joint and then divides into three branches, to supply one branch to each anterior toe. These tendons of flexor muscles are attached to the upper surfaces of the phalanges. The contraction of tendons of extensor muscles serves to open the toes, when the bird raises its shank while taking off the perch.

3. Dermal muscles. The dermis of pigeon has some muscles which derive from parietal muscles and move individual feathers.

ENDOSKELETON OF FOWL (*GALLUS*)

The endoskeleton of birds (fowl or *Gallus*) is quite light and completely ossified and contains no marrow. They are filled with air and are **pneumatic**. The endoskeleton comprises the usual two divisions, *viz.*, the **axial skeleton** and the **appendicular skeleton**. The **axial skeleton** forms the thin axis of the body and includes the main **skull**, **vertebral column**, **sternum** and **ribs**. The **appendicular skeleton** includes the bones of the **limbs** and **girdles** by means of which they are connected to the axial skeleton.

In pigeon, the bones of the forearm and hand, and of the leg, are non-pneumatic. In very good fliers (such as some birds of prey and albatross), wing and leg bones are pneumatic. The long bones have no epiphysis at the ends.

The skeleton of backbone and limb girdles allow to carry the weight of the body on the wings or on the legs. For this purpose there are two plate-like girders, the sternum and synsacrum, curved in opposite directions.

Axial Skeleton

Characteristics of Skull

1. The skull is large but very light due to spongy bones.
2. There has been a complete fusion of bones so that it has no sutures, but it is not deviated from the reptilian type of skull.
3. Number of bones is reduced.
4. Jaw bones form a toothless beak.
5. Monocondylic, *i.e.,* having a single occipital condyle.
6. Cranium is rounded and large for accommodating the well-developed brain.

7. Cranium does not extend forward into orbital region (**tropibasic**).
8. Large orbits separated by a thin membranous interorbital septum.
9. Autostylic jaw suspensorium.
10. In fowl, palate is **schizognathous** (short vomers and palatines meet together).

Regions of skull. The skull of fowl consists of **cranium, sense capsules** and **visceral skeleton** including the upper and lower jaws and hyoid apparatus.

1. Cranium (Brain case). It encloses the brain. For the study of the cranium it can be further divided into the following regions :

***(i)* Occipital region.** It is the back part of the skull. It is formed by the usual bones, the **basioccipital** (basal), the two **occipitals** (lateral sides) and the **supraoccipital** (above), all of which enclose a large rounded opening, the **foramen magnum**. Immediately below the foramen is a single rounded **occipital condyle,** mostly formed by the basioccipital. The supraoccipital anteriorly joins the parietals.

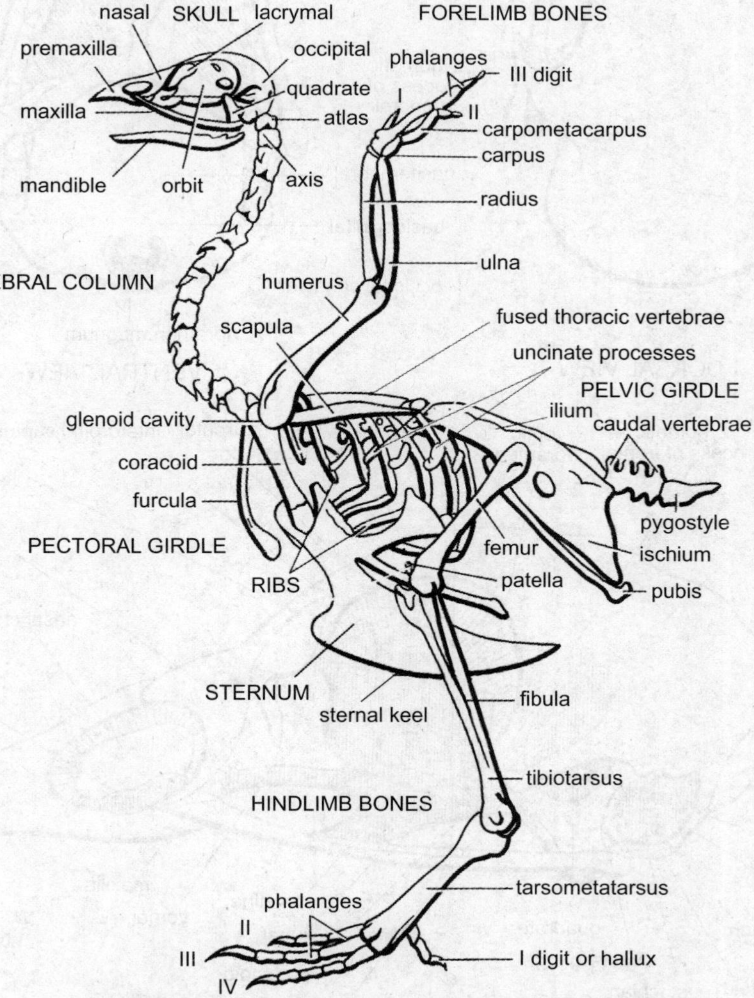

Fig. 26.14. Complete skeleton of a fowl in lateral view.

***(ii)* Parietal region.** It lies in front of the occipital region. Its roof is formed by two fused square-shaped **parietals**. The lateral sides of parietal region is formed by the **squamosals** and

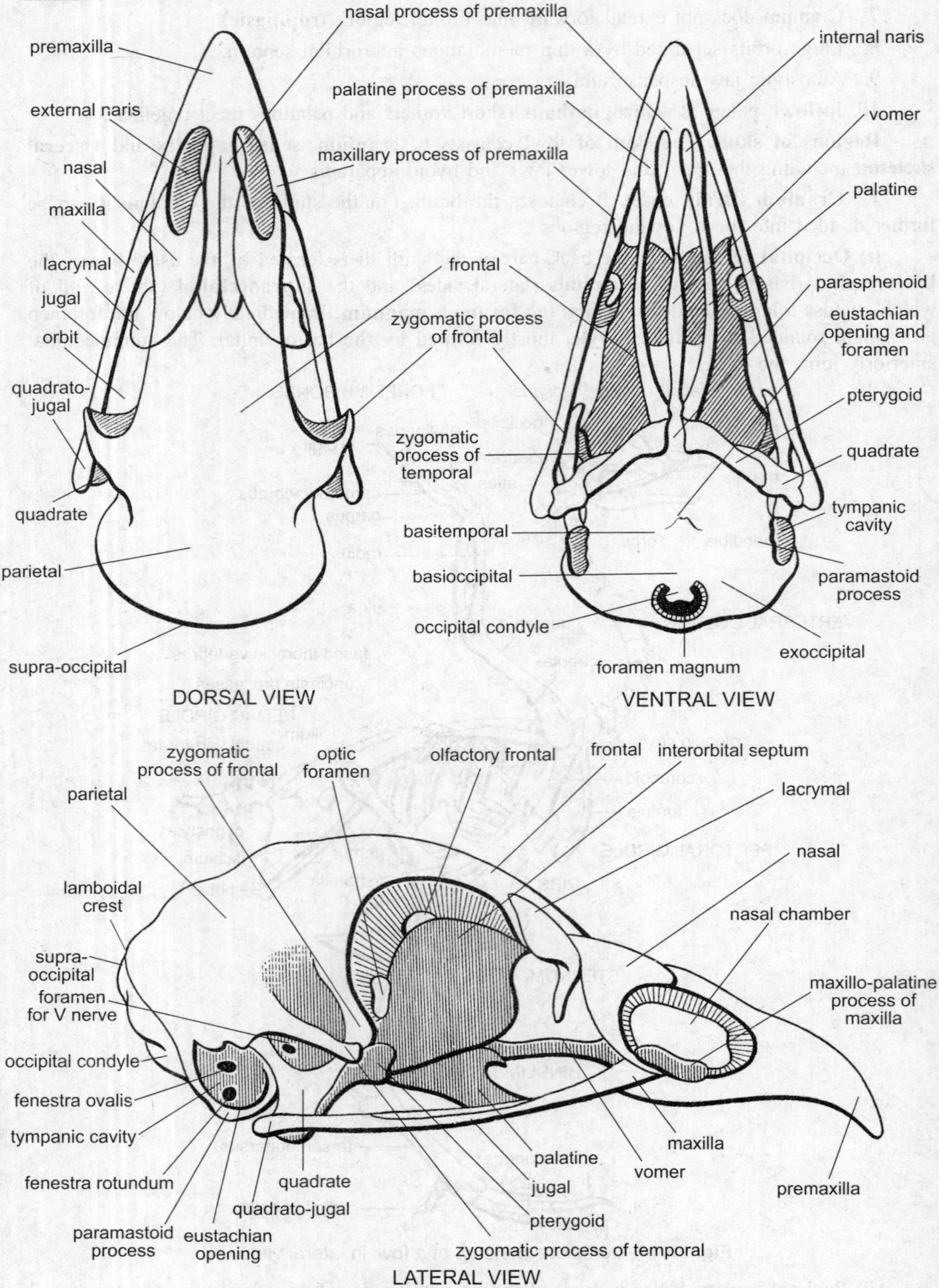

Fig. 26.15. Fowl. Skull in dorsal, ventral and lateral view.

alisphenoids and the floor by a large **basisphenoid**. Ventrally the basisphenoid is covered over by a paired broad membrane bone, the **basitemporal** or **parasphenoid**. The basisphenoid continues forwards as in lizards, by a slender **parasphenoid rostrum** which represents the anterior portion of parasphenoid. Each squamosal is produced outward and forwards into a zygomatic process. The squamosal bounds the tympanic cavity and firmly united to other cranial bones.

***(iii)* Frontal region.** It is anterior to parietal region. The roof of the frontal region consists of a pair of large **frontal** bones. The lateral posterior end of each frontal is drawn out into a strong **zygomatic** or **post-orbital process** which joins the zygomatic process of the squamosal. The alisphenoids are continued forward into the **orbitosphenoids** forming the sides of the frontal region, while its base is formed by a poorly developed **presphenoid**, lying above the parasphenoid rostrum.

2. Sense capsules. These are closely united with the cranium and lodge the various sense organs. The sense capsules include **auditory**, **optic** and **olfactory capsules**.

***(i)* Auditory capsules.** A pair of auditory capsules, enclosing the organs of hearing, are attached behind one on either lateral side of the occipital region of the skull. Each capsule is mainly formed by the **prootic bone**. (The small opisthotic of early embryo unites with the exoccipital and the epiotic with the supraoccipital.) A large cup-like tympanic cavity of the middle ear is present on each posterior lateral side of the skull bounded by the squamosal above and the basitemporal below. The tympanic cavity contains an **tympanic membrane** attached just within its prominent outer edge, and a slender rod-like **columella** made of bone and cartilage. Its inner portion is like a slender rod-like bone, called **columella auris**, and its outer part is called **extra-columella** which is a triradiate cartilage. Three rays of extra columella are the **supra-stapedial**, **extra-stapedial** and **infra-stapedial**. Within the tympanic cavity are several openings; two of these lie about the middle of the cavity. The upper one is **fenestra ovalis** and the lower one is **fenestra rotunda**. The inner end of columella auris fits into the fenestra ovalis, while its outer end touches the tympanic membrane. The tympanic cavity communicates with the mouth through an **eustachian canal**, the posterior funnel-shaped opening of which lies in the anterior ventral part of the tympanic cavity. The eustachian canal runs forwards and inwards between the basitemporal and the basisphenoid and joins with its fellow of the opposite side to open by a common median aperture in the roof of the posterior part of the mouth cavity.

***(ii)* Optic capsules.** The two orbits are very large to accommodate the relatively massive eyes. They are separated from one another by a narrow vertical partition, the **inter-orbital septum**, which is formed by a **mesoethmoid** together with **presphenoid, parasphenoid rostrum** and **orbitosphenoids**. Each orbit is bounded anteriorly by the frontal and large lacrymal, dorsally by the frontal and posteriorly by the post-orbital process of frontal and alisphenoid. The lacrymal is produced below in a curved pointed process hanging freely. Ventrally the orbit is incomplete. The interorbital septum is perforated by many openings. A large median **optic foramen** lies in the posterior part of the orbit. It actually perforates the orbito-sphenoid and communicates the two orbits with one another.

***(iii)* Olfactory capsules.** The olfactory capsules are cartilaginous and greatly reduced in size in accordance with the poor sense of smell. They are present in front of cranium. The **ecto-ethmoid** or **turbinals** are comparatively poorly developed. The nasals are a pair of thin, triangular plate-like bones forming the roof and sides of the nasal chambers. Each nasal sends out three processes, a posterior process uniting with the frontal and also with lacrymal. Other two anterior processes furnish the posterior boundary of the external nare and uniting with the process of premaxilla. The two vomers fuse into a small, median, slender bone, lying in continuation with the parasphenoid rostrum at the base of nasal chambers, and separating the two nasal chambers at the base. In pigeon vomers are absent.

Visceral skeleton. The visceral skeleton includes the upper jaw, lower jaw and hyoid apparatus. Jaws are not toothed and form boundaries of the mouth gape. Hyoid apparatus supports the tongue.

(*i*) Upper jaw. The upper jaw on each side forms two bony arcades : the **outer arcade** or **infra-orbital arcade** or **sub-orbital bar** consisting of premaxilla, maxilla, jugal and quadrato-jugal and the **inner arcade** or **palato-pterygo-quadrate bar** is formed by pterygoid, palatine and quadrate.

Premaxillae. Premaxillae are united in front into a large triradiate bone which forms practically the whole of the upper beak. Posteriorly it is produced into three paired processes : (*i*) The nasal processes are the longest, run backwards closely along the inner side of nasals to unite the mesethmoid and frontals forming the upper margin of the external nares. (*ii*) The maxillary processes run backwards and slightly outwards forming anterior part of the upper jaw. (*iii*) The inner palatine processes are the smallest, run horizontally backwards below maxillae and join ultimately with palatines. They form the anterior part of roof of buccal cavity (palate).

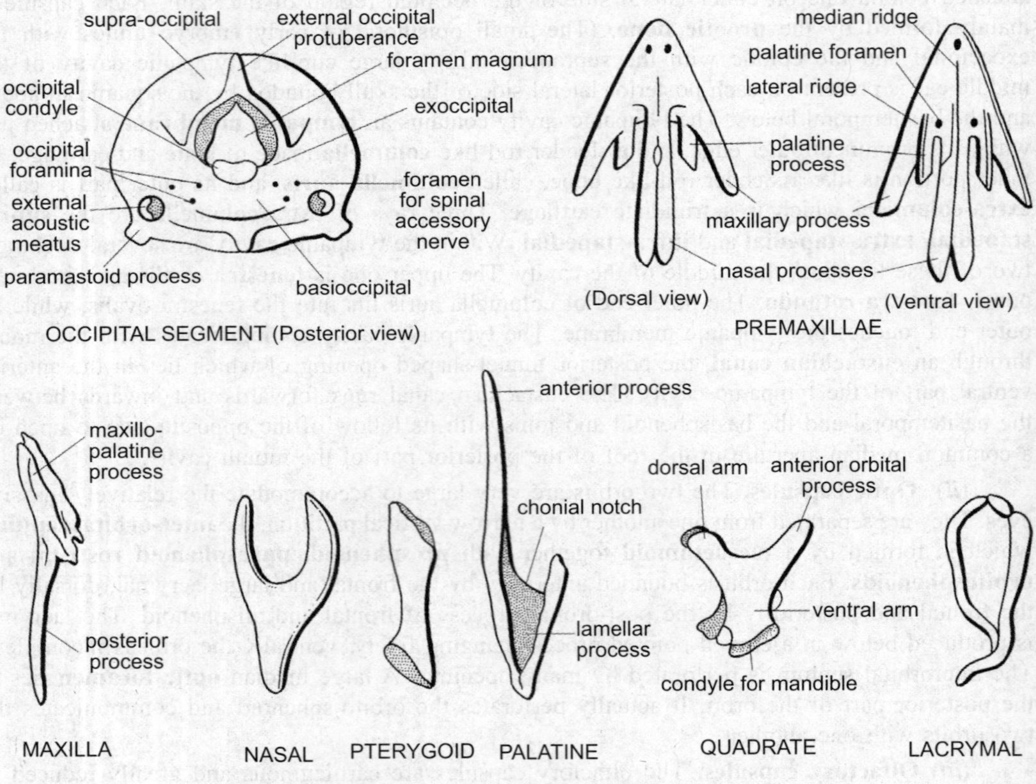

Fig. 26.16. Fowl. Disarticulated bones of skull.

Maxilla. Maxilla is a slender rod-shaped bone. Its anterior end articulates with maxillary process of maxilla and produced inwards into spongy **maxillo-palatine process**. The slender posterior end of maxilla is continued backwards by a slender jugal and quadrato-jugal to the quadrate. The posterior zygomatic process of the maxilla forms the anterior part of the outer arcade or infraorbital arcade or suborbital bar or zygomatic arch.

Jugal. The jugal is a very slender rod which forms the middle part of the sub-orbital bar lying dorsally upon the other two components of the bar, *i.e.,* maxilla and quadrato-jugal.

Quadrato-jugal. The quadrato-jugal is also a rod-like slender bone and forms the posterior part of the sub-orbital bar. Its hind end is thickened and articulates with quadrate.

Quadrate. The quadrate is a stout, triradiate bone articulating by two facets on its otic process with the roof of the tympanic cavity. Its **dorsal arm** articulates with the squamosal slightly above the anterior margin of the tympanic cavity. Its **ventral arm** forms a condyle for articulation with the mandible. Condyle also articulates on the outer side with the quadrato-jugal and on the inner side with the pterygoid. Its anterior arm (**orbital process**) runs anteriorly parallel to the pterygoid and ends blindly.

Pterygoid. It is a short and stout, rod-shaped bone. It lies obliquely behind the palatine between the presphenoid rostrum and quadrate of its own side. It articulates behind with the quadrate and in front with the parasphenoid rostrum and palatines.

Palatine. It is slender bony bar lying in front of parasphenoid rostrum. Anteriorly it is attached with the processes of premaxilla and maxilla. Its posterior end articulates with the pterygoid. Posteriorly it is also expanded inwards into a broad lamella, the sides of which articulate with the rostrum.

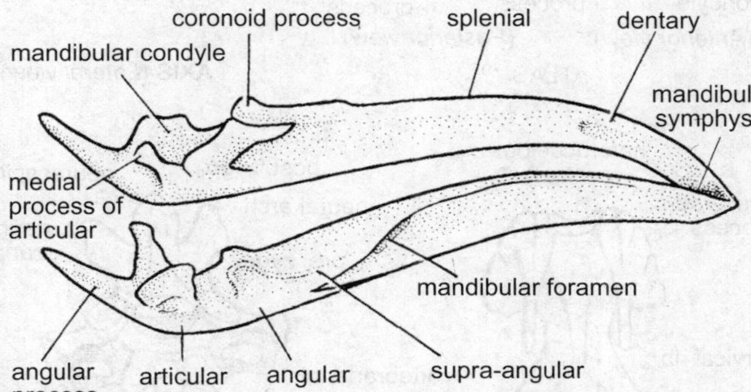

Fig. 26.17. Fowl. Mandible (Complete).

Lower jaw. The V-shaped lower jaw consists of two long laterally compressed rami which are firmly united in front. Each ramus is composed of five bones, *i.e.,* one replacing bone, the **articular**, and four investing bones, the **angular**, **supra-angular**, **dentary** and **splenial**, developing around a cartilaginous bar, **Meckel's cartilage**. All these bones are fused together.

Articular forms the posterior expanded part of each ramus. On the dorsal surface it bears an elongated mandibular condyle for the articulation with the quadrate. Posteriorly it is produced into a medial and angular process. **Angular** is a slender bone and lies beneath the articular along the lower inner posterior third border of each ramus. **Supra-angular** forms the dorsal part of about one-third of the posterior upper part of ramus. It bears a **coronoid process**. **Dentary** is the largest bone and forms about the complete distal half of the ramus. Anteriorly it is united with its counterpart of the other side forming the **mandibular symphysis**. It is devoid of teeth. **Splenial** is a thin splint-like bone lying on the inner surface of the middle part of ramus.

Hyoid apparatus. Hyoid apparatus lies in the floor of the mouth and supports the tongue. It is composed of three bones : an anterior **entoglossal** or **glossohyal** or **ceratohyal**, a median bony **basihyal** and a posterior **urohyal**. Entoglossal bears a cartilaginous process in front. Entoglossal gives off a pair of backwardly directed short **anterior cornua** which represents the free ends of ceratohyals. From the junction of basihyal and urohyal arise a pair of long backwardly directed **posterior cornua** or **thyrohyals.** Each thyrohyal consists of two rod-like pieces, the proximal ceratobranchial and distal epibranchial.

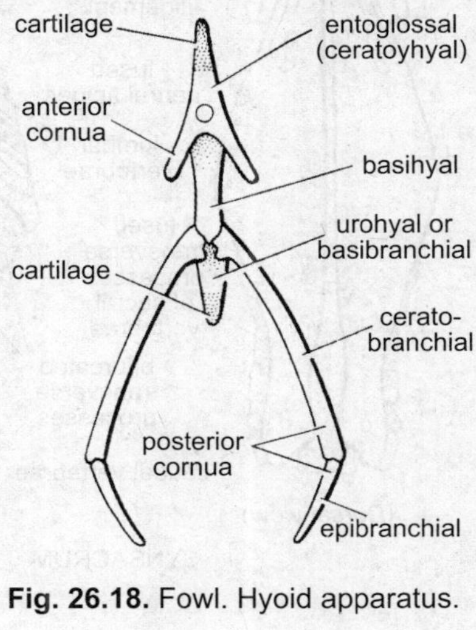

Fig. 26.18. Fowl. Hyoid apparatus.

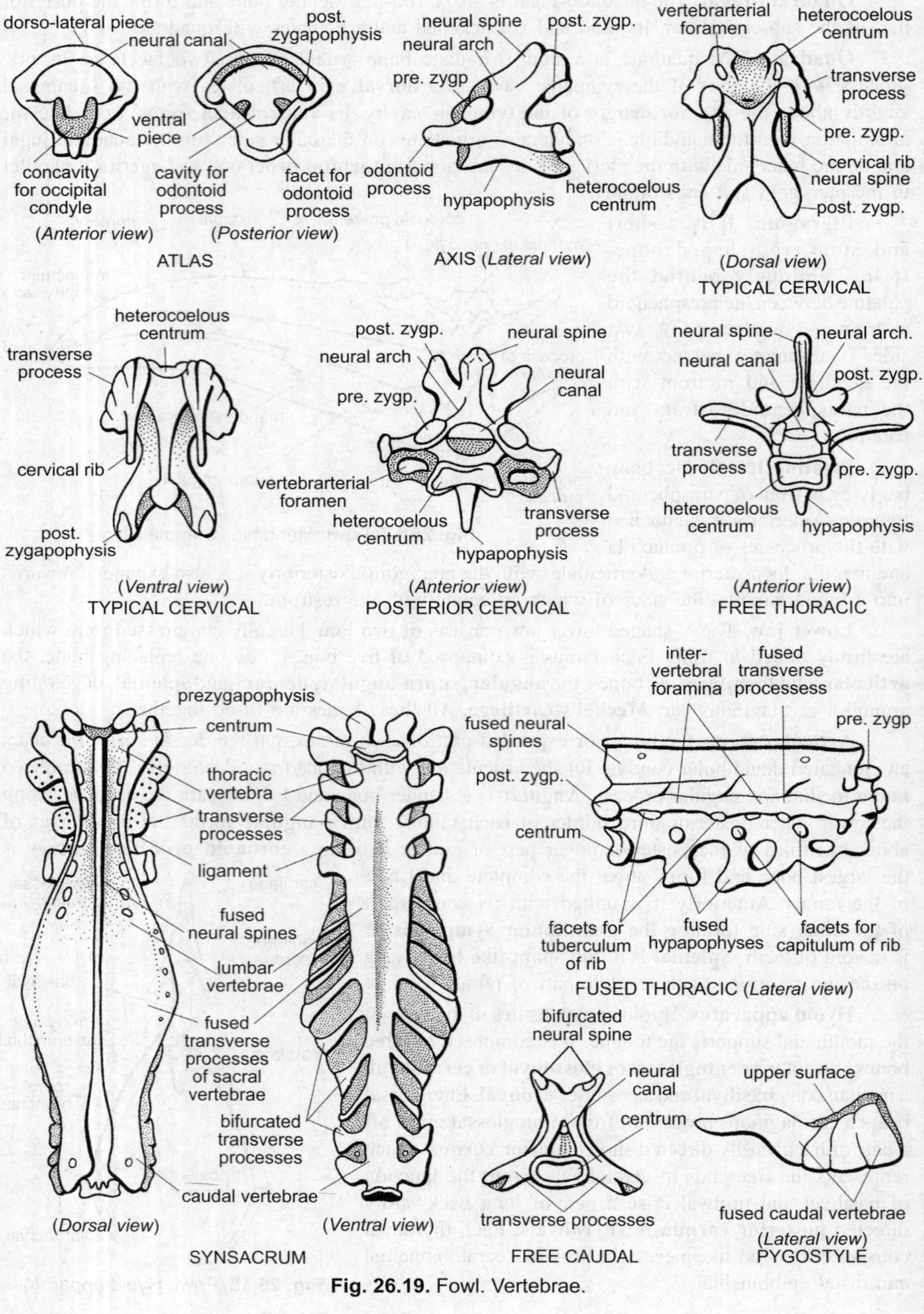

Fig. 26.19. Fowl. Vertebrae.

Vertebral column. The vertebral column of birds is peculiar in having a large number of cervical vertebrae due to long mobile neck; thoracic, lumbar and sacral vertebrae are fused giving rigidity which is advantageous in flight. Caudal vertebrae are fused forming the pygostyle which supports the tail feathers. Epiphysis is lacking in long bones. Vertebral column of *Gallus* is divided into the following regions, *i.e.,* **cervical**, **thoracic**, **sacral** and **caudal**.

1. Cervical vertebrae. These are 14 to 16 in number. Their articulation with each other allows greater movement of the long neck and head.

(*a*) Atlas vertebra. Atlas is the first vertebra of the cervical region. It is small ring-like in appearance. Centrum, neural spine, transverse processes, ribs and prezygapophyses are absent. It is formed of three pieces, a ventral and two dorso-lateral uniting mid-dorsally to form the **neural arch**. Its ventral piece is thick like the centrum and notched above to receive the **odontoid process** of the axis. Its anterior side bears a cup-like cavity to provide articulation to the single occipital condyle of the skull. In life, neural canal is divided by a transverse ligament into an upper spinal canal and a lower odontoid canal for odontoid process of axis. The posterior border of the neural arch is produced into a pair of postzygapophyses for the articulation with the prezygapophyses of axis vertebra.

(*b*) Axis vertebra. Axis is the second vertebra of the cervical region. Transverse processes, ribs and vertebrarterial canal are absent. Centrum is produced in front into a peg-like **odontoid process** by which it articulates within notch present on the dorsal side of the ventral piece of the atlas vertebra. Neural spine is blunt. Pre- and post-zygapophyses are present. A small **hypapophysis** is present below the posterior face of centrum. Head and atlas get mobility over the axis.

(*c*) Typical cervical vertebra. The typical cervical vertebra (3rd to 10th) has a long body or centrum. The centrum is **heterocoelous** or saddle-shaped. Anterior surface is concave from side to side and convex from above downwards, while the posterior surface is convex from side to side and concave from above downwards. The neural arch is short and notched both anteriorly and posteriorly. The neural spine is rudimentary. Transverse processes are short and irregular, and arise laterally from the anterior end. These are fused with the rudimentary posteriorly directed spine-like cervical ribs. The bases of cervical ribs have vertebrarterial canal for vertebral artery. Prezygapophyses are flat articular facets on the anterior side of transverse processes facing upwards and inwards and postzygaphyses are projecting backwards from the posterior side of the neural arch and facing downwards. Posterior carvical vertebrae (*i.e.,* behind Xth) are shorter. Their centrum bears a ventral spine-like process, the **hypapophysis**. Neural spine is also distinct. Cervical ribs are no longer visible and vertebrarterial canal is also open groove-like, not closed.

2. Thoracic vertebrae. These are seven in number. First and sixth are free, while 2nd to 5th are fused and seventh is fused with the synsacrum.

(*a*) Fused thoracic vertebrae. Fused thoracic vertebrae are formed by the fusion of 2nd to 5th thoracic vertebrae. The centrum is heteroceleous. Their neural spines, transverse processes and hypapophyses are fused with each other forming a continuous plate-like ridges perforated by intervertebral gaps. Fused hypapophyses form a ventral bridge. Fused transverse processes and centra of vertebrae possess on either side the tubercular and capitular facets for the articulation of thoracic ribs. Pre- and post-zygapophyses are present on the anterior and posterior side of the fused thoracic mass respectively.

(*b*) Free thoracic vertebra. First and sixth free thoracic vertebra are slightly smaller than the typical cervical. The centrum is heterocoelous and neural spine is elongated, pointed and well-developed. Transverse processes are also well-developed and outwardly directed. Pre- and post-zygapophyses are present on the anterior and posterior surfaces respectively. Hypapophysis is prominent beneath the centrum. It also carries capitular and tubercular facets for the articulation of the thoracic ribs.

3. Synsacrum. Synsacrum is a long compound bone formed by the fusion of 14 to 16 vertebrae, supporting the long ilium of the pelvic girdle on both sides. The fused vertebrae of synsacrum are last thoracic having a pair of free thoracic ribs and has no hypapophysis, six lumbars, two sacrals and about seven caudal vertebrae. All the fused vertebrae of lumbar region have free well-developed transverse processess which are applied against long ilia. The neural spines of these vertebrae are also fused to form a vertical crest which is continuous in front with that of the last thoracic and attached with the dorsal margins of the ilia. Two sacral vertebrae lie behind the lumbars and fused with them. Their transverse processes are also united to form bony plates which support the ilia along their outer margins. Sacral vertebrae are fused with their transverse processes. The hindermost part of the synsacrum is formed by the remaining seven anterior caudal vertebrae. The centra of few anterior caudal vertebrae are laterally compressed, while those of two or three posterior vertebrae are dorso-ventrally compressed. The transverse processes in each except the last one is divided into dorsal and ventral portions. The dorsal portions of these fuse to form bony plates continuous with those of the vertebrae in front. The ventral portion is rod-shaped in the anterior four or five vertebrae and is regarded as ribs but in the remaining are smaller. The components of synsacrum are so intimately fused that it is difficult to distinguish them separately.

4. Caudal vertebrae. Tail is short and has a few free vertebrae and pygostyle (fused caudal vertebrae).

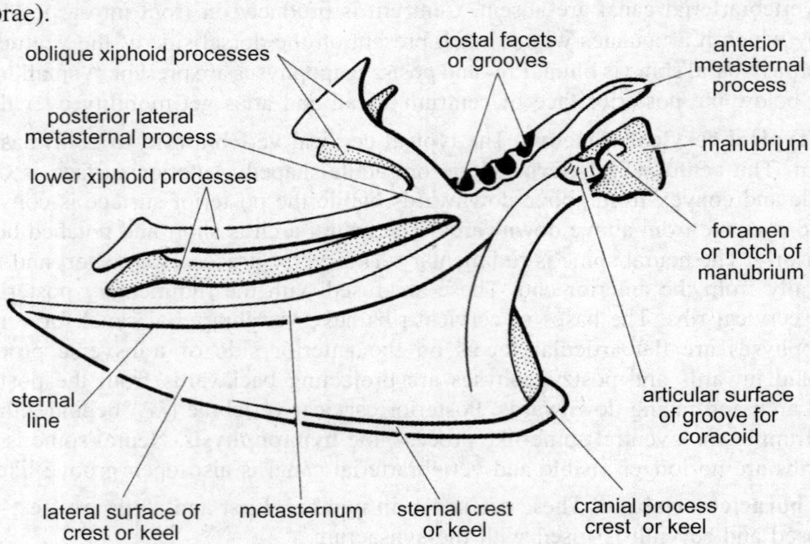

Fig. 26.20. Fowl. Sternum (Lateral view).

(a) Free caudal vertebra. Four or five free caudal vertebra are small. Their centra are heterocoelous, and neural spines are small and bifid. Transverse processes are somewhat cylindrical and directed downwards and outwards. Pre- and post-zygapophyses are absent. Movement of tail feathers is due to these vertebrae.

(b) Pygostyle. Pygostyle is the last fused vertebra of the caudal region. It is commonly known as **ploughshare**. It is formed by the fusion of four or more of hindermost caudal vertebrae. It looks like a verticaly triangular and laterally compressed plate. Centrum, neural spine, pre- and post-zygapophyses are absent. It supports the rectrices (tail feathers) and provides attachment to caudal muscles.

Sternum

Sternum (Fig. 26.20) is a broad bone lying in the breast region, therefore, it is also known as breast bone. It is a boat-shaped bone present in the ventral thoracic region and consists of the following parts :

1. Metasternum. It is a broad plate-like bone forming the body proper. It is concave dorsally and convex ventrally. It is broader towards the anterior and posterior sides and narrow in the middle.

2. Keel. Ventral convex surface of metasternum is produced ventrally into a median, triangular ridge called the **sternal crest** or **carina** or **keel**. The great pectoral muscles are attached with the keel.

3. Manubrium. Manubrium is a small plate-like structure arising vertically from the anterior end of the ventral surface of the metasternum and called the **manubrium** or **presternum**. It also bears a large foramen at its base called the **notch of manubrium**.

4. Metasternal processes. There are a pair of large processes on either side of the metasternum, a large posterior and a small anterior metasternal processes. Anterior processes are also known as **costal processes** and posterior processes are also called **xiphoid processes**. Each posterior metasternal process is divided again just at its origin in two large laterally compressed processes, the upper is oblique lateral xiphoid processes and posterior lateral xiphoid processess. The upper oblique processes terminate in a broad end just behind the last sternal rib. Lower larger xiphoid processes run downward and backward.

5. Costal surfaces. These are four cup-like depressions in the dorso-lateral margins of the anterior part of the metasternum. These depressions serve for the attachment of sternal ribs.

6. Coracoid grooves. There is a deep transverse groove on either lateral side of metasternum, at the base of manubrium which provide articular surface for the coracoid bones. Coracoid grooves communicate with each other through the foramen of manubrium.

Ribs

There are seven pairs of thoracic ribs in fowl. One pair of ribs articulates with each thoracic vertebra. Each thoracic rib consists of a dorsal **vertebral** and a ventral **sternal** portions meeting at an angle. By the sternal portion the ribs are united with sternum. First, second and sometimes the seventh ribs do not have sternal parts and do not reach the sternum. The vertebral portion of the ribs leaving first and last pairs bear backwardly directed **uncinate processes** which overlap the next rib behind. The proximal end of vertebral portion is bifurcated into a dorsal **tuberculum** and ventral **capitulum**. The tuberculum is attached to the transverse process, while the capitulum is attached to the centrum of the corresponding vertebra.

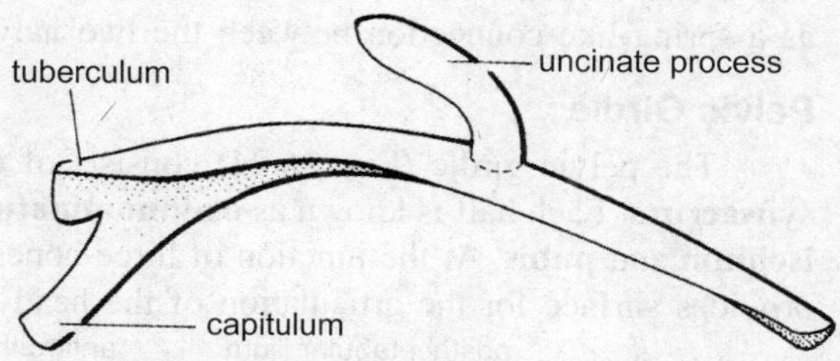

Fig. 26.21. Fowl. A vertebral thoracic rib.

Appendicular Skeleton

Pectoral Girdle

The pectoral girdle (Fig. 26.22 & 26.23) is very stout bony structure connected with the sternum on either side to support the wings. On either side it consists of a scapula, coracoid and a clavicle.

Scapula. Scapula is long, flattened, slightly curved, sabre-shaped bone lying above the thoracic ribs and parallel to the vertebral column. The expanded anterior end of scapula is firmly united with the coracoid by ligament. Its anterior end bears a shallow depression forming a part of **glenoid cavity**. The glenoid end of scapula is produced into an **acromian process** to provide articular surface to the clavicle and to form a part of the **foramen triosseum**. Suprascapula is wanting.

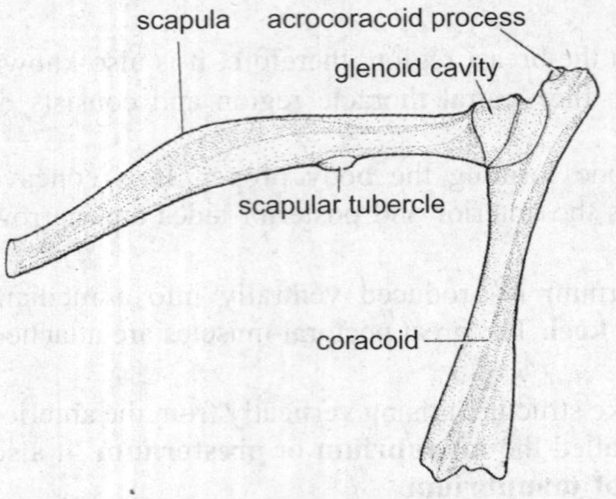

Fig. 26.22. Fowl. Pectoral girdle (Right half). Outer view,

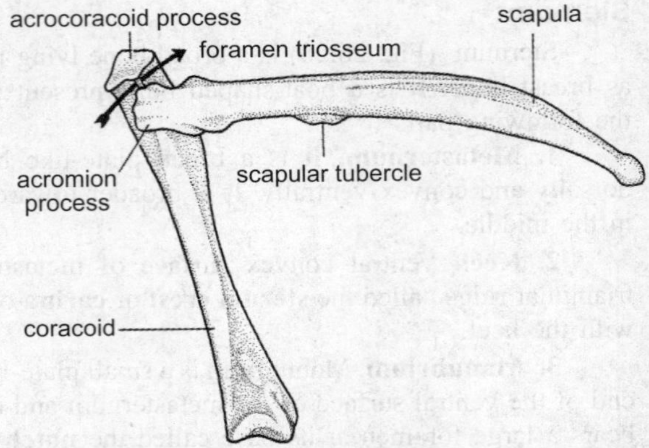

Fig. 26.23. Fowl. Pectoral girdle. Inner view.

Coracoid. Coracoid is a stout, straight rod-like bone directed downwards and articulates with the **coracoid groove** on the antero-lateral edge of the sternum at the base of manubrium. The upper end of coracoid on its inner side articulates with the scapula and on its outer side it bears a deep cup-shaped depression which forms the greater part of the glenoid cavity. The upper end of coracoid is also produced into a hook-like **acrocoracoid** process which articulates with the clavicle.

Clavicles. Clavicles are a pair of slender, curved delicate rod-like bones connected by their expanded upper ends with the scapula and acrocoracoid process of coracoid to enclose a circular **foramen triosseum**.

Furcula. Ventrally the two clavicles are fused with a small interclavicle to form a laterally compressed disc or **hypocleidium**. It is connected with the rostrum of the sternum by a ligament. The V-shaped bone, thus, formed is known as **furcula** or **merry thought** or **wish bone**. It works as a spring-like connection between the two halves of shoulder girdle.

Pelvic Girdle

The pelvic girdle (Fig. 26.24) consists of two separate halves lying on either side of the **synsacrum**. Each half is known as **os-innominatum**. Each os-innominatum is composed of **ilium, ischium** and **pubis**. At the junction of three bones is present a concavity, the **acetabulum**, which provides surface for the articulation of the head of femur.

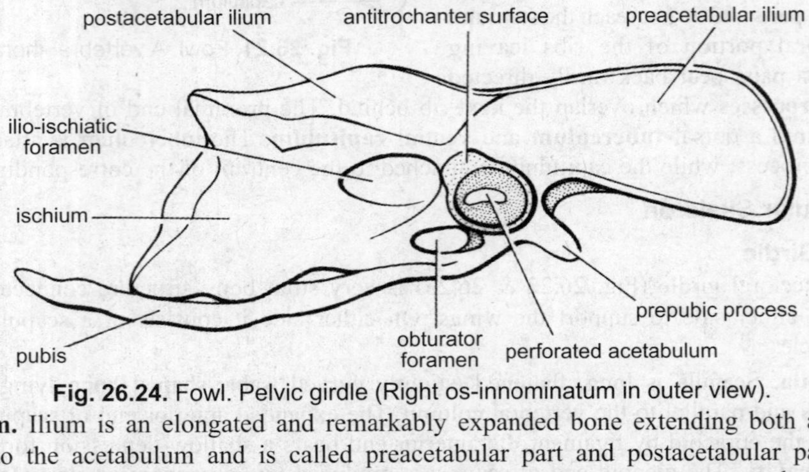

Fig. 26.24. Fowl. Pelvic girdle (Right os-innominatum in outer view).

Ilium. Ilium is an elongated and remarkably expanded bone extending both anterior and posterior to the acetabulum and is called preacetabular part and postacetabular part of ilium respectively. The inner margin of ilium is fused with the transverse processes and neural spines of the synsacrum. The outer surface of its anterior part is concave and posterior part is fused with the ischium. Ilium forms the dorsal part of acetabulum. On the outer surface above the acetabulum is a projection, the **antitrochanter**, which articulates with the great trochanter of femur.

Ischium. Ischium is also dorso-ventrally flattened bone projecting backwards behind the acetabulum and parallel to the posterior part of the ilium. Ischium is fused posteriorly with the ilium but separated anteriorly from it by a large, oval **ilio-ischiatic foramen**.

Pubis. Pubis is a long, thin, curved, slender bone directed backward parallel to the ventral margin of ischium with which it is usually fused. It forms ventral part of acetabulum. Behind the acetabulum the pubis and ischium are separated by an oval opening, the **obturator foramen**. Just in front and outside the acetabulum the pubis gives off a short and blunt process, the **prepubic** or **preacetabular process**.

Bones of Forelimb

Forelimbs of birds are modified as wings and correspond to the forelimbs of a mammal.

1. Humerus. Humerus (Fig. 26.25) is the bone of upper arm of forelimb. It is an elongated stout bone expanded at both the ends. The proximal end is expanded into smooth convex surface, the **head** which fits into the glenoid cavity of pectoral girdle. The head is bordered by two tuberosities, **preaxial** and **postaxial**. From the smaller preaxial tubersity extends in front a short crest or **deltoid ridge** for the insertion of great pectoral and deltoid muscles. The postaxial tuberosity is larger and has a **pneumatic foramen** on the dorsal side of the proximal end communicating with its air cavity. The distal end possesses trochlea or condyles for the articulation with the radius and ulna.

2. Radius and ulna. Radius and ulna are the bones of the forearm.

Radius (Fig. 26.25) is shorter, slender and straight bone. Proximally it has a cup-shaped articular surface for the outer trochlea of humerus. Distally it bears a knob which articulates with the radiale carpal.

Ulna (Fig. 26.25) is stouter and larger than the radius and also slightly curved. Proximally it bears a small projection, the **olecranon process**, and a large articular facet for the inner condyle of the humerus. Distally it articulates with the ulnare carpal and radius.

3. Carpals. Carpals are two irregular bones of wrist. Preaxial **radiale** is smaller and articulates with the radius and the postaxial **ulnare** is larger and articulates with the ulna.

4. Carpometacarpus. Carpometacarpus (Fig. 26.25) is the bone of manus or palm. It is a compound bone formed by the fusion of three metacarpals with the distal row of carpals. The **first metacarpal** is very short, stumpy projection at the proximal end. The **second metacarpal** is strong and somewhat straight bone. The **third metacarpal** is thinner, slightly curved outwardly on the postaxial side and fused at both ends with the second metacarpal.

5. Phalanges. Three metacarpals bear three clawless digits (fingers) which possess small rod-like bones called **phalanges. First digit** or **pollex** is preaxial and consists of a single phalanx, **second digit** (index) has two phalanges and **third** postaxial digit has a single phalanx.

Bones of Hindlimb

In birds, hindlimbs are modified for bipedal locomotion. Tarsus is absent.

1. Femur. Femur (Fig. 26.26) is the bone of thigh of hindlimb. It is short and powerful bone flattened at both the ends. Proximally on the inner side it bears a rounded **head** which fits into **acetabulum**. On the outer side of the head is an irregular prominent process, the **great trochanter**. Between the great trochanter and head is the articular surface for the **antitrochanter** of ilium. Distally it has a deep **groove** or **intercondylar fossa** bounded laterally by two condyles for the articulation with the tibiotarsus. Outer condyle receives the upper end of fibula.

Patella is a sesamoid bone formed by the ossification of tendons. It is found in the tendon in front of femur-tibiotarsal joint.

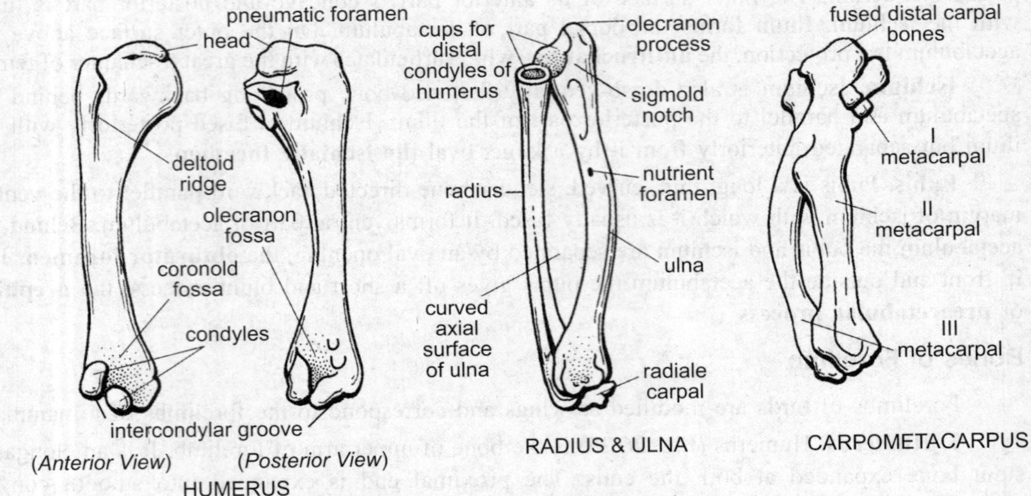

Fig. 26.25. Fowl. Forelimb bones.

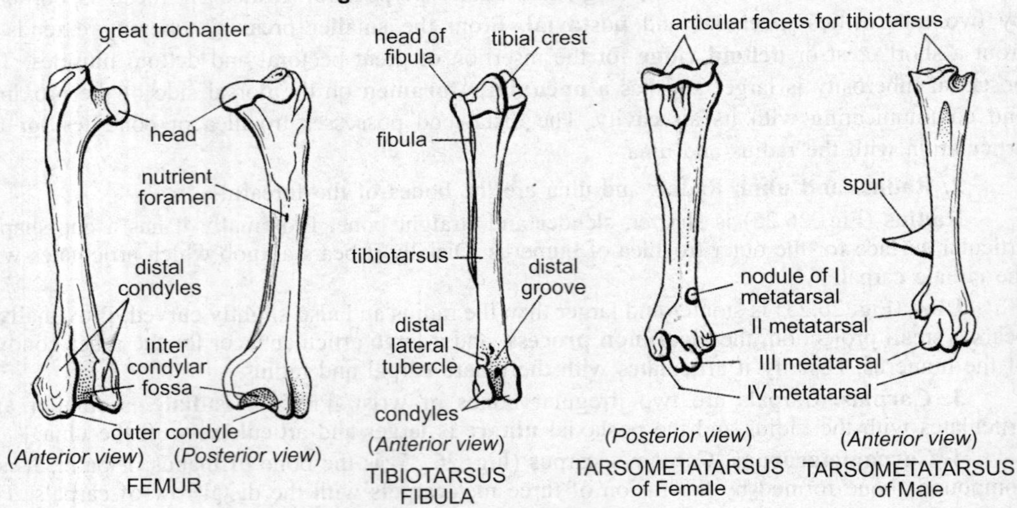

Fig. 26.26. Fowl. Hindlimb bones.

2. Tibiotarsus and fibula. Tibio-tarsus and fibula (Fig. 26.26) are the bones of **shank** of hindlimb. Tibiotarsus is a long and stout nearly straight bone found on the inner side of shank. It is formed by the fusion with the proximal row of tarsals, *i.e.,* astragalus and calcaneum. The proximal end of tibiotarsus is expanded and has on its anterior face a median **cnemial** or **tibial crest** and two articular surfaces for the condyles of the femur. Distally it articulates with the tarso-metatarsus by a pulley-like articular surface, *i.e.,* a groove bounded by two condyles.

Fibula (Fig. 26.26) is a smaller, slender bone closely applied to the outer surface of the tibiotarsus. Proximally it articulates with the outer condyle of femur. Its distal end does not reach up to the lower end of tibiotarsus and terminates in a sharp point.

3. Tarsals. In adult birds, the proximal row of tarsals are fused with tibia and distal row with metatarsus. In between the two rows is present the **intertarsal ankle joint**.

4. Tarsometatarsus. Tarsometatarsus (Fig. 26.26) is a compound bone of foot or **pes** and formed by the fusion of distal row of tarsals with the second, third and fourth **metatarsals**. It is a stout, long bone and proximally it bears two cup-like facets for the articulation with the tibio-tarsus. First **metatarsal** is a small nodule-like bone attached to the inner and posterior surface of tarso-metatarsus near its distal end. At its distal end, the three metacarpals become free, each has a pulley-like articular surface for the corresponding digit. In male, the tarsometatarsus often bears a stout pointed spinous fighting **spur** on its medial side.

5. Phalanges. These are small, slender bones. Metatarsus carries four toes : first or **hallux** is directed backward and has two phalanges, and the remaining three are directed forward. Second, third and fourth toes bear 3, 4 and 5 phalanges respectively. Terminal phalanx of each toe has a horny claw.

COELOM AND VISCERA

The coelom of birds (pigeon) is well developed and divided into two cavities, anterior thoracic and posterior abdominal cavities, by a membranous oblique septum or **diaphragm**. The thoracic cavity has two dorso-lateral **pleural cavities**, each enclosing a lung. The two pleural cavities are separated from each other by the antero-ventral **pericardial cavity** enclosing the heart lying in the middle between the ventral portions of the pleural cavities. The abdominal cavity lies posterior to the diaphragm and is called a **peritoneal** or **abdominal cavity** which contains the rest of the visceral organs such as digestive, excretory, reproductive organs suspended by mesenteries. Coelom is filled with a watery coelomic fluid and is bounded by peritoneum.

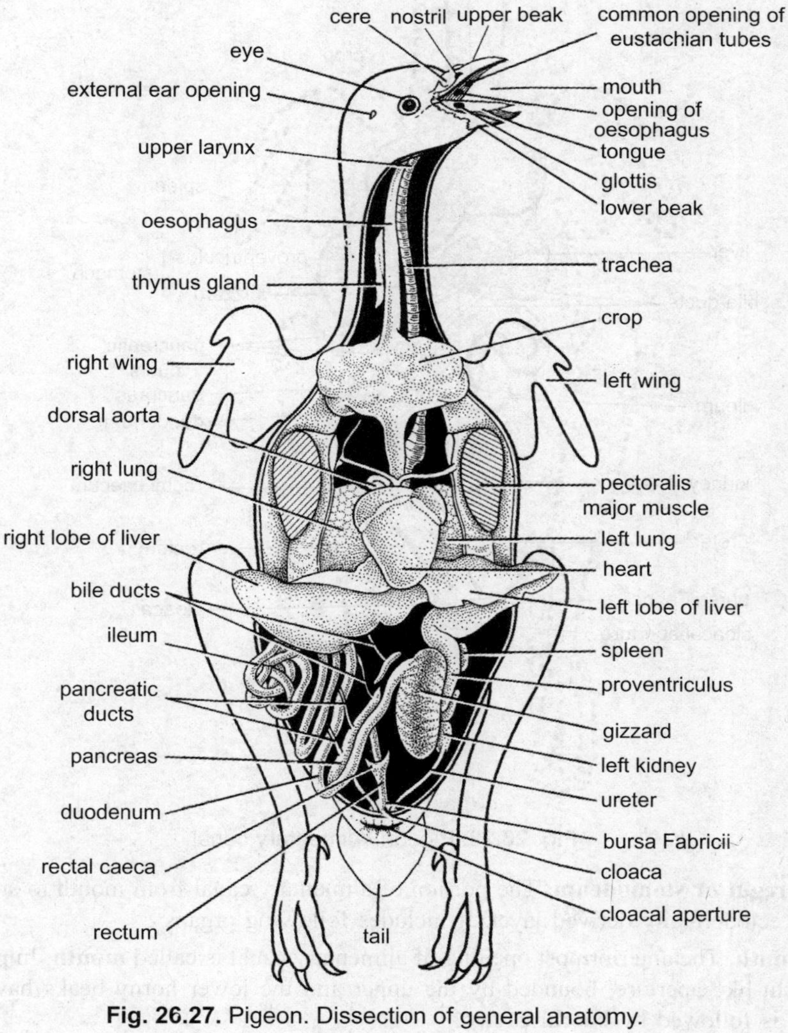

Fig. 26.27. Pigeon. Dissection of general anatomy.

DIGESTIVE SYSTEM

The digestive system of pigeon is well developed and includes an alimentary canal and the digestive glands.

Alimentary Canal

The alimentary canal of pigeon (Fig. 26.28) is long, tubular and coiled. It comprises mouth, buccal cavity, pharynx, oesophagus, stomach, small intestine and large intestine which opens to the exterior by cloacal aperture. All the segments of alimentary canal fall into following three categories : foregut or stomodeum, midgut or mesenteron and hindgut or proctodeum.

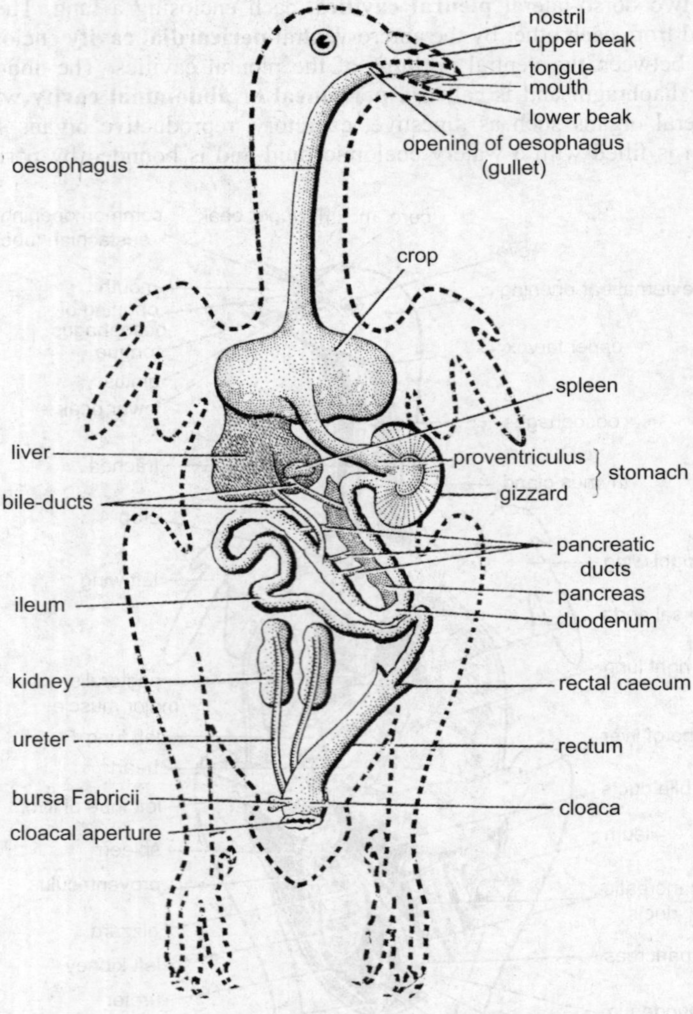

Fig. 26.28. Pigeon. Alimentary canal.

1. Foregut or stomodeum. The portion of alimentary canal from mouth to stomach region is lined by ectodermally derived layer. It includes following organs :

***(i)* Mouth.** The anteriormost opening of alimentary canal is called **mouth**. In pigeon, mouth is a wide slit-like aperture, bounded by the upper and the lower horny beaks having no teeth. The mouth is followed by buccal cavity.

***(ii)* Buccal cavity.** The featureless buccal cavity has a large, narrow, triangular and pointed at the tip, the **tongue** at its floor. The tongue has few taste buds and mucous glands, and has the function of manipulation of food. The buccal cavity is followed by pharynx.

***(iii)* Pharynx.** The posteriormost part of buccal cavity may be called the **pharynx**. A pair of elongated apertures, the **posterior nares**, opens in the roof of the pharynx. They are covered by palatal folds of skull roof. Just behind the posterior nares opens a single median aperture of the **pharyngotympanic** or **eustachian tubes**. At the floor of the pharynx occurs an oval aperture with tumid lips which is called **glottis**. The glottis opens into the trachea. Posteriorly, the pharynx opens into the oesophagus.

***(iv)* Oesophagus and crop.** Oesophagus or **gullet** is a long, wide, distensible and thick-walled tube which runs backward through the neck to join a large dilated reservoir or **crop**. It is a large, thin-walled, bilobed, elastic and non-glandular sac.The crop enables the bird to store quickly swallowed food for later digestion. It is specially large in **graminivorous** (grain-eating) birds such as pigeons, finches, buntings, parrots, etc. The grains are moistened and softened in it. The epithelial lining of the crop in both sexes thickens and sheds a white, slimy, proteinous and fatty "**crop milk**" during the breeding season on which the youngs are fed by both parents. Pigeon milk is produced by the degeneration of the epithelial cells lining the crop, under the control of **prolactin** hormone, secreted by the anterior lobe of the pituitary gland. This hormone stimulates and controls the formation of pigeon's milk. Pigeon's milk contains water (65–81%), protein (casein) (13.3–18.8%), fat (6.9–12.7 %) and lactose (1.5%). It is more nourishing than the cow's milk. Pigeon's milk contains 35 per cent of fat in comparison to 3 to 5 per cent in cow's milk. The young ones fed on it double their weight in two days. The pigeon's milk is regurgitated to the youngs.

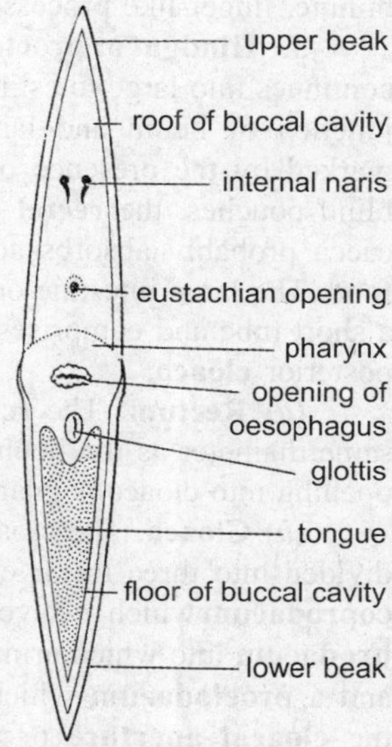

Fig. 26.29. Pigeon. Bucco-pharyngeal region.

The crop also contains some mucus secreting glands. Beyond the crop, before opening into the stomach, the oesophagus become thick-walled and narrow tube.

***(v)* Stomach.** The stomach is differentiated into an anterior glandular **proventriculus** and a posterior muscular **gizzard**.

***(a)* Proventriculus.** The proventriculus is a small, thick-walled, and glandular structure appearing externally as a slight dilation of the oesophagus but it is a gastric structure. Its thick mucus lining secretes the gastric juice. The **spleen** is a small, oval red body and remains attached to the right side of the proventriculus by peritoneum.

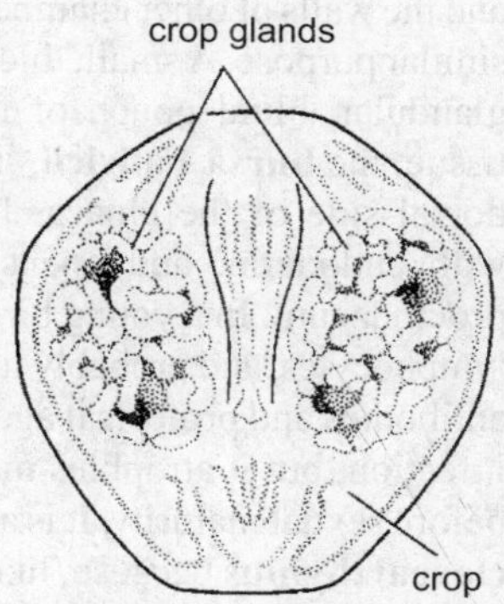

Fig. 26.30. Pigeon. Crop cut open.

***(b)* Gizzard.** The gizzard represents the pyloric region. It is large, hard, muscular and laterally compressed having the shape of biconvex lens. The thick walls of gizzard have thick muscles radiating from two tendons. Its narrow lumen is lined by an epithelium in which are present numerous minute tubular glands, which secrete a fluid (koilin) which becomes thick, horny and of a yellow or green colour and lines the gizzard. Its yellow or green colour is due to regurgitated bile. The cavity of gizzard always contains small stones or grit swallowed by the bird. These stones help the gizzard in grinding the food. The gizzard continues into small intestine and the opening of gizzard into small intestine is guarded by a sphincter, called the **pyloric valve** or **pylorus**. In carnivorous birds the gizzard is not so muscular.

2. Midgut or mesenteron. The midgut, mesenteron or **small intestine** is a narrow tube and has a lining of endodermally derived epithelium. The small intestine is divided in an anterior duodenum and a posterior ileum. The bile and pancreatic ducts usually open into the distal limb of duodenum. In pigeon the left bile duct enters close to the pylorus.

***(a)* Duodenum.** The duodenum arises from the dorsal side of gizzard so that the **pyloric opening** of gizzard into duodenum lies close to the **cardiac opening** of proventriculus into gizzard. The duodenum forms a U-shaped loop enclosing the pancreas between its two limbs.

***(b)* Ileum.** The portion of the small intestine behind the duodenum is called **ileum**. Ileum is a long and coiled tube of uniform diameter. Its inner epithelial lining is thrown into numerous, minute, finger-like processes or **villi**, which greatly increase its area of secretion and absorption.

3. Hindgut or proctodaeum. The slender ileum continues into large intestine of similar diameter. The juncticn of ileum and large intestine is externally marked by the presence of a pair of small conical, blind pouches, the **rectal** or **colic caeca**. The rectal caeca probably absorbs some water from digestive food. The large intestine or hindgut (proctodaeum) is a short tube and comprises an anterior **rectum** and a posterior **cloaca**.

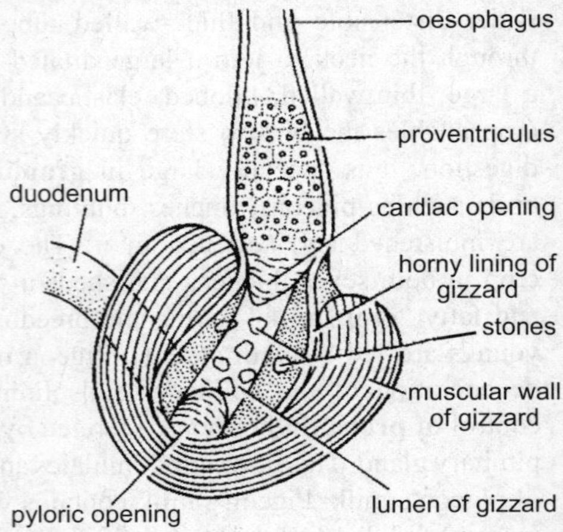

Fig. 26.31. Pigeon. Stomach in sagittal section.

***(a)* Rectum.** The rectum is narrow and is of same diameter as the ileum. It opens into cloaca. Its opening into cloaca is guarded by an **anal sphincter**.

***(b)* Cloaca.** The cloaca is a large chamber and divided into three linear compartments : an anterior **coprodaeum** which receives the rectum, a short middle **urodaeum** into which urinary and genital ducts open, and a **proctodaeum** which opens to the outside by the **cloacal aperture** or **vent**. The urinary products are made solid by absorption of water in the urodaeum and the walls of other chambers serve a similar purpose. A small, thick-walled, glandular, blind pouch of lymphatic tissue, the **bursa Fabricii**, lies on the dorsal side of the cloaca. It is lined with endoderm, and opens into the proctodaeum. In a young bird it forms lymphocytes, and probably it produces antibodies and protects it against local infection, but it atrophies in the adult before sexual maturity. It is also called **cloacal thymus** because, like thymus, it produces lymphocytes. Like thymus, it is prominent in young animals and usually much reduced in the adult.

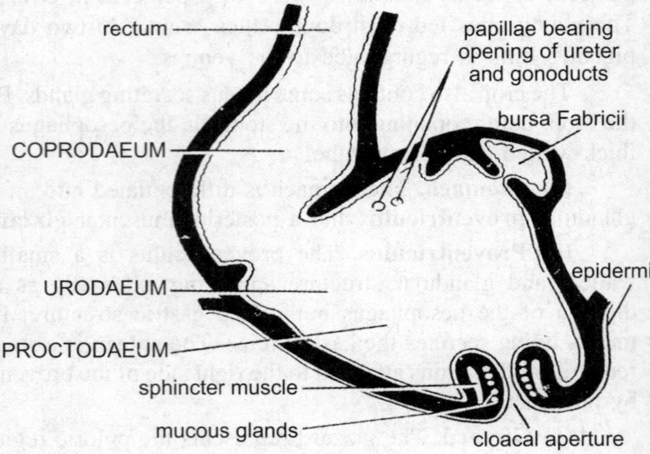

Fig. 26.32. Pigeon. Cloaca in longitudinal section.

Digestive Glands

In pigeon, following digestive glands occur :

***(i)* Buccal glands.** Pigeon has no salivary gland in the buccal cavity, but has a median **sub-lingual gland** and a pair of angle **glands** near the corners. These glands are the only buccal glands which secrete mucus to moisten the food and probably the amylase enzyme. The tongue also bears few mucous glands.

(*ii*) Gastric glands. The epithelial lining of proventriculus contains many gastric glands which secrete certain gastric juices containing peptic enzymes.

(*iii*) Liver. The liver of pigeon is large, compact, dark red, bilobed, containing a large right lobe and a small left lobe. From each lobe of the liver arises a **bile duct**, the left bile duct opens into the proximal limb near the pylorus and the right duct opens into the distal limb of duodenum. In pigeon, gall bladder is lacking, but it is present in *Gallus* and many other species. The liver secretes **bile juice**.

(*iv*) Pancreas. Between two limbs of the duodenum is a large, compact, reddish digestive gland, the **pancreas**. Pancreas gives off three separate **pancreatic ducts** which pour the pancreatic secretions containing many enzymes into the distal limb of duodenum.

(*v*) Intestinal glands. The epithelial linings of intestine have many glands which secrete many enzymes. The duodenum is lined with villi and single or branched crypts of Lieberkuhn. Goblet cells are also present.

Physiology of Digestion

Food and feeding. The pigeon is **graminivorous** and it feeds chiefly on **cereals**, pulses, seeds, etc. These seeds are picked up with the **beak**, manipulated by tongue and lubricated inside the buccal cavity with the secretion of buccal glands. The food is swallowed as such, because beaks lack teeth.

Digestion. The food which is lubricated and moistened by the secretions of buccal glands inside the buccal cavity passes through the oesophagus into the crop, where it is stored. Inside the crop, the food is softened by joint action of body warmth, water, mucus, amylase enzyme of buccal secretions and also due to bacterial autolytic action. From crop the food passes into proventriculus where it is chemically acted by peptic enzyme of gastric juice. From proventriculus the food enters the gizzard where mechanical breakdown of food takes place by muscular contractions of gizzard wall and grinding action of small stones inside the lumen of gizzard. In gizzard little digestion of food occurs and it is made acidic by secretion of its walls which have many hydrogen ions. The partly digested food, the **chyme**, passes through the pylorus into the duodenum. In the duodenum the chyme is mixed with bile, pancreatic juices and intestinal secretions. The chemical conversion of fats, proteins and carbohydrates into glycerol and fatty acids, amino acids and monosaccharides, respectively, takes place in duodenum like other vertebrates. The bile salts help in the digestion and absorption of fats. Thus, in duodenum, all the three classes of foodstuffs (*i.e.,* proteins, fats and carbohydrates) are converted into readily diffusible micromolecules (*i.e.,* amino acids, glycerol and fatty acids and monosaccharides). These micromolecules readily diffuse into the thin walls of ileum and mix into the blood. The bile salts help in the digestion and absorption of fat. Thus, digestion, absorption and assimilation of food occurs in the small intestine.

The undigested portion of food passes to the rectum, where absorption of remaining water takes place. The faecal matter which is almost dry due to absorption of water is finally ejected through the cloacal aperture.

The large surface area, high temperature and great activity of birds necessitate a high food intake, especially in the smaller types. This is due to rapid passage of food through the gut. The amount of food taken per day may reach nearly 30 per cent of the body weight (6g) in the very small goldcrest and about 12 per cent in starling weighing 75g. The food is very efficiently utilised inspite of the rapid passage.

RESPIRATORY SYSTEM

The flight activity requires a continuous and abundant supply of oxygen, therefore, the respiratory system of birds is highly developed and well differentiated. Respiration is **pulmonary** and respiratory organs are simple. Muscular **diaphragm** in birds is lacking. The respiratory system includes the **respiratory tract**, the **respiratory organs** or **lungs** and the **air sacs**.

1. Respiratory Tract

The respiratory tract includes nares, nasal sacs, glottis, larynx, trachea and syrinx.

***(i)* Nares, nasal sacs and glottis.** The **external nares** or **nostrils** are a pair of slit-like oblique apertures occurring at the base of upper beak and are overlapped by a swollen sensitive skin, the **cere**. The external nares open into short **olfactory** or **nasal sacs** which remain communicated to the pharynx by **internal nares** or **choanae**. A median slit-like **glottis** lies behind the base of the tongue and it opens into the trachea. At the anterior end of trachea is present an expanded chamber, the **larynx**.

***(ii)* Larynx.** The larynx is an expanded, voiceless chamber which occurs at the anteriormost margin of trachea and is greatly reduced in birds. The larynx is supported by a triangular **cricoid** cartilage, formed of four pieces, two of which lying above are called **procricoids** present only in some birds. The larynx also has a pair of partly ossified curved arytenoids cartilages. Thyroid cartilage found in mammals is absent. There are no **vocal cords** which are the characteristics of mammals, therefore, the larynx of birds does not function as a sound producing organ. Larynx opens into a long trachea.

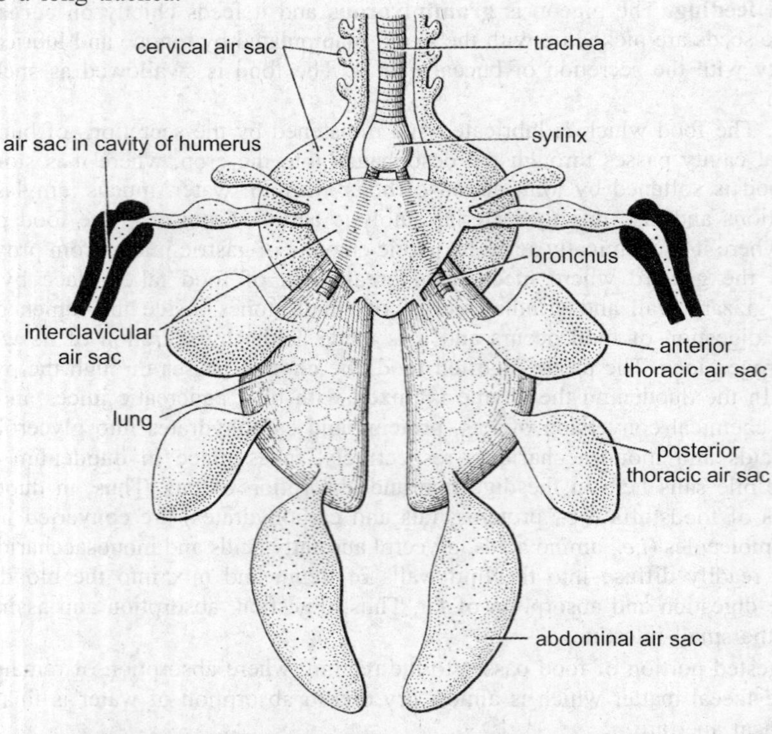

Fig. 26.33. Pigeon. Respiratory system (lungs and air sacs) in ventral view.

***(iii)* Trachea.** The trachea is a long, cylindrical and flexible tube running backward through the neck. It occurs beneath the oesophagus and is displaced to the left in the middle region by the crop. On entering the thoracic cavity it expands into a **syrinx** and then divides into two **bronchi**, one for each lung. The walls of tracheal and bronchial tubes are supported by a series of closely-set **rings**. All the tracheal and the first bronchial rings are bony and completely surround the tube. While the bronchial rings are incomplete mesially and cartilaginous.

***(iv)* Syrinx.** The syrinx is a vocal organ found only in birds. It is present at the junction of trachea and paired bronchi. It is an expanded chamber called **tympanum**. Its wall is supported by the last three or four tracheal rings and the first half-rings of the two bronchi. Its inner mucous membrane forms a cushion-like thickening on each side. At the junction of bronchi a bar of cartilage, called the **pessulus**, extends dorso-ventrally and supports an inconspicuous fold of mucous membrane, the **membrana semilunaris**. The mucous membrane of inner walls of each bronchus forms an **internal tympaniform membrane** and an **external tympaniform membrane** associated with the outer wall. These tympaniform membranes enclose an **interclavicular air sac** just beneath the semilunar membrane. The movement or functioning of tympaniform membranes is controlled and regulated by a pair of **intrinsic syringeal muscles** which arise from the sides of the trachea and are inserted into the syrinx, and a pair of **sternotracheal muscles** arise from the sternum and are inserted into the trachea. The sound (voice) is produced by the vibrations of the semilunar membrane as the air is forcibly expelled from the lungs, during expiration, passes between them. The pitch of the voice is altered by sternotracheal and intrinsic syringeal muscles which change the tension on the semilunar tympaniform membranes and so the pitch of the sound. The syrinx is absent in ostriches and storks, etc.

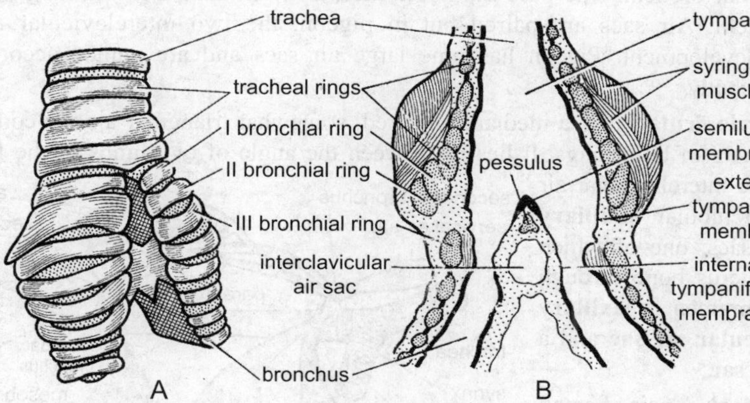

Fig. 26.34. Pigeon. Syrinx. A–External view; B–In longitudinal section.

2. Respiratory Organs or Lungs

The lungs of pigeon are bright red, small, compact, slightly distensible, being solid, spongy, highly vascular organs. They differ from those of other vertebrates in that they do not store air. They lie in the **pleural cavities** which are separated from the peritoneal or abdominal coelom by an oblique septum. The dorsal surface of the lungs is closely applied to the ribs and thoracic vertebrae and has no peritoneal covering. Their ventral surface is free and covered by a special fibrous peritoneal membrane called **pleura** or **pulmonary aponeurosis**. Small fan-like muscles called **costopulmonary muscles** arise from the junction of vertebral and sternal ribs and are attached to the pulmonary aponeurosis.

The bronchus entering a lung from the antero-ventral side does not branch immediately as in other tetrapoda and is called **primary bronchus**. Within the lung, it loses its cartilaginous rings and continues up to its distal or posterior end, and is called **mesobronchus**.The mesobronchus gives off branches to air sacs and also gives off several **secondary bronchi** which may be distinguished as dorsal, ventral and lateral according to their position. The secondary bronchi (dorsal and ventral) then branch into **tertiary bronchi** or **parabronchi**. The parabronchi form an intricate system of branching and anastomosing thin-walled **air capillaries** forming loops and also join **recurrent bronchi**. The lung parenchyma has hexagonal areas, each has a central parabronchus, surrounded by a system of air capillaries and interlacing blood capillaries. Thus, parabronchi and air capillaries do not end blindly. There is no dead space in the lungs. Each air capillary opens by both ends into a parabronchus. Alveoli are not found in the lungs. The air capillaries are lined with a vascular respiratory membrane and they come into contact with blood capillaries for an exchange of gases. In birds, thus, the branches do not have blind endings in alveoli as in mammals. They form an inter-communicating system of bronchi, parabronchi, and air capillaries forming complete air circuits in which air circulates.

3. Air Sacs

From the mesobronchus arise secondary bronchi (dorsal, ventral and lateral), dorsal and ventral divide into parabronchi, but in each lung lateral secondary bronchi do not divide but pass through walls of the lung and enter the air sacs except the abdominal air sacs which are connected with the mesobronchus. The air sacs are, thus, dilations of the bronchi. The air sacs are large, thin-walled membranous, non-muscular and non-vascular structures. They do not increase the respiratory surface and lie among the viscera and also extend into some of the larger bones. Their total volume is more than that of lungs and fill up much of the body cavity. Openings of bonchi into air sacs are called **ostia**.

From the anterior ends of air sacs arise small **recurrent bronchi** or **sacco-bronchi**, which connect the sacs with parabronchi and the air capillaries of the lungs. Cervical air sacs have no recurrent bronchi. The pure air is returned from air sacs into the lungs through these recurrent bronchi. Air sacs are paired but in pigeon, the two interclavicular air sacs fuse during early development. Pigeon has nine large air sacs and are named according to their position in the body.

(*a*) Interclavicular. It is a median, unpaired, somewhat triangular air sac connected to the secondary bronchi of both lungs. It lies in between the angle of two limbs of the furcula. From each side of the interclavicular air sac arises two tubular **auxiliary** (**axillary**) air sacs, one of which enters the humerus bone through a pneumatic foramen as **axillary** or **extra-clavicular air sac** and a **clavicular air sac**.

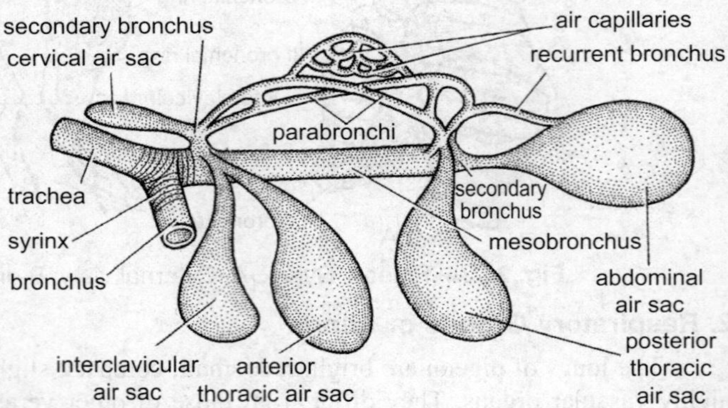

Fig. 26.35. Pigeon. Scheme of histological structure of a lung and origin of air sacs.

(*b*) Cervical. A pair of small cervical air sacs arises anteriorly, one from each lung. They lie at the base of the neck, dorsal to the interclavicular and alongside the vertebral column. They give out minor saccular branches in the neck.

(*c*) Anterior thoracic. From the side of each lung arises an anterior thoracic air sac which lies at ventral side of lung in the anterior portion of the thorax, in close contact with the ribs and the pericardium. Each extends back and overlaps the posterior thoracic air sacs. Its ventral wall remains covered by the oblique septum.

(*d*) Posterior thoracic. A pair of small, posterior thoracic air sacs is found in the posterior part of the thoracic cavity just in front of abdominal sacs. Each sac overlaps the posterior end of its lung and communicates with its outer posterior angle.

(*e*) Abdominal. From the distal end of each lung arises a large abdominal air sac. Each abdominal air sac lies along the dorsal wall of the abdomen, ventral to the kidneys, amongst the coils of the small intestine.

Functions of Air Sacs

The air sacs are thin reservoirs of air which communicate with bronchi on the one hand and with the pneumatic cavities of the bones, on the other hand. They serve the following functions :

1. Accessory respiratory organs. The air sacs are not respiratory organs but they help in respiration. They act as bellows forcing their air into the lungs for ventilation at each expiration to compelety renew the air in the lungs, thus, there is no "dead space" of unrespired air in the lungs. But, it is claimed that the anterior air sacs are **expiratory** (interclavicular, cervical and anterior thoracic) and they are more active during flight. The posterior air sacs are **inspiratory** (posterior thoracic and abdominal), they are more active when the bird is not flying.

2. Lightness. The air sacs also act as **balloons** giving buoyancy in flight and reducing the specific gravity of the bird due to the contained warm air. This view is no longer supported now.

3. Temperature regulation. The air sacs also help to maintain and regulate body temperature acting as a cooling device by losing body heat through internal evaporation, *i.e.,* water vapours diffuse from the blood into cavities of air sacs and pass out through the lungs, accompanied by loss of body heat.

4. Cardiac movements. According to **Muller** (1908), the air sacs allow movement to the heart in a very rigid thorax.

5. Flight. The best flying birds possess most highly developed air sacs which confirm their association with flight.

4. Respiratory Mechanism

In pigeon and other birds the expiration is an active process, and not inspiration as in other vertebrates.The lungs have a large internal respiratory surface.

1. Breathing at rest. In a resting pigeon during inspiration, the respiration is brought about by **costopulmonary** or **intercostal muscles** (inspiratory) which lowers the sternum, the air sacs expand and the lungs are compressed, thereby increasing the thoracic and abdominal cavities. It reduces the pressure on the lungs. Air is drawn into the posterior air sacs through the mesobronchus. At the same time air present in the lungs enters into the anterior air sacs. Thus, in inspiration a large amount of air goes from secondary bronchi into the air sacs.

Expiration is brought about by movements of abdominal muscles (expiratory) which raises the sternum compressing the air sacs and expansion of lungs. By compression of posterior air sacs their air is forced into the lungs through recurrent bronchi, and the stale air of anterior air sacs enters the mesobronchus, trachea and finally out through the nostrils.

2. Breathing during flight. During flight, the sternum and ribs become immovable to brace the wings, and the ordinary method of inspiration and expiration by intercostal and abdominal muscles respectively is abandoned. The air movements in and out of the lungs is done by the elevation and depression of the back with the help of the wing's strokes. The faster a bird flies, the more rapid is the air circulation and gaseous exchange in the lungs.

In birds the air is constantly renewed and passes several times over the respiratory epithelium of air capillaries, because of intercommunicating system of tubules. The aeration of blood is absolutely complete, allowing a high degree of muscular efficiency and maintaining a high body temperature which ranges from 102° to 111°F (passerine birds).

CIRCULATORY SYSTEM

The birds have an efficient circulation, a high rate of metabolism and a high and constant temperature. The birds and mammals are the only vertebrates that have a complete separation of the respiratory and systemic circulations, making possible a high arteriolar pressure, which allows materials to reach the tissues rapidly.

The circulatory system of pigeon includes heart, arteries, veins, the blood and lymphatic system.

1. Heart

The heart is four chambered, 2 auricles and 2 ventricles. The sinus venosus is absent.

External structure. The heart of pigeon is large-sized, reddish in colour, triangular and compact. It lies midventrally in the thorax. It is enclosed in a thin, transparent membranous sac, the **pericardium**, which is a derivative of coelomic epithelium. The pericardial wall is made of an outer parietal layer and an inner visceral layer. The two layers enclose a narrow **pericardial** cavity, filled with a watery **serous** or **pericardial fluid.**The pericardial fluid protects the heart from shocks and injuries and permits movements during beating.The auricles remain well marked from ventricles by an external deep, shallow, transverse groove, the **coronary sulcus** or the **auriculo-ventricular groove** which separates the two anterior, smaller, darker auricles from the posterior, larger, lighter in colour, the ventricles. The two auricles are also demarcated from each other by a faint, **inter-auricular groove**. The sinus venosus has disappeared being absorbed into the wall of the right auricle.

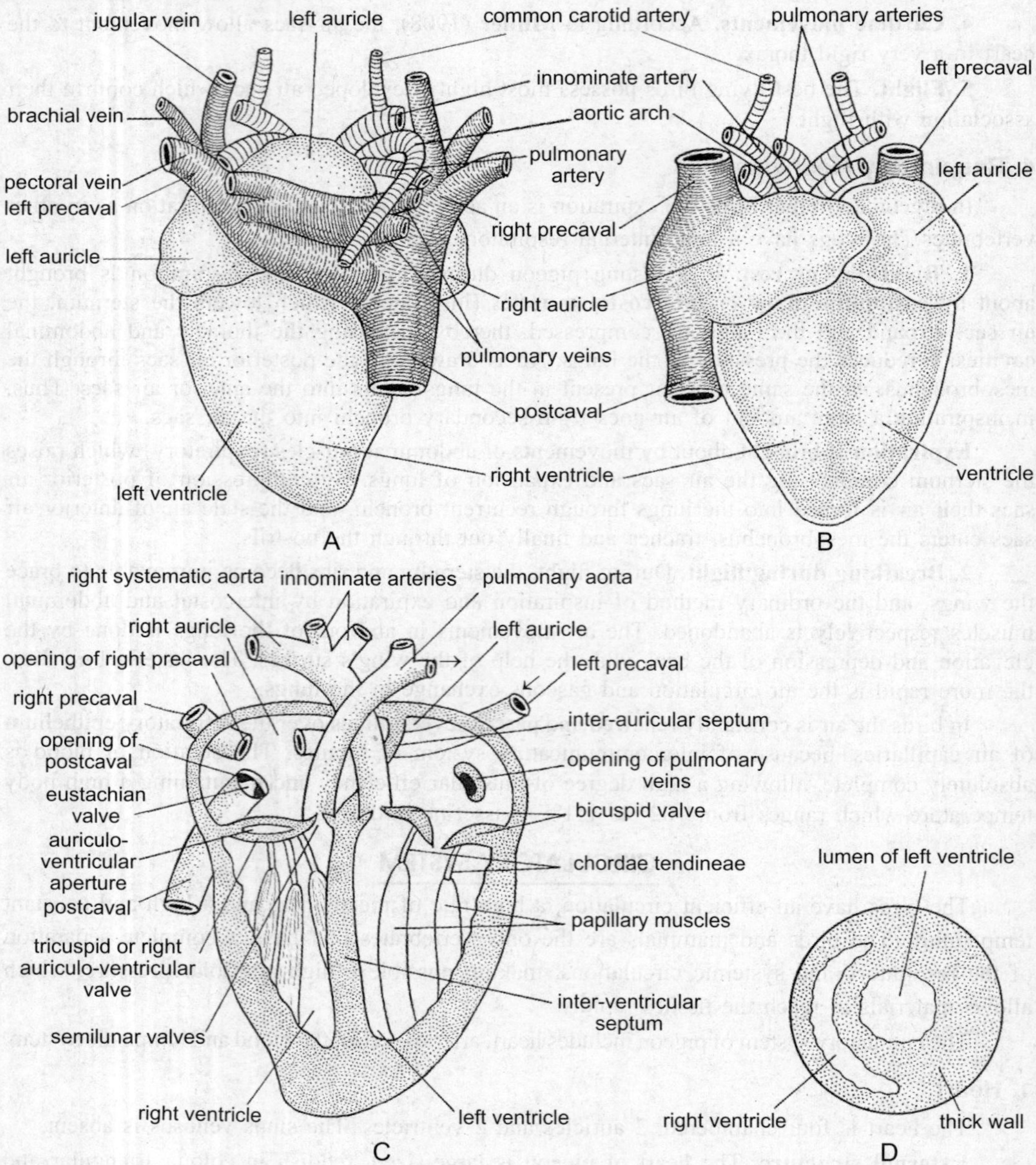

Fig. 26.36. Pigeon. Structure of heart. A–Dorsal view; B–Ventral view; C–Internal structure in ventral dissection; D–Ventricle in T.S.

Internal structure. Internally, the two thin-walled auricles are separated by a complete, thin, membranous partition, the **inter-auricular septum**. It has a small oval area in the middle, called the **fossa ovalis,** representing the foramen ovale in the embryo. The two thick-walled muscular ventricles are also completely separated by a thick and muscular **inter-ventricular septum**. The **left ventricle** is larger with powerful muscular walls. It has a circular cavity. Its propulsive force has great power of pumping blood. The **right ventricle** is smaller with a crescentic cavity in section. It partly encircles the left ventricle. The ventricles have bundles of muscle fibres arranged longitudinally to form **columnae carneae**.

The **right auricle** is larger than left auricle and it receives the three large veins, the right and left precavals and the postcaval, in its dorsal wall. The opening of the postcaval vein is gurarded by a muscular **eustachian valve**. The **right auricle** opens into the right ventricle by a crescentic aperture, the **right auriculo-ventricular aperture**. This aperture is guarded by a pair of muscular flaps, having no chordae tendinae, forming the right auriculo-ventricular valve. It represents the tricuspid valve of the mammals.The right ventricle gives off a single trunk, the **pulmonary aorta**, which bifurcates into two pulmonary arteries, each going to a lung. The opening of the pulmonary trunk is guarded by three **semilunar valves**.

The **left auricle** is smaller and receives four pulmonary veins from the lungs. It opens into the left ventricle by a circular **left auriculo-ventricular aperture**. It remains guarded by two membranous flaps forming the **left auriculo-ventricular valve**, which represents the bicuspid or mitral valve of mammals. The flaps are connected to the thick **papillary muscles** which arise from the wall of the left ventricle, by the **chordae tendinae**.The left ventricle gives rise to the single **right systemic** or **aortic arch** which crosses over to the right side of the heart and continues as dorsal aorta. The **left aortic arch**, which is present in the frog and lizard, is absent in birds. The opening of right aortic arch is guarded by three semilunar valves. The cavities of both ventricles are traversed by bars of muscles, called **trabeculae** or **columnae carneae**.

Functioning of heart. The heart is a pumping station of circulatory system which pumps the blood to arteries. In pigeon, the venous and arterial blood never mixes in heart as in frog, but remains well separated like mammals since it is four chambered. The right half receives and discharges only venous (unoxygenated) blood, while the left half only arterial or oxygenated blood. Thus, birds possess a complete **double circulation** which includes a **pulmonary circulation** and a **systemic circulation**.

(*i*) Pulmonary circulation. The right ventricle pumps venous blood into the lungs through the pulmonary aorta and its two pulmonary arteries. The oxygenated blood from the lungs is returned into the left auricle through four pulmonary veins.

(*ii*) Systemic circulation. The left ventricle sends blood into the right aortic arch, the dorsal aorta and through smaller arteries to the various organs and tissues of the body, where they break up into capillaries. These capillaries unite to form veins, which finally form three great veins, which return blood into the right auricle. From right auricle, the blood enters the right ventricle through right auriculo-ventricular aperture.

Heart rate. The heart gives a rapid heartbeat to ensure rapid blood circulation, high body temperature (102° to 112° F) and rapid metabolism, for production of energy. The size and rate of heartbeats vary with the size and activity of birds. The larger birds have in general relatively smaller heart and less heart beats. Rates increase with exercise and a maximum of 102 per minute has been recorded in a humming bird. The larger heart allows a greater stroke output and higher pressure than in mammals.

Heartbeats. The heart of pigeon has two pacemakers, a **sinu-auricular node** in the wall of the right auricle, and an **auriculo-ventricular node** in the inter-auricular septum. The muscles of auricles and ventricles are not continuous (unlike amphibians and reptiles). A wave of excitation starts from the sinu-auricular node and goes to the auriculo-ventricular node from which fibres of the **bundle of His** convey the excitation wave to the ventricles. The branches of the fibres break up into numerous **Purkinje fibres**, which spread as a network into the ventricle. When left auricle contracts, the oxygenated blood from the left auricle passes to the left ventricle. From the left ventricle the blood is taken by the right aortic or systemic arch and distributed to the body. Thus, systemic circulation completes. Likewise, when right auricle contracts, its unoxygenated or venous blood passes to right ventricle, the contraction of which forces the blood into pulmonary artery for oxygenation in the lungs.

2. Arterial System

The heart pumps its unoxygenated blood to **pulmonary aorta** and oxygenated blood to **aortic** or **systemic arch**; both of which collectively constitute the arterial system of pigeon.

1. Pulmonary aorta. From right ventricle arises a single **pulmonary aorta** which passes ventral to aortic or systemic arch and bifurcates into pulmonary arteries, each of which enters into a lung.

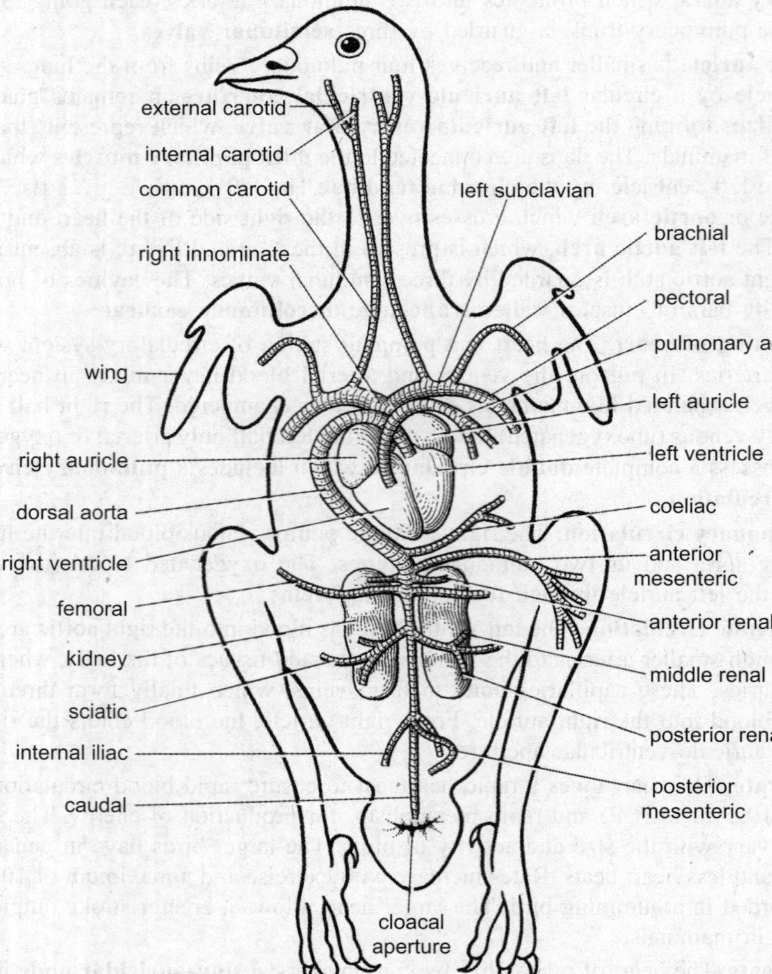

Fig. 26.37. Pigeon. Arterial system.

2. Aortic arch or systemic aorta. The right aortic or systemic arch arises from the left ventricle, the left aortic arch being absent in adult birds. The right aortic arch gives out two small **coronary arteries** to the heart, just at the point of its origin. It then gives off a pair of short, stout and thick **right** and **left innominate arteries**. Each innominate artery divides into an anterior **common carotid artery** and a lateral **subclavian artery**. The common carotid artery in turn divides into a **verterbral** to neck vertebrae, an **internal carotid** to the brain, an **external carotid** to the head and neck, and a **comes nervi vagi**, which joins the external carotid.The subclavian artery divides into a **brachial artery** going to the wing, and a very large **pectoral artery** which gives off a thin **internal mammary artery** to the ventral thoracic wall, then it forms three main branches going to the pectoral muscles of the flight.

Dorsal aorta. The right aortic arch now curves over dorsally toward the right side of heart and passes backwards mid-dorsally as **dorsal aorta**. The dorsal aorta gives off following branches to different body organs :

(*i*) A single, median, strong **coeliac artery** which supplies branches to proventriculus, gizzard, duodenum, liver and spleen.

(*ii*) An unpaired **anterior mesenteric artery** to the greater part of small intestine.

(*iii*) Two **anterior renal arteries** to the first lobes of kidneys. In male, each renal artery gives off a branch to the testis; but in female, only the left renal artery gives off a branch to the ovary (the right ovary is absent in birds).

(*iv*) A pair of **femoral arteries** runs outwards dorsal to the anterior lobes of kidneys and supplies blood to pelvic muscles and outer muscles of the thighs.

(*v*) A pair of **sciatic** or **ischiatic arteries** passes outwards to supply blood to inner muscles of the thighs and the legs. Each sciatic artery gives off a **middle renal artery** to the middle lobe of the kidney and a **posterior renal artery** to the posterior lobe of the kidney.

(*vi*) A pair of **iliac arteries** to the hinder part of pelvic region.

(*vii*) An unpaired **posterior mesenteric artery** to the large intestine (rectum) and cloaca

(*viii*) At last the dorsal aorta continues into the tail as **caudal artery.**

3. Venous System

The venous system of pigeon includes : **(*i*)** the **caval veins**, **(*ii*)** the **pulmonary veins**, **(*iii*)** the **hepatic portal system** and **(*iv*)** the **renal portal system**.

1. Caval veins. The non-oxygenated or venous blood of entire body of pigeon is collected and conveyed to the right auricle of the heart by three large veins, two **precavals** from the anterior part of the body and one **postcaval** from the posterior region of the body.

A. Precavals. The precaval vein of each side (smaller right and larger left) is formed by the confluence of a large common **jugular vein** and a **subclavian vein** as in lizard. The common jugular vein is formed by the union of **external** and **internal jugular veins** from the head and a **vertebral vein** from the spinal cord and neck muscles.The jugular veins of the two sides are joined just beneath the base of the skull by a transverse loop called **jugular anastomosis**. The jugular anastomosis prevents blocking of the flow of blood from the head when one vein is momentarily shut off by twisting the neck.

The subclavian vein receives a slender **brachial vein** from the wing and a large **pectoral vein** from the pectoral muscles. A small **internal mammary** from the ventral thoracic walls also joins the pectoral vein.

B. Postcaval. The postcaval vein is formed by the fusion of all the veins from the posterior part of the body. A short, unpaired **caudal vein** from the tail after entering the abdomen bifurcates into right and left **hypogastric** or **renal portal veins**. From the bifurcation of the caudal vein arises a median **coccygeo-mesenteric vein** which is characteristic of birds. It receives the blood from cloaca and rectum, and runs forward to join the hepatic portal vein. Each renal portal vein instead of breaking up into capillaries in the kidney, gives off only a few small branches, **afferent renal veins**, which carry blood to the kidney. The main renal portal vein passes forwards through the substance of the kidney and receives the **femoral** and **sciatic veins** from the leg to form the **iliac vein**. At the junction of iliac and renal veins is present a valve which probably regulates the flow of blood into the renal portal vein. The femoral vein also receives the **efferent renal veins** from the kidney. As soon as it leaves the kidney, it is called a **common iliac vein**. The common iliac veins of the both sides unite to form a short **postcaval vein**, which runs forwards through the substance of right lobe of liver and receives three pairs of **hepatic veins**, and then enters the right auricle.

Thus, the main part of blood from the caudal and pelvic regions is taken directly to the heart, and not through the renal capillaries as in most fishes, and all amphibians and reptiles.

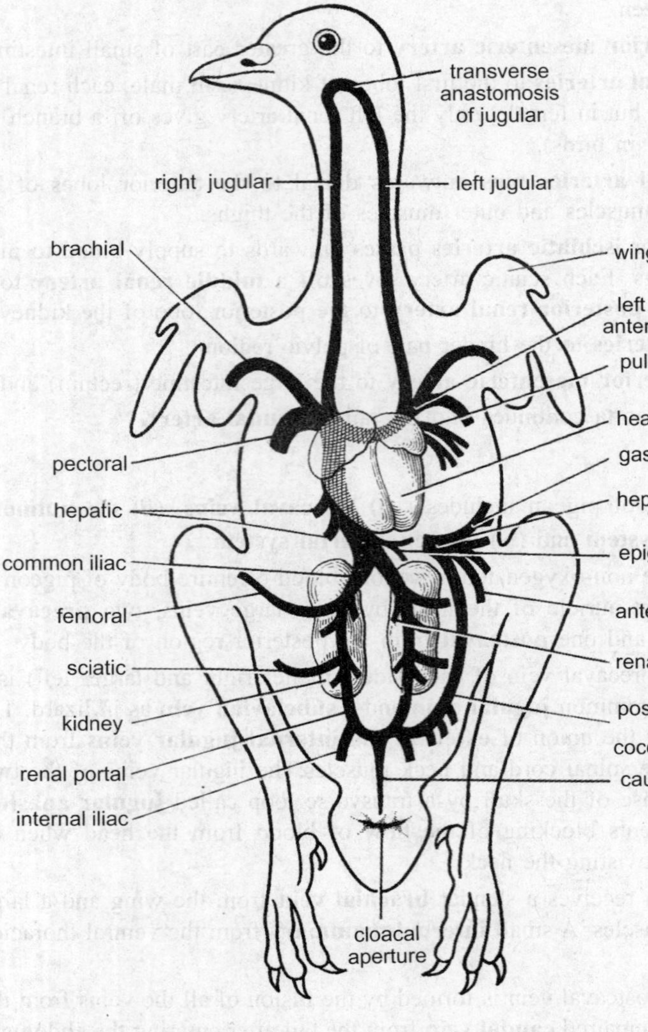

Fig. 26.38. Pigeon. Venous system.

2. Pulmonary veins. From lungs two pulmonary veins return oxygenated blood into the left auricle. Both pulmonary veins before opening into auricle unite togther and open by a single aperture in left auricle.

3. Hepatic portal system. The blood of various parts of alimentary canal is returned to liver by a **hepatic portal vein** which is formed by the union of **gastro-duodenal, anterior mesenteric, posterior mesenteric,** and **coccygeo-mesenteric veins. Coccygeo-mesenteric vein** arises from the point of bifurcation of the caudal vein, runs parallel with the rectum and joins with the hepatic portal vein. An **epigastric vein** returns blood from the fat-laden peritoneum (**great omentum**) and joins one of the hepatic veins. The **hepatic portal vein** collects blood from rectum, ileum, duodenum and gizzard, and it divides into two branches, one entering each liver lobe into which they form capillary network. The blood of hepatic portal vein is filtered through the liver and then conveyed to the postcaval vein through the hepatic veins.

Because the **coccygeo-mesenteric vein** connects the two portal systems (*i.e.,* renal portal and hepatic portal), it is considered homologous with the anterior abdominal vein of amphibians and reptiles. But this homology is not accepted by some authorities. They regard the **epigastric vein** to be the homologous of the abdominal vein of amphibians and reptiles.

4. Renal portal system. The renal portal system of pigeon is greatly reduced and includes two **hypogastric** or **renal portal veins**, each of which gives off only a few **afferent renal veins** to each kidney and do not form capillary system in the kidney. Thus, the blood from the posterior region of body reaches directly to the heart, and not through the renal capillaries, as in amphibians and reptiles.

4. Blood

The blood of birds is rich in red blood corpuscles. They are oval and nucleated. They are large 10–15 μm × 5–8 μm and numerous 2.5–3.5 million per cmm. They carry a large amount of haemoglobin that gives up its oxygen suddenly at a relatively high oxygen tension. The red corpuscles are smaller in actively flying birds than in the larger flightless birds.

Haemopoietic tissue is widespread in the young, but is restricted mainly to the marrow in the adult, although it may also be found in liver and spleen. The white blood corpuscles are more numerous than in mammals. Platelets are not found in pigeons, but it clots quickly. Thrombocytes are present. Every part of the body is nourished by the blood.

Bird's temperature is considerably higher than that of mammals, usually about 42°C to nearly 45°C in some birds. In the absence of sweat glands, how it is kept constant is not known. Heat loss is minimised by the absence of vascularised extremities (feet). The formation of wing from large avascular surfaces is the secret of the success of birds.

Air sacs may serve to conserve heat by providing an air cushion for the viscera, which might otherwise lose heat. There is a system of direct arterio-venous connections in the feet, and elsewhere. These anastomotic regions have powerful muscles, whose contraction closes them and forces the blood through the capillary system. There must have been a system of nervous pathways for the control of upward and downward temperature regulation. To avoid heat loss one species (a nightjar) is known to hibernate. Certain humming birds and sunbirds have a serious heat loss problem so they become poikilothermic in night.

5. Lymphatic System

A lymphatic system with a few lymphatic glands is associated with the blood vascular system. Lymph glands are found in some aquatic and wading birds. There is a pair of lymph hearts in the sacral region of the embryo and these may persist in the adult. The lymph vessels unite to form two **thoracic ducts** (homologous with subvertebral trunk) opening into the precaval veins. Adult pigeons lack any lymph hearts. The spleen is a small, oval and reddish gland attached to the right side of the proventriculus by a peritoneal fold.

Spleen. It is a small, ovoid red body attached by peritoneum to the right side of the proventriculus.

NERVOUS SYSTEM

The well developed nervous system of pigeon consists of the **central nervous system** comprising brain and spinal cord, **peripheral nervous system** including cranial and spinal nerves and **autonomic nervous system** comprising sympathetic and parasympathetic systems.

A. Central Nervous System

1. Brain. The brain of birds is much more highly developed than that of reptiles and is relatively largest, next to those of mammals. The hemispheres are without superficial convolutions.

External structure. The brain of pigeon is relatively larger than in lower forms, but in shape it is short, broad, rounded in form and whitish in colour. It completely fills the roomy cranial cavity. It is covered by two protective membranes or **meninges** (singular, *meninx*), an outer **duramater** of connective tissue and an inner **pia-arachnoid** rich in blood supply. The narrow space between two meninges has connective and fatty tissues, veins and **cerebro-spinal fluid**. The brain is divisible into forebrain, midbrain and hindbrain.

A. Forebrain. The **olfactory bulbs** or **lobes** are extremely small and degenerate due to poorly developed organs of smell. But is *Apteryx* olfactory chamber is highy developed extending from tip of the beak to the level of the **optic foramina**. The **cerebral hemispheres** are pyriform, large-sized, convex, and separated by a deep sagittal fissure. The cerebral hemispheres are relatively smooth and lack characteristic fissures of mammalian brain and reach the cerebellum posteriorly. The ventro-lateral portions are enormously developed and usually called the "**corpus striatum**", whereas the medial ventral walls are thin and the pallium is quite small, thin and not folded. The **diencephalon** is covered completely on the dorsal side by the posterior extension of the cerebral hemispheres and anterior extension of the cerebellum. A small pineal body lies dorsally at the posterior part of sagittal fissure of the cerebrum and an anterior choroid plexus. On the ventral surface of the visible diencephalon, the two prominent and large optic nerves cross each other forming the **optic chiasma**. Behind the chiasma projects a median process, the **infundibulum**, bearing a large **hypophysis** or **pituitary gland**, which lacks the intermediate lobe.

B. Midbrain. The midbrain is highly developed due to keen sight of birds, with two large very well developed **optic lobes.** These are lateral in position because they have been pressed outwards by the cerebral hemispheres which meet posteriorly with the cerebellum. These are of large and rounded form. Both optic lobes remain connected together by a transverse **optic commissure**.

C. Hindbrain. The **cerebellum** is highly developed, large, folded and extends forwards. It is divisible into a large median lobe or **vermis** and two small, conical, lateral lobes or **flocculi**. The vermis has transverse grooves forming folds and ridges, so that the superficial grey matter is greatly increased in extent. The convoluted cerebellum indicates the delicate sense of equilibrium and the great power of muscular coordination belonging to birds. The cerebellum extends far backwards covering a large part of the medulla oblongata, which descends almost vertically from it, to join the spinal cord. At the point of junction of medulla oblongata and spinal cord occurs a well-marked ventral flexure, as in lizards.

Internal structure. The brain is a hollow organ and its cavities are called **ventricles**. Each major division of pigeon's brain has greatly reduced ventricles. The two cavities of cerebral hemispheres are called **lateral ventricles** or the **paracoels**. Their roof or **pallium** is thin and not folded, and there is very little development of cerebral cortex like the reptiles. The ventro-lateral walls of cerebral hemispheres are very thick and formed of **corpora striata** or **hyperstriatum**. The corpora striata of both the lobes are interconnected by a transverse **anterior**

commissure. The corpora striata are characteristic of birds, their size makes the lateral ventricle small. They receive fibres from the thalamus and send fibres (**crura cerebri**) to the midbrain, cerebellum, medulla, and indirectly to the spinal cord. Posteriorly, each lateral ventricle or paracoel communicates by a small common aperture, the **foramen of Monro,** with the narrow, vertical cavity of the diencephalon, termed the **third ventricle** or **diacoel**. The thin roof of diacoel is called **epithalamus**. It contains a richly vascular tissue, the **anterior choroid plexus** located just in front of pineal body. The diacoel remains bounded laterally by the thick **optic thalami,** while its thick floor is termed **hypothalamus**.

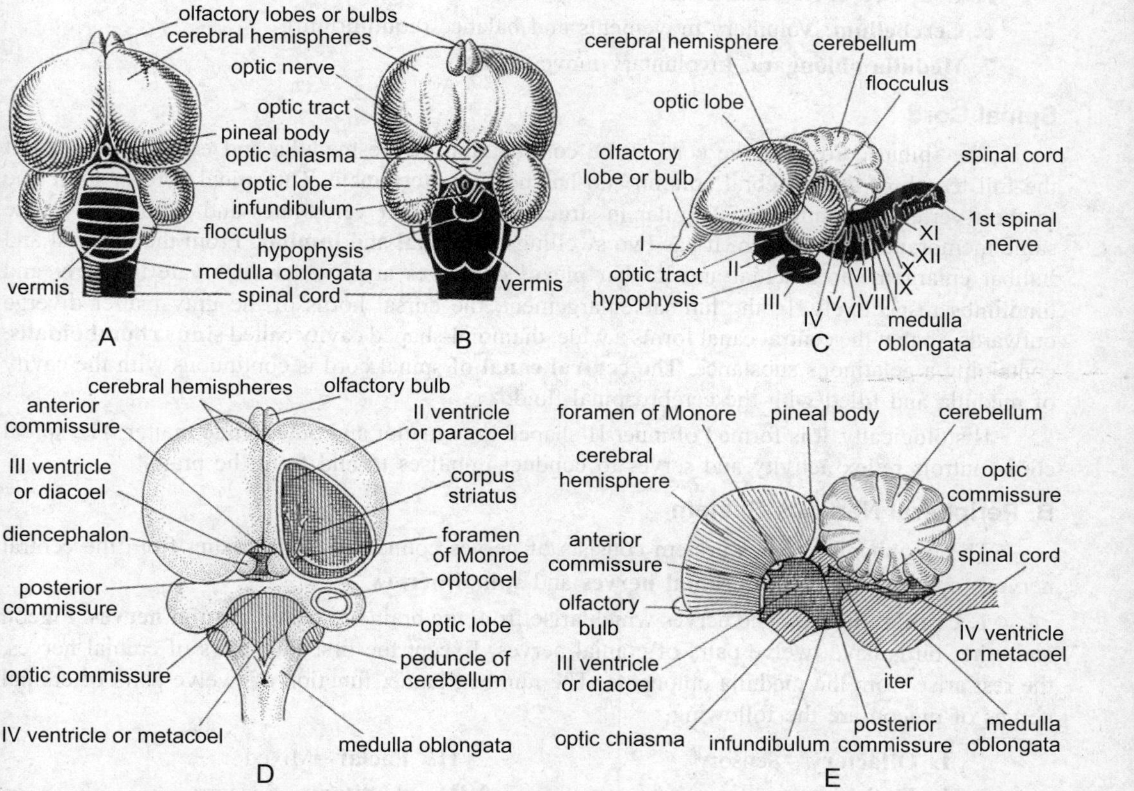

Fig. 26.39. Pigeon. Structure of brain. A–Dorsal view; B–Ventral view; C–Left lateral view; D–Cavities in dorsal dissection; E–In sagittal section.

The diacoel remains separated from the narrow cavity of the midbrain, called **iter**, by **posterior commissure**. The iter communicates dorsally with the cavities of the optic lobes, called the **optocoels**. Thick longitudinal bands of nerve fibres forming the **crura cerebri** run beneath the optic lobes and connect the diencephalon with the medulla.

The cavity of the medulla oblongata is called **fourth ventricle** or the **metacoel** (or **myelocoel**). It sends a branch to cerebellum. Its roof is thin, highly vascular, called **posterior choroid plexus**, and its sides and floor are thick-walled. The anterior and posterior choroid plexes secretes the cerebrospinal fluid which fills in the ventricles of brain.

Histologically, the brain has a layer of **grey matter,** containing nerve cells, and a layer of **white matter**, containing nerve fibres. The grey matter is outer in the olfactory lobes, cerebrum, diencephalon and cerebellum, while in optic lobes and medulla it is internal.

Functions. The functions of different parts of brain are following :

1. Olfactory lobes. Smell.

2. Cerebral hemispheres. Voluntary movement or conditioned reflexes, intelligence, will, memory, emotions, etc. The corpora striata are concerned in the control of the reflex behaviour governing the lives of birds, such as eating, locomotion, copulation, nest building and care of the young.

3. Diencephalon. Relay impulses to cerebral hemispheres, integrating centre of autonomic system and perception of extreme cold, heat, pain, etc.

4. Optic lobes. Sight.

5. Crura cerebri. Conduction of sensory impulses.

6. Cerebellum. Voluntary movements and balance (equilibrium).

7. Medulla oblongata. Involuntary movements.

Spinal Cord

The spinal cord or nerve cord is the continuation of the medulla and extends throughout the full length of the vertebral column and has no filum terminale. The spinal cord is short due to short vertebral column. It is similar in structure as in other vertebrates and is covered by the same membranes as the brain. It has two swellings, **brachial** and **lumbar**. From the brachial and lumbar enlargements brachial and lumbar plexes of nerves arise which innervate the fore- and hindlimbs respectively. In the lumbar enlargement, the dorsal horns of the gray matter diverge outwards, so that the central canal forms a wide, diamond-shaped cavity called **sinus rhomboidalis,** containing a gelatinous substance. The **central canal** of spinal cord is continuous with the cavity of medulla and filled with the cerebrospinal fluid.

Histologically, it is formed of inner H-shaped grey matter and outer white matter. The spinal cord controls reflex activity and serves to conduct impulses to and from the brain.

B. Peripheral Nervous System

The peripheral nervous system consists of nerves connected to or arising from the central nervous system. It includes **cranial nerves** and **spinal nerves**.

1. Cranial nerves. The nerves which arise from the brain are called **cranial nerves**. Pigeon and other birds have twelve pairs of cranial nerves. Except the first four pairs of cranial nerves, the rest arise from the medulla oblongata. The number, name, function of twelve pairs of cranial nerves of pigeon are the following :

I. **Olfactory**—Sensory
II. **Optic**—Sensory
III. **Oculomotor**—Motor
IV. **Trochlear**—Mixed
V. **Trigeminal**—Mixed
VI. **Abducens**—Motor
VII. **Facial**—Mixed
VIII. **Auditory**—Sensory
IX. **Glossopharyngeal**—Mixed
X. **Vagus**—Mixed
XI. **Spinal accessory**—Motor
XII. **Hypoglossal**—Motor

2. Spinal nerves. The nerves which arise from the spinal cord are called **spinal nerves**. The number of spinal nerves of pigeon corresponds to the number of its vertebrae. Each spinal nerve arises as a **dorsal** and **ventral** root. The dorsal root has a ganglion and carries only **somatic sensory** and **visceral sensory fibres**. The ventrai root has **somatic motor** and **visceral motor fibres**. The dorsal and ventral roots unite inside the vertebral column (inside the central canal) and come outside the column as spinal nerves. Each spinal nerve divides into **dorsal ramus**, long **ventral ramus** and **ramus communicans**.

In pigeon, the **cervical spinal nerves** pass out to the neck region. The last three **cervical nerves** and first **thoracic nerve** form a **brachial plexus** going to the wing. Next five thoracic spinal nerves pass between the ribs. Following these is a **lumbosacral plexus** which may be separated into **lumbar**, **sacral** and **pudendal plexuses**. The lumbar plexus of three nerves supplies the thighs. Five nerves of the sacral plexus unite to form a **sciatic nerve** which passes through the thigh and goes to the lower leg. The pudendal plexus sends branches to the cloaca and the tail.

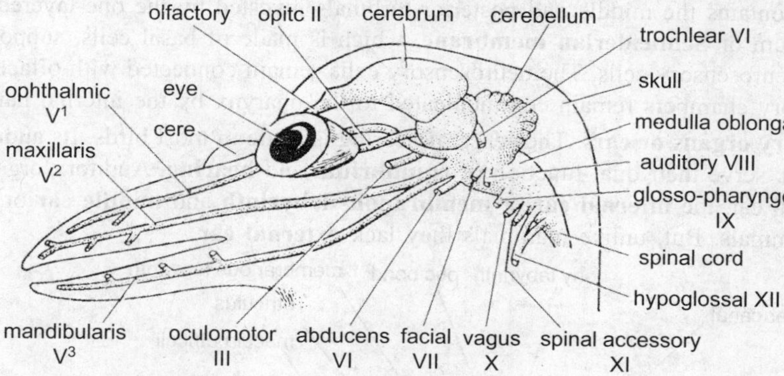

Fig. 26.40. Pigeon. Cranial nerves in left lateral dissection.

C. Autonomic Nervous System

The autonomic nervous system of pigeon includes **sympathetic** and **parasympathetic systems** and is very much like that of mammals. It contains nerves and ganglia. Fibres link up the spinal nerves with the ganglia and the spinal cord. The sympathetic nerves supply chiefly the urinogenital, alimentary, respiratory and circulatory systems.

SENSE ORGANS

The pigeon, like other vertebrates, has receptors or sense organs for touch, smell, taste, sight and hearing which are stimulated by the environment. These sense organs are termed external receptors or **exteroceptors**. The pigeon possesses following exteroceptors :

1. Tactile organs. These are poorly developed in birds due to feathery covering of the body. Tactile organs of pigeon remain confined to the bill and tongue of pigeon. The **cere** is a sensitive soft fold of skin at the base of the upper beak in pigeons, is said to have a stimulating effect during love making. The **corpuscles of Grandry** in the bill of ducks and other birds are probably tactile receptors. They are composed of cells with a flattened nerve ending between cells. These are comparable to Meissner's corpuscles in mammals. **Merkel's corpuscles** are also found in many birds.The **corpuscles of Herbst**, resembling Pacinian corpuscles of mammals, are found in the dermis, are vibration receptors, sensitive to mechanical deformation by rapid pressure changes. They are found in great numbers in the beaks of all birds and also in the feather follicles, between tibia and fibula and in the tip of the tongue of woodpecker. Tactile nerves are also present at the base of the feathers, especially those of the wings and tail.

2. Gustatory organs (Chemoreceptors). Sense of taste and smell are little developed. The sense organs of taste, the **taste buds**, occur in limited number on the dorsal surface of tongue. The sense of taste is poorly developed in pigeons.

3. Olfactory organs. Birds are usually unable to distinguish delicate odours, and on the whole their **sense of smell** is very poor, as flying animals cannot depend on smell. The nasal cavity is large but the olfactory epithelium is restricted. Birds use the nose to test air coming from the internal nostrils. In kiwis, olfactory sense is well developed. These birds are nocturnal and terrestrial.

Structure of olfactory organs. The nostrils, overhung by cere, lead into the small, paired **olfactory sacs** or **nasal chambers** in the base of upper beak.The two chambers are separated medially by mesethmoid and bounded externally by ectoethmoid. The ectoethmoid produces inwards three scroll-like **turbinal processes** to increase the olfactory surface. The nasal chamber has an anterior **non-sensory respiratory** part and a posterior **sensory** part. The non-sensory part or **vestibulae** contains the anterior turbinal covered by laminated epithelium. The sensory or **olfactory** part contains the middle and posterior turbinals invested by the one-layered olfactory mucous epithelium or **Schneiderian membrane**, which is made of basal cells, supporting cells and elongated neurosensory cells. The neurosensory cells remain connected with olfactory nerve. Both the olfactory chambers remain communicated to the pharynx by the internal nares.

4. Auditory organs or ears. The sense of hearing is acute in most birds. Its auditory sense organs, the **ears**, serve their dual function of **equilibrium** and **hearing**. Auditory organs consist of a fundamental ear, the **internal ear** or **membranous labyrinth** and **middle ear** or **tympanic cavity,** like mammals. But, unlike mammals they lack **external ear**.

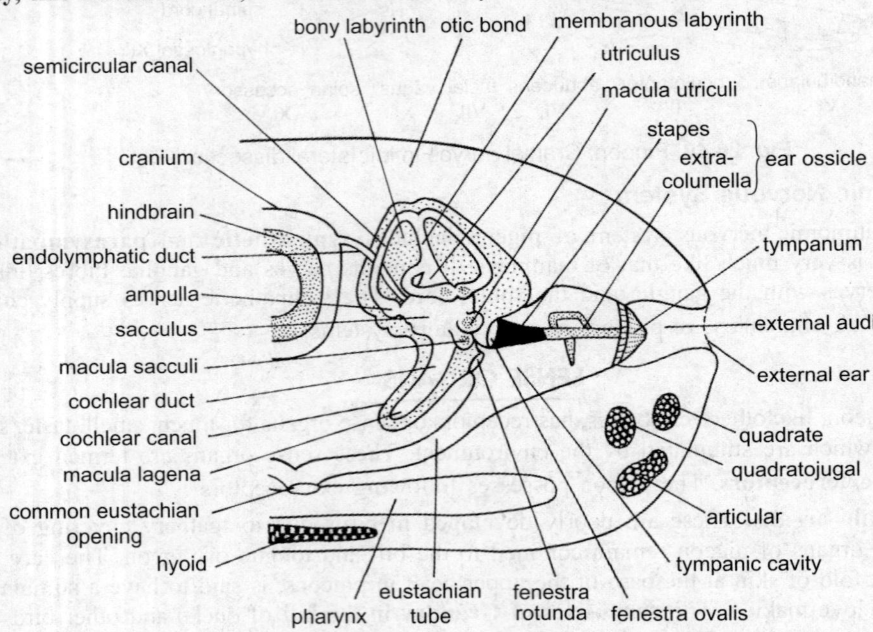

Fig. 26.41. Pigeon. T.S. head through internal ear.

Internal ear. The **internal ear** or **membranous labyrinth** lies embedded in a dense ivory-like bone in the side wall of the skull and is surrounded with a **perilymph** fluid. Membranous labyrinth is filled with a dense fluid, called **endolymph**. It contains a high concentration of potassium as in mammals, maintained by the activity of a tegmentum vasculosum, which also absorbs sodium. It consists of three **semicircular canals** (one horizontal, one anterio-vertical and one posterio-vertical), relatively small **sacculus** and **utriculus** and a short blind tube, the **cochlear duct** or **lagena**. Lagena is larger than in reptiles and less developed than in mammals. The proximal limbs of anterior and posterior semicircular canals unite to form a **crus commune**. The sacculo-utricular connection is narrow. From the sacculus arises an endolymphatic duct which ends in the duramater.

The cochlear duct, with its surrounding bony labyrinth or **cochlear canal**, is called **cochlea**. The cochlear duct is a slightly curved tube, surrounded by a perilymphatic space within a bony tube. In a transverse section, the cochlea shows three chambers : an upper **scala vestibuli,** a middle **scala media** and a lower **scala tympani**. The scala media is the actual cochlear duct and contains endolymph, while other two scalae are filled with perilymph. At the apex of the cochlea (scala media) the scala vestibuli and scala tympani are continuous, this junction is known as the **helicotrema**, near it the scala media ends blindly. At the inner end of cochlea, scala vestibuli is continuous with fenestra ovalis, while the scala tympani with the fenestra rotunda. The floor of scala media is formed by the **basilar membrane** and the roof by the **Reissner's membrane**. The basilar membrane consists of tall auditory hair cells (40), together constituting an **organ of Corti**. The organ of Corti is sensitive to sounds of higher frequencies. The free ends of the auditory cells bear hairs, which are embedded in a **tectorial membrane**, lying above the organ of Corti. From the basal ends of the cells arise nerve fibres which unite to form the **cochlear branch** of the auditory nerve. The cochlear duct at its apical end has another set of auditory hair cells, their hairs embedded in a gelatinous **cupola terminalis** having minute calcareous **otoliths**. At the tip of the cochlear duct is the **lagena**. Groups of auditory cells are called **cristae** and **maculae**. Those present in the ampullae of three semicircular canals are called **cristae ampullares** which possess the sense of direction and equilibrium. One macula is present in utriculus and one in sacculus. One macula is present in the lagena (tip of cochlear duct). Macula lagena is separated from the other maculae by a long cochlear duct. It perceives low frequency sounds.

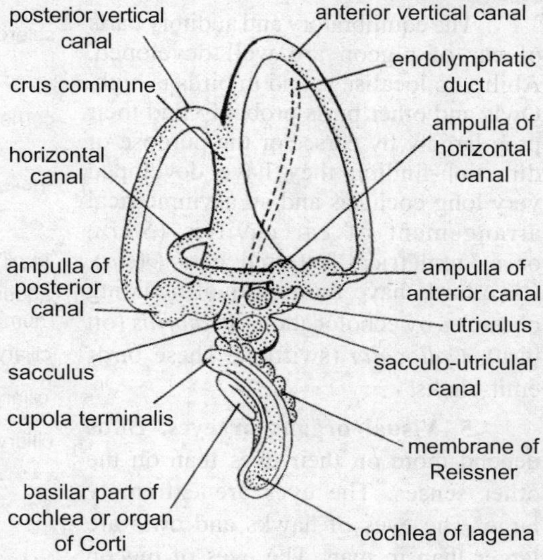

Fig. 26.42. Pigeon. Internal ear (membranous labyrinth).

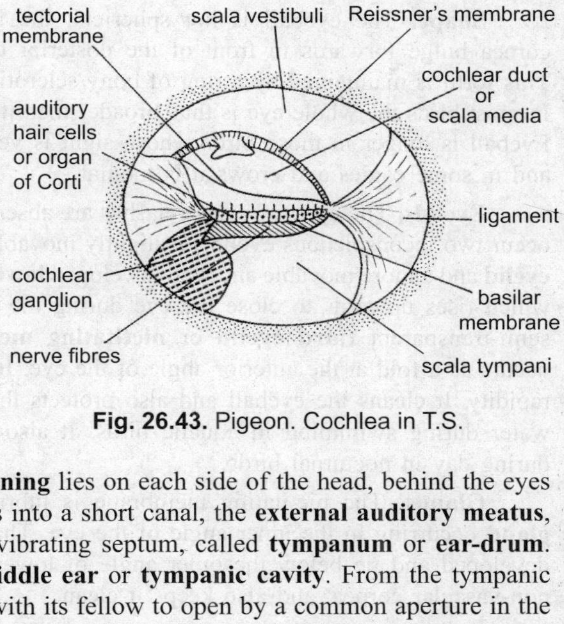

Fig. 26.43. Pigeon. Cochlea in T.S.

Middle ear. A circular **external ear opening** lies on each side of the head, behind the eyes concealed beneath **auricular feathers.** It leads into a short canal, the **external auditory meatus**, at the base of which lies a thin, transparent, vibrating septum, called **tympanum** or **ear-drum**. The ear-drum is followed by the cavity of **middle ear** or **tympanic cavity**. From the tympanic cavity arises a **eustachian tube** which unites with its fellow to open by a common aperture in the pharynx. The eustachian tube serves to equalise the air pressure on both sides of the tympanum.

Across the tympanic cavity lies a single rod-like bone, the **columella auris**, which transmits sound waves from tympanum across the tympanic cavity to the **fenestra ovalis** of the inner ear. The outer end of columella auris has three rayed cartilaginous processes called **extra columella** connected to the inner surface of tympanic membrane, and its inner disc-like bony part, the **stapes**, is wholly occluded the fenestra ovalis. Columella auris (stapes) is derived from the hyoid arch. Sound is transmitted from the tympanum by the columella auris (stapes). Extra columella reduce the amplitude and increase the force of vibrations. There is single middle ear muscle attached to columella and tympanum and innervated by facial nerve. Acoustic vibrations are transmitted to an oval window and so around cochlea to a round window, as in mammals.

The equilibratory and auditory parts of ear of pigeon are well developed. Ability to localise sound in birds is high. Owls and other birds probably find their prey largely by ears. For the purpose of direction-finding they have developed very long cochleas and an asymmetrical arrangement of ear cavities (*Strix*) or asymmetrical external ears (*Asio*). Cave birds have the power of avoiding obstacles by echolocation, *Steatornis* (oil bird), *Collocalia* (swiftlet). These birds emit clicks.

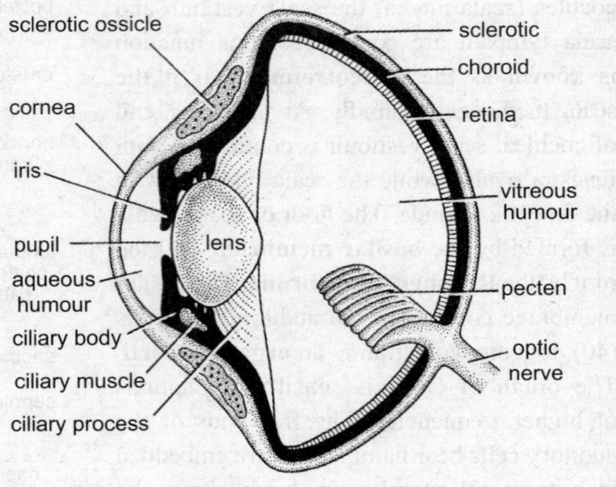

Fig. 26.44. Pigeon. Eye in sagittal section.

5. Visual organs or eyes. Birds depend more on their eyes than on the other senses. The eyes are extremely large. The eyes of hawks and owls are larger than in man. The eyes of pigeon are well developed and are very large in correlation with an aerial life for a precise vision over considerable distances.

Shape. The eyeball is not spherical, the lens and cornea bulge forwards in front of the posterior chamber. This form is maintained by a ring of bony sclerotic plates. In most birds the whole eye is thus broader than it is deep. Eyeball is longer in those birds whose sight is very acute and in some eagles and crows it is tubular.

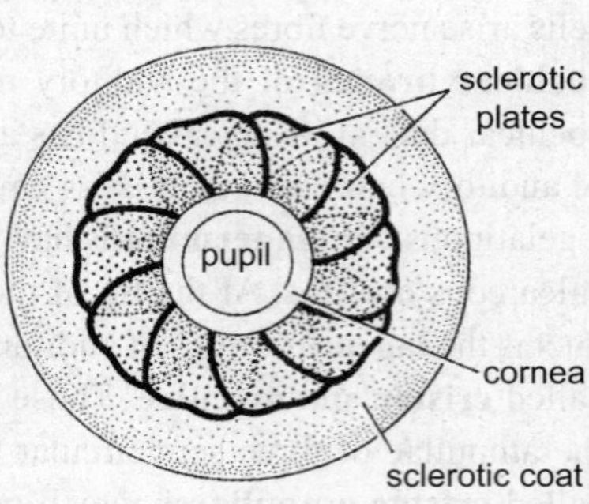

Fig. 26.45. Pigeon. Eye in outer view showing sclerotic plates.

Eyelids. The eyebrows or eyelashes are absent. There occur two inconspicuous eyelids, a slightly movable **upper eyelid** and a more movable and well developed **lower eyelid** which rises upwards to close the eye during the sleep. A semi-transparent **third eyelid** or **nictitating membrane** occurs as a fold at the anterior angle of the eye. It can be drawn posteriorly over eye with great rapidity. It cleans the eyeball and also protects the eyes from wind and during flight and from water during swimming in aquatic birds. It also protects the eyes from glare of the sunlight during day in nocturnal birds.

Glands. The nictitating membrane is lubricated by the oily secretion of a **Harderian gland** occurring in the inner angle of the eye. The tear gland or **lacrymal glands** are also well developed and lie below the outer angle of lower eyelid. Their watery secretion nourishes the non-vascular cornea, and also keeps it clean.

The wall of hollow eyeball consists of three usual layers, namely, an outer **sclerotic**, a middle **choroid** and an inner **retina**.

Sclerotic. The external coat of the eyeball is sclerotic. In the posterior hidden part of the eye, it is opaque, white, dense and cartilaginous. In front in the exposed part of the eye, it bulges out to form a convex, transparent and horny **cornea** of connective tissue. The cornea is externally covered by a thin, transparent, sensitive and vascular epithelial membrane, the **conjunctiva**, which is formed by modified epidermis and is continuous with the mucous lining of the eyelids. Anteriorly, at the junction of cornea and sclerotic coat, the latter is strengthened by a ring of 10–12 small overlapping bony, **sclerotic plates** or **ossicles**.

Choroid. The sclerotic coat is followed by the middle layer, the **choroid**, which is thin, dark pigmented and highly vascular. The choroid closely lines the sclerotic, but it separates in front to form a circular, pigmented diaphragm, the **iris**, perforated by a rounded aperture, the **pupil**. The iris regulates the amount of the entering light. It contains intrinsic circular and radial smooth muscles, the circular muscles contract the pupil, while radial muscles dilate it. Along the peripheral margin of the iris, the choroid forms a ring-like **ciliary body** which is a thickened fold containing smooth ciliary muscles. From ciliary body arises striated **ciliary processes** or **suspensory ligaments** and attached to the lens. The ciliary muscles are divided into anterior **Crampton muscles** and posterior **Brucke muscles**. The Brucke muscles draws the lens forward into the anterior chamber so that since the shape of the eye is fixed by the sclerotic plates, the lens becomes more curved and, hence, accommodated for near vision. Contraction of the iris sphincter assists in this process. At the same time the Crampton muscles pull the cornea reducing its radius and further assist in accommodation.The circular and radial muscles of the iris and ciliary muscles are under the control of the autonomic nervous system receiving sympathetic and parasympathetic fibres.

Retina. The innermost coat of the eyeball is a thin, light sensitive nervous layer called **retina**. It is transparent, devoid of blood vessels, thick and consists of nerve fibres, nerve cells and minute **rods** and **cones**. Pigeon, being a diurnal bird, has largely more cones than rods. The high resolving power and high powers of discrimination and of movement detection depend on the great density of the cones, as many as 1 million per square millimetre in the fovea of a hawk. Nocturnal bird's retina is composed mainly or completely of rods.

Sensitive spots. The retina has two sensitive spots or **fovea**. The **central fovea**, which lies near the centre of retina as a slight depression, is more sensitive and used for lateral or **monocular vision**. The second fovea, the **temporal fovea**, lies more towards the outer side of the eye and is used for forward or **binocular vision.** The foveae have comparatively more cones and give more distinct vision. The cones of birds often contain carotenoid oil droplets, which are also found in frogs, turtles and marsupials. In diurnal birds, they are red, yellow, orange or colourless, but in nocturnal ones pale yellow or colourless. Colour droplets may produce narrow-band sensitivity channels for the mediation of colour discrimination. The lower part of retina contains yellow droplets and upper part (dorso-posterior) contains red droplets. These droplets serve as filters and increase the colour vision of birds up to great degree. The red may provide finer colour discrimination of objects in feeding.

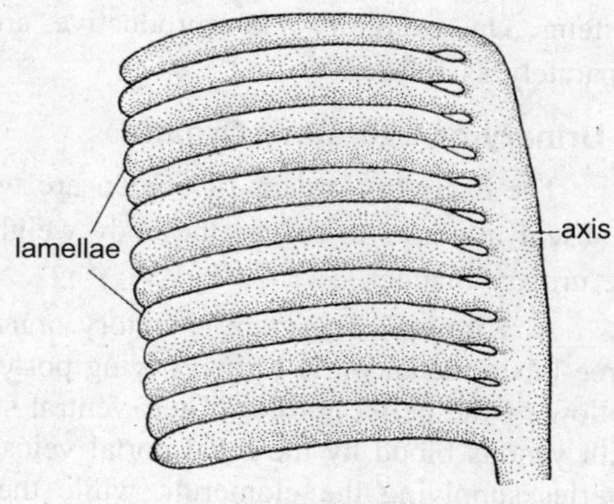

Fig. 26.46. Pigeon. Pecten.

Lens. The lens is biconvex, soft, pliable, crystalline, colourless, transparent and is surrounded by a fibrous capsule. It remains suspended just behind the iris by suspensory ligaments. It divides the eye cavity into a small anterior **aqueous chamber** and a large posterior **vitreous chamber**. The aqueous chamber is filled with a colourless watery fluid, the **aqueous humour**, while vitreous chamber contains a thick colourless, gelatinous **vitreous humour**. The two humours keep the eyeball tight or tense and also serve to focus the light rays on the retina.

Pecten. Pecten (Fig. 26.46) is a pleated, strongly pigmented vascular fold projecting into the cavity of the eye from the entrance of the optic nerve. It is large and much pleated in predatory birds, which detect minute movements at great distances, and is small and smooth in nocturnal birds. It is also well developed in diurnal birds (pigeon). Except kiwi, all birds have pecten.

Functions of pecten. There are many speculations about the function of pecten but none is known definitely. Probably its main function is to bring oxygen and nourishment to the retina, which in birds has no capillary circulation. It helps in accommodation, it is not likely that it actually assists in focussing, for instance, by pressing forward the lens, and no changes have been seen in it during accommodation. However, it might possibly assist by adjusting the intraocular pressure, which must be increased by the extensive changes in the lens during accommodation.

Vision. In pigeon, because the cornea projects outwards and the posterior part is expanded, so the eye is broader than deep. Due to expansion of retina over the broad posterior portion, the distant objects are sharply focussed on it. Further, though the eyes are lateral in position, there is an overlapping of the two visual fields to some extent. This is called **binocular vision**. It is worth noting that there occurs little movement in pigeons' eyes due to ill-development of extrinsic eyeball muscles and that is compensated by flexibility of neck which turns the neck very quickly.

ENDOCRINE GLANDS

Pigeon has all the vertebrate endocrine glands such as **thyroid**, **parathyroid**, **pituitary**, **thymus**, **adrenal glands**, **testes** and **ovaries**. They have similar functions as in other vertebrates.

URINOGENITAL SYSTEM

The excretory and reproductive organs of pigeon are very closely connected with each other structurally, so are studied together under the heading **urinogenital system**. But, both systems, *viz.*, excretory and reproductive, are physiologically independent, so should be studied separately as follows :

1. Urinary or Excretory Organs

The excretory organs of pigeon are two **kidneys** and two **ureters**. The **urinary bladder** is absent in adults to reduce the body weight (flight adaptation). Among birds, ostrich only has the urinary bladder.

A. Kidneys. The main excretory organs of pigeon (birds) are two, dark brown, flattened, three-lobed metanephric **kidneys** lying posterio-dorsally in the body cavity and embedded in the hollow of the pelvic girdles. Their ventral surface is covered by peritoneum. They are provided with venous blood by the renal portal veins and arterial blood from the renal arteries.The renal arteries supplying the glomeruli, while the tubules are supplied both by efferent glomerular arterioles and branches of the portal veins. A small yellowish, elongated and streak-like **adrenal body** lies attached ventrally to the anterior lobe of each kidney. It is an endocrine gland.

Each kidney consists of masses of numerous tightly packed, convoluted **uriniferous tubules**. Birds have comparatively more tubules than those of mammals. Each uriniferous tubule has small **glomerulus** and a specialised portion, called the **loop of Henle**. Loops of Henle are long. The glomerulus filters the blood and filtrate passes through the long loop, where much of its water is reabsorbed. Thus, much concentrated urine passes down the ureter.

B. Ureter. Uriniferous tubules of each kidney unite to form a ureter or metanephric duct. The ureter is a narrow, straight tube, arising ventrally from the anterior lobe of the kidney and running backwards to open into the middle compartment of the cloaca, the **urodaeum**, through its dorsal wall.

Physiology of excretion. The excretory system is highly specialised for water-saving. The birds are **urecotelic** like reptiles, because in them the end product of urinary excretion is relatively insoluble uric acid which is synthesised in liver. Semisolid viscous urine comes from the kidneys into the urodaeum from where it passes up into the coprodaeum and large intestine, where further water is resorbed and the mixed faeces and urinary products are then excreted as the characteristic semisolid white **guano**. Urine is hyperosmotic to the blood. Birds also excrete salt by nasal glands, especially well developed in marine forms.

2. Male Reproductive Organs

Sexes are separate and sexual dimorphism is absent in pigeon. The male reproductive organs of pigeon (Fig. 26.47) are **testes** and **vasa deferentia** and there occurs no penis like mammals. Among birds, a definite penis (and also clitoris) is found in ratites, anseriformes and a few other birds. Ostrich and ducks only have an erectile penis.

1. Testes. Two oval and white testes, each attached to the anterior end of a kidney by **mesorchium** (fold of peritoneum). The right testis is slightly smaller than the left.The testes increase many times in size during the breeding season, usually under regulation by changes in illumination. The weight of testis is about 1000 times greater in the breeding season than it is in the non-breeding period, when it contains only spermatogonia. Both testes have masses of coiled **seminiferous tubules** with groups of **interstitial cells** between them. The cells of lining epithelium of seminiferous tubules produce **spermatozoa** by undergoing spermatogenesis. The interstitial cells are also called **Leydig cells** and produce sex hormone testosterone. The seminiferous tubules join to form a long epididymis. The testes are permanently retained within the body cavity. There is no temperature-regulating scrotum, as found in mammals. It is supposed that the abdominal air sacs may reduce testis temperature and be concerned indirectly with spermatogenesis, but this is not yet proved.

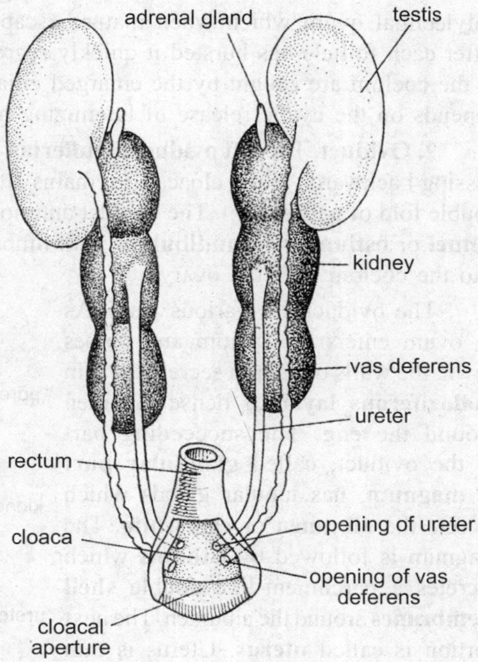

Fig. 26.47. Pigeon. Male urinogenital system.

2. Vasa deferentia. From the inner border of each testis arises a convoluted Wolffian or mesonephric duct, its anterior end is an **epididymis** and the rest is a vas deferens**.** The epididymis is connected with the seminiferous tubules of the testis by extremely fine tubules, the **vasa efferentia**. Vas deferens runs backwards along the outer side and parallel to the ureter of that side to open into the urodaeum on a very small **erectile papilla** posterior to the ureter. It is the only copulatory organ of most birds. For temporary storage of spermatozoa, the hind end of each vas deferens becomes swollen, called the **seminal vesicle**.There is no copulatory organ in pigeon.

3. Female Reproductive Organs

The adult female has only the left **ovary** and left **oviduct** as their main reproductive organs. In the embryo there are two ovaries and two oviducts, but during development, the ovary and oviduct of right side become more or less completely atrophied.

1. Ovary. A large-sized, irregular-shaped left ovary occurs at the ventral side of anterior lobe of left kidney. It remains attached to the kidney by **mesovarium**, a double-fold peritoneum. During breeding season, the size of the ovary increases considerably due to the influence of **FSH** and **LH** of anterior pituitary. The ovary secretes **oestrogen** which modifies the accessory sexual organs and behaviour. The surface of ovary remains studded with numerous **follicles** or **ovisacs**

of different sizes and each contains a single **ovum** or **oocyte**. The ova are at various stages of development. These follicles project from the surface of the ovary in birds. Of the large number of oocytes only few ripen to make the large follicles. Each follicle produces a large yolky, polylecithal ovum which when mature escapes by the rupture of the follicle into the coelom. After each follicle has bursted it quickly regresses. There is no corpus luteum. The released ova in the coelom are caught by the enlarged ciliated and muscular funnel of the oviduct. Ovulation depends on the cyclic release of luteinizing hormone (LH).

2. Oviduct. The **left oviduct** (**Mullerian duct**) is a long, broad, thick-walled, convoluted tube passing backwards to the cloaca. It remains attached to the dorsal body wall by **mesotubarium**, a double fold of peritoneum. The oviduct anteriorly has an expanded muscular and ciliated **coelomic funnel** or **ostium** or **infundibulum** with fimbriated margin. It opens by a wide, slit-like aperture into the coelom near the ovary.

The oviduct has various parts. As an ovum enters the ostium and passes down, the walls of ostium secrete the thin **chalaziferous layer** of dense albumen around the egg. The succeeding part of the oviduct, called **glandular part** or **magnum**, has tubular glands which secrete the **albumen** or **egg white.** The magnum is followed by **isthmus** which secretes a parchment-like double **shell membranes** around the albumen. The next portion is called **uterus**. Uterus is thin walled and is lined by **nidamental glands** which form an outer hard **calcareous shell** around the shell membranes. The last portion of oviduct is **vagina** which is muscular, thick-walled and contains mucus secreting unicellular glands which secrete pigment, external cuticular layer of the cell and mucus for expelling the egg. The oviduct opens into urodaeum.

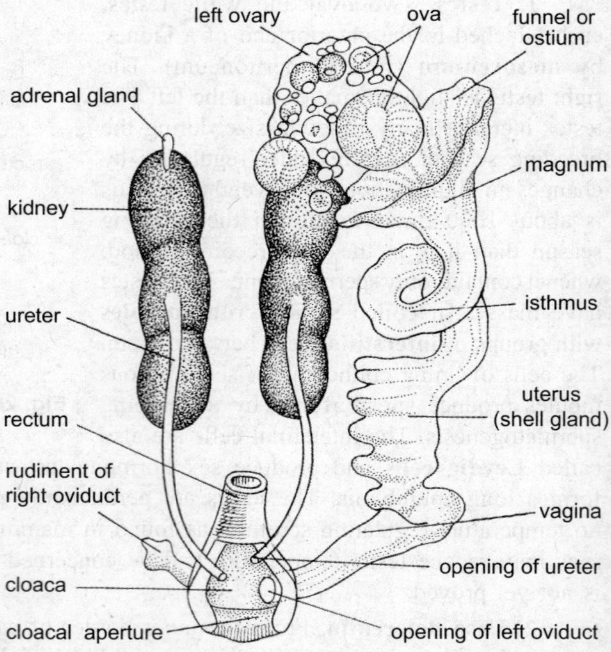

Fig. 26.48. Pigeon. Female urinogenital system.

As much as one-third of the weight of calcium in the whole skeleton is needed by the pigeon for its two eggs. A reserve is collected as the ovarian follicles mature. The oestrogen they produce increases the uptake of calcium from the food and stimulates its deposition in the bones. After ovulation the oestrogen level falls, the calcium is then mobilised from the bones, probably by parathyroid hormone, and its concentration in the blood becomes very high, until used by the eggs.

In many birds the egg shells are coloured, pigments derived from bile by breaking down haemoglobin are deposited in the shells by nidamental glands of the oviduct. In pigeon, there occurs a rudimentary right oviduct close to the right side of cloaca and sometimes vestiges of right ovary.

Significance of One Ovary

Loss of an ovary and oviduct in birds reduce weight as well as decrease the number of eggs produced, which is an advantage in flight. Further, the retention of a single ovary also helps in the safe manipulation of large eggs with breakable shells.

COPULATION

In copulation the proctodaea of male and female are everted and pressed together so that the sperm is ejaculated direct into the urodaeum of the female and travel up the oviduct. The **fertilisation** takes place in the upper part of the oviduct, before the coverings (albumen, shell membranes and calcareous shell) are formed around the egg.

The female pigeon being **oviparous** lays two eggs in a rough nest. The eggs are incubated by both parents for 14 days at a temperature of 38° to 40°C. At the end of incubation the youngs called **squabs** hatch by breaking the shell with a **caruncle**. The squabs are **altricial** being naked and unable to fend for themselves. They are fed for a time by both parents on **pigeon milk** formed by the epithelium of the crop. Crop gland activity is controlled by **prolactin** hormone.

IMPORTANT QUESTIONS

Long Answer Questions

1. Describe the general characteristics of Aves. What features in the anatomy of *Columba livia* (pigeon) are adapted to its habits?
2. Give an account of structure and development of a typical feather.
3. Describe the various types of feathers found in *Columba livia* (pigeon).
4. Give an account of exoskeleton in birds.
5. Describe the skeleton of fowl (*Gallus*) in detail.
6. Give an illustrated account of flight muscles of pigeon.
7. Give an account of perching mechanism in birds.
8. What is synsacrum? Name the component bones of synsacrum. Draw and label the parts. Mention its functional significance.
9. Describe the food, feeding habits and digestive system in the common rock pigeon.
10. Describe the respiratory organs and respiratory mechanism in pigeon. Justify the modifications in respiratory organs in relation to the aerial mode of life in birds.
11. Give the structure of heart in birds. Discuss how blood circulation takes place in birds.
12. Describe the arterial and venous system of a pigeon.
13. Describe the structure of brain in pigeon and discuss its peculiarities in detail.
14. Sense of vision in birds is better developed than in other animals. Support this view giving suitable details.
15. Respiratory system in pigeon is ideally suited to its aerial mode of life. Discuss.
16. Describe how lungs of birds are adapted to aerial mode of life.
17. Explain how the pectoral girdle of birds is suited for flight.
18. Describe the urinogenital system of a bird. Discuss the significance of presence of single ovary in birds.

Short Answer Questions

1. Briefly give an account of exoskeleton in pigeon.
2. Mention the functions of feathers in birds.
3. Explain the structure of quill feather in birds.
4. Explain the sternum in birds.
5. Draw a neat and labelled diagram of pectoral girdle of a bird.
6. Give a brief account of flight muscles of birds.
7. Describe briefly the flight mechanism in a bird.
8. Answer briefly a pigeon can have sustained flight, while a fowl cannot. Explain.
9. Explain the process of digestion in a bird.
10. Write short notes on the following : (*i*) Air sacs in birds; (*ii*) Feathers; (*iii*) Flight muscles; (*iv*) Perching mechanism; (*v*) Syrinx.
11. Write a note on the long trachea in birds.
12. Distinguish between pecten and ectodermal conus.

Very Short Answer Questions

1. What is filoplume?
2. What is synsacrum?
3. What is pygostyle?
4. How many quill feathers are present in one forelimb of pigeon?
5. Distinguish between remiges and rectrices.
6. Discuss the pubic symphysis in birds.
7. What is pterolysis?
8. How is avian eye suited for its special problems?
9. Show how alimentary canal of birds is adapted for their mode of feeding.
10. Name the air sacs in a bird.

11. State two important functions of air sacs in birds.
12. What is pecten?
13. What is ala spuria?

Multiple Choice Questions

1. The study of birds is known as
(*a*) Ornithology (*b*) Herpetology
(*c*) Ichthyology (*d*) Oology.
2. Study of bird's eggs is
(*a*) Ornithology (*b*) Herpetology
(*c*) Oology (*d*) Ichthyology.
3. Which of the following is not true for birds?
(*a*) Flying (*b*) Feathered
(*c*) Quadruped (*d*) Possess wings.
4. The common rock pigeon does not nest on
(*a*) Trees (*b*) Fissures of rocks
(*c*) Forts (*d*) Houses.
5. Jaws in *Columba livia* are covered by
(*a*) Nictitating membrane
(*b*) Cere
(*c*) Rhamphotheca
(*d*) Operculum.
6. Nostrils in pigeon are overhung by
(*a*) Feathers
(*b*) Operculum or cere
(*c*) Rhamphotheca
(*d*) Nictitating membrane.
7. The foot of birds is covered by
(*a*) Epidermal glands (*b*) Dermal plates
(*c*) Dermal scales (*d*) Feathers.
8. The arrangement and distribution of feathers on birds is known as
(*a*) Plumage (*b*) Pterolysis
(*c*) Pterylae (*d*) Aptera.
9. The remiges attached to second digit of hand of birds are
(*a*) Digitals (*b*) Manuals
(*c*) Metacarpals (*d*) Cubitals.
10. Down feathers are present in
(*a*) Newly hatched pigeons
(*b*) Male pigeons
(*c*) Adult pigeons
(*d*) Female pigeons.
11. Bastard wing is attached to
(*a*) Hand (*b*) Second digit
(*c*) Forearm (*d*) Pollex or thumb.
12. In some birds, at the junction of quill and rachis, there appears to be a structure very similar to another feather
(*a*) Rachis (*b*) Hyporachis
(*c*) Inferior umbilicus
(*d*) Superior umbilicus.
13. Which of the following types of feather is seen in the young ones of birds?
(*a*) Quill (*b*) Contour
(*c*) Filoplume (*d*) Down.
14. The only cutaneous glands in pigeons are
(*a*) Sweat glands
(*b*) Mammary glands.
(*c*) Sebaceous glands
(*d*) Preen glands or uropygial gland.
15. The uropygial gland of bird is a simple
(*a*) Branched saccular gland
(*b*) Saccular gland
(*c*) Branched tubular gland
(*d*) Coiled tubular gland.
16. The type of vertebrae seen in pigeon is
(*a*) Acoelous
(*b*) Procoelous
(*c*) Amphicoelous (*d*) Heterocoelous.
17. Cervical vertebrae of birds are
(*a*) Heterocoelous (*b*) Amphicoelous
(*c*) Procoelous (*d*) Acoelous.
18. The number of caudal vertebrae seen in the synsacrum of pigeon is
(*a*) 11 (*b*) 7
(*c*) 5 (*d*) 2.
19. Select the movable bones present in the skull of pigeon
(*a*) Quadrate (*b*) Upper jaw
(*c*) Frontal (*d*) Basioccipital.
20. Which of the following is found to constitute the roof of the skull in pigeon?
(*a*) Jugal (*b*) Nasal
(*c*) Basioccipital (*d*) Rostrum.
21. The coelom in birds is divided into thoracic and abdominal cavity by
(*a*) Transverse septum
(*b*) Oblique septum
(*c*) Diagonal septum
(*d*) Longitudinal septum.
22. The foramen triosseum is an aperture between
(*a*) Clavicle and coracoid
(*b*) Coracoid and scapula
(*c*) Clavicle and scapula
(*d*) Clavicle, coracoid and scapula.
23. Which of the following is not a flight muscle?
(*a*) Pectoral muscles
(*b*) Gastrocnemius
(*c*) Coraco-brachialis longus
(*d*) Coraco-brachialis.

24. Pectoralis minor is a
(*a*) Perching muscle
(*b*) Elevator muscle
(*c*) Visceral muscle
(*d*) Depressor muscle.

25. The pigeon's milk is manufactured in
(*a*) Buccal cavity (*b*) Oesophagus
(*c*) Crop (*d*) Stomach.

26. In birds gizzard is used for
(*a*) Chewing food
(*b*) Grinding food
(*c*) Mixing food
(*d*) Digesting food.

27. What fraction of foodstuff is converted into energy by birds?
(*a*) Half (*b*) One-third
(*c*) Two-thirds (*d*) One-fourth.

28. The number of air sacs present in bird is
(*a*) 6 (*b*) 8
(*c*) 9 (*d*) 10.

29. The voice box of birds is situated in the
(*a*) Larynx (*b*) Bronchus
(*c*) Syrinx (*d*) Pharynx.

30. The voice box in birds is known as
(*a*) Larynx (*b*) Syrinx
(*c*) Pharynx (*d*) Lungs.

31. The heart of pigeon is
(*a*) 2 chambered
(*b*) 3 chambered
(*c*) Incompletely divided four chambered
(*d*) Completely divided four chambered.

32. Coccygeo-mesenteric vein collects blood from
(*a*) Kidney
(*b*) Tail
(*c*) Pelvis
(*d*) Cloaca and rectum.

33. Which of the following animal groups has richest blood?
(*a*) Amphibia (*b*) Reptiles
(*c*) Birds (*d*) Mammals.

34. The meninges covering the brain of a bird are
(*a*) Duramater and piamater
(*b*) Duramater and pia-arachnoid layer
(*c*) Duramater and arachnoid layer
(*d*) Piamater and arachnoid layer.

35. In pigeons gray matter is internal in
(*a*) Olfactory lobes
(*b*) Cerebral hemisphere
(*c*) Cerebellum
(*d*) Optic lobe and medulla oblongata.

36. The cavity of midbrain is called
(*a*) Diacoel (*b*) Optocoel
(*c*) Iter (*d*) IV Ventricle.

37. In birds voluntary movements are controlled by
(*a*) Cerebellum (*b*) Optic lobes
(*c*) Medulla oblongata (*d*) Crura cerebri.

38. Which of the following parts of the eye is characteristic of birds?
(*a*) Lens (*b*) Retina
(*c*) Pecten (*d*) Iris.

39. In pigeons retina possesses
(*a*) Only cones
(*b*) More rods than cones
(*c*) Rods and cones equally
(*d*) More cones than rods.

40. Pecten is absent in which bird?
(*a*) Peacock (*b*) Pigeon
(*c*) Crow (*d*) Kiwi.

41. The retina of nocturnal birds is composed of
(*a*) Mainly cones (*b*) Mainly rods
(*c*) Cones and rods in equal proportion
(*d*) Larger proportion of cones than rods.

42. Owls have well developed cochlea, hence, they are able to
(*a*) See even in dark
(*b*) Smell the presence of prey
(*c*) Locate sound
(*d*) Taste the food.

ANSWERS

1. (*a*)	2. (*c*)	3. (*c*)	4. (*a*)	5. (*c*)	6. (*b*)	7. (*a*)	8. (*b*)
9. (*a*)	10. (*a*)	11. (*d*)	12. (*b*)	13. (*d*)	14. (*d*)	15. (*b*)	16. (*d*)
17. (*a*)	18. (*b*)	19. (*a*)	20. (*b*)	21. (*b*)	22. (*d*)	23. (*b*)	24. (*b*)
25. (*c*)	26. (*b*)	27. (*b*)	28. (*c*)	29. (*c*)	30. (*b*)	31. (*d*)	32. (*d*)
33. (*c*)	34. (*b*)	35. (*d*)	36. (*c*)	37. (*a*)	38. (*c*)	39. (*d*)	40. (*d*)
41. (*b*)	42. (*c*)						

27 Chapter

Class 8. Aves (The Birds) : General Characters, Classification and Representative Types

Contents of this Chapter

The birds get a living by moving in the air, which is the most difficult medium of all and also the most rewarding for freedom of movement. The birds are active animals. Flight enables them to avoid their enemies and to seek new habitats. They have low weight and high power. The lightness and strength are achieved by using thin sheets and tubes of bone and extending air sacs into them. There are no teeth or heavy jaws and the food eaten is quickly digested and of high energy value. Accumulation of fat is very little. Youngs develop outside the body. They have high temperatures and metabolic rates, rapid heartbeats, and high performance of respiratory and muscular systems. They have more elaborate social and family habits. The distinctive characters of birds are as follows :

GENERAL CHARACTERS

1. Birds are feather-clad, warm-blooded, air-breathing, oviparous, bipedal flying vertebrates.

2. **Body** is roughly boat-shaped and divisible into head, neck, trunk and tail. Jaw bones projected to form toothless **beak** (**bill**). Tail is short and stumpy.

3. **Limbs** are two pairs. The forelimbs are modified into **wings**, each bearing three clawless digits and provided with feathers for flight. The hindlimbs are adapted for walking, running, scratching, perching, swimming or wading, and bearing four toes. First or hallux directed backward.

4. **Exoskeleton** is epidermal and horny. Feathers provide warmth. Legs are covered by **scales**. **Claws** on toes and sheath on beaks. No skin gland except the oil glands at the base of tail.

5. **Skin** is dry.

6. **Pectoral muscles** are well developed used for flight.

7. **Endoskeleton**. The bones, forming the skeleton, are spongy, light in weight containing air cavities. Long bones pneumatic and without marrow.

8. **Skull** is **monocondylic** having single rounded occipital condyle. Cranium large and sutures indistinct.

9. **Lower jaw** is composed of five or six bones.

10. Modern birds have no teeth. A horny beak is present.

11. Vertebral column short. Centra of vertebrae are **heterocoelous**. Cervical vertebrae with reduced pointed cervical ribs, first thoracic vertebrae free and some are fused together, last thoracic + sacral vertebrae, lumbars and anterior caudals are fused to form a **synsacrum**. The posterior caudal vertebrae are usually fused to form a **pygostyle**.

12. **Vertebral ribs** are double headed, provided with **uncinate process**.

13. **Sternum** is broad, usually with a longitudinal mid-ventral **keel** for the attachment of flight muscles.

14. Both clavicles and a single interclavicle are fused to form a U- or V-shaped **furcula**.

15. In pectoral girdle, the coracoid is usually more or less pillar-like and the scapula is sabre-shaped.

16. **Pelvic girdle** is large, elongated and fused with the synsacrum along its entire length. Pubic and ischiatic symphysis absent and acetabulum is perforated.

17. Proximal carpals are free and distal carpals and metacarpals are united to form a **carpo-metacarpus**.

18. Proximal tarsals are fused with the tibia to form a **tibio-tarsus** and metatarsals with the distal tarsals to form the **tarso-metatarsus**.

19. In the leg, the **ankle joint** is inter-tarsal.

20. The gullet is dilated into a **crop** for the storage of food. Stomach is divided into a glandular **proventriculus** and a muscular **gizzard.** A pair of small **rectal caeca** are present at the junction of small intestine and rectum. **Cloaca** is three chambered.

21. Lungs are spongy and non-distensible with thin-walled air sacs.

22. The voice is produced by the **syrinx** situated at the junction of trachea with the bronchii. Lyrinx is without vocal cords.

23. The **heart** is four chambered. Sinus venosus and truncus arteriosus are absent. Only right aortic arch is present in the adult. Renal portal system is vestigial. RBCs are oval, nucleated and biconvex. Homoiothermous (warm blooded).

24. Kidneys are **metanephric**, three lobed and ureters open into the cloaca. Urinary bladder is absent. Nitrogenous wastes are uric acid **(urecotelic)**.

25. Brain is smooth and large. The cerebral hemispheres and cerebellum are well developed and the optic lobes are displaced laterally. Olfactory lobes small with poorly developed sense of smell.

26. Eyes are large with nictitating membrane, pecten and surrounding sclerotic plates.

27. Sexes are separate and **sexual dimorphism** is well marked. Male has a pair of abdominal testes. Copulatory organ is absent except ratites and Anseriformes. Only left ovary is present. The right ovary and oviduct are more or less completely atrophied.

28. **Fertilisation** is internal. **Oviparous**. Ovum large containing much yolk invested with albumen, shell membranes and shell.

29. The embryo has an amnion, chorion, allantois and a large yolk sac. Newly hatched young is either **precocious** or **nudifugous** (able to run or swim and can obtain their own food) or **altricious** or **nudicolous** (more or less naked and depend on parents for their food).

30. **Parental care** is highly developed.

CLASSIFICATION

It is estimated that the total number of bird species that have existed is 154,000, of which 9,600 are alive today (**Brodkorb**, 1971). **Wetomore** (1960) described the 34 orders, out of which 27 orders are of living birds (2 orders of these have recently become extinct) and 7 orders of fossil birds. Class Aves is divided into 2 subclasses. The classification is adopted from **Storer** and **Usinger** (1957).

SUBCLASS I. ARCHAEORNITHES

(Gr., *archios* = ancient + *ornithos* = bird)

1. Extinct Jurassic birds of Mesozoic age, about 155 million years ago.
2. Wings with little power of flight.
3. Teeth embedded in sockets were present in both jaws.
4. Tail long with more than 13 (18–20) caudal vertebrae bearing rectrices arranged in two lateral rows.
5. Pygostyle was absent.
6. Vertebrae were amphicoelous. Thoracic ribs without uncinate process. Abdominal ribs present.
7. Sternum without keel.
8. Three carpals and metacarpals were free, and three clawed digits.

Order Archaeopterygiformes

Examples : *Archaeopteryx lithographica* and *Archaeornis siemensi* found in the Upper Jurassic limestone of Solenhofen in Bavaria. Each was about the size of a crow.

SUBCLASS II. NEORNITHES

(Gr., *neos* = modern + *ornithos* = *birds*)

1. Living as well as extinct birds.
2. Wings usually well developed used for flight with a few exceptions.
3. Tail is short, ends in a pygostyle. Rectrices are arranged in a semicircle around the pygostyle.
4. Teeth are absent except in few fossil birds.
5. Vertebrae are heterocoelous in living forms.
6. Caudal vertebrae are 13 or less. Few caudal vertebrae free and rest fused into a pygostyle.
7. The sternum is well developed and usually provided with a keel or carina.
8. The metacarpals are fused with the distal carpals to form a carpo-metacarpus.
9. Except in one case, not more than two digits of the hand bear claws. In nearly all cases claws are absent in the hand.
10. Thoracic ribs with uncinate processes. Abdominal ribs are absent.

The subclass is divisible into four superorders :

SUPERORDER I. ODONTOGNATHAE

(Gr., *odontos* = teeth)

1. Extinct upper Cretaceous birds.

2. Teeth present on jaws, advantageous for catching fish.
3. Brain is of avian type rather reptilian (**Edinger**).

Order 1. Hesperornithiformes

1. Large marine diving and swimming birds.
2. Sharply pointed teeth present in grooves and not in sockets.
3. Premaxillae without teeth.
4. Vertebrae amphicoelous.
5. Sternum without keel.
6. Shoulder girdle was much reduced.

Examples : *Hesperornis*, *Enaliornis*, *Baptornis* and *Neogoeornis*.

Order 2. Ichthyornithiformes

1. Wings well **developed**.
2. Brain was **of** avian type.
3. Whether **teeth** were present is not definite.
4. Neck vertebrae amphicoelous.
5. Sternum with well developed keel.

Examples : *Ichthyornis*, *Apatornis* belonged to upper Cretaceous of North America.

SUPERORDER II. PALAEOGNATHAE OR RATITAE

(Gr., *palaios* = old + *gnathos* = jaw; L., *ratis* = raft)

1. Large sized, flightless, walking or running birds.
2. Wings are reduced or absent. Feathers without interlocking mechanism.
3. Tail feathers irregularly arranged or absent. Teeth are absent.
4. Except Tinamus and Kiwi, oil gland is lacking.
5. Skull is palaeognathous (dromaeognathous), *i.e.,* vomer is broad and large and located in between palatines.
6. Quadrate is single headed.
7. Sternum devoid of keel or carina, or flat, raft-like.
8. Uncinate processes in ribs are absent or vestigial.
9. Tail vertebrae are free. Pygostyle small or absent.
10. Coracoid and scapula are small and fused at an obtuse angle.
11. Clavicles are small or absent.
12. Except Rhea and Emu, ilium and ischium are not united posteriorly.
13. Pectoral muscles not well developed.
14. Syrinx is absent.
15. Penis in male is large and erectile and female possesses a clitoris.
16. Youngs are precocious.
17. Ratites are not found in India. They are classified into 7 orders.

Order 1. Struthioformes

(Gr., *struthio* = ostrich + form)

1. Large flightless terrestrial birds.
2. Feathers without aftershaft.
3. Head, neck and legs sparsely feathered.
4. Head is comparatively small and the neck is very long and flexible.
5. The beak is short and broad.
6. Only two toes (3rd and 4th) on each foot.
7. Pubic symphysis present.

Examples : *Struthio camelus* (True ostriches of Africa and Western Asia).

Order 2. Casuariiformes

(NL. *casuarius* = genus of cassowary + form)

1. Large flightless terrestrial birds.
2. Head is sparsely feathered, while neck and body densely feathered. Head bears a comb-like structure.
3. Feathers with aftershaft nearly equal to shaft.
4. Wings small or rudimentary.
5. Three toes with typical claws on each foot.

Examples : *Dromaeus* (Emus of New Zealand), *Casuarius* (Cassowaries of Australia and New Guinea).

Order 3. Aepyornithiformes

1. Flightless, terrestrial **extinct** elephant birds.
2. Wings vestigial or tiny.
3. Legs powerful with 4 toes.
4. Eggs were surprisingly large.

Examples : *Aepyornis titan*, *Mulleornis* (Giant elephant birds of Africa).

Order 4. Dinornithiformes

1. Giant flightless, terrestrial, **extinct** birds (extinct about 700 years ago).
2. Feathers with large aftershaft.
3. Wings almost absent.
4. The beaks were short.
5. Legs massive bearing 4 toes.

Examples : *Dinornis maximus* (Moas of New Zealand).

Order 5. Apterygiformes

(Gr., *a* = not + *pteryx* = wing + form)

1. Flightless terrestrial birds.
2. Feathers fluppy hair-like without aftershafts.
3. Wings are rudimentary or degenerate.
4. The beak is long and slender with nostrils at the tip of the maxilla.
5. The eyes are small.
6. The neck and legs are comparatively short.
7. Eggs are relatively largest of all the living birds.
8. Nocturnal, feeding on worms and insects.

Examples : *Apteryx* (Kiwis of New Zealand).

Order 6. Rheiformes

(Gr., *Rhea* = mother of Zeus + form)

1. Flightless terrestrial and swift-running birds.
2. Head, neck and thighs are feathered.
3. Wings are better developed.
4. Legs bear 3 toes with heavy and typical claws.
5. Ischia form a ventral symphysis.

Examples : *Rhea americana* (American ostrich found in South America).

Order 7. Tinamiformes

(NL. *tinamus* = genus of tinamou)

1. Small, terrestrial, essentially cursorial with poor power of flight.
2. Wings are short and rounded.
3. Sternum is keeled.
4. Vegetarian.

Examples : *Tinamus* (Tinamou), *Rhyncotus.*

SUPERORDER III. IMPENNAE

Order 1. Sphenisciformes

(Gr., *spheniscus* = wedge + form)

1. Modern, aquatic, flightless birds.
2. Wings paddle-like or flippers.
3. Feet are webbed.

Examples : *Aptenodytes* (Penguins).

SUPERORDER IV. NEOGNATHAE

(L., *carina* = a keel)

1. Modern, usually small, flying birds.
2. Wings well developed.
3. Feathers having interlocking mechanism.
4. Rectrices are present and regularly arranged.
5. Pterylae are regular, while irregular in ratites.
6. Oil gland is present.
7. Skull is **neognathus**, *i.e.*, vomer short and palatines meet with each other.
8. Quadrate is double-headed.
9. Sternum with a well developed keel.
10. Ribs with uncinate processes.
11. Pygostyle present.
12. Scapula and coracoid unite at a right angle.
13. Ilium and ischium are posteriorly united.
14. Pectoral muscles well developed.

This superorder includes several orders which are as follows :

Order 1. Gaviiformes

(L., *gavia* = sea mew + form)

It includes marine birds.

1. Legs are short and at the posterior end of the body.
2. Tail comprises 18–20 short stiff feathers.
3. Toes are fully webbed.
4. Patella reduced.
5. Powerful in flight.

Examples : *Gavia immer* (Common loon).

Order 2. Podicipeformes or Colymbiformes

(Gr., *kolymbos* = diving bird)

1. Freshwater divers.
2. Feet are lobed.
3. Tail comprises a tuft of down feathers.

4. Patella is large.
5. Tarsus is compressed.

Examples : *Podicipes*, *Podilymbus*.

Order 3. Procellariiformes

(L., *procella* = a tempest + form)

1. Nostrils are tubular.
2. Horny sheath of bill consisting of several plates.
3. Skull with large nasal glands.
4. Hind toe is vestigial or none.
5. Feathers compact and oily in texture.
6. Wings are long and narrow.
7. Strictly marine, nest usually on islands.

Examples : *Diomedea* (Albatrosse), *Puffinus*, *Procellaria* (Petrels), sheawaters.

Order 4. Pelecaniiformes

(L., *pelicanus* = pelican + form)

1. Aquatic fish-eating and colonial nesters.
2. Nostrils are vestigial or absent.
3. Presence of a gular pouch on the throat.
4. Four toed webbed feet.
5. Beaks are long with wide gapes for catching and swallowing the struggling prey.

Examples : It includes pelicans, darters, gannets, and cormorants. *Pelecanus* (Pelican), *Phalacrocorax* (Cormorant), *Anhinga* (Indian darter).

Order 5. Ciconiiformes

(L., *ciconia* = a stork + form)

1. Long snake-like neck and long-legged birds.
2. Feathers are decorative.
3. Beaks abruptly decurved at middle in flamingos.
4. Toes are without web.
5. Young born naked.
6. Food comprises fish and other aquatic animals.

Examples : It includes marshy wading birds: *Ardea* (Great blue heron), *Phenicopterus* (Flamingo), *Bubulcus* (Egret), *Nycticorax* (Night Heron), *Platalea* (Spoonbill), *Ciconia* (Stork).

Order 6. Anseriformes

(L., *anser* = goose + form)

1. Beak or bill is broad, covered with soft cornified epidermis.
2. Margins of the beak containing many transverse horny ridges (lamellae).
3. Tongue is fleshy.
4. Legs are short and feet are webbed.
5. Tail is usually short, consisting of many feathers.

Examples : Aquatic birds such as Ducks, Geese, Swans. *Anas* (Wild duck), *Nettion crecca* (Common teal), *Anser indica* (Bar-headed goose), *Cygnus* (Swan).

Order 7. Falconiformes

(L., *falco* = falcon + form)

1. Beak is short and curved at the tip.
2. Mandibles are sharp edged.
3. Feet usually with sharp curved claws adapted for grasping and holding the prey.
4. Diurnal and strong fliers.

Examples : Vultures, Kites, Hawks and Falcons. *Milvus migrans* (Pariah kite), *Haliaster indus* (Brahminy kite), *Astur badius* (Sparrow hawk), *Pseudogyps* (Vulture), *Falco pelegrinus* (Falcon), *Aquila* (Eagle).

Order 8. Galliformes

(L., *gallus* = a cock + form)

1. Terrestrial game birds with short and powerful flight.
2. Beak is short.
3. Feathers with aftershaft.
4. Feet usually massive adapted for scratching and running.
5. Graminivorous.

Examples : Game birds. *Gallus* (Jungle fowl), *Pavo cristatus* (Peacock), *Coturnix* (Quail), *Francolinus* (Grey partridge), *Phasianus* (Pheasants).

Order 9. Gruiformes

(L., *grus* = crane + form)

1. Small, medium and large-sized birds.
2. Flightless or weak fliers or strong fliers.
3. Feathers with aftershaft.
4. Legs long and beaks heavy.

Examples : Crane-like wading birds. *Grus* (Crane), *Fulica atra* (Common coot), *Antegone antegone* (Sarus crane), *Choriotis* (bustard), rail.

Order 10. Diatrymiformes

1. Large flightless **extinct** birds.
2. Wings atrophied.
3. Massive head, beak and neck.
4. Foot bears 4 toes.

Example : *Diatryma* (Extinct).

Order 11. Charadriiformes

(NL., *charadrius* = genus of plovers + form)

1. Shore dwelling aquatic birds.
2. Fore- and hindlimbs are slender and enormously elongated.
3. Toes usually webbed.
4. Beaks mud probing.
5. Feathers are dense and firm.

Examples : *Lobivanellus indicus* (Red wattled lapwing), *Hydrophasianus chirugus* (Pheasant tailed jacana), *Tringa glariola* (Sand piper), *Capella* (Snipe), *Larus* (Gull), etc.

Order 12. Columbiformes

(L., *columba* = dove + form)

1. Skin thick and soft.
2. Beak usually short and slender.
3. Tarsus usually shorter than toes.
4. Crop large producing "pigeon milk" to feed young ones.
5. Graminivorous and frugivorous.

Examples : *Columba livia* (Blue rock pigeon), *Goura cristata* (Crowned pigeon), *Streptopelia risoria* (Ring turtle dove), *Streptopelia chinensis* (Spotted dove), *Raphus* (extinct dodo), etc.

Order 13. Cuculiformes

(L., *cuculus* = cuckoo + form)

1. Toes 2 in front and 2 behind, out hind toe reversible.
2. Feet not adapted for grasping.
3. Tail long and beak moderate.
4. Many cuckoos are parasitic, the female laying eggs in nest of other birds.

Examples : *Cuculus canorus* (Cuckoo), *Eudynamis scolopaceous* (Koel), *Centropus sinensis* (Crow pheasant).

Order 14. Psittaciformes

(L., *psitacus* = parrot + form)

1. Feathers are green, blue, yellow or red.
2. Beak stout, narrow, sharp-edged and hooked at the tip.
3. Upper jaw is movably articulated with the skull.
4. Feet zygodactylus, *i.e.,* two toes in front and two toes behind.
5. Outer toe is not reversible.

Examples : *Psittacula eupatria* (Large Indian Parakeet), *Psittacula krameri* (Green parrot), *Melopsittacus* (Budgeriger).

Order 15. Strigiformes

(Gr., *strix* = owl + form)

1. Head large and rounded.
2. Eyes large and directed forwards. Each in a disc of radial feathers.
3. Retina contains principally the rods.
4. Ear opening large, often with flap-like cover.
5. Beak short, feet with sharp claws adapted for grasping.
6. Nocturnal and predators.

Examples : *Bubo bubo* (Great horned owl), *Ketupa zeylonensis* (Brown fish owl), *Tylopus*.

Order 16. Caprimulgiformes

(L., *caprimulgus* = goat sucker + form)

1. Beak is small and delicate.
2. Mouth is wide and margined with long bristle-like sensory feathers.
3. Feathers are soft and lax.
4. Legs and feet small and weak, not adapted for grasping.
5. Nocturnal and insectivorous.

Examples : *Caprimulgus* (Goat sucker), *Chordeiles* (Night hawks), *Phaeloenoptilus* (Whippoor wills).

Order 17. Apodiformes (Micropodiformes)

1. Usually small birds.
2. Legs are very short and feet very small.
3. Wings very long formed of short humerus and long distal segments.
4. Beak is small and weak or slender with long tubular tongue.

Examples : Swifts (*Apus*) are insectivorous, Humming birds (*Trochilus*).

Order 18. Coliformes

1. Small passerine birds.
2. First and fourth toes are reversible.
3. Tail is very long.
4. Insectivorous and frugivorous.

Example : *Colius* (Mouse birds or Colies) is confined to Africa.

Order 19. Trogoniformes

1. Beak short and stout with bristles at the base.
2. Feet small and weak.
3. Feathers brilliant in colour, often green.

Example : *Trogon.*

Order 20. Coraciformes

(Gr., *korax* = crow or raven + form)

1. Beak strong.
2. Third and fourth toes fused at the base.

Examples : Kingfishers and their allies. *Halcyon smyrnensis* (White-breasted kingfisher), *Ceryle rudis* (Pied kingfisher), *Dichoceros bicornis* (Great hornbill), *Tochus birostris* (Grey hornbill), *Upupa epops* (Hoopoe).

Order 21. Piciformes

(L., *picus* = woodpecker + form)

1. Tail feathers are stiff with pointed tips.
2. Beaks stout and powerful.
3. Protrusible tongue is roughened or with barbs at the tip.
4. Two toes in front and two or one behind, not reversible.

Examples : Woodpeckers and their allies. *Dryobates* (Yellow front pied woodpecker), *Brachypternus bengalensis* (Golden beak woodpecker), *Dendrocopos*, etc.

Order 22. Passeriformes

(L., *passer* = sparrow + form)

1. Toes three in front and one behind, adapted for perching.
2. Beaks are adapted for cutting.

Examples : *Passer domesticus* (Common house sparrow), *Corvus splendens* (Common house crow), *Corvus macrorhynchos* (Indian jungle crow), *Acridotheres tristis* (Common myna), *Saxicoloides fulicata* (Indian robin), *Oriolus oriolus* (Golden oriole), *Dicrurus macrocercus* (Black drongo or King crow), *Ploceus philippinus* (Weaver bird or Baya), *Muscicapa* (Flycatchers), Bulbul (*Molpestes*), Swallos (*Hirundo*), *Pica* (Magpie), *Loxia* (Crossbill), *Sturnus* (Starling), *Alauda* (Larks), *Fringilla* (Finches), *Anthus* (Pipits), *Motacilla* (Wagtails), *Certhia* (Tree-creepers), *Parus* (Tits), *Lanius* (Shrikes), *Sylvia* (Warblers).

REPRESENTATIVE TYPES

1. *Archaeopteryx*. Two perfectly preserved specimens of *Archaeopteryx* (Fig. 27.1) were found in Bavaria in the fine grained lithographic limestone of the Upper Jurassic about 140 million years ago. It was discovered by **Andreas Wagner** in 1861. Another skeleton was found in 1877 and a third was also found in 1956 from the same place. Had not the impression of feathers been found, they would have been taken for some bipedal diapsid reptile. *Archaeopteryx*, like *Seymouria*, is an example of mosaic evolution. It is of the size of a crow with a long tapering tail and dwelt in forests; though the body was covered with feathers, its body axis was elongated like that of a lizard. It is an extinct bird and is a transitional stage showing characters of both reptiles and birds.

Reptilian Characters of *Archaeopteryx*

Archaeopteryx presents striking affinities with reptiles which are given here :

1. The bones are solid with no air spaces.

2. The jaws have small equal conical enamel-crowned and thecodont teeth like those of reptiles. There are 26 teeth in the upper jaw and 6 in the lower.

3. The forelimbs are reptilian. There are three well formed digits with separate phalanges, there are 2 phalanges in the first, 3 in the second and 4 in the third digit. The digits bear no feathers but project from the anterior margin of the wing end in curved claws.

4. There are three free metacarpals, there being no carpometacarpus.

5. The sternum is flat without keel.

6. The vertebral column is elongated and vertebrae are free and amphicoelous with simple concave articulation as in reptiles.

7. There is a sacrum of only 5 or 6 vertebrae as in dinosaurs.

8. The vertebrae bear thin, single-headed ribs with no uncinate processes.

9. Besides the cervical and thoracic ribs, there are abdominal ribs or gastralia in the ventral wall of the abdomen, as in *Sphenodon* and crocodilians.

10. The bones of the pelvic girdle are not fused, the ilium is long, the pubis is directed posteriorly and probably there is an ischiac symphysis.

11. Hindlimb has separate tibia and fibula of about equal size with no reduction in size of fibula.

12. Proximal tarsals are free.

13. There is an elongated tail with 20 free caudal vertebrae tapering gradually to the distal end, there being no pygostyle.

14. The cerebral hemispheres were smooth, long and narrow and cerebrum was small, thus, the brain is reptilian.

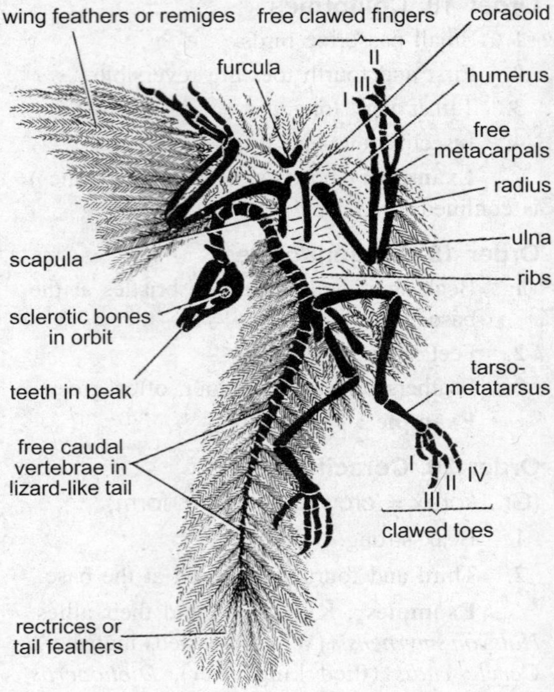

Fig. 27.1. Fossil *Archaeopteryx*. Berlin specimen.

Avian Characters of *Archaeopteryx*

1. The skull is large, monocondylic and bones are fused.

2. There is short blunt beak formed by elongation of both the jaws.

3. The eyes are large and avian with a ring of sclerotic bones.

4. Flight feathers are attached only to the back of ulna and manus, there are 6 to 7 primaries and 10 secondaries.

5. In the pectoral girdle the coracoid is small, scapula is slender and curved and they are at right angles as in birds.

6. There is U-shaped furcula formed by fused clavicles.

7. Distal tarsals are fused with the metatarsals to form a slender tarsometatarsus.

8. There are four clawed digits with 2, 3, 4, 5 phalanges, with the hallux small, directed backwards and opposable.

9. Rectrices are attached to both sides of the tail unlike any bird.

10. Recent studies have shown that *Archaeopteryx* was endothermal or warm-blooded as other birds.

Archaeopteryx could not fly efficiently but could only volplane, from tree to tree in forests, because the sternum is small, flat and without a keel, the wing spread is small, and it has an elongated, feathered lizard-like tail. It shows that birds arose from bipedal, arboreal ancestors and it also shows the gliding origin of flight in birds.

Significance of *Archaeopteryx*

Many characteristics of birds show close resemblance to the archosaurian reptiles. *Archaeopteryx* had achieved some powers of flight or gliding, but they were less specialised for the purpose than are modern birds. The whole body axis was still elongated and lizard-like. Its fossil and other like fossils suggest that the birds arose from a race of bipedal arboreal reptiles, living in forests and accustomed to running, jumping and gliding among the branches. There has been much controversy about the origin of flight, some maintaining that the earliest birds were terrestrial and used the wings to assist in running, leading eventually to a take-off, perhaps at first for short distances. A recent theory is that the enlarged feathers on the back of the hands may have served as nets to assist in catching insects (**Osborn,** 1979).

Archaeopteryx showed both reptilian and avian characters. It was certainly not a bird like the modern birds and also it was not a true reptile. Except feathers it resembled to archosaurian reptiles. It was presumably not the immediate ancestor of modern birds. Its immediate descendants are also not known. It was distinctly separate from the birds due to presence of non-pneumatised bones, forelimb ended in 3 clawed digits, with separate metacarpals and phalanges and elongated tail. *Hesperornis* and *Ichthyornis*, cretaceous birds, were distinctly birds, except they had reptile-like teeth. Hence, *Archaeopteryx* was placed in a separate subclass **Archaeornithes** since it had certain different characters not found in fossil birds and modern birds placed in **Neornithes**.

COMMON BIRDS OF INDIA

The Indian continent, including India, Pakistan, Bangladesh, Nepal, Sri Lanka and Myanmar, contains one of the richest and most varied bird fauna. It includes about 2,500 species and subspecies out of which nearly 500 are winter visitors from North.

Some of the largest Indian birds include the Sarus Crane (*Antigone antigone*) standing one and a half metres, and the Himalayan Bearded Vulture with a wingspread of nearly two and a half metres. Among the smallest Indian birds are Tickell's Flower pecker (*Dicaeum erythrorhynchos*) and the sun birds, about half the size of the House Sparrow or even less. The pheasants and the sun birds are noted for their glistening colours and brilliancy of plumage. The most accomplished Indian songsters are the Greywinged (*Tardus boulboul*) of the Himalayas, the Malabar Whistling Thrush (*Myiophoneus horsfieldii*) and the Shama (*Copsychus malabaricus*). The best talker is certainly the Hill Myna (*Gracula religiosa*).

1. House Sparrow. *Passer domesticus* (Fig. 27.2) is commonly known as **house sparrow** and in Hindi it is called **gauriyya**. It is a small bird growing up to 10 cm to 16 cm in length. The upper surface is earthy brown streaked with blackish and fulvous and underparts are whitish. Sexual dimorphism is distinct. The female is ash white and the male is earthy brown with blackish throat and breast and white abdomen. Beak is conical and eyes small. Feet are adapted for perching with three toes anterior and the first toe of hallux posterior. Sparrows are omnivorous but chiefly granivorous. Food consists chiefly of seeds and grain gleaned on the ground, but they also raid ripening crops of wheat and other cereals. It is an unfailing commensal on man. In winter these birds collect in large flocks to feed in and around cultivation. Breeding occurs almost throughout the year. 3 to 5 pale white greyish eggs are laid at a time and successive broods are often raised. These are useful to agriculture by destroying large number of insect pests. Nest is a collection of straw and rubbish stuffed into a hole in a wall. *Passer domesticus* is distributed world-wide except the Andamans and Nicobars.

2. Crow. *Corvus splendens* (Fig. 27.2) is commonly known as **house crow** in English and in Hindi it is called **kowwa** or **kag**. It is a most familiar bird of Indian towns and villages, growing in length from 32 to 42 cm. The body is covered with a more or less black plumage, while the neck and breast are grey coloured. Each leg bears 4 clawed toes, three directed forwards and one backwards. Feet are adapted for perching. Omnivorous. It eats almost anything–dead rat, kitchen refuse, fish, eggs, locusts, termites, fruits and grain, etc. Nest is a platform of twigs with a cup-like depression lined with tow, coir fibre, etc. The koel commonly lays its eggs in the nests of crows. It is a useful scavanger and an unfailing commensal of man. Although they destroy locusts and other injurious insects when they are swarming but also raid ripening crops such as wheat and maize and do considerable damage to fruits in orchards. *Corvus splendens* is distributed throughout India, Pakistan, Sri Lanka and Myanmar.

3. Weaver bird or Baya. *Ploceus philippinus* (Fig. 27.2) is commonly known as weaver bird or baya. It looks like sparrow and lives in flocks. It is among the world's best nest builders. Elaborately woven bottle-shaped nests are commonly seen hanging from tree branches in the country-side. Weaver birds damage cereal crops.

4. Common myna. *Acriodotheres tristis* (Fig. 27.2) is commonly known as common myna. It measures about 22.5 cm in length. The colour is perky dark brown getting paler on the underparts. Head, throat and breast are glossy black. Beak and legs are yellow. There is a small patch of bare yellow skin behind each eye. When it flies, a white patch or bar becomes conspicuous on each wing. Sexes are alike. The 'bank myna' (*Acridotheres ginginianus*) or the 'ganga myna' in Hindi, is similar to common myna or 'deshi myna' (*A. tristis*), but its colour is pale bluish grey instead of brown and the naked skin around eyes is brick-red instead of yellow. Mynas are carnivorous eating fruits, insects and kitchen scraps. *A. tristis* is said to be the grasshopper hunter. It is common throughout Indian Region. It is a confirmed associate of man and follows him wherever he opens new settlements.

5. Hill myna. *Gracula religiosa* (Fig. 27.2) is commonly known as **pahari myna** or **bolti myna** in Hindi. The bird is about 30 cm in length. A glossy jet black myna with a conspicuous white patch on the wings, yellow bill (beak) and legs, and bright orange-yellow patches of naked skin and wattles on the head. Both the sexes are similar. Hill myna is arboreal and seen in pairs or noisy flocks in hill forest. It feeds on the various wild figs in company with green pigeons, hornbills and other fruit-earing birds. It has a habit of settling on bare tops and dead trees in forest clearings at sunset and uttering its loud, sharp, creaky shrieks. In flight the wings produce a metallic whirring noise, as in green pigeons. An accomplished mimic and talker, and much prized as a cage bird. Nesting season is March to October. Nest is a collection of grass, leaves, feathers, etc., stuffed into natural hollows in lofty tree-trunks in forest, usually 10 metres to 20 metres up. Eggs are 2 or 3, deep blue sparsely spotted and blotched with reddish brown or chocolate. Hill myna is restricted and patchy in distribution. In India proper in three distinct areas : (*i*) Himalayan foothills to about 800 metres elevation from Almora to Assam, (*ii*) Chhota Nagpur, Orissa and South East Madhya Pradesh, (*iii*) Western Ghats north to about Mumbai. Also in Andamans, Sri Lanka and South Myanmar.

6. Hoopoe. Hoopoe or *Upupa epops* (Fig. 27.2) is known as 'hudhud' in Hindi. It is about the size of a myna. A reddish fawn coloured bird with black and white zebra markings on back, wings and tail. A conspicuous fan-shaped crest, and long, slender, gently curved bill (beak) which is about 5 cm long. Both the sexes are alike or similar. Singly or pairs usually on the ground in lightly-wooded country. The hoopoe is found practically all over India. It is fond of lawns, gardens and grooves in and around villages and towns. It walks and runs with a quail-like but waddling gait, probing into the soil for food with bill partly open like forceps. When digging, the crest is folded back and projects in a point behind the head. It is flicked open and erected fanwise from time to time. The hoopoe is often found pecking and feeding on insects on the ground. Its name echoes its distinctive calls '*hoo-po*' and *hud-hud-hud*. It emits a soft musical, penetrating *hoo-po* or *hoo-po-po* repeated in runs, often intermittently for 10 minutes at a stretch. It feeds on insects, grubs and pupae, hence, it is beneficial to agriculture. Nesting season principally February to May. It nests in natural tree hollow or hole in wall or ceiling of a building. Nest is lined with straw, rags and rubbish. Eggs 5 or 6, white in colour. Both sexes share in feeding the young. Hoopoe is found throughout India, Bangladesh, Pakistan, Sri Lanka and Myanmar.

7. Koel. Koel or Kokila (Fig. 27.2) is known as *Eudynamis scolopacea*. Its body is slender with a longer tail measuring up to 42 cm in length about the size of a crow. Sexual dimorphism is well marked. **Male** is glistening metallic black all over with a striking yellowish green bill or beak and crimson or blood red eyes. **Female** is brown profusely spotted and barred with white.

Fig. 27.2. Some common Indian birds.

It is very common and widespread bird of gardens and groves, and is perhaps better known by its melodious voice than by its appearance. It is the male bird whose voice is often heard in summer during mango season. The female koel does not sing but utters a sharp and quick repeated 'kik-kik-kik'. Food consists chiefly of banyan and peepal figs, various berries and hairy caterpillars. Koel is a nest parasite and does not build a nest of its own but deposits its eggs in crow's nest leaving them to be hatched, and the young to be reared, by the foster parents. It is entirely arboreal and never descends to the ground. During winter it is silent and, thus, often overlooked and presumed to have migrated and becomes noisy with the advance of hot weather. Koel is distributed throughout India, Pakistan, Sri Lanka and Myanmar.

8. Large Indian parrot. The Alexandrine or the large Indian parakeet, *Psittacula eupatria* (Fig. 27.2), is known as 'hiraman tota' in Hindi. It is about the size of a pigeon, with a slender body and a long pointed tail. The body is covered by a brilliant grass green plumage. The beak is short but stout, sharp-edged, deeply hooked and red. There is a conspicuous maroon patch on each shoulder. The female is green all over, but the male has a rose-pink collar and black throat. Upper mandible is movable on the frontal bone of skull. Feet are adapted for grasping, holding and climbing. There are 2 front toes and 2 hind toes. The outer hind toe is not reversible. Food consists chiefly of fruits. Parrots are gregarious with loud voices. Parrots can copy and speak some words like human being but not as distinctly as the hill myna. Even then they are popular domestic cage birds as they are easily procured. Parrots affect wooded country, orchards and cultivation. Occasionally collects in large flocks which do considerable damage to ripening fruits and standing crops of maize and jowar. Voice deeper and more powerful than that of the commoner Rose-ringed species. Nesting season is chiefly December to April, varying locally. Nest is an unlined hollow in a tree-trunk at moderate heights and up to 30 metres up. Eggs 2 to 4, white, blunt ovals. Both sexes share all domestic duties. Parrots are distributed practically throughout India, Bangladesh, Sri Lanka and Myanmar. In Pakistan apparently only in the environment of Karachi.

9. Kites. *Milvus migrans* (Fig. 27.2) is commonly known as **pariah kite** and in Hindi it is called **cheel**. It is quite large in size and grows up to 60 cm in length. It is brown in colour. The tail is forked, a distinguishing character from all other similar birds. No sexual dimorphism. Both sexes are alike and share in the domestic duties. Food chiefly comprises offal and garbage, earthworms, winged termites, lizards, mice and young birds, etc. It lives in the neighbourhood of human habitations. Nesting season from September to April. Nest untidy platform of twigs, iron wire, tow rags and rubbish up in a large tree or on roof or cornice of a building. Eggs 2 to 4, dirty pinkish white, lightly spotted and blotched with reddish brown. *Milvus* (kite) is found throughout India, Bangladesh, Pakistan, Sri Lanka and Myanmar.

In India, we have two kinds of kites, the common pariah kite (*Milvus migrans*) and the brahminy kite (*Haliaster indus* Fig. 27.2). The brahminy kite is smaller than the other and much handsomer. Its head, neck and breast are pure white, while the rest of the plumage is rich chest-nut brown. Its tail is rounded and not forked or wedged. Largely a scavanger in sea ports. Food chiefly comprises offal, fish, frogs, small snakes, etc. Winged termites emerging from rain-sodden ground are hawked in the air. Nesting season principally December to April. Nest is a loose platform of twigs, lined with green leaves, etc., built up in a large peepul, banyan or similar tree growing near water. Eggs 2, greyish white, speckled and blotched with dingy reddish brown. Both sexes share in the domestic duties. The brahminy kite is found throughout India, Bangladesh, Sri Lanka and Myanmar.

10. Vultures. Two species of vultures are common. The black Pondichery or king vulture *Sarcogyps calvus* (Fig. 27.3) is commonly known as 'raj gidh'. A huge, black turkey-like vulture measuring about 2 metres across the wings. Its deep black colour is relieved by two white patches on upper thighs, a white band on underside of wings, and a white collar at the base of neck.

The naked head, neck and legs are blood red in colour. The stout bill (beak) is hooked at the tip and bears a soft naked cere at the base. The feet are adapted for grasping with sharp claws. Both the sexes are alike. A carrion-feeder, it is timid, cowardly and repulsive-looking. Nesting season December to April. Nest is a massive platform of twigs in the top of a large tree 10 to 13 metres up often close to a village. Egg is single, white, fine textured roundish oval. It is found throughout India, Bangladesh, Pakistan and Myanmar. It is not found in Sri Lanka.

The commonest Indian vulture is the whitebacked or Bengal vulture *Pseudogyps bengalensis* (Fig. 27.3) called 'gidh' in Hindi. A heavy, dirty blackish brown vulture with scrawny, naked head and neck. At rest and while banking in the air, the white back is diagnostic. In overhead flight a whitish band stretching along the underside of wings. Sexes are alike. It is our commonest vulture. A carrion-feeder and useful scavenger on the countryside and in the environs of towns and villages. It collects in large gatherings to demolish animal carcases with astonishing promptness and incredible speed. Nesting season October to March. Nest is a large untidy platform of sticks in the top of a banyan, tamarind or similar tree, often along roadsides or near villages. Egg single, white occasionally speckled and spotted with reddish brown. It is found throughout India, Bangladesh, Pakistan and Myanmar. It is not found in Sri Lanka.

11. Owls. There are many species of owl. Two important species are described here. *Bubo bubo* (Fig. 27.3) is commonly known as '**great horned owl**' and in Hindi it is called **ghughu**. It is a large, heavy and robust bird, measuring up to 60 cm in length. The colour is dark brown streaked and mottled with tawny buff and black. Head is large and bears two conspicuous black ear tufts or horns. Two prominent feather tufts above head look like ears, hence, the name 'great horned owl'. The beak is short, parrot-like. The eyes are large, round, yellow and directed forwards. The legs are fully feathered, which are diagnostic of horned owl. Food consists mainly of small mammals, birds, lizards and other reptiles; also large insects and occasionally even fish and crabs. The bird spends the day resting on the ground under shelter of a bush or on some shady rock projection in a ravine or river. It is mainly nocturnal but it is frequently seen on move during the day time. By maintaining a constant check on the rodents and other destructive vermin they are of great economic value to agriculture and deserve the strictest protection. Nesting season principally November to April. Eggs 3 or 4, creamy white, broad roundish ovals with a smooth texture. Laid without nest on bare soil, on ledge of cliff, or under shelter of bush on level ground. *Bubo bubo* is distributed throughout India, Bangladesh, Pakistan and Myanmar. It is not found in Sri Lanka.

Brown fish owl (*Bubo zeylonensis*) or 'ullo' in Hindi. It is similar to the horned owl in all respects, but its legs are unfeathered which is diagnostic. Sexes are alike. A similar variety is the barn owl or screech owl (*Tyto alba*) about the size of jungle crow. It has a large round head with a monkey-like face, but no ear projections. Sexes are alike. The barn owl is also found throughout India, Bangladesh, Pakistan, Sri Lanka and Myanmar.

12. Peacock. *Pavo cristatus* (Fig. 27.3) is commonly known as common peafowl or peacock. In Hindi it is called **mor** or **mayur**. It is the **national bird of India** and to kill it is penal crime. Sexual dimorphism present. The male bird is about the size of a vulture, very gracious and beautifully coloured. Its ornamented gorgeous ocellated tail is about 1 or 1.5 metres long. The gorgeous ocellated or eyed trail of cock is actually not his tail but abnormally elongated upper tail coverts. Cock has also fan-shaped crest, brilliant metallic blue head, neck and breast. The hen is crested like cock, but lacks the trail and is a sober mottled brown with dull metallic green on the lower neck. Hen lays 3 to 5 eggs. Head bears eyes, nares at the base of short bill (beak) and ear openings. The feet are adapted for scratching and running. Peacock inhabits crop fields, dense scrubs, forests having streams and rivers. Peacocks are shy and polygamous birds usually living in a party or drove of one cock and 4 to 5 hens. They feed on grain, vegetable shoots, insects, lizards, snakes, etc. At night they roost up in lofty trees, and at early dawn the jungle resounds with loud ugly screaming *may-awe* calls of the cocks. The dance of male peacock with its gorgeous tail coverts spread like a fan is very famous. The peacock dances specially on cloudy and rainy days to attract hen. Peacock is found throughout India, Bangladesh and Sri Lanka.

13. Flamingo. *Phoenicopterus* (Fig. 27.3) is commonly called 'Flamingo' and 'Boghans' or 'Charaj baggo' or 'Rajhans' in Hindi. It is a pale rosy-white bird, stands one and a half metres high. Wing coverts are brilliant scarlet with black wing-border. It has elongated bare pink legs, long sinuous neck and heavy pink bill (beak) turned down at an angle (broken) from about half of its length. The toes are webbed like ducks. Head bears eyes, nostrils at the base of beak and decurved beak. The fleshy tongue works like a plunger sieving out the water through the comb-

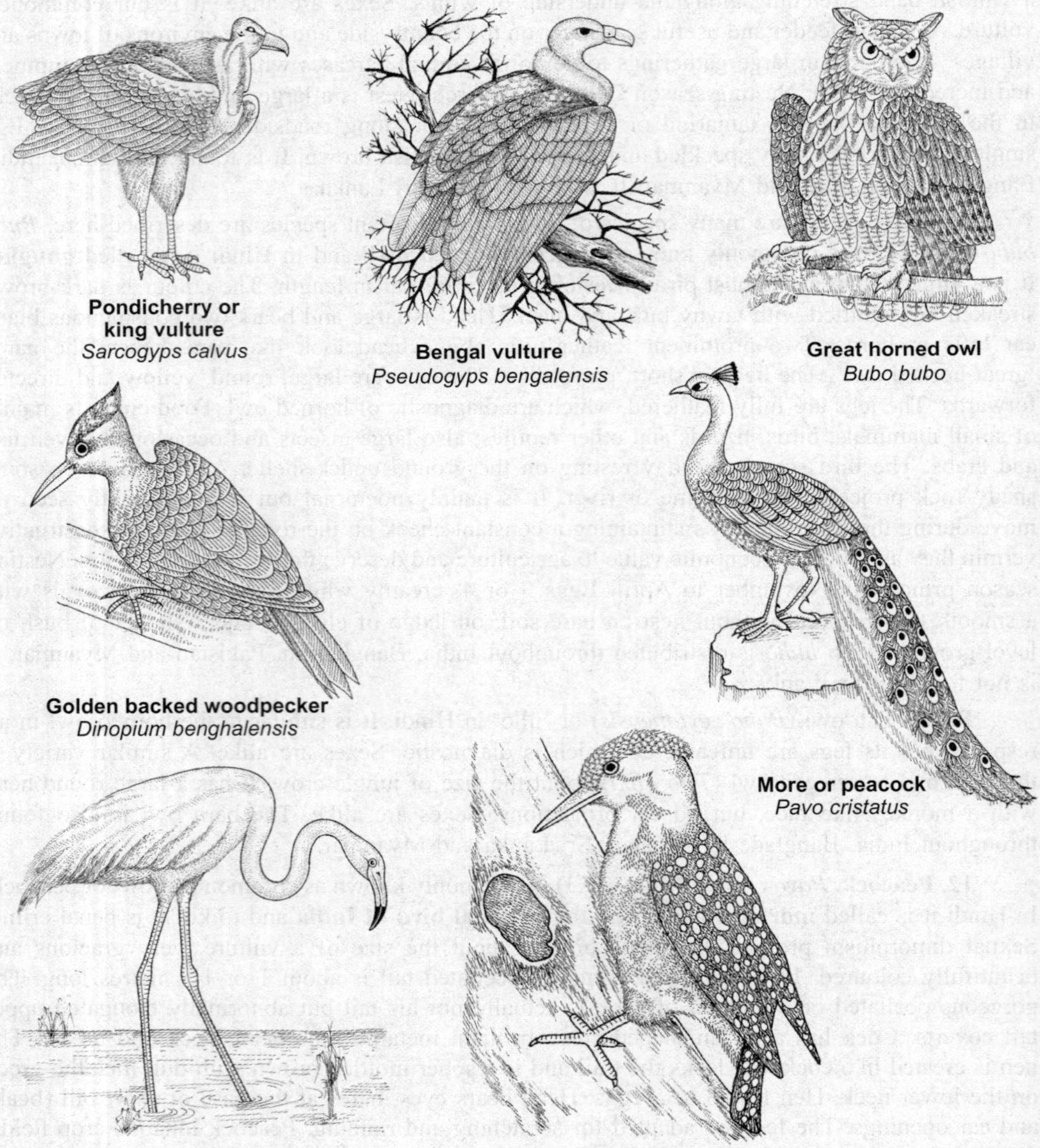

Fig. 27.3. (Cont). Some common Indian birds.

like fringes or lamellae along the edge of the mandibles, leaving the minute food particles behind. The method of procuring food is to wade into shallow water and feed with their long necks bent down and head completely immersed. Flamingos fly with fairly rapid wing beats in V-formation like geese. The slender neck is stretched in front, while the long red legs trail behind. The nest is cone-shaped mound of scrapped-up and plastered semi-liquid mud which becomes hard and sun-baked and has an average height of about 30 cm. A flat pancake-like depression is tamped on the top in which the eggs–2 or only 1– are laid. The incubating flamingo sits on this with its legs folded under and not standing astride the mound. The only known breeding place of flamingo is the Great Rann of Kutch whose vast numbers collect between October and March when water conditions are favourable. Flamingos live in flocks at jheels, brackish lagoons, on tidal mudflats, ponds and lakes. They wade in shallow water and feed on worms, small molluscs, crustaceans, insect larvae and seeds, etc. They fly rapidly. Flamingos are resident more or less throughout India, Pakistan, Bangladesh, Sri Lanka, Afghanistan, France, East Africa, Siberia, etc.

14. Woodpeckers. *Dinopium benghalense* (Fig. 27.3) is commonly known as **golden-backed woodpecker** and in Hindi it is called **kath-phorwa**. It is a small bird and is about 22 cm in length. Upper plumage is distinctive golden yellow and black, while lower is buffy white streaked with black. The entire crown and occipital crest is crimson in male only, partly so in the female. Head bears large eyes, nostrils, ear openings and long stout pointed beak adapted for wood-chiselling and piercing. The tongue is protrusible and barb tipped. Toes are four, two directed in front and two behind. The tail is stiff and wedge-shaped. *Dinopium* (Woodpecker) inhabits the grooves, gardens and light scrub in villages. It ascends to tree trunks tapping on the bark and peer into cracks for ants, grubs, wood-boring beetles and other insects injurious to trees for its food. It is found in village groves, gardens and in lightly wooded country. *Dinopium* is distributed throughout India, Pakistan, Sri Lanka and Myanmar.

The golden-backed woodpecker (*Brachypternus* or *Dinopium*) is about the size of a myna. The yellow fronted pied or Mahratta woodpecker (*Dryobates* or *Dendrocopos mahrattensis* Fig. 27.3) is much smaller, about the size of a bul bul.

15. Nilkanth. *Coracias benghalensis* (Fig. 27.4) is commonly known as **Indian roller** or **blue jay** and in Hindi it is called **nilkanth** or **sabzak**. It is an Oxford and Cambridge blue bird about the size of a pigeon. Head is big with heavy and black bill. The breast is rufous brown and pale blue abdomen and under tail. The dark and pale blue portions of the wings show up as brilliant bands in flights. Sexes alike. Food chiefly comprises crickets, grasshoppers, beetles, and other insects but occasionally lizards, mice and frogs are also eaten. Nesting season chiefly March to July. Eggs 4 or 5, glossy white roundish ovals. The nest is a collection of straw, feathers and rubbish in a natural tree hollow. It is essentially an inhabitant of open cultivated countryside and avoids dense forests. It is usually seen perched on an exposed tree stump or telegraph wires. The bird does great service to agriculture by destruction of some injurious pests. *Coracias benghalensis* (nilkanth) is distributed throughout India, Pakistan, Sri Lanka and Myanmar.

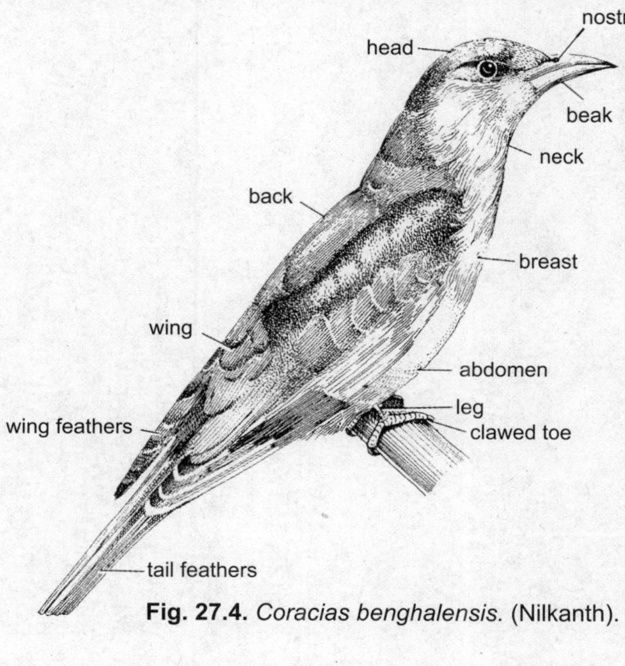

Fig. 27.4. *Coracias benghalensis.* (Nilkanth).

IMPORTANT QUESTIONS

Long Answer Questions

1. Describe the distinctive features or general characters of class Aves.
2. Classify birds giving important characters and examples of each group.

Short Answer Questions

1. Write short notes on :
   (*i*) Crow; (*ii*) Koel;
   (*iii*) Peacock; (*iv*) Parrot;
   (*v*) Hoopoe; (*vi*) Owl;
   (*vii*) Flamingo.

Multiple Choice Questions

1. In birds, the exoskeleton is epidermal and horny, represented by
   (*a*) Feathers, forming a body covering
   (*b*) Scales, similar to those of fishes
   (*c*) Claws, on wings
   (*d*) Sheaths on toes.
2. Centra of vertebrae in birds is
   (*a*) Procoelous (*b*) Heterocoelous
   (*c*) Amphicoelous (*d*) Acoelous.
3. In birds, the last 3 or 4 tail vertebrae are fused to form
   (*a*) Synsacrum (*b*) Pygostyle
   (*c*) Furcula (*d*) Wishbone.
4. In birds, sound is produced by
   (*a*) Air sacs (*b*) Trachea
   (*c*) Larynx (*d*) Syrinx.
5. In birds, middle ear contains
   (*a*) 1 ossicle (*b*) 2 ossicles
   (*c*) 3 ossicles (*d*) 4 ossicles.
6. Order Archaeopterygiformes includes
   (*a*) *Hesperornis* (*b*) *Archaeopteryx*
   (*c*) *Icthyornis* (*d*) *Enaliornis*.
7. Flightless birds are included in
   (*a*) Odontognathae (*b*) Impennae
   (*c*) Carinatae (*d*) Ratitae.

Answers

1. (*a*) 2. (*b*) 3. (*b*) 4. (*d*) 5. (*a*) 6. (*b*) 7. (*d*)

28

Chapter

Aves (Birds) : General Account

Contents of this Chapter

FLIGHTLESS BIRDS (RATITAE)

The living wingless or flightless birds belong to the superorder Palaeognathae or Ratitae of the subclass Neornithes. They are characterised by having a palaeognathous palate, a keelless raft-like sternum and curly feathers. Ratitae have persisted on earth over a long period of geological time. Ratitae are represented by several species on earth today. These birds are much alike and all are ground-dwelling. They are usually very large, some reaching as much as 4 metres in height. They have well developed powerful legs, small heads and rudimentary wings with fluppy feathers which are useless for flight. An interesting fact is that all modern flightless birds live in parts of the world where there are no carnivores, *i.e.,* no members of the cat or wolf tribes. Table 28.1 gives the differences between Ratitae and Carinatae.

TABLE 28.1. DIFFERENCES BETWEEN RATITAE AND CARINATAE.

	Structure	Ratitae (Palaeognathae)	Carinatae (Neognathae)
1.	Distribution	Restricted, discontinuous. Except ostrich, they are confined to Southern continents and islands.	Cosmopolitan, found all over world.
2.	Habitat	Terrestrial–usually living in deserts.	Arboreal, terrestrial or aquatic.
3.	Locomotion	Flightless running birds.	Flying birds.
4.	Size	Usually large-sized.	Usually small-sized.
5.	Wings	Vestigial or absent.	Well developed.
6.	Feathers	Without interlocking mechanism. Devoid of hooks and hooklets. Barbs free.	Barbs interlocked due to barbules and barbicels.
7.	Pterylae	Irregular, continuous.	Regular, separated by apteria.
8.	Rectrices	Absent or irregularly arranged.	Regularly arranged in a fan-like manner.
9.	Down feathers	Absent.	Present.
10.	Aftershaft	Well developed.	Poorly developed.
11.	Uropygial gland	Absent except kiwi.	Present.
12.	Rhamphotheca (horny sheath to the beak)	Compound, composed of several separate pieces.	Usually undivided, may be compound in Albatross.
13.	Oil gland	Usually absent.	Present.
14.	Nature of skull	Large and strongly built.	Small, light and fragile.
15.	Skull sutures	Remain distinct for a long time.	Disappear very early.
16.	Type of skull	Dromaeognathous or palaeognathous.	Neognathous, never dromaeognathous.
17.	Quadrate	Y-shaped bone, articulates with the squamosal by a single or partly divided facet.	Never Y-shaped, articulates by two facets.
18.	Vomer	Large, broad and separates the palatines.	Small, narrow, allows palatines to meet.
19.	Palatines	Short, do not articulate with rostrum.	Long, articulate with parasphenoidal rostrum.
20.	Pterygoids	Immovable.	Movable.
21.	Basipterygoid processes	Well developed.	Small or absent.
22.	Maxillopalatine process	Small, do not articulate with one another.	Unite with one another to form large secondary palate.
23.	Sacral vertebrae	Not firmly ankylosed.	Most of them are firmly ankylosed.
24.	Pygostyle	Small or absent.	Well developed.
25.	Sternum	Flat, raft-like due to vestigial or no keel.	Well developed mid-ventral keel.
26.	Uncinate process	Vestigial or absent.	Present.
27.	Coracoid and scapula	Comparatively small, and fused at an obtuse angle.	Comparatively large, not fused but meeting at an acute angle.

(Contd.)

	Structure	Ratitae (Palaeognathae)	Carinatae (Neognathae)
28.	Clavicles	Small or absent.	Both clavicles and interclavicle.
29.	Furcula	Absent.	Present and well developed.
30.	Acromian and acrocoracoid process	Absent or vestigial.	Well developed.
31.	Ilium and ischium	Usually remain free posteriorly.	Become fused posteriorly.
32.	Pelvic symphysis	Pubic symphysis present in ostrich and ischiatic symphysis found in Rhea.	Pubic and ischiatic symphyses absent.
33.	Limb bones	Not pneumatic.	Pneumatic.
34.	Digits in forelimbs	Usually three digits are present, two with claws.	Usually three digits are found and claws are absent in adult.
35.	Hindlimbs	Usually elongated, adapted to cursorial life.	Usually adapted for cursorial/ arboreal/aquatic life.
36.	Pectoral muscles	Poorly developed.	Well developed.
37.	Caecum	Large.	Small.
38.	Syrinx	Usually absent.	Present.
39.	Air sacs	Poorly developed.	Well developed.
40.	Penis in male	Present.	Absent.
41.	Ovary in female	Both the ovaries may be present.	Usually right ovary is absent.
42.	Clittoris in female	Present.	Absent.
43.	Eggs	Shell thick, hard. Pore canal branched.	Shell thin, fragile. Pore canal unbranched.
44.	Newly hatched young.	Usually precocial, *i.e.*, independent.	Altricial, *i.e.*, helpless and dependent on parents.

Orders of Flightless Birds

The living flightless birds belong to the following 4 orders of the superorder Palaeognathae :

Order 1. Sruthioformes. Ostriches of Southwest Asia and Africa.

Order 2. Rheiformes. Rheas of South America.

Order 3. Casuariiformes. Cassowaries and Emus of Australia and adjacent islands.

Order 4. Apterygiformes. Kiwis of New Zealand.

Discontinuous Distribution

Flightless birds (Ratitae) furnished a remarkable example of discontinuous distribution. Taking into consideration the living forms only, each of the great Southern landmasses contains one order of Ratitae (or Palaeognathae) not found elsewhere. They are distributed in far-flung areas such as South America (rheas), Africa and Asia (ostriches), Australia (cassowaries and emus) and New Zealand (kiwi). It is indeed very remarkable that two genera so closely allied as are *Rhea* and *Struthio*, should occupy areas so distantly separated from each other as South America and Africa. The living members are entirely wanting in Europe and North America. But the existence of their fossils in North America and Europe only proves that the flightless birds were once not confined to their present limits alone and possibly pervaded the greater part of the earth.

Origin

The origin and ancestry of flightless birds (Ratitae) have been a very puzzling problem and different views have been put forward. According to the most accepted view, these birds evolved from flying ancestors and only secondarily re-adapted for a flightless terrestrial mode of life. This is supported by the fact that the ratites possess many morphological characters showing adaptation to flight. It seems hardly conceivable that their ancestors had acquired those characters and were not able to fly. One reason for the acquisition of flight by the ancestral birds probably was to be better able to flee from their enemies on the ground. If the enemies are removed, then the need for flight is no longer present, hence why fly. Therefore, in the areas where there were abundant food supply upon the ground and practically no terrestrial predators and competitors, the ancestral ratites simply ceased flying and remained on the ground all of the time, thus, conserving their energies for procurement of food. The above view propounded by **Romer** is also supported by present-day geographical distribution of the ratites. All of them live in the regions that are now, or were in the past free from formidable bird enemies on the ground.

The ancestry of certain ratites can be traced back to the early Cenozoic Era. A number of large ground-dwelling neognathous birds also lived then. There might have been a keen competition at that time between birds and early mammals for the conquest of the land surface which had been vacated by then by the large reptiles. While the mammals won in general, only a few ground-dwelling birds survived and these gave rise to the modern flightless birds.

Examples of Ratitae (Flightless Birds)

1. Ostrich. African ostrich (*Struthio*) (Fig. 28.1) is represented by at least 4 species occurring naturally in arid regions of Africa and Arabia. It is the largest living bird attaining a height of more than 2.4 metres and a weight of 150 kg. Feathers are without aftershaft. Wings are small and nearly kept folded during running, but may be expanded to work as a brake or steering. Head is small with a broad beak, neck is long, body is heavy, and tail is short. Head, neck and legs are sparsely feathered. The skin is so thick that it can be made into leather, curly feathers do not lie in pterylae. Two wing fingers are clawed which are used for defence. Tail feathers are replaced by long tail coverts. Hindlimbs are long and powerful, they kick very hard and are not covered by feathers. Only the 3rd and 4th toes are present bearing stunted nails, the 3rd toe is larger and clawed. Thus, ostrich is a flightless but very swift running bird. Male and female birds differ in the colour of their plumage. The tail and wing feathers are white, while the rest are black in the male. Female is clad in a sombre brownish livery. The long sturdy legs are admirably suited to match the speed of zebras and antelopes in whose company it feeds. It can run at a speed of 80 kilometres per hour, covering more than 8 metres in a single stride. Both beak and feet serve as weapons of defence. Two beaks are more or less round and enclose a broad space. Prominent eyes have lashes on the upper eyelid. The food consists chiefly of herbage including seeds and fruits. Contrary to popular belief, an ostrich does not hide its head in sand when in danger. Ostriches travel about in groups. They are polygamous as one male attends several females. The male has a single solid penis which can be retracted into a pouch of the cloaca. Female has a clitoris. They lay large eggs in depressions of sand which are incubated by both parents. The young ones run about as soon as they are hatched. The egg of the ostrich is very large and next in size to that of mackerel shark. Each egg weighs nearly 1.5 kg and requires 50 minutes to boil it. Ostrich meat is coarse and of little use for food. The feathers of adult bird are used by women as decorative plumes. Ostriches are easily domesticated and there are a number of ostrich farms in Africa and United States.

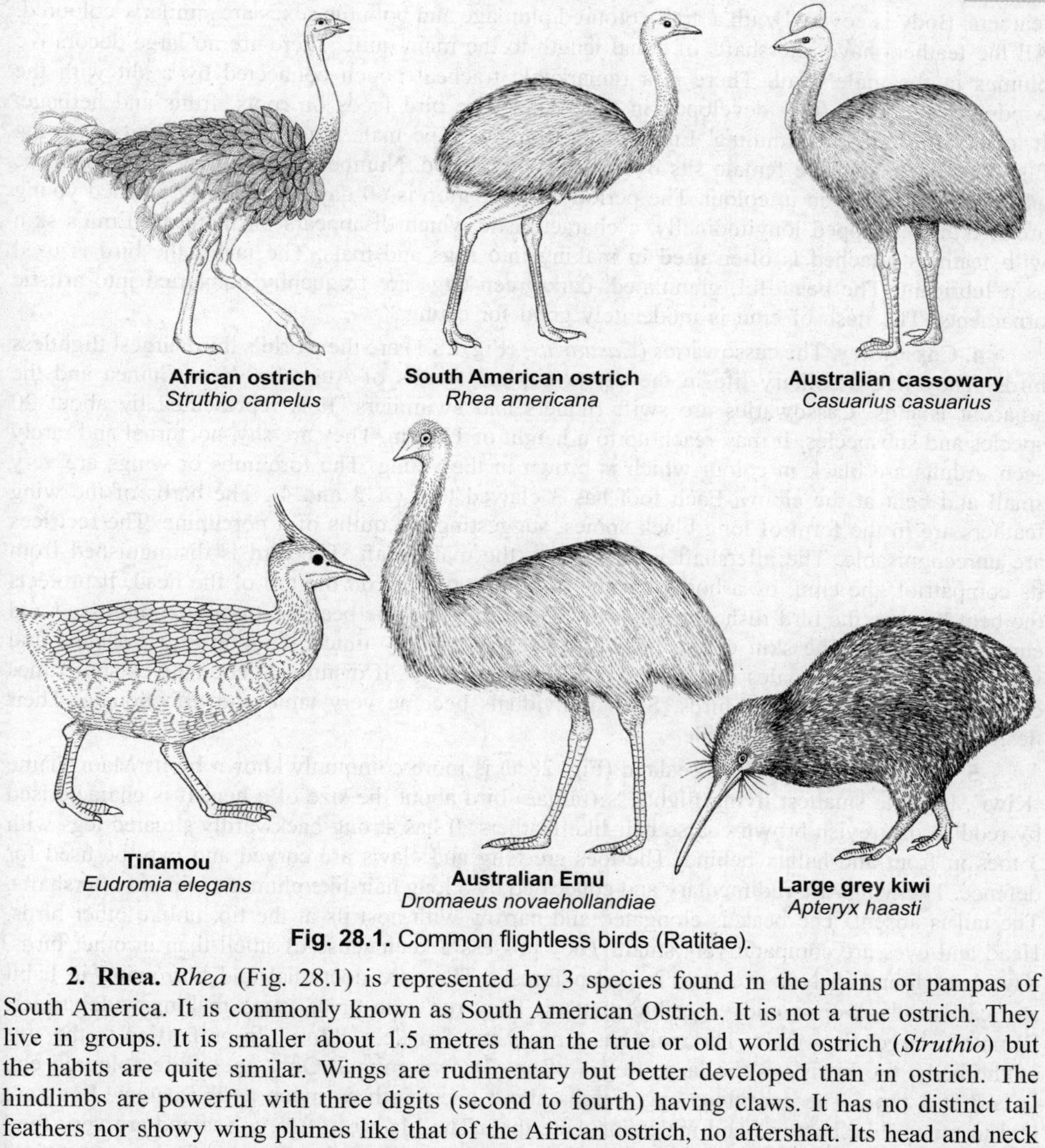

Fig. 28.1. Common flightless birds (Ratitae).

2. Rhea. *Rhea* (Fig. 28.1) is represented by 3 species found in the plains or pampas of South America. It is commonly known as South American Ostrich. It is not a true ostrich. They live in groups. It is smaller about 1.5 metres than the true or old world ostrich (*Struthio*) but the habits are quite similar. Wings are rudimentary but better developed than in ostrich. The hindlimbs are powerful with three digits (second to fourth) having claws. It has no distinct tail feathers nor showy wing plumes like that of the African ostrich, no aftershaft. Its head and neck are feathered, not naked as in ostrich. Beak flat and broad. Rheas are very fond of bathing and are even able to swim. Food varies from vegetable products to insects. The male is polygamous. During breeding season, the males select and segregate with 3–7 females to form a harem. The nest is dug by the male and all the females use to lay eggs in the same nest. Nearly 50 eggs are laid in a season and male incubates them. Freshly laid eggs are lemon yellow in colour and hatching requires 40 days incubation. Though the plumes of Rhea are of an inferior quality to those of ostrich, they are used in the manufacture of feather dusters. There are two species of *Rhea*, *viz.*, *Rhea americana* (American Rhea) and *Pteronemia pennata* (Darnicus Rhea).

3. Emu. Next to ostrich, emu (*Dromaeus*) (Fig. 28.1) is the largest living bird, standing more than 1.8 metre high.There are 2 species confined to Australia, inhabiting sandy plains or open forests. Emu has more slender built than an ostrich. It has a shorter neck covered with feathers. Body is covered with a drab-coloured plumage and both the sexes are similarly coloured. All the feathers have aftershafts of equal length to the main quill. There are no large decorative plumes in the male's tail. There is a remarkable tracheal pouch connected by a slit with the windpipe, and only fully developed in the adults. The bird feeds on roots, fruits and herbage. It loves bathing and swimming. Emu is monogamous. The male digs the nest and sits over the first brood of eggs. The female sits over the second brood. Number of eggs may be 15 or more and the eggs are green in colour. The period of incubation is 60 days. The newly hatched young are strikingly stripped longitudinally, a characteristic which disappears on maturity. Emu's skin with feathers attached is often used in making into rugs and mats. The fat of the bird is used as a lubricant. The beautiful, granulated, dark-green eggs are frequently converted into artistic ornaments. The flesh of emu is moderately good for eating.

4. Cassowary. The cassowaries (*Casuarius*) (Fig. 28.1) are the world's third largest flightless birds. They live a solitary life in the dense tropical forests of Australia, New Guinea and the adjacent islands. Cassowaries are swift runners and swimmers. It is represented by about 20 species and subspecies. It may reach up to a height of 135 cm. They are shy, nocturnal and rarely seen. Adults are black in colour which is brown in the young. The forelimbs or wings are very small and bent at the elbow. Each foot has 3 clawed toes (2, 3 and 4). The barbs of the wing feathers are in the form of long black spines, suggesting the quills of a porcupine. The rectrices are unrecognisable. The aftershaft is as long as the main shaft. The bird is distinguished from its compatriot, the emu, by a horny casque or helmet present on the top of the head. It protects the bent head as the bird rushes through the dense jungles. The beak is strongly compressed and curved at the tip. The skin of the head and neck is brightly tinted. They are quarrelsome and vicious birds and old males may attack even human beings, if disturbed. The natives sometimes capture and rear the young birds. Such individuals become very tame and affectionate. Their flesh is considered very palatable.

5. Kiwi. *Apteryx* of New Zealand (Fig. 28.1) is more commonly known by its Maori name 'Kiwi'. It is the smallest living flightless (ratitae) bird about the size of a hen. It is characterised by reddish or greyish brown coarse hair-like feathers. It has strong backwardly situated legs with 3 toes in front and hallux behind. The toes are long and claws are curved and may be used for defence. The wings are rudimentary and concealed by a long hair-like plumage without aftershafts. The tail is absent. The beak is elongated and narrow with nostrils at the tip, unlike other birds. Head and eyes are comparatively small. They possess a keen sense of smell than in other birds. Kiwis are found in New Zealand in damp forests. They are nocturnal and burrowing in habit and eat insects, worms and vegetable matters. They run with great speed making wide strides. Female is larger than male. The male has a penis and female a clitoris. In proportion to the size of the body, the kiwi lays the largest eggs of any known animal. Only one white coloured egg is laid at a time. The enormous egg weighs about one-fourth as much as the parent. Eggs are laid in tunnel-like burrows and males incubate them.The Maories of New Zealand are very fond of kiwi flesh, either roasted or boiled, while their chiefs utilised the plumage for ornamentation. In order to protect kiwi from extermination the Government of New Zealand has set aside some sanctuaries where kiwis may remain unmolested. *Apteryx* has three species, *Apteryx australis* (common brown kiwi), *Apteryx hoasti* (greater spotted kiwi) and *Apteryx oweni* (little spotted kiwi).

ORIGIN OF BIRDS

Birds constitute a very specialised group of vertebrates which have evolved from reptiles during Mesozoic era. The actual documentary records of the stages showing how such reptiles were transformed into birds are rare because of the non-availability of abundant avian fossils. Luckily a transitional stage between reptiles and aves is represented by *Archaeopteryx* and *Archaeornithes,* the two fossils from Germany. These combine in themselves many of the reptilian features with undoubted avian features, so that the reptilian origin of birds is obvious. The gap between this stage and the actual reptilian ancestor is yet to be filled. Two fossil specimens of *Archaeopteryx* discovered from the Upper Jurassic beds added the strongest evidence to this contention.

The exact nature of hypothelical ancestor '**Proaves**' which formed the link between *Archaeopteryx* and the reptile has been a matter of speculation. Some of the theories are mentioned here as they are of interest.

According to **Osborne**, the ancestral bird was an arboreal four-legged parachuting animal which glided from tree to tree like a flying squirrel. The parachute became an active wing with the development of feathers and strengthening of wing muscles. In other flying animals like bats and pterodactyls with patagium, the hindlimbs are incapacitated for walking, whereas in birds, they are prominent.

Nopsca put forward the theory that the 'Proaves' was a cursorial long-tailed, bipedal creature which ran about and leaped along the ground with their strong hindlimbs, flapping their forelimbs in the air as an aid to increased speed. The forelimbs which were the wings became broadened by the development of scales in the hinder part of the arm and the scales became sprayed out to form quill feathers.

Steiner suggests that the 'Proaves' should have been an arboreal form with four legs, the hindlegs being modified for springing, and forelegs for climbing and steadying the body after a leap. The quill feathers developed along the hinder aspect of the forelimbs and the tail which helped the animal to buoy up the body in the air. The transformation of the forelimbs into wings was accompanied by the development of bipedalism.

Beebe suggests that the Proaves should have been an arboreal tetrapod with backwardly directed feathers in all the legs, both sets serving as parachutes in gliding. This view is supported by indications of tufts of feathers in the hinder aspect of the limbs of *Archaeopteryx*.

Gregory is of the view that birds have a dual origin, some have evolved from cursorial and others from arboreal ancestors, which will be seen to be a compromise of several theories.

Presently the origin of birds is considered mainly on two lines : diphyletic origin and monophyletic origin.

Diphyletic Origin of Birds

The earliest known fossil birds include both flying (*Archaeopteryx, Ichthyornis*) as well as flightless (*Hesperornis, Diatryma*) types. The recently extinct Moas and Elephant birds were also flightless. The most primitive living birds or Ratitae (Ostrich, Rhea, Cassowary, Emu and Kiwi) and Penguins are also flightless. This led some authors, notably **P. R. Lowe**, to believe in the **diphyletic** (two-lines-of-descent) origin of birds. They maintain that the flightless and flying birds of today have descended from flightless ancestors. According to Lowe, the present-day flightless birds were never capable of flight, and their wings are not degenerate now, but better developed than any time in their past history.

Monophyletic Origin of Birds

In Ratitae, the legs are well-developed and powerful, the wings vestigial, and the feathers are fluffy. But a recently discovered fossil of *Eleuherornis*, a probable ancestor of the present-day ostrich from the Eocene of Switzerland, shows closer affinities to flying forms than does the present-day ostrich and poses a serious blow to the concept of diphyletic origin of birds. Presently most palaeontologists believe that the Carinatae are more primitive. Presumably the Ratitae evolved from flying ancestors but readapted to a terrestrial mode of life in areas with abundant food and few competitors or enemies. The more usually accepted view today maintains that birds have a monophyletic (one-line-of-descent) origin, *i.e.*, all birds have evolved from a single ancestor, perhaps close to *Archaeopteryx*. Accordingly, the flightless birds have evolved by loss of flight from flying ancestors. The weight of the known evidence also favours this view.

ORIGIN OF FLIGHT IN BIRDS

The evolution of the birds is associated with the changes from terrestrial to aerial life.The reptilian origin of birds is a self-evident fact, but detailed evidence of the steps how cold-blooded terrestrial reptiles were converted into warm-blooded flying birds are insufficiently known except the discovery of a fossil bird, *Archaeopteryx*, in the Jurassic bed. We know nothing about the evolution of feathers from reptilian scales, although intermediate structures between scales and feathers are present on the legs of ostrich and fowl. We, of course, do not know just how flight evolved in the ancestors of *Archaeopteryx*. It is presumed that the ancestors were becoming more active and possibly warm-blooded and feathers developed from scales primarily to conserve their body heat. Later, the feathers enlarged on the limbs and that the tail probably to provide stability in fast running on ground or in rudimentary gliding from branches.

Different theories have been proposed to explain the origin of flight in birds, starting from either a terrestrial, bipedal and cursorial ancestor (**Nopcsa**) or an arboreal ancestor (**Osborn**, **Steiner**, **Beebe**, etc).

1. Theory of cursorial origin of flight. Nopcsa has advocated the cursorial origin of flight. According to him, the proaves was a running reptile. While running fast on the ground, this reptile lifted the forelimbs off from the ground. The scales over the margin of the limbs became expanded as feathers. **Nopcsa** believed that birds arose from bipedal dinosaur-like reptiles which used to run on the ground by hindlimbs. According to his theory, the ancestors of birds were long-tailed, **cursorial**, bipedal animals. They were fast runners who leaped on their strong hindlimbs and flapped their forelimbs in air to help them along, as do many modern birds that run fast. Gradually the forelimbs enlarged due to fraying out or elongation of scales forming quill-feathers through the process of mutation and selection. In the end, the forelimbs became organs of flight or wings rather than accessories to rapid running.

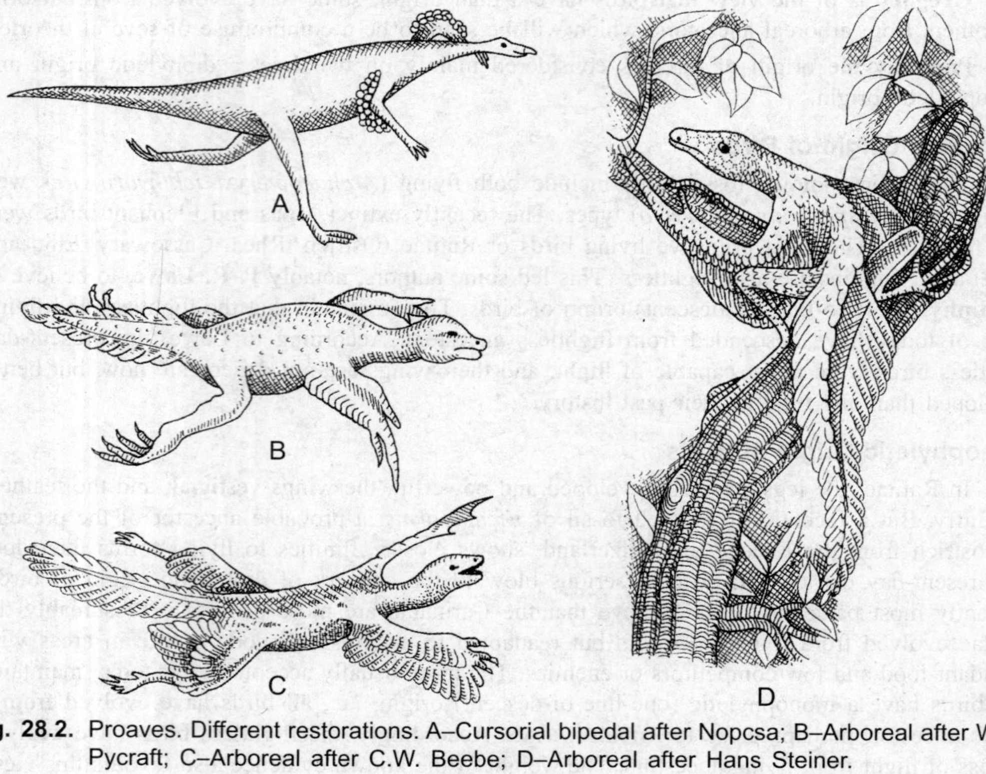

Fig. 28.2. Proaves. Different restorations. A–Cursorial bipedal after Nopcsa; B–Arboreal after W.P. Pycraft; C–Arboreal after C.W. Beebe; D–Arboreal after Hans Steiner.

2. Theory of arboreal origin of flight. This theory of arboreal origin of flight postulates that the ancestral birds were arboreal creatures. This concept is supported by many ornithologists. Proaves was arboreal form which used to live on trees. They climbed trees from which they glided to the ground or to other trees, like the modern flying squirrels. **Osborn** has put forward the **pair-wing theory** and advocated that only the forelimbs were utilised in climbing the trees and also during jumping from one tree to another. This has led to the origin of patagia between the limbs and the body. **Beebe** and **Heilmann** established the **four-wing theory** which signifies that both the forelimbs and hindlimbs were need during the process. Both the pairs of the limbs were utilised for climbing the trees and also during parachuting from tree to the ground and from tree to tree. As time went on, the forelimbs gradually enlarged and eventually converted into wings which were able to support the animal in flight. The feathers of the hindlimbs vanished and the tail shortened as seen in the modern birds. Thus, flight was an outgrowth of climbing and gliding.

Steiner maintains that the hindlimbs of the ancestor were modified for springing, while the forelimbs were used for climbing and balancing the body after a leap in the air.

3. Compromise theory of origin of flight. Gregory advanced a compromise idea and believed in the dual origin of birds. Believers in the dual origin of birds maintain that some birds evolved from arboreal and others from cursorial ancestors.

4. Diving theory of origin of flight. Newmann regarded the proaves as aquatic reptile. Flight might have started in connection with soaring over the water during diving for fishes.

BIRDS ARE GLORIFIED REPTILES

Nearly a century ago, **T.H. Huxley** called birds 'glorified reptiles', thereby meaning that birds have evolved from some reptilian ancestor and that they are better evolved in their organisation. The evidence from comparative anatomy, embryology and palaeontology with fossil *Archaeopteryx* as the transitional form suggests that bird is a highly specialised descendant of some reptilian ancestor.

The steps and the impulse under which evolution of birds took place from a reptilian stock are not properly understood at present. Whatever may be mode and direction of origin of birds from ancestral reptiles, the modern birds show a marked superiority over their reptilian ancestors.

Some of the main changes leading to the glorification or superiority of birds over reptiles are as follows :

1. The reptiles are generally sluggish, **cold-blooded** (**poikilothermous**) and earth-bound, the birds are active, **warn-blooded** (**homoiothermous**) and alert, having more life in them than any other living creature. They have a higher rate of metabolism and correlated carefully regulated high body temperature. They are warm-blooded and due to this they are equally active throughout the year.

2. Birds show a more **rapid locomotion** due to power of flight, perhaps also the most efficient in the animal kingdom.

3. The development of avian **feathers** from reptilian scales is definitely an advance. It has made it possible for birds to fly. The insulating feather covering preserves body heat and makes the birds warm-blooded. The elongation of feathers on forelimbs has modified them into **wings** for flight. Thus, the variety and colouration of various types of feathers show an advance over the simple uniform type of scales in reptiles.

4. Birds, like mammals, have a completely **four-chambered heart** with **double circulation**, in which there is no mixing of pure and impure bloods. Only a single systemic arch, that of the right side, persists in birds, whereas both the systemic arches are present in reptiles.

5. The **renal portal system** is well developed in reptiles, while vestigial in birds. It is absent in mammals.

6. In birds respiratory system is highly evolved. The lungs, with the development of **air sacs**, bring about complete aeration and are many times more effective than those of reptiles.

7. Birds have developed the **voice** with all the varied features of calls and songs, in contrast to the silent reptiles.

8. Kidneys are metanephric, but show two specialisations not found in reptiles. First, the uriniferous tubules are relatively longer, and second, a U-shaped piece, called the **loop of Henle**, is usually inserted in their middle. This condition is similar to that of mammals.

9. The endoskeletal, muscular, alimentary and reproductive organs also show an advance over those of reptiles.

10. The **brain** is proportionately bigger with better development of cerebrum and cerebellum. The sense of sight and hearing are better developed. **Eyes** are more efficient with better power of accommodation. Ears possess cochlea with an organ of Corti and, therefore, are better developed.

11. Bird's **eggs** resemble those of reptiles but differ in that many of them are coloured.

12. Birds possess higher grades of **intelligence** and **behaviour** practically unknown in reptiles, such as : (*i*) Parental care and extreme emotion in preparation of more or less elaborate nest, (*ii*) While most reptiles leave their eggs to hatch in ordinary environmental conditions, all birds incubate their eggs, thus ensuring hatching and better survival, (*iii*) Periodic migrations or wandering to and fro their winter and summer abodes, (*iv*) Monogamous life with selected mating, courtship behaviour and affection for the mate and (*v*) Singing and mating calls during breeding season.

Thus, Huxley was fully justified in commenting that 'birds are glorified reptiles'.

BIRDS AS A FLYING MACHINE

The flight is a spontaneous action. At take-off the bird has to acquire sufficient forward momentum to provide lift. In many birds the jump is provided by the legs, which is adequate for the take off. Larger or heavier birds such as peacock run or swim rapidly to gather enough forward momentum for a take-off. Many large birds nest on a cliff or tree which gives them an up-current for the take-off. Small-sized birds such as pigeons usually take a quick jump by means of their legs followed by the beating of their wings.

The pigeons and other birds on the principle of aeroplane or heavier than air machine. Its body is well adapted for aerial mode of life. Such flight adaptations include the following ways :

1. Centralisation. The internal organs are well centralised and well adapted for their mode of life. For example, presence of light horny beak without teeth and muscular gizzard having stones for grinding the seeds. Tail is short and provided with tail feathers as rudder during flight. The head is small and light, and the neck is long which can be retracted during flight. The body axis is short and the body is streamlined and perfectly balanced with centre of gravity well below the wings.

2. Lightness and rigidity. The bones are light, strong and pneumatic built on a hollow girder principle.

3. Wings for support. The large, broad wings and tail provide support during flight. Birds are able to adjust their wings and tail during flight in the fast flowing air.

4. Sustained power. Birds have a great sustained power of flight. This is attained by the well developed flight muscles, large heart allows a greater stroke output and higher pressure than in mammals, large-sized red blood corpuscles which carry a large amount of haemoglobin that gives up its oxygen suddenly at a relatively high oxygen tension and high basal metabolic rate and high body temperature (42°C–45°C). High body temperature and great activity of birds necessitate a high food intake. This is made possible by rapid passage of food through the gut.

5. Steering and balancing. This is done by the help of wings and tail. Wings are the active means of flight. The wings support the body weight in air. Depression of the wing elevates and propels the bird through air. Rotational movements of the wing also help the bird during flight. Thus, wings act as balancing and flying organs. The tail with tail feathers acts as a rudder for steering during flight, it suddenly checks flight and as a counterbalance in perching. It also directs the course of flight by spreading or folding and elevating or depressing the tail.

Mechanism of Flight

The pigeon and other birds fly on the principle of **indirect movement**. They move the air, which, by its displacement, moves the birds. Air displaced by the beating of wings, sets up currents that keep the animal aloft and move forward, resulting in flight. A plane surface moved through the air in a direction inclined at an angle to its plane is called **aerofoil**. The forces generated can be resolved into a **lift force** acting upwards and a **drag force** to stop the motion. On this fact depends the power of supporting weight in the air that is possessed by birds. Both lift and drag forces are proportional to the square of the speed and the requirement for sustained flight in still air is that the object shall have sufficient speed to generate a lift force equal to its weight (**Lighthill,** 1975, **Pennycuick**, 1972, 1975).

The flow of air over the upper surface of the wing reduces the pressure there and provides the main portion of the lift.

By tilting the wing the pressure on the underside can be increased, but the air flow becomes turbulent especially at the hind edge destroying the lift. When an aerofoil fails below the critical speed, it stalls (drops suddenly) being no longer supported. The smooth flow of air over the wing disturbs the wings' hinder (trailing) edge and by eddies round the end (tip vortex). A small wing area is necessary for fast flight, at least for high speeds. Wing provides the forward momentum as well as the lift. For **flapping flight** the wings must be moved relatively fast and for this small size is an advantage. While a large wing area allows slow flight. Although a small wing area reduces drag, many fast flying birds have a large wing-span. It allows a slow rate of descent when gliding, reducing the expenditure of energy necessary to sustain flight.

A pointed wing tends to stall first at its tip and is only suitable for fast fliers. Whereas birds with shorter broader wing possess slower flight.

The condition of the air around the wing is of first importance for maintenance of lift. If there is not a smooth stream over the upper and under surfaces, the air becomes turbulent, and the aerofoil stalls. This tends to happen either if the speed falls too low or if the angle of the wing relative to the line of motion increases above 20°. Turbulence is mitigated by the openings (slots) which let through part of the air and provide the necessary smooth stream. The spaces that occur between the feathers, especially towards the wing tip junction as slots. It provides a very efficient high life device. Such slots are conspicuous in slow fliers (rooks) and especially in those that soar on thermal up-currents (vultures). Slots are also found in the wings of large fast-flier birds (pheasants). In most birds there is a thick loading edge and a thinner trailing edge. Nearly all wings are convexly curved in cross section, especially in the region of forearm. This arrangement directs the air stream over the upper surface of the wing in such a way as to provide an extra lift by creating a **suction zone** of reduced pressure. Highly convexly curved wings and less wing area reduces the speed of the bird.

In some birds feathers are allowed to twist only when the wing is being raised and the barbs of the feathers open like the vanes of a blind when under pressure from above, but close when the pressure is from below. In fast flying birds with a slow wing beat (gulls and swans) the wing is probably rigid on the up as well as on down strokes, and is twisted so as to produce forward and upward components on the upstroke.

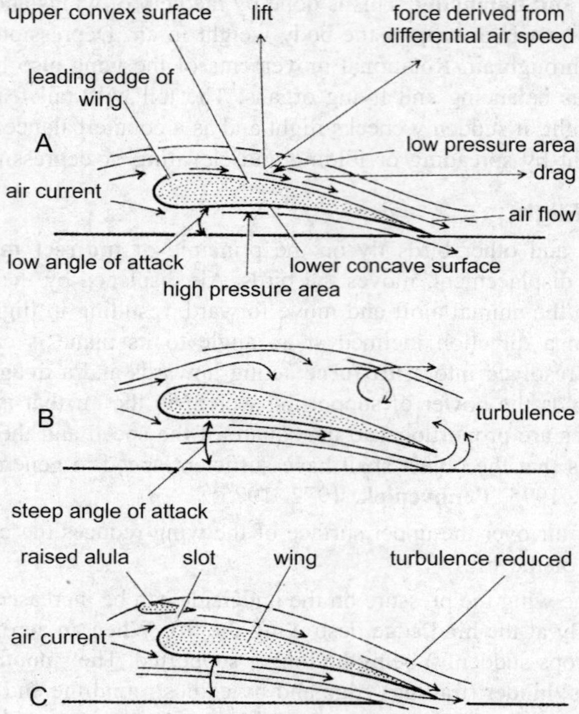

Fig. 28.3. Effect of a wing on air stream in flight. A–Wing tilted favourably so that air flows smoothly over its upper surface to increase lift; B–Wing lifted sharply so that air becomes turbulent above destroying lift and producing a stall; C–Raising of alula forms a slot through which air moves rapidly and smoothly, reducing turbulence.

The whole upward movement is usually faster than the downward one. Before the wing tip has reached its highest point, the upper arm is already beginning to descend and in this way the line of flight is maintained almost straight and does not follow a wavy path as it would do if the parts of the wing vibrated together. In small birds the wing works more nearly as a whole and the flight differs in several respects from that of larger birds. In general the wing is a very labile system and regulates itself automatically with changes in the aerodynamical forces. This regulation is produced partly by feather plasticity and joint mobility, with participation of reflex muscular adjustments that are little understood.

MODES OF FLIGHT

The modes of flight in birds are of three or four kinds. All these kinds of flight may be used at different times by the same bird. In all the types of flight, the tail helps to support and balance the body, serving as a rudder. The tail feathers of pigeon act as rudder for vertical and lateral steering, elevating the tail produces an upward slant and lowering the tail a downward turning. Landing is achieved by lowering and fanning out the tail feathers which act as a break, then the legs are lowered and the pigeon drops on a perch. The chief kinds of flights are as follows :

1. Gliding or skimming. The simplest and probably the most primitive mode of flight is gliding (Fig. 28.4A). Birds hold their wings spread, motionless and glide for a considerable distance without flapping them. Gliding depends for its movement on the velocity acquired by previous strokes, or by descending from a higher to a lower level or by making use of air-currents. The gliding flight can be shown for a short time, bird soon loses velocity or height. Gliding flight can be readily observed in shore birds coming in for a landing; in ducks, gulls and herons over water, in swallows and swifts in the air, in pigeons gliding from their loft to the ground or in falcon swooping upon its quarry.

2. Soaring and sailing. This is the most remarkable and highly specialised mode of flight (Fig. 28.4B). This type of flight is illustrated by birds with a large wing-span, such as albatross, vulture, falcon, stork, crow, etc. The bird usually at a high level, describes great circles without any movement of the wings, whatsoever. The bird rises without loss of kinetic energy.

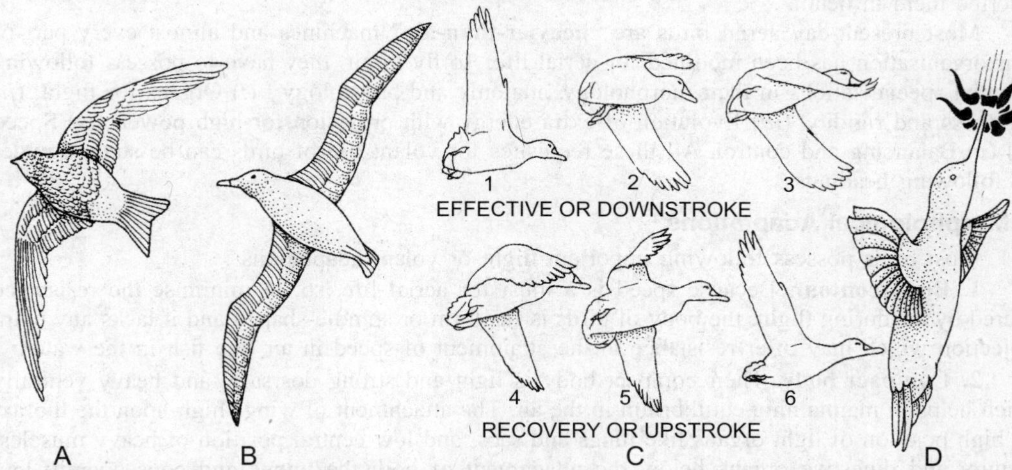

Fig. 28.4. Modes of flights in birds. A–Gliding (swift); B–Soaring (albatross); C–Flopping (duck); D–Hovering (humming bird).

3. Flapping. This is most common or ordinary mode of flight (Fig. 28.4C). All birds fly by flapping their wings up-and-down. Each flapping includes an effective downstroke and a recovery upstroke of the wings. To start with, the wings are held vertically and fully spread. In the **downstroke**, they move obliquely forward, downward and backward, their distal portions tilted upwards. Thus, there is both lift and thrust. In the **upstroke**, the wings are partly folded and their primary feathers spread out for the air to slip through, thus making it easier to lift them. They move up and backwards. As a net result, the bird will be propelled forwards and sustained in the air. The pigeon flies by flapping method. The flight includes the quick take-off by jump and flapping of wings in downward and backward stroke, which propel the bird forward, and by forward and upward movement of recovery stroke, this is done by bringing back the wing edgewise to the resistance of the air. The downward stroke of wings works as a lift force acting upwards. The lift force is proportional to the speed, and the requirement for sustained flight in the air is that the bird should have sufficient speed to generate a lift force equal to its weight. The flow of air over the upper surface of wings reduces the pressure and provides a portion of the lift. By tilting the wing the pressure on the underside can be increased. Pigeons can beat their wings at least eight times in one second.

4. Hovering. This is a peculiar kind of flopping flight. Support of weight without horizontal movement requires very much greater consumption of energy than forward flight and the largest humming birds weigh only 20 gm. The principle involved is also that used for short times by larger birds in take-off and landing (**Weiss-Fogh**, 1973). Lift for weight support requires horizontal movements and since the wings beat in a plane perpendicular to the body, hovering is achieved essentially by holding the body vertically (Fig. 28.4D). In this way the humming birds can remain in one place in the air or even move backwards. The wings beat backwards and forwards, often as fast as 200 times a second.

FLIGHT OR AERIAL ADAPTATION OF BIRDS

Birds are the only amniotes whose bodies are highly specialised for aerial mode of life. Besides aerial mode of life, most birds are also well adapted for terrestrial mode of life. However, during their evolutionary course, a marked adaptive radiation has occurred in them, as certain birds either became exclusively terrestrial and flightless (*e.g.,* ostrich, kiwi, etc.), and certain other birds became well adapted for aquatic as well as terrestrial mode of life (*e.g.,* ducks). Here we are mainly concerned with **aerial** or **volant adaptations** of birds, therefore, we will describe them in detail.

Most present-day aerial birds are **"heavier-than-air"** machines and almost every part of their organisation has been modified for aerial life. To fly in air, they have to possess following essential specialisations in their morphology, anatomy and physiology : (*i*) Organs for flight, (*ii*) Lightness and rigidity, (*iii*) Evolution of extra energy with provision for high power, (*iv*) Speed and (*v*) Balancing and control. All these requisites for volant life of birds can be studied under the following headings :

A. Morphological Adaptations

Most birds possess following important flight or volant adaptations :

1. Body contour. Because speed is a must for aerial life, so, to minimise the resistance offered by air during flight, the body of birds is fusiform or spindle-shaped and it lacks any extra projection which may offer resistance in the attainment of speed in air like fish in the water.

2. Compact body. Their compact body is light and strong dosrsally and heavy ventrally which helps in maintaining equilibrium in the air. The attachment of wings high upon the thorax, the high position of light organs like lungs and sacs, and low central position of heavy muscles, sternum and digestive organs below the attachment of both the wings and consequently low centre of gravity are other morphological facts of great significance.

3. Body-covering of feathers. Body of all birds is covered by special integumentary derivatives called **feathers**. Feathers are diagnostic of birds, since no other group of animal kingdom has ever developed them. Feathers have following advantages for birds :

(*a*) The smooth, closely fitting and backwardly directed contour feathers make the body streamlined helping them to pass through the air by reducing the friction to the minimum.

(*b*) The feathery covering makes the body light and at the same time protects from the hazards of environmental temperature.

(*c*) The feathers hold a considerable blanket of enveloping air around the body and add much to its buoyancy.

(*d*) The non-conducting covering of feathers insulates the body perfectly and prevents loss of heat which enables the bird to endure intense cold at high altitudes and also to maintain a constant temperature.

(*e*) Feathers of wings form a broad surface for striking the air.

4. Forelimbs modified into wings. The forelimbs have transformed into unique and powerful propelling organs, the **wings**. The wings

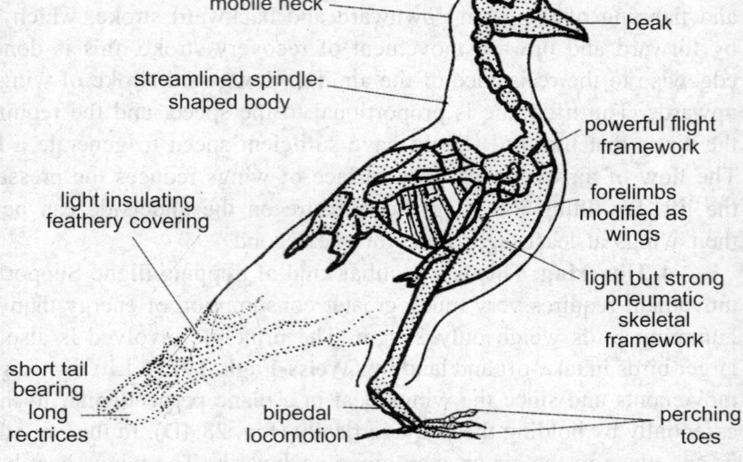

Fig. 28.5. Diagram showing the streamlined body of bird.

are the sole organs of flight. These organs have complicated structural constructions consisting of a framework of bones, muscles, nerves, blood vessels, feathers, etc. Both wings spring from the anterior region of trunk. During rest they remain folded against the sides of the body, but during flight they become expanded. The surface area of the wings is increased by the development of elongated flight-feathers, the **remiges**. The **vane** of each remix forms a flexible and continuous surface for striking the air in flight. The flight feathers of a wing also form a broad surface for supporting the bird in air. The particular shape of the wing, with thick strong leading edge, convex upper surface and concave lower surface, causes reduction in air pressure above and increase below, with minimum turbulence behind. This helps in driving the bird forwards and upwards during flight.

5. Mobile neck and head. The transformation of forelimbs into wings is duly compensated by the presence of beak or bill used for feeding, nest building, preening, and offence and defence. The mouth is drawn out into a horny beak which acts as a pair of forceps in picking up the things and in various other activities such as nest building, pruning, etc., which are normally done by forelimbs in other animals. The neck in birds is also very long and flexible for the movement of head necessary for various functions.

6. Bipedal locomotion. As anterior part of the body of birds becomes concerned with flight, the posterior part of body becomes modified for movement on land. For locomotion on the ground and to support the entire body weight, the hindlimbs occupy a somewhat anterior position on the trunk and become more stouter in case of ratites which are running birds.

7. Perching. The hindlimbs of a bird are well specialised for an arboreal life. Their muscles are developed in such a manner that when a bird sits on a branch of the tree, the toes close round the twig automatically. This happens due to so called **perching mechanism**. When the bird settles on the branch of a tree, the legs are bent and put the **flexor tendons** on the stretch. With the exertion of the pull, the toes are bent spontaneously around the perch. A bird can go to sleep in this position without any fear of falling off.

8. Short tail. The short tail of a bird bears a tuft of long tail feathers or **rectrices**, which spread out in a fan-like manner and serve as a rudder during flight. They also assist in steering, lifting and counterbalancing during flying and perching.

B. Anatomical Adaptations

Flying birds have following anatomical modifications for volant life :

1. Flight muscles. The action of the wings is controlled by the flight muscles which are greatly developed, weighing about one-sixth of the entire bird, while the muscles of the back remain greatly reduced. The muscle fibres comprising the flight muscles are of striated type and well vascularised to withstand fatigueness after prolonged activity. The wings are depressed by large muscles, **pectoralis major** and elevated by **pectoralis minor**. Other muscles are small and help the above muscles in their functioning.

2. Lightness and rigidity of endoskeleton. The skeletal framework of flying birds is very stout and is lightly built on the "**hollow-girder principle**". Most of the bones are **pneumatics**, filled with air sacs and provided with a secondary plastering to make them rigid. Bone marrow is lacking in bones of birds. Further skeletal framework becomes compact, centralised, rigid due to **fusion of bones**. Endoskeleton of birds thus contains the following characteristics :

(*i*) The **skull bones** are paper-like thin and show a tendency towards the reduction in their number. These bones are firmly fused with each other. The posterior portion of the skull is spongy. **Teeth** are lacking.

(*ii*) All the thoracic vertebrae except the last are fused into a single mass giving rigidity to the dorsal part of vertebral column. Fusion of vertebrae provides a firm fulcrum for the action of wings in striking air. The uncinate processes of thoracic ribs help in providing compactness, necessary for flight by concentrating the mass. The arched clavicles fused with interclavicle and powerful pillar-like coracoids of pectoral girdle are well suited to resist the inward pressure of the down-stroke.

(*iii*) The **heterocoelous** vertebrae confer great flexibility and birds can move their neck through 180°, which help in preening feathers in all parts of the body.

(*iv*) The shortening of caudal vertebrae and formation of **pygostyle** has assisted stability in air.

(*v*) Sternum or breast bone is expanded having a median ridge or **keel** for the attachment of major flight muscles in flying birds, while it is without a keel in running birds, like ostrich. Sternum is also hollow in which viscera are located.

(*vi*) The fusion of the pelvis with synsacram (*viz.,* fused mass of last thoracic vertebra + lumbar vertebrae + sacral vertebrae + few anterior caudal vertebrae) not only supports the weight of the body when the bird is walking, but also counteracts the effect of shocks as the bird alights. The absence of a ventral symphysis of ischia and pubis permits laying of large eggs.

(*vii*) The fusion of distal tarsals with the metatarsals to form a **tarsometatarsus**, and that of proximal tarsals with the lower end of tibia to form a **tibiotarsus**, help to strengthen the legs for bipedal gait.

(*viii*) The skeleton of forelimbs is completely modified for the attachment of feathers (remiges) and flight muscles. There are only three digits, which are more or less fused.

3. Digestive system. The rate of metabolism in birds is very high, so the food requirements are great and digestion is rapid. Most birds are very selective in their diet and accordingly their beaks are variously modified. Further, because undigested waste is minimum and is immediately got rid of, consequently the rectum becomes much reduced in length and never stores the undigested food. The ill-development of rectum of flying birds indicates towards the fact that the flying animals cannot afford to bear the weight of faeces. The absence of gall bladder in birds minimises the bodyweight to some extent.

4. Respiratory system. As a flying bird requires great and sustained power, therefore, its respiratory system is specialised in such a fashion that the food is combusted (oxidised) rapidly and completely to liberate large amount of energy. To meet the extensive rate of metabolism greater amount of oxygen molecules is needed by the body tissues. For this purpose, the dense, compact, and inelastic lungs are supplemented by a remarkable system of **air sacs**, which grow out from lungs and occupy all available space between internal organs, even extending to the cavities of hollow bones. The air sacs primarily reduce the specific gravity of the bird and also facilitate complete aeration of the lungs. The avian lungs are aerated twice at each breath which secures perfect oxygenation of blood. The air sacs help in regulating body temperature by internal perspiration. Further, insertion of air sacs in between the flight muscles like pads reduces mechanical friction and increases the mobility in muscular action.

5. Circulatory system. Rapid metabolism requires large oxygen supply to the tissues, which can be achieved by an efficient circulatory system. Accordingly, avian heart is large-sized, four-chambered, powerful and efficient. Due to double circulation in it, the oxygenated and deoxygenated bloods remain completely separated. Further, red blood cells of birds contain large amount of haemoglobin which is responsible for quick and perfect aeration of body tissues.

6. Warm-bloodedness. Due to perfect aeration of blood, the body temperature remains high (40°– 46°C) and does not change with change of environmental temperature. For this reason birds are called warm-blooded or homeothermal animals. The high and constant body temperature enables the bird to take flights at high altitudes and also facilitates activeness in every season.

7. Excretory system. The avian excretory system becomes specialised in three ways : (*i*) For the retention of water, the uriniferous tubules with **Henle's loops** are efficient in water absorption. The coprodaeum of cloaca is another efficient water-absorbing organ of birds.

(*ii*) For reducing the weight of body, there occurs no urinary bladder and the semi-solid urine is immediately excreted out, not retained for long in the body. (*iii*) The metabolic nitrogenous wastes are converted into less toxic and insoluble organic compounds such as **uric acid** and **urates**, which is an important physiological volant adaptation.

8. Brain and sense organs. The avian brain is highly developed consisting of well-developed centres of equilibrium, muscular coordination and instinct. The cerebellum is much developed and convoluted controlling the sense of equilibrium and muscular coordination. The cerebrum is also large and relatively smooth controlling voluntary movements, behaviour, intelligence and memory. Birds have to depend mostly on the sense of sight, so, the eyes are large and optic lobes are well developed due to acute vision. Sense of smell is poorly developed corresponding the ill-development of olfactory lobes.

9. Reproductive organs. In female birds, the presence of a single functional ovary of left side also leads to reduction of body weight which is essential for flight.

Thus, it becomes evident that birds are fully developed for terrestrial, arboreal and aerial environments.

BEAKS OF BIRDS

The entire modern world of Aves (birds) is characterised by the absence of teeth. Their absence is compensated by the development of horny sheaths called **rhamphotheca** for the jaws which constitute the beak. The beak is also referred to as bill. The upper and lower jaws are often described by naturalists as the upper and lower mandibles. The sheath of each jaw in the majority of birds is entire, but in some it is composed of several pieces. The beak is essentially a structure to oblain food, to preen feathers, to collect nest materials to build the nest and also to act as the organ of defence.

The shape of the beak varies in different birds according to their feeding habits. Because of the functional diversities, the beaks have undergone extensive range of modifications in different birds. Figure 28.6 shows a few varieties of beaks in birds. Some of the most important and common types are described here.

Types of Beaks in Birds

1. Seed-eating beak (Seed-crushing beak). (*a*) The seed-eating beaks are short, stout, conical and pointed at the tip. (*b*) This type of beaks are characteristic of small seed-eating or graminivorous birds such as **sparrows, finches** and **cardinals**, etc. (*c*) The weaker beaks are used to pierce the small seeds, while powerful beaks crush large, hard-shelled seeds, fruit stones, etc.

2. Cutting and biting beak. (*a*) The cutting and biting type beaks are long and strong and provided with sharp horny edges. (*b*) This type of beaks can be used for various purposes. (*c*) These are found in **crows** (*Corvus*) and **rovens**, etc.

3. Fruit-eating beak. (*a*) The fruit-eating beaks are large, sharp, powerful and hooked. (*b*) The upper beak is movable on the skull and is well adapted for tearing fruits and gnawing hard nuts and seeds. (*c*) Such beaks are found in **parrots** and **cockatoos**. In hornbills beak is large and heavy having cellular structure in its inner region that act as resonators to produce loud sound.

4. Piercing and tearing beak. (*a*) The piercing and biting beaks are short, pointed, sharp-edged and hooked at the tip and are adapted for piercing and tearing the flesh into pieces. (*b*) They are operated by well developed mandibular muscles. (*c*) These are commonly found in carnivorous birds shuch as **vultures**, **hawks**, **eagles**, **kites** and **owls**.

5. Insectivorous beak. (*a*) The insectivorous beaks are found in **swifts, swallows, flycatchers, night jars** and **hoopoe**, etc. (*b*) In **hoopoe** the beak is long, slender and slightly curved, adapted for turning the leaves and probing into soil for insect larvae, pupae and mites, etc. In swallows,

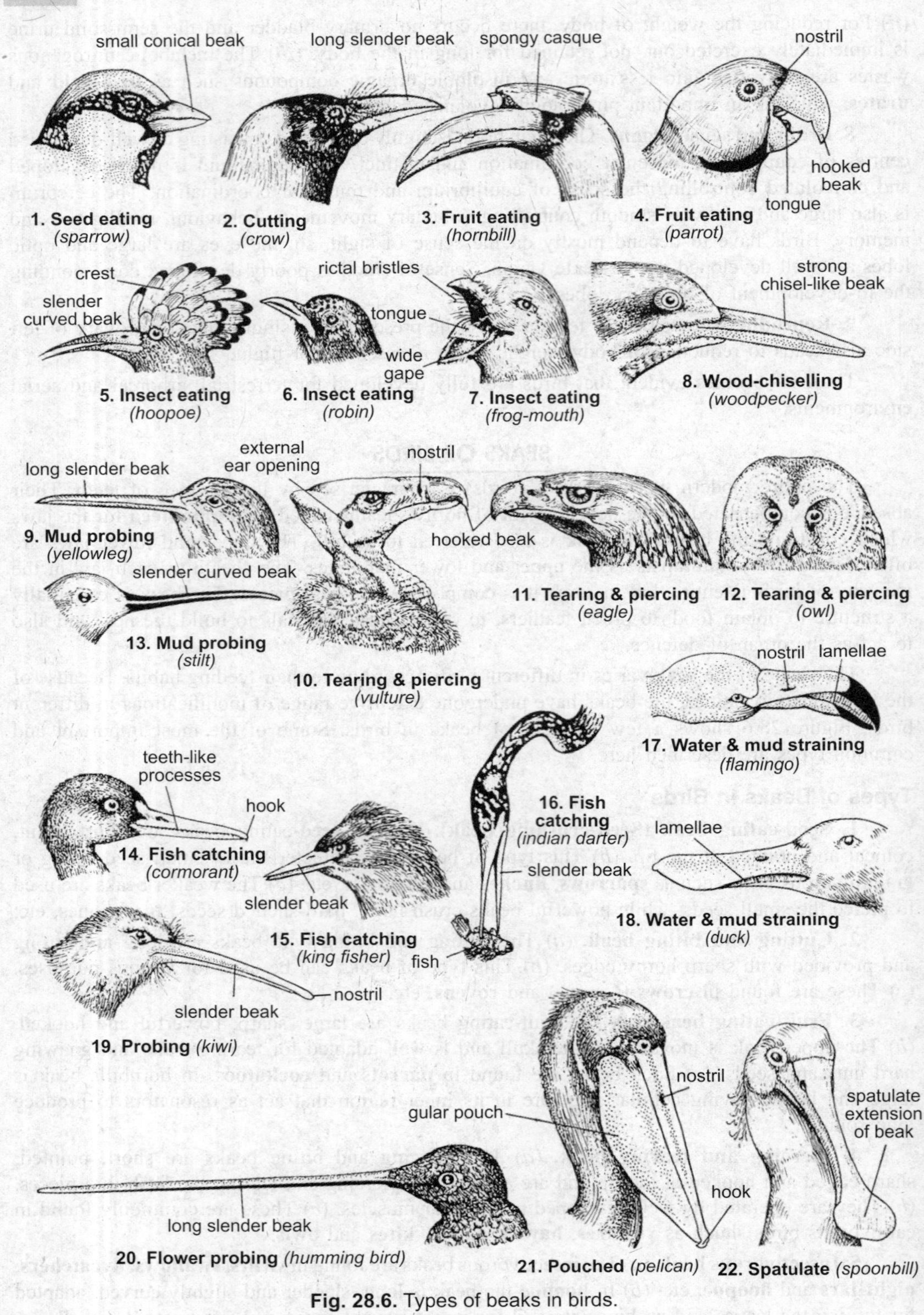

Fig. 28.6. Types of beaks in birds.

frog-mouth and swifts, the beak is small, wide and delicate to prey upon flying insects. In flycatchers, the beak is small, strong with notched mandibles at the tip end. Rictal bristles are also present at the base of the beak.

6. Fish-catching beak. (*a*) The fish-catching beaks are long, powerful and sharply pointed found in **storks**, **herons**, **kingfishers**, etc. (*b*) In **cormorant** the beak is slightly curved at the tip and have lateral tooth-like processes adapted for capturing the fish. In snake birds (Indian darters), these lateral tooth-like processes over the beaks are in the form of needle-like processes. (*c*) In **kingfisher** the beak is long, powerful and sharply pointed to capture fish, frogs, tadpoles, molluscs and other aquatic animals.

7. Wood-chiselling beak. (*a*) The wood-chiselling beaks are long, straight, stout and chisel-like and are strongly articulated with the skull. (*b*) The skull bones are also thick and shock absorbent, Neck muscles are also strong. (*c*) This type of beaks are used for chiselling and drilling into bark or wood of trees for the prey, *e.g.*, insects and their larvae and for nest construction. (*d*) It is commonly found in **woodpeckers**.

8. Mud-probing beak. (*a*) This type of beaks are very long, slender and slighly curved to probe the mud deep for searching the aquatic worms and larvae. (*b*) These are commonly found in **stilts**, **snipes**, **sandpipers**, **jacanas**, **lapwings** and **curlews**, etc.

9. Mud and water straining beak. (*a*) This type of beaks are broad and flat, and the margins of the jaws are provided with transverse lamellae or horny serrations making it an efficient filter or sieve. (*b*) Because of the presence of transverse lamellae on the margin of jaws, the mud and water pass out leaving the food into the mouth. (*c*) This type of beaks are commonly found in **ducks**, **teals**, **geese** and **flamingos**. In flamingo, the beak is distally decurved. The two halves of lower jaw are enlarged to form a wide cavity that is covered over by upper jaw.

10. Flower-probing beak. (*a*) This type of beaks are long, narrow and pointed for probing flowers and for sucking honey. (*b*) It is found in **humming birds**, which become stationary in air over the flowers for sucking honey. Their beaks are slightly decurved adapted for such function.

11. Pouched beak. (*a*) It is found in **pelicans**. (*b*) The beak is large having a large gular pouch. (*c*) The gular pouch is an extension of skin and attached to the mandible for storing engulfed fish.

12. Spatulate beak. (*a*) It is a characteristic of **spoonbill**. (*b*) It is specially flattened throughout the length but ends in a broad spatulate or spoon-like expansion. (*c*) The spatulate expansion is used for dabbling in mud and water in search of fish, tadpoles, worms, insects and other small aquatic animals.

TYPES OF FEET OR CLAWS IN BIRDS

The feet of birds are also modified variously in accordance with the character of the environment and the manner of locomotion (Fig. 28.7).

1. Running or cursorial feet. (*a*) In running birds, the legs are powerful and number of toes are reduced. (*b*) The hind toe is elevated, reduced or absent. (*c*) Ratites, *e.g.,* **Rhea, Emu** and **Cassowary** and **coursers**, **bustards** are short-toed birds and having 3 toes directed forward. In **ostrich** the legs have only two toes, *i.e.,* 3rd or middle toe and 4th or outer toe is smaller and devoid of nail.

2. Perching feet. (*a*) The perching type of feet are found in majority of birds (passer birds), *e.g.,* **sparrows, crows, bulbuls, koels, mynas** and **robins**, etc. (*b*) In this type of foot there are four movable toes, three toes in front are slender and the hind or **hallux** is long, strong and sharply clawed. (*c*) Because of opposable toes these birds can fasten the foot to a branch of tree.

3. Climbing feet. (*a*) The climbing type of feet are found in **parrots, woodpeckers** and **hoopoes**. (*b*) The feet are zygodactylous. The second and third toes are directed forwards,

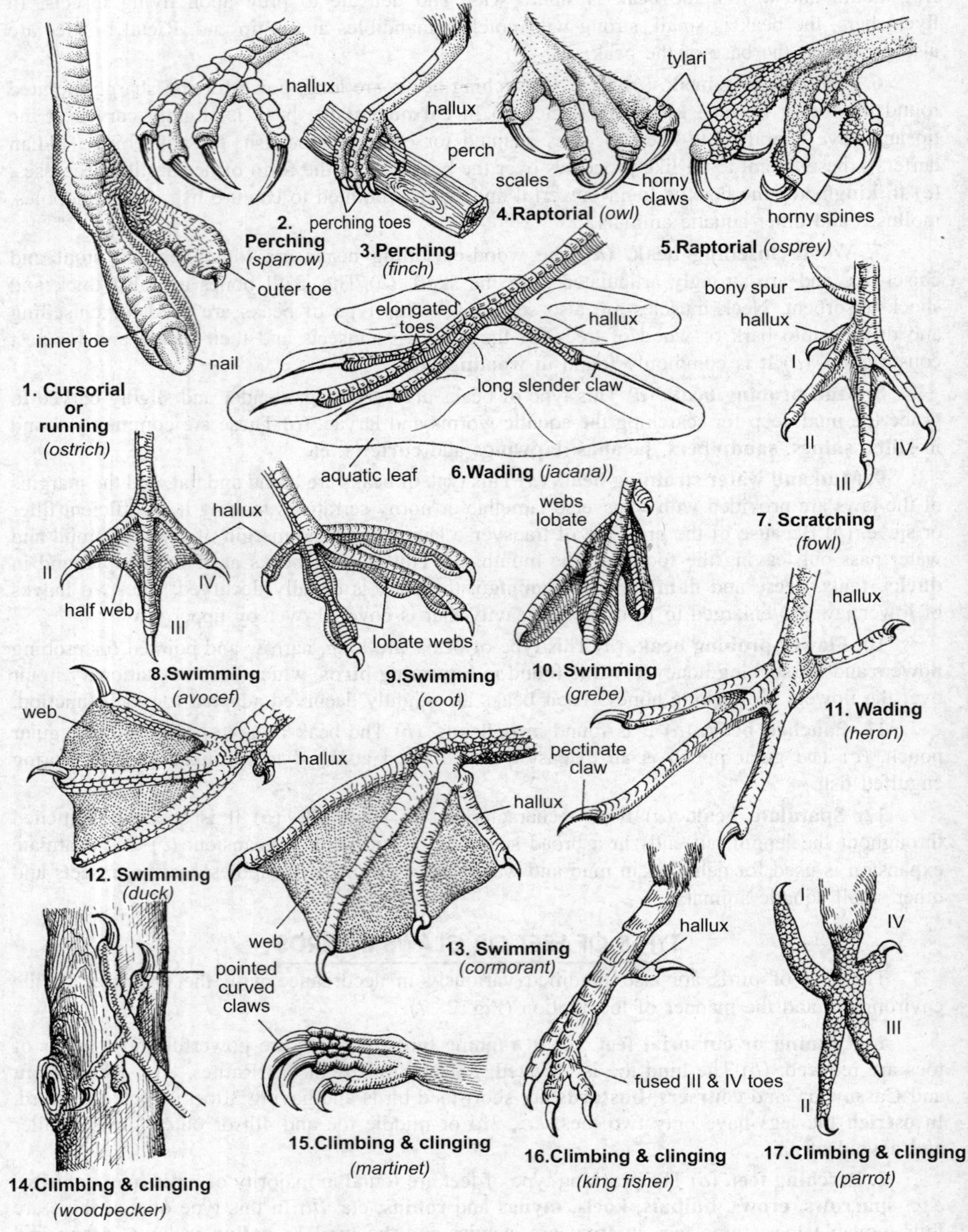

Fig. 28.7. Types of feet or claws in birds.

while the first and fourth toes are directed backwards. (*c*) The foot is modified for grasping and especially adapted for climbing on vertical surfaces of trees and walls.

4. Clinging feet. (*a*) The clinging type of feet are found in **martinets, swifts** and **humming birds**. (*b*) All the four toes are directed forwards and have long, sharp and curved claws. (*c*) The foot is modified for clinging on the trees or any vertical surface while feeding.

5. Scratching feet. (*a*) The scratching type of feet are found in **fowls, quails** and **pheasants**. (*b*) The feet are stout and provided with strong clawed toes. (*c*) The second, third and fourth toes are directed forwards, while the first toe is directed backwards. (*d*) The foot is modified for running as well as scratching the earth. (*e*) In the male, the foot is usually provided with a pointed bony **spur** for fighting and for treading the females.

6. Raptorial feet. (*a*) The raptorial type of feet are found in predatory or carnivorous birds such as **ospreys**, **hawks**, **kites**, **eagles**, **owls** and **vultures**, etc. (*b*) All the four toes are large and the hallux is strongly developed. The claws are large, strong, sharp and curved. The undersurface of all toes possesses large fleshy bulbs called **tylari**. These are well developed in hawks. (*c*) In **osprey** and **ketupa**, tylari are absent but sharp denticulated spines are present which serve to the slippery creatures (fish). (*d*) The foot is modified for grasping and holding the prey.

7. Swimming feet. In swimming birds the feet are modified in two ways, *i.e.*, **swimming** and **diving type** and **swimming** and **paddling type**.

(*a*) Swimming and diving feet. (*i*) In swimming and diving type of feet, toes are partially or completely webbed. This type of feet are found in **coots** and **grebes**. In these birds the web is lobate on the undersurface and the toes are free. (*ii*) In grebes each toe is paddle-like bordered on either side by fleshy lobes of integument. (*iii*) The nails are flat in a grebe and in coot slightly curved. (*iv*) This type of feet are modified for swimming and diving in water.

(*b*) Swimming and paddling feet. (*i*) Swimming and paddling type of feet are found in **cormorants**, **pelicans**, **ducks** and **teels**. Here the toes are webbed. (*ii*) In **cormorants** and **pelicans** all the four toes are connected with tough web all along their length. (*iii*) In **ducks** and **teals** the first toe in comparatively small and free, while three anterior toes are completely webbed.

8. Wadding feet. (*a*) The wadding type of feet are found in **herons**, **jacanas**, **lapwing** and **snipes**, etc. (*b*) In this type the legs and toes are very long and slender and the web is absent or feebly developed. (*iii*) The feet are modified for walking over the aquatic leaves (vegetation) or marshes.

PALATE IN BIRDS

In general, the skull in various birds does not show much variation. Amongst some of the universal skull characters of birds are : an early ankylosis of bones except in Ratitae, rounded and spacious cranium, single small rounded occipital condyle formed by basioccipal, the upper beak composed mainly by the two premaxillae united into a large triradiate bone, the slender maxillo-jugal arch, the large parasphenoidal rostrum, the freely articulated quadrate and the reptilian post-frontals. Orbits are large, separated by a thin interorbital septum formed by the mesethmoid which is continued anteriorly with cartilaginous internasal septum.

Importance of palate in classification. There are, however, marked differences in the structure of the palatal region of skull in birds which provides an important character for their classification. The living birds are classified in two suborders : Palaeognathae and Neognathae. In the **Palaeognathae** or **Ratitae** are included all the flightless birds. They have a **dromaeognathous** type of palate in skull in which the large vomer is extended posteriorly, so that the two palatines do not meet with one another and with the rostrum. On the other hand, in **Neognathe** or **Carinatae** are included all the modern flying birds. They have a **neognathous** type of palate with three subtypes (**schizognathous, desmognathous** and **aegithognathous**), in which the vomer is small or absent, so that the palatines meet the rostrum.

Kinds of Palate in Birds

Huxley, in 1877, pointed out the following four types of palate in birds (Fig. 28.8), based on the relations of vomer, palatines, pterygoids and maxillo-palatine processes.

1. Palaeognathous or dromaeognathous palate. This type of palate is characteristic of Ratitae, such as the ostrich, rhea, kiwi and tinamous, etc.

(*i*) Vomer is large and broad behind connected with the palatines.

(*ii*) Palatines do not articulate with the parasphenoidal rostrum, because the vomer intervenes between the two.

(*iii*) Maxillo-palatine processes are small and do not unite with one another or with the vomer.

(*iv*) Basipterygoid processes of basisphenoid are well developed and they articulate with the hinder part of the pterygoids.

(*v*) Pterygoids are immovably fixed to vomer and are reptilian in form.

This type of palate is primitive and occurs in the Palaeognathae.

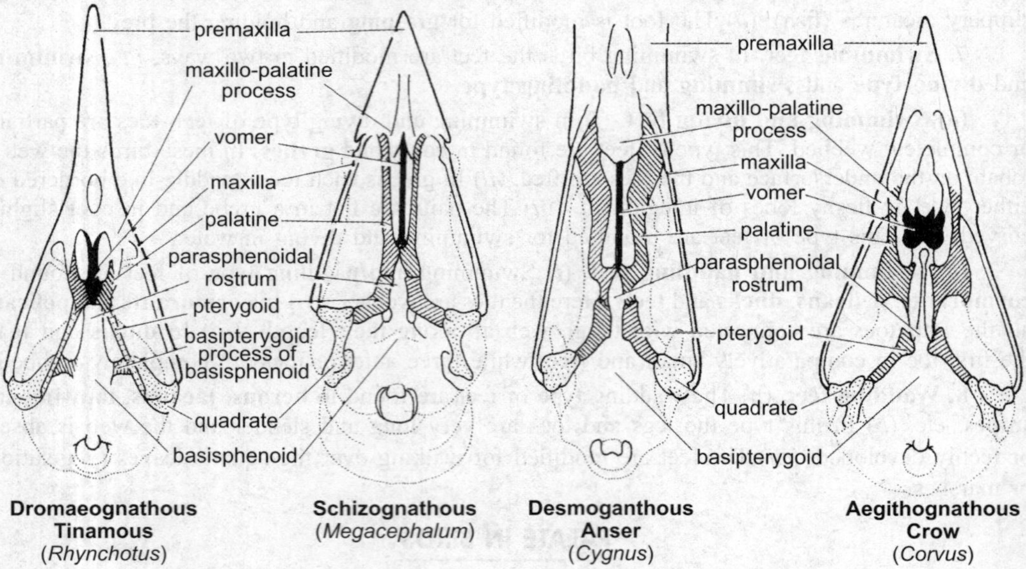

Fig. 28.8. Types of skull palate in birds.

2. Schizognathous palate. This type of palate is common in a variety of birds, such as pigeons, fowls, gulls, plovers, cranes, woodpeckers, trogons, etc.

(*i*) Vomer is small and pointed in front, or absent.

(*ii*) Palatines and pterygoids articulate with the parasphenoidal rostrum at the point where they join one another.

(*iii*) Maxillo-palatine processes do not unite with one another or with the vomer.

(*iv*) Basipterygoid processes may be absent or small and arising at the base of the rostrum.

(*v*) Pterygoids are movably articulated.

3. Desmognathous palate. This type of palate is common in most of the wading and swimming birds such as storks, herons, ducks and geese, besides parrots, birds of prey, cuckoos, etc.

(*i*) Vomer is often abortive or so small that it disappears in the skeleton. When present, it is always narrow, slender and tapers to a point in front.

(*ii*) Palatines and pterygoids articulate with the rostrum.

(*iii*) Maxillo-palatines are large and united with one another across the middle line, often forming a flat, spongy palate ventral to the vomer.

(*iv*) Basipterygoid processes are absent.

(*v*) In parrots, a special type of desmognathous type of sliding palate occurs so that the depression of the lower jaw automatically raises the upper jaw.

4. Aegithognathous palate. This is similar to the schizognathous type of palate. It occurs in passerine birds such as crows, swifts, bulbuls, etc. Vomer is short and broad and truncated instead of being pointed in front. Posteriorly the vomer is deeply cleft embracing the rostrum.

MIGRATION OF BIRDS

Individual organisms often move about in their environment in response to a varied group of stimuli. When such movements result in the temporary or permanent absence from one home range and the establishment of a residence in another, the resulting movement is said to be a **migration** (L., *migrare* = to travel). Migration is a periodic passing of animals from one place to another (**Cahn**). Bird's migration is a two-way journey, *i.e.,* a regular, periodic, to-and-fro movement of some birds between their summer and winter homes or from a breeding and nesting place to a feeding and resting place.

Migratory and resident birds. Majority of birds have the inherent quality to move from one place to another to obtain the advantages of the favourable condition. Such birds are called **migratory birds**. Some birds such as bobwhite and ruffled sand grouse, however, do not migrate at all and they remain throughout the year in a country. Such birds are called **resident birds**. Migratory birds cover thousands of miles in their periodic seasonal journeys. Migration is a remarkable device to obtain the advantage of the favourable conditions in more northerly regions only during the summer. Migration occurs during the particular period of the year and the birds usually follow the same route.

Kinds of Migrations

The migratory birds migrate in a variety of ways and accordingly following kinds of avian migrations have been recognised by different ornithologists :

1. Latitudinal migration. The latitudinal migration usually means the movement from north to south and *vice versa*. Most birds live in the landmasses of the northern temperate and subarctic zones where they get facilities for nesting and feeding during summer. During winter, when the northern hemisphere becomes covered with ice and snow, these birds move towards South for shelter. Several North American and Eurasian birds cross the equator to spend winter in deeper and warmer parts of South America and Africa. The American golden plover (*Pluvialis dominica*) passes the nine months of winter 8,000 miles south in the plains of Argentina, thus, enjoying two summers each year and knows not a hint of winter. Ruff breeds at Siberia and travels to Great Britain, Africa, plains of India and Sri Lanka, thus, travelling a distartce of 6,000 miles.

An opposite but lesser movement also occurs in the southern hemisphere when the seasons are changed. Some tropical birds migrate to breed during rainy season to the outer tropics and return to the central tropics in dry season. Marine birds may also make extensive migrations. The Arctic tern (*Sterna*) breeds in the north temperate zone, and migrates to the Antarctic along both sides of the Atlantic. Penguins make migrations by swimming.

2. Altitudinal migration. Altitudinal or vertical migrations from high mountains in the summer to low valleys in the winter occur in many Indian and foreign mountaineer birds. In India a number of species during summer migrate from planes to the slopes of Himalayas ascending thousands of feet above sea level and return to planes on the commencement of winter, *e.g.,* common wood-cock, Bush chat and *Scolopax ruticola.* Such altitudinal migration also occurs in the grebes and coots of Andes in Argentina, violet green swallows of Great Britain, and the willow ptarmigan of Siberia. The brown plumage of willow ptarmigan turns white in winter, and diet of insects shifted to buds and twigs of alders and willows.

3. Longitudinal migration. Longitudinal or east-west and vice-versa migration also occur in some birds. The Patagonian plover visits the Falkland Island, and south Patagonian in September and October for breeding.

4. Partial migrations. Many species, such as blue birds and many blue jays of Canada and northern United States and barn owls (*Tyto alba*) migrate southwards to mingle with the sedentary populations of the southern states, are only **partial migrants**. In such cases, all the birds of a group of migratory birds do not leave the native land, visible throughout the year. Actually these are partial migrants, because the birds visible in winter are not the same as seen in summer. Songthrush, redbreast, titmouse, finch, etc., are partial migrants.

5. Erratic migration. The **erratic**, **vagrant**, **irregular** or **wandering** migration occurs in great blue heron, cuckoos, thrushes and warblers. In such birds, after breeding, the adults and the young may stray from their home to disperse in all directions over many or a few hundred miles in search of food and safety from enemies. Sometimes hurricanes take the sea birds as far as 2000 miles away from home seas and there they die either due to exhaustion or due to unknown shores.

6. Seasonal migration. Some ornithologists of temperate countries have classified migratory birds according to seasons. Thus, in Britain, swifts, swallows, nightingales and cuckoos are **summer visitors**, because they arrive in spring from the south, remain there to breed and leave for the south in autumn. Some birds, such as fieldfare, snow bunting and redwing, are **winter visitors**, as they arrive in autumn, chiefly from north, stay throughout the winter, and fly northwards again in spring. While some birds such as snipes and sandpipers are the **birds of passage**, seen for short time twice a year on their way to colder or warmer countries in spring and autumn.

Besides these kinds of avian migrations, following three kinds of migrations can also be recognised in different birds :

A. Climatic migrations occur as a result of daily or seasonal changes in the climate of the environment. The well known north-south migration of many ducks and geese is a good example of climatic migration.

B. Alimental migrations occur as a result of food or water shortages and may occur at any time in a year.

C. Gametic migrations result from a need to occupy some special region or environment for some part of the reproductive process. Most migratory birds perform gametic migration.

Modes of Flight in Migration

During their migration, most migratory birds display following significant features :

1. Time of migratory flights. The most migratory birds either fly during daytime or night-time and accordingly following two types of migration can be recognised :

(*a*) Diurnal migration. Some birds fly mainly by day, such as crows, swallows, robins, blackbirds, hawks, blue-birds, jays, cranes, loons, pelicans, geese, ducks, swans, and other shore birds. They may stop to forage in suitable places, but swallows and swifts capture their insect food in the air as they travel. These diurnal migratory birds often travel in flocks, which may be well-organised (ducks, geese and swans) or loose (swallows).

(*b*) Nocturnal migration. A vast majority of birds are nocturnal migrants. These include mostly small-sized birds, such as sparrows, warblers, thrushes, etc. These birds prefer to fly at night, under the protective cover of darkness, to escape their enemies. By flying at night, they arrive at the daybreak, take rest, procure food during daytime and then start again at the approach of night.

2. Range of migration. The range of migration commonly varies from one or few miles to thousands of miles in different groups of migratory birds but it is almost constant for a particular species. For example, the Himalayan snow partridges descend a few hundred feet and cover a distance of about one or two miles, *Chicades* travel (descend) about 8,000 feet, while golden plovers, sandpipers, boblinks and swallows cover 6,000 to 9,000 miles from the Arctic to the grassy plains of Argentina or the Patagonian beaches in South America. *Arctic tern* breeds along the coast of Labrador in summer and then migrates to its home destination to the edges of Antarctica in winter which is about 11,000 miles. It again returns in the summer to the coast of Labrador travelling about the same distance. White stork of Europe migrates to South Africa in winter taking a journey of about 8,000 miles. It is calculated that 5,000 million land birds migrate from Europe to Africa each autumn and half of them succeed in returning next spring (**Moreau**, 1972).

3. Altitude of flight. Some birds fly quite close to the earth, while most routine migration occurs within 3,000 feet .of the earth. Certain small land birds have been reported to fly in night at 5,000 to 14,000 feet altitudes. Further, some avian species even cross the Andes and the Himalayas at altitude of 2,000 feet or more.

4. Velocity during migration. The speed or velocity of flight of migratory birds varies from individual to individual and species to species. It is affected by the speed of air and its direction, and it has been reported that birds travel faster during migration than at any other time. The flight speeds of certain migratory birds has been tabulated in the following table :

TABLE 28.2. FLIGHT SPEEDS OF BIRDS (AFTER WETTY, 1962).

	Species	Distance (miles)	Elapsed time of flight (Days)	Miles per day
	A. Speed of return to nest after being experimentally displaced.			
1.	Manx shear water	3293	12.9	263
2.	Laysan albatross	3200	10.0	320
3.	White stork	1404	11.9	118
4.	Noddy tern	1081	5.0	216
5.	Homing pigeon	1007	1.5	671
6.	Alpine swift	1007	3.0	336
7.	Herring gull	870	4.1	212
	B. Speed in moving from one point to another during migration.			
8.	Arctic tern	8699	114	76
9.	Slender-billed shear water	6835	53	129
10.	Short-billed dowitcher	2299	20	115
11.	Lesser yellow legs	1926	7	275
12.	Peregrine falcon	994	21	47
13.	Mallard	553	2	276
14.	Turnstone	497	25	19

During migration, thus, cranes, carrians, crows and finches fly with the speed of 30 miles per hour, while cross-bill flies with the speed of 30 miles per hour. The maximum speed recorded so far by **E.C. Stuart** is about 170–200 miles per hour (*e.g.,* Indian swifts). Birds cover hundreds of miles nonstop in a day or a night with an average of about 500 miles. Migratory birds usually fly five or six hours per day and also take rest in between for drink and food. Golden plovers fly nonstop from Hudson Bay and Alaska to South America covering a distance of 2,400 miles.

5. Routes of migration. The migratory birds usually follow definite lines of flight. The route followed by them may be the same while going and returning back or may be different.

The nocturnal small birds migrate with the general air flow. In spring it takes place from South to North along warm air currents, and in autumn from North to South with the cool wind of North. Change in their course occurs due to configuration of land, coastline, path of great rivers and intervening mountain ranges, etc. Different migratory birds may follow the following routes during their migration :

(*a*) Sea routes. Marine birds follow sea routes. The land birds are known to cross as much as 400 miles of ocean in a stroke but if there are intermediate islands the distance covered may be more. Certain birds have been seen crossing the Atlantic Ocean between Azores and Portugal (900 miles) and the ocean between the continents of North America and Bermuda, etc.

(*b*) Coastal routes. The coastal routes afford migration for a large number of migrants. Certain important migratory coastal routes are : East Atlantic coastline, West Atlantic coastline, East Pacific coastline, West Pacific coastline, East Indies coastline and West Indies coastline.

(*c*) River-valley routes. While migrating from planes to hills and from hills to planes, the migratory birds cross rivers and river-valleys falling in the way.

(*d*) Mountain ranges. Very rarely the birds cross mountain ranges.

The river valleys, mountain ranges and coastal routes provide good landmarks for the migrating birds, which enable the birds to recognise and remember the routes and entrances to the countries. Deviations in path sometimes occur due to configuration of land, coastline, course of great rivers or intervening mountain chains.

6. Segregation during migration. Some birds such as kingfishers, swifts, and night-hawks travel in separate companies, but certain other birds such as swallows, vultures, blue birds, turkeys, etc., usually travel in mixed companies of several species due to similarity in their size, method of search of food, etc. In some avian species, the male and female individuals travel separately. Males arrive first to build the nests. The young birds usually accompany their mothers.

7. Order of migration. During migration the birds follow a definite order which is strictly followed. Normally the adults migrate first and they are followed by youngs. It has been found that urge of migration occurs due to the maturity of gonads which instigate them to migrate towards their breeding, grounds. Adult birds return to the same general and even detailed places at both ends of the journeys. Young birds mostly do not learn from their elders, indeed may leave before them, flying off in a direction that is presumably genetically determined. Hence, the adults with ripe gonads start the migration. During the return flight the order becomes reversed—the young birds start first and follow the same path which their parents had followed while coming from that place. In adult precedence, there is always a definite sequence, the adult males take the lead, adult females next in order and the birds of the year follow them and in the end come the weak and wounded birds.

8. Regularity of migration. Several species of migratory birds show a striking regularity, year after year, in their timings of arrival and departure. In spite of long distances travelled or vagaries of weather, they are often punctual within a day or two in their time of arrival. Further, most migratory birds come back to the same breeding place year after year.

Problems of Migration

The phenomenon of avian migration has remained an ornithological riddle and its various aspects are still little understood. Certain scientific explanations have been forwarded from time to time to explain certain problematical issues concerning the avian migration and they are the following :

1. Problem of way-finding or navigation. It is still a riddle that what guides the young ones of a migratory bird to migrate and follow the same course which their parents had taken. Probably it is genetically determined. Various explanations have been given for what determines the direction and course of migration.

***(i)* Topographic cues.** Topographic cues can give long range guidance in daylight. The migratory birds utilise various natural structures or topographical features, such as great rivers, river valleys, coastal lines, chains of oceanic islands, mountain ranges, etc., as the landmarks during their flight. Landmarks may be learned. Ducks and geese travelling in family groups transmit the information from generation to generation. Most birds make little use of landmarks except for the last stages of homing. They navigate mainly by the sun and stars.

***(ii)* Experience.** A few naturalists have suggested that the birds learn by experience. Some older members, benefited by a tradition of following a path in past several years, become leaders to guide the younger generations. However, birds certainly do not learn their migratory route from elders, as some of them do not fly in flocks at all.

(iii) Some ornithologists such as **Von Middendroff** and **H.L. Yeaglev** advanced the idea that birds navigate through responses to the **earth's magnetic field** and their inner ear reacts to the mechanical **Coriolis effect** produced by the rotation of the earth. But there is no evidence that birds have some magnetic sense, etc.

(iv) The east-west and north-south gradients of gravity and the magnetic intensity are supposed to have some function in avian migration.

(v) **Mathews** (1955) and **Sauer** (1957) have emphasised the position of the sun (during day time) or stars (during night) helps the birds to navigate along definite route. Experimental evidences by creating artificial planetarium have shown that the shifting of the position of sun causes a change of the migration during night.

Kramer (1951) experiments on caged starlings show restlessness at times when migration is due and flutter in the appropriate direction in relation to the sun. Kramer showed that birds have a clock that does this. It has been experimentally proved that pigeons return home after release from a distant place even if they have never been there before, nor have had any previous training in returning from situations out of sight from the loft. **Mathews** (1955) and others have shown that upon release the birds fly off towards home provided that they can see the sun.

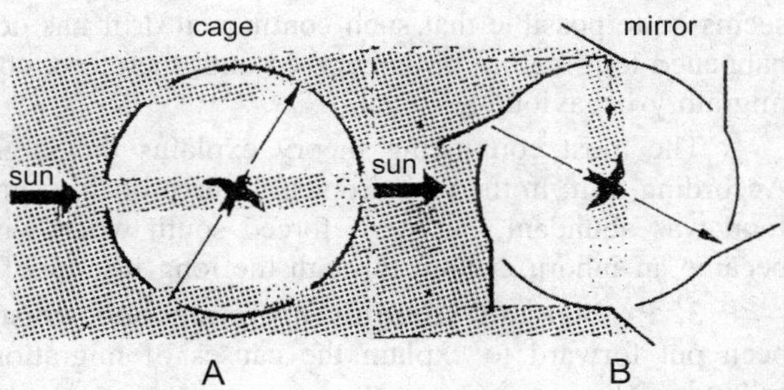

Fig. 28.9. Gustav Kramer's experiment with starlings to show that day migrants use sun as a compass.

Navigation by starlight has been studied by following the direction of restlessness movements of birds in a planetarium. Warblers and others altered their directions when the stars were shifted. The stars change and presumably it is a pattern that is recognised.

***(vi)* Telluric currents.** According to some naturalists, homing instinct and telluric currents enable the migratory birds to migrate along definite routes. It is probable that much migration in fact makes use of the wind. The economy of energy moving down wind has obvious selective value and this has made birds into very good meteorologists. Birds often fly at the height where the wind is most favourable.

Inspite of the evidence for navigation by the sun and stars much remain uncertain. It is not clear what determines the initial compass bearing and what is responsible for the reversal of migration between spring and autumn. Birds usually avoid flying in clouds. There is some evidence that birds can use magnetic cues. Pigeons released under conditions of total overcast 50 km from home take up an initial bearing towards home. Robins (*Erithacus*) oriented their nocturnal activity in the migratory direction even without visual cues. Particles of magnetite have been found in the brain, and muscles of the head and trunk of pigeons and sparrows. Magnetic forces may influence the cells and/or receptors in the muscles (**Presti** and **Pettigrew**, 1980).

2. Problem of origin of migration. This has remained a problem for biologists that what started migration or how and why phenomenon of avian migration is originated. **One theory** which seeks to explain the origin of bird migration assumes that the original home of birds was in the north at a time when these regions were warmer than they are now. With the advent of colder winters the birds were driven farther and farther south to find adequate supplies of food and escape the freezing weather. Each summer, however, they returned to their homelands to breed.

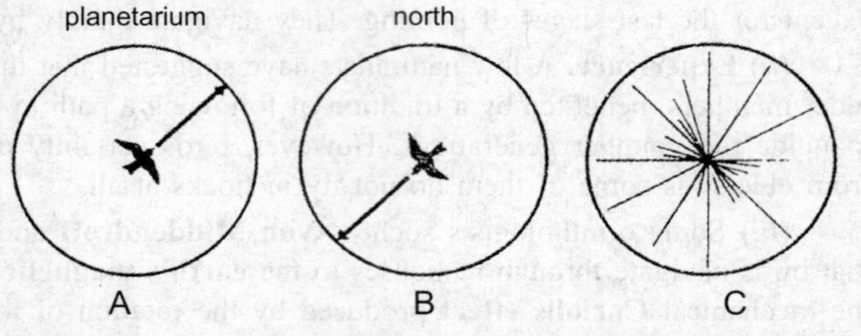

Fig. 28.10. Franz Sauer's experiment with warblers in a planetarium to show that night migrants use sun as a compass.

Another theory assumes that birds originated in more equatorial regions at a time when all the landmasses of the world were located there. Gradually, according to this theory, continents broke apart and drifted (the solid continental crust presumably "floats" upon the deeper, liquid basalt layers of the earth) to their present locations. Birds attempted to reach their original homelands to breed each summer, and these homelands gradually drifted farther and farther to the north. It seems quite possible that such continental drift has occurred, but most geologists believe that it happened long before birds evolved and, therefore, could not have had any influence upon their migratory behaviour.

The most convincing theory explains the migratory behaviour in **evolutionary terms**. According to it, in the past, tropical birds probably spread out in colder northern latitudes, where food was abundant, but were forced south when winter came. The event or series of events became an inborn custom through the long history of the race.

3. Problem of stimulus or immediate causes for migration. Several theories have been put forward to explain the causes of migration in birds. A few important theories are following :

(*a*) Environmental stimuli. External ecological stimuli such as growing scarcity of food, decrease of day-length. or sun glare, increased cold and stormy weather, increase of atmospheric pressure, etc., are found to excite birds to migrate to the better suited places. The heat glare and drought also provoke migration among birds.

Migrations are obviously correlated with the seasons and changes in day-length certainly play a central part in initiating migration. It was first shown by **Rowan** in 1925 that juncos (*Junco*) or crows submitted to artificially increased day-length in winter show precocious migratory movements if released. The longer days also caused growth of the gonads, which regress in autumn.

The increase of food availability in spring and its decrease in autumn may be the actual triggers to the migration change.

(*b*) Gonadial stimuli. It has been suggested that the ripening of sex organs in birds causes a physiological change which leads to an impulse for migration and the bird is evoked to leave its winter quarters and reach the summer breeding grounds. According to **Rowan** (1925), the spring migration is stimulated by the hormones secreted by gonads (*i.e.,* testes and ovaries) of birds and reduction of sex hormones produces the autumn migration. Later work has shown that this is only partly true and that other endocrine influences are also at work. Many nocturnal migrants make characteristic calls while in flight and these noises overhead may stimulate birds on the ground to join in the movement.

(*c*) Thyroid hypothesis. According to this hypothesis certain thyroid hormones produce certain necessary changes in the metabolism of migratory birds during the time of migrations and these changes compel the bird to migrate.

(*d*) Antipituitary hormone. The antipituitary hormone is found to regulate the migration and instinct to migrate.

(*e*) Metabolic hypothesis. Experimentally produced artificial prolonged day length in several migrant species can produce hyperphagia, increased body weight, deposition of fat and restlessness even in late autumn. Non-migrants showed no such effects. The temporal programme of migration is an inherited species-specific character.

The pattern of migration of many species must have been evolved quite recently. At the end of last glaciation 18,000 years ago there can have been hardly any breeding territories north of Mediterranean. The maximum of northern breeding was probably about 5,000 years ago and human clearance has probably reduced it by more than half. However, swallows (*Hirundo*), house martins (*Delichon*) and swifts (*Apus*) have taken advantage of human habitation.

Physiology of Migration

Because all long-distance migrants have to fly for several hundreds of miles without food or drink, therefore, in them deposition of fats takes place just prior to migration. The fat has a physiological advantage over the glycogen, which is the usual storage product of animal tissues, that fat produces more energy per unit weight. It also produces more water to cope with high rate of metabolism and breathing.

Disadvantages of Avian Migration

Migration is expensive in terms of food and energy requirements and it is dangerous in terms of risk of predation and exposure to climatic factors. Every year, hundreds of millions of migratory birds never reach their destination. The following factors create great problem to the migratory birds : cold weathers, snowstorms, hurricanes, shortage of food, strong current of wind, fog, man-made lighthouses, mountains, big buildings, television towers, ceilometer beams at the airports, etc.

Advantages of Avian Migration

If advantages of avian migration have not outweighed its disadvantages, it could not be undertaken. Indeed, such behaviour would never have been established through evolution had it not strong survival value for the species. Many species do not migrate, since there is no need for them to do so; other means of survival have been exploited instead. Advantages to be expected from a change of living range (or migration) include better climate, new or more plentiful food supply, increased space for breeding and nesting, longer daylight, working hours for gathering food to feed offspring, and the opportunity for gene exchange with individuals of other parts of the range.

ECONOMIC IMPORTANCE OF BIRDS

Birds occupy a very significant position in the human society. They provide us food, medicine, fertilizers, beautiful decoratives, give us enchanting songs and bring about pollination. In addition, they form various sources of amusement and exercise biological control over injurious crop pests. Besides their many beneficial aspects, certain birds are harmful for mankind. All sorts of economic significances of birds can be best studied under following headings :

I. Beneficial Birds

Some of the important beneficial birds to man are as follows :

1. As food. Both the flesh and eggs of birds are used as food by man. Certain animals such as snakes, cats, civets, mongoose, etc., also prey upon small birds and their eggs. The flesh of certain birds such as chicks, ducks, geese, praise chickens, pigeons, turkeys and shore birds is much relished and is regarded as delicacy. Several species of these birds have been domesticated and hybridised by man for many centuries to produce better qualities of feathers, flesh and messengers, etc. Domestic fowl is supposed to be a delicacy and its eggs are considered to be the best standard food, second only to milk. The eggs of fowl are also used in tofees, pastries, cakes and biscuits. The great demand of flesh and eggs of fowl has given rise to the paying poultry industry.

Besides flesh and eggs, even nests of certain birds have food value. For instance, Chinese use the nests of a particular species of swift (*e.g., G. collocalia*) as food. These are said to have tonic property and are formed purely of salivary secretion.

2. In medicine. Feathers and muscles of certain birds are utilised in certain **Ayurvedic** and **Unani** medicines. In Unani aystem, patients of chest diseases and tuberculosis are prescribed to keep pigeons' bodies in contact with the chest. Flesh of pigeon is said to be good for the patients of paralysis. The eggs of fowl are used in tonics and various medicines. The eggs of fowl have great experimental value, as they are used as a medium to grow cultures of viruses, bacteria, fungi, nematode larvae, etc.

3. Decorative and commercial value. The simple or multicoloured and festidious feathers of different birds have great economic importance for man. They have been used extensively for pillows, quilts, blankets, clothing and sleeping bags. The down feathers of aquatic birds, such as ducks and geese, provide the warmest insulation and much used for Arctic clothing, ski attire and for sub-zero. sleeping bags.

Feathers with beautiful and distinctive colours and shapes have been used in decoration of houses and ornamentation by native peoples and modern women. Red Indians of America use wing and tail feathers of golden eagles for ceremonial headdresses and feathers of flicker and valley quail for decoration. Domesticated poultry and ostriches, rheas, etc., provide ornamental feathers. Badminton shuttlecocks are manufactured out of feathers. Many natives use feathers in their arrows so that they may go farther. In Argentina and Brazil, the feathers of rheas are used as feather dusters. In India, the tail feathers of peacocks are used in house decoration, ornamentation and in making a variety of toys, fans, etc. In the Mughal period, the long feathers of birds were used as writing pens.

4. Agricultural value. The birds are of great agricultural value since they act as pollinator, fertilizer and help in biological control in the following manner :

(*a*) As pollinator. Many small birds like humming birds, living in grooves and meadows on flowering plants bring about their pollination. Certain flowers are specially designed for bird-pollination and respond to them only. The fruit-eating birds help in the dispersal of seeds far and wide.

(*b*) As fertiizer. The faecal matter of birds contains nitrogen, phosphates, calcium and iron, etc., and is called **guano**. It is extensively used as fertilizer. On islands of the coast of Chile, where little rains fall, and which are breeding places of numerous migratory and sea-birds, the guano accumulates in enormous quantities and is exported to different countries.

(*c*) In biological control. Birds prove one of the best friends of farmers as they exercise **biological control** over injurious crop pests. The seed-eating birds control the harmful weeds of agricultural fields by consuming thousands of tons of weed seeds annually. Likewise, the insectivorous birds devour millions of tons of harmful insects which live as pests on the crop plants. Certain carnivorous birds such as owls, eagles and hawks prey upon mice, rats, moles, squirrels, hares, rabbits and other rodents and, thus, protect the crops from damage. Birds also pick up larvae of insects which devour tbe plant leaves and also feed up worms of harmful nature.

5. As scavengers. Carrion-eating (scavenger) birds, like vultures, hawks, eagles and crows, etc., are of great sanitary value, as they feed upon the dead bodies and decaying organic substances. These birds check the spreading of such diseases, which might be caused due to decay and putrefaction of dead bodies. Certain other scavengers like cattle-egrets, ox-peckers, rhinoceros-bird, African starlings and crocodile birds, etc., remove parasites from the wild animals on which they live.

6. Miscellaneous value. Besides above described values, birds have following uses for man :

(*i*) As predators. Certain birds such as peacocks, eagles, kites and other carnivorous birds are destroyers of several venomous and injurious creatures, such as snakes, scorpions, etc.

(*ii*) As messengers. Some species and varieties of pigeons have been trained and used as messengers in wars and love affairs from early days.

(*iii*) As signals. Many birds profess changes in seasons. In India, the appearance of cuckoo suggests the onset of spring. The dance and peculiar voice of peacocks in cloudy weather indicate rainfall. Certain birds help hunters by giving **danger signals** to them. Some sparrows cry before carnivorous animals such as tigers hidden in the bushes. A typical path sparrow cry along the path of king cobra.

(*iv*) Entertainment value. Game birds like ducks, quails, herons, pheasants, etc., are used for entertainment. These birds are hunted by persons as a recreation.

(*v*) Singer birds like cuckoo, bulbuls and canary amuse persons by their sweet and melodious voice.

(*vi*) Parrots, Mynas, and Parakeets amuse man by repeating their voices and are, thus, good sources of recreation and so they are kept in cages and aviaries as pets. Pigeons, cocks and partridges are tamed for play. Parrots are used by the roadside astrologers.

II. Injurious Birds

In spite of their great positive economic values, birds are not all beneficial to mankind. Many birds, in fact, cause tremendous damage to mankind. Some of the injurious acts of birds are the following :

1. Menace to agriculture. Some birds damage crops in various ways. They feed upon newly planted seeds, young seedlings, destroy leaves and flowers and spoil fruits and seeds. In Malva, crows do great damage to groundnut crops by eating groundnuts. In Australia sparrows have become serious crop pests. In India, sparrows, parrots, etc., do great harm to agriculturists and horticulturists. As parrots destroy fruits either ripe or unripe, sparrows destroy granaries. Further, the insectivorous birds may feed on beneficial insects and worms along with injurious ones.

2. Destroyer of game birds. The hawks, owls, etc., cause great damage to poultry yards by preying upon young chickens. They also attack upon the game birds and kill them. *Nestor notabilis* (New Zealand parrot) dig out the kidneys and fat of sick and weak sheep.

3. Damage to fisheries. Certain fish-eating birds such as herons, etc., feed upon fishes and cause great damage to fish industry.

4. Carriers of diseases. Many birds act as carriers to large number of harmful larval stages and transmit them to their new hosts. They also spread the noxious germs of various fatal diseases by feeding upon infected insects and worms and depositing their excreta in human food.

5. Pests of honey bees. Birds destroy and feed upon honey bees and their nests. Certain birds (honey-guides) attack on the nests of honey bees by the cooperation of little badger and feed upon their grubs and on themselves, while badger feeds upon the honey. Badger has a sting-proof hide and coat of coarse hairs and, hence, it is protected from the stings of bees.

Whatever might be the injurious effect of birds to man, their uses overweigh them and, hence, it can be safely concluded that birds are the best companions of mankind.

IMPORTANT QUESTIONS

Long Answer Questions

1. Give an account of Ratitae (flightless birds).
2. Describe the origin of birds. Mention the important theories regarding the origin of birds.
3. 'Birds are glorified reptiles'. Critically analyse the statement.
4. 'Birds as a flying machine'. Critically analyse the statement.
5. Give an account of flight mechanism in birds.
6. Describe various modes of flight in birds.
7. Give an account of various modifications in the structures of birds that are associated with flight or aerial adaptations.
8. Give an account of palate in birds.
9. Discuss migration of birds in detail.
10. Discuss the economic importance of birds.
11. Write an essay on : (*i*) Flight adaptations; (*ii*) Origin of birds; (*iii*) Ratitae (Flightless birds); (*iv*) Migration of birds; (*v*) Economic importance.

Short Answer Questions

1. Differentiate between Ratitae and Carinatae in a tabular form.
2. Discuss birds as a flying machine.
3. Describe the various types of beaks or feet found in birds.
4. Mention any five adaptations in birds for aerial mode of life.
5. Describe the prinicipal types of palate found in birds.
6. Write short notes on :
   (*i*) Kiwi (*Apteryx*);
   (*ii*) Migratory birds;
   (*iii*) Fossil birds; (*iv*) *Tinamous*.
7. Explain reptilian characters found in birds.

Multiple Choice Questions

1. Select the type of palate seen in *Rhea*
   (*a*) Dromaeognathous
   (*b*) Schizognathous
   (*c*) Desmognathous
   (*d*) Aegithognathous.
2. Probable ancestors of present-day ostrich is
   (*a*) *Ichthyornis*
   (*b*) *Archaeopteryx*
   (*c*) *Hesperornis*
   (*d*) *Eleuherornis*.
3. Fossil record of *Archaeopteryx* have been discovered from
   (*a*) Britain (*b*) France
   (*c*) Germany (*d*) U.S.A.
4. Birds are glorified
   (*a*) Reptiles (*b*) Vertebrates
   (*c*) Amphibians (*d*) Fishes.
5. The most common mode of flight is
   (*a*) Gliding (*b*) Flapping
   (*c*) Hovering (*d*) Soaring.
6. Beak in sparrow is
   (*a*) Seed eating type
   (*b*) Cutting type
   (*c*) Tearing and piercing type
   (*d*) Fruit eating type.
7. Long distance migrant is
   (*a*) Snow parrtidges
   (*b*) Arctic tern
   (*c*) Chicades
   (*d*) European white stork.
8. Which of the following reptiles possess gizzard?
   (*a*) Lizards (*b*) Crocodiles
   (*c*) Snakes (*d*) Turtles.

ANSWERS

1. (*a*) 2. (*d*) 3. (*c*) 4. (*a*) 5. (*b*) 6. (*a*) 7. (*b*) 8. (*b*)

29 Type 11. *Oryctolagus* (The Rabbit)

Chapter

The mammals (Gr., mammae = mammary glands) are at the top of the animal kingdom and are undoubtedly derived from the reptilian subclass **Synapsida**, most probably from mammal-like reptiles, the **Therapsids**. The transition from primitive reptiles to early mammal-like forms occurred in between Upper Carboniferous and the end of the Triassic but they suffered a setback in Permian. In Jurassic only a few small mammals existed but in Cretaceous these small forms became abundant and later on from them in early of Tertiary a number of different types of mammals arose which became perfect. Therefore, the Tertiary period is known as the **age of mammals**. Thereafter, mammals have radiated, adapted and colonised almost all parts of the globe in different ecological conditions. However, the mammals are characterised by the following features.

Contents of this Chapter

DISTINGUISHING CHARACTERS

1. The body is usually covered with **hairs** which are epidermal in origin.

2. **Warm-blooded** or **ectothermal** or **homoiothermal** animals.

3. The skin is provided with **sebaceous**, **sweat**, **scent**, and **milk-glands**. The milk-glands produce milk for the nourishment of the young for some time after birth. Only due to the possession of **milk-glands** and **mammae** the name mammalia has been derived for this group of animals.

4. Tail is generally long and mobile.

5.Two pairs of pentadactyle limbs present which are adapted variously for walking, running, climbing, burrowing, swimming or flying.

6. Skull is provided with two **occipital condyles**, *i.e.*, **dicondylic** skull. The toes are usually provided with horny **claws**, **nails**, **hoofs** or **fleshy pads** in aquatic forms.

7. **Cervical vertebrae** usually seven. Usually external ear or **pinna** with external auditory meatus present. The pinna may be reduced or absent in burrowing and aquatic forms.

8. The lower jaw is reduced to a single bone, the **dentary**, which articulates with the squamosal.

9. Both the jaws are provided with teeth embedded in sockets, *i.e.*, **thecodont**. Only two sets of teeth are recognised during the lifetime, the **deciduous** or **milk teeth** which are replaced by the other set called **permanent teeth**. This condition is known as **diphyodont**.

10. The teeth are usually differentiated into four types depending upon their feeding habits, *i.e.*, **heterodont**. However, teeth are rarely absent.

11. The **nasal passage** is usually long and mobile.

12. The middle ear is provided with three ear ossicles, the **malleus**, **incus**, and **stapes**.

13. The **tongue** is usually mobile.

14. The eyes are provided with mobile lids.

15. Heart four chambered, aortic arch asymmetrical, *i.e.*, only **left aortic arch** present.

16. Respiration occurs by lungs and usually a larynx with **vocal cords** present. A muscular transverse partition, the **diaphragm**, separates the body cavity into an anterior thoracic cavity and a posterior abdominal cavity.

17. The erythrocytes are more, spherical and non-nucleated (except in camel).

18. Brain well developed and specialised with different centres like memory, learning, etc.

19. Males are with a copulatory organ, the **penis**, and the testis situated outside the abdominal cavity in a **scrotal sac** (except in aquatic mammals and elephant).

20. Fertilisation internal. Usually viviparous.

21. The development of the embryo occurs in the uterus of the mother where a **placenta** is formed by the association of uterine and foetal tissues. The placenta helps the embryo in the physiological exchange of materials from the maternal blood and, thus, it brings about nutrition, excretion and respiration, etc., of the embryo in the womb.

SYSTEMATIC POSITION

Phylum	....	Chordata
Subphylum	....	Vertebrata
Superclass	....	Tetrapoda
Class	....	Mammalia
Subclass	....	Theria
Infraclass	....	Eutheria

Order	....	Lagomorpha
Family	.....	Leporidae
Genus	....	*Oryctolagus*
Species	....	*cuniculus* (Rabbit)

HABIT AND HABITAT

Oryctolagus cuniculus (European grey rabbit) is worldwide (cosmopolitan) in distribution. Originally it belonged to Western Europe and Africa, but later on it was carried by men to various parts of the world including India. It was wild and from it various domestic varieties have been developed, which differ in colour, size and fur texture. *Oryctolagus cuniculus* or the Rabbit is usually found in the fields, grass lands and open wood lands. It leads a **fossorial** life, *i.e.*, lives in burrows, which it digs into the ground with the help of its clawed forelimbs. The burrows are in the form of spacious branched **tunnels**, which open in various directions. These burrows are known as **warrens**. The burrows are usually used by the rabbits for retiring or at the approach of danger and to rear their offsprings. The rabbits are nocturnal (crepuscular) in habit, usually feed in twilight at dawn or dusk and they feed on green vegetation (**herbivorous**), usually near their warren. They are **coprophagous**, *i.e.*, eat their soft night droppings without mastication.

The rabbits are **gregarious** in habit, *i.e.*, they live in groups and even several families consisting of hundreds of rabbits may live in one burrow. They are defenceless but clever and agile and fast runners, and cover nearly 32–40 km in an hour. The average life span of a rabbit is seven to eight years. A rabbit breeds nearly four times a year. A rabbit gives nearly five to eight young ones in each litter or breed which become sexually mature in 4 to 6 months. The young ones when born are blind, deaf, naked (without hairs) and helpless. Later on when mother starts feeding the young ones on her milk, they grow, their eyes and ears are opened, hairs develop on the body and they become mature. Thus, **parental care** is shown by the female rabbit. The rabbits are **polygamous**, *i.e.*, one male lives in the company of many females.

DISTRIBUTION

The rabbits are cosmopolitan in distribution but their original home was Western Europe and Africa, from where they spread or were carried by men to various parts of the world. They are not found widely in the wild state in India but their domesticated varieties are common in all parts.

ENEMIES

The natural enemies of rabbits are men, snakes, fox, wolf, jackal, wild cats, owls and other preying birds. They are without organs of defence but they are clever enough to hide in the burrows on seeing the enemies.

ECONOMIC IMPORTANCE

The rabbits are of great economic importance to the mankind because they are both useful and harmful. Their enemies are snakes (pythons), owl, hawks, buzzards (predatory birds) and mammals such as foxes, wolves, cats, etc. Man uses its fur for the manufacture of purses, gloves, caps,etc. Their meat is palatable which is very delicious. The rabbits are also used for various biological researches in the hospitals and educational institutions. They are pet animals and are domesticated for entertainment. The rabbits are harmful as they damage crops, vegetables and the grass of the lawns. They also befoul the grass of the lawns and pastures with their excreta. Consequently sheep and other animals grazing on these places do not like to eat the grass. Rabbits also carry germs of some human diseases.

IMPORTANT ADAPTATIONS

The rabbits are defenceless and have several enemies, even then they are well suited to protect themselves from their enemies due to several adaptations found in them. Some of the important adaptations are :

1. They are fast runners and always remain alert.
2. The protective colouration and burrowing mode of life.
3. The sense organs of smell and hearing are keen.
4. They possess a wide range of vision.
5. They can easily live in the different environmental conditions.
6. They possess 13 pairs of ribs which facilitate respiration during fast running.
7. The fecundity of reproduction is very high.
8. The mother cares the young ones, etc.

RABBIT AND HARE

The rabbits and hares are closely related to one another and belonged to the order **Logomorpha** (Gr., *lagos* = hare). Picas (*Ochotona*), sometimes called conies, of Asia and Western North America are mountain hares with the habit of storing hay, which they eat in the winter. They form colonies, but each animal guards a territory on which he makes a pile of cut vegetation, turning it to dry in the sun. Therefore, rabbits and hares both were kept in the genus *Lepus* for a long time. But recently they are distinguished from each other on several points and separate generic names have been given to them, *Lepus* for hare and *Oryctolagus* for rabbit. A number of species of *Lepus* are found in the different parts of the world, *e.g.*, *L. cuniculus* (common hare), *L. europeans* (European hare), *L. ruficaudatus* (North Indian hare), *L. nigricollis* (Black napped South Indian hare), *L. dayanus* (desert hare), *L. americanus* (snow shore hare), *L. arcticus* (Arctic islands), etc. Fossil logomorphs are found back to the Oligocene. Few remains are known from the Eocene, but this was also distinct in the Palaeocene. However, both of them can be distinguished on the following points :

TABLE 29.1. DIFFERENCES BETWEEN RABBIT AND HARE.

Characters	Rabbit *(Oryctolagus)*	Hare *(Lepus)*
1. Habits	Gregarious (living in groups) fossorial (burrowing), crepuscular in nature (*i.e.*, comes out of the burrow after sunset and before sunrise for feeding). It can be domesticated.	Nocturnal, remains active during night for feeding. Solitary, does not live in groups. It cannot be domesticated.
2. Burrow	Digs burrow like branching tunnel having several exits. Leads fossorial life.	Vagabond, does not burrow but lies hidden in a depression in the vegetation.
3. Size	Smaller, 30–45 cm long.	Larger, 50–70 cm long.
4. Nature	Docile, can be domesticated.	Wild, cannot be domesticated.
5. Speed	Fast runner, about 30–40 km per hour.	Faster runner, about 50–60 km per hour.
6. Danger signal	With sound produced by striking ground with hindlimbs.	With sound produced by grinding incisor teeth against one another.
7. Body colour	Brownish grey above, white below.	Brown above, white below.
8. Snout	Narrower	Broader
9. External ears or pinnae	Comparatively shorter and of same colour.	Comparatively longer with black tips
10. Forelimbs	Shorter than hindlimbs.	Fore-and hindlimbs almost equal.
11. Hindlimbs	Comparatively smaller.	Comparatively larger.

(Contd.)

Characters	Rabbit *(Oryctolagus)*	Hare *(Lepus)*
12. Claws	Strong for digging.	Not so strong.
13. Radius	Shorter than humerus.	Larger than humerus.
14. Posterior nasal openings	Narrower than palatal bridge.	Wider than palatal bridge.
15. Interparietal bone of skull	Distinct.	Fused with supra-occipital.
16. Upper front incisor teeth	Have longitudinal grooves on outer surface.	Longitudinal grooves absent.
17. Nasal bones, chambers and passages	Narrower.	Broader.
18. Young ones in one litter	5–8.	Only 2.
19. Newly born young	Weak, naked and blind.	Strong, covered with fur and eyes open.

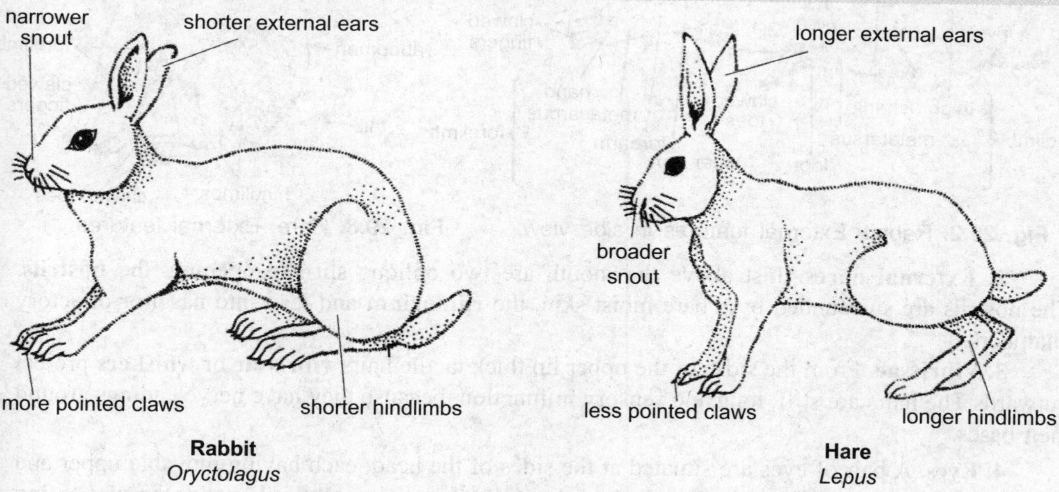

Fig. 29.1. Comparison of rabbit and hare.

EXTERNAL FEATURES

Shape, Size and Colour

The rabbit is about sixteen inches (40 cm) in length from mouth to anus and weighs two to four pounds. Its body is pointed anteriorly and broad posteriorly, which is covered with soft uniform fur or hairs. It keeps the body temperature constant, *i.e.,* 38.8°C., acting as heat insulator. The colour of the European wild rabbit, *Oryctolagus cuniculus*, is dusty brown above but the ventral side and lower part of tail remains white. Its colouration is protective camouflaging with the surroundings. The colouration of the domestic varieties of rabbits vary greatly, *e.g.,* it may be pure white, pure black or white with brown or black patches, etc.

Divisions of Body

The rabbit shows a typical mammalian form of the body which consists of **head, neck, trunk** and **tail**. The trunk is further divisible into **thorax** and **abdomen**.

Head. The head is large and spherical posteriorly but produced anteriorly into a large pointed blunt **snout** or **muzzle**. The head bears the following structures :

1. Mouth. The snout has a terminal transverse slit-like **mouth**, which is surrounded by two soft, fleshy movable **lips**. The upper lip is divided in the middle into right and left halves due to a vertical cleft extending up to the nostrils. Such a divided lip is known as the **hare lip**, due to which the upper front incisors are exposed.

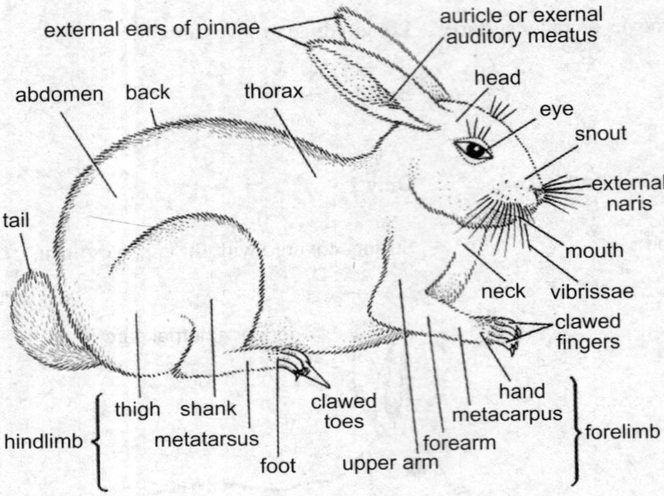

Fig. 29.2. Rabbit. External features in side view.

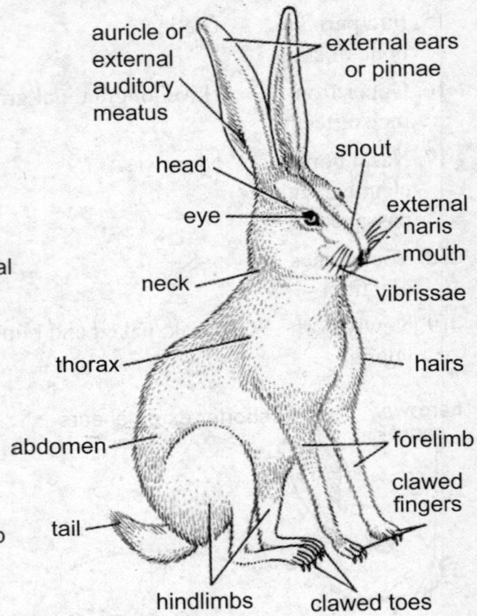

Fig. 29.3. Hare. External features.

2. External nares. Just above the mouth are two oblique slit-like opening, the **nostrils**. The nostrils are surrounded by a bare moist skin, the **rhinarium** and lead into nasal or olfactory chambers.

3. Vibrissae. From the sides of the upper lip thick tactile hairs **vibrissae** or **whiskers** project outwards.The hairs are stiff, long and sensory in function, because they have nerve endings around their bases.

4. Eyes. A pair of eyes are situated at the sides of the head, each having movable upper and lower **eyelids** with very fine, short **eyelashes**. A small white coloured third eyelid, the **nictitating membrane**, is also present in the inner anterior corner of the eye. The nictitating membrane is also movable and stretched across the cornea and used for cleaning cornea.

5. External ears or pinna. A pair of large, movable trumpet-shaped **external ears** or **pinnae** are found situated at the posterior lateral side of the head. The long pinnae are movable in all directions to receive the sound waves. Each pinna has an **external auditory meatus** at its base, which is closed below by the **tympanic membrane**. Both the pinnae remain upright when the rabbit is on the alert and laid back on frightening and running.

Neck. The neck is an extension of the body which connects the head with the trunk at a slight angle. It enables the head to move in all directions. The neck of the rabbit is short and flexible. Its short neck is advantageous in its burrowing and fast running habits.

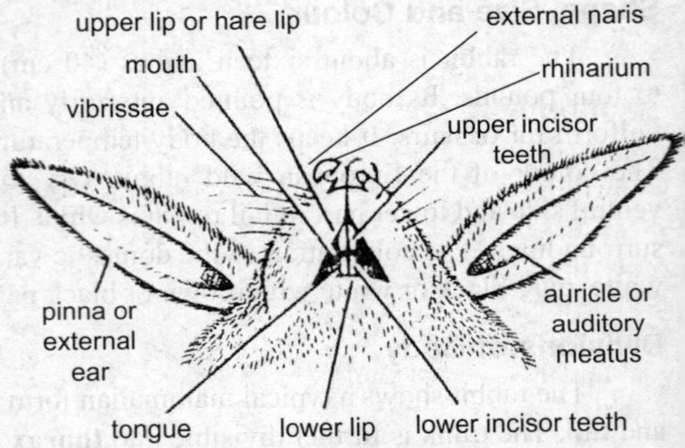

Fig. 29.4. Rabbit. Head in ventral view.

Trunk. The neck is followed by large, cylindrical trunk, which is divided into anterior **thorax** and posterior broad, soft-bellied **abdomen**. The thorax or the chest forms a bony cage having ribs at the sides and sternum ventrally. The abdomen is without ribs and sternum. The cavity of the thorax is known as **thoracic cavity** in which tender body parts like heart and lungs are protected.

1. Teats. On the ventral side between the thorax and abdomen are situated 4–5 pairs of well developed and functional **teats** or **nipples** in females and rudimentary in males. The 4 or 5 pairs of **mammary glands** open on the outside at the teats.

2. Anus. At the posterior end of the abdomen, at the base of tail is found the **anus**, which is the external opening of the digestive tract. A pair of hairless depressions are found in both the sexes, one on each side of the anus, called the **perineal pouch** into which the ducts of the **perineal glands** open. The secretion of the perineal glands has a strong odour, characteristic of rabbit.

3. Urinogenital aperture. Urinogenital opening is situated at the tip of the **penis** in male in front of anus. The penis is cylindrical, muscular and covered by the skin. In the male a pair of **scrotal sacs** are situated, one on either side of the penis, in which testes are lodged. The scrotal sacs are thin-walled bags of skin.

In the female the slit-like urinogenital aperture or **vulva** is present beneath the anus and at the anterior margin of **vulva** is present a rod-like **clitoris** which is similar to the penis of the male.

Limbs or Appendages

The trunk bears two pairs of **pentadactyle limbs**. Both pairs of limbs take part in locomotion and support the weight of the body. Forelimbs are shorter than the hindlimbs.

1. Forelimbs. The forelimbs are short and are held rigid to take the shock at the end of a leap. Each forelimb has a proximal upper arm or **brachium**, a middle forearm or **antebrachium** and the distal hand or **manus** with a wrist or **carpus**, palm or **metacarpals** and five fingers or **digits** with sharp, curved claws. The forelimbs are used for digging the burrow. The palm is hairy.

2. Hindlimbs. The hindlimbs are longer and more powerful than the forelimbs. Each hindlimb has a proximal **thigh**, a middle **shank** or **crus** and a distal **foot** or **pes** with an ankle or **tarsus**, **metatarsals** and four clawed digits. Hallux (first toe) is absent. The hindlimbs are main locomotory organs. The sole is hairy.

Tail. A short, bushy tail is found at the hind end of the trunk. The lower surface of the tail has a white hairy patch in wild rabbit, which is used for warning signal to other rabbits when danger approaches.

LOCOMOTION

The mode of locomotion in rabbits may be of three types : *viz.,* **walking**, **running** and **leaping**. For walking, firstly, the forelimbs are moved forward as a whole nearly through 90 degrees, therefore, the hands are directed in the same direction as the head and the pre-axial end of the limb remains situated near and parallel to the side of the body. Then the upper arm is bent backwards, the forearms forward, while the hand rests with palms downward on the ground. Exactly in the same way, the hindlimbs are moved forward, hence, the preaxial side is near the body but the thigh remains directed forwards, while the shank is bent backwards and the sole of the foot comes in contact with the ground. So in such movements in rabbit, the limbs lie under the body forming vertical levers. However, during running the body is not stable. When the speed of walking increases, each foot is raised off the ground before the previous one is kept. At the time of leaping only two feet touch the ground, actually the hind feet hit the ground at a point in front of which the forelimbs have just left. Thus, its locomotion is well adapted for its terrestrial mode of life.

INTEGUMENT

The bodies of all vertebrates are invested by an outer protective covering, the **integument** or **skin**, which is partly responsible for retaining the body shape. The skin is comparatively thick and covered with hairs. It consists of two main layers : the **epidermis** which is ectodermal in origin and the **dermis** which is mesoderamal in origin. Several other structures which are derived from the skin are found associated with it, *e.g.,* hairs, claws, nails, various glands, etc.

1. Epidermis. The epidermis is the outermost, thick and more complex layer formed of stratified epithelium.The epidermis is mainly formed of two layers, the superficial outer cornified layer and deeper inner Malpighian layer.

Cornified layer. The outermost layer of cells are hard, scale-like, dead, fully keratinized and flattened. They form the **straturm corneum**. It is thick on palms and soles. This layer contains hard **keratin** which prevents the passage of water and solutes. In places where there is considerable friction this layer may become very thick. This layer is cast off in shreds periodically which is called dandruff. New cells are constantly added from below when granular cells become horny and dead. Epidermis contains openings of sweat glands and hair follicles.

***(i)* Stratum germinativum.** It is the innermost layer based on basement membrane secreted by the dermis. It is formed of a single layer of columnar cells. These cells continuously divide mitotically and, thus, new cells are added into the epidermis. All the glands are derived from this layer.

***(ii)* Stratum spinosum.** It lies above the stratum germinativum layer. Its cells are polyhedral in shape and arranged in several layers. These cells are gradually pushed upwards and become flattened and keratinized due to deposition of **keratin** (scleroprotein).

***(iii)* Stratum granulosum.** It is present above the stratum spinosum. Its cells contain granules of keratohyalin. This layer is especially developed in palms and soles, etc., where the epidermis is thick.

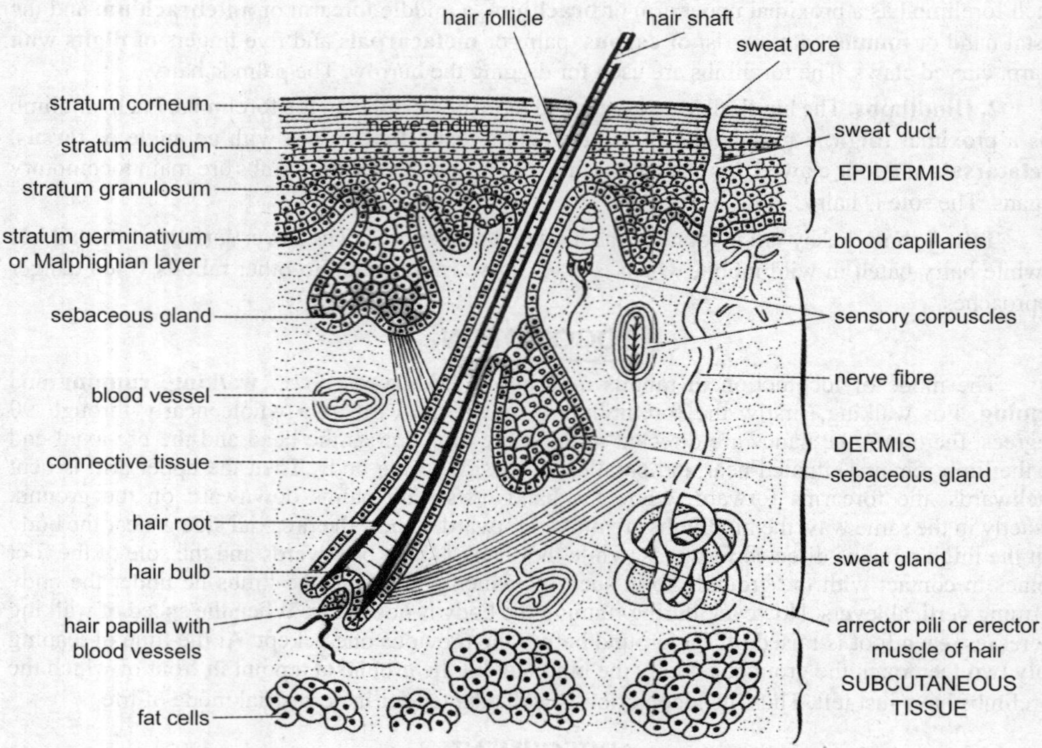

Fig. 29.5. Rabbit. V.S. of skin.

***(iv)* Stratum lucidum.** This layer is formed of thin transparent and hard shiny and refractile cells. A chemical substance **eleiden** is found in this layer. The stratum lucidum is transformed into stratum corneum when it is removed. It is the innermost layer of the epidermis which lie over the dermis. The cells of this layer are very active and divide to form new cells. As the new cells are formed, the older ones are pushed upwards and at the same time they become flattened, their protoplasm becomes granular and horny due to the formation of **keratin** which is a protein. Finally they come to the surface and become perfectly flat and horny.

2. Dermis. The epidermis is followed by the **dermis**, which is thick, tough, flexible and elastic. The dermis is formed of dense connective tissue fibres, unstriped muscles, blood capillaries, lymph vessels, nerves, pigment cells, and fat cells. The nerve endings, receptors and blood capillaries are distributed throughout the dermis. Nerves and sensory organs of touch, pressure, temperature and pain are abundantly found in dermis. A few nerve endings penetrate the epidermis. Pigment cells are present in the outer layer of dermis. These cells contain **melanin**. Some of the epidermal glands, *e.g.,* sweat gland, sebaceous gland, etc., are found in the dermis. The deeper part of the dermis is formed of fatty or adipose tissue consisting of groups of fat cells. Fat is found stored in the lobules of the fat cells in the deeper parts of dermis and in the subcutaneous tissue. The fatty layer acts as a reserve food supply and as a heat insulator. The functions of the dermis are to provide firmness and flexibility to the skin, protecting and supporting the body and carrying blood to the general body surface.

Skin Derivatives

1. Hair. The hairs are epidermal in origin and found in the mammals only. A hair develops as a thickening of the stratum germinativum of epidermis. Later on, it is differentiated into two parts : the basal part deeply situated in the dermis is called the **root**, which is bulb-like and the upper part that traverses the dermis and projects out of the epidermis is called the **shaft**.The hair is enclosed in a tube-like invagination of the epidermis formed by stratum Malpighii, the **hair follicle**. At the base of the follicle in the dermis a small swelling of dermal tissues is found which is evaginated to form the **dermal papilla** or **hair papilla**. The hair papilla is richly supplied by nerves and fine blood vessels carrying nourishment which diffuses into the root of the hair. The epidermal cells situated above the hair papilla multiply actively to form the **hair bulb**. From the hair bulb develops a cylindrical shaft of cornified cells, the base or root of the hair. Constant addition of cells causes the hair base to extend through the follicle and pierce the skin as a column of cells. The hair base is only living, while the cells of hair become horny and die beyond the skin. The hair is made oily by the secretion of **sebaceous gland** opening into the hair follicle. With each hair follicle is attached a small smooth

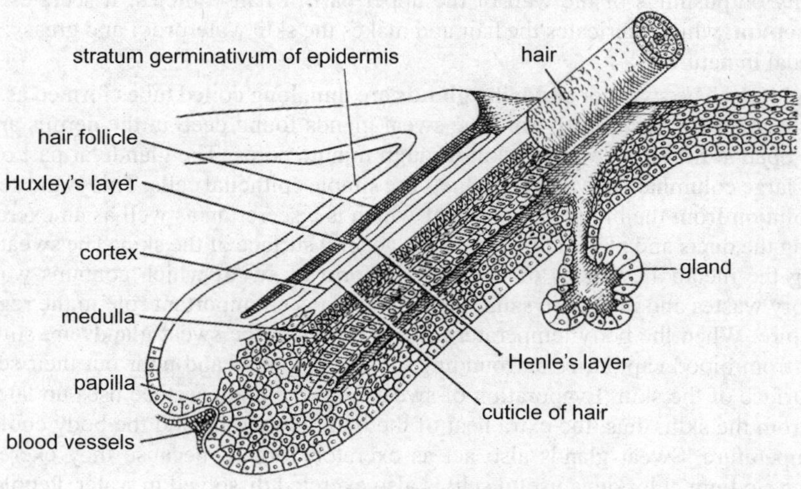

Fig. 29.6. Rabbit. Structure of hair.

muscle, the **erecter pili,** for erecting the hair due to fear, excessive cold, etc. In a cross section the hair has three regions. The outer is a **cuticular layer** formed of overlapping microscopic scales, which is followed by a pigmented fibrous **cortex** consisting of elongated horny epidermal cells. In the centre of the hair is the **pith** or **medulla** formed of rounded cells. In larger hairs it contains air spaces.

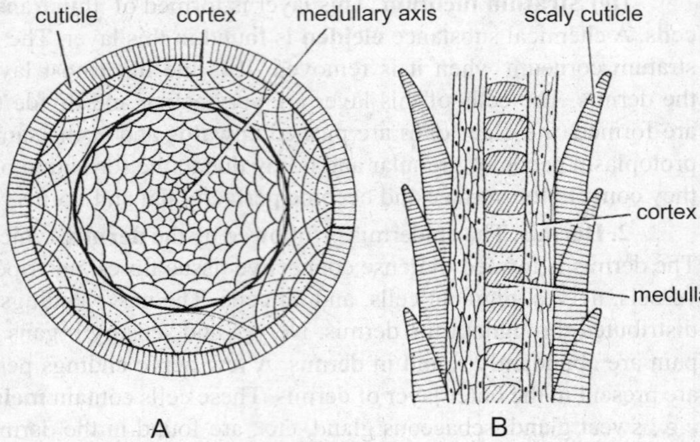

Fig. 29.7. Rabbit shaft. A – In T.S.; B – In LS.

Functions of hairs. The principal function of hair is the part it plays in temperature regulation, but it can be tactile, protective and posseses sexual functions as well. In man and certain other mammals (cetaceans, sirenians) that have become relatively hairless, heat loss is retarded by a compensatory layer of hypodermal fat (**blubber**) that arises in the superficial fascia.

Vibrissae on the facial region and sometimes near carpus and tarsus are richly innervated and tactile in nature.

Spines in spiny anteaters, hedgehogs and porcupines are the modified hairs and used as defensive organs.

Bristles (hairs) in the nostrils and auditory meatus of many mammals, including man, serve to block the entrance of foreign substances and insects.

Hairs in specialised areas assist in the retention of animal odours significant in courtship.

2. Cutaneous or epidermal glands. These are abundant and peculiar in the skin of mammals. The epidermal glands are sebaceous, sweat, mammary, lacrimal, Meibomian, scent and wax glands. These glands are multicellular and present in the dermis.

***(a)* Sebaceous glands.** The sebaceous glands are large glandular, branched alveoli-like saccular glands found attached with the upper part of the hair follicle by a small duct. The sebaceous glands are actually outpushings of the wall of the upper part of hair follicles. It secretes an oily secretion called **sebum**, which lubricates the hair and makes the skin waterproof and greasy. Sebum is also bacteriocidal in nature.

***(b)* Sweat glands.** The sweat or **sudorific glands** are thin, long coiled tubes formed as tubular invagination of the **stratum germinativum**. The sweat glands found deep in the dermis and their long spiral ducts open at the surface of the skin through minute pores. The glandular part of sweat glands formed of large columnar cells and their ducts are simple epithelial cells. This gland separates a salty, watery solution from the blood called **sweat** which is a secretion as well as an excretion. It is conducted along the ducts and passed outside on the general surface of the skin. The sweat glands help in removing the metabolic wastes of the body in form of **sweat** which contains water and dissolved excretory wastes and inorganic salts. This gland plays an important role in the regulation of body temperature. When the body temperature rises too much, the sweat glands are stimulated to take up water from blood capillaries surrounding the sweat glands and pour out their secretion on the general surface of the skin. Evaporation of sweat from the body surface uses up latent heat of vaporisation from the skin, thus, the extra heat of the body is used up and the body cools down reducing the temperature. Sweat glands also act as excretory organs because they excrete urea dissolved in water. Sodium chloride, a useful salt, is also excreted dissolved in water. People living in Tropical countries, thus, need more salt in their diet since they perspire profusely in summers. In rats sweat glands are not found. In arboreal animals they occur on plantar and palmar surfaces.

(*c*) Mammary glands. These are modified sweat glands and one of the most important characteristics of placental mammals. The mammary glands consist of branching system of tubules and alveoli secreting milk, and their ducts open on small projection of the skin surface, called **nipples** or **teats**. The mammary glands arise along the mammary lines which run from axilla to inguinal region on each side of the trunk. They differ considerably in number and distribution from one pair (*e.g.,* most primates, giraffes) to eleven pairs occur in eutherians. Rabbit generally possesses four pairs of mammary glands. Mammary glands may be restricted to the thoracic region (*e.g.,* Chiroptera, Sirenia, and all primates except lemurs) or to the inguinal area (*e.g.,* cetaceans, ungulates). In females they secrete milk for the nourishment of the young after birth. In males they remain undeveloped and functionless. Milk is cellularly synthesised from precursors in the blood stream and passes through the membranes as alveolar cells.

Other glands. A number of other glands are also found associated with the skin of mammals. They are as follows :

(*i*) Perinaeal glands. These glands are situated under the skin near the anus and open into hairless depressions, called **perinaeal pouches**. These glands are in fact modified sebaceous glands which secrete a secretion with characteristic odour of rabbit.

(*ii*) Meibomian glands. These glands are found on the margins of eyelids just beneath the conjunctiva and open on the free edges of the eyelids. These are also modified sebaceous glands having long ducts into which several alveoli open. An oily secretion is secreted by these glands for the lubrication of the eyelashes which make them soft and flexible.

(*iii*) Scent glands. These glands with strongly smelling secretions open on various parts of the integument are regarded as modified sebaceous glands or more rarely sweat glands. Examples of such glands are the **occipital glands** of camel, **cerumen glands** of antelopes and deer, **temporal glands** of elephant, **pedal glands** of ruminants, **preputial glands** of musk deer and beaver, etc. These glands are often found near anus or in inguinal region and open into special cutaneous pits.

(*iv*) Lacrymal gland. It has several openings on the conjunctival surface beneath the upper eyelid towards the outer side of the eye.

(*v*) Harderian gland. It lies at the inner side of eyeball and mainly on its lower surface. It opens in connection with the nictitating membrane at the inner angle. Harderian gland is absent in primates.

(*vi*) Wax glands. These are present in the auditory canal and secrete fatty **earwax** or **cerumen**. It lubricates and protects the tympanum.

Other Derivatives of the Skin

Claws, nails, horns and hooves are the other derivatives of the skin found in most of the mammals.

Functions of the Integument

The integument or the skin performs the following functions :

1. Shape. It helps in maintaining a characteristic **shape** of the body.

2. Protection. It **protects** the body organs from mechanical injuries due to friction and blows. It is a **germ-proof** covering of the body which prevents the entry of bacteria and the other germs. It prevents an undue **evaporation** of water.

3. Defence. Claws, horns, nails and hooves are the derivatives of the skin which serve as useful tools, such as digging, weapons of offence and defence.

4. Body temperature. The sweat glands of the skin help in regulating the **body temperature** by evaporating the sweat. The dermis gives rise to the dermal bones of the skull. In some reptiles it forms dermal bony plates.

5. Excretion. The skin also helps in removing the **metabolic wastes** of the body along with the sweat. The shedding of the **keratinized** dead cells of the epidermis is a kind of excretion.

6. Secretion. The skin serves as an organ of **secretion**. The milk is produced in the **mammary glands** and the **sebaceous gland** secretes an oily secretion, the **sebum**, which is responsible for water-proofing of the skin and hairs.

7. Sexual function. It helps in **sexual selection**. The colour of the integument and various integumental structures serve to attract the opposite sex.

8. Sensation. The fine **nerve endings** and special **tactile corpuscles** in the integument provide the **sensory function** to the skin. It receives stimuli of touch, pain, pressure, temperature, etc.

9. Synthesis of vitamin D. The skin is also capable of synthesising vitamin D in the presence of sunlight.

10. Other uses. (*i*) In some mammals like bats, the skin forms **patagium** to help the animal in flight. The patagium looks like a wing which is merely fold of skin. The flying-squirrels are also provided with such patagium.

(*ii*) It also serves to absorb oils and ointments to some extent, thus, serving as an organ of absorption and checks excessive absorption of injurious substances.

(*iii*) The dermis gives rise to dermal bones of the skull. In some reptiles it forms dermal bony plates.

(*iv*) It acts as an organ of **storage of reserve food**. Fats are usually stored in the liver and muscles, but a considerable amount of fats are stored in the form of subdermal fat in the skin. In **whales** such a fatty layer, called **blubber**, is several centimetres thick..

ENDOSKELETON

The endoskeleton of rabbit is chiefly formed of bone and cartilaginous part is very little. Exactly like those of other vertebrates, the skeleton of rabbit can also be divided into two parts : (*i*) the **axial skeleton** is present along the longitudinal axis of the body and consists of the bones of **skull**, the **vertebral column**, the **ribs** and the **sternum**; (*ii*) the **appendicular skeleton** lies at right angle to the longitudinal axis of the body and consists of the bones of **limbs** and the **girdles**.

Axial Skeleton

Characteristics of Skull

Some important characteristic points in the mammalian skull are as follows :

1. Since there is a general tendency to increase in the size of the brain, the skull has a short posterior **cranial part** for lodging the brain and the long anterior **facial part** comprising mainly the jaws. In higher mammals the facial part lies below the cranial part.

2. The number of bones in the skull is much reduced, many of them are fused intimately so that their separating boundaries are marked only by the **sutures**.

3. Skull is **dicondylic**, *i.e.,* 2 occipital condyles. Each exoccipital bears an occipital condyle.

4. **Tropibasic** skull–a vertical interorbital septum is present in between two orbits. Cranium does not extend into orbital region.

5. The food passage is well separated from the nasal passage due to the development of **palate** which is formed of premaxillae, maxillae and palatines.

6. A **zygomatic arch** on either side of the skull is formed by squamosal, jugal and maxillary bones.

7. The **auditory capsules** are formed by the union of **periotic** and **tympanic** forming a swollen **tympanic bulla**.

8. The **articular** and **quadrate** of the jaws become separated and free, and form **malleus** and **incus** respectively (two ear-ossicles of the three). **Stapes** forms the columella.

9. Otic bones, prootic, epiotic and opisthotic, are fused to form a single **periotic**.

10. Turbinal bones are much folded and, thus, increases the olfactory surface of nasal chambers.

11. Only a single bone, the **dentary**, forms one half of the lower jaw.

12. Jaws suspensorium is **craniostylic**, *i.e.*, dentary, articulates with the cranium (skull) by **squamosal**.

13. Prefrontal, postfrontal, parasphenoid and quadratojugal are lacking. Pterygoids scale-like.

14. Premaxillae, maxillae and dentaries bear the **thecondont** teeth (teeth embedded in sockets). Teeth are **diphyodont** (milk and permanent) and **heterodont** (different types). Canines are absent leaving a space, called **diastema**. Dental formula is as follows :

$$i\frac{2}{1}, c\frac{0}{0}, pm\frac{3}{2}, m\frac{3}{3} = 28$$

Parts of Skull of the Rabbit

The skull of the rabbit can be divided into : **1. cranium** which encloses the brain, **2. sense capsules** (olfactory, auditory and orbital) closely attached with the cranium, and **3. visceral skeleton** includes the upper and lower jaws and hyoid apparatus.

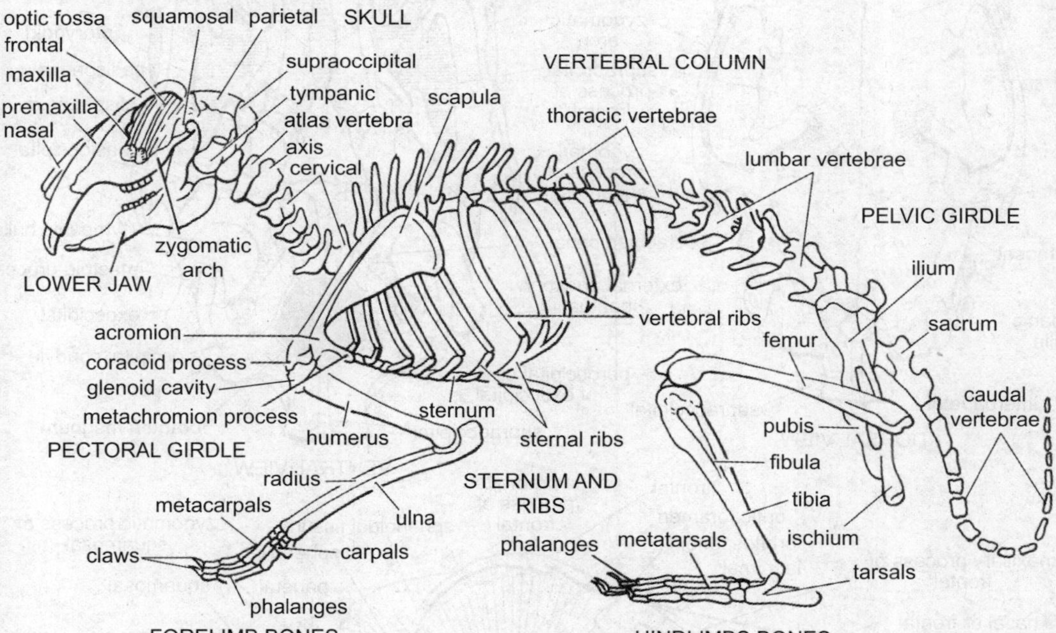

Fig. 29.8. Rabbit. Complete skeleton in lateral view.

1. Cranium. The various bones constituting the cranium can be grouped into three segments, a posterior **occipital segment**, middle **parietal segment** and anterior **frontal segment**.

***(i)* Occipital segment.** It is the hindermost part of the cranium. It consists of four cartilaginous bones completely fused together and encircling a large passage, the **foramen magnum**, through which spinal cord comes out. The bone making the dorsal boundary of the foramen magnum is flat, shield-like square-shaped **supraoccipital**, the ventral boundary is marked by a flat bone, the **basioccipital**, and the lateral boundaries are marked by a pair of **exoccipitals**. Each ex-occipital bears a prominent oval or a somewhat elongated **occipital condyle** and also gives off a downwardly directed paroccipital process which lies in close contact with postero-ventral surface of tympanic bulla.

Supraoccipital articulates in front with the parietals, and on lateral sides with the squamosals and periotics.

Basioccipital anteriorly articulates with basisphenoid and laterally with periotics.

The occipital condyles articulate with the facets of the first vertebra, the **atlas**.

(*ii*) **Parietal segment.** It is present in front of occipital segment. On lateral sides, both are separated from each other by auditory capsules and squamosals. It consists of five bones, a triangular **basisphenoid** on the mid-ventral side, **alisphenoids** on lateral sides and two **parietals** and **interparietal** on the dorsal side.

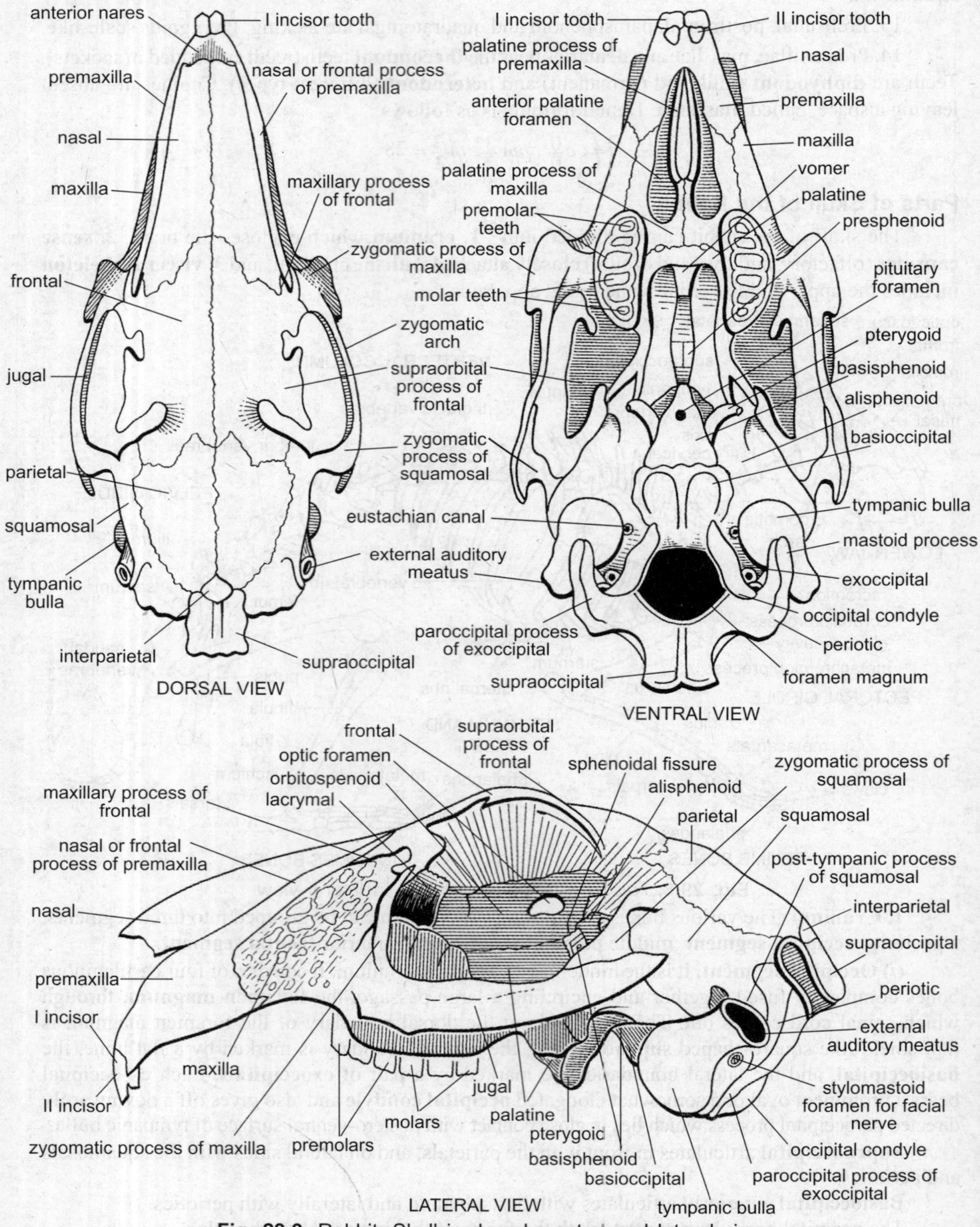

Fig. 29.9. Rabbit. Skull in dorsal, ventral and lateral view.

Basisphenoid and alisphenoid. Basisphenoid is a flat, median triangular cartilage bone. Its apex joins anteriorly with the presphenoid and its broad posterior end is connected with the basioccipital by a thin plate of cartilage. Dorsal surface of basisphenoid has a depression, the **sella turcica**, in which pituitary gland is placed. Basisphenoid also bears a **pituitary foramen** at about its middle. Basisphenoid laterally articulates with the alisphenoids.The alisphenoids are broad wing-like and firmly united with the basisphenoid.

Each alisphenoid is produced below into a bilaminate process, the **pterygoid process**, connected with the palatine and bounded in front by a slit-like aperture, the **sphenoidal fissure** (foramen lacerum anterius) opening into the cranial cavity.

Interparietal. It is a small, median wedge-like, triangular bone on the dorsal side in between parietals and supraoccipital.

Parietal. The parietal bones are a pair of thin, slightly arched bones protecting the brain dorsolaterally. Both the parietals are firmly united with each other by a suture along the mid-dorsal line. They remain separated from alisphenoids by the squamosals. Each parietal gives off a **ventral process** from its posterior outer border which extends beneath the squamosal.

Squamosal. It is roughly a rectangular bone situated ventral to parietal on either side. Outer surface of squamosal is produced into a strong **zygomatic process** which forms the posterior part of **zygomatic arch**. It unites anteriorly with **jugal** in the formation of a complete zygomatic arch. Below the root of the process is a hollow **glenoid fossa** for articulation with the condyle of mandible. Squamosal from its posterior side gives off a slender process, the **post-tympanic process**, which becomes applied to the outer surface of the **periotic**.

(*iii*) Frontal segment. This segment also consists of five bones, a **presphenoid** ventrally, two **orbitosphenoids** one on either side and two **frontals** dorsally.

Presphenoid and orbitosphenoids. Presphenoid forms the floor of the frontal region of cranium. It is a median, thin, small bone situated in front of the basisphenoid and connected with it by a cartilage. The presphenoid bone forms the lower and anterior boundary of the optic foramen. Orbitosphenoids are connected with the lateral sides of the presphenoids. These form the sides of the frontal region. In rabbit orbitosphenoids are partially fused and form a thin, vertical median interorbital septum, which surrounds the **optic foramen**. These bones form the lateral wall of cranium and orbits. Orbitosphenoids articulate with palatine in front, with frontals above and behind with the squamosals and alisphenoids.

Frontals.The frontals are large, and strong bones protecting the brain from the dorsal and lateral sides. Both the frontals are united with each other by a suture along the mid-dorsal line. Outer middle part of each frontal forms a prominent ridge over the orbit, known as the **supra-orbital process.** Frontals unite posteriorly with the parietals and anteriorly with the nasals. A slender **maxillary process** arises from the anterior side of each frontal running in between the premaxilla and maxilla. On the ventral side each frontal articulates with the orbitosphenoid.

(*iv*) Ethmoidal region. It lies anterior to the cranial cavity and not well separated from olfactory capsules. It includes a single **mesethmoid**.

Mesethmoid. It is a median veritcal bone which extends forwards and downwards in front of presphenoid. It is called **nasal septum** which separates the two olfactory chambers from each other. Posterior part of mesethmoid forms a narrow plate of bone, the **cribriform plate**, which forms the anterior boundary of the brain case. It is perforated by small foramina through which the fibres of the olfactory nerves enter the cranial cavity.

2. Sensory capsules. Sensory capsules are **olfactory** lodging the organs of smell, **auditory** containing organs of hearing and orbits for eyes. These are closely fused with the cranium.

(*i*) Olfactory Capsules

Nasal. Roof of olfactory chambers is formed by two flat bones, the **nasals** which unite with each other in the mid-dorsal line. Each nasal has on its inner surface a very thin, hollow process the **nasoturbinal**. Anterior end of each nasal forms the upper boundary of external nostril. Each nasal articulates laterally with the premaxilla.

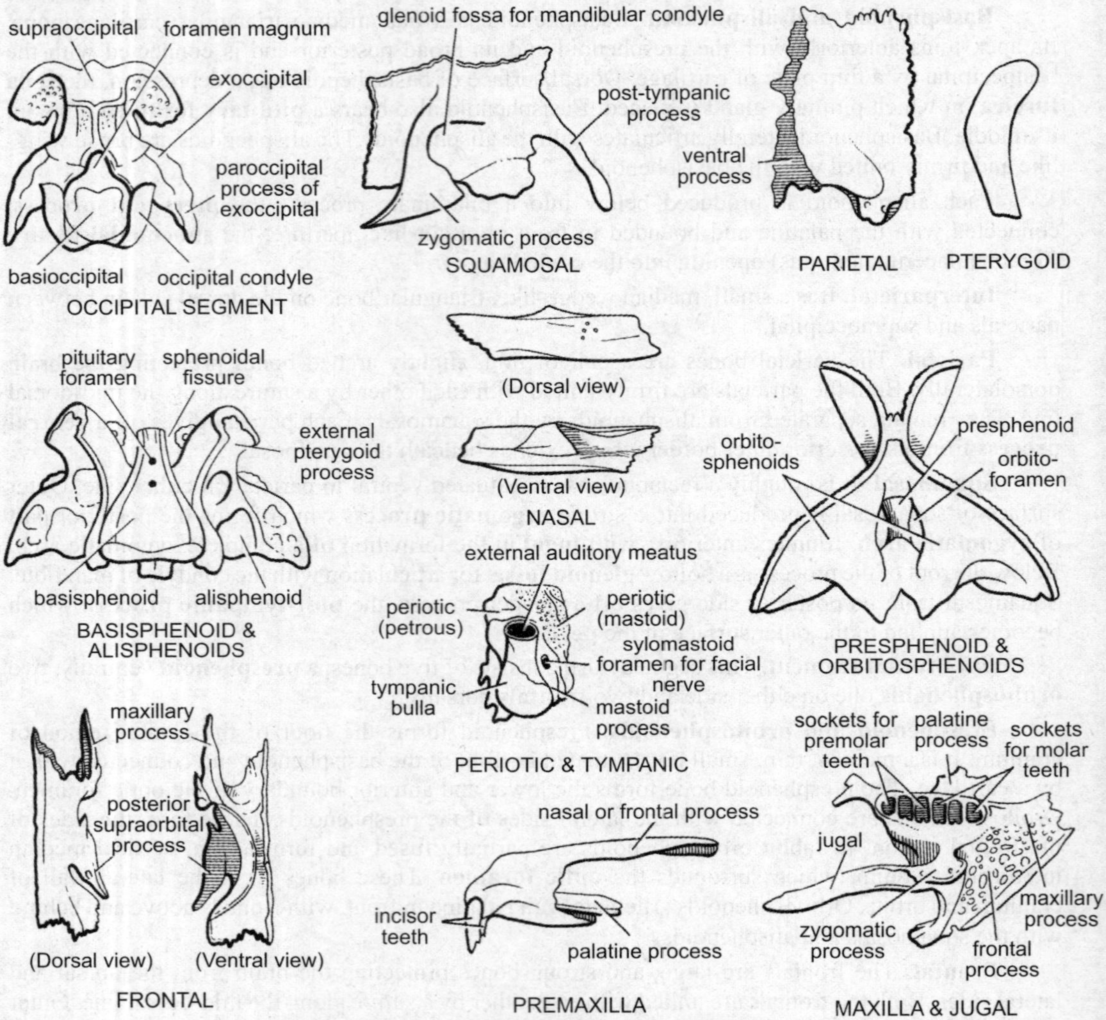

Fig. 29.10. Rabbit. Disarticulated bones of skull.

Vomer. Two vomers are long slender fused bones forming the floor of olfactory capsules. These dorsally articulate with ventral (inferior) edge of the mesethmoid (nasal septum). Vomers possess delicate lateral wings.

Turbinals. Each nasal or olfactory chamber or cavity separated from other cavity by a median septum, partly cartilaginous, partly bony by the **mesethmoid**, contains the **turbinals** or turbinate bones of its side. These are ethmoturbinals, maxilloturbinal and nasoturbinal.

***(ii)* Auditory capsules.** The auditory capsules are found closely fused with the postero-lateral region of the cranial region of the skull. Each capsule is formed by the fusion of two bones, the periotic and tympanic, and encloses the internal ear.

Periotic. Each **periotic** is an irregular cartilage bone formed by the fusion of prootic, epiotic and opisthotic bones. It is found in between squamosal and occipital segment and externally visible forming a bulge and perforated in the adults. The periotic consists of two parts : (*i*) an internal, hard, bony **petrous part** and (*ii*) an outer posterior porous **mastoid part** which is produced downwards into a **mastoid process** which is closely applied to the posterior side of tympanic. Petrous part bears two apertures–an anterior **fenestra ovalis** and another posterior **fenestra rotunda**. It also bears a ventral rounded swelling called the **promontary** for lodging the cochlea.

Tympanic. It is a flask-shaped membrane bone attached to the outer surface of the periotic. Its lower, flask-shaped swollen part is known as **tympanic bulla**. It encloses the tympanic cavity of middle ear containing three ear ossicles arranged linearly–the **malleus, incus** and **stapes**. The neck of the flask constitutes the **external auditory meatus** closed at the base by the tympanic membrane in life.Tympanic on its posterior side bears a **stylomastoid foramen** for the facial nerve.

(*iii*) Orbits. The orbits are situated on the sides of the frontal segment of the cranial region. A median bony inter-orbital septum separates the two orbits. Each orbit has a median wall formed of dorsal frontal which project above the orbit as **supra-orbital ridge**, median ventral **orbitosphenoid**, **alisphenoid** and **ethmoid** bones. The outer side of the orbit is protected by the **zygomatic arch** and the lower side of the orbit is unossified.

3. Visceral skeleton. It includes the upper and lower jaws and hyoid apparatus.

(*i*) Upper jaw. Bones of the upper jaw are premaxillae, maxillae, jugals, pterygoids and palatines. These are closely associated with the cranium and olfactory capsules.

Premaxillae. These are large bones and form the anteriormost part of the upper jaw. Both the premaxillae unite in front in the median line. Each premaxilla ventrally bears two sockets for incisor teeth. Each presmaxilla also bears two processes–a **nasal** process extending backward up to frontal in between nasal and maxilla; a **palatine process (**ventral**)** which meets with its fellow of other side in the mid-ventral line and both extend up to palatine processes of maxillae.

Maxilla. Maxilla is a large, irregular-shaped bone forming the greater part of the upper jaw and side of the face. The ventrat surface of **maxillae** bears 6 sockets for the cheek teeth. Each maxilla gives off a number of processes. A median flat **palatine process** is given out from the ventral surface of each maxilla. Both the palatine processes are united by a suture in the middle line forming the anterior part of hard palate. A flat stout process from the outer surface of each maxilla, called the **zygomatic process**, passes backwards and joins with a similar forwardly directed process of the squamosal, thus, forming the **zygomatic arch**. Each maxilla anteriorly gives off a **maxillary process** which joins with the premaxilla.

Palatine. The palatines are thin, more or less vertically situated bones in the mid-ventral side. Both palatines form an inner process in front, which meet together to form the posterior part of hard palate. Each palatine posteriorly unites with the alisphenoid and pterygoid and encloses a pterygoid fossa between it and basisphenoid.

Pterygoid. The pterygoids are small, irregular, bony plates situated behind the palatines and joined with the pterygoid process of alisphenoids behind.

Jugal. Molar or jugal is a laterally compressed bone which extends in between zygomatic processes of maxilla and squanosal. It forms the greater part of the zygomatic arch.

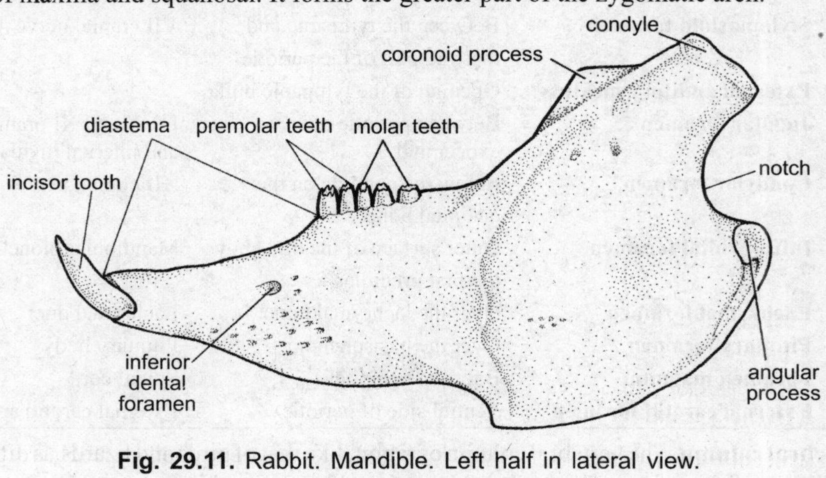

Fig. 29.11. Rabbit. Mandible. Left half in lateral view.

***(ii)* Lower jaw.** The lower jaw consists of two halves or rami which are united in front by the **mandibular symphysis**. Each ramus is formed of a single bone, called **dentary** or **mandible**. Each dentary has a horizontal part which bears sockets for the attachment of teeth like incisors, premolars and molars, and a gap without teeth, called the diastema.The posterior vertical or ascending part of dentary carries the articular surface or **condyle** for articulation with the glenoid cavity of squamosal. In front of the condyle is the compressed **coronoid process**. On the postero-lateral surface of the mandible is present a slight depression, called **messenteric fossa,** in which are attached large **masseter** muscle. Temporal muscle is attached with the medial side of coronoid process.The posterior, lower border of the dentary gives off a rounded and inwardly directed **angular process.** On the outer anterior surface of dentary has a **inferior dental** or **mental foramen** and on the inner side is present the **mandibular foramen.**

***(iii)* Hyoid.** The hyoid is bony in rabbit and consists of a small stout thick body, called **basihyal**, placed transversely between the two halves of mandible and bears two pairs of **cornua** or **horns**. The small **anterior cornua** or ceratohyals are situated below the tongue and consists of a series of small bones such as **ceratohyal, epihyal, stylohyal** and **tympanohyal**. The last one is fused with the tympanic and periotic bones. The long and posteriorly directed **posterior cornua** consists of a single **thyrohyals.**

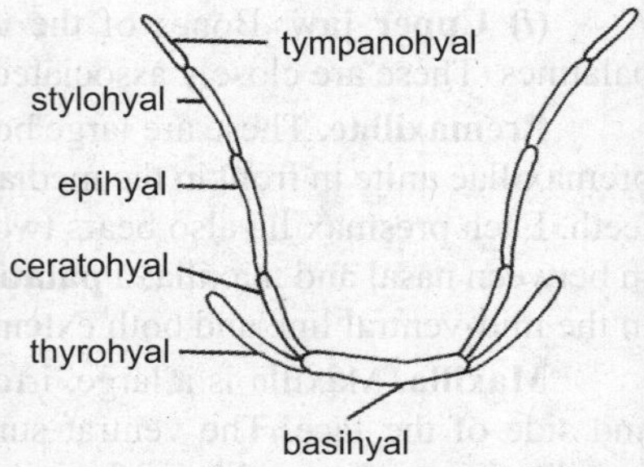

Fig. 29.12. Rabbit. Hyoid apparatus.

Important foramina in the skull of rabbit. The important foramina in the skull of rabbit can be easily explained with the help of the following table :

TABLE 29.2. IMPORTANT FORAMINA OF SKULL OF RABBIT.

S.No.	Name of the foramen	Situation	Name of nerve passing through the foramen
1.	**Optic foramen**	Orbitosphenoid	Both optic nerves
2.	**Sphenoidal fissure or superior orbital fissure**	Between orbito-and alisphenoid	III, IV, ophthalmic and maxillary branches of V and VI nerves
3.	**Foramen lacerum**	In front of periotic	Mandibular branch of V cranial nerve and internal carotid artery
4.	**Stylomastoid foramen**	Between the tympanic and mastoid part of the periotic	VII cranial nerve
5.	**External auditory meatus**	Opening of the tympanic bulla	—
6.	**Jugular foramen**	Between periotic and exoccipital	IX, X and XI oranial nerves and internal jugular vein
7.	**Condylar foramen**	Below the condyle on the occipital bone	XII cranial nerve
8.	**Infra-orbital foramen**	Outer surface of maxilla above the first premolar	Mandibular branch of V
9.	**Lachrymal foramen**	Over the lachrymal bone	Lachrymal duct
10.	**Pituitary foramen**	Over the basisphenoid	Pituitary body
11.	**Foramen magnum**	Between occipitals	Spinal cord
12.	**External carotid foramen**	Ventral side of periotic	External carotid artery

Vertebral column. The vertebral column of rabbit, like that of birds and lizards, is differentiated into the following five regions. The total number of vertebrae in rabbit is about 45–47.

1. Cervical
2. Thoracic
3. Lumbar
4. Sacral
5. Coccygeal or caudal

Vertebral formula of rabbit : C_7, $T_{12\text{-}13}$, $L_{6\text{-}7}$, S_4, Cd_{15} where C, T, L, S, Cd stand for cervical, thoracic, lumbar, sacral and caudal respectively.

Vertebral column of mammals is distinguished from other vertebrates in the following respects:

Important Characteristics of Mammalian Vertebrae

1. The centra are more a less flattened on both the surfaces, *i.e.*, **amphiplatyan** type. The centra, on either side, is provided with small bony plate, **epiphysis**.

2. A thin **epiphysial cartilage** separates the centrum and epiphysis in the embryonic condition, which later become fused with the centrum. Thus, in adults no epiphysial cartilage is found.

3. In between the adjacent centra are present **intervertebral discs** of fibrocartilage. The central portion of the intervertebral disc is called **nucleus pulposus**, which represents the remnant notochord. The discs are shock-absorbing cushions and probably represent the hypocentrum.

A. Cervical vertebrae. Out of the seven cervical vertebrae, first and the second are highly modified, known as **atlas** and **axis** respectively. Remaining 3rd to 7th are more or less alike and can be called **typical cervicals**.

1. Atlas vertebra. It is ring-like without any solid centrum and zygapophyses. It consists of a large neural arch but a reduced neural spine (The centrum is, however, present in the embryonic condition which later fuses with the axis vertebra and known as **odontoid process**). The neural canal is large and divided into two parts in living condition. The upper part is the **neural canal** for the passage of spinal cord and lower part is occupied by the odontoid process of the axis. Two large concave **occipital facets** are present in the anterior face of the altas for articulation with the occipital condyles of the skull. This **atlanto-occipital joint** allows movements of the head in sagittal plain, as in nodding the head. The transverse processes arising from the sides are broad, long and wing-like for the attachment of muscles that hold and rotate the head and neck. These are not transverse processes but are enlarged flattened cervical ribs, perforated basally by the **vertebrarterial canals**. A pair of lateral and a mid-ventral articular facets are present on the posterior face of the atlas for the odontoid process of axis.

2. Axis vertebra. It is the second cervical vertebra. Its neural spine is high, ridge-like, laterally compressed and elongated antero-posteriorly. Its centrum bears a peg-like **odontoid process** in the anterior face which is developmentally the centrum of atlas. This process forms the **atlanto-axial joint**, which allows the rotation of the skull and atlas on the axis. This movement is facilitated by a pair of smooth articular surfaces on the anterior face of the axis one on either side of the odontoid process. The transverse processes are short and perforated basally by **vertebrarterial canals** for vertebral artery. A pair of post-zygapophyses are found but pre-zygapophyses are absent.

3. Typical cervical vertebra. A typical cervical vertebra is broad and has a small flattened centrum, a large neural arch and a small neural spine. Pre-and post-zygapophyses are well developed. Transverse processes are bifurcated into dorsal and ventral lamellae perforated basally by a **foramen transversaria** or **vertebrarterial canal.**The cervical ribs are much reduced and more or less incorporated in the vertebra. The transverse processes and reduced ribs provide surface for the attachment of neck muscles.

The cervical vertebrae from 3rd to 6th are similar in structure. But the 7th cervical differs from others in having a more elongated neural spine, in having its transverse processes simple and imperforate and in the presence of a small concave semilunar facet at the posterior edge of the centrum for the articulation of thoracic ribs.

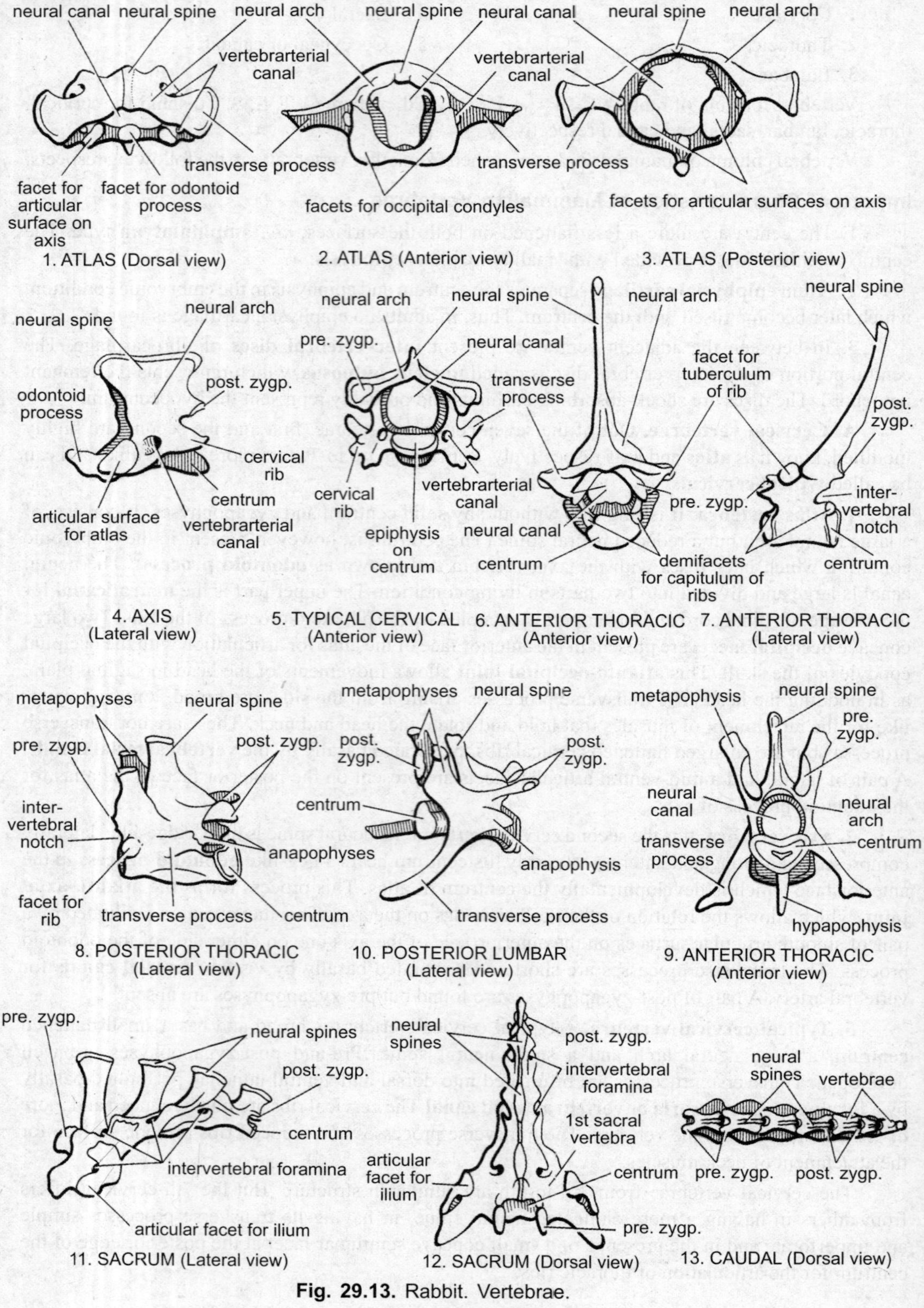

Fig. 29.13. Rabbit. Vertebrae.

B. Thoracic vertebrae. The thoracic vertebrae are 12–13 and each vertebra is provided with a well developed centrum, a neural arch, long neural spine and pre- and post-zygapophyses. The transverse processes are short and stout. Each bears near its extremity a small smooth articular surface or **tubercular facet** for the tubercle of a rib. On the anterior and posterior borders of each vertebra is a little semilunar facet, the **capitular facet**, situated at the junction of the centrum and neural arch. The neural spines of the anterior thoracic vertebrae are more or less straight and directed backwards. Poterior 4 or 5 thoracic vertebrae are little different from the anterior thoracic vertebrae. They possess longer and stout **centra**, short **neural spines**, reduced transverse processes with no tubercular facets, and **zygapophyses** are distinct. Capitular facets are present near the anterior of centrum. Neural spine is directed upward. Metapophyses and anapophyses are present on the anterior end of neural arch and the posterior end of neural arch below post-zygapophyses respectively.

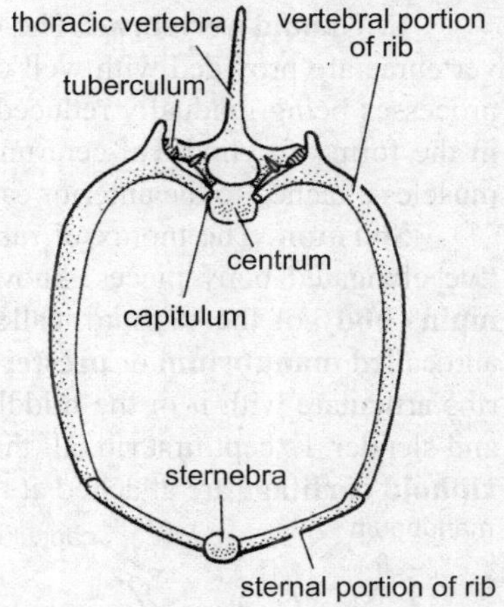

Fig. 29.14. Rabbit. A thoracic vertebra with its ribs.

Ribs. Thoracic ribs are 12 or 13 pairs and found articulating with the thoracic vertebrae. Each rib is a bony curved rod divided into a dorsal longer bony **vertebral rib** and a ventral smaller cartilaginous **sternal rib**. Vertebral part is **biramus (double headed)**, the heads are known as **tuberculum** and **capitulum** which articulate with the transverse process and with the demi-facets on the centra of two adjacent vertebra respectively.The sternal part of the rib articulates with the sternum. The sternal parts of all the thoracic ribs, except the last five, meet the sternum below and are called **true ribs**.The last three or four ribs have no sternal parts and they along with one anterior to them are not connected with the sternum and, hence, known as **floating ribs**.

C. Lumbar vertebrae. They are 7 and out of which the first two are called **anterior lumbars**. Each anterior lumbar has a large, stout, strongly built **centrum**, a **neural arch** enclosing a broad **neural canal** and well developed forwardly directed **neural spine**. The **transverse processes** are large, distally expanded beside pre- and post-zygapophyses and directed downwards and forwards. There are two pairs of bony processes, called **mammilary processes**, *i.e.,* **metapophysis** is present at the anterior end of neural arch and it is sloping forward. Beneath it lies the median pre-zygapophysis; **anapophysis** is present at the posterior end of neural arch beneath the post-zygapophysis and it is a small backwardly directed process. Each anterior lumbar also has a median ventral process beneath the centrum, called **hypapophysis**.

Posterior lumbar. From third to seventh lumbars are called posterior lumbars. These resemble the anterior lumbars in all details but the hypapophysis is absent, only a small ridge is present in its place.

D. Sacral vertebrae (sacrum). There are four vertebrae in the sacral region of rabbit which fuse together to form a compound bone, the **sacrum** but only the first articulates with ilium of the pelvic girdle. It is believed that out of these four vertebrae constituting sacrum, only first vertebra is the sacral vertebra and the remaining three vertebrae are the anterior caudals. Each vertebra of the sacrum is provided with a long neural spine, tubercle-like projections on the upper side representing zygapophyses, and intervertebral foramina. The first vertebra is the largest and has large, flattened, stout transverse processes, probably fused sacral ribs. These articulate with the ilia of pelvic girdle. This joint is called **sacro-iliac joint**. Hypapophyses and anapophyses are absent and the metapophyses are relatively small. The sacro-iliac joint provides strength to the pelvic girdle and the vertebral column at the time of throwing the body forward when the hindlimbs are straightened.

E. Caudal vertebrae. The caudal vertebrae are sixteen in rabbit. Only the anterior caudal vertebrae are provided with well developed neural spines, neural arches and zygapophyses. These processes being gradually reduced until the terminal vertebrae near the end of the tail is only left in the form of cylindrical centrum. The transverse processes are absent in caudal vertebrae. The muscles attached to the anterior caudal vertebrae provide movements of the tail in many planes.

Sternum. The thorax of rabbit is bounded mid-ventrally by the sternum which consists of five elongated bony pieces, known as **sternebrae.** Thus, the sternebrae together constitute the **main body** of the sternum, called **mesosternum.**The first anteriormost sternebra is the longest and called **manubrium** or **presternum**. It is ventrally produced into a **keel**. The first pair of sternal ribs articulate with it in the middle. Sixth sternebra is the smallest of all and the last one is long and slender. Except first rib, all the sternal ribs called **xiphisternum** terminating into an expanded **xiphoid cartilage** are attached at the intersternebral junctions.

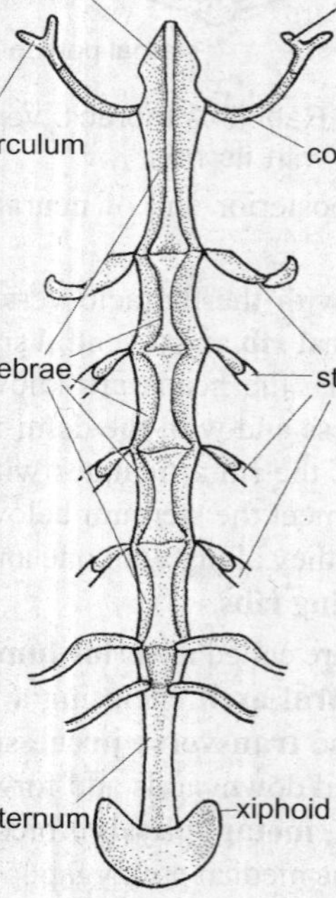

Fig. 29.15. Rabbit. Sternum.

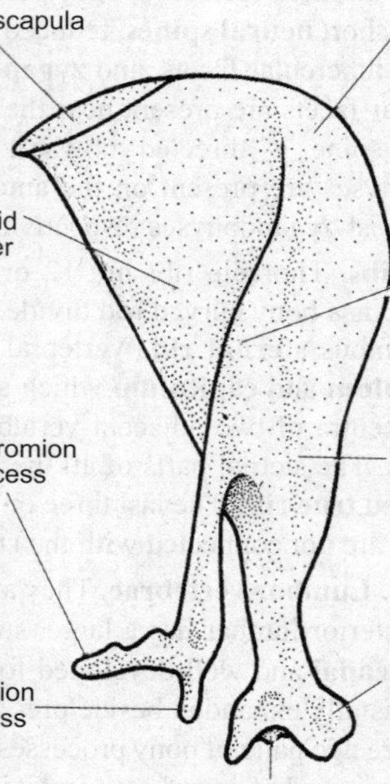

Fig. 29.16. Rabbit. Pectoral girdle. Right half in outer view.

Appendicular Skeleton

Pectoral Girdle

It consists of two separate halves placed dorsal to the anterior thoracic ribs in between the forelimbs.It supports forelimbs and protects the soft parts of the body from the ventral side. Each half of the pectoral girdle is known as **innominate**.Thus, one half of the pectoral girdle is formed of a broad, more or less triangular bony plate, called **scapulo-coracoid** and a small **clavicle** bone.

Scapulo-coraciod. It is mainly formed of **scapula** which is a thin, flat, and more or less triangular bony structure. Its outer surface bears a prominent ridge, called the **spine** which divides its surface into antero-dorsal and postero-dorsal portions to which are attached muscles. The spine terminates ventrally into an expanded knob-like structure, the **acromion process** which posteriorly bears backwardly directed **metacromion process**. The apex of scapulo-coracoid is directed downwards and forwards terminating below into a concave **glenoid cavity** for the head of humerus. Above this cavity is present a small hook-like coracoid process, the rudimentary coracoid. The

suprascapula is in the form of a thin strip of cartilage situated along the dorsal or vertebral border of the scapula.

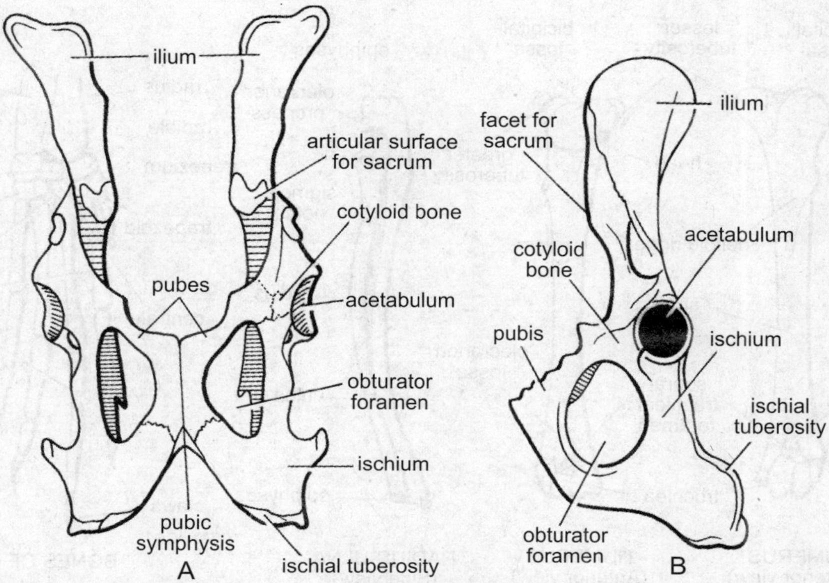

Fig. 29.17. Rabbit. Pelvic girdle. A – Complete girdle in ventral view; B – Left half.

Clavicle is a thin, slightly curved bone extending between the acromion processes and manubrium of the sternum.

Pelvic Girdle

The pelvic girdle is also formed of two equal halves or **os-innominate**. Both the halves are joined together mid-ventrally by **pubic symphysis** to form a stout and strong **girdle** situated in the pelvic region between the two hindlimbs.

Each half or os-innominatum is formed of three bones, the **ilium**, **ischium**, and **pubis**. All these bones are fused together to form a single hip bone.

The **ilium** is the antero-dorsal longest bone, which bears a rough flat articular surface roughly at about the middle of its length for sacrum. The anterior and dorsal edge of the ilium is raised into **iliac-crest**. The ilium extends posteriorly up to the **acetabulum**. The postero-dorsal part of the os-innominate is formed by the **ischium**. The posteriormost part of ischium is broad and projects outwards into a prominent **ischial tuberosity**. The **pubis** is a narrow bone and forms the ventro-median portion of innominate. Both the pubes unite with each other on the mid-ventral line to form a **pubic-symphysis**. The pubis do not take part directly in the formation of acetabulum, because a **cotyloid** bone is present in between the acetabulum and pubis. A big **obturator foramen** is present between the **ischium** and **pubis** which is always covered by the obturator membrane and muscles in the lifetime. **Acetabulum** is only formed by the ilium and ischiam on both sides of the girdle and into it articulates the head of humerus.

Forelimbs Bones

Humerus. The bone of the **upper-arm** is the **humerus**, which is a long bone having a proximal large rounded head for the articulation in the glenoid cavity of scapula. The proximal end of the humerus near the head is provided with a **bicipital groove** in between the two tuberosities (greater and lesser) for the attachment of muscles.The anterior surface of the humerus below the head has a projection called **deltoid ridge** for the attachment of muscles. The distal end of the humerus is provided with a pulley-like **trochlea** for articulation with ulna. Just above trochlea are present two fossae (depressions)–anterior smaller is **coracoid fossa** and posterior larger is **olecranon fossa**. Both the fossae communicate with each other through a **supra-trochlear foramen**.

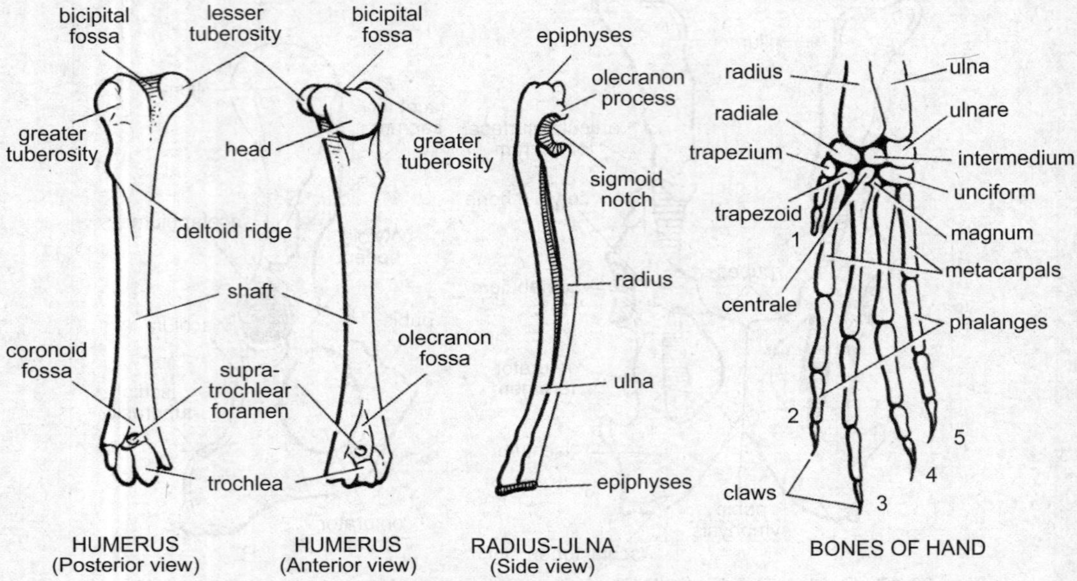

Fig. 29.18. Rabbit. Forelimb bones.

Radius and ulna. The bones of the **forearm** are **radius** and **ulna** which are closely held together at the two ends, so that they cannot move over each other. The **ulna** is the long bone and proximally bears a prominent projection called the **olecranon process**. It articulates with the **olecranon fossa** of humerus. Olecranon process is basally notched called the **sigmoid notch** into which fits the trochlea of humerus. The **radius** is the smaller bone than ulna and situated towards the inner side. The radius and ulna are distally provided with epiphysis and articulate with the wrist bones.

Bones of hand. The wrist **bones** or **carpals** are 9 small bones arranged in two rows. **Proximal row** has three carpals, called **radiale, ulnare** and **intermedium**. The radiale is situated below the radius, ulnare below the ulna, while the intermedium is situated between them. The **distal row** consists of **trapezium** and **trapezoid** situated below the **radiale, centrale** and **magnum** situated below the intermedium and **unciform** situated below the ulnare. The unciform is actually fused two carpals. Besides these, a sesamoid **pisciform** is present on the ventral side of carpus.

The bones of the **palm** or **manus** are five long **metacarpals**, which support the five digits having 2, 3, 3, 3, 3 **phalanges** respectively. The terminal phalanx of each digit bears a horny **claw**.

Hindlimb Bones

Femur. The bone of the **thigh** is the **femur** which is a long bone with a flattened proximal end. Its flattened proximal end bears a rounded smooth **head** towards the inner side for the articulation with the acetabulum. The proximal end of the femur also bears three **trochanters** for the attachment of muscles. The first or **greater trochanter** is situated above the head, the second or **lesser trochanter** is situated below the head, while the **third trochanter** is stiuated below the greater trochanter. The deep groove below the head and greater trochanter is the **digital fossa**. The main body of the femur or **shaft** terminates distally into a pair of expanded **condyles** enclosing the **intercondylar groove** for the articulation with tibio-fibula. On the anterior side of this groove is called the **patellar groove** into which moves the patella bone. These condyles are provided with articular facets for articulating with the tibia.

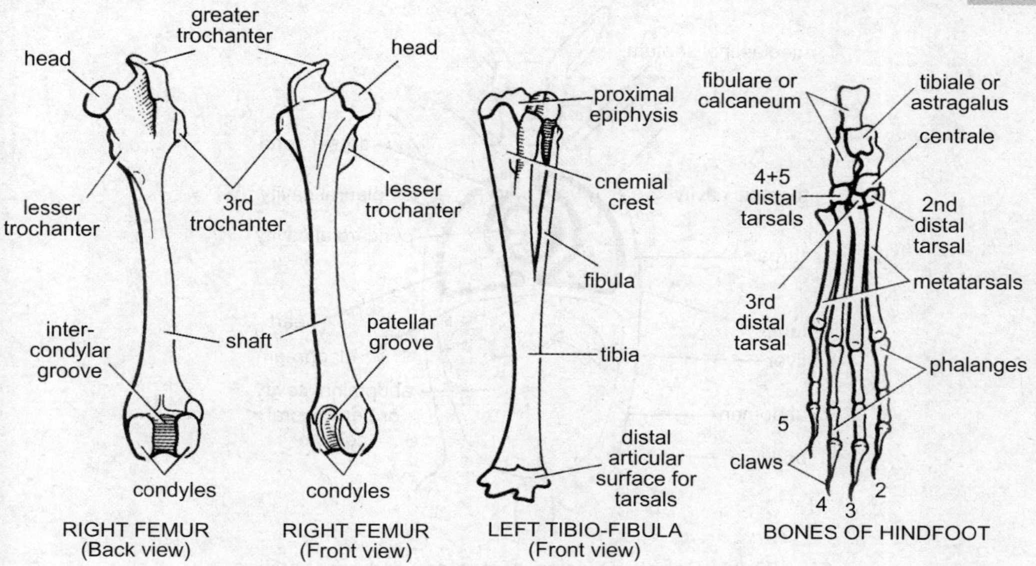

Fig. 29.19. Rabbit. Hindlimb bones.

Tibio-fibula. The **shank** of the hindlimb is provided with long stout and straight **tibia** and a small slender **fibula** bones. The fibula is a reduced slender bone which is fused with the tibia distally, while proximally it is free. The proximal end of these bones are provided with **proximal epiphysis** which articulates with the condyles of the fenur. Tibio-fibula articulate distally with the bones of the **ankle** or **tarsus**. A **cnemial crest** is present on the proximal dorsal end of tibia whose two depressions articulate with two condyles of femur.

Bones of foot. The **ankle** is formed of six **tarsus** arranged in three rows. The first proximal row is formed of two tarsals, a **tibiale** and **intermedium**, both fused to form the **astragalus** located on the preaxial side. The other largest is the **calcaneum** which is produced into a process behind its articulation with the tibio-fibula.The astragalus bears a pulley-like surface for articulation with tibia. The middle row has a single bone, the **centrale,** just in front of astragalus. The distal row has three tarsals which are **mesocuneiform, ectocuneiform** and **cuboid**.

The bones of **sole** or **foot** are 4 long **metatarsals** (the first is absent). Four digits or toes are only present, each formed of three **phalanges**. The last phalanx of each digit is provided with a **claw**. **Hallux** (first toe) is absent.

COELOM AND VISCERA

Coelom

The body cavity of rabbit is divided into an anterior thoracic and posterior abdominal cavities which are separated from each other by a transverse muscular partition, the **diaphragm**. Both these cavities are derived from the original **perivisceral body cavity** or coelom of the rabbit. Unlike frog, the thoracic cavity is further divided into three cavities, two of them are situated laterally and lodge soft spongy lungs known as **pleural cavities**, while median one situated between the two pleural cavities is known as **pericardial cavity** lodging the heart and the roots of the great vessels. Thoracic cavity also contains the trachea, bronchi and posterior part of oesophagus. The oesophagus perforates the diaphragm and enters the abdomen. The **abdominal** or the **peritonial cavity** is large and contains all the remaining visceral organs, *e.g.,* stomach, intestine, liver and pancreas, spleen, urinogenital organs, etc.

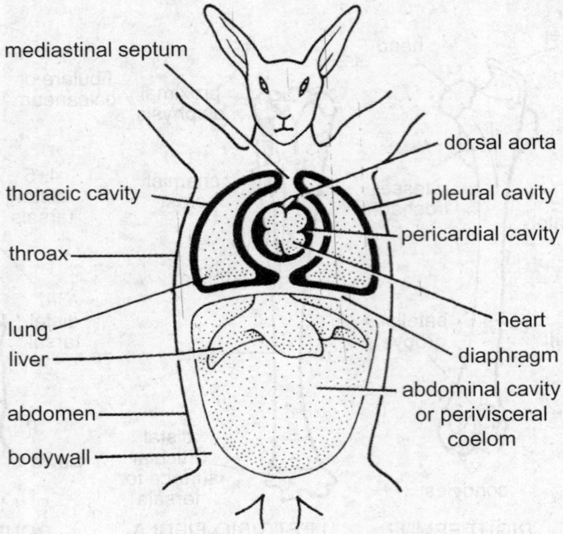

Fig. 29.20. Rabbit. Division of coelom or body cavities.

The coelom is lined by outer **parietal epithelium** and inner **visceral epithelium**, which covers and in fact encloses all the organs. The pleural cavity is lined by a moist membrane, the **pleura**, which can be divided into **parietal** and **visceral** parts. The **parietal pleura** lines the inner side of the pleural cavity, while the **visceral pleura** is closely applied to the lungs. The pleural cavities are separated from each other by a **mediastinal space** which contains oesophagus and trachea. The viseral pleura of two sides are closely applied in the ventral part of the mediastinal cavity to form a median **mediastinal septum**.

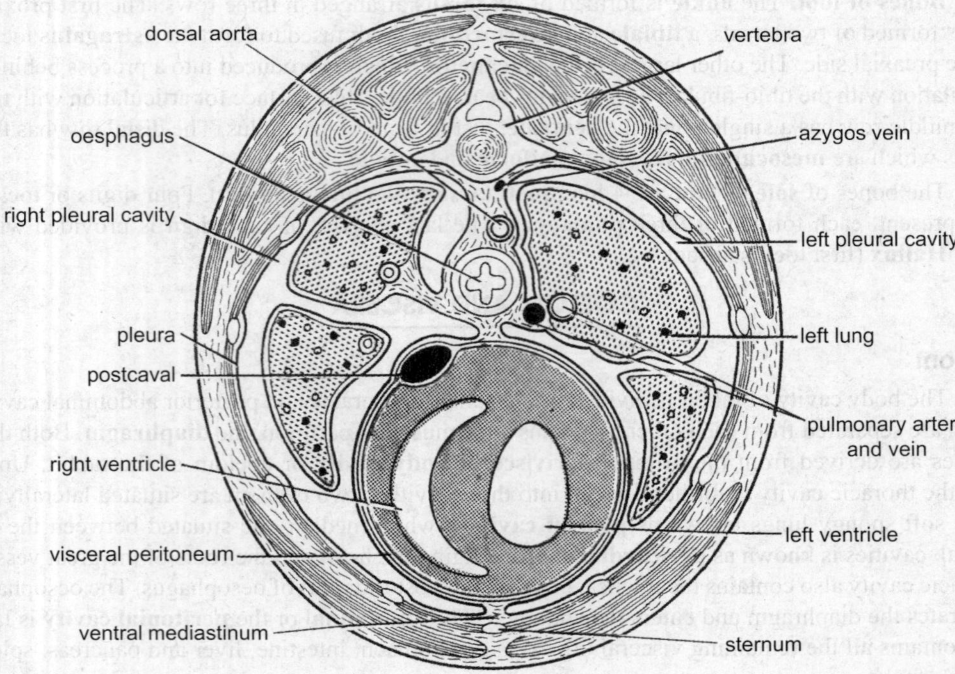

Fig. 29.21. Rabbit. Cross-section of the thorax.

In the **pericardial cavity** the outer parietal membrane **(parietal pericardium)** is transparent and the inner visceral membrane (**visceral pericardium** or **epicardium**) is closely applied to the heart. The parietal pericardium is separated from the heart by the pericardial cavity.

The membranes of the abdominal cavity are **parietal** and **visceral peritoneum**. The peritoneum forms double layered **mesenteries** or **ligaments** which support the alimentary canal and organs of the viscera.

Viscera

It is clear from the above description that the heart, lungs, oesophagus and trachea are contained in the **thoracic cavity**. The **abdominal cavity** has a dark brown five-lobed **liver** fitted into the concavity of the **diaphragm**, and a saccular dark blue **gall bladder** is situated in one of the lobes of the liver. The oesophagus opens into the **stomach** which lies dorsally to the left lobe of the liver. The **spleen** is attached to the left border of the stomach.The stomach is continued into a long coiled **intestine**, occupying largely the whole abdominal cavity. A pinkish irregular structure, the **pancreas**, is found in the 'U'-shaped duodenum. Two bean-shaped **kidneys** are found attached to the dorsal wall of the abdomen outside the coelom. The right kidney is situated a bit anterior to the left one. Ureters are seen coming out from the kidneys to join a thin-walled sac, the **urinary bladder**. A small rounded yellow coloured **adrenal gland** is found in front of each kidney. The adrenal glands are endocrine in nature. In a male rabbit the two **testes** are situated in two pouches of the bodywall known as the **scrotal sacs** lying outside the abdominal cavity. The scrotal sacs are situated one on either side of the **penis**. The abdominal cavity is continued into the cavity of the scrotal sacs by **inguinal canals**. The vas deferens of the testes unite with the neck of the urinary bladder to form a common **urinogenital canal** or **urethra** which passes in the penis. The penis is the copulatory organ of the male rabbit. In a female rabbit, two small oval **ovaries** are situated behind the kidneys. From behind each ovary a convoluted duct, the **oviduct**, arises which dilates to form a **uterus**. The two uteri thus formed unite together to form a **vagina**. The neck of the uinary bladder joins the vagina to form a **urinogenital canal** or **vestibule** which opens out by a slit-like **vulva**. The vulva is the external aperture of the female reproductive system.

DIGESTIVE SYSTEM

The digestive system of the rabbit consists of the **alimentary canal** and the **digestive glands** associated with the alimentary canal.

A. Alimentary Canal

The alimentary canal of rabbit is a long coiled tube of variable diameter starting from mouth and terminatings at anus. It consists of mouth, vestibule, pharynx (forming bucco-pharyngeal cavity), oesophagus, stomach, small intestine, caecum, large intestine and anus.

1. Bucco-pharyngeal Cavity

1. Mouth. The mouth is a transverse slit-like terminal aperture situated a little below the anterior tip of snout. The mouth is bounded by two soft mobile and muscular lips. These are externally covered by a hairy skin and internally lined with mucous membrane. The upper lip is divided by a median cleft extending up to the nostrils. Through this cleft the upper incisors remain exposed outwardly even when mouth is closed.

2. Vestibule. The mouth opens into a narrow vertical space, called the **vestibule**, present in between lips and cheeks and gums of the jaws. Its mucous membrane contains mucous secreting glands.

3. Buccal cavity. The vestibule opens into a large spacious **buccal cavity** in between the jaws. It is lined by the mucous membrane and contains the palate, tongue and teeth.

(*i*) Palate. It forms the roof of buccal cavity. The anterior end of the palate is hard, supported by the premaxillae, maxillae and palatine bones forming the **hard palate**. The mucous membrane lining the palate (ectodermal in orgin) forms transverse ridge, known as **palatal rugae**. The posterior end of the palate is soft, smooth, fleshy and formed of connective tissue, called **soft palate**. The soft palate posteriorly hangs down freely into pharynx as a small flap, the **uvula**. The palate divides the original buccal cavity into a dorsal **nasal passage** and ventral **food passage**. The separation of the buccal cavity from the nasal passage by the palate enables the rabbits to retain their food in the mouth during chewing and to breathe at the same time. There is no clear cut demarcation between the buccal cavity and the pharynx, therefore, both these structures are considered together as **bucco-pharyngeal cavity**. There is a pair of small openings of **naso-palatine ducts** in the anterior part of hard palate, which lead into the **nasal** or **olfactory cavities**. A tubular **Jacobson's organ** opens into each naso-palatine duct whose function is probably to recognise different kinds of food. These are enclosed in a cartilage and situated on the floor of the nasal cavities.

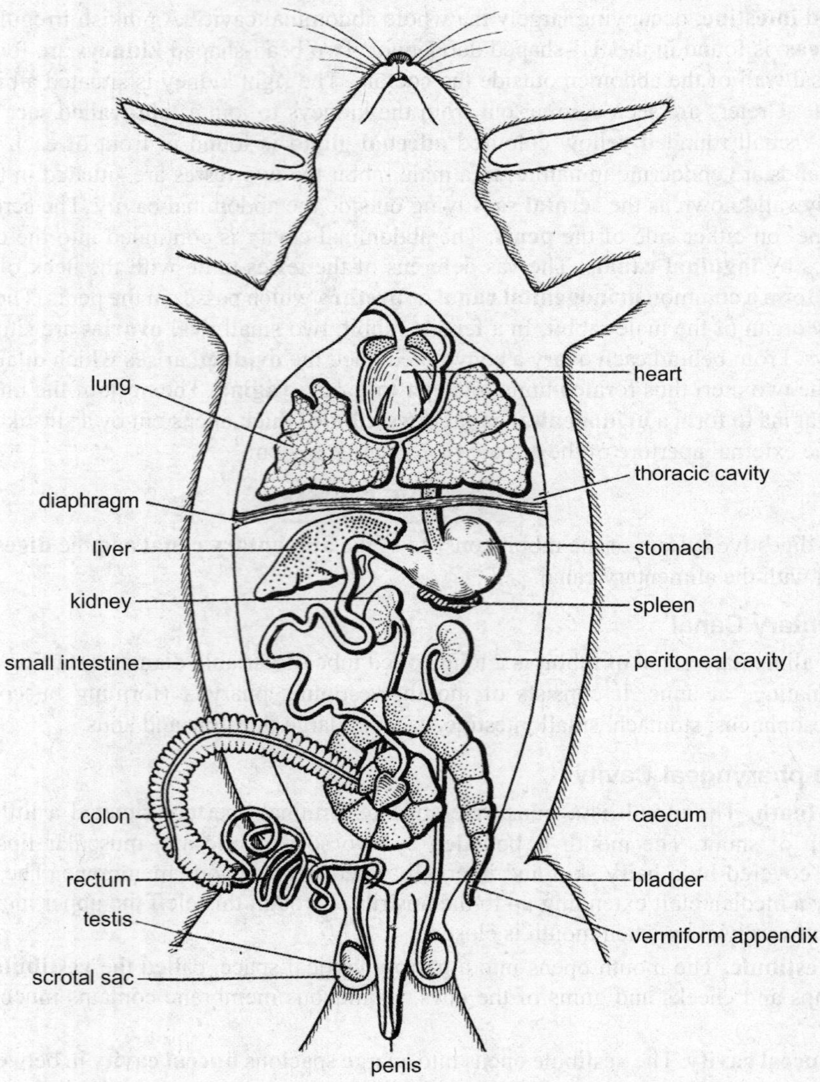

Fig. 29.22. Rabbit. Coelom and viscera.

(*ii*) Tongue. The tongue is situated on the floor of the buccal cavity and the greater part of its length is attached to the floor of the buccal cavity with a free rounded tip in front. It is a fleshy, muscular and movable organ.The tongue is covered with mucous membrane beset with numerous papillae. On certain papillae are situated the **taste buds**. On the basis of shape and size, these papillae are of four types : **(*i*) fungiform**, which are numerous mushroom-shaped situated at the margins of the tongue, **(*ii*) filiform**, which are numerous filamentous, situated at the upper surface of the tongue, **(*iii*) circumvallate** are large-sized, few in number situated at the base of the tongue, and **(*iv*) foliate** are broad, leaf-like situated at the sides of the base of the tongue. The function of the tongue is to manipulate the food and mixing of saliva with the food.

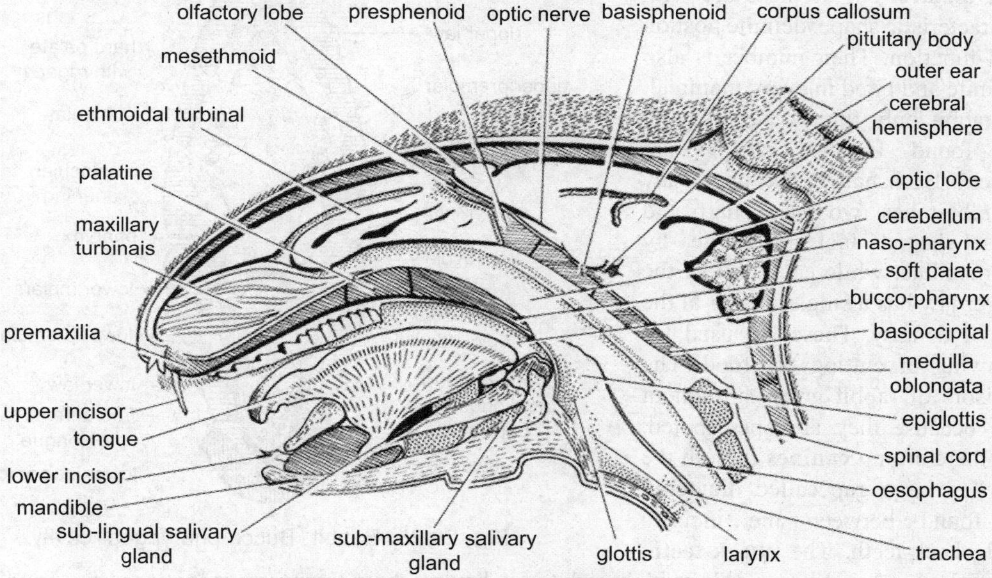

Fig. 29.23. Rabbit. Sagittal section of the head.

(*iii*) Teeth. Both the jaws of rabbit bear teeth. These are situated on the premaxillae and maxillae of the upper jaw and dentary bones of the lower jaw. The teeth are situated in the cup-like

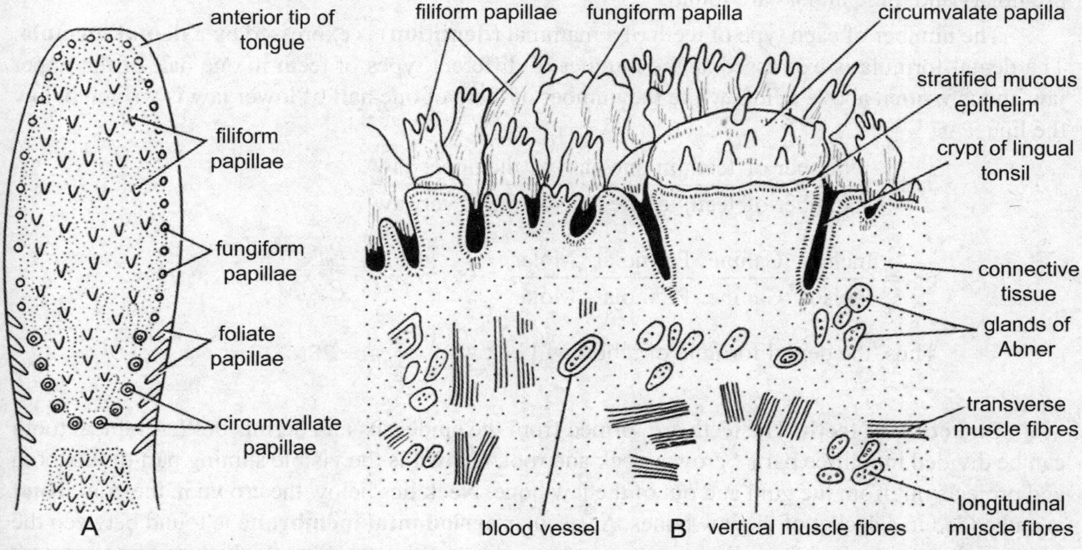

Fig. 29.24. Rabbit. Tongue showing different types of papillae. A – Entire tongue in dorsal view; B – V. S. of tongue.

sockets of the jaw bones, *i.e.,* **thecodont** (L., *theca* = a case or sheath + G. *odous* = teeth) and are of differnet shapes, *i.e.,* **heterodont (**Gr., *heteros* = different**).** Teeth in rabbit are also **diphyodont**, *i.e.*, two sets of teeth are seen during the lifetime of rabbit like majority of other mammals. The first set is known as **milk-teeth** which are **deciduous** in the young condition. The milk-teeth are replaced by the other set of **permanent teeth** in the adults which are not replaced.

Heterodont dentition. The teeth in mammals are four different types: **incisors**, **canines**, **premolars** and **molars**. All of these are with characteristic shape, definite positon and function. Their number is also definite and fixed in every mammal. In rabbit only three types of teeth are found, because **canines** are absent. Each half of the upper jaw of rabbit has two long, sharp and chisel-shaped incisors borne by premaxillae, while each half of the lower jaw has a single incisor at the tip of dentary. These are used for gnawing or cutting the food. The incisors of rabbit grow throughout life because they are open-rooted. As there are no **canines** in both the jaws, a large gap called **diastema** is found between the incisors and cheek-teeth. The **cheek-teeth**, premolars and molars are alike with broad crown having sharp transverse ridges used for grinding the food. So they are also known as **grinders**. Such types of cheek-teeth are known as **lophodont**. Each half of the upper jaw has three premolars and three molars, while in the lower jaw only two premolars and three molars are found.

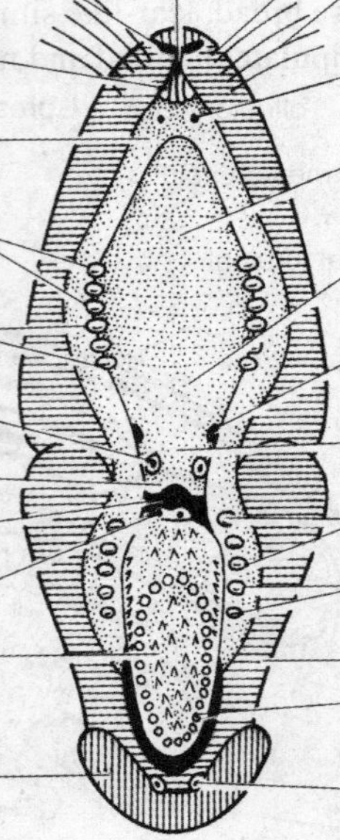

Fig. 29.25. Rabbit. Bucco-pharyngeal cavity.

The number of each type of teeth of a mammal (**dentition**) is expressed by a **dental formula**. The dental formula is expressed as the number of different types of teeth in one half of the upper jaw and is written above a line, while the number of teeth of one half of lower jaw is written below the line, *e.g.,*

$$\text{D.f.} = \frac{\text{Number of teeth in one half of the upper jaw}}{\text{Number of teeth in one half of the lower jaw}}$$

$$\text{or } = \frac{\text{Incisor}}{\text{Incisor}}, \frac{\text{Canine}}{\text{Canine}}, \frac{\text{Premolar}}{\text{Premolar}}, \frac{\text{Molar}}{\text{Molar}} \quad \text{or } = \frac{I}{I}, \frac{C}{C}, \frac{P_m}{P_m}, \frac{M}{M}$$

Thus, the dental formula of rabbit will be : $\frac{2}{1}, \frac{0}{0}, \frac{3}{2}, \frac{3}{3} = 28.$

Structure of teeth. The teeth are formed from the epidermis and dermis both. A typical tooth can be divided into three parts : crown, neck and root. **Crown** is the visible shining part of the tooth and projects out from the gum and out of the jaw bone. **Neck** lies below the crown in the gum. **Root** is embedded in a socket of the jaw bones. A vascular **periodontal membrane** is found between the root and the socket. It is formed of connective tissue fibres (Sharpey fibres) which run from cement layer of root to bone. The bulk of the tooth is formed by hard **dentine** (dermal bony substance). In the crown it is covered over by a thin, hard and shining enamel layer. It is not found over root. Inside the tooth a **pulp cavity** is found which contains connective tissue, blood vessels and nerves. The pulp cavity is lined by a layer of **odontoblast** or **bone cells**. The dentine is pierced by many fine **canaliculi** in which the protoplasmic processes of odontoblasts run. Further, the attachment of teeth on the jaw bones is strengthened by the gum. Furthermore, the teeth are of two types : open rooted and close rooted. In the incisors of rodents, lagomorphs and elephants, pulp cavity is basally opened so these teeth grow continuously throughout life, and are called **open rooted**. In **close rooted** the basal aperture of pulp cavity becomes closed at a certain age, due to which teeth could not get nourishment and, thus, further growth ceases.

4. Pharynx. The buccal cavity indistinguishably merges posteriorly into the **pharynx** which is small and narrow cavity. The soft palate divides the pharynx incompletely into three parts : (*i*) **nasopharynx** is dorsal to the soft palate, (*ii*) **oropharynx** beneath the soft palate, and (*iii*) **laryngopharynx** around the freely suspended **uvula** or **velum palati**. It communicates with the first two cavities. The nasopharynx communicates anteriorly with the nasal chambers through **internal nares**. It has also a pair of oval eustachian openings on lateral sides. Eustachian openings lead to the tympanic cavities through the eustachian tubes. Just behind the tongue, the floor of laryngopharynx contains a median vertical slit-like **glottis**, which leads into larynx. The glottis is guarded by a thin, bilobed cartilaginous flap called **epiglottis**. It develops from its anterior margin. When the animal swallows the food, the soft palate closes the internal nares and the

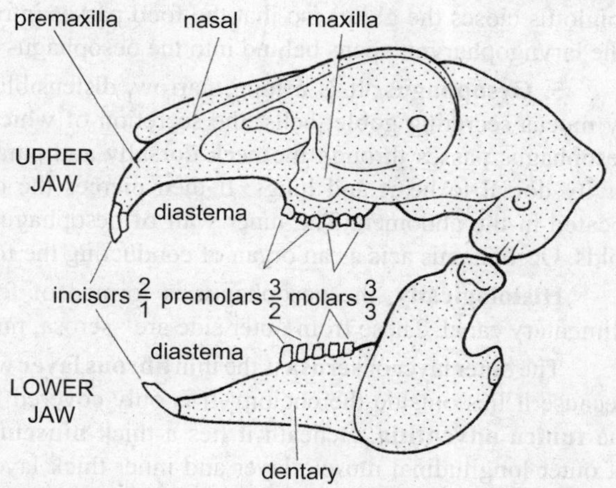

Fig. 29.26. Rabbit. Dentition. Arrangement of teeth on jaws.

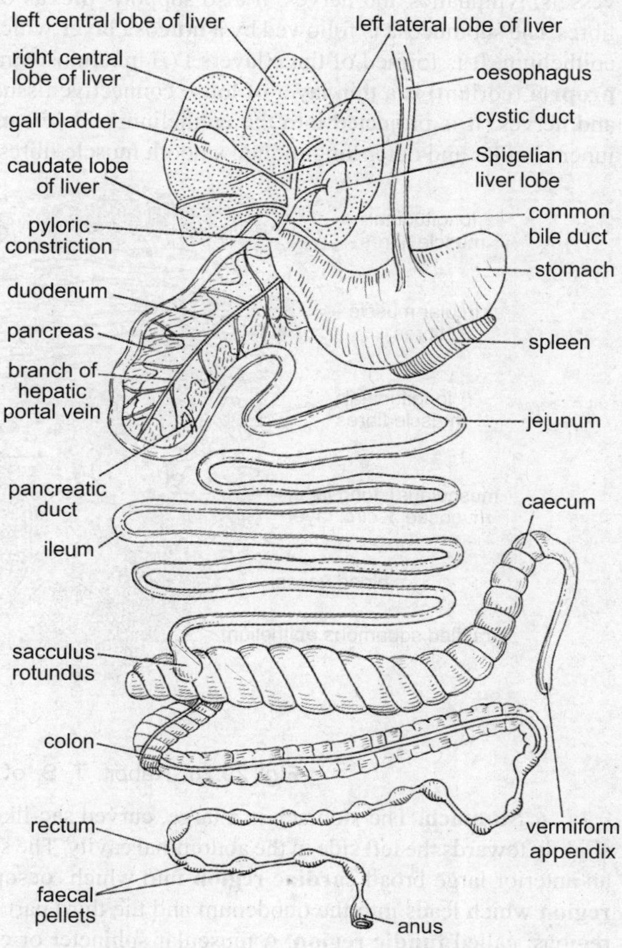

Fig. 29.27. Rabbit. Alimentary canal with associated glands.

epiglottis closes the glottis, so that the food passes into the oesophagus and never into the trachea. The laryngopharynx opens behind into the oesophagus through a wide **gullet**.

5. Oesophagus. It is a long, narrow, distensible muscular tube. The oesophagus is joined by mucus secreting goblet cells, the secretion of which smoothens the oesophageal passage. The oesophagus passes through the neck dorsally and parallel to trachea and passes through thoracic cavity dorsal to heart and lungs. It then pierces the diaphragm to open into a sac-like stomach located in the abdomen. The inner wall of oesophagus is elevated into a number of longitudinal folds. Oesophagus acts as an organ of conducting the food and no digestion occurs here.

Histologically, the oesophagus is formed of four layers which are found throughout the alimentary canal. These from outer side are : serosa, muscular coat, submucosa and mucosa.

The outer layer or **serosa** is the thin **fibrous layer** which is absent in the upper part of oesophagus because it lies outside the coelom. It is only covered by a thin layer of fibrous connective tissue, the **tunica adventitia**. Beneath it lies a thick **muscularis layer**. The muscularis layer is formed of outer longitudinal muscle layer and inner thick layer of **circular muscles**. In the upper half of oesophagus the muscle fibres are of striped type and the lower half contains the unstriped type. The muscularis layer is followed by a **submucosa** layer. It is a connective tissue layer having blood vessels, lymphatics and nerves. It also supports **plexus of Meissner** composed of nerve cells and fibres.The submucosa is followed by a **mucosa layer** which is mainly formed of stratified equamous epithelium. It is formed of three layers : (*i*) innermost stratified squamous **epithelium**, (*ii*) **lamina propria** (corium) is a thin layer of loose connective tissue having blood capillaries, lymph vessels and nerves. It is projected into the epithelium,and (*iii*) **muscularis mucosae** is a narrow layer of inner circular and outer longitudinal smooth muscle fibres. It separates mucosa from submucosa.

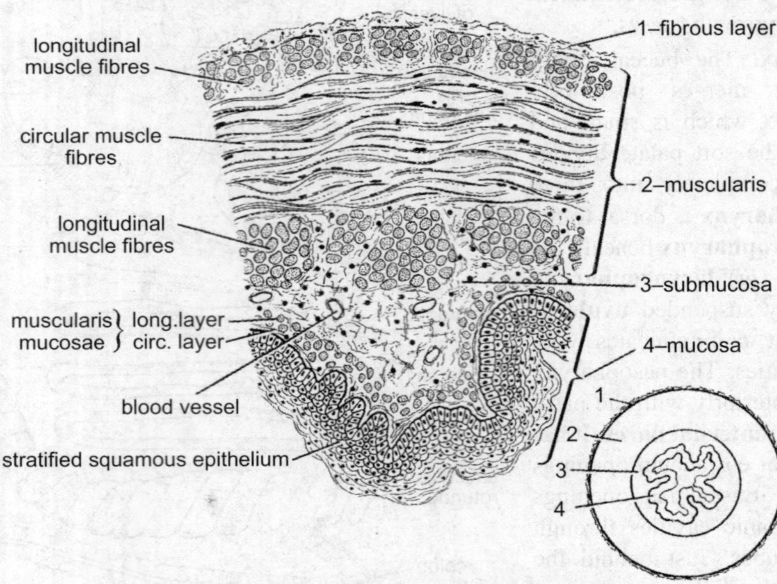

Fig. 29.28. Rabbit. T. S. of oesophagus.

6. Stomach. The stomach is a large, curved sac-like structure situated behind the diaphragm slightly towards the left side in the abdominal cavity. The stomach is differentiated into three regions, an anterior large broad **cardiac region** into which oesophagus opens, a narrow posterior **pyloric region** which leads into the duodenum and the third part is situated in between cardiac and pyloric regions, called **fundic region**. A muscular sphincter or **cardiac valve** is present at the opening of the oesophagus in cardiac region which prevents the backward passage of food from stomach to oesophagus. Similarly a circular **pyloric sphincter** is present at the opening of pyloric stomach into duodenum called **pylorus**. It prevents the backward movement of food from duodenum to pyloric stomach. The end of pyloric stomach is externally marked by a circular groove, the **pyloric constriction**. The cardiac and pyloric openings are near to each other. Near the outer curve of stomach is present a red elongated **spleen**. The function of stomach is to store food, churning of food and partial digestion of food.

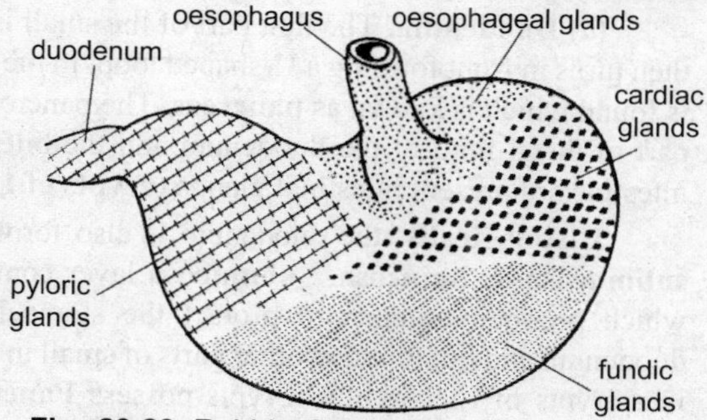

Fig. 29.29. Rabbit. Stomach showing distribution of gastric glands.

Histologically, the stomach has the same construction as the oesophagus and is formed of outer **serosa, muscularis, submucosa** and **mucosa** layers. The distinguishing features of stomach are : The **muscularis mucosae** layer is not well developed. The inner **mucous membrane** lining is well developed and thrown out into large number of **folds** and equal number of alternating deep **gastric pits,** into the bases of which open the ducts of long tubular simple or branched **gastric glands**. These are of three types : (*i*) Gastric glands of the cardiac region, called **cardiac glands**, contain only mucus secreting cells. (*ii*) Gastric glands of the pyloric region, called **pyloric glands**, also produce mucus only. (*iii*) Gastric glands of the fundic region, called **fundic glands**, contain (*i*) **mucous secreting cells**, (*ii*) **zymogen cells** or **propepsin**, (*iii*) **prorenin secreting cells** and (*iv*) **oxyntic cells** or **hydrochloric acid producing cells**. The fundic region of the stomach opens in the intestine. In stomach villi are absent. **Submucosa** is like that of oesophagus. Muscular coat is well developed and has three layers : inner oblique muscles, middle circular muscles and outer longitudinal muscles. Outermost layer of stomach is **serosa** formed by single layer of simple squamous cells.

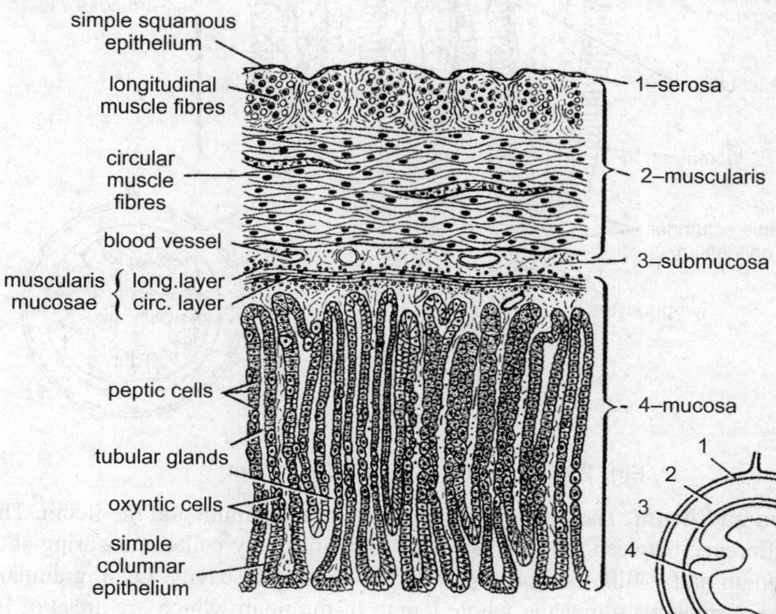

Fig. 29.30. Rabbit. T.S. of stomach.

7. Intestine. The intestine of the rabbit is very much coiled and long, nearly 1.2 metres in length, and can be divided into **small** and **large intestines.**

(*i*) Small intestine. It is formed of three parts : duodenum, jejunum and ileum.

(*a*) **Duodenum.** The first part of the small intestine is the **duodenum** which runs backwards, then turns in front forming a U-shaped loop. In the loop of the duodenum a pinkish diffused structure is found which is known as **pancreas**. The pancreatic duct from the pancreas opens in the proximal part of distal limb of the duodenum and the bile duct opens in the proximal limb. In the wall of intestine are present intestinal glands (**crypts of Lieberkuhn**) and branched **Brunner's glands**.

Histologically, the duodenum is also formed of four layers, outermost **serosa**, **mucularis**, **submucosa** and **mucosa**. **Submucosa** layer contains branched glands, called **Brunner's glands**, which produce **mucous** to protect the stomach lining from acid. These are characteristic of duodenum and not found in other parts of small intestine. Mucus secreted by these glands discharged into crypts of Lieberkuhn. Crypts possess **Paneth** and **argentaffin cells** which secrete digestive enzymes. **Crypts of Lieberkuhn** are pocket-like structures between the villi. Muscularis mucosae of mucosa is reduced. Mucosa layer is formed of columnal cells and contains **goblet cells** which secrete mucus. Muscular layer has no oblique muscle fibres. Circular muscle layer is thicker than the longitudinal muscle layer.

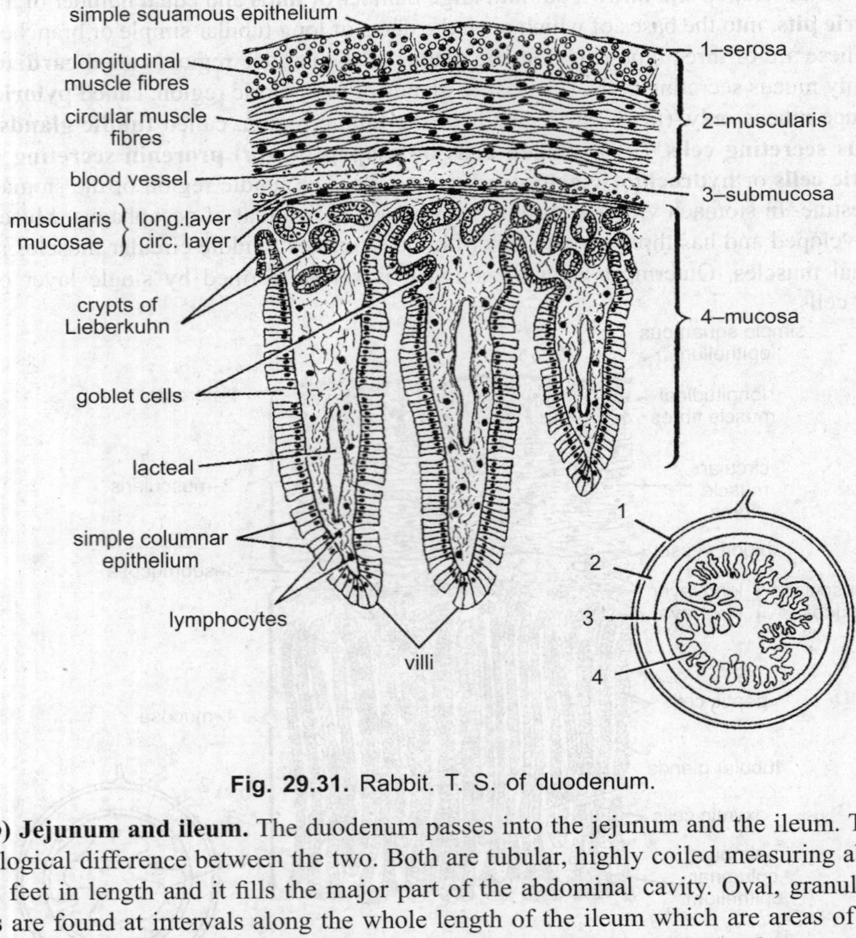

Fig. 29.31. Rabbit. T. S. of duodenum.

(*b*) **Jejunum and ileum.** The duodenum passes into the jejunum and the ileum. There is no morphological difference between the two. Both are tubular, highly coiled measuring about seven to eight feet in length and it fills the major part of the abdominal cavity. Oval, granular **Peyer's patches** are found at intervals along the whole length of the ileum which are areas of **lymphoid tissue**. Jejunum and ileum are held with abdominal wall by fan-shaped folds of **mesenteries**. Their inner lining is elevated into a number of small finger-like process, called the **villi**. These increase the inner absorptive surface. The elevations of submucosa, villi and crypts of Lieberkuhn (intestinal glands) are larger in jejunum than the ileum.

Histologically, the ileum resembles with the duodenum. The **mucosa** of the ileum is thrown out into large number of finger-like folds, the **villi**, which are less in ileum, while more and taller in jejunum. Villi in jejunum are tongue-shaped with swollen ends. Only crypts of Lieberkuhn are present in ileum, and Brunner's glands are absent. Lamina propria of mucosa has yellowish, oval, granular masses, called **Peyer's patches** (lymph nodules). These produce lymphocytes. Crypts of Lieberkuhn are lined by columnar cells and goblet cells.

The ileum before opening into the large intestine dilates to form a round **sacculus rotundus** which opens into the **caecum**. The sacculus rotundus has an **ileo-caecal valve** by which the contents of the ileum can be directed into the caecum before passing into the **colon**.

Caecum. It is found at the junction of ileum and colon. It is a thin-walled tube, about 50 cm long. Its external surface shows spiral constrictions. The caecum leads into a blindly ending finger-like narrow tube, about 15 cm long and thick-walled, known as **vermiform appendix**. Cellulose is digested within the caecum by the action of bacteria.

(*ii*) Large intestine. The large intestine proceeds from the junction of caecum and sacculus rotundus. The opening of the caecum into the large intestine is much wider. The large intestine can be divided into two parts, a **colon** and the **rectum**.

(*a*) Colon. The colon is the first part of large intestine which is about 45 cm long and bears longitudinal muscle bands, the **taeniae**. The wall of the colon is constricted at the sides of the taeniae at regular intervals to form pocket-like pouches, the **haustra**.

(*b*) Rectum. The rectum is about 75 cm long and at intervals it is swollen to form rounded bead-like structures and finally opens outside by the **anus**. The inner wall of the large intestine is devoid of villi but there are small folds containing glands of Lieberkuhn and many mucous secreting goblet cells. Mucosa is formed of stratified epithelium near anus. Muscularis mucosae and muscular layer are thicker than in colon. Muscle fibres are of striated type. Villi and Brunner's glands are absent in large intestine. Crypts of Lieberkuhn are present, formed by the invagination of mucous membrane. Peyer's patches are present in colon and projected into submucosa. Muscularis mucosae and muscular layers are thin in colon.

Anus. It is the external opening of rectum located at the base of tail. Anus is guarded by **anal sphincter**.

B. Digestive Glands

In addition to the glands found in the wall of the alimentary canal, there are certain other glands outside it but in its close association that produce secretions essential for the process of digestion. These glands are the **salivary glands**, **pancreas** and **liver**.

1. Salivary glands. Four pairs of salivary glands are found around the buccal cavity that open in it by their salivary ducts. In addition to the salivary glands, mucous glands are also found on the palate and tongue. The various salivary glands of rabbit on the basis of their position are as follows :

(*i*) Parotid glands. These glands are situated at the base of pinnae and their long fine ducts open behind the upper incisors.

(*ii*) Infra-orbital glands. These glands are situated below the orbits and their ducts open near the upper molars.

(*iii*) Submandibular or submaxillary glands. These glands are situated on the inner side of the angles of lower jaws. Their long ducts open behind the lower incisors.

(*iv*) Sub-lingual glands. These glands are situated below the tongue. Their several short ducts open below the free part or tip of the tongue.

Saliva is watery, alkaline and secreted by these glands, which contain slippery **mucin** and watery enzyme **ptyalin**. The mucin lubricates the food and the food-passage, while ptyalin helps in the digestion of starch in the form of maltose. The salivary glands are stimulated to produce saliva by reflex actions caused due to the presence of food in the buccal cavity.

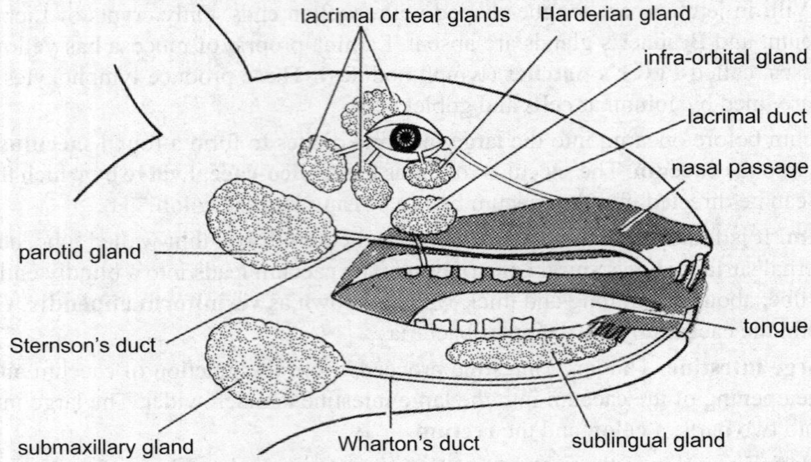

Fig. 29.32. Rabbit. Glands in the head.

2. Pancreas. It is an irregular, pinkish gland situated in 'U'–shaped duodenum supported by mesentery. Its duct opens into the distal limb of duodenum. **Histologically**, it consists of a large number of branching tubules, called **acini,** embedded in connective tissue containing blood and

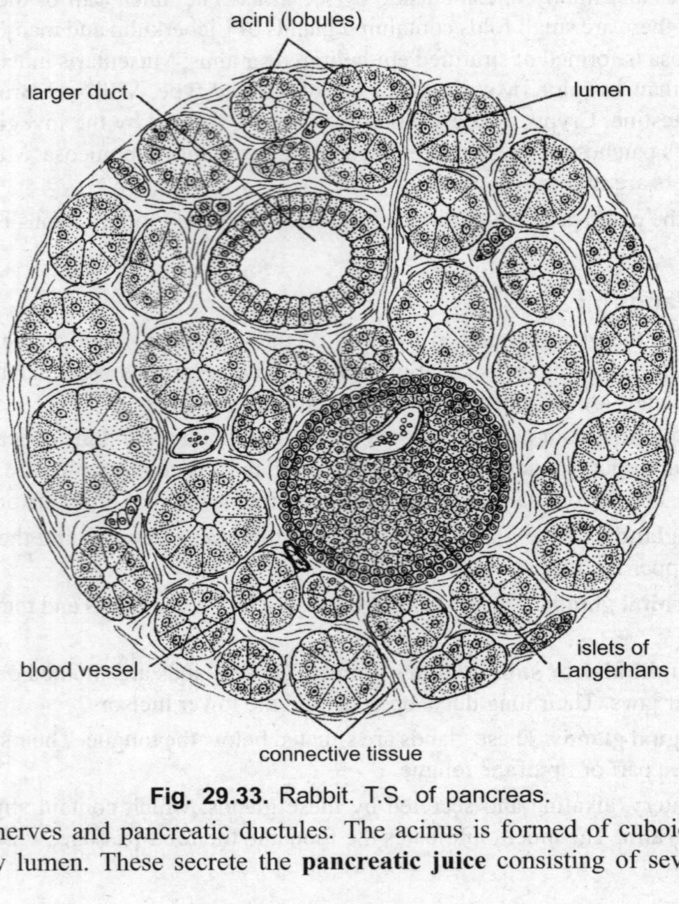

Fig. 29.33. Rabbit. T.S. of pancreas.

lymph vessels, nerves and pancreatic ductules. The acinus is formed of cuboidal epithelial cells around a narrow lumen. These secrete the **pancreatic juice** consisting of several enzymes like

trypsinogen, **amylase** and **lipase**, which act on proteins, starches and lipids respectively. The small ducts arising from the acini unite together to form a common **pancreatic duct**, which opens at the distal end of duodenum.

Several groups of cells, called **islets of Langerhans**, are found situated in between the acini of pancreas constituting an **endocrine gland**. Their secretion is a hormone, called **insulin**, produced by beta cells, which plays an important role in the metabolism of carbohydrates. It regulates the amount of sugar in blood. The deficiency of insulin causes a disease, called **diabetes**. Alpha cells of islets of Langerhans secrete **glucagon** hormone which increases the concentration of glucose in blood. Its deficiency causes **hypoglycemia**.

3. Liver. The liver is the largest gland of the body and remains attached to the posterior concavity of diaphragm by a fold of peritoneum, called **faciparum ligament.** It consists of **five lobes**, three are on the left and two on the right sides. The left lobes are **left lateral**, **left central** and a **Spigelian lobe**, while the right lobes are **right central** *or* **cystic** and **caudate lobe.** The **gall bladder** is a large saccular thin-walled and dark green structure, situated in a groove on the posterior surface of right central lobe. A large **cystic duct** arises from the gall bladder which receives several small **hepatic ducts** (equal to the number of lobes) from the different lobes of the liver carrying bile to the gall bladder, where it is stored and condensed. Thus, a large **common bile duct** is formed which opens in the proximal limb of the duodenum near pylorus guarded by sphincter.

Histological structure. Liver is a spongy, racemose type of structure. It is externally covered by a fibro-elastic serous coat (liver capsule). Each lobe of liver consists of a number of **hepatic lobules** separated from one another by a thin layer of connective tissue called septa or **Glisson's capsule** but the separation of hepatic lobules in rabbit and man is incomplete. But in pig's liver it is complete. Each lobule is polygonal and consists of a mass of **polyhedral hepatic cells** arranged in radial rows, called the **hepatic cords**, around a central or intralobular vein. Radial rows of hepatic cells are 1 or 2 cells thick and anastomosing. Liver cells contain fine granular cytoplasm, large central nuclei and deposits of fat and glycogen. Between the hepatic cords are present narrow irregular spaces, the **lacunae**, through which run blood capillaries, **hepatic sinusoids** having broken walls of endothelial cells. Hepatic sinusoids are formed from interlobular branches of hepatic portal vein and hepatic artery. Sinusoids empty into the intralobular veins which unite to form the hepatic vein. Some specialised **Kupffer cells** are also found amongst liver cells which generally help in the destruction of many germs. At the corners of the adjacent lobules is found the **interlobular blood vessels** (branch of hepatic artery and hepatic portal vein), while the **intralobular blood vessels** are lying in the centre of the lobules. Interlobular

Fig. 29.34. Rabbit. T.S. of liver.

blood vessels and bile duct lie within interlobular connective tissue called the **portal areas** or **canals**. The intercellular spaces are left in between the heptic cells which receive the secretions of these cells, the **bile**. A number of such spaces unite together to form many **interlobular bile ductules** ultimately uniting together to form **hepatic ducts** which open in the **common bile duct**.

Functions of liver. The important functions of liver can be summarised as below :

1. It secretes **bile** which is an alkaline dark green coloured fluid, having several organic and inorganic salts (bile salts), bile pigments, cholesterol, lecithin, water as well as some waste substances. The bile performs following functions :

(*i*) It makes the chyme **alkaline** better suited for the action of pancreatic juice.

(*ii*) It is responsible for the **emulsification** of fats.

(*iii*) It helps in removing the excretory products like bile pigments, inorganic salts, toxins, etc., from the body.

(*iv*) It stimulates **peristalsis**.

(*v*) Some bile salts are necessary for the **absorption** of fat-soluble vitamin K and other vitamins soluble in fats.

(*vi*) Bile acts as an **antiseptic**, therefore, it does not allow the growth and multiplication of bacteria.

2. The excess of sugars are stored in the liver cells as **glycogen** by the process of glycogenesis.

3. It helps in maintaining a constant sugar level in the blood, *i.e.*, at the time of need glycogen is converted into glucose (glycogenolysis).

4. It is a storage organ for the salts of iron and copper, and ribouncleoproteins.

5. Vitamin K is manufactured and stored in liver. It also stores vitamin D.

6. Lymph is formed from the liver.

7. It decomposes proteins and amino acids to ammonia from which urea is formed by a cyclical chain of reactions.

8. It produces **fibrinogen** and **prothrombin** responsible for clotting of blood.

9. It produces **heparin** which checks the coagulation of blood in the blood vessels.

10. It helps in removing various unwanted substances like carbolic acid, cresol, etc., from the blood coming from the alimentary canal.

11. It helps in changing amino-acids into albumin which regulate the salt-water balance in the body.

12. It produces **red blood corpuscles** in the foetus of mammals.

13. The Kupffer cells of the liver destroy the germs of many diseases and worn out RBCs & by phagocytosis.

14. It is the main heat producing centre of the body.

15. It produces certain substances which check anaemia.

Besides these three types of glands, **intestinal glands** are found in the mucosa of small villi secreting intestinal juice, and **gastric glands** are found in the mucosa of stomach which secrete mucus, pepsinogen, renin and hydrochloric acid

Food and Feeding

Food. The rabbit feeds on green vegetables, leaves of plants, roots, grasses, etc., *i.e.*, **herbivorous.**

Gnawing. It cuts food into small pieces by its sharp incisors which are transferred into the buccal cavity by movable lips.

Mastication. The food is masticated in the buccal cavity by premolar and molar teeth where saliva mixes with the food.

Swallowing. Since saliva contains mucilaginous mucin which helps in lubricating and binding the masticated food into a **food-bolus**. The food-bolus is then swallowed into the oesophagus from where it passes into the stomach by peristaltic contractions. During swallowing the glottis is closed by the epiglottis and the nasal passage also becomes closed by the upward movement of soft palate.

Physiology of Digestion

Digestion is the process of hydrolysis of complex foodstuffs into simpler soluble forms in which they can be absorbed. The hydrolysis occurs with the help of organic catalysts, called **enzymes**. The food generally contains **proteins**, **fats**, **carbohydrates** and **inorganic salts**, **vitamins** and **water.** From these foodstuffs, proteins, fats and carbohydrates require digestion and the rest substances are more or less directly absorbed in the gastro-intestinal tract without undergoing the process of digestion. Digestion of food starts from the buccal cavity itself, continues in the stomach and completed in the intestine.

1. Digestion of food by ptyalin in the buccal cavity. As soon as food comes in the buccal cavity mastication occurs during which saliva from different salivary glands mixes with the food. Thus, during mastication the food is turned into a pulpy mass or **bolus** with the help of tongue and cheek muscles. The saliva contains largely water, **mucin** and an enzyme, **ptyalin**. Ptyalin acts on starch or polysaccharides turning them into compound sugars like maltose and dextrin. Mucin of the saliva lubricates the food and the food passage.

$$\text{Starch + Ptyalin} \longrightarrow \text{Maltose and Dextrin.}$$

The food in the form of bolus is swallowed with the help of tongue which pushes bolus behind and the pharyngeal muscles force it into the oesphagus which is known as **deglutition**. The epiglottis prevents it from entering into the larynx. Once the bolus comes in the oesophagus it is pushed into the stomach due to the wave of contraction (**peristalsis**) in oesophagus which is facilitated by the mucus secreted from the mucous glands of the oesophageal wall.

2. Digestion of food by pepsin in stomach. In the stomach food remains for a longer time. It is churned, broken into smaller fragments by the contraction of stomach muscles. It is also mixed with the gastric juice. A hormone, **gastrin,** secreted in the stomach stimulates the gastric glands to produce **gastric juice** which mixes with the food. The gastric juice contains nearly 90% water, 0.5% HCl, proenzymes like **prorenin** and **propepsin** and **gastric lipase**.

The hydrochloric acid (*i*) checks bacterial decomposition of the food, (*ii*) stops the action of ptyalin on starch, (*iii*) brings the food in acidic medium, (*iv*) acts as an activator for the proenzymes–prorenin and propepsin, converting them into **renin** and **pepsin** respectively, and (*v*) stops contractions of the stomach. The renin is only found in the young rabbits.

The **renin** acts on milk protein **casein** in presence of Ca ions and converts it into **curd** which is **calcium paracasein**. The **pepsin** acts on proteins including calcium paracasein to change them into **peptones** and **proteoses**. The **gastric lipase** acts on fats to convert them into **fatty acids** and **glycerol**. The amount of gastric lipase is very less, therefore, it is a negligible reaction.

$$\text{Renin + Milk protein (casein)} \xrightarrow{\text{Ca++}} \text{Curd (cal.paracasein)}$$

$$\text{Pepsin + Proteins including curd} \longrightarrow \text{Peptones and Proteoses}$$

$$\text{Lipase + Fats} \longrightarrow \text{Fatty acids and glycerol.}$$

The presence of **pyloric sphincter** helps in retaining the food in the stomach for a longer time during which the food is churned and mixed thoroughly with the gastric juice until it takes the form of a semidigested thick pasty solution, called **chyme**. The pyloric sphincter only opens when the chyme becomes acidic. The food in the form of chyme passes into the duodenum by the peristaltic contractions of the stomach wall.

3. Digestion of food in intestine. Duodenum secretes a hormone, **cholecystokinin**, which stimulates gall bladder to pour **bile-juice** in the duodenum. Another hormone, **secretin**, is also produced in duodenum which stimulates pancreas to produce **pancreatic-juice** which is also poured in the duodenum. At the same time from the glands of crypts of Lieberkuhn of duodenum and small intestine a juice called **succus-entericus** is produced. The **chyme**, thus, meets with the bile juice, pancreatic juice and juice from intestine, the succus-entericus, as soon as it comes in the duodenum.

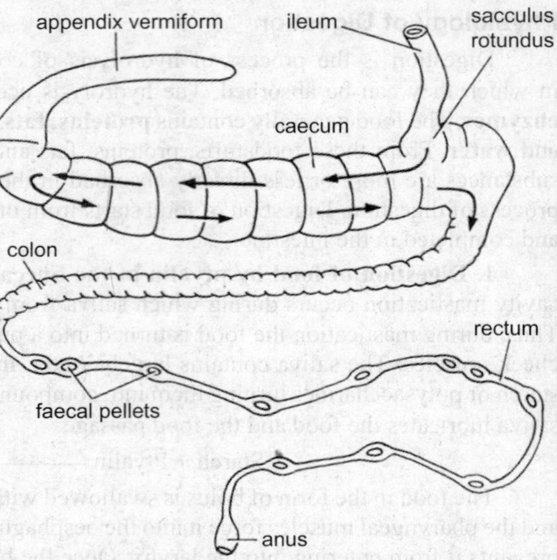

Fig. 29.35. Rabbit. Large intestine showing pasage of food in caecum.

***(i)* Bile-juice.** It is an alkaline, dark green coloured fluid containing **water**, **bile salts**, and **bile pigments**. It has no digestive enzymes, therefore, it does not take part directly in digestion but plays an important role in the process of digestion. (*i*) It contains important bile salts such as **glycocholate** and **taurocholate** of sodium and **cholesterol**, (*ii*) Sodium glycocholate, taurocholate and cholesterol break down fats into smaller globules to form **emulsion** which can be easily acted upon by the enzymes of pancreas. Sodium bicarbonate neutralises the acid of chyme and makes it alkaline. This is essential because pancreatic juice and intestinal juice operate only in alkaline medium. (*iii*) Bile juice also prevents **putrefaction** of the food. (*iv*) The bile pigments present in the bile juice are **biliverdin** and **bilirubin** which are formed by the breakdown of haemoglobin of worn-out red blood corpuscles in the liver and are excretory waste products. The green colour of the bile juice is only due to the presence of these pigments and they are passed out with faeces as such. (*v*) Bile salts also help in the absorption of fats and fat soluble vitamins – A, D and K.

***(ii)* Pancreatic juice.** It is also an alkaline fluid containing largely **water**, proenzymes **trypsinogen** and **chymotrypsinogen**, and enzymes **amylase, lipase,** etc. (*a*) **Trypsinogen** is activated by the **enterokinase** of the succus entericus and converted into **trypsin**. **Trypsin** activates chymotrypsinogen to change into **chymotrypsin**. Trypsin and chymotrypsin are proteolytic enzymes acting upon peptones, proteoses and fresh remaining protein if any, to convert them into amino acids. (*b*) **Amylase** acts upon starch and glycogen to change them into the molecules of **maltose**, a disaccharide sugar. (*c*) **Lipase** acts on fatty emulsion formed by the bile and converts it into **fatty acids** and **glycerol.**

$$\text{Trypsin + Peptones, Proteoses} \longrightarrow \text{Amino acids}$$

$$\text{Amylase + Starch} \longrightarrow \text{Maltose}$$

$$\text{Lipase + Fatty emulsion} \longrightarrow \text{Fatty acids and glycerol.}$$

***(iii)* Succus entericus.** It is also alkaline watery fluid containing an activator, **enterokinase**, and several enzymes like **peptidases** (proteolytic enzymes, *e.g.*, **erepsin**), **lipase**, **invertase**, **maltase** and **lactase**. (*a*) **Enterokinase** activates pancreatic trypsinogen to trypsin.(*b*) The **peptidases** (erepsin) act on polypeptides to change them into **amino-acids**. (*c*) **Lipase** acts on remaining fat emulsion converting it into **fatty acids** and **glycerol**. (*d*) **Maltase** hydrolyses maltose to **glucose**. (*e*) **Invertase** and **lactase** act on sucrose and lactose respectively to change them into glucose.

$$\text{Peptidases + Polypeptides} \longrightarrow \text{Amino acids}$$

$$\text{Lipase + Fat emulsion} \longrightarrow \text{Fatty acids and Glycerol}$$

$$\text{Maltase + Maltose} \longrightarrow \text{Glucose}$$

$$\text{Invertase + Sucrose} \longrightarrow \text{Glucose and Fructose}$$

$$\text{Lactase + Lactose} \longrightarrow \text{Glucose and Galactose.}$$

Thus, the digestion of various foodstuffs is completed and as the food passes through the intestine it receives more water and forms a fluid emulsion, called **chyle**, which is slowly pushed onwards by peristalsis. The enzymatic activity remains continous throughout the process and finally the **proteins**, **carbohydrates** and **fats** are hydrolysed into **amino-acids**, **glucose** or similar **monosaccharide sugars** and **fatty acids** and **glycerol** respectively.

4. Digestion of cellulose in caecum. The **cellulose** in the diet of rabbit (herbivores) remains undigested, for which it passes in the caecum. Thus, the digestion of cellulose occurs in the caecum which is well developed in rabbit and herbivorous mammals. Cellulose decomposing symbiotic bacteria and protozoans are found in the caecum, where it is decomposed to soluble sugars and bacteria in return get nitrogenous food. The process of digestion and absorption of cellulose is very slow, therefore, food remains for a long time in the caecum. However, only partial decomposition of cellulose has been reported so far in the case of rabbit. The digested foodstuffs are absorbed in the intestine and the remaining fluid content of the alimentary canal passes into the colon and rectum, where water is absorbed and semisolid **faeces** remains. Some **putrefying bacteria** of the colon help in the formation of faeces which passes out through the **anus**.

5. Coprophagy. It has been observed that rabbit feeds on night excreta (**coprophagy**) which is moist and soft for the incomplete digestion of cellulose. Rabbit only feeds on the faeces produced during night. Thus, the decomposed and simplified cellulose in the excreta is again subjected for digestion during its passage through the gut. The digested cellulose does not enter the caecum again but passes directly in the colon where water is absorbed and dry faecal pellets are formed. It has also been reported that if the rabbits are not allowed to feed on night excreta, they die soon. Faeces produced during the daytime is not eaten because it is dry and hard. The various minerals, water and vitamins, etc., are directly absorbed more or less without undergoing the process of digestion.

Absorption

The end products of digestion, *e.g.*, amino acids, glucose, other monosaccharides, fatty acids and glycerol are absorbed in the intestinal wall (mainly in ileum) through the intestinal villi. The

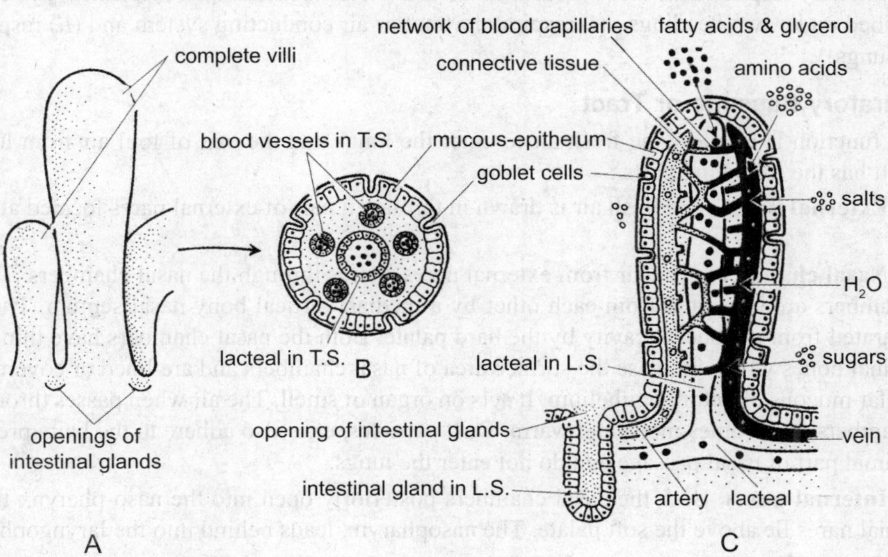

Fig. 29.36. Rabbit. Absorption of food through villi. A – Entire villi; B – A villus in T. S.; C – A villus in L. S.

amino acids, glucose and other monosaccharides are easily absorbed in the blood capillaries of intestinal villi by simple process of diffusion. They are finally carried to the liver through the portal circulation. The **fatty acids** and **glycerol** pass in the **intestinal lacteals** which are lymph capillaries in the **intestinal villi**. These are taken in the blood circulation through lymphatic system. Recent electron microscopic studies have revealed that it is not necessary for fats to change into fatty acids and glycerol for being absorbed, because the fat globules coming in contact with the **microvilli** of the intestinal cells are passed directly through their cell membrane into the lymphatic capillaries of the villi.

Egestion or Defaecation

After absorption of digested food in the ileum, the remaining undigested solid residue like vegetable fibres and cellulose, etc., enters the large intestine. Water is absorbed in the colon leaving the solid humid faeces. In the **rectum**, faeces is converted into small **pellets** which are periodically discarded through the anus.

Assimilation

The absorbed amino-acids and monosaccharides first come into the liver through portal circulation and then go to the heart for circulation in the different parts of the body where tissues pick up their requirements. The **amino acids** are building material and constantly needed for the formation of protoplasm. The excess of amino acids are subjected to the process of **deamination** in liver forming **ammonia** which is soon converted into **urea**, which is excreted by the kidneys. The excess of monosaccharides are stored in the form of **glycogen** usually in the liver cells. Whenever glucose level falls down in the blood then glycogen is soon converted into glucose (**glycogenolysis**) to maintain its normal level in the blood. Liver is also concerned with fat metabolism and the fatty globules are found filled in the liver cells. Energy is also released from the fats. The fats are also converted into amino acids and carbohydrates by the liver.

RESPIRATORY SYSTEM

Pulmonary respiration is the rule since rabbit is a terrestrial animal. Its respiratory system may be deseribed under two headings : (*i*) respiratory tract or air conducting system and (*ii*) respiratory organs (lungs).

I. Respiratory Passage or Tract

Its function is to allow the fresh air to enter the lungs and the exit of foul air from lungs to outside. It has the following parts :

1. External nares. The fresh air is drawn in through a pair of external nares located at the tip of snout.

2. Nasal chambers. The air from external nares passes through the nasal chambers. The two nasal chambers are separated from each other by a median vertical bony nasal septum. These are also separated from the buccal cavity by the hard palate. Both the nasal chambers have thin scroll-like turbinal bones which increase the surface area of nasal chambers and are lined or covered over by vascular mucous olfactory epithelium. It acts on organ of smell. The air when passes through the nasal chambers, it becomes moist and warm. Dust particles, etc., also adhere to the hairs present in the proximal part of nasal passage and do not enter the lungs.

3. Internal nares. Both the nasal chambers posteriorly open into the naso-pharynx through the internal nares lie above the soft palate. The nasopharynx leads behind into the laryngopharynx.

4. Glottis. The floor of laryngopharynx bears a median, vertical, slit-like aperture, the **glottis**, which leads into larynx. From its anterior margin arises a bilobed cartilaginous flap covered by mucous membrane, called the **epiglottis**. During swallowing of food it covers the glottis.

5. Larynx. The **larynx** or **voice-box** is a small dilated chamber which is the sound producing organ of rabbit. It is a modified anterior part of the trachea. The wall of larynx is supported by four cartilages : thyroid, cricoid and a pair of arytenoids. (*i*) The largest shield-shaped or girdle-like and most anterior is the **thyroid cartilage** which supports the larynx ventrally and laterally. It is incomplete dorsally. (*ii*) Just posterior to it is a ring-like **cricoid cartilage** which is broad on the dorsal side but narrow ventrally. (*iii*) A pair of small **arytenoid cartilages** are situated at the dorsal side of the larynx, anterior to cricoid. There is also a pair of small nodule-like **cartilages** of **Santorini**, situated at the anterior tips of arytenoid cartilages. A pair of membranous folds, the **vocal chords**, are found extending between the thyroid above and aryteniod cartilages below. The vocal cords are the folds of the lateral wall of the larynx and close the glottis except for a slit-like aperture between their free edges. These are formed of elastic connective tissue covered by mucous membrane. In resting condition the vocal cords lie at an acute angle to one another and the glottis remains widely open to allow free passage of air.

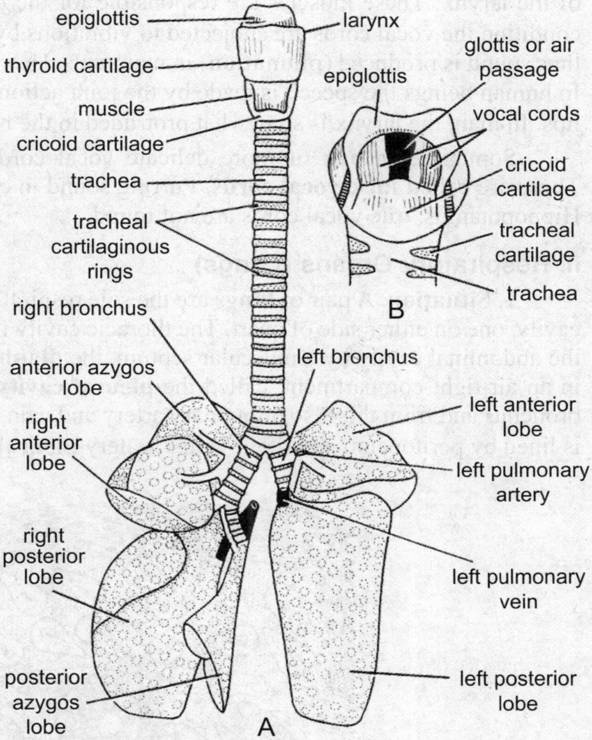

Fig. 29.37. Rabbit. A – Respiratory system in ventral view; B – Vocal cords in dorsal view of

Sound Production

At the time of sound production the vocal cords lie close to each other. The vocal cords are regulated by the action of a set of muscles (intrinsic laryngeal muscles) attached to the cartilages

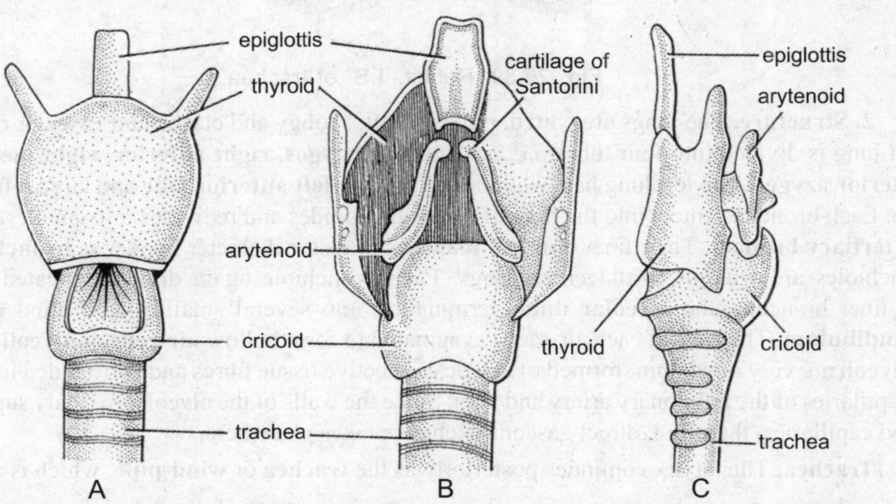

Fig. 29.38. Rabbit. Trachea and larynx. A–Dorsal view; B–Internal structure; C–Lateral view.

of the larynx. These muscles are responsible for the pitch and quality of sound produced. In this condition the vocal cords are subjected to vibrations by the current of air passing through them and thus sound is produced (**phonation**) in mammals. However, loud sounds are not produced in rabbit. In human beings the speech is made by the joint action of vocal cords, tongue, pharynx, mouth and lips. In man, the larynx is somewhat protruded in the neck and called **Adam's apple**.

Sometimes a pair of more delicate vocal cords are present anterior to true vocal cords. These are called **false vocal cords**. Purring sound in cat is produced by these false vocal cords, In Hippopotamus, true vocal cords are not found.

II. Respiratory Organs (Lungs)

1. Situation. A pair of **lungs** are the sole respiratory organs, which are situated in the thoracic cavity, one on either side of heart. The thoracic cavity in the mammals is completely separated from the abdominal cavity by a muscular septum, the **diaphragm**. Each lung is enclosed and suspended in an air-tight compartment, called the **pleural cavity**. Its suspension is from the site of entry of bronchus and from this also pulmonary artery and vein, and nerves enter the lung. The pleural cavity is lined by peritoneum and filled with a watery **pleural fluid**.

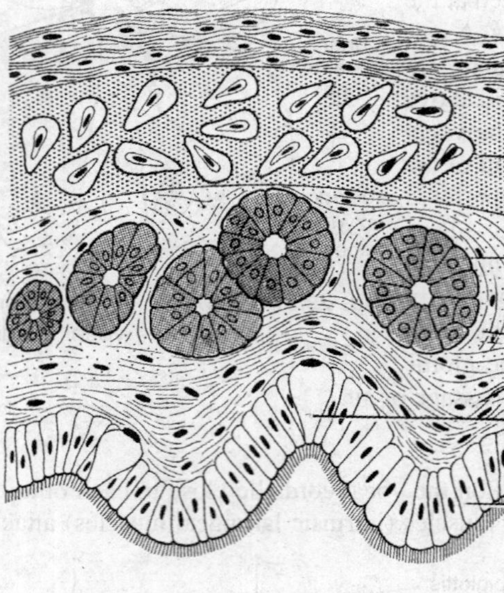

Fig. 29.39. Rabbit. T.S. of trachea.

2. Structure. The lungs are paired, pinkish, soft, spongy and elastic and vascular organs. The right lung is divided into four lobes, *i.e.,* an **anterior azygos, right anterior**, **right posterior** and **posterior azygos**. The left lung has two lobes : a smaller **left anterior lobe** and large **left posterior lobe**. Each bronchus enters into the lung of their side, divides and redivides into smaller **secondary** and **tertiary bronchi**. Their finer terminal branches of lesser diameter known as **bronchioles**. The bronchioles are without cartilagenous rings. Each bronchiole again divides repeatedly to form still finer branches, the **alveolar ducts** terminating into several small dilated blind **air-sacs** or **infundibulum**. The wall of each air-sac is evaginated to form hollow **air-cells** or **alveoli**. The walls of alveoli are very much thin, formed of elastic connective tissue fibres and surrounded in a network of capillaries of the pulmonary artery and vein. Since the walls of the alveoli are richly supplied with blood capillaries, therefore, direct gaseous exchange takes place here.

Trachea. The larynx continues posteriorly as the **trachea** or **wind-pipe**, which is a long tube running through the neck just ventral to the oesohagus. After entering the thoracic cavity, dorsal to the heart, its lower end divides into two **primary bronchi** – right and left. Each brochus enters the lung of its side. The wall of the trachea and primary bronchi are supported by numerous C-shaped cartilaginous **tracheal rings**, which are incomplete dorsally but united by smooth muscles. The trachea is a flexible tube and its cartilaginous rings prevent the trachea from collapse and allow a free passage of air from both sides, *i.e.*, outside the body into the lungs and vice-versa. Internally the trachea and bronchl are lined by **ciliated** and **mucous epithelium**. Mucus keeps the inner surface moist and slimy. It also holds the bacteria and dust particles coming in with air. These are finally swept back to the pharynx by the upward beating of the cilia.

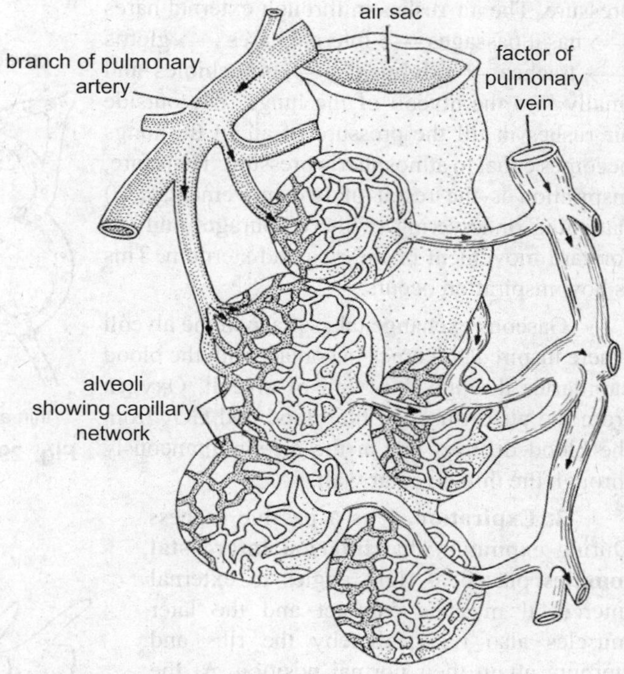

Fig. 29.40. Rabbit. Diagram showing the arrangement of blood vessel in alveoli.

III. Mechanism of Breathing or Respiration

The mechanism of breathing consists of intake of fresh air into the lungs (**inspiration**) and elimination of respired air from the lungs (**expiration**). The thoracic cavity with air-tight pleural cavities having lungs, is a basket or box-like structure formed dorsally by the **vertebral column**, ventrally by the **sternum** and by the **ribs** on the sides. The ribs are movably articulated with the vetebral column dorsally and sternum ventrally. The basket is closed behind by a dome-shaped muscular sheet, the **diaphragm**, which is attached with the vertebral column on one hand and sternum on the other. Two sets of muscles, the **internal intercostal** and **external intercostal**, are situated between the two successive ribs. The mechanism of breathing can be divided into two stages :

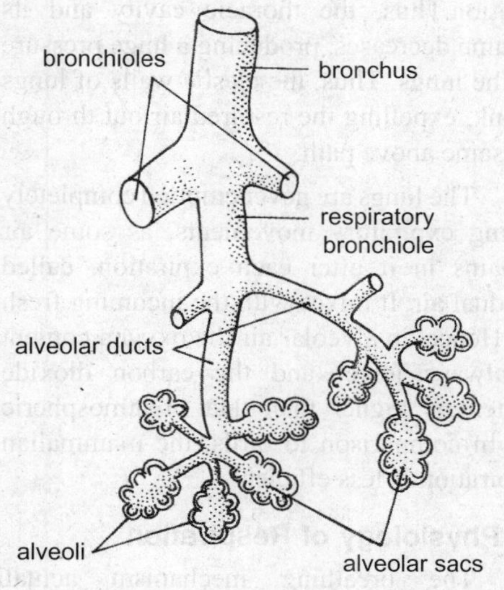

Fig. 29.41. Rabbit. Terminal branching of bronchiole in a lung.

1. Inspiration. The respiration is an active process during which fresh air is drawn in the lungs. It is done by the contraction of the obliquely arranged **external intercostal muscles** dragging the ribs and the sternum forwards and downwards. At the same time the radially arranged muscles of the **diaphragm** contract, thereby it becomes flattened. This results in the increase of the internal capacity of the thoracic cavity, thus, decreasing the pressure within the thoracic cavity. It causes the lungs to expand, which causes the air to rush into the lungs due to more outside atmospheric pressure. The air rushes in through external nares ⟶ nasal passages ⟶ internal nares ⟶ glottis ⟶ trachea ⟶ bronchi ⟶ bronchioles and finally into the alveoli of the lungs. The outside air rushes in till the pressure of air in the lungs becomes equal to atmospheric pressure. Therefore, inspiration is the result of two movements : (*i*) flattening or lowering of the diaphragm and (*ii*) forward movement of the ribs and sternum. This is how inspiration occurs.

Gaseous exchange takes place in the alveoli where inspired air comes in contact with the blood capillaries through thin wall of alveoli. Oxygen from the air diffuses in the blood and CO_2 from the blood diffuses out in the air simultaneously through the thin alveolar wall.

2. Expiration. It is a passive process. During expiration, the **internal intercostal muscles** placed at right angles to external intercostal muscles contract and the later muscles also relax, thereby the ribs and sternum attain their normal position. At the same time the muscles of the diaphragm relax, bringing it in its normal dome-shaped position.Thus, the thoracic cavity and its volume decreases, producing a huge pressure on the lungs. Thus, the elastic walls of lungs shrink, expelling the respired air out through the same above path.

The lungs are never emptied completely during expiratory movements, as some air remains in it after each expiration, called residual air. It mixes with the incoming fresh air. Hence, in alveolar air the oxygen content is always lower and the carbon dioxide content is higher than that of atmospheric air. In comparison to birds, the mammalian respiration is less efficient.

IV. Physiology of Respiration

The breathing mechanism actually involves the intake of oxygen and passing out of CO_2, which constitute the **external respiration**. The actual use of oxygen in the body tissues for the oxidation of foodstuffs liberating energy and other metabolic wastes constitute the **tissue respiration** or **internal respiration**. Actually, the oxygen which diffuses in the blood combines with the hamoglobin of RBCs to form **oxyhaemoglobin**, such blood is known

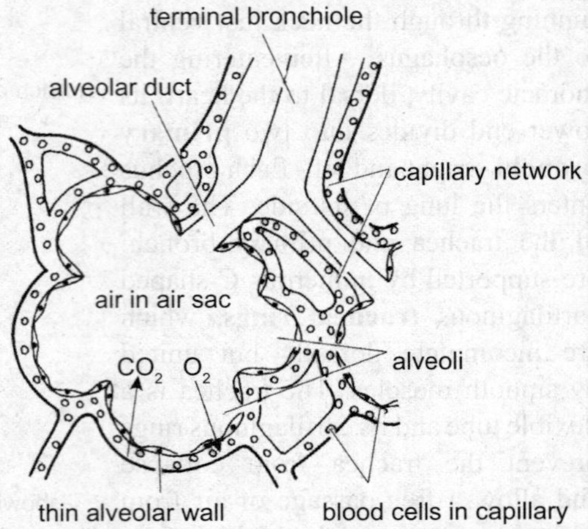

Fig. 29.42. Rabbit. Section through alveoli to show gaseous exchange between lung and blood capillaries.

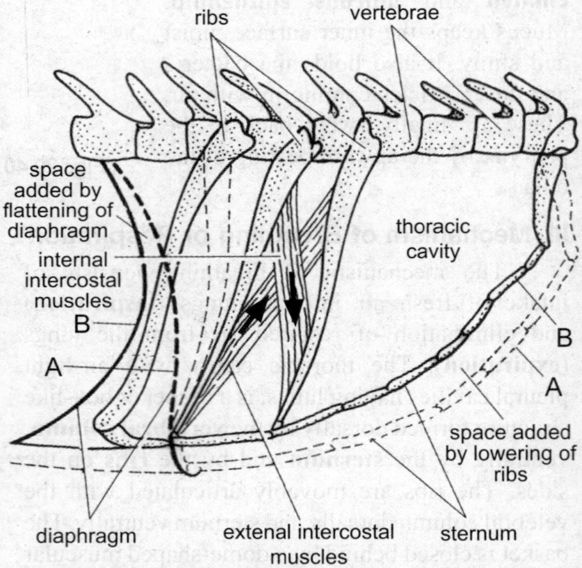

Fig. 29.43. Rabbit. Mechanism of breathing. A–Normal position; B–During inspiration.

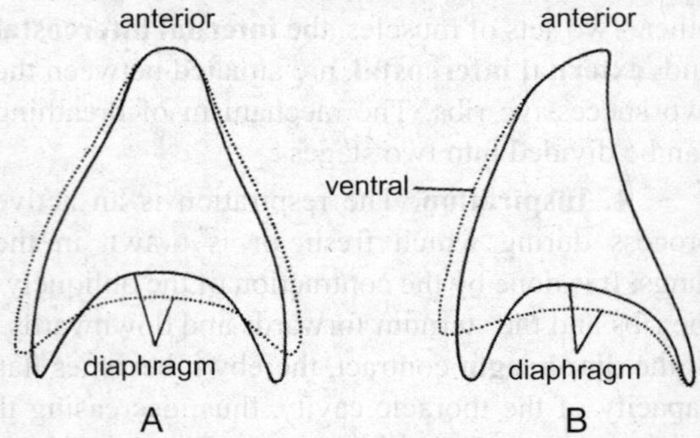

Fig. 29.44. Rabbit. Diaphragm during breathing. A–From below; B–From side.

as **oxygenated**.Thus, the oxygenated blood goes to the heart through the pulmonary veins from where it is circulated in the different parts of the body, finally reaching to the tissues of the organs. Here, the oxyhaeomoglobin gives up its oxygen and changes into haemoglobin.This oxygen diffuses into the tissues which is used in the oxidation of foodstuffs, finally resulting into the formation of energy, CO_2 and water. CO_2 thus produced diffuses in the blood which is carried partly in the RBCs and partly in the plasma.

BLOOD VASCULAR SYSTEM

The blood vascular system of rabbit (mammal) consists of a circulatory media, called the **blood**, channels through which the blood flows, called **blood vessels**, and a central pumping organ, the **heart**, which pumps the blood in the blood vessels. However, like that of the other vertebrates, the blood vascular system of rabbit is of **closed** type. This system is generally concerned with the distribution of materials (digested food, water, oxygen, hormones, excretory products, etc.) from one part of the body to the other.

Heart

The heart lies in between the two lungs within the median space, the **mediastinum** slightly to the left side of the thoracic cavity.

1. External features. The heart is a pear-shaped, muscular, four-chambered pumping organ. The pointed **apex** of the heart is directed posteriorly and broad **base** towards the anterior side.

(*i*) Pericardium. Heart is found completely enclosed within a thin-walled, double layered membranous sac, the **pericardium**. It is connected to the ventral thoracic wall and posterior diaphragm. In between the two layers of pericardium, outer parietal and inner visceral layer is a narrow space, known as **pericardial cavity** filled with a water **pericardial fluid**. The pericardial fluid protects the heart from external injuries and allows its free movement.

External divisions. A transverse groove distinctly separates the heart into an anterior smaller **auricular part** and posterior larger **ventricular part**.This groove is called **coronary sulcus** or auriculo-ventricular groove.

(*ii*) Auricles. The auricular part consists of **right** and **left auricles**.The left auricle is smaller than the right. The sinus venous is absent and merged in the right auricle. The posterior part of each auricle is swollen flap-like, called **auricular appendix**, slightly covering the corresponding ventricles.

(*iii*) Ventricles. The ventricular part consists of **right** and **left ventricles**. The ventricle is divided into right and left parts by an oblique **interventricular groove**, extending from the top of the heart backwards towards the right but not reaching the apex. Therefore, left ventricle is bigger than the right ventricle and includes the apex of the heart. The **conus** or **truncus arteriosus** is absent and merged with the ventricles.

2. Internal structure. The internal structure of the heart can be easily seen by dissecting it in a longitudinal fashion. Internally the heart is divided into four chambers–anterior two auricles and posterior two ventricles.

(*i*) Auricles. The auricles are thin-walled, separated from each other by a thin, muscular and vertical **inter-auricular septum.** A small oval area, called **fossa ovalis**, is found on the septum, which has an opening, the **foramen ovale** in the embryo through which the two auricles are communicated with each other. The blood goes from right auricle to left auricle without entering in the lungs, which are not yet functional. The foramen ovale is closed in the adults when lungs become fully functional. The inner lining of the auricular wall forms a network of low muscular ridges, which are called **musculi pectinati.**

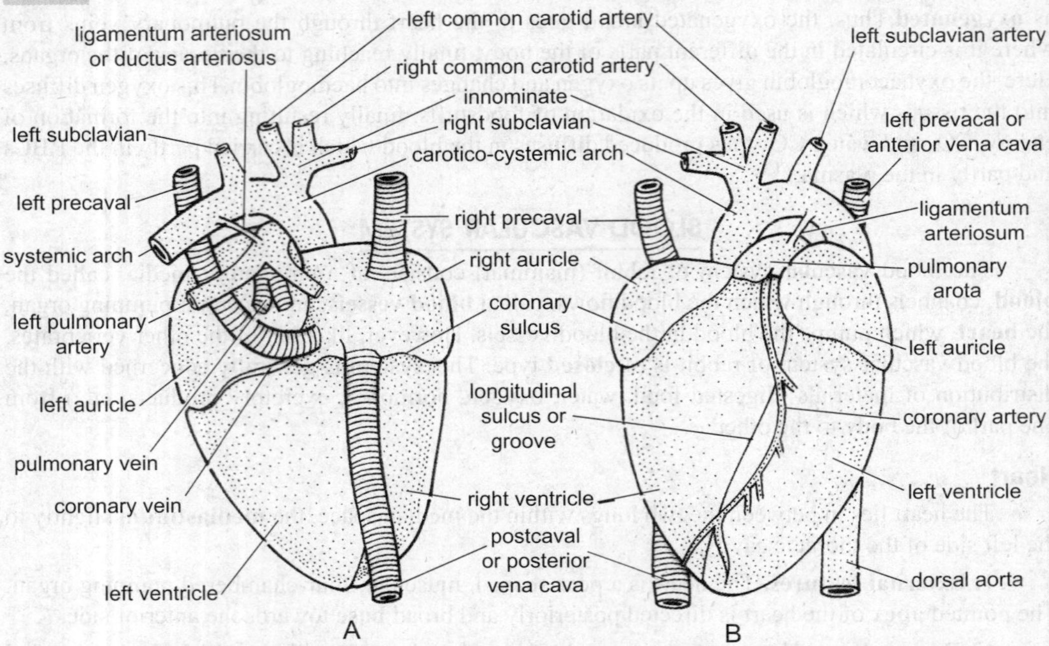

Fig. 29.45. Rabbit. External features of heart. A – Dorsal view; B – Ventral view.

The right auricle receives **venous blood** from all parts of the body except lungs by two superior vena cavae (**precavals**) and an inferior vena cava (**postcaval**). All open separately into the right auricle. The opening of postcaval is guarded by a rudimentary eustachian valve. The left auricle receives oxygenated (aerated) blood through the **pulmonary veins** through a common pulmonary opening. Both the auricles open behind into the ventricle of their side through the **auriculo-ventricular aperture**. The right auriculo-ventricular aperture is guarded by **right auriculo-ventricular valve**, called **tricuspid valve** made of three triangular flaps or cusps. The left auriculo-ventricular aperture is guarded by a **bicuspid** or **mitral valve** made of two flaps.

(*ii*) Ventricles. Like the auricles, the ventricle is also divided completely by an oblique vertical **inter-ventricular septum** into two chambers, the right and left ventricles. The left ventricle is thick-walled and larger than the right and is circular in shape in transverse section. It also forms the apex of the heart. The smaller right ventricle is crescentic in transverse section. The inner wall of the ventricles is raised into irregular small ridges, called **columnae carneae**. Besides these, there are large conical ridges, known as **papillary muscles**. The free edges of tricuspid and bicuspid valves hanging freely in the ventricle are attached with the **papillary muscles** or **columnae carneae** of the ventricle through long, tough connective tissue strands, the **chordae tendineae**. The tricuspid and bicuspid valves between right and left auriculo-ventricular apertures respectively allow the passage of blood only in one direction, *i.e.,* from the auricles into the ventricles and not from the ventricles to the auricles.

A **pulmonary aorta** arises from the right ventricle which carries venous blood to the lungs for oxygenation. The opening of pulmonary aorta into the right ventricle is guarded by three **semilunar valves** which allow the passage of blood from the ventricle into the aorta and prevent the reflux of blood into the ventricle. Similarly, a **carotico-systemic aorta** arises from the left ventricle which carries oxygenated blood to the different parts of the body. Its opening into the left ventricle is also guarded by three semilunar valves allowing the passage of blood only in one direction, *i.e.*, from left ventricle to aorta and prevents the backward flow of blood in the ventricle. At the place where carotico-systemic aorta crosses the pulmonary aorta, a muscular strand **ligamentum arteriosum**, is present in adults but during embryonic condition both the **aortae** are connected with each other by an artery called **ductus arteriosus** or **Botalli**.

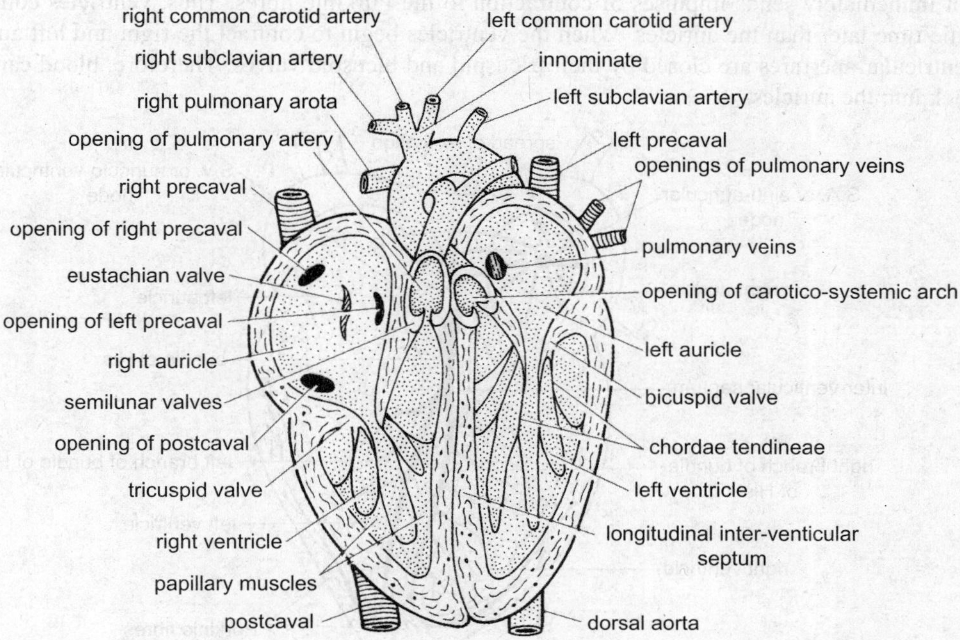

Fig. 29.46. Rabbit. Internal structure of heart in ventral view.

Working of the heart. The heart performs the function of a **force pump** beating continuously throughout life. It pumps blood into the arteries from the ventricles and also acts as a **suction pump** by drawing blood into the auricles. During heartbeats, once the auricles contract and the ventricles relax or dilate and then auricles relax and the ventricles contract. Contraction and relaxation goes on simultaneously. Contraction is called **systole** and the relaxation is called **diastole**. The heart muscles are of cardiac type which have the inherent tendency for continuous beating.

The right auricle receives venous or impure blood from different parts of the body through precaval and postcaval veins. The left auricle receives oxygenated or pure blood from the lungs through pulmonary veins. The venous and oxygenated blood are forced into their respective chambers of ventricles by the contraction of the auricles. The wave of contraction in the heart originates from **sinu-auricular node (SA-node)** or **pacemaker**, which is a group of specialised cardiac muscle cells with nerve fibres. It is situated in the wall of the right auricle between the openings of right precaval and postcaval. It represents the remnants of sinus venosus of lower vertebrates.The wave of contraction spreads like concentric rings into the two auricles and, thus, both the auricles contract at the same time. At the same time the tricuspid and bicuspid valves of the right and left auriculo-ventricular apertures respectviely are forced to open towards the ventricles and, thus, forcing the venous blood from right auricle into the right ventricle and oxygenated blood from left auricle into the left ventricle. The auricles then relax and another wave of contraction begins from the proximal end of the ventricles. This wave of auricular contraction travels over a bridge called the **auriculo-ventricular node** (AV-node). It is situated on the lower end of inter-auricular septum and is made of specialised cardiac muscle fibres. From AV-node arises the **bundle of His** which bifurcates in the inter-ventricular septum into auriculo-ventricular bundles. Their fibres called **Purkinje fibres** ramify into the ventricular wall. Wave of contraction from SA-node stimulates the AV-node, which not immediately sends impulses of contraction to the Purkinje fibres. Thus, ventricles contract, a little time later than the auricles. When the ventricles begin to contract the right and left auriculo-ventricular apertures are closed by their tricuspid and bicuspid valves. Therefore, blood cannot go back into the auricles.

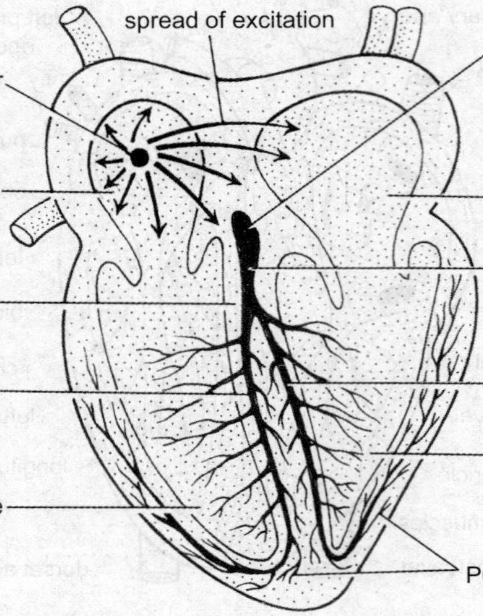

Fig. 29.47. Rabbit. Conduction of heartbeat. Nervous control.

The semilunar valves of the aortic arches are still closed and the blood is blocked in all directions, finally resulting into an increase in the blood pressure inside the ventricles. When this pressure is more than those of the arteries, the semilunar valves of the aortic arches open and the blood is forced in them. Thus, the oxygenated blood from the left ventricle goes into the **carotico-systemic aorta** for the distribution in the different parts of the body and venous blood from right ventricle goes into the pulmonary aorta carrying it into the lungs for oxygenation. Then the ventricles relax and the auricles are again filled with blood, similar wave of contraction starts again after a very short rest or **pause**. This action goes on continuously as long as the animal is alive.

A complete contraction **(systole)** and relaxation **(diastole)** of the heart constitutes a **heartbeat** or heart cycle.

Double circulation. Now it is clear from the above descriptions that there remains a complete separation of venous and oxygenated blood in a mammalian heart. The right part of the heart is concerned with venous blood and the left part with the oxygenated blood. In one complete circuit in the body, the blood passes twice through the heart, once through its right side and then through its left side. This type of circulation of blood is known as **double circulation**. The complete separation of ventricles into two chambers enables the venous and oxygenated blood not to mix.

Course of blood circulation. The circulation of blood in the body parts from the heart can be represented as given below :

Oxygenated blood from lungs ⟶ left auricle ⟶ left ventricle ⟶ carotico-systemic aorta ⟶ arteries ⟶ arterioles and ⟶ arterial capillaries ⟶ various organs of the body ⟶ venous capillaries and venules **(venous blood)** ⟶ veins ⟶ precavals and postcaval ⟶ right auricle ⟶ right ventricle ⟶ pulmonary aorta ⟶ pulmonary arteries ⟶ lungs (for **purification**) ⟶ pulmonary veins ⟶ left auricle (oxygenated blood). Thus, the blood passes through the heart twice – once through pulmonary circuit and second time through systemic circulation.

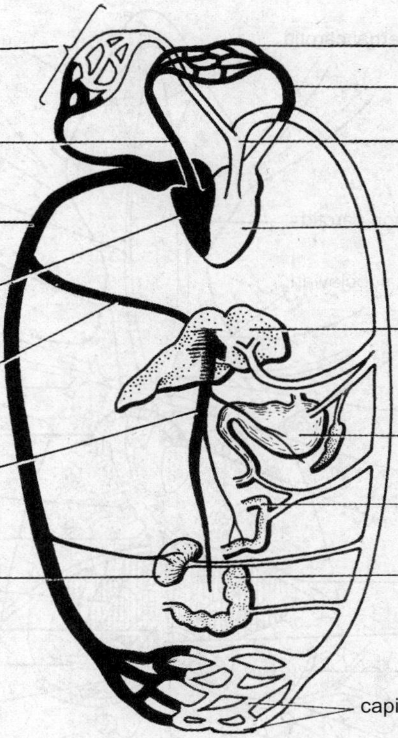

Fig. 29.48. Diagram showing double circulation of blood in a mammal.

BLOOD VESSELS

The blood vessels in rabbit is a system of closed channels through which blood flows. The blood vessels are the **arteries** and **veins**. The arteries are those vessels which carry blood **away** from the heart and veins are those which carry blood **towards** the heart. The arteries, thus, supply blood to the various parts of the body and constitute the **arterial system**. The veins similarly collect impure or deoxygenated blood from the different parts of the body and together constitute the **venous system**. In the closed blood vascular system the artery supplying a particular organ ramify to form **arterioles** finally forming **arterial capillaries**, which transform into venous capillaries. Both types of capillaries are united with each other. The **venous capillaries** unite together to form **venules** so as to form **veins**.

1. Arterial System

In rabbit, the arterial system consists of two large vessels, called the **aortic arches** directly originating from the ventricles, known as (*i*) the **pulmonary aorta** and the (*ii*) the **left carotico-systemic aorta**.

1. Pulmonary aorta. The pulmonary aorta originates from the right ventricle and runs dorso-posteriorly and then divides into two : **right** and **left pulmonary arteries**, each of them going to the lungs of their sides. The pulmonary aorta close to its bifurcation is connected with the carotico-systemic aorta by a vessel called **ductus arteriosus** in the embryo, but in the adult these are connected by a cord of connective tissue, the **ligamentum arteriosum.** It is remnant of ductus arteriosus.

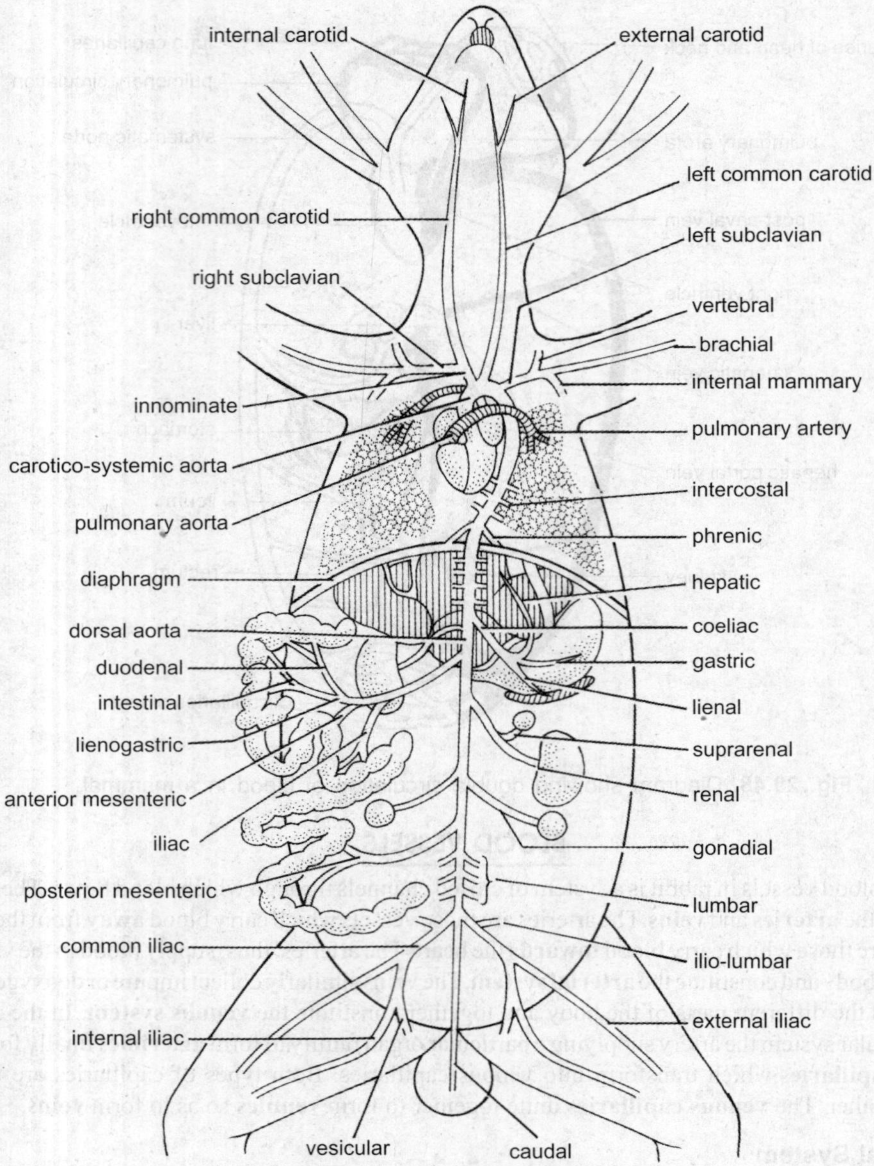

Fig. 29.49. Rabbit. Arterial system.

2. Carotico-systemic aorta. The carotico-systemic aorta originates from the left ventricle and arches towards the left side within pericardium. It gives off two small right and left **coronary arteries** supplying blood to the wall of the heart. The carotico-systemic aorta now arches to the left and passes ventral to trachea and gives off two main arteries : innominate and left suclavian.

***(i)* Innominate artery.** Innominate artery soon divides into **right** and **left carotids** which travel anteriorly through the neck along each side of trachea and near the angle of jaws each divides into **external** and **internal carotids**. The internal carotid supplies blood to the different parts of the brain and the external carotid supplies to the back of the head, tongue, lips, salivary glands, pinna and muscles of the jaw.

***(ii)* Right subclavian.** The right subclavian artery originates from the base of the right carotid and soon divides into three main branches : (*i*) **vertebral artery** supplying to the vertebral column and brain, (*ii*) **internal mammary artery** supplying to the diaphragm, pericardium, mammary glands, etc., and (*iii*) **brachial artery** supplying blood in the forelimbs.

***(iii)* Left subclavian.** The left subclavian artery originates from a little left of the innominate from the aorta and divides into three branches like that of the right subclavian.

3. Dorsal aorta. The carotico-systemic aorta now turns to the left and runs posteriorly above the heart and then runs backward mid-dorsally below the vertebral column as a **dorsal aorta**. The dorsal aorta pierces the diaphragm and continues in the abdomen up to the tip of the tail. The part of the dorsal aorta in the thorax is known as **thoracic aorta**, while its part in the abdomen is known as **abdominal aorta**. The following arteries arise from the dorsal aorta in the order given below : **(*i*)** A series of paired (5–7) **intercostal arteries** which supply blood to the ribs. **(*ii*** A pair of small **phrenic arteries** supplying blood to the muscles of the diaphragm. These two arise from the thoracic aorta. The later ones arise from the abdominal aorta. **(*iii*)** A median **coelic artery** originates just behind the diaphragm which goes to the left side and divides into two : a **hepatic artery** supplying to the liver and a **lineogastric artery** supplying to the stomach and spleen. **(*iv*) Anterior mesenteric artery.** It arises from the dorsal aorta just behind the coelic artery and divides to form a number of branches supplying to the duodenum, pancreas, ileum, caecum and colon. **(*v*) Renal arteries.** A pair of **renal arteries** arise from the dorsal aorta supplying to the kidneys. The **right renal artery** may be situated a bit below to the anterior mesentric. Each renal artery gives off a **suprarenal** branch to the suprarenal gland. **(*vi*) Gonadial arteries.** The paired **gonadial arteries** arise from the dorsal aorta which supply the gonad of its side. The gonadial artery in male is known as **spermatic** and in female the **ovarian**. **(*vii*) Posterior mesenteric artery.** The median posterior mesenteric artery originates from the dorsal aorta behind the gonadial and supplies blood to the lower part of the colon and the rectum. **(*viii*) Lumbar arteries.** A series of lumbar arteries originating from the dorsal aorta behind the posterior mesenteric supply blood to the dorsal body wall. **(*ix*) Common iliac arteries.** The dorsal aorta then divides into two common iliacs each supplying blood to the hindlimb of its side. Each common iliac gives off an **iliolumbar** for the dorsal abdominal wall. Then, each common iliac divides into **internal iliac** and **vesicular** supplying to the organs of the pelvis, *e.g.*, urinary bladder, uterus, rectum, anus, etc., and an **external iliac** which continues as **femoral artery** supplying to the hindlimbs. **(*x*) Caudal artery.** The caudal artery supplies blood to the tail region. It may arise mid-dorsally from the dorsal aorta, a little in front of its bifurcation.

2. Venous System

The blood supplied from the heart in the different parts of the body is collected by the veins as deoxygenated blood and finally poured into the heart. The venous system consists of a pair of **pulmonary veins**, **right** and **left precavals** and a **postcaval** vein, **coronary** veins and **hepatic portal system**.

1. Pulmonary veins. The pulmonary veins bring oxygenated blood from both the lungs. Both of them unite to form a common pulmonary vein which opens in the left auricle by a single opening.

2. Precaval veins or right and left anterior vena cavae. The anterior right and left precavals collect blood from the anterior regions of the body, *e.g.*, head, neck, shoulders, forelimbs and thoracic wall. Each precaval is formed by the union of following branches :

(*i*) External jugular veins. The external jugular veins collect blood from the head, tongue and shoulder region through anterior and posterior facials, deep external jugular and cephalic veins. Both the external jugulars are long veins running through the neck and are connected together by a **transverse jugular anastomosis**.

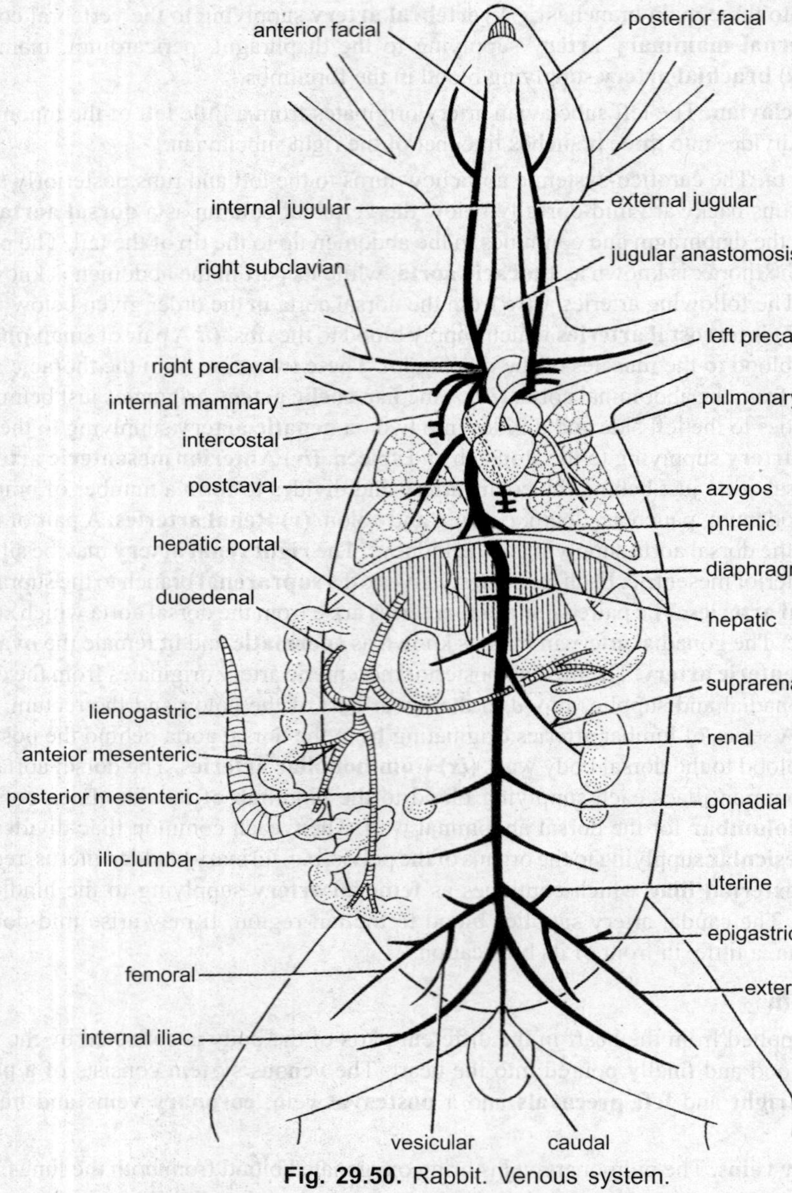

Fig. 29.50. Rabbit. Venous system.

(*ii*) **Internal jugular veins.** The internal jugular veins collect blood from the brain and occipital region of skull. The internal jugulars are connected with the external jugulars posteriorly.

(*iii*) **Subclavian veins.** The subclavian veins collect blood from the shoulder and forelimbs, and are connected with the external jugulars.

(*iv*) **Internal mammary vein.** It drains blood from the mammary glands and joins the precavals.

(*v*) **Anterior intercostal vein.** The right precaval also receives an **anterior intercostal vein** collecting blood from the anterior intercostal spaces.

(*vi*) **Azygos vein.** An azygos vein brings blood from the posterior intercostal spaces (*i.e.*, posterior 8 or 9 ribs) and lumbar region. The azygos and intercostal veins are not found in the left precaval but a few small veins on the left side form a **hemiazygos vein** which meets with the azygos vein of the right side through transverse anastomosis.

(*vii*) **Coronary veins.** A pair of small **coronary veins** collecting blood from the wall of the heart open in the left precaval only.

3. Postcaval vein. The postcaval vein is a large, median vein formed by the union of a number of veins from the abdominal region, caudal region and hindlimbs. It receives a number of veins running along the corresponding arteries.

(*i*) **Caudal vein.** It drains blood and marks the beginning of postcaval.

(*ii*) **External iliac veins.** A pair of large veins and each is formed by the union of **femoral vein** collecting blood from the outer side of hindleg, a **vesicular vein** from urinary bladder, seminal vesicle in male and uterus in female, and a **posterior epigastric vein** from pubic region.

(*iii*) **Internal iliac veins.** These collect the blood from the inner region of thighs.

(*iv*) **Ilio-lumbar veins.** They are paired **veins** bringing blood from the muscles of the back.

(*v*) **Genital veins.** A pair of **gonadial veins** bringing blood from the gonads (**spermatic** and **ovarian** in male and female respectively).

(*vi*) **Renal veins.** A pair of **renal veins** bringing blood from the kidneys (renal portal system being absent in rabbit). Each renal vein receives a suprarenal vein from the corresponding suprarenal gland.

(*vii*) **Hepatic veins.** Two pairs of **hepatic veins** arise from the liver lobes.

(*viii*) **Phrenic veins.** A pair of **phrenic veins** bringing blood from the diaphragm. These pierce the diaphragm and join the postcaval vein. Thus, the postcaval receiving these veins during its course, travels towards the anterior sides and finally opens in the right auricle.

3. Hepatic Portal System

The blood from the different parts of the alimentary canal are not directly carried to the heart but collected by the **hepatic portal vein** which brings blood into the liver. The hepatic portal vein is formed by :

(*i*) **Lineo-gastric vein.** Collecting blood from stomach and spleen.

(*ii*) **Duodenal vein.** Collecting blood from duodenum.

(*iii*) **Anterior mesentric vein.** Collecting blood from the ileum, caecum and colon.

(*iv*) **Posterior mesentric vein.** Collecting blood from the rectum and anus.

The **hepatic portal vein**, thus, enters in the left lobe of the liver where it branches into capillaries. The blood from the liver is collected by **hepatic veins** which meet with the postcaval.

Blood

The blood is a fluid connective tissue which flows in the blood vessels. It consists of a fluid matrix, the **plasma**, and formed elements, the **blood corpuscles** which are found floating in the fluid matrix. These are red blood corpuscles (erythrocytes), white blood corpuscles (leucocytes), and platelets. The red colour of the blood is due to red blood corpuscles containing **haemoglobin**, a red coloured iron pigment. It carries oxygen. The blood of rabbit (mammal) resembles with that of frog in all the structural and functional details but, however, it exhibits certain differences :

(*a*) The **erythrocytes** of rabbit are small, nearly 7.5 μ in diameter, circular, biconcave, non-nucleated and more in number, but in frog they are large, nearly 23 –26 μ in diameter, oval, nucleated, biconvex and less in number.

(*b*) The number of erythrocytes are more in rabbit, *e.g.,* 5×10^6/cu mm but in frog they are less, nearly 4×10^6/cu mm.

(*c*) The **thrombocytes** or **blood platelets** of rabbit are irregular, non-nucleated and more in number, while they are spindle-shaped, nucleated and less in number in frog.

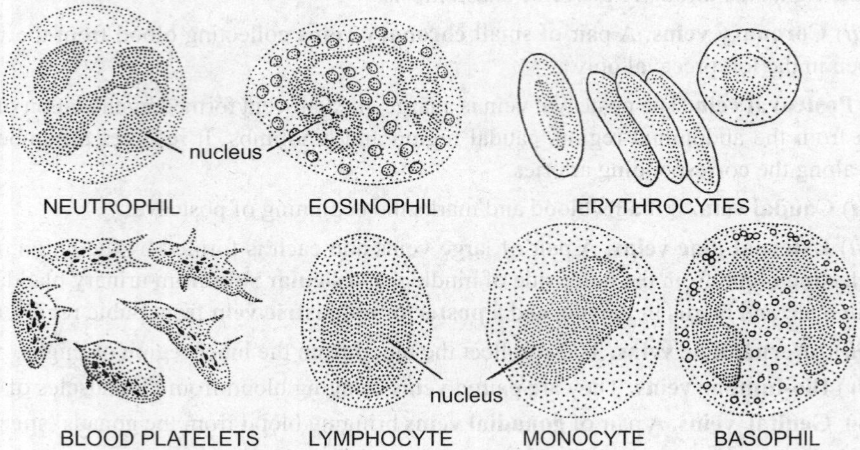

Fig. 29.51. Blood of mammal.

Leucocytes are of two types on the basis of presence of granules in their cytoplasm : granulocytes and non-granulocytes.

Granulocytes are of three types on the basis of the reaction of granules to dyes : (*i*) **acidophil** (eosinophil), (*ii*) **basophil** and (*iii*) **neutrophil**. Their nuclei are also many lobed. **Agranulocytes** are lymphocytes with large nucleus and monocytes with saucer-shaped nucleus and abundant cytoplasm. These are phagocytic, destroy the bacteria, etc.

LYMPHATIC SYSTEM

Besides blood vascular system, the lymphatic system is also found in chordates. It comprises lymph, lymph vessels and lymph nodes.

1. Lymph. Blood including RBCs never leaves the blood vessels, but plasma and leucocytes (WBCs) of the blood escape into the intercellular space of the tissues from blood capillaries. This colourless part of the blood (without erythrocytes and plasma-proteins) is called the **lymph** or tissue fluid. The lymph is also a transportable media in the body like that of the blood. It carries food and oxygen from the blood to the tissues and receives substances from the tissues to re-enter blood by means of **lymphatic vessels**. Some lymph also enters the venous capillaries by osmosis.

2. Lymph vessels. The lymphatic vessels are formed by the union of **lymph capillaries**. Lymph capillaries are formed of simple epithelium permeable even to protein molecules. Fluid pressure in capillaries is nearly zero. Lymph capillaries are not united with blood capillaries. Unlike blood capillaries their diameter is not uniform throughout. These are found in the intercellular spaces. Lymph vessels are thin-walled like veins and provided with valves. More lymph flows due to squeezing pressure of the surrounding muscles since lymph heart is not found in mammals. Valves present in the lymph vessels allow the lymph flow only in one direction. All the lymph vessels of the posterior side of the body and from the left side of the head, neck and thorax communicate with the main lymphatic trunk called **thoracic duct** situated below the vertebral column. The thoracic duct finally opens in the **left subclavian** at the place where it meets with the **left external jugular vein**. The lymph vessels from the right side of the head, neck, thorax, right arm open in the **right lymphatic duct** which opens in the right subclavian vein where it meets with the **right external jugular vein**.

3. Lymph nodes. Lymph nodes are formed of a meshwork of connective tissue fibres encased in a capsule. The lymph vessels pass through a lymph node at certain places and these are found only in mammals. In rabbit the lymph nodes are found in the head, neck, arm pits, near large blood vessels, groins, on the tonsils and Peyer's patches on the intestine are formed by these nodes.

The **lymph hearts** are not found in rabbit, therefore, the lymph flows in the lymph vessels with the help of body muscles, by pressure exerted to the small vessels by osmosis and absorption at tissue fluids. Though the lymphatic system is a **closed system** but it does not form a complete circuit of flow in the body.

4. Functions. (*i*) Lymphatic system returns the excessive tissue fluid produced in the body to blood. It absorbs fat through lymph capillaries (**lacteals**) present in the intestinal villi.

(*ii*) Lymph carries digested food, oxygen, hormones, etc., from the blood to the tissue cells. Similarly, the waste products, carbon dioxide, etc., are transported from cells to the blood.

(*iii*) Lymph nodes produce lymphocytes of the blood.

(*iv*) Lymph nodes act as filters of poisonous and foreign substances like bacteria, dust and debris, etc., and, thus, purify the blood. Tonsils, adenoids and appendix are lymph nodes and become swollen due to infection.

NERVOUS SYSTEM

The nervous system of rabbit consists of three divisions :

1. The **central nervous system** including brain and spinal cord.

2. The **peripheral nervous system** including nerves coming out from the brain and spinal cord, *i.e.,* **cranial nerves** and **spinal nerves**.

3. The **autonomic nervous system** including **sympathetic nervous system** and **parasympathetic nervous system**.

1. Central Nervous System

1. Brain. The brain of rabbit is essentially made on the same plan as that of the frog and other vertebrates but in rabbit is more complicated. The brain is found enclosed in the **cranial cavity** of the skull.

Meninges. Brain is surrounded by 3 membranes which are collectively known as **meninges**. In rabbit, these membranes are **piamater**, **arachnoid layer** and **duramater**. The **piamater** is a thin highly vascularised membrane closely applied with the brain. The **duramater** is the outermost tough membrane applied with the inner surface of the cranial cavity. The **arachnoid layer** is situated in between piamater and duramater. Both above and below the arachnoid layer there are spaces called **subdural cavity** and **subarachnoid cavity** respectively. These are crossed by fibres.

Cerebrospinal fluid. The subdural and subarachnoid spaces are filled with the watery cerebrospinal fluid. The arachnoid layer at some places is provided with vascular **arachnoid villi** which absorb the cerebrospinal fluid which then reaches the venous system. The brain is hollow whose various cavities are called ventricles and are also filled with the above fluid. The cerebrospinal fluid is lymph-like and secreted by the anterior and posterior choroid plexuses.

Parts of brain. The brain of rabbit can be divided into three regions : **(*a*)** the **forebrain** or **prosencephalon, (*b*)** the **midbrain** or **mesencephalon**, and **(*c*)** the **hindbrain** or **rhombencephalon**. Their parts are the same in all the vertebrates.

1. Forebrain. It consists of a pair of **olfactory lobes**, a pair of **cerebral hemispheres** and an unpaired **diencephalon.**

(***a***) **Olfactory lobes.** The olfactory lobes are the anteriormost part of the forebrain and are club-shaped, projecting in front of the cerebral hemispheres. Each olfactory lobe consists of an

olfactory bulb and an **olfactory tract**. The olfactory tract is clearly seen from the ventral side, as it is found concealed beneath the frontal lobes of cerebral hemispheres and connected with the hippocampal lobes. Each olfactory bulb gives off an olfactory nerve. Each olfactory lobe has a narrow cavity, called the **rhinocoel (I ventricle)**. These lobes are responsible for controlling the sense of smell.

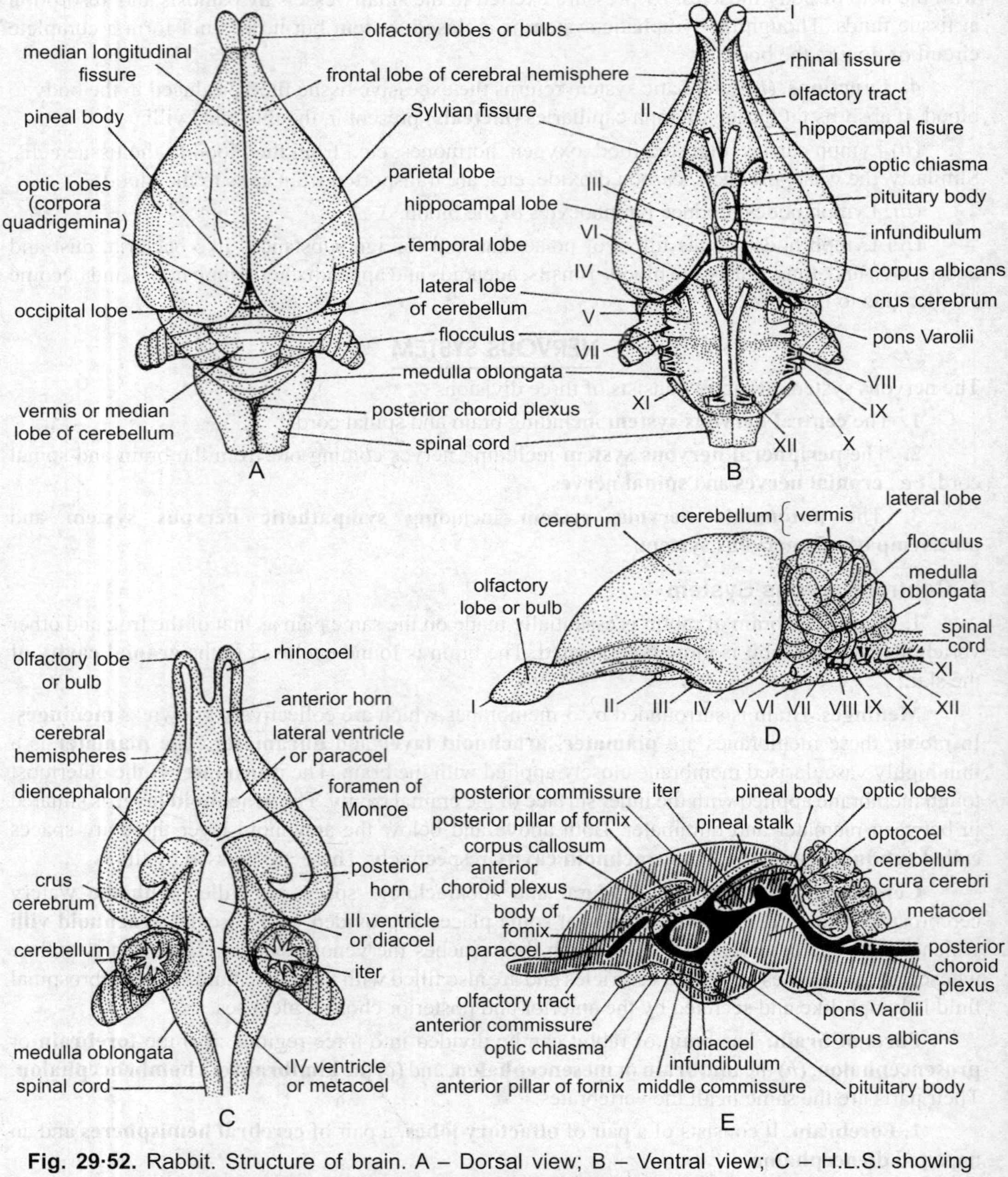

Fig. 29·52. Rabbit. Structure of brain. A – Dorsal view; B – Ventral view; C – H.L.S. showing ventricles; D – Lateral view; E – Sagittal section.

(*b*) **Cerebral hemispheres or cerebrum.** The **cerebral hemispheres** or **telencephalon** are well developed and form about 2/3 of the whole brain. They are narrow in front and broad behind, and their surface is nearly smooth. They extend backwards, covering the diencephalon and mid-brain, and almost touch the cerebellum. They partly cover the olfactory lobes anteriorly. Dorsally a deep longitudinal median fissure separates the two cerebral hemispheres. The hemispheres are further divided into many lobes by fissures. A shallow lateral and oblique **Sylvian fissure** divides each hemisphere into a larger, **frontal lobe** and a smaller postero-lateral **temporal lobe**. Ventrally a longitudinal **hippocampal sulcus** or **fissure** on each hemisphere separates it into an outer lobe and an inner lobe, called the **hippocampal** or **pyriform** lobe. A shallow ventral longitudinal **rhinal fissure** separates the olfactory duct from the hippocampal lobe. Internally both the cerebral hemispheres are connected by a dorsal broad transverse band of nervous tissue, called the **corpus callosum** (characteristic of placental mammals). The cavity of each cerebrum is called the **paracoel** or **lateral ventricle** or **II ventricle.** Both the cavities are connected with each other and also with the diacoel through a narrow passage, called the **foramen of Monro**. The roof of paracoel is formed of corpus callosum and floor, and lateral walls by **corpus striatum**. Both the corpora striata are connected by a transverse **anterior commissure**.

Cerebral hemispheres control all the involuntary actions of the body and the centres of intelligence, memory, thought, etc., are found in these hemispheres. The cerebral hemispheres are not well developed in the lower animals.

(*c*) **Diencephalon.** The **diencephalon** or **thalamencephalon** is narrow, rectangular, lies below and covered by the posterior extension of cerebral hemispheres. From the dorsal side of diencephalon arises a slender **pineal-stalk** having a terminal rounded **pineal body**. Remaining part of the roof (dorsal side) is non-nervous and highly vascular. It is called the **anterior choroid plexus**. On the ventral side the floor of diencephalon gives off a median rounded **infundibulum**, which carries an oval **Rathke's pouch** or **hypophysis**. Behind it is attached a small rounded **corpus albicans**. Infundibulum and hypophysis together form the **pituitary body**. In front of the infundibulum is present the **optic chiasma** formed by the crossing of two optic nerves.

The cavity of diencaphalon is called the **diacoel** or **III ventricle.** The roof of diacoel is called **epithalamus**, floor is **hypothalamus** and the lateral thickened walls are **optic thalami** (singular is thalamus). Optic thalami are connected with each other by a transverse **middle commissure** occupying the greater part of diacoel. It is formed of only nerve cells without fibres.

The centre of autonomic nervous system lies in **diencephalon**. It controls the different chemical activities regarding the metabolism of carbohydrates, fats and proteins. The temperature of the body and reproductive activities are controlled by this part of the brain.

2. Midbrain. The midbrain is largely covered by the cerebral hemispheres. It consists of the **optic lobes** which are four in number, two on each side situated behind the pineal body. All the optic lobes are collectively called **corpora quadrigemina**. The anterior pair of optic lobes are larger called **superior colliculi**, and the posterior lobes are smaller called the **inferior colliculi**. Superior colliculi are associated with the sight, while inferior colluculi are concerned with hearing. The cavity of midbrain is narrow, called the **iter**. Its floor is formed of thick longitudinal bands of nerve fibres called the **crura cerebri**. These connect the forebrain with the hindbrain. The optic lobes are the centre of sense of sight. Crura cerebri control the activity of muscle fibres.

3. Hindbrain or rhombencephalon. It consists of the **cerebellum** or **metencephalon** and **medulla oblongata** or **myelencephalon**.

(*a*) **Cerebellum.** The cerebellum of rabbit is well developed and consists of a large median lobe, called **vermis**, two **lateral lobes** situated on the sides of the vermis and two **floccular lobes** situated on the ventro-lateral sides of the lateral lobes. The surface of the carebellum is thrown out into a number of folds, therefore, increasing the area of grey matter. Cerebellum has no cavity. It is responsible for maintaining the equilibrium and it coordinates the voluntary muscular movements. The ventral surface of cerebellum has strong transverse bands of fibres connecting the right and left halves of cerebellum. It is called **pons Varolii.**

(*b*) **Medulla oblongata.** The **medulla oblongata** is the posteriormost part of the brain which is broad and triangular anteriorly but gradually narrows posteriorly so as to form the **spinal cord**. Its cavity is called the **metacoel** or **IV ventricle**. It continues into the central canal of spinal cord. The roof of the metacoel is non-nervous and vascular which forms the **posterior choroid plexus**. Medulla oblongata and pons Varolii control the involuntary activities of the body, such as digestion, respiration, excretion and circulation, etc.

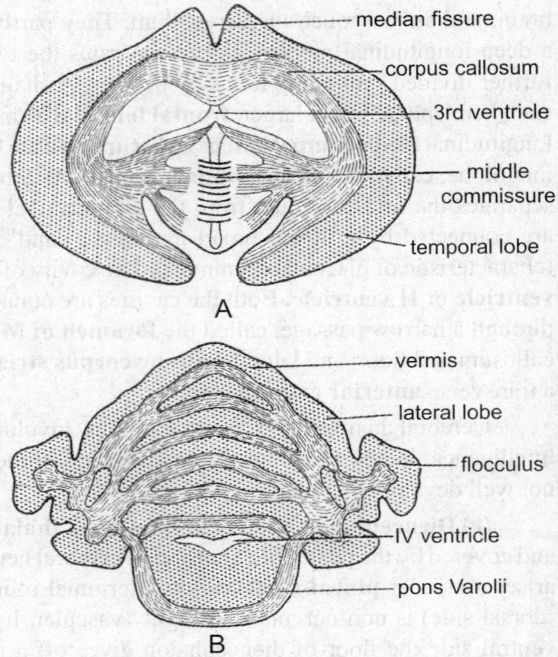

Fig. 29.53. Rabbit. A – T.S. of cerebrum; B – T.S. of cerebellum.

Histology. The brain is formed of two types of nervous tissues. The outer part of brain is made of **grey matter** having only nerve cells. It is called **cortex**. The inner part of brain is the **medulla** formed of **white matter** having only nerve fibres and neuroglia (supporting structure).

Ventricles of the brain. Each olfactory lobe has a narrow cavity, the **rhinocoel**. The rhinocoel of each side is connected with the cavity of the cerebral hemisphere, **lateral ventricle** of its side. The diencephalon which is situated just below the cerebral hemispheres has the third ventricle or **diacoel**. The lateral ventricles of the hemispheres and diacoel are connected with each other through **foramen of Monro**. A narrow passage, the **iter** or **Sylvian aqueduct**, is found in the midbrain which connects the diacoel with the 4th ventricle or **metacoel**. The 4th ventricle is the cavity of the medulla oblongata which is broader anteriorly but gets narrower posteriorly.

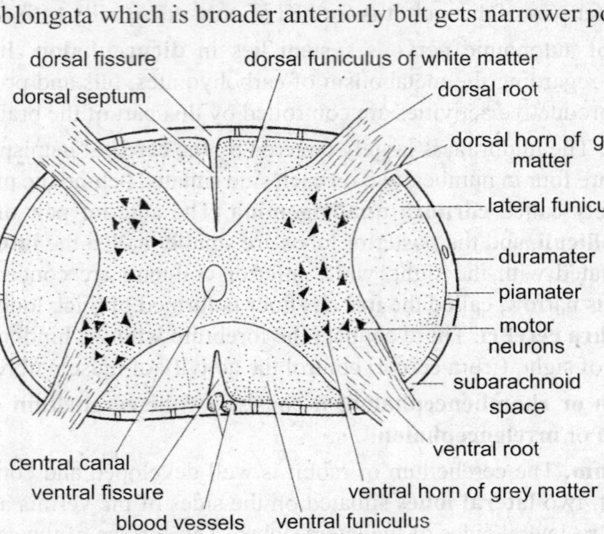

Fig. 29.54. Rabbit. T.S. of spinal cord.

2. Spinal cord. The spinal cord is the continuation of the medulla oblongata after coming out of the cranium through foramen magnum. It runs through the **neural canal** of the vertebrae and terminates behind in the lumbar region as **filum terminale**. The spinal cord is white, thick-walled tube formed of nervous tissue. The filum terminale is a thread-like non-nervous structure. The spinal cord is swollen at the levels of fore-and hindlimbs, which are called **brachial** and **sciatic swellings** respectively. These swellings give off nerves to the fore-and hindlimbs. There are two median longitudinal fissures, one on its dorsal surface, called **mid-dorsal fissure**, and the other on its ventral surface, called **mid-ventral fissure**. The 4th ventricle of the medulla is continued in the spinal cord as **central canal**, which is filled with the cerebro-spinal fluid.

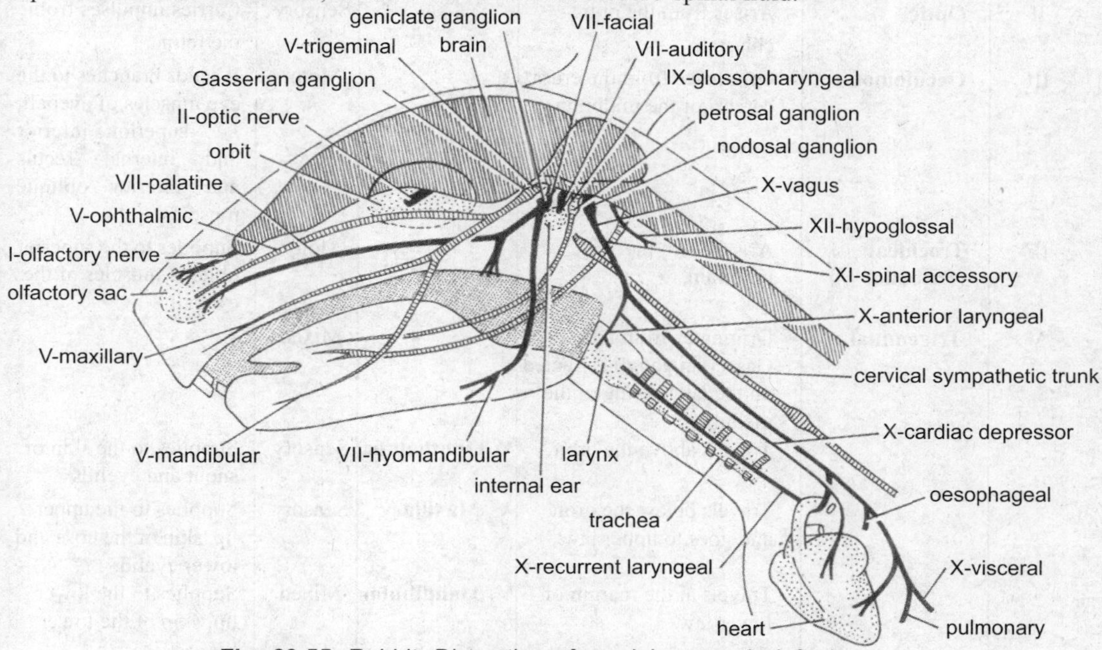

Fig. 29.55. Rabbit. Dissection of cranial nerves in left view.

Meninges. Like brain, the spinal cord is also enclosed in three membranes : outer **duramater**, middle **arachnoid** layer and inner **piamater**. The subarachnoid space is filled with cerebrospinal fluid.

Histology. The structure of spinal cord can be well explained with the help of its transverse section. In T.S. it appears to be a more or less spherical structure having a **central canal** in its centre. It is composed of a central **grey matter** with nerve cells (neurons), which seems to be a butterfly-like or H-shaped structure. The grey matter is projected into the **dorsal** and **ventral horns**. This grey matter is surrounded by the **white matter** formed of medullated nerve fibres. The white matter is distinctly divided into two compartments by a dorsal fissure and a narrow **dorsal septum** extending up to central canal and a broad, deep **ventral fissure**. The white matter of each compartment can be further distinguished into dorsal, lateral and ventral columns, *i.e.,* **dorsal funiculus, lateral funiculus** and **ventral funiculus** respectively, due to the characteristic shape of the grey matter. The spinal cord provides passage for the impulses to and from the brain. It plays an important role in **reflex actions.**

2. Peripheral Nervous System

1. Cranial nerves. The nerves coming out from the brain constitute the **cranial nerves**. In rabbit (mammals) twelve pairs of cranial nerves are found. First ten pairs of cranial nerves are like that of fishes and amphibians. The two additional pairs are the **spinal accessory** (XI) and the **hypoglossal** (XII). The **facial** (VII) in mammals lacks ophthalmic, buccal and lateral branches. Each cranial nerve originates from a distinct separate root. The disposition, origin, nature and distribution of cranial nerves of rabbit can be easily explained with the help of following table 29.3.

TABLE 29·3. CRANIAL NERVES OF RAT OR RABBIT.

Sl. No.	Name of the nerve	Origin and disposition	Branch if any	Nature	Distribution and supply
1	2	3	4	5	6
I	**Olfactory**	Arises from the ventral surface of the anterior end of the olfactory lobe.	—	Sensory	Brings impulses from the nasal sacs to the olfactory lobe.
II	**Optic**	Arises from the optic-chiasma.	—	Sensory	Carries impulses from the retina.
III	**Occulomotor**	Originates from the crura cerebri of the midbrain.	—	Motor	It sends branches to the eye muscles of eyeball, *i.e.,* superior, inferior and internal rectus and inferior oblique muscles.
IV	**Trochlear (Pathetic)**	Arises from the midbrain.	—	Motor	Supplies to the superior oblique muscles of the eyeball.
V	**Trigeminal**	Originates from the Gasserian ganglia situated on the lateral side of the medulla.		Mixed	
		Travels above the orbit.	**V_1 Ophthalmic**	Sensory	Supplies to the skin of snout and eyelids.
		Travels below the orbit and goes to upper jaw.	**V_2 Maxillary**	Sensory	Supplies to the upper lip, skin of the nose and lower eyelid.
		Travels at the margin of lower jaw.	**V_3 Mandibular**	Mixed	Supplies to the lower lip, skin of the lower jaw, pinna.
VI	**Abducens**	Originates from the ventral side of the medulla.	—	Motor	Supplies to the posterior rectus muscles of the eyeball.
VII	**Facial**	Originates from the geniculate ganglia situated below the Gasserian ganglia on the sides of medulla.		Mixed	
			VII_1 Palatine	Sensory	Supplies to the roof of the buceal cavity and nasal chambers (palate).
		Travels near the tympanic membrane.	**VII_2 Chorda tympani**	Sensory	Supplies to the salivary glands and taste-buds, the tongue.
		Travels at the outer surface of lower jaw.	**VII_3 Hyomandibular**	Mixed	Supplies to the muscles of the lower jaw, neck and pinna.

(Contd.)

Sl. No.	Name of the nerve	Origin and disposition	Branch if any	Nature	Distribution and supply
1	2	3	4	5	6
VIII	**Auditory**	Originates from the sides of the medulla		Sensory	
			VIII_1 Vestibular		Supplies to the utriculus, sacoulus and semicircular canals of the internal ear.
			VIII_2 Cochlear		Supplies to the cochlea.
IX	**Glossopharyngeal**	Originates from the sides of medulla	**Laryngeal**	Mixed	Supplies to the tongue, pharynx and salivary glands.
			Pharyngeal	Mixed	Supplies to the pharynx salivary glands.
X	**Vagus**	Originates from the sides of medulla just behind to glassopharyngeal.		Mixed	
			X_1 Superior laryngeal	Motor	Supplies to muscles of the laryngeal region.
			X_2 Recurrent laryngeal	Motor	Supplies to muscles of the laryngeal region.
			X_3 Cardiac	Motor	Supplies to muscles of the heart.
			X_4 Pneumogastric	Sensory and Motor	Supplies to the lungs, oesophagus and stomach.
			X_5 Depressor	Motor	Supplies to muscles of diaphragm.
XI	**Spinal accessory**	Arises from the sides of the medulla and spinal cord	—	Motor	Supplies to muscles of the pharyngeal, laryngeal and cervical regions.
XII	**Hypoglossal**	Arises from the ventral surface of medulla oblongata.	—	Motor	Supplies to the muscles of the neck and tongue.

2. Spinal nerves. The spinal nerves in rabbit are 37 pairs which can be divided into five zones, *viz.*, 8 pairs **cervical**, 12 pairs **thoracic**, 7 pairs **lumbar**, 4 pairs **sacral** and 6 pairs **caudal** nerves. Each arises from the cord by two roots : dorsal and ventral. The sensory nerve fibres are found in the dorsal root and motor nerve fibres are found in the ventral root. The dorsal root after coming out from the spinal cord bears a **dorsal root ganglion**. The **dorsal** and **ventral roots** of the spinal nerves originate from the dorsal and ventral horns of the grey matter of the spinal cord respectively. The dorsal and ventral roots unite in the neural canal and come outside the vertebral column through intervertebral foramen as a single nerve of mixed nature. It now immediately divides into three branches (rami) : dorsal, ventral and visceral. The **dorsal ramus** innervates the skin and muscles of back. The **ventral ramus** goes to the ventro-lateral parts of the body. The visceral or **ramus communicans** unites with the sympathetic system and innervates the viscera. The 1st, 2nd and 3rd spinal nerves of the cervical region supply the skin by their ramus dorsalis and to the muscles of the neck by their ramus ventralis.

Plexes. The ventral rami of 4th, 5th, 6th cervical nerves of the neck region meet to form a network, called **cervical plexus**. It sends a **phrenic nerve** to muscles of diaphragm. The 5th, 6th, 7th and 8th cervical nerves and the first thoracic nerve in the region of forelimbs anastomose together to form a **brachial plexus**. The 4th, 5th, 6th and 7th lumbar nerves and the 1st, 2nd and 3rd sacral nerves in the region of hindlimbs unite to form **sacral plexus** or **lumbo-sacral plexus**.

3. Autonomic Nervous System

The autonomic nervous system is mainly responsible for controlling the involuntary visceral cardiac and glandular muscles of the body. It is of two types : **sympathetic nervous system** and **parasympathetic nervous system**.

1. Sympthetic nervous system. There is a pair of ganglionated sympathetic nerves or cords which arise from the brain and pass out the cranium through carotid canals and extend from the neck to the end of the abdomen situated at the lateral sides of the vertebral column and enter the tail as a fine filament. In rabbit, the sympathetic nerve bears three ganglia in the neck region, ten in the thorax, six in lumbar, four sacral and one caudal, total 24 ganglia on one side. The ganglia of sympathetic nerves are connected with the ramus communicans of the corresponding spinal nerves, *i.e.,* one ganglia for each spinal nerve. Syumpathetic nerves innervate the heart, glands and visceral organs.

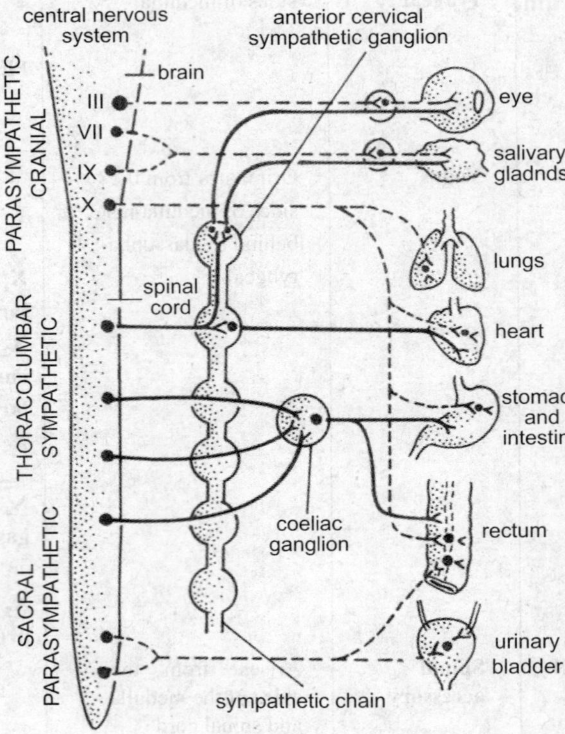

Fig. 29.56. Rabbit. Sympathetic (solid lines) and parasympathetic (broken lines) autonomic nervous system (Diagrammatic).

2. Parasympathetic nervous system. Parasympathetic nervous system is formed by the fibres of IIIrd, VIIth, IXth and Xth cranial nerves and from 2nd, 3rd and 4th sacral nerves. Ganglia of parasysmpathetic nerves lie in the head, neck and sacral regions. They innervate all those organs which are innervated by the sympathetic nerves.

The IIIrd cranial nerve runs to the **ciliary ganglion** in the orbit and from these nerve fibres supply to the muscles of the iris. The parasympathetic fibres from the VII and X cranial nerves supply to the lacrimal gland, salivary gland, lungs, heart, liver and different parts of alimentary canal. The parasympathetic nerve fibres of the sacral region unite to form a **pelvic ganglia** on each side and the nerve fibres from this ganglion supply to the bladder, kidney, rectum and gonads.

Working. The sympathetic and parasympathetic nervous systems work independently but opposite to one another. Thus, each visceral organ innervated by a **sympathetic fibre** stimulates the organ to start the function and **parasympathetic fibre** inhibits or slows down the function of that organ. For example, sympathetic stimulation causes increased heartbeat, rise in blood pressure, bronchial dilation, peripheral vascular contraction, sweat gland activity, dilation of pupil, contraction of arrector pili; inhibition of saliva, contraction of certain sphincters, splenic contraction and increased blood-sugar content. In short the animal exhibits symptoms of fear or anger and is made ready to fight or flight. Sympathetic fibres are **adrenergic** or sometimes (as in sweat glands) **cholinergic**. Opposite to sympathetic action, parasympathetic stimulation leads to dilation of blood vessels (except intrinsic coronary circulation of heart), increased salivary secretion, constriction of bronchi, increased peristalsis, relaxation of sphincter, constriction of pupil. Parasympathetic fibres are **cholinergic**.

SENSE ORGANS

The **sense organs** or **receptor organs** of rabbit are essentially similar with that of the frog except in some details. The sense organs detect the changes in the external and internal environments. The changes in the environment are known as **stimuli**. The various sense organs of rabbit are :

1. Organs of touch — Cutaneous (Skin) receptor.
2. Organs of taste — Gustatoreceptors (Tongue)
3. Organs of smell — Olfactoreceptors (Olfactory sacs or nasal chamber)
4. Organs of sight — Photoreceptors (Eyes)
5. Organs of hearing and equilibrium — Statoacoustic receptors (Ears)

1. Organs of Touch (Skin)

The skin of mammals is highly sensitive which is provided with several types of receptors. These receptors are microscopic and are of various types situated beneath the epidermis. Each receptor is concerned with a particular stimulus.

1. Free nerve endings. Sense of touch, cold, warmth, pressure and pain, etc., are perceived by free nerve endings. The fine branching fibres of sensory nerves lie just beneath the epidermis of skin. These fibres also extend into the epidermis.

2. Basket nerve endings. These are concerned with sense of touch. A fine network of branching sensory nerve fibres are found around the hair follicle. Sense of touch is perceived by touching the hair.

3. Encapsulated nerve endings. These are in the form of capsules, *e.g.,* Meissner's and Pacinian capsules. **Meissner's capsules** are found just beneath the epidermis. There are formed of a naked axon surrounded by a sheath of connective tissue capsule. These are sensitive to touch. **Pacinian capsules** are found in dermis and several internal organs. Each capsule is formed of a core of single axon ending in an ovoid bulb, which is surrounded by a sheath of connective tissue. These detect pressure.

4. Neuromuscular bundles or spindles. These are found inside the body, *i.e.,* in skeletal muscles, tendons and joints. These are called **kinaesthetic** receptors and apprise the central nervous system about the degree of muscle tension. It is formed of nerve ending in bulb or end organ, called the **neuromuscular bundle** or **neuromuscular spindle**.

The tactile organs are characteristically known as **corpuscles**. These corpuscles are ovoid bodies supplied with nerve-endings. Each corpuscle is formed of a sheath of connective tissue arranged in concentric rings around the **core of a nerve fibre**. The corpuscles are **tangoreceptors**, **frigidoreceptors**, **caloreceptors** and **algesireceptors**, *i.e.,* sensitive to touch, cold, warmth or heat and pain respectively. However, the pain receptors (algesireceptors) are without special corpuscles and only in the form of fine nerve fibres. These corpuscles are also responsible for sensitiveness to humidity and pressure, etc.

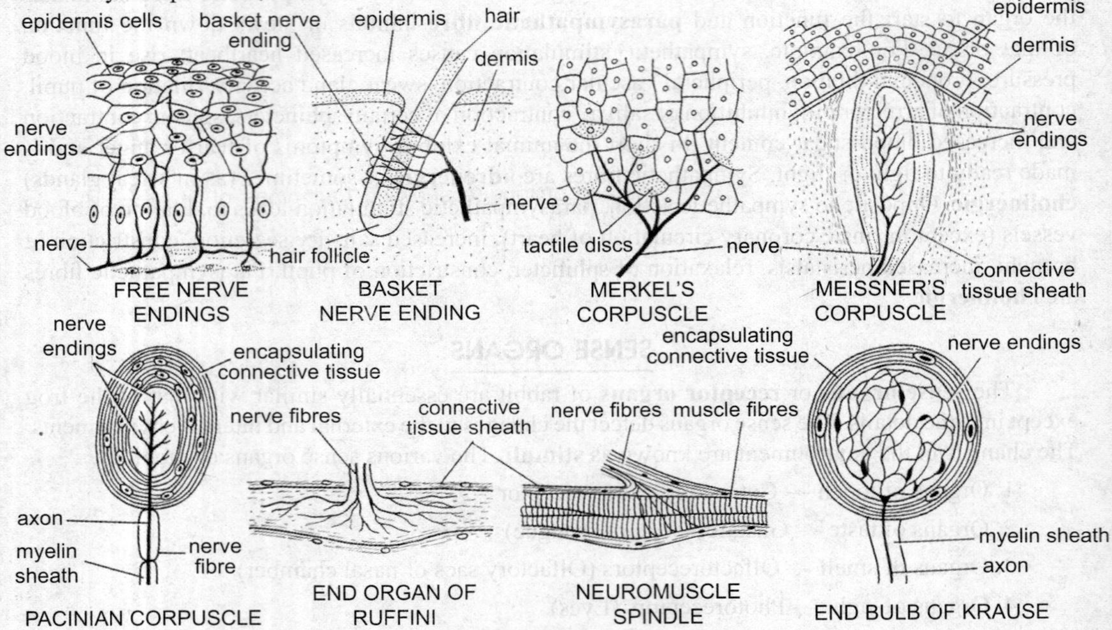

Fig. 29.57. Rabbit. Various types of cutaneous receptors.

2. Organs of Taste (Gustatoreceptors)

These are the **taste-buds** found on the tongue papilla and on the soft palate (roof of the buccal cavity). The taste bud is barrel-shaped and made of two kinds of cells, the **neurosensory cells** (gustatory cells) and **supporting cells**. The gustatory cells are elongated, spindle-shaped and provided with sensory hairs projecting in a depression between tongue-papillae called **taste-pore**. The inner ends of neurosensory cells are supplied by the nerve fibres from VII and IX cranial nerves. These cells are stimulated by the substances dissolved in the mucous and saliva and the sensation of bitter, sweet, salty, sour, etc., finally reaches to the brain.

3. Organs of Smell (Olfactoreceptors)

The nasal passage is provided with scroll-like turbinal bones, known as **ethmoturbinals**, **maxilloturbinals** and **nasoturbinals** on the basis of their origin. These turbinals and roof of nasal chambers are covered over with **olfactory epithelium** or **Schneiderian epithelium**. The olfactory organs have spindle-shaped olfactory cells, mucous cells and columnar supporting cells. The olfactory cells externally bear several delicate olfactory hairs and their inner ends are connected with nerve fibres which enter the olfactory lobes of the brain. The mucous cells produce mucus which keeps back particles of dust and it also dissolvs **odoriferous** substances which are in gas form and stimulate the neurosensory cells or olfactory cells. The sense of smell is well developed in rabbits. The rabbit in embryonic condition bears **Jacobson's organ** in the roof of buccal cavity, in the form of a groove on the lower medial aspect of each nasal cavity. Its duct is separate from the nasal apparatus and opens into the mouth cavity. It is not found in adults.

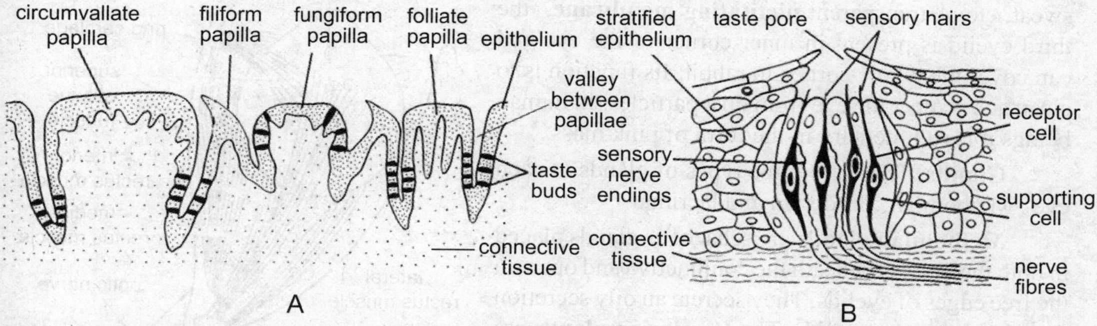

Fig. 29·58. Rabbit. A – V. S. of upper surface of tongue showing four types of papillae; B – A single taste bud in V.S.

The epithelium of ethmoturbinals actually constitutes the olfactory region which is sensory to smell. The epithelium of maxilloturbinal moistens and warms the air on its way to lungs and is not sensory. The epithelium of nasoturbinals has no olfactory cells.

Each nasal passage has an ovoid sensory pad having minute papillae and ridges. These pads are tactile and serve as distance receptors for testing air.

4. Organs of Sight or Eyes

The eyeballs are spherical photoreceptor organs, situated one on either side of the head in the **orbits**. The eyes of mammals are similar in their structures.

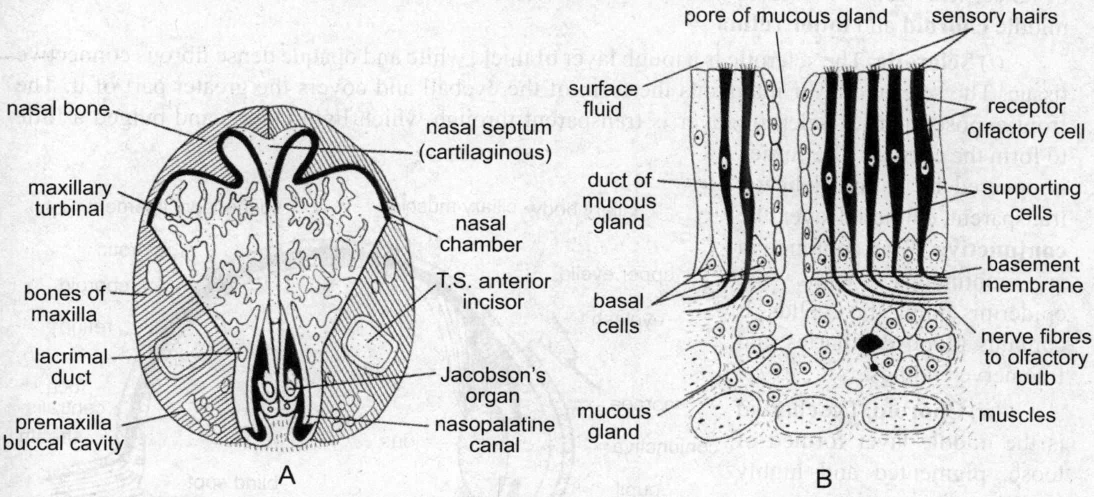

Fig. 29·59. Rabbit. A – Snout in V. S. through nasal chambers and Jacobson's organs; B – Olfactory epithelium with mucous glands in section.

External structure. The eyeball is spherical and hollow. Its about one-fifth part is visible externally and the remaining part remains hidden within the body orbit. Each eyeball is provided with six sets of muscles which move the eyeball in the orbit. Out of the six sets, four sets are **rectus muscles** and two sets are **oblique muscles**. These muscles are attached on one side with the eyeball and on the other side with the orbit. The exposed part of the eyeball is known as **cornea** which is well protected by two movable **eyelids** the upper and lower. Both the eyelids are provided with stiff hairs or **eyelashes** on their margins. These eyelashes protect the eyes from dust particles, rain water and sweat, etc. A transparent **nictitating membrane**, the third eyelid is present in inner corner of the eye and can cover the whole cornea in rabbit. Its function is to clean and protect the eye from dust particles. In human beings it is rudimentary in the form of pink mass.

Glands. There are three types of glands in each eye : Meibomian, Harderian and lacrimal.

Meibomian glands are sebaceous glands placed in both the eyelids beneath the conjunctiva and open on the free edges of eyelids. They secrete an oily secretion for lubricating the eyelids. The **Harderian glands** are situated at the inner side below the lower eyelids and open in connection with nictitating membrane at the inner angle, whose secretion keeps the conjunctiva moist. The **lacrimal gland** has several openings on the conjunctival surface below the upper eyelid, towards the outer side of eye. These glands secrete a saline watery fluid, the **tears**, on the surface of the eye (conjunctiva). It keeps the eye moist, soft, clean and free from bacteria. The excess of tears accumulate towards the inner corner of eye and are drained by a **naso-lacrimal duct** into the nasal chamber. Harderian glands are absent in primates.

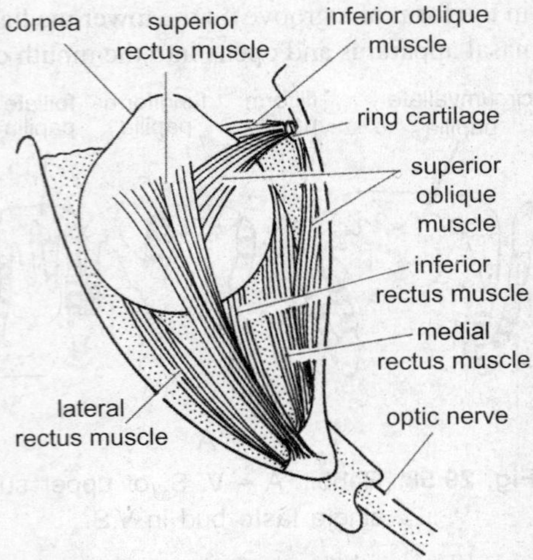

Fig. 29.60. Rabbit. The extraocular muscles of eye.

Internal structure. The internal structures of the eyeball can be well explained with the help of its vertical section as shown in the diagram. Its wall consists of three coats : an outer **sclerotic**, middle **choroid** and **inner retina**.

(*i*) **Sclerotic.** The sclerotic is a tough layer of thick, white and opaque dense fibrous connective tissue. The sclerotic layer maintains the form of the eyeball and covers the greater part of it. The front exposed part of sclerotic layer is transparent through which light enters, and bulged a little to form the **cornea**. The cornea is covered by a thin delicate, transparent, epithelial layer, the **conjunctiva**. The conjunctiva is continuous with the epidermis lining the eyelids. It contains blood capillaries and free nerve endings.

(*ii*) **Choroid.** The choroid is the middle layer formed of loose, pigmented and highly vascular connective tissue. This layer helps in darking the cavity of the eyeball to check the internal reflection of the light. Its blood capillaries provide nourishment to the retina. The choroid is found beneath the posterior part of sclera. In front at the junction

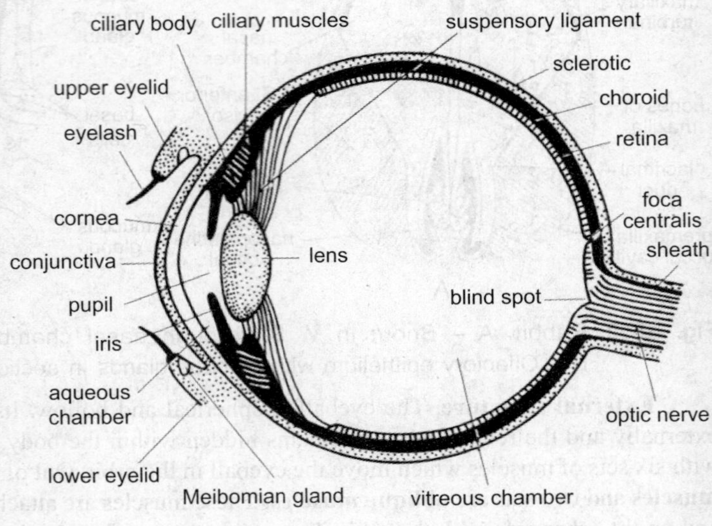

Fig. 29.61. Rabbit. Diagrammatic V.S. of eyeball.

of sclerotic and cornea the choroid swells and bends inwards into the cavity of the eyeball forming the **ciliary body** which is provided with **ciliary muscles** which project into vascular **ciliary processes** or **folds**. These secrete aqueous humour. In front of ciliary body, the choroid becomes separate from the cornea or at the corneal margin the choroid is continued as **iris**, which is separated by a space (**anterior chamber of eye**) from the cornea. In this region its pigmentation gives the characteristic colour to the eye. It is perforated by a circular aperture, the **pupil**. In the iris are present two sets of smooth muscle fibres : radiating fibres for dilation and circular fibres for contraction of the pupil. Thus, the diameter of the pupil is controlled by the contraction and relaxation of the muscles of the iris.

Lens. Just behind the iris is a solid biconvex lens enclosed in a lens capsule. Lens is formed of concentrically arranged transparent fibres. Lens capsule is thin, transparent and elastic. Lens is supported by the **suspensory ligaments** attached with the ciliary bodies.

(*iii*) **Retina.** The retina is the innermost transparent layer of the eyeball whose outer layer is pigmented and closely applied with the choroid and the inner nervous layer which is sensory layer. The nervous layer is provided with an outer layer of **rods** and **cones** which are photoreceptor cells. The outer pigmented layer is of simple cuboidal or columnar epithelium. The inner layer is also a simple cuboidal epithelium in the **pars iridica** (beneath the iris) and in the **pars ciliaris** (in the region of ciliary body). Retina beneath the iris is also pigmented. This part of retina is not light sensitive and relatively simple in structure. Only the posterior part of retina (**pars optica**) is light sensitive and highly complex, containing light-sensitive elements and nerve cells and fibres. Light-sensitive elements are rods and cones. Their inner ends connect with small bipolar nerve cells which in turn connect with large nerve cells **(ganglion cells)**. From the ganglion cells arise fibres, all of which converge on a small circular area, towards nasal side of the posterior pole of eyeball, called **optic disc**, and here they pierce the coats of eyeball to become optic nerve.

The **rods** are more sensitive to low intensity of light and contain a pigment **rhodopsin**. They are suitable for dark (night vision). The **cones** are sensitive to high intensity light and more suited for day vision. They produce a sharp image. Cones are also sensitive to light of relatively narrow frequency bands. Thus, they are associated with recognition of colours, which is only found in primates.

Rods are more towards the periphery, while cones are more concentrated towards the centre. The site where all the nerve fibres converge and unite to form the optic nerve and then leave the eye is called the **blind spot**. It is without any photoreceptor cells and, therefore, does not produce any impression. Just above the blind spot in the retina in line with the optical axis is a slight, oval, depression called **yellow spot** or **area centralis**. **Moghe** (1957), however, has described that yellow spot is not found in the eyes of rabbit. The oval depression in the area centralis, called **fovea**, is the region for most distinct vision. Here rods are absent and only cones are present.

Chambers. The cavity of the eyeball is divided into an anterior and a posterior chamber by the iris, lens and suspensory ligaments. The anterior chamber

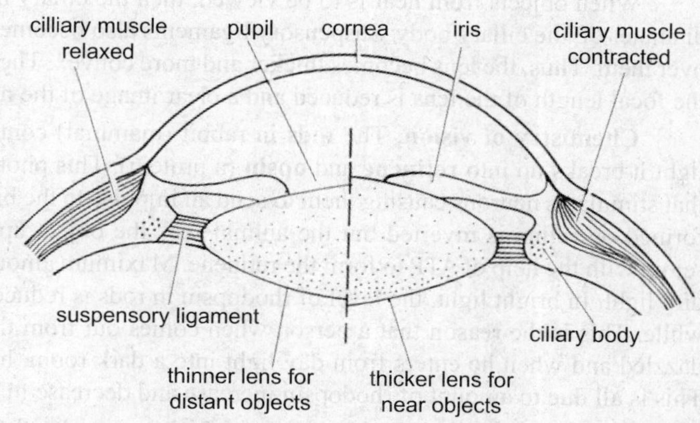

Fig. 29.62. Changes in curvature of lens during accommodation in mammalian eye.

between lens and cornea is the small **aqueous chamber** filled with a watery fluid, the **aqueous humour**. The posterior chamber lies between lens and retina is large, called **vitreous chamber**. The vitreous chamber is filled with a gelatinous fluid, the **vitreous humour**. The aqueous humour is secreted by the ciliary body and carried through **canal of Schlemm** at the base of cornea. It nourishes cornea and lens and maintains intraocular pressure. The disease **glaucoma** is resulted due to imbalance of intraocular pressure. It damages the retina.

Working of the Eye

Image formation. The cornea, aqueous humour, lens and vitreous body together constitute the dioptric apparatus, which focuses an image of external objects on the retina. The iris is a diaphragm by which amount of light which enters can be regulated. The eye works exactly like that of a photographic camera. The light rays reflected from an object are refracted by the cornea and the rays are then focussed by the lens on retina, so that a sharp, inverted image is formed. The inverted retinal image is interpreted by the brain and the real sensation of sight arises and the animal sees the object in an upright way. However, the inverted retinal image is never reinverted by the brain.

The type of vision which is found in higher mammals is called **binocular (stereoscopic) vision**. In such a vision the fields of vision of the two eyes overlap or even coincide each other and, hence, two images of the same object are not seen. Thus, both the eyes can see a single object in three dimensions, and the animal can estimate the distance also. This type of vision in which distance is also estimated is called **binocular vision**.

But in the lower mammals like rabbit, both the eyes have divergent axes of vision because the fields of vision do not overlap each other. Therefore, the rabbit can see all around by its two eyes, each eye covers a different field of vision. Such type of vision is called **monocular vision**.

Accommodation or focussing. The power of **accommodation** is well exhibited by the eyes of rabbit for the objects situated at various distances. In other words, it can be said that the power of accommodation is an adaptation of seeing the objects situated at various distances. The accommodation is brought about by changing the convexity of the lens because the distance between lens and retina is fixed and unchangeable. Thus, change in convexity of lens is essential for the formation of distinct images on the retina.

In **normal condition** when the eyes are at rest, the lens is kept flattened by the suspensory ligaments and, thus, it is adjusted for seeing distant objects. It is due to relaxed condition of the circular muscles of ciliary body. The diameter of the ciliary body is also much increased due to outward pressure of the fluid of eyeball. It, thus, increases the tension of the suspensory ligament, which pulls the lens capsule making the lens thinner or flattened.

When objects from near is to be viewed, then the ciliary muscles contract, thus, reducing the diameter of the ciliary body. Suspensory ligaments also become relaxed due to lessening of tension over them. Thus, the lens becomes thicker and more convex. The cornea also arches outwards. Thus, the focal length of the lens is reduced and a clear image of the near object is formed on the retina.

Chemistry of vision. The rods in rabbit (mammal) contain **rhodopsin (visual purple)**. In light it breaks up into **retinene** and **opsin** (a protein). This photochemical reaction releases energy that stimulates neurons causing them to send an impulse to the brain through optic nerve. The image formed on retina is inverted but the animal sees the object upright. Later the retinene and opsin rejoin with the help of ATP to form the retinene. Maximum amount of rhodopsin is present is rods in dim light. In bright light, the level of rhodopsin in rods is reduced not immediately but after a short while. This is the reason that a person when comes out from the dark into the day light, becomes dazzled and when he enters from day light into a dark room, he becomes like a blind for a while. This is all due to amount of rhodopsin increase and decrease in rods.

The cones contain iodopsin pigment which is stimulated only during bright light. In dim light one cannot see the colours, they are only distinguishable in bright light.

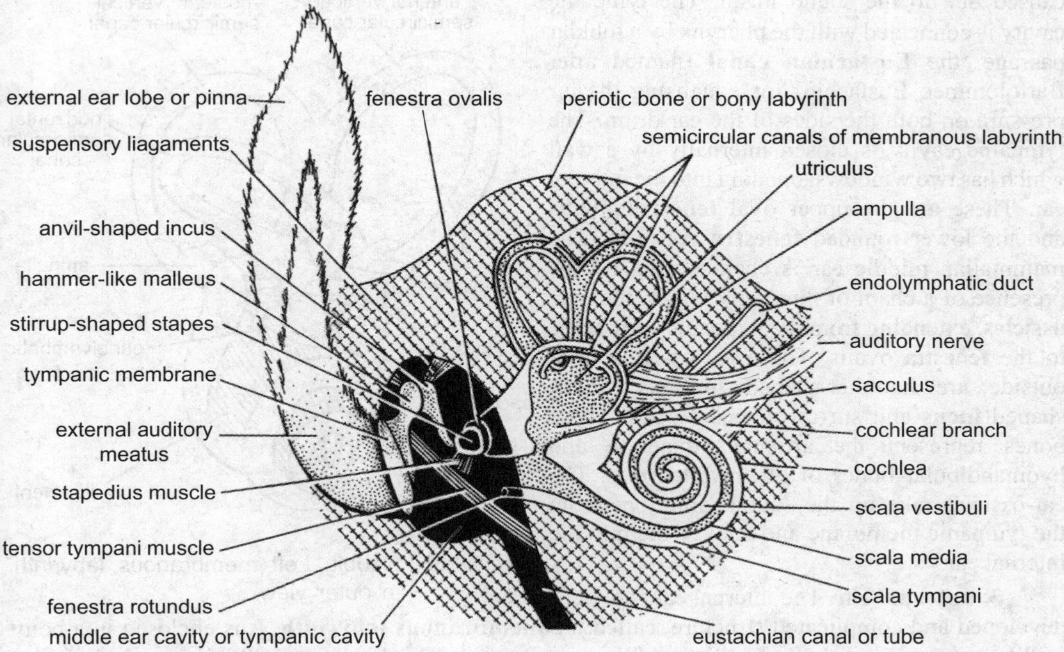

Fig. 29.63. Rabbit. Diagrammatic V.S. of head passing through ear region.

5. Organs of Hearing or Ears

The ears are **stato-acoustic** organs of rabbit, performing the function of **hearing** and **equilibrium** or the ears are sensitive to the frequencies of sound waves and to changes in relation to gravity.

The mammalian ear has three parts, an **external**, a **middle** and an **internal**. The internal ear transforms these sound vibrations into nerve impulses which are communicated to the brain through VIII cranial nerve.

1. External ear. Generally in all mammals a large **external ear** or **pinna** or **auricle** is found which is movable in most animals. It is a skin covered elastic cartilaginous projection from the lateral sides of head. The pinna is a wide-mouthed funnel usually capable of being moved or turned in different directions by its muscles. The opening of the funnel-shaped pinna leads into a tubular passage, called the **external auditory meatus**. It ends into a cone-shaped membrane, called the **tympanum** or **ear drum**. The walls of auditory meatus are covered by skin containing hairs, oil glands and wax glands. The hairs, oil and wax protect the ear drum from the harm caused due to dust and small insects. The odour of the wax inhibits the insects to enter the canal. The function of the external ear is to collect the sound vibrations and to detect the direction of **sound vibrations** and send them to middle ear.

2. Middle ear. The middle ear is the tympanic cavity which starts from the **tympanic membrane** or **ear drum**. It is an air-filled cavity enclosed in the tympanic bone. The ear drum is exposed to the impact of sound vibrations

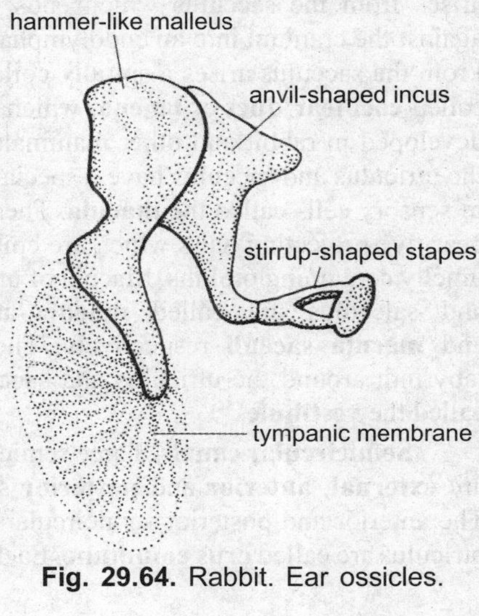

Fig. 29.64. Rabbit. Ear ossicles.

caused due to the sound in air. The tympanic cavity is connected with the pharynx by a tubular passage, the **Eustachian canal** (named after Bartolommeo Eustachio) for equalising the air pressure on both the sides of the ear drum. The tympanic cavity is closed internally by a wall which has two windows, opening into the internal ear. These are the upper oval **fenestra ovalis** and the lower rounded **fenestra rotunda**. The mammalian middle ear is characterised by the presence of a chain of three tiny bones, the **ear-ossicles**, extending from the tympanic membrane to the fenestra ovalis. These ear-ossicles from outside are hammer-shaped **malleus**, anvil-shaped **incus** and stirrup-shaped **stapes**. These bones represent the articular, quadrate and hyomandibular bones of other vertebrates. The ear-ossicles take up the sound vibrations from the tympanic membrane and convey them to the internal ear.

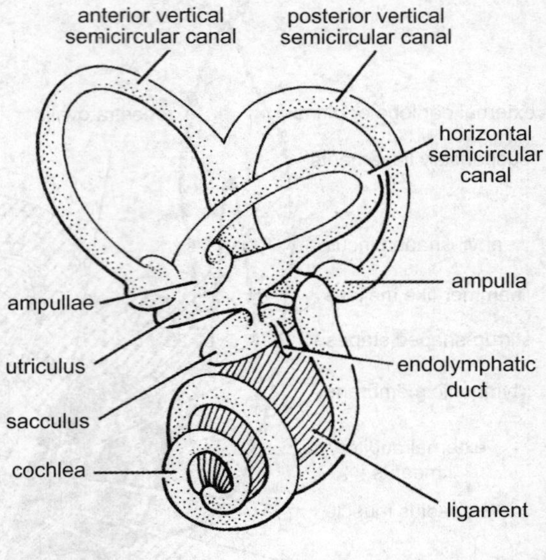

Fig. 29.65. Rabbit. Left membranous labyrinth in outer view.

3. Internal ear. The internal ear is a well-developed and complicated structure, called the **membranous labyrinth**. It is enclosed in a bony **auditory capsule** of the same shape as the membranous labyrinth, formed by the **periotic bone**. The space between the bony capsule and the membranous labyrinth is filled with a liquid, called **perilymph**. A similar fluid is found within the membranous labyrinth, called the **endolymph** having tiny calcareous particles, the **otoliths**. The membranous labyrinth is formed of a larger dorsal **utriculus** and a smaller ventral **sacculus** forming the body proper, and semicircular canals and cochlea.

(*i*) **Utriculus and sacculus.** The utriculus and sacculus are connected together by a small duct, called **sacculo-utricular canal**. A small, narrow endolymphatic duct arises from the sacculus which ends blindly against the cranium into an endolymphatic sac. From the sacculus arises a spirally coiled tube called **cochlear duct** or **lagena**, which is well developed in rabbit and other mammals. Both the utriculus and sacculus have a special group of sensory cells called the **macula**. These cells bear five projecting hairs which are embedded in jelly containing otoliths. Macula of utriculus and salcculus are called **macula utriculi** and **macula sacculi** respectively. The bony labyrinth around the utriculus and sacculus is called the **vestibule**.

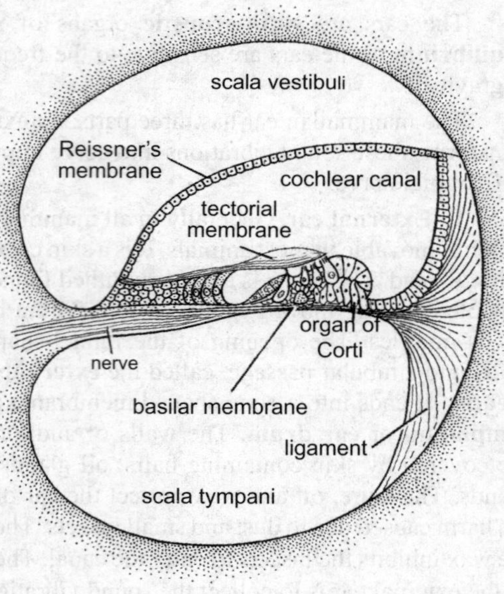

Fig. 29.66. Rabbit. T.S of cochlea.

Semicircular canals. Three semicircular canals connect at both the ends with utriculus. These are **external**, **anterior** and **posterior** semicircular canals, situated at right angles to each other. The anterior and posterior semicircular canals arise from a common canal, after their origin from utriculus are called **crus commune**. Each semicircular canal has a swollen **ampulla** at its lower end.

Each ampulla has a sensory area known **crista ampullaris** which is formed by sensory **hair cells** and **supporting cells**. The hairs are embedded in a jelly cone having no otoliths.

The VIII cranial or auditory nerve supporting to ear divides into two branches, **vestibular branch** supplying to the semicircular canals, utriculus and sacculus and a **cochlear branch** supplying to the cochlea.

(*iii*) **Cochlea.** The spirally coiled watch spring-like **cochlear duct** or **lagena** arising from sacculus is enclosed within the similarly spirally coiled cochlear canal of the periotic bone, which together constitute the **cochlea**. It is the organ of hearing. Thus, the cochlea is a bony tube lined with connective tissue. In cross section the cochlea shows three chambers or canals : Middle chamber or **scala media** is the cochlear duct arising from the sacculus and is filled with endolymph. It terminates blindly at the apex of spiral or cochlea. The upper and lower canals of the cochlea are **scala vestibuli** and **scala tympani** respectively. These are parts of bony labyrinth or cochlear canal and are filled with perilymph. Both the canals are longitudinally separated from each other by a **spiral lamina** but communicate with each other at the tip of spiral by a small opening, called the **helicotrema**. The scala vestibuli and scala tympani communicate basally with the fenestra ovalis and fenestra rotunda respectively.

The epithelial wall of the scala media rests above on the **Reissner's membrane** and below on the **basilar membrane**. Basillar membrane is formed of tightly stretched transverse connective tissue fibres. It contains a series of sensory receptor **hair cells** and long columnar supporting cells. Hair cells are basally connected with the nerve fibres of cochlear nerve of auditory nerve. Over the hair cells touching the hairs is found a thin, gelatinous, ribbon-shaped sheet of connective tissue, called the **tectorial membrane**. Both of these are collectively called the **organ of Corti**.

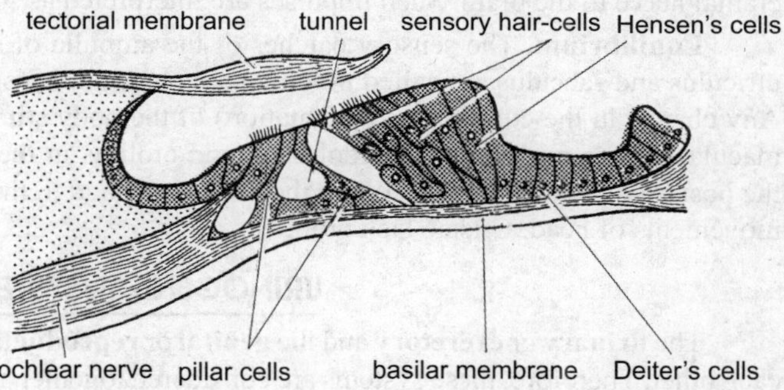

Fig. 29.67. Rabbit. Organ of Corti.

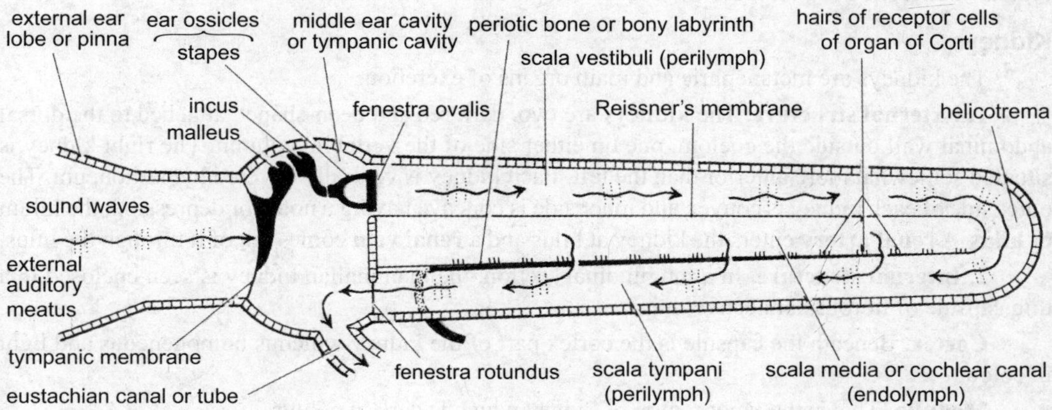

Fig. 29.68. Rabbit. Diagrammatic representation of the mechanism of hearing in the mammalian ear. Arrows indicate pathway and vibrations.

Working of ear. Ear performs two functions : hearing and equilibrium. The cochlear duct of sacculus of the membranous labyrinth is responsible for **hearing**, while maculi of sacculus, utriculus and cristae of semicircular canals help in **equilibrium**.

(*i*) **Hearing.** The sound waves are collected by the movable pinna which travel through the external auditory meatus and cause the ear drum to vibrate. The vibrations are then transmitted through the ear-ossicles of the middle ear and fenestra ovalis into the perilymph of internal ear. The vibrations of the membrane of fenestra ovalis cause alternate increase and decrease in the pressure of the perilymph of scala vestibuli which is transmitted to the scala tympani through helicotrema and escape through the fenestra rotunda back into the middle ear. The membrane of fenestra rotunda due to the pressure bulges out into the middle ear. Thus, it acts as a **pressure-relief valve**.

The vibrations of the perilymph of scala tympani and scala vestibuli cause endolymph of the scala media and basilar membrane to vibrate. These vibrations cause the tectorial membrane floating in the endolymph of scala media to brush the sensory hairs of organ of Corti. Finally the stimulated hair cells of the organ of Corti sends a message through nerve impulses which are carried by the VIII cranial nerve to the brain. Such impulses are interpreted as sound by the brain.

Equilibrium. The sensory patches of the ampulla of semicircular canals called **cristae** and utriculus and sacculus are called **maculae** are responsible for maintaining equilibrium of the body. Any change in the equilibrium (orientation) of the body stimulates the hair cells of the cristae and maculae due to movement in endolymph and otoliths in them. Maculae respond to the change in the posture of head and body. While the cristae respond to the changes in the direction or rotational movements of head. Cristae lack otoliths.

URINOGENITAL SYSTEM

The **urinary** or **excretory** and the **genital** or **reproductive systems** are closely associated with each other. Therefore, these systems are considered together and collectively known as **urinogenital system** because they are somewhat morphologically inerrelated. Certain urinogenital ducts in male also discharge the gametes. But functionally, both the systems are separate.

EXCRETORY SYSTEM

This system is generally responsible for the elimination of metabolic waste products (nitrogenous waste products in the form of urea, etc.), excess of salts and water from the body. The organs of excretion in rabbit are a pair of **kidneys, ureters** and a **urinary bladder**.

Kidneys

The kidneys are metanephric and main organs of excretion.

1. External structure. The **kidneys** are two, dark red and bean-shaped attached to the dorsal abdominal wall outside the coelom, one on either side of the vertebral column. The right kidney is situated somewhat more anterior than the left. Each kidney is ventrally covered by peritoneum. The outer side of each kidney is convex and inner side is concave having a notch or depression, the hilum or **hilus**. A **renal artery** enters the kidney at hilus and a **renal vein** comes out of it through the hilus.

2. Internal structure. In a longitudinal section, the mammalian kidney is seen enclosed in a thin capsule of fibrous connective tissue.

Cortex. Beneath the capsule is the cortex part of the kidney which is homogeneous and light in colour.

Medulla. This is the inner region of kidney which is dark in colour.

Pelvis. It is a large, funnel-shaped space in the centre of kidney towards the hilus. The urine is collected here and later drained off by the ureter.

Pyramids. In rabbit and man, the medulla is formed of a number of lobes, called **pyramids**, and the cortex is continued inside between the pyramids to form **renal columns of Bertini**. In rat the conical pyramid is single. The pyramids are projected into the pelvis. The kidney is formed of a mass of fine, long, convoluted tubules, called the **uriniferous tubules** or **nephrons** embedded in connective tissue having blood and lymph vessels, nerves and smooth muscle fibres.

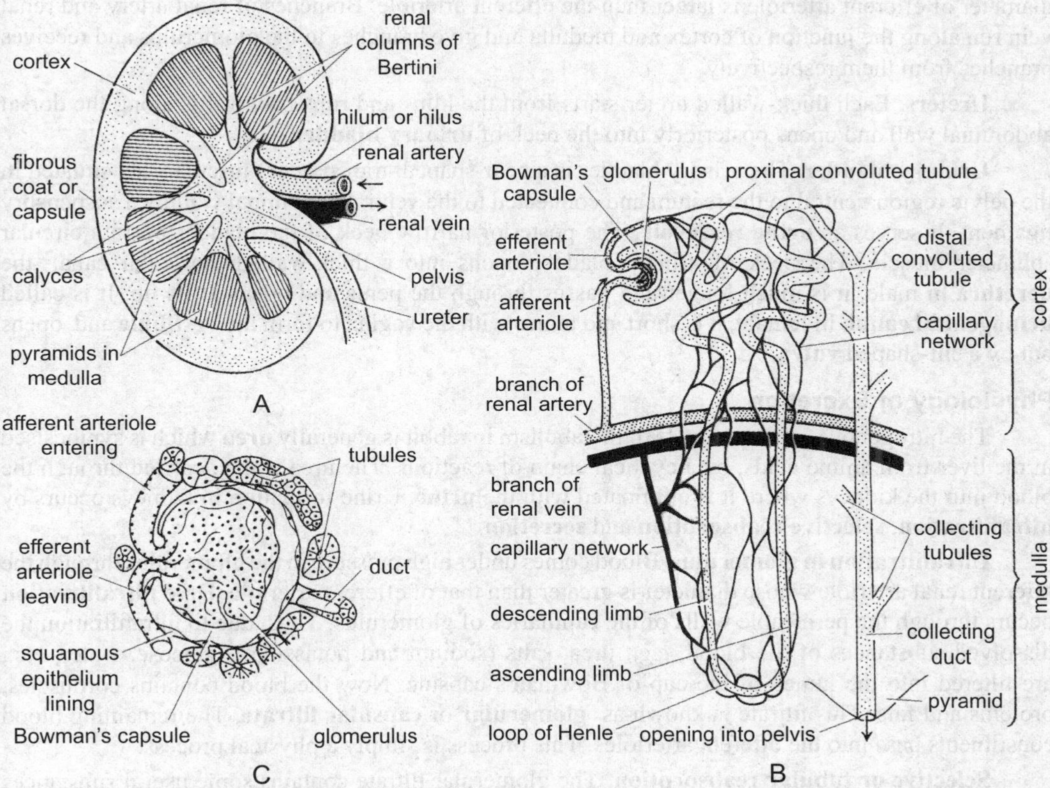

Fig. 29·69. Rabbit. Excretory organs. A–Structure of a kidney in sagittal section; B–Structure and blood supply of a single nephron or uriniferous tubule; C–A Malpighian capsule magnified.

Each uriniferous tubule is formed of a **Malpighian capsule** and **convoluted duct**.

The Malpighian capsule consists of a proximal cup-shaped **Bowman's capsule** in the cortex and in its lumen is present a tuft of blood capillaries, called **glomerulus**. The Bowman's capsule enclosing the glomerulus is collectively known as **Malpighian capsule** or **renal corpuscle**.

The convoluted tubule behind the Malpighian capsule is divisible into three regions : (*a*) proximal convoluted tubule, (*b*) Henle's loop and (*c*) distal convoluted tubule.

The **proximal convoluted tubule** starts from the Bowman's capsule and makes a few coils in the cortex and then proceeds downwards as **descending limb** in the medulla to form a thin **loop of Henle**. It again proceeds towards the cortex as **ascending limb** and again extends towards the cortex as asending limb and forms a few coils in the cortex, called the **distal convoluted tubule**. The distal convoluted tubule joins with one of the larger collecting ducts. The collecting ducts collectively form a pyramid and finally open into the pelvis. The wall of the tubule is made of cuboidal epithelial cells which are ciliated at interval, whereas the wall of Bowman's capsule is formed of single layer of squamous epithelium.

Blood supply of the kidney. The renal artery after entering into the kidney divides and redivides forming several **arterioles**. Each arteriole enters a Bowman's capsule as an **afferent**

arteriole which capillarises to form the **glomerulus** and then an **efferent arteriole** emerges from the glomerulus and again capillarises around the convoluted tubule to distribute the blood in the remaining part of the tubule. These capillaries unite together to form a **venule** and a number of such venules unite together to form a **renal vein** which comes out of the kidney from the hilus. The diameter of efferent arteriole is larger than the efferent arteriole. Branches of renal artery and renal vein run along the junction of cortex and medulla and give branches to the glomerulus and receives branches from them respectively.

Ureters. Each thick-walled ureter starts from the hilus and runs backwards along the dorsal abdominal wall and opens posteriorly into the neck of **urinary bladder**.

Urinary bladder. The urinary bladder is a pear-shaped, transparent muscular sac, situated in the pelvic region ventral to the rectum and connected to the ventral abdominal wall by a suspensory ligament. It serves as **urine reservoir**. The posterior narrow neck of the bladder bears a circular sphincter muscle. The neck of urinary bladder opens into a thick-walled, muscular canal, the **uerethra**.In male, it is much longer and passes through the penis and opens at its tip. It is called **urinogenital canal**. In female, it is short and unites with the vagina to form the **vestibule** and opens out by a slit-shaped **vulva**.

Physiology of Excretion

The nitrogenous waste products of metabolism in rabbit is generally **urea** which is synthesised in the liver from amino acids, by a cyclical chain of reactions. The urea is then carried through the blood into the kidneys where it is eliminated with the **urine**. Urine formation in kidneys occurs by **ultrafiltration**, **selective reabsorption** and **secretion**.

Ultrafiltration in glomerulus. Blood comes under high pressure in the glomerulus through the afferent renal arteriole whose diameter is greater than that of efferent arteriole. Here **ultrafiltration** occurs through the permeable walls of the capillaries of glomerulus. Thus, due to ultrafiltration the dissolved substances of the blood, *e.g.,* urea, salts (sodium and potassium), glucose, water, etc., are filtered into the lumen of the cup of Bowman's capsule. Now the blood contains corpuscles, proteins and fats. This filtrate is known as **glomerular** or **capsular filtrate**. The remaining blood constituents pass into the efferent arterioles. This process is simply a physical process.

Selective or tubular reabsorption. The glomerular filtrate contains some useful substances like water, glucose, amino acids and some salts. These useful substances are **reabsorbed** in the proximal convoluted tubules from the glomerulur filtrate. The absorbed substances then pass into the blood circulation through capillaries of the efferent arterioles found around the convoluted tubule which finally joins to form renal venules. The maximum water is reabsorbed in the loop of Henle and no reabsorption occurs in the distal convoluted tubule. Glucose, amino acids and some urea are reabsorbed in the proximal convoluted tubule, while chloride and bicarbonate of sodium are reabsorbed in the proximal and distal convoluted tubules. The water and urea are reabsorbed by passive diffusion. Amino acids, salts and sugars, etc., are reabsorbed by active transport.

Thus, as the glomerular filtrate passes through the convoluted tubule, the useful substances are reabsorbed and the concentration of waste products including urea increases.

Tubular secretion. In this process some substances like remnant urea, creatine, creatinine, ammonia, hydrogen, potassium ions, various drugs, etc., are secreted into the lumen of tubule from the efferent capillaries surrounding the tubules.

After tubular secretion the substance which is left in the tubule is rich in excretory wastes and finally known as **urine** which contains nearly 95% water, 2% urea, 6% Ca and Na chlorides, some uric acid, creatinine, ammonia, etc. It also contains a pigment, **urochrome**, formed by the breakdown of haemoglobin. It gives yellow colour to the urine.

The urine, thus, formed passes into the collecting tubule and finally reaches the pelvis and ureter from where it is collected in the urinary bladder. The bladder contracts and forces the urine outside through the urinogenital canal and urinogenital aperture.

Besides excretion, kidney regulates the body fluids (blood and water) within the body, thus, also acting as **homeostatic organ**. Homeostasis is the method for regulating concentrations and contents within the body to preserve equilibrium in any animal.

Other organs of excretion. Besides kidney, some other organs also perform excretory functions. The CO_2 evolved during tissue respiration is removed by the lungs. Some amount of water is removed in the form of vapours with the expired air and some salts are also removed with the sweat. Bile salts and bile pigments are removed by the alimentary canal which are produced in the liver.

REPRODUCTIVE SYSTEM

The **sexes** are separate and **sexual dimorphism** is well marked in rabbit.

Male Reproductive System

Male reproductive system (Fig. 29.70) consists of a pair of **testes**, a pair of **vasa deferentia**, **uterus masculinus** or **seminal vesicle**, **urethra**, **penis** and a number of **accessory glands** like **prostate**, **Cowpers**, **perineal**, and **rectal glands**.

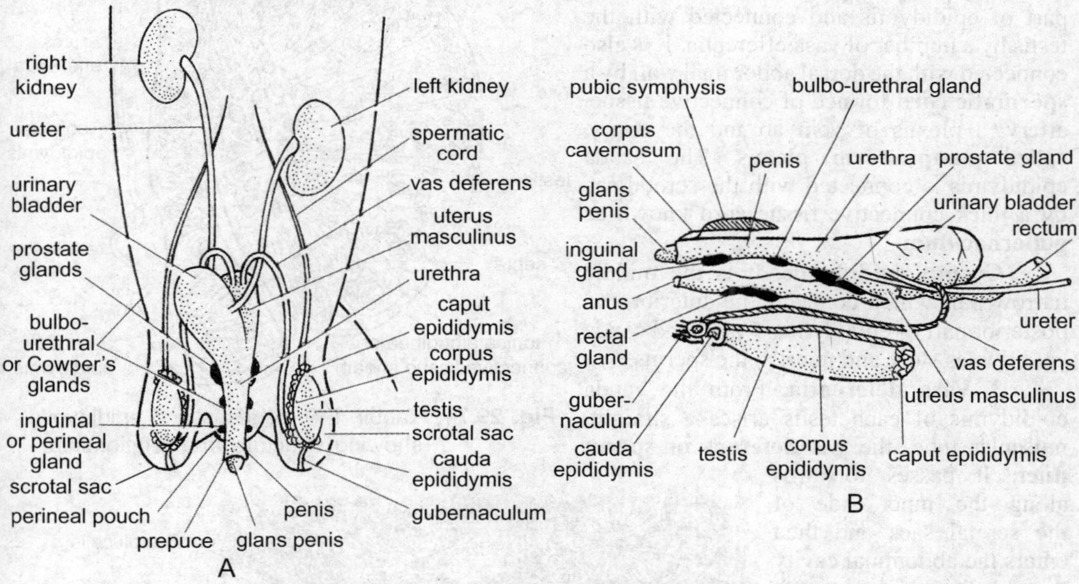

Fig. 29.70. Rabbit. Male urinogenital system. A–Ventral dissection; B–In side view.

1. Testes. The testes are paired, light pink-coloured, ovoid bodies. These lie in the foetus in the abdominal cavity near the kidneys. But at puberty these descend outside from the body cavity through **inguinal canals** into a special part of the coelom, the **scrotal sacs**. The scrotal sacs are covered externally by a thin hairy skin, situated ventrally in the pubic region on either side of the penis. In most mammals, testes remain within scrotal sacs throughout life. But in rabbit, rat and other rodents, they remain within scrotum only during breeding season. In non-breeding season they are withdrawn into the abdominal cavity through the inguinal canals, thus, these canals remain open throughout life in these animals. It is due to temperature difference, abdominal cavity temperature is higher than that of scrotal sacs. Development of sperms need a slight lower temperature, so they descend in scrotum during breeding season. The cavity of the scrotal sacs is connected with the abdominal cavity by a narrow passage, known as **inguinal canal**.

Histologically, each testis is covered by a tough coat of fibrous connective tissue called **tunica albuginea**, which is projected into the substance of testis in the form of a number of septa forming a number of lobules within the kidney. Each lobule is occupied by long convoluted **seminiferous tubules** which are bound together by connective tissue. The seminiferous tubules are lined by **germinal epithelial cells** which are of two types : smaller, more numerous are the **spermatogonial cells** which produce **spermatozoa** after **spermatogenesis**, and others are a few large-sized columnar supporting cells, called **Sertoli cells**. These nourish the spermatozoa before they leave the tubule. The connective tissue having **interstitial cells** or **cells of Leyding** are found in between the seminiferous tubules. These cells are endocrine in function and secrete hormones, which control the secondary sexual characters.

In each testis, all the seminiferous tubules open into a networks, called the **rete testis**. It communicates by many fine ductules, the **vasa efferentia** into the epididymis. The vasa efferentia are lined internally by ciliated epithelium.

2. Epididymis. It is a long, narrow and highly convoluted tubule forming a compact protrusion all along the inner surface of the testis. The epididymis is divisible into three regions : anterior **caput epididymis** or **globus major**, middle narrow **corpus epididymis** and posterior **cauda epididymis** or **globus minor**.

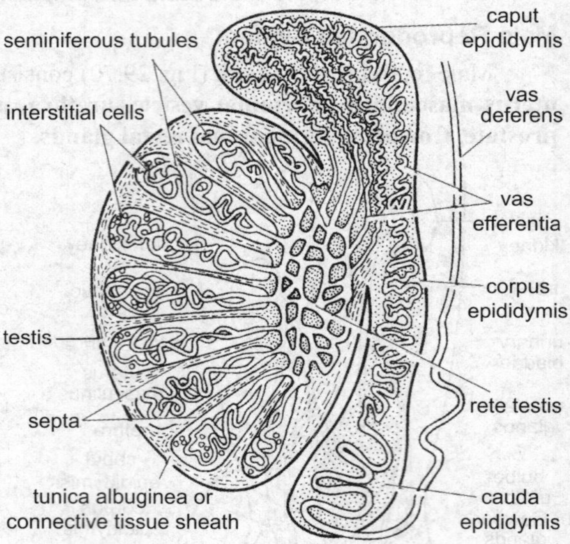

Fig. 29.71. Rabbit. L.S. of testis and epididymis showing structure and relationship.

Caput epididymis. It is the anterior part of epididymis and connected with the testis by a number of vasa efferentia. It is also connected with the dorsal abdominal wall by a **spermatic cord** formed of connective tissue, artery, a plexus of vein around the artery, called **pampiniform plexus**. The cauda epididymis is connected with the scrotal sac by a thick connective tissue cord known as **gubernaculum**.

Corpus epididymis. It is the middle narrow part which connects the anterior and posterior part of epididymis. Caput and cauda epididymes store and nourish the sperms.

3. Vasa deferentia. From the cauda epididymis of each testis arises a straight, muscular tube, the **vas deferens** or **sperm duct**. It passes forwards along the inner side of the scrotal sacs and then enters the abdominal cavity through the inguinal canal. Then it curves over the ureter and passes backwards to open in a small median bag, called **uterus masculinus** or **seminal vesicle** situated above the neck of the bladder. The seminal vesicle opens dorsally into urethra. Urethra is the distal narrow part of the urinary bladder. The urethra, thus, serves as a passage for **urine** and the **spermatic fluid**. It opens at the tip of penis.

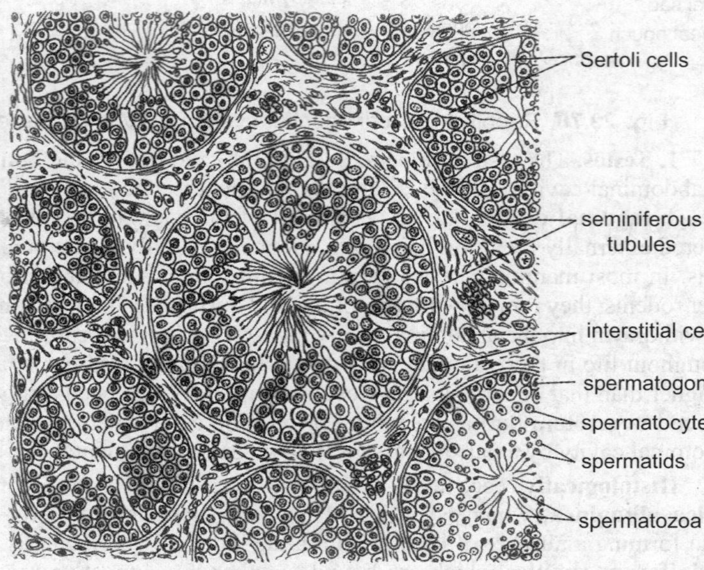

Fig. 29.72. Rabbit. T.S. of testis.

4. Urethra. The neck of urinary bladder

and the seminal vesicle open into a thick-walled muscular duct, called the **urethra** or **urinogenital canal**.

5. Penis. The penis is a muscular, cylindrical, and vascular erectile organ, which hangs in front of the anus. The penis is covered by a loose sheath of skin which hangs freely over and beyond the tip of the penis as a **prepuce**. The cap-like tip of the penis covered by prepuce is called **glans penis**. The penis consists of three spongy, vascular longitudinal columns around urethra, called **corpora spongiosum** and above it two **corpora cavernosa**. These spongy muscles are surrounded individually by tunica albuginea, a fibrous connective tissue layer. Around these three columns is present elastic areolar tissue layer containing blood vessels and nerves. The blood spaces of the penis get filled with the blood at the time of sexual excitement to make the penis stiff and enlarged.

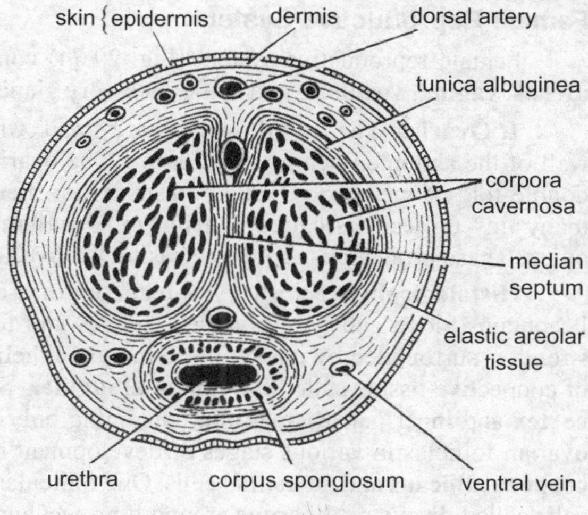

Fig. 29.73. Rabbit. T.S. of penis.

6. Accessory glands. A number of accessory sex glands open into the urethra. Their secretions, along with those of epididymis, seminal vesicles form the **seminal fluid** (semen). These sex glands are as follows :

(*i*) **Prostate gland.** A large prostate gland is situated dorsally at the base of uterus masculinus which opens into the urethra by numerous ducts. Its secretion is white and alkaline, which makes the passive sperm active.

(*ii*) **Cowper's glands or bulbo-urethral glands**. A pair of Cowper's glands are situated just anterior to the prostate gland on urethra. Their secretion neutralises the acidity and, thus, protects the sperms.

(*iii*) **Perineal glands.** A pair of elognated scent glands lie anterior to the Cowper's glands. They open into the hairless perineal depressions one on either side of anus. Their secretion gives a characteristic odour of rabbit.

(*iv*) **Rectal glands.** A pair of rectal glands lie on the dorsal side of rectum. Their function is not known. These two last glands are not associated with the male reproductive organs.

The secretions of prostate and Cowper's glands activate and nourish the sperms, provide a fluid media for the flow of sperms and neutralise the acidity of the urethral passage.

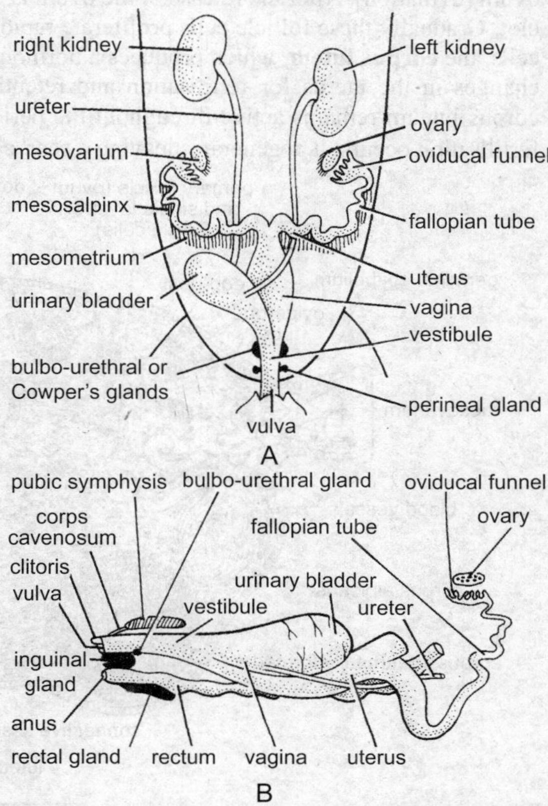

Fig. 29.74. Rabbit. Female urinogential system. A – Ventral dissection; B – In side view.

Female Reproductive System

Female reproductive system (Fig. 29.74) consists of a pair of **ovaries**, a pair of **oviducts**, **uterus**, **vagina**, **vestibule** and some accessory glands.

1. Ovaries. The paired **ovaries** are small, whitish, oval bodies found attached to the dorsal wall of the abdomen behind the kidneys. The ovaries are attached with the dorsal body wall by a double fold of mesentries, called **mesovarium**. Each ovary when ripe shows on its outer surface many tiny projections called **ovarian** or **Graafian follicles**. Each follicle contains a developing ovum. These projections rupture releasing the **ova**.

Histologically, each ovary is made up of a dense fibrous connective tissue called **stroma**. It contains blood and lymph vessels, nerves and follicles in various stages of development. The stroma is surrounded by a layer of germinal epithelial cells and internal to it is present a thin layer of connective tissue, called the **tunica albuginea**. Stroma is further differentiated into outer broad **cortex** and inner narrow **medulla** containing only blood and lymph vessels. The cortex contains ovarian follicles in various stages of development and regressing follicles like **corpora lutea** and **corpora albicans** and interstitial cells. One follicular cell develops to become an **ovum** and the other cells called **discus proligerous** around it protect and nourish the ovum. It is attached to one side of the follicle. In a mature Graafian follicle, a fluid-filled cavity appears which separates an outer layer of cells, the **membrana granulosa**, from the cells of discus proligerous. The follicular cells around the ovum secrete a thick membrane called **zona pellucida**. It is covered by another membrane of striated columnar cells, called **corona radiata**. Such a mature follicle is known as a **Graafian follicle**. The Graafian follicle migrates towards the surface of the ovary and ruptures to release the ovum (ovulation). After the release of the ovum is left a mass of follicular cells surrounding a blood clot. Gradually these follicle cells proliferate rapidly and transform into a yellow coloured mass of cells, the **corpus lutem**, which produces a hormone, progesterone. It is responsible for preparatory changes in the uterus for fertilisation and retention of the embryo during **gestation**. Thus, the corpus luteum remains active throughout the period of embryonic development. If, somehow, no fertilisation occurs, it degenerates leaving a scar, called **corpus albicans**.

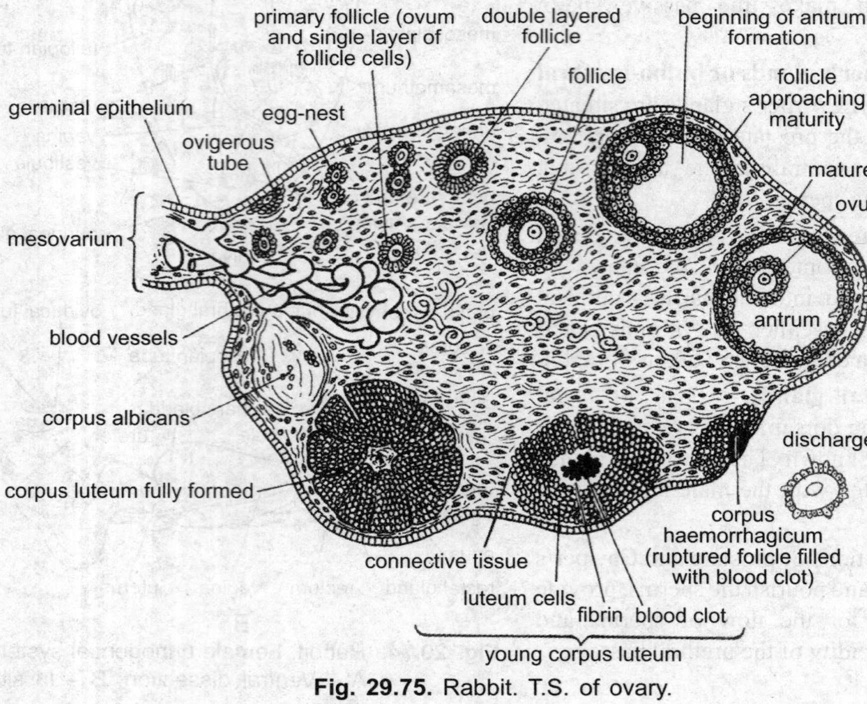

Fig. 29.75. Rabbit. T.S. of ovary.

2. Oviducts. There is a **fimbriated** (lined by cilia) **funnel** or **ostium** close to the outer side of each ovary which leads into the **oviduct**. The fimbriated funnel receives the ovum after their discharge from the ovary. Each oviduct behind the funnel is differentiated into two parts : first part, the **follopian tube**, is coiled, narrow, internally ciliated and glandular and the second part is long, thick, muscular, glandular and coiled with wider diameter, called **uterus**. Both the uteri are attached with the dorsal abdominal wall by a mesentery. Fertilised ova are implanted on the uterine wall for the development. The developing embryo remains attached with the uterine wall by a placenta.

3. Vagina and vestibule. The uteri of both the sides join together along the median line to form a common wide passage, called the **vagina**. It lies over the urinary bladder. The vagina passes backwards and joins with the neck of the urinary bladder, forming a common passage, called **urinogenital canal** or **vestibule**. It extends backwards ventral to rectum and opens outside by a slit-like aperture, the **vulva**.

4. Clitoris. It is small, erectile, and knob-like, and attached to the anterior wall of vulva. It is homologous to the **penis** of male because it consists of a pair of erectile tissue (corpora cavernosa) which is highly sensitive.

5. Accessory glands. A pair of small **Bartholin's glands** are found embedded in the dorsal wall of vestibule. The secretion of these glands is discharged in the vestibule which lubricates the vaginal passage. They are similar to the Cowper's glands of the males. The **prostate glands** are not found in female rabbit. A pair of **perineal glands** and **rectal glands** are also found like those of the male rabbit.

Copulation. The domestic rabbits can copulate at any time but wild rabbits copulate from January to June. Usually a female in most mammals copulates only when she is in **oestrus** condition or **on heat**, of which rabbit may be an exception. At the time of copulation the penis of the male is introduced in the vagina of the female where semen is ejaculated.

Fertilisation and Development. After about 8–10 hours of copulation, the Graafian follicles mature and **ovulation** occurs. The ova discharged from the Graafian follicle are immediately caught by the ostium and then passes down the fallopian tubule where fertilisation occurs. The ova undergo first maturation division before their discharge from the ovary. The second maturation division occurs at the time of fertilisation after the entry of sperm in the ovum. At the time of fertilisation the entire sperm enters the ovum but its tail degenerates. The penetration of sperm into ovum is helped by an enzyme **hyaluronidase** secreted by the acrosome of the sperm. Usually only one sperm enters the ovum. After the entry of sperm in the ovum, the head of sperm rotates through 180 degrees and finally the male and female pronuclei fuse together resulting into the formation of **zygote nucleus**. The zygote then passes down the fallopian tube and reaches the uterus where it gets attached to the wall of the uterus. After next few days a special connection **placenta** is formed between the developing embryo and uterine wall. The placenta helps the embryo in all the physiological exchange of materials with the mother until birth. The **gestation** period in rabbit is about 28–32 days, after which mother gives birth to the young ones.

ENDOCRINE SYSTEM

Generally two types of glands are found in the body of all the vertebrates–the **exocrine** and **endocrine**.

1. Exocrine glands. The exocrine glands are those which liberate their secretions through the **ducts** in an organ, such as **salivary glands**, **pancreas**, **liver**, etc.

2. Endocrine glands. The endocrine glands are those which liberate their secretions directly into the blood and are without any duct. Therefore, these glands are also known as **ductless glands**. The secretions of these glands are known as **hormones** which assist the nervous system in coordinating or integrating various body activities. Their secretions are of **vital importance** for various metabolic processes and physiological activities to continue in the body.

Hormones. The term hormone was given by **E.H. Starling (1905)**, a British physiologist. The hormones are the chemical agents or messengers secreted by the ductless glands and are carried by the blood to the different parts of the body called target organ where they perform definite functions. Many endocrine glands cannot secrete their hormones unless they are stimulated to do so by other hormones secreted by other glands. The amount of the various hormones released in blood are maintained at a proper level. Hormones are chemical messengers and bring about chemical coordination of different parts of the body. It is opposite to nervous coordination which is done by the nervous system. The action of endocrine system is slow and of long duration, whereas the action of nervous system is quite fast. The hormones produced by the endocrine glands are classified on the basis of their activities, such as :

(*a*) **Metabolic hormones** are those which stimulate the metabolic activities in the body, *e.g.*, **insulin** stimulate carbohydrate and fat metabolism.

(*b*) **Regulatory hormones** are those which control and regulate the rate of secretion of other endocrine glands, *e.g.,* the hormones of **pituitary gland**.

(*c*) **Morphogenetic hormones** are those which affect the rate of development of the parts of the body, *e.g.,* the hormones of the **thyroid gland**.

The position of endocrine glands in the body and their secretions are roughly the same in all vertebrates. The various **endocrine glands** in rabbit are as follows :

1. **Pituitary** or **hypophysis**
2. **Thyroid**
3. **Parathyroid**
4. **Adrenal**
5. **Islets of Langerhans** of pancreas
6. **Gonads**
7. **Placenta** during pregnancy
8. **Gastro-intestinal tract**.

In addition to these, the **thymus** and **pineal body** were previously considered as endocrine, but now they are not considered as endocrine glands.

1. Pituitary gland. It is a small nut-like gland situated on the ventral side of the diencephalon of the brain, attached with the hyopthalamus by infundibulum. The hormones secreted by this gland regulate the functions of many other endocrine glands and also play a vital role in the chemical coordination of the body. Therefore, this gland is also known as **master gland** or **bandmaster of endocrine orchestra**. Mammalian pituitary consists of three lobes of different embryological origin. The **anterior** and **intermediate lobes** arise from the roof of the mouth and together known as **adenohypophysis**. The **posterior lobe** develops from hypothlamus of diencephalon and is called **neurohypophysis.**

The **adenohypophysis** secretes many chemically distinct hormones which are named by adding the suffix **"trophic"** or **"tropic"** or **'in'** to the name of the part of the body affected by them, such as growth hormone (GH) is called **somatotrophic hormone** (STH) or **somatotropin**.

(*i*) **Somatotrophic hormone (STH) or Growth hormone (GH)**. It controls the protein metabolism and growth of the skeleton. An excess secretion of this hormone results

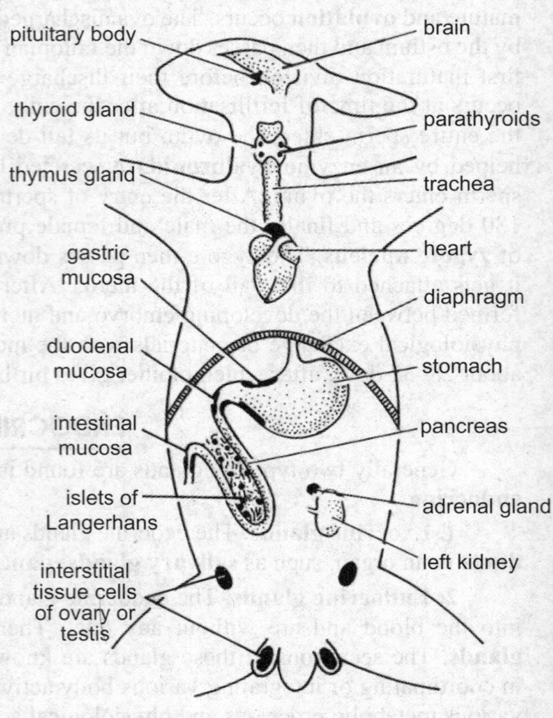

Fig. 29.76. Rabbit. Position of main endocrine glands.

in **gigantism** (production of giants with abnormal heights). Its deficiency results in **dwarfism**. Sometimes this gland does not begin over-secretion till maturity, then instead of giants, gorilla-like man with huge head, hands, feet and protruding jaws are the result, which is known as **acromegaly**.

***(ii)* Thyroid stimulating hormone (TSH)**. This hormone stimulates the thyroid gland to secrete **thyroxine**.

***(iii)* Adreno-corticotropic hormone (ACTH)**. It controls and stimulates the **adrenal cortex** to secrete its hormone.

***(iv)* Gonadotropic hormone (GTH).** Three types of gonadotropic hormones are known, *e.g.,* **follicle stimulting hormone (FSH), luteinizing hormone** (LH) and **luterotropic hormone** or **prolactin** (LTH). **FSH** hormone affects the development of gonads and formation of gametes. **LH** stimulates the production of sex hormones from the gonads. **LTH** stimulates the mammary glands to secrete the milk after the birth of the young ones.

***(v)* Melanocyte stimulating hormone (MSH) or chromatotropic hormone (CTH) or intermedin**. This hormone stimulates the distribution and concentration of the **melanin** pigment in the chromatophores of the skin.

***(vi)* Diabetogenic hormone or pancreotropin hormone.** This hormone affects the metabolism of carbohydrates. It converts reserved glycogen into glucose, *i.e.,* reverse to insulin.

The neurohypophysis or posterior lobe of pituitary secretes only two important hormones : **(*a*) Oxytocin.** It stimulates the contraction of some smooth muscles of the uterus at the time of childbirth and also stimulates the secretion of milk from the mammary glands. **(*b*) Vasopressin** or **antidiuretic hormone**. This hormone regulates the reabsorption of water by the renal tubules. It also causes the construction of small arteries of the body which causes the general increase in the arterial blood pressure. Its deficiency causes a disease, called **diabetes insipidus**, which is an abnormality characterised by exeretion of large amount of urine and, thus, the animal feels great thirst due to which it drinks more and more water.

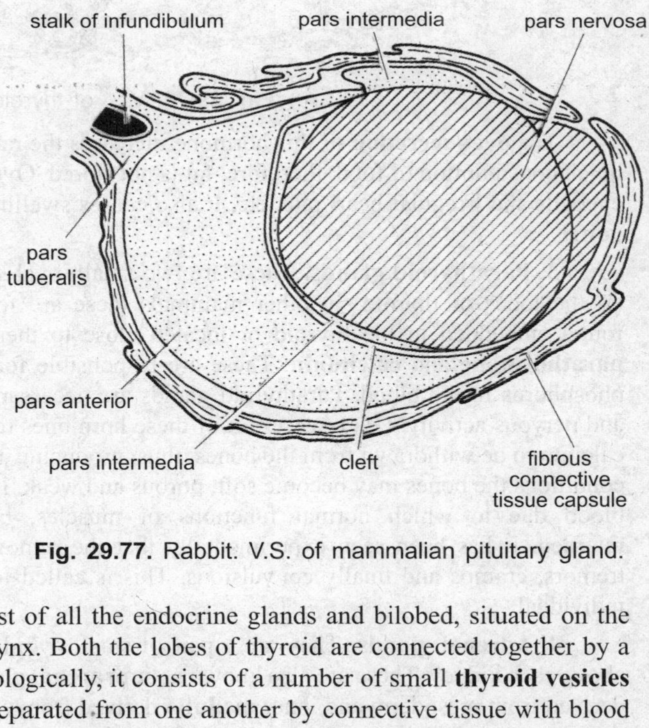

Fig. 29.77. Rabbit. V.S. of mammalian pituitary gland.

2. Thyroid gland. It is the largest of all the endocrine glands and bilobed, situated on the either side of the trachea below the larynx. Both the lobes of thyroid are connected together by a connective tissue, called **isthmus**. Histologically, it consists of a number of small **thyroid vesicles** or **follicles** in the form of closed sacs separated from one another by connective tissue with blood and lymph vessels. Each follicle is lined by cuboidal epithelial cells. These cells are richly provided with active secretory agents which secrete yellow colloidal secretion into the cavities of follicles. This colloidal secretion is a globular protein, known as **thyroglobulin**, which is also known as **iodothyronines** or thyroid hormones. It consists of nearly four most important hormones of which **thyroxine** is one.

Thyroxine regulates metabolism, *i.e.,* heat production and energy liberation in all the organ systems and also controls the rate of growth. The **under-secretion** of this hormone during childhood lowers the rate of metabolism and causes the animal to become feeble minded, growth is retarded, heart rate is also lowered and sexual development is retarded. This condition is known as **cretinism**. The **deficiency** of thyroxine in adults stimulates the formation of new gland tissue resulting in an enlargement of the thyroid called **goitre**. The goitre is actually caused by the deficiency of iodine in the diet. **Hypothyroidism** causes **myxoedema** in which person becomes fatty and sluggish due to poor rate of oxidation.

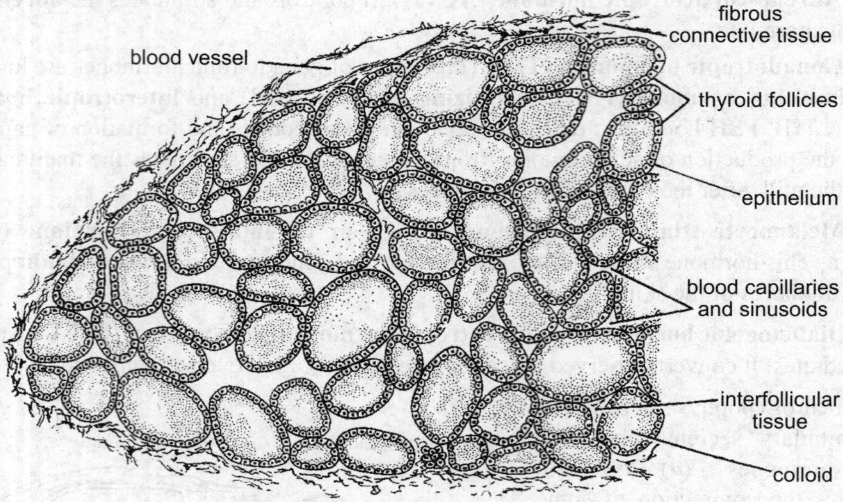

Fig. 29.78. V.S. of thyroid gland.

The **over-secretion** of this hormone increases the rate of oxidation and is responsible for the quick consumption of food, therefore, fat is not stored. Over-secretion also results in the bulging of eyeballs and irregular heart rate, etc. It also causes swelling of the thyroid gland resulting into the **goitre**.

3. Parathyroid glands. These are two small, oval glands in rabbit, situated antero-laterally on the lobes of thyroid. In other mammals these are four in number. These glands are either found embedded in the thyroid or located close to them. Parathyroid secretes two hormones, **parathormone** and **calcitonin**. These are responsible for regulating the amount of calcium and phosphorus in the blood. Parathyroid glands are necessary for the growth of bones, muscle tone and nervous activity. **Over-secretion** of these hormones results in tumors on the gland and causes calcium to be withdrawn from the bones, thus, producing a high calcium level in the blood. In such conditions the bones may become soft, porous and weak. Its **deficiency** lowers calcium level in the blood due to which normal functions of muscles, bone growth and teeth formation are retarded. It has been seen experimentally that the removal of **parathyroid** results in muscular tremors, cramps and finally convulsious. This is called **tetany** which results in the death of the individual.

4. Adrenal glands. They are a pair of small ovoid white or yellowish glands situated just above each kidney. These are also known as **suprarenal glands**. This gland consists of outer **cortex** derived from mesoderm and inner **medulla** derived from ectoderm.

(*i*) **Cortex.** The cortex secretes a group of hormones collectively known as **cortin**. These all are steroids and synthesised from cholesterol. **Corticosterone** and **cortisol** constitute the major portion of the total corticosteroid production. These two hormones are mainly active in carbohydrate and protein metabolism and are physiologically classified as **glucocorticoids**. The third is **aldosterone** which is mainly concerned with electrolyte balance, *i.e.*, retention of sodium and release of potassium by the kidney. It is physiologically classified as **mineralocorticoid**. Their production is under the control of ACTH with the usual negative feedback from adrenal cortex to pituitary *via* hypothalamus.

Production of aldosterone is not dependent upon ACTH. **Deoxycorticosterone** and **aldosterone** are highly potent in correcting the imbalance in the metabolism of sodium and potassium. **Cortisone** and **cortisol** (hydrocortisone) have profound influence on protein and carbohydrate metabolism, and suppression of inflammatory reaction.

Over-production of **aldosterone** causes hypertension with sodium retention, low serum potassium, and excess aldosterone in urine. The **deficiency** of cortin results in muscular weakness, nervous depression, low blood pressure, appetite is lost, the metabolic rate falls below normal, etc. This condition is known as **Addison's disease**. **Over-production** of **androgens** leads to virilism, hirsutism and precocious masculinisation with enhanced protein synthesis.

(*ii*) **Medulla.** The medulla produces a very important hormone known as **adrenaline** which is responsible for controlling the involuntary muscles and blood pressure. It increases the heart beat, increases the blood sugar level, thus, increasing the metabolic rate, dilates the bronchi and pupil. The adrenaline is, however, used in the treatment of **asthma** and similar respiratory troubles. **Noradrenaline** regulates the blood pressure under normal conditions and constricts the blood vessels.

5. Islets of Langerhans. Clusters of specialised cells, (Fig. 29.33) described by **Paul Langerhans** (1869). These are found distributed in the **pancreas**, which are endocrine in nature. These clusters are formed of two types of cells : (*i*) **beta-cells**, which constitute the maximum part of the cluster and secrete the hormone **insulin**, and (*ii*) the **alpha-cells** constituting nearly 20% of the cluster and believed to secrete a second hormone called **glucagon**. The C-cells and D-cells are in very low percentage and their function is not known. **Insulin** hormone is necessary for the conversion of excess sugars into glycogen in liver and muscles and **glucagon** function is conversion of glycogen back into glucose, when the sugar contents of the blood fall below the normal level. Deficiency of insulin results in **diabetes mellitus** in human beings.

6. Gonads. The gonads also perform endocrine function by producing **sex-hormones** in addition to their normal function of producing gametes.

(*i*) Testis. Testis secretes **testosterone** or male sex-hormone. The hormone testosterone is actually secreted by the **interstitial cells** or **cells of Leydig** situated in the connective tissue of the testes between seminiferous tubules. This hormone is responsible for the development of secondary sexual characters in male, *e.g.,* deep voice, beard on the face and the distribution of hairs on the body in man, etc.

(*ii*) Ovary. Ovary secretes **estrogen (oestradiol)**, a female sex-hormone produced by the cells surrounding the Graafian follicles. It is responsible for the mature growth, development and maintenance of the female reproductive system. It stimulates the secondary sexual characters and also contributes to the sexual desire of the female.

Progesterone. It is produced by the corpus luteum of the ovary formed after ovulation. It suspends further ovulation during pregnancy. It maintains pregnancy and prepares the uterine wall for the attachment of embryo with the uterine wall.

7. Placenta. The placenta also produces **estrogen** and **progestin** (chorionic gonadotropin). These hormones augment the action of pituitary luteinising hormone in maintaining the corpus luteum during pregnancy. These hormones work along with the hormones of ovary. In rabbit, etc., the placenta as well as the ovary secrete **relaxin** hormone. This relaxes pelvic ligaments for the easy birth of the child.

8. Gastro-intestinal tract. Some hormones like **gastrin**, **secretin**, **enterogastrone** and **cholecystokinin** are also produced in the different parts of gastro-intestinal tract which stimulate gastric glands to produce gastric juice, **secretin** stimulates pancreas to produce pancreatic juice into duodenum, **enterogastrone** inhibits the gastric secretion and **cholecystokinin** causes contraction of gall bladder to expel bile into duodenum.

9. Thymus. It is a large bilobed gland in the upper chest region in front of heart and below the base of trachea. In many mammals it remains as such throughout the life, while in some it disappears altogether at the time of puberty. Its secretion stimulates metabolism, growth and helps in attaining sexual maturity. Its removal from the embryo hampers the development of reproductive system. It is regarded as lymphatic gland and, hence, its endocrine nature is not definitely known.

10. Pineal body. It is situated on the roof of the diencephalon of the brain on a small pineal stalk. The nature and physiological role of the secretion produced by pineal body is not definitely known. However, its removal retards growth, and quickens the maturity or has no effect.

TABLE 29.4. SUMMARY OF VERTEBRATE HORMONES.

S. No.	Source (Endocrine gland)	Hormone	Physiologic effects (Functions)
1.	**Pineal body**	?	?
2.	**Pituitary**		
	(*a*) ***Anterior lobe***	1. *Somatotropin*	Controls growth and metabolism of protein. Deficiency causes dwarfism, excess causes gigantism and acromegaly
		2. *Thyrotropin* (TSH)	Stimulates thyroid
		3. *Adrenocorticotropin* (ACTH)	Stimulates adrenal cortex
		4. *Pancreatotropin* (PTH)	Suppresses secretion of insulin
		5. *Follicle-stimulating hormone* (FSH)	Affects growth of Graafian follicles in ovary and seminiferous tubules in testis
		6. *Luteinizing hormone* (LH)	Affects production and release of gonadal hormones
		7. *Prolactin* (LTH)	Affects mammary glands and milk production
	(*b*) ***Intermediate lobe***	*Intermedin*	Affects pigments in skin chromatophores
	(*c*) ***Posterior lobe***	1. *Oxytocin*	Stimulates contraction of uterine muscles and secretion of milk
		2. *Vasopressin*	Regulates blood pressure and water loss by kidneys
3.	**Thyroid**	*Thyroxine*	Increases O_2 consumption or basal metabolic rate. Deficiency leads to cretinism in children and myxoedema in adults. Hyperthyroidism causes goitre formation and exophthalmia.
4.	**Parathyroids**	*Parathormone*	Regulates calcium and phosphorus metabolism
5.	**Thymus**	?	Regulates growth, sexual maturity and immunity
6.	**Islets of Langerhans**		
	(*a*) ***Beta cells***	1. *Insulin*	Oxidation of glucose and its conversion into glycogen. Deficiency causes diabetes mellitus
	(*b*) ***Alpha cells***	2. *Glucagon*	Conversion of liver glycogen into blood
7.	**Adrenal**		
	(*a*) ***Cortex***	1. *Cortisone*	Conversion of proteins to carbohydrates
		2. *Aldosterone*	Metabolism of sodium and potassium
		3. *Adrenosterone*	Stimulates male characteristics
	(*b*) ***Medulla***	1. *Adrenaline*	Supplements action of sympathetic nervous system during emergencies
		2. *Noradrenaline*	Regulates normal blood pressure
8.	**Gonads**		
	(*a*) ***Testes***	*Testosterone*	Development and maintenance of male sex characters
	(*b*) ***Ovaries***	1. *Estrogen*	Development and maintenance of female characteristics

(*Contd.*)

S. No.	Source (Endocrine gland)	Hormone	Physiologic effects (Functions)
		2. *Progesterone*	Changes associated with pregnancy
	(*c*) ***Placenta***	1. *Chorionic gonadotropin*	Acts with other hormones to maintain pregnancy
		2. *Relaxin*	Relaxes pelvic ligaments at childbirth
9.	**Alimentary canal**		
	(*a*) ***Stomach***	*Gastrin*	Induces gastric secretion
	(*b*) ***Duodenum***	1. *Enterogastrone*	Slows down gastric secretion
		2. *Secretin*	Stimulates pancreatic secretion
		3. *Cholecystokinin*	Stimulates bile secretion

IMPORTANT QUESTIONS

Long Answer Questions

1. Describe the structure of mammalian skin and enumerate its functions.
2. Describe the food, feeding habits, digestive organs and process of digestion in rabbit.
3. Discuss the mammalian respiratory system and illustrate it with labelled figures.
4. What do you understand by tissue respiration? Describe the physiology of respiration in mammal.
5. Describe the structure and function of mamalian heart, with a note on its automacity.
6. Give an account of arterial or venous system of rabbit.
7. Describe the structure of mammalian blood and mention how many kinds of corpuscles are found in mammalian blood.
8. Describe in detail the structure of brain of rabbit.
9. Discuss various sense organs of rabbit.
10. Describe fully the internal ear of a mammal and discuss its advancement over an amphibian ear.
11. Discuss the mechanism of hearing in rabbit. Draw a well labelled diagram of T.S. of ear of rabbit.
12. Describe the structure and working of mammalian eye. Also give the structure of retina of eye.
13. Describe the structure and physiology of mamnalian kidney.
14. Describe the male and female urinogenital organs of rabbit.
15. Write an essay on endocrine system of rabbit.

Short Answer Questions

1. Draw a neat and well labelled diagram of the V.S. of skin of rabbit.
2. Describe the structure of hair of rabbit.
3. Describe the thoracic framework of rabbit.
4. Sketch and name the parts of the axis in *Oryctolagus*.
5. Draw the cross section of spinal cord, roots and ganglia of a mammal and name all the parts.
6. Distinguish between sacculus and utriculus.
7. Differentiate between corpus callosum and corpus striatum.
8. Describe the structure of internal ear of rabbit and the mechanism of hearing.
9. Describe in tabular form the origin, distribution and nature of cranial nerves of rabbit.
10. Draw well labelled diagrams of the following organs of mammals: (*i*) V. S. of skin, (*ii*) T. S. of tongue, (*iii*) T. S. through the region of thorax; (*iv*) T. S. of ear region of rabbit, (*v*) Ventral and dorsal view of brain, (*vi*) T. S. of cochlea, (*vii*) Male or female urinogenital system of rabbit. (*viii*) T. S. of ovary of mammal.
11. Write short notes on : (*i*) Differences between hare and rabbit, (*ii*) Corpus callosum, (*iii*) Sylvian fissure, (*iv*) Graafian follicle.
12. Write short note on pineal eye.
13. Write short note on organ of Corti.

Very Short Answer Questions

1. What is erector pili?
2. What is the type of dentition found in rabbit?

3. Give the dental formula of *Oryctolagus*.
4. The incisors of rodents grow throughout their life. Explain it in not more than two lines.
5. How many types of teeth are found in mammals?
6. How many pairs of salivary glands are found in rabbit?
7. Write the name of duct arising from submaxillary (submandibular) gland.
8. What is uvula? Mention its functions.
9. How many certilages give support to the wall of larynx?
10. Write the position of mitral valve in the heart.
11. Write the name of pacemaker in the heart.
12. Which systemic artery is found in rabbit?
13. What is corpora quadrigemina?
14. Why the eyes of albino rat look pink in colour?
15. Where image is formed in the eye?
16. Write the names of ear ossicles of rabbit.
17. What is cochlea? Give its function.
18. How does the arrangement of white and grey matter different in brain and in spinal cord.
19. What is function of Sertoli cells in testes?
20. Which hormone is secreted from corpus luteum? Mention its function.

Multiple Choice Questions

1. The zoological name of common rabbit is
   (*a*) *Lepus ruficaudatus*
   (*b*) *Camelus dromadarius*
   (*c*) *Oryctolagus cuniculus*
   (*d*) *Orycteropus*.
2. The burrow of rabbit is called
   (*a*) Form (*b*) Barrel
   (*c*) Warren (*d*) Barren.
3. The burrow of hare is called
   (*a*) Form (*b*) Barrel
   (*c*) Warren (*d*) Barren.
4. The external nares in rabbit are surrounded by
   (*a*) Vibrissae (*b*) Hare lip
   (*c*) Nictitating membrane (*d*) Rhinarium.
5. The perineal pouch bears
   (*a*) Cloaca (*b*) Anus
   (*c*) Vulva
   (*d*) Perineal scent glands.
6. The tail in rabbit serves as
   (*a*) Balancing organ (*b*) Sensory organ
   (*c*) Decorative structure (*d*) Hearing aid.
7. The outer thin cornified epidermis forming a sheath over the growing hair is called
   (*a*) Periderm (*b*) Torus
   (*c*) Epitrichium (*d*) Unguis.
8. The glands and keratin structures in rabbit are derived from
   (*a*) Stratum spinosum
   (*b*) Stratum granulosum
   (*c*) Stratum lucidum
   (*d*) Stratum germinativum.
9. Sweat glands are absent in the skin of
   (*a*) Rabbit (*b*) Cat
   (*c*) Rat (*d*) Man.
10. The secretion of sebaceous glands is
    (*a*) Sebum (*b*) Sweat
    (*c*) Tears (*d*) Ear wax.
11. Mammary glands are modified
    (*a*) Sabaceous glands (*b*) Sweat glands
    (*c*) Tear glands
    (*d*) Meibonian glands.
12. Buccal cavity is divided into nasal and food passage by
    (*a*) Teeth (*b*) Tongue
    (*c*) Palate (*d*) Vestibule.
13. Dental formula of rabbit is
    (*a*) 2123 / 2033 (*b*) 2103 / 2103
    (*c*) 2033 / 1033 (*d*) 2234 / 2033.
14. The wide toothless gap due to the absence of canines is
    (*a*) Epiglottis (*b*) Glottis
    (*c*) Diastemma (*d*) Cheek.
15. The duct of infraorbital salivary glands opens near
    (*a*) Upper incisors (*b*) Upper molars
    (*c*) Tongue
    (*d*) Angles of mandibles.
16. The large intestine of mammals is divided into
    (*a*) Colon and rectum
    (*b*) Colon and aterial
    (*c*) Aterial and haustra
    (*d*) Anus and colon.
17. Mucin is a mixture of
    (*a*) Carbohydrates (*b*) Proteins
    (*c*) Fatty acids (*d*) Lipids.

18. Ptylin is present in
    (*a*) Gastric juice (*b*) Bile juice
    (*c*) Saliva
    (*d*) Succus entericus.
19. Bilirubin and biliverdin are derived by the disintegration of
    (*a*) Globulin (*b*) Haemoglobin
    (*c*) Myosin (*d*) DNA.
20. Hormone insulin is secreted by
    (*a*) Gastric glands
    (*b*) Pancreatic glands
    (*c*) Islets of Langerhans
    (*d*) Cystic lobe of liver.
21. Rabbits re-eat the faeces produced during
    (*a*) Morning (*b*) Evening
    (*c*) Night (*d*) Noon.
22. Adam's apple in human neck is
    (*a*) Pharynx (*b*) Glottis
    (*c*) Larynx (*d*) Trachea.
23. True vocal cords are absent in
    (*a*) Man (*b*) Dog
    (*c*) Lion
    (*d*) Hippopotamus.
24. Blood vascular system of rabbit is
    (*a*) Open type (*b*) Closed type
    (*c*) Semiclosed type (*d*) Haemocoel.
25. The mammalian heart is enclosed in a protective covering known as
    (*a*) Epicardium (*b*) Myocardium
    (*c*) Mesocardium (*d*) Pericardium.
26. The left auriculo-ventricular valve is known as
    (*a*) Semilunar valve
    (*b*) Eustachian valve
    (*c*) Mitral valve
    (*d*) Tricuspid valve.
27. Fossa ovalis represents
    (*a*) Sinus venosus
    (*b*) Musculi pectinati
    (*c*) Auricular appendix
    (*d*) Foramen ovale.
28. Veins carrying oxygenated blood
    (*a*) Pulmonary veins (*b*) Caudal veins
    (*c*) Caval veins (*d*) Aorta.
29. Funnel-shaped space towards the concavity of kidney is
    (*a*) Capsule (*b*) Cortex
    (*c*) Pelvis (*d*) Medulla.
30. Renal columns of Bertini are formed by
    (*a*) Capsule (*b*) Cortex
    (*c*) Medulla (*d*) Pelvis.
31. The volume of urine is controlled by
    (*a*) Thyroxine (*b*) Adrenalin
    (*c*) Insulin (*d*) ADH.
32. Involuntary activities of body are controlled by
    (*a*) Cerebrum (*b*) Cerebellum
    (*c*) Spinal cord
    (*d*) Pons and medulla oblongata.
33. Contraction of melanophores is regulated by
    (*a*) Aldosterone (*b*) Noradrenaline
    (*c*) Melatonine (*d*) Serotonine.
34. Hormone controlling the contraction of uterine muscles at the time of childbirth
    (*a*) Prolactin (*b*) Vasopressin
    (*c*) Oxytocin (*d*) Gonadotropin.
35. Cells of islets of Langerhans responsible for the secretion of glucagon
    (*a*) Alpha cells (*b*) Beta cells
    (*c*) Gamma cells (*d*) T-cells.
36. Removal of parathyroid glands leads to
    (*a*) Diabetes mellitus
    (*b*) Diabetes insipidus
    (*c*) Myxoedema (*d*) Tetany.
37. Addison's disease results from the underactivity of
    (*a*) Adrenal medulla
    (*b*) Adrenal cortex
    (*c*) Thymus
    (*d*) Parathyroid.

ANSWERS

1. (*c*)	2. (*c*)	3. (*a*)	4. (*d*)	5. (*d*)	6. (*a*)	7. (*a*)	8. (*d*)
9. (*c*)	10. (*a*)	11. (*a*)	12. (*c*)	13. (*c*)	14. (*c*)	15. (*b*)	16. (*a*)
17. (*b*)	18. (*c*)	19. (*b*)	20. (*c*)	21. (*c*)	22. (*c*)	23. (*d*)	24. (*b*)
25. (*d*)	26. (*c*)	27. (*d*)	28. (*a*)	29. (*c*)	30. (*b*)	31. (*d*)	32. (*d*)
33. (*c*)	34. (*c*)	35. (*a*)	36. (*d*)	37. (*b*)			

30 Chapter Class 9. Mammalia : General Characters, Classification and Representative Types

Contents of this Chapter

GENERAL CHARACTERS

1. Mammals are hair-clad (except Cetacea), air-breathing, warm-blooded oviparous (prototherians) or viviparous animals.

2. The skin has sudoriporous (sweat) glands, sebaceous (oil) glands and sometimes scent glands. Females have mammary glands with teats for nourishing the youngs.

3. Limbs two pairs and are either plantigrade or digitigrade or unguligrade. Each limb with 5 or fewer digits. Cetaceans and sirenians have no hindlimbs.

4. External ears (pinnae) are present.

5. Exoskeleton includes horny, epidermal hairs, spines, scales, claws, nails, hoofs, horns and bony dermal plates.

6. Endoskeleton completely ossified. Skull with two occipital condyles, and is without prefrontals, postfrontals, quadrato-jugal, supraorbitals and basipterygoids. Otic bones fused into periotic which with tympanic forms tympanic bulla.

The lower jaw is composed of a single bone, on each side, the **dentary** which articulates with squamosal of its side. Vertebrae with terminal epiphyses and centra is acoelous (flat). With few exceptions, mammals possess seven cervical vertebrae. Ribs bicephalous, *i.e.,* with two heads.

7. Presence of a muscular diaphragm between thoracic and abdominal cavities.

8. Alimentary canal without cloaca. Buccal cavity is separated from nasal passage by a hard palate formed of premaxillae, maxillae and palatines. Teeth are heterodont (differentiation into incisors, canines and molars), thecodont (embedded in the alveolar pockets of jaw bone) and diphyodont (only two sets of teeth, a milk set replaced by permanent set).

9. Respiration by lungs (pulmonary). Glottis is guarded by epiglottis. Larynx with vocal cords.

10. Heart is four-chambered with only a left aortic arch. Renal portal system is absent. RBCs are small and non-nucleated, homoiothermous.

11. Brain with four optic lobes. Cerebrum and cerebellum are large and convoluted. Corpus callosum connecting both the cerebral hemispheres is present. Cranial nerves 12 pairs.

12. The kidney is metanephros. Ureters open into urinary bladder. Ureotelic.

13. Eyes are protected by lids. Upper eyelids are movable. Middle ear has 3 ear ossicles–malleus, incus and stapes. Cochlea spirally coiled.

14. Sexes separate. Male with erectile penis. Eggs small with very little yolk.

15. Fertilisation is always internal, after copulation.

16. Except monotremes, all are viviporous. The young, except in monotremes, develops in the uterus for some time and born alive.

CLASSIFICATION

Mammals include about 5000 living species and several fossil forms. Mammals are classified on the basis of (1) mode of caring for their youngs, (2) dentition, (3) foot posture, (4) exoskeleton (nails, claws and hoofs), (5) complexity of brain and (6) systematics. For the purpose of this text, we shall refer only 18 living orders of mammals which are first divided into 2 subclasses : Prototheria and Theria. **G. G. Simplon** provided a complete review of the group in a publication entitled "The Principles of Classification and a Classification of Mammals" (*Bull. Am. Mus. Nat. Hist.* 1945 Vol. 85). He recognised 18 living and 14 extinct orders of mammals.

SUBCLASS I. PROTOTHERIA

(Gr., *protos* = first + *therios* = beast)

1. No pinna on the ear.
2. In the skull, tympanic bulla and lacrimals are absent; the jugal is reduced or absent.
3. Teeth are present only in young, adults with horny beak.
4. The pectoral girdle possesses large coracoid bones and interclavicle. The pelvic girdle possesses epipubic bones extending from the pelvis. Vertebrae without terminal epiphyses. The ribs have only a single head, the tuberculum being absent. The cervical ribs are present.
5. A cloaca is present into which ureters and urinogenital sinus open.
6. Mammary glands without nipples.
7. There is no corpus callosum in the brain.
8. The cochlea is a simple process of the sacculus and not coiled.
9. Testes are abdominal.
10. Females are oviparous.
11. Found in Australia, Tasmania and New Guines.

Order 1. Monotremata

(Gr., *monos* = single + *trema* = opening)

The general characters are same as that of Prototheria.

Examples : *Echidna* (Spiny ant-eater), *Ornithorhynchus* (Platypus).

SUBCLASS II. THERIA

(Gr., *ther* = animal)

1. Includes modern viviparous marsupial and placental mammals.
2. Ear usually with external pinna.
3. Mammary glands with nipples or teats.
4. Teeth are usually present both in young and adults. Vertebrae with terminal epiphyses.
5. Cloaca is usually absent.
6. Testes usually in the scrotal sac. Vasa deferentia and bladder opening through a common urethra in penis. Ureters open into the base of bladder.
7. Oviducts open into a vagina.

Subclass Theria is divided into two living infraclasses :

INFRACLASS I. METATHERIA

(Gr., *meta* = between or after)

1. Marsupium or brood pouch is present in the females.
2. Mammary glands bearing nipples are sebaceous.
3. Epipubic (marsupial) bones are usually present and attached with the pubis. Separate coracoid and interclavicle are absent.
4. Corpus callosum is feebly developed or absent.
5. Vagina and uterus are double (didelphic condition).
6. Placenta is rudimentary yolk sac type.
7. They are confined to Australian region.

Order 1. Marsupialia

(Gr., *marsypion* = pouch)

Young ones are born immature and develop in mursupium attached with nipples or teats. In dentition 3 premolar and 4 molars are usually present. Other characters are same as that of Metatheria.

Examples : *Didelphis* (Opossum), *Dasyurus* (Tiger cat), *Caenolestes*, *Notoryctes* (Marsupial mole), *Parameles* (Bandicoot), *Macropus* (Kangaroo), *Phascolarctos* (Koala).

INFRACLASS II. EUTHERIA

(Gr., *eu* = true + *therios* = beast)

1. Higher viviparous placental mammals. Marsupium (marsupial pouch) is entirely absent.
2. Epipubic bones are absent.
3. Cloaca is absent.
4. Corpus callosum is present.
5. Testes are usually contained in scrotal sacs.
6. Vagina is single.
7. The young always nourished for a considerable time in the uterus by means of allantoic placenta and born in a relatively advanced state.

Order 1. Insectivora

(L., *insectum* = insect + *vorare* = to eat)

1. Small and furry animals.
2. Snout usually long and tapering.
3. Feet usually five-toed with claws, inner toe not opposable.
4. Molars sharp and pointed for insects feeding. Dental formula : *i* 3/3, *c* 1/1, *p* 4/4, *m* 3/3.
5. Placenta discoidal.
6. Nocturnal and terrestrial.

Examples : *Erinaceus, Paraechinus* (Hedgehog), *Scapanus, Talpa* (Mole), *Sorex, Echinosorex* (Shrews), *Solenodon* (Solenodon).

Order 2. Dermoptera

(Gr., *derm* = skin + *pteron* = wing)

1. Commonly known as "flying lemurs."
2. Furry skin stretched between equal-sized fore- and hindlimbs and the tail, called patagium.
3. Incisor teeth 2/3.
4. Nocturnal in habit.

Examples : *Galeopithecus* or *Cynocephalus* having two species from South-eastern Asia.

Order 3. Chiroptera

(Gr., *cheiros* = hand + *pteron* = wing)

1. Commonly known as true flying mammals or bats.

2. Forelimbs are modified into membranous wings.
3. Hindlimbs are weak and have sharp clawed digits which are included in wing membrane.
4. Eyes are small with vision weak and ears with large pinnae.
5. Teeth are small and sharp. Dental formula : *i* 2/3, *c* 1/1, *p* 3/3, *m* 3/3.
6. Sternum has a keel.
7. Mostly nocturnal in habit.

It is divided into two suborders

Suborder I. Megachiroptera	Suborder II. Microchiroptera
1. Large fruit eating bats (frugivorous).	1. Small insectivorous bats.
2. Nocturnal often present in flocks.	2. Nocturnal, found solitary or in flocks.
3. Snout is elongated without nose leaf and eyes large.	3. Snout short with nose leaf, and eyes small.
4. Pinna small without any appendage.	4. Pinna large with leaf-like appendages (**tragus**).
5. Thumb and second digits are clawed.	5. Only thumb is clawed.
6. Tail is absent or small without interfemoral membrane.	6. Tail large and enclosed in an interfemoral membrane.
7. Molars without sharp cusps and longitudinal grooves. Dental formula : *i* 2/2, *c* 1/1, *p* 3/3, *m* 2/3. **Examples** : Flying foxes. *Pteropus, Xanthorpyia Cynopterus.*	7. Molars with sharp cusps and transverse grooves. **Examples** : *Magaderma, Rhinolophus, Desmodus* (Vampire bat), *Eptesicus* and *Myotis* (Brown bat), *Vespertilio.*

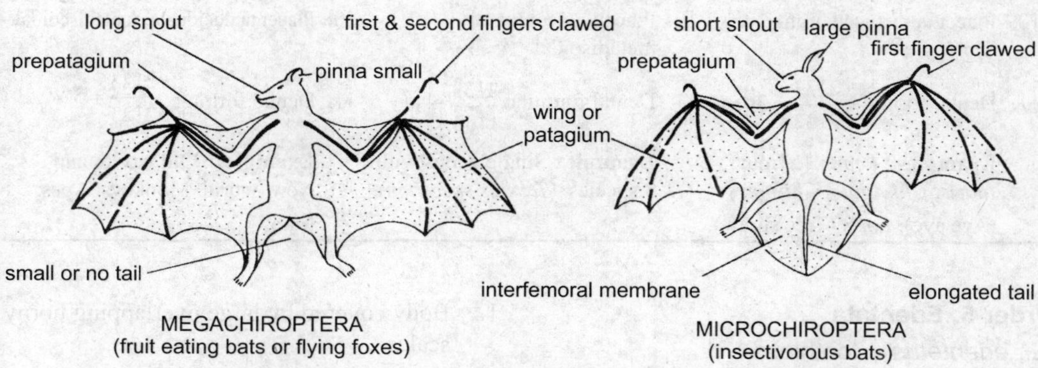

Fig. 30.1. Comparison of megachiroptera and microchiroptera.

Order 4. Primates

(L., *primus* = of the first rank).

Primitive mammals with well developed brain. Cerebrum is large and convoluted.

1. Five digits of hand and foot with flat nails.
2. Thumb and great toe are shorter and opposable for grasping.
3. Eyes are directed forward and surrounded by bony ring.
4. Femur has no third trochanter.
5. Mode of walking is plantigrade.
6. Placenta discoidal or metadiscoidal and haemochorial.

This order is divided into three suborders: **Lemuroidea**, **Tarsioidea** and **Anthropoidea**. Anthropoidea is further subdivided into two infraorders : **Platyrrhina** and **Catarrhina**.

Suborder I. Lemuroidea (*L., lemures* = ghost) Primitive	Suborder II. Tarsioidea (*Gr., tarsus* = first foot) Slightly advanced	Suborder III. Anthropoidea (*Gr., anthropos* = man) Advanced
1. Exclusively arboreal, nocturnal, omnivorous and solitary.	1. Arboreal, nocturnal, insecti–vorous and solitary.	1. Terrestrial or arboreal, diurnal, and gregarious.
2. Head with elongated muzzle (snout).	2. Head with shorter snout.	2. Head with short face.
3. Eyes lateral with poor vision.	3. Eyes large, forward with good vision.	3. Eyes forward with fine vision (stereoscopic).
4. Orbits confluent with temporal foesae.	4. Orbits communicate with temporal fossae by a wide fissure.	4. Orbits are separated from temporal fossae by a bony partition.
5. Long tail is never prehensile.	5. Tail is longer with a tuft at the tip and non-prehensile.	5. Tail when present, prehensile.
6. Forelimbs are shorter than hindlimbs.	6. Limbs are extremely long, hindlimbs longer than the fore-limbs.	6. Hands and feet are prehensile except in man.
7. Second toe or finger with claw but the rest have flat nails.	7. Digits have adhesive pads and flats nails except the second and third toes which have claws.	7. Fingers and toes have flat nails except in few forms.
8. Cerebellum is not covered by cerebrum.	8. Cerebellum not covered by cerebrum.	8. Cerebral hemispheres are more highly developed and covering the cerebellum.
9. Duplex uterus. Clitoris with urethra.	9. Bicornuate uterus. Clitoris traversed by urethra.	9. Uterus simplex. Clitoris not traversed by urethra.
10. Placenta diffused, non-deciduate.	10. Placenta deciduate, metadisoidal.	10. Placenta deciduate, metadiscoidal.
11. Dental formula $\frac{2133}{2133} = 36$.	11. Dental formula $\frac{2133}{1133} = 34$.	11. Dental formula $\frac{2123}{2123} = 32$.
Examples : *Lemur* (Lemur), *Indris*, *Chirogale*, *Chiromys* (Aye aye), *Loris* (Tree shrew).	**Example** : Single genus with 3 species. (*Tarsius spectrum*).	**Examples** : Old world and New world Monkeys, Apes, Man.

Order 5. Edentata

(L., *edentetus* = toothless)

1. Teeth present or absent, if present, incisors and canines are always absent. Only molars are present.
2. Feet have well developed strong and curved claws.
3. The testes are abdominal.

Examples : Sloth (*Bradypus*), Armadillos (*Dasypus*), American ant-eaters, (*Myrmecophaga*).

Order 6. Pholidota

(Gr., *pholis* = a horny scale)

1. Body covered by large overlapping horny scales.
2. Hairs are sparsely found between the scales.
3. Snout is elongated.
4. Tongue is long, sticky and protrusible.
5. Teeth are entirely absent.
6. Ears are reduced.
7. Limbs are short bearing five digits. Forelimbs have well developed claws.
8. Nocturnal and burrowing in habits.

Example : Single genus *Manis* (Pangolin or scaly ant-eater).

Fig. 30.2. Some modern primates. Only heads.

Comparison between Platyrrhina and Catarrhina.

Platyrrhina	*Catarrhina*
1. Arboreal.	1. Arboreal or terrestrial.
2. Comparatively smaller in size.	2. Relatively larger in size.
3. Nose flat. Nostrils widely separated.	3. Nose raised. Nostrils near to each other.
4. Cheek pouches absent.	4. Cheek pouches present except apes.
5. Tail long and prehensile.	5. Tail, if present, non-prehensile.
6. Ischial callosities on buttocks absent.	6. Ischial callosities on buttocks red.
7. Dental formula : 2, 1, 3, 3 / 2, 1, 3, 3	7. Dental formula : 2, 1, 2, 3 / 2, 1, 2, 3
8. Offsprings more than one.	8. Gave birth to one.
9. Found in Central and South America.	9. Found in Asia and Africa.
Examples : *Ateles* (Spider monkey), *Aloutta* (Howling monkey), marmosets, etc.	**Examples** : *Macaca* (Rhesus monkey), *Presbytis* (Langur), *Hylobates* (Gibbon), *Papio* (Baboon), *Pango* (Orangutan), *Pan* (Chimpanzee), *Gorilla* (Gorilla), *Homo sapiens* (Man).

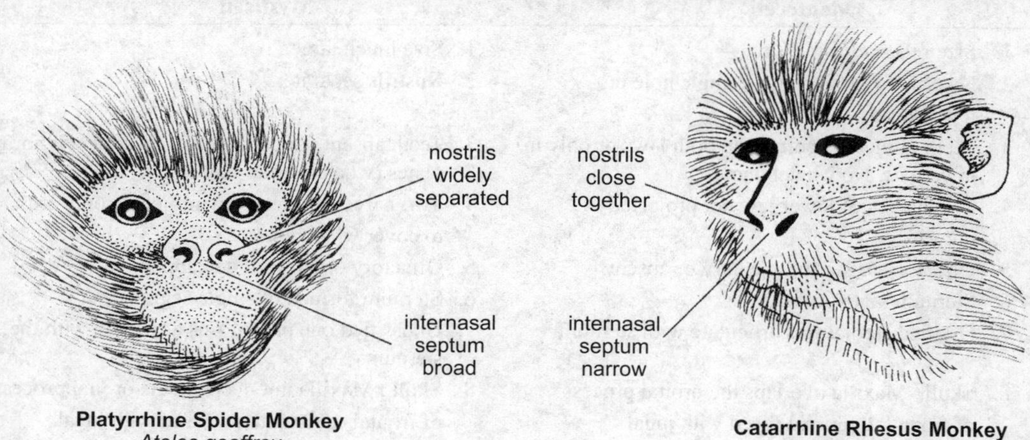

Fig. 30.3. Comparison between New World Monkeys (Platyrrhina) and Old World Monkeys (Catarrhina). Only heads.

Order 7. Lagomorpha

(Gr., *logos* = hare + *morphe* = form)

1. Two pairs of incisors are present in the upper jaw, first pair large and chisel-like found in front of smaller second pair.
2. Tibia is fused with the fibula. Soles of feet are hairy. Toes bear claws.
3. Testes are always external in the scrotum.

Examples : *Oryctolagus* (Rabbit) and *Lepus* (Hare), *Ochotona* (Pika).

Order 8. Rodentia

(L., *rodo* = gnaw)

1. Usually small gnawing mammals.
2. Incisors single pair, long chisel-like, rootless and persistently growing with enamel, confined chiefly to the anterior end. Canines are entirely absent.
3. Space known as diastemma present between the incisors and molars, *i.e.*, at the place of canine.

Examples : *Funambulus* (Squirrel), *Rattus rattus* (House Rat), *Mus* (Mouse), *Hystrix* (Porcupine), *Cavia* (Guinea pig), *Castor* (Beaver), *Cynomys* (Prairie dog).

Order 9. Cetacea

(Gr., *ketos*, or L., *cetus* = a whale)

1. Medium sized to very large aquatic mammals.
2. Body is usually fish-like.
3. Hairy covering of skin is reduced to few bristles on the muzzle.
4. Head long often pointed without neck.
5. Eyes are minute. Nostrils on the top of the head and ear openings are minute, without external ears.
6. Forelimbs are modified as paddles, digits are enclosed in a common integument devoid of claws. Hindlimbs are entirely absent.
7. Tail is long ending in two broad, transverse, fleshy flukes notched in the mid-line. It is used in swimming.
8. Fleshy dorsal fin is present in some forms. Dorsal and tail fins have no skeletal elements.
9. A thick layer of fat (**blubber**) is present beneath the skin.
10. Bones of skull are spongy and contain fat.
11. Stomach is complex and consists of three or four chambers.
12. Testes are abdominal.
13. Placenta is non-deciduate and diffuse.

Examples : Living catacea is divided into two suborders : **Odontoceti** (toothed whales) and **Mysticeti (Mystacoceti)** (wholebone whales).

Comparison between Odontoceti and Mysticeti.

Odontoceti	Mysticeti
1. Size relatively smaller.	1. Size much larger.
2. Nostrils fused forming a single hole on top of head.	2. Nostrils separate.
3. Teeth numerous pointed in both jaws or only in lower jaw. No whalebone.	3. Teeth absent. Upper jaw bears a series of horny plates (whalebone or baleen).
4. Two halves of lower jaw not projected outwards.	4. Two halves of lower jaw projected outwards to cover whalebones.
5. Olfactory organ rudimentary or absent.	5. Olfactory organ well developed.
6. Sternum many jointed.	6. Sternum formed of single segment.
7. Ossified sternal ribs articulate with sternum.	7. Unossified one pair of ribs articulate with the sternum.
8. Skull : Maxilla overlaps the orbital process of frontal. Lacrimal fused with jugal. Tympanic not fused with periotic.	8. Skull : Maxilla not overlaps the orbital process of frontal. Lacrimal not fused with jugal. Tympanic fused with periotic.

(*Contd.*)

Odontoceti	Mysticeti
9. Body oil not edible.	9. Body oil edible.
10. Feed on fish.	10. Feed on planktons.
Examples : *Physeter catodon* (Sperm whale), *Orcinus* (Killer whale), *Phocaena* (Porpoise), *Platanista gangeticus* (Ganges dolphin), *Delphinus delphis* (Common dolphin).	**Examples :** *Balaenoptera musculus* (Blue whale), *B. physalus* (Common fin whale), *Balaena* (Right whale).

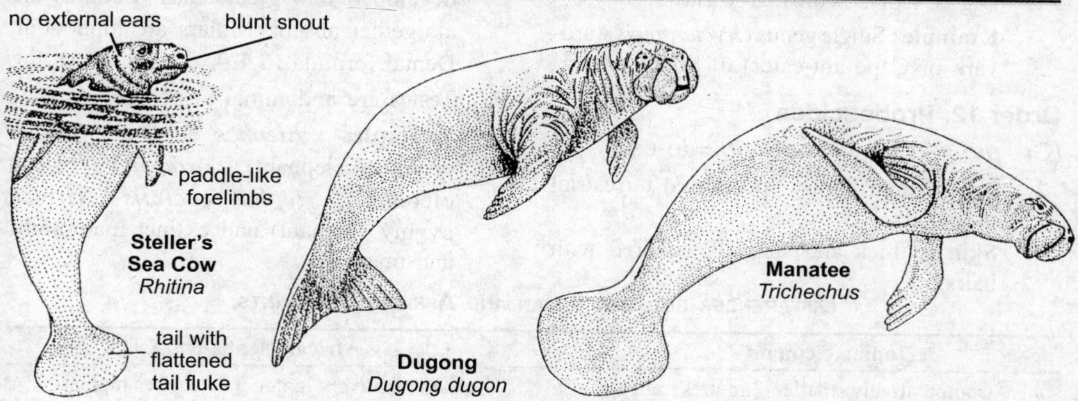

Fig. 30.4. Examples of Sirenia.

Order 10. Carnivora

(L. *caro* = flesh + *vorare* = to eat)

1. Small to large quadrepels, terrestrial, arboreal or aquatic, predatory, flesh-eating mammals.
2. Teeth generally have sharp and cutting edges. Canines are large and fang-like for tearing flesh. The incisors are small and always 3 on each side in each jaw. Molars are cutting type.
3. Toes or digits never less than four and all have strong and sharp claws.

Living carnivores are divided into two suborders : **Fissipedia** and **Pinnipedia**.

Suborder I. Fissipedia

1. Terrestrial carnivores having separate toes. Incisors are always six in each jaw and are feebly developed in many cases. Canines are always very large and strong.
2. The last premolar in the lower jaw known as **carnassial** teeth and the last two molars are crushing teeth.
3. The digits are mostly five and never less than four. All the digits are separate and have claws.

Examples : *Panthera leo* (Lion), *Panthera tigris* (Tiger), *Prionalurus bengalensis* (Leopard cat), *Acinonyx* (Hunting leopard), Civet cat, *Herpestes* (Mongoose), *Hyaena* (Hyaena), *Canis lupus* (Wolf), *Vulpes* (Fox), *Canis familiaris* (Dog), *C. aureus* (Jackal), *Lutra* (Otter), *Meles* (Badger), *Acinonyx* (Cheetah), *Felis domesticus* (Domestic cat), *Ursus* (Bear), *Procyon* (Racoon), *Mustela* (Mink), *Ailuropoda* (Panda), *Mephites* (Shunk).

Suborder II. Pinnipedia

1. Marine carnivores, with torpedo-shaped body.
2. The greater part of the limbs are enclosed within the body skin and hands and feet are fully webbed. Limbs modified into paddles or flippers.
3. Tail is very short, useful in swimming.
4. External ear small or degenerated.

Examples : *Otario jubata* (Sea-lion), *Trichecus* (Walrus), *Phoca* (Common seal), *Callorhinus* (Fur seal).

Order 11. Tubulidentata

(L., *tubulus* = tube-like + *dens* = tooth)

1. Body stout, somewhat pig-like.

2. Skin is very thick, covered with sparse hairs.
3. Snout is long and tubular with round nostrils at the tip.
4. Tongue slender and protrusible.
5. Ears are long, erect, and pointed.
6. Milk teeth numerous and permanent teeth only 4 or 5 without enamel and perforated by numerous tubules of pulp.
7. Digits 4 or 5 with heavy claws.

Example : Single genus *Orycteropus* (Aard-vark or Cape ant-eater) of South Africa.

Order 12. Proboscidea

(Gr., *pro* = in front + *boskein* = to eat)

1. Largest and highly specialised terrestrial mammals.
2. Skin is thick and sparsely covered with hairs.
3. Long proboscis or trunk which is an elongation of the nose and upper lip, bearing the nostrils at the tip.
4. Eyes are small and ears very large.
5. Legs are large, pillar-like and bear five functional digits with small, nail-like hoofs on both fore-and hindlimbs.
6. Two incisor teeth of the upper jaw developed into great tusks. Canines are altogether absent. Molars are lophodont. Dental formula : *i* 1/0, *c* %, *p* 3/3 *m* 3/3.
7. Testes are abdominal in position.

Examples : *Elephas maximus* (Indian or Asian elephant), *Loxodonta* (African elephant), *Elephas cyclotis* (African pygmy elephant) and extinct mammoths and mastodon.

DIFFERENCES BETWEEN INDIAN AND AFRICAN ELEPHANTS.

Indian elephant	African elephant
1. Comparatively smaller, 3 metres high at shoulder. Weight over 6 tons.	1. Comparatively larger, 3·45 metres high at shoulder. Weight over 7 tons.
2. Forehead high, dome-shaped and clefted.	2. Forehead low, flat and not clefted.
3. Back slopping.	3. Back arched.
4. External ears smaller.	4. External ears very much large.
5. Tip of proboscis with one lip.	5. Tip of proboscis with two lips.
6. Male with tusks, smaller in size.	6. Tusks present in both sexes. Relatively larger in male.
7. Distributed in South-East Asia.	7. Distributed throughout Africa, South of Sahara desert.

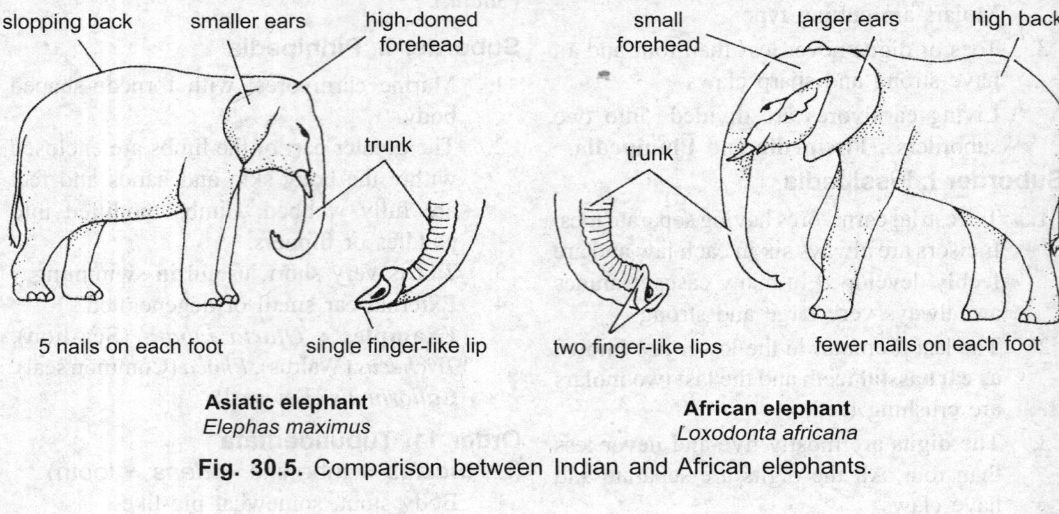

Fig. 30.5. Comparison between Indian and African elephants.

Order 13. Hyracoidea

(Gr., *hyrax* = shrew + *eidos* = form)

1. Small guinea pig-like animals with a split snout.
2. Ears are short and the tail is much reduced.
3. Legs short with four digits in the forelimbs and three in hindlimbs.
4. Testes are abdominal.
5. Dental formula : *i* 1/2, *c* 1/0, *pm* 4/4, *m* 3/3.

Example : *Procavia* or *Hyrax* (Conies of South Africa).

Order 14. Sirenia

(Gr., *siren* = sea nymph)

1. Large, herbivorous, aquatic mammals commonly known as **sea cows**.
2. Body spindle-shaped with paddle-like fore-limbs. Hindlimbs are absent.
3. Tail flattened with horizontal lateral flukes, with or without notch.
4. Nostrils are on the upper surface of the blunt snout.
5. Few hairs over the body.
6. Stomach complex, consisting of several chambers.
7. Testes are abdominal.

Examples : *Halicore* = *Dugong* (Dugong), *Trichechus* (Manatees).

Order 15. Perissodactyla

(Gr., *perissos* = odd + *dactyloss* = *toes*)

1. Large-sized odd-toed hoofed mammals or ungulates. Toes 1 or three.
2. The middle (3rd) digit of fore-and hind-limbs is pre-eminent and carries most of the weight.
3. Teeth are lophodont. Incisors present in both jaws.
4. Stomach is simple.

Examples : *Equas cabalus* (Horse), *Equus zebra*, (Zebra) *Tapirus* (Tapir), *Rhinoceros* or *Diceros* (Rhinoceros).

Order 16. Artiodactyla

(Gr., *artios* = even + *dactylos* = digit)

1. Terrestrial and semi-aquatic even-toed hoofed mammals.
2. Fore-and hindlimbs bear two, rarely four digits.
3. Teeth are **selenodont** or **bunodont**. Incisors and canines in upper jaw are usually absent.
4. Stomach is complicated, 4-chambered.
5. Except pig and peccaries, all chew or ruminate their cud.

Examples : *Hippopotamus* (Horse of river), *Sus* (Pig), *Dicotyles* (Peccaries), *Camelus* (Camel), *Cervus* (Deer), *Moschus* (muskdeer), *Giraffa* (Giraffe), *Antelope* (Blackbuck), Cow, Buffalo, *Ovis* (Sheep), *Capra* (Goat), *Bos indicus* (Ox), *Bos grunniens* (Yak).

REPRESENTATIVE TYPES

A. Monotremes or Egg-laying Mammals

1. *Tachyglossus*. *Tachyglossus aculeatus,* also called *Echidna* (Fig. 30.6), is a spiny ant-eater found in Australia and Tasmania. Another genus, *Zaglossus* (*Proechidna*), is found in New Guinea. The body is covered above with strong pointed spines and coarse hairs, ventral side has no spines but only hairs. The animal can roll into a ball for protection. It is terrestrial and burrowing though it does not make tunnels. It is nocturnal and specialised for ant-eating. The

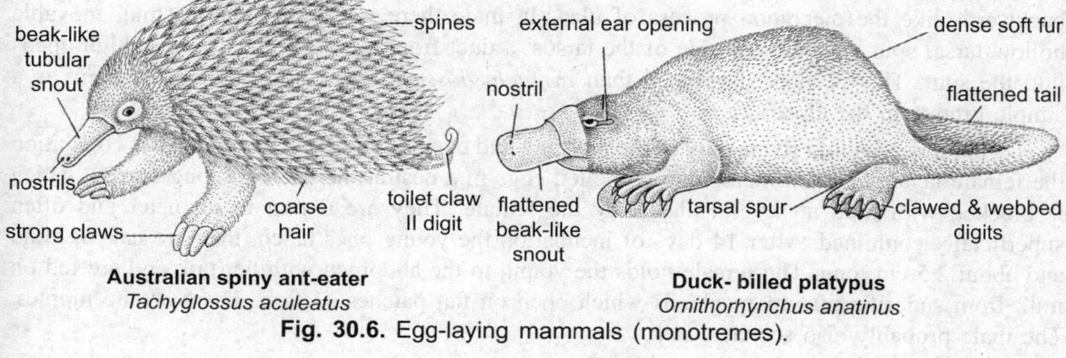

Fig. 30.6. Egg-laying mammals (monotremes).

jaws form a long, tubular snout or rostrum which is tactile and has a terminal rounded mouth. There are no teeth at any stage. There is a long sticky, protrusible tongue, used for capturing ants and insects. The dorsal surface of the tongue has horny serrations which grind the insects against ridges of the palate. The saliva neutralises the formic acid of ants. Eyes are small with no nictitating membrane, they are reptilian with little accommodation. Ears have incospicuous pinna. The tail is vestigial. The girdles and limbs are reptilian. The limbs are held laterally, and have five digits with sharp claws used for burrowing and tearing up ant-hills. In the hindlimbs the second digit has a long, curved **toilet-claw** to clean the spines and hair. In the male there is a grooved horny **poison-spur** on the tarsus and a poison gland in the thigh from which poison passes into the spur. Cervical vertebrae bear separate ribs, in all there are 16 ribs. There are no sweat glands. Penis is a simple groove on the floor of the cloaca, it is used only for the passage of sperms, urine passes through a special urinary canal. The female lays one or two leathery eggs in a nest, then it places them in a pouch of skin on the abdomen, which develops during breeding season. This pouch is supported by two epipubic bones arising from the pelvic girdle. The young hatches in the pouch in an immature condition and is nourished on milk from mammary glands which are modified sweat glands and have no nipples. The young suck milk from moistened hair. The youngs are carried in the pouch till they are 7.5 to 10 cm long. The male also has mammary glands and secretes milk, this condition is known as **gynaecomastism** in which both parents share in feeding the young. It is also found in Platypus.

2. *Zaglossus* or *Proechidna*. It is confined to New Guinea. It is larger in size than *Echidna*. Body is covered over by short spines concealed by long fur. Head is larger and beak is twice longer than the head. Limbs are three-toed with claws.

3. *Ornithorhynchus*. *Ornithorhynchus anatinus* (Platypus or duck-mole) (Fig.30.6) differs widely from *Tachyglossus* in external appearance, but in a few minor details only, the two resemble closely in anatomical structure. Platypus lives in long burrows on river banks in Australia and Tasmania. It is about 20 inches (50–60 cm) long. The flat body is covered with soft brown waterproof under fur and harsher outer fur. Platypus can roll up into a ball. Upper jaw forms a broad, flat muzzle like the bill of a duck. It is covered with soft hairless skin and has pitted tactile organs. The **muzzle** is moist and flexible having dorsal nostrils, used for capturing worms and molluscs from mud of river beds which can be stored in cheek pouches. **Teeth** are found in the young but are replaced in the adult by horny epidermal plates used for breaking shells of molluscs, sand taken in with food assists in mastication. **Eyes** are small and have nictitating membranes, ears have no pinnae. When under water, the eyes and ears are covered by skin pouches and the animal fends by touch and smell. There is a broad flat 5 inches (12.5 cm) long tail, used for swimming. Front limbs are short and powerful having five digits with strong claws for digging. Lying below and projecting beyond the claws is a web having leathery extensions; the web is used for swimming, when walking or digging the web can be folded back. Hindlimbs have five digits with sharp claws and also webbed. The proximal end of the fibula projects backwards like the olecranon process of ulna. In male there is a horny sharp, stout, movable hollow tarsal spur on the inner side of the tarsus, a duct from a poison gland in the thigh opens into the spur. This is more developed than in *Tachyglossus*. Testes are internal, the penis is a simple groove in the cloaca.

Platypus goes into rivers only in the morning and evening in search of food. After copulation the female lays two small membranous shelled eggs in a nest at the end of a long burrow which is blocked with mud at several places by the female. They are 6 mm in diameter and often superficially conjoined. After 14 days of incubation the young ones hatch, they are naked, blind and about 2·5 cm long. The female holds the young to the abdomen with her tail, and are fed on milk from sudorific mammary glands which open on flat patches of skin. There are no nipples. The male probably also secretes milk.

B. Marsupials or Pouched Mammals

1. *Macropus.* *Macropus* (kangaroo) (Fig. 30.7) is found in Australia, Tasmania and New Guinea on grassy plains. It has about 25 species. *Macropus* is the largest kangaroo. It is herbivorous and terrestrial and causes much damage to crops. It is up to 6 feet (1.80 metre) standing. The front part of the body including small head, neck, and small forelimbs and hind part of the body, legs, and tail are large. The small head has a small flat brain case, teeth are numerous with a dental formula 3,1,2,4/1,0,2,4 = 34. The small forelimbs have five clawed digits and are not used in locomotion. They are used as hands and to support the body while feeding. Hindlimbs are long and powerful, the foot has four digits, the hallux being absent, second and third metatarsals and digits are thin and small, these two digits are united together by integument so that the foot appears three-toed. The fourth metatarsal and digit are very large with a strong claw which is used in fighting, the fifth toe is small. The tail is long, round and thick, it serves as third leg in locomotion and at rest. Locomotion is bipedal with modifications in ilia and thigh muscles. The kangaroo moves rapidly by long erect springs of over 25 feet by jumping. The tail is used for balancing.

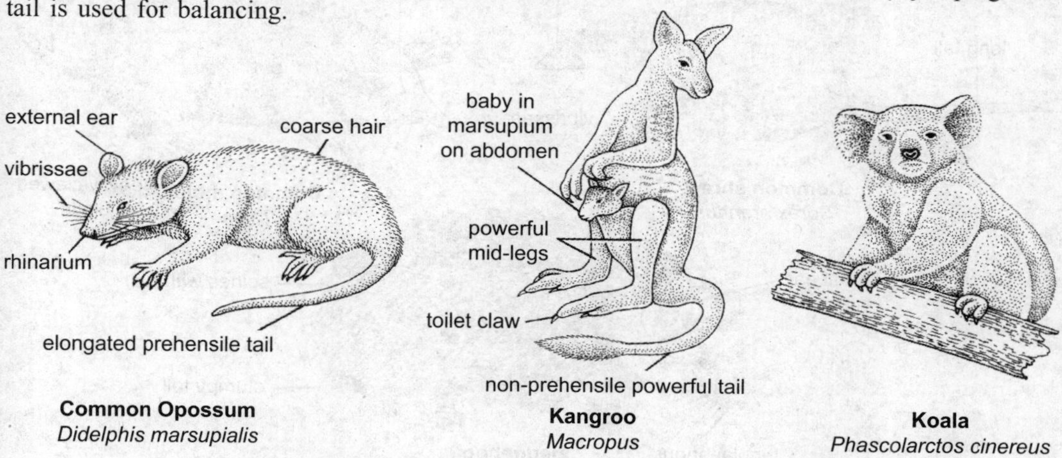

Fig. 30.7. Pouched mammals (marsupials).

The young develop in the uterus with a yolk sac placenta. They are born immature and naked, grub-like, about one inch (2.5 cm) long and are transferred into a large marsupium of the mother supported by epipubic or marsupial bones of the pelvic girdle. In the marsupium the young are fed on milk from true mammary glands having nipples. The young venture out when they are four to six months old. Kangaroos are terrestrial, gregarious and herbivorous, and cause damage to standing crops. They are hunted for meat, fur and sport.

2. *Didelphis.* *Didelphis* (Opossum) (Fig. 30.7) is found in Australia and North and South America. It is a small, arboreal, nocturnal and insectivorous marsupial. *Didelphis marsupialis* is the common opossum found in Eastern U.S.A. to South America. It is about 50 cm long and its body is covered over by a coarse fur. It has a long prehensile tail. Its legs are smaller, snout is elongated and eyes are large and black over its white face. It has a long scaly prehensile tail.

3. *Phascolarctos* (Koala). *Phascolarctos* (Fig. 30.7) is arboreal and feeds upon eucalyptus forests of Eastern Australia. It is about 75 cm long. Its body is covered over by woolly fur and has large ears, a soft pad on nose, and rudimentary tail. Female carries its young in a posteriorly opening abdominal pouch. First digit on both limbs is oppsable, used for grasping the objects. Incisor 10/8 in the upper and lower jaws respectively. It is arboreal, nocturnal and omnivorous. Gestation period is 12 to 14 days. Female give birth to 7–13 youngs, each about 1 cm long.

C. Insectivorous Mammals (Insectivora)

1. *Sorex.* *Sorex* (shrew) (Fig. 30.8) is rat-like. It is 3 inches (7.50 cm) long excluding the tail. The tail is about 2 inches. It is reddish-brown and pale. Snout is elongated and body is covered with silky fur. Eyes and ears are small. First pair of incisors is very long, lower canines are absent, all teeth are red, the dentition is 3, 1, 3, 3 / 2, 0, 1, 3 = 32. Limbs short and have five clawed digits. It lives in tunnels (burrows) and comes out in the night and feeds on insects, worms, and mice. Shrews have a higher rate of metabolism than other mammals. They cannot live without food for more than 10 hours. Because of their great expenditure of energy, shrews daily consume twice as much food as their weight. They are untamable, hence the name. Cats do not eat shrews because of their rancid odour. Shrews are found in Asia, Europe, and North America. The Indian shrew belongs to genus *Soriculus*, commonly called "chhuchhunder."

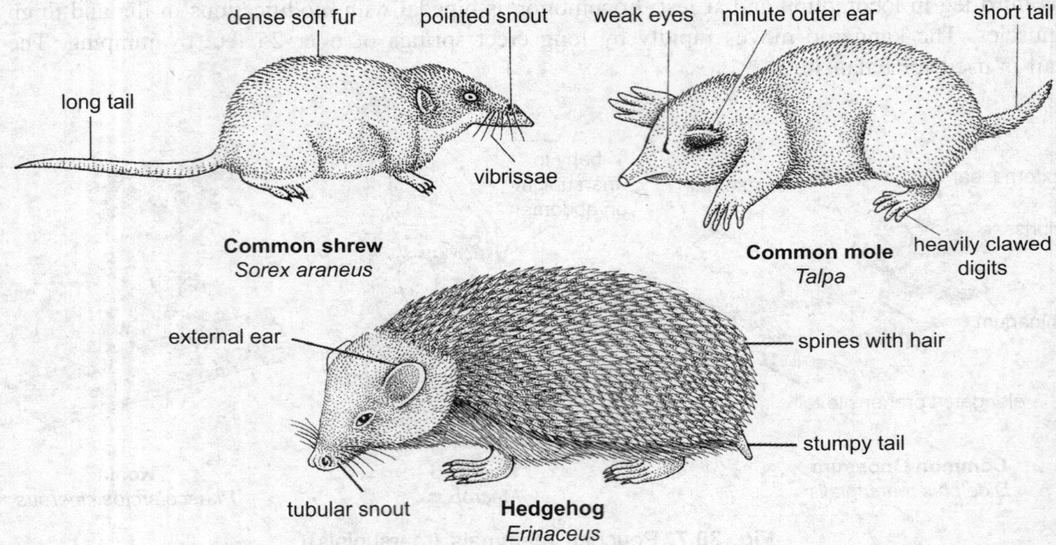

Fig. 30.8. Insect-eating mammals (Insectivora).

2. *Crocidura.* *Crocidura* (musk-shrew) is like a rat in body. The colour is light gray and the fur is silky. It has a pointed snout, eyes are small, dentition is complete, pinnae are small and depressed. Tail is comparatively short. The snout, pinnae, feet and tail are pinkish in colour with scanty hair. It is characterised by a strong obnoxious odour due to secretion of a pair of musk glands behind the arm pits. Their secretion is stronger during the breeding season. Musk glands are absent in females of some species. It is insectivorous and comes into houses at night in search of insects. It is found in Asia, Africa, Europe and N. America.

3. *Erinaceus.* *Erinaceus* (Fig. 30.8) is commonly known as hedgehog. Their body is small and globular with pointed snout, small stumpy tail and short legs. Eyes are small and black and pinnae are short. On the dorsal side hairs are modified to form sharp, outwardly projecting spines, including hairs and the ventral side has soft fur. A band of muscles passes along the sides and over the neck and base of the tail forming a sphincter for moving the spiny area. The animal rolls up into a ball when in danger and the spines project outward causing harm to the attacking animal. It is nocturnal and feeds on insects, slugs, fruits, birds and snakes. It is immune to snake bite. The hedgehog hibernates in winter. It is found in Asia, Africa, and Europe.

4. *Talpa.* *Talpa* (Fig. 30.8) is commonly known as **short-tailed mole** and is also called "chhuchhunder." Moles are related to shrews but are larger in size and have enormous shoulder girdle and muscles. They are profoundly specialised for a subterranean burrowing life. Body is thick, cylindrical and about 6 inches (15 cm) long. The head is wedge-shaped with a prenasal bone in the snout, eyes are rudimentary, and pinnae are minute. Tail is short with no hair, and is sensory. It has a thick velvety fur. Forelimbs are strong, manus is large and broad, five digits have long, broad nails, there is an extra sickle-shaped bone of the carpus called **radial sesamoid** or falciform bone. The hand is used for digging. The mole lives underground digging long tunnels rapidly. It almost never comes to the surface and has a remarkable sense of touch. It feeds on earthworms. It is found in Asia and Europe, the golden mole is African.

D. Flying Mammals

1. *Pteropus* or *Cynopterus* (Frugivorous bats). *Pteropus* (Fig. 30.9) is commonly known as **flying fox** due to fox-like head with long snout and large eyes. They belong to the suborder Megachiroptera. It is a large bat with a wing spread over five feet (150 cm) though the body is only about a foot (30 cm) in length. The body is covered with brown fur. Ears are oval and the two edges of the ear are in contact at the base. Tail is rudimentary. The thumb (first digit) and second digit are clawed. Dental formula is *i* 2/2, *c* 1/1, *pm* 3/3, *m* 2/3. The molars are not tubercular but marked with a longitudinal groove. Social in habit and move about in droves of considerable size. *Pteropus* occurs in India, Sri Lanka, Australia, Africa, and Madagascar.

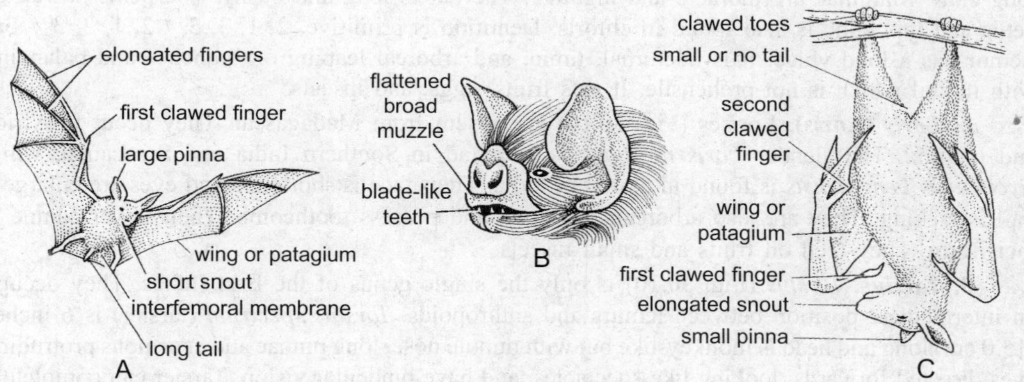

Fig. 30.9. Flying mammals or bats. A – An insect eating bat (Microchiroptera); B – Head of a Vampire bat, *Desmodus*; C – A fruit eating bat (Megachiroptera).

2. *Rhinolophus* (Insectivorous bats). They belong to suborder Microchiroptera. They are small in size and their body is covered with soft fur. Snout is short with or without nose leaf. Pinnae are large and often provided with special flaps in front of ears used for navigation or echolocation. Eyes are small and the vision is weak as the visual rods are poorly developed. Tail is included in the interfemoral membrane provided with a distinct flap. Only the thumb (first digit) is clawed in the forelimbs. Hindlimbs are weak and have five clawed digits. Nocturnal in habit. During the day they spend their time sleeping in caves or other dark sheltered places and come out at dusk for feeding. Gregarious, living in colonies of thousands. They produce ultrasonic sound waves which after striking on the solid objects are reflected back and are picked up by their ears, guiding in their flight. *Rhinolophus* is commonly known as horse-shoe bat and *Rhinopoma* is mouse-tailed bat. They are cosmopolitan in distribution.

3. *Desmodous* (Vampire bat). Tree vampire bats occurs in tropics and are famous because of their sanguinivorous habit. They often feed on fresh blood of sleeping domestic animals, even human beings. Their incisors are razor-sharp with which they slit the skin of the prey, where hairs and feathers are scanty, and the oozing blood is lapped up. The vampires are very sly so that the sleeping victim often remains unaware. They transmit paralytic rabies in their victims.

Associated with the blood sucking habit is the very small lumen of their oesophagus through which solid food cannot pass. The first digit of the forelimb bears claw. Tail is present. Pinnae are large with a lobe called **tragus**. Molars bear transverse grooves. Vampires do not have a nose-leaf. Instead, they have a naked pad wearing U-shaped grooves. Common genera are *Desmodus* and *Megaderma*.

E. Primates

The order Primates is interesting because it includes man, besides lemurs, tarsiers, monkeys and apes. They inhabit chiefly warmer parts of the world. They stand first in the animal kingdom in brain development. However, they are relatively unspecialised and mostly tree dwelling (arboreal). Limbs have 5 digits and generally bear nails instead of claws. Great toe and thumb are opposable to other digits for grasping. Their eyes are in front of head and meant for stereoscopic vision. Some primates are as follows :

1. *Lemur*. *Lemur* (lemur) (Fig. 30.10) is found today only in Madagascar and neighbouring islands. Their fossils of the Palaeocene and Eocene are found in Africa, Europe, and America. These are supposed to be the lowest primates. The head and skull are long, face is fox-like with cleft and moist upper lip, the large eyes are directed sideways, there is no binocular vision. Hands are broad. Thumb and great toe are opposable. Digits with nails but second toe bears a long claw. Mammae are thoracic and inguinal. The tail is long and bushy. The penis has an os penis and this bone is also found in clitoris. Dentition is primitive, 2, 1, 3, 3, / 2, 1, 3, 3 = 36. Lemur has a loud voice. It is nocturnal, timid, and arboreal leaping on branches and balancing with its tail which is not prehensile. It eats fruits, eggs, and insects.

2. *Loris* (Loris). Lorises (Fig. 30.10) are absent from Madagascar. They occur in Africa and in Asia. The slender *Loris tardigradus* is found in Southern India and Sri Lanka, while *Nycticabus bengalensis* is found in Northern India. Their face is shortened and eyes are enlarged. Tail is lacking. They are also arboreal, nocturnal and possess toothcomb. Their lower canine is incisiform. They feed on fruits and small insects.

3. *Tarsius*. *Tarsius* (Fig. 30.10) is only the single genus of the East Indies. They occupy an intermediate position between lemurs and anthropoids. *Tarsius spectrum* (tarsier) is 6 inches (15.0 cm) long and head is monkey-like but with minute nose, long pinnae and enormous protruding eyes directed forwards, looking like spectacles, and have binocular vision. Tarsier can completely turn its head around without moving the body. Limbs are long, the hindlimbs being longer with elongated tarsals, hence the name. They have five thin digits with round adhesive pads and flat nails, except the second and third digits of the foot which have retained the ancestral claws. Tail is long with a bushy tuft at the tip. Ventral side of tail is scaly. It is nocturnal, insectivorous and arboreal living in pairs in hollows of trees. It also feeds on small amphibians and reptiles.

4. *Daubentonia*. *Daubentonia* (= *Cheiromys*) *madagascariensis* is commonly known as Aye-aye (Fig. 30.10). It is about the size of a cat. It is found in Madagascar. It is like lemurs in many ways. There is one incisor on each side in each jaw continually growing as in rodents and used for gnawing. It has a very long and wiry middle finger for impaling insect grubs.

5. *Ateles*. *Ateles* (Fig 30.10) is commonly called spider monkey. It is also a New world monkey with a long prehensile lail used in hanging over the tree branches or for picking up small objects. *Ateles paniscus* is found in tropical forests from Mexico to Brazil.

6. *Alouatta*. *Alouatta* (howling monkey) is known for its loud howling due to an ossified bag-like outgrowth of the larynx which acts as a resonator, and a much enlarged bell-shaped hyoid having pharyngeal air sacs. In howling the throat is inflated. Howling is to assert the territorial rights of the clan. These are the largest of the New World monkeys. Body and limbs are slender. It is an ugly looking monkey with protruding snout, naked face, hair on the forehead is directed backwards and that on the head forwards. The colour is black or brown, thumb is opposable, nose is flat, and the long tail is prehensile. It lives in the forests of South America in families, and cooperation is by sound language.

7. *Macaca* and *Semnopithecus*. *Macaca* (rhesus monkey or macaque) (Fig. 30.10) has several species in Asia and North Africa; *Semnopithecus* or *Presbytis* (langur) (Fig. 30.10) of North India has no cheek pouches, hindlimbs are longer than forelimbs, and stomach is divided into three chambers. The colour is silvery gray with a black face. *Macaca mulatta* is common in Northern India, China and Vietnam. It has simple stomach and has cheek pouches for food storage. *Rh blood factor* was first discovered in this monkey. These are used in biological investigations for testing the effects of newly discovered medicines, etc. They live in groups in open country and not arboreal. *Mandrillus* (mandrill) has red and purple stripes on the face. It is found in West Africa.

8. *Papio*. *Papio* (Fig. 30.10) is commonly known as baboon. The large baboons are the Old World monkeys from Africa. They have a dog-like face with a long muzzle and large canine teeth. They are the most ferocious and perhaps the most despised. They are not arboreal but live in open country in a family groups with hierarchial organisation.

9. *Hylobates*. *Hylobates* (gibbon) (Fig. 30.10) is a small ape. It has several species in Assam and South-east Asia. It is about 3 feet (90 cm) long with slender body and limbs. The males are black but females are yellowish. It is peculiar among apes in having ischial callosities. Skull is rounded and has no crest, canines are large and sabre-like. Voice of gibbons is very powerful. Arms are long and muscular used for locomotion in trees by swinging in the branches. Arms touch the ground in standing posture, it is erect in walking and sitting but it balances itself in walking by knuckles touching the ground. Feet are used for holding objects. Gibbons live in trees and rarely descend to the ground. It is frugivorous and arboreal, living in hilly forests where it swings along the branches by its arms with great speed.

10. *Pongo*. *Pongo* or *Simia satyrus* or *Pongo pygmaeus* (orang-utan) (Fig. 30.10) has only one species found in Borneo and Sumatra. It is about 4 feet (120.0 cm) high with long arms and short legs. It has reddish-brown hair. The head is short and broad with eyes very close together. The skull has a sagittal crest for attachment of strong neck muscles. Male has a big throat, sac-like goitre and fatty swellings in the cheeks. Body has a heavy build, consequently it is a poor climber. On ground it runs on all fours. It builds a platform between two branches of a tree on which it lives and feeds on buds and fruits.

11. *Anthropithecus*. *Anthropithecus troglodytes* or *Pan troglodytes* (chimpanzee) (Fig. 30.10) is an African ape of black colour. It is about 5 feet (150 cm) high but the body is not heavy and it is a good climber. It is most intelligent among apes and can be tamed in young age. Head is large with prominent brow ridges, skull has a prominent sagittal crest for neck muscles, jaws are heavy and project forwards (prognathous). Chimpanzees are limited to central equatorial Africa. They build nests in trees and are largely frugivorous.

12. *Gorilla*. *Gorilla* (gorilla) (Fig. 30.10) has a single species restricted to tropical African forests. It is about five and a half feet (165 cm) high and has a massive build weighing up to 6 maunds (250 kg). The skin and coarse hair are black and chest is broad. Powerful arms reach the middle of the legs in standing. Skull has a heavy sagittal crest for very powerful neck muscles, brow ridges are prominent, jaws are heavy and project forwards, canines are large. It is a very intelligent and fierce ape. All these characters are more pronounced in old males than in females or young males. On ground locomotion is bipedal and plantigrade, but they often walk on all fours. It is transitional between an arboreal and terrestrial life. It builds nests in trees and eats fruits and plants. It lives in families of several females and young males with an old male as the head. It is very strong and fights fiercely with its hands and teeth.

Fig. 30.10. Modern Primate mammals.

F. Toothless Mammals

1. ***Bradypus***. *Bradypus* (3-toed sloth) (Fig. 30.11) is specialised for an arboreal life, slow moving on the trees and can hardly walk on the ground. Head is short, rounded with a flat face, forward looking eyes, pinnae are small and tail is rudimentary. Limbs are long and slender, the forelimbs being longer. There are three digits in each limb having long, curved, hook-like claws by which it hangs upside down from branches and also sleeps in this position. The hair is heavy, coarse, and longitudinally grooved and algae grow on the hair imparting a greenish-blue colour to the animal – a protective adaptation. Teeth are very few, only 4 or 5 similar molars in each half of each jaw. Incisors are absent. It is a nocturnal, vegetarian, and sotitary animal of slow habits. It is found in the rain forests of Central and South America. *Choloepus*, the two-toed sloth of South America, has two clawed digits in the forelimbs and three in the hindlimbs. The sloths and sea cows (Sirenia) are the only mammals without the typical seven cervical vertebrae.

2. *Dasypus*. *Dasypus novemcinctus* (armadillo) (Fig. 30.11) is nocturnal, fossorial, and omnivorous, found in Central and South America. It is about 75 cm long. The head is short, broad, and depressed. Armadillos are unique among mammals in having an armour of bony dermal plates or **osteoderms** which are covered with horny epidermal scales or scutes, this armour forms two large shields, a scapular and a pelvic shield, between the shields are 3 to 9 transverse bands of the armour. Bony osteoderms and scutes also cover the head and tail. Embedded in the armour are hairs. Ventral side has no armour. There is a long protrusible tongue for feeding on insects, but it also eats dead bodies. In the adult incisors and canines are absent but there are numerous continuously growing molars with no enamel. Eyes are small, pinnae are long and close together. Limbs are short with 3 to 5 digits in forelimbs for digging, and 5 digits in hindlimbs, digits bear strong claws. It can roll into a ball for defence. The animal is unique in producing four or eight young ones at one time which are of the same sex because they develop from a single fertilised egg. This is because of **polyembryony** in which the single fertilised egg divides at an early stage into 4 or 8 secondary embryos, all of which develop within a single amnion.

Related to armadillos were the extinct glyptodonts (*Glyptodon*) of the Pleistocene. Their skull had a hemispherical carapace made of many small polygonal pieces fused together. The carapace was supported by blocks of fused vertebrae. The thick tail had armoured rings often with spines. There was no clavicle and zygomatic arch had a long downgrowth.

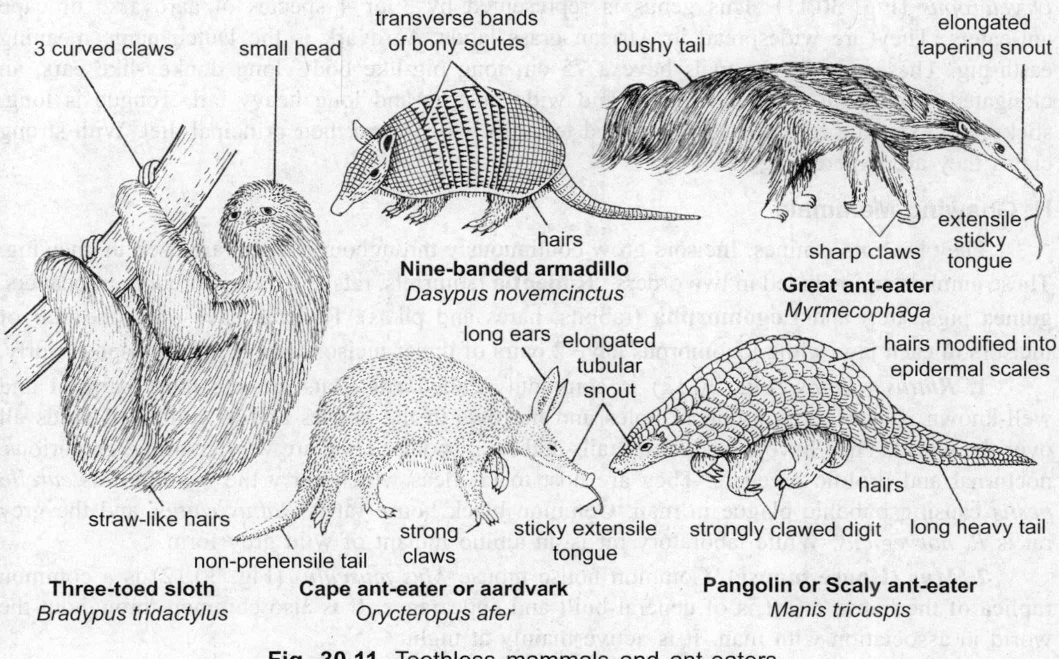

Fig. 30.11. Toothless mammals and ant-eaters.

3. *Myrmecophaga*. *Myrmecophaga* (giant ant-eater) (Fig. 30.11) is found in South America. It is over 6 feet (180.0 cm) long with coarse hair and a long and very bushy tail. Across the shoulder is a broad black and brown stripe. The head is long and narrow drawn out into a long narrow snout with nares at the tip. The mouth is a small aperture at the tip of the snout and eyes are small. There are no teeth. There is a very long protrusible sticky tongue for capturing ants and termites. Forelimbs are powerful with 5 clawed digits, third claw is very large, curved and sharp used for digging ant-hills and for defence. Hindlimbs have 4 digits with normal claws. In walking the digits of the manus are bent and their dorsal surface is placed on the ground.

G. Other Ant-eaters

Scaly ant-eaters or pangolins are found in Old World belonging to the order Pholidota, and the cape ant-eaters (aardvarks) belong to the order Tubulidentata. They resemble with edentates (upper three animals) only due to similar mode of life.

1. *Manis.* *Manis* (scaly ant-eater or pangolin) (Fig. 30.11) is about 5 feet (150.0 cm) long. The head, body, and tail except snout, sides of face and ventral surface of body are covered with large, rounded, horny epidermal overlapping scales and between the scales are some coarse hair. Head is small with a short pointed snout and eyes and pinnae are small. There are no teeth. There is long sticky tongue retracted into a special sac used for feeding white ants. Limbs are short and strong with 5 digits. Forelimbs have long curved claws for opening termite hills and for burrowing. Tail is long and broad. Stomach is reduced but lined with a thick layer of keratin for grinding insects. Pangolin also climbs trees. When in danger, it rolls into a ball for self-defence. It is said that it often feigns death, erects the scales and traps the ants which crawl over by closing the scales, then it enters water and devours the ants as they float on water. Pangolin is a burrowing terrestrial animal found in India, Africa, and Asia. *Manis crassicaudata* is found in South India. *M. pentadactyla* is distributed in India and Sri Lanka and *M. aurita* in Nepal and China.

2. *Orycteropus* (Cape ant-eater). Order Tubulidentata is represented by a single genus *Orycteropus* (Fig. 30.11). This genus is represented by 3 or 4 species of aardvarks or cape ant-eaters. They are widespread in African grass lands. Aardvark is the Dutch name meaning earth-pig. These peculiar animals have a 75 cm long pig-like body, long donkey-like ears, an elongated snout with a tubular mouth and wide nostrils and long heavy tail. Tongue is long, sticky and extensile for collecting ants and termites which form their principal diet. With strong claws they are expert diggers.

H. Gnawing Mammals

They have no canines. Incisors grow continuously throughout life and are used in gnawing. These animals are included in two orders ; **Rodentia** (squirrels, rats and mice, porcupines, beavers, guinea pigs, etc.) and **Logomorpha** (rabbits, hares and pikas). Rodents have only one pair of incisors in each jaw, while logomorphs have 2 pairs of upper incisors arranged antero-posteriorly.

1. *Rattus.* *Rattus* (Fig. 30.12) is commonly called **rat**. Rats are the most harmful and well-known rodent pests, living in holes and burrows in the houses and in cultivated fields all over the world. They have naked, scaly tails and long snouts. They are very cunning, gregarious, nocturnal and prolific breeders. They are host to rat fleas which carry the bacillus *Pasteurella pestis* causing bubonic plague in man. Common black house rat is *Rattus rattus*, and the grey rat is *R. norvegicus.* White laboratory rat is an albino mutant of wild grey form.

2. *Mus* (House mouse). Common house mouse *Mus musculus* (Fig. 30.12) is a common raplica of the house rat. It is of general built and appearance. It is also common throughout the world in association with man. It is active mainly at night.

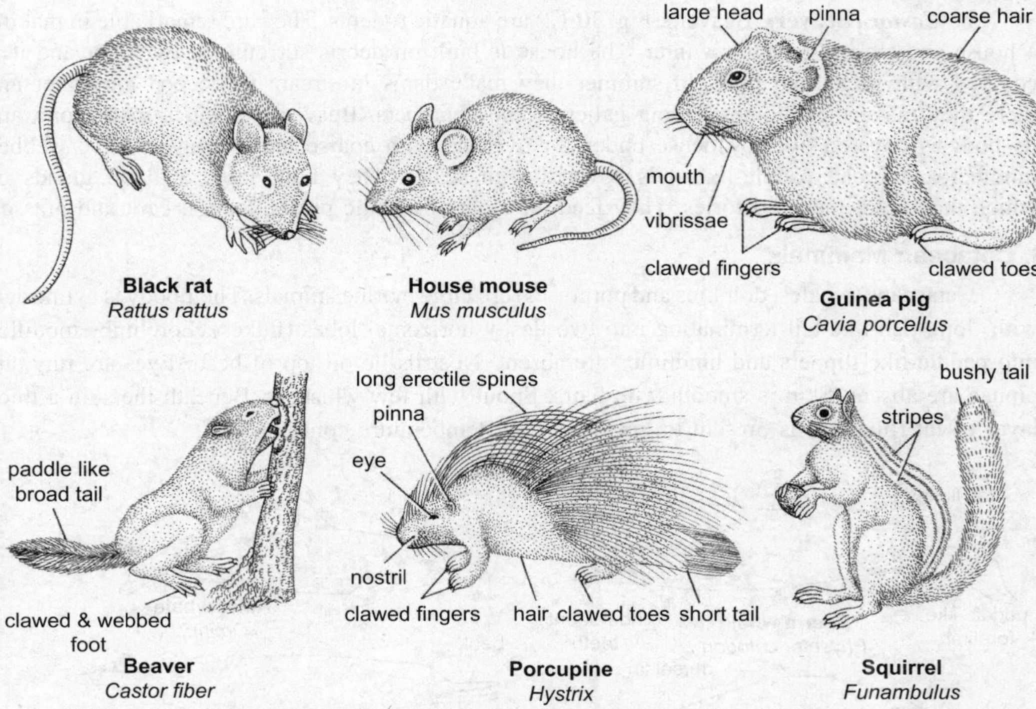

Fig. 30.12. Gnawing mammals.

3. *Cavia.* *Cavia* (Fig. 30.12) is commonly called Guinea pig. Like white rats, guinea pigs (*Cavia*) are also used extensively for research. In the wild, they occur in rocky or swampy areas near forests or grasslands. Body is about 20 cm long, with fairly coarse hairy coat, large head, long thin limbs and rudimentary tails. It digs its own burrow or occupies deserted burrows of other animals. Like rabbit, it is crepuscular, vegetarian, coprophagous and easily domesticated.

4. *Funambulus*. *Funambulus* (Fig. 30.12) is commonly called **squirrel** or **gilhari** in Hindi. Body is elongated and covered with fur. Three to five longitudinal stripes of dark colour are present on the back. Tail is long and bushy. The eyes and pinnae are large. The forelimbs have an inconspicuous thumb and hindlimbs have four clawed digits. The soles of forelimbs are naked and those of hindlimbs are hairy. Arboreal and active climber. Diurnal and builds nests of twigs and leaves. Feeds on nuts, seeds and fruits. Squirrels are world-wide in distribution with the exception of Australian region and islands of Madagascar.

5. *Hystrix.* (Fig. 30.12) is commonly called porcupine or syahi in Hindi. Body is covered with long erectile spines which are modified hairs. The spines are solid in the middle of the body but on the tail they are expanded into hollow quills which make much rattling. The spines are black and white, the middle of the spine being banded with black. A great crest of stiff long hairs is present on the head. Eyes and pinnae are small. Tail is short and never prehensile. The soles of the feet are smooth and not covered with rough tubercles. The tongue has serrated scales arranged in transverse rows which are directed backwards. Nocturnal and herbivorous feeding on crops, vegetables and plant roots. During day time, it mostly remains in the burrow. The spines present on the body are, however, excellent weapons of offence and the animal charges somewhat backwards to make best use of them against the enemy. *Hystrix* is found from East Indies to Africa, Europe and Asia. *Hystrix leucura* and *H, bengalensis* are found in India. *Hystrix indica* is also found in Indian forests, rocky hills, etc.

6. *Castor (*Beaver*)*. Beavers (Fig. 30.12) are aquatic rodents. They are remarkable in making a house and storing food for winter. The house is built on debris surrounded by water and it is covered with sticks and mud. In summer they make dams in stream to deepen the water and there they make their homes having galleries and chambers. Beaver is about a metre long and its body is covered over by dense under fur overlaid with coarse hairs. Beavers have webbed hindfeet and broad paddle-like tails covered with scales. They have large anal oil glands by which they mark their territories. They feed on roots of aquatic plants, leaves, bark and grasses.

I. Cetacean Mammals

Cetaceans (whales, dolphins and porpoises) are large marine animals. Their body is cylindrical with long fish-like tail terminating into two fleshy horizontal lobes (flukes). Forelimbs modified into paddle-like flippers and hindlimbs are absent. Nostrils lie on top of head. Eyes are tiny and pinnae are absent. Skin is smooth with hairs. Snout with few whiskers. Beneath the skin a thick layer of fat (blubber) is present to keep the body temperature constant.

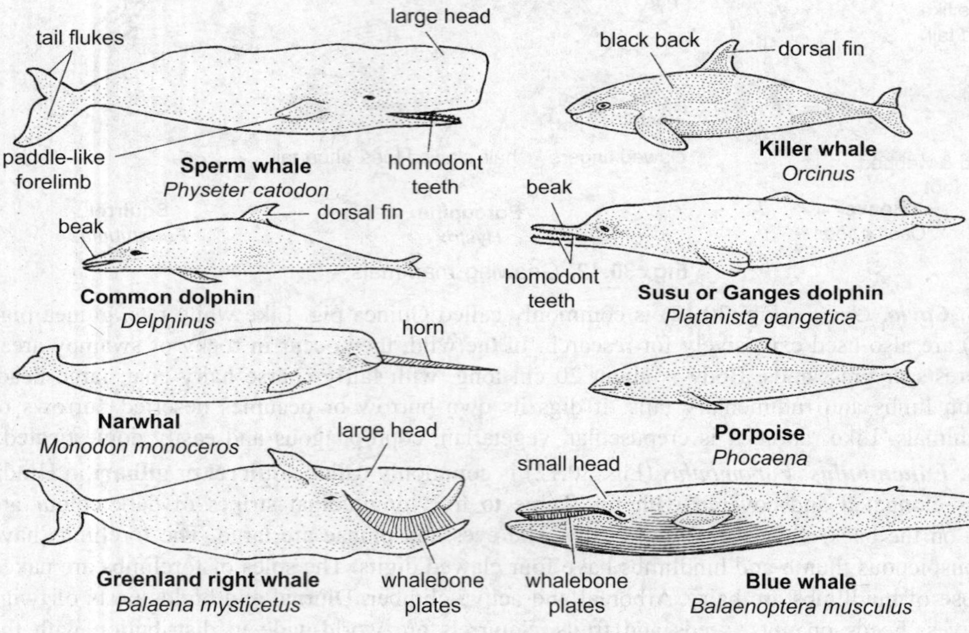

Fig. 30.13. Cetacean mammals.

1. *Physeter*. *Physeter catodon* (sperm whale) (Fig. 30.13) is an enormous heavy-bodied whale over 60 feet (up to 25 metres) long. It is the largest toothed whale. Head is large and produced into a swollen rostrum which is a reservoir of **spermaceti** or liquid oil contained in a large cavity. Spermaceti makes valuable whale oil which solidifies in air, it makes the animal hydrostatic. It is non-edible and used as lubricant and in making candles. Dorsal fin is absent. Nostril is single. Lower jaw is shorter, narrow and armed with numerous conical teeth which fit into grooves of the upper jaw. Upper jaw is longer and has vestigial teeth in the gums. The skull is highly modified, maxillae and nasals of the two sides are unequal. All seven cervical vertebrae are fused together. Intestinal secretion forms a stone-like **ambergris** which is used in making perfumes. *Physeter* is a fast swimmer. It feeds on squids and is found in the Southern Seas.

2. *Orcinus*. *Orcinus orca* (killer whale) (Fig. 30.13) has a fish-like body. It is black with white and yellow marks. The male is 30 feet (9 metres) long and has a long, triangular and pointed dorsal fin. The female is 18 feet (6 metres) and has a recurved dorsal fin. It is a fierce cunning hunter with conical teeth in both jaws, feeding on fish, birds, and marine mammals. It even attacks large whales.

3. *Platanista.* *Platanista gangetica* (Ganges dolphin) (Fig. 30.13) is a river dolphin, found in the Ganges, Indus, Brahmaputra, and their tributaries. It is more primitive than other whales and is related to forms found in South America and China. It is a small toothed whale 6 to 9 feet (2–3 metres) long, commonly called "susu". Males are slightly smaller with shorter beak or snout. It is dark gray in colour and is highly specialised. The beak is fairly long with thin sharp teeth (about 200) on both jaws. Head is round and skull has a high fan-shaped bony process called a **maxillary crest**, neck short in which the cervical vertebrae are not fused and in this respect it differs from other whales. It has rudimentary eyes. Forelimbs are modified into simple paddles with no increase in the number of digits or phalanges. Dorsal fin is absent represented by only a ridge. Nipples are close to the anus. It feeds on fish and crustaceans groping in the river mud. *Platanista* differs from marine dolphins and porpoises which have well formed eyes, no neck, and have a dorsal fin.

4. *Delphinus.* *Delphinus* (Fig. 30.13) (common marine dolphin) is very common, having a large number of teeth. It is 8 feet (2.5 metres) long with a rounded head, distinct short snout or beak with 130 conical teeth on the two jaws. Eyes are well developed. There is a large dorsal fin. Neck with free cervical vertebrae, but atlas and axis are fused. It has a wide distribution in all seas. All dolphins are very intelligent.

5. *Phocaena.* *Phocaena* (porpoise) (Fig. 30.13) is a small, 6 to 8 feet (about 2 metres) long whale. The body is black above and light below. The head is rounded, no distinct beak but jaws have numerous teeth. Dorsal fin is large with a row of tubercles. Paddles (forelimbs) are broad with the number of phalanges reduced. There is no neck and six cervical vertebrae are fused. At regular intervals porpoises arch their backs above water to breathe air. It feeds on fish, crustaceans and molluscs. It is found in the Atlantic and the Pacific Oceans and often swims up in rivers. It is gregarious and often follows ships. Because of its intelligence it can be easily trained.

6. *Monodon.* *Monodon* (Fig. 30.13) is commonly called narwhal. *Monodoon monoceros* is found in Arctic waters. It grows up to 5 metres in length. In male, one of the two uppper incisors grows into a spirally twisted tusk or horn, up to 2 metres long, and of unknown function. The female retains undeveloped incisors, buried in premaxillae.

7. *Balaenoptera* or *Sibbaldus.* *Balaenoptera* or *Sibbaldus musculus* (blue whale) (Fig. 30.13) is a whalebone whale. It is the largest whale being 100 feet (up to 35 metres) long and weighing 120 tons (an elephant is 6 tons). The torpedo-shaped body is blue above and yellow below, hence, it is often called the sulphur-bottom whale. The skin is smooth and naked but deeply grooved below the throat. Dorsal fin is small and far back. Head is small and the mouth cavity is provided with a series of parallel horny plates hanged from the palate, called the **baleen** or **whalebone**, acting as sieve or strainer for planktons. Cervical vertebrae are not fused. It is found in the Antarctic and Arctic Oceans. It migrates to temperate waters for breeding.

J. Carnivorous Mammals

Carnivores are swift and strongest mammals. Their jaws are powerful with large canines for killing and tearing the prey and cheek teeth for cutting. Limbs provided with strong claws for gripping. Suborder **Fissipedia** includes terrestrial carnivores having strong curved claws on the limbs, *e.g.,* cats, dogs, hyaenas, wolves, foxes, badgers, racoons, lions, tigers, bears, etc. Suborder **Pinnipedia** are aquatic carnivores having flipper-shaped limbs with webbed digits, *e.g.,* seals, walrus, sea lions, etc.

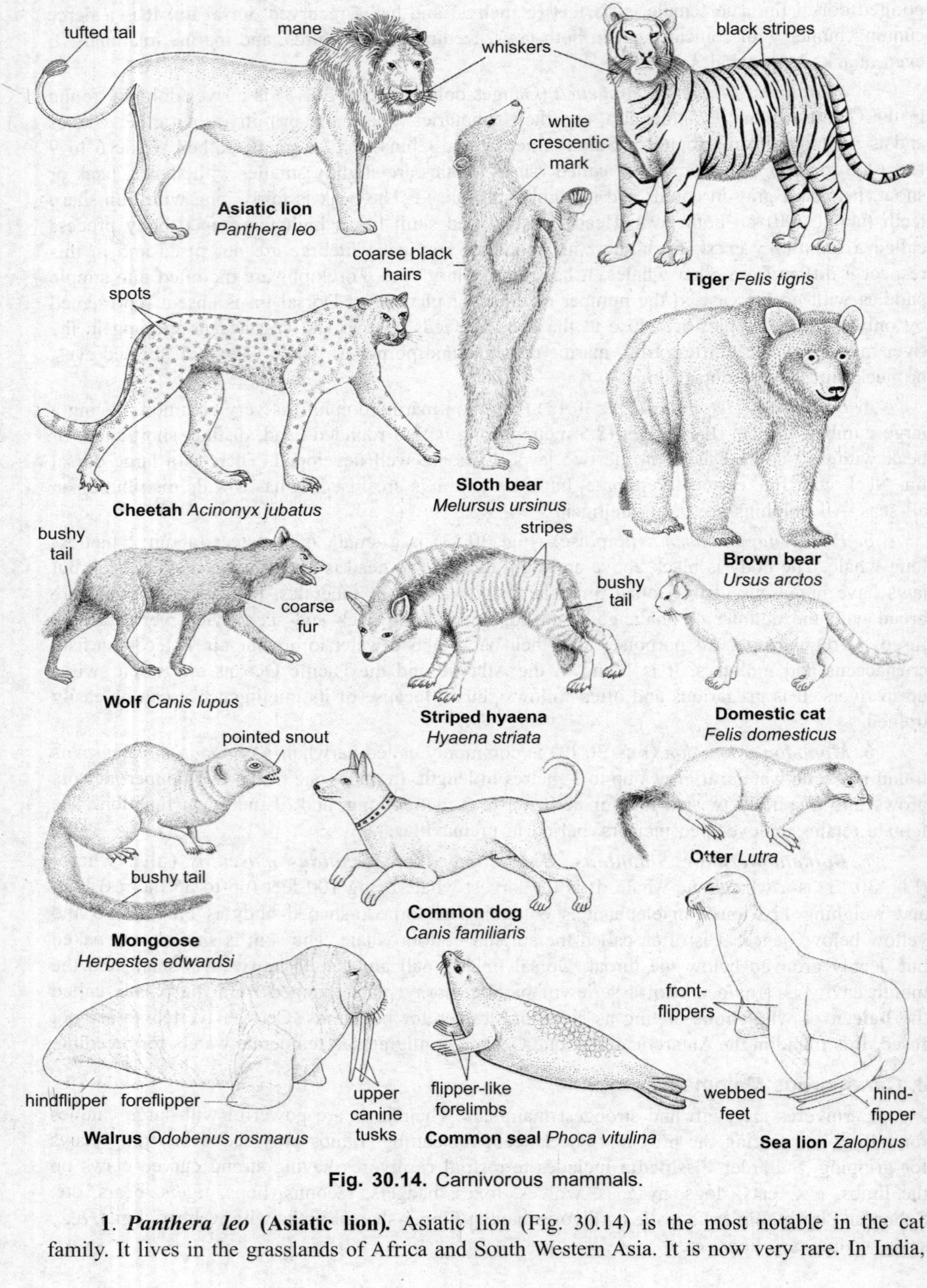

Fig. 30.14. Carnivorous mammals.

1. ***Panthera leo* (Asiatic lion).** Asiatic lion (Fig. 30.14) is the most notable in the cat family. It lives in the grasslands of Africa and South Western Asia. It is now very rare. In India, it is preserved in a sanctuary in Gir forests in Gujarat. Body is about 2 metres long and uniform tawny yellow in colour. Tail is tufted. The male is easily distinguished having a ruff of long hairs called **mane** round the neck and shoulders. Digits have sharp curved claws, which are retractile into sheaths and pads below digits form compact cushions. They hunt large grazing animals (antelope, zebra, etc.) and live in a group called **pride**. They secure their prey with a mighty spring after stalking or lying in wait. Teeth are highly specialised for shearing and tearing but not for crushing, carnassials are highly developed, premolars and molars are small. It is very robust, famous for its strength, courage and specialised roar.

2. *Felis tigris* (Tiger). *Panthera tigris* or *Felis tigris* (Fig. 30.14) is a large and much dreaded carnivore, found in forests of Asia including India. The colour of its coat is orange-yellow, striped with black. Tiger is solitary and male has no mane. It can kill the largest herbivorous animals and attacks even man. Since, 1972, the tiger has become the national animal of India. Under save tiger project, it is protected in several sancturies such as Sariska, Ranthambore, Simlipal, Corbett, etc.

3. *Acinonyx* (Cheetah). Cheetah or hunting leopard, *Acinonyx jubatus* (Fig. 30.14), inhabits grasslands and semideserts of Africa and South-Western Asia. It is 1.5 metres long, leanly built with non-retractile claws. It is the fastest running land's carnivorous animal reaching a top speed of 70 miles per hour. They hunt in groups. They are easily tamed and they make docile pets. Cheetah became extinct in India early in the present century.

4. *Canis lupus* (Wolf). *Canis lupus* (Fig. 30.14) is restricted to relatively uninhabited areas of Northern Hemisphere. They have coarse brown or grey fur, bushy tail and pointed ears. They have slender limbs and are generally good runners, moving on the tips of their toes (digitigrade). They hunt in packs of about 30 and prey on small mammals. It is a great menace to stock raisers, killing calves and other young domestic animals.

5. *Felis domesticus* (Cat), *Panthera pardus* (Leopard), *Uncia* (Snow leopard). They are terrestrial, nocturnal hunters with short but high skull and large eyes. Digits have sharp curved claws, which are retractile into sheaths and pads below digits form compact cushions. Tail is long and tufted in lion. They secure their prey with a mighty spring after stalking or lying in wait. Teeth are highly specialised for shearing and tearing but not for crushing, carnassials are highly developed, premolars and molars are small. They occur mostly in the tropics but have a wide range of distribution, except in Australia and Madagascar. Lion is preserved in Gir forest in Gujarat. Male is distinguishable by the presence of a ruff of long hairs round the neck and shoulders, called **mane**. It is very robust with a specialised roar.

6. *Hyaena*. Hyaenas (Fig. 30.14) are scavengers, feeding on carrion though they are also carnivorous. They are spotted (*Crocuta*–African hyaena) or striped (*Hyaena striata* – Indian hyaena) is also called "lakad baggha." Its hind part is low and front part is heavy. Pinnae are large. Jaws are massive with massive cheek teeth used for crushing. There is only one molar in each jaw but the upper molar is vestigial and placed transversely. Carnassials are well developed. Limbs end in four digits with non-retractile claws. Voice is almost human. They are found in Asia and Africa.

7. *Canis familiaris* (Dog), *Canis aureus* (Jackal), *Vulpes* (Fox) and *Cyon* (Wild dog) (Fig. 30.14). They are universally distributed except in Australia, New Zealand and Madagascar. Dentition is complete except for the absence of last upper molar. Carnassials are well developed. Limbs are long and strong for fast running, forelimbs have 4 to 5 digits, hindlimbs have 4 digits. Claws are non-retractile. Pollex and hallux are reduced. Locomotion is digitigrade (move on the tips of their toes). They prey over cattles (sheep, goat and calves, etc.) and even children. Thus, they are a great menace to farmers.

8. *Meles* (Badger) and ***Lutra* (Otter).** In both badger and otter, the body is slender and neck long. Carnassials are well developed, last upper molar is expanded on its inner side. Claws are non-retractile. Badger is omnivorous, the large claws of forelimbs and snout are modified for digging roots. Otter (Fig. 30.14) is semi-aquatic, found in rivers and lakes being adapted for an aquatic life. The short, dense fur over the body is oily. The digits in both the limbs are webbed. Tail somewhat flattened and pinnae small. Food consists of fish, crabs and frogs.

9. *Herpestes*. *Herpestes* (Fig. 30.14) is commonly known as mongoose or nyola in Hindi. Body is elongated and covered with yellowish grey fur. Head is elongated and possesses pointed snout. Eyes are small and pinnae are small and rounded. Pinnae possess complicated folds so as to close the ear openings when the animal is buried in the mud. Tail is long. Fore-and hindlimbs have five digits with fossorial claws. Carnivorous and highly predaceous. It also takes vegetable food. *Mongoose* is famous for its fight with the snakes. It kills the snakes and feeds on their blood. *Herpestes* is distributed between the Ethiopian and Oriental regions. *Herpestes* is very common in India and represented by *H. bengalensis, H. edwardsi, H. fuscus and H. vitticolis.*

10. Bear. Bears have a wide distribution. They are large with long coarse hair. Limbs have five clawed digits and locomontion is plantigrade. Bears can climb trees. Tail is short. They are partly carnivorous, but in sloth-bear the carnassials are not formed, molars are large, elongated and bunodont, *i.e.,* crowns with blunt and rounded tubercles. The **sloth-bear** (*Melursus ursinus*) (Fig. 30.14) is found in the hilly forests from India to Sri Lanka. It has long coarse black hair and a white crescentic mark on its chest. It has very large claws. Snout is long with an elongated lower lip, the tongue is extensible, and some teeth are rudimentary. The sloth-bear lives mostly on fruits, honey and insects, and also sucks up white ants from their hills which are broken by its claws with great speed. *Selenarctos tibetanus*, the black Himalayan bear, is larger than a sloth-bear. It has a white V-shaped mark on the chest. It is a carnivorous form with well-developed carnassials.

***Ursus arctos* (brown bear)** (Fig. 30.14) is found in Asia, Europe and North America. *Thalarctos maritimus* (polar bear) is found in the Arctic region. It is the largest and strongest amongst other types of bears. Body and legs are covered with dense white fur. It feeds on fishes and seals.

11. *Odobenus.* *Odobenus* (Fig. 30.14) is commonly known as Walrus. Walrus is a largest seal found in the Arctic Ocean. The wrinkled body is large and heavy, there is very little hair, but there is a thick layer of subcutaneous fat. On the upper lip are thick bushy sensory bristles. There are short cartilaginous pinnae. The upper canines form long downwardly directed tusks in both sexes, tusks are used for digging up food and for locomotion on land. Forelimbs form paddles which are the primary swimming organs, hindlimbs can be turned forwards for locomotion on land and for swimming. They are aquatic, carnivorous animals which feed on fish, crustaceans and mostly molluscs. Walrus breeds once in three years for which it comes on land.

12. *Phoca.* *Phoca* (seal) (Fig. 30.14) is found in all seas, and is gregarious. Each adult male controls about 50 to 60 females. Body is fish-like, streamlined and covered with hairs which are also found on soles and palms. Proximal parts of limbs are short but distal parts are lengthened and the digits are bound together by the skin to form paddles. Hindlimbs are turned backwards, and up to the ankles they are bound with the short tail. Both limbs are used for swimming. The pinnae have been lost completely though they are present in fur seals. Incisors are small but canines are well developed for eating fish. Their blubber is rich in oil. They come on land for their annual breeding. Locomotion on land is mainly by forelimbs and wriggling of the abdomen. *Phoca vitulina* (Fig. 30.14) is the common seal having no pinnae. Its hindflippers cannot be directed forward. *Zalophus* (sea lion) (Fig. 30.14) and *Callorhinus* (fur seal) are with pinnae and can move on land by all the four limbs.

K. Proboscidean Mammal

Elephas. *Elephas indicus-maximus,* the Indian elephant, and *Loxodonta,* the African elephant. **Indian elephant** is smaller, pinnae are broad and fan-like but not very large. Head is rounded, tip of the trunk has only one sensory projection, forehead is concave, tusks are large in males but just project beyond the lips in females. Molars have transverse parallel ridges. **African elephant** is larger, up to eleven and a half feet high with enormous upcurved. Tusks in both sexes, very large pinnae, head is sloping, trunk has two sensory projections, forehead is convex, and molars have transverse diamond-shaped ridges. The African elephant has never been trained or domesticated.

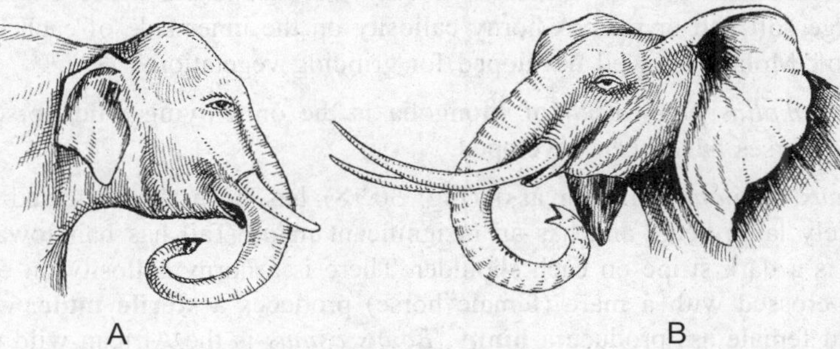

Fig. 30.15. A – *Elephas indicus*; B – *Elephas africana.*

L. Hyracoidean Mammal

Procavia. *Procavia* or *Hyrax* (coney) (Fig. 30.16), belongs to Hyracoidea, is a rabbit-like animal with a split snout. Pinnae are short, and tail is so small that it appears to have no tail. Testes are abdominal lying close to kidneys. Uterus paired and a single pair of mammae are pectoral. Incisors grow continuously from persistent pulp. It resembles ungulates in structure of molars, absence of clavicle and acromian, and in the reduction of digits to 4 in the manus and 3 in the pes. In addition to the normal colic caecum, there is a pair of caeca arising from the large intestine. It has a dorsal gland covered with hair of a different colour to that covering the body. It is herbivorous and lives in colonies in rocks. Its colour blends with that of rocks. It is found from Ethiopia to Arabia.

Fig. 30.16. *Procavia syriacus* (Coney).

M. Sirenian Mammal

Halicore. *Halicore* (dugong) (Fig. 30.17) belongs to the order Sirenia, is found in the Red Sea and along the shores of Indian and Pacific Oceans. The body is heavy with a fish-like tail having a tail fluke like that of a whale. Nostrils are far up on the head. Teeth are much reduced. The jaws have horny pads for chewing. The upper jaw of the male has a pair of tusks pointing downwards, which are enlarged incisors. There are 7 cervical vertebrae, while manatees have 6. The sea cows swim by the body and tail. Young are born in water and nursed by pectoral mammae. The mother often holds the young by one paddle.

Fig. 30.17. *Halicore* (Dugong).

N. Ungulates (Hoofed Mammals)

These are divided into two orders : (*i*) **Perisodactyla** having odd number of toes, *e.g.*, horse, ass, zebra, tapir, rhinoceros), and (*ii*) **Artiodactyla** having even number of toes, *e.g.*, pig, hippopotamus, camel, deer, musk deer, sheep, goat, giraffe, blackbuck, ox, water buffalo, yak, bison.

1. ***Equus.*** *Equus caballus* (domestic horse) (Fig. 30.18) has been a useful commensal of man having been domesticated for centuries. It is a swift running animal feeding on grasses. Head is proportionately small. There is a well developed mane on the neck, and the tail is bushy with hair from the root downwards. Limbs have only the third digits which bears a hoof of keratin having a large soft pad or heel. A horny callosity on the inner side of each limb represents the vestigial hoof. Molars are well developed for grinding vegetation.

Equus caballus przewalskii of Mongolia is the only living wild horse. From this other domesticated horses have been developed.

Equus hemionus (Indian ass) (Fig. 30.18) has been domesticated by man. It has a proportionately large head, and has an insignificant mane. Tail has hair towards the distal half only. There is a dark stripe on each shoulder. There is a horny callosity on each forelimb only. A male ass crossed with a mare (female horse) produces a **sterile mule**, while a male horse (stallion) and female ass produce a **hinny**. *Equus asinus* is the African wild ass.

Equus zebra (Zebra) (Fig. 30.18) is African. It lives in herds on grassy plains. It has conspicuous black and yellow stripes. This is a concealing colour which makes the animal invisible in strong sunlight in an open plain, also at dusk and in moonlight, and in the cover of thickets.

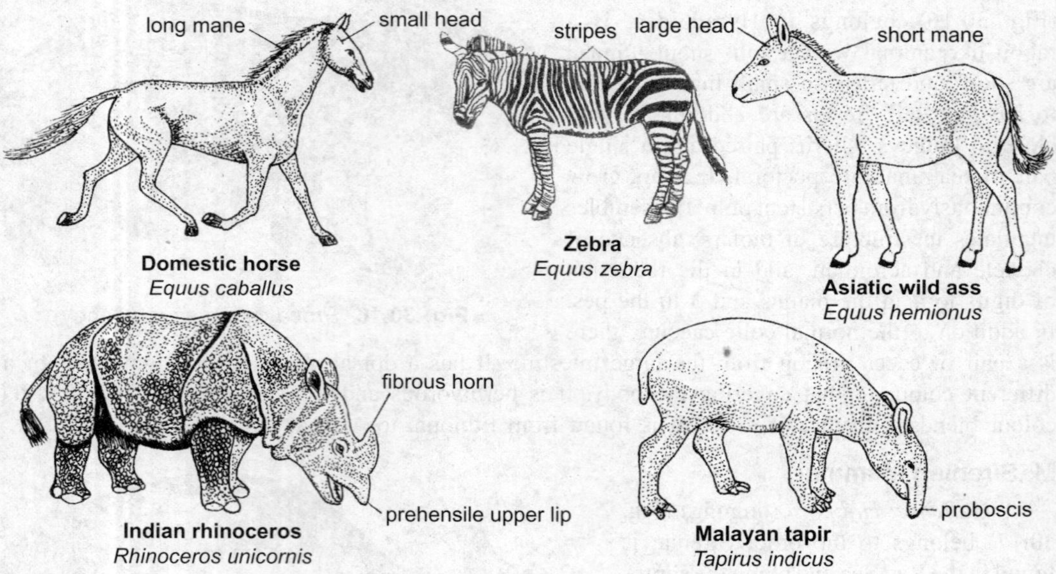

Fig. 30.18. Odd-toed hoofed mammals.

2. Tapirus. *Tapirus* (tapir) (Fig. 30.18) shows discontinuous distribution and is found now in Central and South America, Malaya, and Indonesia. It is brownish-black with thick hair but the young ones have white spots. It lives in dense forests close to water into which it retires. They are nocturnal, herbivorous and heavy ungulates. Their legs are short, forefeet bear 4 hoofed toes and hindfeet have 3 hoofed toes. Tail is very short. The upper lip and nose form a flexible proboscis.

3. Rhinoceros. *Rhinoceros unicornis* (Indian rhinoceros) (Fig. 30.18) is found in Assam and Nepal. It is about 12 feet long and 6 feet high, and has a single horn on the snout. The horn is made of keratinised skin and matted hair, it is said to have aphrodisiac and other medicinal values, so that large sums are paid for it, this results in indiscriminate shooting and the animal is on the verge of extinction in India. It is massive with thick, strongly tuberculated skin which is thrown into enormous fold on the neck, shoulders, hip, and over the limbs. There is practically no hair except on pinnae and at the tip of a thin tail. Massive limbs end in three digits each. The upper lip is long and prehensile. Upper canines are absent. Testes are permanently abdominal. It is herbivorous, nocturnal and lives singly or in pairs. When annoyed, it is very dangerous and charges fiercely. *Dicerorhinus* is the two-horned rhinoceros of Indonesia. *Diceros bicornis* is the two-horned black rhinoceros of Africa. White rhinoceros, *Rhinoceros simus* is the largest African rhinoceros. *Chaeropsis liberiensis* is the pigmy rhinoceros and is about 60 cms in height at shoulder level.

Fig. 30.19. Even-toed ungulate mammals.

4. *Sus*. *Sus scrofa* (pig) (Fig. 30.19) has a long head, mobile snout with terminal nostrils. Colour is dark gray with rusty and white stiff hairs, feet and ears are black. Coat is sparse with a mane of black bristles on the neck and back. Limbs are short and narrow with 4 digits but only the third and fourth reach the ground and used in walking. Dentition is complete, 3, 1, 4, 3/3,

1, 4, 3 = 44. Canines of both jaws grow continuously and form long triangular and upcurved tusks for defence and digging roots, they are better developed in males. Pigs are omnivorous, they feed on carrion, snakes, insects, roots, tubers and cultivated crops causing much damage. They are gregarious living in families in grass or bush of marshy places and are found in Asia, Europe and North Africa. *Sus cristatus* is the Indian wild boar which is also found in Myanmar and Malaysia.

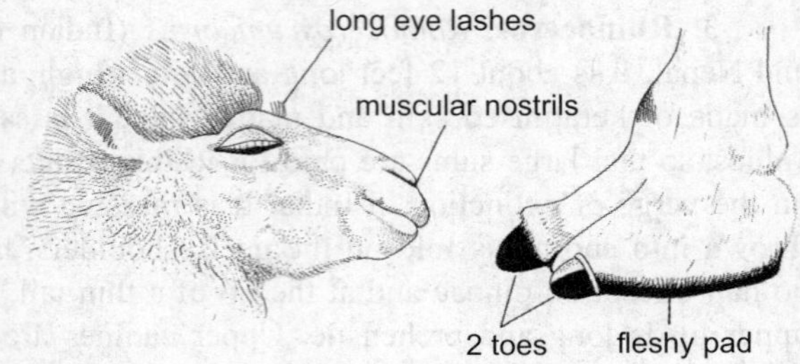

Fig. 30.20. Camel. Head and foot showing adaptations to sand.

5. ***Hippopotamus***. *Hippopotamus amphibius* (hippopotamus) (Fig. 30.19) is an amphibious, herbivorous and gregarious animal of most of the rivers of Africa. It has a large heavy body up to about 14 feet long and up to 4 tons in weight with a barrel-like thorax. There is a broad muzzle with nostrils, eyes and ears are on the dorsal side of the head as a specialisation for a life in water. It can remain submerged for only 5 minutes, after which it comes to the surface to breathe air. During day they remain in water but come out on land at night to feed on vegetation. The skin is thick and smooth, having almost no hair. A thick layer of fat is found beneath the skin. Sweat is red coloured. Tail is short and limbs with 4 hoofed digits, all digits reach the ground. Dentition is complete, incisors and canines are worn off but grow continuously. Lower canines are tusk-like and used for fighting. The enormous stomach has three chambers, and the animal is a non-ruminant.

6. ***Camelus***. *Camelus dromedarius* (Fig. 30.19) is the Arabian or Dromedary camel with one hump on the back. It is no longer found in the wild state but is entirely domesticated. *Camelus bactrianus* (Fig. 30.19) is the cold Gobi desert Bactrian camel with two humps. The hump is a reservoir of food and also provides water and heat. Body has a small area to conserve water. Limbs with only 2 toes (didactyl) with large supporting pads for walking on desert sand. Digits have nails. In locomotion both limbs of one side are moved at the same time, and not alternately. Snout is slender with a clefted upper lip. There are heavy fringes of hair around the eyes and nostrils to protect them from sand. Stomach is four-chambered. Food passes into the sacculated rumen of the stomach where it is softened and is called **cud** which is regurgitated and re-chewed. The large rumen has sacculated wall forming pouches or **water cells**, each has a sphincter muscle and is separated from others by muscular walls, reticulum has similar water cells, it was supposed that the water cells were for storing water for use in deserts, but this is not true because pouches cannot hold enough water, and the walls of rumen and reticulum are glandular and their function is digestive. Camels are adapted for desert conditions, they can do without water for 17 days, almost no water is lost with urine, and they can endure loss of water by evaporation of sweat without harm. Some water is stored in the muscles and connective tissue, but most of this water, and that in the water cells, comes indirectly from the breakdown of glycogen in muscles and of fat in the hump. Camels live up to an age of 40 years. Camel is called the ship of the desert. They provide transport, hide, flesh and also wool.

7. ***Lama***. *Lama* (Llama) (Fig. 30.19) resembles a camel but it is smaller, has long woolly hairs and has no hump. It dwells in mountains of South America but it is now domesticated. It has a habit of spitting the contents of the stomach on attacker or when annoyed (protective device). Testes are abdominal and descend into scrotal sacs only during the breeding season. It provides fur, flesh and fat (tallow).

8. ***Axis axis***. *Axis axis* (spotted deer or cheetal) is found in the forests of Sri Lanka and Indian plains and in hills up to 4000 feet, but it is not found in Assam, Punjab and Rajasthan.

It lives where there is good grazing and plenty of water and often comes near villages. It lives in herds of 20 to 30. It is 2·5 to 3 feet high, but South Indian forms are smaller. The colour is bright fawn with white spots with a black line along the spine. Throat and belly are white. Tail is as long as the head. The males have branched antlers up to 34 inches long, each has 3 tines or branches. The antlers are shed annually. It generally gives birth to two young ones at a time which have the colouration of adults.

9. *Moschus*. *Moschus moschiferus* (musk deer) is found in Central Asia and the Himalayas from 8000 to 12,000 feet. It is about 20 inches high and has thick, long bristly hairs. Colour is grayish brown with gray spots on the back. Hindlegs are much longer than the forelegs, so that in walking the hindlegs are bent and all four long pointed hoofs touch the ground. There are no horns. The ears are long. There are no suborbital glands on the face which are generally present in Cervidae. Tail is very small and completely hidden in the long hairs of the anal region. The male has downwardly projecting canines of the upper jaw which are 3 inches long and used for fighting. The male has a musk pouch below the skin of the abdomen which contains musk in the form of a paste, when dry it is used for perfumes and is a valuable product. The male also has caudal glands near the tail on the thighs, which open on the ventral side of the tail. Their secretion is for attracting the female. The musk deer has a gall bladder which is absent in other Cervidae. *Moschus* holds a place between deer and antelopes, and is regarded as an undeveloped deer in having no antlers, no glands of the face and foot, and in possessing a gall bladder.

10. *Cervulus*. *Cervulus muntjac* (barking deer) is found in India, Malaysia, China and Japan in thickly wooded hills up to 6000 feet. It is 2 feet high and colour is chestnut with black streaks on the face. Antlers are 3 inches long, each with one tine (branch) on the main stem. Antlers are borne on bony hair-covered pedicels which are 3 inches long and extend down forming bony ridges on the face. Does have tufts of bristly hair in place of antlers. Tail is as long as the head. It lives in pairs or small families in thickets, and is diurnal. Its call sounds like a short bark of a dog. It breeds in all seasons producing one or two youngs which are spotted.

11. *Giraffa*. *Giraffa camelopardalis* (giraffe) (Fig. 30.19) is confined to tropical Africa and lives in tree-studded grasslands. It is the tallest living animal, being 16 feet high. The body is relatively small with an enormously long neck which has seven cervical vertebrae. The colour is yellowish with large brownish spots separated by white spaces. Forelimbs are longer and bear the main bulk of the body and neck. Both sexes have small unbranched 3 to 5 horns as bony outgrowths of frontals covered with unmodified skin continuous with that of the head which are never shed. Between the horns is a median prominence. Upper canines are absent. Forelimbs are longer than hindlimbs. Limbs have only two digits ending in hoofs. The long neck is used for browsing on trees and as a look-out tower from long grasses. In locomotion the limbs of each side are used simultaneusly producing a rocking motion. It lives in herds, feeding on leaves and branches by its long sensitive prehensile tongue.

12. *Boselaphus*. *Boselaphus tragocamelus* (blue bull or nilgai) (Fig. 30.21) is found in the plains of India but not found in Bengal and Assam. They live in small herds of 4 to 10 in undulating plains with grass and bushes, and avoid thick forests. It both grazes and browses on grass, fruits and leaves of trees, and causes much damage to crops. *Boselaphus* is 52–56 inches high, the females

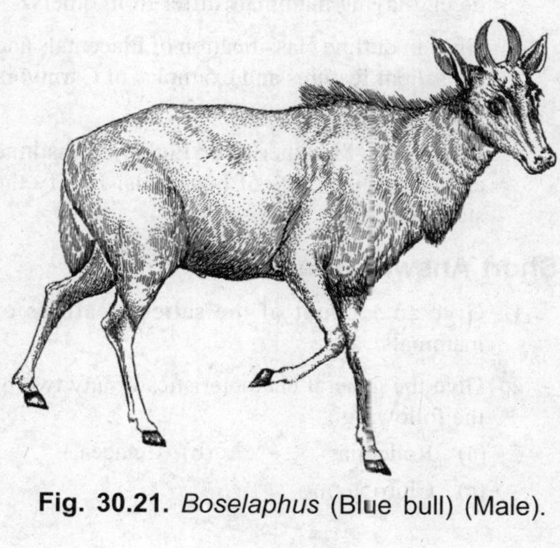

Fig. 30.21. *Boselaphus* (Blue bull) (Male).

being smaller. It is an ungainly antelope with high withers and low rump. Colour is iron gray in old males, young bulls and cows are tawny. Both sexes have a short dark mane on the neck and males have a tuft of stiff black hairs on the throat which is present at times in females also There is a white ring below each fetlock and two white spots on each cheek. Males have small cone-shaped curved horns which are 8 inches long. They reproduce in all seasons.

13. *Antilope.* *Antilope cervicapra* (black buck) (Fig. 30.22) is not a deer as is commonly supposed but an antelope of the grassy plains of India. It avoids forests. In antelopes unbranched horns are present in both sexes generally, which are permanent and are never shed. It is about 32 inches high and tail is 7 inches long. Body is covered with short hairs. The upper part is brown which in old males becomes dark brown or black, lower part is white with a ring around each eye. The does and young are yellowish fawn. Males have spirally ringed horns which at times may be present in does also. There is a large preorbital gland below each eye. It lives in herds with an old male, a few young bucks and several females. They feed on grass and crops near villages. Eyesight is very keen, and they run with great speed taking long leaps. They breed in all seasons, producing 1 or 2 young ones at one time. It is very graceful antelope.

Fig. 30.22. *Antilope cervicapra* (Black buck).

IMPORTANT QUESTIONS

Long Answer Questions

1. Give an account of distinguishing characters and outline classification of class Mammalia.
2. Give a brief classification of class Mammalia stating important characters and examples of each order.
3. Give a brief classification of Mammalia. How do egg-laying mammals differ from others?
4. Give an outline classification of Placentals and the salient features and examples of Carnivora and Primates.
5. Prototheria, Metatheria and Eutheria constitute an ascending series of mammals. Justify the statement.

Short Answer Questions

1. Give an account of the salient features of mammals.
2. Give the general characteristics of any two of the following :

   (a) Rodentia; (b) Cetacea;

   (c) Chiroptera.
3. Explain the differences between Old World monkeys and New World monkeys.
4. Distinuguish between the following in the tabular form :

   (*a*) Indian and African elephants; (*b*) Horse and ass; (*c*) Lemuroidea, Tarsioidea and Anthropoidea; (*d*) Tiger, Lion and Leopard; (*e*) Megachiroptera and Microchiroptera; (*vi*) New World monkeys and Old World monkeys; (*vii*) Spiny and scaly ant-eaters; (*ix*) Odontoceti and Mysticeti.
5. Write short notes on : (*i*) Ant-eaters; (*ii*) Bats; (*iii*) Duckbilled *Platypus*; (*iv*) *Echidna;* (*v*) Kangaroo; (*vi*) Pangolin; (*vii*) Primates; (*viii*) Cetacea.
6. How are whales adapted to their mode of life.

Very Short Answer Questions

1. Write the systematic position of whales.
2. What is Chiroptera?
3. Classify and write a very short note on *Felis leo*.

4. Write a brief account of Cetacea.
5. Compare sea cow and whales.
6. Write a few lines on radar of bats.
7. Write short note on Duckbilled *Platypus*.

Multiple Choice Questions

1. Monotremates are confined to
   (*a*) Oriental region
   (*b*) Ethiopian region
   (*c*) Australian region
   (*d*) Neotropical region.
2. Marsupium is present in
   (*a*) Marsupials (*b*) Monotremes
   (*c*) Insectivores (*d*) Chiropterans.
3. In mammals hindlimbs are absent in
   (*a*) Rodents (*b*) Cetaceans
   (*c*) Morsupials (*d*) Chiropterans.
4. Teeth in mammals are
   (*a*) Acrodont,homodont, polyphyodont
   (*b*) Acrodont, homodont, monophyodent
   (*c*) Thecodont, homodont, diphyodont
   (*d*) Thecodont, heterodont, diphyodont.
5. The dental formula of suborder Lemuroidea is
   (*a*) $\frac{2133}{2133}$ (*b*) $\frac{2133}{1133}$
   (*c*) $\frac{2123}{2123}$ (*d*) $\frac{2023}{2023}$.
6. Largest mammalian suborder is
   (*a*) Lagomorpha (*b*) Rodentia
   (*c*) Sirenia (*d*) Cetacea.
7. Largest living land animals are included in the suborder
   (*a*) Artiodactyla (*b*) Perissodactyla
   (*c*) Hyracoidea (*d*) Proboscidea.
8. The zoological name of kangaroo is
   (*a*) *Petaurus*
   (*b*) *Phascolomys*
   (*c*) *Macropus*
   (*d*) *Hotoryctes.*
9. The only marsupial that has a placenta is
   (*a*) *Perameles*
   (*b*) *Macropus*
   (*c*) *Petaurista*
   (*d*) *Phalangista.*
10. Manatees or seacows are aquatic mammals included under the order
    (*a*) Lagomorpha (*b*) Sirenia
    (*c*) Cetacea (*d*) Pinnipedia.
11. The subcutaneous fat of whales is called
    (*a*) Blubber (*b*) Fat
    (*c*) Soluble fat (*d*) Cholesterol.
12. Whales have been hunted for their
    (*a*) Blubber (*b*) Meat
    (*c*) Bones (*d*) Blood.

ANSWERS

1. (c) 2. (a) 3. (b) 4. (d) 5. (a) 6. (b) 7. (d) 8. (c)
9. (a) 10. (c) 11. (a) 12. (b)

31

Chapter

Subclass Prototheria (Order Monotremata)

Contents of this Chapter

The Prototheria is the subclass of class Mammalia and includes egg-laying mammals distributed only in Australia and the neighbouring islands of Tasmania and New Guinea. Prototheria includes a single order, Monotremata (= one aperture or Ornithodelphia). These include the duck-billed platypus (*Ornithorhynchus*) and one species of echidna or spiny ant-eater, *Tachyglossus* (swift tongue) from Australia and two species of *Zaglossus* from New Guinea. Both are basically similar to each other, but too different from other mammals. Monotremes derived from an early stock on the principal mammalian line of descent.

DISTINCTIVE CHARACTERS

1. **Habits and habitat.** The platypus is highly modified for aquatic life and spiny ant-eaters are terrestrial. Both are burrowing in habit, both burrow and make nests for rearing their youngs. Both are insectivorous, nocturnal, warm-blooded, quadrupedal and oviparous.

2. External features. Body of platypus is flattened and covered with a dense, soft under-fur protected by harsher outer fur. Body of spiny ant-eaters is covered above with strong pointed spines, between which are coarse hairs, while the lower surface of the body has only hairs. In platypus jaws are produced into a depressed beak covered with soft, but tough skin. In spiny ant-eaters jaws are produced into a long sensitive pointed rostrum. External ears or pinnae are absent. Tail flat and elongated in platypus and vestigial in spiny ant-eaters. Mammary glands devoid of nipples. Mammary pouch on the abdomen of female is present in which milk is secreted through a large number of (120) specialised sweat glands, but the ducts of these are not united to open on central nipples. In platypus, forelegs are short and powerful. The five digits are webbed and end in strong claws. Hindlimbs are less specialised. In the male a hollow tarsal spur is connected with a small crural gland whose secretion is poisonous. This gland is well developed in platypus and less developed in spiny ant-eaters.

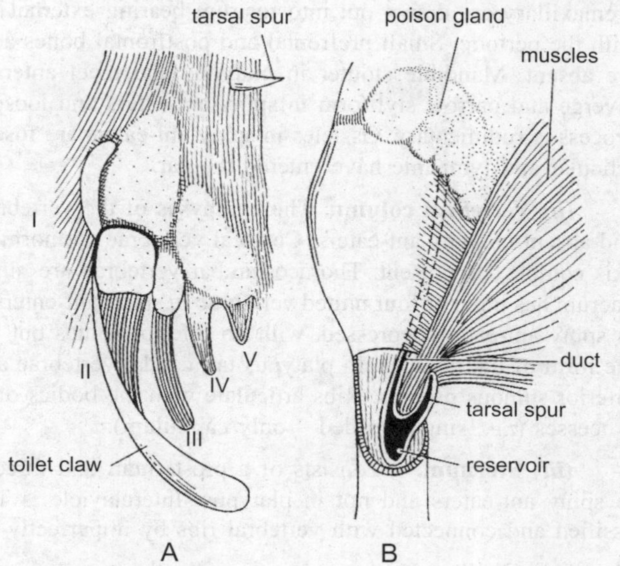

Fig. 31.1. Tarsal spur. A–Foot of male echidna; B–Poison apparatus of male platypus dissected.

3. Exoskeleton. The platypus has five short dark brown epidermal hairs, whereas spiny ant-eater's back is covered by a mixture of spines and hairs, and the belly has hairs only. Skin is glandular.

4. Body cavity. It is divided by a typical mammalian diaphragm into an ant-erior thoracic and a posterior abdominal cavities.

5. Endoskeleton

(*i*) **Skull.** Dicondylic. The bones of skull early become fused together showing no trace of their boundaries. Brain case is larger and much more rounded in spiny ant-eaters than in platypus. Unlike higher mammals, pterygoid bones are separate. Parasphenoid is also present like lower vertebrates. Temporal canal perforates the posterior part of zygomatic arch, absent in higher mammals. Zygomatic arch is stouter in platypus than that of spiny ant-eaters. Postorbital process formed by jugal is present in platypus and absent in spiny ant-eaters. The nasal and

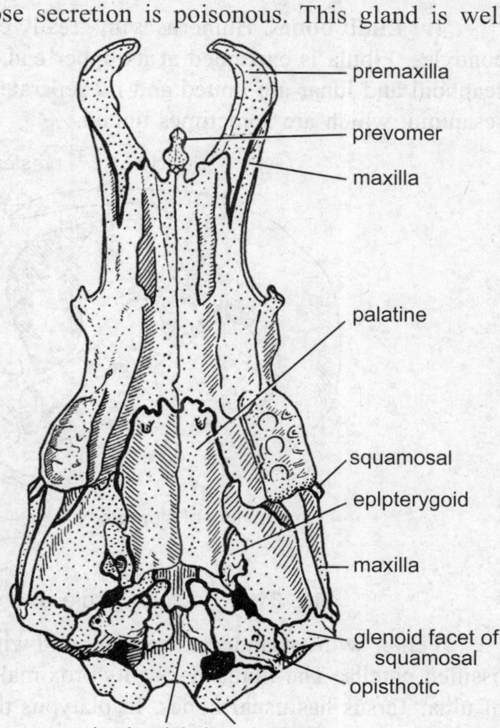

Fig. 31.2. Ventral view of skull of platypus.

premaxillary are drawn out into rostrum bearing external nares in both. Tympanic is not united with the periotic. Small prefrontal and postfrontal bones are present. Lacrimals and alisphenoids are absent. Mandible stouter in platypus and meet anteriorly a short distance and then again diverge and narrow styliform in spiny ant-eaters and loosely united at the symphysis. Coronoid process is rudimentary. Ossicles in spiny ant-eaters are fused, and separate in platypus. Quadrate, articular and tympanic have entered the ear.

(*ii*) **Vertebral column.** The epiphyses of the vertebrae are not well developed in platypus and absent in spiny ant-eaters. Cervical vertebrae are normal and zygapophyses are absent. Atlas-axis complex is present. Thoraco-lumbar vertebrae are nineteen with short transverse processes. Sacrum has three or four united vertebrae in spiny ant-enters, and two in platypus, caudal vertebrae in spiny ant-eaters depressed, with no inferior spines but with about five sub-vertebral bones in the form of flat nodules. In platypus tail caudal vertebrae are twenty or twenty one, each with an inferior spinous process. Ribs articulate with the bodies of the vertebrae and not with transverse processes (*i.e.*, single-headed – only capitulum).

(*iii*) **Sternum.** It consists of a presternum and three keeled sternbrae. Xiphisternum only in spiny ant-eaters and not in platypus. Interclavicle is T-shaped like reptiles. Sternal ribs are ossified and connected with vertebral ribs by imperfectly ossified intermediate ribs.

(*iv*) **Girdles.** In the pectoral girdle there are separate clavicles and a median T-shaped interclavicle. Coracoid region has two separate bones, coracoid and precoracoid. In front of coracoid is the epicoracoid. The scapula is unlike that of other mammals and with well developed acromian process and without spine. In pelvic girdle pubes and ischia form a long symphysis. With the anterior border of pubes are articulated a pair of large epipubic or marsupial bones, presumably for the support of the marsupial pouch.

(*v*) **Limb bones.** Humerus with greatly expanded extremities and prominent tubrosities and condyles. Fibula is expanded at its upper end like the ulna of other mammals. In the carpus the scaphoid and lunar are united and no separate centrale. There are a radial and two large palmar sesamoid, which are sometimes united.

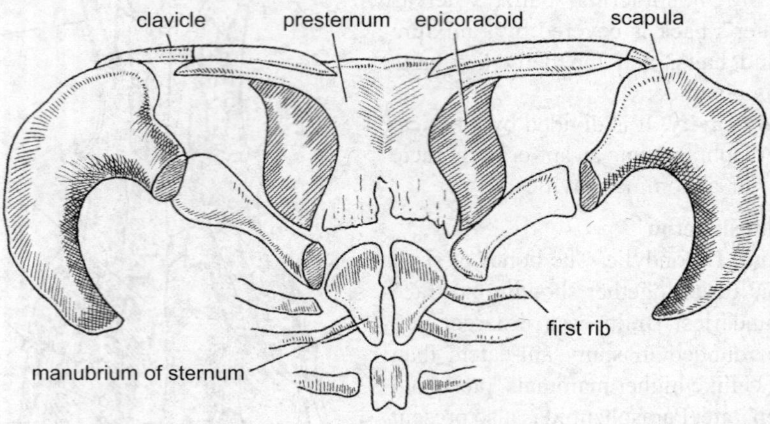

Fig. 31.3. Pectoral girdle and sternum of platypus (Ventral view).

Femur with expanded extremities and with external and internal trochanters. There is a large ossified patella. The fibula has at its proximal end a compressed process like olecranon process of ulna. Tarsus has usual bones. In platypus the astragalus and calcaneum are firmly united and in male there is an accessory ossification on the inner side for the tarsal spur. Metatarsals are as usual.

6. Digestive system. In spiny ant-eaters, teeth are absent in all stages of development. In young platypus flattened, rigid teeth are present and these are replaced by horny structures, formed by an ingrowth of epidermis beneath them and used for breaking the shells of molluscs. Spiny ant-eaters have a tiny round mouth at the end of long beak. In spiny ant-eaters posterior part of tongue is beset with horny serrations which grind the food against corresponding ridges on the palate. Submaxillary, parotid and sublingual glands are present whose secretions help to neutralise formic acid discharged by their preys, ants.

The stomach of monotremes is almost globular, intestine is differentiated into small and large, and caecum is also present. Rectum opens into the cloaca posterior to the ureters and vasa deferentia. Liver is large and there is a gall bladder with bile duct which traverses the pancreas and joins the pancreatic duct.

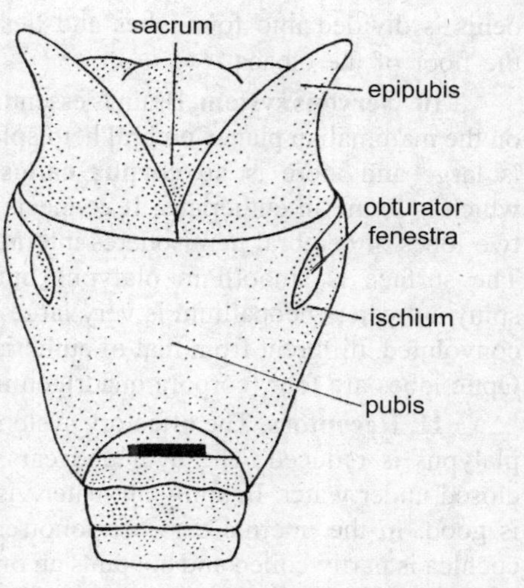

Fig. 31.4. Pelvic girdle of platypus (Ventral view).

7. Respiratory system. Respiration is pulmonary, by lungs. Larynx is also present.

8. Circulatory system. Heart is 4-chambered. Right auriculo-ventricular valve is incomplete and partly muscular. There are no chordae tendinae. Left auriculo-ventricular aperture is guarded by tricuspid valve, not by bicuspid as in other mammals. Single left aortic arch resembles those of other mammals. There are no renal portal veins. In spiny ant-eaters an anterior abdominal vein, arising on the bladder and distributed to the liver. Erythrocytes are small, circular and non-nucleated. In monotremes heat regulation is imperfect though far advanced over that of living reptiles. In the absence of sweat-glands, heat-reduction by vasodilation occurs. The temperature of spiny ant-eaters rises with that of the environment above 30ºC and the animal dies of heat stroke at about 37°C, if unable to cool itself by burrowing. The platypus has few sweat-glands but it lives mostly in water or in a cool burrow.

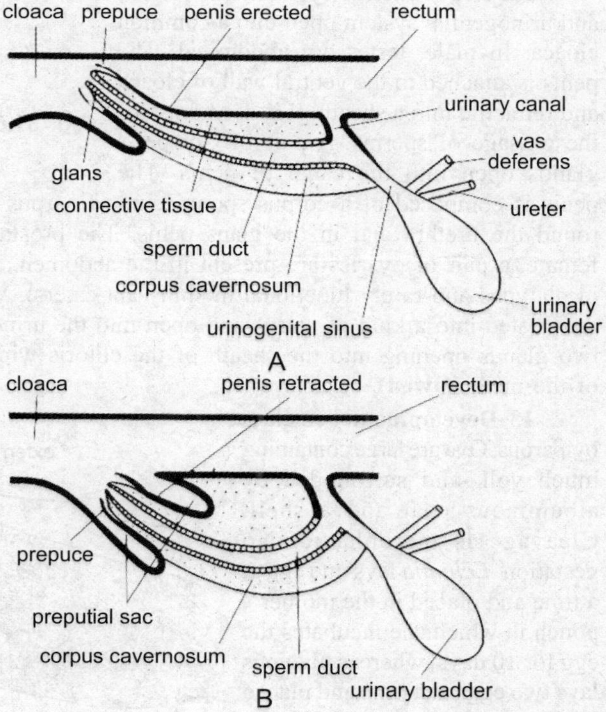

Fig. 31.5. Diagrammatic L.S. through cloacal region of a male monotreme. A – Penis erected; B – Penis retracted.

9. Excretory system. Kidneys are metanephric. Ureters do not empty directly into the bladder but open into the urethra through a common aperture over a papilla. The urine flows through a short urinary canal into the large urinogenital sinus. Therefore, it does not traverse the penis which carries only sperms. The glans penis is divided into four lobes and lies on the floor of the cloaca.

10. Nervous system. Brain is essentially on the mammalian plan. Cerebral hemisphere is large and there is no **corpus callosum** which is found in eutherians. It connects the two lobes of cerebral hemispheres internally. The surface is smooth in platypus but in spiny ant-eaters neopallium is very large and convoluted, different from that of eutherians. Optic lobes are four (corpora quadrigemina).

11. Receptors. The olfactory system of platypus is reduced. The eyes and ears are closed under water. In spiny ant-eater vision is good. In the internal ears of monotremes cochlea is partly coiled and contains an organ of Corti. Like reptiles it also carries lagena cochlea at its tip.

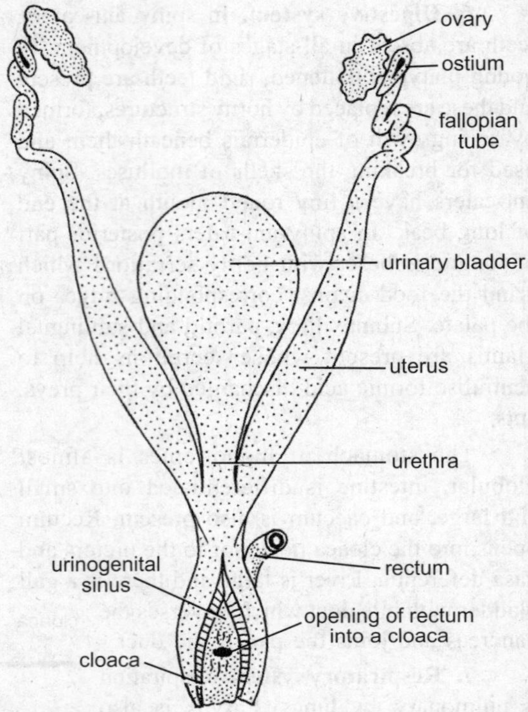

Fig. 31.6. Female reproductive system of platypus (Diagrammatic).

12. Reproductive system. The rectum and urinogenital system open into a common cloaca. In male, testes are abdominal. The penis is attached to the ventral wall of cloaca and retractile into a sheath. It is used only for the passage of sperms. A pair of Cowper's glands open into the canal of penis. The penis is composed of a corpus spongiosum, a corpus fibrosum and cavernous tissue is present round the urethra and in the glans penis. The prostate and vesicula seminales are absent. In female, a pair of ovaries are present in the abdomen. The right ovary is smaller (not functional in platypus and rarely functional in spiny ant-eaters). Vagina absent. The lower ends of oviducts are dilated into a kind of uterus and open into the urinogenital sinus in front of ureter. There are two glands opening into the sheath of the clitoris which are compared to the Cowper's glands of the male **(Owen)**.

13. Development. Female is oviparous. Ova are large containing much yolk and surrounded by albuminous coat and a shell. Cleavage is meroblastic. No gestation. *Echidna* lays one egg at a time and placed in the mother's pouch in which she incubates the egg for 10 days, whereas platypus lays two eggs at a time and placed in the nest in the burrow.

Monotremes possess a caruncle on the head as well as the egg-tooth for breaking out the shell. After incubation and hatching, the young enters the pouch of mother and is fed by milk.

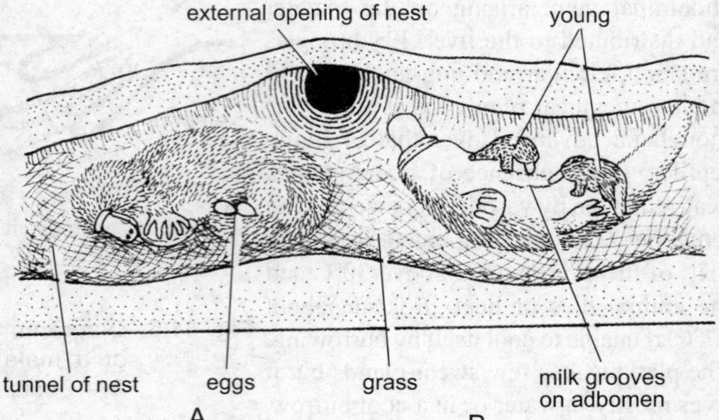

Fig. 31.7. Female platypus. A – Incubating her two eggs; B – Newly hatched young lapping up milk on the belly of mother.

CLASSIFICATION OF PROTOTHERIA

Subclass Prototheria includes only one living order Monotremata which includes two families, the Ornithorhynchidae and Tachyglossidae. Both differ widely from each other in external appearance and in relatively minor anatomical characters. But still they are bound together in a single order.

Family 1. Ornithorhynchidae. Aquatic and burrowing in habit. Body is covered with dense soft fur. Facial portion of snout broad and elongated and covered with a leathery skin. Feet webbed with 5 clawed toes. The web on the forefoot extends beyond the claws. Pinnae are absent and tail is flattened and elongated. In females, mammary pouch is present over the abdomen in which milk is secreted through a large number of pores. Nipples are absent. Teeth are absent in adults but present in embryos. **Example :** Duck-billed platypus (*Ornithoryhuchus*).

Family 2. Tachyglossidae. Terrestrial and burrowing in habit. Body is covered with spines and intermingled coarse hairs. Snout is elongated and edentulous. Tongue is long and protractile. Tail is vestigial. Limbs with 5 spade-like clawed digits. Pinnae are absent. Teeth are absent in all stages of development. In females, abdominal pouch develops during sexual season in which a single egg develops. Milk is secreted in the pouch through numerous pores. **Examples :** *Tachyglossus* and *Zaglossus* (echidna or spiny ant-eater).

AFFINITIES OF PROTOTHERIA

Prototherians are fully mammalian in some characters but they also possess a number of reptilian characters. Thus, this group is a connecting link between the reptilian ancestors from which the mammals evolved and the modern mammals. The prototherians are also sometimes referred to as living fossils as they represent a very primitive level of mamalian evolution.

1. Reptilian affinities. Although there is no definite connecting link known between Prototheria and Reptilia, there is enough evidence to show that mammals had a reptilian ancestry. This view is further supported by the following resemblances between the monotremes and the living reptiles :

1. Skull with epipterygoids, pterygoids, palatines, dumb bell-shaped prevomers and annular tympanic bone. Tympanic bulla is not developed. Lacrymal, alisphenoid are absent, while jugal is reduced or absent. Prefrontals and postfrontals are also present.

2. Vertebrae without epiphyses except in tail region of platypus. Cervical vertebrae bear free ribs. Caudal vertebrae with inferior spines.

3. Ribs are single headed (*i.e.,* capitulum).

4. Sternum. A median T-shaped interclavicle present.

5. Pectoral girdle. Scapula with developed acromion process and without spine. Coracoids and precoracoids and in front of coracoids are epicoracoids. In pelvic girdle, epipubic or marsupial bones are articulated with the anterior border of pubes.

6. Cloaca is present.

7. Anterior abdominal vein is present.

8. Nervous system. Corpus callosum is absent, which connect the two cerebral hemispheres in mammals.

9. Body temperature regulation is imperfect.

10. Cochlea of internal ear is with lagena.

11. Ureters open into the urinogenital sinus and do not open into the bladder.

12. Testes are abdominal. Penis is erectile and only used for carrying sperms.

13. Oviducts separately open into the urinogenital sinus (cloaca). Vagina absent.

14. Females are oviparous and no uterine gestation.

15. Eggs are large macrolecithal and cleidoic. Cleavage is meroblastic.

16. Young hatches out of the egg by breaking the egg shell with the help of caruncle over the head and egg tooth.

2. Avian affinities. The relationship of prototherian with aves (birds) does not have solid facts. The similar characters present in theria are chiefly due to the fact that both possess common reptilian ancestry. The important points of resemblance are as follows :

1. Shape of beak of *Platypus* resembles with birds.
2. Teeth in both are absent.
3. Feet of both are webbed.
4. Presence of obliterated sutures of skull.
5. Tarsal region bears spur.
6. Oil gland is present.

3. Mammalian affinities. The prototherians are essentially mammals as they possess the following typical mammalian characters :

1. Body is covered with hairs. Pinna (external ear) is present.
2. Skin is richly glandular and has sweat and sebaceous glands.
3. Mammary glands are present which open within marsupial pouch through numerous ducts and pores. Nipples are absent.
4. A typical mammalian diaphragm is present.
5. Skull is dicondylic. Chondrocranium is typically mammalian.
6. Lower jaw is formed of a single bone, the dentery.
7. Middle ear has three ear ossicles. Malleus is large, incus small and stapes elongated. Cochlea is slightly coiled.
8. Cervical vertebrae are typically seven.
9. Sternum is segmented with a presternum, 3 keeled sternebrae and xiphisternum present only in spiny ant-eaters.
10. Salivary glands present in the buccal cavity.
11. Lobes of liver typically mammalian.
12. Heart 4-chambered.
13. Only left aortic arch present.
14. Erythrocytes (RBCs) circular and non-nucleated.
15. Optic lobes are four (corpora quadrigemina).
16. Penis with glans penis is present and its canal is surrounded by corpus spongiosum.

4. Special characters (Peculiar characters). The following characters are special and peculiar to Prototheria :

1. Hollow tarsal spur is present over the tarsus bone of male, which is connected to crural gland whose secretion is poisonous.
2. Milk glands derived from sweat glands (not sebaceous as in other mammals) and without teats.
3. Development of temporary abdominal pouch (or mammary glands) in female in breeding season for the development of young.
4. Jaws elongated forming a beak or rostrum, bearing external nares.
5. Teeth present in young platypus are replaced by horny plates in adults.
6. Presence of epipubic or marsupial bones for the support of marsupium.
7. Like ulna, fibula is expanded at its upper end in platypus for the attachment of muscles used in swimming.

8. Ear ossicles relatively large showing transition.
9. Imperfectly warm blooded, body temperature varies from 25° to 28°C.
10. Right ovary is smaller and usually not functional.

Conclusion. Peculiar affinities of Prototheria lead us to some definite conclusions. Prototheria are peculiar in having the following characters :

1. Peculiar blending of reptilian and mammalian characters suggests an intermediate stage between two groups.

2. Presence of primitive, degenerate and highly specialised characters suggests that they represent an early separate side line from main mammalian stock. This justifies their placement in a separate subclass Prototheria from the rest of mammals which are placed in the subclass Theria.

3. Monotremes show that reptiles, birds and mammals together constitute a natural group more homogeneous than the group Ichthyopsida (fishes + amphibians) or even the superclass Pisces.

According to phylogenetic consideration two hypotheses have been held :

1. It is said that monotremes evolved independently from some early mammal-like reptiles and continued to survive in isolation.

2. It is believed that monotremes have been derived from very early marsupials with peculiar characters and divergent specialisation.

Among mammals, the position of monotremes is very controversial. The monotremes are without doubt mammals but in view of their many reptilian features they may be designated as unfinished mammals which have failed to their evolution into typical mammals. They are not to be regarded as the ancestors of higher mammals but as a side line of mammalian evolution, having probably arisen from some different mammal-like reptilian stock than that from which the other mammalian groups evolved. The monotremes have retained those characters through which higher mammals have passed.

IMPORTANT QUESTIONS

Long Answer Questions

1. Give an account of general characters and affinities of Prototheria.
2. What are egg-laying mammals? Describe their general characters, habits and habitat and distribution. How do monotremes resemble the reptiles?
3. Monotremes exhibit connecting link between reptiles and mammals. Discuss it briefly.

Short Answer Questions

1. Discuss the statement that monotremes show a peculiar mixture of reptilian and mammalian characters.
2. Give two examples of monotremes and state where they are found.
3. Write short notes on :
   (*i*) Prototheria; (*ii*) Egg-laying mammals.

Multiple Choice Questions

1. Typical prototherian character is
   (*a*) 4-chambered heart
   (*b*) Presence of cloaca
   (*c*) Tarsal spur in males
   (*d*) Segmented sternum.
2. Spiny ant-eaters are included in the family
   (*a*) Ornithorhynchidae
   (*b*) Tachyglossidae
   (*c*) Notoryctidae
   (*d*) Didelphidae.
3. Which of the following is a reptilian affinity of monotremes?
   (*a*) T-shaped interclavicle
   (*b*) Body covered with hair
   (*c*) Dicondylic skull
   (*d*) Presence of pinnae.

ANSWERS

1. (*c*) 2. (*b*) 3. (*a*)

32 Infraclass Metatheria (Order Marsupialia)

Chapter

Contents of this Chapter

The subclass Theria constitute the "modern" mammals. This group includes three infraclasses, *viz.*, the probably ancestral Jurassic Pantotheria (= Trituberculata), marsupial Metatheria and placental Eutheria, the most highly organised and advanced mammals. Metatheria and Eutheria are presumed to have arisen from some untraced branch of Jurassic pantotheres and both diverged along their separate lines of evolution during early Cretaceous period. In the upper Cretaceous period, the marsupials were more numerous than the eutherians. It was possibly during Cretaceous that marsupials entered New Guinea, Australia and adjacent islands, which were isolated from Asia probably in the late Cretaceous. Here they were able to survive free from competition with eutherians (except for bats and later rodent invaders). It is probable that the basic stock was arboreal, though many have become terrestrial. Tree-kangaroos (*Dendrolagus*) have become secondarily arboreal from a fully terrestrial offshoot. Metatheria were widely distributed over many parts of the world as recently as the Miocene period. But they are now confined to the Australian region (except New Zealand), to South America and a few species to North America.

DISTINCTIVE CHARACTERS

1. Distribution. Almost entirely confined to the Australian region with the exception of the American opossums.

2. Habits and habitat. Marsupials are terrestrial and carnivorous (*e.g.,* native-cats, tasmanian Devil, marsupial wolf), terrestrial and herbivorous kangaroos; arboreal and insectivorous *Didelphys, Marmosa, Chironectes,* etc.; arboreal phalangers or opossums and semi-arboreal Phascogale and pouched rats and mice, and pouchless Jerboa-like marsupials and banded ant-eaters; burrowing desert marsupial mole and bandicoots; and flying and climbing phalangers.

Marsupials are herbivorous (*e.g.,* kangaroos), insectivorous, carnivorous (thylacines and dasyures), and omnivorous, diurnal or nocturnal, warm-blooded, air-breathing, viviparous pouched animals.

3. External features. Body is covered over with hairs. Pinna (external ear) is well developed, but absent in moles. Tail is generally long and prehensile, and an important organ of balance in kangaroo rats and jumping mice. Mammary glands are modified sebaceous glands and have elevated nipples. Females usually have a marsupial pouch or marsupium but it is absent in *Didelphys* and the Dasyuridae (*e.g., Dasyurus, Thylacinus,* etc.). Marsupium encloses the nipples. The number of teats bears a relation to the number of young produced at a birth.

4. Exoskeleton. It includes ectodermal structures, the hairs derived from the Malpighian layer of epidermis and claws. Skin also contains sweat gland, sebaceous glands and scent glands.

5. Body cavity. Typical muscular diaphragm divides the body cavity into thorax and abdominal cavities.

6. Endoskeleton

(*i*) **Skull.** Skull is dicondylic. Brain case small. Orbit and temporal fossa fully confluent and no postorbital bar. Bony palate is incomplete posteriorly. Jugal bone reaches back so as to participate in the formation of glenoid cavity for articulation of lower jaw. Lacrymal bone extends outside the orbit. Tympanic bulla is absent. Tympanic, periotic and squamosal remain separate. Alisphenoid assists in forming the bulla. In kangaroos alisphenolid extends backward to join the greatly elongated paroccipital process. The lower jaw, consisting of a single dentary, has inflected or inturned inner angle to which large lateral pterygoid muscle is attached.

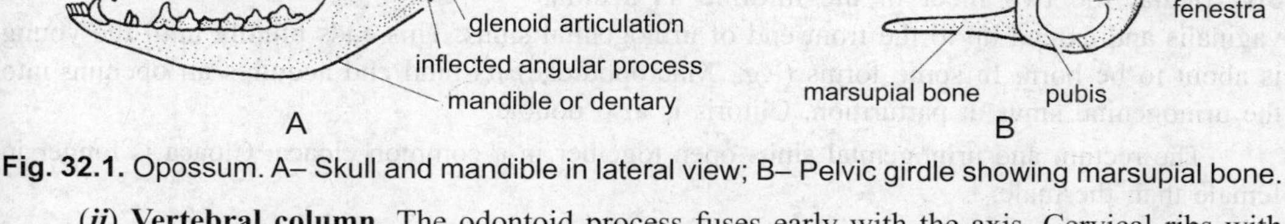

Fig. 32.1. Opossum. A– Skull and mandible in lateral view; B– Pelvic girdle showing marsupial bone.

(*ii*) **Vertebral column.** The odontoid process fuses early with the axis. Cervical ribs with their respective vertebrae. Seventh cervical vertebra is pierced by a foramen for the vertebral artery. Presence of 13 rib-bearing thoracic vertebrae, as in placentals, and 7 lumbars as usual. Sacrum has two vertebrae, but it may be reinforced by caudal vertebrae. Caudal region is short in koala and wombats, long in opossums, dasyures, phalangers and kangaroos. Chevron bones are generally present in caudal vertebrae, except in koala and wombats.

(*iii*) **Girdles.** In pectoral girdle, coracoid is reduced as in placentals, and scapula enlarged and provided with a spine. Clavicle is present except in bandicoots. There is no interclavicle. In pelvic girdle, epipubic bones project forward from the pubes in all. In *Thylacinus* these are represented by small cartilages.

(*iv*) **Limbs.** Forearm bones are separate and generally adapted for pronation and supination. Thumb is not opposable, but the two inner digits of the hand can frequently be opposed to the three outer in grasping. Hand with five clawed digits except in *Chaeropus* and the carpus is without centrale. In hindlimb five digits, fibula is free and can be rotated on the tibia. In some cases first digit can be used as a thumb. Hallux is frequently absent. Second and third digits are in most cases slender and united by the skin (syndactyly).

7. Digestive system. There is only one set of teeth functional throughout life with the exception of a milk molar which is replaced by the last premolar. Incisors are more numerous than in placentals. 3 premolars and 4 molars as against 4 P and 3 M in placentals. Dentition in opossum is I 3/4, C 1/1, P 3/3, M 4/4. Stomach is usually simple, but in kangaroos it is much elongated and sacculated. Caccum is usually present, absent in dasyures.

8. Respiratory, circulatory and excretory systems. Respiratory, circulatory and excretory systems are typically mammalian. Respiration by lungs. Heart is without fossa ovalis. Auriculo-ventricular valves are membranous and attached to the papillary muscles by chordae tendineae.

9. Nervous system. Brain is relatively smaller. Olfactory bulbs large. Corpus callosum is absent and anterior commissure connects the two corpora striata (nerve band present in the lateral walls and floor of paracoels, lateral ventricle of cerebrum) is large. Cerebral hemisphere surface is convoluted. Cerebellum is small and simple. Cochlea of internal ear is spirally coiled.

10. Reproductive system. In male, there are no vesiculae seminales. Testes descend into scrotal sacs placed in front of the penis. Glans penis (tip of the penis) is bifid in several species.

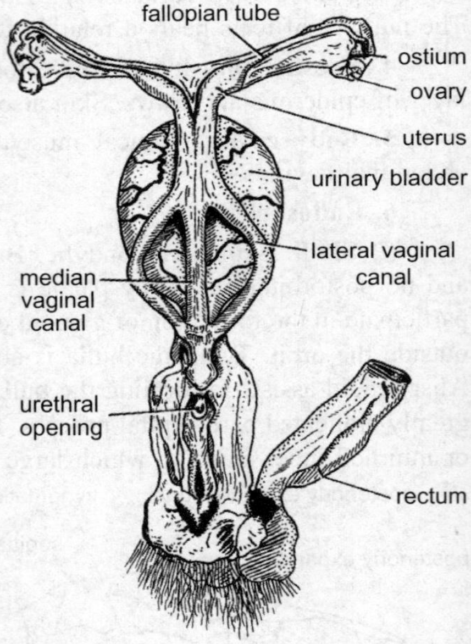

Fig. 32.2. Female reproductive system of kangaroo.

In females, the Mullerian ducts are separate posteriorly and open separately into the long urinogenital sinus. They are differentiated into oviduct, uterus and vagina on each side (*e.g., Didelphis*). In other forms the anterior part of vagina gives off a backwardly directed diverticula, the two meet in the midline as a sinus vaginalis and extend up to the front end of urinogenital sinus. This ends blindly until the young is about to be born. In some forms (*e.g.,* Macropodidae) its hind end acquires an opening into the urinogenital sinus at parturition. Clitoris is also double.

The rectum and urinogenital sinus open together in a common cloaca. Cloaca is longer in female than the male.

11. Development. Females are viviparous. The egg is yolky and covered with albumen and a shell membrane. Cleavage is unequal. Placenta is of yolk sac type, vascular wall of yolk sac is in contact with the somewhat hypertrophied uterine wall. In *Perameles* (bandicoot) allantois develop a nutritive function to some extent. Embryos are born very young as little as 8 days from conception in opossum. They are developed in the pouch attached to the teats. The young stays in the pouch for 8–9 months and the mother remains in lactationally induced anoestrus.

Population control. The males of marsupial mouse (*Antechinus*) die one week after mating. It is due to increased adrenocortical activity in the last weeks of life due to which they become more aggressive in nature. Females survive for two or three seasons.

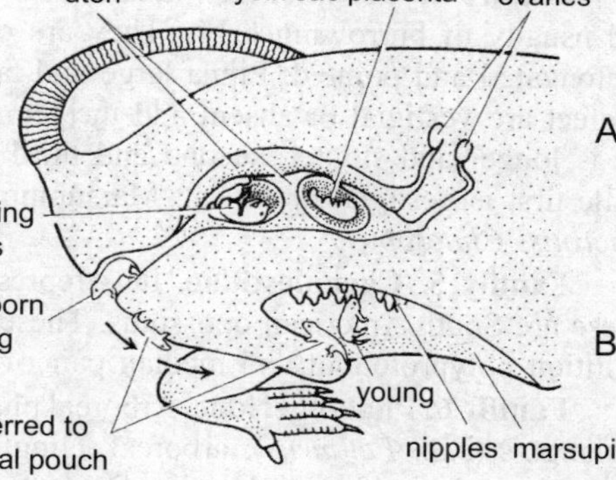

Fig. 32.3. Female opossum. A – Showing uterine development; B – Showing marsupial development.

CLASSIFICATION OF MARSUPIALIA

Marsupials are mammals basically similar to the Eutheria, but their youngs are born in a rudimentary condition and are generally sheltered during their later development within an integumentary pouch, the **marsupium**. However, pouch is lacking in the Australian Numbat (marsupial ant-eater, *Myrmecobius fasciatus*) in which the youngs are dragged along, clinging to the nipples. Marsupium is also lacking in American *Chironectes, Marmosa* and *Monodelphis*.

Metatheria are now confined to the Australian region except New Zealand, to South America and to North America. Their classification is based on dentition.

Group 1. Polyprotodontia having more than three incisors in each half of the upper jaw.

Group 2. Diprotodontia having never more than three incisors in the upper jaw, and one pair in the lower jaw.

Another classification is based on foot-structure (toes), *i.e.,* syndactyly and didactyly.

1. Syndactyly. Diprotodonts are syndactylous, *i.e.,* second and third digits of the hindfoot are united by a common sheath.

2. Didactyly. Polyprotodontia are didactylous, *i.e.,* all toes are separate.

Classification of Order Marsupialia.

Scheme 1. On the basis of incisors		*Scheme 2. On the basis of toes*	
Groups	**Families**	**Groups**	**Families**
Group 1. Polyprotodontia	1. Didelphidae	**1. Didactyly**	**1.** Didelphidae
More than 3 incisors in each	2. Dasyuridae	All the toes	**2.** Dasyuridae
half of upper jaw.	3. Notoryctidae	separate.	**3.** Notoryctidae
	4. Peramelidae		**4.** Peramelidae
Group 2. Diprotodontia	1. Caenolestidae	**2. Syndactyly**	**1.** Peramelidae
Less than 3 incisors in each	2. Phalangeridae	2nd and 3rd toes	**2.** Phalengeridae
half of upper jaw.	3. Phascolomidae	of hindfoot	**3.** Phascolomidae
	4. Macropodidae	united.	**4.** Macropodidae

Family 1. Didelphidae. These are polyprotodont opossums of North, Central and South America (*e.g., Didelphys, Marmosa, Chironectes*). They are small, arboreal, generally insectivorous animals with an elongated, naked muzzle, a well-developed nailless opposable hallux and long prehensile tail. An incomplete pouch is present or absent in most living forms. *Chironectes* (South American opossum) has webbed toes.

Family 2. Dasyuridae. Terrestrial, carnivorous, pollex rudimentary, foot four-toed, hallux when present, small and clawless and tail non-prehensile. The marsupium is well-developed [*e.g.,* Native cat (*Dasyurus*), marsupial wolf (*Thylacinus*), banded ant-eaters (*Myrmecobius*)].

Family 3. Notoryctidae. *Notoryctes,* the Marsupial mole, is a small, burrowing desert animal having short, powerful, furred 5-toed limbs. Third and fourth toes of the forefoot are provided with large, flat, triangular claws. Pinna is absent, eyes vestigial and the fur is silky. Well developed pouch opens backwards.

Family 4. Peramelidae. Bandicoots are Australian terrestrial grass and forest dwellers, and usually of burrowing habit. They are of the size of a large rat to that of a rabbit. Muzzle is elongated and pointed. Pinna large and tail is short and also long. First and fifth digits of the forefeet are vestigial or absent and the remaining three are nearly equal. In hindfoot, the fourth toe is longer and stouter, second and third are small and slender and united by a web of skin, while first is vestigial or absent. Marsupium opens backwards. Examples : *Perameles, Thylacis, Macrotis, Chaeropus.*

Family 5. Caenolestidae. It is represented by *Caenolestes, Orolestes* and *Rhyncholestes.* These are South American opossums. These are small rodent-like forms. Toes didactylous, upper dentition polyprotodont and median pair of lower incisors are enlarged.

Family 6. Phalangeridae. Arboreal phalangers (*e.g., Trichosurus, Pseudocheirus*) and flying-phalangers (*e.g., Petaurus*); arboreal, clinging sloth-like koala (*Phascolarctos*). Phalangers have both fore-and hindfeet prehensile. Second and third toes of hindfoot are slender and united by a web, but the hallux is nailless and opposable. Tail is well-developed. Flying phalangers have lateral folds of skin extending from the fore-to the hindlimbs and acting as a parachute during gliding from tree to tree. Koala is sluggish and inoffensive. It eats the leaves of *Eucalyptus*. It never drinks, but licks up earth. Gestation period is of only 35 days.

Family 7. Phascolomidae. Wombat (*Phascolomys*) are large, heavy, thick-bodied vegetarian animals living in burrows and emerging in night. Head short and flat, limbs short, thick with strong claws on all digits except hallux. Tail is very short.

Family 8. Macropodidae. Kangaroos (*Macropus*) are herbivorous, adapted for swift terrestrial locomotion. They have small head and neck. The forelimbs are small and five-toed. Hindlimbs are long and powerful. Foot is narrow with four toes, hallux is absent. Second and third toes are small and united by integument. Middle toe is long and powerful. Tail is very long, usually thick and helps to support the body while at rest. Marsupium is large. Tree-kangaroos (*Dendrolagus*) have relativery shorter and thicker hindlimbs.

DISTRIBUTION OF METATHERIA

Metatheria were widely distributed over many parts of the world as recently as the Miocene period. They are now confined to the Australian region (with the exception of New Zealand), to South America and (a few species) to North America. During Tertiary period they were widely distributed in South America. The Australasian region, excluding New Zealand, contains more than four times as many marsupials as occur in the Americas, where their congregation is chiefly in the southern continent. Two families, Didelphyidae and Epanorthidae (*Caenolestes*) are found in the Neotropical region and the genus *Phalanger* extends into the island of Celebes, where it is represented by two peculiar species.

Geological history of the group is interesting. In Australia, which is now its chief home, no fossil remains are found before Pleistocene. In America, remains of both polyprotodonts and diprotodonts are found as far back as Eocene, and in the case of Epanorthidae (Genus *Cimolestes, Telacodon* and *Batodon*) in the upper Cretaceous. Didelphyds are found in the Tertiaries of Europe, and there is a number of fossil forms in the Jurassic Formation of Europe and N. America which belong to the group polyprotodont. These continents were probably connected about 100 million years ago. By the time placentals appeared Australia and South America were no longer connected with the other continents and the marsupials became extinct in Europe and North America which were still connected. South America remained isolated until the formation of the Panama bridge in the Pliocene and most of its marsupial fauna became extinct after this time. In North America also there were no marsupials between the middle Miocene and the Pleistocene, when opossums (*Didelphis*) appeared, probably from South America. Australian marsupials only became threatened by the advent of Man from Europe and the other placentals he brought with him. However, large *Macropus* have thrived with the spread of agriculture. Marsupials diversified in Australia due to isolation, and differentiated structurally into numerous types, arboreal, fruit-eating, grazing, gnawing, digging, burrowing, ant-eating, insectivorous or carnivorous.

AFFINITIES OF MARSUPIALS

Marsupials show a mixture of primitive and advanced features.

1. Affinity with Prototheria

1. Presence of cloaca. The rectum and urinogenital sinus open together in a common cloaca.
2. Presence of clavicles, epipubic bones in pelvic girdle and tympanic bone is ring-like.
3. Tympanic bulla is absent. Tympanic is not united with the periotic
4. Brain is relatively smaller with large olfactory bulbs and anterior commissure. Corpus callosum is absent.
5. True allantoic placenta is absent.
6. Vesicula seminales are absent in males.

Differences with the Prototherians

1. Metatherians have well develope pinna, absent in prototherians.
2. Metatherians have well-developed marsupium.
3. Vertebrae with epiphysis, Ribs bicephalous (monocephalous in prototherians). No interclavicle. Coracoids reduced as in placentals.
4. Teeth present in the the adults (only one set of teeth).
5. Cochlea is spirally coiled, whereas it is partly coiled in Prototheria.
6. Vivparous.
7. Uterine gestation (No gestation in monotremes).
8. Testes in the scrotal sacs, whereas in prototherians these are abdominal. Glans penis is bifid.

2. Affinity with Eutheria

Metatherians have many advanced features similar to eutherians or higher placental mammals :

1. Presence of hairs over the body.
2. Presence of external ears (pinnae).
3. Sebaceous mammary glands with teats.
4. Coracoids reduced, interclavicle absent and ribs bicephalous.
5. Teeth heterodont : incisors, canines, premolars and molars.
6. Brain with 4 optic lobes. Cochlea spirally coiled.
7. Testes in scrotal sacs. Presence of erectile penis in male.
8. In female uterus and vagina are present in the oviduct.
9. Viviparous. Ova microlecithal. Uterine gestation and placenta present.

Differences with the Eutherians

1. Metatherians have restricted distribution, found only in Australia and Americas (North and South America).
2. Presence of marsupial pouch.
3. Number of incisors is more in both the jaws.
4. Tympanic bone absent. Alisphenoid assists in forming the tympanic bulla. Periotic and squamosal remain separate. Jugal extending back and participates in forming the glenoid cavity for the articulation of lower jaw. Well-developed epipubic bones. Ventral posterior border of lower jaw is inwardly inflected.
5. Corpus callosum is absent in the brain.
6. In *Perameles*, a true chorio-allantoic placenta is present but it is absorbed after the birth of the young. Others have yolk sac placenta.
7. A pair of uteri and vaginae are present in females. Each vagina has a separate opening into the urinogenital canal.

8. Ureters pass between the genital ducts, whereas in eutherians they pass outside the genital ducts.

9. In male penis is sometimes bifid and behind the scrotal sacs.

10. Gestation period is brief.

Systematic Position

It is obvious that the Metatheria are more advanced than the primitive, reptile-like, oviparous Prototheria. They are more closely related with the Eutheria, but do not belong to the same grade of evolution. Therefore, they are put under a separate infraclass **Metatheria**, while the rest of the higher and truly placental mammals are placed in the infraclass **Eutheria** and both combined in the subclass **Theria**.

Phylogenetic Consideration of Metatheria (Marsupials)

Comparative study of biological organisation of marsupials and placentals reveals that the marsupials appear to be 'second grade of mammals.' For this reason, it was the common practice for the zoologists to regard the marsupials as the transitional step in the evolution of mammals between the ancestral Jurassic mammals and the Coenozoic placentals. But now it is believed that placental mammals and marsupials evolved independently from some common pantotherian ancestor in the late Jurassic period and both of them evolved side by side. Fig. 32.4 will give an idea about the phylogenetic relationship of marsupials with the monotremes and placentals.

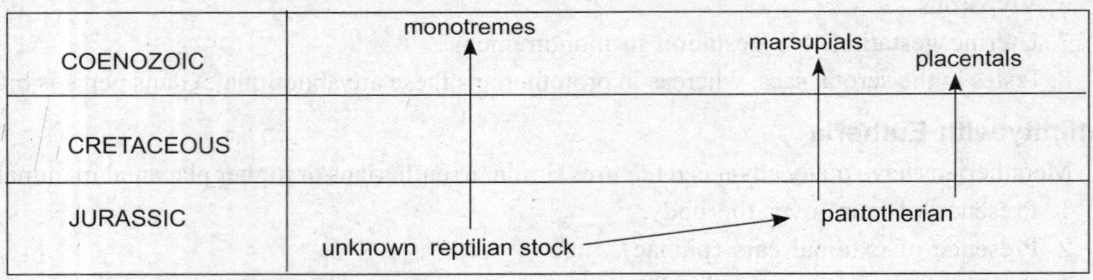

Fig. 32.4. Showing phylogenetic relationship between three surviving mammalian groups.

IMPORTANT QUESTIONS

Long Answer Questions

1. Give the general characters, distribution and affinities of Metatheria.
2. What are marsupials? What are their chief differences from Eutheria? Which of the group is of higher grade of organisation and why?
3. How are the marsupials more advanced than the monotremes, yet more primitive than the other mammals?

Short Answer Questions

1. Write the general characters of Metatheria.
2. Write the affinities of Metatheria.
3. Write short notes on : (*i*) Distribution of Metatheria; (*ii*) Differences between Prototheria and Metatheria; (*iii*) Eutherian characters in Metatheria. (*iv*) Marsupium.

Multiple Choice Questions

1. Marsupials present in America
   (*a*) Opossum (*b*) Kangaroo
   (*c*) Bandicoots (*d*) Wombat.
2. Metatherian teeth are more than
   (*a*) 44 (*b*) 32
   (*c*) 40 (*d*) 36.
3. Marsupials with the allantoic placenta
   (*a*) Opossum (*b*) Bandicoots
   (*c*) Petaurus (*d*) Kangaroo.

ANSWERS

1. (*a*) 2. (*a*) 3. (*b*)

33

Mammalia : General Account

Chapter

Contents of this Chapter

DENTITION IN MAMMALS

Teeth and Dentition

The hard and usually pointed structures connected to the jaw bones in the buccal cavity of vertebrates are known as **teeth**. The structure, kind, number and arrangement of teeth are collectively called **dentition**. Although teeth are found among fishes, amphibians and reptiles and are also known to have been present in ancestral birds, but they are most highly specialised in mammals. Teeth are as a rule present in the foetal as well as adult conditions of mammals. In all such cases they are never found on the palatal bones but are present on the premaxillae, maxillae and mandibles and are, unlike those in most lower vertebrates, differentiated in form and function.

Functions of Teeth

Teeth play an important role in everyday life of an animal. The functions of teeth are as follows : (*i*) The primary function of teeth is to **grasp and hold the prey or food** in the mouth cavity (buccal cavity). (*ii*) Teeth are modified to serve as a **grinding mill** for chewing food. (*iii*) Teeth may serve as **weapons** for offence and defence by working as tearing organs.

Significance of Teeth

(*i*) Teeth are so characteristic of mammals that their classification is based largely on their dentition. Thus, study of dentition is important for taxonomic work on mammals. (*ii*) The number of teeth present gives an idea of the approximate age of the mammal. (*iii*) Study of dentition has helped in deciding the pedigree or ancestry of certain mammals.

Toothless Mammals

Not all mammals possess teeth. In some monotremes, *Echidna* and in some American ant-eaters, however, they are entirely absent at any time. In some spiny ant-eaters (*Tachyglossus*) no teeth are found in any stage. A secondary toothless condition occurs in some mammals. In Platypus (*Ornithorhynchus*), embryonic teeth are replaced in the adult by horny epidermal plates but no true teeth are present. The true or great ant-eater (*Myrmecophaga*) also has no teeth. Also the whalebone whale and many edentates lack teeth in the adult condition only. In most of these mammals, teeth make their appearance temporarily in the foetus and are soon absorbed. In whalebone whales foetal teeth are replaced before or soon after birth by baleen plates for straining the planktonic food. Although other ant-eaters, armadillos and certain other forms possess teeth, they are imperfect structures differing from typical mammalian teeth.

Differentiation (Shape) of Teeth

Morphologically, teeth can be distinguished into two types such as **homodont** and **heterodont**.

1. Homodont teeth. In vertebrates other than mammals, all the teeth present are similar in shape and size. They are said to be **homodont** or **isodont**. Among mammals only certain cetaceans have homodont dentition, *i.e.,* teeth which are all similar in shape. Further the number of teeth in these homodont mammals varies between 2 and 200. In certain mammals such as toothed whales, dolphins, porpoises and armadillos, teeth become secondarily uniform or homodont.

2. Heterodont teeth. Mammalian teeth are characteristically **heterodont**, *i.e.,* dissimilar in shape, size and functions. The heterodont dentition commonly includes four kinds of teeth such as **incisors**, **canines**, **premolars** and **molars**. The differentiation depends upon the nature of food eaten and the manner of procuring or securing it.

Attachment of Teeth

The manner of attachment of teeth at their bases with the jaw bones varies throughout the vertebrates. These are of the following three types :

1. Acrodont type. This condition occurs in most vertebrates in which teeth are attached on the crest of jaw bone or attached to the free surface or summit of the jaw bone as in shark or frog. Such teeth are apt to break off easily but are replaced.

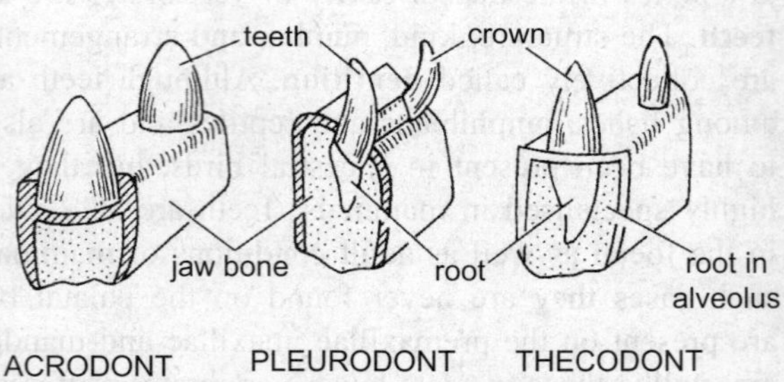

Fig. 33.1. Three methods of attachment of teeth to jaws.

2. Pleurodont type. This condition occurs commonly in urodels and lizards. Teeth are attached to the shelf-like indentations on the inner margin of jaw bone by their bases as well as one side. Acrodont and pleurodont teeth are rootless, so that nerves and blood vessels enter the pulp cavity along lateral side at the base of the tooth.

3. Thecodont type. Thecodont type of teeth are found in some fishes, crocodilians and mostly in mammals. In this type the teeth have roots (one or more) and the roots are embedded in sockets called **alveoli** or **theca** of jaw bones, a crown projects above the socket. In mammals the roots of thecodont teeth are longer, they may be open or closed below. In the open type the pulp cavity has a wide opening or root canal, such teeth continue to grow throughout life by addition of dentition, *e.g.,* incisors of rodents, tusks of elephants. Most teeth are of the closed type in which the opening of the pulp cavity is very small and serves only for passage of blood vessels and nerves, such teeth do not grow after reaching a definite size.

Succession of Teeth

According to their permanence or replacement (succession), teeth fall into three categories : polyphyodont, diphyodont and monophyodont.

1. Polyphyodont. In most of the lower vertebrates, the teeth are being constantly replaced an indefinite number of times during life, *i.e.,* successions are numerous and continuous throughout life. Such a condition is known as **polyphyodont** which is not found in mammals.

2. Diphyodont. In most mammals, there are only two successions, *i.e.,* teeth develop during life in two successive sets. This condition is known as **diphyodont**. Teeth of the first set are called **deciduous**, **lacteal** or **milk teeth**. They usually erupt after birth and they are replaced by permanent dentition. In many mammals such as guinea pigs milk teeth develop and are lost before birth. Milk dentition has no molars included. Later, milk teeth are replaced in the adult by the **permanent teeth** which last throughout life. If lost they are not replaced. In cape ant-eaters or aardvarks, milk teeth out number the permanent teeth.

3. Monophyodont. In some mammals such as platypus, marsupials, moles, sirenians, cetaceans (toothless whales), etc., only one set of teeth develops, known as **monophyodont** condition.

Kinds of Teeth (Variation of Form and Structure)

In different mammals the different kinds of teeth may undergo variations of form and structure. In many cases these variations are found to be correlated with the feeding habits and the type of food eaten. As stated earlier, when the growth of the root ceases, as it does in the majority of mammals, the opening of the pulp cavity becomes reduced to a narrow foramen. In some cases, however, the root grows indefinitely and in these forms the opening of the pulp cavity remains permanently wide such roots are said to be 'rootless'. In such cases the crown of the tooth is always long and such long-crowned teeth are called hypsodont in contradistinction to the ordinary short-crowned teeth, which are called brychydont. These and other variations may be found in many type of teeth. For the sake of convenience the different variations of the different types of teeth and their functional adaptation are studied hereunder under separate heads. There are four types of teeth are found in mammals. These are as follows : incisors, canines, premolars and molars.

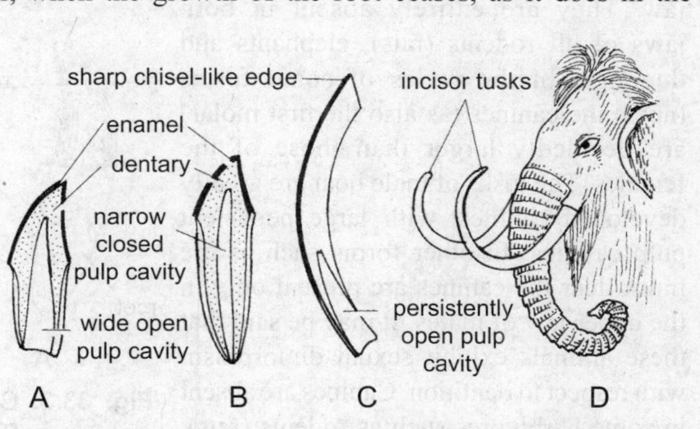

Fig. 33.2. Various forms of incisor teeth. A–Immature open-rooted; B–Mature close-rooted; C–Open-rooted rodent incisor; D–Open-rooted upper incisors in mastodon.

1. Incisors. These are the front teeth borne by the premaxillae in upper jaw and tips of dentaries in lower jaw. They are single-rooted, monocuspid and long, curved and sharp-edged. They are adapted for seizing, cutting and biting (Fig. 33.2).

Incisors are sometimes totally absent as in sloths; sometimes they are partially absent and in the upper jaw of many artiodactyles (cow, giraffe), where the incisors of the lower jaw bite against a callous pad in the corresponding region of the upper jaw. The incisors are small in carnivorous animals (cat, tiger, dog), where they are used in a limited way. In the insectivorous mammals the upper and lower incisors are elongated and work like a pair of forceps to catch hold of the active prey.

In rodents (squirrel and rat) and lagomorphs the incisors assume great functional importance in that they are constantly used for gnawing at hard-vegetable matter. The gnawing incisors of such animals are rootless and can, therefore, grow throughout life, thereby compensating for rapid wear. In such cases only the anterior surface of the incisor is covered with enamel. The incisors of the two jaws oppose each other and are kept sharp by whetting against each other, since the dentine wears away more rapidly than the harder enamel. The tusks of the elephant also are rootless incisors. In some elephants (Loxodonta) the enormous up-curved tusks measure about eleven and half inches (11.5 inches) and weigh nearly 350 pounds. The tusks are composed of hard dentine except for temporary cap of enamel at the tip. In lemurs, incisors are denticulate like a comb, serving for cleaning fur.

2. Canines. Canines are present in the majority of mammals. They always have conical pointed elongated single rooted crowns and are used for piercing and tearing. A single canine tooth occurs in each half of each jaw, just outside the incisors. Upper canines are the first teeth on maxillae.

In carnivorous animals the canine teeth assume great functional importance. In such animals they tend to become larger and much stronger to serve an efficient organs for piercing and tearing the flesh of the prey. In walruses only the upper canines develop into enormous offensive tusks. In the herbivorous animals, on the other hand, the canines are of relatively less importance and, if present, they are much less developed than those of carnivores. In some of the artiodactyla (cow and giraffe) the canines are absent only in the upper jaw. They are entirely absent in both jaws of all rodents (rats), elephants and dugongs. In the males of some forms (apes) the canines (as also the first molar) are decidedly larger than those of the females. The tusks of male boar are greatly developed canines, with large persistent pulp cavities. In other forms such as the musk deer the canines are present only in the upper jaw of males. It may be said that these animals exhibit sexual dimorphism with respect to dentition. Canines are absent in some herbivores such as rodents (rats), lagomorphs (rabbits) and some ungulates (ox) leaving a wide toothless space called **diastema**.

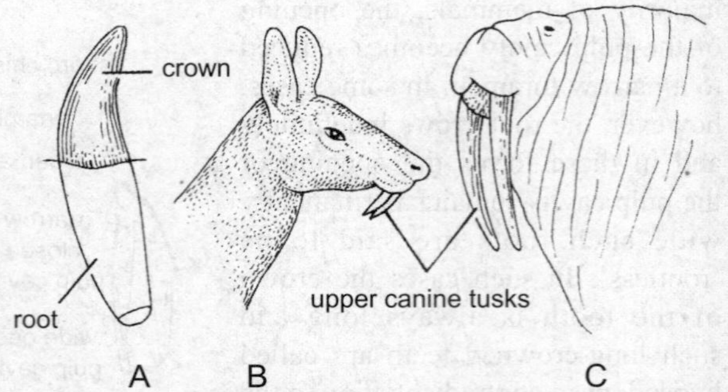

Fig. 33.3. Canine teeth. A — Of carnivore (jaguar); B — Of male musk deer; C — Of walrus.

3. Cheek teeth (premolars and molars). Canines are followed by **premolars** followed by **molars**. Both types are collectively called **cheek teeth**. Their crowns have broad surfaces with ridges and tubercles meant for crushing, grinding and chewing. Premolars usually have two roots and two cusps and are represented in milk dentition. Molars generally have more than two roots and several cusps and do not have milk predecessors (Fig. 33.4). The cheek teeth or the grinding teeth are the teeth which show the greatest amount of variation in their form and structure. This is shown by the variety of the number and arrangement of cusps and ridges at the apex of the crown. The number of cusps of a cheek teeth may vary from one to many.

In carnivores, last premolars in upper jaw and first molars in lower jaw may have very sharp cusps for cracking bones and shearing tendons. These are called **carnassial teeth**. In lemurs, first premolars are like canines. In crab-eater seal molars bear denticulate processes to strain plankton. In higher primates (man), last molar is called **wisdom tooth**. Its eruption may be delayed, or it is imperfectly formed or absent in man.

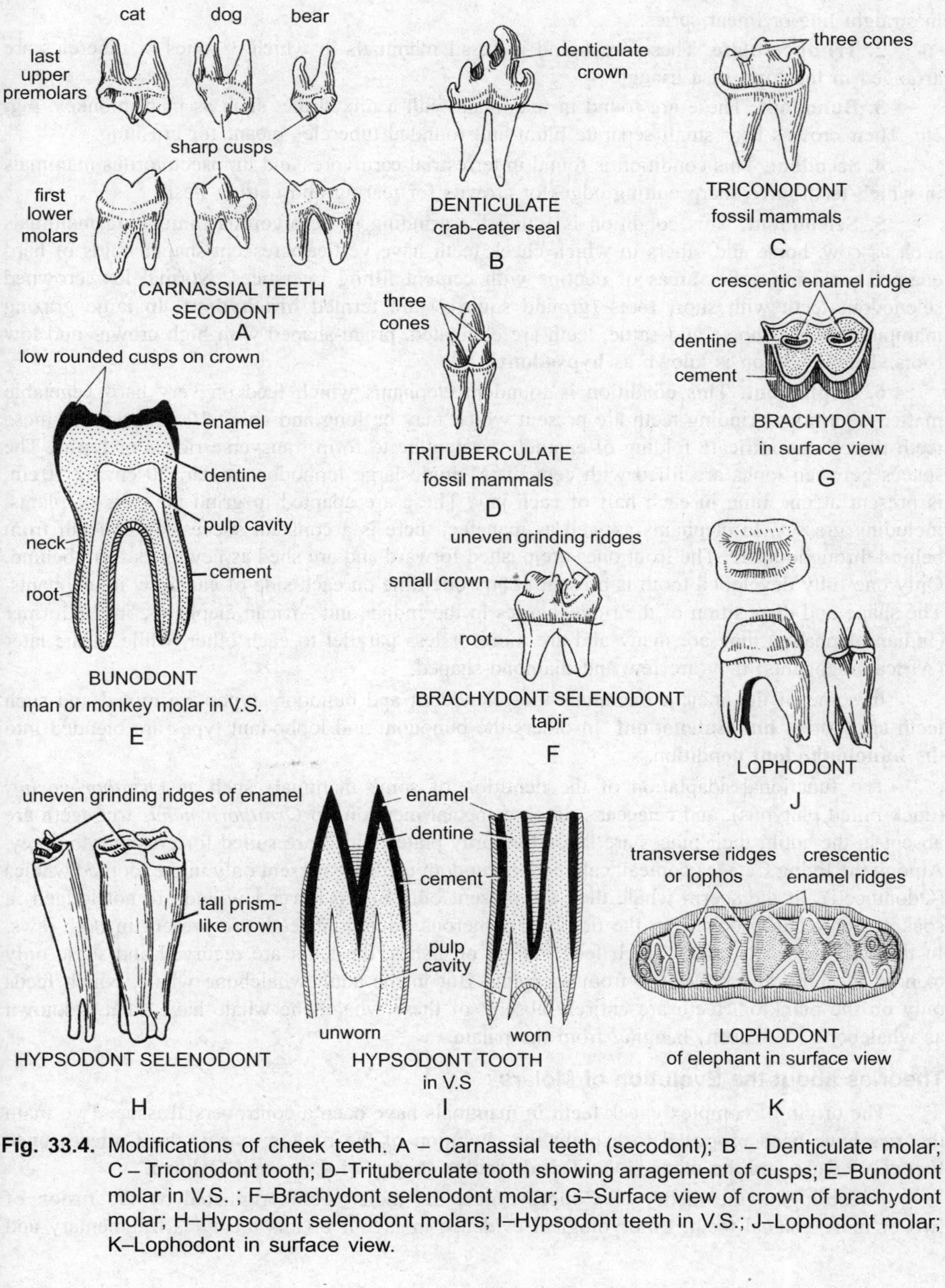

Fig. 33.4. Modifications of cheek teeth. A – Carnassial teeth (secodont); B – Denticulate molar; C – Triconodont tooth; D–Tritubercullate tooth showing arragement of cusps; E–Bunodont molar in V.S. ; F–Brachydont selenodont molar; G–Surface view of crown of brachydont molar; H–Hypsodont selenodont molars; I–Hypsodont teeth in V.S.; J–Lophodont molar; K–Lophodont in surface view.

Types of Cheek Teeth

Cheek teeth are of various types depending on the number, shape and arrangement of cusps. These are of the following types :

1. Triconodont. These are found in fossil Mesozoic mammals in which 3 cones are arranged in straight line or linear series.

2. Trituberculate. These are found in fossil mammals in which 3 cones or tubercles are arranged in the form of a triangle.

3. Bunodont. These are found in mammals with a mixed diet such as man, monkey, pig, etc. Their crowns bear small separate blunt and rounded tubercles meant for crushing.

4. Secodont. This condition is found in terrestrial carnivores and in insectivorous mammals in which teeth have sharp cutting edges or crowns for tearing and cutting flesh.

5. Selenodont. This condition is found for grinding in herbivorous ruminating mammals such as cow, horse and others in which cheek teeth have vertical crescent-shaped folds of hard enamel enclosing softer areas of dentine with cement filling interstices. Normal low crowned selenodont teeth with short roots (ground squirrel) are termed **brachydont.** In large grazing mammals such as horse and cattle, teeth are elongated, prism-shaped with high crowns and low roots. This condition is known as **hypsodont**.

6. Lophodont. This condition is found in elephants which feed on very hard vegetable matter, enormous grinding teeth are present which may be long and about 10 cm wide. In these teeth there is an intricate folding of enamel and dentine to form transverse ridges or **lophs**. The spaces between lophs are filled with cement. A single large lophodont molar, 30 cm by 10 cm, is present at one time in each half of each jaw. These are adapted to grind all sorts of plants, including grasses. In elephants as well as manatee, there is a constant succession of teeth from behind throughout life. The front ones are pushed forward and are shed as new ones form behind. Only one fully developed tooth is present at any one time on each side of each jaw in elephants. The shape and disposition of the ridges varies in the Indian and African elephants; in the former (Indian elephants) they are many and are more or less parallel to each other, while in the later (African elephants) they are few and diamond-shaped.

In some of the ancient mammals the selenodont and bunodont types are mixed and such teeth are termed **bunosalenodont**. In others the bunodont and lophodont types are blended into the **bunolophodont** condition.

The functional adaptation of the dentition of some mammals such as *Ornithorhynchus* (duck-billed platypus), and cetaceans deserve special mention. In *Ornithorhynchus* true teeth are absent in the adult; their places are taken by horny plates which are suited for crushing the prey. Among the living Cetacea conical, calcified, homodont teeth are present only in the toothed whales (Odontoceti). In the sperm whale they are present only in the lower jaw and are not lodged in sockets but set in grooves. In the dolphin, numerous homodont teeth are present in both jaws. In these toothed cetaceans, which feed mainly on fishes, the teeth are recurved and serve only to hold the prey and prevent it from escaping. But in the adult whalebone whale, which feeds only on the plankton, teeth are entirely absent; of these whalebone whale have what is known as whalebone (or baleen) hanging from the palate.

Theories about the Evolution of Molars

The origin of complex cheek teeth in mammals have been a controversial issue. Two main theories have been proposed to explain the evolution of the molars; one is the **Concrescence theory** of **Rose** and the other is the **Tritubercular theory** of **Cope** and **Osborn**.

According to the theory of **Concrescence**, the molar teeth originated by the fusion of two or more conical teeth. Embryological evidence in favour of this theory is fragmentary and inconclusive. This is now abandoned in favour of the second theory which is now accepted with certain modifications.

The **Cope-Osborn** theory of Tritubercular, based on palaeontological evidence postulates that the mammalian molar has been derived from a simple reptilian conical tooth by the development of additional cusps. In the Mesozoic mammals (Pantotheria or Trituberculata) from which modern mammals have evolved, the crown of the molar tooth has three cusps or tubercles arranged more or less in the form of a triangle. This pattern of tooth has been **tritubercular**, and it is assumed that the molars of the modern mammals have arisen from this type. The tritubercular tooth itself has arisen from the primitive cone by the development of lateral cusps. Later from these cusps or folds additional cusps or folds develop which finally give rise varied mammalian molars or cheek teeth.

Dental Formula

One of the important characteristics of mammalian dentition is that the number of teeth in any particular species is so surprisingly constant that it is considered to be of great significance in classification. The number of each type of teeth of a mammal or dentition is expressed by a **dental formula** (D.F.). The dental formula is expressed as the number of different types of teeth in one half of the upper jaw and is written above a line, while the number of teeth of one half of the lower jaw is written below the line, *e.g.*,

$$\text{D.F.} = \frac{\text{Number of teeth in one half on the upper jaw}}{\text{Number of teeth in same half on the lower jaw}}$$

$$\text{or} \quad = \frac{\text{incisor}}{\text{incisor}}, \frac{\text{canine}}{\text{canine}}, \frac{\text{premolar}}{\text{premolar}}, \frac{\text{molar}}{\text{molar}} \times 2$$

$$\text{or} \quad \frac{i}{i}, \frac{c}{c}, \frac{pm}{pm}, \frac{m}{m} = 2$$

Since two halves of each jaw are identical, only the teeth of one side are recorded. Those of the upper and lower jaws are separated by a horizontal line. Kinds of teeth are denoted by their initial letters *i, c, pm* and *m* representing incisors, canines, premolars and molars, respectively. Number of teeth shown in the formula multiplied by 2 gives the total number of teeth of a species. A typical mammalian dentition includes 44 permanent teeth which are shown by the dental formula as follows :

$$i\frac{3}{3}, c\,\frac{1}{1}, pm\,\frac{4}{4}, m\,\frac{3}{3} \times 2 = 44$$

To simplify further, it is customary to omit letters, so that the same formula may be written as 3, 1, 4, 3 / 3, 1, 4, 3 = 44. When a certain type of teeth is lacking, it is indicated by a zero. Dental formula of some common mammals are as follows :

Dog	$\frac{3,1,4,2}{3,1,4,3} = 42$	Horse, Pig and Mole	$\frac{3,1,4,3}{3,1,4,3} = 44$
Cat	$\frac{3,1,3,1}{3,1,2,1} = 30$	Squirrel	$\frac{1.0.2,3}{1,0,1,3} = 22$
Rat	$\frac{1,0,0,3}{1,0,0,3} = 16$	Cow, Sheep and Goat	$\frac{0,0,3,3}{3,1,3,3} = 32$
		Monkey	$\frac{2,1,2,3}{2,1,2,3} = 32$

Rabbit	$\frac{2,0,3,3}{1,0,2,3} = 28$	Lemur	$\frac{2,1,3,3}{2,1,3,3} = 36$
Kangaroo	$\frac{3,1,2,4}{1,0,2,4} = 34$	Man	$\frac{2,1,2,3}{2,1,2,3} = 32$
Opossum	$\frac{5,1,3,4}{4,1,3,4} = 50$	Elephant	$\frac{1,0,0,3}{0,0,0,3} = 14$

Structure of Teeth

Structure of teeth is similar in all the vertebrates. A typical mammalian tooth is differentiated into 3 parts : crown, neck and root. **Crown** is the visible part of the tooth projecting out from the gum and out of the jaw bone. **Root** is the basal part, embedded in a socket or alveolus of the jaw bone. The junction of crown and root is called the **neck**. A tooth is made largely of a hard dermal bony substance the **dentine** (ivory). The crown is externally covered by a thin very hard and shining layer of **enamel**. Enamel is the hardest substance in the body of vertebrates. The root of tooth is surrounded and fixed to the jaw bone by a layer of **cement** and a vascular **periodontial membrane**. It is made of strong connective tissue fibres (Sharpey's fibres) running from cement to bone. Inside the tooth is narrow **pulp cavity**. It contains connective tissue or **pulp** with blood vessels and nerves. The blood vessels and nerves enter the pulp cavity through its opening present at the apex of the root. Lining the pulp cavity is a layer of bone cells, the **odontoblasts**, which send fine protoplasmic processes into **canaliculi** of dentine. The main mass of the tooth is formed of a very hard substance, called **dentine**, which encloses the pulp cavity. The dentine is covered over by enamel in the crown region and by cement in the root region. Further, the attachment of tooth on the jaw bones are strengthened by the gum. In the incisors of lagomorphs, rodents and elephants, the pulp cavity remains open basally so that these teeth continue to grow throughout life and are termed **open-rooted**. In the majority of mammals, including man, the basal aperture of pulp cavity becomes closed at a certain age so that nourishment stops and further growth ceases. Such teeth are termed **close-rooted**.

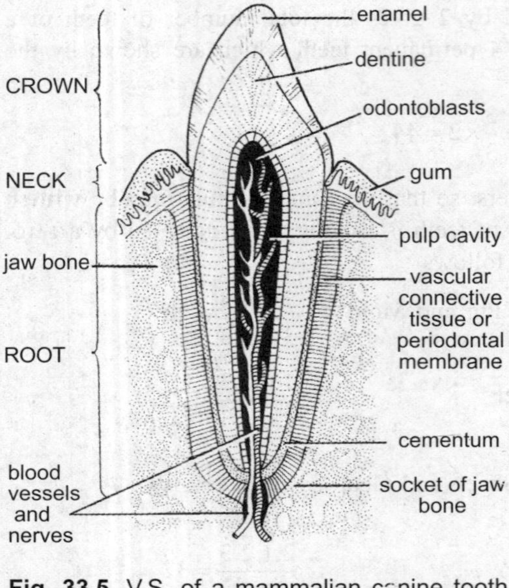

Fig. 33.5. V.S. of a mammalian canine tooth.

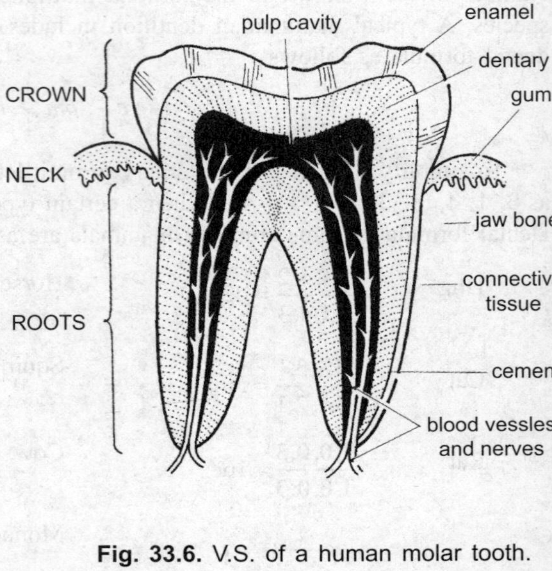

Fig. 33.6. V.S. of a human molar tooth.

Development of a Tooth

In mammals, teeth develop in the gum or the soft tissue, covering the borders of premaxillae, maxillae and denteries (Fig. 33.7). As in all vertebrates, development of a tooth is similar to that of a placoid scale. Enamel of tooth is derived from epidermis, while the rest of tooth from dermis or mesenchyme. In the beginning of development of a tooth, there is a thickening of ectoderm along the margin of jaw bone. The basal layer of ectoderm, the **Malpighian layer**, forms a continuous ridge-like vertical invagination into the underlying dermis. This forms the **dental lamina** which retains its connection with the outer epidermis. Mesodermal cells multiply rapidly beneath the ectodermal ingrowth or dental lamina forming a series of solid bud-like outgrowths at intervals, called **tooth germs**. The number of tooth germs is as many as the number of milk teeth. In each tooth germ, the inverted cup-like epithelial cap will secrete the enamel, hence, termed the **enamel organ**. This gives rise to the enamel of the tooth. The outer columnar cells of enamel organ become differentiated into **odontoblasts** which secrete a layer of **dentine** on their outer surface. The cells of inner epithelial layer of enamel organ similarly become **ameloblasts** which form a cap of hard enamel around the top and sides of dentine. No enamel is deposited in the root. Dental papilla is retained as **pulp**. The central cavity of pulp goes on increasing to become **pulp cavity**. The nerves and blood vessels enter the pulp cavity through the basal opening. Up to this stage the tooth remains inside the tissue (gum). In the later stage, its eruption (the eruption of dental papilla) through the overlying epidermis is known as **cutting of tooth**. Around the root of tooth appears **cement** or **cementum** which is a modified bone. Odontoblasts become inactive when tooth is fully formed. However, in rodents, lagomorphs, etc., the odontoblasts remain active throughout life and teeth continue to grow.

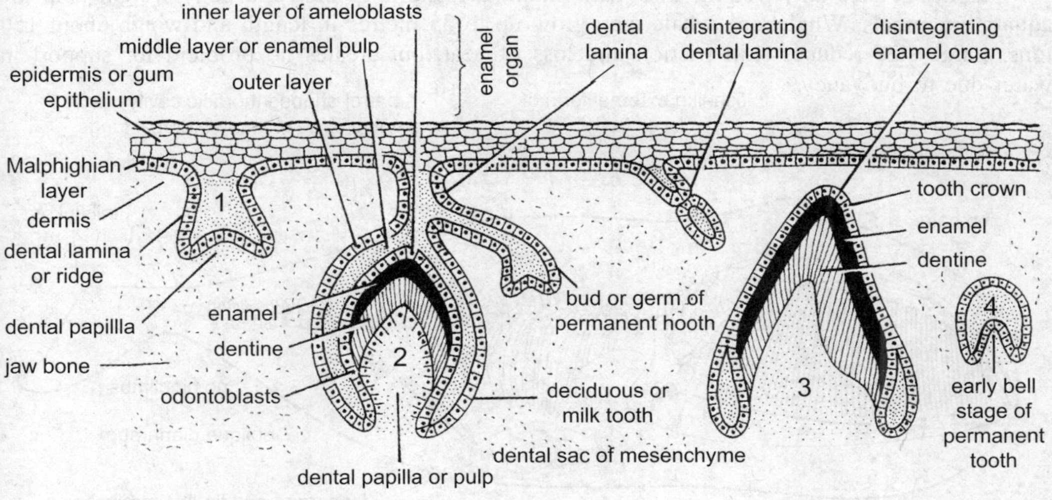

Fig. 33.7. Development of a mammalian tooth.

AQUATIC MAMMALS AND ADAPTATIONS

Mammals are primarily terrestrial animals. However, some of them have adopted an aquatic mode of life. The aquatic mammals have evolved from terrestrial mammals. The fact that all of them are not gill-breathers but breathe air through lungs, indicate their original terrestrial mode of life. All the aquatic mammals are really terrestrial lung-breathing forms which have reverted to an aquatic life, and they have done so with remarkable success, the whales being the most successful. They have reverted to water probably because of extreme competition on land for food and shelter.

[I] Aquatic Mammals

There are several aquatic mammals. Aquatic mammals belong to several orders of Mammalia. Depending on the degree for aquatic adaptation the aquatic mammals have been divided into the following categories :

1. Amphibious mammals. These mammals do not live permanently in water. They live on land but go into water for food and shelter. They show only partial aquatic adaptations such as (*i*) small external ears, (*ii*) webbed feet, (*iii*) flattened nails, (*iv*) subcutaneous fat. The mammals of this category include the beaver (*Castor*), musk rat (*Ondatra*), nutria (*Myocaster*), otter (*Lutra*), mink (*Mustela*) and many others. The amphibious mammals belong to several orders of mammalia such as Carnivora, Rodentia, Artiodactyla, Marsupialia, Monotremata, etc.

2. Aquatic mammals. The mammals under this category spend most of the time in water and usually come to land for reproduction. The typical examples are seals and hippopotamus.

3. Marine mammals. These mammals never come to land and are perfectly at home in water. The typical examples are whales.

[II] Aquatic Adaptations

The adaptations or specialisations of truly aquatic mammals (Cetacea and Sirenia) are divided into 3 major categories : (*i*) modifications of original structures, (*ii*) loss of structures, and (*iii*) development of new structures (Fig. 33.8).

A. Modifications of Original Structures

1. Body shape. In aquatic mammals, body shape is of prime importance. The external fish-like form, elongated head, indistinct neck and tapering streamlined body offers little resistance and swims rapidly through water.

2. Large size and weight. In aquatic mammals, the large size and body weight help the aquatic mammals. Whalebone whale may grow up to 35 metres in length and weigh about 150 tons. Large size reduces skin friction and loss of heat, but creates no problem for support in water due to buoyancy.

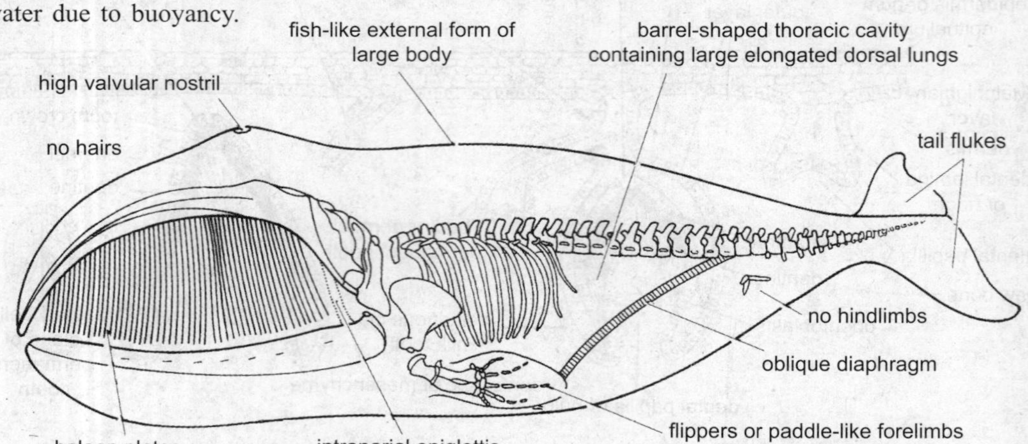

Fig. 33.8. Body outline and skeleton of Greenland right whale *Balaena mysticetus* showing aquatic adaptations.

3. Flippers. In aquatic mammals, the forelimbs are transformed into skin-covered, unjointed paddles or flippers, having no separate indication of fingers. These paddles or flippers can move as a whole only at the shoulder joint. The broad and flattened paddles or flippers serve as balancers and provide stability during swimming.

4. Hyperdactyly and hyperphalangy. In aquatic mammals, extra digits (hyperdactyly) and extra phalanges (hyperphalangy) up to 14 or more in some forms, serve to increase the surface area of flippers for greater utility for swimming in water.

5. High and valvular nostrils. In aquatic mammals, the nostrils are placed far back on the top of head so that animal can breathe air without raising head much out of water. The nostrils can also be closed by valves during diving under water.

6. Mammary ducts. In aquatic mammals, during lactation, ducts of mammary glands dilate to form large reservoirs of milk, which is pumped directly into mouth of young by the action of a special compressor muscle. This arrangement facilitates suckling of young under water.

7. Oblique diaphragm. In aquatic mammals, oblique diaphragm makes the thoracic cavity larger, dorsal and barrel-shaped for providing more space to lungs for expansion.

8. Large lungs. In aquatic mammals, large unlobulated and highly elastic lungs ensure taking down maximum air before submergence. Like swim bladders of fishes, the dorsal lungs also serve as hydrostatic organs in maintaining a horizontal posture during swimming.

9. Intra-narial epiglottis. In aquatic mammals, elongated, tubular and intra-narial epiglottis, when embraced by the soft palate, provides a continuous and separate air-passage, thus, allowing breathing and feeding simultaneously.

10. Endoskeleton. In aquatic mammals, the cranium becomes small but wider to accommodate the short and wide brain. The facial part of skull projects forming elongated snout or rostrum (porpoise). The zygomatic arches are reduced. Due to reduced neck, the cervical vertebrae are fused into a solid bony mass. Zygapophyses are reduced. Sacrum is also reduced. Ribs become arched dorsally to increase thoracic cavity. Bones are light and spongy. In Cetacea, bones are filled with oil.

11. Teeth. In toothed whales, teeth are monophyodont, homodont and numerous, as many as 250. This helps in capturing or seizing prey, prevent its escape and swallowing it without mastication. Usually, the mobility of jaws is reduced as they have no function in mastication.

B. Loss of Structures

In aquatic mammals, there is a loss of a few structures which are usually present in other mammals. These are as follows :

1. There is a loss of hairs. Skin surface usually remains **smooth** and **glistening** due to **loss of hairs** except for a few sensory bristles on snout or lips in some cases.
2. **Pinnae** are also absent. Presence of hairs and pinnae may obstruct or impede the ever flow of water over body surface and interfere with the speed and elegance of movement through water.
3. **Nictitating membranes**, **eye cleansing glands**, **lacrimal glands** and all kinds of **skin glands** (sweat and sebaceous) are also absent because they would have been useless under water.
4. Skin losses its muscles and nerves due to thickening and immobility.
5. **Hindlimbs** are represented only by button-like knobs in the foetus but disappear in the adult.
6. **Pelvis** is also rudimentary.
7. **Fingernails** are absent except for traces in foetus.
8. **Scrotal sacs** are also absent and testes remain inside abdomen.

C. Development of New Structures

1. Tail flukes. In aquatic mammals, some large, lateral or horizontal expansions of the skin develop on tail. These expansions are called tail flukes. These are not supported by fin-rays. Their up and down strokes not only propel the body through water but enable rapid return to the surface for breathing after prolonged submersion.

2. Dorsal fin. In most Cetacea develop an unpaired adipose dorsal fin without internal skeletal support. It serves as a rudder or keel during swimming.

3. Blubber. In aquatic mammals, the blubber is the thick subcutaneous layer of fat, which compensates for the lack of hairy covering. Blubber acts as a heat insulator. It not only retains the warmth of the body but also provides a ready reservoir of food and water during emergency. The fat also reduces the specific gravity of the animal, thus, imparting buoyancy. Blubber also provides an elastic covering to allow changes in body volume during deep diving and also counteracts the hydrostatic pressure.

4. Baleen. In whalebone whales, teeth are absent. Instead, the upper jaw carries two transverse rows of numerous triangular fringed horny plates of **baleen** or **whalebone**. These serve as an effective sieve for straining plankton (mostly kril) which forms their chief food.

5. Foam. Foam is a fine emulsion of fat, mucus and gas. Each middle ear cavity sends an inner pneumatic prolongation, which meets with the fellow on the other side below the skull. These extensions contain foam. It probably serves to insulate sound and improves audition or hearing under water.

6. Melon. In some aquatic mammals, the melon is a receptor present in front of nostrils. It consists of a fatty mass traversed by muscle fibres. It possibly serves to detect pressure changes in water.

7. Harderian glands. In aquatic mammals, eyes under water remain protected by a special fatty secretion of Harderian glands.

FLYING (VOLANT) MAMMALS AND ADAPTATIONS

Many mammals, specially living in forest areas, have become modified to live on trees. Their mode of living is named as **arboreal**. Arboreal mammals are able to climb the trees and use their branches as the highways. There are three possible modes of aerial locomotion or flight–gliding, soaring and flapping. **Gliding** is the rudimentary flight displayed by certain fishes (*Exocoetus*), amphibians (*Rhachophorus*), lizards (*Draco*), and among some mammals by some phalangers, flying lemurs and squirrels. **Soaring** and **flapping** are more efficient and need more profound morphological and physiological adaptations. Soaring flight is attained by certain birds and flapping flight is accomplished both by birds and bats (Fig.33.9).

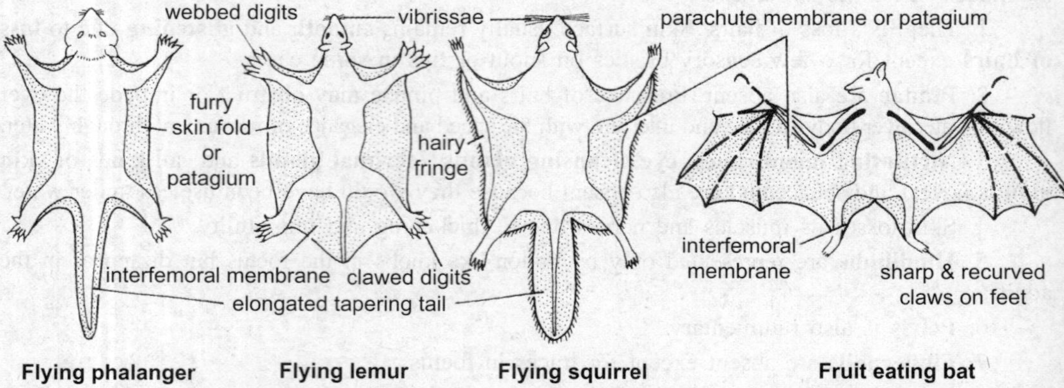

Fig. 33.9. Gliding and flying mammals (diagrammatic).

I. Gliding Mammals and Adaptations

1. Gliding mammals. Common examples are : (*i*) **Order Marsupialia.** Flying phalangers (*Petaurus*) and feather tails (*Acrobates*).

(*ii*) **Order Dermoptera.** Flying lemurs (*Cynocephalus*).

(*iii*) **Order Rodentia.** Flying squirrels. Kashmir wooly flying squirrel (*Eupetaurus cinereus*) and brown flying squirrel (*Petaurista phillipensis*).

2. Gliding adaptations. In gliding mammals, body is elongated, flattened and streamlined. Limbs are long and equal. Tail is long and gradually tapering. There is a double fold of furred skin, called **patagium** or **parachute membrane** stretched on either lateral side between neck, limbs, body and tail. It is sometimes also provided with a cartilaginous rod springing from the elbow or wrist. When the animal is at rest, the parachute is scarcely visible as it remains folded close to the body by its own elasticity.

3. Nature of gliding flight. Gliding is not a true continuous flight but merely a prologed aerial leap, covering 10 to 20 metres at the most. This type of movement involves no propulsion other than the initial force of jumping. Gliding is characterised by leaping or jumping from a high point and held up by some sustaining organs, then to glide to lower level. Thus, there is no locomotive force other than gravity. The gliding mammals glides from one tree to another, smoothly and swiftly downwards, supported by the outstretched parachute and limbs. Aerial progress is in a straight line, rapidly losing height and with little manoeuvrability. However, the flight can be guided to some degree by changing the position of the limbs, by twisting the parachute membrane, and by using the tail. Before alighting the animal raises the front body part to check speed and to soften its impact on the landing target.

II. Flapping or Flying Mammals and Adaptations

1. Bats. Bats belong to the order **Chiroptera**. They are the only mammals with true and sustained flight. This flight is effected by the flapping or up and down strokes of wings. Bats are built for flight and compete very well with birds.

Flight adaptations of bats. Bats possess many adaptations which are infinitely more profound. The modifications are not so conspicuous in their external features. But radical changes have taken place internally in their skeleton and musculature. The skull of bats has been altered in the most exaggerated manner for special purpose.

(*a*) Wings. Wings or **patagia** of bats are thin like a paper and elastic membranes. These are extensions of leathery skin from the lateral sides of body, legs and tail. Forearm is greatly elongated, carrying a hand with 5 fingers. In small insectivorous **Microchiroptera**, the first finger or **pollex** is short, free and sharply clawed. The other 4 fingers are dawless, enormously lengthened and embedded in the web of wing to support it. Like the ribs of an umbrella, the fingers open and close the wing and keep it taut when expanded. In large fruit-eating bats or **Megachiroptera**, the 2nd finger also ends in a claw. The 3rd finger is the largest, corresponding to the leading edge of the wing membrane.

In most bats, an **inter-femoral membrane**, which also enclose the tail wholly or partially between the hindlegs. It is also supported by a spur of bone, the **calcar**, projecting from the tarsus of each foot. A similar **ante-brachial membrane** connects the neck with the humerus bone of upper arm. Thus, there is a continuous and uninterrupted parachute of skin around the body of bat.

The muscles and other body structures are specially designed in bats to support and operate the wings. Bat possess a capacious **thorax** containing a remarkably large **heart** and **lungs**. The keeled sternum offers space for the attachment of great **pectoral muscles** which sustain the arms in flight.

In bats, though converted into wings for flight, the arms and hands are also used for walking, climbing, holding food and killing prey in flight.

(*b*) Legs. In bats, the hindlegs are small, weak and with sharp curved claws on toes used for suspending the bat head down from a tree branch or perch, while resting or sleeping. Knee-joints are directed backwards, instead of forwards, which helps in maximum spread of wing membranes but is of little help in other movements.

(*c*) Tail. In bats, the tail is variable in size. It may be large, small or scarcely visible. When well developed, the tail supports the inter-femoral membrane which can act as a brake to flight, used as a large pouch for holding prey or food, or even as an aerial cradle for the reception of a new born baby.

(*d*) Teeth. In bats, the milk dentition is curious in adaptation to flight. The young are often born with small hook-like or needle-like teeth, similar and sometimes forked. These are supposed to serve the young for gripping firmly the maternal teats while the mother is flying.

(*e*) Senses. Bats are extremely modified for nocturnal flight. They are able to fly even in total darkness with ease and swiftness avoiding obstacles. They possess special sense organs which allow them to perceive prey and surrounding obstacles at a distance even in total darkness.

Bats are supersensitive to sound and possess a natural **echo-apparatus** or **radar system** of their own. In flight bats keep their mouth open and continually emit tremendous bursts of supersonic or ultrasonic sounds which are quite inaudible to human ear. The warning echos reflected back enable bats to locate and evade obstacles in their path. In numerous insectivorous bats sensitive accessory lobes on ears called **tragus**, help to pick up warning echos. Besides, certain bats have skin flaps encircling nose, called **nose-leaf**. The exact use of nose-leaf is not well-understood, but probably it acts as an **antenna** for the perception of air vibrations. The **wing webs** are abundantly supplied with scattered nerves and blood vessels and probably act as tactile organs which also help in avoiding the obstacles in their course during nocturnal flights. Bats are so incredibly sensitive that thousands of them may swirl around for hours in total darkness in a cave without a single collision. Engineers are not able to understand the aerodynamics of bats. It is said that the aerodynamics of bats is not easily understandable and beyond our knowledge at present.

ADAPTIVE RADIATION IN MAMMALS

During Mesozoic era, the age of reptiles (dinosaurs), mammals were small, generalised and rare. By the end of Mesozoic or beginning of Coenozoic, the dinosaurs vanished and mammals suddenly expanded into varied evolutionary patterns. Early in Cretaceous period, placental mammals became distinct from marsupials. During Eocene and Oligocene, most of the orders of mammals originated moving into habitats and ecological niches vacated by the extinct dinosaurs. This evolution from a single ancestral species to a variety of forms which occupy different habitats is called **adaptive radiation** or **divergent evolution**. The concept of adaptive radiation in evolution was developed by **H.F. Osborn** in 1898. Examples often given as evidence include Darwin's finches of the Galapagos Islands, varied limb structure of mammals, Australian Marsupials, etc.

Figure 33.10 shows adaptive radiation in mammals. It is based on limb structure.

1. Radiation in limb structure of mammals. Mammalian limbs are the modifications of the pentadactyl limb. Primitive, ancestral mammals are believed to have been short legged, five fingered creatures living on the ground. Their limbs were not modified for any particular type of locomotion. These animals were **terrestrial**. These terrestrial ancestors formed the ancestors of modern mammals. These mammals having primitive limb structure are placed in the centre of the figure 33.10. Therefore, adaptive radiation occurred in five different lines or habitats with modifications in their limb structure. From these terrestrial mammals the different lines radiated in the following manner :

1. One evolutionary line radiates to form **arboreal forms** which have adapted limbs for life in trees (*e.g.,* squirrels, sloths, monkeys, etc.).

2. Another line leads to **aerial** representing mammals adapted for flight (*e.g.,* bats). Only bats occupy the position at the terminus of this line, since they are the only truely flying mammals. Somewhere along this line we can place for gliding mammals such as "flying squirrel." The arboreal and aerial forms not arose independently from the terrestrial forms as shown in the diagram. It is believed that the ancestral aerial forms were previously lived in trees having gliding type of locomotion which later gave rise to true flight. Hence, perhaps the gliding formed transitional type of locomotion between climbing and true flight.

3. Third line of radiation gave rise to **cursorial forms** (*e.g.*, horses and antelopes). They have developed limbs suitable to rapid movements over the surface of the ground. Along this line also developed other mammals with less strongly modified limbs, such as wolves, foxes, hyaenas, lions.

4. Fourth line of radiation formed the burrowing mammals, the **fossorial mammals**. Some of the fossorial mammals, like the moles, have modified their forelimbs for digging but they are poorly adapted for locomotion on the ground. While others like pocket gophers and badgers are expert diggers but they have retained structures enabling them to move readily on the surface of ground.

5. Fifth line of radiation leads to the **aquatic mammals**. (*i*) Whales and porpoises having limbs strongly adapted for aquatic life, but they cannot move about on land. (*ii*) While seals, sea lions and walruses have also strongly modified limbs for aquatic life but they are also able to move about on land. (*iii*) The third group includes accomplished swimmers such as others and polar bears which are equally at home in water or on land.

Thus, all the mammals of different radiating lines have limbs more or less adapted for some particular mode of locomotion. All the lines start from a common centre representing the short, pentadactyl limbs of terrestrial mammals. From the centre, evolutionary lines radiate out in various directions. Hence, adaptive radiation is evolution in several directions starting from a common ancestral type.

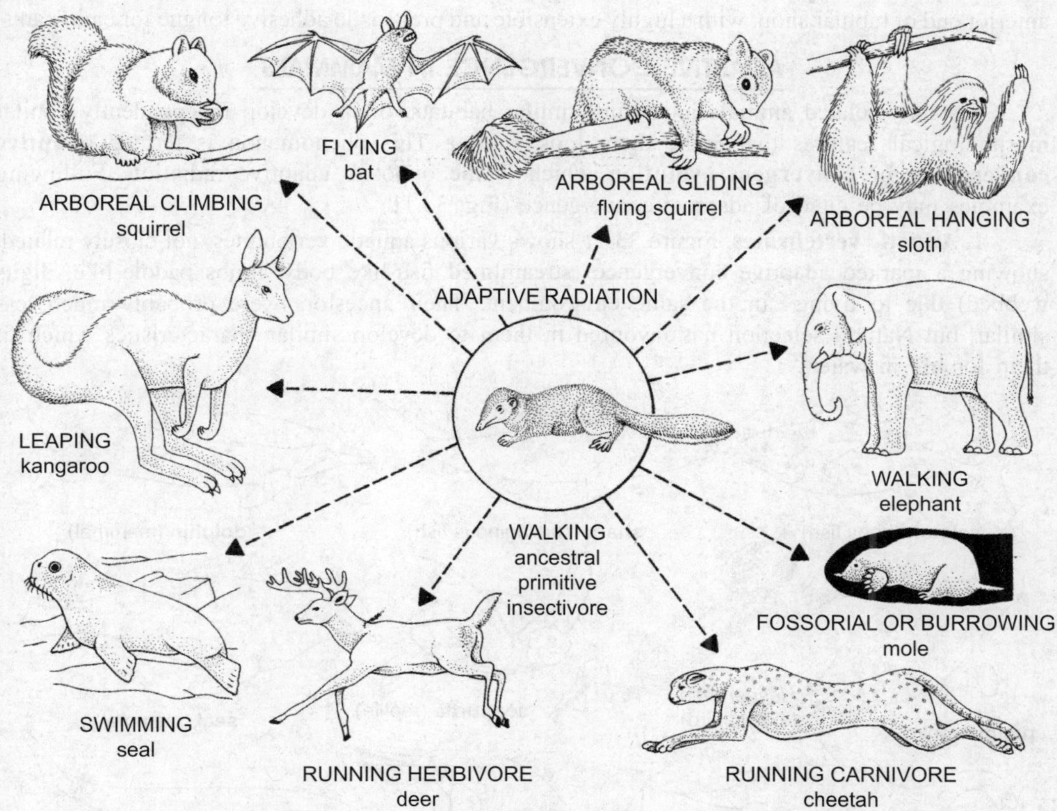

Fig. 33.10. Adaptive radiation or divergent evolution in mammals, based on locomotion.

Evolutionary relationships. All the mammals included in the diagram have modified pentadactyl limbs, hence, they must be related to each other. Possession of this common limb pattern indicates close relationship not only among mammals but also by birds, reptiles and amphibians – by all vertebrates except fishes.

2. Tooth radiation in mammals. The mammals with few exceptions (Cetacea, ant-eaters, etc.) possess heterodont dentition, *i.e.,* the teeth in contrast to those of reptiles, are differentiated into different forms with very distinct functions. The different types of teeth are **incisors** for biting, **canines** for grasping, tearing or for defence or offence, **premolars** and **molars** are for grinding. The premolars and molars show greatest structural modification for different types of food.

In **insectivorous type** (stem form in mammals), premolars and molars are low-crowned, simple, with few cusps, generally sharp pointed and suitable for crushing feeble prey. In **carnivorous type**, premolars and molars are high crowned, trenchant, shearing structures (carnassial). The jaws have little or no lateral movement. Cats have no grinding teeth, while dogs have more of grinders. In Odontoceli (*e.g.,* toothed whales) there is no tooth differentiation and the teeth are practically alike. In sperm whale, *Physeter*, no teeth in the upper jaw, but germs of upper teeth are present, while in whalebone whales (Mystaceti) upper teeth are totally absent and their place is taken by curious baleen or whalebone which hangs from the palate. In **herbivorous types**, incisors are for seizing and cutting the vegetation. In ruminants, they are absent in the upper jaw, but a horny pad is present there. Canine teeth are of little importance for herbivores, but in musk deer they are used for defence and in swine they are used for uprooting the vegetation. Grinders (premolar and molars) may be **short-crowned** and **brachydont** adapted for succulent leaves and twigs, or **long-crowned** and **hypsodont** adapted for harsh grasses. In **myrmecophagous type** teeth have disappeared, jaws reduced and mouth opens at the extreme anterior end of tubular snout with a highly extensible and prehensile adhesive tongue for eating ants.

ADAPTIVE CONVERGENCE IN MAMMALS

Distantly related animals inhabiting similar habitats, often develop independently similar morphological features that make them look similar. This phenomenon is termed **adaptive convergence** or **convergent evolution** which is the opposite adaptive radiation. Following examples may be cited of adaptive convergence (Fig. 33.11) :

1. Aquatic vertebrates. Figure 33.11 shows various aquatic vertebrates, not closely related, showing a marked adaptive convergence (streamlined fish-like body, limbs paddle-like, digits webbed) due to living in the same environment. Their ancestors were probably much less similar, but Natural selection has favoured in them to develop similar characteristics which fit them for life in water.

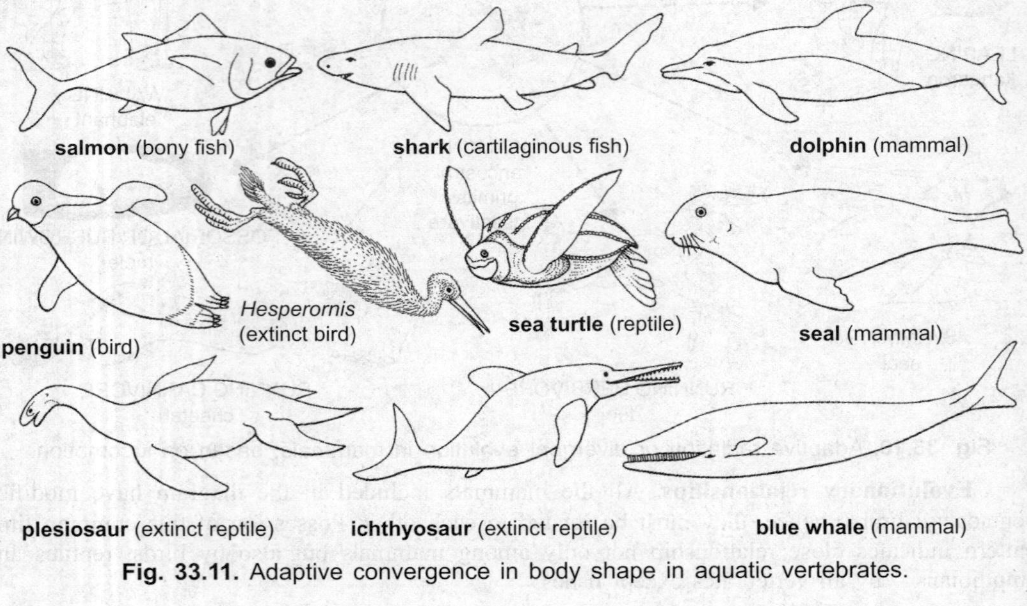

Fig. 33.11. Adaptive convergence in body shape in aquatic vertebrates.

2. Placental and marsupial mammals. During Palaeocene, Eocene and Oligocene, the marsupials in Australia and the placentals in the rest of the world passed through much adaptive radiation. They moved into habitats vacated due to extinction of dinosaurs. Marsupials and placental mammals are not closely related, but convergent evolution has resulted in similar looking members in both the groups occupying similar ecological niches and leading similar ways of life. Thus, there are native or marsupial wolves, mice, cats, ant-eaters, moles, sloths, flying phalangers and wombats. These are not related to their true counterparts in placental mammals but resemble them due to convergent evolution (Fig. 33.12).

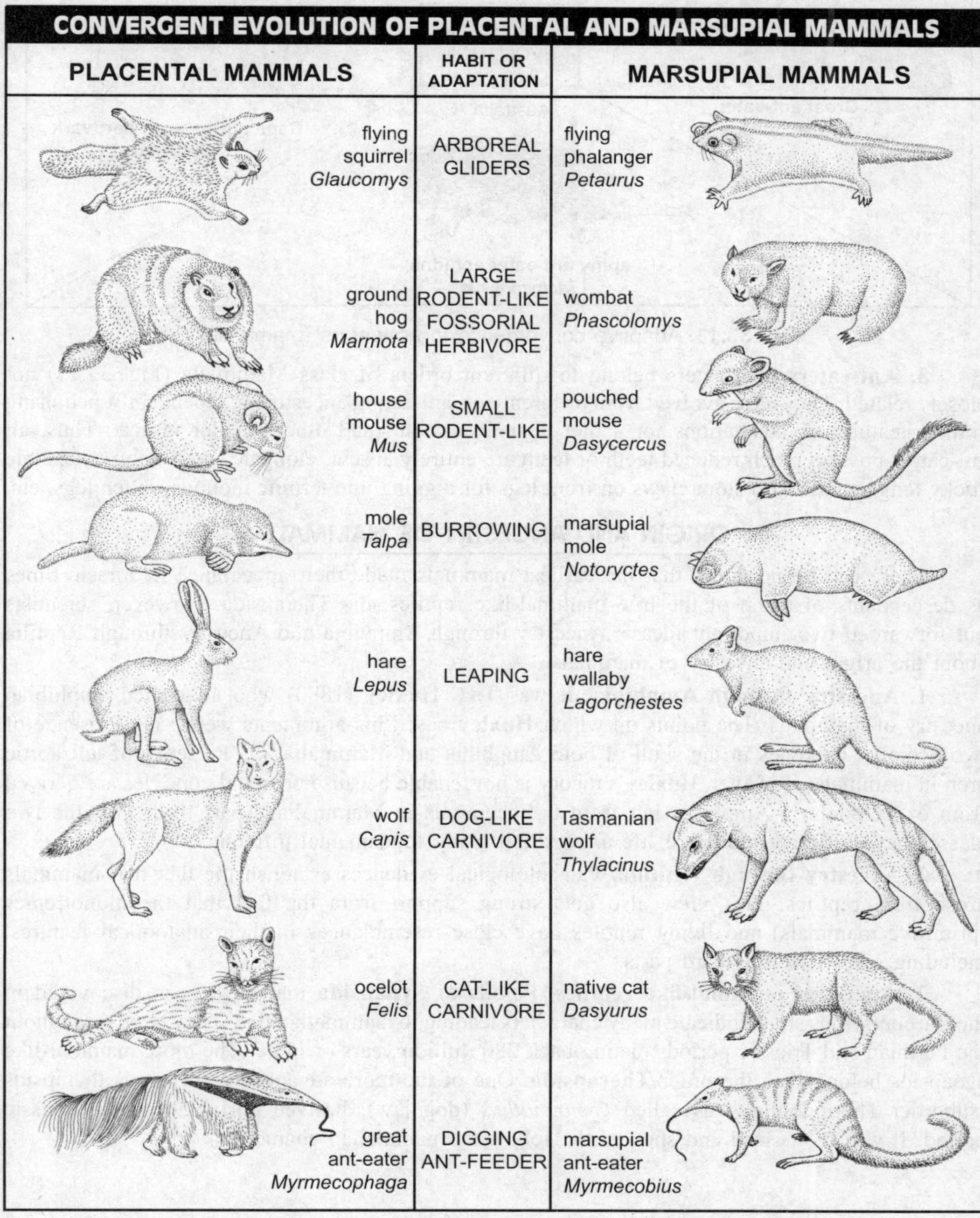

Fig. 33.12. Convergent evolution in placental and marsupial mammals.

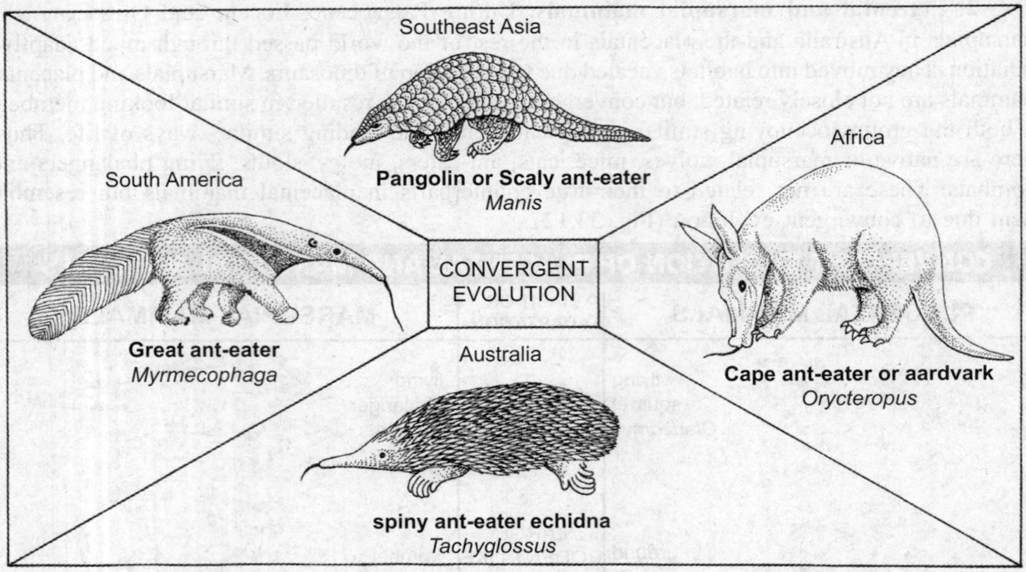

Fig. 33.13. Adaptive convergence in ant-eaters (mammals).

3. Ant-eaters. Ant-eaters belong to different orders of class Mammalia (Fig. 33.13) not closely related. They have evolved from different non-ant-eating ancestors independently, acquiring similar features or adaptations for a diet of ants, termites and other smaller insects. Thus, all ant-eaters possess much reduced teeth or teeth are entirely absent, elongate snout, long extensible sticky tongue and sharp stout claws on front legs for digging into termite mounds, rotten legs, etc.

ORIGIN AND ANCESTRY OF MAMMALS

There can be no doubt that the earliest mammals made their appearance in Jurasic times as descendants of some of the true mammal-like reptiles, the Therapsida. However, scientists put forwarded two important ideas : Ancestry through Amphibia and Ancestry through Reptilia about the origin and ancestry of mammals.

1. Ancestry through Amphibia. It was **T.H. Huxley** (1880) who advocated amphibian ancestry of mammals. The points on which **Huxley** based his arguments were : (1) Presence of two occipital condyles in the skull of both Amphibia and Mammalia, (2) Presence of left aortic arch in mammals. However, Huxley's theory is not tenable because occipital condyles are derived from exoccipitals in Amphibia but from basioccipitals in Mammalia. Apart from this, the two classes have different modes of life and exhibit many fundamental differences.

2. Ancestry through Reptilia. Palaentological evidences establish the fact that mammals arose from reptiles. This view also gets strong support from the fact that the monotremes (primitive mammals) and living reptiles have close resemblances in their anatomical features, including soft as well as hard parts.

3. Ancestral mammal-like reptiles. Fossils of **Synapsida** that have been discovered in the Carboniferous strata indicate many characters leading to mammalian line. They lived throughout the Permian and Triassic periods, dating back 280 million years or more. The more mammal-like synapsids belonged to the order **Therapsida**. One of the more advanced carnivorous therapsids (suborder Theriodontia) was called *Cynognathus* (dog jaw). It lived during the early Triassic period. It was wolf-sized and showed the following mammalian characters :

(*i*) The presence of two occipital condyles which develop from two exoccipitals, and enlarged lateral temporal fossa.

(*ii*) A well developed secondary or false palate which separates the nasal passages from the mouth cavity.

(*iii*) Heterodont dentition with teeth embedded in sockets and differentiated into small peg-like incisors, elongated canines, and cusped molars.

(*iv*) The quadrate of the skull and the articular of the lower jaw forms a joint, but both of them are reduced in size, thus, foreshadowing the beginning of squamosal-dentary joint of the mammals.

(*v*) The dentary of lower jaw is greatly enlarged and other reptilian bones of the lower jaw reduced or altogether absent.

(*vi*) The scapula at its front border was everted or turned out which is the beginning of the scapular spine of the mammals.

(*vii*) The number of phalanges in the digits are reduced to two in the thumb and great toe, and three in all others.

(*viii*) The sacrum and consequently the ilia are elongated, the latter in the form of anterior iliac blades.

(*ix*) Typical upright mammalian limbs capable of generating considerable speed.

All the aforesaid characters point unmistakably to the fact that the ancestors of the mammals are the Therapsid reptiles.

Therapsids retained several reptile-like features also. Their skull was intermediate between that of reptiles and mammals, having small cranium, parietal foramen, single middle ear bone, reduced quadrate and quadratojugal, many lower jaw bones, etc. It is also not known whether therapsids were warm-blooded, had hairs instead of scales and nursed their young. They were not necessarily the direct ancestors of present day mammals.

4. First true mammals. There is very little evidence from the fossil remains, mainly teeth and jaws, which reveal very little about the first true mammals. They were mostly tiny creatures no longer bigger than rats and mice and insignificant. But they could still manage to survive by exploiting different ways of life from those practiced by the contemporary gigantic reptilian enemies. They were nocturnal, thus, avoiding direct conflict and competition with the mostly diurnal reptiles. They were either burrowing hunting for insects, or arboreal in contrast to their ground-dwelling herbivorous or carnivorous contemporaries. They had a regulated high body temperature

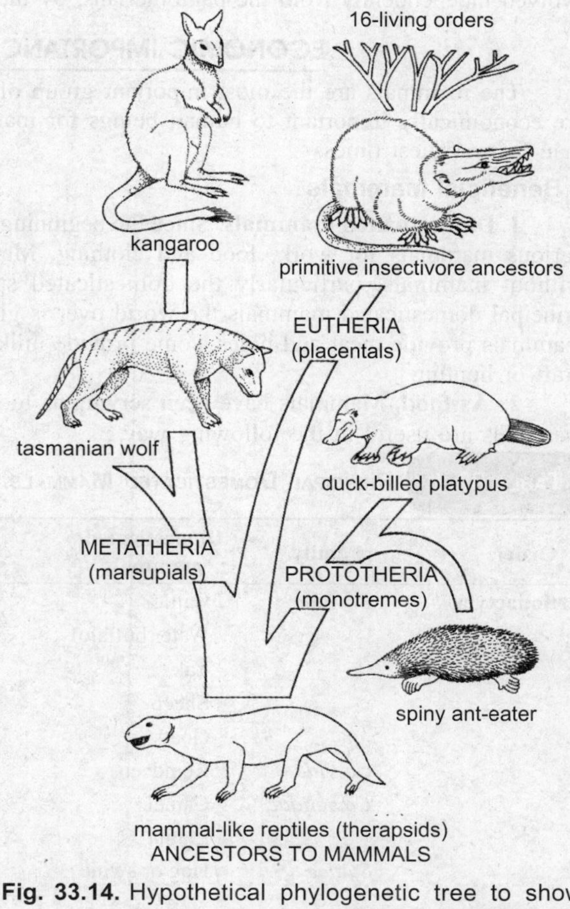

Fig. 33.14. Hypothetical phylogenetic tree to show evolution of major groups of class Mammalia.

(endothermic), hairy integument, and probably carried their young in pouches for further development after birth, and safety. They were provided with larger brains and greater intelligence.

By the end of Cretaceous period (Mesozoic age) the vast majority of dominant reptiles became extinct for reasons which are still not understood well. Many ecological niches were now left open to mammals who started their great adaptive radiation. By the close of Cretaceous period, placental mammals became extinct from marsupials. During Coenozoic age, there were established all the orders of placental mammals, so that it is called the **Age of Mammals**.

5. Polyphyletic origin. Living mammals are divided into two different reproductive subclasses. The primitive reptile-like egg-laying (oviparous) monotremes are in the subclass **Prototheria**. All other mammals give birth to living young (viviparous) and form the subclass **Theria**. The living therians are divided into marsupials or infraclass **Metatheria**, and placentals or infraclass **Eutheria** (Fig. 33.14).

Whether mammalian evolution is monophyletic or polyphyletic? Nothing is known about the origin of primitive Prototheria, for fossils older than Pleistocene are unknown. They have remained primitive probably due to isolation from rest of World in Australia and New Zealand. During Jurassic period, there were atleast 5 well-established orders of mammals, now all extinct. According to the phylogeny given in Fig. 33.14, origin of mammals is polyphyletic because they derive from atleast two Triassic reptilian stocks, cynodonts and ictidosaurs. It is generally assumed that the living Prototheria possibly evolved from the docodonts, while Metatheria and Eutheria evolved independently from the pantotherians, by the end of Cretaceous period.

ECONOMIC IMPORTANCE OF MAMMALS

The mammals are the most important group of animals from human point of view. They are economically important to human beings for many reasons as they have served or harmed man from earliest times.

I. Beneficial Mammals

1. Domesticated mammals. Since the beginning of civilization, man has been domesticating various mammals for work, food and clothing. Man would probably not be able to survive without mammals, particularly the domesticated species. The useful information about the principal domesticated mammals the world over is given in the table 33.1. All the domesticated mammals provide meat and hides, some provide milk, wool and hair; others serve for transport, draft or hunting.

2. As food. Mammals have been serving as food for man since very ancient time. Various mammals are useful in the following way :

TABLE 33.1. PRINCIPAL DOMESTICATED MAMMALS.

Order	Family	Domesticated mammals	Source or wild stock whence derived	Region of origin
Artiodactyla	*Bovidae*	Cattle	*Bos*	India, Europe
		Water buffalo	*Bubalus bubalis*	India
		Yak	*Poephagus grunniens*	Tibet
		Sheep	*Ovis*	Asia
		Goats	*Copra*	Western Asia
	Cervidae	Reindeer	*Rangifer*	Northern Europe
	Camelidae	Camels	*Camelus*	Mangolia
		Llama	*Auchenia*	South America
	Suidae	Hog or swine	*Sus*	Asia, Europe

(*Contd.*)

Order	Family	Domesticated mammals	Source or wild stock whence derived	Region of origin
Perissodactyla	*Equidae*	Horse	*Equus caballus*	Western Asia
		Grey ass	*Equus onager*	Western Asia
		White ass	*Equus asinus*	North Africa
Carnivora	*Canidae*	Dogs	*Canis familiaris*	Asia or Europe
		Cats	*Felis*	Asia or Africa
Rodentia	*Muridae*	Rats	*Rattus*	Western Asia
		Mice	*Mus*	Asia or Europe
Logomorpha	*Leporidae*	Rabbits	*Oryctolagus*	Europe, N. Africa
Proboscidea		Indian elephant	*Elephas*	Asia
		African elephant	*Loxodonta*	Africa

(*i*) Meat. The domesticated mammals provide enormous quantities of various kinds of meat and meat products. Large even-toed ungulates, such as cattle, buffaloes, pigs, goats, sheep, reindeer, etc., provide the bulk of meat supply of the world. The big game mammals, such as deers, provide the necessary proteins for many people. Seals, whales, rabbits and even dogs, rats and bears are also used as food in many parts of the world.

(*ii*) Milk. Cows, buffaloes, goats, sheep, llamas, camels, reindeers, etc., provide milk which forms an important product for human food. Ghee, butter, cheese and curd are prepared from milk.

3. Commercial products. Countless items of commercial value are produced by different products of mammals. These are as follows :

(*i*) Hides. The hides of various domesticated as well as wild mammals provide **leather** and **fur** for the manufacture of clothing and for a variety of other purposes. Animal **pelts** have been used as clothing for many centuries. Eskimos in the Arctic use reindeer and bear skins for clothing, seal hides for footwear and skin of other species for boats. Much of the early explorations of North America were made by trappers in search of furs. Various mammals such as otter, mink, weasel, marten, ermine, beaver, fur seal, muskrat, skunk, fox, reacoon, lynx, bear, rabbit, kangaroo, house cat and innumerable other mammals provide natural furs of soft texture and pleasing colours. Leather and fur are also used for the manufacture of suitcases, bags, shoes, harness, saddles, belts, purses, garments, etc., and support big industries. Muskrat is the most important and leading fur-producing mammal in U.S.A. Many mink and fox farms running have proved quite profitable.

Most of the fur-bearing mammals are fast becoming scarce so that man-made acrylic fibres, designed to look-like fur, now compete with and replacing the natural furs.

(*ii*) Hair and wool. The **wool** of sheep is the most important animal fibre. It is extensively used for weaving into cloth for winter. Goats, alpaca, camels, etc., also produce wool. **Hairs** of camels, pigs, sabbles and a host of others are made into brushes, etc.

(*iii*) Perfume. Ambergris, used as a base in fine perfume, is a product of intestinal canal of sperm whales. **Musk**, another substance used in perfumes, is obtained from the glands of muskdeer of Central and Eastern Asia.

(*iv*) Oil and fat. Blubber of baleen whale and seal are the source of edible fat or oil. Eskimos use seal blubber as fuel oil. Oil of sperm whale is not edible. It is used as a base for cosmetics and in manufacture of candles and lubricants. Pigs provide cooking fat.

(*v*) Ivory and baleen. Ivory is obtained mainly from the tusks of elephants and walrus. It is used for making toys and in decorations. **Baleen** or **whalebone** is obtained from toothless whales. It was formerly an important commercial article.

(*vi*) Glue and gelatine. These are obtained from the skin, horns, hoofs and bones of cattle.

4. Beasts of burden. Various mammals have been domesticated by man to serve as beasts of burden for transport and draft of heavy loads. In the more progressive countries, they have been greatly replaced by automotive vehicles. However, in the underdeveloped countries, cattle, horses, mules, sheep and goats are still being used for work in plains and mountains, elephants in jungles, camels and llamas in deserts, and reindeers and dogs on ice in cold countries. Of all the beasts of burden, perhaps the horse has played the greatest role in the development of human civilisation.

5. In agriculture. Bullocks, buffaloes, yak, horses and camels are employed for **ploughing** fields. They provide **manure** in the form of dung and urine. The manure nitrify soil and yield fuel biogas. **Bone meal** is prepared by crushing bones of cattle, sheep and goats, etc., is a good fertilizer. Some carnivorous predators, such as mangooses, prey upon rats and mice which are serious pests of crops. Ant-eaters feed upon harmful insects. Skunks destroy grubs and cutworms, while moles and shrews consume enormous number of insect larvae.

6. As pollinators. Arboreal mammals such as bats, squirrels, monkeys , etc., visit one tree after another in search of fruits may incidently help in the pollination of flowers. In East Indies, bats are said to pollinate bananas.

7. As scavengers. Hyaenas and jackals are of great utility as scavengers in nature. The common pig also clears away the refuge in the Indian villages and towns.

8. In recreation. In amusement no animal is safe.

(*i*) Pets. Since time immemorial, man has tried to tame a large number of mammals as pets such as monkeys, bears, mangooses, rabbits, etc. But only dogs and cats have become popular and of general interest.

(*ii*) Zoos. Lions, tigers, horses, elephants and monkeys are generally employed in sports and circuses. **Bull fights** are by no means confined to spain only. Larger and rarer varieties of mammals are exhibited in zoological gardens and national parks.

(*iii*) Hunting. Hunting is generally done by civilised man. It is actually for pleasure and recreation. Dogs, leopards, elephants and horses are used in the hunting. Important game animals are deer, elk, moose, mountain sheep, bear, lion, tiger, rabbit, raccoon, opossum, tree squirrel, etc. They are sought for their heads, hides and meat. Various countries now protect their game animals to prevent their extermination.

(*iv*) Decoration. Antlers heads, and pelts of various mammals are displayed as trophies or momentos and for decoration of houses.

(*v*) Laboratory animals. Common mammals such as rats, guinea pigs, rabbits, cats, pigs, monkeys, apes, dogs, etc., are extensively used in laboratories for research and scientific studies.

II. Harmful Mammals

There are some mammals which are decidedly harmful or injurious to man. These are as follows :

1. Pests. Many mammals act as pests. Most dangerous mammals to crops are the rodents. They destroy crops in fields. They are harmful in houses and buildings also. They consume all kinds of food stuffs and destroy property. Rabbits, mice, wood chucks and some others forage in vegetable fields and gardens and gnaw at bark to trees. Pocket gophers eats roots of plants, while squirrels damage grain and other crops. Besides, bigger herbivorous ungulates such as deer, bear, boar, hippopotamus, elephant, kangaroo, etc., also damage and eat up crops. Cats in kitchen forage on food besides rats.

2. Predators. Several wild carnivorous or predatory mammals are constant menace to the lives of people and livistock. Lions, tigers, wolves and bears prey at times on domestic cattle, sheep, hogs, poultry, etc. Wolves and hyaenas sometimes attack sleeping persons. In tropical Africa, the sanguivorous vampire bats feed on blood of various mammals, including man.

3. Disease carriers. Rodents are the worst carriers and distributors of the germs of certain diseases. Rats, mice, pigs, cats, dogs, etc., transmit many diseases to humans such as bubonic plague, typhus, relapsing fever, spotted fever, tularemia, trichinia, hydrophobia, etc.

4. Economic importance of rats. Rats occur in almost all the terrestrial habitats that support life. They multiply rapidly and soon establish their colony. Rats are omnivorous, feeding on seeds, fruits, grains, roots, insects, snails and invertebrate and vertebrate bodies. They are mainly fossorial, nesting and living much of the time in burrows. They live freely in competition with humans.

Damage caused by rats includes grazing of young trees, root destruction and seed eating. Generally rats prefer very young plants having a high nutritional value than mature ones. These injured plants remain weak and small. Various species of rats are distributed in different parts and cause damage to different crops. The short tailed bandicoot rat *Nesokia indica* is found across Egypt and India. It is mainly a subterranean feeder. The lesser bandicoot, *Bandicota bengalensis* is found alongwith *Nesokia indica* in Bangladesh and Myanmar. It causes 20–40% losses in cereals. The soft furred rat, *Millardia meltada,* a pest of cereals, is found in India and Pakistan. *Bandicota bengalensis* and *Millardia meltada* cause 10% or more losses to various crops and wheat. *Rattus losea* and *Bandicota indicata* are principal pests of crops in China. In Australia, various crops including seedling fruits are damaged by bark stripping. In Australia, the number of rats increases tremendously reaching nearly 3000 holes in about 0.5 in one hectare. When the food is exhausted, a large number of rats die and enormous external parasites grow on their dead bodies. Rice field rat, *Rattus agriventor* and other species of rat are serious pests of rice. They gnaw the small plants and eat grain in fields. The population of rat damaging rice has one or two peaks correlated with single or double crops per year. *Bandicota bengalensis* causes 7–22% loss of rice in North India and 5–18% in South India and at some places the loss is up to 60%.

In sugarcane fields, the populations of rats are established. They find suitable shelter and food in sugarcane fields. Rats cause direct loss of cane by eating into internodes. This permits entry of harmful microorganisms and cause physiological stress, which can reduce weight of sugar content. Rats also damage the oil palm crop by eating upon their ripe oil bearing fruitlets and carrying them with them. Unripe fruitlets are also gnawed by the rats *Rattus rattus diandi*, which is the commonest oil palm pest in Malayasia. Rat species like *Rattus praetor, Rattus exulans* attack coconuts. Rats climb the palms of all ages to feed on developing nuts when then fall prematurely. In some regions the loss due to coconut fall is up to 50%.

Rats also damage the crops of potato, groundnut, sweet potato, maize, casava and others to large extent. Up to 16% of ripening pine apples are damaged by *Bandicota bengalensis* and *Rattus rattus*.

Rats as carriers of disease. Some rodent species are reservoirs of a large number of infectious organisms of man. If these infectitious organisms are transmitted to human and domestic animal populations may cause outbreak of diseases often leading to high mortality. Some of the diseases for which rats are the reservoirs are given in the following table.

Disease (agent)	(Reservoir)
Lana fever	*Mastonys netalensis*
Venenzuelen haemorhagic fever	*Sigmodon, Oryzomys*
Tick borne encephalitis	*Microtus*
Scrub typhus	*Rattus* spp.
Murine typhus	*Rattus* spp.
Rocky mountain spotted fever	*Sigmodon*
Leptospirosis	*Rattus* spp.
Salmonellosis	*Rattus* spp.
Toxoplasmosis	*Rattus norvagicus, Sigmodon, Rattus rattus*
Leishmaniasis	*Rattus* spp., *Oryzomys*
Chagas disease	*Rattus* spp.

Control of Rats

For the control of rat population various methods are employed which are given below :

1. Repellants and attractants. In this method chemicals like cyclohexamide and tributylin acetate are used as repellants. The repellants may be incorporated in the packing material to keep the rats away from the packed food stuffs. Various types of attractants such as sugar, archis oil are used to attract the rats to the traps laid for them in the fields. Sometimes the poisons are also mixed into baits used which are placed in rat burrows. Selective rodenticides like Norborne carboximide, Quintos and Brodifacoum are also used.

2. Fumigation. In this method fumes of chemicals like Fomafog, are introduced into the rat burrows by suitable apparatus.

Rats as food. In India and Central Asian, African and some European countries some species of rats are used as food. The rats, *Ondalria zibethica* is served as a delicacy in restaurants in Holand and Belgium. The cane rat, *Jhryononys swinderianus* forms an important constituent of the food of majority of the population in West Africa.

Use of rat skin. Skin of larger species of rats are tanned. It can be used in preparation of gloves, purses and other small items. This helps in providing jobs to people and also controls the rat population.

IMPORTANT QUESTIONS

Long Answer Questions

1. Write an essay on dentition in mammals.
2. Give an account of aquatic adaptations.
3. Discuss how adaptive radiation in mammals could have helped in their evolutionary success.
4. Write an essay on economic importance of mammals.
5. Write an essay on origin and ancestry of mammals.

Short Answer Questions

1. If you were shown the teeth of a rodent, an ungulate and a carnivore, how could you differentiate one from the other?
2. State the difference between horn and antler.
3. Describe the dentition and dental formula of a mammal in relation to its diet.
4. Explain how bats are adapted for an aerial mode of life.
5. Write short notes on : (*i*) Flying adaptations in bats; (*ii*) Dental formula; (*iii*) Development of tooth, (*iv*) Flying mammals; (*v*) Adaptive convergence in mammals; (*vi*) Domesticated mammals; (*vii*) *Cynognathus*.

Very Short Answer Questions

1. What is the characteristic type of dentition in mammals?
2. What is diphyodont dentition?
3. Name the mammal in which teeth are lacking.
4. What are cheek teeth?
5. Explain polyphyodont dentition.
6. Name the animal in which epidermal horny plate and baleen plates are found over jaws instead of teeth.
7. Advantages of binocular vision is primates. Explain it in not more than two lines.
8. Role of ears in the flight of a bat.

Multiple Choice Questions

1. Mammary glands are modified
   (*a*) Sebaceous glands
   (*b*) Sudorific glands
   (*c*) Cutaneous glands
   (*d*) Scent glands.
2. Milk dentition in mammals lack
   (*a*) Molars (*b*) Premolars
   (*c*) Canines (*d*) Incisors.
3. First teeth on maxillae are
   (*a*) Lower canines (*b*) Upper canines
   (*c*) Upper molars
   (*d*) Lower premolars.
4. Baleen is present in
   (*a*) Whales (*b*) Elephants
   (*c*) Frogs (*d*) Lizards.

ANSWERS

1. (*a*) 2. (*a*) 3. (*b*) 4. (*a*)

34

Chapter

Introduction to Vertebrate Embryology

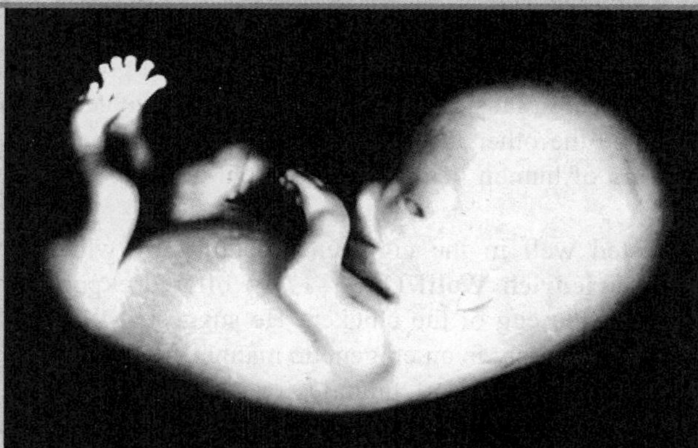

Contents of this Chapter

Embryology (GK., *embryon* = embryo; + *logia* = discourse) is a study of the origin and development of animals dealing with changes through which a fertilised egg must pass before it assumes the adult state. Fertilisation of an ovum by a spermatozoon results in the formation of a zygote. Development of a single-celled zygote into an adult involves a series of steps or stages resulting in a gradual increase in the complexity of structure. The stages of embryonic development differ in various chordates, yet the chief phases are basically similar in all. The differences are related primarily to the amount and distribution of yolk present in an egg. The inert yolk or vitellin furnishes nourishment for the developing embryo. The yolk also influences on the pattern of cleavage, on the morphogenetic movements of the blastomeres during gastrulation and on the type of development, *i.e.,* indirect with larval forms or direct with juvenile stages.

Embryogenesis or **embryogeny** may be defined as the formation and development of embryos. In fact it includes all the changes by which a fertilised ovum or zygote is transformed into an adult. So long as the developing individual remains in the egg, it is called an embryo. In some lower animals the amount of yolk is less in egg, so that the embryo hatches in earlier stages of development, called a **larva**.

Usually, it is very different in form and structure from the adult. Examples are **catterpillars** of insects and **tadpoles** of frogs. The larva undergoes transformation into the adult by the process of **metamorphosis**. In higher vertebrates like reptiles, birds and mammals, the eggs are richly supplied with yolk. Their embryos continue development until they attain a form resembling the adult. Examples are **chicks** of birds and foetuses of mammals.

HISTORICAL REVIEW

Aristotle (384–322 B.C.) was the first Greek philosopher who described the ontogenetic development of chick and many other forms. The doctrines of Aristotle about the development were accepted for a very long time. **William Harvey** (1578 – 1657) and **Marcello Malpighi** (1628–1694) contributed information on the various stages of the development of chick on the basis of their studies with the help of simple lens. With the discovery of the microscope, **Leeuwenhock** (1632–1723) described the sperm of man and other mammals.

Some ovists namely **Swammerdam** and **Bonnet** advocated an extreme form of preformation theory called **encasement** or "**emboitment**" **theory**. This theory holds that successive generations of individual organisms pre-existed one inside the other in the germ cells of the mother. It was estimated that, as many as, 200 million years of human beings were present, already delineated in the ovaries of Eve.

Such theories of preformation persisted well in the eighteenth century by which time (in 1759) the German investigator **Caspar Friedrich Wolff** (1733–1794) offered experimental evidence that no preformed embryo existed in the egg of the chicken. He suggested that during embryonic development the organs formed successively in an epigenetic manner. **Wolff** advocated that the future embryonic regions of an egg first consist of granules or "globules" (*viz.,* cells or their nuclei) lacking in any arrangement, *i.e.,* these globules do not reveal any resemblance with the form or structure of the future embryo. Only gradually did these "globules" organise into rudiments (germ layers) which, in turn, took on the characteristics of the various organs of the embryo. This method of progressive development from the simpler to the more complex, through the utilisation of building units (globules or cells) is called **epigenesis**. Today this theory is accepted in a modified form.

K.E. Von Baer (1792–1876), the **father of modern embryology**, was the first embryologist who first of all, presented the embryological data in a coherent form, made various land mark embryological investigations and made certain very important generalisations. He forwarded the **germ layer theory** which states that "various structures of the body arise from the same germ layers in different species of animals". His most important generalisation is known as **Baer's law** which states that "more general features that are common to all the members of a group of animals are, in the embryo, developed earlier than the more special features which distinguish the various members of the group".

Baer's law was formulated before the recognition of evolutionary theory, therefore, later on it is reinterpreted in the light of evolutionary theory by **Muller** and **Haeckel** (1864) and named as **biogenetic law**.

In 1824, **Prevost** and **Duman** described cleavage or segmentation of the egg. **Hertwig** in 1875 observed the main events taking place in fertilisation of an egg by a sperm. **Von Bender** (1883) proved that the male and female sex cells contribute the equal number of chromosomes to the fertilised egg. During the last days of Nineteenth and early days of Twentieth century, embryologists like **Weismann** (1883), **Endres** (1885), **Spermann** (1901 and 1903) and **Morgan** (1908) made experimental and analytical investigations and, thus, a new branch of embryology gave way for the initiation of experimental embryology. In 1883, **A. Weismann** (1834–1913) suggested convincingly that a child in no way inherits its characters from the bodies of the parents but from the sex cells alone. These germ cells, in turn, acquired their characters directly from the pre-existing germ cells of the same kind.

Wilhelm Roux (1850–1924) in 1881, performed a classical experiment which may be viewed as marking the beginning of the science of experimental embryology. He took a frog's egg at the two cell stage of cleavage and touched one of the two cells with a hot needle, thus, destroying the nucleus. He observed that the uninjured cell continued dividing and developed into what he interpreted to be a one-half blastula, a one-half gastrula, and ultimately a one-half embryo. He, thus, concluded that certain areas of the egg are already destined in the ovary to develop into special region. Thus, the pigmented cytoplasm of animal pole of frog's unfertilised egg chiefly develops into the head region of the animal, while yolky cytoplasm of vegetal pole of egg forms posterior region.

In 1891, the German Scientist **Hans Driesch** performed experiment on sea urchin eggs, similar to **Roux**. He suggested that early cleavages of the egg are equational and have a "quantitative division of homogeneous material", therefore, the blastomeres have equal potentialities and their fate is determined by their position. Development as observed by Driesch in sea urchin eggs was to be called regulative development and the eggs which were capable of performing such regulative development were called regulative eggs.

Various operative and chemical procedures have been employed in attempts to analyse the developmental processes leading to or involved in the formation of the blastula, the gastrula and the actively swimming larva. Such an experimental approach of **T. Boveri** (1910), **J. Runnstrom** (1928), **S. Horstadius** (1928) and **C.M. Child** (1936) on sea urchin eggs has contributed a most important theory the **gradient theory**. The two gradients are, therefore, the **animal gradient** with a centre of activity at the animal pole, and the **vegetal gradient** with a centre of activity at the vegetal pole.

C.M. Child (1936), while recognising the physico-chemical nature of the two gradients, proposed the existence of a single physiological or oxidative metabolic gradient in sea urchin egg.

It was in 1969 and 1972 that **Horstadius** and **Josefsson** succeeded in isolating animalising and vegetalising substances from the mature unfertilised egg and early cleavage stages of sea urchin. **Arnold** (1976) has suggested that the egg cortex by controlling the displacement of membrane receptors and enzyme systems, modulates metabolism in growth, division and cell surface interaction.

In 1924, **Spermann** and **Hilde Mangold** published a classical paper providing definitive proof of the organising action of transplanted dorsal lip in the production of secondary embryos, establishing firmly the concept of induction as a basic mechanism in embryonic development. Spermann, thus, has recognised a primary organiser in the form of archenteron in amphibian gastrula and got Nobel Prize of 1935, for such a landmark discovery in experimental embryology.

In modern terms induction can be defined as a type of intercellular communication which is required for differentiation, morphogenesis and maintenance (**Graham** and **Wareing**, 1976). It is also found that during induction some chemical is transmitted from one tissue to the other and this chemical acts on the genes of the cells being induced to develop into a particular manner. What the substance is has not been still determined, but it appears to be a relatively larger molecule (**Pai**, 1985).

MODERN EMBRYOLOGY

With the discovery of the chrozmosomes, genes and genetic code, it has become evident that all the properties of any organism are determined by the sequence of the triplets in the DNA molecule. The sequence of the base triplets can directly determine what kind of proteins can be produced by an organism. All the morphological and physiological manifestations of an organism depend on the assortment of proteins, coded for by the hereditary DNA. The modern embryology is heading towards analytical embryology on the basis of the analysis through molecular biology techniques.

SCOPE OF EMBRYOLOGY

Embryology is the most important biological science. It explains the details of the ontogenetic development of an animal from a single fertilised cell. It gives basic information about the physiology, genetics, sex determination, various diseases and organic evolution. Embryology plays a key role in human welfare. It helps in understanding the causes of congenital malformations, cancer, ageing and in improving the breeds of domestic animals, in controlling pests and vectors of diseases and in the formation of test-tube babies. Some of the latest phenomena such as teratogenesis, cancer, animal breeding, test-tube babies and cloning, and pest control are most important fields in animal embryology. With the success in cloning experiments of **Ian Wilmut** (1996) a new concept of cloning without involving germ cells has originated which is useful for the biological resources. The advantages of cloning plants and animals are numerous. High-yield food plants such as wheat, corn and rice can be selected and abundantly reproduced. Cloning would give animal breeders a tool for exactly reproducing highly desirable animals; for example, cloning would make it possible to create 1000 copies of prize dairy cow to help feed growing populations. Endangered species might be saved by cloning numerous replicas of the best of the few remaining individuals.

GAMETOGENESIS

The embryogenesis (embryonic development) of a sexually reproducing multicellular animal is prefaced by the gametogenis, *i.e.,* the formation and ripening of two highly dissimilar and specialised **sex-cells** or **gametes**, namely a large-sized, non-motile, nutrient filled cell the **ovum** or **egg** and a small-sized, motile, sex-cell, the **spermatozoon** or **sperm**, both of which unite and give origin to a diploid **zygote**. Formation of sex-cells or gametes is termed **gametogenesis**. It is accompanied by a special type of nuclear division, called **meiosis**. As a result, the nuclei of gametes formed contain only half or **haploid** number of chromosomes. When male and female sex-cells (sperms and ova) unite at the time of fertilisation, the resulting cell or **zygote** again has the full or **diploid** number of chromosomes.

The production of male germ cells, the **sperms** or spermatozoa occurs in the male gonads, the testes, by a process called **spermatogenesis**. Each sperm consists of a head, middle piece and tail. It is preferable to call them sperm cells or simply sperms.

The production of female germ cells, the **ova** takes place in female gonads, the **ovaries**, and the process is called **oogenesis**. The word 'egg' is often loosely used for ova or secondary oocytes. It may be reserved for more complex structures such as the hen's egg which may even contain early embryonic stages.

The processes of **meiosis** and **gemetogenesis** are described in full details in our Textbook of Cell Biology, Genetics, Evolution and Ecology. This is not necessary to describe them in detail in this Textbook.

EMBRYONIC DEVELOPMENT

The stages of embryonic development differ in various chordates, yet the chief phases are basically similar in all. The differences are related primarily to the amount and distribution of yolk present in an egg. The inert yolk or vitellin furnishes nourishment for the developing embryo. The yolk also influences on the pattern of cleavage, on the morphogenetic movements of the blastomeres during gastrulation and on the type of development, *i.e.,* **indirect** with larval forms or **direct** with juvenile stages.

Yolk. The amout of yolk varies in the eggs of different chordates, it determines the size of the egg and the pattern of early development (cleavage and blastulation, etc.). The eggs are classified according to the distribution of yolk they contain into two main types, namely, isolecithal and telolecithal eggs.

Types of Eggs

A. Isolecithal or **homolecithal eggs** have very little yolk which is uniformly distributed evenly in the cytoplasm. Such eggs are found in various chordates, *e.g.*, *Amphioxus*, tunicates and marsupial and eutherian mammals.

B. Telolecithal eggs contain a considerable amount of yolk, which has a polarised distribution. Due to its gravity, it is concentrated more in vegetal hemisphere than that of animal hemisphere. Such polarised distribution of yolk is found in mesolacithal and macrolecithal eggs. In fact, in macrolecithal eggs, the amount of yolk is so massive that it almost occupies the whole space of the egg, except a small space at the animal pole where the nucleus or germinal vesicle lies in the form of cap over the yolk. The telolecithal eggs may be either moderately telolecithal (*e.g.,* eggs of Amphibia, *Petromyzon* and Dipnoi) or highly telolecithal, (*e.g.,* cartilaginous and bony fishes, reptiles, birds and egg-laying mammals). All eggs are enclosed in one or two vitelline membranes.

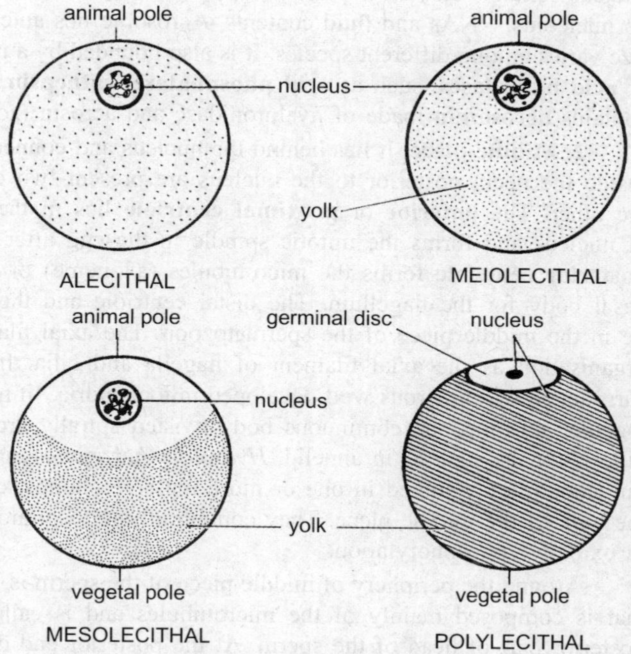

Fig. 34.1. Types of chordate ova. 1 – and 2–Isolecithal; 3 – and 4 –Telolecithal.

C. Centrolecithal eggs found in insects and some hydrozoa, contain a large amount of yolk concentrated in the centre of the egg surrounded by thin peripheral layer of active cytoplasm.

Classification of Eggs on the Basis of Amount of Yolk

1. Microlecithal or oligolecithal eggs are small sized, containing a small amount of yolk. Such eggs are found in *Amphioxus*, tunicates, and marsupial and eutherian mammals, and also in certain invertebrates such as *Hydra* and sea urchin.

2. Mesolecithal eggs contain moderate amount of yolk, *e.g.,* annelid worms, molluscs, Petromyzontia, Dipnoi and Amphibia.

3. Macrolecithal, megalecithal or polylecithal eggs contain massive amount of yolk such as eggs of insects, *Myxine*, elasmobranch fishes, reptiles, birds and prototherian mammals.

The egg or ovum is surrounded by a thin plasma membrane and around it is present a **vitelline membrane**, which is non-cellular and transparent layer of mucoprotein. It is often much thicker and stronger than the underlying fine plasma membrane. It is differently named in various groups of animals such as **chorion** in fishes and **zona pellucida** in reptiles and mammals.

Spermatozoa. A **spermatozoon** (Gr., *sperma* = seed + *zoon* = animal) or male gamete of vertebrates despite its small size is an exceedingly complex cell. It has a head, a middle piece, and a tail, all of these are contained by a continuous plasma membrane, like other living cell.

1. Head. The **head** has a nucleus invested by a thin layer of cytoplasm which projects in front as a pointed acrosome, both performing two basic functions of the sperm – genetic and activating, respectively. The nucleus occupies most of the space of the sperm head. It is enveloped by a typical double nuclear membrane, which lacks the nuclear pores except the lower part. The nucleus contains only its haploid complement of DNA bound by basic proteins. The nucleus has no nucleolus, RNAs and fluid contents. Acrosome lies anterior to the nucleus and its shape and size varies among different species. It is also bounded by a unit membrane and contains a number of **acid hydrolases**, such as **acid phosphatase**, **cathepsin**, **hyaluronidase**, etc. In mammals, it contains **acrosomin** made of hyaluronidase and acrosin (zona lysin).

2. Middle piece. It lies behind the nucleus and connected with the head by a narrow **neck**. Inside the neck, posterior to the nucleus are present two centrioles, both lie at right angles to the other. The **anterior** or **proximal centriole** lies in the depression in the posterior surface of nucleus and forms the mitotic spindle in the egg after fertilisation. The **distal centriole** or **posterior centriole** forms the microtubules (axoneme) of the sperm tail (flagellum). It acts as basal body for the flagellum. The distal centriole and the proximal part of the axial filament lie in the middle piece of the spermatozoon. The axial filament of the sperm tail has the same organisation as the axial filament of flagella and cilia. In middle piece the axial filament is surrounded by numerous well developed mitochondria. In mammals, the mitochondria are joined together forming one continuous body twisted spirally around the axial filament. However, in other animals, such as in annelid, *Hydroides hexagonus*, and in sea urchin, *Arbacia punctulata*, mitochondria are joined in one or more massive clumps, called **mitochondrial bodies** forming the bulk of the middle piece. They contain all the respiratory enzymes and are extremely active in oxidative phosphorylation.

Around the periphery of middle piece of the sperm is found a condensed layer of cytoplasm that is composed mainly of the microtubules and is called **manchette**. It also surrounds the posterior part of head of the sperm. At the posterior end of middle piece occurs a dark ring or fibrous thickenings beneath the plasma membrane, forming the boundary between the middle piece and tail. It is called **ring centriole** or **Jensen's ring**.

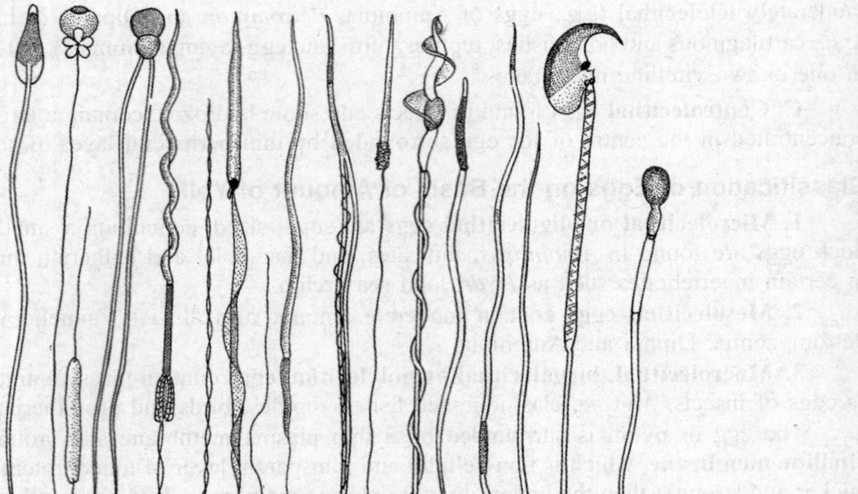

Fig. 34.2. Forms of flagellate spermatozoa. 1. *Arbacia* (sea urchin); 2. *Amphioxus*; 3. *Petromyzon*; 4. *Opeanus* (toad fish); 5. *Raja* (elasmobranch); 6. *Bufo* (toad); 7. *Rana* (frog); 8. *Amphiuma* (salamander); 9. *Anguis* (lizard); 10. *Frigilla* (bird); 11. *Fundulus* (domestic fowl); 12. *Echidna* (monotrematous mammal); 13. *Mus* (eutherian mammal); 14. man.

3. Tail. The tail is a long vibratile flagellum containing an axial filament along its whole length and projecting behind the cytoplasm of the tail as an **end piece**. Tail has two main parts : principal piece and end piece. The principal piece constitutes most of the tail length, consists of a **central core**, comprising the axial filament. Surrounding this core is a microtubular fibrous **tail sheath** which some time appear as semicircular ribs oriented perpendicular to the long axis of the filament or as helical coils. In human sperms, out of nine coarse fibres found around axial filament, of the tail two coarse fibres are fused with the surrounding ribs so as to form **anterior** and **posterior columns** extending throughout the length of the principal piece.

The **end piece** is merely a short tapering portion of tail containing only the axial filament covered with cytoplasm and plasma membrane.

Spermatozoa are discharged from the body floating in a seminal fluid or semen secreted by the seminiferous tubules and accessory reproductive glands. Spermatozoa are always produced in very large numbers.

FERTILISATION

Fertlisation (L., *fertilis* = to bear**)**. It is the fusion of two gametes (spermatozoa and ova) and so their nuclei to form a diploid zygote. It activates the egg to form fertilisation membrane outside the egg plasma membrane to start its metabolism and to start its cleavage.

During fertilisation process the jelly coats and egg membraes such as vitelline membrane and plasma membrane secrete the **fertilizin** and the sperms tip secrete **antifertilizin**, both interact with each other and sperms, thus, agglutinated. It occurs in the female genital tract. The membrane of the acrosomal vesicle of the acrosome and the plasma membrane of the sperm breakdown and the severed edges of the two membranes fuse to form an opening through which the contents of the acrosomal vesicle are released. The inner acrosomal membrane grows into one or many **acrosomal tubules** which come in contact with the vitelline membrane and plasma membrane. In mammals, the plasma membrane and outer acrosomal membrane break and fuse to give rise to extensive vesiculations and the sperm is possibly phagocytosed by the egg. The acrosome now releases the lytic enzymes or lysins (**acrosomin** in sea urchins) which help the sperm to penetrate the egg envelopes by liquefying them locally, without affecting the plasma membrane. In mammals including human females, the sperms first penetrate the multiple layers of follicular cells (granulosa cells) which are held together by an adhesive substance **hyaluronic acid**. The acrosome releases the **hyaluronidase** and proteolytic enzymes for penetrating the follicle cell layers, corona radiata and zona pellucida. Hyaluronidase is supposed to dissolve the cement between cells of corona radiata. **Zona lysins** or proteolytic acrosomal enzymes are responsible for the passage of the sperms through zona pellucida.

The apical part of sperm plasma membrane (originally the inner acrosomal membrane) extends forward to form an acrosomal tubule. It projects through the egg membranes to reach the egg plasma membrane or oolemma. The shape and size of acrosomal tubule varies with species and is entirely absent in mammals. The tip of acrosomal tubule fuses with egg plasma membrane, while in mammals sperm come in contact with the egg surface by its lateral aspects. After the fusion, the plasma membrane of the egg and tip of acrosomal tubule dissolve at the point of contact. In teleost fishes acrosome is lacking and so the plasma membrane of sperm head fuses directly with the plasma membrane of ovum. After fusion of the both the plasma membranes, the plasma membrane of ovum becomes permeable to sodium, potassium and calcium ions. Calcium is essential for the fertilisation process. pH of egg cytoplasm also increases due to inflow of Na^+ and outflow of H^+ ions. Within seconds after membranes contact, changes occur in egg cortex. In bony fishes and frogs, the cortical granules are broken down after sperms' penetration into the egg cytoplasm and their contents become liquefied and extruded on the surface plasma membrane of the egg. They gradually fill up the perivitelline space in between chorion and egg plasma membrane in bony fishes, and the space between vitelline membrane and egg plasma membrane in frogs. Thus, **fertilisation membrane** is formed by the rupture of cortical granules outside the plasma membrane. This is due to the cortical reaction stimulated by the penetrating sperm. Fertilisation membrane blocks the entrance of other living sperms. The vitelline membrane or chorion does not transform into fertilisation membrane. In some mammals (*e.g.,* man, rabbit and hamster) the cortical granules burst open and release their contents in the space betwen egg plasma membrane and zona pellucida. Cortical granules are not found in urodele amphibians and, hence, no fertilisation membrane formation occurs.

In most species only one sperm enters the egg and this is called **monospermic fertilisation**. When many sperms penetrate the single ovum (*e.g.,* in polylecithal eggs of some insects, elasmobranchs, urodeles, reptiles and birds, and also in microlecithal eggs of bryozoans), it is called **polyspermic fertilisation**. In this case, the genetic material of only one sperm is incorporated in the zygote nucleus, and other sperm nuclei are degenerated.

After the penetration of sperm inside the egg cytoplasm, its nucleus moves inward, swells and its chromatin which is very closely packed becomes finely granular. It finally becomes vesicular and is called **male pronucleus**. Similarly the egg nucleus after second meiotic division undergoes changes and becomes **female pronucleus**, which swells, increases in volume and becomes vesicular. Later on male and female pronuclei fuse together, that is, the nuclear membrane of both pronuclei are broken at the point of contact and their contents unite into one mass, which is finally bounded by a common nuclear envelope, forming a **zygote nucleus**. This type of fusion of both pronuclei (male and female) is called **amphimixis**.

Significance of fertilisation. 1. The male and female nucleus possess haploid (n) number of chromosomes. The fertilisation restores the specific parental diploid chromosome number. 2. Fertilisation brings together the chromosomes and genes from two different parents, resulting into a new genetic recombination. 3. Fertilisation activates the egg to undergo cleavage.

Types of fertilisation. According to place and nature of fluid media, fertilisation is of two types : 1. External fertilisation, 2. Internal fertilisation.

1. External fertilisation. When the fertilisation occurs in the aquatic medium outside the body of female, it is called **external fertilisation**. Aquatic medium may be sea water or freshwater. In marine animals, sexually mature adults shed eggs and sperms freely into the surrounding water. The sperms and eggs are laid in water in astronomical numbers, and also in close proximity.

2. Internal fertilisation. In terrestrial forms, where eggs are completely enclosed in impermeable envelopes before being laid such as oviparous animals or where they are retained within maternal body throughout development such as ovo-viviparous and viviparous animals (*e.g.,* elasmobranchs and mammals) the sperms are transmitted internally, *i.e.,* in the body of female, by the intromittent organ of male. In these forms the fertilisation may occur in the lower portion of oviduct (*e.g.,* urodela) or in the upper portion of the oviduct such as salamanders, reptiles, aves and most mammals. In viviparous fishes such as *Gambusia affinis* and *Heterandria formosa*, and certain eutherian mammals such as *Ericulus*, fertilisation occurs in the ovarian haploid follicles.

The results of fertilisation are : (*a*) an activation of the egg to undergo its second maturation division for preparing a haploid female nucleus; (*b*) an introduction of a centriole by the sperm which divides to form two centrioles, since a centriole is lacking in a mature ovum; (*c*) a restoration of a diploid number of chromosomes in the zygote; (*d*) a change in the periphery of the egg which precludes the entry of other sperms; (*e*) separation of the vitelline membrane from the egg to allow the zygote to rotate.

Cleavage

The division of an activated egg (zygote) by a series of mitotic cell divisions into a multitude of cells which become the building units of future organism, is called **cleavage** or

segmentation (Ger., *kleiben* = to cleave). During cleavage, cells do not grow in size and early cleavage divisions occur synchronously, which is lost during late cleavage. During cleavage, there is no growth in the resulting blastomeres and the total size and volume of the embryo remains the same. The blastomeres do not move so the general shape of the embryo remains the same except the formation of a cavity, the **blastocoel** in the interior. During cleavage, chemical conversion of reserve food material (yolk, glycogen and ribonucleotides) into active cytoplasm takes place. Thus, a steady increase of respiration occurs throughout cleavage. During cleavage, nucleo-cytoplasmic ratio in cells is reduced, which permits the cells to be more metabolically active, because such nuclei have less cytoplasm to control. Thus, the cleavage converts the egg into a compact mass of cells or blastomeres called **morula**.

Types of cleavage. The type of cleavage taking place depends largely on the amount of yolk present. Following types of cleavages occur :

1. Holoblastic or total cleavage. In this type of cleavage, the entire egg divides by each cleavage furrow. It is subdivided into two types :

(*i*) Complete or equal holoblastic cleavage occurs in microlecithal and isolecithal eggs, the entire zygote divides completely to produce a number of almost equal-sized cells, *e.g.,* eutherian mammals, *Amphioxus*, tunicates.

(*ii*) Unequal holoblastic cleavage occurs in mesolecithal and telolecithal eggs, the zygote divides completely to form unequal-sized blastomeres, *i.e.,* small-sized cells towards the animal pole which has almost no yolk, larger cells towards the yolky vegetal pole, *e.g.,* cyclostomes, elasmobranchs, Dipnoi and Amphibia.

2. Meroblastic or incomplete cleavage occurs in polylecithal eggs in which only the small germinal disc lying at the animal pole consisting of clear cytoplasm and a nucleus, undergoes a series of incomplete divisions forming an area of cells at the animal pole, the large yolky portion beneath the germinal disc remains unsegmented, *e.g.,* toleosts, reptiles, birds and egg-laying mammals. Here the germinal disc is of disc-shape, so the cleavage is also called **discoidal**.

3. Superficial cleavage. This type of incomplete cleavage is found in **centrolecithal eggs**, *e.g.,* insects and many arthropods. The nucleus lying in the centre of the egg yolk surrounded by an island of cytoplasm undergoes cleavage, and each nuclei is surrounded by small amount of cytoplasm. They later move towards the periphery in the peripheral cytoplasm. Here their cytoplasm fuses with the peripheral cytoplasm. Later the peripheral cytoplasm becomes subdivided by furrows extending inward from the surface, thus, a layer of peripheral or superficial cells is formed which surrounds the central undivided yolk.

Pattern of cleavage. The pattern of cleavage due to organisation of egg may be of following types :

1. Radial cleavage. When successive cleavages extend through the egg, at right angles to one another and the resulting blastomeres become symmetrically arranged around the polar axis. Such type of cleavage is called radial cleavage, and is found in echinoderms (*e.g., Synapta* and *Paracentrotus*, etc).

2. Biradial cleavage. When first three cleavage planes are not arranged at right angles to each other, it is called biradial cleavage, *e.g.,* Acoela like *Polychoerus* and Ctenophora.

3. Spiral cleavage. The rotational movement of cells around the egg axis during cleavage is due to spiral cleavage. The spiral cleavage results due to oblique positions of mitotic spindles in the blastomeres. Thus, it is also called **oblique cleavage**. In successive cleavages, the rotational movements alternate in clockwise direction or anticlockwise direction. It is found in Turbellaria, Nematoda, Rotifera, Annelida and molluscs except cephalopods.

4. Bilateral cleavage. In this type of cleavage the mitotic spindles and cleavage planes remain bilaterally arranged with reference to a plane of symmetry which coincides with the median plane of the embryo. It is found in Tunicata, *Amphioxus*, Amphibia and higher mammals.

5. Determinate and Indeterminate cleavage. The cleavage in nematodes is of a special type of bilateral cleavage in which definite blastomeres give rise to specific parts of the embryo. This type of cleavage is called **determinate** or **mosaic cleavage**. In vertebrates, the plane of cleavage is less rigid, the cleavage pattern has no definite relation to the embryo. This type of cleavage is called **indeterminate** or **regulative** and is found in echinoderms, *Balanoglossus*, coelenterates and amphibians. A first cleavage blastomere of a sea urchin or an amphibian or a mammal, when isolated can alter its usual destiny and develop into a perfect (but small) embryo. Similarly, when two fertilised eggs, made to adhere like a two-cell stage, they produce a single giant embryo. This is **regulative development**.

STAGES OF EMBRYOGENY

1. Morula

During early cleavages, the blastomeres tend to assume a spherical shape and their mutual pressure flattens the surfaces of the blastomeres in contact with each other, but their free surfaces remain spherical. Thus, cleavage process develops a multicellular body with loosely arranged blastomeres with in fertilisation membrane, called **morula** (Latin word for mulberry) resembling mulberry, *e.g.,* amphibian and coelenterates. In macrolecithal eggs, morula is a cellular flattened disc at the animal pole.

2. Blastula

As cleavage proceeds the cells increase in number but become smaller. The cells withdraw from the centre and arrange themselves towards the surface to form a true epithelium, which may be single cell thick as in *Amphioxus*, echinoderms, etc., or many cell thick as in most vertebrates. Due to rearrangement of cells to form the epithelium or **blastoderm** a fluid-filled space or **blastocoel** or segmentation cavity is formed. This stage is called blastula and the process of formation is called **blastulation**.

Types of Blastulae

1. Coeloblastula. It is in the form of a hollow sphere formed of a single layer of blastoderm and the blastocoel is filled with mucopolysaccharides. Examples, echinoderms and *Amphioxus*.

2. Stereoblastula. In spirally cleaving eggs of annelids, molluscs, nemerteans and some planarians, blastula is solid, having no blastocoelic cavity. In them micromeres accumulate as cluster of cells over macromeres of vegetal hemisphere.

3. Periblastula or superficial blastula. In superficially cleaving eggs of insects, the blastocoelic cavity is not found. The central yolk is surrounded by peripherally arranged cells.

4. Discoblastula. In large yolky eggs of fishes, reptiles and birds discoblastula is found. It is a small multilayered flat disc separated from the yolk by a narrow subgerminal cavity.

5. Amphiblastula. It is found in amphibians. The blastula contains micromeres in the animal hemisphere and macromeres in the vegetal hemisphere, and a small fluid-filled eccentric blastocoel in the animal hemisphere.

6. Blastocyst. It is found in mammals. Cleavage is regular and a small cavity appears inside the dividing cells, which gradually increases in volume. This is the blastocoel. The cells surrounding the blastocoel are the trophoblast cells or nutritive cells and an inner cell mass of formative cells displaced to one pole of the blastocyst.

3. Gastrula

A rearrangement of the cells of the blastula occurs in which some cells are differentiated and come to lie inside, while the other cells enclose them, this stage is **gastrula** and the processes converting the blastula into a gastrula are known as **gastrulation**. Gastrulation process (morphogenetic movements of cells) converts a simple one-layered blastula into a two-layered (*e.g., Amphioxus*) or a three-layered (*e.g.,* all vertebrates) gastrula (Gr., *gaster* = stomach or gut). The single layer of blastula is called **blastoderm**, **ectoblast** or **proctoderm**. The three layeres (ectoderm, mesoderm and endoderm) are called **germinal layers**. The blastocoel is generally obliterated and the inner layer of cells (endoderm) of the gastrula enclose a new cavity called **archenteron** which opens on one side to the exterior by a **blastopore**. During gastrulation embryo acquires antero-posterior polarity and bilateral symmetry.

4. Organogeny

After gastrulation the continuous masses of cells of the three germ layers split up into smaller groups of cells, called primary organ rudiments, each of which produces a certain organ or part of the animal body. These organ rudiments further develop simple organs and parts and, thus, embryo develops into either larval form or a miniature adult. Thus, the formation of organs from the germ layers is called **organogenesis.**

Derivatives of Germ Layers

1. Ectoderm. The ectoderm forms a neural tube which gives rise to the **brain, spinal cord**, and **nerves**. The forebrain forms the **retina**, and part of the **iris**. The ectoderm forms the **lens**, **conjunctiva**, and a part of the **cornea**, the **membranous labyrinth** and the lining of the **nose**. In fishes and aquatic amphibians, the sensory parts of the **lateral line system** arise from the ectoderm. The neural crest cells lying between the outer ectoderm and on both sides of the neural tube give rise to **ganglia** of the spinal nerves and autonomic nervous system, the **neurilemma** of peripheral nerves, **chromatophores** of the skin, some neural crest cells give rise to mesenchyme which produces the visceral arches, and some neural crest cells wander inwards and form the **suprarenal gland** near the kidneys, but in mammals they form the **medulla** of adrenal glands. Supporting part of the central nervous system called **neuroglia** is derived from the neural tube. The ectoderm forms the **epidermis** of the skin and many epidermal derivatives, such as **skin glands**, epidermal **scales**, **nails**, **claws**, **hoofs**, **horns**, **feathers** and **hairs**. Ectodermal invaginations form the **stomodaeum** and **proctodaeum** which meet the archenteron, the ectoderm of the stomodaeum forms the lining of the mouth and lips, **glands** of buccal cavity, **enamel** of teeth, covering of tongue, and anterior and intermediate lobes of the **pituitary gland** (the posterior lobe of the pituitary is formed from the forebrain). The ectoderm of proctodaeum forms the lining of the cloaca and some anal and cloacal glands. From the dorsal side of the forebrain one or two evaginations take place, the anterior one is an eye-like **parietal body** which is present in lower forms only, the posterior one is the **pineal body** found in all vertebrates.

2. Endoderm. The archenteron is formed from endoderm, it becomes the lining of the adult alimentary canal, except in the buccal cavity and cloaca. Two outgrowths of the digestive tract form the **liver** and **pancreas**, the endoderm forming their epithelial lining only, and also of the gall bladder and bile duct. From the pharynx the endoderm pushes out to form several pairs of **pharyngeal pouches**. In cyclostomes, fishes, and amphibians, the pharyngeal pouches meet the ectoderm to form gill-clefts which open to the exterior. In amniotes, the pharyngeal pouches do not perforate to the exterior, in tetrapoda, the first pair is modified to form the cavity of the **middle ear** and **Eustachian tube**. An evagination of the pharynx along with some pharyngeal pouches form a **thyroid gland**. In air-breathing vertebrates the endoderm of pharynx forms the lining of the **larynx**, **trachea**, and **lungs**. Endoderm of some pharyngeal pouches form part of the **tonsils**, **thymus**, **parathyroid glands** and **ultimobranchial bodies**. In amniotes, the archenteron forms a large bag, the **allantois**, its lining is endodermal. Endoderm cells of the archenteron grow out in embryos developing from polylecithal eggs to form the lining of the **yolk sac** to enclose the yolk, the yolk sac disappears in the adult. It must be noted that organs arising from the archenteron have only their lining and epithelial cells formed from endoderm, the supporting tissues of these organs are mesodermal.

3. Mesoderm. The mesoderm becomes differentiated into three major parts : a dorsal **epimere** which is segmented, a median **mesomere**, and a ventral **hypomere**. Further development of mesoderm forms **mesenchyme** which is not a germinal layer but a primitive kind of embryonic connective tissue with branching cells forming a network. Nearly all mesenchyme comes from mesoderm though other germinal layers may also contribute to its formation.

***(i)* Epimere.** The epimere is differentiated into sclerotome, dermatome, and myotome. The middle parts of epimeres form mesenchyme which gathers around the neural tube and notochord to form the **sclerotome**. The mesenchymatous sclerotome forms the **vertebral column**. The **dermatome** transforms into mesenchyme which migrates to lie below the ectoderm and gives rise to the **dermis** of the skin. The remaining portion of the epimere is called **myotome**, the adjacent myotomes are separated by myocommata which are connective tissue partitions. The myotomes of the two sides grow down between the skin and somatic layer of mesoderm to meet midventrally, they give rise (with some exceptions) to **voluntary muscles** of the body and body wall.

***(ii)* Mesomere** forms the **urogenital organs** and their ducts, the terminal parts of these ducts may have ectodermal or endodermal lining.

***(iii)* Hypomere** splits into somatic and splanchnic layers of mesoderm enclosing the **coelom**. The splanchnic layer forms mesenchyme which gives rise to **involuntary muscles** and **connective tissue** of the alimentary canal and of the organs formed as outgrowths of the archenteron. The splanchnic mesoderm forms the **heart**. The remainder of the splanchnic mesoderm together with the somatic mesoderm forms the lining of the **coelom**, **pericardium** and **lung pleura** or peritoneum. Splanchnic mesoderm also forms the **mesenteries** and **omenta**.

***(iv)* Mesenchyme** (Gk., *mesos* = middle + *enchyma* = infusion) gives rise to all the **connective tissue**, **blood vessels**, **lymph vessels**, **lymph nodes**, **blood corpuscles**, all **involuntary muscles**, parts of the **eye**, **dentine** of teeth, and to cartilage and bones of the entire **skeleton**, except the vertebral column. It is claimed that voluntary muscles of limbs are formed from mesenchyme and not from myotomes.

TABLE 34.1 GERM LAYERS AND THEIR DERIVATIVES IN VERTEBRATES.

Germ layer	In Embryo	In Adult
1. Ectoderm	1. Somatic ectoderm	Epidermis, cutaneous derivatives, cutaneous sense organs, olfactory organs, lateral line, eye lens, membranous labyrinth, anterior pituitary, lining of stomodaeum and proctodaeum and enamel.
	2. Neural ectoderm	Central and peripheral nervous system, retina and posterior pituitary.
	3. Neural crest	Branchial skeleton, nervous ganglia and sensory nerves, adrenal medulla and chromatophores.
2. Endoderm	Archenteron (enteron)	Lining of alimentary canal except that of stomodaeum, and proctodaeum, lining of glands, liver, pancreas, respiratory organs, tonsils, urinary bladder, thyroid, parathyroid and thymus glands, accessory reproductive glands and auditory tube.
3. Mesoderm	1. Notochord	
	2. Epimere	(a) Dermatome–dermis (b) Sclerotome–vertebral column (c) Myotome–skeletal muscles and appendicular skeleton.
	3. Mesomere	Excretory organs and reproductive tracts.
	4. Hypomere	(a) Somatic or parietal layer (b) Splanchnic layer–visceral layer, heart, blood vessels and cells, gonads, spleen, visceral muscles (c) Coelom–Body cavities.

IMPORTANT QUESTIONS

Long Answer Questions

1. Give an account of gametogenesis.
2. Describe various types of eggs and cleavage found in chordates.
3. Describe the main types of cleavages, blastulation and gastrulation in vertebrates.
4. Describe the process of blastulation in chordates and comment upon the importance of blastula.
5. What do you understand by germinal layers? Discuss the formation of germinal layers with suitable illustrations and just outline the roles played by germinal layers in the organogenesis in amniota.
6. Show how three germinal layers develop in vertebrates and name the adult body parts that arise from each layer. Review the process of mesoderm and coelom formation in *Amphioxus*.

Short Answer Questions

1. What are monoestrous and polyestrous?
2. Distinguish between irregular and spiral cleavage.
3. What are the factors that control the pattern of cleavage?
4. Distinguish between epiblast and hypoblast.
5. Write short notes on the following : (*i*) Blastula; (*ii*) Blastopore; (*iii*) Fertilisation; (*iv*) Oogenesis; (*v*) Parthenogenesis; (*vi*) Types of cleavage; (*vii*) Planes of cleavage; (*viii*) Germ layers and their fate; (*ix*) Scope of embryology.

Very Short Answer Questions

1. Define the following terms : (*a*) Placenta; (*b*) Mesonephros; (*c*) Epiboly; (*d*) Coelom; (*e*) Fertilisation.

Multiple Choice Questions

1. The spermatid undergoes some transformation before it is liberated as an active sperm. These changes are known as
   (*a*) Spermatogenesis (*b*) Spermiolysis
   (*c*) Spermiogenesis (*d*) Sperm maturation.
2. Process of sperm formation is called
   (*a*) Oogenesis (*b*) Gametogenesis
   (*c*) Spermatogenesis (*d*) Meiosis.
3. The acrosome of sperm is derived from
   (*a*) Mitochondria (*b*) Nucleus
   (*c*) Golgi apparatus (*d*) Centrosome.
4. Yolk is the main reserve food in the ovum. It is mainly made up of two ingredients
   (*a*) Fats and carbohydrates
   (*b*) Lipids and proteins
   (*c*) Carbohydrates and minerals
   (*d*) Lipids and minerals.
5. Union of sperm and ovum is called
   (*a*) Parthenogenesis (*b*) Fertilisation
   (*c*) Polyspermy (*d*) Karyokinesis.
6. Eggs of insects are
   (*a*) Microlecithal (*b*) Centrolecithal
   (*c*) Oligolecithal (*d*) Macrolecithal.
7. Which of the following produce centrolecithal egg?
   (*a*) Fishes (*b*) Amphibians
   (*c*) Mammals (*d*) Insects.
8. Which membrane prevents polyspermic fertilisation?
   (*a*) Fertilizin membrane
   (*b*) Antifertilizin membrane
   (*c*) Fertilisation membrane
   (*d*) Vitelline membrane.
9. Cleavage in which furrows are at right angle to each other is called
   (*a*) Radial (*b*) Spiral
   (*c*) Indeterminate (*d*) Determinate.
10. Meroblastic and discoidal cleavage is characteristic of egg type
    (*a*) Centrolecithal (*b*) Mesolecithal
    (*c*) Polylecithal (*d*) Alecithal.
11. Gastrulation results in the formation of
    (*a*) Blastocoel
    (*b*) Three germinal layers
    (*c*) Archenteron
    (*d*) Archenteron and germinal layers.
12. Formation of archenteron occurs in which of the following stage of development
    (*a*) Blastulation (*b*) Gastrulation
    (*c*) Organogeny (*d*) Cleavage.
13. Coelom is formed from which of the following germ layer
    (*a*) Ectoderm (*b*) Endoderm
    (*c*) Mesoderm (*d*) All of the three.
14. Which of the following is mesodermal in origin
    (*a*) Archenteron (*b*) Notochord
    (*c*) Neural tube (*d*) None of these.

15. In the two layered gastrula the inner invaginative layer is
(*a*) Ectoderm (*b*) Endoderm
(*c*) Mesoderm (*d*) All three.
16. Embryo becomes triploblastic during
(*a*) Cleavage (*b*) Blastulation
(*c*) Gastrulation (*d*) Organogeny.
17. Mass of cells formed after cleavage is known as
(*a*) Morula (*b*) Blastula
(*c*) Gastrula (*d*) Both (*a*) and (*b*).
18. Which of the following layer is formed at last during gastrulation?
(*a*) Ectoderm (*b*) Endoderm
(*c*) Mesoderm (*d*) None of these.
19. Mesoderm is derived from
(*a*) Ectoderm (*b*) Endoderm
(*c*) Both of these (*d*) None of these.
20. When cleavage furrow bisects both the poles passing through centre of the egg and median axis runs between centres of animal and vegetal pole. It is called
(*a*) Meridional plane of cleavage
(*b*) Vertical plane of cleavage
(*c*) Latitudinal plane of cleavage
(*d*) Equatorial plane of cleavage.
21. Superficial cleavage is found in
(*a*) Cockroach (*b*) Starfish
(*c*) Frog (*d*) Man.
22. In honeybee drones are formed by
(*a*) Fertilisation (*b*) Organogenesis
(*c*) Parthenogenesis (*d*) All of these.
23. Brain is formed from
(*a*) Neural crest
(*b*) Embryonic neural tube
(*c*) Coelom
(*d*) None of these.
24. Sperms in animals are produced as a result of complete meiotic division and subsequent morphological and cytological changes in the meiotic divisions. The metamorphosis is called
(*a*) Spermatogenesis (*b*) Spermiogenesis
(*c*) Gametogenesis (*d*) Oogenesis.
25. Sperms usually originate in
(*a*) Vas deferens
(*b*) Seminiferous tubules
(*c*) Spermatocytes
(*d*) Kidney.
26. The head of mature sperm is composed of
(*a*) Two centrioles and axial filament
(*b*) Mitochondrial sheath and cytoplasm
(*c*) Elongated nucleus covered by acrosomal material
(*d*) Acrosome without nucleus.
27. Sperms produce substances of enzymatic nature called sperm lysins. In mammals these are known as
(*a*) Androgens (*b*) Hyaluronic acid
(*c*) Hyaluronidase (*d*) Estrogens.
28. The number of chromosomes present in the oogonium is
(*a*) Haploid (*b*) Diploid
(*c*) Euploid (*d*) Polyploid.
29. The early division of zygote is termed as
(*a*) Mitosis (*b*) Meiosis
(*c*) Cleavage (*d*) Metamorphosis.
30. Complete holoblastic cleavage occurs in
(*a*) Alecithal eggs
(*b*) Microlecithal eggs
(*c*) Telolecithal eggs
(*d*) Megalecithal eggs.
31. Development an egg without fertilisation is called
(*a*) Metamorphosis (*b*) Oogenesis
(*c*) Parthenogenesis (*d*) Pedogenesis.
32. Which of the following is concerned with asexual reproduction?
(*a*) Zygote (*b*) Spores
(*c*) Gametes (*d*) Ovaries.
33. In teleosts, reptiles and birds, the cleavage pattern is
(*a*) Incomplete meroblastic type
(*b*) Unequal holoblastic type.
(*c*) Equal holoblastic type
(*d*) Complete meroblastic type.

ANSWERS

1. (*a*)	2. (*c*)	3. (*c*)	4. (*b*)	5. (*b*)	6. (*b*)	7. (*d*)	8. (*c*)
9. (*b*)	10. (*c*)	11. (*d*)	12. (*b*)	13. (*c*)	14. (*b*)	15. (*b*)	16. (*c*)
17. (*a*)	18. (*c*)	19. (*c*)	20. (*a*)	21. (*a*)	22. (*c*)	23. (*b*)	24. (*b*)
25. (*b*)	26. (*c*)	27. (*c*)	28. (*b*)	29. (*c*)	30. (*b*)	31. (*c*)	32. (*b*)
33. (*a*)							

35 Development of *Herdmania*

Chapter

Contents of this Chapter

Development of *Herdmania* has been described by **Sebastian** (1953) and **Das** (1957). Development is indirect involving a free swimming tadpole larva. *Herdmania* (a tunicate) is the most appropriate one of the best examples of **retrogressive metamorphosis** in which the highly developed tadpole larva undergoes retrogressive changes to become the most degenerated and sedentary adult. The phenomenon of retrogressive metamorphosis is peculiar to urochordates and provides a clue to their chordate nature.

GAMETES

Spermatozoa

The sperm of *Herdmania* is flagellated and about 4 μm in length. Each sperm has an anterior head bearing a middle mucleus and capped with a beak-like **acrosome**, a middle piece with neck and a very long tail. According to **Das** (1936), sperms of *Herdmania* are polymorphic, *i.e.*, three types of sperms having acrosome either shorter, equal or longer than the head.

Ova

The ova or unfertilised eggs are microlecithal having distinct animal and vegetal poles. Mature ovum is about 0.3 mm in diameter containing an eccentric nucleus with a prominent nucleolus. The egg is surrounded by an inner vitelline membrane secreted by the ovum which is closely applied to it, and outer to it are two chorion membranes inner and outer secreted by follicular cells of the ovary. The space between inner chorion and vitelline membrane is the **perivitelline space** filled with **perivitelline fluid**. The perivitelline space contains a large number of small-sized **inner follicle cells** or **test cells**, some of which float in the fluid and most of them are attached to vitelline membrane. These cells nourish the ovum and secrete a chorion digesting enzyme to facilitate hatching. Follicle cells are also found attached to the outer surface of chorion membrane. These cells keep the ovum floating in outer perivitelline fluid. Maturation of ovum occurs in sea water.

Fertilisation

Fertilisation is external and cross-fertilisation. Spermatozoon penetrates the ovum nearer to vegetal pole. Polyspermy is not allowed by the chorion layers. After fertilisation, various cytoplasmic regions are rearranged by a process known as **ooplasmic segregation**, *i.e.*, yellow and clear cytoplasm segregates in the lower hemisphere with a cap of clear cytoplasm over it and gray yolk-laden endoplasm occupies the upper two-thirds of the egg.

Presumptive Areas

The entry of sperm into ovum results in movements of ovum's cytoplasm (ooplasm), so that the fertilised egg or zygote is **mosaic**, *i.e*, it shows **presumptive areas** with predictable future.

EMBRYONIC OR PRELARVAL DEVELOPMENT

Cleavage

Cleavage is determinate. The zygote undergoes holoblastic, somewhat unequal and determinate cleavage, *i.e.*, the fate of blastomeres is predetermined. The first cleavage is vertical, meridional and divides the gray crescent into two unequal parts. Second cleavage is also vertical but at right angle to the first. The resultant four blastomeres are different. Two smaller blastomeres are at posterior side and two larger blastomeres are at front. Only two of these four contain parts of gray crescent. The third cleavage is horizontal and passes little above the equator which divides the blastomeres into 8 cells arranged in two tiers. Fourth cleavage results into 16 blastomeres. Subsequently, cleavage becomes irregular and finally results into a solid ball of cells called **morula**. A single-layered flat **coeloblastula** with an internal fluid-filled segmentation cavity or blastocoel is formed at 64-cell stage. In all it takes about 110 minutes.

Thus, a round 64-celled compact blastula is formed having a blastocoel in the interior. The upper smaller cells are micromeres or ectoderm, and lower larger cells are macromeres or endoderm.

Gastrulation. It takes place by **emboly** or invagination of macromeres, which now enclose an **archenteron**. The wide ventral opening of the archenteron is the blastopore, the archenteron increases obliterating the blastocoel. On the completion of gastrulation, the blastopore closes and the embryo begins to lengthen along the embryonic axis. The upper surface or neural plate of the elongating gastrula invaginates along the mid-line to form a groove which subsequently forms the **neural tube**. This stage of the embryo is

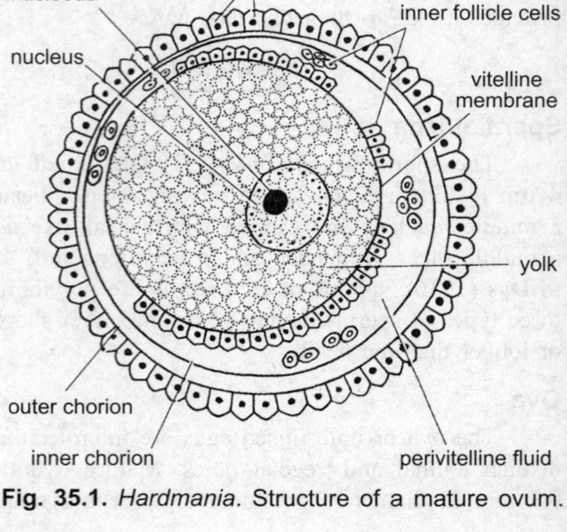

Fig. 35.1. *Hardmania*. Structure of a mature ovum.

neurala. The anterior portion of neural tube remains wider and develops into **brain vesicle**, and the posterior narrow portion of neural tube forms the **spinal cord**. For a short time, the brain-vesicle communicates to the exterior through **neuropore**. The hinder part of neurula becomes extended into a tail, into which neural tube, notochord, mesoderm and mesenchyme are extended. With further development, the neurula develops into an ascidian **tadpole larva**, which ruptures the egg membranes and becomes free-swimming.

LARVAL DEVELOPMENT

The blastopore closes and develops a rudiment of tail. The embryo elongates and forms a tailed larva. The presumptive notochordatal cells separated from the roof of the archenteron,occupy the central core of larval tail. Archenteron produces presumptive mesoderm as solid bands and not as hollow coelomic sacs as in *Balanoglossus* or *Branchiostoma*. About 8 hours after fertilisation, the chorions rupture, probably dissolved by an enzyme secreted by the inner follicle cells, and the fully formed larva hatches out to become free-swimming.

Tadpole Larva

The fully developed ascidian tadpole is minute, about 1.5 mm long and motile. Its body is covered by a thin **tunic** or **test** secreted by the ectoderm. It has an anterior short oval **head** or **trunk** and a posterior long **tail**.

1. Head or trunk. At the anterior end of the trunk are three **adhesive papillae** made of ectodermal cells. The nervous system consists of an anterior enlarged hollow **sense vesicle** or **brain vesicle**. The sense vesicle is continued posteriorly into a solid, thick mass of nerve cells, the **visceral ganglion**, it is continued posteriorly into a **nerve cord** which is hollow and continued into the tail. It lies mid-dorsally. The sense vesicle contains a pigmented **statocyst (otocyst)** and two pigmented unequal **ocelli** or simple **eyes** as sense organs.

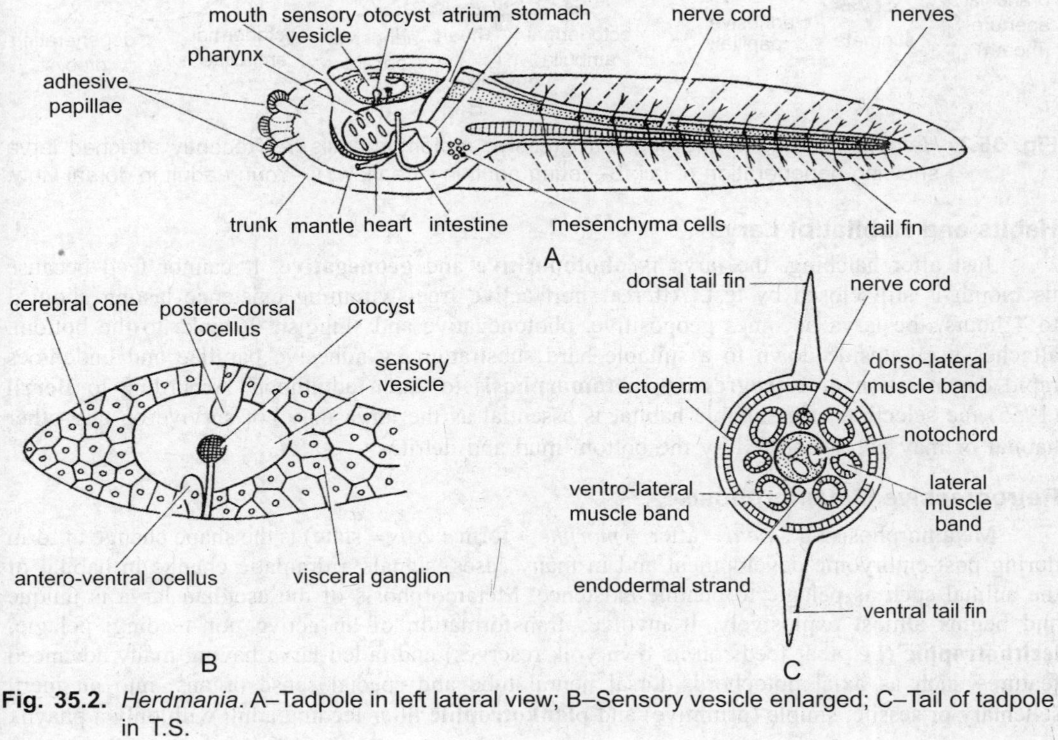

Fig. 35.2. *Herdmania*. A–Tadpole in left lateral view; B–Sensory vesicle enlarged; C–Tail of tadpole in T.S.

The **alimentary canal** begins from antero-dorsal mouth leading into narrow branchial siphon, a large sac-like pharynx, short oesophagus, stomach, intestine and a small rectum. The **anus** opens into the left side of atrium. The pharynx has a ventral **endostyle** and also a pair of large gill-slits or **stigmata**. Each gill-slit finally splits to form six stigmata on either lateral side An **atrial cavity** is formed around the pharynx laterally and dorsally and opens to the exterior by an **atrial aperture** lying dorsally. The **heart** and **pericardium** is located postero-ventrally in between pharynx and stomach. **Mesenchyme cells** found scattered all over the body beneath the ectoderm and in a mass in the posterior region of trunk.

2. Tail. It is a powerful locomotory organ being laterally compressed and pointed terminally. Tail is about 0.9 mm long and is fringed with a vertical continuous **tail fin** formed by the test and marked with oblique striations looking like fin rays. It contains a dorsal tubular nerve cord, a notochord beneath the nerve cord and lateral muscle bands. The anterior end of notochord slightly projects into the trunk region.

The ascidian tadpole larva is unquestionably a true chordate, having a dorsal hollow nerve tube, a notochord in the tail and paired gill-slits in the pharynx and after about three hours of active, non-feeding and free-swimming existence, it sinks to the bottom.

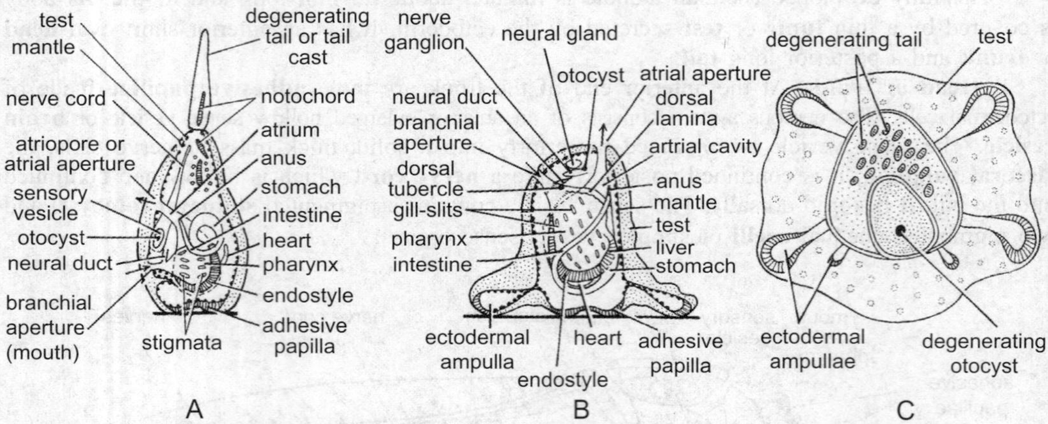

Fig. 35.3. *Herdmania.* Stages showing retrogressive metamorphosis. A–Recently attached larva showing degeneration of tail; B–Young adult in side view; C–Young adult in dorsal view.

Habits and Habitat of Larva

Just after hatching, the larva is **photopositive** and **geonegative**. It cannot feed because its mouth is still closed by test. After a short active free-swimming existence lasting about 3 to 4 hours, the larva becomes geopositive, photonegative and sluggish. It sinks to the bottom, attaches itself upside down to a suitable hard substratum by adhesive papillae and undergoes rapid degeneration or **retrogressive metamorphosis** to attain adulthood. According to **Berril** (1955) the selection of a suitable habitat is essential as the larva may not survive on any other habitat or may get suffocated by the bottom mud and detritus.

Retrogressive Metamorphosis

Metamorphosis Gr., *meta* = after + *morphe* = form + *osis* = state) is the shape change in form during post-embryonic development and in many cases, signals a dramatic change in habitat of the animal such as pelagic to benthic existence. Metamorphosis of the ascidian larva is unique and begins almost explosively. It involves transformation of an active non-feeding, pelagic, **lecithotrophic** (*i.e.*, that feeds on its own yolk reserves) and tailed larva having many advanced features such as axial notochord, dorsal neural tube and special sense organs, into an inert, sedentary or sessile, simple (primitive) and **plankotrophic** filter feeding adult with only a phaynx with stigmata and endostyle, both indicating the chordate features of adult ascidian. This type of metamorphosis which shows degenerative or retrogressive changes from larva to adult is called **retrogressive metamorphosis**. It involes the following two types of changes : (*i*) **retrogressive** and (*ii*) **progressive**.

1. Retrogressive changes. These changes involve destruction and disappearance of some of the larval structures such as follows :

1. Long tail of larva with caudal fin shortens and finally disappears.
2. Caudal muscles, nerve cord and notochord disappear as they break down and are consumed by phagocytes.
3. Larval sense organs (the ocellus and the otolith) are lost and sensory vesicle is transformed into an adult cerebral ganglion.
4. Adhesive papillae disappear completely.
5. Anterior region between point of attachment (adhesive papillae) and mouth shows rapid growth, while original dorsal side with atriopore stops growth. This causes shifting of mouth through 90°. Therefore, the final branchial and atrial apertures in the adult represent the original anterior and dorsal sides of the larva.

2. Progressive changes. Some larval structures necessary for survival become more elaborated and specialised in each adult, such as :

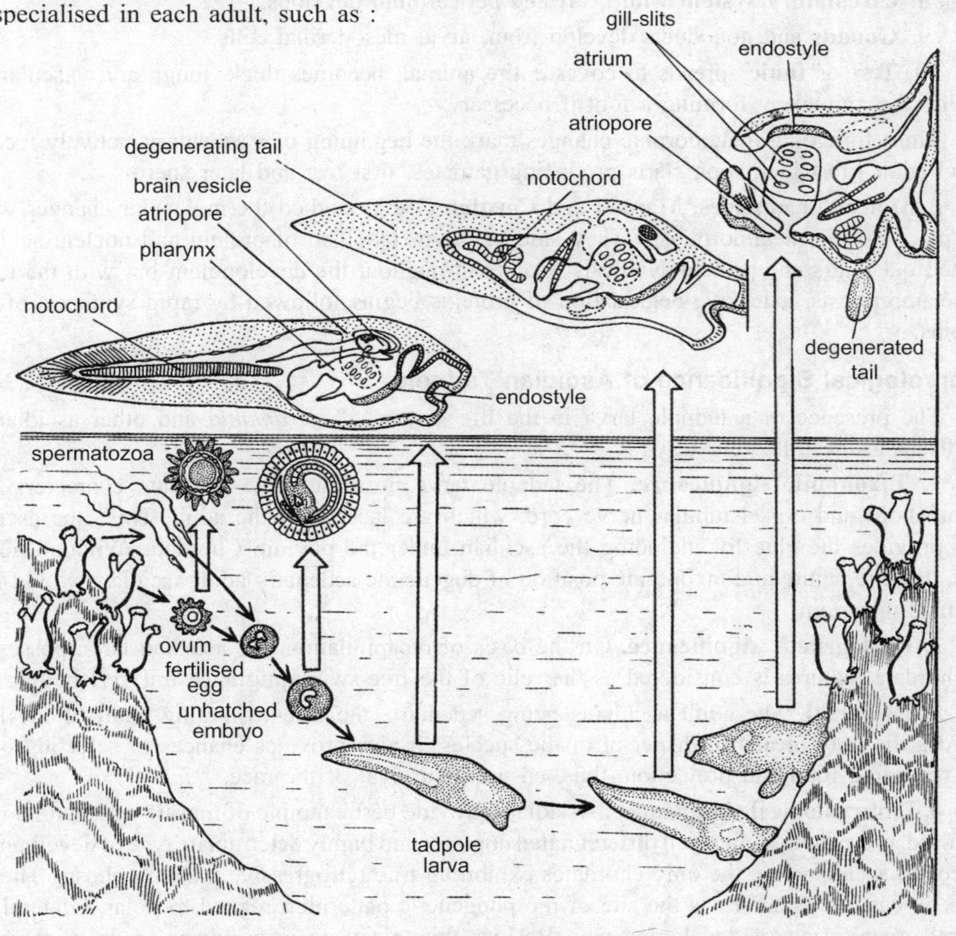

Fig. 35.4. Life cycle of a simple ascidian, showing retrogressive metamorphosis of tadpole larva, later its attachment with the solid substratum.

1. Due to loss of tail, the trunk becomes pear-shaped and four **larger ectodermal ampullae** grow out of its four corners. These ampullae firmly anchor the metamorphosing tadpole to the substratum and also serve for respiration as a blood-like fluid keeps circulating through them. Soon two more **smaller ectodermal ampullae** appear dorso-laterally.

2. Anterior region between point of attachment (adhesive papillae) and mouth exhibits rapid growth, while original dorsal side with atriopore stops growth. This causes shifting of mouth through 90°. The body too rotates so that general form of the adult sessile organism is assumed.

3. Adult **neural glands** and **nerve** or **cerebral ganglion** are formed by the neural tube and trunk ganglion come to lie mid-dorsally between mouth and atriopore. The trunk ganglion itself persists as **visceral nerve**.

4. With the absorption of its test covering, the **mouth** becomes functional and filter-mode of feeding by incoming ciliary water currents.

5. **Pharynx** greatly enlarges, develops blood vessels and **stigmata** multiply rapidly, forming the **branchial sac**.

6. **Stomach** enlarges, intestine elongates and gets curved and liver develops.

7. **Atrial cavity** becomes more extensive.

8. **Circulatory system** with heart and pericardium develops.

9. **Gonads** and gonoducts develop from larval mesodermal cells.

10. **Test** or **tunic** spreads to cover entire animal, becomes thick, tough and vascular and attaches the animal by forming a foot if necessary.

Thus, foregoing metamorphic changes mark the beginning of a sedentary, actively feeding, sexual adult life which soon starts producing gametes, first ova and later sperms.

3. Molecular changes. Manket and **Cowden** (1965) studied the molecular changes which take place during metamorphosis. They studied the metabolism of protein and nucleic acid and pointed out that some protein synthesis occurs throughout the development but with the outset of metamorphosis, extensive degradation of proteins begins followed by rapid synthesis of new proteins.

Embryological Significance of Ascidian Tadpole

The presence of a tadpole larva in the life history of *Herdmania* and other ascidians is significant in the following ways :

1. Taxonomic significance. The tadpole larva possesses true chordate characters such as notochord and dorsal tubular nerve cord, which are lacking in the adult. Thus, the ascidian larva provides the clue for including the ascidian under the phylum Chordata. Without tadpole larva, the true nature and taxonomic position of degenerate sedentary adult ascidians would have remained uncertain.

2. Phylogenetic significance. On the basis of recapitulation, the ascidian larva possessing the chordate features is considered as the relic of the free-swimming ancestral vertebrates.

3. Dispersal. The adult ascidians being sedentary, the free-swimming habit of the larva provides the only means of dispersal of the species. It also provides chances of selecting better sites regarding food and protection, thus, ensuring survival of the race.

4. Embryological significance. Ascidians provide best example of mosaic eggs with a well-organised, pre-patterned and well differentiated ooplasm and highly determinate type of development. Moreover, ascidians are the only chordates exhibiting true retrogressive metamorphosis. The egg cortex in case of ascidians is the site of morphogenetic patterning related to polar, bilateral and general organisation of developing egg. Besides this, cleavage in ascidians tends to segregate cytoplasmic territories, having different biological, histochemical and biochemical properties.

IMPORTANT QUESTIONS

Long Answer Questions

1. Give a detailed account of metamorphosis of *Ascidia*.
2. Describe asexual reproduction in ascidians.
3. Describe the development of tadpole larva of *Herdmania* and discuss various changes it has to undergo during retrogressive metamorphosis.
4. What is retrogressive metamorphosis? Discuss with special reference to the life history of an ascidian.

Short Answer Questions

1. Describe the structure of tadpole larva of *Herdmania* along with a labelled diagram.
2. Write short notes on : (*i*) Ascidian tadpole, (*ii*) Retrogressive metamorphosis.

Very Short Answer Questions

1. What is protogyny in *Herdmania*?
2. What is the type of cleavage in *Herdmania*?
3. Blastocoel is formed in which stage of embryo during development.
4. What is retrogressive metamorphosis?
5. Why tadpole larva is supposed to be more advanced than the adult animal?

Multiple Choice Questions

1. Development of *Herdmania* is
   (*a*) Indirect (*b*) Direct
   (*c*) Partly indirect (*d*) Both (a) and (b).
2. Fertilisation in *Herdmania* is
   (*a*) External (*b*) Internal
   (*c*) Both
   (*d*) None of the above.
3. In *Herdmania* metamorphosis is
   (a) Progressive
   (b) Retrogressive
   (c) Incomplete metamorphosis
   (d) None.
4. Function of otolith (otocyst) in *Herdmania* tadpole is
   (*a*) Photoreceptor
   (*b*) Tactile receptor
   (*c*) Maintenance of equilibrium
   (*d*) All these.
5. Larva of *Herdmania* shows
   (*a*) True nonchordate characters
   (*b*) True chordate characters
   (*c*) Echinoderm characters
   (*d*) Hemichordate characters.
6. Number of adhesive papillae on trunk of *Herdmania* larva is
   (*a*) 5 (*b*) 3
   (*c*) 8 (*d*) 2.
7. Fertilised eggs of *Herdmania* are
   (*a*) Mosaic showing presumtive areas
   (*b*) Nonmosaic
   (*c*) Sometimes mosaic
   (*d*) All these.
8. Opening of archenteron in *Herdmania* is called
   (*a*) Blastopore (*b*) Gastropore
   (*c*) Archenteron (*d*) Coelopore.
9. Archenteron in *Herdmania* produces
   (*a*) Presumptive ectoderm
   (*b*) Presumptive mesoderm
   (*c*) Coelomic sacs
   (*d*) Tail.

ANSWERS

1. (*a*) 2. (*a*) 3. (*b*) 4. (*c*) 5. (*b*) 6. (*b*) 7. (*a*) 8. (*a*)
9. (*b*)

36 Chapter

Development of *Branchiostoma* (*Amphioxus*)

Contents of this Chapter

Development of *Branchiostoma* (*Amphioxus*) is indirect involving a larval stage. Early embryology of *Branchiostoma* is simple. Therefore, the transformation of egg, which having less yolk, into a complex and differentiated animal is far easier to follow than in any other vertebrate. The early development of *Branchiostoma* (*Amphioxus*) is of great phylogenetic significance because it resembles with those of invertebrates like echinodermates on one hand and vertebrates on the other. Development of *Amphioxus* was described by many scientists such as **Hatscheck** (1882, 1888), **Wilson** (1883), **Cerfontaine** (1906), and **Conklin** (1932). The work of **Conklin** (1932) is the most recent and accepted one.

GAMETES

Spermatozoa

Spermatozoon of *Amphioxus* has a typical structure of flagellate spermatozoon, *i.e.,* it contains a spherical head, a very short middle piece and a long tail.

Ova

The unfertilised ovum or secondary oocyte of *Amphioxus* is 0.10 mm to 0.12 mm in diameter. It is **microlecithal** and **isolecithal**. It is surrounded by a plasma membrane, which is further surrounded by a thin **vitelline membrane** of mucopolysaccharides. Its nucleus is eccentric lying at the animal pole.

The egg undergoes its first maturation division before it leaves the ovary. The animal pole is free from yolk and the cytoplasm towards the vegetal pole contains some yolk distributed evenly.

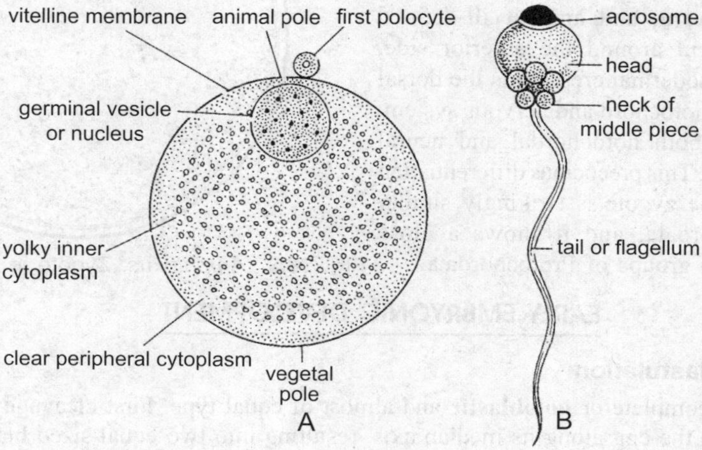

Fig. 36.1. *Amphioxus*. A–Unfertilised ovum; B–Sperm.

FERTILISATION

Fertilisation is external, taking place in the surrounding sea water where eggs and spermatozoa are shed. Before fertilisation, the ovum has outer thin **vitelline membrane** enclosing a **peripheral cytoplasmic layer**, central **yolky cytoplasm** mainly towards the vegetal pole and a fluid-filled **germinal sac** or **nucleus** towards the animal pole. During fertilisation, the sperm enters the egg near the vegetal pole which provides a stimulus for the egg (secondary oocyte) to undergo its second maturation division, a polar body is extruded which comes to lie near the animal pole inside the fertilisation membrane, within perivitelline space. It is a space formed by the elevation of fertilisation membrane from the plasma membrane. It marks the animal pole of the egg. The nuclei of sperm and egg migrate towards the equator of the egg and fuse to form **zygote nucleus**. The site of fusion is generally above the equator of the egg and slightly toward the side which will eventually be the posterior end of the embryo.

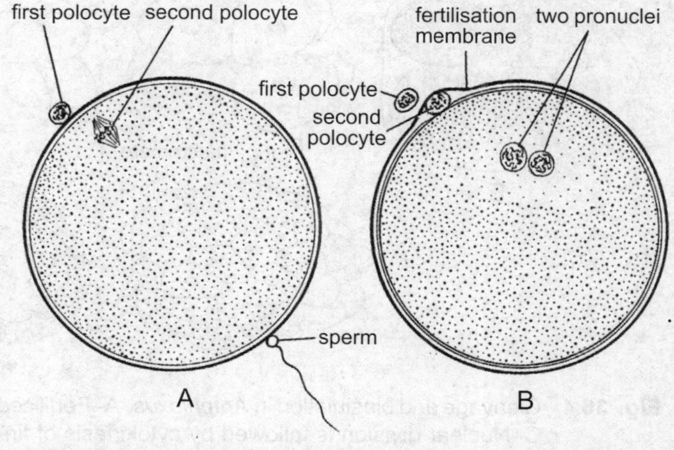

Fig. 36.2. *Amphioxus* egg. A–At fertilisation; B–At karyogamy.

Presumptive Areas

Immediately after fertilisation the cytoplam of the zygote gets so arranged as to foreshadow the future parts of the embryo. The clear cytoplasm occupies the animal hemisphere develops mainly into the **ectoderm**, the granular yolky cytoplasm at the vegetal pole gives rise to **endoderm**, a basophilic cytoplasmic crescent at the posterior end forms **mesoderm**, and another ill-defined cytoplasmic crescent around the anterior side, opposite to the mesodermal crescent is the dorsal crescent forms the notochord and nervous system. Thus, it includes both notochordal and neural ectodermal material. This precocious differentiation of cytoplasm of the zygote is strikingly similar to that of Urochordata, and it shows a close relationship of two groups of Protochordata.

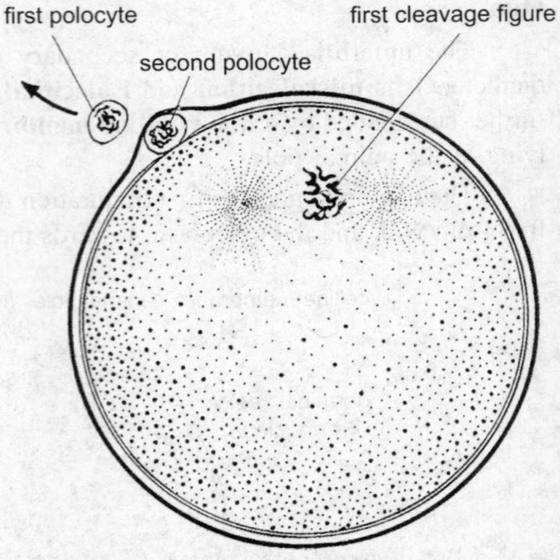

Fig. 36.3. *Amphioxus*. Zygote in first cleavage.

EARLY EMBRYONIC DEVELOPMENT

Cleavage and Blastulation

Cleavage is complete or **holoblastic** and almost of equal type. First cleavage is meridional which cuts through the egg along its median axis, resulting into two equal sized blastomeres, the right and left, which forms the right side and left side of the adult animal. The second cleavage is again meridional, but at the right angle to the first and divides the first two blastomeres into four

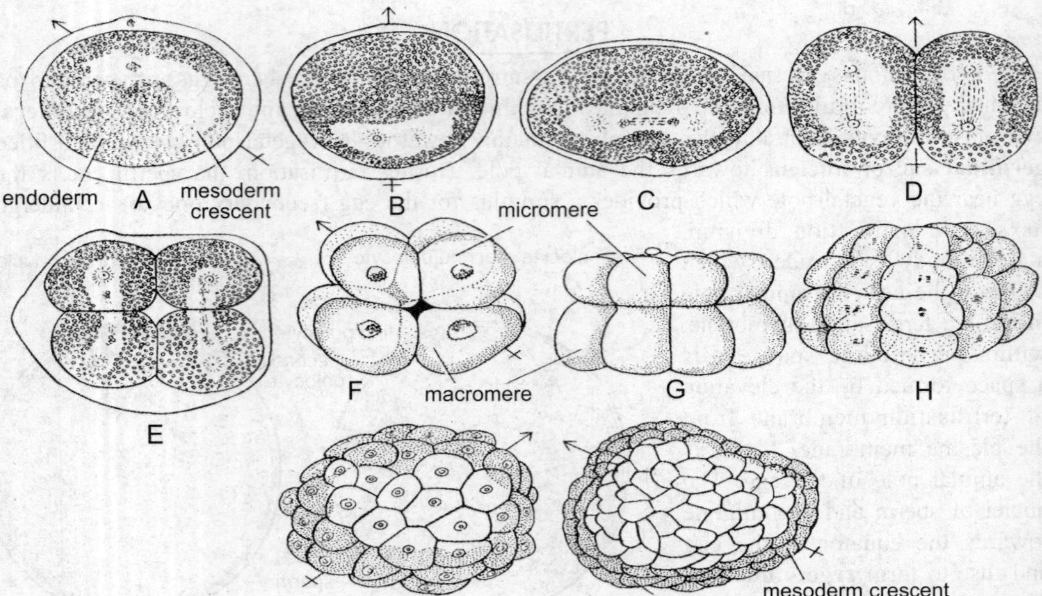

Fig. 36.4. Cleavage and blastulation in *Amphioxus*. A–Fertilised egg; B–Mitosis of first cleavage started; C–Nuclear division is followed by cytokinesis of first cleavage–mitosis; D–Two-cell stage; E–Four-cell stage; F–Eight-cell stage; G–Sixteen-cell stage; H–Thirty-two cell stage; I–Morula stage; J–Blastula stage.

equal-sized blastomeres. The third division is horizontal passing just above the equator forming four upper smaller **micromeres** near the polar body, and four lower large **macromeres** at the vegetal pole. Fourth cleavage is again meridional, cutting all the eight blastomeres synchronously. Sixteen-cell stage is formed having eight micromeres and eight macromeres. The fifth cleavage is latitudinal or horizontal and synchronous, resulting into thirty-two cells arranged in four tiers. The sixth cleavage is meridional and synchronous producing sixty four blastomeres. Later cleavages will not be synchronous up to this stage, the blastomeres remain loosely packed and form the **morula**. Meanwhile, a semifluid material accumulates in the centre of the morula. It pushes all the blastomeres outward during further cleavage of the morula, so that they become arranged in a single layer, called **blastoderm**, enclosing a central fluid-filled cavity the **blastocoel**. This embryonic stage is called the **blastula** and a fully developed blastula of *Amphioxus* is called the **coeloblastula**.

The cells at the vegetal pole of blastula are larger than at the animal pole and the blastoderm is, therefore, thicker at the vegetal pole. The cells of ventral or mesodermal crescent are small sized, spherical amoebocytes.

Gastrulation

In *Amphioxus*, gastrulation is started when there are about 800 cells in the embryo, *i.e.*, between 9th and 10th cleavage. It involves two basic types of morphogenetic movements of the embryonic cells : epiboly and emboly. Both the processes go on side by side.

1. Epiboly. It is the expansion of ectodermal cells of the embryo.

2. Emboly. The blastoderm at the vegetal pole (endodermal plate) becomes flat and subsequently bends inwards (invagination). Thus, the embryo, instead of spherical becomes converted into a cup-shaped structure, having a large cavity, the **archenteron**, opening outside by a wide **blastopore**. The cup has double walls, an external and internal epidermis, both of which are continuous with each other over the rim of the cup-shaped embryo. In this embryo, the prospective mesoderm lies in the ventral lip of blastopore and the prospective notochord lies in the dorsal lip of blastopore.

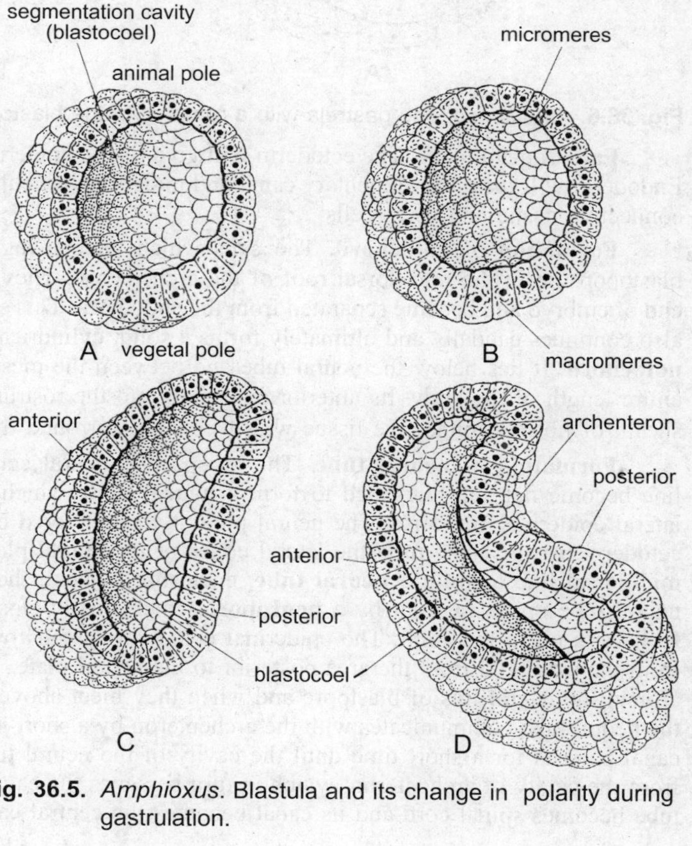

Fig. 36.5. *Amphioxus*. Blastula and its change in polarity during gastrulation.

As the endodermal plate moves inward dorsal notochordal cells involute, move inward along with the endoderm due to which prospective notochordal cells come to lie beneath the prospective neural ectoderm. Similarly, the prospective mesoderm in the ventral lip of the blastopore also moves inward and side by side its lateral horns converge toward the dorsal side of the embryo, so that they come to lie on both sides of the presumptive notochord. Now the embryo also elongates in antero-posterior direction and in this elongation all the prospective cells take part. The blastopore subsequently narrows due to contraction of the rim of blastopore. The blastopore leads into a newly formed cavity, the **archenteron**. Thus, a two layered **gastrula** is formed, in which the outer layer of cells becomes ciliated but it is enclosed in the vitelline membrane. Growth continues and the gastrula becomes further elongated antero-posteriorly. Growth of the dorsal lip reduces the blastopore and shifts it from a nearly dorsal to a posterior position. Gastrula has an outer layer of ciliated **ectoderm** cells, in which mid-dorsally are cells of **neural plate** and ventro-laterally are cells of **epidermal ectoderm**; the inner cells of the gastrula have a mid-dorsal strip of **notochord** cells, on the two sides of which are **mesoderm** cells, and the lateral and ventral inner cells are **endoderm**.

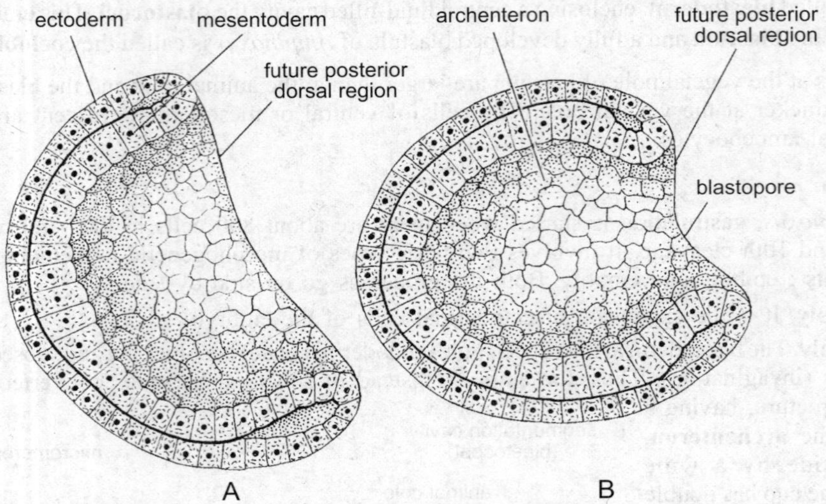

Fig. 36.6. *Amphioxus.* The gastrula with a remnant of the blastocoel still present at the blastopore.

Fate of germ cells. The ectoderm will give rise to epidermis, nervous system and receptors. Endoderm will form the alimentary canal and midgut diverticulum. Mesoderm will form muscles, connective tissue and germ cells.

Formation of notochord. The cells which were invaginated from the dorsal lip of the blastopore lie in the mid-dorsal roof of the archenteron. They evaginate dorsally at the anterior end of embryo and become separated from the endoderm. This evagination of notochordal material also continues caudally and ultimately forms a solid, cylindrical cord of cells, which is called the **notochord**. It lies below the neural tube and between the mesodermal somites. It extends in the entire length of the body. Its anterior extension into the rostrum takes place later. A notochordal sheath of fibrous connective tissue will eventually surround the notochord.

Formation of neural tube. The prospective neural ectoderm cells along the mid-dorsal line become flat and thickened to form a **neural plate** which sinks inwards below the level of lateral epidermal ectoderm. The neural plate is then covered by the free edges of the epidermal ectoderm. At the same time the lateral edges of the neural plate grow upwards and meet in the mid-dorsal line to from a **neural tube**, enclosing a canal, the **neural canal** or **neurocoel**. The neural tube opens in front by a **neuropore** which finally closes and its place is indicated by a Kolliker's pit in the adult. The epidermal ectoderm spreads over the surface of the neural plate from the sides and from the area posterior to the neural plate. Posteriorly the neural folds extend around the lateral lips of blastpore and when they meet above they close the blastopore, so that the neural tube communicates with the archenteron by a short **neurenteric canal**. The neurenteric canal persists for a short time until the cavity of the neural tube becomes completely separated from the cavity of archenteron, which in turn becomes the cavity of alimentary canal. The neural tube becomes spinal cord and its canal becomes the central canal of the central nervous system.

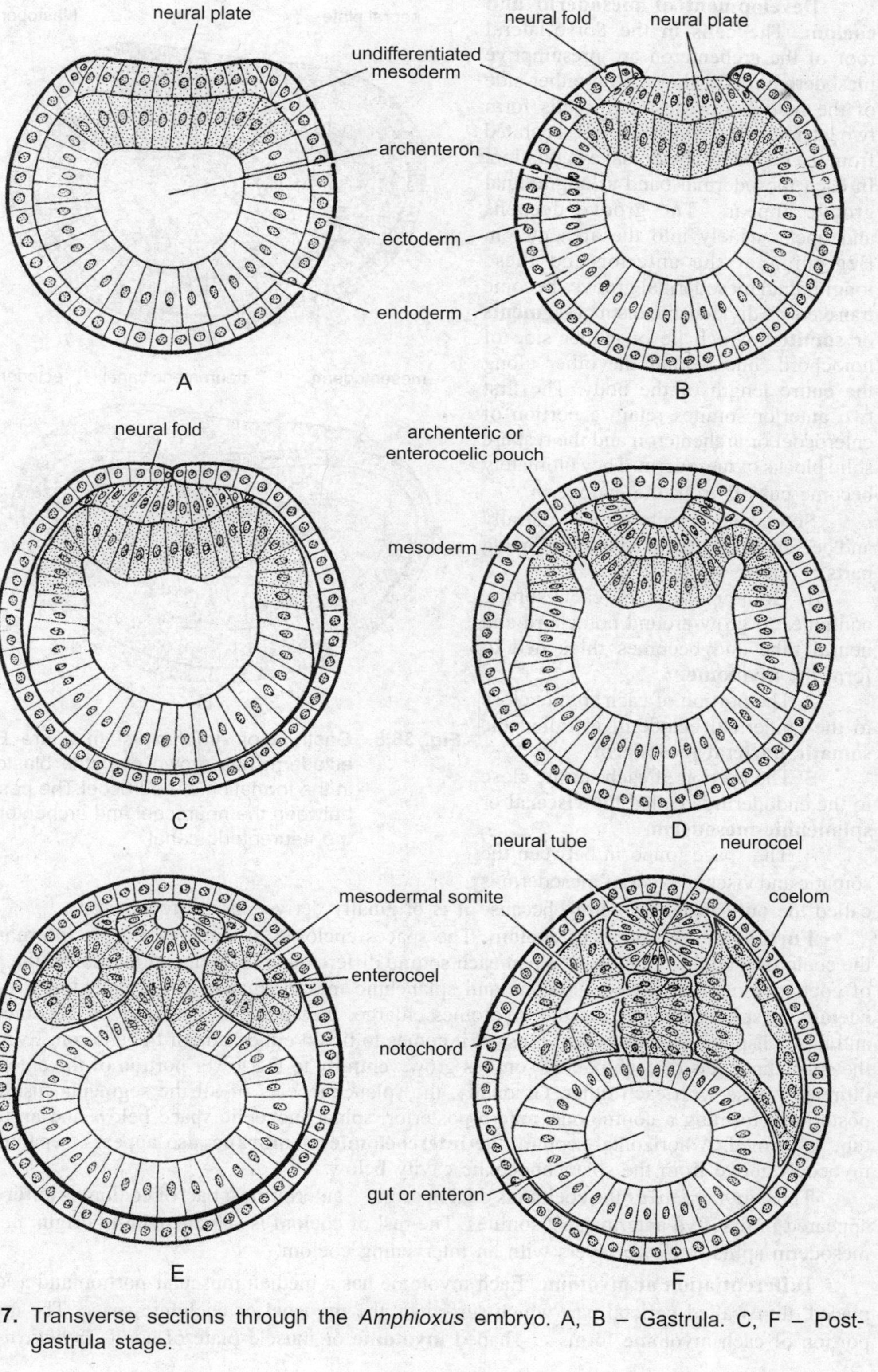

Fig. 36.7. Transverse sections through the *Amphioxus* embryo. A, B – Gastrula. C, F – Post-gastrula stage.

Development of mesoderm and coelom. The cells in the dorso-lateral roof of the archenteron are presumptive mesodermal cells, they lie on either side of the notochord. Mesoderm cells form two lateral bands, which also get separated from the endoderm by dorsal evagination. In each mesodermal band a longitudinal groove appears. The groove deepens and opens widely into the archenteron. Beginning at the anterior end these longitudinal mesodermal grooves become transversely divided into distinct **segments** or **somites**, which lie on either side of notochord, one behind the other along the entire length of the body. The first two anterior somites retain a portion of enterocoel or archenteron and the rest are solid blocks of mesoderm. They ultimately become cut off from the endoderm.

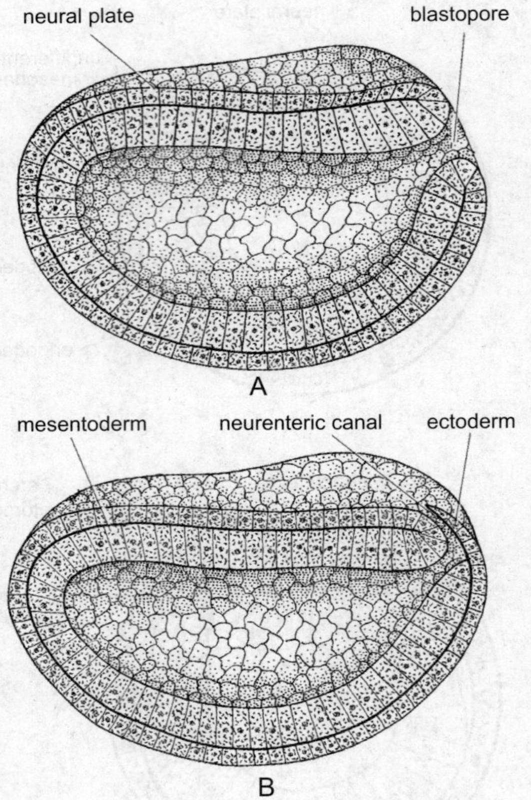

Fig. 36.8. Gastrula of *Amphioxus*. In figure B the ectoderm has grown over the blastopore in the formation of neurocoel.The passage between the neurocoel and archenteron is the neurenteric canal.

Soon, each somite grows ventrally and becomes differentiated into following parts :

1. Some portion of each somite continues to grow around notochord and neural tube and becomes thickened to form the **myotome**.
2. The portion of each somite close to the epidermal ectoderm is called the **somatic parietal mesoderm**.
3. The portion of each somite close to the endoderm is called the **visceral** or **splanchnic mesoderm**.
4. The space found in between the somatic and visceral layer of mesoderm is called the **enterocoelic coelom**, because it is originally derived from archenteron.

Further development of coelom. The spaces enclosed by somites are the forerunners of the coelom of adult *Amphioxus*. When each somite differentiates into dorsal myotome, the portion of coelom found in between somatic and splanchnic mesodermal layers is called ventro-lateral **coelom** or **splanchnocoel**. As the myotomes enlarge, the splanchnocoelic space becomes more and more displaced ventrally and most of it comes to lie on either side of the enteron. Eventually, the splanchnocoels of each pair of somites grow ventrally to the lower portion of the enteron and ultimately fuse with each other. Gradually, the splanchnocoels of all the segments fuse antero-posteriorly forming a continuous, antero-posterior, splanchnocoelic space below and around gut tube is formed. A horizontal septum, the **intercoelomic membrane** also appears, separating the myocoels above from the splanchnocoelic cavity below.

Thus, in *Amphioxus* the coelom is dual in origin – anteriormost part of coelom is **enterocoelic** appeared in first two anteriormost somites. The rest of coelom is **schizocoelic** in origin, *i.e.,* solid mesoderm splits into two layers with an intervening coelom.

Differentiation of myotome. Each myotome has a median muscular portion and a laterally placed, thin-walled parietal part which surrounds the **myocoel** or coelomic space. The muscular portion of each myotome forms > -shaped **myotome** or muscle plate of adult *Amphioxus*.

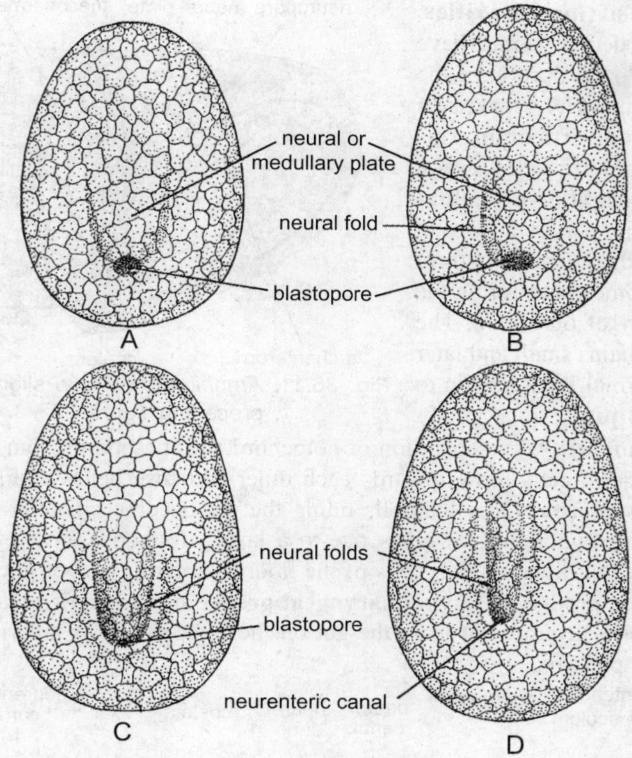

Fig. 36.9. Gastrula of *Amphioxus* showing progressive overgrowth of the ectoderm over the blastopore in the formation of the neural folds.

The myocoelic portion in each segment gives rise : (*i*) lower sclerotomic diverticula which extend between myotome and notochord, and nerve cord. Its inner wall wraps around the notochord and nerve cord and forms a **skeletogenous sheath** of connective tissues which ensheath these structures. Whereas its outer wall covers inner side of myotome with a **connective tissue covering**. (*ii*) ventral diverticulum extends between lateral wall of the splanchnocoel and epidermal layer of the body wall and separates the parietal mesoderm of the splanchnocoel from the epidermal wall. This ventral diverticula together with parietal wall of the myocoel above, forms the dermatome. Both the inner and outer layers of ventral diverticulum fuse to form the dermis of the skin. (*iii*) the dorsal slerotomic diverticala forms the **fin-ray cavities** in dorsal fin.

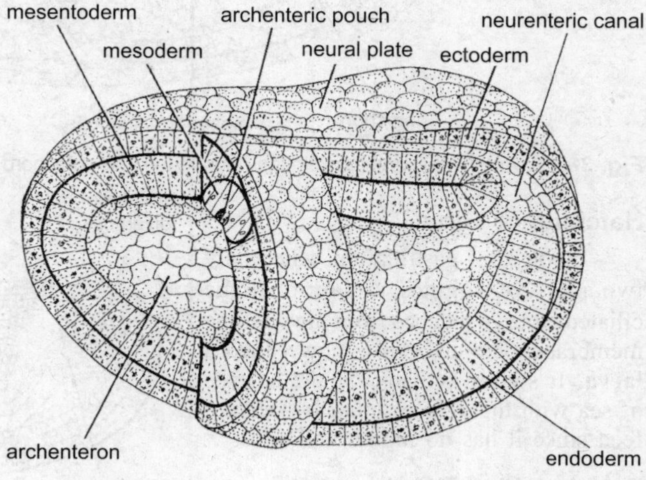

Fig. 36.10. *Amphioxus* embryo with an archenteric pouch in formation.

Development of head cavities. After about eight mesodermal somites have been separated, the dorsal portion of the anterior end of the gut cavity gives rise a **dorsal** or **anterior gut diverticulum**. This diverticulum splits to form a left and a right diverticula. The **right diverticulum** extends into the anterior part of embryo in between ectoderm and notochord, in front of first pair of myotomes, and forms the **head cavity** or cavity of oral hood. The left **diverticulum** remains small and later fuses with an ectodermal invagination to form the **Hatschek's pit**.

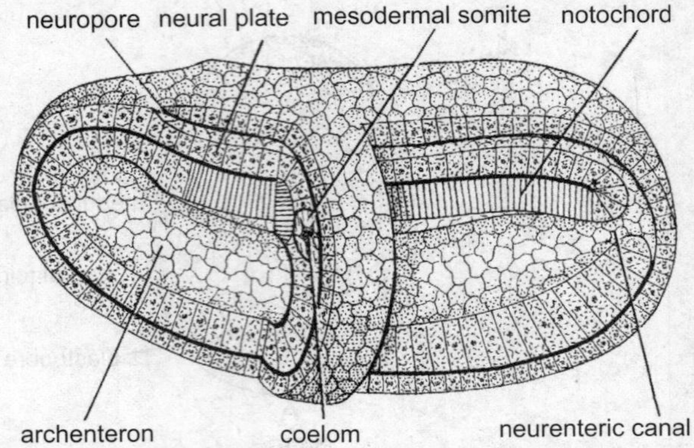

Fig. 36.11. *Amphioxus* embryo slightly older than in the preceding figure.

Development of gut. After separation of notochord and mesoderm from the endoderm, the edges of these endoderm cells grow towards each other and fuse below the notochord to form a tubular **gut** or **mesenteron**. Later the cells lining the gut become ciliated, and the splanchnic layer of mesoderm surrounding them gives rise to a layer of smooth muscles of the gut. After separation of anterior gut diverticulum, cells of the floor of anterior part of gut become columnar and form an elongated groove, the **hypopharyngeal groove** or **endostyle**. Much later a **midgut diverticulum** or **liver** will grow out from the gut on the right side. The anterior part of gut also enlarges to form the pharynx.

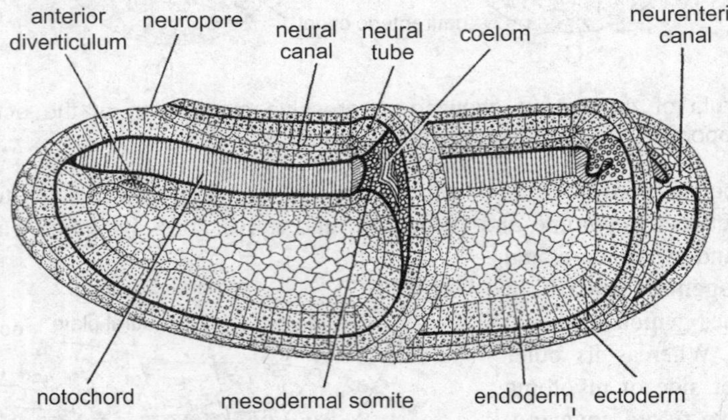

Fig. 36.12. *Amphioxus* embryo with the neural tube, the notochord and the mesodermal somites differentiated.

Hatching of Larva

Soon after gastrulation, when only two pairs of somites are formed, the ciliated embryo hatches from the vitelline membrane. The embryo is now called a **larva**. It swims actively on the surface of sea with the help of cilia. It does not feed since it has no mouth and anus.

Post-hatching Development

The ciliated larva swims actively and elongates gradually, becomes laterally

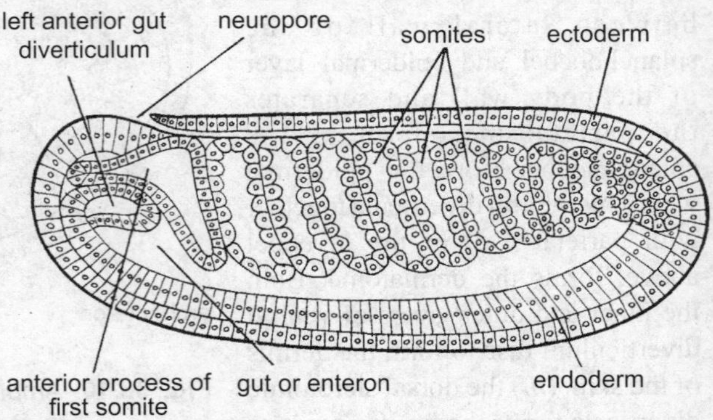

Fig. 36.13. Embryo of *Amphioxus* with the anterior gut diverticulum.

compressed, the number of somites increases with elongation. Later on cilia disappear. Each larva undergoes following changes to become an adult.

1. Mouth. It develops when the larva acquires about 16 to 18 pairs of mesodermal somites. At about the level of second mesodermal somite, on the left side of the ventral surface of head, the ectoderm fuses with the endoderm and then there a perforation occurs forming the mouth opening. The mouth gets bordered with cilia.

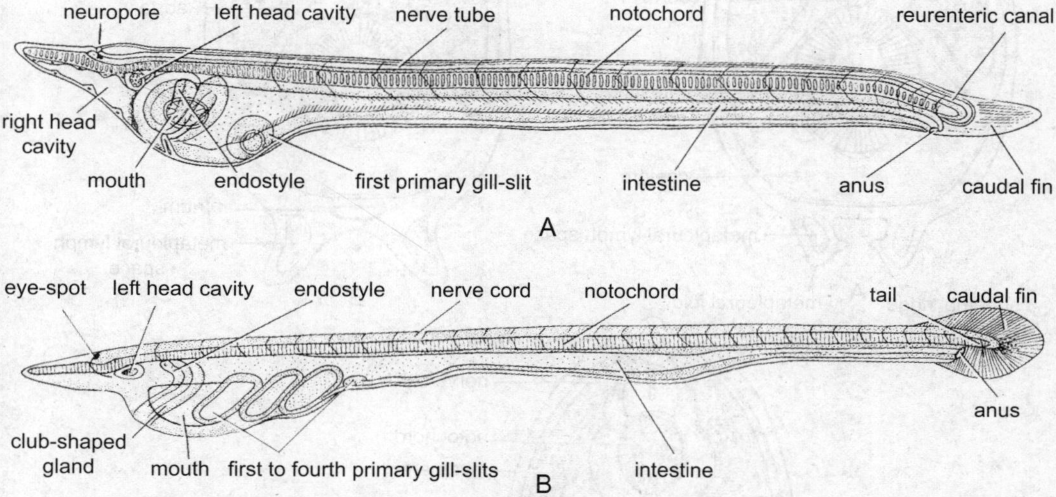

Fig. 36.14. Young larva of *Amphioxus*. A–Larval stage after opening of mouth and first gill-slit; B–Larval stage in which four primary gill-slits are present.

2. Anus. At the caudal end of the larva on the ventral side, a small area of endoderm fuses with the ectoderm and then a perforation occurs at this site to form the **anus**. Actually it lies slightly on the left side of the mid-ventral line. The larva begins to feed on plankton. Dorsal to the anus a tail grows out posteriorly.

3. Gill-slits. At the time of mouth formation, the endoderm opposite to the first pair of somites pushes ventrally and fuses with the ectoderm. This site perforates to form the **first gill-slit**. This gill-slit later on moves up on the right side of the body. Later fourteen more gill-slits also appear in a series and eventually shift to the right side of the larval body. Out of these fourteen gill-slits, the first and the last five gill-slits degenerate or close up. These remaining eight gill-slits are called **primary gill-slits**. Soon another set of eight gill-slits called **secondary gill-slits** appear at the dorsal side of the primary gill-slits. Thus, two rows of gill-slits are formed on the right side of the body. Eventually, first set of primary gill-slits is shifted to the left side of the pharynx. These gill-slits of both sides, later on, are subdivided by tongue bars and new gill-slits are added at the posterior end of both sets of gill-slits.

The first opening or cleft arises as a diverticulum from the floor, it lies on the right of the pharynx and is formed before the mouth, it becomes a tube opening at one end into the pharynx and at the other on the ventral side of the body, eventually it forms a coiled, glandular **club-shaped gland** which is a larval organ and disappears during metamorphosis.

4. Club-shaped gland. From the anterior end of the gut floor grows out a coiled tube, which acquires an opening at its one end with the gut on the right side and at the other end with the exterior, slightly to the left side, just ventral to mouth. This is called **club-shaped gland**. Its cells are glandular. It begins to disappear when first set of eight gill-slits are formed.

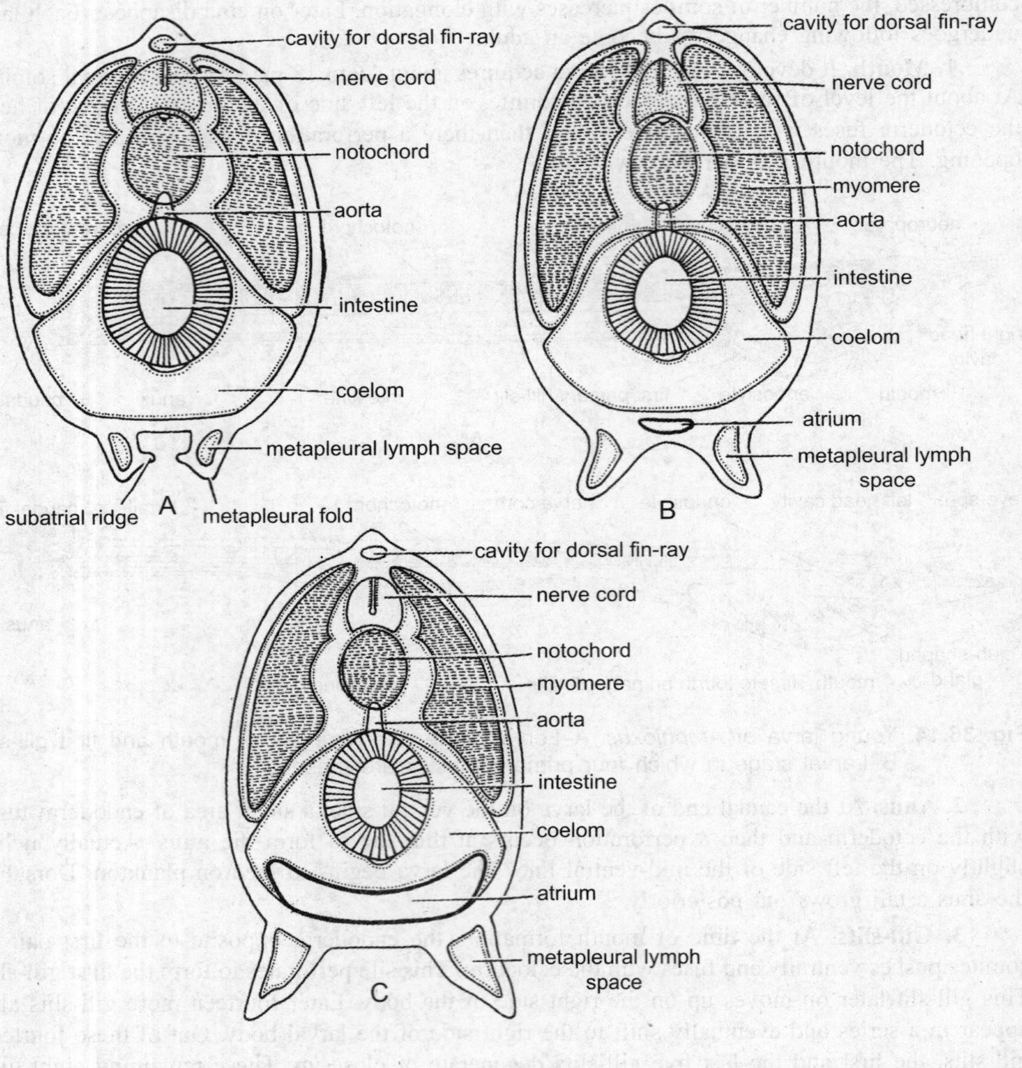

Fig. 36.15. *Amphioxus*. A, B and C stages in the development of atrium.

5. Endostyle. A V-shaped thickening arises from the wall of the gut, this is the rudiment of the endostyle. It is made of ciliated and glandular cells and is formed on the right side at the anterior end of the pharynx in front of club-shaped gland. The two limbs of the V fuse and grow back along the mid-ventral line to become the endostyle.

6. Myotomes on either side of body increase in number.

7. Nephridia. They also develop during larval life and are ectodermal in origin. Hatschek's nephridium develops as a small tube behind the pre-oral pit.

8. Atrium. Shortly before the secondary gill-clefts of the right side are formed, the body wall forms two longitudinal **metapleural folds** on the ventral side. The atrial folds start appearing from posterior end where they are quite close to each other. Anteriorly they grow and enclose the gill-slits in between them. From the inner side of each metapleural fold arises a shelf called the **epipleur**. The epipleurs grow towards each other and fuse together to enclose a space, the **atrium**. The atrium is lined with ectoderm. It becomes closed anteriorly but remains open posteriorly by an atriopore. The atrium gradually enlarges displacing the body wall and reducing the splanchnocoel around the anterior part of the gut.

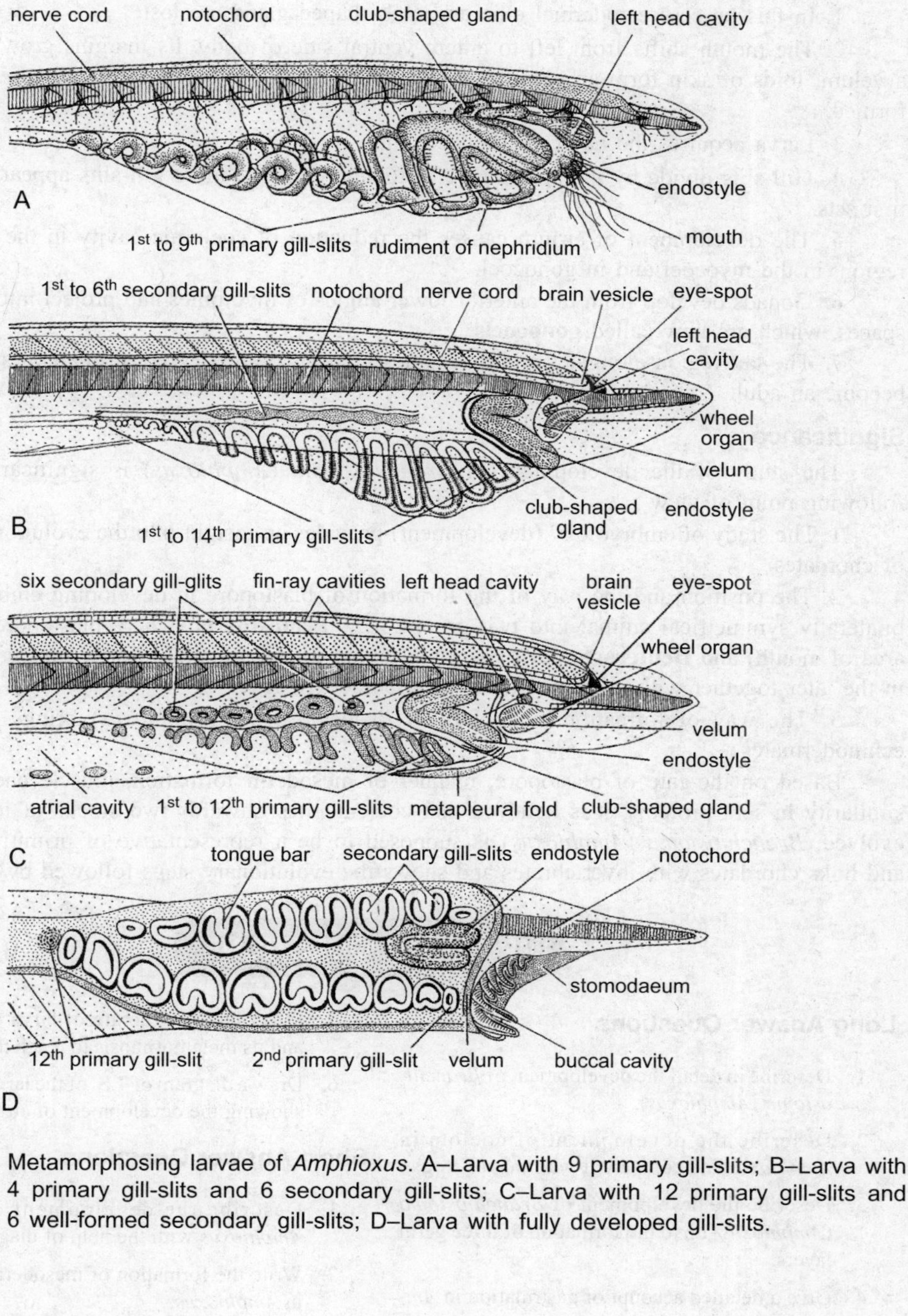

Fig. 36.16. Metamorphosing larvae of *Amphioxus*. A–Larva with 9 primary gill-slits; B–Larva with 4 primary gill-slits and 6 secondary gill-slits; C–Larva with 12 primary gill-slits and 6 well-formed secondary gill-slits; D–Larva with fully developed gill-slits.

METAMORPHOSIS

For some time the larva continues to feed on plankton of the surface of the sea for about three months, and sinks to the floor of the sea where it grows and undergoes a slow metamorphosis to become an adult in several months. Some very significant metamorphic changes are the following :

1. In this process ectodermal cilia and club-shaped gland are lost.
2. The mouth shifts from left to antero-ventral side of body. Its margins grow in to form a velum, folds of skin form an **oral hood** and oral cirri inside which a ciliated **wheel organ** is formed.
3. Larva acquires the habit of filter-feeding.
4. Gill-slits divide by the appearance of tongue bars and more gill-slits appear behind the first sets.
5. The development of atrium causes the reduction of coelomic cavity in the pharyngeal region, in the myocoel and in gonocoel.
6. Gonads develop from the anterior lower angles of myotomes and project into myocoelic spaces, which are now called **gonocoels**.
7. The tail region elongates and it assumes a considerable degree of bilateral symmetry to become an adult.

Significance

The study of the development of *Branchiostoma* (*Amphioxus*) is significant from the following point of view :

1. The study of embryology (development) provides an insight into the evolutionary history of chordates.
2. The position and the way of the formation of blastopore in developing embryos divide bilaterally symmetrical animal into two groups, *viz.,* **Protostomia** (where blastopore marks the area of mouth) and **Deuterostomia** (where blastopore marks the anus). The *Branchiostoma* falls in the later together with other chordates and echinodermates.
3. The way of formation of coelom in *Branchiostoma* (*Amphioxus*) brings it closer to echinodermates.

Based on the fate of blastopore, manner of mesoderm formation, muscle chemistry and similarity in sera proteins, it is believed that coelenterates onwards two stocks of invertebrates evolved. *Branchiostoma* (*Amphioxus*) is supposed to be a representative of primitive chordate and links chordates with invertebrates and shows the evolutionary steps followed by vertebrates.

IMPORTANT QUESTIONS

Long Answer Questions

1. Describe in detail the development of *Branchiostoma* (*Amphioxus*).
2. Describe the development of coelom in *Amphioxus* and compare it with that of frog.
3. Describe the development of *Branchiostoma* (*Amphioxus*) up to the formation of three germ layers.
4. Give a detailed account of gastrulation in *Amphioxus*.
5. Describe the structure of larva of *Amphioxus* and its metamorphosis to reach the adult stage.
6. Draw a diagram of T.S. of the larval *Amphioxus* showing the development of atrium.

Short Answer Questions

1. Describe the development of atrium in *Amphioxus* with the help of diagrams.
2. Write the formation of mesoderm and coelom in *Amphioxus*.

3. Write short notes on : (*i*) Cleavage in *Amphioxus*; (*ii*) Gastrula in *Amphioxus*; (*iii*) Organogenesis in *Amphioxus*; (*iv*) Formation of mesoderm and coelom in *Amphioxus*.

Very short Answer Questions

1. What is involution and which cells involute during gastrulation in *Amphioxus*?
2. What is notogenesis and neurogenesis?
3. Write about neurentic canal.
4. What is the type of coelom formation in *Amphioxus*?
5. What type of development is in *Amphioxus*?

Multiple Choice Questions

1. Development of *Amphioxus* resembles with those of
   (*a*) Coelenterates (*b*) Annelids
   (*c*) Echinodermates (*d*) All these.
2. Number of chromosomes in the fertilised eggs of *Amphioxus* are
   (*a*) 12 (*b*) 20
   (*c*) 21 (*d*) 24.
3. Cleavage in *Amphioxus* is
   (*a*) Holoblastic (*b*) Meroblastic
   (*c*) Discoidal (*d*) Superficial.
4. Yolky cytoplasm of *Amphioxus* develops into
   (*a*) Ectoderm (*b*) Endoderm
   (*c*) Mesoderm (*d*) (*a*) and (*b*).
5. Gastrulation in *Amphioxus* is formed by
   (*a*) Invagination (*b*) Epiboly
   (*c*) Convergence (*d*) Divergence.
6. Emboly occurs in *Amphioxus* development during
   (*a*) Cleavage (*b*) Blastulation
   (*c*) Gastrulation (*d*) Organogenesis.
7. Which amongst the following is olfactory pit in *Amphioxus*?
   (*a*) Hatschek's pit (*b*) Kolliker's pit
   (*c*) Hasel's pit (*d*) Larval pit.
8. At the time of hatching of *Amphioxus* larva, the number of formed mesodermal somites are
   (*a*) One pair (*b*) Two pairs
   (*c*) Three pairs (*d*) Four pairs.

ANSWERS

1. (*c*) 2. (*d*) 3. (*a*) 4. (*b*) 5. (*a*) 6. (*c*) 7. (*b*) 8. (*b*)

37

Chapter

Development of Frog

Contents of this Chapter

Frogs lay their eggs in water in early spring. In pseudocopulation or mating, the male frog firmly clasps the body of the female frog by his forelegs and enlarged thumb pads (**nuptial pads**). These nuptial pads help in clasping the body of female. This sexual embrace is called **amplexus**. As the eggs are extruded through the cloaca of female (**oviposition**), the male deposits sperm cells over them (**insemination**). Thus, **fertilisation** is **external**, taking place in water.

Spawning

The mesolecithal eggs of frog enclosed in a protective gelatinous albumen are laid in water. The cluster or masses of eggs which remain stick together is called **spawn**. A spawn of *Rana tigrina* may have 3000 to 4000 ova. The spawn is laid during pseudocopulation or **amplexus**.

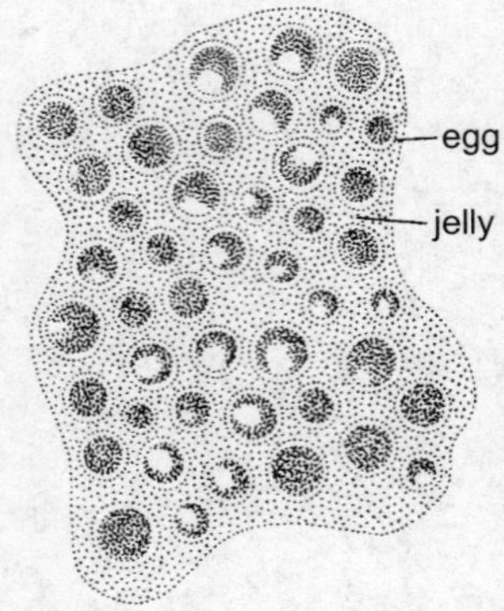

Fig. 37.1. Frog. Spawn of frog.

1. Fertilisation

In frog, fertilisation is external and occurs at once in water outside the body of the oviparous female.

The entire sperm penetrates the ovum anywhere around the animal hemisphere. If fertilisation is delayed, the albumen layers around the ovum become too thick for the sperm to pass through them and the ovum also starts to show degeneration. In the fertilisation process, vesicular sperm nucleus and vesicular female nucleus (or pronuclei) fuse together to form **zygote nucleus**. The fusion of both male and female pronuclei is called **amphimixis**. The fertilised egg or zygote is about 1.6 mm in diameter, it rotates within the vitelline membrane so that the animal pole becomes dorsal. The upper half of the zygote or **animal hemisphere** is pigmented black and it contains the cytoplasm and a nucleus, the lower **vegetal hemisphere** is white and full of yolk. On one side between the black and white areas is a **gray crescent** region which marks the future dorsal side. At this region cortex becomes thin and this area is crescent-shaped. The plane passing through the centre of grey crescent and the animal pole defines the median plane of bilateral symmetry. It coincides with the embryonic axis and is the only plane which separates the egg into two eqivalent parts, each containing half the crescent material.

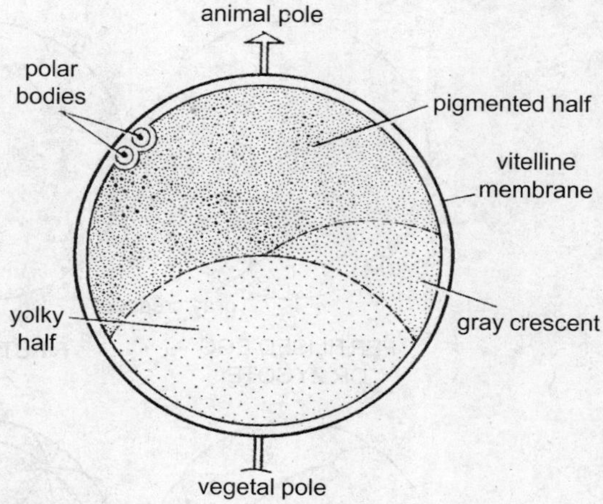

Fig. 37.2. A fertilised egg of frog (after Huettner, 1949).

2. Cleavage and Blastulation

Cleavage or segmentation is **holoblastic** and **unequal**. A vertical furrow from the animal to the vegetal pole divides the zygote completely into two equal-sized cells. A second vertical furrow at right angles to the first divides the zygote into four cells. The third cleavage is horizontal and above the equator which segments the zygote into upper four smaller, black-coloured cells, and lower four larger, white-coloured cells. The cells formed by cleavage are **blastomeres**, the upper black blastomeres are called **micromeres**, and lower white ones are **macromeres**. Further cleavages divide the micromeres more rapidly than the lower macromeres whose division is hindered by yolk. The blastomeres' mutual pressure flattens their surfaces in contact with each other but free surfaces of each blastomere remain spherical. At this stage the whole embryo acquires a characteristic appearance reminiscent of a mulbery and so it is called **morula**.

About fourth and fifth cleavage stages a small space, the **blastocoel** appears between the blastomeres of morula. In the beginning it is like a narrow crevices between blastomeres of morula, which gradually increases as the cleavage goes on. The blastocoel shifts more and more towards the animal pole due to more rapid multiplication of the micromeres, and infiltrated by water and albuminous fluid secreted by the surrounding cells. The blastocoel also enlarges due to uptake of more water. As cleavage proceeds, the blastomeres arrange themselves into a true epithelium called **blastoderm**. It is two-cell thick towards animal pole of the egg and forms the roof of blastocoel, while the sides and floor of the blastocoel is occupied by multilayered blastoderm of large yolky blastomeres. Thus, the resulting embryo having fluid-filled blastocoel is called blastula.

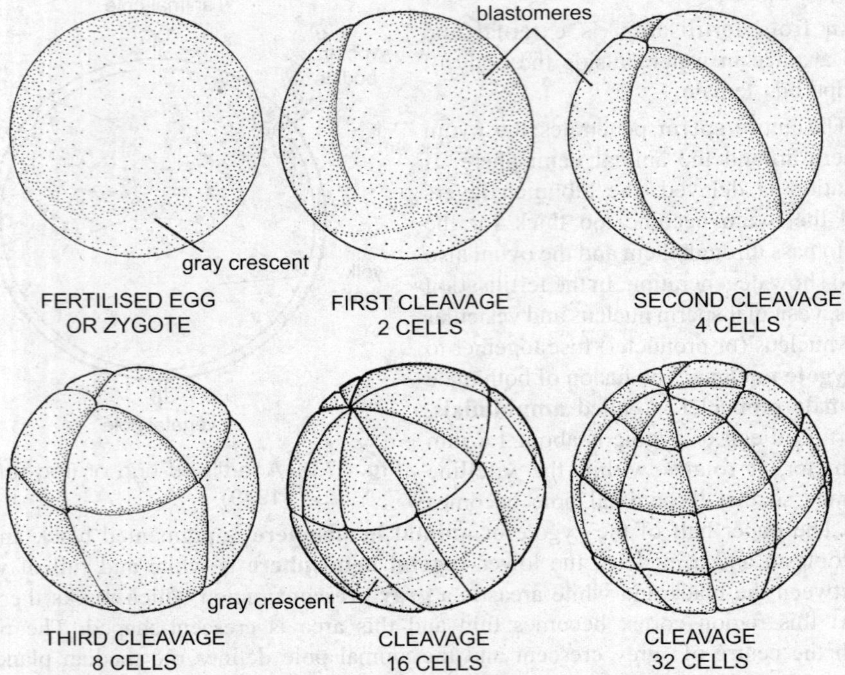

Fig. 37.3. Frog. Cleavage in fertilised egg or zygote.

3. Presumptive Areas

In the blastula, the blastomeres which have to form different germinal layers and different organs of the adult frog can be observed by artificial-vital staining methods of **Vogt** (1925) and prospective organ region maps or fate maps have been prepared. According to the fate map studies, the whole surface of blastula can be divided into the following three areas :

***(i)* Prospective ectoderm area** is present on and around the animal pole and it is pigmented black. The **neural ectoderm** occurs largely on the future dorsal side of blastula, while the **epidermal ectoderm** occupies the antero-ventral side of the blastula. Inside the neural ectoderm occurs a small sub-area that develops into the eye of the embryo. The sub-area of nose, sucker, ears and mouth are present inside the epidermal ectoderm.

***(ii)* Prospective notochord and mesoderm area** is present behind the pigmented animal hemisphere. It is crescentic gray area, the **marginal zone** along the equator of blastula. It has blastomeres for the formation of notochord and mesoderm of the embryo. The large area of dorsal side of the gray crescent is occupied by notochordal cells. Beneath the notochordal area, toward the vegetal pole lies a narrow strip of cells which form the pre-chordal plate of the embryo. On either sides of notochordal area, the part of grey crescent forms the segmental muscles (somites) and tail mesoderm is a narrow strip of cells on the dorsal side, toward animal hemisphere. Lateral and ventral parts of grey crescent give rise to ventro-lateral mesoderm.

***(iii)* Prospective endodermal area** is the entire non-pigmented area of the vegetal hemisphere, which give rise to endodermal lining of the mouth, gill region, pharynx, midgut and hindgut, and other organs such as liver, pancreas, urinary bladder and certain endocrine glands.

4. Gastrulation

Gastrulation is a process of migration and re-arrangement of prospective organ forming cells already present in the blastula. It is brought about by several types of morphogenetic movements taking place at the same time.

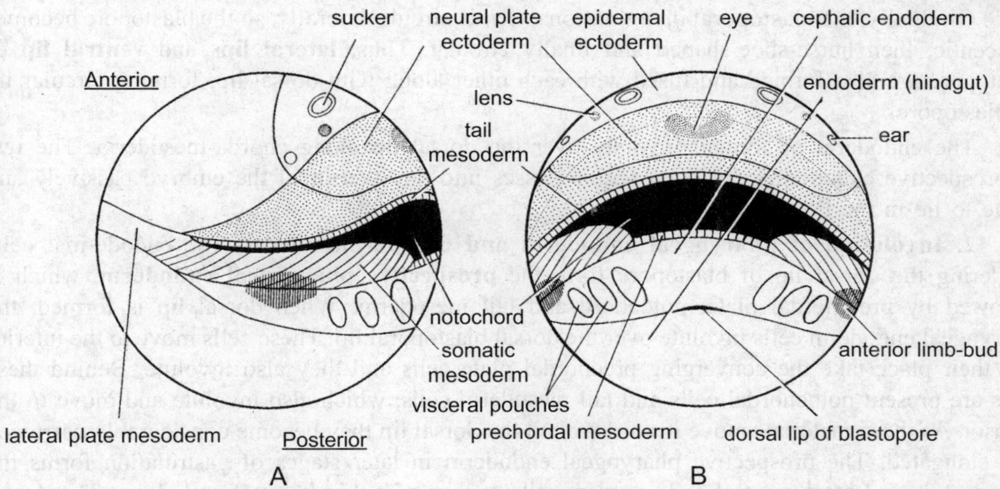

Fig. 37.4. Frog. Fate maps of blastula. A–Ventral view from the lateral side; B–Viewed from the dorsal side.

1. Invagination of endoderm. Certain prospective endodermal cells just beneath the mid-dorsal point of gray crescent of blastula assume the elongate shape of a bottle and move toward the interior of the blastula. Their steadily elongating necks remain attached to the surface of the blastula with the outermost cementing layer. Thus, as the bulky-cell bodies move inward, a pull exerted along their attenuated necks and creates an indentation at the surface. With continued multiplication and attenuation of bottle cells, the invagination deepens, and expands internally to form the **archenteron** or **gastrocoel** and its outer opening (original indentation) is called the **blastopore** lying at the future posterior end. The area immediately above the blastopore is the **dorsal lip** of blastopore.

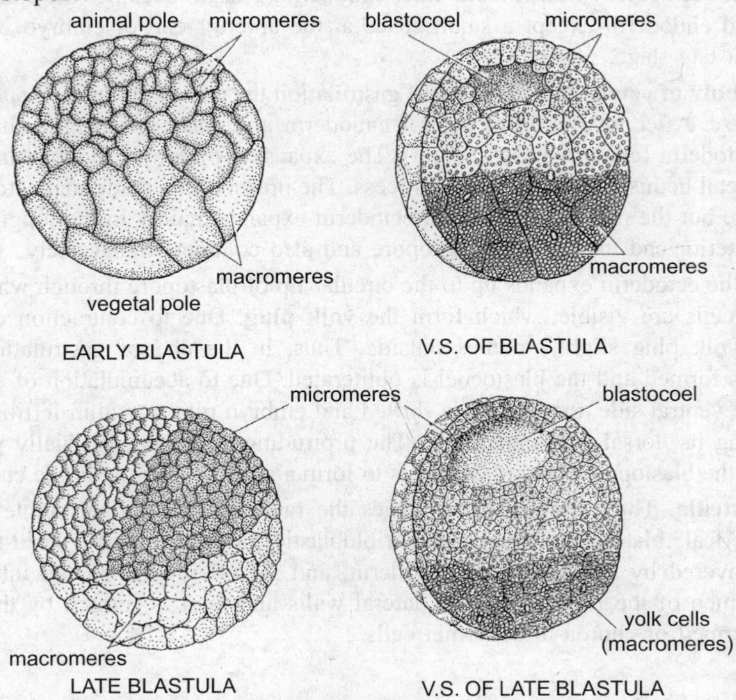

Fig. 37.5. Frog. Early and late blastula.

Gradually, the blastoporal invagination extends circulo-laterally, so the blastopore becomes crescentic, then horse-shoe-shaped and finally circular. Thus, **lateral lips** and **ventral lip** of blastopore are also formed and fused with each other along with dorsal lip, forming circular lip of blastopore.

The endoderm of foregut involute over the dorsal lip along chorda-mesoderm. The rest of prospective endoderm of vegetal region passes into the interior of the embryo passively and come to lie in the floor of gastrocoel.

2. Involution of pharyngeal endoderm and chorda-mesoderm. The endodermal cells bordering the dorsal lip of blastopore form the **prospective pharyngeal endoderm**, which is followed by pre-chordal plate, notochord and tail mesoderm. When dorsal lip is formed, the pharyngeal endoderm cells involute over the dorsal blastoporal lip. These cells move to the interior and their place take the converging prechordal plate cells and they also involute. Behind these cells are present notochordal cells and tail mesoderm cells, which also involute and move to the interior. As these materials move inward around the dorsal lip they become considerably narrowed and elongated. The prospective pharyngeal endoderm in later stages of gastrulation forms the **foregut** whose lateral, ventral and anterior walls consist of a thin layer of endoderm. The dorsal wall of foregut consists of prechordal plate and anterior tip of notochord. The notochord cells of the posterior region also involute and move anteriorly over the dorso-lateral lips of blastopore. Thus, the notochord forms the mid-dorsal wall of the archenteron, which is in the form of strip. The prechordal plate also forms the dorsal wall of the archenteron in front of notochord. The tail mesoderm remains near the blastopore, and marks the posterior end of the embryo.

The mesoderm (*i.e.*, trunk somites and ventro-lateral mesoderm) rolls over the lateral and ventral lips of blastopore and then invaginates. After rolling inside the entire mesoderm (*i.e.*, notochord, prechordal plate mesoderm, somites and ventro-lateral mesoderm) move from posterior side (blastoporal end) towards the anterior side as a single unit called **chorda-mesodermal mantle** in between outer ectoderm and inner endoderm. Thus, it occupies the entire space between ectoderm and endoderm except a small space at the anterior end of embryo where mouth will be formed in late stage.

3. Epiboly of ectoderm. Throughout gastrulation the embryo retains its spherical shape and a uniform size. After involution of gastral endoderm and entire mesoderm, this space is taken up by the ectoderm (epidermal and neural). The expansion of ectoderm from animal hemisphere over the vegetal hemisphere is an active process. The presumptive epidermal ectoderm expands in all directions, but the presumptive neural ectoderm expands mainly in the longitudiual direction, *i.e.*, from anterior end towards the blastopore and also contracts transversely.

Thus, the ectoderm expands up to the circular lip of blastopore through which unpigmented endodermal cells are visible, which form the **yolk plug**. Due to contraction of circular lip of blastopore, yolk plug slightly comes outside. Thus, in the end of gastrulation a new cavity **gastrocoel** is formed and the blastocoel is obliterated. Due to accumulation of endodermal mass on the future ventral side, the gravity is shifted and embryo rotates within fertilisation membrane so as to bring its dorsal side uppermost. The protruding yolk plug gradually withdraws to the interior, and the blastopore steadily contracts to form a slit-like opening in the end of gastrulation.

4. Gastrula. Thus, gastrulation changes the radially symmetrical single layered blastula into a spherical, blaterally symmetrical, triploblastic gastula having a head-to-tail axis. It is externally covered by ectoderm and endoderm, and mesoderm lies in the interior. Gastrocoel forms the lumen of the forming gut. Its lateral walls and floor is formed by the endoderm and its roof is formed of chorda-mesodermal cells.

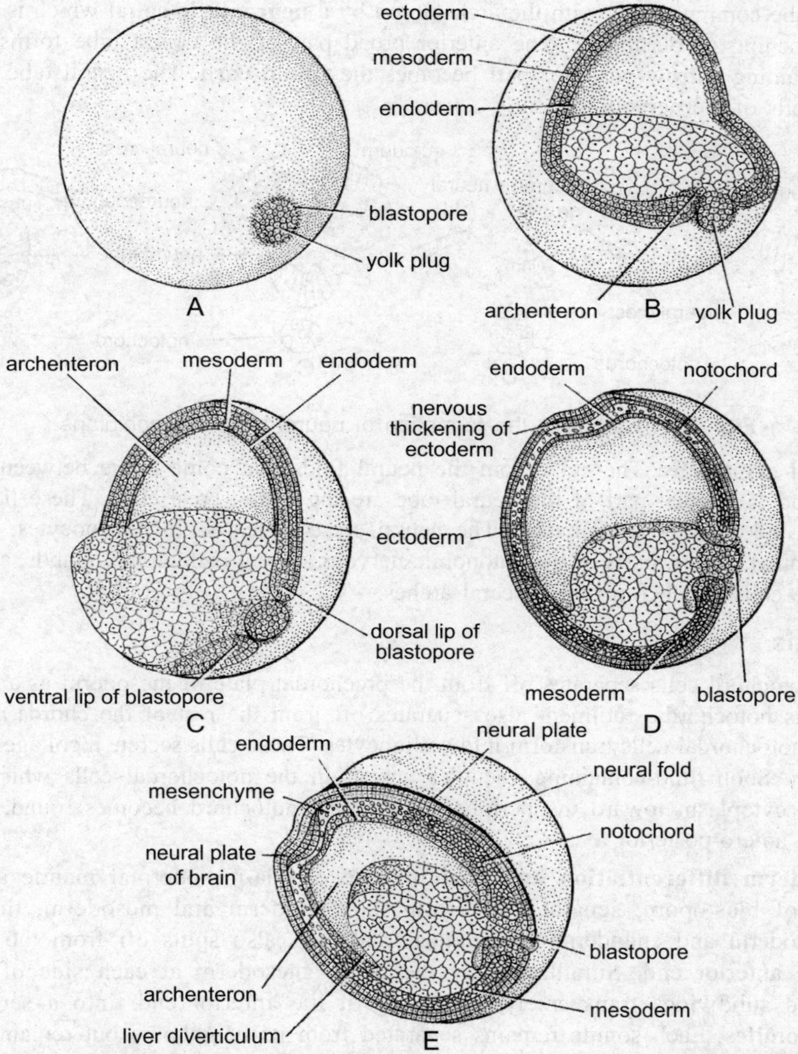

Fig. 37.6. Frog. A–External appearance of gastrula; B, C–Partial sections of early gastrula showing delaminations of endoderm and mesoderm: D, E–Sections of late gastrula showing the three germ layers.

Neurulation

By the time gastrulation is being completed, the ectoderm along the mid-dorsal side of the embryo thickens to form a pear-shaped **medullary** or **neural plate**. The neural plate cells change in shape and become elongated and arranged themselves into a columnar epithelium. The epidermal cells remain more or less flat and arranged as a stratified epithelium usually two cells thick. The edges of the neural plate become thickened and slightly raised above the general level as ridges called **neural folds**. The neural plate narrows transversely especially in its posterior parts and the neural folds raised higher due to which a **neural groove** is formed along its length. The neural folds grow and fuse with each other in the mid-dorsal line to form a neural tube. The lateral epidermal ectoderm of either side also meet and fuse at the mid-dorsal line above the neural tube, thus, enclosing it. The neural tube remains open in front for a time as a **neuropore**, posteriorly the neural folds cover and fuse over the blastopore so that the cavity of the neural tube communicates with the archenteron by a **neurenteric canal** which is the narrow canal-like opening of blastopore. The anterior broad part of the neural tube forms the **brain** and the remaining narrow posterior part becomes the **spinal cord**. The neural tube also forms **neuroglia** cells of the nervous system.

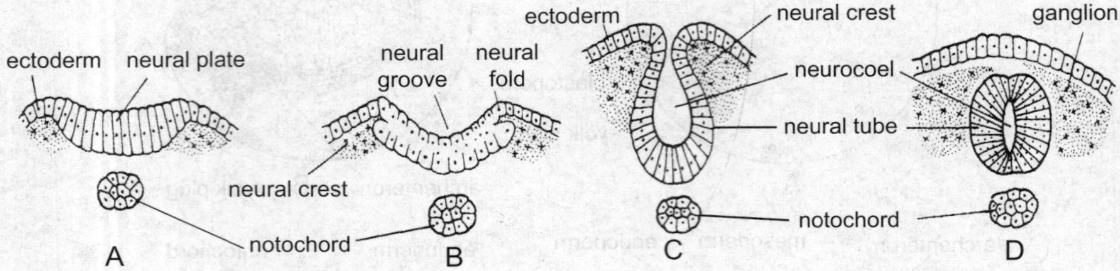

Fig. 37.7. Stages in the formation of neural tube in amphibians.

Neural crest cells. The cells from the neural folds that come to lie between the dorsal epidermis and the dorsal part of the neural tube are the neural crest cells, These lie along the dorso-lateral sides of the neural tube. The neural crests give rise to melanocytes, dorsal root ganglia of spinal nerves, parts of the autonomic nervous system and adrenal glands, and to some mesenchyme cells which form the visceral arches.

Notogenesis

The notochord cells separates off from the prechordal plate of mesoderm as a narrow rod of cells. This notochordal rudiment also separates off from the rest of the chorda-mesodermal mantle and notochordal cells transform into colligocytes. These cells secrete a collagenous sheath around them. Soon fluid-containing vacuoles appear in the notochordal cells which push the nucleus and cytoplasm toward the periphery. Thus, the notochord becomes round, turgid and elongated in antero-posterior axis.

Mesoderm differentiation and coelom. The chorda-mesodermal mantle at the time of closure of blastopore, separates off from the endoderm and mesoderm, thus, lies in between ectoderm and endoderm. The notochordal cells also splits off from the prechordal plate at the anterior end. Simultaneously the tip of mesoderm at each side of notochord thickens and subdivides transversely, beginning at the anterior end, into a series of cell masses or **somites**. Each somite remains separated from its neighbours but remains joined to the lateral and ventral parts of the mesodermal mantle on each side by strands of cells. These lateral and ventral mesodermal mantles are called **lateral plates** and these strands of cells in between somites and lateral plates are called **mesomeres** or **nephrotomes**, which later on forms the kidney tissue.

Each somite or **epimere** (dorsal part of mesoderm) on either side of notochord splits into three layers : (*i*) inner is the **sclerotome** or skeleton-forming tissue around the notochord, (*ii*) middle is the **myotome** whose cells differentiate to form the striated muscle fibres of the somatic muscles, and (*iii*) outermost narrow strip is the **dermatome** which forms the dermis of skin.

The **hypomere** or **lateral plate** of mesoderm of each side is divided by a split which passes downwards on each side to separate the hypomere into an outer **somatic** or **parietal layer**, and an inner **splanchnic** or **visceral layer** and the space between these two layers is a **splanchnocoel** or **perivisceral coelom**. The inner visceral layer gives rise to smooth muscles of the intestine and to the blood and blood vessels, and outer somatic layer with the ectoderm forms the **somatopleure**. Anteriorly the coelom is restricted to the ventral side only as a **pericardial cavity** below the pharynx which gets separated from the splanchnocoel by a transverse septum.

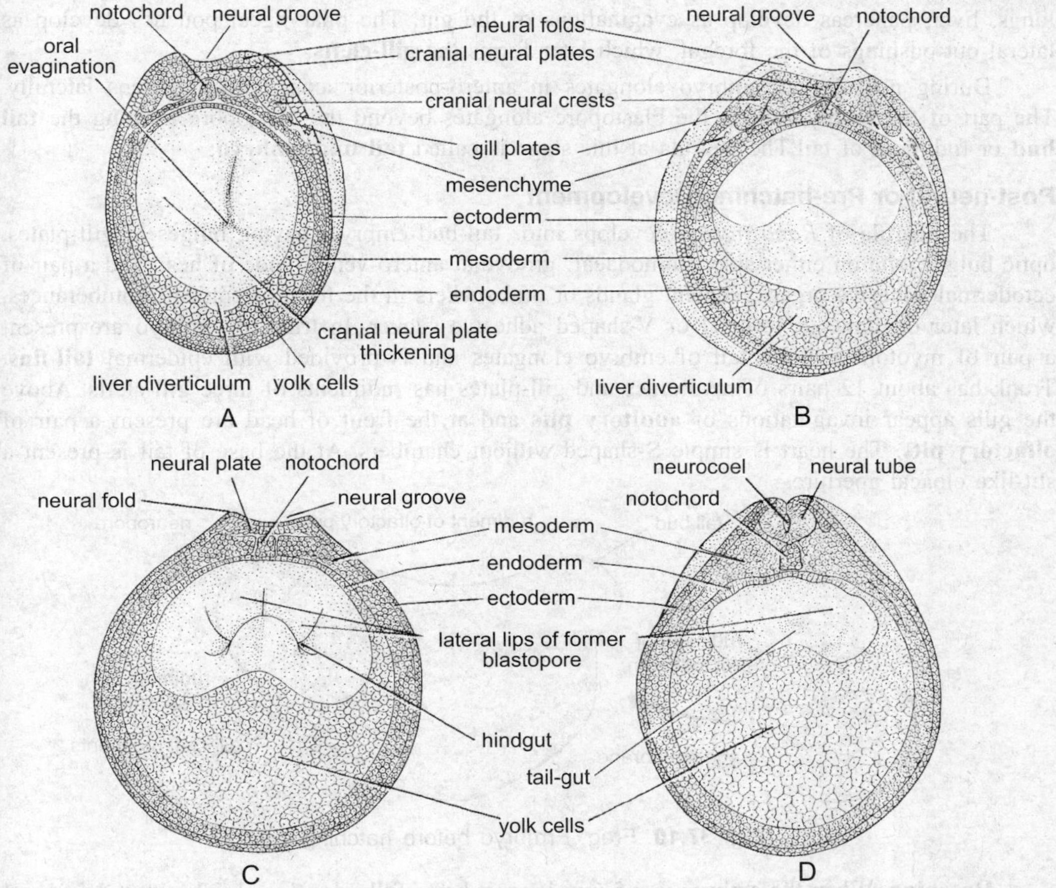

Fig. 37.8. Frog. A,B–Anterior sections of early neurula; C,D–Posterior sections of late neurula.

Primitive gut. The endoderm differentiates slowly and eventually forms the enteron or primitive gut. At the end of gastrulation it is an open trough. After the formation of notochord and separation of mesoderm, the free margins of endoderm fuse in the mid-dorsal line beneath the notochord to form the definitive **gut** or **enteron**. The floor of enteron persists as thick layer of large yolk-ladden cells. At the antero-ventral end the enteron makes contact with the ectoderm, where later on mouth invagination occurs which communicates with the enteron. Later, the

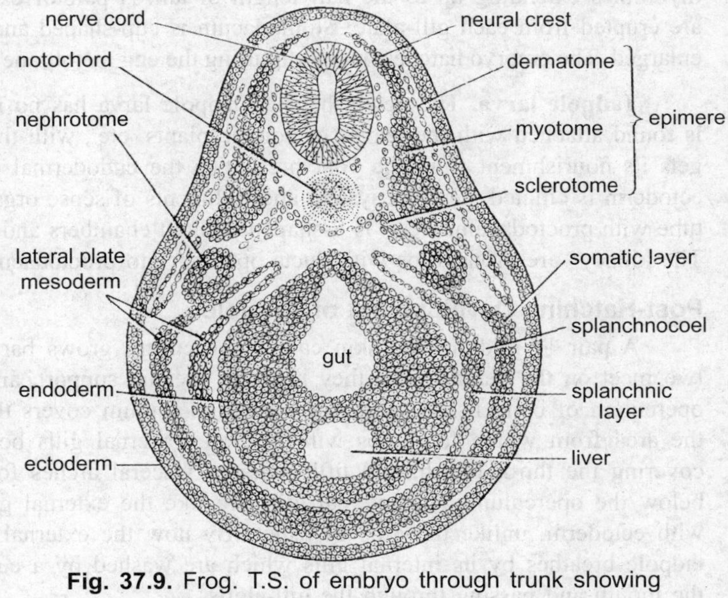

Fig. 37.9. Frog. T.S. of embryo through trunk showing mesoderm differentiation.

lungs, liver, pancreas develop as evaginations of the gut. The pharyngeal pouches develop as lateral out-pushings of the foregut, which later forms the **gill-clefts**.

During neurulation, embryo elongates in antero-posterior axis, it also flattens laterally. The part of the embryo above the blastopore elongates beyond the blastopore forming the **tail bud** or rudiment of tail.The neurula at this stage is called **tail bud embryo**.

Post-neural or Pre-hatching Development

The neurula of *Rana tigrina* develops into tail-bud embryo, having bulges of gill-plates, optic bulges, one on either side, stomodaeal groove at antero-ventral side of head and a pair of ectodermal adhesive organs, cement glands or oral suckers in the form of conical protuberances, which later on unite to form U or V-shaped adhesive sucker. In trunk of embryo are present a pair of myotomes. Soon tail of embryo elongates and is provided with epidermal **tail-fins.** Trunk has about 12 pairs of myotomes and gill-plates has rudiments of three gill-clefts. Above the gills appear invaginations of **auditory pits** and at the front of head are present a pair of **olfactory pits**. The heart is simple S-shaped without chambers. At the base of tail is present a slit-like cloacal aperture.

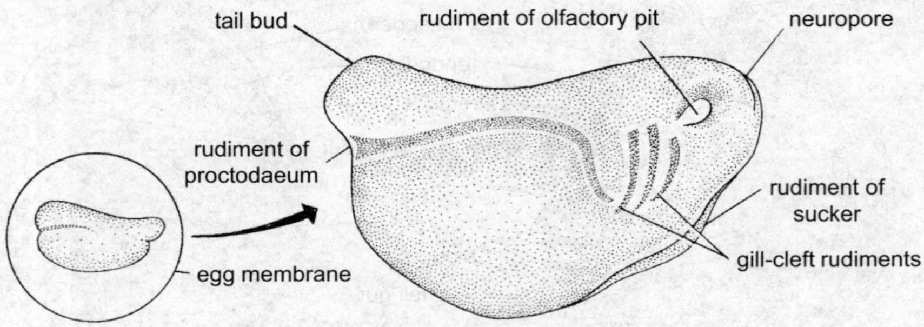

Fig. 37.10. Frog. Ermbryo before hatching.

Hatching. When the embryo is 4.5 mm long, it has a fully developed tail with tail-fins and myotomes extending up to the half length of tail. A pair of external gills in the form of buds are erupted from each gill-plate. Stomodaeum is cup-shaped and V-shaped oral sucker becomes enlarged. The embryo hatches out by rupturing the egg membrane and is now called **tadpole larva**.

Tadpole larva. The newly hatched tadpole larva has no mouth, and is unable to feed. It is found attached with the leaves of aquatic plants, etc., with the help of its adhesive sucker. It gets its nourishment from the yolk present in the endodermal cells of the floor of midgut. Its ectoderm is ciliated, nervous system and rudiments of sense organs are present. Gut is a straight tube with proctodaeum. Heart is S-shaped without chambers and blood vessels are being formed. The kidneys are pronephros with ducts opening into proctodaeum.

Post-Hatching Development of Tadpole

A pair of folds of the skin called **operculum** grows backwards and downwards till the two meet on the ventral side, they have no skeletal support, and are not homologous with the operculum of bony fishes. Posteriorly each operculum covers the gill-clefts, external gills, and the area from which forelimbs will develop. External gills begin to degenerate and the skin covering the third, fourth, and fifth pairs of visceral arches forms paired **internal gills** lying below the operculum. The internal gills are like the external gills, their filaments are covered with ectoderm, unlike the gills of fishes. By now the external gills have disappeared and the tadpole breathes by its internal gills which are washed by a current of water entering through the mouth and passing through the gill-clefts.

1. External gill-stage of tadpole. Tadpole at this stage is about 5.5 mm long and the following structures are visible in it :

1. Five pairs of branchial, pharyngeal or visceral pouches are formed in the gill-plate area due to outpushing of the endodermal lining of pharynx. The epidermal ectoderm also invaginate and fuses with the outpushing of pharynx and then a slit-like opening is formed which is called **gill-slit** or **gill-cleft**, by which pharynx communicates with the outside. In frog only four pairs of gill-slits are perforated.

2. Soon from the sides of head in the pharyngeal region three pairs of external gills are projected out, which are feathery extensions of the integument above the gill-slits. There are bathed by the surrounding water.

3. Stomodaeum develops in the form of pit, whose outer opening is the mouth. The mouth is surrounded by fringed lips and also acquires a pair of horny jaws. The fringed lips has two rows of tiny, needle-like horny teeth. A 10.5 mm long tadpole contains 3 rows of fringed teeth.

4. Adhesive suckers disappear.

5. The larval gut is differentiated into pharynx, oesophagus, stomach and intestine. Liver and gall bladder also develop. The intestine is very long and coiled like a watch spring due to herbivorous mode of feeding.

6. Myotomes extend up to the tip of tail.

7. Melanophores appear in the skin of dorso-lateral surface of head, trunk and tail.

8. The cornea becomes transparent and eye lens is visible.

9. Lateral line sensory system is visible on either lateral side of the tail.

10. A pair of giant neurons called **Mauthner cells** appear in the hindbrain. It is an interneuron that transmits impulses from the lateral line and auditory receptors to the motor output system of spinal cord.

11. A pair of pronephric kidneys becomes functional and excretes ammonia.

12. On either side of cloacal aperture, at the junction of trunk and tail, a pair of hindlimb buds appear.

The tadpole swims actively with the help of tail and feeds on algal and other aquatic vegetation.

2. Internal gill-stage of tadpole. 1. The opercular folds grow backward from the hyoid arch of each side covering the external gills and gill-slits and finally fuse with each other ventrally and with the belly wall. Thus, an operculum or gill-cover is formed enclosing the external gills and gill-slits and open outside by a ventro-lateral opening, the **spiracle** located on the left side of the body.

2. External gills later fall off and four pairs of filamentous internal gills develop on the walls of gill-slits.

3. Intestine is still coiled and long.

4. Different parts of hindlimbs such as thigh, shank, ankle, foot and five toes become well formed in the tadpole of 40 mm long.

5. A pair of forelimb buds appear behind the head but remain hidden within operculum. As development proceeds, the left forelimb emerges through the spiracle. The right forelimb appears later.

6. In a mature tadpole, a pair of lungs develop from the pharynx. Now the larva breathes by both, the internal gills and lungs.

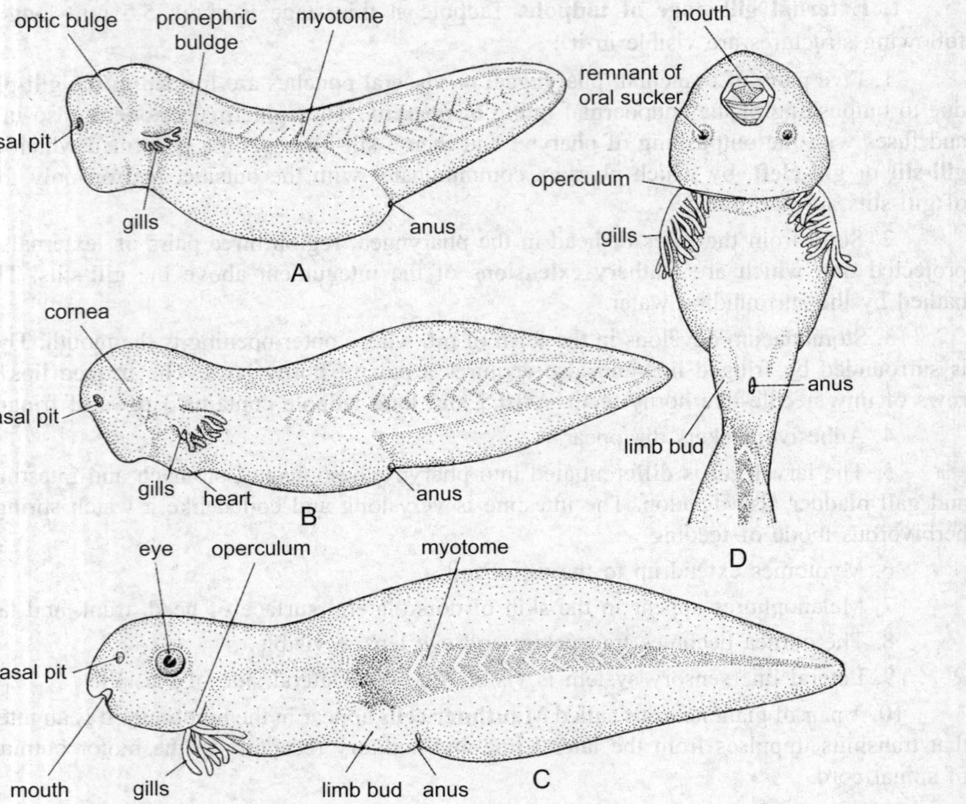

Fig. 37.11. Frog. Different stages of tadpole after hatching.

Thus, a **fully formed tadpole larva** is a fish-like creature. It has a well-developed locomotory tail with tail fin and muscles. It has a pair of eyes, nostrils, mouth, long spirally coiled intestine, cloacal opening and spiracle. Adhesive glands have lost. It respires with the help of internal gills but in later stage lungs develop, so it breathes by both internal gills and lungs. Soon gill-slits close, internal gills absorbed and opercular cavity disappeared and, thus, it breathes only with the help of lungs. It frequently comes to surface to gulp air. Lateral line system is well developed. Mesonephric kidneys are formed. Hindlimbs appear first and later the forelimbs, which are hidden within operculum in the beginning.

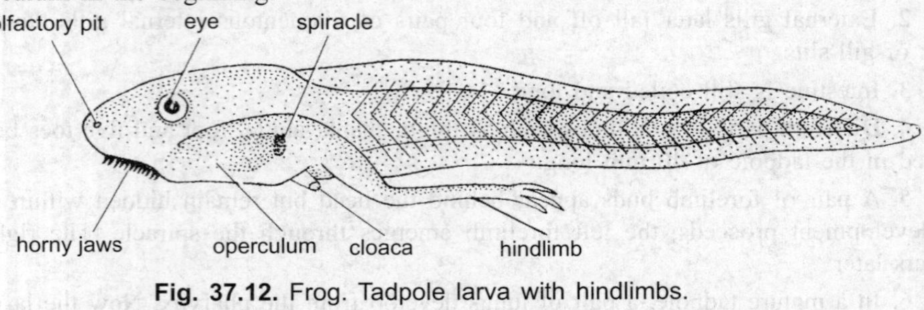

Fig. 37.12. Frog. Tadpole larva with hindlimbs.

Metamorphosis

Metamorphosis (Gk., *metamorphoun* = to transform) is the abrupt transition from larval to adult form. It includes morphological, anatomical, physiological and behavioural, hormonally regulated changes in the larval form to transform it into the adult form. In frog, metamorphosis is of progressive type. In frog, it is associated with a transition from an aquatic to a terrestrial mode of life and from a herbivorous to carivorous mode of feeding. The metamorphosis involves numerous structural, biochemical and physiological changes. These changes include the degeneration of exising structure, construction of new structures and modification of larval structures. Metamorphosis is controlled by hormones such as thyroxine of thyroid gland. The following changes occur during metamorphosis :

1. The first sign of metamorphosis is that tadpoles frequently come to the water surface to take air into their lungs through the mouth because internal gills are resorbed, gill-slits are closed, opercular fold also falls off and lungs develop for aerial respiration.

2. The tadpoles stop feeding, they are nourished on the substance of the tail, the tail becomes shorter and resorbed. The ciliated larval skin along with the epidermal horny jaws with horny teeth and labial fringes are cast off.

3. The vascular system of larva having aortic arches for internal gills are reduced and becomes modified for aerial respiration.

4. The larval haemoglobin in RBCs having higher affinity for oxygen and independent from pH is replaced by adult haemoglobin which has lower affinity for oxygen and highly sensitive to acid.

Fig. 37.13. Frog. Stages in metamorphosis.

5. The intestine shortens because of the change from a herbivorous to a carnivorous diet. The stomach and liver increase in size and start secretion of **ceruloplasmin**, a copper-binding serum protein.

6. Lateral line system and Mauthner cells of brain degenerate and disappear.

7. The limbs increase in size and differentiate.

8. The mouth widens, true jaws develop and tongue becomes long. Eyes become prominent on the dorsal surface of head with eyelids and nictitating membrane. Rhodopsin visual pigment appears.

9. Middle ear develops in connection with the pharyngeal pouch located between mandibular and hyoid arches. Tympanic membrane also develops.

10. Larval pronephros change into mesonephros of adult.

11. Erythropoesis occurs in spleen instead of liver.

12. Brain also becomes highly differentiated.

13. The synthesis of melanin and serotonin (a local vasoconstrictor) begins in the skin.

14. In stomach peptic activity starts for the digestion of animal tissue.

At cellular level, the cell modifications are evident in eye and eyelids, limbs, skin, operculum, tongue, liver, pancreas, intestine and lungs. Probably no cell or tissue or organ remain unaffected.

Endocrine function of pancreas also starts at metamorphosis to turn over the carbohydrates in liver. Thus, a young frog is formed, it still has a stump of the tail and it leaves water for some damp land, it feeds on insects and continues growing till it assumes the adult form and colour.

Hormonal control of metamorphosis. The metamorphosis is under control of hormones secreted by hypothalamus, hypophysis and thyroid. Thyrotropin-releasing factor (TRF) and prolactin-releasing-inhibiting factor (PIH) of hypothalamus, and thyroid-stimulating hormone (TSH) and prolactin hormone (PH) of hypophysis regulate the secretion of thyroid hormone (TH) such as throxine from thyroid gland. Thyroxine causes the degeneration and necrosis of some cells or tissues and organs in the larva, and stimulates the growth and diffentiation of organs needed in the adult frog. Iodine is an essential component of thyroxine hormone and it has been found that iodine alone can cause metamorphosis in frogs' larva.

Significance

The study of embryology of frog is practically useful to us in a variety of ways :

1. It helps in interpretation of avian and mammalian development.
2. It explains the evolutionary transition of lower chordates into higher chordates.
3. It explains the evolution of lung-breathing animals from gill-breathing animals.
4. It also explains the evolution of various physiological requirements present in air-breathing and land-living animals.

IMPORTANT QUESTIONS

Long Answer Questions

1. Describe the life history of Indian bull-frog, *Rana tigrina*.
2. Discuss the development of frog up to the end of gastrulation.
3. Describe the development of frog up to the formation of three geriminal layers and also mention their fate.
4. Describe the structure of tadpole of frog. Discuss the metamorphic changes it undergoes to become the adult.
5. What do you understand by metamorphosis? Explain with reference to the life history of frog.
6. With suitable diagrams explain the stages of development of post-larval and anterior abdominal veins of frog.
7. Describe the phenomenon of metamorphosis in frog.
8. What are the biochemical changes taking place during fertilisation of frog?
9. Give an account of formation of notochord, nerve cord, coelom and enteron in the development of frog.
10. Describe the enterocoelic method of coelom formation.

Short Answer Questions

1. Explain the effects of yolk on gastrulation in the development of frog.
2. Discuss the importance of gastrulation in frog's egg.
3. Write short notes on : (*i*) Neurogenesis, (*ii*) Notogenesis, (*iii*) Presumptive areas, (*iv*) Fate maps in frog, (*v*) Formation of heart in frog, (*vi*) Formation of coelom and mesoderm in frog, (*vii*) Formation of eyes in frog, (*viii*) Epiboly.

Very Short Answer Questions

1. What is amplexus?
2. What is spawn?
3. To embrace the female frog, what structures are developed in male frog during breeding season?
4. What is the type of development in frog?
5. What is the purpose of cone reception in frog during development ?
6. What is perivitelline space?
7. How is fertilisation membrane formed in the egg of frog?
8. How polyspermy is checked?
9. What is the type of egg and cleavage in frog?

10. What is gastrulation?

11. Define fate maps in frog.

Multiple Choice Questions

1. On forelimbs of male frog, well developed pads are formed for clasping the female frog at the time of copulation are called
   (*a*) Nuptial pads (*b*) Tuberosities
   (*c*) Suchial buds (*d*) None of these.
2. Development of frog is
   (*a*) Direct (*b*) Indirect
   (*c*) Both of these (*d*) None of these.
3. Fertilisation in frog is
   (*a*) External (*b*) Internal
   (*c*) Both of these (*d*) None of these.
4. Diploid number of chromosomes in the zygote of frog is
   (*a*) 22 (*b*) 24
   (*c*) 26 (*d*) 25.
5. The type of egg of frog is
   (*a*) Alecithal (*b*) Centrolecithal
   (*c*) Isolecithal (*d*) Telolecithal.
6. Fertilisation membrane in frog is formed
   (*a*) Before fertilisation (*b*) After fertilisation
   (*c*) Before oviposition (*d*) After oviposition.
7. First cleavage in frog is
   (*a*) Holoblastic equal
   (*b*) Holoblastic unequal
   (*c*) Horizontal (*d*) Meroblastic.
8. Blastula of frog is called
   (*a*) Holoblastula (*b*) Coeloblastula
   (*c*) Amphiblastula (*d*) Discoblastula.
9. Coelom of frog is
   (*a*) Acoelous (*b*) Pseudocoelous
   (*c*) Enterocoelous (*d*) Schizocoelous.
10. Coelom of frog is derived from
    (*a*) Splitting of mesoderm
    (*b*) From archenteron
    (*c*) Both of these (*d*) None of these.
11. Gastrulation is a process of
    (*a*) Migration of mesoderm
    (*b*) Migration and rearrangement of cells already formed in blastula
    (*c*) Formation of ectoderm
    (*d*) Differentiation of parts from layers already formed.
12. During metamorphosis of tadpole of frog following changes occur
    (*a*) Morphological (*b*) Biochemical
    (*c*) Molecular (*d*) All these.
13. Name the germ layer from which the internal ear is developed
    (*a*) Ectoderm only (*b*) Endoderm only
    (*c*) Mesoderm only
    (*d*) Ectoderm and endoderm.
14. What is Rathke's pouch?
    (*a*) It is a rudimentary pouch of ectoderm posterior to the last gill
    (*b*) It is a rudiment of the infundibulum
    (*c*) It is a rudiment of the pars nervosa
    (*d*) It is a rudiment of the pars anterior.
15. On the basis of pigmentation, surface of the blastula of frog can be divided into
    (*a*) Two regions (*b*) Three regions
    (*c*) Four regions (*d*) Five regions.
16. Innermost compartment of **myocoel** called **sclerotome** contribute to the formation of
    (*a*) Dermis (*b*) Muscles
    (*c*) Vertebral skeleton (*d*) All these.
17. What is the structure normally differentiated from optic cup?
    (*a*) Only sensory pigment layers of retina
    (*b*) Sensory layers of retina and lens
    (*c*) Sclerotic coat and optic nerve
    (*d*) Iris and sensory pigment layers of retina.
18. The cytoplasm is concentrated in frog's egg in
    (*a*) Middle region (*b*) Peripheral region
    (*c*) Vegetal pole (*d*) Animal pole.

Answers

1. (*a*)	2. (*b*)	3. (*a*)	4. (*c*)	5. (*d*)	6. (*b*)	7. (*b*)	8. (*b*)
9. (*d*)	10. (*a*)	11. (*b*)	12. (*d*)	13. (*a*)	14. (*b*)	15. (*b*)	16. (*c*)
17. (*d*)	18. (*d*)						

38 Development of Chick (Fowl)

Chapter

Contents of this Chapter

The embryology of chick or the common fowl (*Gallus gallus*) is preferable to study to that of pigeon (*Columba livia*) because of several advantages. The embryology of chick has been worked out more extensively only because of the following reasons :

1. Eggs of fowl are large or of convenient size, easily available throughout the year and can be incubated artificially.

2. The various developmental stages can be easily available in laboratory conditions for experimental purposes.

3. Moreover, the process of development has been most thoroughly worked out in fowl.

4. The embryology of chick bears many resemblances with those of reptiles and mammals. A comparative study of embryology of different birds exhibits that it is essentially

similar in all the birds with only minor unimportant differences. Despite these reasons, the embryology of chick is important due to phylogenetic significance. In chick, development is direct without a larval stage.

Fertilisation

The ova are released from the ovary in the form of primary oocytes. The second maturation division occurs after ovulation in the oviduct. The released primary oocyte in the body coelom is grasped and swallowed by the ostium of the oviduct. The secondary oocyte or **ovum** after second maturation division is surrounded by several (5 or 6) sperms which enter in it (polyspermy), but only one sperm succeeds in the fertilisation process. The nucleus of one sperm fuses with the female nucleus (**amphimixis**) and the nuclei of other sperms degenerate. As the fertilised egg passes downward inside the oviduct, it rotates, undergoes cleavage and various accessory egg membranes are laid down over the developing egg. Thus, fertilisation is internal. The various egg membranes secreted around egg are albumen, shell membranes and shell.

Structure of Egg of Hen

The egg is about 3.0 cm in diameter and is **polylecithal**. It is entirely filled up by the yolk and over it, *i.e.,* in animal pole lies a small cytoplasmic disc with a nucleus. This disc is called **blastodisc**. The **yolk** has a central mass of white yolk around which are alternate concentric layers of yellow and white yolk. The central flask-shaped white yolk called **latebra** runs from the centre to the lower side of the blastodisc and there it spreads to form the **nucleus of Pander**. The yolk and blastodisc are bounded by a plasma membrane and outer **vitelline membrane**, which is a converted form of zona radiata. The vitelline membrane is of double origin. The inner layer of this membrane is produced in the ovary between the oocyte and follicle cells and it is composed of very tough fibres. The outer layer of it is formed in the upper part of fallopian tube. The yolk contains 48.7% water, 32.6% phospholipids and fats, 16% proteins, 1% carbohydrates and 1.1% other chemical molecules. Its proteins are in the form of phosvitin and lipovitelline or livetin. The fat is predominantly neutral fat (50%), rest is phosphatids, cerebrosides and cholesterol.

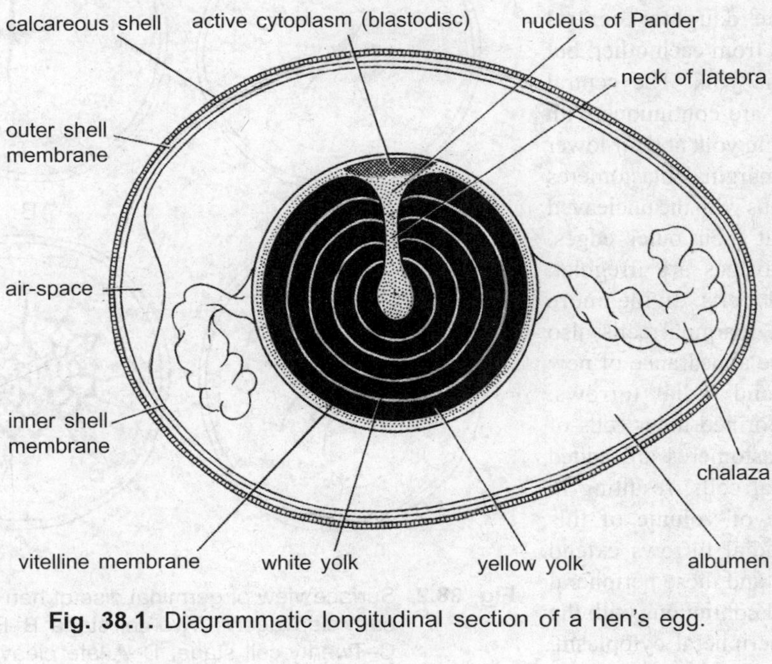

Fig. 38.1. Diagrammatic longitudinal section of a hen's egg.

The fertilised egg or zygote is covered by a layer of dense viscous albumen which forms a thin **chalaziferous layer** around the vitelline membrane. This dense albumen forms two twisted cords or **chalazae**, one at each end of the zygote. They are formed by rotation of the egg during its movement through the oviduct. Around the chalaziferous layer is a thick layer of watery albumen. All albumen is secreted by the upper glandular walls or magnum of the oviduct. The functions of albumen are to provide nutrition to the ambryo, serves as a water store and also acts as protective envelope for protecting the embryo from mechanical and chemical injuries. The albumen and yolk also contains a variety of enzymes, vitamins, pigments and phophorus. The isthmus part of the oviduct secretes two **shell membranes** made of tough keratin fibres matted together. The two shell membranes are closely applied except at the blunt end of the egg where they are separated by an **air space** formed after the egg is laid. The nidamental glands of the oviduct secrete a porous, **calcareous shell** which soon hardens. It is pierced by a large number (about 7,000) of fine pores filled with a protein related to collagen. The diameter of the pores varies from 0.04 to 0.05 mm. These pores allow exchange of gases (oxygen and carbon dioxide) during respiration of the developing embryo. The egg is laid 24 hours after fertilisation, and its further development takes place, when the egg is incubated by the female. **Incubation** must continue steadily for 21 days at a temperature of 103°F.

Cleavage and Blastulation

Cleavage is **meroblastic** or **discoidal** and is confined only to be germinal disc or **blastodisc**. It does not segment the yolk of the polylecithal egg and it is later eventually surrounded by the growing tissues of the embryo. **First two cleavages** are at right angles to each other in the centre of blastodisc. These cleavage furrows do not cut the germinal disc completely through in the vertical plane. The **third set of cleavage** furrows is vertical, cutting across the second set of vertical furrows. The **fourth cleavage** furrow is also vertical and circular cutting across all the cleavage furrows, forming eight **central blastomeres** which are surrounded by eight **marginal blastomeres**. Thus, these cleavage furrows separate the daughter central blastomeres from each other, but not from the yolk. The central blastomeres are continuous with the underlying yolk at their lower ends. The marginal blastomeres are continuous with the uncleaved cytoplasm at their outer edges. Further cleavages are irregular. The central cells divide more rapidly. The marginal cells also divide by the appearance of new horizontal and radial furrows. The newly formed inner cells of marginal blastomeres are added to the central cells, resulting in the increase of volume of this area. The radial furrows extend peripherally and these peripheral cells are still continuous with the uncleaved peripheral cytoplasm.

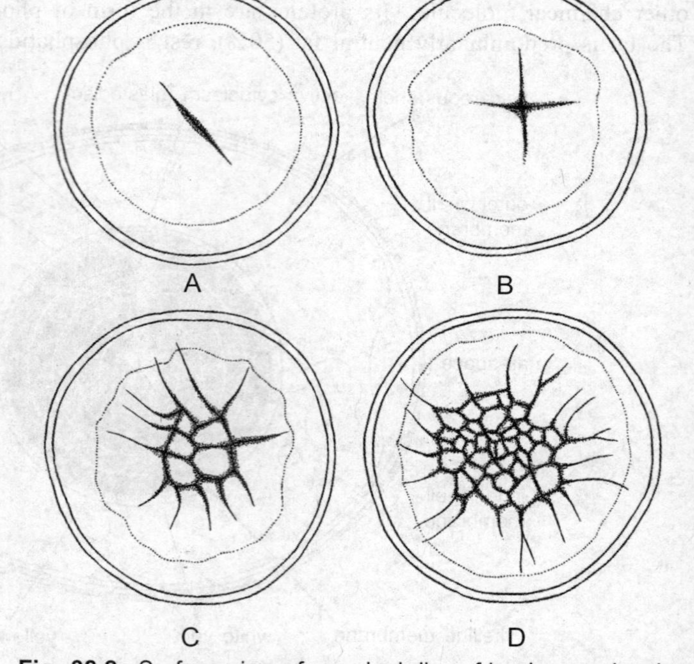

Fig. 38.2. Surface view of germinal disc of hen's egg showing early cleavage. A–Two-cell stage; B–Four-cell stage; C–Twenty-cell stage; D–A late cleavage stage.

In later stage of cleavage, the blastomeres of the central area become separated from the underlying yolk due to the appearance of a horizontal cleavage in these cells. This cleavage extends peripherally cutting the inner ends of the blastomeres. Thus, a space also appears in the beginning beneath the central cells which also extends peripherally as the horizontal cleavage extends outward.

The cavity beneath the central cells, *i.e.,* in between central cells and yolk is called the **subgerminal cavity**, which is filled with a fluid diffused from the albumen through vitelline membrane. Thus, due to further cleavage the blastodisc becomes cellular, called the **blastoderm**– a round disc, 5 to 6 cells deep in the centre but only 1 to 2 cells deep at the periphery.

The appearance of subgerminal cavity separates the blastoderm from the underlying yolk, but the marginal cells remain overlapping the yolk. The embryo is now called the **blastula** stage. At blastula stage the embryo reaches in the uterus. It may be compared to the blastula of *Amphioxus* and frog, its sub-geminal cavity is equivalent to the blastocoel, the blastoderm is the animal pole, and the yolk is the vegetal pole.

During later part of cleavage, about 12 to 14 hours after the egg reaches the uterus or 6 to 8 hours before the egg is laid, some cells on the inner or under side of the blastoderm become detached or delaminated from the blastoderm and fall on the floor of subgerminal cavity due to pιesence of relatively more yolk. The delamination of these yolky cells from the blastoderm starts at the posterior edge and spreads forward until whole blastoderm becomes free from yolky cells. As a result, the epithelial layer in the central region of blastoderm becomes thinner (few layers of cells) and transparent. Thus, this region is called the **area pellucida** because it seems to be transparent when viewed from the upper side. The peripheral part of blastoderm, the yolky cells are not delaminated (shed), so this part of the blastoderm seems to be opaque, because beneath these cells blastocoel is not present. This region of blastoderm is, thus, called **area opaca**. These delaminated cells at the posterior edge of area pellucida gradually link up with each other, forming a continuous layer of flattened cells, which extends anteriorly. This layer is called the **hypoblast** and the upper layer is the **epiblast** containing ectoderm and mesoderm cells. Hypoblast is exclusively composed of endoderm cells.

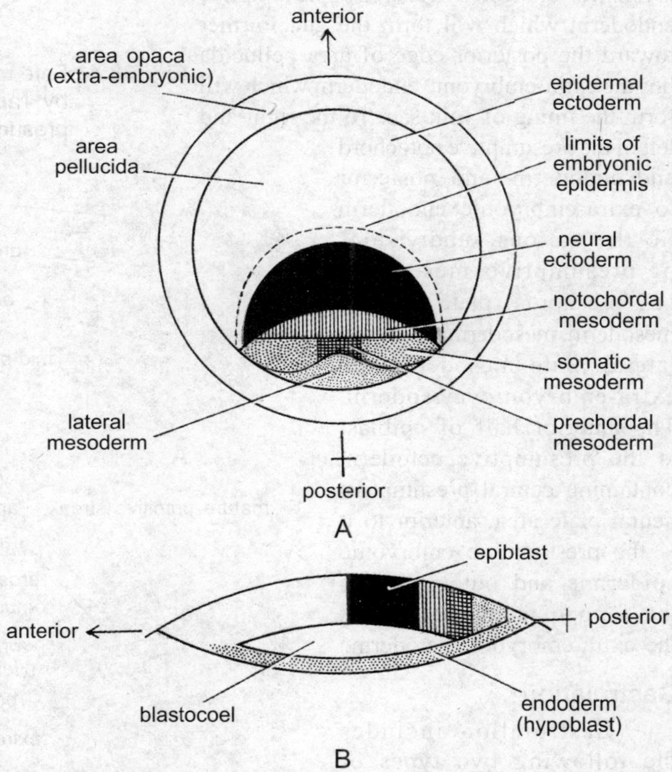

Fig. 38.3. Fate maps of chick blastoderm. A–Surface view of epiblast; B–Diagrammatic section of discoblastula showing the fate map of epiblast and hypoblast at a glance.

The egg is laid by the female about the time the blastula is formed or even a little later.

Presumptive Fate Maps of Blastula

Fate maps of the blastula of chick has been prepared by using the vital stains such as carmine or carbon (charcoal) particles or radioactive thymidine. It shows that blastomeres of area opaca do not form any part of the embryo proper, they form only extra-embryonic membranes. The epiblast and hypoblast of area pellucida have different fates in the course of embryonic development. The fate maps prepared by the use of tritiated thymidine has shown the following structures in epiblast : In the centre of area pellucida lies a small area destined to produce the notochord. Posterior to it, in the median plane is found an elongated oval area of presumptive endoderm which will form the gut. Further toward the posterior edge of area pellucida lies the extra-embryonic endoderm which will form the lining of yolk sac. To the right and left of the presumptive notochord and endoderm, and posterior to extra-embryonic endoderm lie the various subdivisions of presumptive mesoderm, *i.e.,* prechordal plate or head mesoderm, mesodermal somites, lateral plate mesoderm and extra-embryonic mesoderm. The anterior half of epiblast is the presumptive ectoderm containing central presumptive neural plate area, anterior to it is the presumptive embryonic epidermis and outer to it in the form of complete ring is the extra-embryonic ectoderm.

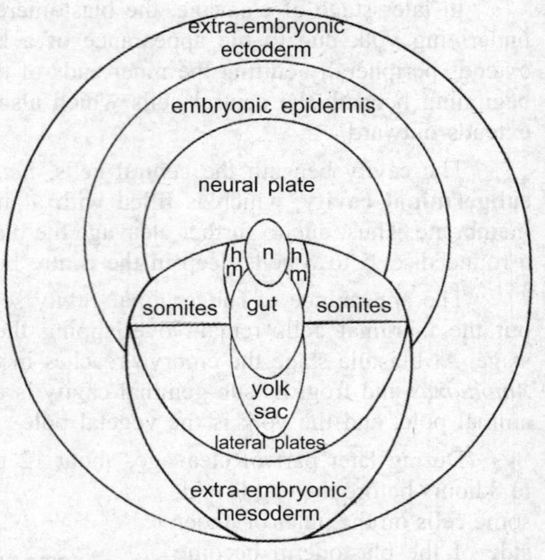

Fig. 38.4. Fate map of the epiblast of a bird prepared by radioactive thymidine marking : hm, presumptive head mesoderm; n, notochord.

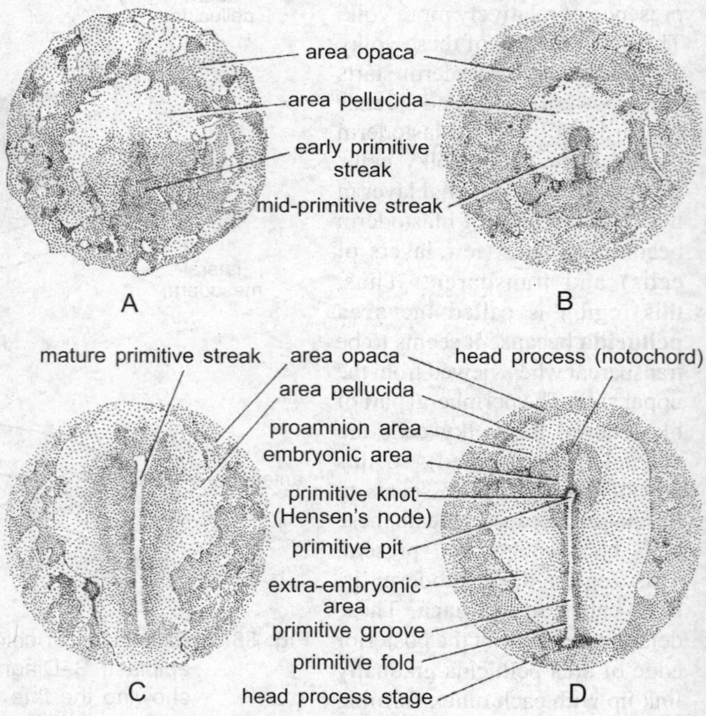

Fig. 38.5. Surface view of chick blastoderm showing development of primitive streak (gastrulation) and head process (neurulation). A–Initial streak (stage 2); B–Intermediate streak (stage 3); C–Definitive streak (stage 4); D–Head process stage (stage 5 embryo of 19 to 22 hours of incubation).

Gastrulation

Gastrulation includes the following two types of morphogenetic movements :

1. Emboly. It involves only the epiblast which contains cells of ectoderm, mesoderm and notochordal cells. It includes convergence, invagination and involution. Formation of primitive streak and head process is due to emboly.

2. Epiboly. It includes the overgrowth of ectoderm or epiblast and also of hypoblast.

1. Formation of Primitive Streak

Various prospective mesodermal and endodermal cells forming notochord of the epiblast converge toward the posterior edge of the area pellucida and form a conical thickening in the midline, called the **initial primitive streak**. It appears after 6 to 7 hours of incubation. The primitive streak grows anteriorly because of proliferation of its own cells as well as of the addition of cells that migrate to it from anterior and lateral parts of area pellucida. The elongated axis of the primitve streak marks the antero-posterior axis of the future embryo. It, thus, eventually extends to about three fifths of the entire length of area pellucida. This is the fully developed **definitive primitive streak** and it is usually completed after 18 to 19 hours of incubation. The area pellucida also becomes pear-shaped. Along the middle of the primitive streak, when it is fully developed runs a narrow furrow, the **primitive groove**. At the anterior end of the primitive streak there is a thickening, the **primitive knot** or **Hensen's node**. The centre of Hensen's node is excavated to form a funnel-shaped depression.

The movements in the blastoderm leading to the final placement of cells in the hypoblast and to the formation of the primitive streak in the epiblast may be called **pregastrular movements**.

Invagination and involution. At the stage of short primitive streak, the cells of the blastoderm already begin to migrate (invaginate and involute) into the blastocoel cavity between epiblast and hypoblast. Immigrating cells are replaced by more epiblast cells converging toward the streak area. The inward migrating cells also spread out side ways and forward from the anterior end of primitive streak. The notochordal cells immigrate through primitive pit. Endodermal cells invaginate through that part of the streak which lies just behind primitive pit. The mesodermal cells of somites just follow the path of endodermal cells. Whereas the lateral plate mesoderm cells invaginate through the middle section of primitive streak, but only after the disappearance of endoderm from the area pellucida. The extra-exbryonic mesoderm (of the yolk sac) immigrate through the posterior part of primitive streak.

Meanwhile, some hypoblast cells expand into the area opaca to become extra-embryonic endoderm (the lining of yolk sac), while other hypoblast cells attach to mesodermal and notochordal cells are carried along by the latter's migration.

Formation of head process. Prospective notochoral cells converge on the node, sinks through it and then passes directly forward as a tongue of tissue known as **head process** or **notochord process**.

Disappearance of primitive streak. With the gradual disappearane of endodermal, notochordal and mesodermal cells from the primitive streak, it begins to shrink from anterior towards posterior side and its remains are partly included in the tail bud and partly into the cloacal region of the embryo.

Development of head process. The midline area of notochordal tissue develops into a rigid rod, anterior to the receding primitive streak. As the streak regresses posteriorly, the embryo develops anterior to it. The head process consists of a thick central mass of cells and more diffuse lateral wings. In the beginning it is also blended in the midline with the hypoblast. The thicker central portion forms the definitive notochord, whereas the lateral wings form the paraxial (somitic) mesoderm. With its differentiation, the notochord becomes detached from the hypoblast below, except at the extreme end. Thus, the head process stage is completed at about 20 to 25 hours of incubation. Gastrulation is also completed at this stage.

Completion of endoderm. The first cells that migrate through the anterior part of streak form the endoderm. As the Hensen's node recedes backward, and the notochordal process elongates, the presumptive endoderm of the middle and posterior part of the gut, located just behind the node, migrate inside as an endodermal strip beneath the notochord. The original hypoblast at the floor of the blastocoel contribute a very less amount to the gut, the upper migrated endodermal cells form the major part of the gut. In chick no archenteron is formed during gastrulation.

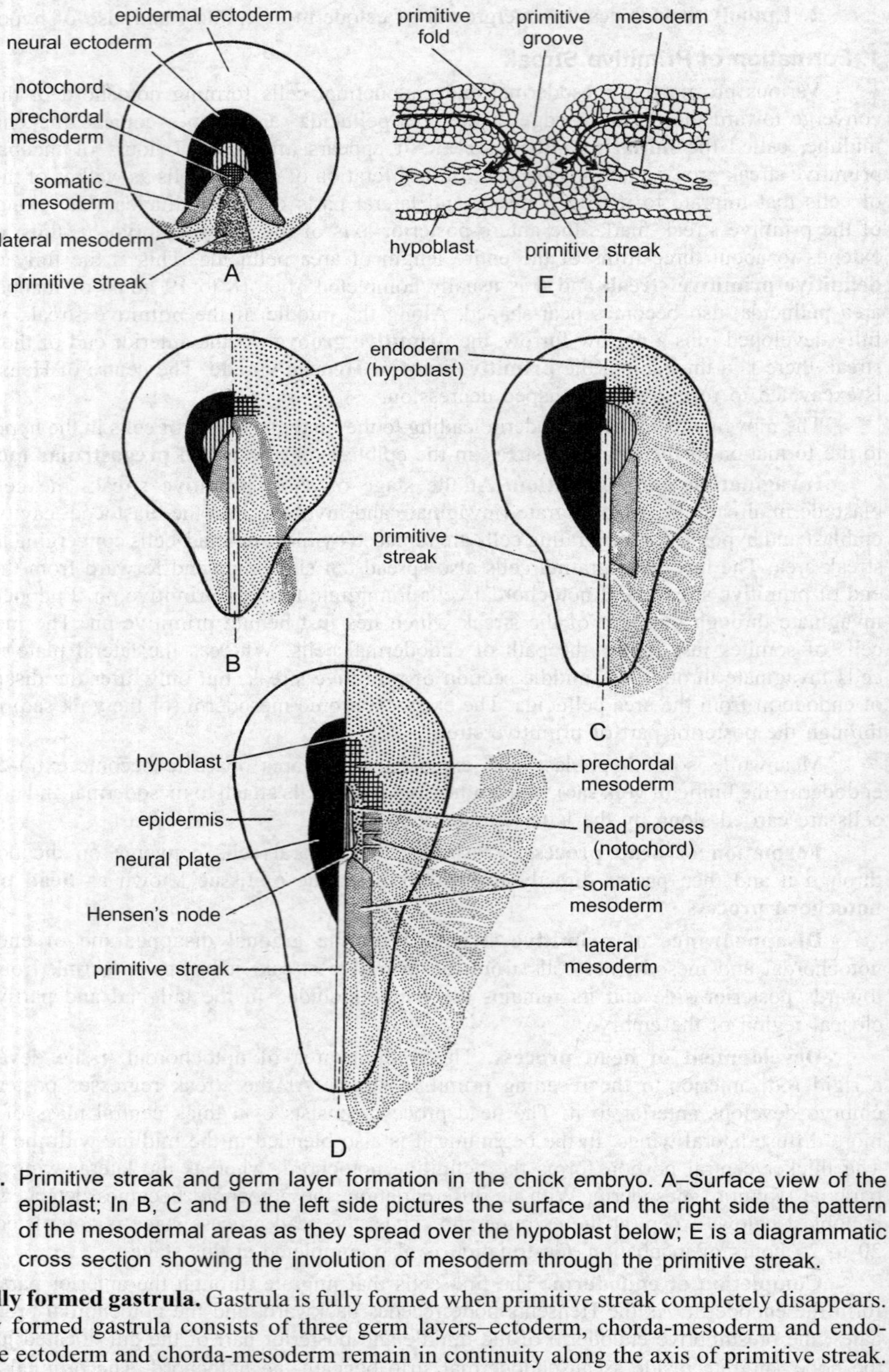

Fig. 38.6. Primitive streak and germ layer formation in the chick embryo. A–Surface view of the epiblast; In B, C and D the left side pictures the surface and the right side the pattern of the mesodermal areas as they spread over the hypoblast below; E is a diagrammatic cross section showing the involution of mesoderm through the primitive streak.

Fully formed gastrula. Gastrula is fully formed when primitive streak completely disappears. The fully formed gastrula consists of three germ layers–ectoderm, chorda-mesoderm and endoderm. The ectoderm and chorda-mesoderm remain in continuity along the axis of primitive streak. The endoderm is also united with the mesoderm and ectoderm at the anterior and posterior end of streak.

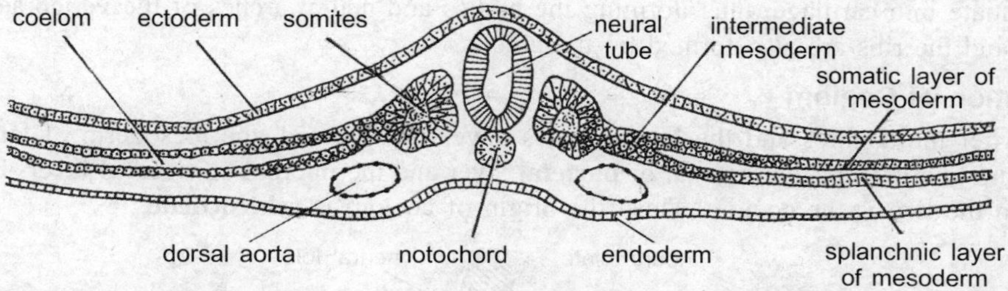

Fig. 38.7. Cross section of an early chick embryo, illustrating the relationship of the germ layers.

Formation of Neural Tube (Neurogenesis)

The ectoderm, anterior and lateral to the head process, becomes thickened to form the neural plate while the gastrulation process is going on. The neural plate appears in the brain region. As Hensen's node recedes farther and farther, parts of the neural plate become differentiated, and the anterior parts of the neural plate proceed to close into a tube, the **neural tube**. The formation of neural tube occurs due to sinking in of the neural plate due to which a **neural groove** is formed along its longitudinal axis. Its elevated margins are called **neural folds**, which rise up and grow toward the midline and fuse to form the **neural tube**.

Formation of Notochord and Mesoderm

While the neural plate is folding into the neural tube, the chorda-mesoderm is also differentiating. Its most anterior part, the **prechordal plate mesoderm** gives rise to the mesenchyme of head, and behind it the notochordal cells become separated from the rest of the adjoining sheets of mesoderm and differentiated into **notchord**. On either side of notochord are three longitudinal strands or sheets of mesoderm.

1. Epimere, axial or somatic mesoderms are the thicker dorsal medial strands, which merge anteriorly into the mesenchyme of head.

2. Mesomeres or intermediate mesodermal strands are located lateral to the axial mesoderm. They are thin in an early stage.

3. Hypomeres or lateral plate mesoderm lies beneath the mesomere.

Formation of Somites

The axial or somatic mesoderm begings to differentiate after at about 21 hours of incubation. It becomes a thick band on either side of notochord and nerve cord. A short distance anterior to the primitive knot a transverse cleft appears across each band, marking the division between first and second somites. About one hour later a second cleft appears posterior to first cleft and, thus, second somite is formed. This process goes on approximately after each hour up to 20 hours. After that somites formation become somewhat slower.

The first three pairs of mesodermal somites are formed in the paraxial mesoderm in the presumptive hind brain region, and other mesodermal somites appear as the primitive knot regresses. Somites, when first formed, are masses of mesodermal cells with a small cavity in the centre. The cells are arranged radially around the central cavity. With further development, they become extended in the dorso-ventral direction and flattened medio-laterally. The cavity (**myocoel**) of each somite also changes, from spherical to a narrow vertical slit. An inner wall and an outer wall become clearly distinguishable, corresponding to the parietal and visceral layers of the lateral plates. The inner wall of the somites becomes very much thicker and produces skeletogenous tissue and the voluntary striated muscles of the body. The outer of the somites is thinner and forms the connective tissue layer of the skin and is, therefore, called the **dermatome**. Furthermore, the dorsal part of the inner wall of the somite which is the source of somatic muscles of the body is called **myotome**, whereas the lower edge of the inner wall is called **sclerotome** which is responsible for skeletogenous tissue. The sclerotome breaks up into a mass of mesenchymal cells. The cells migrate into the spaces surrounding the notochord and spinal cord, envelop these organs and later differentiate into cartilage, thus, forming the bodies and neural arches of the vertebrae. Haemal arches and the ribs are also formed by these cells.

Formation of Coelom

After the somites and the lateral plates have been formed, the mesoderm of lateral plate splits into two layers : the external or **parietal** layer and the internal or **visceral** layer. The cavity between the two is the coelom. Thus, the origin of coelom is **schizocoelic**.

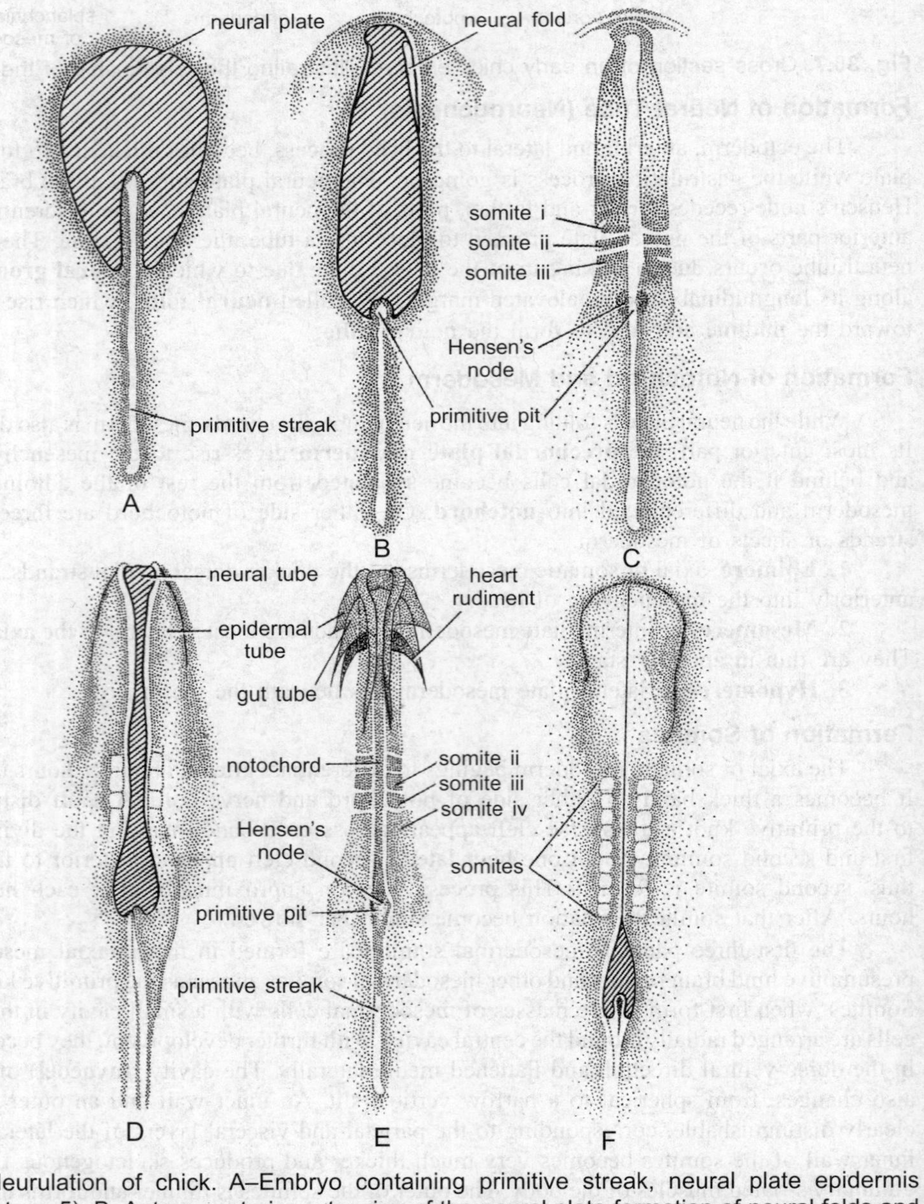

Fig. 38.8. Neurulation of chick. A–Embryo containing primitive streak, neural plate epidermis (stage 5); B–Antero-posterior extension of the neural plate formation of neural folds and shortening of the primitive streak (Stages 7 to 8); C–Internal anatomy of embryo of stages 7 to 8; D–Surface view of embryo of 3 to 4 pair somite stage (stage 8, 26 to 28 hours of incubation); E–Internal anatomy of stage 8 embryo; F–Surface view of an emrbyo of 9 to 10 pairs somite stage (Stage 10, 33 to 38 hours of incubation).

Folding of Embryo

As stated earlier, the entire blastoderm does not form the embryo. Its central portion, called the area pellucida, only forms the embryo, and its outer portion, the area opaca forms the extra-embryonic membranes, such as yolk sac, amnion, chorion and allantois. The body of the embryo becomes separated from the yolk sac. This is achieved by the formation of folds which appear all around the body of embryo. The folds involve all three germ layers and are directed downward and inward, undercutting the body of the embryo proper. They are known as **body folds**. The various folds do not appear simultaneously. The first to appear is the **head fold** just in front of the head. It undercuts the head and anterior part of the trunk of the embryo, so that these parts project freely over the surface of yolk sac. The lateral and posterior parts of the body fold develop soon after. The **posterior fold** undercuts the tail end and the posterior part of trunk which also project freely over the surface of yolk sac. These body folds gradually contract underneath the embryo and eventually the body of embryo is connected with the yolk sac and other extra-embryonic membranes by a narrow stalk, the **umbilical cord**.

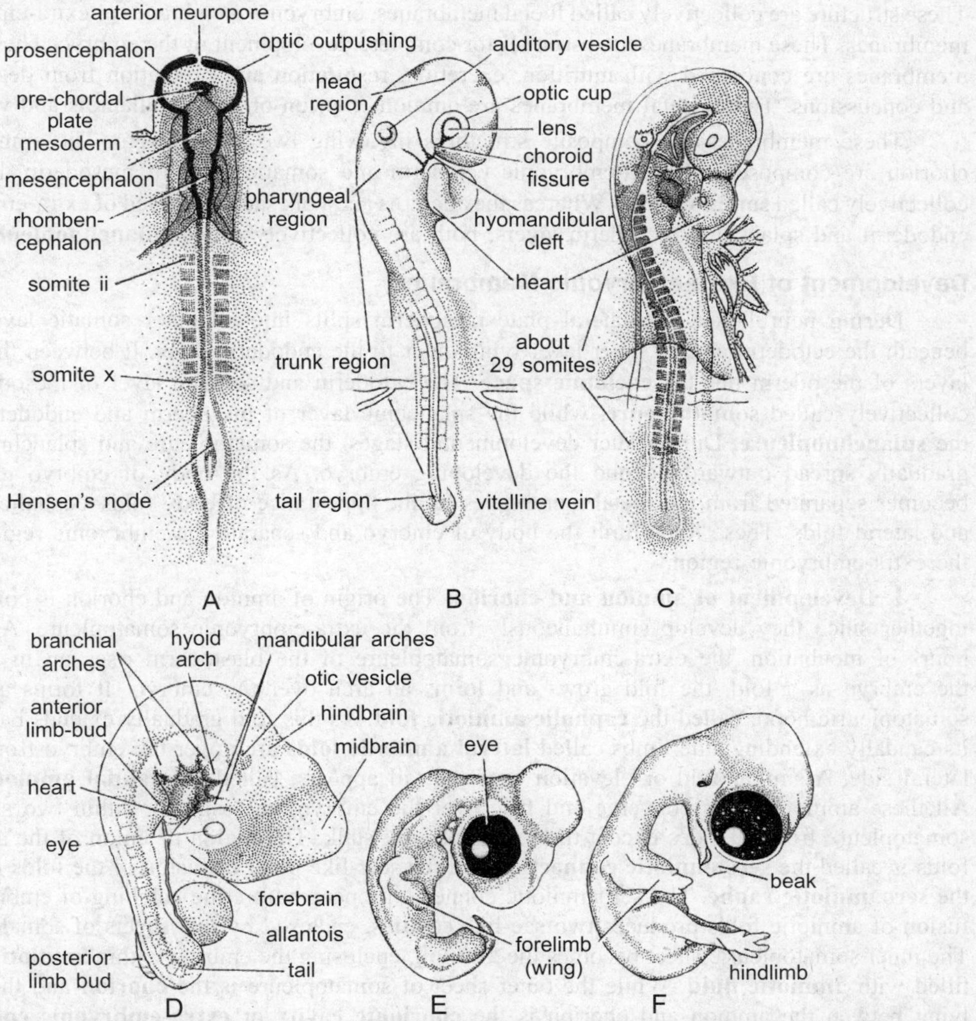

Fig. 38.9. Post-neurular development. A–Stained preparation of the stage 10; B–External view of embryo of about 27–28 pairs of somites (51 to 56 hours of incubation, stage 16); C–Internal anatomy of stage 16 (of B); D–Embryo of about 72 to 75 hours of incubation (Stage 20); E–Embryo of stage 27; F–Embryo of stage 23.

Flexure and Torsion

When the head is formed it gets bent down ventrally with respect to the main axis of the embryo, this bending is called **cranial flexure**. Then the head at first, and gradually the entire body turn sideways so that the embryo comes to lie on its left side over the yolk, this dextral twist is knows as **torsion**. The anterior end in its cervical region becomes still more bent over towards its ventral surface. Various organs are formed on the third and fourth days of incubation, such as the nervous system, sense organs, four pairs of pharyngeal pouches out of which only the first three pairs meet the ectoderm to open as gill-clefts, five pairs of visceral arches, heart and blood vessels, and paired segmental somites.

EXTRA-EMBRYONIC (FOETAL) MEMBRANES OF CHICK

The blastoderm besides forming the embryo, gives rise to certain other structures which do not take part in the formation of embryo proper, but are external to the developing embryo. These structure are collectively called foetal membranes, embryonic membranes or extra-embryonic membranes. These membranes are essential for complete development of the embryo. These foetal membranes are concerned with nutrition, excretion, respiration and protection from desiccation and concussions. These foetal membranes are amnion, chorion or serosa, allantois and yolk sac.

These membranes are composite structures involving two germ layers. The amnion and chorion are composed of extra-embryonic ectoderm and somatic layer of mesoderm, both are collectively called **somatopleure**. Whereas the allantois and yolk sac are formed of extra-embryonic endoderm and splanchnic mesoderm layers, both are collectively called **splanchnopleure**.

Development of Extra-embryonic Membranes

During neurulation, the lateral plate mesoderm splits into an outer somatic layer lying beneath the ectoderm and an inner layer lying outer to the endoderm layer. It between these two layers of mesoderm lies the **coelomic space**. The ectoderm and somatic layer of mesoderm are collectively called **somatopleure**, while the splanchnic layer of mesoderm and endoderm form the **splanchnopleure**. During later developmental stages, the somatopleure and splanchnopleure gradually spread outward beyond the developing embryo. As the body of embryo grows, it becomes separated from the foetal membranes by the appearance of body folds–cephalic, caudal and lateral folds. These folds limit the body of embryo and separate the embryonic region from the extra-embryonic region.

1. Development of amnion and chorion. The origin of amnion and chorion is considered together since they develop simultaneously from the extra-embryonic somatopleure. About 30 hours of incubation, the extra-embryonic somatopleure of the blastoderm rises up in front of the embryo as a fold, the fold grows and forms an arch over the embryo. It forms a double somatopleuric hood, called the **caphalic amniotic fold**. As this fold gradually extends backward, its caudally extending side limbs called **lateral amniotic folds** arch over the embryo from either lateral side. A similar fold or elevation over the tail appears, called the **caudal amniotic fold**. All these amniotic folds converge and fuse over the embryo, enclosing it within two sheets of somatopleure from all sides except the region of yolk stalk. The region of union of the amniotic folds is called the **seroamniotic connection** and the scar-like point of fusion of the folds is called the **seroamniotic raphe**. The seroamniotic connection opens before the hatching of embryo. The fusion of amniotic folds produces two sac-like cavities, enclosed by two sheets of somatopleure. The inner somatopleuric sheet becomes the **amnion**, enclosing the embryo within **amniotic cavity** filled with **amniotic fluid**. While the outer sheet of somatopleure is the **chorion** and the cavity lying betwen the amnion and chorion is the **chorionic cavity** or **extra-embryonic coelom**. It is lined by mesoderm of somatopleure and splanchnopleure. The somatopleure of amnion and chorion grows continuously peripherally and laterally and ultimately encloses the embryo and also the remaining two foetal membranes.

Functions of amnion and chorion. The amnion, chorion and amniotic cavity serves the following functions for the embryo :

1. The **amnion** protects the embryo from desiccation and sudden temperature changes.

2. The amniotic fluid is an efficient shock absorber and protects the embryo from the mechanical shocks.

3. The shell and shell membranes are protective and prevent desiccation of the embryo.

4. The mesoderm of amnion during later development forms muscle cells which contract rhythmically, rocking the embryo with in the amniotic fluid so as to prevent it from adhesion to the embryonic membranes.

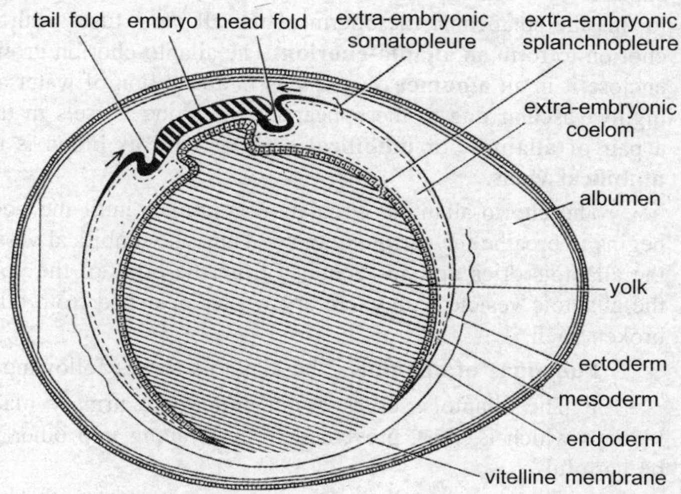

Fig. 38.10. An early chick embryo showing body folds delimiting from extra-embryonic areas.

2. Development of allantois. About the third day of incubation the floor of the endodermal hindgut begins to bulge as a bladder, called the **allantois**. The allantoic evagination is formed from the splanchnopleure, that is, inner layer of endoderm and an outer layer of splanchnic mesoderm. It grows rapidly and spreads into the extra-embryonic coelom, the space between the yolk sac, the amnion and the chorion. The distal part of the allantois expands and remains connected with the hindgut of the embryo by means of a narrow **allantoic stalk**. When the body folds contract, separating the embryo from the extra-embryonic parts, the allantoic stalk is enclosed together with the stalk of yolk sac, forming an **umbilical cord**.

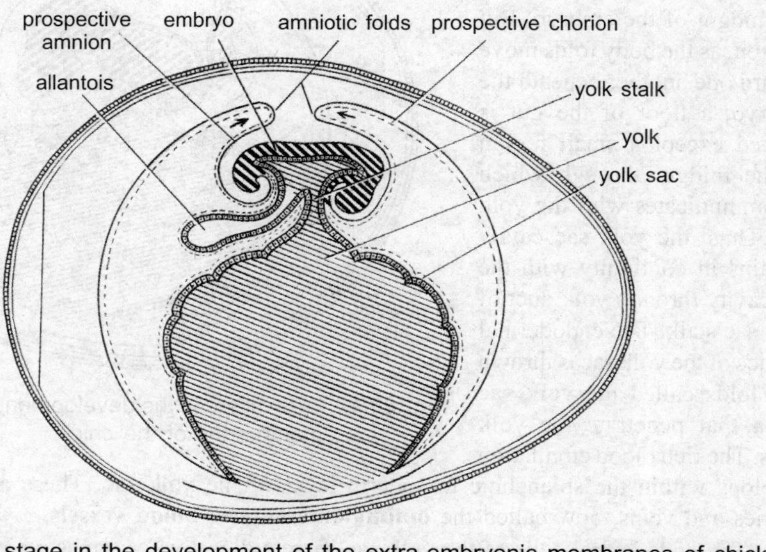

Fig. 38.11. Early stage in the development of the extra-embryonic membranes of chick.

As the allantois vesicle enlarges and spreads outward, its distal part becomes flattened and expands between the amnion and yolk sac on one side and the chorion on the other side. Due to this the splanchnic mesoderm of the allantois fuses with the inner somatic mesoderm of the chorion to form an **allanto-chorion**. The allanto-chorion grows around the albumen of the egg to enclose it in an **albumen sac**. It aids in absorption of water and albumen. The allantois becomes highly vascular due to the appearance of blood vessels in the splanchnopleure. It also receives a pair of **allantoic** or **umbilical arteries**, and its blood is returned into a pair of **allantoic** or **umbilical viens.**

The chorio-allantoic circulation continues until the young chick breaks the egg-shell and begins to breathe the surrounding air. Thus, the umbilical vessels close, the circulation ceases and the allantois dries up and separates from the body of the young chick. At the time of hatching, the allantoic vesicle with excretory wastes detached from allantoic stalk and left attached to the broken shell.

Functions of allantois. The allantois serves following functions for the embryo :

1. The allantois serves as an embryonic urinary bladder collecting uric acid from the kidneys which is, thus, prevented from escaping into other parts of the embryo where it might be harmful.

2. The vascular allanto-chorion is in contact with shell membranes and it brings about the respiration of the embryo by an exchange of oxygen and carbon dioxide through the porous shell.

3. Development of yolk-sac. During neurulation, the gut region of embryo has a roof and side but no floor. The ventrally open gut rests on the yolk mass. The formation of yolk sac first among foetal membranes is essential since yolk is supplied to the developing embryo with the help of yolk sac. The extra-embryonic splanchnopleure grows over the yolk and eventually surrounds the entire yolk to form the **yolk sac**. It has an inner layer of endoderm and an outer layer of splanchnic mesoderm. The yolk sac in the beginning is joined by a wide yolk stalk to the midgut of the embryo. But later on, as the body folds move toward one another beneath the embryo, a floor of the gut is formed except a small region of the midgut through which it communicates with the yolk sac. Thus, the yolk sac cavity remains in continuity with the gut cavity through yolk duct of yolk sac stalk. The endodermal surface of the yolk sac is thrown into folds called the **yolk sac septa** that penetrate the yolk mass. The rich blood circulation develops within the splanchnic mesoderm layer of the yolk sac. These are the paired vitelline arteries and veins, now called the **omphalomesenteric blood vessels**.

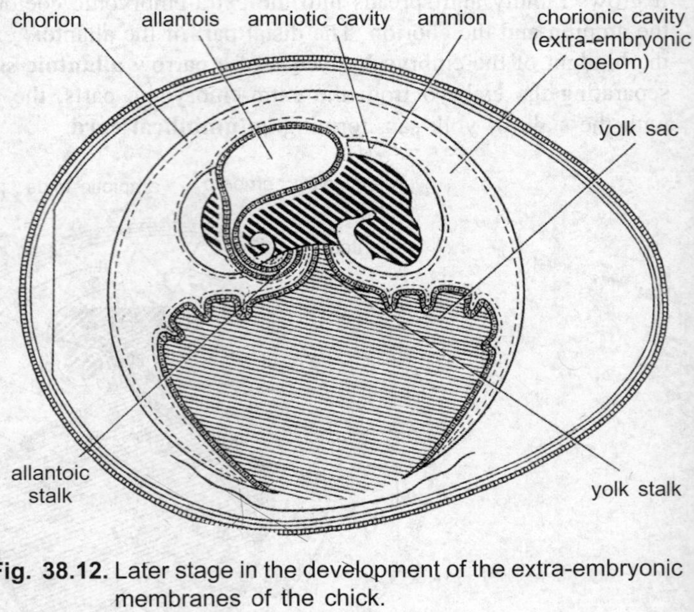

Fig. 38.12. Later stage in the development of the extra-embryonic membranes of the chick.

The endodermal cells of the yolk sac secrete digestive enzymes which digest the yolk. The digested yolk is collected by left and right vitelline veins, both of which open into an unpaired

ductus venosus which open into **sinus venosus** of the heart. Thus, the yolk sac serves as a digestive and absorptive surface by which yolk is made available to the embryo. Shortly before hatching the shrivelled up remains of the yolk sac is retracted into the abdominal cavity of the embryo, and the walls of the abdominal cavity close behind it.

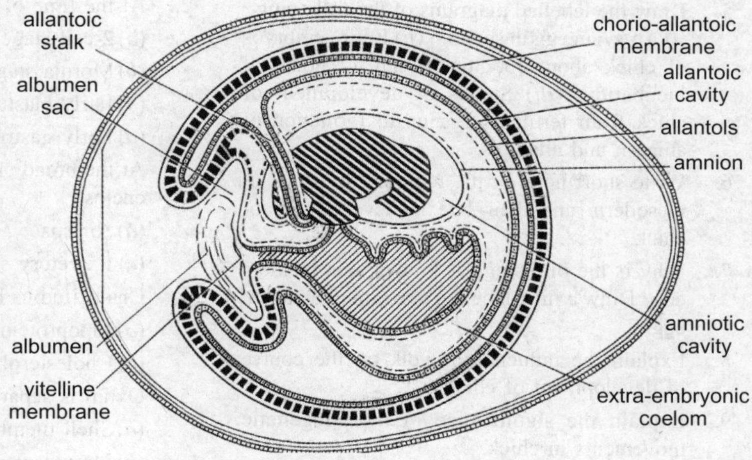

Fig. 38.13. The fully matured extra-embryonic membranes of the chick.

The amnion, chorion, and allantois are devices by which the embryo can develop on dry land (as also in reptiles and mammals). The yolk sac (also found in fishes, reptiles and mammals) develops for the absorption of yolk.

Hatching

The early embryo was at right angles to the long axis of the egg, later it comes to lie along the long axis with its head near the blunt end. The beak of the chick is covered by a horny **caruncle** by which it first pierces the inner shell membrane and air is taken into the lungs from the air space, soon after the caruncle breaks the egg shell and the chick emerges after 21 days of incubation at 103°F.

Since the inside surface of shell is concave, it is easier for the chick to break open the shell and shell membrane as all the force is concentrated at one point. The outer surface of the shell is convex, hence, the force exerted on the outside of the shell is distributed on all sides and the shell does not break easily. The chick breaks open the shell by repeated blows of the egg-tooth and hatches out leaving behind the allanto-chorion and amnion in the shell. The newly hatched chick is **precocious** or **nidifugous**, *i.e.,* it is fully clad with wet feathers which soon dry up on the exposure to air, and is able to walk and feed.

IMPORTANT QUESTIONS

Long Answer Questions

1. Describe in detail the structure of hen's egg and explain the functions of its various parts.
2. Give an account of development of chick till the formation of germinal layers.
3. Describe the development of chick up to the formation of primitive streak.
4. Discuss the process of gastrulation in chick and compare it with that of frog.
5. What do you understand by foetal membranes? Describe their formation and functions in chick.

Short Answer Questions

1. Describe the structure of a freshly laid egg of hen.
2. Describe the mesoderm formation and its differentiation during three days of incubation.
3. Differentiate between : (*i*) Blastulation and gastrulation; (*ii*) Development and growth; (*iii*) Area opaca and area pellucida; (*iv*) Embryo and larva.
4. Write short notes on the following : (*i*) Egg of fowl; (*ii*) Primitive streak; (*iii*) Hensen's node; (*iv*) Extra-embryonic membranes; (*v*) Allantois.

5. Draw the labelled diagrams of the following : (*i*) An avian egg in section; (*ii*) Entire embryos of chick about 33, 36, 48 and 72 hours of incubation; (*iii*) Stages of development of chick from fertilised egg up to formation of amnion and allantois.
6. Write short note on the regional divisions of mesoderm and their fate in development of chick.
7. Why is the bird's egg is known as a cleidoic egg? Draw a fully labelled diagram of an avian egg.
8. Explain the influence of yolk on the course of development of chick.
9. Explain the significance of morphogenetic movements in chick.
10. Explain how the Hensen's node of chick is homologous to the dorsal lip of frog.

Very Short Answer Questions

1. What is the type of egg in birds?
2. Write the type of cleavage found in hen's egg.
3. What is blastodisc?
4. In birds, what type of blastula is?
5. Define delamination and ingression.
6. What is the aim of incubation in birds?
7. Define primitive streak.
8. What is concrescence that occurs during formation of primitive streak.
9. At what stage primitive streak is fully formed?
10. What is neuropore?
11. At what stage first pair of somites is formed?
12. Write the number of somites formed in 24 hours chick embryo.
13. Write total number of somites formed in chick embryo.
14. What is the type of coelom in chick?
15. What are area vasculosa and area vitellina?
16. Why foetal membranes develop in amniotes?
17. Write about area pellucida and area opaca.
18. Write the names of foetal membranes.
19. What structures develop from area opaca and area pellucida?
20. What are the functions of yolk sac and allantois?
21. What is umbilical cord?

Multiple Choice Questions

1. Viability period of cock's sperm is
   (*a*) About 3 hours (*b*) About 3 days
   (*c*) About 3 weeks (*d*) About 3 months.
2. At the time of laying the chick embryo is in
   (*a*) 2-cell stage
   (*b*) Morula stage
   (*c*) Early blastula stage
   (*d*) Early gastrula stage.
3. At the broad end of egg the shell membranes enclose
   (*a*) Air space (*b*) Nutritive space
   (*c*) Excretory space (*d*) Circulatory space.
4. Chalaziferous layer of albumen is made up of
   (*a*) Lipoprotein (*b*) Ovomucoid
   (*c*) Cholesterol (*d*) Lipoid.
5. Ovum is separated from albumin by
   (*a*) Shell membrane
   (*b*) Plasma membrane
   (*c*) Fertilisation membrane
   (*d*) Dense albumin.
6. Body of embryo proper of chick is formed by
   (*a*) Area pellucida
   (*b*) Area opaca
   (*c*) Latebra
   (*d*) Nucleus of Pander.
7. White yolk beneath the blastodisc is called
   (*a*) Latebra (*b*) Neck of Latebra
   (*c*) Germinal disc (*d*) Nucleus of Pander.
8. Blastodisc is united with the yolk mass by
   (*a*) Epiblast (*b*) Endoblast
   (*c*) Mesoblast (*d*) Periblast.
9. Hensen's node is formed of cells of presumptive
   (*a*) Notochord (*b*) Mesoderm
   (*c*) Gut (*d*) Heart.
10. Extra-embryonic structures are formed by
    (*a*) Area opaca (*b*) Area pellucida
    (*c*) Latebra (*d*) Nucleus of Pander.
11. In chick, the incubation period is
    (*a*) 14 days (*b*) 21 days
    (*c*) 28 days (*d*) 35 days.
12. Primitive streak begins after
    (*a*) 12 hours (*b*) 15 hours
    (*c*) 18 hours (*d*) 24 hours.
13. The first sign of metameric segmentation in chick embryo is the appearance of
    (*a*) Ectoblastic somites
    (*b*) Endoblastic somites
    (*c*) Mesoblastic somites
    (*d*) Meroblastic somites.

14. Anterior region of area pellucida in front of neural plate is without mesoderm and is called
(*a*) Proamnion (*b*) Preamnion
(*c*) Postamnion (*d*) Mesoamnion.

15. 4 pairs of somites are formed after
(*a*) 12 hours of incubation
(*b*) 15 hours of incubation
(*c*) 18 hours of incubation
(*d*) 24 hours of incubation.

16. Cleavage in birds produces a cellular cap known as
(*a*) Ectoderm (*b*) Endoderm
(*c*) Blastoderm (*d*) Chordamesoderm.

17. The function of chalaza in the hen's egg is to
(*a*) Nourish the embryo
(*b*) Balance the egg
(*c*) Keep the blastoderm on the top
(*d*) None of the above.

18. One of the following is responsible for the protection of the embryo
(*a*) Amnion (*b*) Allantois
(*c*) Chorion (*d*) Placenta.

19. Blastoderm or germinal disc refers to the protoplasm of the egg, this is restricted in a small area in
(*a*) Birds egg (*b*) Amphibian egg
(*c*) Reptilian egg (*d*) Mammalian egg.

20. The significance of the greater amount of yolk in bird eggs as compared to the amout of yolk in the mammalian egg is that
(*a*) Birds need more energy for development
(*b*) Mammals do not depend totally on yolk for development
(*c*) Birds develop slower and, therefore, need a greater supply of yolk
(*d*) Yolk of mammals is more concentrated.

21. Which of the following embryonic membranes is excretory in function?
(*a*) Amnion
(*b*) Allantois
(*c*) Yolk sac
(*d*) Vitelline.

22. Amnion and chorion refers to
(*a*) Blastoderm of the egg
(*b*) Protoplasm of the egg
(*c*) Extra-embryonic membranes
(*d*) Yolk present in the egg.

23. The embryonic respiratory organ is
(*a*) Amnion (*b*) Chorion
(*c*) Allantois (*d*) Yolk sac.

24. The embryonic urinary bladder is
(*a*) Amnion (*b*) Chorion
(*c*) Allantois (*d*) Yolk sac.

ANSWERS

1. (*c*)	2. (*d*)	3. (*a*)	4. (*b*)	5. (*c*)	6. (*a*)	7. (*d*)	8. (*d*)
9. (*a*)	10. (*a*)	11. (*b*)	12. (*c*)	13. (*c*)	14. (*a*)	15. (*d*)	16. (*c*)
17. (*b*)	18. (*a*)	19. (*a*)	20. (*a*)	21. (*b*)	22. (*c*)	23. (*c*)	24. (*c*)

39

Chapter

Development of Rabbit

Contents of this Chapter

Both birds and mammals have evolved from reptilian ancestors, but the developmental features of mammals have become greatly modified. The embryo develops within a specialised part of the oviduct called **uterus** or womb, the embryo establishing an intimate contact with the uterine wall and the young are born alive so that the eutherian mammals are **viviparous**.

Egg

The egg of rabbit is 0.1 mm in diameter and is a secondary oocyte when shed from the Graafian follicle of the ovary, it is surrounded by a non-cellular, transparent striated secondary membrane, the **zona pellucida** secreted by the surrounding follicle cells. Around the zona pellucida are some follicle cells (cumulus cells surrounding the oocyte) forming a **corona radiata** and several layers of granulosa cells from the cumulus oophorus, which breaks free from the follicle and passes with the fluid into the oviduct at the time of ovulation. In rabbit, the ovulation takes place after the copulation (*i.e.*, female rabbit is an induced ovulator). The ovulation of rabbit and other mammals is found to be controlled by both the hypothalamus and pituitary gland. In rabbit, several ova mature and ovulate simultaneously. The egg is **alecithal** with almost no yolk, the nucleus lies in the upper half or animal hemisphere, but the lower half or vegetal hemisphere has some particles of yolk distributed evenly in the cytoplasm.

Fertilisation

The fertilisation is internal and the egg is **fertilised** in the upper part of the oviduct *i.e.*, in the fallopian tube, and it immediately undergoes a second maturation division to become a mature ovum. Two polar bodies lie within the zona pellucida near the animal pole. In mammals, the entire spermatozoon penetrates the egg but the tail soon degenerates, penetration of sperms into an egg is brought by an enzyme **hyaluronidase** present in the semen. The zygote passes down the oviduct and is covered by a coat of dense albumen.

EARLY EMBRYONIC DEVELOPMENT

Cleavage

Cleavage is **holoblastic** but unequal. Unlike the lower animals, cleavage in mammals is not always synchronous. The number of blastomeres do not increase by a regular doubling sequence but tend to show arithmetical progression. The first cleavage furrow is vertical dividing the zygote completely into two slightly unequal **blastomeres**, one being smaller and more opaque than the other. The second cleavage is also vertical and at right angle to the first, resulting into four cells. The third cleavage first occurs in two larger blastomeres and then in the remaining two smaller blastomeres, producing eight blastomeres. All the cleavages are at right angles to one another, thus, showing asynchrony of cleavage. After this rate of mitotic divisions increases and the 16-cell stage is reached. This 16-cell stage is a solid mass of blastomeres, representing the **morula**. Thus, zygote, during its journey in the fallopian tube, cleaves to form the morula by the time it reaches the upper part of uterus. It lies for about 2 days free in the uterine lumen, bathed by the secretion of the uterine glands. Soon an eccentrically placed fluid-filled cavity, the **blastocoel** appears within the morula which is then known as a **blastocyst** or **blastula**. The blastocoelic cavity gradually enlarges due to accumulation of the fluid that passes in the morula from the uterine cavity through the zona pellucida.

Blastocyst

The embryo swells rapidly and the liquid food collects in a cavity which separates an outer layer of smaller flat **trophoblast cells**, enclosing the fluid-filled blastocoel and the knob of round-cells, called **inner cells mass**, at one side of the blastocoel. The embryo is now called a **blastocyst**, which appears like a blastula but is not comparable to it. Those trophoblast cells which lie above the inner cell mass are called **cells of Rauber**. As the blastocyst grows the inner cell mass becomes a knob-like thickening at one pole, and it is called an **embryonal knob** because all parts of the embryo will be derived from it. The embryonal knob is comparable to the area pellucida of the chick, the trophoblast is equivalent to the extra-embryonic ectoderm, the cavity of the embryo to the subgerminal cavity together with the space filled by the yolk in chick, and the blastocyst is equivalent to the blastoderm of the chick.

During the first few days, it is still enclosed by the zona pellucida but it later ruptures due to the blastocyst expansion and thinning of the zona pellucida. Loss of blastocyst layer exposes the trophoblast cells to the uterine wall.

Implantation

The trophoblast cells opposite the embryonal knob form small papillae, **trophoblastic villi** which penetrate into depressions or **crypts** in the uterine wall, so the blastocyst becomes attached or is **implanted** in the uterus of the mother. These villi absorb food from the uterus. The process of implantation is initiated and controlled by a protein called **blastokinin** secreted by the uterus into the uterine cavity. The trophoblast cells not in contact with inner cell mass later increase in size and invade the uterine tissue, and those trophoblast cells overlying the inner cell mass continues dividing and form the extra-embryonic ectoderm (*i.e.,* primary chorionic villi).

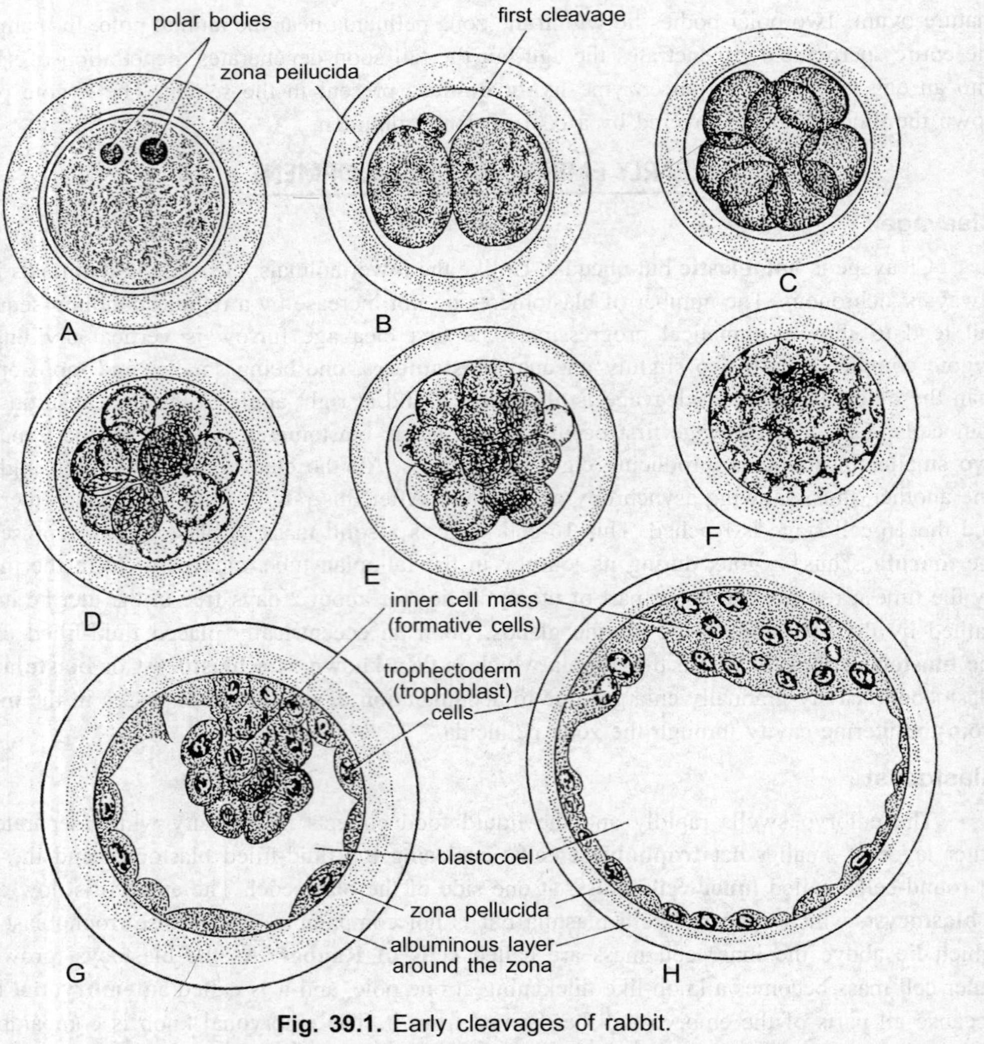

Fig. 39.1. Early cleavages of rabbit.

Inner cell mass cells facing the blastocoel become flattened and rich in rough endoplasmic reticulum. These are the **primary endodermal cells**, which separate from the inner cell mass and start migrating on the inner surface of the trophoblast or trophectoderm. The primary endoderm, thus, forms the lining of the blastocoel, now called the **yolk sac**. It does not seem to contribute to the definite endoderm of the embryo. As the hypoblast (or endoderm) spreads out to enclose the cavity of the yolk sac, the inner cell mass also spreads out and becomes arranged into a plate, resembling the epiblast of the blastodisc of reptiles and birds.

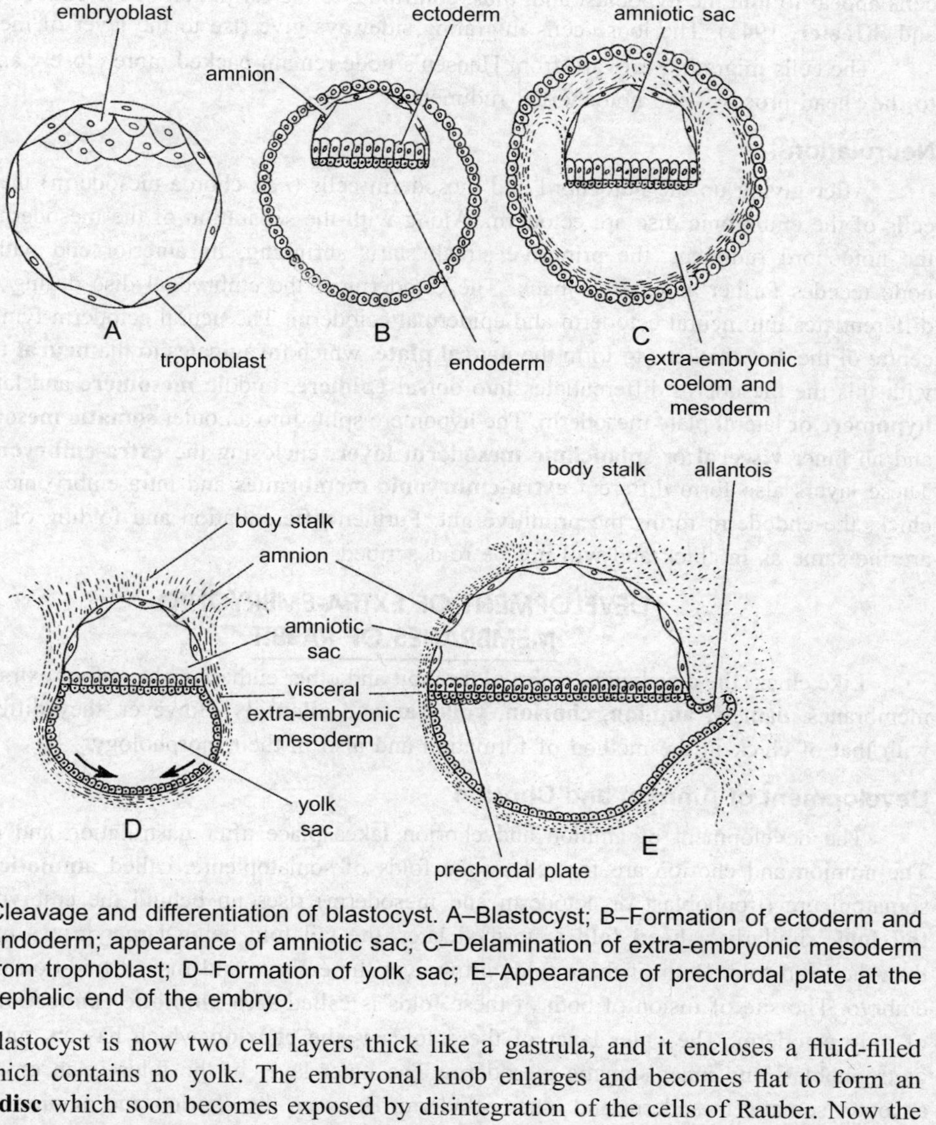

Fig. 39.2. Cleavage and differentiation of blastocyst. A–Blastocyst; B–Formation of ectoderm and endoderm; appearance of amniotic sac; C–Delamination of extra-embryonic mesoderm from trophoblast; D–Formation of yolk sac; E–Appearance of prechordal plate at the cephalic end of the embryo.

The blastocyst is now two cell layers thick like a gastrula, and it encloses a fluid-filled **yolk sac** which contains no yolk. The embryonal knob enlarges and becomes flat to form an **embryonal disc** which soon becomes exposed by disintegration of the cells of Rauber. Now the epiblast becomes temporarily superficial. The blastodisc, consisting of an epiblast and a hypoblast, becomes quite sharply delimited from the remainder of the embryo. The epiblast consists of a thick plate of columnar cells, clearly distinguishable from the flat and more irregularly arranged cells of the trophoblast. The hypoblast cells beneath the blastodisc may become cubiodal or even columnar, and thus, they differ from the cells of extra-embryonic endoderm lining the internal surface of the trophoblast.

Gastrulation

The gastrulation of rabbit resembles with that of reptiles and birds. A primitive streak is formed, and a Hensen's node is seen at the anterior end of the primitive streak. The primitive streak is, however, much shorter than in birds, and when fully formed, does not surpass half the length of the blastodisc, being confined to its posterior part. The cells of the primitive streak migrate downward and sideways between the epiblast and the hypoblast. Some of the migrating cells appear to join the hypoblast and, thus, contribute to the formation of the endoderm (**Heuser** and **Streeter**, 1941). The loose cells migrating sideways give rise to the layer of mesoderm.

The cells migrating forward from Hensen's node remain packed more closely and give rise to the "head process"–the notochordal rudiment.

Neurulation

After giving up the notochord and mesoderm cells (*i.e.,* chorda-mesoderm) the remaining cells of the embryonic disc are ectoderm. Along with the separation of the mesoderm layer and the notochord rudiment, the primitive streak starts shrinking, its anterior end with Hensen's node recedes farther and farther back. The ectoderm of the embryonal disc during neurulation differentiates into neural ectoderm and epidermal ectoderm. The neural ectoderm lying along the centre of the disc thickens to form the **neural plate**, which invaginates to the **neural tube**. Along with this the mesoderm differentiates into dorsal **epimere**, middle **mesomere** and latero-ventral **hypomere** or lateral plate mesoderm. The hypomere splits into an outer **somatic mesoderm layer** and an inner **visceral** or **splanchnic mesoderm layer**, enclosing the **extra-embryonic coelom**. These layers also form different **extra-embryonic membranes** and intra-embryonic parts. Like chick, the endoderm forms the primitive gut. Further differentiation and folding of the embryo are the same as in chick and will not be re-described.

DEVELOPMENT OF EXTRA-EMBRYONIC MEMBRANES OF RABBIT

Like chick, the developing embryo of rabbit and other eutherians have four extra-embryonic membranes, namely, **amnion**, **chorion**, **yolk sac** and **allantois**. However, they differ markedly with that of chick in the method of formation and also in their morphology.

Development of Amnion and Chorion

The development of amnion and chorion takes place after gastrulation and neurulation. The amnion and chorion are formed by the folds of somatopleure, called **amniotic folds**. The somatopleure (trophoblast or ectoderm and mesoderm) rises up behind the embryo to form a **tail fold**, similarly a **head fold** is formed later, the tail fold being larger grows more rapidly than the head fold. Both of these folds grow over the embryo and fuse at mid-dorsal end of the embryo. The site of fusion of both of these folds is called **sero-amniotic connection** consisting of only ectoderm. The outer layer of these folds is the **chorion** which has an outer ectoderm or trophoblast and inner somatic mesoderm. The inner layer of the folds is an **amnion** having an outer somatic mesoderm and inner ectoderm. Between the chorion and amnion is an **extra-embryonic coelom**. Thus, the embryo is enclosed in a fluid-filled **amniotic cavity** lined with ectoderm which acts as a protective cushion and prevents desiccation of the embryo.

Functions of amnion and chorion. The **amnion** completely encloses the developing embryo and it contains a fluid, the amniotic fluid in which the embryo bathes. The embryo freely moves within the amnion and it also protects the embryo from abrasion and outward concushions.

The **chorion** is the outer surface through which exchange of substances occurs between embryonic tissue and the maternal tissue. The chorion also develops more or less complex **villi**, which penetrate into the uterine tissue and, thus, greatly increase the absorptive surface of the chorion to absorb nourishment.

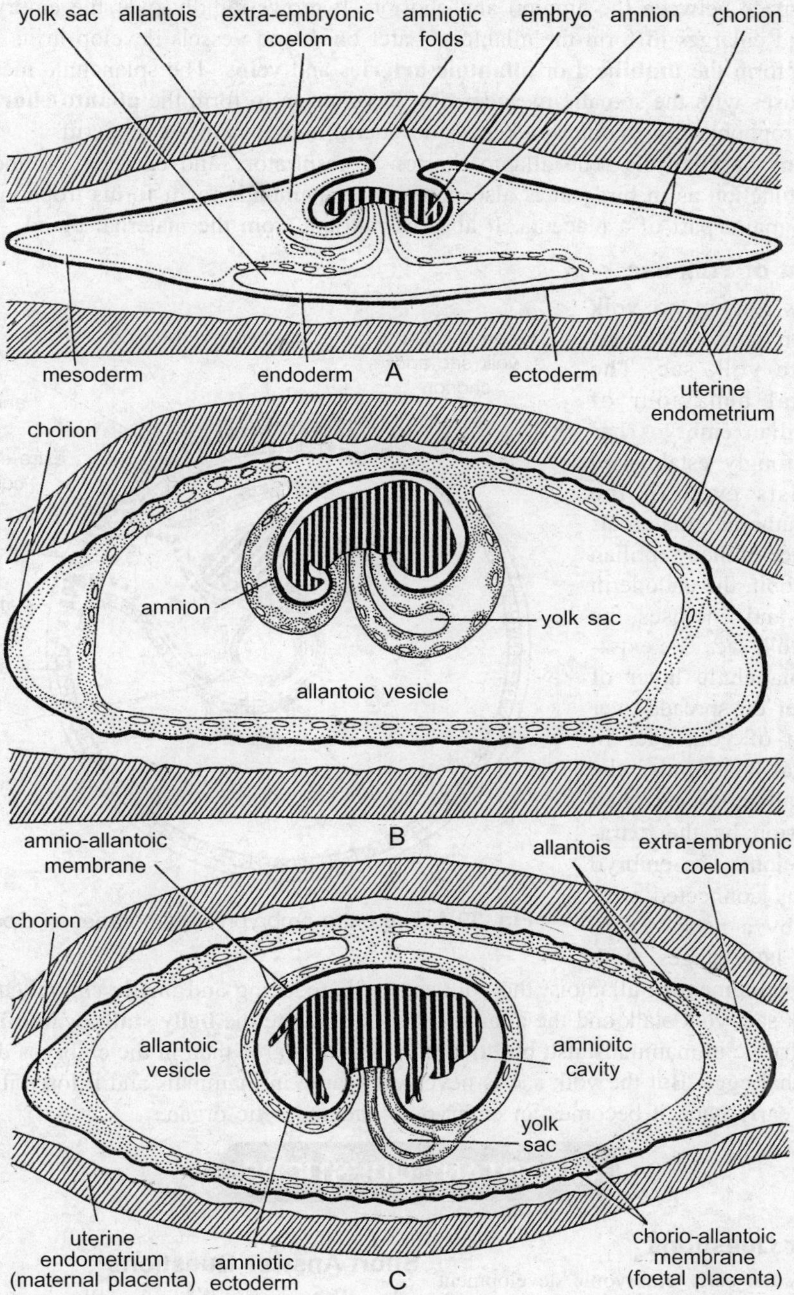

Fig. 39.3. Development of extra-embryonic membranes of pig. A, B–Early stages; C–Fully developed extra-embryonic membranes.

Development of Allantois

The allantois of rabbit and other mammals is formed of splanchnopleure, *i.e.,* outer endoderm and inner splanchnic mesoderm. Even before the closure of the amniotic cavity the hindgut of the embryo grows out in the form of a sac which gradually increases in the extra-embryonic coelom, *i.e.,* space between the amnion and chorion. It grows rapidly over the embryo outside the amnion and enlarges to form the allantois. Later on, blood vessels develop in its wall, later some of them form the **umbilical** or **allantoic arteries** and **veins**. The splanchnic mesoderm of the allantois fuses with the somatic mesoderm of the chorion to form the **allanto-chorion** which has an outer trophoblast, a middle thickened mesoderm, and an inner endoderm.

Functions of allantois. The allanto-chorion is respiratory and excretory (removes CO_2, urea, etc.) in function as in birds. It is also nutritive in mammals for it forms trophoblastic villi which are the major part of a placenta. It absorbs nutrient from the maternal blood.

Development of Yolk-sac

There is almost no yolk in mammalian eggs, yet their embryos have yolk sac. The developmental behaviour of ancestral reptilian embryos has become so strongly established that it persists even in the present-day eutherian mammals as an inheritance of their reptilian ancestry. In rabbit, the endoderm spreads over and encloses the cavity of the yolk sac. The extra-embryonic splanchnic layer of mesoderm later on spreads over the endoderm of yolk sac. In most placental mammals, the yolk sac is small and is separated from the chorion by the extra-embryonic coelom. The embryo proper remains connected with the yolk sac by a narrow **yolk stalk**. During later stages, with the gradual development of allantois, the yolk sac starts reducing and finally shrivelled out. This shrivelled yolk sac, yolk stalk and the allantoic stalk constitute the **belly stalk** or **umbilicus**. The yolk sac is nutritive in mammals also but to a much lesser degree than in the embryos developing from polylecithal eggs. But the yolk sac is never very large in mammals and it loses its nutritive function very early, then it becomes an embryonic haemopoietic organ.

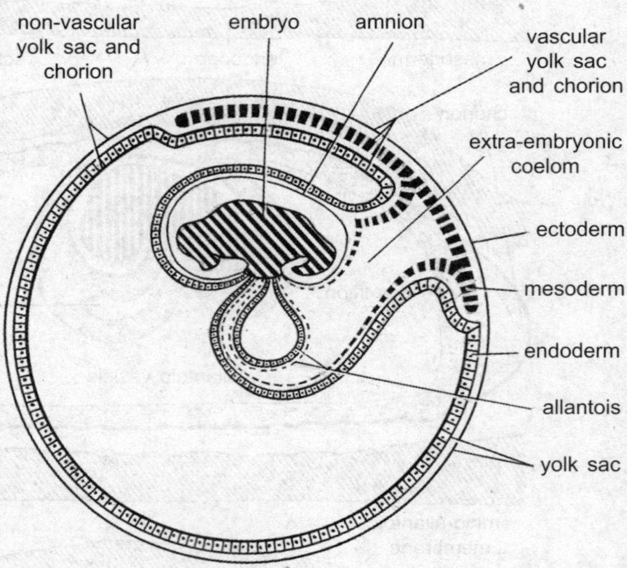

Fig. 39.4. The extra-embryonic membranes of *Opossum.*

IMPORTANT QUESTIONS

Long Answer Questions

1. How does the early embryonic development (up to gastrula stage) of a mammal differ from that of *Amphioxus* and that of frog?
2. Describe the development of foetal membranes in a mammal (rabbit) and write their functions.
3. Draw neatly labelled diagrams to illustrate the development of foetal membranes of rabbit.

Short Answer Questions

1. Write short notes on the following : (*i*) Structure of egg of a rabbit; (*ii*) Early embryonic development of rabbit; (*iii*) Amnion and chorion in rabbit; (*iv*) Yolk sac; (*v*) Allantois in rabbit.

Very Short Answer Questions

1. What is blastocyst or blastosphere?
2. Write about implantation.

3. Which foetal membrane develops placenta?
4. Write about yolk sac in eutherian mammals.

Multiple Choice Questions

1. The eggs of mammals are
   (*a*) Alecithal (*b*) Mesolecithal
   (*c*) Isolecithal (*d*) Telolecithal.
2. The cleavage in mammals is
   (*a*) Holoblastic equal (*b*) Holoblastic unequal
   (*c*) Meroblastic (*d*) Apiblastic.
3. Gestation period of rabbit is
   (*a*) 28 to 32 days (*b*) 48 to 50 days
   (*c*) 10 to 20 days (*d*) 5 to 6 days.
4. In mammals embryo develops within a specialised part of the oviduct is known as
   (*a*) Ovisac (*b*) Yolk sac
   (*c*) Uterus (*d*) Vagina.
5. Fertilisation in mammals occur in the
   (*a*) Ovary (*b*) Fallopian tubes
   (*c*) Uterus (*d*) Vagina.
6. After the main organs of embryo and its embryonic membranes have been formed, then the mammalian embryo is known as
   (*a*) Endometrium (*b*) Morula
   (*c*) Gastrula (*d*) Foetus.
7. The cells destined to become the embryo proper in a mammal constitute the
   (*a*) Trophoblast
   (*b*) Cells of Rauber
   (*c*) Inner cell mass
   (*d*) Cellular capsule.
8. The chorion is composed of
   (*a*) Ectoderm and mesoderm
   (*b*) Syncytial trophoblast
   (*c*) Cytotrophoderm
   (*d*) All the above.
9. During development of human embryo, the blood tissue starts developing in the 15th week. This development starts by the formation of
   (*a*) Blood plasma (*b*) Blood platelets
   (*c*) RBCs and WBCs (*d*) Blood islands.
10. In rabbit, morula develops into the fluid-filled ball called
    (*a*) Coeloblastula (*b*) Blastocyst
    (*c*) Trophoblastula (*d*) Gastrula.
11. The acrosome of sperm produces an enzyme
    (*a*) Hyaluronidase (*b*) Androgammone
    (*c*) Gynogammone (*d*) Acrozin.
12. Cells of Rauber are found in
    (*a*) *Amphioxus* (*b*) Frog
    (*c*) Chick (*d*) Rabbit.
13. Yolk sac placenta where yolk sac wall is intimately connected to uterine wall is found in
    (*a*) Rabbit (*b*) Opossum
    (*c*) Rat (*d*) Man.
14. Chorio-allantois in case of mammals perform following functions
    (*a*) Nutrition (*b*) Respiration
    (*c*) Excretion (*d*) All these.

Answers

1. (*a*)	2. (*c*)	3. (*a*)	4. (*c*)	5. (*b*)	6. (*d*)	7. (*c*)	8. (*a*)
9. (*d*)	10. (*b*)	11. (*d*)	12. (*d*)	13. (*b*)	14. (*d*)		

40

Chapter

Placenta and Placentation in Mammals

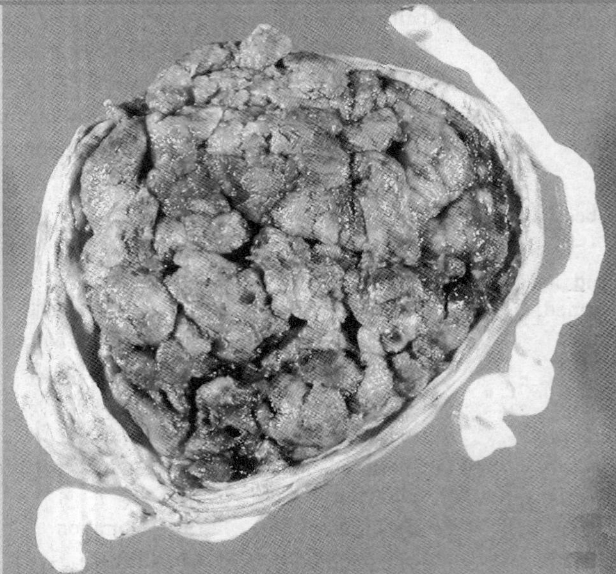

Contents of this Chapter

The **placenta** is a Greek word and it means a "*flat cake*". This name is derived from the human placenta which is flat, rounded mass, shaped more or less like a pancake. The term placenta in its broadest sense refers to any region in a viviparous organism where meternal and embryonic tissues of any kind are closely apposed, and which serves as a site for physiological exchange between parent and embryo (**Kent**, 1987). The placenta can be defined as a temporary organ which is formed jointly by the extra-embryonic membranes of the foetus and maternal tissue. The developing embryo or foetus of viviparous mammals obtains its nourishment from the maternal uterine tissue.

The term placenta applies to any type of organ built up of maternal and foetal tissues jointly and which serves for the transport of nutrient substances from the tissues of mother into those of embryo. The mode of formation and fusion of the placenta to the uterine wall is called **placentation**.

The phenomenon of placentation is invariably related with the viviparity of the organism. In all viviparous animals, the development of the young takes place inside the uterus of the mother because the amount of stored yolk of egg does not remain sufficient for the development of an embryo which can lead an independent mode of existence, therefore, the developing embryo has to depend on the mother in lesser or greater degree for nourishment, oxygen supply and other physiological assignments. Such embryos which develop inside the uterus of the mother often get attached with uterine wall to draw necessary substances from the maternal (uterine) circulation by means of a organ called placenta.

The placentae are not found exclusively in mammals but appear also in animals belonging to various groups of the animal kingdom such as in *Peripatus* (Protracheata), *Salpa* (Tunicata), *Mustelus laevis* (Elasmobranchii) and certain lizards (Reptilia). The nature of the tissues entering into the formation of the placenta is not the same in all cases. For example, in placental fishes and reptiles, the yolk sac comes in close relation with the maternal blood stream and vitelline circulation carries the materials to and from the embryo. However, the placentae of mammals have following mode of origin, structure, type and physiology.

MODE OF ORIGIN AND TYPES OF MAMMALIAN PLACENTAE

A mammalian placenta, typically, is a structure produced by the apposition or fusion of the extra-embryonic membranes (*i.e.*, chorion) with the endometrium of uterus for the purpose of physiological exchange. It, therefore, follows that the placenta from the point of view of its origin, consists of two parts : a **foetal placenta**, furnished by the extra-embryonic membranes and a **maternal placenta**, furnished by the uterine endometrium.

Now, it would be obvious that while on the maternal side a single component, the **endometrium** is involved; on the foetal side, one has to consider the prospective roles of four elements – **amnion**, **chorion**, **yolk sac** and **allantois**. The first of these, the amnion, may be ruled out immediately, as it is making no direct contribution to the placenta. This leaves the other three, of which the chorion, because of its most external position, is the membrane making immediate contact with the endometrium. In **chick embryo**, the chorion plays its role by way of a vascular supply, which it acquires from the allantois. In **mammals**, there are two possible sources of chorionic vascularisation – the **vitelline circulation** provided by the yolk sac and **allanotic circulation**, provided by the allantois. Thus, it can be said that in mammals, there exists two essentially different main types of placentae – the **chorio-vitelline placenta** and the **chorio-allantoic placenta**.

A. Chorio-vitelline Placenta or Chorionic Placenta or Yolk-sac Placenta

In some marsupials (*Didelphys, Macropus*), the allantois remains relatively small and never makes contact with the chorion. Whereas the yolk sac becomes very large and fuse broadly with the chorion. In these forms, the chorion gains its blood supply from the network of vitelline blood vessels of yolk sac. Such a placenta is called **yolk sac placenta** or **chorio-vitelline placenta**. In such a foetal placenta, the chorion never advances beyond a smooth membrane is close apposition with the vascular uterine lining, the endometrium. The chorio-vitelline placenta also occurs as a temporary structure in insectivores, rodents and horse (eutherians).

B. Chorio-allantoic Placenta or Allantoic Placenta

In some marsupials (*e.g., Parameles, Dasyurus*), and all eutherians the yolk sac remains rudimentary and the allantois becomes well developed and vascularised to fuse with chorion and to furnish the latter the blood supply. Such a foetal placenta is called **chorio-allantoic placenta**. In this kind of placenta, the chorion is not smooth, but bears root-like, vascular processes, the **villi**, which grow out from the chorion into the adjacent maternal tissue. Thus, in chorio-allantoic placenta, the chorion lined with allantois becomes associated with the uterine wall during the gestation period. In it the allantoic (umbilical) vessel transports the materials. As the embryonic body develops, it separates from various membranes until only the **umbilicus cord** connects the foetus with the placenta. The final part of the placenta consists of chorionic **villi** formed of outer chorionic ectoderm, inner allantoic endoderm and in between these is a highly vascularised mesodermal connective tissue of double origin–somatic or chorionic and allantoic or splanchnic. The ectodermal cells have mircovilli to increase the absorptive surface. The trophoblast that give rise to villi often becomes syncytial on its outer border towards endometrium (syncytiotrophoblast). The maternal part of placenta consists of the epithelium of endometrium and the stroma of endometrium with its glands and the uterine blood vessels.

CLASSIFICATION OF CHORIO-ALLANTOIC PLACENTA

The chorio-allantoic placenta has been classified into various types on the basis of its morphology, arrangement of villi, histology and electron microscopy.

A. Morphological Classification of Placenta

On the basis of closeness of foetal and maternal tissues, placentae may be of the following three types :

1. Non-deciduous placenta or semiplacenta. In most mammals, the implantation is superficial, *i.e*, the blastocyst lies in the cavity of the uterus in contact with the uterine wall. The contact may be made more intimate by the surface of the blastocyst by forming finger-like outgrowths which penetrate into depressions in the wall of the uterus. Such outgrowths are intially formed by the trophoblast (*i.e.,* the epithelial layer covering the blastocyst), but later on the connective tissue and blood vessels invade the outgrowths. These outgrowths are called **chorionic villi**, the blood vessels of chorionic villi are the branches of allantoic blood vessels in case of chorio-allantoic placenta. (In chorio-vitelline placenta, vitelline blood vessels give their branches to chorionic villi).

At the time of birth, when **parturition** (the separation of the foetus and its membranes from the mother's body) occurs, the chorionic villi are simply drawn out from the depressions in the wall of the uterus and, thus, maternal and foetal tissues are separated without further damage to the uterine wall and no bleeding occurs. This type of placenta is called **non-deciduate** or **non-deciduous placenta** and is found in pigs, cattle and some other mammals. Further, the chorionic villi of a non-deciduate placenta, because lie in apposition with the endometrium, but, do not fuse with it, so such a placenta is also called **semiplacenta**.

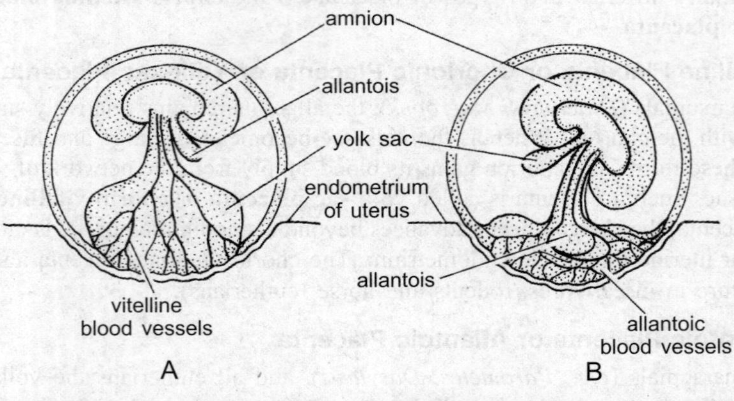

Fig. 40.1. Comparison of two types of mammalian placentae. A–Chorio-vitelline placenta; B–Chorio-allantoic placenta.

2. Deciduous placenta or placenta vera. In other mammals, however, the degree of intimacy between maternal and foetal tissues becomes further increased. The wall of the uterus becomes eroded to various degress through the action of the trophoblast and the embryonic tissues penetrate into the uterine wall, establishing a more intimate contact and facilitating the passage of substances from the mother to the foetus and from the foetus to the mother. Here because the chorionic villi fuse with the eroded uterine mucosa, such placenta is called **placenta vera** (true placenta). At the end of pregnancy the uterine wall is no longer intact and when the foetus with its membranes including the chorion is removed, more or less extensive haemorrhage from the uterine wall ensues (*i.e.,* at birth, when such placenta is discharged, the uterine lining also tears away with some bleeding). Such a type of placenta found in higher eutherian mammals and is called **deciduate** or **deciduous placenta**.

The maternal tissues which are expelled at birth in the case of deciduate placenta are called **deciduae**. The haemorrhage at parturition is normally stopped by the same mechanism as serves for the expulsion of the newborn, the contraction of the muscular wall of the uterus constricts the blood vessels and, thus, slows down the flow of blood, until clotting of the blood stops the haemorrhage altogether.

3. Contra-deciduate placenta. In *Perameles* and *Talpa* (mole), somewhat modified type of deciduate placenta occurs, which is called **contra-deciduate placenta**. In such case, not only there is a loss of maternal tissue but also of the foetal portion of the placenta, both of which absorbed in situ by maternal leucocytes.

B. Classification of Placentae According to the Distribution of Villi on Chorion

In different mammals the pattern of distribution of villi varies from species to species and accordingly following kinds of placentae have been recognised :

1. Diffuse placenta. In some mammals (*e.g.*, ungulates, pig, sow, mare, horse, lemur, etc.) the chorionic villi remain scattered all over the surface of the chorion and their placentae are correspondingly expensive. Such placentae are called diffuse placentae.

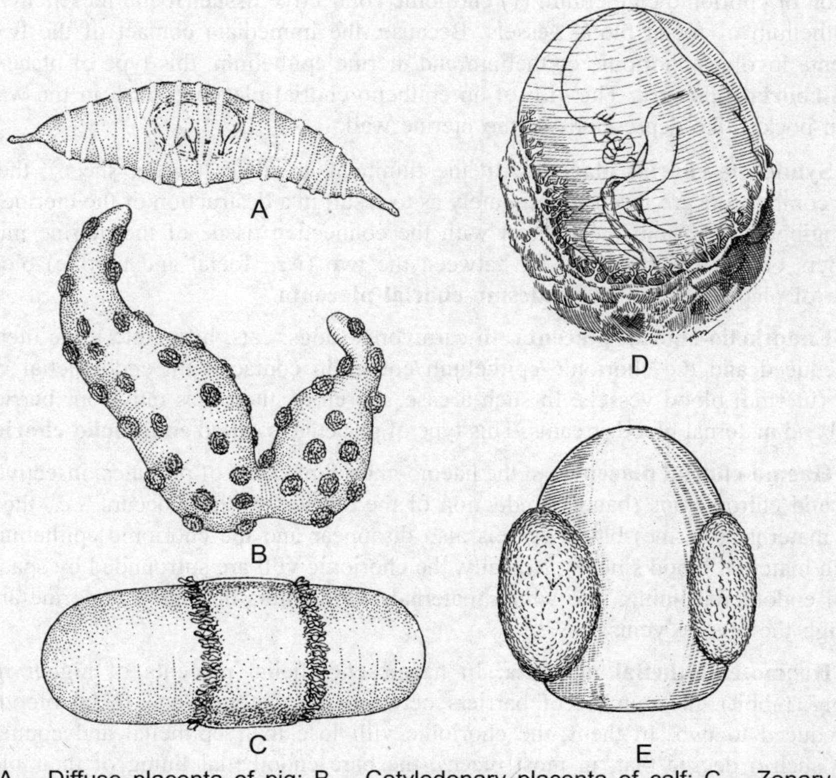

Fig. 40.2. A – Diffuse placenta of pig; B – Cotyledonary placenta of calf; C – Zonary placenta of dog; D – Monodiscoidal placenta of man; and E – Bidiscoidal placenta of monkey.

2. Cotyledonary placenta. In a cotyledonary placenta, the villi are found in groups or patches, while the rest of the chorion surface is smooth. The rosettes or patches of villi are called **cotyledons**, and the placenta of this type is found in ruminants (cud-chewing), ungulates such as cattle, sheep and deer. Camel and giraffe have an intermediate type of placenta in which villi are arranged in cotyledons as well as scattered.

3. Zonary placenta. In a zonary placenta, the villi are developed in the form of a belt or girdle-like band around the middle of their blastocyst or chorionic sac, which is more or less elliptical in shape. Such a placenta occurs in carnivores (*e.g.*, cats, dogs, etc.). Raccoon has incomplete zonary placenta.

4. Discoidal placenta. In insctivores, bats, rodents (mouse, rat, rabbit, etc.) and bear, the villi are restricted to a circular disc or plate on the dorsal surface of blastocyst.

5. Metadiscoidal placenta. In primates also discoidal placenta is found but of special type, *i.e.,* chorionic villi are at first scattered but later on become restricted to one or two discs. Thus, in man the placenta has a single disc-shaped villous area and is called **monodiscoidal placenta**. In the monkeys, the placenta consists of two disc-shaped villous areas and such a placenta is called **bidiscoidal placenta**.

C. Classification of Placenta According to Histology

On histological basis, following types of mammalian placentae have been recognised :

1. Epithelio-chorial placenta. The epithelio-chorial type placenta is most primitive type and it is found in marsupials, ungulates (pig, horse, sow, cattle, etc.) and lemurs. In this case, placenta is formed of six tissue or membranes : (*i*) the endothelium of the maternal blood vessels; (*ii*) endometrial connective tissue (mesenchyme); (*iii*) uterine epithelium; (*iv*) the ectoderm of the chorion or chorionic epithelium; (*v*) chorionic connective tissue (foetal mesenchyme) and (*vi*) the endothelium of foetal blood vessels. Because, the immediate contact of the two halves of the placenta involves chorionic epithelium and uterine epithelium, this type of placenta is called **epithelio-chorial placenta**. The villi of an epithelio-chorial placenta, push in the wall of uterus and lie in pocket-like depressions of the uterine wall.

2. Syndesmo-chorial placenta. In the ruminant ungulates (cattle, sheep), the foetal and maternal components are fused so intimately as to result in a destruction of the uterine epithelium, thus, bringing the chorion into contact with the connective tissue of the uterine mucosa. Only five barriers or tissues, therefore, lie between the two (*viz.*, foetal and uterine) blood streams. This type of placenta is called **syndesmo-chorial placenta**.

3. Endothelio-chorial placenta. In carnivores (dogs, cats, bears, etc.), the uterine mucosa is also reduced and the chorionic epithelium comes in contact with endoethelial walls of the maternal (uterine) blood vessels. In such a case, therefore, there lies only four barriers between the foetal and maternal blood streams. This type of placenta is called **endothelio-chorial placenta**.

4. Haemo-chorial placenta. In the haemo-chorial placenta of primates, insectivores (moles, shrews), and chiropterans (bats), a reduction of the barriers to three occurs, *i.e.,* the endothelial walls of maternal (uterine) blood vessels also disappear and the chorionic epithelium is bathed directly in maternal blood sinuses. Actually, the chorionic villi are surrounded by spaces (sinuses) devoid of endothelial lining, into which maternal blood enters through the uterine arteries flows out through the uterine veins.

5. Haemo-endothelial placenta. In haemo-endothelial placenta of higher rodents (rat, guinea pig, rabbit), the number of barriers between the maternal and foetal blood streams is further reduced to two. In them, the chorionic villi lose their epithelial and connective tissue layers to such a degree that, in most places, the bare endothelial lining of their blood vessels alone separates the foetal blood from the maternal blood sinuses.

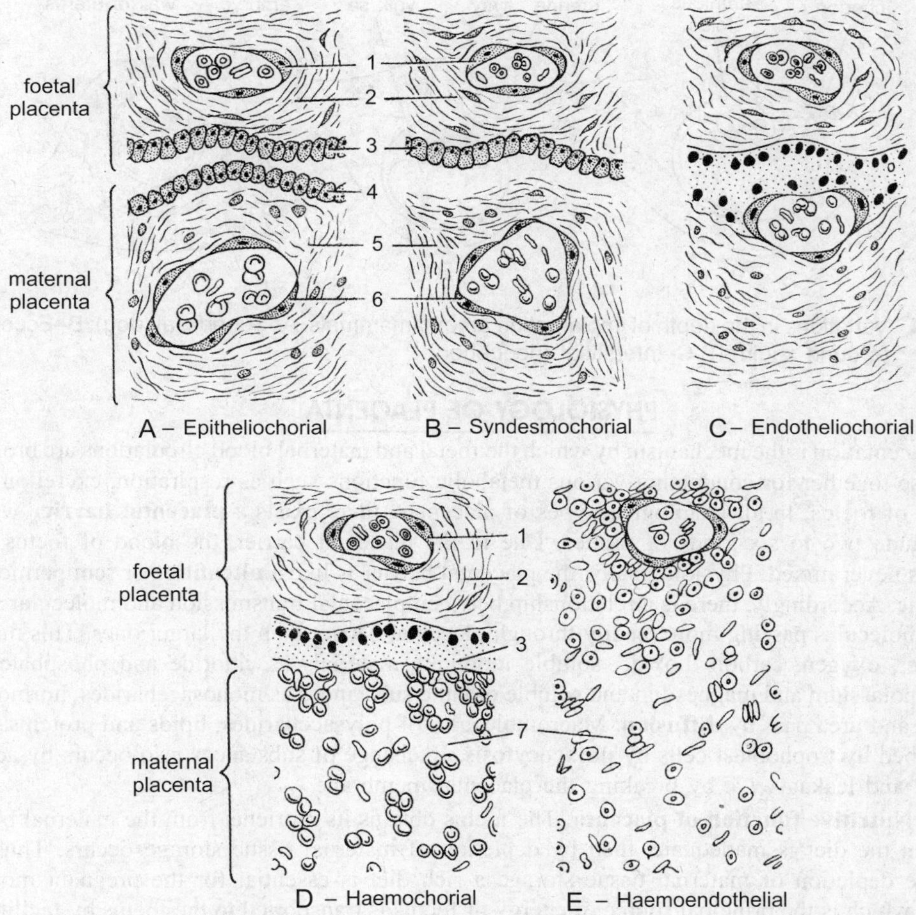

Fig. 40.3. Mammalian placental types arranged in a series to show the progressive elimination of barriers between the maternal and foetal circulation. 1. Endothelium of foetal blood vessel; 2. Chorionic connective tissue; 3. Chorionic epthelium; 4. Uterine epithelium; 5. Endometrial connective tissue (mucosa); 6. Endothelium of maternal blood vessel; 7. Maternal blood pool.

D. Classification of Placentae According to the Mode of Implantation

The relation of the chorionic sac to the uterine wall varies greatly among placental mammals. In general, following three types of implantation may be distinguished, although transitional conditions occur :

1. Superficial implantation. Growth of the chorionic sac brings it into contact with the lining of the main uterine cavity. This type of implantation is called **central implantation**, *e.g.*, ungulates, carnivores, monkey.

2. Eccentric implantation. The chorionic sac lies for a time in a fold or pocket which looses off from the main cavity, *e.g.,* beaver, rat, squirrel.

3. Interstitial implantation. The chorionic sac penetrates into the substance of the uterine lining, *e.g.*, hedgehog, guinea pig, some bats, ape and man.

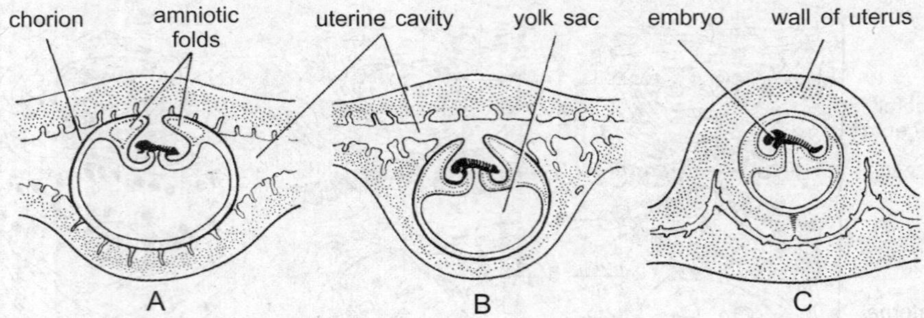

Fig. 40.4. Variations in the depth of implantation among mammals. A–Superficial (dog); B–Eccentric (ground squirrel). C–Interstitial (hedgehog).

PHYSIOLOGY OF PLACENTA

Placentation is the mechanism by which the foetal and maternal blood circulations are brought very close together for couducting various metabolic functions such as respiration, excretion and nutrition of foetus. In all histological types of placentae, there exists a **placental barrier** which may include two to six kinds of tissues. Due to the placental barrier, the blood of foetus and mother is never mixed. Physiologically, the placental barrier is like a **ultrafilter** or semipermeable membrane. Accordingly, there is a relationship between placental transmission and molecular size, smaller molecules passing more readily through placental barrier than the larger ones. This means that water, oxygen, carbon dioxide, soluble inorganic materials as chloride and phosphates of sodium, potassium and magnesium and soluble organic substances as monosaccharides, hormones, vitamins and urea pass by **diffusion**. Macromolecules of polysaccharides, lipids and proteins may be absorbed by trophoblast cells by **pinacocytosis**. Exchange of substances also occurs by active transport and leakage, *i.e,* by breaking the placental membrane.

1. Nutritive function of placenta. The foetus obtains its nutrients from the maternal blood and when the diet is inadequate, then by depletion of maternal tissue storage occurs. Thus, to avoid the depletion of maternal tissue storage a rich diet is essential for the pregnant mother. **Glucose** which is the principal source of energy of foetus is transported to the foetus by facilitated diffusion. **Lipids** needed for foetal growth are transported from the mother to the foetus or synthesised in the foetus (*e.g.,* triglycerides and fatty acids). **Amino acids** are transferred to the foetus by active transport and then proteins are synthesised in the foetal body. **Water**, **sodium**, **potassium** and **chloride** enter the foetal blood by simple diffusion. Whereas **calcium**, **phosphorus** (phosphate), **iron** and **iodine** enter the foetal blood by active transport. **Water soluble vitamins** are transported by active transport but the **fat soluble vitamins** enter the maternal blood by simple diffusion. **Hormones** like insulin, steroids from the adrenals, thyroid hormones and chorionic gonadotropin or placental lactogen cross the placenta at a very slow rate.

2. Respiratory function of placenta. Intake of oxygen and output of carbon dioxide occurs by simple diffusion, across the foetal membrane, *i.e.*, oxygen from maternal blood diffuses into the foetal blood, whereas carbon dioxide from the foetal blood diffuses into the maternal blood.

3. Excretory function of placenta. Excretory wastes such as urea, uric acid and creatinine from the foetus are excreted to the maternal blood through placenta by diffusion. The kidneys of the mother eliminate these wastes of foetus along with her own waste products.

4. The placenta during first half of pregnancy is impermeable to bacteria and macromolecules. Therefore, it reduces the chances of appearance of maternal ill health on the foetus. But in the later half of pregnancy, the placenta becomes more permeable and chances of passing the germs from mother to foetus become more. For example, syphilitic spirochetes enter the foetus during this period and causes congenital syphilis in the foetus.

5. The antibodies, which have developed in the blood of a mother who has acquired immunity to certain diseases, such as **diphtheria**, **scarlet fever**, **small pox** and **measles**, are passed to the foetus, which, thus, becomes passively immunised and unsusceptible to these illnesses in the first period after birth. Certain **pathogenic organisms** and **viruses** can penetrate through the placental barrier and can infect the foetus, if the mother is infected by those pathogens. This is known to happen with **syphilis**, **small pox**, **chicken pox**, **measles**, and **rubella** or German measles.

Many drugs used medically may penetrate the placental barrier and sometimes cause most adverse effects on the embryo. Thus, it is believed that, the drug **thalidomide**, which was used as a sedative, when taken by women in early pregnancy (25 to 44 days), caused very extensive deficiencies in the development of limbs, the alimentary canal (non-perforation of the anus) and the heart. The children born to such mothers had flipper-like limbs (**phocomelia**) and were called **thalidomide babies**. Drugs like quinine and aspirin, exposure to X-rays and radium, alcoholic drinks and cigarette smoking, etc, are found to be potent teratogens. In horses, it also acts like an endocrine gland. It secretes many hormones such as **estradiol**, **progesterone**, **chorionic gonadotropin** and **placental lactogen**. In some animals, such as rabbit, the placenta is a significant source of **relaxin**, which is a protein hormone for the **relaxation** of the pelvic ligaments to facilitate the birth of the young. Human placenta produces a variety of protein and steroid hormones such as oestrogens, and progesterones. Protein hormones are chorionic gonadotropin, placental lactogen, chorionic thyrotropin and corticotropin and relaxin, etc.

IMPORTANT QUESTIONS

Long Answer Questions

1. What is placentation? Deseribe various types of placenta found in mammals.
2. Describe structure, formation and functions of placenta in mammals.
3. Give an illustrated account of different types of allantoic placenta.
4. Write an essay on placentation in mammals.
5. Describe the placenta on the basis of histology.

Short Answer Questions

1. Distinguish between the terms yolk sac placenta and chorio-allantoic placenta.
2. On the basis of shape and distribution of villi, describe the various types of placenta.
3. Describe the yolk sac placenta formed in marsupials.
4. Write the functions of placenta.
5. Describe labelled diagrams of :
   (*i*) Epitheliochorial type of placenta;
   (*ii*) Endotheliochorial type of placenta;
   (*iii*) Syndesmochorial type of placenta;
   (*iv*) Haemochorial type of placenta.
6. Write short notes on : (*i*) Implantation of mammalian embryo; (*ii*) Parturition; (*iii*) Placenta; (*iv*) Physiological role of placenta.

Very Short Answer Questions

1. Define placenta.
2. What is the position of placenta in monotremes?
3. What is yolk sac placenta and in which animal is it found?
4. What is allantoic placenta?
5. Write the name of metatherian in which allantoic placenta is found alongwith yolk sac.
6. What is chorionic placenta? Give its examples.
7. Define non-deciduate placenta.
8. Write about contra-deciduous placenta.
9. Name the type of placenta found in man and apes.

Multiple Choice Questions

1. The egg-laying mammals are
   (*a*) Prototherians
   (*b*) Metatherians
   (*c*) Eutherians
   (*d*) Both prototherians and metatherians.
2. Yolk sac placenta is found in
   (*a*) *Macropus* and *Didelphis*
   (*b*) *Rattus rattus*
   (*c*) *Perameles*
   (*d*) *Talpa.*

3. Yolk sac placenta is derived from
(*a*) Yolk sac and allantois
(*b*) Yolk sac and chorion
(*c*) Yollk sac and amnion
(*d*) Yolk sac only.

4. The type of placenta found in primates is
(*a*) Yolk sac placenta
(*b*) Allantoic placenta
(*c*) Chorionic placenta
(*d*) Allanto-chorionic placenta.

5. Cotyledonary placenta is found in
(*a*) Cow and camel (*b*) Cow and deer
(*c*) Camel and giraffe (*d*) Pig and horse.

6. Contra-deciduous placenta is characteristic of
(*a*) Elephant (*b*) Tiger
(*c*) Cow (*d*) Bandicoot.

7. The primitive type of placenta in mammals is referred to
(*a*) Epithelio-chorial placenta
(*b*) Syndesmo-chorial placenta
(*c*) Endothelio-chorial placenta
(*d*) Haemo-endothelial placenta.

8. The structure which helps in the nutrition of the embryo when the embryo is within uterus in mammals is known as
(*a*) Yolk sac (*b*) Amniotic membrane
(*c*) Crypts (*d*) Placenta.

9. Parturition refers to
(*a*) Degree of intimacy between foetal tissue and maternal tissue
(*b*) The separation of foetus and its membranes from the mother's body
(*c*) Endometrium
(*d*) Placental tissue.

10. The placenta in monkey is known as
(*a*) Bidiscoidal placenta
(*b*) Discoidal placenta
(*c*) Zonary placenta
(*d*) Diffuse and cotyledonary placenta.

11. Placentation of pig is of the
(*a*) Contradeciduate type
(*b*) Indeciduate type
(*c*) Deciduate type
(*d*) Intermediate non-deciduate type.

12. Placentation is not exclusively exhibited by mammal, but also found in other groups of animals like
(*a*) Bony fishes
(*b*) Cartilaginous fishes
(*c*) Amphibians
(*d*) Reptiles (some lizards).

13. Hormone secreted from the placenta at the time of parturition
(*a*) Progesterone (*b*) Relaxin
(*c*) Lactogen (*d*) Estradiol.

14. Haemochorial placenta is found in
(*a*) Cat and camel (*b*) Man and monkeys
(*c*) Giraffe and horse (*d*) Lion and tiger.

15. Eutherian mammals are
(*a*) Oviparous (*b*) Viviparous
(*c*) Ovoviviparous (*d*) None of the above.

16. Attachment of developing embryo to the wall of uterus
(*a*) Implantation (*b*) Ovulation
(*c*) Placentation (*d*) Orientation.

17. In chorionic placenta allantois is limited to
(*a*) Chorionic villi (*b*) Uterine wall
(*c*) Yolk sac (*d*) Umbilical cord.

18. Syndesmo-chorial placenta lacks
(*a*) Chorionic epithelium
(*b*) Uterine connective tissue
(*c*) Chorionic + allantoic mesoderm
(*d*) Uterine epithelium.

19. Functions of placenta include
(*a*) Exchange of substances between mother and foetus
(*b*) Ultrafilter
(*c*) Endocrine gland
(*d*) All of the above.

20. Viral or bacterial infection of placenta is known as
(*a*) Dermatitis (*b*) Appendicitis
(*c*) Syphilis (*d*) Placentitis.

Answers

1. (*a*)	2. (*a*)	3. (*b*)	4. (*c*)	5. (*b*)	6. (*d*)	7. (*a*)	8. (*d*)
9. (*b*)	10. (*a*)	11. (*b*)	12. (*b* & *d*)	13. (*b*)	14. (*b*)	15. (*b*)	16. (*a*)
17. (*d*)	18. (*d*)	19. (*d*)	20. (*d*)				

41

Chapter

Integument and its Derivatives in Vertebrates

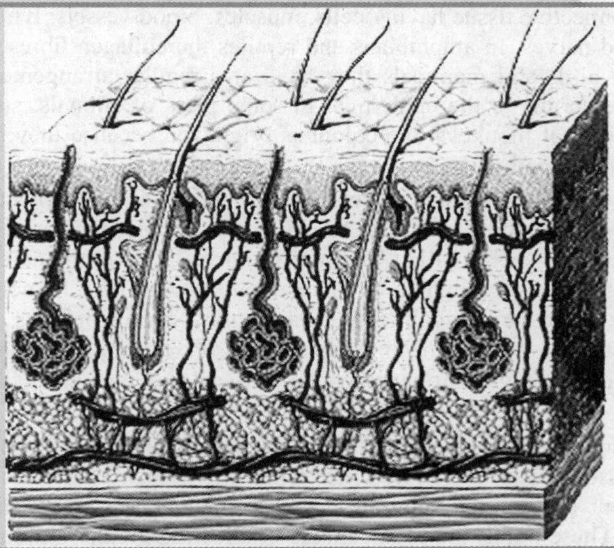

Contents of this Chapter

The integument or the skin forms the external protective covering of an animal (vertebrate). It is continuous with the mucous membrane lining the mouth, rectum, nostrils, eyelids, and the opening of urinogenital ducts. Conjunctiva of eyeball and outer surface of eardrum is also skin. The skin comes in contact with the environment, terrestrial or aquatic, and it influences the character of the skin. The evolution of the vertebrate integument is correlated with the transition of vertebrates from an aquatic to a terrestrial environment.

Structure of skin. The vertebrate skin differs from that of invertebrates in having two layers, an outer **epidermis** derived from the ectoderm, an and inner **dermis** or **corium** formed from the mesenchyme of the dermatomes (mesoderm). The relative amount of the two layers varies with the environment.

1. Epidermis. The epidermis is made of stratified epithelium (several layers of columnar epithelial cells).

These cells are held together tightly by minute **intercellular bridges** found on the surface of cells. The innermost layer of the epidermis is the **stratum Malpighii** or **stratum germinativum** placed over a thin basement membrane. The cells divide constantly to produce new cells. These cells move upwards and tend to become flattened and their protoplasm becomes horny, this is called **keratinisation**. In aquatic fishes and amphibians the keratinised layer forms a cuticle, but in amniotes the outermost keratinised layer forms a layer of cells called **stratum corneum** of hard, horny, flat, and cornified cells made largely of **keratin** which is a tough, waterproof and insoluble protein. It affords protection against mechanical injuries, fungal and bacterial attacks, and prevents desiccation. In many tetrapoda the dead stratum corneum is shed periodically in pieces or all at once. There is no stratum corneum in cyclostomes and fishes since they are completely aquatic but it is found in animals exposed to air. In aquatic vertebrates the epidermis has mucous glands, secreting the mucus which keeps the skin slimy and protects it from bacteria. There are no blood vessels in the epidermis.

2. Dermis. The dermis below the epidermis has an outer **loose layer** and an inner **dense layer**. The dermis is made of dense connective tissue having cells, muscles, blood vessels, lymph vessels, collagen and elastic fibers, and nerves. In amphibians and reptiles the collagen fibres lie at right angles in their planes, but in birds and mammals they have an irregular arrangement. Substances pass by diffusion from the dermis to the epidermis. In some parts of animals, such as the comb of a cock, face and buttocks of monkeys, the colour is bright red because they are intensely vascularised and contain much **blood**.

The skin also contains pigment. If the pigment is present in the epidermis, then it occurs as a diffuse substance or as granules and if the pigment is found in the dermis, then it is in the form of granules in special branching cells called **chromatophores**. The pigment can either collect as a central ball making the skin lighter, or spread out into all the branches making the skin darker, thus, chromatophores bring about colour variations. Chromatophores are of three kinds, **melanophores** contain brown to black melanin, **lipophores** or **xanthophores** have yellow red fatty pigments, and **iridocytes** or **guanophores** contain crystals of guanine which refract light.

Under the dermis the skin has loose subcutaneous areolar tissue which separates the skin from the underlying muscles. It may also contain fat and muscles, especially in mammals. During the course of evolution the integument of anamnia shows a decrease in thickness and also a decrease in the degree of ossification. These trends are of advantage in allowing greater mobility. In amphibians they permit respiration by the skin. But in amniota the skin becomes progressively thicker to prevent loss of water and to retain the body heat.

FUNCTIONS OF THE INTEGUMENT

1. Protection. The integument forms a covering of the body and is protective. It protects the body against entry of foreign bodies and against mechanical injuries. It protects the tissues against excessive loss of moisture, this is very important because both aquatic and terrestrial animals are dependent upon water in their bodies for various metabolic activities. The integument forms protective derivatives, such as scales, bony plates, layer of fat, feathers and hair which reduce the effect of injurious contacts. In some animals the skin shows protective colouration which makes the animals resemble their environment, thus, making them almost invisible to their enemies. Poison glands of toads, slippery skin of aquatic animals and an armour of spines of some mammals are protective devices of the integument. The skin forms a covering which prevents the passage of water and solutes in one of the following ways : (*a*) by formation of **cuticle** in Protochordata and embryos of fishes and amphibians, (*b*) by secreting a coat of **mucus** in fishes and aquatic amphibians, and (*c*) by formation of **keratin** layers in the epidermis of tetrapoda. Keratin is formed from the cytoplasm of degenerating cells of the epidermis which finally form a layer of horny stratum corneum.

2. Temperature control. Heat is produced constantly by oxidation of food stuffs in tissues. This heat is distributed evenly by the circulating blood. The body heat is lost constantly with expired breath, with faeces and urine, and from the surface of the skin. The integument regulates heat and maintains a constant temperature in endothermal animals. In birds the heat is regulated by adjustment of feathers which retain a warm blanket of air, when feathers are held close to the body they remove warm air and body cooled, when feathers are fluffed out they keep the warm air enclosed. In mammals constant evaporation of sweat regulates the body heat. In cold weather contraction of skin's blood capillaries reduces the loss of body heat. In some animals fat in the skin prevents loss of heat because it is a non-conductor of heat.

3. Food storage. The skin stores fat in its layers as reserve food material which is used for nourishment in times of need. In whales and seals the fat of the skin forms a thick layer, called **blubber** which is not only reserve food but also maintains the body temperature.

4. Secretion. The skin acts as an organ of secretion. Glands of the skin are secretory. In aquatic forms there are secretory mucous glands whose secretion keep the skin moist and slippery. In mammals sebaceous glands secrete oil which lubricates the skin and hairs. Mammary glands produce milk for nourishment of the young. In birds uropygial glands secrete oil for preening the feathers. Odours of scent glands attract the opposite sex. Lacrymal glands' secretion wash the conjunctiva of eyeball in mammals. Ear wax (cerumen) secreted by the glands of auditory meatus greases the eardrums and avoids insects to enter the canal.

5. Excretion. The integument acts as an organ of excretion. Shedding of the corneal layer during ecdysis removes some waste substances. In mammals metabolic waste (salts, urea and water) is removed from the blood by means of sweat. Chloride secreting cells are found in gills of marine fishes.

6. Sensation. The skin is an important sense organ because it has various kinds of tactile cells and corpuscles which are sensory to touch, temperature changes, heat, cold, pressure and pain.

7. Respiration. In amphibians, the moist skin acts as an organ of respiration, in frogs the respiratory function of the skin is greater than that of the lungs.

8. Locomotion. Derivatives of the integument bring about locomotion in some animals, such as the fins of fishes aid in locomotion in water, the web of skin in the feet of frogs and aquatic birds aid in swimming, feathers of the wings and tail of birds are used for flying, and extensions of the integument forming "wings" of flying lizards, extinct pterodactyls, flying squirrels and bats.

9. Dermal endoskeleton. The skin contributes to the endoskeleton. It forms the dermal bones of vertebrates and also forms parts of the teeth. Endoskeleton of head protects the brain and sense organs. In the body it protects the soft, tender viscera.

10. Sexual selection. The skin acts as an organ of sexual selection. It provides the feathers of birds which often have brilliant colours which are for sexual attraction. Some integumentary glands of mammals produce odours far attracting the opposite sex. Antlers of male deer distinguishes it from female.

Besides the above functions, mammalian skin synthesises the vitamin D with the help of sebum of sebaceous glands. Brood pouches beneath skin in some fishes and amphibians protect unhatched eggs. Nasal glands of tetrapods keep the nostrils free of dirt and water. Skin also have the power of absorption of oils, ointments, etc.

DERIVATIVES OF THE INTEGUMENT

Both layers of integument have given rise to various types of derivatives. The epidermis gives rise to integumentary glands, epidermal scales, horns, digital structures, different corneal structures, feathers, and hairs. The dermis forms dermal scales of fishes and of some reptiles, plates or scutes in reptiles, fin rays in fishes and antlers in mammals.

I. EPIDERMAL DERIVATIVES

Epidermal derivatives are epidermal glands (unicellular and multicellular), epidermal scales and scutes, horns, digital structures (claws, nails and hoofs), feathers and hairs.

1. Epidermal Glands

Epidermal glands are formed from the Malpighian layer of the epidermis. They arise from the epidermis and often penetrate the dermis. According to their structure they are unicellular or multicellular, tubular or alveolar and simple or compound (branched) glands. These are lined by cuboidal or columnar cells.

(*a*) Unicellular glands are single modified cells found among other epithelial cells, they are present in amphioxus, cyclostomes, fishes and larvae of amphibians. Unicellular glands are known as **mucous cells** or **goblet cells**. They secrete a protein **mucin** which combines with water to form mucus which lubricates the surface of the body. Other unicellular glands are **granular cells** and large **beaker cells** of cyclostomes and fishes, they also secrete mucus.

(*b*) Multicellular glands are of two types :

(*i*) Tubular glands are multicellular tubes of uniform diameter formed as ingrowths of the Malpighian layer into the dermis, *e.g.,* glands of Moll on the margin of the human eyelids. Tubular glands may become coiled at the base deep in the dermis, *e.g.,* sweat or sudoriferous glands of mammals, or they may divide into many tubules which are then called **compound tubular glands**, *e.g.,* mammary glands of females and of males in monotremes and primates, etc., and gastric glands in stomach.

(*ii*) Alveolar or saccular glands are multicellular downgrowths of the Malpighian layer into the dermis, having a tubular duct whose terminal parts form a rounded expansion to become flask-shaped, *e.g.,* mucous and poison glands of amphibians. Alveolar glands may branch into many lobules which finally open into a common duct, they are then called **compound alveolar glands**, *e.g.,* mammary glands of eutherians, and salivary glands.

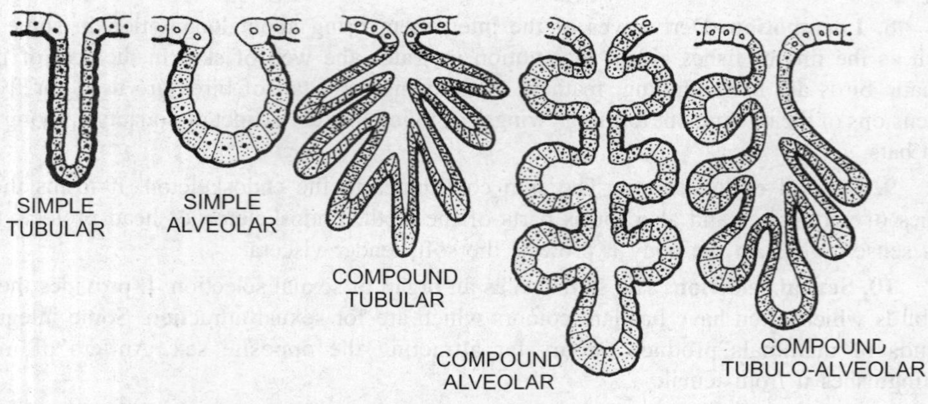

Fig. 41.1. Different types of epidermal glands.

Kinds of epidermal glands. According to function, the epidermal glands of vertebrates are of the following types :

1. Mucous glands. They may be unicellular or multicellular. The unicellular glands are mucous gland cells, granular cells and beaker cells of amphioxus, cyclostomes and fishes. They secrete mucus which keeps the skin moist and slippery, and also affords protection against bacteria and fungi. Mucous cells and granular cells lie near the surface, but the beaker cells lie more deeply and extend from the Malpighian layer to the surface.

Multicelluar mucous glands are alveolar found in some fishes and amphibians. They occur all over the surface of the body and produce mucus for lubricating the skin and in amphibians they keep the skin moist to aid in respiration.

2. Poison glands. Amphibians also have alveolar **poison glands** which are larger but less numerous than mucous glands. In toads masses of poison glands form **parotoid glands** behind the head. The secretion of poison glands has a burning taste and is used as a defence. Caecilians have giant poison glands. Some tubular glands are found on the feet and suctorial discs of tree frogs which aid in climbing. Tubular glands are also found on the swollen glandular thumb pads of male frogs and toads during the breeding season. They aid in clasping the female during amplexus.

3. Luminescent organs or photophores. They are found in longitudinal rows near the ventral side of the body in those fishes which live in deep sea where no light penetrates. Each photophore is a group of epidermal cells lying in the dermis. Each photophore has a lower layer of luminous cells below which is a layer of reflecting pigment cells, and the upper layer of mucous cells forms a lens. The glandular cells produce phosphorescent light which is transmitted to the outside by other cells. The light helps to attract the prey of deep sea fishes.

4. Femoral glands. Femoral glands are found in male lizards (*e.g., Uromastix*) below the thighs in a row from the knee to the cloaca. They secrete a sticky substance which hardens into short spines that are used for holding the female during copulation.

5. Uropygial glands. These are the only glands in birds, and they are best developed in aquatic birds. Uropygial glands are branched alveolar glands located on the dorsal side at the base of the tail or uropygium in the form of swelling. They secrete an oil which is odoriferous and attracts the opposite sex. The oil contains **pomatum** which is picked up with the beak and used for preening and water-proofing the feathers.

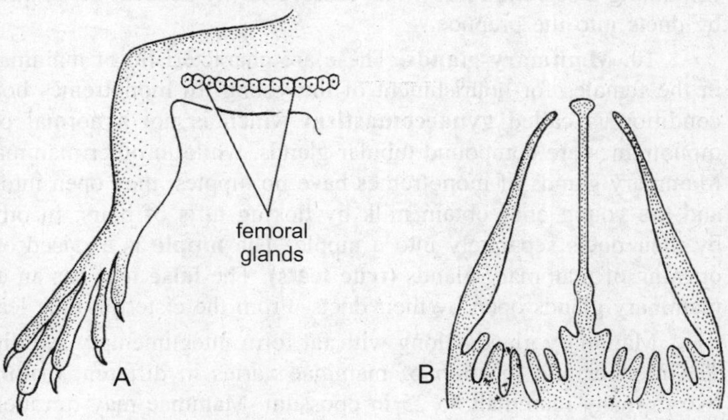

Fig. 41.2. Epidermal glands. A–Femoral glands of lizard; B – Uropygial glands of bird.

6. Sweat glands. The largest number and variety of epidermal glands are found in the skin of mammals. They are tubular or alveolar and multicellular. **Sudorific** or **sweat glands** are long, coiled tubular glands embedded deep in the dermis. Their upper part forms a duct which opens on the surface by a pore and the lower coiled part lies in the dermis surrounded by a network of blood capillaries. Sweat is secreted by sweat glands continuously, may or may not be visible. Sweat contains a large amount of water having dissolved salts of sodium, potassium and urea. Sweat removes some metabolic wastes and regulates the body temperature by evaporation.

Sweat glands are not uniformly distributed. In man they are more numerous on palms, soles, and arm pits. In cats, rats, and mice they are confined to the soles of the feet. In rabbit they are around the lips; in bats on the sides of head; in ruminants on the muzzle and the skin between the digits and in hippopotamus they are found only on the pinnae. There are no sweat glands in *Tachyglossus*, Sirenia, and Cetacea. In some mammals the secretion of sudorific glands is red-coloured. In hippopotamus it is red and spreads on the head and back. The male giant kangaroo *Macropus rufus* secretes red sweat. Modified sudorific glands form **glands of Moll** in the margins of the human eye in connection with eye-lashes. **Ceruminous glands** in the external ear passages of mammals are modified sweat glands and secrete a waxy substance which combines with the secretion of sebaceous glands to form earwax which catches dust. Oil glands form ceruminous glands in the external ears of some gallinaceous birds.

7. Sebaceous glands. Sebaceous glands are alveolar glands opening in hair follicles containing hairs. They also independently open at the skin surface around the genital organs, tip of the nose, and edges of the lips. Sebaceous glands secrete an oil (**sebum**) to lubricate the hairs and also covers the skin with a film of oily coating. The oily secretion of sebaceous glands contain waxes, fatty acids, and cholesterol, which makes the skin **pliable**.

Sebaceous glands are absent in *Manis* (pangolin), and Sirenia, and Cetacea which practically have no hairs. Modified sebaceous glands form **Meibomian glands** in the eyelids, each has a long straight duct into which separate alveoli open. They produce an oily secretion which forms a film over the lacrimal fluid or tears holding them evenly on the surface of the eyeball for keeping the eye moist, in weeping the oily film is broken and tears flow out. **Ceruminous glands** of external auditory meatus are modified sebaceous glands. Their greasy or waxy secretion, called the **cerumen** traps the insects and dust particles.

8. Scent glands. Scent glands are modified sebaceous glands or sweat glands. Their secretion is an allurement to the opposite sex. Scent glands are located in the deer family on the head near the eyes. Skunks and carnivores have scent glands around the anus, and pigs and goats have scent glands between their toes.

9. Preputial glands. These are found in many kinds of mammals. In beaver, large sacs containing a secretion known as **castoreum** lie beneath the skin on either side of penis and open by ducts into the prepuce.

10. Mammary glands. These are characteristic of mammals. They secrete milk generally in the females for nourishment of the young. In monotremes both sexes may secrete milk, this condition is called **gynaecomastism** which is not a normal condition. Mammary glands of monotremes are compound tubular glands, while in other mammals they are compound alveolar. Mammary glands of monotremes have no nipples, they open into pits on the surface of the skin, and the young ones obtain milk by licking tufts of hairs. In others the mammary glands open by their ducts separately into a nipple. The **nipple** is a raised outgrowth of the breast, bearing opening of mammary glands **(true teats)**. The **false teat** has an elongated cistern into which the mammary glands open by their ducts. From the cistern a tube leads to the surface of the nipple.

Mammary glands along with fat form integumentary swellings called **mammae** or breasts. The number and location of mammae varies in different mammals. The number ranges from two in many mammals to 25 in opossum. Mammae may run along two ventral **milk lines** from the arm pits to the groin, or they may be axillary, thoracic, abdominal or inguinal in position.

2. Epidermal Scales and Scutes

1. Epidermal scales. After the evolutionary loss of dermal scales of fishes, amniotes developed an entirely new type of scale derived from the epidermis. The skin of vertebrates is rarely naked, and it is usually provided with protective scales, bony plates, feathers or hairs. There are no epidermal scales in fishes and amphibians. They appear for the first time in reptiles. They are cornified derivatives of the Malpighian layer and are generally shed and replaced. The scleroprotein called the **keratin** is continually accumulated in the permanently growing layer of the epidermis called the **stratum germinativum**. These cells are called cornified and they become dead. Thus, the stratum corneum cells become cornified and form hard horny structures, such as scales and scutes in reptiles, beaks, claws, horns, nails, hoofs, feathers and hairs in different vertebrates.

Scales are most noticeable on lizards and snakes. They are continually being produced by the permanently growing layer of the **stratum germinativum** and are generally folded so as to overlap one another. When they are fully grown they become separated from the stratum germinativum and appear as non-living, cornified structures.

The epidermal scales form a protective covering of the body (a continuous armour). The scales in snakes and lizards are continuous and undergo ecdysis periodically. Before ecdysis, the new scales that will replace the old ones are formed. The entire corneal layer of scales is shed as whole in snakes. The old epidermal covering becomes loosened first in the head region. This skin, including even the transparent plates over the eyes, is turned back and the snake finally crawls out of the old covering, leaving the old covering inside out. The scales on the ventral side in most snakes differ from their other scales in being long and transversely arranged, they aid in locomotion. Turtles and crocodiles have different kinds of scales which do not overlap, nor undergo periodic ecdysis, but the scales are gradually worn off and replaced. **Large epidermal scales**, such as those on the shell of turtles and on the head of snakes, are generally called **scutes**. In birds the scales are confined to the shanks and feet and some at the base of the beak. They generally overlap as in snakes and lizards. In mammals epidermal scales are found on the tail and paws of rats, mice, beavers, musk rats and shrews. These scales are not much cornified, nor do they undergo ecdysis; hairs project from beneath the scales. In scaly ant-eaters large epidermal scales cover the entire body, except the ventral side, the scales are reptilian and undergo ecdysis singly. In armadillos there are large scales which fuse to form plates on the head, shoulders and hips; in the middle of the body, except mid-ventrally, the scales fuse to form ring-like bands; these scales do not undergo ecdysis but are gradually worn off and replaced.

Some epidermal scales of the tail of a rattle-snake are modified to form a rattle. It consists of a series of old dried scales. Rattles of rattle snakes represent horny remnants of the skin which adhere to the base of the tail and are not lost during ecdysis. During ecdysis the scale at the tip of the tail is not shed and it forms a ring. Thus, after several ecdysis a series of rings form a rattle, each new ring being larger than its predecessor. In older rattle snakes the terminal rattles are often lost. In tortoises and turtles, the dermal plates forming the carapace and plastron are covered externally with cornified epidermal scales. These are not shed regularly like lizards and snakes. A few turtles lack scales and have a leathery skin. The bodies of alligators and their relatives are also covered with epidermal scales which gradually wear off and are replaced.

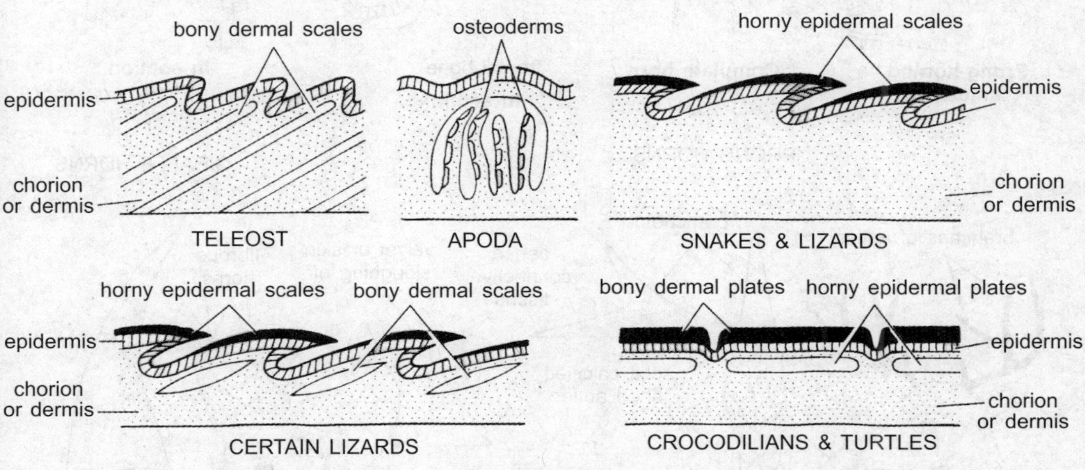

Fig. 41.3. Diagrammatic V.S. through skin of various vertebrates showing relationship of various types of scales.

In turtles, tortoises and modern birds there are no teeth. Each jaw bone being covered with horny sheath formed of several fused plates or scales which form a beak. In monotremes there is a soft bill which differs from the beak of birds is not being covered with modified epidermal scales.

3. Horns

Horns are found in ungulate (even-toed hoofed) mammals only. **True horns** of the hollow type are found in pronghorn, cattle, antelope, sheep and goats consist of an inner core of bone which is an outgrowth from the frontal bone. It is encased in a keratinised, epidermal covering. True horns continuously grow throughout life and are not shed. Five types of horns are recognised :

(*a*) Rhinoceros or **keratin fibre horn** has no skeletal element. It is made by keratinised cells of the epidermis and consists of matted keratin fibres bound together, but its fibres are not true hair. It is a permanent epidermal structure and if broken it grows again. There is one horn in the Indian rhinoceros and two in the African species.

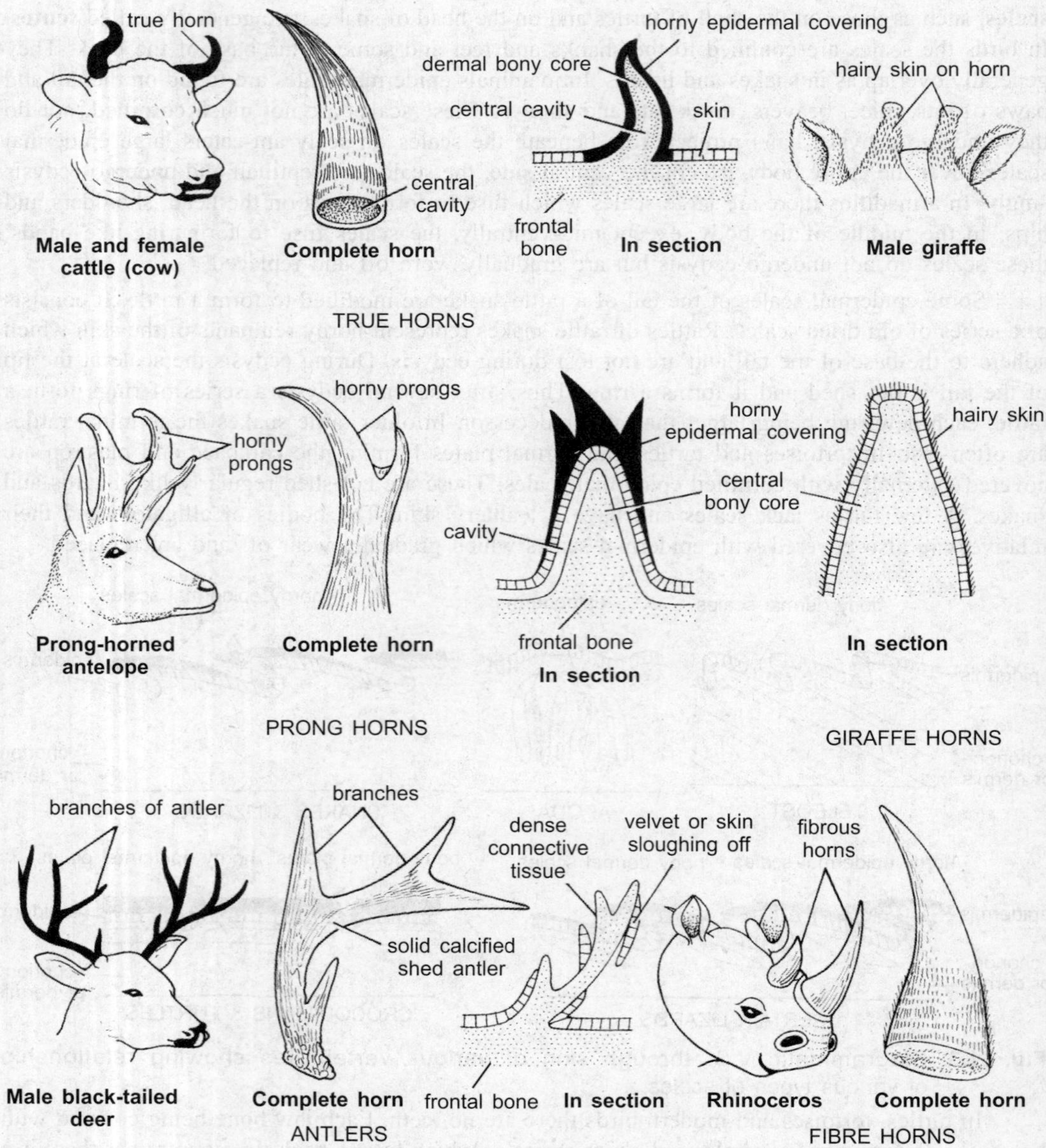

Fig. 41.4. Types of mammalian horns and antlers.

(*b*) **Pronghorn** is a **true horn**, consists of a permanent projection of the frontal bone covered by a hard, horny epidermal sheath. The sheath is forked bearing one to three prongs made only of horny sheath. The horny sheath is shed annually and is replaced by another which grows from the skin that surrounds the core. It is found in Russian antelope *Antilocapra.*

(*c*) **Giraffe horns.** They develop from cartilaginous protrusions which are present at birth. They ossify and fuse at the top of the skull, where they appear as knobs permanently covered with living skin and hair. Giraffe possesses three of these knobs, one is median and anterior to the other two. These horns are short, unbranched and are permanent, and are present in both sexes.

TABLE 41.1. DISTINCTION BETWEEN ANTLERS AND HORNS.

Antler	Horn
1. Anters are outgrowths over the head of artiodactyls.	1. Same
2. Antlers are found only in Cervidae, but absent in Asiatic genera *Moschus* and *Hydropotes*.	2. Horns (hollow) are found in Bovidae and in Antilocapridae in a modified form.
3. Normally present in males except *Rangifer.*	3. Commonly occur in both sexes–larger in males.
4. Shed annually.	4. No shedding and if occurs only outer sheath is shed.
5. Antler is made of bone and covered by velvet (skin + hairs).	5. Horn consists of inner core of bone, an outgrowth of frontal bone covered over by keratinised epidermal covering.

(*d*) **Antlers** are found in the males of deer family, but they are present in both sexes in reindeer and caribou. An antler consists of a branching solid outgrowth of the frontal bone formed of dense connective tissue. It is covered during growth by hairy, vascular skin called 'velvet'. The velvet is shed exposing the antler naked when the antler reaches full growth. Thus, the antler consists only of dermal bone. The bony antler is also shed annually after the breeding season, and a new antler develops. Antlers are solid mesodermal bone, but they are formed under the influence of the integument. Formation of antlers is controlled by the hormones of testes and anterior lobe of the pituitary.

4. Digital Structures

In amniota the distal ends of digits have claws, nails or hoofs formed from the horny layer of the epidermis. They grow parallel to the surface of the skin and are built on the same plan.

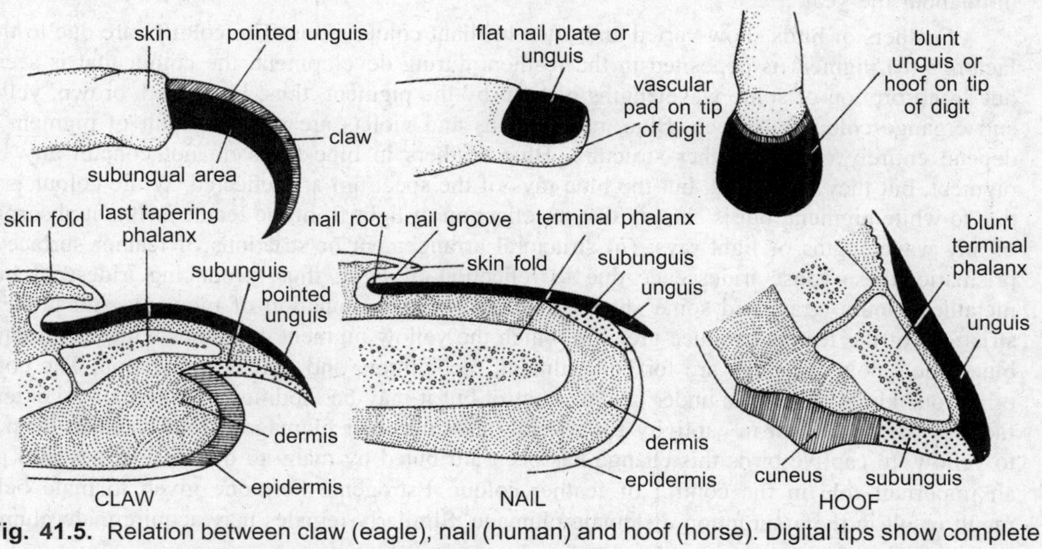

Fig. 41.5. Relation between claw (eagle), nail (human) and hoof (horse). Digital tips show complete above and in sagittal section below.

1. Claws. Claws made their appearance first in the reptiles. A claw is made of a hard horny dorsal scale-like plate called **unguis** and a relatively soft ventral **subunguis**, both converge terminally and cover the terminal part of the last phalanx. Claws of reptiles and birds are similar but in mammals the subunguis is much reduced and is continuous with a pad at the end of a digit. In cat family claws are retractile.

2. Nails. They are found in primates. The dorsal unguis is large and flat and subunguis is soft and much reduced. Tip of the digit forms a sensitive and vascular pad over which the nail groove is present. It is formed by the invagination of epidermis. Growth of the unguis takes place from the nail root lying below the skin in the nail groove.

3. Hoofs. They are found in ungulates. The horny unguis is thick and around the end of the digit, and encloses the thickened subunguis which touches the ground. Subunguis surrounds the soft, horny **cuneus**. Tip of digit, thus, forms a pad containing a blunt phalanx. Nails and hoofs of mammals are modified claws. Whalebone plates of toothless whales are also the modification of stratum corneum.

5. Feathers

Feathers are found only in birds and are formed from the epidermis in which the stratum corneum is highly specialised. Feathers are light, strong, elastic, waterproof and show many colours due to pigments and structural arrangement. The pigments are carotenoids and melanins. Carotenoids are frequently called lipochromes which are soluble in fat solvents like methanol, ether or carbon disulphide, and insoluble in water. Animal red is zooerythrin and animal yellow is zooxanthin are the two groups of carotenoids. Melanins are soluble only in acids. Eumelanin granules vary from black to dark brown and phaeomelanin granules may be almost colourless to reddish brown. Many of the feather colours are the product of both carotenoids and melanin. They form a protective covering (conserve body heat), provide buoyancy to the water birds during swimming and at rest, regulate body temperature and support the body in flight.

There are three kinds of feathers in birds, they are contour feathers, down feathers, and filoplumes. The development of feather is like that of scales.

Shedding and replacement of feathers is **moulting** which takes place gradually, one moulting usually takes an average time of six weeks. At the base of each feather follicle a dermal papilla persists from which new feathers will form, so that there is a continuous replacement of feathers throughout life. The replacement in some birds is seasonal, while in others it is gradual throughout the year.

Feathers of birds show varied and often brilliant colourations. The **colours** are due to three factors : (*a*) pigment is deposited in the feathers during development, the colour that is seen is due to absorption of some wavelengths of light by the pigment, thus, black, red, brown, yellow, and organge colours are seen. Blue, most greens and violets are not the result of pigment but depend entirely on the feather structure. Blue feathers in blue birds do not contain any blue pigment, but they absorb all, but the blue rays of the spectrum are reflected. White colour is not due to white pigment, but is caused, due to reflection of light from the feather without absorption of any wavelengths of light rays; (*b*) structural arrangement or striations of feather surface are prismatic, these cause iridescence due to reflection of light, thus, producing iridescent hues, metallic colours, gray, and some shades of blue; (*c*) a combination of pigments and prismatic striations of the feather produce green in which the yellow pigment combines with the structural blue. The colours of birds are for concealment, recognition, and sexual stimulation. The colour of various kinds of birds is under genetic control but it may be modified by internal and external factors. The birds kept in captivity for several years get their plumage colour changed from red to yellow. In captive birds this change has been attributed by many to diet. Hormones also play an important role in the control of feather colour. Estrogenic hormone given to male before moult result in the assumption of female plumage. Similarly females may acquire male plumage by giving the testosterone.

Oxidation and abrasion are external factors that change in colour to a lesser or greater degree in most species of birds. Carotenoids fades in sunlight.

6. Hair

Hair is found only in mammals. It projects at an acute angle from the skin. Hair covers the entire integument in most (furred mammals), but in others only traces are left, such as whales have only a few coare hairs on the snout. But during development the body of the embryos of all mammals is covered with a coating of fine hair called **lanugo** which is usually shed before birth and replaced by a new one. Hair is entirely epidermal in origin.

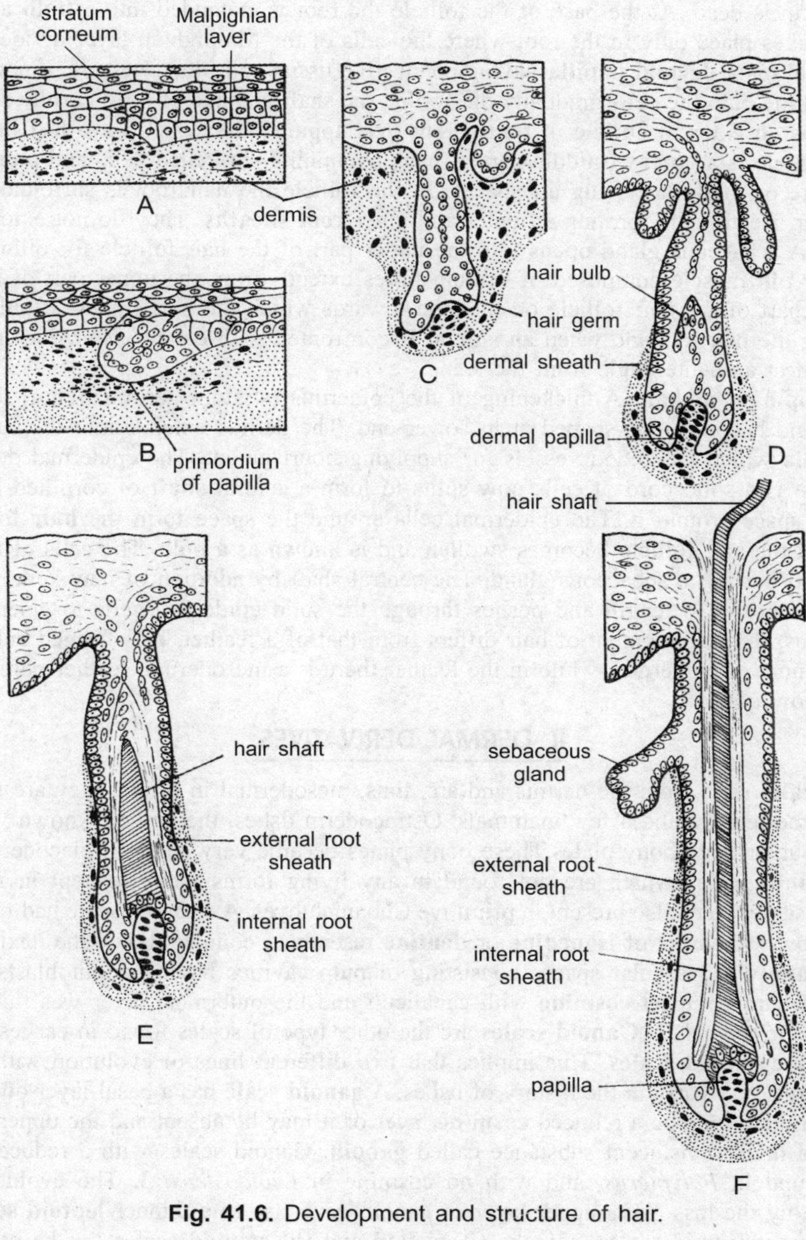

Fig. 41.6. Development and structure of hair.

The hairs trap air which does not transmit body heat and, thus, act as insulators for the body. In some animals the colours of hair are protective. Hair in nostrils and ears prevent entry of dust, eyebrows and eyelashes protect the eyes, vibrissae are delicate organs of touch (act as tactile organs), hairs on the tail are used to drive away insects. In some animals, such as lions and some monkeys the **mane** distinguishes the male. Hairs are also modified in spines, scales, horns, etc., in some mammals.

Structure. Hairs are not modified scales but are new outgrowths of the epidermis only. A hair has an upper projecting **shaft** and a lower **root** lying in a **hair follicle** which is a sunken pit in the dermis. The **shaft** is made of only dead, keratinised cells. The part of the hair protruding above the skin is dead. At the base of the follicle the root is expanded into a **bulb** and growth of the hair takes place only in the root where the cells of the Malpighian layer divided actively. Below the bulb is a **dermal papilla** having connective tissue and blood vessels, it nourishes the hair. Beyond the bulb the cells gradually die so that the shaft is made of dead cornified cells. The hair shaft has an external **cuticle** of transparent overlapping cells which have lost their nuclei, inside the cuticle is a **cortex** (middle part of hair) containing shrivelled cells and pigments, and a central core or **medulla** having air spaces. In the follicle the hair root is surrounded by two layers of hair sheath cells forming an outer and inner **root sheaths**. They do not extend beyond the follicle. A sebaceous gland opens into the upper part of the hair follicle for oiling the hair. An **arrector pili** muscle composed of smooth fibres extends from the upper part of the dermis to the basal part of the hair follicle on the side towards which the hair slopes. It pulls the hair base causing the hair to stand when an animal is confronted with danger. Hair does not project vertically but at an acute angle from the skin.

Development of hair. A thickening of the epidermis (stratum germinativum) pushes into the dermis and becomes cup-shaped at its lower end. The dermis extends into the cup forming a **hair papilla** which has blood vessels for supplying nourishment. The epidermal downgrowth which at first is a solid cord of cells now splits to form a central **shaft** of cornified keratinised cells, and a space around it. The epidermal cells around the space form the **hair follicle**. The lower part of the hair follicle becomes swollen and is known as a bulb. The cells of the follicle thickened and bud off a sebaceous gland. The central shaft by addition of new keratinised cells from the root grows in length and pushes through the solid epidermal cells to emerge outside the skin. Thus, the development of hair differs from that of a feather. It is formed entirely from the solid column of epidermis, while in the feather there is a mesodermal feather pulp extending into the hollow quill.

II. DERMAL DERIVATIVES

The **scales** arise from the dermis and are, thus, mesodermal in origin, they are found only in fishes, some reptiles and a few mammals. Ostracoderm fishes, the earliest known vertebrates, had an armour of large bony plates.These bony plates became very small in placoderms to give rise to **cosmoid scales** which are not found in any living forms today (except in *Latimeria*). **1. Cosmoid scales** were also present in primitive Choanichthyes. A cosmoid scale had four distinct layers, the lowest layer is of **isopedine** or **dentine** resembled compact bone, the next layer was a **spongy bone** with vascular spaces consisting of pulp cavities having odontoblasts, the third layer was of hard compact **cosmine** with canaliculi and the outermost layer was thin but hard **vitrodentine** or enamel. **2. Ganoid scales** are the other type of scales found in earliest primitive bony fishes are **ganoid scales**. This implies that two different lines of evolution with regard to scales appeared very early in the history of fishes. A **ganoid scale** has a basal layer of **isopedine**, above which there may be a reduced **cosmine** layer or it may be absent and the uppermost layer is made of a hard, translucent substance called **ganoin**. Ganoid scales with a reduced cosmine layer are found in *Polypterus*, and with no cosmine in *Lepidosteus*. **3.** The evolution of the ganoid scale by the loss of its upper layer of ganoin gave rise to a thinner **leptoid scale**. There are two types of leptoid scales, namely, **(*i*) cycloid** and **(*ii*) ctenoid** scales. In the other line of evolution the cosmoid scale lost its three lower layers, only the fourth enamel-like dentine layer was retained and somewhat elaborated to form the **placoid scale**. Thus, in the present fishes there are four types of dermal scales : placoid, ganoid, ctenoid and cycloid scales.

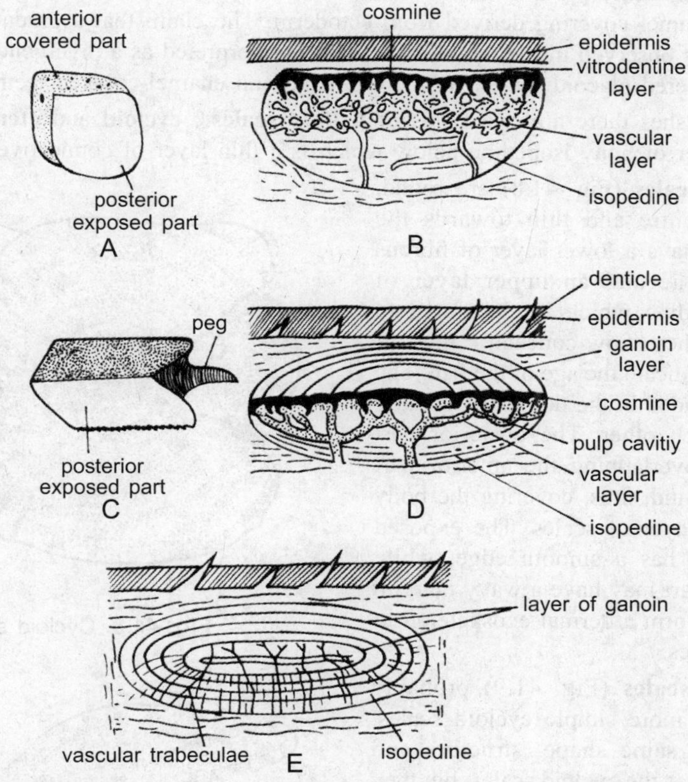

Fig. 41.7. Structure of scales of fishes. A–An isolated scale; B–T.S. of cosmoid scale; C–An isolated ganoid scale; D–T.S. of Palaeoniscoid scale; E–T.S. of lepidostoid scale.

Embedded in the dermis of elasmobranchs in oblique rows and projecting from the epidermis are dermal denticles of placoid scales forming an exoskeleton. The placoid scales were evolved by loss of some layers from the cosmoid scales. Placoid scales are found only in elasmobranchs, in sharks they are very small and give the skin a rough texture, but in skates they are large.

Structure. A **placoid scale** has a flat bony **basal plate** bearing a trident **spine** which projects above the epidermis and points backwards. Inside the spine is a **pulp cavity** containing pulp made of connective tissue, blood vessels, and a layer of **odontoblast** cells. The basal plate is made of calcified dentine, and the spine has mostly dentine which is covered with a cap of hard modified dentinę called **vitrodentine** and not enamel as often stated erroneously. The basal plate and spine are both of mesodermal origin. The cured skin of sharks containing placoid scales is called **shagreen** which is used for polishing and for handle covers.

Placoid scales are the fore-runners of vertebrate teeth because the two have essentially the same form and structure and a gradation from placoid scales to teeth is seen in the mouth of shark. Shark teeth are enlarged placoid scales formed in the skin of jaws. But there are objections to this supposition. A more recent view is that both placoid scales and teeth are modified remnants of the bony dermal plates found in the ancestral ostracoderms and placoderms, so that teeth and placoid scales are lomologous strutures. Moreover, it has been shown that there is no epidermal enamal layer in placoid scales, but it is a layer of **vitrodentine** formed from dermal cells. Enamel is the hardest substance in the body, whereas vitrodentine is only a hardened outer layer of dentine. The enamel organ does not secrete enamel, it only plays a role in shaping the spine, so that the entire placoid scale is mesodermal like the scales of bony fishes, whereas a tooth has an enamel covering derived from ectoderm. The claim that a gradation from placoid scales to teeth is observed in the mouth of a shark is interpreted as a divergence shown between vitrodentine-covered placoid scales, on the one hand and enamel-covered teeth on the other.

In bony fishes there are two kinds of **leptoid scales** : **cycloid** and **ctenoid scales**. They have a thin layer of bony isopedine below which is a thin layer of connective tissue.

Cycloid scales (Fig. 41.8) are round, thick in the centre and thin towards the margins. They have a lower layer of fibrous connective tissue and an upper layer of bone-like isopedine which is elaborated to form dentine. They show concentric lines of growth which indicate the age of the fish. The scales lie embedded in the dermis diagonally overlapping each other. The posterior part of each scale overlapping the anterior part of the scale behind, thus, covering the body with a double layer of scales. The exposed part (posterior) has a smooth edge, while the concealed part may have a wavy margin. Cycloid scales form a dermal exoskeleton in many bony fishes.

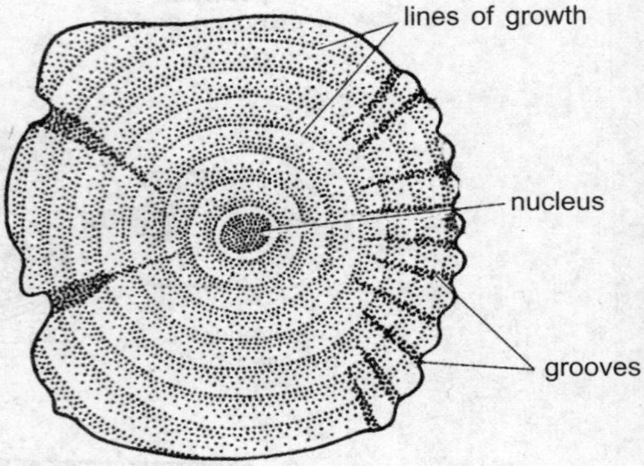

Fig. 41.8. Cycloid scale.

Ctenoid scales (Fig. 41.9) probably arose from the more simple cycloid scales. They have the same shape, structure and concentric lines as the cycloid scales, but they differ in having small teeth or **cteni** on their free posterior part. Their anterior concealed part may have notched or scalloped margin. Ctenoid scales form a dermal exoskeleton of most bony fishes. The cycloid or ctenoid scales covering the lateral line canals are perforated by a vertical tube of the lateral line opening on the surface. In some flat fishes both cycloid and ctenoid scales are present. In some catfishes there are no scales, in eels scales are very small and embedded in the dermis. In sea-horses scales form a continuous armour covering the body. Scales of bony fishes arise only from dermis, they form a protective exoskeleton but do not hamper movement.

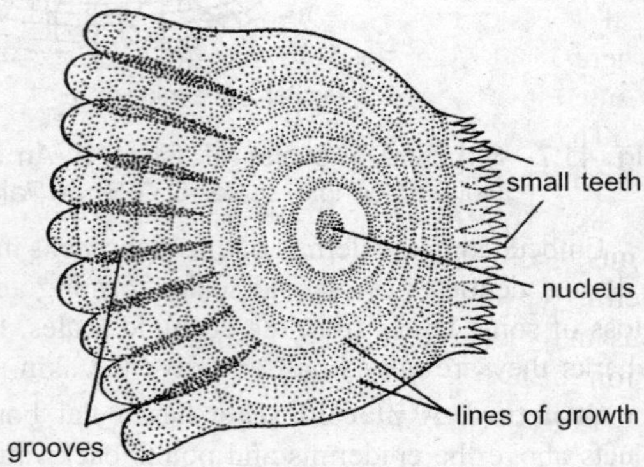

Fig. 41.9. Ctenoid scale.

In **tetrapoda** dermal scales are known as **bony plates** or **osteoderms**. In **Apoda** (Amphibia) only vestiges of osteoderms are present. They are found in pockets below the epidermis and are not visible externally. Two most ancient groups of reptiles, the turtles and crocodiles, have retained bony dermal plates. Turtles have continuous osteoderms below the epidermal plates of the carapace and plastron. These osteoderms form a rigid dermal skeleton which becomes connected with the endoskeleton. In crocodiles there are osteoderms below the epidermal plates only on the back and the throat.

In birds and mammals there is a tendency for elaborating epidermal structures with an accompanied reduction or loss of dermal derivatives. In **mammals**, osteoderms are found in armadillos lying below the epidermal scales. They are bony plates of spongy texture. Extinct glyptodons had a rigid bony armour of osteoderms. In some whales bony osteoderms may be present on the back and the dorsal fin.

INTEGUMENT IN DIFFERENT CLASSES OF CHORDATES

The fundamental structure of skin in all the vertebrates is the same but there are certain variations in different classes.

1. Protochordata. In *Branchiostoma* the skin is simple without keratin. The **epidermis** is single layered made of tall or columnar cells. These are ciliated in *Balanoglossus.* Epidermis has numerous unicellular gland cells which secrete a thin cuticle in *Branchiostoma.* **Dermis** (corium) is gelatinous in *Amphioxus*.

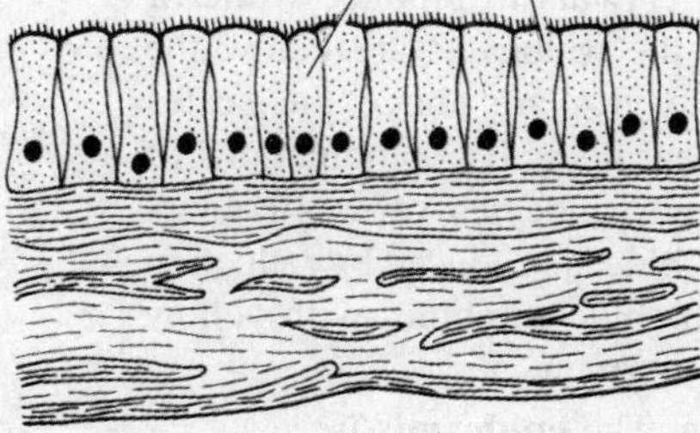

Fig. 41.10. V.S. of skin of a young *Amphioxus*.

2. Cyclostomata. Epidermis is multilayered (stratified) but has no keratin. It has three types of unicellular gland cells : **mucous glands** secrete mucous, **club cells** probably scab-forming cells, and **granular cells** are of unknown function. Below the epidermis is the **cutis** formed of collagen and elastic fibres. Star-shaped pigment cells are also present in the cutis.

3. Pisces. The epidermis has several layers of simple and thin cells, but there is no dead stratum corneum. The outermost cells are nucleated and living. The stratum Malpighii replenishes the outer layers of cells which have some keratin. Unicellular goblet or **mucous gland cells** are found in the epidermis, as in all aquatic animals. The mucus makes the skin slimy reducing friction between body surface and water, protects the skin from bacteria and fungi, and assists in the control of osmosis. Multicellular epidermal glands like **poison glands** and light producing organs (photophores) may also be found. The epidermis rests on a delicate basement membrane.

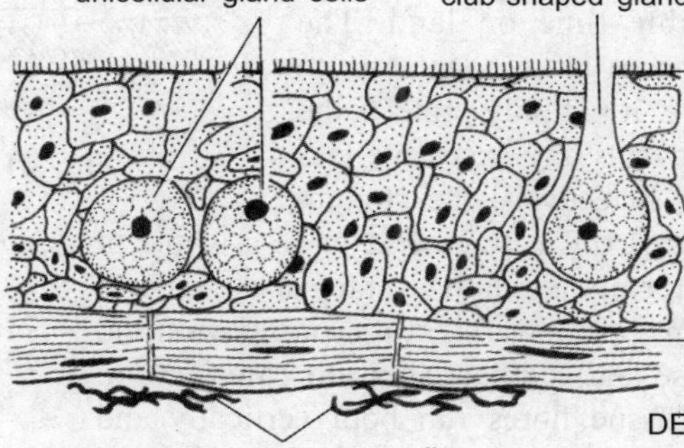

Fig. 41.11. V.S. of skin of a larval cyclostome.

The dermis contains connective tissue, smooth muscles, blood vessels, nerves, lymph vessels, and collagen fibres. The connective tissue fibres are generally not arranged at right angles, but run parallel to the surface. Scales are embedded in the dermis and projected above the epidermal surface. These are of five types. The elasmobranchs have **placoid scales**, Chondrostei and Holostei have **ganoid scales**, while most Teleostei have **cycloid** and **ctenoid** scales lodged in pouches of the dermis. Extinct Crossopterygii had **cosmoid** scales.

Many bony fishes show more brilliant colours than any other group of animals. The colours of fishes are due to **chromatophores** and **iridocytes**.

(*a*) Chromatophores in the dermis are derived from neural crest cells. They contain pigments which not only produce colours but also cause variations of colours. Chromatophores containing brown or black pigment are known as **melanophores** and those containing red, yellow, or orange pigment are collectively called **lipophores**.

(*b*) Iridocytes or **guanophores** are reflecting cells. They have no pigment but contain crystals of guanin. They lie in the dermis and cause irides-cence. Iridocytes reflect light from guanin crystals to produce white or silvery colours if the iridocytes are below the scales, if the iridocytes are above the scales they cause iridescent hues. By combinations of chromatophores and iridocytes various colours are produced, *e.g.*, blue is produced by reflection from iridocytes, the blue combines with yellow pigment to produce green.

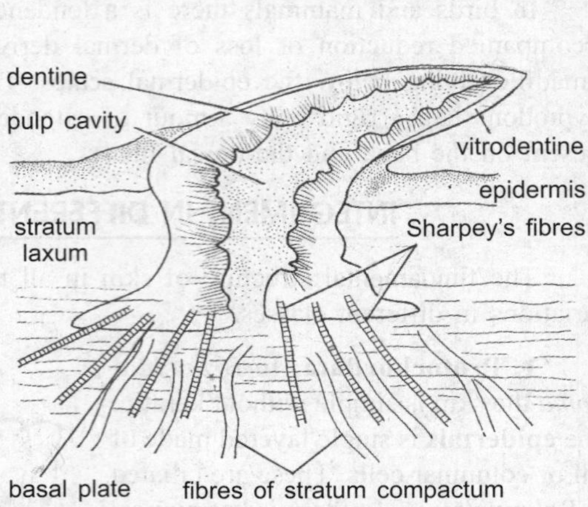

Fig. 41.12. V.S. of integument of shark.

4. Amphibia. The **epidermis** is multilayered, the outermost layer is a **stratum corneum** made of flattened, highly keratinised cells. Such a dead layer appears first in amphibians, and is best formed in those which spend a considerable time on land. The stratum corneum is an adaptation to terrestrial life, it not only protects the body but prevents any excessive loss of moisture. In ecdysis, the stratum corneum is cast off in fragments or as a whole in some. The **dermis** is relatively thin, it is made of two layers, an upper loose **stratum spongiosum** and a lower dense and compact **stratum compactum**. Connective tissue fibres run both vertically and horizontally. There are two kinds of glands, they are multicellular **mucous glands** and **poison glands** in the dermis, but they are derivatives of the epidermis. The mucous glands produce mucus which not only forms a slimy protective covering but also helps in respiration. The poison glands found in toads as parotid glands produce a mild but unpleasant poison which is protective, keep the enemies away. In the upper part of the dermis are **chromatophores** which have black melanophores and yellow lipophores, these produce the colour of the skin. The ability of the skin for changing colour to blend with the environment is well developed. Skin of labyrinthodontia, the stem Amphibia had a armour of dermal seales. Bony dermal scales are found embedded in the skin of some Gymnophiona (Apoda) and a few tropical toads. These scales are absent in modern Amphibia.

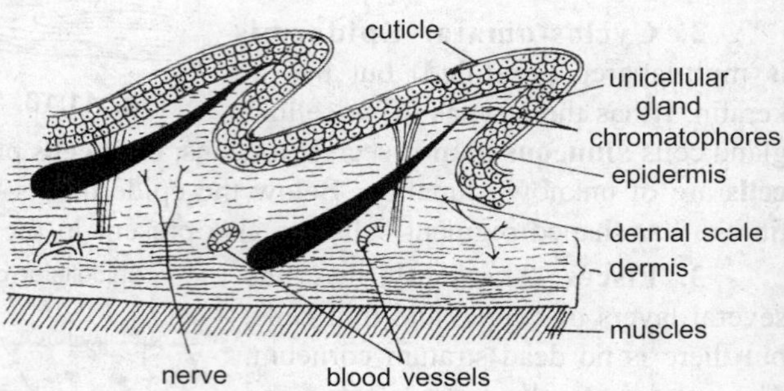

Fig. 41.13. V.S. of skin of bony fish.

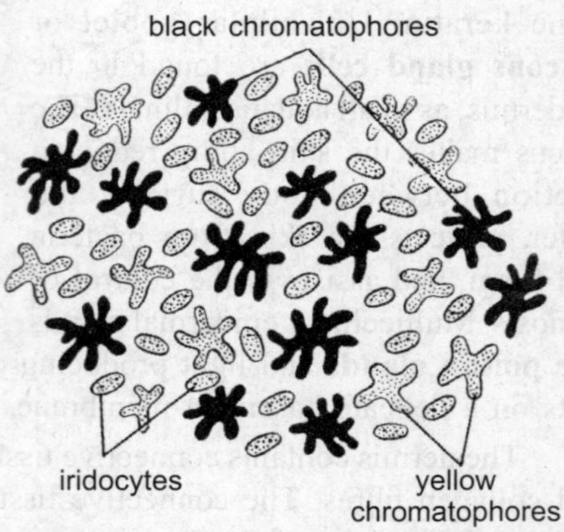

Fig. 41.14. Chromatophores and iridocytes in skin of fishes.

The skin is sensitive to light in amphibians, especially in cave-dwelling forms. It **is** an important organ of respiration, **and** also enables the frog to respire **under** water for long periods, during hiber**nation** or aestivation it is the only organ of **respiration**.

The skin **is** loose being attached to muscles **only** at certain places by connective tissue septa which mark the boundaries of subcutaneous lymph spaces.

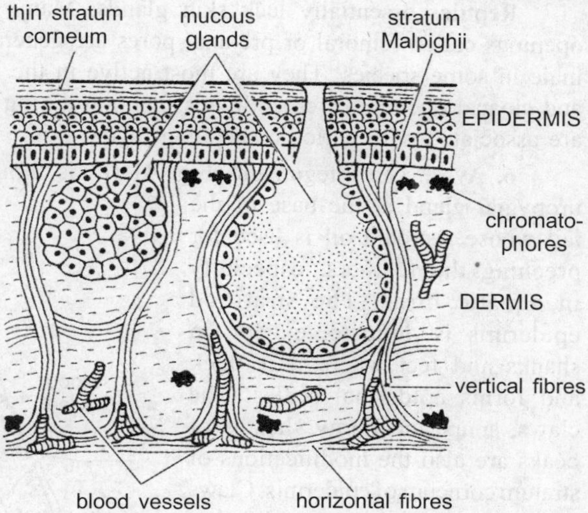

Fig. 41.15. V.S. of skin of frog.

5. Reptilia. The integument (Fig. 41.17) is thick and dry. It prevents any loss of water. It has almost no glands, this is an adaptation to prevent evaporation of water.

The **epidermis** has a well developed **stratum corneum** which make the skin dry and prevent any loss of body moisture, thus, well adapted to a terrestrial life. The epidermis produces horny scales. Scales are shed periodically in fragments or cast in a single slough, as in snakes and some lizards. The scales often form **spines** or **crests**. Below the epidermal scales are dermal bony plates or **osteoderms** in tortoises, crocodiles, and some lizards (*Heloderma*). These are retained for life and are not shed off. These may form dermal bones in the skull lying superficially or they may be found in the dermis.

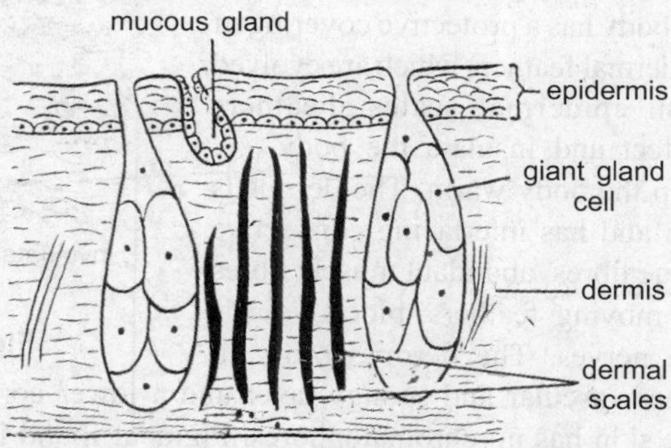

Fig. 41.16. V.S. of skin of *Ichthyophis* showing dermal scales.

The **dermis** is thick and has an upper loose connective tissue layer and a lower layer tella subjunctiva and separating the two is a horizontal layer of fibrous connective tissue. The upper layer has an abundance of chromatophores in snakes and lizards like fishes and ampibians. Leather of high commercial value is made from the skin of lizards, snakes, and crocodiles.

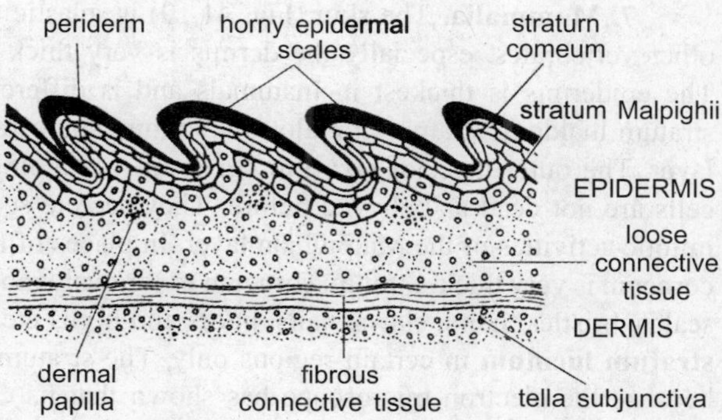

Fig. 41.17. V.S. of skin of lizard.

Many lizards and snakes have elaborate colour patterns, for concealment or as warning colours. There is marked colour change in certain lizards, such as *Chameleon*, the colour may change with the environment for concealment, or it may change in courtship or threat. In *Calotes,* the chromatophores have no nerve fibres, they are controlled by hormones of the posterior lobe of the pituitary. In *Chameleon*, chromatophores are controlled by the autonomic nervous system.

Reptiles essentially lack skin glands. Many lizards have glands near the cloaca. Their openings called femoral or pre-anal pores are generally smaller in female and found only in the male in some species. They are most active in the breeding season. **Musk glands** in the throat and cloacal opening of crocodilians function during courtship. Generation glands found recently are associated with periodic shedding of the skin.

6. Aves. The integument (Fig. 41.18) is thin, loose, dry and devoid of glands except a uropygial gland at the base of the tail whose secreted oil is used for preening the feathers, especially in aquatic birds. The stratified epidermis is delicate, except on shanks and feet where it is thick and forms epidermal scales. The claws, spurs and horny sheaths of beaks are also the modifications of stratum corneum of epidermis. Claws and beaks are variously modified in birds according to habitat. The rest of the body has a protective covering of epidermal feathers which are evolved from epidermal scales. Feathers protect and insulate the body, *i.e,* keep the body warm. The dermis is thin and has interlacing connective tissue fibres, abundant muscle fibres for moving feathers, blood vessels and nerves. The dermis forms an upper vascular and spongy layer and a lower compact layer. The dermis also contains fat cells. The skin has no chromatophores. Pigment found in melanocytes migrate into feathers and scales. Colour patterns of birds are vivid, they are for concealment, recognition, and sexual stimulation. The colours are mainly produced by reflection and refraction of light from surface of feathers.

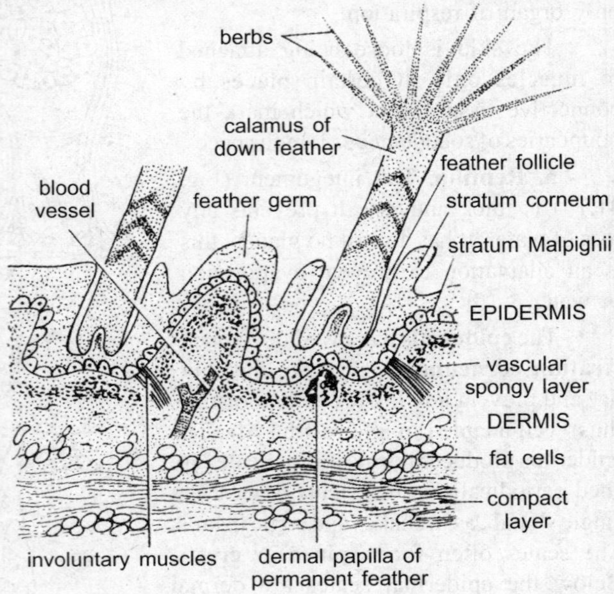

Fig. 41.18. V.S. of skin of a bird.

7. Mammalia. The skin (Fig. 41.19) is elastic and waterproof and is much thicker than in other vertebrates, especially the dermis is very thick and tough and is used for making leather. The epidermis is thickest in mammals and is differentiated into five layers : stratum corneum, stratum lucidum, stratum granulosum, stratum spinosum and stratum germinativum or Malpighian layer. The outer layer of **stratum corneum** containing keratin, its cells lose their nuclei, but the cells are not dead as believed before. They secrete several hormones, one of which represses the mitotic activities of the Malpighian layer. In places of friction, such as soles and palms, the stratum corneum is very thick. Stratum corneum is variously modified in various mammals to form epidermal scales, bristles, hairs, claws, nails, hoofs and horns etc., Below the stratum corneum is a refractive **stratum lucidum** in certain regions only. The stratum lucidum is now known as a **barrier layer** because the electron microscope has shown that its cells become compact and closely united to form a region which prevents passage of substances into or out of the body. Stratum lucidum contains a chemical known as **eleidin**. Keratohyalin and eleidin are intermediate products in the formation of **keratin**. Below this is a stratum granulosum which is having darkly-staining granules of keratohyalin. Below the stratum granulosum is a **stratum spinosum** whose cells are held together by spiny intercellular bridges, each bridge has two arms in close contact, one arm arising from each cell. Lastly there is a stratum germinativum or Malpighian layer which rests on a thin basement membrane. The Malpighian layer forms new cells continously which move towards the surface and become flat and keratinised till the stratum corneum has flat, cornified cells made only of keratin. This layer is sloughed off continuously and replaced by new cells. There are no mucous glands in the epidermis of mammals. The keratin from the epidermis at ends of digits forms claws, nails or hoofs.

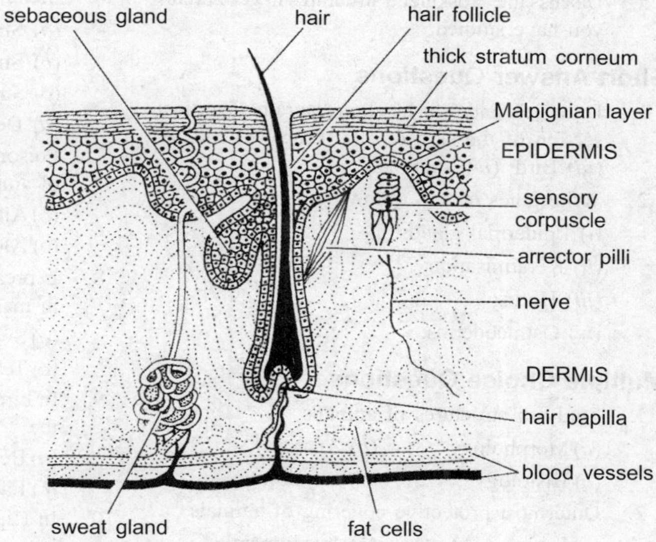

Fig. 41.19. V.S. of skin of a mammal.

The **dermis** is best developed in mammals. The upper part of the dermis in contact with the epidermis is the **papillary layer** which is made of elastic and collagen fibres with capillaries in between. It is thrown into folds to form rows of **dermal papillae**, especially in areas of friction. The greater lower part of the dermis is a **reticular layer** having elastic and collagen fibres. In both layers there are blood vessels, nerves, smooth muscles, certain glands, tactile corpuscles, and connective tissue fibres extending in all directions. Below the dermis the subcutaneous tissue has a layer of fat cells forming **adipose tissue** which helps to maintain body heat. In making leather only the dermis is used. Dermal scales are not found in mammals except armadillos.

In the lowest layer of the epidermis are **pigment** granules but there are no pigment-bearing chromatophores in mammals. In man some branching dendritic cells or **melanoblasts** lie between the epidermis and dermis, they contain pigment.

The epidermis forms hairs, sudorific glands, sebaceous glands and mammary glands. **Hairs** form an epidermal covering. Shafts of hair project above the skin and their roots are embedded in hair follicles, into each of which opens a branching sebaceous gland. Hairs form an insulating layer which prevents a loss of body heat, thus, hairs keep up the body temperature. **Sebaceous glands** are outpushings of the wall of hair follicle and produce an oily substance which keeps the hair supple and prevents its wetting in water. It also lubricates the skin. In the dermis are present coiled **sudorific** or **sweat glands**, which occur all over except lips and glans penis. **Mammary glands** are modified sebaceous glands, but in monotremes they are modified sudorific glands. They are functional only in females for producing milk for the young. **Mucous glands** are not found in mammals.

IMPORTANT QUESTIONS

Long Answer Questions

1. What is integument? Describe the integument and its derivatives in vertebrates.
2. Give an account of the integument in a vertebrate you have studied. Explain the functions of integument.
3. Discuss how the integument of reptiles and birds are adapted to their respective modes of life.
4. Give a comparative account of integument of reptiles, birds and mammals.

5. Discuss the exoskeletal structures in vertebrates you have studied.

Short Answer Questions

1. Draw neat and well-labelled diagram of vertical section of the skin of : (*i*) Frog; (*ii*) Lizard; (*iii*) Bird; (*iv*) Mammal.
2. Write notes on the following :
   (*i*) Epidermal glands;
   (*ii*) Keratinisation;
   (*iii*) Horns;
   (*iv*) Ostracoderms.

Multiple Choice Questions

1. Study of structures of animals
   (*a*) Morphology (*b*) Anatomy
   (*c*) Histology (*d*) Physiology.
2. Outermost protective covering of animals
   (*a*) Scales (*b*) Integument
   (*c*) Feathers (*d*) Hairs.
3. Which is not a function of skin?
   (*a*) Protection (*b*) Locomotion
   (*c*) Secretion (*d*) Digestion.
4. Integumentary gland secreting tears
   (*a*) Sebaceous gland (*b*) Mammary gland
   (*c*) Lacrymal gland (*d*) Uropygial gland.
5. Dermis of vertebrate integument is derived from
   (*a*) Ectoderm (*b*) Endoderm
   (*c*) Mesoderm (*d*) Ecto-mesoderm.
6. Melanocytes are located in
   (*a*) Stratum corneum
   (*b*) Stratum germinativum
   (*c*) Stratum lucidum
   (*d*) Dermis.
7. Poison secreted by parotoid (parotid) glands of amphibians is
   (*a*) Alkaloid (*b*) Fatty acid
   (*c*) Alcohol (*d*) Carbohydrate.
8. In presence of sunlight vitamin D is synthesised in mammalian skin from
   (*a*) Sweat (*b*) Sebum
   (*c*) Tears (*d*) Earwax.
9. In birds, uropygial gland is present just above the
   (*a*) Beak (*b*) Eye
   (*c*) Ear (*d*) Tail.
10. In Hippopotamus, sweat glands are restricted in
   (*a*) Lips (*b*) Ears
   (*c*) Muzzle (*d*) Soles of feet.
11. Typical cornification are modification of
   (*a*) Stratum corneum
   (*b*) Stratum lucidum
   (*c*) Stratum germinativum
   (*d*) Dermis.
12. Reflecting pigment cells in luminescent glands contain
   (*a*) Guanine crystals (*b*) Adenine crystals
   (*c*) Cytosine crystals (*d*) Thymine crystals.

ANSWERS

1. (*b*)	2. (*b*)	3. (*d*)	4. (*c*)	5. (*c*)	6. (*d*)
7. (*d*)	8. (*b*)	9. (*d*)	10. (*b*)	11. (*a*)	12. (*a*)

42 Endoskeleton in Vertebrates

Chapter

Contents of this Chapter

- Bone
- Axial Skeleton
- Appendicular Skeleton
- Muscles

In vertebrates the skeleton forms a framework of the body to which muscles are found attached. The skeleton is made of cartilage or bone or of both cartilage and bone. The presence of an endoskeleton is a characteristic of vertebrates, though many vertebrates also have a dermal exoskeleton.

Dermal bony skeleton. The **dermal exoskeleton** consists of dermal **scales** and **fin rays** in fishes. The dermis forms bony **dermal plates** of fossil ostracoderms, placoderms and amphibians. It also forms the bony armour (plates and scutes) of sturgeons, turtles, crocodilians, and some mammals (armadillos). Thus, it is somewhat artificial to make a distinction between dermal exoskeleton and endoskeleton because dermal bones contribute to the formation of some parts of the endoskeleton.

Endoskeleton. It lies deep in the body and develops from mesenchyme. In embryonic stage it is formed of cartilage but in adults it is replaced by bones, *i.e.*, bones developed in place of existing cartilages are called **cartilage** or **replacing bones**. If the bones develop in the dermis then these are called **dermal** or **membrane bones**. But both the above formed bones are similar in structure. The **notochord** forms the endoskeleton in *Amphioxus*, but in a majority of living vertebrates the notochord is only embryonic and in the adult it is replaced largely by a vertebral column.

Subdivisions of endoskeleton. The endoskeleton of vertebrates has an **axial skeleton** made of skull, vertebral column, ribs, and sternum, and an **appendicular skeleton** consisting of pectoral and pelvic girdles and the skeleton of paired limbs and unpaired appendages. These two types are included in **somatic skeleton**. A **visceral skeleton** forms the support of jaws and gills in fishes, but in higher forms it contributes to the skull and hyoid and is included in the axial skeleton.

Functions of Endoskeleton

1. The endoskeleton forms a framework for the support of the body.
2. It permits growth because it is living and grows in size with the rest of the body.
3. It maintains a definite shape and form of an animal.
4. It provides protection for delicate vital organs of the body.
5. It furnishes a firm and adequate surface for attachment of muscles by means of tendons.
6. In bone marrow blood corpuscles are manufactured.
7. Ear ossicles aid in hearing.
8. Tracheal rings and ribs help in breathing.

There are joints in the endoskeleton so that movement of parts is possible. In transition from water to land levers in the form of limbs were provided which not only bring about locomotion, but also lift the body above the ground. The vertebral column became a rigid axis to support the heavy head and to resist the thrust exercised by the limbs.

The long bones of vertebrates contain **yellow marrow** which stores fats, but in mammals the long bones and skull bones also have **red marrow** or **myeloid tissue** besides yellow marrow. In amphibians, reptiles, and birds the bone marrow forms all kinds of corpuscles, while in mammals the red marrow forms erythrocytes, blood platelets, and granular leucocytes only, the other white corpuscles are formed in the lymphatic tissues. Embryos of mammals have only red marrow, later some of this is replaced by yellow marrow.

BONE

In a vertebrate embryo the endoskeleton is at first, cartilaginous, which is replaced by bone in most adult vertebrates. In cyclostomes and elasmobranchs, the skeleton does not pass beyond the cartilage stage. In higher vertebrates the endoskeleton is largely bony.

Types of Bones

Bone exists in cancellous and compact form.

(*i*) **Spongy** or **cancellous** or endochondral bone consists of small pieces joined together irregularly, the small spaces between the pieces contain bone marrow.

(*ii*) **Compact** or **periosteal bone** is a hard, solid mass of bone without any apparent spaces.

The broad ends of **long bones** are made of spongy bone underlying a layer of compact bone. The shafts of long bones are made of compact bone enclosing a marrow

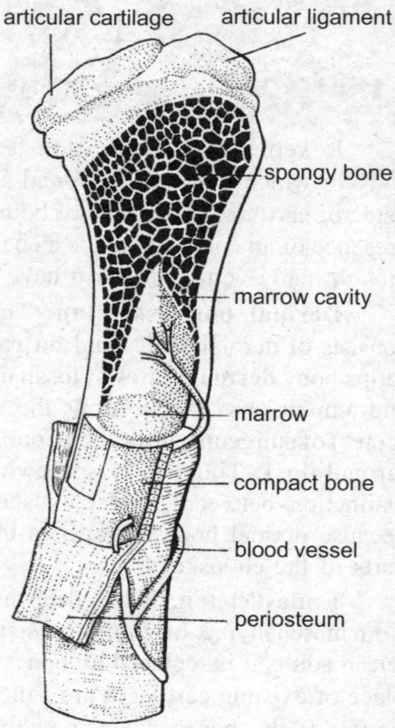

Fig. 42.1. Long bone showing compact and spongy parts.

cavity. The **skull bones** have a layer of spongy bone between two layers of compact bone called inner and outer tables.

Bone formation. Bone is formed in two ways, by a recplacement of pre-existing cartilage with bone, called **cartilage bone**, and secondly by direct **ossification of mesenchyme** or connective tissue without an intervention of cartilage, in which case the bone is known as **membrane** or **dermal bone**. In both types of bones the method of formation is similar.

1. In formation of cartilage bones known as **endochondral bone formation**, the original cartilage cells become arranged in rows and then they die. The matrix becomes calcified and invaded by blood vessesls and connective tissue eroding the cartilage to make intercommunicating channels between bars of calcified cartilage. The connective tissue fibroblasts become differentiated into **osteoblasts**. The osteoblasts secrete ossein of bones in concentric layers around bars of cartilage. The original perichondrium becomes the periosteum. Several osteoblasts fuse to form new cells called **osteoclasts**, which reconstruct the forming bone and dissolve away unwanted portions.

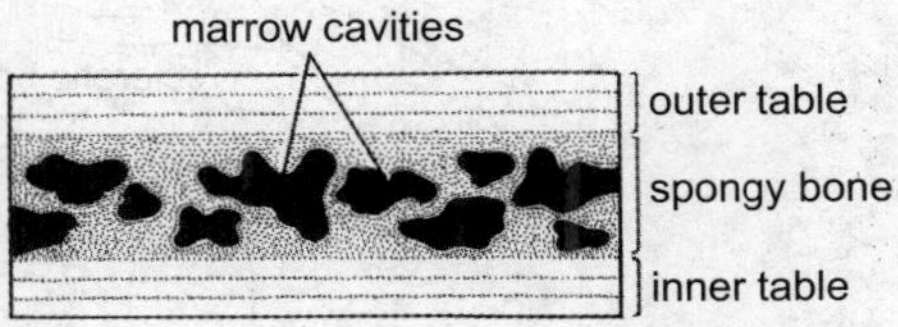

Fig. 42.2. T.S. of dermal bone of skull.

2. In formation of membrane bones mesenchyme cells collect and form an interlacing network and in between the cells of the network are present collagen fibres. An amorphous matrix is formed between the cells and fibres. Bone is secreted by cells which are now called **osteoblasts**, and the matrix gets impregnated with calcium. The osteoblasts now known as **osteocytes** lie in spaces in the matrix. Later the bony tissue matures.

The cartilage bones and membrane bones have the same structure. They differ only in embryonic origin, and many bones contain tissues of both kinds. Dermal bones are laid down layer by layer forming compact bone, while cartilage bones generally form cancellous bone, though an admixture of the two is found in many bones.

Heterotopic endoskeleton. In some animals certain bones may be present which do not belong either to the axial skeleton or to the appendicular skeleton, they are spoken of as **heterotopic bones.** These are :

1. Sesamoid bones are formed by ossification of a tendon where the tendon moves over a bony surface, such as patella or knee cap which appears first in some lizards, and two sesamoid bones are commonly present in the wrist (pisciform bones) of tetrapoda. They are a radial sesamoid anterior to the radiale, and a pisciform posterior to the unlnare.

2. Os cordis is a bone in the interventricular septum of deer and bovines.

3. Os penis or **os priapi** is a bone giving rigidity to the penis in marsupials, insectivores, rodents, Carnivora, Chiroptera, Cetacea, walrus and lower primates.

4. Os rostralis is a bone in the snout of some ungulates, such as pigs.

5. Os falciforme is an additional bone in the palm of moles which helps in digging.

6. Sclerotic bones form a ring in the eyeball of lizards and birds.

7. Os clitoridis, present in the clitoris of otters, rabbits and several rodents.

8. Pessulus is found in syrinx of birds.

9. Epipubic is found in ventral abdominal wall of monotremes and murspials.

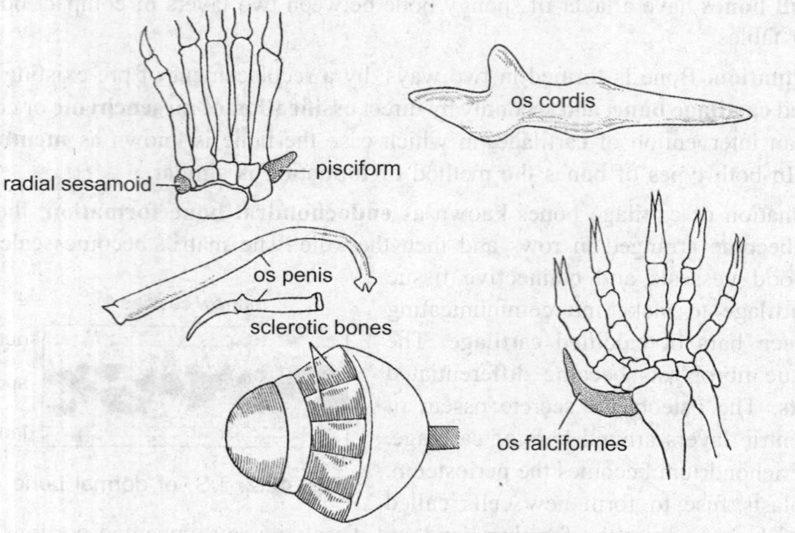

Fig. 42.3. Heterotopic bones.

Structure and formation of long bones. In long bones, such as the femur, the main shaft is **diaphysis** made of cartilage. A ring of bony tissue around the diaphysis form the compact bone which surrounds spongy bone. Ossification begins in the middle of the diaphysis and extends towards the ends. The ends remain cartilaginous for some time and continue to grow, increasing the length of the bone.

In mammals, and to some extent in reptiles, the long bones and vertebrae are formed from three centres of ossification : a central **diaphysis** and two terminal **epiphyses**. Between the diaphysis and each epiphysis arises a new cartilaginous pad, the **epiphyseal plate** where increase in length takes place. Eventually epiphyseal plates are replaced by cancellous bone and epiphyses are fused to the diaphysis, after which no further growth in length can take place. Ends of long bones are covered with articular cartilage.

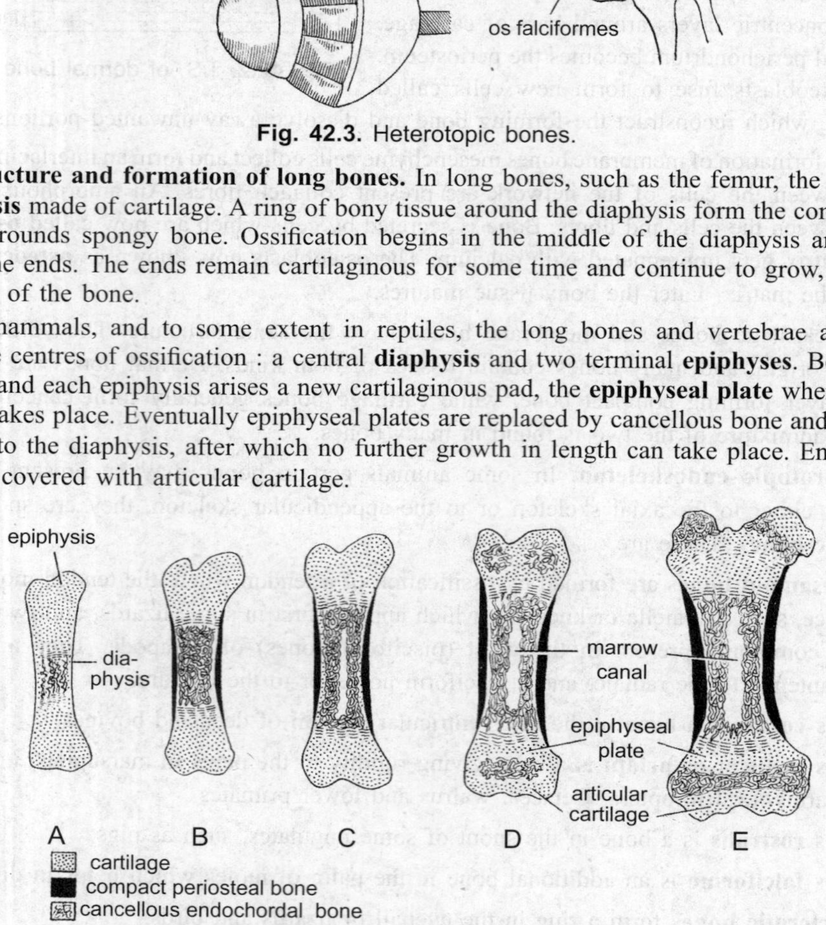

Fig. 42.4. Stages in ossification and growth of long bone.

AXIAL SKELETON

Skull

The skeletal framework of a vertebrate head is called a **skull**. A lamprey has a skull consisting of a braincase and cartilages of the tongue. A shark has a skull made of a braincase and isolated upper and lower jaw bars. But commonly in animals ranging from fishes to mammals, the skull has a brain case to which the upper jaws are welded by a series of bones, the lower jaw is not included. However, the skull of vertebrates is derived from three differnt embryonic components.

1. Chondrocranium. It includes cartilaginous brain case or neurocranium and cartilaginous capsules of olfactory, optic, and otic sense organs. The chondrocranium is replaced by bones in most vertebrates.

2. Splanchnocranium. It is derived from the visceral or pharyngeal skeleton which is cartilaginous but becomes largely replaced or invested by bones in higher forms. It provides support to the gills and forms the jaws and suspensorium in gnathostomes.

3. Dermatocranium. It consists of dermal bones which become attached to the chondrocranium and splanchnocranium in bony fishes and tetrapoda. A dermatocranium is absent in cyclostomes, elasmobranchs, and a few higher fishes because the entire skull is cartilaginous.

Development of chondrocranium. After the formation of central nervous system and notochord in the embryo, the mesenchyme cells form a membranous covering around the brain and the anterior part of the notochord. Cartilage is formed in this membrane. It gives rise to a pair of cartilaginous plates, the **parachordals** lying below the midbrain and hindbrain and anterior part of the notochord. In front of the parachordals forms a pair of curved cartilaginous rods, the **trabeculae** or **prechordals** lying below the anterior part of the brain. At the same time cartilaginous capsules arise around the developing sense organs. They are a pair of **olfactory capsules** around the organs of smell, a pair of **optic capsules** around the eyes, and a pair of auditory or **otic capsules** around the auditory organs (internal ears).

The parachordals grow larger and fuse in the middle line forming a **basilar plate** leaving a small opening or **basicranial fenestra**. The two prechordals or trabeculae grow towards each other and fuse in the midline to form an **ethmoid plate**. The ethmoid plate and the basilar plate grow towards each other and fuse to form a single **basal plate** forming the floor of the brain. A large opening, the **hypophyseal fenestra** is present in the basal plate, which lodges the pituitary gland.

The olfactory and otic capsules join the basal plate to form a cartilaginous **chondrocranium.** The optic capsules are sometime cartilaginous but more usually are fibrous, and never fuse with the chondrocranium but remain free and form the sclerotic of the eye, thus, they permit mobility of eyes.

The floor of the brain, thus, formed by the basal plate begins to grow upwards on the sides. In elasmobranchs the side walls grow and meet above the brain to enclose it in a brain-box or **neurocranium**. A few openings are left uncovered for the cranial nerves and blood vessels. But in most other vertebrates there is no roof of cartilage, only the posterior or occipital region gets roofed over by cartilage, and the rest of the brainbox has only a dorsal membrane. Later on membrane or dermal bones form the roof of brain. The formation of chondrocranium leaves a large opening or foramen magnum behind through which the spinal cord emerges.

Development of splanchnocranium. A visceral skeleton is formed partly from the neural crest cells and from splanchnic mesoderm around the pharynx between gill-clefts for their support. It consists of series of a paired visceral bars (usually seven pairs) of cartilage which become united with one another ventrally by an unpaired carilage to form **visceral arches.** The visceral arches are horse-shoe-shaped and encircle the pharynx all around except dorsally. Typically there are seven visceral arches in fishes, though this number varies form four to nine in differnt groups. The **first** visceral arch is the **mandibular arch**, the **second** is the **hyoid arch**, the remaining ones are **branchial arches** because they generally support gills in lower aquatic vertebrates.

In all vertebrates, except Agnatha, the **mandibular arch** forms jaws for supporting the mouth. This arch on either side is divided into a dorsal **palatopterygoquadrate** cartilage and a ventral **Meckel's cartilage**. The palatopterygoquadrate or palatoquadrate forms the **upper**

jaw, while the Meckel's cartilage forms the **lower jaw**. The **second** or **hyoid arch** on either side gives off a dorsal **hyomandibular** cartilage to support and connect jaws to chondrocranium beneath auditory region and a ventral cartilage which forms the **hyoid apparatus** which gives support to the tongue. The remaining or branchial arches form support for the gills or larynx. The branchial arches support the gills in fishes and in tetrapoda these are much reduced, and form the hyoid apparatus and cartilages of the larynx.

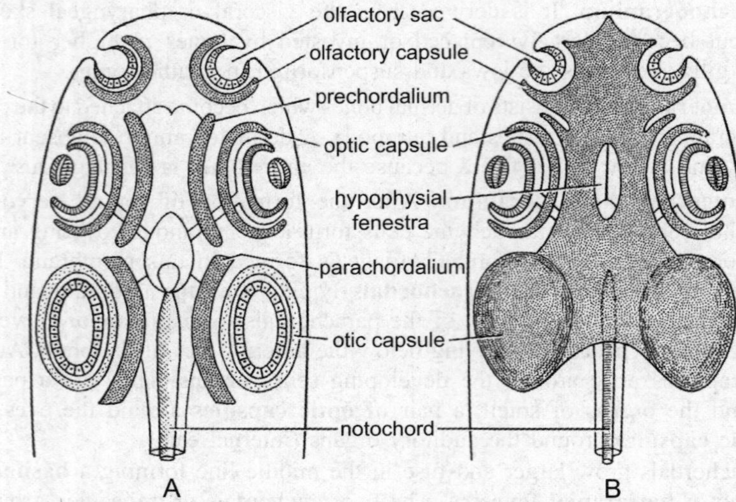

Fig. 42.5. Development of chondrocranium of skull.

Development of dermatocranium. The formation of the skull stops at the cartilaginous stage in cyclostomes and Chodrichthyes, but from bony fishes upwards cartilage bones and dermal or membrane bones become incorporated with the skull.

The dermal bones constitute the dermatocranium, they appear in the head region of bony fishes as scales. These dermal scales are actually parts of the exoskeleton which sink inwards and fuse with the roof of the chondrocranium to complete a protective envelope around the brain. The dermal bones may remain independent of the rest of the skull so that they can be removed easily by boiling, or they may become fused with cartilage and cartilage bones so that all distinction between replacing and dermal bones is lost in the adult.

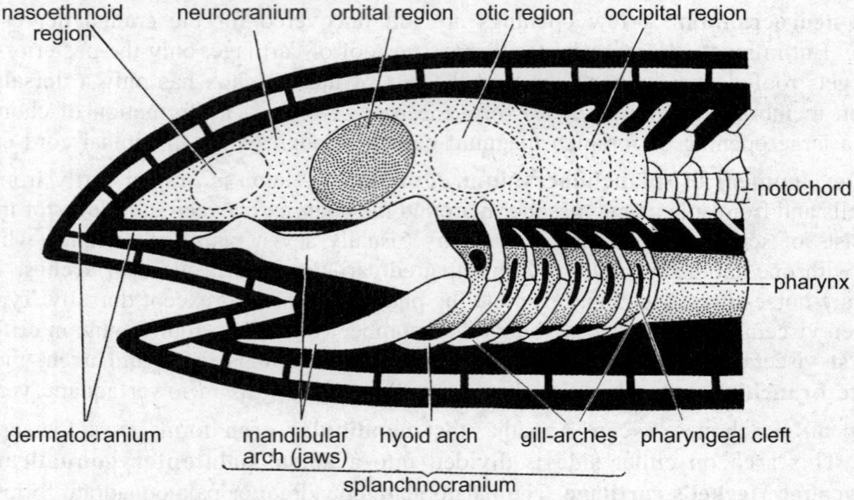

Fig. 42.6. Chondrocranium and splanchnocranium of elasmobranch.

TABLE. 42.1. GENERAL DIVISIONS OF ENDOSKELETON IN A LAND VERTEBRATE.

I. Axial skeleton (median)			*II. Appendicular skeleton (lateral, paired)*		*Heterotopic bones (miscellaneous)*
Skull	**Vertebral column**	**Thoracic basket**	**Girdles**	**Limb bones**	**Develop in association with certain organs**
A. Neurocranium 1. Cranium or brain box surrounding brain 2. Sense capsules (*i*) Olfactory-nose (*ii*) Optic-eyes (*iii*) Auditory ears	**Vertebrae** 1. Cervical-neck 2. Thoracic-chest 3. Lumbar-lower back 4. Sacral-hip 5. Caudal-tail	**A. Ribs** Paired; bony cartilaginous B. Sternum-Breast bone	**A. Pectoral** Anterior or shoulder girdle. Includes : 1. Scapula-dorsal 2. Clavicle-anterior 3. Coracoid-posterior	**A. Forelimb** 1. Humerus -upper arm 2. Radius and ulna-forearm 3. Carpals-wrist 4. Metacarpals-palm 5. Phalanges-fingers	**1. Os cordis**-Inter ventricular septum of heart in deer and bovines. 2. Rostral-Pig's snout.
B. Dermatocranium Membrane or dermal bones of skull **C. Splanchnocranium** Includes visceral arches or pharyngeal skeleton 1. Upper jaw 2. Lower jaw 3. Hyoid 4. Larynx			**B. Pelvic** Posterior or hip girdle Includes : 1. Ilium-dorsal 2. Pubis-anterior 3. Ischium-posterior	**B. Hind limb** 1. Femur-thigh 2. Tibia & fibula-shank 3. Tarsals-ankle 4. Metatarsals-sole 5. Phalanges-toes	**3. Os penis-** Penis of bats, rodents, marsupials, carnivores, insectivores, whales, lower primates. **4. Os clitoridis-** Clitoris of otters, rabbits several rodents. **5. Pessulus-** Syrinx of bird. **6. Epipubic-** Ventral abdominal wall of monotremes and marsupials. **7. Sesamoid-** Pisciform in hand, patella (kneecap), etc.

TABLE 42.2. TYPES OF BONES IN SKULL OF VERTEBRATES.

Region of skull	Cartilage or replacement bones	Membrane or dermal bones	Bones of mixed origin
A. Chondrocranium 1. Occipital	* Supraoccipital Exoccipital * Basioccipital	Parietal * Interparietal Postparietal	
2. Parietal	* Basisphenoid Pleurosphenoid		
3. Frontal	Orbitosphenoid * Presphenoid	Frontal Postfrontal Lacrimal	Prefrontal
4. Olfactory capsule	* Mesethmoid Turbinals } (Ethmoid) Cribriform } (Ethmoid) Ectethmoid	Nasal Vomer Septomaxillary	
5. Otic capsule	Epiotic Prootic Opisthotic	Squamosal Supratemporal	Sphenotic Pterotic
6. Optic capsule	Sclerotic		
7. Palate		* Parasphenoid Vomer (in mammals) Endopterygoid or Pterygoid, Ectopterygoid	Palatine
B. Splanchnocranium 1. Upper jaw	Quadrate (incus) Epipterygoid Alisphenoid Metapterygoid	Premaxilla Maxilla Jugal Quadratojugal	
2. Lower jaw	Articular (malleus) Mentomeckelian	Dentary (mandible) Coronoid Splenial Angular Supraangular	
3. Hyoid arch	Hyomandibular Columella (stapes) Symplectic Inter-epi-, hypo-, cerato- * basihyal		
4. Gill-cover		Preopercular Opercular Subopercular Interopercular Gular	

N.B. – All bones are paired. Single bones are marked with an asterisk (*).

Jaw suspension or suspensoria. The method by which the upper and lower jaws are suspended or attached from the chondrocranium is known as **jaw suspension** or **suspensorium**. Amongst the visceral arches, the **first** (mandibular) **arch** consists of a dorsal **palatopterygoquadrate bar** forming the upper jaw, and ventral **Meckel's cartilage** forms the lower jaw. The **second** (hyoid) **arch** consists of a dorsal **hyomandibular** supporting and suspending the jaws with the cranium, and a ventral **hyoid**. The remaining visceral arches support the gills and are, hence, called **branchial arches**. Thus, splanchnocranium forms the jaws and suspends them with the chondrocranium. Suspensoria are of five types :

1. Autodiastylic. The jaws are attached by ligaments (anterior and posterior) to the chondrocranium. The hyoid arch does not support the jaws but remain completely free as the posterior branchial arches. The gill-cleft in front of the hyoid arch does not form a spiracle but forms a complete gill, *e.g.*, early bony fishes (acanthodians).

2. Amphistylic. The upper jaw (mandibular arch) has basal and otic processes which are attached by ligaments to the chondrocranium. Besides this, the hyomandibular of the hyoid arch is also attached to the chondrocranium. At the other end both jaws are suspended from it. Thus, it is a double suspension in which both the mandibular and hyoid arches are attached to the chondrocranium. This type of suspensorium is found in Crossopterygii and in some primitive sharks *Heptanchus* and *Hexanchus*.

3. Hyostylic. The upper jaw is (palatoquadrate) is loosely articulated with the cranium by anterior ethmopalatine ligament and posterior spiracular ligament. Both jaws are suspended from the hyomandibular which is attached to the otic region of the skull. Thus, only hyoid arch binds both the jaws with the cranium and, hence, it is called **hyostylic**. It is found in most elasmobranchs and bony fishes. These fishes are able to swallow large preys.

In fishes the autodiastylic suspension is the most primitive, the amphistylic condition was derived from it, while the hyostylic condition is the most recent having been arrived independently.

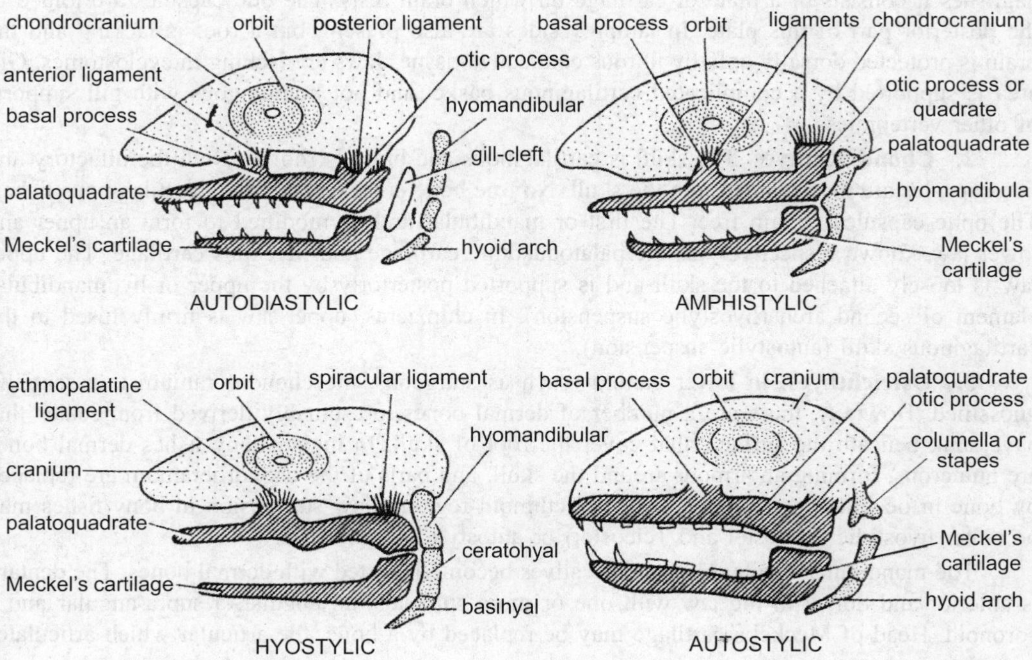

Fig. 42.7. Types of jaw suspensoria in vertebrates.

4. Autostylic. The upper jaw (palatoquadrate) is completely fused by its processes to the bony skull and the articular of lower jaw is suspended from the quadrate of the upper jaw. The hyomandibular do not take part in suspensorium and modified into columella or stapes of the middle ear. Some authorities use the term **autosystylic** for autostylic. It is found in extinct placoderms, chimaera, lung fishes and tetrapods. *i.e*, amphibians, reptiles and birds. In these, the **quadrate** of the upper jaw articulates with the **articular** of the lower jaw.

The autostylic suspension is divided into three varieties :

(*a*) Holostylic. The upper jaw is fused to the skull and the lower jaw is suspended from it. The hyoid arch is complete and not attached to the skull, *e.g.,* Holocephali (chimaera).

(*b*) Monimostylic. In many tetrapoda, except mammals, hyomandibular forms columella (middle ear bone) and the articular of lower jaw articulates with the quadrate of the upper jaw. The quadrate becomes an immovable part of the skull.

(*c*) Streptostylic. In lizards, snakes and birds the articulation is between quadrate and articular, but the quadrate is not firmly fused with the skull and is movable at both ends. This autostylic suspension is distinguished as streptostylic.

5. Crainostylic. The upper jaw is fused with the cranium in its entire length. Hyomandibular forms the stapes of middle ear bone. The quadrate and articular also modified into malleus and incus respectively. Thus, squamosal of the skull and dentary of lower jaw articulate with each other and both are dermal bones. It is found in mammals. Some consider it as modification of autostylic type.

Skull in Various Classes of Vertebrates

In all the vertebrates neurocranium, dermatocranium and splanchnocranium are the same. But still there are differences in form and detailed structure of skull.

1. Cyclostomata. In cyclostomes the skull is cartilaginous and very incomplete. In hagfishes it consists of a plate of cartilage on which brain rests. The otic capsules are joined to the posterior part of this plate. In lamprey sides are also present, but a roof is lacking and the brain is protected dorsally only by fibrous connective tissue. Jaws are lacking in cyclostomes. Gill area is supported by a complicated cartilaginous basket and not homologous with gill supports of other vertebrates.

2. Chondrichthyes. The skull is cartilaginous and brain is roofed. Both the olfactory and otic capsules are incorporated into the skull. No true bone is present in living cartilaginous fishes. The optic capsules remain free. The first or mandibular arch is modified to form an upper and lower jaw, known respectively as the palatoquadrate cartilage and Meckel's cartilage. The upper jaw is loosely attached to the skull and is supported posteriorly by the upper or hyomandibular element of second arch (hyostylic suspension). In chimaeras, upper jaw is firmly fused to the cartilaginous skull (autostylic suspension).

3. Osteichthyes. In lower ganoids such as sturgeons, the chondrocranium is essentially unossified. However, there are a number of dermal bones, presumably derived from scales that have sunk beneath the surface, that cover the root of skull. In higher bony fishes dermal bones are numerous, forming an armour around the skull, and parts of the chondrocranium are replaced by bone in occipital, sphenoidal, otic, and ethmoid regions. Jaw suspension in bony fishes may be either hyostylic (Holostei and Teleostei) or autostylic (Dipnoi).

The mandibular cartilage or its derivatives become invested with dermal bones. The dentary is anterior and dorsal in the jaw with one or more splenials, an angular, a supra-angular and a coronoid. Head of Meckel's cartilage may be replaced by a bone, the articular which articulates with the quadrate.

4. Amphibia. The skull of amphibians shows a reduction in the number of bones and a flattening of the skull. In the modern amphibian skull, the principal bones are paired premaxillaries,

maxillaries, nasals, frontals, parietals and squamosals. Much of the upper surface of the head lacks a bony covering especially anurans. Parts of the chondrocranium still remain unossified. The underside of brain case is covered by dermal bone, the parasphenoid. The length of the occipital region is much reduced. The embryonic chondrocranium persists considerably in the adult, only some of it is replaced by cartilage bones. The basioccipital supraoccipital, basisphenoid and presphenoid are absent. The otic capsule has a ventral membrane covered aperture, the **fenestra ovalis** into which the stapedial plate of the columella fits. The dorsal surface of the skull is solidly covered by dermal bones leaving opening for nares and eyes. Exoccipitals are ossified and bear two occipital condyles one in each exoccipital. The basisphenoid, presphenoid and alisphenoid are cartilaginous. Epipterygoid is never ossified in amphibians and incompletely ossified in urodeles and fused with the otic region. Articular of lower jaw is bony in caudates and cartilaginous in anurans. A dentary, splenial and angular may be present.

5. Reptilia. In the skull two more segments have been incorporated than in Amphibia, hence, the hypoglossal nerve emerges from the skull. The entire chondrocranium is ossified except the naso-ethmoidal region. There are more dermal bones than in amphibians. The parietals are fused together (except Chelonia) and an interparietal foramen is prominent in lizards. Between the orbits is an interorbital septum which is imperfect in lizards. Behind the orbits are one or two **temporal fossae** (except Chelonia) for muscles of the jaws. In Chelonia, the roof of skull is solid having no temporal openings. Parasphenoid begins to disappear and is fused with the basisphenoid. In the occipital region all four bones surround the foramen magnum, and there is a single occipital condyle. The prootic, epiotic, and opisthotic remain separate in the otic region. A stout **ectopterygoid** or transverse bone is added between the maxilla and pterygoid (except Chelonia), and a vertical **epipterygoid** runs from the prootic to the pterygoid. The lower jaw in each half has one cartilage bone and five dermal bones, the Meckel's cartilage becomes very slender. The **quadrate** forms a movable union with the squamosal on one side and on the other side the quadrate articulates with the lower jaw, such a suspension is known as **streptostylic**. In turtles, crocodiles and tuatara, quadrate is fused with the skull. A quadratojugal is absent The hyoid arch forms a columella of the middle ear from its hyomandibular, the lower part of the hyoid arch and third and fourth visceral arches form the hyoid. The remaining visceral arches form cartilages of larynx and trachea. Secondary palate is well developed in crocodiles, less developed in turtles and absent in others.

6. Aves. The skull is made on the same plan as in reptiles and it resembles the skull of lizards. The bird skull differs in being larger and very light due to pneumatic bones. Its bones are fused together so that there are practically no sutures. There is a large, arched cranium due to a greater development of the brain which lies in the posterior part. The orbits are immense and lie in front of the cranium. They are separated by a thin **interorbital septum**, thus, pushing the brain case posteriorly, developed better than in reptiles. Orbits are very large and each is continuous with a temporal fossa. Each orbit is surrounded by a ring of thin dermal bones called **sclerotics**. There is a large pointed **beak** formed by the premaxillae and dentaries. There are no teeth in modern birds. There is a single occipital condyle formed from basioccipital and exoccipitals, shifted to the ventral side so the skull is at right angles to vertebral column. Foramen magnum is surrounded by all four occipital bones. Auditory capsules are sunk inwards, their bones fuse together and also with adjoining bones. The Meckel's cartilage is much reduced, it has one cartilage bone and four ensheathing dermal bones. There is a freely movable quadrate articulating with the lower jaw, the suspension is **streptostylic** in which the lower jaw comes to have two movable joints, one at each end of the quadrate. Like reptiles, hyomandibular forms the columella of the middle ear, the rest of the hyoid arch, along with the first branchial arch forms the hyoid apparatus.

7. Mammalia. There is much variation in the skulls of different mammals, yet there are certain common features. In the evolution of the skull from fishes to mammals there has been a considerable reduction in the number of skull bones by **loss** and **fusion**, (usually 35 in number). The prefrontals, postfrontals, postorbitals and quadratojugal are lacking. In mammals four occipital bones are fused together. Presence of two occipital condyles, one in each exoccipital. Various degrees of fusion occur in the sphenoidal area. The presphenoid, orbitosphenoids, basisphenoid and alisphenoids may be separate or all fused into a single bone. Due to the formation of a large brain the cranium is very large having expanded dorsally and laterally. Its cavity is closed in front by a **cribriform plate**. The petrosal may fuse with the squamosal to form the temporal bone. The tympanic encloses the **ear ossicles**, the quadrate of the upper jaw forming the **incus**, articular of the lower jaw forming the **malleus** and hyomandibula or columella of reptiles forms the **stapes**.

The **bony palate** in mammals is formed from the premaxillae, maxillae and palatines, in some the pterygoids also contribute to the bony palate.

The lower jaw in each half is made of a single **dentary**, with no trace of Meckel's cartilage in the adult. It articulates with a glenoid fossa in the squamosal. The major part of the hyoid arch forms a **hyoid** apparatus. The remaining visceral arches form the thyroid, epiglottis, arytenoids, cricoid in the larynx and also the rings of the trachea.

Vertebral Column

Notochord. The primitive axial skeleton is a notochord present in all chordate embryos. It is formed from chordamesoderm cells. It is a stiff rod below the nerve cord and above the alimentary canal running from the infundibulum to the hind end of the body. At first the cells of the notochord are packed closely but later they fuse together and become vacuolated, except a layer of peripheral cells. The notochord is enclosed in an inner and outer elastic fibrous sheath of connective tissue called elastica interna and elastica externa respectively.

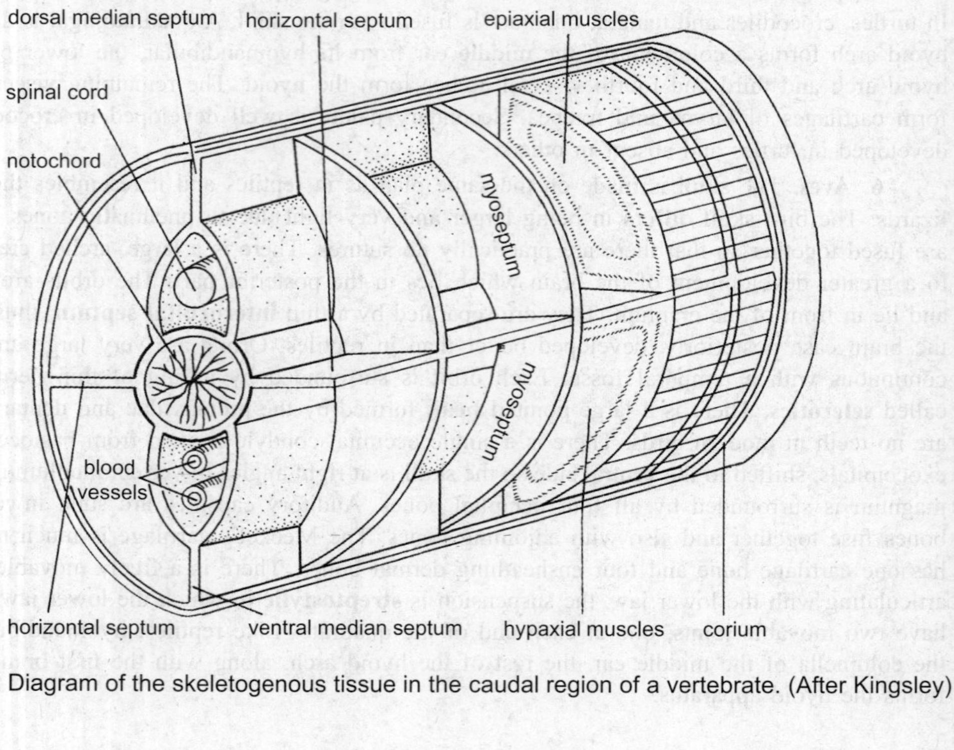

Fig. 42.8. Diagram of the skeletogenous tissue in the caudal region of a vertebrate. (After Kingsley)

TABLE: 42.3. MAJOR SKELETAL DERIVATIVES OF VISCERAL ARCHES IN REPRESENTATIVE VERTEBRATES.

Visceral arch	Dogfish (*Scoliodon*)	Teleost (Bony fish)	Amphibian (*Necturus*)	Amphibian (Frog)	Reptile & Bird (*Uromastix* & Pigeon)	Mammal (Rat or rabbit)
I.	Meckel's cartilage Pterygoquadrate	Articular Quadrate Epipterygoid Metapterygoid	Articular Quadrate Cartilage in lateral roof of mouth	Articular, Mentomecklian in some species Quadrate Annulus tympanicus	Articular Quadrate Epipterygoid – –	Malleus Incus Alisphenoid –
II.	Hyomandibula Ceratohyal Basihyal	Hyomandibula Symplectic Interhyal Epihyal Ceratohyal Hypohyal Entoglossal	Rudimentary – Ceratohyal – – Hypohyal	Columella – Anterior horn – – Body of hyobranchial apparatus	Columella – Anterior horn – – Body of hyobranchial apparatus Enoglossus	Stapes – Anterior horn – – Body of hyoid
III.	Pharyngobranchial Epibranchial Ceratobranchial Hypobranchial	Pharyngobranchial Epibranchial Ceratobranchial Hypobranchial	Epibranchial Ceratobranchial	Body of hyobranchial apparatus	Second horn Body of hyobranchial apparatus	Posterior horn Body of hyoid
IV.	Branchial elements (as in III)	Branchial elements (as in III)	Branchial elements (as in III)	Posterior horn. Body of hyobranchial apparatus	Posterior horn	Thyroid cartilage
V.	Branchial elements	Branchial elements	Epibranchial only. Homologies of laryngeal cartilages remain in doubt	Not clearly delineated		Thyroid Cricoid Arytenoid
VI.	Branchial elements	Branchial elements	Missing or not clearly delineated			
VII.	Bears no gill. Some reduction.	Reduced	Missing			

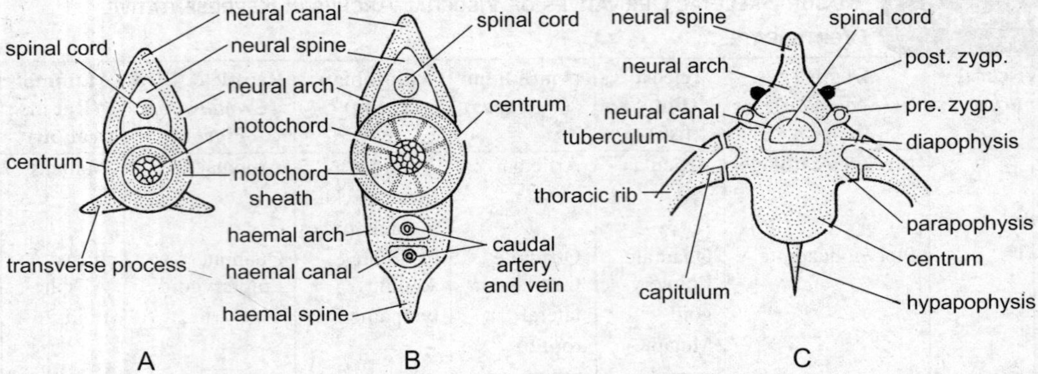

Fig. 42.9. Structure of a vertebra showing processes in cephalic view, A – Trunk vertebra of shark; B – Caudal vertebra of shark; C – Typical tetrapod vertebra.

In protochordates (*Amphioxus*) and cyclostomes (lamprey) the axial skeleton is primarily the notochord, which persists throughout life of the animal. In hagfishes small cartilaginous elements are present in the caudal region, and in lampreys two pairs of cartilaginous elements occur on either side of the spinal cord in each body segment. They do not meet above to form a neural arch. In some of lower bony fishes (sturgeon), the notochord persists unchanged and, although neural and haemal arches are present, a centrum is lacking. In chimaeras and lungfishes, cartilage begin to invade the notochordal sheath. In most fishes centrum is well developed. In fishes and higher animals, notochord is later surrounded by cartilaginous or bony rings called **vertebrae**. Notochord is practically obliterated in tetrapods.

The vertebral column is made of a metameric series of vertebrae extending longitudinally from the skull to the tip of the tail.

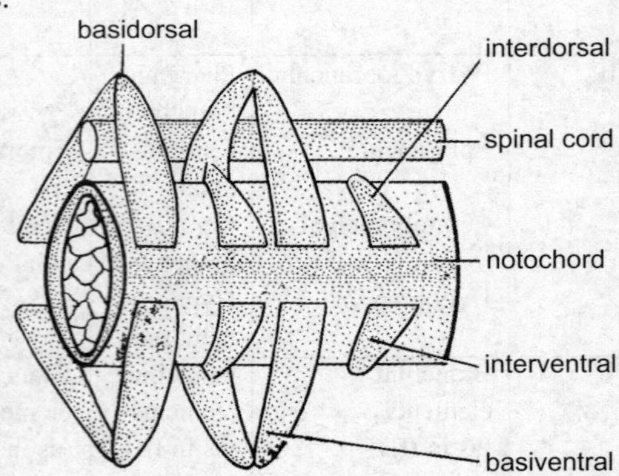

Fig. 42.10. Cartilaginous arcualia formed from sclerotomes (show for two vertebrae).

Vertebrae. Vertebrae of various animals are different, even in the same vertebral column there are differences, but all vertebrae conform to a basic plan. Generally fish vertebrae are divisible into trunk or preanals and the caudal or postanals. The caudals are recognisable by the presence of a haemal arch on the underside of centrum or notochonrd. Centrum in fishes is amphicoelous. In tetrapods vertabral column has five regions : cervical, thoracic, lumbar, sacral and caudal, each have several vertebrae. Amphibians have a single cervical (atlas) and one sacral (9th) vertebra.

Basic structure of vertebra. A **typical vertebra** has a cylindrical body or **centrum** which encloses or replaces embryonic notochord. Above the centrum is a **neural arch** enclosing a **neural canal** through which the nerve or spinal cord passes. The neural arch is produced dorsally into a **neural spine**. In the caudal region of fishes, below the centrum is a **haemal arch** enclosing a **haemal canal** through which the caudal artery and vein pass. The haemal arch is produced below into a **haemal spine**.

Types of processes. Various kinds of projections or **apophyses** arise from a vertebra. They are : ***(i)*** **zygapophyses** present over anterior and posterior faces of neural arch are articulations between successive vertebrae. **Prezygapophyses** are anterior projections facing upwards and **postzygapophyses** are posterior projections facing downward. ***(ii)*** **Transverse processes.** Lateral transverse processes arise from centrum, ***(iii)* diapophyses** arising from the neural arch or the centrum and serve for attachment of the upper head (tuberculum) of a two-headed (bicipital) rib, ***(iv)* parapophyses** arise laterally from the centrum for attachment of the ventral head (capitulum) of a bicipital rib, ***(v)* basapophyses** are ventro-lateral processes of the centrum or haemal arch, they are remnants of a haemal arch or articulate with the haemal arch when present, ***(vi)* pleurapophyses** are lateral projections to which short ribs are fused, ***(vii)* hypapophysis** is a mid-ventral projection from the centrum.

Development of vertebrae. Mesenchyme cells are produced from metameric sclerotomes. A sclerotome becomes differentiated into a dense posterior or **caudal half** and a less dense anterior or **cranial half**. The caudal and cranial halves separate, then the caudal half of each sclerotome fuses with the cranial half of the following sclerotome. Thus, definite sclerotomes are formed and they come to lie inter-segmentally alternating with myotomes. This is an adaptation by which each myotome will become connected with two successive vertebrae and their ribs, by this arrangement movement of the vertebral column occurs. The mesenchyme cells of the definite sclerotomes spread to form a skeletogenous layer around the notochord and nerve cord.

Arcualia. Four pairs of cartilages, called **arcualia** are laid down in each sclerotome lying on the two sides of the notochord. The dorsal arcualia are two **interdorsals** and two **basidorsals**, the ventral arcualia are two **interventrals** and two **basiventrals**. The basidorsals and basiventrals are derived from the caudal half of the sclerotome and will form the anterior parts of a vertebra, while the interdorsals and interventrals arise from the cranial half of the sclerotome and will form the posterior parts of a vertebra. Generally the basidorsals and basiventrals are larger than the other arcualia.

The two basidorsals extend above the nerve cord and unite to form a neural arch.

1. Formation of centrum. The basiventrals, interdorsals, and interventrals form a **centrum** around the notochord. In the tail region the basiventrals also extend below to form a **haemal arch**.

2. There is another method of formation of centra, in some elasmobranchs cells from the basidorsals and basiventrals penetrate and spread into the secondary notochordal sheath and secrete cartilage to form a ring-shaped centrum around the constricted notochord, such a centrum is called a **chordal centrum**. But in other elasmobranchs, bony fishes, and tetrapoda, the arcualia grow around the notochord outside the notochordal sheaths to form a centrum known as a **perichordal centrum** which later grows and obliterates the notochord. In both cases the centrum becomes fused to the neural and haemal arches.

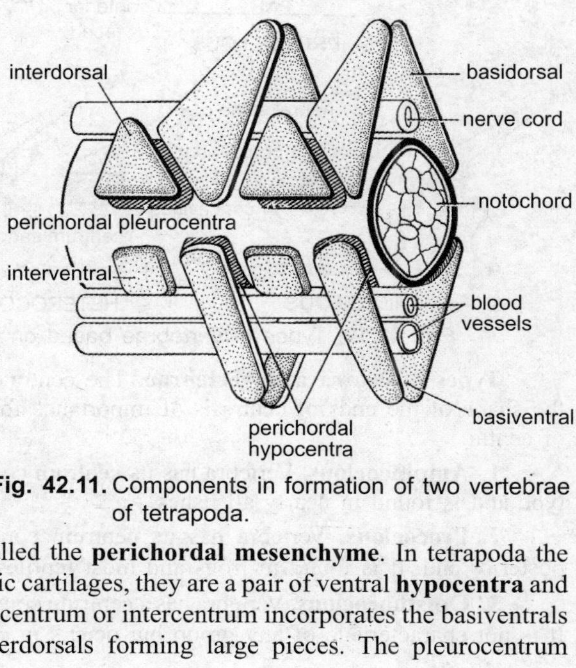

Fig. 42.11. Components in formation of two vertebrae of tetrapoda.

3. The centrum is not always formed from arcualia only, there is another method of centrum formation. The mesenchyme lying between the definite sclerotomes and notochord, internal to the arcualia, is called the **perichordal mesenchyme**. In tetrapoda the perichordal mesenchyme forms four embryonic cartilages, they are a pair of ventral **hypocentra** and a pair of dorsolateral **pleurocentra**. The hypocentrum or intercentrum incorporates the basiventrals and the pleurocentrum incorporates the interdorsals forming large pieces. The pleurocentrum piece increases and the hypocentrum piece decreases to form the centrum of amniotes, but in amphibians the hypocentrum forms the centrum, while the pleurocentrum decreases or disappears.

Types of centrum. The parts forming a centrum are not always homologous and different lines of evolution are seen.

1. In crossopterygians and some fossil amphibians the two hypocentra fuse to form a large piece incomplete dorsally, but the two pleurocentra remain separate, the neural arch rests on the hypocentrum and pleurocentra, such a vertebra is **rachitomous**.

2. In some other extinct amphibia the hypocentrum and pleurocentrum are equal, the former is anterior and the latter posterior and each forms a disc-like centrum, but on both there is single neural arch, this is an **embolomerous vertebra**, it was evolved from a rachitomous vertebra.

3. On the other hand, the rachitomous vertebra gave rise to a **stereospondylous vertebra** of modern amphibians in which the hypocentrum enlarges to form a centrum and the pleurocentrum disappears.

4. The embolomerous vertebra gave rise to **gastrocentrous vertebra** of reptiles and mammals in which the pleurocentrum enlarges to form the centrum, while the hypocentrum is reduced to a small piece (reptiles) or it is changed to form an intervertebral disc (mammals).

Generally there is one verterbra per segment, this is known as **monospondyly**. In some cases the arcualia form two centra and only one neural arch in one segment, this is known as **diplospondyly**. It is found in the tail of some fishes (*Amia*), some extinct amphibians, and some lizards. A second type of diplospondyly is seen in the tail region of some elasmobranchs in which each segment has two complete vertebrae with two centra, two neural arches, and two haemal arches.

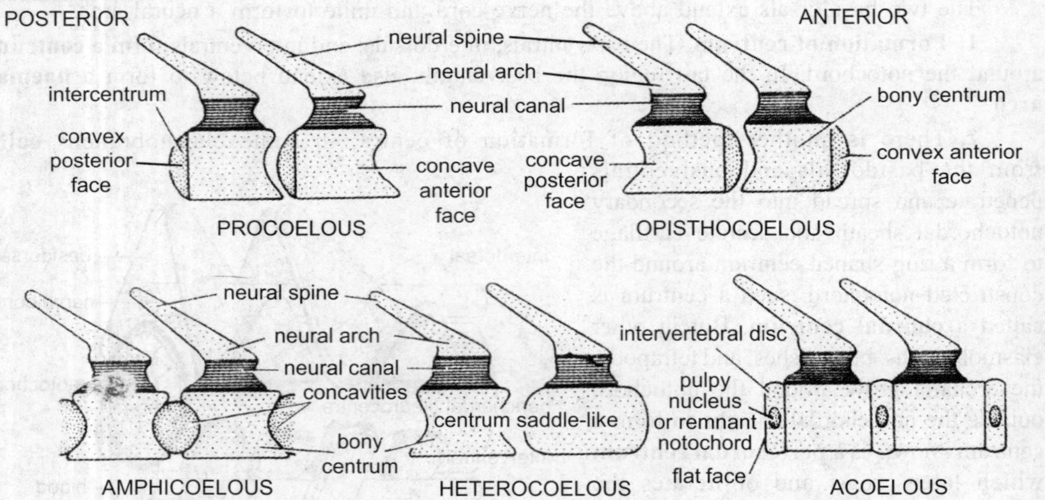

Fig. 42.12. Types of vertebrae based on shape of centra in sagittal section.

Types of centra and vertebrae. The centra of vertebra are placed end to end in a row, the shape of the ends of centra is of importance for articulation. There are six principal shapes of centra.

1. Amphicoelous. Vertebra has its centrum concave at both ends. This is the most primitive type and is found in nearly all fishes.

2. Procoelous. Vertebra has its centrum concave at the anterior end and convex at the posterior end. It is found in frogs and most reptiles.

3. Opisthocoelous. Vertebra has a centrum convex at the anterior end and concave posteriorly. It is not characteristic of any group but occurs in all classes.

4. Heterocoelous. Vertebra has the centrum saddle-shaped transversely. The anterior end is convex dorsoventrally and concave sideways, while the posterior end is opposite of the anterior end. This is the most specialised vertebra and is found in neck region of birds.

5. Amphiplatyan or acoelous. Vertebra has its centrum flat at both ends and is found in mammals.

6. Biconvex. Centrum is convex at both the ends. It is found in the 9th (sacral) vertebra of frog.

Ribs. Ribs are cartilaginous or bony structures, which are either fused to or articulated with vertebrae. Typically there is one pair of ribs to each vertebra. In **fishes** there are two kinds of ribs (*i*) **dorsal** or **intermuscular ribs** extend from transverse processes into a skeletogenous septum between two successive myotomes. (*ii*) The second kind are **ventral** or **pleural ribs** which arise from the centrum and lie between the body wall muscles and parietal peritoneum. Most fishes have either dorsal or ventral ribs, but some have both kinds, *e.g.*, *Polypterus* and many teleosts. In elasmobranchs only dorsal ribs are present, while most teleosts have only ventral ribs. In **tetrapoda** there is one pair of ribs to each vertebra, it is now believed that they represent the ventral ribs of fishes. All ribs are short in tetrapoda, except those in the thorax. In amniotes the ribs of the thoracic region unite ventrally with the sternum and dorsally articulate with the vertebrae by two heads. In these two-headed or **bicipital** ribs the **tuberculum** or upper head articulates with a process of the neural arch called **diapophysis**, while the **capitulum** or lower head articulates with a process of the centrum called **parapophysis** (Fig. 42.9). In the cervical region where ribs are fused with veretebrae, the space between the two heads of a rib is a **vertebrarterial canal** through which a nerve and blood vessels pass.

Abdominal ribs are also called **gastralia** or ventral ribs. There are V-shaped dermal bones in the ventral region of the abdomen extending postero-dorsally with their apices pointing in front. Gastralia were present in some fossil amphibians, in *Sphenodon* and crocodilians. They are dermal elements, while true ribs are endoskeleton structures.

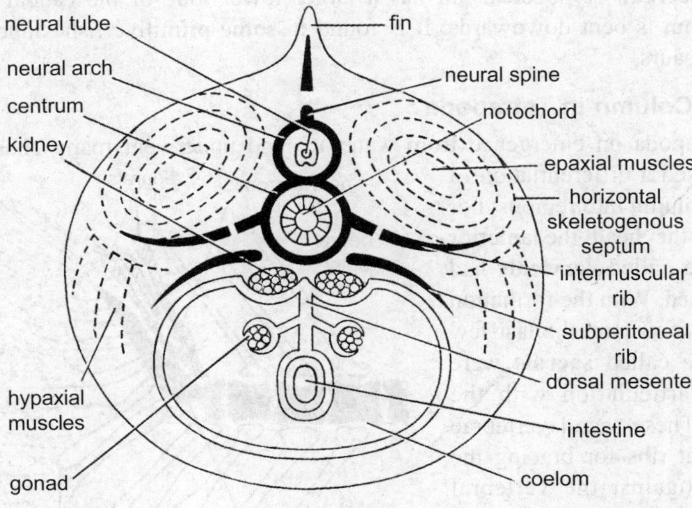

Fig. 42.13. T.S. of trunk region of a vertebrate showing dorsal and ventral ribs.

1. Vertebral Column in Fishes

The **vertebral column** of fishes has two types of vertebrae, **trunk vertebrae** and **caudal vertebrae**, the anus marks the point of transition between the two types of vertebrae. The vertebrae are **amphicoelous**.

Types of tails or caudal fins: In fishes the shape of the tail is correlated with the direction taken by the terminal portion of the notochord or vertebral column.

1. Protocercal. When the notochord extends straight to the tip of the tail and the caudal fin is divided equally then the tail is **protocercal**, as in embryonic fishes, adult cyclostomes, and some amphibians.

2. Heterocercal. Heterocercal tail has the vertebral column bent upwards in the posterior region so that the caudal fin is divided into a narrow dorsal lobe and a broader ventral lobe. It is typical of elasmobranchs (*Scoliodon*) and some lower bony fishes. The heterocercal tail made its appearance after the protocercal tail, and it gave rise to the next two types of tails.

3. Diphycercal. Diphycercal tail has lost the terminal upturned part of the vertebral column, and the vertebral column does not reach the end. Externally the tail has become symmetrical secondarily having two equal lobes of the caudal fin. It is found in Dipnoi, *Polypterus* and *Latimeria.*

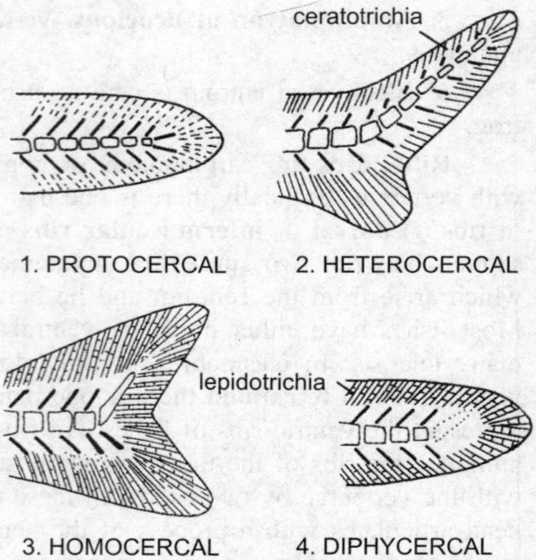

Fig. 42.14. Types of tails or caudal fins in fishes.

4. Homocercal. In homocercal tail, the terminal part of the vertebral column is bent upwards, but the dorsal lobe of the caudal fin is lost during development and the ventral lobe divides into two equal parts, so that the caudal fin appears symmetrical externally. This tail is common in bony fishes.

5. Hypocercal. Hypocercal tail has a larger lower lobe of the caudal fin into which the vertebral column is bent downwards. It is found in some primitive fishes and in extinct reptiles called ichthyosaurs.

2. Vertebral Column in Tetrapoda

The tetrapoda on emergence from water to land underwent many changes. One of these changes involved a differentiation of the vertebral column into regions. For movement of the head the anterior trunk vertebrae called **cervicals** had their ribs reduced. With the formation of limbs for use on land the last few trunk vertebrae called **sacrals** were modified for articulation with the pelvic girdle. These sacral vertebrae bear short stout ribs for bracing the pelvic girdle against the vertebral column. The **caudal** vertebrae are reduced in number, and the terminal ones are reduced to small cylindrical centra only with no arches. Thus, tetrapoda have cervical, trunk (dorsal), sacral and caudal vertebrae. In many reptiles and all birds and mammals the

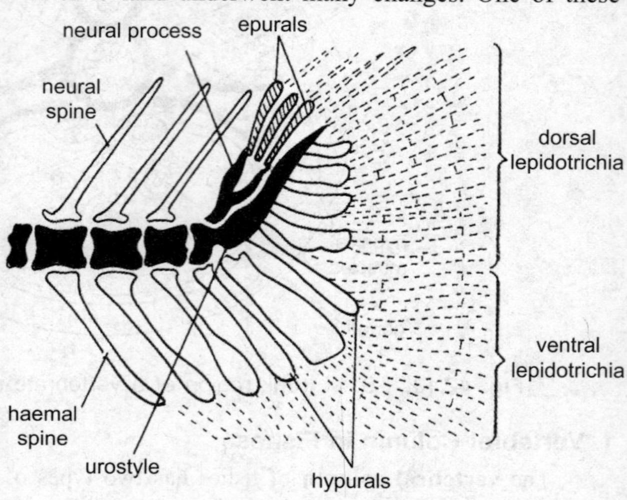

Fig. 42.15. Skeleton of homocercal tail.

trunk vertebrae are further differentiated into **thoracic** vertebrae having ribs and **lumbar** vertebrae with reduced or no ribs.

***(i)* Amphibia.** The centra of vertebrae are **stereospondylous** being formed from hypocentra, the pleurocentra having disappeared. There is a single **cervical** vertebra (**atlas**) which articulates with the skull. This is followed by trunk vertebrae, then there is a single **sacral** vertebra connected to pelvic girdle. Caudal vertebrae are found only in tailed forms. All **caudal** vertebrae, except the first, bear haemal arches with haemal spines.

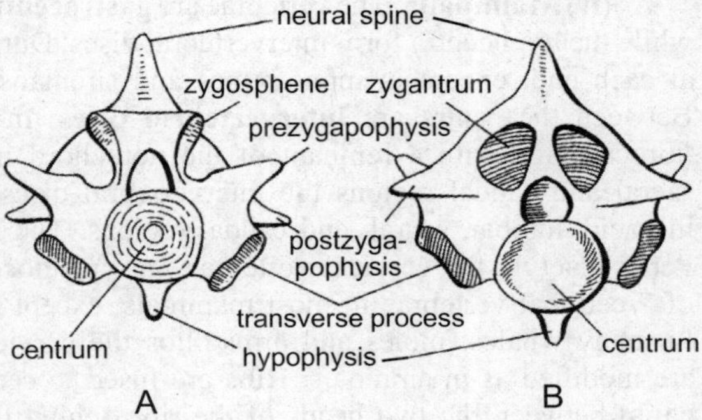

Fig. 42.16. Vertebra of snake. A–Anterior view; B–Posterior view.

***(ii)* Reptila.** The vertebrae are **gastrocentrous**, *i.e.,* the centra are formed from pleurocentra, the hypocentrum being lost, but in *Sphenodon* it persists as a small ventral remnant. The vertebral column in most lizards and crocodilians is differentiated into cervical, thoracic, lumbar, sacral, and caudal regions. In all living forms two sacral vertebrae fuse to form a **sacrum** whose transverse processes and ribs are attached to the pelvic girdle. Caudal vertebrae have small V-shaped **chevron bones** on the ventral side of centrum, they are formed from the **hypocentra** and are not homologous with haemal arches of lower forms in which haemal arches are formed from **basiventrals**.

All types of centra are found in reptiles, in turtles and *Sphenodon*. They are **amphicoelous**, some have **opisthocoelous** vertebrae. Most lizards, snakes and crocodilians have **procoelous** vertebrae. Sacral vertebrae of crocodilians are **amphiplatyan**. In snakes and some lizards in addition to zygapophyses, there is a special pair of apophyses called **zygosphenes** at the anterior end fitting into concavities called **zygantra** located on the posterior side of each vertebra.

***(iii)* Aves.** The vertebrae are formed in typical manner from four pairs of arcualia. The chief characteristic of the vertebral column is its rigidity due to an ankylosis between vertebrae. The rigidity is an advantage in flight. The cervical region is very mobile but the other regions are capable of little movement. The vertebral column provides a fulcrum for the wing strokes. In the cervical region the number of vertebrae is variable from 8 to 25. These vertebre are **heterocoelous** or saddle-shaped, the anterior face of the centrum is convex dorsoventrally, but concave from side to side, the posterior face has its curves in just the opposite direction so that there is proper articulation. Thus, in vertical section the centrum appears opisthocoelous and in horizontal section it is procoelous. But parrots have **opisthocoelous** cervical vertebrae, and in *Archaeopteryx* they were **amphicoelous**. Cervical vertebrae bear short ribs, ventrally each vertebra has a **hypapophysis**. Thoracic vertebrae are more or less fused together, though some are free, their centra are produced below into hypapophyses, their broad transverse processes bear ribs. Ribs are attached to thoracic vertebrae and are united ventrally to the sternum. The posterior thoracic, lumbar, sacral, and the first few caudal vertebrae are all fused together to form a single **synsacrum**, which becomes fused with the pelvic girdle. The number of vertebrae making up the synsacrum varies from 13 to 20 in different birds. On examining the ventral side of the synsacrum the separate transverse processes of vertebrae are seen which reveal the number of vertebrae forming the synsacrum. Only two of the vertebrae of the synsacrum are **sacral**, except in ostrich which has three. The synsacrum provides a rigid framework valuable in flight and the girdle with its longitudinal curvature distributes the weight of the body evenly on the legs in standing or walking The number of caudal vertebrae is small, the first few after the synsacrum are free but the posterior ones are fused to form a **pygostyle** which supports the tail. The **free caudal** vertebrae are amphicoelous.

(iv) **Mammalia.** The vertebrae are **gastrocentrous**. The centra being formed from pleurocentra, while the hypocentra form intervertebral discs. During development each centrum has an **epiphysis** at each end, except in monotremes and sirenians, the epiphyses finally fuse with the centrum. Between the centra are **intervertebral discs**. In front of the sacrum these discs are made of fibrocartilage with a remnant of the notochord in the centre, called **nucleus pulposus**. In the sacral and caudal regions the intervertebral discs are bony. The vertebral column has cervical, thoracic, lumbar, sacral, and caudal regions. The vertebrae are **amphiplatyan** (acoelous), but in Perissodactyla the cervical vertebrae are opisthocoelous. An outstanding feature is the presence of 7 cervical vertebrae in most mammals, except in some Edentata which have 6 to 9 vertebrae. In many whales, moles and armadillos the cervical vertebrae fuse together. The atlas and axis are modified as in amniotes. **Ribs** are fused to cervical vertebrae. There being a **vertebrarterial canal** between the two heads of the rib. Behind the neck the number of vertebrae is variable in mammals. **Thoracic vertebrae** bear ribs which connect directly or indirectly with the sternum, but some posterior ribs do not reach the sternum and are called **floating ribs**. **Lumbar vertebrae** are strong and longer than others. They have prominent transverse processes directed forward. The anterior lumbar vertebrae bear two additional processes called **mammillary porcesses** or **metapophyses**. **Sacrum** has several vertebrae fused together : 3 in dog, 4 in rabbit, 5 in horse, and 5 in man. There is no sacrum in whales because of a reduced pelvic girdle. **Caudal vertebrae** are variable in number : 16–18 in rabbit, 20–23 in dog, 15–21 in horse, and 3–4 in man which unite to form a single **coccyx**. Usually the most posterior caudal vertebrae are reduced to only centra, whereas the more anterior ones possess all the parts.

Sternum

The sternum is an alongated structure in the midvental region of the thorax. It is absent in fishes but occurs in tetrapoda. Some tetrapoda such as *Proteus*, *Amphiuma,* caecilians, snakes, turtles and limbless lizards have no sternum. Primitively the sternum is a plate of cartilage, but in later forms it has a series of cartilages or bones or both. In tetrapoda the sternum articulates with the pectoral girdle. In amniotes (but not in amphibians) it also articulates with ribs to form a **thoracic basket** which is protective and aids in breathing.

The origin of a sternum is obscure. It is explained by three main theories :

1. The median portions of two pectoral girdles united midventrally to form a sternum.

2. The ventral ends of the more anterior ribs fuse to form a rod of cartilage on each side. The two rods then unite to form a sternum. Such a sternum is called a **neosternum** and is found in amniotes. **Objection** to this theory is based on the presence of a sternum in most amphibians, but in no case are ribs associated with it.

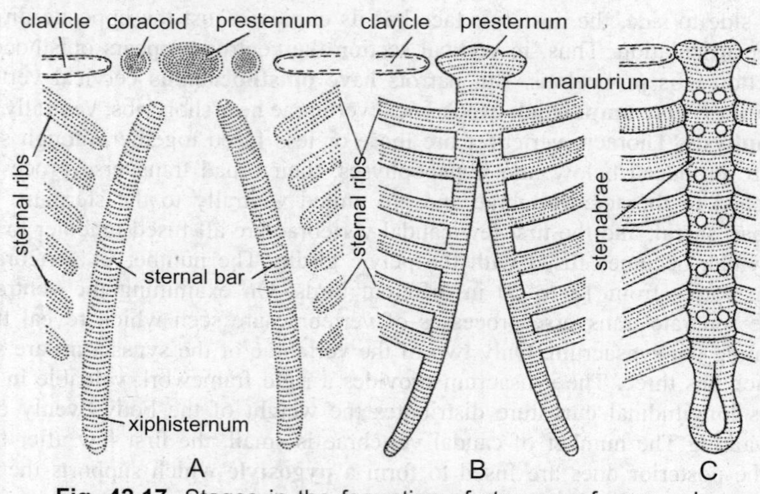

Fig. 42.17. Stages in the formation of sternum of mammal.

3. Cartilages called **parasternals**, which are the same as abdominal ribs, develop in the ventral myocommata of the thoracic region in many amphibians and lizards. Some of the anterior parasternals fuse to form a sternum, such a sternum is called an **archisternum**. This theory is regarded as the most probable.

In mammals the mesenchyme condenses mid-ventrally between the pectoral girdles. This mesenchyme chondrifies to form a **presternum** cartilage. Posterior to it two elongated bars of cartilage arise, they unite with each other medially and fuse with the presternum. They also unite with the ribs laterally. The central part of the bars joining the ribs becomes the **mesosternum**, while its free posterior end is the **xiphisternum**, later the cartilage ossifies.

1. Pisces. Sternum is lacking.

2. Amphibia. The sternum is in two parts : anterior to the clavicles lies an inverted Y-shaped bony **omosternum** with expanded cartilaginous **episternum** in front. Posterior to the coracoids is a bony **sternum proper** or **mesosternum** having an expanded cartilaginous **xiphisternum** posteriorly. Sternum is absent in caecilians, *Amphiuma* and *Proteus*. The homology of such a sternum with that of amniotes is doubtful.

3. Reptilia. The sternum is a rhomboidal plate of cartilage between the coracoids. The sternal parts of three pairs of thoracic ribs are attached to it. A T-shaped dermal bone, the **episternum (interclavicle)** lies ventral to the sternum, but it is not a part of the sternum. A sternum is absent in turtles, snakes, and most limbless lizards.

4. Aves. The sternum has a broad triangular bone extending far posteriorly below the abdomen. The mid-ventral portion is drawn out into a **carina** or **keel** in flying birds for insertion of large muscles of flight. It forms a T-shaped girder to bear the weight of the heavy abdominal viscera. Projecting anteriorly is a pair of **costal** processes to which thoracic ribs are attached. Posterolaterally are two pairs of **xiphoid** processes. **Running birds** have a rounded sternum with no keel. Extinct flying reptiles have a carinate sternum.

5. Mammalia. The sternum is made of a series of bones fixed together and arranged in a row lying mid-ventrally in the thorax. It has become segmented secondarily and has three regions : an anterior presternum **(manubrium)**, middle **mesosternum** consisting of five pieces or **sternebrae**, the fifth is very small, and a posterior region called **metasternum** (**xiphisternum**), which is long and has a terminal expanded **xiphoid cartilage**. Seven pairs of ribs are joined to the sternum in such a way that each rib articulates with two sternebrae. Bats have carinate sternum.

APPENDICULAR SKELETON

The appendicular skeleton consists of skeletal elements of pectoral and pelvic girdles, and of the appendages. The appendages are median and paired. Median appendages are found in fishes and aquatic tetrapoda, and paired appendages are found in all vertebrates except cyclostomes. In fishes the paired appendages are pectoral and pelvic fins, in tetrapoda they are fore- and hindlimbs.

Origin of Appendages

1. Fin-fold theory. According to the **fin-fold theory** of **Balfour**, the ancestral vertebrate (fish) had a pair of lateral fin-folds, one on each side of the trunk, they fused behind the anus with a median ventral fin which extended along the tail to the mid-dorsal side of the body. This hypothetical condition resembles the median fins and metapleural folds of *Amphioxus*. Persistence of centain parts and disappearance of others of the fin-folds and median fin resulted in the formation of paired and median fins respectively in fishes. Later on, the limbs of tetrapoda evolved from the paired fins of fishes. Thus, there are two transitions in the evolution of appendages, first from fin-folds to fins, and second from fins to limbs. This theory is supported by certain evidences : (*i*) Basic skeletal structures of paired and median fins are similar. This shows a common mode of origin. (*ii*) In embryos of certain elasmobranchs, a continuous series of muscle buds appear, certain of which later disappear leaving only paired and median fins. (*iii*) In some extinct Deveonian acanthodian sharks were present a row of numerous small accessory spiny fins on either side between pectoral and pelvic fins. These are supposed to be the remnants of fin folds. (*iv*) In *Cladoselache* (extinct shark) pectoral and pelvic fins are broad, having no basal notches and are supported by parallel cartilaginous rods, the pterygiophores. Their origin may be like primitive fin-fold. But no primitive fossil fish possessing continuous fin folds is known. Thus, this theory is not widely accepted now.

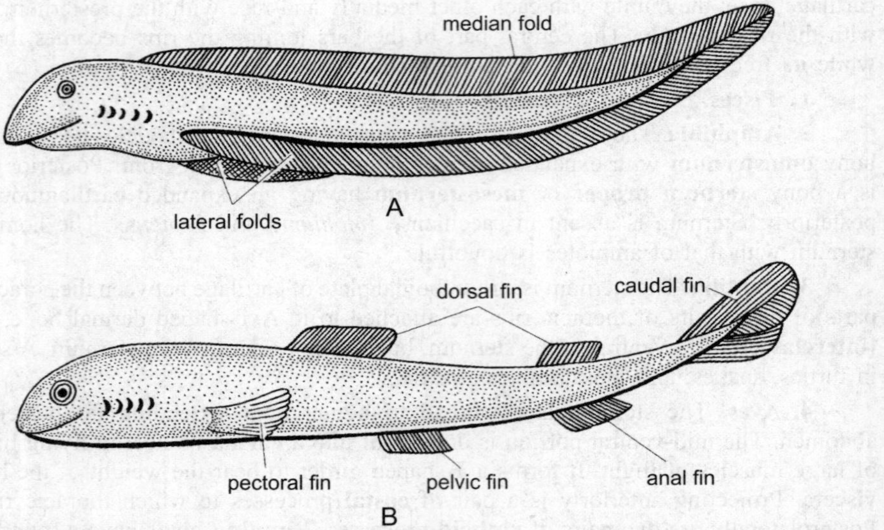

Fig. 42.18. Origin of appendages. A – Continuous fin-folds; B – Formation of permanent fins from continuous folds.

2. Fin-spines theory. According to **fin-spines theory**, probably the first fins are formed of metamerically arranged spines with fleshy webs, like extinct acanthodian sharks. Later on, all spines with webs may have been disappeared leaving only an anterior and a posterior pair, that later gave rise to pectoral and pelvic fins.

3. Ostracoderm theory. Certain ostracoderms had lateral fleshy lobes on either lateral side. They may be supposed to be the precursors of pectoral fins. Certain other ostracoderms possessed ventro-lateral rows of dermal spines like acanthodians. Leaving certain spines in the pectoral and pelvic regions, all were lost. This may be supposed to be the origin of paired fins and later limbs.

The paired and median fins are supported by a series of endoskeletal **pterygiophores**; each divided into pieces: those in the basal part are **basals**, while the distal pieces are **radials**. Muscles are attached to basals and radials for moving the fins. The distal part of a fin is vane-like and is supported by integumentary **fin rays**. The horny unjointed fin rays are **ceratotrichia** found in elasmobranchs, bony jointed fin rays are **lepidotrichia** found in most bony fishes, and fibrous and jointed fin rays are **campotrichia** found in Dipnoi. In elasmobranchs and Dipnoi the paired fins have basal lobes and a vane-like part, these are called **lobed fins**. In most bony fishes the basal lobe is missing and paired fins have only a vane-like part projecting from the body, they are called **rayed fins**.

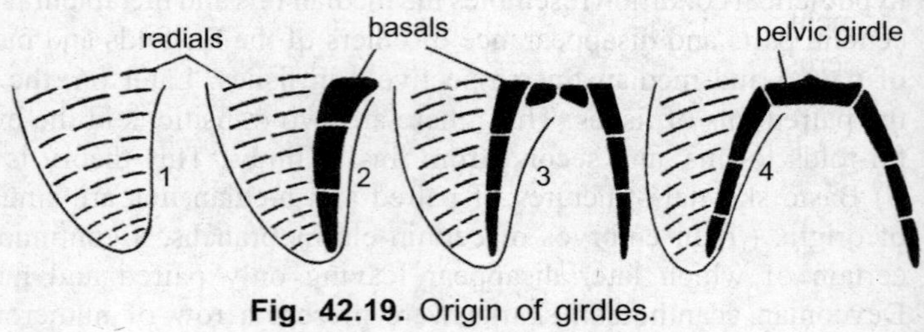

Fig. 42.19. Origin of girdles.

Origin of Girdles

The basal pterygiophores fuse into larger pieces. The large anterior-most basal of each side extends towards the mid-ventral line, and fuse together to form a girdle made of a bar of cartilage.

Girdles and Limbs in Tetrapoda

Girdles. The girdles and paired limbs of tetrapoda are remarkably similar and have homologous parts.

Pectoral girdle has several cartilage bones. They are a **scapula**, a **suprascaphla, precoracoid**, and a **coracoid** in each half. Between the scapula and coracoid a glenoid cavity appears for the first time. Two dermal bones, the clavicle and interclavicle are added in each half. But there is a gradual reduction or loss of some bones.

Pelvic girdle has three cartilage bones in each half. They are **ilium**, **ischium** and **pubis**. There are no dermal bones like pectoral girdle. All three bones participate in the formation of an **acetabulum** cavity. A **cotyloid** or **acetabular** bone often forms a part of the acetabulum in mammals. Generally a large **obturator foramen** lies between the ischium and pubis. The two halves of the pelvic girdle often meet midventrally at the pubic or ischiac symphysis. In most tetrapoda the pelvic girdle is attached to the axial skeleton through the ilia articulating with the sacral vertebrae.

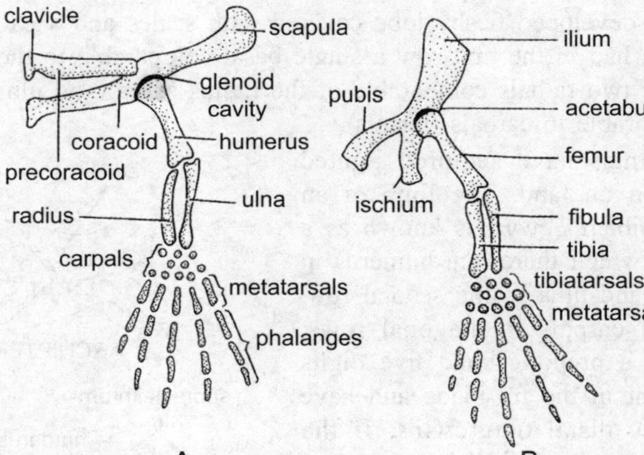

Fig. 42.20. Girdles and limb-bones of tetrapoda. A – Pectoral girdle and forelimb; B – Pelvic girdle and hindlimb.

Homology. The bones of the pectoral and pelvic girdles are homologous, the scapula with the ilium, the precoracoid with the pubis and cotyloid, and the coracoid with the ischium. The clavicle is a dermal bone and is not homologous with the pubis which is a cartilage bone.

Limbs

The **limbs** of tetrapoda are pentadactyl. They are used not only for locomotion but also to support the weight of the body above the ground. Thus, they are provided with joints. Each has three segments, the **stylopodium**, **zeugopodium** and **autopodium** with joints between the segments. The **stylopodium** is the upper arm (or thigh), having a single humerus (or femur). The humerus joins the pectoral girdle at the glenoid cavity, while the femur joins the pelvic girdle at the acetabulum cavity. The **zeugopodium** is the forearm (or shank). It contains two parallel bones, the **radius** and **ulna** (or tibia and fibula). The radius and tibia are lateral, while ulna and fibula are medial. Between the stylopodium and zeugopodium is an elbow joint (or knee joint). The **autopodium** has three divisions, a carpus or wrist (tarsus or ankle), metacarpus or palm (metatarsus or sole), and digits forming fingers or toes. The **carpus** or **tarsus** in the primitive condition has 10 bones in 3 rows, the first row has a radiale (or tibiale), an intermedium, and an ulnare (or fibulare), in the second row are generally two centralia, while the third row has five distal carpals (or distal tarsals). The **metacarpus** (or metatarsus) has five long metacarpals (or metatarsals). The **digits** are generally five in number or **pentadactyl**, with the first digit on preaxial side and the fifth on the postaxial side. Each digit has a linear row of phalanges, they are 2, 3, 4, 5, 4 beginning from the first to the last. There is some evidence that early tetrapoda had seven digits, one on the preaxial side was **prehallex** (or **prepollex**), and the postaxial digit was a **postminimus**. But these were probably additional bones of the wrist or ankle.

Origin of Pentadactyl Limbs

1. Gegenbaur's theory. It was erroneously believed by **Gegenbaur** that Dipnoi were the ancestors of tetrapoda. The skeleton of their lobed fins, as seen in *Neoceratodus*, consists of a jointed axis of numerous bones from which biserial segmented radials and fin rays arose on both sides for supporting the fin. This type of fin is called an **archipterygium**. Gegenbaur's theory deriving the tetrapod limb from the archipterygium is now discarded.

2. From crossoptergians fins. The link between the fins of fishes and limbs of tetrapoda is found in fossil crossopterygians. The lobed fins of extinct *Sauripterus* and *Eusthenopteron* were like paddles. They are called **ichthyopterygium** and suggest the beginnings of tetrapod limbs. Their fin had a well-developed fleshy lobe covered with scales and a paddle-like fin blade. In the lobe the skeleton had in the first row a single basal comparable to the humerus (femur), in the second row were two radials comparable of the radius (tibia) and ulna (fibula). The rest of the radials are comparable to carpals (tarsals).

Terrestrial animals have acquired jointed limbs for locomotion on land. The limb of an ancient extinct amphibian *Eryops* is known as a **cheiropterygium** in which there is a humerus in the first row, radius and ulna in the second row, followed by several carpals in diagonal rows, after which there is a prepollex and five digits which are not present in the fin blade and have been formed as new distal outgrowths. If the skeleton of an ichthyopterygium is compared to that of a cheiropterygium, it is easy to see how the transformation of a fish fin into an amphibian limb took place. The former shows that even before locomotion on land was adopted the arrangement of its skeleton had already come to resemble that of land animals. Thus, the crossopterygian fin or ichthyopterygium gave rise to a pentadactyl limb by an elongation of the basal lobe and its skeleton, by a loss of the fin blade and supporting fin rays and by formation and separation of terminal radials to form digits.

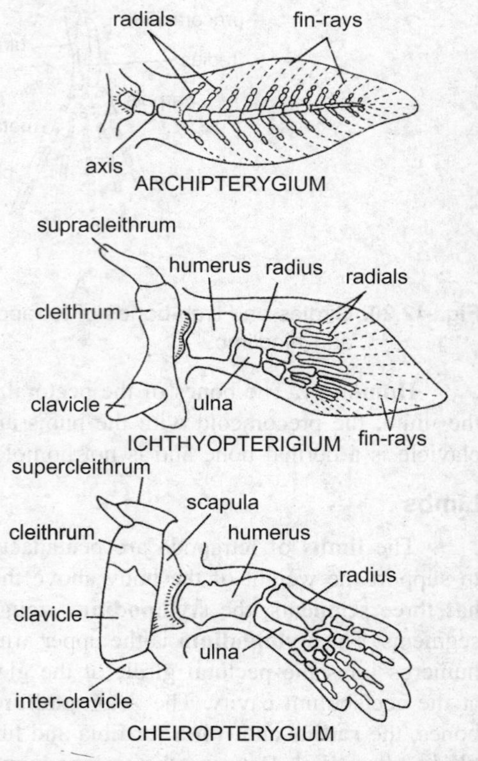

Fig. 42.21. Origin of tetrapod pentdactyl limb. Archipterygium of *Neoceratodus*; Ichthyopterygium of *Sauripterus*; Cheiropterygium of *Eryops*.

Girdles and Limbs in Tetrapoda

1. Amphibia (frog). **Pectoral girdle** is an arch enclosing the chest. The two halves are united midventrally and closely related to the sternum. (In toads the two halves overlap). Each half has a **coracoid**, **precoracoid**, **clavicle**, **paraglenoid cartilage**, **scapula** and **suprascapula**. The two precoracoid catilages fuse ventrally. The clavicle is the only dermal bone. It is fused to and reinforces a cartilaginous precoracoid. Glenoid cavity is bordered by coracoid, paraglenoid, and scapula. A partly bony suprascapula reinforces the scapula. On the inner margin of the coracoid is an epicoracoid cartilage. Anura have two kinds of pectoral girdles : (*a*) **firmisternal** in which the two epicoracoid cartilages are fused mesially along their entire lengths, (*b*) **arciferal** in which the two epicoracoids are fused only along their anterior edges. A **sternum** having two separate parts is fused to the middle of the pectoral girdle. The pectoral girdle is reinforced anteriorly by an omosternum and posteriorly by a mesosternum and their cartilaginous expansions, the episternum and xiphisternum respectively.

Forelimb. Humerus has a prominent deltoid ridge. Radius and ulna fuse into a **radioulna**. There are five carpals in two rows, in the first row the intermedium and ulnare are fused together, and there is a radiale. There is a single centrale. The three carpals of the second row are fused together. Thus, the number of carpals is reduced. Only four digits are present besides a rudimentary **pollex** enclosed in the skin.

Pelvic girdle. It is V-shaped and has a long ilium on each side. It articulates with the transverse processes of the sacral (9th) vertebra. The ilia act as shock absorbers when the animal lands after a jump. Pubis and ischium are small. The pubis is only partly ossified. The two pubes and two ischia are fused together to form a solid fulcrum. All three bones join at the acetabulum cavity. Pelvic girdle is different from that of any other vertebrate, with very long ilia it forms a long lever for transferring the force from the hindlimbs to the vertebral column in jumping.

Hindlimb. The tibia and fibula fuse into a **tibiofibula** or crural bone. There are five tarsals in two rows, the promixal row has two long bones, an inner astragalus (ulnare) and outer calcaneum (fibulare) which make the ankle very long for jumping. There are five digits of which the first is called **hallux**. There is an additional digit called **calcar** or **prehallux** enclosed in the skin. It probably represents an additional tarsal and may be regarded as a remainder of a fin ray of the ancestral fin. It helps in stretching the web of the foot.

2. Reptilia (*Varanus*). In reptiles the girdles and limbs are well-developed, except in snakes and limbless lizards, but vestigial pelvis and hindlimbs are found in some snakes.

Pectoral girdle. It is well formed and joined to the sternum. Each half has a bony coracoid and scapula, between which is a glenoid cavity. There is an expanded suprascapula of calcified cartilage. Between the coracoids of the two sides is an expanded epicoracoid cartilage formed by the fusion of two. In each half there are three fenestrae between the coracoid and epicoracoid. A T-shaped unpaired dermal bone called interclavicle (episternum) lies midventrally. On the arms of the interclavicle are two clavicles bracing the girdle, but they do not meet ventrally. Clavicles and interclavicle are dermal bones. Attached to the coracoids is a sternum made of cartilage.

In **turtles** the coracoids and scapulae are slender rods, while the clavicle and interclavicle form dermal bony plates of the plastron.

Forelimb. It is typical with no unusual feature. Humerus is broad, radius and ulna are separate, carpals are in two rows : four in the first row including a centrale, and five in the second row. There are five digits ending in claws.

Pelvic girdle. All three bones are fused inseparably to form a single **innominate bone** in each half. Ilium points backwards and is firmly united with two sacral ribs of the sacrum so that the articulation is post-acetabular. Both pubis and ischium meet their fellows to form pubic and ischiac symphyses. At the pubic symphysis is a **epipubis cartilage** and at the ischiac symphysis is a **hypoischium cartilage**. The pubis has a small foramen for the obturator nerve. Between two innominate bones is a large **cordiform** (or ischiopubic) **foramen** divided by a ligament into two obturator foramina. Acetabulum is bordered by all three bones.

Hindlimb. It is typical, but the tibia begins to become larger than the fibula. The tarsals are reduced by fusion, the proximal row has two bones and distal row has three, between the two rows of tarsals is an **intratarsal ankle** joint where bending takes place. There are five digits ending in claws.

In some reptiles, but not in *Varanus*, two structures appear for the first time in the hindlimbs, first the calcaneum bears a posterior **calcaneal process** or heel, secondly a sesamoid bone forms a **patella** or knee cap in front of the tibia.

3. Aves (*Gallus*). The appendicular skeleton of birds is remarkably uniform. The pectoral girdles and forelimbs are modified for flight. The pelvic girdles and hindlimbs are modified to support the weight of the body in standing and walking. Bones are tubular, spongy, and pneumatic so that they are light, but lightness is achieved without any sacrifice of strength. The marrow cavity has no marrow but there are supporting bony struts which make the bones strong to bear stresses.

Pectoral girdle. It lies far backwards towards the centre of the body. It has a large, stout coracoid attached to the sternum at the coracoid groove by means of a **synarthrosis** or immovable articulation. Coracoid is joined at its other end at right angle with a sword-like scapula lying over and bracing the ribs. A glenoid cavity is formed by an imperfect union of scapula and coracoid. It is greatly displaced above the centre of gravity, consequently the coracoids are elongated. In front of and attached to each coracoid is a thin clavicle. The two clavicles are joined ventrally to a small, flat interclavicle forming a **furcula** or "wishbone". In flying birds the furcula is joined to coracoids and is well developed. In non-flying birds it is reduced or absent. The arched clavicles and strong coracoids brace the sternum and resist the inward pressure of the poweful down-stroke of the wings. At the junction of coracoid, scapula, and clavicle is a **foramen triosseum** through which the tendon of a pectoralis minor muscle passes and is inserted on the humerus.

Forelimb and its skeleton are modified for flight. The wings are attached high up on the thorax towards the centre of the body for balancing the weight in flying. The forelimb has a stout humerus with a deltoid ridge. Humerus has a pneumatic foramen through which an air sac enters making the bone spongy and light. Air in pneumatic bones oxygenates blood and adjusts air pressure when a bird descends from a great height. Such bones have no marrow. Radius and ulna are well formed. The ulna is more strongly developed than the radius, the ulna is large for attachment of flight feathers. Carpus has four carpals in two rows–in the first row are radiale and ulnare, but the carpals of the second row fuse to three fused metacarpals to form a **carpometacarpus**. Of the three metacarpals the first is much reduced, all three are fused proximally, but distally the second and third are free forming two rods. There are three digits: the first has one phalanx, second has two phalanges, and third has one phalanx. The digits are reduced to almost a one-fingered condition. Muscles of the upper arm are large and strong, those of the forearm are reduced, and those of the hand have atrophied.

Pelvic girdle. It has a longitudinal curvature which distributes the body weight evenly in bipedal locomotion. Ilia and ischia have become long and wide and lie longitudinally. Large expanded ilium extends in front and behind the acetabulum, and is fused to the synsacrum along its entire length. Ischium is broad and extends posteriorly. It lies parallel to and is fused with the ilium, an **ilioischiatic foramen** lies between the two bones. A slender pubis extends backwards and ends freely. Between the ischium and pubis is an **obturator foramen**. All three bones are fused together to form an **innominate bone** which has an incompletely ossified acetabulum with a foramen. Pubic and ischiac symphyses are absent so that large-shelled eggs can be laid. But in ostrich (*Struthio*) there is a pubic symphysis, while in *Rhea*, the South American ostrich, there is an ischiac symphysis.

Hindlimb. These are modified for bipedal locomotion. A strong femur articulates at the acetabulum. Fibula is much reduced and is short. The tibia has a **cnemial crest**, in front of the junction of femur and tibia. A sesamoid bone forms a **patella**. The tibia is fused with the proximal **tarsals** (astragalus and calcaneum) to form a **tibiotarsus**. The distal tarsals fuse with the second, third, and fourth fused metatarsals to form a compound bone, the **tarsometatarsus**. The ankle joint is between the two rows of tarsals is **intratarsal** in position. The first metatarsal is free and projects near the lower end of the tarsometatarsus. In the fowl and some other birds a bony **spur** projects backwards from the tarsometatarsus, it has a horny epidermal covering. The spur is more developed in male birds for fighting. The digits are never more than four, having 2, 3, 4, 5 phalanges. The last phalanx of each digit bears a claw. The first digit (hallux) usually points backward, while the other three digits are directed forward.

4. Mammalia. The appendicular skeleton of mammals shows much variation from a reptile-like condition to a high degree of specialisation.

Pectoral girdle of *Ornithorhynchus*, (a monotreme), has a scapula, coracoid, and a cartilaginous precoracoid (epicoracoid) on each side and there is an expanded suprascapula. Between coracoid and scapula is a glenoid cavity on each side. Ventrally the two coracoids are connected with the manubrium of the sternum. Between the precoracoids is a median T-shaped episternum (interclavicle) forming a connection between the girdle and sternum. Two clavicles rest on the episternum and each is attached to a simple acromion of the scapula. Clavicles and episternum are dermal bones. Such a pectoral girdle is like that of a reptile.

In **most mammals** the precoracoids and interclavicle are lost, and the coracoids are much reduced, the clavicle may be a strong bony arch from the scapula to the sternum or it may be reduced or even lost. When the clavicles are lost all connection between the pectoral girdle and axial skeleton (sternum) disappears.

Pectoral girdle (Rabbit)**.** Each half of the girdle has a triangular scapula or shoulder blade, at the end of which is a glenoid cavity. Near the glenoid cavity there is a **coracoid process** formed by the reduced coracoid. On the outer surface of the scapula is a spine which has two projections, an **acromion process** and a **metacromion process** for attachment of muscles. There is a clavicle, a dermal bone forming a slender rod attached by ligaments to the acromion process and manubrium of the sternum. Clavicles are retained only in those mammals which have an extensive movement of forelimbs, in others they are reduced or lost.

Forelimb. The limbs of most mammals are capable of extensive rotation, extension and bending. **Humerus** is stout with a deltoid ridge, the distal end has two expanded condyles between which is a pulley or trochlea for articulation with ulna. Above the trochlea is an olecranon fossa perforated by a **supratrochlear** (supracondyle) **foramen** for the passage of a nerve and brachial artery. **Radius** and **ulna** are separate but tightly bound together. In primates, with prehensile forelimbs, the radius and ulna are not fixed but the radius can rotate over the ulna to turn the hand to either a prone or a supine position. The ulna is produced into an elbow or **olecranon process**, in front of which is a **sigmoid notch** for articulation with the humerus. **Carpus** has eight bones, three in the first row, then a small centrale and four distal carpals. There are five metacarpals having five digits with 2, 3, 3, 3, 3 phalanges ending in claws.

In primates the first digit or **pollex** is independent of the others. It can be brought opposite the other digits and palm. This opposable thumb can handle minute objects. This factor has led to a superior position of primates.

Pelvic girdle. In each half the ilium, ischium, and pubis fuse to form an innominate bone. The two halves meet by a midventral **pubic symphysis**. In rabbit and some others there is a small **acetabular** (**cotyloid**) bone at the lower margin of the acetabulum. The acetabulum lies where all three bones meet (except the pubis). There is a large obturator foramen between the pubis and ischium on each side. The two ilia articulate with the transverse processes of the first vertebra of the sacrum.

Hindlimb. Femur is strong with a head for articulation with the acetabulum. There are three trochanters for attachment of muscles, distal end of femur has two smooth condyles. **Tibia** is strong and has a cnemial crest, above is a sesamoid bone or **patella** formed in the tendon of triceps femoralis muscle. It forms a knee cap which reinforces the ligments of the knee joint. **Fibula** is reduced and fused distally to the tibia but separate proximally. **Tarsus** has six bones, proximal row has astragalus and calcaneum, its postior projection, the calcaneal process forms a heel. Centrale lies in front of the astragalus, distal row has three tarsals. There are four metatarsals bearing digits. Each digit has three phalanges ending in claws. First digit or hallux is absent. In some primates the hallux is opposable in the same way as the pollex of the hand.

Postures of limbs. In terrestrial mammals there are three kinds of postures of hands and feet in locomotion.

1. Plantigrade. Plantigrade locomotion is the most primitive, in which the entire hand and foot are in contact with the ground, as in bears, rabbit, some insectivores, and man.

2. Digitigrade. Digitigrade locomotion is used for increased speed. In it only the digits are placed on the ground. The wrist and ankle are lifted above, as in dogs, cats and wolves.

3. Unguligrade. Unguligrade locomotion is an extreme condition in which only the tips of the digits are placed on the ground and the tips are covered with hoofs. This is the most specialised condition, as in deer, horse, and cattle.

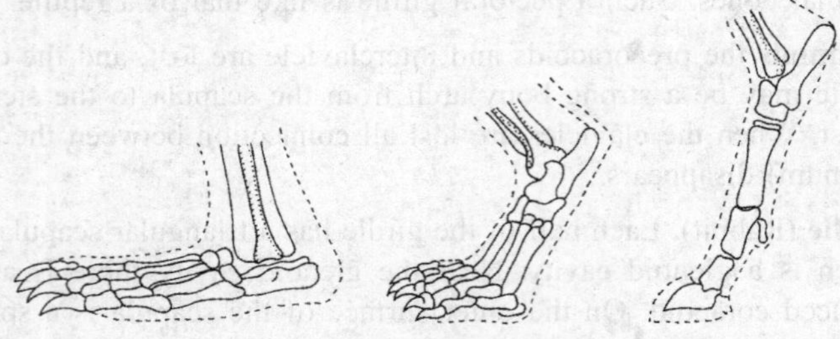

Fig. 42.22. Foot postures in locomotion of mammals.

MUSCLES

Divisions of mesoderm. During development, the mesodern cells grow between the ectoderm and endoderm and gradually differentiate into three different regions, a dorsal **epimere**, a middle **mesomere**, and a lower **hypomere**. The epimere and hypomere give rise to body muscles, while the mesomere forms excretory and reproductive organs. The **epimere** becomes marked off into somites to form segmentally arranged **myotomes** which give rise to voluntary or striated muscles of the body. The **hypomere** is unsegmented and forms an outer somatic (parietal) and inner splanchnic (visceral) muscle layers, enclosing the coelom. The splanchnic layer forms smooth or involuntary muscles of the gut and also the cardiac muscles of the heart.

In lower forms (fishes) there is a linear series of muscle segments or myotomes, which are paired striated muscle fibres run longitudinally in the myotomes. Each myotome is sharply bent four times from the dorsal to the ventral side, hence, in section the myotomes appear as concentric bundles. Each myotome lies opposite the point of articulation of two vertebrae and is supplied by a spinal nerve. Between successive myotomes are tough connective tissue partitions, known as **myocommata** which are joined on the outer side to the skin on the inner side to the vertebral column. When fibres of myotomes contract, they pull the myocommata flexing the body and vertebral column. The muscles of the head and gill region are concerned with movements of the jaws and closing and opening of gill-apertures (branchial muscles), eyes, and pharynx, they do not show any segmental arrangement and are also of striated type.

In cyclostomes there is no separation of the body or axial musculature into dorsal and ventral groups. In cartilaginous and bony fishes the axial musculature is separated by a lateral or horizontal septum into dorsal epaxial muscles and ventral hypaxial muscles. Epaxial muscles are innervated by dorsal branches of spinal nerves, while hypaxial muscles are innervated by ventral branches of spinal nerves.

In the head region, during embryonic development of fishes as well as higher vertebrates, there are three pairs of somite-like mesodermal blocks called the premandibular, mandibular and hyoid somites. Premandibular somite gives rise to four of the six extrinsic muscles of the eye, *i.e.*, superior, inferior, anterior rectus and inferior oblique muscles. Superior oblique muscle arise from the mandibular somite and posterior or external rectus muscle arise from hyoid somite.

In **tetrapoda** the primitive metameric condition becomes more and more obscure. This is brought about by splitting of myotomes into various layers, disappearance of myocommata, degeneration of parts of some myotomes, and a displacement and fusion of other parts of myotomes. The dorsal parts of myotomes are reduced and split into short and long systems of longitudinally directed muscles, the shorter ones are connected to vertebrae and longer ones form dorsal musculature. The ventral parts of myotomes split into layers, some of which move the ribs in respiration.

Appendicular muscles. In elasmobranchs hollow buds grow out from some myotomes. Their metameric arrangement is lost and they give rise to dorsal (elevates or extends) and ventral muscles (lower or adducts) for moving the paired fins. In tetrapoda these dorsal and ventral muscles differentiate into separate extrinsic and intrinsic muscles, which provide for a greater freedom of movements of limbs on land.

Dermal muscles. Dermal muscles of the integument generally appear in only amniotes. They are derived from parietal muscles. In snakes dermal muscles move the ventral scales in locomotion. In birds they cause movement of feathers, in mammals they are highly developed, their contractions move the skin in lower mammals, some move the hair. In higher mammals dermal muscles bring about facial expressions.

Swimming in Bony Fishes

The bony fishes are the masters of waters at all depths. The body has become shorter and flattened laterally, increasing the vertical axis, so that the length is much smaller in ratio to the depth by which flexibility is reduced. Thus, the body form provides advantages for swimming over the more elongated elasmobranchs.

Bony fishes also swim by passing waves of curvature or bending alternately on each side of the body from the head to tail. The myotomes contract one after the other causing **metachronal contractions** from the head to the tail along alternate sides of the body. The large caudal fin of a homocercal tail resists transverse movement, still the tail does move transversely rotating towards and away from the axis of locomotion. The effect of the caudal fin combined with the shortness of the body and reduced flexibility is that the front part of the body makes only small transverse movements so that the head keeps nearly straight.

In most bony fishes conditions are different from elasmobranchs due to the presence of an **air bladdar** as a hydrostatic organ which maintains the bony fish steadily at any depth, it makes the specific gravity of the fish the same as the water it displaces. In elasmobranchs the specific gravity is more than that of water.

The fins have changed correspondingly, and very small movement of fins is sufficient to keep the fish at a constant depth or to change its depth. The pectoral fins are placed high up on the body, they prevent rolling and act as brakes in stopping, but they do not act as elevators, and are essentially used for steering. While the pelvic as well as dorsal and aual fins effect stability and assist in maintaining the body upright in water. The pelvic fins are small and have shifted forward, when the fish stops the head does not rise because the pelvic fins tilt the head downward. The dorsal fin is well developed, in elasmobranchs it is fixed but in bony fishes it can be folded up and down, in turning the dorsal fin is raised.The dorsal and caudal fins prevent pitching in a transverse plane.

Turning is produced either by passing of waves on one side of the body only or by asymmetrical braking with pectoral fins. In turning, the dorsal fin is raised and the symmetrical caudal fin is used to increase the efficiency by which the front part of the body is bent on the tail which acts as a fulcrum, this enables the fish to turn through 180° within a circle having a diameter of the length of the fish.

Rapid forward progress is generally produced by lateral movement of the posterior part of the body and caudal fin. In **fast-swimming fishes** the tail is crescent-shaped, the caudal peduncle is narrow and the body streamlined for speed. The **slow-moving fishes** entirely rely upon their fins for locomotion. Such fishes are usually protected by a heavy integumental covering or by spines or concealing colouration. The heavily armoured sea horses swim principally by movement of the dorsal fin with some assistance by the pectoral fins. Pipe fishes, puffers and turkey fishes swim by means of fins. Many of the rays swim by means of undulatory movements of the large pectoral fins. Most of the eel-like fishes rely on lateral undulations of the entire body for locomotion as does a water snake.

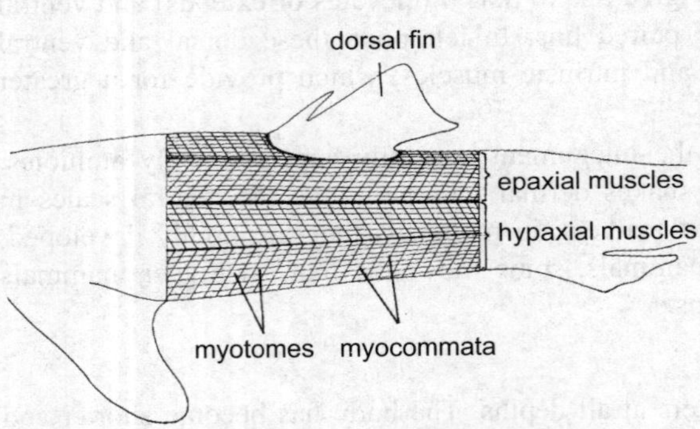

Fig. 42.23. Myotomes of *Scoliodon*.

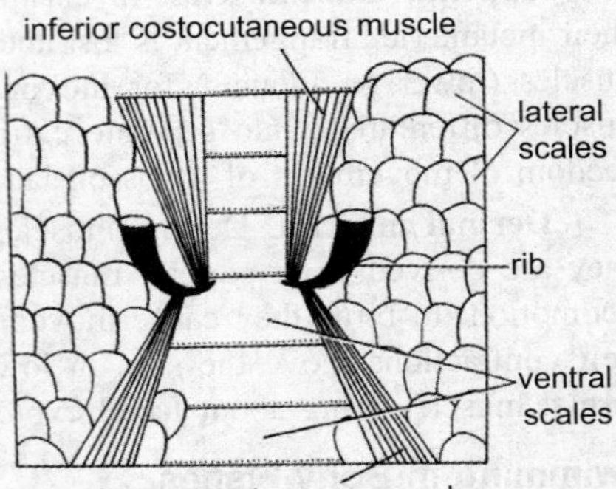

Fig. 42.24. Locomotory muscles from ribs to ventral scales in snake.

Swimming in Snakes

In **snakes** there are no limbs but locomotion is brought about by lateral muscular undulations of the body and by the large ventral scales, these scales form broad transverse bands. Two pairs of small **costocutaneous parietal muscles** pass from the ribs to each ventral scale. These muscles alternately raise and lower the ventral scales. When the scales are raised they hold the ground and push the body forward. The ground surface should be rough for scales to work, a snake cannot move on a smooth surface. Snakes crawl in four different ways : 1. Horizontal undulatory; 2. Rectilinear; 3. Concertina; 4. Sidewinder progression.

1. Horizontal undulatory movement. The track led by a snake in such movement is a wavy line. The body glides along in a series of waves and each part of the body passes along the same track. In soft dust or sand it is evident that the dirt or sand is raised or pushed up on the side opposite the axis of travel so as to form a pivot.

2. Rectilinear motion. In this type of motion the axis of the body is essentially straight and movement is affected by alternate movement of the ventral scutes and the body itself. An undulatory movement raises the scutes, carries them forward and permits them to become anchored to the ground. Following this the body slides forward on the part of the ventral skin and the body. The snake appears to move along a straight path without any lateral motion.

3. Concertina progression. It consists of alternate curving and straightening of the body. After the body is drawn up in a series of curves, the tail serves as an anchor and movement is effected by pushing the rest of the body forward until it is straight. The head and neck are then used as anchors and the remainder of the body is drawn up in a series of curves. This mode of progression is not as rapid as rectilinear motion. It is used when snakes are stalking their prey.

4. Sidewinding. It is used by certain desert snakes. It is essentially a series of lateral looping movements in which only a vertical force is applied and no more than two parts of the body contact the ground at any one time. The resulting tracks are a series of parallel, diagonal, J-shaped marks. The curve of J is made by the head and neck. Its example is sidewinder rattle snake (*Crotalus cerastes*) of North America.

IMPORTANT QUESTIONS

Long Answer Questions

1. Give an account of different types of jaw suspensorium in vertebrates.
2. Compare the skull of a lizard with that of a bird.
3. Describe the pectoral and pelvic girdles of frog, *Varanus*, (lizard), *Gallus* (bird) and rabbit. Discuss how the structure of girdle is suited to the mode of life in these animals.

Short Answer Questions

1. Give a brief account of the following :

(*i*) Basic structure of a vertebra; (*ii*) Development of chondrocranium; (*iii*) Jaw suspensorium; (iv) Pectoral girdle; (*v*) Pelvic girdle.

Multiple Choice Questions

1. In living amphibian, exoskeleton is
   (*a*) Hairy (*b*) Horny
   (*c*) Nails (*d*) Absent.
2. Cartilage bone is
   (*a*) Parietal (*b*) Prefrontal
   (*c*) Quadrate (*d*) Lacrimal.
3. Which of the following is a pelvic bone?
   (*a*) Ilium (*b*) Scapula
   (*c*) Coracoid (*d*) Clavicle.
4. Cranium is a constituent of
   (*a*) Splanchnocranium (*b*) Chondrocranium
   (*c*) Neurocranium (*d*) Visceral skeleton.
5. Skull in cyclostomes is made of
   (*a*) Cartilage (*b*) Replacing bones
   (*c*) Membrane bones (*d*) Dermal plate.
6. Jaw suspensorium in elasmobranchs is
   (*a*) Autostylic (*b*) Hyostylic
   (*c*) Autodiastylic (*d*) Amphistylic.
7. A single prominent mid-ventral piece of centrum in vertebrates is
   (*a*) Hypapophysis (*b*) Zygapophysis
   (*c*) Diapophysis (*d*) Parapophysis.
8. In modern amphibians, the columella of middle ear is a modified
   (*a*) Mandibular (*b*) Hyomandibular
   (*c*) Sphenoid (*d*) Basisphenoid.
9. Foramen magnum in birds faces
   (*a*) Downwards (*b*) Upwards
   (*c*) Right (*d*) Left.
10. Pituitary gland is housed in
    (*a*) Olfactory capsule
    (*b*) Ethmoid plate
    (*c*) Hypophysial fenestra
    (*d*) Basal plate.

ANSWERS

1. (*d*) 2. (*c*) 3. (*a*) 4. (*c*) 5. (*a*) 6. (*b*) 7. (*a*) 8. (*b*)
9. (*a*) 10. (*c*)

43

Chapter

Coelom and Digestive System in Vertebrates

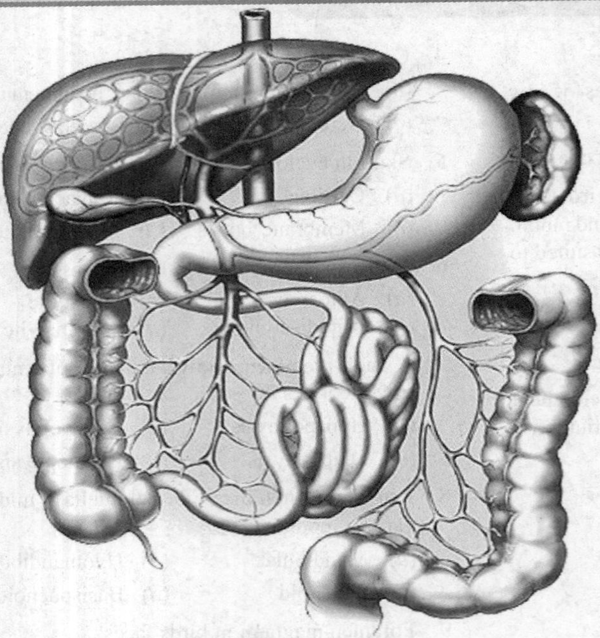

Contents of this Chapter

COELOM

In vertebrates the dorsal part of the mesoderm is **epimere** which is segmented into somites, the middle part is **mesomere** which forms the urinogenital organs, and the lower part is the **hypomere** or **lateral plate mesoderm** which is never segmented and its cavity is the **coelom** or **splanchnocoel**. The outer wall of the lateral plate is the **somatic** or **parietal** layer, which becomes the lining of the body wall and is known as **parietal peritoneum**. The inner wall of the lateral plate is the **splanchnic** or **visceral layer**, which becomes the covering of the alimentary canal and other viscera, and is called the **visceral peritoneum**. The kidneys lie outside the coelomic cavity and so only ventrally covered by the peritoneum. The coelom is the space enclosed within the parietal and visceral peritonium.

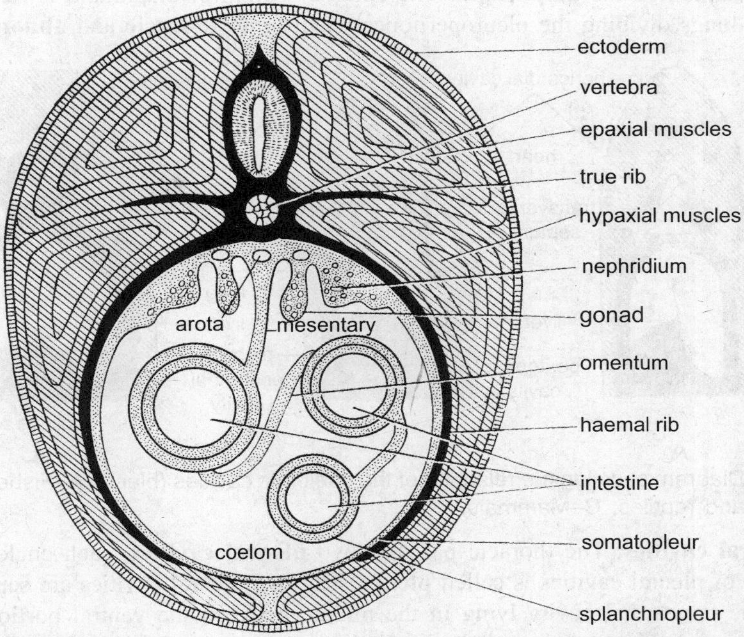

Fig. 43.1. Diagrammatic cross section of the body of a vertebrate to show the different parts of the viscera and how the digestive tract is suspended in the body cavity.

Mesentery

The parietal and visceral peritoneum are in contact with each other above and below the intestine forming a double-layered **mesentery**. The dorsal mesentery which is above the intestine persists, but the ventral mesentery disappears leaving only a few ligaments. In some regions the alimentary canal is joined to an organ by a remnant of the ventral mesentery, which is then known as an **omentum** or ligament. The coelom is divided longitudinally by the dorsal mesentery.

1. Pericardial cavity or coelom. Behind the heart a partition, called **transverse septum** is formed, which divides the coelom into a small anterior **pericardial cavity** containing the heart. The peritoneum lining the pericardial cavity is a pericardial membrane inside which is a clear pericardial fluid. The peritoneal cavity is lined with parietal and visceral peritoneum which secrete a coelomic fluid.

2. Abdominal or peritoneal cavity. The large posterior cavity which contains the rest of the viscera is called the abdominal or peritoneal cavity.

Position of Pericardial Cavity in Various Vertebrates

In **fishes** and **urodeles** the pericardial cavity is anterior to the peritoneal cavity, the transverse septum lying straight transversely. In **anurans** and **amniotes** the pericardial cavity comes to lie ventral to the anterior part of the peritoneal cavity, because the transverse septum becomes oblique. In that part of the peritoneal cavity which lies dorsal to the pericardial cavity lungs are present in tetrapoda, consequently the peritoneal cavity containing both lungs and the viscera is now called a **pleuroperitoneal cavity**. This condition is found in anurans and most reptiles.

Separation of Thoracic and Abdominal Coelomic Cavities

In **crocodilians**, **birds** and **mammals** a new fold descends from the dorsal body wall and fuses with the transverse septum to form a new partition. This partition in birds is membranous with some muscles and is called an **oblique septum**, but in mammals it is a highly muscular dome-shaped, callled the **diaphragm**. The oblique septum or diaphragm is formed immediately behind the lungs dividing the pleuroperitoneal cavity into **thoracic** and **abdominal portions**.

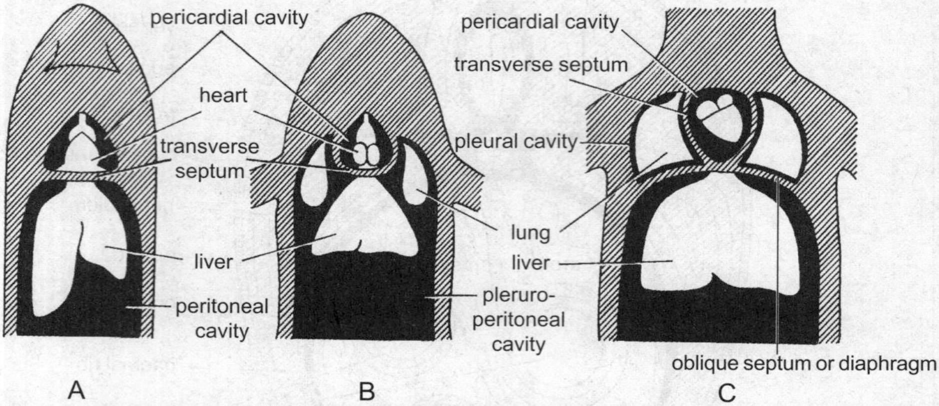

Fig. 43.2. Diagram showing the relations of the coelomic cavities (black). A–Fishes; B–Amphibians and reptiles; C–Mammals.

Pleural cavities. The thoracic part has two **pleural cavities**, each enclosing a lung. The peritoneum of pleural cavities is called **pleura**. The two pleural cavities are separated from each other by the pericardial cavity lying in the middle between the ventral portions of the pleural cavities. The abdominal portion of the cavity lies posterior to the oblique septum or diaphragm and is called a **peritoneal** or **abdominal cavity**, which contains the rest of the viscera. Thus, crocodilians, birds, and mammals have four compartments of the coelom, a pericardial cavity, two pleural cavities, and a peritoneal cavity. This arrangement greatly increases the efficiency of lungs for respiration.

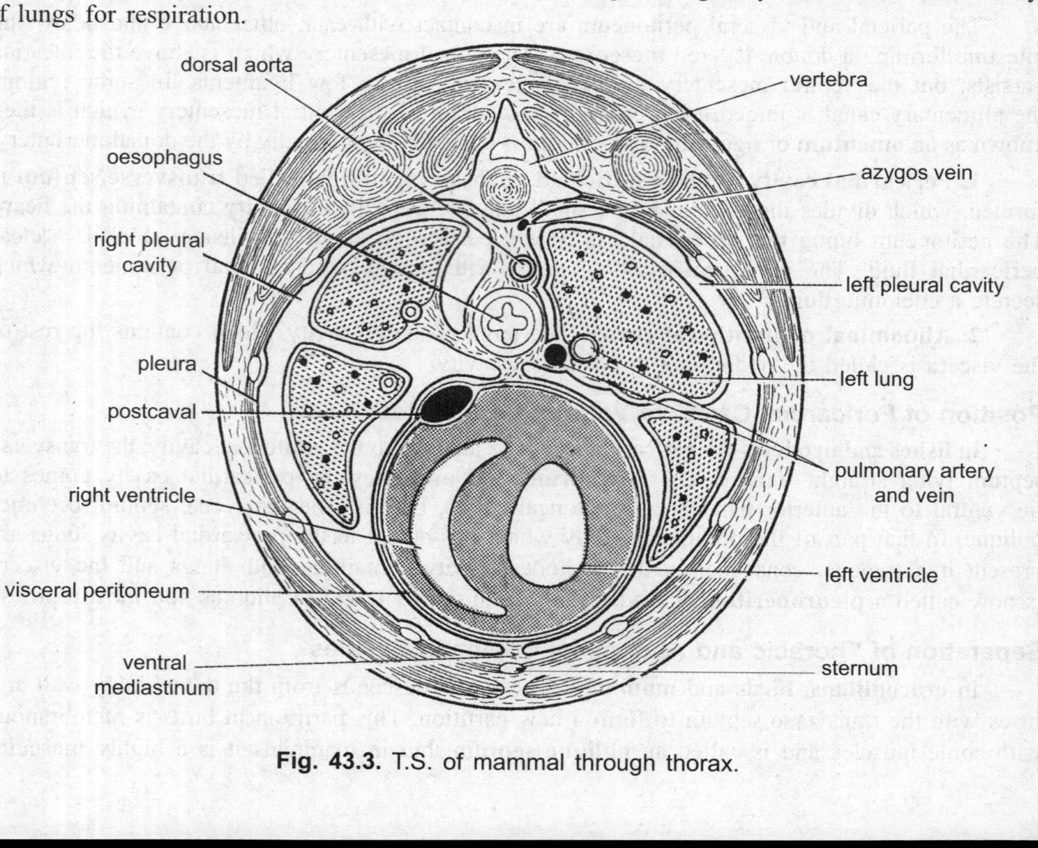

Fig. 43.3. T.S. of mammal through thorax.

Mediastinal Space

In mammals the two pleural cavities grow ventrally below the pericardium, separating it from the ventral body wall. They grow inwards so that their median walls meet below the pericardial cavity to form a **mediastinal septum**. The two parts of the mediastinal septum spread apart dorsally to form a **mediastinal space** which contains the pericardial cavity and the enclosed heart, major blood vessels, oesophagus, and trachea.

Greater omentum. In many mammals the mesentery supporting the stomach is drawn out into a double-walled bag, called **greater omentum**, which protects the abdominal viscera and has fat deposited in it.

It is convenient to speak of the viscera as lying inside the coelom, but actually the visceral organs and alimentary canal are covered by visceral peritoneum and are outside the coelom (because coelom is the space between parietal and visceral peritoneum). They have pushed into the coelom carrying the peritoneum of the coelomic wall before them, thus, they appear inside the coelom.

DIGESTIVE SYSTEM

Embryonic Digestive Tract

Archenteron. The embryonic archenteron becomes the lining of the adult digestive tract and of all its derivatives. Splanchnic mesoderm adds layers of connective tissue and smooth muscles around the archenteron. Ectodermal invagination of the head forms the **stomodaeum** leading into oral cavity, and a similar mid-ventral ectodermal invagination forms **proctodaeum**, which leads into the hindgut. The stomodaeum becomes the adult buccal cavity and gives rise to teeth enamel, epithelial covering of tongue, glands, *e.g.*, mucous, poison and salivary, etc., and Rathke's pouch of anterior pituitary gland. The proctodaeum forms either a small terminal part of the cloaca in lower vertebrates and rectum in mammals.

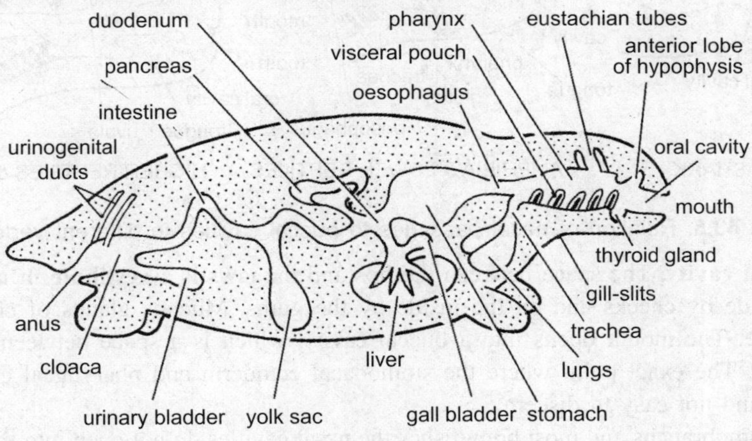

Fig. 43.4. Alimentary canal and its chief derivatives in a vertebrate.

The alimentary canal in embryos from stomach to cloaca is attached to the dorsal body wall by a double fold of peritoneum, called the **dorsal mesentery**, and to ventral body wall by a **ventral mesentery**. In adults, dorsal mesentery persists but the ventral mesentery dispappers leaving only in the region of liver and urinary bladder.

Digestive Tract of Adult

The digestive tract differentiates for different functions into the following regions : mouth, buccal cavity, pharynx, oesophagus, stomach, small intestine, large intestine and cloaca.

Following outgrowths arise from the digestive tract : oral glands, Rathke's pouch, thyroid gland, gill-clefts, tympanic cavity, thymus and other glands of gill-clefts, trachea, lungs, swim bladder, liver, pancreas, yolk sac, and urinary bladder.

Histology. The wall of the alimentary canal is made of four concentric layers. (*i*) An outermost **visceral peritoneum** or **serous coat** is made of mesothelial cells and thin layer of connective tissue. It is lacking in the oesophagus. (*ii*) Below this is a **muscular layer** formed of smooth muscle fibres arranged in outer longitudinal and inner circular muscle fibres. Between the two layers of muscles is a network of nerve cells and nerve fibres of the autonomic nervous system, known as **myenteric plexus** or **plexus of Auerbach**. (*iii*) Beneath the muscle layer is a **submucosa** made of connective tissue having elastic fibres, fat, blood and lymph vessels, nerve cells and fibres glands. (*iv*) The innermost layer is a **mucosa** composed of three regions : (*a*) outer-most narrow **muscularis mucosa** of outer longitudinal and inner circular smooth muscle fibres, (*b*) middle thin layer of **lamina propria** of connective tissue, blood vessels, nerves and nodules of lymphatic tissue, and (*c*) a **basement membrane** supporting a layer of columnar **epithelial cells** which are often glandular and ciliated.

Mouth. Mouth is the opening leading into buccal cavity. In lampreys (cyclostomes) it is a circular opening at the base of buccal funnel and remains permanently open due to lack of jaws, etc. In gnathostomes it is terminal. Mouth is bounded by lips which are immovable and formed of cornified skin in fishes, amphibians and reptiles. In mammals these are fleshy and muscular.

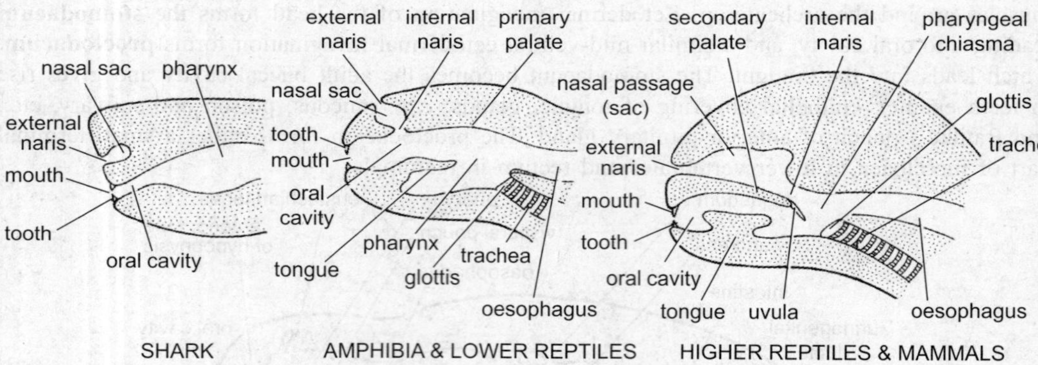

Fig. 43.5. Relation of nasal passages to buccal cavities in different vertebrates.

Buccal cavity. The space between the lips and the jaws is a **vestibule**. It may be bounded on the outside by cheeks and on the inside by the gums. Mucous glands of cheeks open into the vestibule. The mouth opens into a buccal cavity, which is a space between the mouth and the pharynx. The exact point where the stomodaeal ectoderm and pharyngeal endoderm merge is variable and not easy to discern.

In elasmobranchs and most bony fishes the nasal cavities do not open into the buccal cavity. In Chondrichthyes and tetrapoda (amphibians and most reptiles) the nasal cavities open into the buccal cavity by **choanae** or **internal nares**, which are primitively placed anteriorly, but in crocodiles, birds and mammals they become posterior in the pharynx due to the formation of a **secondary palate**, which effectively separates the respiratory nasal passage from the mouth cavity or food passage. In birds, this palate is cleft due to which nasal and buccal cavities communicate with each other. In mammals, secondary palate is continued posteriorly as a membranous **soft palate**. In human beings soft palate hangs into the laryngeal pharynx in the form of fleshy process, called **uvula**.

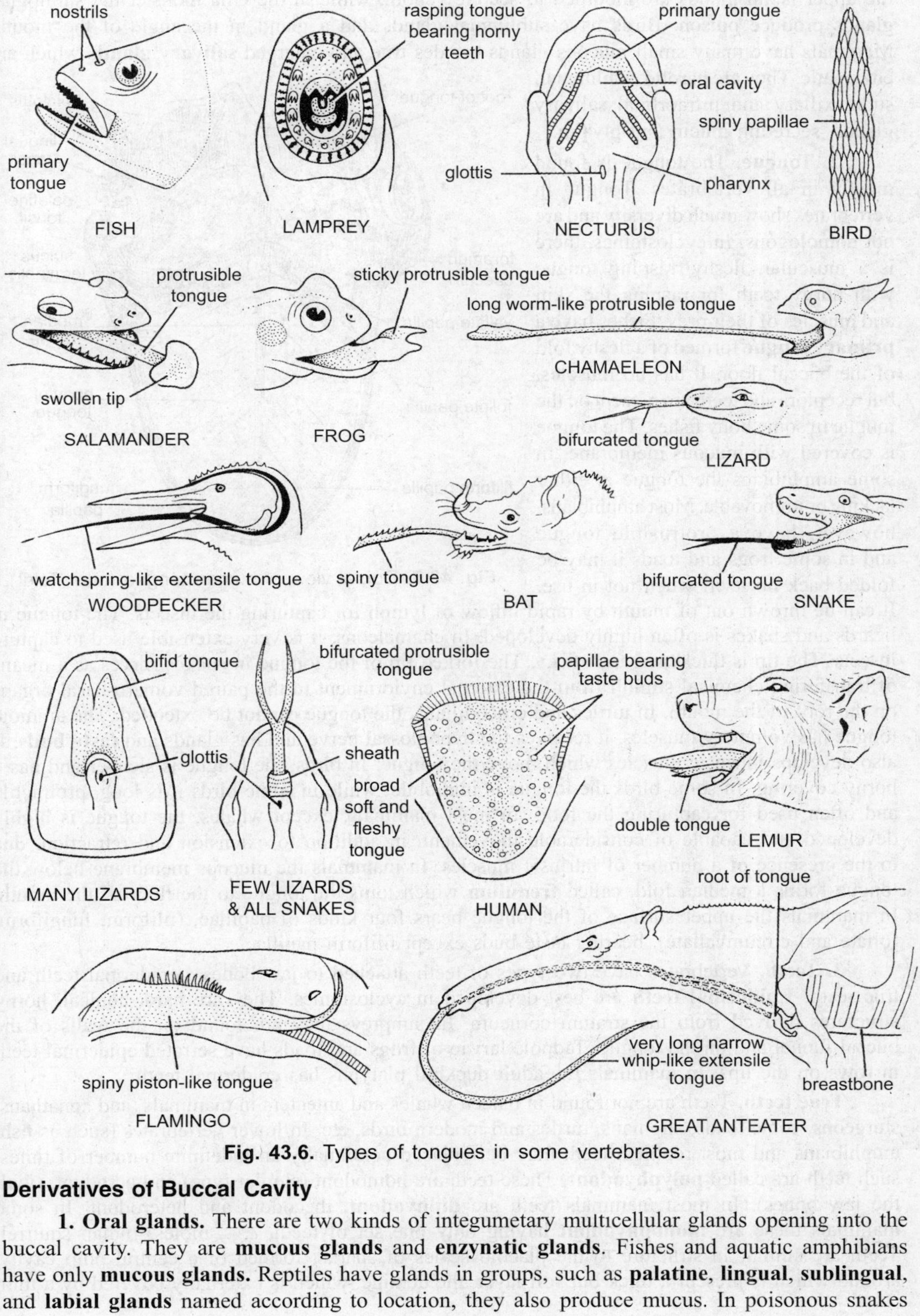

Fig. 43.6. Types of tongues in some vertebrates.

Derivatives of Buccal Cavity

1. Oral glands. There are two kinds of integumetary multicellular glands opening into the buccal cavity. They are **mucous glands** and **enzynatic glands**. Fishes and aquatic amphibians have only **mucous glands.** Reptiles have glands in groups, such as **palatine, lingual, sublingual,** and **labial glands** named according to location, they also produce mucus. In poisonous snakes the upper labial glands are modified to secrete venom, while in the Gila monster the sublingual glands produce poison. Birds have sublingual glands and a gland in the angle of the mouth. Mammals have many small mucous glands besides true and enlarged **salivary glands** which are enzymatic. They are parotid, sublingual, submaxillary and infraorbital salivary glands, secreting mucin and ptyalin.

2. Tongue. The tongue is found mostly in all vertebrates. Tongue in vertebrates show much diversity and are not homologous. In cyclostomes, there is a muscular, fleshy, rasping tongue with horny teeth for rasping the skin and muscles of their prey. Fishes have a **primary tongue** formed of a fleshy fold of the buccal floor. It has no muscles, but receptors and teeth are present on the tongue in some bony fishes. The tongue is covered with mucous membrane. In some amphibians the tongue is either lacking or immovable. Most amphibians, however, have a protrusible tongue and in some frogs and toads it may be folded back on itself when not in use. It can be thrown out of mouth by rapid inflow of lymph for capturing the insects. The tongue in lizards and snakes is often highly developed. In chameleons, it is very extensible used to capture insects. The tip is thickened and sticky. The forked tip of the tongue in snakes serves as a means of transfering chemical stimuli from the external enviornment to the paired vomero-nasal organs on the roof of the mouth. In turtles and crocodilians, the tongue cannot be extended. The amniote tongue has voluntary muscles, it receives the hypoglossal nerve and has glands and **taste buds**. It also develops intrinsic muscles which move the tongue. In birds, the tongue is slender and has a horny covering. In some birds the tongue is immobile, while in some birds it is long, protractile and often used for capturing the food. In most mammals, except whales, the tongue is highly developed and capable of considerable movement, in addition to extension and retraction, due to the presence of a number of intrinsic muscles. In mammals the mucous membrane below the tongue forms a median fold, called **frenulum** which joins the tongue to the floor of the mouth. In mammals the upper surface of the tongue bears four kinds of papillae, (filiform, fungiform, foliate and circumvallate), bearing taste buds except filiform papillae.

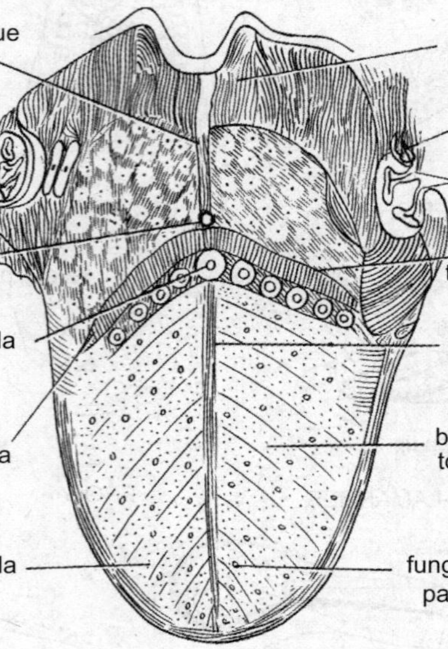

Fig. 43.7. Dorsal view of human tongue (after Toldt).

3. Teeth. Vertebrates have two types of teeth attached to jaw bones : epidermal teeth and true teeth. **Epidermal teeth** are best developed in cyclostomes. They are hard, conical, horny structures derived from the stratum corneum. In lampreys they are found on the walls of the buccal funnel and on the tongue. Tadpole larvae of frogs and toads have serrated epidermal teeth in rows on the lips. In mammals the adult duckbill platypus has epidermal teeth.

True teeth. Teeth are not found in baleen whales and anteaters in mammals, and agnathans, sturgeons, some toads, sirenians, turtles and modern birds, etc. In lower vertebrates (such as fish, amphibians and most reptiles) teeth may be replaced continually an indefinite number of times, such teeth are called **polyphyodont**. These teeth are homodont (similar type) and acrodont. (with the jaw bones). In most mammals teeth are **diphyodont**, thecodont and heterodont. In some mammals these are **monophyodont** having only one set of teeth, *e.g.*, moles, Indian squirrel. Teeth are similar in structure to the placoid scales of sharks, formed of a central pulp cavity, around which is present a thick but soft layer, the dentine, which is externally covered by a thin, extremely hard enamel. These are supposed to have derived from bony scales of ostracoderms and placoderms. For details readers may see the Dentition in Mammals.

4. Adenohypophysis. The anterior lobe of pituitary gland develops as a dorsal evagination of stomodaeum, called **Rathke's pouch**, which constricts off to form the anterior and middle lobes of pituitary gland (adenohypophysis). The posterior lobe of pituitary or neurohypophysis is the ventral evagination of diencephalon, called infundibulum. Thus, it is nervous part.

Pharynx. The part of the alimentary canal immediately behind the buccal cavity is a pharynx, lined with endoderm. It is a common passage serving both for the digestion and respiration. As a part of the digestive system it is used as a passage for food from the buccal cavity to the oesophagus, its muscles initiate swallowing. In fishes, the pharynx is large and laterally perforated for gill-slits, while in tetrapoda, it is short and bears openings of nostrils. In embryos, the wall of pharynx gives off a number evaginations which develop into spiracles, gill-clefts, air bladders, lungs, tonsils and a few endocrine glands (*e.g.*, thymus, thyroid and parathyroids).

Oesophagus. The oesophagus is short in most fishes and amphibians because they lack neck, but it is longer in amniotes due to presence of neck. The oesophagus of reptiles is more elongate than that of fishes and amphibians. In granivorous and carnivorous birds, a portion of the oesophagus is enlarged into a sac-like pouch called **crop** which serves to store food which has been eaten quickly. The crop is essentially lacking in digestive glands although in pigeons the crop has 2 crop glands in both sexes, they are really not glands but cell-forming structures, the cells form 'pigeon milk' which is fed to the young. In mammals, the oesophagus is long, lacks glands and varies in relation to the length of neck. It passes through the diaphragm, the portion below the diaphragm is covered with visceral peritoneum which is lacking from the upper part. Oesophagus has mucous glands. Its lining forms longitudinal folds, or finger-like fleshy papillae (elasmobranchs) or horny papillae in marine turtles.

Histologically, the oesophagus differs from the rest of the alimentary canal in three facts : (*i*) It has no visceral peritoneum because it lies outside the coelom, its outermost covering layer is a thin **tunica adventitia**. (*ii*) The muscle fibres in its anterior part are **striped**, middle part has both striped and unstriped muscles, and the posterior part has only unstriped muscles. But there are exceptions in ruminating mammals, all the muscles are striped or voluntary. (*iii*) The mucous membrane lining is made of stratified squamous epithelial cells and not of columnar cells.

Stomach. There is practically no stomach in cyclostomes, chimeras, lung fishes and some primitive teleost fishes, since it has no gastric glands, but in most fishes and tetrapoda it is dilated for storage and maceration of solid food, and digestion of food because it contains gastric glands. The first part of the stomach, next to the oesophagus, is the **cardiac** region and the lower end near the intestine is the **pyloric** region, which has a **pylorus** or **pyloric valve** in which the mucous membrane lining is surrounded by a thick sphincter muscle which regulates the opening and closing of the pyloric stomach into the intestine.

Stomach is straight in cyclostomes, gar, *Belone*, etc., and spindle-shaped in *Proteus, Necturus*, some lizards and snakes. In turtles and tortoises, it is a wide curved tube, and in elasmobranchs the stomach is J-shaped. In crocodiles and birds the stomach has two parts, a **proventriculus** with gastric glands, and a highly muscular **gizzard**, which represents the pyloric region and has a hard, cornified lining for grinding food. In mammals the stomach lies transversely and may be a simple sac or divided into 3 regions, namely **cardiac**, **fundic** and **pyloric** and each region has its gastric glands. In many ruminants the stomach has four chambers : a **rumen, reticulum, omasum** and an **abomasum**. It is claimed that the first three chambers are modifications of the oesophagus, and abomasum is the true stomach representing the cardiac, fundic, and pyloric parts of the stomach. It has been shown embryologically that all four chambers are modified regions of the stomach. In camels, there is no omasum, the rumen and reticulum have pouch-like **water cells** which were once believed to store water, but they are probably digestive.

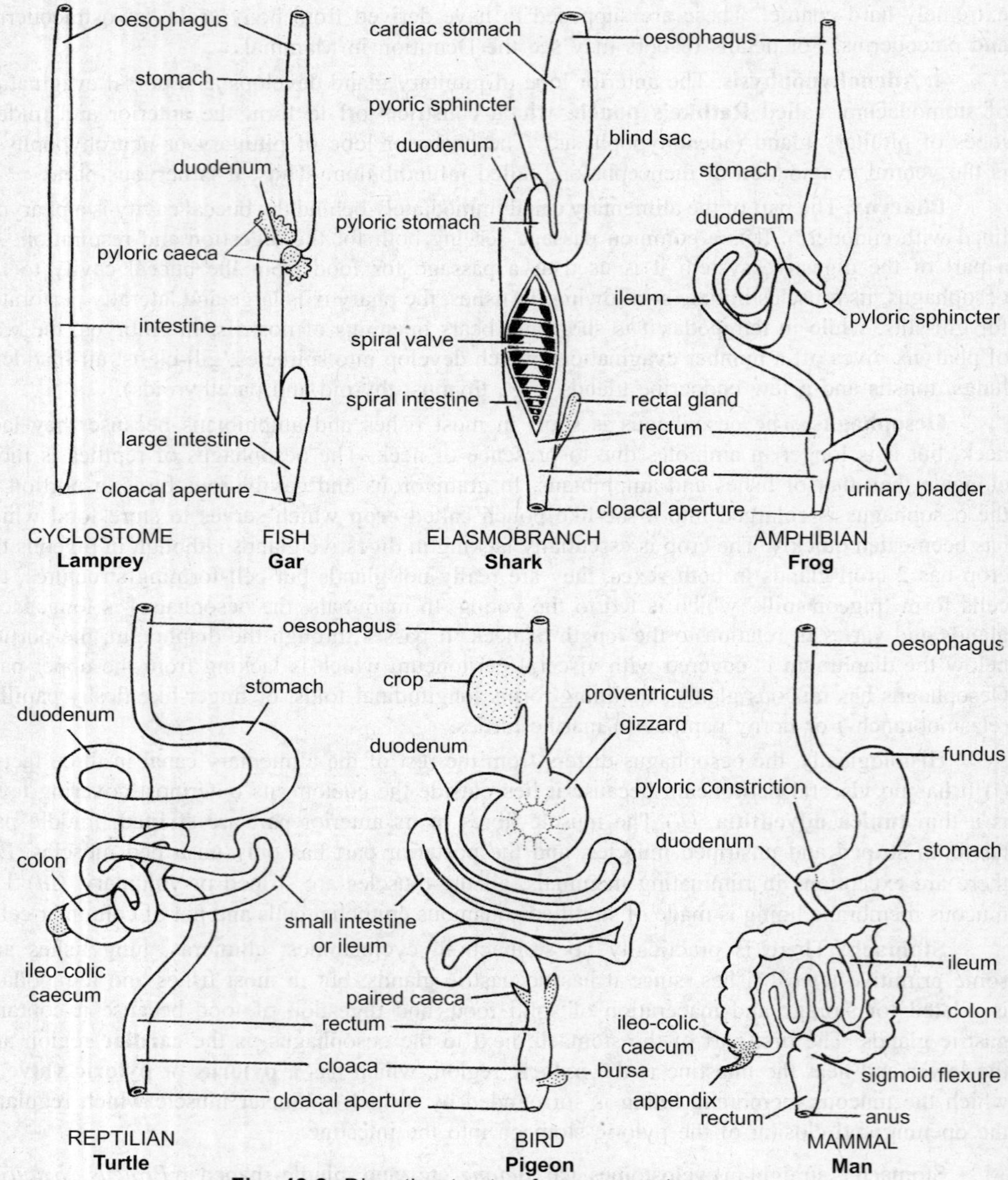

Fig. 43.8. Digestive tracts of some vertebrates.

Histologically, the stomach has the typical parts of the alimentary canal, but it has two peculiarities, the muscularis mucosa is made of an outer longitudinal layer and an inner circular layer of muscles. The epithelium lining is thick with several types of glandular cells forming **gastric glands** of three types called cardiac, fundic, and pyloric gastric glands. The cardiac and pyloric glands secrete only mucus from their surface cells. Fundic glands (or cardiac glands in some) have three kinds of cells, **mucous neck cells** produce mucus, **oxyntic cells** produce hydrochloric acid, they may be present in the cardiac region also, **zymogen cells** or peptic cells secrete pepsin. In most animals the zymogen cells also secrete two **proenzymes** called propepsin and prorennin which are converted by hydrochloric acid into pepsin and rennin respectively. The secretions of all stomach cells form a mixture called gastric juice.

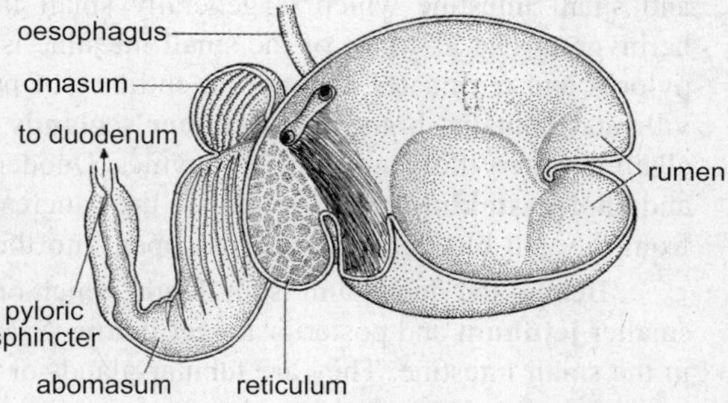

Fig. 43.9. Diagram of a ruminant stomach.

Small intestine. Small intestine is long, narrow and coiled tube after the pylorus. It is the most important part of the digestive tract because the digestion and absorption of food take place in it. In cyclostomes, the intestine is a short straight tube with a spirally arranged longitudinal flap extending into it. In elasmobranchs, it is divided into small and large portions, and the small portion has a spiral valve which greatly increases the absorptive surface. A spiral valve is also present in the small intestine of a few more primitive bony fishes, but is lacking in higher forms in which the intestine is long and coiled. In caecilians, it is little coiled and not differentiated into a small and large tract. In frogs and toads it is relatively long and coiled. In reptiles, it is more coiled than in amphibians. For the first time in vertebrates a caecum or blind diverticulum arises at the junction of small and large intestines. However, this is not permanent in all reptiles. In birds, the small intestine is coiled or looped and one or two colic caeca are also present at the junction of small and large intestines. In most mammals also the small intestine is proportionately long and coiled. Its length, however, is correlated with food habits. In herbivores it is relatively more longer in comparison to insectivores and carnivores. There is a blind pocket or caecum at the junction of the colon

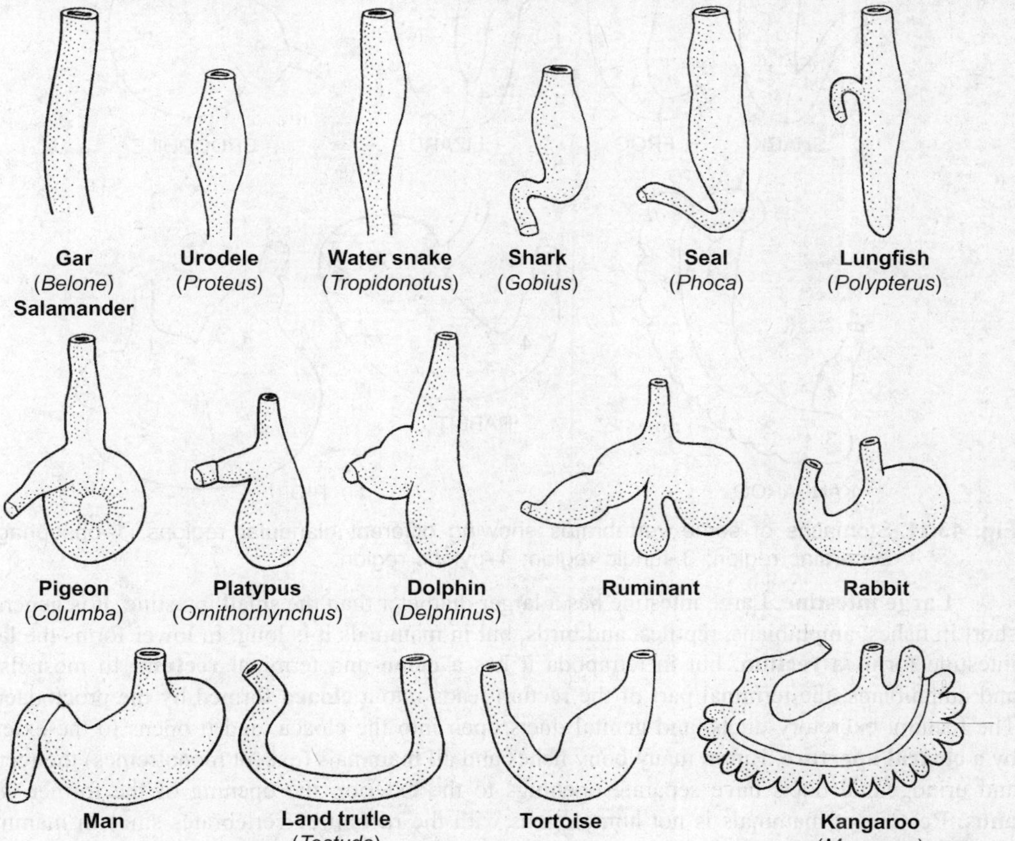

Fig. 43.10. Different shapes of vertebrate stomachs including human.

and small intestine which is generally small in carnivorous species and quite long in many herbivores. The first part of the small intestine is the **duodenum**, which is short beginning from pylorus and terminates beyond the entrance of pancreatic and hepatic ducts. It has many folded villi and contains branching **Brunner's glands** in the submucosa which secrete mucus, some alkaline watery-fluid, and a little enzymes. Duodenum also produces two hormones called **secretin** and **cholecystokinin** which stimulate the pancreas and gall bladder to liberate their juices. Ducts from the gall bladder and pancreas open into the duodenum.

Behind the duodenum is an ileum, which only in mammals is differentiated into an anterior smaller **jejunum** and posterior longer **ileum**. A large number of small digestive glands are present in the small intestine. They are tubular glands or **crypts of Lieberkuhn** found through the entire length, they secrete mucus and a **succus entericus** which has several enzymes. The lining of the small intestine is folded to form small **villi**, which increase the surface area for secretion and absorption. The villi are covered densely by minute finger-like projections, called **microvilli** which assist in absorption into the villi. In mammals nodules of lymphoid tissue called **Peyer's patches** are found on the ileum.

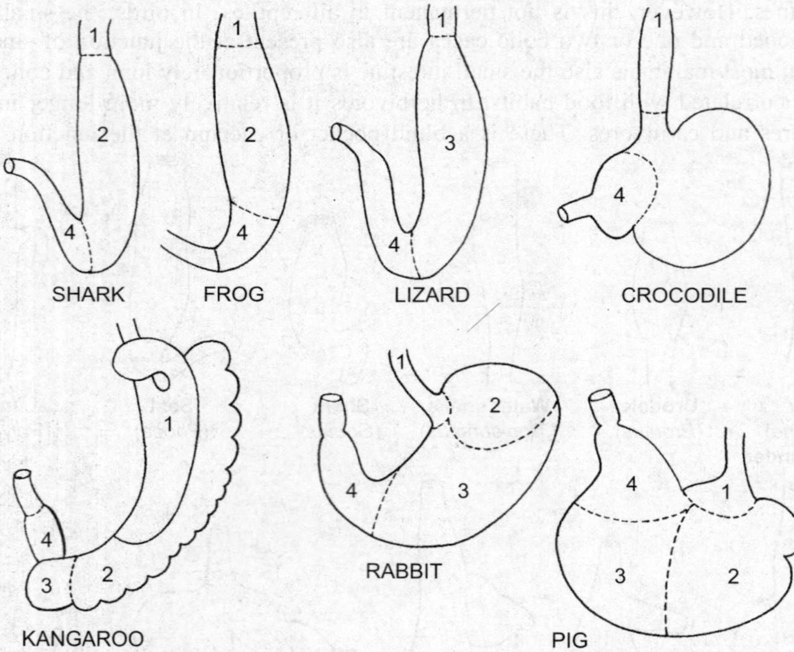

Fig. 43.11. Stomachs of some vertebrates showing different glandular regions. 1–oesophagus; 2–cardiac region; 3–fundic region; 4–pyloric region.

Large intestine. Large intestine has a larger diameter than the small intestine. It is generally short in fishes, amphibians, reptiles, and birds, but in mammals it is long. In lower forms the large intestine forms a **rectum**, but in tetrapoda it has a **colon** and terminal **rectum**. In most fishes and amphibians, the terminal part of the rectum leads into a **cloaca** formed by the proctodaeum. The rectum, excretory ducts, and genital ducts open into the cloaca, and it opens to the exterior by a **cloacal aperture**. But in many bony fishes and all mammals (except monotremes) the rectum and urinogenital ducts have separate openings to the exterior, the opening of the former is an **anus.** Rectum of mammals is not homologous with the rectum of vertebrates since in mammals it is derived by partitioning of embryonic cloaca. In most vertebrate embryos there is a **postanal gut** as an extension of the intestine into the tail, but it disappears later.

In elasmobranchs the large intestine bears a pair of **rectal glands** which secrete mucus and sodium chloride. In amniotes there is an **ileocolic valve** at the junction of small and large intestines, which is absent in fishes. It prevents bacteria to enter into ileum from colon. In amniotes from this junction arises an **ileocolic caecum** which are two in birds. It contains cellulose digesting bacteria. It is very long in herbivorous mammals (rabbit, horse, cow, etc.). In primates caecum is small having a vestigial **vermiform appendix**.

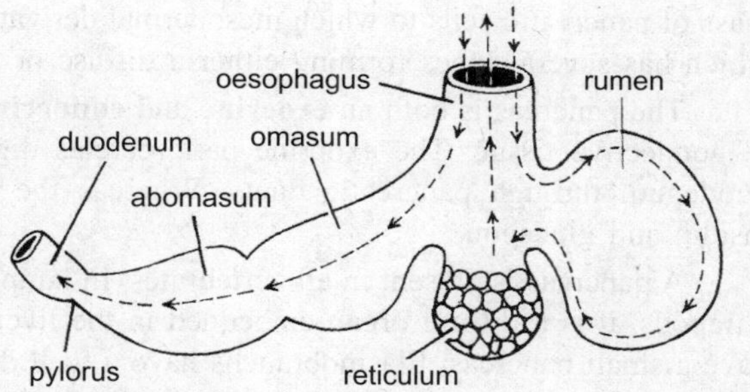

Fig. 43.12. Stomach of a ruminant or cud-chewing mammal. Arrows indicate course of food.

Digestive Glands

1. Liver. The liver arises as a single or double outgrowth from the ventral wall of the embryonic archenteron. This outgrowth forms a hollow hepatic diverticulum, which soon differentiates into an **anterior part**, which proliferates to become the liver and its bile ducts, and a **posterior part**, which gives rise to the gall bladder and cystic duct. The bile ducts join to form a hepatic duct which unites with the cystic duct to form a **common bile duct** or **ductus choledochus**. The region of the archenteron from which the liver arises becomes the duodenum.

The liver is the largest lobed gland in the body, suspended by a double layer of peritoneum from the transverse septum or its representative.

A **gall bladder** is for storage of bile secreted by the hepatic cells, lies in the liver and drains into the duodenum through common bile duct formed by the union of cystic duct and hepatic duct. A gall bladder is not indispensable and is lacking in many birds and mammals.

A liver is present in all vertebrates. In cyclostomes, it is small, single lobed (lampreys) and two lobed in hagfishes. It is bilobed in elasmobranchs, two or three lobed in bony fishes, amphibians, reptiles and birds and many lobed in mammals. Liver is long, narrow and cylindrical in fishes, urodeles and snakes. It is short, broad and flattened in birds and mammals. A gall bladder and bile duct are present in larval cyclostomes but they are absent in the adult. Fishes, amphibians, and reptiles generally have a gall bladder, but it is lacking in many birds. Most mammals have a gall bladder, but it is absent in Cetacea and Ungulata.

The liver secretes a watery, alkaline bile but have no enzymes. It neutralises the acidity of food entering the duodenum. It aids in the digestion of fats.

2. Pancreas. Pancreas is formed from the endoderm of the embryonic archenteron. A single dorsal diverticulum from embryonic duodenum and one or two ventral outgrowths from the liver form pancreatic diverticula. The proximal parts of diverticula form pancreatic ducts, but these ducts undergo reduction or fusion so that only one or two pancreatic ducts remain in the adult, they open into the duodenum either separately or after uniting

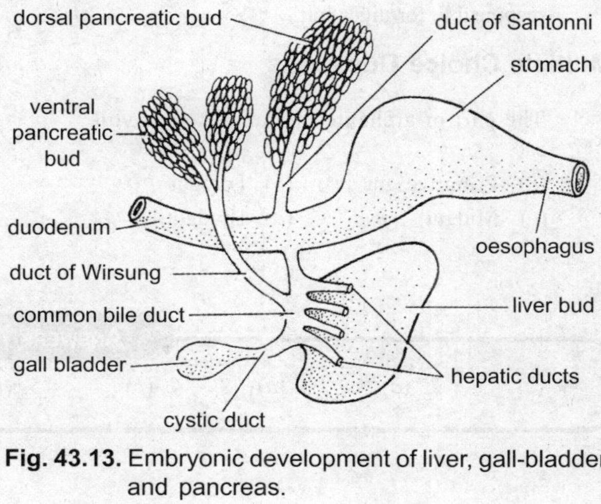

Fig. 43.13. Embryonic development of liver, gall-bladder and pancreas.

with the common bile duct. The distal parts of diverticula undergo budding to form the main mass of pancreatic cells to which mesodermal derivatives are added. Thus, a single gland is made which has several lobes forming either a diffuse or a compact pancreas.

The pancreas is both an **exocrine** and **endocrine gland**, bound together by delicate strands of connective tissue. The exocrine part secretes digestive enzymes which are poured into the duodenum through pancreatic ducts. Whereas the endocrine part secretes hormones such as insulin and glucagon.

A pancreas is present in all vertebrates. In lampreys, some bony fishes, lungfishes and lower tetrapods, it is a diffuse organ embedded in the liver, mesenteries and intestinal wall. Hagfishes have a small pancreas. Elasmobranchs have a well defined bilobed pancreas. In higher tetrapoda it is generally a compact gland. One or two pancreatic ducts open into the duodenum.

IMPORTANT QUESTIONS

Long Answer Questions

1. Describe the coelom in vertebrates.
2. Give a comparative account of digestive system of *Uromastix* (lizard) and *Columba* (bird).
3. Compare the digestive system of *Columba* (bird) with *Oryctolagus* (rabbit) and give reasons for their differences.
4. Give a comparative account of stomach in different vertebrates you have studied.
5. Give a brief account of teeth.

Short Answer Questions

1. Write the notes on the following : (*i*) Gizzard; (*ii*) Ileum; (*iii*) Colon; (*iv*) Pancreas; (*v*) Liver; (*vi*) Ruminant stomach; (*vii*) Digestive glands responsible for digestion.

Multiple Choice Questions

1. The part of archenteron connected to yolk sac
   (*a*) Primitive gut (*b*) Foregut
   (*c*) Midgut (*d*) Hindgut.
2. True muscular lips are found in
   (*a*) Cyclostomes (*b*) Fishes
   (*c*) Amphibians (*d*) Mammals.
3. Largest oral glands are found in
   (*a*) Reptiles (*b*) Birds
   (*c*) Amphibians (*d*) Mammals.
4. In frogs tongue is
   (*a*) Primary tongue (*b*) Definite tongue
   (*c*) Immobile tongue (*d*) Horny tongue.
5. Oesophagus in adult vertebrates is derived from
   (*a*) Primitive gut (*b*) Foregut
   (*c*) Midgut (*d*) Hindgut.
6. Large oesophagus is found in
   (*a*) Dogfish (*b*) Frog
   (*c*) Sparrow (*d*) Giraffe.
7. Epithelial lining in a true stomach contains
   (*a*) Salivary glands (*b*) Mucous glands
   (*c*) Goblet glands (*d*) Gastric glands.
8. In ruminants, true stomach is represented by
   (*a*) Abomasum (*b*) Omasum
   (*c*) Rumen (*d*) Reticulum.

ANSWERS

1. (*c*) 2. (*d*) 3. (*d*) 4. (*b*) 5. (*b*) 6. (*d*) 7. (*d*) 8. (*a*)

44

Chapter

Respiratory System in Vertebrates

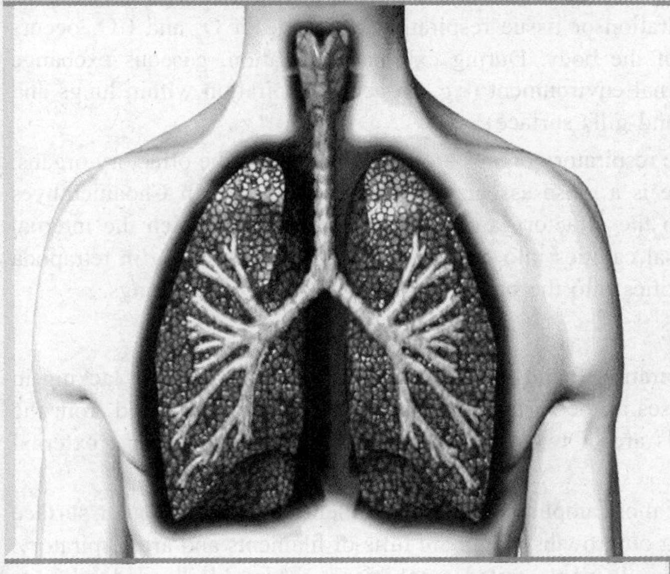

Contents of this Chapter

In vertebrates the blood takes up oxygen from the respiratory organs (*e.g.,* skin, gills, lungs) and transports it to each and every cell of tissues. The red blood corpuscles contain **haemoglobin**, a respiratory pigment. Haemoglobin combines readily with oxygen to form oxyhaemoglobin which is carried by blood from the respiratory organs to the tissue cells where it readily gives up its oxygen in the tissues where it oxidises the food stuffs to form CO_2, heat and energy. Carbon dioxide formed in the tissues enters the blood and red blood corpuscles and is carried away as carbonic acid and carbonates of the sodium and potassium from the tissues to the respiratory surface where free carbon dioxide is liberated.

RESPIRATORY ORGANS

In vertebrates the **skin** may be respiratory (*e.g.,* anurans), while in some fishes and aquatic turtles, the vascular **rectum** or **cloaca** is respiratory. But there are two main types of respiratory organs : **gills** for aquatic respiration and **lungs** for aerial respiration. Both gills and lungs may occur in the same animal.

Accessory respiratory organs are also present in some vertebrates. In both kinds of respiration two conditions are essential, firstly the respiratory organs must have a rich blood supply with very thin moist epithelium covering the blood vessels so that these blood vessels are brought into close contact with the environment (water or air). Secondly in the organs of respiration the blood vessels should be reduced to thin capillaries which expose a large surface area to the environment, so that blood is brought into close contact with the water or air in the respiratory organs.

Exchange of oxygen and carbon dioxide occurs at two places, *i.e.,* in the respiratory organs and in tissues. During internal respiration or tissue respiration exchange of O_2 and CO_2 occurs between blood and tissues (cells) of the body. During external respiration, gaseous exchange takes place between blood and external environment (*e.g.,* in aerial respiration within lungs and in aquatic respiration within water and gills surface).

In lower aquatic vertebrates the respiratory organs are not connected to the olfactory organs, but in air-breathing vertebrates there is a close association between the two. In Choanichthyes there is a direct connection between the olfactory and respiratory organs in which the internal nares or **choanae** open from the nasal cavities into the buccal cavity, but it is only in tetrapoda that air enters through the nasal cavities into the buccal cavity and then into the lungs.

A. Gills

Gills are used for aquatic respiration found in fishes and amphibians. Gills are lacking in anamniotes. Besides exchange of gases at the surface of gills, salts are also eliminated from the gills surface in marine teleosts. Gills are of two types on the basis of their position : external gills and internal gills.

1. External gills. In larvae of most amphibians the integument covering the outer surface of visceral arches gives off branching outgrowths which are tufts of filaments and are respiratory. Thus, these are of ectodermal origin. In some tailed amphibians, external gills and gill-slits are retained throughout life, but in some tailed and all tailless amphibians they are lost during metamorphosis and, hence, called **larval gills**. In the embryo usually five pairs of pharyngeal pouches arise, of which only second, third, and, fourth perforate and open to the exterior. In most amphibians these gill-clefts are closed during metamorphosis. The third, fourth, and fifth visceral arches bear external gills. External gills are in direct contact with water and an exchange of gases occurs through their surface epithelium. Later an operculum arises and covers these gill-clefts and gills externally in tadpoles so that the gills become enclosed in an opercular chamber lined with ectoderm. These external gills soon degenerate and a new set of gills, called **internal gills** develop from the same visceral arches. However, these internal gills and operculum are not homologous with those of fishes. External gills, though rare in fishes, are found in some larval forms of lampreys, *Polypterus* (bichir) has one pair of external gills. Dipnoi (*Lepidosiren*) have four pairs of filamentous external gills attached to the outer edges of the branchial arches. The external gills of fishes disappear in the adult. Gills may be pectinate, bipinnate, dendritic or leaf-like. A gill has a narrow central axis bearing double row of filaments. These are richly vascularised by aortic arches. Gill-slits are not found.

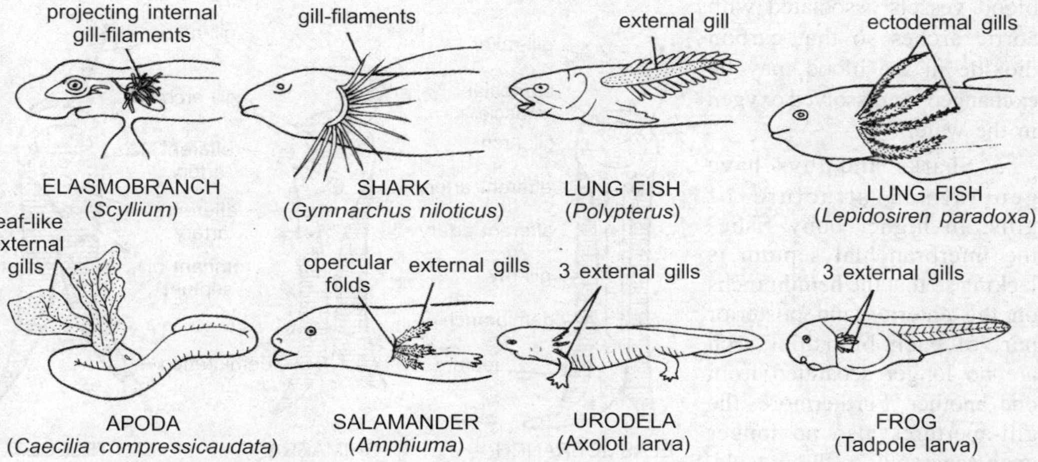

Fig. 44.1. Larval external gills of fishes and amphibians.

2. Internal or true gills. Internal gills are characteristic of fishes. These are placed in the gill-slits and attached to the visceral arches. In amniotes embryonic pharyngeal pouches do not communicate to the outside through gill-slits in the adults. Gills are not found in them.

Gill-slits. Embryologically gill-cleft develops as a result of a series of evaginations from the pharynx which grow outward and meet corresponding ectodermal invaginations from the outside. At each point of junction a cleft or pouch is formed so that water coming into the pharynx from the mouth may pass to outside. The partitions between these passage ways contain cartilaginous or bony supports for the gill-filaments which are located on each side. These gill-slits persist in the adults of cyclostomes, fishes and certain amphibians, but become abolished in higher vertebrates. The number of gill-slits are 6–14 pairs in cyclostomes, 5 to 7 pairs in sharks and rays, 4 pairs in most bony fishes.

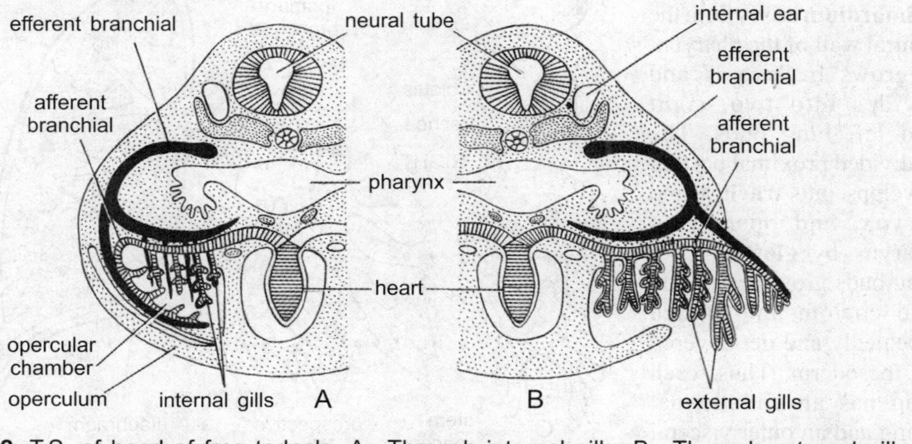

Fig. 44.2. T.S. of head of frog tadpole. A –Through internal gills; B –Through external gills.

Structure of a true gill. These gills develop on the walls of gill-pouches or gill-arches. A gill is formed of two rows of a series of **gill-filaments or gill-lamellae** develop from the epithelium covering the **interbranchial septum** on both sides. Interbranchial septum extends outward from the cartilaginous arch. A single row of gill-filaments on one side of interbranchial septum forms a half gill called a **hemibranch** or **demibranch**. An interbranchial septum with two hemibranchs (anterior and posterior) forms a complete gill or **holobranch**. These are richly supplied with blood vessels associated with aortic arches so that carbon dioxide in the blood may be exchanged for dissolved oxygen in the water.

Sharks and rays have generalised structure of gills. In higher bony fishes the interbranchial septum is lacking so that the hemibranchs on the anterior and posterior part of each branchial arch are no longer separated from one another. Furthermore, the gill-apertures also no longer open separately to the outside. Instead, the gills are enclosed in a single chamber and covered externally by a large **bony operculum** which opens and closes posteriorly to permit water to pass to the outside. In most bony fishes there are four pairs of functional gills.

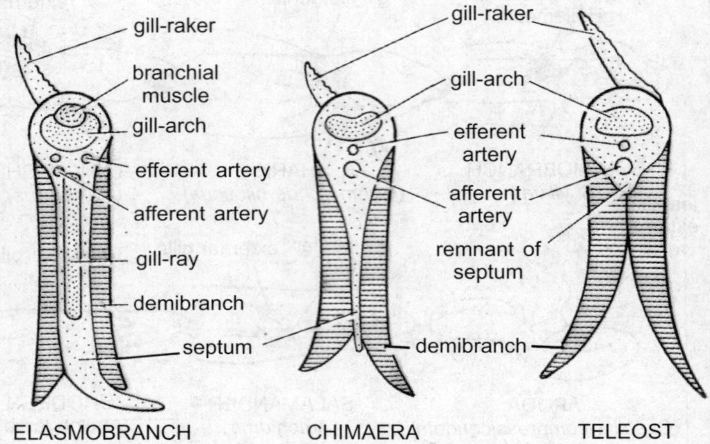

Fig. 44.3. Types of gills in fishes in section.

Spiracles. In sharks and rays an anterior pair of non-respiratory openings, one on either side between mandibular and hyoid arches are called **spiracles** or **pseudobranchs**. These are internally closed. These become closed or lost in lung fishes and bony fishes.

B. Lungs

Most adult amphibians and all amniotes breathe by means of **lungs**, though lungs are also present in lung fishes. In an embryo a hollow outpushing, called **lung primordium** arises from the ventral wall of the pharynx. It grows backwards and divides into two, right and left lung buds. The undivided proximal portion develops into **trachea** and **larynx**, and opens into pharynx by **glottis**. Later lung buds grow posteriorly into coelom and branch repeatedly and get covered by mesoderm. Thus, each lung has an endodermal lining and an outer visceral peritoneum and in between the two mesodermal mesenchyme having blood and lymph vessels, nerves, and smooth muscle fibres and connective tissue. Inner endodermal epithelium of lungs is raised into a network of ridges to increase the vascularised surface exposed to the action of air.

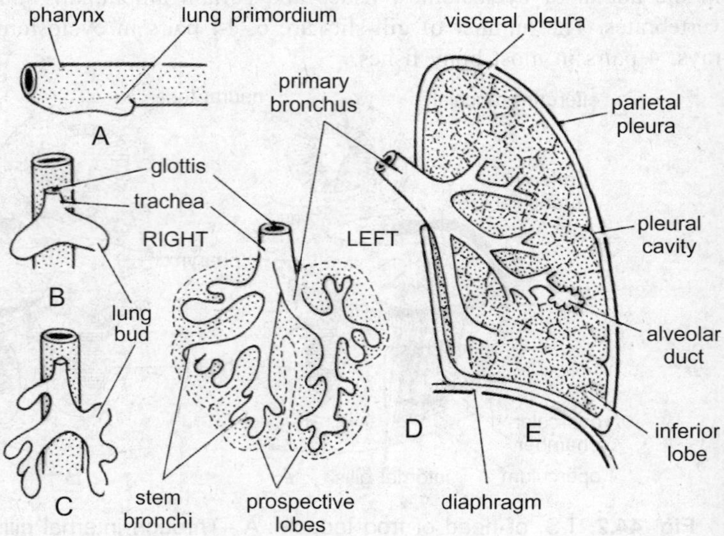

Fig. 44.4. Development of a vertebrate lung in embryo. A – Lung primordium; B & C – Lung buds; D – Embryonic lung; E – Lung at birth.

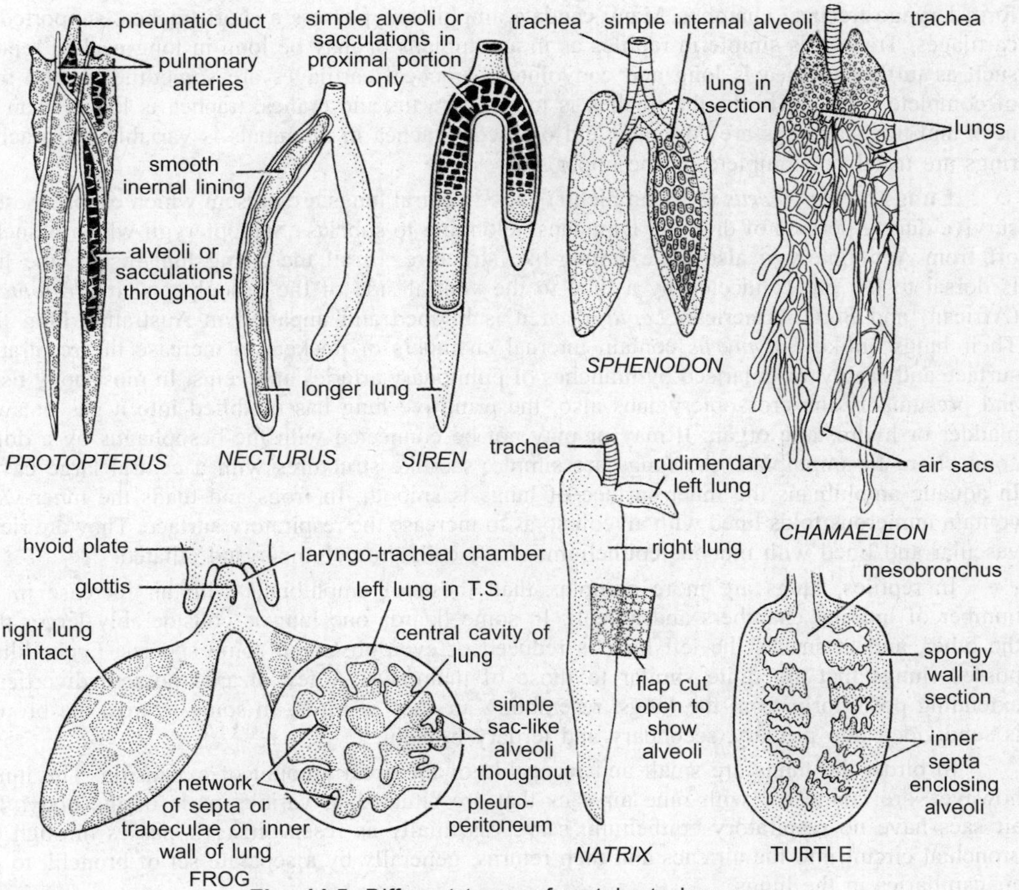

Fig. 44.5. Different types of vertebrate lungs.

In lower forms, the lungs are hollow bags, but in higher forms the ridges increase in number and unite with one another across the lumen of the lung to convert it into a solid but spongy structure with innumerable air spaces. In mammals, the internal surface area of lungs may be thirty times that of the external surface area of the body. The original duct of the lung sac connecting the pharynx to the lungs becomes a **trachea** in most. Trachea is absent in anurans. In many tetrapoda the anterior end of the trachea becomes modified into a **larynx** or sound box which opens into the pharynx by a **glottis**. At its lower end, the trachea divides into two bronchi, each of which enters a lung. The bronchi divide to form an immense system of bronchioles carrying air into minute bags or **alveoli**. The alveoli have very thin walls invested with blood capillaries, an exchange of gases occurs in the alveoli.

Larynx. The upper end of trachea is enlarged, especially in frogs and toads, to form the larynx or sound box in which the vocal cords are located. In *Necturus*, it is supported by a pair of lateral cartilages bounding the glottis. In other amphibians, each lateral cartilage is divided into a dorsal **arytenoid** and a ventral **cricoid** cartilage. In frog, both the cricoids fuse to form a cartilaginous ring. Larynx is not more developed in reptiles. Larynx is not sound producing organ in birds, but serves to modulate tones that originate in the **syrinx**. Syrinx lies at the lower end of trachea where it divides into two bronchi. It is the sound producing organ. Larynx is greatly developed in mammals. Its wall is supported by a pair of **arytenoid**, single **cricoid** and a single **thyroid** cartilage on the ventral surface. Glottis may be closed at the time of swallowing of food by a flap of muscular **epiglottis**.

Trachea. Trachea is extremely short or absent in Anura. It is merged with the larynx to form laryngo-tracheal chamber. Many caudate amphibians possess a short trachea, supported by cartilages. Trachea is simple in reptiles as in amphibians or may be long in long-necked reptiles such as turtles, trachea is long and convoluted. Tracheal cartilages are sometimes in the form of complete rings. In birds, the trachea is long. In swans and cranes, trachea is longer than the neck and tracheal rings are complete and ossified. Trachea in mammals is variable and tracheal rings are usually incomplete on the upper side.

Lungs. In *Polypterus* (African bichir) paired ventral lungs are present which enable these to survive during periods of draught. Dipnoans belonging to subclass Sarcopterygii which branched off from Actinopterygii also have a lung-like structure. In all the living lung fishes, the lung is dorsal to the gut connected by a tube to the ventral side of the oesophagus. In *Protopterus* (African) and South American *Lepidosiren* it is bilobed and unpaired in Australian lung fish. Their lungs unlike *Polypterus* contain internal chambers or pockets to increase the respiratory surface and highly vascularised by branches of pulmonary arteries and veins. In most bony fishes and presumably the crossopterygians also, the primitive lung has modified into a gas or swim bladder or hydrostatic organ. It may or may not be connected with the oesophagus by a dorsal connection. In amphibians the lungs are simple, sac-like structures with a central large cavity. In aquatic amphibians the inner surface of lungs is smooth. In frogs and toads the inner walls contain numerous folds lined with alveoli so as to increase the respiratory surface. They are richly vascular and lined with mucous epithelium whose cells are columnar and ciliated.

In reptiles, lungs are more complex than those of amphibians with an increase in the number of internal chambers and alveoli. In some lizards one lung is considerably larger than the other, and in snakes the left lung is reduced or even absent in some species. Crocodilians possess lungs that are quite similar to those of mammals. A few lizards possess diverticula, extending posteriorly from the lungs, resembling air sacs of birds. In some lizards, the bronchi is subdivided into primary, secondary and tertiary bronchi.

In birds, the lungs are small and incapable of the great amount of expansion. The lungs, however, are connected with nine air sacs that are situated in various parts of the body. The air sacs have no respiratory epithelium, serve essentially as reservoirs. Air passes through the bronchial circuit into the air sacs and then returns, generally by a separate set of bronchi, to the air capillaries in the lungs.

The respiratory system of the mammal is much less complicated than that of the bird. The primary bronchi after entering the lung into secondary bronchi which divide into smaller and smaller bronchioles, finally terminating in tiny alveoli or blind pockets in which there is an exchange of gases. In most mammals, lungs are subdivided externally into lobes, *i.e.,* left lung has two lobes and right lung has three lobes in man and four lobes in rabbit. Lungs are simple and without lobes in whales, sirenians, elephants, hyrax and several perissodactyles. Right lung is lobulated in monotremes and rats. In sirenians, the lungs are elongated.

Mechanism of Respiration

In fishes and amphibians the mechanism of respiration is the same, the floor of the buccal cavity is lowered and water (fishes) or air (amphibians) is taken in, then the mouth is closed and the floor of the buccal cavity is raised which forces the water into gill-clefts in fishes or the air into the lungs in amphibians. In amniotes air is taken in by increasing the volume of the lungs by an expansion of the thorax, this is done by movement of the ribs (and by movements of the diaphragm in mammals). In turtles where ribs are fixed to the carapace, the volume of lungs is increased by movements of the neck and limbs.

C. Accessory Respiratory Organs

Gills are the chief respiratory organs in aquatic venrtebrates, like fishes and some aquatic urodeles, etc. The land vertebrates have the lungs for respiration. There are also other accessory structures for respiration, *i.e.,* for taking oxygen directly from water or air.

1. Skin. Some fishes are able to survive outside water. The common eel, *Anguilla* can travel by wriggling on damp grass though it has no special respiratory organs, but it has vascular areas in the **skin** by which it can breathe both in water and on land. Secondly the opening of the operculum is small and rounded so that the eel can retain water in the branchial chamber and journey on land. In amphibians also, the moist skin is highly vascular. Lungless salamanders (plethodonts) respire only through skin. Their larvae loses gills at metamorphosis and lungs do not develop in adults. African male hairy frog, *Astylosternus* have vascular hairy cutaneous outgrowths which act as respiratory surface. Vascular caudal fin of *Periophthalmus* (mud-skipper) acts as respiratory organ during submergence. In the mud-skipper *Periophthalmus* of the Indian and Pacific Oceans the caudal fin is highly vascular, the head and trunk of the fish project above water when it perches on a rock, only the caudal fin remains submerged and acts as a respiratory organ.

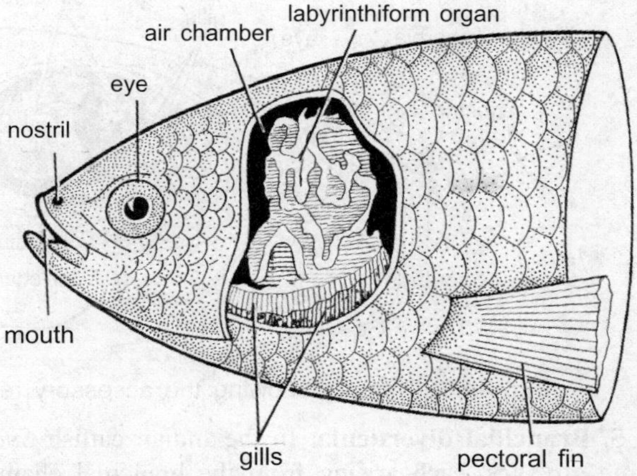

Fig. 44.6. Diagram showing the accessory respiratory organs of *Anabas*.

2. Swim-bladders. (*i*) The Indian climbing perch *Anabas* ***scandens*** **(Fig. 44.6) has** special air chambers above the gills, where three concentrically folded **bony laminae**, called **labyrinthiform organs** are developed from the first epibranchial bone on **each** side. Their covering vascular mucous membrane brings about respiration. *Anbas* is so **depend**ent on air that even in water it comes to the surface to gulp air and it is asphyxiated if prevented from doing so. It can survive for long periods on land and makes excursions by means of its many long spines on the operculum and ventral fins. (*ii*) In *Ophiocephalus* there is an **accessory branchial cavity** on each side above the gills.

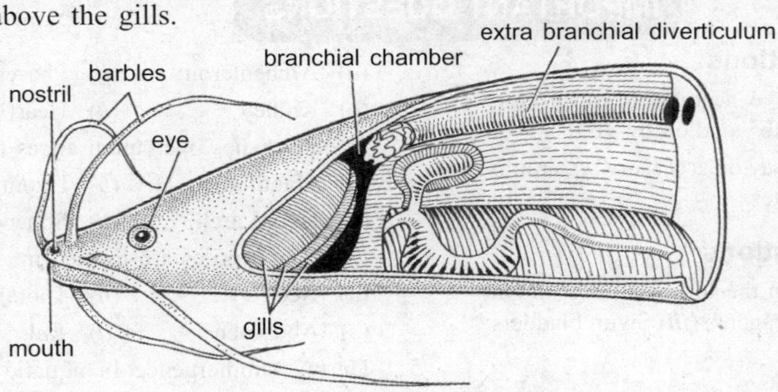

Fig. 44.7. Diagram showing the accessory respiratory organs of *Saccobranchus*.

3. Epithelial lining. The loach *Misgurnus* swallows air which passes through the intestine and is voided by the anus, the highly vascular mucous membrane absorbs oxygen from the air, carbon dioxide is also passed through the anus.

There may be other special organs for gaseous interchange. In *Calichthys* rectal respiration takes place, the rectum is highly vascular into which water is alternately taken in and pumped out.

4. Pharyngeal diverticula. The Indian 'Cuchia eel' *Amphipnous* has poorly developed gills, but on each side of the body there is a **vascular sac** as an outgrowth of the pharynx which opens anteriorly into the first gill-cleft. These sacs are respiratory.

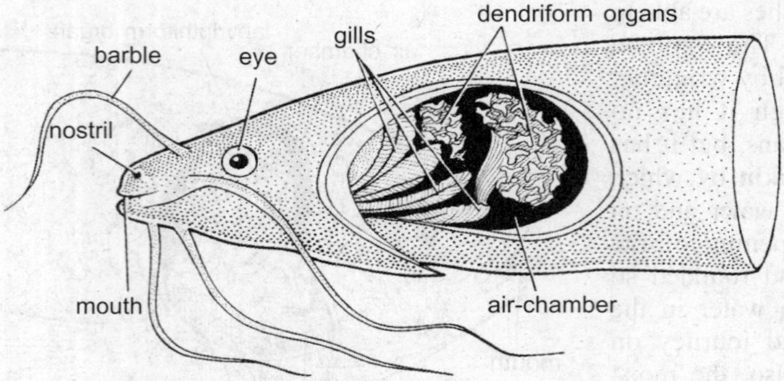

Fig. 44.8. Diagram showing the accessory respiratory organs of *Clarias*.

5. Branchial diverticula. In the Indian catfish *Saccobranchus* (Fig. 44.7) there is a pair of large **air sacs**, each arising from the branchial chamber and extending laterally backwards into the trunk muscles. They can be filled with air for respiration. The catfish *Clarias* (Fig. 44.8), found in Indian and African rivers, has a pair of **supra-branchial organs**, each lying on one side and divided into two parts, a highly branched **arborescent organ** formed from second and fourth branchial arches, and a **vascular sac** of the branchial chamber which encloses the arborescent organ. Several **gill-fans** formed by coalescing of gill-filaments close the entrance of the suprabranchial organ. Air is taken into the organ through the mouth continuosly, and *Clarias* cannot only live outside water for several hours but it can move along on damp grass. Accessory respiratory organs are found generally in tropical fishes of amphibious habit, they are devices for sustaining life out of water.

6. Swim or Air-bladders. Swim or air-bladder arises as a diverticulum from the pharynx or oesophagus in bony fishes. It is originally lateral in positon but becomes dorsal. It usually lies below or lateral to the vertebral column outside the coelom.

IMPORTANT QUESTIONS

Long Answer Questions

1. Give a comparative account of respiratory organs in vertebrates studied by you.
2. Write a detailed essay on accessory respiratory organs in vertebrates.

Short Answer Questions

1. Write short note on the following : (*i*) Larval gills ; (*ii*) Pseudobranchs; (*iii*) Swim bladders; (*iv*) Air bladder.

Multiple Choice Questions

1. A complete gill is called
   (*a*) Holobranch (*b*) Demibranch
   (*c*) Hemibranch (*d*) Pseudobranch.
2. Gills and lungs of vertebrates are the derivatives of embryonic
   (*a*) Archenteron (*b*) Pharynx
   (*c*) Kidney (*d*) Heart.
3. In bony fishes operculum arises from
   (*a*) Mandibular (*b*) Hyomandibular
   (*c*) Hyoid arch (*d*) Sphenethmoid.
4. Lungs of reptiles are located in
   (*a*) Neck (*b*) Thorax
   (*c*) Abdomen (*d*) Tail.
5. During submergence in aquatic turtles the accessory respiratory organs are
   (*a*) Skin (*b*) Gills
   (*c*) Air sacs (*d*) Cloacal bladder.
6. Respiratory organs of embryonic vertebrates
   (*a*) Amnion (*b*) Chorion
   (*c*) Egg membrane (*d*) Yolk sac.

ANSWERS

1. (*a*) 2. (*b*) 3. (*c*) 4. (*c*) 5. (*d*) 6. (*d*)

45

Chapter

Circulatory System in Vertebrates

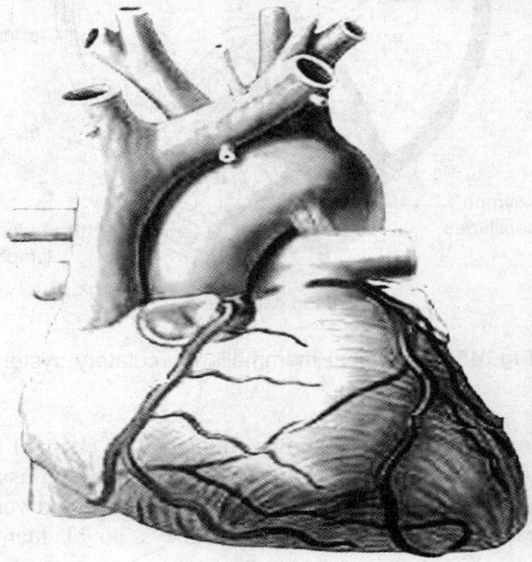

Contents of this Chapter

In the circulatory system of vertebrates there are two systems of elaborately branching tubes, which ramify throughout the body and carry fluids to the tissues. They are a **blood vascular** system and a **lymphatic** system. When the term circulatory system is used, it refers only to the blood vascular system.

The **blood vascular system** is a closed system in vertebrates, it has a contractile heart and continuous tubes called vessels. The **lymphatic system** is an open system with lymph spaces. The blood vessels which carry blood away from the heart are arteries, which divide into thinner **arterioles**, branching into extremely thin and small **capillaries**. The wall of a capillary is made of a single layer of tasselated endothelial cells. Every blood vessel, including the heart, has a lining of endothelial cells or endothelium. Blood does not come in contact directly with tissue cells. Substances pass from and into capillaries through the tissue fluid contained in tissue spaces between cells. Exchange of substances between the blood and tissue cells of the body takes place through the capillary walls. This is brought about by **pinocytosis** in the endothelial cells of capillaries.

Under an electron microscope the endothelial cells show many small vesicles which are invaginations of the plasma membranes. These vesicles move from one side of the cell and release their contents on the opposite side. Thus, the blood gives up oxygen, nutrient material and hormones to the cells for metabolism, and the cells give out water, carbon dioxide, and nitrogenous wastes into the blood to carry them to the excretory organs for their quick elimination.

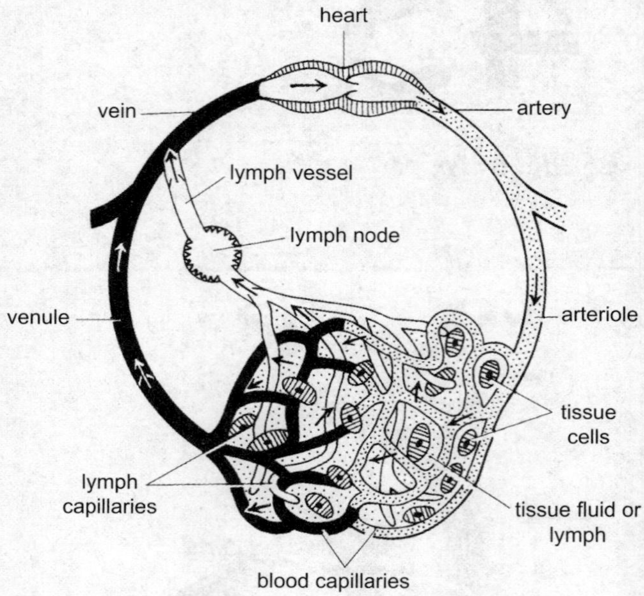

Fig. 45.1. A typical mammalian circulatory system.

The capillaries form a network in all body tissues except cartilage and epithelium. From the capillaries the blood passes into thin **venules** which combine to form **veins** which carry blood towards the heart. But some veins (portal veins, renal veins, and hepatic vein) have capillaries which are just like those of arteries. But all blood does not pass through capillaries into venules. There are also some **through channels** between arterioles and venules lying in some organs, such as the skin. There are also **arteriovenous anastomoses** between arterioles and venules found in the digits. The function of such connections is not clearly known, though it is claimed that they regulate blood pressure and circulation. At some places exchange of material between tissues and blood takes place through thin-walled spaces or **sinusoids**. In certain organs a blood vessel may form a coiled network of tiny blood vessels called **rete mirabile** (kidneys, air bladder).

Parts of Circulatory System

In chordates, the circulatory system is of **closed type**. This type is also found in annelids (invertebrates). Another type is of **open type** in which capillaries are not found. It is found in molluscs and arthropods. Here the blood flows through arteries into various organs, which pass through blood spaces or sinuses and then again into vessels (veins) to the heart.

The circulatory system includes the heart, arteries, veins, capillaries and blood. **Heart** is a modified blood vessel with muscular walls, which periodically contracts to pump the blood into the various parts of the body through definite vessels. **Arteries** and their branches form an arterial system which carry the blood from the heart. **Veins** and their tributaries constitute a venous system which takes the blood from the capillaries of arteries or arterioles and carry that to the heart.

Portal system. In the portal system, the blood is not returned directly to the heart, but there is an interposing organ (liver or kidney) in the course of the returning blood. The vein bringing blood starts in capillaries and ends in capillaries, the vein concerned acting both as afferent and efferent vessel, the afferent vessels end in capillaries just like arteries, then the blood is collected into systemic veins. All vertebrates have a hepatic portal system in which blood passes into two sets of capillaries in the liver. Lower vertebrates and embryos of higher vertebrates have a renal portal system also in which blood passes through two sets of capillaries in the kidney before reaching the heart. The capillaries in the pituitary gland form a pituitary portal system which is a small but an important system.

Lymphatic system. It is found in chordates except cyclostomes and cartilaginous fishes. It includes lymph and lymph vessels. **Lymph** is a tissue fluid found amongst the body cells. It is blood plasma minus red blood corpuscles and some proteins. **Lymph capillaries** forming a network of thin blind-ending vessels, which collect lymph. **Lymph vessels** are thin-walled vessels formed by the union of blood capillaries. These empty into veins. **Lymph nodes** are found on lymph vessels in mammals. These form lymphocytes of blood used for body defence against diseases.

EVOLUTION OF HEART IN VERTEBRATES

The heart is an unpaired organ but its origin is bilateral. In an embryo the mesenchyme forms a group of **endocardial cells** below the pharynx. These cells become arranged to form a pair of thin **endothelial tubes**. The two endothelial tubes soon fuse to form a single **endocardial tube** lying longitudinally below the pharynx. The splanchnic mesoderm lying below the endoderm gets folded longitudinally around the endocardial tube. This two-layered tube will form the heart in which the splanchnic mesoderm thickens to form a **myocardium** or muscular wall of the heart and an outer thin **epicardium** or visceral pericardium. The endocardial tube becomes the lining of the heart known as **endocardium**. Folds of splanchnic mesoderm meet above to form a dorsal **mesocardium** which suspends the heart in the coelom. Soon a transverse septum is formed behind the heart which divides the coelom into two chambers, an anterior **pericardial cavity** enclosing the heart and a posterior **abdominal cavity**. The heart is a straight tube but it increases in length and becomes S-shaped because its ends are fixed. Appearance of valves, constriction, partitions in the heart, and differential thickenings of its walls form three or four chambers in the heart.

1. Single-chambered Heart

In *Amphioxus* (primitive chordate), a true heart is not found. A part of ventral aorta beneath the pharynx is muscular and contractile and acts as heart.

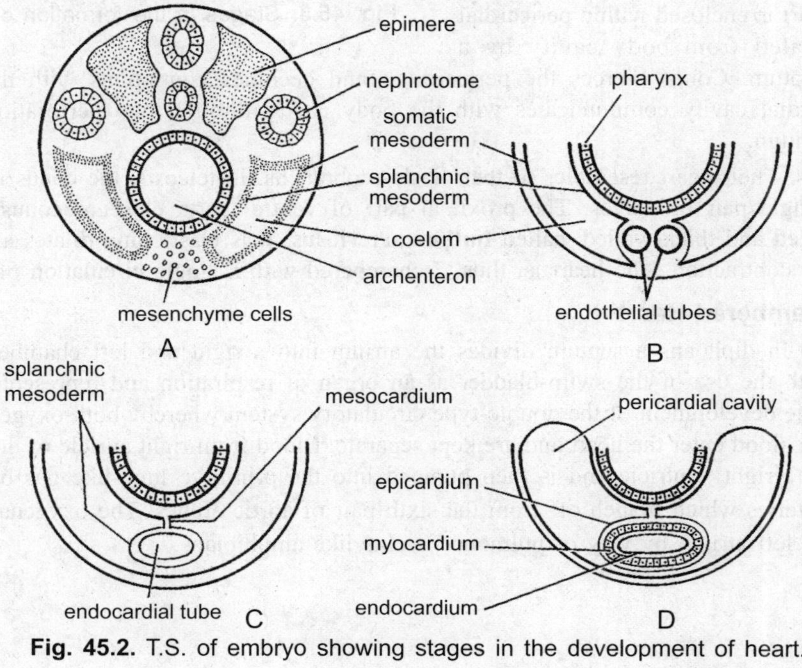

Fig. 45.2. T.S. of embryo showing stages in the development of heart.

2. Two-chambered Heart

In cyclostomes, there are four chambers arranged in a linear order : a thin-walled **sinus venosus**, a slightly muscular **atrium** (auricle), a muscular **ventricle** and a muscular **conus arteriosus** or **bulbus cordis**. It lies in the body cavity in which other visceral organs are also present. Out of four chambers, only atrium and ventricle correspond to the four chambers (paired atria and paired ventricles) of the higher vertebrates. In the evolution of heart many changes have taken place.

Elasmobranchs. Except Dipnoi, the circulatory system in fishes from cyclostomes to teleosts, only unoxygenated blood goes to the heart. From there it is pumped to the gills, aerated and then distributed to the body. The heart of cartilaginous dogfish is muscular and dorsoventrally bent S-shaped tube with four compartments in a linear series. They are sinus venosus and atrium for receiving venous blood, and a ventricle and conus arteriosus for pumping this blood. The heart is a **branchial venous heart**. The sinus venosus and conus arteriosus are **accessory chambers**. Atrium and ventricle are true chambers, thus, it is a 2-chambered heart. The sinus venosus opens anteriorly into atrium through sinu-atrial aperture guarded by a pair of valves. Atrium lies dorsal to ventricle and opens ventrally into ventricle through an atrio-ventricular aperture guarded by a pair of valves. The thick-walled, muscular ventricle opens into a narrow **conus arteriosus** containing valves in two series.

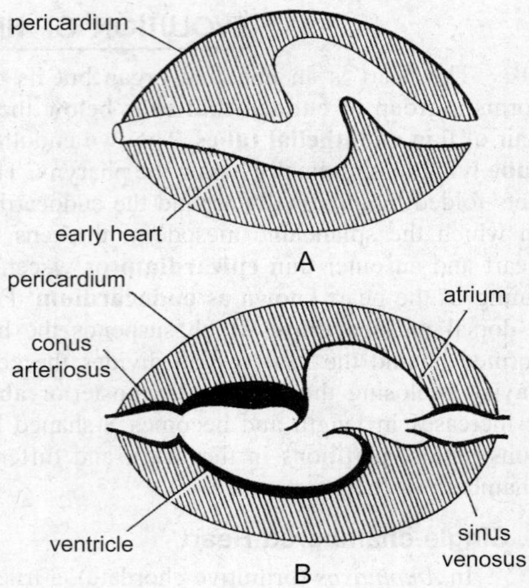

Fig. 45.3. Stages in the formation of heart.

The heart is enclosed within pericardial cavity separated from body cavity by a transverse septum. Conus pierces the pericardium and becomes continuous with the ventral aorta. Pericardial cavity communicates with the body cavity through two perforations in the transverse septum.

Teleosts. Their heart resembles to that of clasmobranchs. In teleosts, the conus is reduced and has a single pair of valves. The proximal part of ventral aorta close to conus becomes greatly enlarged and thick-walled, called **bulbus arteriosus**. It is elastic and dilates at the time of ventricular contraction. The heart is, thus, 2-chambered with a single circulation of blood.

3. Three-chambered Heart

Dipnoi. In diphoans a septum divides the atrium into a right and left chamber. This is correlated with the use of the swim-bladder as an organ of respiration and represents the first step toward the development of the double-type circulatory system whereby both oxygenated and unoxygenated blood enter the heart and are kept separate. Blood from right auricle of the lungfish passes into the right ventricle and is then pumped into the primitive lung-like gas bladder by pulmonary arteries which branch off from the sixth pair of aortic arches. The oxygenated blood returns to the left atrium by way of pulmonary veins like amphibians.

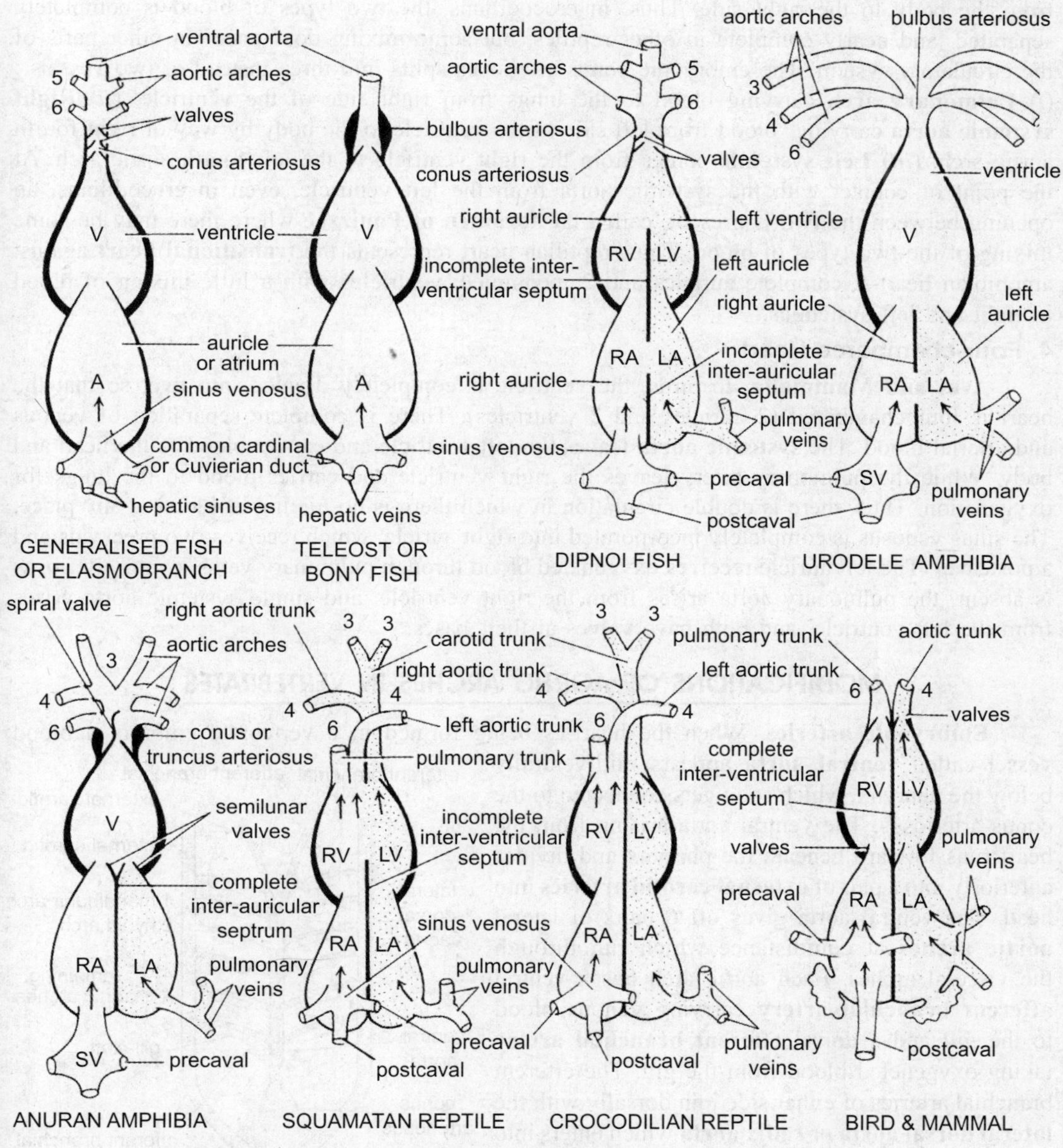

Fig. 45.4. Evolution of heart in different classes of vertebrates.

Amphibia. In amphibians, the dorsal atrium shifts anterior to ventricle. The sinus venosus opens into right atrium dorsally and not posteriorly. The atrium is completely divided into right and left chambers and has no foramen ovale in the inter-auricular septum, which remains open in dipnoans. Deep pockets develop in the ventricular cavity. The conus arteriosus divides into systemic and pulmonary vessels by a spiral valve. In lungless salamanders, the inter-atrial septum is incomplete and pulmonary veins are absent.

Reptilia. In reptiles, the heart is further advanced. The atrium is always completely separated into a right and left chamber, and in many forms the sinus venosus is incorporated into the wall of the right atrium. The ventricle is also partly divided by a septum in most reptiles, and in the alligators and crocodiles is completely two-chambered. This means that oxygenated blood coming from the lungs to the left side of the heart is essentially separted from the non-oxygenated blood from the body to the right side. Thus, in crocodilians, the two types of blood is completely separated, and nearly complete in other reptiles, but some mixing does occur in other parts of the circulatory system. The embryonic conus arteriosus splits into three instead of two vessels : (*i*) **Pulmonary arch** carrying blood to the lungs from right side of the ventricle. (*ii*) **Right systemic aorta** carrying blood from left side of the ventricle to the body by way of right fourth aortic arch. (*iii*) **Left systemic** comes from the right ventricle to the left fourth aortic arch. At the point of contact with the systemic aorta from the left ventricle, even in crocodilians, an opening between the two is present, called the **foramen of Panizzae** where there may be some mixing of the two types of blood. Thus, reptilian heart represents the **transitional heart** against amphibian heart–2 complete auricles and 2 incomplete ventricles with a little mixing of blood in right and left systemics.

4. Four-chambered Heart

Aves and Mammalia. In birds, the ventricle is completely divided into two, so that the heart is four chambered (2 auricles and 2 ventricles). There is complete separation of venous and arterial blood. The **systemic aorta** leaves the left ventricle and carries blood to the head and body. While the pulmonary artery leaves the right ventricle and carries blood to the lungs for oxygenation. Thus, there is double circulation in which there is no mixing of blood at any place. The sinus venosus is completely incorporated into right auricle, which receives two precavals and a postcaval. The left auricle receives oxygenated blood through pulmonary veins, conus arteriosus is absent, the pulmonary aorta arises from the right ventricle, and single systemic aorta arises from the left ventricle, and both have valves at their bases.

MODIFICATIONS OF AORTIC ARCHES IN VERTEBRATES

Embryonic arteries. When the heart is being formed in a vertebrate embryo, a blood vessel called **ventral aorta** appears mid-ventrally below the pharynx, which soon gets connected to the conus arteriosus. The ventral aorta arising from the heart runs forward beneath the pharynx and divides anteriorly into a pair of **external carotid arteries** into head. The ventral aorta gives off 6 pairs of lateral **aortic arches** at equidistance which run through the visceral arches. Each aortic arch has a ventral **afferent branchial artery** carrying venous blood to the gill and a dorsal **efferent branchial artery** taking oxygeneted blood from the gill. The efferent branchial arteries of either side join dorsally with the **lateral dorsal aorta** or **radix aorta** which enters into the head as **internal carotid artery**. The first aortic arch is **a mandibular aortic arch**, the second is a **hyoid aortic arch**, the remaining ones are called third, fourth, fifth and sixth aortic arches. The lateral dossal aorta fuse behind the pharynx to form a **dorsal aorta** which is continued mid-dorsally into the tail as a **caudal artery**. From the dorsal aorta paired and unpaired arteries arise which supply various organs of the body. In an embryo with a yolk sac, a pair of **vitelline ariteries** arise from the dorsal aorta and supply the yolk sac. In embryos of amniotes a pair of **umbilical** or **allantoic arteries** arise from the

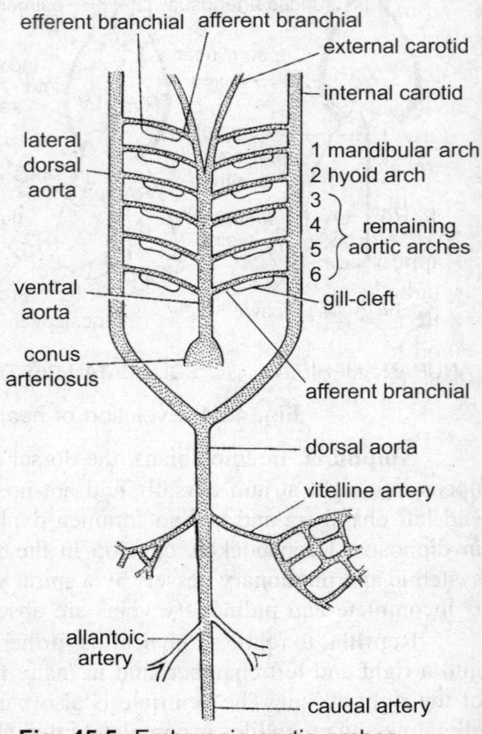

Fig. 45.5. Embryonic aortic arches and arteries.

dorsal aorta supplying blood to the allantois. In an adult the vitelline arteries fuse to form the main mesenteric artery, the major part of the allantoic arteries is lost, but their remnants form hypogastric or internal iliac arteries.

Aortic Arches in Vertebrates

In various adult vertebrates, the arterial system appears to be different, but they are built on the same basic fundamental plan. The difference is due to increasing complexity of heart due to a change in respiration from gills to lungs. There is a progressive reduction in the number of aortic arches in the vertebrate series.

Cyclostomata. In *Petromyzon*, there are 7 pairs of aortic arches. In other cyclostomes these vary from 6 pairs in *Myxine* and 15 pairs in *Eptatretus*.

Pisces. Although six is considered to have been the basic number of aortic arches for fishes. This number is reduced to five even in sharks and rays with the loss of the first pair, the mandibular aortic arch or it is represented by an efferent pseudobranchial artery. In most bony fishes both the mandibular (*i*) and hyoid (*ii*) aortic arches disappear or are much reduced. In *Polypterus* and Dipnoi (lung-fishes), gills are not well developed. So the **pulmonary artery** arises from the efferent part of 6th arch on either side which supplies blood to the air bladder or lung. In elasmobranchs and Dipnoi each arch has one afferent and two efferent branchial arteries (formed by splitting) in each gill. In bony fishes each gill has one afferent and one efferent artery.

In tetrapoda aortic arches do not break up into afferent and efferent parts because true internal gills are absent. In all tetrapoda the first and second arches disappear.

Amphibia. Here the aortic arches show modification due to loss of gills and appearance of the lungs. In **urodeles** there are external gills present as respiratory organs in addition to lungs. The III, IV, V and VI aortic arches are present, though the fifth pair is much reduced in *Siren*, *Amphiuma* and *Necturus*. The aortic arches are not broken in the external gills into afferent and efferent portions, because branches arising from IV, V, and VI aortic arches form capillaries in the external gills. The lateral dorsal aortae between the III and IV aortic arches persist as a vascular connection, the **ductus caroticus**. The VI aortic arch forms the pulmo-cutaneous arch or artery on either side taking blood to the lung and skin. It also retains a connection with the lateral dorsal aorta known as a **ductus arteriosus** (duct of Botalli).

In the larva of anuran (frog tadpole), arrangement of aortic arches are like an adult urodele due to presence of gills. At metamorphosis, with the loss of gills, I, II, and V aortic arches disappear completely, only the IIIrd, IVth and VIth aortic arches are present. The lateral dorsal aorta between the third and fourth aortic arches (ductus caroticus) also disappears. Thus, the third aortic arch along with a part of the ventral aorta becomes the carotid arch carrying oxygenated blood to the head region. The fourth aortic arch along its lateral dorsal aorta forms the systemic arch. The sixth aortic arch becomes the pulmocutaneous arch supplying venous blood to lungs and skin. The ductus arteriosus disappears during metamorphosis. Thus, adult anurans have only III, IV and VI aortic arches. These are also retained by amniotes.

Reptilia. In reptiles, the gills are fully replaced by lungs. Only III, IV and VI aortic arches are present. With the partial separation of the ventricle into two parts, the distal portion of the conus arteriosus and the entire ventral aorta are split into three vessels, *i.e.,* two aortic or systemic and one pulmonary. **Right systemic arch** (IV) arise from the left ventricle carrying oxygenated blood to the carotid arch (III). The **left systemic** (IV) and pulmonary aortae (VI) take their origin from the right ventricle. The left systemic carries deoxygenated or mixed blood to the body through dorsal aorta. While the pulmonary artery takes deoxygenated blood to the lungs. The ductus caroticus disappears, but it persists in snakes and some lizards (*Uromastix*). The ductus arteriosus disappears in most reptiles though it persists in a reduced form in *Sphenodon* and some turtles. Due to mixing of blood, reptiles are cold blooded animals like fishes and amphibians.

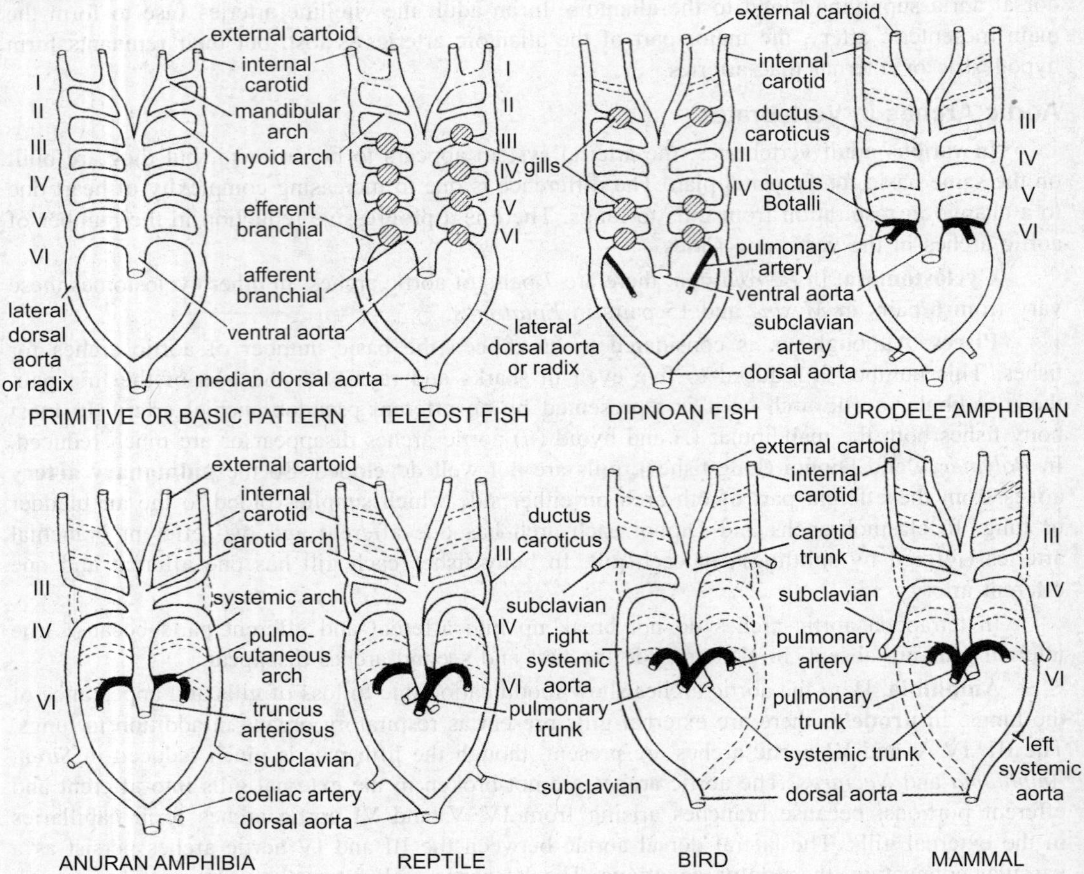

Fig. 45.6. Modification of aortic arches in representative vertebrates.

Aves. In birds, the III, IV and VI aortic arches are present. They follow the general pattern of reptiles with some differences. With the complete division of the ventricle into two parts, the conus arteriosus and ventral aorta have split to form two vessels, **systemic aorta** arising from the left ventricle and a **pulmonary aorta** from the right ventricle. **Third aortic arch** with remnants of lateral and ventral aortae form the carotids which arise from systemic aorta. **Fourth aortic arch** forms the **systemic aorta** on the right side only. It unites with the lateral aorta of its own side and forms the dorsal aorta. Part of the fourth aortic arch of the left side forms the **left subclavian artery**, the rest along with its lateral dorsal aorta disappears. The **sixth aortic arch** forms the pulmonary aorta. **Ductus caroticus** and **ductus arteriosus** disappear.

Mammalia. In mammals also the III, IV and VI aoric arches persist. The ventricle is divided completely into two parts. The conus arteriosus and ventral aorta split to form two vessels : (*i*) a **systemic aorta** arising from the left ventricle, and (*ii*) a **pulmonary aorta** from the right ventricle. **Third aortic arch** with remnants of lateral and ventral aortae forms the carotid arch. **Fourth aortic arch** forms the systemic aorta on the left side only, while on the right side its proximal portion forms an **innominate** and **right subclavian artery**, the rest along with its lateral dorsal aorta disappears. **Sixth aortic arch** forms the pulmonary aorta. The ductus arteriosus degenerates but it persists in some until hatching or birth in a reduced form on the left side as a thin **ligamentum arteriosum**.

VENOUS SYSTEM

Embryonic veins. In all the vertebrate embryos, the venous system is simple and similar. The veins are mostly paired and symmetrically arranged. In embryos without a yolk sac, a **sub-intestinal vein** is formed in the splanchnic mesoderm below the gut. It loops around the anus and is continued posteriorly as (*i*) a **caudal vein** into the tail.

In all embryos having a yolk sac (whether containing yolk or not), a pair of **vitelline veins** arises from the yolk sac and joins the posterior part of the developing heart which becomes the sinus venosus. In fact the fusion of these vitelline veins is responsible for the heart formation in bony fishes, reptiles and birds. Each vitelline vein at its posterior end joins the subintestinal vein formed in the same way as in the embryos having no yolk sac.

(*ii*) A pair of **subcardinal veins** arises between the kidneys and joins the caudal vein.

(*iii*) Paired **anterior** and **posterior cardinal veins** are formed, which carry the blood from the head and posterior parts of the body respectively. The anterior and posterior cardinal veins of each side unite to form a **ductus Cuvieri** or **common cardinal** vein which passes inwards through the transverse septum to enter the sinus venosus.

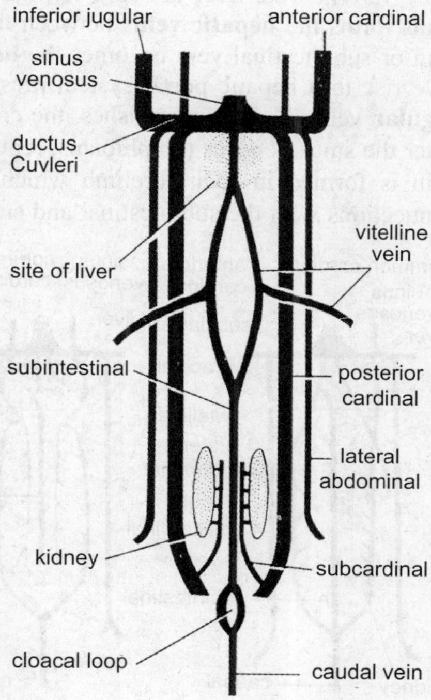

Fig. 45.7. Primitive embryonic veins.

In **fishes** and **salamanders** (urodeles) an **inferior jugular** vein comes from the ventral side of the head to join the common cardinal vein. It has no homologue in other vertebrates. In **anamniotes** a pair of **lateral** or **ventral abdominal veins** come from the body wall to enter the common cardinal veins.

Pisces. Common cardinal vein (duct of Cuvier) enter the sinus venosus from each side and is formed by the fusion of anterior and posterior cardinals. Blood from the head is collected by the anterior cardinals, and blood from the kidneys and gonads is collected by the posterior cardinals. Also entering the ducts of Cuvier are the paired lateral abdominal veins which receive blood from the body wall and paired appendages. The **renal portal system** consists of the caudal vein and two renal portal veins situated lateral to the kidneys which capillarised in the kidneys. **Hepatic portal system** carries blood from the stomach and intestine and returns it to the liver, where, after passing through series of sinusoids, it enters the sinus venosus by way of paired hapatic veins.

In **teleosts**, the lateral abdominal veins are lacking so that blood from the **subclavians**, draning the pectoral appendages, enters the sinus venosus directly, and the blood from **iliac veins**, draining the pelvic appendages, passes into the postcardinals.

In **dipnoans**, a single ventral **abdominal vein** is present, presumably derived from a fusion of lateral abdominals. This vein receives blood from the ilacs by way of paired pelvic veins and enters the right duct of Cuvier. Furthermore, from the right postcardinal system, a new vein, the postclaval makes its appearance that is of major importance in higher vertebrates. It is connected with the caudal vein and passes forward through the liver to the sinus.

In embryos of **amniotes** the lateral abdominal veins are known as **umbilical** or **allantoic** veins because they drain the allantois. The vitelline and umbilical veins are lost at birth, only their remnants persist.

MODIFICATIONS OF VEINS

As in the case of arteries, the veins of different vertebrates are arranged on the same fundamental plan. Any variations shown by them follow a sequence in the vertebrate series. In its development the venous system of higher vertebrates passes through the stages seen in embryos of lower forms.

1. When the liver is being formed the proximal part of the vitelline veins or subintestinal veins forms the **hepatic veins** between the liver and the heart, the distal part of the left vitelline vein or subintestinal vein becomes the **hepatic portal vein** which forms sinusoids in the liver to give rise to a hepatic portal system present in all. 2. Anterior cardinal veins persist as **internal jugular** veins. 3. Except in fishes, the common cardinal veins become the **precaval** veins which enter the sinus venosus (amphibians, reptiles) or the right auricle (birds, mammals). A **subclavian** vein is formed in each forelimb which joins the precaval vein. 4. The caudal vein loses its connections with the subintestinal and subcardinal veins, its anterior part splits into two branches

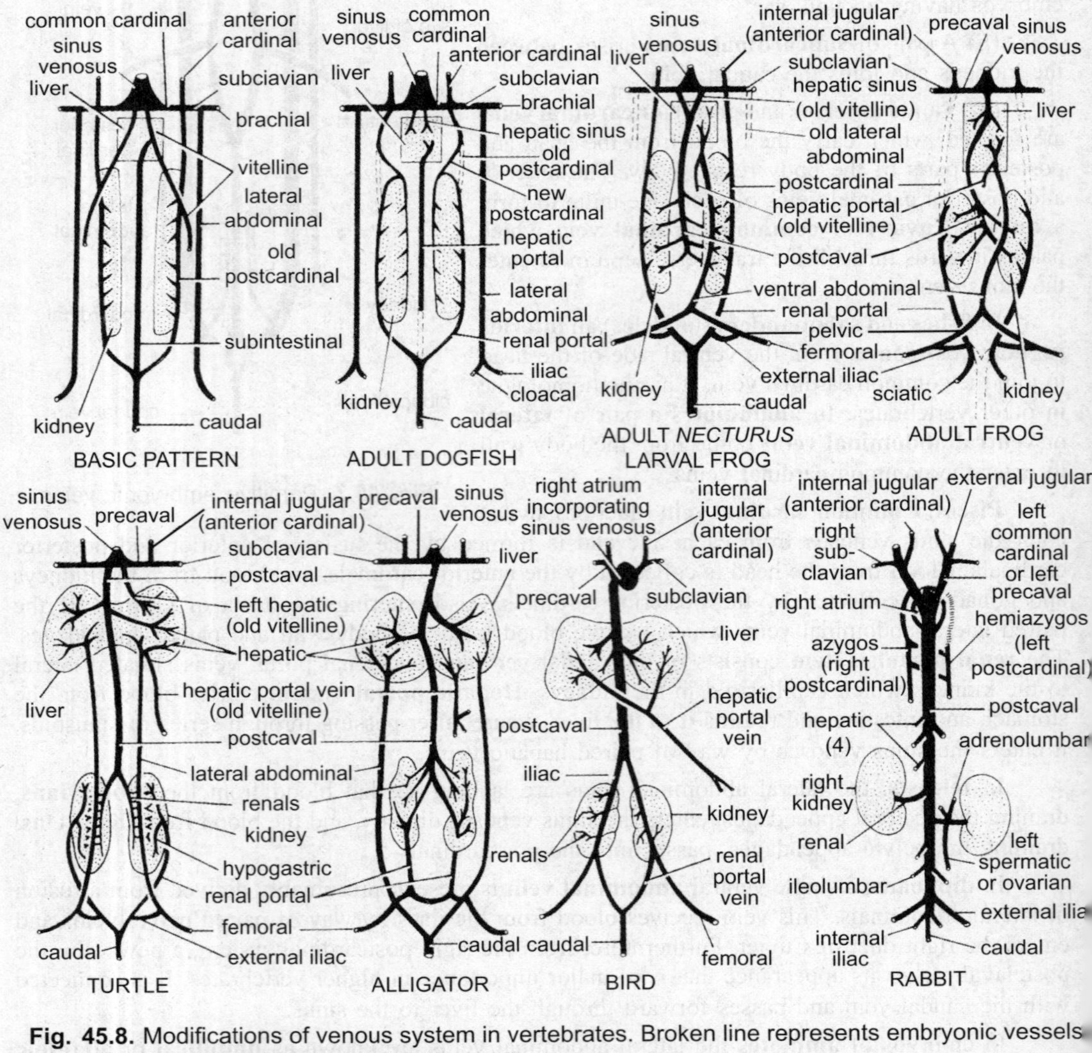

Fig. 45.8. Modifications of venous system in vertebrates. Broken line represents embryonic vessels that have disappeared in the adult.

which join the posterior cardinal veins in all except mammals. 5. Posterior cardinal veins persist as such in fishes, but in others each breaks up into two portions, an anterior portion which disappears in amphibians, reptiles, and birds, but in mammals the right anterior portion forms an **azygos** vein, the left one may form a **hemiazygos** vein which loses its connection with the precaval and joins the azygos vein by a transverse anastomosis; the posterior portions of the posterior cardinal veins which have joined the caudal vein, become the **renal portal** veins. The renal portal vein of each side forms capillaries in the kidney to form a renal portal system which is complete in fishes, amphibians, and reptiles, but becomes much reduced in birds, and is absent in mammals because the posterior portions of posterior cardinal veins disappear in mammals. 6. The renal portal veins of fishes collect blood only from the tail, but in amphibians, reptiles, and birds they also usurp the veins from the legs. 7. In some fishes and all tetrapoda the vitelline veins join the subcardinal vein to form a **postcaval** vein. In amphibians and reptiles the postcaval vein extends to the posterior ends of the kidneys, in birds it joins the renal portal veins, thus, reducing the renal portal system, in mammals the postcaval vein joins the veins from the legs and tail, so that the renal portal system is completely eliminated. 8. The two lateral abdominal veins persist as such in fishes, but from Dipnoi upwards they fuse to form an **anterior abdominal** vein which joins the hepatic portal vein near the liver, the anterior abdominal vein, thus, connects the renal portal and hepatic portal systems. An **iliac** vein is formed in each hindlimb which joins the anterior abdominal vein. The anterior abdominal vein begins to lose its importance in reptiles, in birds it is modified to form the epigastric and coccygeo-mesenteric veins, in mammals it disappears, except in *Tachyglossus*. 9. In air-breathing forms, **pulmonary veins** grow out from the left auricle and enter the lungs. 10. The umbilical (allantoic) veins of the **embryo** disappear when the lungs become functional as respiratory organs.

Tetrapoda

The venous system of amphibians very much resembles that of lungfishes except that the abdominal vein enters the **hepatic portal system** rather than the sinus venosus. Anterior cardinals persist as **internal jugular veins** in all adult tetrapods. **Inferior jugular veins** are absent. Common cardinals become the anterior vena cavae or **precavals**, which join sinus venosus in amphibians and reptiles. In birds and mammals, precavals directly enter the right auricle of heart, because sinus venosus is absent. In man, cat, etc., left precaval is lost so the blood of left side enters the right precaval through a **brachio-cephalic** branch. In lungless salamanders, the pulmonary veins are absent since inter-atrial septum is incomplete.

The **post cardinal veins** join the caudal vein posteriorly and precavals anteriorly to form the common cardinals in larval frog and *Necturus*.

In **reptiles**, there is greater development of pulmonary veins and postcaval vein and reduction in the importance of **renal portal system** which brings blood to the kidneys from the posterior part of the body.

In **birds**, there are two functional **precaval veins** formed by the union of the jugular and subclavian on each side, and a complete post caval. It receives blood from the limbs by way of the **renal portals**, which pass through kidneys, but do not break up into capillaries and are not comparable to the renal portals of lower vertebrates.

In **mammals**, there may be one or two precaval veins. There is a single postcaval vein, whose embryological development is very complicated. All the caval veins enter the right auricle directly, as the sinus venosus is absorbed into the wall of this chamber of the heart in embryonic life. There is no renal portal system although the hepatic portal system very much resembles that of other vertebrates.

LYMPHATIC SYSTEM

Development of lymph vessels. Lymph vessels or **lymphatics** arise as spaces in the mesenchyme of an embryo. They arise near the large veins but considerably later than and independently of the blood vessels. The lymph spaces unite and then branch to form anastomosing lymph vessels running throughout the body. The bordering mesenchyme cells form the flat endothelial cells lining the lymph vessels. In certain places in some vertebrates connective tissue covers plexuses of lymph capillaries to lymph nodes. Connections are made between the lymphatics and certain veins.

The **lymphatic system** differs from the blood vescular system in being an open system having lymph spaces between tissue cells, and it does not form a complete circuit. Lymphatic system includes the **lymph capillaries**, **lymph vessels** and **lymph nodes**. Since lymph flows only in one direction, that is from the tissues towards the heart, it has only capillaries and lymph vessels which are equivalent to veins. It has blind **lymph capillaries** lying between the tissue cells. They are interwoven with but independent of blood capillaries. Lymph capillaries form an anastomosing network in all the body organs except in the nervos system. The diameter of lymph capillaries is larger and less uniform than that of blood capillaries. At their commencement the closed ends of lymph capillaries have small swollen knobs which help in collecting lymph from tissues. The lymph capillaries unite to form larger lymph vessels or **lymphatics** which finally discharge their contents into veins. Thus, lymphatics are efferent vessels draining from lymph capillaries into veins.

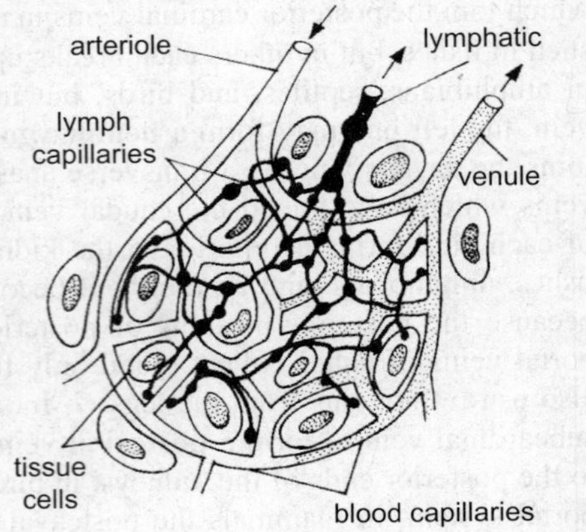

Fig. 45.9. Blood and lymph capillaries in tissue cells.

Structure. Lymphatics are made of **connective tissue fibres** with some unstriped muscles and are lined with **endothelial cells**. The largest lymphatics have three-layered walls: (*i*) a **tunica externa**, (*ii*) a **tunica media**, and (*iii*) a **tunica interna** just as in blood vessels. The lymphatics have valves arranged in pairs. The valves are more numerous than in veins, and prevent a backflow. Lymphatics in the intestine are called **lacteals** which absorb emulsified fats.

Lower vertebrates. Many lower vertebrates, except cyclostomes and cartilaginous fishes, have large **lymph sinuses**, the lymphatics communicate not only with lymph sinuses but also with the coelom by minute apertures or **stomata**. In lower vertebrates there are muscular **lymph hearts** which pump the lymph into veins. In the path of some lymphatics of higher vertebrates (mammals) are some **lymph nodes** (wrongly called lymph glands). They are masses of lymph capillaries enclosed in connective tissue and contain lymphocytes and macrophages. Lymph nodes produce lymphocytes whose function is not clear, probably they transport DNA to those body cells which are unable to form this substance. The **macrophages** are phagocytic destroying foreign bodies and bacteria. Tonsils, thymus, spleen and Peyer's patches are also lymph nodes.

Lymph. Filtration pressure causes the plasma and leucocytes of blood to escape from the blood capillaries, into tissues. Most of this fluid is absorbed by blood capillaries, but some of it passes by diffusion into lymph capillaries and is then known as **lymph**. Lymph is a colourless fluid, it is a part of the blood which has filtered out from blood capillaries. Actually it is blood minus its red blood corpuscles and plasma proteins. There is less calcium and phosphorus in lymph than in blood. Lymph has the power to coagulate slowly. Lymph fills body spaces and bathes the tissue cells where it is known as **tissue fluid**. It picks up colloids, broken tissue cells and bacteria. It is a medium of exchange between the blood and tissue cells. It carries food and oxygen to the cells and takes away water and waste substances from the cells. The lymph re-enters the blood through the lymphatics opening into veins, but some lymph enters the venous capillaries by osmosis.

Pisces. Lymph vessels are well developed in fishes. They are below the skin, in the muscles and viscera. They run along the larger veins and extend into the head, tail and fins. The lymph vessels open into veins in the anterior, middle and posterior parts of the body. **Lymph hearts** are present only in some fishes but lymph nodes are lacking.

Amphibia (Frog). In the tissues are lymph spaces from which arise lymph capillaries. The lymph capillaries unite to form lymph vessels, some of which dilate to form large **lymph sinuses**. Between the skin and muscles are spacious **subcutaneous** lymph sinuses separated from each other by fibrous membranes. Around the dorsal aorta is a **subvertebral sinus** which also encloses the kidneys. From this diffuse lymphatic system, lymph is pumped into veins by two pairs of **lymph hearts** : (*i*) an anterior pair just behind the transverse processes of the third vertebra, opening into the **subscapular veins** and (*ii*) a **posterior pair** of lymph hearts on either side at the end of the urostyle, which can be seen if the skin is removed. They open into the **femoral veins**.

Reptilia. The lymphatic system has lymph spaces in tissues from which lymph capillaries arise and form numerous lymphatics. The main lymphatic, called **subvertebral trunk** divides in front into two branches, each of which enters a precaval vein. There is only one pair of **lymph hearts** opening into the iliac veins posteriorly.

Aves. There is an extensive distribution of lymph vessels, which finally form two **thoracic ducts** (homologous with subvertebral trunk) opening into precaval veins. In the embryo, there is a pair of lymph hearts in the sacral region. They are almost always lost so that the adult has no lymph hearts.

Mammalia. The lymphatic system is highly developed having an extensive network of lymph capillaries and lymph vessels in the entire body. The larger lymphatics are interrupted at different places by **lymph nodes**.

Lymph nodes occur in the head, neck, arm-pits, groins, close to large blood vessels, and they also form the tonsils and Peyer's patches on the intestine. Lymph nodes contain anastomosing lymph capillaries, lymphocytes, reticular cells, and macrophages, which are phagocytic engulfing bacteria and foreign bodies. Thus, they are responsible for bodily protection.

The **lymph capillaries** coming from intestinal villi are called **lacteals**, which carry emulsified fats so that their lymph becomes milky-white in appearance and is called **chyle**.

The lymphatics coming from the left side of the head, neck, arm, and thoracic region open into a **thoracic duct** lying below the vertebral column. It runs in front and opens into the left external jugular vein near its junction with the subclavian vein. The lower end of the thoracic duct, below the diaphragm, is expanded to form a **receptaculum** or **cisterna chyli**. The lymphatics coming from the legs and lower trunk region, and the lacteals from the intestine pour their lymph into the **cisterna chyli**. In some mammals, but not in all, there is a **right lymphatic duct** which receives the lymphatics from the right side of the head and neck, right arm, and right side of the thorax. It opens into the right external jugular vein, where it joins the right subclavian vein.

There are no **lymph hearts** in mammals but lymph is propelled slowly through the lymph vessels and nodes by the movements of the body muscles and by the osmotic pressure in the lymph capillaries.

Spleen

A spleen is a dark red organ found in all vertebrates except cyclostomes and Dipnoi. The spleen is haemolymphatic in character and it is interposed in the blood stream and blood, not lymph, filters through it. In an embryo the spleen is formed from mesenchyme cells as a thickening in the dorsal mesentery suspending the stomach. Blood vessels arise in it and bring about a differentiation of mesenchyme cells to form splenic cells and pulp.

Structure. The spleen is covered by an outer **capsule** from which **trabeculae** pass into the substance of the spleen, dividing it into compartments or **lobules**. The capsule and trabeculae are made of dense connective tissue containing elastic fibres and unstriped muscles, they also contain blood vessels, lymph vessels and nerves. Only blood vessels pass beyond the trabeculae to enter the parenchyma of the spleen where they form capillary network of arteries and veins, and many sinusoids connecting the arteries and veins directly. The splenic lobules have splenic cells of different sizes which may be pigmented. They also have red and white pulp. The **red pulp** is made of diffuse lymphatic tissue of reticular cells and fibres, macrophages, lymphocytes and blood corpuscles. The **white pulp** consists of compact lymphatic tissue around small arteries. In section, it appears as isolated circular massses, called **splenic nodules** or corpuscles. The spleen is regarded as a haemolymph organ.

Functions of spleen. 1. The spleen functions as a **haemopoietic organ**. In an ordinary embryo it forms both erythrocytes and leucocytes. It continues to do so in the adult, except in mammals where it forms only non-granular leucocytes.

2. The red pulp acts as an efficient blood filter. The old erythrocytes coming into the spleen are destroyed by phagocytic macrophages. Their iron is sent to the bone marrow for being used again, and the haemoglobin goes to the liver to form bile pigment.

3. The spleen acts as a storehouse for blood, one-sixth of the total blood can be stored in the spleen. Large volumes of erythrocytes are stored in the splenic sinusoids. Contractions of the spleen force out the red blood corpuscles into the blood stream when there is a demand for them.

4. The spleen produces antibodies from the cells received from the thymus gland. These cells serve as a defence mechanism against bacteria and some diseases. Though the spleen is extremely important, it is not essential for life. If it is removed then its functions are taken over by other haemopoietic tissues.

IMPORTANT QUESTIONS

Long Answer Questions

1. Describe the evolution of heart in vertebrates.
2. Give a comparative account of hearts in fishes, amphibians, reptiles, birds and mammals.
3. Give a comparative account of hearts and aortic arches in reptiles, birds and mammals.
4. Describe the aortic arches of vertebrates and discuss their modifications in different groups of vertebrates.

Short Answer Questions

1. Write short notes on the following :
(*i*) Branchial or venous heart; (*ii*) Truncus arteriosus; (*iii*) Ductus Botalli; (*iv*) Ductus caroticus; (*v*) Foramen of Pannizzae; (*vi*) Renal portal system; (*vii*) Hepatic portal system.

Multiple Choice Questions

1. In vertebrates, circulatory system does not
(*a*) Transport nutrients
(*b*) Transport excretory products
(*c*) Transport impulses
(*d*) Transport hormones.
2. Smallest arteries are connected to smallest veins by
(*a*) Capillaries (*b*) Arterioles
(*c*) Venules (*d*) Muscles.
3. In urodeles, conus is replaced by
(*a*) Aorta (*b*) Vena cava
(*c*) Foramen ovale
(*d*) Bulbus arteriosus.
4. The opening of coronary sinus is guarded by
(*a*) Mitral valve (*b*) Tricuspid valve
(*c*) Semilunar valve (*d*) Besian valve.
5. The union of sinus venosus with right auricle in mammals is marked externally by
(*a*) Sulcus terminalis
(*b*) Sinus venarum
(*c*) Crista terminalis
(*d*) Appendix auriculae.
6. In man, renal portal system is
(*a*) Well developed
(*b*) Poorly developed
(*c*) Primitive than that of frog
(*d*) Absent.
7. Biliverdin and bilirubin are the bile pigments but are the waste products formed by breaking down of
(*a*) WBCs (*b*) Platelets
(*c*) Haemoglobin (*d*) Thrombin.

ANSWERS

1. (*d*) 2. (*a*) 3. (*d*) 4. (*d*) 5. (*a*) 6. (*a*) 7. (*c*)

46 Chapter

Nervous System in Vertebrates

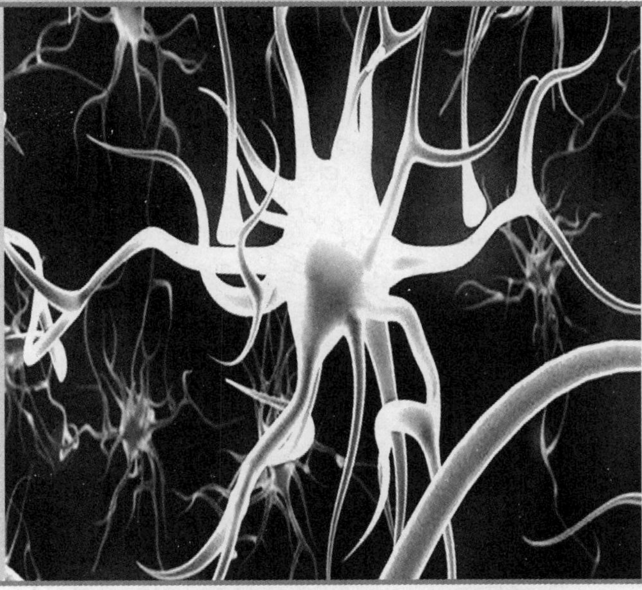

Contents of this Chapter

NERVOUS SYSTEM AND ITS FUNCTIONS

The nervous system is meant to perceive stumuli detected by the receptors and transmit them to the concerning parts of the body and finally respond to responses through effectors. The nervous system is highly specialised in vertebrates and plays at least three vital roles.

1. Response to stimuli. Responding to all sorts of stimuli, it acquaints the organism with them so that the organism may react and orient itself favourably in the surrounding environment.

2. Coordination. In association with endocrine system, the nervous system also serves to coordinate and integrate the activities of various parts of the body so that they act harmoniously as a unit. This makes possible the integrated control of the internal body environment (**homeostasis**). However, the nervous system brings about rapid coordination by means of nerves, whereas the endocrine system does so gradually and slowly by secreting hormones into blood.

3. Learning. In higher vertebrates at least, the nervous system serves as a centre for learning by accumulating memories from past experiences.

ANATOMY OF NERVOUS SYSTEM : THE NEURON

The nervous system is formed of nerve cells, **neurons**, which are surrounded by **neuroglia**, a tender network of connective tissue. Neurons are the distinct units of nervous tissue and are of ectodermal origin.

Structure of neuron. A neuron has a **cell body** or **cyton** from which arise two types of **fibres**.

***(i)* Cyton.** A cyton has a nucleus and numerous basophilic **Nissl's granules**. These granules are made of ribonucleic acid. They produce new cytoplasm in neurons and are also concerned with protein synthesis in nerve cells. The cytoplasm of cyton has fine threads or **neurofibrils**, forming a network. Mass of nerve cell bodies found in the brain or spinal cord is called nucleus and a similar mass found outside the central nervous system is known as a **ganglion**.

***(ii)* Nerve fibres.** A neuron has a number of processes or fibres projecting from the cell body. These fibres on the basis of movement of nervous impulse are of two types : (*a*) **Dendrites.** These are generally several short branching processes having Nissl's granule, and close to the cell body. They are **afferent** because the neuron receives impulses through them. (*b*) **Axon.** Axon or **axis cylinder** is a single process which is generally long and without Nissl granules. It has terminal branches, called **axon endings**, which have small swellings at their ends. Axons are **efferent** because impulses are sent out through them.

Dendrites lie in contact either with receptors or with the axon endings of another neuron, from either of which they receive stimuli. Neurofibrils found in the cytoplasm of cyton extend both into the dendrites and axons and their branches. They probably transmit impulses.

Nerve fibre. An axon is covered with one or two sheaths and is then called a **nerve fibre**. A nerve fibre has a central thin cytoplasmic strand, the **axis cylinder**. It is continuous with the cell body or cyton. The nerve

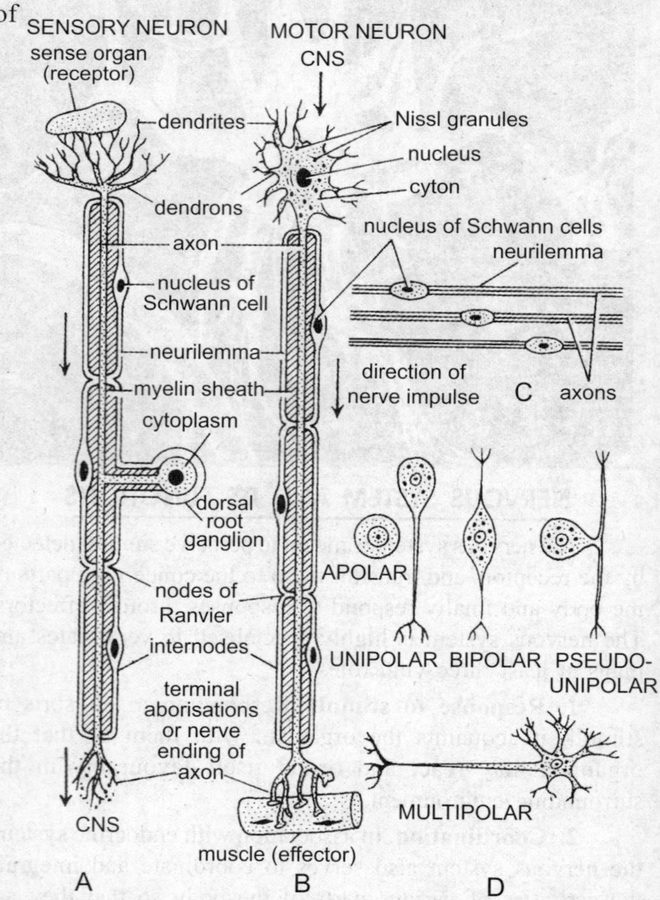

Fig. 46.1. Structure of neurons and nerve fibres. A–Sensory neuron; B–Motor neuron; C–Non-medullated nerve fibres; D–Kinds of neurons.

fibres outside the brain and spinal cord are covered by a thin continuous sheath, known as a **neurilemma** which is a tough membrane of protein and is made of flat sheath cells or **Schwann cells**. Each sheath cell has a thin layer of cytoplasm with a flat nucleus. In nerve fibres there is a second sheath, called a **myelin sheath** or **medullary sheath** between the axis cylinder and its neurilemma. Myelin is a shining white substance made of the bodies of sheath cells wrapped around the axon many times. It has alternate layers of proteins and fats. The myelin sheath, unlike neurilemma, is interrupted at intervals by circular constrictions, called **nodes of Ranvier**. The part of a nerve fibre between two nodes of Ranvier is called an **internode**, a single Schwann cell covers an internode. Nerve fibres having a thick layer of a myelin sheath are called **myelinated** or **medullated fibres** and they appear white in colour. While the nerve fibres having a thin layer of a medullary sheath are **non-medullated fibres** and they appear gray in colour. Medullated fibres conduct impulses much faster than non-medullated ones.

Collateral branches may be given off by the axon at right angles. Nerve fibres lying in bundles are known as **fibre tracts** when present in the central nervous system, but when the bundles of nerve fibres are found outside the brain or spinal cord, then they are called **nerves**.

Synapse. Nerve impulses are relayed from the axon endings of one neuron to the dendrites of another, they are arranged in such a way that the axon endings of a neuron rest on the dendrites or even on the cyton of the next neuron. This linking of neurons is called a **synapse** or **synapsis**. The electron microscope has shown that each axon terminates in a tiny swelling, called a **terminal button**, which is full of mitochondria. They are very close to each other but have no organic connection with the terminal branches of another neuron. This very small gap between these two opposite neurons or terminal branches of axon of one neuron and dendrites of another neuron is called the **synapse**. Thus, the whole nervous system forms a chain of neurons with a synapse.

Neuroglia. Besides the neurons in the brain and spinal cord, other cells of ectodermal origin are called **neuroglia**. They are also present as packing cells between the neurons. They have branching processes and they take the place of connective tissue, but true connective tissue is not present. Lining the cavities of the brain and spinal cord are some **ependymal cells**, which are like epithelial cells and are often ciliated.

In the brain and spinal cord the nerve cell bodies, their dendrites, neuroglia cells and the proximal unmyelinated portions of axons form a **gray matter**, which appears gray in colour, while the bundles of myelinated nerve fibres form the **white matter**.

Divisions of Nervous System

The vertebrate nervous system has three divisions : (*i*) A **central nervous system** comprising the brain and spinal cord. Its function is to receive the stimulus from the receptors and transmit its response to the effectors. Thus, it coordinates all the functions of the body. (*ii*) A **peripheral nervous system** consisting of cranial and spinal nerves arising from the brain and spinal cord respectively. It forms a connecting link between the receptors, central nervous system (CNS) and effectors. (*iii*) An **autonomic nervous system** made of two ganglionated sympathetic nerves, ganglia in the head and viscera, and their

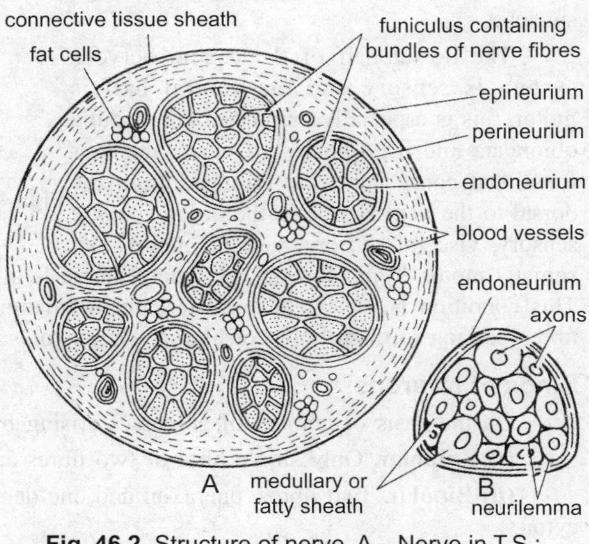

Fig. 46.2. Structure of nerve. A – Nerve in T.S.; B – A bundle of nerve fibres in T.S.

connecting nerves. The autonomic nervous system is often regarded as a part of the peripheral nervous system because the two are connected. But all the three divisions of the nervous system are connected intimately both structurally and functionally.

Nerve. A nerve is formed of a number of fibres (axons or dendrites) outside the central nervous system. These are bound together in a smaller bundle by white connective sheaths, called **perineurium**. A number of such smaller bundles of nerve fibres are externally surrounded by another coat of fibrous connective tissue, called **epineurium**, which has blood vessels.

Functional Divisions of the Nervous System

The body functions of the nervous system are somatic or visceral. **Somatic functions** are those which are carried out by the voluntary muscles, skeleton, and the skin and its derivatives. **Visceral functions** are those which are performed by the digestive, circulatory, excretory, endocrine, and the urinogenital system. Both somatic and visceral functions are divided into two components each : (*i*) **afferent** or **sensory**, and (*ii*) **efferent** or **motor**. Thus, the nervous system has four functional divisions :

1. Somatic sensory. The somatic sensory fibres carry impulses from the skin, sense organs, voluntary muscles, and joints to the central nervous system.

2. Somatic motor. The somatic motor fibres carry impulses from the central nervous system to the voluntary muscles.

3. Visceral sensory. The visceral sensory component carries sensations from the viscera and organs of taste and smell to the central nervous system.

4. Visceral motor. The visceral motor component carries impulses from the central nervous system to the involuntary muscles, branchial muscles, and glands.

Association or adjustor neurons. These are located in the brain and spinal cord. They link the afferent and efferent fibres through synapses.

The dorsal half of the central nervous system is sensory and the ventral half is motor, this is especially marked in the medulla oblongata and the spinal cord where the four functional components are arranged from the dorsal to the ventral side in the order, somatic sensory, visceral sensory, visceral motor, and somatic motor. Typically each peripheral nerve is also made of all four functional components. This condition is typical of amniotes, but in anamniotes the visceral motor component has its fibres passing through both dorsal and ventral roots (Fig. 46.3).

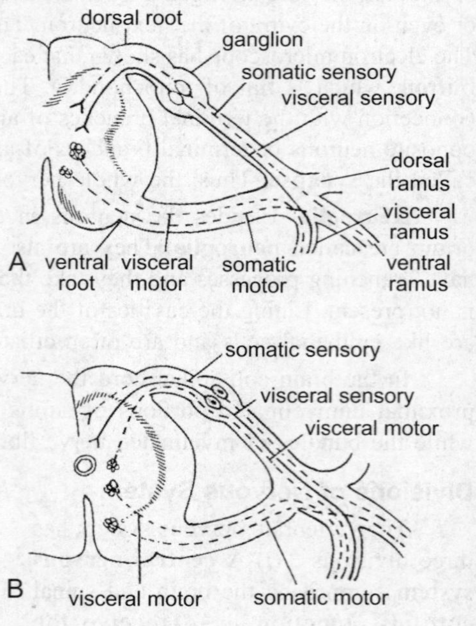

Fig. 46.3. Functional divisions of the nervous system; A – In amniota; B – In anamniota.

Kinds of Neurons

On the basis of number of processes arising from a neuron, these are of 3 types :

***(i)* Unipolar.** Only single fibre or two fibres arising very closely from the cyton.

***(ii)* Bipolar.** Two fibres, one axon and one dendrite, arising from the opposite sides of the cyton.

(*iii*) **Multipolar.** A number of processes, *i.e.,* one axon and a number of dendrites arise from the cyton.

Neurons are the longest cells in the body. Axons of the cytons may reach every part of the body, near or distant.

Nerve impulse. Impulses do not travel across synapses but a fresh impulse is set up on the other side of a synapse. When an impulse reaches the terminal buttons, they give off a small amount of **acetylcholine**, while terminal buttons of postganglionic fibres of the autonomic nervous system give off **sympathin**. On being released the acetylcholine or sympathin contact the dendrites of the next neuron in which they cause a fresh impulse. After causing an impulse the acetylcholine or sympathin are immediately destroyed by enzymes. The impulses travel in only one direction in neurons, from the dendrites to the cell body and then through the axon.

Development of the Central Nervous System

In a vertebrate embryo the mid-dorsal strip of ectoderm form a longitudinal thickening, known as a **medullary** or **neural plate**. The neural plate sinks downwards but its edges grow upwards to form **neural folds**. The neural folds grow towards each other and eventually fuse to form a hollow **neural tube** of several layers of cells above the notochord. The lateral ectoderm re-forms above the neural tube. The cells of the neural tube produce two kinds of cells, the **neuroblasts** and **spongioblasts**. The neuroblasts develop into neurons, while the spongioblasts form the neuroglia and ependymal cells lining the neural canal. The neural tube at first is open at both ends, having a **neuropore** in front communicating with the exterior, posteriorly the neural tube communicates with the archenteron cavity by a very short **neurenteric canal**, but both openings are soon closed. The anterior end of the neural tube enlarges forming the brain, while the remaining neural tube will give rise to the spinal cord. The cavity of the neural tube will become the ventricles of the brain and central canal of the spinal cord.

Development of brain. The anterior thickened, enlarged end of the neural tube known as **encephalon** is the embryonic brain. It undergoes differential growth and acquires two constrictions, which divide it into a series of three lobes, the **primary cerebral vesicles**. The primary cerebral vesicles are a **forebrain** or prosencephalon, a **midbrain** or mesencephalon, and a **hindbrain** or rhombencephalon. This embryonic brain is bent downwards between the forebrain and midbrain in the region of the mesencephalon and this curvature is known as **cranial flexure** or cephalic flexure. In later development the cranial flexure is lost in cyclostomes, fishes, amphibians and reptiles, but persists in birds and mammals.

1. Forebrain. The forebrain or **prosencephalon** is further sub-divided by a constriction into an anterior **telencephalon** and a posterior **diencephalon**. The anterior wall of the telencephalon is known as **lamina terminalis** which marks the original anterior boundary of the brain. The telencephalon grows beyond the lamina terminalis and forms paired **cerebral hemispheres**. The roof of the cerebral hemispheres becomes the **pallium** and its ventro-lateral thicker walls form the **corpora striata**. Each cerebral hemisphere has a cavity, the **lateral ventricle**. From the antero-ventral part of the telencephalon paired **olfactory lobes** (rhinencephalon)

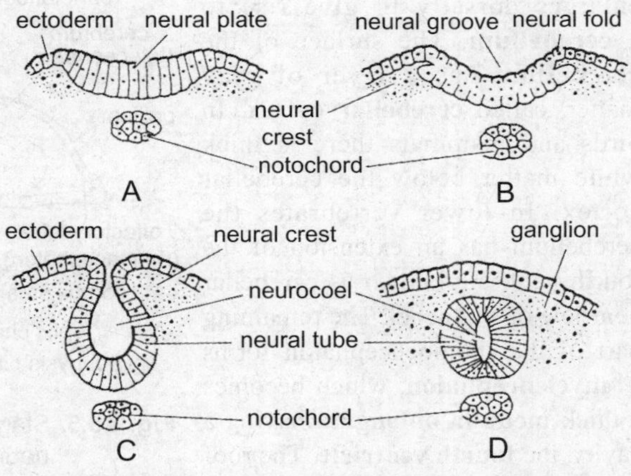

Fig. 46.4. Stages in the embryonic development of central nervous system in T.S.

grow out towards the nose. Each olfactory lobe has a cavity, the **rhinocoel**, which communicates with the lateral ventricle of its side. The **diencephalon** or thalamencephalon has an upper **epithalamus**, a middle **thalamus**, and a lower **hypothalamus**. Dorsally the diencephalon gives out a median unpaired **paraphysis**, which is well developed in fishes but is much reduced in reptiles and birds and is absent in mammals. Behind the paraphysis the diencephalon gives out two more unpaired evaginations, an anterior **parietal body** and a posterior **pineal body** or **epiphysis**. The parietal body forms an eye in lampreys, some fishes, and many reptiles, in others it is lost in the adult. The pineal body forms a simplified eye in lampreys but in all other vertebrates it becomes glandular and is believed to be an endocrine gland, but there is no real evidence for this, most probably the pineal body is a vestigial organ of no functional significance. Between the thalami of diencephalon is a **third ventricle** which communicates with each lateral ventricle by an aperture called a **foramen of Monro**. The thalami give rise to optic vesicles which form parts of the eyes. The roof of the diencephalon is a thin ependymal layer which fuses with the vascular piamater to form a **tela choroidea**. Vascular folds of the tela choroidea project into the third ventricle forming an **anterior choroid plexus**. The lower part of the diencephalon known as hypothalamus forms a ventral **infundibulum**. An ectodermal evagination of the roof of the stomodaeum forms a **Rathke's pouch** or **hypophysis**, which fuses with the infundibulum to form a **pituitary body** or **hypophysis cerebri**, which loses its nervous character and becomes an important endocrine gland. Except in a few lower forms the Rathke's pouch loses its connection with the stomodaeum. The fibres of the optic nerves cross in front of the infundibulum to form an **optic chiasma**.

2. Midbrain. The midbrain or **mesencephalon** has a thick roof. Its dorso-lateral walls form two **optic lobes** (four in mammals). The optic lobes are hollow, each having an **optocoel** except in mammals where they have no cavity and are solid. The ventral wall of mesencephalon forms thick **crura cerebri**, which are tracts of nerve fibres joining the diencephalon with the hindbrain. Passing through the mesencephalon from the third to the fourth ventricle is a narrow **iter** or **aqueduct of Sylvius**, which in non-mammals is also connected to the optocoels.

3. Hindbrain. The hindbrain or **rhombencephalon** forms a **metencephalon** from its anterior dorsal part. The metencephalon enlarges dorsally to give rise to a **cerebellum**. The surface of the cerebellum has a layer of gray matter, called **cerebellar cortex**. In birds and mammals there is thick white matter below the cerebellar cortex. In lower vertebrates the cerebellum has an extension of the fourth ventricle known as cerebellar ventricle or **metacoel**. The remaining part of the rhombencephalon forms a **myelencephalon**, which becomes a thick **medulla oblongata** having a cavity, the **fourth ventricle**. The roof of the medulla has a **posterior choroid plexus** formed in the same way as the

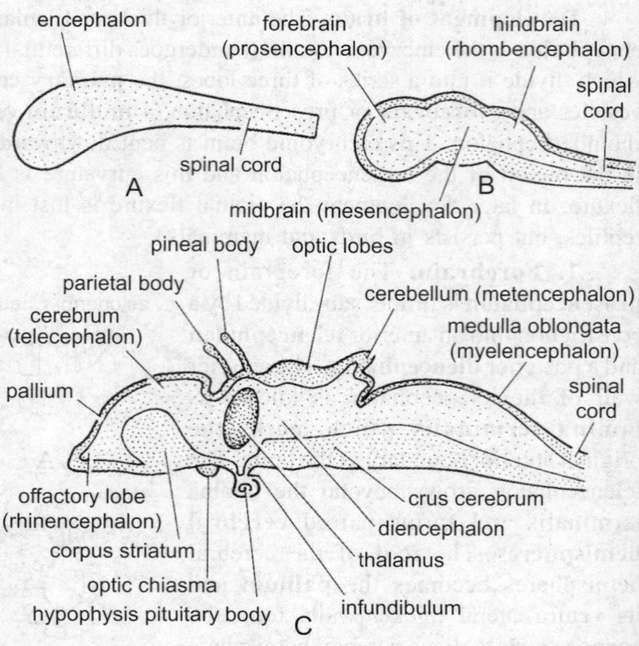

Fig. 46.5. Stages in development of brain. A–Anterior end of neural tube in lateral view; B–M.L.S. of embryonic brain to show three primary cerebral vesicles; C–Differentiation of brain from three vesicles.

anterior choroid plexus. In lower vertebrates the floor of the metencephalon and myelencephalon does not differ, but in higher forms the floor of the metencephalon thicknes due to many tracts of nerve fibres, and in mammals it forms a conspicuous **pons Varolii**.

Commissures. The two similar halves of the brain are connected by transverse tracts of nerve fibres, called **commissures**. In most vertebrates there are only anterior and posterior commissures, the **anterior commissure** originates in the lamina terminalis and lies in front of the third ventricle, it joins the two corpora striata. The **posterior commissure** lies in the roof of the diencephalon at its junction with the midbrain. In mammals there are several other commissures. The commissures make bilateral integration possible.

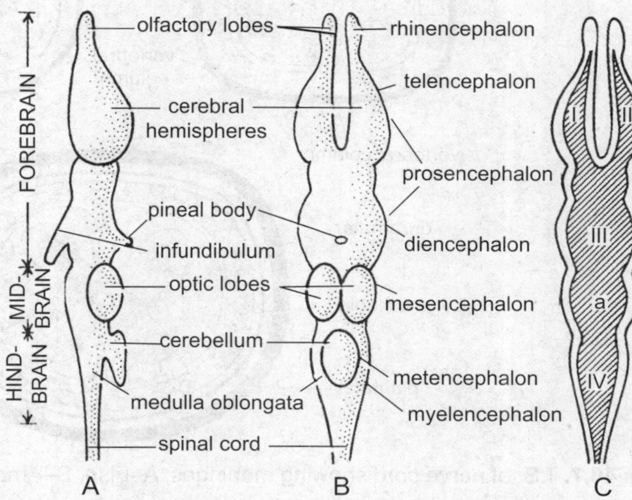

Fig. 46.6. Pattern of generalised vertebrate brain. A–Lateral view; B–Dorsal surface; C–H.L.S. showing ventricles.

Spinal cord. Nerve cord or spinal cord is formed from the neural tube behind the brain. The nerve cord is a cylindrical tube somewhat flattened dorso-ventrally. Its anterior end is wide where it is continuous with medulla, the posterior end generally tapers to a fine thread, the **filum terminale**. In fishes it extends to the posterior end of the tail. It extends the full length of the vertebral column in amphibians, reptiles and birds, but in mammals it is short and does not extend into the tail. In cyclostomes, fishes and limbless amphibians the nerve cord has a uniform diameter, but in tetrapoda it has two enlargements called **cervical (brachial)** and **lumbar** enlargements. They are formed due to a larger number of nerve cell bodies whose fibres form nerves going to limbs.

Histology. The nerve cord has a thin **central canal** lined with ciliated ependymal cells. On its mid-ventral side the nerve cord acquires a deep **ventral fissure**, mid-dorsally is a slight dorsal **sulcus** from which a **dorsal septum** extends into the interior. The dorsal septum and ventral fissure divide the nerve cord incompletely into two halves connected across the middle by commissures. There is an outer **white matter** with neurogila and nerve fibres, and an inner **gray matter** having nerve cell bodies, non-medullated fibres, and neuroglia. In **fishes** and **amphibians** the gray matter is rounded or quadrangular in section, but in **amniotes** it is H-shaped forming dorsal and ventral horns or columns. The gray matter divides the white matter into dorsal, lateral, and ventral **funiculi**.

Meninges. The brain and nerve cords are surrounded by protective connective tissue membranes called meninges. Their complexity increases with the advance in the scale of evolution.

In **cyclostomes** and **fishes** there is generally a single meninx, the **meninx primitiva** around the brain and spinal cord, it has a network of blood vessels. Between the skull or vertebral column and the meninx primitiva is a **perimeningeal space** filled with fat and connective tissue fibres.

In **amphibians**, **reptiles** and **birds** the meninx primitiva splits to form two meninges : an outer tough fibrous **duramater** and a thin vascular **piamater** covering the brain and nerve cord, between these two meninges is a **subdural space** having a small amount of fluid. Between the enclosing skeleton and the duramater is an **epidural space** having connective tissue, fat, and blood vessels.

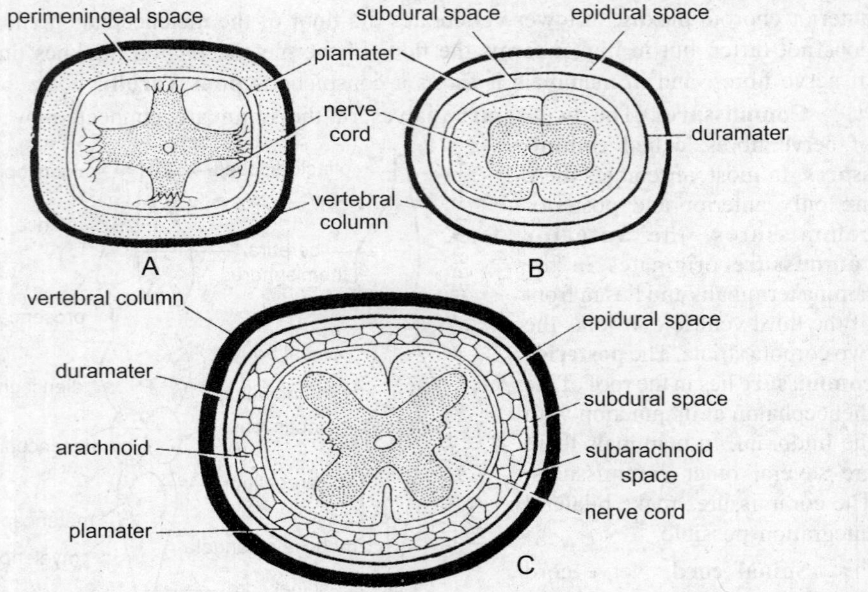

Fig. 46.7. T.S. of nerve cord showing meninges. A–Fish; B–Amphibian, reptile, and bird, C–Mammal.

In **mammals** there are three meninges, an outer duramater, a middle non-vascular **arachnoid membrane**, and an inner highly vascular **piamater**. Between the enclosing skeleton and the duramater is an epidural space in the region of the vertebral column only, but in the brain region the duramater fuses with the lining of the skull (endorachis) so that there is no epidural space. Between the duramater and the arachnoid is a subdural space, between the arachnoid and piamater is a **subarachnoid space** having a network of fine fibres, filled with **cerebrospinal fluid**. The piamater and the arachnoid being joined by fibres are together known as **leptomeninges**. The arachnoid forms **arachnoid villi** which return the cerebrospinal fluid to the venous blood. The cerebrospinal fluid is formed by the choroid plexuses, which fills the ventricles, central canal, subdural space, and subarachnoid space.

In the roof of the medulla are three apertures, a median **formen of Magendie**, and two lateral **foramina of Luschka**, through these foramina the cerebrospinal fluid passes into the spaces between the meninges so that the surface of the central nervous system is bathed by the fluid.

Neural crest. After the formation of the neural tube, masses of ectoderm cells appear on each side between the neural tube and ectoderm, these cells are **neural crests** which extend on each side along the whole length of the neural tube. The neural crests form segmental **dorsal root ganglia** of spinal nerves and ganglia of the autonomic sympathetic nerves. They also form parts of the ganglia of sensory cranial nerves. Some neural crest cells form those mesenchyme cells which give rise to **visceral arches**. While other neural crest cells form the **piamater** and **arachnoid** membrane of the central nervous system. Still other neural crest cells form **chromatophores** in lower vertebrates. **Chromaffin cells** of the medulla of adrenal glands arise from neural crest cells.

COMPARATIVE ACCOUNT OF BRAIN IN VERTEBRATES

1. Cyclostomata. The cerebral hemispheres are barely indicated. Two olfactory lobes are prominent. Cavities of cerebral hemispheres (lateral ventricles) are rudimentary. From the dorsal wall of diencephalon two median eye-like structures may arise. The anterior one is the parietal body and the posterior one is the pineal body. These two are vestigial in *Eptatretus* and absent in *Myxine*. The vental part of diencephalon gives rise to the infundibulum of the hypophysis as well as the posterior lobe. The optic lobes are not well differentiated.

Cerebellum is poorly developed due to sluggish and slow moving habit. Medulla oblongata is very well developed in all vertebrates since it serves as a centre for many vital body activities (respiration, heart action, metabolism, etc.). It also serves as a relay centre and in fishes is a centre for lateral line system and inner ear.

2. Elasmobranchii. The brain is large and different from that of cyclostomes and bony fishes. In fishes, telencephalon is primarily olfactory in function and olfactory lobes are large in elasmobranchs. Each olfactory lobe has a rhinocoel cavity. The cerebral hemispheres (cerebrum) are better developed. Cerebrum in fishes, however, consists of a basal ganglionic mass known as **corpus striatum** and a thin, dorsal, epithelial layer called the **pallium** composed essentially of non-nervous tissue and the centre of brain activity is back in the mesencephalon. The **diencephalon** is narrow with a thin roof having an anterior choroid plexus below which is the third ventricle. Its lateral walls form thick thalami which serves primarily as a relay centre for olfactory and visual impulses and is associated with several sensory and glandular structures. Dorsally the diencephalon has a long-tubular **pineal body** and the floor of the diencephalon give rise to the stalk or infundibulum of the hypophysis as well as the posterior lobe. A thin-walled vascular sensory organ, **succus vasculosus** is attached to hypophysis. It is connected with the cerebellum by fibre tracts. The midbrain in fishes is the centre of nervous coordination. Dorsally it develops two prominences known as the optic lobes concerned with visual reception. Cerebellum is a centre of muscular coordination, it is larger than that of cyclostomes and dipnoans. Medulla is well developed and at its antero-lateral margins are present ruffle-like **restiform bodies,** which assist the cerebellum in the maintenance of equilibrium.

TABLE 46.1. SUBDIVISIOS, PARTS AND ASSOCIATED STRUCTURES OF A VERTEBRATE BRAIN.

Divisions	Subdivisions	Parts	Cavity	Associated Structure
I. Prosencephalon (Forebrain)	**1. Telencephalon**	Rhinencephalon	I Ventricle (*Rhinocoel*)	Olfactory bulbs Olfactory tracts Olfactory lobes Palaeocortex on pallium
		Cerebral hemispheres	II or Lateral Ventricles (*Paracoels*) ↓ Foramen of Monro ↓	Corpora striata or basal ganglia Corpus callosum Neocortex on pallium Paraphysis
	2. Diencephalon	Epithalamus (roof)	III Ventricle (*Diacoel*)	Habenulae Pineal apparatus Parapineal or parietal
		Thalamus (sides)		
		Hypothalamus (floor)		Hypothalmic nuclei Optic chiasma Median eminence Infundibular stalk Pituitary Saccus vasculosus Mamillary bodies Anterior choroid plexus
II. Mesencephalon (Midbrain)	—	Crura cerebri (floor)	*Iter* or *cerebral aqueduct*	Optic lobes } Auditory lobes } Tectum Cerebral peduncles
III. Rhombencephalon (Hindbrain)	**1. Metencephalon** **2. Myelecephalon**	Cerebellum Medulla oblongata	IV Ventricle (*Metacoel*)	Trapezoid body Pons Restiform bodies Pyramids

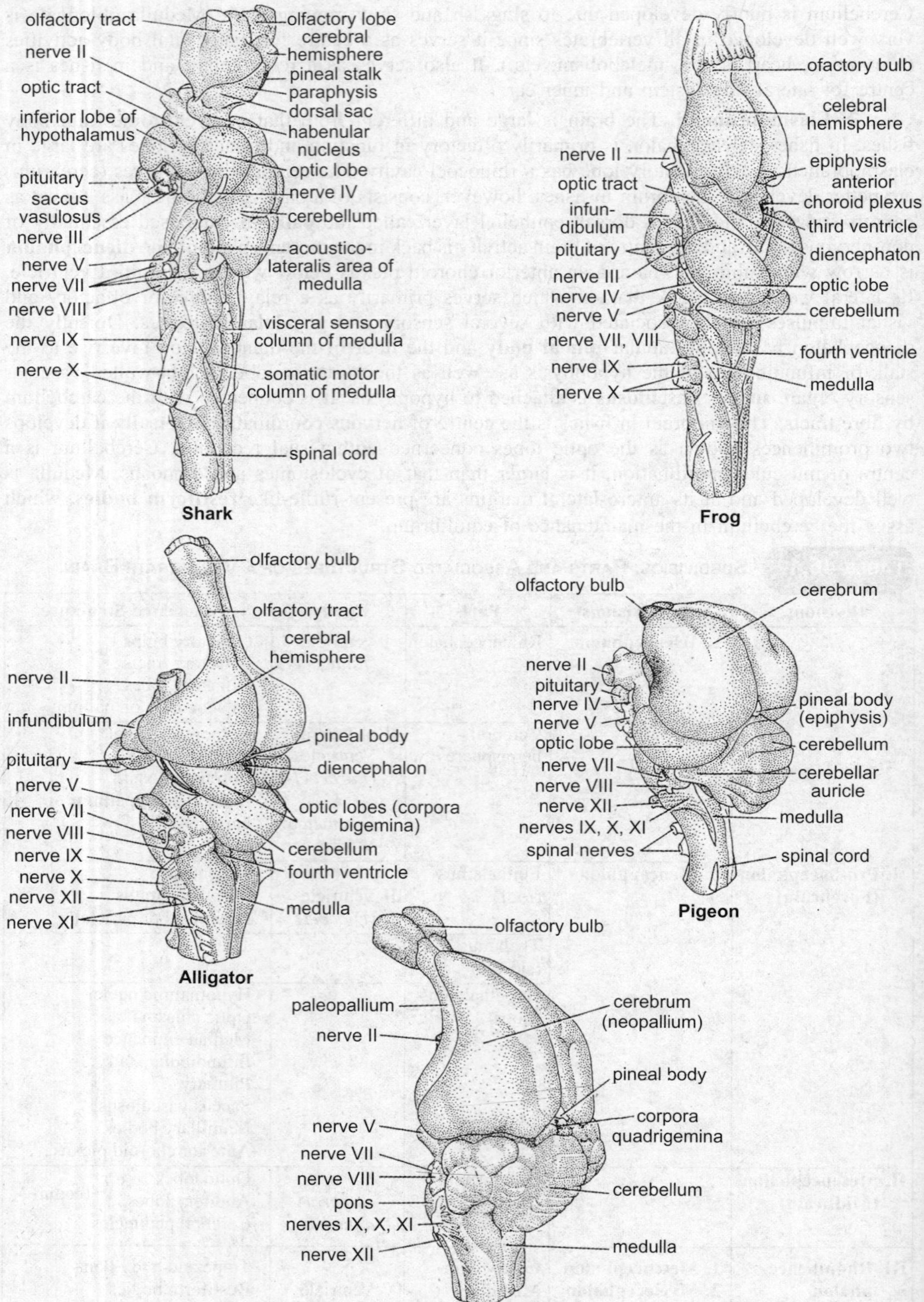

Fig. 46.8. Comparative anatomy of the vertebrate brain in lateral view.

3. Osteichthyes (Bony fish). The brain is more specialised than in elasmobranchs. Olfactory lobes are large, having no olfactory peduncles and are closely applied in front of the cerebral hemispheres. Each contains a rhinocoel continuous with the lateral ventricle. The cerebral hemispheres are better developed. The roof or pallium of the cerebral hemispheres is thin and non-nervous, consequently it is not equivalent to the pallium of other animals. The corpus striatum is very thick. The lateral ventricle is undivided. Diencecephalon is small hidden dorsally by the midbrain. Only the pineal body with stalk is present. Parapineal body is not found in modern bony fishes. Melatonin has been identified in the pineal gland of certain fishes. It has an inhibitory effect on gonadial activity. In immature Pacific salmon (genus *Oncorhynchus*) melatonin content is 6 times more than that of adults. On the ventral side the diencephalon has an infundibulum to which the pituitary gland is attached below. Optic lobes are very large. Cerebellum is larger than that of elasmobranchs. **Restiform bodies** are also present in some bony fishes. **Vagal lobes** are the bulgings from the antero-lateral sides of medulla, and are found in bottom feeders which have taste buds on their body surface.

4. Amphibia. The nervous system of amphibians is still basically like that of fishes. The pallium of cerebral hemispheres is invaded for the first time in vertebrates by nerve cells due to which the cerebral hemispheres become enlarged in accordance with more complex activities of locomotion, hibernation, breeding, etc. The olfactory lobes are fused together in the median plane. The corpora striata form a thick floor of the cerebrum. The **diencephlaon** is short but optic fibres end in it. The pineal body is found in all amphibians, but only anurans possess a parietal body or a pineal end organ. The optic lobes are large and are the centre of brain activity like fishes. The members of this class are notably slow and sluggish in their movement, so the cerebellum is very small, like a simple narrow transverse band. Medulla is also small. Except in caecilians, flexures are lacking in amphibian brain.

5. Reptilia. The brain shows many features which are characteristic of the mammalian brain. In all anamniotes the midbrain is the centre of brain activity, but in reptiles for the first time there is a shift in the nerve centre to the cerebrum. This is correlated with a marked increase in the size of the cerebral hemisphere as a result of invasion of the pallium by many nerve cells so as to form the **neopallium**. Two long olfactory lobes are larger than that of amphibians due to greater thickness and enlargement of corpora striata. **Parietal eye** is found in *Sphenodon* and some modern lizards, but absent or vestigial in some other reptiles. Behind the optic lobes are present a pair of **auditory lobes** which are not hollow. The third ventricle is reduced and form a narrow **cerebral aqueduct**. Cerebellum is relatively larger than that of amphibians. However, it does not attain the size of sharks, and not near the size of birds or mammals. This is corelated with relatively limited locomotive powers of most reptiles.

6. Aves. The brain of birds shows a considerable advance over that of reptiles. In shape it is short, broad, and rounded. The olfactory lobes are extremely small and degenerate due to poor sense of smell. The **cerebral hemispheres** are very large and covers the diencephalon and optic lobes. Its size is due to enlargement of the corpus striatum rather than the cerebral cortex. The cerebral cortex is smooth. The size of corpora striata makes the lateral ventricles small. The **diencephalon** is covered dorsally by the cerebral hemispheres and cerebellum. It has a small dorsal pineal body, an anterior choroid plexus and a narrow third ventricle. The **thalami** are highly developed and are connected by fibres to the corpora striata and spinal cord.

The **midbrain** is highly developed with exceptionally large rounded optic lobes. This seems to be correlated with the remarkable powers of sight that birds possess.

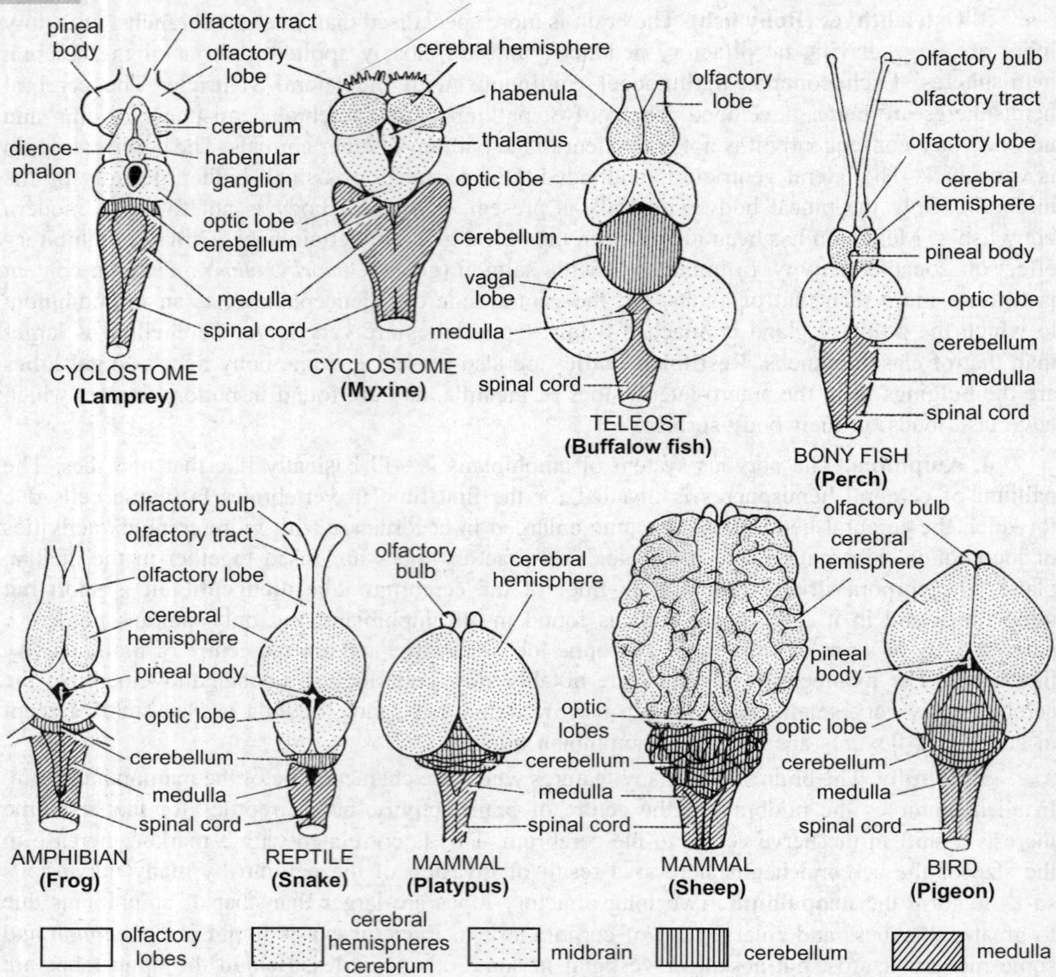

Fig. 46.9. Comparative anatomy of brains of representative vertebrates in dorsal view.

The **cerebellum** is larger than in reptiles and is deeply fissured, although it is not as large as in mammals. Ventral to cerebellum, the avian brain shows the beginning of the development of a pons. In the cerebellum masses of nerve cell bodies form nuclei so that it has outer layers of gray matter and an inner mass of white matter, forming a branching pattern called **arbor vitae**. A small projection of the fourth ventricle extends into the cerebellum. It controls equilibrium and movements in all places during flight. The medulla oblongata has a posterior dorsal choroid plexus and a fourth ventricle, which does not extend into the cerebellum.

7. Mammalia. The brain reaches its highest development in mammals with better integration and mastery over the environment. The cerebral hemispheres reaching the status of a dominant integrating part of the brain and acting as coordinating centres of the brain. The cerebral hemispheres in Prototheria are smaller and smooth like that of reptiles. In Metatheria, they are larger and smooth. In most eutherians, the cerebral hemispheres are immense, projecting forwards above the olfactory lobes and backwards above the diencephalon and midbrain and divided into lobes. They have convolutions on the surface so that there are ridges or **gyri** and depressions or **sulci**. The outer layer or cortex of the cerebrum is composed of gray matter. The right and left cerebral hemispheres are connected with one another by a broad white commissure, called the **corpus callosum**. The olfactory lobes are small compared with those of lower vertebrates.

The diencephalon consists of a dorsal **epithalamus,** a lateral **thalamus**, and a ventral **hypothalamus**. A pineal gland is present on the roof of the diencephalon, but it shows no eye-like structure. Thalamus is an important relay centre. Hypothalamus is very important in mammals and consists of four parts. These are the **infundibulum** forming the stalk and posterior lobe of the pituitary, the **optic chiasma**, where right and left optic nerves cross enroute to the brain; and the **mammillary bodies**, which integrate olfactory impulses. The hypothalamus controls a great many mammalian functions including blood pressure, sleep, water content, fat and carbohydrate metabolism, body temperature and possibly rhythmic activities such as moulting, migration and pituitary secretion.

The **midbrain** in mammals is of less importance than in lower vertebrates, in which it is really the brain centre. It is divided into four prominences, called the **corpora quadrigemina**. The two superior lobes are concerned with sight, whereas the two inferior lobes are probably concerned with hearing. The III ventricle or **iter** of midbrain is laterally compressed vertical passage called the **cerebral aqueduct**.

The **cerebellum**, which is the centre of control of body movement, is most highly developed in mammals. Its surface is convoluted and it is divided into a number of lobes : central lobe or **vermis,** two lateral lobes, and two outer floccular lobes. Beneath the cerebellum is a typical mammalian structure the **pons**. This relay centre is a conspicuous feature of the ventral metencephalon. **Pons Varolii** have crossing or decusating fibres connecting opposite sides of the cerebellar cortex. These are the transverse nerve fibres.

Besides pons, there is a short tract of transverse fibres, called **corpus trapezoideum** relaying impulses for sound.

The **medulla oblongata** lies ventrally and is much thickened. It has a dorsal posterior choroid plexus below which is the fourth ventricle joined in front to the iter and behind to the central canal. The medulla has centres for regulation of digestion, respiration and circulation.

PERIPHERAL NERVOUS SYSTEM

The peripheral nervous system consists of nerves connected to or arising from the central nervous system, it has cranial and spinal nerves. All the nerves arise in pairs. **Cranial nerves** originate from brain and emerge through foramina of the skull. Except the first four pairs of cranial nerves, the rest arise from the medulla oblongata. There are ten pairs of cranial nerves in anamniota (cyclostomes, fishes and amphibians), and twelve pairs in amniota (reptiles, birds and mammals). There is a paired **nervus terminalis** or number zero nerve arising from the cerebral hemisphere in all vertebrates except birds, it goes to the organ of Jacobson.

The cranial nerves do not show a clear metameric arrangement, yet they represent a regular series of segmental dorsal and ventral roots of the segments of the head, but the dorsal and ventral roots do not join. Table 46.2 summarises the cranial nerves of vertebrates showing their serial numbers, names, origin from brain, distribution, nature and functions. The head segments and their dorsal and ventral roots are given below.

Head segment	Dorsal root	Ventral root
1. Premandibular	Ophthalmic profundus of V	Oculomotor III
2. Mandibular	Ophthalmic superficialis, maxillary and mandibular of V	Trochlear VI
3. Hyoid	Facial VII, Auditory VIII	Abducens VI
4. First branchial	Glossopharyngeal IX	Nil
5. Second to fifth branchials	Vagus X and Spinal accessory XI	Hypoglossal XII

TABLE 46.2. CRANIAL NERVES (PAIRED) OF VERTEBRATES.

S.No.	Name	Origin	Distribution	Nature	Functions
I.	**Olfactory**	Olfactory lobe or bulb	Olfactory epithelium in nasal cavity	*Sensory*	Smell
II.	**Optic**	Optic lobe on midbrain	Retina of eye	*Sensory*	Sight
III.	**Oculomotor**	Floor of midbrain	Eye, 4 muscles of eyeball	Motor	Movement of eyeball, iris, lens, eyelid
IV.	**Trochlear**	Floor of midbrain	Eye, superior oblique muscles of eyeball	Motor	Rotation of eyeball
V.	**Trigeminal**	Side of medulla	Head, face, jaws, teeth	*Sensory* Motor	Forehead, scalp, upper eyelid, side of nose, teeth, Movement of tongue, jaw muscles for chewing
VI.	**Abducens**	Side of medulla	External rectus muscle of eyeball	Motor	Rotation of eyeball
VII.	**Facial**	Side and floor of medulla	Anterior 2/3 tongue. Muscles of face, neck and chewing.	*Sensory* Motor	Taste, Facial expression, chewing, movement of neck
VIII.	**Auditory (acoustic)**	Side of medulla	Organ of Corti in cochlea Semicircular canals	*Sensory*	Hearing Equilibrium
IX.	**Glossopha-ryngeal**	Side of medulla	Posterior 1/3 tongue, mucous membrane and muscles of pharynx	*Sensory* Motor	Taste & touch, Movements (swallowing) of pharynx
X.	**Vagus (pneumo-gastric)**	Side and floor of medulla	Muscles of pharynx, vocal cords, lungs, heart, oesophagus, stomach, intestine	*Sensory* Motor	Vocal cords, lungs Respiratory reflexes, peristaltic movements, speech, swallowing, secretion of gastric glands, inhibition of heart beat.
XI.	**Spinal accessory**	Floor of medulla	Muscles of palate, larynx, vocal cords, neck, shoulder	Motor	Muscles of pharynx, larynx, neck, shoulder movements
XII.	**Hypoglossal**	Floor of medulla	Muscles of tongue, neck	Motor	Movements of tongue.

Spinal nerves. The **spinal nerves** arise from the spinal cord. They are also paired segmental structures emerging through intervertebral foramina of the vertebral column. Their number corresponds approximately to the number of vertebrae present. Each spinal nerve is formed by two roots : (*i*) A **dorsal root** (sensory) attached to the dorsal horn of the gray matter. The dorsal root has always a ganglion containing nerve cell bodies of sensory fibres. (*ii*) A **ventral root** (motor) arising from the ventral horn of the gray matter. The nerve cell bodies of motor fibres are always located in the brain or spinal cord. The roots have different embryonic origin : the dorsal roots arise from neural crests, while the ventral roots arise from the gray matter of the spinal cord. In **anamniota**, the dorsal roots contain **somatic sensory, visceral sensory** and **visceral motor fibres**.

In **amniota** the dorsal roots have only somatic sensory and visceral sensory fibres. The **ventral roots** have visceral motor and somatic motor fibres in all. Except in some cyclostomes, a dorsal and a ventral root of each side unite to form a spinal nerve. In cyclostomes, the sensory and motor roots do not join together to form a common trunk. In **lampreys**, they remain separate and emerge alternately from the spinal cord, whereas in **hagfishes** there may be an incomplete union. In all other fishes the dorsal sensory root and ventral motor root unite with each other outside the vertebral column to form a common trunk. In **amphibians** dorsal and ventral roots of the spinal nerves unite in their passage through the intervertebral foramen rather than outside as in most fishes or within the neural canal as in amniotes (reptiles, birds and mammals). In mammals, complicated plexuses, resulting from the intermixing of fibres from the ventral branches of spinal nerves are found. These are differentiated into cervical, brachial, lumbar and sacral plexuses. Each spinal nerve divides into three branches or rami, a **dorsal ramus** supplying the skin and muscles of the back, a **ventral ramus** going to body muscles and skin of ventral side, and a communicating **ramus communicans** or **visceral ramus** going to the viscera and a ganglion of the autonomic nervous system. A ramus communicans has two parts, a **white ramus** and a **gray ramus**. There are no gray rami in elasmobranchs. The white ramus has medullated visceral sensory and visceral motor fibres. The axons of visceral motor fibres go to autonomic ganglia and form synapses with non-medullated fibres of the gray ramus. The medullated visceral motor fibres are called **preganglionic fibres** because they terminate in autonomic ganglia where they form synapses with non-medullated **postganglionic fibres** of the gray ramus. These fibres enter the spinal nerves and pass into the dorsal or ventral rami to supply structures under involuntary control. In brief the white ramus of ramus communicans carries medullated visceral sensory fibres and medullated preganglionic motor fibres, the gray ramus carries only non-medullated postganglionic autonomic motor fibres (Fig. 46.10).

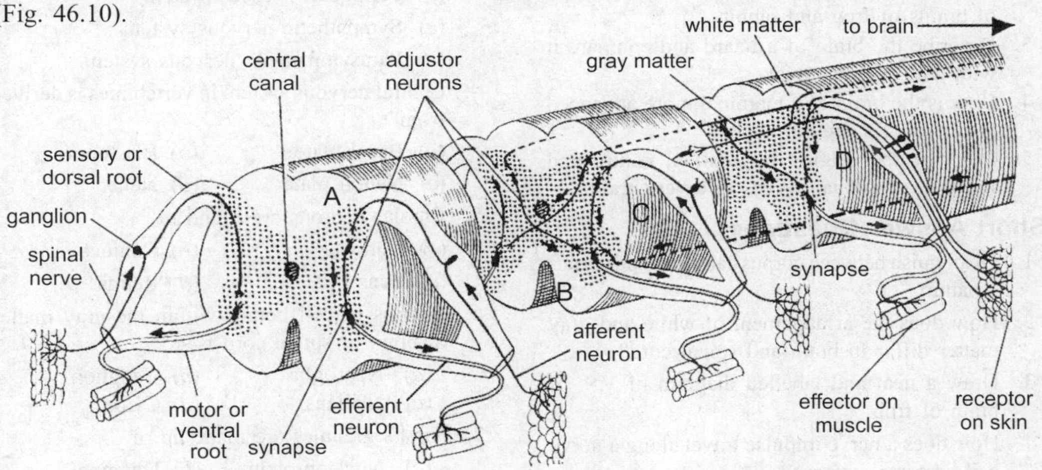

Fig. 46.10. Simplified stereogram of the vertebrate spinal cord and nerves to show relations of neuron involved in reflex arcs. A–Simple reflex arc; B–Reflex arc with one adjustor neuron; C–Reflex arc with cross connections; D–Reflex arc with cross connections and also others to and from the brain.

AUTONOMIC NERVOUS SYSTEM

The autonomic nervous system is so called because it is partly independent and not under voluntary control, though it is involuntarily controlled by the nerve centres located in the central nervous system. It is also connected to spinal nerves and some cranial nerves. Cranial and spinal somatic nerves mainly innervate the voluntary (skeletal) muscles of the animal. While the **autonomic nerves** and **ganglia** innervate the smooth or involuntary muscles of the viscera (alimentary canal, heart, etc,) and glands, and, thus, control the internal environment of the body. Little is known about the autonomic nervous system in **cyclostomes** although

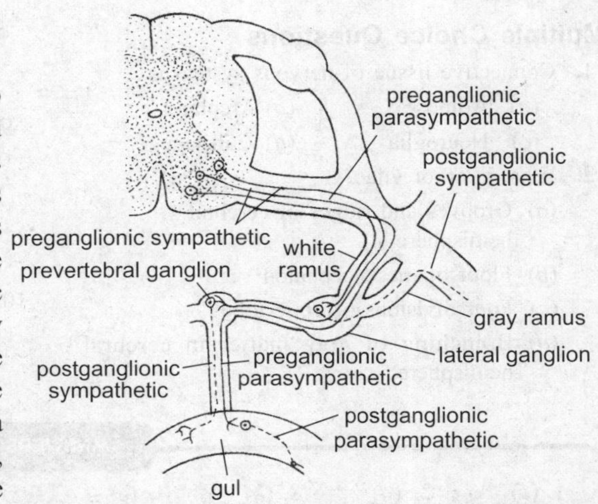

Fig. 46.11. T.S. of spinal cord showing distribution of autonomic nerve fibres.

sympathetic ganglia are present. The digestive system and heart are supplied by the vagus nerve.

In **elasmobranchs**, an irregular series of sympathetic ganglia lies along the body wall. Fibres from these ganglia connect both to the spinal cord and to the smooth muscles of the digestive and circulatory systems. In **bony fishes** autonomic nervous system is more advanced *i.e.,* the sympathetic ganglia are arranged in a chain extending forward to the trigeminal nerve.

IMPORTANT QUESTIONS

Long Answer Questions

1. Discuss the comparative account of brain of vertebrates.
2. Give a comparative account of the structure of brains of frog and rabbit.
3. Describe the brain of a lizard and compare it with that of a mammal.
4. How is the brain of a mammal more advanced than that of a reptile?
5. Describe the origin, distribution, nature and function of cranial nerves in a vertebrate.

Short Answer Questions

1. Distinguish between corpus callosum and corpus striatum.
2. How does the arrangement of white and gray matter differ in brain and spinal cord?
3. Draw a neat and labelled diagram of V.S. of brain of frog.
4. How does a nerve impulse travel along a nerve cell across a synapse?
5. Write short note on the following :
   (*i*) Pineal eye; (*ii*) Autonomic nervous system; (*iii*) Neuron; (*iv*) Synapse; (*v*) Conduction of nerve impulse.

Multiple Choice Questions

1. Connective tissue of nervous system is
   (*a*) Adipose (*b*) Areolar
   (*c*) Neuroglia (*d*) Cartilage.
2. What is arbor vitae?
   (*a*) Grooves and ridges of cerebral hemisphere
   (*b*) Floor of mesencephalon
   (*c*) Four divisions of optic lobes
   (*d*) Branching of gray matter in cerebral hemisphere.
3. Connecting link between the receptors, central nervous system and effectors is
   (*a*) Central nervous system
   (*b*) Peripheral nervous system
   (*c*) Sympathetic nervous system
   (*d*) Parasympathetic nervous system.
4. Central nervous system in vertebrates is derived from
   (*a*) Basal plate (*b*) Blastopore
   (*c*) Neural plate (*d*) None.
5. Bipolar neurons are found in
   (*a*) Cornea (*b*) Conjuctiva
   (*c*) Lens (*d*) Retina.
6. A mass of cell bodies within the gray matter of brain or spinal cord is
   (*a*) Neuroglia (*b*) Ganglion
   (*c*) Nucleus (*d*) Cyton.
7. Nissl's granules are made up of
   (*a*) Nucleoprotein (*b*) Thymine
   (*c*) DNA (*d*) RNA.
8. Cerebral hemispheres in eutherians are connected internally by
   (*a*) Corpus albicans
   (*b*) Corpus callosum
   (*c*) Corpus luteum
   (*d*) Aqueduct of Sylvius.
9. Function of restiform bodies in elasmobranchs
   (*a*) Swimming
   (*b*) Steering
   (*c*) Maintenance of equilibrium
   (*d*) Sound production.
10. Terminal nerves are numbered
    (*a*) 0 (*b*) 11
    (*c*) 12 (*d*) 13.

ANSWERS

1. (*c*) 2. (*d*) 3. (*b*) 4. (*c*) 5. (*d*) 6. (*c*) 7. (*d*) 8. (*b*)
9. (*c*) 10. (*a*)

47

Chapter

Receptor Organs in Vertebrates

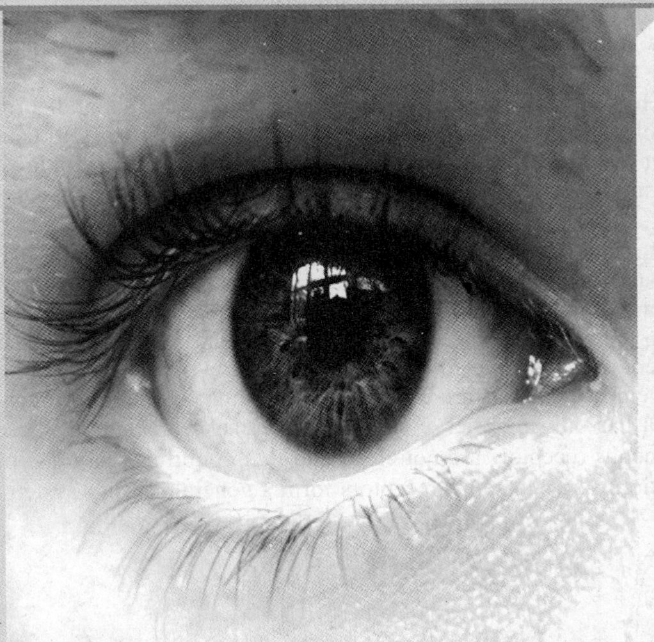

Contents of this Chapter

Sensory systems consist of peripheral receptor cells and integrating neurons in the brain that provide a biologically relevant image of the environment to an organism. The receptor elements range from single cell, such as tactile corpuscles, to complicated structures, such as the eye. A vertebrate has receptors or sense organs for touch, smell, taste, sight, and hearing, which are stimulated by the environment. These sense organs are termed external receptors or **exteroceptors**. There are other sense organs found in the body, which detect temperature, pain, hunger, thirst, fatigue, and muscle position. They are spoken of as internal receptors or **interoceptors**. Besides these two, third is **proprioceptors**, which are stretch receptors found in the muscles, joints, tendons, connective tissue and skeletons. All receptors are closely associated with the nervous system and respond to external or internal stimuli. **Impulses** are transmitted from receptors by sensory fibres to the central nervous system where they are interpreted as sensations or messages, which are sent to effector organs through efferent or motor nerve fibres, for responding in an appropriate manner.

List of Common Senses

1. **Touch.** It includes contact, pressure, heat and cold, etc.
2. **Taste.** Receive stimulus by chemicals in solution.
3. **Smell.** Receive volatile chemicals and gases in air.
4. **Hearing.** Receive sound vibrations.
5. **Sight.** Receive light waves.

CUTANEOUS (GENERAL) RECEPTORS

The entire body surface of aquatic vertebrates, and the moist parts of the skin of terrestrial vertebrates are sensitive to chemicals. Certain parts of the skin are sensitive to contact and temperature changes. The cutaneous receptors are free nerve endings of peripheral nerves, and some encapsulated nerve endings or tactile corpuscles.

1. Free nerve endings. Branches of sensory nerves terminate in the epidermis and form patches, cornea of the eye, hair follicles, and mucous membrane. These parts are sensory to pain (**algesireceptors**), or to touch (**tangoreceptors**). The nerve endings do not reach the outermost layer of the epidermis so that the stimulus first travels through the outer cells before it reaches the nerve endings.

2. Tactile corpuscles of various kinds are located in the dermis or in the sub-epidermal connective tissue. They have a basic structure of nerve endings or tactile cells enclosed in layers of connective tissue. Tactile corpuscles are found all over the skin, in birds they are found also on the beak, in mammals they occur also on palms, soles, and external genital organs. The encapsulated corpuscles may be tactile (tangoreceptors), sensitive to pressure, to heat (**caloreceptors**), or to cold (**frigidoreceptors**). But even free nerve endings are capable of discerning sensation of touch, heat, cold, and, pain, *e.g.*, the cornea has only free nerve endings yet it can appreciate all these sensations. The parts played by cutaneous receptors are not clear, and it is doubtful if any one of them is sensory to a particular sensation. The cutaneous receptors form a general sensory system.

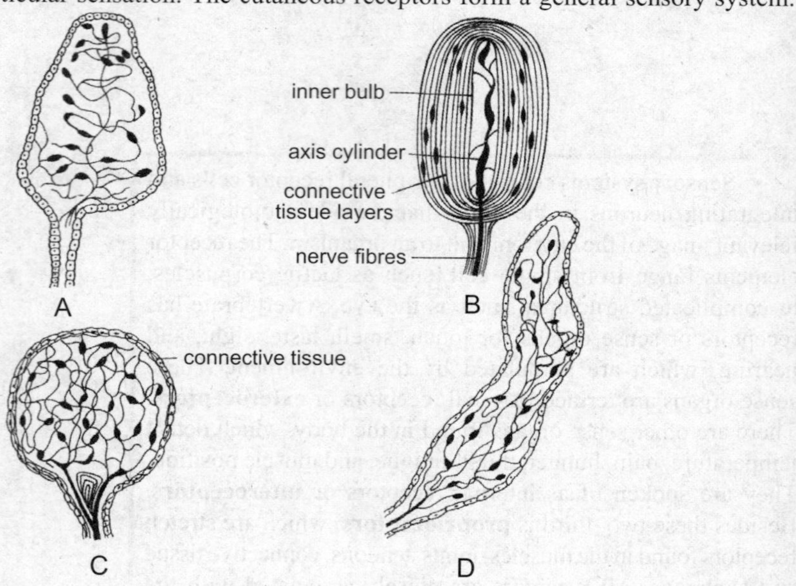

Fig. 47.1. Cutaneous sensory receptors. A–Meissner's corpuscle from the dermal papilla; B–Pacinian corpuscle; C–End bulb of Krause; D–Meissner's corpuscle.

CHEMORECEPTORS

Chemoreceptors or organs of chemical sense consist of **olfactory organs** and **organs of taste**. Both these organs are stimulated only by chemical substances or odours in air (nostrils) and in solution (tongue). The medium for dissolving substances for taste is water for aquatic animals and mucus for land animals. The **olfactory organs** can respond to a low concentration of the dissolved substance, whereas **organs of taste** need a higher concentration of the dissolved substance for a response.

Olfactory Organs in Vertebrates

Olfactory organs (olfacto-receptors) are a pair of invaginations of the ectodermal cells of the skin forming olfactory sacs on the anterior end of head. Their external openings are called **nostrils** or **nares**. In most **fishes** the olfactory organs consist of a pair of pits lined with folds or ridges of sensory epithelium. The **cyclostomes** have a single median olfactory organ. This is a blind pit in the lampreys, but in hagfishes it opens into the pharynx. Dipnoans resemble higher vertebrates in possessing paired nasal passages that open by means of choanae into pharynx. The nasal passages, therefore, have both internal and external openings. The olfactory epithelium within canals appears in the form of folds. The olfactory epithelium in **amphibians** is generally smooth and is restricted to the upper part of the nasal passages Within each of the nasal passages, which are elongate in **higher reptiles** because of the development of a secondary palate, there is a shelf or concha which serves to increase the surface for the olfactory epithelium. In **birds**, the nasal passages have three shelves or conchae. In most birds there are external nostrils leading into the nasal passages, but in certain members of Pelecaniformes these are closed. Generally the nostrils are separated internally from one another by a septum, but in New World vultures as well as some of the Gruiformes, a partition is lacking. Olfactory epithelium in most birds is restricted to the surface of the upper most or superior concha. This is correlated with the small size of olfactory lobes and poor sense of smell. But kiwi has relatively more developed sense organs. In **mammals**, the sense of smell is highly developed in many mammals. It is due to the development of the nasal conchae into elaborate scroll-like structures, which greatly increase the surface available for olfactory epithelium. In whales, the olfactory organs are essentially non-functional.

Organs of Jacobson or vomeronasal organ. In many tetrapoda there is a pair of vomeronasal organs which are sac-like chambers lying below the nasal cavities but above the buccal cavity. They have a pigmented epithelial lining like that of the olfactory organs. Each opens by a short duct into the olfactory organ in amphibians, but in others the duct opens into the buccal cavity. The organ of Jacobson receives nerves from the nervus terminalis, a branch from the olfactory nerve, and a branch from the trigeminal nerve. The organ is believed to aid by smelling the recognition of food held in the mouth, and in lizards and snakes it appreciates the scent introduced into it by the tip of the tongue.

The organ of Jacobson first appears in amphibians as evaginations from the nasal passage and is lined with olfactory epithelium. It is believed to be an aid in tasting food. They may also be important in reproductive behaviour in case of *Ensatina* (lungless plethodontid salamander) since the first act of male in courtship to nose the females' head and neck. It is best developed in *Sphenodon*, lizards and snakes and is connected with the roof of mouth rather the nasal canal. It is also well formed in monotremes, marsupials, insectivores, and rodents. But in turtles, crocodiles, birds, and many mammals such as Primates and Cetacea, it is found only in the embryo and is absent in the adult.

Gustatory or Organs of Taste in Vertebrates

Organs of taste (gustatoreceptors) consist of **taste buds** which perceive the sense of taste from the dissolved substances. Their structure is fairly uniform in vertebrates. Each taste bud consists of cluster of **neurosensory cells** and **supporting cells** arranged to form a barrel-like structure embedded in stratified epithelium. Each neurosensory cell is long and narrow with a thin **taste hair** at its free tip and a sensory nerve fibre at its base. Taste hairs project into a depression or **taste pore** in mammals, in others they project above the surface, there being no taste pore. In man, there are four fundamental sensations of taste : sweet, salty, bitter, and sour.

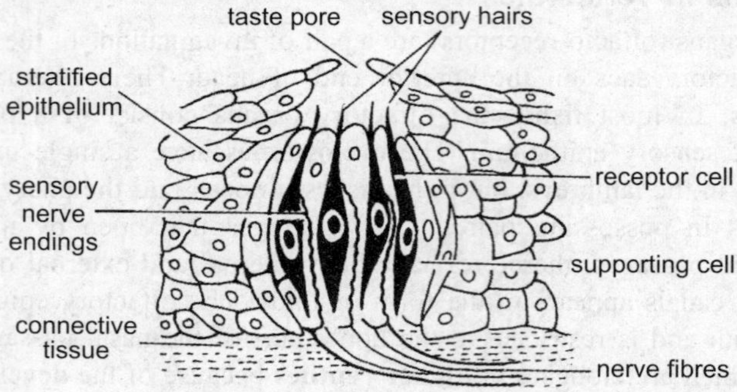

Fig. 47.2. Diagrammatic representation of a taste bud.

Taste buds are widely distributed in fishes since they live in a liquid medium, being found in the mouth, pharynx and outer surface of the head. In some fishes they occur on the entire body surface including even the fins. The taste buds are innervated by the V, VII, IX and X cranial nerves. In amphibians, the taste buds are restricted to the roof of the mouth, the tongue and mucosa that lines the jaws. In most reptiles, the taste buds are restricted largely to the pharyngeal region. Taste buds are lacking on the tongue of most birds, although they are found on the lining of the mouth and pharynx. In mammals, there are various kinds of **papillae** on the tongue which possess taste buds except filiform papillae. Some taste buds of vertebrates are not for tasting but for testing substances in the pharynx to cause reflexes which prevent solid particles from entering air passages.

EYES OR PHOTORECEPTORS

Eyes are of two types : (*i*) Median and (*ii*) Lateral.

1. Median eye. In amphibians, both the parietal and pineal bodies are known to function as photoreceptors, being sensitive to wavelength and intensity of light. It has been suggested that they may play a role in thermoregulation, compass orientation and circadian rhythm. Pineal eye is also present in tuatara and a few lizards. It aids in regulating the amount of exposure to sunlight (**Stebbins** and **Eakin**, 1958).

2. Lateral eyes. The paired eyes of vertebrates, except for small differences, are remarkably constant in structure, innervation, and extrinsic muscles. The differences occur in eyelids, glands of the eyes, and in the mechanism of accommodation. Fishes have no movable eyelids and tear glands. Cornea remains flat in lower vertebrates and buldging in higher vertebrates.

There is no basic difference between the eyes of fishes and those of other vertebrates. Differences are in the methods of accomodation or special adaptations to a particular mode of life or are the result of degeneration. Accomodation or the ability of the eye to adjust itself to near and far vision, is accomplished in fishes by moving the lens forward or backward so as to change its distance from the sensitive retina. Sharks are fast moving predatory fishes, usually have distant vision, they move the lens forward to see near objects. In most teleosts, the eyes are normally adjusted to near vision so the lens is moved back toward the retina when distant vision is required. The eyes of some fishes are highly specialised, as an adaptation to a particular mode of life. South American "four-eyed fish" (*Anableps*) inhabits quiet water where it floats with the upper half of the eyes above the surface. So that it may see both in the air and in water, presumably at the same time. Its lens is divided into two parts, each part being a different distance from the retina. Another four-eyed fish" *Dialommus fuscus* (Galapagos blenny) spends considerable time out of water. Its lens is not divided but it has a divided cornea–anterior section for air vision and posterior section for vision under water. Many deep fishes have very large eyes for vision in regions of low-light intensity. Other deep sea fishes and some cave inhabiting species have degenerated eyes, have lost the power of sight.

In various lower vertebrate groups accomodation to near and distant vision is accomplished by moving the lens forward or backward so as to change the distance between the lens and the sensitive retina. Accomodation in **reptiles** and most other amniotes is accomplished by changing the shape of lens. It may be flattened for distant vision or rounded for near vision through the action of the muscles of the ciliary body. **Eyelids** are present in many reptiles and generally are more movable than those of amphibians. In snakes and some of the burrowing lizards, movable lids are absent, and the eye, which itself is capable of moving, is covered with immovable transparent skin. These so called "eye plates" are believed to have been derived from the lids and are like scales which shed periodically. Many reptiles possess the **nictitating membrane**, situated beneath the upper and lower lids and closes the eye by moving backward or outward from the anterior or median margin of the eye aperture.The blood that a horned toad squirts from the eye comes from blood vessels in the nictitating membrane which rupture easily as a result of muscular contraction.

The amphibian eye is basically like that of other vertebrates. The lens is fixed for relatively distant vision but may be moved forward toward the cornea for seeing close objects, by means of small muscles of accomodation. The pupillary aperture may be vertical, horizontal, three-cornered or four-cornered. Eyelids are poorly represented in aquatic forms but are well developed in many terrestrial species. The lower lid is usually more movable than the upper lid. **Lacrimal** or **tear glands** are present in many amphibians, although they are poorly developed. The eye, however, is kept moist by the oily secretion from the **Harderian gland**.

In **birds**, eye are highly developed and quite large in proportion to body size. Accomodation is accomplished by action of the ciliary muscles which change the shape of the lens. An ossified sclerotic ring is present and surrounds the ciliary region. A fan-shaped **pecten** is present in the eye, which extends into the posterior chamber from the point at which the optic nerve emerges from the retina. Pecten provides nourishment to avascular parts of the eye. As in other groups of vertebrates, in the retina of diurnal birds, cones predominate, whereas rods are much more numerous in nocturnal species.

The **mammalian eye** is basically like that of most other vertebrates, with certain modifications correlated with habits. As in birds, nocturnal mammals have a preponderance of rods in the retina, while in diurnal species cone cells predominate. Land mammals have an emmetropic condition in air but become hypermetropic in water. Aquatic or marine mammals have developed emmetropic vision under the water. In pinnipeds this is accomplished by a permanent change in the curvature of the lens. They also have special dioptric mechanisms for air vision. In terrestrial mammals, the strong sphincter muscles in eye are present in the iris that change the shape of the lens to preserve visual acuity when under water.

In fossorial species such as moles (Talpidae), the eye may be of little importance. This is also true of the Gangetic dolphin (*Platanista gangetica*) living in very turbid water. Retina is more sensitive to light, lens is devoid of a crystalline body and probably serves no function.

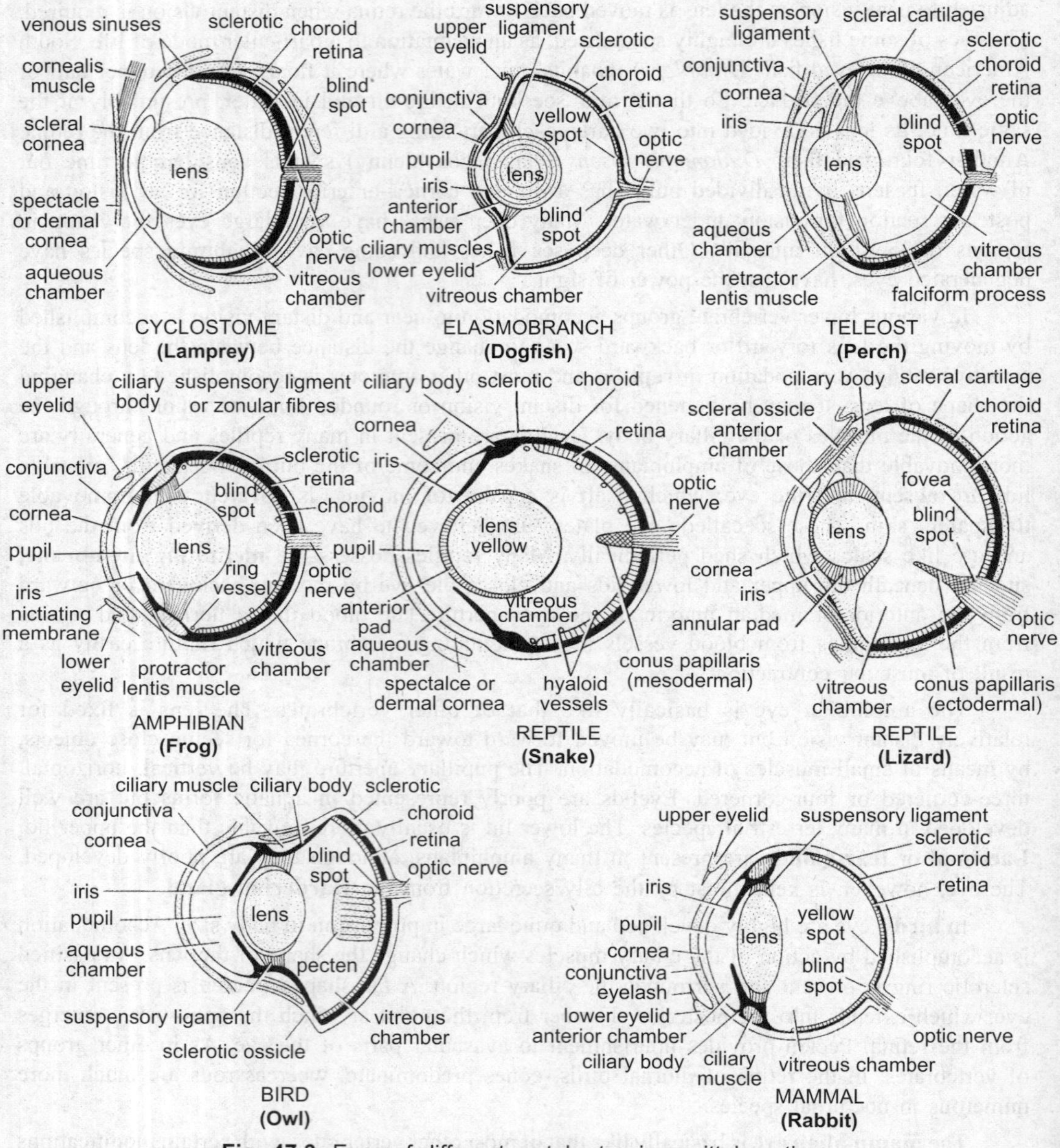

Fig. 47.3. Eyes of different vertebrates in sagittal section.

Lateral Line System

Lateral line system is found in cyclostomes, fishes, aquatic amphibians and aquatic larval stages of terrestrial amphibians. It is not found in terrestrial vertebrates. It is related to an aquatic mode of life. The receptor organs are sensory papillae or **neuromasts** which are arranged in rows on the body. In most fishes there is a row on each side of the body which branches into three parts on each side of the head. The neuromasts may be on the surface of the skin (cyclostomes and a few fishes) or they may be in covered canals which open to the outside through small pores (cartilaginous and bony fishes). A lateral line system, similar to that of fishes, is present in larval amphibians and even in the adult stage of certain aquatic species. Lateral line organs are lacking in reptiles and other amniotes. Lateral line receptors detect pressure changes and currents in the water.

EARS OR STATOACOUSTIC ORGANS

The vertebrate ear is commonly regarded only as an organ of hearing (**phonoreceptor**), but at least in higher vertebrates, it serves the dual function of equilibrium and hearing.

The **ear** of a **fish**, is very different from the ear of a mammal. In mammals ear consists of an external pinna which funnels sound waves into the external auditory canal. At the inner end of this canal these waves strike the ear drum or tympanic membrane and are then transmitted across the middle ear or tympanic cavity by means of a chain of small bones to the cochlea or ventral part of the inner ear. From this complicated structure sensory impulses pass to the brain by means of the auditory nerve. The dorsal part of the inner ear consists primarily of three semicircular canals which are essentially organs of balance or orientation. The ear of fish, however, is not suited for the reception of sound. There is no external or middle ear, nor is a cochlea present.

The inner ear of a fish consists of a dorsal **utriculus**, which is connected with the semicircular canals, and a ventral enlargement known as the **sacculus**. It is from the sacculus that the lagena of amphibians, reptiles and birds as well as the cochlea of mammals arises to permit hearing in true sense. While certain fishes are definitely able to detect vibrations in the water. The ear of a fish is primarily an organ of balance, enabling it to maintain its equilibrium.

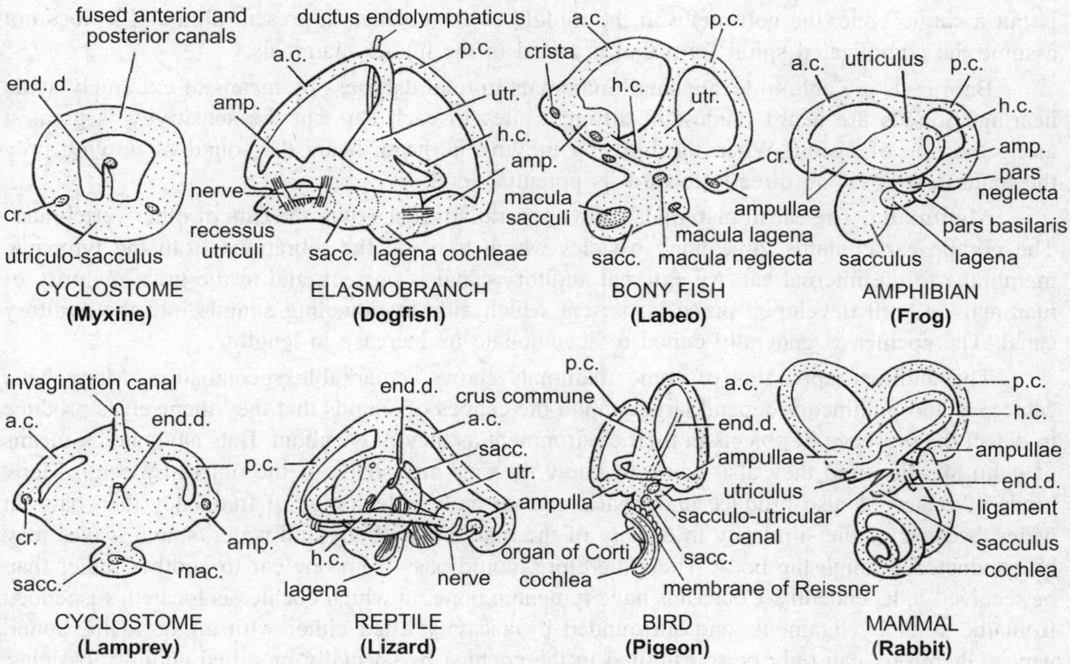

Fig. 47.4. Internal ears of representative types of vertebrates. a.c.–anterior vertical semicircular canal; amp.–ampulla; cr.–crista; end.d.–endolymphatic duct; h.c.–horizontal canal; mac.–macula; p.c.–posterior vertical canal; sacc.– sacculus; utr.–utriculus.

In **cyclostomes**, the membranous labyrinth is degenerate with only one (*Myxine*) or two (*Petromyzon*) semicircular canals. In fishes the membranous labyrinth is complete with three semicircular canals, and in many teleosts it is connected to the air bladder by a duct or a chain of bony Weberian ossicles.

Amphibia. There is considerable variation in the auditory apparatus of amphibians. Salamanders and their relatives lack any middle ear, although it is believed that they can detect vibration. Most frogs and toads, however, have a middle ear and an external ear drum. Sounds are transmitted from the ear drum across the tympanic cavity to the inner ear by means of a bone called the **columella** which is homologous with the hyomandibular element of the gill-arches of cartilaginous fishes. The tympanic cavity, since it is derived eubryologically from the first pharyngeal or gill-pouch, is concerned with the pharynx by means of the eustachian tube. This permits equalisation of pressure on both sides of the tympanic membrane. From the ventral part of the sacculus of the inner ear there is a ventral outpocketing called the **lagena** which is the forerunner of cochlea of mammals. This is believed to be concerned with the reception of sound vibrations.

Reptilia. There is considerable variation in the structure of the ear in reptiles. Lizards have a well developed middle ear and in some forms the eardrum has sunk into a depression. Sound waves, thus, pass through a short canal known as the external auditory meatus in order to impinge upon the tympanic membrane. This is the first indication of an external ear in vertebrates. In snakes the tympanic membrane, middle ear and eustachian tube are lacking. Vibrations are received by snakes are transmitted by way of the quadrate to the columella and then to the inner ear. The lagena is more elongate than in amphibians and in the crocodilians actually forms a cochlear duct rather similar to birds.

Aves. The ear in most birds lacks an external pinna and has only an internal ear and a middle ear. In the barn owl (*Tyto alba*), the external pinna is well developed. As in reptiles there is but a single bone, the **columella** in the middle ear. A cochlea is present although it does not assume the complicated spiral form that is found in the higher mammals.

Barn owls are able to locate small mammals in total darkness by means of extremely acute hearing. Sounds are heard loudest in different sites in each ear, but the sensitivity is greatest along the line of vision. When the bird, by moving its head, hears the sound of moving prey the loudest, it is facing directly toward its potential food.

Mammalia. The ear in mammals shows several advances over the ears of other vertebrates. The middle ear contains three bony ossicles which transmit the vibrations from the tympanic membrane to the internal ear. An external auditory canal is present, and in the great majority of mammals, a well developed pinna is present which aids in funneling sounds into the auditory canal. The cochlea is generally coiled to accomodate its increase in length.

The auditory apparatus of some mammals shows remarkable specialistion. Many bats, cetaceans and pinnipedes depend largely upon the echoes of sounds that they themselves produce to detect the presence of objects in their environment as they move about. Bats navigate by means of **echo-location** and they also come to know the size and shape of the object by echo. Many kinds of cetaceans also produce underwater sounds with wider range of frequency than bats. In water, because of the similarity in density of the mammalian body and water, sound waves may be conducted through the body. They, therefore, could pass from one ear to another rather than be received independently. Cetaceans have tympanic bone, in which cochlea is located, suspended from the skull by ligaments and surrounded by a cavity filled either with air or foam. Sound waves, therefore, can only be transmitted in the cochlea by specially modified auditory ossicles. The sound itself is produced in nasal sacs just inside the blowhole, rather than in a larynx as in terrestrial mammals. Recent experiments also indicate that sound reception from the water to the middle and inner ear may be by way of lower jaw. The mandibles of porpoises, dolphins and other toothed cetaceans are hollow and each is composed of an outer covering of very thin bone around an oil-filled cavity, which would favour sound transmission. The back of the mandible is very close to the middle ear.

INTERNAL RECEPTORS OR INTEROCEPTORS

Interoceptors are directly affected by stimuli arising within the body itself. Those found in the walls of the alimentary canal and viscera are called **interoceptors**, and those located in the tendons, joints, and voluntary muscles are spoken of as **proprioceptors**.

1. Interoceptors are connected to the visceral sensory nerves but they do not form definite sense organs. They are stimulated by conditions in the digestive system and supply information to the central nervous system. Interoceptors detect appetite, nausea, hunger, and thirst. **Hunger** is not to be confused with **appetite**. **Hunger** is a nutritional deficiency in the blood, while **appetite** is a memory of food causing internal changes in conjunction with sight, colour, or taste. **Thirst** is a condition of the mucous membrane of the throat due to decrease of water and an increase of salt content in the blood. Some **genital corpuscles** are present in the dermis of nipples and external genital organs, they are associated with sexual sensations.

2. Proprioceptors are connected to the somatic sensory nerves. They generally form spindle-shaped bodies consisting of free nerve endings enclosed in layers of connective tissue or **encapsulated corpuscles** enclosing tactile cells. They are found in tendons, joints and voluntary muscles. Proprioceptors are stimulated only indirectly by the environment. They supply information to the central nervous system regarding visceral pains, position of the body, and coordinated movements of parts of the body. They give an idea of weight of objects, and provide a "muscle sense" making one aware of the degree of tension in a muscle.

IMPORTANT QUESTIONS

Long Answer Questions

1. What are receptors? Classify them.
2. Describe the structure of eye and explain the process of focussing mechanism in vertebrates.
3. Describe the structure of ear in vertebrates.
4. Explain the mechanism of hearing and equilibrium in vertebrates.

Short Answer Questions

1. Distinguish between sacculus and utriculus.
2. Write a note on organ of Corti.
3. Draw a detailed labelled diagram showing the structure of mammalian eye.
4. What are neuromast organs? How do they function?
5. Write short notes on the following :
   (*i*) Accomodation of eye; (*ii*) Cochlea;
   (*iii*) Mechanoreceptors; (*iv*) Organ of Jacobson;
   (*v*) Proprioreceptors; (*vi*) Taste buds;
   (*vii*) Parietal eye; (*viii*) Nerve endings;
   (*ix*) Chemoreceptors; (*x*) Eustachian tube.

Multiple Choice Questions

1. What are the receptors of touch and pressure, vibrations and balance?
   (*a*) Thermoreceptors
   (*b*) Chemoreceptors
   (*c*) Mechanoreceptors
   (*d*) Photoreceptors.
2. Which of the following is a visual receptor?
   (*a*) Proprioceptor
   (*b*) Thermoreceptor
   (*c*) Exteroceptor
   (*d*) Interoceptor.
3. In vertebrates taste buds are present on
   (*a*) Palate (*b*) Tongue
   (*c*) Pharynx (*d*) All.
4. Muscles are
   (*a*) Mechanoreceptors (*b*) Thermoreceptors
   (*c*) Chemoreceptors (*d*) Photoreceptors.
5. Pecten is found in the eye of
   (*a*) Fish (*b*) Frog
   (*c*) Lizard or Snake (*d*) Bird.

ANSWERS

1. (*c*) 2. (*d*) 3. (*d*) 4. (*a*) 5. (*d*)

48

Chapter

Urinogenital System in Vertebrates

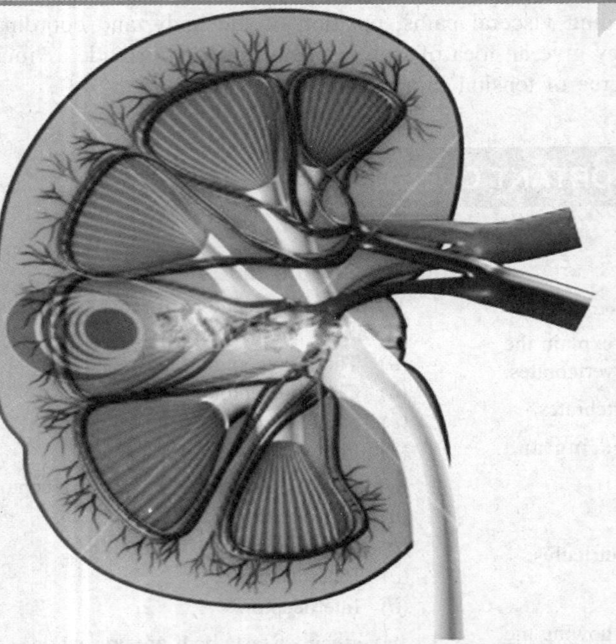

Contents of this Chapter

The end products of metabolism are waste substances, they are carbon dioxide, ammonia, urea, uric acid, pigments, creatinine and some inorganic salts. Carbon dioxide is removed through the skin and gills or lungs, but the others are excreted through the kidneys, though in some fishes gills also excrete some waste substances. Most excretory substances are in solution in water, and any excess of water is also eliminated by kidneys. The excretory and reproductive systems are closely associated structurally, particularly in the males in which the male urinary ducts are also used to discharge sperms. Hence, it is customary to consider the two systems together as a **urogenital** or **urinogenital system**. But the evolutionary tendency is towards a complete separation of the two systems.

VERTEBRATE KIDNEYS AND DUCTS

Basic Structure

In vertebrates, there is a pair of compact kidneys, lying dorsal to the coelom in trunk region, one on either side of vertebral column. A kidney is made of a large number of **uriniferous tubules** or **nephrons**. Their number, complexity and arrangement is different in different groups of vertebrates.

The uriniferous tubules arise in an embryo from a special part of the mesoderm, called **mesomere** or **nephrotome** extending on each side along the entire trunk in between upper segmental mesoderm or **epimere** and lower lateral plate mesoderm or **hypomere**. Primitively the uriniferous tubules develop from the nephrotome in a sequence commencing from the anterior end. They are segmental in arrangement with one pair of uriniferous tubules for each trunk segment. A uriniferous tubule has three parts : (*i*) A ciliated **peritoneal funnel** near the proximal end of the tubule which opens into the splanchnocoel by a **nephrostome** or **coelomostome** (often the peritoneal funnel itself is called a nephrostome). (*ii*) A convoluted ciliated tubule opening into a **longitudinal collecting duct**, and a **Malpighian body** or **renal corpuscle**. The Malpighian body has a double-walled **Bowman's capsule**, enclosing a network of inter-arterial blood capillaries, called **glomerulus** where filtration of blood takes place. Bowman's capsule and glomerulus together form the **Malpighian body** or **renal corpuscle**. An **afferent arteriole** brings blood into the glomerulus and an **efferent arteriole** takes blood away from it. Then the efferent arteriole breaks up into capillaries along the entire course of the uriniferous tubule and finally the blood goes to a renal vein. Encapsulated glomerulus is **internal glomerulus** which is common. The glomerulus without a capsule suspended freely in the coelomic cavity is called **external glomerulus** which is found in embryos and larvae. Malpighian bodies with glomeruli are lacking in some fishes, embryos and larvae and their kidneys are called **aglomerular**.

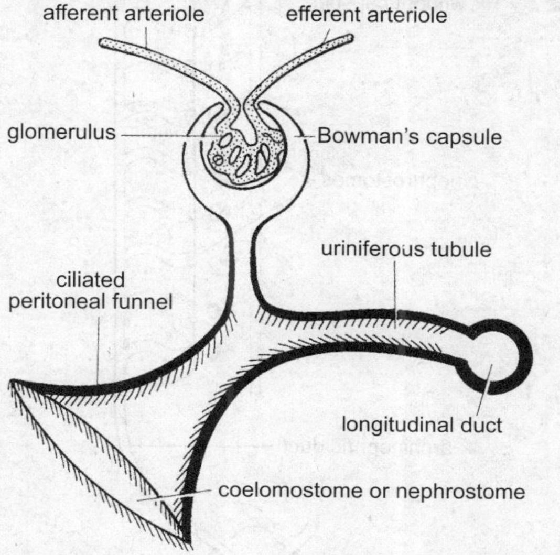

Fig. 48.1. Structure of an embryonic kidney tubule.

In an adult the uriniferous tubules are elongated and coiled, so that their segmental arrangement is lost and they become enclosed in a connective tissue capsule to form a **kidney**.

1. Archinephros. The ancestral vertebrates had a pair of kidneys running through the entire length of the coelom. Each had segmentally arranged tubules; one pair per body segment. Each tubule opened separately into the coelom by a **peritoneal funnel** and **nephrostome**. Near each funnel was an external glomerulus (without capsule) suspended in coelom. All the tubules of each kidney opened into a common longitudinal duct (Wolffian or archinephric duct) which joined the cloaca. This kidney is called an **archinephros** or **holonephros** and its duct is an **archinephric duct**.

Among the living vertebrates an archinephros is found only in the larval *Myxine* and some apodan amphibians. In present-day vertebrates the uriniferous tubules develop antero-posteriorly in two or three stages in succession, these stages are pronephros, mesonephros and metanephros. These stages have evolved from the original archinephros. Thus, in present vertebrates, kidneys are of three types : **pronephros**, **mesonephros** and **metanephros**. All three are fundamentally alike, differing principally in their relationship to the blood system, in degree of complexity, and in efficiency.

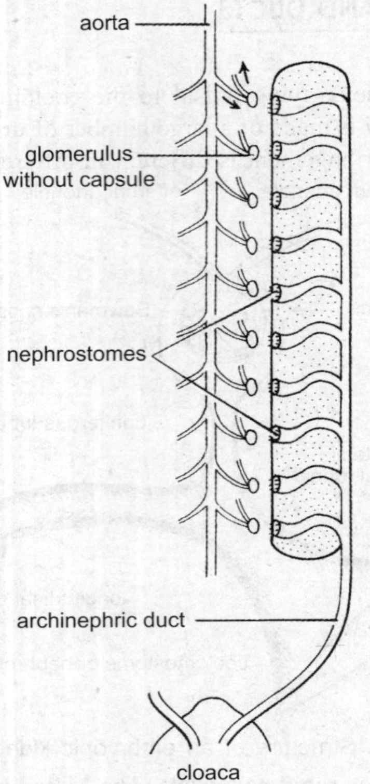

Fig. 48.2. Primitive ancestral vertebrate kidney or archinephros.

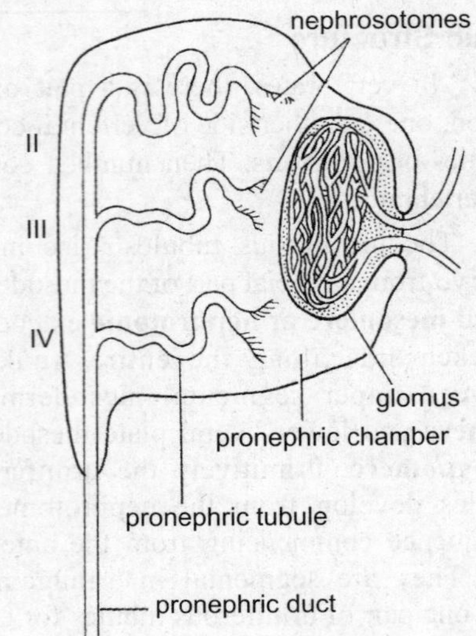

Fig. 48.3. Encapsulated pronephric kidney of 15 mm frog larva.

2. Pronephros. Pronephros is the most primitive and while present in the embryonic development of all vertebrates, is functional in the adult of none. Some larval cyclostomes, however, have a kidney, part of which appears to be homologous to the embryonic pronephros of higher vertebrates. Pronephros develops in the anterior-most part of the nephrotome. There are only 3 to 15 uriniferous tubules in each, one pair to each segment. Near each tubule is a glomerulus projecting into the coelom, which is not connected with the tubule is the external glomerulus. Each tubule opens into coelom by a funnel or nephrostome. The uriniferous tubules of each pronephros open into a common **pronephric duct** which grows back to enter the embryonic cloaca. In some, there is a large **pronephric chamber** which surrounds the glomus (all glomeruli) and tubules. All glomeruli project into the pronephric chamber where they may unite to form a single compound glomerulus called **glomus**. Pronephric chamber is derived from pericardial or pleuroperitoneal cavity. All the tubules of a pronephros open into a common pronephric duct opening posteriorly into the embryonic cloaca.

A pair of pronephroi appear in all vertebrate embryos but they become functional kidneys only in some cyclostomes and embryos of all anamniotes. In others they degenerate during

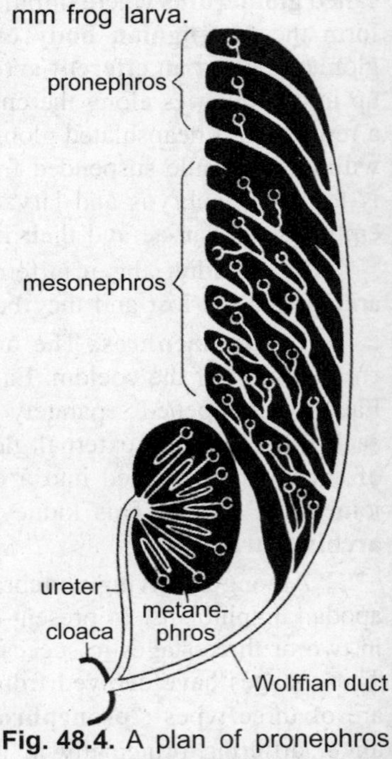

Fig. 48.4. A plan of pronephros mesonephros, and metanephros.

development but the pronephric ducts persist. Pronephros is replaced by mesonephros. In those vertebrates in which pronephroi become adult kidneys, they are called **head kidneys** due to its anterior position behind the head.

3. Mesonephros. Mesonephros develops from the middle part of the nephrotome behind the pronephros which degenerates. At first it consists of paired segmental uriniferous or mesonephric tubules, each with a peritoneal funnel opening into the coelom and internal glomerulus enclosed in a Bowman's capsule both are collectively called renal corpuscle. These mesonphric uriniferous tubules join the existing **pronephric duct** or **archinephric duct** on each side, which with the disappearance of pronephros is called a **mesonephric duct** or **Wolffian duct**. Later the mesonephric

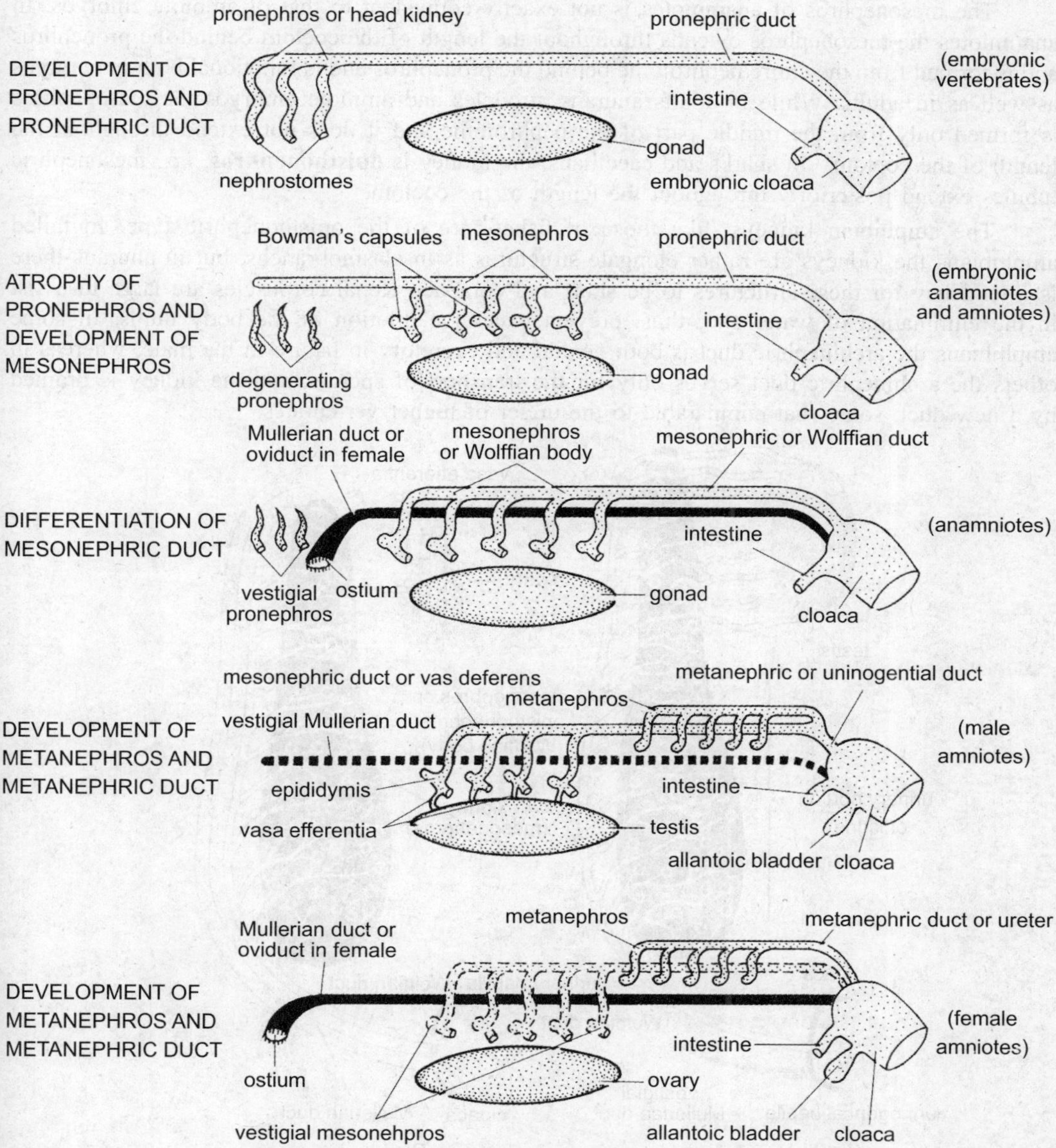

Fig. 48.5. Evolution of kidney in vertebrates.

tubules undergo budding to form hundreds of tubules, so that their segmental arrangement is lost. The mesonephric tubules are coiled or convoluted both proximally and distally and lead into a common longitudinal collecting duct the archinephric duct. This in turn leads to the outside, usually by way of the cloaca. Water, salts and waste products from the blood stream pass into the capsule and then may flow through the tubule to the mesonephric duct (original archinephric duct) and ultimately out of the body. In sharks it functions as a gonadial duct and the kidneys have developed new accessory urinary ducts. The mesonephric tubules have no peritoneal funnels. Mesonephroi form the adult functional kidneys in some cyclostomes, fishes, amphibians and the embryos of amniotes in which they degenerate in the adult. The mesonephroi of amniote embryos lack peritoneal funnels, except in monotremes.

The mesonephros of anamniotes is not exactly equivalent to that of amniote embryos. In anamniotes the mesonephros extends throughout the length of the coelom behind the pronephros and is formed from the entire nephrotome behind the pronephros and is functional both in embryos as well as in adults. While in adults anurans, urodeles and amniote embryos the mesonephros is formed only from the middle part of the nephrotome and it does not extend throughout the length of the coelom. In sharks and caecilians, the kidney is **opisthonephros**, *i.e.,* mesonephric tubules extend posteriorly throughout the length of the coelom.

The amphibian kidneys, like those of fishes, are of the opisthonephric type. In tailed amphibians the kidneys are rather elongate structures as in elasmobranchs, but in anurans there is a tendency for these structures to be short and compact. Renal corpuscles are large to assist in the elimination of water and, thus, prevent excessive dilution of the body fluids. In some amphibians the archinephric duct is both genital and excretory in nature in the male, whereas in others the archinephric duct serves only for the transport of sperms, and the kidney is drained by a new duct, somewhat comparable to the ureter of higher vertebrates.

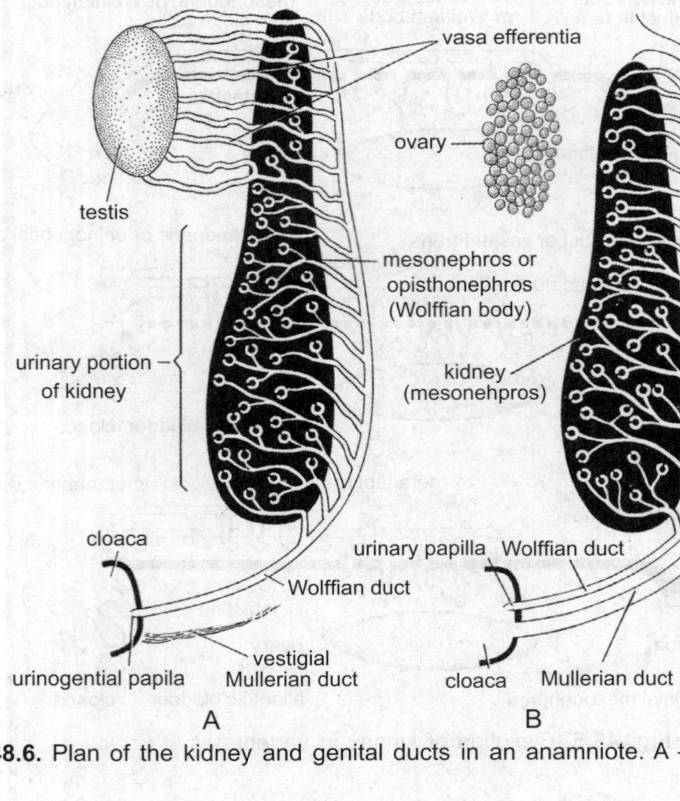

Fig. 48.6. Plan of the kidney and genital ducts in an anamniote. A – Male; B – Female.

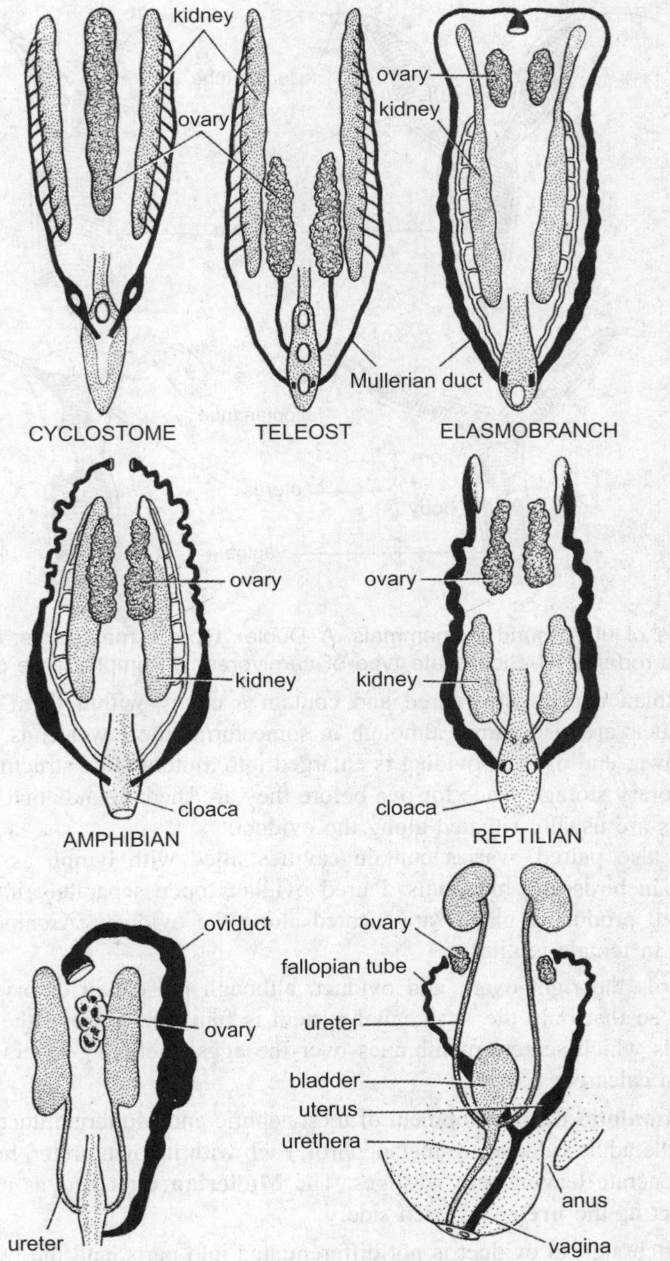

Fig. 48.9. Reproductive and urinary ducts of female vertebrates.

In most vertebrates the ovary is not directly connected with the oviduct so that the eggs enter the coelom and then pass into the ostium. In some bony fishes, however, where huge number of eggs are produced during the short breeding season, the ovaries are continuous with the oviducts so as to prevent the ova from escaping into the coelom. In many teleosts the lower parts of the paired oviducts are fused. Most bony fishes are oviparous, but there are some in which eggs hatch internally.

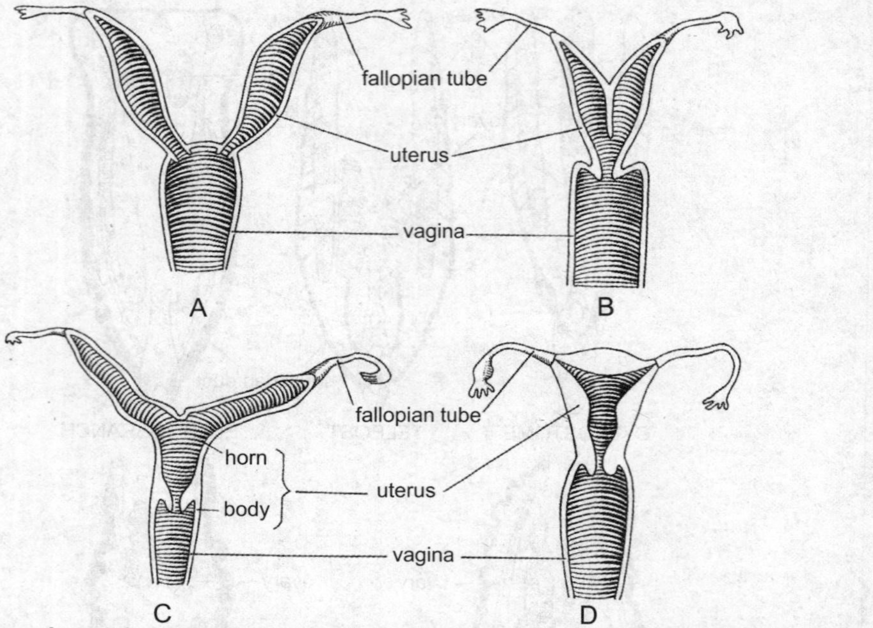

Fig. 48.10. Types of uteri found in mammals. A–Duplex type in marsupials; B–Bipartite type of some rodents; C–Bicornuate type of carnivores; D–Simplex type of primates.

The amphibian ovaries are paired and contain a cavity within them that is filled with lymph. The oviducts are also paired although in some forms the lower ends are fused together. Frequently the lower end of each oviduct is enlarged into a uterus-like structure or ovisac which serves as a temporary storage space for ova before they are shed. Glands that secrete a jelly-like covering for eggs are usually situated along the oviduct.

In reptiles also paired ovaries contain cavities filled with lymph as in amphibians, or may be solid as in birds and mammals. Paired oviducts open separately into the cloaca. The albumen and shell producing glands are located along the oviducts. Archinephric or Wolffian duct degenerates in female reptiles.

In most birds the right ovary and oviduct, although present in embryonic development, become vestigial so that only the left genital system is functional. Along the course of oviduct are several glands which secrete membranes over the eggs, including layers of albumen, shell membranes and a calcareous shell.

In female **amniotes** the development of mesonephric and Mullerian ducts is the same as in anamniotes, but the adult kidneys are metanephroi, each with its own ureter, hence, mesonephros and its duct degenerate leaving only vestiges. The **Mullerian duct** acts as an **oviduct** and the **metanephric duct** as the **ureter** on each side.

In most vertebrates an oviduct is not differentiated into parts, and the two oviducts remain separate and open into the cloaca, but in mammals each oviduct is differentiated into two or three parts : (*i*) a narrow anterior **Fallopian tube**, (*ii*) wide posterior **uterus** whose posterior part may be differentiated as a **vagina**. In **monotremes** the two oviducts are separate so that there are two separate **uteri** opening into the cloaca. **Monotremes** have no vagina. With the separation of urinary and genital systems from the rectum, the cloaca is eliminated in marsupials and eutherians. In **marsupials** there are two uteri (hence, Didelphia), and the terminal portions of oviducts are differentiated as **vaginae**. The two vaginae are separate but joined along the median line. They receive the penis of the male. But in **eutherian mammals** the two vaginae fuse into a single tube (hence, Monodelphia). The two uteri may fuse partly or completely, thus, four kinds of uteri are recognised in Eutheria :

1. Duplex uterus found in rodents, elephants, and some bats. The two uteri do not fuse but enter and open separately in the vagina. **2. Bipartite uterus** found in carnivores, pigs and cattle. The lower ends of the two uteri fused together so that they open by a single aperture (os uterus) into the vagina. **3. Bicornuate uterus** found in ungulates, whales and most bats. The lower ends of the two uteri fused more completely with no partition wall between the fused portions and this part has a single cavity. **4. Simplex uterus** found in Primates. The two uteri fused completely forming a single body with one cavity. There is no correlation between the development of various types of uteri and the phylogenetic relationship between the orders of mammals.

IMPORTANT QUESTIONS

Long Answer Questions

1. Give a general account of the evolution of urinogenital system in the vertebrates.
2. Give a comparative account of the structures of kidney in the vertebrates you have studied.
3. Trace the fate of pro-, meso-, and metanephros in vertebrates.
4. Describe the evolution of genital ducts in different vertebrates.
5. Give a comparative account of the male reproductive organs of frog and rabbit.
6. Compare the female genital system of a bird with that of a mammal.

Short Answer Questions

1. Enumerate the basic functions of mammalian kidney.
2. Write a comparative account of male copulatory organs of dogfish, frog, lizard, pigeon and rabbit.
3. Distinguish between Cowper's glands and perineal glands.
4. Write short note on the following :
   (*i*) Archinephron; (*ii*) Mesonephros; (*iii*) Metanephros (*iv*) Nephron; (*v*) Mulierian duct; (*vi*) Opisthonephros; (*vii*) Renal corpuscle; (*viii*) Pronephros.

Multiple Choice Questions

1 Functional unit of kidney is
   (*a*) Neuron (*b*) Nephron
   (*c*) Glomeruli (*d*) Renal corpuscle.
2. Hypothetical primitive kidney is
   (*a*) Archinephros (*b*) Pronephros
   (*c*) Metanephros (*d*) Nephron.
3. Archinephros is found in
   (*a*) Fishes (*b*) Amphibia
   (*c*) Reptiles
   (*d*) Larval cyclostomes.
4. Wolffian body is
   (*a*) Archinephros (*b*) Pronephros
   (*c*) Mesonephros (*d*) Metanephros.
5. Urinary bladder opens to exterior through
   (*a*) Pelvis (*b*) Tubule
   (*c*) Urethra (*d*) Ureter.
6. The functional adult gonad is derived from which part of genital ridge?
   (*a*) Pregonal (*b*) Gonal
   (*c*) Epigonal (*d*) Endogonal.
7. In mammals, testes are found in
   (*a*) Scrotal canal (*b*) Inguinal canal
   (*c*) Bidder's organ (*d*) Peyer's patches.
8. Oviduct in vertebrates is modified
   (*a*) Urinary duct (*b*) Inguinal canal
   (*c*) Wolffian duct (*d*) Mulierian duct.
9. In human, each kidney has nearly
   (*a*) 10,000 nephrons
   (*b*) 100,000 nephrons
   (*c*) 50,000 nephrons
   (*d*) 1 million nephrons.
10. Henle's loop and major portion of the collecting tubules are situated in the
   (*a*) Cortex (*b*) Pelvis
   (*c*) Medulla (*d*) Pyramid.

Answers

1. (*b*) 2. (*a*) 3. (*d*) 4. (*c*) 5. (*c*) 6. (*b*) 7. (*a*) 8. (*d*)
9. (*d*) 10. (*c*)

49

Chapter

Endocrine System in Vertebrates

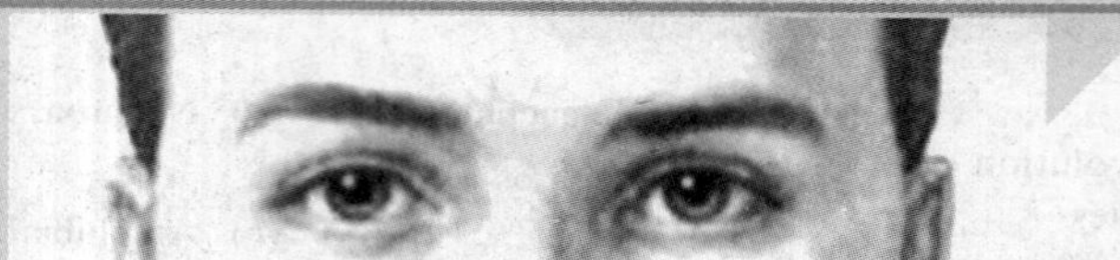

Contents of this Chapter

ENDOCRINE AND EXOCRINE GLANDS

Glands are the groups of cells which are specialised in structure and function. They produce chemical substances required for the body for various metabolic activities. These glands are of two types : endocrine and exocrine.

1. Endocrine glands. There are secretory glands found in various parts of the body which have no ducts of their own. Their secretions called **hormones** or internal secretions are carried by diffusion into blood vessels. These are chemical regulators of the body. They function essentially as catalytic agents by stimulating other glands, regulating growth, controlling metabolic activity, and maintaining chemical equilibrium in the body without undergoing change themselves. Such hormone secreting these glands are known as **endocrine glands** or ductless glands. All endocrine glands are derived from embryonic epithelial tissues.

2. Exocrine glands. Those glands which liberate their secretions into ducts are called **exocrine glands** or glands of external secretion. They may discharge their secretions into an internal organ, such as ducts of salivary glands and liver discharge their secretion into alimentary canal. Besides these, sweat and mammary glands produce their secretion to the exterior, *i.e.*, over the body surface.

A few glands like pancreas and gonads are exocrine as well as endocrine in nature. Pancreas secretes enzymes which are discharged into the duodenum through duct, whereas islets of Langerhans, endocrine part, secretes hormones which diffuses into the blood of the capillaries.

Hormones (Gr., hormon = arouse or exite). These are the secretions of endocrine glands. The term hormone was coined by **E.H. Ṣtarling** (British physiologist) in 1905.

Functions. 1. The hormones of endocrine glands coordinate the function of the body. Some of them act on muscles and glands due to secretion of chemicals in the same way as nerve impulses. Thus, endocrine glands assist the nervous system in integrating body activities. 2. Some hormones regulate metabolic activities, maintaining chemical equilibrium of the body and controlling growth. 3. Some endocrine glands play an important part in the development processes. 4. Some endocrine glands are influenced by or they influence other endocrine glands. 5. Some exert a restraining influence over the others and their secretions are called chalones in distinction from hormones or autocoids. 6. There are some endocrine glands which are under the control of the nervous system.

Properties of hormones. 1. The hormones are of a chemical nature which may influence the body generally or only certain specific organs. 2. Hormones diffuse rapidly through tissues and produce their effects quickly. 3. Their actions may be of long duration or they may induce immediate and short-term responses. 4. Hormones are destroyed rapidly after their release, so that they do not produce unsuitably long responses. If the balance of hormones in the body is upset, it results in various pathological manifestations.

Though the endocrine glands are widely separated, they are an integrated group linked together functionally by nervous and circulatory systems, so that collectively they constitute an **endocrine system** which coordinates and dominates various activities by the hormones, but the endocrine and nervous systems are not sharply separated from each other, and some of their activities overlap.

Classification of hormones. They may be classified on the basis of their biochemical nature as follows :

1. **Phenolic hormones** such as thyroxine and adrenaline.
2. **Steroid hormones** like estradiol, progesterone, testosterone.
3. **Proteinous hormones** like parathormone, insulin, prolactin.

ENODOCRINE GLANDS

The following diagram (Fig 49.1) shows the approximate position of the important endocrine glands in the body of a man and woman. The chief endocrine glands of vertebrates are : (i) thyroid, (ii) parathyroid, (iii) adrenal, (iv) pituitary, (v) thymus, (vi) pineal, (vii) pancreas, (viii) mucous membrane, and (ix) gonads.

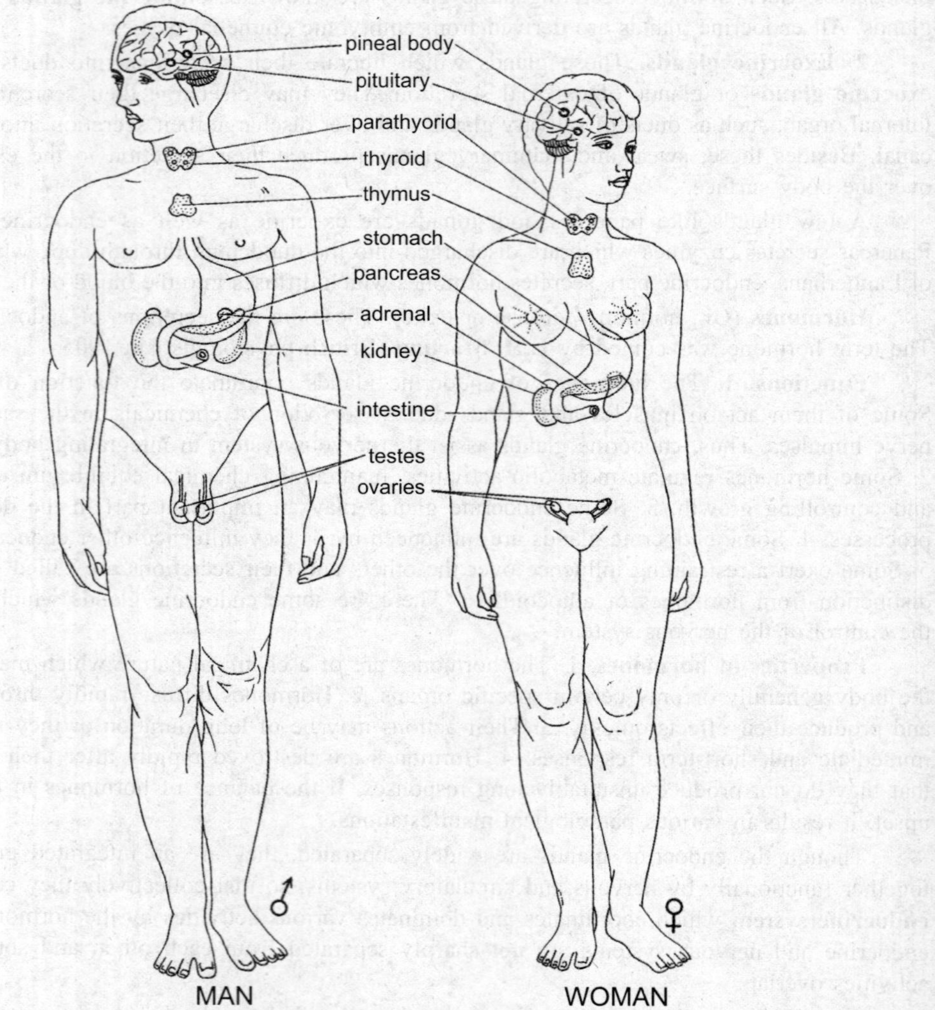

Fig. 49.1. Diagram showing the location of various human endocrine glands in man and woman.

1. Thyroid Gland

Development. Thyoid gland arises in an embryo as a diverticulum of the floor of the pharynx which proliferates to form follicles and loses its connection with the pharynx. In lower vertebrates it arises mid-ventrally from between the second and fourth visceral-clefts, but in higher forms it develops between the first and second clefts.

Structure. The thyroid has small follicles of cubical cells. The lumen of follicles is filled with an iodine-rich colloid. The follicles are enclosed in a capsule of connective tissue.

Fish-like vertebrates have well developed thyroid glands. It is unpaired in elasmobranchs but paired in adult amphibians, not only regulates metabolic activity in the body but is believed to be important in influencing the periodic moulting of the outer layer of the skin. In amniotes the two lobes of the thyroid are joined by a narrow **isthmus** of thyroid tissue. The thyroid is homologous with **endostyle** of lower chordates. In the ammocoete larva of *Petromyzon* and in *Amphioxus* there is an endostyle as an open groove in the floor of the pharynx and its cells contain iodine, but in gnathostomes its major part atrophies and the remaining part becomes the thyroid gland of the adult.

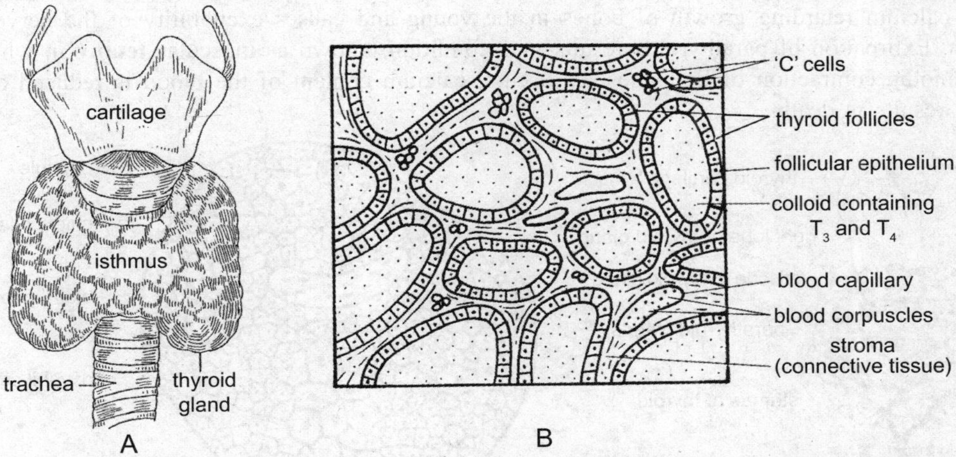

Fig. 49.2. A – Human thyroid gland seen from front; B –T.S. of thyroid gland.

Hormones and their Functions

1. Thyroxin. The hormone of thyroid is a protein, called **thyroglobulin** which breaks down into a number of molecules of an amino acid known as **thyroxin** which contains 65% iodine. It increases the rate of metabolism for release of energy, also affects other endocrine glands and is affected by them. In young animals it stimulates growth and in lower vertebrates it is responsible for metamorphosis. Tadpoles fed on thyroid extract quickly metamorphose into frogs without growing up. **Deficiency of thyroxin** in a young child causes a condition known as **cretinism** in which growth is retarded, intelligence is low, sex organs do not develop, legs are bent and the tongue and abdomen protrude out. If the thyroid degenerates in adult life, it results in a disease called **myxoedema** in which there is sluggishness, swelling of subcutaneous tissue mainly of the face and hands, body temperature is low, and the mind becomes dull.

Hypersecretion. Excessive production of thyroxin causes **exophthalmic goitre** in which thyroid is enlarged, eyes protrude, metabolic rate is high, heart beat is rapid and the body is wasted away. **Deficiency of iodine** in thryoxin causes simple goitre by a pathological enlargement of thyroid.

2. Calcitonin. Secretion of thyroid also contains a hormone, called **calcitonin** which keeps the bones healthy and prevents an undue loss of calcium from them.

2. Parathyroid Glands

Development. Parathyroid glands arise as ventral or dorsal outgrowths of the third and fourth visceral-clefts.

Position. In cyclostomes, cartilaginous and bony fishes parathyroid glands have not been found as compact glands but are probably represented by diffuse cells behind the last pair of visceral clefts. In tetrapoda they form one or two pairs of compact glandular masses lying close to the thyroid glands or partially embedded in the thyroid or thymus glands. In reptiles the parathyroid glands are usually situated posterior to the thyroid gland, which is unpaired, at least in adult reptiles. In man they form four minute yellowish glands attached to the dorsal side of thyroid.

Hormones. 1. Parathormone. The hormone of parathyroids is a protein, called **parathormone**. Its chemical nature is not known but it is concerned with maintaining the normal amount of calcium and phosphorus in the blood. It plays an important part in bone formation, a function it shares with vitamin D. In birds it plays an active role in regulating the formation of the calcareous shell of eggs. **Hyperactivity**. Over-secretion of parathormone causes calcium and prcsphorus to go out of bones and teeth, making them soft and weak. **Hypoactivity.** Under-secretion lowers blood calcium retarding growth of bones in the young and causes excitability of the nervous system. Extirpation of parathyroids results in a condition, known as muscular **tetany** in which a spasmodic contraction of limb muscles occurs, calcium content of the blood is reduced and finally results in death.

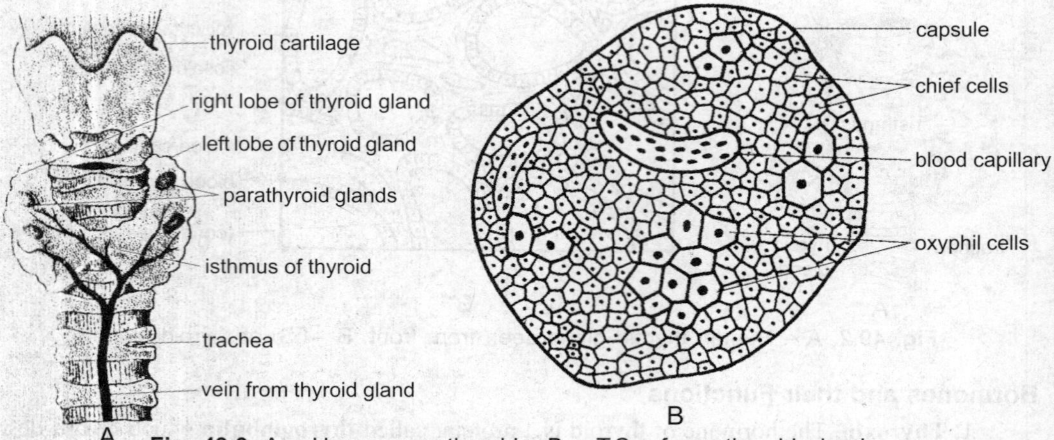

Fig. 49.3. A – Human parathyroids; B – T.S. of parathyroid gland.

3. Adrenal Glands

Adrenal glands occur in fishes as two separate sets of tissues called suprarenal which are homologous with the mammalian adrenal medulla and inter-renal which are homologous with the cortex in mammals. The **suprarenal bodies** or **chromaffin bodies** are bright yellow, small, paired masses of cells along the sympathetic ganglia. The **inter-renal bodies** are pink-coloured paired or unpaired masses of cells between the kidneys. In amphibians, reptiles and birds the two components of the adrenals, the cortex and medulla are joined together rather being separate in fishes. Adrenal glands lie close to the kidneys. In mammals the **adrenal glands** have two portions, an outer mesodermal **adrenal cortex** homologous with the inter-renal bodies, and an inner ectodermal **adrenal medulla** homologous with the suprarenal bodies or chromaffin bodies. The adrenal glands are attached in front of the kidneys.

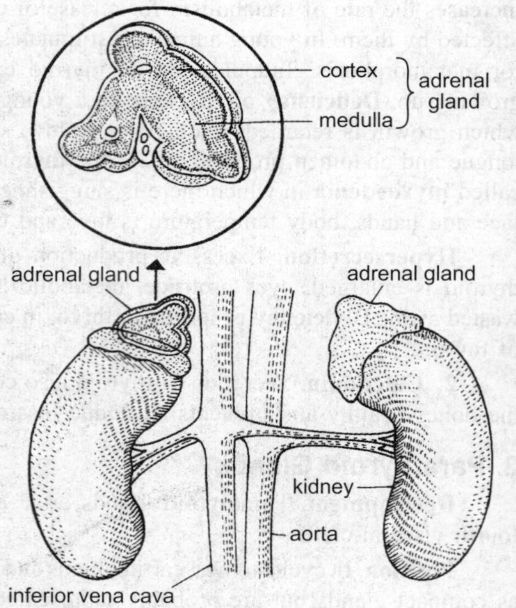

Fig. 49.4. Diagram showing the location of adrenal gland and its parts in human.

Hormones of cortex. Secretions of the cortex or inter-renal bodies are essential for life. They are concerned with carbohydrate metabolism, balance of salts in the blood, balance of sodium and fluids in the body, maintenance of the volume of circulating blood, control of sexual maturity, and with a resistance to many diseases. **1. Cortin.** Several hormones secreted by the cortex are collectively called **cortin** which maintains a balance between the salts in the blood and those in the bones. **2. Cortisone.** A steroid **cortisone** from the cortex of mammals is concerned with carbohydrate and protein metabolism. It is now used in treating arthritis and certain types of leukemia.

3. Aldosterone. The cortex also produces **aldosterone**, which controls sodium and potassium excretion.

Hyperactivity. Over-secretion of the cortex causes marked increase of male characters in both sexes. Removal of adrenal cortex results in death.

Hormones of medulla. The hormone of the medulla or suprarenal bodies is **epinephrine** or **adrenalin** which is regularly present in the blood. It maintains the tone of involuntary muscles keeping the blood vessels contracted and blood pressure normal. It controls the conversion of glycogen from the liver and muscles into blood sugar. The total effect of epinephrine is to provide the animal with a suitable respiratory rate and heart beat ensuring sufficient oxygen and blood sugar to meet the needs of the body in time of stress. Epinephrine is also secreted in response to emotional factors, such as anger, fear or pain, then it prepares the animal to fight or to retreat. When extra epinephrine is injected into the blood it increases the rate of heart beat, enhances blood pressure, dilates the air passages to the lungs, increases the amount of blood sugar by breaking down glycogen, and dilates the pupils. These effects of epinephrine are the same as those produced by stimulation of the autonomic nervous system, both affecting the same organs. A substance called **sympathin** is similar to epinephrine and is produced by the ends of the sympathetic nerve fibres.

4. Pituitary Gland

Pituitary gland lies below the diencephalon in a depression of the basisphenoid bone, called **sella turcica**.

Development. The pituitary is a compound organ. It is formed in the embryo by an ectodermal evagination of the stomodaeum, called **Rathke's pouch** which finally fuses with a ventral downgrowth of the infundibulum to form a **pituitary gland** or **hypophysis cerebri**. Except in a few lower forms the Rathke's pouch loses its connection with the stomodaeum. Some authorities use the term **hypophysis** for Rathke's pouch, while others use it for the entire pituitary gland, here the latter nomenclature is followed.

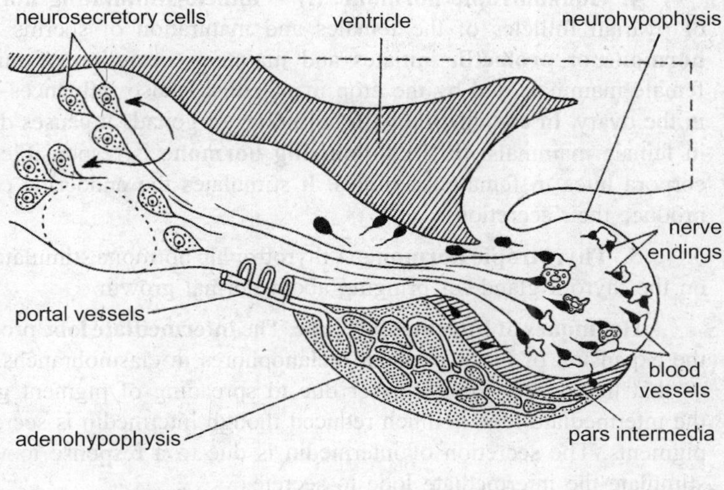

Fig. 49.5. Diagrammatic sagittal section of the hypophysis (pituitary) of a mammal.

Structure. The pituitary gland is well developed in fishes and other vertebrates. The pituitary gland or hypophysis is generally made of four parts : called **pars anterior**, **pars tuberalis,** **pars intermedia** and **pars nervosa**. The pars anterior and pars tuberalis are together known as the **anterior lobe** of pituitary, pars intermedia as **intermediate lobe**, and pars nervosa as the **posterior lobe**. The anterior and intermediate lobes are formed from the Rathke's pouch, while the posterior lobe comes from the infundibulum. The lobes of the pituitary are made of compact masses of cells with a hypophysial portal system, but the different lobes are unrelated functionally. Each lobe producing different hormones. The hormones of the pituitary gland not only affect other organs of the body but they also influence other endocrine glands. The pituitary is also influenced by other endocrine glands so that the members of the endocrine system cooperate in an adaptive fashion for the general welfare of the animal.

Hormones of Anterior lobe (Adenohypophysis). The anterior lobe of the pituitary gland produces about ten hormones which are as follows :

1. Growth hormone or somatotropic hormone. The growth hormone causes normal growth of the bones and the body of the animal. It probably stimulates the thyroid. **Over-secretion** of the growth hormone during young stages before bones have stopped growing leads to **gigantism** in which the bones and body become excessively large. If over-secretion of the growth hormone occurs after the long bones stopped growing then **acromegaly** occurs in which the bones of the head, jaws, hands and feet become abnormally large. Long bones become thick, and the spine is curved.

2. Diabetogenic hormone. A diabetogenic hormone maintains the balance between the blood sugar and liver glycogen. If this hormone is secreted in excess, then glucose will not be stored in the liver as glycogen but will remain in the blood to raise the sugar level which will cause **diabetes**. But the diabetogenic hormone is controlled by insulin of the pancreas.

3. Adrenotropic hormone. An adrenotrophic hormone stimulates and exerts a nutritive influence on the adrenal cortex or inter-renal bodies so that the body responds to counteract the harmful effects of stressess, such as starvation, burns and haemorrhage. The hypothalamus secretes a potent substance which acts on the anterior lobe of the pituitary to produce its hormone.

4. Gonadotropic hormone. (i) A **follicle-stimulating hormone** controls the development of ovarian follicles of the females and maturation of sperms in the males. **(ii)** A **lactogenic hormone** or **prolactin** initiates and maintains secretion of milk by the mammary glands of female mammals and by the crop in pigeons. It also influences the activity of the corpora lutea in the ovary. In conjunction with hormones of gonads it causes development of mammary glands in female mammals. **(iii)** A **luteinizing hormone** is responsible for ovulation and formation of corpora lutea in female mammals. It stimulates the endocrine cells of the ovaries and testes to produce their secretions.

5. Thyrotropic hormone. A thyrotrorhic hormone stimulates and exerts a nutritive influence on the thyroid gland for bringing about normal growth.

Hormones of Intermediate lobe. The **intermediate lobe** produces **intermedin** which controls the expansion of black pigment melanophores in elasmobranchs, amphibians, and many reptiles so that the skin becomes darker due to spreading of pigment granules. In birds and mammals the intermediate lobe is much reduced though intermedin is secreted but it has no effect on their pigments. The secretion of intermedin is due to a response to visual stimuli of the eyes which stimulate the intermediate lobe to secrete.

Hormones of Posterior lobe (Neurohypophysis). The **posterior lobe** of the pituitary produces **pituitrin** made of two hormones : **(i) vasopressin** or pitressin and **(ii) oxytocin** or pitocin. But it has been shown recently that the hormones of the posterior lobe are actually produced by the hypothalamus, and the posterior lobe only stores these hormones and is not a true endocrine gland.

(i) Vasopressin regulates blood pressure and acts on urinary tubules to reduce the amount of water production in urine.

(ii) Oxytocin causes contraction of muscles of the uterus, to hasten child birth. Injury to the posterior lobe leads to excessive secretion of dilute urine and a consequent thirst. The posterior lobe increases in size in females during pregnancy.

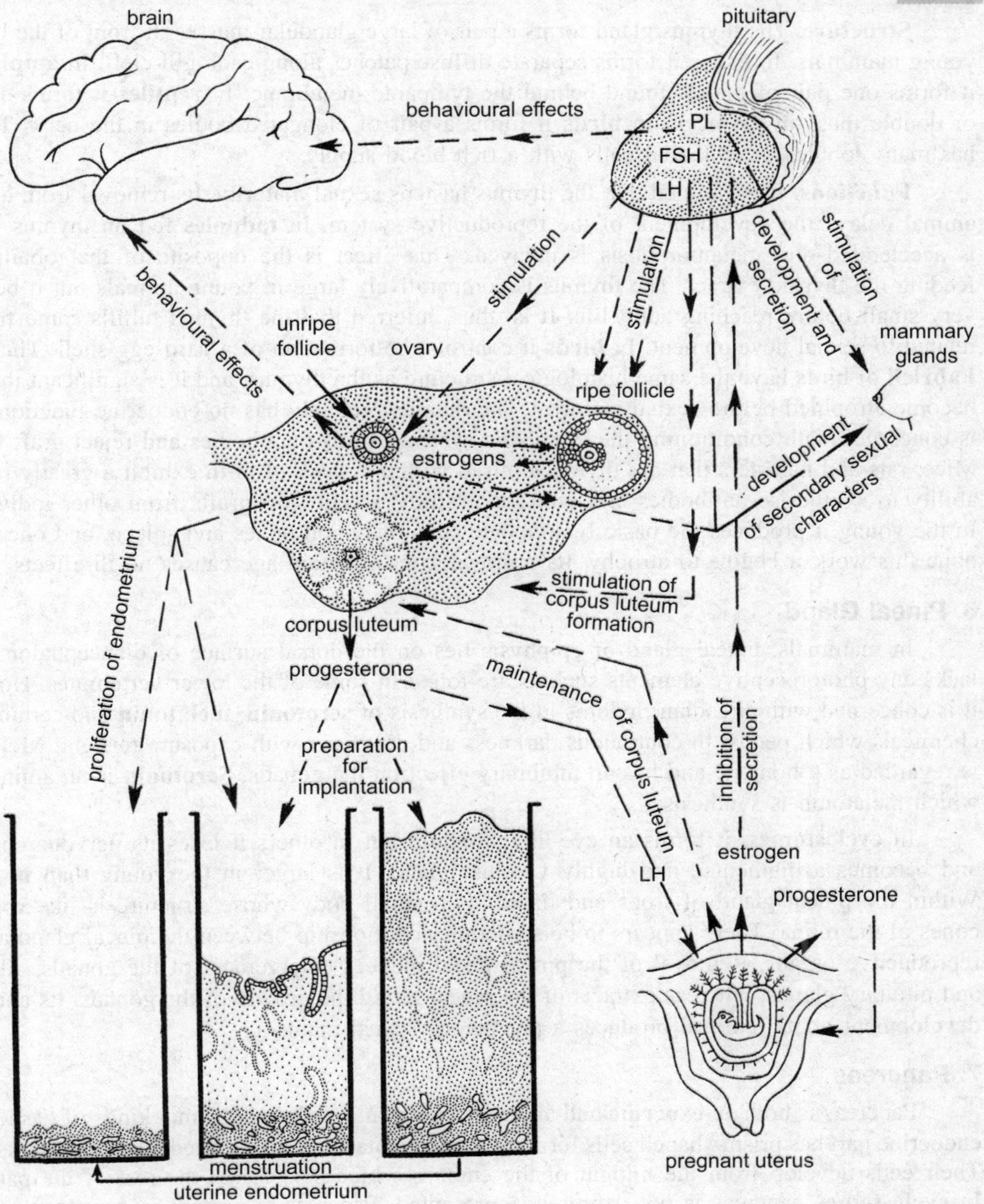

Fig. 49.6. Diagram showing the functions of various gonadotropin hormones secreted by pituitary glands.

5. Thymus Gland

Thymus gland, like the parathyroids and the thyroid, is derived embryologically from the foregut. In mammals this gland migrates posteriorly during development close to the anterior part of the heart. Thymus gland is formed from one to five pairs of visceral-clefts in different vertebrates. In mammals it is formed largely from the third pair. The second pair of visceral-clefts form **tonsils** which are found only in mammals as masses of lymphoid tissue. The fifth pair of visceral-clefts form **ultimobranchials** or **postbranchial bodies** in tetrapoda. Only the left ultimobranchial is present in lizards, and their occurrence in birds and mammals is doubtful. Ultimobranchials may be lymphoid tissue but their function is unknown.

Structure. The thymus gland forms a pair of large glandular masses in front of the heart in young mammals. In **fishes** it forms separate diffuse patches along each gill-cleft. In **amphibians** it forms one pair of glands found behind the tympanic membrane. In **reptiles** it forms a single or double mass in the neck. In **birds** it forms a pair of elongated bodies in the neck. Thymus has many lobules of epithelial cells with a rich blood supply.

Functions. It is claimed that the thymus hastens sexual maturity. Its removal from a young animal delays the development of the reproductive system. In **tadpoles** fed on thymus growth is accelerated and metamorphosis is delayed. This effect is the opposite of that obtained by feeding the thyroid extract. The thymus is comparatively large in young animals but it becomes very small before reaching adult life. It is, thus, inferred that the thymus fulfills some function related to sexual development. In **birds** it controls the formation of a hard egg shell. The **bursa Fabricii** of birds have the same histological structure as the thymus, and it is significant that both become atrophied before sexual maturity. The thymus probably has no endocrine function and it is concerned with conditioning the lymphatic system to form antibodies and reject graft tissues. Mice, rats and hamsters that are thymectomised within a week of birth exhibit a greatly reduced ability to synthesise antibodies and will successfully accept skin grafts from other individuals. In the young it produces the basic lymphocyte cells of lymph nodes and spleen, and once it has done this work it begins to atrophy. Its removal after a certain age causes no ill-effects.

6. Pineal Gland

In mammals, pineal gland or epiphysis lies on the dorsal surface of diencephalon, but it lacks any photoreceptive elements such as are found in some of the lower vertebrates. However, it is concerned with circadian rhythms in the synthesis of **serotonin**, **melatonin** and certain other chemicals which persist in continuous darkness and disappear with exposure to light. **Melatonin** is regarded as a homone and has an inhibitary effect on the gonads. **Serotonin** is an amine from which melatonin is synthesised.

In **cyclostomes**, it forms an eye-like structure but in others it loses its nervous character and becomes a pigmented and highly vascular gland. It is larger in the young than in adults. Within the pineal gland of frogs and lizards is a small body whose structure is like rods and cones of the retina. There appears to be some inter-relationship between the pineal gland and the reproductive organs. Removal of the pineal gland induces enlargement of the gonads, adrenals, and pituitary glands, whereas extracts of the pineal retard the growth of the gonads. Its abnormal development or its removal produces a precocious sexual maturity.

7. Pancreas

Pancreas is both an **exocrine** and an **endocrine** gland with two distinct kinds of tissues. The endocrine part has prism-shaped cells forming several isolated groups, called **islets of Langerhans**. Their cells develop from the midgut of the embryo which also forms the rest of the pancreas. In **cyclostomes** pancreas is not definitely represented. Pancreas is present in **cartilaginous** and **bony** fishes and possesses well marked islands of Langerhans. In some bony fishes the islets of Langerhans are grouped together outside the pancreas but in others they lie within the pancreas as anastomosing cords alternating with sinusoids forming rounded masses with no ducts.

Hormones. The islets of Langerhans have two types of cells known as alpha-cells and beta-cells. The **beta-cells** produce a hormone called **insulin**. In a normal animal insulin controls carbohydrate metabolism and stimulates the liver and striated muscles to store glycogen and make the tissues oxidise glucose as a source of energy. **Deficiency of insulin** causes a disease called **diabetes mellitus** in which there is an excess of unused sugar in the blood which is excreted with urine, so that the sugar level becomes low in the blood. The level of glycogen in liver and muscles decreases, and there is also an increased formation of carbohydrates from body proteins and fats so that the body is wasted away. There is muscular weakness and finally coma sets in.

The **alpha-cells** of the islets of Langerhans produce another hormone, known as **glucagon** which is controlled by a hormone from the anterior lobe of the pituitary and acts as an antagonistic agent to insulin increasing the amount of sugar in the blood by breaking down glycogen of liver into glucose. A **diabetogenic hormone** of the pituitary maintains the balance between blood sugar and liver glycogen. If the diabetogenic hormone is produced in excess, then sugar is not stored in the liver as glycogen but remains in the blood and causes diabetes. Thus, diabetes can be caused in two ways, either by under-production of insulin by the pancreas, or by over-production of diabetogenic hormone by the pituitary. But the diabetogenic hormone is controlled by insulin.

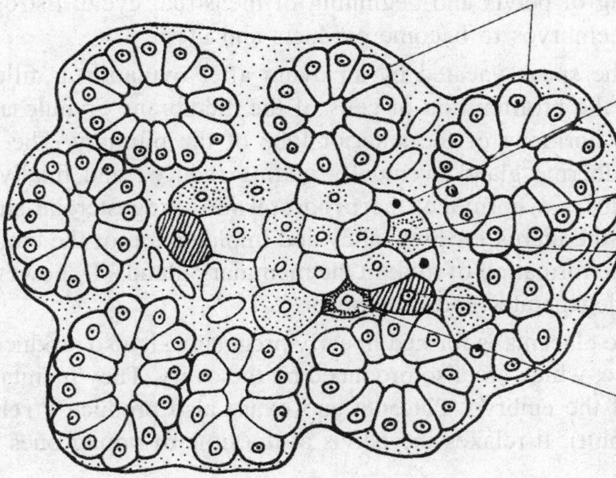

Fig. 49.7. A part of T.S. of pancreas.

8. Mucous Membrane

Mucous membrane of the alimentary canal contains digestive glands which produce digestive juices and also certain hormones which are as follows : **1. Gastrin.** The mucous membrane of the stomach is stimulated by gastric juice to produce a hormone gastrin which passes into the blood and then returns to the stomach causing the gastric glands to secrete gastric juice. **2. Enterogastrone.** The wall of duodenum secretes enterogastrone which inhibits gastric secretion. **3. Secretin.** The mucous membrane of the duodenum produces a hormone secretin which goes through the blood to the pancreas and causes secretion of pancreatic juice. **4. Cholecystokinin.** The mucous membrane of the duodenum also produces a hormone cholecystokinin which makes the gall bladder to release its stored bile.

9. Gonads

Gonads are both cytogenic (producing sex cell) and endocrine in function. The gonads of fishes, like those of other backboned animals are partly endocrine in function and are responsible for various secondary sexual characters that usually manifest themselves during the breeding season.

1. Testes are made of seminiferous tubules between which are found groups of interstitial cells or **cells of Leydig** which are endocrine in function and produce **testosterone**, the male hormone. Substances having properties of testosterone are called **androgens**. Some androgens are found in the male urine. Testosterone is a liquid and it brings about normal development and functioning of **accessory sex organs**, such as intromittent organs, reproductive ducts, and associated glands. Androgens prevent the development of Mullerian ducts in the males. Testosterone is responsible for development of **secondary sexual characters** which distinguish the male from the female such as antlers of stags, brighter plumage of male birds, hair on the face, deeper voices, larger body size and strength, and particularly male behaviour in mating. If a mammal is castrated before sexual maturity then its masculine characters and structures fail to develop, if it is castrated after attaining sexual maturity then its accessory sex organs are reduced. Testes mature under the influence of a hormone from the anterior lobe of the pituitary.

2. Ovaries. The hormones produced by the ovary are as follows :

(i) Estradiol or estrogens. The follicle cells around developing ova are responsible for secreting a female sex hormone estradiol. Substances having properties of estradiol are called estrogens. Androgens and estrogens are present in both sexes but androgens predominate in the males and estrogen in the females. Ovarian estrogens are partly responsible for development of female **accessory sex organs**, such as development of genital organs, *i.e.*, uterus, fallopian tubes, associated glands, and development of mammary glands, fat deposition beneath the skin, changes in body contour, broadening of pelvis and beginning of menstrual cycle. Estrogens differentiate the Mullerian ducts of the embryos to become oviducts and uteri.

(ii) Progesterone. The space vacated by an ovum after ovulation is filled with a yellow **corpus luteum** formed by the proliferation of cells of the membrane granulosa. Its formation is stimulated by a luteinising hormone of the anterior lobe of the pituitary. The corpus luteum is a true but a temporary endocrine gland eventually getting reduced and finally absorbed in the ovary. Corpus luteum produces a hormone, the **progesterone.** Progesterone suspends ovulation during pregnancy, prepares the uterus for pregnancy and implantation of the embryo. It maintains pregnancy and completes the formation of milk in the mammary glands. If, however, no pregnancy occurs the corpus luteum degenerates and disappears.

3. Placenta. When the placenta is formed in early pregnancy, it also produces two hormones, **estrogens** and **progesterone**, which are also produced by the ovary. They maintain the pregnancy, and proper development of the embryo. Placenta and ovary also produce a **relaxin** hormone in some mammals (*e.g.* in rabbit). It relaxes the pelvis at the time of young ones birth for its easy passage through vagina.

ENDOCRINE GLANDS IN VERTEBRATES

Elasmobranchii. Elasmobranchs have most of the vertebrate endocrine glands. They are thyroid pituitary, adrenal tissue, pancreas, mucous membrane of stomach and duodenum, and gonads. There are no thymus and parathyroids as such. The functions of **pituitary glands** are not known except that it stimulates and keeps the glands healthy and it causes expansion of melanophores. The single **thyroid** lies below the anterior end of ventral aorta, it often has small ciliated cavity as a remainder of its origin from the endostyle. The adrenal gland has two widely separate sets of tissues, the **suprarental bodies** or chromaffin cells are a segmental series of small glands along the autonomic ganglia some in the posterior cardinal sinus, some on each side of the oesophagus, and some in the kidneys. The **inter-renal bodies** lie between the kidneys as single or paired glands. Each **islet of Langerhans** has prism-shaped cells around a cavity. **Gonads** contain endocrine glands which produce steroid hormones, testes have interstitial cells which secrete testosterone, ovaries have follicle cells producing estrogen and progesterone.

Amphibia. All the vertebrate endocrine glands are present. The relationship of the various lobes of the **pituitary gland** differs from that of higher vertebrates because the pars anterior and pars tuberalis do not form an anterior lobe, and pars anterior lies posterior to other lobes. Hormones of the pituitary cause maturation of the sex cells and control colour changes of the skin, intermedin causing expansion of melanophores. **Thyroid** is paired and lies deeply, each thyroid lateral to the hyoid, its secretion besides controlling normal growth also brings about periodic ecdysis. **Parathyroids** are two pairs of small, red, round glands, one pair lying on each side of the posterior part of the hyoid. **Thymus gland** is formed from the second pair of visceral cleft. The two components of the **adrenal glands** are closely intermixed. They form a yellow band on the ventral surface of each kidney. The endocrine part of **gonads** is definitely present but is a matter of conjecture because there are almost no interstitial cells in the testes. There are a few follicle cells in the ovaries around developing ova. The hormone of the testes during the breeding season causes formation of certain secondary sexual characters, such as changes in reproductive ducts and a swelling or copulatory pad on the thumb. There are no corpora lutea in ovaries.

Reptilia. All vertebrate endocrine glands are present. **Thyroid** is a single gland in snakes and turtles but is paired in lizards and lies halfway below the trachea. It controls ecdysis. **Parathyroids** are two pairs at some distance posterior to the thyroid lying near the aortic arches. **Pituitary gland** lies somewhat posteriorly because the infundibular stalk is long, the anterior lobe is loosely connected in front to the other lobes, lizards and snakes have no pars distails. **Adrenals** lie close to gonads. The **pineal gland** registers solar radiations in lizards, and it probably secretes a homone which regulates the behaviour with regard to temperature and plays a part in the control of reproduction. **Gonads** control the appearance of secondary sexual characters during the breeding season, such as changes in the skin pigmentation, and in males the formation of femoral glands and hemipenes. Interstitial cells are present in testes as endocrine glands. In those females which give birth to young ones corpora lutea appear in the ovaries for a successful completion of pregnancy. Follicle cells are present in the ovaries.

Aves. All vertebrate endocrine glands are present. **Thyroid glands** are paired, one on each side of the trachea where it branches into bronchi. Thyroid secretes thyroxin which not only controls the general metabolic function but is also concerned with periodic moulting of feathers. **Parathyroids** are one or two small pairs in different birds, they lie slightly posterior to the thyroids. **Pituitary gland** has no intermediate lobe but intermedin is secreted by the anterior lobe. **Thymus** is a pair of elongated glands, one on each side of the trachea. They are large in a young bird but much reduced in the adult. **Adrenal glands** are yellow-coloured and lie close to the gonads anterior to the kidney, the suprarenal tissue lies in crevices between masses of itnerrenal tissue. **Testes** have extremely few interstitial cells but hormones are produced which control the functioning of the reproductive system and formation of secondary sexual characters, such as brighter colour of beaks and plumage, and in some the formation of a comb in the male. **Ovaries** have follicle cells as endocrine glands. But the formation of corpora lutea is doubtful in the single ovary of most birds.

The endocrine glands of **mammals** have been described earlier in the chapter.

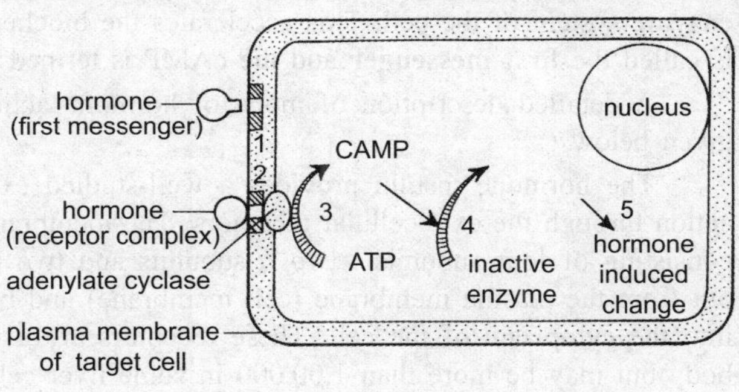

Fig. 49.8. Action of peptide hormone.

MOLECULAR MECHANISM OF HORMONE ACTION

The hormones are released in very small quantities, yet they cause wide spread responses in cells or tissues all over the body. These reponses can be quite specific and selective in different cells. All vertebrate hormones belong to one of the four chemical groups : (i) Hormones such as adrenaline and thyroid hormone are small molecules derived from amino acid tyrosine. (ii) Hormones such as vasopressin and oxytocin are short peptides. (iii) Hormones such as insulin and glucagon are longer polypeptide chains. (iv) Hormones such as testosterone and estrogen are steroid.

All hormones do not act in the same way. Catecholamines (epinephrine and norepinephrine), peptide and polypeptide (protein) are not liquid soluble, therefore, they cannot enter their target cells through the bilipid layer of plasma membrane (cell membrane). Instead, these water soluble hormones interact with a surface receptor, usually a glycoprotein, and, thus, initiate a change of events within it. The hormone insulin provides a well-studied example of how this happens.

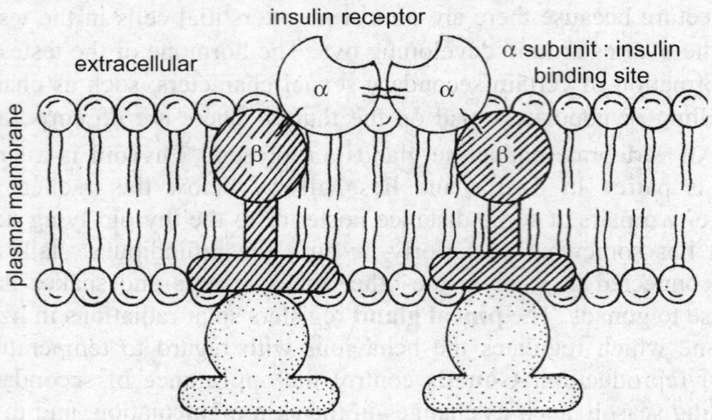

Fig. 49.9. Insulin receptor.

The molecular mechanism of hormone action is described under two separate headings : (i) mode of hormone action through the extracellular receptors and (ii) mode of hormone action through the intracellular receptors.

Mode of Hormone Action through Extracellular Receptors

The molecules of amino acid derivatives, peptides or polypeptides (protein) hormones bind to specific receptor molecules located on the plasma membrane. The hormone-receptor complex causes the release of an enzyme **adenylate cyclase** from the receptor site. This enzyme forms **cyclic adenosine monophosphate (cAMP)** from ATP of the cell. The **cAMP** activates the existing enzyme system of the cell. This accelerates the biochemical reactions in the cell. The hormone is called the **first messenger** and the cAMP is termed the **second messenger**.

A detailed description of mode of hormone action through the extracellular receptors is given below.

The hormone insulin provides a well-studied example to explain the mode of hormone action through the extracellular receptors. The membrane bound receptors of insulin is a protein consisting of four subunits : two α-subunits and two β-subunits. The two α-subunits protrude out from the plasma membrane (cell membrane) and bind insulin. The two β-subunits protrude into the cytoplasm of the cell. These receptors are usually less than 100 in most cells in our body but may be more than 1,00,000 in some liver cells.

(i) Binding to the receptor. Binding of insulin to the outer α-subunits of the receptor causes a structural change in the β-subunits, which is also an enzyme, a tyrosine kinase. The activated β-subunits add phosphate groups of specific tyrosine residues located in cytoplasmic domain of the receptor, as well as a variety of insulin receptor substrates.

(ii) Second messengers–the mediator. A second example which is widely used in hormonal control of cell function is through adenylate cyclase. The hormone-receptor complex does not directly stimulate adenylate cyclase. It is done through a transducer G-protein. **Alfred Gilmans** has shown that the G-protein is a peripheral membrane protein consisting of α, β and γ subunits. It interconverts between a GDP form and a GTP form. In muscle or liver cells, the hormones such as adrenaline bind receptor to form the hormone-receptor complex in the plasma membrane. The hormone-receptor complex induces the release of GDP from the G-protein. The α-subunit bearing GTP dissociates from βγ subunit of G-protein. The activated βγ-subunit of G-protein activates adenylate cyclase. The activated adenylate cyclase catalyses the formation of cAMP from ATP.

$$\text{ATP} \xrightarrow[\text{Mg}^+]{\text{Adenylate cyclase}} \text{cAMP} + \text{PPi} + \text{H}^+$$

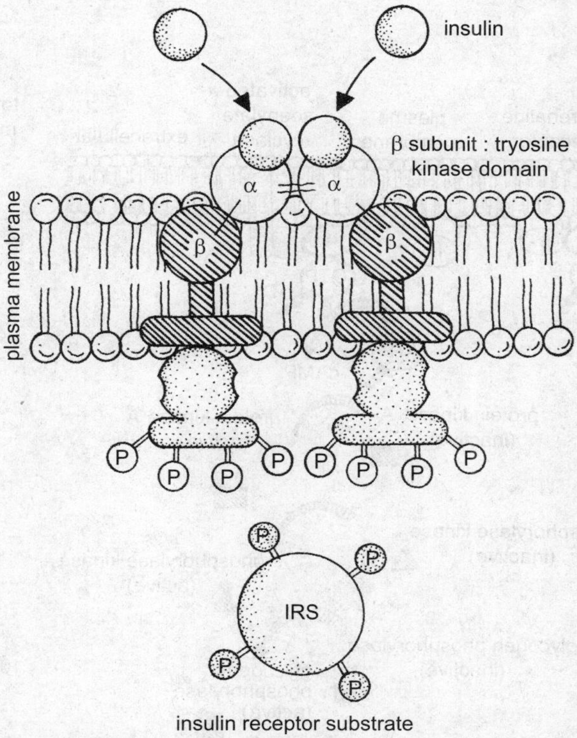

Fig. 49.10. Insulin hormone binding to its receptor.

G-protein activates enzyme **phosphodiesterase.** This enzyme makes **phosphatidylinositol 4,5 biphosphate** (PIP_2) into a pair of mediators : **inositoltriphosphate** (IP_3) and **diacylglycerol** (DG). IP_3 and DG are examples of second messengers. IP_3 is water-soluble, therefore, diffuses into cytoplasm to release another messenger Ca^{2+} ions from intracellular endoplasmic reticulum activating many calcium-mediated processes. While DG remains in the plasma membrane where it activates an enzyme **protein kinase C**, which in turn, activates many other enzymes such as pyruvate dehydrogenase to bring about the physiological effects.

(iii) Amplification of signal. The increased level of cAMP activates the enzyme **protein kinase** A. Activated protein kinase A activates the enzyme phosphorylase kinase. Each molecule of protein kinase activates about 100 molecules of enzyme, phosphorylase kinase and soon. As a result, a single molecule of adrenaline releases as many as 100 million molecules of glucose within only 1 or 2 minutes. However, a very small quantity of hormone is needed.

In a few instances, **cyclic guanosine monophosphate** (cGMP) which is only slightly different from cAMP acts in similar manner as a "second messenger".

(iv) Antagonistic effect. The effect of hormones which act against each other are called antagonistic effect. Many body cells use more than one second messenger. In heart cells, cAMP acts as a second messenger that increases muscle cell contraction in response to adrenaline, while **cyclic guanosine monophosphate** cGMP acts as another second messenger which decreases muscle contraction in response to acetylcholine. Thus, the sympathetic and parasympathetic nervous systems achieve antagonistic effect on heart beat. Another example of antagonistic effect is insulin and glucagon. Insulin lowers blood sugar level, while glucagon raises blood sugar level.

(v) Synergistic effect. When two or more hormones complement each other's actions and they are needed for full expression of the hormone effects, are called synergistic effects. For example, the production, secretion and ejection of milk by mammary glands require the synergistic effects of estrogens, progesterone, prolactin and oxytocin hormones.

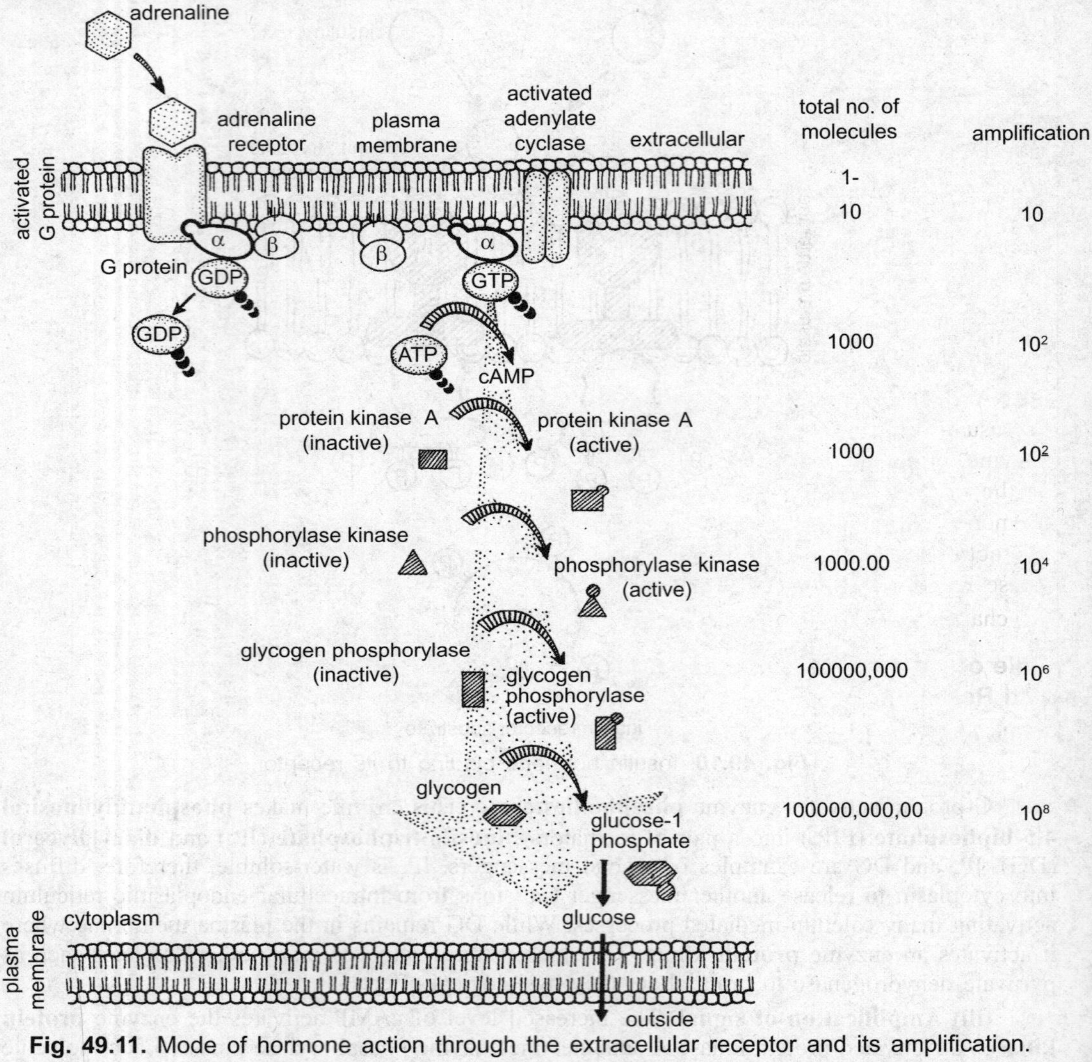

Fig. 49.11. Mode of hormone action through the extracellular receptor and its amplification.

Thus, it is evident from the above description that cAMP is not the only second messenger utilised by different hormones. Besides cAMP, certain other intracellular second messengers, as mentioned above, are cyclic guanosine monophosphate (cGMP), diacylglycerol (DG), inositol triphosphate (IP_3) and Ca_2^+

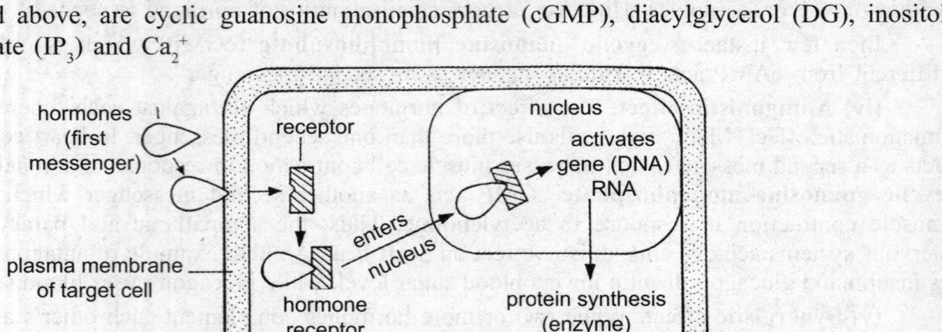

Fig. 49.12. Action of steroid hormone.

Mode of Hormone Action though Intracellular Receptors

Steroid and thyroid hormones are lipid soluble and easily pass through the plasma membrane (cell membrane) of a target cell into the cytoplasm. In the cytoplasm, they bind to specific **intracellular receptor proteins** forming a hormone-receptor complex that enters the nucleus. In the nucleus, the hormone-receptor complex binds to specific regulatory sites on the chromosomes and activates certain genes (DNA). The activated gene transcribes mRNA which directs the synthesis of proteins and usually enzymes in the cytoplasm. The enzymes promote the metabolic reactions in the cell. The actions of lipid-soluble hormones are slower and last longer than the actions of water-soluble hormones. These cause physiological responses that are characteristic of the steroid hormones.

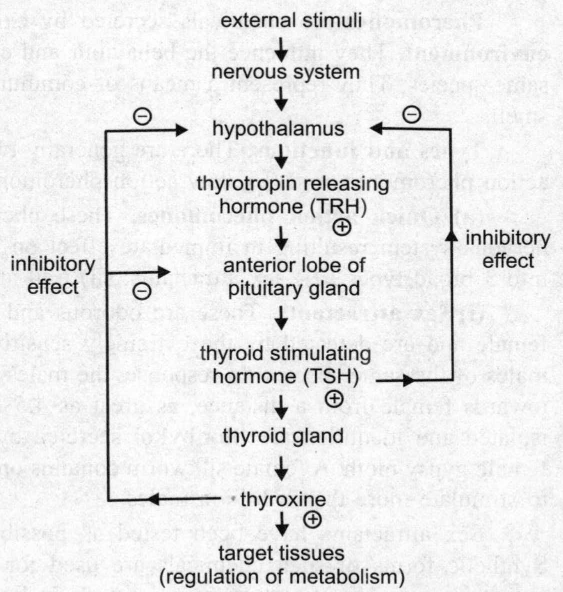

Fig. 49.13. Feedback control involving the hypothalamus, anterior lobe of pituitary gland, thyroid gland, target tissues.

Role of Hormones as Messengers and Regulators

(Role of Hormones in Homeostasis)

Hormones as messengers [Hypothalamo-hypophysial (pituitary) axis]. Hypothalamus is a part of the forebrain. Its hypothalamic nuclei-masses of gray matter containing neurons, are located in the white matter in the floor of the third ventricle of the brain. The neurons (neuronsecretory cells) of hypothalamic nuclei secrete some hormones called **neurohormones** (releasing hormones or releasing factors) into the blood. The neurohormones are carried to the anterior lobe of the pituitary gland (hypophysis), by a pair of hypophysial portal veins (Fig. 49.12). In the pituitary gland (hypophysis) the neurohormones stimulate it to release various hormones. Hence, the neurohormones are also called "releasing hormones or releasing factors".

Hormones as regulators (Feedback control). Homeostasis means maintenance of static or constant conditions in the internal environment of the body. Essentially all the organs and tissues of body perform functions. Hormones help to regulate cellular functions. When the secretion of hormones is under the control of factors or other hormones it is called **feedback control**. The regulation of secretion of thyroxine from the thyroid gland is an example of such feedback control mechanism. Feedback control is of two types.

(i) Positive feedback control. If the level of thyroxine is less than the normal limits in the blood, thyroxine level stimulates the hypothalamus to secrete more of TRH which results in the increased secretion of thyroxine. Such regulatory effect is called **positive feedback control**.

(ii) Negative feedback control. The thyroxine releasing hormone (TRH) from the hypothalamus stimulates the anterior lobe of the pituitary gland to secrete the thyroid stimulating hormone (TSH). The TSH in turn stimulates the thyroid gland to secrete **thyroxine**. A high level of thyroxine in the blood exerts an inhibitory effect on hypothalamus in such a way that less of TRH and TSH is produced respectively. This eventually results a decrease in thyroxine. This is called **negative feedback control.**

PHEROMONES

Pheromones are chemicals secreted by exocrine glands and released into the **external environment**. They influence the behaviour and cause specific reactions in other animals of the same species. They represent a means of communication, a means of imparting information by smell.

Types and functions. There are generally two major categories of pheromones : (a) quick action pheromones, and (b) slow action pheromones.

(a) Quick action pheromones. These pheromones act in some way on the recipient's nervous system resulting in immediate effect on its behaviour. These pheromones are grouped into 3 broad types : (i) sex attractants; (ii) trail attractants; and (iii) alarm pheromones.

(i) Sex attractants. These are odorous and volatile substances. These are exuded by the female and are detected by the extremely sensitive chemoreceptors present in the antennae of males of the same species. In response, the males become sexually stimulated and can be drawn towards female from a distance, as great as 4.5 km in some cases. Among the sex attractants isolated and identified are **bombykol** secreted by female silkworms, and **gyplure** secreted by female gypsy moth. A female silkworm contains on an average 0.01 mg of sex attractant sufficient to stimulate more than a billion males.

Sex attractants have been tested as possible specific insecticides in biological control. Synthetic forms of such chemicals are used for attracting insect pests which are killed with potent insecticides. If attractants are spread in large quantities over an insect-infested field, the confused males might never find females and mate with them.

(ii) Trail pheromones. Trail pheromones are secreted by the fire ants returning to their nest after finding food, so that other ants can also follow the trail to reach food. These are highly volatile and evaporate within two minutes only so that there is no risk of confusion created by the old trails.

Similarly, worker bees on finding food, secrete **geraniole**, to guide other worker bees to the source of food, in addition to their wagging dance.

(iii) Alarm pheromones. These pheromones are released by ants when disturbed. Alarm pheromones are less specific than sex attractants. They have a lower molecular weight. This causes several different species of ants responding to the same alarm substance.

(b) Slow action pheromones. These pheromones act more slowly and affect growth and differentiation of the recipients. They play an important role in regulating the composition and activities of colonial insects such as ants, bees and termites. Queen bees secrete **9-ketodeconic acid**. When ingested by worker bees, it retards development of their ovaries and ability to build royal cells rearing new species. It also serves as a sex attractant for males during nuptial flight.

In termites, special castes such as queens, kings and soldiers secrete inhibitory pheromones. These pheromones affect the corpus allatum of the nymphs and prevent their development into these specialised castes. If a queen dies, there is no longer any "anti-queen" pheromone released so that one or more nymphs develop into queens. Same happens in case of other specialised castes. The males of migratory locusts secrete pheromones which accelerate the growth of young locusts. In mice, if several females are put together in a cage their estrous cycle become erratic. However, if one male mouse is also placed with them, his odour synchronises the estrous cycles of all the females **(Whitten effect)**.

The existence of human pheromones remains doutful. According to French biologist **J. Le Magnen**, the odour of **15-hydroxy pentadeconic acid** is clearly perceived only by sexually mature women, while males and young girls are relatively insensitive to it. The observations also show that among room mates and close friends in girls, there is more synchronisation of menstrual cycles. Whether it involves any pheromones is not very well understood.

IMPORTANT QUESTIONS

Long Answer Questions

1. What is endocrine system? Discuss the role of this system in the life of man.
2. What are endocrine glands? Give their names and describe the structure and function of any four of them.
3. Write an essay on endocrine system in vertebrates.
4. What are hormones? Describe their chemical nature, properties and functions.
5. Describe the mechanism of hormone action.
6. Justify the statement "Pituitary is the master gland". Give an account of the hormones of the adenohypophysis and their functions.
7. Give a brief account of hormonal control of ovulation and pregnancy in mammals.
8. Give an account of hormones of neurohypophysis of pituitary. What is diabetes insipidous?
9. Give an account of the location, structure and origin of parathyroid glands. What is the role of parathormone in bone remodelling?
10. Describe in detail the role of pancreas in carbohydrate metabolism in the body.
11. Give an account of the endocrine role of gastrointestinal mucosa, skin and kidneys.

Short Answer Questions

1. Define endocrine glands, exocrine glands, hormone and pheromone with suitable example.
2. Write short notes on the following:
   (i) Acromegaly; (ii) Exophthalmia; (iii) Goitre; (iv) Islets of Langerhans; (v) Neurosecretion; (vi) Target organ; (vii) Pheromone.
3. What are the differences between hormones and vitamins?
4. What is goitre? What are its causes?
5. What are the causes of gigantism and acromegaly? What are differences between two?
6. What is the role of adrenaline hormone on human beings.
7. What do you know about Addison's disease?
8. Describe the hormones of thyroid gland. Give an account of their functions.

Multiple Choice Questions

1. What is the function of endocrine system?
   (*a*) Muscular coordination
   (*b*) Mechanical coordination
   (*c*) Chemical coordination
   (*d*) Nervous coordination.
2. Ductless glands producing inhabiting chemicals are
   (*a*) Exocrine glands (*b*) Endocrine glands
   (*c*) Receptors (*d*) Chalone.
3. Master of endocrine orchestra is
   (*a*) Thyroid (*b*) Hypothalamus
   (*c*) Pituitary (*d*) Thymus.
4. Thyroxine is a
   (*a*) Phenolic hormone (*b*) Steroid hormone
   (*c*) Proteinous hormone (*d*) Fatty acid.
5. Progesterone is secreted by the
   (*a*) Corpus luteum (*b*) Placenta
   (*c*) By young follicles (*d*) Both (*a*) & (*b*).
6. Menstrual cycle is controlled by
   (*a*) Testosterone
   (*b*) Testosterone and progesterone
   (*c*) Estrogens
   (*d*) Estrogens and progesterone.
7. In mammals the placenta secretes a hormone
   (*a*) Relaxin (*b*) Insulin
   (*c*) Toxin (*d*) Pepsin.
8. Corpus luteum is responsible for the production of
   (*a*) Progesterone (*b*) FSH
   (*c*) LH
   (*d*) Lactogenic hormone.
9. An endocrine structure that progressively degenerates later disappears during development
   (*a*) Testes (*b*) Ovary
   (*c*) Thymus
   (*d*) Reticuloendothelial system.
10. Thyroxine is
    (*a*) An enzyme (*b*) A hormone
    (*c*) A vitamin
    (*d*) An excretory product.
11. Growth hormone is produced by
    (*a*) Pituitary gland (*b*) Thyroid gland
    (*c*) Adrenal gland (*d*) Thymus gland.
12. Islets of Langerhans produce
    (*a*) Rennin (*b*) Ptylin
    (*c*) Insulin (*d*) HCl.

13. Integrative system in the body are
    (*a*) Nervous
    (*b*) Endocrine
    (*c*) Nervous and endocrine
    (*d*) Blood vascular.

14. Oestrogens are secreted by
    (*a*) Ovaries (*b*) Spleen
    (*c*) Testes (*d*) Adrenals.

15. The hormone which plays a role in converting glucose into glycogen is produced by
    (*a*) Pancreas (*b*) Pituitary
    (*c*) Parathyroids (*d*) Thymus.

16. When there is a marked decrease in the blood pressure which hormone could be administered to the patient?
    (*a*) Adrenaline (*b*) Insulin
    (*c*) Thyroxine (*d*) Prolactin.

17. If thyroid extract is added to water containing tadpole of frog it will
    (*a*) Kill the tadpoles
    (*b*) Hasten their metamorphosis
    (*c*) Slow down their metamorphosis
    (*d*) Keep them in larval stage.

18. Antidiuretic hormone produced by the pituitary gland helps in the absorption of
    (*a*) NaCl (*b*) Water
    (*c*) Sugar (*d*) HCl.

19. Which endocrine gland is absent in elasmobranch?
    (*a*) Pituitary
    (*b*) Islets of Langerhans
    (*c*) Thymus and parathyroid
    (*d*) Thyroid.

20. Hormonal activities of pituicytes are regulated by
    (*a*) Hypothalamus (*b*) Thyroid
    (*c*) Thymus
    (*d*) Negative feedback.

21. Pheromones are released into
    (*a*) Target cells (*b*) Blood
    (*c*) Coelom
    (*d*) External environment.

22. Bombykol is secreted by
    (*a*) Female silkmoth (*b*) Male silkmoth
    (*c*) Female gypsy moth
    (*d*) Male gypsy moth.

23. Gyplure is secreted by
    (*a*) Female gypsy moth (*b*) Male gypsy moth
    (*c*) Female silkmoth (*d*) Male silkmoth.

Answers

1. (*c*)	2. (*b*)	3. (*c*)	4. (*a*)	5. (*a*)	6. (*d*)	7. (*a*)	8. (*a*)
9. (*c*)	10. (*b*)	11. (*a*)	12. (*c*)	13. (*c*)	14. (*a*)	15. (*a*)	16. (*a*)
17. (*b*)	18. (*b*)	19. (*c*)	20. (*a*)	21. (*d*)	22. (*a*)	23. (*a*)	

50 Some Comparative Charts of Vertebrate Animal Types

Chapter

In this chapter a comparative survey of the anatomy of the typical vertebrate animal types, such as *Scoliodon* (dogfish), *Rana* (frog), *Uromastix* (lizard), *Columba* (pigeon) and *Oryctolagus* (rabbit) is given for the readers to have a cursory glance of vertebrate anatomy.

Contents of this Chapter

TABLE 50.1. COMPARISON : DISTRIBUTION, HABITS, AND HABITAT OF VERTEBRATE TYPES.

Characters	*Scoliodon* (Dogfish)	*Rana* (Frog)	*Uromastix* (Spiny-tailed lizard)	*Columba* (Pigeon)	*Oryctolagus* (Rabbit)
1. Distribution	Wide geographical distribution; recorded from Indian, Pacific and Atlantic Oceans along sea-coasts.	Found throughout India.	Found in sandy tracts in North-West India and Kathiawar (Gujarat).	Found throughout India. Widely distributed in Asia, Europe and Northern Africa.	Cosmopolitan in distribution.
2. Habitat	Marine like most sharks inhabiting an open sea; sometimes found in estuaries and rivers.	It lives in fresh-water or near water in damp places on land; **amphibious**.	It lives in dry, soft, sandy tracts with scant vegetation in burrows; **terrestrial**.	It lives in fissures and holes of rocks, old buildings and side walls of wells; **aerial**.	It lives in fields, grass-lands and open wood-lands and in burrows under ground; **terrestrial** and **fossorial**.
3. Habit					
(*i*) Mode of locomotion	Active swimming in water.	Jumping and leaping on land, swimming and floating in water.	Creeping and running on land.	Flying in air and walking on land.	Walking, running and leaping on land.
(*ii*) Nature	Active throughout, gregarious, and predaceous.	**Nocturnal**, harmless and solitary.	Gentle and quiet, can be non-poisonous, tamed in captivity, **diurnal** and solitary.	Inoffensive, harmless, timid, **diurnal** and gregarious.	**Crepuscular**, timid, can be domesticated and gregarious.
(*iii*) Food and feeding	Feeds on crabs, lobsters, worms, fishes, etc.; **carnivorous**.	Feeds on insects, worms, molluscs and small frogs; **carnivorous**, cannibalism also occurs.	Chiefly feeds on grasses, leaves and fruits; **herbivorous** and may also prey on insects.	Mainly feeds on grains, pulses, seeds of fruits and grasses; **herbivorous**. Sometimes on insects.	Feeds on green vegetation; **herbivorous** and also **coprophagous**.
(*iv*) Sound	Silent.	Croaking sound during mating season but usually silent.	Rarely very slow squeaking sound but usually silent.	Deep gootur-goon sound.	Usually silent, sometimes squeaking sound.
(*v*) Body temperature	Cold-blooded (poikilothermal).	Cold-blooded (poikilothermal).	Cold-blooded (poikilothermal)	Warm-blooded (homoio-thermal).	Warm-blooded (homoio-thermal).

(Contd.)

Characters	*Scoliodon* (Dogfish)	*Rana* (Frog)	*Uromastix* (Spiny-tailed lizard)	*Columba* (Pigeon)	*Oryctolagus* (Rabbit)
(*vi*) Hibernation and aestivation	None.	Hibernates in winter and aestivates in summer.	Hibernates in winter.	None.	None.
(*vii*) Breeding season	Throughout the year.	Rainy season from July to September.	Spring season from March to April.	Throughout the year.	Throughout the year.
(*viii*) Courtship and mating	Courtship occurs before mating (copulation). During mating male twines around female, inserts its claspers into female's cloacal aperture and transfers sperms.	Courtship occurs: during courtship male croaks loudly to attract its female. Mating (copulation) is called **amplexus** during which male mounts upon the back of female and clasps its abdomen with the help of copulatory pads on its hands so that eggs are expelled through female's cloacal aperture and then male sheds sperms over them.	Courtship takes place before mating (copulation). During mating male and female lie side by side opposing their cloacae, then male everts its hemipenes and inserts into the cloaca of female to transfer the sperms.	Courtship takes place before mating (copulation) which is much pronounced. During mating male rides on the back of female, their cloacae are everted and pressed against each other during which sperms are transferred into the cloaca of female from male.	Courtship does not take place. During mating (copulation) male inserts its penis into the vulva of female to transfer the sperms.
(*ix*) Fertilisation	Internal, takes place in a part of oviduct between oviducal funnel and shell gland.	External, takes place in open water.	Internal, takes place in oviduct.	Internal, takes place in oviduct.	Internal, takes place in oviduct.
(*x*) Oviparity and viviparity	**Viviparous** or **oviparous**; gives birth to 3-7 young ones.	**Oviparous**; female spawns several hundred eggs without shell in water.	**Oviparous**; female lays 10-15 dirty white coloured shelled eggs in burrow.	**Oviparous**; female lays 2 large-sized white shelled eggs at a time in the nest.	**Viviparous**; female gives birth to 5-8 young ones in a litter and protected in burrow.

(Contd.)

(*xi*) **Development**	Direct, takes place in uterus. No metamorphosis.	Indirect, takes place in water to produce tadpole larva which under goes metamorphosis to become adult.	Direct, takes place inside the egg shell. No metamorphosis.	Direct, takes place inside the egg shell. No metamorphosis.	Direct, takes place in uterus. No metamorphosis.
(*xii*) **Extra-embryonic membranes and placenta**	Anamniota, no embryonic membranes. Only yolk sac placenta present.	Anamniota, neither embryonic membranes nor placenta.	Amniota, embryonic membranes–chorion, amnion and allantois develop. No placenta.	Amniota, embryonic membranes–chorion, amnion and allantois develop. No placenta.	Amniota, embryonic membranes–chorion, amnion and allantois develop. Yolk-sac and allantoic placenta develop.
(*xiii*) **Parental care**	Does not exist.	Does not exist.	Does not exist. Egg develops with the help of sun heat.	Both parents incubate the eggs and feed youngs with pigeons' milk after their hatching till they are able to lead an independent life.	Mother feeds the young ones on its milk and also defends them from enemies.
(*xiv*) **Family life**	Not noticed.	Not noticed	Not noticed	Noticed, **monogamous**, *i.e.*, one male lives with one female throughout the life.	Noticed, but **polygamous**, *i.e,* one male may live with several females.
(*xv*) **Change of body colour**	Does not occur.	Change in body colouration takes place matching to the surroundings.	Does not occur.	Not noticed.	Not noticed.

TABLE 50.2 COMPARISON : EXTERNAL FEATURES OF VETEBRATE TYPES.

Characters	***Scoliodon* (Dogfish)**	***Rana* (Frog)**	***Uromastix* (Spiny-tailed lizard)**	***Columba* (Pigeon)**	***Oryctolagus* (Rabbit)**
1. **Shape and size**	Body spindle-shaped, elongated, stream-lined and bilaterally	Body stream-lined, pointed, triangular,	Body elongated, depressed. Measures from 20 -30	Body spindle-shaped or fusiform. Measures	Body elongated, pointed anteriorly

(Contd.)

Characters	*Scoliodon* (Dogfish)	*Rana* (Frog)	*Uromastix* (Spiny-tailed lizard)	*Columba* (Pigeon)	*Oryctolagus* (Rabbit)
	compressed. Measures from 30–60 cm in length. Adapted for swimming.	flattened dorso-ventrally. Measures from few to many cm (12-18 cm). Adapted for swimming as well as for living on land.	cm. Adapted for burrowing in the ground.	about 33 cm in length. Adapted for rapid flight.	and broad posteriorly. Measures about 40 cm in length. Adapted for leaping, walking and fossorial life.
2. Body colouration	Dorsal surface dark grey, ventral surface pale white, while the portion of caudal fin is more or less dark.	Dorsal surface is green with black spots and streaks, while ventral surface is paler. Colour changes to harmonise with the surroundings.	Dorsal surface is yellow-brown with dark spots, while the ventral surface is pale.	Body is slaty gray with glistening green and purple sheen on upper breast and around neck. Two black bars bars on the wings.	Wild rabbit dusty brown but colour of domestic rabbit varies greatly; they may be white, black or white with brown or black patches.
3. Exoskeleton and Integument	Dermal placoid scales embedded in skin which is rough.	No exoskeleton, skin surface smooth and slimy.	Horny over-lapping epi-dermal scales over the body and claws on digits. Skin surface dry and rough.	Body covered with horny epidermal feathers, claws on the digits, scales on feet and and beak. Skin soft and dry.	Soft fur of horny epider-mal hairs covering the body, claws on the digits. Skin soft and dry.
4. Body divisions	Body divisible into head, trunk and tail.	Body divisible into head and trunk only.	Body divisible into head, neck trunk and tail.	Body divisible into head, neck, trunk and tail.	Body divisible into head, neck trunk and tail.
			1 - HEAD		
(*i*) Position	Head lies in line with rest of the body.	Head lies in line with rest of the body.	Head lies in line with rest of the body.	Head lies at an angle with rest of the body.	Head lies at an angle with rest of the body.
(*ii*) Shape and size	Large, depress-ed and triang-ular, not distinct from trunk and produced anteriorly in the form of a	Large, pointed, triangular, flattened and forms a short, blunt anterior snout; not completely	Small, slightly triangular and depressed, forms a short blunt snout.	Small, round-ed, distinct and produced into an elon-gated and tough beak.	Large, pear-shaped, distinct and forms a large blunt snout or or muzzle.

(Contd.)

Characters	*Scoliodon* (Dogfish)	*Rana* (Frog)	*Uromastix* (Spiny-tailed lizard)	*Columba* (Pigeon)	*Oryctolagus* (Rabbit)
	wedge-shaped snout or rostrum.	marked off from the trunk.			
(*iii*) **Mouth**	Small, ventral, crescentic opening bounded by jaws without lips.	Large terminal, wide semicircular opening bounded by hard immovable lips.	Large, wide, terminal transverse slit-like opening bounded by hard immovable lips provided with scales.	V-shaped, terminal wide opening bounded by horny upper and lower beaks. Lips absent.	Subterminal transverse slit-like bounded by movable fleshy lips. Upper lip divided into two halves by a vertical cleft and bears tactile vibrissae.
(*iv*) **Nostrils**	Ventral, cresentic partly covered by skin fold and exclusively olfactory in nature.	Dorso-lateral at the tip of snout, small circular openings and only respiratory in nature.	Antero-dorsal on the snout, small oval with a valve, olfactory and respiratory in nature.	Dorsal, oblique slit-like at the base of upper beak covered by soft skin called cere and respiratory in nature.	Just above mouth, large, oblique slit-like, without valves, respiratory as well as olfactory in nature.
(*v*) **Ear openings**	Neither ear openings nor tympanum.	Ear openings absent. Behind each eye a circular tympanic membrane covered by skin present.	Behind each eye is a sunken pit. At the base of the pit tympanic membrane present.	Below and behind eye is the ear opening covered by feathers. Tympanic membrane sunken into external auditory meatus.	Ear opening is surrounded by movable pinna. Tympanic membrane is sunken at the base of external auditory meatus.
(*vi*) **Eyes**	Small, lateral and circular. Pupil narrow and vertical. Eyelids immovable. Nictitating membrane movable.	Large, dorso-lateral and bulging, upper eyelid fleshy; opaque immovable and lower eyelid thin, transparent movable. Nictitating	Small, dorso-lateral in the middle of head and elliptical. Pupil round, eyelids movable. Nictitating membrane movable.	Large, lateral, round with circular pupil, eyelids. movable. Nictitating membrane movable.	Large, lateral oval with bulging cornea with round pupil. Eyelids movable, bear eyelashes. Nictitating membrane movable.

(Contd.)

Characters	*Scoliodon* (Dogfish)	*Rana* (Frog)	*Uromastix* (Spiny-tailed lizard)	*Columba* (Pigeon)	*Oryctolagus* (Rabbit)
		membrane movable.			
(*vii*) Gill-slits	Five vertical and oblique gill-slits behind each eye.	Absent in adults but present in larva.	Absent.	Absent.	Absent.
(*viii*) Brow spot	Absent.	Median faint brow spot on the head between eyes.	Absent.	Absent.	Absent.
(*ix*) Vocal sacs	Not found	One on either side of ventro-lateral surface of head as bluish wrinkled skin patch in males only.	Absent.	Absent.	Absent.
2 - NECK					
(*i*) Shape and size	Absent.	Absent.	Short having prominent folds of skin on the ventro-lateral side.	Long, cylindrical, flexible and well demarcated from head and trunk.	Short, flexible, vertical and well demarcated from head and trunk.
3 - TRUNK					
(*i*) Shape and size	Large, almost elliptical in transverse section, gradually tapers behind into tail.	Short, broad and somewhat dorso-ventrally flattened with a characteristic sacral hump on back.	Large, elongated, broad, convex dorsally and flattened ventrally.	Compact, spindle-shaped with a characteristic keel at its midventral line.	Large, cylindrical, differentiated into anterior thorax and posterior abdomen.
(*ii*) Integument or skin	Rough due to embedded dermal scales, no wrinkles or folds and attached firmly with muscles.	Smooth, slimy, with wrinkles as dermal plicae and attached loosely with the muscles due to lymphatic space filled with lymph.	Rough with scales, dry and wrinkled.	Dry, loose, covered with feathers and thin.	Glandular, thick, loose and covered with hairs.
(*iii*) Limbs	Trunk with fins for locomotion; 2 dorsal and 1 ventral median	Paired limbs present; forelimbs smaller than hindlimbs. Hand	Paired fore- and hindlimbs present. Both limbs with 5 clawed digits.	Forelimbs modified into wings bearing long feathers	Paired fore- and hindlimbs. Hand bears 5 clawed digits adapted for

(Contd.)

Characters	*Scoliodon* (Dogfish)	*Rana* (Frog)	*Uromastix* (Spiny-tailed lizard)	*Columba* (Pigeon)	*Oryctolagus* (Rabbit)
	fins, and paired pectoral and pelvic fins on the trunk.	bears 4 fingers and foot bears 5 toes connected by thin transparent skin (web). Claws not found, adapted for jumping on land and swimming in water.	Web not found. Adapted for creeping and burrowing on land.	for flying and 3 clawless digits. Hind legs stout, foot bears 4 digits with claws, adapted for walking and perching.	digging the burrow. Foot bears 4 clawed digits, adapted for walking and leaping.
(*iv*) Teats or nipples	Not found.	Not found.	Not found.	Not found.	On the ventral side between thorax and abdomen 4–5 pairs of teats or nipples present which are well developed and functional in females only. Rudimentary in males.
(*v*) Copulatory organs and scrotal sac	Copulatory organs are a pair of rod-like grooved **claspers** in male. Testes abdominal scrotal sac absent.	Not found.	Copulatory organs are a pair of **hemipenes** in male projecting out from cloaca. Testes abdominal. Scrotal sac absent.	Not found.	Copulatory organ is a cylindrical erectile **penis** in male. Testes migrate out of the abdomen in a special scrotal sac at the base of penis.
(*vi*) Apertures	**Cloacal aperture** situated between two pelvic fins on the ventral side as longitudinal slit-like opening. A pair of **abdominal pores,** one on either side	**Cloacal aperture** small, circular and terminal at the vent. No abdominal pores.	Thick-lipped, small, transverse **cloacal aperture** at the junction of trunk and tail on the ventral surface. Abdominal pores absent, but 2-18 **preanofemoral**	Large, transversely elongated **cloacal aperture** on the ventral surface at the junction of uropygium with the trunk. Neither abdominal nor	Cloaca absent. Separate anus and urinogenital openings present at the hind end.

(Contd.)

Characters	*Scoliodon* (Dogfish)	*Rana* (Frog)	*Uromastix* (Spiny-tailed lizard)	*Columba* (Pigeon)	*Oryctolagus* (Rabbit)
	of cloaca, communicating coelom to the exterior.		**pores** present in a single row on the ventral surface of thigh from cloaca to knee joint.	preanofemoral pores.	
(*vii*) Lateral lines and dorsal line.	A pair of lateral lines run on either side of the body from the head to the posterior end of tail.	Lateral lines absent in adults. A mid-dorsal line runs from the tip of the snout to the vent on the dorsal surface.	No such lines are marked.	Such lines absent.	Such lines are not found.
			4 - TAIL		
(*i*) Shape and size	Laterally compressed, long, tapers gradually bearing a caudal fin which is heterocercal. Used for swimming.	Absent in adult but tail and caudal fin present in larva.	Long, thick and broad at the base and tapers posteriorly bearing transverse rows of strongly keeled scales. Used as an organ of offence and defence.	Short, stumpy and conical bearing oil gland and feathers, used for steering . during flight.	Short, bushy, curved upwards. Used for giving warning signals when danger approaches and as balancer while leaping.

TABLE 50.3. **COMPARISON : INTEGUMENT OF VERTEBRATE TYPES.**

Characters	*Scoliodon* (Dogfish)	*Rana* (Frog)	*Uromastix* (Spiny-tailed lizard)	*Columba* (Pigeon)	*Oryctolagus* (Rabbit)
(*i*) Skin surface and attachment	Rough, stiff and leathery. Firmly attached to body muscles.	Thin, smooth, slimy and moist. Loosely attached to body due to subcutaneous lymph spaces below dermis.	Thick, dry and rough. Forms loose folds along the sides of neck and trunk.	Thin, dry, hard and flexible. Attached loosely for free movement during flight.	Thick, dry, elastic, glandular, loosely attached and modified variously.
(*ii*) Colour	Dark grey dorsally and pale white ventrally. More darker in the	Dark green on back and limbs with black and brown patches and streaks but	Yellow-brown on the dorsal surface with dark spots, while the ventral	Colour of rock pigeon is slaty grey with metallic green and	Colour of rabbit is dusty-brown.

(Contd.)

Characters	*Scoliodon* (Dogfish)	*Rana* (Frog)	*Uromastix* (Spiny-tailed lizard)	*Columba* (Pigeon)	*Oryctolagus* (Rabbit)
	tail region. In fishes colour patten is great.	on the ventral side pale yellow.	surface is pale. In reptiles pattern is elaborate for concealment and warning.	purple sheen on the upper breast and around neck. 2 black bars on each wing.	
(*iii*) Ability to change in colour	Skin colour does not change. However, some fishes have protective colouration.	Skin colour changes to harmonise with the surroundings – a protective measure.	Skin colour does not change. Colour change occurs in *Calotes* and chamaeleons.	Body colouration does not change.	Body colouration does not change.
(*iv*) Epidermis	Simple, thin, composed of many layers of epithelial cells. Cornified stratum corneum not found. No moulting.	Many layered epithelial cells which are stratified. Stratum corneum of thin, flat and dead keratinized cells which sheds in patches.	Stratified stratum corneum well developed and composed of thin, flattened dead cells which is sloughed off periodically in bits or in one piece.	Stratified, comparatively thinner and periodically sloughed off and replaced.	Highly stratified. Stratum corneum is composed of hard, scale-like, dead, fully keratinized flattened cells and cast off in shreds. No regular moulting.
(*v*) Dermis	Consists of connective tissue fibres which run parallel to the surface, blood and lymph vessels and also pigment cells.	Consists of an outer loose stratum spongiosum having loose connective tissue, blood vessels and lymph spaces, and an inner stratum compactum having dense connective tissue layer with horizontal and vertical strands. Pigment cells also present in dermis.	Thick layer being composed of muscle-fibres, blood vessels, nerves, connective tissue and also pigment cells.	Thin layer, consists of connective tissue, muscle fibres, blood capillaries, nerves and without pigment cells.	Thickest, consists of connective tissue fibres, unstriped muscles, blood capillaries, nerves, fat cells, tactile receptors and pigment cells. Deeper part of dermis is composed of subdermal adipose tissue.
(*vi*) Pigmentation	Pigment cells– chromatophores and iridophores containing	Chromatophores are located in the upper	Chromatophores are found in epidermis.	Pigment granules in the feathers, neither in epider-	Pigment granules present in the hairs and

(Contd.)

Characters	*Scoliodon* (Dogfish)	*Rana* (Frog)	*Uromastix* (Spiny-tailed lizard)	*Columba* (Pigeon)	*Oryctolagus* (Rabbit)
	guanin are present in dermis.	portion of dermis.		mis nor in dermis. Colour appears due to reflection and refraction of light by feathers.	epidermis. Pigment cells found in the dermis.
(*vii*) Epidermal glands	Numerous unicellular mucous secreting goblet gland cells in epidermis. Multicellular poison glands and luminous glands (photophores) are present in some fishes.	Multicellular mucous gland cells in stratum spongiosum which open at the surface of epidermis. Poisonous parotoid gland present in toads.	Mucous glands not found in skin except a few femoral glands on the ventral surface of the thigh of male individuals. Some reptiles possess scent or musk glands.	Glands not found in skin except a single preen or oil-gland on tail. Mucous glands absent.	Skin highly glandular but no mucous glands. These are sweat glands, sebaceous glands, mammary glands, and scent glands.
(*viii*) Dermal scales	In the form of placoid scales derived from dermis.	Absent in frog. In some caecilians scales are found embeddend in skin.	Dermal scales absent in *Uromastix*. In some lizards, crocodilians and turtles dermal scales or scutes or plates are found.	Dermal scales absent in birds.	Dermal scales or plates found in armadillos and whales.
			FUNCTION		
	Protective, sensory, gives texture to body. Not related to respiration because impermeable to water.	Protective, sensory, gives texture to body. Respiratory structure because permeable to water.	Protective, sensory, water-proof, hence, not related to respiration.	Protective, sensory, insulating, water-proof covering. Not related to respiration.	Protective, sensory, insulating and water-proof. Not related to respiration.

TABLE 50.4. COMPARISON : EXOSKELETAL STRUCTURES OF VERTEBRATE TYPES.

Characters	*Scoliodon* (Dogfish)	*Rana* (Frog)	*Uromastix* (Spiny-tailed lizard)	*Columba* (Pigeon)	*Oryctolagus* (Rabbit)
1. Kind	Exoskeleton present in the form of dermal placoid scales.	Exoskeleton not found.	Exoskeleton present in the form of epidermal horny	Exoskeleton present in the form of feathers, claws,	Exoskeleton present in the form of hairs, scales and

(Contd.)

Characters	*Scoliodon* (Dogfish)	*Rana* (Frog)	*Uromastix* (Spiny-tailed lizard)	*Columba* (Pigeon)	*Oryctolagus* (Rabbit)
			scales and claws.	scales and sheaths of beak.	claws.
2. Epidermal scales	Epidermal scales absent. Dermal placoid scales, mesodermal in origin, found embedded in skin and arranged in oblique rows all over the body.	Epidermal and dermal scales absent.	Scales are epidermal in origin. These are thin, small overlapping covering the whole body.	Scales are very small on lower legs, feet and base of beak which are epidermal in origin.	Scales are not found in rabbit. But found on feet and tails of rats, beavers, etc.
3. Dermal scales	Dermal scales of 4 types (cosmoid, ganoid, ctenoid and cycloid) are present in bony fishes.	Absent.	Dermal scutes and plates are found in snakes, crocodilians and turtles. Absent in *Uromastix*.	Exoskeleton in the birds includes scales, claws, horny sheaths of beaks and feathers.	Exoskeleton in the mammals includes hoofs, nails, horns, etc.
4. Feathers and hairs	Absent.	Absent.	Absent. But horns are present in horned toad, rattle in rattle snakes and horny beak in turtles.	Hairs not found. Feathers present; these are quill feathers (remiges, rectrices and coverts), contour feathers, filoplumes and down feathers which are dermal in origin. Horny sheaths of beaks present in all birds.	Feathers absent. Hairs present, epidermal in origin. Hairs are modified into hair horns in rhino, scales in pangolins spines in echidna, quills in porcupines. Claws become nails in primates or hoofs in ungulates. Horns and antlers found in ungulates.
5. Moulting	Scales do not moult but grow in size with age.	Moulting of stratum corneum occurs in the form of small pieces.	Scales shed off periodically in small bits.	Feathers shed off and replaced seasonally.	Shedding of hairs does not not take place. But horny covering of pronghorns and skin of antlers shed seasonally.

TABLE 50.5. COMPARISON : SKULLS OF VERTEBRATE TYPES.

Characters	*Scoliodon* (Dogfish)	*Rana* (Frog)	*Uromastix* (Spiny-tailed lizard)	*Columba* (Pigeon)	*Oryctolagus* (Rabbit)
1. Shape	Dorso-ventrally flattened and elongated.	Dorso-ventrally flattened, triangular and broad.	Dorso-ventrally flattened and elongated.	Dorsally convex and elongated.	Dorsally convex and elongated.
2. Surface, weight and pneumacity	Smooth without sutures, heavier and solid.	Rough with distinct sutures and light in weight.	Rough with distinct sutures, heavier and solid.	Smooth, sutures not seen due to complete fusion of bones, lighter and pneumatic.	Rough having distinct sutures, light and few bones spongy.
3. Nature of skull skeleton	Cartilaginous only, chondrocranium persists throughout life. Dermatorcranium absent.	Poorly ossified, chondocranium much cartilaginous present. Dermatocranium present.	Entire chondrocranium is ossified except nasoethmoidal region. Dermatocranium present.	Ossified chondrocranium is replaced by neurocranium. Dermatocranium present.	Ossified completely, chondrocranium is replaced by neurocranium. Dermatocranium present.
4. Occipital condyles	Dicondylic.	Dicondylic.	Monocondylic.	Monocondylic	Dicondylic.
5. Number of skull bones	The whole chondrocranium is made of a single piece of cartilage; bones are not found.	Few bones; basisphenoid, alisphenoid, orbitosphenoid. Presphenoid, supraorbitals, postorbital, absent.	Large number of bones; alisphenoids, orbitosphenoids, presphenoids absent; supraorbitals, postorbitals, prefrontals, epipterygoid present.	Few and thin dermal bones; alisphenoids, orbitosphenoids and presphenoids present.	Prefrontal, postfrontal, parasphenoid, quadratojugal absent; alisphenoids and basisphenoids present.
6. Cranium	Simple, small.	Small and narrow due to small brain.	Small cranium.	Large and rounded,	Larger, rounded and dome-shaped.
7. Roof of cranium	Flat arched cartilaginous roof. Dermal bones absent.	Roof formed by fusion of paired fronto-parietal bones. No parietal fossa.	Flattened roof formed by frontal and parietal bones and a median parietal foramen present.	Roof formed by completely fused parietal and frontal bones and no parietal foramen.	Roof formed by distinct paired frontals and parietals and no parietal foramen.
8. Floor of cranium	Flattened floor; cartilaginous.	Floor by T-shaped parasphenoid.	Floor by basisphenoid and reduced parasphenoid bones.	Floor by basitemporal and much reduced parasphenoid bones.	Floor by basisphenoid and presphenoid bones.

(Contd.)

Characters	*Scoliodon* (Dogfish)	*Rana* (Frog)	*Uromastix* (Spiny-tailed lizard)	*Columba* (Pigeon)	*Oryctolagus* (Rabbit)
9. Occipital region and foramen magnum	Cartilaginous. Posteriorly directed foramen maganum.	Consists of paired exoccipitals. Posteriorly directed foramen magnum.	Consists of supraoccipital, paired exoccipitals and basioccipital bones. Posteriorly directed foramen magnum.	Consists of usual 4 bones fused inseparably. Ventrally directed foramen magnum.	Consists of usual 4 bones completely fused. Ventrally directed foramen magnum.
10. Interorbital septum	Absent. Skull platybasic.	Absent. Skull platybasic.	Thin septum. Skull tropibasic.	Thin septum. Skull tropibasic.	Thin septum. Skull tropibasic.
11. Sense capsules **(*i*) Auditory bones**	Absent.	Auditory capsule formed of prootic bone.	Epiotic and opisthotic bones fused with exoccipital and supraorbitals respectively and free bone prootic.	Prootic.	Auditory capsule is formed of periotic bone by the fusion of prootic, epiotic and opisthotic.
(*ii*) Tympanic bone	Absent.	Tympanic bone ring-like.	Tympanic bone annular.	Tympanic bone annular.	Tympanic bone flask-shaped forms tympanic bulla with periotic bones.
(*iii*) Ear ossicles	No. Hyomandibular present.	Only columella is the ear-ossicle which is modified hyomandibular.	Ear ossicle is columella.	Columella present which is modified hyomandibular.	Ear ossicles are malleus, incus and stapes derived from articular, quadrate and hyomandibular bones respectively.
(*iv*) Eye orbits	Eye orbits are formed by preorbital and post-orbital cartilages. having supra-orbital ridges.	Eye orbits are bordered by pterygoids, palatines and fronto-parietal.	Eye orbits are bordered by pre-frontal, lacrimal, supra-orbital, post-orbital and they are separated by interorbital septum.	Eye orbits are separated by inter-orbital septum and bounded anteriorly by frontal, posteriorly by post-orbital process of frontal and alisphenoid.	Eye orbits are bordered dorsally by frontals and laterally by zygomatic arch.

(Contd.)

Characters	*Scoliodon* (Dogfish)	*Rana* (Frog)	*Uromastix* (Spiny-tailed lizard)	*Columba* (Pigeon)	*Oryctolagus* (Rabbit)
(*v*) Olfactory capsule	Two olfactory sacs are large oval separated by internasal septum.	Two olfactory sacs are roofed by nasals, floored by vomers separated by mesethmoid.	Two olfactory capsules are formed by separate nasals from the dorsal side and a vomer from the ventral side.	Olfactory capsules greatly reduced covered by Y-shaped nasal bones, internasal septum cartilaginous, ectoethmoids or turbinals poorly developed.	Olfactory capsules are roofed by nasal and floored by vomers which are fused with internasal septum. Olfactory capsule encloses scroll bones formed of ethmo-, maxillo-and naso-turbinals.
12. Upper jaw	It is formed by two palatoquadrate cartilages joined in front, derived from mandibular arch.	It is formed by premaxilla, maxilla, quadratojugal bones and cartilaginous quadrate.	Formed by premaxilla, maxilla, jugal and quadrate bones from outer side, while pterygoid, palatine, transpalatine, epipterygoid and squamosal from inner side.	Formed by premaxilla, maxilla, jugal; quadrato-jugal and quadrate from outer side, while pterygoid and palatine from inner side.	It consists of premaxilla maxilla, jugal, pterygoid and palatine.
13. Lower jaw	Formed by two Meckel's cartilages united anteriorly, derived from mandibular arch.	Formed of Mentomeckelian, dentary and angulo-splenial bones and Meckel's cartilage.	Formed of articular, angular, supra-angular, coronoid, splenial and dentary bones around an axial Meckel's cartilage.	Formed of articular, angular, supra-angular, dentary and splenial bones surrounding a Meckel's cartilage.	Formed of a single dentary or mandible bone.
14. Teeth	Homodont, homologous to placoid scales, not attached to jaw cartilages.	Homodont and acrodont; situated on premaxilla, maxilla and vomer bones.	Homodont and pleurodont; situated on premaxilla, maxilla and dentaries.	Teeth not found.	Heterodont and thecodont; situated on premaxilla, maxilla and dentaries.

(Contd.)

Characters	*Scoliodon* (Dogfish)	*Rana* (Frog)	*Uromastix* (Spiny-tailed lizard)	*Columba* (Pigeon)	*Oryctolagus* (Rabbit)
15. Jaw suspensorium	Hyostylic; both jaws suspended from chondocranium through hyomandibular.	Autostylic; upper jaw fused with cranium, lower jaw articulates with quadrate.	Autostylic; upper jaw fused with cranium, lower jaw articulates with quadrate with its articular cartilage.	Autostylic; upper jaw fused with cranium; articular of lower jaw articulates with quadrate.	Craniostylic; upper jaw fused, lower jaw (dentary) articulates with squamosal of skull.
16. Miscellaneous **(*i*) Lacrimal**	Lacrimal not found.	Lacrimal absent.	It is small perforated.	It is large perforated.	Small, not perforated.
(*ii*) Sphenethmoid	Sphenethmoid absent.	Present.	Not found.	Not found.	Not found.
(*iii*) Quadrate	Quadrate absent.	Present. Cartilaginous.	Rod-like, small and not movable.	Y-shaped, movable, stout bone.	Modified into incus–an ear ossicle.
(*iv*) Zygomatic arch	Zygomatic arch not found.	Absent.	Absent.	Absent.	It is characteristic formed by the fusion of the processes from squamosal and maxilla.
(*v*) Rostrum or snout	Rostrum or snout is formed of 2 dorso-lateral and 1 ventro-median cartilaginous bars.	Jaws form the snout.	Jaws form the snout.	Snout not found; **beak** is present, formed by premaxilla, maxilla and dentaries.	No beak; snout is formed by jaws.
(*vi*) Secondary palate	Absent.	Absent.	Absent.	It is incomplete, formed by palatines and pterygoids.	It is complete, formed by palatines, maxillae and premaxillae.

TABLE 50.6. COMPARISON : PECTORAL GIRDLES (SHOULDER GIRDLES) OF VERTEBRATE TYPES.

Characters	*Scoliodon* (Dogfish)	*Rana* (Frog)	*Uromastix* (Spiny-tailed lizard)	*Columba* (Pigeon)	*Oryctolagus* (Rabbit)
1. Nature	Not much developed; cartilaginous.	Well developed; bony and cartilaginous both.	Well developed; bony and cartilaginous both.	Well developed for flight; stout; bony only and light.	Well developed for various purposes; largely bony.
2. Location	Found embed-	Found in	Located in the	Located at the	Located at the

(Contd.)

Characters	*Scoliodon* (Dogfish)	*Rana* (Frog)	*Uromastix* (Spiny-tailed lizard)	*Columba* (Pigeon)	*Oryctolagus* (Rabbit)
	ded in the ventro-lateral body wall posterior to gill-clefts and supports pectoral fins.	thoracic region, provides attachment to forelimbs and their muscles, protects the heart and lungs.	ventro-lateral thoracic wall; protects the heart and supports forelimbs.	antero-dorsal sides of the trunk and supports the wings.	antero-lateral sides of trunk, protects the heart and supports forelimbs.
3. General Shape	U-shaped.	Inverted arch-like in shape.	Inverted arch-like.	Somewhat V-shaped.	Somewhat triangular in shape.
4. Gross structure	Composed of right and left halves fused mid-ventrally.	Composed of two identical halves which are united mid-ventrally.	Composed of two halves which are united mid-ventrally.	Composed of two halves separated widely from each other.	Composed of two halves separated widely from each other.
5. General attachments	Neither attached to vertebral column nor to sternum which is absent.	Both halves united mid-ventrally with sternum.	Both halves unite ventrally with T-shaped interclavicle and rhomboidal sternal plate.	Both halves connected firmly with sternum through V-shaped furcula formed of two clavicles and an interclavicle.	Both halves neither connected to sternum nor to vertebral column.
6. Components	Each half is made of a scapulocoracoid cartilage being differentiated into scapula and coracoid parts.	Each half is made of two bones, scapula and coracoid.	Each half is composed of two bones, scapula and coracoid.	Each half is made of two bones, scapula and coracoid.	Each half is composed of a single coraco-scapula bone.
7. Scapula	Rod-like tapering structure directed dorsally.	Scapula is flat, stout bony plate-like and broader at both the ends.	It is bony, oblong and flat plate-like which is narrow in the middle and broader at the outer end and inner end is narrow.	It is long, flattened sabre-shaped dorsal bone lying above the thoracic ribs and parallel to the vertebral column.	The coraco-scapula is mainly formed of scapula which is thin, flat and somewhat triangular bony plate-like.
8. Scapular process	Absent.	Absent. At the the anterior end of scapula is present mesoscapular	Scapula gives out an anterior ossified process, **mesoscapula**.	It is connected with coracoid by ligaments. Anterior end of scapula	At its outer surface a prominent ridge called scapular

(Contd.)

Characters	*Scoliodon* (Dogfish)	*Rana* (Frog)	*Uromastix* (Spiny-tailed lizard)	*Columba* (Pigeon)	*Oryctolagus* (Rabbit)
		process.		bears an scapular tubercle and acromian process.	spine present, bearing at its tip acromian and metacromian processes.
9. Supra-scapula	Suprascapular cartilage present at the dorsal end of scapula.	It is thin, broadly flattened plate-like calcified cartilage at the dorsal end of scapula and covers the first 4 vertebrae of the vertebral column.	It is more or less rectangular, thin plate of calcified cartilage attached to the dorsal end of scapula. Its distal border is free and does not cover the vertebrae.	Not found.	It is in the form of thin strip of calcified cartilage situated along the dorsal border of scapula.
10. Coracoid	Ill-developed ventral flat-plate-like part of scapulocoracoid cartilage; supports the floor of pericardial cavity.	Broad and dumb-bell-shaped. A rod-like cartilaginous precoracoid is separated anteriorly from it forming a coracoid foramen. Both the coracoids meet mid-ventrally by an X-shaped catilaginous epicoracoid.	It is a large, flat and fenestrated bone; two large fenestrae divide the coracoid into outer procoracoid, middle mesocoracoid and inner coracoid proper. Epicoracoid irregular cartilaginous plate-like along its anterior border covering the fenestrae.	It is a stout, straight, pillar-shaped bone; its lower end articulates with the sternum and anteriorly it has an acrocoracoid process. Epicoracoid not found.	It is reduced, present in the form of small knob-like coracoid process of scapula above the glenoid cavity. Epicoracoid absent.
11. Glenoid cavity	Not found; pectoral fins are directly attached to pectoral girdle.	Present at the junction of scapula and coracoid; articulates the head of humerus.	Present at the junction of scapula and coracoid; both bones share in its formation and receives the head of humerus.	Present as a cup-shaped depression at postero-lateral joint of scapula coracoid bones and articulates with the head of humerus.	Present at postero-ventral apex of scapula only for the articalation of the head of humerus.
12. Clavicles	Present, placed ventrally.	Slender rod-like transverse bone on either side attached in front of precoracoid cartilage.	Small, slender, curved bone on either side separated in the middle by interclavicle.	Clavicles are long, slender, rod-like bones, attached dorsally to scapula and	Slender, thin curved bony rod, articulates with manubrium of sternum on

(Contd.)

Characters	*Scoliodon* (Dogfish)	*Rana* (Frog)	*Uromastix* (Spiny-tailed lizard)	*Columba* (Pigeon)	*Oryctolagus* (Rabbit)
				coracoid, while fused ventrally with interclavicle.	one hand and acromian process of scapula on the other.
13. Interclavicle	Not found.	Not found.	Present, T-shaped bone between clavicles and two halves of pectoral girdle.	Present, the two clavicles fused ventrally with a small interclavicle forming a disc called **hypocleidium** so that a V-shaped composite bone is formed, called **furcula** or merry thought or wishbone.	Interclavicle is not found in mammals except egg-laying mammals.
14. Foramen triosseum	Not found.	Not found.	Not found.	Found at the place where dorsal end of clavicle, acromian process of scapula and acrocoracoid process of coracoid meet. It is for the passage of the tendon of pectoralis minor muscle.	Not found.
15. Adaptation	Adapted to support the pectoral fins which help in swimming.	Adapted to support the stout forelimbs.	Adapted to support the forelimbs which help in creeping.	Adapted to support the strong wings which help in flight.	Adapted to support the strong forelimbs which help in running and burrowing, etc.

TABLE 50.7. COMPARISON : PELVIC GIRDLES (HIP GIRDLES) OF VERTEBRATE TYPES.

Characters	*Scoliodon* (Dogfish)	*Rana* (Frog)	*Uromastix* (Spiny-tailed lizard)	*Columba* (Pigeon)	*Oryctolagus* (Rabbit)
1. Nature	Not much developed, simple flattened and cartilaginous rod called ischio-pubic bar.	Well developed, bony and carti-laginous both.	Well built, bony, strong and solid. adapted for tetrapod locom-tion.	Well built, bony, large and light, adapted for bipedal locomotion.	Well built, solid, bony and heavy, adapted for running.
2. Location	Found embed-ded in ventral abdominal wall in front of cloaca to support pelvic fins.	Found in the posterior region of trunk to support pelvic region and hindlimbs.	Found located at the hind end of trunk in pelvic region to support the hindlimbs.	Found in the pelvic region lying on either side of the synsacrum for the support of hindlimbs.	Found located in the pelvic region to support the hindlegs.
3. General shape	Transverse rod-like.	V-shaped	Each half triradiate in shape.	Flattened plate-like.	Each half somewhat triradiate in shape.
4. Gross structure	Composite rod-like structure; not divided into two halves; pelvic fins are directly attached to it.	Composed of two identical halves (each called os-innomi-natum) which are united posteriorly into a median disc.	Composed of two identical halves (each called os-innominatum) which meet mid-ventrally but not united.	Composed of two identical halves (each called os-innominatum) which remain separated from one another.	Composed of two similar halves (each called os-innominatum) which are firmly united together mid-ventrally at pubic symphysis.
5. Attachment	Not attached to vertebral column.	Both halves run parallel with vertebral column. Median disc supports urostyle.	Iliac bones attached to first sacral vertebra.	Attached to synsacrum firmly.	Ilia articulates with sacrum.
6. Components (bones)	Not differenti-ated into separate bones.	Each half is composed of 3 bones– ilium, ischium and pubis.	Each half is formed of ilium, ischium, and pubis.	Usual 3 bones- ilium, ischium and pubis are the components of one half of the girdle.	In addition to usual ilium, ischium and pubis bones there is a cotyloid bone in each half of the girdle.
7. Ilium	Ilium is represented by a small iliac	Ilium is greatly elongated which runs forwards	Ilium is stout, rod-like; its tip articulates with	Ilium is an elongated and expanded	Ilium is the most anterior and longest

(Contd.)

Characters	*Scoliodon* (Dogfish)	*Rana* (Frog)	*Uromastix* (Spiny-tailed lizard)	*Columba* (Pigeon)	*Oryctolagus* (Rabbit)
	process having a foramen for the passage of nerve.	to meet the transverse process of 9th vertebra. There is a prominent vertical ridge on its dorsal surface called iliac crest. Both ilia unite posteriorly with median disc forming iliac symphysis.	the notch in the transverse process of 1st sacral vertebra. It bears preacetabular process. No iliac symphysis since both ilia are wide apart from each other.	bone, differentiated into preacetabular and postacetabular parts as it extends both anterior and posterior to acetabulum and fused with synsacrum and posterior part with the ischium. No iliac symphysis.	bone with a flat articular surface for articulation with the sacrum; its anterior and dorsal edge is raised into iliac crest. Neither iliac symphysis nor antitrochanter process.
8. Antitrochanter process	Absent.	Absent.	Absent.	Antitrochanter process above the acetabulum for articalation with greater trochanter of femur.	Absent.
9. Ischium	Ischium not distinct from pubis.	Ischium is small and slightly oval bone, meets with its fellow by ischiatic symphysis in postero-dorsal part of disc.	Ischium is flat, oblong bone, runs inward to meet its fellow of the other side at the ischiatic symphysis. From the symphysis is projected a small cartilaginous preischium in front and a hypoischium behind. It articulates with pubis and ilium of its side.	Ischium is a dorsoventrally flattened bone projecting backwards behind the acetabulum. It is fused posteriorly with ilium but separated anteriorly from it by an ilio-ischiatic foramen. Neither hypoischium nor ischiatic symphysis present.	Ischium is a posterodorsal broad, slightly curved bone having a prominent projection called ischial tuberosity. Neither hypoischium nor ischiatic symphysis present.
10. Ilio-ischiatic foramen	Absent.	Absent.	Absent.	Present at the junction of anterior part of ischium and postacetabular part of ilium.	Absent.

(Contd.)

Characters	*Scoliodon* (Dogfish)	*Rana* (Frog)	*Uromastix* (Spiny-tailed lizard)	*Columba* (Pigeon)	*Oryctolagus* (Rabbit)
11. Pubis and prepubis	Pubis not distinct from ischium. **Prepubis** and **epipubis** absent.	Pubis of each half is somewhat reduced, triangular piece of calcified cartilage situated posteriorly on on the ventral side. It is fused with ischium and ilium to form the disc. **Epipubis** and **prepubis** absent.	Pubis is a long slender bone. It meets with its fellow at pubic symphysis. **Epipubis** as a nodule of calcified cartilage at the anterior end of pubic symphysis. **Prepubis** is a small rod-like process arising from the middle of pubis.	Pubis is a long slender bone directed parallel to the outer margin of ischium backwardly to which it is usually fused. Pubic symphysis and epipubis absent. Prepubis is absent in pigeon. In fowl prepubis projects in front of acetabulum as a process.	Pubis is a narrow bone which joins with its fellow of the opposite side on the mid-ventral line at pubic symphysis. **Epipubis** and **prepubis** absent.
12. Obturator foramen	Not found.	Not found.	A small obturator foramen in pubis for obturator nerve present. Ischiopubic fenestrae present between ischium and pubis.	A small obturator foramen present where pubis meets ischum. In place of ischiopubic fenestrae, ilioischiatic foramen present.	Obturator foramen large between pubis and ischium. No other fenestrae or foramen.
13. Acetabulum	Not found.	Acetabulum is cup-like depression on either side of the disc for articulation of the head of femur; formed by all the three bones.	Acetabulum is concave depression, directed laterally at the junction of all the three bones for articulation of the head of femur.	Acetabulum is deep lateral depression formed by all the three bones, perforated at the base for articulation of the head of femur.	Acetabulum is deep cup-like depression at the junction of ilium, ischium and cotyloid bones for articulation of the head of femur; pubis plays no part in its formation.

(Contd.)

Characters	*Scoliodon* (Dogfish)	*Rana* (Frog)	*Uromastix* (Spiny-tailed lizard)	*Columba* (Pigeon)	*Oryctolagus* (Rabbit)
14. Cotyloid bone	Absent.	Absent.	Absent.	Absent.	Present; it connects pubis with acetabulum.
15. Adaptation	Adapted to support the pelvic fins which help in swimming.	Adapted to support the pelvic region and long, stout hindlimbs used for leaping, jumping and swimming.	Adapted to protect the structures of pelvic region and hindlimbs for tetrapod locomotion.	Adapted for flight and bipedal locomotion.	Adapted to protect the hip region and hindlimbs for running and leaping.

LIMBS OF VERTEBRATE TYPES

The limbs of vertebrate types are much varied. In cyclostomes paired appendages are absent; in fishes, *i.e.*, cartilaginous sharks and bony fishes, median and paired **fins** are present for locomotion in water, while in tetrapods (Amphibia, Reptilia, birds and mammals) these are **pentadactyle limbs** adapted for varied modes of locomotion.

In *Scoliodon*, the median fins are dorsal and ventral, while paired fins are pectoral and pelvic. The skeleton of median fins consists of a series of cartilaginous rods called **somactids** or **pterygiophores** bearing a double series of numerous horny fin fays or **ceratotrichia**. Each somactidia has three segments, the proximal, mesial and distal segments. First dorsal fin is supported by twenty-one somactidia. The second dorsal fin and median ventral fin are built on the same plan as the first dorsal. The pectoral fin is supported by a series of endoskeletal and exoskeletal elements. The endoskeleton has **pterygiophores** of cartilage to which muscles are attached for moving the fin. The proximal pterygiophores are **basals** and distal ones are **radials**. There are three thick basals, outer **proptetrygium**, middle **mesopterygium** and inner **metapterygium**. Radials are numerous arranged in three rows, proximal, mesial and distal radiating cartilage called **somactidia** from which unjointed exoskeletal elements called **ceratotrichia** radiate to support the fin blade. Pelvic fin is also made of same skeletal elements but there is a single basal in it called **basipterygium**. In bony fishes, basals are absent and radials are reduced to small ossicles lying within the body wall.

However, as referred to, the tetrapods possess limbs in place of paired fins of fishes. The limbs are pentadactyle (5-toed). The skeletal elements of limbs in tetrapods are basically built on the same pattern. The limbs (both forelimb and hindlimb) can be divided into proximal, middle and distal parts. Thus, the skeletal elements of these parts for forelimbs are **humerus**, **radius-ulna** and **carpals**, **metacarpals** and **phalanges**. Similarly, the skeletal elements of hindlimbs are **femur**, **tibio-fibula** and **tarsals, metatarsals** and **phalanges**. The distal part of forelimb is called **manus** or **hand**, while that of hindlimb is called **pes** or **foot**. As mentioned, the skeletal elements of limbs in tetrapods exhibit great variations, though built on the same pattern, due to adaptations for different modes of locomotion.

Table 50.8. Comparison : Bones of Forelimbs of Vertebrate Types.

Characters	*Rana* (Frog)	*Uromastix* (Spiny-tailed lizard)	*Columba* (Pigeon)	*Oryctolagus* (Rabbit)
1. **Skeletal components**	These are: humerus, radioulna, carpals, metacarpals and phalanges.	These are: humerus, radioulna, carpals, metacarpals and phalanges.	These are: humerus, radioulna, carpals, metacarpals and phalanges.	These are: humerus, radioulna,carpals, metacarpals and phalanges.
2. **Humerus**	Humerus is the bone of upper arm. It is short, stout, cylindrical bone havng slightly curved **shaft**. Its proximal end has a **head** of calcified cartilage which fits into glenoid cavity of pectoral girdle. There is a **deltoid ridge** below head on the shaft and at its distal end rounded **trochlea** or **capitulum** with a **condylar ridge** on either side. The trochlea articulates with the proximal groove of radio-ulna. No other tuberosity.	Humerus is the bone of upper arm. It has an elongated **shaft** whose both ends are expanded. The proximal end bears rounded **head, medial process** and crest-like **deltoid ridge** which fits into glenoid cavity of pectoral girdle. There is a **bicipital fossa** between medial process and head. The distal end bears pulley-like **trochlea** divisible into **radial epicondyle** and **ulnare epicondyle** by which it articulates with radius and ulna respectively.	Humereus is the bone of upper arm. It has an elongated, slightly curved, flat **shaft** expanded at both the ends. The proximal ends bears an expanded head bordered with two tuberosities, preaxial and postaxial. Short **deltoid ridge** and a pneumatic foramen in the postaxial tuberosity. The distal end bears two condyles with an inter-condylar groove for articulation with radius and ulna. There is a coronoid fossa anterior to groove.	Humerus is the bone of upper arm. It has a long, stout and rod-like **shaft** having a rounded **head** at its proximal end for articulation with the glenoid cavity of pectoral girdle. The proximal end also has a **bicipital groove, lesser** and **greater tuberosities** and a **deltoid ridge**. The distal end is expanded into epicondyles and a pulley-like **trochlea** for articulation with the ulna bone of forearm. There is an **olecranon fossa** with a **supratrochlear foramen** for the passage of brachial artery.
3. **Radius-ulna**	Radius-ulna are a **compound bone** of forearm. Both fused lengthwise to form a compound skeletal element. Its proximal end has a **concavity** to receive the trochlea of humerus. Its ulna projects into a short, conical **olecranon process** to form elbow joint. The distal end of radius-ulna is flat covered by calcified	Radius-ulna are **separate bones**, placed somewhat parallel to one another in the forcearm. **Radius** is smaller and slender having a shaft and two **epiphyses** ; distally it bears a concave articular facet for carpals and a styloid process. **Ulna** is a stouter, comparatively longer bone lying on the outer side of	Radius-ulna are **separate bones** of forearm. **Radius** is slender, shorter and straight bone bearing a cup-shaped articular surface proximally for articulation with the trochlea of humerus, while distal end is knob-like. **Ulna** is stouter, longer and outwardly curved bone.Proximally it bears small	Radius-ulna are separate bones of forearm. These are elongated and slightly curved. bones closely held together. **Radius** is smaller and inner bone. **Ulna** is longer bone, proximally it bears a deep **sigmoid notch** for trochlea of of humerus and an **olecranon process** to articulate with olecranon fossa of humerus. Both these

(Contd.)

Characters	*Rana* (Frog)	*Uromastix* (Spiny-tailed lizard)	*Columba* (Pigeon)	*Oryctolagus* (Rabbit)
	cartilage having a **groove** to form two **articular facets** for articulation with the carpals.	the radius, proximally it bears **olecranon process**, while distally a concave articular facet for carpals.	**olecranon process,** a large articular facet **sigmoid notch** for articulation with the inner condyle of the humerus, and distally articulates with ulnare carpal and radius. Ulna also bears a **nutrient foramen** in it.	bones distally bear **epiphyses** to articulate with carpals.
4. Carpals	Carpals are wrist bones. These are 6 small bones arranged in two rows of 3 each; bones of proximal rows are **ulnare**, **inter-medium** and **radiale**, while those of distal rows are **capitohamatum**, **trapezoid** and **trapezium**. Pisciform bone is lacking.	Carpals are wrist bones. Carpals are arranged in two rows of 3 and 5 each, having a **centrale** in between and a **pisciform** (total 10 carpals). The proximal row of 3 carpals are **radiale**, **ulnare** and **intermedium**, while 5 carpals of distal row are small bones.	Only 2 free carpals in proximal row; these are **radiale** and **ulnare** which articulate with **radius** and **ulna** respectively. Distal carpals fused with metacarpals. No **pisciform.**	Carpals are wrist bones arranged in three rows. The proximal row of 3 bones **radiale**, **ulnare** and **intermedium**. The distal row of carpals are 4–**trapezium**, **trapezoid, unciform** and **magnum**. One **centrale** in between 2 rows. The **pisciform** present below carpus.
5. Metacarpals	Metacarpals are 5 bones of palm or manus. These are rod-like, elongated bones.First metacarpal very much reduced belongs to pollex or thumb.	Metacarpals are the bones of palm, consisting of 5 enlongated bones.	Palm has a single bone, the **carpome-tacarpus** formed by the fusion of 3 metacarpals and distal carpals. First rudimentary, while second and third elongated and fused at the two ends. At their proximal end the distals are fused.	Metacarpals are the bones of palm, consisting of 5 elongated, rod-like bones. The first metacarpal of thumb or pollex is reduced.
6. Phalanges	Phalanges are the bones of fingers. There are only 4 fingers, pollex or thumb absent. Hence, **phalangeal formula** 0,2, 2, 3, 3. Claws absent.	Phalanges are the bones of 5 fingers which are short and clawed. **Phalangeal formula** 2, 3, 4, 5, 3.	Phalanges belong to 3 short clawless fingers which are 1, 2, 1, respectively.	Phalanges are the short bones of 5 clawed fingers. **Phalangeal formula** 2, 3, 3, 3, 3.

(Contd.)

Table 50.9. Comparison : Bones of Hindlimbs of Vertebrate Types.

Characters	*Rana* (Frog)	*Uromastix* (Spiny-tailed lizard)	*Columba* (Pigeon)	*Oryctolagus* (Rabbit)
1. Skeletal components	These are : femur, tibiofibula, tarsals, metatarsals and phalanges.	These are : femur, tibia and fibula, tarsals, metatarsals and phalanges.	These are : femur, tibiotarsus and fibula, tarsometatarsus and phalanges.	These are : femur, tibiofibula, tarsals, metatarsals and phalanges.
2. Femur and patella	Femur is the bone of thigh. It has a long, slender, slightly curved **shaft** whose both ends are expanded and covered by calcified cartilage. Its proximal end bears a swollen **head** which fits into acetabulum of pelvic girdle— forming ball and socket joint, while distal end has a condyle to articulate with tibiofibula. No **trochanters** and **patella**.	Femur is the bone of thigh. It has a long, stout, elongated **shaft** having expanded ends with epiphysis. The proximal end bears a rounded **head** which fits into acetabulum of pelvic girdle. The pre-axial side of shaft bears a **lesser trochanter**, and on its post-axial side a reduced **greater trochanter**. The distal end is pulley-like having two condyles for articulation of tibia and fibula. **Patella** absent.	Femur is the bone of thigh. It has a long, cylindrical, slightly curved powerful **shaft** having broad ends. Proximally the shaft has a rounded **head** which fits into acetabulum. On the outer side of head **greater trochanter** present: **lesser trochanter** absent. Between greater trochanter and head, there is an articular surface for antitrochanter of ilium. Distally the shaft is pulley-like with two condyles and a deep groove for articulation with tibia and fibula. **Patella**, a sesamoid bone, present in front of groove forming knee-joint.	Femur is the bone of thigh. It has a long, elongated. cylindrical, stout **shaft** whose both ends are expanded. Proximal end of shaft bears a rounded smooth **head** which fits into acetabulum. 3 trochanters : **greater**, **lesser** and a **3rd**. Distally the shaft has a pair of expanded condyles encloing the pattelar groove. The condyles are provided with articular facets for articulation with tibia. **Patella** present at the knee-joint.
3. Tibio-fibula	Tibia and fibula have fused to form a **compound bone** of shank, called **tibio-fibula**. It is the largest bone, shaft slightly curved and flattened at the two ends covered by cartilage. A longitudinal median groove	Tibia and fibula are the **separate bones** of the shank. **Tibia** is stout, slightly curved bone and bears a ridge, **cnemial crest**, along the dorsal surface. Proximal end of tibia bears concave facets for	**Tibiotarsus** and **fibula** are the two **separate bones** of the shank. Tibiotarsus is a compound bone formed by the fusion of tibia and proximal row of tarsals. It is a long and stout, nearly	Tibia and fibula are the bones of the shank. **Tibia** is inner, well-built, stout and strong bone. Proximally, tibia bears a cnemial crest and two concave facets for articulation with the condyles of femur. **Fibula** is reduced, small slender bone

(Contd.)

Characters	*Rana* (Frog)	*Uromastix* (Spiny-tailed lizard)	*Columba* (Pigeon)	*Oryctolagus* (Rabbit)
	demarcates the boundary between tibia and fibula. Near the proximal end of tibia a **cnemial** or **tibial crest** present; proximal end articulates with the condyles of femur. The distal end articulatea with the astragalus-calcaneum.	articulation with the condyles of femur. **Fibula** is comparatively slender bone; proximally articulates with femur. The distal end of both these bones bear convex surfaces for articulation with the tarsals. A **wide gap** between tibia and fibula present.	straight bone. Its proximal end is expanded and has **cnemial crest** and two concave articular facets for the condyles of femur. Distally, tibiotarsus bears pulley-like articular surface for articulation with tarsometatarsus. **Fibula** is slender, reduced bone. It is swollen proximally forming head, while tapers distally, closely applied to the outer surface of tibiotarsus and terminates without reaching the ankle.	which is fused with tibia distally but free proximally and attached to epiphysis having a narrow **gap** between the two bones. After fusion, tibia and fibula form a compound bone **tibiofibula**, the longest bone. Distally tibio-fibula bears articular surface for tarsals.
4. Tarsals	Tarsals are the bones of ankle. They consist of 4 bones in two rows of 2 each. The proximal row consists of two long bones fused together at their proximal and distal ends enclosing a wide gap between them; the inner bone is thinner, slightly curved called **astragalus** or **tibiale**, while the outer one is thicker and straight called **calcaneum** or **fibulare**. Distal row of tarsals are 2 very small bones.	Tarsals are the bones ankle; consist of 5 bones in two rows. Two tarsals of proximal row are fused to articulate with tibia and fibula. Distal row of tarsals are three which articulate with the metacarpals.	Free tarsal bones are not found. The proximal tarsals are fused with tibia, while distal tarsals with metatarsals.	Tarsals are the bones of ankle; consist of 6 bones arranged in 3 rows. The first or proximal row has 2 tarsals, **astragalus** or **tibiale** and **calcaneum** or **fibulare** ; calcaneum bears a **calcaneal process** which forms the **heel**. The middle row has a single bone **centrale**, while the distal row has 3 bones **mesocuneiform**, **ectocuneiform** and **cuboid** for artculating with metatarsals.
5. Metatarsals	Metatarsals are elongated rod-like bones of the foot.	Metatarsals are 5 rod-like bones of the foot supporting	Metatarsals 2nd, 3rd and 4th, and distal tarsals are	Metatarsals are 4 elongated, rod-like bones of the foot

(Contd.)

Characters	*Rana* (Frog)	*Uromastix* (Spiny-tailed lizard)	*Columba* (Pigeon)	*Oryctolagus* (Rabbit)
	These are 5 in number to support the five toes.	the five toes.	fused to form a stout, long compound bone of the foot called **tarsometarsus**. Proximally tarsometatarsus bears two cup-like facets to articulate with tibiotarsus, while distally it bears 3 pulley-like structures, each representing one metatarsal. In male, it bears a **spur**. The 1st metatarsal is reduced nodule-like at the inner posterior side of tarsometatarsus.	forming sole for supporting 4 toes. The first metatarsal is absent because there is no first toe or hallux.
6. Phalanges	Phalanges are the bones of the toes; phalangeal formula 2, 2, 3, 4, 3. Claws absent.	Phalanges are the bones of the toes; phalangeal formula 2, 3, 4, 5, 3. Distal phallax of each toe with a claw.	Phalanges are the bones of the toes; phalangeal formula 2, 3, 4, 5. Distal phallax of each toe with a claw. Fifth toe not found.	Phalanges are the bones of the toes; phalangeal geal formula 3, 3, 3, 3. Distal phallax of each toe with a claw. Hallux is absent.

TABLE 50.10. COMPARISON : VERTEBRAL COLUMN OF VERTEBRATE TYPES.

Characters	*Scoliodon* (Dogfish)	*Rana* (Frog)	*Uromastix* (Spiny-tailed lizard)	*Columba* (Pigeon)	*Oryctolagus* (Rabbit)
1. Gross structure	Cartilaginous, long, consists of about 130 **vertebrae**, develop around a persistent beaded notochord.	Bony, exceptionally short, consists of 10 vertebrae. Notochord does not persist.	Completely ossified, long, consists of about 83–85 vertebrae. Notochord does not persist.	Ossified and bony, long, consists of 45–48 vertebrae. Notochord does not persist.	Ossified and bony, long, consists of 45–47 vertebrae. Notochord does not persist.
2. Division	Divisible only into **trunk vertebrae** and **caudal vertebrae**.	Due to absence of neck and tail, vertebral column is not differentiated into regions	Vertebral column is divisible into the vertebrae of **cervical**, **thoraco-lumbar, sacral**	Vertebral column is divisible into **cervical, thoracic, synsacral**	Vertebral column is divisible into **cervical**, **thoracic**, **lumbar**,

(Contd.)

Characters	*Scoliodon* (Dogfish)	*Rana* (Frog)	*Uromastix* (Spiny-tailed lizard)	*Columba* (Pigeon)	*Oryctolagus* (Rabbit)
		but 1st vertebra represents the vertebra of neck region, 9th represents the vertebra of sacral region, 10th represents vertebra of tail region and from 2nd to 8th to that of trunk region.	and **caudal** regions.	(lumbar and sacral) and **caudal** regions.	**sacral** and **caudal** regions.
3. Peculiarity	Centrum **amphicoelous** which is pierced by thread-like notochord; bears **Maltese cross**. Vertebra **asterospondylous**. Caudal vertebrae with **haemal arch** enclosing **haemal canal**.	First nine vertebrae ring-like, 10th one long, rod-like called **urostyle** Centrum **procoelous, amphicoelous** (VIII) and biconvex (IX).	Long, ossified. **Epiphyses** absent, centrum **procoelous**. Thoracolumbar vertebrae with monocepthalous ribs.	Long but many vertebrae fused; thoracic, lumbar and sacral vertebrae fused, caudal vertebrae fused. **Epiphyses** absent, centrum **heterocoelous**. Thoracic ribs present.	**Epiphyses** present, centrum **amphiplatyan**. Cartilaginous **intervertebral discs** present between adjacent vertebrae. Thoracic ribs present.
4. Cervical vertebrae	Not differentiated.	Neck absent, hence, cervical vertebrae not differentiated.	Consists of 10 vertebrae, *i.e.*, atlas, axis and 8 typical cervical.	Consists of 14—16 vertebrae.	Consists of 7 vertebrae.
(*i*) Atlas	—	Small ring-like centrum and neural spine reduced, and transverse processes and prezygapophysis absent.	Small ring-like. Centrum not distinct. Neural spine, pre- and post-zygapophyses and transverse process absent.	Small ring-like. Centrum, neural spine, transverse process, prezygapophyses, vertebrarterial foramina absent. A pair of postzygapophyses present for articulation with prezygapophysis of axis vertebra.	Comparatively large, signet-ring-like with no centrum and zygapophyses but broad wing-like transverse processes perforated basally by vertebrarterial canal present and neural spine rudimentary.
(*ii*) Axis	—	Not found.	Slightly larger than typical,	Centrum produced	Odontoid process peg-

(Contd.)

Characters	*Scoliodon* (Dogfish)	*Rana* (Frog)	*Uromastix* (Spiny-tailed lizard)	*Columba* (Pigeon)	*Oryctolagus* (Rabbit)
			bears a backwardly directed spine-like process below the odontoid process, neural spine crest-like, pre-zygapophysis rudimentary and post-zygapophysis and a hypophysis below the posterior end of centrum present. Transverse processes absent.	in front into odontoid process. Neural spine blunt, pre- and post-zygapophyses and a hypophysis present. Transverse process present.	like (part of centrum), neural spine flat, ridge-like laterally fattened and elongated. Post-zygapophysis present, hypophysis not found. Transverse processes present basally perforated.
(*iii*) Typical cervical			Stout, elongated procoelous centrum, neural spine vertical crest-like, pre- and postzygapophyses present, hypophyses present, and centrum bears lateral articular facets for cervical ribs.	Centrum longer, heterocoelous. Neural spine rudimentary, pre-zygapophysis flat and post-zygapophysis projects backwards. Cervical ribs fused with transverse processes and possess vertebrarterial canal. Transverse processes arise laterally at the anterior end. Last cervical fused with thoracic.	Broad with short, flattened centrum, neural canal large, neural spine small zygapophyses well developed transverse processes. short, bifurcated having vertebrarterial canal. Cervical ribs much reduced. But 7th cervical bears elongated neural spine and vertebrarterial canal absent.
5. Trunk vertebrae	Amphicoelous centrum with Maltese cross. Neural spine attached to two neural arches enclosing neural canal, transverse	Vertebrae from 2nd to 8th represent trunk vertebrae. Typical vertebra (2nd to 7th) procoelous, neural arch ring-like, neural	Trunk vertebrae are thoraco-lumbar (28) and sacral. Similar to cervicals but slightly larger, centrum procoelous, neural spine vertical	Trunk vertebrae are fused. Last thoracic vertebra, lumbar vertebrae, 2 sacrals and 7 caudal vertebrae fused to form **synsacrum**.	Trunk vertebrae are formed of thoracic, lumbar and sacral. **Thoracic** vertebrae are 6–7 anterior thoracic and 4–5 poste-

(Contd.)

Characters	*Scoliodon* (Dogfish)	*Rana* (Frog)	*Uromastix* (Spiny-tailed lizard)	*Columba* (Pigeon)	*Oryctolagus* (Rabbit)
	processes broad, plate-like directed ventro-laterally, monospondylous.	spine small, blunt spinous directed backward. Transverse processes, pre- and post-zygapophyses present. 8th vertebra amphicoelous, transverse processes long, slender directed outwardly; pre- and post-zygapophyses prominent.	crest-like, pre- and post-zygapophyses present, hypophysis absent, capitular facet present at the junction of neural arch and centrum for articulation with thoracic rib.	Centrum heterocoelous. Neural spines of thoracic and lumbar vertebrae fused to form dorsal crest. Their hypophyses absent. Transverse processes stout, free. Transverse processes of sacral vertebrae fused to form bony plates. Sacral ribs fused with their transverse processes. Transverse processes of caudal vertebrae except the last are bifurcated into dorsal and ventral processes. Dorsal processes united to form bony plates. Hypophyses fused to form ventral bridge. First three thoracic and last cervical have fused to form **fused thoracic vertebrae**. 4th thoracic free having well developed neu-	rior thoracic. **Anterior thoracic** vertebrae bear tall, slender, backwardly directed neural spines, neural arches bear pre- and post-zygapophyses, transverse processes short, centrum short and thick. **Posterior thoracic** vertebrae bear long and stouter centra, short neural spines, more prominent zygapophyses and reduced transverse processes without tubercular facets but bear a complete capitular facet, and possess metapophyses and anapophyses.

(Contd.)

Characters	*Scoliodon* (Dogfish)	*Rana* (Frog)	*Uromastix* (Spiny-tailed lizard)	*Columba* (Pigeon)	*Oryctolagus* (Rabbit)
				ral arch and long neural spine, transverse prcesses and hypophysis.	
(*i*) **Lumbar vertebrae**	—	—	—	Six lumbars are fused with other vertebrae to form synsacrum already described.	7 in number; **2 anterior lumbar** and **5 post-lumbar**. Anterior lumbars bear hypophyses, crest-like neural spine, large distally expanded transverse processes, metapophyses and anapophyses. **Posterior lumbars** are without hypophyses.
(*ii*) **Sacral vertebrae**	—	9th vertebra represents sacral vertebra. Centrum biconvex, transverse processes stout, long, backwardly directed, neural spine greatly reduced and post-zygapophysis absent.	Two, centrum procoelous, neural spine small crest-like, pre- and post-zygapophyses well developed, hypophyses absent, transverse processes of 1st sacral are strong expanded at the tips and strongly notched. Transverse processes of 2nd sacral not notched at the tips.	Last thoracic, six lumbars, two sacrals and about seven anterior caudal vertebrae have fused to form a compound bone called **synsacrum**.	Three or four, all fused to form **sacrum**. Each bears neural spine, zygapophyses, and intervertebral foramina. First sacral largest.
6. Caudal vertebrae	Many, neural spine much larger and point-	Caudal vertebra represented by 10th vertebra-	Many, gradually become smaller posteriorly.	4–5 free anterior caudal vertebrae	16, gradually become smaller with decrea-

(Contd.)

Characters	*Scoliodon* (Dogfish)	*Rana* (Frog)	*Uromastix* (Spiny-tailed lizard)	*Columba* (Pigeon)	*Oryctolagus* (Rabbit)
	ed, neural arch short, transverse processes meet and fuse below centrum to form haemal arch enclosing haemal canal. Haemal spine and neural spine long.	the **urostyle**. It is long, triangular having pointed tip, bears a vertical ridge. Transverse processes pre- and post-zygapophyses absent. It represents several fused vertebrae of the tail of the tadpole.	Centrum procoelous, neural spine long, pre- and post-zygapophyses well developed, transverse processes long and slender. They bear Y-shaped **cheveron bone** attached to ventral surface of the centrum, but posterior caudal vertebrae without cheveron bone and are smaller.	short with centrum heterocoelous, neural spine bifid and transverse processes, while posterior 4 or more caudal vertebrae fused to form **pygostyle**. It is large, vertical plough-shaped bone.	sing processes and finally remain rod-like centra only.

TABLE 50.11. COMPARISON : DIGESTIVE SYSTEM OF VERTEBRATE TYPES.

Characters	*Scoliodon* (Dogfish)	*Rana* (Frog)	*Uromastix* (Spiny-tailed lizard)	*Columba* (Pigeon)	*Oryctolagus* (Rabbit)
			ALIMENTARY CANAL		
1. Parts of alimentary canal	It consists of buccal cavity, pharynx, oesophagus, stomach, intestine and cloaca.	It consists of buccal cavity, pharynx, oesophagus, stomach, intestine and cloaca.	It consists of buccal cavity, pharynx, oesophagus, stomach, intestine and cloaca.	It consists of buccal cavity, pharynx, oesophagus, stomach, intestine and cloaca.	It consists of buccal cavity, pharynx, oesophagus, stomach, intestine and anus. Cloaca is absent.
2. Mouth opening	Small, crescentic at the ventral side of head, bounded by jaws and leads into buccal cavity.	Wide, terminal, horizontal, semicircular along the anterior end of head, bounded by jaws and leads into buccal cavity.	Wide, terminal along the anterior end of head, semicircular, bounded by jaws and leads into buccal cavity.	Terminal, wide slit-like aperture bounded by jaws and horny beaks and leads into buccal cavity.	Transverse, slit-like subterminal aperture at the snout, bounded by jaws and leads into buccal cavity.
3. Jaws and lips	Lower jaw movable, lips absent. Jaws covered by spi-	Lower jaw movable, lips hard and immovable, scaleless.	Lower jaw movable, lips hard, immovable and scales	Lower jaw movable, lips absent, beaks cover the	Lower jaw movable, lips present, fleshy and hairy.

(Contd.)

Characters	*Scoliodon* (Dogfish)	*Rana* (Frog)	*Uromastix* (Spiny-tailed lizard)	*Columba* (Pigeon)	*Oryctolagus* (Rabbit)
	nous skin.		present.	jaws	Upper lip cleft bearing vibrissae.
4. Buccal cavity	Dorsoventrally flattened and spacious.	Wide and large.	Narrow anteriorly and broad posteriorly.	Somewhat triangular and narrow.	Large, spacious and wide.
(*i*) **Vestibule**	Absent.	Absent.	Absent.	Absent.	Narrow space between lips, cheeks and jaws is called vestibule in which mouth opens.
(*ii*) **Teeth**	Homodont; sharp, directed backwardly, several rows on the skin covering jaws and are replaced several times during life time (polyphyodont). These are modified placoid scales. used in grasping prey.	Small, conical teeth present on upper jaw only in one row, attached to jaw bones (acrodont) homodont and polyphyodont, used in holding the prey.	Teeth small, conical, acrodont and pleurodont present on both the jaws in a single row, used in grasping and holding the prey.	Teeth not found.	Teeth of several types on both the jaws (heterodont) embedded with their roots in the sockets of jaws (thecodont) and are replaced once in lifetime (diphyodont), used in cutting, holding and masticating the food.
(*iii*) **Diastema**	Absent.	Absent.	Absent.	Absent.	Diastema, a toothless space between incisor and premolar.
(*iv*) **Palate**	Absent, skull forms the roof of buccal cavity.	Absent, skull forms the roof of buccal cavity.	Absent, skull forms the roof of buccal cavity. A false palate is present in crocodilians.	Palate incompletely developed.	Palate developed, separates the nasal passage from food passage, hence, forms the roof of buccal cavity.
(*v*) **Tongue**	Tongue is thick, flat, non-muscular, non-glandular and non-	Large, muscular, sticky, attached anteriorly and free posteriorly	Large, muscular, glandular, attached midventrally, free anteriorly	Large, narrow, triangular attached ventrally and non-pro-	Large, muscular, attached mid-ventrally and grooved

(Contd.)

Characters	*Scoliodon* (Dogfish)	*Rana* (Frog)	*Uromastix* (Spiny-tailed lizard)	*Columba* (Pigeon)	*Oryctolagus* (Rabbit)
	protrusible fold of mucous membrane at the base of buccal cavity. Not used in food capture and taste buds also absent.	which is notched. Protrusible, used in capturing prey and bears few taste buds.	which is bifid and protrusible covered with papillae having taste buds.	trusible covered with fine horny processes and few taste buds.	mid-dorsally. Anterior tip free, protrusible and covered with four kinds of papilae having taste buds.
(*vi*) **Internal nares**	Absent.	Two small openings on the roof of buccal cavity in front of vomerine teeth.	Two small rounded openings near the anterior end on the roof of buccal cavity.	Two small slit-like openings situated at the posterior end of buccal cavity or pharynx.	Both the nostrils open into a nasal passage which opens posterior to buccal cavity into the roof of the pharynx.
(*vii*) **Pharynx**	Posterior region of buccal cavity represents pharynx. Both are inseparable.	Posterior region of buccal cavity represents short pharynx.	Posterior broad region of buccal cavity represents the pharynx.	Buccal cavity merges behind into pharyngeal cavity.	Pharynx is short at the posterior end of buccal cavity and differentiated into **nasopharynx, oropharynx** and **laryngopharynx.**
(*viii*) **Eustachian opening**	Absent.	A pair of wide eustachian openings lie on the roof, one on either side laterally near jaw angles.	A pair of eustachian openings lie, one on either side of roof of pharynx.	Single eustachian opening in the middle of roof of pharynx. behind internal nares.	Sides of nasopharyngeal wall are pierced by a pair of oval eustachian openings.
(*ix*) **Gill-slits**	A pair of pit-like spiracle is present on the lateral wall of pharynx and 5 vertical gill-slits are present on either lateral side of pharynx.	Absent.	Absent.	Absent.	Absent.
(*x*) **Glottis**	Since there is neither trachea	Median slit-like opening in the	Median slit-like opening at the	Oval opening on the floor	Median verical slit-like

(Contd.)

Characters	*Scoliodon* (Dogfish)	*Rana* (Frog)	*Uromastix* (Spiny-tailed lizard)	*Columba* (Pigeon)	*Oryctolagus* (Rabbit)
	nor lungs, hence it is absent.	floor of bucco-pharyngeal cavity leads into laryngo-tracheal chamber.	floor of pharynx leads into trachea.	of pharynx which leads into trachea.	opening in the floor of laryngopharynx, leads into larynx.
(*xi*) Epiglottis	Absent.	Absent.	Absent.	Absent.	Bilobed cartilaginous flap or epiglottis guards glottis against food entering into it.
(*xii*) Other bucco-pharyngeal structures	Mucous lining of the pharynx contains dermal denticles.	Floor of pharynx in males only contains an opening of vocal sac on either lateral side near jaw angles. Eye-balls bulge internally on the roof into buccal cavity.	Mucous lining of the pharynx is thrown out into distensible longitudinal folds.	Nasal passages open through internal nares into the roof of the pharynx.	Nasal passages open through internal nares into roof of the laryngo-pharynx.
5. Oesophagus	Short, wide tube with thick muscular wall having longitudinal mucous folds. Opens into cardiac stomach with a sphincter or oesophageal valve.	Short, wide, highly distensible with prominent longitudinal folds and not demarcated from pharynx and stomach.	Long, narrow, muscular tube with mucous folds and highly distensible.	Long, wide, distensible, muscular, thick-walled tube. At the base of neck, it expands into a thin-walled bilobed elastic sac called **crop** or food-reservoir.	Long, narrow, elastic, muscular tube of uniform diameter. No crop end opens into stomach.
6. Stomach	Long, muscular, U-shaped, divisible into proximal long, broad cardiac part and short, narrow distal pyloric part.	Large, broad, curved, muscular sac on the left side in the body cavity. Proximal cardiac and distal pyloric parts	Long, tubular, curved, muscular tube on the left side in the body cavity. Not demarcated into cardiac and pyloric parts	Represented by an anterior narrow tube-like glandular **proventriculus** and a posterior broad, thick-walled mus-	Large, bean-shaped on the left side in the abdominal cavity lying transversely and differentiated

(Contd.)

Characters	*Scoliodon* (Dogfish)	*Rana* (Frog)	*Uromastix* (Spiny-tailed lizard)	*Columba* (Pigeon)	*Oryctolagus* (Rabbit)
	The junction is marked by a blind sac and a sphincter valve. Cardiac part has well developed longitudianl mucous folds.	not marked off externally. Blind sac and sphincter valve absent.	externally but cardiac part possesses well developed longitudinal muscular folds.	cular **gizzard**. It is internally lined by thick horny epithelium.	into cardiac, fundic and pyloric parts.
7. Bursa entiana and gizzard	Bursa entiana is a small, thick-walled muscular sac at the distal end of pyloric stomach through which it opens into intestine. No gizzard.	Neither bursa entiana nor gizzard.	Neither bursa entiana nor gizzard.	Bursa entiana absent but a well developed, muscular gizzard present which contains stone grits to help in grinding the food.	Bursa entiana and gizzard absent.
8. Intestine	Straight, short and wide tube. Not differentiated into small and large intestines.	Coiled, long and narrow tube, differentiated into small and large intestines.	Coiled, long and narrow tube, differentiated into small and large intestines.	Coiled, long and narrow tube, differentiated into small and large intestines.	Coiled, long and narrow tube, differentiated into small and large intestines.
(*i*) Small intestine	Not differentiated into duodenum and ileum.	Differentiated into duodenum and ileum.	Duodenum and ileum well marked.	Duodenum and ileum well differentiated.	Duodenum and ileum well marked.
(*a*) Duodenum	Absent.	Straight tube, forms "U" with stomach, receives hepatopancreatic duct.	Straight tube, receives separate ducts from pancreas and gall bladder.	U-shaped tube, receives two ducts from liver and 3 ducts from pancreas.	U-shaped loop-like, receives one duct each from pancreas and gall bladder.
(*b*) Ileum and valve	Not distinct. Internal mucous lining is folded into a longitudinal spiral or scroll valve. Villi absent.	Small and coiled. Mucous lining forms several longitudinal folds. True villi absent. No spiral valve.	Long and coiled. Mucous lining forms folds but spiral valve and true villi absent.	Long and coiled. Inner mucous lining projects into several small finger-like villi. No spiral valve.	Very long and coiled. Villi numerous and well developed but no spiral valve.
(*c*) Accessory structures					
(*i*) Sacculus rotundus	Absent.	Absent.	Absent.	Absent.	Ileum at its distal end expanded to form sacculus rotundus.

(Contd.)

Characters	*Scoliodon* (Dogfish)	*Rana* (Frog)	*Uromastix* (Spiny-tailed lizard)	*Columba* (Pigeon)	*Oryctolagus* (Rabbit)
(*ii*) **Caecum**	Absent.	Absent.	Junction of small and large intestines bear a large caecum and an internal ileocolic valve.	Junction is marked by the presence of a pair of short rectal caeca.	Large thin-walled, tubular, spirally constricted caecum present into which sacculus rotundus opens through ileo-caecal valve.
(*iii*) **Vermiform appendix**	Absent.	Absent.	Absent.	Absent.	Caecum ends distally into a blind vermiform appendix.
9. Large intestine	Distal end of intestine forms a short, narrow rectum which opens into **cloaca**.	Short and broad **rectum** opens into **cloaca**.	Represented by thin-walled narrow **colon** and thick-walled, broad rectum opening into **cloaca.**	Represented by a short but broad rectum only which leads into **cloaca**.	Long and consists of anterior sacculated **colon** and a posterior beaded **rectum**.
10. Rectal glands	Rectum receives a tubular **rectal gland** of unknown function dorsally.	Absent.	Absent.	Absent.	Absent.
11. Cloaca and associated structures	Rectum opens into simple cloaca through anus guarded by **anal sphincter**. It contains urinogenital apertures, a pair of abdominal pores from peritoneal coelom. **Bursa Fabrici** is lacking.	Single sac-like cloaca into which rectum opens by anus. It contains urinogenital apertures. and **bursa Fabrici**. Abdominal pores not found. Anal sphincter present.	Rectum opens into 3 linearly arranged chambers forming cloaca; **coprodaeum**, **urodaeum** and **proctodaeum**. Anal sphincter present but **abdominal pores** and **bursa Fabrici** not found.	Rectum opens into cloaca by anus guarded by **anal sphincter**. Cloaca 3 chambered as in lizard. Abdominal pores not found. In young birds only a thick-walled small pouch called **bursa Fabrici** present dorsally on proctodaeum.	Cloaca absent. Rectum opens directly to outside by anus having **anal sphincter**. Abdominal pores and **bursa Fabrici** not found.

(Contd.)

Characters	*Scoliodon* (Dogfish)	*Rana* (Frog)	*Uromastix* (Spiny-tailed lizard)	*Columba* (Pigeon)	*Oryctolagus* (Rabbit)
12. **Cloacal aperture**	Cloaca opens out ventrally between pelvic fins by a longitudinal slit-like cloacal aperture.	Cloaca opens out through a circular cloacal aperture at the hind - end of trunk between hindlimbs.	Cloaca opens midventrally at the junction of trunk and tail by a transverse slit-like cloacal aperture.	Cloaca opens at the base of tail ventrally by a transverse slit-like cloacal aperture with tumid lips.	Cloacal aperture absent.
			DIGESTIVE GLANDS		
1. **Salivary glands**	Absent.	Absent.	A few mucus-secreting salivary glands are said to be present in the buccal cavity.	Buccal glands of mouth cavity probably secrete **mucus** but they are not salivary glands.	Salivary glands of different kinds found which open into buccal cavity by separate ducts. They secrete saliva containing an enzyme ptyalin.
2. **Gastric glands**	Stomach contains gastric glands to secrete gastric juice containing HCl and **pepsin**.	Found in stomach lining secreting gastric juice containing HCl and **pepsinogen**.	Found in the mucous lining of stomach.	Found in the lining of proventriculus secreting gastric juice containing HCl and enzymes.	Present in the lining of stomach and secrete gastric juice containing HCl and enzymes.
3. **Mucous glands**	Aquatic forms have no mucous glands.	Lining of bucco-pharyngeal cavity and oesophagus contains mucous glands secreting mucus.	Scanty mucous glands.	Mucous glands few, present in buccal cavity and on tongue.	Mucous glands present in the entire inner lining of alimentary canal.
4. **Pancreas and pancreatic duct**	Compact, bilobed, whitish gland between cardiac and pyloric stomachs. A single **pancreatic duct** opens into proximal end of intestine.	Highly branched, irregular, cream-coloured gland lying between stomach and duodenum. Small **pancreatic ducts** open into bile duct.	Elongated, narrow, white-coloured gland between pyloric stomach and duodenum. A single **pancreatic duct** opens directly into duodenum.	Band-like, narrow, pinkish gland between the two limbs of duodenum. Three **pancreatic ducts** open separately into distal limb of duodenum.	Irregular, diffused, pinkish gland surrounded by duodenal loop. A single **pancreatic duct** leads into the distal end of duodenum.

(Contd.)

Characters	*Scoliodon* (Dogfish)	*Rana* (Frog)	*Uromastix* (Spiny-tailed lizard)	*Columba* (Pigeon)	*Oryctolagus* (Rabbit)
(*iv*) Larynx and trachea	Absent.	Both fused to form laryngo-tracheal chamber containing vocal cords. Thyroid cartilage absent.	Both separate. Vocal cords and thyroid cartilage absent. Trachea long cylindrical tube.	Both separate. Larynx without vocal cords and thyroid cartilage. Trachea is a, long, cylindrical flexible tube.	Both separate. Larynx with vocal cords and thyroid cartilage. Trachea long, cylindrical tube.
2. Syrinx	Absent.	Absent.	Absent.	Sound producing organ at the lower end of trachea, characteristic of birds.	Absent.
RESPIRATORY ORGANS					
1. Adult's respiratory organs	Only gills.	Lungs and skin. Gills in tadpoles.	Only lungs.	Only lungs.	Only lungs.
2. Gills and gill-clefts	5 pairs of pharyngeal gill-pouches containing 5 pairs of lamelliform gills.	Tadpole larva with 5 pairs of gill-pouches and 4 pairs of gill-clefts. 3 pairs of external gills which later become internal as in fishes but disappear during metamorphosis.	Temporary pharyngeal gill-pouches and gill-cleft develop during embryogeny which disappear later.	Temporary 4 pairs of pharyngeal gill-pouches and 3 pairs of gill-slits develop during embryogeny only. Adults without gills.	Temporary only 4 pairs of gill-pouches develop during embryogeny. Gill-slits do not develop.
3. Lungs and peritoneum	Not found.	Paired, unlobed, thin-walled, elastic, ovoid, hollow sacs of pinkish colour, and found suspended freely in the peritoneal cavity.	Paired, unlobed, thin-walled, elongated, fusiform, elastic hollow sacs of orange colour, and lie freely in the thoracic cavity and surrounded by peritoneum.	Paired, unlobed, small, ovoid,compact, inelastic, bright red coloured and found attached dorsally to thoracic vertebrae and ribs. Covered by peritoneum except the dorsal attached side.	Paired, large, soft, elastic and spongy, pink coloured, each lying in the lateral pleural cavity of the thorax. Lungs lobed; right of 4 lobes and left of 2 lobes. Lungs surrounded by peritoneum on all sides.

(Contd.)

Characters	*Scoliodon* (Dogfish)	*Rana* (Frog)	*Uromastix* (Spiny-tailed lizard)	*Columba* (Pigeon)	*Oryctolagus* (Rabbit)
4. Alveoli	—	Few in number. Central cavity large.	Alveoli shallow due to low inner septa. Central cavity smaller.	Alveoli absent. Air capillaries form the substance of the lung. These are the final branches of tertiary bronchi or parabronchi.	Alveoli present. Terminal branches of bronchioles end into a cluster of small, blind alveoli.
5. Air sacs	Absent.	Absent.	Absent.	Different types of large and small thin-walled, elastic, non-vascular air sacs are found connected to the lungs.	Absent.
6. Skin as respiratory organ	No.	Moist and richly vascularised skin serves efficiently in the exchange of gases. This is **cutaneous respiration**.	No.	No.	No.
7. Bucco-pharyngeal cavity as respiratory organ	No.	The epithelial lining of bucco-pharyngeal cavity is highly vascularised, hence, helps efficiently in gaseous exchange.	No.	No.	No.
MECHANISM OF RESPIRATION					
1. Breathing movements	Buccal floor is lowered by a series of hypobranchial muscles causing **inspiration**. Buccal floor is raised by constr-	Lowering of bucco-pharyngeal floor is caused by sternohyal muscles resulting into **inspiration** and the	Movements in the ribs and intercostal muscles cause **inspiration** as well as **expiration**.	Joint action of the movements of sternum, intercostal muscles and abdominal muscles cause **inspiration**	The combined action of the movements of diaphragm and intercostal muscles cause **inspiration** and

(Contd.)

Characters	*Scoliodon* (Dogfish)	*Rana* (Frog)	*Uromastix* (Spiny-tailed lizard)	*Columba* (Pigeon)	*Oryctolagus* (Rabbit)
	ictor and inter-branchial muscles causing **expiration**.	raising of bucc-opharyngeal floor is caused by petrohyal muscles resulting into **expiration**.		and **expiration** both.	**expiration** both.
2. Efficacy	Well vascularised gills serve as efficient organ for gaseous exchange in water.	The inefficient lung respiration is well supplemented by cutaneous and bucco-pharyngeal respirations.	Less number of alveoli in lungs cause poor aeration of blood.	Most efficient due to less residual foul air as gaseous exchange takes place both during inspiration as well as during expiration.	Less efficient due to more residual foul air than the birds.

TABLE 50.13. COMPARISON : HEARTS OF VERTEBRATE TYPES.

Characters	*Scoliodon* (Dogfish)	*Rana* (Frog)	*Uromastix* (Spiny-tailed lizard)	*Columba* (Pigeon)	*Oryctolagus* (Rabbit)
1. Location	Located mid-ventrally beneath the pharynx in pericardial cavity separated from peritoneal cavity by a partition called **septum transversum**, perforated by **pericardio-peritoneal canal**.	Located mid-ventrally beneath the oesophagus in thoracic cavity. Septum transversum not found.	Located mid-ventrally above sternum in thoracic cavity. Septum transversum not found.	Located mid-ventrally in thoracic cavity surrounded by lobes of liver.	Located mid-ventrally between two pleural cavities enclosing lungs in a median pericardial cavity of the thorax.
2. Pericardium	Heart remains protected within a 2-layered membranous **pericardium**.	Heart remains enclosed in a thin, transparent, 2-layered sac called **pericardium**.	Heart remains enclosed within a thin, transparent, 2-layered sac called **pericardium**.	Heart remains protected by a thin, transparent, 2-layered membranous sac called **pericardium**.	Heart remains protected by a transparent, 2-layered membranous sac called **pericardium**.
3. Size and shape	Small, S-shaped and dorso-ventrally curved.	Small, conical or triangular.	Small, roughly triangular.	Comparatively larger and conical in shape.	Larger and pear-shaped.

(Contd.)

Characters	*Scoliodon* (Dogfish)	*Rana* (Frog)	*Uromastix* (Spiny-tailed lizard)	*Columba* (Pigeon)	*Oryctolagus* (Rabbit)
4. Colour	Reddish-brown.	Reddish.	Reddish.	Reddish.	Reddish.
5. Chambers	2-chambered but consists of a linear series of 4 compartments : sinus venosus, auricle, ventricle and conus arteriosus which are all distinguished externally. From these auricle and ventricle are the true chambers.	3-chambered but consists of 5 compartments: 2 auricles, ventricle, sinus venosus and truncus arteriosus. Only 2 auricles and ventricle are the true chambers.	3-chambered but consists of 4 compartments: 2 auricles, an incompletely divided ventricle and sinus venosus. All are faintly demarcated externally.	4-chambered, consists of 2 auricles and 2 ventricles only. Ventricles not demarcated externally.	4-chambered, consists of 2 auricles and 2 vertricles. All are demarcated externally.
6. Sinus venosus	Triangular, thin-walled, transversely elongated, lying at the base of pericardial cavity. Receives blood vessels from the body laterally by two **ducti Cuvieri** and two **hepatic sinuses** posteriorly.	Triangular, thin-walled, dark coloured at the dorsal surface of heart. Receives 2 anterior precavals and a posterior postcaval veins from the body.	Large, thin-walled on the dorsal side of heart covering both the auricles, bilobed; right larger and left smaller. Receives 2 precavals and one postcaval veins from the body.	Absent, said to be incorporated in right auricle. The 2 precavals and one postcaval veins from the body open directly into right auricle.	Absent, merged into right auricle; their union represents a groove called **sulcus terminalis** externally and internally by a ridge **crista terminalis**. The three vena cavae open directly into right auricle.
7. Sinu-atrial aperture	Sinus venosus opens into posterior end of auricle by a sinu-atrial aperture guarded by a pair of membranous **valves**.	Sinus venosus opens into dorsal wall of right auricle by a large, oval sinuatrial aperture and guarded by a pair of flap-like **valves**.	Sinus venosus opens into dorsal wall of right auricle through an oval transverse aperture with muscular lip-like **sinu-aricular valves**.	Sinus venosus not found. But the opening of postcaval into right auricle is guarded by a muscular **Eustachian valve**.	Sinus venosus absent. But the opening of postcaval into right auricle is guarded by a muscular flap of **Eustachian valve**.
8. Auricles or atria	Single chambered, somewhat	2-chambered, somewhat	2-chambered, divided interna-	2-chambered, divided inter-	2-chambered, divided inter-

(Contd.)

Characters	*Scoliodon* (Dogfish)	*Rana* (Frog)	*Uromastix* (Spiny-tailed lizard)	*Columba* (Pigeon)	*Oryctolagus* (Rabbit)
	triangular auricle or atrium. Inter-auricular septum absent.	rectangular, internally divided into two by inter-auricular septum, *i.e*, right and left auricles.	lly into right and left auricles by inter-auricular septum. Right auricle is projected into a diverticulum at its dorsal antero-median end.	nally by an inter-auricular septum into right and left auricles. Diverticulum not found.	nally by an inter-auricular septum into right and left auricles. Diverticulum absent.
9. Auricular wall	Thin-walled, moderately muscular and spongy.	Thin-walled, no muscular processes.	Thin-walled, inner lining in the form of slightly elevated muscular ridges.	Comparatively thick-walled and inner surface is thrown out into short muscular ridges.	Comparatively thick-walled, inner surface is thrown out into muscular ridges called **musculi pectinati**.
10. Auricular appendix	Auricle projected laterally at both sides beyond ventricle in the form of ear-like **auricular appendages**.	No such projection.	Right auricle projected as diverticulum.	No such projection.	Both auricles are projected behind into a swollen flap-like **auricular appendix** covering the ventricles slightly.
11. Ventricle	Small, conical, thick-walled chamber situated ventral to sinus venosus and auricle. Interventricular septum not found, hence, undivided.	Small, conical, thick-walled chamber situated posterior to the auricles. No interventricular septum, hence, undivided.	Small, conical, thick-walled chamber situated behind auricles. An oblique musclar ridge or septum divides it incompletely into two chambers–a large dorsal **cavum dorsale** and a small ventral **cavum pulmonale**.	Larger, conical thick-walled situated posterior to auricles. A vertical interventricular septum divides it into right and left ventricles.	As in birds.
12. Chordae tendineae	Ventricular cavity is traversed by several muscular	Thread-like chordae tendineae arise from the flaps	Fine chords like chordae tendineae arise from free edges of aur-	The papillary muscles are attached by chordae ten-	Chordae tendineae are long, tough, connective tissue

(Contd.)

Characters	*Scoliodon* (Dogfish)	*Rana* (Frog)	*Uromastix* (Spiny-tailed lizard)	*Columba* (Pigeon)	*Oryctolagus* (Rabbit)
	strands called **chordae tendineae**, making it spongy in nature.	of auriculo-ventricular valve and attached to ventricular wall.	iculo-ventricular valves and attached to inner wall of ventricle.	dineae with the auriculoventricular valves.	strands connecting the papillary muscles to the free edges of auriculoventricular valves.
13. Columnae carneae	Not found.	Inner wall of ventricle gives rise to several irregular ridges called **columnae carneae**.	Comparatively prominent ridges from the inner wall of ventricle.	Ventricular cavity is traversed by several bars of muscular strands.	Inner wall of ventricle is projected in the form of several irregular muscular ridges.
14. Papillary muscles	Not found.	Not found.	Not found.	Present in the form of prominent muscular projections from the wall of ventricles.	Found as large, conical nipple-shaped muscular processes from the inner wall of ventricles.
15. Pulmonary veins	Not found.	Right and left pulmonary veins unite to form a common pulmonary vein which opens into left auricle.	A common pulmonary vein opens into left auricle.	Four pulmonary veins unite to open by a common aperture into left auricle.	Two pulmonary veins unite to open into left auricle by a common aperture.
16. Auriculo-ventricular aperture and valves	Auricle opens into ventricle through single auriculo-ventricular aperture guarded by a pair of membranous valves.	Both auricles open into single ventricle by a common large auriculo-ventricular aperture guarded by two pairs of flap-like auriculo-ventricular valves.	Both auricles open into ventricle through separate right and left auriculo-ventricular apertures due to backward extension of inter-auricular septum into the ventricle. Each auriculo-ventricular aperture is guarded by	Right and left auricles open into their respective ventricles through circular right and left auriculo-ventricular apertures. Right opening is guarded by a strong muscular flap, *i.e.*, right auriculo-ventricu-	Both auricles open into their respective ventricles by separate right and left auriculo-ventricular apertures. Right aperture is guarded by 3 triangular flaps or cusps forming **tricuspid ventricular**

(Contd.)

Characters	*Scoliodon* (Dogfish)	*Rana* (Frog)	*Uromastix* (Spiny-tailed lizard)	*Columba* (Pigeon)	*Oryctolagus* (Rabbit)
			a single semilunar flap forming left auriculoventricular valve.	lar valve, while left opening is guarded by two membranous flaps forming left auriculo-ventricular valve that correspond to the bicuspid valve of mammals.	**valve**, while left aperture is guarded by 2 flaps forming left **bicuspid** or **mitral valve**.
17. Conus or truncus arteriosus	Ventricle continues anteriorly into an undivided muscular tube forming **conus arteriosus**. It has 2 transverse rows of 5 semilunar valves, each row having 3 larger and 2 smaller valves. **Spiral valve** is not found.	Truncus arteriosus is a tubular chamber arising anteriorly from the right ventral side of the ventricle. Internally its cavity is divided by 3 semilunar valves into a distal **synangium** and a proximal **pylangium**. Pylangium is further divided into **cavum pulmocutaneum** and **cavum aorticum** by a spiral valve.	Conus or truncus arteriosus not found.	Conus or truncus arteriosus not found.	Conus or truncus arteriosus not found.
18. Aortic arches	Conus arteriosus continues anteriorly into ventral aorta which gives off 5 pairs of lateral afferent branchial arches.	Truncus arteriosus bifurcates anteriorly into 2 aortic trunks; each trunk further gives rise to 3 aortic arches: common carotid, pulmocutaneous and syste-	3 aortic arches arise directly from the ventricle; pulmonary from cavum pulmonale, right systemic from cavum dorsale and left systemic from between	Only 2 aortic arches arise directly, one from each ventricle; pulmonary from right ventricle and right systemic from left ventricle	Exactly like bird but left systemic arising from left ventricle, rounds the heart from left side. Right systemic and

(Contd.)

Characters	*Scoliodon* (Dogfish)	*Rana* (Frog)	*Uromastix* (Spiny-tailed lizard)	*Columba* (Pigeon)	*Oryctolagus* (Rabbit)
		mic. Ventral aorta is not found.	the two. Ventral aorta absent.	which rounds the heart from right side. Left systemic and ventral aorta absent.	ventral aorta absent.
19. Foramen Panizzae	Absent.	Absent.	Found as a bridge connecting two systemic aortae at the point of their crossing.	Absent.	Absent.
20. Condition of heart	Since it receives only venous blood and sends it to gills only, hence, it is called **venous heart** and **branchial heart**. There is **single circulation**.	Heart receives venous and oxygenated blood both and supplies mixed blood to body parts, hence, it is a **transitional heart** and there is **single circulation**.	Heart receives venous and oxygenated blood which are mixed in the incompletely divided ventricle. This is also a **transitional heart** and there is **single circulation**.	Since heart is completely 4-chambered, hence, no mixing of venous and oxygenated blood occurs. Therefore, venous blood goes to the lungs and oxygenated blood to the body parts, *i.e.*, **double circulation**.	Exactly like birds but more efficient.

TABLE 50.14. COMPARISON : BRAINS OF VERTEBRATE TYPES.

Characters	*Scoliodon* (Dogfish)	*Rana* (Frog)	*Uromastix* (Spiny-tailed lizard)	*Columba* (Pigeon)	*Oryctolagus* (Rabbit)
1. Shape, size and main parts	Elongated, flattened, 3 times longer than width, simple and consists of 3 main parts; forebrain (prosencephalon), midbrain (mesencephalon) and hindbrain (rhombencephalon).	Elongated, flattened, 3 times longer than width, simple and consists of 3 main parts ; forebrain midbrain and hindbrain.	Elongated like frog but larger and broader. Made of usual 3 main parts; forebrain, midbrain and hindbrain.	Comparatively larger, more complex, broader and somewhat rounded. Made of usual three parts as in earlier cases.	Comparatively largest, complex, most advanced and specialised. Consists of usual 3 main parts.

(Contd.)

Characters	*Scoliodon* (Dogfish)	*Rana* (Frog)	*Uromastix* (Spiny-tailed lizard)	*Columba* (Pigeon)	*Oryctolagus* (Rabbit)
2. Meninges	Brain is enclosed and protected by a single membrane called **meninx primitiva**.	Brain is enclosed and protected by two membranes : a thin inner **piamater** and a thick outer **duramater**.	Meninges are like that of frog.	Brain is enclosed and protected by two meninges : inner **pia-archnoid** and outer **duramater**.	Brain is enclosed and protected by three membranes : inner **piamater**, middle **arachnoid** and outer **duramater**.
			FOREBRAIN		
3. Parts	Consists of olfactory lobes, cerebral hemispheres and diencephalon.	Consists of 3 components as in fish.	Consists of 3 components as in frog.	Consists of usual 3 components as in earlier cases.	Consists of usual 3 components.
4. Olfactory Lobes					
(*i*) Location	Placed at the antero-lateral sides of cerebral hemispheres, therefore, widely separated.	Placed in front of cerebral hemispheres from which separated by slight constrictions.	Placed in front of cerebral hemispheres without constriction side by side on a long olfactory peduncle.	Placed in front of cerebral hemispheres by which olfactory lobes are largely covered.	Placed distinctly at the anterior end of cerebral hemispheres.
(*ii*) Shape and size	Large, bilobed and highly developed; sense of smell well developed.	Small, spherical, not much developed due to poor sense of smell.	Small due to poor sense of smell.	Very small, conical, ill-developed due to poor sense of smell.	Small, elongated, club-shaped. Sense of smell well developed.
(*iii*) Parts	**Olfactory tract** or **peduncle** and a bilobed **olfactory bulb** closely applied to large olfactory sac.	Olfactory tract and bulb not differentiated. Olfactory bulb not closely applied to olfactory sac.	**Olfactory tract** long, slender bearing a small nut-like **olfactory bulb** which is not closely applied to olfactory sac.	Olfactory tract and bulb not differentiated. Olfactory bulb not closely applied to sac.	**Olfactory tract** hidden beneath anterior end of cerebral hemispheres. Only club-shaped **olfactory bulb** visible from dorsal side and closely applied to olfactory sac.

(Contd.)

Characters	*Scoliodon* (Dogfish)	*Rana* (Frog)	*Uromastix* (Spiny-tailed lizard)	*Columba* (Pigeon)	*Oryctolagus* (Rabbit)
(*iv*) **Rhinocoel**	Spacious cavity of the olfactory lobe is **rhinocoel**.	Small and narrow.	Narrow passage.	Absent, hence, olfactory lobe solid.	Present.
5. Cerebral Hemispheres					
(*i*) **Shape and size**	Large, somewhat rectangular, not divided into right and left hemispheres due to absence of median groove.	Large, long, oval, narrow in front and broad behind. Divided into right and left hemispheres due to longitudinal median fissure.	Oval, narrow in front and broad behind. Divided into two hemispheres due to dorsal median longitudinal fissure.	Very large, pyriform and divided into two hemispheres of right and left sides by a deep sagittal fissure. Covers olfactory lobes in front and diencephalon behind.	Large, narrow in front and broad behind. Divided into two hemispheres of right and left sides due to median longitudinal fissure. Covers olfactory lobes in front and midbrain behind.
(*ii*) **Surface**	Smooth, no fissures, folds and lobes.	Smooth, no fissures, folds and lobes.	Smooth, no fissures and lobes.	Smooth, folds, fissures and lobes not found.	**Sylvian fissure** divides into frontal and temporal lobes, **rhinal fissure** demarcates olfactory tract, while **hippocampal fissure** or **sulcus** demarctes **hippocampal lobe**.
(*iii*) **Neuropore**	Present on the mid-ventral side for terminal nerve.	Absent.	Absent.	Absent.	Absent.
(*iv*) **Cerebral cortex**	Not developed. Lining of lateral ventricles is formed by gray matter.	Beginning of cerebral cortex starts.	Present but poorly developed.	Present but more developed than lizard and less than mammal.	Present, well developed.
(*v*) **Roof (Pallium)**	Roof of cerebral hemispheres, *i.e.*, pallium poorly developed.	Little more developed than in fishes.	Still more increased than in amphibians.	More developed than in lizards.	Well developed.

(*Contd.*)

Characters	*Scoliodon* (Dogfish)	*Rana* (Frog)	*Uromastix* (Spiny-tailed lizard)	*Columba* (Pigeon)	*Oryctolagus* (Rabbit)
(*vi*) Corpora striata	Corpora striata, *i.e.*, ventro-lateal walls of cerebral hemispheres very poorly developed.	Better developed than in fishes.	Still better developed than in amphibians and thickened.	Thickened, becomes well conspicuous.	Relatively less developed. Present in the ventro-lateral walls.
(*vii*) Corpus callosum	Not found.	Not found.	Not found.	Not found.	Present in the form of special dorsal transverse band of nervous tissue interconnecting two hemispheres internally.
(*viii*) Lateral ventricles (Paracoel)	Spacious but unbranched.	Present, not branched.	Present, not branched.	Present, unbranched.	Well developed and branched.
6. Diencephalon					
(*i*) Shape and size	Small, narrow, covered dorsally by anterior extension of cerebellum.	Short, rhomboidal but cerebellum does not cover dorsally.	Small, rounded, covered dorsally by posterior part of cerebral hemispheres and optic lobes.	Covered dorsally by cerebral hemispheres and cerebellum.	Covered completely by backward extension of cerebral hemispheres.
(*ii*) Epiphyseal apparatus	A long pineal stalk having a small rounded pineal body arising from the dorsal roof of diencephalon. Parietal organ absent.	Short pineal stalk having small rounded pineal body but in adults pineal body separates and lies above skull. Parietal organ absent.	Epiphyseal apparatus consists of an anterior parietal organ and a posterior pineal body.	Short pineal stalk which is vertical and pineal body small, spherical. No parietal organ.	Slender pineal stalk directed posteriorly, pineal body small and rounded. Parietal organ absent.
(*iii*) Infundibulum	In the form of ventral hollow projection behind **optic chiasma**; consists of a large **median lobe** and 2 late-	In the form of a large median bilobed projection; **inferior lobes** and **saccus vasculosus** not found.	In the form of a small projection behind optic chiasma; **inferior lobes** and **saccus vasculosus** absent.	In the form of small projection without inferior lobes and saccus vasculosus.	In the form of small projection without inferior lobes and saccus vasculosus.

(Contd.)

Characters	*Scoliodon* (Dogfish)	*Rana* (Frog)	*Uromastix* (Spiny-tailed lizard)	*Columba* (Pigeon)	*Oryctolagus* (Rabbit)
	ral small **inferior lobes** having thin-walled **saccus vasculosus**.				
(*iv*) **Pituitary body**	Posterior infundibular part bearing sac-like **hypophysis** together constitute the **pituitary body**.	Infundibulum with flattened oval hypophysis together constitute pituitary body.	Infundibulum and posterior hypophysis together constitute pituitary body.	Ventral infundibulum and posterior hypophysis form pituitary body but without intermediate lobe.	Infundibulum and hypophysis along with intermediate lobe consitute pituitary body.
(*v*) **Corpus albicans**	Not found.	Not found.	Not found.	Not found.	**Corpus albicans** or corpus mammillare is a small rounded body behind pituitary.
(*vi*) **Middle commissure**	Optic thalami present on the lateral sides of diacoel and middle commissure absent.	Optic thalami present and middle commissure absent.	Like that of amphibians.	Not found.	Middle commissure present; it connects optic thalami across diacoel.
			MIDBRAIN		
1. Optic lobes	Two large, dorsal, hollow optic lobes or **corpora bigemina** present and covered mostly by cerebellum, control vision.	Two large, dorso-lateral, rounded, hollow optic lobes or **corpora bigemina** present with optocoels but uncovered.	Two median, oval, dorso-laterally placed hollow optic lobes or **bigemina** present and uncovered by cerebrum.	Two very large, spherical, hollow, laterally displaced optic lobes or corpora bigemina present; laterally displaced due to extensions of cerebral hemispheres and cerebellum but remain	Four, almost solid spherial ball-like **corpora quadrigemina** present and mostly covered by posterior extensions of cerebral hemispheres.

(Contd.)

Characters	*Scoliodon* (Dogfish)	*Rana* (Frog)	*Uromastix* (Spiny-tailed lizard)	*Columba* (Pigeon)	*Oryctolagus* (Rabbit)
				connected by transverse **optic commissure**.	
2. Crura cerebri	Floor of midbrain or **crura cerebri** poorly developed and hidden ventrally by inferior lobes of infundibulum and saccus vasculosus.	Situated beneath optic lobes longitudinally connecting diencephalon and medulla oblongata.	Thickened floor forms **crura cerebri**, which are relatively less developed.	Crura cerebri thickened as in *Uromastix*.	Much developed than in lower vertebrates.
			HINDBRAIN		
1. Cerebellum					
(*i*) Shape and size	Large, elongated and rhomboidal. Dorsal extension covers midbrain and posterior extension to medulla.	Small and narrow dorsal transverse band behind optic lobes.	Small, flat, semicircular, ridge covering anterior dorsal surface of medulla.	Very large and antero-posteriorly elongated. Covers midbrain anteriorly and medulla posteriorly.	Comparatively larger, elongated transversely. Covers partially midbrain in front and medulla behind.
(*ii*) Division	2 transverse furrows divide cerebellum into 3 lobes.	Undivided.	Undivided.	Divided into 2 small **lateral flocculi** and one median **vermis lobe**.	Divided into 5 lobes, a median **vermis**, **2 lateral lobes** terminating into **flocculi**.
(*iii*) Surface	Irregular folds at the dorsal surface.	Smooth because no folds.	Smooth externally.	Rough due to several folds.	Very rough due to much folds.
(*iv*) Ventricle	Large epicoel.	Small epicoel.	Small epicoel.	No ventricle, hence, solid.	Narrow and branched epicoel.
(*v*) Pons Varolii	Absent.	Absent.	Absent.	Absent.	Two lateral sides of cerebellum connected ventrally by a stout, nervous band.

(Contd.)

Characters	*Scoliodon* (Dogfish)	*Rana* (Frog)	*Uromastix* (Spiny-tailed lizard)	*Columba* (Pigeon)	*Oryctolagus* (Rabbit)
(*vi*) Arbor vitae	Absent.	Absent.	Absent.	Absent.	White matter looks tree-like in gray matter called arbor vitae.
2. Medulla oblongata					
(*i*) Shape and size	Large, hollow, tapering behind, partly covered anteriorly beneath cerebellum.	Small, conical, hollow and uncovered.	Small, triangular, hollow and uncovered.	Small, hollow and covered beneath cerebellum.	Broad, triangular, hollow, tapering and concealed partly by cerebellum.
(*ii*) Restiform bodies	A pair of irregular thin-walled outgrowths at the antero-lateral border of medulla, called restiform bodies.	Not found.	Not found.	Not found.	Not found.
(*iii*) Ventral flexure	Absent.	Absent.	Junction of medulla and spinal cord has well marked ventral flexure.	Well marked like lizards	Absent.
(*iv*) Cranial nerves	10 pairs.	10 pairs.	12 pairs. XI is spinal accessory and XII is hypoglossal.	12 pairs as in lizards.	12 pairs.

TABLE 50.15. COMPARISON : EXCRETORY SYSTEMS OF VERTEBRATE TYPES.

Characters	*Scoliodon* (Dogfish)	*Rana* (Frog)	*Uromastix* (Spiny-tailed lizard)	*Columba* (Pigeon)	*Oryctolagus* (Rabbit)
1. Excretory organs	Consist of a pair of **kidneys**, a pair of **ureters**, and **urinogenital sinus**. Urinary bladder absent.	Consist of a pair of **kidneys**, a pair of **ureters**, **urinary bladder** and **cloaca**.	Consist of a pair of **kidneys**, a pair of **ureters**, **urinary bladder** and **cloaca.**	Consist of a pair of **kidneys**, a pair of **ureters** and **cloaca**. Urinary bladder not found.	Consist of a pair of **kidneys**, a pair of **ureters**, **urinary bladder**, and **urethra**.

(Contd.)

Characters	*Scoliodon* (Dogfish)	*Rana* (Frog)	*Uromastix* (Spiny-tailed lizard)	*Columba* (Pigeon)	*Oryctolagus* (Rabbit)
2. Kidneys	Elongated, flattened, ribbon-like, attached to dorsal abdominal wall; each kidney has two distinct parts : anterior narrow **non-excretory** part (genital in male but non-genital in female) and **posterior** broader **excretory part**. Such a kidney is called **opisthonephros**.	Elongated, oval, flat, attached dorsally one on either side of the vertebral column in posterior abdominal cavity. Not differentiated into non-excretory and excretory parts. Whole kidney is excretory. It is **mesonephric**.	Small, irregular, attached dorsally and lie in pelvic region of abdominal cavity. Each kidney bilobed; anterior broad lobes remain free, while posterior narrow lobes united to form V-shaped structure. It is **metanephric** kidney.	Small, flat, dorsally attached in pelvic region of abdominal cavity and embedded in hollows of pelvis. Each kidney is made of anterior, middle and posterior lobes. It is **metanephric** kidney.	Small, bean-shaped, placed asymmetrically in the anterior part of abdominal cavity attached dorsally with the abdominal wall. Not divided into lobes. Kidneys are **metanephric**.
(i) **Histology**	Composed of a compact mass of uriniferous tubules, not differentiated into cortex and medulla and ventrally covered by **peritoneum**.	Similar to fish.	Similar to fish and amphibia.	Differentiated into **cortex** and **medulla** and made of a large number of **uriniferous tubules**. Covered ventrally by **peritoneum**.	Differentiated in outer **cortex** and inner **medlla**, made of much convoluted **uriniferous tubules** and covered ventrally by **peritoneum**.
(ii) **Nephrostomes (peritoneal funnels)**	Present.	Present.	Absent.	Absent.	Absent.
(iii) **Uriniferous tubules or nephrons**	Bear special urea absorbing area and **loop of Henle** absent.	Urea absorbing area and loop of Henle not present. Glomerulus helps in urine filtration.	Similar to amphibians.	Urea absorbing area not found but loop of Henle helpful in water absorption present. Glomerulus helps in filtering	Urea filtered through glomerulus and tubular reabsorption, loop of Henle present.

(Contd.)

Characters	*Scoliodon* (Dogfish)	*Rana* (Frog)	*Uromastix* (Spiny-tailed lizard)	*Columba* (Pigeon)	*Oryctolagus* (Rabbit)
				excretory wastes.	
3. Ureters	Ureters or opisthonephric ducts or kidney ducts of both kidneys run over ventral sides of kidneys and open into urinogenital sinus that leads into cloaca. In female ureters unite to open by a common aperture on a urinary papilla but in male both open separately.	Ureters or mesonephric ducts run along outer surface of kidneys and open separately directly into cloaca. Urinogenital sinus not found.	Ureters or metanephric ducts run ventrally to kidneys and open separately into middle part (**urodaeum**) of cloaca. Urinogenital sinus not found.	Like that of lizards.	Ureters or metanephric ducts originate from the concavity (**hilus**) of kidneys and open into urinary bladder. At the origin the ureter is funnel-like expansion to form **pelvis**. Cloaca not found.
4. Urinary bladder	Absent.	Large, bilobed, thin-walled elastic urinary bladder opens ventrally into cloaca by a **sphinctered aperture.**	Small, thin-walled inelastic sac-like urinary bladder opens ventrally into coprodaeum of cloaca.	Absent.	Large, median, pear-shaped muscular sac-like; its neck forms urethra to open at the tip of penis in male and into the vestibule of female that opens outside through **vulva**.
5. Nature of excretion	**Ammonotelic** because nitrogenous wastes are excreted mainly as ammonia than anything else.	**Ammonotelic** in tadpole stage but **ureotelic** in adult because nitrogenous wastes excreted mainly are urea dissolved in water as urine.	**Uricotelic** because nitrogenous wastes are excreted chiefly as semisolid **uric acid** and **urate granules** which are not much soluble in water.	Similar to reptiles.	**Ureotelic** because urea is the main nitrogenous excretory waste excreted in the form of urine.

(Contd.)

TABLE 50.16. COMPARISON : REPRODUCTIVE SYSTEMS OF VERTEBRATE TYPES.

Characters	*Scoliodon* (Dogfish)	*Rana* (Frog)	*Uromastix* (Spiny-tailed lizard)	*Columba* (Pigeon)	*Oryctolagus* (Rabbit)
Sexual dimorphism	Distinct; the inner portion of pelvic fins in males only develop into claspers for transferring spermatozoa during copulation.	Indistinct but during breeding season male frog develops **nuptial pad** at the base of first finger to clasp the female frog in amplexus.	Somewhat distinct; in male a pair of **hemipenes** present midventrally behind cloaca and it also possesses **preanofemoral pores**.	Not distinct in pigeon.	Well marked; male bears penis and scrotal sacs bearing testes.
1. Male Reproductive System	Consists of a pair of testes, vasa efferentia, paired vasa deferentia, epididymes, seminal vesicles, urinogenital sinus, sperm sac and claspers.	Consists of a pair of testes, several vasa efferentia, paired urinogenital ducts and cloaca.	Consists of a pair of testes, a pair of vasa deferentia, 2 hemipenes and cloaca.	Consists of a pair of testes, a pair of vasa deferentia, a pair of seminal vesicles and cloaca.	Consists of a pair of testes, 2 vasa deferentia, epididymis and a urethra, penis and some accessory glands.
***(i)* Testes**	Elongated, very long, cylindrical and attached to dorsal body wall by **mesorchium** (double fold of peritoneum). Posterior ends of testes are attached to a rectal gland by a non-glandular tissue.	Small, rod-like, **oval bodies, yellowish in colour** attached to antero-ventral surface of kidneys by **mesorchium.** Rectal gland absent.	Small, oval, white coloured bodies, situated much anterior to kidneys and attached to dorsal abdominal wall by mesorchium, right testis is placed a little anterior to the left testis. Rectal gland absent.	Comparatively large, oval, white coloured, right one is a little smaller than the left, situated antero-ventral to kidneys and attached by **mesorchium**. Rectal gland absent.	Smaller, oval, pinkish bodies, situated outside from the abdominal cavity in **scrotal sacs**. Testis descends out in scrotal sac through **inguinal canal**. Rectal gland absent.
***(ii)* Scrotal sacs**	Absent.	Absent.	Absent.	Absent.	Present.
***(iii)* Gubernaculum and spermatic cord**	Absent.	Absent.	Absent.	Absent.	Present; **gubernaculum** attaches the posterior end of testis with scrotal sac

(Contd.)

Characters	*Scoliodon* (Dogfish)	*Rana* (Frog)	*Uromastix* (Spiny-tailed lizard)	*Columba* (Pigeon)	*Oryctolagus* (Rabbit)
					and **spermatic cord** attaches the anterior end of testis to dorsal abdominal wall. These structures are comparable to mesorchium of other vertebrates.
(*iv*) **Vasa efferentia**	Emerge from anterior end of testis to join **Wolffian duct** or vas deferens or mesonephric duct.	Emerge from inner end of testis and enter kidney to join **Bidder's canal** which opens into urinogenital duct or ureter.	Emerge from outer posterior surface of testis and form a highly convoluted mass, the **epididymis**.	Emerge from anterior end of testis and form **epididymis**.	Emerge from inner border of testis and form **epididymis**.
(*v*) **Epididymis**	Extremely coiled anterior part of vas deferens at the ventral surface of anterior genital part of kidney.	Not found. Epididymis is not formed by the vas deferens.	Anterior part of vas deferens forms thickened, convoluted, undifferentiated mass of tubules at the outer surface of testis.	Anterior part of vas deferens forms a coiled mass of tubules at the inner surface of testis but poorly developed.	Well developed, irregular narrow, compact, ridge-like mass of convoluted tubules at the inner surface of testis. Differentiated into caput epididymis, cauda epididymis and corpus epididymis.
(*vi*) **Vasa deferentia**	Each Wolffian duct or vas deferens functions only as genital duct and opens separately into urinogenital sinus.	Vas deferens of each side unite with the ureter of its side to form **urinogenital duct** which opens separately in the roof of cloaca.	Vas deferens of each side travel separately to ureter and open into the middle chamber of cloaca (urodaeum).	As in lizards.	Each vas deferens arising from **cauda epididymis** of their side, open into the neck of urinary bladder of **urethra**.

(Contd.)

Characters	*Scoliodon* (Dogfish)	*Rana* (Frog)	*Uromastix* (Spiny-tailed lizard)	*Columba* (Pigeon)	*Oryctolagus* (Rabbit)
(*vii*) **Seminal vesicle**	Each vas deferens expands terminally to form wide tube-like sperm storing organ, called **seminal vesicle**.	Terminal part of urinogenital duct enlarges to form a temporary seminal vesicle.	Absent.	Seminal vesicle small for temporary storage of sperms during breeding season.	Small sac-like structure called **uterus masculinus** or **seminal vesicle** opens directly into dorsal end of **urethra**.
(*viii*) **Sperm sacs**	A pair of blind, club-shaped **sperm sacs** of doubtful function open into urinogenital sinus.	Not found.	Not found.	Not found.	Not found.
(*ix*) **Copulatory organs**	A pair of siphons for flushing sperm with sea water and a pair of claspers help in transferring spermatozoa into cloaca of female.	No copulatory organs.	A pair of hemipenes found midventrally at the base of tail help in transferring spermatozoa into the cloaca of female.	No copulatory organs.	Penis present in front of anus transfers spermatozoa into vagina of female.
(*x*) **Accessory reproductive glands**	Not found.	Not found.	Preanofemoral glands produce a horny secretion which forms temporary spines for grasping females during copulation.	Not found.	A pair of each of prostrate glands, Cowper's gland, perineal and rectal glands. Their secretions either contribute in the formation of semen or help in attracting female.
2. Female Reproductive System	Consists of a pair of ovaries, oviducts, shell glands, uteri, epigonal organs, a vagina and cloaca.	Consists of a pair of ovaries, oviducts and a cloaca.	Consists of a pair of ovaries, oviducts, vagina, shell glands, and a cloaca.	Consists of only left ovary, left oviduct including uterus, shell gland and vagina and a cloaca.	Consists of a pair of ovaries, oviducts, uteri, single vestibule, vagina, clitoris and some accessory reproductive glands.

(Contd.)

Characters	*Scoliodon* (Dogfish)	*Rana* (Frog)	*Uromastix* (Spiny-tailed lizard)	*Columba* (Pigeon)	*Oryctolagus* (Rabbit)
(*i*) **Ovaries**	Paired, irregular, small, lobulated structures, situated behind liver attached to dorsal abdominal wall by a double peritoneal fold called **mesovarium**.	Paired, large, irregular, hollow, lobulated structures situated near kidneys attached to dorsal abdominal wall by. **mesovarium**.	Paired, small oval structures situated a little anterior to kidneys attached to dorsal body wall by **mesovarium**. Right ovary is somewhat anterior to left ovary.	Only left ovary present which is large, irregular structure situated ventral to anterior lobe of left kidney and attached by **mesovarium**.	Paired, small, oval white strucutres situated symmetrically and found behind the kidneys attached to dorsal abdominal wall by **mesovarium**.
(*ii*) **Epigonal organs and rectal glands**	Epigonal organs are elongated structures connected anteriorly to ovaries and posteriorly to the rectal gland of unknown function.	Both absent.	Both absent.	Both absent.	Both absent.
(*iii*) **Oviducts**	**Mullerian ducts** or **oviducts** are a pair of large and straight ducts originating from a **common oviducal funnel** situated near septum transversum. Soon from the point of origin oviducts descend on both the sides and open into the cloaca.	Oviducts are very long and much coiled glandular tubes opening behind into cloaca. Their anterior ends possess separate wide **oviducal funnel** having openings called **ostia** at the base of lungs.	Oviducts are long, wide, thin-walled coiled tubules opening behind separately into cloaca (**urodaeum**). Anterior end of each oviduct forms a wide oviducal funnel having an **ostium**.	Left oviduct is well developed, long, wide tube which is thick-walled and opens behind into cloaca (**urodaeum**). Anteriorly it widens near ovary to form **oviducal funnel** having a fimbriated ostium. Right oviduct may present as vestiges.	Oviducts are large, coiled tubules opening behind within the uterus; these are called **Fallopian tubes**. Anteriorly each oviduct opens near ovary by a small **oviducal funnel** having fimbriated **ostium**.
(*iv*) **Shell glands**	The oviduct swells a little behind the funnel to form **shell gland**. It secretes a thin membrane around descending fertilised egg.	Not found.	Each oviduct dilates a little to form shell gland slightly before opening into cloaca. It secretes shell over the fertilised egg.	Behind ostium the left oviduct is differentiated into **magnum, isthmus, uterus** and **vagina**. Shell is secreted in the uterus part of oviduct.	Absent.

(Contd.)

Characters	*Scoliodon* (Dogfish)	*Rana* (Frog)	*Uromastix* (Spiny-tailed lizard)	*Columba* (Pigeon)	*Oryctolagus* (Rabbit)
(*v*) **Uterus**	Each oviduct forms a wide uterus in the region of kidneys in which embryos develop.	Uterus is absent but each oviduct is dilated to form **ovisac** before opening into cloaca.	Not found.	True uterus not found but the hind part of oviduct forms the so-called uterus.	True uterus present where embryos develop; the fallopian tube is followed by a much wider, thick-walled muscular tube called **uterus**.
(*vi*) **Vagina and vestibule**	The uteri of both sides unite posteriorly to form a common passage called **vagina** which opens into cloaca. Vestibule not found.	Vagina and vestibule absent, **Ovisacs** directly open into cloaca.	Terminating part of each oviduct forms the so-called **vagina** which opens into cloaca (**urodaeum**). Vestibule absent.	The last part of single left oviduct is vagina which opens in cloaca (urodaeum). Vestibule absent.	The uteri of both sides meet into a common, long, wide median **vagina**. It opens with urethra into common vestibule or urinogenital sinus.
(*vii*) **Vulva**	Absent but cloaca opens outside through a large median shallow groove.	Absent. Cloaca opens out through a small circular cloacal aperture.	Absent but cloaca opens out directly through a transverse slit-like aperture.	Vulva not present but cloaca opens midventrally by a transverse aperture.	Vestibule opens outside ventral to anus through a longitudinal slit-like aperture called **vulva**.
(*viii*) **Accessory reproductive glands**	Not found.	Not found.	Not found.	Not found.	Accessory female reproductive glands are **Bartholin's** or **Cowpers' glands**, **perineal glands** and **rectal glands**.

(*Contd.*)

Characters	*Scoliodon* (Dogfish)	*Rana* (Frog)	*Uromastix* (Spiny-tailed lizard)	*Columba* (Pigeon)	*Oryctolagus* (Rabbit)
(*ix*) **Milk glands**	Absent.	Absent.	Absent.	Crop glands secrete pigeon's milk to feed the young ones.	Mammary glands secrete milk and open on 4 or 5 pairs of ventral teats or nipples.
3. Life History					
(*i*) **Breeding season**	Throughout the year.	From July to September (rainy season).	During March and April in North India.	Throughout the year.	Entire year specially from January to June.
(*ii*) **Fertilisation**	Internal inside oviduct after copulation, **ovo-viviparous**.	External in water, **oviparous**.	Internal inside oviduct, **oviparous**.	Internal inside oviduct, **oviparous**.	Internal, **viviparous**.
(*iii*) **Eggs**	3 to 7 small, kept in uteri.	Indefinite small laid in water in a common jelly to form spawn.	10 to 15 large shelled eggs laid in burrow.	Usually 2 large, oval shelled laid in nest.	Microscopic, kept in uteri.
(*iv*) **Incubation**	No.	No.	No.	Incubation by both parents.	No.
(*v*) **Development**	Uterine 3 to 7 youngs develop in each uterus and nourished by yolk sac placenta.	In water. Tadpoles develop which undergo metamorphosis to form adults.	On land, inside egg. Newly hatched youngs are like adults.	On land, inside egg. Newly hatched young is naked and helpless. Fed on pigeon's milk.	Inside uteri. Gestation period is of 30 days. Newborn young is fed on milk.

51

Chapter

Wildlife in India

Contents of this Chapter

Wildlife comprises all living organisms (plants, animals, microorganisms) in their natural habitats which are neither cultivated/domesticated nor tamed. But in its strictest sense, it includes uncultivated mammals, birds, reptiles, amphibians and fishes, etc., which are generally hunted.

Wildlife includes "Animals like mammals, birds, reptiles, amphibians and fishes and their youngs and also eggs of birds and reptiles" **(Wildlife (Protection) Act, 1972)**. It also includes habitat like land, water or vegetation which is the natural home of any wild animal.

India has a rich heritage of wildlife. In India wildlife has a long history and tradition of conservation. The conservation ethic was observed in Ashrams of our sages which were the seats of learning in the country's ancient past.

The Vedas include hymns in praise of animals and the

Indian Mythology is full of references to several animal-like Gods such as the monkey-headed *Hanuman*, elephant-headed *Ganesh*, boar-headed *Varahavatar*, lion-headed *Narasinghavatar*, turtle-like *Koormavtar*, fish-like *Matsyavatar*, snake-like *Shesh Nag*, etc. We also learn about snake worship (*Nag pooja*), eagle worship (*Garud pooja*), cow-worship (*Gau pooja*), and vehicle of Lord Shiva (*Bull Nandi*), vehicle of Saraswati (*Swan*), vehicle of Lord Ganesh (*rat*), etc.

King Pandu in Mahabharata was cursed by the hind deer for shooting her stag in mating. Rama was helped by Hanuman against Ravana, the flying deer hunted by Kalidas to the safety of rishi's ashram, eight celestial points of compass being supposed to be guarded by Indira's elephant, lion is one of the many incarnations of God Vishnu, the mongoose noticed in the Mahabharata as a teacher of wisdom of King Yudhishthira, the deer is always associated with God Brahma and the constant companion of Lord Mahadeva, the wild boar is referred to as the "Boar of Heaven". About thirty different mammals are mentioned by name in the Samhitas (four principal Vedas).

The first recorded game laws were promulgated by **Kautilya** in his *Arthashastra* in the third century B.C. He proclaimed severe punishment for killing, entrapping animals in protected areas (sanctuaries). The **Great Ashoka** enacted laws for the protection of animals in his kingdom in third century B.C. Even in Christianity, Francis of Assisi loved and spoke to protect wild animals. The condition of our wildlife deteriorated in the Moghul rule and later in British rule when slaughter of animals became the fashion of the day. Babar and Temur killed thousands of rhinoceroses in Kashmir and Northern India. Colonel Pollock, military engineer of British East India Company in Assam, shot a rhinoceros or buffalo almost daily for breakfast. Former Kings and Nawabs of various states of India were also very fond of hunting tigers.

Some wild animals are so characteristic that they become symbols of their home countries. Thus, tiger is associated with India, white bear with Russia, Giant Panda with China, Kangaroo with Australia, Kiwi with New Zealand and Springbok with South Africa.

IMPORTANCE OF WILDLIFE

Wildlife is important for the human beings, animals and plants. These are so closely interlinked that disturbance in one produces imbalance in the others.

1. Ecological balance. Wildlife maintains balance of nature through regulation of population of different species by self-regulation and feedback, passage of food and energy through series of different populations (food chain/food web), circulation of inorganic nutrients between abiotic and biotic environments, and prevention of leaching and runoff by plants. Terrestrial and aquatic animals play their role in the maintenance of balance of nature. Thus, it preserves the environment as a self-sustaining system. It balances populations and maintains food-chain and natural cycles.

2. Gene bank. Wildlife serves as a gene bank for breeding improved varieties in agriculture, animal husbandry and fishery. Plant breeders have been able to produce high-yielding and disease- and stress-resistant varieties from wild relatives of crop plants which possess useful genes for the purpose of breeding. Scientists have been constantly examining the wild relatives of crop plants for the presence of useful genes that can be introduced to breeding programmes. Hence, gene bank maintenance is essential. Some old rice varieties from Kerala saved rice cultivation when brown plant hopper (*Nila-parvata lugens*) attacked all modern rice varieties.

The production of high-yielding, disease-resistant crops, livestock and fish cannot continue without the wild relatives of the cultivated varieties. This is because of pests and pathogens evolve new strains; climate change; soils vary; and consumers' demands change with time. Man is aware that the rich diversity of organisms today is the product of natural evolution occurring through 3.5 billion years. A species once lost cannot be retrieved. Therefore, it would be unethical to be responsible for the destruction of a species.

3. Plant propagation. Insects and birds, etc., are useful in pollinating certain plants which is very essential in plant propagation. Cross pollination increases the diversity of genetic recombination and viability of the plant vegetation.

4. Cleaning of environment. Cleaning of environment and keeping it in hygienic state is carried by scavengers like vultures, eagles, jackals, hyaenas, etc. Microorganisms feed upon dead animals, convert them into different nutrients and, thus, release energy back to the soil making it fertile.

5. Soil erosion. Soil erosion is prevented by plant cover, litter, mixing of litter by movement of wild animals. Microorganisms convert litter into spongy humus.

6. Scientific importance. Scientific studies of many of wildlife species are of direct value to humans. These studies and researches in anatomy, physiology, ecology, evolutionary aspects, etc., are of direct value in saving human life. Sea urchins have helped greatly in the understanding of human embryology. A desert toad has helped in early determination of pregnancy. Rhesus monkeys have contributed a lot to the present knowledge of human blood groups. Antlers of deer are useful in determining the degree of radioactive contamination of natural environment. Armadillos contributed to the development of vaccine for leprosy. We do not know when some obscure wild animal species may be put to prominence by providing a clue to human health and survival. When any drug of any disease is invented, research of its effects and side effects are carried on wild animals before applying on human body. Rh factor in human blood was found due to Rhesus monkey. Chimpanzee helped us in conducting serological protein tests. Studies on animal behaviour helped the psychiatrist to read the human mind. Some wild flora and fauna have medicinal values. Muskpod of musk deer and rhinoceros' horn are of great medicinal value. Fat of tiger or hornbill is used for curing rheumatism. Snake venom is used for preparing antivenom for snake bite.

7. Economic importance. Wildlife forms an important natural resource. Various useful products are obtained from plants such as timber, firewood, paper, gum, resins, tannins, several drugs, essential oils, spices, lac, silk, honey, hair, feathers, guano (the faecal matter of sea-fishes used as manure or the manure made from fish), leather, musk, ivory, etc., are obtained from wildlife. Besides these, benefits such as tendu leaf, cork, rudraksha, etc., are also obtained from wildlife. Tribals living in forests also get food (tubers, roots, leaves and fruits of plants and meat from animals), medicines and other commercial products. The commercial value of wildlife is best seen in the world's marine fisheries with an annual output of about 100 million tons of sea food worth billions of rupees. Freshwater fish and other aquatic creatures also provide large amount of food for people. Fish are as big as any other modern industry in respect of employment and income. An entire industry for fur trade is supported by fur-bearing animals. Trade in live and dead animals, though illegal and banned throughout the world, supports thousands of people and also earns foreign exchange. For example, an Indian rhinoceros may fetch equivalent of ₹ 1,25,000 in the world market. Similarly, the ivory of elephants, the horns of rhinoceros, the glands of musk deer, the antlers of deer, etc., all provide high prices. Wildlife of a country may even attract people from abroad and earn foreign exchange. Wildlife tourism can be made a big source of income. Thus, it is important to develop more biological reserves, parks, safaris and zoos for tourism from economical point of view. The tourist industry of Kenya (East Africa), based on its wildlife, ranks third after coffee and sisal.

8. Potential importance. Just as all present-day cultivated/domesticated plants and animals are derived from wildlife; new foods, beverages such as tea, coffee, cocoa, etc., drugs and other useful products may be obtained from wildlife.

9. Game importance. Wildlife has its importance in game also. In several European and American countries, millions of people hunt or fish for recreation, spending billions of dollars on these sports.

10. Aesthetic importance. There is a great aestheic value of wildlife throughout the world because of sheer beauty, tranquillity and appeal to the humans. A world without colourful and melodious birds, graceful beasts and deep forests would be a poorer place for humans to live. Without wild animals, a countryside would be static and monotonous. People feel pleasure, satisfaction and happiness in the presence of wildlife. Pandit Jawaharlal Nehru said, "Life would become very dull and colourless if we did not have these magnificient animals and birds to look at and to play with." Smt. Indira Gandhi also said, "A threat to any species of plant and animal life is a threat to man himself."

11. Ethical importance. Generally civilised people think that they have no right to destroy wildlife (plants and animals), rather they feel an obligation for the conservation of nature and protection of wildlife. In fact, all religions preach respect and reverence for life and consider it wrong and unethical to take the life of an animal.

BIODIVERSITY

Biodiversity encompasses all species of plants, animals, and microorganisms and the ecosystems and ecological processes of which they are parts. It is an umbrella term for the degree of nature's variety including both the number and frequency of ecosystems, species or genes in a given assemblage. Thus, the term '**biodiversity**' refers to the totality of 'genes, species and ecosystems' of a region. It is usually considered at three different levels : (*i*) genetic diversity, (*ii*) species diversity, and (*iii*) ecosystem diversity.

(*i*) Genetic diversity is the sum total of genetic information, contained in the genes of individuals of plants, animals and microorganisms that inhabit the earth.

(*ii*) Species diversity refers to the variety of living organisms on earth and has been variously estimated to be between 5 and 50 million or more, though only about 1.4 million have actually been described.

(*iii*) Ecosystem diversity refers to the variety of habitats, biotic communities, and the ecological processes in the biosphere as well as the tremendous diversity within ecosystems in terms of habitat differences and the variety of ecological processes.

Biodiversity is defined as "the variability among living organisms from all sources, including terrestrial, marine and other aquatic ecosystems and the ecological complexes of which they are a part; this includes diversity within species between species and of ecosystems."

Conservation and sustainable use of biodiversity is fundamental to ecologically sustainable development. Biodiversity is part of our daily lives and livelihood, and constitutes resources upon which families, communities, nations and future generations depend.

Every country has the responsibility to conserve, restore and sustainably use the biological diversity within its jurisdiction. Biological diversity is fundamental to the fulfilment of human needs. An environment rich in biological diversity offers the broadest array of options for sustainable economic ability, for sustaining human welfare and for adapting to change. Loss of biodiversity has serious economic and social costs for any country.

The experience of the past few decades has shown that as industrialisation and economic development in the classical sense have taken place, patterns of consumption, production and needs change, straining, altering and even destroying ecosystems.

India, a megabiodiversity country while following the path of development, has been sensitive to the needs of conservation and, hence, is rich in biological resources. Ethos of conservation and harmonious living with nature is very much ingrained in the lifestyies of India's people.

Magnitude of Biodiversity

The foundation for assessing the importance of biodiversity is an inventory of how many species exist and which species exist where. At the global level the plants and animals are relatively well known. **Erwin** (1982), for example, suggests as many as 30 million species in total, with most undescribed species living in tropical forests. The known and described number of species of all organisms on the earth is between 1.7 and 1.8 million which is fewer than 15 per cent of the actual number. About 61 per cent of the known species are insects. Only 4,650 species of mammals are known to biological science. A large number of plant species (2,70,000) and vertebrates are known. But the fact remains that basic knowledge of the organisms that make up most ecosystems, especially in the tropics, is inadequate. Information about bacteria, viruses, protists and Archaea is only fragmentary. For convenience, many assume that about 10 million species exist though the final figure is likely to be 30–50 millions.

Status of Biodiversity in India

India occupies only about 2.4 per cent of the world's land area but its contribution to the world's biodiversity is approximately 8 per cent of the total number of species which is estimated to be 1.75 million. Of these, 126,188 have been described in India. The species recorded includes flowering plants (angiosperms), mammals, birds, reptiles, amphibians and fishes, constituting 17.3 per cent of the total, whereas nearly 60 per cent of India's biowealth is contributed by fungi and insects. Based on the available data, India ranks tenth in the world and fourth in Asia in plant diversity, and ranks tenth in the number of endemic species of higher vertebrates in the world.

There are about 410 species of mammals (8.86% species of the world), about 1,250 species of birds (about 12% of the world species), about 2,546 species of fishes (about 11% of the world species), about 197 species of amphibians (4.4% of the world species), and more than 408 species of reptiles (6% of the world species) are found in India. Among these groups the highest levels of endemism are found in amphibians.

Uses of Biodiversity

Humans derive many direct and indirect benefits from biological diversity. All our food comes from wild species brought into domestication. Most of our medicines, pharmaceuticals, fibres, rubber and timber come from biological resources. The biodiversity also provides many ecological services free of charge that are responsible for maintaining ecosystems. Our water is supplied by one of nature's most important processes, called the 'hydrological cycle'. Forested watersheds provide clear, high-quality water for domestic or agricultural use, while healthy rivers provide water, transport and fish.

Threats to Biodiversity

Major biodiversity threats are as follows :

1. Habitat destruction and fragmentation;
2. Extension of agriculture;
3. Filling up of wetlands;
4. Conversion of rich biodiversity site for human settlement and industrial development;
5. Destruction of coastal areas;
6. Uncontrolled commercial exploitation;
7. Disturbance and pollution; and
8. Introduction of non-native (exotic) species.

Vanishing Wildlife

Before the arrival of man about two million years ago, animal species were subjected to natural causes of extinction. A few examples are of extinct ammonites, large cephalopods and brachiopods of Devonian period and dinosaurs of Mesozoic Era. Siwaliks (Himalayan range) were formed in the last and in it are found the largest number of fossils of Tertiary period. Siwalik fossils include mastodons, hippopotamus, rhinoceros, Sivatherium (large four-horned ruminant), giraffe, horses, camels, bison, deer, antelope, pigs, chimpanzees, orangutans, baboons, langurs, macaques, cheetals, sabre-toothed cats, lions, tigers, sloth bear, Aurochs, leopards, wolves, sholes, porcupines, rabbits, etc.

Many fossil tree species have also been found in the intertrappean beds including *Grewioxylon* (Eocene) and *Heritieroxylon keralensis* (middle Miocene) from Kerala and *H. arunachalensis* (Mio-Pliocene) from Arunachal Pradesh. *Grossopteris* (fern fossil) was discovered from India and Antarctica led to the discovery of Gondwanaland. Fossil cycads are also found in India. From Narmada valley was discovered the first dinosaur, *Titanosaurus indicus* (**Richard Lydekker**, 1877). Another dinosaur fossil from the same area was *Rajasaurus narmadensis*. Whale fossil *Himalayacetus subathuensis* was found in Simla hills (Eocene). This area was under water during Tertiary period. Another whale fossil was *Remingtonocetus* about 43–46 million years old.

During Indus valley civilisation species of wild cattle *Bos primegenius nomadicus* was vanished. Other mammals which became extinct in India are Indian/Asiatic cheetah, Javan Rhinoceros and Sumatran rhinoceros.

Pink-headed duck (*Rhodonessa caryophyllacea*) and Himalayan Quail (*Ophrysia superciliosa*) have became extinct.

The number of species of wild fauna disappeared during the last four centuries are 7 species, 11 species, 27 species and 67 species in seventeenth, eighteenth, nineteenth and twentieth centuries respectively.

Causes of Destruction

It is very necessary to access the factors which are hampering in the growth of population of the species such as decimating (death) factors like diseases, predators, etc., or scarcity of food, water and shelter.

Human population. Man can easily destroy but cannot create a new species. Forests are unlocked property, it is vast and open and, hence, vulnerable for theft, hunting and killing, poaching, etc. Enormous growth in human population results in conversion of forest into agricultural fields, hunting of wild animals for food, etc. Increase in human population and enhancement in cattle (livestock population) goes side by side which causes adverse effect on wild papulation. Pollutants as insecticides, byproducts of industries in the form of gases and water pollutants, quarry of mines, etc., inside forest and such other activities resulting into habitat shrinkage as well as death and disease among the wild animals. Hunting, poaching, killing, smuggling, etc., of wild animals have been hampering and suffering the wild animals. Majority of people are poor and illiterate and are unaware about the importance of wildlife and its role in ecosystem. Wildlife is very much essential for ecobalance and is ultimately essential for human life.

For the existence of an organism/wildlife, food, water and shelter are required. These basic components are exclusively related with wildlife habitats. Wildlife habitats are presently undergoing tremendous changes primarily by the man for his needs, encroachment of forest land for grazing and agricultural purposes due to mushroom growth in human population, constructing roads and cities, etc. Therefore, man's manipulation of environment for his needs is the most prevalent factor affecting wildlife habitat and wildlife populations. For instance, in U.S.A., the dense old growth forests may be disastrous to the spotted owl's nesting and feeding requirements, while it greatly increases preferred forage food for elk.

Destruction of habitat. The most serious depletion of wildlife is due to habitat destruction. Habitats which protect wildlife are being converted for human settlements, harbours, dams, reservoirs, croplands, grazing grounds, mining operations, etc. Environmental pollution and deforestation have also caused the degradation of important habitats.

Migratory animals are also vulnerable to the destruction of habitats because disturbance at any point of their migratory routes affects them. Some of the dams are blocking spawning, migration of fishes by inundating habitats and by altering the physical environment. Many species of whales, sea cows and sea turtles are facing total extinction as they are caught by mechanical devices for the sea industry.

Vultures and kites feed on carcases. Since carcases are being buried or burnt now, the population of large flying bird, California condor (*Gymnogyps californianus*) has started declining.

Asiatic wild ass (*Asinus hemionus khur*) in Rann of Kutch, Indian lion, musk deer, etc., are facing danger of imminent extinction due to reduction of habitat.

Wildlife diseases. In wild animals, there are a number of diseases which are decimating factors and cause a lot of loss of wild animals. It is also very difficult to diagnose and treat the disease in wild animals because they are free ranging animals. **Khera** (1980) pointed out that the prevalence of epizootic diseases among wildlife has also been a major factor associated with decline in number of some species of wildlife population. Important diseases found in wild animals are as follows :

1. Foot and Mouth disease (FMD) caused by Picorna virus mostly occurring in wild boar, cheetal, sambhar and gaur (hoofed mammals).

2. Rinderpest caused by Paramyxo virus found in ruminants such as cattle, cheetal, sambhar, gaur and wild buffalo, etc. It also infects wild animals through the cattle.

3. Anthrax caused by *Bacillus anthrasis* (bacteria) found in mammals (enlargement of spleen).

4. Brucellosis caused by *Brucella* bacteria found in cattle, cheetal, sambhar, etc. It causes tumor in the joints.

5. Botulism a kind of paralysis caused by *Clostridium botulinum* (bacteria). It makes the food poisonous. It occurs in aquatic birds.

6. Tuberculosis caused by *Bacterium tuberculosis.* It is found in wild animals like human beings. It is also common in monkeys.

7. Rabies caused by Rhabdo virus. It is very common in pet dogs and street dogs, wolf, jackal, fox, jungle cat, mongoose, etc.

8. Distemper caused by Paramyxo virus. It is found in carnivorous mammals. Fever affects central nervous system.

Adverse climate. Adverse climatic conditions also cause casualty in wild animals such as very hot weather, excessive rainfall and cold, flood, earthquake, volcano, etc. The population having excessive number than the carrying-capacity is more liable to death. Such deaths can be minimised by increasing carrying capacity, food, water and shelter.

Accidents. Fire, flood, earthquake and road crossing, etc., also cause their accidental death. In forest fire, many wild animals, their young ones, eggs, shelter are burnt. Such accidents can be avoided or minimised by proper management.

Wild animals are also killed by poisoning by man to save agricultural crops and cattle, etc.

Illegal hunting. Illegal hunting of wild animals by man is also to be checked. It is done by smugglers and local people living inside forest or neighbouring forest. It causes heavy loss in wild populations.

CONSERVATION, PRESERVATION, PROTECTION AND CATEGORIES OF WILD ANIMALS

1. Conservation

Conservation means the maintenance of a reasonable number of members of every species from the largest mammals to the smallest invertebrates in their own habitat without destroying that habitat (**Noel Simon**).

According to Indian Forest Record (1965), conservation means, planned management and wide use of natural resources, so as to prevent over-exploitation, destruction or neglect, *i.e.*, wildlife conservation.

2. Preservation

Preservation is the protection of a species without regard to the consequences. Preservation is to save and maintain the wild animals against injury or destruction as well as keeping them safe and undisturbed from private or public use.

3. Protection

Guarding the wild species against danger or injury is protection. **Partial protection** of certain species is to enforce closing of hunting, fishing, etc., and restricted shooting. **Total protection** is achieved by forming sanctuaries or by legally prohibiting the killing or minimising of a particular species in any place.

4. Extinction of a Species

Extinction of a species is a part of a natural process. Extinction is "biological reality" because no species has as yet existed for more than a few million years without evolving into something different or dying out completely. But with the gradual emergence of human beings being a major evolutionary force, people have been increasingly exploiting the wildlife. As many as 500 million kinds of plants, animals, and microorganisms have made this planet as their home since life began over 3.5 billion years ago. Today, there are more than 30 million species alive. The rate of decline has been rapid in the last one hundred years. It is estimated that about 25,000 plant species and 1000 vertebrate species and subspecies are threatened with extinction. Several invertebrates are also invariably vulnerable. At least about 10% of the living species are supposed to be in danger.

5. Extinct Species

Extinct species are those which were found in the past but became disappeared and not found now. They are no longer known to exist in wild, though it may survive in cultivation. According to CITES, a species is said to be extinct, if it is not definitely known in the wild during the past 50 years. Extinction results ecological hazards, imbalances in ecosystem and food chain/food-web, ultimately affecting present and future generations of human beings. For example, extinction of cheetah, two-horned rhinoceros, mountain quail, Pin-headed duck from India and Dodo bird from Mauritius, etc.

Causes of extinction. Causes of extinction are as follows :

(*i*) Hunting. Hunting is large-scale destruction of wildlife for food, safety and pleasure. It was started with the use of fire as a means of hunting. Man-made forest fires have caused the extinction of several species in the past. Hunting as a sport and needless killing of wild animals is dangerous. Disappearance of dodo (*Didus ineptus*) of Mauritius and cheetah (*Acinonyx jubatus*) from India are recent calamities. A number of orchids and medicinal plants have also disappeared.

(*ii*) Destruction of habitats. Destruction of habitats of wildlife is also a threat to them. It has been destroyed due to establishment of new human settlements, croplands, grazing grounds, quarry in mining sites, etc. Deforestation caused by jhuming, cutting of trees for timber and overgrazing, conversion of forest into agricultural land, building of roads and rails, construction of dams/reservoirs, etc. Damage to the forest and grasslands are caused by acid rain. Hence, the most serious depletion of wildlife comes from habitat destruction.

6. Endemic Species

Endemic species is that which is found in a particular natural habitat beyond which it is unknown. The plant and animal species confined to a given region and having originated there are called endemic species. India has a large number of endemic species like *Ficus religiosa*, *Butea monosperma*, etc.

THE IUCN RED LIST CATEGORIES

The World Conservation Union (WCU), formerly known as International Union for the Conservation of Nature and Natural Resources (IUCN), has recognised eight Red List Categories of species. These are as follows :

1. Extinct (Ex). A taxon is **Extinct** where there is no reasonable doubt that the last individual has died.

2. Extinct in the Wild (EW). A taxon is **Extinct in the Wild** when exhaustive surveys, in known and/or expected habitats, have failed to record an individual.

3. Critically Endangered (CE). A taxon is **Critically Endangered** when it is facing an extremely high risk of extinction in wild in the immediate future.

4. Endangered (E). A taxon is **Endangered** when it is not critically Endangered, but is facing a high risk of extinction in the wild in the near future.

5. Vulnerable (V). A taxon is **Vulnerable** when it is not Critically Endangered or Endangered, but is facing a high risk of extinction in the wild in the medium-term future.

6. Lower Risk (LR). A taxon is **Lower Risk** when it has been evaluated and does not satisfy the criteria for Critically Endangered, Endangered or Vulnerable.

7. Data Deficient (DD). A taxon is **Data Deficient** when there is inadequate information to make a direct, or indirect, assessment of its risk of extinction.

8. Not Evaluated (NE). A taxon is **Not Evaluated** when it has not yet been assessed against the above criteria.

The IUCN Red List is a catalogue of those taxa which are facing the risk of extinction.

The species which are threatened with extinction are included in **Vulnerable, Endangered** or **Critically Endangered** categories.

Rare Species (R). Taxa with small world population that are not at present "Endangered" or "Vulnerable" but are at risk.

In practice, 'Endangered' and 'Vulnerable' categories may include, temporarily, taxa whose populations are beginning to recover as a result of remedial action, but whose recovery is insufficient to justify that transfer to another category. These taxa are usually localised within restricted geographical areas or habitats or are thinly scattered over a more extensive range.

Indeterminate (I). Taxa known to be 'Endangered', 'Vulnerable' or 'Rare', but where there is not enough information to say which of the three categories is appropriate.

The IUCN Red List System was initiated in 1963, and since then evaluation of the conservation status of species and subspecies is continuing on a global scale.

The 2000 IUCN Red List contains assessments of more than 18,000 species, of which 11,000 are threatened.

Status of Threatened Species. According to 2000 IUCN Red List, there are 11,046 species listed as threatened, *i.e.*, Critically Endangered, Endangered or Vulnerable. Of these 5,485 are animals and 5,611 are plants. 1,939 species are listed as Critically Endangered, of which 925 are animals and 1,014 are plants.

PROTECTED INDIAN WILDLIFE

There is a great deal of protected Indian wildlife. Tables 51.1, 51.2 and 51.3 provide a list of some common Indian mammals, birds and reptiles which are threatened or endangered with extinction. They have been declared as protected species so that their hunting (capturing, poisoning, injuring, killing, etc.) is illegal under the provisions of the Wildlife (Protection) Act, 1972.

TABLE 51.1. IUCN RED LIST OF SOME INDIAN THREATENED MAMMALS.

Common name	Scientific name	Threatened category
Asiatic golden cat	*Felis temmincki*	I (Indeterminate)
Asiatic lion	*Panthera leo persica*	E (Endangered)
Asiatic wild dog/dhole	*Cuon alpinus*	V (Vulnerable)
Asiatic black bear	*Selenarctos thibetanus*	E (Endangered)
Banteng	*Bos javanicus*	V (Vulnerable)
Bengal fox	*Vulpes bengalensis*	I (Inderminate)
Bharal	*Ovis nahura*	T (Threatened)
Bison or gaur or mithun	*Bos gaurus*	T (Threatened)
Black buck	*Antelope cervicapra*	T (Threatened)
Brow-antlered deer	*Cervus eldi eldi*	V (Vulnerable)
Brown bear	*Ursus arctos*	V (Vulnerable)
Brown palm civet	*Paradoxurus jerdoni*	V (Vulnerable)
Capped langur	*Presbytis pileatus*	T (Threatened)
Caracal	*Felis caracal*	T (Threatened)
Chinkara or Indian gazelle	*Gazella gazelle bennetti*	T (Threatened)
Chital	*Axis axis*	T (Threatened)
Clawless otter	*Anonyx cinerea*	K (Insufficiently known)
Clouded leopard	*Neofelis nebulosa*	V (Vulnerable)
Common otter	*Lutra lutra*	V (Vulnerable)
Crab eating macaque	*Macaca irus umbrosa*	T (Threatened)
Desert cat	*Felis libyca ornate*	E (Endangered)
Dugong/seacow	*Dugong dugon*	V (Vulnerable)
Fishing cat	*Felis viverrina*	T (Threatened)
Flying squirrels	*All species*	T (Threatened)
Four-horned antelope	*Tetraceros quadricornis*	T (Threatened)
Ganges river dolphin	*Platanista gangetica*	V (Vulnerable)
Giant squirrels	*Ratufa macroura, R. indicus*	T (Threatened)
Golden langur	*Trachypithecus geei*	R (Rare species)
Goral	*Nemorhaedus goral*	V (Vulnerable)
Great Indian rhinoceros	*Rhinoceros unicornis*	E (Endangered)
Grey/Indian wolf	*Canis lupus*	V (Vulnerable)
Himalayan ibex	*Capra ibex*	T (Threatened)
Himalayan crestless procupine	*Hystrix hodgsoni*	T (Threatened)
Himalayan W-toothed shrew	*Crocidura attenuata*	V (Vulnerable)
Himalayan marten	*Martes flavigula*	I (Inderminate)
Himalayan musk deer	*Moschus chrysogaster*	V (Vulnerable)
Himalayan shrew	*Soriculus nigrescens*	V (Vulnerable)
Himalayan tahr	*Hemitragus jemlahicus*	T (Threatened)
Hoolock gibbon	*Hylobates hoolock*	E (Endangered)

(Contd.)

Common name	Scientific name	Threatened category
Hyaena	*Hyaena hyaena*	T (Threatened)
Indian elephant	*Elephas maximus*	E (Endangered)
Indian pangolin	*Manis crassicaudata*	T (Threatened)
Indian wild ass	*Equus hemionus khur*	T (Threatened)
Jackal	*Canis aureus*	V (Vulnerable)
Kashmir stag/hangul	*Cervus elaphus hanglu*	E (Endangered)
Leopard	*Panthera pardus*	T (Threatened)
Lesser panda	*Ailurus fulgens*	V (Vulnerable)
Lion-tailed macaque	*Macaca silenus*	T (Threatened)
Loris	*Loris tardigradus*	T (Threatened)
Malabar civet	*Viverra megaspila*	T (Threatened)
Markhor	*Capra falconeri*	T (Threatened)
Musk deer	*Moschus moschiferus*	T (Threatened)
Nilgai	*Boselaphus tragocamelus*	E (Endangered)
Nilgiri langur	*Presbytis johni*	T (Threatened)
Nilgiri tahr	*Ilemitragus hylocrius*	T (Threatened)
Otter	*Lutra perspicillata*	T (Threatened)
Pallas's cat	*Felis manul*	T (Threatened)
Pig-tailed macaque	*Macaca nemestrina*	T (Threatened)
Pigmy hog	*Sus sulvanius*	T (Threatened)
Red fox	*Vulpes vulpes*	T (Threatened)
Sambar	*Cervus unicolor*	T (Threatened)
Sloth bear	*Melursus ursinus*	T (Threatened)
Slow loris	*Nycticebus coucang*	T (Threatened)
Snow leopard	*Panthera uncial*	T (Threatened)
Swamp deer or gond	*Cervus duvauceli*	T (Threatened)
Tibetan antelope or chiru	*Panthelope hodgsoni*	T (Threatened)
Tibetan fox	*Vulpes ferrilatus*	T (Threatened)
Tibetan gazelle	*Procapra picticaudata*	T (Threatened)
Tibetan wild ass	*Equus heminonus kiang*	T (Threatened)
Tiger	*Panthera tigris*	T (Threatened)
Wild buffalo	*Bulbalus bubalis*	T (Threatened)
Wild pig	*Sus scrofa*	T (Threatened)
Wild yak	*Bos grunniens*	T (Threatened)

Indian threatened mammals. There are nearly 372 species of mammals found in India including various species listed in threatened categories (IUCN[1] Red List of Threatened Animals Table 51.1). This database provides information on the distribution of mammalian species in various protected areas of India.

Indian threatened reptiles. There are 447 species of reptiles found in India (**Murthy**, 1994) including a number of species listed in threatened categories (Table 51.2). This database provides information on the distribution of reptiles in Indian protected areas. Large number of information related to species distribution on various protected areas has been linked with various protected areas including tiger reserves.

TABLE 51.2. IUCN RED LIST OF SOME INDIAN THREATENED REPTILES.

English name	Scientific name	Threatened category
Asian Giant Tortoise	*Manouria emys*	V (Vulnerable)
Assam Roofed Turtle	*Kachuga sylhetensis*	T (Threatened)
Black Pond Turtle	*Geoclemys hamiltonii*	I (Indeterminate)
Central Asian/Oxus Cobra	*Naja naja oxiana*	K (Insufficiently known)
Cochin Forest Cane Turtle	*Geoemyda silvatica*	V (Vulnerable)
Elongated Tortoise	*Indotestudo elongata*	K (Insufficiently known)
Estuarine Crocodile	*Crocodylus porosus*	V (Vulnerable)
Gharial (Long-snouted Crocodile)	*Gavialis gangeticus*	E (Endangered)
Green Sea Turtle	*Chelonia mydas*	E (Endangered)
Hawksbill Turtle	*Eretmochelys imbricata*	E (Endangered)
Indian Egg-eating Snake	*Elachistodon westerwanii*	R (Rare species)
Indian Python	*Python molurus molurus*	V (Vulnerable)
Leathery Turtle	*Dermochelys coriacea*	E (Endangered)
Loggerhead	*Caretta caretta ??*	V (Vulnerable)
Mugger (Marsh Crocodile)	*Crocodylus palustris*	V (Vulnerable)
Monitor Lizards	*Varanus griseus, V. flavescens, V. salvator*	T (Threatened)
Olive Ridley Turtle	*Lepidochelys olivacea*	E (Endangered)
Pythons	*Python molurus, P. reticulatus*	T (Threatened)
Red-crowned Roofed Turtle	*Kachuga kachuga*	I (Indeterminate)
Three-keeled land Tortoise	*Melanochelys tricarinata*	I (Indeterminate)
Travancore Tortoise	*Indotestudo forstenii*	R (Rare species)
Yellow Monitor Lizard	*Varanus flavescens*	I (Indeterminate)
?	*Oligodon nikhili*	R (Rare species)

TABLE 51.3. IUCN RED LIST OF THREATENED INDIAN BIRDS.

English name	Scientific name	Threatened category
Baer's Pochard	*Aythya baeri*	CR(Critically endangered)
Bengal Florican	*Houbaropsis bengalensis*	CR(Critically endangered)
Christmas Island Frigatebird	*Fregata andrewsi*	CR(Critically endangered
Forest Owlet	*Heteroglaux blewitti*	CR(Critically endangered)
Great Indian Bustard	*Ardeotis nigriceps*	CR(Critically endangered)
Himalayan Quail	*Ophrysia superciliosa*	CR(Critically endangered)
Indian Vulture	*Gyps indicus*	CR(Critically endangered)
Jerdon's Courser	*Rhinoptilus bitorquotus*	CR(Critically endangered)
Pink-headed Duck	*Rhodonessa caryophyllacea*	CR(Critically endangered)
Red-headed Vulture	*Sarcogyps calvus*	CR(Critically endangered)
Siberian Crane	*Leucogeranus leucogeranus*	CR(Critically endangered)
Slender-billed Vulture	*Gyps tenuirostris*	CR(Critically endangered)
Sociable Lapwing	*Vanellus gregarius*	CR(Critically endangered)
Spoon-billed Sandpiper	*Eurynorhynchus pygmeus*	CR(Critically endangered)
White-bellied Heron	*Ardea insignis*	CR(Critically endangered)
White-rumped Vulture	*Gyps bengalensis*	CR(Critically endangered)
Black-bellied Tern	*Sterna acuticauda*	EN (Endangered)
Black -chinned (Nilgiri) Laughingthrush	*Strophocincla cachinnans*	EN (Endangered)
Egyptian Vulture	*Neophron percnopterus*	EN (Endangered)
Greater Adjutant	*Leptoptilos dubius*	EN (Endangered)
Green Peafowl	*Pavo muticus*	EN (Endangered)
Lesser Florican	*Sypheotides indicus*	EN (Endangered)

(Contd.)

English name	Scientific name	Threatened category
Manipur Bush -quail	*Perdicula manipurensis*	EN (Endangered)- Upgraded from VU
Masked Finfoot	*Heliopais personatus*	EN (Endangered)
Narcondam Hornbill	*Aceros narcondami*	EN (Endangered)
Nilgiri Blue Robin	*Myiomela major*	EN (Endangered)
Oriental Stork	*Ciconia boyciana*	EN (Endangered)
Red-breasted Goose	*Branta ruficollis*	EN (Endangered)
Saker Falcon	*Falco cherrug*	EN (Endangered)
Spotted Greenshank	*Tringa guittifer*	EN (Endangered)
White-bellied Blue Robin	*Myiomela albiventris*	EN (Endangered)
White-headed Duck	*Oxyura leucocephala*	EN (Endangered)
White-winged Duck	*Cairina scutulata*	EN (Endangered)
Yellow-breasted Bunting	*Emberiza aureola*	EN (Endangered)-Upgraded from VU
Beautiful Nuthatch	*Sitta formosa*	VU (Vulnerable)
Black-brested Parrotbill	*Paradoxornis flavirostris*	VU (Vulnerable)
Black-necked Crane	*Grus nigriocollis*	VU (Vulnerable)
Blyth's Tragopan	*Tragopan blythii*	VU (Vulnerable)
Bristled Grassbird	*Chaetornis striata*	VU (Vulnerable)
Broad-tailed Grassbird	*Schoenicola platyurus*	VU(Vulnerable)
Bugun Liocichla	*Liocichla bugunorum*	VU (Vulnerable)
Cheer Pheasant	*Catreus wallichi*	VU (Vulnerable)
Chestnut-breasted Patridge	*Arborophila mandellii*	VU (Vulnerable)
Dalmatian Pelican	*Pelecanus crispus*	VU (Vulnerable)
Dark-rumbped Swift	*Apus acuticauda*	VU (Vulnerable)
Eastern Imperial Eagle	*Aquila helica*	VU (Vulnerable)
Great Knot	*Calidris tenuirostris*	VU (Vulnerable)
Great Slaty Woodpecker	*Mulleripicus pulverulentus*	VU (Vulnerable)
Greater Spotted Eagle	*Aquila clanga*	VU (Vulnerable)
Green Munia (Avadavat)	*Amandava formosa*	VU (Vulnerable)
Grey-sided Thrush	*Turdus feae*	VU (Vulnerable)
Grey-crowned Prinia	*Prinia cinereocapilla*	VU (Vulnerable)
Hooded Crane	*Grus monacha*	VU (Vulnerable)
Houbara Bustard	*Chlamydotis undulata*	VU (Vulnerable)
Indian Skimmer	*Rynchops albicollis*	VU (Vulnerable)
Indian Spotted Eagle	*Aquila hastata*	VU (Vulnerable)
Jerdon's Babbler	*Chrysomma altirostre*	VU (Vulnerable)
Kashmir Flycatcher	*Ficedula subrubra*	VU (Vulnerable)
Lesser Adjutant	*Leptoptilos javanicus*	VU (Vulnerable)
Lesser White-fronted Goose	*Anser erythopus*	VU (Vulnerable)
Long-tailed Duck	*Clangula hyemalis*	VU (Vulnerable)
Marbled Teal	*Marmaronetta angustirostris*	VU (Vulnerable)
Marsh Babbler	*Pellorneum palustre*	VU (Vulnerable)
Nicobar Megapode	*Megapodius nicobariensis*	VU (Vulnerable)
Nicobar Sparrowhawk	*Accipiter butleri*	VU (Vulnerable)
Nilgiri Pipit	*Anthus nilghiriensis*	VU (Vulnerable)
Nilgiri Wood-pigeon	*Columba elphinstonii*	VU (Vulnerable)
Pale-backed Pigeon	*Columba eversmanni*	VU (Vulnerable)
Pale-capped Pigeon	*Columba punicea*	VU (Vulnerable)
Pallas's Fish-eagle	*Haliaeetus leucoryphus*	VU (Vulnerable)
Red-faced Malkoha	*Phaenicophaeus pyrrhocephalus*	VU (Vulnerable)
Rufous-necked Hornbill	*Aceros nipalensis*	VU (Vulnerable)

(Contd.)

English name	Scientific name	Threatened category
Rusty-throated Wren-babbler	*Spelaeornis badeigularis*	VU (Vulnerable)
Sarus Crane	*Grus antigone*	VU (Vulnerable)
Sclater's Monal	*Lophophorus sclateri*	VU (Vulnerable)
Slender-billed Babbler	*Turdoides longirostris*	VU (Vulnerable)
Snowy-throated Babbler	*Stachyris oglei*	VU (Vulnerable)
Socotra Cormorant	*Phalacrocorax nigrogularis*	VU (Vulnerable)
Swamp Francolin	*Francolinus gularis*	VU (Vulnerable)
Tawny-breasted Wren-babbler	*Spelaeornis lognicaudatus*	VU (Vulnerable)
Western Tragopan	*Tragopan melanocephalus*	VU (Vulnerable)
White-browed Buschat	*Saxicola macrorhynchus*	VU (Vulnerable)
White-naped Tit	*Parus nuchalis*	VU (Vulnerable)
White-throated Bushchat	*Saxicola insignis*	VU (Vulnerable)
Wood Snipe	*Gallinago nemoricola*	VU (Vulnerable)
Yellow Weaver	*Ploceus megarhynchus*	VU (Vulnerable)
Yellow-throated Bulbul	*Pycnonotus xantholaemus*	VU (Vulnerable)
Alexandrine Parakeet	*Psittacula eupatria*	NT (Near threatened)-Upgraded from least concern
Andaman Crake	*Rallina canningi*	NT (Near threatened)
Andaman Cuckoo-dove	*Macropygia rufipennis*	NT (Near threatened)
Andaman Drongo	*Dicrurus andamanesis*	NT (Near threatened)
Andaman Hawk-owl	*Ninox affinis*	NT (Near threatened)
Andaman Scops-owl	*Otus balli*	NT (Near threatened)
Andaman Serpent-eagle	*Spilornis elgini*	NT (Near threatened)
Andaman Treepie	*Dendrocitta bayleyi*	NT (Near threatened)
Andaman Woodpecker	*Dryocopus hodgei*	NT (Near threatened)
Andaman Wood-pigeon	*Columba palumboides*	NT (Near threatened)
Asian Dowitcher	*Limnodromus semipalmatus*	NT (Near threatened)
Austen's Brown Hornbill	*Anorrhinus austeni*	NT (Near threatened)
Beach Thick-knee	*Esacus giganteus*	NT (Near threatened)
Black-and-rufous Flycatcher	*Ficedula nigrorufa*	NT (Near threatened)
Black-headed Ibis	*Threskiornis melanocephalus*	NT (Near threatened)
Blackish-breasted Babbler	*Sphenocichla humei*	NT (Near threatened)
Black-necked Stork	*Ephippiorhynchus asiaticus*	NT (Near threatened)
Black-tailed Godwit	*Limosa limosa*	NT (Near threatened)
Blossom-headed Parakeet	*Psittacula roseata*	NT (Near threatened)-Upgraded from least concern
Blyth's Kingfisher	*Alcedo hercules*	NT (Near threatened)
Brown-winged Kingfisher	*Pelargopsis amauroptera*	NT (Near threatened)
Buff-breasted Sandpiper	*Tryngites subruficollis*	NT (Near threatened)
Chestnut-backed Laughingthrush	*Garrulax nuchalis*	NT (Near threatened)
Chevron-breasted Babbler	*Sphenocichla roberti*	NT (Near threatened)
Cinereous Vulture	*Aegypius monachus*	NT (Near threatened)
Derbyan Parakeet	*Psittacula derbiana*	NT (Near threatened)
Eurasian Curlew	*Numenius arquata*	NT (Near threatened)
European Roller	*Corcacias garrulus*	NT (Near threatened)
Falcated Duck	*Anas falcata*	NT (Near threatened)
Ferruginous Duck	*Aythya nyroca*	NT (Near threatened)
Firethroat	*Luscinia pectardens*	NT (Near threatened)
Giant Babax	*Babax waddelli*	NT (Near threatened)
Great Pied Hornbill	*Buceros bicornis*	NT (Near threatened)

(Contd.)

English name	Scientific name	Threatened category
Great Snipe	*Gallinago media*	NT (Near threatened)
Great Stone-Plover (Thick-knee)	*Escaus recurvirostris*	NT (Near threatened)-Upgraded from least concern
Grey-headed Fish-eagle	*Ichthyophaga ichthyaetus*	NT (Near threatened)
Grey-headed Bulbul	*Pycnonotus priocephalus*	NT (Near threatened)
Grey-headed Parakeet	*Psittacula finschii*	NT (Near threatened)-Upgraded from least concern
Hume's Pheasant	*Syrmaticus humiae*	NT (Near threatened)
Japenese Quail	*Coturnix japonica*	NT (Near threatened)
Jouanin's Petrel	*Bulweria fallax*	NT (Near threatened)
Kerala Laughinghthrush	*Strophocincla fairbanki*	NT (Near threatened)
Laggar Falcon	*Falco jugger*	NT (Near threatened)
Lesser Fish-eagle	*Ichthyophaga humilis*	NT (Near threatened)
Lesser Flamingo	*Phoeniconaias minor*	NT (Near threatened)
Little Bustard	*Tetrax tetrax*	NT (Near threatened)
Long-billed Bush-warbler	*Bradypterus major*	NT (Near threatened)
Long-tailed Parakeet	*Psittacula longicauda*	NT (Near threatened)
Long-tailed Wren-babbler	*Spelaeornis chocolatinus*	NT (Near threatened)
Malabar Pied Hornbill	*Anthracoceros coronatus*	NT (Near threatened)
Nicobar Bulbul	*Hypsipetes nicobariensis*	NT (Near threatened)
Nicobar Parakeet	*Psittacula caniceps*	NT (Near threatened)
Nicobar Pigeon	*Caloenas nicobarica*	NT (Near threatened)
Nilgiri Flycatcher	*Eumyias albicaudatus*	NT (Near threatened)
Oriental Darter	*Anhinga melanogaster*	NT (Near threatened)
Painted Stork	*Mycteria leucocephala*	NT (Near threatened)
Pallid Harrier	*Circus macrourus*	NT (Near threatened)
Red Kite	*Milvus milvus*	NT (Near threatened)
Red-breasted Parakeet	*Psittacula alexandri*	NT (Near threatened)-Upgraded from least concern
River Lapwing	*Vanellus duvaucelii*	NT (Near threatened)
River Tern	*Sterna aurantia*	NT (Near threatened)
Rufous-rumped Grassbird	*Graminicola bengalensis*	NT (Near threatened)
Rufous-throated Wren-babbler	*Spelaeornis caudatus*	NT (Near threatened)
Rufous-vented Prinia	*Prinia burnesii*	NT (Near threatened)
Rusty-bellied Shortwing	*Brachypteryx hyperythra*	NT (Near threatened)
Satyr Tragopan	*Tragopan satyra*	NT (Near threatened)
South Nicobar Serpent-eagle	*Spilornis klossi*	NT (Near threatened)
Spot-billed Pelican	*Pelecanus phillippensis*	NT (Near threatened)
Swinhoe's Storm-petrel	*Oceanondroma monorhis*	NT (Near threatened)
Tibetan Eared-pheasant	*Crossoptilon harmani*	NT (Near threatened)
Treepie	*Dendrocitta bayleyi*	NT (Near threatened)
Tytler's Leaf-warbler	*Phylloscopus tytleri*	NT (Near threatened)
Ward's Trogon	*Harpactes wardi*	NT (Near threatened)
White-cheeked Partridge	*Arborophila atrogularis*	NT (Near threatened)
Yellow-rumped Honeyguide	*Indicator xanthonotus*	NT (Near threatened)
Large-billed Reed-warbler	*Acrocephalus orinus*	DD (Data deficient)
Nicobar Scops-owl	*Otus alius*	DD (Data deficient)
Sillem's Mountain-finch	*Leucosticte sillemi*	DD (Data deficient)

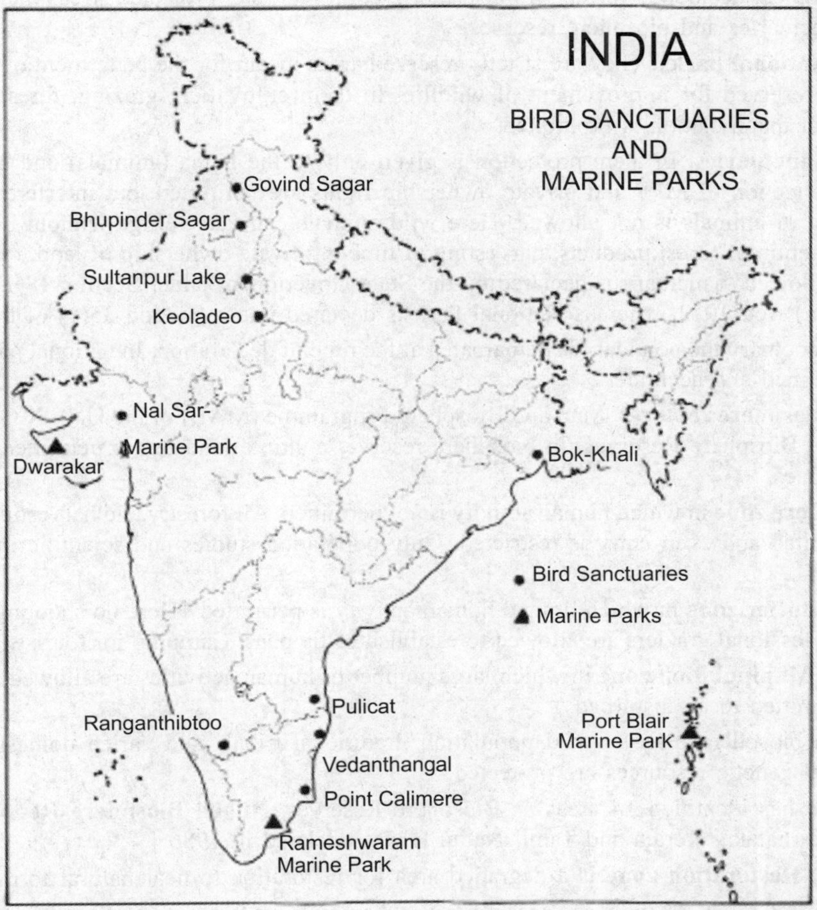

Fig. 51.1. Bird sanctuaries and marine parks in India.

Indian threatened birds. According to IUCN, there are 1,228 species of birds found in India including 65 species listed in various threatened categories (Table 51.3). This database provides information on the distribution of birds in various protected areas of India. Large number of information related to species distribution on various protected areas have been collected and documented in the form of check lists and linking with each PAs of India is under process.

PROTECTED AREAS IN INDIA

India is rich in various biogeographical provinces, ranging from the cold deserts of Ladakh and Spiti to the hot deserts of Thar, temperate forests in the Himalayas to the lush green tropical rain forests of the low lands. India has also large freshwater bodies such as Wular and Manasbal lakes in Kashmir, Chilka lake in Orissa and Kolleru lake in Andhra Pradesh and the rugged and rich coastline and coral reefs of Deccan.

Protected Areas are ecological/biogeographical areas where wildlife is conserved. There habitats and natural resources are conserved and poaching is prevented. They are delimited to protect biological diversity. They are cold desert (Ladakh and Spiti), hoı desert (Thar), wetland (Assam and N.E. States), saline swampy areas (Sunderbans, Rann of Kutch), mangroves, temperate forests, subtropical forests, tropical forests, tropical wet evergreen forests, tropical moist deciduous forests, tropical deciduous forests, tropical thorn, coral reef, etc. Protected Areas include national parks, sanctuaries and biosphere reserves.

1. National parks. They are strictly reserved areas meant for the betterment of the wildlife. They are reserved for improvement of wildlife. In them cultivation, grazing, forestry operation and habitat manipulation is prohibited.

2. Sanctuaries. In them protection is given only to the fauna (animals) and harvesting of timber, collection of MFP and private ownership rights are permitted, but interference with the well-being of animals is not allowed. Here wild animals can take refuge without being hunted. Here collection of forest products, harvesting of timber, private ownership of land, tilling of land, etc., are allowed. Sanctuary is declared by the State Government under Section 18(1) of Wildlife (Protection) Act, 1972, whereas National Park is declared under Section 35(1) of the Act.

In sanctuary the boundary is demarcated at the time of declaration. In national park boundary is well-defined and accurate.

3. Biosphere reserves. Man and Biosphere Programme (MAB) of the UNESCO evolved the concept of Biosphere Reserves. In biosphere reserve, multiple land use is permitted designating various zones.

(*i*) Core zone in which human activity is not permitted. All forestry and harvesting operations are prohibited and even entry is restricted. Only population studies and scientific investigations are allowed.

(*ii*) Buffer zone in which limited human activity is permitted. Here no shooting is allowed, but no professional graziers are allowed to establish cattle pens. Camping for tourists are allowed.

(*iii*) Manipulation zone in which large number of human activities are allowed, but ecology is not permitted to be disturbed.

In a biosphere reserve, wild population, traditional tribals and varied domesticated plant and animal genetic resources are protected.

India has identified 14 areas as Biosphere Reserves. **Nilgiri Biosphere Reserve** includes parts of Karnataka, Kerala and Tamil Nadu. It was declared in 1986.

(*iv*) Restoration zone is a degraded area for restoration to near natural form.

4. Safari park. An enclosed park where wild animals are kept uncaged in the open providing natural habitat for viewing to the public.

5. Zoo. An area set aside for exhibiting the wild animals kept in cages and artificial enclosures. Here animals' freedom is restricted.

6. Zoological garden. A place where a large number of mammals, birds, reptiles, fishes, etc., are shown in a confined area in or near a city. Animals are usually kept in small enclosures or in cages. It is used for recreation and education of the public.

7. Zoological park. It is a zoo where animals are comparatively free and are shown in the natural surrounding with barriers and restrictions hidden from view. It is best located on the outskirts of cities where enough land is available.

8. Sanctum sanctorum. It is a sanctuary within the sanctuary or inner portion of a wild sanctuary or national park in which no forest operation or management is allowed. Even visitors are not allowed to prevent any kind of disturbance to wildlife.

Conservation of Wildlife

The management of human use of the biosphere so that it may yield the greatest sustainable benefit to present generation and to maintain its potential to meet the needs and aspirations of future generations is called the conservation. It is scientific management of wildlife to maintain it at its optimum level.

The conservation of wildlife is directly related to healthy and better forests. Wildlife conservation includes protection, preservation, and perpetuation of rare species of plants and animals in their natural habitats. Conservation of living resources has three specific objectives :

1. To maintain essential ecological processes and life supporting systems.
2. To preserve diversity of species.
3. Sustainable utilisation of species and ecosystems which support rural communities and major industries.

Conservation Strategies

For wildlife conservation and its propagation, proper management techniques should be employed. Sanctuaries, national parks, biosphere reserves, projects, etc., have been created for exclusively protecting the wild flora and fauna in India as well as in other countries of the world. Scientists of 100 countries of the world have evolved comprehensive "World Conservation Strategies" for the judicious use of resources. To save the existing species of wildlife they proposed some steps which are as follows :

1. Efforts should be made to preserve the endangered species. Species that are sole representative of their family or genus should receive special attention. Endangered species should be given priority over a vulnerable one, a vulnerable species over a rare one and a rare species over other categories.

All the threatened species should be protected. Priority be given belonging to monotypic genera, endangered overvulnerable, vulnerable over rare and rare over other species.

2. Wildlife should be protected in their natural habitat *in situ* and in zoo and botanical gardens (*ex situ*). The threatened species should be conserved *in situ* as well as in *ex situ*.

3. Identify the habitats of wild relatives of the economically valuable and useful plants and animals and preserve them in protected areas like sanctuaries, national parks and biosphere reserves.

4. The critical habitats of the species like feeding, breeding, nursery and resting areas should be protected (safeguarded).

5. In case of migratory or wide ranging animals, protected areas should be established to preserve their habitats.

6. For migratory or wide ranging animals, pollution and exploitation of the environment along their migration routes should be controlled.

7. Unique ecosystem (national parks and biospheres) should be protected as a matter of priority.

The national protection programmes have to be coordinated with international programmes like biosphere reserve programme of UNESCO. Man and Biosphere Project and National Parks and Protected Areas of International Union for Conservation of Nature and Natural Resources (IUCN).

National Wildlife (Protection) Act was enacted in India in 1972. Wildlife protection strategies were formulated in India in 1983. Biosphere reserves have also been put into practice since 1986. Wildlife Institute of India is located at Dehradun (Uttaranchal). Indian Board for Wildlife (IBWL) was established in 1952.

8. The productive capacities of exploited species and ecosystems have to be determined and their utilisation should not exceed from those capacities.

9. International trade in wild plants and animals has to be regulated by appropriate legislative and administrative measures.

India is a signatory to the Convention on International Trade in Endangered Species of Wild Flora and Fauna (CITES).

Smt. Indira Gandhi (Late Prime Minister of India) addressed in her inaugural speech of World Conservation Strategies in 1980. She said that Indian tradition teaches us that all forms of life – human, animal and plant – are so closely interlinked that disturbance in one gives rise to imbalance in the other ... Nature is beautifully balanced. Any disturbance creates a chain reaction which may not be visible for some time.

IMPORTANT INDIAN WILDLIFE

The science of zoogeography has both ecological and historical aspects and the two are intimately interwoven. Animals and plants are living indicators of the characteristics of their environment. Their ranges mark the places in which environmental conditions are the same or similar. The evolution and distribution of species throws light upon the geological evolution of various parts of earth and upon the course of global changes in climate and vegetation. Based mainly on historical-cum-geographical factors, **Philip Lutley Schaler** (1825–1913) and **Alfred Russel Wallace** (1823–1913) have divided the world into six zoo-geographical regions, namely–Neoarctic, Palaearctic, Ethiopian, Oriental, Australian and Neotropical. India is of recent origin and it is a part of Oriental region. North Indian fauna during tertiary period were mastodons, eleven species of elephants, Siwalik bison, buffalo, ox, tamarau as well as the recent African animals like hippopotamus, giraffe, chimpanzee, rhinoceros and four-horned ruminant *Sivatherium*. Area was covered with savannah and woodlands. Asiatic lion, striped hyaena and antelopes can be the relics of the past. The dhole, most endangered top predator, is on the edge of extinction. Less than 2500 members of species are alive in the world. From Siwalik were discovered fragments of jaw of *Ramapitheus* (primitive hominid ape).

India has **three sub-regions** on the basis of physiography and climate :

1. Himalayan mountain systems which has three distinct sub-zones : (*i*) Himalayan foothills (from eastern frontiers of Kashmir to Assam), (*ii*) High altitudes in the Western Himalayas (from Kashmir including Ladakh to Kumaon), and (*iii*) Eastern Himalayan sub-region.

2. Peninsular-Indian sub-region which shares the animals of North Africa, such as lion, cheetah, leopard, hyaena, jackal and antelopes.

3. Indo-Malayan sub-region (Tropical evergreen forests) has similarities with Indian, Malayan and Indo-Chinese fauna. This sub-region contains red panda, gibbon, tree shrew, tapir, giant squirrel, and flying lemur.

1. Lion (*Panthera leo persica*). It is gregarious carnivorous animal. It prefers open scrub forest mixed with thorny deciduous forest. The temperature of its habitat should not be more than 45°C in summer and not below 5°C in winter. It is found only in Gir forest of Gujarat State as well as in the whole Asian continent. Its habitat should be improved to raise the carrying-capacity for the prey species to meet the full demand of lion. For herbivorous prey species, there should be total control in grazing of the livestock. Moghul emperor Jahangir was fond of hunting male lions and tigers. Col. Smith during Sepoy Mutiny in India in 1857 killed 300 lions of which 50 were from Delhi alone. By 20th century, the lions population was only 15. The last lion was killed at Anadra and Jaswantpura (Rajasthan) in 1876. Gir Lion Sanctuary Project was started in 1972. Due to conservation measures, lion population increased gradually to 250 in 1977.

2. Indian Tiger (*Panthera tigris tigris*). Tiger is a solitary carnivorous animal having apparent territory. It is nocturnal predatory inhabiting dense forest such as thorny forest, dry and moist deciduous forest, evergreen and semi-evergreen forest. For raising its population, pasture lands should be improved for raising the carrying-capacity of the habitat for herbivorous preys. Since 1972, tiger has been declared India's National Animal. In the beginning of 20th century tiger population was about 40,000. In early 1970s their population was reduced to about

1800 due to unrestricted killing for skin, flesh and fat, etc. Indian tiger census conducted in 2008 showed their existence of only 1411 tigers. To save them from extinction, Project Tiger was started on April 1, 1973, by the Government of India with the help of W.W.F. (Worldwide Fund for Nature). Today there are 39 Project Tiger wildlife reserves in India covering an area of 37,761 sq. km.

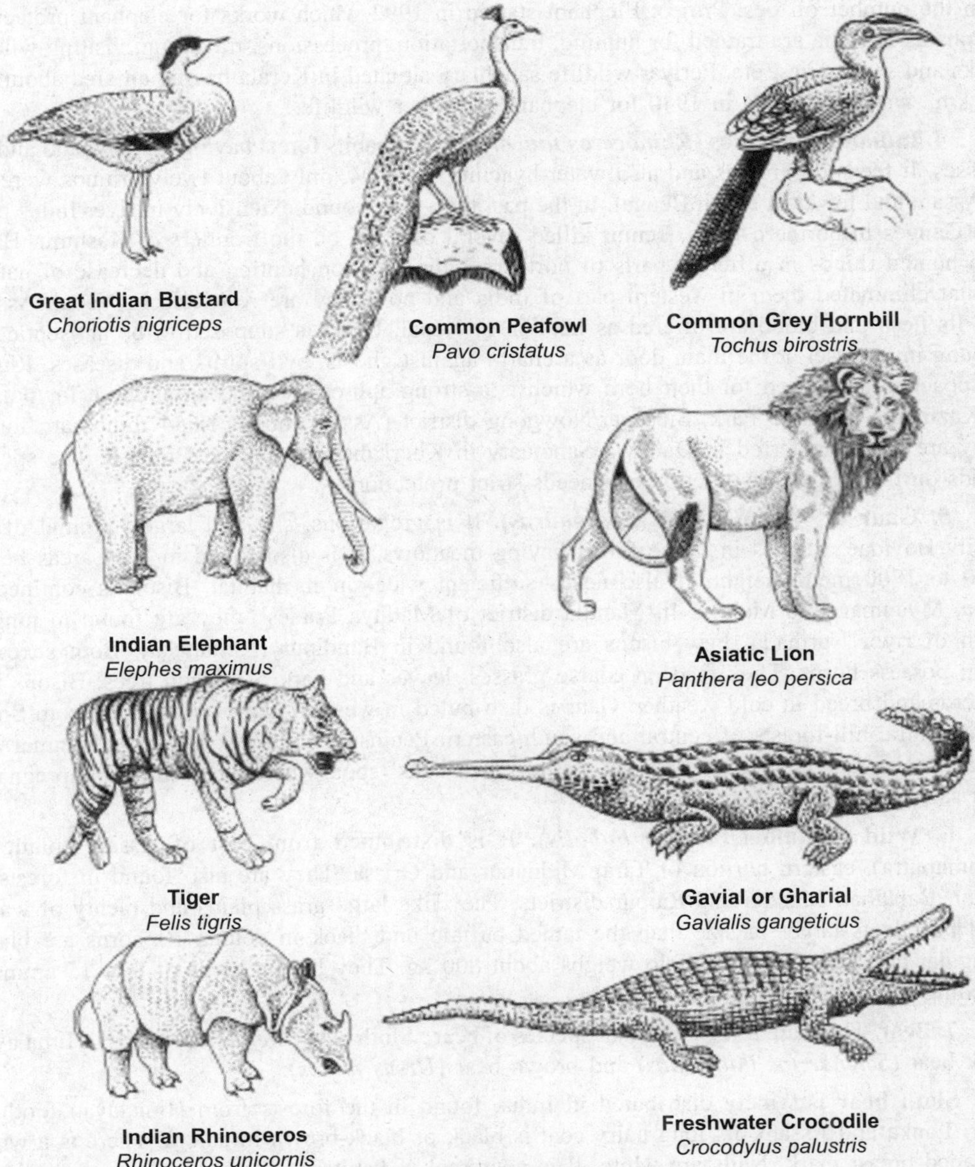

Fig. 51.2. Important wildlife of India.

3. Elephant (*Elephas maximns*). It is found in plains and hilly forest up to 1500 metre elevation. It needs a lot of water for drinking and bathing, so there must be perennial river, lake, etc., in the habitat. Its food is bamboo and grass, which should be in sufficient quantity. An adult elephant needs about three quintals green fodder daily. It is distributed throughout India except Madhya Pradesh, Andhra and Maharashtra. Generally only the males have large tusks, which are the extension of second pair of incisors. Canines and all incisors have lost. Lengthened nose and upper lip forms the trunk. In some males, tusks are no longer than females (a few inches long) and called tuskless or Makhana. Elephants have very poor sight, but smell and hearing are acute. Limbs are pillar-like. Toes are embedded in a common mass of foot and encased in a common skin. Their position is indicated externally by broad flat nails which may be fewer than the number of toes. Project Elephant started in 1992 which works for elephant protection. Elephants in India are trained for hunting, transportation, processions, travelling, visiting wildlife parks and sanctuaries, etc. Periyar wildlife sanctuary situated in Kerala having an area about 777 sq. km., was established in 1940 for elephant and other wildlife.

4. Indian Rhinoceros (*Rhinoceros unicornis*). It inhabits forest having marshy land and tall grasses. It feeds on grasses and also water-hyacinth. In 1904, only about twelve rhinos were left in Assam and fewer in North Bengal. In the past they were found extensively in river Indus plain and Ganges in northern India. Temur killed several of them on the frontiers of Kashmir. Babar also hunted rhinos in different parts of northern India. Wanton hunting and decrease of natural habitat eliminated them in western part of India and now they are restricted in Nepal, Assam, etc. Its flesh and blood are offered as libation in Nepal. Urine is supposed to be antiseptic and is hung in a vessel at the main door as a charm against ghosts, evil spirits and diseases. Rhinos are poached and killed for their horn which is a strong aphrodisiac and an antidote for poison. In Kaziranga National Park, Sibsagar/Nowgong district (Assam) about 1,654 rhinos are found. They are also transferred in Dudhwa Sanctuary in Kheri district of Uttar Pradesh. The species stands on the verge of extinction and needs strict protection.

5. Gaur or Indian Bison (*Bos gaurus*). It is gregarious, shy and largest animal of the family Bovidae. It lives in dense forest having meadows. It is also found in hilly areas below 1500 to 1800 metre height. It also needs sufficient water in its habitat. Bison is confined in India, Myanmar, and Malaya. In Mandla district of Madhya Pradesh they are found in jungles north of river Narmada. Few bisons are also found in Bandipur (Karnataka). Both sexes of bison possess horns. They feed on coarse grasses, leaves and bark of certain trees. Bisons live in herds and breed in cold weather. Gaur is distributed in western ghats southwards from South Maharashtra, hill-forests of central and south-eastern Peninsula and West Bengal, Myanmar and Malay Peninsula. Gaur has the habit of visiting "salt-licks" spots where the ground is impregnated with salts and other minerals.

6. Wild Buffalo (*Bubalus bubalis*). It is distributed from east of Assam (plane of Brahmaputra), eastern portion of Tara, Midnapur and Orissa. They are also found in forests of Bastar, Balaghat, Mandla and Raipur districts. They like large grass plains and plenty of water. Wild buffalo is larger in size than the tamed buffalo and black in colour. Its horns are black, triangular and large. Adult buffalo weighs about 800 kg. They live in herds of 8 to 15 animals. Breeding starts in autumn.

7. Bear. There are three important species of bear : Sloth bear (*Melursus ursinus*), Himalayan black bear (*Selenarctos thibetanus*) and brown bear (*Ursus arctos*).

Sloth bear is widely distributed in India, found in the forests from Himalayan foothills to Sri Lanka and Assam. Its long hairy coat is black or black-brown in colour. There is a white V-shaped breast mark. Nails are white. It is nocturnal in habit and feeds on honey, insects and fruits. At the time of hunger, it takes carrion. Sloth bear shares habitat of tiger and elephant.

Himalayan black bear is found in forests of Himalayas up to 3,000 metre height. Its hairy black coat is shorter and smoother than the sloth bear. V-shaped breast mark is white or yellowish. Nails are black. It is carnivorous and kills sheep, goat, fowl and even cattle, though its main food is honey and fruit. It is nocturnal in habit and climbs the trees. It hibernates in winter.

Brown bear is found in Himalayas above the tree line (treeless mountain areas). It is large and heavy having hair coat of reddish brown colour. It mainly feeds on grass, roots, insects, fruits and grain but also eats meat of goat, sheep, cattle and fowl.

8. Black buck (*Antilope cervicapra*). It is called Indian antelope, lives outside the forest in herds. It is gregarious and male has its territory. It is found in 13 states of India but the largest number is found in Rajasthan. About 10,000 are found in Jodhpur district. In Thar desert in Taal Chhaper Sanctuary in Churu about 1400 live in 7 sq. km. Black buck is found in 8 wildlife sanctuaries and nearly 14 other areas in Rajasthan. Black buck prefers open grassy fields and is dependent on water. It is not found in arid areas. Black buck is not found in Jaisalmer, parts of Bikaner and parts of Jodhpur with less than 200 mm annual rainfall. It feeds on small grasses.

9. Cheetal (*Axis axis*). Sambhar (*Cervus unicolor*), **Four-horned antelope** (*Tetracerus quadricornis*). **Cheetal** lives in moist deciduous forest, evergreen and thorny forests. It feeds on grasses, leaves, flowers and fruits.

Sambhar lives in open dry deciduous forest, and also found in dry and moist deciduous forest and evergreen forest. It is the largest among all deer species. It lives in small herds.

Four-horned antelope prefers dry and bushy savannah but also lives in open grassy field near the forest. It is solitary animal and generally lives in pair.

Cheetal and sambhar both are prey of top carnivore.

10. Swamp deer/Barasingha (*Cervus duvauceli*). It is found in open grassy lands with marshy lands. Its habitat should have sufficient water. It lives in marshy tracts of Tarai and Duars from northern parts of upper Gangetic plains east to Assam. Its other subspecies (*C. d. branderi*) thrives on the open grassy land of Madhya Pradesh. They are now mostly confined to the Kanha National Park in Madhya Pradesh.

11. Musk deer (*Moschus moschiferous*). Himalayan musk deer is a small primitive deer, was widespread throughout Himalayas from Pakistan, through India, Nepal, Bhutan, Myanmar, Tibet and South-west China. Its head is dark grey with slight orange-brown patches above and below the eyes with white tipped ears. Throat is whitish around a central grey oval patch. Along the back, pelage is dark grey or brown, becoming orange-brown around the anal region. Male deer lacks antlers but possess tusks (canines of upper jaw) for fighting. A unique feature of this is presence of musk gland in male, which is chiefly responsible for its decline. The gelatinous, brown musk secreted by preputial gland has been used in medicines and cosmetics for centuries. Its cost now is 40,000 to 59,000 U.S. Dollar per kg., in the international market. Its habitat destruction due to destruction of forest for livestock, timber and fuelwood is the main cause for the decline of species. The species was listed as vulnerable in Red Data Book (IUCN, 1974). IUCN and WWF Project was launched with the cooperation of India to conserve the musk deer. The project is based in Kedarnath Sanctuary, Uttar Pradesh.

12. Indian Crocodiles. In India, there are three species of crocodilians such as : (*i*) The Mugger or Freshwater Swamp Crocodile (*Crocodylus palustris*). It has average adult size of 3.5 metres, inhabiting rivers, pools, ponds, village tanks, lakes, swamps and reservoirs. (*ii*) The Saltwater or Estuarine Crocodile (*Crocodylus porosus*). It grows more than 7 metres and is restricted to the coastal mangrove area in the saltwater (sea). (*iii*) The Gharial (*Gavialis gangeticus*). It is the sole living member of the family Gavialidae. It has unique long snout. It has large size of more than 7 metres. It is a fish-eating, riverine species found in large rivers such as Ganga, Brahmaputra, Mahanadi, Kosi, Gandak, etc., of North Indian Himalayan-fed river system.

Once abundant in all the major rivers and even ponds, they are among threatened animals today. Their population declined because of uncontrolled and all-season hunting for skin, flesh and sport. Loss of habitat due to construction of dams, diversion of rivers and human interference were other factors. In the wake of declining population, 'Save the Crocodile' projects were launched in 1974 under the guidance of Dr. H.R. Bustard as the chief technical adviser.

13. Great Indian Bustard (*Aredotis nigricaps*). It is one of the rarest birds of the world. In mid-1980s, bustard population was estimated to be between 500 and 1500, of which half of the birds surviving in Rajasthan. Bustard survived in nearly 200,000 sq. km. of the Thar desert. Surveys of 1993 and 1994 indicate that bustard numbers have almost half in the whole Rajasthan. In mid-1980s survey of bustard indicated that it is found in Jaisalmer, Barmer, Jodhpur, Bikaner, Pali, Jalore, Ajmer, Bhilwara, Tonk, Kota and Sawai Madhopur (11 districts). Bustard is distributed from central Punjab to central Tamil Nadu, western Orissa into eastern Pakistan.

Great Indian Bustard is under Schedule 1 of Wildlife (Protection) Act, 1972. Rajasthan Government has declared it as State Bird.

Bustard population decreased due to the development of 649-km Indira Gandhi Nahar Project (IGNP). It has resulted in expansion of agriculture, land colonisation, development of new towns and their expansion and change in natural vegetation due to extensive plantation of exotic trees, Shooting also played a major role in decimation of bustards of the Thar desert.

Bustards live in flocks. It was listed as globally threatened in 1966. During last 10 to 12 years the bustard population has crashed in many areas and now the total population could be as low as 500. It is extinct in Karera and Sorsan bustard areas.

Bustard is about one metre tall with long sturdy yellow legs without hind toe. Its plumage is dull brown above and white below. One male lives with 3 to 5 hens. It feeds on arthropods, lizards, snakes, mice and also grains and young shoots of plants. Their breeding season is from July to October and lay one or two eggs.

14. Common Peafowl (*Pavo cristatus*). It is also called peacock and in Hindi "Mor" or "Mayur". It is found throughout India up to 1650 metres in Himalayas. It displays sexual dimorphism, male has a gorgeous occellated tail feathers, which are not found in female bird. The bird is not threatened but it is the National Bird of India.

Green peafowl (*Pavo muticus*). It was distributed widely from north-east India to southern China, Myanmar, Thailand, Laos, Vietnam, Cambodia and Indonesia. It is not found in Malaysia and Bangladesh. It has declined rapidly and now occurs in fragmented, greatly reduced populations. In India it is restricted to far north-eastern states in a very reduced number. Its number is declining due to hunting for meat, exploitation for trade and persecution by farmers, habitat conversion to farmland. It is a vulnerable species and receives legal protection in India and also in other countries.

15. Hornbilis. Rufous-Necked Hornbill (***Aceros ripalensis***) is found in mountainous regions between eastern Nepal and Vietnam. It is now absent from or very rare in this range, but still occurs in southern China, north-eastern India, Myanmar, Thailand, Laos and Vietnam. In India, it has been recorded in West Bengal, Arunachal Pradesh, Assam, Meghalaya, Manipur, Mizoram, Nagaland and Sikkim. Recently it has been scarce in West Bengal and Assam and perhaps disappeared from its previous range. It is threatened by the combination of habitat loss and hunting.

Narcondam Hornbill (*Aceros narcondami*). It is found in Narcondam, a small area of the Andaman Islands in the Bay of Bengal. In 1972 their number was 400, while in 1998 their number reduced to 295 to 320 birds. Its population is susceptible to climatic disasters and disease. The primary threats arise from establishment of police outpost on the island manned by 17 persons in 1969. In 1976 police introduced pairs of goats which increased to 130–150 in 1998 and further increased to 250 due to which natural woodland regeneration is reduced. Each year at least 10–12 live standing trees are cut down for fuelwood and for poles to make fences to check goats not to enter the vegetable plots. Hunting for meat was also a threat to these birds. It is also listed in Schedule 1 of Wildlife (Protection) Act, 1972. Narcondam Island has been protected since February 1977 as a wildlife sanctuary.

TABLE 51.4. ENDANGERED SPECIES OF INDIAN HORNBILLS.

S.No.	Common name	Zoological name	Distribution
1.	Common grey hornbill "Dhanesh" in Hindi	*Tochus birostris*	In semi-open forests of North India.
2.	Malabar grey hornbill	*Tochus griseus*	Western Ghats
3.	White-throated brown hornbill	*Ptilolaemus tickelli*	Everngreen forests of Eastern India from 300–5000 metre.
4.	Rufous-necked hornbill	*Aceros ripalensis*	Deciduous evergreen forests of Sikkim, Arunachal Pradesh, Manipur, and Mizoram.
5.	Wreathed hornbill	*Rhyticeros undulatus*	Evergreen forests of Himalayan foothills south of Brahmaputra.
6.	Narcondam hornbill	*Rhyticeros (undulatus) narcondami*	Narcondam Island of Andaman Islands.
7.	Giant hornbill	*Buceros bicornis homri*	Heavy forests of Terai up to 2000 metre height. Western Ghats.
8.	Indian pied hornbill	*Anthracoceros malabaricus*	Terai and lower foothill forests.
9.	Malabar pied hornbill	*Anthracoceros coronatus*	*Plains to 300 metre altitude forests of South India.*

Project Tiger

Tiger is our National Animal. It is found in diverse habitats and in different parts of the country. Tiger is at the apex position as top carnivore of the complex food-chain in most of our forest ecosystems. Over the years, the over-exploitation of the forest areas, merciless hunting, unscientific management, etc., reduced the habitat of tiger as well as leading to a rapid decline has, therefore, been in India not only as effort to save an endangered species but also with equal importance as a means of preserving biotopes. Project Tiger has been inspired by such an approach. The Indian Board for Wildlife (IBWL) set up a Task Force for studying the condition of tiger population and its status. On the recommendation of this Task Force, Project Tiger was initiated a Central Sector Scheme in 1973 with 9 Tiger Reserves (total area : 13,017 sq. km.) located in different habitat types in 9 diffferent states, but two more reserves have since been subsequently added to it constituting 11 Tiger Reserves in 10 different states of the country (total area : 15,800 sq. km.). But this number is not last and increasing always.

The main aim of the project was to conserve and improve the natural habitat of the tiger under different habitat types. The management practices and strategies in the reserves are controlled in such a way that all the limiting factors of habitats are removed. The steps to be taken in this direction are intensive anti-poaching drive, fire prevention, elimination of cattle-grazing, soil conservation, water management, eradication of weeds, non-interferance by human activities, relocation of human habitation and so on.

The project started initially as a Central Sector Scheme and expenditure incurred by the States was provided by the Central Governement till 1979–80. After this, the project had been given the status of Centrally Sponsored Scheme and the Centre and States are sharing cost on a 50:50 basis. The Worldwide Fund for Nature and Natural Resources (WWF) is also extending financial and technical help.

TABLE 51.5. DETAILS OF TIGER RESERVES OF INDIA.

S.No.	Tiger Reserves	Year of Establishment	Total Area (in sq. km.)	State
1.	Bandhavgarh	1993–1994	1162	Madhya Pradesh
2.	Bandipur	1973–1974	866	Karnataka
	Nagarhole (extension)	1999–2000	643	Karnataka
3.	Bhadra	1998–1999	492	Karnataka
4.	Bori Satpura	1999–2000	1486	Madhya Pradesh
5.	Buxa	1982–1983	759	West Bengal
6.	Corbett	1973–1974	1316	Uttaranchal
7.	Dampo	1994–1995	500	Mizoram
8.	Dudhwa Katerniaghat	1987–1988	811	Uttar Pradesh
9.	Indravati	1982–1983	2799	Chhattisgarh
10.	Kalakad Mundanthurai	1988–1989	800	Tamil Nadu
11.	Kanha	1973–1974	1945	Chhattisgarh
12.	Manas	1973–1974	2840	Assam
13.	Melghat	1973–1974	1677	Maharashtra
14.	Nagarjunsagar-Srisailam	1982–1983	3568	Andhra Pradesh
15.	Namdapha	1982–1983	1985	Arunachal Pradesh
16.	Pakhui-Nameri	1999–2000	1206	Arunachal Pradesh
17.	Palamau	1973–1974	1026	Jharkhand
18.	Panna	1994–1995	542	Madhya Pradesh
19.	Pench (M.P.)	1992–1993	758	Madhya Pradesh
20.	Pench	1992–1993	257	Maharashtra
21.	Periyar	1978–1979	777	Kerala
22.	Ranthambhore	1973–1974	1334	Rajasthan
23.	Sariska	1978–1979	866	Rajasthan
24.	Simplipal	1973–1974	2750	Orissa
25.	Sunderbans	1973–1974	2585	West Bengal
26.	Tadoba-Andhari	1993–1994	620	Maharashtra
27.	Valmiki	1989–1990	540	Bihar

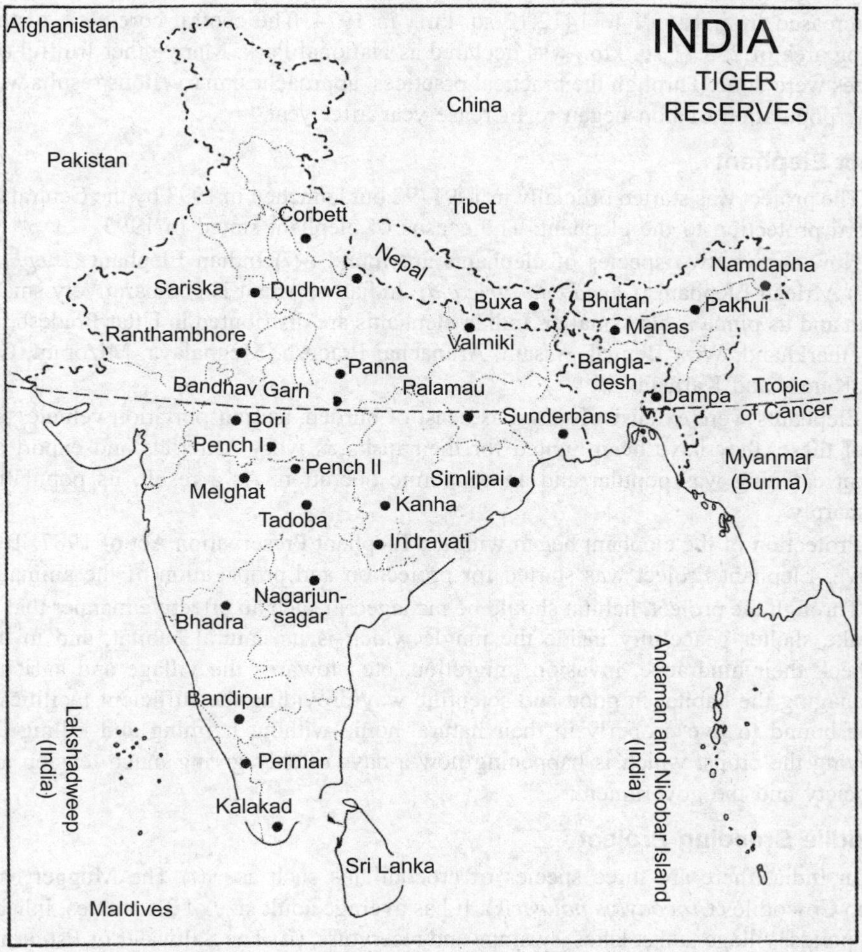

Fig. 51.3. Tiger reserves in India.

Project Lion (Gir Lion Project)

The lion stands as top carnivore in the food-chain of the ecosystem. The Indian race of lion (*Panthera leo persica*) is found only in the Gir Forest of Junagarh district in the Saurashtra peninsula of Gujarat State in whole Asian continent and, hence, also called **"Asiatic Lion"**. Another race of lion (*Panthera leo*) is found in Africa and called **"African Lion"**.

Today the Asiatic lion is restricted only in the Gir Forest of Gujarat State. Even in this very forest; due to merciless hunting, conversion of forest into agricultural land, uncontrolled cattle-grazing and spreading of infectious diseases through them, pressure of maldharis inside the forest, etc., hampered the habitat of this magnificent creature too much and subsequently declined its population reaching to the endangered stage.

Observing the situation of Gir Forest, it was discussed at the technical session of International Union for Conservation of Nature and Natural Resources (IUCN) in New Delhi in November, 1969 towards the conservation of Asiatic Lion and its habitat. A number of wildlifers and ecologists all over the world were consulted in 1972. The State Government gave guidelines to Forest Department for the management of this project.

The guidelines of the project were implemented in the same year. The area of sanctuary was increased from 1265.1 to 1412.12 sq. km., in 1974. The central core-zone of the sanctuary covering area of 258.71 sq. km., was declared as National Park. Many other fruitful and effective practices were done. Through the practical practices, approaches, marvellous results were achieved and the population of lion began to increase year after year.

Project Elephant

The project was started officially in 1991–92 but launched in 1993 by the Central Government to afford protection to the elephant. The census of elephant stated in 1993.

Now-a-days, two species of elephants are found : (*i*) Indian Elephant (*Elephas maximus*) and (*ii*) African Elephant (*Loxodonta africana*). Indian elephant is comparatively smaller than the African and its pinna is also smaller. Indian elephants are distributed in Uttar Pradesh, Uttaranchal, Bihar, Jharkhand, West Bengal, Assam, Arunachal Pradesh, Meghalaya, Mizoram, Orissa, Tamil Nadu, Kerala and Karnataka.

Elephants were extensively used as beast of burden, as transportation vehicle, and so on. In spite of these, they have been hunted for their tusks as ivory materials and export also. Hence, elephant catching was popular and an economic operation. As a result, its population declined very sharply.

Protection of the elephant began with the Elephant Preservation Act of 1987. To make more effective, Elephant Project was started for protection and propagation of the animal.

Through the project, habitat should be managed in such an effective manner that the elephant may take shelter peacefully inside the jungle which is its natural habitat, and in this way, we can check their hindrance, invasion, migration, etc., towards the village and agricultural-fields. By managing the habitat in good and scientific way providing the sufficient facilities, the animal will be bound to live properly in their natural-home without harming and killing the man and destroying the crops, which is happening now-a-days oftenly giving much tension and worry to our society and the government.

Crocodile Breeding Project

In India, there are three species of crocodilians such as : (*i*) The Mugger or Freshwater Swamp Crocodile (*Crocodylus palustris*). It has average adult size of 3.5 metres, inhabiting rivers, pools, ponds, village tanks, lakes, swamps and reservoirs. (*ii*) The Saltwater or Estuarine Crocodile (*Crocodylus porosus*). It grows more than 7 metres and is restricted to the coastal mangrove areas in the saltwater (sea). (*iii*) The Gharial (*Gavialis gangeticus*). It is the sole living member of the family Gavialidae. It has unique long smout. It has large size of more than 7 metres. It is a fish eating riverine species found in large rivers such as Ganga, Brahmaputra, Mahanadi, Kosi, Gandak, etc., of the North Indian Himalayan-fed river systems.

Crocodiles have catastrophically declined worldwide in the post-war period largely as a result of sophisticated hunting methods for their hides by the luxury leather market. The problem continued around poaching and destruction of the remaining crocodile resource by fishery activities either direct or indirect, lethal effect of set nylon nets being used in fishing, disappearance of habitat of the sand-banks (which is essential for gharial nesting) replaced by concrete embankments, loss of riverine-habitats by dam-construction for irrigation or hydroelectric schemes. As a result, these crocodilian species came on the verge of extinction.

Crocodile hunting is now legally banned in India. The Wildlife (Protection) Act, 1972 lists both species of crocodile and gharial under Schedule I which affords total protection at all times. Similarly, Export I Instruction No 46/73 forbids the export of crocodiles and gharials, their hides or products therefrom.

Project Crocodile Breeding and Management was started as the report given by FAO expert, Dr. H.R. Bustard in 1974, stating "only management will restore the crocodile quickly and it appears that without management, the gharial will become extinct". The actual project was started on April 1, 1975 in Orissa. Gharial eggs were collected and hatched for the first time in captivity anywhere in the world at Tikerpada, District Dhenkanal, Orissa, in June 1975. A small batch was also hatched at Kukrail near Lucknow the same year. The Tikerpada hatched gharials were successfully reared for subsequent release back into the wild. At the same time, a Saltwater Crocodile Project was initiated in the tidal mangrove forests at Bhitar Kanika, District Cuttack, Orissa, and a Mugger Project and a Captive Breeding Project were also initiated, the latter for capitive-breeding of all three species was located at Nandankanan Biological Park, Orissa. All these projects were started by the Government of India under the help and guidance of FAO and UNDP. The Head Office of the Project is at Hyderabad. The project has main functions such as : (*i*) Conservation and Management of Crocodiles and Development of Sanctuary; (*ii*) Rehabilitation of Crocodiles.

Conservation and management of crocodiles and development of sanctuary. For the conservation of crocodile/mugger, their eggs are collected and hatched and reared in sanctuaries and released in rivers after attaining proper length of 1.2 metre. With the development of husbandry centres, steps have been taken to gazette and manage sanctuaries in ideal habitat areas for all three crocodilian species into which individuals were reared in. The various husbandry centres could be released when they attain a length of 1.2 metres. The first sanctuaries to be gazetted in the country were Satkoshia Gorge Sanctuary and Bhitar Kanika Sanctuary, both in Orissa, Tristate Chambal Sanctuary of Madhya Pradesh–Rajasthan–Uttar Pradesh and the Katerniaghat Sanctuary in northern Uttar Pradesh (Uttaranchal). With the exception of Bhitar Kanika, declared for the saltwater crocodile, these sanctuaries were all for gharial, which due to its critically endangered status, was given prime attention during the early stages of the project.

The management of sanctuaries is, of course, a long-term task, but immediate steps were taken to try to cut down disturbances which resulted in the loss of many animals (for instance, fishing with nylon gill-nets was immediately banned in all four sanctuaries mentioned above), to implement protection and to make a start on implementing management plans (the first of which was ready in December, 1980).

Release. Young crocodiles of size 1.2 metre are released into ideal areas of the natural habitat in small batches in early spring. Selected areas are specifically managed in the sanctuaries. This includes following cares : (*a*) Location of release–ideal habitat areas should be located where they will be free from disturbances. (*b*) Timing of release–it is important that the release be carried out at a time when water levels are low so that the young crocodiles can gain an intimate knowledge of their future home-range prior to the onset of the monsoon floods. The ideal time in South India is early February, this may be delayed by 4–6 weeks in the extreme North of the country.

Rehabilitation of crocodiles. Crocodiles breed in the end of winter. In the end of March or in the first week of April, the pregnent female makes 10–15 circular ditches of radius 30 – 50 cm in the sand of the river's bank. Out of these 10–15 ditches, she lays whole of its eggs in one ditch and covers all the ditches by sands. The aim to construct more than one ditch is to protect the eggs from predators like jackals, etc. At a time, the female lays up to 100 eggs. Male and female crocodiles also watch the ditch in the night. Under the effects of heat and moisture of the sand, the embryos develop and hatch in 60–70 days (incubation period) duration. The timing of hatching of youngs from the eggs is acknowleded by the female by hearing special sound emitted by them. On hearing such sound, the female takes out the youngs by removing sand of the ditch. The young are removed by the female from the unhatched eggs also by breaking it. After this, the female brings the youngs into water keeping them on its back and looks after them till the next breeding season. But even after taking so much care, large number of youngs are eaten by fishes, jackals, birds and crocodiles themselves and very less number of youngs survive in nature.

***(i)* Gharial.** Gharial rehabilitation started in 1977 with release of 26 individuals into Mahanadi river, Orissa. By January, 1980, 107 individuals had been released into Mahanadi where the wild population had been reduced to 5 individuals. 3 individuals were also released in Rajasthan in a trial release in October, 1977. Large scale releases by Uttar Pradesh (into the National Chambal Sanctuary) commenced in May, 1979, and between then and late March, 1980, 185 were released. The total number of gharial released by the end of March, 1980, totalled 324.

***(ii)* Saltwater Crocodile.** The first saltwater crocodile release occurred in Bhitar Kanika Wildlife Sanctuary, also in Orissa, on April 27, 1977. By January, 1980, a total of 125 had been introduced into this one sanctuary. The total release up to May, 1980, including 40 released in West Bengal and 3 in Andhra Pradesh, was 168.

***(iii)* Mugger.** The first release of Indian mugger took place on February 7, 1977, with 4 individuals being released at Ethipothalla, Andhra Pradesh. This group has now been strengthened to 8 individuals and a further 33 have been released into Kinnersani Wildlife Sanctuary of Andhra Pradesh (March, 1980). On March 9–10, 1979, 130 were released in Tamil Nadu followed by 47 on May 29, 1979, hence, a total of 218 individuals have been introduced.

TABLE 51.6. SCHEMES OF THE GOVERNMENT OF INDIA PROJECT, CROCODILE BREEDING AND MANAGEMENT, OPERATING WITH FAO/UNDP TECHNICAL ASSISTANCE.

Name of Project	Nature of Project	Species	Location	Date of Commencement
Satkoshia Gorge Gharial Scheme	Husbandry/rehabilitation	Gharial	Satkoshia Gorge Wildlife Sanctuary, Orissa	April 1975
Bhitarkanika Saltwater Crocodile Scheme	Husbandry/rehabilitation	Saltwater Crocodile	Bhitarkanika Wildlife Sanctuary, Orissa	May 1975
Kukrail National Scheme	Husbandry/rehabilitation	Gharial (extended to mugger)	Kukrail, Lucknow	1975
Chambal Sanctuary Gharial Scheme, Rajasthan	Husbandry/rehabilitation	Gharial	Kotah, Chambal River	1975
Nandankanan Captive Breeding Project	Captive breeding & husbandry	All three species	Nandankanan Biological Park, Orissa	1976
Katerniaghat Gharial Scheme	Husbandry/rehabilitation	Gharial	Katernia Ghat Wildlife Sanctuary, Bahraich, Northen Uttar Pradesh	1976
Sunderban Saltwater Crocodile Scheme	Husbandry/rehabilitation in association with Tiger Project	Saltwater Crocodile	Sunderbans, West Bengal	March 1976
Madras Project	Husbandry/rehabilitation for release elsewhere	Mugger	Guindy Park, Madras	1976
Nehru Zoological Park Rehabilitation & Captive Breeding Project	Husbandry for sanctuaries release and captive breeding	All three species	Nehru Zoological Park, Hyderabad.	1977
Sathanur Reservoir Mugger Scheme	Husbandry/rehabilitation	Mugger	Sathanur, District North Arcot, Tamil Nadu	1977
Amravati Reservoir Mugger Scheme	Husbandry/rehabilitation	Mugger	Amravati, Coimbatore District, Tamil Nadu	January 1977

(Contd.)

Name of Project	Nature of Project	Species	Location	Date of Commencement
Hoggenakal Mugger Scheme	Husbandry/rehabilitation	Mugger	Hoggenakal, Dharmapuri District, Tamil Nadu	1977
Gir Mugger Scheme	Husbandry/rehabilitation	Mugger	Gir National Park, Gujarat	January 1977
Neyyar Mugger Scheme	Husbandry/rehabilitation	Mugger	Neyyar, Kerala	1977
Parambikulam Mugger Scheme	Husbandry/rehabilitation	Mugger	Kerala	1977
Coringa Saltwater Crocodile Project	Rehabilitation	Saltwater Crocodile	Coringa Wildlife Sanctuary, A.P.	March 1978
Krishna Sanctuary Mugger Scheme	Rehabilitation	Mugger	Krishna Wildlife Sanctuary, A.P.	1978
Lanjumadugu Mugger Scheme	Egg collection & rehabilitation	Mugger	Godavari River, A.P.	May 1978
Manjira Mugger Scheme	Egg collection & rehabilitation	Mugger	Manjira, Medak Dist., A.P.	May 1978
Kinnersani Mugger Scheme	Rehabilitation	Mugger	Kinnersani Wildlife Sanctuary, Khammam District, A.P.	April 1980
Pappikonda Gharial Scheme	Rehabilitation	Gharial	Pappikonda Wildlife Sanctuary, East Godavari Dist., A.P.	1978
National Chambal Gharial Project	Protection & rehabilitation	Gharial	National Chambal Sanctuary, U.P./M.P./Rajasthan	1978

Note : Additional schemes have commenced/are about to commence in a number of States/Union Territories including Karnataka, the Andamans, Maharashtra, Arunachal Pradesh, Assam and Nagaland.

SANCTUARY AND NATIONAL PARK

Sanctuaries and national parks are the final refuge of wildlife and constitute an insurance against the total disappearance of species.

Sanctuary

"It is an area where killing and capturing of any species of birds or animals is prohibited except under orders of competent authority and whole boundaries and characters should be sacrosanct (free from outrage) as for as possible" (IBWL, 1952).

The IBWL (Indian Board for Wildlife) has further clarified the position by stating that while the management of sanctuaries does not involve suspension or restriction of normal forest operations, it is desirable to set aside a completely sacrosanct area within a sanctuary to be known as 'Abhayaranya'. It is also indicated that sanctuaries should be made accessible to the public. In these "Ashrams" of wildlife, destruction of the wild animals is prohibited by law. These areas will be from nurseries of wild animals to replenish areas with depleted stock from time to time.

According to Wildlife (Protection) Act, 1972, Section 18, "The State Government may by notificatiopn declare any area to be a sanctuary if it considers that such area is of adequate ecological, faunal, floral, geomorphological, natural or zoological significance, for the purpose of protecting, propagating or developing wildlife, or its environment." No person is allowed to move freely inside the sanctuary without the permission of the authority. Inside the sanctuaries, carrying of weapons without permission, setting fire or candle any fire or leave any fire burning, use of explosives and chemicals are strictly prohibited. The competent authority can take proper steps from management point of view for the security of wild animals and may regulate, control or prohibit grazing or movement of cattle or fishing in the interest of wild animals.

National Park

"A national park is a relatively extensive area : (1) in which one or more ecosystems have not been physically altered by human exploitation and occupation, where plant and animal species, geomorphological sites and habitats are of special scientific, educational and recreational interest, or where the natural scenery is of great beauty; (2) or eliminate exploitation or occupation of entire area in the briefest possible time and to effectively complete respect for the ecologic, geomorphologic or aesthetic features which lead to its establishment; (3) where visitors are allowed and special condition for inspirational, educational, cultural and recreational purposes" (IUCN).

"An area dedicated by statute for all time to conserve scenery, natural and historical objects of national significance and wildlife, and where provision is made for the enjoyment of the same by the public" (IBWL).

According to Wildlife (Protection) Act, 1972, Section 35, "Whenever it appears to the State Government that an area, whether within a sanctuary or not, is by the reason or its ecological, faunal, floral, geomorphological or zoological association or importance, needed to be constituted as national park for the purpose of propagating or developing wildlife therein or its environment it may, by notification, declare its intention to constitute such area as National Park."

From the definition, the difference between sanctuary and national park shows that sanctuary should be natural or of zoological significance, whereas a national park should be of zoological association or importance.

TABLE 51.7. DIFFERENCES BETWEEN SANCTUARY AND NATIONAL PARK.

Sanctuary	National Park
1. Sanctuary is declared by the State Government under Section 18(1) of Wildlife (Protection) Act, 1972.	1. National Park is declared by State Government under Section 35(1) of Wildlife (Protection) Act, 1972.
2. The boundary of a sanctuary is demarcated approximately at the time of its declaration.	2. The boundary of a national park is well-defined and accurate at the very time of its declaration.
3. In a sanctuary, the claim of rights of land of the people is settled after its declaration.	3. In national park, the claim of rights of land of people is settled before its declaration.
4. The alteration in boundary of the sanctuary may be done by the order of State Government.	4. The alteration of the boundary of a national park may be done only after passing through the Legislative Assembly of the State Government.
5. In sanctuary, grazing or movement of the cattle may be permitted for the benefit of wild animals.	5. In national park, grazing or movement of cattle is not permitted.
6. In sanctionary, fishing may be allowed.	6. In national park, fishing is not permissible.

Indian Sanctuaries

There are 493 existing wildlife sanctuaries in India, covering an area of 117,291.03 sq. km., which is 3.57% of the geographical area of the country. Another 217 sanctuaries are proposed in the Protected Area Network Report covering an area of 16,669.44 sq. km. Maximum number of sanctuaries exist in size class less than 10 sq. km., and there are only 2 sanctuaries, having more than 5,000 sq. km. The network of sancturies will go up 709 after full implementation of the above report (Rodgers *et. al.*, 2002) (Table 51.8 and Figure 51.4).

TABLE 51.8. STATEWISE BREAK-UP OF WILDLIFE SANCTUARIES.

State/UTs	Area of state (sq. km.)	No. of WLS	Area covered (sq. km.)	% of state area
Andhra Pradesh	275,068	22	12600.09	4.58
Arunachal Pradesh	83,743	11	7606.37	9.08
Assam	78,438	15	883.16	1.13
Bihar	94,163	11	2949.17	3.13
Chhattisgarh	135,194	10	3419.46	2.53
Goa	3702	6	647.96	17.50
Gujarat	196,024	21	16422.72	8.38
Haryana	44,212	9	279.92	0.63
Himachal Pradesh	55,673	32	5770.85	10.37
Jammu & Kashmir	222,235	15	10312.25	4.60
Jharkhand	79,714	10	1862.75	2.34
Karnataka	191,791	21	3888.22	2.03
Kerala	38,863	12	2143.36	5.52
Madhya Pradesh	308,252	25	7158.40	2.32
Maharashtra	307,690	35	14376.66	4.67
Manipur	22,327	3	393.30	1.76
Meghalaya	22,429	3	34.20	0.15
Mizoram	21,081	4	771.00	3.66
Nagaland	16,579	3	20.34	0.12
Orissa	155,707	18	6969.15	4.48
Punjab	50,362	10	316.73	0.63
Rajasthan	342,239	24	5712.83	1.67
Sikkim	7,096	5	265.10	3.74
Tamil Nadu	130,058	19	2539.82	1.95
Tripura	10,486	4	603.62	5.76
Uttar Pradesh	240,926	23	5222.47	2.17
Uttaranchal	53,485	6	2395.52	4.48
West Bengal	88,752	15	1203.28	1.36
		Union Territories		
Andaman & Nicobar	8249	96	377.95	4.58
Chandigarh	114	1	25.42	22.30
Dadra & Nagar Haveli	491	1	92.16	18.77
Daman & Diu	112	1	2.18	1.95
Delhi	1483	1	13.20	0.89
Lakshadweep	32	1	0.01	0.03
Pondicherry	493	0	0.00	0.00
India	3,287,263	493	117,291.03	3.56

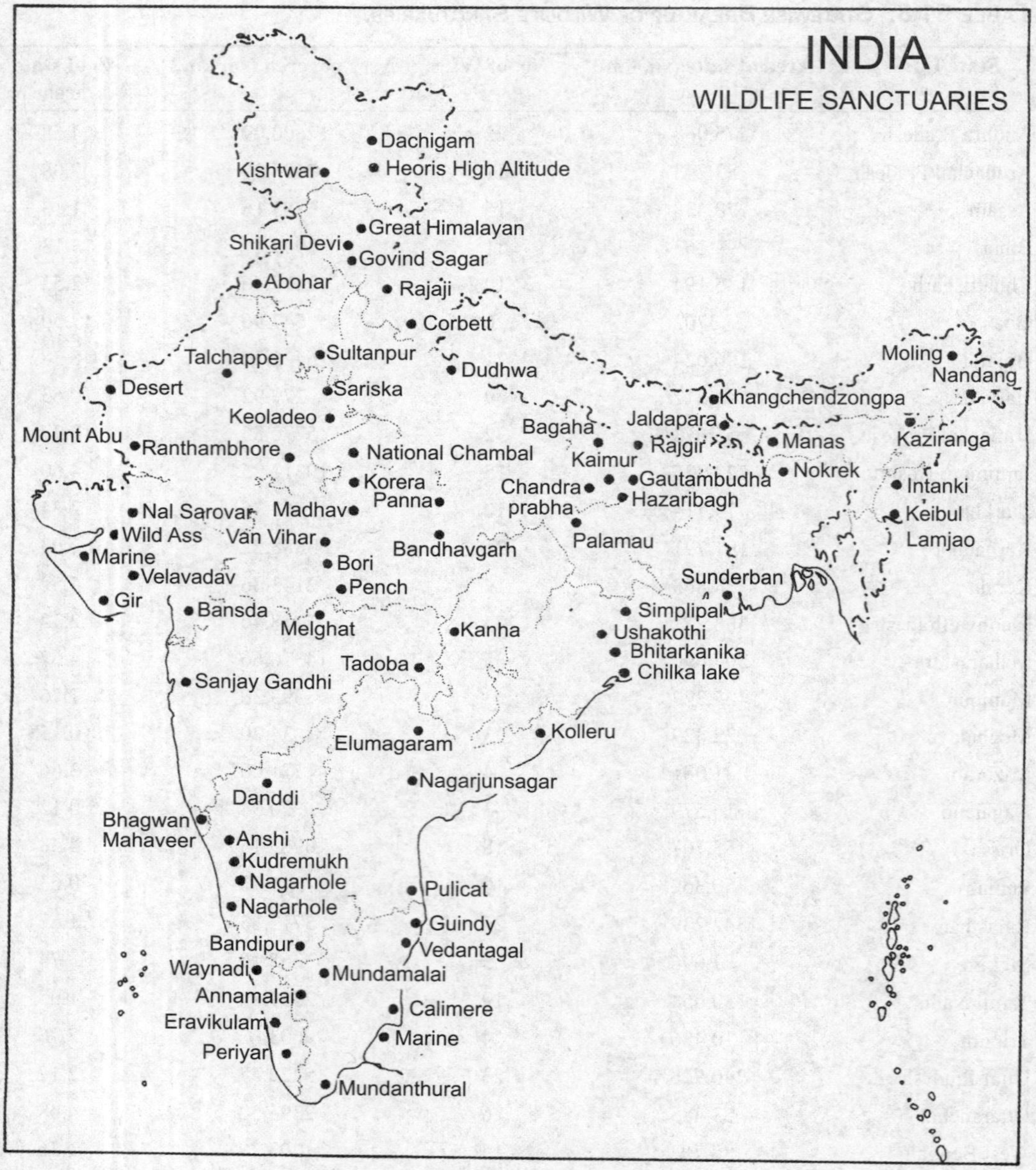

Fig. 51.4. Wildlife sanctuaries in India.

National Parks

There are 89 existing National Parks in India covering an area of 37,526.90 sq. km., which is about 1.14% of the geographical area of the country (National Wildlife Database, 2003). In addition to the above, 74 National Parks covering an area of 16,630.08 sq. km., are proposed in the Protected Area Network Report (Rodgers, *et. al.*, 2002). The network of parks will go up to 164 after full implementation of the above report. Maximum number of parks exist in size about 100–500 sq. km., and there are 7 parks having more than 5,000 sq. km. (Table 51.9 and Figure 51.5).

TABLE 51.9. STATEWISE BREAK-UP OF NATIONAL PARKS OF INDIA.

State/UTs	Area of state (sq. km.)	No. of National Parks	Area covered (sq. km.)	% of state area
Andhra Pradesh	275,068	4	373.23	0.14
Arunachal Pradesh	83,743	2	2290.82	2.74
Assam	78,438	5	1968.60	2.51
Bihar	94,163	1	335.65	0.36
Chhattisgarh	135,194	3	2929.50	2.17
Goa	3702	1	107.00	2.89
Gujarat	196,024	4	479.67	0.24
Haryana	44,212	1	1.43	0.00
Himachal Pradesh	55,673	2	1429.40	2.57
Jammu & Kashmir	222,235	4	4680.25	2.09
Jharkhand	79,714	1	231.67	0.29
Karnataka	191,791	5	2435.14	1.27
Kerala	38,863	3	536.52	1.38
Madhya Pradesh	308,252	9	3656.36	1.19
Maharashtra	307,690	5	955.93	0.31
Manipur	22,327	1	40.00	0.18
Meghalaya	22,429	2	267.48	1.19
Mizoram	21,081	2	250.00	1.19
Nagaland	16,579	1	202.02	1.22
Orissa	155,707	2	990.70	0.64
Punjab	50,362	0	0.00	0.00
Rajasthan	342,239	4	3856.53	1.13
Sikkim	7096	1	1784.00	25.14
Tamil Nadu	130,058	5	307.84	0.24
Tripura	10,486	0	0.00	0.00
Uttar Pradesh	240,926	1	490.00	0.20
Uttaranchal	53,485	6	4077.00	7.62
West Bengal	88,752	5	1693.25	1.91
Union Territories				
Andaman & Nicobar	8249	9	1156.91	14.02
Chandigarh	114	0	0.00	0.00
Dadra & Nagar Haveli	491	0	0.00	0.00
Daman & Diu	112	0	0.00	0.00
Delhi	1483	0	0.00	0.00
Lakshadweep	32	0	0.00	0.00
Pondicherry	493	0	0.00	0.00

However, a comprehensive list of various wildlife found in different sanctuaries and national parks are appended in Table 51.11.

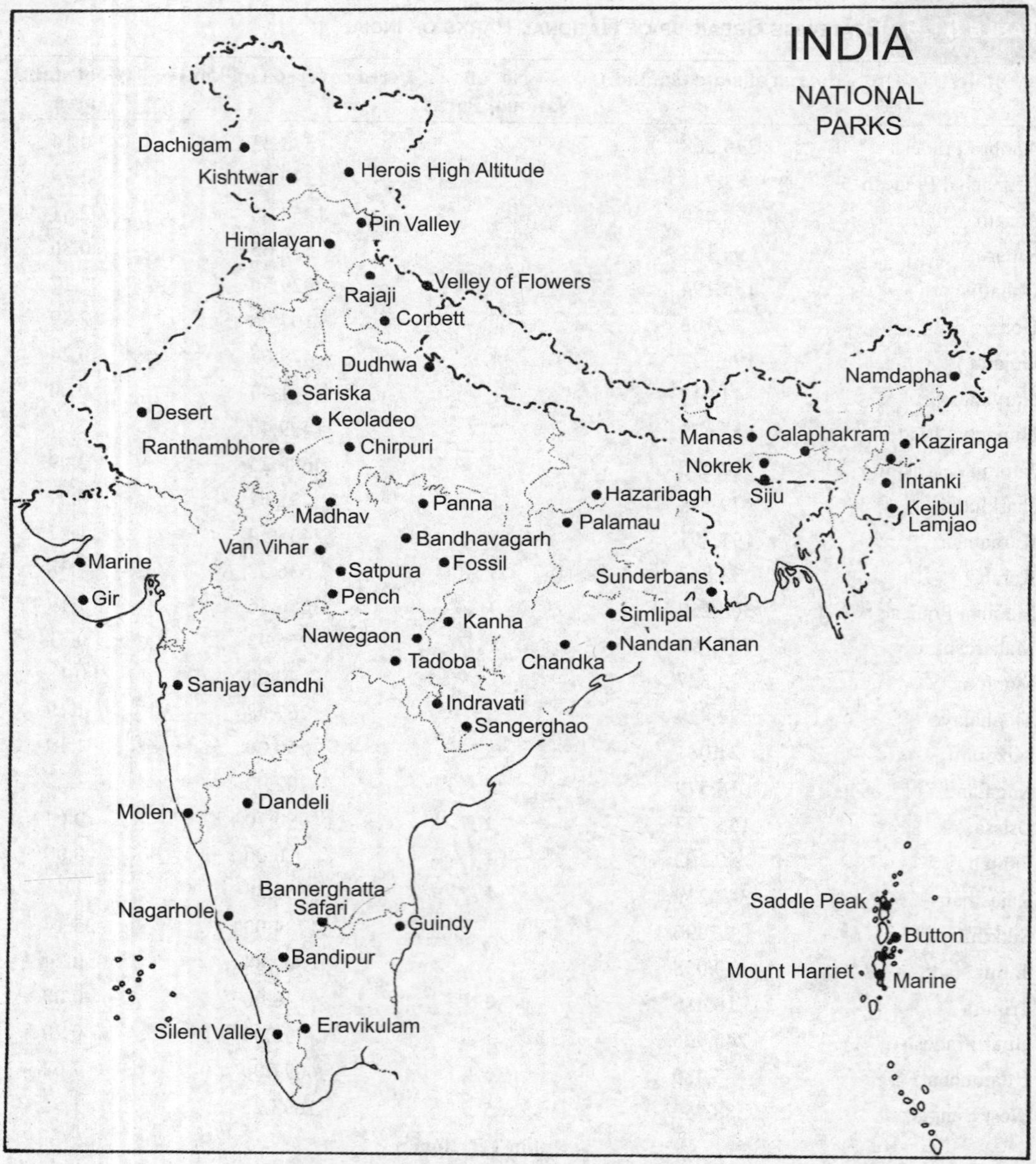

Fig. 51.5. National parks in India.

SOME NATIONAL PARKS OF INDIA

1. Corbett National Park (Project Tiger)

Total area : 520.80 sq. km.

Year of establishment : 1936

Location : Himalayan foothills in Nainital and Garhwal districts, Uttaranchal.

Climate : Annual temperature varies between 4°C and 46°C.

This is the first National Park of this subcontinent whose first name was Hailey National Park in the name of the then U.P. Governor, Mr. Malcom Hailey. After independence, it was named as Ram Ganga National Park in the name of Ram Ganga River. Further, to honour Jim Corbett (famous hunter and naturalist) its name was changed as Corbett National Park in 1956. In 1973, National Park was declared as Tiger Reserve to give special protection to the tigers of this part.

Fauna : Mammals : Herbivores : Elephant, Sambhar, Nilgai, Goral, Cheetal, Para, Barking deer, Wild boar, etc.

Carnivores : Tiger, Panther, Wilddog, Jackal, Red fox, Black bear, Sloth bear, Common otter, Indian civet, Palm civet, Mongoose, Long-eared hedgehog, Shrew, etc. Bats, Hare, Northern Palm Squirrel and Porcupine are also common.

Birds. Peacock, Jungle fowl, Partridges, Water birds like Ducks, Falcons, Coots, Dab chicks, Naktas, Kingfishers, besides some carnivorous birds like Hawks, Vultures, etc. Many migratory birds also visit the Ram Ganga River during winter.

Reptiles. Crocodile (Mugger) and Gharial, Common monitor lizard, some lesser lizards and soft shelled tortoises. Python, Cobra, Viper, Water-snake, Boa, Cat-snake and Wolf-snake, etc.

Tiger is our national animal and found in diverse habitats and in different parts of the country. Tiger is a top carnivore of the complex food chain in our forest ecosystems. From several years, over-exploitation of the forest areas, merciless hunting, unscientific management, etc., reduced the habitat of tiger that leads to a rapid decline in its population. Thus, tiger conservation in India is very necessary to save this endangered species. IBWL set up a task force for studying the tiger population and its status. On recommendations of this task force, Project Tiger was initiated in 1973 with eleven Tiger Reserves located in ten different states.

2. Betla National Park (Tiger Reserve)

Total area : 1026 sq. km.

Climate : Summer maximum temperature 48ºC and winter temperature 3ºC.

It is located in the western part of Chotanagpur plateau in Jharkhand state. Tiger reserve was declared on 4th June 1974 and in 1986 it was declared as National Park. Tiger reserve comprises most of the forests of Daltonganj South Forest Division. These forests were very rich in wildlife such as Tigers, Leopards, Bear, Wolf, Jungle cat, Wild dog, Mongoose, Small civet, Palm civet, Jackal and Hyaena. In 1934, tiger population was 32 (one tiger per 9.3 sq. km.) and when it was declared as Tiger Reserve its population density was one tiger per 33 sq. km. Main river in this area is **Koel** which has its three tributaries, Auranga, Burha and Pandra.

Main rocks of this reserve are laterite, quartzite, gneiss, gondwana and amphibolites, etc.

3. Kanha National Park (Project Tiger)

Total area : 940 sq. km.

It is located in Sylvan Maikal Hills in Mandla and Balaghat districts on the Satpura ranges in Madhya Pradesh. Its name Kanha was given from the old Kanha village. Forest is rich in sal trees, bamboos, etc. Important wild animals are Tiger, Panther, Cheetal, Sambhar, Nilgai, Gaur (Indian bison), Langur (*Presbytis entellus*), Barasingha (Swamp deer) *Cervus duvauceli branderi*, Barking deer, Blackbuck, Wild pig, Wild dog, Jackal, etc. Swamp deer named "Branderi" was saved from extinction.

4. Sunderbans National Park (Tiger Reserve)

Total area : 2585 sq. km.

It is located in West Bengal. The mangrove forests occurring at the mouth of Ganga and Brahmaputra river system (river divides itself into hundreds of streams which join Bay of Bengal) is known as Sunderbans. Mangrove vegetation or "Mangals" have high degree tolerance for saline environment. They possess breathing-root, (pneumatophores), stilt-roots for support, salt excretory glands, vivipary, etc. The mangrove fauna is supreme. Tiger, birds (Storks, Herons, Domestic ducks, etc.), reptiles (Olive Ridley Turtle and Estuarine Crocodile–both are endangered species), Fishes (Mud-skippers and Semiterrestrial gobies both are independent of water) and crustaceans (Land crabs and Fibber crabs are most famous). Spotted deer, Pigs, Monitor lizards and Monkeys also inhabit the forest.

5. Gir National Park (Gir Lion Project)

Total area : 258.71 sq. km.

It is situated in Saurashtra peninsula of Junagarh district of Gujarat. It was declared as sanctuary in 1965 and the project was started in 1972. The fauna of this park is Asiatic lion (*Panthera leo persica*), Panther, Striped hyaena, Sambhar, Nilgai, Cheetal, Four-horned antelope, Chinkara, Wild boar and Crocodiles. Its habitat mainly includes dry teak forests and open scrub jungle.

Lion stands as top carnivore in the food chain of the ecosystem. It is found only in Gir Forest in the whole Asian Continent and, hence, called Asiatic lion. Even in this forest, due to merciless hunting, conversion of forest into agricultural land, uncontrolled cattle grazing and spreading infectious diseases, etc., reduced its habitat and subsequently declined population to the endangered state.

6. Bandipur National Park (Tiger Reserve)

Total area : 874.20 sq. km.

It is located in Mysore district (80 km from Mysore city) of Karnataka. Project Tiger Plan Scheme was initiated in 1973–74 in Bandipur National Park. Rivers of this area are Kabim, Nagu and Moyar. The reserve includes southerm tropical moist deciduous and dry deciduous forests. Here teak trees are abundant and other trees are rosewood, sandalwood, laurel, yellow teak, kydia and bamboos. The fauna of this reserve is Cheetal, Muntjac, Sambhar, Chausingha, Gaur, Wild boar, Elephant, Moue deer, Black-naped, Hare, Sloth bear, Langur, Bonnet macaque, Giant squirrel, Flying squirrel, Tiger, Leopard, Wild dog, Jungle cat, Small Indian civet, Toddy cat, Striped-neck and common mongoose, Crocodiles, Snakes and Monitor lizards. Avian fauna is Pied hornbill, Green pigeon, Woodpeckers, Drongoes, Bee-eaters, Kingfishers, Peafowl, Jungle fowl, Partridge and Quails.

7. Bannerghatta National Park

Total area : 104.27 sq. km.

It is situated in Karnataka, just 22 km from Bangalore. It was created in 1974 to protect local fauna and flora. Area is covered by dry thorny scrub and dry deciduous scrub jungle. Lions and Tigers are found roaming free in an enclosed area of 15 ha. Visitors can move inside the Safari in a closed minibus of the park. Safari also have Gaur, Cheetal, Sambhar, Barking deer, poisonous and non-poisonous snakes, rare varieties of tortoises and lizards.

8. Kaziranga National Park

Total area : 849.79 sq. km.

It is located in Sibsagar district of Assam. It is about 217 km from Guwahati. Its fauna is Great Indian One-horned Rhinoceros (*Rhinoceros unicornis*), Wild buffalo, Tiger, Leopard, Elephant, Gaur, Gat, Mongoose, Otter, Civet cat, Wild boar, Swamp deer, Hog deer, Sambhar, Langur, Python, Pelican, Partridge, Floricans, Hoolock and a number of migratory birds during winter. Painted storks, Adjutant stork, Sarus, Cranes, Darter, Brahminy duck and Bar-headed goose are also kept here in the small island surrounded by water and some shore land.

9. Keibul Lamjao National Park

Total area : 40 sq. km.

It is situated at the south-eastern corner of Loghtak Lake in Manipur State. It was named earlier as Sanctuary in 1968 but in 1977 it was declared as a National Park. Its fauna is Brow-antlered deer or Thamin (*Cervus eldi eldi*) locally called "Sangai" (also called Dancing Deer), Hog deer, Wild goat, Water birds, etc.

10. Dachigam National Park (Project Hangul)

Total area : 141.00 sq. km.

It is situated in Jammu and Kashmir State. Hangul Project was taken up by IUCN/WWF in 1970. It was declared as National Park in 1981. Its main animal is Hangul (Kashmir Stag) (*Cervus eiaphus hunglu*), which is a large deer roaming singly or in herds of 2 to 18. It is also found in north of Chamba (Himachal Pradesh). Habitat of Dachigam includes scrubs and savannah forests. Along Hangul are also found Black bear, Brown bear, Leopard and Snow leopard, Musk deer, Serow and Pheasants, etc.

Fomous National Parks of World

The first national park in the world, the **Yellowstone National Park**, was founded in 1872 in U.S.A. Since then, about 2,000 parks have been established all over the world. These offer protection to thousands of endangered species in their natural habitats. Some parks have been created for specific and very rare endangered species to be saved from extinction. Table 51.10 provides a list of some of such famous parks of the world.

TABLE 51.10. FAMOUS NATIONAL PARKS OF WORLD CREATED FOR SPECIFIC ENDANGERED SPECIES.

S. No.	Name of National Park	Country	Specific Endangered Animal
1.	Alberta National Park	Congo, Africa	Mountain Gorilla
2.	Mt. Simien National Park	Ethiopia, Africa	Abyssinian Ibex or Steinbok
3.	Bontebok National Park	South Africa, Africa	Bontebok antelope
4.	Everglades National Park	Florida, U.S.A.	Puma
5.	Rifugio di Los Padres	California, U.S.A.	Californian Condor
6.	Ordesa National Park	Spain, Europe	Steinbok of Pyrenees
7.	Bialowieska National Park	Poland, Europe	Europian Bison
8.	Corbett National Park	Uttar Pradesh, India	Indian Tiger
9.	Dachigam National Park	Kashmir, India	Kashmir Stag
10.	Gir National Park	Gujarat, India	Asiatic Lion
11.	Kaziranga National Park	Assam, India	One-horned Rhinoceros
12.	Ghana Bird Sancturay (Keoladeo National Park)	Bharatpur (Rajasthan) India	Avifauna

TABLE 51.11. SOME WILDLIFE SANCTUARIES AND NATIONAL PARKS OF INDIA.

S.No.	Name and Location	Area in sq. km.	Important Animals Found
1.	**Nagarjuna Sagar (Ikshwaka Sanctuary)** Guntur, Prakasham, Karnool, Mahbubnagar and Nalgonda Distt., Andhra Pradesh	3568	Tiger, panther, slothbear, wild bear, nilgai, chital, sambar, black buck, jackal, fox, wolf, muggar crocodile.
2.	**Pulicat (Lake) Sanctuary** Nellore Distt., Andhra Pradesh	500	Flamingo, pelican, duck, teal, stork, crane, heron
3.	**Kaziranga National Park** Sibsagar, Jorhat Distt., Assam	430	Rhinoceros, elephant, wild buffalo, gaur, sambar, swamp deer, hogdeer, wild boar, tiger, leopard, gibbon, python, pelican, stork, florican
4.	**Manas Sanctuary** Barpeta Distt., Assam	80	As in Kaziranga. Also wild dog, panther, golden langur, water monitor, great pied hornbill
5.	**Hazaribagh Sanctuary** Hazaribagh, Bihar	186	Tiger, leopard, hyaena, wild boar, gaur, samabar, chital, nilgai, peafowl

(*Contd.*)

S. No.	Name and Location	Area in sq. km.	Important Animals Found
6.	**Palamau Sanctuary** Daltongunj, Bihar	980	Elephant, panther, leopard, wild boar, barking deer, gaur, chital, sambar, peafowl
7.	**Kaimur Sanctuary** Rohtas, Bihar	1342	Tiger, leopard, chinkara, sambar, nilgai, crocodiles
8.	**Gir National Park** Sasan-Gir, Junagarh Distt., Gujarat	1412	Asiatic lion, panther, striped hyaena, sambar, nilgai, chital, 4-horned antelope, chinkara
9.	**Sultanpur (Lake) Bird Sanctuary** Gurgaon, Haryana	1.2	Wild boar, crocodiles, python, green pigeon Sarus crane, spot bill, ducks, ruddy shell
10.	**Dachigam Sanctuary** Srinagar, Jammu and Kashmir	89	Snow leopard, black & brown bears, hangul, musk deer, serow
11.	**Shikari Devi Sanctuary** Mandi, Himachal Pradesh	213	Black bear, panther, snow leopard, goral, barking & musk deers, serow, flying fox, monal, chir, chukor, partridge
12.	**Govind Sagar Bird Sanctuary** Bilaspur, Himachal Pradesh	100	Teal, ducks, goose, crane
13.	**Bandipur National Park** Mysore Distt., Karnataka	874	Elephant, tiger, panther, wild boar, wild dog, sloth bear, gaur, barking deer, 4-horned antelope, sambar, chital, malabar squirrel, green pigeon
14.	**Periyar Sanctuary** Idukki Distt., Kerala	777	Elephant, tiger, panther, wild boar, wild dog, sloth bear, gaur, nilgai, sambar, barking deer. black nilgiri langur, grey hornbill, egret
15.	**Kanha National Park** Mandla & Balaghat, Madhya Pradesh	940	Tiger, panther, wild boar, wild dog, gaur, barasingha, sambar, chital, black buck, nilgai, barking and mouse deers
16.	**Tadoba National Park** Chandrapur, Maharashtra	116	Tiger, panther, sloth bear, gaur, sambar, chital, nilgai, chinkara, crocodiles
17.	**Pench National Park** Nagpur, Maharashtra	257	Tiger, panther, sloth bear, gaur, sambar, chital, nilgai, chinkara, barking deer, peafowl
18.	**Bhitar Kanika Sanctuary** Cuttack, Orissa	170	Salt water crocodile, leopard, hyaena, chital, sambar, giant squirrel, water monitor, king cobra, python, storks, ibis
19.	**Chilka Lake Bird Sanctuary** Balagaon, Orissa	900	Flamingo, pelican, egret, ibis, cormorant crane, duck, sandpiper, curlew
20.	**Simlipal Sanctuary** Baripad, Mayurbhanj, Orissa	303	Tiger, leopard, elephant, wild boar, gaur, sambar, mouse deer, flying squirrel, mugger
21.	**Sariska (Project Tiger)** Alwar, Rajasthan	195	Tiger, panther, wild boar, hyaena, sambar, chinkara, nilgai, 4-horned antelope, langur
22.	**Keoladeo Ghana Bird Sanctuary** Bharatpur, Rajasthan	29	Siberian crane, storks, herons, cormorant, spoon bill, egret, ibis, etc., sambar, chital, boar, python
23.	**Desert National Park** Jaisalmer, Barmer, Rajasthan	3000	Great Indian bustard, black buck, chinkara
24.	**Annamalai Sanctuary** Coimbatore, Tamil Nadu	958	Elephant, tiger, panther, sloth bear, wild dog, gaur, chital, sambar
25.	**Corbett National Park** Nainital Distt., Uttar Pradesh	525	Elephant, tiger, panther, sloth bear, wild boar, nilgai, sambar, chital, crocodiles, python, king cobra, peafowl, partridge
26.	**Jaldapara Sanctuary** Madarihat, West Bengal	115.5	Rhino, elephant, tiger, leopard, gaur, deers, sambar, variety of birds

(Contd.)

S.No.	Name and Location	Area in sq. km.	Important Animals Found
27.	**Sajnakhali Sanctuary** 24-Parganas, West Bengal	362	Tiger, wild boar, chital, storks, cormorant, herons, ibis, pelican, bittern, darter
28.	**Sundarbans (Tiger Reserve)** 24-Parganas, West Bengal	2585	Tiger, wild boar, deers, estuarine crocodile, gangetic dolphin

BIOSPHERE RESERVES

The concept of Biosphere Reserves was evolved under UNESCO's Man and Biosphere (MAB) Programme in 1971 with the following objectives :

1. To conserve, for present and future human use, the diversity and integrity of biotic communities of plants and animals within natural ecosystems, and to safegaurd the genetic diversity of species on which their continuing evolution depends.

2. To provide areas for ecological and environmental research including, particularly baseline studies both within and adjacent to these reserves, such research to be consistent with objective (1) above.

3. To provide facilities for education and training.

4. To promote international cooperation.

The concept of Biosphere Reserves is of immense value to conserve the gene-pool resources of flora and fauna in the country and to serve as bench-marks for future studies. This requires a detailed survey and classification of natural ecosystems for proper identification of their biological diversity and pristine (ancient) attributes (consider as belonging). The areas identified should be large enough to be viable. In view of the virtual disappearance of many genuinely natural ecosystems in the country, those that remain in such natural or near natural condition need to be identified with utmost urgency. The recent Silent Valley controversy highlights the urgency of the problem.

In a biosphere reserve, multiple land use is permitted by designating various zones. There is the Core Zone (where no human activity is permitted), the Buffer Zone (where limited human activity is allowed) and the Manipulation Zone (where a large number of human activities would go on). In a biosphere reserve, wild population as well as traditional lifestyles of tribals and varied domesticated plant and animal genetic resources are protected.

Biosphere Reserves are multipurpose protected areas which are meant for preserving genetic diversities in representative ecosystems by protecting wild populations, traditional lifestyle of tribals and domesticated plant/animal genetic resources. There are some 243 biosphere reserves in 65 countries of the world. In India 14 potential sites were identified in 1979 by Core Advisory Group but only 13 biosphere reserves have been set up by now. Each biosphere reserve has a :

(*a*) **Core Zone:** No human activity is allowed.

(*b*) **Buffer Zone:** Limited human activity is permitted.

(*c*) **Manipulation Zone:** Human activity is allowed but ecology is not permitted to be disturbed.

(*d*) **Restoration Zone:** Degraded areas for restoration to near natural form.

India has identified 14 areas to be declared as Biosphere Reserves. Of these the first biosphere reserve, **Nilgiri Biosphere Reserve** having an area of 5520 sq. km., was established in 1986. The second biosphere reserve was established in 1988, **Nanda Devi Biosphere Reserve**. In this very year (1988) third biosphere reserve, **Nokrek Biosphere Reserves** was set up. Till 2001 following biosphere reserves were established as shown in Table 51.12 and Figure 51.6.

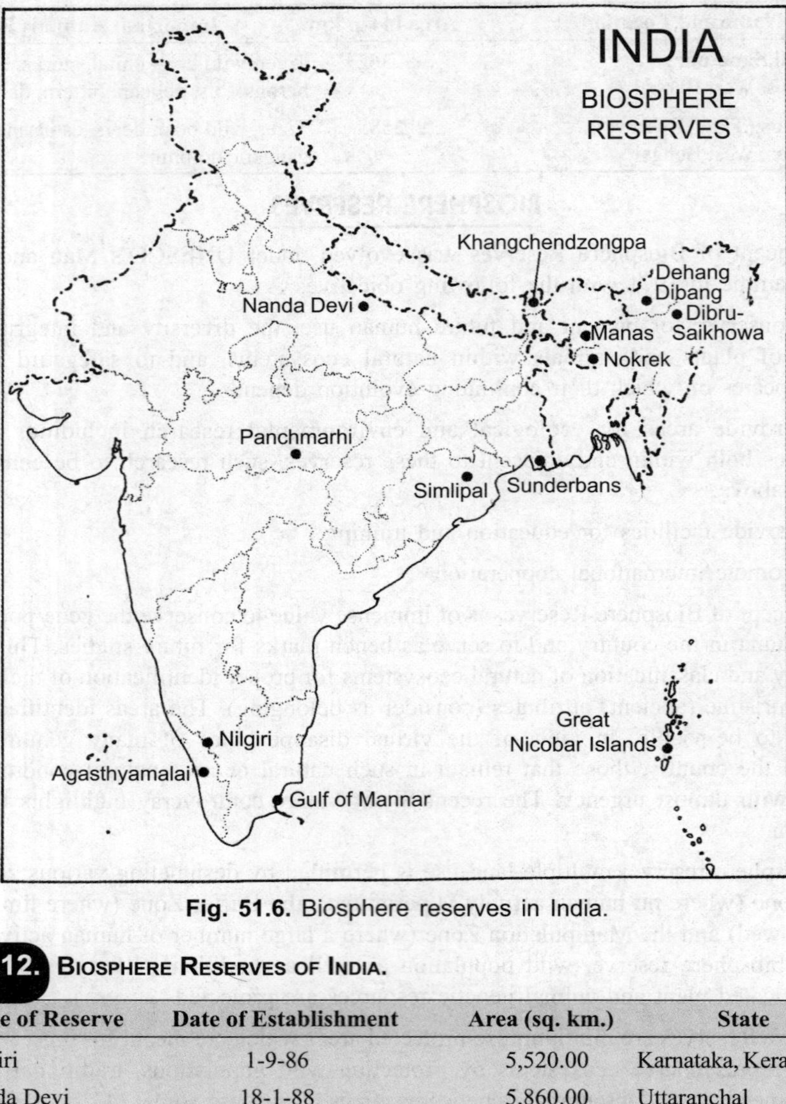

Fig. 51.6. Biosphere reserves in India.

TABLE 51.12. BIOSPHERE RESERVES OF INDIA.

S.No.	Name of Reserve	Date of Establishment	Area (sq. km.)	State
1.	Nilgiri	1-9-86	5,520.00	Karnataka, Kerala, Tamil Nadu
2.	Nanda Devi	18-1-88	5,860.00	Uttaranchal
3.	Nokrek	1-9-88	80.00	Meghalaya
4.	Great Nicobar Islands	6-1-89	885.00	Andaman & Nicobar Islands
5.	Gulf of Mannar	18-2-89	10,500.00	Tamil Nadu
6.	Manas	14-3-89	2,837.00	Assam
7.	Sunderbans	29-3-89	9,630.00	West Bengal
8.	Simlipal	22-6-94	4,374.00	Orissa
9.	Dibru-Saikhowa	28-7-97	765.00	Assam
10.	Dehang-Dibang	2-9-98	5,111.50	Arunachal Pradesh
11.	Panchmarhi	3-3-99	4,926.00	Madhya Pradesh
12.	Khangchendzongpa	7-2-2000	2,619.92	Sikkim
13.	Agasthyamalai	12-11-2001	1,701.00	Kerala

ZOOLOGICAL GARDENS AND MUSEUMS

Several zoological gardens and museums were also established to conserve the species in their wild state, to provide scientific, educational and recreational opportunities and to earn revenue.

The zoological gardens are somewhat different from sanctuaries and wildlife national parks as the animals are kept in cages in zoological gardens for show. The animals, thus, do not get natural habitat in zoological gardens.

Some important zoological gardens and museums of India are shown in Table 51.13 and Figure 51.7.

1. Natural History Museum, **Mumbai**, has excellent collection of animals, particularly fishes and birds.

2. Tarapore Varsova Aquarium, **Mumbai**, attached to sea and has beautiful collection of marine fishes and crustaceans.

3. Indian Museum, **Kolkata**, is the biggest museum in India.

4. Zoological Garden, **Kolkata**, is famous for talking parrots. It has good species collection.

5. Zoological Garden, **Delhi**, **Lucknow**, **Kanpur**, have some rare species of animals.

6. Geological Museum, **ISI, Kolkata**, are famous for complete skeleton of Indian Dinosaurs.

7. Zoological Garden, **Hyderabad**, has a rich collection of lions, tigers, monkeys and crocodiles.

TABLE 51.13. SOME INDIAN ZOOLOGICAL GARDENS AND MUSEUMS.

Zoo	Locality	Town
Kamla Nehru Zoological Gardens	Kankaria	Ahmedabad
Nandankanan Zoological Park	Shaeed Nagar	Bhubaneswar
Veermata Jijabai Bhosle Udyan	Byculla	Mumbai
Natural History Museum	Byculla	Mumbai
Zoological Garden	Byculla	Mumbai
Tarapore Varsova Aquarium	Byculla	Mumbai
Kolkata Zoological Garden	Alipore	Kolkata
Indian Museum	Alipore	Kolkata
Geological Museum	Alipore	Kolkata
M.C. Zoological Park	Chhat-Bir	Chandigarh
Padmaja Naidu Himalayan Zoological Park	Darjeeling	Darjeeling
Delhi National Zoological Park	New Delhi	New Delhi
Assam State Zoo	Guwahati	Assam
Nehru Zoological Park	Hyderabad	Hyderabad
Jaipur Zoological Gardens	Jaipur	Jaipur
Sakkarbaug Zoo, Junagadh	Junagadh	Junagadh
Kanpur Zoological Park	Azadnagar	Kanpur
Arignar Anna Zoological Park	Chennai	Chennai
Zoological Garden	Chennai	Chennai
Sri Chamarajendra Zoological Gardens	Indiranagar	Mysore
Peshawe Park Zoological Gardens	Peth	Pune (Poona)
Trivandrum Zoological Gardens	Trivandrum (Thiruvananthapuram)	Trivandrum (Thiruvananthapuram)

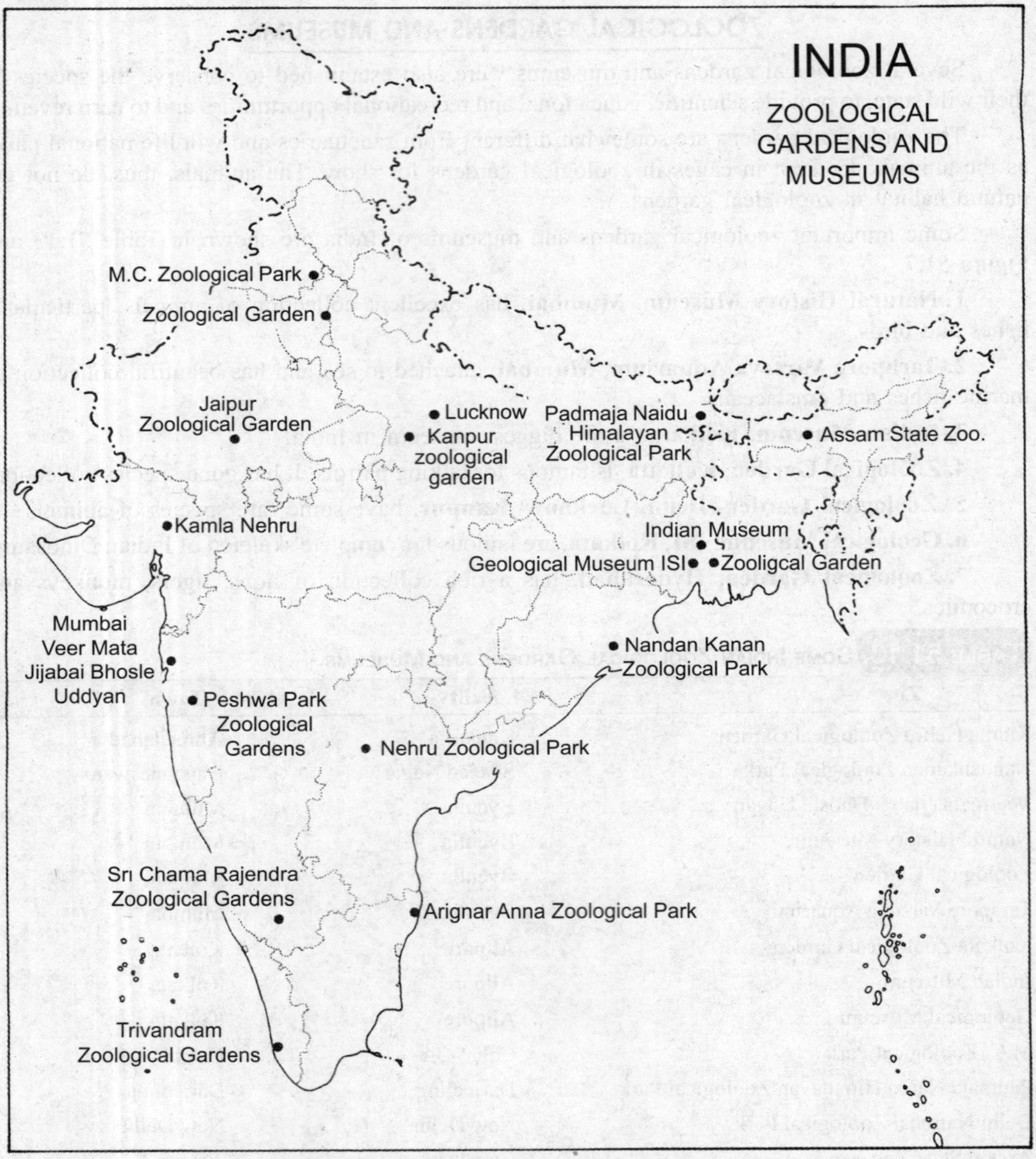

Fig. 51.7. Zoological gardens and museums of India.

WILDLIFE ORGANISATIONS

There are several organisations which are engaged in conservation of wildlife. Some of these are mentioned below :

1. Convention on International Trade in Endangered Species of Wild Fauna and Flora (CITES)

Habitat destruction and unregulated commercial exploitation are the principal causes of the decline of wildlife. In India, jackal exploited for the fur, musk deer for musk pod and rhinoceros for its horn, crocodile for its skin, etc., for fantastic prices in the world market by poachers, traders and smugglers. In order to prevent over-exploitation of wild fauna and flora through international trade, a Convention on International Trade in Endangered Species of Wild Fauna and Flora (CITES) was signed at Washington on 3rd March, 1973. After its ratification by ten countries, it came into effect on 1st July, 1975. Government of India signed this convention in July 1974 and became a party to it from 18 October 1976. Under this convention, Inspector General of Forests and Director of Wildlife Preservation are the Management Authorities in India supported by Director, Zoological Survey of India (ZSI), Director, Botanical Survey of India (BSI) and Director, Central Marine Fisheries Research Institute.

2. Worldwide Fund for Nature (WWF)

It is the world's largest voluntary organisation set up in 1961. Its above name was given in 1990 and its previous name was World Wildlife Fund (WWF). Its headquarter is at Gland, Switzerland, since 1979. Giant Panda is the official symbol of WWF. An executive committee comprises one president and ten members of the Board of Trustees of WWF. They are responsible to the Board for the administration of the fund and also supervise raising and disbursement of fund, etc. WWF depends entirely upon voluntary contributions for raising its fund. WWF works in collaboration with IUCN and is supported by UNO, UNESCO, UNEP and FAO.

WWF India was formed in 1969 having a Board of 8 Trustees. Its headquarter is in Bombay and three Regional offices in Kolkata, Delhi and Chennai.

3. Bombay Natural History Society (BNHS)

It is a private organisation, set up in 1983 by seven residents of Bombay. The society collects information and specimens of flora and fauna from India, Myanmar and Sri Lanka. The society focuses official and public attention to the need for proper understanding, conserving and development of the rich and varied biodiversity of wildlife of India through publications, lectures, films and field visits, etc. It surveyed mammals and birds of India from 1911 to 1923. The society is under the affiliation of University of Bombay and provides M.Sc. and Ph.D. degrees in Botany and Zoology.

4. Wildlife Preservation Society of India (WPSI)

It is a non-government organisation established in 1958 at Dehradun by a group of wildlife conservationists. Its aim is to protect, conserve and propagate wildlife in the country. The Society has to advise and help the government and wildlife administrators in the formation, maintenance and protection of sanctuaries and national parks. It also supports in enforcing Wildlife (Protection) Act. It also promotes wildlife tourism. The Society is publishing a journal named "Cheetal" together with BNHS.

5. Indian Board for Wildlife (IBWL)

It is the advisory body that advises Government of India in regard to wildlife policy in the country. It was founded in 1952 as an Advisory Body under the chairmanship of Inspector General of Forests in the name of "Central Board for Widlife". Its function was to suggest how India's wildlife could be safeguarded. Later it was renamed as Indian Board for Wildlife (IBWL). A Zoological Park was set up in Delhi in 1955 to exhibit animals in open air enclosures. It was due to the efforts of IBWL. India became a party of CITES in 1976 on the recommendation of IBWL. National Environmental Conservation Policy and National Forest Policy have also been formulated on recommendation of IBWL.

6. International Union for Conservation of Nature and Natural Resources (IUCN)

IUCN formation started after International Conference for the Protection of Nature held at Paris in 1931 which led to formation of the IUCN. After the last war (Second World War) and in collaboration with the UNESCO (United Nations Economic, Social and Cultural Organisation), IUCN General Assembly met at Paris in 1948 and at Finland in 1952.

IUCN is the international, non-government organisation was founded in 1948. Its aim is wildlife conservation. IUCN has more than 475 member organisations and 116 governmental agencies in 110 countries. IUCN prepares Red Data Book for endangered species. It is modified every year and addition or deletion of animal is done in the list of Book on the basis of animal's status.

7. Red Data Book

Special Survival Commission (SSC) comes under IUCN. Its aim is to conserve biological diversity, save, restore and manage species and their habitats. SSC prepares Red Data Book for endangered species and it is modified every year on the basis of the species status. The animal whose status becomes improved and comes out of endangered category is removed from the list of the Book.

WILDLIFE INSTITUTE OF INDIA (WII)

A separate Directorate of Wildlife Environmental Education and Research was established at the Forest Research Institute and Colleges, Dehradun, in the Fifth Plan. A post-graduate diploma course in Wildlife Management, of one academic year's duration, for people with forestry qualifications has been started by the Directorate since 1977. The Directorate has also sponsored fellowship grants for conducting research on wildlife topics in the universities. Research in field oriented management techniques is also being conducted at the Directorate. Proposals for strengthening the research base and organising short-term orientation course for decision makers, administrators and professional groups are being considered for the Sixth Plan.

To train people and expand the activities in wildlife, Government of India has established a Wildlife Institute of India (WII) at Dehradun. This institute is engaged in management and extension of wildlife. It runs several courses at post-graduate level along with diploma and other short-term courses in specialised fields. As a result of training by this institute a number of trained persons are now engaged in conservation and extension of wildlife in India.

WILDLIFE ACT

Wild animals are the integral part of the ecosystem performing a vital role in maintaining food chain and foodweb. Thus, their existence and survival on the earth planet is so important and essential for the natural balance and also for the existence of human life.

India is an underdeveloping country facing many problems like tremendous growth in human population, poverty, unemployment, illiteracy and unawareness towards wildlife and nature. As a result, there is lot of pressure on forest and wild animals. Encroachment of forest land converting it into agricultural field, illicit felling of trees, illegal hunting and poaching, smuggling and merciless killings of wild animals have resulted in dwindling population of wild animals and shrinkage of their habitats. Several species have been brought to the endangered/threatened category, many are on the verge of extinction surpassing the stage of critical limit, and extinction of many important species has taken place. Therefore, the forest personnel as well as people should have holistic view and approach in forest activity whether it is tree and/or wild animal.

In the year 1800, for the first time in India, the exploitation of wildlife was started on scientific basis. In 1806, first Forest Conservator was appointed at Madras Presidency. In 1847, an eminent botanist was appointed as Forest Conservator at Bombay. In 1855, the administrative structure of Forest Department was framed. In 1864, Sir Brandis became the first Forest Inspector General. In 1887, first systematic law in India was enacted by the British called "Wild Birds' Protection Act of 1887". But due to wanton killing of wild animals and birds, a comprehensive legislation was passed, *i.e.*, "Wild Birds and Animals (Protection) Act, 1912". Again in 1935, the Act was amended calling it "Wild Birds and Animals (Protection) Act, 1935". All these laws were aimed at not hunting the game animals and trade in products derived from wild animals. Much attention was not taken in conserving and propagating the wild animals throughout the country.

Before independence, the protection of wildlife was the responsibility of native rulers and local forest officers under the Indian Forest Act, 1927. There was rapid depletion in the population of wild animals. British rulers realised the importance of conservation of wildlife and enforced various laws like Rhino Protection Act, Elephant Protection Act, Arms Act, etc., from time to time to save them. During Second World War the wild animals were mercilessly hunted by the soldiers and army officers wherever they halted. Bombay Natural History Society (BNHS), a private agency, established in 1883 took the first step in formulating a comprehensive Act for Bombay in 1951 called Bombay Wild Animals and Wild Birds Protection Act, 1951. After independence, in 1952 an advisory board calling Central Board for Wildlife was set up, which was renamed as Indian Board for Wildlife (IBWL).

Later Wildlife (Protection) Act, 1972 was formulated to conserve and propagate the wild animals with the salient provisions of creating sanctuaries and national parks as well as penalties against illegal hunting, poaching and trading, etc. The Act provides legal protection to the wildlife and to the endangered species of fauna listed in different Schedules.

This Act has been amended in 1982, 1986, 1991 and 1993. Amendment in 1991 has imposed total ban on hunting, constituted Central Zoo Authority, recognition of zoos, emphasising protection of plants, etc., through insertion of Chapter IIIA and Chapter IVA, amendment of certain sections, omission of words such as Special Game, Big Game and Small Game from the Schedules, addition of one more Schedule VI enlisting certain plants.

Need of Wildlife Management / Conservation

In ancient years, the wildlife in India were found sufficient; but due to enormous growth in human population, deforestation started for converting forest land into agricultural purposes and subsequently the habitat of wildlife started shrinking and their population depleted and still depleting. The period of Second World War was the most devastating period for wildlife as the wild animals were killed mercilessly wherever the militaries got shelter. Following are the main reasons causing depletion in wildlife population :

(*i*) Conversion of forests into agricultural fields.

(*ii*) In ancient time, hunting was done rudely (wantonly), recklessly, mercilessly and unscientifically.

(*iii*) Creation of forest roads as well as allowing there the hunters with their vehicles.

(*iv*) Use of firearms in hunting and hunting were not done selflessly.

(*v*) Increase in livestock, grazing by them in forest areas resulting in loss of food and shelter of wild animals as well as infection of diseases.

(*vi*) Establishment of industries in forest areas, quarrying of mines, etc.

(*vii*) Pollution of environment, rivers, water reservoirs and ocean by insecticides, polluted gases and water from industries as byproducts.

(*viii*) Use of insecticides on the large scale in agricultural fields.

(*ix*) Misuse of licensed firearms.

(*x*) Lighting of fire by the local people inside the forest.

These factors not only depleted the number of wildlife but some became extinct, some are threatened and some are on the verge of extinction. Their number has come to such a stage that the benefits derived from them is negligible.

Problems of Widlife Management in India

Mitra (1980) pointed out that the management of wildlife in India is beset with several problems. The shrinkage and degradation of forest and non-forest wildlife habitats as well as serious decline in the status of wild animals; conservation efforts started gathering momentum in the country in the early 1970s. The National Wildlife Action Plan was drawn up against this background. It is a document setting out an agenda of countrywide activities in the field of wildlife conservation, which was released by the late Prime Minister Indira Gandhi in November, 1983. The Action Plan covers a period of 5 years and contains ten sections, each representing a major area of activity in the field of wildlife which are as follows : 1. Establishment of Representative Network of Protected Areas (PA = Sanctuary and National Park); 2. Management of Protected Areas and Habitat Restoration; 3. Wildlife Protection in Multiple-use Areas; 4. Rehabilitation of Endangered and Threatened Species; 5. Captive Breeding Programmes; 6. Wildlife Education and Interpretation; 7. Research and Monitoring; 8. Domestic Legislation and International Conventions; 9. National Conservation Strategy; 10. Collaboration with Voluntary Bodies.

The National Wildlife Action Plan has provided a vital framework for conservation and management of India's diverse wildlife resources and a yardstick by which progress can be measured.

Methods of Wildlife Conservation

In wildlife conservation, the objective is that the wildlife is to be conserved in such a way that it may remain in the nature in peak status. The most commonly adopted methods for the conservation of wildlife, in brief, are as follows :

1. Habitat management. Habitat management includes ecological study of the habits and habitats of wildlife species, protection, preservation and improvement of habitats, census and statistical data regarding species to be conserved, etc.

2. Establishment of national parks, wildlife reserves, and wildlife sanctuaries. The establishment of national parks, wildlife reserves and wildlife sanctuaries and zoological gardens serves many purposes such as : (*i*) to conserve the species in their wild state; (*ii*) to provide scientific, educational and recreational opportunities; and (*iii*) to earn revenue from tourists by attracting them.

3. Breeding in captivity. Tree species, such as ginko and metasequoia, have survived only in captivity. Tiger, White Tiger and Indian Lion today live in National Parks. Gorilla is protected in the National Park of Alberta. The European Bison, saved at the eleventh hour, survives in the National Park of Biolowieska in Poland. Similarly Pere David's Deer, Whooping Crane, Hawaiian Goose, Parma Wallaby and Arabian Gazzelle, once threatened with extinction, are considered to be already 'on the way back' as a result of captive breeding.

4. Reintroduction of species. Several animal species which were almost extinct, such as Arabian Oryx, Vicuna of High Andes, Russian Antelope or Saiga, Trumpeter Swan, Black Buck, Flamingos, etc., were allowed to reproduce and flourish in suitable places similar to the original ones. Later these were reintroduced in several parks and sanctuaries and areas of their original natural habitats.

5. Mass education. For any conservation programme, there is a great need of education to the people to achieve their participation. To achieve this objective, the methods adopted are as follows : (*i*) celebration of wildlife week every year; (*ii*) publicity through media and film shows; (*iii*) holding essay competitions, lectures, seminars, conducted tours, etc.; (*iv*) setting up nature clubs in educational institutions; (*v*) publication of wildlife books and journals, etc.; (*vi*) establishment of Natural History Museum, etc.

6. Promulgation of laws. All the countries have promulgated laws for the protection and conservation of wildlife. In India, Wildlife (Protection) Act, 1972, a comprehensive Central Government Legislation, was enacted in 1972. Killing, capturing and hunting of wildlife without prior permission from competent authority, and poaching have become punishable under Wildlife (Protection) Act, 1972. In 1976, 42nd amendment in the Indian Constitution, protection of wildlife have been included in Concurrent List. Further amendment in law is required for severe and deterrent punishment to poachers.

IMPORTANT QUESTIONS

Long Answer Questions

1. What is wildlife? What do you understand by wildlife? What are its values for mankind?
2. What are sanctuaries and national parks? Give a detailed account of wildlife of India.
3. "Wildlife is an asset to be protected and preserved to your own advantage and to benefit future generations." Discuss.
4. Write an essay on any one of the following : (*i*) Vanishing Wildlife of India; (*ii*) Methods of Conservation of Wildlife; (*iii*) Wildlife Organisations.

Short Answer Questions

1. Write notes on any two or one of the following :
   (*i*) Importance of wildlife in India;
   (*ii*) Vanishing wildlife;
   (*iii*) Importance of wildlife;
   (*iv*) Biosphere reserves;
   (*v*) Zoological gardens and museums;
   vi) Project Tiger;
   (*vii*) Musk deer.

Multiple Choice Questions

1. Which of the following is national animal of India?
   (*a*) Elephant (*b*) Deer
   (*c*) Tiger (*d*) Bear.
2. Project Tiger was launched in India in the year
   (*a*) 1973 (*b*) 1975
   (*c*) 1978 (*d*) 1990.
3. In India first biosphere reserve was established in
   (*a*) Nanda Devi
   (*b*) Corbett National Park
   (*c*) Nilgiri
   (*d*) Manas.
4. In which of the following states Periyar Wildlife Sanctuary is situated?
   (*a*) Uttar Pradesh (*b*) Kerala
   (*c*) Meghalaya (*d*) Manipur.
5. In which of the following states Crocodile Project is situated?
   (*a*) Madhya Pradesh (*b*) Uttar Pradesh
   (*c*) Assam (*d*) Bengal.
6. In which of the following national parks Indian rhino is conserved?
   (*a*) Corbett National Park
   (*b*) Kaziranga National Park
   (*c*) Manas Sanctuary
   (*d*) Kaimur Sanctuary.
7. Armadillo contributed to the development of vaccine for
   (*a*) Tuberculosis (*b*) Hepatitis B
   (*c*) Leprosy (*d*) AIDS.

ANSWERS

1. (*c*) 2. (*a*) 3. (*c*) 4. (*b*) 5. (*b*) 6. (*b*) 7. (*c*)

GLOSSARY

Abdomen. Body region that contains the viscera; in mammals, limited to the part of the body extending from the diaphragm to the pelvis.

Abductor. A muscle that draws a part away from the median line or centre.

Aboral. Opposite to mouth.

Absorption. Taking in of fluids or other substances by cells or tissues.

Abyssal. The bottom waters of the deep sea marked by the absence of light and, hence, of plants; where carnivores live.

Abysal zone. Ocean zone below 2000 metres; all aquatic environment below bathyal zone.

Acclimatise. To become habituated to a different environment.

Acetabulum. Cup-shaped socket of the hip into which the femur fits.

Acetylcholine. A chemical substance produced at or near parasympathetic nerve endings which is thought to be involved in conduction of nerve impulses across synapses.

Acinus. One of the small terminal sacs in a lung or multicellular gland.

Acquired character. Modification of the body that occurs during the life of an individual as a result of environmental conditions.

Acromion. Pertaining to the prolongation of the spine of the scapula forming the point of the shoulder.

Adaptation. A process by which an organism becomes better suited to its environment.

Adaptive radiation. Evolution and spread of a single line of descent of organisms into several ecological niches, resulting in a series, sometimes of a strikingly different form, each adapted to a particular habitat.

Adductor. A muscle that draws a structure toward the median line.

ADH. Antidiuretic hormone. Hormone secreted by posterior hypophysis; presence decreases water reabsorption in the kidney.

Adipose tissue. Fat tissue, composed of polygonal or rounded cells. Mostly for energy storage.

ADP. Adenosine diphosphate. ATP with one less phosphate. Of lower energy than ATP.

Adrenal gland. A double endocrine gland located on or near the kidneys; consists of the cortex and the medulla, each secreting different hormones.

Adrenaline. A hormone produced by the medullary portion of the adrenal glands.

Aerial. Inhabiting or frequenting the air.

Aerobic. Able to live and grow only in the presence of free oxygen.

Aestivate. To pass the summer in a quiet, dormant state.

Afferent. A vessel or structure leading to or toward a certain region, *e.g.*, an afferent artery carries blood to the lungs.

Agglutinin. A substance in the blood which causes cells or microorganisms to clump.

Agglutination. Clumping or cohesion of cells.

Air-bladder. In fishes, a thin-walled sac, situated dorsally, which serves as a float; a hydrostatic organ.

Albinism. A condition in which the normal pigment of the skin, hair and the eyes is lacking.

Albumen. White portion of the reptiles' or birds' egg, surrounding the yolk or zygote, and supplying food for the embryo.

Alimentary. Pertaining to digestion or the digestive tract.

Allantois. An extra-embryonic membrane arising as an outgrowth of the cloaca in reptiles, birds and mammals.

Allergy. A condition in which substances that are normally harmless cause a marked hypersensitivity or reaction.

Altitudinal migration. Migration perpendicular to the earth's surface, *e.g.*, from the foot to the apex of a mountain.

Alveolus. A small cavity, such as the tiny air-sac in the mammalian lung, the secreting portion of an alveoler gland or the socket of a tooth.

Amnion. Innermost membrane which encloses the embryo in reptiles, birds and mammals.

Amniota. A group of vertebrates possessing an amnion and an allantois – reptiles, birds and mammals.

Amphibia. A class of vertebrates that hatch as gill-breathing larval and usually metamorphose into lung-breathing adults; *e.g.*, the frog, toad and salamander.

Amphibious. Pertaining to animals capable of living either in water or land, such as frog and toad.

Amphicoelous. A structure that is concave at both ends, applied to some vertebrae.

Amylase. A carbohydrate-splitting enzyme.

Amylopsin. A pancreatic amylase. An enzyme produced by the pancreas which acts on carbohydrates.

Anabolism. The constructive phase of metabolism in which the cells build protoplasm from food materials.

Anaerobic. Able to live and grow without the presence of air or free oxygen.

Analogous. Body parts with similar functions but usually genetically dissimilar.

Anamniota. A group of vertebrates without an amnion, *e.g.*, cyclostomes, fishes and amphibians.

Anastomosis. The union of two or more blood vessels, nerves or other structures.

Anatomy. The study of the structure of animals and plants.

Animal pole. In telolecithal eggs, the end where cytoplasm congregates.

Anion. Negatively charged ion.

Ankylosis. A union or knitting together of two or more bones or parts of the bones.

Anterior. Pertaining to the front or head end of an animal.

Antibody. A protein produced in an animal body which neutralises or reacts with a foreign substance, *i.e.*, with an antigen.

Antigen. A foreign substance, usually protein, that causes the production of specific antibodies within an animal body.

Antitoxin. A substance in the body fluids naturally occurring or induced which neutralises specific poisons or toxins.

Antrum. A cavity or chamber, referring specially to one within a bone.

Anura. The order of tailless amphibians to which frogs and toads belong.

Anus. Posterior opening of the digestive tract.

Aorta. The large artery carrying blood away from the heart.

Aortic arch. One of the paired arteries which connects the ventral and dorsal aortae in the pharynx or gill-region.

Aperture. Orifice or opening.

Apical. The top or apex.

Appendage. Portion of the body that projects and has a free end, such as limbs.

Appendicular skeleton. Part of the vertebrate skeleton that consists of limbs and limb girdles.

Apteryla. Area of bird skin on which feathers do not grow.

Aquatic. Pertaining to water. Living in water.

Aqueous humor. Liquid within eye (jelly-like) in the area between the cornea and the lens.

Arachnoid. Arachnoid membrane; the central of the three meninges covering the vertebrate brain and spinal cord; it is very fine and delicate.

Arboreal. Pertaining to trees or tree-living.

Archenteron. Primitive digestive tract of a metazoan embryo, formed during gastrulation.

Arteriole. Small artery.

Arteriosclerosis. Reduction of elasticity of small arteries.

Artery. A blood vessel that carries blood away from the heart.

Articulation. A joint as between two segments or structures.

Artificial classification. Classification based on characters of convenience without relation to phylogenetic significance.

Artiodactyla. Order of the even-toed ungulates to which cattle, swine and deer belong.

Arytenoid. Pertaining to a pair of laryngeal cartilages that lie posterior to the thyroid cartilage.

Asexual reproduction. Reproduction without involving gametes or sex cells.

Assimilation. The conversion of digested foods and other materials into protoplasm.

Asymmetry. Without symmetry; condition in which opposite sides of an animal are not alike.

Atlas. First cervical vertebra; first vertebra upon which the skull rests.

ATP. Adenosine triphosphate, an energy-rich molecule formed in internal respiration.

Atrial cavity. Cavity in protochordates and some chordate larvae (*e.g.*, tunicates, *Amphioxus*, frog tadpoles) through which water passes after moving out of the gill-slits.

Atriopore. The exterior opening from the atrial cavity in the *Amphioxus*.

Atrioventricular node. In the heart, lying within the interventricular septum, a group of cells which, when stimulated, sends an impulse to the base of the ventricles initiating a contractile response from the ventricles.

Atrioventricular valves. Heart valves between the atria and the ventricles. Includes the tricuspid and bicuspid valves.

Atrium. Chamber or cavity; may refer to the atrium of the heart or to the specialised outer cavity of the *Amphioxus*.

Atrophy. The gradual regression of the whole body or of its parts.

Auditory. Pertaining to hearing or ears.

Auricle. Heart chamber into which the veins empty.

Autonomic. Independent; self-governing.

Autonomic nervous system. System of ganglia and nerves controlling involuntary actions of the body; the action of the ductless glands, viscera, blood vessels and all organs containing involuntary muscle, and communicating with the central nervous system by way of the roots of the spinal and cranial nerves.

Autotomy. The automatic breaking off of a part of the body of an animal; self-mutilation.

Aves. A class of vertebrate to which the birds belong.

Axial skeleton. Part of the vertebrate skeleton that consists of the skull, vertebrae, sternum and ribs.

Axis. Second cervical vertebra.

Axolotl. Pedogenic larvae of salamander of the genus *Ambystoma*.

Axon. Fibre of a nerve cell that conducts impulses away from the cell body of which it is a part.

Azygos. Unpaired anatomical structure, such as the azygos vein of mammals.

Backbone. The vertebral column, especially if composed of bone.

Barbule. Processes emanating from barbs of a feather. Have hooklets.

Barrier. Any type of obstruction – physical, chemical or biological – that prevents the migration of animals or gradual extension of their territories.

Basement membrane. Any membrane upon which an epithelium lies.

Basophil. Type of leucocyte; multinucleate.

Bathyal. Ocean zone encompassing upper 200 metres of water.

Behaviour. Reactions of the whole organisms to its environment.

Benthos. Organisms living on or in the bottom, either of the ocean or freshwaters, from the edge of water down to the greatest depths.

Biscuspid valve. In mammalian heart, the bilobed valve between the left atrium and ventride.

Ectoderm. Outermost germ layer of the gastrula which gives rise to the epidermis, sense organs and nervous system.

Ectotherm. Individual which obtains heat directly from the environment.

Effector. Any part of the body, such as a muscle or gland, transforming motor impulses into motor action.

Efferent. Conveying outward or away from a structure; opposite of afferent.

Egest. To discharge unusable food or residues from the digestive tract.

Egg. Non-motile female gamete.

Ejaculatory duct. Duct conveying semen to the urethra; especially in mammals.

Elasmobranch. Cartilaginous fish such as a shark or ray.

Electrical potential. Separation of charge.

Electrocardiogram. Mechanical tracing based on heart contractile movements. Also called ECG.

Electroencephalogram. Graphic record of the variations in the electric potential of brain. Also called EEG.

Electron micrograph. Enlarged photograph of an object taken by attaching a camera to an electron microscope.

Embolus. Blood clot within a blood vessel.

Emboly. Invaginative gastrulation.

Embryo. The young animal that is passing through its developmental stages, usually within the egg membranes or within the maternal uterus.

Embryogeny. Development of an organism.

Embryonic membranes. Cellular membranes formed during embryonic development, *e.g.*, amnion, chorion, allantois.

Enamel. Dense white covering of the crown of a tooth, the hardest substance produced in the animal body.

Endemic. Ecologically found only in a particular region in which it originated.

Endocrine. Pertaining to the ductless glands.

Endoderm. Innermost layer of the early embryo which gives rise to the lining of the digestive tract. Also called endoderm.

Endogenous. Coming from within.

Endoparasite. A parasite living within the body of the host.

Endoskeleton. A supporting structure on the inside of an animal, whether it be cartilaginous, bony, or of other material.

Endostyle. Ciliated groove in the ventral surface of the pharynx in *Amphioxus* and some other protochordates.

Endothelium. Cellular membrane that lines the blood vessels, heart and lymphatic vessels of vertebrates.

Endotherm. Individual which produces its own heat metabolically.

Enterokinase. Vertebrate intestinal enzyme converting trypsinogen to trypsin.

Eosinophil. Type of leucocyte, non-phagocytic, uni- or binucleate.

Epiboly. Posterior growth of a fold of the blastoderm over the surface of an embryo in the process of forming the enteron during gastrulation.

Epididymis. Coiled structure containing efferent tubules of mammalian testis.

Epiglottis. Flap of cartilaginous tissue in mammals that closes the air passage to lungs in swallowing.

Epinephrine. Hormone secreted by the medulla of the adrenal glands.

Epiphysis. End of a long bone; also the pineal body.

Epithelium. Usually a sheet of cells covering either external or internal body surface.

—, simple. An epithelium, unilaminar usually composed of squamous (flat) cells.

—, cuboidal. Epithelium composed of cells shaped like cubes. Found in glands and ducts.

—, columnar. Epithelium which is relatively tall and found in stomach and intestine.

—, stratified. Epithelium composed of more than one layer. The layers are usually even. Found in skin, urethra, etc.

—, pseudo-stratified. Epithelium composed of more than one layer. Layers are not even. Cells may be of various shapes.

—, cuboidal ciliated. Epithelium of the cuboidal variety, with surface having short cilia. Frequently found in ducts.

—, flagellated. Epithelium with long flagella. Found in digestive cavity.

Erythrocyte. Red blood corpuscle.

Estrogen. A hormone produced by the ovarian follicle which together with one of the pituitary hormones, influences estrus.

Ethmoid. A small bone forming the upper wall of the nasal passageway.

Eustachian tube. Tube leading from middle ear to pharynx in higher vertebrates.

Evagination. An outpocketing from any surface by growth especially a cavity.

Evolution organic. Process by which organisms have changed through time, both structurally and functionally; hence, descent with change has made possible the change of organisms from simple to complex forms.

Excretion. Discharge of metabolic wastes, also the substance discharged.

Exocrine. Type of gland that releases its secretions through ducts.

Expiration. The expulsion of water or air from the lungs in a vertebrate.

External respiration. Consists of two phases: (*i*) breathing which brings air and blood together in the lungs and (*ii*) the transportation of oxygen and carbon dioxide between the lungs and the cells.

Exteroceptors. Sense organs capable of receiving external stimuli.

Family. The principal subdivision of an order.

Facia. A band of connective tissue that covers and supports or binds parts together.

Fauna. Referring to animal life of a given period or region.

Faeces. The indigestible unabsorbed residue of digestion.

Femur. The thigh-bone.

Fertilisation. Union of a mature ovum and a mature sperm to form a zygote.

Fibre. Thread or filament produced by cells but external to them.

Fibril. Thread or filament produced by cells and located within them.

Fibrin. A blood protein and major component of vertebrate blood clots.

Fibrous joint. Joint without a cavity; bones are directly apposed.

Fibula. Narrower of two bones in vertebrate hindlimb between knee and ankle.

Filoplume. Hair-like feather with relatively few barbs at tip.

Fin. Locomotor or steering structure and extension of body of aquatic animal.

Fissure. Furrow, cleft or slit.

Flexor. A muscle whose function is to bend or flex a joint, thereby decreasing the angle between the two component parts.

Floating ribs. The two most caudal ribs which have no ventral attachment.

Follicle. A cellular sac or covering.

Follicle stimulating hormone. FSH, Hormone produced by anterior pituitary; stimulates Graafian follicles to mature; also stimulates nurse cells to supply nutrients.

Foramen. A natural opening in a bone or membrane through which blood vessels and/or nerves usually run.

Foramen of Monro. A passageway between the lateral ventricles and third ventricle of the brain.

Foramen magnum. Cranial foramen through which spinal cord enters.

Fossa. Pit or depression in a bone.

Fossil. Remains or other indications of prehistoric forms of life.

Fovea centralis. Optic centre of mammalian retina where only cone cells are present.

Frontoparietal. Wall of the long flat bones forming the roof of the cranium as in the frog.

Function. The action of any part of an animal.

Gamete. A mature reproductive or germ cell, either ovum (egg), sperm or other types of reproductive cell.

Gametogenesis. Process of development of gametes.

Ganglion. Group of mass nerve cell bodies, usually located outside the central nervous system in vertebrates.

Ganoid. Pertaining to shiny fish scale, enamel covered.

Gastric. Pertaining to the stomach, as the gastric glands.

Gastrin. Mammalian hormone produced by stomach wall in presence of food.

Gastrocnemius. Large muscle on the posterior side of the lower leg of a vertebrate.

Gastrula. Stage in development in which the embryo usually consists of two germ layers (ectoderm and endoderm) with a cavity surrounded by endoderm.

Gastrulation. Process by which a gastrula is formed; the invagination of the blastula.

Genus. The taxonomic subdivision of a family. A genus is usually composed of several species.

Gestation. Period between fertilisation and birth of a mammal.

Gill. Type of respiratory organ for aquatic organism.

Gill-arches. Walls adjacent to the gill-slits; wall bearing gills.

Gill-rakers. Spiny ossifications attached to first four gill-arches serving to filter large particles from incurrent water.

Gill-slit. Series of paired openings in the wall of the pharynx and body of chordates.

Gizzard. Muscular part of the digestive tract as in the birds, used for grinding ingested food.

Glomerulus. Small coiled mass of capillaries contained in the Bowman's capsule of the kidney.

Glottis. Opening from the pharynx into the larynx of a vertabrate with lungs.

Goiter. Enlargement of thyroid gland related to either hypersecretion or hyposecretion of thyroid hormone.

Gonad. Reproductive organ, either ovary, testis or ovotestis, in which gametes are produced.

Gorilla rib. In some humans, an extra rib; present normally in gorillas and chimpanzees.

Graafian follicle. Ovarian follicles containing eggs.

Gray crescent. The primary organiser. Portion of the gastrula which induces mesoderm, endoderm and ectoderm. Becomes dorsal lip of blastopore during gastrulation.

Habitat. Environment in which an animal lives.

Hair. Slender filamentous skin accessory of mammals.

Harem. Group of females, normally kept for copulatory purposes.

Hallux. First digit of avian foot; directed backwards.

Haversian canal. One of the canals of bone that permits the passage of blood vessels and nerves.

Halicotrema. At apical end of cochlea, area from where basilar partition (basilar membrane) ends to apex.

Haemol. Referring to the blood or blood vascular system.

Haeme. Iron containing red pigment.

Haemocyanin. Bluish respiratory pigment.

Haemoglobin. Pigment in red blood corpuscles of vertebrates capable of carrying oxygen.

Haemophilia. Abnormal inherited condition in man in which there is a delayed clotting of blood.

Hemichordata. Deuterostomial enterocoelomate animal phylum.

Hensen's node. The anterior end of the primitive streak.

Heparin. Protein-containing molecule capable of preventing blood clotting.

Hepatic. Pertaining to the liver.

Hepatic portal system. System of veins from digestive tract to capillaries of vertebrate liver.

Herbivorous. Feeding chiefly on plants.

Hermaphrodite. An individual possessing both male and female reproductive organs.

Heterocercal. Pertaining to the type of tail that is asymmetrical internally as well as externally. Examples shark and sturgeon tails.

Heterocrine gland. Gland with a combination of exocrine and endocrine activity.

Heterodonts. Having different types of teeth.

Hibernation. Passing of the winter in dormant inactive state.

Histogenesis. Origin, development and differentiation of the tissues of an organism.

Histology. Branch of anatomy that deals with the microscopic structure of tissue and organs.

Holoblastic egg. An egg that divides completely into cells during cleavage.

Homeotherm. Individual which maintains a constant body temperature.

Homocercal. Pertaining to the type of the tail that is externally symmetrical but internally asymmetrical.

Homodonts. Having the same type of teeth.

Homologous. Corresponding in structure and derived from a common primitive origin.

Homology. Basic similarity; structural likeness of an organ or part of one kind of animal with the comparable unit in another, resulting from descent from a common ancestry. These organs may or may not have the same function.

Hooklet (barbicel). Hooked structure which holds together feather barbs.

Hormone. A chemical regulator or coordinator secreted by cells or a ductless gland and having a specific action on some organ or organs at a distance from the gland. Carried by the blood or other body fluids.

Humerus. Bone of the vertebrate's upper arm.

Humoral. Pertaining to the body fluids such as hormones or other chemical agents.

Hypertrophy. Abnormal increase in the size of an organ or other body part.

Hypoblast. Germinal layer which spreads down, under trophoblast, forming the yolk sac.

Hypocercal. Caudal fin in which notochord or vertebral column extends into the ventral portion. Ventral portion extends further than dorsal portion.

Hypophysis. Pituitary gland.

Ileum. Posterior and longest part of the small intestine of a mammal.

Iliocaecal valve. Valve between the ileum and the colon.

Incus. Second ear bone, lying between malleus and stapes.

Infundibulum. Stalk-like down-pushing of the diencephalon of the brain, which, along with the embryonic hypophysis, will give rise to the pituitary gland of the adult.

Inspiration. Intake of air to the lungs.

Integument. Outer covering, especially the skin, of a vertebrate.

Inter-cartilaginous bone. Bone formed from cartilage, which was originally formed from mesenchyme.

Intercellular. Between the cells.

Intermedin. In some vertebrates; a hormone secreted by intermediate lobe of pituitary gland. Major activity is chromatophore control.

Inter-ventricular septum. Partition between the ventricles in the heart.

Intestine. Part of the digestive tract posterior to the stomach.

Intracellular. Within cells.

Invaginate. To infold; folding or inpushing of a layer of cells into a cavity, as of a blastula to form gastrula.

Invertase. Sucrase. Enzyme acting upon sucrose; hydrolyses sucrose to glucose and fructose.

Involution. The process of rolling or turning in of cells over a rim.

Isolecithal. Pertaining to the eggs that have yolk distributed evenly throughout the cytoplasm.

Jaw. Structure normally used for chewing or holding food.

Jejunum. Part of the intestine extending from the duodenum to the ileum in a vertebrate.

Joint. Junction of two separate bones or other hard structures.

Jugular. Pertaining to the throat, as jugular vein.

Keel. Distinct median ridge of avian sternum.

Keratin. Protein formed by certain epidermal tissues; such as vertebrate skin.

Kidney. Chief organ for the excretion of liquid nitrogenous wastes in vertebrates.

Labial. Pertaining to the lips.

Labyrinthodont. Fossil amphibian of the late Palaeozoic.

Lacrimal. Pertaining to tears.

Lacteal. Pertains to milk; often refers to the lymph vessels of the small intestine of a vertebrate.

Lactogenic. Milk producing, *e.g.*, milk producing hormone of vertebrate pituitary gland.

Larynx. Organ situated between the trachea and the base of the tongue into which the glottis opens and which contains the vocal folds. A larynx is typically found in all lunged vertebrates above fishes, except birds.

Lateral line. Sense organs in fishes and some amphibians extending bilaterally and longitudinally.

Latitudinal migration. Migration in a north-south plane.

Lecithality. Pertaining to oocyte yolk concentration.

Leucocyte. White blood cell or corpuscle.

Ligament. Tough, fibrous band of tissue connecting bones and supporting viscera.

Lingual. Pertaining to the tongue, as lingual artery.

Lipase. A fat splitting enzyme.

Lipoid. Of fatty nature.

Loop of Henle. Kidney structure in medulla of kidney; majority of water resorption occurs here.

Lumbar. The region just posterior to the ribs of vertebrates.

Lumen. Internal cavity within a body or structure such as the lumen of the intestine, a gland duct or blood vessel.

Lung. Organ for aerial respiration.

Lutenizing hormone. Hormone secreted by anterior pituitary; induces Graafian follicles to burst open and allow egg to escape.

Lymphatic system. System of vessels and nodes in vertebrates which lead from tissue spaces to large veins entering heart, a part of the circulatory system.

Lymph node. Area of lymphatic system which produce antibodies.

Lymphocyte. Vertebrate white blood cell with round or kidney-shaped nucleus.

Macromere. The relatively large, yolk-laden cells that are present during the cleavage stage of certain animals.

Malleus. First ear bone, attached to tympanum.

Malpighian body. Structure in the vertebrate kidney composed of Bowman's capsule and the glomerulus.

Mandible. A jaw; the lower jaw in vertebrates.

Marsupial. Mammal that possesses a pouch.

Maxilla. The large bone of the upper jaw in vertebrates.

Meatus. Small passage or canal, as in the external meatus of the middle ear.

Medulla. Inner region of a gland or other structure in animals.

Medulla oblongata. Most posterior of the principal divisions of the vertebrate brain.

Medullated. Used principally for certain nerve fibres covered with a myelin sheath.

Meisner's corpuscle. A light touch receptor.

Melanin. Black pigment found in cytoplasmic granules of chromatophore cells.

Membranous labyrinth. The semicircular canals responsible for balance, and associated sacculus and utriculus.

Meninges. The three membranes (duramater, arachnoid and piamater) covering the brain and spinal cord.

Mentomeckelian. Refers to the anterior bones or cartilages of vertebrate lower jaw.

Meroblastic. Egg cleavage in which yolk is left undivided as only part of the cytoplasm divides.

Mesencephalon. Midbrain of vertebrates.

Mesentery. A thin double-walled sheet of peritoneum that supports organs in the abdominal cavity of vertebrates.

Mesoblast. Blastula cells from which mesoderm develops.

Mesoderm. The middle layer of embryonic cells, between the ectoderm and endoderm.

Mesonephric duct. Duct leading from mesonephric type of kidney to cloaca.

Mesonephros. Type of vertebrate kidney, present in embryos of most vertebrates.

Metabolism. The sum total of the reactions, mainly chemical, that occur within the protoplasm of organism.

Metacarpals. Proximal bones of the hand.

Metamorphosis. Marked structural changes or transformation during development; for example, as from larva to adult.

Metapleural folds. Two folds on the anterior, ventral surface of the *Amphioxus*, extending posteriorly from the mouth to the atriopore.

Metatarsals. Proximal bones of the foot.

Metencephalon. Anterior part of the hindbrain in vertebrates.

Migration. Movement of a part of population of a species (usually in groups) from one region to another.

Mimicry. Adaptive resemblance of certain animals to others or surroundings resulting in protection for the mimic from predators.

Molar. Permanent mammalian posterior teeth.

Monocyte. Type of leucocyte; large phagocytic, uninucleate.

Monophyletic. Evolved from a single known ancestral type.

Morphogenesis. Development of form and size.

Morula. Solid ball of cells resulting from egg cleavage.

Mucosa. A membrane that secretes mucus.

Mucous. Pertaining to mucus.

Mucous membrane. Consists of an epithelium with some subepithelial connective tissue which lines cavities that communicate with the exterior.

Mucus. Viscous secretion containing mucin; a product of the mucous glands.

Muscularis. Thin layer of smooth muscle between the tunica propria and the submucosa in the wall of the oesophagus, stomach and intestine.

Myelencephalon. Posterior part of vertebrate hindbrain.

Myelin. The fat-like substance surrounding the axis cylinder of a medullated nerve fibre.

Myocardium. The body of the heart, as distinguished from the pericardium.

Myofibril. One of the numerous longitudinal small fibrils contained within the protoplasm of the muscle cell or fibre.

Myomere. Muscle somite or segment.

Myosin. A protein of muscle that contributes to contraction.

Nares. Opening of the air passages, both external and internal, in the head of a vertebrate.

Nasal. Pertaining to the nose or nostrils.

Necrosis. Death.

Neoteny. Sexually mature, larval state which is permanent.

Nephron. The functional unit of the vertebrate kidney.

Nerve. Bundle of nerve fibres located outside the central nervous system.

Nerve cord. Cord of neurons and (usually) ganglia forming part of a central nervous system.

Neural. Pertaining to the nervous system.

Neural canal. Canal through the vertebrae; the canal formed by the neural arches.

Neural plate, groove, and tube. Three successive phases in the development of central nervous system in vertebrates.

Neural spine. The dorsal projection of a vertebra.

Neurilemma. Outermost sheath of a nerve fibre.

Neurocoel. Cavity in a chordate nerve cord.

Neuron. A nerve cell, including the cell body and all its processes.

Neutrophil. Type of leucocyte; phagocytic fragmented nuclei. Make up 65 to 75 per cent of leucocytes.

Neuromast. Aggregation of sensory cells, sensory hairs and cupulae in the lateral line organ of fishes and some amphibians.

Nictitating membrane. Transparent, thin membrane possessed by many vertebrates which opens and closes laterally, across the cornea.

Nocturnal. Active at night.

Node of Ranvier. Place on a nerve fibre where the medullary sheath is constricted.

Notochord. Characteristic cylindrical rod of supporting cells in chordates, found dorsal to the digestive tract and ventral to the nerve cord; either surrounded or supplemented by the vertebrae in most vertebrates.

Nutrient. Any substance used in metabolism either inorganic or organic.

Nutrition. Sum of the processes concerned in the growth, maintenance, and repair of the living body as a whole or of its constituent parts.

Occipital. Pertaining to the base of the skull of a vertebrate.

Olfactory. Pertaining to the sense of smell.

Oligodendrocyte. Myelin-containing neurons within the central nervous system.

Oligolecithal. Egg with yolk in low concentration and dispersed rather evenly throughout the oocyte.

Omnivorous. Eating all kinds of food, both plants and animals.

Ontogeny. Entire developmental history of the individual organism.

Oocyte. Egg mother cell from which are produced by the first maturation division, the secondary oocyte and the first polar body.

Oogenesis. Process of formation of ova.

Oogonium. In animals, the primordial egg cell prior to the process of maturation (meiosis); cell from which primary oocytes are produced.

Operculum. Structure covering the gills of fishes and tadpoles.

Ophthalmic. Pertaining to the eye.

Optic. Pertaining to the eye or to sight.

Optic lobes. Thickenings on the dorsal surface of the midbrain (mesencephalon).

Optic nerves. Nerves from the eye to the brain.

Oral. Pertaining to the mouth.

Orbit. Region of the skull of the vertebrate surrounding the eye.

Organ of Corti. Sensory organs within the cochlea.

Organogeny. Process of the formation of specialised tissue and organ system during embryonic development.

Osseus. Pertaining to the bone.

Ossicle. A small bone or bone-like supporting structure.

Osteichthyes. Class of bony fishes.

Osteoblast. Mesenchyme cell which has differentiated into protein-matrix secreting cell.

Osteology. Study of bones.

Ostium. A small mouth-like opening such as the anterior end of the oviduct in the frog and other vertebrates.

Otolith. A limy particle in the inner ear of vertebrates.

Ovary. Female gonad in which the eggs (ova) multiply and develop.

Oviduct. A tube that conveys eggs from ovary to uterus or to exterior.

Oviparous. Producing eggs that hatch outside the body of the mother; egg laying animals.

Ovoviviparous. Producing eggs that hatch within the parent's body but are not nourished by the mother's blood stream through a placenta.

Ovulation. Release of eggs from the ovary.

Ovum. An egg; a non-motile female gamete.

Pacinian corpuscle. Heavy touch receptors.

Palatine. Bone that serves as an anterior brace of the upper jaw in the skull of the frog and other vertebrates.

Palaeontology. The science that deals with ancient life of the earth as revealed by fossils, impressions and other remains found in the earth's surface layers.

Palaeozoic. Geological era occurring between the Pre-Cambrian and the Mesozoic and dating approxinately from 500 to 200 millions years ago.

Pancreas. Digestive gland that discharges into the intestine, produces digestive enzymes and insulin.

Papilla. Any small nipple-shaped elevation.

Parasphenoid. Bone forming the floor of the vertebrate cranium.

Parasympathetic. Subdivision of the vertebrate autonomic nervous system with centres in the brain and anterior portion of spinal cord.

Parathormone. A secretion of the parathyroid glands. Influences calcium and phosphorus metabolism.

Parathyroid. One of several small ductless glands, closely associated with the thyroid gland of vertebrates.

Parietal. Pertaining to the outer wall of the coelom.

Parietal bone. One of a pair of bones located just posterior to the frontal bones.

Patella. Knee-cap.

Pectoral. Pertaining to the upper thoracie area.

Pectoral girdle. Group of bones and cartilage connecting the forelimb to the axial skeleton in vertebrates.

Pedogenesis. Reproduction by larvae.

Pelvic girdle. Group of bones to which the bones of hindlimb are attached, connecting the hindlimb to the axial skeleton in the vertebrates.

Penis. Male organ of copulation for conveying sperms to the female genital tract.

Pentadactyl. Having five fingers toes or digits.

Pepsin. An enzyme concerned with protein digestion in animals.

Periblast. In fishes, the layer of cells under the blastocoel.

Pericardium. Closed membranous sac surrounding the heart.

Periosteum. Connective tissue sheath that covers the surface of a bone.

Peripheral nervous system (PNS). Nervous system excluding the central nervous system (brain and spinal cord).

Perissodactyla. Odd-toed ungulates including horses and rhinoceroses.

Peristalsis. Type of smooth muscle contraction in which the wave of contraction follows a wave of relaxation, passing along a hollow organ, especially the digestive tract.

Peritoneum. Thin membrane (mesoderm) that lines the coelom and covers the viscera in many animals, especially vertebrates.

Perkinje fibres. Myocardial conducting tissue forming a network in the ventricular walls.

Phagocyte. Type of white blood cell that engulfs and digests bacteria and other foreign materials.

Phalanges. Bones of digits. Singular phalanx.

Pharyngeal-cleft. Slit in the wall of the pharynx; same as gill-slit, gill-cleft or branchial-slit.

Pharynx. Anterior portion of the digestive tract between the mouth cavity and the oesophagus.

Photoreceptor. Light sensitive cell or organ.

Phrenic. Pertaining to the diaphragm.

Phylogeny. Ancestral or evolutionary history.

Phylum. Any of the main taxonomic divisions into which the animal kingdom is divided.

Physiology. Science dealing with function in organism.

Piamater. Thin, innermost, membranous covering of the brain and spinal cord.

Pineal. A structure located on the roof of the vertebrate animals.

Pinna. External ear.

Pituitary body. An endocrine gland located on the ventral surface of the brain and composed of anterior, posterior and intermediate lobes.

Pituitrin. Combined secretions of the posterior lobe of the hypophysis (pituitary gland).

Placenta. Organ for attaching the foetus of mammals to the uterine wall; serves in foetal nourishment, respiration and excretion. Made of maternal and foetal tissues.

Placoid scale. Scale found in cartilaginous fishes, with basal dentine and forward pointing spine.

Plantigrade. Walking with the whole sole of the foot bearing on the ground as does the bear or man.

Plastron. Ventral plate of, *e.g.*, a turtle shell.

Platelet. In blood, the corpuscles, when ruptured, begin the clotting sequence. Discoidal form.

Pleura. Membrane that covers lungs and lines inner wall of thorax.

Pleural. Pertaining to the cavity, that portion of the coelom which contains the lungs. The membrane covering the lung and lining the pleural cavity.

Plexus. Network chiefly of nerves or blood vessels.

Poikilothermic. Cold blooded animals, without internal temperature controls.

Polyembryony. One egg producing many larvae.

Pons Varolii. Midbrain, a routing station for nervous impulses passing between spinal cord and cerebrum.

Porphyropsin. In freshwater fishes, a retinal pigment based on vitamin A_2.

Portal vein. Vein that branches into capillaries before reaching the heart.

Posterior. The tail or toward the hind or rear end. Opposite of anterior.

Predaceous. Capturing living animals for food.

Predator. Animal that preys upon other animals for its food.

Prehensile. Adapted for holding or grasping.

Premaxillae. A pair of anterior bones of the upper jaw in vertebrates.

Primitive streak. Structure developing on dorsal edge of blastodisc; amoebocytic migration of blastodermal cells occurs along this line.

Proboscis. Tubular extension of the nose, lips or pharynx.

Progesterone. Vertebrate hormone secreted by the corpus luteum and by the mammalian placenta during pregnancy.

Pronephros. The first kidney embryologically formed in the vertebrate.

Proprioceptor. Sense receptor for internal stimuli.

Prostate gland. Gland surrounding the neck of the bladder and urethra in the male mammal.

Protandry. Production of sperm, and then ova, by the same gonad.

Protocercal fin. Straight caudal fin.

Proventriculus. Anterior portion of avian stomach.

Proximal. Nearer the point of attachment of an organ; for example, the shoulder is at the proximal end of the arm.

Pteryla. Area of bird skin on which feathers grow, feather tract.

Ptyalin. An enzyme of the saliva in some vertebrates which acts on starch.

Pubis. Antero-ventral area of pelvic girdle.

Pulmonary. Pertaining to the lung.

Pupil. Hole in the eye through which light travels. Appears black because of the lack of illumination from within the eye.

Pygostyle. Fused tail vertebrae of birds.

Pyloric. Pertaining to the pylorus.

Pyloric caeca. Digestive glands.

Pyloric valve. Valve at the posterior end of the stomach.

Pylorus. Opening to the intestine from stomach.

Pyriformis. A muscle arising from the urostyle in the frog.

Quadratojugal. Posterior bone of the upper jaw in some vertebrates.
Quadruped. Four-footed animals.

Radial cleavage. Cleavage in which successive tiers of equal-sized blastomeres lie directly above and below one another; cleavage is indeterminate.
Radio-ulna. The fused radius and ulna bones as in frogs and toads.
Radius. The bone of the lower arm, located on the thumb side of vertebrates.
RBC. Red blood corpuscle (erythrocyte).
Receptor. A sensory end organ.
Rectum. Terminal portion of the large intestine in vertebrates.
Regeneration. Replacement by growth of a part of the body that has been lost.
Renal. Pertaining to the kidney.
Renin. A stomach enzyme that acts on casein, a protein of milk.
Reproduction. Production by an organism of others of its kind.
Reproductive migration. Migration to an area where copulatory activity is expected.
Respiration. Use of oxygen by the cell; this is usually termed cellular respiration or internal respiration.
Retina. Light sensitive layer of an eye.
Rudimentary. Pertaining to a body part not completely developed.
Rugae. Ridges or folds, as in the lining of the vertebrate stomach.
Ruminant. Cud-chewing mammal as the cow.

Sacculus. Portion of membranous labyrinth of the internal ear.
Sacrum. Posterior portion of vertebral column attached to the pelvic girdle.
Salivary. Pertaining to the saliva-secreting glands of the mouth.
Sarcolemma. Thin membrane surrounding a skeletal or striated muscle cell.
Sarcomere. Muscle unit containing A, I, Z and H bands.
Sarcoplasm. Protoplasmic substance of skeletal muscle cells.
Scapula. Shoulder blade in a vertebrate.
Sclerotic coat. The white of the eye or outside layer.
Scrotum. External skin pouch in most mammals, containing testes.
Scute. Horny plates arranged symmetrically but not correspondingly to plastron and carapace.
Sebaceous glands. Skin glands usually found in connection with hair follicles of the mammals. They produce an oily secretion which lubricates the skin and hair.
Sebum. The fatty secretion of the sebaceous glands.
Secondary sexual characters. Characters other than the sex-glands (gonads) and related organs in which males and females of a species differ.
Self fertilisation. Fertilising an egg by a sperm from the same individual.
Semen. Fluid that carries the sperms in the males of most animals.
Semicircular canals. Canals in the vertebrate ear, associated with the sense of equilibrium.
Seminal. Pertaining to the spermatozoa.
— , receptacles. Sac-like organ in several types of animals, which receive and store sperms after their release, considered a part of female genital system.
— , vesicles. Sac-like organ in several types of animals, which receive and store sperms after their release, considered a part of the female genital system.
Seminiferous tubule. A tube to conduct seminal fluid.
Sense organ. An organ sensitive to a particular type of stimulus.
Sensory cell. Sensory cell that is very sensitive to stimuli, a receptor.
Serosa. Chorion of reptiles and birds.
Serous coat. The peritoneal covering of the visceral organs.
Sertoli cells. In vertebrates, seminiferous, non-gametic cells which seem to provide nourishment for developing sperm.

Serum. Blood plasma that separates from a clot and is without cells or fibrin.

Sexual dimorphism. Phenomenon of two sexes of a given species differing in secondary sexual characters.

Sinu-atrial node. In the heart, the "pacemaker" cells near the tricuspid valve. Receiving stimulus from a vagus nerve. The sinu-atrial node (SA node) initiates a wave of contraction across the atria.

Sinus venosus. In fishes, amphibians, and reptiles, a large chamber receiving blood from veins and transmitting it to the right atrium.

Sinusoid. Large, irregular. tortuous, blood spaces which are comparable to capillaries in that their walls form an endothelial tube, but they differ from capillaries in that their lamina are of greater size.

Skeleton. External or internal hardened framework of an animal body, giving support and protection to the soft parts.

Somatic nervous system. System controlling voluntary acts.

Species. Group of actually or potentially inter-breeding natural populations that resemble one another closely and are reproductively isolated from other such groups.

Sperm. A mature male reproductive cell or gamete.

Spermatogenesis. Process of formation of spermatozoa.

Spermatogonium. Male sex cell before it begins differentiating as sperm. Immature male gamete.

Spermatozoa. Mature male gametes.

Sphincter. Ring of smooth muscle surrounding a tube or opening, which by its contraction closes the lumen.

Spinal column. A series of vertebrae in the vertebrates, which encloses the spinal cord.

Spinal cord. Part of the central nervous system extending the length of spinal column and continuing anteriorly into the brain of vertebrates.

Spiral valve. In some amphibians, flattened twisted structure in truncus arteriosus.

Splanchnic. Pertaining to the viscera.

Spleen. A large organ, characteristic of most vertebrates, lying near the stomach and known to destroy red blood corpuscles in mammals.

Squamosal. T-shaped bone of the skull of a vertebrate, bracing posterior end of the jaws.

Stapes. Third ear bone, attached to incus and oval window of cochlea.

Steapsin. Enzyme of the pancreas that acts on fats.

Sternum. Bone or bones of the ventral side of the thorax of a vertebrate; breast bone.

Stimulus. Change in the external or internal environment of an animal that brings about a response.

Stolon. A branch upon which colonial individuals may bud upright.

Stratified. Arranged in layers, one above the other.

Stratum compactum. Inner layer of the corium or dermis or dermis of the skin.

Stratum corneum. Outer layers of cells in the epidermis of the vertebrate skin.

Stratum germinativum. Mitotic layer of cells in the epidermis of skin of a vertebrate.

Stratum spongiosum. Outer layer of connective tissue in the dermis (corium) of a vertebrate.

Striated. Cross striped as one of the principal types of muscle cells.

Subcutaneous. Just beneath the skin, as subcutaneous connective tissue.

Symmetry. Characteristic of geometric equivalence of two or more parts of an organism.

—, radial. From where any plane through longitudinal axis will divide organism into equivalent parts.

—, biradial. From where two specific different radial planes will divide organism into different sets of halves.

—, spherical. From where any plane passing through centre of organism will divide it into equivalent halves.

—, bilateral. From where one unique plane (mid-sagittal) will divide organism into equivalent halves.

Syrinx. Vocal organ of birds.

System. Group of organs concerned with the same general function, as circulation or digestion.

Systemic. That part of the circulatory system not involved directly in respiration.

Systemic arch. Any large artery of a vertebrate, carrying blood from the heart to the dorsal aorta.

Systole. Contraction phase of the auricles or ventricles of the heart.

Tactile. Pertaining to the sense touch.

Tarsals. Bones of the ankle.

Tarsus. An ankle bone of vertebrates.

Taxon. Category of animal classification.

Taxonomy. Science that deals with the classification of organisms.

Telencephalon. Forebrain of vertebrates.

Telolecithal. Pertaining to eggs that have a large amount of yolk in the vegetal portion of the cell.

Tendon. Tough, cord-like connective tissue extension at the end of a muscle which attaches it to other structures.

Testis. Male reproductive gland (gonad) in which sperms are formed.

Testosterone. Male hormone secreted by the testes.

Tetrapod. Four-limbed vertebrate.

Thalamus. Portion of vertebrate forebrain.

Thorax. Major division of an animal just posterior to the neck or to the head if no neck is present.

Thrombin. Substance formed from prothrombin during vertebrate blood clotting; converts fibrogen to fibrin.

Thymus. Two-lobed endocrine gland located just below the thyroid of mammals.

Thyroid. Endocrine gland in the neck of vertebrates which exerts an influence on growth and metabolism.

Thyroxine. Hormone secreted by the thyroid gland.

Tibia. The larger medial bone of the vertebrate lower leg.

Tissue. Group of cells of similar structure, with intercellular substances, if any, which perform a specialised function.

Tornaria. Larva of particular enteropneust hemichordates.

Trachea. Wind-pipe of vertebrates.

Transverse process. One of the lateral projections of a vertebra.

Tricuspid valve. In mammalian heart, the trilobed valve between the right atrium and ventricle.

Trigeminal. Fifth cranial nerve.

Triploblastic. Derived from three primary germ layers – ectoderm, mesoderm and endoderm.

Trophoblast. Germinal layer from which chorion develops.

Truncus arteriosus. In lower vertebrates, base of aorta.

Trypsin. Enzyme promoting digestion of protein in vertebrates.

Tympanic membrane. Ear drum.

Tympanum. Vibrating membrane involved in the sense of hearing.

Ulna. The bone of the little finger side of the forearm in vertebrates from amphibians to man.

Umbilical cord. The cord-like connection between the embryo or foetus of a mammal and the placenta, composed mainly of blood vessels and connective tissue.

Ungulate. Manmals with hoofs.

Unguligrade. Walking or adapted for walking on hoofs.

Ureter. Tube that carries urine away from the kidney to the urinary bladder or cloaca.

Urethra. Duct that carries urine from the bladder to the outside in mammals.

Urine. Liquid waste excreted by the kidneys.

Uriniferous tubule. One of the excretory tubules within the kidney of higher animals, consisting of a coiled tube and a capsule.

Urodela. Amphibians with tails such as salamanders and newts.

Urogenital. Pertaining to organs of both the urinary and the reproductive systems taken collectively.

Utriculus. Portion of the membranous labyrinth of the inner ear, contains maculae.

Urostyle. Terminal rod-like bone of the frog spinal column.

Uterus. Enlarged portion of the oviduct in which at least part of the development of an animal takes place.

Vagina. Posterior part of the female reproductive tract. It often receives the copulatory organ of the male in mating.

Vagus. Tenth cranial nerve in vertebrates.

Valve. Any structure that limits or closes an opening.

Vasa efferentia. Small ducts carrying sperms from the testes to the kidney in the frog and to the ductus epididymis or to other similar tubules in higher animals.

Vas deferens. Duct that carries sperms away from testis.

Vascular. Pertaining to vessels in animals, they usually carry blood or lymph.

Vasomotor centre. In the medulla of the brain, an area responsible for arterial tonus.

Vasomotor nerves. Nerves that control the contraction and expansion of blood vessels.

Vein. Blood vessel that carries blood toward the heart.

Velar tentacles. In cephalochordates, tentacles projecting from the mouth region; act as strainers.

Velum. Membranous band of tissue.

Ventral. Pertaining to the belly; away from the back.

Ventricle. Any of the small chambers in the anatomy of animals; specifically in the heart, a chamber from which blood is distributed; in the brain any of the several large subdivisions of the central space.

Vermiform appendix. A slender, tubular pouch projecting from the caecum of the large intestine of some mammals.

Vertebra. One unit of the vertebrate spinal column.

Vertebral column. Series of vertebrae in a vertebrate animal, Backbone.

Vertebrate. Animals having a vertebral column.

Vessel. Tube-like structure conveying fluid, such as blood or lymph.

Vestibule. An outer cavity with an entrance to a (usually) larger, deeper cavity.

Vestigial. Degenerate structure that was better developed or functional in more primitive or ancestral groups.

Vibrissae. Stiff hair in or at the nostril.

Villus: A minute, finger-like projection, especially those on the intestinal lining of vertebrates which increase the absorptive surface.

Viscera. Organs enclosed within the great cavities of the body, especially the abdomen.

Visceral skeleton. Vertebrate supporting framework of jaws and gill-arches.

Visual. Pertaining to the sense of sight.

Vitreous. Glassy in appearance.

Vitreous humor. Liquid within the eye in the area between the lens and retina.

Vitelline. Pertaining to yolk.

Viviparous. Giving birth to living young that develop from egg within the body of the mother and are nourished from her blood stream as with most mammals.

Vulva. External portions of female reproductive organs.

Warm blooded. Animals (birds and mammals) whose body temperature remains rather constant regardless of external temperature.

White blood corpuscle. Colourless blood cell (leucocyte).

Wolffian ducts. Mesonephric ducts.

Yolk. Stored nutritive materials for nourishment of the embryo.

Yolk plug. Near the end of gastrulation, the circular spot of yolk left visible, around which the blastopore lies.

Yolk sac. Area in which yolk is enclosed or in the case of, for example, humans, area of enclosed cytoplasm (there is no yolk).

Zoogeography. Branch of zoology dealing with the geographic distribution of animals.

Zoology. Science of animal life.

Zygapophysis. One of the four projections, two anterior and two posterior, extending from the neural arch of a vertebra.

Zygote. Cell that results from the fertilisation of one gamete by another. The product of the union of two gametes of any type, not necessarily an egg and sperm.

Selected References

Adams, L.A., 1946, *Introduction to the Vertebraes*, John Wiley and Sons, New York, U.S.A.

Adams, L.A. and Eddy, S., 1949, *Comparative Anatomy*, John Wiley and Sons, New York, U.S.A.

Alexander, R.M., 1982, *Locomotion in Animals*, Blackie, London, U.K.

Al-Hussaini, A.H. and Demian, E.S., 1967, *Practical Animal Morphology*, Macmillan and Company Ltd., London, U.K.

Allen, A., 1963, *The Book of Bird Life*, D. Van Nostrand Company, Inc. Princeton, New Jersy, U.S.A.

Ali, Salim, 1979, *The Book of Indian Birds*, Bombay Natural History Society, Bombay, India.

Ali, Salim and Fatehally, L., 1975, Common Birds, National Book Trust, New Delhi, India.

Anonymous, 1980, *Wild Life Ecology*, Zoological Survey of India, Kolkata, India.

Arey, L.B., 1965, *Developmental Anatomy,* W.B. Saunders Company, Philadelphia and London.

Austin, Oliver, L., 1968, *Birds of the World*, Hamlyn, London, U.K.

Balinsky, B.I. 1970, *An Introduction to Embryology*, W.B. Saunders Company, Philadelphia and London.

Barrington, E.J.W., 1965, *The Biology of Hemichordata and Protochordata*, Oliver and Boyd, London, U.K.

Beddard, F.E., 1920, *Mammalia* (Cambridge Natural History), Macmillan and Company Ltd., London, U.K.

Bellairs, A. and Carrington R., 1966, *The World of Reptiles*, Chatto and Windus, London, U.K.

Bellairs, A., 1969, *The Life of Reptiles*, Vol. I and II, Weidenfield and Nicholson.

Berrill, N.J., 1950, *The Tunicata*, Ray Society, London.

Berrill, N.J., 1955, *The Origin of Vertebrates*, Oxford University Press, New York, U.S.A.

Berrill, N.J. 1966, *Biology in Action,* Heinemann Educational Books Ltd., London, U.K.

Bhatia, B.L., 1951, *An Elementary Text Book of Zoology*, Macmillan and Company Ltd., London, U.K.

Boolootian, R.A. and Stiles, K.A., 1976, *College Zoology*, Macmillan and Company Inc. New York, U.S.A.

Boolootian, R.A., 1979, *Zoology : An Introduction to the Study of Animals*, Macmillan and Company Inc. New York, U.S.A.

Bridge, T.W. and Boulenger, G.A., 1904, *Fishes* (Cambridge Natural History) Macmillan and Company Ltd., London, U.K.

Buckles, Mary Parker, 1977, *Mammals of the World*, Bantam Books, London, U.K.

Buffaloe, N.D.,1964, *Principles of Biology*, Prentice Hall of India (P) Ltd., New Delhi, India.

Carr, A., 1963, *The Reptiles*, Time-Life International, Neederland.

Carrington, R. 1968, *The Mammals,* Time-Life International, Neederland.

Chapman, G. and Baker, W.B., 1964, *Zoology*, Longmans Green, London, U.K.

Chiasson, R.B. 1959, *Laboratory Anatomy of Pigeon,* W.M.C. Brown Company, Iowa, U.S.A.

Colbert, E.H., 1955, *Evolution of Vertebrates*, Wiley, New York. U.S.A.

Colbert, E.H.1961, *Dinosaurs : Their Discovery and Their World,* E.P., Dutton, New York, U.S.A.

Das, S.M. 1936, *Herdmania,* The Indian Zoological Memoirs, Lucknow Publishing House, Lucknow, India.

Dasmann, R.E., 1982, *Wild Life Biology,* Wiley Eastern Ltd., New Delhi, India.

Day, Francis, 1958, *The Fishes of India,* London, U.K.

Dehaan, R.L. and H. Ursprung, 1965, *Organogenesis,* Holt Rinehart and Winston, New York, Chicago, U.S.A.

Deoras, P.J., 1965, *Snakes of India,* National Book Trust, India, New Delhi, India.

Ditmars, R.L., 1940, *Reptiles of the World*, Macmillan Company, New York, U.S.A.

Ditmars, R.L., 1952, *Snakes of the World*, Macmillan Company, New York, U.S.A.

Dorst, Jean, 1962, *The Migration of Birds,* Williams Heinemann, London, U.K.

Eaton, Jr., T.H., 1960, *Comparative Anatomy of the Vertebrates,* Harper and Row Publishers, New York, U.S.A.

Eimeri, S. and Irven, D., 1972, *The Primates,* Time-Life International, Neederland.

Elliot, A.M., 1964, *Zoology,* Peter Owen, London, U.K.

Evans, A.H., 1935, *Birds,* (Cambridge Natural History), Macmillan and Company Ltd., London. U.K.

Evans. D.H., 1997, *The Physiology of Fishes*, CRC Press.

Fring, H. and Fring, M. 1970, *Concepts of Zoology*, Macmillan and Company Ltd., Toronto, U.S.A.

Gadow, H.,1935, *Amphibia and Reptiles,* (Cambridge Natural History), Macmillan and Company Ltd, London, U.K.

Gee, E.P., 1964, *The Wildlife of India*, Collins, London, U.K.

Gharpurey, K.G., 1954, *The Snakes of India and Pakistan*, Bombay Natural History Society, Bombay, India.

Goodnight, C.J., M.L. Goodnight and Gray, P., 1970, *General Zoology,* Oxford I.B.H. Publishing Company, New Delhi, India.

Goodrich, E.S.,1958, *Studies on the Structure and Development of Vertebrates,* Vol. I & Vol. II, Dover Publications Inc., New York, U.S.A.

Grasse, P.P. (Ed.) 1950, *Traite de Zoology*, Oiseaux, 15, Masson, Paris, France.

Grasse, P.P. (Ed.) 1955, *Traite de Zoology*, Mammalia, 17, Masson, Paris, France.

Groove, A.J. and Newell, G.E.1955, *Animal Biology*, University Tutorial Press Ltd., London, U.K.

Guyer, M.F. and Lane, C.E.,1964, *Animal Biology*, Harper and Row Publishers, New York, U.S.A.

Hall, B.K., 1992, *Evolutionary Developmental Biology*, Chapman and Hall, London, U.K.

Harmer, S.F., 1904, *Hemichordata* (Cambridge Natural History) Macmillan & Company Ltd., London, U.K.

Herdman, W.A., 1904, *Ascidians and Amphioxus* (Cambridge Natural History) Macmillan and Company Ltd., London, U.K.

Herman O. and Owen, J.A. 1909, *Birds Useful and Birds Harmful,* The University Press, Manchester, U.K.

Hickmann, C.P. 1961, *Principles of Zoology*, Mosby, St. Louis, U.S.A.

Hildebrand, M. 1974, *Analysis of Vertebrates*, John Wiley and Sons, New York, U.S.A.

Holmes, S.J. 1930, *The Biology of the Frog,* The Macmillan Company, New York, U.S.A.

Huettner, A.F., 1949, *Fundamentals of Comparative Embryology*, The Macmillan Company, New York, U.S.A.

Hyman, L.H., 1942, *Comparative Verbtebrate Anatomy,* The University of Chicago Press, Chicago, U.S.A.

Hyman, L.H., 1959, *The Invertebrates, Vol. V*, McGraw Hill Book Company, Inc. New York, U.S.A.

Jollie, M., 1968, *Chordate Morphology,* Affiliated East-West Press, Pvt. Ltd., New Delhi, India

Kent, G C., 1969, *Comparative Anatomy of Vertebrates,* The C.V. Mosby Company, Saint Louis, U.S.A.

Khanna, S.S., 1996, *An Introduction to Fishes*, Central Book Depot, Allahabad, India

Kingsley, J.S., 1957. *Outlines of Comparative Anatomy of Vertebrates*, Central Book Depot, Allahabad, India.

Kluge, A.G., 1977, *Chordate Structure and Function*, Macmillan Company, Inc., New York, U.S.A.

Lagler, F. K., 1967, *Ichthyology,* John Wiley, New York, U.S.A.

Lane, N.G., 1992, *Life of the Past,* 3rd Edn., Macmillan, New York, U.S.A.

Long, J.A., 1995, *The Rise of Fishes : 500 Million Years of Evolution,* Johns Hopkins University Press, Baltimore, Maryland, U.S.A.

Lillie, F.R., 1952, *Development of Chick,* Henry Holt, New York, U.S.A.

Lincoln, F.C., 1950, *Migration of Birds*, U.S. Govt., Washington, U.S.A.

Majupuria, T.C.,1954, The anatomy of the hemipenes in *Uromastix hardwickii* Gray. Agra Univ. J. Res. 3 (2) : 539-543.

Majupuria, T.C.,1961, Morphology of cloaca of *Uromastix hardwickii* Gray. *Proc. Indian Sci. Congr.* III : 400.

Manter, H.W. and Miller D.D., 1959, *Introduction to Zoology*. New York.

Marshall, A.M., 1939, *The Frog,* Macmillan and Company Ltd., London, U.K.

Marshall, A.J. (Ed.), 1960-61, *The Biology and Comparative Physiology of Birds,* Academic, Press, New York and London.

McFarland, W.N., 1979, *Vertebrate Life,* Macmillan and Company Ltd., Inc., New York, U.S.A.

Neal, H.V. and Rand, H.W., 1946, *Chordate Anatomy,* The Blakiston Company, Philadelphia, U.S.A..

Nelson, O.E., 1953, *Comparative Embryology of the Vertebrates,* The Blakiston Company, New York, U.S.A.

Newman, H.H. 1958, *The Phylum Chordata*, The Macmillan Company, New York, U.S.A.

Noble, G.K., 1954, *The Biology of the Amphibia,* Dover Publications Inc., New York, U.S.A.

Norman, J.R., 1963, *A History of the Fishes,* Ernest Benn Ltd., London, U.K.

Orr, R.T., 1976, *Vertebrate Biology* , Toppan Company Ltd., Tokyo, Japan.

Parker, T.J. and Haswell, W.A. (Revised by **Marshall A.J.**), 1964, *A Text Book of Zoology*, Vol. II, Macmillan and Company Ltd., London, U.K.

Patten, B.M.,1958, *Foundations of Embryology,* McGraw-Hill Book Company, Inc, New York, U.S.A.

Peterson, R.T., 1965, *Thr Birds,* Time-Life International, Neederland.

Pope, C.H., 1955, *The Reptiles of World*, A Knopf., New York, U.S.A.

Prater, S.H., 1965, *The Book of Indian Mammals*, Bombay Natural History Society, Bombay, India.

Romer, A.S. 1941, *Man and the Vertebrates,* University of Chicago Press, Chicago, U.S.A.

Romer, A.S. 1945, *Vertebrate Palaeontology*, University of Chicago Press, Chicago, U.S.A.

Romer, A.S. 1962, The *Vertebrate Body*, W.B. Saunders Company, Philadelphia and London.

Rugh, R., 1951, *The Frog : Its Reproduction and Development*, The Blackiston, Toronto.

Saharia, V.B., 1982, *Wildlife in India*, Natraj Publishers, Dehradun, India.

Sarabhai, D.S., 1932. *The Endoskeleton of Labeo rohita* (Ham. Buch.) , Journal and Proceedings of Asiatic Society of Bengal, Vol. 28 No. 1, Pages, 337-341, Calcutta, India.

Sedgewick, A., 1927, *A Student's Text Book of Zoology*. Vol. III, George Allen and Unwin Ltd., London, U.K.

Sedgewick, A., 1966, *A Student's Text Book of Zoology*. Vol. II, Central Book Depot, Allahabad, India.

Seshadari, B., 1969, *The Twilight of India's Wildlife,* Oxford University Press, Bombay, India.

Sibely, C.G and Ahiquist, J.E., 1990, *Phylogeny and Classification of Birds : A Study in Molecular Evolution,* Yale University Press, New Haven , Connecticut, U.S.A.

Singh, S., 1986, *Conserving India's Natural Heritage,* Natraj Publishers, Dehradun, India.

Singh, S.K., 2005, *Textbook of Wildlife Management,* International Book Distributing Company, Lucknow, India.

Srivastava, M.D.L.,1952, *An Introduction to the Comparative Anatomy of Vertebrates*, Pothishala Limited, Allahabad, India.

Storer, T.I. and Usinger, R.L., 1965, *General Zoology*, Tata-McGraw-Hill Publishing Company Ltd., Bombay-New Delhi, India.

Thillayampalam, E.M., 1952, *Scoliodon*, The Indian Zoological Memoirs, Lucknow Publishing House, Lucknow, India.

Tilak, Raj, 1982, *Uromastix hardwickii* Gray, Kedar Nath Ram Nath, Meerut, India.

Thomson, J.A., 1923, *The Biology of Birds,* Sidgwik and Jackson Ltd., London, U.K.

Torrey, T.W., 1971, *Morphogenesis of Vertebrates,* John Wiley and Sons, Inc., New York, U.S.A.

Turner, C.D., 1960, *General Endocrinology*, W.B. Saunders, London, U.K.,

Villee, Walker and Barnes, 1973, *General Zoology*, W.B. Saunders, Philadelphia, U.S.A.

Van Tyne, J. and Berger, A.J., 1959, *Fundamentals of Ornithology*, John Wiley and Sons, Inc. New York, U.S.A.

Wallace, G.J., 1969, *An Introduction to Ornithology*, The Macmillan Company, London, U.K.

Walter, H.E. and Sayles, L.P., 1965, *Biology of the Vertebrates*, The Macmillan Company, New York, U.S.A.

Weichert, C.K., 1965. *Anatomy of the Chordates*, McGraw Hill Book Company, Inc. New York, U.S.A.

Wetmore, A., 1951, *A Revised Classification of the Birds of the World*, Smith. Misc. Coll. 117, 1.

Weisz, P.B., 1968, *Elements of Zoology*, Tata McGraw-Hill Publishing Company Ltd., Bombay and New Delhi, India.

Whinchester, A.M., 1964, *Biology and Its Relation to Mankind*, Affiliated Press, New York, U.S.A.

Woodsedalek, J.E., 1963, *General Zoology,* W.C. Brown, U.S.A.

Wright, E.P., *Mammalia*, Casel and Company Ltd., London, U.K.

Yapp, W.B., 1965, *Vertebrates, Their Structure and Life*, Oxford University Press, New York, U.S.A.

Yapp, W.B., 1970, *The Life and Organization of Birds*, E. Arnold, London, U.K.

Young, J.Z., 1957, *The Life of Mammals*, Oxford University Press, London, U.K.

Young, J.Z. 1962, *The Life of Vertebrates*, Oxford University Press, London, U.K.

INDEX

A

B

C

D